W9-BCW-627

ADVANCED ENGINEERING MATHEMATICS

ADVANCED ENGINEERING MATHEMATICS

Peter V. O'Neil
University of Alabama in Birmingham

Wadsworth Publishing Company
Belmont, California

A division of Wadsworth, Inc.

Mathematics Editor: Richard Jones

Production: Greg Hubit Bookworks

Designer: Rich Chafian

Copy Editor: Trevor Grayling

Technical Illustrator: Renaissance Studios

Printed in the United States of America

1 2 3 4 5 6 7 8 9 10—87 86 85 84 83

Library of Congress Cataloging in Publication Data

O'Neil, Peter V.
Advanced engineering mathematics.

Includes bibliographical references and index.
1. Engineering mathematics. I. Title.
TA330.053 515 82-6927
ISBN 0-534-01136-5 AACR2

Preface

This book is intended to cover many of the postcalculus topics in mathematics which are useful or necessary to the modern engineer or physicist. The subject matter separates naturally into six basic areas:

Part One Ordinary Differential Equations

Part Two Vectors and Matrices

Part Three Vector Analysis

Part Four Fourier Analysis and Boundary Value Problems

Part Five Complex Analysis

Part Six Numerical Methods

Certain other topics (such as probability and statistics, or calculus of variations) could be included. However, these six areas seem to constitute the core of engineering mathematics curricula, and every effort has been made to include in them the topics which instructors feel are necessary and important.

In writing this text, I have tried to keep the two basic users in mind—the student and the instructor. First, the book is intended to serve the student. Hopefully, a student using the book for a course or reference will find it to be a genuinely helpful aid in mastering and using the material. At the same time, the book should serve the instructor's needs. It should be convenient to teach from, with the right topics in the right order, appropriate illustrations and examples, and problems that are sufficient in number and variety.

In attempting to realize these objectives, I have used the following as guidelines:

1. Topics and their order of presentation have been dictated by existing courses in engineering mathematics curricula and by suggestions of reviewers.
2. Each new concept and technique is illustrated, and examples usually include sufficient details to enable students to easily follow them.
3. Most sections end with a large number of routine drill exercises, followed by more challenging problems. Some examples and problems require the use of calculators. These can easily be omitted, but I felt that any attempt at presenting this kind of

material in a realistic setting should take into account the computational aspects which will be encountered by most practicing engineers and scientists.

4. A large number of applications are included. These begin with the construction of a mathematical model of the phenomenon under consideration, followed by the mathematics needed for the solution.

The six parts of the book are for the most part independent of one another; so an instructor may teach a portion of an engineering mathematics sequence without having taught the other parts or read many other sections of the book. To facilitate this, I now include some remarks on organization and prerequisites for each part.

PART ONE DIFFERENTIAL EQUATIONS

Chapters 0 through 7 constitute a first course in ordinary differential equations, following the usual first year of calculus. Chapter 0 is optional; Chapters 1 and 2 are basic, although certain parts, such as Bernoulli and Riccati equations, may be omitted or assigned as separate reading. Chapter 3 is not required for subsequent work, with the exception that differential operator notation developed in Section 3.6 is used in Section 7.1 for solving systems and may be delayed until then. Chapters 4 and 5 are independent and may be done in either order. Because of the importance of transform methods in modern engineering (particularly electrical engineering), more material is included on Laplace transforms than is customary in a first course. This material is needed later only for a short section in Chapter 7 on solving systems. The material on Bessel functions in Chapter 6 depends upon the series solution methods of Chapter 5. With the exception of several examples, Sturm-Liouville theory and oscillation theory covered in Sections 6.4 and 6.5 do not require anything beyond Chapter 2. Finally, in Chapter 7, one can go directly to a consideration of phase plane, critical points, and stability after treating linear systems in Section 7.1.

Part One assumes familiarity with differentiation, including implicit differentiation and the chain rule, and techniques of integration. These are usually covered in a first-year calculus course. Students are also encouraged to make efficient use of integration tables. Also needed are improper integrals (for Laplace transforms and, briefly, for the gamma function in Section 6.2 in connection with Bessel functions); power series (for series solutions in Chapter 5 and special functions in Chapter 6—power series are briefly reviewed in Section 5.1); and partial derivatives (used briefly in connection with exact differential equations and integrating factors in Sections 1.4 and 1.5).

PART TWO VECTORS AND MATRICES

In this book, vectors and matrices are covered in two chapters. Chapter 9 is devoted to the algebra and geometry of vectors. The first topics are 3-vectors as points and arrows, followed by the algebraic operations dot and cross product and scalar triple product. The approach is geometric, with some attention given to using vectors to find equations of lines and planes. These considerations are followed by n-vectors, the vector space R^n, subspaces in R^n, linear independence, bases, dimension, and, finally, a brief introduction to abstract vector spaces. The latter may be omitted in treating matrices in Chapter 10, which is organized as follows:

Algebra of matrices: Sections 10.1–10.6, 10.9

Solution of systems of linear equations: 10.7, 10.8

Determinants: 10.10, 10.11

Applications of determinants: 10.12–10.14

Eigenvalues and their applications: 10.15–10.22

Some special matrices: 10.23

Careful proofs are given for many results. However, problems are solved algorithmically, giving the student step-by-step rules for such things as solving systems of linear equations, finding a matrix inverse, or finding eigenvalues. The instructor may choose to emphasize one or both of these aspects.

Most students taking a course from Part Two will have had a year of calculus. Perhaps this is not a logical necessity, but students seem to do better with some calculus behind them. The student should have seen the notions of straight line and plane and should have had elementary trigonometry (at least sine and cosine) to understand dot and cross products. Complex arithmetic is needed for eigenvalues.

PART THREE VECTOR ANALYSIS

This part consists of just one chapter and covers vector differential and integral calculus. The organization is as follows:

Vector differential calculus, including gradient, divergence, curl, velocity, and acceleration: Sections 11.1–11.5

Vector integral calculus, including the theorems of Gauss, Green, and Stokes: 11.6–11.10, 11.14

Applications: 11.11, 11.12

Curvilinear coordinates, including divergence, curl, gradient, and the Laplacian operator written in orthogonal curvilinear coordinates: 11.13

For this material, the student needs partial differentiation, including the chain rule, and should have some experience visualizing and sketching simple three-dimensional surfaces such as the sphere, cone, and cylinder. Double and triple integrals are needed for the vector integral theorems. The student should be able to use polar coordinates to evaluate double integrals and cylindrical and spherical coordinates to evaluate triple integrals. Finally, basic vector algebra from Sections 9.1 and 9.3 is needed.

PART FOUR FOURIER ANALYSIS AND BOUNDARY VALUE PROBLEMS

Part Four consists of two chapters, one devoted to Fourier series, integrals, and transforms, and the other to their use in solving boundary value problems. The organization is as follows:

Fourier series (including multiple series) and forced oscillations: Sections 12.1–12.4, 12.8

Fourier integrals: 12.5, 12.6

Fourier transforms: 12.9, 12.10

Solution of boundary value problems by separation of variables, using Fourier series and integrals: 13.1–13.8

Fourier and Laplace transform solution of boundary value problems: 13.9, 13.10.

Most of the sections in Chapter 13 are independent of one another and may be covered in any order or omitted individually.

For Chapter 12, the student must have some experience with infinite series of functions and with improper integrals. Most students have not encountered multiple series, but they can handle multiple Fourier series if one maintains an intuitive approach.

For the most part, prerequisites for Chapter 13 consist of those parts of Chapter 12 being used for the particular problem under consideration.

PART FIVE COMPLEX ANALYSIS

Part Five is the second-longest part of the book, consisting of six chapters. The material proceeds in the standard way, roughly as follows: complex arithmetic; complex functions; differentiation and analyticity; complex integration; Cauchy's theorem and some of its consequences; Taylor and Laurent series; residues and the residue theorem; evaluation of real integrals and summation of series; conformal mappings; and additional applications.

The student needs to be familiar with the calculus of real-valued functions of two real variables, plus line integrals and Green's theorem in the plane. The student should also have seen real series of constants and real power series expansions of functions of one real variable.

PART SIX NUMERICAL METHODS

This part consists of just one chapter and outlines methods for approximating solutions of certain commonly encountered problems. The sections are independent of one another and may be treated in any order.

For Sections 20.1 through 20.6, only elementary calculus is needed. Sections 20.7 through 20.9 require some knowledge of ordinary differential equations, and Section 20.20 treats the Dirichlet problem. Section 20.11 requires that the student be familiar with matrix operations, eigenvectors, and eigenvalues.

As a further aid to the reader, Chapter 20 is followed by a list of references for supplementary reading, a collection of facts and formulas commonly needed, a guide to the theorems to help in their location when needed, and solutions to many of the odd-numbered problems. I have also included sections covering the historical development of various topics such as Fourier series and ordinary differential equations. These help put a subject in perspective and provide a sense of its development which frequently is lost in the normal modern textbook organization.

In bringing this book from conception to completion, many people have made significant contributions. Among these were the following reviewers who pointed out many errors and slips and suggested improvements:

Mathematics

Stuart Black
California State University
Long Beach

J. Michael Bossert
California State University
Sacramento

Jerald P. Dauer
University of Nebraska

Robert Fennel
Clemson University

Charles N. Friedman
University of Texas

Michael Gregory
University of North Dakota

Euel Kennedy
California State Polytechnic
University

Myren Krom
California State University
Sacramento

Doug Moore
University of California
Santa Barbara

Thomas O'Neil
California State Polytechnic
University

Raymond D. Terry
California State Polytechnic
University

Joseph Verdina
California State University
Long Beach

Michael Williams
Virginia Polytechnic Institute
and State University

Electrical Engineering

Arwin A. Dougal
University of Texas

Charles P. Newman
Carnegie-Mellon University

Mechanical Engineering

Kenneth Krieger
University of New Orleans

J. P. Vanyo
University of California
Santa Barbara

Eileen Schauer typed parts of the manuscript and helped with the photocopying, and the editorial staff at Wadsworth, particularly mathematics editor Rich Jones and Marta Kongsle, gave me good assistance throughout the project. I owe an enormous debt of gratitude to Tom O'Neil for his outstanding contribution to this book.

No doubt some mistakes and misprints will manage to survive the various proofreadings and corrections. I would appreciate having these called to my attention. I would also welcome suggestions for changes or improvements, both from instructors and from students who have used the book.

Contents

PART ONE

DIFFERENTIAL EQUATIONS

CHAPTER ZERO

Introductory Remarks

Many problems in the physical sciences and engineering are posed mathematically in terms of equations involving derivatives of the function which expresses the relationship one seeks. Such an equation is called a *differential equation* and is termed *ordinary* if the derivatives are of functions of one variable, *partial* if partial derivatives occur.

Basically, differential equations arise because our interaction with nature is initially in the form of observations of objects or systems in motion. Hence, we observe and measure rates of change, or derivatives; formulate conjectures about how the various quantities involved relate to one another; test these conjectures by experiment; and finally derive the differential equations describing the process of interest. The problem then is to solve the differential equation. Here are several examples of differential equations as descriptions of natural phenomena.

EXAMPLE 1 Electrical Circuits

Suppose we are interested in flow of current through a series circuit, such as shown in Figure 1. Customarily, current is thought of as a function $I(t)$ of time t. Suppose the resistance R, the capacitance C, and the inductance L are all constant, and that at time t the capacitor has total charge $Q(t)$. (If I is measured in amperes, then R is in ohms, C in farads, L in henrys, and Q in coulombs.)

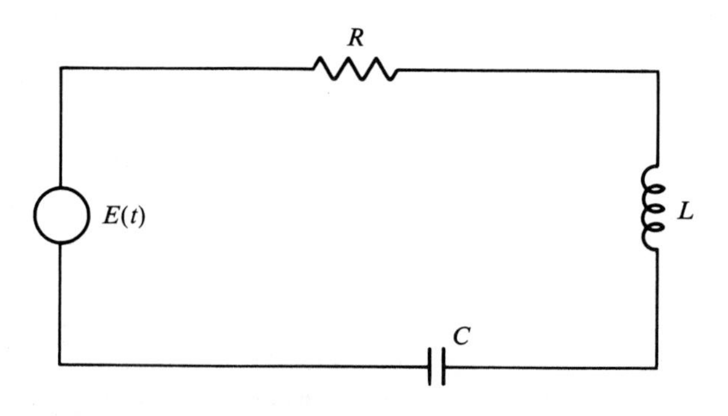

Figure 1. *RLC* circuit.

Experiment tells us that

$$\text{voltage drop across the resistance} = IR$$

$$\text{voltage drop across the capacitor} = \frac{Q}{C}$$

and

$$\text{voltage drop across the inductance} = L\frac{dI}{dt}.$$

Further, Kirchhoff's second law for circuits says that the impressed voltage $E(t)$ is the sum of the voltage drops in the circuit. Thus,

$$L\frac{dI}{dt} + RI + \frac{Q}{C} = E(t),$$

a differential equation involving I and Q. Usually we know E, R, L, and C, and attempt to solve for Q. To eliminate I and have an equation just for Q, we have

$$I = \frac{dQ}{dt}.$$

Substituting this into the differential equation gives us

$$L\frac{d^2Q}{dt^2} + R\frac{dQ}{dt} + \frac{1}{C}Q = E,$$

a differential equation involving only Q and its derivatives as unknowns. We would solve this for Q, then obtain I as dQ/dt.

This is an ordinary differential equation (no partial derivatives), and is of second order (this is the order of the highest derivative occurring in the differential equation).

EXAMPLE 2 Free Oscillations

Suppose we suspend a ball from a spring, pull down on the ball, and then release it. Can we describe the resulting motion?

To analyze a problem such as this, we must consider the forces acting on the ball. Imagine the spring as in Figure 2(a). In its unstretched state, it has a length L. If we hang a ball of mass m from the spring and leave the system in equilibrium, the spring stretches an amount d, so that the ball is now $L + d$ from the ceiling [Figure 2(b)]. We measure vertical displacement of the ball from $L + d$ as a convenience [Figure 2(c)]. Thus, $L + d$ is $y = 0$, and y is chosen positive in the downward direction, negative upward. At $L + d$, the spring is in *static equilibrium.*

Here are the forces acting on the ball:

1. The magnitude of the attraction of gravity is mg, where g is the constant magnitude of acceleration due to gravity, and is about 980 centimeters/second^2 or 9.8 meters/second^2.
2. Hooke's law says that the magnitude of the restoring force of the spring is proportional to the distance stretched. Experiments verify this. The constant of proportionality k is called the *spring modulus,* and varies from one spring to another. The larger the value of k, the stiffer the spring.

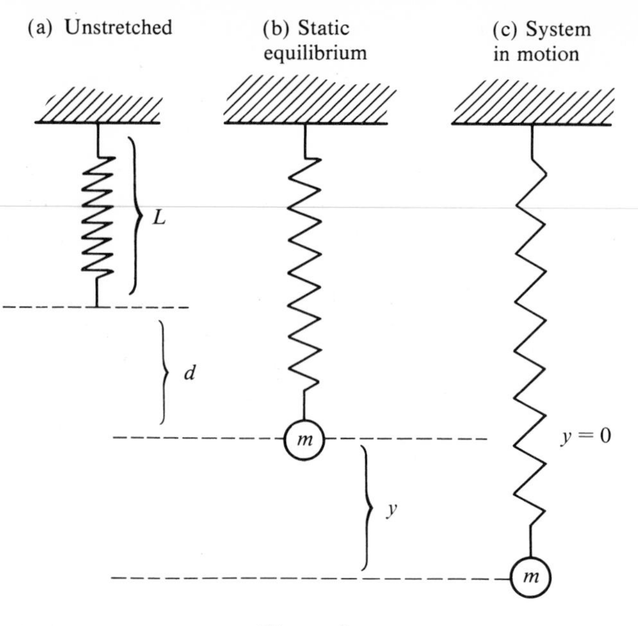

Figure 2

At static equilibrium, this force is $-kd$ (minus because the spring tends to pull the ball upwards). If the ball is pulled down a distance y from static equilibrium, an additional force $-ky$ is exerted on the ball. Thus, the total force on the ball from the spring is

$$-kd - ky.$$

Adding the forces due to gravity and the spring, we have a total force of

$$mg - kd - ky.$$

In static equilibrium, $y = 0$, and all forces balance. Hence,

$$kd = mg.$$

Thus, the total force becomes $mg - mg - ky$, or just $-ky$.

Now consider two cases:

1. Undamped system Here we imagine that damping effects (such as air resistance) are negligible. Using Newton's second law (force equals mass times acceleration), with t = time and d^2y/dt^2 the acceleration, we have

$$m\frac{d^2y}{dt^2} = -ky.$$

Thus, the behavior of y with time is governed by the second order differential equation

$$m\frac{d^2y}{dt^2} + ky = 0.$$

As we shall see later, this has solutions of the form

$$y(t) = A\cos\left(\sqrt{\frac{k}{m}}\,t\right) + B\sin\left(\sqrt{\frac{k}{m}}\,t\right),$$

where A and B are constants which must be determined from additional data of the problem (such as initial position and velocity of the ball). Thus, in the undamped case, the ball exhibits a periodic, up and down motion, often called a *harmonic oscillation.*

2. Damped system If the ball is connected to a dashpot, as in Figure 3, then a new force comes into play, tending to damp out the motion. Experiment shows that the damping force is proportional to the velocity dy/dt. Thus, the total force is now $-ky - c(dy/dt)$ for some constant c, called the *damping constant*. The equation for y, from Newton's second law, is now

$$m\frac{d^2y}{dt^2} = -ky - c\frac{dy}{dt}$$

or

$$m\frac{d^2y}{dt^2} + c\frac{dy}{dt} + ky = 0.$$

Solutions of this are more complicated than in the undamped case, and the resulting motion will be overdamped, underdamped, or critically damped, depending on the relative sizes of m, c, and k. We shall treat these in detail in Section 2.6 when we have learned to solve certain second order differential equations.

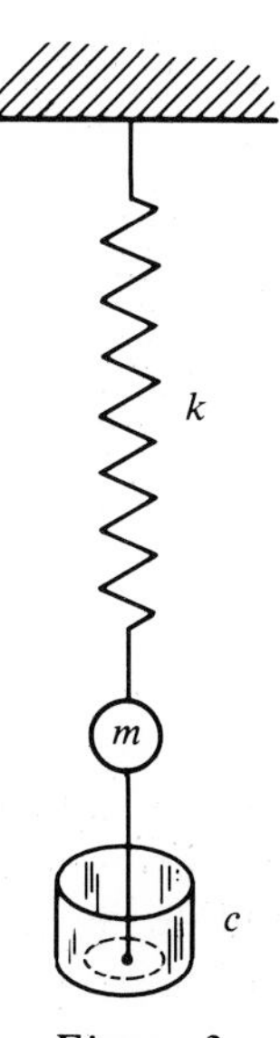

Figure 3. Damped spring system.

EXAMPLE 3 Radioactive Decay

Experiments indicate that radioactive elements, such as radium, decay at a rate proportional to the mass present at any given time. We would like to write a formula to compute this mass at any time.

Let $M(t)$ be the mass at time t. If the mass decreases at a rate proportional to itself, then for some constant k, which depends on the element,

$$\frac{dM}{dt} = -kM.$$

This is a first order differential equation for M, and happens to be an easy one to solve. Write it as

$$\frac{dM}{M} = -k\,dt$$

and integrate* to get

$$\ln(M) = -kt + c,$$

* Actually $\int dM/M = \ln|M|$, but mass is positive and so $\ln|M| = \ln(M)$.

where c is a constant to be determined. Then

$$M(t) = e^{-kt+c} = e^c e^{-kt} = Ae^{-kt},$$

with e^c written as A for convenience.

How do we find A? We must have measured the mass at some time t_0. Suppose $M(t_0) = m$. Then

$$M(t_0) = Ae^{-kt_0} = m;$$

so

$$A = me^{kt_0}.$$

Then

$$M(t) = me^{kt_0}e^{-kt} = me^{-k(t-t_0)}$$

gives the mass at any time t.

EXAMPLE 4 Motion of a Falling Body

Consider the problem of describing the motion of an object falling under the influence of gravity. Assuming that such things as wind resistance are negligible, we begin with the experimentally observed fact that acceleration due to gravity is a constant, which we denote g. Let $y(t)$ be the height of the object at time t, as in Figure 4. Then

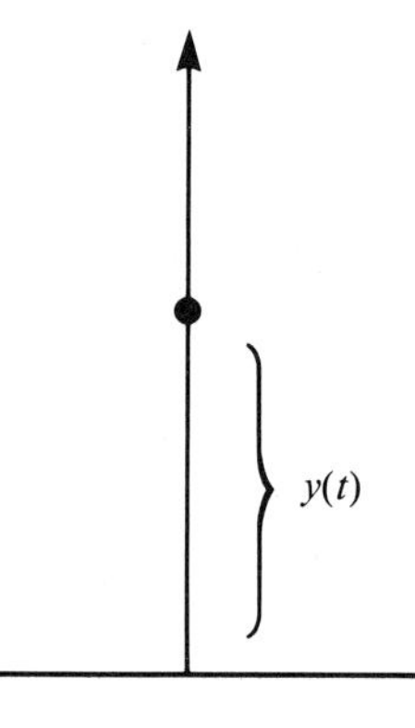

Figure 4

$$\text{acceleration} = \frac{d^2y}{dt^2} = g.$$

Integrating gives us

$$\frac{dy}{dt} = gt + k$$

for some constant k. Integrating again gives us

$$y(t) = g\frac{t^2}{2} + kt + C,$$

where C is a second constant of integration.

To determine the integration constants, note that

$$y(0) = C.$$

Thus, C is the height of the object at time 0.

Next,

$$y'(0) = k,$$

so that k is the velocity at time 0.

Thus, the position of the object at any time $t > 0$ is given by

$$y(t) = g\frac{t^2}{2} + y'(0)t + y(0),$$

which is completely known if we know the initial position $y(0)$ and the initial velocity $y'(0)$.

EXAMPLE 5 Simple Harmonic Motion of a Pendulum

Suppose we suspend a ball of mass m at the end of a rod of length L and set it in motion swinging back and forth. We would like to write an equation describing the resulting motion of the ball.

Let θ be the angle of displacement from the vertical, as in Figure 5. Here, θ varies with time t and is measured in radians. The tangential component of the weight mg is $-mg\sin(\theta)$ (minus because this force tends to restore the pendulum to equilibrium). The displacement s from the equilibrium position is given by $s = L\theta$. Thus, the acceleration d^2s/dt^2 is equal to

$$L\frac{d^2\theta}{dt^2}.$$

We assume that air resistance and the mass of the rod are negligible. Then Newton's second law tells us that

$$mL\frac{d^2\theta}{dt^2} = -mg\sin(\theta)$$

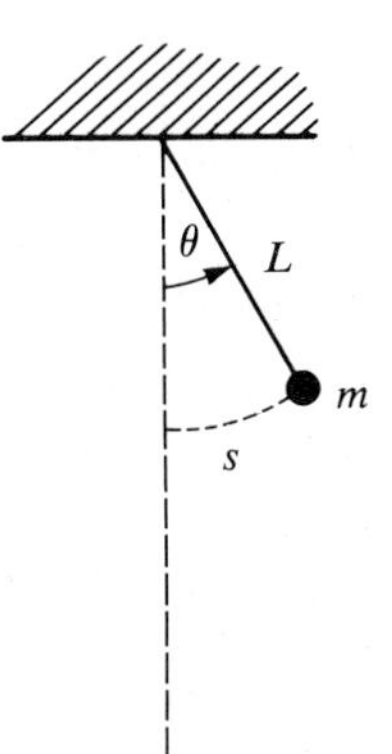

Figure 5.
Simple pendulum.

or

$$\frac{d^2\theta}{dt^2} + \frac{g}{L}\sin(\theta) = 0,$$

a second order differential equation for θ. Note that the mass m has cancelled out, implying that the resulting motion does not depend on the mass of the ball. This surprising result was observed by Galileo in the early seventeenth century. (Incidentally, Galileo was also the first to observe that falling objects accelerate at a constant rate. The story—probably apocryphal—is that he did this by observing objects dropped from the leaning tower of Pisa.)

These are just five simple but important examples of differential equations, but they are enough to suggest several questions. Given a differential equation, how do we know whether there is a solution? How many solutions might there be, and how are they related? How do we find a solution? If we cannot find a solution, can we approximate one numerically? Or, can we find by other means some properties of solutions? (For example, are some solutions periodic; do some solutions grow or decay with time?)

Our object in the next seven chapters is to begin answering some of these questions. In the course of this, we shall solve the differential equations arising in

Examples 1 through 4, and also see that the pendulum equation of Example 5 is astonishingly difficult, considering the apparent simplicity of the phenomenon it describes. We shall apply series methods to its solution in Section 5.2.

We shall now ease into the subject by beginning with some first order differential equations we can solve in a fairly straightforward manner, and then work toward some of the more difficult equations arising in physical applications.

PROBLEMS FOR CHAPTER 0

In each of the following problems, a physical process is described, together with certain observations which have been experimentally verified, and, in some cases, simplifying assumptions. In the spirit of Examples 1 through 5, derive a differential equation governing the process. Be clear in labeling and defining the variables and in explaining the basis for each step of the derivation. Use diagrams where possible to illustrate the variables.

1. *Velocity of a bullet.* A bullet weighing one ounce is shot vertically upward from the earth's surface. The muzzle velocity is 1 500 feet/second. The forces acting on the bullet are (1) air resistance, which is about $v^2/1000$ (where v is velocity), and (2) gravitational pull of the earth. Neglecting other forces (such as wind), find a differential equation for the velocity $v(t)$ at time t.
2. *Newton's law of cooling.* Newton verified experimentally that the surface temperature of an object changes at a rate proportional to the difference between the temperature of the object and that of the surrounding medium, say air. Assuming that the air temperature is constant, write a differential equation for the temperature of an object cooling in air according to Newton's law.
3. *Compound interest.* A man invests \$4000 at 6% interest, compounded continuously. Write a differential equation for the amount he will have in his account at any later time (assuming no withdrawals). Solve the equation by integrating. At what time t will he have become a millionaire?
4. *Motion on an inclined plane.* A 50-pound block is released from rest at the top of an inclined plane making an angle of 30° with the horizontal. Air resistance is $\frac{1}{2}v$, where v is the velocity, and the coefficient of friction is 0.34. The force of friction on the object is then of magnitude 0.34 N, where N is the force exerted by the surface on the object perpendicular to the slide. Assuming that weight, air resistance, and friction are the only forces acting on the block, write a differential equation for its velocity $v(t)$ at time t.
5. *Population modeling.* Assume that, at any time t, both the birth rate and the death rate in the United States are constant multiples of the population at that time (not necessarily the same constants). Write a differential equation for the population at time t, assuming zero immigration and emigration.
6. *Mixing of salt in water.* At an initial time $t = 0$, S_0 pounds of salt were dissolved in 500 gallons of water stored in a tank. At each time $t > 0$, 2 pounds/gallon of salt are poured into the tank at a rate of 6 gallons/minute. The solution is stirred to dissolve the salt, and the brine solution is drawn off from a tap at the rate of 6 gallons/minute. Find a differential equation for the amount of salt in the tank at any time $t > 0$. Solve the resulting differential equation by direct integration. Using this solution, describe what happens to the amount of salt in the tank as $t \to \infty$.
7. *Acid drainage.* A hemispherical tank of radius R is filled with acid, which is pouring under the influence of gravity out of a circular hole of radius r at the bottom of the tank. Assume Torricelli's law,* that the velocity $v(t)$ of fluid flowing from the opening is proportional to $\sqrt{2gh(t)}$, where g = constant acceleration due to gravity and $h(t)$ = height at time t of fluid level above the opening. Derive a differential equation for the depth

* Evangelista Torricelli (1608–1647) was a student of Galileo. His "law" holds if the opening is not "too small."

of the acid at any time. Solve this differential equation by integrating, and determine (in terms of R and r) how long it will take the tank to drain completely.

8. Redo Problem 7, with the additional assumption that a float valve has been installed at the bottom opening, making the opening at any time t of radius equal to one-fourth the depth of the acid at time t.
9. *Boyle-Mariotte gas law.* At constant temperature and low pressure p, the rate of change of the volume V of a gas with the pressure has been observed to be proportional to $-V/p$. Determine a differential equation for V in terms of p; then solve the differential equation by integrating.
10. A ball of mass m is thrown upward from the surface of the earth. The forces acting on the ball are the constant acceleration due to gravity, and air resistance, which is proportional to the velocity. Write a differential equation for the motion of the ball.
11. A light beam directed downward into the ocean is partially absorbed as it passes through the water. Its intensity decreases at a rate proportional to its intensity at any given depth. Write a differential equation for the intensity as a function of depth.
12. The thickness of the ice on a frozen lake increases at a rate proportional to the square root of the time. Write a differential equation for the thickness at any time t.
13. A spherical raindrop falling through a cloud accumulates moisture as it falls. Assume that its volume increases with respect to distance fallen at a rate proportional to its cross-sectional area at that time. Derive a differential equation for the radius of the drop.
14. Derive a differential equation for the motion of a block of wood floating in water, assuming that it bobs up and down so that its top and bottom remain horizontal, and the sides vertical.*
15. Derive the differential equation

$$\frac{d^2\theta}{dx^2} + \frac{2}{x}\frac{d\theta}{dx} + \frac{g}{v}\frac{1}{x}\theta = 0$$

for a simple pendulum whose length is increased at a steady rate of v inches per minute. Here, t is time, L_0 is the initial length, and

$$x = \frac{L_0 + vt}{v}.$$

16. A block of mass m is falling through an oil mixture in which the forces acting are acceleration due to gravity, and a resistant force proportional to v, where v is the velocity at time t. Derive a differential equation for v.
17. A spherical body of radius R and mass m falls in a dense grease in which the forces acting on the sphere are (1) acceleration due to gravity, (2) a buoyant force equal to the weight of the grease displaced by the sphere, and (3) a resistive force given by $6\pi\mu Rv$, where v is the velocity of the sphere and μ is a constant called the viscosity coefficient of the grease. Find a differential equation for v.

* It will be useful here to use Archimedes' principle: An object submerged in a fluid is buoyed up by a force equal to the weight of the fluid displaced. This was used by Archimedes to determine the proportions of gold and silver in King Hiero's crown.

CHAPTER ONE

First Order Differential Equations

1.0 INTRODUCTION

A *first order differential equation* is one of the form

$$F(x, y, y') = 0,$$

involving only the dependent and independent variables and a first derivative. The example of radioactive decay in the introductory remarks involved a first order differential equation. Other examples are

$$y' + 2xy^2 - 4x^3 = 0$$

and

$$(y')^{3/2} + x^2 - \cos(xy') = 0.$$

A *solution* of a differential equation is a function which satisfies the equation. Thus, $y = f(x)$ is a solution of $F(x, y, y') = 0$ if $F(x, f(x), f'(x)) = 0$ for all x in, say, some interval. For example, $y = e^{2x}$ is a solution of $y' - 2y = 0$, since by direct substitution $(e^{2x})' - 2e^{2x} = 2e^{2x} - 2e^{2x} = 0$, in this case for all x.

A *general solution* of $F(x, y, y') = 0$ is a solution containing one arbitrary constant. For example, $y = Ce^{2x}$ is a solution of $y' - 2y = 0$ for any real number C, and hence is a general solution of $y' - 2y = 0$. In this context, we often call the solution obtained by making a specific choice of constant a *particular solution*. Thus, $y = 3e^{2x}$ and $y = -9e^{2x}$ are two particular solutions of $y' - 2y = 0$.

In many applications, we solve the problem under consideration by finding a general solution of the given differential equation and then choosing the constant to find the particular solution having the desired property. Two illustrations of this are given in Examples 6 and 7; and we shall see many more throughout the chapter.

Even with first order differential equations, unless the equation has a fairly simple form, solution may be very difficult or even impossible in terms of elementary functions. In the next seven sections we treat special kinds of first order equations for which we can often find solutions. The remainder of the chapter will then be devoted to applying these results to the electrical circuit equation of Example 1, to filling in some theoretical gaps, and to looking at the geometry inherent in first order differential equations.

PROBLEMS FOR SECTION 1.0

In each of Problems 1 through 10, determine whether or not the given function is a solution of the given differential equation; C always denotes a constant.

1. $y' = -\dfrac{2y + e^x}{2x}$; $y = \dfrac{C - e^x}{2x}$

2. $y' + y = 1$; $y = 1 + Ce^{-x}$

3. $2yy' = 1$; $y = \sqrt{x - 1}$

4. $x^2yy' = -1 - xy^2$; $y = \dfrac{4 - x^2}{2x}$

5. $y' + y = 0$; $y = Ce^{-x}$

6. $y' = \dfrac{2xy}{2 - x^2}$; $y = \dfrac{C}{x^2 - 2}$

7. $xy' = -y + x$; $y = \dfrac{x^2 - 3}{2x}$

8. $y' + 2y = 0$; $y = \sin(3x) - 4$

9. $\sinh(x)y' + y\cosh(x) = 0$; $y = \dfrac{-1}{\sinh(x)}$

10. $y' - 3y = x$; $y = \dfrac{-x}{3} + \dfrac{1}{3} + Ce^{3x}$

1.1 SEPARABLE EQUATIONS

A differential equation $F(x, y, y') = 0$ is called *separable* if it can be written in the form

$$A(x)\,dx = B(y)\,dy.$$

In this event, we can integrate the left side with respect to x and the right side with respect to y to obtain a general solution. The arbitrary constant required in a general solution is exactly the constant of integration.

One example of a separable equation was the radioactive decay equation of Example 3. Here are some others.

EXAMPLE 6

$$\frac{dy}{dx} = 3x^2 + 1$$

is separable. Write the differential equation as

$$dy = (3x^2 + 1)\,dx$$

and integrate both sides to get

$$y = x^3 + x + C,$$

where C is a constant of integration. For any constant C, $y = x^3 + x + C$ is a solution of the differential equation. Thus, $y = x^3 + x + C$ is a general solution.

Given more information, we can assign C a value. For example, if we want a solution $y(x)$ such that $y(1) = 4$, then we need

$$y(1) = 2 + C = 4,$$

and hence must choose $C = 2$. The particular solution satisfying $y(1) = 4$ is

$$y(x) = x^3 + x + 2.$$

EXAMPLE 7

$$\frac{dy}{dx} = 8x^3y^2$$

is separable if $y \neq 0$, since it can be written

$$\frac{1}{y^2}\,dy = 8x^3\,dx.$$

Integrating gives us

$$-\frac{1}{y} = 2x^4 + C,$$

where C is any constant. If we want to solve for y, we have a general solution

$$y(x) = \frac{-1}{2x^4 + C}.$$

For a solution taking on a particular value of y at a given value of x, we would have to choose C appropriately. For example, if we want a solution satisfying $y(2) = 3$, then we need

$$y(2) = \frac{-1}{32 + C} = 3$$

and we must choose $C = -\frac{97}{3}$.

EXAMPLE 8

$$\frac{dy}{dx} = \frac{3x^2 + y}{x - y}$$

is not separable. We can write

$$(x - y)\,dy = (3x^2 + y)\,dx,$$

but no amount of algebra will isolate all the x's on one side and all the y's on the other.

There is no nice test to determine easily whether or not a first order equation is separable. In practice, we usually try to separate the variables until we either succeed or become convinced that the equation is not separable.

PROBLEMS FOR SECTION 1.1

In each of Problems 1 through 20, determine whether or not the equation is separable. If it is, find the general solution; if it is not, leave it alone. Check each solution by substituting back into the differential equation.

1. $\frac{y^2}{x}\frac{dy}{dx} = 1 + x^2$

2. $x\frac{dy}{dx} = \frac{e^y}{yx}$

3. $\cos(y)\frac{dy}{dx} = \sin(x + y)$

4. $\sec(x)\frac{dy}{dx} = \tan(y)$

5. $e^{x+y}\frac{dy}{dx} = 2x$

6. $\frac{dy}{dx} = \frac{(x + 1)^2 - 2y}{y}$

7. $(x + y)\frac{dy}{dx} = 2x$

8. $\frac{dy}{dx} = \frac{\cos(x)\cos(y)}{e^x}$

9. $\ln(y^x)\frac{dy}{dx} = 3x^2y$

10. $(x + y)^2\frac{dy}{dx} = 2xy$

11. $\frac{dy}{dx} = \frac{8x + y^2 - 1}{2yx}$

12. $\frac{x}{y}\frac{dy}{dx} = \frac{2y^2 + 1}{x + 1}$

13. $(2x + y)\frac{dy}{dx} = \frac{4x^2}{y} + y + 4x$

14. $(2y + xy)\frac{dy}{dx} = 1 + \frac{4}{x} + \frac{4}{x^2}$

15. $\frac{dx}{dy} = x^3 + 3y$

16. $(x + y)^2\frac{dy}{dx} = x^2 + y^2$

17. $x\sin(y)\frac{dy}{dx} = \frac{1}{\cos(y)}$

18. $2e^xy^2\frac{dy}{dx} = x + 2$

19. $(1 - x^2)\frac{dy}{dx} = y - 1$

20. $\frac{dy}{dx} = y\frac{(x^2 - 2x + 1)}{y + 3}$

In each of Problems 21 through 30, you are given a separable differential equation, together with an extra condition the solution is to satisfy. Find the general solution of the differential equation; then use the condition to solve for C, as in Examples 6 and 7. Check your solutions.

21. $\frac{dy}{dx} = 3x^2(y + 2)$; $y(4) = 8$

22. $x\frac{dy}{dx} = y^2$; $y(3) = 5$

23. $\frac{dy}{dx} = \frac{x^2 + 2}{y}$; $y(1) = 7$

24. $y\frac{dy}{dx} = \frac{x^2}{y + 4}$; $y(3) = 2$

25. $\frac{dy}{dx} = \frac{x - 1}{y + 2}$; $y(-1) = 6$

26. $x^2\frac{dy}{dx} = \frac{1}{y}$; $y(4) = 9$

27. $\frac{dy}{dx} = \frac{e^y}{x}$; $y(1) = 4$

28. $\frac{dy}{dx} = \frac{-2\sin(x + 1)}{y^2}$; $y(\pi) = 4$

29. $\frac{dy}{dx} = \frac{-3x}{y + 4}$; $y(2) = 7$

30. $y\frac{dy}{dx} = \frac{8x + 1}{y^2}$; $y(1) = -5$

31. Radioactive elements tend to decay at a rate proportional to the amount of the element present at any given time. If $x(t)$ is the amount of the element present at time t, write a separable differential equation for $x(t)$.

(a) Suppose that P_0 pounds are present at time 0, and P_1 pounds at time $t_1 > 0$. Solve for $x(t)$ in terms of time and the constants P_0, P_1, and t_1.

(b) Find an expression for the time T such that $x(T) = \frac{1}{2}P_0$. T is called the *half-life* of the element.

(c) At what time τ will only 10% of the original amount be left? Here the original amount is the amount at time 0.

32. A bank compounds interest continuously on its savings accounts. If \$7000 is invested at time 0, and three days later the account has \$7200 in it from principal plus interest, what is the interest rate?

33. Pleasantdale Graft and Embezzlement gives 7.5% interest on savings accounts, compounded quarterly. Pleasantdale National Trust also gives 7.5%, but compounds continuously.

(a) Derive formulas for the total amount in each account after n years, assuming that \$10,000 is deposited on the first of the year.

(b) How much better off are you with PNT after five years?

34. A spherical raindrop evaporates at a rate proportional to its surface area. Suppose it was originally $\frac{1}{8}$ inch in radius, and 40 minutes later it is $\frac{1}{24}$ inch in radius.

(a) Find an expression for the radius at any time t.

(b) Find an expression for the surface area and volume at any time t.

(c) When is the raindrop reduced to $\frac{1}{1000}$ inch in radius?

(d) At what time is the surface area reduced by half from its original value?

(e) At what time is the volume reduced by $\frac{1}{2}$ from its original volume?

35. Population P is assumed to increase at a rate proportional to the population at any time t. If at time $t = 1$ there are 10,000 people, and at $t = 4$ there are 20,000 people, how many people will there be at $t = 8$? At what time T will the population be triple its value at $t = 0$?

36. From Newton's law of cooling (see Problem 2, Chapter 0), a body's heat energy changes at a rate proportional to the difference between its temperature and that of its surroundings. Assuming an environment of constant temperature u_0, find a formula for the temperature $u(t)$ at any time t, assuming that at time 0 the temperature was 12 degrees Celsius, and at time 2 it was 11 degrees Celsius.

1.2 SOME APPLICATIONS OF SEPARABLE DIFFERENTIAL EQUATIONS

Here are three examples in which separable differential equations model real-world phenomena.

EXAMPLE 9 Fall of a Parachutist

Imagine a parachutist falling toward earth at a speed $v(t)$, with t = time. We would like an explicit expression for v, assuming that the parachute opens at the beginning of the fall.

The forces acting on the parachutist are primarily the downward pull of gravity, and the drag caused by the parachute. The gravitational pull is mg, where m is the mass and g is the acceleration due to gravity. To determine the drag of the chute, we turn to experiments on air resistance which tell us that the resistant force is proportional to v^2, and hence equal to kv^2 for some constant k. Thus, assuming other factors to be negligible, the total force acting on the chutist is

$$mg - kv^2.$$

In this equation, we chose down as the positive direction; the drag then operates in the opposite direction, hence the plus sign before mg and minus sign in front of kv^2.

Now apply Newton's second law:

$$\text{mass} \times \text{acceleration} = \text{force}.$$

Here, acceleration is dv/dt, the rate of change of velocity with time. Thus, Newton's law becomes in this case

$$m\frac{dv}{dt} = mg - kv^2.$$

Thus far we have obtained a differential equation for v. Now we want to solve for v. The equation is separable, since we can write it as

$$\frac{m\,dv}{mg - kv^2} = dt$$

(remember that m, g, and k are constants).

The rest is integration technique. To make the integration easier, write the equation as

$$\frac{m\,dv}{k[(mg/k) - v^2]} = dt$$

or

$$\frac{dv}{v^2 - (mg/k)} = -\frac{k}{m}\,dt.$$

Now, by using partial fractions (found in the "techniques of integration" section of most calculus books), we have

$$\frac{1}{[v^2 - (mg/k)]} = \frac{1}{(v + A)(v - A)} = \frac{1}{2A}\left(\frac{1}{v - A} - \frac{1}{v + A}\right),$$

where $A = \sqrt{mg/k}$. Thus, we have

$$\frac{1}{2A}\left(\frac{dv}{v - A} - \frac{dv}{v + A}\right) = -\frac{k}{m}\,dt.$$

Integrating yields

$$\frac{1}{2A}[\ln(v - A) - \ln(v + A)] = -\frac{k}{m}t + C,$$

where C is a constant of integration. This is the same as

$$\frac{1}{2A}\ln\left(\frac{v - A}{v + A}\right) = -\frac{k}{m}t + C.$$

To solve for v, write

$$\ln\left(\frac{v-A}{v+A}\right) = \frac{-2Ak}{m}t + 2AC;$$

so

$$\frac{v-A}{v+A} = e^{-2Akt/m} \cdot e^{2AC} = Be^{-2Akt/m},$$

where $B = e^{2AC}$ is also constant.

Solving for v finally gives us

$$v(t) = \frac{A(1 + Be^{-2Akt/m})}{1 - Be^{-2Akt/m}}.$$

Putting $A = \sqrt{mg/k}$ back into this solution gives us

$$v(t) = \frac{\sqrt{mg/k}\,(1 + Be^{-2\sqrt{gk/m}t})}{1 - Be^{-2\sqrt{gk/m}t}}.$$

Even before we determine the constants B and k, this formula yields an interesting conclusion: As $t \to \infty$, $e^{-2\sqrt{gk/m}t} \to 0$; hence $v(t) \to \sqrt{mg/k}$ as a limiting value, regardless of the initial velocity of the parachutist. Practically speaking, after sufficiently long, the parachutist will settle into a constant-velocity fall. This limiting velocity does depend upon m, however, and so will vary from one person to the next.

Now for the constants. We know that g is about 9.8 meters/second^2. For the earth's atmosphere and a standard parachute, k is about 30 kilograms/meter. Let us suppose that the mass (parachutist plus equipment) is 80 kilograms, just to be specific. This leaves the constant of integration B, and we must have more information to determine it. Suppose that the velocity at time 0 is 15 meters/second. Then

$$v(0) = 15 = \sqrt{\frac{mg}{k}}\,\frac{(1 + B)}{1 - B}.$$

Hence,

$$B = \frac{15 - \sqrt{mg/k}}{15 + \sqrt{mg/k}},$$

about 0.492. Thus, under the given conditions,

$$v(t) = (5.112)\left(\frac{1 + 0.492e^{-3.83t}}{1 - 0.492e^{-3.83t}}\right).$$

This person's limiting velocity is about 5.112 meters/second.

EXAMPLE 10 An Electrical Circuit Equation

Electrical circuit equations (see Example 1) are generally not separable. For a circuit such as in Figure 6, however, the equation for I is

$$L\frac{dI}{dt} = E(t),$$

and this can be written

$$\frac{dI}{dt} = \frac{1}{L} E(t).$$

A fairly common form for the impressed voltage is

$$E(t) = A \cos(\omega t),$$

where A and ω are constants. Then $dI/dt = (A/L) \cos(\omega t)$, and an integration yields

$$I(t) = \frac{A}{\omega L} \sin(\omega t) + C.$$

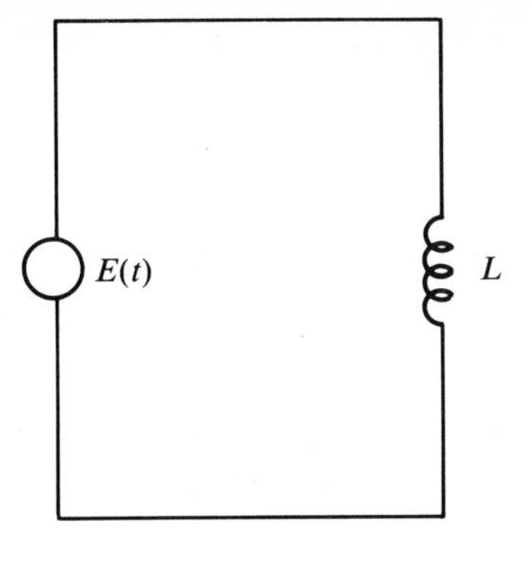

Figure 6

We could determine C if we knew, say, $I(0)$. If, for example, $I(0) = 0$, then $C = 0$ and $I(t) = (A/\omega L) \sin(\omega t)$.

EXAMPLE 11 Melting

Assume that a sphere of ice melts at a rate proportional to its surface area. We would like an expression for the volume at any time t.

Let $V(t)$ be the volume at time t. We interpret melting as the rate of change of volume with time, i.e., dV/dt. If $r(t)$ is the radius at time t, then for some constant of proportionality k (which must be found experimentally),

$$\frac{dV}{dt} = k \cdot (\text{surface area}) = k(4\pi r^2).$$

Now, this equation involves two unknowns, V and r, so we try to eliminate r. We know that, at any time t,

$$V = \tfrac{4}{3}\pi r^3.$$

(This assumes symmetrical melting so that the ice retains a spherical shape.) Then,

$$r = \left(\frac{3V}{4\pi}\right)^{1/3}.$$

Then,

$$\frac{dV}{dt} = 4\pi k \left(\frac{3V}{4\pi}\right)^{2/3} = (4\pi)^{1/3} 3^{2/3} k \cdot (V)^{2/3}.$$

This is a separable equation, since we can write it as

$$\frac{dV}{V^{2/3}} = (4\pi)^{1/3} 3^{2/3} k \, dt.$$

Then,

$$3V^{1/3} = (4\pi)^{1/3} 3^{2/3} kt + C,$$

where C is a constant of integration. This can be solved for V to yield

$$V = \left[\left(\frac{4\pi}{3} \right)^{1/3} kt + C^* \right]^3,$$

where C^* is $C/3$ and is still an arbitrary constant. Approximately, we have

$$V = (1.612kt + C^*)^3.$$

To determine C^*, we would need to know V at some time t_0.

These are but three examples of separable differential equations arising under natural circumstances. There are many others, some of which we pursue in the problems.

PROBLEMS FOR SECTION 1.2

1. Redo Example 9 (falling parachutist), assuming that the air resistance is proportional to v instead of v^2. Derive the differential equation, solve it (using the same constants as in Example 9), and discuss the significance of the solution [is there still a limiting velocity; how does T_0 influence the fall; compare the velocity with which the person strikes the earth under the two extremes $T_0 = 0$ (chute opens at point of jump) and T_0 = time of descent (chute never opens)].
2. In Example 10, solve for and discuss the behavior of the current under the conditions that the electromotive force is constant and the initial current $I(0)$ is 4 amperes. Redo the problem with $E(t) = 2\cos(\omega_1 t) + 4\sin(\omega_2 t)$ and $I(0) = 3$ amperes.
3. *Population equation.* Suppose that the population P of bacteria in a culture at time t changes at a rate proportional to $P - P^3$. Solve for P at any time t, assuming that $P > 1$ for all t.
4. *Moment of inertia of a crankshaft.* Suppose that the crankshaft in a diesel engine revolves at R radians per second. The torque applied to the crankshaft by the gases is called the *throttle torque* and is of the form kx, where x is the coordinate of throttle position and k is constant. If I is the moment of inertia of the crankshaft, the inertial torque acting *against* the throttle torque is $I(dR/dx)$. Also opposing the throttle torque is a damping torque of the form cR. Thus, using Newton's law in torsional form,

$$(\text{moment of inertia}) \cdot (\text{angular acceleration}) = \text{torque},$$

we have

$$I \frac{dR}{dx} + cR = kx.$$

If $c \neq 0$, this is not separable (it is, however, linear, and can be solved by the method of Section 1.6). If $c = 0$, solve the equation. Evaluate the constant of integration by assuming that $R(0) = 0$.
5. *Pressure in a perfect gas.* For an adiabatic process involving a perfect gas, pressure p is related to volume v by

$$\frac{dp}{dv} = -\left(\frac{c_p}{c_v} \right) \frac{p}{v},$$

where c_p = specific heat at constant pressure and c_v = specific heat at constant volume. Solve for pressure as a function of volume.

6. *Blackbody radiation.* Consider thermal energy radiated by a blackbody. If $u =$ energy density and $T =$ absolute temperature, then it was found experimentally that

$$\frac{du}{dT} = \frac{4u}{T}.$$

Solve for u as a function of T. (The Stefan-Boltzmann law says that the rate at which thermal energy is radiated by a blackbody of surface area A and Stefan-Boltzmann constant σ is $\sigma A T^4$. How is this result related to the solution for u above?)

7. Solve the differential equation for the velocity of a bullet fired vertically (see Problem 1, Chapter 0). Assume that the force due to gravity is negligible.
8. Solve for the temperature of an object cooling in air according to Newton's law (see Problem 2, Chapter 0).
9. Solve for the volume V of a gas as a function of the pressure p if the rate of change of the volume with respect to the pressure is proportional to $-V/p^2$.
10. Suppose, in Example 9, the resistant force is proportional to v^4, not v^2. Write the new differential equation for the falling parachutist and find a general solution. How would the parachutist's motion differ, if at all, from that found in Example 9 where a constant terminal velocity was reached? Is this a reasonable model?
11. Solve the current equation of Example 10 if the electromotive force is given by $E(t) = 1 - e^{-t}$. Discuss what happens to the current as time increases.
12. Solve the current equation of Example 10 if the electromotive force is $e^{-t} \cos(4t)$. What happens to the current as time increases?
13. Solve the differential equation for the radius of a falling raindrop (Problem 13, Chapter 0).
14. Solve the differential equation for the light intensity in Problem 11, Chapter 0.
15. Solve the differential equation for the thickness of ice discussed in Problem 12, Chapter 0.
16. Realistic models of population growth usually assume a saturation value, that is, some value beyond which the population cannot grow, due to limitations in space, food, natural resources, and other factors. Let $P(t)$ be the population at time t, and suppose that P satisfies

$$\frac{dP}{dt} = kP(A - P),$$

in which k and A are constants.

(a) Solve for $P(t)$.

(b) Show that $\lim_{t\to\infty} P(t) = A$. Thus, A is the saturation constant.

(c) What is the significance of k?

17. Let H and K be chemicals which react together. At time t, let $x(t)$ and $y(t)$ denote the concentrations (say, in moles per liter) of H and K. Assume that one molecule of H combines with one molecule of K to form a new product, and let $z(t)$ be the amount by which $x(t)$ and $y(t)$ have decreased in time t.

(a) Explain why $x(t) = x(0) - z(t)$ and $y(t) = y(0) - z(t)$.

(b) Under constant temperature, the rate of reaction dz/dt satisfies

$$\frac{dz}{dt} = kxy$$

for some constant k. Use this, together with (a), to solve for $z(t)$ and hence also for $x(t)$ and $y(t)$.

(c) Find expressions for the times T_x and T_y it takes for the original amounts of H and K, respectively, to be halved.

18. Suppose, in Problem 17, two molecules of K combine with one of H to form a new product.

(a) Explain why $x(t) = x(0) - z(t)$ and $y(t) = y(0) - 2z(t)$.

(b) Experiment shows that

$$\frac{dz}{dt} = kxy^2$$

for some constant k. Use this and the result of (a) to solve for z, and hence for x and y at any time t.

(c) Find an expression for the time it takes for the original amount of H to be halved, and for the time it takes for the original amount of K to be halved.

19. Suppose, in Problem 17, α molecules of H combine with β of K to form a new product. Now we find by experiment that

$$\frac{dz}{dt} = kx^\alpha y^\beta.$$

Write a differential equation for $z(t)$, noting that $x(t) = x(0) - \alpha z(t)$ and $y(t) = y(0) - \beta z(t)$. What prevents you from solving this differential equation, as you did in Problem 17, where $\alpha = \beta = 1$, and Problem 18, where $\alpha = 1$ and $\beta = 2$?

20. Suppose that a chemical H breaks down when an electric current is passed through it. Assume that the reaction occurs according to the following rule: The amount $x(t)$ of H decreases with time at a rate proportional to the α power of x. Write and solve a differential equation for the amount $x(t)$ present at time t.

21. In Example 9, we integrated $dV/(v - A) - dV/(v + A)$ to obtain $\ln(v - A) - \ln(v + A)$. This tacitly assumed that $v - A$ and $v + A$ are positive. Is this assumption justified? Carry out the details of solving for v if we carry out the integration to get $\ln|v - A| - \ln|v + A|$.

22. Write a differential equation for the velocity of the falling parachutist in the following model: The parachutist falls for T seconds, experiencing a drag proportional to v, then opens the parachute, experiencing from then on a drag proportional to v^2. Try to solve the resulting differential equation.

1.3 HOMOGENEOUS AND "NEARLY HOMOGENEOUS" EQUATIONS

Most first order equations we encounter are not separable. In some cases, a change of variables can be chosen which results in an equation which is separable. Here is one such case, together with the change of variables to use.

Suppose the differential equation is of the form

$$\frac{dy}{dx} = f\left(\frac{y}{x}\right).$$

That is, dy/dx is isolated on one side, and on the other is some expression of (y/x)'s. Such an equation is called *homogeneous*. Homogeneous differential equations can be approached as follows:

Put $u = y/x$, and use this to eliminate y. Since $y = ux$, $dy/dx = x(du/dx) + u$, and the differential equation becomes

$$x\frac{du}{dx} + u = f(u).$$

This is separable, since we can write it as

$$\frac{du}{f(u) - u} = \frac{1}{x}\, dx.$$

We can (in principle) solve this for u, and then obtain y as ux. In practice, the integration on the left can turn out to be difficult, or even impossible, in terms of elementary functions.

EXAMPLE 12

Solve

$$x \frac{dy}{dx} = \frac{y^2}{x} + y.$$

This is not separable, but we recognize it as

$$\frac{dy}{dx} = \left(\frac{y}{x}\right)^2 + \frac{y}{x} = u^2 + u,$$

with $u = y/x$. The equation in terms of u and x is

$$u + x \frac{du}{dx} = u^2 + u.$$

This is separable, and can be written as

$$\frac{du}{u^2} = \frac{dx}{x}.$$

Integration gives us

$$-\frac{1}{u} = \ln|x| + C.$$

Hence,

$$u = \frac{-1}{\ln|x| + C},$$

with C an arbitrary constant. A general solution for y can be written

$$y = xu = \frac{-x}{\ln|x| + C},$$

whenever $\ln|x| + C \neq 0$.

EXAMPLE 13

Solve

$$x^3 \frac{dy}{dx} = x^2 y - 2y^3.$$

Upon dividing by x^3, this becomes

$$\frac{dy}{dx} = \frac{y}{x} - 2\left(\frac{y}{x}\right)^3.$$

With $u = y/x$, this is

$$x\frac{du}{dx} + u = u - 2u^3.$$

Then,

$$\frac{du}{-2u^3} = \frac{dx}{x},$$

and we integrate to obtain

$$\frac{1}{4u^2} = \ln|x| + C.$$

This gives us

$$\frac{x^2}{4y^2} = \ln|x| + C$$

or

$$y = \sqrt{\frac{x^2}{4(\ln|x| + C)}} = \frac{|x|}{2\sqrt{\ln|x| + C}},$$

a general solution for y. If we want to satisfy an additional condition, then we must choose C accordingly. For example, if we want a solution such that $y(1) = 6$, then we must have

$$6 = \frac{1}{2\sqrt{C}},$$

and hence must choose $C = \frac{1}{144}$. The solution satisfying $y(2) = 6$ is then

$$y = \frac{|x|}{2\sqrt{\ln|x| + \frac{1}{144}}}.$$

It is not always immediately apparent when an equation is of the form $dy/dx = f(y/x)$. In practice, a few algebraic manipulations usually tell us one way or the other. Problem 17 at the end of this section gives a test which is sometimes useful.

We are now also able to handle differential equations of the form

$$\frac{dy}{dx} = f\left(\frac{ax + by + c}{dx + ey + h}\right),$$

in which a, b, c, d, e, and h are constants. This can be written

$$\frac{dy}{dx} = f\left(\frac{a + b(y/x) + (c/x)}{d + e(y/x) + (h/x)}\right),$$

which is homogeneous when $c = h = 0$ and nonhomogeneous when $c \neq 0$ or $h \neq 0$. In the latter case, the equation is "near enough" to being homogeneous in that a change of variables will still enable us to solve it. Consider, then, the differential equation

$$\frac{dy}{dx} = f\left(\frac{ax + by + c}{dx + ey + h}\right),$$

and look at two cases.

Case 1 Suppose that $ae - bd \neq 0$. Now change variables from x and y to X and Y by setting

$$x = X + A \quad \text{and} \quad y = Y + B,$$

with A and B to be chosen later. Then

$$\frac{dy}{dx} = \frac{dY}{dX}$$

and the differential equation becomes

$$\frac{dY}{dX} = f\left(\frac{a(X + A) + b(Y + B) + c}{d(X + A) + e(Y + B) + h}\right) = f\left(\frac{aX + bY + (aA + bB + c)}{dX + eY + (dA + eB + h)}\right).$$

Now choose A and B so that

$$aA + bB + c = 0$$
$$dA + eB + h = 0.$$

Since a, b, c, d, e, and h are known constants, and $ae - bd \neq 0$, there exist A and B satisfying these equations. With these choices, we have

$$\frac{dY}{dX} = f\left(\frac{aX + bY}{dX + eY}\right),$$

a homogeneous equation in X and Y which we can solve by the method given above. This solution in terms of X and Y gives a solution of the original differential equation upon putting $X = x - A$ and $Y = y - B$.

Before going on to case 2, we shall look at an illustration of case 1.

EXAMPLE 14

Solve

$$\frac{dy}{dx} = \left(\frac{2x + y - 1}{x - 2}\right)^2.$$

This is not homogeneous but is of the form under consideration, with $a = 2$, $b = 1, c = -1, d = 1, e = 0, h = -2$, and $f(t) = t^2$. Set

$$x = X + A \quad \text{and} \quad y = Y + B$$

to get

$$\frac{dY}{dX} = \left(\frac{2X + Y + (2A + B - 1)}{X + (A - 2)}\right)^2.$$

Choose A and B so that

$$2A + B - 1 = 0$$

and

$$A - 2 = 0.$$

Then $A = 2$ and $B = -3$, and we have

$$\frac{dY}{dX} = \left(\frac{2X + Y}{X}\right)^2 = \left(2 + \frac{Y}{X}\right)^2.$$

This is homogeneous, and setting $u = Y/X$ gives us (after a little manipulation)

$$\frac{du}{u^2 + 3u + 4} = \frac{dX}{X}.$$

This can be written

$$\frac{du}{(u + \frac{3}{2})^2 + \frac{7}{4}} = \frac{dX}{X},$$

and an integration yields

$$\ln|X| = \frac{2}{\sqrt{7}} \arctan\left[\frac{2}{\sqrt{7}}\left(u + \frac{3}{2}\right)\right] + C.$$

Since $X = x - 2$, $Y = y + 3$, and $u = Y/X = (y + 3)/(x - 2)$, the original differential equation has general solution

$$\ln|x - 2| = \frac{2}{\sqrt{7}} \arctan\left\{\frac{2}{\sqrt{7}}\left[\left(\frac{y + 3}{x - 2}\right) + \frac{3}{2}\right]\right\} + C.$$

Now we turn to case 2 of the general consideration.

Case 2 Suppose that $ae - bd = 0$. In this case, set

$$v = \frac{ax + by}{a}.$$

Observe that, since $ae = bd$, then also

$$v = \frac{dx + ey}{d}.$$

Now,

$$\frac{dv}{dx} = 1 + \frac{b}{a}\frac{dy}{dx},$$

so that

$$\frac{dy}{dx} = \frac{a}{b}\left(\frac{dv}{dx} - 1\right).$$

The differential equation now becomes

$$\frac{a}{b}\left(\frac{dv}{dx} - 1\right) = f\left(\frac{av + c}{dv + h}\right)$$

or

$$\frac{dv}{dx} = 1 + \frac{b}{a} f\left(\frac{av + c}{dv + h}\right).$$

This can be written

$$\frac{dv}{1 + \frac{b}{a} f\left(\frac{av + c}{dv + h}\right)} = dx,$$

which is separable and subject to the method of Section 1.1.

EXAMPLE 15

Solve

$$\frac{dy}{dx} = \frac{2x + y - 1}{4x + 2y - 4}.$$

Here $a = 2$, $b = 1$, $c = -1$, $d = 4$, $e = 2$, $h = -4$, and $f(t) = t$. Note that $ae - bd = 0$. Thus, let

$$v = \frac{2x + y}{2}.$$

The differential equation becomes

$$\frac{dv}{dx} = 1 + \frac{1}{2}\left(\frac{2v - 1}{4v - 4}\right)$$

or

$$\left(\frac{8v - 8}{10v - 9}\right) dv = dx,$$

a separable equation which we can solve by integration to get

$$\frac{4v}{5} - \frac{2}{25} \ln|10v - 9| + C = x.$$

Recall now that $v = (2x + y)/2$. Thus, the original differential equation has general solution

$$\tfrac{2}{5}(2x + y) - \tfrac{2}{25} \ln|10x + 5y - 9| + C = x$$

or
$$-\frac{x}{5} + \frac{2y}{5} - \frac{2}{25}\ln|10x + 5y - 9| + C = 0.$$

PROBLEMS FOR SECTION 1.3

In each of Problems 1 through 16, determine whether or not the equation is homogeneous. If it is, find the general solution and check the answer by substituting back into the equation. If it is not homogeneous, do not attempt a solution.

1. $\frac{dy}{dx} = \frac{2y^2 + x^2}{x^2}$ **2.** $\frac{dy}{dx} = \frac{y}{x + y}$ **3.** $x\frac{dy}{dx} = y^2$ **4.** $\frac{dy}{dx} = \frac{x + y}{x}$

5. $\frac{dy}{dx} = \frac{2x + y}{x - y}$ **6.** $\frac{dy}{dx} = \frac{(x + y)^2}{x^2}$ **7.** $\frac{dy}{dx} = 3xy$ **8.** $x^2\frac{dy}{dx} = x^2 + y^2$

9. $y\frac{dy}{dx} = x^2 - y^2$ **10.** $x^3\frac{dy}{dx} = x^2y - y^3$ **11.** $\frac{dy}{dx} = \frac{y^2}{x^2} + \frac{y}{x}$ **12.** $x\frac{dy}{dx} = ye^{y/x}$

13. $\frac{dy}{dx} = \frac{x}{y} - 1$ **14.** $\frac{dy}{dx} = \frac{x - 2y}{x + 2y}$ **15.** $xy' = x - 3y$ **16.** $\frac{dy}{dx} = \frac{x^2 - y^2}{x^2 + y^2}$

17. Show that $dy/dx = g(x, y)$ is homogeneous if $g(x, tx) = g(1, t)$ for t real. Use this to test each of the differential equations in Problems 1 through 16 above.

In each of Problems 18 through 35, solve the given homogeneous differential equation; then determine the constant C to satisfy the other given condition.

18. $\frac{dy}{dx} = \frac{y^2 - x^2}{x^2}$; $y(2) = 6$

19. $x^2\frac{dy}{dx} = y^2 + x^2$; $y(1) = 4$

20. $y^2x^2\frac{dy}{dx} = x^4 - y^4$; $y(0) = 2$

21. $\frac{dy}{dx} = \frac{x - y}{x + y}$; $y(2) = 7$

22. $3x\frac{dy}{dx} = 2x + y$; $y(1) = 8$

23. $\frac{dy}{dx} = \frac{8x - 4y}{x - 2y}$; $y(2) = -7$

24. $\frac{dy}{dx} = 3y + \frac{y^2}{x}$; $y(1) = 4$

25. $(x + y)\frac{dy}{dx} = 4x - 3y$; $y(2) = 6$

26. $(x - 2y)\frac{dy}{dx} = x - y$; $y(1) = 4$

27. $\frac{dy}{dx} = \frac{x^2 - y^2}{x^2}$; $y(2) = 8$

28. $\frac{dy}{dx} = \frac{x - y}{3x + 2y}$; $y(1) = -5$

29. $-y\frac{dy}{dx} = x$; $y(2) = 3$

30. $2x^2\frac{dy}{dx} = y^2 + x^2$; $y(3) = 5$

31. $y^2\frac{dy}{dx} = x^2 + \frac{y^3}{x}$; $y(3) = 4$

32. $2x^2\frac{dy}{dx} = y^2 - 3x^2$; $y(1) = -2$

33. $(x + 2y)\frac{dy}{dx} = y - 3x$; $y(1) = 6$

34. $\frac{dy}{dx} = \frac{x + y}{3x - 2y}$; $y(4) = 3$

35. $\frac{dy}{dx} = 2 + \frac{y^2}{x^2}$; $y(4) = -7$

In each of Problems 36 through 50, the differential equation is not homogeneous but can still be solved by the methods of this section. In each case, find the general solution.

36. $\dfrac{dy}{dx} = \dfrac{2x + y - 2}{x - y + 4}$ **37.** $\dfrac{dy}{dx} = \dfrac{x - 3y + 1}{x + y}$ **38.** $\dfrac{dy}{dx} = \dfrac{y + 4y - 3}{2x + 6y - 2}$ **39.** $\dfrac{dy}{dx} = \dfrac{x - y + 2}{x - y + 3}$

40. $\dfrac{dy}{dx} = \dfrac{3x + y - 1}{6x + 2y - 3}$ **41.** $\dfrac{dy}{dx} = \dfrac{2x + 3y - 4}{8x + y - 2}$ **42.** $\dfrac{dy}{dx} = \left(\dfrac{x - y + 1}{2x - 2y}\right)^2$ **43.** $\dfrac{dy}{dx} = \dfrac{x - 2y}{3x - 6y + 4}$

44. $\dfrac{dy}{dx} = \dfrac{x - 2y + 1}{6x - y + 3}$ **45.** $\dfrac{dy}{dx} = \left(\dfrac{x - y + 2}{x + 1}\right)^2$ **46.** $\dfrac{dy}{dx} = \dfrac{x + y - 4}{2x + y}$ **47.** $\dfrac{dy}{dx} = \dfrac{3x + 2y + 4}{x - y + 6}$

48. $\dfrac{dy}{dx} = \dfrac{x - 2y + 8}{x + y + 1}$ **49.** $\dfrac{dy}{dx} = \dfrac{4x - y + 3}{x + 2y - 2}$ **50.** $\dfrac{dy}{dx} = \dfrac{x - y + 6}{3x - 3y + 4}$

In order to get some practice in using these techniques when the variables have names other than x and y, solve the following.

51. $\dfrac{dr}{d\theta} = \dfrac{2r - \theta + 1}{r + \theta + 7}$ **52.** $\dfrac{dw}{dt} = \dfrac{5w + t - 2}{w - 2t}$ **53.** $\dfrac{dz}{du} = \dfrac{2z - 4u + 3}{5z - 10u + 1}$ **54.** $\dfrac{dx}{dw} = \dfrac{x + 3w + 3}{w - 2}$

55. $\dfrac{dp}{dx} = \dfrac{3p + 2x - 1}{p - 4x + 2}$ **56.** $\dfrac{dq}{dr} = \dfrac{r - q + 5}{r - q + 4}$ **57.** $\dfrac{de}{dq} = \dfrac{q - 4e + 5}{3q + 12e + 2}$ **58.** $\dfrac{dc}{dv} = \dfrac{v + c - 4}{3v + 3c - 8}$

59. $\dfrac{dx}{dt} = \dfrac{2t - x - 5}{t - 5x + 11}$ **60.** $\dfrac{dk}{ds} = \dfrac{s - 3k}{s + k}$

1.4 EXACT DIFFERENTIAL EQUATIONS

We can sometimes write a first order differential equation $F(x, y, y') = 0$ as

$$\frac{dy}{dx} = \frac{-M(x, y)}{N(x, y)}$$

or, equivalently,

$$M(x, y)\,dx + N(x, y)\,dy = 0.$$

This differential equation may turn out to be separable or homogeneous, in which case we know how to solve it, at least in principle. If it is neither, one strategy is to try to find a function $F(x, y)$ whose total differential dF is $M\,dx + N\,dy$. If such a function exists, we call the differential equation $M\,dx + N\,dy = 0$ *exact*, and $F(x, y)$ is called a *potential function* for the differential equation.

Note that, when $dF = M\,dx + N\,dy$, the equation $M\,dx + N\,dy = 0$ becomes just

$$dF = 0,$$

with general solution $F(x, y) = C$. Thus, finding $F(x, y)$ is tantamount to solving the differential equation.

How can we find such an F, or else know that one does not exist? If such F exists, then

$$dF = \frac{\partial F}{\partial x}\, dx + \frac{\partial F}{\partial y}\, dy = M\, dx + N\, dy.$$

Then,

$$\frac{\partial F}{\partial x} = M \quad \text{and} \quad \frac{\partial F}{\partial y} = N.$$

In practice, we try to reconstruct F by integrating these equations. Here are three examples of this method.

EXAMPLE 16

Solve

$$\frac{dy}{dx} = \frac{-2xy^3 - 2}{3x^2y^2 + e^y}.$$

This equation is not separable, but can be written

$$(2xy^3 + 2)\, dx + (3x^2y^2 + e^y)\, dy = 0.$$

We want $F(x, y)$ such that

$$\frac{\partial F}{\partial x} = 2xy^3 + 2$$

and

$$\frac{\partial F}{\partial y} = 3x^2y^2 + e^y.$$

Choose one of these to begin, say

$$\frac{\partial F}{\partial x} = 2xy^3 + 2.$$

To reverse the partial derivative with respect to x (in which y is held constant), integrate with respect to x, holding y constant. We get

$$F(x, y) = \int \frac{\partial F}{\partial x}\, \partial x = \int (2xy^3 + 2)\, \partial x = x^2y^3 + 2x + k(y).$$

Here the "constant of integration" $k(y)$ may involve y, which was constant in the integration.

Now we need

$$\frac{\partial F}{\partial y} = 3x^2y^2 + k'(y) = 3x^2y^2 + e^y.$$

Then $k'(y) = e^y$, and we may choose $k(y) = e^y$.

Thus, with $F(x, y) = x^2y^3 + 2x + e^y$, the differential equation is

$$dF = \frac{\partial F}{\partial x}\,dx + \frac{\partial F}{\partial y}\,dy = 0,$$

with general solution

$$x^2y^3 + 2x + e^y = C.$$

This equation implicitly defines y as a function of x satisfying

$$\frac{dy}{dx} = \frac{-2xy^3 - 2}{3x^2y^2 + e^y},$$

or x as a function of y satisfying

$$\frac{dx}{dy} = \frac{3x^2y^2 + e^y}{-2xy^3 - 2}.$$

Explicitly solving for y in terms of x is difficult here.

Just for illustration, let us show that $F(x, y) = C$ implicitly defines $y = y(x)$ satisfying the differential equation. Differentiate

$$x^2y^3 + 2x + e^y = C$$

implicitly with respect to x to get

$$2xy^3 + 3x^2y^2\frac{dy}{dx} + 2 + e^y\frac{dy}{dx} = 0.$$

Then

$$\frac{dy}{dx} = \frac{-2xy^3 - 2}{3x^2y^2 + e^y},$$

which is the original differential equation.

EXAMPLE 17

Solve

$$\frac{dy}{dx} = \frac{-\cos(xy) + xy\sin(xy)}{-x^2\sin(xy) + 2y}.$$

This can be written

$$[\cos(xy) - xy\sin(xy)]\,dx + [-x^2\sin(xy) + 2y]\,dy = 0.$$

We want $F(x, y)$ such that

$$\frac{\partial F}{\partial x} = \cos(xy) - xy\sin(xy)$$

and

$$\frac{\partial F}{\partial y} = -x^2\sin(xy) + 2y.$$

Choose one of these, say the second, and integrate with respect to y, holding x fixed:

$$F(x, y) = \int \frac{\partial F}{\partial y}\, \partial y = \int (-x^2 \sin(xy) + 2y)\, \partial y = x \cos(xy) + y^2 + k(x).$$

Then,

$$\frac{\partial F}{\partial x} = -yx \sin(xy) + \cos(xy) + k'(x) = \cos(xy) - xy \sin(xy).$$

Then $k'(x) = 0$, and we may choose $k(x) = 0$ (or any other constant).

Then $F(x, y) = x \cos(xy) + y^2$ is a potential function for the differential equation, and the general solution is given implicitly by

$$x \cos(xy) + y^2 = C.$$

Again, explicitly solving for x or y here is not possible in elementary terms.

If we wanted a solution satisfying an additional condition, we would have to choose C appropriately. For example, suppose we want a solution satisfying $y(2) = 6$. Even though $y(x)$ is given implicitly by $x \cos(xy) + y^2 = C$, we can still obtain C by setting $x = 2$ and $y = 6$ to obtain $C = 2 \cos(12) + 36$ (approximately 37.6877). Thus, the solution $y(x)$ satisfying $y(2) = 6$ is given implicitly by

$$x \cos(xy) + y^2 = 2 \cos(12) + 36.$$

The method just illustrated will also tell us if the differential equation is not exact, and hence if a potential function does not exist.

EXAMPLE 18

Consider

$$\frac{dy}{dx} = \frac{-2xy^3 - 2xy}{3x^2y^2 + e^y},$$

a slight variation of Example 16. This differential equation can be written

$$(2xy^3 + 2xy)\, dx + (3x^2y^2 + e^y)\, dy = 0.$$

We want $F(x, y)$ such that

$$\frac{\partial F}{\partial x} = 2xy^3 + 2xy \quad \text{and} \quad \frac{\partial F}{\partial y} = 3x^2y^2 + e^y.$$

Choose one, say the first, and integrate:

$$F(x, y) = \int \frac{\partial F}{\partial x}\, \partial x = \int (2xy^3 + 2xy)\, \partial x = x^2y^3 + x^2y + k(y).$$

Then we need

$$\frac{\partial F}{\partial y} = 3x^2y^2 + x^2 + k'(y) = 3x^2y^2 + e^y.$$

This requires that

$$k'(y) = e^y - x^2,$$

which is impossible if k is a function of y alone (arising as a "constant" of integration with respect to x). Thus, this differential equation is not exact.

It is possible to test for exactness without going through the search procedures of these examples, but a technical condition is required of M and N and the region over which they are defined. This condition is a test for exactness of $M\,dx + N\,dy = 0$:

> If M, N, $\dfrac{\partial N}{\partial x}$, and $\dfrac{\partial M}{\partial y}$ are continuous over a rectangular region R, then $M\,dx + N\,dy = 0$ is exact for (x, y) in R if and only if $\dfrac{\partial M}{\partial y} = \dfrac{\partial N}{\partial x}$ in R.

In practice, many of our functions M and N are continuous with continuous partials over the whole plane, and so the above condition about the rectangle automatically holds for any rectangle you choose. A proof of this result is given in the section on potential theory in the vector analysis chapters.

EXAMPLE 19

In Example 16,

$$M = 2xy^3 + 2 \quad \text{and} \quad N = 3x^2y^2 + e^y.$$

Here, M and N are defined over the entire plane, and

$$\frac{\partial M}{\partial y} = 6xy^2 = \frac{\partial N}{\partial x}.$$

Thus, $M\,dx + N\,dy = 0$ is exact (as we found before by determining $F(x, y)$).

In Example 17,

$$M = \cos(xy) - xy\sin(xy)$$

and

$$N = -x^2\sin(xy) + 2y.$$

Then

$$\frac{\partial M}{\partial y} = -x\sin(xy) - x\sin(xy) - x^2y\cos(xy)$$

and

$$\frac{\partial N}{\partial x} = -2x\sin(xy) - x^2y\cos(xy).$$

These agree (over any rectangle you choose); hence $M\,dx + N\,dy = 0$ is exact (over any rectangle). Again, we found this explicitly before by deriving F.

In Example 18,

$$M = 2xy^3 + 2xy \quad \text{and} \quad N = 3x^2y^2 + e^y.$$

Then

$$\frac{\partial M}{\partial y} = 6xy^2 + 2x \quad \text{and} \quad \frac{\partial N}{\partial x} = 6xy^2.$$

These are different unless $x = 0$ (the y-axis), which cannot contain a rectangle. Thus, there is no region over which $M\,dx + N\,dy = 0$ is exact, which agrees with our conclusion in Example 16.

It is important for the student to realize two things in connection with exact differential equations:

1. If $F(x, y)$ is a potential function for $M\,dx + N\,dy = 0$, then the general solution of the differential equation is $F(x, y) = C$ (*not*, for example, $y = F(x, y)$, a commonly seen mistake).
2. Solutions $F(x, y) = C$ are in implicit form, and not in the customary state $y = f(x)$. This may seem unusual at first, but is inherent in the method and the student must come to be at ease with it. It is important to remember that one can still check the solution by differentiating implicitly, solving for dy/dx, and seeing if this is the original differential equation.

In the next section we offer a partial remedy for the circumstance in which $M\,dx + N\,dy = 0$ is not exact.

PROBLEMS FOR SECTION 1.4

In each of Problems 1 through 25, use the test of this section to determine whether or not the given equation is exact. If it is, solve the equation by finding a potential function and check your answer by implicit differentiation.

1. $(2y^2 + ye^{xy})\,dx + (4xy + xe^{xy} + 2y)\,dy = 0$
2. $[2\cos(x + y) - 2x\sin(x + y)]\,dx - 2x\sin(x + y)\,dy = 0$
3. $(y + e^x)\,dx + x\,dy = 0$
4. $(4xy + 2x)\,dx + (2x^2 + 3y^2)\,dy = 0$
5. $(4xy + 2x^2y)\,dx + (2x^2 + 3y^2)\,dy = 0$
6. $\cos(y)e^{x\cos(y)}\,dx - x\sin(y)e^{x\cos(y)}\,dy = 0$
7. $\dfrac{x}{x^2 + y^2}\,dx + \dfrac{y}{x^2 + y^2}\,dy = 0$
8. $(3x^2 + 3y)\,dx + (2y + 3x)\,dy = 0$
9. $\sinh(x)\sinh(y)\,dx + \cosh(x)\cosh(y)\,dy = 0$
10. $[e^x\sin(y^2) + xe^x\sin(y^2)]\,dx + [2yxe^x\sin(y^2) + e^y]\,dy = 0$
11. $\left(\dfrac{1}{x} + y\right)dx + (3y^2 + x)\,dy = 0$

12. $(6x - ye^{xy})\,dx + (e^y - xe^{xy})\,dy = 0$

13. $yx^{y-1}\,dx + x^y \ln(x)\,dy = 0$

14. $[16x^3y - 3\cos(x)]\,dx + [4x^4 + 3\cos(y)]\,dy = 0$

15. $\left(\dfrac{2x^3 + 2xy^2 - y}{x^2 + y^2}\right) dx + \left(\dfrac{x}{x^2 + y^2}\right) dy = 0$

16. $[-y\sin(xy) + 2x]\,dx + [3y^2 - x\sin(xy)]\,dy = 0$

17. $(2xy - e^y)\,dx + (x^2 - y^2)\,dy = 0$

18. $[2x\sin(2xy) + 2x^2y\cos(2xy)]\,dx + 2x^3\cos(2xy)\,dy = 0$

19. $\dfrac{x}{y^2}\,dx - \dfrac{x^2}{y^3}\,dy = 0$

20. $\left[1 - \dfrac{y}{x^2}\sec^2\left(\dfrac{y}{x}\right)\right] dx + \dfrac{1}{x}\sec^2\left(\dfrac{y}{x}\right) dy = 0$

21. $[3x^2 - y\cos(x)]\,dx - \sin(x)\,dy = 0$

22. $9x^2\,dx + 3y^2\,dy = 0$

23. $-y^2\sin(x)\,dx + [2y\cos(x) + 2]\,dy = 0$

24. $y\cosh(xy)\,dx + [1 + x\cosh(xy)]\,dy = 0$

25. $2xye^{x^2}\,dx - e^{x^2}\,dy = 0$

In each of Problems 26 through 32, solve the exact differential equation and then find a choice of the constant to satisfy the additional condition.

26. $dx + (-3y^2 + 2e^y)\,dy = 0; \quad y(1) = 4$

27. $(-8x + ye^{xy})\,dx + (2y + xe^{xy})\,dy = 0; \quad y(2) = -6$

28. $[\cosh(x - y) + x\sinh(x - y)]\,dx - x\sinh(x - y)\,dy = 0; \quad y(0) = 4$

29. $[2y - y^2\sec^2(xy^2)]\,dx + [2x - 2xy\sec^2(xy^2)]\,dy = 0; \quad y(1) = 2$

30. $(1 + e^{y/x} - (y/x)e^{y/x})\,dx + e^{y/x}\,dy = 0; \quad y(1) = -5$

31. $y^3\,dx + (3xy^2 - 1)\,dy = 0; \quad y(-2) = 5$

32. $\dfrac{-1}{xy^2}\,dx - \dfrac{1}{xy^2}\,dy = 0; \quad y(1) = 3$

In each of Problems 33 through 40, write the given differential equation in the form $M\,dx + N\,dy = 0$ and test for exactness. If the equation is exact, solve it.

33. $\dfrac{dy}{dx} = \dfrac{1 - 3y^4}{12xy^3}$

34. $\dfrac{dy}{dx} = \dfrac{x\cos(2y - x) - \sin(2y - x)}{2x\cos(2y - x)}$

35. $\dfrac{dy}{dx} = \dfrac{e^y}{1 - xe^y}$

36. $\dfrac{dy}{dx} + \dfrac{y}{x - 3y^2} = 0$

37. $\dfrac{dy}{dx} = \dfrac{3y + 3x^2y^2}{2x^3y - 3x}$

38. $\dfrac{dy}{dx} = \dfrac{6e^y - 4x^3y}{x^4 - 6xe^y}$

39. $\dfrac{dy}{dx} = \dfrac{2y - y\cos(x)}{2x + \sin(x)}$

40. $\dfrac{dy}{dx} = \dfrac{e^{x-y}}{e^{x-y} - 1}$

41. Show that every separable differential equation (Section 1.1) is exact. Are there exact differential equations which are not separable?

42. Consider the differential equation

$$\frac{x}{x^2 + y^2}\,dx - \frac{y}{x^2 + y^2}\,dy = 0.$$

(a) Show that $\partial M/\partial y = \partial N/\partial x$, except at the origin.

(b) Show by finding a potential function $F(x, y)$ that the differential equation is exact over any region in the plane which does not contain the origin.

(c) Show that the differential equation is not exact if the region includes the origin (e.g., on the inside of the square bounded by $x = 1, x = -1, y = 1, y = -1$).

(d) In view of (a), how are (b) and (c) consistent with the test for exactness?

(e) Solve the differential equation for any rectangle not containing the origin.

To get some practice with other variables, test each of the following for exactness and solve the differential equation if it is exact.

43. $[2r\cos(r\theta) - r^2\theta\sin(r\theta)]\,dr - r^3\sin(r\theta)\,d\theta = 0$

44. $2r\theta^2\,dr + [2r^2\theta + \sin(\theta)]\,d\theta = 0$

45. $(2u - 2v)\,du - 2u\,dv = 0$

46. $(2z + 2t^2)\,dz + (4zt - 3)\,dt = 0$

47. $(1/u)\,du + (1/v)\,dv = 0$

48. $(2x - 2s)\,dx + (-2x + 7s)\,ds = 0$

49. $[3x^2 + 3u\cosh(ux)]\,dx + 3x\cosh(ux)\,du = 0$

50. $e^{yt}\,dy + te^{yt}\,dt = 0$

1.5 INTEGRATING FACTORS AND BERNOULLI EQUATIONS

Suppose we have a first order differential equation

$$\frac{dy}{dx} = \frac{-M(x, y)}{N(x, y)},$$

which we write as

$$M\,dx + N\,dy = 0.$$

If this is exact, we can proceed to a solution as in the previous section. If not, we can sometimes find a nonzero function $u(x, y)$ such that

$$(uM)\,dx + (uN)\,dy = 0$$

is exact. We call u an *integrating factor* for $M\,dx + N\,dy = 0$.

What good is an integrating factor? To understand this, note that exactness of $(uM)\,dx + (uN)\,dy = 0$ means that this equation has a potential function $F(x, y)$ and hence a general solution $F(x, y) = C$. Then $F(x, y) = C$ is a general solution of $u(M\,dx + N\,dy) = 0$, and this is the same as $M\,dx + N\,dy = 0$ if $u \neq 0$. To sum up:

> If $M\,dx + N\,dy = 0$ is not exact, we seek nonzero $u(x, y)$ such that $(uM)\,dx + (uN)\,dy = 0$ is exact. If $F(x, y)$ is a potential function for $(uM)\,dx + (uN)\,dy = 0$, then $F(x, y) = C$ is a general solution of $M\,dx + N\,dy = 0$.

EXAMPLE 20

Consider

$$(y^2 - 6xy)\,dx + (3xy - 6x^2)\,dy = 0. \qquad \text{(a)}$$

Here,

$$M = y^2 - 6xy \quad \text{and} \quad N = 3xy - 6x^2.$$

Since $\partial M/\partial y \neq \partial N/\partial x$ in general, (a) is not exact. Now multiply (a) by y to obtain

$$(y^3 - 6xy^2)\,dx + (3xy^2 - 6x^2y)\,dy = 0. \tag{b}$$

This is exact, since

$$\frac{\partial}{\partial y}(y^3 - 6xy^2) = 3y^2 - 12xy = \frac{\partial}{\partial x}(3xy^2 - 6x^2y).$$

Thus, $u(x, y) = y$ is an integrating factor for (a). We shall explain later how we got it. For now, just verify our assertions about what $u(x, y)$ does for us.

Since (b) is exact, it has a potential function. One such function (with details of finding it omitted) is

$$F(x, y) = xy^3 - 3x^2y^2.$$

Then $xy^3 - 3x^2y^2 = C$ is a general solution of (b). We claim that this is also a general solution of (a). To see this, differentiate $xy^3 - 3x^2y^2 = C$ implicitly with respect to x:

$$y^3 + 3xy^2\frac{dy}{dx} - 6xy^2 - 6x^2y\frac{dy}{dx} = 0.$$

This is the same as

$$(y^3 - 6xy^2)\,dx + (3xy^2 - 6x^2y)\,dy = 0$$

or

$$y[(y^2 - 6xy)\,dx + (3xy - 6x^2)\,dy] = 0.$$

Assuming that $y \neq 0$, this in turn gives us

$$(y^2 - 6xy)\,dx + (3xy - 6x^2)\,dy = 0,$$

the original differential equation (a).

Now, how do we find integrating factors? If u is an integrating factor for $M\,dx + N\,dy = 0$, then $(uM)\,dx + (uN)\,dy = 0$ is exact, so that we must have

$$\frac{\partial(uM)}{\partial y} = \frac{\partial(uN)}{\partial x}.$$

Then,

$$u\frac{\partial M}{\partial y} + M\frac{\partial u}{\partial y} = u\frac{\partial N}{\partial x} + N\frac{\partial u}{\partial x},$$

a partial differential equation for $u(x, y)$. If M and N are reasonably well behaved, this always has a solution for u, at least in theory. In practice, however, this equation can be at least as difficult to solve as the original differential equation $M\,dx + N\,dy = 0$. This means that in the real world we can find integrating factors only under special circumstances. Here are two things to try:

1. Sometimes we can find an integrating factor which is a function of x only or y only (as in Example 20, where $u = y$). In particular, if $u = u(x)$, then $\partial u/\partial y = 0$, and the above partial differential equation for $u(x, y)$ is

$$\frac{1}{u}\frac{du}{dx} = \frac{1}{N}\left(\frac{\partial M}{\partial y} - \frac{\partial N}{\partial x}\right).$$

If, in addition, the right side is a function just of x, then we have a separable differential equation which we can solve for $u(x)$ by integrating.

By the same token, if there is a u which depends only on y, then $\partial u/\partial x = 0$, and the partial differential equation for u becomes

$$\frac{1}{u}\frac{du}{dy} = \frac{1}{M}\left(\frac{\partial N}{\partial x} - \frac{\partial M}{\partial y}\right).$$

If the right side is just a function of y, we again have a separable differential equation which we can solve for $u(y)$.

2. If (1) fails, try a function of the form $u = x^a y^b$ and attempt to solve for a and b to make this an integrating factor. We try this in Example 23 below.

   If (1) and (2) both fail, you will have to be more clever, and perhaps a little lucky. Experience is also helpful, and one can acquire some skill in constructing integrating factors if one works with them often enough.

EXAMPLE 21

Consider

$$(y^2 - 6xy)\,dx + (3xy - 6x^2)\,dy = 0$$

as in Example 18. Here,

$$\frac{\partial M}{\partial y} - \frac{\partial N}{\partial x} = (2y - 6x) - (3y - 12x) = -y + 6x.$$

Since $\partial M/\partial y - \partial N/\partial x \neq 0$ in general, the equation is not exact. However, observe that

$$\frac{1}{M}\left(\frac{\partial N}{\partial x} - \frac{\partial M}{\partial y}\right) = \frac{y - 6x}{y^2 - 6xy} = \frac{y - 6x}{y(y - 6x)} = \frac{1}{y}.$$

Thus,

$$\frac{1}{M}\left(\frac{\partial N}{\partial x} - \frac{\partial M}{\partial y}\right)$$

is independent of x, and the equation for u is

$$\frac{1}{u}\frac{du}{dy} = \frac{1}{y}.$$

This has solution $u = y$, the integrating factor we used in Example 20.

EXAMPLE 22

Solve

$$\left(3x^2y + 6xy + \frac{y^2}{2}\right) dx + (3x^2 + y)\, dy = 0.$$

Here,

$$\frac{\partial M}{\partial y} - \frac{\partial N}{\partial x} = (3x^2 + 6x + y) - (6x) = 3x^2 + y.$$

In general, this is nonzero and so the equation is not exact. But

$$\frac{1}{N}\left(\frac{\partial M}{\partial y} - \frac{\partial N}{\partial x}\right) = 1,$$

which is independent of y. Thus, we may solve for u by solving

$$\frac{1}{u}\frac{du}{dx} = \frac{1}{N}\left(\frac{\partial M}{\partial y} - \frac{\partial N}{\partial x}\right) = 1.$$

Then $u = e^x$ is an integrating factor, and

$$\left(3x^2y + 6xy + \frac{y^2}{2}\right) e^x\, dx + (3x^2 + y)e^x\, dy = 0$$

is exact. A potential function for this equation is

$$F(x, y) = 3x^2ye^x + \frac{y^2}{2}\, e^x.$$

Thus,

$$\left(3x^2y + \frac{y^2}{2}\right) e^x = C$$

is a general solution of the original differential equation.

EXAMPLE 23

Solve

$$(2y^2 - 9xy)\, dx + (3xy - 6x^2)\, dy = 0.$$

Here,

$$\frac{\partial M}{\partial y} - \frac{\partial N}{\partial x} = (4y - 9x) - (3y - 12x) = y + 3x,$$

which, in general, is nonzero. Thus, the differential equation is not exact. To attempt suggestion (1) above, calculate

$$\frac{1}{N}\left(\frac{\partial M}{\partial y} - \frac{\partial N}{\partial x}\right) = \frac{y + 3x}{3xy - 6x^2}$$

and note that this is not a function of x alone. Next, compute

$$\frac{1}{M}\left(\frac{\partial N}{\partial x} - \frac{\partial M}{\partial y}\right) = \frac{-y - 3x}{2y^2 - 9xy},$$

which is not a function of y alone. Thus, suggestion (1) above will not work. Try (2): Attempt to find an integrating factor of the form $u = x^a y^b$. Remember that u must satisfy

$$u\frac{\partial M}{\partial y} + M\frac{\partial u}{\partial y} = u\frac{\partial N}{\partial x} + N\frac{\partial u}{\partial x}.$$

Upon substituting $u = x^a y^b$ into this, we get

$$4x^a y^{b+1} - 9x^{a+1}y^b + 2bx^a y^{b+1} - 9bx^{a+1}y^b$$
$$= 3x^a y^{b+1} - 12x^{a+1}y^b + 3ax^a y^{b+1} - 6ax^{a+1}y^b.$$

Divide by $x^a y^b$ to obtain

$$4y - 9x + 2by - 9bx = 3y - 12x + 3ay - 6ax.$$

This can be written

$$(1 + 2b - 3a)y + (3 - 9b + 6a)x = 0.$$

Since x and y are independent, this can hold for all x and y only if a and b are chosen so that

$$1 + 2b - 3a = 0$$

and

$$3 - 9b + 6a = 0.$$

Solve these to get $a = b = 1$. Thus, $u = xy$ is an integrating factor.

To check this out, multiply the original differential equation by xy to get

$$(2xy^3 - 9x^2y^2)\,dx + (3x^2y^2 - 6x^3y)\,dy = 0.$$

This is exact, with potential function $F(x, y) = x^2y^3 - 3x^3y^2$. Thus, the original equation has general solution

$$x^2y^3 - 3x^3y^2 = C.$$

Occasionally one can identify a particular kind of differential equation for which integrating factors can be found. One such is the *Bernoulli equation*,

$$P(x)y' + Q(x)y = R(x)y^\alpha,$$

in which α is a constant (not necessarily an integer). Here we shall suppose that $\alpha \neq 0$ and $\alpha \neq 1$, because these cases can be treated by other means. In fact, when $\alpha = 1$, the Bernoulli equation is separable, and when $\alpha = 0$, it is linear and can be solved by the method to be given in Section 1.6.

The Bernoulli equation can be written in differential form as

$$(Q(x)y - R(x)y^\alpha)\,dx + P(x)\,dy = 0.$$

In general, this is not exact. However, an integrating factor is

$$u(x, y) = \frac{1}{y^\alpha P(x)} \exp\left[(1-\alpha)\int \frac{Q(x)}{P(x)}\,dx\right].$$

This can be verified directly, but it is instructive to see how one obtains such an integrating factor. Here is the basic line of reasoning. We seek a function $u(x, y)$ such that

$$u(Qy - Ry^\alpha)\,dx + uP\,dy = 0$$

is exact. Thus, we require that

$$\frac{\partial}{\partial x}(uP) = \frac{\partial}{\partial y}[u(Qy - Ry^\alpha)]. \tag{a}$$

In the previous section we saw two things to try, namely, u as a function of x or y alone, or $u = x^a y^b$ for some a and b. The reader can check that neither of these will work here for all choices of P, Q, R, and α. The next thing to try for u is something more general than $x^a y^b$, but still simple enough to be tractable. Try $u = f(x)y^b$, and attempt to choose $f(x)$ and b to make this an integrating factor. Substituting u into (a) gives us

$$f'(x)P(x)y^b + f(x)P'(x)y^b = f(x)Q(x)(b+1)y^b - R(x)f(x)(b+\alpha)y^{b+\alpha-1}.$$

Now observe that, if we choose $b = -\alpha$, then the last term on the right cancels out and every remaining term has a common factor of $y^{-\alpha}$. Specifically, we then have

$$f'(x)P(x)y^{-\alpha} + f(x)P'(x)y^{-\alpha} = f(x)Q(x)(1-\alpha)y^{-\alpha}.$$

Upon dividing by $f(x)P(x)y^{-\alpha}$, we obtain the separable differential equation for $f(x)$:

$$\frac{f'(x)}{f(x)} = -\frac{P'(x)}{P(x)} + (1-\alpha)\frac{Q(x)}{P(x)}.$$

Integrate to get

$$\ln[f(x)] = -\ln[P(x)] + (1-\alpha)\int \frac{Q(x)}{P(x)}\,dx$$

or

$$f(x) = \frac{1}{P(x)} \exp\left[(1-\alpha)\int \frac{Q(x)}{P(x)}\,dx\right].$$

Then $u(x, y) = y^{-\alpha} f(x)$ is an integrating factor, as stated before.

EXAMPLE 24

Solve

$$y' + y = y^4.$$

This is a Bernoulli equation with $P(x) = Q(x) = R(x) = 1$ and $\alpha = 4$. Calculate

$$u(x, y) = \frac{1}{y^4} e^{-3\int dx} = y^{-4} e^{-3x}.$$

Multiply the differential equation $(y - y^4)\,dx + dy = 0$ through by u to get

$$(y^{-3} - 1)e^{-3x}\,dx + y^{-4}e^{-3x}\,dy = 0,$$

an exact equation with general solution

$$-\tfrac{1}{3}e^{-3x}y^{-3} + \tfrac{1}{3}e^{-3x} = C.$$

This is also a general solution of $y' + y = y^4$.

EXAMPLE 25

Solve

$$xy' + 2y = xy^3.$$

This is a Bernoulli equation with $P = x$, $Q = 2$, $R = x$, and $\alpha = 3$. Here,

$$u(x, y) = \frac{1}{xy^3}\exp\left[-2\int \frac{2}{x}\,dx\right] = \frac{1}{xy^3}\,e^{-4\ln(x)} = \frac{1}{x^5y^3}.$$

Write the differential equation as

$$(2y - xy^3)\,dx + x\,dy = 0.$$

Multiply through by u to get the exact equation

$$\left(\frac{2}{x^5y^2} - \frac{1}{x^4}\right)dx + \frac{1}{x^4y^3}\,dy = 0.$$

A general solution is found to be

$$-\frac{x^{-4}y^{-2}}{2} + \frac{x^{-3}}{3} = C.$$

It is sometimes useful to remember that one can interchange the traditional roles of x and y as independent and dependent variables. Here is an example.

EXAMPLE 26

Solve

$$y^2\,dx + (2yx - x^4)\,dy = 0.$$

This can be written

$$y^2x' + 2yx = x^4,$$

a Bernoulli equation in which $x' = dx/dy$ and x and y are interchanged in the standard form of the Bernoulli equation. Here, we have

$$P(y)\frac{dx}{dy} + Q(y)x = R(y)x^4,$$

with $P = y^2$, $Q = 2y$, and $R = 1$. An integrating factor is

$$u(x, y) = \frac{1}{x^4y^2}\exp\left[-3\int \frac{2}{y}\,dy\right] = \frac{1}{x^4y^8}.$$

Multiplying the differential equation by u gives us

$$\frac{1}{x^4y^6}\,dx + \left(\frac{2}{x^3y^7} - \frac{1}{y^8}\right) dy = 0,$$

which has general solution

$$\frac{-1}{3x^3y^6} + \frac{1}{7y^7} = C.$$

Finally, here is an example in which α is not an integer.

EXAMPLE 27

Solve

$$x^2y' + xy = -y^{-3/2}.$$

This can be written

$$(xy + y^{-3/2})\,dx + x^2y\,dy = 0.$$

An integrating factor is

$$u = \frac{1}{y^{-3/2}x^2}\exp\left[\frac{5}{2}\int \frac{1}{x}\,dx\right] = y^{3/2}x^{1/2}.$$

Thus, we must solve the exact equation

$$(x^{3/2}y^{5/2} + x^{1/2})\,dx + x^{5/2}y^{3/2}\,dy = 0.$$

We obtain the general solution

$$\tfrac{2}{5}x^{5/2}y^{5/2} + \tfrac{2}{3}x^{3/2} = C.$$

In the next section we shall solve the important class of equations obtained by setting $\alpha = 0$ in the Bernoulli equation.

PROBLEMS FOR SECTION 1.5

In each of Problems 1 through 12, (a) show that the equation is not exact by using the test of Section 1.4, (b) find an integrating factor, (c) find the general solution, and (d) check the solution by implicit differentiation.

1. $(3xy + y + 4)\,dx + \frac{1}{2}x\,dy = 0$
2. $(\frac{3}{2}y^2 + 3y + 4x^2)\,dx + x(y + 1)\,dy = 0$
3. $3x^2y\,dx + (2x^3 - 2)\,dy = 0$
4. $(90xy + 4y^4)\,dx + (72x^2 + 7xy^3)\,dy = 0$
5. $[7x^5y^5 + 2y\sin(x) + xy\cos(x)]\,dx + [6x^6y^4 + 2x\sin(x)]\,dy = 0$
6. $(3xy^2 + 12y)\,dx + (3x^2y + 12x)\,dy = 0$
7. $(6xy + 5y^4)\,dx + (4x^2 + 7xy^3)\,dy = 0$
8. $(x^2 + 1)y^3\,dx + (\frac{4}{3}x^3y^2 + 4y^2x - \frac{1}{3}x^3 - x)\,dy = 0$
9. $(1 + x + y^2)\,dx + 2y\,dy = 0$
10. $(4xy + 6y^2)\,dx + (2x^2 + 6xy)\,dy = 0$

11. $(2x - 2y - x^2 + 2yx)\,dx + (2x^2 - 4yx - 2x)\,dy = 0$ (*Hint:* Try $u = e^{ax}e^{by}$.)
12. $(x^3 + 1)(\frac{1}{3}y^3 + \frac{1}{2}y^2)\,dx + xy(y + 1)\,dy = 0$
13. Given the differential equation $M\,dx + N\,dy = 0$, prove the following:
    (a) If $1/N(\partial M/\partial y - \partial N/\partial x) = f(x)$, a function of x only, then $e^{\int f(x)\,dx}$ is an integrating factor of $M\,dx + N\,dy = 0$.
    (b) If $1/M(\partial M/\partial y - \partial N/\partial x) = g(y)$, a function of y only, then $e^{-\int g(y)\,dy}$ is an integrating factor of $M\,dx + N\,dy = 0$.
14. Suppose that $dy/dx = -M(x, y)/N(x, y)$ is homogeneous (Section 1.3). Prove that $1/(Mx + Ny)$ is an integrating factor of $M\,dx + N\,dy = 0$, provided that $1/(Mx + Ny) \neq 0$.
15. Suppose that $u(x, y)$ and $v(x, y)$ are integrating factors of $M\,dx + N\,dy = 0$, and that u is not a constant multiple of v. Show that

$$\frac{u(x, y)}{v(x, y)} = C$$

is a general solution of $M\,dx + N\,dy = 0$.

In each of Problems 16 through 30, use an integrating factor to find the general solution and then find a particular solution satisfying the given condition.

16. $dx + x\,dy = 0$; $y(2) = -3$
17. $3y\,dx + 4x\,dy = 0$; $y(1) = 6$
18. $y(1 + x)\,dx + 2x\,dy = 0$; $y(1) = -3$ (*Hint:* Try $u = y^a e^{bx}$.)
19. $2xy\,dx + 3\,dy = 0$; $y(3) = 5$ (*Hint:* Try $u = y^a e^{x^2}$.)
20. $2y\,dx + 3x\,dy = 0$; $y(3) = -4$
21. $2y(1 + x^2)\,dx + x\,dy = 0$; $y(2) = 3$ (*Hint:* Try $u = x^a y^b e^{x^2}$.)
22. $[\sin(x - y) + \cos(x - y)]\,dx - \cos(x - y)\,dy = 0$; $y(\pi) = 4$
23. $2(y^3 - 2)\,dx + 3xy^2\,dy = 0$; $y(3) = -5$
24. $2y(2y + 1)\,dx + 2x(3y + 1)\,dy = 0$; $y(1) = 7$
25. $y(2x^2 + 1)\,dx + x\,dy = 0$; $y(1) = -2$
26. $(x - 2y^2)\,dx - 2xy\,dy = 0$; $y(-2) = 4$
27. $3y\,dy + 4x\,dx = 0$; $y(1) = 9$
28. $9y\,dx + 15x\,dy = 0$; $y(-2) = -3$
29. $y(3x + 2)\,dx + 3x\,dy = 0$; $y(1) = -7$
30. $2y(1 - x)\,dx + x\,dy = 0$; $y(2) = 6$
31. Go through the last fifteen problems and see which you can do by other methods covered thus far in this chapter. For example, in Problem 16, $dx + x\,dy = 0$ can be written $dx/-x = dy$ and hence is separable.

In each of Problems 32 through 41, you are given a Bernoulli equation. Find the general solution.

32. $x^3y' + x^2y = 2y^{-4/3}$
33. $y' + xy = xy^2$
34. $2x^2y - xy = y^{3/2}$
35. $x^3y' + 2x^2y = y^{-3}$
36. $x^2y' - x^2y = y^3$
37. $x^2y' - xy = e^x y^3$
38. $x^4y' + x^3y = y^{-3/4}$
39. $x^2y' - 3xy = -2y^{5/3}$
40. $2x^3y' + 4x^2y = -5y^4$
41. $xy' + 2xy = 3y^2$

In each of Problems 42 through 51, determine if the given equation is a Bernoulli equation. Do not attempt to solve these equations.

42. $2xyy' + x^2y = 4y^3$
43. $(y^2 - xy^{3/2})\,dx - 2xy\,dy = 0$
44. $(2y - x^3y^4)\,dx + x^2\,dy = 0$
45. $x^2y' - 3y^3 = 2xy$
46. $4y\dfrac{dx}{dy} - 2y^2x = 8y^2x^{4/3}$
47. $xy' - 4x^2y^2 = 18y$

(*Hint:* Remember Example 26.)

48. $(x^3 - 3)y\,dx + 4x^3\,dy = 0$

49. $4y^{3/2}\,dx + (y^2x - y^3x^3)\,dy = 0$

50. $y' + y^{4/3} = x^2y$

51. $4y^3\,dx + 8y^{3/2}\,dy = 0$

1.6 LINEAR FIRST ORDER DIFFERENTIAL EQUATIONS

A number of important first order equations have the form

$$y' + p(x)y = q(x).$$

This is called a *linear* first order differential equation, and can be recognized as a Bernoulli equation with $\alpha = 0$. An explicit formula for its solution can be obtained by first multiplying through by $e^{\int p(x)\,dx}$ to get

$$y'e^{\int p(x)\,dx} + pye^{\int p(x)\,dx} = qe^{\int p(x)\,dx}.$$

This can be written

$$(ye^{\int p(x)\,dx})' = qe^{\int p(x)\,dx}.$$

Integrating then gives us

$$\int (ye^{\int p(x)\,dx})'\,dx = ye^{\int p(x)\,dx} = \int qe^{\int p(x)\,dx}\,dx + C.$$

Then

$$y = e^{-\int p(x)\,dx}\int qe^{\int p(x)\,dx}\,dx + Ce^{-\int p(x)\,dx}.$$

To sum up:

> A general solution of $y' + p(x)y = q(x)$ is
>
> $$y = e^{-\int p(x)\,dx}\int q(x)e^{\int p(x)\,dx}\,dx + Ce^{-\int p(x)\,dx}.$$

Example 28

Solve

$$y' + y = \sin(x).$$

This is a linear first order differential equation, with $p(x) = 1$ and $q(x) = \sin(x)$. We have

$$\int p(x)\,dx = x.$$

Hence, a general solution is

$$y = e^{-x} \int \sin(x)e^{x}\, dx + Ce^{-x}$$

$$= e^{-x}\left(\frac{e^{x}[\sin(x) - \cos(x)]}{2}\right) + Ce^{-x}$$

$$= \tfrac{1}{2}[\sin(x) - \cos(x)] + Ce^{-x}.$$

If we want a solution satisfying an additional condition, say $y(\pi) = -4$, then we must solve for C in the equation

$$-4 = \tfrac{1}{2}[-\cos(\pi)] + Ce^{-\pi}$$

to obtain $C = -9e^{\pi}/2$. The solution satisfying $y(\pi) = -4$ is then

$$y = \frac{1}{2}[\sin(x) - \cos(x)] - \frac{9e^{\pi}}{2}e^{-x}.$$

(Here, $-9e^{\pi}/2$ is approximately -104.1331.)

Example 29

Solve

$$y' + \frac{1}{x}y = 3x^2.$$

Here, $q(x) = 3x^2$, $p(x) = 1/x$, and $\int p(x)\, dx = \ln(x)$. A general solution is

$$y = e^{-\ln(x)} \int 3x^2 e^{\ln(x)}\, dx + Ce^{-\ln(x)}.$$

Now recall that $e^{\ln(x)} = x$ and $e^{-\ln(x)} = 1/x$. Thus, the general solution can be written

$$y = \frac{1}{x}\int 3x^3\, dx + \frac{C}{x} = \frac{3x^4}{4} + \frac{C}{x}.$$

Again, if we want a solution satisfying, say, $y(1) = 5$, then we solve for C in

$$5 = \tfrac{3}{4} + C,$$

obtaining $C = 17/4$. The solution satisfying $y(1) = 5$ is then

$$y = \frac{3x^3}{4} + \frac{17}{4x}.$$

One problem encountered with this method is in performing the required integrations. Here is a typical example.

EXAMPLE 30

Solve

$$y' + xy = \cos(x).$$

Here $p(x) = x$, $q(x) = \cos(x)$, and $\int p(x)\,dx = \frac{1}{2}x^2$. Thus, the general solution is of the form

$$y = e^{-x^2/2} \int \cos(x) e^{x^2/2}\,dx + Ce^{-x^2/2},$$

and this integral cannot be evaluated in terms of elementary functions. The answer must be left in this form, or we must go to infinite series or some other representation.

In the next section, we discuss Riccati equations, which contain the general linear equation as a special case.

PROBLEMS FOR SECTION 1.6

In each of Problems 1 through 12, solve the given linear differential equation. Become accustomed to using integral tables to help with the integrations. In cases where integrations are impossible in elementary form, solutions may contain integral signs, as in Example 30. Check solutions by substituting back.

1. $y' + y = 2\cos(x)$ **2.** $y' - y = \sinh(x)$ **3.** $y' - \frac{1}{x}y = x^2 + 2$

4. $y' - xy = 3x + e^x$ **5.** $y' + y = 5x + e^{2x}$ **6.** $-y' + 2y = 8x^2$

7. $y' - \frac{3}{x}y = 2x^2$ **8.** $xy' + \frac{2}{x}y = 4$ **9.** $3y' - y = 4x$

10. $2y' + 3y = e^{2x}$ **11.** $y' - 4y = e^x - \sin(x)$ **12.** $y' - 6y = x - \cosh(x)$

13. Write the linear equation $y' + p(x)y = q(x)$ as

$$[p(x)y - q(x)]\,dx + dy = 0.$$

Show that $e^{\int p(x)\,dx}$ is an integrating factor.

14. *Permanent magnet DC motor.* Imagine a permanent magnet, direct-current motor with an armature moving between the poles of the magnet and a coil wound around the armature. Suppose that the coil has n turns, and that the armature has $2n$ conductors of length s parallel to the axis of the armature and at a distance R from the axis. The total torque on the armature is $2nRBsI$, where B = average flux density of the magnetic field and I = current. If N = speed of the rotor, then

$$M\frac{dN}{dt} + kN = 2nRBsI, \tag{a}$$

where k is some constant of proportionality and M = moment of inertia of the armature (M = constant). The term kN accounts for friction. Let E = voltage, L = inductance, and R = resistance (of the coil), respectively. Then I satisfies

$$L\frac{dI}{dt} + RI = E - E_c, \tag{b}$$

where E_c = counter electromotive force generated in the coil by the conductors moving in the magnetic field. Assuming that $E - E_c$, L, and R are constant, solve the linear equation (b) for I. Substitute this into the linear equation (a) and solve for N, assuming B to be constant.

15. Solve the linear equation for revolution of a crankshaft in a diesel engine (Problem 4, Section 1.2) when $c \neq 0$.

16. *Rotation of a propellor.* Consider rotation of a propellor on a diesel-driven ship at sea. If M is the moment of inertia of the rotating shaft of the engine, N is the number of revolutions per minute, and x the throttle coordinate, then N satisfies an equation of the form

$$M\frac{dN}{dx} + AN^a = f(x),$$

where A and a are constant. Solve this when $M =$ constant, $A = 10$, $a = 1$, and $f(x) = x^2$.

In each of Problems 17 through 27, find a solution of the linear differential equation which also satisfies the given condition, as in Examples 28 and 29.

17. $y' - y = 2e^{4x}$; $y(0) = -3$
18. $y' + 3y = 8x^3$; $y(1) = 4$
19. $y' + \frac{3}{x}y = x\sin(x)$; $y(\pi) = 0$
20. $y' - 3y = xe^{2x}$; $y(0) = 2$
21. $y' + \frac{4}{x}y = 2$; $y(1) = -4$
22. $y' + 6y = \cosh(2x) - x$; $y(3) = -2$
23. $y' + \frac{3}{2x}y = x^3 - 1$; $y(3) = 7$
24. $y' - 4y = x + \sin(x)$; $y(0) = 3$
25. $y' - \frac{5}{9x}y = 3x^3 + x$; $y(-1) = 4$
26. $y' + \sqrt{3}\,y = e^{2x} - 5$; $y(0) = 2$
27. $y' - 5y = e^{5x}\sin(x)$; $y(0) = 4$
28. Show that the Bernoulli equation

$$y' + q(x)y = r(x)y^\alpha$$

reduces to a linear equation when we set $z = y^{1-\alpha}$.

In each of Problems 29 through 33, use the method of Problem 28 to reduce the Bernoulli equation to a linear equation. Solve the resulting linear equation and hence obtain a general solution of the Bernoulli equation.

29. $y' + \frac{1}{x}y = 3x^2y^3$
30. $y' - 2y = 4xy^2$
31. $y' + \frac{3}{x}y = -2xy^{5/2}$
32. $y' + 8y = 2x^3y^{3/4}$
33. $y' + \frac{4}{x}y = xy^4$
34. Will the transformation defined in Problem 28 reduce the general Bernoulli equation $P(x)y' + Q(x)y = R(x)y^\alpha$ to a linear equation?
35. Suppose that the rate of change of price $P(t)$ with time is proportional to the difference between supply $S(t)$ and demand $D(t)$. Solve for $P(t)$ when D and S are given by $D(t) = -BP(t)$ for some positive constant B, and $S(t) = k[1 - \cos(at)]$ for some constants k and a.
36. Solve for price $P(t)$, as in Problem 35, when supply and demand are given by

$$D(t) = A - BP(t) \quad \text{and} \quad S(t) = k[1 - \cos(at)].$$

Here, A and B are positive constants, and k and a are constants.

37. Solve for $P(t)$ in Problem 35 if $D(t)$ is constant, and $S(t) = k[1 - \cos(at)]$.
38. Solve for the price $P(t)$ of Problem 35 if $D(t)$ is constant and $S(t) = A + BP(t)$, with A and B positive constants.

1.7 RICCATI EQUATIONS

A *Riccati equation* is one of the form

$$y' = P(x)y^2 + Q(x)y + R(x).$$

This is linear if $P(x) = 0$, and not linear otherwise. In this section we consider the problem of solving such an equation.

Usually we seek a general solution, which will contain one arbitrary constant. Depending upon $P(x)$, $Q(x)$, and $R(x)$, there may be no apparent way of obtaining such a solution. However, *if we can somehow* (often by observation, guessing, or trial and error) *produce one specific solution* $y = S(x)$, *then we can obtain a general solution* as follows:

Change variables from y to z by setting

$$y = S(x) + \frac{1}{z}.$$

Then,

$$y' = S'(x) - \frac{1}{z^2}z',$$

and substitution into the Riccati equation gives us

$$S'(x) - \frac{1}{z^2}z' = P(x)\left(S(x) + \frac{1}{z}\right)^2 + Q(x)\left(S(x) + \frac{1}{z}\right) + R(x).$$

This can be written

$$S'(x) - \frac{1}{z^2}z' = [P(x)S(x)^2 + Q(x)S(x) + R(x)]$$
$$+ \left[P(x)\frac{1}{z^2} + 2P(x)S(x)\frac{1}{z} + Q(x)\frac{1}{z}\right].$$

Now, $S(x)$ is assumed to be a solution of the Riccati equation; hence the first term on the left cancels the terms in the first brackets on the right, leaving

$$-\frac{1}{z^2}z' = P\frac{1}{z^2} + 2PS\frac{1}{z} + Q\frac{1}{z}.$$

After multiplying through by $-z^2$, this can be written

$$z' + (2PS + Q)z = -P.$$

This is a linear equation for z which we can solve by the method of Section 1.6, obtaining

$$z = \frac{C}{u(x)} + \frac{1}{u(x)}\int -P(x)u(x)\,dx,$$

where

$$u(x) = \exp\left[\int [2P(x)S(x) + Q(x)]\, dx\right].$$

Then,

$$y = S(x) + \frac{1}{z}$$

is a general solution of the Riccati equation.

Usually this method of solving Riccati equations presents two difficulties:

1. One must first find a specific solution $y = S(x)$.
2. One must be able to perform the necessary integrations.

Failing in either of these, we must go to some other method, such as infinite series which are treated in Chapter 5.

EXAMPLE 31

Solve

$$y' = \frac{1}{x} y^2 + \frac{1}{x} y - \frac{2}{x}.$$

This is a Riccati equation with $P(x) = Q(x) = 1/x$ and $R(x) = -2/x$. Here we can observe that $y = 1$ is a solution. Substituting $S(x) = 1$ into the method described above, we change variables by setting

$$y = 1 + \frac{1}{z}.$$

The differential equation transforms to

$$z' + \frac{1}{x} z = -\frac{1}{x},$$

a linear equation whose general solution is easily found to be

$$z = -\frac{1}{3} + \frac{C}{x^3}.$$

A general solution of the Riccati equation is then

$$y = 1 + \frac{1}{\dfrac{-1}{3} + \dfrac{C}{x^3}} = \frac{2x^3 + K}{-x^3 + K}$$

in which $K = 3C$ is still an arbitrary constant.

Often we want a solution satisfying an additional condition. Suppose, for example, we want a solution satisfying $y(1) = 3$. Then solve for K in

$$3 = \frac{2 + K}{-1 + K}$$

to get $K = \frac{5}{2}$. The solution satisfying $y(1) = 3$ is then

$$y = \frac{2x^3 + \frac{5}{2}}{-x^3 + \frac{5}{2}}.$$

EXAMPLE 32

Solve

$$y' = y^2 - 2xy + x^2 + 1.$$

This is a Riccati equation with $P = 1$, $Q = -2x$, and $R = x^2 + 1$. By inspection, we find that one solution is $y = x$. Thus, we change variables by letting

$$y = x + \frac{1}{z}.$$

The Riccati equation becomes

$$z' = -1,$$

with general solution

$$z = -x + C.$$

The original differential equation then has general solution

$$y = x + \frac{1}{-x + C}.$$

As an example in which not everything works out so neatly, consider the following:

EXAMPLE 33

Solve

$$y' = \frac{1}{x} y^2 + xy + 2x(1 - x^2).$$

One solution is $y = x^2$. Let

$$y = x^2 + \frac{1}{z}.$$

Then the Riccati equation becomes

$$z' + 3xz = \frac{-1}{x}.$$

This has general solution

$$z = e^{-3x^2/2} \int \frac{-1}{x} e^{3x^2/2}\, dx + Ce^{-3x^2/2}.$$

The Riccati equation then has general solution

$$y = x^2 + \left[e^{-3x^2/2} \int \frac{-1}{x} e^{3x^2/2}\, dx + Ce^{-3x^2/2} \right]^{-1}.$$

Since the integrations cannot be carried out in elementary terms, this is the best we can do with this equation at this point in our treatment of differential equations.

PROBLEMS FOR SECTION 1.7

In each of Problems 1 through 10, find the general solution; then find a particular solution satisfying the given condition.

1. $y' = \dfrac{1}{x^2} y^2 - \dfrac{1}{x} y + 2 - \dfrac{1}{x^2}$; $y(1) = 3$ [Try $S(x) = x$]

2. $y' = \dfrac{1}{2x} y^2 - \dfrac{1}{x} y - \dfrac{4}{x}$; $y(2) = -6$ [Try $S(x) = 4$]

3. $y' = -\dfrac{1}{x} y^2 + \dfrac{2}{x} y$; $y(1) = 4$

4. $y' = y^2 + \left(\dfrac{1}{x} - 2\right) y - \dfrac{1}{x} + 1$; $y(2) = 1$

5. $y' = -e^{-x}y^2 + y + e^x$; $y(0) = 6$ [Try $S(x) = -e^x$]

6. $y' = e^{2x}y^2 - 2y - 9e^{-2x}$; $y(0) = 4$ [Try $S(x)$ of the form ae^{bx}]

7. $y' = \dfrac{1}{16x^2} y^2 - y + 4x(4 + x)$; $y(1) = -3$ [Try $S(x) = ax^b$]

8. $y' = y^2 - \dfrac{6}{x^2} y - \dfrac{6}{x^3} + \dfrac{9}{x^4}$; $y(2) = 4$

9. $y' = xy^2 + \left(-8x^2 + \dfrac{1}{x}\right) y + 16x^3$; $y(2) = 6$

10. $y' = \dfrac{e^{-3x}}{x} y^2 - \dfrac{1}{x} y + 3e^{3x}$; $y(3) = -4$

In each of Problems 11 through 15, find an expression for the general solution. It may be necessary to leave some integrals in the final result, as in Example 33.

11. $y' = \dfrac{1}{x^3} y^2 + y - 2x^2$ [Try $S(x) = ax^b$]

12. $y' = \dfrac{1}{3x} y^2 + y - 3$

13. $y' = (1 - 2x)y^2 + x^2y - 14(1 - 2x) - \sqrt{14}\,x^2$

14. $y' = y^2 + xy - 1$

15. $y' = \frac{1}{9}y^2 + xy + 3 - 4x^2$

16. Write the Riccati differential equation in differential form as

$$[P(x)y^2 + Q(x)y + R(x)]\, dx - dy = 0.$$

Suppose that $y = S(x)$ is one solution. Show that

$$u(x, y) = \frac{1}{[y - S(x)]^2} \exp\left\{\int [2P(x)S(x) + Q(x)]\, dx\right\}$$

is an integrating factor. Use this result to redo Problems 1, 2, 3, and 4.

17. Let $S_1(x)$ and $S_2(x)$ be particular solutions of the Riccati equation $y' = P(x)y^2 + Q(x)y + R(x)$. Let

$$u_1(x, y) = \frac{1}{[y - S_1(x)]^2} \exp\left\{\int [2P(x)S_1(x) + Q(x)]\, dx\right\}$$

and

$$u_2(x, y) = \frac{1}{[y - S_2(x)]^2} \exp\left\{\int [2P(x)S_2(x) + Q(x)]\, dx\right\}.$$

Show that the general solution of the Riccati equation is then

$$\frac{u_1(x, y)}{u_2(x, y)} = C.$$

(*Hint:* Use the results of Problem 16 above, and Problem 15, Section 1.5.)

18. Use the result of Problem 17 to show that the general solution of a Riccati equation always has the appearance

$$y = \frac{F(x) + CG(x)}{H(x) + CJ(x)}.$$

Go back over Problems 1 through 5 and show how in each case the general solution can be put into this form.

19. Consider the special Riccati equation

$$y' = ay^2 + by^\alpha,$$

with a and b constant. Show that the general solution can be obtained in closed form when

$$\alpha = \frac{-4n}{2n + 1} \quad \text{or} \quad \alpha = \frac{-4n}{2n - 1},$$

with $n = 0, 1, 2, 3, \ldots$. Here, *closed form* means that all integrations can be explicitly performed in terms of elementary functions. (It is true, but do not try to prove it, that when α is not of the above form, then the general solution of this Riccati equation cannot be written in closed form. To appreciate this, try solving

$$y' = 2y^2 + 4y^{7/3},$$

in which $\alpha = \frac{7}{3}$ does not have the required form.)

20. Riccati equations, which are first order, are related to second order equations, which we shall study in the next chapter. As one example of this, suppose that $F(x)$ is a solution of

$$y'' - \left(Q(x) + \frac{P'(x)}{P(x)}\right)y' + P(x)R(x)y = 0.$$

Show that

$$y = \frac{F'(x)}{-P(x)F(x)}$$

is a solution of $y' = P(x)y^2 + Q(x)y + R(x)$.

1.8 *RL* AND *RC* CIRCUITS

In Example 1 we derived a differential equation for the current in a series circuit containing resistance, inductance, and capacitance. When the circuit contains only resistance, we obtain a separable differential equation, which we solved in Example 10 for the case of an alternating electromotive force. We are now in a position to solve the current equation in two more important cases (though still not the completely general one).

CASE 1: *RL* CIRCUITS

A circuit such as in Figure 1, but with no capacitance, is called an *RL circuit*. A typical *RL* circuit is shown in Figure 7. From Example 1, current $I(t)$ in such a circuit is governed by the first order differential equation

$$L\frac{dI}{dt} + RI = E(t).$$

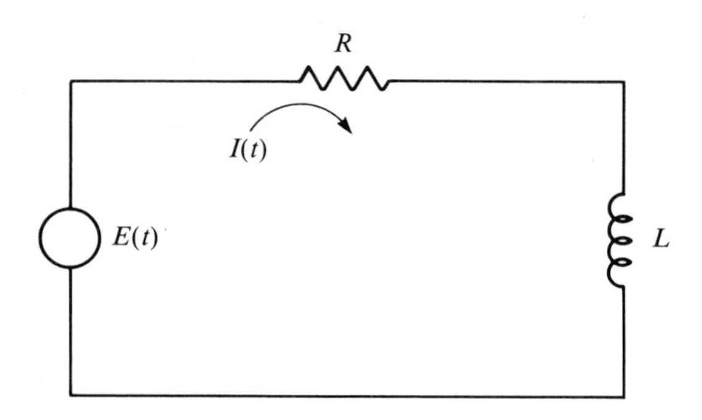

Figure 7. *RL* circuit.

This can be written in linear form

$$\frac{dI}{dt} + \frac{R}{L}I = \frac{1}{L}E(t).$$

Here, t is the independent variable instead of x, and I the dependent variable instead of y. From Section 1.6, we immediately have the general solution

$$I(t) = \frac{1}{L}e^{-Rt/L}\int e^{Rt/L}E(t)\,dt + ke^{-Rt/L},$$

with k used instead of C because in this context C is used to denote capacitance. Given $E(t)$, this equation yields $I(t)$. Two subcases are of special interest.

Case 1(a) If E = constant (constant electromotive force *RL* circuit), we obtain the general solution

$$I(t) = \frac{1}{L}e^{-Rt/L}\left(E\frac{L}{R}e^{Rt/L}\right) + ke^{-Rt/L} = \frac{E}{R} + ke^{-Rt/L},$$

with k a constant we would need more information [e.g., the value of $I(0)$] to determine. Note that, regardless of the value of k,

$$I(t) \to \frac{E}{R}$$

as $t \to \infty$. Thus, with increasing time, current in a constant electromotive force RL circuit tends toward a constant value equal to the electromotive force divided by the resistance. This is shown in the graph of Figure 8.

Putting $t = 0$ into the solution

$$I(t) = \frac{E}{R} + ke^{-Rt/L},$$

we obtain

$$I(0) = \frac{E}{R} + k.$$

Then, in terms of initial current, resistance, and the constant electromotive force, the constant of integration is

$$k = I(0) - \frac{E}{R},$$

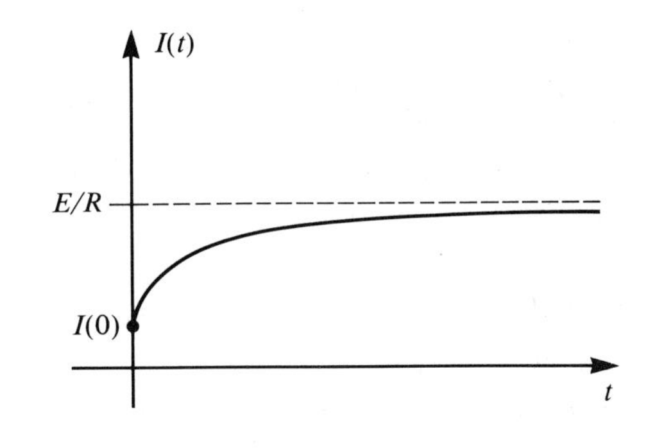

Figure 8

and the solution can be written

$$I(t) = \frac{E}{R} + \left(I(0) - \frac{E}{R}\right)e^{-Rt/L} = \frac{E}{R}(1 - e^{-Rt/L}) + I(0)e^{-Rt/L}.$$

Note that, if $I(0) > E/R$, then $I(t)$ decreases with time, while if $I(0) < E/R$, then $I(t)$ increases with time. The latter is shown in Figure 8.

Case 1(b) If $E(t) = A\sin(\omega t)$, for some positive constant A, then we have a *periodic electromotive force* (of period $2\pi/\omega$). The general solution for $I(t)$ is

$$\begin{aligned} I(t) &= \frac{1}{L}e^{-Rt/L}\int e^{-Rt/L}A\sin(\omega t)\,dt + ke^{-Rt/L} \\ &= \frac{A}{R^2 + \omega^2 L^2}[R\sin(\omega t) - \omega L\cos(\omega t)] + ke^{-Rt/L}. \end{aligned}$$

By a trigonometric identity, this is customarily written as

$$I(t) = \frac{A}{\sqrt{R^2 + \omega^2 L^2}}\sin(\omega t - \delta) + ke^{-Rt/L},$$

where

$$\delta = \arctan\left(\frac{\omega L}{R}\right).$$

Here, δ is called the *phase shift*. As in case 1(a), we can draw a conclusion before determining k. As $t \to \infty$,

$$I(t) \to \frac{A}{\sqrt{R^2 + \omega^2 L^2}}\sin(\omega t - \delta),$$

so that the current tends to behave like a sine wave, as shown in Figure 9.

As in case 1(a), we can determine k if we know, for example, the initial current. Putting $t = 0$ into the solution gives us

$$I(0) = \frac{-A\sin(\delta)}{\sqrt{R^2 + \omega^2 L^2}} + k,$$

so that

$$k = I(0) + \frac{A\sin(\delta)}{\sqrt{R^2 + \omega^2 L^2}}.$$

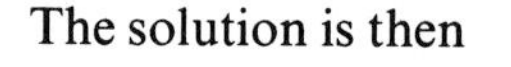

The solution is then

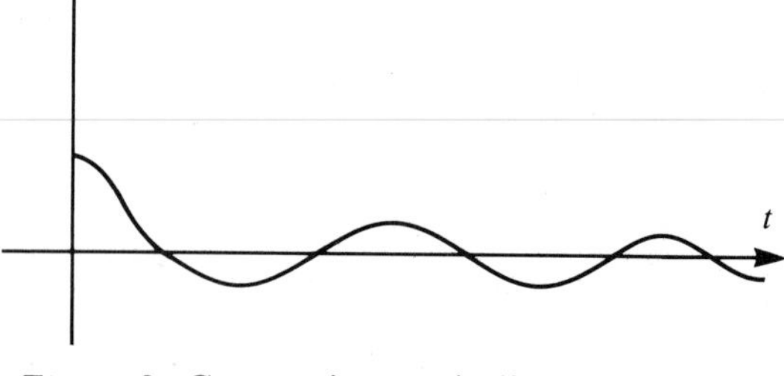

Figure 9. Current in a periodic electromotive force RL circuit.

$$I(t) = \frac{A}{\sqrt{R^2 + \omega^2 L^2}}\sin(\omega t - \delta) + \left(I(0) + \frac{A\sin(\delta)}{\sqrt{R^2 + \omega^2 L^2}}\right)e^{-Rt/L}$$

for current in a periodic electromotive force RL circuit.

CASE 2: *RC* CIRCUITS

A circuit such as in Figure 1, but with no inductance, is called an *RC circuit*. A typical RC circuit is shown in Figure 10. From Example 1, current flow in an RC circuit is governed by

$$RI + \frac{1}{C}Q = E(t),$$

where the charge $Q(t)$ is related to $I(t)$ by

$$I = \frac{dQ}{dt}.$$

Recalling that R and C are constant by assumption, differentiate the circuit equation to get

$$R\frac{dI}{dt} + \frac{1}{C}\frac{dQ}{dt} = \frac{dE}{dt}.$$

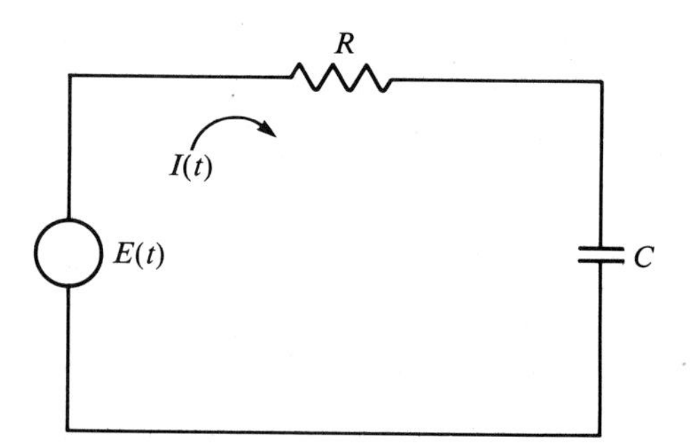

Figure 10. RC circuit.

Replacing dQ/dt with I, we obtain the linear differential equation

$$\frac{dI}{dt} + \frac{1}{RC}I = \frac{1}{R}\frac{dE}{dt}.$$

As with RL circuits, we shall examine this in some detail for two kinds of electromotive force.

Case 2(a) If $E = \text{constant}$, then $dE/dt = 0$, and the equation for the current in a constant electromotive force RC circuit is simply

$$\frac{dI}{dt} + \frac{1}{RC} I = 0.$$

This is separable, since we can write it as

$$\frac{1}{I}\, dI = -\frac{1}{RC}\, dt.$$

The general solution is

$$I(t) = ke^{-t/RC}.$$

As $t \to \infty$, $I(t) \to 0$, so current in a constant electromotive force RC circuit dies out with time.

Case 2(b) If $E = A \sin(\omega t)$ for some positive constant A, we have a *periodic electromotive force*. Now $dE/dt = \omega A \cos(\omega t)$, and the current equation is

$$\frac{dI}{dt} + \frac{1}{RC} I = \frac{\omega A}{R} \cos(\omega t).$$

This linear differential equation has general solution

$$I(t) = \frac{\omega AC}{1 + \omega^2 R^2 C^2} [\cos(\omega t) + \omega RC \sin(\omega t)] + ke^{-t/RC}.$$

As in case 1(b), this can be written

$$I(t) = \frac{\omega AC}{\sqrt{1 + (\omega RC)^2}} \sin(\omega t - \delta) + ke^{-t/RC},$$

where $\delta = \arctan(-1/\omega RC)$. As $t \to \infty$,

$$I(t) \to \frac{\omega AC}{\sqrt{1 + (\omega RC)^2}} \sin(\omega t - \delta).$$

Hence, $I(t)$ tends to behave like a sine wave as t increases. As before, we can find k if we know $I(t)$ at some time.

PROBLEMS FOR SECTION 1.8

1. Obtain an expression for the current $I(t)$ (in amperes) in an RL circuit having a resistance of 7 ohms, an inductance of 5 henrys, and an electromotive force of $\sin(\omega t) + \cos(\omega t)$ volts. Assume that the current at time 0 is 2 amperes.
2. Find the current $I(t)$ in an RL circuit when the resistance is 6 ohms, the inductance 2 henrys, and the electromotive force is $\sin(\omega_1 t) + \sin(\omega_2 t)$ volts. The current at time 0 is $\frac{1}{2}$ ampere.
3. Find the current in an RC circuit having a resistance of 6 ohms, capacitance of 8 farads, and an electromotive force of $t \sin(\omega t)$. The initial current is 2 amperes.
4. Find the current in an RC circuit having a resistance of 18 ohms, a capacitance of 12 farads, and an electromotive force of $\cos(2t) + \sin(3t)$ volts. The initial current is 3 amperes.

5. In a constant electromotive force RL circuit, we got

$$I(t) = \frac{E}{R}(1 - e^{-Rt/L}) + I(0)e^{-Rt/L}.$$

Assume for convenience that $I(0) = 0$.

(a) Show that $I(t)$ is increasing with time.

(b) Find t_0 such that $I(t_0)$ is about 63% of its limiting value E/R. This value of t_0 is called the *inductive time constant* of the circuit.

(c) Would the inductive time constant change if $I(0)$ were nonzero? If so, in what way?

6. Using the fact that $I = dQ/dt$ (see Example 1), derive the equation

$$\frac{dQ}{dt} + \frac{1}{RC}Q = \frac{1}{R}E(t)$$

for charge $Q(t)$ in an RC circuit.

(a) Solve this for Q in the case of constant electromotive force. To evaluate the constant of integration, let the charge at time 0 be Q_0.

(b) Show that, as $t \to \infty$, $Q(t)$ has a limiting value which is independent of Q_0. Find this limiting value.

(c) Graph Q as a function of t. For what value of time does $Q(t)$ have a maximum? What is the maximum value of Q?

(d) Determine at what time Q has been reduced to 1% of its original value Q_0. Does this value of t depend on Q_0? If so, in what way?

7. Using the differential equation of Problem 6, determine the charge in an RC circuit with electromotive force $E(t) = A\cos(\omega t)$, where A and ω are positive constants. To evaluate the constant of integration, let $I(0) = I_0$.

8. Solve for the current in an RL circuit in which $R = 2$ ohms, $L = 25$ henrys, and the electromotive force is a series of two impulses, say

$$E(t) = \begin{cases} 1, & 0 \le t \le 10 \\ 0, & 10 < t < 20 \\ 1, & 20 \le t \le 30 \\ 0, & t > 30 \end{cases}$$

Discuss the behavior of $I(t)$ with increasing time. Graph I as a function of time.

9. Solve for the current in an RL circuit in which $R = 2$ ohms, $L = 25$ henrys, and the electromotive force is of the form $E(t) = Ae^{-t}$, with A any positive constant. Graph $I(t)$ as a function of time.

10. Solve for the current in an RL circuit in which the electromotive force has the form $E(t) = A\cos(\omega t) + Be^{-t}$, with A and B positive constants. Treat the general case, in which R is any resistance and L any inductance. Graph I as a function of time, and determine what happens as t increases.

11. Repeat Problem 10, with an RC circuit in place of the RL circuit of that problem.

12. Obtain an expression for the current $I(t)$ in an RL circuit having a resistance of R ohms, an inductance of L henrys, and an electromotive force of $E(t) = A + B/(t + 1)$ volts. The current at time $t = 0$ is 3 amperes.

13. Find the current $I(t)$ in an RL circuit if the electromotive force is given by $E(t) = A + B/(t + 1)^\alpha$, where A and B are positive and α is a positive integer. Discuss how various choices of α influence the resulting current. Assume that $I(0) = 0$.

14. Find the current in an RL circuit having electromotive force $E(t) = A\sin(\omega_1 t) + B\cos(\omega_2 t)$. How is the solution affected by choices of ω_1 and ω_2? Assume that $I(0) = 0$.

15. Solve for the current in an RC circuit having a resistance of R ohms, capacitance of C farads, and electromotive force $E(t) = 1 - \cos(2t)$. Assume that $I(0) = 0$.

16. Solve for the current in an RC circuit if $I(0) = 0$ and the electromotive force is $E(t) = 1 + \sin(3t)$.
17. Find the charge $Q(t)$ in an RL circuit if $I(0) = 0$, $Q(0) = 0$, and $E(t) = 1 + e^{-t}$.
18. Find the charge $Q(t)$ in an RC circuit if $I(0) = 0$, $Q(0) = 0$, and $E(t) = e^{-t} - e^{-2t}$.

1.9 EXISTENCE, UNIQUENESS, AND THE PICARD ITERATION SCHEME

Thus far we have developed methods for solving some (but not all) first order differential equations. We shall now confront two issues which have been avoided up to now.

One issue is *existence*. It is easy to make up a differential equation which cannot possibly have any solution, for example,

$$(y')^2 + y^2 = -1.$$

The second issue is *uniqueness*. All our experience to date suggests that first order equations often have infinitely many solutions, but only one satisfying an additional condition.

In order to formulate a precise statement concerning existence and uniqueness, we introduce the following terminology: A first order differential equation

$$y' = f(x, y)$$

together with the condition

$$y(x_0) = y_0,$$

with x_0 and y_0 given constants, is called an *initial value problem*. For example,

$$y' = 2xy + x^2; \qquad y(1) = 4$$

is an initial value problem, as is

$$y' = x^3 + 1; \qquad y(0) = -1.$$

In practice, we have actually solved many initial value problems. Here is the theorem underlying what we were doing:

THEOREM 1

Suppose that $f(x, y)$ and $\partial f/\partial y$ are continuous in a rectangle R centered at the point (x_0, y_0). Then, there is an interval $x_0 - h \le x \le x_0 + h$ about x_0 on which there exists a unique solution of the initial value problem

$$y' = f(x, y); \qquad y(x_0) = y_0.$$

In words, Theorem 1 says that there exists exactly one solution of $y' = f(x, y)$ passing through a given point (x_0, y_0), provided that $f(x, y)$ and its first partial derivative $\partial f/\partial y$ are continuous on some rectangle about (x_0, y_0). This unique solution is defined on *some* interval of length $2h$ about x_0. In general, the value of h will depend upon $f(x, y)$ and the given point (x_0, y_0), and one cannot make a general statement about how large it will be. These ideas are depicted in Figure 11.

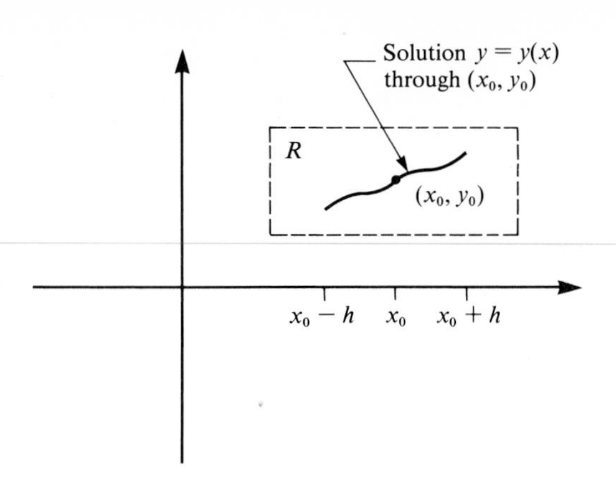

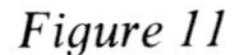

Figure 11

EXAMPLE 34

Consider the initial value problem

$$y' = 2x^3y + y^2x^2 - e^{x\cos(y)}; \qquad y(1) = 4.$$

Here,

$$f(x, y) = 2x^3y + y^2x^2 - e^{x\cos(y)}$$

and

$$\frac{\partial f}{\partial y} = 2x^3 + 2yx^2 + x\sin(y)e^{x\cos(y)}.$$

Both $f(x, y)$ and $\partial f/\partial y$ are continuous on any rectangle you choose about (1, 4). Thus, there is a unique solution $y = y(x)$ defined on *some* interval about $x_0 = 1$. We are not able to produce this solution by elementary means because $f(x, y)$ is too complicated in this example.

We do not intend to prove Theorem 1 in this treatment, but we should mention the *Picard iteration scheme* used in one of the standard proofs.

Consider the initial value problem

$$y' = f(x, y); \qquad y(x_0) = y_0.$$

Define a sequence of functions as follows: Let

$$y_1(x) = y_0 + \int_{x_0}^{x} f(t, y_0)\,dt$$

$$y_2(x) = y_0 + \int_{x_0}^{x} f(t, y_1(t))\,dt$$

$$y_3(x) = y_0 + \int_{x_0}^{x} f(t, y_2(t))\,dt$$

$$\vdots \qquad\qquad \vdots$$

$$y_n(x) = y_0 + \int_{x_0}^{x} f(t, y_{n-1}(t))\,dt.$$

Thus, we generate $y_1(x)$ from y_0 and $f(x, y)$, $y_2(x)$ from $y_1(x)$ and $f(x, y)$, and so on, determining each function in the sequence from the one immediately preceding it and from $f(x, y)$. One means of proving Theorem 1 is to show that this sequence converges, and that in fact its limit is the solution we seek. Here is an example.

EXAMPLE 35

Consider the initial value problem (with linear differential equation)

$$y' + y = 2; \qquad y(0) = 1.$$

We can solve this problem exactly and then see how the Picard iteration scheme does. The solution (unique by Theorem 1) is

$$y(x) = 2 - e^{-x}.$$

Now generate some of the functions in Picard's sequence. In the initial value problem,

$$f(x, y) = 2 - y; \qquad x_0 = 0 \quad \text{and} \quad y_0 = 1.$$

(It just happens here that f is independent of x.) Calculate

$$y_1 = 1 + \int_0^x f(t, 1)\, dt = 1 + \int_0^x (2 - 1)\, dt = 1 + x,$$

$$y_2 = 1 + \int_0^x f(t, 1 + t)\, dt = 1 + \int_0^x (2 - (1 + t))\, dt = 1 + x - \frac{x^2}{2},$$

$$y_3 = 1 + \int_0^x f(t, y_2(t))\, dt = 1 + \int_0^x \left[2 - \left(1 + t - \frac{t^2}{2}\right)\right] dt = 1 + x - \frac{x^2}{2} + \frac{x^3}{6},$$

$$y_4 = 1 + \int_0^x f(t, y_3(t))\, dt$$

$$= 1 + \int_0^x \left[2 - \left(1 + t - \frac{t^2}{2} + \frac{t^3}{6}\right)\right] dt = 1 + x - \frac{x^2}{2} + \frac{x^3}{6} - \frac{x^4}{24},$$

and so on. There is a clear pattern emerging, and we conjecture that in fact

$$y_n(x) = 1 + x - \frac{x^2}{2} + \frac{x^3}{2 \cdot 3} - \frac{x^4}{2 \cdot 3 \cdot 4} + \frac{x^5}{2 \cdot 3 \cdot 4 \cdot 5} - \cdots + (-1)^n \frac{x^n}{2 \cdot 3 \cdot 4 \cdots n}$$

$$= 1 - \sum_{j=1}^{n} (-1)^j \frac{x^j}{j!}.$$

Here we have used the *factorial notation*: $j!$ (read "j factorial") is the product of the integers from 1 to j if j is a positive integer, and $0!$ is defined to be 1. Thus,

$$2! = 2, \qquad 3! = 2 \cdot 3 = 6, \qquad 4! = 2 \cdot 3 \cdot 4 = 24,$$

and so on.

Those familiar with power series will recall that

$$e^{-x} = \sum_{j=0}^{\infty} \frac{(-1)^j x^j}{j!} = 1 - x + \frac{x^2}{2} - \frac{x^3}{2 \cdot 3} + \frac{x^4}{2 \cdot 3 \cdot 4} - \frac{x^5}{2 \cdot 3 \cdot 4 \cdot 5} + \cdots$$

Thus, as $n \to \infty$, $y_n(x) \to 2 - e^{-x}$, the solution of the initial value problem.

This example was chosen to illustrate the idea with a minimum of detail. The student should not, however, conclude that Picard iteration will often yield the solution in closed form. Two unusual features of Example 35 were (1) we could guess $y_n(x)$ after calculating just a few of the functions of the sequence, and (2) we could recognize $\lim_{n\to\infty} y_n(x)$ in closed form as a familiar function. Usually both of these fortunate circumstances fail to materialize. Further, in most cases even the first few y_n's are quite tedious to calculate. Thus, the Picard scheme is a useful theoretical tool in proving Theorem 1, but is of relatively little value in finding, or even approximating, solutions of specific initial value problems. Better approximation methods can be found in the chapter on numerical methods.

PROBLEMS FOR SECTION 1.9

In each of Problems 1 through 10, verify by Theorem 1 that the given initial value problem has a unique solution. Then use the methods of the preceding sections to find the solution. In applying Theorem 1, you may assume continuity of elementary functions involving polynomials, exponentials, trigonometric functions, and the like.

1. $y' = x - 2y;\ \ y(1) = 3$

2. $(3x^2y + 3y^2)\,dx + (x^3 + 6xy)\,dy = 0;\ \ y(2) = 4$

3. $y' = \dfrac{y^2}{x^2} + 2\dfrac{y}{x};\ \ y(1) = 6$

4. $y' + 3y = \cos(x);\ \ y(\pi) = 4$

5. $y' - \dfrac{1}{x}y = 2x^3;\ \ y(1) = 1$

6. $[\cos(xy) - xy\sin(xy)]\,dx - x^2\sin(xy)\,dy = 0;\ y(4) = -1$

7. $y' - 4y = e^{2x};\ \ y(0) = -4$

8. $y' = \dfrac{x - y}{x};\ \ y(-2) = 6$

9. $y' = \sin(x) - 4y;\ \ y(\pi) = -3$

10. $2xy' + y = 4;\ \ y(1) = 1$

In each of Problems 11 through 20, find $y_1(x)$ through $y_5(x)$ in the Picard iteration scheme.

11. $y' = 4 - y;\ \ y(0) = 0$

12. $y' = 2x^2;\ \ y(1) = 3$

13. $y' = xy;\ \ y(0) = 1$

14. $y' = ye^x;\ \ y(1) = 1$

15. $y' = 2 - x;\ \ y(2) = -2$

16. $y' = x^2y;\ \ y(0) = 1$

17. $y' = y + x;\ \ y(1) = 4$

18. $y' = (1 - x^2)y;\ \ y(0) = -2$

19. $y' = y\sin(x);\ \ y(\pi/2) = 1$

20. $y' = 2y + x^2y;\ \ y(0) = 3$

In each of Problems 21 through 30, use Theorem 1 to show that the given initial value problem has a unique solution in some interval about the given value of x_0. Do not attempt to find the solutions of these problems.

21. $y' = 2y^2 + 3xe^y\sin(xy);\ \ y(2) = 4$

22. $y' = 4xy + \cosh(x);\ \ y(1) = -1$

23. $y' = (xy)^3 - \sin(y);\ \ y(2) = 2$

24. $y' = x^5 - y^5 + 2xe^y;\ \ y(3) = \pi$

25. $y' = (2x + 3y)^2 - e^x;\ \ y(1) = 1$

26. $y' = x^2e^{-2x}y + y^2;\ \ y(3) = 8$

27. $y' = xy^2 + e^y;\ \ y(1) = 4$

28. $y' = 2xy + \sinh(x + y);\ \ y(0) = -3$

29. $y' = y^2 - x^2 + \cos(xy);\ \ y(1) = 2$

30. $y' = 2xy^2 + e^{xy}\sin(x);\ \ y(3) = -4$

31. Find two different solutions of the initial value problem

$$y' = y^{1/3}; \qquad y(0) = 0.$$

Why does this not contradict Theorem 1?

1.10 ISOCLINES, DIRECTION FIELDS, AND SKETCHING SOLUTIONS

In many (but not all) instances, a first order differential equation can be written

$$\frac{dy}{dx} = f(x, y) \tag{a}$$

with a single derivative term isolated on one side. In such a case we can often get some idea of what solutions look like geometrically without actually solving the equation.

Suppose that $f(x, y)$ is defined over some region D of the xy-plane. At any point (x_0, y_0) in D, the graph of a solution $y = y(x)$ of (a) has slope $y'(x_0) = f(x_0, y_0)$, where $y_0 = y(x_0)$. Thus, we know the *slope* of each solution curve through each point of D. Knowing this, we can construct a picture of the solution curves themselves, as follows:

First, draw the curves

$$f(x, y) = k$$

for each constant k. (Of course, in practice, we draw only a few such curves to get an idea of their general appearance.) These curves are called *isoclines* of (a). Along each isocline, draw a number of short line segments of slope k, as in Figure 12. Each such

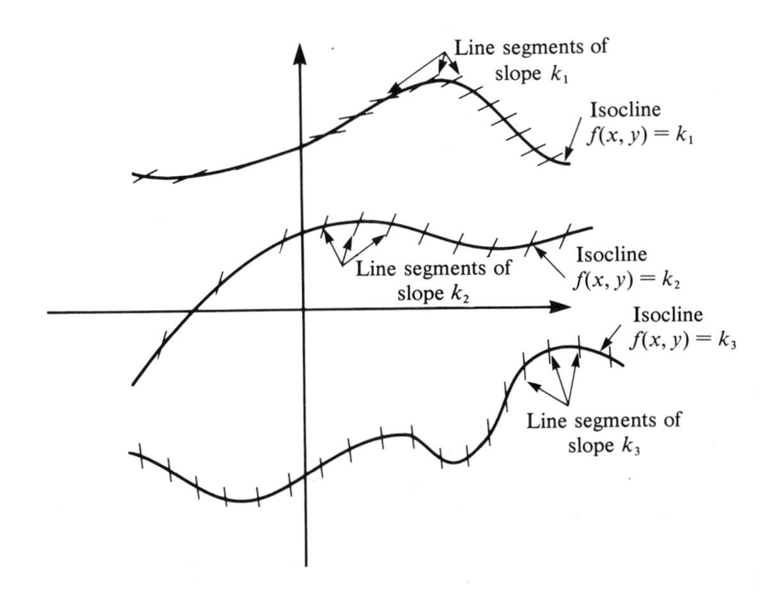

Figure 12. Isoclines and direction field.

line segment shows the direction of the solution passing through that point of that isocline. Using these segments as guides, we can sketch solutions passing through any given point of D. Appropriately enough, these short line segments constitute the *direction field* of the differential equation.

Here are two examples of isoclines, direction fields, and their use in sketching solutions.

EXAMPLE 36

Consider the differential equation

$$y' = x^2 y.$$

This is separable, and can be solved exactly by the method of Section 1.1 to yield

$$y(x) = Ce^{x^3/3},$$

where C is any constant.

Now look at $y' = x^2 y$ from a purely geometric point of view. The equation is $y' = f(x, y)$, with $f(x, y) = x^2 y$. This is defined everywhere, so we let D be the entire xy-plane. The isoclines are curves

$$x^2 y = k$$

or

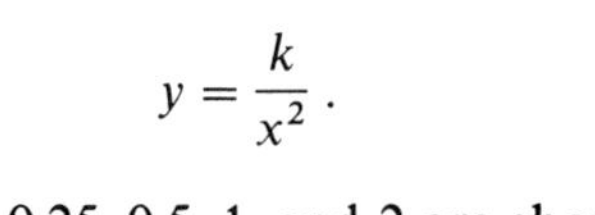

$$y = \frac{k}{x^2}.$$

Isoclines corresponding to $k = 0.25$, 0.5, 1, and 2 are shown in Figure 13. The short line segments on each isocline make up the direction field of the differential equation.

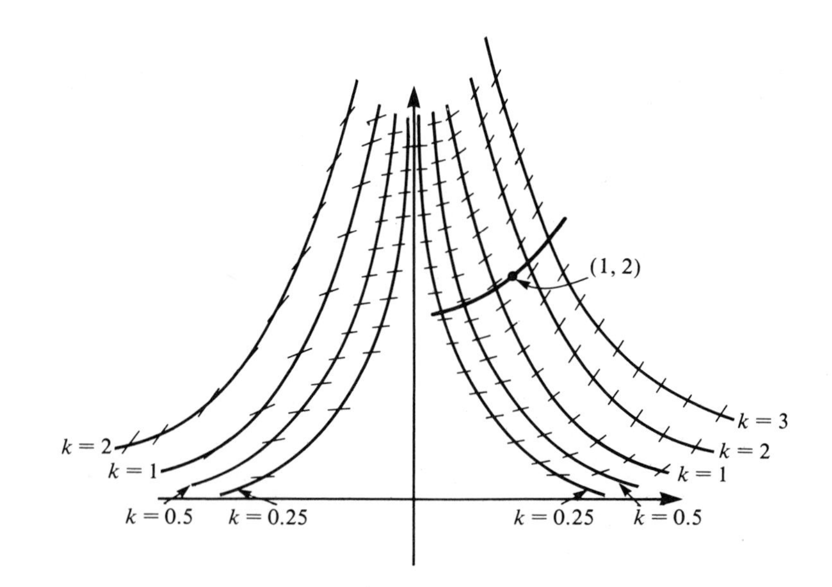

Figure 13. Isoclines and direction field for $y' = x^2 y$; solution through (1, 2).

Each segment on the isocline $y = k/x^2$ has slope k, and the totality of line segments gives a qualitative picture of the solutions.

For example, suppose we want a solution passing through (1, 2). We sketch in Figure 13 through (1, 2) a curve whose direction coincides with that of the direction field on each isocline it crosses. This does not give the exact solution, but rather a general idea of its behavior near (1, 2). We can see, for example, that it is increasing rapidly as x increases from values just below 1 to values greater than 1. To see how this conclusion holds up, we can in this example obtain the solution through (1, 2) explicitly; it is $y = 2e^{(x^3-1)/2}$, whose graph does in fact resemble the sketch drawn from the direction field in Figure 13.

EXAMPLE 37

Consider

$$y' = x - y.$$

This is a linear equation, since it can be written

$$y' + y = x.$$

Using the method of Section 1.6, we get solutions

$$y(x) = x - 1 + Ce^{-x}.$$

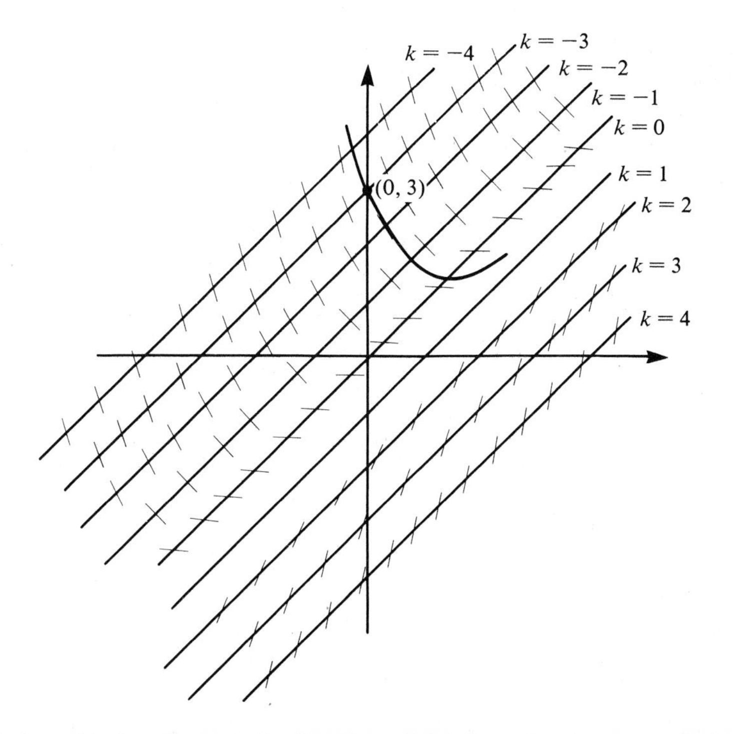

Figure 14. Isoclines and direction field for $y' = x - y$; solution through (0, 3).

Now look at the equation geometrically. The isoclines are curves

$$x - y = k,$$

and these are straight lines, as shown in Figure 14. The short line segments constitute the direction field of $y' = x - y$. Each line segment on $x - y = k$ has slope k (the segments on $x - y = 1$ are of slope 1, and so are on this isocline and don't show up separately).

As an illustration, we have used the direction field to sketch a solution through (0, 3) in Figure 14. The exact solution through (0, 3) is $y = x - 1 + 4e^{-x}$.

The method we have discussed here is sometimes useful for getting a quick idea of the behavior of the solution of $y' = f(x, y)$ in the vicinity of the point (x_0, y_0). It is, however, just a rough guide which can be used to determine a further course of action or perhaps draw some conclusion about qualitative behavior of solutions. In a later chapter we shall discuss methods of approximating solutions numerically to a high degree of accuracy.

PROBLEMS FOR SECTION 1.10

In each of Problems 1 through 20, (a) draw some isoclines of the given equation near the given point; (b) draw in some of the direction field elements; (c) use the direction field to sketch a solution through the given point; (d) make some conclusions about qualitative behavior of the solution through the given point from the sketch; and (e) obtain the exact solution through the given point and compare conclusions from (d) with those you can draw from the solution itself.

1. $y' = x + y$; (2, 2)

2. $y' + 2xy = x$; (1, 3)

3. $y' = x^2 + 2y$; (2, 4)

4. $y' = e^x - y$; (1, 1)

5. $y' = xy$; (0, 4)

6. $y' = y(1 + x)$; (2, 3)

7. $y' + \frac{1}{x} y = 2x$; (3, 3)

8. $y' - xy = 2x$; (3, 1)

9. $y' + 2y = \sin(x)$; $(\pi, 2)$

10. $y' + 2xy = x^2$; (3, 4)

11. $y' = 3x - y$; (1, −4)

12. $y' + 3x = \sin(x)$; (1, 2)

13. $y' + \frac{1}{x} y = e^{-x}$; (−1, 3)

14. $y' - \frac{2}{x} y = e^x$; (1, 1)

15. $y' + y = 2 - x$; (2, −5)

16. $y' - 2y = x^2 + 1$; (−1, 7)

17. $y' - \frac{3}{x} y = x^4 + 2$; (3, 2)

18. $y' + y = 1 - 3x$; (2, 2)

19. $y' = 2y - x$; (1, 3)

20. $y' - 8y = 1$; (1, 4)

In each of Problems 21 through 30, use the direction field to sketch the solution in the vicinity of the given point. Do not solve the differential equation exactly.

21. $y' = xy^2 - 2$; (1, 3)

22. $y' = 8x^2y + 4e^{-x}$; (0, 1)

23. $y' = x\cos(2x) - y$; (1, 1)

24. $y' = 2x + e^xy$; (−3, 1)

25. $y' = 3x^2y - 2y$; (0, 1)

26. $y' = \sin(x)y - 3x^2$; (0, 1)

27. $y' - y^2x = e^{-2x} + 1$; (1, 2)

28. $y' = 7x^3y^2 - 2x$; (−1, 1)

29. $y' = 3x^2y - e^{-x}$; (2, −4)

30. $y' = 4x^3 - 3y + x\cos(2x)$; (−1, 2)

1.11 ORTHOGONAL AND OBLIQUE TRAJECTORIES

In many contexts we encounter two families of curves, in which each curve of one family is *orthogonal* (or *perpendicular*) to each curve of the other family. Examples are the parallels and meridians on a globe, or the equipotential lines and electric force lines in electrostatics. In this section we show how differential equations enter into such considerations.

As background, consider the notion of a *family of curves*. By this we mean simply an expression

$$F(x, y, K) = 0,$$

in which K is a parameter (constant) left to our choice. For each value of K, we obtain an equation in x and y which graphs as a curve in the xy-plane. The totality of such curves comprises the family.

EXAMPLE 38

Let

$$F(x, y, K) = x^2 + y^2 - K^2.$$

The equation

$$F(x, y, K) = 0$$

is then just

$$x^2 + y^2 = K^2.$$

When $K = 0$, this degenerates to a single point, (0, 0); if $K \neq 0$, we get a circle of radius $|K|$ about the origin. Thus, the family of curves in this example consists of all circles about the origin, together with the origin itself.

EXAMPLE 39

Let

$$F(x, y, K) = x^2 - y^2 - K^2.$$

The family $F(x, y, K) = 0$ consists of all hyperbolas $x^2 - y^2 = K^2$ when $K \neq 0$, and the straight lines $y = x$ and $y = -x$ corresponding to $K = 0$.

As we have seen in many examples, a general solution of a first order differential equation contains one arbitrary constant, and so may be thought of as a family of curves. For example, the family $y = x^3 + x + C$ is the general solution of $y' = 3x^2 + 1$.

Conversely, a family of curves can often be thought of as a general solution of a differential equation, which is then called the *differential equation of the family*. To find this differential equation from a family $F(x, y, K) = 0$, differentiate to eliminate the parameter K, as in these two examples.

EXAMPLE 40

Consider the family of circles $x^2 + y^2 - K^2 = 0$. Implicitly differentiating with respect to x gives us

$$2x + 2yy' = 0$$

or

$$y' = -\frac{x}{y}.$$

Here, $y' = -x/y$ is the differential equation of the family $x^2 + y^2 - K^2 = 0$. Conversely, the family of circles is a general solution of the differential equation.

EXAMPLE 41

Consider the family

$$y = \frac{3x^3}{4} + \frac{K}{x}.$$

To eliminate K, calculate

$$y' = \frac{9x^2}{4} - \frac{K}{x^2}.$$

Then,

$$K = x^2\left(\frac{9x^2}{4} - y'\right).$$

Substituting this for K into the family gives us

$$y = \frac{3x^3}{4} + \frac{1}{x}x^2\left(\frac{9x^2}{4} - y'\right),$$

which simplifies algebraically to

$$y' + \frac{1}{x}y = 3x^2.$$

This is the differential equation of the family; and the family is a general solution of the differential equation.

In many applications we are given a family of curves, and we want to find the *orthogonal trajectories*, that is, a second family of curves, each of which is orthogonal (perpendicular) to each curve of the first family. We can do this as follows: Suppose that the given family is $F(x, y, K) = 0$. Eliminate K to find the differential equation $y' = f(x, y)$ of this family. At each (x_0, y_0), $f(x_0, y_0)$ is the slope of the curve in family $F(x, y, K) = 0$ passing through (x_0, y_0). A curve through (x_0, y_0) orthogonal to this curve must have slope*

$$\frac{-1}{f(x_0, y_0)}.$$

* Recall that, for two lines to be perpendicular, the slope of one is the negative reciprocal of the slope of the other.

Thus, the family orthogonal to $F(x, y, K) = 0$ has differential equation

$$y' = -\frac{1}{f(x, y)},$$

and the general solution of this differential equation is the family of orthogonal trajectories.

EXAMPLE 42

Suppose we want the orthogonal trajectories of the family of circles $x^2 + y^2 - K^2 = 0$. We found in Example 40 that this family has differential equation $y' = -x/y$. Thus, the orthogonal trajectories have differential equation

$$y' = \frac{y}{x}.$$

This is a separable differential equation, and we find that the general solution is

$$y = Cx.$$

This family of straight lines constitutes the orthogonal trajectories of the given family of circles. Figure 15 shows some of the circles and lines of these families, and one can see that each line is orthogonal to each circle.

EXAMPLE 43

Find the orthogonal trajectories of the family $x^2 + Ky^2 = 4$.

First, we need the differential equation of this family. Differentiating with respect to x gives us

$$2x + 2Kyy' = 0$$

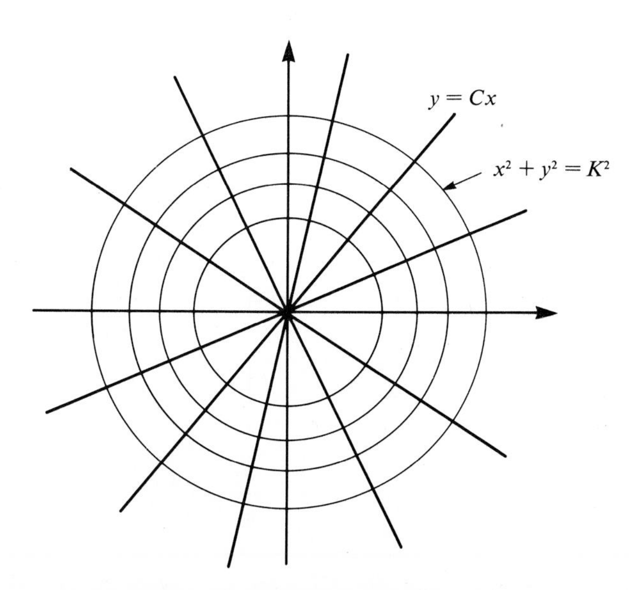

Figure 15. Orthogonal trajectories.

or

$$y' = \frac{-x}{Ky}.$$

From $x^2 + Ky^2 = 4$, we have

$$K = \frac{4 - x^2}{y^2}.$$

Then,

$$y' = \frac{-x}{y}\frac{1}{K} = \frac{-xy}{4 - x^2}.$$

The differential equation of the orthogonal trajectories is

$$y' = \frac{4 - x^2}{xy}.$$

This is a separable differential equation whose general solution is

$$\frac{y^2}{2} = 4 \ln|x| - \frac{x^2}{2} + C,$$

and this is the family of orthogonal trajectories of $x^2 + Ky^2 = 4$.

Once the idea behind orthogonal trajectories is understood, it is fairly easy to digest the notion of oblique trajectories. Suppose we are given a family of curves $F(x, y, K) = 0$. Any curve which intersects each curve of this family at a given angle α ($\alpha \neq 90°$) is called an *oblique trajectory* of the family. The totality of oblique trajectories of $F(x, y, K) = 0$ is called the *family of oblique trajectories of* $F(x, y, K) = 0$. We shall now describe how to find this family of oblique trajectories.

Suppose that the differential equation of the family $F(x, y, K) = 0$ is $y' = f(x, y)$. The solution curve C of this differential equation passing through a given point (x, y) has slope $f(x, y)$ at (x, y). Thus, the tangent to C at (x, y) makes an angle $\theta = \arctan[f(x, y)]$ with the x-axis, as shown in Figure 16. An oblique trajectory $C^{\#}$, intersecting C at an angle α at (x, y), will have a tangent line making an angle α with the tangent to C at (x, y). The slope of $C^{\#}$ at (x, y) is then $\tan(\theta + \alpha)$ or

$$\tan\{\arctan[f(x, y)] + \alpha\}.$$

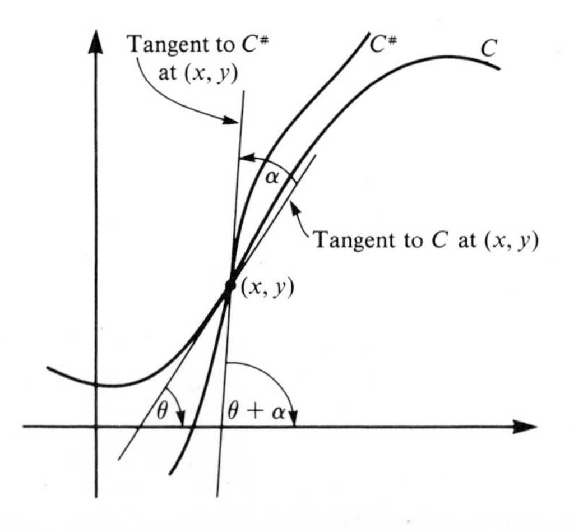

Figure 16

Thus,* at (x, y) the oblique trajectory has slope

$$\frac{\tan\{\arctan[f(x, y)]\} + \tan(\alpha)}{1 - \tan\{\arctan[f(x, y)]\}\tan(\alpha)},$$

which is the same as

$$\frac{f(x, y) + \tan(\alpha)}{1 - f(x, y)\tan(\alpha)}.$$

The differential equation of the family of oblique trajectories is then

$$y' = \frac{f(x, y) + \tan(\alpha)}{1 - f(x, y)\tan(\alpha)}.$$

The family of oblique trajectories is the general solution of this differential equation.

EXAMPLE 44

Find the family of oblique trajectories intersecting the circles $x^2 + y^2 = K^2$ at an angle of 45°.

First, we know from Example 40 that the family of circles has differential equation

$$y' = -\frac{x}{y}.$$

Thus, $f(x, y) = -x/y$ in the above discussion. Since $\tan(45°) = 1$, the differential equation of the oblique trajectories is

$$y' = \frac{-\dfrac{x}{y} + 1}{1 + \dfrac{x}{y}}$$

or

$$y' = \frac{y - x}{y + x}.$$

This is a homogeneous differential equation with general solution

$$\frac{1}{2}\ln\left|\frac{y^2}{x^2} + 1\right| + \arctan\left(\frac{y}{x}\right) = -\ln|x| + C,$$

and this is the family of oblique trajectories.

EXAMPLE 45

Find a family of oblique trajectories intersecting the lines $y = Kx$ at 30°.

The differential equation of the family $y = Kx$ is found to be

$$y' = \frac{y}{x}.$$

* Here we use the trigonometric identity $\tan(A + B) = [\tan(A) + \tan(B)]/[1 - \tan(A)\tan(B)]$.

Thus, $f(x, y) = y/x$. The oblique trajectories have differential equation

$$y' = \frac{\frac{y}{x} + \tan(30^\circ)}{1 - \frac{y}{x}\tan(30^\circ)}.$$

This is a homogeneous equation with general solution

$$\frac{1}{\tan(30^\circ)} \arctan\left(\frac{y}{x}\right) - \frac{1}{2}\ln\left(1 + \frac{y^2}{x^2}\right) = \ln|x| + C.$$

This is the family of oblique trajectories.

These considerations of orthogonal and oblique trajectories can be recast in polar coordinates. To illustrate, we shall do this for orthogonal trajectories.

Suppose we have a family of curves

$$F(\theta, r, K) = 0,$$

in which θ and r are polar coordinates. Let the differential equation of this family be $f(\theta, r, r') = 0$, in which $r' = dr/d\theta$. We claim that the family of orthogonal trajectories has differential equation

$$f\left(\theta, r, \frac{-r^2}{r'}\right) = 0,$$

obtained by replacing r' by $-r^2/r'$ in the differential equation of the given family.

To see this, recall the following fact from calculus: If C is a curve given in polar coordinates, and ψ is the angle from the radius to the tangent at a point P, as shown in Figure 17, then

$$\tan(\psi) = \frac{r}{r'}.$$

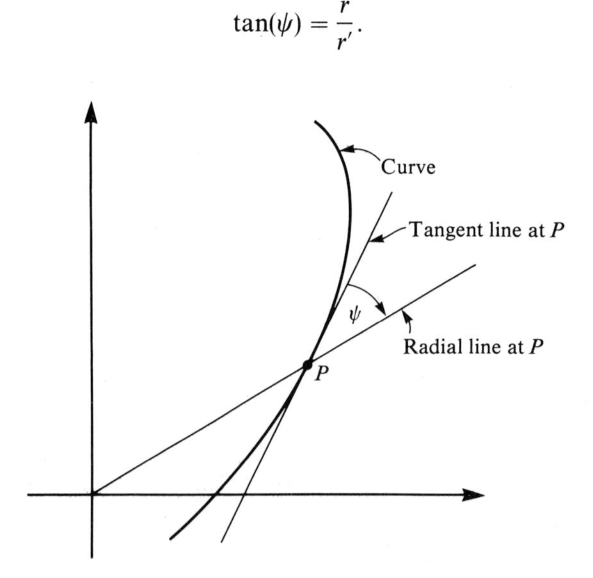

Figure 17

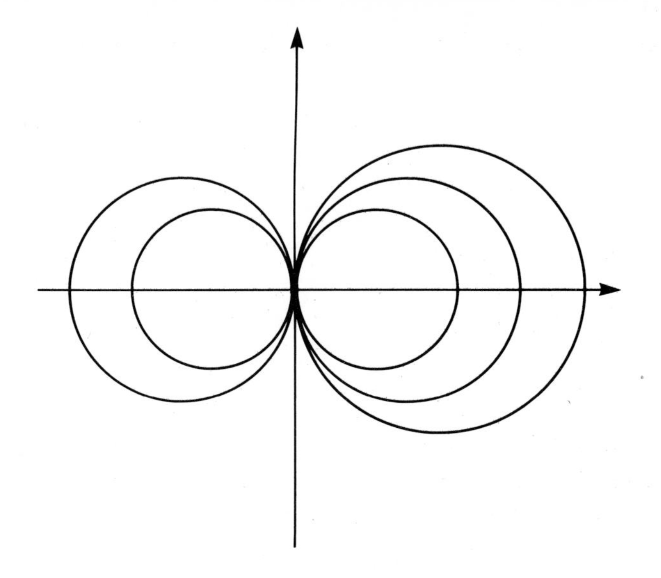

Figure 18. Circles in the family $r = K\cos(\theta)$; centers are $(0, \frac{1}{2}K)$, radii $\frac{1}{2}|K|$.

Then, for a curve orthogonal to C at P, the tangent of the angle from the radius to the tangent line is

$$\frac{r}{-1/\tan(\psi)}$$

or

$$\frac{-r^2}{r'}.$$

EXAMPLE 46

Consider the family $r = K\cos(\theta)$ of circles tangent to the y-axis at the origin, as shown in Figure 18. Here, K must be positive as it represents a diameter of a circle. Differentiating, we get

$$r' = -K\sin(\theta),$$

so that

$$K = \frac{-r'}{\sin(\theta)}.$$

The differential equation of the family is then

$$\frac{r'}{r} = \frac{-\sin(\theta)}{\cos(\theta)}.$$

For the orthogonal trajectories, replace r' by $-r^2/r'$ to get

$$\frac{r}{r'} = \frac{\sin(\theta)}{\cos(\theta)}.$$

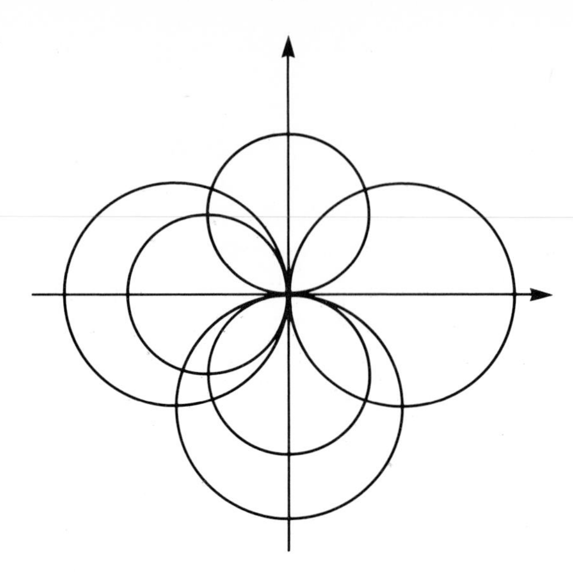

Figure 19. Orthogonal families $r = K\cos(\theta)$ and $r = C\sin(\theta)$.

This is separable, with general solution

$$r = C\sin(\theta),$$

where C is any positive constant. This is a family of circles tangent to the x-axis at the origin. The two families are shown in Figure 19.

PROBLEMS FOR SECTION 1.11

In each of Problems 1 through 20, find the orthogonal trajectories of the given family of curves.

1. $x + 2y = K$
2. $2x^2 - 3y = K$
3. $\frac{1}{2}x^2 + y^2 = K$
4. $x + Ky = 1$
5. $x^2 - 3y^2 = K$
6. $y = 2Kx^2 + 1$
7. $x + 2y^2 = K$
8. $x^2 - Ky^2 = 1$
9. $x + y = Kx^2$
10. $y = e^{Kx}$
11. $y = Ke^x$
12. $y = (1 + Kx)/(1 - Kx)$
13. $y^2 + (x - K)^2 = 1$
14. $x - 2y = K$
15. $y = (x - K)^2$
16. $y^2 - x^2 = K$
17. $2x - 3(y + K) = 4$
18. $y^2 = Kx^3$
19. $x^2 - Ky = 1$
20. $2x - 3yK = 1$

In each of Problems 21 through 30, find oblique trajectories making the given angle with the given family of curves.

21. $x^2 - y^2 = K^2$; $45°$
22. $x^2 + y^2 = K^2$; $60°$
23. $y + Kx = 1$; $30°$
24. $y = 2x + K$; $45°$
25. $Ky - x = 1$; $30°$
26. $x^2 + (y - K)^2 = 1$; $30°$
27. $y = Ke^x$; $45°$
28. $x^2 + y^2 = Ky$; $60°$
29. $x = Ke^y$; $30°$
30. $x^2 = Ky$; $60°$

31. Let $F(\theta, r, K) = 0$ be a family of curves in polar coordinates. Derive a method for finding the oblique trajectories of this family, with an angle α ($\alpha \neq 90°$).

In each of Problems 32 through 36, find the orthogonal trajectories of the given family in polar coordinates.

32. $r = K\theta$ **33.** $r = Ke^{-\theta}$ **34.** $r = K[1 - \cos(\theta)]$
35. $r = K\theta^2$ **36.** $r = K\ln(\theta)$

CHAPTER SUMMARY

This chapter was devoted to specific types of first order differential equations for which some method may enable us to produce solutions, either explicit or implicitly defined. In general, the easiest kind of first order equation to solve is the separable one, which reduces immediately to an exercise in integration. Homogeneous equations $y' = f(y/x)$ can be rendered separable by a change of variables $u = y/x$. Equations $y' = f[(ax + by + c)/(dx + ey + h)]$, while usually not homogeneous or separable, can similarly be handled by appropriate changes of variables.

The method available for solving exact differential equations leads us to consider the notion of an integrating factor. Often, finding an integrating factor presents a serious problem, but in some cases proves tractable. In particular, there is a formula for an integrating factor for any Bernoulli equation, as there is for any linear equation. In fact, one can write a formula for the general solution of any linear equation, and the only obstacle to obtaining an explicit solution remains in performing the necessary integrations. Riccati equations represent an important class which can be approached through linear equations and a change of variables.

In all of these considerations, one must distinguish between a general solution, which in the first order case contains an arbitrary constant, and particular solutions. Geometrically, we may think of a general solution of a first order differential equation as a family of curves, while an initial value problem has as solution a particular curve of this family. The geometric aspects of first order equations are also apparent in consideration of orthogonal and oblique trajectories.

For first order equations, one can think of our treatment as a collection of techniques aimed at finding solutions of particular kinds of equations. In this context, the linear differential equation is simply one of the types we can solve, but most of the first order equations considered were nonlinear. As we go on to higher order differential equations, linearity will become more important. In fact, we shall restrict our attention in the next chapter to linear second order differential equations. For these, there is a substantial body of theory and some effective techniques, while nonlinear second order equations present difficulties we shall not be prepared to treat until Chapter 5.

Here is an outline of some of the methods and ideas of this chapter.

Type of Differential Equation	*Method of Solution*
Separable: $A(x)\,dx = B(y)\,dy$	Integration

Type of Differential Equation	*Method of Solution*
Homogeneous: $\frac{dy}{dx} = f\left(\frac{y}{x}\right)$	Let $u = y/x$ and obtain the separable equation $\frac{du}{f(u) - u} = \frac{dx}{x}$.
$\frac{dy}{dx} = f\left(\frac{ax + by + c}{dx + ey + h}\right)$	If $ae - bd \neq 0$, let $x = X + A$, $y = Y + B$, where $aA + bB + c = 0$ and $dA + eB + h = 0$. Then solve the homogeneous equation $\frac{dY}{dX} = f\left(\frac{aX + bY}{dX + eY}\right)$. If $ae - bd = 0$, let $v = (ax + by)/a$ and solve the separable equation $\frac{dv}{1 + \frac{b}{a} f\left(\frac{av + c}{dv + h}\right)} = dx$.
$M(x, y)\, dx + N(x, y)\, dy = 0$	Exact if $\partial M/\partial y = \partial N/\partial x$; then integrate to find $F(x, y)$ such that $dF = M\, dx + N\, dy$ and obtain general solution $F(x, y) = C$. If inexact, try to find an integrating factor; the first things to try are $u = u(x)$, $u = u(y)$, or $u = x^a y^b$.
Bernoulli equation: $P(x)y' + Q(x)y = R(x)y^\alpha$, $\alpha \neq 0, 1$	An integrating factor is $u = \frac{1}{y^\alpha P(x)} \exp\left[(1 - \alpha) \int \frac{Q(x)}{P(x)}\, dx\right]$.
Linear equation: $y' + p(x)y = q(x)$	General solution is $y = e^{-\int p(x)\, dx} \int q(x) e^{\int p(x)\, dx}\, dx + Ce^{-\int p(x)\, dx}$.

Type of Differential Equation	*Method of Solution*
Riccati equation: $y' = P(x)y^2 + Q(x)y + R(x)$	Find (by some means) a particular solution $S(x)$; let $y = S(x) + 1/z$, and solve the linear equation $z' + (2PS + Q)z = P$.

OTHER TOPICS

Isoclines of $y' = f(x, y)$ are curves $f(x, y) = k$; the direction field of $y' = f(x, y)$ consists of line segments of slope k drawn across the isoclines $f(x, y) = k$.

Trajectories: If $F(x, y, K) = 0$ has differential equation $y' = f(x, y)$, then the orthogonal trajectories have differential equation $y' = -1/f(x, y)$; and the oblique trajectories making an angle α with $F(x, y, K) = 0$ have differential equation

$$y' = \frac{f(x, y) + \tan(\alpha)}{1 - f(x, y)\tan(\alpha)}.$$

In polar coordinates, if $f(\theta, r, r')$ is the differential equation of a family $F(\theta, r, K) = 0$, then the orthogonal trajectories have differential equation $f(\theta, r, -r^2/r') = 0$.

Picard iterates: Given $y' = f(x, y), y(x_0) = y_0$, let

$$y_n = y_0 + \int_{x_0}^{x} f(t, y_{n-1}(t))\, dt.$$

Then the solution of the initial value problem is $y(x) = \lim_{n\to\infty} y_n(t)$, on some interval about x_0.

SUPPLEMENTARY PROBLEMS

In each of Problems 1 through 20, find the general solution. Note that these problems represent all types covered in the chapter, and so identifying the method to use is the first step. In doing this, it is often useful to write the equation in various different ways.

1. $y' = \dfrac{8x - 2y + 1}{6x - 4y}$

2. $y' - \dfrac{1}{x}y = \sin(2x)$

3. $(3x^2 - 4y)\,dx - 4\,dy = 0$

4. $y' - y^2 = 2xy - (4x + 4)$

5. $6x\,dx - 2y\,dy = 0$

6. $xy' + y = 2y^{3/2}$

7. $(x + 2y)\,dx = (y + 1)\,dy$

8. $\dfrac{dy}{dx} = -\dfrac{1}{x}y + e^{-x}$

9. $[12x^2e^y - \sin(x - y)]\,dx + [4x^3e^y + \sin(x - y)]\,dy = 0$

10. $(18x - 2y + 4)\,dx = (y - 2x + 1)\,dy$

11. $(2x - 2y + 4)\,dx = (4x - 4y + 6)\,dy$

12. $(x^2 - 2)\,dy = (y + 3)\,dx$

13. $y' = \dfrac{4x - 2y}{x + y - 6}$

14. $xy' = x - y$

15. $y' = xy^2 - 2x^2y + x^3 + 1$

16. $(4x - y)\,dx + (8y - 2x + 1)\,dy = 0$

17. $\dfrac{dy}{dx} = y - \cos(2x)$

18. $[e^{-y}\cos(x) - xe^{-y}\sin(x)]\,dx - xe^{-y}\cos(x)\,dy = 0$

19. $y' = \dfrac{4x - 2y + 1}{2x - y + 6}$

20. $\dfrac{dy}{dx} = \dfrac{x - 2y + 4}{3x + y - 1}$

In each of Problems 21 through 30, solve the initial value problem.

21. $x^2\dfrac{dy}{dx} = y^2 + x^2;\quad y(1) = 4$

22. $x\dfrac{dy}{dx} = y + \ln(x);\quad y(4) = 21$

23. $\dfrac{dy}{dx} = \dfrac{-8x - 9x^2y^2}{6x^3y};\quad y(2) = 40$

24. $\dfrac{dy}{dx} = \dfrac{1 - x^2}{y};\quad y(6) = 3$

25. $x\dfrac{dy}{dx} + 3y = \sin(x);\quad y(\pi) = 0$

26. $\dfrac{dy}{dx} = \dfrac{4x^2y + 4xy + y^2}{-y - 2x^2};\quad y(4) = 8$

27. $[24x^2y + e^x\sin(y)]\,dx + [8x^3 + e^x\cos(y)]\,dy = 0;\quad y(1) = 1$

28. $x^2\dfrac{dy}{dx} = x^2 - y^2;\quad y(2) = -4$

29. $\dfrac{dy}{dx} = 8y - e^{-x};\quad y(-1) = 6$

30. $2\dfrac{dy}{dx} - 3y = e^{-x}\ln(x);\quad y(2) = 8$

31. A physician's waiting room contains 2000 cubic feet of air which is initially free of carbon monoxide. From time 0, carbon monoxide from people smoking in the building is drawn into the room at a rate of 0.2 cubic feet per minute. The circulating fan mixes the carbon monoxide with air in the room, and also blows out mixed air–carbon monoxide at a rate of 0.2 cubic feet per minute.
(a) Find the concentration of carbon monoxide in the room at time t.
(b) When air in a room is 0.013% carbon monoxide, the occupants are in trouble. At what time will this concentration be reached?

32. Determine values of C and K so that the following equation is exact:

$$[Cx^2ye^y + 2\cos(y)]\,dx + [x^3e^yy + x^3e^y + Kx\sin(y)]\,dx = 0.$$

33. Find the orthogonal trajectories of the family

$$c^2x^2 + 2y^2 = y + 2.$$

34. *Clairaut's equation* is of the form

$$y = xy' + f(y').$$

Show that, for any constant k such that $f'(k)$ is defined, $y = kx + f(k)$ is a solution. Use this fact to solve
(a) $y = xy' + 3x^{-x^2}$ (b) $y = xy' + \cos(2x^2 + 1)$
(c) $y = y'\tan(x) - (y')^2\sec^2(x)$ (*Hint:* This is not a Clairaut equation, but transforms into one if you put $z = \sin(x)$.)

35. A certain radioactive substance decays at a rate proportional to the mass present at any time t. At time zero there was a mass of two grams. In eighty years there will be 1.7 grams.
(a) How many grams will there be in 200 years?
(b) In how many years will there be exactly one gram left?

36. A six-pound ball falls from rest from the top of the Empire State Building. As it falls, it is pulled downward by gravity, but the fall is also impeded by air resistance. This force is proportional to 0.6 times the velocity, which is measured in feet per second. Find a general formula for the velocity as a function of time.

37. In each of the following, sketch some isoclines and some elements of the direction field of the differential equation. Use these to sketch a solution through the given point.
(a) $y' - 4xy = e^{-2x}$; $(2, -3)$
(b) $y' = \ln(1 + x) + y$; $(0, 1)$
(c) $y' = 2x^2y - 3e^x$; $(1, 2)$
(d) $y' - 2xy = x^2$; $(1, 1)$
(e) $y' + 4xy^2 = 2y$; $(1, 3)$
(f) $y' = 1 + y^2$; $(2, 4)$

38. In each of the following, find the orthogonal trajectories of the given family.
(a) $2xy = K(x + 1)$ (b) $y^2 - K^2 = 2x + 1$ (c) $y^3 = Kx^3$
(d) $2x - 4y^2 = K$ (e) $(x - K)^2 = y^2 + 1$

39. In each of the following, find the oblique trajectories of the given family at the given angle.
(a) $y^2 = Kx$; $30°$ (b) $x^2 + y^2 = K^2$; $60°$ (c) $y = Kx$; $45°$ (d) $2xy = K$; $50°$

40. In each of the following, find the orthogonal trajectories in polar coordinates.
(a) $r = K[1 - \sin(\theta)]$ (b) $r = K\theta e^{\theta}$ (c) $r = K \sin(4\theta)$

41. Find the current in an RL circuit if $I(0) = 0$ and $E(t) = e^{-2t} \cos(\omega t)$.

42. Find the current in an RC circuit if $I(0) = 1$ and $E(t) = 1 + \cos(2\pi t)$.

43. In each of the following, use the Picard iteration scheme to find $y_1(x)$ through $y_5(x)$.
(a) $2y' - 4y = 8$; $y(1) = -3$
(b) $y' = 2y - x^2$; $y(0) = 1$
(c) $y' = xy$; $y(0) = 2$
(d) $y' - 2y = xy$; $y(2) = 4$
(e) $y' = y + 2$; $y(0) = 1$

In each of Problems 44 through 55, find a general solution of the differential equation.

44. $y' - \dfrac{3}{x} y = \cos(2x) - e^{4x}$

45. $(8x - 4x^2)y\,dx + 4x^2dy = 0$

46. $2xyy' = x + 8y$

47. $y' = \dfrac{4x - 2y}{x + 6y}$

48. $y' = \dfrac{2}{x} y - e^{2x}$

49. $[2 + y \cos(xy)]\,dx + x \cos(xy)\,dy = 0$

50. $3 \sin(y)\,dx + 3x \cos(y)\,dy = 0$

51. $xy' + y = x^3y^{4/3}$

52. $y' - \dfrac{2}{x} y = \sin(3x) - 1$

53. $y' = y - 3x^2$

54. $y' = \dfrac{3x - 2y + 1}{x + 2y - 4}$

55. $y' = y + 2x^2$

56. An oil tanker of mass M is sailing in a straight line. At time 0 it shuts off its engines and coasts. Assume that the water tends to slow the tanker with a force proportional to v^{α}, where v is the velocity at time t.
(a) Derive a differential equation for v as a function of time.
(b) Show that the tanker moves in a straight line and eventually comes to a full stop if $0 < \alpha < 2$. What happens if $\alpha \geq 2$?

57. Find the shape a mirror should have so that light rays from a point O in the axis are reflected into beams

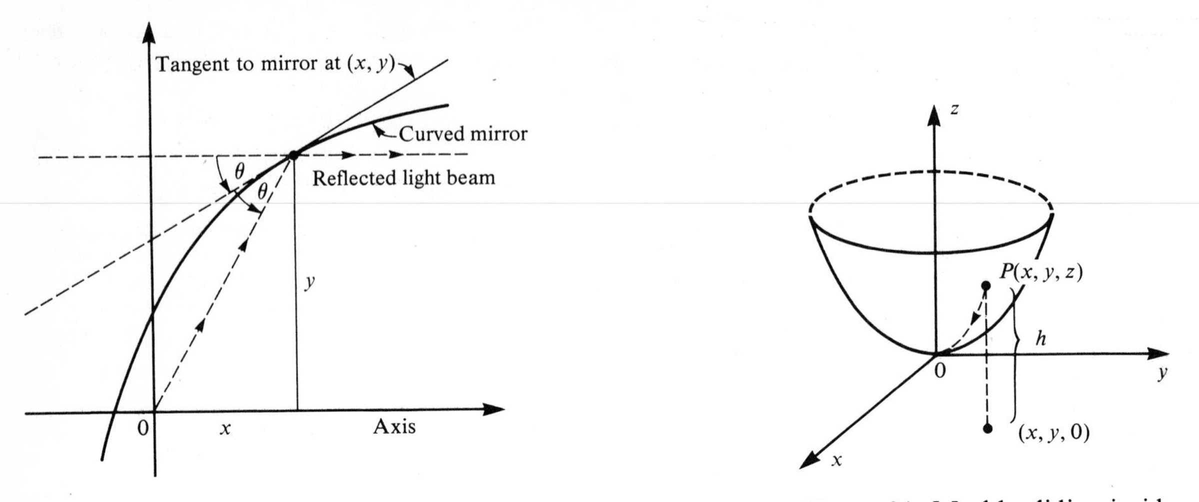

Figure 20. Light beam from 0 reflected parallel to the axis.

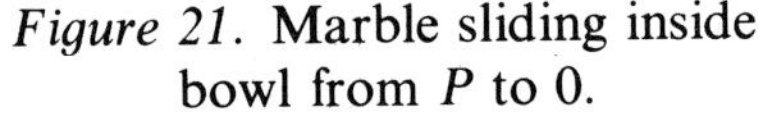

Figure 21. Marble sliding inside bowl from P to 0.

parallel to this axis. [*Hint*: Show from Figure 20 that $\tan(2\theta) = y/x$. Hence, find dy/dx, which is equal to $\tan(\theta)$.]

58. Here is a problem involving a marble released to roll in a bowl. Imagine a bowl in the shape of $z = x^2 + y^2$ (as shown in Figure 21), resting on a table taken as the xy-plane. A marble of mass m is initially at point P in the yz-plane, and is released from rest to roll down the sloping sides of the bowl. If h is the vertical height of P above the table, then one can show that

$$\frac{dy}{dt} = -\sqrt{2g}\sqrt{\frac{h - y^2}{1 + 4y^2}}.$$

Using this, we can find the time T it takes the marble to reach the bottom of the bowl as follows:

(a) Argue that at $t = 0$, $y = \sqrt{h}$, and at time $t = T$, $y = 0$.

(b) Change variables by letting $y = \sqrt{h}\cos(\theta)$. From (a), show that

$$T = \sqrt{\frac{1 + 4h}{2g}} \int_0^{\pi/2} \sqrt{1 - k^2 \sin^2(\theta)}\, d\theta.$$

This is an elliptic integral, and for most values of k cannot be evaluated in elementary terms. Values of this integral for various values of k can be found in tables of elliptic integrals. Elliptic integrals also appear in consideration of the motion of a simple pendulum.

59. Find the orthogonal trajectories of the family $y = x + 2 - Ke^y$. Sketch some of the curves of the family and those of the family of orthogonal trajectories.

60. Consider the following RC circuit:

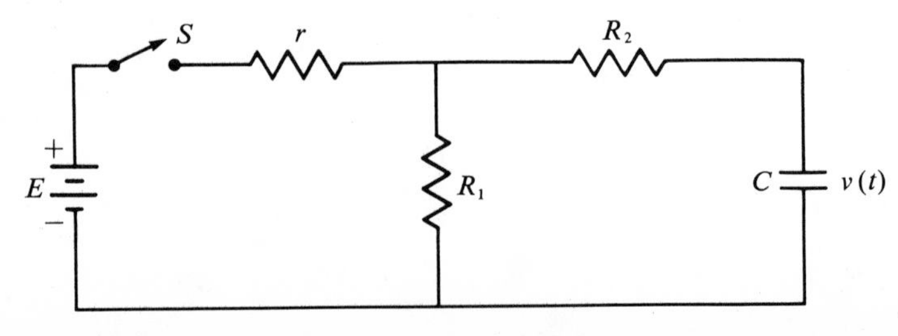

(a) Develop a complete mathematical model of this circuit. That is, write the differential equations which completely characterize the transient behavior of the capacitor voltage and state the requisite initial conditions when the switch S is closed.

(b) Find the circuit time-constant τ.

(c) Find the steady-state capacitor voltage. Assume as circuit parameters that $r = 2\ \Omega$, $R_1 = \Omega$, $R_2 = 3\ \Omega$, $C = 30\ \mu F$, and $E = 9$ V.

CHAPTER TWO

Linear Second Order Differential Equations

2.0 INTRODUCTION

A *second order differential equation* is an equation of the form

$$F(x, y, y', y'') = 0,$$

involving a second, but no higher, order derivative. Some examples are

$$y'' + 2xy' = e^x \cos(y),$$
$$y'' - y = x,$$

and

$$(y')^2 - 2y'' + y = 1.$$

Some examples encountered in Chapter 0 were

$$L\frac{d^2Q}{dt^2} + R\frac{dQ}{dt} + \frac{1}{C}Q = E,$$

for charge in an RLC circuit (see Example 1); the spring equation

$$m\frac{d^2y}{dt^2} + ky = 0$$

of Example 2; and the pendulum equation

$$\frac{d^2\theta}{dt^2} + \frac{g}{L}\sin(\theta) = 0$$

of Example 5.

A *solution* of $F(x, y, y', y'') = 0$ is a function $y = f(x)$ satisfying the equation. For example, $y = \cos(2x)$ is a solution of $y'' + 4y = 0$, and $y = e^x - x$ is a solution of $y'' - y = x$. These are verified by direct substitution of the function into the differential equation.

Sometimes we must restrict ourselves to a given interval J. We say that $y = f(x)$ is a solution of $F(x, y, y', y'') = 0$ on J if $F(x, f(x), f'(x), f''(x)) = 0$ for all x in J. For example, it is easy to verify by substitution that

$$y = \sqrt{\frac{2}{\pi x}}\sin(x)$$

is a solution of

$$y'' + \frac{1}{x}y' + \left(1 - \frac{1}{4x^2}\right)y = 0$$

on the interval (0, 1). Actually, this is a solution on any interval J containing only positive numbers.

As previous examples indicate, second order differential equations model a number of phenomena for which first order differential equations are inadequate. There is also a somewhat richer theory and collection of techniques associated with second order differential equations, and we shall begin to study these now.

Recall that, in the first order case, we were only able to explicitly solve equations having certain forms. Similarly, we are not able to treat the general second order differential equation $F(x, y, y', y'') = 0$, but must restrict ourselves to several tractable forms. For most of this chapter we shall consider *linear* second order differential equations, which are defined in the next section.

2.1 LINEAR SECOND ORDER DIFFERENTIAL EQUATIONS: EXISTENCE AND UNIQUENESS OF SOLUTIONS

A second order differential equation is *linear* if it has the form

$$y'' + P(x)y' + Q(x)y = F(x).$$

Thus,

$$y'' - 2xy' + 3y = 4e^{-2x} \quad \text{and} \quad y'' - 2y = 0$$

are linear, while

$$y'' - yy' + 6xy = 0 \quad \text{and} \quad (y'')^2 - 2y = 4x$$

are not linear.

The following is the basic theorem on linear second order differential equations.

THEOREM 2

Let $P(x)$, $Q(x)$, and $F(x)$ be continuous on an interval J. Let x_0 be any point in J, and let A and B be any real numbers. Then there exists exactly one solution $y(x)$ of $y'' + P(x)y' + Q(x)y = F(x)$ satisfying

$$y(x_0) = A \quad \text{and} \quad y'(x_0) = B.$$

In general, the problem of finding $y(x)$ such that

$$y'' + P(x)y' + Q(x)y = F(x); \qquad y(x_0) = A, \qquad y'(x_0) = B$$

is called an *initial value problem.* Theorem 2 says that an initial value problem has one, and only one, solution over any interval where the coefficient functions are continuous. We shall not attempt to prove Theorem 2 in this treatment, but will make use of it in deriving properties of solutions.

Compare Theorem 2 with Theorem 1. Although the latter is not restricted to the linear case, the two theorems are similar in that the number of conditions given at x_0 equals the order of the differential equation. In the second order case, solutions involve two constants of integration because of the second derivative term; hence two pieces of data; $y(x_0) = A$ and $y'(x_0) = B$; are needed to solve for the constants and determine a unique solution.

EXAMPLE 47

Consider the initial value problem

$$y'' - 2y = x^2 - 1; \qquad y(1) = 3, \qquad y'(1) = -5.$$

It is easy to check that, for any constants c_1 and c_2,

$$y = c_1 e^{\sqrt{2}x} + c_2 e^{-\sqrt{2}x} - \tfrac{1}{2}x^2$$

is a solution. Thus, this differential equation has infinitely many solutions. (Don't worry for now about how we found them—that comes later.)

Now look at the initial conditions. We must choose c_1 and c_2 so that

$$c_1 e^{\sqrt{2}} + c_2 e^{-\sqrt{2}} - \tfrac{1}{2} = 3$$

and

$$\sqrt{2}c_1 e^{\sqrt{2}} - \sqrt{2}c_2 e^{-\sqrt{2}} - 1 = -5.$$

This can be done, and in only one way. We obtain

$$c_1 = \frac{e^{-\sqrt{2}}}{2\sqrt{2}}\left(\frac{7\sqrt{2}}{2} - 4\right) \quad \text{and} \quad c_2 = \frac{e^{\sqrt{2}}}{2\sqrt{2}}\left(4 + \frac{7}{2}\sqrt{2}\right).$$

(To four decimals, $c_1 = 0.0816$ and $c_2 = 13.0152$.)

In this example, the coefficient functions, -2 and $x^2 - 1$, are continuous everywhere, and so we need not worry about restricting ourselves to any particular interval.

Another kind of problem commonly encountered is the *boundary value problem,* in which one seeks to determine a solution of a differential equation which is valid on a given interval, and also to satisfy certain conditions prescribed at the end points of the interval. This is quite different from an initial value problem, in which the solution and its derivative must take on given values at the same point. Boundary value problems arise, for example, in consideration of the vibrations of a string pegged at both ends and set in motion. The boundary conditions reflect the assumption that the ends of the string are held fixed. We shall see specific examples of this kind of application in Chapter 13.

There is no general result like Theorem 2 for boundary value problems, since one can construct apparently simple examples in which both existence and uniqueness fail. For example, the boundary value problem

$$y'' + y = 0; \qquad y(0) = y(\pi) = 0$$

has infinitely many solutions $y = C\sin(x)$. By contrast, the boundary value problem

$$y'' + y = 0; \qquad y(0) = y'(\pi) = 0$$

has only the trivial solution $y = 0$; and the boundary value problem

$$y'' + y = x; \qquad y(0) = 0, \qquad y'(\pi) = 0$$

has no solution at all. (The assertions about the last two examples will be clear after we have discussed general solutions of second order linear differential equations in Sections 2.3, 2.5, and 2.7.)

In Section 6.4 we shall study a special class of boundary value problems, called Sturm-Liouville problems, which are particularly important in engineering applications. For the remainder of this chapter we shall concentrate on initial value problems and general solutions of second order linear differential equations.

PROBLEMS FOR SECTION 2.1

In each of Problems 1 through 15, you are given an initial value problem and a function which involves two arbitrary constants, c_1 and c_2. (a) Show that the function is a solution of the differential equation for any choices of the constants. (b) Determine how to choose c_1 and c_2 to solve the initial value problem, as in Example 47.

1. $y'' - 5y' + 4y = 0;\ \ y(0) = 1,\ \ y'(0) = -4;\ \ y = c_1e^x + c_2e^{4x}$
2. $y'' - y' - 2y = 0;\ \ y(0) = 3,\ \ y'(0) = 2;\ \ y = c_1e^{-x} + c_2e^{2x}$
3. $y'' + 8y' + 16y = 0;\ \ y(0) = y'(0) = 3;\ \ y = e^{-4x}(c_1 + c_2x)$
4. $y'' + 16y = 0;\ \ y(\pi) = y'(\pi) = 1;\ \ y = c_1\cos(4x) + c_2\sin(4x)$
5. $y'' - 16y = 0;\ \ y(0) = 1,\ \ y'(0) = -2;\ \ y = c_1e^{4x} + c_2e^{-4x}$
6. $x^2y'' + 2xy' - y = 0;\ \ y(1) = y'(1) = 3;\ \ y = (1/\sqrt{x})(c_1x^{\sqrt{5}/2} + c_2x^{-\sqrt{5}/2})$
7. $x^2y'' + 4xy' + 2y = 0;\ \ y(2) = 0,\ \ y'(2) = 4;\ \ y = c_1/x + c_2/x^2$
8. $y'' + 4y = 2x;\ \ y(\pi/4) = 0,\ \ y'(\pi/4) = 1;\ \ y = c_1\cos(2x) + c_2\sin(2x) + x/2$
9. $y'' + 2y' + 2y = 0;\ \ y(\pi/2) = 1,\ \ y'(\pi/2) = -1;\ \ y = e^{-x}[c_1\cos(x) + c_2\sin(x)]$
10. $x^2y'' + 3xy' + 3y = 0;\ \ y(1) = 2,\ \ y'(1) = 0;\ \ y = (1/x)\{c_1\cos[\sqrt{2}\ln(x)] + c_2\sin[\sqrt{2}\ln(x)]\}$
11. $y'' - 9y = 3;\ \ y(0) = 1,\ \ y'(0) = 4;\ \ y = c_1e^{3x} + c_2e^{-3x} - \frac{1}{3}$
12. $y'' - 7y' + 12y = 2e^x;\ \ y(0) = -3,\ \ y'(0) = 1;\ \ y = c_1e^{3x} + c_2e^{4x} + \frac{1}{3}e^x$
13. $y'' + (2/x)y' - (3/x^2)y = 0;\ \ y(1) = y'(1) = 1;\ \ y = (1/\sqrt{x})[c_1x^{\sqrt{13}/2} + c_2x^{\sqrt{13}/2}]$
14. $y'' + 5y' + 8y = 0;\ \ y(1) = 0,\ \ y'(1) = 1;\ \ y = (1/x^2)\{c_1\cos[2\ln(x)] + c_2\sin[2\ln(x)]\}$
15. $y'' + y' - 12y = x^2 - 1;\ \ y(0) = 1,\ \ y'(0) = -3;\ \ y = c_1e^{3x} + c_2e^{-4x} - \frac{1}{12}x^2 - \frac{1}{72}x + \frac{59}{864}$

2.2 THEORY OF LINEAR HOMOGENEOUS SECOND ORDER DIFFERENTIAL EQUATIONS

A linear second order differential equation

$$y'' + P(x)y' + Q(x)y = F(x)$$

is *homogeneous** on an interval J if $F(x) = 0$ for all x in J. If $F(x) \neq 0$ for at least one x in J, then the differential equation is *nonhomogeneous*.

EXAMPLE 48

Let

$$F(x) = \begin{cases} 0 & \text{if } 0 \le x \le 1 \\ 1 & \text{if } 1 < x \le 2. \end{cases}$$

Then, $y'' + xy' - 2y = F(x)$ is homogeneous on $[0, 1]$, but nonhomogeneous on $[0, 2]$.

In many instances, J is the whole real line. In this case we omit reference to J. This is common, for example, when $P(x)$ and $Q(x)$ are defined for all x and no restrictions are needed.

We shall now determine some properties of solutions of $y'' + P(x)y' + Q(x)y = 0$. These properties will help us know what to look for later when we begin actually solving such equations.

THEOREM 3

Let $y_1(x)$ and $y_2(x)$ be solutions of $y'' + P(x)y' + Q(x)y = 0$ on an interval J. Then,

1. $y_1(x) + y_2(x)$ is also a solution on J.
2. $cy_1(x)$ is a solution on J for any real number c (similarly, so is $cy_2(x)$).

Proof of (1) Substitute $y_1(x) + y_2(x)$ into the differential equation to get, for any x in J,

$$[y_1(x) + y_2(x)]'' + P(x)[y_1(x) + y_2(x)]' + Q(x)[y_1(x) + y_2(x)]$$
$$= [y_1'' + P(x)y_1' + Q(x)y_1] + [y_2'' + P(x)y_2' + Q(x)y_2] = 0 + 0 = 0,$$

by the assumption that $y_1(x)$ and $y_2(x)$ are themselves solutions on J.

Proof of (2) Substitute $cy_1(x)$ into the differential equation to get

$$[cy_1(x)]'' + P(x)[cy_1(x)]' + Q(x)[cy_1(x)] = c[y_1'' + P(x)y_1' + Q(x)y_1] = c \cdot 0 = 0.$$

Note that (1) and (2) can be combined to give us an alternate statement of Theorem 3:

THEOREM 3′

If $y_1(x)$ and $y_2(x)$ are solutions of

$$y'' + P(x)y' + Q(x)y = 0$$

---

* The term *homogeneous*, as used here, must not be confused with its different meaning in connection with first order differential equations in Chapter 1.

on J, and c_1 and c_2 are any real numbers, then

$$c_1 y_1(x) + c_2 y_2(x)$$

is also a solution on J.

Generally, any function of the form $c_1 y_1(x) + c_2 y_2(x)$, with c_1 and c_2 real, is called a *linear combination* of $y_1(x)$ and $y_2(x)$. Thus, Theorems 3 and 3′ may be phrased:

For a linear homogeneous differential equation, any linear combination of solutions is again a solution.

EXAMPLE 49

It is easy to check that $\sin(2x)$ and $\cos(2x)$ are both solutions of

$$y'' + 4y = 0.$$

Thus, so is any linear combination

$$c_1 \sin(2x) + c_2 \cos(2x)$$

for c_1 and c_2 real. The reader should verify this by direct substitution into $y'' + 4y = 0$.

Similarly, e^{2x} and e^{-2x} are solutions of

$$y'' - 4y = 0.$$

Hence, so is any linear combination

$$c_1 e^{2x} + c_2 e^{-2x}.$$

For example, with $c_1 = c_2 = \frac{1}{2}$, we get $\cosh(2x)$, and with $c_1 = \frac{1}{2}$ and $c_2 = -\frac{1}{2}$, we get $\sinh(2x)$. It is easy to check that in fact $\cosh(2x)$ and $\sinh(2x)$ are solutions of $y'' - 4y = 0$.

Finally,

$$x^{2+\sqrt{2}} \quad \text{and} \quad x^{2-\sqrt{2}}$$

are solutions of

$$y'' - \frac{3}{x} y' + \frac{2}{x^2} y = 0$$

on any interval J containing only positive numbers. Thus, any linear combination

$$c_1 x^{2+\sqrt{2}} + c_2 x^{2-\sqrt{2}}$$

is also a solution on J.

EXAMPLE 50

The conclusion of Theorem 3 *fails* for nonhomogeneous differential equations. For example, $y_1(x) = e^x + x$ and $y_2(x) = -e^x + x$ are solutions of $y'' - y = -x$. But $y_1(x) + y_2(x) = 2x$ is not a solution, as one can verify by substitution.

If $y_2(x)$ is a constant multiple of $y_1(x)$ on an interval J, then the linear combination $c_1 y_1(x) + c_2 y_2(x)$ is again just a constant multiple of $y_1(x)$. In such a case, $y_2(x)$ does not make a significant contribution to the solution. We call $y_1(x)$ and $y_2(x)$ *linearly dependent* on J if one is a constant multiple of the other *for all* x in J.

If $y_1(x)$ and $y_2(x)$ are not constant multiples of each other for all x in J, we call them *linearly independent* on J. In this case, linear combinations $c_1 y_1(x) + c_2 y_2(x)$ give functions that are honestly different from both $y_1(x)$ and $y_2(x)$ (if c_1 and c_2 are both nonzero).

EXAMPLE 51

The functions

$$y_1(x) = e^{2x} + e^{-2x} \quad \text{and} \quad y_2(x) = 4\cosh(2x)$$

are solutions of $y'' - 4y = 0$ for all x. Here, $y_1(x)$ and $y_2(x)$ are linearly dependent (on any interval), since $y_2(x) = 2y_1(x)$, and, similarly, $y_1(x) = \frac{1}{2}y_2(x)$. Thus, given one of these functions, we really already know the other is a solution of $y'' - 4y = 0$ by Theorem 3(2).

On the other hand, e^{2x} and e^{-2x} are also solutions of $y'' - 4y = 0$, and are linearly independent on any interval. The linear combination $c_1 e^{2x} + c_2 e^{-2x}$ is also a solution for any choice of the constants, and is different from both e^{2x} and e^{-2x} if the constants are chosen nonzero.

Usually one can tell by looking at two functions whether or not they are linearly independent on a given interval. In the case of very complicated functions (for example, infinite series to be encountered in Chapter 5), and also for some theoretical considerations, the following test is important:

THEOREM 4

Let $y_1(x)$ and $y_2(x)$ be solutions of

$$y'' + P(x)y' + Q(x)y = 0$$

on an interval $[a, b]$. Let

$$W(y_1, y_2) = y_1(x)y_2'(x) - y_1'(x)y_2(x).$$

Then,

1. Either $W(y_1, y_2) = 0$ for all x in $[a, b]$, or $W(y_1, y_2) \neq 0$ for all x in $[a, b]$. (That is, $W(y_1, y_2)$ either vanishes identically for $a \leq x \leq b$, or is never zero for $a \leq x \leq b$.)
2. $y_1(x)$ and $y_2(x)$ are linearly independent on $[a, b]$ if and only if $W(y_1, y_2) \neq 0$ for some x in $[a, b]$.

The function $W(y_1, y_2)$ is called the *Wronskian* of $y_1(x)$ and $y_2(x)$, and can be remembered as the determinant

$$W(y_1, y_2) = \begin{vmatrix} y_1(x) & y_2(x) \\ y_1'(x) & y_2'(x) \end{vmatrix}.$$

The value of conclusion (1) is in applying (2); we need only test $W(y_1, y_2)$ at *one* value of x in $[a, b]$. If $W(y_1, y_2) = 0$ at x_0 in $[a, b]$, then by (1), $W(y_1, y_2) = 0$ for all x in $[a, b]$; and by (2), $y_1(x)$ and $y_2(x)$ are linearly dependent on $[a, b]$. If $W(y_1, y_2) \neq 0$ for *some* x_0 in $[a, b]$, then $W(y_1, y_2) \neq 0$ for all x in $[a, b]$; and $y_1(x)$ and $y_2(x)$ are linearly independent on $[a, b]$.

Proof of (1) Begin by observing that the derivative of the Wronskian is given by

$$W' = y_1'y_2' + y_1y_2'' - y_1''y_2 - y_1'y_2' = y_1y_2'' - y_1''y_2.$$

Now recall that both $y_1(x)$ and $y_2(x)$ are solutions of $y'' + P(x)y' + Q(x)y = 0$. Thus,

$$y_1'' + P(x)y_1' + Q(x)y_1 = 0$$

and

$$y_2'' + P(x)y_2' + Q(x)y_2 = 0.$$

Multiply the first of these equations by $-y_2$ and the second by y_1 to get

$$-y_2 y_1'' - P(x)y_2 y_1' - Q(x)y_2 y_1 = 0$$
$$y_1y_2'' + P(x)y_1y_2' + Q(x)y_1y_2 = 0.$$

Add these equations to get

$$-y_2y_1'' + y_1y_2'' + P(x)(-y_2y_1' + y_1y_2') = 0.$$

This is exactly

$$W' + P(x)W = 0.$$

Thus, W satisfies a linear first order differential equation, which we easily solve (see Section 1.6) to get

$$W(y_1, y_2) = Ce^{-\int P(x)\,dx},$$

where C is any constant. If $C = 0$, then $W(y_1, y_2) = 0$ for all x; if $C \neq 0$, then $W(y_1, y_2) \neq 0$ for all x for which $\int P(x)\,dx$ is defined, since the exponential function is never zero. This proves (1).

Proof of (2) We shall prove that $y_1(x)$ and $y_2(x)$ are linearly dependent on $[a, b]$ if and only if $W(y_1, y_2) = 0$ for all x in $[a, b]$.

First, suppose that $y_1(x)$ and $y_2(x)$ are linearly dependent on $[a, b]$. Then $y_2(x) = cy_1(x)$ for some constant c. Then

$$W(y_1, y_2) = y_1y_2' - y_1'y_2 = cy_1y_1' - cy_1y_1' = 0$$

for all x in $[a, b]$.

Conversely, suppose that $W(y_1, y_2) = 0$ for all x in $[a, b]$. We want to show that $y_1(x)$ and $y_2(x)$ are linearly dependent, that is, that one is a constant multiple of the other for $a \leq x \leq b$.

If $y_2(x) = 0$ for all x in $[a, b]$, then $y_2(x) = 0 \cdot y_1(x)$ and the two functions are linearly dependent.

Hence, suppose that $y_2(x_0) \neq 0$ for some x_0 in $[a, b]$. By continuity of $y_2(x)$, there is some interval $[c, d]$ inside $[a, b]$ such that $y_2(x) \neq 0$ for $c \leq x \leq d$. Then we can divide W by $y_2^2(x)$ to get:

$$\frac{W(y_1, y_2)}{y_2^2(x)} = \frac{y_1(x)y_2'(x) - y_2(x)y_1'(x)}{y_2^2(x)},$$

and this is zero because $W(y_1, y_2) = 0$ for $c \leq x \leq d$. Recognizing the right side as the derivative of a quotient, we have

$$\left(\frac{y_1}{y_2}\right)' = 0 \quad \text{for} \quad c \leq x \leq d.$$

Then, for some constant K,

$$\frac{y_1(x)}{y_2(x)} = K;$$

hence, $y_1(x) = Ky_2(x)$ for $c \leq x \leq d$.

We must now show that $y_1(x) = Ky_2(x)$ for the perhaps larger interval $a \leq x \leq b$. Recall that x_0 is in $[c, d]$, and hence also in $[a, b]$, and that $y_1(x)$ and $Ky_2(x)$ both satisfy the initial value problem

$$y'' + P(x)y' + Q(x)y = 0; \qquad y(x_0) = y_1(x_0), \qquad y'(x_0) = y_1'(x_0)$$

defined on $[a, b]$. By Theorem 2, $y_1(x)$ and $Ky_2(x)$ must be identical on all of $[a, b]$, completing the proof of (2) of Theorem 4.

EXAMPLE 52

The functions

$$y_1(x) = e^{2x} \quad \text{and} \quad y_2(x) = e^{-2x}$$

are solutions of $y'' - 4y = 0$ on any interval. The Wronskian is

$$W(y_1, y_2) = \begin{vmatrix} e^{2x} & e^{-2x} \\ 2e^{2x} & -2e^{-2x} \end{vmatrix} = -4,$$

which is nonzero everywhere. Of course, e^{2x} and e^{-2x} are linearly independent on any interval, since neither is a constant multiple of the other.

The next theorem is the reason for this discussion of linear independence. It tells us that we know *all* solutions of $y'' + P(x)y' + Q(x)y = 0$ on a given interval if we know just *two* linearly independent solutions.

THEOREM 5

Let $y_1(x)$ and $y_2(x)$ be linearly independent solutions of

$$y'' + P(x)y' + Q(x)y = 0$$

on an interval J. Then, every solution of $y'' + P(x)y' + Q(x)y = 0$ on J is of the form

$$c_1y_1(x) + c_2y_2(x)$$

for constants c_1 and c_2. That is, every solution of $y'' + P(x)y' + Q(x)y = 0$ is a linear combination of any two linearly independent solutions.

The expression

$$c_1 y_1(x) + c_2 y_2(x)$$

is called the *general solution* of $y'' + P(x)y' + Q(x)y = 0$ when $y_1(x)$ and $y_2(x)$ are linearly independent; and the functions $y_1(x)$ and $y_2(x)$ are said to constitute a *fundamental system of solutions* of $y'' + P(x)y' + Q(x)y = 0$.

Here is an example, after which we prove Theorem 5.

EXAMPLE 53

We have observed that e^{2x} and e^{-2x} are solutions of $y'' - 4y = 0$, on any interval. Since e^{2x} and e^{-2x} are linearly independent, they form a fundamental system of solutions of $y'' - 4y = 0$. The general solution of $y'' - 4y = 0$ is

$$c_1 e^{2x} + c_2 e^{-2x}.$$

Similarly, $\sin(2x)$ and $\cos(2x)$ are linearly independent solutions of $y'' + 4y = 0$, on any interval, and hence form a fundamental system of solutions for $y'' + 4y = 0$. The general solution of $y'' + 4y = 0$ is

$$c_1 \sin(2x) + c_2 \cos(2x).$$

Proof of Theorem 5 Let $y(x)$ be any solution of $y'' + P(x)y' + Q(x)y = 0$. We must show that there are constants c_1 and c_2 such that $y(x) = c_1 y_1(x) + c_2 y_2(x)$.

Choose any x_0 in the given interval J. Let

$$y(x_0) = A \quad \text{and} \quad y'(x_0) = B.$$

We first attempt to solve the algebraic equations

$$y_1(x_0)X + y_2(x_0)Y = A$$
$$y_1'(x_0)X + y_2'(x_0)Y = B.$$

Here, $y_1(x_0)$, $y_2(x_0)$, $y_1'(x_0)$, $y_2'(x_0)$, A, and B are constants, and X and Y denote the unknowns. From elementary algebra, this system of two equations in two unknowns has a solution exactly when the determinant of coefficients is nonzero, i.e.,

$$\begin{vmatrix} y_1(x_0) & y_2(x_0) \\ y_1'(x_0) & y_2'(x_0) \end{vmatrix} \neq 0.$$

But this is exactly the Wronskian evaluated at x_0, and this is nonzero by Theorem 4(2) and the hypothesis that $y_1(x)$ and $y_2(x)$ are linearly independent. Thus we have solutions, say, $X = c_1$ and $Y = c_2$. Now observe that $c_1 y_1(x) + c_2 y_2(x)$ is a solution of the initial value problem

$$y'' + P(x)y' + Q(x)y = 0; \qquad y(x_0) = A, \qquad y'(x_0) = B.$$

Since $y(x)$ is also a solution of this problem, then $y(x) = c_1 y_1(x) + c_2 y_2(x)$ by the uniqueness part of Theorem 2. This completes the proof of Theorem 5.

Theorem 5 suggests a strategy for finding the general solution of $y'' + P(x)y' + Q(x)y = 0$, namely, develop a method for producing two linearly independent solutions. It also suggests a strategy for solving the initial value problem $y'' + P(x)y' + Q(x)y = 0$; $y(x_0) = A$, $y'(x_0) = B$. We first find the general solution, which has the form $c_1 y_1(x) + c_2 y_2(x)$, and then use the two pieces of data to solve for the constants c_1 and c_2. We begin developing this strategy in the next section, and conclude this section with an example involving an initial value problem.

EXAMPLE 54

Solve the initial value problem

$$y'' + 4y = 0; \qquad y(\pi) = 3, \qquad y'(\pi) = -2.$$

First, we know from Example 53 that the general solution of $y'' + 4y = 0$ is

$$y = c_1 \sin(2x) + c_2 \cos(2x).$$

Now we must choose the constants to fit the data. We need

$$y(\pi) = 3 = c_2$$

and

$$y'(\pi) = -2 = 2c_1.$$

Thus, we must choose $c_1 = -1$ and $c_2 = 3$, and the solution of the initial value problem is

$$y = -\sin(2x) + 3\cos(2x).$$

PROBLEMS FOR SECTION 2.2

In each of Problems 1 through 15, (a) verify that the given functions are solutions of the differential equation; (b) use the Wronskian to show that they are linearly independent; and (c) write the general solution.

1. $y'' - k^2 y = 0$; $y_1 = \cosh(kx)$, $y_2 = \sinh(kx)$; all x $(k \neq 0)$
2. $y'' + k^2 y = 0$; $y_1 = \sin(kx)$, $y_2 = \cos(kx)$; all x $(k \neq 0)$
3. $y'' + 4y' - 12y = 0$; $y_1 = e^{2x}$, $y_2 = e^{-6x}$; all x
4. $y'' + 11y' + 24y = 0$; $y_1 = e^{-8x}$, $y_2 = e^{-3x}$; all x
5. $y'' + \dfrac{3}{x} y' + \dfrac{2}{x^2} y = 0$; $y_1 = \dfrac{1}{x}\cos[\ln(x)]$, $y_2 = \dfrac{1}{x}\sin[\ln(x)]$; $1 \leq x \leq 4$
6. $y'' + 5y' - 24y = 0$; $y_1 = e^{3x}$, $y_2 = e^{-8x}$; all x
7. $y'' - 10y' + 25y = 0$; $y_1 = e^{5x}$, $y_2 = xe^{5x}$; all x
8. $y'' - y' - 6y = 0$; $y_1 = e^{-2x}$, $y_2 = e^{3x}$; all x
9. $x^2 y'' + 3xy' + y = 0$; $y_1 = 1/x$, $y_2 = (1/x)\ln(x)$; $x > 0$
10. $x^2 y'' + 6xy' - y = 0$; $y_1 = x^{(-5+\sqrt{29})/2}$, $y_2 = x^{(-5-\sqrt{29})/2}$; $x > 0$
11. $y'' + 11y' - 42y = 0$; $y_1 = e^{3x}$, $y_2 = e^{-14x}$; all x
12. $y'' + 11y' + 28y = 0$; $y_1 = e^{-4x}$, $y_2 = e^{-7x}$; all x
13. $x^2 y'' - 7xy' + 16y = 0$; $y_1 = x^4$, $y_2 = x^4 \ln(x)$; $x > 0$

14. $y'' + \frac{1}{x}y' + \left(1 - \frac{9}{4x^2}\right)y = 0;\quad y_1 = \sqrt{\frac{2}{\pi x}}\left(\frac{\sin(x)}{x} - \cos(x)\right),\quad y_2 = \sqrt{\frac{2}{\pi x}}\left(\frac{\cos(x)}{x} - \sin(x)\right);$ all $x > 0$

15. $y'' + \frac{1}{x}y' + \left(1 - \frac{1}{4x^2}\right)y = 0;\quad y_1 = \sqrt{\frac{2}{\pi x}}\sin(x),\quad y_2 = \sqrt{\frac{2}{\pi x}}\cos(x);$ all $x > 0$.

16. Let $y_1 = x^2$ and $y_2 = x^3$ for all x.
(a) Show that $W(y_1, y_2) = x^4$.
(b) Why does it not contradict Theorem 4(1) that $W(y_1, y_2)$ is zero for $x = 0$, but nonzero for all other values of x?

17. Use the results of Problem 1 to solve the initial value problem

$$y'' - k^2y = 0;\qquad y(\pi) = 1,\qquad y'(\pi) = 0.$$

18. Use Problem 3 to solve the initial value problem

$$y'' + 4y' - 12y = 0;\qquad y(0) = 1,\qquad y'(0) = -1.$$

19. Use Problem 4 to solve the initial value problem

$$y'' + 11y' + 24y = 0;\qquad y(0) = 2,\qquad y'(0) = -4.$$

20. Verify that $y_1(x) = x$ and $y_2(x) = x^2$ are linearly independent solutions of $x^2y'' - 2xy' + 2y = 0$ on the interval $[-1, 1]$, but that $W(y_1, y_2) = 0$ at $x = 0$. Why does this not contradict Theorem 4(1)?

21. Use Problem 8 to solve the initial value problem

$$y'' - y' - 6y = 0;\qquad y(-1) = 3,\qquad y'(-1) = 6.$$

22. Use Problem 11 to solve the initial value problem

$$y'' + 11y' - 42y = 0;\qquad y(0) = -5,\qquad y'(0) = 0.$$

23. Use Problem 12 to solve the initial value problem

$$y'' + 11y' + 28y = 0;\qquad y(2) = 3,\qquad y'(2) = 0.$$

24. Use Problem 13 to solve the initial value problem

$$y'' - \frac{7}{x}y' + \frac{16}{x^2}y = 0;\qquad y(1) = 2,\qquad y'(1) = 4.$$

25. Use Problem 15 to solve the initial value problem

$$y'' + \frac{1}{x}y' + \left(1 - \frac{1}{4x^2}\right)y = 0;\qquad y(\pi) = -5,\qquad y'(\pi) = 8.$$

26. Give an example to show that a product of two solutions of $y'' + P(x)y' + Q(x)y = 0$ is not necessarily a solution.

2.3 GENERAL SOLUTION OF $y'' + Ay' + By = 0$ FOR $A^2 - 4B \geq 0$

Even the linear differential equation $y'' + P(x)y' + Q(x)y = 0$ is too difficult to tackle without some assumptions about $P(x)$ and $Q(x)$. In this section we shall concentrate on the constant coefficient differential equation

$$y'' + Ay' + By = 0 \tag{a}$$

in which A and B are constant. The basic method of solution is to substitute an exponential function

$$y = e^{rx}$$

and see if we can choose r so as to obtain a solution. Upon putting $y = e^{rx}$ into (a), we get

$$r^2 e^{rx} + rAe^{rx} + Be^{rx} = 0.$$

Since e^{rx} is never zero, we can divide out the common exponential factor to get

$$r^2 + Ar + B = 0.$$

This is called the *characteristic equation* for (a), and is a quadratic equation for r. Solve it to obtain

$$r = \frac{-A \pm \sqrt{A^2 - 4B}}{2}.$$

This leads to three possibilities:

1. Two distinct, real values for r (when $A^2 - 4B > 0$).
2. Only one real value for r (when $A^2 - 4B = 0$).
3. Two distinct, complex values for r (when $A^2 - 4B < 0$).

Case 1 Suppose that the characteristic equation has two distinct, real roots. This happens when $A^2 - 4B > 0$, and the roots are then

$$r_1 = \frac{-A + \sqrt{A^2 - 4B}}{2} \quad \text{and} \quad r_2 = \frac{-A - \sqrt{A^2 - 4B}}{2}.$$

Corresponding to these we get two solutions:

$$y_1(x) = e^{r_1 x} \quad \text{and} \quad y_2(x) = e^{r_2 x}.$$

These are linearly independent on any interval J by Theorem 4(2), since the Wronskian is

$$y_1 y_2' - y_2 y_1' = r_2 e^{r_1 x} e^{r_2 x} - r_1 e^{r_1 x} e^{r_2 x} = (r_2 - r_1) e^{(r_1 + r_2)x}$$

which is never zero, by the assumption that $r_1 \neq r_2$.

Thus, $e^{r_1 x}$ and $e^{r_2 x}$ form a fundamental system of solutions of (a), and the general solution in case 1 is

$$y = c_1 e^{r_1 x} + c_2 e^{r_2 x},$$

with c_1 and c_2 arbitrary constants.

EXAMPLE 55

Solve

$$y'' + 4y' - 2y = 0.$$

Here $A = 4$ and $B = -2$. The characteristic equation is

$$r^2 + 4r - 2 = 0,$$

with two real solutions $r_1 = -2 + \sqrt{6}$ and $r_2 = -2 - \sqrt{6}$. Thus, the general solution is

$$y = c_1 e^{(-2+\sqrt{6})x} + c_2 e^{(-2-\sqrt{6})x}.$$

This could also be written

$$y = e^{-2x}(c_1 e^{\sqrt{6}\,x} + c_2 e^{-\sqrt{6}\,x}),$$

where c_1 and c_2 are arbitrary constants.

Case 2 Suppose that the characteristic equation has only one root. This happens when $A^2 - 4B = 0$, and the root is then $r = -A/2$. This gives us just one solution

$$y_1(x) = e^{-Ax/2},$$

but we need another to form a fundamental system of solutions. To find $y_2(x)$, we apply a technique called *reduction of order*, which is treated more generally in Section 2.11. The idea is to try to produce $u(x)$ such that $u(x)y_1(x)$ is a solution. Substituting $u(x)e^{-Ax/2}$ into $y'' + Ay' + By = 0$ gives us

$$u''e^{-Ax/2} - Au'e^{-Ax/2} + \frac{A^2}{4}ue^{-Ax/2} + Au'e^{-Ax/2} - \frac{A^2}{2}ue^{-Ax/2} + Bue^{-Ax/2} = 0.$$

We can divide out the nonzero $e^{-Ax/2}$. Further, many terms cancel, since $A^2 - 4B = 0$, and we are left with

$$u'' = 0.$$

Then $u(x) = c_1 x + c_2$. Since we only need *one* $u(x)$ such that $u(x)y_1(x)$ is a solution which is linearly independent from $y_2(x)$, we need not maintain c_1 and c_2 in complete generality. In fact, if we choose $c_1 = 1$ and $c_2 = 0$, then we get $u(x) = x$ and $y_2(x) = xe^{-Ax/2}$. Further, $y_1(x)$ and $y_2(x)$ are linearly independent, by Theorem 4(2), since the Wronskian is

$$\begin{aligned} y_1 y_2' - y_2 y_1' &= e^{-Ax/2}\left(e^{-Ax/2} - \frac{A}{2}xe^{-Ax/2}\right) - xe^{-Ax/2}\left(-\frac{A}{2}e^{-Ax/2}\right) \\ &= e^{-Ax} \neq 0. \end{aligned}$$

Thus, in case 2 the general solution of (a) is

$$y = c_1 e^{-Ax/2} + c_2 xe^{-Ax/2}$$

or

$$y = e^{-Ax/2}(c_1 + c_2 x),$$

where c_1 and c_2 are arbitrary constants.

Note: In practice, one does not repeat the above derivation every time $A^2 - 4B = 0$; just remember in this case that the general solution is $e^{-Ax/2}(c_1 + c_2 x)$.

EXAMPLE 56

Solve

$$y'' + 4y' + 4y = 0.$$

Here the characteristic equation is

$$r^2 + 4r + 4 = 0$$

with only one solution, $r = -2$. The general solution is then

$$y(x) = e^{-2x}(c_1 + c_2 x).$$

EXAMPLE 57

Solve the initial value problem

$$y'' - 12y' + 6y = 0; \qquad y(0) = 1, \qquad y'(0) = 4.$$

First, find the general solution of the differential equation. The characteristic equation is

$$r^2 - 12r + 6 = 0,$$

with roots $6 + \sqrt{30}$ and $6 - \sqrt{30}$. Thus, the general solution is

$$y = c_1 e^{(6+\sqrt{30})x} + c_2 e^{(6-\sqrt{30})x}.$$

We now need

$$y(0) = c_1 + c_2 = 1$$

and

$$y'(0) = (6 + \sqrt{30})c_1 + (6 - \sqrt{30})c_2 = 4.$$

Solve these to obtain

$$c_1 = \frac{\sqrt{30} - 2}{2\sqrt{30}} \quad \text{and} \quad c_2 = \frac{\sqrt{30} + 2}{2\sqrt{30}}.$$

Thus, the solution of the initial value problem is

$$y = \frac{\sqrt{30} - 2}{2\sqrt{30}} e^{(6+\sqrt{30})x} + \frac{\sqrt{30} + 2}{2\sqrt{30}} e^{(6-\sqrt{30})x}.$$

If we wanted only an approximate solution, say to four decimals, then we could approximate $\sqrt{30}$ as 5.4772 and obtain

$$y_{\text{app}} = 0.3174e^{11.4772x} + 0.6826e^{0.5228x}.$$

EXAMPLE 58

Solve

$$y'' - 8y' + 16y = 0; \qquad y(1) = 3, \qquad y'(1) = -2.$$

First, find the general solution of $y'' - 8y' + 16y = 0$. The characteristic equation is

$$(r - 4)^2 = 0.$$

Hence, the general solution is

$$y = e^{4x}(c_1 + c_2 x).$$

Now we need

$$y(1) = e^4(c_1 + c_2) = 3$$

and

$$y'(1) = 4e^4 c_1 + 5e^4 c_2 = -2.$$

Solve these to obtain

$$c_1 = 17e^{-4} \quad \text{and} \quad c_2 = -14e^{-4}.$$

Thus, the solution is

$$y = e^{4x}(17e^{-4} - 14e^{-4}x) = e^{4(x-1)}(17 - 14x).$$

We now return to the general discussion and the remaining case.

Case 3 Suppose that the indicial equation has complex roots. This happens when $A^2 - 4B < 0$. The roots are then

$$r_1 = \frac{-A + i\sqrt{4B - A^2}}{2} \quad \text{and} \quad r_2 = \frac{-A - i\sqrt{4B - A^2}}{2}.$$

Formally, we would expect $e^{r_1 x}$ and $e^{r_2 x}$ to comprise a fundamental system of solutions (and they do). However, this raises two issues: First, what does e^{rx} mean when r is complex? Second, since $y'' + Ay' + By = 0$ does not explicitly involve complex numbers, one is perhaps a little surprised to find complex numbers in the solutions. To resolve the first difficulty, we pause in the next section to examine the complex exponential function. For the second issue, we return in Section 2.5 to solutions of $y'' + Ay' + By = 0$ when the characteristic equation has complex roots.

PROBLEMS FOR SECTION 2.3

In each of Problems 1 through 25, find the general solution of the differential equation.

1. $y'' - 4y' + y = 0$
2. $y'' + 6y' - 40y = 0$
3. $y'' - 16y' + 64y = 0$
4. $y'' + 3y' - 18y = 0$
5. $y'' + 6y' - 3y = 0$
6. $y'' - 7y' + 2y = 0$
7. $y'' + 2y' - y = 0$
8. $y'' + 16y' + 64y = 0$
9. $y'' + 2y' - 16 = 0$
10. $y'' - 3y' - 3y = 0$
11. $y'' - 14y' + 49y = 0$
12. $y'' + y' - 3y = 0$
13. $y'' + 7y' - 5y = 0$
14. $y'' - 6y' + 3y = 0$
15. $y'' + 12y' + 36y = 0$
16. $y'' + 11y' - 11y = 0$
17. $y'' + 3y' - 2y = 0$
18. $y'' - 14y' + y = 0$
19. $y'' - 10y' + 25y = 0$
20. $y'' - 6y' + 7y = 0$
21. $y'' - 14y' + 2y = 0$
22. $y'' + 2y' - 13y = 0$
23. $y'' + 14y' - 2y = 0$
24. $y'' - 7y' - 2y = 0$
25. $y'' + 18y' + 81y = 0$

In each of Problems 26 through 35, solve the initial value problem.

26. $y'' + 3y' - 2y = 0; \quad y(0) = 1, \quad y'(0) = 0$

27. $y'' - 2y' + y = 0; \quad y(1) = 1, \quad y'(1) = -3$

28. $y'' - 4y' + 4y = 0; \quad y(0) = 3, \quad y'(0) = 5$

29. $y'' + 2y' - 3y = 0; \quad y(0) = 1, \quad y'(0) = 1$

30. $y'' - 2y' - 5y = 0; \quad y(0) = 0, \quad y'(0) = 3$

31. $y'' + 12y' + 36y = 0; \quad y(0) = -2, \quad y'(0) = -3$

32. $y'' - 2y' + y = 0; \quad y(1) = y'(1) = 0$

33. $y'' + 3y' - 2y = 0; \quad y(0) = 2, \quad y'(0) = -3$

34. $y'' - 7y' + 2y = 0; \quad y(2) = 1, \quad y'(2) = 0$

35. $y'' - 6y' + 9y = 0; \quad y(-1) = 1, \quad y'(-1) = 7$

2.4 BACKGROUND ON COMPLEX EXPONENTIAL FUNCTIONS

The key to solving $y'' + Ay' + By = 0$ in the case $A^2 - 4B < 0$ is the complex exponential function. Here we define this function and give some of its properties.

For a complex number $x + iy$ (x and y are real and $i^2 = -1$), we define*

$$e^{x+iy} = e^x[\cos(y) + i\sin(y)] = e^x\cos(y) + ie^x\sin(y).$$

Since x and y are both real, so are e^x, $\cos(y)$, and $\sin(y)$; hence e^{x+iy} is well defined. For example,

$$e^{2+i\pi} = e^2\cos(\pi) + ie^2\sin(\pi) = -e^2 \qquad \text{(a real number!)},$$

$$e^{2+i(\pi/2)} = e^2\cos\left(\frac{\pi}{2}\right) + ie^2\sin\left(\frac{\pi}{2}\right) = ie^2 \qquad \text{(pure imaginary)},$$

and

$$e^{2+3i} = e^2\cos(3) + ie^2\sin(3).$$

Note that, when $y = 0$, $x + iy = x + i0 = x$, and

$$e^{x+iy} = e^x[\cos(0) + i\sin(0)] = e^x;$$

so our definition is consistent, and reduces to the real exponential when the complex number is in fact real (i.e., has 0 as coefficient of i).

Complex exponentials obey the usual rules for exponentials, but these will be treated later in a chapter on complex analysis. The property we need for differential equations is this: For x real, and a and b real constants,

$$\frac{d}{dx}e^{(a+ib)x} = (a+ib)e^{(a+ib)x}.$$

* A motivation for this definition is given in Problem 15 at the end of the section.

To derive this, note that

$$\frac{d}{dx} e^{(a+ib)x} = \frac{d}{dx}[e^{ax}\cos(bx) + ie^{ax}\sin(bx)]$$

$$= \frac{d}{dx}[e^{ax}\cos(bx)] + i\frac{d}{dx}[e^{ax}\sin(bx)]$$

$$= ae^{ax}\cos(bx) - be^{ax}\sin(bx) + iae^{ax}\sin(bx) + ibe^{ax}\cos(bx)$$

$$= ae^{ax}[\cos(bx) + i\sin(bx)] + ibe^{ax}[\cos(bx) + i\sin(bx)]$$

$$= (a+ib)e^{ax}[\cos(bx) + i\sin(bx)]$$

$$= (a+ib)e^{(a+ib)x}.$$

We shall also, in the course of calculations with complex exponentials, assume that the usual rule

$$e^{\alpha+\beta} = e^{\alpha}\cdot e^{\beta}$$

holds for complex as well as real α and β. A proof of this is indicated in the exercises.

PROBLEMS FOR SECTION 2.4

In each of Problems 1 through 10, use the definition to write e^{x+iy} in the form $a+ib$ [as with $e^{2+i\sqrt{2}} = e^2\cos(\sqrt{2}) + ie^2\sin(\sqrt{2})$].

1. $e^{1-i\pi}$ **2.** $e^{\pi+2i\pi}$ **3.** $e^{4i\pi}$ **4.** $e^{1-i\sqrt{2}}$ **5.** $e^{\pi i}$

6. e^{8-2i} **7.** e^{1+4i} **8.** e^{2-i} **9.** e^{i} **10.** $e^{1-i(\pi/2)}$

11. Use the definition of e^{x+iy} to show that

$$e^{x-iy} = e^x[\cos(y) - i\sin(y)].$$

12. Use the result of Problem 11, and the definition of e^{x+iy}, to show that

$$\cos(y) = \frac{1}{2}(e^{iy} + e^{-iy}) \quad\text{and}\quad \sin(y) = \frac{1}{2i}(e^{iy} - e^{-iy}).$$

13. Use the result of Problem 12 to derive

$$\cos(y) = \cosh(iy) \quad\text{and}\quad i\sin(y) = \sinh(iy).$$

14. Use the definition of e^{x+iy} and trigonometric identities to show that

$$e^{x_1+iy_1}\cdot e^{x_2+iy_2} = e^{x_1+x_2+i(y_1+y_2)}.$$

15. Here is a motivation for the definition of e^{x+iy} for the special case $x=0$. Fill in the details.

(a) Use the Taylor series about 0 for the exponential function to obtain

$$e^{iy} = \sum_{n=0}^{\infty}\frac{(iy)^n}{n!}.$$

(b) Show from (a) that

$$e^{iy} = \sum_{n=0}^{\infty}\frac{(-1)^n y^{2n}}{(2n)!} + i\sum_{n=0}^{\infty}\frac{(-1)^n y^{2n+1}}{(2n+1)!}.$$

(c) Deduce from (b) that

$$e^{iy} = \cos(y) + i\sin(y).$$

2.5 GENERAL SOLUTION OF $y'' + Ay' + By = 0$ FOR $A^2 - 4B < 0$

Continuing from the end of Section 2.3, we are attempting to solve $y''+Ay'+By=0$ in the case that $A^2-4B<0$. The characteristic equation $r^2+Ar+B=0$ then has complex roots

$$r_1 = \frac{-A + i\sqrt{4B - A^2}}{2} \quad \text{and} \quad r_2 \frac{-A - i\sqrt{4B - A^2}}{2}.$$

As a convenience, write

$$p = \frac{-A}{2} \quad \text{and} \quad q = \frac{\sqrt{4B - A^2}}{2}.$$

Then,

$$r_1 = p + iq \quad \text{and} \quad r_2 = p - iq.$$

Now, $e^{r_1 x}$ and $e^{r_2 x}$ are complex exponential solutions of $y'' + Ay' + By = 0$. Remember, however, that any linear combination of solutions is also a solution. Write

$$e^{r_1 x} = e^{px}\cos(qx) + ie^{px}\sin(qx)$$

and

$$e^{r_2 x} = e^{px}\cos(qx) - ie^{px}\sin(qx).$$

Then

$$\tfrac{1}{2}(e^{r_1 x} + e^{r_2 x}) = e^{px}\cos(qx)$$

is a solution, and

$$\frac{1}{2i}(e^{r_1 x} - e^{r_2 x}) = e^{px}\sin(qx)$$

is also a solution. Furthermore, it is easy to calculate the Wronskian of these solutions and obtain e^{2px}, which is never zero. Thus,

$$y_1(x) = e^{px}\cos(qx) \quad \text{and} \quad y_2(x) = e^{px}\sin(qx)$$

form a fundamental system of solutions of $y'' + Ay' + By = 0$ when $A^2 - 4B < 0$; and the general solution is

$$y = c_1 e^{px}\cos(qx) + c_2 e^{px}\sin(qx) = e^{px}[c_1\cos(qx) + c_2\sin(qx)].$$

Here, $p + iq$ and $p - iq$ are the complex roots of $r^2 + Ar + B = 0$.

This form of the general solution is usually preferred to $c_1 e^{r_1 x} + c_2 e^{r_2 x}$, although this complex exponential form is correct also.

EXAMPLE 59

Solve

$$y'' + 2y' + 6y = 0.$$

The characteristic equation is

$$r^2 + 2r + 6 = 0,$$

with complex roots

$$r_1 = -1 + i\sqrt{5} \quad \text{and} \quad r_2 = -1 - i\sqrt{5}.$$

Here, $p = -1$, $q = \sqrt{5}$, and the general solution is

$$y = c_1 e^{-x} \cos(\sqrt{5}\,x) + c_2 e^{-x} \sin(\sqrt{5}\,x).$$

EXAMPLE 60

Solve the initial value problem

$$y'' + 2y' + 3y = 0; \qquad y(1) = 2, \qquad y'(1) = -3.$$

First, find the general solution of $y'' + 2y' + 3y = 0$. The characteristic equation is $r^2 + 2r + 3 = 0$, with roots $-1 + i\sqrt{2}$ and $-1 - i\sqrt{2}$. Thus, the general solution is

$$y = c_1 e^{-x} \cos(\sqrt{2}\,x) + c_2 e^{-x} \sin(\sqrt{2}\,x).$$

Now we need

$$y(1) = c_1 e^{-1} \cos(\sqrt{2}) + c_2 e^{-1} \sin(\sqrt{2}) = 2$$

and

$$y'(1) = c_1[-e^{-1} \cos(\sqrt{2}) - \sqrt{2}\,e^{-1} \sin(\sqrt{2})]$$
$$+ c_2[-e^{-1} \sin(\sqrt{2}) + \sqrt{2}\,e^{-1} \cos(\sqrt{2})] = -3.$$

Solve these to get

$$c_1 = \frac{e}{\sqrt{2}}[2\sqrt{2} \cos(\sqrt{2}) + \sin(\sqrt{2})]$$

and

$$c_2 = \frac{e}{\sqrt{2}}[2\sqrt{2} \sin(\sqrt{2}) - \cos(\sqrt{2})].$$

Thus, the initial value problem has solution

$$y = \frac{e}{\sqrt{2}} e^{-x}\{[2\sqrt{2} \cos(\sqrt{2}) + \sin(\sqrt{2})]\cos(\sqrt{2}\,x)$$
$$+ [2\sqrt{2} \sin(\sqrt{2}) - \cos(\sqrt{2})]\sin(\sqrt{2}\,x)\}.$$

If we want to approximate to, say, four decimal places, then we compute $\sqrt{2} \approx 1.4142$, $c_1 \approx 2.7464$, and $c_2 \approx 5.0703$. The approximate solution is then

$$y_{\text{app}} = e^{-x}[2.7464 \cos(1.4142x) + 5.0703 \sin(1.4142x)].$$

Writing the general solution of $y'' + Ay' + By = 0$ should become quite automatic with practice. Here is a summary which covers this section as well as Section 2.3.

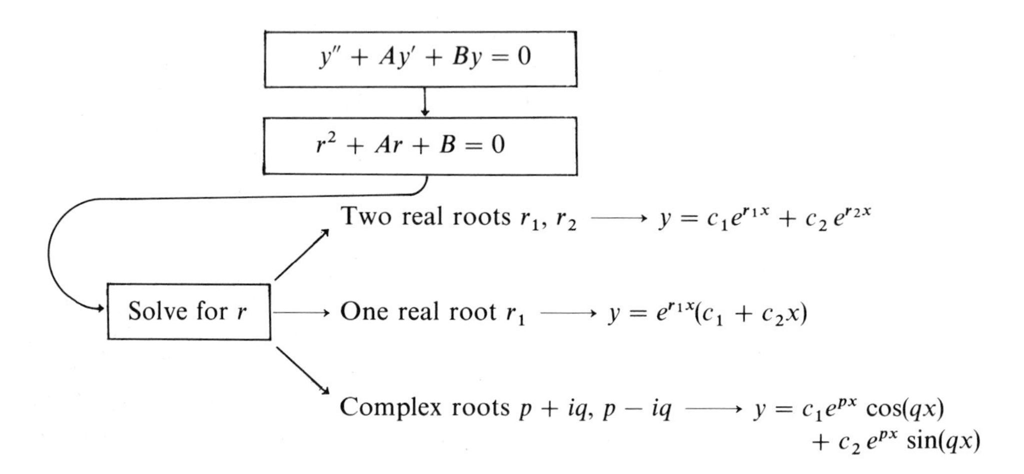

EXAMPLE 61

(1) Solve

$$y'' + 6y' - 3y = 0.$$

The characteristic equation is $r^2 + 6r - 3 = 0$, with roots $-3 + 2\sqrt{3}$, $-3 - 2\sqrt{3}$. The general solution is

$$y = c_1 e^{(-3+2\sqrt{3})x} + c_2 e^{(-3-2\sqrt{3})x}.$$

(2) Solve

$$y'' - 3y' + 8y = 0.$$

The characteristic equation is $r^2 - 3r + 8 = 0$, with complex roots $(3 + i\sqrt{23})/2$ and $(3 - i\sqrt{23})/2$. The general solution is

$$y = c_1 e^{3x/2} \cos\left(\frac{x\sqrt{23}}{2}\right) + c_2 e^{3x/2} \sin\left(\frac{x\sqrt{23}}{2}\right).$$

(3) Solve

$$y'' - 12y' + 36y = 0.$$

Here the characteristic equation is $r^2 - 12r + 36 = 0$, with just one root, $r = 6$. The general solution is

$$y = e^{6x}(c_1 + c_2 x).$$

Now we are in a position to treat nonhomogeneous linear second order differential equations. We begin doing this in Section 2.7, after devoting the next section to an important application.

PROBLEMS FOR SECTION 2.5

In each of Problems 1 through 20, find the general solution of the differential equation. *Note:* These problems mix all three cases of Sections 2.3 and 2.5.

1. $y'' - 4y' + 8y = 0$
2. $y'' - 4y' + y = 0$
3. $y'' + y' + y = 0$
4. $y'' + 2y' + 3y = 0$
5. $y'' + 22y' + 121y = 0$
6. $y'' + 2y' - y = 0$
7. $y'' - 4y' = 0$
8. $y'' + 8y' - 3y = 0$
9. $y'' - 4y' + 2y = 0$
10. $y'' + 6y' + 9y = 0$
11. $y'' + 10y' - y = 0$
12. $y'' + 3y' + 5y = 0$
13. $y'' - 14y' + 49y = 0$
14. $y'' + 2y' - 4y = 0$
15. $y'' - 3y' + 8y = 0$
16. $y'' + y' + 4y = 0$
17. $y'' + 10y' + 25y = 0$
18. $y'' + 3y' - 4y = 0$
19. $y'' + 11y' + 2y = 0$
20. $y'' + 3y' - 5y = 0$

In each of Problems 21 through 30, solve the initial value problem.

21. $y'' + 2y' + 4y = 0;\quad y(0) = 1,\quad y'(0) = 0$
22. $y'' - 2y' + 3y = 0;\quad y(0) = 0,\quad y'(0) = 3$
23. $y'' + 2y' - 3y = 0;\quad y(0) = 0,\quad y'(0) = -2$
24. $y'' + y' + y = 0;\quad y(0) = 2,\quad y'(0) = 0$
25. $y'' - 4y' + 2y = 0;\quad y(0) = y'(0) = 1$
26. $y'' + y' + 3y = 0;\quad y(0) = -3,\quad y'(0) = 2$
27. $y'' + 2y' + 6y = 0;\quad y(1) = -3,\quad y'(1) = 4$
28. $y'' + 2y' + y = 0;\quad y(-2) = 3,\quad y'(-2) = -1$
29. $y'' - 4y' + 7y = 0;\quad y(\pi) = 1,\quad y'(\pi) = 1$
30. $y'' - 6y' + 9y = 0;\quad y(3) = 0,\quad y'(3) = 2$

31. The simple pendulum equation is, from Example 5,

$$\theta'' + \frac{g}{L}\sin(\theta) = 0.$$

If θ is small, then $\sin(\theta)$ is approximately θ, and we have

$$\theta'' + \frac{g}{L}\theta = 0.$$

Solve this to get approximate solutions for motion of the pendulum through small angles. Show that the motion is periodic, with period $2\pi\sqrt{L/g}$.

32. The second order differential equation

$$A(x)y'' + B(x)y' + C(x)y = 0$$

is called *exact* if it can be written in the form

$$[A(x)y']' + [F(x)y]' = 0$$

for some $F(x)$.

Show that, if $A(x)y'' + B(x)y' + C(x)y = 0$ is exact, then necessarily

$$A''(x) - B'(x) + C(x) = 0.$$

The point to exactness for second order differential equations is that $[A(x)y']' + [F(x)y]' = 0$ can be integrated once, reducing the problem to a first order linear differential equation for y. Write a formula for the general solution in this case.

33. Using the results of Problem 32, solve each of the following exact second order differential equations.

(a) $4xy'' + x^2y' + 2xy = 0$

(b) $(2x^2y')' + (xy)' = 0$
(c) $(xy')' + (3xy)' = 0$
(d) $(x^2y')' + (xy)' = 0$
(e) $(2x^2y')' + (4xy)' = 0$

34. A ball of mass m is thrown upward from the earth with initial velocity v_0. Acceleration due to gravity is constant, g, and air resistance is proportional to the velocity v.
(a) Write a differential equation for the height $x(t)$ of the ball (from the earth's surface) at any time t.
(b) Solve for $x(t)$.
(c) Find the maximum height attained.
(d) Show that the time required for the ball to reach maximum height is less than the time required for the ball to fall back to earth from this height.

35. Show that

$$y'' + y = 0; \qquad y(0) = y'(\pi) = 0$$

has only the trivial solution. [*Hint:* Find the general solution of $y'' + y = 0$, and then solve for the constants so that $y(0) = 0$ and $y'(\pi) = 0$.]

In each of Problems 36 through 45, find a solution of the boundary value problem by first finding the general solution of the differential equation and then solving for the constants to satisfy the boundary conditions.

36. $y'' - 4y = 0; \quad y(0) = 1, \quad y(\pi) = 0$
37. $y'' + 3y' - 2y = 0; \quad y(0) = -1, \quad y(2) = 3$
38. $y'' - 6y' + 9y = 0; \quad y(1) = 1, \quad y(2) = -4$
39. $y'' - 10y' + 24y = 0; \quad y(2) = 0, \quad y(4) = -1$
40. $y'' - 4y = 0; \quad y(\pi/2) = 1, \quad y(\pi) = -1$
41. $y'' + 4y = 0; \quad y(0) = 0, \quad y(1) = 1$
42. $y'' - 3y' + y = 0; \quad y(0) = -2, \quad y(1) = 0$
43. $y'' + 3y' + 2y = 0; \quad y(0) = y(3) = 5$
44. $y'' - 2y' + 4y = 0; \quad y(0) = 1, \quad y(1) = 0$
45. $y'' + 6y' - 2y = 0; \quad y(2) = 3, \quad y(5) = -2$

2.6 DAMPED AND UNDAMPED FREE MOTION OF A MASS ON A SPRING

In Chapter 0, Example 2, we derived the equation

$$m\frac{d^2y}{dt^2} + ky = 0.$$

This governs the motion of a body with mass m attached as in Figure 2 to a spring having spring constant k, and in the absence of damping or external driving forces.

Rewriting the equation as

$$y'' + \frac{k}{m}y = 0,$$

we now recognize it as a second order, constant coefficient, homogeneous, linear differential equation—in short, the kind we have just learned to solve. We have only to keep our notation straight. In this setting, t (for time) replaces x, and y' means dy/dt.

The characteristic equation is

$$r^2 + \frac{k}{m} = 0,$$

with complex roots

$$i\sqrt{\frac{k}{m}} \quad \text{and} \quad -i\sqrt{\frac{k}{m}}.$$

Thus, the solution is of the form

$$y(t) = c_1 \cos\left(\sqrt{\frac{k}{m}}\, t\right) + c_2 \sin\left(\sqrt{\frac{k}{m}}\, t\right).$$

This is often written

$$y(t) = d \cos(\omega_0 t - \delta),$$

where $d = \sqrt{c_1^2 + c_2^2}$, $\delta = \arctan(c_2/c_1)$, and $\omega_0 = \sqrt{k/m}$. This motion is called *harmonic oscillation*, and it has a frequency of $\omega_0/2\pi$.

We can determine c_1 and c_2 for specific cases if we know more information about the system. For example, observe that

$$y(0) = c_1 \quad \text{and} \quad y'(0) = c_2\,\omega_0.$$

Thus, c_1 is the position of the mass at time 0, and c_2 is $(1/\omega_0)$ times the initial velocity. Given these, the position y is predictable at all future times (under the ideal conditions assumed in the derivation in Chapter 0). For example, if $y(0) = 5$ and $y'(0) = 0$ (i.e., the spring is extended 5 units and the mass released from rest), then for all times $t > 0$,

$$y(t) = 5 \cos(\omega_0 t).$$

A more interesting scenario is one in which the mass is connected to a dashpot, as in Figure 3. In Example 2 we assumed that this produced a damping force of cy', where c is the damping constant. The resulting differential equation for y was

$$my'' + cy' + ky = 0.$$

This can be written

$$y'' + \frac{c}{m}\,y' + \frac{k}{m}\,y = 0,$$

an equation we are now in a position to analyze completely.

The characteristic equation is

$$r^2 + \frac{c}{m}\,r + \frac{k}{m} = 0,$$

with roots

$$-\frac{c}{2m} \pm \frac{1}{2m}\sqrt{c^2 - 4km}.$$

We now have three cases, all of which can occur, but resulting in quite different motion of the mass attached to the spring.

Case 1 $c^2 - 4km > 0$.

This is called *overdamping*. Putting

$$r_1 = -\frac{c}{2m} + \frac{1}{2m}\sqrt{c^2 - 4km} \quad \text{and} \quad r_2 = -\frac{c}{2m} - \frac{1}{2m}\sqrt{c^2 - 4km},$$

the solution is of the form

$$y = c_1 e^{r_1 t} + c_2 e^{r_2 t}.$$

Now, c and k are positive constants. Thus,

$$\sqrt{c^2 - 4km} < c.$$

Then

$$\frac{1}{2m}\sqrt{c^2 - 4km} < \frac{c}{2m}.$$

Hence,

$$r_1 = -\frac{c}{2m} + \frac{1}{2m}\sqrt{c^2 - 4km} < 0.$$

Obviously $r_2 < 0$ also. Thus, $y(t) \to 0$ as $t \to \infty$, and the motion dies out with time. Eventually the mass will simply assume the static equilibrium position in this case. (See Problem 5 for a follow-up on this.)

Case 2 $c^2 - 4km = 0$.

This is called *critical damping*. Now the solution is of the form

$$y(t) = e^{-ct/2m}(c_1 + c_2 t).$$

As in case 1, $y(t) \to 0$ as $t \to \infty$, and the motion dies out with time.

To see the significance of c_1 and c_2, note first that $y(0) = c_1$, that is, c_1 measures initial position. Next,

$$y'(0) = c_2 - \frac{cc_1}{2m}.$$

Hence,

$$c_2 = y'(0) + \frac{cy(0)}{2m},$$

where $y'(0)$ = initial velocity.

Since $e^{-ct/2m} \neq 0$, $y(t)$ can be zero only when $c_1 + c_2 t = 0$, and this can happen only if

$$t = \frac{-c_1}{c_2} = \frac{-y(0)}{y'(0) + \dfrac{cy(0)}{2m}}.$$

If the number on the right is positive, then the mass passes through the equilibrium position $y = 0$ exactly once, at that time. If the number on the right is negative, the mass never passes through the equilibrium position.

Case 3 $c^2 - 4km < 0$.

This is called *underdamping*. The indicial equation now has complex roots

$$-\frac{c}{2m} + \frac{i}{2m}\sqrt{4km - c^2} \quad \text{and} \quad -\frac{c}{2m} - \frac{i}{2m}\sqrt{4km - c^2},$$

and the solution is of the form

$$y = e^{-ct/2m}\left[c_1 \cos\left(\sqrt{4km - c^2}\,\frac{t}{2m}\right) + c_2 \sin\left(\sqrt{4km - c^2}\,\frac{t}{2m}\right)\right].$$

This is often written

$$y = de^{-ct/2m}\cos(\bar{\omega}t - \delta),$$

where

$$d = \sqrt{C_1^2 + C_2^2}, \qquad \delta = \arctan\left(\frac{c_2}{c_1}\right) \quad \text{and} \quad \bar{\omega} = \frac{\sqrt{4km - c^2}}{2m}.$$

Because c and m are positive, $e^{-ct/2m} \to 0$ as $t \to \infty$, and the underdamped motion decays to zero with increasing time. As it dies out, it oscillates with a frequency $\bar{\omega}/2\pi$. The motion is not, however, periodic.

Note that, as $c \to 0$,

$$y'' + \frac{c}{m}y' + \frac{k}{m}y = 0 \quad \to \quad y'' + \frac{k}{m}y = 0$$

and

$$de^{-ct/2m}\cos(\bar{\omega}t - \delta) \to d\cos(\omega_0 t - \delta).$$

That is, as the damping constant is taken smaller, approaching zero, then

damped equation $\to$ undamped equation

and

underdamped motion $\to$ harmonic oscillation.

This is consistent with intuition.

If we allow an external driving force $F(t)$ in addition to damping, the motion is governed by

$$my'' + cy' + ky = F(t)$$

or

$$y'' + \frac{c}{m}y' + \frac{k}{m}y = \frac{1}{m}F(t),$$

a nonhomogeneous linear differential equation. We shall spend the next two sections learning how to solve such equations and then return to forced spring motion in Section 2.9.

PROBLEMS FOR SECTION 2.6

1. We obtained the solution

$$y = c_1 \cos\left(\sqrt{\frac{k}{m}}\, t\right) + c_2 \sin\left(\sqrt{\frac{k}{m}}\, t\right)$$

of $y'' + (k/m)y = 0$. Write out the details of showing that this solution can be written

$$y = d\cos(\omega_0 t - \delta),$$

where $\delta = \arctan(c_2/c_1)$, $\omega_0 = \sqrt{k/m}$, and $d = \sqrt{c_1^2 + c_2^2}$.

Obtain a particular solution of the problem meeting the requirements

$$y(0) = 1, \qquad y'(0) = \tfrac{1}{2}$$

by solving for d and δ in the above equation. Graph the resulting solution, showing both amplitude and frequency of the harmonic oscillation.

2. Using the solution $y = d\cos(\omega_0 t - \delta)$, determine how d and δ must be chosen to have an initial ($t = 0$) displacement y_0 and an initial velocity v_0.

3. Solve the pendulum equation

$$\frac{d^2\theta}{dt^2} + \frac{g}{L}\sin(\theta) = 0$$

for small displacements by replacing $\sin(\theta)$ by θ for θ close to zero. What analogy can you draw between pendulum motion through small angles and undamped motion of a mass suspended from a spring?

4. Graph the general solution of $my'' + cy' + ky = 0$ in the case of overdamping.

5. How many times can the mass pass through the static equilibrium ($y = 0$) position in the case of overdamping?

What condition can you place on initial displacement $y(0)$ to guarantee that the mass never passes through the static equilibrium position?

How does initial velocity $y'(0)$ influence whether the mass passes through $y = 0$?

6. Graph the general solution of the damped equation $my'' + cy' + ky = 0$ in the case of critical damping.

7. Find solutions corresponding to

$$y(0) = A, \qquad y'(0) = 0 \qquad (A \text{ positive, constant})$$

in all four cases: undamped, overdamped, critically damped, and underdamped. Graph all four solutions on the same set of axes for comparison purposes, clearly marking each curve according to the case it represents.

8. Find solutions corresponding to

$$y(0) = 0, \qquad y'(0) = A \qquad (A \text{ positive, constant})$$

in the cases undamped, overdamped, critically damped, and underdamped. Graph all four solutions on the same set of axes to compare behavior, clearly labeling each curve according to the case it represents.

9. In order to gauge comparative effects of initial position and initial velocity on the resulting motion, find solutions in the overdamped case

(a) when $y(0) = A$ and $y'(0) = 0$,

(b) when $y(0) = 0$ and $y'(0) = A$.

Graph both solutions on the same set of axes. What conclusions do you draw about the two cases?

10. Find the critical damping solutions when

(a) $y(0) = A, \quad y'(0) = 0;$

(b) $y(0) = 0, \quad y'(0) = A.$

Graph both solutions on the same set of axes. What conclusions do you draw about the influence of initial displacement and initial velocity on the resulting motion?

11. Do Problem 10 with underdamping replacing critical damping.

12. Do Problem 10 with harmonic oscillations replacing critical damping solutions.

13. In underdamped motion, what effect on the frequency of the motion would result from increasing the damping constant?

14. Suppose that $y(0) = y'(0)$. Determine the maximum displacement in the critical damping solution, and show that the time at which this maximum occurs is independent of the actual value of $y(0)$ [assuming that $y(0)$ is not zero].

15. If we let m get larger in the underdamping case, the solution dies out more quickly. Intuitively, however, it would seem that a heavier mass would stretch the spring more and cause more oscillation. Which is correct?

16. In the treatment of $y'' + (k/m)y = 0$ at the beginning of this section, we obtained the general solution

$$y = c_1 \cos\left(\sqrt{\frac{k}{m}}\, t\right) + c_2 \sin\left(\sqrt{\frac{k}{m}}\, t\right),$$

and claimed that this can be written in the form

$$y = d \cos(\omega_0 t - \delta),$$

where $\omega_0 = \sqrt{k/m}$. Show in detail how this solution is obtained from the above general solution.

17. In the case of underdamping, we obtained the general solution

$$y = e^{-ct/2m}\left[c_1 \cos\left(\sqrt{4km - c^2}\,\frac{t}{2m}\right) + c_2 \sin\left(\sqrt{4km - c^2}\,\frac{t}{2m}\right)\right].$$

Show in detail how this solution can be written in the form

$$y = de^{-ct/2m} \cos(\bar{\omega} t - \delta).$$

18. Consider underdamped motion, with general solution written as

$$y = de^{-ct/2m} \cos(\bar{\omega} t - \delta).$$

(a) Graph this solution, recalling that $\bar{\omega}$, d, and δ are constants.

(b) The natural period of an unforced undamped system, whose differential equation was $my'' + ky = 0$, is

$$T = 2\pi \sqrt{\frac{m}{k}}.$$

In the overdamped case, the motion does not exhibit this periodicity. Define the *quasi period* as the time between successive maxima and minima of $y(t)$. Show that the quasi period is given by

$$T_q = \frac{4m\pi}{\sqrt{4km - c^2}}.$$

(c) Show that

$$T_q = \frac{T}{\sqrt{1 - \dfrac{c^2}{4km}}}.$$

Hence, conclude that, when $c^2/4km$ is very small, effects of damping are negligible in computing the quasi period, and T_q is approximately T.

19. Prove the principle of conservation of energy for the undamped harmonic oscillator. The principle says that the total energy of the system remains constant throughout the motion. [*Hint:* The energy is a sum of the kinetic energy $\frac{1}{2}mv^2$(v = velocity) and the potential energy $\frac{1}{2}ky^2$. To show that this sum is a constant, show that the differential equation $my'' + ky = 0$ can be written

$$mv\frac{dv}{dy} = -ky,$$

and integrate.]

20. Consider free, undamped motion of a mass on a spring. Suppose that the acceleration at distance d from the equilibrium position is a. Prove that the period of the motion is

$$2\pi\sqrt{\frac{d}{a}}.$$

21. A mass m_1 is attached to a spring and allowed to vibrate with free, undamped motion having period p. Suppose at some time a second mass m_2 is instantaneously glued onto m_1. Prove that the total mass $m_1 + m_2$ now executes simple harmonic motion with period

$$p\sqrt{1 + \frac{m_2}{m_1}}.$$

In Problems 22 through 27, you will need a scientific calculator to compute the requested numbers. Compute to four decimal places.

22. Using the solution $y = d\cos(\omega_0 t - \delta)$, solve $y'' + (k/m)y = 0$ if $k = 1.3$, $m = 2$, $y(0) = 1.72$, and $y'(0) = 2.4$.

23. Let $m = 3$, $c = 2.4$, and $k = \sqrt{3}$. Here $c^2 - 4km = -15.0246$, so we have underdamping. Find the solution corresponding to $y(0) = 3.72$ and $y'(0) = 1.6$.

24. Let $m = 1$, $c = 3.45$, and $k = 2.4$. Here we have overdamping. Find the solution satisfying $y(0) = 1.2$ and $y'(0) = 2.74$.

25. Let $c = 4.069$, $k = 1.8$, and $m = 2.2996$. Here $c^2 - 4km = 0.00036$, which we take to be approximately zero. With this assumption, find a critical damping solution with $y(0) = 3.24$ and $y'(0) = 1.52$.

26. Find a solution satisfying $y(0) = 1.8$, $y'(0) = 0.32$ in the case that $c = \sqrt{5}$, $k = 2.46$, and $m = 1.92$.

27. Find a solution satisfying $y(0) = 2.36$ and $y'(0) = 1.4$ if $m = 2.31$, $c = \sqrt{7}$, and $k = 3.02$.

2.7 THEORY OF LINEAR NONHOMOGENEOUS SECOND ORDER DIFFERENTIAL EQUATIONS

We now have the idea of a general solution of $y'' + P(x)y' + Q(x)y = 0$, and a method for producing this general solution when $P(x) = A$ and $Q(x) = B$, where A and B are constants. This case was useful in analyzing free motion of a mass attached to a spring.

Often, however, we encounter a *nonhomogeneous* differential equation

$$y'' + P(x)y' + Q(x)y = F(x). \tag{a}$$

For example, current in an RLC circuit (Example 1) is governed by

$$LI'' + RI' + \frac{1}{C} I = E(t),$$

and damped motion of a mass on a spring subjected to a driving force $F(t)$ is described by

$$my'' + cy' + ky = F(t).$$

In this section we lay the theoretical groundwork needed to get some insight into solving (a). The key is in the following theorem, which says that we can find *all* solutions of (a) if we can find just *one* solution, together with the general solution of $y'' + P(x)y' + Q(x)y = 0$.

THEOREM 6

Let $y_1(x)$ and $y_2(x)$ form a fundamental system of solutions of $y'' + P(x)y' + Q(x)y = 0$ on an interval J. Let $y_p(x)$ be any solution of $y'' + P(x)y' + Q(x)y = F(x)$ on J. Then, every solution of $y'' + P(x)y' + Q(x)y = F(x)$ is of the form

$$y(x) = c_1 y_1(x) + c_2 y_2(x) + y_p(x)$$

for some choice of the constants c_1 and c_2.

Proof Let $y(x)$ be any solution of (a) on J. Since $y_p(x)$ is also a solution by hypothesis, then

$$[y'' + P(x)y' + Q(x)y] - [y_p'' + P(x)y_p' + Q(x)y_p]$$
$$= (y - y_p)'' + P(x)(y - y_p)' + Q(x)(y - y_p) = F(x) - F(x) = 0.$$

Hence, $y(x) - y_p(x)$ is a solution of the homogeneous equation $y'' + P(x)y' + Q(x)y = 0$. Since $y_1(x)$ and $y_2(x)$ form a fundamental system of solutions of the homogeneous equation on J, then $y(x) - y_p(x)$ must be a linear combination of $y_1(x)$ and $y_2(x)$. Then, for some constants c_1 and c_2,

$$y(x) - y_p(x) = c_1 y_1(x) + c_2 y_2(x).$$

This is the same as

$$y(x) = c_1 y_1(x) + c_2 y_2(x) + y_p(x),$$

as we wanted to prove.

Thus, solving the homogeneous linear differential equation $y'' + P(x)y' + Q(x)y = 0$ is a major step in solving the nonhomogeneous differential equation $y'' + P(x)y' + Q(x)y = F(x)$. In view of Theorem 6, we adopt the following terminology:

$$y'' + P(x)y' + Q(x)y = 0 \qquad \text{(b)}$$

is called the *associated homogeneous equation* of (a). Often we denote the general solution of (b) as $y_h(x)$, and any particular solution of (a) as $y_p(x)$. The expression

$$y_h(x) + y_p(x)$$

then contains all solutions of (a), and is called the *general solution* of $y'' + P(x)y' + Q(x)y = F(x)$.

EXAMPLE 62

Consider

$$y'' - 4y = x.$$

The associated homogeneous equation is

$$y'' - 4y = 0,$$

with fundamental system of solutions e^{2x} and e^{-2x}. The general homogeneous solution of $y'' - 4y = x$ is

$$y_h = c_1 e^{2x} + c_2 e^{-2x}.$$

One solution of $y'' - 4y = x$ is

$$y_p = -\tfrac{1}{4}x.$$

(We describe a method for finding such a solution in the next section—for now take y_p as given for the sake of illustration.) Then, the general solution of $y'' - 4y = x$ is

$$y = y_h + y_p = c_1 e^{2x} + c_2 e^{-2x} - \tfrac{1}{4}x,$$

with c_1 and c_2 arbitrary constants. This expression contains every solution of $y'' - 4y = x$ (on any interval J, because the functions involved in this example are defined everywhere).

Note that $\cosh(2x)$ and $\sinh(2x)$ also form a fundamental system for $y'' - 4y = 0$. Thus, the general solution of $y'' - 4y = x$ could also be written

$$y = A\cosh(2x) + B\sinh(2x) - \tfrac{1}{4}x.$$

With $c_1 = (A + B)/2$ and $c_2 = (A - B)/2$, these two expressions for the general solution become identical.

EXAMPLE 63

Consider

$$y'' + 4y = e^x.$$

The associated homogeneous equation $y'' + 4y = 0$ has fundamental system of solutions $\cos(2x)$ and $\sin(2x)$. Thus,

$$y_h = c_1 \cos(2x) + c_2 \sin(2x).$$

A particular solution of $y'' + 4y = e^x$ is

$$y_p = \tfrac{1}{5}e^x.$$

(Again, we shall describe how to find this in the next section.) Thus, the general solution of $y'' + 4y = e^x$ is

$$y = y_h + y_p = c_1 \cos(2x) + c_2 \sin(2x) + \tfrac{1}{5}e^x,$$

with c_1 and c_2 arbitrary constants.

We can summarize this section as follows:

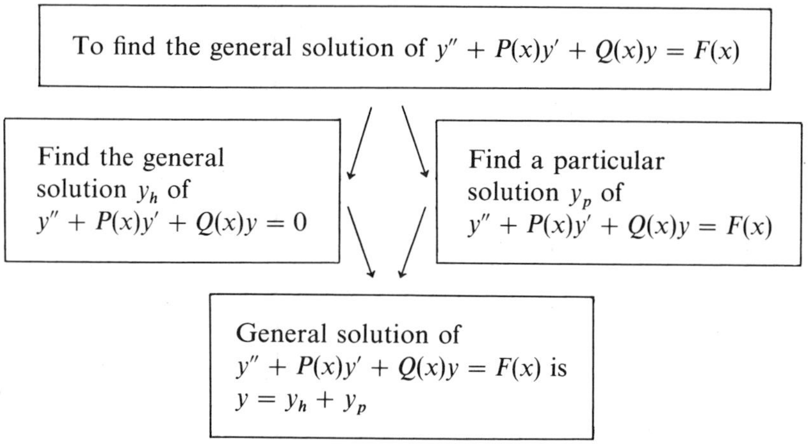

To find the general solution of $y'' + P(x)y' + Q(x)y = F(x)$

Find the general solution y_h of $y'' + P(x)y' + Q(x)y = 0$

Find a particular solution y_p of $y'' + P(x)y' + Q(x)y = F(x)$

General solution of $y'' + P(x)y' + Q(x)y = F(x)$ is $y = y_h + y_p$

As usual (at least until Chapter 4), one solves an initial value problem by finding the general solution and then solving for the constants to satisfy the initial data.

EXAMPLE 64

Solve

$$y'' - 4y = x; \qquad y(1) = 3, \qquad y'(1) = -2.$$

From Example 62, the general solution of $y'' - 4y = x$ is

$$y = c_1 e^{2x} + c_2 e^{-2x} - \tfrac{1}{4}x.$$

We must choose c_1 and c_2 so that

$$y(1) = c_1 e^2 + c_2 e^{-2} - \tfrac{1}{4} = 3$$

and

$$y'(1) = 2c_1 e^2 - 2c_2 e^{-2} - \tfrac{1}{4} = -2.$$

Solve these to get

$$c_1 = \frac{19}{16e^2} \quad \text{and} \quad c_2 = \frac{33e^2}{16}.$$

The initial value problem then has solution

$$y = \frac{19}{16e^2} e^{2x} + \frac{33e^2}{16} e^{-2x} - \frac{1}{4} x.$$

To four decimals, we have approximately

$$y_{\text{app}} = 0.1607e^{2x} + 15.2399e^{-2x} - 0.2500x.$$

PROBLEMS FOR SECTION 2.7

In each of Problems 1 through 15, verify that the given function is a solution of the differential equation, and then find the general solution.

1. $y'' - 4y' + 2y = 3x + 2$; $y_p = \frac{3}{2}x + 4$
2. $y'' + y = x^3$; $y_p = x^3 - 6x$
3. $y'' + 2y' + y = e^x$; $y_p = \frac{1}{4}e^x$
4. $y'' - 4y' - 3y = x - 1$; $y_p = -\frac{1}{3}x + \frac{7}{9}$
5. $y'' - 6y' - y = 3\cos(2x)$; $y_p = -\frac{15}{169}\cos(2x) - \frac{36}{169}\sin(2x)$
6. $y'' + 3y' - 4y = 8e^{2x} + 1$; $y_p = \frac{4}{3}e^{2x} - \frac{1}{4}$
7. $y'' + 8y' - 2y = \cos(x) - \sin(x)$; $y_p = \frac{5}{73}\cos(x) + \frac{11}{73}\sin(x)$
8. $y'' + 2y' + 2y = x^2$; $y_p = \frac{1}{2}x^2 - x + \frac{1}{2}$
9. $y'' + 16y' - 2y = 4$; $y_p = -2$
10. $y'' + 2y' + 8y = 3e^{4x} + 1$; $y_p = \frac{3}{32}e^{4x} + \frac{1}{8}$
11. $y'' - 4y' - 2y = x^3 - 4x + 3$; $y_p = -\frac{1}{2}x^3 + 3x^2 - \frac{23}{2}x + \frac{49}{2}$
12. $y'' - y' + 4y = -2\sin(3x)$; $y_p = -\frac{3}{17}\cos(3x) + \frac{5}{17}\sin(3x)$
13. $y'' - 3y' + 2y = x$; $y = \frac{1}{2}x + \frac{3}{4}$
14. $y'' + 2y' - 5y = 20$; $y_p = -4$
15. $y'' - y' + 10y = 4x^2 + 2x$; $y_p = \frac{2}{5}x^2 + \frac{7}{25}x - \frac{13}{250}$

In each of Problems 16 through 30, use the result of the problem cited to solve the initial value problem.

16. $y'' - 4y' + 2y = 3x + 2$ (Problem 1); $y(0) = 1$, $y'(0) = 0$
17. $y'' + y = x^3$ (Problem 2); $y(0) = -2$, $y'(0) = 3$
18. $y'' + 2y' + y = e^x$ (Problem 3); $y(0) = y'(0) = 0$
19. $y'' - 4y' - 3y = x - 1$ (Problem 4); $y(0) = 1$, $y'(0) = 0$
20. $y'' - 6y' - y = 3\cos(2x)$ (Problem 5); $y(\pi/2) = 0$, $y'(\pi/2) = 1$
21. $y'' + 3y' - 4y = 8e^{2x} + 1$ (Problem 6); $y(0) = -4$, $y'(0) = 0$
22. $y'' + 8y' - 2y = \cos(x) - \sin(x)$ (Problem 7); $y(\pi) = 3$, $y'(\pi) = 0$
23. $y'' + 2y' + 2y = x^2$ (Problem 8); $y(\pi) = y'(\pi) = 0$
24. $y'' + 16y' - 2y = 4$ (Problem 9); $y(3) = 2$, $y'(3) = 0$
25. $y'' + 2y' + 8y = 3e^{4x} + 1$ (Problem 10); $y(0) = y'(0) = 0$
26. $y'' - 4y' - 2y = x^3 - 4x + 3$ (Problem 11); $y(0) = y'(0) = 0$
27. $y'' - y' + 4y = -2\sin(3x)$ (Problem 12); $y(0) = 0$, $y'(0) = 1$
28. $y'' - 3y' + 2y = x$ (Problem 13); $y(0) = -5$, $y'(0) = 6$
29. $y'' + 2y' - 5y = 20$ (Problem 14): $y(1) = y'(1) = 3$
30. $y'' - y' + 10y = 4x^2 + 2x$ (Problem 15); $y(3) = 1$, $y'(3) = 0$

2.8 FINDING PARTICULAR SOLUTIONS OF $y'' + P(x)y' + Q(x)y = F(x)$

In general, finding even one solution of $y'' + P(x)y' + Q(x)y = F(x)$ can be very difficult. In this section we shall describe two methods which are often useful.

The first method is applicable to the special case where $P(x)$ and $Q(x)$ [but not necessarily $F(x)$] are constant. Since we can always find the general homogeneous solution y_h in this case, finding a particular solution y_p of $y'' + Ay' + By = F(x)$ will enable us, by Theorem 6, to write the general solution $y_h + y_p$.

METHOD 1 UNDETERMINED COEFFICIENTS

Another name for this method is "educated guessing." Sometimes the form of $F(x)$ suggests what $y_p(x)$ should look like in general appearance, usually involving one or more unknown constants. We substitute the conjectured "solution" into the differential equation, and then try to solve for the constants to assure that it is indeed a solution. Here are three examples to illustrate the basic idea, after which we shall discuss the method in more detail.

EXAMPLE 65

Find a solution of

$$y'' - 4y = 8x^2 - 2x.$$

Observe that $8x^2 - 2x$ is a polynomial. Since derivatives of polynomials are polynomials, we would guess that some polynomial can be inserted into $y'' - 4y$ to get $8x^2 - 2x$. Further, the polynomial we try should not contain powers x^3 or higher because $y'' - 4y$ must have x^2 as its highest power term. Thus, we are led to try

$$y_p = ax^2 + bx + c.$$

Note that, although $8x^2 - 2x$ has a zero constant term, we cannot assume that y_p will also have a zero constant term. It may or may not—in fact, we shall see that in this example, $c \neq 0$.

Calculate

$$y_p' = 2ax + b \quad \text{and} \quad y_p'' = 2a.$$

Upon substitution into the differential equation, we get

$$y_p'' - 4y_p = 2a - 4(ax^2 + bx + c) = -4ax^2 - 4bx + 2a - 4c = 8x^2 - 2x.$$

The only way these two polynomials can be equal (for all x) is for the coefficients of each power of x to match. Thus, we need

$$-4a = 8, \qquad -4b = -2, \quad \text{and} \quad 2a - 4c = 0.$$

Solve these to get

$$a = -2, \qquad b = \tfrac{1}{2}, \quad \text{and} \quad c = -1.$$

Thus, a particular solution is

$$y_p = -2x^2 + \tfrac{1}{2}x - 1,$$

as is easily verified by substituting back into the differential equation.

It is easy to solve the associated homogeneous equation $y'' - 4y = 0$ in general to obtain

$$y_h = c_1e^{2x} + c_2e^{-2x}.$$

Thus, the general solution of $y'' - 4y = 8x^2 - 2x$ is

$$y = c_1 e^{2x} + c_2 e^{-2x} - 2x^2 + \tfrac{1}{2}x - 1.$$

EXAMPLE 66

Find a solution of

$$y'' + 2y' - 3y = 4e^{2x}.$$

Since derivatives of e^{2x} are always constant multiples of e^{2x}, we conjecture that there is some number k such that $y_p = ke^{2x}$. Substituting this into the differential equation gives us

$$4ke^{2x} + 4ke^{2x} - 3ke^{2x} = 4e^{2x}$$

or

$$5k = 4.$$

Then $k = \frac{4}{5}$, and

$$y_p = \tfrac{4}{5}e^{2x}.$$

Again, once we have y_p, it is easy to find the general solution of $y'' + 2y' - 3y = 4e^{2x}$, should we want it. The general homogeneous solution of $y'' + 2y' - 3y = 0$ is easily found to be

$$y_h = c_1 e^x + c_2 e^{-3x}.$$

The general solution of $y'' + 2y' - 3y = 4e^{2x}$ is then

$$y = c_1 e^x + c_2 e^{-3x} + \tfrac{4}{5}e^{2x}.$$

EXAMPLE 67

Solve

$$y'' - 3y' + 7y = x^3 - 3x - 4\cos(2x).$$

Notice that $F(x)$ here involves two kinds of functions: a polynomial, and a cosine. Derivatives of polynomials are polynomials, so we can try to account for the $x^3 - 3x$ term by putting $ax^3 + bx^2 + cx + d$ into y_p, reasoning as in Example 65. However, cosine can have both sines and cosines as derivatives, depending on how many derivatives are taken. Thus, we try to account for the $-4\cos(2x)$ term by inserting $h\cos(2x) + k\sin(2x)$ into y_p. Thus, we try

$$y_p = ax^3 + bx^2 + cx + d + h\cos(2x) + k\sin(2x),$$

and attempt to solve for a, b, c, d, h, and k so that this is a particular solution. Calculate

$$y_p' = 3ax^2 + 2bx + c - 2h\sin(2x) + 2k\cos(2x)$$

and

$$y_p'' = 6ax + 2b - 4h\cos(2x) - 4k\sin(2x).$$

Substitute into the differential equation:

$$6ax + 2b - 4h\cos(2x) - 4k\sin(2x) - 9ax^2 - 6bx - 3c + 6h\sin(2x) - 6k\cos(2x) + 7ax^3 + 7bx^2 + 7cx + 7d + 7h\cos(2x) + 7k\sin(2x) = x^3 - 3x - 4\cos(2x).$$

Group terms:

$$7ax^3 + (-9a + 7b)x^2 + (6a - 6b + 7c)x + (2b - 3c + 7d) + (-4h - 6k + 7h)\cos(2x) + (-4k + 6h + 7k)\sin(2x) = x^3 - 3x - 4\cos(2x).$$

Now equate coefficients of like terms on both sides of the equation:

$$\begin{aligned} 7a &= 1 && [\text{coefficients of } x^3] \\ -9a + 7b &= 0 && [\text{coefficients of } x^2] \\ 6a - 6b + 7c &= -3 && [\text{coefficients of } x] \\ 2b - 3c + 7d &= 0 && [\text{constant term}] \\ 3h - 6k &= -4 && [\text{coefficient of } \cos(2x)] \\ 3k + 6h &= 0 && [\text{coefficient of } \sin(2x)]. \end{aligned}$$

Solve these to get

$$a = \tfrac{1}{7}, \quad b = \tfrac{9}{49}, \quad c = -\tfrac{135}{343}, \quad d = -\tfrac{531}{2401}, \quad h = -\tfrac{12}{45}, \quad k = \tfrac{24}{45}.$$

Thus, a particular solution is

$$y_p = \tfrac{1}{7}x^3 + \tfrac{9}{49}x^2 - \tfrac{135}{343}x - \tfrac{531}{2401} - \tfrac{12}{45}\cos(2x) + \tfrac{24}{45}\sin(2x).$$

Note that, although $F(x)$ in the differential equation does not contain an x^2 term, a constant term, or a $\sin(2x)$ term, y_p does. Such terms may not be omitted from the guessed form of y_p.

It is now easy to check that the general solution of the given differential equation is $y_h + y_p$, where

$$y_h = e^{-3x/2}\left[c_1\cos\left(\sqrt{19}\,\frac{x}{2}\right) + c_2\sin\left(\sqrt{19}\,\frac{x}{2}\right)\right].$$

With these three examples as introduction, we shall now examine the method more carefully. Two problems immediately come to mind:

1. The differential equation must have constant coefficients [although of course $F(x)$ need not be constant].
2. Guessing a form for y_p may not be easy. For example, what would you guess to solve $y'' + 4y = \tan(x)$?

These disadvantages are offset by the frequency of occurrence in practical problems of differential equations to which the method applies. In fact, we shall use it in the next section to analyze forced oscillations of a mass on a spring.

There is, in addition, a fine point of crucial importance, and it is this:

> If the conjectured solution y_p satisfies the associated homogeneous differential equation, then the method as illustrated thus far fails.

For example, given $y'' + 2y' - 3y = 4e^x$, we would be tempted by our success in Example 66 to try $y_p = Ae^x$. However, Ae^x satisfies $y'' + 2y' - 3y = 0$; hence substitution of Ae^x into the differential equation yields the equation $0 = 4e^x$, an impossibility.

> In such a case, we can still succeed by introducing a factor of x or x^2 into the proposed solution.

Here are two illustrations of this procedure.

EXAMPLE 68

Find a particular solution of

$$y'' + 2y' - 3y = 4e^x.$$

Trying $y_p = Ae^x$ won't work, as we have just pointed out. Thus, try $y_p = Axe^x$. We have

$$y_p' = Ae^x + Axe^x \quad \text{and} \quad y_p'' = 2Ae^x + Axe^x.$$

Substituting into the differential equation gives us

$$2Ae^x + Axe^x + 2(Ae^x + Axe^x) - 3Axe^x = 4e^x.$$

Then,

$$4Ae^x = 4e^x;$$

so we must choose $A = 1$. A solution is then $y_p = xe^x$.

EXAMPLE 69

Find a particular solution of

$$y'' - 6y' + 9y = 8e^{3x}.$$

Here we might try $y_p = Ae^{3x}$. But e^{3x} turns out to be a solution of the associated homogeneous equation, and so won't work. Next we try $y_p = Axe^x$. However, in this example, xe^{3x} is also a solution of the homogeneous equation. [In fact, the general solution of $y'' - 6y' + 9y = 0$ is $y = e^{3x}(c_1 + c_2 x)$.] Thus, we try $y_p = Ax^2e^{3x}$. Substitute into the differential equation and cancel terms where possible to obtain

$$2Ae^{3x} = 8e^{3x}.$$

Then $A = 4$, and a particular solution is $y_p = 4x^2e^{3x}$.

Based on the ideas illustrated thus far, it is possible to draw up a table of things to try corresponding to the form of $F(x)$.

$F(x)$	$y_p(x)$
$c_0 + c_1x + \cdots + c_nx^n$	$d_0 + d_1x + \cdots + d_nx^n$
$ce^{\alpha x}$	$de^{\alpha x}$
$c\sin(\beta x)$ or $c\cos(\beta x)$	$a\sin(\beta x) + b\cos(\beta x)$
$(c_0 + c_1x + \cdots + c_nx^n)e^{\alpha x}$	$(d_0 + d_1x + \cdots + d_nx^n)e^{\alpha x}$
$(c_0 + c_1x + \cdots + c_nx^n)\sin(\beta x)$ or $(c_0 + c_1x + \cdots + c_nx^n)\cos(\beta x)$	$(d_0 + d_1x + \cdots + d_nx^n)\sin(\beta x) + (e_0 + e_1x + \cdots + e_nx^n)\cos(\beta x)$
$(c_0 + c_1x + \cdots + c_nx^n)e^{\alpha x}\sin(\beta x)$ or $(c_0 + c_1x + \cdots + c_nx^n)e^{\alpha x}\cos(\beta x)$	$(d_0 + d_1x + \cdots + d_nx^n)e^{\alpha x}\sin(\beta x) + (e_0 + e_1x + \cdots + e_nx^n)e^{\alpha x}\cos(\beta x)$

One now proceeds as follows in applying undetermined coefficients:

1. Check that the differential equation has constant coefficients.
2. Determine if $F(x)$ is a sum of terms from the left side of the table.
3. As a first guess, try for y_p a sum of corresponding terms on the right.
4. If *any term contained in* this y_p is a solution of the associated homogeneous equation, replace y_p by xy_p; if *any term contained in* xy_p is also a homogeneous solution, use x^2y_p.

In view of (4), one often obtains y_h first to check for occurrence of homogeneous solutions in the terms of y_p.

EXAMPLE 70

Find a particular solution of

$$y'' + 3y' - y = 7e^{2x}\cos(4x).$$

Looking down the left side of the table, the last entry fits $F(x)$ with $n = 0$, $\alpha = 2$, and $\beta = 4$. Thus, we attempt $y_p = ae^{2x}\sin(4x) + be^{2x}\cos(4x)$. Since neither $e^{2x}\sin(4x)$ nor $e^{2x}\cos(4x)$ is a solution of $y'' + 3y' - y = 0$, this choice of y_p will work. Compute y_p', y_p'', and substitute into the differential equation to get

$$(-7a + 28b)e^{2x}\cos(4x) + (-28a - 7b)e^{2x}\sin(4x) = 7e^{2x}\cos(4x).$$

Equating coefficients of like terms on both sides of the equation gives us

$$-7a + 28b = 7$$
$$-28a - 7b = 0.$$

Solve these to get

$$a = -\tfrac{49}{833} \quad \text{and} \quad b = \tfrac{196}{833}.$$

A particular solution is then

$$y_p = \tfrac{1}{833}[-49e^{2x}\cos(4x) + 196e^{2x}\sin(4x)].$$

EXAMPLE 71

Find a particular solution of

$$y'' + 9y = -e^{-4x} + x^2\cos(3x).$$

We can break this into two problems:

(1) $y'' + 9y = -e^{-4x}$

and

(2) $y'' + 9y = x^2\cos(3x)$.

The solution we seek will be a sum of the solutions of (1) and (2).

Problem (1) is routine. We try $y_p = Ae^{-4x}$ and find that $A = -\frac{1}{25}$.

Problem (2) illustrates item 4 after the table. We are tempted to try

$$y_p = (d_0 + d_1x + d_2x^2)\cos(3x) + (e_0 + e_1x + e_2x^2)\sin(3x).$$

However, $\cos(3x)$ and $\sin(3x)$, which occur as terms of y_p, are solutions of $y'' + 9y = 0$. Thus, multiply by x and try

$$y_p = (d_0x + d_1x^2 + d_2x^3)\cos(3x) + (e_0x + e_1x^2 + e_2x^3)\sin(3x).$$

Substitute this into Problem (2) and equate coefficients of like terms on both sides of the equation to obtain:

$$\begin{aligned} 2d_1 + 6e_0 &= 0 && \text{[from } \cos(3x) \text{ terms]} \\ -6d_0 + 2e_1 &= 0 && \text{[from } \sin(3x) \text{ terms]} \\ 6d_2 + 12e_1 &= 0 && \text{[from } x\cos(3x) \text{ terms]} \\ -12d_1 + 6e_2 &= 0 && \text{[from } x\sin(3x) \text{ terms]} \\ 18e_2 &= 1 && \text{[from } x^2\cos(3x) \text{ terms]} \\ -18d_2 &= 0 && \text{[from } x^2\sin(3x) \text{ terms].} \end{aligned}$$

Terms involving $x^3\cos(3x)$ and $x^3\sin(3x)$ cancel out. Solving these equations, we get

$$d_1 = \tfrac{1}{36}, \qquad d_0 = d_2 = e_1 = 0, \qquad e_0 = -\tfrac{1}{108}, \quad \text{and} \quad e_2 = \tfrac{1}{18}.$$

Thus, a particular solution of Problem (2) is

$$\tfrac{1}{36}x^2\cos(3x) + (-\tfrac{1}{108}x + \tfrac{1}{18}x^3)\sin(3x).$$

A particular solution of $y'' + 9y = -e^{-4x} + x^2\cos(3x)$ is then

$$y_p = -\tfrac{1}{25}e^{-4x} + \tfrac{1}{36}x^2\cos(3x) + (-\tfrac{1}{108}x + \tfrac{1}{18}x^3)\sin(3x).$$

METHOD 2 VARIATION OF PARAMETERS

Given $y'' + P(x)y' + Q(x)y = F(x)$, in which $P(x)$ and $Q(x)$ need not be constant, the variation of parameters method begins with finding a fundamental system of solutions $y_1(x)$ and $y_2(x)$ of $y'' + P(x)y' + Q(x)y = 0$. We now attempt to find a particular solution y_p of $y'' + P(x)y' + Q(x)y = F(x)$ of the form

$$y_p = u(x)y_1(x) + v(x)y_2(x).$$

In order to find two such functions $u(x)$ and $v(x)$, we will need two equations. One will arise from the requirement that $uy_1 + vy_2$ is to be a solution of the differential equation. We can impose a second requirement at our convenience. In fact, we will soon set a condition which will simplify y_p', with the intention of ultimately obtaining equations for $u'(x)$ and $v'(x)$ which we can integrate to find $u(x)$ and $v(x)$.

To begin, compute

$$y_p' = u'y_1 + v'y_2 + uy_1' + vy_2'.$$

To simplify this expression (and hence also that for y_p''), we impose the condition

$$u'y_1 + v'y_2 = 0. \tag{a}$$

This gives us one equation for u' and v', and also simplifies y_p' to just

$$y_p' = uy_1' + vy_2'.$$

Then,

$$y_p'' = u'y_1' + v'y_2' + uy_1'' + vy_2''.$$

Substitute y_p, y_p', and y_p'' into the differential equation to obtain

$$u'y_1' + v'y_2' + uy_1'' + vy_2'' + P(x)(uy_1' + vy_2') + Q(x)(uy_1 + vy_2) = F(x).$$

Rewrite this as

$$u[y_1'' + P(x)y_1' + Q(x)y] + v[y_2'' + P(x)y_2' + Q(x)y_2] + u'y_1' + v'y_2' = F(x).$$

Recalling that y_1 and y_2 are solutions of $y'' + P(x)y' + Q(x)y = 0$, this reduces to

$$u'y_1' + v'y_2' = F(x), \tag{b}$$

a second equation for u' and v'. Solve (a) and (b) to obtain

$$u' = \frac{-y_2 F(x)}{y_1 y_2' - y_2 y_1'} \quad \text{and} \quad v' = \frac{y_1 F(x)}{y_1 y_2' - y_2 y_1'}.$$

In each expression, the denominator is exactly the Wronskian $W(y_1, y_2)$, which is nonzero by the assumption that y_1 and y_2 form a fundamental system of solutions of $y'' + P(x)y' + Q(x)y = 0$. We finally have

$$u(x) = \int \frac{-y_2 F(x)}{W(y_1, y_2)}\, dx \quad \text{and} \quad v(x) = \int \frac{y_1 F(x)}{W(y_1, y_2)}\, dx.$$

These give u and v such that $y_p = uy_1 + vy_2$ is a particular solution of $y'' + P(x)y' + Q(x)y = F(x)$.

Note: In carrying out the integrations to get u and v, the integration constants can be chosen as zero since we are looking for some particular choice of $u(x)$ and $v(x)$, not every possible one.

EXAMPLE 72

Find a particular solution of

$$y'' + 4y = \tan(2x).$$

The homogeneous differential equation is $y'' + 4y = 0$, with fundamental system of solutions $y_1 = \cos(2x)$ and $y_2 = \sin(2x)$. The Wronskian is easily computed to be $W(y_1, y_2) = 2$. Compute

$$\begin{aligned} u(x) &= \int \frac{-\sin(2x)\tan(2x)}{2}\,dx \\ &= \frac{1}{4}\sin(2x) - \frac{1}{4}\ln\left|\tan\left(\frac{\pi}{4} + x\right)\right| \end{aligned}$$

and

$$v(x) = \int \frac{\cos(2x)\tan(2x)}{2}\,dx = -\frac{1}{4}\cos(2x).$$

Thus, a particular solution is

$$\begin{aligned} y_p(x) &= \frac{1}{4}\sin(2x)\cos(2x) - \frac{1}{4}\cos(2x)\ln\left|\tan\left(\frac{\pi}{4} + x\right)\right| - \frac{1}{4}\sin(2x)\cos(2x) \\ &= -\frac{1}{4}\cos(2x)\ln\left|\tan\left(\frac{\pi}{4} + x\right)\right|. \end{aligned}$$

We now have easily that the general solution of $y'' + 4y = \tan(2x)$ is

$$y(x) = c_1\cos(2x) + c_2\sin(2x) - \frac{1}{4}\cos(2x)\ln\left|\tan\left(\frac{\pi}{4} + x\right)\right|.$$

EXAMPLE 73

Find a particular solution of

$$y'' - \frac{4}{x}y' + \frac{4}{x^2}y = x^2 + 1.$$

On any interval J containing only positive numbers, $y_1 = x$ and $y_2 = x^4$ form a fundamental system of solutions of the homogeneous equation $y'' - (4/x)y' + (4/x^2)y = 0$. We shall explain how we got these in Section 2.12, where we treat Euler equations. For now, consider them as given for the sake of illustration.

To find y_p using variation of parameters, first compute

$$W(y_1, y_2) = y_1 y_2' - y_2 y_1' = 3x^4.$$

Then,

$$u(x) = \int \frac{-x^4(x^2 + 1)}{3x^4}\,dx = -\frac{x^3}{9} - \frac{x}{3}$$

and

$$v(x) = \int \frac{x(x^2 + 1)}{3x^4}\,dx = \frac{1}{3}\ln|x| - \frac{1}{6}x^2.$$

A particular solution is then

$$y_p(x) = \left(-\frac{x^3}{9} - \frac{x}{3}\right)x + \left(\frac{1}{3}\ln|x| - \frac{1}{6}x^2\right)x^4$$

$$= -\frac{x^4}{9} - \frac{x^2}{3} + \frac{x^4}{3}\ln|x| - \frac{1}{6}x^6.$$

If one wishes, it is now easy to write the general solution. It is

$$y(x) = c_1 x + c_2 x^4 - \frac{x^4}{9} - \frac{x^2}{3} + \frac{x^4}{3}\ln|x| - \frac{1}{6}x^6.$$

One difficulty with variation of parameters is that the formulas for $u(x)$ and $v(x)$ require integrations which can be difficult. The student should review the techniques of integration portion of beginning calculus, and should also become handy with the use of integral tables.

We still have a long way to go with second order differential equations. However, the theory and methods we have now will enable us to continue the analysis of electrical circuits and spring motion begun in previous sections.

PROBLEMS FOR SECTION 2.8

In each of Problems 1 through 20, find the general homogeneous solution, and then use undetermined coefficients to find a particular solution. Finally, find the general solution of the given equation.

1. $y'' + 6y' - 5y = 3x^2 - 4$

2. $y'' - 2y' + 11y = 3e^{2x}$

3. $y'' + y' - 7y = x^3 - 4$

4. $y'' - 3y' + 5y = e^x + \cos(2x)$

5. $y'' - y' + 14y = x - 2\sin(3x)$

6. $y'' + y' - 6y = e^{-3x} + 7e^{-2x}$

7. $y'' + 2y' - 12y = x^2 - x + 2e^{-3x}$

8. $y'' - 7y' + 2y = xe^x$

9. $y'' + y = x^2 - 4 + \sinh(3x)$

10. $y'' - 6y' + 7y = \sin(2x) + \sin(3x)$

11. $y'' - 4y' + 6y = e^{2x} - 3e^{4x}$

12. $y'' + 8y' - 3y = x^3 - \cos(2x)$

13. $y'' + 2y' + y = -3e^{-x} + 8xe^{-x} + 1$

14. $y'' + 4y' - 12y = \cos(3x) + x$

15. $y'' + 8y' + 7y = 6\cosh(2x) - 4$

16. $y'' - 4y' + 7y = -x^2 + 3x$

17. $y'' - y' - 6y = 8x^3 - 5$

18. $y'' + 4y = -3\cos(3x) + \sin(2x)$

19. $y'' - 4y = 5\sinh(2x)$

20. $y'' + 7y' - 6y = x + 1$

In each of Problems 21 through 35, find the general homogeneous solution, and then use variation of parameters to find a particular solution. Finally, find the general solution of the given equation.

21. $y'' + y' - 2y = x$

22. $y'' + y = \tan(x)$

23. $y'' - y' + 2y = e^{x/2}$

24. $y'' + 2y' + 4y = 1/x$

25. $y'' - 6y' + 2y = \cos(x)$

26. $y'' + \frac{2}{x}y' - \frac{6}{x^2}y = x^2 - 2$

(*Hint:* Use $y_1 = x^2$, $y_2 = x^{-3}$.)

27. $y'' + y' - 6y = x$

28. $y'' - 5y' + 6y = 8\sin^2(4x)$

29. $y'' + 9y = 3\sec(3x)$

30. $y'' - 16y = -\tan(x)$

31. $y'' - y' - 12y = 2\sinh^2(x)$

32. $y'' + 8y' + 15y = 7x^2 - \cos^2(x)$

33. $y'' + 2y' - 8y = e^{4x} - 1$

34. $y'' - 4y' + 3y = -3\sin(x + 2)$

35. $y'' + y' - 7y = 2x - 3$

In each of Problems 36 through 45, solve the initial value problem by first finding the general solution of the differential equation and then solving for the constants to fit the given conditions. Use undetermined coefficients or variation of parameters, as you choose, in finding particular solutions.

36. $y'' + 11y' + 28y = x^3 - 4;\quad y(0) = y'(0) = 0$

37. $y'' + 2y' + y = -3\sin(2x);\quad y(0) = -3,\quad y'(0) = 0$

38. $y'' + 8y' + 12y = e^{-x} + 7;\quad y(0) = 1,\quad y'(0) = 0$

39. $y'' + y = 5\sin(2x) - 4;\quad y(0) = 0,\quad y'(0) = -3$

40. $y'' - 4y = -7e^{2x} + x;\quad y(0) = 1,\quad y'(0) = 3$

41. $y'' + y' - 12y = -x^2 + 2;\quad y(0) = 1,\quad y'(0) = 0$

42. $y'' + 10y' + 24y = 1;\quad y(2) = -3,\quad y'(2) = 5$

43. $y'' + 3y' - 28y = -\cos(3x);\quad y(0) = y'(0) = 0$

44. $y'' + 4y' + 5y = -4\cos^2(3x);\quad y(0) = 1,\quad y'(0) = -4$

45. $y'' + 16y = 8\cos(4x) y(\pi) = 0,\quad y'(\pi) = 7$

46. In treating variation of parameters, we considered a differential equation of the form $y'' + P(x)y' + Q(x)y = F(x)$, in which the coefficient of y'' is 1. Go through the derivation of the method and determine what happens if the coefficient of y'' is not 1, but a function of x, say $R(x)$. That is, apply the derivation to the differential equation $R(x)y'' + P(x)y' + Q(x)y = F(x)$.

47. What happens in the variation of parameters method if $y_1(x)$ and $y_2(x)$ are chosen as solutions of the associated homogeneous equation, but are not linearly independent?

2.9 ANALYSIS OF FORCED OSCILLATIONS OF A MASS ON A SPRING

We have derived the differential equation

$$my'' + cy' + ky = 0$$

for motion of a body of mass m suspended from a spring with modulus k and attached to a dashpot producing a damping force with damping constant c. This equation was analyzed in Section 2.6. We now extend this analysis by allowing a variable driving force $F(t)$ to act on the system. The resulting motion is called *forced motion* [as opposed to *free motion* when $F(t) = 0$], and the governing differential equation is

$$my'' + cy' + ky = F(t).$$

Writing this as

$$y'' + \frac{c}{m}y' + \frac{k}{m}y = \frac{1}{m}F(t), \tag{a}$$

we have a constant coefficient nonhomogeneous equation we are now equipped to attack, given $F(t)$. We shall consider the commonly encountered case of a periodic driving force, say

$$F(t) = A\cos(\omega t),$$

where A and ω are constants. We then have

$$y'' + \frac{c}{m} y' + \frac{k}{m} y = \frac{A}{m} \cos(\omega t).$$

The general solution is of the form $y_h + y_p$. Here y_h, the general homogeneous solution, was obtained in all cases in Section 2.6 in considering free oscillations. Thus, we concentrate on finding y_p. The undetermined coefficients method is convenient for this choice of the driving force, so we shall try

$$y_p = \alpha \cos(\omega t) + \beta \sin(\omega t).$$

Substituting into the differential equation gives us

$$-\alpha\omega^2 \cos(\omega t) - \beta\omega^2 \sin(\omega t) + \frac{c}{m} [-\alpha\omega \sin(\omega t) + \beta\omega \cos(\omega t)]$$
$$+ \frac{k}{m} [\alpha \cos(\omega t) + \beta \sin(\omega t)] = \frac{A}{m} \cos(\omega t).$$

This can be written

$$\left[-\alpha\omega^2 + \frac{\beta\omega c}{m} + \frac{k\alpha}{m}\right] \cos(\omega t) + \left[-\beta\omega^2 - \frac{\alpha\omega c}{m} + \frac{\beta k}{m}\right] \sin(\omega t) = \frac{A}{m} \cos(\omega t).$$

Then,

$$-\alpha\omega^2 + \frac{\beta\omega c}{m} + \frac{k\alpha}{m} = \frac{A}{m}$$

and

$$-\beta\omega^2 - \frac{\alpha\omega c}{m} + \frac{\beta k}{m} = 0.$$

These can be written

$$(k - m\omega^2)\alpha + \omega c\beta = A$$
$$-\omega c\alpha + (k - m\omega^2)\beta = 0.$$

Solving for α and β gives us

$$\alpha = \frac{A(k - m\omega^2)}{(k - m\omega^2)^2 + \omega^2 c^2} \quad \text{and} \quad \beta = \frac{A\omega c}{(k - m\omega^2)^2 + \omega^2 c^2}.$$

Put $\omega_0^2 = k/m$, as in Section 2.6. Then a particular solution is

$$y_p(t) = \frac{Am(\omega_0^2 - \omega^2)}{m^2(\omega_0^2 - \omega^2)^2 + \omega^2 c^2} \cos(\omega t) + \frac{A\omega c}{m^2(\omega_0^2 - \omega^2) + \omega^2 c^2} \sin(\omega t),$$

assuming that $m^2(\omega_0^2 - \omega^2) + \omega^2 c^2 \neq 0$. (This is true if $c \neq 0$ *or* $\omega_0 \neq \omega$.)

We now consider several cases, drawing on the results of Section 2.6. If $c \neq 0$, we have three possibilities.

Case 1 Overdamping ($c^2 - 4km > 0$)

The general solution is then $y_h + y_p$, with y_h as in case 1 of Section 2.6 and y_p as just derived. Then, $y(t) = c_1 e^{r_1 t} + c_2 e^{r_2 t} + y_p$, where

$$r_1 = -\frac{c}{2m} + \frac{1}{2m}\sqrt{c^2 - 4km} \quad \text{and} \quad r_2 = -\frac{c}{2m} - \frac{1}{2m}\sqrt{c^2 - 4km}.$$

We saw before that r_1 and r_2 are negative. Hence, as t increases, the term $c_1 e^{r_1 t} + c_2 e^{r_2 t}$ exerts less influence on the solution, which tends toward the oscillatory and periodic solution y_p. We call y_h the *transient* solution and y_p the *steady-state solution.* The frequency of the steady-state solution is the same as that of the input force $A\cos(\omega t)$.

Case 2 Critical damping ($c^2 - 4km = 0$)

Here we obtain

$$y(t) = e^{-ct/2m}(c_1 + c_2 t) + y_p(t).$$

Again, as $t \to \infty$, $y(t) \to y_p(t)$, the steady-state solution.

Case 3 Underdamping ($c^2 - 4km < 0$)

Now

$$y(t) = de^{-ct/2m}\cos(\bar{\omega}t - \delta) + y_p(t),$$

where d is an arbitrary (nonnegative) constant, $\bar{\omega} = \sqrt{4km - c^2}/2m$, and δ is constant. Here, $y_h(t)$ again decays to 0 as $t \to \infty$, and $y(t) \to y_p(t)$.

In all cases of damped forced oscillation, then, the motion approaches the steady-state motion with increasing time.

If $c = 0$, so that we have forced undamped motion, we have to assume that $\omega_0 \neq \omega$. Now

$$y_p(t) = \frac{A}{m(\omega_0^2 - \omega^2)}\cos(\omega t),$$

and so

$$y(t) = d\cos(\omega_0 t - \delta) + \frac{A}{m(\omega_0^2 - \omega^2)}\cos(\omega t).$$

This is similar to the underdamped solution when $c \neq 0$, except that the factor $e^{-ct/2m}$ is missing here. Thus, y_h does not decay with time in the undamped case, and the motion is a sum of two harmonic oscillations, one with frequency $\omega_0/2\pi$, the other with frequency $\omega/2\pi$. Sometimes $\omega_0/2\pi$ is called the *natural frequency* and $\omega/2\pi$ the *input frequency*.

While ω must not equal ω_0 for this solution to be valid, an interesting thing happens if natural and input frequencies are taken closer. As $\omega \to \omega_0$, the amplitude of y_p becomes larger. Thus, a system with no damping (or even c very close to zero) will experience larger and larger oscillations as natural and input frequencies become closer. In the real world there is always some damping. Nevertheless, when c is small (say, compared to the mass), and the natural and input frequencies are very close, oscillations can build up to a large enough magnitude to damage the system. An extreme example was the Tacoma bridge in Washington, which experienced vibrations of increasing violence until it finally collapsed.* Many physics and engineering departments have films showing the dramatic thrashing of the bridge up to the end.

A phenomenon called *resonance* occurs when natural and input frequencies actually coincide, and there is no damping. We cannot use the above solution in this case, as the denominator would vanish; so we go back to the governing differential equation. With $\omega = \omega_0$ and $c = 0$, it is

$$my'' + ky = A\cos(\omega_0 t)$$

or

$$y'' + \omega_0^2 y = A^* \cos(\omega_0 t),$$

with $A^* = A/m$. We find that the general solution is

$$y = c_1 \cos(\omega_0 t) + c_2 \sin(\omega_0 t) + \frac{A^*}{2m\omega_0} t \sin(\omega_0 t).$$

The factor of t in the last term occurs because the forcing function $A^* \cos(\omega_0 t)$ is itself a solution of $y'' + \omega_0^2 y = 0$ (recall Examples 68 and 71). It is this factor which causes the amplitude to increase with time in the case of resonance. As above, a system in resonance will eventually experience too great a strain and be damaged by the unbounded oscillations.

We conclude this section with some examples involving forced, damped motion. These will also illustrate some of the practical aspects of calculating results. You may imagine the units in any system you like: for example, in the cgs system, length is in centimeters, mass in grams, and force in dynes; in the mks system, length is in meters, mass in kilograms, and force in newtons. Both systems use seconds for time.

EXAMPLE 74

Determine the motion of the mass when the spring constant k is 4, damping constant c is 2, mass m is 3, and the external force is $F(t) = 2\cos(5t)$, if the initial displacement is 5 and the initial velocity is 2.

We must solve the initial value problem

$$3y'' + 2y' + 4y = 2\cos(5t); \qquad y(0) = 5, \qquad y'(0) = 2.$$

---

* Technically, the Tacoma bridge experienced something engineers call *strut flutter*, with the energy supplied by the wind. Another classic example of resonance occurred near Manchester, England, in 1831, when a column of soldiers marching across the Broughton bridge set up a periodic force whose frequency very closely approximated the natural frequency of the bridge, causing its collapse.

It is routine to find the general solution of the differential equation. It is

$$y(t) = c_1 e^{-t/3} \cos\left(\sqrt{11}\,\frac{t}{3}\right) + c_2 e^{-t/3} \sin\left(\sqrt{11}\,\frac{t}{3}\right)$$

$$-\frac{142}{5141}\cos(5t) + \frac{20}{5141}\sin(5t).$$

To satisfy the initial conditions, we need to choose c_1 and c_2 so that

$$y(0) = 5 = c_1 - \tfrac{142}{5141}$$

and

$$y'(0) = 2 = -\tfrac{1}{3}c_1 + \tfrac{\sqrt{11}}{3}c_2 + \tfrac{100}{5141}.$$

Solve these to get

$$c_1 = \frac{25{,}847}{5141} \quad \text{and} \quad c_2 = \frac{56{,}393}{5141\sqrt{11}}.$$

Thus, the solution of the problem is

$$y = \frac{25{,}847}{5141} e^{-t/3} \cos\left(\sqrt{11}\,\frac{t}{3}\right) + \frac{56{,}393}{5141\sqrt{11}} e^{-t/3} \sin\left(\sqrt{11}\,\frac{t}{3}\right)$$

$$-\frac{142}{5141}\cos(5t) + \frac{20}{5141}\sin(5t).$$

To four decimals, we have the approximate solution

$$y_{\text{app}} = 5.0276e^{-t/3}\cos(1.1055t) + 3.3074e^{-t/3}\sin(1.1055t)$$
$$-0.0276\cos(5t) + 0.0039\sin(5t).$$

Note: One can always write the general solution of a differential equation in many different ways. For example, $3y'' + 2y' + 4y = 2\cos(5t)$ also has general solution (see case 3, underdamping, above)

$$y = de^{-t/3}\cos\left(\frac{\sqrt{11}}{3}\,t - \delta\right) - \frac{142}{5141}\cos(5t) + \frac{20}{5141}\sin(5t).$$

However, the equations for d and δ resulting from the initial conditions $y(0) = 5$, $y'(0) = 2$, are more difficult to solve than those for c_1 and c_2 above. Thus, one attempts to choose a form for the general solution which makes the rest of the problem as simple as possible.

EXAMPLE 75

Solve for the motion of the body when the spring constant k is 2.9, damping constant c is 4.31, mass m is 1.2, and the driving force is given by $F(t) = 3.6e^{-t}$. The initial displacement is 3.7, and the initial velocity is 1.4.

We must solve the initial value problem

$$1.2y'' + 4.31y' + 2.9y = 3.6e^{-t}; \qquad y(0) = 3.7, \qquad y'(0) = 1.4.$$

In this, and in the next example, we have made the numbers a little messier than in previous examples in order to illustrate the computations one encounters in practice. (Actually, one usually sees much worse than this.) You should use a calculator to do the arithmetic involved. We shall carry out calculations to four decimal places; thus, " = " in the following should be read "equal up to four places."

The differential equation can be written

$$y'' + 3.5917y' + 2.4167y = 3e^{-t}.$$

It is routine to obtain the general solution. It is

$$y = c_1e^{-0.8968t} + c_2e^{-2.6949t} - 17.1429e^{-t}.$$

We now need

$$y(0) = 3.7 = c_1 + c_2 - 17.1429$$

and

$$y'(0) = 1.4 = -0.8968c_1 - 2.6949c_2 + 17.1429.$$

Solve these to obtain

$$c_1 = 22.4830 \quad \text{and} \quad c_2 = -1.6401.$$

Thus, the solution is

$$y = 22.4830e^{-0.8968t} - 1.6401e^{-2.6949t} - 17.1429e^{-t}.$$

The rapidly damping motion is accounted for by the fact that the damping constant in this example is large compared to the mass and the spring constant.

EXAMPLE 76

Suppose that the damping constant is 2.46, the mass 8.35, and the spring constant 4.72. Let the driving force be $F(t) = 2.7\cos(4t) + 1.43\sin(4t)$. The initial displacement is 2.6, and the initial velocity is 1.73.

Again carrying calculations to four places, we can write the differential equation as

$$y'' + 0.2946y' + 0.5653y = 0.3234\cos(4t) + 0.1713\sin(4t).$$

The general solution is found to be

$$y = c_1e^{-0.1473t}\cos(0.7373t) + c_2e^{-0.1473t}\sin(0.7373t) \\ - 0.0217\cos(4t) - 0.0094\sin(4t).$$

We now need

$$y(0) = 2.6 = c_1 - 0.0217$$

and

$$y'(0) = 1.73 = -0.1473c_1 + 0.7373c_2 - (4)(0.0094).$$

These yield

$$c_1 = 2.6217 \quad \text{and} \quad c_2 = 2.9212.$$

Thus, the solution is

$$y = 2.6217e^{-0.1473t}\cos(0.7373t) + 2.9212e^{-0.1473t}\sin(0.7373t) - 0.0217\cos(4t) - 0.0094\sin(4t).$$

As usual, the student should check this result by substituting the solution into the differential equation, and also checking that $y(0) = 2.6$ and $y'(0) = 1.73$.

The student may have noticed that the differential equations for *RLC* circuits and for damped, driven spring motion are very similar. In the next section we shall discuss this similarity and some of its ramifications, and then return in Section 2.11 to the business of solving differential equations.

PROBLEMS FOR SECTION 2.9

In each of Problems 1 through 10, find the transient and steady-state solutions for the given values of m, c, k, and the driving force.

	m	c	k	$F(t)$		m	c	k	$F(t)$
1.	1	6	1	$3\sin(\omega t)$	**2.**	1	4	2	$3\cos(\omega t) + 2\sin(\omega t)$
3.	2	5	2	$4\cos(2t) + 2\cos(3t)$	**4.**	2	5	2	$\sin(4t) - \sin(2t)$
5.	2	6	3	$\cos(t) - \sin(3t)$	**6.**	6	1	4	$3\sin(2t)$
7.	5	2	1	$2\cos(4t) - \sin(3t)$	**8.**	4	2	2	$3e^{-t} + \sin(2t)$
9.	8	2	4	$4\sin^2(3t)$	**10.**	4	3	6	$e^{-3t} + 2\cos(5t) - t$

In each of Problems 11 through 20, find the general solution for the displacement y with the given values of m, c, k, and $F(t)$.

	m	c	k	$F(t)$		m	c	k	$F(t)$
11.	1	1	1	e^{-2t}	**12.**	1	2	1	$\sin(4t)$
13.	2	4	2	$\cos(2t) + \sin(4t)$	**14.**	3	6	3	e^{-t}
15.	2	4	2	$e^{-t}\sin(t)$	**16.**	3	6	3	$\cos(t) + \sin^2(2t)$
17.	6	1	1	$e^{-t}\sin(2t)$	**18.**	4	2	6	$t^2\sin(t)$
19.	5	7	2	$t + e^{-t}$	**20.**	4	4	3	$\cos(3t) - 4\sin(t)$

In each of Problems 21 through 30, solve the given initial value problem and describe the resulting motion of the mass-spring system it represents.

21. $2y'' + y' + 6y = 3\sin(4t);\quad y(0) = 0,\quad y'(0) = 1$

22. $y'' + 2y' + 4y = \cos(2t);\quad y(0) = 2,\quad y'(0) = 0$

23. $y'' + 6y' + 2y = \cos(t) + \cos(3t);\quad y(0) = y'(0) = 1$

24. $2y'' + 4y' + y = \sin(2t) - \sin(t);\quad y(0) = y'(0) = 2$

25. $y'' + 2y' + y = 2\sin(t) - \cos(3t);\quad y(0) = 2,\quad y'(0) = 1$

26. $y'' + 4y' + 10y = 2e^{-t}\sin(t + 1);\quad y(0) = 1,\quad y'(0) = 0$

27. $y'' + 3y' + 7y = 4\cos^2(3t);\quad y(0) = 1,\quad y'(0) = 4$

28. $5y'' + y' + 3y = 3\sin^2(4t) - e^{-t};\quad y(0) = 2,\quad y'(0) = -2$

29. $3y'' + y' + 5y = 2;\quad y(0) = y'(0) = 3$

30. $8y'' + y' + 2y = 3\cosh(t);\quad y(0) = 1,\quad y'(0) = 5$

In each of Problems 31 through 40, use a calculator to solve the initial value problem, carrying out the calculations to four decimal places, as in Examples 75 and 76.

31. $2.34y'' + 0.5y' + 8.22y = 0.2\cos(3t);\quad y(0) = 0.23,\quad y'(0) = 3.45$

32. $4.35y'' + 2.2y' + 7y = 3.24\sin(3t) - 4.2\cos(3t)$; $y(0) = 0.5$, $y'(0) = 7.3$
33. $21.7y'' + 3y' + 2.75y = 4e^{-t}$; $y(0) = 3$, $y'(0) = 2.1$
34. $5.73y'' + 4.6y' + 2y = 3.4te^{-t}$; $y(0) = 0.5$, $y'(0) = 0$
35. $3y'' + 3.4y' + 2y = 6.88\cos^2(2t)$; $y(0) = -1.2$, $y'(0) = 3.5$
36. $2.6y'' + 4.8y' + 8y = 0.2\sin(3.5t)$; $y(0) = 0$, $y'(0) = 7.3$
37. $6.38y'' + 4.4y' + 9.22y = 0.5t^2e^{-t}$; $y(0) = 3$, $y'(0) = 0$
38. $0.2y'' + 4.6y' + y = 8.42\sin^2(2.1t)$; $y(0) = 3.6$, $y'(0) = 0$
39. $6.88y'' + 2y' + 4.62y = 0.2t^2$; $y(0) = 0.23$, $y'(0) = 1$
40. $5.77y'' + 2y' + 1.56y = 4.65e^{-2t}\sin(t)$; $y(0) = 0.27$, $y'(0) = 1$
41. In the case of overdamped forced motion (case 1 above) and $F(t) = A\cos(\omega t)$, find a solution subject to

$$y(0) = B, \quad y'(0) = 0; \tag{1}$$

then (independently) find a solution subject to

$$y(0) = 0, \quad y'(0) = B. \tag{2}$$

(a) On the same set of axes, sketch the solutions in (1) and (2) to compare influence of initial displacement and initial velocity on the resulting motion. What conclusions do you draw?
(b) On the same set of axes, sketch the graphs of the solution in (1), the transient solution in (1), and the steady-state solution in (1).
(c) On the same set of axes, sketch the graphs of the solution in (2), the transient solution in (2), and the steady-state solution in (2).

42. Repeat the project of Problem 41 for the case of critical damping.
43. Repeat the project of Problem 41 for the case of underdamping.
44. Completely analyze the motion of a body of mass m suspended from a spring with constant k and attached to a dashpot with damping factor c, with an external driving force $F(t) = e^{-t}$. Consider all the cases that arise, depending upon the relative sizes of m, k, and c.
45. Do the analysis of Problem 44, but with $F(t) = t$.
46. Do the analysis of Problem 44, but with $F(t) = Ae^{-t}\sin(\omega t)$.
47. Consider undamped forced motion of a mass on a spring, governed by

$$my'' + ky = A\cos(\omega t).$$

(a) Assuming that $y(0) = y'(0) = 0$, obtain the solution

$$y = \frac{A}{m(\omega_0^2 - \omega^2)}[\cos(\omega t) - \cos(\omega_0 t)].$$

In this, $\omega_0 = \sqrt{k/m}$, and we assume that $\omega \neq \omega_0$.

(b) Use trigonometric identities to write this solution as

$$y = \frac{2A}{m(\omega_0^2 - \omega^2)}\sin\left(\frac{(\omega_0 - \omega)t}{2}\right)\sin\left(\frac{(\omega_0 + \omega)t}{2}\right).$$

(c) Suppose that $\omega_0 - \omega$ is much smaller than $\omega + \omega_0$. Explain from (b) why the resulting motion is a rapid sinusoidal oscillation with a periodic variation of amplitude. Such a motion is called a *beat*, and the variation of amplitude with time is called *amplitude modulation*.
(d) Show that the distance between successive maxima of $y(t)$ is

$$\frac{4\pi}{\omega_0 + \omega}.$$

(e) Sketch a graph of $y(t)$.

48. Show that damped forced motion of a mass on a spring, with forcing function $A\cos(\omega t)$, is always bounded.

49. Consider damped, forced motion governed by

$$my'' + cy' + ky = A\cos(\omega t).$$

Show that the amplitude of the steady-state solution is a maximum if ω is chosen so that

$$\omega^2 = \frac{k}{m} - \frac{c^2}{2m^2}.$$

The value of this is that, in building a seismic detector, we would try to choose k, c, and m so that ω^2 is as near to this value as possible to maximize the response.

2.10 COMPARISON OF *RLC* CIRCUITS AND FORCED, DAMPED SPRING MOTION

In Chapter 0 we derived the differential equation governing charge $Q(t)$ in an RLC circuit. With inductance L, resistance R, and capacitance C all constant, and an electromotive force $E(t)$, the differential equation is

$$LQ'' + RQ' + \frac{1}{C}Q = E(t).$$

Assuming that $E'(t)$ exists, differentiate this equation with respect to t to get

$$LQ''' + RQ'' + \frac{1}{C}Q' = E'(t).$$

Since current $I(t)$ is related to charge by $I(t) = Q'(t)$, then we have a second order differential equation for I:

$$LI'' + RI' + \frac{1}{C}I = E'(t).$$

Now recall that the equation for forced, damped spring motion is

$$my'' + cy' + ky = F(t).$$

Observe that the differential equation for the circuit is virtually identical to that for the mechanical system, with the following comparisons:

Spring System	Electrical Circuit
$my'' + cy' + ky = F(t)$	$LI'' + RI' + \frac{1}{C}I = E'$
displacement $y(t)$	current $I(t)$
driving force $F(t)$	derivative E' of electromotive force
mass m	inductance L
damping constant c	resistance R
spring modulus k	reciprocal $1/C$ of capacitance

This kind of analogy has two major uses: First, we need not solve the current equation, because we have already just done all the work [at least in the important case $E(t) = A\sin(\omega t)$, $E' = \omega A\cos(\omega t)$]. By taking the solutions of the spring problem and substituting L for m, I for y, R for c, and so on, we obtain solutions of the current equation in the various cases.

Perhaps more important are the practical ramifications. An experimenter can continue this analogy in the laboratory, where circuits are generally easier and cheaper to construct than spring systems. Furthermore, accurate measurements are taken more easily of electrical circuits than of spring systems.

Finally, awareness of this particular analogy should alert us to be on the lookout for other analogies to be drawn between apparently unrelated physical phenomena. Scientists, engineers, and mathematicians have often benefitted from realizing that one phenomenon (little understood) is "like" another (better understood) if one makes the appropriate substitutions and analogies.

Here are two examples involving *RLC* circuits. You should use a calculator to do the computations.

EXAMPLE 77

Solve for the current in an *RLC* circuit with resistance 25 ohms, capacitance 0.001 farad, inductance 0.3 henry, and voltage source $E(t) = 121.7\sin(42t)$. Assume that $I(0) = I'(0) = 0$.

We must solve the initial value problem

$$0.3I'' + 25I' + \frac{1}{0.001}I = E'(t) = 5111.4\cos(42t); \qquad I(0) = I'(0) = 0.$$

Writing the differential equation as

$$I'' + 83.3333I' + 3333.3333I = 17{,}038\cos(42t),$$

we find the general solution

$$I(t) = c_1 e^{-41.6667t}\cos(39.9653t) + c_2 e^{-41.6667t}\sin(39.9653t) + 1.8173\cos(42t) + 4.0531\sin(42t).$$

We now need

$$I(0) = 0 = c_1 + 1.8173$$

and

$$I'(0) = 0 = -41.6667c_1 + 39.9653c_2 + (42)(4.0531).$$

Then,

$$c_1 = -1.8173 \quad \text{and} \quad c_2 = -6.1541.$$

The solution is then

$$I(t) = -1.8173e^{-41.6667t}\cos(39.9653t) - 6.1541e^{-41.6667t}\sin(39.9653t) + 1.8173\cos(42t) + 4.0531\sin(42t).$$

EXAMPLE 78

Find the current in an RLC circuit with $R = 72$, $C = 0.003$, $L = 0.2$, and $E(t) = 150\sin(25t)$. Assume that the initial current and initial charge are both zero.

This is quite a different kind of problem from the one just completed. The differential equation is

$$0.2I'' + 72I' + \frac{1}{0.003}I = E'(t) = 3750\cos(25t)$$

or

$$I'' + 360I' + 1666.6667I = 18{,}750\cos(25t).$$

However, we do not have an initial value problem yet, because we do not know $I'(0)$; in its place we are given $Q(0)$. We shall proceed in the usual way, and attempt to find $I'(0)$ from $Q(0)$.

First, solve the differential equation, obtaining the general solution $I(t) = c_1e^{-4.6907t} + c_2e^{-355.3093t} + 0.2379\cos(25t) + 2.0558\sin(25t)$. Now, $I(0) = 0$; so we have

$$c_1 + c_2 + 0.2379 = 0.$$

We need a second equation for c_1 and c_2. To get this, we shall determine $I'(0)$ from $Q(0)$, which is given. Remember that, from Example 1,

$$LQ'' + RQ' + \frac{1}{C}Q = E \quad \text{and} \quad I = Q'.$$

Then,

$$LI' + RI + \frac{1}{C}Q = E.$$

In particular,

$$LI'(0) + RI(0) + \frac{1}{C}Q(0) = E(0).$$

In this particular problem, $I(0) = 0$ was given, $Q(0) = 0$ was given, and $E(0) = 0$. Hence, $I'(0) = 0$ also. Then,

$$I'(0) = 0 = -4.6907c_1 - 355.3093c_2 + (25)(2.0558).$$

We now have two equations to solve for c_1 and c_2. We obtain

$$c_1 = -0.3877 \quad \text{and} \quad c_2 = 0.1498.$$

The solution is then

$$I(t) = -0.3877e^{-4.6907t} + 0.1498e^{-355.3093t} + 0.2379\cos(25t) + 2.0558\sin(25t).$$

Note: Remember that the solutions in the last two examples are approximate because of the numerical calculations. In both examples, the student should substitute the approximate solution back into the initial value problem to get some feeling for how good the solution is.

PROBLEMS FOR SECTION 2.10

1. Let $E(t) = A\sin(\omega t)$, and consider the current equation

$$LI'' + RI' + \frac{1}{C}I = E'(t) = A\omega\cos(\omega t).$$

(a) Obtain the general solution, delineating the three cases which occur in finding the general homogeneous solution.

(b) Independent of (a), find the general solution in each case by using the table of correspondences between electrical circuit terms and mechanical terms used in the preceding section and making the appropriate substitutions in the solutions derived there.

(c) Compare the results of (a) and (b) in each case. (*Note:* The constant A used in Section 2.9 for the driving force coefficient is replaced above by $A\omega$. Don't forget to take this into account in your replacements.)

In each of Problems 2 through 10, solve for the current in the *RLC* circuit under the given conditions.

	R	L	C	$E(t)$	$I(0)$	$I'(0)$
2.	400	0.12	0.004	$120\sin(20t)$	0	0
3.	200	0.10	0.006	$200\cos(40t)$	0	0
4.	250	0.30	0.015	$30\sin(10t)$	0	0
5.	300	0.35	0.002	$20\cos(150t)$	4	0
6.	350	0.20	0.002	$10\cos(5t) + 50\sin(20t)$	0	18
7.	150	0.15	0.001	$e^{-t}\cos(3t)$	0	4
8.	200	0.55	0.015	te^{-t}	2	1
9.	160	0.80	0.055	$4\sin^2(2t)$	3	1
10.	450	0.95	0.007	$e^{-t}\sin^2(3t)$	1	5

In each of Problems 11 through 20, solve for the current in the *RLC* circuit under the given conditions.

	R	L	C	$E'(t)$	$I(0)$	$Q(0)$
11.	50	0.55	0.022	$4\cos(3t)$	2	0
12.	300	0.40	0.005	$3e^{-2t}$	1	1
13.	150	0.20	0.05	$2\sin(4t) - e^{-t}$	2	1
14.	340	0.45	0.025	$\cos(4t) + \sin(5t)$	0	0
15.	220	0.35	0.040	$e^{-t} + 4e^{-5t}$	1	4
16.	130	0.40	0.003	$\cos^2(4t) + 2e^{-4t}$	2	1
17.	210	0.65	0.001	$\sin(2t + 1)$	1	0
18.	230	0.90	0.004	$e^{-t}\cos(3t)$	2	4
19.	190	0.85	0.003	te^{-4t}	1	3
20.	300	0.50	0.045	$t\cos(4t)$	2	6

21. In an *RLC* circuit with $E(t) = A\sin(\omega t)$, for what value of C will the steady-state current have its maximum value? For what value of L will this amplitude be a maximum?

22. Solve for $I(t)$ in an *RLC* circuit if $E(t) = Ae^{-t}$. Consider all cases that arise.

23. Solve for $I(t)$ in an *RLC* circuit if $E(t) = Ate^{-t}$. Consider all cases that arise.

24. Suppose $R = 0$ and $E(t) = 0$. Show that the impressed charge $Q(t)$ on the capacitor is periodic in time with frequency $1/\sqrt{LC}$. (This is called the *natural frequency* of the system.)

25. Suppose that $R = 0$ and $E(t) = A\cos(\omega t)$. Show that resonance occurs; that is, show that the charge on the capacitor becomes infinite as $t \to \infty$ when $\omega = 1/\sqrt{LC}$.

26. Show that, if $E(t) = A\cos(\omega t)$, then the charge $Q(t)$ must remain bounded if $R \neq 0$.

2.11 REDUCTION OF ORDER

We now return to the general problem of finding solutions of second order equations. Recall that for $y'' + P(x)y' + Q(x)y = F(x)$, the general solution is of the form $y_h + y_p$, that is, the general solution of the associated homogeneous equation plus any particular solution of the given nonhomogeneous equation [if $F(x)$ is not identically zero].

In this section we introduce a set of techniques which are often grouped under the single heading *reduction of order*. The general idea is to reduce the problem of solving a second order differential equation to a problem of solving one or more first order differential equations. There are three circumstances in which this method of attack is effective.

A. ABSENT DEPENDENT VARIABLE

In general, a second order differential equation is of the form $F(x, y, y', y'') = 0$, with x independent and y dependent upon x. If y does not appear explicitly in the differential equation (although y' may appear, and y'' must if the equation is second order), then we have an equation of the form $f(x, y', y'') = 0$. In this event, let $u = y'$ to obtain $f(x, u, u') = 0$, a first order differential equation. Solve this for u; then integrate to obtain y.

EXAMPLE 79

Solve

$$xy'' + 2y' = 4x^3.$$

Here y does not appear explicitly. Let $u = y'$ to obtain

$$xu' + 2u = 4x^3,$$

which can be written

$$u' + \frac{2}{x}u = 4x^2.$$

This is a linear first order differential equation for u, which we easily solve by the method of Section 1.6 to obtain the general solution

$$u = \frac{4x^3}{5} + \frac{C}{x^2}.$$

Then,

$$y(x) = \int u(x)\,dx = \frac{x^4}{5} - \frac{C}{x} + K,$$

with C and K arbitrary constants. This solves the problem.

B. ABSENT INDEPENDENT VARIABLE

If x does not occur explicitly in $F(x, y, y', y'') = 0$, then the differential equation has the form $f(y, y', y'') = 0$. As in (A), let $u = y'$. But, unlike (A), we now attempt to put the original differential equation in terms of y, u, and u'. To do this, write $y' = u$, and, by the chain rule,

$$y'' = \frac{d}{dx}\left(\frac{dy}{dx}\right) = \frac{d}{dx}(y') = \frac{dy'}{dy}\frac{dy}{dx} = \left(\frac{du}{dy}\right)u.$$

Thus,

$$f(y, y', y'') = 0 \quad \text{becomes} \quad f\left(y, u, u\frac{du}{dy}\right) = 0,$$

a first order differential equation in which y is thought of as independent, and u as a function of y. We attempt to solve this for u and then use $u = y'$ to solve for y.

EXAMPLE 80

Solve

$$y'' - 2yy' = 0.$$

Here, x does not appear explicitly. Let

$$u = y', \qquad u\frac{du}{dy} = y''.$$

We get

$$u\frac{du}{dy} - 2yu = 0.$$

Assuming that $u \neq 0$, this is the same as

$$\frac{du}{dy} = 2y \quad \text{or} \quad du = 2y\,dy.$$

The latter is a separable first order differential equation. One integration gives us

$$u = y^2 + C,$$

where C is any constant. Since $u = dy/dx$, we now have a second differential equation

$$\frac{dy}{dx} = y^2 + C.$$

This also happens to be separable, and can be written as

$$\frac{dy}{y^2 + C} = dx.$$

If $C > 0$, an integration gives us

$$\frac{1}{\sqrt{C}}\arctan\left(\frac{y}{\sqrt{C}}\right) = x + K,$$

where K is any constant. This implicitly defines solutions of $y'' - 2yy' = 0$. If we wish, we can solve explicitly for y as a function of x and obtain

$$y(x) = \alpha\tan(\alpha x + \beta),$$

in which $\alpha = \sqrt{C}$ and $\beta = K\sqrt{C}$. We leave it for the student to derive the solution when $C = 0$ or $C < 0$.

C. USING ONE SOLUTION TO FIND A SECOND SOLUTION

The methods of (A) and (B) do not assume that the given differential equation is either linear or homogeneous. In (C), we describe a method for finding a second solution $y_2(x)$ of $y'' + P(x)y' + Q(x)y = 0$, given that we can find one solution $y_1(x)$. The method will actually ensure that $y_2(x)$ is linearly independent of $y_1(x)$. Here, $P(x)$ and $Q(x)$ need not be constant.

Suppose then that $y_1(x)$ satisfies $y'' + P(x)y' + Q(x)y = 0$. Look for a second solution of the form

$$y_2(x) = v(x)y_1(x).$$

We now show how to choose $v(x)$. Substitute y_2 into the differential equation to get

$$vy_1'' + v''y_1 + 2v'y_1' + P(x)(vy_1' + v'y_1) + Q(x)vy_1 = 0.$$

This can be written

$$v''y_1 + v'[2y_1' + P(x)y_1] + v[y_1'' + P(x)y_1' + Q(x)y_1] = 0.$$

The last term in brackets vanishes by assumption that y_1 is a solution of $y'' + P(x)y' + Q(x)y = 0$, and we are left with

$$v''y_1 + v'[2y_1' + P(x)y_1] = 0.$$

This can be solved by the method of (A) above. In particular, let $u = v'$ to get

$$u'y_1 + u[2y_1' + P(x)y_1] = 0$$

or, if $y_1 \neq 0$,

$$u' + \left(\frac{2y_1'}{y_1} + P(x)\right)u = 0.$$

This is a linear, first order differential equation for u (remember that y_1 and $P(x)$ are presumed known). Solve it by the method of Section 1.6, then find $v(x) = \int u(x)\,dx$. Having found $v(x)$, then $y_2(x) = v(x)y_1(x)$ is a second solution of $y'' + P(x)y' + Q(x)y = 0$.

We leave it for the student to use the Wronskian to show that $y_1(x)$ and $y_2(x)$ are linearly independent, and hence form a fundamental system of solutions of $y'' + P(x)y' + Q(x)y = 0$, at least on any interval where $y_1(x) \neq 0$.

Here is a summary of the method:

> Suppose that $y_1 \neq 0$ is a solution of $y'' + P(x)y' + Q(x)y = 0$. Solve
>
> $$u' + \left(\frac{2y_1'}{y_1} + P(x)\right)u = 0$$
>
> and let $v(x) = \int u(x)\,dx$. Then $y_2(x) = v(x)y_1(x)$ is a second, linearly independent solution of $y'' + P(x)y' + Q(x)y = 0$.

We have already used this method to produce a second solution of $y'' + Ay' + By = 0$ when the characteristic equation has repeated roots (see Section 2.3, case 2). Here is a different example.

EXAMPLE 81

Consider the differential equation

$$y'' + \frac{3}{x}y' + \frac{1}{x^2}y = 0.$$

Take as given one solution $y_1 = 1/x$ (we explain how we got this in the next section). First, solve

$$u' + \left(\frac{2y_1'}{y_1} + P(x)\right)u = u' + \frac{1}{x}u = 0.$$

Using the method of Section 1.6, we get $u = 1/x$. Then

$$v(x) = \int \frac{1}{x}\,dx = \ln|x|.$$

Thus, a second solution is $y_2(x) = (1/x)\ln|x|$.

Note: In this example we let the constants of integration be zero in finding both $u(x)$ and $v(x)$. This was because we did not want all possible choices of u or v, but simply one choice which would yield a second solution $y_2(x) = v(x)y_1(x)$. Choosing the integration constants to be zero was thus a convenience we could afford.

PROBLEMS FOR SECTION 2.11

In each of Problems 1 through 10, use the method of (A) or (B) to find a solution of the differential equation. Answers may be left in integral form when they cannot be done in terms of elementary functions.

1. $xy'' = 2 + y'$ **2.** $x^2y'' - 2x = y'$ **3.** $2y'' = 1 + y$
4. $1 - y'' = 2y' + x^2$ **5.** $xy'' - 2y' = 1$ **6.** $y'' = 3y - 2$
7. $-3y'' - 2y' = 8x + 2$ **8.** $1 - y' = 4y''$ **9.** $xy'' + 2y' = x$
10. $-3y'' + 2y' = 4x - 5$

In each of Problems 11 through 18, find a second solution, given one solution. Solutions may be left in integral form when they cannot be done in terms of elementary functions.

11. $x^2y'' + xy' + (x^2 - \frac{1}{4})y = 0;\quad y_1(x) = \cos(x)/\sqrt{x}$
12. $(1 - x^2)y'' - 2xy' + 2y = 0;\quad y_1(x) = x$
13. $y'' - xy' + y = 0;\quad y_1(x) = x$
14. $xy'' + (1 - x)y' + 2y = 0;\quad y_1(x) = x^2 - 4x + 2$
15. $xy'' - xy' + y = 0;\quad y_1(x) = x$
16. $y'' - xy' + 3y = 0;\quad y_1(x) = x^3 - 3x$
17. $y'' - \dfrac{2x}{1 - x^2}y' + \dfrac{6}{1 - x^2}y = 0;\quad y_1(x) = \dfrac{3}{2}x^2 - \dfrac{1}{2}$

18. $y'' - \dfrac{2x}{1+x^2}y' + \dfrac{2}{1+x^2}y = 0; \quad y_1(x) = x$

2.12 EULER EQUATIONS

In this section we continue to consider differential equations having some form that enables us to find a solution by some fairly simple trick. One such class was the constant coefficient equation

$$y'' + Ay' + By = 0,$$

where we always found at least one exponential solution. We now consider *Euler's equation**

$$x^2y'' + Axy' + By = 0,$$

where A and B are constants. We shall discuss two ways of solving such an equation.

The first way is to seek solutions $y = x^r$ for appropriate choices of r. Substitute $y = x^r$ into the differential equation to get

$$r(r-1)x^r + Arx^r + Bx^r = 0.$$

If $x \neq 0$, then this reduces to the *characteristic equation*

$$r(r-1) + Ar + B = 0$$

or

$$r^2 + (A-1)r + B = 0,$$

with at least one solution for r guaranteed. We now have three cases to examine.

Case 1 $(A-1)^2 - 4B > 0$

Here we have two real roots for r:

$$r_1 = \frac{(1-A) + \sqrt{(A-1)^2 - 4B}}{2} \quad \text{and} \quad r_2 = \frac{(1-A) - \sqrt{(A-1)^2 - 4B}}{2}.$$

Two solutions are x^{r_1} and x^{r_2}, and these are linearly independent on any interval over which they are defined. Thus, the general solution here is

$$y = c_1x^{r_1} + c_2x^{r_2}.$$

EXAMPLE 82

Solve

$$x^2y'' + 5xy' - 2y = 0.$$

---

* Sometimes called the Euler-Cauchy equation.

Putting $y = x^r$ into the equation gives us the characteristic equation

$$r(r-1) + 5r - 2 = 0$$

or

$$r^2 + 4r - 2 = 0.$$

This has roots $-2 + \sqrt{6}$ and $-2 - \sqrt{6}$. Thus, the general solution is

$$y = c_1 x^{-2+\sqrt{6}} + c_2 x^{-2-\sqrt{6}} = \frac{1}{x^2}(c_1 x^{\sqrt{6}} + c_2 x^{\sqrt{6}}),$$

which is defined for any $x > 0$. (Recall that $x^{\sqrt{6}} = e^{\sqrt{6}\ln(x)}$ and $x^{-\sqrt{6}} = e^{-\sqrt{6}\ln(x)}$; hence we need $x > 0$.)

Case 2 $(A-1)^2 - 4B = 0$

Here we have one real root for r, and it is

$$r = \frac{1-A}{2}.$$

One solution is then $x^{(1-A)/2}$, and we can find a second solution by reduction of order. We look for a solution of the form

$$y = x^{(1-A)/2}v,$$

where v is some function of x. It is routine to find v using method (C) of the preceding section, and we omit the details. One finds that $v(x) = \ln(x)$; hence a second solution is

$$y_2(x) = x^{(1-A)/2}\ln(x),$$

for $x > 0$. The general solution in this case is then

$$y(x) = x^{(1-A)/2}[c_1 + c_2 \ln(x)].$$

Note: It is not necessary to rederive this result each time. Just remember that when the characteristic equation for a Euler equation has just one real root r_1, then the general solution is

$$y = x^{r_1}[c_1 + c_2 \ln(x)].$$

EXAMPLE 83

Solve

$$x^2y'' + 5xy' + 4y = 0.$$

The characteristic equation is

$$r^2 + 4r + 4 = 0,$$

with just one root, $r = -2$. The general solution is then

$$y = x^{-2}[c_1 + c_2 \ln(x)].$$

Case 3 $(A-1)^2 - 4B < 0$

Here we obtain two complex roots for r,

$$r_1 = \frac{(1-A) + i\sqrt{4B-(A-1)^2}}{2} \quad \text{and} \quad r_2 = \frac{(1-A) - i\sqrt{4B-(A-1)^2}}{2}.$$

Write these as

$$r_1 = p + iq \quad \text{and} \quad r_2 = p - iq,$$

with

$$p = \frac{1-A}{2} \quad \text{and} \quad q = \frac{1}{2}\sqrt{4B-(A-1)^2}.$$

Formally, the two solutions are

$$x^{p+iq} \quad \text{and} \quad x^{p-iq}.$$

But what do these mean? Write

$$x^{p+iq} = x^p \cdot x^{iq},$$

and recall that, for real α, $x^\alpha = e^{\alpha \ln(x)}$. Thus, we can interpret

$$x^{iq} = e^{iq \ln(x)} = \cos[q \ln(x)] + i \sin[q \ln(x)],$$

using the results on complex exponentials from Section 2.4.

Thus, if $x > 0$,

$$x^{p+iq} = x^p \cdot x^{iq} = x^p\{\cos[q \ln(x)] + i \sin[q \ln(x)]\}$$

and, by similar reasoning,

$$x^{p-iq} = x^p\{\cos[q \ln(x)] - i \sin[q \ln(x)]\}.$$

These form a fundamental system of solutions for $x > 0$. However, just as in Section 2.5, case 3, these solutions are unwieldy, and we can exploit the fact that linear combinations are again solutions to find a simpler fundamental system. Write the general solution as

$$y = c_1 x^p\{\cos[q \ln(x)] + i \sin[q \ln(x)]\} + c_2 x^p\{\cos[q \ln(x)] - i \sin[q \ln(x)]\}.$$

With $c_1 = c_2 = \frac{1}{2}$, we obtain the solution

$$x^p \cos[q \ln(x)].$$

With $c_1 = -c_2 = 1/2i$, we get the solution

$$x^p \sin[q \ln(x)].$$

It is easy to check by Theorem 4 that these also form a fundamental system of solutions when $x > 0$. Thus, we shall write the general solution in case 3 as

$$y = c_1 x^p \cos[q \ln(x)] + c_2 x^p \sin[q \ln(x)], \qquad x > 0.$$

Note: As in case 2, it is not necessary to go through the above derivation each time. Just remember that for an Euler equation $x^2y'' + Axy' + By = 0$, when the characteristic equation $r^2 + (A - 1)r + B = 0$ has complex roots $p + iq$ and $p - iq$, then the general solution is

$$y = c_1 x^p \cos[q \ln(x)] + c_2 x^p \sin[q \ln(x)].$$

EXAMPLE 84

Solve

$$y'' + 4y' + 6y = 0.$$

Here, the characteristic equation is

$$r^2 + 3r + 6 = 0,$$

with complex roots

$$\frac{-3 + i\sqrt{15}}{2} \quad \text{and} \quad \frac{-3 - i\sqrt{15}}{2}.$$

Thus, the general solution, for $x > 0$, is

$$y = c_1 x^{-3/2} \cos\left(\frac{\sqrt{15}}{2} \ln(x)\right) + c_2 x^{-3/2} \sin\left(\frac{\sqrt{15}}{2} \ln(x)\right).$$

The second method of solving Euler equations is also a good illustration of the use of transformations. Change variables from x and y to z and y by setting

$$z = \ln(x), \qquad x > 0.$$

By the chain rule, we compute

$$\frac{dy}{dx} = \frac{dy}{dz}\frac{dz}{dx} = \frac{1}{x}\frac{dy}{dz}$$

and

$$\frac{d^2y}{dx^2} = \frac{d}{dx}\left(\frac{dy}{dx}\right) = \frac{d}{dx}\left(\frac{1}{x}\frac{dy}{dz}\right) = -\frac{1}{x^2}\frac{dy}{dz} + \frac{1}{x}\frac{d}{dx}\left(\frac{dy}{dz}\right)$$

$$= -\frac{1}{x^2}\frac{dy}{dz} + \frac{1}{x}\frac{d}{dz}\left(\frac{dy}{dz}\right)\frac{dz}{dx} = -\frac{1}{x^2}\frac{dy}{dz} + \frac{1}{x^2}\frac{d^2y}{dz^2}.$$

Now substitute into $x^2y'' + Axy' + By = 0$ to get

$$x^2\left(-\frac{1}{x^2}\frac{dy}{dz} + \frac{1}{x^2}\frac{d^2y}{dz^2}\right) + Ax\left(\frac{1}{x}\frac{dy}{dz}\right) + By = 0.$$

Then,

$$\frac{d^2y}{dz^2} + (A - 1)\frac{dy}{dz} + By = 0,$$

a constant coefficient equation which we easily solve for y in terms of z. Then obtain y in terms of x by recalling that $z = \ln(x)$.

It is not necessary to go through the above derivation each time you want to use this method. Just remember that $z = \ln(x)$ transforms

$$x^2y'' + Axy' + By = 0 \quad \text{to} \quad \frac{d^2y}{dz^2} + (A - 1)\frac{dy}{dz} + By = 0,$$

and solve the latter for y in terms of z. Then substitute $z = \ln(x)$ to obtain y in terms of x.

EXAMPLE 85

Solve

$$x^2y'' - 4xy' + 6y = 0.$$

Setting $z = \ln(x)$ transforms this differential equation into

$$\frac{d^2y}{dz^2} - 5\frac{dy}{dz} + 6y = 0,$$

which has general solution

$$y = c_1e^{3z} + c_2e^{2z}.$$

Since $z = \ln(x)$, then the general solution in terms of x is

$$y = c_1e^{3\ln(x)} + c_2e^{2\ln(x)} = c_2x^3 + c_2x^2.$$

As usual, one solves an initial value problem by finding the general solution of the differential equation and then solving for the constants to match the initial data.

EXAMPLE 86

Solve

$$x^2y'' + 2xy' + 4y = 0; \qquad y(1) = 4, \qquad y'(1) = -7.$$

For the differential equation, the characteristic equation is

$$r^2 + r + 4 = 0,$$

with roots

$$\frac{-1 + i\sqrt{15}}{2} \quad \text{and} \quad \frac{-1 - i\sqrt{15}}{2}.$$

The general solution is

$$y = c_1x^{-1/2}\cos\left(\frac{\sqrt{15}}{2}\ln(x)\right) + c_2x^{-1/2}\sin\left(\frac{\sqrt{15}}{2}\ln(x)\right).$$

Now we need

$$y(1) = 4 = c_1$$

and

$$y'(1) = -7 = -\frac{1}{2}c_1 + \frac{\sqrt{15}}{2}c_2.$$

Solve these to get

$$c_1 = 4 \quad \text{and} \quad c_2 = \frac{-10}{\sqrt{15}}.$$

The solution is then

$$y = \frac{4}{\sqrt{x}}\cos\left(\frac{\sqrt{15}}{2}\ln(x)\right) - \frac{10}{\sqrt{15}}\frac{1}{\sqrt{x}}\sin\left(\frac{\sqrt{15}}{2}\ln(x)\right).$$

To four decimal places, we have

$$y_{\text{app}} = \frac{1}{\sqrt{x}}\{4\cos[1.9365\ln(x)] - 2.5820\sin[1.9365\ln(x)]\}.$$

PROBLEMS FOR SECTION 2.12

In each of Problems 1 through 25, find the general solution of the Euler equation (assume $x > 0$). In each problem, use both methods of this section.

1. $x^2y'' - 2xy' + y = 0$
2. $x^2y'' + 3xy' + y = 0$
3. $x^2y'' - 2xy' + 4y = 0$
4. $x^2y'' - 3xy' + 2y = 0$
5. $x^2y'' + 4xy' + 10y = 0$
6. $x^2y'' - 8xy' + 4y = 0$
7. $x^2y'' + 2xy' - 5y = 0$
8. $x^2y'' - 3xy' + 2y = 0$
9. $x^2y'' + 5xy' + 4y = 0$
10. $x^2y'' - 12xy' + 2y = 0$
11. $x^2y'' - 4xy' + 6y = 0$
12. $x^2y'' + 7xy' - 2y = 0$
13. $x^2y'' - 4xy' + 12y = 0$
14. $x^2y'' + 7xy' + 6y = 0$
15. $x^2y'' - 2xy' + 8y = 0$
16. $x^2y'' + xy' + y = 0$
17. $x^2y'' + 4xy' + 5y = 0$
18. $x^2y'' + 15xy' + 49y = 0$
19. $x^2y'' - 8xy' + 10y = 0$
20. $x^2y'' + 25xy' + 144y = 0$
21. $x^2y'' + 2xy' - y = 0$
22. $x^2y'' + 2xy' + y = 0$
23. $x^2y'' + xy' - 4y = 0$
24. $x^2y'' - 2xy' + 8y = 0$
25. $x^2y'' - 3xy' + 4y = 0$

In each of Problems 26 through 35, solve the initial value problem.

26. $x^2y'' + 3xy' + 2y = 0;\quad y(1) = y'(1) = 3$
27. $x^2y'' + 5xy' + 20y = 0;\quad y(1) = 0,\quad y'(1) = 2$
28. $x^2y'' + 5xy' - 21y = 0;\quad y(2) = 1,\quad y'(2) = 0$
29. $x^2y'' + 25xy' + 144y = 0;\quad y(1) = -3,\quad y'(1) = 0$
30. $x^2y'' + xy' - y = 0;\quad y(2) = 1,\quad y'(2) = -3$
31. $x^2y'' - 3xy' - 5y = 0;\quad y(2) = 3,\quad y'(2) = -7$
32. $x^2y'' + 10xy' + 24y = 0;\quad y(1) = 0,\quad y'(1) = 4$
33. $x^2y'' + 3xy' + 37y = 0;\quad y(1) = 1,\quad y'(1) = 0$
34. $x^2y'' - 6xy' - 12y = 0;\quad y(3) = y'(3) = -2$
35. $x^2y'' - 8xy' + 10y = 0;\quad y(1) = 4,\quad y'(1) = 2$

2.13 SUMMARY OF METHODS

Thus far we have collected a number of methods for solving certain second order differential equations. In this section we display them together to try to give the student an overall picture, as well as a review device. The key ideas are repeated in the notation of the chapter.

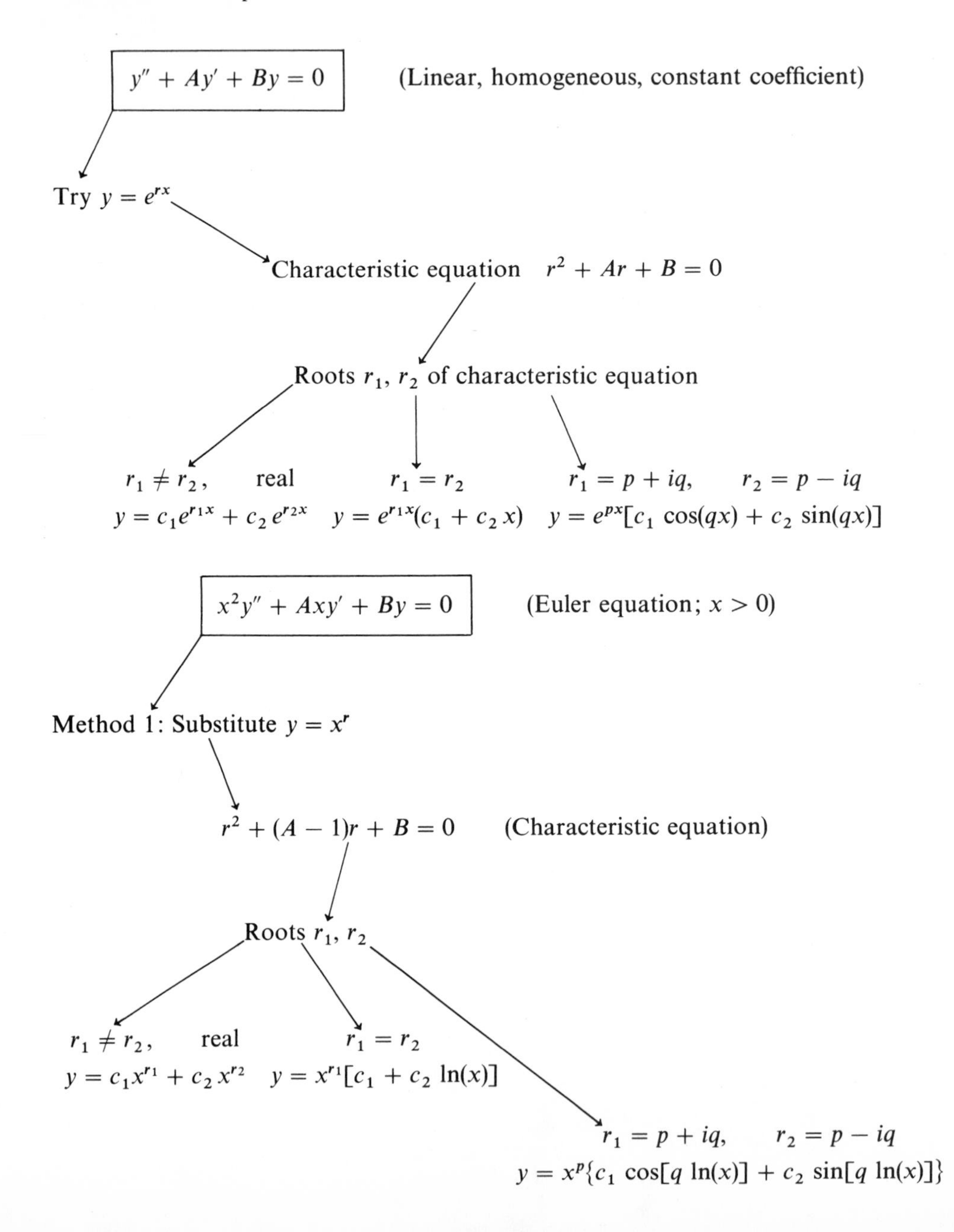

Method 2 for $x^2y'' + Axy' + By = 0$:

Solve $d^2y/dz^2 + (A - 1)\,dy/dz + By = 0$ for y in terms of z; then substitute $z = \ln(x)$ to solve the Euler equation in terms of x.

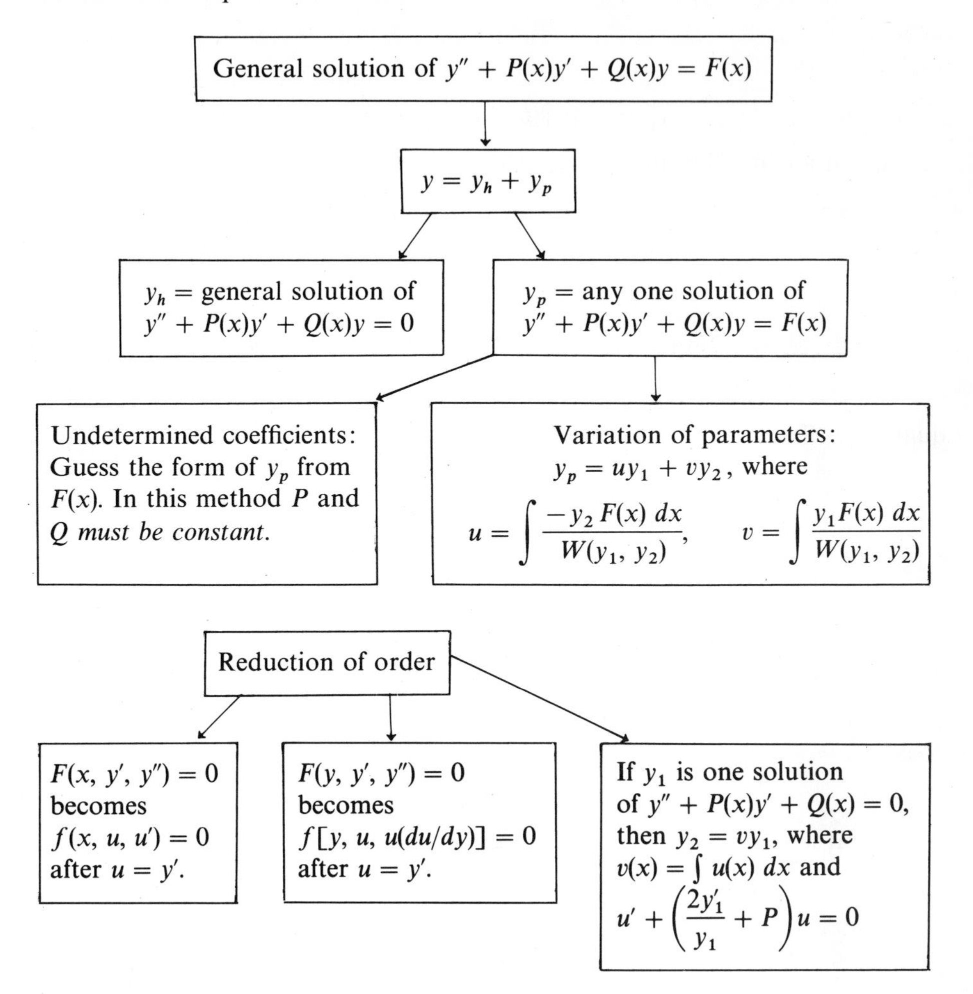

Here are some additional examples.

EXAMPLE 87

Find the general solution of

$$y'' - 3y' + 2y = 2x + 8\sin(2x).$$

First we find y_h. The characteristic equation is

$$r^2 - 3r + 2 = 0,$$

with roots $r_1 = 1$ and $r_2 = 2$. Thus,

$$y_h = c_1e^x + c_2e^{2x}.$$

For a particular solution y_p, try undetermined coefficients. We would guess a solution of the form

$$y_p = \alpha x + \beta + \gamma \sin(2x) + \delta \cos(2x).$$

Then,

$$y'_p = \alpha + 2\gamma \cos(2x) - 2\delta \sin(2x),$$
$$y''_p = -4\gamma \sin(2x) - 4\delta \cos(2x),$$

and substitution into the differential equation gives us

$$-4\gamma \sin(2x) - 4\delta \cos(2x) - 3\alpha - 6\gamma \cos(2x) + 6\delta \sin(2x)$$
$$+ 2\alpha x + 2\beta + 2\gamma \sin(2x) + 2\delta \cos(2x) = 2x + 8 \sin(2x).$$

This can be written

$$2\alpha x + (-3\alpha + 2\beta) + (-4\delta - 6\gamma + 2\delta) \cos(2x) + (-4\gamma + 6\delta + 2\gamma) \sin(2x)$$
$$= 2x + 8 \sin(2x).$$

Equating coefficients of like terms, we have

$$2\alpha = 2$$
$$-3\alpha + 2\beta = 0$$
$$-2\delta - 6\gamma = 0$$
$$6\delta - 2\gamma = 8.$$

Then,

$$\alpha = 1, \quad \beta = \tfrac{3}{2}, \quad \gamma = -\tfrac{2}{5}, \quad \text{and} \quad \delta = \tfrac{6}{5}.$$

Then,

$$y_p = x + \tfrac{3}{2} - \tfrac{2}{5} \sin(2x) + \tfrac{6}{5} \cos(2x).$$

The general solution of $y'' = 3y' + 2y = 2x + 8 \sin(2x)$ is

$$y = c_1 e^x + c_2 e^{2x} + x + \tfrac{3}{2} - \tfrac{2}{5} \sin(2x) + \tfrac{6}{5} \cos(2x).$$

EXAMPLE 88

Find the general solution of

$$x^2 y'' - 2xy' + 9y = 0.$$

This is an Euler equation. If we use method 1 for solving it, then the characteristic equation is $r^2 - 3r + 9 = 0$, with roots

$$r_1 = \frac{3 + 3i\sqrt{3}}{2} \quad \text{and} \quad r_2 = \frac{3 - 3i\sqrt{3}}{2}.$$

The general solution is

$$y = x^{3/2}\left\{c_1 \cos\left[\frac{3\sqrt{3}}{2} \ln(x)\right] + c_2 \sin\left[\frac{3\sqrt{3}}{2} \ln(x)\right]\right\}.$$

If you want to use method 2, set $z = \ln(x)$ and solve the constant coefficient linear homogeneous equation

$$\frac{d^2y}{dz^2} - 3\frac{dy}{dz} + 9y = 0.$$

In terms of z, we get

$$y = e^{3z/2}\left[c_1 \cos\left(\frac{3\sqrt{3}}{2}z\right) + c_2 \sin\left(\frac{3\sqrt{3}}{2}z\right)\right].$$

Upon putting $z = \ln(x)$ and recalling that $e^{3\ln(x)/2} = x^{3/2}$, this yields the same general solution as obtained by method 1.

EXAMPLE 89

Find the general solution of

$$y'' - 4y' + 4y = x\cos(x).$$

To find y_h, first solve the characteristic equation

$$r^2 - 4r + 4 = (r - 2)^2 = 0.$$

Then $r = 2$ and

$$y_h = e^{2x}(c_1 + c_2 x).$$

To find y_p, we can use undetermined coefficients or variation of parameters. Variation of parameters leads us to $y_p = ue^{2x} + vxe^{2x}$, where

$$u = -\int x^2 e^{-2x}\cos(x)\,dx \quad \text{and} \quad v = \int xe^{-2x}\cos(x)\,dx.$$

These can certainly be done, but only after some work; so undetermined coefficients is probably a little easier here. Try a solution

$$y_p = (Ax + B)\cos(x) + (Cx + D)\sin(x).$$

After calculating the derivatives and substituting into the differential equation, we get

$$\begin{aligned}y_p'' - 4y_p + 4y_p &= -2A\sin(x) + 2C\cos(x) - (Ax + B)\cos(x) - (Cx + D)\sin(x)\\ &\quad - 4A\cos(x) - 4C\sin(x) + 4(Ax + B)\sin(x) - 4(Cx + D)\cos(x)\\ &\quad + 4(Ax + B)\cos(x) + 4(Cx + D)\sin(x) = x\cos(x).\end{aligned}$$

Upon grouping terms, we get

$$\begin{aligned}&(-2A + 4B - 4C + 3D)\sin(x) + (-4A + 3B + 2C - 4D)\cos(x)\\ &\qquad + (-5A - 4C)x\cos(x) + (4A + 3C)x\sin(x) = x\cos(x).\end{aligned}$$

Then,

$$\begin{aligned}-2A + 4B - 4C + 3D &= 0\\ -4A + 3B + 2C - 4D &= 0\\ -5A - 4C &= 1\\ 4A + 3C &= 0.\end{aligned}$$

Solve these to get

$$A = 3, \qquad B = \tfrac{4}{5}, \qquad C = -4, \quad \text{and} \quad D = -\tfrac{22}{5}.$$

A particular solution can then be written

$$y_p = \left(\frac{75x + 20}{25}\right)\cos(x) + \left(\frac{-100x + 110}{25}\right)\sin(x).$$

The general solution is

$$y = e^{2x}(c_1 + c_2 x) + \left(\frac{75x + 20}{25}\right)\cos(x) + \left(\frac{-100x + 110}{25}\right)\sin(x).$$

PROBLEMS FOR SECTION 2.13

In each of Problems 1 through 20, find the general solution.

1. $x^2y'' - 2xy + 7y = 0$
2. $y'' - 4y' + 4y = 2\sin(3x) - 4$
3. $y'' - y' + 2y = -e^{3x} + x$
4. $x^2y'' - xy' + 2y = 0$
5. $x^2y'' + 2xy' - y = 0$
6. $y'' + 5y' - 2y = 8x^2 + 2x - 1$
7. $y'' - 30y' + 225y = e^x - \sin(2x)$
8. $y'' + 10y' - 2y = 2x^3 - \cos(3x)$
9. $y'' + 4y' + 9y = x^2 - 3\cos(2x) + e^x$
10. $y'' - 8y' - 9y = e^{-x} + 2e^{9x}$
11. $x^2y'' - 2xy' + \frac{1}{2}y = 0$
12. $x^2y'' - 4xy' + 7y = 0$
13. $y'' - 2y' + 22y = \cosh(x) - x^2 + 3$
14. $x^2y'' - 4xy' + 6y = 0$
15. $y'' + 5y' - 2y = \sin(2x) - 5\cos(4x)$
16. $x^2y'' + 5xy' + 4y = 0$
17. $y'' + 2y' + 5y = x^3 - 3x + 2$
18. $x^2y'' - 4xy' + 2y = 0$
19. $y'' - 10y' + 9y = x^2 + e^x$
20. $y'' - 3y' - 28y = \sec^2(2x)$

In each of Problems 21 through 25, solve the initial value problem.

21. $x^2y'' + 5xy' + 29y = 0;\quad y(1) = y'(1) = 4$
22. $y'' - 10y' + 25y = \sin(5x) + 2e^{5x};\quad y(0) = y'(0) = 0$
23. $y'' - 4y' - 12y = 2x;\quad y(0) = 1,\quad y'(0) = 4$
24. $y'' + 64y = 3\cos(8x);\quad y(\pi) = 0,\quad y'(\pi) = -1$
25. $y'' - 6y' - 7y = x^3 - 3x;\quad y(0) = 1,\quad y'(0) = -4$

CHAPTER SUMMARY

This chapter was primarily concerned with the theory and techniques of solution of specific kinds of second order differential equations which are encountered frequently in practice. It also provides the essential basis from which we can consider more advanced methods. The theory of second order linear differential equations extends in a very direct way to the higher order case, as do the methods we have learned for equations having a particular appearance.

Even with all this behind us, however, we are still quite limited in the kinds of differential equations we can solve. Note, for example, that we are as yet unable to solve the harmless looking pendulum equation $\theta''(t) + (g/L)\sin(\theta) = 0$. In the next chapter we shall consider differential equations of arbitrary order n, and then proceed to transform, series, and special function techniques which are called upon when methods developed thus far fail for practical reasons or do not apply to the problem at hand.

SUPPLEMENTARY PROBLEMS

In each of Problems 1 through 25, find the general solution of the differential equation.

1. $x^2y'' - 16xy' + 2y = 0$

2. $y' + 2xy'' = e^{-x}$

3. $y'' + 2y' - 4y = \cos(3x) - e^x$

4. $y'' + 8y' + 16y = x\cos(2x)$

5. $y'' + y' - y = e^{2x} - 2e^x$

6. $yy' - 4y'' = 0$

7. $y'' - 8y' + 16y = \cosh(4x)$

8. $y'' + 9y' + 16y = x^5 - 2x^2 + 1$

9. $x^2y'' - 3xy' + y = 0$

10. $y'' + 9y = \sin(3x) - \sinh(2x)$

11. $2y' - y'' = x^2$

12. $y'' + 18y = \cos(2x)$

13. $y'' - \dfrac{3}{x}y' + \dfrac{2}{x^2}y = 0$

14. $y'' - 18y' + 81y = 3x^3 - 5$

15. $y'' - 5y' + 9y = x^2 - \sin(x)$

16. $y'' + 10y' - 4y = e^x - 5$

17. $x^2y'' - 2xy' + 14y = 0$

18. $y'' - 6y' = 8e^{2x}$

19. $y'' + 13y' - 8y = x^2 + 2$

20. $y'' + 5y' - 12y = \tan(3x)$

21. $2y'' + y'\ln(y') = 0$

22. $y'' + 7y' - 5y = x^3 - 3x + 1$

23. $y'' - y' - 8y = 3x^2 + 5$

24. $x^2y'' - 9xy' + 2y = 0$

25. $y'' + 3y' + 10y = x^3$

26. Obtain a solution for the displacement $y(t)$ in the spring system of Figure 22. Assume that $y(0)$ and $y'(0)$ are initially zero.

27. Determine the displacement in the spring system of Figure 23. Assume that $y(0) = -1$ and $y'(0) = 2$.

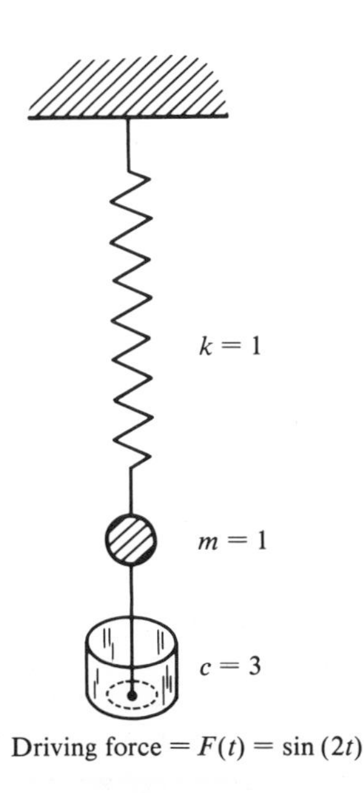

Driving force $= F(t) = \sin(2t)$

Figure 22

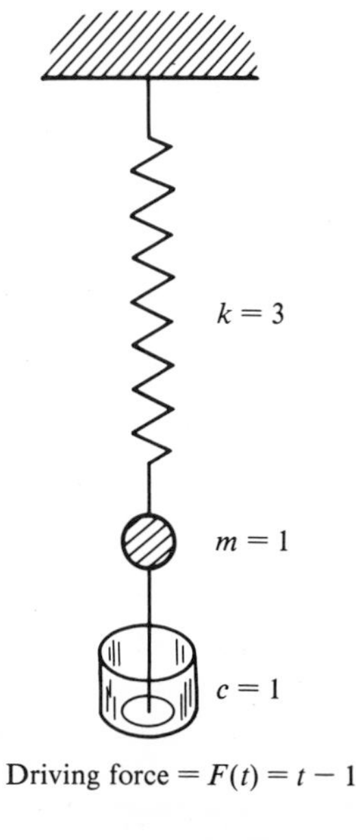

Driving force $= F(t) = t - 1$

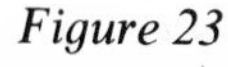

Figure 23

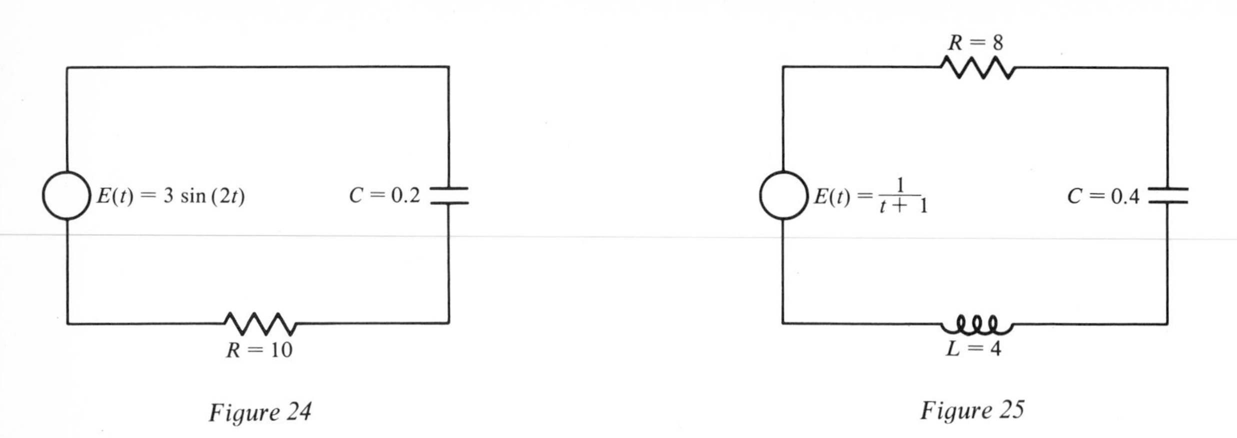

Figure 24 *Figure 25*

28. Determine the charge $Q(t)$ and current $I(t)$ in the circuit of Figure 24. Assume that I and I' are initially zero.

29. Determine the charge $Q(t)$ and current $I(t)$ in the circuit of Figure 25. Assume that I and I' are initially zero.

In each of Problems 30 through 37, solve the initial value problem.

30. $x^2y'' + 13xy' + 45y = 0$; $y(1) = y'(1) = 2$

31. $y'' + 4y' - 96y = 3e^{8x} - \cos(2x)$; $y(0) = y'(0) = 0$

32. $y'' - y' - 20y = x^2 - 1$; $y(0) = y'(0) = 0$

33. $y'' - 2y' + y = 2\sin(3x)$; $y(0) = 2$, $y'(0) = 1$

34. $y'' - 4y' + 2y = x^2 + 1$; $y(0) = 4$, $y'(0) = 0$

35. $y'' - 2y' + 7y = e^x$; $y(0) = y'(0) = -2$

36. $y'' + 8y' - 2y = x + 5$; $y(-1) = -3$, $y'(-1) = 2$

37. $y'' + 22y' + 121y = 28$; $y(0) = -5$, $y'(0) = 2$

38. Suppose that y_1 and y_2 are solutions of $y'' + P(x)y' + Q(x)y = 0$ on $[a, b]$, and that y_1 and y_2 both have a relative maximum or relative minimum at x_0, $a < x_0 < b$. Prove that y_1 and y_2 do *not* form a fundamental system of solutions of $y'' + P(x)y' + Q(x)y = 0$ on $[a, b]$.

39. A particle of mass m moves along a straight line, having traveled a distance $x(t)$ in t seconds. The total force acting on the particle is $F(x)$.
(a) Use Newton's laws of motion to write a differential equation for x.
(b) Show that the energy of the particle is a constant of the motion.
(*Hint:* The kinetic energy is $\frac{1}{2}m(x')^2$, and the potential energy is $-\int_0^x F(z)\,dz$.

40. A ball of mass m is thrown upward from the surface of the earth. The initial velocity is v_0, and the forces acting on the ball are air resistance, which is proportional to the square of the velocity, and gravity.
(a) Write and solve the differential equation for the height of the ball at time t.
(b) Find the maximum height attained, and the time it takes to reach this height.
(c) Is it true that the time it takes the ball to reach its maximum height is less than the time it takes it to fall back to earth?

41. Repeat Problem 40, assuming that air resistance is proportional to v^n, where v is the velocity and n is a positive integer. What influence does n have on the resulting motion?

42. An object is thrown vertically upward from the surface of the earth with initial velocity v_0. Neglecting air resistance, show that the body will return to earth if $v_0 < \sqrt{2gR}$, and will continue moving away if $v_0 > \sqrt{2gR}$. Here, R is the radius of the earth, and $\sqrt{2gR}$ is called the *escape velocity*. What happens if $v_0 = \sqrt{2gR}$? (*Hint:* Use Newton's inverse square law of gravitational attraction.)

43. For a simple pendulum, we obtained

$$\theta'' + \frac{g}{L}\sin(\theta) = 0.$$

Suppose that the pendulum is released from rest at time $t = 0$ from the position $\theta = -\alpha$, $0 < \alpha < \pi/2$. Show that, on the first half-swing of the pendulum,

$$t = \sqrt{\frac{L}{2g}} \int_{-\alpha}^{\theta} \frac{d\phi}{\sqrt{\cos(\phi) - \cos(\alpha)}}.$$

44. Prove the conservation of energy for the simple pendulum. (*Hint:* Show that the kinetic energy is $\frac{1}{2}m(ds/dt)^2$, where $s = L\theta$, and that the potential energy is $mgL[1 - \cos(\theta)]$.

45. If θ is small, then $\sin(\theta)$ is almost θ. In this case, the pendulum equation is just $\theta'' + (g/L)\theta = 0$, with sine and cosine solutions. The period of the motion turns out to be $2\pi\sqrt{L/g}$.

Suppose now that we do not assume small θ, so that the differential equation is $\theta'' + (g/L)\sin(\theta) = 0$.

(a) Letting $u = d\theta/dt$, show that

$$u\frac{du}{d\theta} + \frac{g}{L}\sin(\theta) = 0.$$

(b) Integrate the result of (a) to get

$$\frac{1}{2}u^2 - \frac{g}{L}\cos(\theta) = C,$$

where C is constant.

(c) Suppose that, at $t = 0$, $\theta = \theta_0$ and $\theta' = 0$. Show that, as the pendulum moves from its initial position θ_0 to $\theta = 0$ (vertical position), then

$$\frac{d\theta}{dt} = -\sqrt{\frac{2g}{L}}\sqrt{\cos(\theta) - \cos(\theta_0)}.$$

(d) Suppose that θ_0 is such that, as θ moves from θ_0 to 0, the time consumed is $\frac{1}{4}$ of the period p. Show that

$$p = 4\sqrt{\frac{L}{2g}} \int_0^{\theta_0} \frac{d\theta}{\sqrt{\cos(\theta) - \cos(\theta_0)}}.$$

(e) Use trigonometric identities to show that

$$p = 2\sqrt{\frac{L}{g}} \int_0^{\theta_0} \frac{d\theta}{\sin^2(\theta_0/2) - \sin^2(\theta/2)}.$$

(f) Letting $\sin(\theta/2) = \sin(\theta_0/2)\sin(\alpha)$, show that

$$p = 4\sqrt{\frac{L}{g}} \int_0^{\pi/2} \frac{d\alpha}{\sqrt{1 - k^2\sin^2(\alpha)}},$$

where $k = \sin(\theta_0/2)$. The integral in this expression for the period is an elliptic integral of the first kind. It cannot be evaluated in terms of elementary functions, but can be approximated for different values of k by using tables which have been constructed for this purpose.

(g) If θ is small, show that k is near zero. Hence, obtain from (f) that $p = 2\pi\sqrt{L/g}$, the period for small oscillations (see Problem 31, Section 2.5).

46. Using the result of Problem 45(f), show that the period p of a simple pendulum is given by

$$p = 2\pi\sqrt{\frac{L}{g}}\left[1 + \frac{1^2}{2^2}k^2 + \frac{1^2 3^2}{2^2 4^2}k^4 + \frac{1^2 3^2 5^2}{2^2 4^2 6^2}k^6 + \cdots\right],$$

in which $k = \sin(\theta_0/2)$ and $\theta_0 = \theta(0)$, as in Problem 45.
[*Hint:* Recall from the binomial theorem that

$$(1 + \alpha)^r = 1 + r\alpha + \frac{r(r-1)}{2!}\alpha^2 + \frac{r(r-1)(r-2)}{3!}\alpha^2 + \cdots$$

if $|\alpha| < 1$. Use this to expand the integrand in Problem 45(f), and integrate the resulting series term by term, using the standard integral result

$$\int_0^{\pi/2} \sin^{2n}(\alpha)\,d\alpha = \frac{1 \cdot 3 \cdot 5 \cdots (2n-1)}{2 \cdot 4 \cdot 6 \cdots (2n)}\frac{\pi}{2}$$

for $n = 1, 2, 3, \ldots$. This formula is easily derived using the gamma (or beta) function, as we will see in Chapter 6.]

Use this infinite series for p to argue again that, for small θ, the period is approximately $2\pi\sqrt{L/g}$.

47. A particle of mass m inside the earth at a distance r from the center feels a gravitational force of $F = -mgr/R$, where R is the radius of the earth (assume that the earth is a sphere). Show that a particle in an evacuated tube through the earth's center will execute simple harmonic motion, and determine the period of this motion.

48. A stone of mass m is thrown downward toward the earth from a height of 1500 feet and at an initial velocity of 12 feet/second. The forces acting on it are gravity and air resistance, which is assumed to be proportional to the velocity. Find the velocity as a function of time, and determine the time T when the stone hits the earth. Also determine the impact velocity.

49. A ball of mass m is thrown vertically downward from a stationary dirigible h feet off the ground. The initial velocity of the ball is v_0. Neglecting air resistance, show that the ball will impact on the ground at time $(\sqrt{v_0^2 + 2gh} - v_0)/g$.

50. A four-pound brick is dropped from a building h feet high. After it has fallen k feet ($k < h$), a six-pound brick is dropped from the same point. Neglecting air resistance,

(a) Show that, when the first brick hits the ground, the second still has $(2\sqrt{kh} - h)$ feet to go.

(b) Of what importance is the weight of each brick in this problem?

CHAPTER THREE

Higher Order Differential Equations

3.0 INTRODUCTION

Thus far we have seen some practical and theoretical results on first and second order differential equations. Many of these results generalize in a straightforward way to higher order differential equations. If for convenience we denote the kth derivative $d^k y/dx^k$ by $y^{(k)}$, then an nth order differential equation is of the form

$$F(x, y, y', \ldots, y^{(n)}) = 0,$$

involving $x, y, y', y^{(2)}, \ldots, y^{(n-1)}$, and $y^{(n)}$. We assume in an nth order equation that $y^{(n)}$ is explicitly present, but no higher derivative; x, y, y', and lower derivatives may or may not appear explicitly. For example,

$$y^{(4)} - 3y' + 2y = 0$$

is a fourth order equation (x, y'', $y^{(3)}$, do not appear explicitly);

$$y^{(6)} - 2y^{(4)} + x^2 y + 2x = 0$$

is a sixth order equation, and so on. As usual, a solution of $F(x, y, y', \ldots, y^{(n)}) = 0$ is a function satisfying the equation.

Differential equations of higher than second order arise in many contexts. Here is one illustration of this.

EXAMPLE 90 Deflection of a Uniformly Loaded Beam

Consider a horizontal beam, fixed or clamped at one end, and supporting a weight suspended from the other end, as in Figure 26. We would expect the beam to experience some bending, due to the suspended weight and also the stress caused by the extension of the beam from its supported end. Our objective here is to derive an equation for the deflection of the beam from the horizontal (i.e., "amount of bending").

Suppose that the beam has length L, is clamped at the left end, and that a weight W has been suspended from the right end (e.g., by a pulley system). Assume that the beam has a uniform cross section and is much longer than it is thick. Also assume a constant density and a weight (of the beam) of w pounds per foot in the x-direction.

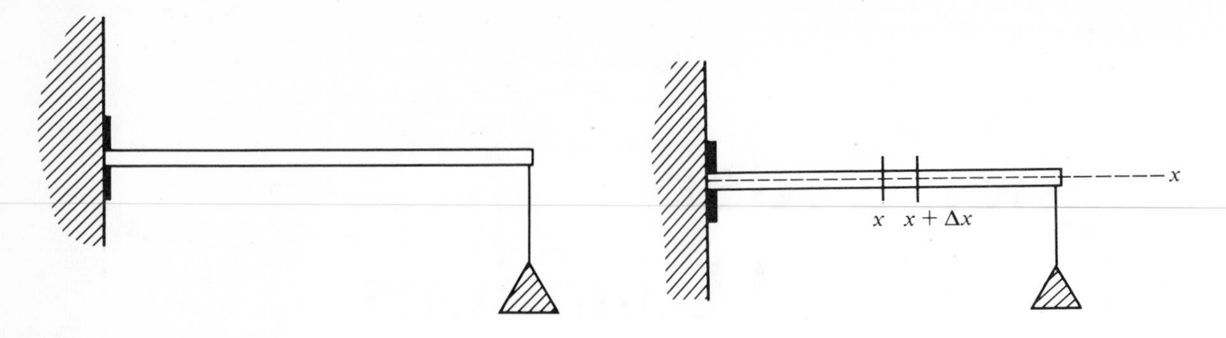

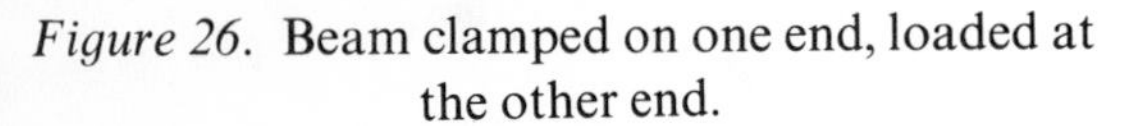

Figure 26. Beam clamped on one end, loaded at the other end.

Figure 27

Look at the piece of the beam between x and $x + \Delta x$, as in Figure 27. The part of the beam to the right of x applies a stress to the part to the left of x. This stress induces a force which can be decomposed into an x-component $X(x)$ and a y-component $Y(x)$. The horizontal stresses also produce a moment $\mu(x)$ (called the *bending moment*) about a line through the origin perpendicular to the xy-plane. In sum, then, the part of the beam to the left of x applies forces $-X(x)$, $-Y(x)$, and $-\mu(x)$ to the part to the right of x.

Now apply the same analysis at $x + \Delta x$. The part of the beam to the right of $x + \Delta x$ applies stresses to the part to the left of $x + \Delta x$, yielding forces $X(x + \Delta x)$, $Y(x + \Delta x)$, and $\mu(x + \Delta x)$.

Now, the segment between x and $x + \Delta x$ is in equilibrium; hence the horizontal and vertical components of force on the segment cancel out. Thus,

$$X(x + \Delta x) - X(x) = 0$$

and

$$Y(x + \Delta x) - Y(x) = w\,\Delta x.$$

Divide these by Δx to get

$$\frac{X(x + \Delta x) - X(x)}{\Delta x} = 0$$

and

$$\frac{Y(x + \Delta x) - Y(x)}{\Delta x} = w.$$

In the limit as $\Delta x \to 0$, we get

$$X' = 0 \tag{1}$$

and

$$Y' = w. \tag{2}$$

Further, the total moment on the segment between x and $x + \Delta x$ about the x-axis must vanish. Hence,

$$\mu(x + \Delta x) - \mu(x) + x[Y(x + \Delta x) - Y(x)] + Y(x + \Delta x)\,\Delta x = \left(x + \frac{\Delta x}{2}\right) w\,\Delta x.$$

Divide by Δx to get

$$\frac{\mu(x + \Delta x) - \mu(x)}{\Delta x} + x\left(\frac{Y(x + \Delta x) - Y(x)}{\Delta x}\right) + Y(x + \Delta x) = \left(x + \frac{\Delta x}{2}\right)w.$$

Letting $\Delta x \to 0$, we get

$$\mu' + xY' + Y = xw.$$

Since $Y' = w$ from (2), this reduces to

$$\mu' = -Y. \tag{3}$$

Assuming that the weight W acts at the end of the bar, then (1), (2), and (3) would be solved subject to

$$Y(L) = -W, \qquad \mu(L) = -LW.$$

With this as preparation, we can now attack the problem of deriving an explicit equation for deflection $y(x)$ of the beam from the horizontal (Figure 28). The Euler-Bernoulli law states that the bending moment at any point is proportional to the curvature* of the beam there. If y' is much smaller than y'', then this radius of curvature is approximately y''. Thus, the Euler-Bernoulli law becomes

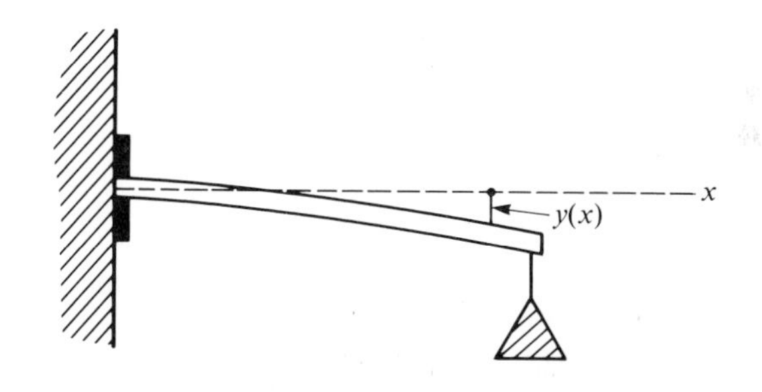

Figure 28. Deflection $y(x)$ of a loaded beam from the horizontal.

$$\mu = (EI)y'', \tag{4}$$

where E is Young's modulus and I is the moment of inertia of a cross section of the beam with respect to the axis of the beam.

---

* Here is a brief review of the notion of curvature: In Figure 29 we show $\phi(s)$ as the angle between the tangent and the horizontal at a point on the curve $y = y(x)$. Here, ϕ is measured as a function of arc length s. The curvature is the rate of change of ϕ with s, $d\phi/ds$. Intuitively, curvature measures how fast the "bending" changes. The greater the curvature, the greater the "rate of bending." Many elementary calculus texts include a formula for the curvature in terms of $y(x)$. It is

$$\text{curvature} = \frac{y''}{[1 + (y')^2]^{3/2}}.$$

Thus, for a straight line $y = ax + b$, curvature $= 0$, which is consistent with intuition. Note that the curvature is approximately y'' if y' is much smaller than y''.

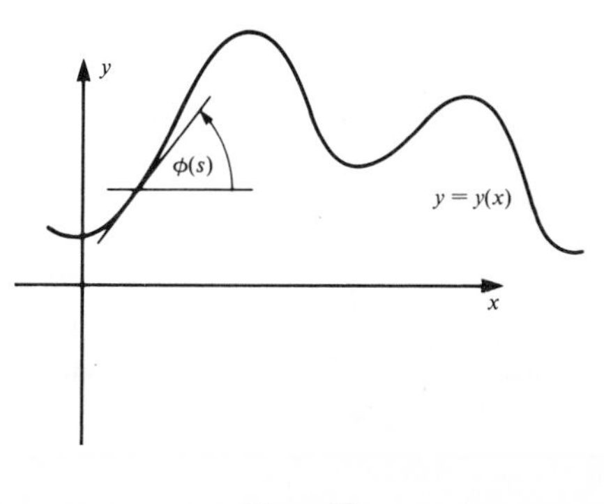

Figure 29

To derive a differential equation for y, differentiate (4) and use (3) to get

$$\mu' = (EI)y^{(3)} = -Y.$$

Differentiate again and use (2) to get

$$(EI)y^{(4)} = -Y' = -w.$$

Thus, we have

$$y^{(4)} = -\frac{w}{EI},$$

a fourth order differential equation for y.

As we shall see in later chapters, RLC circuits connected in series, and spring systems involving more than one spring, can result in differential equations of arbitrarily high order.

In the next section we begin the process of identifying some higher order equations we can solve, and then learning how to solve them.

PROBLEMS FOR SECTION 3.0

1. The fourth order differential equation derived in Example 90 for the deflection of a beam is separable and can be solved by four integrations. What kinds of information should you specify to determine a unique solution in this problem? (*Hint*: You should look at conditions at $x = 0$ and $x = L$.)
2. An nth order differential equation

$$y^{(n)} = F(x, y, y', \ldots, y^{(n-1)})$$

can always be converted to a *system* of *first order* differential equations as follows:

Define new variables by

$$z_1 = y, \qquad z_2 = y', \qquad z_3 = y'', \qquad \ldots, \qquad z_n = y^{(n-1)}.$$

We now have a system of first order equations in the new variables $z_1, \ldots, z_n$ (all functions of x):

$$\begin{aligned} z_1' &= z_2 \\ z_2' &= z_3 \\ &\vdots \\ z_{n-1}' &= z_n \\ z_n' &= F(x, z_1, z_2, \ldots, z_n). \end{aligned}$$

If we can solve this system, obtaining

$$z_1 = f_1(x), \qquad z_2 = f_2(x), \qquad \ldots, \qquad z_n = f_n(x)$$

satisfying all the equations of the system, then $y = z_1 = f_1(x)$ should satisfy the nth order equation. Prove this.
3. The method of Problem 2 converts the problem of solving an nth order equation into one of solving a system of first order equations. These arise frequently in consideration of electrical and mechanical systems. Later we shall develop matrix and transform methods for solving systems of differential equations. For now, use the method of Problem 2 to convert each of the following differential equations to a system of first order differential equations.

Do not solve the resulting system.

(a) $y^{(4)} = 2xy' + y'' - 3y^{(3)}$

(b) $y^{(5)} = 8y' - 4x^3 + 2xy' + y^{(3)}$

(c) $y^{(4)} = 2y^{(3)} - 6xy + xe^x$

3.1 THEORETICAL CONSIDERATIONS

The general nth order differential equation

$$F(x, y, y', \dots, y^{(n)}) = 0$$

is not approachable by elementary means. Hence, we look for special cases which are important and which we have some hope of solving. In this, we shall imitate to a large extent our results and experience from Chapters 1 and 2.

An nth order differential equation is called *linear* if it is of the form

$$y^{(n)} + P_{n-1}(x)y^{(n-1)} + P_{n-2}(x)y^{(n-2)} + \cdots + P_1(x)y' + P_0(x)y = F(x). \qquad \text{(a)}$$

We call $P_{n-1}(x), \dots, P_1(x), P_0(x)$, and $F(x)$ the *coefficient functions.* Often these functions are defined only on some interval J, in which case the discussion is restricted to J.

A *solution* of (a) is a function $y = f(x)$ such that

$$f^{(n)}(x) + P_{n-1}(x)f^{(n-1)}(x) + \cdots + P_1(x)f'(x) + P_0(x)f(x) = F(x).$$

For example, it is easy to check that e^{2x}, e^x, and e^{-x} are solutions of

$$y^{(3)} - 2y'' - y' + 2y = 0$$

on the whole real line.

The following theorem gives the higher order version of the existence and uniqueness results of Theorem 2.

THEOREM 7

Let n be an integer greater than or equal to 2. Let $P_0(x), P_1(x), \dots, P_{n-1}(x)$, and $F(x)$ be continuous on an interval J, and let x_0 be in J. Let $A_0, \dots, A_{n-1}$ be given constants. Then the initial value problem

$$y^{(n)} + P_{n-1}(x)y^{(n-1)} + \cdots + P_1(x)y' + P_0(x)y = F(x);$$

$$y(x_0) = A_0, \qquad y'(x_0) = A_1, \qquad \dots, \qquad y^{(n-1)}(x_0) = A_{n-1}$$

has a solution on J, and this solution is unique.

As in the case $n = 2$, one now breaks the study of the linear nth order differential equation into the homogeneous and nonhomogeneous cases. We call equation (a) *homogeneous* if $F(x)$ is identically zero; otherwise (a) is *nonhomogeneous.*

THE HOMOGENEOUS CASE

Here we consider the homogeneous linear equation

$$y^{(n)} + P_{n-1}(x)y^{(n-1)} + \cdots + P_1(x)y' + P_0(x)y = 0. \tag{b}$$

Theorem 8 generalizes Theorem 3 of Chapter 2.

THEOREM 8

1. If $y_1(x), \ldots, y_k(x)$ are solutions of (b), then $y_1(x) + \cdots + y_k(x)$ is a solution of (b).
2. If $y(x)$ is a solution of (b), then so is $cy(x)$ for any real number c.

That is, for an nth order linear homogeneous equation,

sums of solutions are solutions

and

constant multiples of solutions are solutions.

Given functions $y_1(x), \ldots, y_k(x)$, and real numbers $c_1, \ldots, c_k$, we call

$$c_1 y_1(x) + \cdots + c_k y_k(x)$$

a *linear combination* of $y_1(x), \ldots, y_k(x)$. In this terminology, Theorem 8 may be rephrased:

Any linear combination of solutions of an nth order linear homogeneous equation is also a solution.

Functions $y_1(x), \ldots, y_k(x)$ are called *linearly dependent* over an interval J if there is a linear combination $c_1 y_1(x) + \cdots + c_k y_k(x)$ which is zero *for all x in J*, with not all the c_i's zero. If, however, $c_1 y_1(x) + \cdots + c_k y_k(x) = 0$ can hold *for all x in J* only by choosing each c_i to be zero, then we call $y_1(x), \ldots, y_k(x)$ *linearly independent* on J. If J is the whole real line, then reference to J is omitted, or we may use the phrase "for all x."

For example, $5\cos(2x)$, $3\cos^2(x)$, and $-4\sin^2(x)$ are linearly dependent, since for all x,

$$\tfrac{1}{5}[5\cos(2x)] - \tfrac{1}{3}[3\cos^2(x)] - \tfrac{1}{4}[-4\sin^2(x)] = 0.$$

This follows from the trigonometric identity $\cos(2x) = \cos^2(x) - \sin^2(x)$.

However, $\sin(x)$, $\cos(x)$, and e^x are linearly independent. To see this, suppose that there is a linear combination of these functions which is zero for all x:

$$c_1 \sin(x) + c_2 \cos(x) + c_3 e^x = 0 \qquad \text{for all } x.$$

Putting $x = 0$, we then have

$$c_2 + c_3 = 0.$$

Setting $x = \pi$, we have

$$-c_2 + c_3 e^{\pi} = 0.$$

Finally, setting $x = \pi/2$, we get

$$c_1 + c_3 e^{\pi/2} = 0.$$

These are easily solved to get

$$c_1 = c_2 = c_3 = 0.$$

Thus, the only linear combination of $\sin(x)$, $\cos(x)$, and e^x which can be zero for all x is the linear combination having all the constants zero. From the definition, these functions are linearly independent.

There is an nth order version of the Wronskian test (Theorem 4 for the case $n = 2$) which is often useful. Given n functions $y_1(x), \ldots, y_n(x)$, their Wronskian is the $n \times n$ determinant*

$$W(y_1, \ldots, y_n) = \begin{vmatrix} y_1 & y_2 & \cdots & y_n \\ y_1' & y_2' & \cdots & y_n' \\ \vdots & \vdots & & \vdots \\ y_n^{(n-1)} & y_2^{(n-1)} & \cdots & y_n^{(n-1)} \end{vmatrix}.$$

THEOREM 9

Let $y_1(x), \ldots, y_n(x)$ be solutions of

$$y^{(n)} + P_{n-1}(x)y^{(n-1)} + \cdots + P_1(x)y' + P_0(x)y = 0$$

on an interval J. Then, $y_1(x), \ldots, y_n(x)$ are linearly independent on J if and only if there is some x_0 in J such that $W(y_1, \ldots, y_n) \neq 0$.

This theorem is of mainly theoretical value, since for large n the practical difficulties of evaluating an $n \times n$ determinant are considerable. For $n = 2$ or $n = 3$ the labor is not prohibitive, and here is an example.

EXAMPLE 91

We have already noted that $\cos(x)$, $\sin(x)$, and e^x are linearly independent. Further, these functions are solutions of

$$y^{(3)} - y'' + y' - y = 0.$$

The Wronskian is

$$\begin{vmatrix} \cos(x) & \sin(x) & e^x \\ -\sin(x) & \cos(x) & e^x \\ -\cos(x) & -\sin(x) & e^x \end{vmatrix},$$

which works out to $2e^x$. This is nonzero, consistent with Theorem 9.

We can now generalize Theorem 5:

---

* Determinants are treated in detail in Chapter 9. For $n = 3$, the 3×3 determinant is given by

$$\begin{vmatrix} a & b & c \\ d & e & f \\ g & h & k \end{vmatrix} = a(ke - fh) + b(fg - dk) + c(dh - eg).$$

THEOREM 10

Let $y_1(x), \ldots, y_n(x)$ be linearly independent solutions of

$$y^{(n)} + P_{n-1}(x)y^{(n-1)} + \cdots + P_1(x)y' + P_0(x)y = 0$$

on an interval J. Then, every solution on J is of the form

$$y(x) = c_1 y_1(x) + c_2 y_2(x) + \cdots + c_n y_n(x).$$

A proof of this result can be modeled after that of Theorem 5, and we leave the details to the student.

Thus, to solve the homogeneous linear equation

$$y^{(n)} + P_{n-1}(x)y^{(n-1)} + \cdots + P_1(x)y' + P_0(x)y = 0,$$

we seek n (the same as the order of the differential equation) linearly independent solutions $y_1, \ldots, y_n$. *Every* solution can then be written as a linear combination $c_1 y_1(x) + \cdots + c_n y_n(x)$ of these n solutions. For this reason, we call the expression

$$c_1 y_1(x) + \cdots + c_n y_n(x)$$

the *general solution* of (b) when $y_1(x), \ldots, y_n(x)$ are linearly independent.

EXAMPLE 92

e^{2x}, e^x, and e^{-x} are linearly independent solutions of

$$y^{(3)} - 2y'' - y' + 2y = 0.$$

(If you calculate the Wronskian, you get $-6e^{2x}$.) The general solution is therefore

$$y = c_1 e^{2x} + c_2 e^x + c_3 e^{-x}.$$

THE NONHOMOGENEOUS CASE

Suppose we now want to solve the nonhomogeneous linear equation

$$y^{(n)} + P_{n-1}(x)y^{(n-1)} + \cdots + P_1(x)y' + P_0(x)y = F(x). \tag{a}$$

We call

$$y^{(n)} + P_{n-1}(x)y^{(n-1)} + \cdots + P_1(x)y' + P_0(x)y = 0 \tag{b}$$

the *associated homogeneous equation*; its general solution is the *general homogeneous solution* of (a).

Here is the analogue of Theorem 6:

THEOREM 11

Let y_p be any solution of

$$y^{(n)} + P_{n-1}(x)y^{(n-1)} + \cdots + P_1(x)y' + P_0(x)y = F(x). \tag{a}$$

Let $y_1(x), \ldots, y_n(x)$ be linearly independent solutions of

$$y^{(n)} + P_{n-1}(x)y^{(n-1)} + \cdots + P_1(x)y' + P_0(x)y = 0.$$

Then, given *any* solution $y(x)$ of (a), there are constants $c_1, \ldots, c_n$ such that

$$y(x) = c_1 y_1(x) + \cdots + c_n y_n(x) + y_p(x).$$

In short, the expression

[general solution of associated homogeneous equation (b)]
+ [any particular solution of (a)]

contains *every* solution of (a); hence is called the *general solution* of (a).

As in Chapter 2, we usually denote the general homogeneous solution y_h, and a particular solution of (a) y_p. The general solution of (a) is then

$$y = y_h + y_p.$$

EXAMPLE 93

One solution of

$$y^{(3)} - 2y'' - y' + 2y = 2x - 1$$

is

$$y_p = x.$$

(This can be found by inspection or by the methods developed in Section 3.3.) Further,

$$y_h = c_1 e^{2x} + c_2 e^x + c_3 e^{-x}$$

from Example 92.

The general solution of $y^{(3)} - 2y'' - y' + 2y = 2x - 1$ is then

$$y = c_1 e^{2x} + c_2 e^x + c_3 e^{-x} + x.$$

Theorems 10 and 11 tell us what to look for in finding general solutions of both the homogeneous and nonhomogeneous nth order linear differential equation. The remainder of this chapter is devoted to the practical matter of actually finding solutions.

In concluding this section, we wish to emphasize the role of the initial data in the initial value problem of Theorem 7. The general solution of

$$y^{(n)} + P_{n-1}(x)y^{(n-1)} + \cdots + P_1(x)y' + P_0(x)y = F(x)$$

involves n constants (all contained in y_h). One can think of an nth order differential equation as requiring n integrations for solution, and hence n constants of integration. If we specify values for $y, y', \ldots, y^{(n-1)}$ at a point x_0, then we are giving n pieces of data, which give us n equations for the integration constants. Solving these for the constants produces the unique solution of the initial value problem. Here is an example.

EXAMPLE 94

Solve the initial value problem

$$y^{(3)} - 2y'' - y' + 2y = 2x - 1; \qquad y(0) = 1, \qquad y'(0) = -3, \qquad y''(0) = 4.$$

We know from Example 93 that the general solution of the differential equation is

$$y = c_1 e^{2x} + c_2 e^x + c_3 e^{-x} + x.$$

To satisfy the initial conditions, we need

$$y(0) = c_1 + c_2 + c_3 = 1$$
$$y'(0) = 2c_1 + c_2 - c_3 + 1 = -3$$
$$y''(0) = 4c_1 + c_2 + c_3 = 4.$$

Solve these to get

$$c_1 = 1, \qquad c_2 = -3, \quad \text{and} \quad c_3 = 3.$$

Thus, the solution is

$$y = e^{2x} - 3e^x + 3e^{-x} + x.$$

PROBLEMS FOR SECTION 3.1

1. Prove that $e^{k_1 x}$, $e^{k_2 x}$, and $e^{k_3 x}$ are linearly independent* if k_1, k_2, and k_3 are distinct real numbers.
2. Prove Theorem 8.
3. Prove that e^x, $\sin(x)$, and $\cos(x)$ are linearly independent.*

In each of Problems 4 through 10, show that the given functions are solutions of the given equation, and show that these solutions are linearly independent. Then write the general solution of the differential equation.

4. $y^{(3)} - 5y'' + 2y' + 8y = 0$; e^{4x}, e^{2x}, e^{-x}
5. $y^{(3)} - 2y'' - 7y' - 4y = 0$; e^{4x}, e^{-x}, xe^{-x}
6. $y^{(4)} - y^{(3)} - 9y'' - 11y' - 4y = 0$; $e^{4x}, e^{-x}, xe^{-x}, x^2e^{-x}$
7. $y^{(3)} + y' = 0$; $1, \cos(x), \sin(x)$
8. $y^{(3)} + y'' - 7y' - 15y = 0$; $e^{3x}, e^{-2x}\cos(x), e^{-2x}\sin(x)$
9. $y^{(3)} + 4y'' - 3y' - 18y = 0$; $e^{-3x}, xe^{-3x}, e^{2x}$
10. $y^{(4)} + 4y^{(3)} + 6y'' + 4y' + y = 0$; $e^{-x}, xe^{-x}, x^2e^{-x}, x^3e^{-x}$

In each of Problems 11 through 17, verify that the given function is a particular solution of the nonhomogeneous differential equation. Next, use the information from the problem cited to write the general solution of the nonhomogeneous differential equation.

11. $y^{(3)} - 5y'' + 2y' + 8y = x^2$; $\frac{1}{64}(8x^2 - 4x + 11)$; Problem 4
12. $y^{(3)} - 2y'' - 7y' - 4y = 3\cos(2x)$; $\frac{1}{250}[6\cos(2x) - 33\sin(2x)]$; Problem 5
13. $y^{(4)} - y^{(3)} - 9y'' - 11y' - 4y = -x^3 + 2$; $\frac{1}{4}x^3 - \frac{33}{16}x^2 + \frac{255}{32}x - \frac{1729}{128}$; Problem 6
14. $y^{(3)} + y' = -4e^{2x}$; $-\frac{2}{5}e^{2x}$; Problem 7
15. $y^{(3)} + y'' - 7y' - 15y = 8\sin(3x)$; $\frac{2}{15}\cos(3x) - \frac{1}{15}\sin(3x)$; Problem 8
16. $y^{(3)} + 4y'' - 3y' - 18y = 2x - 5$; $-\frac{1}{9}x + \frac{8}{27}$; Problem 9
17. $y^{(4)} + 4y^{(3)} + 6y'' + 4y' + y = 4\cosh(3x)$; $\frac{1}{128}[17\cosh(3x) - 15\sinh(3x)]$; Problem 10

---

* You may proceed directly from the definition, or you may use Theorem 9 *if* you first show that the functions in question are solutions of an appropriate differential equation.

In each of Problems 18 through 24, solve the differential equation of the cited problem subject to the given initial conditions.

18. Problem 11; $y(0) = 0, \quad y'(0) = 1, \quad y''(0) = 0$

19. Problem 12; $y(0) = y'(0) = 0, \quad y''(0) = 1$

20. Problem 13; $y(0) = 1, \quad y'(0) = y''(0) = 0, \quad y^{(3)}(0) = 2$

21. Problem 14; $y(0) = y'(0) = 1, \quad y''(0) = -2$

22. Problem 15; $y(0) = 1, \quad y'(0) = -1, \quad y''(0) = -4, \quad y^{(3)}(0) = 2$

23. Problem 16; $y(0) = -4, \quad y'(0) = 6, \quad y''(0) = -2$

24. Problem 17; $y(-1) = 3, \quad y'(-1) = -2, \quad y''(-1) = 0, \quad y^{(3)}(-1) = 5$

25. Prove part of Theorem 9 by showing that the Wronskian is zero if the solutions $y_1(x), \ldots, y_n(x)$ are linearly dependent.

26. Prove Theorem 10.

27. Prove Theorem 11.

3.2 SOLVING $y^{(n)} + A_{n-1}y^{(n-1)} + \cdots + A_1 y' + A_0 y = 0$

We shall tackle the constant coefficient nth order linear homogeneous equation first. The idea is the same as when $n = 2$, but the execution is complicated by the practical difficulty of finding roots of nth degree polynomials.

Suppose we want to solve

$$y^{(n)} + A_{n-1}y^{(n-1)} + \cdots + A_1 y' + A_0 y = 0,$$

with $A_{n-1}, \ldots, A_1, A_0$ constant. Try for a solution $y = e^{rx}$. Substituting this in and then dividing out e^{rx} gives us the characteristic equation

$$r^n + A_{n-1}r^{n-1} + \cdots + A_1 r + A_0 = 0.$$

This will have n roots, but some may be complex and some may be repeated. In any event, finding them may be difficult, and we do not have the clearly distinguished three cases that occurred when $n = 2$.

EXAMPLE 95

Solve

$$y^{(4)} + 3y^{(3)} - 16y'' + 12y' = 0.$$

The characteristic equation is

$$r^4 + 3r^3 - 16r^2 + 12r = 0,$$

with roots $r = 0, 1, 2$, and -6. Thus, four solutions are

$$e^{0x} = 1, \quad e^x, \quad e^{2x}, \quad \text{and} \quad e^{-6x}.$$

These are linearly independent, and the general solution is

$$y = c_1 + c_2 e^x + c_3 e^{2x} + c_4 e^{-6x}.$$

Suppose we have initial conditions

$$y(0) = 2, \qquad y'(0) = -3, \qquad y''(0) = 4, \qquad y^{(3)}(0) = -8.$$

Then we must have

$$\begin{aligned} y(0) &= c_1 + c_2 + c_3 + c_4 = 2 \\ y'(0) &= c_2 + 2c_3 - 6c_4 = -3 \\ y''(0) &= c_2 + 4c_3 + 36c_4 = 4 \\ y^{(3)}(0) &= c_2 + 8c_3 - 216c_4 = -8. \end{aligned}$$

Solve these to get

$$c_1 = \frac{4256}{672}, \qquad c_2 = \frac{-4224}{672}, \qquad c_3 = \frac{1260}{672}, \quad \text{and} \quad c_4 = \frac{52}{672}.$$

Thus, the unique solution of the initial value problem is

$$y = \tfrac{1}{672}(4256 - 4224e^{x} + 1260e^{2x} + 52e^{-6x}).$$

EXAMPLE 96

Solve the initial value problem

$$y^{(4)} - 5y^{(3)} + 3y'' + 19y' - 30y = 0;$$
$$y(0) = -2, \qquad y'(0) = 0, \qquad y''(0) = 1, \qquad y^{(3)}(0) = 14.$$

The characteristic equation is

$$r^4 - 5r^3 + 3r^2 + 19r - 30 = 0,$$

with roots $r = -2, 3, 2 + i, 2 - i$. Four solutions are

$$e^{-2x}, \quad e^{3x}, \quad e^{(2+i)x} \quad \text{and} \quad e^{(2-i)x}.$$

Recall that

$$e^{(2-i)x} = e^{2x}[\cos(x) + i \sin(x)]$$

and

$$e^{(2+i)x} = e^{2x}[\cos(x) - i \sin(x)].$$

Since linear combinations of solutions are solutions, we use these to get "nicer" solutions

$$e^{2x}\cos(x) \quad \text{and} \quad e^{2x}\sin(x)$$

to be used in place of $e^{(2+i)x}$ and $e^{(2-i)x}$, as in Section 2.5.

The general solution is

$$y = c_1e^{-2x} + c_2e^{3x} + c_3e^{2x}\cos(x) + c_4e^{2x}\sin(x).$$

For the initial conditions, we need

$$\begin{aligned} y(0) &= c_1 + c_2 + c_3 = -2 \\ y'(0) &= -2c_1 + 3c_2 + 2c_3 + c_4 = 0 \\ y''(0) &= 4c_1 + 9c_2 + 3c_3 + 4c_4 = 1 \\ y^{(3)}(0) &= -8c_1 + 27c_2 + 2c_3 + 11c_4 = 14. \end{aligned}$$

Solve these to get

$$c_1 = \frac{-74}{170}, \quad c_2 = \frac{-136}{170}, \quad c_3 = \frac{-130}{170}, \quad \text{and} \quad c_4 = \frac{520}{170}.$$

Thus, the solution of the initial value problem is

$$y = \tfrac{1}{170}[-74e^{-2x} - 136e^{3x} - 130e^{2x}\cos(x) + 520e^{2x}\sin(x)].$$

EXAMPLE 97

Solve

$$y^{(4)} - y = 0.$$

The characteristic equation is $r^4 - 1 = 0$, with only one root, 1 (repeated four times). Thus, our method gives us only one solution, $y = e^x$.

Here we take a cue from case 2 of Section 2.3. The reader should check that xe^x, x^2e^x, and x^3e^x are also solutions. Further, these four solutions are linearly independent. Thus, the general solution is

$$y = e^x(c_1 + c_2x + c_2x^2 + c_3x^3).$$

In general, when the characteristic equation has a real root r, occurring k times as a root, then this root generates solutions

$$e^{rx}, \quad xe^{rx}, \quad x^2e^{rx}, \quad \ldots, \quad x^{k-1}e^{rx},$$

and these are linearly independent.

EXAMPLE 98

Solve

$$y^{(5)} + 2y^{(4)} - 3y^{(3)} - 4y'' + 4y' = 0.$$

The characteristic equation is

$$r^5 + 2r^4 - 3r^3 - 4r^2 + 4r = 0,$$

with roots 1 (twice), -2 (twice), and 0 (once). Thus, the general solution is

$$\begin{aligned} y &= e^x(c_1 + c_2x) + e^{-2x}(c_3 + c_4x) + c_5e^{0x} \\ &= e^x(c_1 + c_2x) + e^{-2x}(c_3 + c_4x) + c_5. \end{aligned}$$

In theory, then, we can solve the linear, constant coefficient, homogeneous nth order equation. If we can overcome the *algebraic* problem of finding the roots of the characteristic equation, then the general solution of

$$y^{(n)} + A_{n-1}y^{(n-1)} + \cdots + A_1 y' + A_0 y = 0$$

is easily written. The Appendix at the end of this chapter addresses the question of finding roots of polynomials.

PROBLEMS FOR SECTION 3.2

In each of Problems 1 through 25, find the general solution.

1. $y^{(5)} - 32y = 0$
2. $y^{(3)} - 3y' + 2y = 0$
3. $y^{(3)} - 4y'' - y' + 4y = 0$
4. $y^{(4)} + 5y'' + 4y = 0$
5. $y^{(4)} - 4y^{(3)} + 7y'' - 8y' + 10y = 0$
6. $y^{(3)} - 8y'' + 5y' + 50y = 0$
7. $y^{(3)} - 2y' + 4y = 0$
8. $y^{(3)} - 8y = 0$
9. $y^{(3)} + 4y'' - 11y' + 6y = 0$
10. $y^{(5)} - 8y^{(3)} + 16y = 0$
11. $y^{(4)} + 3y^{(3)} - 18y'' + 4y' + 24y = 0$
12. $y^{(8)} - 256y = 0$
13. $y^{(4)} - 5y'' + 4y = 0$
14. $y^{(3)} + 4y'' + 4y' + 16 = 0$
15. $y^{(3)} - 6y'' + 21y' - 26y = 0$
16. $y^{(3)} + 4y'' + 11y' + 14y = 0$
17. $y^{(4)} + 2y^{(3)} - 59y'' - 60y' + 900y = 0$
18. $y^{(3)} - y'' + 2y' - 2y = 0$
19. $y^{(3)} - 4y'' - 19y' - 14y = 0$
20. $y^{(3)} + 49y' = 0$
21. $y^{(3)} + 6y'' - 69y' - 154y = 0$
22. $y^{(3)} + 5y'' - 31y' + 21y = 0$
23. $y^{(4)} - 2y^{(3)} - 12y'' - 14y' - 5y = 0$
24. $y^{(3)} - 2y'' - y' + 14y = 0$
25. $y^{(3)} + 6y'' + 6y' + 36y = 0$

In each of Problems 26 through 35, solve the initial value problem.

26. $y^{(3)} - 8y = 0$; $y(0) = 1$, $y'(0) = y''(0) = 0$
27. $y^{(3)} - y'' + 2y' - 2y = 0$; $y(0) = -1$, $y'(0) = 2$, $y''(0) = 5$
28. $y^{(3)} + y'' + 4y' + 4y = 0$; $y(0) = 0$, $y'(0) = y''(0) = 6$
29. $y^{(3)} + 2y'' + 29y' + 148y = 0$; $y(\pi) = y'(\pi) = 0$, $y''(\pi) = 8$
30. $y^{(4)} - 256y = 0$, $y(1) = 0$; $y'(1) = 2$, $y''(1) = 0$, $y^{(3)}(1) = -1$
31. $y^{(3)} - 11y'' + 7y' + 147y = 0$; $y(0) = 4$, $y'(0) = 0$, $y''(0) = 6$
32. $y^{(3)} - 14y'' + 69y' - 90y = 0$; $y(0) = y'(0) = 0$, $y''(0) = -4$
33. $y^{(3)} - 8y'' + y' + 42y = 0$; $y(0) = -2$, $y'(0) = y''(0) = 3$
34. $y^{(3)} - 8y'' + 5y' + 50y = 0$; $y(-2) = 3$, $y'(-2) = 1$, $y''(-2) = -6$
35. $y^{(3)} + 17y'' + 40y' - 300y = 0$; $y(0) = y'(0) = y''(0) = -1$

3.3 SOLVING $y^{(n)} + A_{n-1}y^{(n-1)} + \cdots + A_1y' + A_0 y = F(x)$

We shall now tackle the nth order linear, constant coefficient, nonhomogeneous equation. As we know from Theorem 11, we need only one solution y_p, together with the general homogeneous solution y_h, for the general solution. In the last section we considered how to find y_h. The two methods we consider here for finding y_p are undetermined coefficients and variation of parameters, direct generalizations from the $n = 2$ case.

UNDETERMINED COEFFICIENTS

As in the case $n = 2$, the method consists of formulating a good conjecture for y_p based upon the appearance of $F(x)$. The discussion of Section 2.8 carries over almost verbatim to the higher order case, because higher derivatives of exponentials, polynomials, and sines or cosines remain, respectively, exponentials, polynomials, and sines or cosines. Thus, the table of things to try given in Section 2.8 (p. 117) applies here, with one important exception. Note (4) at the end of the table must be amended to allow for roots of the characteristic equation of multiplicity greater than 2. For the nth order case, the appropriate statement is:

4′. Suppose a term $r(x)$ contained in the proposed y_p is a solution of the homogeneous equation. Find the smallest integer s such that $x^s r(x)$ is not a solution of the homogeneous equation; then try $x^s y_p$ as a proposed solution of the nonhomogeneous equation.

In practice, one can determine s from the characteristic equation of the associated homogeneous equation. Remember that each term in y_h comes from a root of the characteristic equation. For example, if 2 is a root, then e^{2x} is a homogeneous solution; if 2 is a root twice [i.e., $(r - 2)^2$ is a factor of the characteristic equation], then e^{2x} and xe^{2x} are homogeneous solutions. Continuing in this way, if $(r - 2)^k$ is a factor of the characteristic equation, then $e^{2x}, xe^{2x}, \ldots, x^{k-1}e^{2x}$ are homogeneous solutions. In general, then, in Note 4′, s is the largest integer such that (r minus the root)^s is a factor of the characteristic equation, or, equivalently, s is the multiplicity of the root.

EXAMPLE 99

Find the general solution of

$$y^{(3)} - 4y'' + y' + 6y = x^3 - 4x + 2.$$

It is easy to solve the associated homogeneous equation and get

$$y_h = c_1e^{2x} + c_2 e^{-x} + c_3 e^{3x}.$$

Now we want a particular solution y_p of the nonhomogeneous equation. From $F(x) = x^3 - 4x + 2$, we are led to try a solution

$$y_p = Ax^3 + Bx^2 + Cx + D.$$

Substituting this into the differential equation gives us

$$6A - 24Ax - 8B + 3Ax^2 + 2Bx + C + 6Ax^3 + 6Bx^2 + 6Cx + 6D = x^3 - 4x + 2.$$

Equating coefficients of like powers of x on both sides of the equation gives us

$$\begin{aligned} 6A &= 1 && \text{(from } x^3\text{)} \\ 3A + 6B &= 0 && \text{(from } x^2\text{)} \\ -24A + 2B + 6C &= -4 && \text{(from } x\text{)} \\ 6A - 8B + C + 6D &= 2 && \text{(from the constant term).} \end{aligned}$$

Solve these to get

$$A = \tfrac{1}{6}, \qquad B = -\tfrac{1}{12}, \qquad C = \tfrac{1}{36}, \quad \text{and} \quad D = \tfrac{11}{216}.$$

Then,

$$y_p = \frac{x^3}{6} - \frac{x^2}{12} + \frac{x}{36} + \frac{11}{216}.$$

The general solution is

$$y = c_1 e^{2x} + c_2 e^{-x} + c_3 e^{3x} + \frac{x^3}{6} - \frac{x^2}{12} + \frac{x}{36} + \frac{11}{216}.$$

EXAMPLE 100

Find the general solution of

$$y^{(3)} + 2y'' - y' = 4e^x - 3\cos(2x).$$

First find y_h. The characteristic equation for $y^{(3)} + 2y'' - y' = 0$ is $r^3 + 2r^2 - r = 0$, with roots 0, $-1 + \sqrt{2}$, and $-1 - \sqrt{2}$. Thus,

$$y_h = c_1 + c_2 e^{(-1+\sqrt{2})x} + c_3 e^{(-1-\sqrt{2})x}.$$

From $F(x) = 4e^x - 3\cos(2x)$, we guess a particular solution

$$y_p = Ae^x + B\cos(2x) + C\sin(2x).$$

Substituting into the differential equation gives us

$$Ae^x + 8B\sin(2x) - 8C\cos(2x) + 2Ae^x - 8B\cos(2x) - 8C\sin(2x) - Ae^x + 2B\sin(2x) - 2C\cos(2x) = 4e^x - 3\cos(2x).$$

Then

$$2A = 4$$
$$-10C - 8B = -3$$
$$-8C + 10B = 0.$$

Then

$$A = 2, \qquad B = \tfrac{6}{41}, \quad \text{and} \quad C = \tfrac{15}{82}.$$

A particular solution is

$$y_p = 2e^x + \tfrac{6}{41}\cos(2x) + \tfrac{15}{82}\sin(2x).$$

Then the general solution is

$$y = c_1 + e^{-x}(c_2 e^{\sqrt{2}x} + c_3 e^{-\sqrt{2}x}) + 2e^x + \tfrac{6}{41}\cos(2x) + \tfrac{15}{82}\sin(2x).$$

EXAMPLE 101

Find the general solution of

$$y^{(3)} - y'' - 8y' + 12y = 7e^{2x}.$$

We find that the characteristic equation is

$$(r-2)^2(r+3) = 0.$$

This has a double root $r = 2$; hence e^{2x} and xe^{2x} are homogeneous solutions. Thus, instead of trying $y_p = Ae^{2x}$, we try $y_p = Ax^2e^{2x}$. Substituting into the differential equation gives us

$$10A = 7.$$

Hence, $y_p = \frac{7}{10}x^2e^{2x}$.

The general solution is

$$y = c_1e^{-3x} + e^{2x}(c_2 + c_3 x) + \tfrac{7}{10}x^2e^{2x}.$$

EXAMPLE 102

Find the general solution of

$$y^{(4)} + 11y^{(3)} + 36y'' + 16y' - 64y = -3e^{-4x} + 2\cos(2x).$$

The characteristic equation is

$$(r + 4)^3(r - 1) = 0.$$

Hence,

$$y_h = (c_1 + c_2 x + c_3 x^2)e^{-4x} + c_4 e^x.$$

Since -4 is a root of multiplicity 3 of the characteristic equation, we put in y_p the term Ax^3e^{-4x} to correspond to $-3e^{-4x}$ in $F(x)$. Corresponding to $2\cos(2x)$ we put in terms $B\cos(2x) + C\sin(2x)$. Thus, we try

$$y_p = Ax^3e^{-4x} + B\cos(2x) + C\sin(2x).$$

After substitution into the differential equation, equate coefficients of like terms on both sides to get

$$\begin{aligned} -30A &= -3 \\ -192B - 120C &= 2 \\ 56B - 192C &= 0. \end{aligned}$$

Solve these to get

$$A = \frac{1}{10}, \qquad B = \frac{-6}{681}, \quad \text{and} \quad C = \frac{-7}{2724}.$$

The general solution is

$$y = (c_1 + c_2 x + c_3 x^2)e^{-4x} + c_4 e^x + \tfrac{1}{10}x^3e^{-4x} - \tfrac{6}{681}\cos(2x) - \tfrac{7}{2724}\sin(2x).$$

VARIATION OF PARAMETERS

The idea here is a direct generalization from the case $n = 2$ treated in Section 2.8. We consider a differential equation of the form

$$y^{(n)} + P_{n-1}(x)y^{(n-1)} + \cdots + P_1(x)y' + P_0(x)y = F(x), \qquad \text{(a)}$$

in which the coefficient of $y^{(n)}$ is 1 and $P_{n-1}(x), \ldots, P_0(x)$ are functions of x alone. Unlike undetermined coefficients, these coefficients need not be constant.

Suppose that $y_1(x), \ldots, y_n(x)$ are linearly independent solutions of the associated homogeneous equation

$$y^{(n)} + P_{n-1}(x)y^{(n-1)} + \cdots + P_1(x)y' + P_0(x)y = 0.$$

We seek a particular solution of (a) of the form

$$y_p(x) = u_1(x)y_1(x) + u_2(x)y_2(x) + \cdots + u_n(x)y_n(x). \tag{b}$$

The strategy is to obtain n equations for $u_1'(x), \ldots, u_n'(x)$, and then solve and integrate to obtain $u_1(x), \ldots, u_n(x)$. The condition that (b) is to be a solution of (a) will give us one equation. This leaves $(n - 1)$ other conditions for us to impose. We will set these $(n - 1)$ conditions in such a way as to simplify $y_p', \ldots, y_p^{(n-1)}$, giving us $(n - 1)$ additional equations for $u_1', \ldots, u_n'$.

Specifically, begin with

$$y_p' = u_1'y_1 + u_2'y_2 + \cdots + u_n'y_n + u_1y_1' + \cdots + u_ny_n',$$

and set

$$u_1'y_1 + \cdots + u_n'y_n = 0. \tag{1}$$

Then

$$y_p'' = u_1'y_1' + \cdots + u_n'y_n' + u_1y_1'' + \cdots + u_ny_n''.$$

Set

$$u_1'y_1' + \cdots + u_n'y_n' = 0. \tag{2}$$

Continuing, we now have

$$y_p^{(3)} = u_1'y_1'' + \cdots + u_n'y_n'' + u_1y_1^{(3)} + \cdots + u_ny_n^{(3)}.$$

Set

$$u_1'y_1'' + \cdots + u_n'y_n'' = 0. \tag{3}$$

Continue in this way, calculating $y_p^{(4)}, \ldots, y_p^{(n-1)}$, at each step setting the sum of terms involving $u_1', \ldots, u_n'$ equal to zero. This gives us $(n - 1)$ equations (1), (2), (3), ..., $(n - 1)$ for $u_1', \ldots, u_n'$. Finally, calculate $y_p^{(n)}$ and substitute everything into the differential equation, giving one more equation for $u_1', \ldots, u_n'$. The n resulting equations are

$$\begin{aligned}
y_1u_1' + y_2u_2' + \cdots + y_nu_n' &= 0\\
y_1'u_1' + y_2'u_2' + \cdots + y_n'u_n' &= 0\\
y_1''u_1' + y_2''u_2' + \cdots + y_n''u_n' &= 0\\
&\vdots\\
y_1^{(n-2)}u_1' + y_2^{(n-2)}u_2' + \cdots + y_n^{(n-2)}u_n' &= 0\\
y_1^{(n-1)}u_1' + y_2^{(n-1)}u_2' + \cdots + y_n^{(n-1)}u_n' &= F(x).
\end{aligned}$$

These must be solved for $u_1', \ldots, u_n'$, and then each u_j' integrated to yield u_j for $j = 1, 2, \ldots, n$.

EXAMPLE 103

Solve

$$y^{(3)} - 4y'' + y' + 6y = \cos^2(x).$$

As we saw in Example 99, three linearly independent solutions of $y^{(3)} - 4y'' + y' + 6y = 0$ are $y_1 = e^{2x}$, $y_2 = e^{-x}$, and $y_3 = e^{3x}$. We try for a particular solution of the form

$$y_p = u_1e^{2x} + u_2e^{-x} + u_3e^{3x}.$$

The equations for u_1', u_2', and u_3' are

$$\begin{aligned} e^{2x}u_1' + e^{-x}u_2' + e^{3x}u_3' &= 0 \\ 2e^{2x}u_1' - e^{-x}u_2' + 3e^{3x}u_3' &= 0 \\ 4e^{2x}u_1' + e^{-x}u_2' + 9e^{3x}u_3' &= \cos^2(x). \end{aligned}$$

Solve these to get

$$\begin{aligned} u_1' &= -\tfrac{1}{3}e^{-2x}\cos^2(x) \\ u_2' &= \tfrac{1}{12}e^{x}\cos^2(x) \\ u_3' &= \tfrac{1}{4}e^{-3x}\cos^2(x). \end{aligned}$$

Integrate these to get

$$u_1 = \frac{e^{-2x}}{24}[1 - 2\sin(x)\cos(x) + 2\cos^2(x)]$$

$$u_2 = \frac{e^{x}}{60}[2 + 2\sin(x)\cos(x) + \cos^2(x)]$$

$$u_3 = \frac{e^{-3x}}{52}\left[2\sin(x)\cos(x) - 3\cos^2(x) - \frac{2}{3}\right].$$

Thus, a particular solution is

$$\begin{aligned} y_p &= u_1e^{2x} + u_2e^{-x} + u_3e^{3x} \\ &= \tfrac{1}{24}[1 - 2\sin(x)\cos(x) + 2\cos^2(x)] \\ &\quad + \tfrac{1}{60}[2 + 2\sin(x)\cos(x) + \cos^2(x)] \\ &\quad + \tfrac{1}{52}[2\sin(x)\cos(x) - 3\cos^2(x) - \tfrac{2}{3}] \\ &= \tfrac{97}{1560} - \tfrac{3}{260}\sin(x)\cos(x) + \tfrac{17}{195}\cos^2(x). \end{aligned}$$

As one can easily imagine, the labor involved with variation of parameters increases greatly as n becomes larger, and usually the undetermined coefficients method is to be preferred when it can be applied.

PROBLEMS FOR SECTION 3.3

In each of Problems 1 through 20, find the general solution of the differential equation. Use undetermined coefficients to find y_p.

1. $y^{(4)} - y = -4\cosh(2x)$

2. $y^{(3)} - 4y'' + y' + 6y = \cos(3x) - 4e^{-2x} + 6e^{2x}$

3. $y^{(3)} - 2y'' + y' - 2y = x^2 - 2x + 4 - 3\cos(x)$

4. $y^{(3)} - 12y' - 16y = 3x^2 - \cos(4x)$

5. $y^{(3)} + 6y'' + 12y' + 8y = e^x - 3\sin(x) - 8e^{-2x}$
6. $y^{(4)} - 2y^{(3)} - 3y'' + 4y' + 4y = 3x^4 - 2x^3 + x - 5$
7. $y^{(3)} - 4y'' + 20y' = x^2 + 4x - 10$
8. $y^{(3)} + y'' - 14y' - 24y = x^3 - 2\cos(x) + 7e^{4x}$
9. $y^{(3)} + 5y'' + 19y' - 25y = x - \cosh(3x)$
10. $y^{(3)} + 12y'' + 36y' = \sin(2x) - \cos(3x)$
11. $y^{(4)} - 10y^{(3)} + 25y'' = -4$
12. $y^{(3)} - 9y'' + 16y' - 4y = x - 1$
13. $y^{(3)} - 8y'' + 25y' - 26y = -7e^{3x}\cos(2x)$
14. $y^{(3)} - 5y'' - 13y' - 7y = 8x^2 - 4$
15. $y^{(3)} - 7y' - 6y = \sin(2x) - 5$
16. $y^{(3)} - 6y'' + 18y' - 40y = 2\sinh(4x) - x^3 + 2$
17. $y^{(3)} - 4y'' + 13y' + 50y = -4\cos(2x)$
18. $y^{(3)} - y'' - 40y' + 112y = x^2 + 8e^{-3x}$
19. $y^{(3)} + y'' - 4y' + 6y = 7e^{3x} - 2x + 4$
20. $y^{(3)} + 2y'' - 20y' + 24y = 4\cos(3x)$

In each of Problems 21 through 25, find the general solution. Use variation of parameters to find y_p.

21. $y^{(3)} - 2y'' - y' + 2y = \sin^2(x)$
22. $y^{(3)} - 2y'' - y' + 2y = e^{-2x}$
23. $y^{(3)} - 13y' + 12y = \cosh(2x)$
24. $y^{(3)} + 5y'' - 8y' + 48y = 3\sin^2(2x)$
25. $y^{(3)} + 9y'' + 26y' + 24y = x^3$

In each of Problems 26 through 35, solve the initial value problem. Use undetermined coefficients or variation of parameters to find y_p.

26. $y^{(4)} - 16y = 0;\quad y(0) = -2,\quad y'(0) = y''(0) = 0,\quad y^{(3)}(0) = 3$
27. $y^{(3)} + 9y'' + 15y' - 25y = x^2 + 2;\quad y(0) = y'(0) = -3,\quad y''(0) = 1$
28. $y^{(3)} + 6y'' + y' + 6y = -3\sin(x) + 4;\quad y(\pi) = y'(\pi) = 0,\quad y''(\pi) = 6$
29. $y^{(3)} - 6y'' + 25y' = -3e^{-2x};\quad y(0) = 1,\quad y'(0) = 0,\quad y''(0) = -2$
30. $y^{(3)} - 5y'' + y' + 7y = -\cos(3x);\quad y(0) = -2,\quad y'(0) = 1,\quad y''(0) = 0$
31. $y^{(3)} - 4y'' - 3y' + 18y = x - e^{2x};\quad y(0) = -1,\quad y'(0) = y''(0) = 0$
32. $y^{(4)} + 4y^{(3)} + 6y'' + 4y' + y = 3e^{-x};\quad y(0) = y'(0) = y''(0) = 0,\quad y^{(3)}(0) = 1$
33. $y^{(3)} - 5y'' + 9y' - 5y = 5y(0) = y'(0) = y''(0) = 0$
34. $y^{(3)} + 2y'' - 4y' - 8y = x^3 - 3x;\quad y(2) = 1,\quad y'(2) = -3,\quad y''(2) = 0$
35. $y^{(3)} + y'' - 5y' + 3y = 2e^{-3x} + e^x;\quad y(0) = -2,\quad y'(0) = 1,\quad y''(0) = 1$
36. In the variation of parameters method, it is assumed that the n solutions $y_1, \ldots, y_n$ of the homogeneous equation are linearly independent. What would happen in the method if these solutions were not linearly independent?

3.4 *N* TH ORDER EULER TYPE EQUATIONS

Another nth order differential equation one should learn to recognize is the *Euler type equation*

$$x^n y^{(n)} + A_{n-1}x^{n-1}y^{(n-1)} + \cdots + A_1 xy' + A_0 y = 0 \qquad (x > 0),$$

in which the coefficient of the jth derivative is a constant times x^j. When $n = 2$, this is exactly the Euler equation of Section 2.12. As in the case $n = 2$, we try a solution $y = x^r$, substituting into the differential equation to obtain an algebraic equation for r. This algebraic equation is called the *characteristic equation*, and yields values of r for which x^r is a solution.

EXAMPLE 104

Solve

$$x^3y^{(3)} + x^2y'' - 2xy' + 2y = 0.$$

Putting $y = x^r$, we have

$$r(r-1)(r-2)x^r + r(r-1)x^r - 2rx^r + 2x^r = 0.$$

Upon dividing out x^r and doing a little algebra, this becomes

$$r^3 - 2r^2 - r + 2 = 0,$$

the characteristic equation for the given differential equation. The roots are -1, 1, and 2; hence solutions are x^{-1}, x, and x^2, for $x > 0$. The general solution is

$$y = c_1x^{-1} + c_2x + c_2x^2.$$

EXAMPLE 105

Solve

$$x^3y^{(3)} + 9x^2y'' + 19xy' + 8y = 0.$$

Putting $y = x^r$ gives us

$$r^3 + 6r^2 + 12r + 8 = 0.$$

This has one root -2 (three times). Thus, the method has given us just one solution, $y = 1/x^2$.

Taking a cue from the $n = 2$ case when r had repeated roots (Section 2.12, case 2), we try two more solutions of the form

$$(\ln(x))\frac{1}{x^2} \quad \text{and} \quad (\ln(x)^2)\frac{1}{x^2}.$$

If you substitute these into the differential equation, you will find that they work. Further, on $x > 0$ all three solutions are linearly independent. Thus, the general solution is

$$y = c_1\frac{1}{x^2} + c_2\ln(x)\frac{1}{x^2} + c_3[\ln(x)]^2\frac{1}{x^2}, \qquad (x > 0).$$

The last example told us something that is useful to remember:

> If the equation for r has r_1 as a real root k times, then this root yields linearly independent solutions
>
> $$x^{r_1}, \quad \ln(x)x^{r_1}, \quad [\ln(x)]^2x^{r_1}, \quad \ldots, \quad [\ln(x)]^{k-1}x^{r_1}.$$

Complex roots of the characteristic equation are handled as in case 3 in Section 2.12. In particular:

> Corresponding to complex roots $p + iq$ and $p - iq$ of the characteristic equation, one obtains linearly independent solutions
>
> $$x^p \cos[q \ln(x)] \quad \text{and} \quad x^p \sin[q \ln(x)].$$

EXAMPLE 106

Solve

$$x^3y^{(3)} - 5x^2y'' + 18xy' - 26y = 0.$$

Putting $y = x^r$, we get

$$r^3 - 8r^2 + 25r - 26 = 0,$$

with roots 2, $3 - 2i$, and $3 + 2i$. Thus, the general solution is

$$y = c_1x^2 + c_2 x^3 \cos[2 \ln(x)] + c_3 x^3 \sin[2 \ln(x)].$$

As in the case of constant coefficient linear differential equations, Euler type equations represent another instance in which we can reduce the problem of solving the differential equation to one of solving for the roots of a polynomial equation.

PROBLEMS FOR SECTION 3.4

In each of Problems 1 through 20, find the general solution of the given Euler type equation.

1. $x^3y^{(3)} - 3xy' + 3y = 0$
2. $x^3y^{(3)} + 2x^2y'' - xy' + y = 0$
3. $x^3y^{(3)} - 7x^2y'' + 27xy' - 40y = 0$
4. $x^3y^{(3)} - 9x^2y'' + 37xy' - 64y = 0$
5. $x^3y^{(3)} + 4x^2y'' - 6xy' - 12y = 0$
6. $x^3y^{(3)} + 7x^2y'' + 4xy' - 4y = 0$
7. $x^3y^{(3)} - 2x^2y'' - 8xy' + 60y = 0$
8. $x^3y^{(3)} + \frac{1}{2}x^2y'' - \frac{7}{2}xy' + 6y = 0$
9. $x^3y^{(3)} + 12x^2y'' + 37xy' + 27y = 0$
10. $x^3y^{(3)} + x^2y'' + 9xy' = 0$
11. $x^3y^{(3)} + 2x^2y'' + 17xy' + 87y = 0$
12. $x^4y^{(4)} - 2x^3y^{(3)} + 7x^2y'' - 15xy' + 16y = 0$
13. $x^3y^{(3)} - 8x^2y'' + 31xy' - 51y = 0$
14. $x^3y^{(3)} + 8x^2y'' - 2xy' - 48y = 0$
15. $x^3y^{(3)} - 7x^2y'' + 19xy' + 104y = 0$
16. $x^3y^{(3)} + 7x^2y'' - 24xy' + 14y = 0$
17. $x^3y^{(3)} + xy' - y = 0$
18. $x^3y^{(3)} + 4x^2y'' + 3xy' + 26y = 0$
19. $x^3y^{(3)} - 3x^2y'' - 20xy' - 8y = 0$
20. $x^3y^{(3)} - 5x^2y'' - 41xy' + 300y = 0$
21. Show that the characteristic equation for

$$x^3y^{(3)} + A_2x^2y'' + A_1xy' + A_0 y = 0$$

is

$$r^3 + (A_2 - 3)r^2 + (A_1 - A_2 + 2)r + A_0 = 0.$$

22. Show that the characteristic equation for

$$x^4y^{(4)} + A_3x^3y^{(3)} + A_2x^2y'' + A_1xy' + A_0y = 0$$

is

$$r^4 + (A_3 - 6)r^3 + (-3A_3 + A_2 + 11)r^2 + (2A_3 - A_2 + A_1 - 6)r + A_0 = 0.$$

In each of Problems 23 through 35, solve the initial value problem.

23. $x^3y^{(3)} + 2x^2y'' - xy' + y = 0;\quad y(2) = y'(2) = 0,\quad y''(2) = 4$

24. $x^3y^{(3)} - 3x^2y'' + 44xy' + 130y = 0;\quad y(1) = y'(1) = -3,\quad y''(1) = 0$

25. $x^3y^{(3)} + 4x^2y'' - 2xy' + 6y = 0;\quad y(1) = -2,\quad y'(1) = 3,\quad y''(1) = 0$

26. $x^3y^{(3)} - 3x^2y'' + 6xy' - 12y = 0;\quad y(3) = 0,\quad y'(3) = 1,\quad y''(3) = 0$

27. $x^3y^{(3)} + 5x^2y'' + 2xy' - 14y = 0;\quad y(1) = y'(1) = 4,\quad y''(1) = -6$

28. $x^3y^{(3)} + x^2y'' - 20xy' - 30y = 0;\quad y(1) = y'(1) = y''(1) = -3$

29. $x^3y^{(3)} + 4x^2y'' - 8xy' + 8y = 0;\quad y(2) = 1,\quad y'(2) = -2,\quad y''(2) = 2$

30. $x^4y^{(4)} + 26x^3y^{(3)} + 91x^2y'' + 77xy' + 16y = 0;\quad y(1) = y'(1) = 2,\quad y''(1) = y^{(3)}(1) = -2$

31. $x^3y^{(3)} - 9x^2y'' - 11xy' + 256y = 0;\quad y(1) = y'(1) = 0,\quad y''(1) = -5$

32. $x^3y^{(3)} + 5x^2y'' - 14xy' + 42y = 0;\quad y(1) = y'(1) = 0,\quad y''(1) = 7$

33. $x^3y^{(3)} - 2x^2y'' - 20y = 0;\quad y(1) = 1,\quad y'(1) = -3,\quad y''(1) = 4$

34. $x^3y^{(3)} - 8x^2y'' + 55xy' - 123y = 0;\quad y(1) = -3,\quad y'(1) = y''(1) = 2$

35. $x^3y^{(3)} + 4x^2y'' - 39xy' - 105y = 0;\quad y(1) = 0,\quad y'(1) = -1,\quad y''(1) = 4$

36. In Section 2.12 we used the transformation $z = \ln(x)$ to solve Euler equations $x^2y'' + Axy' + By = 0$ for $x > 0$. Will this transformation work if $n > 2$ in the nth order Euler equation?

37. What will be the form of the solutions of an nth order Euler equation corresponding to repeated complex roots of the characteristic equation?

3.5 SUMMARY OF METHODS

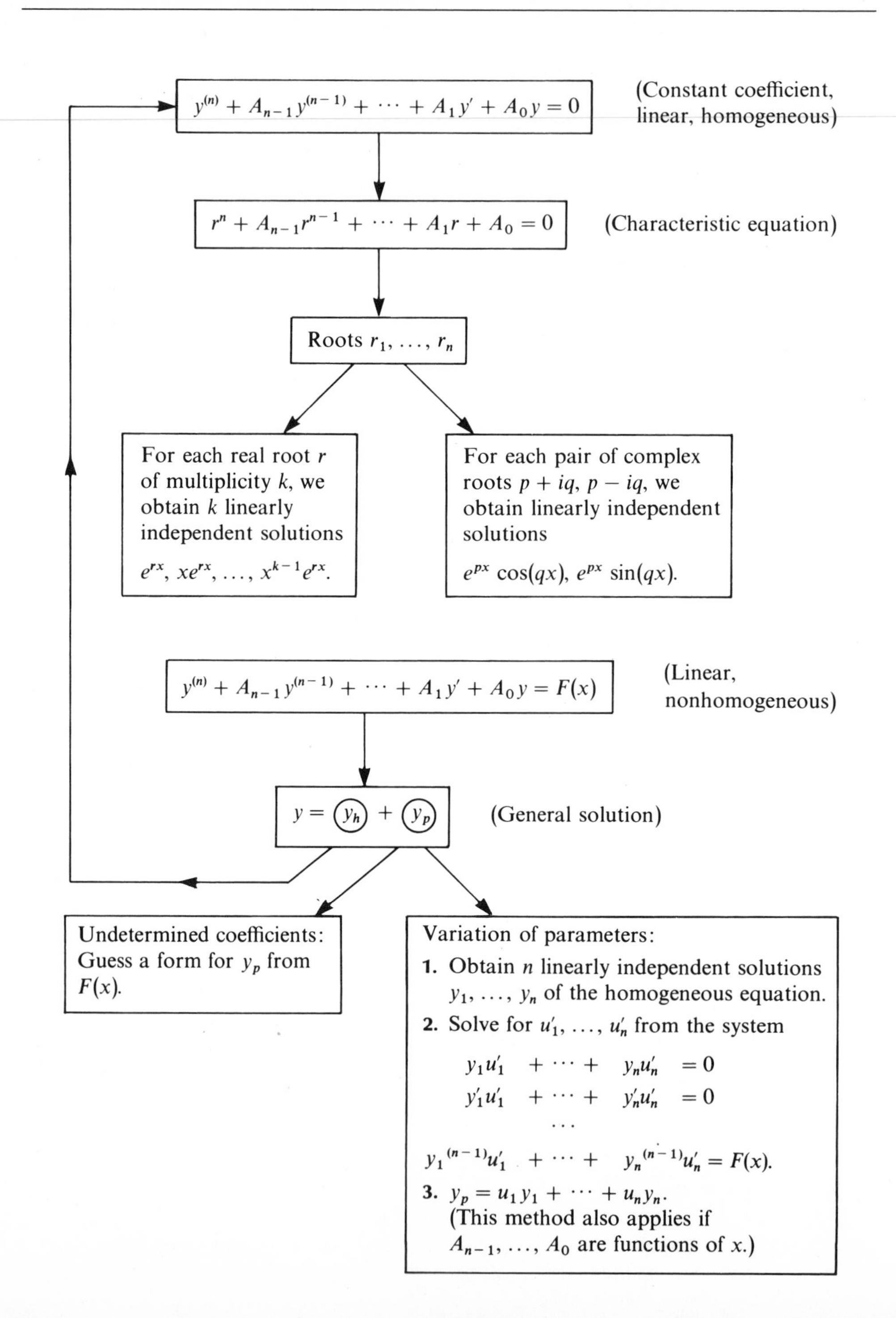

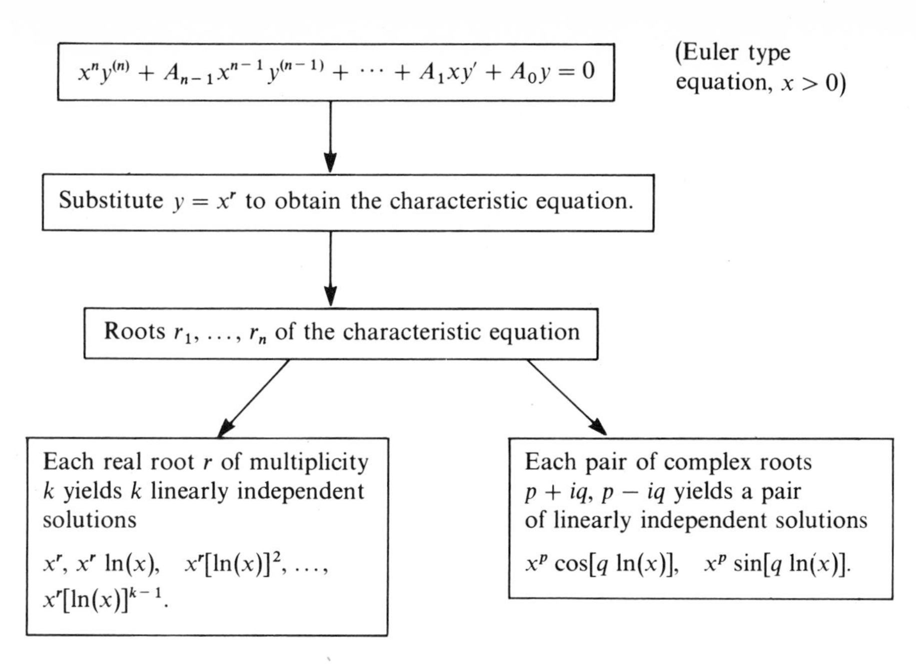

3.6 DIFFERENTIAL OPERATORS

There is a symbolic way of writing differential equations which is sometimes useful. Let D denote the operation of taking a derivative:

$$D = \frac{d}{dx}.$$

Then,

$$D^2 = \frac{d^2}{dx^2}, \qquad D^3 = \frac{d^3}{dx^3},$$

and so on, with D^n denoting the operation of differentiating n times. Often D is called a *differential operator.*

Using the usual rules for differentiating, we can interpret algebraic expressions involving D as instructions for taking derivatives. For example,

$$D^2 - 2D \quad \text{means} \quad \frac{d^2}{dx^2} - 2\frac{d}{dx}.$$

We would then have

$$(D^2 - 2D)(x^3 + 2) = \frac{d^2}{dx^2}(x^3 + 2) - 2\frac{d}{dx}(x^3 + 2) = 6x - 6x^2$$

and

$$(D^2 - 2D)\cos(x) = -\cos(x) + 2\sin(x),$$

and so on.

We can also factor expressions involving D. For example,

$$D^2 - 2D - 8 \quad \text{can be written as} \quad (D + 2)(D - 4).$$

To see that these are in fact the same, observe that they have the same effect on any function of x. We have

$$(D^2 - 2D - 8)y = y'' - 2y' - 8y,$$

while

$$\begin{aligned}(D + 2)(D - 4)y &= (D + 2)(y' - 4y)\\ &= y'' - 4y' + 2y' - 8y = y'' - 2y' - 8y,\end{aligned}$$

which is the same result.

Using this notation, we can often solve differential equations by factoring and looking at each factor separately. Here are two examples.

EXAMPLE 107

Solve

$$y'' - 2y' - 15y = 0.$$

Write this as

$$(D^2 - 2D - 15)y = 0$$

or

$$(D - 5)(D + 3)y = 0.$$

Solutions of this are the same as solutions of $(D - 5)y = 0$ or $(D + 3)y = 0$, giving us $y_1 = e^{5x}$ and $y_2 = e^{-3x}$.

EXAMPLE 108

Solve

$$y^{(3)} - y'' - 10y' - 8y = 0.$$

This can be written

$$(D - 4)(D + 2)(D + 1)y = 0.$$

Solutions are then solutions of $(D - 4)y = 0$, $(D + 2)y = 0$, or $(D + 1)y = 0$, giving us e^{4x}, e^{-2x}, and e^{-x}.

At this point in our treatment, the differential operator notation does not make a significant contribution to our ability to solve differential equations, since factoring the differential equation into operator factors is the same as factoring the characteristic equation. However, differential operators will provide us in Section 7.1 with an important method for solving certain systems of differential equations. They are also important in a variety of advanced topics in differential equations.

PROBLEMS FOR SECTION 3.6

In each of Problems 1 through 10, apply the given operator to the given function.

1. $D^3 - 3D^2;\quad -3\cos(2x)$
2. $D^2 + 4D - 1;\quad x^2 - 4x$
3. $D^2 + 4D - 2;\quad \sin(2x) - 3x$
4. $3D^4 + 2D - 5;\quad e^{-3x} + 2x^3$
5. $(D - 2)^2;\quad 8x^4 + 2x$
6. $(D - 3)(D + 2);\quad \sin(4x) - x$
7. $(2D - 3)(D^2 + 3D - 4);\quad x^4 + 2x - e^x$
8. $8D^3 + 1;\quad \ln(x) - 2$
9. $(D - 3)^4;\quad -3e^{4x}$
10. $(D^2 + 2)(D^2 - 1);\quad x^4 + \cos(3x)$

In each of Problems 11 through 20, find a fundamental system of solutions by writing the differential equation in operator notation and factoring the operator.

11. $2y'' + 11y' + 12y = 0$
12. $y^{(3)} + 3y'' - 4y' - 12y = 0$
13. $y'' - 4y' - 12y = 0$
14. $y^{(3)} + 6y'' - 9y' - 54y = 0$
15. $y'' - 2y' - 35y = 0$
16. $6y'' - 17y' - 14y = 0$
17. $8y'' - 15y' - 2y = 0$
18. $8y'' - 14y' - 15y = 0$
19. $2y^{(3)} + 5y'' - y' - 6y = 0$
20. $9y'' - 4y = 0$
21. Show that

$$(D + 2)(D^2 + 3D + 1) = (D^2 + 3D + 1)(D + 2)$$

in the sense that both sides produce the same result when applied to a function $y(x)$.

22. Show that

$$(D^3 + 3D^2 - 4D + 3)(D^2 - 3D + 1) = (D^2 - 3D + 1)(D^3 + 3D^2 - 4D + 3).$$

23. Using the experience gained from Problems 21 and 22, show that polynomial expressions involving D commute under multiplication. That is, show that

$$(a_n D^n + a_{n-1}D^{n-1} + \cdots + a_1 D + a_0)(b_m D^m + b_{m-1}D^{m-1} + \cdots + b_1 D + b_0)$$
$$= (b_m D^m + b_{m-1}D^{m-1} + \cdots + b_1 D + b_0)(a_n D^n + a_{n-1}D^{n-1} + \cdots + a_1 D + a_0).$$

This result is important in Section 7.1 in applying differential operators to the solution of systems of differential equations.

CHAPTER SUMMARY

This chapter was the natural continuation of Chapter 2 to the higher order case, and also marked the end of the first stage of studying differential equations. In this stage we became generally familiar with the idea of a differential equation and mastered various techniques for solving standard kinds of differential equations which one sees fairly often in practice. We also developed some of the theory behind these methods, and gained some experience in initial value problems which play a central role in applications.

In the next stage we begin to broaden our capabilities by introducing methods which will give us new perspective on problems we already know how to solve and also enable us to solve problems beyond our current means. We begin this program in the next chapter with Laplace transforms, followed by infinite series methods.

APPENDIX: ROOTS OF POLYNOMIALS

In solving nth order, linear, constant coefficient, homogeneous differential equations, or Euler type differential equations, success or failure hinges on our ability to solve for the roots of the polynomial equation

$$r^n + A_{n-1}r^{n-1} + \cdots + A_1 r + A_0 = 0.$$

When $n = 2$, this is a quadratic equation

$$r^2 + A_1 r + A_0 = 0,$$

with roots

$$r = \frac{-A_1 \pm \sqrt{A_1^2 - 4A_0}}{2}.$$

When $n = 3$, there is also a formula for the roots, although it is more complicated.

To solve $r^3 + A_2 r^2 + A_1 r + A_0 = 0$, first put

$$x = y - \frac{A_2}{3}$$

to obtain

$$y^3 + ay + b = 0,$$

where

$$a = \tfrac{1}{3}(3A_1 - A_2^2) \quad \text{and} \quad b = \tfrac{1}{27}(2A_2^3 - 9A_1A_2 + 27A_0).$$

The solutions of $y^3 + ay + b = 0$ are

$$y = \alpha + \beta, \qquad \frac{-(\alpha + \beta)}{2} + \frac{(\alpha - \beta)}{2}\sqrt{3}\,i, \quad \text{and} \quad \frac{-(\alpha + \beta)}{2} - \frac{(\alpha - \beta)}{2}\sqrt{3}\,i,$$

where

$$\alpha = \left\{\frac{-b}{2} + \sqrt{\frac{b^2}{4} + \frac{a^3}{27}}\right\}^{1/3}$$

and

$$\beta = \left\{\frac{-b}{2} - \sqrt{\frac{b^2}{4} + \frac{a^3}{27}}\right\}^{1/3}$$

Having obtained the roots for y, we obtain those for x by recalling that $x = y - (A_2/3)$.

Likewise, there is a formula for the roots when $n = 4$, but we omit it. For $n \geq 5$, there is no general formula like the above which gives the roots in every case.

How do we proceed when faced with solving

$$r^n + A_{n-1}r^{n-1} + \cdots + A_1 r + A_0 = 0,$$

where $n \geq 3$?

First, if we can guess, or find by trial and error, one root, say α, then $r - \alpha$ divides the polynomial, resulting in a polynomial of lower degree which might be easier to solve. For example, suppose we want to solve

$$r^3 + 78.42r^2 - 246.22r + 5.88 = 0.$$

One root is $r = 3$. We can then divide the given polynomial by $r - 3$:

$$\begin{array}{r|l}
 & r^2 + 81.42r - 1.96 \\ \hline
r - 3 & r^3 + 78.42r^2 - 246.22r + 5.88 \\
 & r^3 - 3.00r^2 \\ \cline{2-2}
 & \quad 81.42r^2 - 246.22r \\
 & \quad 81.42r^2 - 244.26r \\ \cline{2-2}
 & \qquad -1.96r + 5.88 \\
 & \qquad -1.96r + 5.88 \\ \cline{2-2}
\end{array}$$

Then,

$$r^3 + 78.42r^2 - 246.22r + 5.88 = (r - 3)(r^2 + 81.42r - 1.96)$$

and the other roots are roots of

$$r^2 + 81.42r - 1.96 = 0,$$

a quadratic we easily solve to get

$$\frac{-81.42 \pm \sqrt{6637.0564}}{2}.$$

Thus, the roots of our third degree polynomial are

$$3, \qquad 0.0240655963, \quad \text{and} \quad -81.4440656$$

(to 7 decimals). The last two roots are, of course, approximations incurred in approximating $\sqrt{6637.0564}$ by 81.46813119.

What do we do if we cannot find enough roots to divide out factors and simplify the problem? Then we must go to numerical techniques—for example, Newton's method for approximating roots of polynomials. This will be treated in the chapter on numerical analysis.* One can also solve for the roots of many polynomials using hand-held calculators. For example, programs are easily written for the Texas Instruments TI-58 and TI-59 models, and the Hewlett-Packard HP-34C has a key called SOLVE which will do the job. The HP-41C also has a program in its Math Pac called ROOTS which finds roots (real or complex) of polynomials of degree five or lower.

---

* Nowadays one can also find Newton's method in some elementary calculus texts. See Thomas, G. B., and Finney, R. L. 1980. *Calculus and analytic geometry*. 5th ed. Reading, Mass.: Addison-Wesley. Also, Swokowski, E. W. 1976. *Calculus with analytic geometry*. 2d ed. Boston: Prindle, Weber and Schmidt.

SUPPLEMENTARY PROBLEMS

In each of Problems 1 through 20, solve the initial value problem. Use any methods of this chapter which seem appropriate to the problem.

1. $y^{(3)} + 18y'' + 108y' + 216y = x - 4;\quad y(0) = y'(0) = 1,\quad y''(0) = -2$
2. $x^3y^{(3)} + 21x^2y'' + 127xy' + 216y = 0;\quad y(1) = y'(1) = 0,\quad y''(1) = -3$
3. $y^{(3)} + 5y'' + 3y' - 9y = 4e^{-3x} - e^x;\quad y(0) = 2,\quad y'(0) = y''(0) = -1$
4. $x^4y^{(4)} + 12x^3y^{(3)} + 148x^2y'' + 216xy' - 216y = 0;\quad y(1) = y'(1) = 0,\quad y''(1) = 2,\quad y^{(3)}(1) = 0$
5. $y^{(3)} - 8y'' + 85y' - 146y = 146;\quad y(0) = y'(0) = -3,\quad y''(0) = 4$
6. $x^3y^{(3)} - 6x^2y'' + 65xy' - 65y = 0;\quad y(1) = -1,\quad y'(1) = 2,\quad y''(1) = -5$
7. $y^{(3)} - 3y' + 2y = 4e^x - 3;\quad y(-1) = 1,\quad y'(-1) = 4,\quad y''(-1) = -3$
8. $y^{(4)} + 2y^{(3)} - 11y'' - 12y' + 36y = 5\sin(3x);\quad y(0) = 1,\quad y'(0) = -4,$ $y''(-1) = 5,\quad y^{(3)}(-1) = 0$
9. $y^{(3)} - y'' + y' - y = 4x;\quad y(\pi) = 1,\quad y'(\pi) = y''(\pi) = 0$
10. $x^3y^{(3)} - 8x^2y'' + 23xy' - 35y = 0;\quad y(1) = y'(1) = y''(1) = -3$
11. $y^{(3)} - 6y'' + 45y' - 148y = 0;\quad y(\pi/6) = -3,\quad y'(\pi/6) = y''(\pi/6) = 0$
12. $y^{(3)} - y'' - 12y' - 40y = \sin^2(x);\quad y(-2) = y'(-2) = 3,\quad y''(-2) = -1$
13. $x^3y^{(3)} - 3x^2y'' + 7xy' - 8y = 0;\quad y(1) = -5,\quad y'(1) = y''(1) = 0$
14. $y^{(3)} + 3y'' - 14y' + 24y = 14;\quad y(2) = -7,\quad y'(2) = 1,\quad y''(2) = 5$
15. $x^4y^{(4)} + 2x^3y^{(3)} + 10x^2y'' - 10xy' + 10y = 0;\quad y(1) = -4,\quad y'(1) = 2,\quad y''(1) = 7,\quad y^{(3)}(1) = 0$
16. $y^{(3)} + 5y'' - 8y' - 48y = \sin(x) - 3;\quad y(0) = 2,\quad y'(0) = -4,\quad y''(0) = 0$
17. $y^{(3)} - 6y'' - 11y' + 116y = 3e^{-4x};\quad y(0) = y'(0) = y''(0) = 0$
18. $y^{(3)} - 6y'' - 59y' + 424y = 9x^2 + 2;\quad y(1) = -5,\quad y'(1) = 1,\quad y''(1) = 6$
19. $y^{(4)} - 2y^{(3)} + 7y'' - 4y' + 10y = 4;\quad y(0) = y'(0) = y''(0) = y^{(3)}(0) = 0$
20. $x^4y^{(4)} + 4x^3y^{(3)} + 8x^2y'' + 2xy' + 10y = 0;\quad y(1) = -5,\quad y'(1) = 2,\quad y''(1) = 6,\quad y^{(3)}(1) = 2$
21. Show that the characteristic equation for the Euler type equation of order n is

$$r(r-1)(r-2)\cdots(r-n+1) + A_1r(r-1)\cdots(r-n+2) + \cdots + A_1r + A_0 = 0.$$

22. Use the Wronskian test (Theorem 9) to show that e^{2x}, $\sin(x)$, and 1 are linearly independent on any interval. *Note:* First produce a differential equation they satisfy.
23. Use the Wronskian test to show that e^{ax}, e^{bx}, e^{cx}, and e^{dx} are linearly independent on any interval, provided that a, b, c, and d are distinct real numbers. *Note:* First you must produce a differential equation which they satisfy.
24. Make the change of variable $x = e^z$ in $x^3y^{(3)} + A_2x^2y'' + A_1xy' + A_0y = 0$ to obtain a third order, linear, constant coefficient differential equation. Use this method to solve the differential equations in Problems 2, 6, 10, and 13 above.
25. Consider the third order differential equation $y^{(3)} + A_2y'' + A_1y' + A_0y = 0$. We call this equation *asymptotically stable* if every solution tends to zero as x goes to infinity. This is important in many engineering and physics applications, in which real-world experience dictates that certain phenomena should decay to zero with time. Prove that the above differential equation is asymptotically stable if A_0, A_2, and $A_1A_2 - A_0$ are positive.

In each of Problems 26 through 30, apply the operator to the function.

26. $(3D^3 + 2D^2 + 1);\quad \ln[\cos(2x)]$
27. $(D^4 + 1);\quad \sin(3x)$
28. $(8D^2 - 4D + 2);\quad x^3 + 5x^2$
29. $(5D^3 + 2D^2);\quad x^4 + 2x - 1$
30. $(2D^2 - 4D + 5);\quad x^5 - e^{3x}$

In each of Problems 31 through 35, factor the differential operator associated with the given differential equation and find a fundamental system of solutions.

31. $2y'' - 5y' - 12y = 0$

32. $y'' - y' - 42y = 0$

33. $y'' - 5y' - 24y = 0$

34. $y^{(3)} + y'' - 34y' + 56y = 0$

35. $y^{(3)} - 8y'' + 5y' + 14y = 0$

CHAPTER FOUR

Laplace Transforms

4.0 INTRODUCTION

Despite the fact that we can now solve some important differential equations, it is easy to imagine situations we cannot yet handle. For example, we can solve for the motion of a spring driven by a periodic or constant force, but what if the force were, say, an impulse? Such a driving force would also be interesting as an electromotive force in an *RLC* circuit. Methods treated up to this point require continuity of $F(x)$, however, and so do not apply. In this section we develop one tool useful for treating these and a variety of other problems. The method is based on the notion of the Laplace transform.

Broadly speaking, a *transform* is a device which changes objects from one form to another. The purpose of such a change is that in a transformed state the object may be easier to work with. Or, a problem may transform into another problem we already know how to solve.

In this broad sense we encounter transforms all the time. For example, consider the problem of cooling a glass of soda. One way is to pour in cold water, but most people do not do this. A more popular way is to use a mold and freezer to transform the water into ice cubes. Another example is the transformation of natural gas to a liquefied state to facilitate transportation.

In mathematics, "transform" usually refers to a device which changes one kind of function or equation into another kind. One attempts to design transforms which change problems we do not know how to solve, or which are difficult, into problems which are in some substantial way easier to solve. This technique has proved very effective in solving differential equations, and many different transforms have been invented. In this chapter we study one of them, the Laplace transform. Eventually we shall apply the Laplace transform to electrical circuits and mechanical systems more complex than those we could handle in the preceding chapters. This will require some ground work, which we now begin.

4.1 DEFINITION OF THE LAPLACE TRANSFORM

The *Laplace transform* $\mathscr{L}[f(t)]$ of a function $f(t)$ is defined to be

$$\mathscr{L}[f(t)] = \int_0^\infty e^{-st} f(t)\, dt = F(s),$$

whenever this integral exists. The integration variable is t; hence the integral defines a function of the new variable s. In order to help keep things straight, we shall usually denote the Laplace transform of a function written in lowercase by the same letter in uppercase. Thus,

$$\mathscr{L}[g(t)] = \int_0^\infty e^{-st} g(t)\, dt = G(s),$$

$$\mathscr{L}[h(t)] = \int_0^\infty e^{-st} h(t)\, dt = H(s),$$

and so on. We shall also customarily use t as the variable of our original function, and s as the variable of its Laplace transform.

The *inverse Laplace transform* of $F(s)$ is a function $f(t)$ such that $\mathscr{L}[f(t)] = F(s)$. If we denote the operation of taking a Laplace transform by $\mathscr{L}$, and of taking an inverse Laplace transform by $\mathscr{L}^{-1}$, then

$$\mathscr{L}[f(t)] = F(s) \quad \text{implies} \quad \mathscr{L}^{-1}[F(s)] = f(t)$$

and, conversely,

$$\mathscr{L}^{-1}[F(s)] = f(t) \quad \text{implies} \quad \mathscr{L}[f(t)] = F(s).$$

In practice, then, we operate algebraically with the symbols $\mathscr{L}$ and $\mathscr{L}^{-1}$, bringing them from one side of an equation to the other just as we would in writing "$ax = b$ implies $x = a^{-1}b$."

EXAMPLE 109

The Laplace transform of $f(t) = t$ is

$$\mathscr{L}(t) = \int_0^\infty e^{-st} t\, dt.$$

To work out the integral on the right, recall that by definition

$$\int_0^\infty A(t)\, dt = \lim_{k \to \infty} \int_0^k A(t)\, dt,$$

whenever this limit exists. Thus, we must first work out

$$\int_0^k e^{-st}t\,dt = \left[\frac{e^{-st}}{s^2}(-st-1)\right]_0^k$$

$$= \frac{e^{-sk}}{s^2}(-sk-1) - \frac{1}{s^2}(-1)$$

$$= \frac{1}{s^2} - \frac{e^{-sk}}{s^2}(1+sk).$$

If $s > 0$, this approaches $1/s^2$ as $k \to \infty$. Thus,

$$\mathscr{L}(t) = \frac{1}{s^2}, \qquad (s > 0).$$

It is easy to show by repeated integration that

$$\mathscr{L}(t^n) = \frac{n!}{s^{n+1}}$$

for any positive integer n. By the same token,

$$\mathscr{L}^{-1}\left(\frac{n!}{s^{n+1}}\right) = t^n.$$

EXAMPLE 110

Let a be constant. We want the Laplace transform of $\cos(at)$.

By definition,

$$\mathscr{L}[\cos(at)] = \int_0^\infty e^{-st}\cos(at)\,dt$$

for all values of s such that this integral exists.

We find that

$$\int_0^k e^{-st}\cos(at)\,dt = \left[\frac{e^{-st}[-s\cos(at) + a\sin(at)]}{s^2+a^2}\right]_0^k$$

$$= \frac{e^{-sk}}{s^2+a^2}[-s\cos(ak) + a\sin(ak)] + \frac{s}{s^2+a^2}.$$

As $k \to \infty$, $e^{-sk}[-s\cos(ak) + a\sin(ak)] \to 0$ if $s > 0$. Thus, for $s > 0$,

$$\mathscr{L}[\cos(at)] = \frac{s}{s^2+a^2}.$$

Conversely,

$$\mathscr{L}^{-1}\left(\frac{s}{s^2+a^2}\right) = \cos(at).$$

We include at the end of the chapter a table of Laplace transforms of commonly encountered functions to save the trouble of integrating each time.

Here are two important properties of Laplace transforms.

THEOREM 12

1. $$\mathscr{L}[f(t) + g(t)] = \mathscr{L}[f(t)] + \mathscr{L}[g(t)]$$

   whenever all three Laplace transforms exist.

2. For any real number a,

   $$\mathscr{L}[af(t)] = a\mathscr{L}[f(t)]$$

   whenever both sides exist.

In words, the Laplace transform of a sum is the sum of the Laplace transforms; and the Laplace transform of a constant times a function is the constant times the Laplace transform of the function.

Proof Both (1) and (2) follow easily from properties of the integral. For (1),

$$\mathscr{L}[f(t) + g(t)] = \int_0^\infty e^{-st}[f(t) + g(t)]\,dt$$

$$= \int_0^\infty e^{-st}f(t)\,dt + \int_0^\infty e^{-st}g(t)\,dt = \mathscr{L}[f(t)] + \mathscr{L}[g(t)],$$

under the assumption that all three integrals involved exist. We leave (2) to the student.

The inverse version of Theorem 12 is the following:

THEOREM 13

1. $$\mathscr{L}^{-1}[F(s) + G(s)] = \mathscr{L}^{-1}[F(s)] + \mathscr{L}^{-1}[G(s)]$$

2. $$\mathscr{L}^{-1}[aF(s)] = a\mathscr{L}^{-1}[F(s)]$$

Theorems 12 and 13 are useful in reading Laplace and inverse Laplace transforms from the table when the function involved is a fairly simple combination of functions covered in the table. Here are some examples.

EXAMPLE 111

Find

$$\mathscr{L}[4\sinh(3t) - 18e^{-5t}].$$

Using Theorem 12 and the table, we write this as

$$4\mathscr{L}[\sinh(3t)] - 18\mathscr{L}(e^{-5t}) = 4\left(\frac{3}{s^2 - 9}\right) - 18\left(\frac{1}{s + 5}\right).$$

The answer can be left like this, or the fractions can be combined to yield

$$\frac{-18s^2 + 12s + 222}{(s^2 - 9)(s + 5)}.$$

EXAMPLE 112

Find

$$\mathscr{L}(t^3 - 8t^2 + 1).$$

This is

$$\begin{aligned}\mathscr{L}(t^3) - 8\mathscr{L}(t^2) + \mathscr{L}(1) &= \frac{3!}{s^4} - 8\left(\frac{2!}{s^3}\right) + \frac{1}{s} \\ &= \frac{6}{s^4} - \frac{16}{s^3} + \frac{1}{s} \\ &= \frac{s^3 - 16s + 6}{s^4}.\end{aligned}$$

EXAMPLE 113

Find

$$\mathscr{L}^{-1}\left(\frac{4}{s^2 + 22}\right).$$

This is

$$4\mathscr{L}^{-1}\left(\frac{1}{s^2 + 22}\right) = \frac{4}{\sqrt{22}}\mathscr{L}^{-1}\left(\frac{\sqrt{22}}{s^2 + (\sqrt{22})^2}\right) = \frac{4}{\sqrt{22}}\sin(\sqrt{22}\,t).$$

EXAMPLE 114

Find

$$\mathscr{L}^{-1}\left(\frac{-3s}{s^2 + 94} + \frac{12}{s - 5}\right).$$

This is

$$\begin{aligned}&-3\mathscr{L}^{-1}\left(\frac{s}{s^2 + 94}\right) + 12\mathscr{L}^{-1}\left(\frac{1}{s - 5}\right) \\ &= -3\mathscr{L}^{-1}\left(\frac{s}{s^2 + (\sqrt{94})^2}\right) + 12\mathscr{L}^{-1}\left(\frac{1}{s - 5}\right) \\ &= -3\cos(\sqrt{94}\,t) + 12e^{5t}.\end{aligned}$$

These examples are simple in the sense that one can, with a minimal amount of algebra, refer the problem directly to the tables. The next three sections are devoted to finding Laplace and inverse Laplace transforms in more complicated circumstances.

Theorem 12 asserted two properties of the Laplace transform under the assumption that all the Laplace transforms involved existed. We shall briefly take up this existence question by considering conditions on $f(t)$ sufficient for $\mathscr{L}[f(t)]$ to exist.

Begin by examining the definition

$$\mathscr{L}[f(t)] = \int_0^{\infty} f(t)e^{-st}\,dt = \lim_{k\to\infty} \int_0^{k} f(t)e^{-st}\,dt.$$

This suggests two ways that $\mathscr{L}[f(t)]$ can fail to exist: (1) $\int_0^k f(t)e^{-st}\,dt$ may fail to exist for some values of k; or (2) the limit on the right may fail to exist or be infinite. These difficulties can be overcome by placing two restrictions on $f(t)$. For the first we need a definition.

We call $f(t)$ *piecewise continuous* on $[a, b]$ if $[a, b]$ can be divided into finitely many subintervals so that $f(t)$ is continuous within each subinterval, and has a finite limit at each end of each subinterval as t approaches the endpoint from within the subinterval. Thus, $f(t)$ must be continuous on $[a, b]$ except possibly at finitely many points, at each of which $f(t)$ experiences at worst a finite jump. An example of a piecewise continuous function defined on the interval $[0, 4]$ is shown in Figure 30, which is the graph of

$$f(t) = \begin{cases} t^2, & 0 \le t \le 2 \\ 1, & 2 < t \le 3 \\ -1, & 3 < t \le 4. \end{cases}$$

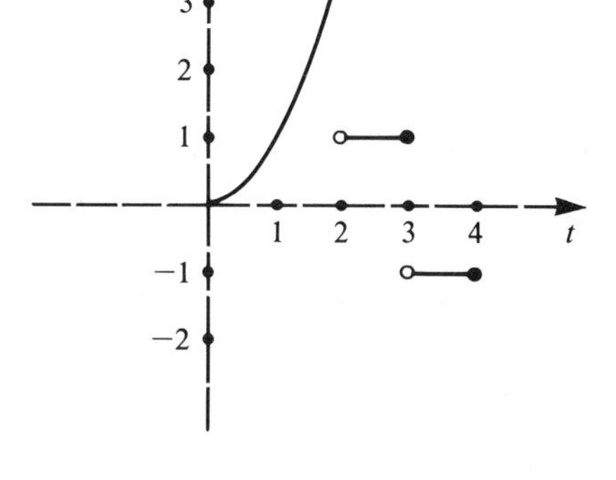

Figure 30. Graph of a piecewise continuous function on [0, 4].

Here $f(t)$ is continuous on (0, 2), (2, 3), and (3, 4), and has finite limits from the left at 2, 3, and 4, and from the right at 0, 2, and 3.

Note that $\int_a^b f(t)\,dt$ exists if $f(t)$ is piecewise continuous on $[a, b]$, since $f(t)$ is then bounded on $[a, b]$ and $\int_a^b f(t)\,dt$ can be written as a sum of integrals of $f(t)$ over subintervals, on each of which $f(t)$ is continuous except possibly at the end points. Further, since e^{-st} is continuous everywhere, $\int_a^b f(t)e^{-st}\,dt$ exists if $f(t)$ is piecewise continuous on $[a, b]$.

There are several ways of addressing the question of convergence of $\int_0^\infty f(t)e^{-st}\,dt$. One condition which is commonly used is to bound $f(t)$ by some function which ensures that $f(t)e^{-st}$ does not grow "too fast" with increasing t. In particular, if for some constants M, b, and t_0, we have

$$|f(t)| \le Me^{bt} \quad \text{for} \quad t \ge t_0,$$

then $|f(t)e^{-st}| \le Me^{(b-s)t}$ for $t \ge t_0$. Convergence of $\int_0^\infty Me^{(b-s)t}\,dt$ for $s > b$ suggests that bounds of the form $|f(t)| \le Me^{bt}$ for t sufficiently large are useful to look at.

Such a bound is easily found for many functions commonly encountered. For example, if $f(t) = t^2$, then we can choose $M = b = 1$ and have $t^2 \le e^t$ for t sufficiently large, since $\lim_{t\to\infty} (t^2/e^t) = 0$. By contrast, e^{t^2} does not have a bound of this form, since e^{t^2} grows faster than Me^{bt} for any choices of M and b.

We can now state an existence theorem.

THEOREM 14

Suppose that $f(t)$ is piecewise continuous on every finite interval $[0, k]$ with $k > 0$. Suppose also that there are constants M, b, and t_0 such that $|f(t)| \leq Me^{bt}$ for $t \geq t_0$. Then, $\mathscr{L}[f(t)]$ exists for $s > b$.

Proof First, the piecewise continuity of $f(t)$ on every interval $[0, k]$ ensures that $\int_0^k f(t)e^{-st}\,dt$ exists for every $k > 0$. Further, $\int_0^\infty Me^{(b-s)t}\,dt$ converges for $s > b$, implying convergence of $\int_0^\infty f(t)e^{-st}\,dt$ for $s > b$.

The hypotheses of Theorem 14 determine a fairly large class of functions for which the Laplace transform will exist. However, these hypotheses do not constitute necessary conditions, only sufficient ones; and it is possible for $\mathscr{L}[f(t)]$ to exist even if one or both of these hypotheses fails. A standard example is $f(t) = t^{-1/2}$, which is not sectionally continuous on any interval $[0, k]$ because of its behavior at zero. Even so, $\int_0^k t^{-1/2}\,dt$ exists for any $k > 0$. Furthermore, one can show (see Problem 51 at the end of this section) that in fact

$$\begin{aligned}\mathscr{L}(t^{-1/2}) &= 2s^{-1/2}\int_0^\infty e^{-z^2}\,dz \qquad (s > 0)\\ &= 2s^{-1/2}\frac{\sqrt{\pi}}{2}\\ &= \left(\frac{\pi}{s}\right)^{1/2} \qquad (s > 0).\end{aligned}$$

In connection with Theorem 14, there is some terminology which the student may see in the literature, and which we will mention but not make much use of. If, for some constants M, b, and t_0, we have $|f(t)| \leq Me^{bt}$ for $t \geq t_0$, then we say that $f(t)$ is of *exponential order* as $t \to \infty$. Sometimes this is written $f(t) = O(e^{bt})$ as $t \to \infty$. For an example of the use of this notation, and a good introduction to the subject, see Rainville, E. 1965. *The Laplace transform.* New York: Macmillan.

In concluding this section, we mention a rationale for investing some effort in learning how to manipulate Laplace transforms. We shall see (beginning with Example 116 of the next section) that the Laplace transform changes an initial value problem into an algebra problem. The strategy is to solve the algebra problem (presumably an easier task than solving the initial value problem), and then use the inverse Laplace transform to retrieve the solution of the initial value problem. Schematically, the situation is as follows:

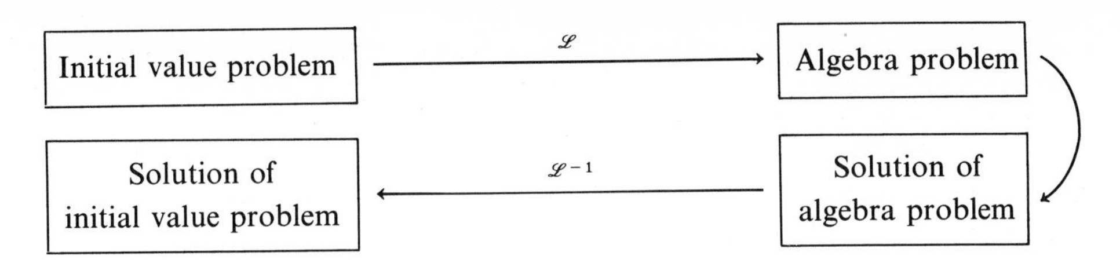

Later this will provide a powerful tool for solving important problems which are beyond the methods of Chapters 1 and 2.

PROBLEMS FOR SECTION 4.1

In each of Problems 1 through 25, compute $\mathscr{L}[f(t)]$ for the given $f(t)$. Use the table at the end of the chapter, in combination with Theorem 12, whenever possible. If you have to use the definition and perform integrations, use integral tables.

1. $2\sinh(3t) - 4$
2. $\cos(t) - \sin(3t)$
3. $4t\sin(2t)$
4. $t^2 - 3t + 5$
5. $t - \cos(5t)$
6. $\cos^2(2t)$
Hint: $\cos^2(2t) = \frac{1}{2}[1 + \cos(4t)]$
7. $2t^2e^{-3t} - 4t + 1$
8. $3e^{-t} + \sin(6t)$
9. $t\sin(3t) - 1 + 2t^2$
10. $(t + 4)^2$
11. $4t^3e^{-t}$
12. $\sinh(4t) - t$
13. $7t^2 - te^{-4t}$
14. $2e^{-t}\cos(3t)$
15. $4t^3\sin(t)$
16. $\frac{1}{4}\sin^3(t) - t + 2$
17. $2\cosh^2(3t)$
18. $5t - 8 + te^{-3t}$
19. $6t\sinh(3t) - 9t^3$
20. $3e^{-2t}\sinh(3t)$
21. $2t^2e^{-t} - \sin^2(t)$
22. $t^4 - 3t^2 + 7$
23. $\frac{1}{2}[\cosh(4t) - \cos(4t)]$
24. $13(t - 2)^2 - 7t^3 - 4$
25. $(t - 1)\cos(3t) + 5e^{-t}$

In each of Problems 26 through 50, use Theorem 13 and the tables, and, in some instances, algebraic manipulation, to find the inverse Laplace transform.

26. $\dfrac{-2}{s + 16}$
27. $\dfrac{4s}{s^2 - 14}$
28. $\dfrac{-1}{s(s^2 + 8)}$

29. $\dfrac{3}{s - 7} + \dfrac{1}{s^2}$
30. $\dfrac{2s - 5}{s^2 + 21}$
31. $\dfrac{3s + 17}{s^2 - 7}$

32. $\dfrac{1}{(s - 3)^2} - \dfrac{2}{s^2} + \dfrac{1}{s}$
33. $\dfrac{2}{(s^2 + 3)^2} - \dfrac{4}{s^7}$
34. $\dfrac{4s - 7}{s^2 + 5} - \dfrac{3}{s - 2}$

35. $\dfrac{1}{(s - 3)(s + 3)} - \dfrac{7s}{s^2 + 15}$
36. $\dfrac{2s}{s^2 - 5} - \dfrac{3s}{s^2 + 5}$
37. $\dfrac{2}{(s + 7)^2}$

38. $\dfrac{1}{s - 4} - \dfrac{2}{(s - 4)^2} + \dfrac{6}{(s - 4)^3}$
39. $\dfrac{1}{s + 7}\left(2 - \dfrac{1}{s + 7}\right)$
40. $\dfrac{-s^2}{(s^2 + 7)^2} - \dfrac{5}{s^4}$

41. $\dfrac{s^2 - 2s + 4}{s^6}$
42. $\dfrac{3s - 6}{(s^2 + 5)^2} - \dfrac{5(s + 1)}{s^6}$
43. $\dfrac{2}{s^3}\left(1 - \dfrac{3}{s} + s^2\right)$

44. $\dfrac{s - 4}{(s^2 + 5)^2} + \dfrac{s}{s^2 + 2}$
45. $\dfrac{s + 1}{(s - 3)(s + 3)} - \dfrac{1}{s}$
46. $\dfrac{s^2 + 2}{3(s^2 + 4)^2}$

47. $\dfrac{s-5}{s^2+6}$

48. $\dfrac{7}{s} - \dfrac{4}{s+2} - \dfrac{5s}{s^2+12}$

49. $\dfrac{-3}{(s+2)^3} + \dfrac{4s}{s^2+6}$

50. $\dfrac{2}{(s-5)^2} + \dfrac{8s}{s^2+2} - \dfrac{4}{s}$

51. Derive the formula

$$\mathscr{L}(t^{-1/2}) = \left(\frac{\pi}{s}\right)^{1/2} \quad \text{for} \quad s > 0.$$

(*Hint:* In the definition of $\mathscr{L}(t^{-1/2})$, change variables by setting $x = t^{1/2}$; in the resulting integral, set $z = s^{1/2}x$. You may assume the fact that $\int_0^\infty e^{-z^2}\,dz = \frac{1}{2}\sqrt{\pi}$.)

In each of Problems 52 through 60, show that the given function is of exponential order as $t \to \infty$.

52. t^3

53. t^x (x any real number)

54. $t^2 \cos(4t)$

55. $t^3 \sin(3t)$

56. $t \cosh(t)$

57. $\sinh(t)$

58. e^{-t}

59. $t^2 e^{3t}$

60. $\dfrac{1-\cos(2t)}{t}$

4.2 CALCULATING LAPLACE TRANSFORMS

In order to carry out the program outlined at the end of the preceding section, we must be able to calculate Laplace transforms and inverse Laplace transforms. This section is devoted to Laplace transforms, and the next two to inverses.

In calculating Laplace transforms, we use tables whenever we can. However, no table can possibly contain every result we are likely to want, and there are certain manipulative rules which the student must master in order to make effective use of the transform. Here are six rules we shall use repeatedly. In deriving them, we assume that s is restricted to values for which the Laplace transform is defined. Along with each rule, we state sufficient conditions for the rule to hold, but again we emphasize that these conditions are not necessary.

A word is in order on the functions $f(t)$ we shall be working with. Since $\mathscr{L}[f(t)] = \int_0^\infty f(t)e^{-st}\,dt$, we are only interested in $f(t)$ when $t \geq 0$. In many applications t denotes time, so this may even be a natural restriction. However, for convenience, we often define $f(t) = 0$ for $t < 0$. This is usually for ease in notation, particularly in connection with the step functions and impulses which we shall soon encounter.

We are now ready to begin developing the rules we shall need.

RULE 1: LAPLACE TRANSFORM OF A DERIVATIVE

If $|f(t)| \leq Me^{bt}$ for $t \geq t_0$, $f(t)$ is continuous for $t \geq 0$, and $f'(t)$ is piecewise continuous on $[0, k]$ for every $k > 0$, then

$$\mathscr{L}(f') = s\mathscr{L}(f) - f(0) \quad \text{for} \quad s > b.$$

Derivation

$$\mathscr{L}[f'(t)] = \int_0^\infty \underbrace{e^{-st}}_{u} \underbrace{f'(t)\,dt}_{dv}$$

Integrate by parts

$$= [e^{-st}f(t)]_0^\infty - \int_0^\infty -se^{-st}f(t)\,dt$$

$$= \lim_{t\to\infty} e^{-st}f(t) - \lim_{t\to 0+} e^{-st}f(t) + s\int_0^\infty e^{-st}f(t)\,dt.$$

Now, $|f(t)| \leq Me^{bt}$ for t sufficiently large. Then $|f(t)e^{-st}| \leq Me^{(b-s)t}$ for t sufficiently large. If $s > b$, then $Me^{(b-s)t} \to 0$ as $t \to \infty$; hence $e^{-st}f(t) \to 0$ as $t \to \infty$. For the second of the above limits, we have by continuity of $f(t)$ at 0 that $\lim_{t\to 0+} e^{-st}f(t) = f(0)$. Thus, we have shown that

$$\mathscr{L}[f'(t)] = -f(0) + s\mathscr{L}[f(t)],$$

which is another way of writing Rule 1.

EXAMPLE 115

We know from the tables that

$$\mathscr{L}[\cos(at)] = \frac{s}{a^2 + s^2}.$$

Now, $\cos(at)' = -a \sin(at)$; so $\sin(at) = -(1/a)\cos(at)'$. Using Rule 1 and Theorem 12(2), we have

$$\mathscr{L}[\sin(at)] = \mathscr{L}\left[-\frac{1}{a}\cos(at)'\right] = -\frac{1}{a}\mathscr{L}[\cos(at)']$$

$$= -\frac{1}{a}\{s\mathscr{L}[\cos(at)] - \cos(0)\}$$

$$= -\frac{1}{a}\left\{\frac{s^2}{a^2 + s^2} - 1\right\}$$

$$= \frac{a}{a^2 + s^2}.$$

Note: This example is strictly for illustration of Rule 1; of course in practice we would find $\mathscr{L}[\sin(at)]$ in the table.

RULE 2: LAPLACE TRANSFORM OF AN NTH DERIVATIVE

$$\mathscr{L}[f^{(n)}] = s^n\mathscr{L}(f) - s^{n-1}f(0) - s^{n-2}f'(0) - \cdots - f^{(n-1)}(0)$$

Derivation Assume that we can apply Rule 1 with $f'(t)$ in place of $f(t)$, and $f''(t)$ in place of $f'(t)$. We then have

$$\begin{aligned}\mathscr{L}(f'') &= \mathscr{L}[(f')'] \\ &= s\mathscr{L}(f') - f'(0) \\ &= s[s\mathscr{L}(f) - f(0)] - f'(0) \\ &= s^2\mathscr{L}(f) - sf(0) - f'(0).\end{aligned}$$

Repeated applications give us Rule 2, assuming at each application that Rule 1 applies to the derivatives involved.

These two rules form the basis for using Laplace transforms to solve initial value problems, as the following shows.

EXAMPLE 116

Solve the initial value problem

$$y'' - 4y = t; \qquad y(0) = 1, \qquad y'(0) = -2.$$

We know how to solve this already, but here is how we would use Laplace transforms. Use Theorem 12 to take the Laplace transform of both sides of the differential equation, obtaining

$$\mathscr{L}(y'') - 4\mathscr{L}(y) = \mathscr{L}(t).$$

By Rule 2, and the initial conditions,

$$\begin{aligned}\mathscr{L}(y'') &= s^2\mathscr{L}(y) - sy(0) - y'(0) \\ &= s^2\mathscr{L}(y) - s + 2.\end{aligned}$$

Further, we read from the table that

$$\mathscr{L}(t) = \frac{1}{s^2}.$$

Thus, the initial value problem has been transformed into

$$s^2\mathscr{L}(y) - s + 2 - 4\mathscr{L}(y) = \frac{1}{s^2}.$$

Solve this *algebraic* equation for $\mathscr{L}(y)$ to find

$$\mathscr{L}(y) = \frac{1}{s^2(s^2 - 4)} + \frac{s - 2}{s^2 - 4}.$$

Some algebraic manipulation (which we describe in detail in the next section) lets us write

$$\mathscr{L}(y) = -\frac{1}{4}\frac{1}{s^2} + \frac{15}{16}\frac{1}{s+2} + \frac{1}{16}\frac{1}{s-2}.$$

Then,

$$y = -\frac{1}{4}\mathscr{L}^{-1}\left(\frac{1}{s^2}\right) + \frac{15}{16}\mathscr{L}^{-1}\left(\frac{1}{s+2}\right) + \frac{1}{16}\mathscr{L}^{-1}\left(\frac{1}{s-2}\right).$$

We find from the table that

$$\mathscr{L}^{-1}\left(\frac{1}{s^2}\right) = t, \quad \mathscr{L}^{-1}\left(\frac{1}{s+2}\right) = e^{-2t}, \quad \text{and} \quad \mathscr{L}^{-1}\left(\frac{1}{s-2}\right) = e^{2t}.$$

Thus, the solution of the initial value problem is

$$y = -\frac{t}{4} + \frac{15}{16}e^{-2t} + \frac{1}{16}e^{2t}.$$

This demonstration was probably not very impressive, since we could have solved the problem easily by previous methods anyway. But it illustrates the program sketched at the end of Section 4.1: The Laplace transform changed an initial value problem for y into an algebra problem for $\mathscr{L}(y)$, which we solved and transformed back into a solution for y. We will see more examples of this later.

RULE 3: LAPLACE TRANSFORM OF A PERIODIC FUNCTION

> If $f(t + \omega) = f(t)$, so that $f(t)$ has period ω, then
>
> $$\mathscr{L}(f) = \frac{1}{1 - e^{-s\omega}} \int_0^\omega e^{-st} f(t)\, dt.$$

Derivation Assuming that the Laplace transform of $f(t)$ exists, we have

$$\begin{aligned}\mathscr{L}(f) &= \int_0^\infty e^{-st} f(t)\, dt \\ &= \int_0^\omega e^{-st} f(t)\, dt + \int_\omega^{2\omega} e^{-st} f(t)\, dt + \cdots \\ &= \sum_{n=0}^{\infty} \int_{n\omega}^{(n+1)\omega} e^{-st} f(t)\, dt.\end{aligned}$$

Putting $t = u + n\omega$, then $u = 0$ when $t = n\omega$, and $u = \omega$ when $t = (n+1)\omega$. Thus,

$$\mathscr{L}(f) = \sum_{n=0}^{\infty} \int_0^\omega e^{-s(u+n\omega)} f(u + n\omega)\, du.$$

Because $f(u + n\omega) = f(u)$ by periodicity of $f(t)$, we have

$$\mathscr{L}(f) = \sum_{n=0}^{\infty} \int_0^{\omega} e^{-s(u+n\omega)} f(u)\, du$$

$$= \sum_{n=0}^{\infty} e^{-sn\omega} \int_0^{\omega} e^{-su} f(u)\, du$$

$$= \left[\sum_{n=0}^{\infty} (e^{-s\omega})^n\right] \int_0^{\omega} e^{-su} f(u)\, du.$$

Since $\sum_{n=0}^{\infty} k^n = 1/(1 - k)$ if $|k| < 1$ (geometric series), then

$$\sum_{n=0}^{\infty} (e^{-s\omega})^n = \frac{1}{1 - e^{-s\omega}} \quad \text{if} \quad e^{-s\omega} < 1.$$

(This is true if $1 < e^{s\omega}$, which holds if $s > 0$ and $\omega > 0$.) Hence,

$$\mathscr{L}(f) = \frac{1}{1 - e^{-s\omega}} \int_0^{\omega} e^{-su} f(u)\, du,$$

as we wanted to show.

EXAMPLE 117

Let $f(t)$ be the square wave of Figure 31, given by

$$f(t) = \begin{cases} 1 & \text{if} \quad 2nc \leq t \leq (2n + 1)c \\ 0 & \text{if} \quad (2n + 1)c < t < (2n + 2)c, \end{cases}$$

for $n = 0, 1, 2, 3, \ldots$, here, c is a given positive constant.

Note that $f(t)$ repeats itself every $2c$ units. That is, $f(t)$ has period $\omega = 2c$. Thus, the Laplace transform of $f(t)$ is

$$\mathscr{L}[f(t)] = \frac{1}{1 - e^{-s\omega}} \int_0^{\omega} e^{-su} f(u)\, du.$$

Now,

$$\int_0^{\omega} e^{-su} f(u)\, du$$

$$= \int_0^{2c} e^{-su} f(u)\, du$$

$$= \int_0^{c} e^{-su}(1)\, du + \int_c^{2c} e^{-su}(0)\, du$$

$$= \left[-\frac{1}{s} e^{-sc}\right]_0^c$$

$$= \frac{1}{s}(1 - e^{-sc}).$$

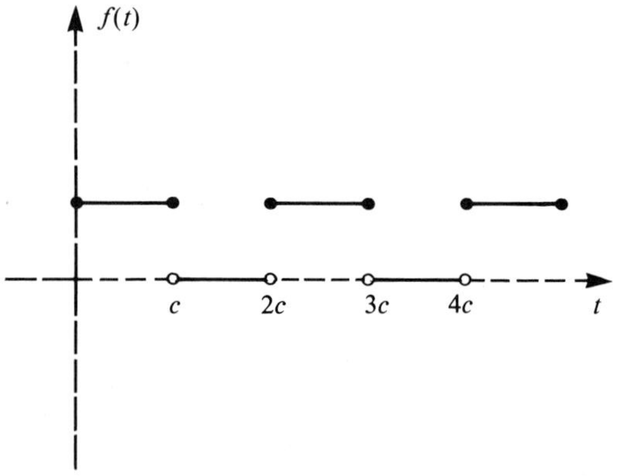

Figure 31. Square wave.

Then, in terms of the period ω (here $c = \omega/2$), we have

$$\mathscr{L}(f) = \frac{1}{s}\left(\frac{1 - e^{-s\omega/2}}{1 - e^{-s\omega}}\right).$$

This is fine as it stands, but could also be written in the handier form

$$\mathscr{L}(f) = \frac{1}{s}\frac{1}{(1 + e^{-s\omega/2})}.$$

RULE 4: SHIFTING IN THE S-VARIABLE

> If $\mathscr{L}[f(t)] = F(s)$, then $\mathscr{L}[e^{at}f(t)] = F(s - a)$ for $s > a$.

In words, we get the Laplace transform of $e^{at}f(t)$ by replacing s by $s - a$ in the Laplace transform of $f(t)$.

Derivation Suppose that $F(s) = \int_0^\infty e^{-st}f(t)\,dt$. Then,

$$\mathscr{L}[e^{at}f(t)] = \int_0^\infty e^{-st}e^{at}f(t)\,dt$$

$$= \int_0^\infty e^{-(s-a)t}f(t)\,dt = F(s - a) \qquad (s > a).$$

EXAMPLE 118

We know that

$$\mathscr{L}[\cos(kt)] = \frac{s}{s^2 + k^2}.$$

Then

$$\mathscr{L}[e^{at}\cos(kt)] = \frac{s - a}{(s - a)^2 + k^2}$$

[replace s by $s - a$ in the Laplace transform of $\cos(kt)$].

Similarly,

$$\mathscr{L}[\sin(kt)] = \frac{k}{s^2 + k^2};$$

hence,

$$\mathscr{L}[e^{at}\sin(kt)] = \frac{k}{(s - a)^2 + k^2}.$$

RULE 5: SHIFTING IN THE *T*-VARIABLE

> Let a be a positive constant. Let $f(t)$ be given, with $f(t) = 0$ if $t < 0$. Define $g(t)$ by
>
> $$g(t) = f(t - a).$$
>
> Then,
>
> $$\mathcal{L}(g) = e^{-as}\mathcal{L}(f).$$

Comparing the graphs in Figure 32, we see that $y = g(t)$ is zero for $t < a$; for $t \geq a$, it is the graph of $y = f(t)$ for $t \geq 0$, but translated a units to the right. Rule 5 says that the Laplace transform of $g(t)$ is just e^{-as} times the Laplace transform of $f(t)$.

Derivation

$$e^{-as}F(s) = e^{-as}\int_0^\infty e^{-st}f(t)\,dt$$

$$= \int_0^\infty e^{-s(t+a)}f(t)\,dt.$$

Let $u = t + a$. Then $u = a$ when $t = 0$ and $u \to \infty$ as $t \to \infty$. Thus,

$$e^{-as}F(s) = \int_a^\infty e^{-su}f(u - a)\,du$$

$$= \int_0^a e^{-su}0\,du + \int_a^\infty e^{-su}f(u - a)\,du$$

$$= \int_0^\infty e^{-su}g(u)\,du = \mathcal{L}(g),$$

since u is a dummy integration variable.

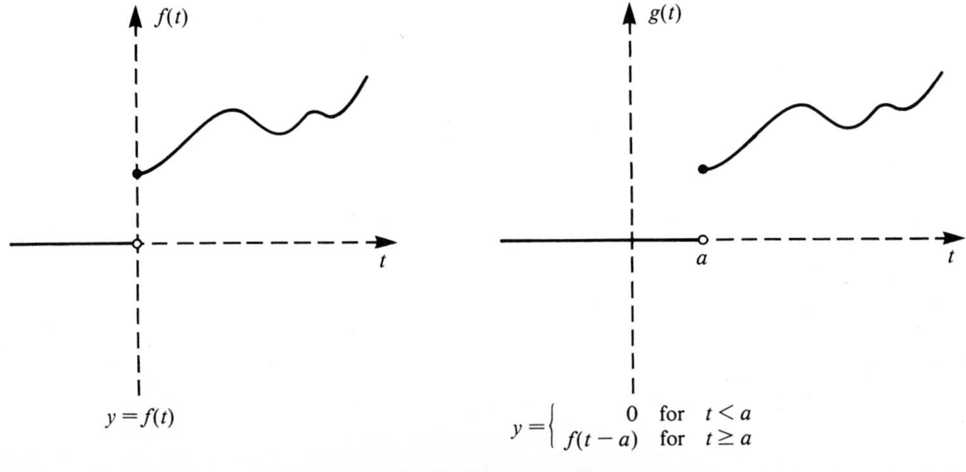

Figure 32. Shifting $y = f(x)$ by a.

EXAMPLE 119

Let

$$g(t) = \begin{cases} 0, & t < 5 \\ t - 5, & t \geq 5. \end{cases}$$

If we let

$$f(t) = \begin{cases} t & \text{for } t \geq 0 \\ 0 & \text{for } t < 0, \end{cases}$$

then we recognize $g(t)$ as the graph of $f(t)$ shifted five units to the right. That is,

$$g(t) = f(t - 5).$$

This is shown in Figure 33.

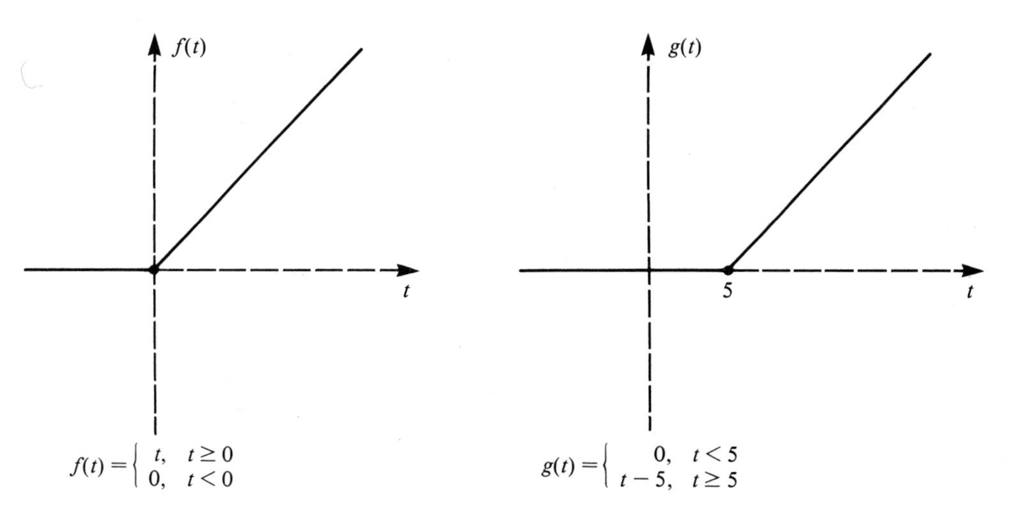

Figure 33

Since $\mathscr{L}(t) = 1/s^2$, then by Rule 5,

$$\mathscr{L}(g) = e^{-5s}\mathscr{L}(t) = \frac{e^{-5s}}{s^2}.$$

Sometimes it is convenient to state Rule 5 in terms of step functions. Define the *unit step function* $u(t)$ by

$$u(t) = \begin{cases} 0, & t < 0 \\ 1, & t \geq 0. \end{cases}$$

This is shown in Figure 34.

Note that $u(t - a)$ is simply $u(t)$ shifted a units to the right:

$$u(t - a) = \begin{cases} 0, & t < a \quad (\text{or } t - a < 0) \\ 1, & t \geq a \quad (\text{or } t - a \geq 0). \end{cases}$$

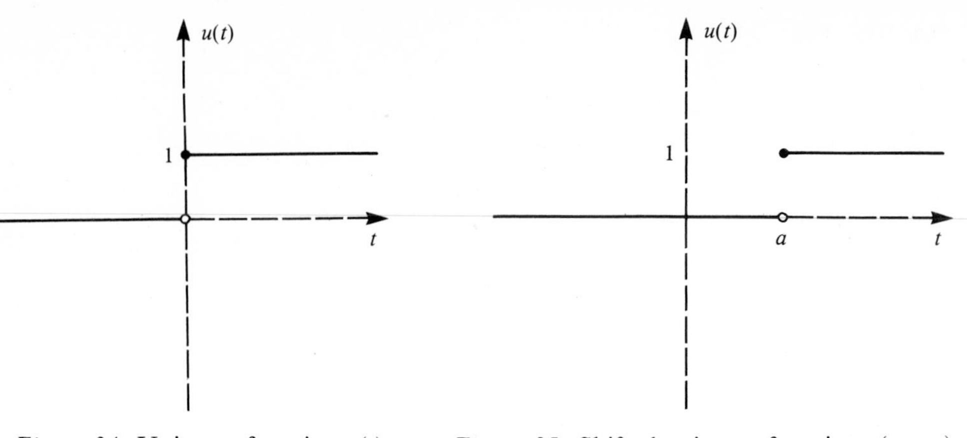

Figure 34. Unit step function $u(t)$.

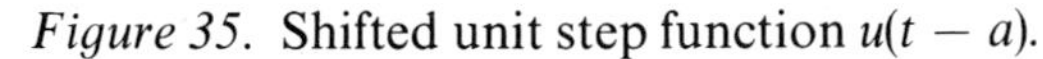

Figure 35. Shifted unit step function $u(t - a)$.

This *shifted unit step function* is shown in Figure 35, and is very useful in writing functions which could otherwise be cumbersome to describe. In particular, given $f(t)$ as shown in Figure 36(a), then

$$u(t - a)f(t - a)$$

is as shown in Figure 36(b). In fact, comparing Figure 36(b) with Figure 32, the function $g(t)$ of Rule 5 is exactly $u(t - a)f(t - a)$. This gives us a convenient alternative way of stating Rule 5:

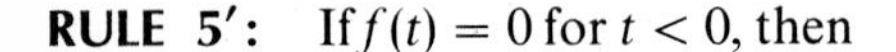

RULE 5′: If $f(t) = 0$ for $t < 0$, then

$$\mathscr{L}[u(t - a)f(t - a)] = e^{-as}\mathscr{L}(f).$$

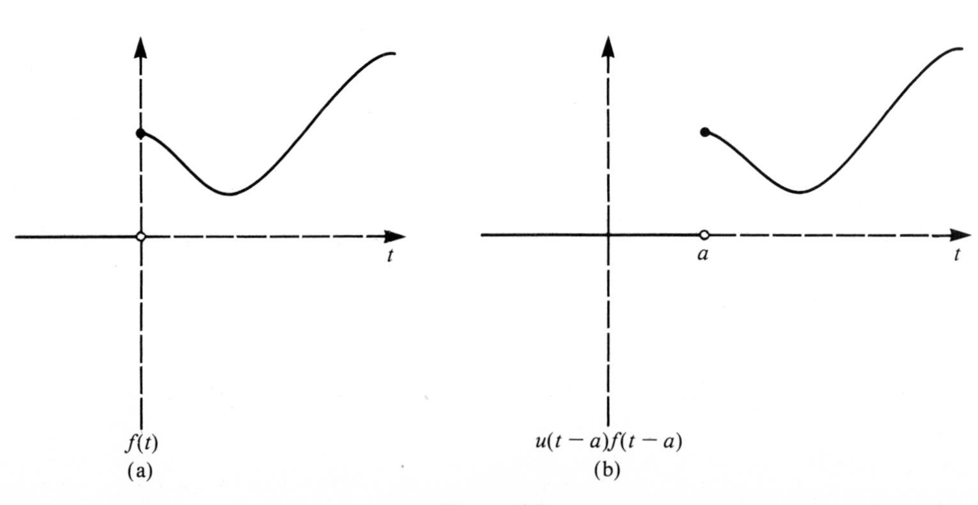

Figure 36

EXAMPLE 120

Let

$$g(t) = \begin{cases} 0, & t < 3 \\ \cos[4(t-3)], & t \geq 3. \end{cases}$$

To find $\mathscr{L}(g)$, note that

$$g(t) = u(t-3)\cos[4(t-3)].$$

Thus, with no additional effort, Rule 5′ gives us

$$\begin{aligned} \mathscr{L}(g) &= \mathscr{L}\{u(t-3)\cos[4(t-3)]\} \\ &= e^{-3s}\mathscr{L}[\cos(4t)] \\ &= e^{-3s}\left(\frac{s}{s^2+16}\right). \end{aligned}$$

As we shall see in Section 4.4, Rule 5′ is particularly useful in solving problems related to mechanical or electrical systems in which the imposed force has jump discontinuities (such as with an impulse).

RULE 6: FUNCTIONS DEFINED BY INTEGRALS

$$\mathscr{L}\left[\int_0^t f(z)\,dz\right] = \frac{1}{s}\mathscr{L}(f).$$

Here, a function $g(t)$ is defined by an integral of another function $f(t)$:

$$g(t) = \int_0^t f(z)\,dz.$$

Rule 6 says that $\mathscr{L}(g)$ is just $(1/s)$ times the Laplace transform of $f(t)$. A derivation of this rule is left to the student (see Problem 26, Section 4.6).

EXAMPLE 121

$$\mathscr{L}\left[\int_0^t \cos(3z)\,dz\right] = \frac{1}{s}\mathscr{L}[\cos(3t)] = \frac{1}{s}\left(\frac{s}{s^2+9}\right) = \frac{1}{s^2+9}.$$

We conclude this section with some additional examples showing how to use the rules just given.

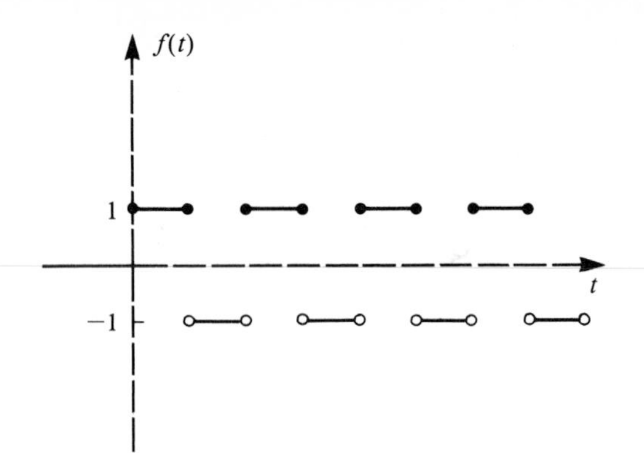

Figure 37. Square wave of Example 122.

EXAMPLE 122

Find $\mathscr{L}[f(t)]$, where

$$f(t) = \begin{cases} 0 & \text{if} \quad t < 0 \\ 1 & \text{if} \quad nc \leq t \leq (n+1)c \\ -1 & \text{if} \quad (n+1)c < t < 2nc, \end{cases}$$

for $n = 0, 1, 2, 3, \ldots$

Thus, $f(t)$ is the square wave shown in Figure 37. Use Rule 3, with $\omega = 2c$. We have

$$\mathscr{L}[f(t)] = \frac{1}{1 - e^{-s\omega}} \int_0^{\omega} e^{-st} f(t)\, dt$$

$$= \frac{1}{1 - e^{-s\omega}} \left[\int_0^c e^{-st}(1)\, dt + \int_c^{2c} e^{-st}(-1)\, dt \right]$$

$$= \frac{1}{1 - e^{-s\omega}} \left[\frac{1}{s}(1 - e^{-sc}) + \frac{1}{s}(e^{-2sc} - e^{-sc}) \right]$$

$$= \frac{1}{1 - e^{-s\omega}} \left[\frac{1}{s}(1 - 2e^{-sc} + e^{-2sc}) \right] = \frac{(1 - e^{-sc})^2}{s(1 - e^{-s\omega})}$$

$$= \frac{(1 - e^{-s\omega/2})^2}{s(1 - e^{-s\omega/2})(1 + e^{-s\omega/2})} = \frac{1}{s} \frac{(1 - e^{-s\omega/2})}{(1 + e^{-s\omega/2})}.$$

EXAMPLE 123

Find $\mathscr{L}[f(t)]$ if

$$f(t) = \begin{cases} 0, & t < 5 \\ t^2 + 1, & t \geq 5. \end{cases}$$

The form of $f(t)$ suggests Rule 5′, but this requires that $f(t)$ first be written in terms not of t itself, but of $(t - 5)$. This is an algebraic manipulation. Write

$$t^2 + 1 = (t - 5)^2 + 10(t - 5) + 26.$$

(This is, incidentally, the Taylor series of $t^2 + 1$ about 5.) Then,

$$f(t) = u(t - 5)h(t - 5),$$

where $h(t) = t^2 + 10t + 26$.

By Rule 5′,

$$\begin{aligned}\mathscr{L}(f) &= e^{-5s}\mathscr{L}(h)\\ &= e^{-5s}[\mathscr{L}(t^2) + 10\mathscr{L}(t) + 26\mathscr{L}(1)]\\ &= e^{-5s}\left(\frac{2}{s^3} + \frac{10}{s^2} + \frac{26}{s}\right).\end{aligned}$$

EXAMPLE 124

Find $\mathscr{L}[e^{-3t}f(t)]$, where

$$f(t) = \begin{cases} 0, & t < 8 \\ t^2 - 4, & t \geq 8. \end{cases}$$

We can write $f(t) = u(t - 8)h(t)$, where $h(t) = t^2 - 4$. Thus, we want

$$\mathscr{L}[e^{-3t}u(t - 8)(t^2 - 4)].$$

Because of the e^{-3t} factor, we look at Rule 4 first and get (with $a = -3$),

$$\mathscr{L}[e^{-3t}u(t - 8)(t^2 - 4)] = F(s + 3),$$

where

$$\mathscr{L}[u(t - 8)(t^2 - 4)] = F(s).$$

To work this out, Rule 5′ seems appropriate, but $u(t - 8)(t^2 - 4)$ is not quite in the right form: We need to write $t^2 - 4$ as some function of $t - 8$. This involves some algebraic juggling. We find* that

$$t^2 - 4 = (t - 8)^2 + 16(t - 8) + 60.$$

Thus,

$$\begin{aligned}u(t - 8)(t^2 - 4) &= u(t - 8)[(t - 8)^2 + 16(t - 8) + 60]\\ &= u(t - 8)g(t - 8),\end{aligned}$$

where

$$g(t) = t^2 + 16t + 60.$$

Now we can apply Rule 5′:

$$\begin{aligned}\mathscr{L}[u(t - 8)(t^2 - 4)] &= F(s)\\ &= \mathscr{L}[u(t - 8)g(t - 8)]\\ &= e^{-8s}\mathscr{L}(g)\\ &= e^{-8s}\left(\frac{2}{s^3} + \frac{16}{s^2} + \frac{60}{s}\right).\end{aligned}$$

* As in the preceding example, this can be done by "intuitive" juggling or, more formally, by expanding $t^2 - 4$ in a Taylor series about 8.

Finally,

$$\mathscr{L}[e^{-3t}f(t)] = F(s+3)$$
$$= e^{-8(s+3)}\left(\frac{2}{(s+3)^3} + \frac{16}{(s+3)^2} + \frac{60}{s+3}\right).$$

This example shows that the rules for finding Laplace transforms are not really disjoint entities, but sometimes must be used in combination.

The next example shows the use of shifted step functions to deal with functions having several jump discontinuities.

EXAMPLE 125

Sometimes we must use combinations of step functions to find a Laplace transform. For example, let $f(t) = 0$ for $t < 0$; and for $t \geq 0$, let $f(t)$ be defined by

$$f(t) = \begin{cases} 2, & 0 \leq t < 3 \\ 2t, & 3 \leq t < 5 \\ \frac{1}{7}t^2, & t \geq 5. \end{cases}$$

The graph of $f(t)$ is shown in Figure 38.

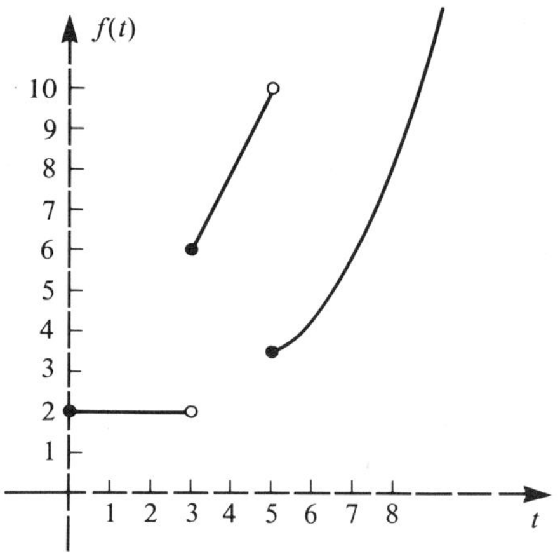

One way to visualize what we are about to do is to think of $f(t)$ as a sum of the three functions graphed in Figure 39. Thus, $f(t) = g_1(t) + g_2(t) + g_3(t)$, where each of $g_1(t)$, $g_2(t)$, and $g_3(t)$ "contains" exactly one of the nonzero pieces of $f(t)$, and their sum, obtained by superimposing the three graphs, is exactly $f(t)$.

Figure 38

Now note that, for $t \geq 0$,

$$g_1(t) = 2[1 - u(t-3)].$$

Check this: For $0 \leq t < 3$, $u(t-3) = 0$, and so $g_1(t) = 2$; for $t \geq 3$, $u(t-3) = 1$, and so $g_1(t) = 0$.

Next, write

$$g_2(t) = 2t[u(t-3) - u(t-5)].$$

Check this: When $t < 3$, $u(t-3)$ and $u(t-5)$ are zero, and so $g_2(t) = 0$; when $3 \leq t < 5$, $u(t-3) = 1$ but $u(t-5) = 0$, and so $g_2(t) = 2t$; and when $t \geq 5$, $u(t-3) = u(t-5) = 1$, and so $g_2(t) = 0$.

Finally, write

$$g_3(t) = \tfrac{1}{7}t^2 u(t-5).$$

Check this: This is $\frac{1}{7}t^2$ when $t \geq 5$, and zero when $t < 5$.

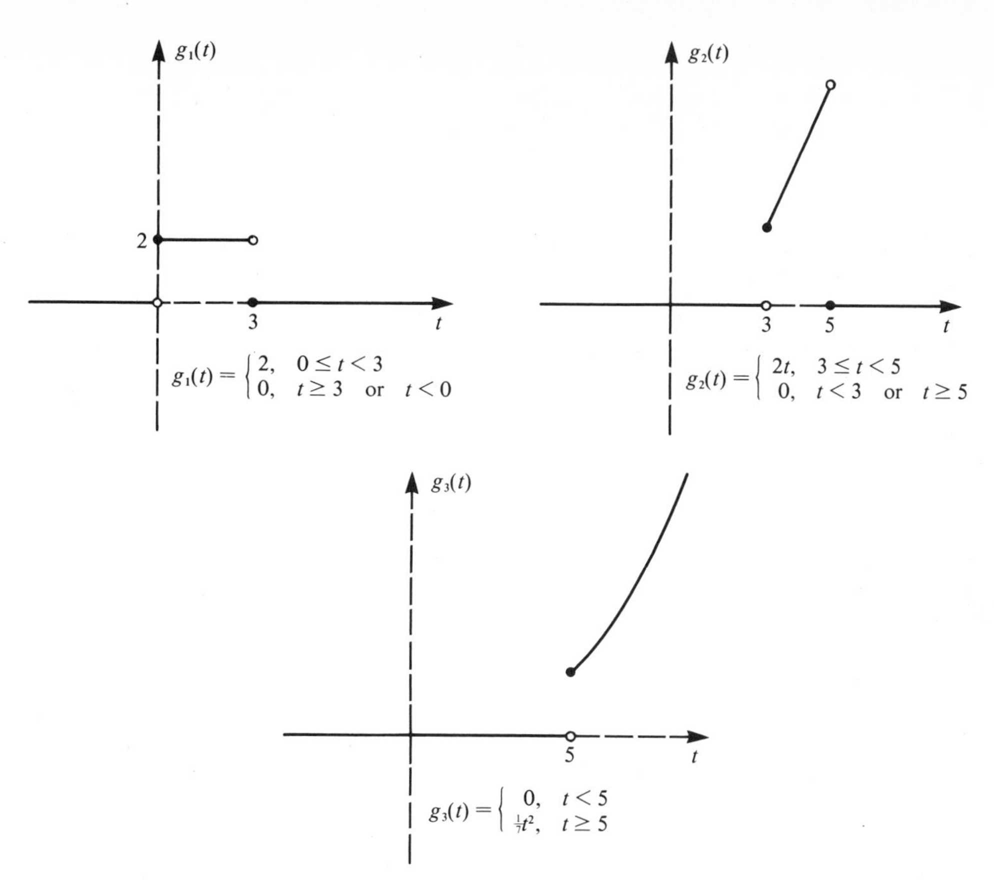

Figure 39

Thus,

$$\begin{aligned} f(t) &= 2[1 - u(t-3)] + 2t[u(t-3) - u(t-5)] + \tfrac{1}{7}t^2u(t-5) \\ &= 2 + (2t-2)u(t-3) + (\tfrac{1}{7}t^2 - 2t)u(t-5). \end{aligned}$$

In order to invoke Rule 5′, rewrite

$$2t - 2 = 4 + 2(t-3)$$

and

$$\tfrac{1}{7}t^2 - 2t = -\tfrac{45}{7} - \tfrac{4}{7}(t-5) + \tfrac{1}{7}(t-5)^2.$$

We now have

$$f(t) = 2 + [4 + 2(t-3)]u(t-3) + [-\tfrac{45}{7} - \tfrac{4}{7}(t-5) + \tfrac{1}{7}(t-5)^2]u(t-5).$$

We can now use the table and Rule 5′ to write

$$\begin{aligned} \mathscr{L}[f(t)] &= \mathscr{L}(2) + e^{-3s}\mathscr{L}(4+2t) + e^{-5s}\mathscr{L}\left(-\frac{45}{7} - \frac{4}{7}t + \frac{1}{7}t^2\right) \\ &= \frac{2}{s} + e^{-3s}\left(\frac{4}{s} + \frac{2}{s^2}\right) + e^{-5s}\left(-\frac{45}{7s} - \frac{4}{7s^2} + \frac{2}{7s^3}\right). \end{aligned}$$

Rules 1 through 6 are all we shall need for a while. In the next two sections we concentrate on inverse Laplace transforms and consider their use in solving initial value problems, particularly those arising from mass-spring systems, electrical networks, and loaded beams.

PROBLEMS FOR SECTION 4.2

In each of Problems 1 through 50, find the Laplace transform of the given function by using the table and applying Rules 1 through 6. In these problems you should not have to perform any integrations, except in carrying out Rule 3 where it applies.

1. $4t^2 - 3t^3$

2. $8 - 5t + 6t^4$

3. $t^2 - \sin(2t) + 1$

4. $\cosh(3t) - 4t^2 + 6u(t-7)$

5. $-e^{-3t}\sin(4t) + 3$

6. $\cos(2t) - \sinh(4t)$

7. $2(t-4)u(t-4)$

8. $5u(t-6) + u(t-7)$

9. $t^2e^{-5t} + 4\cos(2t) - \sinh(6t)$

10. $e^{-5t}\sinh(2t)$

11. $tu(t-4)$

12. $1 - 7u(t-4) + t^2e^{-t}$

13. $\sinh(3t) + \cosh(3t) - 2t^3$

14. $(t-5)^3u(t-5) + \cos(4t)$

15. $2(t-2)^4u(t-2)$

16. $18t^2e^{-3t} + e^{2t}\sin(3t)$

17. $1 - \sinh(6t) + e^{-t}\cosh(t)$

18. $(2t-1)u(t-3)$

19. $e^{2t}\cos(5t) - 3t^3$

20. $t^4e^{-5t} + t$

21. $3e^{-4t}\sin(2t) - \cos(6t) + 6$

22.
$$f(t) = \begin{cases} 0, & t < 0 \\ 2, & 2n \le t < 2n+1 \\ -2, & 2n+1 \le t < 2n+2, \end{cases}$$
for $n = 0, 1, 2, 3, \ldots$

23. $g(t) = e^{3t}f(t)$, with $f(t)$ as in Problem 22.

24. $t^2u(t-5)$

25. $f(t) = \begin{cases} 0, & t < 4 \\ 2t^3, & t \ge 4 \end{cases}$

26. $f(t) = \int_0^t (z^3 + 1)\,dz$

27. $e^{-4t}\cosh(6t)$

28. $f(t) = \begin{cases} 1, & 0 \le t < 3 \\ 2, & t \ge 3 \end{cases}$

29. $f(t) = \int_0^t e^{-4z}\sin(3z)\,dz$

30. $f(t) = \begin{cases} -h, & 2n \le t < 2n+1 \\ 2h, & 2n+1 \le t < 2n+2, \end{cases}$
where $n = 0, 1, 2, 3, \ldots$ and h is any positive constant.

31. $f(t) = \begin{cases} 2t, & 0 \le t < 5 \\ t+2, & t \ge 5 \end{cases}$

32. $e^{3t}\sinh(4t) - 2\cos(5t) + t^2$

33. $(3t-1)u(t-6)$

34.
$$f(t) = \begin{cases} 0, & 0 \le t < 6 \\ 1, & 6 \le t < 8 \\ 0, & t \ge 8 \end{cases}$$

35. $(2t^2 + 3)u(t-7)$

36.
$$f(t) = \begin{cases} 1, & 0 \le t < 3 \\ -5, & 3 \le t < 7 \\ 2t+1, & t \ge 7 \end{cases}$$

37. $f(t) = \begin{cases} h, & 4n \le t < 4n+4 \\ -h, & 4n+4 \le t < 4n+8, \end{cases}$ where h is any positive constant and $n = 0, 2, 4, 6, 8, \ldots$

38. $f(t) = -3\int_0^t e^{-4z}\cosh(3z)\,dz$

39. $(8t^3 - 2)u(t-2)$

40. $(2t+3)u(t-7) - 8(t^2 + 2t)u(t-4)$

41. $f(t) = \begin{cases} -2t^2, & 0 \le t < 9 \\ (t+1)^3, & t \ge 9 \end{cases}$

42. $e^{-2t}u(t-4)$

43. $8t^3u(t-5) - e^{-4t}tu(t-3)$

44.
$$f(t) = \begin{cases} 1, & 0 \le t < 1 \\ -1, & 1 \le t < 3 \\ 8, & t \ge 3 \end{cases}$$

45. $t - (7t^2 + 30)u(t-1)$

46. $2e^{-5t}t^2 - 3tu(t-4)$

47. $(t^4 - 3t^2 + 2)u(t-9)$

48. $f(t) = \begin{cases} -8t, & 0 \le t < 4 \\ t(t+2), & t \ge 4 \end{cases}$

49. $1 - t^2e^{-3t}u(t-2)$

50. $7e^{-4t}\cos(t) - \int_0^t \sin^2(z)\,dz$

In each of Problems 51 through 70, use Rules 1 and 2, as in Example 116, to convert the initial value problem to an algebra problem involving $\mathscr{L}(y)$. Solve for $\mathscr{L}(y)$. Do not attempt at this time to invert and solve for y.

51. $y'' - 6y' + 2y = 0; \quad y(0) = 1, \quad y'(0) = -3$
52. $6y^{(3)} - 2y'' + 3y' - 4y = t; \quad y(0) = 1, \quad y'(0) = 0, \quad y''(0) = -3$
53. $y'' - 3y' - 10y = e^{-t}; \quad y(0) = 2, \quad y'(0) = -4$
54. $y^{(4)} - 3y'' + 2y = 2\cos(4t); \quad y(0) = y'(0) = 1, \quad y''(0) = 0, \quad y^{(3)}(0) = -4$
55. $y'' + 6y' - 18y = e^{-4t}; \quad y(0) = -3, \quad y'(0) = 2$
56. $y'' - 3y = 4\cos(t); \quad y(0) = y'(0) = 0$
57. $y'' + 2y' - 7y = 8; \quad y(0) = 1, \quad y'(0) = 0$
58. $y'' - 7y' + 2y = -e^{-3t}; \quad y(0) = y'(0) = 0$
59. $y'' + 8y' - 2y = 1 - 3\cosh(2t); \quad y(0) = 1, \quad y'(0) = -2$
60. $y^{(3)} - 2y' + y = 14te^{-t}; \quad y(0) = 2, \quad y'(0) = y''(0) = 0$
61. $y^{(3)} - 3y'' + 4y' - 16y = 2u(t-3); \quad y(0) = y'(0) = -1, \quad y''(0) = 0$
62. $y^{(3)} + 8y'' - 10y = 18t - \sin(3t); \quad y(0) = -2, \quad y'(0) = 1, \quad y''(0) = -7$
63. $y^{(4)} + 12y^{(3)} - 2y = 1; \quad y(0) = -4, \quad y'(0) = 1, \quad y''(0) = y^{(3)}(0) = 0$
64. $y'' + y' - 5y = 1 - t^2u(t-3); \quad y(0) = 4, \quad y'(0) = -5$
65. $y^{(3)} - 7y'' + y' + y = 4t^2 - 3e^{-2t}\cos(t); \quad y(0) = y'(0) = 2, \quad y''(0) = 0$
66. $y^{(4)} + 13y^{(3)} - 2y' + 6y = (t-2)u(t-3); \quad y(0) = -3, \quad y'(0) = -4, \quad y''(0) = 1, \quad y^{(3)}(0) = 7$
67. $y^{(3)} + 3y'' - 2y' + y = t^3e^{-4t} + 5t^3; \quad y(0) = y'(0) = 5, \quad y''(0) = -2$
68. $y^{(3)} + 2y'' - 8y = 2t^3 - \cosh(3t); \quad y(0) = 5, \quad y'(0) = y''(0) = 2$
69. $y'' - 10y' + 8y = \sin(3t) - tu(t-4); \quad y(0) = 1, \quad y'(0) = -6$
70. $y^{(3)} - 3y'' + 2y' - y = (t^2 - 2t + 4)u(t-5); \quad y(0) = -3, \quad y'(0) = 2, \quad y''(0) = 0$
71. Let $\mathscr{L}[f(t)] = F(s)$. Prove that, for any constant $c > 0$,

$$\mathscr{L}[f(ct)] = \frac{1}{c}F\left(\frac{s}{c}\right).$$

72. Let $\mathscr{L}[f(t)] = F(s)$. Prove that

$$\mathscr{L}[(-1)^n t^n f(t)] = F^{(n)}(s),$$

recalling that $F^{(n)}(s)$ is the nth derivative of $F(s)$ with respect to s.

4.3 CALCULATING INVERSE LAPLACE TRANSFORMS: PART 1

In the preceding section we concentrated on computing Laplace transforms. When we use Laplace transforms to solve initial value problems, as in Example 116, it will be equally important to be able to take inverse Laplace transforms, i.e., given $F(s)$, find $f(t)$ such that $\mathscr{L}[f(t)] = F(s)$.

Of course, every formula $\mathscr{L}[f(t)] = F(s)$ implies an inverse formula $f(t) = \mathscr{L}^{-1}[F(s)]$, but in practice things are not this easy, and not every result we need will be readily available from the table. Hence, the student must acquire experience and a collection of tricks and techniques. In this section we center upon three techniques: partial fractions and the inverse versions of Rules 4 and 5′. The following section is devoted to the Heaviside expansion formulas, which offer an alternative to partial fractions.

PARTIAL FRACTIONS IN CALCULATING INVERSE LAPLACE TRANSFORMS

As Example 116 showed, it is often useful to be able to write a quotient of polynomials as a sum of simpler quotients. In that example, we wrote

$$\frac{1}{s^2(s^2-4)} + \frac{s-2}{s^2-4} = -\frac{1}{4}\frac{1}{s^2} + \frac{15}{16}\frac{1}{s+2} + \frac{1}{16}\frac{1}{s-2}.$$

The purpose in doing this was that the inverse transform of each function on the right could be read directly from the table, while the left side was not in such recognizable form.

Partial fractions is a name given to the algebraic technique of writing a quotient of polynomials in "simpler" form, as above. The student has no doubt already encountered this technique in elementary calculus (review the section on techniques of integration), where it enables us to write quotients in ways whose integrals are easier to recognize. Such things being easily forgotten, we review the algebra here before applying it to finding inverse Laplace transforms.

Suppose we have a quotient of polynomials,

$$\frac{f(t)}{g(t)} = \frac{a_0 + a_1 t + \cdots + a_p t^p}{b_0 + b_1 t + \cdots + b_r t^r}.$$

We assume that:

1. All coefficients are real;
2. All common factors have been divided out [so that $f(t)$ and $g(t)$ have no common root]; and
3. $g(t)$ is of higher degree than $f(t)$ [otherwise divide $g(t)$ into $f(t)$]. In the above quotient, this means that $p < r$.

Now, $g(t)$ can always be split into a product of factors, each of the form

$$(s-a)^m$$

(linear factors) or

$$(s^2 + ps + q)^n$$

(quadratic factors). We write $f(t)/g(t)$ as a sum of simpler fractions by:

1. Assigning to each $(s-a)^m$ factor of $g(t)$ a sum of terms
$$\frac{A_1}{(s-a)} + \frac{A_2}{(s-a)^2} + \cdots + \frac{A_m}{(s-a)^m};$$
2. Assigning to each $(s^2 + ps + q)^n$ factor of $g(t)$ a sum of terms
$$\frac{B_1 s + C_1}{(s^2+ps+q)} + \frac{B_2 s + C_2}{(s^2+ps+q)^2} + \cdots + \frac{B_n s + C_n}{(s^2+ps+q)^n}; \quad \text{and}$$
3. Solving for the constants $A_1, \ldots, B_1, \ldots, C_1, \ldots$.

Some examples, and their use in finding inverse Laplace transforms, should clarify this process.

EXAMPLE 126

Find

$$\mathcal{L}^{-1}\left(\frac{s^2+2}{s^4-6s^3+32s}\right).$$

Factoring the denominator, the quotient is

$$\frac{s^2+2}{s(s+2)(s-4)^2}.$$

All factors in the denominator are linear, so by (1) of the partial fractions process, we write

$$\frac{s^2+2}{s(s+2)(s-4)^2} = \frac{A}{s} + \frac{B}{s+2} + \frac{C}{s-4} + \frac{D}{(s-4)^2},$$

where A, B, C, and D are constants to be determined. To do this, add the fractions on the right to get

$$\frac{s^2+2}{s(s+2)(s-4)^2} = \frac{A(s+2)(s-4)^2 + B(s)(s-4)^2 + C(s)(s+2)(s-4) + D(s)(s+2)}{s(s+2)(s-4)^2}.$$

This works out to

$$\frac{s^2+2}{s(s+2)(s-4)^2} = \frac{(A+B+C)s^3 + (-6A-8B-2C+D)s^2 + (16B-8C+2D)s + 32A}{s(s+2)(s-4)^2}.$$

The only way for the left side to equal the right for all s is for the numerators to be identical, since the denominators are the same. Equating coefficients of like powers of s in the numerators, we have

$$\begin{aligned} A+B+C &= 0 && \text{(no } s^3 \text{ in } s^2+2) \\ -6A-8B-2C+D &= 1 && \text{(from coefficient of } s^2) \\ 16B-8C+2D &= 0 && \text{(no } s \text{ term in } s^2+2) \\ 32A &= 2 && \text{(from constant term).} \end{aligned}$$

Solve these to get

$$A = \tfrac{1}{16}, \qquad B = -\tfrac{1}{12}, \qquad C = \tfrac{1}{48}, \quad \text{and} \quad D = \tfrac{3}{4}.$$

Then,

$$\frac{s^2+2}{s^4-6s^3+32s} = \frac{1}{16}\frac{1}{s} - \frac{1}{12}\frac{1}{s+2} + \frac{1}{48}\frac{1}{s-4} + \frac{3}{4}\frac{1}{(s-4)^2}.$$

Finally, we have

$$\mathscr{L}^{-1}\left(\frac{s^2+2}{s^4-6s^3+32s}\right)=\frac{1}{16}\mathscr{L}^{-1}\left(\frac{1}{s}\right)-\frac{1}{12}\mathscr{L}^{-1}\left(\frac{1}{s+2}\right)+\frac{1}{48}\mathscr{L}^{-1}\left(\frac{1}{s-4}\right)$$
$$+\frac{3}{4}\mathscr{L}^{-1}\left(\frac{1}{(s-4)^2}\right).$$

We read directly from the table that

$$\mathscr{L}^{-1}\left(\frac{1}{s}\right)=t,\qquad \mathscr{L}^{-1}\left(\frac{1}{s+2}\right)=e^{-2t},\qquad \mathscr{L}^{-1}\left(\frac{1}{s-4}\right)=e^{4t},$$

and $\mathscr{L}^{-1}\left(\frac{1}{(s-4)^2}\right)=te^{4t}.$

Thus,

$$\mathscr{L}^{-1}\left(\frac{s^2+2}{s^4-6s^3+32s}\right)=\frac{t}{16}-\frac{e^{-2t}}{12}+\frac{e^{4t}}{48}+\frac{3te^{4t}}{4}.$$

EXAMPLE 127

Find

$$\mathscr{L}^{-1}\left(\frac{2s+3}{(s^2+2)^2(s-3)}\right).$$

Here the denominator has one linear factor, $s-3$, and a quadratic factor, s^2+2, appearing to the second power. Write

$$\frac{2s+3}{(s^2+2)^2(s-3)}=\frac{A}{s-3}+\frac{B_1s+C_1}{s^2+2}+\frac{B_2s+C_2}{(s^2+2)^2}.$$

If we add the fractions on the right [the greatest common divisor is $(s-3)(s^2+2)^2$] and equate the numerators on both sides, we get

$$2s+3=(A+B_1)s^4+(-3B_1+C_1)s^3+(4A+2B_1-3C_1+B_2)s^2$$
$$+(-6B_1+2C_1-3B_2+C_2)s+(4A-6C_1-3C_2).$$

Equating coefficients of like powers of s on both sides gives us

$$\begin{aligned}A+B_1&=0\\-3B_1+C_1&=0\\4A+2B_1-3C_1+B_2&=0\\-6B_1+2C_1-3B_2+C_2&=2\\4A-6C_1-3C_2&=3.\end{aligned}$$

Solve these to get

$$A=\frac{9}{121},\qquad B_1=\frac{-9}{121},\qquad C_1=\frac{-27}{121},\qquad B_2=\frac{-99}{121},\qquad C_2=\frac{-55}{121}.$$

Hence,

$$\frac{2s+3}{(s^2+2)^2(s-3)} = \frac{9}{121}\frac{1}{s-3} + \frac{\frac{-9s}{121} - \frac{27}{121}}{s^2+2} + \frac{\frac{-99s}{121} - \frac{55}{121}}{(s^2+2)^2}.$$

Then,

$$\mathscr{L}^{-1}\left(\frac{2s+3}{(s^2+2)^2(s-3)}\right) = \frac{9}{121}\mathscr{L}^{-1}\left(\frac{1}{s-3}\right) - \frac{9}{121}\mathscr{L}^{-1}\left(\frac{s}{s^2+2}\right)$$
$$- \frac{27}{121}\mathscr{L}^{-1}\left(\frac{1}{s^2+2}\right) - \frac{99}{121}\mathscr{L}^{-1}\left(\frac{s}{(s^2+2)^2}\right) - \frac{55}{121}\mathscr{L}^{-1}\left(\frac{1}{(s^2+2)^2}\right),$$

and we read directly from the table that

$$\mathscr{L}^{-1}\left(\frac{2s+3}{(s^2+2)^2(s-3)}\right) = \frac{9}{121}e^{3t} - \frac{9}{121}\cos(\sqrt{2}\,t) - \frac{27}{121}\frac{1}{\sqrt{2}}\sin(\sqrt{2}\,t)$$
$$- \frac{99}{121}\frac{t}{2\sqrt{2}}\sin(\sqrt{2}\,t) - \frac{55}{121}\frac{1}{2(\sqrt{2})^3}[\sin(\sqrt{2}\,t) - \sqrt{2}\,t\cos(\sqrt{2}\,t)].$$

EXAMPLE 128

Find

$$\mathscr{L}^{-1}\left(\frac{s^2+2s-4}{s^3-5s^2+2s+8}\right).$$

Write

$$\frac{s^2+2s-4}{s^3-5s^2+2s+8} = \frac{s^2+2s-4}{(s+1)(s-2)(s-4)}$$
$$= \frac{A}{s+1} + \frac{B}{s-2} + \frac{C}{s-4}.$$

This requires that

$$s^2+2s-4 = (A+B+C)s^2 + (-6A-3B-C)s + (8A-4B-2C).$$

Then,

$$A+B+C=1$$
$$-6A-3B-C=2$$
$$8A-4B-2C=-4.$$

Solve these to get

$$A = -\tfrac{1}{3}, \qquad B = -\tfrac{2}{3}, \qquad C = 2.$$

Then,

$$\mathscr{L}^{-1}\left(\frac{s^2+2s-4}{s^3-5s^2+2s+8}\right) = -\frac{1}{3}\mathscr{L}^{-1}\left(\frac{1}{s+1}\right) - \frac{2}{3}\mathscr{L}^{-1}\left(\frac{1}{s-2}\right) + 2\mathscr{L}^{-1}\left(\frac{1}{s-4}\right)$$
$$= -\tfrac{1}{3}e^{-t} - \tfrac{2}{3}e^{2t} + 2e^{4t}.$$

EXAMPLE 129

Find

$$\mathscr{L}^{-1}\left(\frac{2s^2 - s}{(s^2 + 4)^2}\right).$$

Here we have a quadratic factor squared in the denominator. Write

$$\frac{2s^2 - s}{(s^2 + 4)^2} = \frac{A_1 s + B_1}{s^2 + 4} + \frac{A_2 s + B_2}{(s^2 + 4)^2}.$$

Then,

$$(A_1 s + B_1)(s^2 + 4) + A_2 s + B_2 = 2s^2 - s$$

or

$$A_1 s^3 + B_1 s^2 + (4A_1 + A_2)s + 4B_1 + B_2 = 2s^2 - s.$$

Then,

$$\begin{aligned} A_1 &= 0 \\ B_1 &= 2 \\ 4A_1 + A_2 &= -1 \\ 4B_1 + B_2 &= 0. \end{aligned}$$

So $A_1 = 0$, $B_1 = 2$, $A_2 = -1$, and $B_2 = -8$, and we have

$$\frac{2s^2 - s}{(s^2 + 4)^2} = \frac{2}{s^2 + 4} + \frac{-s - 8}{(s^2 + 4)^2}.$$

Then,

$$\begin{aligned} \mathscr{L}^{-1}\left(\frac{2s^2 - s}{(s^2 + 4)^2}\right) &= 2\mathscr{L}^{-1}\left(\frac{1}{s^2 + 4}\right) - \mathscr{L}^{-1}\left(\frac{s}{(s^2 + 4)^2}\right) - 8\mathscr{L}^{-1}\left(\frac{1}{(s^2 + 4)^2}\right) \\ &= \sin(2t) - \tfrac{1}{4}t\sin(2t) - \tfrac{1}{2}[\sin(2t) - 2t\cos(2t)] \\ &= \tfrac{1}{2}\sin(2t) - \tfrac{1}{4}t\sin(2t) + t\cos(2t). \end{aligned}$$

The student should note that use of the table often requires some additional manipulation. In the last example, we needed

$$\mathscr{L}^{-1}\left(\frac{1}{(s^2 + 4)^2}\right).$$

From the table, the closest thing is

$$\mathscr{L}^{-1}\left(\frac{2k^3}{(s^2 + k^2)^2}\right) = \sin(kt) - kt\cos(kt).$$

Choosing $k = 2$ to get $(s^2 + 4)^2$ in the denominator, we then need $2k^3 = 16$ in the numerator; so we put it there and then divide it out, getting

$$\frac{1}{16}\mathscr{L}^{-1}\left(\frac{16}{(s^2 + 4)^2}\right),$$

which we can handle from the table.

EXAMPLE 130

The partial fractions technique depends on our ability to find factors or roots of the polynomial in the denominator. Sometimes this is hard to do (e.g., the coefficients may not be conveniently chosen integers, as in our examples). What do we do then?

As an example, consider the problem of finding

$$\mathscr{L}^{-1}\left(\frac{0.238s^2 - 8.671s}{s^3 - 1.006s^2 - 4.43804s + 1.3758672}\right).$$

This is more typical of what one might actually encounter in an engineering application. It is usually necessary in a case like this to approximate. First, we have to apply some numerical technique (such as Newton's method, discussed in the chapter on numerical analysis), to factor

$$s^3 - 1.006s^2 - 4.43804s + 1.375867 = (s - 0.296)(s + 1.83)(s - 2.54).$$

Then write

$$\frac{0.238s^2 - 8.671s}{(s - 0.296)(s + 1.83)(s - 2.54)} = \frac{A}{s - 0.296} + \frac{B}{s + 1.83} + \frac{C}{s - 2.54}.$$

In the usual way, this gives us equations

$$\begin{aligned} A + B + C &= 0.238 \\ -0.71A - 2.836B + 1.534C &= -8.671 \\ -4.6482A + 0.75184B - 0.54168C &= 0. \end{aligned}$$

Solve these to get (to four decimal places)

$$A = 0.5336, \qquad B = 1.7937, \quad \text{and} \quad C = -2.0894.$$

Then,

$$\mathscr{L}^{-1}\left(\frac{0.238s^2 - 8.671s}{s^3 - 1.006s^2 - 4.43804s + 1.3758672}\right) = 0.5336e^{0.296t} + 1.7937e^{-1.83t} - 2.0894e^{2.54t}.$$

Finding the roots of $s^3 - 1.006s^2 - 4.43804s + 1.375867$, as well as solving the equations for A, B, and C, requires at the least a good hand-held calculator. For polynomials of much higher degree, computer assistance would be needed.

THE INVERSE VERSION OF RULE 4

Rule 4 of the last section said:

$$\text{If} \quad \mathscr{L}[f(t)] = F(s), \quad \text{then} \quad \mathscr{L}[e^{at}f(t)] = F(s - a).$$

The inverse version of this is:

$$\boxed{\text{If} \quad \mathscr{L}^{-1}[F(s)] = f(t), \quad \text{then} \quad \mathscr{L}^{-1}[F(s - a)] = e^{at}f(t).}$$

The tipoff for using this is that we want the inverse transform of a function not just of s, but of $s - a$. Here are two examples.

EXAMPLE 131

Find

$$\mathscr{L}^{-1}\left(\frac{1}{s^2 + 4s + 12}\right).$$

Here, $s^2 + 4s + 12$ has complex roots $-2 \pm i\sqrt{8}$, and so cannot be factored into two linear factors with real coefficients. Thus, partial fractions will not simplify $1/(s^2 + 4s + 12)$.

Try to work the denominator into a function of $s - a$ for some a. Complete the square by writing

$$s^2 + 4s + 12 = (s + 2)^2 + 8.$$

Then,

$$\frac{1}{s^2 + 4s + 12} = \frac{1}{(s + 2)^2 + 8} = F(s + 2),$$

where $F(s) = 1/(s^2 + 8)$. (Thus we chose $a = -2$.)

Now, from the table, we have

$$\mathscr{L}^{-1}[F(s)] = \mathscr{L}^{-1}\left(\frac{1}{s^2 + 8}\right) = \frac{1}{\sqrt{8}} \sin(\sqrt{8}\, t) = f(t).$$

Then,

$$\mathscr{L}^{-1}\left(\frac{1}{s^2 + 4s + 12}\right) = \mathscr{L}^{-1}[F(s + 2)] = e^{-2t} f(t)$$

$$= \frac{1}{\sqrt{8}} e^{-2t} \sin(\sqrt{8}\, t).$$

EXAMPLE 132

Find

$$\mathscr{L}^{-1}\left(\frac{s + 6}{s^2 + 4s + 12}\right).$$

Completing the square in the denominator, we have $s^2 + 4s + 12 = (s + 2)^2 + 8$; hence we center our attention on functions of $s + 2$. This motivates us to write

$$\frac{s + 6}{s^2 + 4s + 12} = \frac{(s + 2) + 4}{(s + 2)^2 + 8} = \frac{s + 2}{(s + 2)^2 + 8} + 4\left(\frac{1}{(s + 2)^2 + 8}\right).$$

Take the last two quotients one at a time. First,

$$\frac{s + 2}{(s + 2)^2 + 8} = F(s + 2),$$

where $F(s) = s/(s^2 + 8)$. Now $\mathscr{L}^{-1}[F(s)] = \cos(\sqrt{8}\,t) = f(t)$. By the inverse of Rule 4,

$$\mathscr{L}^{-1}\left(\frac{s+2}{(s+2)^2+8}\right) = e^{-2t}f(t) = e^{-2t}\cos(\sqrt{8}\,t).$$

We already had

$$\mathscr{L}^{-1}\left(\frac{1}{(s+2)^2+8}\right) = \frac{1}{\sqrt{8}}e^{-2t}\sin(\sqrt{8}\,t)$$

by Example 131, and so

$$\mathscr{L}^{-1}\left(\frac{s+6}{s^2+4s+12}\right) = e^{-2t}\left[\cos(\sqrt{8}\,t) + \frac{4}{\sqrt{8}}\sin(\sqrt{8}\,t)\right].$$

THE INVERSE VERSION OF RULE 5′

Recall Rule 5′:

$$\mathscr{L}[u(t-a)f(t-a)] = e^{-as}\mathscr{L}(f).$$

This inverts as follows:

> If $\mathscr{L}^{-1}[F(s)] = f(t)$, then $\mathscr{L}^{-1}[e^{-as}F(s)] = u(t-a)f(t-a)$.

The tipoff in using this is that we want the inverse Laplace transform of an exponential times a function whose inverse Laplace transform we know. Here are some examples.

EXAMPLE 133

Find

$$\mathscr{L}^{-1}\left(\frac{e^{-3s}}{s+8}\right).$$

By the inverse of Rule 5′,

$$\mathscr{L}^{-1}\left[e^{-3s}\left(\frac{1}{s+8}\right)\right] = u(t-3)f(t-3),$$

where $f(t) = \mathscr{L}^{-1}[1/(s+8)] = e^{-8t}$.

Then $f(t-3) = e^{-8(t-3)}$ and

$$\mathscr{L}^{-1}\left(\frac{e^{-3s}}{s+8}\right) = u(t-3)e^{-8(t-3)}.$$

This function is 0 for $t < 3$, and is the graph of e^{-8t} shifted three units to the right for $t \geq 3$. This is shown in Figure 40.

EXAMPLE 134
Find

$$\mathscr{L}^{-1}\left(\frac{e^{-2s}}{s^2-3s-4}\right).$$

Again, the e^{-2s} factor suggests the inverse of Rule 5′. We have

$$\mathscr{L}^{-1}\left(\frac{e^{-2s}}{s^2-3s-4}\right)=u(t-2)f(t-2),$$

if we can find

$$f(t)=\mathscr{L}^{-1}\left(\frac{1}{s^2-3s-4}\right).$$

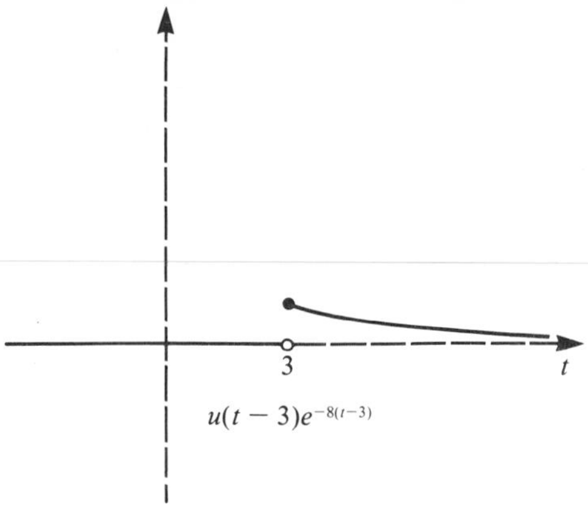

3
t
$u(t-3)e^{-8(t-3)}$

Figure 40

To this end, write $s^2-3s-4=(s-4)(s+1)$, and use partial fractions to write

$$\frac{1}{s^2-3s-4}=-\frac{1}{5}\frac{1}{s+1}+\frac{1}{5}\frac{1}{s-4}.$$

Then,

$$f(t)=-\frac{1}{5}\mathscr{L}^{-1}\left(\frac{1}{s+1}\right)+\frac{1}{5}\mathscr{L}^{-1}\left(\frac{1}{s-4}\right)=-\frac{1}{5}e^{-t}+\frac{1}{5}e^{4t}.$$

Then,

$$\mathscr{L}^{-1}\left(\frac{e^{-2s}}{s^2-3s-4}\right)=u(t-2)f(t-2)$$
$$=-\tfrac{1}{5}u(t-2)e^{-(t-2)}+\tfrac{1}{5}u(t-2)e^{4(t-2)}.$$

EXAMPLE 135

Here is an example involving an initial value problem. Solve

$$y''-2y'-3y=f(t);\qquad y(0)=1,\qquad y'(0)=0,$$

where

$$f(t)=\begin{cases}0, & t<4\\ 3, & t\geq 4.\end{cases}$$

Taking Laplace transforms, we have

$$\mathscr{L}(y'')-2\mathscr{L}(y')-3\mathscr{L}(y)=\mathscr{L}(f).$$

By Rules 1 and 2, and the initial conditions,

$$\mathscr{L}(y'')=s^2\mathscr{L}(y)-sy(0)-y'(0)=s^2\mathscr{L}(y)-s$$

and

$$\mathscr{L}(y') = s\mathscr{L}(y) - y(0) = s\mathscr{L}(y) - 1.$$

Next, note that

$$f(t) = u(t-4)h(t-4),$$

where $h(t) = 3$. Then, by Rule 5′,

$$\mathscr{L}(f) = \mathscr{L}[u(t-4)h(t-4)] = e^{-4s}\mathscr{L}(3) = 3\,\frac{e^{-4s}}{s}.$$

The initial value problem then transforms to

$$s^2\mathscr{L}(y) - s - 2[s\mathscr{L}(y) - 1] - 3\mathscr{L}(y) = 3\,\frac{e^{-4s}}{s}.$$

Solve for $\mathscr{L}(y)$ to obtain

$$\mathscr{L}(y) = \frac{3e^{-4s}}{s(s^2 - 2s - 3)} + \frac{s-2}{(s^2 - 2s - 3)}.$$

Formally, then, the solution of the problem is

$$y(t) = \mathscr{L}^{-1}\left(\frac{3e^{-4s}}{s(s^2 - 2s - 3)}\right) + \mathscr{L}^{-1}\left(\frac{s-2}{(s^2 - 2s - 3)}\right).$$

To write the solution in more explicit form, use a partial fractions decomposition (details of which we omit) to write

$$\begin{aligned} y(t) = {} & -\mathscr{L}^{-1}\left(\frac{e^{-4s}}{s}\right) + \frac{1}{4}\mathscr{L}^{-1}\left(\frac{e^{-4s}}{s-3}\right) + \frac{3}{4}\mathscr{L}^{-1}\left(\frac{e^{-4s}}{s+1}\right) \\ & + \frac{3}{4}\mathscr{L}^{-1}\left(\frac{1}{s+1}\right) + \frac{1}{4}\mathscr{L}^{-1}\left(\frac{1}{s-3}\right) \\ = {} & -u(t-4) + \tfrac{1}{4}u(t-4)e^{3(t-4)} + \tfrac{3}{4}u(t-4)e^{-(t-4)} + \tfrac{3}{4}e^{-t} + \tfrac{1}{4}e^{3t}. \end{aligned}$$

EXAMPLE 136

Solve

$$y^{(3)} - 8y = g(t); \qquad y(0) = y'(0) = y''(0) = 0,$$

where

$$g(t) = \begin{cases} 0, & 0 \le t < 4 \\ 2, & t \ge 4. \end{cases}$$

Here, $g(t) = 2u(t-4)$. Taking the Laplace transform of the differential equation and incorporating the initial conditions gives us

$$s^3\mathscr{L}(y) - 8\mathscr{L}(y) = 2\,\frac{e^{-4s}}{s}.$$

Then,

$$\mathcal{L}(y) = \frac{2e^{-4s}}{s(s^3 - 8)}$$

$$= \frac{2e^{-4s}}{s(s-2)(s^2+2s+4)}$$

$$= -\frac{1}{4}\frac{e^{-4s}}{s} + \frac{1}{12}\frac{e^{-4s}}{s-2} + \frac{1}{6}\frac{e^{-4s}(s+1)}{s^2+2s+4}.$$

Taking inverse Laplace transforms now gives us the solution

$$y(t) = -\tfrac{1}{4}u(t-4) + \tfrac{1}{12}u(t-4)e^{2(t-4)} + \tfrac{1}{6}u(t-4)e^{-(t-4)}\cos[\sqrt{3}(t-4)].$$

Note: In deriving the last term of this solution, we wrote

$$\frac{s+1}{s^2+2s+4} \quad \text{as} \quad \frac{s+1}{(s+1)^2+3}.$$

The last two examples suggest great possibilities for Laplace transforms in solving initial value problems. Section 4.5 illustrates this for problems involving mass-spring systems, electrical networks, and bending of loaded beams.

PROBLEMS FOR SECTION 4.3

In each of Problems 1 through 55, find the inverse Laplace transform of the given function.

1. $\dfrac{1}{s^2 - 4s + 3}$ **2.** $\dfrac{3s+2}{s^2+6s+8}$ **3.** $\dfrac{2s}{s^2-4}$

4. $\dfrac{1}{s+4} - \dfrac{3s}{s^2+16}$ **5.** $\dfrac{s-3}{(s-3)^2+9}$ **6.** $\dfrac{4}{(s-5)^2} - \dfrac{2}{(s-1)^2}$

7. $\dfrac{4}{s^2+9} - \dfrac{1}{(s-3)^2}$ **8.** $\dfrac{s-2}{s^2-4s+8}$

(*Hint:* Complete the square in the denominator.)

9. $\dfrac{4}{s^2-s-2}$ **10.** $\dfrac{3s+1}{s^2+4s}$ **11.** $\dfrac{s-3}{(s-2)^2+2(s-2)+1}$

12. $\dfrac{e^{-3s}}{s^2+4}$ **13.** $\dfrac{2s}{(s-3)^2+4}$ **14.** $\dfrac{s^2-2s+3}{s(s^2-3s+2)}$

15. $\dfrac{s^2+1}{(s-1)(s^2+2)}$ **16.** $\dfrac{s+3}{s^2+6s+15}$ **17.** $\dfrac{e^{-4s}}{s+2}$

18. $\dfrac{3s^2-2}{(s+1)(s^2+6)^2}$ **19.** $\dfrac{3se^{-2s}}{s^2+14}$ **20.** $\dfrac{4s-5}{s^3-s^2-5s-3}$

21. $e^{-2s}\left(\dfrac{s+2}{s^2-4s+8}\right)$ **22.** $\dfrac{s^2+6}{s^3+6s^2+9s}$ **23.** $e^{-5s}\left(\dfrac{2s+1}{s^2-3s+5}\right)$

24. $\dfrac{e^{-4s}}{s^3 - 6s^2 + 5s + 12}$
25. $\dfrac{2s - 3}{s^2 + s - 2}$
26. $\dfrac{s + 2}{s^2 + 2s - 4}$

27. $\dfrac{e^{-2s}}{s^2 + 5s + 6}$
28. $\dfrac{3s^2 + 2s - 1}{(s^2 + 4)(s^2 - 2s + 1)}$
29. $\dfrac{2s - 4}{(s - 1)^4}$

30. $\dfrac{se^{-3s}}{s^2 + 2s + 1}$
31. $\dfrac{e^{-4s}(s^2 + 3s - 2)}{(s + 2)^2(s^2 - 1)}$
32. $\dfrac{8s^2}{(s^2 - 3)^3}$

33. $\dfrac{(-3s + 2)e^{-2s}}{s^2 - 2s + 6}$
34. $\dfrac{se^{-s}}{(s + 1)(s^2 + 3s + 4)}$
35. $\dfrac{2s^2 + 3s - 4}{(s - 3)(s^2 + 4)^2}$

36. $\dfrac{s + 1}{s^2 + 2s}$
37. $\dfrac{(5s - 3)e^{-s}}{s^2 - 7s + 10}$
38. $\dfrac{s^2 - 2s + 4}{s^4 + 4s^3 - 2s^2 - 12s + 9}$

39. $\dfrac{e^{-5s}}{s(s^2 + 9)}$
40. $\dfrac{8s^2 - 3s + 2}{s^3 - 3s^2 - 10s + 24}$
41. $\dfrac{e^{-3s}(s + 4)}{s^2 + 3s - 40}$

42. $\dfrac{2s + 3}{(s^2 + 4)^2(s - 5)}$
43. $\dfrac{se^{-6s}}{(s - 5)^2}$
44. $\dfrac{s + 4}{s^2 + 8s + 21}$

45. $\dfrac{s - 2}{s^2 - 4s + 19}$
46. $\dfrac{e^{-5s}}{s(s^2 + 12)}$
47. $\dfrac{8s^3 - 3s + 2}{s^4 - 3s^3 - 20s^2 + 84s - 80}$

48. $\dfrac{-4e^{-3s}}{s^3 + 2s^2 + s}$
49. $\dfrac{(s - 3)e^{-2s}}{s^2 + 9s - 5}$
50. $\dfrac{8s^3 - s}{(s^2 + 2)^2(s^2 - 4s + 4)}$

51. $\dfrac{se^{-3s}}{s^2 + 4}$
52. $\dfrac{(s + 1)e^{-s}}{(s^2 + 4)(s^2 - 2s + 1)}$
53. $\dfrac{(s - 5)e^{-3s}}{s^2 + 4s + 5}$

54. $\dfrac{-3s^2 + 4s - 9}{s(s^2 + 5)^2}$
55. $\dfrac{-2e^{-2s}(s - 4)}{(s - 5)^3}$

In each of Problems 56 through 65, find an approximate solution for $\mathscr{L}^{-1}[F(s)]$, as in Example 128. Do not attempt these without the help of a computer or calculator to do the arithmetic.

56. $\dfrac{s - 4.73}{s^3 - 3.34s^2 + 6.68s - 5.36}$ (*Hint:* One root of the denominator is 1.34.)

57. $\dfrac{0.37e^{-1.27s}}{s^2 + 0.5s - 0.5}$
58. $\dfrac{s + 1.46}{s^2 + 2.92s + 3.1316}$
59. $\dfrac{2s - 8.4}{s^2 + 0.4s - 0.21}$

60. $\dfrac{e^{-0.2s}}{s^2 - 2.72s - 3.72}$
61. $\dfrac{0.4s - 1}{(s^2 + 0.6)(s + 0.23)}$
62. $\dfrac{0.2e^{-1.56s}}{s^2 - 3s + 2.56}$

63. $\dfrac{0.3s - 5.67}{s^2 + 0.34s - 0.56}$
64. $\dfrac{2.34s^2 + 5.36s - 3.2}{(s^2 - 5.73)(s + 3.84)}$
65. $\dfrac{e^{-4.72s}}{(s^2 + 4.73)^2}$

In each of Problems 66 through 85, solve the initial value problem using Laplace transforms, as in Examples 135 and 136.

66. $y'' + 4y = f(t)$; $y(0) = 1$, $y'(0) = 0$,
where

$$f(t) = \begin{cases} 0, & 0 \le t < 4 \\ 3, & t \ge 4. \end{cases}$$

67. $y'' - 2y' + y = t$; $y(0) = y'(0) = 0$

68. $y'' - 3y' - 4y = e^{-2t}$; $y(0) = 1$, $y'(0) = -1$

69. $y'' + 5y' + 6y = f(t)$; $y(0) = y'(0) = 0$,
where

$$f(t) = \begin{cases} -2, & 0 \le t < 3 \\ 0, & t \ge 3. \end{cases}$$

70. $y^{(3)} - 8y = e^{-3t}$; $y(0) = y'(0) = 0$, $y''(0) = 2$

71. $y^{(4)} - 2y'' + y = 1$; $y(0) = y'(0) = y''(0) = y^{(3)}(0) = 0$

72. $y^{(3)} - y'' + 4y' - 4y = f(t)$; $y(0) = y'(0) = 0$, $y''(0) = 1$,
where

$$f(t) = \begin{cases} 1, & 0 \le t < 5 \\ 2, & t \ge 5. \end{cases}$$

73. $y^{(3)} + 8y = \sin(2t)$; $y(0) = 1$, $y'(0) = y''(0) = 0$

74. $y^{(3)} + 2y' + 3y = 4t$; $y(0) = y'(0) = 0$, $y''(0) = -3$

75. $y'' - 4y' + 4y = f(t)$; $y(0) = -2$, $y'(0) = 1$,
where

$$f(t) = \begin{cases} t, & 0 \le t < 3 \\ t + 2, & t \ge 3. \end{cases}$$

76. $y'' + 8y' - y = 1 - 4t$; $y(0) = y'(0) = -1$

77. $y^{(4)} - 10y'' + 24y = 4$; $y(0) = y'(0) = y''(0) = 0$, $y^{(3)}(0) = 2$

78. $y'' + 7y' - 2y = -2\cos(5t)$; $y(0) = 4$, $y'(0) = 0$

79. $y^{(3)} - 4y'' - 9y' + 36y = 0$; $y(0) = y'(0) = 0$, $y''(0) = 4$

80. $y'' - 4y' + 6y = 3\sinh(t)$; $y(0) = 1$, $y'(0) = 0$

81. $y'' - 4y' + 9y = -\cos(t)$; $y(0) = y'(0) = 0$

82. $y'' + 2y' - 7y = f(t)$; $y(0) = -2$, $y'(0) = 0$,
where

$$f(t) = \begin{cases} 0, & 0 \le t < 5 \\ 2, & t \ge 5. \end{cases}$$

83. $y'' - 3y' + 5y = 7e^{-3t}$; $y(0) = y'(0) = 0$

84. $y'' - 5y' - 2y = t^2 - 1$; $y(0) = 2$, $y'(0) = 0$

85. $y'' + 2y' + y = t^2e^{-4t} - 1$; $y(0) = 1$, $y'(0) = -2$

4.4 CALCULATING INVERSE LAPLACE TRANSFORMS: PART 2 —THE HEAVISIDE EXPANSION FORMULAS

As we have seen, we often need to calculate the inverse Laplace transform of a quotient of polynomials. This can be done using partial fractions. There is, however, an alternative method which utilizes the *Heaviside expansion formulas*. The advantage of this method is that it circumvents much of the algebra involved in finding the coefficients in a partial fractions expansion.

Consider, then, the problem of finding $\mathscr{L}^{-1}(P(s)/Q(s))$, where $P(s)$ and $Q(s)$ are polynomials having no common factor and $Q(s)$ has higher degree than $P(s)$. Let

$$f(t) = \mathscr{L}^{-1}\left(\frac{P(s)}{Q(s)}\right).$$

Heaviside's formulas enable us to write $f(t)$ directly as a sum of terms, each determined by a factor of $Q(s)$. The four cases are as follows:

Case 1 If $Q(s)$ contains an unrepeated linear factor $(s - a)$, then $f(t)$ contains the term

$$H(a)e^{at},$$

where

$$H(s) = \frac{P(s)(s-a)}{Q(s)}.$$

Another expression for $H(a)$ is

$$H(a) = \frac{P(a)}{Q'(a)}.$$

Case 2 If $k \geq 2$ and $Q(s)$ contains the linear factor $(s - a)^k$, but not $(s - a)^{k+1}$, then the corresponding term in $f(t)$ is

$$\left[\frac{H^{(k-1)}(a)}{(k-1)!} + \frac{H^{(k-2)}(a)}{(k-2)!}\frac{t}{1!} + \frac{H^{(k-3)}(a)}{(k-3)!}\frac{t^2}{2!} + \cdots + \frac{H'(a)}{1!}\frac{t^{k-2}}{(k-2)!} + H(a)\frac{t^{k-1}}{(k-1)!}\right]e^{at},$$

in which

$$H(s) = \frac{P(s)(s-a)^k}{Q(s)}$$

and $H^{(j)}(a)$ denotes the jth derivative of $H(s)$ evaluated at a.

Case 3 If $Q(s)$ contains the unrepeated quadratic factor $(s - a)^2 + b^2$, then $f(t)$ contains the terms

$$\frac{1}{b}\left[\alpha_i \cos(bt) + \alpha_r \sin(bt)\right]e^{at},$$

in which α_r is the real part of $H(a + ib)$, α_i is the imaginary part of $H(a + ib)$, and

$$H(s) = \frac{P(s)[(s-a)^2 + b^2]}{Q(s)}.$$

Recall that the real part of a complex number $u + iv$ is u and the imaginary part is v. Both u and v are themselves real numbers.

Case 4 If $Q(s)$ contains the quadratic factor $[(s-a)^2+b^2]^2$, but not $[(s-a)^2+b^2]^3$, then $f(t)$ contains the terms

$$\frac{1}{2b^3}(b\alpha_i-\beta_r)e^{at}\cos(bt)+\frac{1}{2b^3}(b\beta_i-\alpha_r)e^{at}\sin(bt)+\frac{te^{at}}{2b^2}[\alpha_i\sin(bt)-\alpha_r\cos(bt)],$$

where α_r is the real part of $H(a+ib)$, α_i is the imaginary part of $H(a+ib)$, β_r is the real part of $H'(a+ib)$, β_i is the imaginary part of $H'(a+ib)$, and

$$H(s)=\frac{P(s)[(s-a)^2+b^2]^2}{Q(s)}.$$

There is a general formula for terms in $f(t)$ corresponding to factors $[(s-a)^2+b^2]^k$ in $Q(s)$ when $k>2$ but it is extremely complicated. Generally, the Heaviside method is used for quadratic factors only when $k=1$ or 2. We shall give derivations of the above formulas after looking at some examples.

EXAMPLE 137

Find

$$\mathscr{L}^{-1}\left(\frac{3s}{(s^2-2s+5)(s+1)(s-5)}\right).$$

Here $Q(s)$ has unrepeated linear factors $(s-5)$ and $(s+1)$, and an unrepeated quadratic factor (s^2-2s-5), which is $(s-1)^2+4$.

Corresponding to $(s-5)$, $f(t)$ has a term

$$\frac{P(5)}{Q'(5)}e^{5t},$$

which works out to $\frac{1}{8}e^{5t}$.

Corresponding to $(s+1)$, which must be written $[s-(-1)]$, $f(t)$ has a term

$$\frac{P(-1)}{Q'(-1)}e^{-t}$$

or $\frac{1}{16}e^{-t}$.

For the quadratic term $[(s-1)^2+4]$, let

$$H(s)=\frac{3s[(s-1)^2+4]}{[(s-1)^2+4](s+1)(s-5)}=\frac{3s}{(s+1)(s-5)}.$$

Let $a=1$ and $b=2$ in case 3 above, and compute

$$H(1+2i)=\frac{3(1+2i)}{(2+2i)(-4+2i)}=-\frac{3}{8}-\frac{3}{8}i.$$

Then

$$\alpha_r=\text{real part of }(-\tfrac{3}{8}-\tfrac{3}{8}i)=-\tfrac{3}{8}$$

and

$$\alpha_i=\text{imaginary part of }(-\tfrac{3}{8}-\tfrac{3}{8}i)=-\tfrac{3}{8}.$$

The term in $f(t)$ corresponding to $(s-1)^2+4$ in $Q(s)$ is then

$$\tfrac{1}{2}[-\tfrac{3}{8}\cos(2t)-\tfrac{3}{8}\sin(2t)]e^t.$$

Then,

$$\mathscr{L}^{-1}\left(\frac{3s}{(s^2-2s+5)(s+1)(s-5)}\right)=\frac{1}{8}e^{5t}+\frac{1}{16}e^{-t}-\frac{3}{16}e^t[\cos(2t)+\sin(2t)].$$

EXAMPLE 138

Find

$$\mathscr{L}^{-1}\left(\frac{2s^2+7}{(s^2+4)(s+3)^3}\right).$$

Here $Q(s)$ has a cubed linear term and a quadratic term to the first power. To handle the $(s+3)^3$ term, use case 2 above with $k=3$, $a=-3$, and

$$H(s)=\frac{2s^2+7}{s^2+4}.$$

We find that

$$\left[\frac{H''(-3)}{2!}+\frac{H'(-3)}{1!}t+\frac{H(-3)}{2!}t^2\right]e^{-3t}=\left[-\frac{23}{2197}-\frac{6}{169}t+\frac{25}{26}t^2\right]e^{-3t}.$$

For the (s^2+4) term, use case 3 above with

$$H(s)=\frac{2s^2+7}{(s+3)^3}.$$

Write $s^2+4=(s-a)^2+b^2$ with $a=0$ and $b=2$, and compute

$$H(a+ib)=H(2i)=\frac{2(-4)+7}{(3+2i)^3}=\frac{9}{2197}+\frac{46}{2197}i.$$

The term in $f(t)$ corresponding to (s^2+4) is then

$$\frac{1}{2}\left[\frac{46}{2197}\cos(2t)+\frac{9}{2197}\sin(2t)\right].$$

Then,

$$\mathscr{L}^{-1}\left(\frac{2s^2+7}{(s^2+4)(s+3)^3}\right)=\left(-\frac{23}{2197}-\frac{6}{169}t+\frac{25}{26}t^2\right)e^{-3t}+\frac{23}{2197}\cos(2t)+\frac{9}{4394}\sin(2t).$$

EXAMPLE 139

Find

$$\mathscr{L}^{-1}\left(\frac{s-2}{(s^2-4s+13)^2(s+4)(s-6)}\right).$$

To handle the term $(s^2 - 4s + 13)^2$, write it as $[(s - 2)^2 + 9]^2$ and use case 4 with $a = 2, b = 3$, and

$$H(s) = \frac{s - 2}{(s + 4)(s - 6)}.$$

First, compute

$$H(2 + 3i) = \frac{18 - 99i}{1125}.$$

Thus,

$$\alpha_r = \frac{18}{1125} \quad \text{and} \quad \alpha_i = \frac{-99}{1125}.$$

Next, compute

$$H'(2 + 3i) = -\frac{3159}{253{,}125} - \frac{1188}{253{,}125}\, i.$$

Then,

$$\beta_r = -\frac{3159}{253{,}125} \quad \text{and} \quad \beta_i = -\frac{1188}{253{,}125}.$$

By substituting these into the result in case 4, we get the terms in $f(t)$ corresponding to $(s^2 - 4s + 13)^2$. For the term corresponding to $(s + 4)$, use case 1 and compute

$$\frac{P(-4)}{Q'(-4)} = \frac{1}{3375}.$$

Similarly, for the term in $f(t)$ corresponding to $(s - 6)$, compute

$$\frac{P(6)}{Q'(6)} = \frac{2}{3125}.$$

Putting everything together now gives us

$$f(t) = -\frac{1179}{253{,}125} e^{2t} \cos(3t) - \frac{141}{253{,}125} e^{2t} \sin(3t)$$
$$+ \frac{te^{2t}}{54}\left[\frac{-99}{1125} \sin(3t) - \frac{18}{1125} \cos(3t)\right] + \frac{1}{3375} e^{-4t} + \frac{2}{3125} e^{6t}.$$

The Heaviside formulas are particularly convenient when using Laplace transforms to solve differential equations, since this often gives rise to a problem of computing the inverse Laplace transform of a quotient of polynomials. We pursue this in the exercises, concluding this section with derivations of the above results.

Derivation of Case 1 We can write

$$\frac{P(s)}{Q(s)} = \frac{A}{s - a} + G(s),$$

in which A is constant and $G(s)$ has no factor $(s-a)$ in numerator or denominator. Then,

$$f(t) = Ae^{at} + \mathscr{L}^{-1}[G(s)].$$

To solve for A, note that

$$H(s) = \frac{P(s)(s-a)}{Q(s)} = A + (s-a)G(s).$$

Then $H(a) = A$,* and the factor $(s-a)$ in $Q(s)$ gives rise to the term $H(a)e^{at}$ in $f(t)$.

Since $Q(a) = 0$, we can also write

$$\frac{P(s)(s-a)}{Q(s)} = P(s)\left(\frac{s-a}{Q(s)-Q(a)}\right) = A + (s-a)G(s).$$

Letting $s \to a$, we get

$$A = \frac{P(a)}{Q'(a)}.$$

Derivation of Case 2 In theory at least, we can write

$$\frac{P(s)}{Q(s)} = \frac{A_1}{s-a} + \frac{A_2}{(s-a)^2} + \cdots + \frac{A_k}{(s-a)^k} + G(s),$$

where $(s-a)$ does not appear in $G(s)$. Then,

$$H(s) = \frac{P(s)(s-a)^k}{Q(s)} = A_1(s-a)^{k-1} + A_2(s-a)^{k-2}$$
$$+ \cdots + A_{k-1}(s-a) + A_k + (s-a)^k G(s).$$

Immediately† from this,

$$H(a) = A_k.$$

Now compute

$$H''(s) = A_1(k-1)(k-2)(s-a)^{k-2} + A_2(k-2)(k-3)(s-a)^{k-4}$$
$$+ \cdots + 2A_{k-2} + k(k-1)(s-a)^{k-2}G(s) + 2k(s-a)^{k-1}G'(s)$$
$$+ (s-a)^k G''(s).$$

Then,

$$H''(a) = 2A_{k-2} \quad \text{or} \quad A_{k-2} = \tfrac{1}{2}H''(a).$$

* This is an abuse of notation; actually $A = \lim_{s\to a} H(s)$.

† As in the previous derivation, we should, strictly speaking, have $A_k = \lim_{s\to a} H(s)$, $A_{k-2} = \lim_{s\to a} \frac{1}{2}H''(s)$, and so on.

Continuing in this way, we find that

$$H^{(3)}(a) = (3)(2)A_{k-3},$$
$$\vdots$$
$$H^{(k-1)}(a) = (k-1)(k-2)\cdots(3)(2)A_1.$$

In general,

$$A_{k-j} = \frac{H^{(j)}(a)}{j!},$$

for $j = 0, 1, 2, \ldots, (k-1)$. Thus,

$$\frac{P(s)}{Q(s)} = \frac{H^{(k-1)}(a)}{(k-1)!}\frac{1}{s-a} + \frac{H^{(k-2)}(a)}{(k-2)!}\frac{1}{(s-a)^2} + \frac{H^{(k-3)}(a)}{(k-3)!}\frac{1}{(s-a)^3}$$
$$+ \cdots + \frac{H'(a)}{1!}\frac{1}{(s-a)^{k-1}} + H(a)\frac{1}{(s-a)^k} + G(s).$$

Now,

$$\mathscr{L}^{-1}\left(\frac{1}{(s-a)^r}\right) = \frac{t^{r-1}e^{at}}{(r-1)!}.$$

Thus, the factor $(s-a)^k$ in $Q(s)$ corresponds in $f(t)$ to

$$\frac{H^{(k-1)}(a)}{(k-1)!}e^{at} + \frac{H^{(k-2)}(a)}{(k-2)!}\frac{t}{1!}e^{at} + \frac{H^{(k-3)}(a)}{(k-3)!}\frac{t^2}{2!}e^{at}$$
$$+ \cdots + \frac{H'(a)}{1!}\frac{t^{k-2}}{(k-2)!}e^{at} + H(a)\frac{t^{k-1}}{(k-1)!}e^{at},$$

as we wanted to show.

Derivation of Case 3 Write

$$\frac{P(s)}{Q(s)} = \frac{As + B}{(s-a)^2 + b^2} + G(s),$$

where $[(s-a)^2 + b^2]$ does not appear in $G(s)$. Then,

$$H(s) = [(s-a)^2 + b^2]\frac{P(s)}{Q(s)} = As + B + [(s-a)^2 + b^2]G(s).$$

Now,

$$H(a + ib) = aA + B + ibA.$$

Then,

$$\alpha_r = aA + B \quad \text{and} \quad \alpha_i = bA.$$

Solve these for A and B to obtain

$$A = \frac{1}{b}\alpha_i \quad \text{and} \quad B = \frac{b\alpha_r - a\alpha_i}{b}.$$

Then,

$$\frac{P(s)}{Q(s)} = \frac{1}{b}\left[\frac{\alpha_i(s-a)}{(s-a)^2+b^2} + \frac{b\alpha_r}{(s-a)^2+b^2}\right] + G(s).$$

The contribution to $f(t)$ from the factor $[(s-a)^2+b^2]$ in $Q(s)$ is then

$$\frac{1}{b}\mathscr{L}^{-1}\left(\frac{\alpha_i(s-a)}{(s-a)^2+b^2}\right) + \frac{1}{b}\mathscr{L}^{-1}\left(\frac{b\alpha_r}{(s-a)^2+b^2}\right) = \frac{1}{b}[\alpha_i\cos(bt)+\alpha_r\sin(bt)]e^{at},$$

as was to be proved.

The derivation of case 4 is left as Problem 32 of this section.

We are now in a position to move in the next section to some applications of Laplace transforms.

PROBLEMS FOR SECTION 4.4

In each of Problems 1 through 20, use the Heaviside formulas to find the inverse Laplace transform of the given function.

1. $\dfrac{s-3}{(s^2+4)(s+7)}$ **2.** $\dfrac{s^2}{(s+2)^2(s+3)}$ **3.** $\dfrac{3s-4}{(s-1)(s+2)^2}$ **4.** $\dfrac{s^2+1}{(s+2)^3(s-1)}$

5. $\dfrac{-3s-2}{(s+4)^2}$ **6.** $\dfrac{8}{(s-1)^2(s+3)}$ **7.** $\dfrac{-s}{(s-4)^2(s-5)}$ **8.** $\dfrac{s^2+4s+1}{(s-2)^2(s+3)}$

9. $\dfrac{2s-5}{(s^2+2)^2(s-1)}$ **10.** $\dfrac{s+1}{(s^2+3s-1)^2}$ **11.** $\dfrac{4}{(s-2)^2(s+6)}$ **12.** $\dfrac{-5s}{(s^2+s-4)(s+1)}$

13. $\dfrac{2s^2}{(s+2)(s-3)^2}$ **14.** $\dfrac{s-4}{(s+2)(s+1)(s-3)}$ **15.** $\dfrac{4s^2-2}{(s-1)^2(s+2)(s+3)}$ **16.** $\dfrac{1}{(s^2-2s+3)^2(s+5)}$

17. $\dfrac{4s^2+5}{(s+3)(s^2+3s+7)}$ **18.** $\dfrac{s-2}{(s+4)^2(s-2)}$ **19.** $\dfrac{s^3}{(s+3)^2(s+2)^2}$ **20.** $\dfrac{s^2+4s-2}{(s+1)^3(s-2)}$

In each of Problems 21 through 30, solve the initial value problem by using Laplace transforms. Use the Heaviside formulas to find inverse Laplace transforms where applicable.

21. $y'' + 3y' - 4y = e^{-t};\quad y(0) = y'(0) = 0$

22. $y^{(3)} + 4y'' - 2y' + 8y = 1;\quad y(0) = y'(0) = y''(0) = 0$

23. $y'' + 2y' - 3y = e^{-3t};\quad y(0) = y'(0) = 0$

24. $y'' - 4y' + 6y = -2\cos(3t);\quad y(0) = y'(0) = 0$

25. $y'' + 6y' - 8y = 0;\quad y(0) = 1,\quad y'(0) = 0$

26. $y^{(3)} + 8y'' - 2y' + y = 1 - e^{-t};\quad y(0) = y'(0) = y''(0) = 0$

27. $y^{(3)} + 2y'' - 11y' - 12y = 4;\quad y(0) = y'(0) = y''(0) = 0$

28. $y'' + 5y' - 8y = \sin(2t) + 4;\quad y(0) = y'(0) = 0$

29. $y'' - y' - 6y = \cos(2t);\quad y(0) = y'(0) = 0$

30. $y'' - y' + 7y = \cosh(3t);\quad y(0) = 1,\quad y'(0) = 0$

31. Recall that, in case 2 of this section, $P(s)/Q(s)$ contains the factor $(s-a)^k$ with $k > 1$. Show that the term in $f(t)$ corresponding to this factor is the derivative

$$\frac{1}{n!}\frac{\partial^{k-1}}{\partial s^{k-1}}[H(s)e^{st}],$$

evaluated at $s = a$. Here, $H(s)$ is as defined in case 2.

32. Derive case 4 of the Heaviside formulas. [*Hint:* First write

$$\frac{P(s)}{Q(s)} = \frac{As+B}{(s-a)^2+b^2} + \frac{Cs+D}{[(s-a)^2+b^2]^2} + G(s).$$

Now look at $H(s)$, and compute $H(a+ib)$ and $H'(a+ib)$ to solve for A, B, C, and D in terms of real and imaginary parts of $H(a+ib)$ and $H'(a+ib)$.

33. Suppose that $P(s)/Q(s)$ contains a factor $[(s-a)^2+b^2]^3$, but not $[(s-a)^2+b^2]^4$. Determine the terms in $f(t)$ corresponding to this factor.

4.5 LAPLACE TRANSFORMS IN SOLVING TYPICAL ENGINEERING PROBLEMS

This section is devoted to examples of Laplace transforms at work solving initial value problems arising in typical applications.

EXAMPLE 140

Consider the problem of finding the charge $Q(t)$ in the RLC circuit of Figure 41 if the electromotive force is zero until time $t = 2$ and then has constant value k. Analytically,

$$E(t) = k[1 - u(t-2)],$$

as shown in Figure 42.

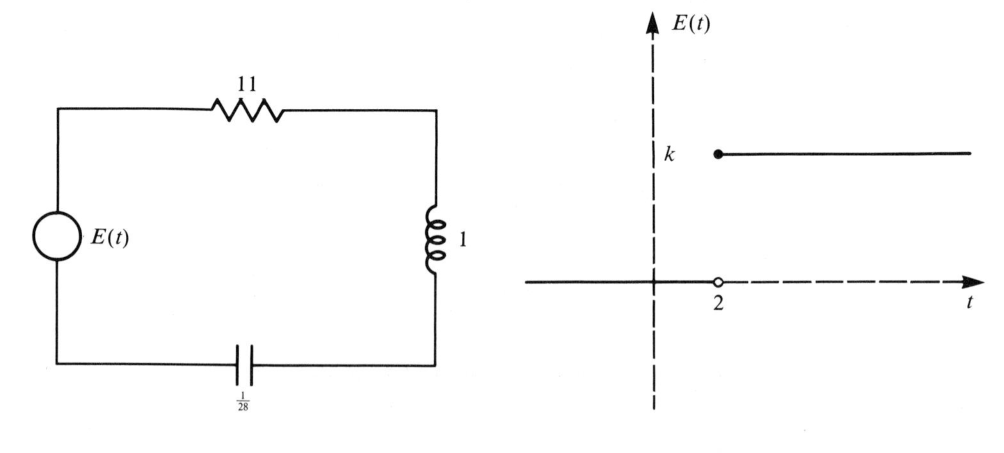

Figure 41

Figure 42

The differential equation for the charge is

$$Q'' + 11Q' + 28Q = E(t).$$

We shall assume initial conditions

$$Q(0) = Q'(0) = 0.$$

Apply the Laplace transform to the differential equation and insert the initial conditions to obtain

$$s^2\mathscr{L}(Q) + 11s\mathscr{L}(Q) + 28\mathscr{L}(Q) = \mathscr{L}[E(t)] = \frac{k}{s} - \frac{k}{s}e^{-2s}.$$

Solve for $\mathscr{L}(Q)$ to obtain

$$\mathscr{L}(Q) = \frac{k}{s(s+4)(s+7)} - \frac{ke^{-2s}}{s(s+4)(s+7)}.$$

It is now routine to invert the Laplace transform to obtain

$$Q(t) = k[1 - u(t-2)]\left[\frac{1}{18} - \frac{e^{-4t}}{12} + \frac{e^{-7t}}{21}\right].$$

EXAMPLE 141

Consider the circuit of Figure 43. We want to solve for the current in each part of the network. Assume that at time 0 all the currents are zero, i.e.,

$$I_1(0) = I_2(0) = I_3(0) = 0.$$

Kirchhoff's current law says that the sum of the currents into or out of any point is zero. Thus, from point P, we have

$$I_1 - I_2 - I_3 = 0.$$

We have three unknowns, I_1, I_2, and I_3, and so need two more equations. Kirchhoff's voltage law says that, at any time t, the sum of the voltage drops around any closed path in the network in a given direction is zero.

Using the closed path on the left, this gives us

$$4I_1 + 2I_2' = 1.$$

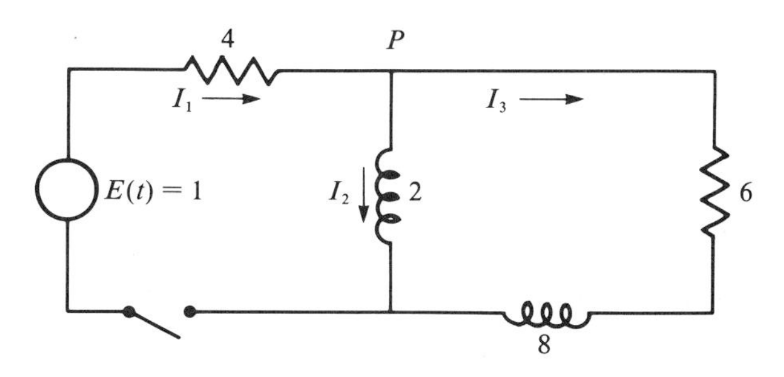

Figure 43

Using the closed path on the right, we get

$$-2I'_2 + 6I_3 + 8I'_3 = 0.$$

This gives us a total of three equations, two of which are first order differential equations. Eliminate I_1 from the last two equations by substituting $I_1 = I_2 + I_3$ to get

$$4I_2 + 4I_3 + 2I'_2 = 1$$
$$-2I'_2 + 6I_3 + 8I'_3 = 0.$$

Taking Laplace transforms of both equations, and recalling that $I_2(0) = I_3(0) = 0$, we have

$$4\mathscr{L}(I_2) + 4\mathscr{L}(I_3) + 2s\mathscr{L}(I_2) = \frac{1}{s}$$

and

$$-2s\mathscr{L}(I_2) + 6\mathscr{L}(I_3) + 8s\mathscr{L}(I_3) = 0.$$

Solve these for $\mathscr{L}(I_2)$ and $\mathscr{L}(I_3)$ to obtain

$$\mathscr{L}(I_2) = \frac{4s + 3}{s(8s^2 + 26s + 12)}$$

and

$$\mathscr{L}(I_3) = \frac{1}{8s^2 + 26s + 12}.$$

It is possible to invert these exactly, but much easier to approximate to, say, four decimals. Write

$$\mathscr{L}(I_2) = \frac{0.5s + 0.375}{s(s + 0.5570)(s + 2.6930)}.$$

Using Heaviside's formula (case 1 of Section 4.4), we get

$$\begin{aligned} I_2 &= \frac{0.375}{(0.5570)(2.6930)} e^{0t} + \frac{0.5(-0.5570) + 0.375}{(-0.5570)(-0.5570 + 2.6930)} e^{-0.5570t} \\ &\quad + \frac{0.5(-2.6930) + 0.375}{(-2.6930)(-2.6930 + 0.5570)} e^{-2.6930t} \\ &= 0.25 - 0.0811e^{-0.5570t} - 0.1689e^{-2.6930t}. \end{aligned}$$

Next, write

$$\mathscr{L}(I_3) = \frac{0.125}{(s + 0.5570)(s + 2.6930)}.$$

Then,

$$\begin{aligned} I_3 &= \frac{0.125}{-0.5570 + 2.6930} e^{-0.5570t} + \frac{0.125}{-2.6930 + 0.5570} e^{-2.6930t} \\ &= 0.0585e^{-0.5570t} - 0.0585e^{-2.6930t}. \end{aligned}$$

Of course, we now also have I_1, since $I_1 = I_2 + I_3$. Note that, as $t \to \infty$, $I_1 \to 0.25$, $I_2 \to 0.25$, and $I_3 \to 0$.

EXAMPLE 142

Consider the mass-spring system of Figure 44, in which y_1 and y_2 are displacements from static equilibrium positions and the positive direction is downward. We neglect the mass of the springs themselves, and also assume that damping is negligible. If there are external driving forces $F_1(t)$ and $F_2(t)$ acting on m_1 and m_2, respectively, then the motion is governed by

$$m_1 y_1'' = -k_1 y_1 + k_2(y_2 - y_1) + F_1(t)$$
$$m_2 y_2'' = -k_2(y_2 - y_1) - k_3 y_2 + F_2(t).$$

This is a system of second order differential equations which can be solved by Laplace transforms when specific data is inserted.

As an illustration, let

$$m_1 = 4, \qquad m_2 = 2,$$
$$k_1 = 2, \qquad k_2 = 2, \qquad k_3 = 1,$$
$$y_1(0) = y_1'(0) = y_2(0) = y_2'(0) = 0$$

and assume driving forces

$$F_1(t) = 1, \qquad F_2(t) = 0.$$

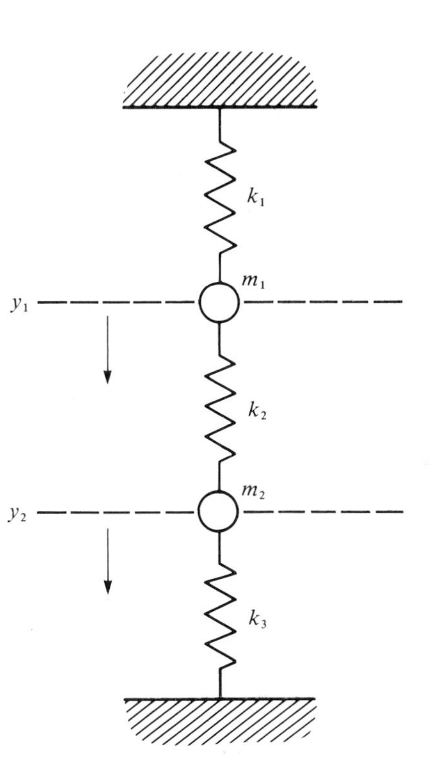

Figure 44

We then have

$$4y_1'' = -2y_1 + 2(y_2 - y_1) + 1$$

and

$$2y_2'' = -2(y_2 - y_1) - y_2.$$

After taking Laplace transforms and rearranging terms, we get

$$(4s^2 + 4)\mathscr{L}(y_1) - 2\mathscr{L}(y_2) = \frac{1}{s}$$
$$-2\mathscr{L}(y_1) + (2s^2 + 3)\mathscr{L}(y_2) = 0.$$

Solve these to obtain

$$\mathscr{L}(y_1) = \frac{2s^2 + 3}{s(8s^4 + 20s^2 + 8)}$$

and

$$\mathscr{L}(y_2) = \frac{1}{s(4s^4 + 10s^2 + 4)}.$$

These are most easily handled by approximation. Write (to four decimal places)

$$\mathscr{L}(y_1) = \frac{0.25s + 0.375}{s(s^2 + 0.5)(s^2 + 2)}$$

$$= \frac{0.375}{s} - \frac{0.3333s}{s^2 + 0.5} - \frac{0.0417s}{s^2 + 2}$$

and

$$\mathscr{L}(y_2) = \frac{0.25}{s(s^2 + 0.5)(s^2 + 2)}$$

$$= \frac{0.25}{s} - \frac{0.1667s}{s^2 + 0.5} - \frac{0.0833s}{s^2 + 2}.$$

Then,

$$y_1 = 0.375 - 0.3333\cos(0.7071t) - 0.0417\cos(1.4142t)$$

and

$$y_2 = 0.25 - 0.1667\cos(0.7071t) - 0.0833\cos(1.4142t).$$

In these expressions, we approximated $\sqrt{0.5}$ by 0.7071 and $\sqrt{2}$ by 1.4142. Note also that we used partial fractions to find the inverse Laplace transforms here; we could have used Heaviside's formulas. The student should do this for the practice.

EXAMPLE 143

In Example 90 (Chapter 3) we considered deflection of a loaded beam. We now consider this problem in more detail.

Suppose we have a beam of length L, as in Figure 45. Distance along the beam is denoted by x, and deflection by y. Given a vertical load $W(x)$, the deflection is governed by

$$y^{(4)} = \frac{1}{EI} W(x),$$

where E = Young's modulus and I = moment of inertia about an axis through the origin and perpendicular to the plane of the diagram.

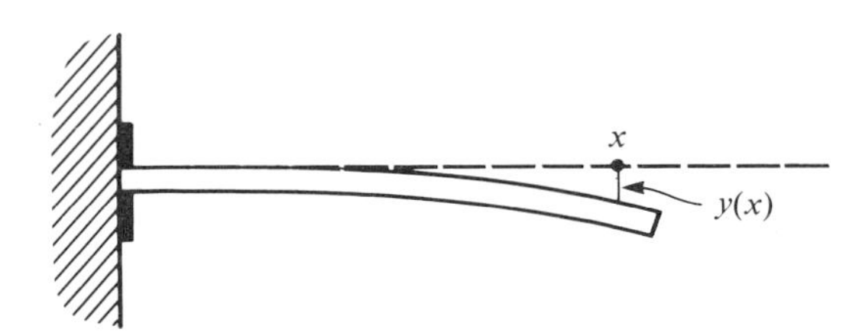

Figure 45. Deflection of a beam.

We shall solve for the deflection, assuming that the weight of the beam is negligible and that $W(x)$ decreases uniformly from a given value at the left end to zero at the midpoint and then remains at 0. This function can be written

$$W(x) = \frac{2w}{L}\left[\frac{L}{2} - x + \left(x - \frac{L}{2}\right)u\left(x - \frac{L}{2}\right)\right], \qquad 0 < x < L,$$

where w = constant = weight at left end. The graph of $W(x)$ is shown in Figure 46.

We assume initial conditions

$$y(0) = 0, \qquad y'(0) = 0,$$

$$y''(0) = \frac{wL^2}{24EI}, \qquad y^{(3)}(0) = \frac{-wL}{4EI}.$$

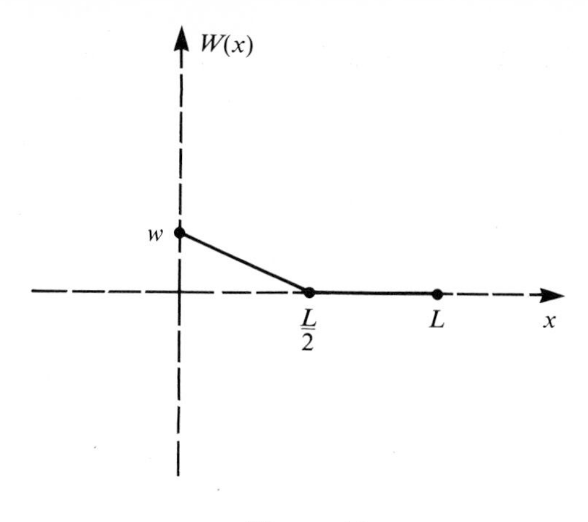

Figure 46

The first two of these conditions correspond to a clamped left end; the other two were chosen in accordance with experience.

Strictly speaking, we cannot yet apply the Laplace transform to the differential equation because $W(x)$ is defined only for $0 \leq x \leq L$. We overcome this difficulty by letting $W(x) = 0$ for $x > L$. We are only interested in the resulting $y(x)$ when $0 \leq x \leq L$, however.

Now calculate

$$\mathscr{L}(y^{(4)}) = s^4\mathscr{L}(y) - s^3(0) - s^2(0) - s\frac{wL^2}{24EI} - \frac{-wL}{4EI}$$

$$= \frac{1}{EI}\mathscr{L}[W(x)] = \frac{1}{EI}\left(\frac{w}{s} - \frac{2w}{L}\frac{1}{s^2} + \frac{2w}{L}\frac{e^{-Ls/2}}{s^2}\right).$$

Then

$$\mathscr{L}(y) = \frac{w}{EI}\frac{1}{s^5} - \frac{2w}{LEI}\frac{1}{s^6} + \frac{2w}{LEI}\frac{e^{-Ls/2}}{s^6} + \frac{wL^2}{24EI}\frac{1}{s^3} - \frac{wL}{4EI}\frac{1}{s^4}.$$

Then

$$y(x) = \frac{w}{12EI}\left[\frac{L^2}{4}x^2 - \frac{L}{2}x^3 + \frac{1}{2}x^4 - \frac{1}{5L}x^5 + \frac{1}{5L}u\left(x - \frac{L}{2}\right)\left(x - \frac{L}{2}\right)^5\right].$$

EXAMPLE 144

We conclude this section with another example from circuit theory. Consider the *RLC* circuit of Figure 47. The electromotive force is an impulse: From time $t = 0$ to $t = 4$ seconds, $E(t) = 0$; thereafter it is 6 volts. We want the charge $Q(t)$, assuming that $Q(0) = Q'(0) = 0$.

The differential equation is

$$20Q'' + 5Q' + 10Q = E(t).$$

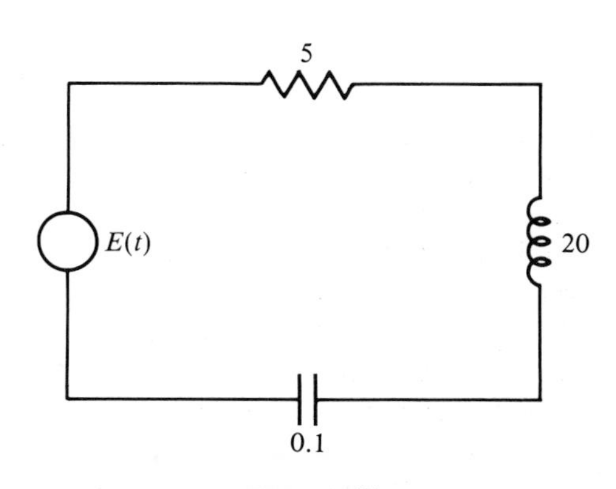

Figure 47

We can write $E(t) = 6u(t-4)$. Taking the Laplace transform of the differential equation gives us

$$20s^2\mathscr{L}(Q) + 5s\mathscr{L}(Q) + 10\mathscr{L}(Q) = 6\mathscr{L}[u(t-4)] = 6\,\frac{e^{-4s}}{s}.$$

Then,

$$\begin{aligned}\mathscr{L}(Q) &= \frac{6}{5}\left(\frac{e^{-4s}}{s(4s^2+s+2)}\right)\\ &= \frac{0.3e^{-4s}}{s(s^2+0.25s+0.5)}\\ &= \frac{0.6e^{-4s}}{s} - \frac{(0.6s+0.15)e^{-4s}}{s^2+0.25s+0.5}.\end{aligned}$$

Now,

$$\mathscr{L}^{-1}\left(\frac{0.6e^{-4s}}{s}\right) = 0.6u(t-4)$$

and, by the inverse version of Rule 5′,

$$\mathscr{L}^{-1}\left(\frac{(0.6s+0.15)e^{-4s}}{s^2+0.25s+0.5}\right) = u(t-4)f(t-4),$$

where

$$f(t) = \mathscr{L}^{-1}\left(\frac{0.6s+0.15}{s^2+0.25s+0.5}\right).$$

To determine $f(t)$, write

$$f(t) = \mathscr{L}^{-1}\left(\frac{0.6s+0.15}{(s+0.125)^2+0.696^2}\right).$$

Using Heaviside's formula (case 3), we let $H(s) = 0.6s + 0.15$ and calculate

$$H(-0.125+0.696i) = 0.075 + 0.4176i.$$

Thus,

$$\begin{aligned}f(t) &= \frac{1}{0.696}(0.4176\cos(0.696t) + 0.075\sin(0.696t))e^{-0.125t}\\ &= [0.6\cos(0.696t) + 0.1078\sin(0.696t)]e^{-0.125t}.\end{aligned}$$

Finally, we have the solution:

$$\begin{aligned}Q(t) &= 0.6u(t-4) - u(t-4)f(t-4)\\ &= u(t-4)\{0.6 - 0.6e^{-0.125(t-4)}\cos[0.696(t-4)]\\ &\quad - 0.1078e^{-0.125(t-4)}\sin[0.696(t-4)]\}.\end{aligned}$$

PROBLEMS FOR SECTION 4.5

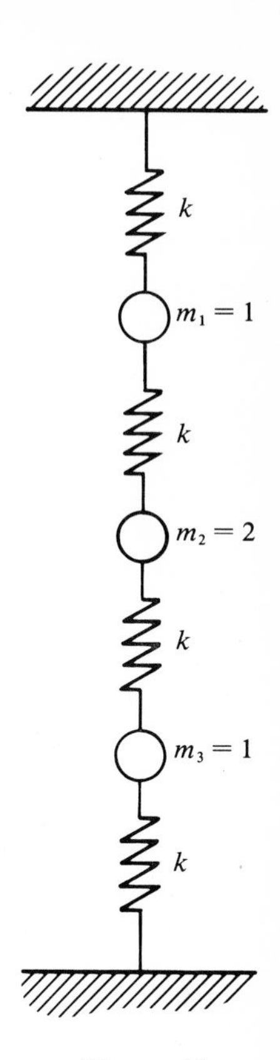

Figure 48

1. Solve for the deflection of a loaded beam (Example 90) when the weight function is given by

$$W(x) = \begin{cases} 0, & 0 \le x < \frac{1}{2}L \\ x - \frac{1}{2}L, & \frac{1}{2}L \le x < L. \end{cases}$$

Assume the same initial conditions as in Example 141.

2. Redo Problem 1 with weight function

$$W(x) = \begin{cases} 0, & 0 \le x < \frac{1}{2}L \\ (x - \frac{1}{2}L)^2, & \frac{1}{2}L \le x < L. \end{cases}$$

3. Redo Problem 1 with weight function $W(x) = x(x - L)$.
4. Derive and solve the differential equations for the mass-spring system of Figure 48. All springs have the same spring constant k. Assume no damping and a zero driving force on each mass. Assume zero initial displacement and velocity of each mass.
5. Redo Problem 4 with a force $F_3(t) = A \sin(2t)$ acting on m_3 and no forces on m_1 or m_2.
6. Redo Problem 4 with force $F_2(t) = u(t - 3)$ acting on m_2 and no forces on m_1 or m_3.
7. Derive and solve the differential equations for the mass-spring system of Figure 48 if all springs have the same spring constant k, spring 1 has a damping constant $c_1 = 1$, the other damping constants are zero, and there is a driving force $F_3(t) = A \cos(2t)$ acting on m_3.
8. Derive and solve the differential equations for the mass-spring system of Figure 48 if springs 1 and 2 have spring constants k, spring 3 has spring constant $2k$, spring 3 has damping constant c_3, and mass m_3 has a driving force $F_3(t) = 1 - u(t - 2)$ acting on it.
9. Solve for the current in the RL circuit of Figure 49 if

$$E(t) = \begin{cases} 0, & 0 \le t < 5 \\ 2, & t \ge 5. \end{cases}$$

Assume that the initial current is zero.

10. Redo Problem 9 for the case where the electromotive force is kept at a constant value k until $t = 5$ seconds and is then switched off.
11. Redo Problem 9 for the case that $E(t)$ is zero until time $t = 4$ and is then Ae^{-t}.

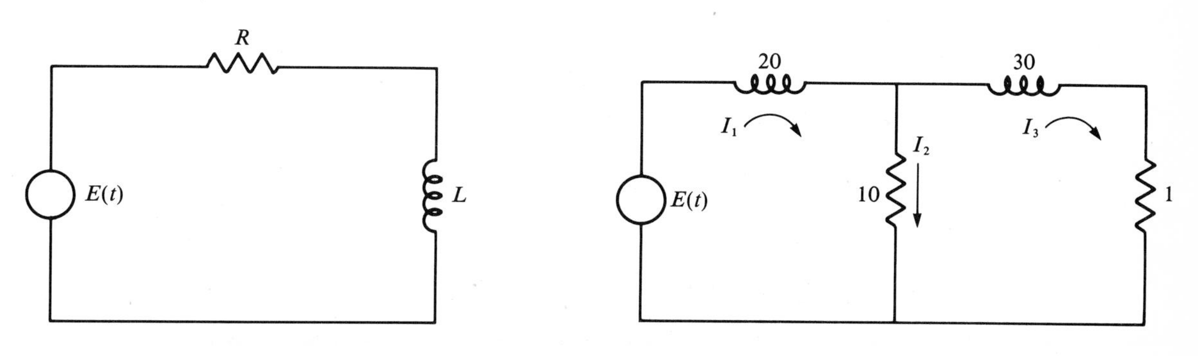

Figure 49 *Figure 50*

12. Solve for the currents in the circuit of Figure 50 if

$$E(t) = \begin{cases} 0, & 0 \le t < k \\ A\sin(\omega t), & t \ge k. \end{cases}$$

Assume zero initial conditions.

13. Solve for the currents in the circuit of Figure 50 if $E(t) = Au(t - k)$. Assume zero initial conditions.

14. Solve for the currents in the circuit of Figure 51 if the initial conditions are zero and $E(t) = k$.

15. Solve for the currents in the circuit of Figure 51 if the initial conditions are zero and $E(t) = A\sin(\omega t)$.

16. Redo Problem 15 for the case that

$$E(t) = \begin{cases} 0, & 0 \le t < 4 \\ A\sin(\omega t), & t \ge 4. \end{cases}$$

17. Solve for the currents in the circuit of Figure 52, assuming zero initial conditions and electromotive force $E(t) = k$.

18. Redo Problem 17 for the case that $E(t) = t[1 - u(t - 3)]$.

19. Redo Problem 17 for the case that $E(t) = A\sin(\omega t)$.

20. Solve for the deflection of the loaded beam of Example 143 if $W(x) = 2u(x - \frac{1}{2}L)$.

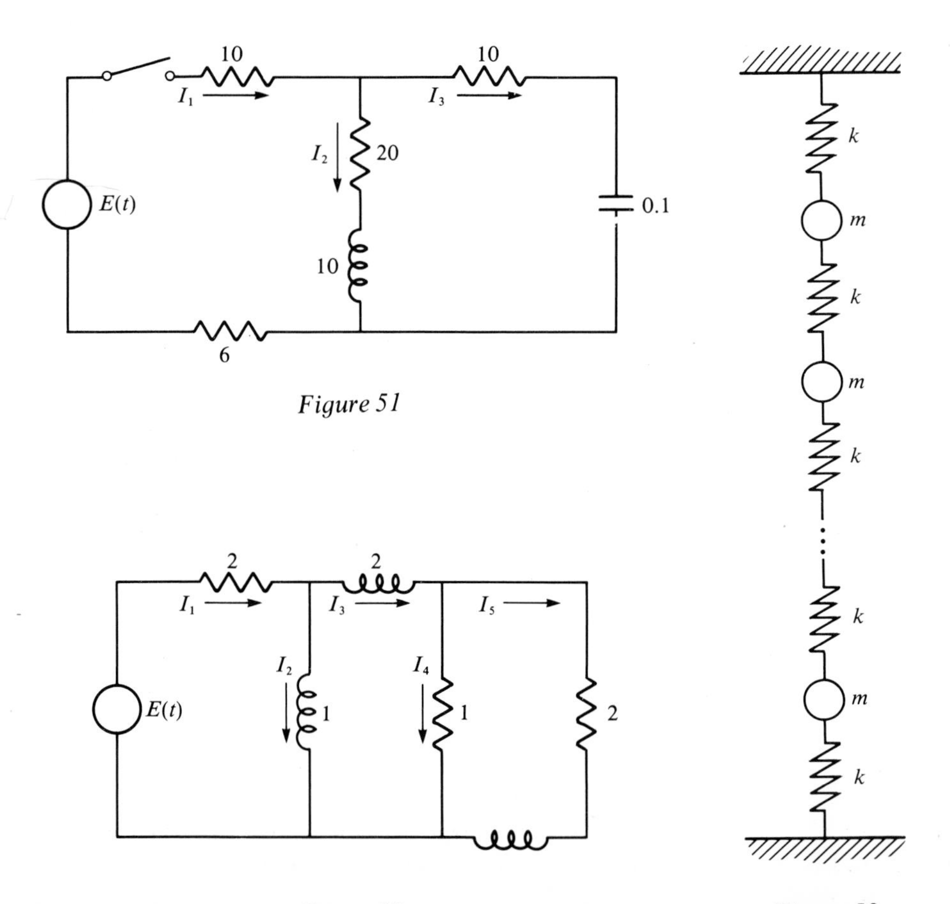

Figure 51

Figure 52

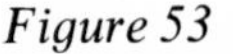
Figure 53

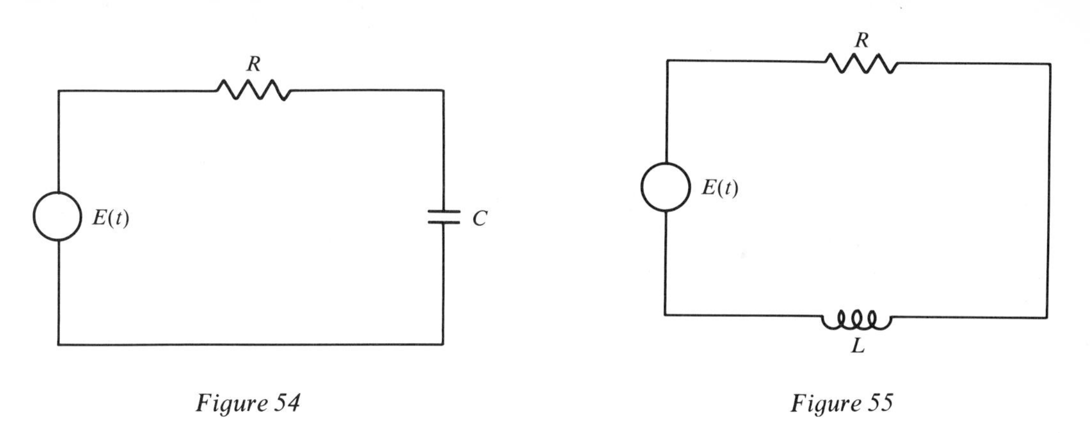

Figure 54

Figure 55

21. Assuming no damping, zero driving forces, and zero initial conditions, solve for the displacements in the mass-spring system of Figure 53. In this system, there are n objects of equal mass m attached by $(n + 1)$ springs as shown. All the spring constants are k. As usual, let y_j be the displacement from static equilibrium of the jth mass (counting from the top), with the positive direction downward.
22. Develop an electrical network which models the mass-spring system of Problem 21 in the sense that the governing differential equations are the same if we make the correct analogies.
23. Solve for the current in the RC circuit of Figure 54 if the electromotive force is given by

$$E(t) = u(t - t_0) - u(t - t_1).$$

Here, t_0 and t_1 are given times and $0 < t_0 < t_1$. Assume zero initial conditions.
24. Solve for the current in the RL circuit of Figure 55 if $E(t)$ is as in Problem 23. Assume zero initial conditions.

4.6 CONVOLUTIONS

It is sometimes useful to have a formula for the inverse Laplace transform of a product $F(s)G(s)$ in terms of the inverse transforms of $F(s)$ and $G(s)$. There is such a formula, and it is known as the *convolution theorem.*

THEOREM 15

If $\mathcal{L}^{-1}[F(s)] = f(t)$ and $\mathcal{L}^{-1}[G(s)] = g(t)$, then

$$\mathcal{L}^{-1}[F(s)G(s)] = \int_0^t f(t - \alpha)g(\alpha)\, d\alpha.$$

Note that, in this integral, α is a dummy variable of integration; hence the result is a function of t.

The function $\int_0^t f(t - \alpha)g(\alpha)\, d\alpha$ is usually called the *convolution of $f(t)$ with $g(t)$* and is denoted $f(t) * g(t)$. In this notation, the convolution theorem says that

$$\mathcal{L}^{-1}[F(s)G(s)] = f(t) * g(t).$$

PROOF Begin from the definition:

$$F(s) = \mathscr{L}[f(t)] = \int_0^\infty e^{-st} f(t)\, dt$$

and

$$G(s) = \mathscr{L}[g(t)] = \int_0^\infty e^{-st} g(t)\, dt.$$

Write

$$F(s)G(s) = F(s) \int_0^\infty e^{-st} g(t)\, dt = \int_0^\infty e^{-st} F(s) g(t)\, dt,$$

since the integration is with respect to t. Replace the integration variable t with the dummy integration variable α:

$$F(s)G(s) = \int_0^\infty e^{-s\alpha} F(s) g(\alpha)\, d\alpha.$$

Now recall that

$$e^{-\alpha s} F(s) = \mathscr{L}[u(t-\alpha) f(t-\alpha)].$$

Hence,

$$e^{-\alpha s} F(s) = \int_0^\infty e^{-st} u(t-\alpha) f(t-\alpha)\, dt.$$

Putting this into the product for $F(s)G(s)$, we get

$$\begin{aligned} F(s)G(s) &= \int_0^\infty [e^{-\alpha s} F(s)] g(\alpha)\, d\alpha \\ &= \int_0^\infty \left[\int_0^\infty e^{-st} u(t-\alpha) f(t-\alpha)\, dt \right] g(\alpha)\, d\alpha \\ &= \int_0^\infty \int_0^\infty e^{-st} g(\alpha) u(t-\alpha) f(t-\alpha)\, dt\, d\alpha. \end{aligned}$$

Now,

$$u(t-\alpha) = \begin{cases} 0, & t < \alpha \\ 1, & t \geq \alpha. \end{cases}$$

Thus,

$$F(s)G(s) = \int_0^\infty \int_\alpha^\infty e^{-st} g(\alpha) f(t-\alpha)\, dt\, d\alpha.$$

Look at Figure 56 which shows the $t\alpha$-plane. The last integration is over $\alpha \le t < \infty$, $\alpha > 0$ (the shaded region). Reverse the order of integration to get

$$F(s)G(s) = \int_0^\infty \int_0^t e^{-st} g(\alpha) f(t-\alpha)\, d\alpha\, dt$$

$$= \int_0^\infty e^{-st}\left[\int_0^t g(\alpha) f(t-\alpha)\, d\alpha\right] dt$$

$$= \mathscr{L}\left(\int_0^t g(\alpha) f(t-\alpha)\, d\alpha\right).$$

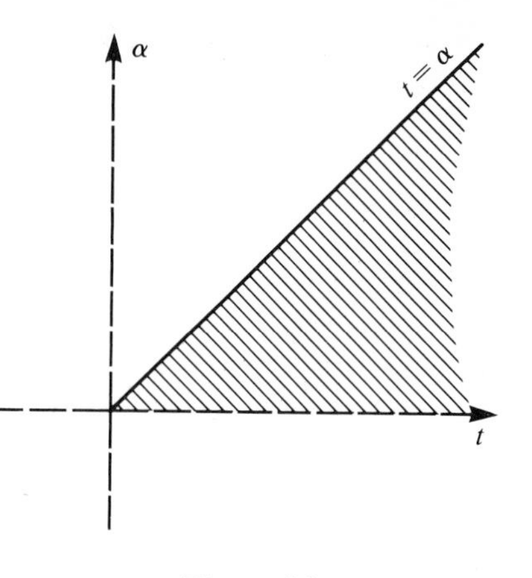

Figure 56

Hence,

$$\mathscr{L}^{-1}[F(s)G(s)] = \int_0^t g(\alpha) f(t-\alpha)\, d\alpha.$$

Before looking at examples and applications, we note that convolution has the following handy property (called the *commutative property*):

$$f(t) * g(t) = g(t) * f(t).$$

To prove this, change variables from α to z by setting $z = t - \alpha$. Then,

$$f(t) * g(t) = \int_0^t f(t-\alpha) g(\alpha)\, d\alpha$$

$$= \int_t^0 f(z) g(t-z)(-dz)$$

$$= \int_0^t g(t-z) f(z)\, dz$$

$$= g(t) * f(t),$$

since z is just a dummy variable of integration.

We now illustrate convolutions in finding inverse Laplace transforms and in solving initial value problems.

EXAMPLE 145

Find

$$\mathscr{L}^{-1}\left(\frac{1}{s(s-4)^2}\right).$$

We could do this by previous methods, but here is a solution by convolution. Choose, say,

$$F(s) = \frac{1}{s} \quad \text{and} \quad G(s) = \frac{1}{(s-4)^2}.$$

Then,

$$f(t) = 1 \quad \text{and} \quad g(t) = te^{4t}.$$

So

$$\mathscr{L}^{-1}[F(s)G(s)] = 1 * te^{4t} = \int_0^t \alpha e^{4\alpha}\, d\alpha = \frac{e^{4t}}{4}\left(t - \frac{1}{4}\right) + \frac{1}{16}.$$

To appreciate the commutative property of convolution, the student should redo this example by computing $te^{4t} * 1$. You will get the same result, but with more effort. Thus, the order in which one does a convolution can be of practical importance.

EXAMPLE 146

Consider a mass-spring system as in Figure 57 with a periodic driving force $A \sin(\omega t)$ and no damping. Assuming that the mass is initially at rest in the static equilibrium position, then the motion is governed by

$$my'' + ky = A \sin(\omega t),$$
$$y(0) = y'(0) = 0.$$

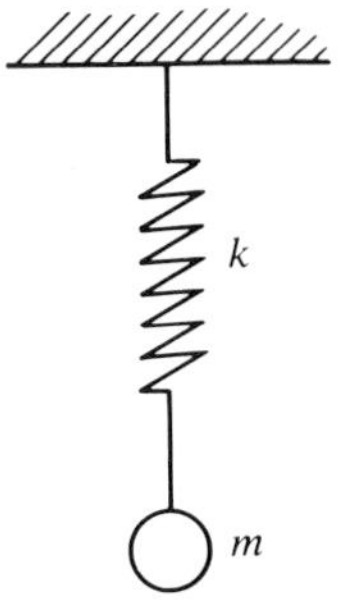

Write the differential equation as

$$y'' + \omega_0^2\, y = B \sin(\omega t),$$

where $\omega_0^2 = k/m$ and $B = A/m$.

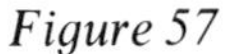

Figure 57

Taking Laplace transforms gives us

$$s^2 \mathscr{L}(y) + \omega_0^2\, \mathscr{L}(y) = \frac{B\omega}{s^2 + \omega^2}.$$

Then,

$$\mathscr{L}(y) = B\omega\left(\frac{1}{s^2 + \omega^2}\right)\left(\frac{1}{s^2 + \omega_0^2}\right).$$

This is a natural place to use the convolution theorem: Since

$$\mathscr{L}^{-1}\left(\frac{1}{s^2 + \omega^2}\right) = \frac{1}{\omega} \sin(\omega t)$$

and

$$\mathscr{L}^{-1}\left(\frac{1}{s^2 + \omega_0^2}\right) = \frac{1}{\omega_0} \sin(\omega_0 t),$$

then the solution for the spring's displacement is

$$y(t) = B\omega \mathscr{L}^{-1}\left(\frac{1}{s^2 + \omega^2} \cdot \frac{1}{s^2 + \omega_0^2}\right)$$

$$= B\omega \int_0^t \frac{1}{\omega} \sin(\omega\alpha) \cdot \frac{1}{\omega_0} \sin[\omega_0(t - \alpha)]\, d\alpha$$

$$= \frac{B}{\omega_0} \int_0^t \sin(\omega\alpha) \sin[\omega_0(t - \alpha)]\, d\alpha.$$

This leads to two cases.

Case 1 If $\omega \neq \omega_0$, then

$$y(t) = \frac{B}{2\omega_0}\left[\frac{-\sin[(\omega-\omega_0)\alpha+\omega_0 t]}{\omega-\omega_0} + \frac{\sin[(\omega+\omega_0)\alpha-\omega_0 t]}{\omega+\omega_0}\right]_0^t$$

$$= \frac{B}{2\omega_0}\left[\frac{\sin(\omega_0 t)-\sin(\omega t)}{\omega-\omega_0} + \frac{\sin(\omega_0 t)+\sin(\omega t)}{\omega+\omega_0}\right].$$

This is the nonresonance case. With no damping, the resulting motion is simply a periodic oscillation.

Case 2 If $\omega = \omega_0$, then

$$y(t) = \frac{B}{\omega}\int_0^t \sin(\omega\alpha)\sin[\omega(t-\alpha)]\,d\alpha = \frac{B}{\omega}\left[\frac{1}{2\omega}\sin(\omega t) - \frac{t}{2}\cos(\omega t)\right].$$

This is the resonance case. Note that, as t grows, so does the amplitude of the oscillation.

EXAMPLE 147

Sometimes the convolution theorem can be used to solve quite general initial value problems. As an illustration, consider the problem

$$y'' - 2y' - 8y = f(t); \qquad y(0) = 1, \qquad y'(0) = 0.$$

Taking Laplace transforms gives us

$$s^2\mathscr{L}(y) - s - 2[\mathscr{L}(y) - 1] - 8\mathscr{L}(y) = \mathscr{L}[f(t)] = F(s).$$

Then,

$$\mathscr{L}(y) = \frac{F(s)}{s^2-2s-8} + \frac{s-2}{s^2-2s-8}$$

$$= \frac{1}{6}\frac{F(s)}{s-4} - \frac{1}{6}\frac{F(s)}{s+2} + \frac{1}{3}\frac{1}{s-4} + \frac{2}{3}\frac{1}{s+2}.$$

Taking inverse Laplace transforms now gives us

$$y(t) = \tfrac{1}{6}f(t) * e^{4t} - \tfrac{1}{6}f(t) * e^{-2t} + \tfrac{1}{3}e^{4t} + \tfrac{2}{3}e^{-2t}.$$

Thus, we can write a solution which holds for any (say continuous) $f(t)$.

In the next section we shall see how convolutions can be used to solve certain kinds of integral equations.

PROBLEMS FOR SECTION 4.6

In each of Problems 1 through 25, use the convolution theorem to find the inverse Laplace transform of the given function. Use integral tables where convenient. Assume throughout that a and b are distinct constants.

1. $\dfrac{1}{(s^2+4)(s^2-4)}$
2. $\dfrac{e^{-2s}}{s^2+4}$
3. $\dfrac{e^{-2s}}{s^2}$
4. $\dfrac{s}{(s^2+a^2)(s^2+b^2)}$
5. $\dfrac{s^2}{(s^2+a^2)(s^2+b^2)}$
6. $\dfrac{1}{(s-3)(s^2+5)}$
7. $\dfrac{s}{(s^2+a^2)(s^2-b^2)}$
8. $\dfrac{s^2}{(s^2+a^2)(s^2-b^2)}$
9. $\dfrac{s}{(s^2-a^2)(s^2-b^2)}$
10. $\dfrac{s^2}{(s^2-a^2)(s^2-b^2)}$
11. $\dfrac{1}{s^2(s^2+2)}$
12. $\dfrac{1}{s^2(s^2-a^2)}$
13. $\dfrac{1}{s(s^2+a^2)^2}$
14. $\dfrac{1}{(s-1)(s^2+4)}$
15. $\dfrac{1}{(s+2)(s^2-9)}$
16. $\dfrac{1}{s^4(s-5)}$
17. $\dfrac{4}{[(s-2)^2+9](s^2-12)}$
18. $\dfrac{-2}{(s^2-5)(s-2)^2}$
19. $\dfrac{e^{-4s}}{s(s+2)}$
20. $\dfrac{2(s-4)}{(s^2+8)[(s-4)^2+1]}$
21. $\dfrac{e^{-3s}}{s^2}$
22. $\dfrac{e^{-4s}}{s(s-2)^3}$
23. $\dfrac{2}{s^3(s^2+5)}$
24. $\dfrac{e^{-3s}}{s(s^2-9)}$
25. $\dfrac{1}{(s-5)[(s-3)^2+8]}$

26. Use the convolution theorem to show that

$$\mathscr{L}^{-1}\left[\frac{1}{s}F(s)\right] = \int_0^t f(\alpha)\,d\alpha,$$

where $f(t) = \mathscr{L}^{-1}[F(s)]$. Thus derive Rule 6 of Section 4.2.

27. Use the convolution theorem to show that

$$\mathscr{L}^{-1}\left[\frac{1}{s^2}F(s)\right] = \int_0^t \int_0^\tau f(\alpha)\,d\alpha\,d\tau,$$

where $f(t) = \mathscr{L}^{-1}[F(s)]$. Thus generalize Rule 6 of Section 4.2.

28. Derive a formula for

$$\mathscr{L}^{-1}\left(\frac{F(s)}{s^2+a^2}\right)$$

in terms of $f(t) = \mathscr{L}^{-1}[F(s)]$.

29. Derive a formula for

$$\mathscr{L}^{-1}\left(\frac{F(s)}{s^2-a^2}\right)$$

in terms of $f(t) = \mathscr{L}^{-1}[F(s)]$.

In each of Problems 30 through 45, use convolutions to write a formula for the solution of the initial value problem, as in Example 147.

30. $y'' - 5y' + 6y = f(t);\quad y(0) = y'(0) = 0$

31. $y'' + 10y' + 24y = f(t);\quad y(0) = 1,\quad y'(0) = 0$

32. $y^{(3)} - y'' - 4y' + 4y = f(t)$; $y(0) = y'(0) = 1$, $y''(0) = 0$
33. $y^{(4)} - 11y'' + 18y = f(t)$; $y(0) = 0$, $y'(0) = 0$, $y''(0) = y^{(3)}(0) = 0$
34. $y'' - k^2 y = f(t)$; $y(0) = 2$, $y'(0) = -4$
35. $y'' - 8y' + 12y = f(t)$; $y(0) = -3$, $y'(0) = 2$
36. $y^{(3)} + 2y'' - 40y' + 64y = f(t)$; $y(0) = 1$, $y'(0) = 0$, $y''(0) = -5$
37. $y'' + y' - 3y = f(t)$; $y(0) = y'(0) = 4$
38. $y'' - 4y' - 5y = f(t)$; $y(0) = 2$, $y'(0) = 1$
39. $y^{(3)} - 3y'' + 6y' - 18y = f(t)$; $y(0) = 0$, $y'(0) = y''(0) = 0$
40. $y'' + 4y' + 8y = f(t)$; $y(0) = y'(0) = 0$
41. $y'' - 8y' - 9y = f(t)$; $y(0) = 0$, $y'(0) = -2$
42. $y^{(4)} - 7y^{(3)} + 14y'' - 14y' + 24y = f(t)$; $y(0) = y'(0) = y''(0) = y^{(3)}(0) = 0$
43. $y'' + 10y' + 24y = f(t)$; $y(0) = 0$, $y'(0) = -2$
44. $y'' + 2y' - 15y = f(t)$; $y(0) = 1$, $y'(0) = -3$
45. $y'' - 3y' + 8y = f(t)$; $y(0) = y'(0) = 0$

4.7 INTEGRAL EQUATIONS, SHIFTED AND MIXED DATA PROBLEMS, AND UNIT IMPULSES

This chapter contains a number of types of problems for which Laplace transforms provide a useful tool.

INTEGRAL EQUATIONS

A differential equation is one involving derivatives with respect to a variable. By the same token, one sometimes encounters equations featuring integrals which are functions of a variable. These are called *integral equations*. Some integral equations are of just such a form that convolution works nicely. Here are two examples in which we use the convolution theorem written as

$$\mathscr{L}\left(\int_0^t f(t-\alpha)g(\alpha)\,d\alpha\right) = \mathscr{L}[f(t)]\mathscr{L}[g(t)].$$

EXAMPLE 148

Find a function $f(t)$ satisfying

$$f(t) = e^{-t} + 2\int_0^t e^{-3\alpha} f(t-\alpha)\,d\alpha.$$

We recognize the integral on the right as being of the right form for the convolution theorem. Take Laplace transforms of both sides to get

$$\mathscr{L}(f) = \frac{1}{s+1} + 2\mathscr{L}(e^{-3t})\mathscr{L}(f).$$

Then

$$\mathscr{L}(f) = \frac{1}{s+1} + \frac{2}{s+3}\mathscr{L}(f).$$

Solve for $\mathscr{L}(f)$:

$$\mathscr{L}(f) = \frac{s+3}{(s+1)^2}.$$

It is now routine to find that

$$\mathscr{L}(f) = \frac{1}{s+1} + \frac{2}{(s+1)^2}.$$

Hence,

$$f(t) = e^{-t} + 2te^{-t}.$$

EXAMPLE 149

Solve

$$f(t) = 2t^2 + \int_0^t \sin(4\alpha)f(t-\alpha)\,d\alpha.$$

We have

$$\mathscr{L}(f) = \frac{4}{s^3} + \mathscr{L}[\sin(4t)]\mathscr{L}(f) = \frac{4}{s^3} + \frac{4}{s^2+16}\mathscr{L}(f).$$

Then,

$$\mathscr{L}(f) = \frac{4(s^2+16)}{s^3(s^2+12)}.$$

Write

$$\mathscr{L}(f) = -\frac{1}{9}\frac{1}{s} + \frac{16}{3}\frac{1}{s^3} + \frac{1}{9}\frac{s}{s^2+12}.$$

Then,

$$f(t) = -\frac{1}{9} + \frac{16}{3}\frac{t^2}{2} + \frac{1}{9}\cos(\sqrt{12}\,t) = -\frac{1}{9} + \frac{8t^2}{3} + \frac{1}{9}\cos(\sqrt{12}\,t).$$

MIXED DATA PROBLEMS

Thus far we have applied Laplace transforms to many initial value problems in which the function and certain of its derivatives were specified at time 0. Sometimes, however, we have various pieces of data given at different times. Such a problem is often called a *mixed data problem*, and may still be solvable by Laplace transform. Here are two examples.

EXAMPLE 150

Solve

$$y'' - y' - 6y = t; \qquad y(0) = 1, \qquad y'(1) = 4.$$

(Note that y and y' are specified *for different values of* t.)

To handle this, let $y'(0) = A$ and proceed as usual (we will have to find A later). We have

$$s^2\mathscr{L}(y) - s - A - [s\mathscr{L}(y) - 1] - 6\mathscr{L}(y) = \frac{1}{s^2}.$$

Solve for $\mathscr{L}(y)$:

$$\mathscr{L}(y) = \frac{1}{s^2(s^2 - s - 6)} + \frac{s + A - 1}{s^2 - s - 6}.$$

This can be written

$$\mathscr{L}(y) = \frac{1}{36}\frac{1}{s} - \frac{1}{6}\frac{1}{s^2} + \frac{1}{45}\frac{1}{s-3} - \frac{1}{20}\frac{1}{s+2} + \left(\frac{A+2}{5}\right)\frac{1}{s-3} + \left(\frac{3-A}{5}\right)\frac{1}{s+2}.$$

Then,

$$y = \frac{1}{36} - \frac{1}{6}t + \frac{1}{45}e^{3t} - \frac{1}{20}e^{-2t} + \left(\frac{A+2}{5}\right)e^{3t} + \left(\frac{3-A}{5}\right)e^{-2t}.$$

To find A, recall that $y'(1) = 4$. Then,

$$4 = -\frac{1}{6} + \frac{1}{15}e^3 + \frac{1}{10}e^{-2} + \frac{3(A+2)}{5}e^3 - \frac{2(3-A)}{5}e^{-2}.$$

Solve this for A to get

$$A = \frac{\frac{25}{6} - \frac{19}{15}e^3 + \frac{11}{10}e^{-2}}{\frac{3}{5}e^3 + \frac{2}{5}e^{-2}}.$$

When substituted for A in $y(t)$, this gives the exact solution.

If we approximate to four decimal places, we have $A = -1.7452$. Using this value, $y(t)$ works out to

$$y_{\text{app}}(t) = 0.0278 - 0.1667t + 0.0732e^{3t} + 0.8990e^{-2t},$$

where the subscript "app" is a reminder that the numbers are approximate values. To get some feeling for how good the approximation is, check to four decimal places that

$$y_{\text{app}}(0) = 1.0000 \quad \text{and} \quad y'_{\text{app}}(1) = 4.0008,$$

which is in good agreement with the required values $y(0) = 1$ and $y'(1) = 4$.

EXAMPLE 151

Solve

$$y'' - 4y' + 4y = \cos(3t); \qquad y(2) = 0, \qquad y'(0) = 0.$$

Let $y(0) = A$ and take Laplace transforms to get

$$s^2\mathscr{L}(y) - sA - 4[s\mathscr{L}(y) - A] + 4\mathscr{L}(y) = \frac{s}{s^2+9}.$$

Then,

$$\mathscr{L}(y) = \frac{s}{(s^2+9)(s^2-4s+4)} + \frac{As-4A}{s^2-4s+4}$$

$$= -\frac{5}{169}\frac{s}{s^2+9} - \frac{36}{169}\frac{1}{s^2+9} + \frac{5}{169}\frac{1}{s-2}$$

$$+ \frac{26}{169}\frac{1}{(s-2)^2} + \frac{A}{s-2} - \frac{2A}{(s-2)^2}.$$

Then,

$$y(t) = -\tfrac{5}{169}\cos(3t) - \tfrac{12}{169}\sin(3t) + \tfrac{5}{169}e^{2t} + \tfrac{26}{169}te^{2t} + Ae^{2t} - 2Ate^{2t}.$$

To solve for A, we have

$$y(2) = 0 = -\tfrac{5}{169}\cos(6) - \tfrac{12}{169}\sin(6) + \tfrac{5}{169}e^4 + \tfrac{52}{169}e^4 + Ae^4 - 4Ae^4.$$

Then,

$$A = -\tfrac{5}{507}e^{-4}\cos(6) - \tfrac{12}{507}e^{-4}\sin(6) + \tfrac{57}{507}.$$

This is about 0.1124, and an approximate solution is

$$y_{\text{app}} = -0.0296\cos(3t) - 0.0710\sin(3t) + 0.1420e^{2t} - 0.0710te^{2t}.$$

DIFFERENTIAL EQUATIONS WITH SHIFTED DATA

In a variation on mixed data type problems, suppose we have a differential equation with data supplied at the same nonzero time. Such a problem is sometimes referred to as having *shifted data*. We can still attempt a solution by Laplace transform, as this example shows.

EXAMPLE 152

Solve

$$y'' + 4y = 1; \qquad y(2) = 0, \qquad y'(2) = 3.$$

To apply Rules 1 and 2, we need $y(0)$ and $y'(0)$. Set $y(0) = A$ and $y'(0) = B$ and proceed. We get

$$s^2\mathscr{L}(y) - sA - B + 4\mathscr{L}(y) = \frac{1}{s}.$$

Then,

$$\mathscr{L}(y) = \frac{1}{s(s^2+4)} + \frac{As+B}{s^2+4}$$

$$= \frac{1}{4}\frac{1}{s} - \frac{1}{4}\frac{s}{s^2+4} + A\frac{s}{s^2+4} + B\frac{1}{s^2+4}.$$

Then,

$$y(t) = \frac{1}{4} - \frac{1}{4}\cos(2t) + A\cos(2t) + \frac{B}{2}\sin(2t).$$

To solve for A and B, we have from the data that

$$y(2) = 0 = \frac{1}{4} - \frac{1}{4}\cos(4) + A\cos(4) + \frac{B}{2}\sin(4)$$

and

$$y'(2) = 3 = \tfrac{1}{2}\sin(4) - 2A\sin(4) + B\cos(4).$$

Solve these for A and B to obtain

$$A = \tfrac{1}{4} - \tfrac{1}{4}\cos(4) - \tfrac{3}{2}\sin(4)$$

and

$$B = 3\cos(4) - \tfrac{1}{2}\sin(4).$$

These can be substituted for A and B in $y(t)$ to give the exact solution.

If we need only an approximation, we can write, say,

$$A = 1.5486 \quad \text{and} \quad B = -1.5825,$$

giving us

$$y_{\text{app}} = 0.25 + 1.2986\cos(2t) - 0.7913\sin(2t).$$

UNIT IMPULSES AND THE DIRAC SYMBOL

Many applications in engineering and physics make use of the notion of an impulse. To express such an idea mathematically, define the function

$$\delta_a(t) = \begin{cases} \dfrac{1}{\varepsilon} & \text{for} \quad a \le t \le a + \varepsilon \\ 0 & \text{for} \quad t < a \quad \text{or} \quad t > a + \varepsilon. \end{cases}$$

If ε is "small," this is called a *unit impulse*, since the area under the nonzero part of the graph is exactly 1 (see Figure 58). We can write $\delta_a(t)$ in terms of the previously defined unit function by

$$\delta_a(t) = \frac{1}{\varepsilon}[u(t-a) - u(t-a-\varepsilon)].$$

Immediately, then,

$$\mathscr{L}[\delta_a(t)] = \frac{1}{\varepsilon}\left(\frac{e^{-as}}{s} - \frac{e^{-(a+\varepsilon)s}}{s}\right)$$

$$= \frac{e^{-as}(1 - e^{-\varepsilon s})}{\varepsilon s}.$$

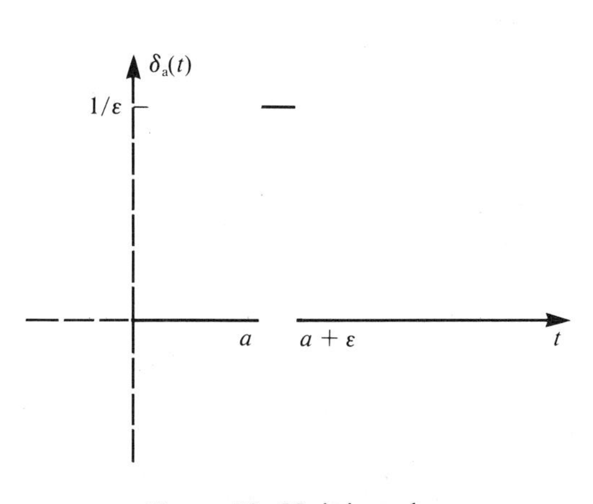

Figure 58. Unit impulse.

The symbol $\delta(t-a)$ denotes the limit of $\delta_a(t)$ as $\varepsilon \to 0$. Strictly speaking, $\delta(t-a)$ is not a function, but is instead an object called a *distribution*. Sometimes $\delta(t-a)$ is called the *Dirac symbol*; more often, for traditional reasons, $\delta(t-a)$ is called the *Dirac delta function*. Roughly speaking, we may think of $\delta(t-a)$ as being 0 for $t \neq a$ and infinity for $t = a$. Thus, $\delta(t-a)$ is useful in representing an instantaneous impulse at time $t = a$. Since

$$\frac{1 - e^{-\varepsilon s}}{\varepsilon} \to s$$

as $\varepsilon \to 0$, then

$$\mathscr{L}[\delta(t-a)] = e^{-as}.$$

The Dirac delta function satisfies the following *filtering property*:

$$\int_0^\infty f(t)\delta(t-a)\,dt = f(a),$$

provided that $f(t)$ is integrable on $[0, \infty)$ and continuous at a. To derive this, note that

$$\int_0^\infty f(t)\delta_a(t)\,dt = \frac{1}{\varepsilon}\int_a^{a+\varepsilon} f(t)\,dt = \frac{1}{\varepsilon}(a + \varepsilon - \varepsilon)f(t_0)$$

for some t_0 between a and $a + \varepsilon$ by the mean value theorem for integrals. Now let $\varepsilon \to 0$. Since t_0 is between a and $a + \varepsilon$, then $t_0 \to a$ and, by continuity, $f(t) \to f(t_0)$. Further, $\delta_a(t) \to \delta(t-a)$, giving us the filtering property.

Note that the formula

$$\int_0^\infty e^{-st}\delta(t-a)\,dt = e^{-as}$$

is just the filtering property applied to $f(t) = e^{-st}$. Further, the expression

$$\int_0^\infty f(x)\delta(x-t)\,dx = f(t)$$

may be written

$$f(t) * \delta(t) = f(t).$$

Thus, in terms of convolution, $\delta(t)$ behaves like the number 1 does in multiplication in the sense that the convolution of a continuous function $f(t)$ with $\delta(t)$ is just $f(t)$ again.

Here is an example of a use of the Dirac delta function.

EXAMPLE 153

Consider bending of a simply supported beam of length L subjected to a load M at a point $x = a$ between its ends. Neglecting the weight of the beam, the differential equation for the deflection $y(x)$ from the horizontal at point x is

$$y^{(4)} = \frac{1}{EI} M\delta(x-a).$$

We assume, as in Example 90, that E is the Young's modulus and I the moment of inertia of the cross section at x with respect to a horizontal line through its centroid. The end conditions are

$$y(0) = y''(0) = y(L) = 0, \qquad y^{(3)}(0) = F_0,$$

where F_0 is the shearing force at the left end $x = 0$.

Taking the Laplace transform of the differential equation gives us

$$s^4Y(s) - s^3y(0) - s^2y'(0) - sy''(0) - y^{(3)}(0) = \frac{M}{EI}e^{-as}.$$

Putting in the values for $y(0)$, $y''(0)$, and $y^{(3)}(0)$, we have

$$Y(s) = \frac{M}{EI}\frac{e^{-as}}{s^4} + \frac{y'(0)}{s^2} + \frac{F_0}{s^4}.$$

(Here we must carry $y'(0)$ along in the calculation, as we do not know its value.) Taking inverse Laplace transforms, we have

$$y(x) = \frac{M}{EI}\frac{(x-a)^3}{6}u(x-a) + y'(0)x + \frac{F_0x^3}{6}.$$

To fill in a value for $y'(0)$, use the data $y(L) = 0$ to write

$$y(L) = 0 = \frac{M}{EI}\frac{(L-a)^3}{6} + y'(0)L + \frac{F_0L^3}{6}.$$

Then,

$$y'(0) = -\frac{M}{LEI}\frac{(L-a)^3}{6} - \frac{F_0L^2}{6}.$$

The solution is then

$$y(x) = \begin{cases} \left(-\dfrac{M}{LEI}\dfrac{(L-a)^3}{6} - \dfrac{F_0L^2}{6}\right)x + \dfrac{F_0x^3}{6} & \text{if } 0 \le x < a \\ \dfrac{M}{EI}\dfrac{(x-a)^3}{6} - \left(\dfrac{M}{LEI}\dfrac{(L-a)^3}{6} - \dfrac{F_0L^2}{6}\right)x + \dfrac{F_0x^3}{6} & \text{if } a \le x \le L. \end{cases}$$

PROBLEMS FOR SECTION 4.7

In each of Problems 1 through 30, solve the integral equation for the unknown function $f(t)$.

1. $f(t) = -1 + \int_0^t f(t-\alpha)e^{-3\alpha}\,d\alpha$

2. $f(t) = -t + \int_0^t f(t-\alpha)\cos(\alpha)\,d\alpha$

3. $f(t) = e^{-t} + \int_0^t f(t-\alpha)\,d\alpha$

4. $f(t) = 2t^2 + \int_0^t f(t-\alpha)e^{-\alpha}\,d\alpha$

5. $f(t) = \cos(t) + \int_0^t f(t-\alpha)e^{-2\alpha}\,d\alpha$

6. $f(t) = -1 + t - 2\int_0^t f(t-\alpha)\sin(\alpha)\,d\alpha$

7. $f(t) = e^{-2t} - 3\int_0^t f(t-\alpha)e^{-3\alpha}\,d\alpha$

8. $f(t) = 2t + 1 + \int_0^t f(t-\alpha)e^{-\alpha}\,d\alpha$

9. $f(t) = -1 - 3\int_0^t f(t-\alpha)\cosh(\alpha)\,d\alpha$

10. $f(t) = 2t^2 - \int_0^t f(t-\alpha)e^{-3\alpha}\,d\alpha$

11. $f(t) = -2t + \int_0^t f(t-\alpha)\alpha\,d\alpha$

12. $f(t) = \sin(2t) - 2\int_0^t (t-\alpha)^2 f(\alpha)\,d\alpha$

13. $f(t) = 3 + \int_0^t \cos[2(t-\alpha)]f(\alpha)\,d\alpha$

14. $f(t) = \sinh(3t) - 4\int_0^t e^{2(t-\alpha)}f(\alpha)\,d\alpha$

15. $f(t) = 3t^2 + \int_0^t \sinh[3(t-\alpha)]f(\alpha)\,d\alpha$

16. $2f(t) + 1 = t - \int_0^t e^{3(t-\alpha)}f(\alpha)\,d\alpha$

17. $f(t) = \int_0^t e^{2(t-\alpha)}\sin[3(t-\alpha)]f(\alpha)\,d\alpha$

18. $f(t) = \int_0^t \alpha e^{-\alpha}f(t-\alpha)\,d\alpha$

19. $f(t) = 2t + e^{-t} + \int_0^t \alpha f(t-\alpha)\,d\alpha$

20. $f(t) = 2t^2 + 3\int_0^t (t-\alpha)^3 f(\alpha)\,d\alpha$

21. $f(t) = \cos(2t) - \int_0^t f(\alpha)\cos[2(t-\alpha)]\,d\alpha$

22. $f(t) = 1 - e^{-2t} + \int_0^t (t-\alpha)f(\alpha)\,d\alpha$

23. $f(t) = 2t^2 + \int_0^t \cosh[3(t-\alpha)]f(\alpha)\,d\alpha$

24. $f(t) = 3 + \int_0^t f(t-\alpha)\alpha^3\,d\alpha$

25. $f(t) = e^{-t} + 3\int_0^t \alpha f(t-\alpha)\,d\alpha$

26. $f(t) = 4 - \int_0^t f(\alpha)\cosh[4(t-\alpha)]\,d\alpha$

27. $f(t) = -2te^{-t} + \int_0^t f(t-\alpha)\,d\alpha$

28. $f(t) = \sin(3t) - \int_0^t f(t-\alpha)\sinh(4\alpha)\,d\alpha$

29. $f(t) = 1 - t + \int_0^t (t-\alpha)f(\alpha)\,d\alpha$

30. $f(t) = te^{-t} + \int_0^t f(t-\alpha)\sin^2(\alpha)\,d\alpha$

In each of Problems 31 through 55, solve the mixed or shifted data problem by Laplace transform methods.

31. $y'' - 4y' + 4y = t;\ \ y(1) = 0,\ \ y'(0) = -2$
32. $y'' - 10y' + 21y = 1;\ \ y(2) = 1,\ \ y'(0) = -3$
33. $y'' + 4y' - 6y = 1;\ \ y(0) = -3,\ \ y'(1) = 1$
34. $y'' - 3y' - 4y = -\sin(2t);\ \ y(1) = y'(1) = 4$
35. $y'' + k^2y = t - 4;\ \ y(3) = -2,\ \ y'(0) = 0\ \ (k \neq 0)$
36. $y'' + 2y' + 5y = 3e^{-t};\ \ y(1) = 0,\ \ y'(0) = -3$
37. $y'' - 12y' + 27y = \cosh(2t);\ \ y(0) = y'(2) = 0$
38. $y'' + 11y' + 30y = 1 + 3t;\ \ y(2) = -3,\ \ y'(2) = 5$
39. $y'' + 7y' + 12y = te^{-2t};\ \ y(2) = -3,\ \ y'(0) = 0$
40. $y'' - 8y = 3t - \sinh(t);\ \ y(1) = 6,\ \ y'(0) = 3$
41. $y'' + 5y' + y = t;\ \ y(0) = y'(1) = -4$
42. $y'' + 14y' + 45y = 1 - 4\sin(t);\ \ y(2) = 0,\ \ y'(0) = 0$
43. $y'' + 8y' + 16y = 2;\ \ y(1) = 0,\ \ y'(1) = -3$

44. $y'' - 3y' + 5y = -2t;\quad y(3) = -4,\quad y'(3) = 0$
45. $y'' - 4y' - 12y = 8e^t;\quad y(0) = 0,\quad y'(4) = 0$
46. $y'' + 3y' - 21y = t^2 + 1;\quad y(2) = y'(2) = -1$
47. $y'' - 2y = 2t - \sinh(t);\quad y(0) = y'(2) = 0$
48. $y'' + 4y' + 9y = 2;\quad y(1) = y'(1) = -5$
49. $y'' - 5y' + 6y = 2t^2e^{-t};\quad y(1) = 1,\quad y'(0) = 0$
50. $y'' + 14y' + 49y = t - \sin(4t);\quad y(0) = -1,\quad y'(0) = 0$
51. $y'' - 2y' + y = t;\quad y(0) = 0,\quad y'(1) = -2$
52. $y'' + 5y' + 6y = e^{-t};\quad y(1) = y'(1) = 3$
53. $y'' + k^2y = \sin(t);\quad y(0) = 1,\quad y'(2) = 0 \qquad (k > 0, k \neq 1, \text{ and } \cos(2k) \neq 0)$
54. $y'' - 4y = \sinh(t);\quad y(0) = 2,\quad y'(1) = 0$
55. $y'' - 8y' - 9y = t;\quad y(0) = 0,\quad y'(2) = 0$
56. Solve for the deflection of the loaded beam of Example 153 if $W(x)$ is as given there, but the boundary conditions are

$$y(0) = y'(0) = 0, \qquad y''(L) = y^{(3)}(L) = 0.$$

(With these conditions, the beam is said to be *imbedded*, or fixed, at the left end $x = 0$ and *free* at the right end $x = L$.)

57. Solve for the deflection of a beam imbedded at the left end, free at the right end, if the load is given by

$$W(x) = \begin{cases} 0, & 0 \leq x < \frac{1}{3}L \\ w_0, & \frac{1}{3}L \leq x \leq \frac{2}{3}L \\ 0, & \frac{2}{3}L < x \leq L. \end{cases}$$

Here, w_0 is a given positive constant.

58. Solve

$$y^{(4)} = \frac{M}{EI}\,\delta(x - a);$$

$y(0) = y'(0) = 0,\ y''(0) = B_0,\ y^{(3)}(0) = F_0,\ y(L) = y''(L) = 0.$
This is for the deflection of a beam horizontally restrained at both ends, with a load at $x = a$ and with the weight of the beam neglected.

4.8 LAPLACE TRANSFORM SOLUTION OF DIFFERENTIAL EQUATIONS WITH POLYNOMIAL COEFFICIENTS

We have seen that Laplace transforms are quite successful in solving initial value problems involving constant coefficient differential equations. In some instances, they are also effective when the coefficients are polynomials. The key lies in the following fact:

$$\text{If} \quad \mathscr{L}[f(t)] = F(s), \quad \text{then} \quad \mathscr{L}[t^n f(t)] = (-1)^n \frac{d^n F(s)}{ds^n}.$$

A derivation of this is surprisingly simple. Begin with

$$F(s) = \int_0^{\infty} e^{-st} f(t)\, dt.$$

By successive differentiation, we obtain

$$F'(s) = \int_0^{\infty} -te^{-st} f(t)\, dt = \mathscr{L}[-tf(t)],$$

$$F''(s) = \int_0^{\infty} t^2 e^{-st} f(t)\, dt = \mathscr{L}[t^2 f(t)],$$

$$F^{(3)}(s) = \int_0^{\infty} -t^3 e^{-st} f(t)\, dt = \mathscr{L}[-t^3 f(t)],$$

$$\vdots$$

$$F^{(n)}(s) = \int_0^{\infty} (-1)^n t^n e^{-st} f(t)\, dt = \mathscr{L}[(-1)^n t^n f(t)].$$

The result now follows upon dividing both sides by the constant $(-1)^n$.

As the following examples show, this formula is often useful in solving differential equations having polynomial coefficients.

EXAMPLE 154

Solve

$$y'' + 2ty' - 4y = 1; \qquad y(0) = y'(0) = 0.$$

Applying the Laplace transform to the differential equation gives us

$$\mathscr{L}(y'') + \mathscr{L}(2ty') - 4\mathscr{L}(y) = \mathscr{L}(1) = \frac{1}{s}.$$

In the usual way,

$$\mathscr{L}(y'') = s^2 Y(s) - sy(0) - y'(0) = s^2 Y(s).$$

But now

$$\mathscr{L}(ty') = -\frac{d}{ds}\mathscr{L}(y') = -\frac{d}{ds}[sY(s) - y(0)] = -Y(s) - sY'(s).$$

Thus, the differential equation for y transforms to

$$s^2 Y - 2Y - 2sY' - 4Y = \frac{1}{s},$$

a differential equation for $Y(s)$. This can be written

$$Y' + \left(\frac{3}{s} - \frac{s}{2}\right) Y = \frac{-1}{2s^2},$$

a linear first order equation which we solve by the methods of Section 1.6 to get

$$Y(s) = \frac{1}{s^3} + C\,\frac{e^{s^2/4}}{s^3}.$$

If $C \neq 0$, then $\lim_{s\to\infty} Y(s)$ is infinite. We shall impose the condition that $\lim_{s\to\infty} Y(s) = 0$ in order for the improper integral defining the Laplace transform to converge. Thus, we must choose $C = 0$, giving us

$$Y(s) = \frac{1}{s^3}.$$

Then,

$$y(t) = \mathscr{L}^{-1}\left(\frac{1}{s^3}\right) = \frac{t^2}{2}.$$

It is easy to check this solution by directly substituting it back into the original problem.

EXAMPLE 155

Solve

$$ty'' + (4t - 2)y' - 4y = 0; \qquad y(0) = 1.$$

Apply the Laplace transform to the differential equation and rearrange terms to get

$$(s^2 + 4s)Y' + (4s + 8)Y = 3$$

or

$$Y' + \left(\frac{4s + 8}{s(s + 4)}\right)Y = \frac{3}{s(s + 4)}.$$

This is a linear first order equation which we solve to obtain

$$Y(s) = \frac{s}{(s + 4)^2} + \frac{6}{(s + 4)^2} + \frac{C}{s^2(s + 4)^2}.$$

Upon taking inverse Laplace transforms, we get

$$y(t) = e^{-4t} + 2te^{-4t} + C[-\tfrac{1}{32} + \tfrac{1}{16}t + \tfrac{1}{16}te^{-4t} + \tfrac{1}{32}e^{-4t}],$$

which is a solution for any constant C.

PROBLEMS FOR SECTION 4.8

In each of Problems 1 through 10, use the Laplace transform to find a solution. Only nontrivial solutions are acceptable.

1. $t^2y'' - 2y = 2$

2. $y'' + 2ty' - 4y = 6; \quad y(0) = y'(0) = 0$

3. $y'' + 4ty' - 4y = 0; \quad y(0) = 0, \quad y'(0) = -7$

4. $y'' - 16ty' + 32y = 14; \quad y(0) = y'(0) = 0$

5. $y'' - 8ty' + 16y = 3; \quad y(0) = y'(0) = 0$

6. $t(1 - t)y'' + 2y' + 2y = 6t; \quad y(0) = 0, \quad y(2) = 0$

7. $y'' - 4ty' + 4y = 0; \quad y(0) = 0, \quad y'(0) = 10$

8. $y'' + 8ty' - 8y = 0; \quad y(0) = 0, \quad y'(0) = -4$

9. $ty'' + (t - 1)y' + y = 0; \quad y(0) = 0$

10. $y'' + 8ty' = 0; \quad y(0) = 4, \quad y'(0) = 0$

11. Use Laplace transform methods to solve the Bessel equation

$$ty'' + y' + ty = 0,$$

subject to $y(0) = 1$. (*Hint:* You will obtain a separable differential equation for $Y(s)$. Use a binomial expansion to express $Y(s)$ as an infinite series; then take the inverse Laplace transform term by term to obtain an infinite series solution. The resulting series is called a Bessel function of order zero, and we will study such functions in detail in Chapter 6.)

12. Solve Bessel's equation of order n:

$$t^2y'' + ty' + (t^2 - n^2)y = 0, \qquad n = 1, 2, 3, \ldots$$

(*Hint:* You will need the condition that $\lim_{s\to\infty} Y(s) = 0$ in order to eliminate one constant of integration. As in Problem 11, your answer will be an infinite series, called a Bessel function of order n.)

13. Use Laplace transforms to solve Laguerre's differential equation

$$ty'' + (1 - t)y' + ny = 0,$$

in which n is a positive integer.

14. In Example 155, we specified $y(0)$ but not $y'(0)$. In fact, $y'(0)$ cannot be specified arbitrarily; no matter how you choose the constant C, you will get $y'(0) = -2$. Why does this not violate the existence theorem for second order initial value problems stated in Chapter 2?

15. Solve $x^2y'' + Axy' + By = 0$ using Laplace transforms.

16. Use Laplace transforms to find a nontrivial solution of $y'' + \alpha ty' + \beta y = \gamma$ in each of the following cases:

(a) $y(0) = y'(0) = 0, \quad \gamma \neq 0, \quad \beta = -2\alpha$

(b) $\gamma = y'(0) = 0, \quad y(0) \neq 0, \quad \beta = 0, \quad \alpha \neq 0$

(c) $y(0) = \gamma = 0, \quad y'(0) \neq 0, \quad \beta = -\alpha$

CHAPTER SUMMARY

In this chapter we concentrated primarily on linear, constant coefficient differential equations, but with two important differences from the earlier treatment.

One difference was in the variety of functions which could appear on the right side of the differential equation. With an initial value problem such as

$$y'' - 4y' + 4y = \sin(t); \qquad y(0) = 1, \qquad y'(0) = -2,$$

the methods of Chapter 2 lead easily to the solution and Laplace transforms are not needed. But with something like

$$y'' - 4y' + 4y = \begin{cases} 0, & 0 \leq t < 10 \\ \sin(t), & t \geq 10 \end{cases}; \qquad y(0) = 1, \qquad y'(0) = -2,$$

we find Laplace transforms the most effective tool to date. Similarly, with

$$y'' - 4y' + 4y = f(t); \qquad y(0) = 1, \qquad y'(0) = -2,$$

where $f(t)$ is the square wave of Figure 31, Laplace transforms succeed where previous methods fail.

A second difference is the effectiveness of Laplace transforms in handling both higher order differential equations and systems of differential equations. The latter is a particularly important feature of Laplace transforms since the mechanical systems and electrical networks which one encounters in practice are often governed by more than one differential equation.

One further importance of Laplace transforms is simply that they introduce the student to the idea of a transform. We shall later develop others, such as the Fourier and finite Fourier transforms. The latter see important action in electrical engineering.

In the next chapter we set out in a different direction and develop series methods for solving differential equations. As with Laplace transforms, we shall find that series methods broaden our scope and enable us to solve exactly, or, in some cases, to approximate, solutions of problems which are at this point still beyond our means.

Table of Laplace Transforms

$f(t)$	$F(s) = \mathscr{L}[f(t)]$
1	$\dfrac{1}{s}$
t	$\dfrac{1}{s^2}$
t^n	$\dfrac{n!}{s^{n+1}}$
$\dfrac{1}{\sqrt{t}}$	$\sqrt{\dfrac{\pi}{s}}$
e^{at}	$\dfrac{1}{s-a}$
te^{at}	$\dfrac{1}{(s-a)^2}$
$t^n e^{at}$	$\dfrac{n!}{(s-a)^{n+1}}$
$\dfrac{1}{a-b}(e^{at}-e^{bt})$	$\dfrac{1}{(s-a)(s-b)}$
$\dfrac{1}{a-b}(ae^{at}-be^{bt})$	$\dfrac{s}{(s-a)(s-b)}$
$\dfrac{(c-b)e^{at}+(a-c)e^{bt}+(b-a)e^{ct}}{(a-b)(b-c)(c-a)}$	$\dfrac{1}{(s-a)(s-b)(s-c)}$
$\sin(at)$	$\dfrac{a}{s^2+a^2}$

Table of Laplace Transforms (Continued)

$\cos(at)$	$\dfrac{s}{s^2 + a^2}$
$1 - \cos(at)$	$\dfrac{a^2}{s(s^2 + a^2)}$
$at - \sin(at)$	$\dfrac{a^3}{s^2(s^2 + a^2)^2}$
$\sin(at) - at\cos(at)$	$\dfrac{2a^3}{(s^2 + a^2)^2}$
$t\sin(at)$	$\dfrac{2as}{(s^2 + a^2)^2}$
$t\cos(at)$	$\dfrac{(s - a)(s + a)}{(s^2 + a^2)^2}$
$\dfrac{\cos(at) - \cos(bt)}{(b - a)(b + a)}$	$\dfrac{s}{(s^2 + a^2)(s^2 + b^2)}$
$e^{at}\sin(bt)$	$\dfrac{b}{(s - a)^2 + b^2}$
$e^{at}\cos(bt)$	$\dfrac{s - a}{(s - a)^2 + b^2}$
$\sinh(at)$	$\dfrac{a}{s^2 - a^2}$
$\cosh(at)$	$\dfrac{s}{s^2 - a^2}$
$\sin(at)\cosh(at) - \cos(at)\sinh(at)$	$\dfrac{4a^3}{s^4 + 4a^4}$
$\sin(at)\sinh(at)$	$\dfrac{2a^2 s}{s^4 + 4a^4}$
$\sinh(at) - \sin(at)$	$\dfrac{2a^3}{s^4 - a^4}$
$\cosh(at) - \cos(at)$	$\dfrac{2a^2 s}{s^4 - a^4}$
$\dfrac{e^{at}(1 + 2at)}{\sqrt{\pi t}}$	$\dfrac{s}{(s - a)^{3/2}}$
$J_0(at)$*	$\dfrac{1}{\sqrt{s^2 + a^2}}$

* $J_n(x)$ is Bessel's function of the first kind of order n (see Chapter 6, Sections 6.1 and 6.2).

Table of Laplace Transforms (Continued)

$a^n J_n(at)$*	$\dfrac{(\sqrt{s^2+a^2}-s)^n}{\sqrt{s^2+a^2}}$
$J_0(2\sqrt{at})$	$\dfrac{e^{-a/s}}{s}$
$\dfrac{1}{t}\sin(at)$	$\arctan\left(\dfrac{a}{s}\right)$
$\dfrac{2}{t}[1-\cos(at)]$	$\ln\left(\dfrac{s^2+a^2}{s^2}\right)$
$\dfrac{2}{t}[1-\cosh(at)]$	$\ln\left(\dfrac{s^2-a^2}{s^2}\right)$
$\dfrac{1}{\sqrt{\pi t}} - ae^{a^2t}\operatorname{erfc}(a\sqrt{t})$†	$\dfrac{1}{\sqrt{s}+a}$
$\dfrac{1}{\sqrt{\pi t}} + ae^{a^2t}\operatorname{erf}(a\sqrt{t})$†	$\dfrac{\sqrt{s}}{s-a^2}$
$e^{a^2t}\operatorname{erf}(a\sqrt{t})$	$\dfrac{a}{\sqrt{s}(s-a^2)}$
$e^{a^2t}\operatorname{erfc}(a\sqrt{t})$	$\dfrac{1}{\sqrt{s}(\sqrt{s}+a)}$
$\operatorname{erfc}\left(\dfrac{a}{2\sqrt{t}}\right)$	$\dfrac{1}{s}e^{-a\sqrt{s}}$
$\dfrac{1}{\sqrt{\pi t}}e^{-a^2/4t}$	$\dfrac{1}{\sqrt{s}}e^{-a\sqrt{s}}$
$\dfrac{1}{\sqrt{\pi(t+a)}}$	$\dfrac{1}{\sqrt{s}}e^{as}\operatorname{erfc}(\sqrt{as})$
$\dfrac{1}{\pi t}\sin(2a\sqrt{t})$	$\operatorname{erf}\left(\dfrac{a}{\sqrt{s}}\right)$
$f\left(\dfrac{t}{a}\right)$	$aF(as)$
$e^{bt/a}f\left(\dfrac{t}{a}\right)$	$aF(as-b)$
$f^{(n)}(t)$	$s^nF(s) - s^{n-1}f(0) - s^{n-2}f'(0) - \cdots$ $\cdots - sf^{(n-2)}(0) - f^{(n-1)}(0)$

* $J_n(x)$ is Bessel's function of the first kind of order n (see Chapter 6, Sections 6.1 and 6.2).

† $\operatorname{erf}(x) = (2/\sqrt{\pi})\int_0^x e^{-r^2}\,dr$ is the *error function*; $\operatorname{erfc}(x) = 1 - \operatorname{erf}(x)$ is the *complementary error function*.

Table of Laplace Transforms (Continued)

$\delta_a(t)$	$\dfrac{e^{-as}(1 - e^{-\varepsilon s})}{\varepsilon s}$
$\delta(t - a)$	e^{-as}

SUMMARY OF COMPUTATIONAL RULES

Rule 1: $\mathscr{L}(f') = s\mathscr{L}(f) - f(0)$

Rule 2: $\mathscr{L}(f^{(n)}) = s^n\mathscr{L}(f) - s^{n-1}f(0) - s^{n-2}f'(0) - \cdots - f^{(n-1)}(0)$

Rule 3: If $f(t)$ has period ω, then

$$\mathscr{L}(f) = \frac{1}{1 - e^{-s\omega}} \int_0^\omega e^{-st}f(t)\, dt$$

Rule 4: (s-shift)

$$\mathscr{L}[e^{at}f(t)] = F(s - a)$$

Rule 5′: (t-shift)

$$\mathscr{L}[u(t - a)f(t - a)] = e^{-as}F(s)$$

Rule 6:

$$\mathscr{L}\left[\int_0^t f(z)\, dz\right] = \frac{1}{s}\mathscr{L}(f)$$

Convolution:

$$\mathscr{L}^{-1}[F(s)G(s)] = f(t) * g(t)$$

$$= \int_0^t f(t - \alpha)g(\alpha)\, d\alpha$$

Polynomial Coefficient: $\mathscr{L}[t^n f(t)] = (-1)^n F^{(n)}(s)$

HEAVISIDE FORMULAS

To compute the term in $\mathscr{L}^{-1}[P(s)/Q(s)]$ corresponding to certain factors in $Q(s)$.

Case 1: For a linear factor $(s - a)$ in $Q(s)$, the term is $H(a)e^{at}$ with

$$H(s) = \frac{P(s)(s - a)}{Q(s)}.$$

Case 2: For a factor $(s - a)^k$ in $Q(s)$, the term is

$$\left[\frac{H^{(k-1)}(a)}{(k-1)!} + \frac{H^{(k-2)}(a)}{(k-2)!}\frac{t}{1!} + \cdots + \frac{H'(a)}{1!}\frac{t^{k-2}}{(k-2)!} + H(a)\frac{t^{k-1}}{(k-1)!}\right]e^{at}$$

with

$$H(s) = \frac{P(s)(s-a)^k}{Q(s)}.$$

Case 3: For an unrepeated linear factor $(s-a)^2 + b^2$ in $Q(s)$, the term is

$$\frac{1}{b}[\alpha_i \cos(bt) + \alpha_r \sin(bt)]e^{at}$$

with

$$H(s) = \frac{P(s)[(s-a)^2 + b^2]}{Q(s)},$$

α_i = imaginary part of $H(a + ib)$, and α_r = real part of $H(a + ib)$.

Case 4: For a quadratic factor $[(s-a)^2 + b^2]^2$ in $Q(s)$, the terms are

$$\frac{1}{2b^3}(b\alpha_i - \beta_r)e^{at}\cos(bt) + \frac{1}{2b^3}(b\beta_i - \alpha_r)e^{at}\sin(bt) + \frac{te^{at}}{2b^2}[\alpha_i \sin(bt) - \alpha_r \cos(bt)],$$

where α_r = real part of $H(a + ib)$, α_i = imaginary part of $H(a + ib)$, β_r = real part of $H'(a + ib)$, β_i = imaginary part of $H'(a + ib)$, and

$$H(s) = \frac{P(s)[(s-a)^2 + b^2]^2}{Q(s)}.$$

SUPPLEMENTARY PROBLEMS

In each of Problems 1 through 30, use Laplace transforms to solve for the unknown function satisfying the given conditions.

1. $f(t) = -2 + \int_0^t \sinh(t-\alpha)f(\alpha)\,d\alpha$
2. $y'' - 3y' - 28y = -3u(t-4);\quad y(0) = 1,\quad y'(0) = 0$
3. $y'' - 5y' - 6y = 1 - 2t;\quad y(1) = 0,\quad y'(0) = -2$
4. $y'' + 8y' + 16y = 2\cos(3t);\quad y(1) = -2,\quad y'(2) = 4$
5. $y'' + 2y' - 8y = 1 - u(t-3);\quad y(0) = y'(0) = 0$
6. $y^{(3)} - 2y'' - 19y' + 20y = 2;\quad y(0) = 1,\quad y'(0) = -3,\quad y''(0) = 0$
7. $y'' + 9y' + 14y = 2\sin(t) + t^2;\quad y(0) = y'(0) = 0$
8. $y'' + 3y' - 4y = e^{-2t};\quad y(3) = 0,\quad y'(0) = -1$
9. $y'' + 10y' + 21y = e^{-t} + e^{-2t};\quad y(1) = 3,\quad y'(0) = 0$
10. $f(t) = 3\sin(2t) - \int_0^t \alpha f(t-\alpha)\,d\alpha$
11. $y^{(4)} - 6y^{(3)} + 13y'' - 12y' + 4y = 2u(t-6);\quad y(0) = y'(0) = y''(0) = y^{(3)}(0) = 0$
12. $y'' + 5y' - 14y = u(t-3) - u(t-4);\quad y(1) = 1,\quad y'(0) = -2$
13. $f(t) = 2t - \int_0^t e^{-\alpha}f(t-\alpha)\,d\alpha$
14. $y'' + 5y' - 36y = 3\sinh(t) - 1;\quad y(0) = y'(0) = 0$
15. $y'' - 4y' - 21y = 8e^{-2t};\quad y(0) = y'(0) = 0$
16. $f(t) - 2 = \int_0^t (t-\alpha)e^{-(t-\alpha)}f(\alpha)\,d\alpha$
17. $y'' - 4y' - 5y = 2\cos(5t) - 1;\quad y(0) = -2,\quad y'(0) = 0$
18. $y'' + 14y' + 48y = 2tu(t-4) - 7;\quad y(0) = -3,\quad y'(0) = 4$

19. $y'' + 3y' - 10y = -3\cos(5t);\quad y(0) = 0,\quad y'(0) = 0$

20. $y'' - 6y' - 7y = \sinh(t) - \cos(2t);\quad y(\pi) = 0,\quad y'(0) = 0$

21. $y'' + 11y' + 18y = u(t-1);\quad y(0) = 0,\quad y'(0) = 2$

22. $f(t) = 2\int_0^t e^{-4(t-\alpha)}f(\alpha)\,d\alpha$

23. $y'' + 4y' - 21y = 4u(t-3) + \sin(3t);\quad y(0) = 0,\quad y'(1) = -2$

24. $y^{(3)} + 10y'' + 12y' - 72y = 1;\quad y(0) = 2,\quad y'(1) = -3$

25. $y'' + 2y' - 15y = 2 - e^{-t};\quad y(0) = y'(0) = 1$

26. $y'' - 3y' + 2y = 8u(t-2) + \sin(3t);\quad y(\pi) = 0,\quad y'(0) = 0$

27. $f(t) = \sinh(t) - \int_0^t \alpha f(t-\alpha)\,d\alpha$

28. $y'' + 13y' + 42y = \sin^2(t);\quad y(0) = y'(0) = 0$

29. $y'' - 4y' - 5y = 1;\quad y(0) = 1,\quad y'(2) = 0$

30. $y'' + 11y' + 24y = 2u(t-\pi);\quad y(0) = 1,\quad y'(0) = -3$

31. Use Rule 3 of Section 4.2 to obtain the Laplace transform of $\cos(at)$.

32. Use Rule 3 of Section 4.2 to derive the Laplace transform of $\sin(at)$.

33. Assuming that $\mathcal{L}[\cos(at)] = s/(s^2 + a^2)$, use Rule 6 to obtain $\mathcal{L}[\sin(at)]$.

34. Derive and solve the system of differential equations governing the mechanical system of Figure 59. The driving force on mass m_2 is $F_2(t) = 2u(t-3)$; that on m_1 is zero. Assume that the masses are initially at rest at their equilibrium points. The damping constant is $c = 1$ on spring three due to the dashpot arrangement. Assume no damping effects on the other springs.

35. Redo Problem 34 with forces $F_1 = 2u(t-4)$ and $F_2 = u(t-3)$ acting on masses m_1 and m_2, respectively.

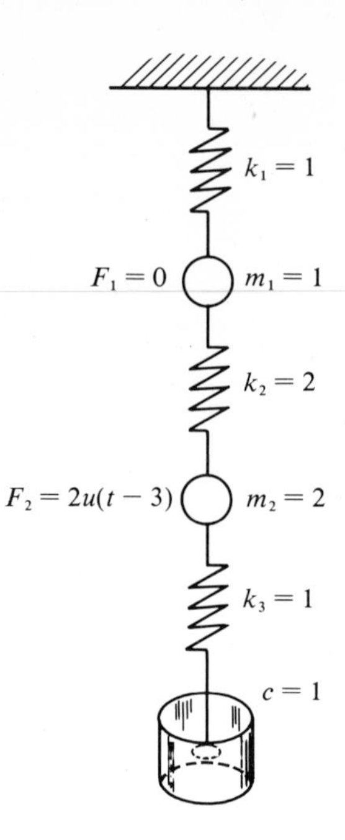

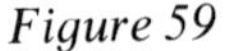

Figure 59

36. Solve for the currents and charge in the network of Figure 60, assuming that the currents are initially zero and that

$$E(t) = 2u(t-4) - u(t-5).$$

37. Redo Problem 36 with

$$E(t) = [1 - u(t-4)]\sin[2(t-4)].$$

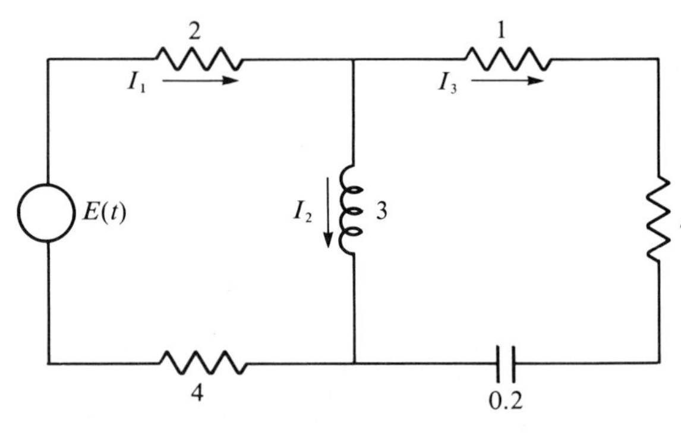

Figure 60

38. Use Laplace transforms to solve the system

$$y_1' - 2y_2' + 3y_3 = 0$$
$$y_1 - 4y_2' + 3y_3' = t$$
$$y_1 - 2y_2' + 3y_3' = -1$$
$$y_1(0) = y_2(0) = y_3(0) = 0.$$

39. Solve the system

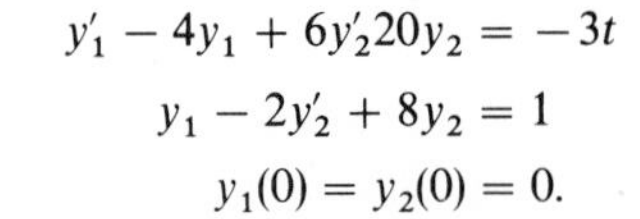

$$y_1' - 4y_1 + 6y_2'20y_2 = -3t$$
$$y_1 - 2y_2' + 8y_2 = 1$$
$$y_1(0) = y_2(0) = 0.$$

40. Solve the system

$$y_1 - 2y_2 + y_1' = 0$$
$$y_2 - 3y_1' + y_2' = t + 1$$
$$y_1(0) = y_2(0) = 0.$$

41. Find the error in the following "solution" of the pendulum equation,

$$\theta'' + \frac{g}{L}\sin(\theta) = 0.$$

Assume that $\theta(0) = \theta'(0) = 0$. Taking Laplace transforms gives us

$$s^2\mathcal{L}(\theta) + \frac{g}{L}\frac{1}{s^2+1} = 0.$$

Then

$$\mathcal{L}(\theta) = -\frac{g}{L}\frac{1}{s^2(s^2+1)} = -\frac{g}{L}\frac{1}{s^2} + \frac{g}{L}\frac{1}{s^2+1}.$$

Then

$$\theta = -\frac{gt}{L} + \frac{g}{L}\sin(t),$$

which implies that $\theta \to -\infty$ as $t \to \infty$!

42. Solve for the currents in the network of Figure 61. Assume that currents are initially zero and that $E(t) = 2u(t-4)$.

43. Solve for the currents in the network of Figure 62. Assume that currents are initially zero and that $E(t) = 3\sin(t)$.

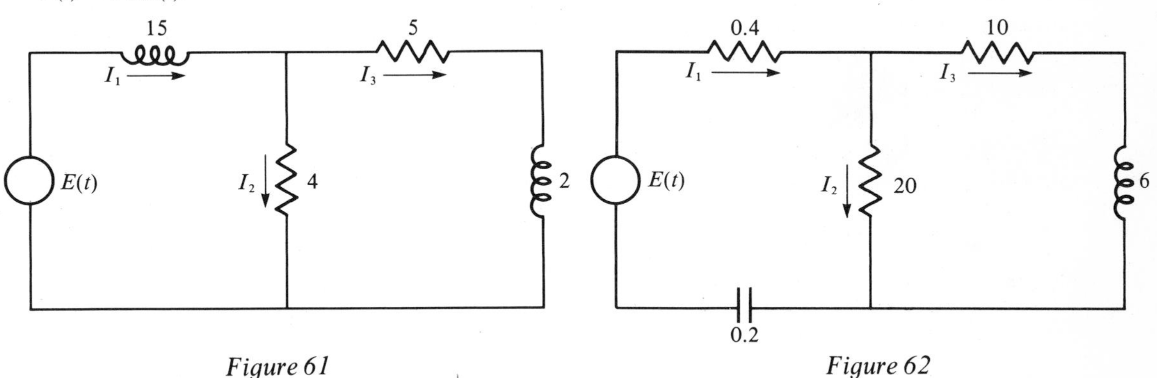

Figure 61   Figure 62

44. Solve for the motion of the masses in Figure 63. Assume no damping, external forces $F_1(t) = 2$, $F_2(t) = 0$, and zero initial displacements and velocities.

45. Redo Problem 44 with $F_1(t) = 1 - u(t-2)$, $F_2(t) = 0$.

46. Suppose that $u(t)$ satisfies

$$au'' + bu' + cu = 0; \qquad u(0) = 0, \qquad u'(0) = \frac{1}{a}.$$

Show that $y(t) = f(t) * u(t)$ satisfies

$$ay'' + by' + cy = f(t); \qquad y(0) = y'(0) = 0.$$

47. Use the idea of Problem 46 to solve each of the following.

(a) $2y'' + 4y' + y = t^2 + 1 \quad y(0) = y'(0) = 0$

(b) $y'' - 6y' + 2y = 4; \quad y(0) = y'(0) = 0$

(c) $y'' - 3y' + y = e^{-t}; \quad y(0) = y'(0) = 0$

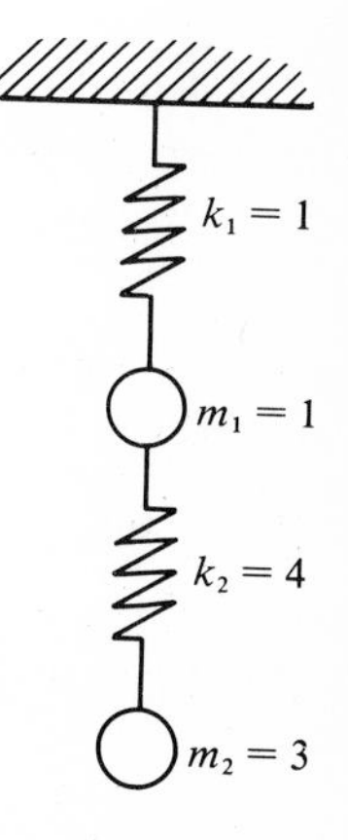

Figure 63

48. Imagine a double mass-spring system as shown in Figure 64. The spring attached to M has spring constant k_1 and damping constant c_1; that attached to m has spring constant k_2 and is undamped. Let M be subjected to a periodic driving force $F(t) = A \sin(\omega t)$. The masses are initially at rest in the equilibrium position.

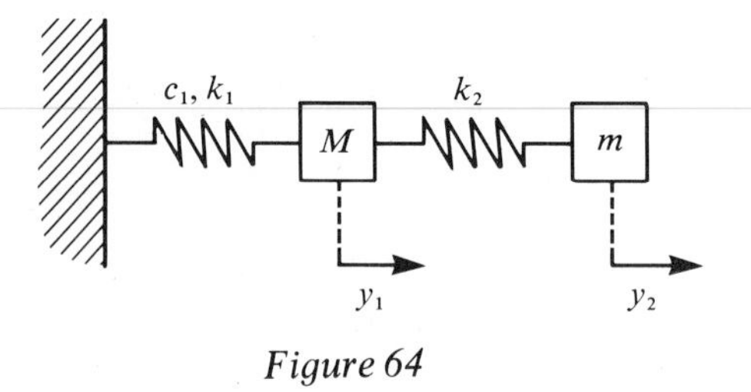

Figure 64

(a) Derive the system of differential equations describing the motion of the masses.
(b) Solve the system.
(c) Show that, if m and k_2 are chosen so that $\omega = \sqrt{k_2/m}$, then the mass m cancels out the forced vibrations of the mass M. In this case, we call m a *vibration absorber*.

49. Imagine a single mass m suspended from a spring having spring constant k. Assume no damping and a driving force $f(t)$. Solve the equation of motion to get

$$y(t) = \frac{1}{\sqrt{mk}} \sin\left(\sqrt{\frac{k}{m}}\, t\right) * f(t),$$

assuming $y(0) = y'(0) = 0$.

50. Two objects of mass m_1 and m_2 are attached at opposite ends of a spring having spring constant k, as shown in Figure 65. The entire apparatus is placed on a highly varnished table. Show that, if stretched and released from

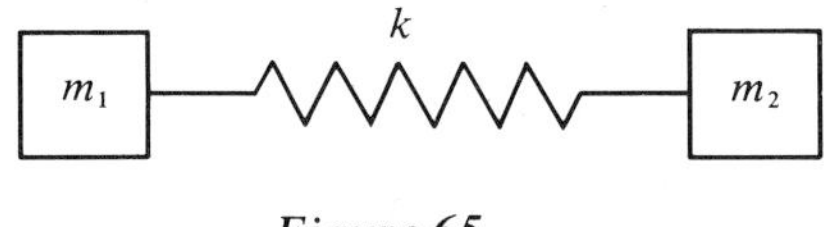

Figure 65

rest, the masses oscillate with respect to each other with period

$$2\pi \sqrt{\frac{m_1 m_2}{m_1 + m_2}}.$$

51. A mass m is suspended from a spring having spring constant k and no damping factor. The mass is at rest in the equilibrium position until, at time t_0, it is suddenly struck sharply by a force acting vertically and lasting τ seconds, after which it stops. Solve for the motion of the mass. (*Hint:* Look at Problem 49 with the appropriate choice of $f(t)$.) Assume that the force has magnitude 1.

52. Do Problem 51 with an impulse at time t_0 substituted for the force described there.

53. In each of (a), (b), and (c), solve the given system.

(a) $x' - 6x + 3y = 8e^t$
$y' - 2x - y = 4e^t$
$x(0) = -1, \quad y(0) = 0$

(b) $x' + 2x - 4y = 5$
$x' - 2x = t$
$x(0) = 0, \quad y(0) = 3$

(c) $x' - 6x + 3y = 0$
$y' - 2x - y = 0$
$x(0) = 14, \quad y(0) = 11$

54. Here is an application to maintenance of equipment in a factory. Suppose a textile mill has $p(t)$ pieces of weaving equipment at time t. Of course, $p(t)$ is integer-valued, but we will think of it as continuous—a ruse which works well for $p(t)$ large. As time increases, various pieces of equipment go out of service for repair or replacement. Let $r(t)$ be the number of replacements made up to time t, and $q(t)$ the number of functional pieces of equipment after time t.

(a) Show that $p(t) = p(0)q(t) + r'(t) * q(t)$.
(b) Solve for $\mathscr{L}[r(t)] = R(s)$ in terms of $\mathscr{L}[p(t)] = P(s)$, $p(0)$, and $\mathscr{L}[q(t)] = Q(s)$.
(c) Solve for $r(t)$ if $p(t) =$ constant and $q(t) = e^{-kt}$.
(d) Solve for $r(t)$ if $p(t) = At + B$ and $q(t) = e^{-kt}$.

55. (Tautochrone) Suppose we are given a point $P_0(x_0, y_0)$ in the xy-plane, as in Figure 66. We seek a curve C from P_0 to $(0, 0)$ having the property that a particle rolling down the curve takes the same amount of time T to reach the origin regardless of where on the curve the particle is initially placed. Such a curve is called a *tautochrone*.

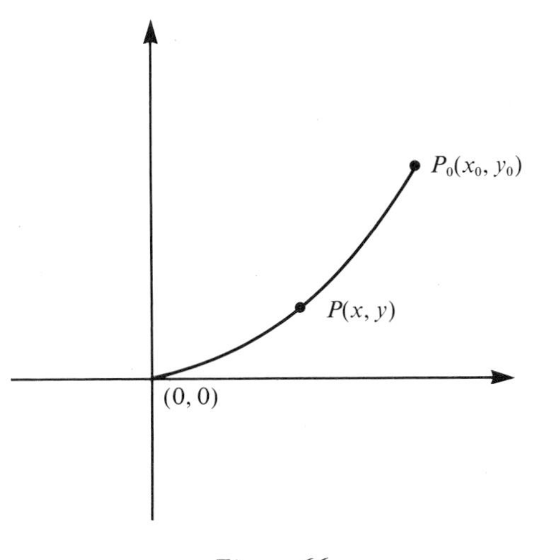

Figure 66

To find T, as well as the equation of the curve, proceed as follows:

(a) Assuming conservation of energy, show that, for a particle at $P(x, y)$ on C,

$$\frac{1}{2} m \left(\frac{ds}{dt}\right)^2 = mg(y_0 - y).$$

Here, friction is neglected, m is the mass of the particle, and s is arc length along C measured from the origin.

(b) Show that

$$T = \frac{1}{\sqrt{2g}} \int_{y=0}^{y=y_0} \frac{ds}{\sqrt{y_0 - y}}.$$

(c) Let $s = f(y)$. Show that

$$T = \frac{1}{\sqrt{2g}} \frac{1}{\sqrt{y}} * f'(y).$$

(d) Explain why $f(0) = 0$, and use (c) to get

$$f'(y) = \frac{\sqrt{2g}}{\pi} \frac{T}{\sqrt{y}}.$$

In this, note that $\mathscr{L}(1/\sqrt{y}) = \sqrt{\pi/s}$.

(e) Using $ds/dy = \sqrt{1 + (dx/dy)^2}$ from elementary calculus, show that

$$dx = \sqrt{\frac{a-y}{y}}\, dy,$$

where $a = 2gT^2/\pi^2$.

(f) Change variables by letting

$$y = a \sin^2\left(\frac{\theta}{2}\right)$$

and obtain from (e) that

$$x = \frac{a}{2}[\theta - \sin(\theta)], \qquad y = \frac{a}{2}[1 - \cos(\theta)].$$

These are parametric equations of a cycloid, and one arch of the cycloid is the tautochrone. A cycloid as given here can be visualized as the curve traced by a point Q on the rim of a wheel of radius $a/2$ rolling along the x-axes, as shown in Figure 67. The parameter θ is the angle through which the line from the center to Q has turned as the wheel rolls. For example, when $\theta = \pi$, Q is at $(\pi a/2, a)$; when $\theta = 2\pi$, Q is at $(\pi a, 0)$, and so on. The tautochrone has the shape of part of the cycloid from $(0, 0)$ to $(\pi a/2, a)$, except that in Figure 66 the tautochrone is turned upside down to form a track for the particle.

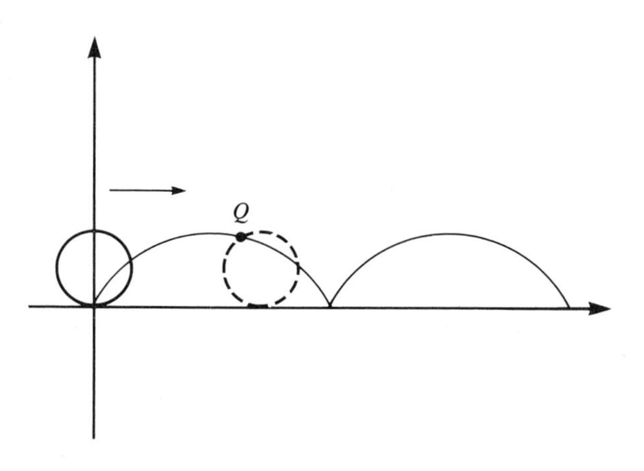

Figure 67. Tautochrone traced by point Q on rim of rolling wheel.

(g) Show by integration that $T = \pi\sqrt{a/g}$ is the time of passage of the particle from *any* point P on the tautochrone to the origin [except, of course, $P = (0, 0)$].

The following five problems require that the student have some experience with circuits beyond that required for previous problems. In particular, some knowledge of transfer functions and frequency response is needed.

56. J. J. Consumer, an electronics devotee, has recently purchased a hobby kit from the Ideal Electronics Company. Among the components in the kit are two resistors (labeled R_1 and R_2), two capacitors (labeled C_1 and C_2), and a switch (labeled S). The battery (marked E) and digital voltmeter are sold separately. Being eager to tinker with the new toy, J. J. fearlessly charges capacitor C_1 to $v_1(0)$ volts and capacitor C_2 to $v_2(0)$ volts and then connects the components as illustrated. But before closing the switch (and possibly destroying some of the components of the kit), J. J. solicits your assistance in accomplishing the following tasks.

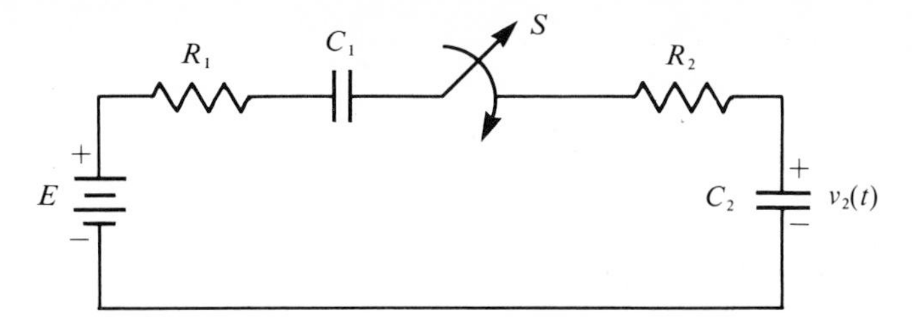

(a) Find the input-output transfer function, $T(s) = V_2(s)/E(s)$, from the source voltage to the voltage across capacitor C_2.
(b) Identify the DC gain K of the circuit as a function of the circuit elements (R_1, R_2, C_1, and C_2).
(c) Identify the time-constant τ of the circuit in terms of the circuit elements.
(d) If the switch S is closed at $t = 0$ and $v_1(0) = v_{10}$ volts and $v_2(0) = v_{20}$ volts, find the steady-state output voltage

$$v_{(2\text{ steady-state})} = \lim_{t\to\infty} v_2(t).$$

(e) If the switch S is closed at $t = 0$ and $v_1(0) = v_{10}$ volts and $v_2(0) = v_{20}$ volts, find the output voltage $v_2(t)$ for all $t > 0$.

57. Consider the electronic circuit shown below:

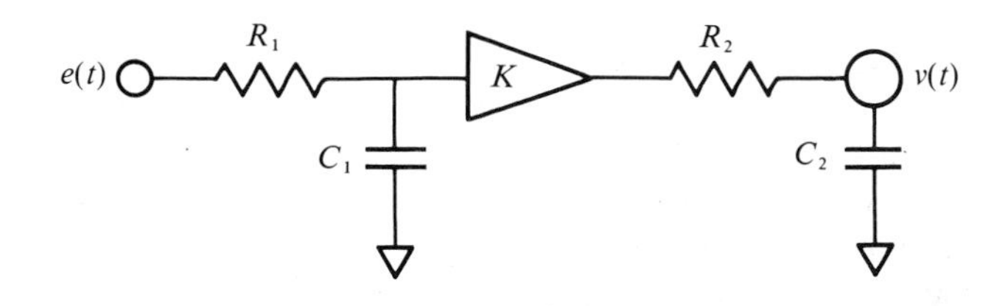

Find the circuit transfer function, the dimensionless damping ratio ζ, and the undamped natural frequency ω_n.

58. A circuit designer proposes the active filter shown below. The amplifier has extremely large input impedance, negligible output impedance, and voltage gain K volts/volt.

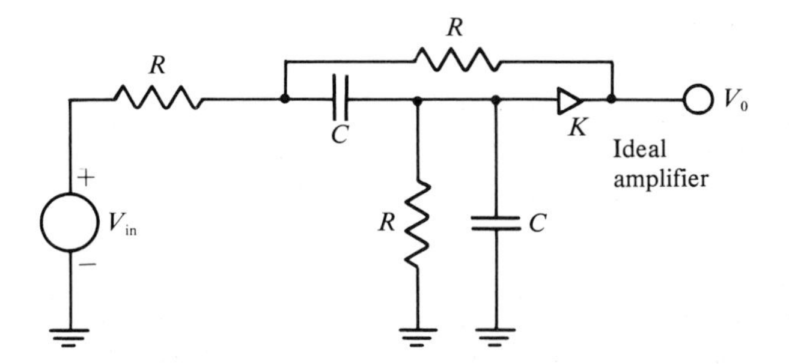

(a) Find the input-output voltage transfer function, $H(s) = V_0(s)/V_{in}(s)$, of the filter.
(b) Find the undamped natural frequency ω_n (second^{-1}), of the filter.
(c) Find the dimensionless damping ratio ζ and quality factor $Q = 1/2\zeta$ of the filter.
(d) Find the center frequency and bandwidth of the filter.
(e) Find and sketch the amplitude response of the filter $|H(j\omega)|$ and thereby characterize the behavior and application of the circuit.

59. An entrepreneur proposes the circuit shown below. A is a quality operational amplifier and $R_1 = 8\ \text{k}\Omega$. The entrepreneur is curious about the workings of the circuit, and for this purpose you are asked to accomplish the following tasks:

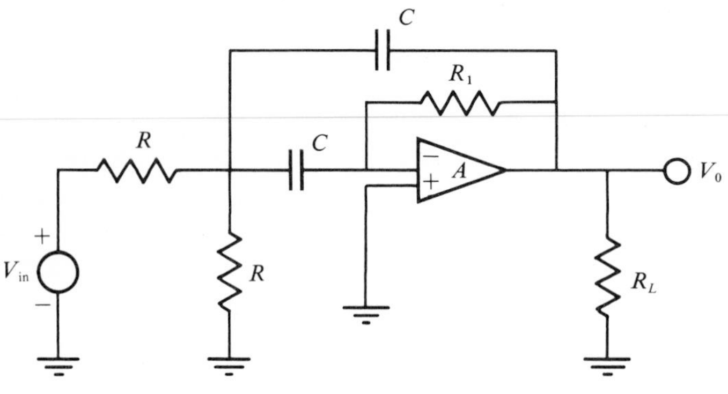

(a) Find the voltage transfer function, $T(s) = V_0(s)/V_{\text{in}}(s)$, in terms of the circuit parameters.
(b) Sketch qualitatively the magnitude $|T(j\omega)|$ of the frequency response of the circuit.
(c) Describe the operation of the circuit.
(d) Find the maximum value of $|T(j\omega)|$.
(e) Find the bandwidth (the full width at one-half power) of the circuit.

60. Upon looking behind an analogue computer patch-board, a technician finds the electronic circuit shown below:

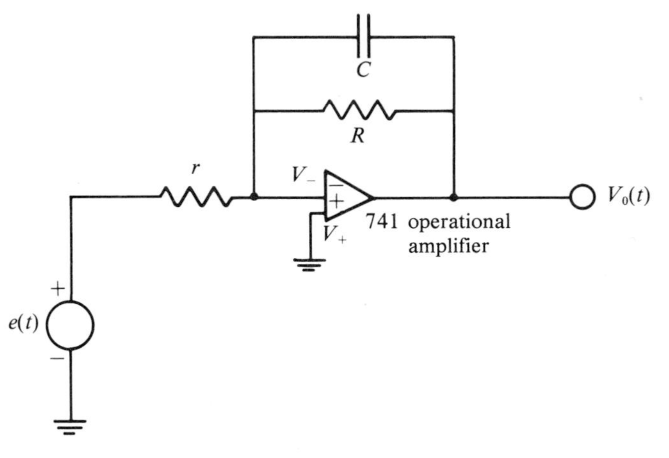

After probing around, the technician discovers that

$$r = 1\ \text{K}\Omega, \qquad R = 1\ \text{M}\Omega, \qquad C = 1\ \mu f,$$

and that the 741 operational amplifier can be characterized by the following three properties:

i. The input impedance exceeds $10^{12}\ \Omega$.

ii. The output impedance is negligible.

iii. The (open-loop) transfer function is

$$A(j\omega) = \frac{V_0(j\omega)}{V_+(j\omega) - V_-(j\omega)} = \frac{K}{j\omega T + 1}. \tag{1}$$

In (1), the gain is $K = 10^6$ volts/volt, or 120 db, and the time-constant is $T = 0.01$ second. Thus, the corner frequency is $\omega_C = 1/T = 100$ radians/second. The technician solicits your assistance in the following:

(a) Obtain the frequency response

$$H(j\omega) \triangleq \frac{V_0(j\omega)}{E(j\omega)}$$

of the circuit, both literally and numerically in terms of the circuit parameters.

(b) Plot and compare the magnitude $|H(j\omega)|$ of frequency response of the electronic circuit with that of the OP Amp.

(c) If the circuit were to be utilized as an analogue filter, would the filter be low pass, band pass, or high pass?

(d) What function(s) can the circuit perform?

CHAPTER FIVE

Series Solutions of Differential Equations

5.0 INTRODUCTION

In this chapter we develop methods of solution of ordinary differential equations based on power series and modifications of power series. Although the methods are fairly general, we shall restrict our attention to first and second order differential equations for two reasons: First, they are the most important for the applications we will develop; and second, the ideas extend naturally from these to higher order equations, and so the student who understands the method of solving second order equations will not be at a loss when faced with a higher order equation.

We begin with a review of the basic tool to be used, the power series.

5.1 REVIEW OF POWER SERIES

A *power series* is a series of the form

$$\sum_{n=0}^{\infty} a_n(x-a)^n .$$

The numbers $a_0, a_1, a_2, \ldots$ are called the *coefficients*, and a is the *center* of the series. Since we could always change variables to $z = x - a$ to get a series $\sum_{n=0}^{\infty} a_n z^n$, we shall for convenience take $a = 0$ unless specific exception is noted.

RADIUS OF CONVERGENCE

We are interested in the values of x for which $\sum_{n=0}^{\infty} a_n x^n$ converges. In most calculus courses it is shown that there are exactly three possibilities:

1. The series converges only if $x = 0$;
2. The series converges for all real x; or
3. For some number R, the series converges if $|x| < R$ and diverges for $|x| > R$.

The number R in (3) is called the *radius of convergence* of $\sum_{n=0}^{\infty} a_n x^n$, and the interval $(-R, R)$, inside of which the series converges, is called the *open interval of convergence.* This is shown in Figure 68.

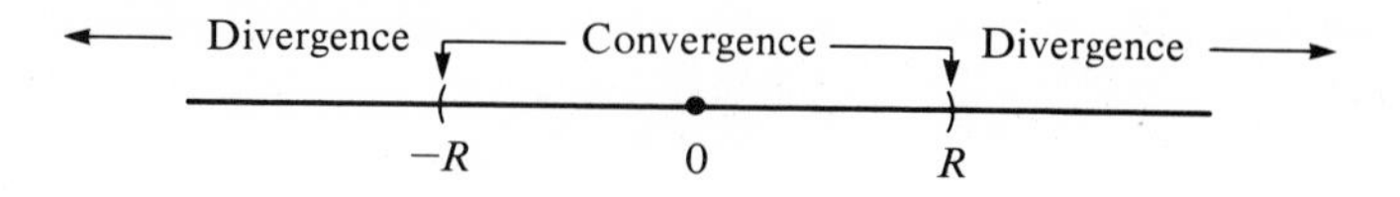

Figure 68. Interval of convergence of a power series.

There are several formulas for R. Two convenient ones are

$$R = \frac{1}{\lim\limits_{n\to\infty} \left| \dfrac{a_{n+1}}{a_n} \right|}$$

and

$$R = \frac{1}{\lim\limits_{n\to\infty} |a_n|^{1/n}},$$

where the a_n's are assumed nonzero.

To derive the first of these, apply the ratio test to $\sum_{n=0}^{\infty} a_n x^n$. Thus, look at

$$\left| \frac{a_{n+1}x^{n+1}}{a_n x^n} \right|,$$

which is the same as

$$|x| \left| \frac{a_{n+1}}{a_n} \right|.$$

If the limit as $n \to \infty$ of this ratio is less than 1, then $\sum_{n=0}^{\infty} a_n x^n$ converges; if the limit is greater than 1, then $\sum_{n=0}^{\infty} a_n x^n$ diverges; and if the limit is 1, we can draw no conclusion. Thus, in summary, $\sum_{n=0}^{\infty} a_n x^n$ will

$$\text{converge if} \quad \lim_{n\to\infty} |x| \left| \frac{a_{n+1}}{a_n} \right| < 1,$$

$$\text{diverge if} \quad \lim_{n\to\infty} |x| \left| \frac{a_{n+1}}{a_n} \right| > 1.$$

Factoring $|x|$ out of the limit, this is the same as saying that

$$\sum_{n=0}^{\infty} a_n x^n \begin{cases} \text{converges if} & |x| < R \\ \text{diverges if} & |x| > R, \end{cases}$$

where

$$R = \frac{1}{\lim\limits_{n\to\infty} \left| \dfrac{a_{n+1}}{a_n} \right|}.$$

This makes R exactly the number we call the radius of convergence of the power series. A similar argument, using the root test in place of the ratio test, yields the second formula given above for R.

It should be noted that these formulas are quite restrictive, assuming as they do that certain limits exist and that none of the a_n's are zero. Even when these formulas do not apply, however, $\sum_{n=0}^{\infty} a_n x^n$ still has a radius of convergence. If the a_n's are very bizarre, then this radius of convergence may be difficult to determine explicitly, but it exists (possibly infinite) nonetheless.

If the denominator is zero in the above formulas for R, we get $R = \infty$, which corresponds to case 2 above [here, $(-R, R) = (-\infty, \infty)$, the whole real line]. If the denominator is infinite, then $R = 0$ and we have case 1, with convergence only at $x = 0$.

EXAMPLE 156

The series $\sum_{n=0}^{\infty} x^n$ has radius of convergence $R = 1$. Here, each $a_n = 1$; so $R = 1/\lim_{n\to\infty} (1) = 1$. Thus, $\sum_{n=0}^{\infty} x^n$ converges if $-1 < x < 1$ and diverges if $x > 1$ or $x < -1$.

By contrast, $\sum_{n=0}^{\infty} (x^n/n!)$ converges for all x (here we find $R = \infty$). Remember that $0! = 1$ and $n! = 1 \cdot 2 \cdots n$ if n is a positive integer.

For purposes of solving differential equations, the question of convergence of $\sum_{n=0}^{\infty} a_n x^n$ at $x = R$ or $x = -R$ is often uninteresting and so we shall not pursue it.

As a more unusual example, consider the power series

$$\frac{1}{2} + \frac{x}{3} + \frac{x^2}{2^2} + \frac{x^3}{3^2} + \frac{x^4}{2^3} + \frac{x^5}{3^3} + \frac{x^6}{2^4} + \frac{x^7}{3^4} + \cdots.$$

Here, $a_{2n} = 1/2^{n+1}$ and $a_{2n+1} = 1/3^{n+1}$ for $n = 0, 1, 2, 3, \ldots$.

In this example, neither $\lim_{n\to\infty} |a_{n+1}/a_n|$ nor $\lim_{n\to\infty} \sqrt[n]{|a_n|}$ exists. However, one can show by means usually discussed in advanced calculus that this power series has radius of convergence $\sqrt{2}$.

OPERATIONS WITH POWER SERIES

Power series can be added and multiplied by constants in the obvious ways:

$$\sum_{n=0}^{\infty} a_n x^n + \sum_{n=0}^{\infty} b_n x^n = \sum_{n=0}^{\infty} (a_n + b_n)x^n$$

(wherever both series on the left converge), and

$$k \sum_{n=0}^{\infty} a_n x^n = \sum_{n=0}^{\infty} k a_n x^n.$$

Further, inside the open interval of convergence, we can differentiate a power series term by term:

$$\frac{d}{dx} \sum_{n=0}^{\infty} a_n x^n = \sum_{n=1}^{\infty} n a_n x^{n-1},$$

and the series on the right has the same radius of convergence as the original series. Note that on the right the differentiated series begins with $n = 1$ since, for $n = 0$, $na_n x^{n-1} = 0$ anyway.

Finally, inside the open interval of convergence we can integrate a power series term by term:

$$\int \sum_{n=0}^{\infty} a_n x^n = \sum_{n=0}^{\infty} \frac{a_n}{n+1} x^{n+1} + C,$$

and the series on the right has the same radius of convergence as the original series. Here, C is the constant of integration.

EXAMPLE 157

$$\frac{d}{dx} \sum_{n=0}^{\infty} x^n = \sum_{n=1}^{\infty} nx^{n-1} \qquad (-1 < x < 1).$$

Similarly,

$$\int \sum_{n=0}^{\infty} x^n = \sum_{n=0}^{\infty} \frac{1}{n+1} x^{n+1} + C.$$

POWER SERIES EXPANSIONS

Given a power series $\sum_{n=0}^{\infty} a_n x^n$, we can define a function $g(x)$ by

$$g(x) = \sum_{n=0}^{\infty} a_n x^n$$

for all x such that the series converges.

Conversely, it is sometimes possible to write a given function $f(x)$ as a power series:

$$f(x) = \sum_{n=0}^{\infty} a_n x^n$$

in some interval $(-R, R)$ about 0 (possibly on the whole real line). When this can be done, we call $f(x)$ *analytic* (at 0). The series $\sum_{n=0}^{\infty} a_n x^n$ is called the *Taylor series** for $f(x)$ about 0, and the a_n's are given by

$$a_n = \frac{f^{(n)}(0)}{n!}.$$

* Often a Taylor series about 0 is called a *Maclauren series*.

Many functions we are familiar with in engineering applications are analytic. For example,

$$e^x = \sum_{n=0}^{\infty} \frac{x^n}{n!} \qquad \text{(for all } x\text{)},$$

$$e^{-x} = \sum_{n=0}^{\infty} \frac{(-1)^n x^n}{n!} \qquad \text{(for all } x\text{)},$$

$$\sin(x) = \sum_{n=0}^{\infty} \frac{(-1)^n x^{2n+1}}{(2n+1)!} \qquad \text{(for all } x\text{)},$$

$$\cos(x) = \sum_{n=0}^{\infty} \frac{(-1)^n x^{2n}}{(2n)!} \qquad \text{(for all } x\text{)},$$

$$\frac{1}{1-x} = \sum_{n=0}^{\infty} x^n \qquad \text{(for } -1 < x < 1\text{)},$$

$$\frac{1}{1+x} = \sum_{n=0}^{\infty} (-1)^n x^n \qquad \text{(for } -1 < x < 1\text{)}.$$

Note that the coefficients in the Taylor series for $f(x)$ about 0 are completely determined by the derivatives of $f(x)$ at 0, i.e.,

$$a_n = \frac{f^{(n)}(0)}{n!}.$$

Conversely, one can show that if $f(x) = \sum_{n=0}^{\infty} b_n x^n$ in some interval about 0, then necessarily this series is exactly the Taylor series about 0, and $b_n = f^{(n)}(0)/n!$. Hence, if $\sum_{n=0}^{\infty} a_n x^n = \sum_{n=0}^{\infty} b_n x^n$ in an interval about 0, then $a_n = b_n$ for $n = 0, 1, 2, \ldots$. This is crucial in using power series to solve differential equations.

ALTERNATING SERIES

In many applications we need to approximate $\sum_{n=0}^{\infty} a_n x^n$ by a finite sum

$$a_0 + a_1 x + \cdots + a_k x^k.$$

How good is such an approximation? Here is one circumstance in which we can make an estimate:

If $\sum_{n=0}^{\infty} a_n x_0^n$ converges for some x_0 with $-R < x_0 < R$; if the terms $a_n x_0^n$ alternate in sign; and if $|a_n x_0^n|$ decreases monotonically to zero as $n \to \infty$, then

$$a_0 + a_1 x_0 + a_2 x_0^2 + \cdots + a_k x_0^k$$

approximates $\sum_{n=0}^{\infty} a_n x_0^n$ to within $|a_{k+1} x_0^{k+1}|$. That is, the error incurred in omitting all terms from $a_{k+1} x_0^{k+1}$ on is not more than $|a_{k+1} x_0^{k+1}|$.

As an example, we know that, for all x,

$$e^{-x} = \sum_{n=0}^{\infty} \frac{(-1)^n x^n}{n!}.$$

Here, $a_n = (-1)^n x^n/n!$ alternates in sign and decreases to zero as $n \to \infty$ for any $x > 0$. Say we want to approximate $1/e$. Put $x = 1$ to get (exactly)

$$e^{-1} = \sum_{n=0}^{\infty} \frac{(-1)^n}{n!}.$$

As an approximation, $1 - 1 + \frac{1}{2} - \frac{1}{6} + \frac{1}{24}$ (summing to $n = 4$) approximates $1/e$ to within $|(-1)^5/5!| = \frac{1}{120}$.

As a check,

$$\tfrac{1}{2} - \tfrac{1}{6} + \tfrac{1}{24} = 0.37500 \ldots,$$

while $1/e = 0.3678794\ldots$. Here, $0.375 - 1/e = 0.00712\ldots < \frac{1}{120} = 0.00833\ldots$.

Similarly, $1 - 1 + \frac{1}{2} - \frac{1}{6} + \frac{1}{24} - \frac{1}{120} + \frac{1}{720}$ approximates $1/e$ to within $|(-1)^7/7!| = 1/5040$. Check this:

$$\tfrac{1}{2} - \tfrac{1}{6} + \tfrac{1}{24} - \tfrac{1}{120} + \tfrac{1}{720} = 0.368055 \ldots$$

and

$$0.368055 \ldots - \frac{1}{e} = 0.001761 \ldots < \frac{1}{5040} = 0.000198 \ldots.$$

SHIFTING INDICES

In applying power series to differential equations, some facility is needed in combining series which involve different powers of x. This can be done by juggling the summation index. Here are two examples.

EXAMPLE 158

Add the power series

$$\sum_{n=0}^{\infty} a_n x^n \quad \text{and} \quad \sum_{n=1}^{\infty} b_n x^{n+2}.$$

In order to factor coefficients of like powers of x in the sum, we want similar appearing exponents on x in the two series. We can, for example, take $\sum_{n=1}^{\infty} b_n x^{n+2}$ and shift indices by putting $m = n + 2$. When $n = 1$, $m = 3$; further, $n = m - 2$, and so

$$\sum_{n=1}^{\infty} b_n x^{n+2} = \sum_{m=3}^{\infty} b_{m-2} x^m.$$

Since m is just a dummy summation variable (as is n), we can use the dummy n in place of dummy m on the right. Finally then,

$$\sum_{n=1}^{\infty} b_n x^{n+2} = \sum_{n=3}^{\infty} b_{n-2} x^n.$$

Check this out:

$$\sum_{n=1}^{\infty} b_n x^{n+2} = b_1 x^3 + b_2 x^4 + b_3 x^5 + \cdots$$

and

$$\sum_{n=3}^{\infty} b_{n-2}x^n = b_1x^3 + b_2x^4 + b_3x^5 + \cdots;$$

so the two sums are indeed the same. Now we can add the series we started with:

$$\begin{aligned}\sum_{n=0}^{\infty} a_n x^n + \sum_{n=1}^{\infty} b_n x^{n+2} &= \sum_{n=0}^{\infty} a_n x^n + \sum_{n=3}^{\infty} b_{n-2}x^n \\ &= \sum_{n=0}^{2} a_n x^n + \sum_{n=3}^{\infty} a_n x^n + \sum_{n=3}^{\infty} b_{n-2}x^n \\ &= a_0 + a_1x + a_2x^2 + \sum_{n=3}^{\infty} (a_n + b_{n-2})x^n.\end{aligned}$$

EXAMPLE 159

Add

$$\sum_{n=0}^{\infty} a_n x^n \quad \text{and} \quad \sum_{n=1}^{\infty} b_n x^{n+1}.$$

As before, we shift indices:

$$\sum_{n=1}^{\infty} b_n x^{n+1} = b_1x^2 + b_2x^3 + \cdots = \sum_{n=2}^{\infty} b_{n-1}x^n.$$

Then,

$$\begin{aligned}\sum_{n=0}^{\infty} a_n x^n + \sum_{n=1}^{\infty} b_n x^{n-1} &= \sum_{n=0}^{\infty} a_n x^n + \sum_{n=2}^{\infty} b_{n-1}x^n \\ &= a_0 + a_1x + \sum_{n=2}^{\infty} a_n x^n + \sum_{n=2}^{\infty} b_{n-1}x^n \\ &= a_0 + a_1x + \sum_{n=2}^{\infty} (a_n + b_{n-1})x^n.\end{aligned}$$

Note that shifting indices to get the powers to look the same often changes the starting index of one of the series. In the last example, we can add terms of $\sum_{n=0}^{\infty} a_n x^n$ to terms of $\sum_{n=2}^{\infty} b_{n-1}x^n$ only from $n = 2$ on since $\sum_{n=2}^{\infty} b_{n-1}x^n$ has no terms below $n = 2$. Thus, we combine $\sum_{n=2}^{\infty} a_n x^n$ and $\sum_{n=2}^{\infty} b_{n-1}x^n$ and separately add on the omitted terms $a_0 + a_1x$. A similar remark applies to Example 158 where we added $\sum_{n=0}^{\infty} a_n x^n$ to $\sum_{n=3}^{\infty} b_{n-2}x^n$ only from $n = 3$ on and then separately put back the omitted terms $a_0 + a_1x + a_2x^2$.

We are now ready to solve some differential equations by series methods.

PROBLEMS FOR SECTION 5.1

In each of Problems 1 through 26, find the radius of convergence of the given series.

1. $\sum_{n=0}^{\infty} \frac{(-1)^n x^n}{n+1}$ **2.** $\sum_{n=1}^{\infty} \frac{2^n}{n!}x^n$ **3.** $\sum_{n=0}^{\infty} \frac{x^n}{n+2}$ **4.** $\sum_{n=0}^{\infty} n^2x^n$

5. $\sum_{n=0}^{\infty} \left(\frac{2n+1}{2n-1}\right)x^n$
6. $\sum_{n=0}^{\infty} n^n x^n$
7. $\sum_{n=0}^{\infty} (-\frac{3}{2})^n x^n$
8. $\sum_{n=0}^{\infty} \left(\frac{n+1}{n+2}\right)x^n$
9. $\sum_{n=0}^{\infty} \left(\frac{n^2-3n}{n^2+4}\right)x^n$
10. $\sum_{n=0}^{\infty} \frac{(-1)^n}{n+2}x^n$
11. $\sum_{n=0}^{\infty} \left(\frac{n}{n^2+1}\right)x^n$
12. $\sum_{n=1}^{\infty} \frac{3^n}{(2n)!}x^n$
13. $\sum_{n=0}^{\infty} \frac{nx^n}{3n+1}$
14. $\sum_{n=1}^{\infty} \left(\frac{n+1}{n}\right)^n x^n$
15. $\sum_{n=0}^{\infty} (-1)^{2n+1}x^{n+2}$
16. $\sum_{n=0}^{\infty} (\frac{2}{3})^n x^{n+1}$
17. $\sum_{n=0}^{\infty} \frac{2^n}{n!}x^n$
18. $\sum_{n=1}^{\infty} \frac{x^n}{n(2n+1)}$
19. $\sum_{n=1}^{\infty} \frac{n!}{2^n}x^{n+3}$
20. $\sum_{n=1}^{\infty} \left(\frac{n-3}{2n+1}\right)x^n$
21. $\sum_{n=1}^{\infty} \frac{(-1)^{n+1}nx^n}{n+4}$
22. $\sum_{n=1}^{\infty} \left(\frac{2}{3}\right)^n \left(\frac{n+1}{n+2}\right)x^n$
23. $\sum_{n=1}^{\infty} \left(\frac{1-2n}{3n}\right)x^n$
24. $\sum_{n=1}^{\infty} \frac{e^n}{n!}x^{n+2}$
25. $\sum_{n=1}^{\infty} (\frac{1}{2})^n (n+1)!\,x^n$
26. $\sum_{n=1}^{\infty} \frac{\ln(n)}{n}x^n$

In each of Problems 27 through 42, find the terms up to and including the x^5 term of the Taylor series about 0 of the given function. *Note:* Some coefficients may be zero.

27. $\ln(x+2)$
28. xe^x
29. $2x\sin(x)$
30. $e^x\cos(x)$
31. $(x+1)^4$
32. $x\tan(x)$
33. $\frac{1}{(1+x)^3}$
34. $\frac{1}{x+1}\sinh(x)$
35. $1-x^2\cos(2x)$
36. $\arctan(x)$
37. $\sec(x+1)$
38. $\cos^2(2x)$
39. $e^x\sin(x)$
40. $4x^2\tan(x)$
41. $\sin(x+1)$
42. $\cos(2x)\sin(2x)$

In each of Problems 43 through 50, find the first six terms (that is, up to and including fifth powers) of the Taylor expansion of the given function about the given point.

43. $x-\cos(2x)$; $\pi/2$
44. $\tan(2x)$; $\pi/6$
45. $(x+1)^2$; $\frac{1}{2}$
46. e^x; -2
47. $\sin(3x)$; $\pi/4$
48. x^2e^x; -1
49. $\ln(1+x)$; $\frac{1}{3}$
50. $\frac{1}{1-2x}$; $\frac{1}{3}$

In each of Problems 51 through 60, shift indices so that all powers of x under the summation are x^n.

51. $\sum_{n=1}^{\infty} \frac{(-1)^{n+1}x^{n+3}}{2n+4}$
52. $\sum_{n=2}^{\infty} \frac{(n+1)^n}{2^n}x^{n+3}$
53. $\sum_{n=2}^{\infty} \frac{(2n+3)}{n}x^{n-1}$
54. $\sum_{n=4}^{\infty} \frac{(-1)^{n+1}x^{n-3}}{n^2+2}$
55. $\sum_{n=1}^{\infty} \frac{2^n x^{n+1}}{n^2-3}$
56. $\sum_{n=1}^{\infty} \frac{2n}{n!}x^{n+1}$
57. $\sum_{n=0}^{\infty} \frac{2n}{n+2}x^{n+3}$
58. $\sum_{n=3}^{\infty} \left(\frac{n-1}{2^n n!}\right)x^{2n-2}$
59. $\sum_{n=0}^{\infty} (n-3)(n-1)x^{n-1}$
60. $\sum_{n=1}^{\infty} \frac{8n^2(-1)^n}{(n+1)^2}x^{n+1}$

In each of Problems 61 through 70, add the given series as in Examples 158 and 159. (There are several ways of writing an acceptable solution.)

61. $\sum_{n=1}^{\infty} 2^n x^{n+1} + \sum_{n=0}^{\infty} (1+n)x^n$
62. $\sum_{n=0}^{\infty} \frac{n!}{2^n}x^{n+3} + \sum_{n=1}^{\infty} \frac{1}{1+n}x^{n-1}$
63. $\sum_{n=1}^{\infty} \frac{n!}{n^2}x^{n-1} + \sum_{n=2}^{\infty} 2^n x^n$

64. $\sum_{n=2}^{\infty} \frac{x^n}{n} + \sum_{n=1}^{\infty} (-1)^n x^{n+2}$

65. $\sum_{n=1}^{\infty} \frac{x^n}{2n} - \sum_{n=3}^{\infty} n^n x^{n-2}$

66. $\sum_{n=1}^{\infty} (2n-1)x^{n+3} + \sum_{n=0}^{\infty} \frac{1}{n+1} x^n$

67. $\sum_{n=2}^{\infty} x^{n-1} + \sum_{n=3}^{\infty} n!\, x^{n+2}$

68. $\sum_{n=1}^{\infty} (-1)^n x^{n+1} + \sum_{n=2}^{\infty} (\frac{3}{2})^n x^{n+2}$

69. $\sum_{n=0}^{\infty} \frac{1}{n+2} x^n + \sum_{n=1}^{\infty} (-2)^n x^{n+1}$

70. $\sum_{n=0}^{\infty} x^{n+3} + \sum_{n=1}^{\infty} (-1)^n \frac{1}{n+1} x^{n+2}$

71. Use the series $e^{-x} = \sum_{n=0}^{\infty} \frac{(-1)^n x^n}{n!}$ to approximate $\sqrt{\frac{1}{e}}$ to within 10^{-5}.

72. Use the series $\cos\left(\frac{x}{4}\right) = \sum_{n=0}^{\infty} \frac{(-1)^n x^{2n}}{(2n)!\, 4^{2n}}$ to approximate cos(2) to within 10^{-5}.

73. Use the series $\ln(1+x) = \sum_{n=1}^{\infty} \frac{(-1)^{n+1} x^n}{n}$ to approximate $\ln(\frac{3}{2})$ to within 10^{-3}.

5.2 POWER SERIES SOLUTIONS OF DIFFERENTIAL EQUATIONS

The power series method consists of substituting $y = \sum_{n=0}^{\infty} a_n x^n$ into the differential equation and attempting to solve for the a_n's to obtain a solution. Success is not guaranteed under all circumstances, but the following theorem gives us cause for optimism under fairly general conditions.

THEOREM 16

First order case If $g(x)$ and $r(x)$ are analytic at 0, then every solution of

$$y' + g(x)y = r(x)$$

is also analytic at 0 and can hence be expressed as a power series $y = \sum_{n=0}^{\infty} a_n x^n$.

Second order case If $P(x)$, $Q(x)$, and $F(x)$ are analytic at 0, then every solution of

$$y'' + P(x)y' + Q(x)y = F(x)$$

is also analytic at 0 and can hence be expressed as a power series $y = \sum_{n=0}^{\infty} a_n x^n$.

Here are some illustrations of the method, beginning with simple problems and proceeding to several difficult ones for which previous methods fail.

EXAMPLE 160

Solve

$$y' + ky = 0,$$

in which k is any constant.

Of course, we already know how to solve this: The general solution is

$$y = Ae^{-kx},$$

with A the constant of integration. But to illustrate the power series method we shall attempt a solution

$$y = \sum_{n=0}^{\infty} a_n x^n.$$

Then*

$$y' = \sum_{n=1}^{\infty} n a_n x^{n-1}$$

and the differential equation becomes

$$\sum_{n=1}^{\infty} n a_n x^{n-1} + \sum_{n=0}^{\infty} k a_n x^n = 0.$$

Shift indices to write

$$\sum_{n=1}^{\infty} n a_n x^{n-1} = \sum_{n=0}^{\infty} (n+1) a_{n+1} x^n.$$

Then the differential equation becomes

$$\sum_{n=0}^{\infty} (n+1) a_{n+1} x^n + \sum_{n=0}^{\infty} k a_n x^n = 0$$

or

$$\sum_{n=0}^{\infty} [(n+1) a_{n+1} + k a_n] x^n = 0.$$

Since this must hold for all x in some interval about 0, then every coefficient in this series must vanish (note the remarks on page 272, Section 5.1). Thus,

$$(n+1) a_{n+1} + k a_n = 0 \quad \text{for} \quad n = 0, 1, 2, 3, \ldots.$$

Then

$$a_{n+1} = -\frac{k a_n}{n+1} \quad \text{for} \quad n = 0, 1, 2, \ldots.$$

This is called a *recurrence relation*, giving each coefficient in terms of (in this case) the previous one. Thus,

$$a_1 = -k a_0,$$

$$a_2 = -k\left(\frac{a_1}{2}\right) = +\frac{k^2 a_0}{2},$$

$$a_3 = -k\left(\frac{a_2}{3}\right) = -\frac{k^3 a_0}{2 \cdot 3},$$

$$a_4 = -k\left(\frac{a_3}{4}\right) = +\frac{k^4 a_0}{2 \cdot 3 \cdot 4},$$

---

* The summation for y' starts at $n = 1$ because the first term is zero when $n = 0$.

and so on. A pattern emerges, and we conjecture in general that

$$a_n = \frac{(-1)^n k^n a_0}{n!} \quad \text{for} \quad n = 1, 2, 3, \ldots,$$

which is a formula easily proved true by induction. (Here, a_0 is arbitrary. When $n = 0$, this formula reduces to $a_0 = a_0$, with the notational convention that $0! = 1$.)

Thus, the series solution is

$$y = \sum_{n=0}^{\infty} a_n x^n = \sum_{n=0}^{\infty} \frac{(-1)^n k^n a_0}{n!} x^n$$
$$= a_0 \sum_{n=0}^{\infty} \frac{(-1)^n (kx)^n}{n!}.$$

Since this series is exactly e^{-kx}, we get $y = a_0 e^{-kx}$ which is in agreement with previous experience (with a_0 in place of A). Note that in this example the solution series converges for all x.

EXAMPLE 161

Solve

$$y'' + k^2 y = 0.$$

Again, we know that the general solution is

$$y = c_1 \sin(kx) + c_2 \cos(kx)$$

but, for illustration, set $y = \sum_{n=0}^{\infty} a_n x^n$. Then

$$y' = \sum_{n=1}^{\infty} n a_n x^{n-1} \quad \text{and} \quad y'' = \sum_{n=2}^{\infty} n(n-1) a_n x^{n-2}.$$

The differential equation becomes

$$\sum_{n=2}^{\infty} n(n-1) a_n x^{n-2} + \sum_{n=0}^{\infty} k^2 a_n x^n = 0.$$

Shift indices in the first summation to get

$$\sum_{n=0}^{\infty} (n+2)(n+1) a_{n+2} x^n + \sum_{n=0}^{\infty} k^2 a_n x^n = 0$$

or

$$\sum_{n=0}^{\infty} [(n+2)(n+1) a_{n+2} + k^2 a_n] x^n = 0.$$

Then, the recurrence relation is

$$(n+2)(n+1) a_{n+2} + k^2 a_n = 0 \quad \text{for} \quad n = 0, 1, 2, \ldots.$$

Then

$$a_{n+2} = \frac{-k^2 a_n}{(n+2)(n+1)} \quad \text{for} \quad n = 0, 1, 2, \ldots.$$

Calculate a few of these terms:

$$a_2 = \frac{-k^2 a_0}{2 \cdot 1}, \qquad a_3 = \frac{-k^2 a_1}{3 \cdot 2},$$

$$a_4 = \frac{-k^2 a_2}{4 \cdot 3} = \frac{+k^4 a_0}{4 \cdot 3 \cdot 2 \cdot 1}, \qquad a_5 = \frac{-k^2 a_3}{5 \cdot 4} = \frac{+k^4 a_1}{5 \cdot 4 \cdot 3 \cdot 2},$$

$$a_6 = \frac{-k^2 a_4}{6 \cdot 5} = \frac{-k^6 a_0}{6 \cdot 5 \cdot 4 \cdot 3 \cdot 2 \cdot 1}, \qquad a_7 = \frac{-k^2 a_5}{7 \cdot 6} = \frac{-k^6 a_1}{7 \cdot 6 \cdot 5 \cdot 4 \cdot 3 \cdot 2},$$

and so on. Note that each a_j with j even depends on a_0, and each a_j with j odd depends on a_1. The emerging pattern is

$$a_{2n} = \frac{(-1)^n k^{2n} a_0}{(2n)!} \quad \text{and} \quad a_{2n+1} = \frac{(-1)^n k^{2n} a_1}{(2n+1)!} \quad \text{for} \quad n = 0, 1, 2, 3, \ldots.$$

Thus, we have the solution

$$\begin{aligned} y &= \sum_{n=0}^{\infty} a_n x^n \\ &= \sum_{n=0}^{\infty} a_{2n} x^{2n} + \sum_{n=0}^{\infty} a_{2n+1} x^{2n+1} \\ &= a_0 \sum_{n=0}^{\infty} \frac{(-1)^n (kx)^{2n}}{(2n)!} + \frac{a_1}{k} \sum_{n=0}^{\infty} \frac{(-1)^n (kx)^{2n+1}}{(2n+1)!}. \end{aligned}$$

Recognizing the Taylor series for $\sin(kx)$ and $\cos(kx)$ about 0, this is the same as $y = c_1 \sin(kx) + c_2 \cos(kx)$ with the purely notational change $c_1 = a_0, c_2 = a_1/k$.

In these two examples, we illustrated the basic technique but only solved problems we already knew how to do anyway. Here are some more impressive examples in which the basic method is the same as above but the details become more complicated because the problems are more difficult.

EXAMPLE 162

Solve

$$y'' + x^2 y = 0.$$

Since x^2 is analytic at 0, we try a solution $y = \sum_{n=0}^{\infty} a_n x^n$. Substituting into the differential equation gives us

$$\sum_{n=2}^{\infty} n(n-1) a_n x^{n-2} + \sum_{n=0}^{\infty} a_n x^{n+2} = 0.$$

An index shift (in both series) gives us

$$\sum_{n=0}^{\infty} (n+2)(n+1) a_{n+2} x^n + \sum_{n=2}^{\infty} a_{n-2} x^n = 0.$$

Then

$$2a_2 + 6a_3 x + \sum_{n=2}^{\infty} [(n+2)(n+1)a_{n+2} + a_{n-2}]x^n = 0.$$

Then

$$2a_2 = 0,$$

$$6a_3 = 0,$$

and, for $n = 2, 3, 4, \ldots,$

$$(n+2)(n+1)a_{n+2} + a_{n-2} = 0.$$

Thus,

$$a_2 = a_3 = 0$$

and, for $n = 2, 3, \ldots,$

$$a_{n+2} = \frac{-a_{n-2}}{(n+2)(n+1)}.$$

Computing some terms, we get

$$a_4 = \frac{-a_0}{4 \cdot 3}, \quad (a_0 \text{ arbitrary, as there is no equation for it})$$

$$a_5 = \frac{-a_1}{5 \cdot 4}, \quad (a_1 \text{ arbitrary})$$

$$a_6 = \frac{-a_2}{6 \cdot 5} = 0,$$

$$a_7 = \frac{-a_3}{7 \cdot 6} = 0,$$

$$a_8 = \frac{-a_4}{8 \cdot 7} = \frac{+a_0}{8 \cdot 7 \cdot 4 \cdot 3} = \frac{a_0}{672},$$

$$a_9 = \frac{-a_5}{9 \cdot 8} = \frac{+a_1}{9 \cdot 8 \cdot 5 \cdot 4} = \frac{a_1}{1440},$$

$$a_{10} = a_{11} = 0,$$

$$a_{12} = \frac{-a_8}{12 \cdot 11} = \frac{-a_0}{12 \cdot 11 \cdot 8 \cdot 7 \cdot 4 \cdot 3} = \frac{-a_0}{88{,}704},$$

$$a_{13} = \frac{-a_9}{13 \cdot 12} = \frac{-a_1}{13 \cdot 12 \cdot 9 \cdot 8 \cdot 5 \cdot 4} = \frac{-a_1}{224{,}640},$$

$$a_{14} = a_{15} = 0,$$

and so on. This gives us the general solution

$$y = a_0 + a_1 x + a_2 x^2 + \cdots$$

$$= a_0\left(1 - \frac{x^4}{12} + \frac{x^8}{672} - \frac{x^{12}}{88{,}704} + \cdots\right) + a_1\left(x - \frac{x^5}{20} + \frac{x^9}{1440} - \frac{x^{13}}{224{,}640} + \cdots\right),$$

where a_0 and a_1 are arbitrary constants. Two independent solutions are

$$y_1(x) = 1 - \frac{x^4}{12} + \frac{x^8}{672} - \frac{x^{12}}{88{,}704} + \cdots \qquad (\text{choose } a_0 = 1,\ a_1 = 0)$$

and

$$y_2(x) = x - \frac{x^5}{20} + \frac{x^9}{1440} - \frac{x^{13}}{224{,}640} + \cdots \qquad (\text{choose } a_0 = 0,\ a_1 = 1).$$

The first solution is an alternating series with coefficients decreasing very rapidly to zero. For an approximation, we have

$$\frac{1}{88{,}704} = 0.00001127\ \ldots.$$

Hence,

$$1 - \frac{x^4}{12} + \frac{x^8}{672}$$

approximates $y_1(x)$ on $(-1, 1)$ to well within 0.00002.

Similarly,

$$\frac{1}{224{,}640} = 0.00000445\ \ldots,$$

and so

$$x - \frac{x^5}{20} + \frac{x^9}{1440}$$

approximates $y_2(x)$ on $(0, 1)$ to within 0.000005.

In this example, a formula for a_n in terms of a_0 and a_1 is quite complicated; hence we were content to write out the solution series for just a few terms. By writing more terms we could increase accuracy to any degree wanted.

EXAMPLE 163

Solve

$$(1 - x^2)y'' - 2xy' + \lambda(\lambda + 1)y = 0,$$

in which λ is constant. This is *Legendre's equation*, and its solutions are called *Legendre functions*. These are important in many engineering problems and will be studied in Chapter 6.

Note that Theorem 16 applies, since the differential equation is

$$y'' - \frac{2x}{1 - x^2}y' + \frac{\lambda(\lambda + 1)}{1 - x^2}y = 0$$

and $-2x/(1 - x^2)$ and $\lambda(\lambda + 1)/(1 - x^2)$ are analytic at 0. To obtain series solutions, substitute $y = \sum_{n=0}^{\infty} a_n x^n$ into the original differential equation to get

$$(1 - x^2)\sum_{n=2}^{\infty} n(n-1)a_n x^{n-2} - 2x\sum_{n=1}^{\infty} na_n x^{n-1} + \lambda(\lambda + 1)\sum_{n=0}^{\infty} a_n x^n = 0$$

or

$$\sum_{n=2}^{\infty} n(n-1)a_n x^{n-2} + \sum_{n=2}^{\infty} -n(n-1)a_n x^n + \sum_{n=1}^{\infty} -2na_n x^n + \sum_{n=0}^{\infty} \lambda(\lambda+1)a_n x^n = 0.$$

Shift indices in the first summation to get

$$\sum_{n=0}^{\infty} (n+2)(n+1)a_{n+2} x^n + \sum_{n=2}^{\infty} -n(n-1)a_n x^n + \sum_{n=1}^{\infty} -2na_n x^n + \sum_{n=0}^{\infty} \lambda(\lambda+1)a_n x^n = 0.$$

Finally we have

$$2a_2 + \lambda(\lambda+1)a_0 + 6a_3 x - 2a_1 x + \lambda(\lambda+1)a_1 x + \sum_{n=2}^{\infty} [(n+2)(n+1)a_{n+2} - n(n-1)a_n - 2na_n + \lambda(\lambda+1)a_n]x^n = 0.$$

Then,

$$2a_2 + \lambda(\lambda+1)a_0 = 0,$$

$$[\lambda(\lambda+1) - 2]a_1 + 6a_3 = 0,$$

and, for $n = 2, 3, \ldots,$

$$(n+2)(n+1)a_{n+2} + [-n^2 - n + \lambda(\lambda+1)]a_n = 0.$$

Then,

$$a_2 = -\frac{\lambda(\lambda+1)a_0}{2},$$

$$a_3 = \frac{[2 - \lambda(\lambda+1)]a_1}{6},$$

and, for $n = 2, 3, 4, \ldots,$

$$a_{n+2} = \frac{[n(n+1) - \lambda(\lambda+1)]a_n}{(n+2)(n+1)}.$$

Here a_0 and a_1 are arbitrary; each a_{2j} depends on a_0, and each a_{2j+1} depends on a_1.

When λ is not a positive integer, the recurrence relation does not yield a simple formula for the coefficients, and we must be content in specific applications to write out just the first few terms of the series. Doing this, we get

$$\begin{aligned} y &= a_0\left[1 - \frac{\lambda(\lambda+1)}{2}x^2 + \left(\frac{6-\lambda(\lambda+1)}{12}\right)\left(\frac{-\lambda(\lambda+1)}{2}\right)x^4 \right. \\ &\quad \left. + \left(\frac{20-\lambda(\lambda+1)}{30}\right)\left(\frac{6-\lambda(\lambda+1)}{12}\right)\left(\frac{-\lambda(\lambda+1)}{2}\right)x^6 + \cdots\right] \\ &\quad + a_1\left[x + \left(\frac{2-\lambda(\lambda+1)}{6}\right)x^3 + \left(\frac{12-\lambda(\lambda+1)}{20}\right)\left(\frac{2-\lambda(\lambda+1)}{6}\right)x^5 \right. \\ &\quad \left. + \left(\frac{30-\lambda(\lambda+1)}{42}\right)\left(\frac{12-\lambda(\lambda+1)}{20}\right)\left(\frac{2-\lambda(\lambda+1)}{6}\right)x^7 + \cdots\right] \\ &= a_0 F(x) + a_1 G(x), \end{aligned}$$

where $F(x)$ contains only even powers of x and $G(x)$ contains only odd powers.

An interesting thing happens when λ is 0, 1, 2, 3, or any nonnegative integer. For example,

$$\begin{aligned}
\lambda = 2 &\quad \text{gives us} \quad F(x) = 1 - 3x^2, \\
\lambda = 3 &\quad \text{gives us} \quad G(x) = x - \tfrac{5}{3}x^3, \\
\lambda = 4 &\quad \text{gives us} \quad F(x) = 1 - 10x^2 + \tfrac{35}{3}x^4,
\end{aligned}$$

and so on. In general, if $\lambda = k$, a nonnegative integer, we can always find a *polynomial* of degree k as one solution of Legendre's equation. These are called *Legendre polynomials* and will be treated in the next chapter.

EXAMPLE 164

Solve

$$y'' - e^x y = 0.$$

Set $y = \sum_{n=0}^{\infty} a_n x^n$ as usual. But here we must also express e^x as a series:

$$e^x = \sum_{n=0}^{\infty} \frac{x^n}{n!}.$$

The differential equation becomes

$$\sum_{n=2}^{\infty} n(n-1)a_n x^{n-2} - \left(\sum_{n=0}^{\infty} \frac{x^n}{n!}\right)\left(\sum_{n=0}^{\infty} a_n x^n\right) = 0.$$

To obtain a few terms of a series solution, write this as

$$\sum_{n=2}^{\infty} n(n-1)a_n x^{n-2} - \left(1 + x + \frac{x^2}{2} + \frac{x^3}{6} + \frac{x^4}{24} + \frac{x^5}{120} + \cdots\right)$$
$$\times (a_0 + a_1 x + a_2 x^2 + a_3 x^3 + a_4 x^4 + a_5 x^5 + \cdots) = 0.$$

After multiplying out the first few terms in the product of the two series, the last equation becomes

$$\begin{aligned}
&2a_2 + 6a_3 x + 12a_4 x^2 + 20a_5 x^3 + 30a_6 x^4 + 42a_7 x^5 + \cdots \\
&\quad - \left[a_0 + (a_0 + a_1)x + \left(\frac{a_0}{2} + a_1 + a_2\right)x^2 + \left(\frac{a_0}{6} + \frac{a_1}{2} + a_2 + a_3\right)x^3\right. \\
&\quad + \left(\frac{a_0}{24} + \frac{a_1}{6} + \frac{a_2}{2} + a_3 + a_4\right)x^4 \\
&\quad \left. + \left(\frac{a_0}{120} + \frac{a_1}{24} + \frac{a_2}{6} + \frac{a_3}{2} + a_4 + a_5\right)x^5 + \cdots\right] = 0.
\end{aligned}$$

Collect coefficients of like powers of x to get

$$(2a_2 - a_0) + (6a_3 - a_0 - a_1)x + \left(12a_4 - \frac{a_0}{2} - a_1 - a_2\right)x^2$$
$$+ \left(20a_5 - \frac{a_0}{6} - \frac{a_1}{2} - a_2 - a_3\right)x^3$$
$$+ \left(30a_6 - \frac{a_0}{24} - \frac{a_1}{6} - \frac{a_2}{2} - a_3 - a_4\right)x^4$$
$$+ \left(42a_7 - \frac{a_0}{120} - \frac{a_1}{24} - \frac{a_2}{6} - \frac{a_3}{2} - a_4 - a_5\right)x^5$$
$$+ \cdots = 0.$$

Since the right side is zero, each coefficient on the left side must vanish, giving us a system of equations:

$$2a_2 - a_0 = 0$$
$$6a_3 - a_0 - a_1 = 0$$
$$12a_4 - \frac{a_0}{2} - a_1 - a_2 = 0$$
$$20a_5 - \frac{a_0}{6} - \frac{a_1}{2} - a_2 - a_3 = 0$$
$$30a_6 - \frac{a_0}{24} - \frac{a_1}{6} - \frac{a_2}{2} - a_3 - a_4 = 0$$
$$42a_7 - \frac{a_0}{120} - \frac{a_1}{24} - \frac{a_2}{6} - \frac{a_3}{2} - a_4 - a_5 = 0$$
$$\vdots$$

Solve these to get

$$a_2 = \frac{a_0}{2},$$
$$a_3 = \frac{a_0 + a_1}{6},$$
$$a_4 = \frac{a_0 + a_1}{12},$$
$$a_5 = \frac{5a_0 + 4a_1}{120},$$
$$a_6 = \frac{13a_0 + 10a_1}{720},$$
$$a_7 = \frac{36a_0 + 29a_1}{5040},$$

and so on, with a_0 and a_1 arbitrary. As far out as x^7, then, a solution is given by

$$y = a_0 + a_1 x + \frac{a_0}{2} x^2 + \left(\frac{a_0 + a_1}{6}\right) x^3 + \left(\frac{a_0 + a_1}{12}\right) x^4$$
$$+ \left(\frac{5a_0 + 4a_1}{120}\right) x^5 + \left(\frac{13a_0 + 10a_1}{720}\right) x^6 + \left(\frac{36a_0 + 29a_1}{5040}\right) x^7 + \cdots.$$

We can find particular solutions by choosing a_0 and a_1. For example, with $a_0 = 1, a_1 = 0$, we get

$$y = 1 + \tfrac{1}{2}x^2 + \tfrac{1}{6}x^3 + \tfrac{1}{12}x^4 + \tfrac{1}{24}x^5 + \tfrac{13}{720}x^6 + \tfrac{1}{140}x^7 + \cdots.$$

And, with $a_0 = 0$ and $a_1 = 1$, we get

$$y = x + \tfrac{1}{6}x^3 + \tfrac{1}{12}x^4 + \tfrac{1}{30}x^5 + \tfrac{1}{72}x^6 + \tfrac{29}{5040}x^7 + \cdots.$$

Again, we can go out as many terms as we wish by doing more work.

EXAMPLE 165

Solve

$$y'' + \sin(x)y = x^2.$$

Let $y = \sum_{n=0}^{\infty} a_n x^n$, and write

$$\sin(x) = \sum_{n=0}^{\infty} \frac{(-1)^n x^{2n+1}}{(2n+1)!}.$$

The differential equation becomes

$$2a_2 + 6a_3 x + 12a_4 x^2 + 20a_5 x^3 + 30a_6 x^4 + 42a_7 x^5 + \cdots$$
$$+ \left(x - \frac{x^3}{6} + \frac{x^5}{120} - \frac{x^7}{5040} + \cdots\right)(a_0 + a_1 x + a_2 x^2 + \cdots) = x^2.$$

Then

$$2a_2 + 6a_3 x + 12a_4 x^2 + 20a_5 x^3 + 30a_6 x^4 + 42a_7 x^5 + \cdots + a_0 x$$
$$+ a_1 x^2 + \left(a_2 - \frac{a_0}{6}\right) x^3 + \left(a_3 - \frac{a_1}{6}\right) x^4 + \left(a_4 - \frac{a_2}{6} + \frac{a_0}{120}\right) x^5 + \cdots = x^2.$$

Equating coefficients of like powers of x on both sides of the equation gives us

$$2a_2 = 0$$
$$6a_3 + a_0 = 0$$
$$12a_4 + a_1 = 1$$
$$20a_5 + a_2 - \frac{a_0}{6} = 0$$
$$30a_6 + a_3 - \frac{a_1}{6} = 0$$
$$42a_7 + a_4 - \frac{a_2}{6} + \frac{a_0}{120} = 0$$
$$\vdots$$

Solve these to get

$$a_2 = 0,$$

$$a_3 = -\frac{a_0}{6},$$

$$a_4 = \frac{1 - a_1}{12},$$

$$a_5 = \frac{a_0 - 6a_2}{120} = \frac{a_0}{120},$$

$$a_6 = \frac{a_1 + a_0}{180},$$

$$a_7 = \frac{10a_1 - a_0 - 10}{5040},$$

and so on. Thus, a solution is

$$y = a_0 + a_1 x - \frac{a_0}{6} x^3 + \frac{(1 - a_1)}{12} x^4 + \frac{a_0}{120} x^5 + \frac{(a_1 + a_0)}{180} x^6 + \frac{(10a_1 - a_0 - 10)}{5040} x^7 + \cdots.$$

As in the previous example, we can get specific solutions by choosing a_0 and a_1, and we can go out as many terms as we like by doing more computation.

The problems we have just done illustrate the basic ideas behind power series solutions. There are two facts to keep in mind:

1. The power series method is usually invoked after previous methods fail. For example, to solve

$$y'' - 4y' + 6y = x^2,$$

we can use easier methods than power series.

2. In most instances where one must use power series, the recurrence relation is usually not solvable for the a_n's in simple terms, and one must often be content to calculate only the first few terms of a solution series. With modern computing devices, one can of course compute many terms fairly easily.

PROBLEMS FOR SECTION 5.2

In each of Problems 1 through 15 (except Problem 3), find two solutions, one with $a_0 = 1$ and $a_1 = 0$, the other with $a_0 = 0$ and $a_1 = 1$. In Problem 3, find the solution with $a_0 = 1$. Write out at least five nonzero terms in each solution.

1. $y'' + xy = 0$

2. $y'' + xy' + 2y = 0$

3. $y' - 2xy = 0$

4. $y'' - 2y' + xy = 0$

5. $y'' - y' + x^2y = 0$

6. $y'' - x^3y = 0$

7. $y'' - x^2y' + 2y = 0$

8. $y'' + (1 - x)y' + 2xy = 0$

9. $y'' + y' + (x - 4)y = 0$

10. $y'' + xy' + 2xy = 0$
11. $y'' + x^2y' + 2y = 0$
12. $y'' + y' - x^3y = 0$
13. $y'' + x^2y' + (x^2 - 1)y = 0$
14. $y'' - 8xy = 2x + 1$
15. $y'' - 4y' + xy = 6x - 4$

In each of Problems 16 through 50, find at least the first five nonzero terms of a general solution. (Thus, each solution has one or more arbitrary constants.)

16. $y'' + 2x^2y' - 3x^2y = 0$
17. $2y'' - 4xy' + 8x^3y = 0$
18. $y'' + 8x^2y' - xy = 0$
19. $y'' + 12y' + x^3y = 0$
20. $y'' + 2y' - 4x^2y = 0$
21. $y'' - 3x^3y' + 4xy = 0$
22. $y'' + 5y' + (x^2 - 2)y = 0$
23. $y'' + 2xy' - y = 0$
24. $y'' - 2x^2y' + (x + 3)y = 0$
25. $y'' + x^2y' + (x^2 - 4)y = 0$
26. $y'' + (2x - 3)y' + (x^2 + 1)y = 0$
27. $y^{(3)} - 4xy'' + 2y' + xy = 0$
28. $y^{(3)} + 2y'' - 4x^2y' + (x - 3)y = 0$
29. $y^{(4)} + x^2y' - 2y = 0$
30. $y^{(3)} + 8xy' - 2y = 0$
31. $y'' + 2xy' - (x^2 + x - 1)y = 0$
32. $y^{(3)} - 4x^2y' + (x - 3)y = 0$
33. $y'' + 2\cos(x)y' = x$
34. $y'' - e^xy' + 2y = 1$
35. $y' + 2\sinh(x)y = 0$
36. $y'' - 2\tan(x)y' + y = 0$
37. $y'' + e^xy = \cos(x)$
38. $y'' - 2y' + e^{-x}y = \sin(x)$
39. $y'' + \tan(x)y = x^3$
40. $y'' - y + \sin(2x)y = x + 1$
41. $2y'' - y' + x^3y = e^x - 3$
42. $-y'' + e^{-3x}y = 2x^2$
43. $y'' + 8x^3y' + 3y = \cosh(x)$
44. $y'' - 4y' + (1/1 + x)y = 2$
45. $y'' - \sin(x)y' + 3y = x^3 - 4$
46. $y'' + x^2y' - 2xy = x^3 - 3x + 2$
47. $y'' + e^{2x}y' - y = \cos(x) + 2$
48. $y'' - 4x^3y = \ln(1 + x)$
49. $y'' + 2xy' - 3y = e^{-x} + 4$
50. $y'' - x^2y' + e^{-2x}y = 1$

51. In Example 162, show that the solution obtained can be written

$$y(x) = a_0\left[1 + \sum_{n=1}^{\infty} \frac{(-1)^n x^{4n}}{\prod_{j=1}^{n}(4j - 1)(4j)}\right] + a_1\left[x + \sum_{n=1}^{\infty} \frac{(-1)^n x^{4n+1}}{\prod_{j=1}^{n}(4j)(4j + 1)}\right],$$

where $\prod_{j=1}^{n} A_j = A_1 \cdot A_2 \cdots A_n$. (Thus, $\prod_{j=1}^{n}$ does for products what $\sum_{j=1}^{n}$ does for sums.)

5.3 THE METHOD OF FROBENIUS

Many important differential equations in physics and engineering have the form

$$P(x)y'' + Q(x)y' + R(x)y = 0.$$

We usually want solutions in some interval about a point x_0.

Suppose that $P(x)$, $Q(x)$, and $R(x)$ are continuous in some interval J about x_0. There are two cases to consider:

Case 1: $P(x_0) \neq 0$ In this case, we call x_0 an *ordinary point* of the differential equation. Then we can divide by $P(x)$ for x in J, obtaining the standard form

$$y'' + q(x)y' + g(x)y = 0,$$

with

$$q(x) = \frac{Q(x)}{P(x)} \quad \text{and} \quad g(x) = \frac{R(x)}{P(x)}.$$

Theorem 2 (Chapter 2) guarantees existence of solutions about x_0, and one can attempt to find them by previous techniques.

Case 2: $P(x_0) = 0$ In this case, x_0 is a *singular point* of the differential equation, and Theorem 2 (Chapter 2, Section 2) does not apply. There may be no nontrivial solution of $Py'' + Qy' + Ry = 0$ about x_0.

For technical reasons which we sketch below, singular points are divided into two categories: A singular point x_0 is *regular* if

$$(x - x_0)\frac{Q(x)}{P(x)} \quad \text{and} \quad (x - x_0)^2\left(\frac{R(x)}{P(x)}\right)$$

both have Taylor series expansions about x_0. Otherwise, x_0 is an *irregular* singular point.

EXAMPLE 166

1. Any Euler equation

$$x^2y'' + axy' + by = 0$$

has a regular singular point at 0. Here $x_0 = 0$, $P(x) = x^2$, $Q(x) = ax$, $R(x) = b$, and

$$(x - x_0)\frac{Q(x)}{P(x)} = x\left(\frac{ax}{x^2}\right) = a,$$

$$(x - x_0)^2\frac{R(x)}{P(x)} = x^2\left(\frac{b}{x^2}\right) = b.$$

2. Legendre's differential equation

$$(1 - x^2)y'' - 2xy' + \lambda(\lambda + 1)y = 0 \qquad (\lambda \text{ constant})$$

has regular singular points at $+1$ and -1. For example, at $+1$,

$$(x - 1)\frac{Q(x)}{P(x)} = (x - 1)\left(\frac{-2x}{1 - x^2}\right) = \frac{2x}{1 + x}$$

is analytic at 1, and

$$(x - 1)^2\frac{R(x)}{P(x)} = (x - 1)^2\left(\frac{\lambda(\lambda + 1)}{1 - x^2}\right) = \frac{(1 - x)\lambda(\lambda + 1)}{1 + x}$$

is also analytic at 1.

3. The equation

$$x^3y'' + 2xy' - y = 0$$

has an irregular singular point at 0.

Now suppose that $P(x)y'' + Q(x)y' + R(x)y = 0$ has a regular singular point at x_0. We will take $x_0 = 0$ for notational simplicity. Rewrite the differential equation as

$$x^2y'' + x[xq(x)]y' + x^2g(x)y = 0 \tag{a}$$

with

$$q(x) = \frac{Q(x)}{P(x)} \quad \text{and} \quad g(x) = \frac{R(x)}{P(x)}.$$

Assume that $x > 0$. (Remember that $P(0) = 0$.)

Now, (a) is an Euler equation if $xq(x)$ and $x^2g(x)$ are constant, and in that case we get solutions $y = x^r$. Here we assume that $xq(x)$ and $x^2g(x)$ are analytic at 0 and not necessarily constant; so we must try something more general for a solution. The *Method of Frobenius* consists of substituting

$$y = \sum_{n=0}^{\infty} a_n x^{n+r}$$

into (a) and choosing the a_n's *and* r to obtain a solution.

The key to the method is selection of r, which we do as follows. Since $xq(x)$ and $x^2g(x)$ are analytic at 0, they have Taylor expansions

$$xq(x) = A_0 + A_1x + A_2x^2 + \cdots$$

and

$$x^2g(x) = B_0 + B_1x + B_2x^2 + \cdots.$$

Substitute these into (a), along with

$$y = \sum_{n=0}^{\infty} a_n x^{n+r},$$

$$xy' = x\sum_{n=0}^{\infty} (n+r)a_n x^{n+r-1} = \sum_{n=0}^{\infty} (n+r)a_n x^{n+r},$$

and

$$x^2y'' = x^2\sum_{n=0}^{\infty} (n+r)(n+r-1)a_n x^{n+r-2} = \sum_{n=0}^{\infty} (n+r)(n+r-1)a_n x^{n+r}.$$

(Note that, in $y' = \sum_{n=0}^{\infty} (n+r)a_n x^{n+r-1}$, we do *not* necessarily have the term corresponding to $n = 0$ vanish because the $n = 0$ term is ra_0x^{r-1}, which need not be zero. Thus, in the Method of Frobenius, the summations for y', y'', ... will still in general begin at $n = 0$, unlike in the power series method, where $\sum_{n=0}^{\infty} na_n x^{n-1}$ has zero first term regardless of a_0.)

Upon making the substitutions into the differential equation, we get

$$\begin{aligned}&[r(r-1)a_0x^r + (1+r)(r)a_1x^{1+r} + \cdots]\\&+ [A_0 + A_1x + A_2x^2 + \cdots][ra_0x^r + (1+r)a_1x^{1+r} + \cdots]\\&+ [B_0 + B_1x + B_2x^2 + \cdots][a_0x^r + a_1x^{1+r} + \cdots] = 0.\end{aligned}$$

After all the multiplications are carried out, the coefficient of each power of x on the left must be zero. Look at the coefficient of x^r to obtain

$$r(r-1)a_0 + A_0a_0r + B_0a_0 = 0.$$

Assuming that $a_0 \neq 0$, then

$$r(r-1) + A_0r + B_0 = 0.$$

This is the *indicial equation* of (a), which we solve for r. Having found r, we obtain a recurrence relation for the a_n's by setting the other coefficients of x^{n+r} equal to zero. This gives us the a_n's (at least in principle), and hence a solution.

The indicial equation is quadratic, and hence gives two roots for r. This leads to three cases, and three different kinds of solutions, which are summarized as follows.

THEOREM 17 Method of Frobenius

Let r_1 and r_2 be roots of the indicial equation of (a), which we assume has a regular singular point at 0. Then,

1. If $r_1 \neq r_2$, and $r_1 - r_2$ is not an integer, then there are two linearly independent solutions

$$y_1 = \sum_{n=0}^{\infty} a_n x^{n+r_1} \quad \text{and} \quad y_2 = \sum_{n=0}^{\infty} b_n x^{n+r_2} \qquad (x > 0).$$

2. If $r_1 - r_2$ is a positive integer, then there are linearly independent solutions of the form

$$y_1 = \sum_{n=0}^{\infty} a_n x^{n+r_1} \quad \text{and} \quad y_2 = Ay_1 \ln(x) + \sum_{n=0}^{\infty} b_n x^{n+r_2} \qquad (x > 0).$$

Here, A is constant and may turn out to be zero.

3. If $r_1 = r_2$, then there are linearly independent solutions

$$y_1 = \sum_{n=0}^{\infty} a_n x^{n+r_1} \quad \text{and} \quad y_2 = y_1 \ln(x) + \sum_{n=0}^{\infty} b_n x^{n+r_1} \qquad (x > 0).$$

Note that, when $r_1 - r_2$ is a positive integer (case 2), then the second solution y_2 may or may not contain a logarithm term; when $r_1 = r_2$ (case 3), y_2 definitely has a logarithm term. We shall see examples of all these possibilities here and in the next section.

Note also that we always begin the Method of Frobenius by assuming a solution $y = \sum_{n=0}^{\infty} a_n x^{n+r}$ and solving the indicial equation for r. The form of the second solution is only apparent after seeing from the roots of the indicial equation which case of Theorem 17 applies.

EXAMPLE 167 Case 1 of Theorem 17

Solve

$$x^2 y'' + x(\tfrac{1}{2} + 2x)y' + (x - \tfrac{1}{2})y = 0 \qquad (x > 0).$$

This is of the form (a) with

$$xq(x) = \tfrac{1}{2} + 2x$$

and

$$x^2 r(x) = x - \tfrac{1}{2}.$$

Since $x^2 = 0$ at $x = 0$, the differential equation has a regular singular point at 0. Setting

$$y = \sum_{n=0}^{\infty} a_n x^{n+r},$$

we get by substitution:

$$\sum_{n=0}^{\infty} (n + r)(n + r - 1)a_n x^{n+r} + \sum_{n=0}^{\infty} \tfrac{1}{2}(n + r)a_n x^{n+r}$$
$$+ \sum_{n=0}^{\infty} 2(n + r)a_n x^{n+r+1} + \sum_{n=0}^{\infty} a_n x^{n+r+1} + \sum_{n=0}^{\infty} -\tfrac{1}{2}a_n x^{n+r} = 0.$$

Shift indices in the third and fourth summations and combine terms to get

$$[r(r - 1)a_0 + \tfrac{1}{2}ra_0 - \tfrac{1}{2}a_0]x^r$$
$$+ \sum_{n=1}^{\infty} [(n + r)(n + r - 1)a_n + \tfrac{1}{2}(n + r)a_n$$
$$+ 2(n - 1 + r)a_{n-1} + a_{n-1} - \tfrac{1}{2}a_n]x^{n+r} = 0.$$

Assuming that $a_0 \neq 0$, we get the indicial equation

$$r(r - 1) + \tfrac{1}{2}r - \tfrac{1}{2} = 0$$

from the coefficient of x^r. From the coefficients of x^{n+r}, we get the recurrence relation

$$(n + r)(n + r - 1)a_n + \tfrac{1}{2}(n + r)a_n + 2(n - 1 + r)a_{n-1} + a_{n-1} - \tfrac{1}{2}a_n = 0,$$

for $n = 1, 2, 3, 4, \ldots$.

Solve the indicial equation to get two roots:

$$r_1 = 1, \qquad r_2 = -\tfrac{1}{2}.$$

Case 1 of Theorem 17 applies: First choose $r = 1$ in the recurrence relation. We get

$$(n + 1)na_n + \tfrac{1}{2}(n + 1)a_n + 2na_{n-1} + a_{n-1} - \tfrac{1}{2}a_n = 0$$

or

$$a_n = \frac{-2(2n + 1)a_{n-1}}{n(2n + 3)} \qquad (n = 1, 2, 3, \ldots).$$

Working backwards to solve for the a_n's, we have

$$\begin{aligned}
a_n &= \frac{-2(2n + 1)}{n(2n + 3)} a_{n-1} \\
&= \frac{-2(2n + 1)}{n(2n + 3)} \frac{(-2)(2n - 1)}{(n - 1)(2n + 1)} a_{n-2} \\
&= \frac{2^2(2n + 1)(2n - 1)}{n(n - 1)(2n + 3)(2n + 1)} a_{n-2} \\
&= \frac{2^2(2n + 1)(2n - 1)}{n(n - 1)(2n + 3)(2n + 1)} \frac{(-2)(2n - 3)}{(n - 2)(2n - 1)} a_{n-3} \\
&= \frac{-2^3(2n + 1)(2n - 1)(2n - 3)}{n(n - 1)(n - 2)(2n + 3)(2n + 1)(2n - 1)} a_{n-3},
\end{aligned}$$

and so on. Observing the trend, we get

$$a_n = \frac{(-1)^n 2^n (2n+1)(2n-1)(2n-3)\cdots(5)(3)}{n(n-1)(n-2)\cdots(2)(1)(2n+3)(2n+1)(2n-1)\cdots(7)(5)}\, a_0$$

$$= \frac{3(-1)^n 2^n a_0}{n!\,(2n+3)} \qquad (n = 1, 2, 3, \ldots).$$

Note that $3(-1)^n 2^n / n!\,(2n+3) = 1$ if $n = 0$, allowing us to write the first solution as

$$y_1 = 3a_0 \sum_{n=0}^{\infty} \frac{(-1)^n 2^n}{n!\,(2n+3)}\, x^{n+1},$$

with a_0 any nonzero number.

For a second solution, put $r = -\frac{1}{2}$ into the recurrence relation. Writing b_n in place of a_n just to avoid confusion, the recurrence relation now becomes

$$(n - \tfrac{1}{2})(n - \tfrac{3}{2})b_n + \tfrac{1}{2}(n - \tfrac{1}{2})b_n + 2(n - \tfrac{3}{2})b_{n-1} + b_{n-1} - \tfrac{1}{2}b_n = 0,$$

for $n = 1, 2, 3, \ldots$.

Then,

$$b_n = \frac{-4(n-1)}{n(2n-3)}\, b_{n-1} \qquad (n = 1, 2, 3, \ldots).$$

Note that $b_1 = 0$; hence $b_2 = b_3 = \cdots = 0$, and $r = -\frac{1}{2}$ gives us the simple solution

$$y_2 = \sum_{n=0}^{\infty} b_n x^{n-1/2} = \frac{b_0}{\sqrt{x}},$$

with b_0 any nonzero number.

EXAMPLE 168 Case 2 of Theorem 17

Solve

$$x^2 y'' + x^2 y' - 2y = 0.$$

Put $y = \sum_{n=0}^{\infty} a_n x^{n+r}$ and obtain (after shifting indices)

$$[r(r-1)a_0 - 2a_0] + \sum_{n=1}^{\infty} [(n+r)(n+r-1)a_n + (n+r-1)a_{n-1} - 2a_n]x^{n+r} = 0.$$

Assuming that $a_0 \neq 0$, we have the indicial equation

$$r^2 - r - 2 = 0,$$

with roots $r_1 = 2$ and $r_2 = -1$. Here $r_1 - r_2 = 3$, so we are in case 2 of Theorem 17. The recurrence relation is

$$(n+r)(n+r-1)a_n + (n+r-1)a_{n-1} - 2a_n = 0.$$

Putting $r = 2$, we get

$$a_n = -\frac{(n+1)}{n(n+3)}\, a_{n-1}.$$

From this we obtain the first solution

$$y_1 = \sum_{n=0}^{\infty} a_n x^{n+2}$$
$$= a_0 x^2\left(1 - \frac{1}{2}x + \frac{3}{20}x^2 - \frac{1}{30}x^3 + \frac{1}{168}x^4 - \frac{1}{1120}x^5 + \frac{1}{8640}x^6 - \frac{1}{75{,}600}x^7 + \cdots\right),$$

in which a_0 is nonzero but otherwise arbitrary.

For a second solution, follow the prescription of Theorem 17(2) with $r_2 = -1$ and b_n in place of a_n to write

$$y_2 = Ay_1 \ln(x) + \sum_{n=0}^{\infty} b_n x^{n-1}.$$

Substitute this into the differential equation and shift indices to get

$$Ax^2 y_1'' \ln(x) + 2Axy_1' - Ay_1 + \sum_{n=0}^{\infty}(n-1)(n-2)b_n x^{n-1} + Ax^2 y_1' \ln(x)$$
$$+ Axy_1 + \sum_{n=1}^{\infty}(n-2)b_{n-1}x^{n-1} - 2\sum_{n=0}^{\infty} b_n x^{n-1} - 2Ay_1 \ln(x) = 0.$$

The terms involving $A \ln(x)$ add up to

$$A \ln(x)(x^2 y_1'' + x^2 y_1' - 2y_1),$$

and this is zero because y_1 is a solution of $x^2 y'' + x^2 y' - 2y = 0$.

Next, substitute in the series for y_1, with $a_0 = 1$ for simplicity (remember that Theorem 17 allows us to use *any* first solution y_1 in seeking a second solution y_2). After some rearranging to gather coefficients of like powers of x, we get

$$(2b_0 - 2b_0)x^{-1} + (-b_0 - 2b_1)x^0 - 2b_2 x + (4A - A + 2b_3 + b_2 - 2b_3)x^2$$
$$+\left(-3A - \frac{A}{2} + 6b_4 + A + 2b_3 - 2b_4\right)x^3$$
$$+\left(\frac{6}{5}A + \frac{3A}{20} + 12b_5 - \frac{A}{2} + 3b_4 - 2b_5\right)x^4$$
$$+\left(-\frac{A}{3} - \frac{A}{30} + 20b_6 + \frac{3A}{20} + 4b_5 - 2b_6\right)x^5$$
$$+\left(\frac{A}{14} + \frac{A}{168} + 30b_7 - \frac{A}{30} + 5b_6 - 2b_7\right)x^6$$
$$+\left(-\frac{A}{80} - \frac{A}{1120} + 42b_8 + \frac{A}{168} + 6b_7 - 2b_8\right)x^7$$
$$+\left(\frac{A}{540} + \frac{A}{8640} + 56b_9 - \frac{A}{1120} + 7b_8 - 2b_9\right)x^8 + \cdots = 0.$$

Set the coefficient of each power of x equal to zero (because the right side is zero). From x^{-1} through x^2, the coefficients give us

$$\begin{aligned} 2b_0 - 2b_0 &= 0, \\ -b_0 - 2b_1 &= 0, \\ -2b_2 &= 0, \\ 3A + b_2 &= 0. \end{aligned}$$

Thus far, b_0 is arbitrary (but nonzero), $b_1 = -\frac{1}{2}b_0$, $b_2 = 0$, and $A = 0$. The x^3 through x^8 terms now give us

$$\begin{aligned} 4b_4 + 2b_3 &= 0 \\ 10b_5 + 3b_4 &= 0 \\ 18b_6 + 4b_5 &= 0 \\ 28b_7 + 5b_6 &= 0 \\ 40b_8 + 6b_7 &= 0 \\ 54b_9 + 7b_8 &= 0 \end{aligned}$$

Then,

$$\begin{aligned} b_4 &= -\tfrac{1}{2}b_3 \\ b_5 &= -\tfrac{3}{10}b_4 = \tfrac{3}{20}b_3 \\ b_6 &= -\tfrac{4}{18}b_5 = -\tfrac{1}{30}b_3 \\ b_7 &= -\tfrac{5}{28}b_7 = \tfrac{1}{168}b_3 \\ b_8 &= -\tfrac{6}{40}b_7 = -\tfrac{1}{1120}b_3 \\ b_9 &= -\tfrac{7}{54}b_8 = \tfrac{1}{8640}b_3 \end{aligned}$$

Thus,

$$\begin{aligned} y_2 &= b_0 x^{-1} - \tfrac{1}{2}b_0 \\ &\quad + b_3(x^2 - \tfrac{1}{2}x^3 + \tfrac{3}{20}x^4 - \tfrac{1}{30}x^5 + \tfrac{1}{168}x^6 - \tfrac{1}{1120}x^7 + \tfrac{1}{8640}x^8 + \cdots) \\ &= \frac{b_0}{x} - \frac{1}{2}b_0 + b_3 y_1. \end{aligned}$$

Since b_3 is arbitrary, we can simplify things by choosing $b_3 = 0$ to drop the term involving y_1. This gives us a particular solution

$$y_2 = b_0\left(\frac{1}{x} - \frac{1}{2}\right),$$

with b_0 nonzero but otherwise arbitrary. It is easy to check by direct substitution into the differential equation that this is indeed a solution.

In this example it happened that $A = 0$; thus the logarithm term did not appear in y_2. We also found a very simple second solution y_2, something which cannot be expected in general.

In the next chapter we shall see an important example of case 2 of Theorem 17 in which the logarithm term is retained in y_2.

EXAMPLE 169 Case 3 of Theorem 17

Solve

$$x^2y'' + 5xy' + (x + 4)y = 0.$$

Setting $y = \sum_{n=0}^{\infty} a_n x^{n+r}$ as usual, we get

$$[r(r - 1) + 5r + 4]a_0 x^r = 0$$

and

$$\sum_{n=1}^{\infty} [(n + r)(n + r - 1)a_n + 5(n + r)a_n + a_{n-1} + 4a_n]x^{n+r} = 0.$$

The indicial equation is $r^2 + 4r + 4 = 0$, with the single root $r = -2$.

To find a solution $y = \sum_{n=0}^{\infty} a_n x^{n-2}$, put $r = -2$ into the coefficient of x^{n+r} in the above equation to get

$$a_n = \frac{-a_{n-1}}{n^2} \qquad (n = 1, 2, 3, \ldots).$$

Then a_0 is arbitrary but nonzero and, for $n \geq 1$,

$$a_n = \frac{(-1)^n a_0}{(n!)^2}.$$

Since $(-1)^n/(n!)^2 = 1$ if $n = 0$, we can write our solution as

$$y_1 = a_0 \sum_{n=0}^{\infty} \frac{(-1)^n}{(n!)^2} x^{n-2}.$$

Now try $y_2 = y_1 \ln(x) + \sum_{n=0}^{\infty} b_n x^{n-2}$. Substituting into the differential equation and rearranging terms gives us

$$4y_1 + 2xy_1' + \sum_{n=0}^{\infty} (n - 2)(n - 3)b_n x^{n-2} + \sum_{n=0}^{\infty} 5(n - 2)b_n x^{n-2}$$

$$+ \sum_{n=0}^{\infty} b_n x^{n-1} + \sum_{n=0}^{\infty} 4b_n x^{n-2}$$

$$+ \ln(x)[x^2y_1'' + 5xy_1' + (x + 4)y_1] = 0.$$

The term in brackets is zero since y_1 is a solution of the differential equation. Substituting the solution for y_1 into the last equation (with $a_0 = 1$ for convenience), we get

$$\sum_{n=1}^{\infty} \left[\frac{4(-1)^n}{(n!)^2} + \frac{2(-1)^n}{(n!)^2}(n - 2) + (n - 2)(n - 3)b_n + 5(n - 2)b_n + b_{n-1} + 4b_n\right]$$

$$\times x^{n-2} + 4x^{-2} - 4x^{-2} + 6b_0 x^{-2} - 10x^{-2} + 4b_0 x^{-2} = 0.$$

Then, $b_0 = 1$ and, for $n = 1, 2, 3, \ldots,$

$$\frac{2(-1)^n}{(n!)^2} n + n^2 b_n + b_{n-1} = 0.$$

Then, for $n \geq 1$,

$$b_n = \frac{-b_{n-1}}{n^2} - \frac{2(-1)^n}{n(n!)^2},$$

and we can compute as many terms of y_2 as we like. Thus,

$$y_2 = y_1 \ln(x) + \frac{1}{x^2} + \frac{1}{x} - \frac{1}{2} + \frac{2}{27}x - \frac{19}{3456}x^2 + \cdots.$$

PROBLEMS FOR SECTION 5.3

In each of Problems 1 through 30, find and solve the indicial equation, determine the recurrence relation, and find the first six terms of each of two linearly independent solutions valid for $x > 0$.

1. $x^2y'' + x(2 - x)y' - 2y = 0$

2. $x^2y'' + x(3 + x)y' + y = 0$

3. $x^2y'' + xy' + x^2y = 0$

4. $x^2y'' + 2xy' + (x - 6)y = 0$

5. $xy'' + y' + x^2y = 0$

6. $x^2y'' + x(x + \frac{1}{2})y' + (x - \frac{1}{2})y = 0$

7. $x^2y'' - xy' + (2x - 8)y = 0$

8. $x^2y'' - xy' + (x^2 - \frac{1}{4})y = 0$

9. $x^2y'' + x(x - \frac{1}{4})y' - \frac{1}{4}y = 0$

10. $x^2y'' + xy' + (x - 2)y = 0$

11. $x^2y'' + x(x^2 - \frac{3}{4})y' + (x - \frac{1}{8})y = 0$

12. $y'' + \frac{1}{x}\left(x - \frac{1}{2}\right)y' - \frac{1}{x^2}y = 0$

13. $y'' + xy' - \frac{2}{x^2}y = 0$

14. $y'' + \frac{1}{x}\left(x^2 + \frac{7}{2}\right)y' + \frac{1}{x^2}(x - 6)y = 0$

15. $y'' + \frac{5}{4x}y' + \left(1 - \frac{9}{4x^2}\right)y = 0$

16. $y'' + \left(1 + \frac{3}{x^2}\right)y' + \left(\frac{1}{x} + \frac{6}{x^2}\right)y = 0$

17. $x^2y'' - 3xy' + (x^2 - 2)y = 0$

18. $x^2y'' + x(x - 2)y + x^2y = 0$

19. $x^2y'' - 4xy' + (x^2 - 3)y = 0$

20. $x^2y'' + 8x^2y' + (2x - 7)y = 0$

21. $x^2y'' + (x^2 - 3x)y' + (x - 4)y = 0$

22. $x^2y'' + x(x^2 - 4)y' + (2x - 3)y = 0$

23. $x^2y'' + 4xy' - 3xy = 0$

24. $x^2y'' + x^2y' - 8y = 0$

25. $2x^2y'' - 2xy' + (3x + 2)y = 0$

26. $x^2y'' - 2xy' + 8xy = 0$

27. $x^2y'' - 4xy' + (x^2 - 3x)y = 0$

28. $x^2y'' - 4xy' + (3x - 2)y = 0$

29. $x^2y'' - 4x^3y' + (x - 6)y = 0$

30. $x^2y'' - 8x^2y' + (x^2 + 4)y = 0$

31. The hypergeometric equation can be written

$$x(1 - x)y'' + [c - (1 + a + b)x]y' - aby = 0.$$

(a) Show that 0 is a regular singular point.

(b) Assuming that $(1 - c)$ is not an integer, use the Method of Frobenius to obtain the first six terms of two linearly independent solutions. (The coefficients will involve the constants a, b, and c.)

CHAPTER SUMMARY

Power series, and the Method of Frobenius, are particularly well suited to nonconstant coefficient equations of the kind generally beyond the reach of previous methods. One disadvantage is that the recurrence relations can be very complicated, with a general expression for a_n in terms of a_{n-1} quite difficult or even impossible to derive in elementary form. In such a case we usually write out as many terms as we need and attempt to compute with these for the problem we are trying to solve.

In several instances, series solutions of certain differential equations are very complicated, but the differential equations themselves arise in such a variety of useful contexts that it is worthwhile developing some facts about these solutions. In this way one enters into the arena of *special functions.* The next chapter serves as a brief introduction to this important area, with particular emphasis on Bessel and Legendre functions.

SUPPLEMENTARY PROBLEMS

In each of Problems 1 through 20, write out the first six nonzero terms of a series solution about 0.

1. $y'' - xy' - x^3y = 0$
2. $y'' - \sin(x)y' + xy = 0$
3. $y'' - 2y' + x^3y = 0$
4. $y'' - 4xy' + 2xy = 0$
5. $y' - \cos(x)y = 0$
6. $y'' + 2x^2y' - y = 0$
7. $y'' + x^3y' - 3y = 0$
8. $2y'' - (x - 4)y' + 7xy = 0$
9. $y'' - x^3y' + 2y = 0$
10. $3y'' + 4y' - (8x^2 + 2)y = 0$
11. $y'' - e^xy' + 2y = 0$
12. $y'' - 8xy' - (x - 3)y = 0$
13. $y^{(3)} + 2x^2y = 0$
14. $y'' + 2x^2y' - (x^2 - \frac{1}{3})y = 0$
15. $2y'' + (x^2 - 6)y' + y = 0$
16. $y^{(3)} + 8x^2y' - 2y = 0$
17. $3y'' + (4x - 7)y' + x^2y = 0$
18. $y' - 3\sin(4x)y = 0$
19. $2y'' + y' - x^4y = 0$
20. $y'' + (x - 3)y' + (x^2 - 4)y = 0$

In each of Problems 21 through 30, write out the first six nonzero terms of a series solution about 0 by expanding the right side in a Taylor series and comparing coefficients of like powers of x on both sides.

21. $y' - 4x^3y = -3e^{2x}$
22. $y'' - xy' + 2y = \sin(\frac{1}{2}x)$
23. $6y'' - x^2y = \cos(x) - \sin(x)$
24. $2y'' + 4x^2y' - 3y = 2\cosh(x)$
25. $y'' - e^xy = \sin(x) + 1$
26. $y'' + 3\sin(x)y' - 4xy = -2\cos(3x)$
27. $y'' - y' + x^3y = x^2 - 7$
28. $y'' + 8x^2y' - y = 2x\sin(x)$
29. $y'' + 4xy' - x^2y = x^3 - 3x + 2$
30. $y'' + e^xy' - y = 2x^2e^{-x}$

31. Find the error in the following "solution" of the pendulum equation.

The differential equation is

$$\theta'' + \frac{g}{L}\sin(\theta) = 0.$$

Write

$$\theta(t) = \sum_{n=0}^{\infty} a_n t^n \quad \text{and} \quad \sin[\theta(t)] = \sum_{n=0}^{\infty} \frac{(-1)^n t^{2n+1}}{(2n+1)!}.$$

Then we need

$$\sum_{n=2}^{\infty} n(n-1)t^{n-2} = \sum_{n=0}^{\infty} \frac{-g}{L}\frac{(-1)^n t^{2n+1}}{(2n+1)!}$$

or

$$\sum_{n=0}^{\infty} (n+2)(n+1)a_{n+2}t^n = \sum_{n=0}^{\infty} \frac{-g}{L} \frac{(-1)^n t^{2n+1}}{(2n+1)!}.$$

From this, find that $a_{2n} = 0$ if $n = 1, 2, 3, \ldots, a_0$ and a_1 are arbitrary, and

$$a_{2n+1} = (-1)^n \frac{g}{L} \frac{1}{(2n+1)!} \qquad (n = 1, 2, 3, \ldots).$$

Then

$$\begin{aligned}\theta(t) &= \sum_{n=0}^{\infty} a_n t^n = a_0 + a_1 t + \sum_{n=1}^{\infty} \frac{(-1)^n}{(2n+1)!} \frac{g}{L} t^{2n+1} \\ &= a_0 + a_1 t + \left[\sum_{n=0}^{\infty} \frac{(-1)^n}{(2n+1)!} \frac{g}{L} t^{2n+1} - \frac{gt}{L}\right] \\ &= a_0 + \left(a_1 - \frac{g}{L}\right)t + \frac{g}{L}\sin(t).\end{aligned}$$

In each of Problems 32 through 50, verify that 0 is a regular singular point, and use the Method of Frobenius to write out the first six nonzero terms of two linearly independent solutions.

32. $y'' + (1 - x^3)y' + \frac{2}{x^2} y = 0$

33. $x^2y'' - xy' + (1 - 3x)y = 0$

34. $y'' - \frac{8}{x} y' + (2 - x)y = 0$

35. $x^2y'' + xy' - (x + 6)y = 0$

36. $x^2y'' - 2xy' - (x - 7)y = 0$

37. $x^2y'' + xy' + (x - 5)y = 0$

38. $x^2y'' + 3x^2y' + (x - 2)y = 0$

39. $x^2y'' + 5xy' - (7x^3 + 2)y = 0$

40. $3x^2y'' - 2xy' + (x^2 - 3)y = 0$

41. $x^2y'' - 4xy' + (x - 5)y = 0$

42. $2x^2y'' - x^2y' + (x + 3)y = 0$

43. $x^2y'' + 2xy' + (x^2 - 7)y = 0$

44. $x^2y'' + x^2y' - y = 0$

45. $-4x^2y'' + xy' + x^2y = 0$

46. $x^2y'' + xy' - (x + 6)y = 0$

47. $x^2y'' - xy' + (x^2 - 2)y = 0$

48. $x^2y'' + 8xy' - (4x - 1)y = 0$

49. $x^2y'' + xy' - (x^2 + 2)y = 0$

50. $x^2y'' + (x^2 - 2x)y' - (x^2 + 5)y = 0$

CHAPTER SIX

Bessel Functions and Legendre Polynomials, Sturm-Liouville Theory, Eigenfunction Expansions, and Oscillation

6.0 INTRODUCTION

Certain differential equations have been observed to occur repeatedly in a wide variety of contexts. One example is *Bessel's equation*,

$$x^2y'' + xy' + (x^2 - \nu^2)y = 0,$$

which may be found (sometimes after a change of variables) in astronomy, heat conduction, electricity, and stress on beams. In fact, the heat energy lost through the sides of a cylindrical hot water tank is emitted as Bessel functions.

Whenever this "frequency of occurrence" phenomenon is noticed, it is convenient and useful to record the differential equation and the properties of its various solutions so that they need not be constantly rederived. Thus was born the area called *special functions*. In this chapter we shall study Bessel's and Legendre's differential equations and their solutions, and then put some of the basic ideas into the broad context of Sturm-Liouville theory, which provides a background for studying many other kinds of special functions as well.

The chapter's final section is an introduction to oscillation theory, and treats a method for analyzing the behavior of solutions of certain differential equations from the form of the equation itself, without explicitly obtaining the solutions. This is part of the *qualitative study* of differential equations, and we will see immediate application in verification of certain properties of Bessel functions. Another aspect of the qualitative study of differential equations is seen in Section 7.3, where we briefly introduce the notion of stability.

6.1 BESSEL FUNCTIONS OF INTEGER ORDER

For $n = 0, 1, 2, 3, \ldots$, the second order differential equation

$$x^2y'' + xy' + (x^2 - n^2)y = 0$$

is called *Bessel's equation of order n.* (The word *order* is thus used in two different senses here, but this is traditional.) One also often sees the equation written in the form

$$y'' + \frac{1}{x}y' + \left(1 - \frac{n^2}{x^2}\right)y = 0.$$

It is routine to check that this differential equation has a regular singular point at zero, suggesting that we try a Frobenius solution

$$y = \sum_{j=0}^{\infty} a_j x^{j+r},$$

in which we use j as summation index because n is a given integer in Bessel's equation. Substitute this into the differential equation and rearrange terms to get

$$a_0[r(r-1) + r - n^2]x^r + a_1[(r+1)r + (r+1) - n^2]x^{r+1}$$
$$+ \sum_{j=2}^{\infty} \{[(j+r)(j+r-1) + (j+r) - n^2]a_j + a_{j-2}\}x^{j+r}.$$

In this, the coefficient of each power of x must be zero. From the coefficient of x^r we get the indicial equation

$$r^2 - n^2 = 0,$$

assuming that $a_0 \neq 0$. This has roots $r = n$ and $r = -n$, and leads to two cases.

Case 1: $n = 0$ Now $r = 0$ is a repeated root of the indicial equation, and we are in case 3 of Theorem 17. To get one solution, set $r = n = 0$ and obtain the recurrence relation

$$j^2a_j + a_{j-2} = 0, \qquad j = 2, 3, 4, \ldots,$$

from the coefficient of x^{j+r} in the above summation. Then,

$$a_j = \frac{-a_{j-2}}{j^2} \quad \text{for} \quad j = 2, 3, 4, \ldots.$$

From this we easily find the even-indexed terms:

$$a_{2j} = \frac{(-1)^j a_0}{2^2 4^2 \cdots (2j)^2} \quad \text{for} \quad j = 1, 2, 3, \ldots .$$

The recurrence relation gives no information about a_1. However, upon setting $r = n = 0$ in the coefficient of x^{r+1} above, we get the coefficient of x^{r+1} to be just a_1, and hence conclude that $a_1 = 0$. From the recurrence relation, then,

$$a_1 = a_3 = a_5 = a_7 = \cdots = 0,$$

and all odd-indexed coefficients are therefore zero. Thus, one solution is

$$y_1 = \sum_{j=0}^{\infty} a_{2j} x^{2j} = a_0 \sum_{j=0}^{\infty} \frac{(-1)^j}{2^2 4^2 \cdots (2j)^2} x^{2j}.$$

This is usually written in a different form. Note that

$$2 \cdot 4 \cdots (2j) = \underbrace{(2 \cdot 2 \cdots 2)}_{j \text{ times}}(1 \cdot 2 \cdot 3 \cdots j) = 2^j j!.$$

Thus,

$$y_1 = a_0 \sum_{j=0}^{\infty} \frac{(-1)^j}{2^{2j}(j!)^2} x^{2j} = a_0 \sum_{j=0}^{\infty} \frac{(-1)^j}{(j!)^2} \left(\frac{x}{2}\right)^{2j}.$$

When we choose $a_0 = 1$, this solution is called *the Bessel function of the first kind of order zero*, and is denoted $J_0(x)$. Thus,

$$J_0(x) = \sum_{j=0}^{\infty} \frac{(-1)^j}{(j!)^2} \left(\frac{x}{2}\right)^{2j} = 1 - \frac{x^2}{2^2} + \frac{x^4}{2^2 4^2} - \frac{x^6}{2^2 4^2 6^2} + \frac{x^8}{2^2 4^2 6^2 8^2} - \cdots$$

is a first solution of $x^2 y'' + xy' + x^2 y = 0$. A graph of $y = J_0(x)$ is shown in Figure 69,

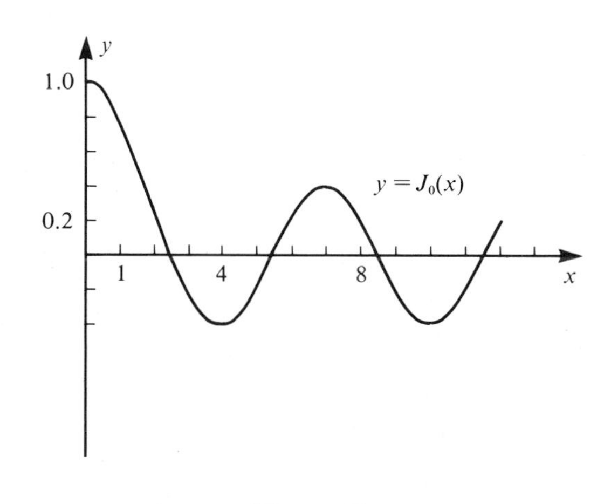

Figure 69

Table 1: Some Values of $y = J_0(x)$ for $0 \leq x \leq 10$

x	$J_0(x)$	x	$J_0(x)$	x	$J_0(x)$	x	$J_0(x)$
0	0	2.5	−0.048384	5.0	−0.177597	7.5	0.266339
0.1	0.997502	2.6	−0.096805	5.1	−0.144335	7.6	0.251602
0.2	0.990025	2.7	−0.142449	5.2	−0.110290	7.7	0.234559
0.3	0.977626	2.8	−0.185036	5.3	−0.075803	7.8	0.215408
0.4	0.960398	2.9	−0.224312	5.4	−0.041210	7.9	0.194362
0.5	0.938470	3.0	−0.260052	5.5	−0.006843	8.0	0.171651
0.6	0.912005	3.1	−0.292064	5.6	+0.026971	8.1	0.147517
0.7	0.881201	3.2	−0.320188	5.7	0.059920	8.2	0.122215
0.8	0.846287	3.3	−0.344296	5.8	0.091703	8.3	0.096006
0.9	0.807524	3.4	−0.364295	5.9	0.122033	8.4	0.069157
1.0	0.765198	3.5	−0.380128	6.0	0.150645	8.5	0.041939
1.1	0.719622	3.6	−0.391769	6.1	0.177291	8.6	0.014623
1.2	0.671133	3.7	−0.399230	6.2	0.201747	8.7	−0.012523
1.3	0.620086	3.8	−0.402556	6.3	0.223812	8.8	−0.039234
1.4	0.566855	3.9	−0.401826	6.4	0.243310	8.9	−0.065253
1.5	0.511828	4.0	−0.397150	6.5	0.260094	9.0	−0.090333
1.6	0.455402	4.1	−0.388670	6.6	0.274043	9.1	−0.114239
1.7	0.397985	4.2	−0.376557	6.7	0.285065	9.2	−0.136748
1.8	0.339986	4.3	−0.361011	6.8	0.293096	9.3	−0.157655
1.9	0.281819	4.4	−0.342257	6.9	0.298102	9.4	−0.176772
2.0	0.223890	4.5	−0.320543	7.0	0.300079	9.5	−0.193929
2.1	0.166607	4.6	−0.296138	7.1	0.299051	9.6	−0.208979
2.2	0.110362	4.7	−0.269331	7.2	0.295071	9.7	−0.221795
2.3	0.055540	4.8	−0.240425	7.3	0.288217	9.8	−0.232276
2.4	0.002508	4.9	−0.209738	7.4	0.278596	9.9	−0.240341
						10.0	−0.245936

and Table 1 gives some values of $J_0(x)$ for $0 \leq x \leq 10$. We shall see later (note Example 171, concerning the displacement of a suspended chain) that it is useful to know some positive roots of the equation $J_0(x) = 0$. Some of these are given in Table 2.

Theorem 17(3) tells us that there is a second solution of Bessel's zero order equation containing a logarithm term. Omitting the details of calculation, we find a second solution

$$J_0(x)\ln(x) + \frac{x^2}{2^2} - \frac{x^4}{2^2 4^2}\left(1 + \frac{1}{2}\right) + \frac{x^6}{2^2 4^2 6^2}\left(1 + \frac{1}{2} + \frac{1}{3}\right) - \frac{x^8}{2^2 4^2 6^2 8^2}\left(1 + \frac{1}{2} + \frac{1}{3} + \frac{1}{4}\right) + \cdots.$$

Table 2: The First Thirty Positive Roots of $J_0(x) = 0$ (In Increasing Order from Zero)

Root number	*Root*	*Root number*	*Root*
1	2.40483	16	49.48261
2	5.52008	17	52.62405
3	8.63573	18	55.76551
4	11.79153	19	58.90698
5	14.93092	20	62.04846
6	18.07106	21	65.18996
7	21.21164	22	68.33146
8	24.35247	23	71.47298
9	27.49348	24	74.61450
10	30.63461	25	77.75602
11	33.77582	26	80.89756
12	36.91709	27	84.03909
13	40.05843	28	87.18063
14	43.19979	29	90.32217
15	46.34119	30	93.46371

This is *Neumann's Bessel function of the second kind of order zero** and is often denoted $Y_0(x)$. A graph is sketched in Figure 70.

Thus, the general solution of Bessel's equation of order zero,

$$x^2y'' + xy' + x^2y = 0,$$

is

$$y = c_1J_0(x) + c_2\, Y_0(x)$$

in some interval $0 < x < \alpha$. In applications, one can often determine c_2 immediately from a boundary condition. For example, if $y(x)$ must remain bounded as x decreases to zero, then we must have $c_2 = 0$, or the logarithm term goes to (minus) infinity.

---

* There is no universal agreement on $Y_0(x)$, and some authors use the second solution

$$\frac{2}{\pi}\left[\ln\left(\frac{x}{2}\right) + \gamma\right]J_0(x) + \frac{x^2}{2^2} - \frac{x^4}{2^24^2}\left(1 + \frac{1}{2}\right) + \frac{x^6}{2^24^26^2}\left(1 + \frac{1}{2} - \frac{1}{3}\right) - \cdots,$$

which is called *Weber's Bessel function of the second kind of order zero.*

In this expression,

$$\gamma = \lim_{n\to\infty}\left[1 + \frac{1}{2} + \frac{1}{3} + \cdots + \frac{1}{n} - \ln(n)\right]$$

is the Euler (or Euler-Mascheroni) constant, which is approximately 0.577215664901533.

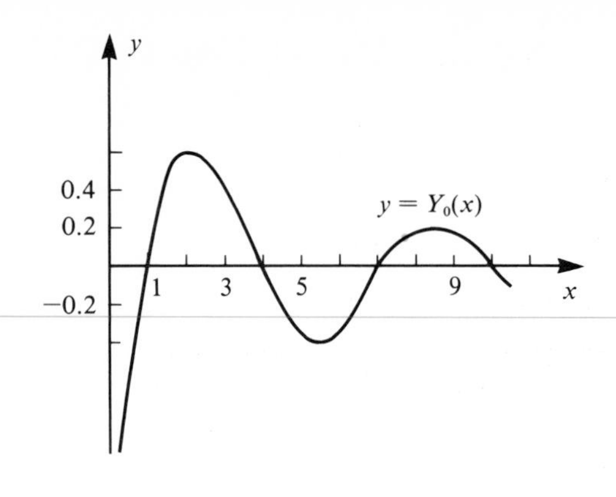

Figure 70

Case 2: $n = 1, 2, 3, \ldots$ Now we are in case 2 of Theorem 17. Again omitting details, which are similar to those encountered in obtaining $J_0(x)$, we get a solution

$$J_n(x) = \sum_{j=0}^{\infty} \frac{(-1)^j}{j!\,(n+j)!}\left(\frac{x}{2}\right)^{n+2j}$$

$$= \frac{x^n}{2^n n!}\left[1 - \frac{x^2}{2^2(n+1)} + \frac{x^4}{2^4 2!\,(n+1)(n+2)} - \frac{x^6}{2^6 3!\,(n+1)(n+2)(n+3)} + \cdots\right].$$

This is a *Bessel function of the first kind of order n.* A second, linearly independent solution is

$$Y_n(x) = J_n(x)\ln(x) - \frac{1}{2}\sum_{j=0}^{n-1} \frac{(n-j-1)!}{j!}\left(\frac{x}{2}\right)^{-n+2j}$$

$$- \frac{1}{2}\sum_{j=0}^{\infty} \frac{(-1)^j}{j!\,(n+j)!}\left(\frac{x}{2}\right)^{n+2j}[\phi(j) + \phi(n+j)],$$

where

$$\phi(m) = 1 + \frac{1}{2} + \cdots + \frac{1}{m} \quad \text{and} \quad \phi(0) = 0.$$

This is *Neumann's Bessel function of the second kind of order n.*

Cases 1 and 2 can now be combined, as substituting $n = 0$ into $J_n(x)$ and $Y_n(x)$ of case 2 yields $J_0(x)$ and $Y_0(x)$.

Thus, for $n = 0, 1, 2, 3, \ldots$, the general solution of

$$x^2y'' + xy' + (x^2 - n^2)y = 0$$

is

$$y = c_1 J_n(x) + c_2 Y_n(x).$$

Again, because of the logarithm term in $Y_n(x)$, c_2 must be chosen as zero any time the solution must remain bounded as x decreases to zero.

For purposes of comparison, Figure 71 shows the graphs of $J_0(x)$, $J_1(x)$, and $J_2(x)$, while Figure 72 shows $Y_0(x)$, $Y_1(x)$, and $Y_2(x)$.

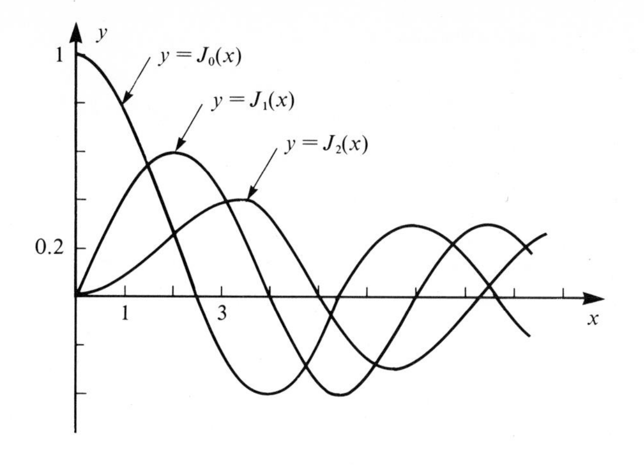

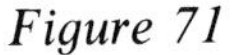
Figure 71

Note from Figure 71 that $J_0(x)$ and $J_1(x)$ oscillate with decreasing amplitude as $x \to \infty$. The oscillations do not, however, occur at fixed intervals, and so Bessel functions are not periodic. Further, the graphs intertwine, as each crosses the axis between two crossings of the other. This behavior is true for any pair $J_n(x)$ and $J_{n+1}(x)$ of successive Bessel functions, and is of importance later in solving certain partial differential equations. A proof of this interlacing of zeros is sketched in Problem 25 at the end of this section. The result will follow more easily from the oscillation theory of Section 6.5.

Bessel's equation occurs in many forms. It is routine to show that

$$y_1 = x^a J_n(bx^c) \quad \text{and} \quad y_2 = x^a Y_n(bx^c),$$

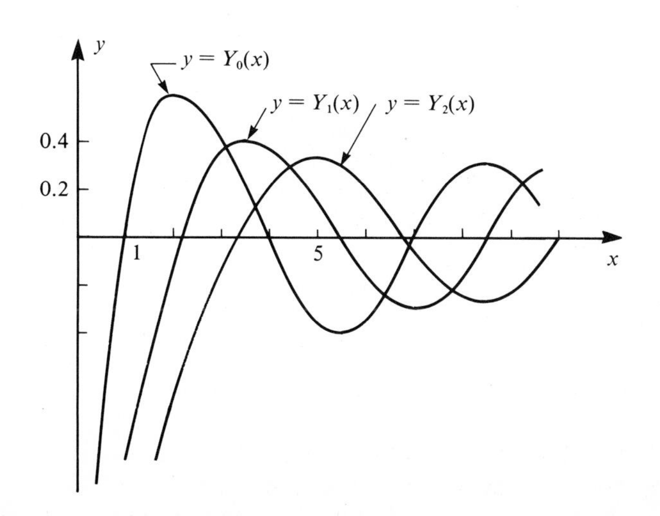

Figure 72

for a, b, and c constant, are solutions of

$$y'' - \left(\frac{2a-1}{x}\right)y' + \left(b^2c^2x^{2c-2} + \frac{a^2 - n^2c^2}{x^2}\right)y = 0. \tag{a}$$

Thus, this differential equation has general solution

$$y = c_1x^aJ_n(bx^c) + c_2x^aY_n(bx^c),$$

provided that $bx^c > 0$ so that $Y_n(bx^c)$ is defined. Here is how this is used:

EXAMPLE 170

Solve

$$y'' + \left(\frac{81x}{4} - \frac{35}{4x^2}\right)y = 0.$$

The strategy is to try to choose a, b, c, and n so that this equation looks like (a). We need

$$2a - 1 = 0$$
$$2c - 2 = 1$$
$$b^2c^2 = \tfrac{81}{4}$$
$$a^2 - n^2c^2 = -\tfrac{35}{4}$$

We find that

$$a = \tfrac{1}{2}, \qquad b = 3, \qquad c = \tfrac{3}{2}, \quad \text{and} \quad n = 2.$$

Thus, the general solution is

$$y = c_1\sqrt{x}\,J_2(3x^{3/2}) + c_2\sqrt{x}\,Y_2(3x^{3/2}).$$

Here is an application of Bessel functions.

EXAMPLE 171 Displacement of a Suspended Chain

We shall now see how Bessel functions enter into the analysis of oscillations in a suspended chain.

Imagine a uniform heavy flexible chain, as shown in Figure 73. The chain is fixed at the upper end and free at the bottom. We consider the problem of describing small oscillations caused by small displacements, in a vertical plane, from the stable equilibrium position.

We assume that each particle of chain oscillates in a horizontal straight line. Let m be the mass of the chain per unit length, L its total length, and let $y(x, t)$ be the horizontal displacement at time t of the particle of chain whose distance from the point of suspension is x.

To derive an equation for y, consider an element of chain of length dx. If the forces acting on the ends of this element are T and $T + dT$, then the horizontal component of Newton's law (force equals mass times

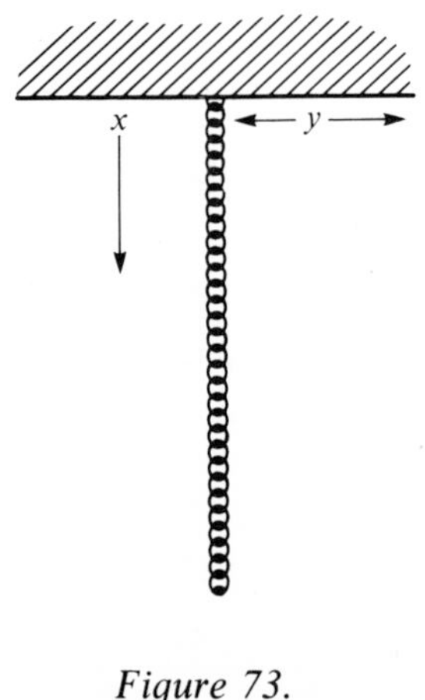

Figure 73.
Suspended chain.

acceleration) is

$$m(dx)\frac{\partial^2 y}{\partial t^2} = \frac{\partial}{\partial x}\left(T\frac{\partial y}{\partial x}\right)dx.$$

Upon dividing out dx, we have

$$m\frac{\partial^2 y}{\partial t^2} = \frac{\partial}{\partial x}\left(T\frac{\partial y}{\partial x}\right).$$

We shall now assume that

$$T = mg(L - x),$$

a reasonable approximation for small disturbances. The equation for y then becomes

$$\frac{\partial^2 y}{\partial t^2} = -g\frac{\partial y}{\partial x} + g(L - x)\frac{\partial^2 y}{\partial x^2}.$$

This is a partial, not ordinary, differential equation, since it involves a function of two variables and its partial derivatives. To solve it, first change variables by putting

$$z = L - x \quad \text{and} \quad u(z, t) = y(L - z, t).$$

Then,

$$\frac{\partial^2 y}{\partial t^2} = \frac{\partial^2 u}{\partial t^2}, \qquad \frac{\partial y}{\partial x} = -\frac{\partial u}{\partial z}; \quad \text{and} \quad \frac{\partial^2 y}{\partial x^2} = \frac{\partial^2 u}{\partial z^2}.$$

The partial differential equation then becomes

$$\frac{\partial^2 u}{\partial t^2} = g\frac{\partial u}{\partial z} + gz\frac{\partial^2 u}{\partial z^2}.$$

We now anticipate a method (to be further developed in Chapter 13) known as *separation of variables*. The idea is to attempt a solution of the form

$$u(z, t) = f(z)\cos(\omega t - \delta).$$

This is motivated partly by the expectation that the oscillations should be periodic in t. Substitute this into the partial differential equation to get

$$-\omega^2 f(z)\cos(\omega t - \delta) = gf'(z)\cos(\omega t - \delta) + gzf''(z)\cos(\omega t - \delta).$$

Dividing out $\cos(\omega t - \delta)$ gives us

$$-\omega^2 f(z) = gf'(z) + gzf''(z)$$

or

$$f''(z) + \frac{1}{z}f'(z) + \frac{\omega^2}{gz}f(z) = 0.$$

This is a Bessel equation for $f(z)$, and can be solved by the method of the preceding example by setting

$$-(2a - 1) = 1, \qquad 2c - 2 = -1, \qquad b^2c^2 = \frac{\omega^2}{g} \quad \text{and} \quad a^2 - n^2c^2 = 0.$$

We get

$$a = n = 0, \qquad c = \frac{1}{2}, \quad \text{and} \quad b = \frac{2\omega}{\sqrt{g}}.$$

Thus,

$$f(z) = c_1 J_0\left(2\omega\sqrt{\frac{z}{g}}\right) + c_2 Y_0\left(2\omega\sqrt{\frac{z}{g}}\right).$$

Since $Y_0(2\omega\sqrt{z/g}) \to -\infty$ as $z \to 0$ (or as $x \to L$), we are forced by the physics of the problem to choose $c_2 = 0$. Then,

$$f(z) = c_1 J_0\left(2\omega\sqrt{\frac{z}{g}}\right).$$

Then,

$$u(z, t) = c_1 J_0\left(2\omega\sqrt{\frac{z}{g}}\right)\cos(\omega t - \delta)$$

and

$$y(x, t) = c_1 J_0\left(2\omega\sqrt{\frac{L - x}{g}}\right)\cos(\omega t - \delta).$$

The frequencies of the normal vibrations are determined by using the condition that the upper end of the chain does not move. Then, for all t, $y(0, t) = 0$. Assuming that $c_1 \neq 0$, this forces us to choose ω so that

$$J_0\left(2\omega\sqrt{\frac{L}{g}}\right) = 0.$$

Admissible values of ω are then obtained by consulting a table of zeros of J_0. From Table 2, we get values

$$\omega_1 = \frac{1}{2}\sqrt{\frac{g}{L}}(2.40483) = 1.2024\sqrt{\frac{g}{L}},$$

$$\omega_2 = \frac{1}{2}\sqrt{\frac{g}{L}}(5.52008) = 2.76\sqrt{\frac{g}{L}},$$

$$\omega_3 = \frac{1}{2}\sqrt{\frac{g}{L}}(8.63573) = 4.3179\sqrt{\frac{g}{L}},$$

and so on. These are the approximate frequencies of the normal modes of vibration. The approximate periods are, respectively,

$$\frac{4\pi}{2.40483}\sqrt{\frac{L}{g}} = 5.2255\sqrt{\frac{L}{g}},$$

$$\frac{4\pi}{5.52008}\sqrt{\frac{L}{g}} = 2.2765\sqrt{\frac{L}{g}},$$

$$\frac{4\pi}{8.63573}\sqrt{\frac{L}{g}} = 1.4552\sqrt{\frac{L}{g}},$$

and so on.

This example is fairly typical of applications of Bessel functions, in that their occurrence may not be immediately apparent but becomes evident only after a change of variables.

The above analysis was essentially carried out by Daniel Bernoulli in about 1732.

Sometimes one encounters modified Bessel functions, which are obtained as follows. First observe that

$$y = c_1 J_0(kx) + c_2 Y_0(kx)$$

is the general solution of

$$y'' + \frac{1}{x} y' + k^2 y = 0.$$

Now set $k = i$, the imaginary unit ($i^2 = -1$). Then,

$$y = c_1 J_0(ix) + c_2 Y_0(ix)$$

is the general solution of

$$y'' + \frac{1}{x} y' - y = 0.$$

We call this differential equation a *modified Bessel equation of order zero*, and $J_0(ix)$ is a *modified Bessel function of the first kind of order zero.*

Usually one denotes

$$I_0(x) = J_0(ix).$$

Since $i^2 = -1$, then

$$I_0(x) = 1 + \frac{x^2}{2^2} + \frac{x^4}{2^2 4^2} + \frac{x^6}{2^2 4^2 6^2} + \cdots.$$

Usually $Y_0(ix)$ is not used, and in its place we set

$$K_0(x) = [\ln(2) - \gamma]I_0(x) - I_0(x)\ln(x) - \frac{x^2}{4} - \cdots.$$

$K_0(x)$ is a *modified Bessel function of the second kind of order zero*, where γ is the Euler constant noted before. Figure 74 shows graphs of $I_0(x)$ and $K_0(x)$.

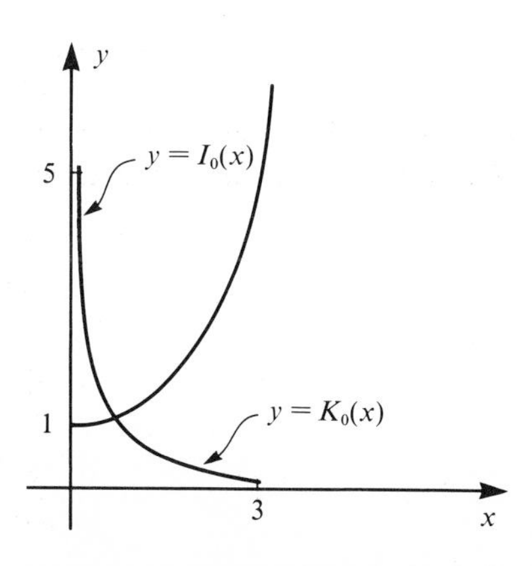

Figure 74

Thus, the general solution of

$$y'' + \frac{1}{x} y' - y = 0$$

is

$$y = c_1 I_0(x) + c_2 K_0(x).$$

One can now easily check that the general solution of

$$y'' + \frac{1}{x} y' - a^2 y = 0$$

is

$$y = c_1 I_0(ax) + c_2 K_0(ax).$$

Here is an application of modified Bessel functions.

EXAMPLE 172 Alternating Current in a Circular Wire. The Skin Effect

In this example we shall show how a modified Bessel function arises in the treatment of alternating current in a circular wire.

Consider an alternating current of period $2\pi/\omega$, given by $D\cos(\omega t)$. Let R be the radius of the wire, ρ its specific resistance, μ the permeability, $x(r, t)$ the current density at radius r and time t, and $H(r, t)$ the magnetic intensity at radius r from the center at time t. To derive an equation for x, we shall need two laws of electromagnetic field theory. They are the following:

Ampere's law: The line integral of the electric force around a closed path equals 4π times the integral of the electric current through the path.

Faraday's law: The line integral of the electric force around a closed path equals minus the partial derivative with respect to time of the magnetic induction through the path.

Now consider a circular path of radius r within the wire, centered about the center of the wire. By Ampere's law,

$$2\pi r H = 4\pi \int_0^r 2\pi r x \, dr.$$

Differentiate both sides with respect to r, giving

$$\frac{\partial}{\partial r}(rH) = 4\pi r x. \tag{1}$$

Next, consider a closed rectangular path in the wire, with two sides along the axis of the cylinder and of length L, and the other sides of length r. By Faraday's law,

$$\rho L[x(0, t) - x(r, t)] = -\frac{\partial}{\partial t}\int_0^r \mu L H \, dr.$$

Differentiate this with respect to r, giving

$$\rho L \frac{\partial x}{\partial r} = \mu L \frac{\partial H}{\partial t}$$

or, upon cancelling L,

$$\rho \frac{\partial x}{\partial r} = \mu \frac{\partial H}{\partial t}. \tag{2}$$

We now want to eliminate H to obtain an equation for x alone. Multiply (2) by r and differentiate with respect to r to get

$$\rho \frac{\partial}{\partial r}\left(r \frac{\partial x}{\partial r}\right) = \mu \frac{\partial}{\partial r}\left(r \frac{\partial H}{\partial t}\right).$$

Now write $r(\partial H/\partial t) = \partial(rH)/\partial t$, since $\partial r/\partial t = 0$, to get

$$\rho \frac{\partial}{\partial r}\left(r \frac{\partial x}{\partial r}\right) = \mu \frac{\partial}{\partial r}\left[\frac{\partial}{\partial t}(rH)\right] = \mu \frac{\partial}{\partial t}\left[\frac{\partial}{\partial r}(rH)\right],$$

assuming that the mixed partials of rH are equal. Substituting (1) into this gives us

$$\rho \frac{\partial}{\partial r}\left(r \frac{\partial x}{\partial r}\right) = \mu \frac{\partial}{\partial t}(4\pi r x),$$

and H has been eliminated. This is usually written

$$\frac{1}{r}\frac{\partial}{\partial r}\left(r \frac{\partial x}{\partial r}\right) = \frac{4\pi\mu}{\rho}\frac{\partial x}{\partial t}.$$

In order to solve this, we employ a trick which is quite standard in treating circuit equations. Set $z(r, t) = x(r, t) + iy(r, t)$, and consider x as the real part of the complex quantity z. Now consider the differential equation

$$\frac{1}{r}\frac{\partial}{\partial r}\left(r \frac{\partial z}{\partial r}\right) = \frac{4\pi\mu}{\rho}\frac{\partial z}{\partial t}.$$

Again anticipating a method for treating such equations, we look for a solution of the form

$$z = f(r)e^{i\omega t}.$$

Substitute this into the differential equation to get

$$\frac{1}{r}\frac{\partial}{\partial r}[rf'(r)e^{i\omega t}] = \frac{4\pi\mu}{\rho} i\omega f(r)e^{i\omega t}.$$

Upon dividing out $e^{i\omega t}$, we have

$$\frac{1}{r}\frac{\partial}{\partial r}[rf'(r)] = \frac{4\pi\mu i\omega}{\rho} f(r).$$

This is

$$f''(r) + \frac{1}{r} f'(r) - \frac{4\pi\mu i\omega}{\rho} f(r) = 0.$$

Now let

$$k = \left(\frac{1+i}{\sqrt{2}}\right)\sqrt{\frac{4\pi\mu\omega}{\rho}}.$$

It is easy to check that $k^2 = 4\pi\mu i\omega/\rho$; hence the differential equation for $f(r)$ is

$$f''(r) + \frac{1}{r} f'(r) - k^2 f(r) = 0,$$

which is a modified Bessel equation with general solution

$$f(r) = c_1 I_0(kr) + c_2 K_0(kr).$$

We require that $f(r)$ remain finite as $r \to 0$, and hence must choose $c_2 = 0$. Thus,

$$f(r) = c_1 I_0(kr).$$

Then,

$$z(r, t) = c_1 I_0(kr) e^{i\omega t},$$

and $x(r, t)$ is the real part of this expression.

To determine c_1, recall that the alternating current in the wire is $D \cos(\omega t)$, a fact we have not used yet (except in putting ω in the exponential term of the solution). Think of $D \cos(\omega t)$ as the real part of $De^{i\omega t}$. Since $De^{i\omega t}$ represents the total current, then

$$De^{i\omega t} = \int_0^R 2\pi r z \, dr = 2\pi c_1 \int_0^R r I_0(kr) e^{i\omega t} \, dr.$$

Upon cancelling out $e^{i\omega t}$ and solving for c_1, we have

$$c_1 = \frac{D}{2\pi \int_0^R r I_0(kr) \, dr}.$$

One can write the integral in the denominator in terms of the derivative of $I_0(kr)$, as follows: Recall that

$$I_0(kr) = 1 + \frac{k^2 r^2}{2^2} + \frac{k^4 r^4}{2^2 4^2} + \frac{k^6 r^6}{2^2 4^2 6^2} + \cdots.$$

Then,

$$\int_0^R r I_0(kr) \, dr = \int_0^R r \, dr + \int_0^R \frac{k^2 r^3}{2^2} \, dr + \int_0^R \frac{k^4 r^5}{2^2 4^2} \, dr + \int_0^R \frac{k^6 r^7}{2^2 4^2 6^2} \, dr + \cdots$$

$$= \frac{R^2}{2} + \frac{k^2 R^4}{2^2 4} + \frac{k^4 R^6}{2^2 4^2 6} + \frac{k^6 R^8}{2^2 4^2 6^2 8} + \cdots.$$

Now differentiate the series for $I_0(kr)$ term by term to get

$$I_0'(kR) = \frac{k^2 R}{2} + \frac{k^4 R^3}{2^2 4} + \frac{k^6 R^5}{2^2 4^2 6} + \cdots.$$

Now observe that

$$\frac{R}{k^2} I_0'(kR) = \frac{R^2}{2} + \frac{k^2 R^4}{2^2 4} + \frac{k^4 R^6}{2^2 4^2 6} + \cdots = \int_0^R r I_0(kr) \, dr.$$

Thus,

$$c_1 = \frac{Dk^2}{2\pi R I_0'(kR)}.$$

The solution for $z(r, t)$ is then

$$z(r, t) = \frac{Dk^2}{2\pi R I_0'(kR)} I_0(kr) e^{i\omega t},$$

and $x(r, t)$ is the real part of this (a nontrivial quantity to extract).

As an application of the above analysis, we shall derive the *skin effect*, which says that, for sufficiently high frequencies, the current flowing through the circular wire at radius r is small compared with the total current, even for r nearly equal to R. Roughly speaking, this says that "most" of the current in a cylindrical wire flows through a thin layer at the outer surface.

To derive this, begin with the above solution for $z(r, t)$. The total current through a coaxial cylinder of radius r is then

$$\int_0^r 2\pi r z(r, t)\, dr = \int_0^r \frac{Dk^2}{2\pi R I_0'(kR)} (2\pi r) I_0(kr) e^{i\omega t}\, dr$$
$$= \frac{Dk^2}{R I_0'(kR)} e^{i\omega t} \int_0^r r I_0(kr)\, dr.$$

By the same reasoning as employed previously in evaluating $\int_0^R r I_0(kr)\, dr$, we get

$$\int_0^r r I_0(kr)\, dr = \frac{r}{k^2} I_0'(kr).$$

Thus, the total current through a coaxial cylinder of radius r is

$$De^{i\omega t}\left(\frac{r I_0'(kr)}{R I_0'(kR)}\right).$$

Now, the total current in the wire is $De^{i\omega t}$. Thus, the ratio of the current in the coaxial cylinder of radius r to the total current is

$$\frac{r I_0'(kr)}{R I_0'(kR)}.$$

We are interested in how this behaves for large k. In Problem 26 at the end of this section we ask the student to show that, for large x, $I_0(x)$ is approximated by

$$\frac{Ae^x}{\sqrt{x}}\left(1 + \frac{1}{8x} + \frac{3^2}{2!(8x)^2} + \frac{3^2 5^2}{3!(8x)^3} + \cdots\right).$$

Here, A is a positive constant whose value does not matter for our purposes. Then, for large x,

$$I_0(x) \approx \frac{Ae^x}{\sqrt{x}},$$

where $\approx$ means "approximately equal to." Then, for large x,

$$I_0'(x) \approx \frac{Ae^x}{\sqrt{x}}\left(1 - \frac{1}{2x}\right) \approx \frac{Ae^x}{\sqrt{x}}.$$

Thus, if the frequency ω of the alternating current is large, then k is large in magnitude and

$$\frac{r I_0'(kr)}{R I_0'(kR)} \approx \frac{r}{R} \frac{e^{kr}}{\sqrt{r}} \frac{\sqrt{R}}{e^{kR}} = \sqrt{\frac{r}{R}}\, e^{-k(R-r)}.$$

Given any $r < R$, $e^{-k(R-r)}$ can be made arbitrarily small in magnitude by taking ω larger. Thus, for ω large, the ratio of the current in the coaxial cylinder of radius r to the total current is near zero, and this is exactly the skin effect. Note that r may be as close as we like to R, and we still obtain this result for sufficiently large frequency.

The next section is devoted to solutions of Bessel's equation in the case that the constant n is not an integer.

PROBLEMS FOR SECTION 6.1

1. Recall the series $\sum_{n=0}^{\infty} z^n/n! = e^z$. Show that, for $n = 0, 1, 2, 3, \ldots, J_n(x)$ is the coefficient of t^n in the series for

$$\exp\left[\frac{x}{2}\left(t - \frac{1}{t}\right)\right].$$

(*Hint:* Write

$$\exp\left[\frac{x}{2}\left(t - \frac{1}{t}\right)\right]$$

as $e^{xt/2}e^{-x/2t}$, expand both of these exponentials in series, and collect the terms which contain t^n in their product. Show that these terms add up exactly to the series for $J_n(x)$ given in the text, for $n = 0, 1, 2, 3, \ldots$.)

2. For $n = 1, 2, 3, \ldots$, define $J_{-n}(x)$ to be the coefficient of t^{-n} in the series expansion of

$$\exp\left[\frac{x}{2}\left(t - \frac{1}{t}\right)\right]$$

(see Problem 1). Prove that

$$J_{-n}(x) = (-1)^n J_n(x) \quad \text{for} \quad n = 1, 2, 3, \ldots.$$

Thus, show that, for n any positive integer, $J_n(x)$ and $J_{-n}(x)$ are linearly dependent.

3. Use the results of Problem 2 to show that, for any integer n,

$$(x^{-n}J_n)' = -x^{-n}J_{n+1}$$

and

$$(x^n J_n)' = x^n J_{n-1}.$$

4. Prove that, for any integer n,

$$xJ_n' = -nJ_n + xJ_{n-1}$$

and

$$xJ_n' = nJ_n - xJ_{n+1}.$$

(*Hint:* Use the results of Problem 3.)

5. Prove that, for any integer n,

$$\frac{2n}{x}J_n = J_{n-1} + J_{n+1}$$

and

$$2J_n' = J_{n-1} - J_{n+1}.$$

6. Use the result of Problem 5 to show that, for any integer n,

$$J_{n-1} = \frac{n}{x} J_n + J'_n$$

and

$$J_{n+1} = \frac{n}{x} J_n - J'_n.$$

7. Prove that

$$\int x^2 J_0(x)\, dx = x^2 J_1(x) + x J_0(x) - \int J_0(x)\, dx$$

and

$$\int x^3 J_0(x)\, dx = x(x^2 - 4) J_1(x) + 2x^2 J_0(x).$$

(*Hint:* Make use of identities derived in previous problems.)

8. Let $u = J_0(\alpha x)$, $v = J_0(\beta x)$.

(a) Show that

$$xu'' + u' + \alpha^2 xu = 0$$

and

$$xv'' + v' + \beta^2 xv = 0.$$

(b) By multiplying the first of these equations by v, the second by u, and subtracting, show that

$$[x(u'v - uv')]' = (\beta^2 - \alpha^2) xuv.$$

(c) Conclude from (b) that

$$(\beta^2 - \alpha^2) \int x J_0(\alpha x) J_0(\beta x)\, dx = x[\alpha J'_0(\alpha x) J_0(\beta x) - \beta J'_0(\beta x) J_0(\alpha x)].$$

This is one of *Lommel's integrals.*

9. Suppose that $J_0(\alpha) = 0$. Prove that

$$\int_0^1 J_1(\alpha x)\, dx = \frac{1}{\alpha}.$$

10. Derive the formula

$$J_0(x) = \frac{2}{\pi} \int_0^{\pi/2} \cos[x \sin(\theta)]\, d\theta.$$

(*Hint:* Begin with $e^{ix \sin(\theta)} = \sum_{n=0}^{\infty} [ix \sin(\theta)]^n / n!$. Integrate both sides from 0 to 2π. Use the formula

$$\int_0^{2\pi} \sin^n(\theta)\, d\theta = \begin{cases} 0 & \text{if } n \text{ is odd} \\ \dfrac{(n-1)(n-3)\cdots(3)(1)(2\pi)}{n(n-2)\cdots(4)(2)} & \text{if } n \text{ is even.} \end{cases}$$

Hence conclude that

$$J_0(x) = \frac{1}{2\pi} \int_0^{2\pi} e^{ix \sin(\theta)}\, d\theta.$$

Now separate $e^{ix \sin(\theta)}$ into real and imaginary parts to derive the requested formula.

In each of Problems 11 through 20, find a solution in terms of Bessel functions for the given equation.

11. $y'' - \frac{1}{x} y' + \left(1 - \frac{3}{x^2}\right) y = 0$

12. $y'' - \frac{3}{x} y' + \left(4 - \frac{5}{x^2}\right) y = 0$

13. $y'' - \frac{3}{x} y' + \left(\frac{1}{4x} + \frac{3}{x^2}\right) y = 0$

14. $y'' - \frac{5}{x} y' + \left(1 - \frac{7}{x^2}\right) y = 0$

15. $y'' - \frac{7}{x} y' + \left(1 + \frac{15}{x^2}\right) y = 0$

16. $y'' - \frac{1}{x} y' + \left(16x^2 - \frac{15}{x^2}\right) y = 0$

17. $y'' + \frac{3}{x} y' + \frac{1}{16x} y = 0$

18. $y'' - \frac{3}{x} y' + \left(1 + \frac{4}{x^2}\right) y = 0$

19. $y'' - \frac{3}{x} y' + \left(4x^2 - \frac{60}{x^2}\right) y = 0$

20. $y'' - \frac{7}{x} y' + \left(36x^4 - \frac{20}{x^2}\right) y = 0$

21. Prove that $(xI_0')' = xI_0$.

22. Using the result of Problem 21, show that

$$\int xI_0(\alpha x)\, dx = \frac{x}{\alpha} I_0'(\alpha x).$$

Use this to justify the conclusion in Example 172 that

$$\int_0^R 2\pi r I_0(kr)\, dr = \frac{2\pi R I_0'(kR)}{k}.$$

23. Prove that

$$(\beta^2 - \alpha^2) \int xI_0(\alpha x) I_0(\beta x)\, dx = x[\beta I_0'(\beta x) I_0(\alpha x) - \alpha I_0'(\alpha x) I_0(\beta x)].$$

(*Hint:* Use the method of Problem 8.)

24. Prove that any solution $y(x)$ of Bessel's equation

$$x^2 y'' + xy' + (x^2 - n^2)y = 0$$

can be written

$$y = Ax^{-1/2}[\cos(x + \alpha) + f(x)],$$

where A and α are constants and $f(x) \to 0$ as $x \to \infty$.

(*Hint:* Put $u = y\sqrt{x}$ into Bessel's equation and examine the resulting differential equation when x is large.)

25. In this problem we ask the student to prove a succession of results leading to conclusions about the roots of the equation $J_n(x) = 0$.

(a) For $k > 1$ and a sufficiently large, there is at least one x in $(a, a + \pi)$ such that $J_n(kx) = 0$. (*Hint:* Show that $u = \sqrt{kx}\, J_n(kx)$ satisfies

$$u'' = -\left(k^2 - \frac{n^2 - \frac{1}{4}}{x^2}\right) u.$$

Next, show that $v = \sin(x - a)$ satisfies

$$v'' = -v.$$

Show from these two differential equations that

$$(uv' - vu')' = \left(k^2 - 1 - \frac{n^2 - \frac{1}{4}}{x^2}\right)uv.$$

From this, show that

$$-u(a + \pi) - u(a) = \int_a^{a+\pi} \left(k^2 - 1 - \frac{n^2 - \frac{1}{4}}{x^2}\right)uv\,dx.$$

Use the mean value theorem for integrals to show that, for some α in $(a, a + \pi)$,

$$-u(a + \pi) - u(a) = u(\alpha) \int_a^{a+\pi} \left(k^2 - 1 - \frac{n^2 - \frac{1}{4}}{x^2}\right)v\,dx.$$

Show that, for a sufficiently large, this integral is positive; hence conclude that $u(a)$, $u(\alpha)$; and $u(a + \pi)$ cannot all have the same sign. From this, prove the result in (a).

(b) The equation $J_n(x) = 0$ has infinitely many positive roots. Use the result of (a).

(c) The equations $J_n(x) = 0$ and $J_n'(x) = 0$ cannot have a common solution if $x \neq 0$.

(d) The equation $J_n(x) = 0$ has no solution in common with solutions of $J_{n-1}(x) = 0$ or $J_{n+1}(x) = 0$ for $x \neq 0$. [*Hint:* Use the results of Problem 5 and part (c).]

(e) The equation $J_{n+1}(x) = 0$ has at least one solution between each pair of positive solutions of $J_n(x) = 0$. [*Hint:* Apply Rolle's theorem to $f(x) = J_n(x)/x^n$ on the interval $[a, b]$, where a and b are successive positive solutions of $J_n(x) = 0$. Remember Rolle's theorem: If f is continuous on $[a, b]$ and differentiable on (a, b), and $f(a) = f(b)$, then for some x_0 in (a, b), $f'(x_0) = 0$.]

(f) Between any pair of positive solutions of $J_n(x) = 0$ is a solution of $J_{n+1}(x) = 0$, and between each pair of positive solutions of $J_{n+1}(x) = 0$ is a solution of $J_n(x) = 0$. Thus, the zeros of J_n and J_{n+1} are said to *interlace*.

26. Show that, for large x, and some positive constant A,

$$I_0(x) = \frac{Ae^x}{x}\left(1 + \frac{1}{8x} + \frac{3^2}{2!\,(8x)^2} + \frac{3^2 5^2}{3!\,(8x)^3} + \cdots\right).$$

[*Hint:* Let $u(x) = I_0(x)/\sqrt{x}$, and show that

$$u'' = \left(1 - \frac{1}{4x^2}\right)u.$$

Note that, for x large, $u(x) \approx Ae^x$, for some constant A. Next, let $u(x) = v(x)e^x$, and show that

$$v'' + 2v' + \frac{v}{4x^2} = 0.$$

Attempt a solution of this differential equation of the form

$$v(x) = 1 + \frac{a_1}{x} + \frac{a_2}{x^2} + \frac{a_3}{x^3} + \cdots.$$

Substitute into the differential equation for v and solve for $a_1, a_2, a_3, \ldots$.]

6.2 NONINTEGER ORDER BESSEL FUNCTIONS

Thus far, we have seen various Bessel functions of integer order as solutions of $x^2y'' + xy' + (x^2 - n^2)y = 0$ or some transformed version thereof. In this section we consider *Bessel's differential equation of order* ν:

$$x^2y'' + xy' + (x^2 - \nu^2)y = 0,$$

where ν is any real number. This requires some preliminary work on the gamma function.

For any $x > 0$, $\Gamma(x)$ (read "gamma of x") is defined by

$$\Gamma(x) = \int_0^\infty t^{x-1}e^{-t}\,dt.$$

This function has many interesting properties, but the important one for us is that

$$\Gamma(x + 1) = x\Gamma(x) \quad \text{for} \quad x > 0.$$

To prove this, integrate by parts:

$$\Gamma(x + 1) = \int_0^\infty t^x e^{-t}\,dt = \lim_{R\to\infty} \int_0^R \underbrace{t^x}_{u}\, \underbrace{e^{-t}\,dt}_{dv}$$

$$= \lim_{R\to\infty} \left\{[t^x(-e^{-t})]_0^R - \int_0^R - e^{-t}xt^{x-1}\,dt\right\}$$

$$= x\int_0^\infty t^{x-1}e^{-t}\,dt = x\Gamma(x).$$

In particular, if n is a positive integer, then

$$\Gamma(n + 1) = n\Gamma(n) = n(n - 1)\Gamma(n - 1) = \cdots = n!\,\Gamma(1).$$

But,

$$\Gamma(1) = \int_0^\infty e^{-t}\,dt = 1.$$

Thus,

$$\Gamma(n + 1) = n!.$$

For this reason, $\Gamma(x)$ is often called the *factorial function.* Similarly, the property $\Gamma(x + 1) = x\Gamma(x)$ is called the *factorial property* of the gamma function, and this is the property we need.

If you apply the method of Frobenius to Bessel's equation of order ν, you get as one solution

$$J_\nu(x) = \sum_{j=0}^\infty \frac{(-1)^j}{j!\,\Gamma(\nu + j + 1)}\left(\frac{x}{2}\right)^{\nu+2j}.$$

This is a Bessel function of the first kind of order v. Note that, when $v = n = 0, 1, 2, \dots$, then $\Gamma(v + j + 1) = \Gamma(n + j + 1) = (n + j)!$, and $J_v(x)$ is exactly $J_n(x)$, which we defined in the previous section.

We can extend the gamma function to negative, noninteger real numbers by using the factorial property.

If $-1 < x < 0$, then $x + 1 > 0$, and we define

$$\Gamma(x) = \frac{1}{x}\,\Gamma(x + 1).$$

This done, we can define $\Gamma(x)$ for $-2 < x < -1$ by

$$\Gamma(x) = \frac{1}{x}\,\Gamma(x + 1),$$

and so on, working backwards from zero. Thus, for example,

$$\Gamma\left(-\frac{1}{2}\right) = \frac{1}{\frac{1}{2}}\,\Gamma\left(\frac{1}{2}\right) = 2\Gamma\left(\frac{1}{2}\right),$$

$$\Gamma\left(-\frac{3}{2}\right) = \frac{1}{\frac{3}{2}}\,\Gamma\left(-\frac{1}{2}\right) = \frac{4}{3}\,\Gamma\left(\frac{1}{2}\right),$$

and so on. Figure 75 shows $y = \Gamma(x)$, and Table 3 gives some values.

In this way, when $v > 0$ (but is not an integer), we can define

$$J_{-v}(x) = \sum_{j=0}^{\infty} \frac{(-1)^j}{j!\,\Gamma(-v + j + 1)} \left(\frac{x}{2}\right)^{-v+2j}.$$

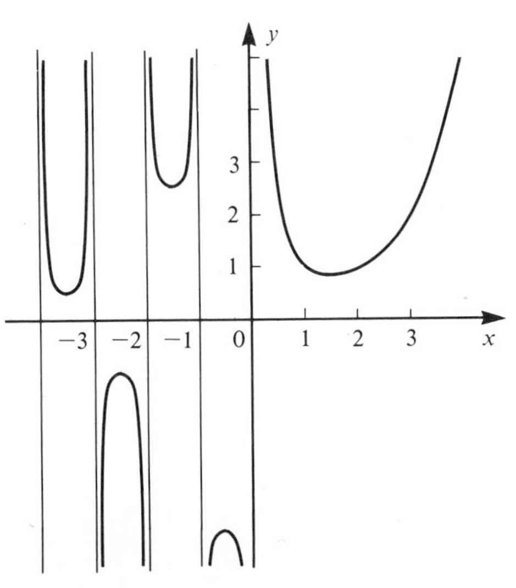

Figure 75. The gamma function $y = \Gamma(x)$.

Table 3: Some Values of $y = \Gamma(x)$

x	$\Gamma(x)$	x	$\Gamma(x)$
1.00	1.00000	1.55	0.88887
1.05	0.97350	1.60	0.89352
1.10	0.95135	1.65	0.90012
1.15	0.93304	1.70	0.90864
1.20	0.91817	1.75	0.91906
1.25	0.90640	1.80	0.93138
1.30	0.89747	1.85	0.94561
1.35	0.89115	1.90	0.96177
1.40	0.88726	1.95	0.97988
1.45	0.88565	2.00	1.00000
1.50	0.88623		

For other values, use the relationship $\Gamma(x + 1) = x\Gamma(x)$. Thus, for example,

$$\Gamma(2.75) = \Gamma(1.75 + 1) = 1.75\Gamma(1.75) = 1.60836.$$

It is easy to check that this satisfies Bessel's equation of order v (where v is not an integer), and is linearly independent from $J_v(x)$. Thus,

> The general solution of
>
> $$x^2y'' + xy' + (x^2 - v^2)y = 0$$
>
> for v not an integer is
>
> $$y = c_1 J_v(x) + c_2 J_{-v}(x).$$

In fact, as in Section 6.1 where v was restricted to be an integer, Bessel's equation of arbitrary order v often appears in different forms. Analogous to (a) of Section 6.1,

> For v not an integer,
>
> $$y = x^a(AJ_v(bx^c) + BJ_{-v}(bx^c)]$$
>
> is the general solution of
>
> $$y'' - \left(\frac{2a-1}{x}\right)y' + \left(b^2c^2x^{2c-2} + \frac{a^2 - v^2c^2}{x^2}\right)y = 0. \qquad \text{(b)}$$

We shall pursue a number of properties of $J_v(x)$ and $J_{-v}(x)$ in the problems at the end of this section. Here is an example showing the use of these functions.

EXAMPLE 173 Stability of a Clamped Circular Wire

Suppose we have an isotropic piece of wire of uniform, circular cross section, clamped at its lower end, and held in a vertical position with upper end free. Intuition suggests that the piece of wire will remain vertical (i.e., be stable) if it is "short enough," while the free end will bend over from the vertical and remain bent (i.e., be unstable) if the wire is "too long." In this example we investigate the longest length at which the wire remains stable. This may also be thought of as a transition length from stability to instability.

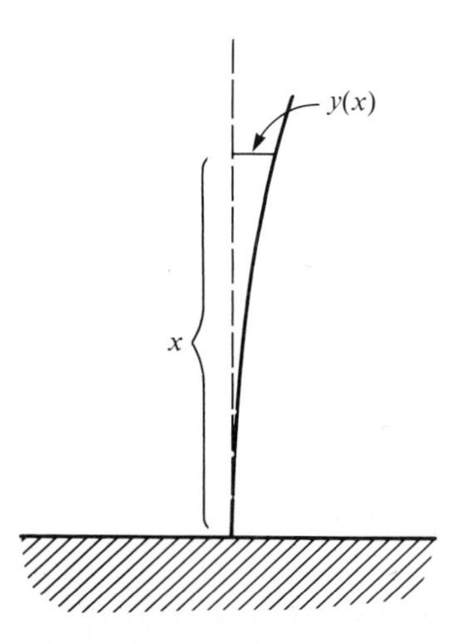

Figure 76. Critical length of a circular wire.

To determine this length, consider the possibility of an equilibrium position varying only slightly from the vertical, as shown in Figure 76. Let x be the distance measured from the clamped end, and let $y(x)$ be the horizontal displacement from the vertical at height x, as shown. Let L be the length of the wire, w its weight per unit length, and

F its flexural rigidity, assumed constant. (Here, $F = \frac{1}{4}\pi R^4 E$, where R is the radius of the wire and E is Young's modulus.) We want to derive an equation for y in terms of x.

Take moments of the portion of wire from x to L to get

$$F\frac{d^2y}{dx^2} = \int_x^L w[y(\bar{x}) - y(x)]\, d\bar{x},$$

in which $\bar{x}$ is a dummy variable of integration. Differentiate this with respect to x to get

$$F\frac{d^3y}{dx^3} = \{w[y(\bar{x}) - y(x)]\}_{x=\bar{x}} - \int_x^L wy'(x)\, d\bar{x} = -wy'(x)(L - x).$$

This gives us a third order differential equation for y:

$$\frac{d^3y}{dx^3} + \frac{w}{F}(L - x)\frac{dy}{dx} = 0.$$

To solve this for $y(x)$, first set $p = dy/dx$ to get

$$\frac{d^2p}{dx^2} + \frac{w}{F}(L - x)p = 0.$$

We shall now make a succession of changes of variables to get this into a recognizable form. First, set

$$z = (L - x)^{3/2}$$

or, equivalently,

$$x = L - z^{2/3}.$$

Substituting this for x in $p(x)$ gives us a function of z:

$$u(z) = p(L - z^{2/3}).$$

By the chain rule,

$$\frac{dp}{dx} = \frac{dp}{dz}\frac{dz}{dx} = -\frac{3}{2}(L - x)^{1/2}\frac{du}{dz} = -\frac{3}{2}z^{1/3}\frac{du}{dz}$$

and

$$\frac{d^2p}{dx^2} = \frac{d}{dz}\left(\frac{dp}{dx}\right)\frac{dz}{dx} = \frac{d}{dz}\left(-\frac{3}{2}z^{1/3}\frac{du}{dz}\right)\left(-\frac{3}{2}z^{1/3}\right)$$

$$= \left(-\frac{1}{2}z^{-2/3}\frac{du}{dz} - \frac{3}{2}z^{1/3}\frac{d^2u}{dz^2}\right)\left(-\frac{3}{2}z^{1/3}\right)$$

$$= \frac{9}{4}\left(\frac{1}{3}z^{-1/3}\frac{du}{dz} + z^{2/3}\frac{d^2u}{dz^2}\right).$$

Substitute these into the differential equation for $p(x)$ to obtain, after some rearrangement,

$$\frac{d^2u}{dz^2} + \frac{1}{3z}\frac{du}{dz} + \frac{4w}{9F}u = 0.$$

Now let $v(z)$ be defined by

$$v(z) = z^{-1/3}u(z).$$

By a calculation similar to that just done, the differential equation for $u(z)$ transforms to

$$\frac{d^2v}{dz^2} + \frac{1}{z}\frac{dv}{dz} + \left(\frac{4w}{9F} - \frac{1}{9z^2}\right)v = 0.$$

This is a Bessel equation with general solution

$$v(z) = c_1 J_{1/3}\left(\frac{2}{3}\sqrt{\frac{w}{F}}\,z\right) + c_2 J_{-1/3}\left(\frac{2}{3}\sqrt{\frac{w}{F}}\,z\right).$$

We leave it for the student to show that $c_1 = 0$. (*Hint:* Use the condition that $dp/dx = 0$ when $x = L$.) Thus,

$$v(z) = c_2 J_{-1/3}\left(\frac{2}{3}\sqrt{\frac{w}{F}}\,z\right).$$

Now note that, at the clamped end, $x = 0$; hence $z = L^{3/2}$. Here, $p = dy/dx = 0$; hence $u(z) = 0$, and so also $v(z) = 0$. Thus,

$$v(L^{3/2}) = c_2 J_{-1/3}\left(\frac{2}{3}\sqrt{\frac{w}{F}}\,L^{3/2}\right) = 0.$$

Assuming that $c_2 \neq 0$, we must have

$$J_{-1/3}\left(\frac{2}{3}\sqrt{\frac{w}{F}}\,L^{3/2}\right) = 0.$$

The smallest positive value of L satisfying this equation is the critical value we seek. Above this value, the wire is unstable; below, it is stable. From tables we find that the smallest positive root of $J_{-1/3}$ is approximately 1.866. Thus, the critical length is approximately given by

$$\frac{2}{3}\sqrt{\frac{w}{F}}\,L^{3/2} = 1.866$$

or

$$L = \left(2.799\sqrt{\frac{F}{w}}\right)^{2/3} = 1.986\left(\frac{F}{w}\right)^{1/3}.$$

There is a vast literature on Bessel functions, including tables for values of Bessel functions of various orders, as well as for the zeros of such functions. References are given under the Special Functions section of the bibliography at the end of the book. In the next section, we introduce another kind of special function called the Legendre polynomial.

PROBLEMS FOR SECTION 6.2

1. Show that $\Gamma(\frac{1}{2}) = \sqrt{\pi}$. Proceed as follows:

(i) Show that $\Gamma(\frac{1}{2}) = 2 \int_0^\infty e^{-x^2}\, dx$.

(ii) Note that $\Gamma(\frac{1}{2})^2 = 4 \int_0^\infty e^{-x^2}\, dx \int_0^\infty e^{-y^2}\, dy = 4 \int_0^\infty \int_0^\infty e^{-x^2-y^2}\, dx\, dy$.

(iii) Switch to polar coordinates and evaluate the resulting double integral by direct integration. *Note:* You will be integrating over the right quarter plane ($r \geq 0, 0 \leq \theta \leq \pi/2$ in polar coordinates). Also, don't forget the Jacobian in your transformation.

2. Prove that, for $n = 1, 2, 3, \ldots,$

$$\Gamma\left(n + \frac{1}{2}\right) = \frac{(2n)!\sqrt{\pi}}{4^n n!}.$$

You may assume that $\Gamma(\frac{1}{2}) = \sqrt{\pi}$.

3. Define the *beta function* by

$$B(x, y) = \int_0^1 t^{x-1}(1-t)^{y-1}\, dt \quad \text{for} \quad x > 0, \qquad y > 0.$$

(a) Prove that $B(x, y) = B(y, x)$.

(b) Prove that

$$B(x, y) = \int_0^{\pi/2} 2 \sin^{2x-1}(\theta)\cos^{2y-1}(\theta)\, d\theta.$$

(*Hint:* Let $t = \sin^2(\theta)$ in the definition.)

(c) Prove that

$$B(x, y) = \frac{\Gamma(x)\Gamma(y)}{\Gamma(x+y)}.$$

(*Hint:* Proceed in the four steps shown below.)

(i) $\Gamma(x) = 2 \int_0^\infty u^{2x-1} e^{-u^2}\, du$

[Put $t = u^2$ in the definition of $\Gamma(x)$.]

(ii) $$\Gamma(x)\Gamma(y) = 4 \int_0^\infty \int_0^\infty u^{2x-1} v^{2y-1} e^{-u^2-v^2}\, du\, dv$$

$$= 4 \lim_{R\to\infty} \iint_{\Delta_R} u^{2x-1} v^{2y-1} e^{-u^2-v^2}\, du\, dv,$$

where Δ_R is the region $x^2 + y^2 \leq R^2, x \geq 0, y \geq 0$.

(iii) Switch to polar coordinates to get

$$\iint_{\Delta_R} u^{2x-1}v^{2y-1}e^{-u^2-v^2}\,du\,dv$$

$$= \int_0^R r^{2x+2y-1}e^{-r^2}\,dr \cdot \int_0^{\pi/2} \cos^{2x-1}(\theta)\sin^{2y-1}(\theta)\,d\theta.$$

(iv) Let $R \to \infty$ to show that $\Gamma(x)\Gamma(y) = \Gamma(x+y)B(x, y)$.

Use the gamma function and the results of Problem 3(b) and (c) to evaluate the following integrals in terms of gamma functions.

4. $\int_0^{\pi/2} \sqrt{\tan(x)}\,dx$ **5.** $\int_0^{\pi/2} \sin^{1/3}(x)\cos^{1/2}(x)\,dx$ **6.** $\int_0^{\pi/2} \sqrt{\sin(x)\cos(x)}\,dx$ **7.** $\int_0^{\infty} \frac{t^{1/2}\,dt}{(1+t)^{5/2}}$

[*Hint:* Put $t = u/(1+u)$ in the definition of the beta function.]

8. Define, for a any real number,

$$(a)_n = \frac{\Gamma(a+n)}{\Gamma(a)}.$$

Thus, $(a)_n = a(a+1)\cdots(a+n-1)$. The series

$$F(a, b, c, x) = \sum_{n=0}^{\infty} \frac{(a)_n(b)_n}{(c)_n\,n!}x^n$$

is called the *hypergeometric series.* It converges for $|x| < 1$ if c is not a negative integer.

(a) Prove that $F(a, b, c, x)$ is a solution of the *hypergeometric differential equation*

$$x(1-x)y'' + [c - (a+b+1)x]y' - aby = 0.$$

(b) Prove that $F(a, b, c, x/b)$ is a solution to the *confluent hypergeometric equation*

$$x\left(1 - \frac{x}{b}\right)y'' + \left[c - \left(1 + \frac{a+1}{b}\right)x\right]y' - ay = 0.$$

9. Show that

$$J_{1/2}(x) = \sqrt{\frac{2}{\pi x}}\sin(x)$$

and

$$J_{-1/2}(x) = \sqrt{\frac{2}{\pi x}}\cos(x).$$

10. Show that

$$J_{3/2}(x) = \sqrt{\frac{2}{\pi x}}\left(\frac{\sin(x)}{x} - \cos(x)\right)$$

and

$$J_{-3/2}(x) = -\sqrt{\frac{2}{\pi x}}\left(\frac{\cos(x)}{x} + \sin(x)\right).$$

11. Show that

$$J_{5/2}(x) = \sqrt{\frac{2}{\pi x}}\left[\left(\frac{3}{x^2} - 1\right)\sin(x) - \frac{3}{x}\cos(x)\right]$$

and

$$J_{-5/2}(x) = \sqrt{\frac{2}{\pi x}}\left[\left(\frac{3}{x^2} - 1\right)\cos(x) + \frac{3}{x}\sin(x)\right].$$

12. Show that

$$x^2 J_\nu'' = (\nu^2 - \nu - x^2)J_n + xJ_{\nu+1}.$$

13. Use the change of variables

$$by = \frac{1}{u}\frac{du}{dx}$$

to transform

$$\frac{dy}{dx} + by^2 = cx^m$$

into

$$\frac{d^2u}{dx^2} - bcx^m u = 0.$$

Thus solve for u in terms of Bessel functions, and from this solve for y.

In each of Problems 14 through 23, use equation (b) of this section to write the general solution of the differential equation.

14. $y'' + \frac{1}{3x}y' + \left(1 + \frac{7}{144x^2}\right)y = 0$

15. $y'' + \frac{1}{x}y' + \left(4x^2 - \frac{4}{9x^2}\right)y = 0$

16. $y'' - \frac{5}{x}y' + \left(64x^6 + \frac{5}{x^2}\right)y = 0$

17. $y'' + \frac{3}{x}y' + \left(16x^2 - \frac{5}{4x^2}\right)y = 0$

18. $y'' - \frac{3}{x}y' + 9x^4 y = 0$

19. $y'' - \frac{7}{x}y' + \left(36x^4 + \frac{175}{16x^2}\right)y = 0$

20. $y'' + \frac{1}{x}y' - \frac{1}{16x^2}y = 0$

21. $y'' + \frac{5}{x}y' + \left(81x^4 + \frac{7}{4x^2}\right)y = 0$

22. $y'' - \frac{5}{x}y' + \left(32x^2 + \frac{8}{x^2}\right)y = 0$

23. $y'' + \frac{1}{x}y' + \left(36x^4 - \frac{81}{25x^2}\right)y = 0$

6.3 LEGENDRE POLYNOMIALS

In Example 163 (Chapter 5) we saw Legendre's differential equation

$$(1 - x^2)y'' - 2xy' + \lambda(\lambda + 1)y = 0.$$

There we obtained linearly independent solutions

$$y_1 = a_0\left[1 - \frac{\lambda(\lambda+1)}{2}x^2 + \left(\frac{6-\lambda(\lambda+1)}{12}\right)\left(\frac{-\lambda(\lambda+1)}{2}\right)x^4 + \left(\frac{20-\lambda(\lambda+1)}{30}\right)\left(\frac{6-\lambda(\lambda+1)}{12}\right)\left(\frac{-\lambda(\lambda+1)}{2}\right)x^6 + \cdots\right]$$

and

$$y_2 = a_1\left[x + \left(\frac{2-\lambda(\lambda+1)}{6}\right)x^3 + \left(\frac{12-\lambda(\lambda+1)}{20}\right)\left(\frac{2-\lambda(\lambda+1)}{6}\right)x^5 + \left(\frac{30-\lambda(\lambda+1)}{42}\right)\left(\frac{12-\lambda(\lambda+1)}{20}\right)\left(\frac{2-\lambda(\lambda+1)}{6}\right)x^7 + \cdots\right].$$

We also remarked that we obtain polynomial or finite series solutions when $\lambda = n$, a nonnegative integer. This section is devoted to such polynomial solutions.

Suppose, then, that $\lambda = n\,(=0, 1, 2, 3, \ldots)$. It is easy to see from y_1 and y_2 that, if n is even, then y_1 is a polynomial of degree n containing only even powers of x; and if n is odd, then y_2 is a polynomial of degree n containing only odd powers of x. For example:

n	$y_1(x)$	$y_2(x)$
0	a_0	
1		a_1x
2	$a_0(1-3x^2)$	
3		$a_1(x-\frac{5}{3}x^3)$
4	$a_0(1-10x^2+\frac{35}{3}x^4)$	
5		$a_1(x-\frac{14}{3}x^3+\frac{21}{5}x^5)$
$\vdots$		$\vdots$

Here, a_0 and a_1 are arbitrary constants. In order to standardize things, it is convenient for later purposes to choose a_0 or a_1 for each n so that each solution has value 1 at $x = 1$. Thus, when $n = 0$, choose $a_0 = 1$; when $n = 1$, let $a_1 = 1$; when $n = 2$, let $a_0 = -\frac{1}{2}$; when $n = 3$, let $a_1 = -\frac{3}{2}$, and so on. The solutions are then called *Legendre polynomials*. The first few are

$$P_0(x) = 1, \quad P_1(x) = x, \quad P_2(x) = \tfrac{1}{2}(3x^2-1), \quad P_3(x) = \tfrac{1}{2}(5x^3-3x),$$
$$P_4(x) = \tfrac{1}{8}(35x^4-30x^2+3), \quad P_5(x) = \tfrac{1}{8}(63x^5-70x^3+15x).$$

Figure 77 shows the graphs of $y = P_0(x)$ through $P_4(x)$.

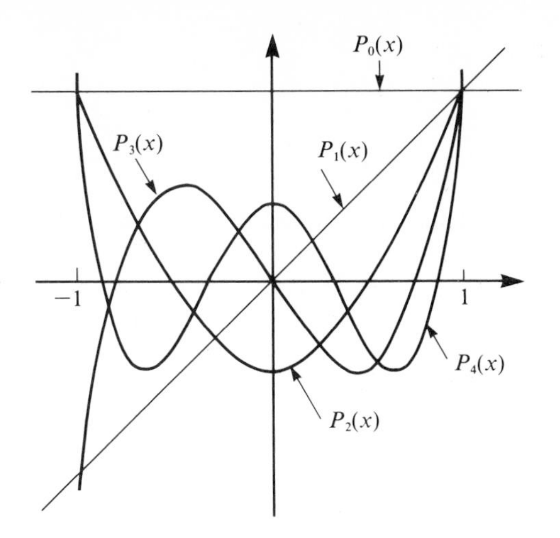

Figure 77. Legendre polynomials.

It is possible to write a summation formula for $P_n(x)$ using the special symbol $\{n\}$, where, for $n = 0, 1, 2, \ldots,$

$$\{n\} = \begin{cases} \dfrac{n}{2} & \text{if} \quad n = 0, 2, 4, 6, 8, \ldots \\[2ex] \dfrac{n-1}{2} & \text{if} \quad n = 1, 3, 5, 7, 9, \ldots. \end{cases}$$

Thus, for example, $\{4\} = 2$, while $\{7\} = 3$. In terms of this, one can show that

$$P_n(x) = \frac{1}{2^n} \sum_{j=0}^{\{n\}} \frac{(-1)^j (2n-2j)!}{j!\,(n-2j)!\,(n-j)!}\, x^{n-2j}.$$

The reader should write out the first few polynomials given by this formula and feel at ease that they are indeed the Legendre polynomials.

Here are some more important properties of Legendre polynomials.

THEOREM 18

$$P_n(x) = \frac{1}{2^n n!} \frac{d^n}{dx^n} [(x^2 - 1)^n] \quad \text{for} \quad n = 0, 1, 2, \ldots.$$

(This is called *Rodrigues's formula.*)

Proof Note that

$$\frac{d^n}{dx^n} x^{2n-2j} = \frac{(2n-2j)!}{(n-2j)!} x^{n-2j}, \qquad j = 0, 1, \ldots, \{n\}.$$

Then,

$$P_n(x) = \frac{1}{2^n n!} \sum_{j=0}^{\{n\}} \frac{(-1)^j}{j!} \frac{n!}{(n-j)!} \frac{(2n-2j)!}{(n-2j)!} x^{n-2j}$$

$$= \frac{1}{2^n n!} \sum_{j=0}^{\{n\}} \frac{(-1)^j}{j!} \frac{n!}{(n-j)!} \frac{d^n}{dx^n} x^{2n-2j}$$

$$= \frac{1}{2^n n!} \frac{d^n}{dx^n} \sum_{j=0}^{\{n\}} \frac{(-1)^j}{j!} \frac{n!}{(n-j)!} x^{2n-2j}$$

$$= \frac{1}{2^n n!} \frac{d^n}{dx^n} \sum_{j=0}^{n} \frac{(-1)^j}{j!} \frac{n!}{(n-j)!} x^{2n-2j}.$$

In the last summation, we extended the range of j. Instead of going from 0 to $\{n\}$, it goes from 0 to n. The additional terms do not matter, since they all involve powers of x less than n, and the nth derivative of each such term is zero. Thus we have added $n - \{n\}$ zeros. Now recall the binomial expansion

$$(x^2 - 1)^n = \sum_{j=0}^{n} \frac{n!}{j!(n-j)!} (-1)^j (x^2)^{n-j}.$$

Then

$$P_n(x) = \frac{1}{2^n n!} \frac{d^n}{dx^n} [(x^2 - 1)^n],$$

as we wanted to show.

THEOREM 19 Recurrence Relations

$$(n + 1)P_{n+1}(x) + nP_{n-1}(x) = (2n + 1)xP_n(x)$$

and

$$P'_{n+1}(x) - P'_{n-1}(x) = (2n + 1)P_n(x)$$

for $n = 1, 2, 3, \ldots$.

A proof of this is sketched in Problem 2 at the end of this section, and the student is asked to fill in the details.

Legendre polynomials and Bessel functions can be placed in a general setting called *Sturm-Liouville theory*, which we discuss in the next section. This theory gives us additional facts about these special functions which are useful later in solving important partial differential equations.

We conclude this section with an application of Legendre polynomials.

EXAMPLE 174 Legendre Polynomials in Potential Theory

The gravitational potential at a point (x, y, z) due to a unit mass placed at the fixed point (x_0, y_0, z_0) is

$$\phi(x, y, z) = \frac{1}{\sqrt{(x - x_0)^2 + (y - y_0)^2 + (z - z_0)^2}}.$$

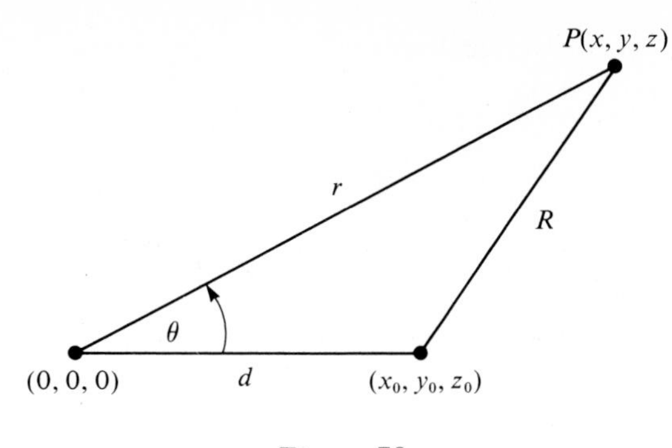

Figure 78

For some problems it is convenient to expand ϕ in powers of r or $1/r$, where $r = \sqrt{x^2 + y^2 + z^2}$ is the distance from P to the origin of the coordinate system. To do this, introduce an angle θ as shown in Figure 78.

Let

$$d = \sqrt{x_0^2 + y_0^2 + z_0^2}$$

and

$$R = \sqrt{(x - x_0)^2 + (y - y_0)^2 + (z - z_0)^2}.$$

By the law of cosines,

$$R^2 = d^2 + r^2 - 2dr\cos(\theta).$$

Then,

$$\phi = \frac{1}{\sqrt{d^2 - 2dr\cos(\theta) + r^2}}.$$

Write this as

$$\phi = \frac{1}{d\sqrt{1 - 2\dfrac{r}{d}\cos(\theta) + \left(\dfrac{r}{d}\right)^2}}.$$

Now, in general, we can (by the binomial theorem) expand as follows:

$$\begin{aligned}\frac{1}{\sqrt{1 - 2at + t^2}} &= 1 + (-\tfrac{1}{2})(-2at + t^2) + \frac{(-\frac{1}{2})(-\frac{3}{2})(-2at + t^2)^2}{2} \\ &\quad + \frac{(-\frac{1}{2})(-\frac{3}{2})(-\frac{5}{2})(-2at + t^2)^3}{6} + \cdots \\ &= 1 + at + \tfrac{1}{2}(3a^2 - 1)t^2 + \tfrac{1}{2}(5a^3 - 3a)t^3 \\ &\quad + \tfrac{1}{8}(35a^4 - 30a^2 + 3)t^4 + \cdots\end{aligned}$$

for $|t| < 1$. We recognize the coefficient of t^n in this expansion as $P_n(a)$. Letting $a = \cos(\theta)$ and $t = r/d$, we have

$$\phi = \frac{1}{d}\frac{1}{\sqrt{1 - 2\dfrac{r}{d}\cos(\theta) + \left(\dfrac{r}{d}\right)^2}} = \frac{1}{d}\sum_{n=0}^{\infty} P_n[\cos(\theta)]\left(\frac{r}{d}\right)^n,$$

where $r/d < 1$ (or $r < d$). Thus, in powers of r, valid when $r < d$,

$$\phi = \sum_{n=0}^{\infty} \frac{1}{d^{n+1}} P_n[\cos(\theta)]r^n.$$

If $r/d > 1$, so that $r > d$, we write

$$\phi = \frac{1}{r\sqrt{\left(\dfrac{d}{r}\right)^2 - 2\dfrac{d}{r}\cos(\theta) + 1}} = \frac{1}{r}\sum_{n=0}^{\infty} P_n[\cos(\theta)]\left(\frac{d}{r}\right)^n,$$

expanding ϕ in powers of $1/r$.

PROBLEMS FOR SECTION 6.3

In Problems 1 and 2, you may assume Leibniz's rule for differentiating a product n times:

$$\frac{d^n}{dx^n}[f(x)g(x)] = \sum_{j=0}^{n} \frac{n!}{j!(n-j)!}\frac{d^j}{dx^j}f(x)\frac{d^{n-j}}{dx^{n-j}}g(x).$$

1. Put $(x^2 - 1)^n = (x - 1)^n(x + 1)^n$ into Rodrigues's formula for $P_n(x)$ to show that

$$P_n(x) = \frac{1}{2^n n!}\sum_{j=0}^{n} \frac{n!}{j!(n-j)!}\frac{d^j}{dx^j}[(x+1)^n]\frac{d^{n-j}}{dx^{n-j}}[(x-1)^n].$$

2. Derive the recurrence relation:

$$(n + 1)P_{n+1} + nP_{n-1} = (2n + 1)xP_n.$$

Hint: Write

$$2^{n+1}(n+1)!\,P_{n+1} = \frac{d^{n+1}}{dx^{n+1}}[(x^2 - 1)^{n+1}] = \frac{d^{n-1}}{dx^{n-1}}\left(\frac{d^2}{dx^2}[(x^2 - 1)^{n+1}]\right).$$

By calculating $(d^2/dx^2)[(x^2 - 1)^{n+1}]$ and substituting into the last equation, show that

$$2^n n!\,P_{n+1} = (2n + 1)\frac{d^{n-1}}{dx^{n-1}}[(x^2 - 1)^n] + 2n\frac{d^{n-1}}{dx^{n-1}}[(x^2 - 1)^{n-1}].$$

Use this and Rodrigues's formula to get

$$P_{n+1} - P_{n-1} = \frac{2n + 1}{2^n n!}\frac{d^{n-1}}{dx^{n-1}}[(x^2 - 1)^n]. \tag{1}$$

Use Leibniz's rule to get

$$P_{n+1} = \frac{1}{2^n n!}\left\{x\frac{d^n}{dx^n}[(x^2 - 1)^n] + n\frac{d^{n-1}}{dx^{n-1}}[(x^2 - 1)^n]\right\}.$$

Hence, conclude that

$$P_{n+1} - xP_n = \frac{n}{2^n n!} \frac{d^{n-1}}{dx^{n-1}} [(x^2 - 1)^n]. \tag{2}$$

Use equations (1) and (2) to derive the desired result.

3. Let $H(x, r) = 1/(1 - 2xr - r^2)^{1/2}$. If expanded in a power series about 0, considered as a function of r, the coefficients will in general be functions of x. That is,

$$H(x, r) = F_0(x) + F_1(x)r + F_2(x)r^2 + F_3(x)r^3 + \cdots.$$

Set $z = 2xr - r^2$ and use the binomial series for $(1 - z)^{-1/2}$ to compute $F_0(x)$ through $F_5(x)$ explicitly. Show that, in fact, $F_j(x) = P_j(x)$ for $j = 0, 1, 2, 3, 4, 5$. (It is true, but we have not asked for a proof, that $H(x, r) = \sum_{n=0}^{\infty} P_n(x)r^n$. That is, $P_n(x)$ is the coefficient of r^n for all nonnegative integers n. For this reason, $H(x, r)$ is called a *generating function* for the Legendre polynomials.)

4. Use the generating function of Problem 3 to give a different derivation of the relation $(n + 1)P_{n+1} + nP_{n-1} = (2n + 1)xP_n$ than that requested in Problem 2. Proceed as follows: First, show by direct calculation that

$$(1 - 2xr + r^2)\frac{\partial H}{\partial r} - (x - r)H = 0.$$

Substitute $H = \sum_{n=0}^{\infty} P_n(x)r^n$ and $\partial H/\partial r = \sum_{n=1}^{\infty} P_n(x)nr^{n-1}$. Collect terms and set the coefficient of r^n in the entire expression equal to zero.

5. Show that

$$\int_{-1}^{1} P_n(x)P_m(x)\, dx = 0 \quad \text{if} \quad n \neq m.$$

(*Hint:* Begin with

$$(1 - x^2)P_n'' - 2xP_n' + n(n + 1)P_n = 0$$

and

$$(1 - x^2)P_m'' - 2xP_m' + m(m + 1)P_m = 0.$$

Multiply the first by P_m, the second by P_n, and subtract. Finally, observe that $(P_n' P_m - P_m' P_n)' = P_n'' P_m - P_n'' P_m$, and integrate the resulting equation between -1 and 1.)

6. Derive the identity

$$nP_{n-1} - P_n' + xP_{n-1}' = 0 \quad \text{for} \quad n = 1, 2, 3, \ldots.$$

(*Hint:* Referring to Problem 3, first show by direct calculation that

$$r\frac{\partial}{\partial r}(rH) - (1 - rx)\frac{\partial H}{\partial x} = 0.$$

Now look at the coefficient of r^n when $H(x, r) = \sum_{n=0}^{\infty} P_n(x)r^n$ is substituted into this equation.)

7. Use the recurrence relation of Problem 2 to show that

$$\int_{-1}^{1} P_n^2(x)\, dx = \frac{2}{2n + 1}, \qquad n = 0, 1, 2, 3, \ldots.$$

8. Prove that

$$P_{2n+1}(0) = 0 \quad \text{and} \quad P_{2n}(0) = (-1)^n \frac{(2n)!}{2^{2n}(n!)^2}.$$

9. Graph $P_5(x)$, $P_6(x)$, and $P_7(x)$. From these and Figure 77, make a conjecture about the roots of the nth Legendre polynomial.

10. Prove that

$$\int_{-1}^{1} x^m P_n(x)\,dx = 0 \quad \text{if} \quad m = 0, 1, \ldots, n-1.$$

11. Prove that

$$P'_{n+1} - P'_{n-1} = (2n+1)P_n.$$

(*Hint:* Use Rodrigues's formula.)

12. Prove that

$$(1 - x^2)P'_n = nP_{n-1} - nxP_n.$$

13. Prove that

$$\int_{-1}^{1} xP_n(x)P_{n-1}(x)\,dx = \frac{2n}{4n^2 - 1} \quad \text{for} \quad n = 1, 2, 3, \ldots.$$

Problems 14 through 23 do not deal with Legendre polynomials, but with other polynomials often treated as special functions. In each problem you may use the results from the preceding problems.

Hermite Polynomials Hermite's differential equation is

$$y'' - xy' + \lambda y = 0,$$

and occurs, for example, in treating the wave mechanics of the harmonic oscillator.

14. Using the method of Frobenius, obtain series solutions of Hermite's equation. Show that, when $\lambda = n = 0, 1, 2, 3, \ldots$, there is a solution which is a polynomial of degree n having these properties:

(i) If n is even, the polynomial contains only even powers of x; if n is odd, it contains only odd powers.

(ii) The coefficient of x^n is 1.

These are called *Hermite polynomials*. The Hermite polynomial of degree n is called $H_n(x)$. Write out $H_0(x)$ through $H_5(x)$.

15. Show that

$$H_n(x) = (-1)^n e^{x^2/2} \frac{d^n}{dx^n} e^{-x^2/2} \quad \text{for} \quad n = 0, 1, \ldots, 5.$$

[Remember in Problem 14 you obtained $H_0(x), \ldots, H_5(x)$.]

16. Using the expression for $H_n(x)$ proved for $n = 0, 1, \ldots, 5$ in Problem 15, show that

$$\int_{-\infty}^{\infty} e^{-x^2/2} H_n{}^2(x)\,dx = \begin{cases} 0 & \text{if} \quad n \neq m \\ \sqrt{2\pi}\, n! & \text{if} \quad n = m. \end{cases}$$

(*Hint:* Use integration by parts.)

17. Show that $H_n(x)$ is the coefficient of r^n in the series for $e^{xr - r^2/2}$.

18. Prove that

$$H_{n+1} - xH_n + nH_{n-1} = 0 \quad \text{for} \quad n = 1, 2, 3, \ldots.$$

Laguerre Polynomials Laguerre's differential equation is

$$xy'' + (1 - x)y' + \lambda y = 0.$$

19. Using the method of Frobenius, show that for $\lambda = 0, 1, 2, \ldots$, there is a polynomial solution of degree n, with coefficient of x^n equal to 1. This is the *Laguerre polynomial* $L_n(x)$. Find $L_0(x), \ldots, L_5(x)$.

20. Show that

$$L_n(x) = (-1)^n e^x \frac{d^n}{dx^n}(x^n e^{-x}) \quad \text{for} \quad n = 0, 1, \ldots, 5.$$

21. Show that

$$\int_0^\infty e^{-x} L_n(x) L_m(x)\, dx = \begin{cases} 0, & n \neq m \\ (n!)^2, & n = m, \end{cases}$$

assuming that

$$L_n(x) = (-1)^n e^x \frac{d^n}{dx^n}(x^n e^{-x}) \quad \text{for} \quad n = 0, 1, 2, \ldots.$$

22. Prove that

$$L_{n+1} + (2n + 1 - x)L_n + n^2 L_{n-1} = 0, \qquad n = 1, 2, 3, \ldots.$$

23. Prove that $L_n(x)$ is the coefficient of r^n in the series for

$$\frac{1}{1-r} e^{-rx/(1-r)}.$$

6.4 STURM-LIOUVILLE THEORY AND EIGENFUNCTION EXPANSIONS

In this and the next section, we shall discuss two ways in which information about solutions of a differential equation can be gleaned from the differential equation itself. This section is devoted to Sturm-Liouville problems, and the next to oscillation theory, which has to do with the frequency with which solutions cross the x-axis.

Often we are interested in a differential equation of the form

$$y'' + R(x)y' + [Q(x) + \lambda P(x)]y = 0, \tag{a}$$

in which we seek to determine values of the constant λ for which there exist nontrivial solutions $y(x)$ on a given interval $[a, b]$, usually satisfying prescribed conditions at (a) and (b). We shall discuss these conditions after putting the differential equation into a form more suitable for the analysis about to be carried out.

Multiply (a) by

$$e^{\int R(x)\,dx}$$

to get

$$y'' e^{\int R(x)\,dx} + R(x)e^{\int R(x)\,dx} y' + [Q(x) + \lambda P(x)]e^{\int R(x)\,dx} y = 0.$$

Since $e^{\int R(x)\,dx}$ is strictly positive, this new differential equation has the same solutions as the original one. Now recognize that

$$y'' e^{\int R(x)\,dx} + R(x)e^{\int R(x)\,dx} y = (e^{\int R(x)\,dx} y)'.$$

Now let

$$r(x) = e^{\int R(x)\,dx}, \qquad q(x) = Q(x)e^{\int R(x)\,dx}, \quad \text{and} \quad p(x) = P(x)e^{\int R(x)\,dx}.$$

Then (a) becomes

$$(ry')' + (q + \lambda p)y = 0. \tag{b}$$

This is called the *Sturm-Liouville differential equation*, and is sometimes referred to as the *Sturm-Liouville form of* (a).

EXAMPLE 175

Consider

$$y'' + \frac{2}{x}\,y' + (x^2 - \lambda x)y = 0.$$

This is (a) with $R = 2/x$, $Q = x^2$; and $P = -x$. Multiply by $e^{\int (2/x)\,dx} = x^2$ to get

$$(x^2 y')' + (x^4 - \lambda x^3)y = 0,$$

a Sturm-Liouville differential equation having the same solutions as the original differential equation.

We now center our attention upon the Sturm-Liouville differential equation, assuming that $r(x)$, $r'(x)$, $q(x)$, and $p(x)$ are continuous on an interval $[a, b]$. We distinguish three kinds of problems.

I. The Regular Sturm-Liouville Problem for $[a, b]$ Assuming that $r(x) > 0$ and $p(x) > 0$ for $a \le x \le b$, find numbers λ and nontrivial functions $y(x)$ satisfying

$$(ry')' + (q + \lambda p)y = 0, \qquad a \le x \le b,$$

and the boundary conditions

$$A_1 y(a) + A_2\, y'(a) = 0, \qquad B_1 y(b) + B_2\, y'(b) = 0.$$

Here, A_1 and A_2 are given, and not both zero, and B_1 and B_2 are given, and not both zero.

II. The Periodic Sturm-Liouville Problem for $[a, b]$ Assuming that $r(x) > 0$ and $p(x) > 0$ for $a \le x \le b$, and $r(a) = r(b)$, find numbers λ and nontrivial functions $y(x)$ satisfying

$$(ry')' + (q + \lambda p)y = 0$$

and the boundary conditions

$$y(a) = y(b), \qquad y'(a) = y'(b).$$

III. The Singular Sturm-Liouville Problem for $[a, b]$ Assuming that $r(x) > 0$ and $p(x) > 0$ for $a < x < b$, find numbers λ and nontrivial functions $y(x)$ satisfying

$$(ry')' + (q + \lambda p)y = 0,$$

and satisfying *one* of the following three sets of boundary conditions:

1. If $r(a) = 0$, then we require that

$$B_1 y(b) + B_2 y'(b) = 0,$$

with B_1 and B_2 given and not both zero.

2. If $r(b) = 0$, we require that

$$A_1 y(a) + A_2 y'(a) = 0,$$

with A_1 and A_2 given and not both zero.

3. If $r(a) = r(b) = 0$, we have no boundary conditions, but require that solutions be bounded on $[a, b]$.

In each of these problems, a number λ for which a nontrivial solution exists is called an *eigenvalue*; such a nontrivial solution is then called an *eigenfunction associated with the eigenvalue* λ. Note that by definition an eigenfunction cannot be the zero function.

Here are some examples of these types of problems and their solution.

EXAMPLE 176 A Regular Sturm-Liouville Problem

Consider the problem

$$y'' + \lambda y = 0; \qquad y(0) = y\left(\frac{\pi}{2}\right) = 0.$$

Here the interval is $[0, \pi/2]$, and we have $r = p = 1$, $q = 0$. Note that $r(x)$ and $p(x)$ are positive on $[0, \pi/2]$, and the boundary conditions are of the form defining a regular Sturm-Liouville problem, with $A_1 = B_1 = 1$, and $A_2 = B_2 = 0$.

In this problem we can explicitly find many eigenvalues and eigenfunctions. Consider cases on λ.

Case 1 If $\lambda = 0$, then $y'' = 0$; so $y = cx + d$. Since $y(0) = 0$, then $d = 0$. Since $y(\pi/2) = 0$, then $c = 0$, and we have only the trivial solution. Thus, $\lambda = 0$ is not an eigenvalue of this problem.

Case 2 If $\lambda > 0$, write $\lambda = \alpha^2$ for $\alpha > 0$, and solve $y'' + \alpha^2 y = 0$ to get

$$y = c\cos(\alpha x) + d\sin(\alpha x).$$

Now, $y(0) = c = 0$; so we are left with $y = d\sin(\alpha x)$. Next,

$$y\left(\frac{\pi}{2}\right) = d\sin\left(\alpha\frac{\pi}{2}\right) = 0.$$

To avoid a trivial solution, we must have $d \neq 0$. Then we need $\sin(\alpha\pi/2) = 0$. This will hold when $\alpha\pi/2$ is an integer multiple of π, say $\alpha\pi/2 = n\pi$ for $n = 1, 2, 3, \ldots$ (remember that $\alpha > 0$). Then, $\alpha = 2n$. Thus, for $n = 1, 2, 3, 4, \ldots$,

$$\lambda = 4n^2$$

is an eigenvalue, with eigenfunctions

$$y = d\sin(2nx).$$

Here, d can be any nonzero real number.

Case 3 If $\lambda < 0$, write $\lambda = -\alpha^2$, with $\alpha > 0$. Then $y'' - \alpha^2 y = 0$, with general solution

$$y = ce^{\alpha x} + de^{-\alpha x}.$$

Now, $y(0) = c + d = 0$; so $c = -d$. Then, $y(\pi/2) = 0 = c(e^{\alpha\pi/2} - e^{-\alpha\pi/2})$. If $\alpha > 0$, then $e^{\alpha\pi/2} - e^{-\alpha\pi/2} = 2\sinh(\alpha\pi/2) > 0$, so $y(\pi/2) = 0$ forces $c = d = 0$, and we have only the trivial solution. Thus, no number $\lambda < 0$ is an eigenvalue of this problem.

In summary,

$$y'' + \lambda y = 0; \qquad y(0) = y\left(\frac{\pi}{2}\right) = 0$$

has eigenvalues

$$\lambda = 4n^2 \quad \text{for} \quad n = 1, 2, 3, 4, \ldots .$$

Each eigenvalue $4n^2$ has associated with it the eigenfunction $\sin(2nx)$, or any nonzero constant multiple thereof. We will show later that there are no complex eigenvalues for this or any other Sturm-Liouville problem as we have defined them.

EXAMPLE 177 A Periodic Sturm-Liouville Problem

Consider

$$y'' + \lambda y = 0; \qquad y(0) = y(\pi); \qquad y'(0) = y'(\pi).$$

This is a periodic Sturm-Liouville problem defined on the interval $[0, \pi]$. Consider cases on λ.

Case 1 If $\lambda = 0$, then $y = cx + d$. Now, $y(0) = d = y(\pi) = c\pi + d$, and $y'(0) = c = y'(\pi)$. But $d = c\pi + d$ means that $c = 0$. Since there are no limitations placed on d by the boundary conditions, we have constant solutions $y = d$ when $\lambda = 0$. Thus, 0 is an eigenvalue, with eigenfunctions $y = \text{constant}\ (\neq 0)$.

Case 2 If $\lambda > 0$, let $\lambda = \alpha^2$, with $\alpha > 0$. Then, $y = c\cos(\alpha x) + d\sin(\alpha x)$. Now, $y(0) = y(\pi)$ gives us

$$c = c\cos(\alpha\pi) + d\sin(\alpha\pi).$$

And $y'(0) = y'(\pi)$ gives us

$$\alpha d = -\alpha c\sin(\alpha\pi) + \alpha d\cos(\alpha\pi)$$

or, since $\alpha > 0$,

$$d = -c\sin(\alpha\pi) + d\cos(\alpha\pi).$$

We cannot have both c and d zero, or else we are left with the trivial solution. Thus, we ask how we can choose α so that the above equations are satisfied. Write them as

$$c[\cos(\alpha\pi) - 1] + d\sin(\alpha\pi) = 0,$$
$$c[-\sin(\alpha\pi)] + d[\cos(\alpha\pi) - 1] = 0,$$

and think of these as two equations in two unknowns, c and d. There are solutions with not both c and d zero when the determinant of the coefficients is zero. Thus, we need

$$\begin{vmatrix} \cos(\alpha\pi) - 1 & \sin(\alpha\pi) \\ -\sin(\alpha\pi) & \cos(\alpha\pi) - 1 \end{vmatrix} = 0.$$

This multiplies out to the equation $\cos(\alpha\pi) = 1$. Thus, we must choose $\alpha\pi = 2n\pi$, $(n = 1, 2, 3, \ldots)$ or, equivalently, $\alpha = 2n$, $(n = 1, 2, 3, \ldots)$. The eigenvalues are then

$$\lambda = \alpha^2 = 4n^2, \qquad n = 1, 2, 3, \ldots .$$

Corresponding to the eigenvalue $\lambda = 4n^2$ are the eigenfunctions

$$y = c\cos(2nx) + d\sin(2nx),$$

with c and d arbitrary but not both zero.

Case 3 If $\lambda < 0$, write $\lambda = -\alpha^2$, with $\alpha > 0$. Now we get $y = ce^{\alpha x} + de^{-\alpha x}$. Here, $y(0) = y(\pi)$ gives us

$$c + d = ce^{\alpha\pi} + de^{-\alpha\pi}.$$

And, $y'(0) = y'(\pi)$ gives us, after dividing out α,

$$c - d = ce^{\alpha\pi} - de^{-\alpha\pi}.$$

These equations can be written

$$c(e^{\alpha\pi} - 1) + d(e^{-\alpha\pi} - 1) = 0,$$
$$c(e^{\alpha\pi} - 1) + d(1 - e^{-\alpha\pi}) = 0.$$

For any choice of $\alpha > 0$, these can hold only if $c = d = 0$, giving us the trivial solution $y = 0$. Thus, there are no negative eigenvalues for this problem.

In summary,

$$y'' + \lambda y = 0; \qquad y(0) = y(\pi), \qquad y'(0) = y'(\pi)$$

has eigenvalues

$$\lambda = 0,$$

with $y =$ constant $(\neq 0)$ as eigenfunctions, and

$$\lambda = 4n^2, \qquad n = 1, 2, 3, \ldots,$$

with corresponding eigenfunctions

$$y = c\cos(2nx) + d\sin(2nx),$$

with not both c and d zero.

EXAMPLE 178 A Singular Sturm-Liouville Problem

Consider

$$[(1 - x^2)y']' + \lambda y = 0,$$

on the interval $[-1, 1]$. Here, $r(x) = 1 - x^2$ vanishes at both -1 and 1, and so we are in case 3 of the boundary conditions for the singular problem. You can check from Section 6.3 that integers of the form $\lambda = n(n+1)$, for $n = 0, 1, 2, 3, \ldots$, are eigenvalues, with corresponding eigenfunctions $y = P_n(x)$, the Legendre polynomials.

Thus far we have defined the kind of problem we want to consider and have looked at some examples. The next theorem gives some basic facts about eigenfunctions and eigenvalues.

THEOREM 20 Sturm-Liouville Theorem

1. For each of the Sturm-Liouville problems defined above, there exists an infinite number of eigenvalues. Further, these can be labeled $\lambda_1, \lambda_2, \ldots$ so that $\lim_{n\to\infty} \lambda_n = \infty$.
2. If λ_m and λ_n are distinct eigenvalues of a Sturm-Liouville problem, with corresponding eigenfunctions $f_m(x)$ and $f_n(x)$, then

$$\int_a^b p(x) f_n(x) f_m(x)\, dx = 0.$$

3. All eigenvalues are real.
4. For a regular Sturm-Liouville problem, all eigenfunctions corresponding to a given eigenvalue are constant multiples of one another.

Conclusion 1 assures us of the existence of nontrivial solutions of a Sturm-Liouville problem, since to each eigenvalue there corresponds at least one nontrivial solution, namely an eigenfunction associated with that eigenvalue. The conclusion that $\lim_{n\to\infty} \lambda_n = \infty$ means roughly that the eigenvalues are spread out and do not accumulate about a finite point (as, for example, the numbers $1 - (1/n)$ accumulate at 1). In many engineering applications, eigenvalues have a physical significance related to the parameters of the problem. For example, in problems dealing with propagation of waves, the eigenvalues may represent fundamental modes of vibration. This is discussed in the chapter on partial differential equations.

In connection with conclusion 2, we define functions $g(x)$ and $h(x)$ to be *orthogonal on* $[a, b]$ *with weight function* $p(x)$ if $\int_a^b p(x)g(x)h(x)\, dx = 0$. We can then rephrase (2) to read: Eigenfunctions associated with distinct eigenvalues are orthogonal on $[a, b]$ with weight function $p(x)$. Recall that $p(x)$ occurs in the differential equation $(ry')' + (q + \lambda p)y = 0$. This conclusion is at the heart of the theory of eigenfunction expansions and Fourier series, which are in turn basic in solving many important partial differential equations.

Conclusion 1 of the theorem requires an analysis beyond what we are prepared to do. Here are proofs of (2) and (3).

Proof of Conclusion 2 We have, from the differential equation,

$$(rf'_n)' + (q + \lambda_n p)f_n = 0$$

and

$$(rf'_m)' + (q + \lambda_m p)f_m = 0.$$

Multiply the first equation by f_m, the second by f_n, and subtract the second from the first to get

$$(rf'_n)'f_m + \lambda_n pf_n f_m - (rf'_m)'f_n - \lambda_m pf_n f_m = 0.$$

Then,

$$(rf'_n)'f_m - (rf'_m)'f_n = \frac{d}{dx}[r(f_m f'_n - f_n f'_m)] = (\lambda_m - \lambda_n)pf_n f_m.$$

Integrating from a to b gives us

$$[r(f_m f'_n - f_n f'_m)]_a^b = (\lambda_m - \lambda_n)\int_a^b p(x)f_n(x)f_m(x)\,dx.$$

Since $\lambda_m \neq \lambda_n$, we will have proved (2) if we can show that the left side of the last equation is zero. We shall do this by considering the boundary conditions. This requires three cases, since we have three different kinds of Sturm-Liouville problems.

Case 1 Suppose we have a regular Sturm-Liouville problem. Then the boundary conditions are of the form

$$A_1 y(a) + A_2 y'(a) = 0,$$
$$B_1 y(b) + B_2 y'(b) = 0,$$

satisfied by both $f_n(x)$ and $f_m(x)$. Thus, looking at just the condition at a,

$$A_1 f_n(a) + A_2 f'_n(a) = 0,$$
$$A_1 f_m(a) + A_2 f'_m(a) = 0.$$

Since A_1 and A_2 are assumed not both zero, then the system of equations

$$f_n(a)X + f'_n(a)Y = 0$$
$$f_m(a)X + f'_m(a)Y = 0$$

has a nontrivial solution. Thus,

$$\begin{vmatrix} f_n(a) & f'_n(a) \\ f_m(a) & f'_m(a) \end{vmatrix} = f_n(a)f'_m(a) - f_m(a)f'_n(a) = 0.$$

Applying the same reasoning at the other endpoint, we get

$$f_n(b)f'_m(b) - f_m(b)f'_n(b) = 0,$$

proving (2) for the regular case.

Case 2 Suppose we have periodic boundary conditions

$$y(a) = y(b), \quad y'(a) = y'(b).$$

These are satisfied by both $f_n(x)$ and $f_m(x)$. The reader can check that

$$r(x)[f_m(x)f'_n(x) - f_n(x)f'_m(x)]_a^b = 0,$$

recalling that $r(a) = r(b)$ in the periodic problem.

Case 3 Suppose we have a singular problem. We leave the details of applying the three kinds of boundary conditions in this case to the student.

Proof of Conclusion 3 To prove that all the eigenvalues are real, suppose that $\lambda = \alpha + i\beta$ is an eigenvalue, with corresponding eigenfunction $f(x)$ which may now also be complex-valued. Write $f(x) = u(x) + iv(x)$. By substituting back into the Sturm-Liouville problem, it is routine to check that $\bar{f}(x) = u(x) - iv(x)$ is an eigenfunction corresponding to the eigenvalue $\bar{\lambda} = \alpha - i\beta$. Now, if $\beta \neq 0$, then $\lambda \neq \bar{\lambda}$; hence, by conclusion 2 of this theorem, we must have

$$\int_a^b p(x)[u(x) + iv(x)][u(x) - iv(x)]\, dx = 0.$$

Then,

$$\int_a^b p(x)[u(x)^2 + v(x)^2]\, dx = 0.$$

Now, $p(x) > 0$ for $a < x < b$; further, $f(x)$ is nontrivial. Then $u^2 + v^2$ cannot be identically zero on $[a, b]$. The above integral is therefore positive. Hence, λ must be equal to $\bar{\lambda}$, forcing $\beta = 0$ and λ to be real.

Here are several examples of Theorem 20, the last two relating it to Bessel functions and Legendre polynomials.

EXAMPLE 179

In Example 176 we had eigenvalues $4n^2$ and corresponding eigenfunctions $\sin(2nx)$ on the interval $[0, \pi/2]$. In keeping with conclusion 1 of the theorem, we would label

$$\lambda_n = 4n^2, \qquad n = 1, 2, 3, \ldots.$$

Note that indeed $\lim_{n\to\infty} \lambda_n = \infty$. For this problem, $p(x) = 1$ and it is easy to verify by explicit integration that

$$\int_0^{\pi/2} \sin(2nx)\sin(2mx)\, dx = 0 \qquad (n \neq m),$$

in keeping with conclusion 2 of the theorem.

EXAMPLE 180

In Example 177 we can label the eigenvalues

$$\lambda_n = 4n^2, \qquad n = 1, 2, 3, \ldots.$$

Corresponding eigenfunctions are

$$f_0(x) = c \text{ (nonzero)}$$

and

$$f_n(x) = c_n \cos(2nx) + d_n \sin(2nx), \qquad n = 1, 2, 3, \ldots,$$

with not both c_n and d_n zero.

Again, one can verify easily that

$$\int_0^\pi [c_n \cos(2nx) + d_n \sin(2nx)][c_m \cos(2mx) + d_m \sin(2mx)]\,dx = 0 \qquad (n \neq m).$$

Conclusion 4 of the theorem fails for this nonregular problem. For example, $\cos(2nx)$ and $\sin(2nx)$ are eigenfunctions associated with $\lambda_n = 4n^2$, but neither is a constant multiple of the other.

EXAMPLE 181

Usually eigenvalues are not as easily found as in the above examples. Consider the regular Sturm-Liouville problem defined on $[0, 1]$:

$$y'' + \lambda y = 0; \qquad y(0) = 0, \qquad 3y(1) + y'(1) = 0.$$

Proceed as before, considering cases on λ.

Case 1 If $\lambda = 0$, then $y = cx + d$. Since $y(0) = 0$, then $d = 0$. Next, $3y(1) + y'(1) = 3c + c = 0$; so $c = 0$, and we have the trivial solution. Thus, 0 is not an eigenvalue.

Case 2 If $\lambda < 0$, write $\lambda = -\alpha^2$, with $\alpha > 0$. Then, $y = ce^{\alpha x} + de^{-\alpha x}$. Now, $y(0) = 0 = c + d$; so $d = -c$. Thus, $y = 2c \sinh(\alpha x)$. Next,

$$3y(1) + y'(1) = 6c \sinh(\alpha) + 2c\alpha \cosh(\alpha) = 0.$$

If $c \neq 0$, this is

$$\frac{\sinh(\alpha)}{\cosh(\alpha)} = -\frac{\alpha}{3}.$$

For $\alpha > 0$, the left side is positive and the right side is negative. Hence the equation can be satisfied by no positive values of α. Thus, $c = 0$, and we again have the trivial solution. This problem has no negative eigenvalues.

Case 3 If $\lambda > 0$, write $\lambda = \alpha^2$, with $\alpha > 0$, and we get $y = c \cos(\alpha x) + d \sin(\alpha x)$. Now, $y(0) = c = 0$; so $y = d \sin(\alpha x)$. Next,

$$3y(1) + y'(1) = 3d \sin(\alpha) + \alpha d \cos(\alpha).$$

If $d \neq 0$, this requires that α be chosen so that

$$\tan(\alpha) = -\frac{\alpha}{3}.$$

This equation cannot be solved algebraically. However, Figure 79 shows a graph of $y = \tan(\alpha)$ and a graph of $y = -\alpha/3$, and one can see that there are infinitely many points of intersection having positive first coordinates. These first coordinates give us values of α satisfying $\tan(\alpha) = -\alpha/3$; hence infinitely many positive eigenvalues. We cannot solve for these eigenvalues exactly, but we can approximate them as closely as

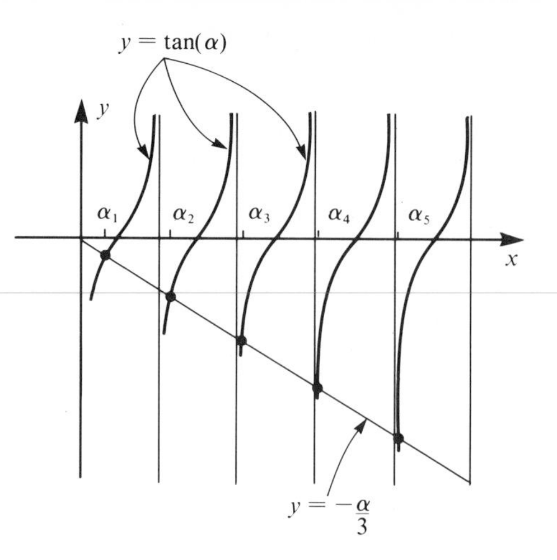

Figure 79. Solutions of $\tan(\alpha) = -\alpha/3$.

we like. To four decimals, we find using Newton's method (see "Numerical Methods," Chapter 20) that

$$\alpha_1 = 2.4556, \qquad \alpha_2 = 5.2329, \qquad \alpha_3 = 8.2045, \qquad \alpha_4 = 11.2560, \qquad \alpha_5 = 14.3434.$$

Thus, the first five eigenvalues are approximately

$$\lambda_1 = 6.0300, \qquad \lambda_2 = 27.3832, \qquad \lambda_3 = 67.3138,$$
$$\lambda_4 = 126.6975, \qquad \lambda_5 = 205.7331.$$

For each number α_n satisfying $\tan(\alpha_n) = -\alpha_n/3$, we get an eigenvalue $\lambda_n = \alpha_n{}^2$ and an eigenfunction $y_n = \sin(\alpha_n x)$, or any nonzero constant multiple thereof. By Theorem 20(2),

$$\int_0^1 \sin(\alpha_n x)\sin(\alpha_m x)\,dx = 0$$

if $n \neq m$. This conclusion is not so obvious by direct integration, since we know the numbers α_n and α_m by defining equations, not as explicitly given numbers. This type of consideration is pursued in the exercises.

EXAMPLE 182 Bessel Functions in Sturm-Liouville Theory

Here we shall see how Bessel functions fit into the framework of Sturm-Liouville theory, and examine what the theory tells us about Bessel functions.

In solving partial differential equations, we shall later encounter the equation

$$x^2y'' + xy' + (\lambda x^2 - n^2)y = 0,$$

in which n is a given positive integer. This can be written in Sturm-Liouville form as

$$(xy')' + \left(\frac{-n^2}{x} + \lambda x\right)y = 0.$$

Typically, we want to solve this on an interval $[0, b]$ subject to

$$y(b) = 0.$$

This is a singular Sturm-Liouville problem, since $r(x) = x$ here, and $r(0) = 0$. For $\lambda > 0$,

$$y = J_n(\sqrt{\lambda}\, x)$$

is a solution. The boundary condition requires that λ be chosen so that

$$J_n(\sqrt{\lambda}\, b) = 0.$$

This has infinitely many solutions, a fact alluded to in Section 6.1 and which will be treated again by oscillation theory in the next section. For example, if $n = 1$, then we get the eigenvalues by solving

$$J_1(\sqrt{\lambda}\, b) = 0.$$

If we wanted, say, the first five eigenvalues, then we would find from a table that the first five zeros of $J_1(x)$ are

3.832, 7.016, 10.173, 13.323, and 16.470.

The first five eigenvalues would then be

$$\lambda_1 = \frac{14.684}{b^2}, \qquad \lambda_2 = \frac{49.224}{b^2}, \qquad \lambda_3 = \frac{103.490}{b^2},$$

$$\lambda_4 = \frac{177.502}{b^2} \quad \text{and} \quad \lambda_5 = \frac{271.261}{b^2}.$$

This list can be continued indefinitely.

For the Sturm-Liouville problem being considered here, $p(x) = x$. Thus, Theorem 20(2) tells us that

$$\int_0^b x J_n(\sqrt{\lambda_k}\, x) J_n(\sqrt{\lambda_m}\, x)\, dx = 0 \quad \text{if} \quad k \neq m.$$

This equation is important in solving certain heat conduction problems in, for example, the radiation of heat energy from a cylindrical tank. It also forms the basis for Fourier-Bessel series, which we discuss briefly in this section and again in the chapter on partial differential equations.

EXAMPLE 183 Legendre Polynomials in Sturm-Liouville Theory

The differential equation

$$(1 - x^2)y'' - 2xy' + \lambda y = 0$$

can be written in Sturm-Liouville form as

$$[(1 - x^2)y']' + \lambda y = 0.$$

Here, $r(x) = 1 - x^2$ is positive on $(-1, 1)$; so we are usually interested in solutions on $[-1, 1]$. Since $r(1) = r(-1) = 0$, we have a singular Sturm-Liouville problem, with no boundary conditions. However, we require that solutions be bounded on $[-1, 1]$.

From Section 6.3, we know that numbers $\lambda_n = n(n+1)$ are eigenvalues for $n = 0, 1, 2, 3, \ldots$. Corresponding eigenfunctions are $y_n = P_n(x)$, the Legendre polynomials. Here, $p(x) = 1$, and the Sturm-Liouville theorem tells us that

$$\int_{-1}^{1} P_n(x)P_m(x)\,dx = 0 \quad \text{if} \quad n \neq m.$$

This fact is important in Fourier-Legendre expansions, which are used in the solution of certain partial differential equations.

We conclude this section with a brief introduction to eigenfunction expansions. Suppose we have a Sturm-Liouville differential equation,

$$(ry')' + (q + \lambda p)y = 0, \qquad a \leq x \leq b,$$

along with regular, singular, or periodic boundary conditions. By Theorem 20(1) there is an infinite sequence $\lambda_1, \lambda_2, \ldots$ of eigenvalues. Let $f_j(x)$ be an eigenfunction associated with λ_j. In many applications to be seen later, we will seek to write a given function $g(x)$ as a series of the eigenfunctions:

$$g(x) = \sum_{n=1}^{\infty} c_n f_n(x).$$

Assuming for the moment that we can do this, how would we choose the c_n's?

The following trick, in conjunction with orthogonality of the eigenfunctions of a Sturm-Liouville problem, lets us solve for the c_n's. Let k be any positive integer, and multiply the last equation by $p(x)f_k(x)$, to get

$$p(x)g(x)f_k(x) = \sum_{n=1}^{\infty} c_n\, p(x)f_n(x)f_k(x).$$

Now integrate, interchanging the sum and integral on the right:

$$\int_a^b p(x)g(x)f_k(x)\,dx = \sum_{n=1}^{\infty} c_n \int_a^b p(x)f_n(x)f_k(x)\,dx.$$

By Theorem 20(2), all terms in the series on the right are zero except the one with $n = k$, giving us

$$\int_a^b p(x)g(x)f_k(x)\,dx = c_k \int_a^b p(x)[f_k(x)]^2\,dx.$$

Then,

$$c_k = \frac{\int_a^b p(x)g(x)f_k(x)\,dx}{\int_a^b p(x)[f_k(x)]^2\,dx}.$$

With this choice of the c_n's, $\sum_{n=1}^{\infty} c_n f_n(x)$ is called a *generalized Fourier series*, or *eigenfunction expansion*, of $g(x)$ on $[a, b]$. Fourier series of sines and cosines, which the reader may have seen, and which are treated in Chapter 12, are a special case of this kind of series.

Computing the coefficients c_n is usually not easy. But even if we succeed here, what relationship will the series have to the function $g(x)$? Rather general theorems can be developed in this connection, relating the series to averages of left and right limits of $g(x)$ at points in $[a, b]$. We put such considerations off until Chapter 12, and state here, without proof, a somewhat restrictive result with a strong conclusion.

THEOREM 21 Convergence of Generalized Fourier Series

Suppose that $\lambda_1, \lambda_2, \ldots$ are the eigenvalues of a regular Sturm-Liouville problem, with $\lambda_1 < \lambda_2 < \cdots$. Let $f_n(x)$ be an eigenfunction associated with λ_n. Let $g(x)$ be continuous on $[a, b]$, and $g'(x)$ piecewise continuous on $[a, b]$. Further, suppose that $g(a) = 0$ if $f_1(a) = 0$, and that $g(b) = 0$ if $f_1(b) = 0$. Then, with the above choice of the c_n's, we have

$$g(x) = \sum_{n=1}^{\infty} c_n f_n(x) \quad \text{for} \quad a \le x \le b,$$

and convergence is both uniform and absolute.

EXAMPLE 184

Go back to the regular Sturm-Liouville problem of Example 176. There, $\lambda_n = 4n^2$ and we may choose $f_n(x) = \sin(2nx)$. The interval if $[0, \pi/2]$.

As an illustration, let $g(x) = x(\pi/2 - x)$. Then, $g(x)$ satisfies the hypotheses of Theorem 21. Compute

$$c_n = \frac{\int_0^{\pi/2} x[(\pi/2) - x]\sin(2nx)\,dx}{\int_0^{\pi/2} \sin^2(2nx)\,dx} = \frac{(1/4n^3)[1 - \cos(n\pi)]}{\pi/4}$$

$$= \frac{1 - \cos(n\pi)}{\pi n^3}.$$

Thus, on $[0, \pi/2]$,

$$x\left(\frac{\pi}{2} - x\right) = \sum_{n=1}^{\infty} \left(\frac{1 - \cos(n\pi)}{\pi n^3}\right)\sin(2nx)$$

$$= \frac{2}{\pi}\sum_{n=1}^{\infty} \frac{1}{(2n-1)^3}\sin[(2n-1)x],$$

since $\cos(n\pi) = (-1)^n$, and $1 - \cos(n\pi) = 1 - (-1)^n = \begin{cases} 0 & \text{if } n \text{ is even} \\ 2 & \text{if } n \text{ is odd.} \end{cases}$

Here is an example in which the eigenfunctions are Legendre polynomials.

EXAMPLE 185 A Fourier-Legendre Expansion

Recall that the Legendre differential equation is

$$[(1 - x^2)y']' + \lambda y = 0, \qquad -1 \le x \le 1,$$

with eigenvalues

$$\lambda_n = n(n+1) \quad \text{for} \quad n = 0, 1, 2, \ldots$$

and eigenfunctions

$$y_n(x) = P_n(x).$$

Let

$$g(x) = \begin{cases} -1 & \text{for} \quad -1 < x < 0, \\ 1 & \text{for} \quad 0 < x < 1. \end{cases}$$

We shall write an eigenfunction expansion of $g(x)$ in terms of Legendre polynomials. The series will be

$$\sum_{n=1}^{\infty} c_n P_n(x),$$

with

$$c_n = \frac{\int_{-1}^{1} g(x)P_n(x)\,dx}{\int_{-1}^{1} [P_n(x)]^2\,dx}.$$

Unlike the previous example, these integrals are not done routinely. In a case such as this, one normally falls back on previously derived properties of the eigenfunctions, such as recurrence relations. Proceed as follows.

First, from the result of Problem 7, Section 6.3, we have

$$\int_{-1}^{1} [P_n(x)]^2\,dx = \frac{2}{2n+1}.$$

Next, recall that $P_n(x)$ is a polynomial containing only even powers of x when n is even, and only odd powers of x when n is odd. Thus,

$$\int_{-1}^{1} g(x)P_n(x)\,dx = \begin{cases} 0 & \text{if } n \text{ is even} \\ 2\int_0^1 P_n(x)\,dx & \text{if } n \text{ is odd.} \end{cases}$$

Thus far, we have

$$c_{2n} = 0 \quad \text{and} \quad c_{2n+1} = \frac{2\int_0^1 P_{2n+1}(x)\,dx}{\dfrac{2}{2(2n+1)+1}} = (4n+3)\int_0^1 P_{2n+1}(x)\,dx.$$

To evaluate this integral, go back to the differential equation for $P_n(x)$,

$$[(1-x^2)P_n']' + n(n+1)P_n = 0.$$

By integrating from 0 to 1, we get

$$-P_n'(0) + n(n+1)\int_0^1 P_n(x)\,dx = 0.$$

Now, by Problem 12, Section 6.3, we have

$$P_n'(0) = nP_{n-1}(0).$$

Hence,

$$\int_0^1 P_n(x)\,dx = \frac{1}{n(n+1)} P_n'(0) = \frac{1}{n+1} P_{n-1}(0).$$

Replacing n by $2n + 1$, we have

$$\int_0^1 P_{2n+1}(x)\,dx = \frac{1}{2n+2} P_{2n}(0) = \left(\frac{1}{2n+2}\right)(-1)^n \frac{(2n)!}{2^{2n}(n!)^2},$$

with the last part coming from Problem 8, Section 6.3.

Thus,

$$c_{2n+1} = (4n+3)\int_0^1 P_{2n+1}(x)\,dx = \frac{(4n+3)}{2(n+1)} \frac{(-1)^n(2n)!}{2^{2n}(n!)^2}.$$

Recalling that each $c_{2n} = 0$, the eigenfunction expansion of $g(x)$ in terms of Legendre polynomials is then

$$\sum_{n=0}^{\infty} \frac{(4n+3)}{2(n+1)} \frac{(-1)^n(2n)!}{2^{2n}(n!)^2} P_{2n+1}(x)$$
$$= \tfrac{3}{2}P_1(x) - \tfrac{7}{8}P_3(x) + \tfrac{11}{16}P_5(x) - \tfrac{75}{128}P_7(x) + \cdots .$$

Since this example does not fall within the hypotheses of Theorem 21, we must ask for what values of x this series converges to $g(x)$. The answer is that the series converges exactly to $g(x)$ for $-1 < x < 0$ and for $0 < x < 1$, but we will put off a careful justification of this until Chapter 12, Section 2.

Eigenfunction expansions in which the eigenfunctions are special functions usually carry the title "Fourier-(name of the functions)." Thus, in the last example we had a Fourier-Legendre series; if we were expanding in a series of Bessel functions, we would have a Fourier-Bessel expansion.

The ideas of Sturm-Liouville theory and eigenfunction expansions form a backdrop for the development of very powerful methods for approaching certain kinds of partial differential equations. We shall develop these ideas in Chapters 12 and 13.

In the next section we introduce some ideas of oscillation theory, with important bearing on some of the ideas we have been discussing.

PROBLEMS FOR SECTION 6.4

In each of Problems 1 through 17, classify the Sturm-Liouville problem as regular, singular, or periodic, and find the eigenvalues and corresponding eigenfunctions. In cases where the eigenvalues are determined by a transcendental equation which cannot be solved explicitly, give a graphical argument for their existence. If you have some computational aid, try approximating some of the eigenvalues in such cases.

1. $y'' + \lambda y = 0;\quad y(-3\pi) = y(3\pi),\quad y'(-3\pi) = y'(3\pi)$

2. $y'' + \lambda y = 0;\quad y'(0) = y'(2\pi) = 0$

3. $y'' + \lambda y = 0;\quad y'(0) = 0;\quad Ay(2\pi) + y'(2\pi) = 0$
(A is a positive constant.)

4. $y'' + \lambda y = 0;\quad y(0) + y'(0) = 0,\quad y(1) = 0$

5. $y'' + \lambda y = 0;\quad y(0) - 2y'(0) = 0,\quad y'(1) = 0$

6. $(xy')' + (\lambda/x)y = 0;\quad y(1) = 0,\quad y'(2) = 0$

7. $(xy')' + (\lambda/x)y = 0;\ y'(1) = 0,\quad y'(2) = 0$
8. $y'' + \lambda y = 0;\quad y(0) = 0,\quad y'(b) = 0$
9. $y'' + \lambda y = 0;\quad y'(0) = y'(b) = 0$
10. $y'' + y' + (1 + \lambda)y = 0;\quad y(0) = y(1) = 0$
11. $y'' + \lambda y = 0;\quad y(0) = 0,\quad y(\pi) + 2y'(\pi) = 0$
12. $x^2y'' + xy' + \lambda y = 0;\quad y(1) = 0,\quad y(x)$ and $y'(x)$ must be bounded as $x \to 0$ from the right.
13. $y'' - 2y' + 2(1 + \lambda)y = 0;\quad y(0) = y(1) = 0$
14. $y'' + \lambda y = 0;\quad y(0) = 0,\quad y(\pi) + 2y'(\pi) = 0$
15. $x^2y'' + 3xy' + \lambda y = 0;\quad y(1) = y(2) = 0$
16. $y'' + \lambda y = 0;\quad y(0) = 0,\quad hy(1) + y'(1) = 0$ (h any positive constant)
17. $y'' + \lambda y = 0;\quad y(0) + y'(0) = 0,\quad y(\pi) + 2y'(\pi) = 0$
18. Write the Hermite differential equation $y'' - xy' + \lambda y = 0$ (see the paragraph preceding Problem 14 in Section 6.3) in Sturm-Liouville form, following the method found at the beginning of this section. The relevant interval for Hermite polynomials is $(-\infty, \infty)$, in which case there are no boundary conditions. Use Theorem 20(2) to prove the result of Problem 16, Section 6.3, for the case $n \neq m$.
19. Write the Laguerre differential equation $xy'' + (1 - x)y' + \lambda y = 0$ (see the paragraph preceding Problem 19 in Section 6.3) in Sturm-Liouville form. Here, the relevant interval is $(0, \infty)$. Use Theorem 20(2) to prove the result given in Problem 21, Section 6.3, for the case $n \neq m$.
20. Find the eigenfunction expansion of $g(x) = x^2$ in terms of the eigenfunctions of Example 181. (*Hint:* You don't know the eigenvalues explicitly, but you can still work out the integrals for the coefficients in the expansion in terms of the λ_n's.)
21. Write a Fourier-Legendre series for

$$g(x) = \begin{cases} 1, & -1 < x < 0, \\ 0, & 0 < x < 1. \end{cases}$$

22. Write a Fourier-Legendre series for $g(x) = 1,\ -1 \le x \le 1$.
23. Write a Fourier-Legendre series for $g(x) = |x|,\ -1 \le x \le 1$. (*Hint:* First show that only even powers of x appear; to evaluate the coefficient of x^{2n}, use the result of Problem 11 of Section 6.3 and integrate by parts.)
24. Write a Fourier-Legendre series for $g(x) = x$.
25. Write a Fourier-Legendre series for $g(x) = x^2 + 1$.
26. Derive the Fourier-Bessel expansion

$$\sum_{n=1}^{\infty} \frac{2J_0(\sqrt{\lambda_n}\, x)}{\sqrt{\lambda_n}\, bJ_1(\sqrt{\lambda_n}\, b)}$$

of $g(x) = 1$ on $0 \le x \le b$. Here the eigenvalues $\lambda_1, \lambda_2, \ldots$ are determined by $J_0(\sqrt{\lambda_n}\, b) = 0$, and the eigenfunctions are $J_0(\sqrt{\lambda_n}\, x)$.
27. Expand $g(x) = x$ on $[0, b]$ in a Fourier-Bessel series in terms of the first order Bessel functions $J_1(\sqrt{\lambda_n}\, x)$, where λ_n is the nth solution of $J_1(\sqrt{\lambda}\, b) = 0$. You should get

$$2 \sum_{n=1}^{\infty} \frac{J_1(\sqrt{\lambda_n}\, x)}{\sqrt{\lambda_n}\, J_2(\sqrt{\lambda_n})}.$$

6.5 THE STURM SEPARATION THEOREM AND THE STURM COMPARISON THEOREM

In this section we address the problem of determining some information about the zeros of solutions of differential equations. Recall that x_0 is a zero of $f(x)$ if $f(x_0) = 0$.

We have already seen instances in which zeros of a solution are important in solving a problem—the suspended chain problem of Example 171, the critical length problem of Example 173, and eigenfunction expansions, for example.

In many instances, such as with Bessel's equation, solutions are so complicated that it is not obvious whether there are zeros or where they occur. We shall now prove two fundamental results which give us information about such things.

THEOREM 22 Sturm Separation Theorem

Let $y_1(x)$ and $y_2(x)$ be linearly independent solutions of

$$y'' + P(x)y' + Q(x)y = 0.$$

Then, $y_1(x)$ and $y_2(x)$ have no common zeros. Further, between each pair of consecutive zeros of one solution is exactly one zero of the other.

Note: This theorem does not assert the existance of zeros, but rather says something about the relative positions of zeros which do exist. The name of the theorem derives from the last part, which asserts that zeros of each function are separated by zeros of the other.

Proof Since y_1 and y_2 are linearly independent, then

$$W(y_1, y_2) = y_1 y_2' - y_2 y_1' \neq 0.$$

This means that y_1 and y_2 cannot have a common zero, since $y_1(x_0) = y_2(x_0) = 0$ would cause W to be zero at x_0, a contradiction.

Now suppose that x_1 and x_2 are consecutive zeros of $y_2(x)$. Then, $y_2(x_1) = y_2(x_2) = 0$; so, at x_1, $W = y_1(x_1)y_2'(x_1)$, and at x_2, $W = y_1(x_2)y_2'(x_2)$. Since W can never be zero, then $y_2'(x_1) \neq 0$ and $y_2'(x_2) \neq 0$. Further, since x_1 and x_2 are *successive* zeros of $y_2(x)$, then $y_2(x)$ has the general appearance of Figure 80(a) or 80(b), which means that $y_2'(x_1)$ and $y_2'(x_2)$ have opposite sign. But W never vanishes, and hence must have the same sign everywhere. Thus, $y_1(x_1)$ and $y_1(x_2)$ must have opposite sign, and so $y_1(x)$ must vanish somewhere between x_1 and x_2.

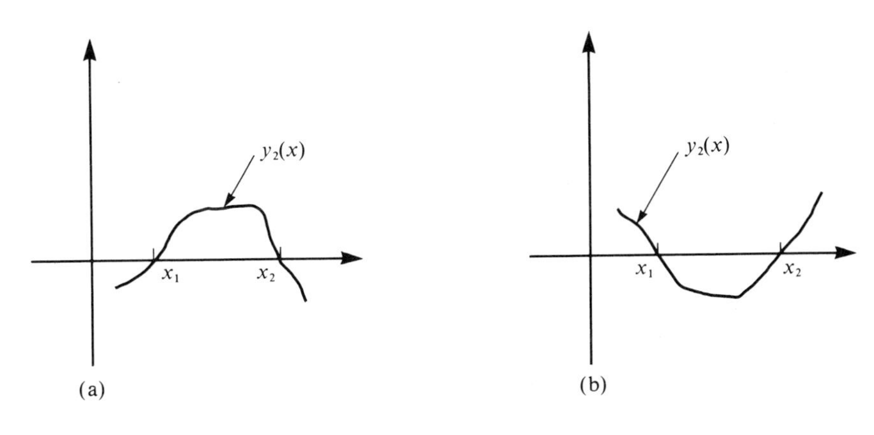

Figure 80

To complete the proof, we must show that $y_1(x)$ vanishes only once between x_1 and x_2. Suppose instead that $y_1(x)$ vanishes at a and b between x_1 and x_2. By reversing the roles of y_1 and y_2 above, we must have a zero of $y_2(x)$ between these two zeros of $y_1(x)$. But then x_1 and x_2 are not consecutive zeros of $y_2(x)$, a contradiction (see Figure 81). This completes the proof.

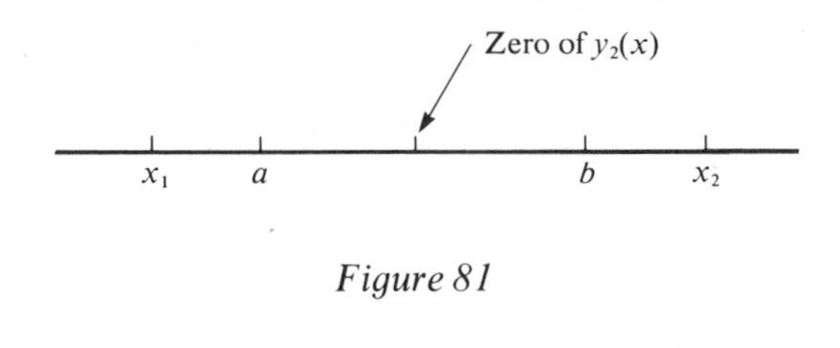

Figure 81

The prototype for the Sturm separation theorem is of course the familiar case of

$$y'' + y = 0,$$

with solutions $\sin(x)$ and $\cos(x)$. The zeros of $\sin(x)$ are at $n\pi$, and those of $\cos(x)$ at $(2n + 1)\pi/2$, with n any integer. One can see in this simple case that each solution has exactly one zero between any pair of consecutive zeros of the other. It should be noted, however, that these functions are periodic, and that the zeros occur at regular intervals, a situation which does not hold in general. The Sturm separation theorem says nothing about the distance between successive zeros in general.

A more impressive example is Bessel's equation of order zero. For $x > 0$, fundamental solutions are $J_0(x)$ and $Y_0(x)$, and the theorem tells us that between any pair of consecutive zeros of one function there is exactly one zero of the other, a fact which is otherwise not obvious. This assumes that these Bessel functions have zeros.

The question of existence of zeros leads us to the notion of oscillation. Again taking a cue from $y'' + y = 0$, we call a nontrivial solution of a differential equation *oscillatory* if it possesses infinitely many zeros, and these zeros are unbounded in the sense that they are not contained in any finite interval. The next theorem is useful in determining from a differential equation whether its solutions are oscillatory.

THEOREM 23 Sturm Comparison Theorem

Let $f(x)$ be a nontrivial solution of

$$(ry')' + q_1 y = 0$$

on $[a, b]$. Let $g(x)$ be a nontrivial solution of

$$(ry')' + q_2 y = 0$$

on $[a, b]$. Suppose that $r(x)$ and $r'(x)$ are continuous on $[a, b]$, and that $r(x) > 0$ for $a \leq x \leq b$. Suppose also that $q_1(x)$ and $q_2(x)$ are continuous on $[a, b]$, and that

$$q_1(x) < q_2(x) \quad \text{for} \quad a \leq x \leq b.$$

Then, between any pair of zeros of $f(x)$ on $[a, b]$ is at least one zero of $g(x)$.

Proof Let x_1 and x_2 be consecutive zeros of $f(x)$ in $[a, b]$, with $x_1 < x_2$, and suppose that $g(x)$ does not vanish on (x_1, x_2). Then the signs of $f(x)$ and $g(x)$ do not change on (x_1, x_2). Suppose, to be specific, that $f(x) > 0$ and $g(x) > 0$ for $x_1 < x < x_2$.

Now,

$$(rf')' + q_1 f = 0$$

and

$$(rg')' + q_2 g = 0.$$

Multiply the first equation by g, the second by $-f$, and add to get $(rf')'g + q_1 gf - (rg')'f - q_2 gf = 0$.
This is the same as

$$(rf')'g - (rg')'f = (q_2 - q_1)fg.$$

But a simple calculation verifies that

$$(rf')'g - (rg')'f = \frac{d}{dx}[r(f'g - fg')].$$

Thus, we have

$$\frac{d}{dx}[r(f'g - fg')] = (q_2 - q_1)fg.$$

Integrating from x_1 to x_2 gives us

$$r(x)[f'(x)g(x) - f(x)g'(x)]_{x_1}^{x_2} = \int_{x_1}^{x_2} [q_2(x) - q_1(x)]f(x)g(x)\,dx.$$

Since $f(x_1) = f(x_2) = 0$, we have

$$r(x_2)f'(x_2)g(x_2) - r(x_1)f'(x_1)g(x_1) = \int_{x_1}^{x_2} [q_2(x) - q_1(x)]f(x)g(x)\,dx.$$

Now look at the sign of each side. Since $q_2(x) - q_1(x) > 0$, and $f(x) > 0$ and $g(x) > 0$ for $x_1 < x < x_2$, then the integral on the right is positive:

$$\int_{x_1}^{x_2} [q_2(x) - q_1(x)]\, f(x)g(x)\,dx > 0.$$

On the left, $r(x_2) > 0$ and $r(x_1) > 0$ by assumption. Since $f(x_2) = 0$ and $f(x) > 0$ for $x_1 < x < x_2$, then $f'(x_2) < 0$ (see Figure 82). Since $g(x) > 0$ on (x_1, x_2) and $g(x)$ is continuous, then $g(x_2) \geq 0$. Thus,

$$r(x_2)f'(x_2)g(x_2) \leq 0.$$

By similar reasoning,

$$r(x_1)f'(x_1)g(x_1) \geq 0.$$

Thus,

$$r(x_2)f'(x_2)g(x_2) - r(x_1)f'(x_1)g(x_1) \leq 0.$$

This is a contradiction. Hence, $g(x)$ vanishes at least once between x_1 and x_2.

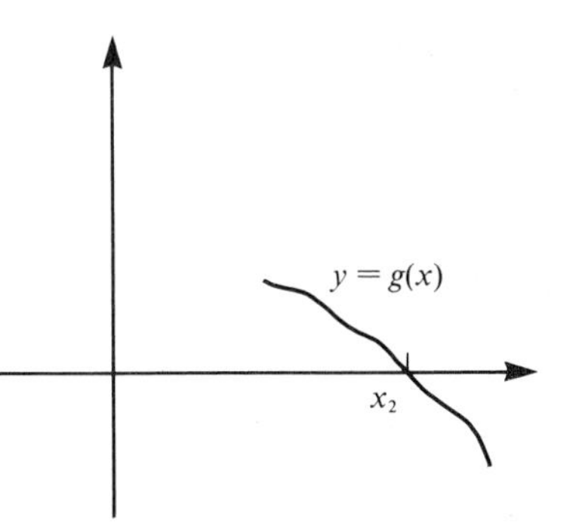

Figure 82

The value of the Sturm comparison theorem is that we can sometimes compare a differential equation whose solutions are complicated functions with a differential equation whose solutions have well-understood properties. In this way we can draw some conclusions about the zeros of the solutions in the more complex case. In actually doing this, it is often useful to reduce a differential equation to what is called *normal form*, in which there is no first derivative term. This is done as follows. Suppose we begin with

$$y'' + P(x)y' + Q(x)y = 0.$$

Let $y(x) = u(x)v(x)$. Then,

$$y' = uv' + u'v \quad \text{and} \quad y'' = u''v + 2u'v' + uv''.$$

Substituting into the differential equation and rearranging terms gives us

$$vu'' + (2v' + Pv)u' + (v'' + Pv' + Qv)u = 0.$$

In order to make the first derivative term disappear, we shall choose v so that

$$2v' + Pv = 0.$$

This separable first order equation has solution

$$v(x) = \exp[-\tfrac{1}{2}\int P(x)\,dx].$$

Then

$$v'' + Pv' + Qv = (Q - \tfrac{1}{2}P' - \tfrac{1}{4}P^2)v$$

and the differential equation becomes

$$vu'' + vqu = 0,$$

with $q = Q - \frac{1}{2}P' - \frac{1}{4}P^2$. Since v is never zero, this is the same as

$$u'' + qu = 0.$$

In summary, the differential equation $y'' + P(x)y' + Q(x)y = 0$ has transformed to $u'' + q(x)u = 0$ under the change of variables

$$y(x) = u(x)\exp\left[-\frac{1}{2}\int P(x)\,dx\right],$$

with

$$q(x) = Q(x) - \tfrac{1}{2}P'(x) - \tfrac{1}{4}P(x)^2.$$

We call $u'' + qu = 0$ the *normal form* of $y'' + P(x)y' + Q(x) = 0$. Clearly, a differential equation and its normal form will generally have different solutions. However, solutions $y(x)$ are related to solutions $u(x)$ by an exponential factor which is never zero. Thus, a solution $u(x)$ of the normal form and a solution $y(x) = u(x)$ $\exp[-\frac{1}{2}\int P(x)\,dx]$ of the original equation *will have the same zeros.* In applying the Sturm comparison theorem, then, we usually reduce the differential equation of interest to normal form, for which comparisons are more easily made. Here is an important illustration.

EXAMPLE 186

What can we say about zeros of Bessel functions of order v?

Begin with Bessel's equation of order v,

$$y'' + \frac{1}{x}y' + \left(1 - \frac{v^2}{x^2}\right)y = 0.$$

This has normal form

$$u'' + qu = 0,$$

in which

$$q = 1 - \frac{v^2}{x^2} - \frac{1}{2}\left(-\frac{1}{x^2}\right) - \frac{1}{4}\left(\frac{1}{x}\right)^2 = 1 + \frac{1 - 4v^2}{4x^2}.$$

We shall compare this with

$$u'' + u = 0,$$

with solution $\sin(x)$ having zeros $x = n\pi$ for $n = 1, 2, 3, \ldots$. [We are only interested in $x > 0$ for Bessel's equation. Note also that we could use $\cos(x)$ instead of $\sin(x)$, or any linear combination of $\sin(x)$ and $\cos(x)$.]

Now, to apply the Sturm comparison theorem, we need to know which is larger, $1 + (1 - 4v^2)/4x^2$ or 1. Clearly,

$$1 + \frac{1 - 4v^2}{4x^2} > 1 \quad \text{exactly when} \quad \frac{1 - 4v^2}{4x^2} > 0,$$

and this happens when $0 \le v < \frac{1}{2}$ (we are interested in $v \ge 0$). Here are three conclusions we can draw.

1. If $0 \le v < \frac{1}{2}$, choose

$$q_1(x) = 1 \quad \text{and} \quad q_2(x) = 1 + \frac{1 - 4v^2}{4x^2}.$$

Then, between any pair of zeros of $\sin(x)$ there is at least one zero of any solution of Bessel's equation of order v [in particular, this applies to $J_v(x)$ and $Y_v(x)$]. Hence, any interval of length π on the positive half of the real line contains at least one zero of $J_v(x)$ and at least one zero of $Y_v(x)$.

2. If $\frac{1}{2} < v < \infty$, choose

$$q_1(x) = 1 + \frac{1 - 4v^2}{4x^2} \quad \text{and} \quad q_2(x) = 1.$$

Then, between any pair of zeros of $J_v(x)$ [or $Y_v(x)$] there is at least one zero of $\sin(x)$. In particular, every interval of length π on $x > 0$ contains at most one zero of $J_v(x)$ or $Y_v(x)$.

3. If $v = \frac{1}{2}$, then the normal form of Bessel's equation of order is exactly $u'' + u = 0$. In this case, successive zeros of $J_{1/2}(x)$ and $Y_{1/2}(x)$ are exactly π units apart. (Note Problem 9, Section 6.2.)

This brief introduction to the qualitative theory of differential equations is supplemented by some of the problems at the end of this section. In the next chapter we shall study systems of differential equations, with the first two sections devoted to methods, and return to qualitative considerations in Section 7.3.

PROBLEMS FOR SECTION 6.5

1. Suppose that $q(x) < 0$ for all x, and that $y(x)$ is a nontrivial solution of $y'' + q(x)y = 0$. Prove that $y(x)$ can have at most one zero.
2. Let $q(x) > 0$ for all $x > 0$, and let $y(x)$ be any nontrivial solution of $y'' + q(x)y = 0$ for $x > 0$. Show that $y(x)$ has infinitely many positive zeros if

$$\int_1^\infty q(x)\, dx = \infty.$$

3. Construct an example to show that the conclusion of Problem 2 fails if we omit the condition that

$$\int_1^\infty q(x)\, dx = \infty.$$

4. Use the Sturm separation theorem to show that the zeros of $y_1(x) = a\cos(x) + b\sin(x)$ and $y_2(x) = d\cos(x) + e\sin(x)$ alternate if $ae - bd \neq 0$. What happens if $ae - bd = 0$?
5. Let $q(x)$ be continuous on $[a, b]$, and suppose that $0 < m < q(x) < M$ for $a \leq x \leq b$. Let $y(x)$ be a nontrivial solution of $y'' + q(x)y = 0$ on $[a, b]$. Prove that

$$\frac{\pi}{\sqrt{M}} < x_2 - x_1 < \frac{\pi}{\sqrt{m}}$$

for any two consecutive zeros x_1 and x_2 of $y(x)$ such that $a \leq x_1 < x_2 \leq b$.
6. One can easily check that $y_1(x) = e^{-x}$ and $y_2(x) = e^{-2x}$ are nontrivial solutions of $y'' + 3y' + 2y = 0$. However, these functions have no zeros. Why does this not violate the Sturm separation theorem?
7. Let $y(x)$ be any nontrivial solution of $y'' + (x^3 + 2x)y = 0$ for $x > 0$. Prove that $y(x)$ must have infinitely many positive zeros. (*Hint:* Use the result of Problem 2.)
8. Let $y(x)$ be any nontrivial solution of $y'' + (1/x)y = 0$ for $x > 0$. Prove that $y(x)$ has infinitely many positive zeros.
9. Obtain a solution of $y'' + (1/x)y = 0$ in some interval $0 < x < a$ by the method of Frobenius. What can you tell from the series solution about the zeros of $y(x)$?

CHAPTER SUMMARY

This chapter was intended as an introduction to special functions, Sturm-Liouville theory, eigenfunction expansions, and oscillation theory. Whole books have been written about each of these important topics, but our purpose has been to familiarize the reader with terminology and some basic facts about these subjects, and to prepare the way for further reading or course work. Some details and applications were included for Bessel functions and Legendre polynomials, two classes of special functions which occur very frequently in engineering and the sciences. These in turn are examples of functions generated as eigenfunctions of Sturm-Liouville problems.

Eigenfunctions and eigenvalues will play an important role in solving partial differential equations later. Then we shall see that the eigenvalues often assume the values of important physical parameters of the problems being considered. Further, the eigenfunctions will form the basis of a general kind of Fourier analysis, giving us series solutions of these problems. For example, solutions of heat conduction problems sometimes involve Fourier-Bessel series. Thus, Sturm-Liouville theory may be viewed as background for the treatment of certain kinds of partial differential equations, as will become apparent in Chapters 12 and 13.

SUPPLEMENTARY PROBLEMS

1. Show that

$$\int x \log(x) J_0(x)\, dx = J_0(x) + x \log(x) J_1(x).$$

2. Show that

$$\int_0^1 x^2 J_0(\alpha x)\, dx = \frac{1}{\alpha} J_1(\alpha) + \frac{1}{\alpha^2} J_0(\alpha) - \frac{1}{\alpha^3} \int_0^\alpha J_0(t)\, dt.$$

3. Show that

$$\int_0^1 x J_0(ax)\, dx = \frac{1}{a} J_1(a),$$

for any positive constant a.

4. Find the general solution of

$$y'' - \frac{5}{x} y' + \left(x^2 - \frac{55}{x^2}\right) y = 0.$$

5. Find the general solution of

$$y'' - \frac{3}{x} y' + \left(\frac{16}{9} x^6 - \frac{12}{x^2}\right) y = 0.$$

6. Find the general solution of

$$y'' - \frac{5}{x} y' + \left(4x^2 + \frac{35}{4x^2}\right) y = 0.$$

7. Find the general solution of

$$y'' - \frac{5}{x} y' + \left(144x^6 + \frac{209}{25x^2}\right) y = 0.$$

8. Lord Kelvin introduced the functions ber(x) and bei(x) by setting

$$I_0\left[\left(\frac{1+i}{\sqrt{2}}\right)x\right] = \text{ber}(x) + i\ \text{bei}(x).$$

Use the fact that

$$\left[\left(\frac{1+i}{\sqrt{2}}\right)x\right]^2 = ix^2,$$

together with the series for $I_0(x)$ given in Section 6.2, to derive the series

$$\text{ber}(x) = 1 - \frac{x^4}{2^2 4^2} + \frac{x^8}{2^2 4^2 6^2 8^2} - \cdots$$

and

$$\text{bei}(x) = \frac{x^2}{2^2} - \frac{x^6}{2^2 4^2 6^2} + \frac{x^{10}}{2^2 4^2 6^2 8^2 10^2} - \cdots .$$

9. Show that

$$\int_0^\infty J_1(x)\,dx = 1.$$

10. Show that

$$\int_0^\infty \frac{1}{x} J_1(x)\,dx = 1.$$

11. Euler's constant γ, which appears in Neumann's Bessel function $Y_n(x)$ of order n, is defined by the expression

$$\gamma = \lim_{n\to\infty}\left[1 + \frac{1}{2} + \frac{1}{3} + \cdots + \frac{1}{n} - \ln(n)\right].$$

Derive the following integral expression for γ:

$$\gamma = \int_0^1 \frac{1-e^{-t}}{t}\,dt - \int_1^\infty \frac{e^{-t}}{t}\,dt.$$

Hint: Begin by showing that

$$1 + \frac{1}{2} + \frac{1}{3} + \cdots + \frac{1}{n} = \int_0^1 (1 + x + x^2 + \cdots x^{n-1})\,dx$$

$$= \int_0^1 \frac{1-x^n}{1-x}\,dx.$$

Change variables by setting $x = 1 - (t/n)$. Then obtain

$$1 + \frac{1}{2} + \cdots + \frac{1}{n} = \int_0^1 \left[1 - \left(1 - \frac{t}{n}\right)^n\right]\frac{1}{t}\,dt + \ln(n) - \int_1^n \left[\left(1 - \frac{t}{n}\right)^n\right]\frac{1}{t}\,dt.$$

Finally, take the limit as $n \to \infty$. You may assume without proof the fact that

$$\lim_{n\to\infty}\left(1 - \frac{t}{n}\right)^n = e^{-t}.$$

12. Use the result of Problem 11 to show that

$$\gamma = -\int_0^\infty e^{-t} \ln(t)\, dt.$$

13. Show that

$$\gamma = \sum_{n=1}^{\infty} \left[\frac{1}{n} + \ln\left(\frac{1}{n}\right)\right].$$

14. Show that

$$\int xI_0(ax)\, dx = \frac{x}{a} I_0'(ax), \quad \text{for} \quad a > 0.$$

15. Show that

$$\int xI_0(ax)^2\, dx = \frac{x^2}{2}\left[I_0(ax)^2 - I_0'(ax)^2\right].$$

(*Hint:* Look at Problem 23, Section 6.1.)

16. Show that

$$K_0(x) = \frac{\pi i}{2}\,[J_0(ix) + iY_0(ix)].$$

17. Show that

$$\int_0^{\pi/2} \tan^n(\theta)\, d\theta = \frac{1}{2}\,\Gamma\left(\frac{n+1}{2}\right)\Gamma\left(\frac{n-1}{2}\right)$$

for $n = 2, 3, 4, \ldots$.

18. Show that

$$\int_0^{\pi/2} \sin^n(\theta)\, d\theta = \int_0^{\pi/2} \cos^n(\theta)\, d\theta = \frac{\sqrt{\pi}}{2}\,\frac{\Gamma\left(\frac{n+1}{2}\right)}{\Gamma\left(\frac{n+2}{2}\right)}$$

for $n = 1, 2, 3, \ldots$.

19. Show that, for $x > 0$,

$$J_2(x) = \frac{2}{x} J_1(x) - J_0(x).$$

20. Show that, for $n = 0, 1, 2, 3, \ldots$,

$$J_{n+2}(x) = \left(2n + 1 - \frac{2n(n^2-1)}{x^2}\right) J_n(x) + (2n+2)J_n''(x).$$

21. Show that, for $x > 0$,

$$\int_0^{\pi/2} J_0[x \sin(\theta)]\sin(\theta)\, d\theta = \frac{\sin(x)}{x}.$$

22. Show that

$$P_{n+1}(x) = \left(\frac{2n+1}{n+1}\right)xP_n(x) - \left(\frac{n}{n+1}\right)P_{n-1}(x),$$

for $n = 1, 2, 3, \ldots$.

23. Show that $P_n(1) = 1$ for $n = 0, 1, 2, 3, \ldots$. (*Hint:* Use the result of Problem 22.)

24. Show that

$$\int_{-1}^{1} xP_n(x)P_n'(x)\,dx = \frac{2n}{2n+1} \quad \text{for} \quad n = 0, 1, 2, \ldots.$$

25. Prove that $P_n(x)$ has no repeated roots. [*Hint:* If α is a repeated root of $P_n(x)$, then $P_n(\alpha) = P_n'(\alpha) = 0$.]

26. Prove that $P_n(x)$ has n distinct roots, all lying between -1 and $+1$, for $n = 0, 1, 2, \ldots$. [*Hint:* First use Problem 25 to note that $P_n(x)$ has no repeated roots. If $n \geq 1$, then $\int_{-1}^{1} P_0(x)P_n(x)\,dx = \int_{-1}^{1} P_n(x)\,dx = 0$ (why?) implies that $P_n(x)$ changes sign at least once in $(-1, 1)$, and hence has at least one root in $(-1, 1)$. Let $\alpha_1, \ldots, \alpha_m$ be the roots of $P_n(x)$ lying in $(-1, 1)$, and let $p(x) = (x - \alpha_1)(x - \alpha_2) \cdots (x - \alpha_m)$. Show that $\int_{-1}^{1} P_n(x)p(x)\,dx \neq 0$. However, show that, if $m < n$, then this integral must be zero; hence conclude that $m = n$ to complete the proof.]

27. Show that

$$P_n(x) = xP_{n-1}(x) + \left(\frac{x^2-1}{n}\right)P_{n-1}'(x) \quad \text{for} \quad n = 1, 2, 3, \ldots.$$

(*Hint:* Use the result of Problem 12, Section 6.3.)

28. Show that

$$P_n'(x) = xP_{n-1}'(x) + nP_{n-1}(x) \quad \text{for} \quad n = 1, 2, 3, \ldots.$$

29. Use the results of Problems 27 and 28 to obtain

$$\left(\frac{1-x^2}{n^2}\right)[P_n'(x)]^2 + P_n(x)^2 = \left(\frac{1-x^2}{n^2}\right)[P_{n-1}'(x)]^2 + P_{n-1}(x)^2,$$

for $n = 1, 2, 3, \ldots$.

30. Use the result of Problem 29 to obtain

$$\left(\frac{1-x^2}{n^2}\right)[P_n'(x)]^2 + P_n(x)^2 \leq 1 \quad \text{for} \quad -1 \leq x \leq 1 \quad \text{and} \quad n = 1, 2, 3, \ldots.$$

Use this result to show that $|P_n(x)| \leq 1$ for $-1 \leq x \leq 1$ and $n = 0, 1, 2, \ldots$.

31. Find the eigenvalues and eigenfunctions of the Sturm-Liouville problem

$$y'' + y' + (\lambda + 1)y = 0; \qquad y(0) = y'(1) = 0.$$

32. Find the eigenvalues and eigenfunctions of the Sturm-Liouville problem

$$x^2y'' + 2xy' + \lambda y = 0; \qquad y(1) = 0, \qquad y(e) = 0.$$

33. The differential equation

$$(1 - x^2)y'' - xy' + \lambda y = 0$$

is known as *Tchebycheff's equation.* Write it in Sturm-Liouville form.

34. Find the eigenvalues and eigenfunctions of the Sturm-Liouville problem

$$y'' + \lambda y = 0; \qquad y(0) + y'(0) = 0, \qquad y'(1) = 0.$$

35. Let α be an eigenvalue of the Sturm-Liouville problem

$$(ry')' + (q + \lambda p)y = 0;$$

$$a_1 y(a) + a_2 y'(a) = 0, \qquad b_1 y(b) + b_2 y'(b) = 0.$$

Let $f(x)$ and $g(x)$ be eigenfunctions corresponding to α. Prove that $f(x)$ and $g(x)$ are linearly dependent on $[a, b]$.

36. Let $y_1(x)$ and $y_2(x)$ be any two solutions of Bessel's equation

$$x^2 y'' + xy' + (x^2 - \nu^2)y = 0.$$

Show that $W(y_1, y_2)$ satisfies $xW' + W = 0$; hence show that $W = c/x$ for some constant c.

37. Show that

$$W(J_\alpha, J_{-\alpha}) = \frac{2}{x\Gamma(\alpha)\Gamma(1-\alpha)} \quad \text{if} \quad \alpha > 0.$$

38. *Tchebycheff polynomials* can be defined by

$$T_n(x) = \sum_{j=0}^{[n/2]} (-1)^j \frac{n!}{(2j)!\,(n-2j)!} (1 - x^2)^j x^{n-2j},$$

for $-1 < x < 1$ and $n = 0, 1, 2, 3, \ldots$. Here, $[a]$ = greatest integer $\leq a$.

(a) Show that $T_0(x) = 1$, $T_1(x) = x$, $T_2(x) = 2x^2 - 1$, $T_3(x) = 4x^3 - 2x$, and $T_4(x) = 8x^4 - 8x^2 + 1$.

(b) Show that

$$(1 - x^2)T_n'' - xT_n' + n^2 T_n = 0;$$

Thus, $y = T_n(x)$ is a solution of Tchebycheff's equation (see Problem 33).

(c) Show that

$$\int_{-1}^{1} (1 - x^2)^{-1/2} T_n(x) T_m(x)\, dx = \begin{cases} 0 & \text{if} \quad n \neq m \\ \pi & \text{if} \quad n = m = 0 \\ \frac{1}{2}\pi & \text{if} \quad n = m \neq 0. \end{cases}$$

(d) Show that $2xT_n(x) = T_{n+1}(x) + T_{n-1}(x)$.

(e) Show that $T_n(x) = \cos[n \arccos(x)]$. *Hint:* Put $x = \cos(\theta)$ and use the binomial theorem to expand the left side of the identity

$$[\cos(\theta) + i\sin(\theta)]^n = \cos(n\theta) + i\sin(n\theta).$$

(f) Show that

$$\frac{1 - xt}{1 - 2xt + t^2} = \sum_{n=0}^{\infty} T_n(x) t^n.$$

For this reason, the function on the left is called a *generating function* for the Tchebycheff polynomials.

39. Define the *associated Legendre functions* $P_q{}^n(x)$ by

$$P_q{}^n(x) = (1 - x^2)^{n/2} \frac{d^n}{dx^n} P_q(x) \quad \text{for} \quad n^2 \leq q^2.$$

Show that $P_q{}^n(x)$ is a solution of

$$(1 - x^2)y'' - 2xy' + \left[q(q + 1) - \frac{n^2}{1 - x^2}\right]y = 0.$$

This differential equation arises in the quantum mechanical treatment of a particle moving on the surface of a sphere.

CHAPTER SEVEN

Linear Systems, Nonlinear Systems, and Stability

7.0 INTRODUCTION

We have seen in previous examples that a physically motivated problem may give rise to a system of differential equations in which several functions $x(t)$, $y(t)$, ... and their derivatives appear. Series electrical networks and mechanical systems involving several masses connected by springs are described by such systems. In this chapter we shall consider two techniques for solving certain kinds of systems and also briefly discuss the important notion of stability.

As with single differential equations, we shall restrict ourselves initially to linear systems in which each differential equation is linear. The two main tools we shall use are differential operators and Laplace transforms. There are very powerful matrix methods as well, but these must wait until Chapter 10, when we shall have developed the necessary matrix machinery.

7.1 SOLUTION OF LINEAR SYSTEMS BY ELIMINATION USING DIFFERENTIAL OPERATORS

To begin, let us briefly review the idea of a differential operator, first introduced in Section 3.6. Using t as an independent variable, we let

$$D = \frac{d}{dt}.$$

Thus, D is the instruction "differentiate with respect to t." Then,

$$Dy = \frac{dy}{dt}, \qquad Dx = \frac{dx}{dt}, \qquad D^2x = \frac{d^2x}{dt^2},$$

$$4Dx = 4\frac{dx}{dt}, \qquad (2D + 1)y = 2\frac{dy}{dt} + y,$$

and so on. Any differential equation, or system of differential equations, can then be written in operator notation. For example,

$$y'' + 2y' + 3y = \cos(2t)$$

becomes

$$(D^2 + 2D + 3)y = \cos(2t);$$

and the system

$$\frac{dx}{dt} + 2\frac{dy}{dt} = 1,$$

$$2\frac{dx}{dt} + 3x\frac{dy}{dt} + 3x - 2y = \sin(t)$$

becomes

$$Dx + 2Dy = 1,$$
$$(2D + 3)x + (3xD - 2)y = \sin(t).$$

Recall that we can perform algebraic operations with the D operator. For example,

$$(D + 3)(D + 2) = D^2 + 5D + 6$$

in the sense that both sides give the same set of instructions. As a verification of this, we have, for any $x(t)$,

$$\begin{aligned}(D + 3)(D + 2)x &= (D + 3)(Dx + 2x) = (D + 3)(x' + 2x) \\ &= Dx' + D(2x) + 3x' + 6x = x'' + 5x' + 6x = (D^2 + 5D + 6)x.\end{aligned}$$

This enables us to "factor" some differential equations. For example,

$$x'' + 2x' + x = 0$$

can be written

$$(D^2 + 2D + 1)x = 0$$

or

$$(D + 1)^2 x = 0.$$

Further, remember (see Problems 21, 22, 23 of Section 3.6) that polynomial expressions in D can be multiplied in either order. For example,

$$(D^2 - 2D + 3)(D + 2) = (D + 2)(D^2 - 2D + 3),$$

since both sides work out to $D^3 - D + 6$. The fact that such products commute is important in applying them to systems of differential equations.

We shall now illustrate how to use this algebra of differential operators to solve systems by what is essentially an elimination technique.

EXAMPLE 187

Consider the system

$$2x' + 3y' + 2x + 7y = 2t,$$
$$10x' + 14y' + 8x + 31y = 3t^2 + 1.$$

It is impossible by simple manipulation (such as adding one equation to the other) to obtain an equation involving just x, x', and t, or one involving just y, y', and t. Rewrite the system in terms of differential operators as

$$(2D + 2)x + (3D + 7)y = 2t,$$
$$(10D + 8)x + (14D + 31)y = 3t^2 + 1.$$

This system involves just x, y, D, and t. Multiply the first equation by $-(14D + 31)$ and the second by $(3D + 7)$, to get

$$-(14D + 31)(2D + 2)x - (14D + 31)(3D + 7)y = -(14D + 31)(2t) = -28 - 62t,$$
$$(3D + 7)(10D + 8)x + (3D + 7)(14D + 31)y = (3D + 7)(3t^2 + 1) = 18t + 21t^2 + 7.$$

Now,

$$(3D + 7)(14D + 31) = (14D + 31)(3D + 7).$$

Adding the two equations in the new system then cancels the y terms, giving us

$$-(14D + 31)(2D + 2)x + (3D + 7)(10D + 8)x = 21t^2 - 44t - 21.$$

This works out to

$$2x'' + 4x' - 6x = 21t^2 - 44t - 21,$$

a second order equation for x. This is easily solved by previous methods, giving us

$$x(t) = c_1 e^t + c_2 e^{-3t} - \tfrac{7}{2}t^2 + \tfrac{8}{3}t + \tfrac{53}{18}.$$

Now we must solve for y. Go back to the original system and begin again. Multiply the first equation by $-(10D + 8)$, the second by $(2D + 2)$, and add to get

$$-(10D + 8)(3D + 7)y + (2D + 2)(14D + 31)y$$
$$= -(10D + 8)(2t) + (2D + 2)(3t^2 + 1).$$

After a little algebra, this works out to

$$y'' + 2y' - 3y = -3t^2 + 2t + 9.$$

The general solution of this is

$$y(t) = c_3 e^t + c_4 e^{-3t} + t^2 + \tfrac{2}{3}t - \tfrac{17}{9}.$$

Now we must understand what we have obtained. In order for x and y to satisfy the given system, $x(t)$ and $y(t)$ must have the above forms. But these involve four arbitrary constants, which is two too many since two first order equations should (at least intuitively) involve two integrations, and hence only two arbitrary constants. It must be that c_1, c_2, c_3, and c_4 are related in some way. Substitute x and y into the first equation of the system to get

$$2(c_1 e^t - 3c_2 e^{-3t} - 7t + \tfrac{8}{3}) + 3(c_3 e^t - 3c_4 e^{-3t} + 2t + \tfrac{2}{3}) + 2(c_1 e^t + c_2 e^{-3t} - \tfrac{7}{2}t^2 + \tfrac{8}{3}t + \tfrac{53}{18}) + 7(c_3 e^t + c_4 e^{-3t} + t^2 + \tfrac{2}{3}t - \tfrac{17}{9}) = 2t.$$

Some terms cancel, and we are left with

$$e^t(4c_1 + 10c_3) + e^{-3t}(-4c_2 - 2c_4) = 0.$$

Now, e^t and e^{-3t} are not constant multiples of each other. Thus, the last equation can hold only if the coefficients vanish:

$$4c_1 + 10c_3 = 0,$$
$$-4c_2 - 2c_4 = 0.$$

Then,

$$c_3 = -\tfrac{2}{5}c_1 \quad \text{and} \quad c_4 = -2c_2.$$

You will get the same result if you substitute x and y into the second equation of the system. Thus, the general solution of the system is

$$x(t) = c_1 e^t + c_2 e^{-3t} - \tfrac{7}{2}t^2 + \tfrac{8}{3}t + \tfrac{53}{18},$$
$$y(t) = -\tfrac{2}{5}c_1 e^t - 2c_2 e^{-3t} + t^2 + \tfrac{2}{3}t - \tfrac{17}{9}.$$

EXAMPLE 188

Solve the system

$$2x' + 3y' + x + 2y = -t,$$
$$3x' + 4y' + 10x + 16y = -e^{2t}.$$

In operator notation, we have

$$(2D + 1)x + (3D + 2)y = -t,$$
$$(3D + 10)x + (4D + 16)y = -e^{2t}.$$

Multiply the first equation by $-(4D + 16)$, the second by $(3D + 2)$, and add to get

$$-(4D + 16)(2D + 1)x + (3D + 2)(3D + 10)x = -(4D + 16)(-t) + (3D + 2)(-e^{2t}).$$

This works out to

$$x'' + 4x = 16t + 4 - 8e^{2t},$$

with general solution

$$x(t) = c_1 \cos(2t) + c_2 \sin(2t) + 4t + 1 - e^{2t}.$$

Now go back to the original system and multiply the first equation by $-(3D + 10)$, the second by $(2D + 1)$, and add the equations to get, after a regrouping of terms,

$$y'' + 4y = -10t - 3 + 5e^{2t}.$$

This has general solution

$$y(t) = c_3 \cos(2t) + c_4 \sin(2t) - \tfrac{5}{2}t - \tfrac{3}{4} + \tfrac{5}{8}e^{2t}.$$

To see how c_1, c_2, c_3, and c_4 are related, substitute $x(t)$ and $y(t)$ into the first differential equation of the system to get

$$\begin{aligned}
&2[-2c_1 \sin(2t) + 2c_2 \cos(2t) + 4 - 2e^{2t}] \\
&\qquad + 3[-2c_3 \sin(2t) + 2c_4 \sin(2t) - \tfrac{5}{2} + \tfrac{5}{4}e^{2t}] \\
&\qquad + [c_1 \cos(2t) + c_2 \sin(2t) + 4t + 1 - e^{2t}] \\
&\qquad\qquad + 2[c_3 \cos(2t) + c_4 \sin(2t) - \tfrac{5}{2}t - \tfrac{3}{4} + \tfrac{5}{8}e^{2t}] = -t.
\end{aligned}$$

After some cancellations, we get

$$(-4c_1 + c_2 - 6c_3 + 2c_4)\sin(2t) + (c_1 + 4c_2 + 2c_3 + 6c_4)\cos(2t) = 0.$$

This can hold for all t only if

$$-4c_1 + c_2 - 6c_3 + 2c_4 = 0$$

and

$$c_1 + 4c_2 + 2c_3 + 6c_4 = 0.$$

For any choices of the constants satisfying these two equations, $x(t)$ and $y(t)$ as given above satisfy the original system. For example, we could solve for c_3 and c_4 in terms of c_1 and c_2 to get

$$c_3 = \frac{-13c_1 - c_2}{20} \quad \text{and} \quad c_4 = \frac{c_1 - 13c_2}{20}.$$

The general solution is then

$$x(t) = c_1 \cos(2t) + c_2 \sin(2t) + 4t + 1 - e^{2t},$$

$$y(t) = -\left(\frac{13c_1 + c_2}{20}\right)\cos(2t) + \left(\frac{c_1 - 13c_2}{20}\right)\sin(2t) - \frac{5}{2}t - \frac{3}{4} + \frac{5}{8}e^{2t}.$$

The differential operator method we have illustrated works in general for systems in which each equation is linear and constant coefficient. It is not necessary that every equation in the system be first order. Here is an illustration of this.

EXAMPLE 189

Solve

$$x'' - 2y = t + 2,$$
$$x' - 3y' + 2y = 3t^2.$$

Write this as

$$D^2x - 2y = t + 2,$$
$$Dx + (-3D + 2)y = 3t^2.$$

Multiply the second equation by $-D$ and add, to get

$$-2y + (3D^2 - 2D)y = t + 2 - 6t$$

or

$$3y'' - 2y' - 2y = -5t + 2.$$

This has general solution

$$y = c_1 e^{at} + c_2 e^{bt} + \tfrac{5}{2}t - \tfrac{7}{2},$$

in which

$$a = \frac{1 + \sqrt{7}}{3} \quad \text{and} \quad b = \frac{1 - \sqrt{7}}{3}.$$

To solve for x, multiply the first equation of the original system by $(-3D + 2)$, the second by 2, and add to get

$$(-3D + 2)D^2x + 2Dx = (-3D + 2)(t + 2) + 2(3t^2).$$

This works out to

$$-3x^{(3)} + 2x'' + 2x' = 6t^2 + 2t + 1,$$

which is easily solved to get

$$x(t) = k_1 e^{at} + k_2 e^{bt} + t^3 - \tfrac{5}{2}t^2 + \tfrac{29}{2}t + k_3.$$

To see how c_1, c_2, k_1, k_2, and k_3 are related, substitute back into the first differential equation of the system to get

$$k_1 a^2 e^{at} + k_2 b^2 e^{bt} + 6t - 5 - 2c_1 e^{at} - 2c_2 e^{bt} - 5t + 7 = t + 2.$$

Then,

$$a^2 k_1 - 2c_1 = 0$$

and

$$b^2 k_2 - 2c_2 = 0.$$

Thus,

$$c_1 = \tfrac{1}{2}a^2 k_1 \quad \text{and} \quad c_2 = \tfrac{1}{2}b^2 k_2.$$

The general solution is then

$$x(t) = k_1 e^{at} + k_2 e^{bt} + t^3 - \tfrac{5}{2}t^2 + \tfrac{29}{2}t + k_3,$$
$$y(t) = \tfrac{1}{2}a^2 k_1 e^{at} + \tfrac{1}{2}b^2 k_2 e^{bt} + \tfrac{5}{2}t - \tfrac{7}{2}.$$

We have already applied Laplace transforms to systems. We shall see more examples of this in the next section.

PROBLEMS FOR SECTION 7.1

In each of Problems 1 through 25, use differential operators and the method of this section to find the general solution of the given system.

1. $2x' + x - y = 0$
$x' + y' + 2x - y = 1$

2. $x' - 2y' + x = 2$
$x' + y' - y = 0$

3. $x' + y' + x - y = 0$
$x' + y' - 2x = 0$

4. $x' + y = e^{-t}$
$x' + y' - 2x = 0$

5. $x' - y' + y = 0$
$x' + y' - x = t$

6. $x' + y' + x = \sin(t)$
$x' - y' + y = 1$

7. $x' + y' - x + 2y = 0$
$x' - 2y' + x + y = 0$

8. $x' + y' = z$
$x' + y' - x = t$

9. $x' - y = e^{-t}$
$x' + y' - 2y = t$

10. $2x' - x + y = 0$
$x' + 2y' + x + y = 1$

11. $x' + y' - x = \cos(t)$
$x' - 3y = 2$

12. $2x' + y' - x = 0$
$x' + y' + 3x - y = 0$

13. $2x' + x - y = 4$
$x' + y' - y = 0$

14. $x' - 2y' - x = e^{-2t}$
$x' + y' - y = 0$

15. $x' + y' - 3x = 0$
$x' - y' + y = t^2$

16. $-2x' + y' - x + 2y = -\cos(3t)$,
$y' + 6y - 2x = 0$

17. $-4x' + 2y' + 8y = e^{-t}$,
$x' - 2y' + 4y - x = 0$

18. $3x' - 2y' + 3x - y = t + 2$,
$x' + 2y' + x + 3y = -t^2 - 3$

19. $3x' - 4y' + x - y = t$,
$3x' - 4y' + 2x - y = -3t^2$

20. $x' - 3y' + x - 2y = 1 - 4t^2$,
$x' - 3y' + 6y - x = 3t + 5$

21. $x' - 3y' + x - 4y = -e^{-t}$,
$2x' - 6y' + x + 2y = 1 - 4t$

22. $3x' + y' - x - y = t$,
$5x' + 2y' + x + y = t^2 + 1$

23. $4x' - 2y' - x = 1$,
$x' + y' + x - y = 0$

24. $2x' - y' + 4x + y = 3$,
$x' + 3y' + x = t^2$

25. $3x' + 4y' - 2x = 0$,
$x' - y' - x + 4y = t$

26. Show that the change of variable $t = e^s$ transforms

$$tx' - ax - by = 0,$$
$$ty' - cx - dy = 0$$

into a linear system with constant coefficients.

In each of Problems 27 through 30, use the method suggested by Problem 26 to find the general solution.

27. $tx' - 3x + 2y = 0$,
$ty' + x + 2y = 0$

28. $tx' + x + 4y = 0$,
$ty' + x + y = 0$

29. $tx' - x + 4y = 0$,
$ty' + x + 3y = 0$

30. $tx' - 4x + 2y = 0$,
$ty' + x + 5y = 0$

In each of Problems 31 through 35, find a solution satisfying the given conditions.

31. $x' - 3y = 1$
$x' + 2y' + x = 0$
$x(0) = y(0) = 0$

32. $x' + x - y = t$
$x' + y' = 0$
$x(0) = y(0) = 0$

33. $x' - 2y' = 4$
$x' - y + x = t^2$
$x(0) = y(0) = 0$

34. $x' - y' + y = 0$
$x' + x - 2y = e^{-t}$
$x(0) = y(0) = 0$

35. $x' + y' + y = 0$
$x' - y' + x = 4$
$x(0) = y(0) = 0$

7.2 SOLUTION OF SYSTEMS BY LAPLACE TRANSFORM

Laplace transforms are particularly effective with systems in which each equation is linear with constant coefficients or certain polynomial coefficients, or which contain a discontinuous forcing function. In such cases, the transform takes the system to an algebraic system, which we then solve and invert. Since Laplace transforms are more suited to problems with prescribed initial data, we provide such information with these examples.

EXAMPLE 190

Solve

$$x' - 2y' + x - y = 1,$$
$$3x' + 4y' - 3x = t;$$
$$x(0) = y(0) = 0.$$

Taking Laplace transforms gives us

$$sX(s) - 2sY(s) + X(s) - Y(s) = \frac{1}{s},$$

$$3sX(s) + 4sY(s) - 3X(s) = \frac{1}{s^2}.$$

Then,

$$(s+1)X(s) + (-2s-1)Y(s) = \frac{1}{s},$$

$$(3s-3)X(s) + 4sY(s) = \frac{1}{s^2}.$$

Solve these to get

$$X(s) = \frac{4s^2 + 2s + 1}{s^2(10s^2 + s - 3)}$$

and

$$Y(s) = \frac{-3s^2 + 4s + 1}{s^2(10s^2 + s - 3)}.$$

These can be inverted using Heaviside's formulas (Chapter 4, Section 4), or we can use partial fractions to write

$$X(s) = -\frac{7}{9}\frac{1}{s} - \frac{1}{3}\frac{1}{s^2} - \frac{31}{99}\frac{1}{s+\frac{3}{5}} + \frac{108}{99}\frac{1}{s-\frac{1}{2}}$$

and

$$Y(s) = -\frac{13}{9}\frac{1}{s} - \frac{1}{3}\frac{1}{s^2} + \frac{62}{99}\frac{1}{s+\frac{3}{5}} + \frac{81}{99}\frac{1}{s-\frac{1}{2}}.$$

Then,

$$x(t) = -\tfrac{7}{9} - \tfrac{1}{3}t - \tfrac{31}{99}e^{-3t/5} + \tfrac{108}{99}e^{t/2}$$

and

$$y(t) = -\tfrac{13}{9} - \tfrac{1}{3}t + \tfrac{62}{99}e^{-3t/5} + \tfrac{81}{99}e^{t/2}.$$

EXAMPLE 191

Solve

$$x'' - 2x' + 3y' + 2y = 4,$$
$$2y' - x' + 3y = 0;$$
$$x(0) = x'(0) = y(0) = 0.$$

Taking Laplace transforms and grouping terms gives us

$$(s^2 - 2s)X + (3s + 2)Y = \frac{4}{s},$$

$$-sX + (2s + 3)Y = 0.$$

Then,

$$X = \frac{4s + 3}{s^2(s + 2)(s - 1)} \quad \text{and} \quad Y = \frac{4}{2s(s + 2)(s - 1)}.$$

Invert these to get

$$x(t) = -3 - 2t + \tfrac{1}{3}e^{-2t} + \tfrac{8}{3}e^t$$
$$y(t) = -1 + \tfrac{1}{3}e^{-2t} + \tfrac{2}{3}e^t.$$

In the next section we introduce some concepts which are useful in analyzing qualitative behavior of nonlinear systems whose solutions may not be obtainable in terms of elementary functions.

PROBLEMS FOR SECTION 7.2

In each of Problems 1 through 15, use Laplace transforms to solve the given system, subject to the given conditions.

1. $x' - 2y' = 1$
$x' + y - x = 0$
$x(0) = y(0) = 0$

2. $2x' - 3y + y' = 0$
$x' + y' = t$
$x(0) = y(0) = 0$

3. $x' + 2y' - y = 1$
$2x' + y = 0$
$x(0) = y(0) = 0$

4. $x' + y' - x = \cos(2t)$
$x' + 2y' = 0$
$x(0) = y(0) = 0$

5. $x' + 2x - y = 1$
$x' + y' - x = 0$
$x(0) = y(0) = 0$

6. $x' - x - y = 4$
$2x' - y = 1$
$x(0) = y(0) = 0$

7. $3x' - y = 2t$
$x' + y' - y = 0$
$x(0) = y(0) = 0$

8. $x' + 4y' - y = 0$
$x' + 2y = e^{-t}$
$x(0) = y(0) = 0$

9. $x' + 2x - y' = 0$
$x' + y + x = t^2$
$x(0) = y(0) = 0$

10. $x' + 4x - y = 0$
$x' + y' = t$
$x(0) = y(0) = 0$

11. $x' + 2y' - y = -t$
$x' + y = 0$
$x(0) = y(0) = 0$

12. $x' + y' - x = 0$
$x' - y' + 2y = 3$
$x(0) = y(0) = 0$

13. $x' - y' - x = 0$
$x' + 4y' + 2y = 1$
$x(0) = y(0) = 1$

14. $x' + y' + x - y = 0$
$x' + 2y' + x = 1$
$x(0) = y(0) = 0$

15. $x' + 2y' - x = 0$
$4x' + 3y' + y = -6$
$x(0) = -1,\ y(0) = 1$

16. Construct and solve the system of differential equations for the currents in the circuit of Figure 83. Assume zero initial conditions.

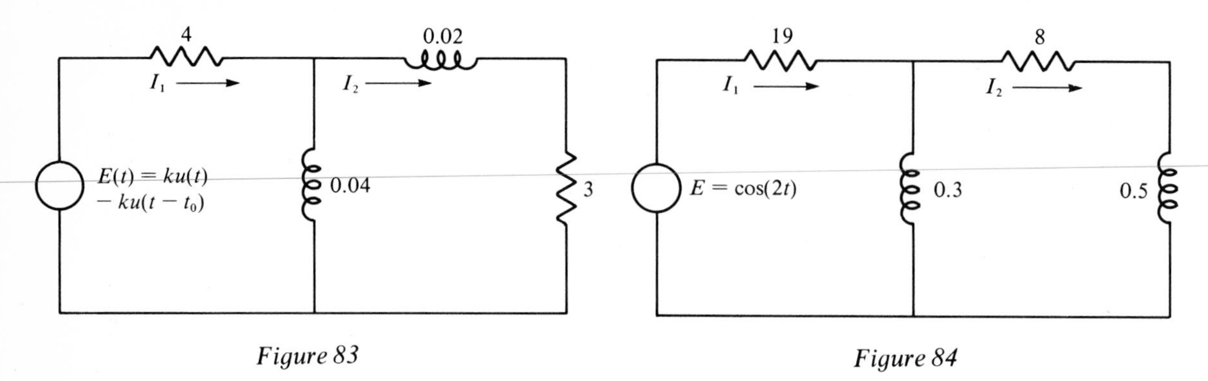

Figure 83

Figure 84

17. Construct and solve the system of differential equations for the currents in the circuit of Figure 84. Assume zero initial conditions.

18. Solve for the motion of the mass-spring system depicted in Figure 85. Assume zero initial displacements.

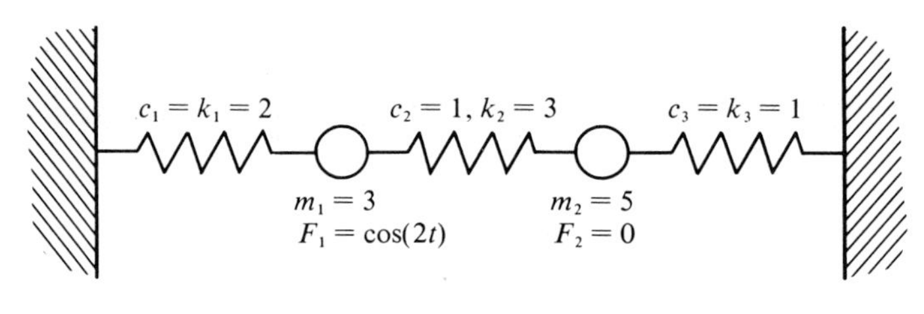

Figure 85

19. Solve for the motion of the mass-spring system of Figure 86. Assume zero initial displacements.

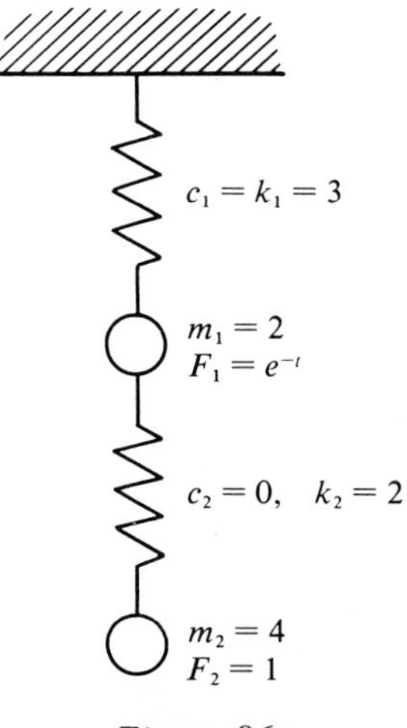

Figure 86

7.3 NONLINEAR SYSTEMS, PHASE PLANE, CRITICAL POINTS, AND STABILITY

At least for linear, constant coefficient systems, we have two effective methods of solution. We shall now sketch some of the ideas behind the notions of phase plane and stability, which enable us to draw certain conclusions about solutions of nonlinear systems, even in those cases where explicit solutions cannot be derived.

We shall consider systems of the form

$$\begin{aligned} \frac{dx}{dt} &= F(x, y), \\ \frac{dy}{dt} &= G(x, y). \end{aligned} \tag{a}$$

Such a system is called *autonomous* because t does not appear explicitly, but only implicitly through $x(t)$, $x'(t)$, $y(t)$, and $y'(t)$. We shall assume the following existence result:

If $F(x, y)$, $G(x, y)$, and their first partial derivatives are continuous in a region containing (x_0, y_0), then there exists a unique solution of (a) satisfying the prescribed conditions $x(t_0) = x_0$, $y(t_0) = y_0$.

For our treatment, $F(x, y)$, $G(x, y)$, and their first partial derivatives will be continuous in the entire plane. Given solutions $x(t)$ and $y(t)$ of (a), the point $(x(t), y(t))$ will generally trace out a path or trajectory in the xy-plane as t varies. In this context we often call the xy-plane the *phase plane.* A solution satisfying $x(t_0) = x_0$, $y(t_0) = y_0$ appears in the phase plane as a trajectory through (x_0, y_0). By the uniqueness part of the existence result, there is only one trajectory through (x_0, y_0). An exceptional case occurs, however, if $F(x_0, y_0) = G(x_0, y_0) = 0$, for then the solution is just $x = x_0$, $y = y_0$, and the trajectory collapses to a single point. We then call (x_0, y_0) a *critical point** of the system (a). In our discussion, we always assume that critical points are *isolated*, that is, in some circle about (x_0, y_0), there are no other critical points.

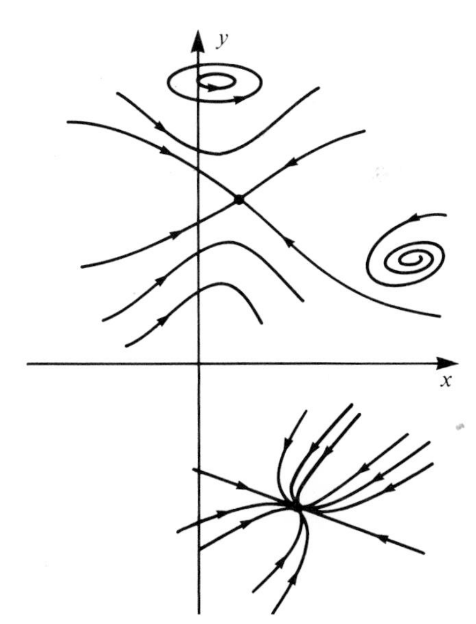

Figure 87. Typical phase portrait.

The *phase portrait* of (a) consists of drawings of trajectories in the phase plane. In such a drawing, we usually attach an arrow to each trajectory to indicate the direction of motion of a point $(x(t), y(t))$ moving along the trajectory as t increases. Thus, one might have a drawing something like that of Figure 87. We shall see below

* In some texts, a critical point is sometimes called an *equilibrium point*.

how the various configurations of trajectories shown might arise. Information about the phase portrait gives us information about solutions of the system. For example, closed trajectories correspond to periodic solutions of the system.

Two basic questions we can ask about systems are the following:

1. If $(x_1(t), y_1(t))$ and $(x_2(t), y_2(t))$ are two trajectories which are "close" at some time, will they remain close as $t \to \infty$? This is called the question of *stability*.
2. How do trajectories behave as $t \to \infty$ (or $t \to -\infty$)? This is treated later in the concept of *asymptotic stability*.

We shall now formulate some definitions which will enable us to pose these questions more carefully, and then give some results and illustrations. Assume in the following that (x_0, y_0) is an isolated critical point of (a).

A trajectory $(x(t), y(t))$ in the phase plane *approaches* (x_0, y_0) if $x(t)$ approaches x_0 and $y(t)$ approaches y_0 as $t \to \infty$ (or $t \to -\infty$). This means that the point $P(x(t), y(t))$ moving along the trajectory approaches $P_0(x_0, y_0)$ at t increases indefinitely or goes to minus infinity.

If $(x(t), y(t))$ approaches (x_0, y_0), and also the ratio

$$\frac{y - y_0}{x - x_0}$$

approaches a finite limit (or plus or minus infinity) as $t \to \infty$ or as $t \to -\infty$, then we say that the trajectory *enters* (x_0, y_0). Since $(y - y_0)/(x - x_0)$ is the slope of the line from (x_0, y_0) to (x, y), this says that the line $\overline{PP_0}$ approaches a definite direction as $t \to \infty$ or as $t \to -\infty$. Figure 88 shows a trajectory approaching (x_0, y_0), while Figure 89 shows a trajectory entering (x_0, y_0). In the latter case, the trajectory approaches a definite direction as it approaches (x_0, y_0).

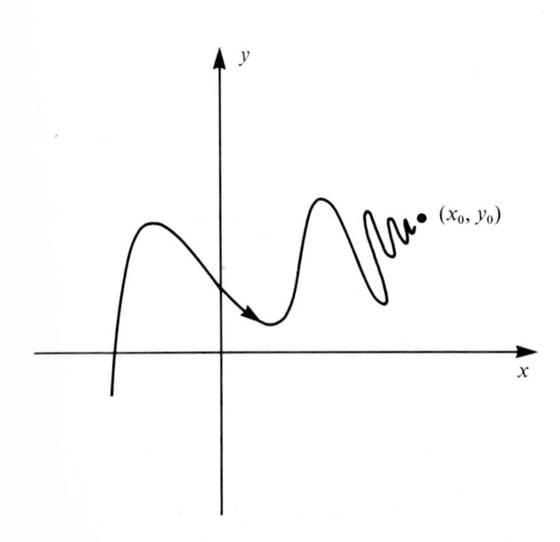

Figure 88. Trajectory approaching (x_0, y_0).

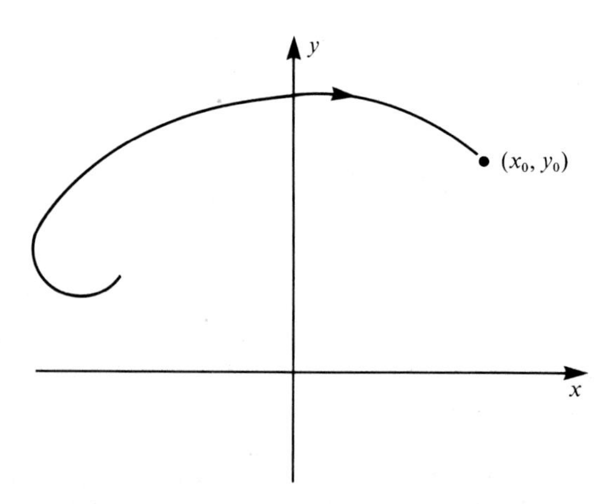

Figure 89. Trajectory entering (x_0, y_0).

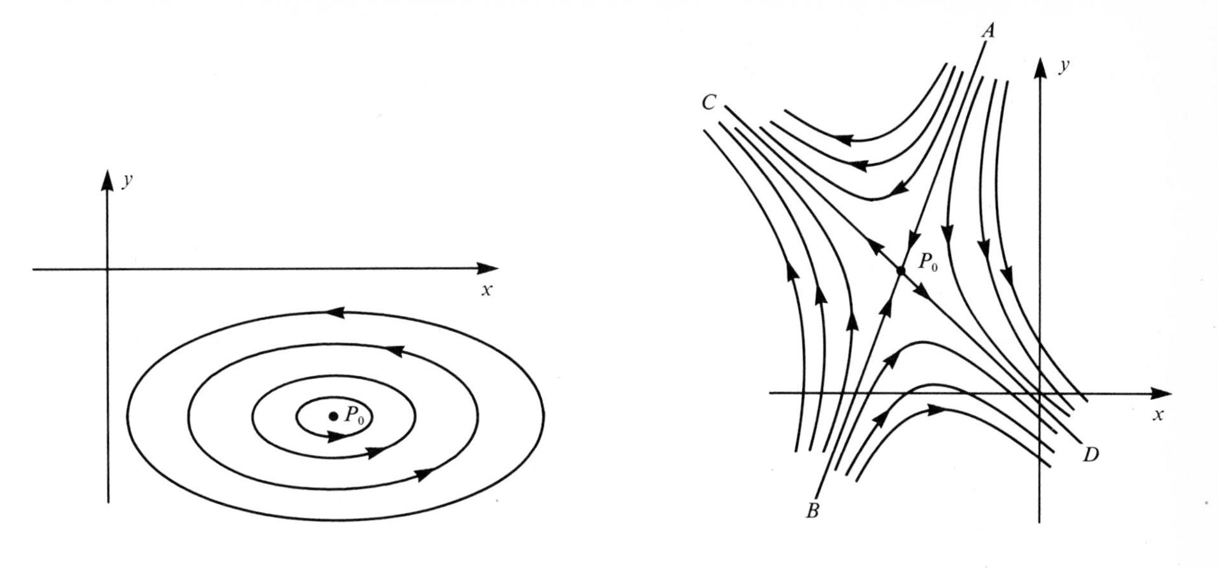

Figure 90. Center at P_0.

Figure 91. Saddle point at P_0.

We shall now distinguish four kinds of critical points:

1. (x_0, y_0) is a *center* if it is surrounded by an infinite family of closed trajectories coming arbitrarily close to (x_0, y_0), but is not approached by any trajectory as $t \to \infty$ or as $t \to -\infty$. This situation is shown in Figure 90.
2. (x_0, y_0) is a *saddle point* if the trajectories are as shown in Figure 91, in which two half-lines AP_0, BP_0 are trajectories approaching P_0 as $t \to \infty$; two half-lines CP_0, DP_0 are trajectories approaching P_0 as $t \to -\infty$; and the other trajectories are hyperbola-like curves having pairs of these half-lines as asymptotes. Note that the half-lines approaching P_0 as $t \to \infty$ have their arrows pointing *toward* P_0, while those approaching P_0 as $t \to -\infty$ have their arrows pointing *away from* P_0.
3. (x_0, y_0) is a *spiral point* if every trajectory within some circle about (x_0, y_0) spirals about (x_0, y_0) infinitely often and approaches (x_0, y_0) as $t \to \infty$ or as $t \to -\infty$. This is shown in Figure 92.
4. (x_0, y_0) is a *node* if it is entered by an infinite family of trajectories as $t \to \infty$ or as $t \to -\infty$. This is shown in Figure 93.

A critical point is called *stable* if, for each $R > 0$, there exists some r, with $0 \le r \le R$, such that every trajectory which is inside the circle $(x - x_0)^2 + (y - y_0)^2 = r^2$ at some time t_0 remains inside the circle $(x - x_0)^2 + (y - y_0)^2 = R^2$ at all later times $t > t_0$. This means that any trajectory coming within r units of (x_0, y_0) at some time t_0 remains at least within R units of (x_0, y_0) at all later times. This is shown in Figure 94.

We call (x_0, y_0) *asymptotically stable* if it is stable and, in addition, there is some circle about (x_0, y_0) such that every trajectory inside the circle at some time t_0 approaches (x_0, y_0) as $t \to \infty$.

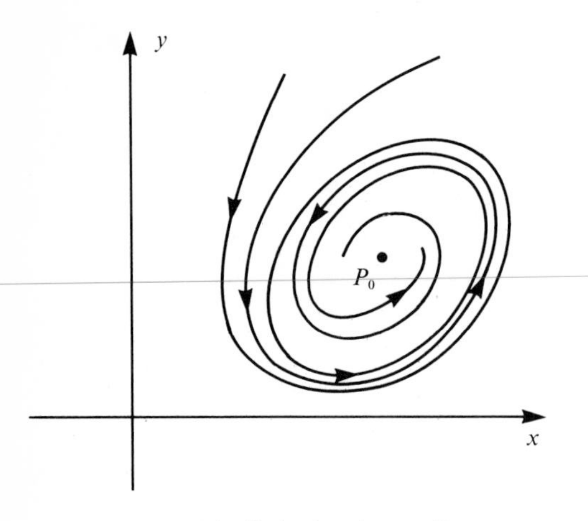

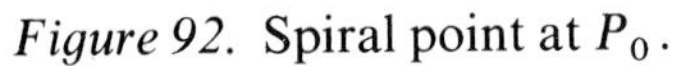
Figure 92. Spiral point at P_0.

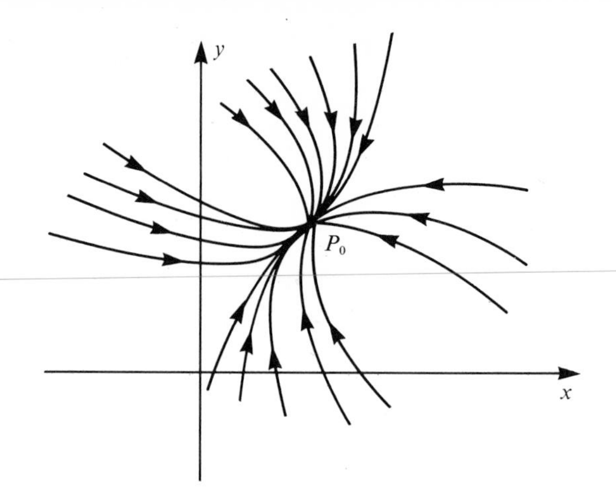

Figure 93. Node at P_0.

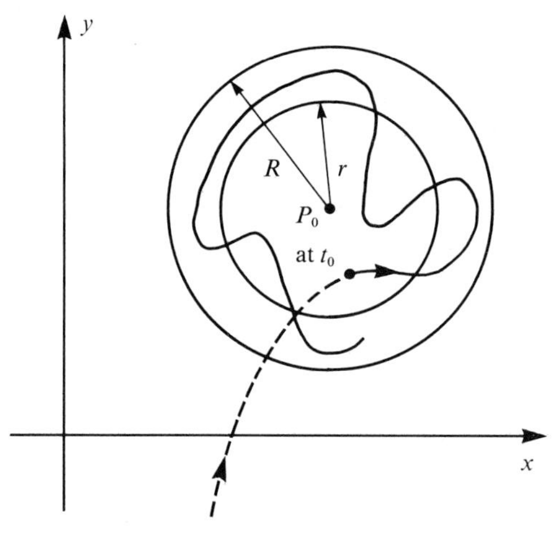

Figure 94. Stability at P_0.

Of course, our aim is to study nonlinear systems. However, if we go back and look at linear systems, we can derive a definitive result which will be useful in the nonlinear case, and which also provides us with some insight and a wealth of instructive examples.

THEOREM 24

Let a, b, c, and d be constants such that $ad - bc \neq 0$. Then, the origin is the only critical point of the system

$$\frac{dx}{dt} = ax + by,$$

$$\frac{dy}{dt} = cx + dy.$$

Further, if λ_1 and λ_2 are the roots of the *auxiliary equation**

$$\lambda^2 - (a + d)\lambda + (ad - bc) = 0,$$

then there are five possible cases:

1. (0, 0) is a *node* if λ_1 and λ_2 are real, distinct, and of the same sign. If λ_1 and λ_2 are both negative, then (0, 0) is asymptotically stable; if both roots are positive, then (0, 0) is unstable (that is, not stable).
2. (0, 0) is a *saddle point* if λ_1 and λ_2 are real, distinct, and of opposite sign. Here, (0, 0) is unstable.
3. (0, 0) is a *spiral point* if λ_1 and λ_2 are complex conjugates with nonzero real parts. Here, (0, 0) is asymptotically stable if λ_1 and λ_2 have negative real parts and is unstable if the real parts are positive.
4. (0, 0) is a *node* if λ_1 and λ_2 are real and equal. In this case, (0, 0) is asymptotically stable if $\lambda_1 < 0$ and unstable if $\lambda_1 > 0$.
5. (0, 0) is a *center* if λ_1 and λ_2 are pure imaginary. In this case, (0, 0) is stable but not asymptotically stable.

We shall prove case 1 and leave the others to the exercises.

Proof of Case 1 Suppose that λ_1 and λ_2 are negative, say $\lambda_1 < \lambda_2 < 0$. Using previous methods, we find that the general solution of the system is

$$x(t) = c_1 e^{\lambda_1 t} + c_2 e^{\lambda_2 t},$$

$$y(t) = c_1\left(\frac{\lambda_1 - a}{b}\right)e^{\lambda_1 t} + c_2\left(\frac{\lambda_2 - a}{b}\right)e^{\lambda_2 t}.$$

To show that (0, 0) is a node, we must produce an infinite family of trajectories entering the origin. To do this, first choose $c_2 = 0$ to get

$$x(t) = c_1 e^{\lambda_1 t},$$

$$y(t) = c_1\left(\frac{\lambda_1 - a}{b}\right)e^{\lambda_1 t}.$$

Thus,

$$y = \left(\frac{\lambda_1 - a}{b}\right)x,$$

which is a half-line (half, since $e^{\lambda_1 t} > 0$ for all t).

---

* The auxiliary equation of the system $dx/dt = ax + by$, $dy/dt = cx + dy$, is actually the characteristic equation of the 2×2 matrix

$$\begin{bmatrix} a & b \\ c & d \end{bmatrix},$$

and the roots are the eigenvalues of this matrix (see Chapter 10, Section 10.15).

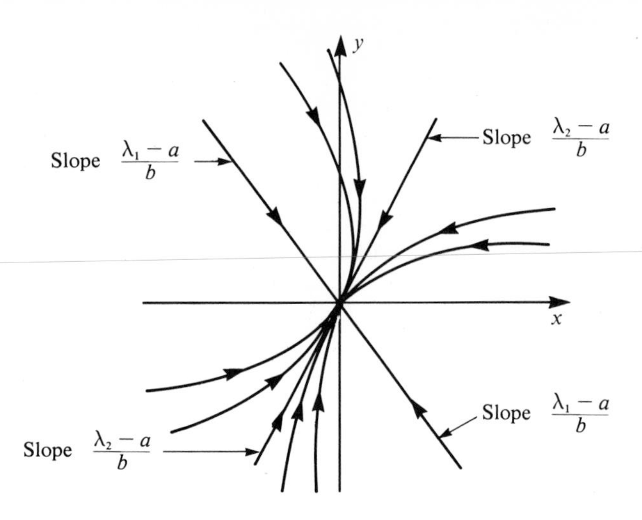

Figure 95. Case 1 of theorem 24: asymptotically stable node at (0, 0).

Now, if $c_1 > 0$, then $x > 0$, and we get a half-line such as in Figure 95. As $t \to \infty$, $x(t) \to 0$ and $y(t) \to 0$ because $\lambda_1 < 0$. Thus, this trajectory enters (0, 0), with slope $(\lambda_1 - a)/b$.

If $c_1 < 0$, then $x < 0$, but we still get a half-line trajectory entering (0, 0) as $t \to \infty$. This half-line trajectory enters (0, 0) with slope $(\lambda_1 - a)/b$.

If $c_1 = 0$ but $c_2 \neq 0$, we realize a similar conclusion, with two half-line trajectories of slope $(\lambda_2 - a)/b$, both entering (0, 0) as $t \to \infty$.

If c_1 and c_2 are both nonzero, then $x(t) \to 0$ and $y(t) \to 0$ as $t \to \infty$ still holds, because λ_1 and λ_2 are both negative. Further,

$$\frac{y}{x} = \frac{c_1\left(\frac{\lambda_1 - a}{b}\right)e^{\lambda_1 t} + c_2\left(\frac{\lambda_2 - a}{b}\right)e^{\lambda_2 t}}{c_1 e^{\lambda_1 t} + c_2 e^{\lambda_2 t}}.$$

This can be written

$$\frac{y}{x} = \frac{c_1\left(\frac{\lambda_1 - a}{b}\right)e^{(\lambda_1 - \lambda_2)t} + c_2\left(\frac{\lambda_2 - a}{b}\right)}{c_1 e^{(\lambda_1 - \lambda_2)t} + c_2}.$$

Since $\lambda_1 < \lambda_2 < 0$, then $\lambda_1 - \lambda_2 < 0$, and as $t \to \infty$ we get

$$\frac{y}{x} \to \frac{\lambda_2 - a}{b},$$

and all these trajectories enter (0, 0) with limiting slope $(\lambda_2 - a)/b$. Thus, (0, 0) is a node, and is asymptotically stable.

If $\lambda_1 > 0$ and $\lambda_2 > 0$, we find the same types of expressions as above, except that now the trajectories enter (0, 0) as $t \to -\infty$, not as $t \to \infty$. Thus, (0, 0) is still a node, but is unstable (see Figure 96).

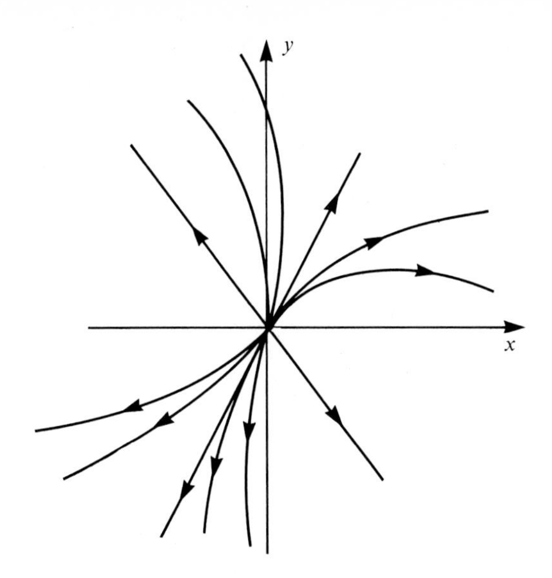

Figure 96. Unstable node at (0, 0).

The same systematic analysis of the general solution (which we have explicitly for the linear case) yields the other conclusions of the theorem. Here are illustrations of cases 1, 2, and 5.

EXAMPLE 192 Case 1 of Theorem 24

Consider the system

$$\frac{dx}{dt} = 3x + y,$$

$$\frac{dy}{dt} = x + 3y.$$

The auxiliary equation is $\lambda^2 - 6\lambda + 8 = 0$, with roots $\lambda_1 = -4$, $\lambda_2 = -2$. Here, $\lambda_1 < \lambda_2 < 0$, and we are in case 1 of Theorem 24. The general solution of the system is found to be

$$x(t) = c_1 e^{-4t} + c_2 e^{-2t},$$
$$y(t) = -7c_1 e^{-4t} - 5c_2 e^{-2t}.$$

If $c_2 = 0$ and $c_1 \neq 0$, we get

$$y(t) = -7x(t).$$

If $c_1 = 0$ and $c_2 \neq 0$, we get

$$y(t) = -5x(t).$$

These give us the half-line trajectories of slope -7 and -5, respectively, entering (0, 0) as shown in Figure 97.

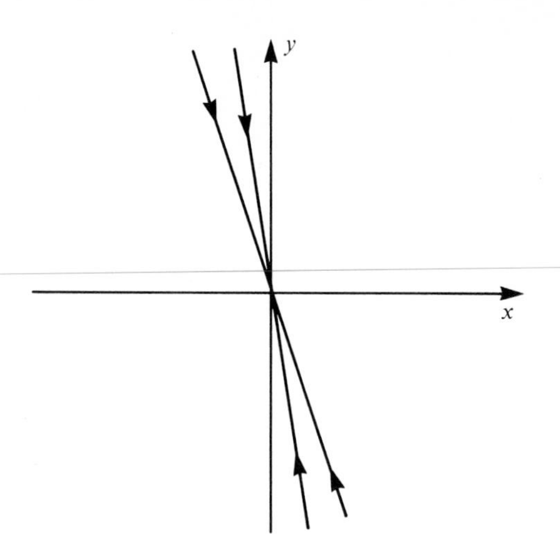

Figure 97. Half-line trajectories entering (0, 0) with slopes -5, -7.

If $c_1 \neq 0$ and $c_2 \neq 0$, then

$$\frac{y}{x} = \frac{-7c_1e^{-4t} - 5c_2e^{-2t}}{c_1e^{-4t} + c_2e^{-2t}} = \frac{-7c_1e^{-2t} - 5c_2}{c_1e^{-2t} + c_2} \to -5 \quad \text{as} \quad t \to \infty.$$

The trajectories formed when c_1 and c_2 are both nonzero thus enter (0, 0) with slope -5. Here, (0, 0) is an asymptotically stable node.

EXAMPLE 193 Case 2 of Theorem 24

Here is an example in which (0, 0) is an unstable saddle point. Consider

$$\frac{dx}{dt} = -x + 3y,$$

$$\frac{dy}{dt} = 2x - 2y.$$

The auxiliary equation is $\lambda^2 + 3\lambda - 4 = 0$, with roots 1 and -4. These are real and of opposite sign. The general solution of the system is

$$x(t) = c_1e^t + c_2e^{-4t},$$
$$y(t) = \tfrac{2}{3}c_1e^t - c_2e^{-4t}.$$

If $c_1 = 0$, then $y = -x = -c_2e^{-4t}$. This trajectory has slope -1, and enters (0, 0) as $t \to \infty$. If $c_2 = 0$, then $y = \frac{2}{3}x = \frac{2}{3}c_1e^t$, which has slope $\frac{2}{3}$ and enters (0, 0) as $t \to -\infty$, *not* as $t \to \infty$.

If c_1 and c_2 are both nonzero, then x does not go to zero as $t \to \infty$ or as $t \to -\infty$. Similarly, y does not go to zero. Thus, we get the trajectories shown in Figure 98. Here, we have a saddle point at (0, 0). Further, (0, 0) is unstable because any trajectories entering a circle about (0, 0) also leave at some later time.

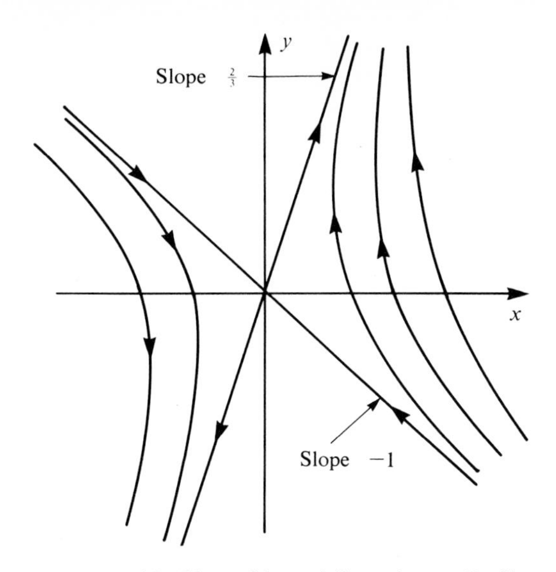

Figure 98. Unstable saddle point at (0, 0).

EXAMPLE 194 Case 5 of Theorem 24

Consider the system

$$\frac{dx}{dt} = 3x + y,$$

$$\frac{dy}{dt} = -13x - 3y.$$

The general solution is

$$x(t) = c_1 \cos(2t) + c_2 \sin(2t),$$
$$y(t) = (2c_2 - 3c_1)\cos(2t) + (-2c_1 - 3c_2)\sin(2t).$$

Here, the trajectories are closed paths about the origin, as shown in Figure 99. Note that neither x nor y approaches zero as $t \to \infty$ or as $t \to -\infty$. Clearly, (0, 0) is stable but not asymptotically stable, and is also a center.

In the exercises we shall ask the reader to analyze other examples. For the remainder of this section we turn our attention back to nonlinear systems, which are the main object of the theory.

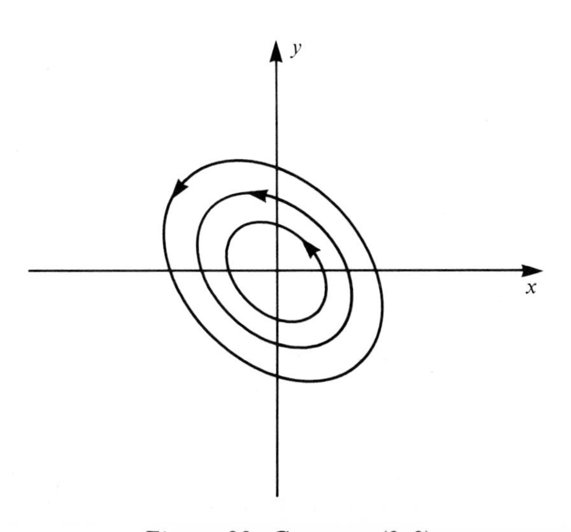

Figure 99. Center at (0, 0).

Our first result in this direction shows how we can sometimes draw conclusions about certain nonlinear systems from associated linear systems which we know how to handle.

THEOREM 25

Suppose we have a nonlinear system of the form

$$\frac{dx}{dt} = ax + by + P(x, y),$$
$$\frac{dy}{dt} = cx + dy + Q(x, y). \tag{a}$$

Assume that $ad - bc \neq 0$, that $P(x, y)$ and $Q(x, y)$ have continuous first partial derivatives for all x and y, and that

$$\lim_{(x, y) \to (0, 0)} \left\{ \frac{P(x, y)}{\sqrt{x^2 + y^2}} \right\} = \lim_{(x, y) \to (0, 0)} \left\{ \frac{Q(x, y)}{\sqrt{x^2 + y^2}} \right\} = 0.$$

Finally, assume that (0, 0) is an isolated critical point [hence, $P(0, 0) = Q(0, 0) = 0$].

Let λ_1 and λ_2 be roots of

$$\lambda^2 - (a + d)\lambda + ad - bc = 0,$$

the auxiliary equation of the associated linear system $dx/dt = ax + by$, $dy/dt = cx + dy$. Then, the nonlinear system (a) and the linear system

$$\frac{dx}{dt} = ax + by,$$
$$\frac{dy}{dt} = cx + dy \tag{b}$$

have the same behavior at (0, 0) in the following cases:

1. If λ_1 and λ_2 are real, distinct, and of the same sign, then (a) and (b) both have a node at (0, 0).
2. If λ_1 and λ_2 are real, distinct, and of opposite sign, then both (a) and (b) have a saddle point at (0, 0).
3. If $\lambda_1 = \lambda_2$ and is real, then (0, 0) is a node of both (a) and (b) *except* if $a = d \neq 0$ and $b = c = 0$. In the latter event, (0, 0) is a node of (b), but may be either a node or spiral point of (a).
4. If λ_1 and λ_2 are complex with nonzero real parts, then (0, 0) is a spiral point of both (a) and (b).

We shall not prove this. However, observe that the form of nonlinear system (a) suggests that it may be viewed as a perturbed version of (b). Further, the hypotheses of the theorem require that, as (x, y) approaches (0, 0), the "differences" $P(x, y)$ and $Q(x, y)$ between (a) and (b) go to zero faster than $\sqrt{x^2 + y^2}$. This is the basis of the connection between the two systems and the behavior of solutions at the origin. We shall also note without proof that, corresponding to case 5 of Theorem 24 in which λ_1 and λ_2 are pure imaginary, (0, 0) is a center of (b), but may be either a center or spiral point of (a).

Here are two illustrations of the power of this theorem.

EXAMPLE 195

Consider the nonlinear system

$$\frac{dx}{dt} = 4x - 2y - 4xy,$$

$$\frac{dy}{dt} = x + 6y - 8x^2y.$$

It is easy to verify that $P(x, y) = -4xy$ and $Q(x, y) = -8x^2y$ satisfy the hypotheses of Theorem 25. Here the auxiliary equation is

$$\lambda^2 - 10\lambda + 26 = 0,$$

with roots $5 + i$ and $5 - i$. By case 4 of the theorem, the nonlinear system has a spiral point at (0, 0). One can check that this spiral point is unstable.

EXAMPLE 196

Consider the system

$$\frac{dx}{dt} = x - 3y + 4xy,$$

$$\frac{dy}{dt} = x + 7y - y^{3/2}x.$$

The auxiliary equation is

$$\lambda^2 - 8\lambda + 10 = 0,$$

with roots

$$4 + \sqrt{6} \quad \text{and} \quad 4 - \sqrt{6}.$$

By case 1 of the theorem, the system has a node at (0, 0).

The following result due to Liapunov discusses stability of (0, 0) for certain nonlinear systems.

THEOREM 26 Liapunov's Theorem

Under the conditions of Theorem 25, we may draw the following conclusions:

1. If λ_1 and λ_2 are either real and both negative, or complex with negative real parts, then (0, 0) is asymptotically stable for both systems (a) and (b).
2. If both λ_1 and λ_2 are positive, or complex with positive real parts, then (0, 0) is an unstable critical point of both systems (a) and (b).

EXAMPLE 197

Consider the system

$$\frac{dx}{dt} = -x + y + x^3y,$$

$$\frac{dy}{dt} = -2x - 3y - x^2y^2.$$

The auxiliary equation is

$$\lambda^2 + 4\lambda + 5 = 0,$$

with roots $-2 + i$ and $-2 - i$. By conclusion 1 of Liapunov's theorem, (0, 0) is an asymptotically stable critical point of the system.

We conclude this section with two remarkable results concerning periodic solutions. For the first result, we need a bit of terminology. A closed curve in the plane

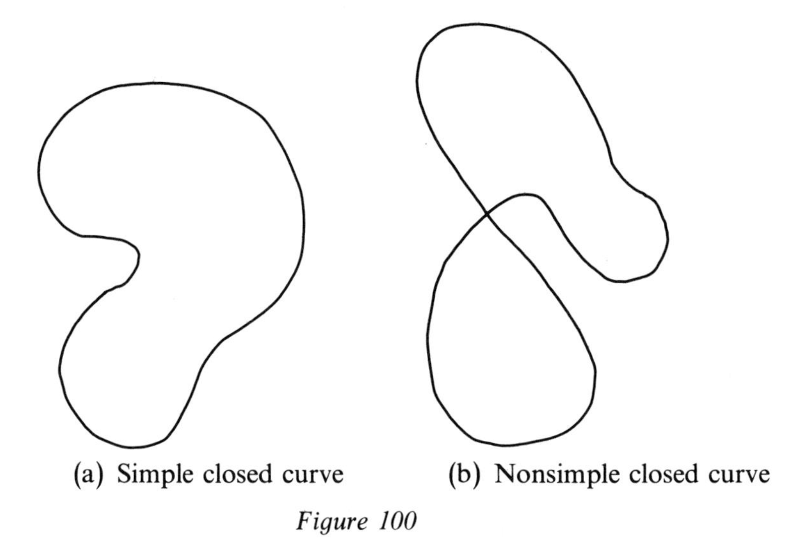

(a) Simple closed curve (b) Nonsimple closed curve

Figure 100

is called *simple* if it does not cross itself. For example, the curve in Figure 100(a) is simple, while that in Figure 100(b) is not. A *region*, for purposes of the next theorem only, consists of all points on and enclosed by a simple closed curve.

THEOREM 27 Poincaré-Bendixson Theorem

Let R be a region in the phase plane, and suppose that R contains no critical point of the system

$$\frac{dx}{dt} = F(x, y),$$

$$\frac{dy}{dt} = G(x, y).$$

Let C be a trajectory that is in R at all times $t > t_0$. Then, either C is a closed trajectory, as shown in Figure 101(a), or is a spiral which approaches a closed trajectory as $t \to \infty$, as shown in Figure 101(b).

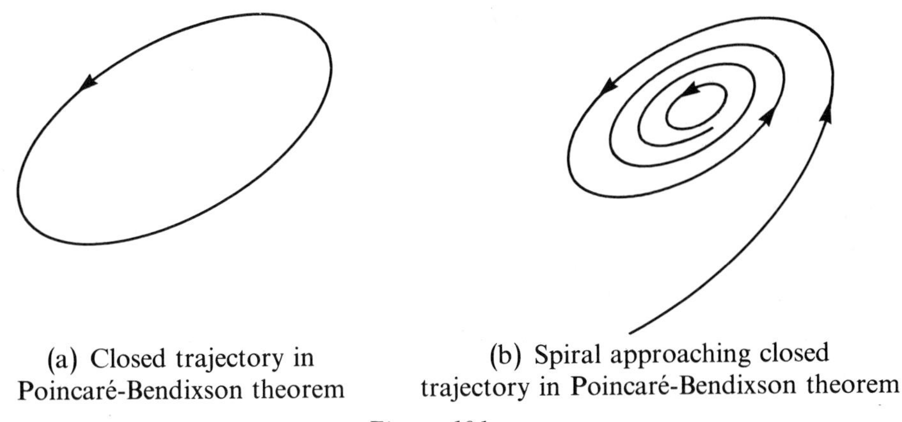

(a) Closed trajectory in Poincaré-Bendixson theorem

(b) Spiral approaching closed trajectory in Poincaré-Bendixson theorem

Figure 101

A much more general version of this theorem can be stated, but one must go into various subtleties in defining a region more carefully.

Lienard's theorem, which we shall state next, applies to the differential equation

$$x'' + p(x)x' + q(x) = 0.$$

This is not a system, but upon setting $y = x'$, we obtain the system

$$\frac{dx}{dt} = y,$$

$$\frac{dy}{dt} = -p(x)y - q(x),$$

whose solutions yield solutions of the second order differential equation.

THEOREM 28 Lienard's Theorem

Let $p(x)$ and $q(x)$ be continuous and have continuous derivatives for all x. Suppose also that

1. $q(x) = -q(-x)$ for all x
2. $q(x) > 0$ for $x > 0$
3. $p(x) = p(-x)$ for all x
4. The function $F(x) = \int_0^x p(x)\,dx$ has exactly one positive zero, say at $x = a$, and $F(x) < 0$ if $0 < x < a$ and $F(x)$ is positive and nondecreasing for $x > a$. Further, $F(x) \to \infty$ as $x \to \infty$.

Then, the system

$$\frac{dx}{dt} = y,$$

$$\frac{dy}{dt} = -p(x)y - q(x)$$

has a unique closed trajectory around the origin in the phase plane. Further, every other trajectory approaches this unique closed trajectory spirally as $t \to \infty$.

As an application of this theorem, consider the *van der Pol equation*

$$x'' + \alpha(x^2 - 1)x' + x = 0,$$

derived by van der Pol in the 1920s in connection with a study of vacuum tubes. Here, α is a positive constant.

One can verify that the hypotheses of Lienard's theorem hold, with $p(x) = \alpha(x^2 - 1)$ and $q(x) = x$. For example,

$$F(x) = \int_0^x \alpha(x^2 - 1)\,dx = \alpha\left(\frac{x^3}{3} - x\right),$$

with a single positive zero at $x = \sqrt{3}$. The system associated with van der Pol's equation is

$$\frac{dx}{dt} = y,$$

$$\frac{dy}{dt} = -\alpha(x^2 - 1)y - x.$$

By Lienard's theorem, this has a unique closed trajectory about the critical point at (0, 0). This closed trajectory represents a periodic solution. That van der Pol's equation has a unique periodic solution is far from obvious by other means.

Due to space limitations, we shall not pursue these topics any further. There are many important ideas and techniques which we have omitted, including limit cycles, Liapunov functions, and the method of Krylov and Bogolyubov.

PROBLEMS FOR SECTION 7.3

For each of the linear systems in Problems 1 through 12, determine the nature of the critical point at (0, 0), and whether (0, 0) is stable, asymptotically stable, unstable, or stable but not asymptotically stable. For each, draw some of the trajectories in the phase plane.

1. $\dfrac{dx}{dt} = x + 2y, \quad \dfrac{dy}{dt} = 2x + 6y$

2. $\dfrac{dx}{dt} = 4x + y, \quad \dfrac{dy}{dt} = -17x - 4y$

3. $\dfrac{dx}{dt} = 2x, \quad \dfrac{dy}{dt} = 8x + 2y$

4. $\dfrac{dx}{dt} = -x + 2y, \quad \dfrac{dy}{dt} = 2x + 3y$

5. $\dfrac{dx}{dt} = 3x - 3y, \quad \dfrac{dy}{dt} = 6x - 3y$

6. $\dfrac{dx}{dt} = x - 4y, \quad \dfrac{dy}{dt} = 4x - 7y$

7. $\dfrac{dx}{dt} = -5x - y, \quad \dfrac{dy}{dt} = 8x + 4y$

8. $\dfrac{dx}{dt} = -2x - 2y, \quad \dfrac{dy}{dt} = 10x + 2y$

9. $\dfrac{dx}{dt} = x + 2y, \quad \dfrac{dy}{dt} = -9x - 5y$

10. $\dfrac{dx}{dt} = x + 4y, \quad \dfrac{dy}{dt} = -x + y$

11. $\dfrac{dx}{dt} = 2x - 3y, \quad \dfrac{dy}{dt} = -8x + 4y$

12. $\dfrac{dx}{dt} = x + y, \quad \dfrac{dy}{dt} = 4x - 3y$

In each of Problems 13 through 24, apply Theorem 25 to determine the nature of the critical point at (0, 0), or else explain why the theorem yields no conclusion.

13. $\dfrac{dx}{dt} = x + 3y + xy^2, \quad \dfrac{dy}{dt} = x + 5y + x^2$

14. $\dfrac{dx}{dt} = -7x + 32y + 3xy, \quad \dfrac{dy}{dt} = 2y + 4x^2y^3$

15. $\dfrac{dx}{dt} = -3x - y, \quad \dfrac{dy}{dt} = 4x - 3y + 2xy^2$

16. $\dfrac{dx}{dt} = 3x - y + x\sin(y), \quad \dfrac{dy}{dt} = 3y - 4x + x^2 - y^2$

17. $\dfrac{dx}{dt} = 2x - 4y + 3x^3y^2, \quad \dfrac{dy}{dt} = x - 2y + 8x^4y^4$

18. $\dfrac{dx}{dt} = x - 2y + 1, \quad \dfrac{dy}{dt} = 3x + 2y + y^{1/2}$

19. $\dfrac{dx}{dt} = 2x + 4y - 3x^2, \quad \dfrac{dy}{dt} = 2x + 6y - xy$

20. $\dfrac{dx}{dt} = x - y + y^{1/3}, \quad \dfrac{dy}{dt} = -x - 2y + 3x^{1/2}$

21. $\dfrac{dx}{dt} = 2x - 4y + xy^2, \quad \dfrac{dy}{dt} = x - 2y + \cos(x - y)$

22. $\dfrac{dx}{dt} = 8x - 2y + x^3y, \quad \dfrac{dy}{dt} = 4x + 6y - x^2 + y^2$

23. $\dfrac{dx}{dt} = -3x + y - 4y^{1/2}, \quad \dfrac{dy}{dt} = 2x + 6y + y^2$

24. $\dfrac{dx}{dt} = 2x - 6y - x^3, \quad \dfrac{dy}{dt} = 4x + 2y + \sqrt{xy}$

25. If you have access to a computer with plotting capability (some microcomputers will do), draw some direction field elements of the van der Pol equation, and sketch solutions for $\alpha = 0.2, 1, 1.3, 2, 4$, and 12.

26. Prove Theorem 24(2).

27. Prove Theorem 24(3).

28. Prove Theorem 24(4).

29. Prove Theorem 24(5).

30. Prove the *Bendixson nonexistence criterion:* Suppose that $G(x, y)$ and $F(x, y)$ have continuous first partial derivatives in a region R of the phase plane, and that $\partial F/\partial x + \partial G/\partial y$ has the same sign at all points of R. Then, the system

$$\frac{dx}{dt} = F(x, y),$$

$$\frac{dy}{dt} = G(x, y)$$

has no closed trajectory in R.

(*Hint:* You need Green's theorem. If you are not familiar with it, put this problem away until you have read the vector analysis chapter.)

31. Recall that the differential equation for free vibrations of a mass on a damped spring is

$$mx'' + cx' + kx = 0.$$

Here, we have used x instead of the previous y in order to be consistent with the notation of this section.

(a) Show that the spring equation is equivalent to the system

$$\frac{dx}{dt} = y,$$

$$\frac{dy}{dt} = -\frac{cy}{m} - \frac{k}{m}x.$$

(b) Show that (0, 0) is the only critical point of this system.

(c) Determine the type of critical point in the cases of no damping, underdamping, critical damping, and overdamping. Also determine the nature of the stability of (0, 0) in each case.

32. Analyze solutions of

$$Ax'' + B(x^2 - 1) + Cx = 0$$

by using Lienard's theorem and a change of variables. Here, A, B, and C are positive constants. (*Hint:* Note that this equation is very similar to the van der Pol equation.)

CHAPTER SUMMARY

This chapter had two main themes. First, for simple systems such as ones with linear, constant coefficient equations, we can work out an explicit solution. If the system is nonlinear, however, we have little hope of solving it explicitly, and must then resort to qualitative methods to learn as much as we can about the solution from the form of the system itself. It is often remarkable how much we can discover in this way. A striking example is the use of Lienard's theorem to show that the van der Pol equation has a unique periodic solution.

In modern engineering and physics research, most of the linear problems have been solved, and such areas as shock waves, boundary layer theory, and the many-body problem involve delicate analysis of nonlinear equations and systems. Stability considerations are even now being invoked to attempt explanations of the red spot on Jupiter (considered as a problem in turbulent fluid flow) and the quite surprising structure of the rings of Saturn, as discovered by the *Voyager* spacecraft. The methods of nonlinear analysis thus assume more practical importance than ever before in applying mathematics to the problem of understanding how nature really behaves.

SUPPLEMENTARY PROBLEMS

In each of Problems 1 through 15, use the method of differential operators to find the general solution of the given system.

1. $x' + 2y' + x = 1$
   $2x' + y' + x = 0$
2. $x' + y' + x - y = 0$
   $2x' + 2y' - y = 0$
3. $x' + x - y = 1$
   $2x' + y' = t$
4. $3x' + 4x - y = t^2$
   $x' - 2y' + y = 0$
5. $2x' + 4x - y' = 0$
   $x' + y' + y = 4$
6. $x' - 3y' + y = e^{-t}$
   $x' + 4y = t$
7. $2x' + 3y' + x = 0$
   $x' - 2y' + y = 0$
8. $2x' + 3y' - x = 0$
   $x' + 2y' = 2t$
9. $x + y' - 2y = 1$
   $x' - 2y' + y = t$
10. $x' + y' - 3x = 0$
    $2x' + 4y' + y = 0$
11. $x' + 2y' - y = 0$
    $x - 2y' + 4y = t^2$
12. $x' + 3y' + x = 4$
    $2x' - y' = -t$
13. $x' + 3y' + 2y = 0$
    $2x' + 3y' + y = t$
14. $2x' + 4x - y = 2e^{3t}$
    $x' - 2y' = 1$
15. $x' + y - x = -2$
    $x - 2y' + y = 0$

In each of Problems 16 through 25, use Laplace transforms to solve the given system, subject to the initial conditions.

16. $x' + 4y = 1$
    $2x' + y - x = 0$
    $x(0) = y(0) = 0$
17. $x' + y' - x = 0$
    $2x' + 4y = t$
    $x(0) = y(0) = 0$
18. $x' + 2y' - x = -3$
    $2x' + 4y' + y = 0$
    $x(0) = y(0) = 0$
19. $x' + y' = -\cos(t)$
    $2x' + y - x = 1$
    $x(0) = y(0) = 0$
20. $x' + 2y + x = 0$
    $y' - 3y + x = 1$
    $x(0) = y(0) = 0$
21. $x' + y' + y = t$
    $x' - 4y' + y = 0$
    $x(0) = y(0) = 0$
22. $2x' + 3y = e^{-t}$
    $x' + y' - x = 0$
    $x(0) = y(0) = 0$
23. $x' + 4y' - y = 1$
    $x' + 2y = -3$
    $x(0) = y(0) = 0$
24. $x' + 6y - x = -t$
    $2x' + 4y' - y = 0$
    $x(0) = y(0) = 0$
25. $2x' + 3y' - x = 0$
    $2x' + 4y' = 3t$
    $x(0) = y(0) = 0$

In each of Problems 26 through 35, find the general solution of the given linear system. Then determine whether the origin is a node, center, spiral point, or saddle point, and whether it is stable, unstable, asymptotically stable, or stable but not asymptotically stable. Finally, draw some of the trajectories in the phase plane.

26. $\dfrac{dx}{dt} = x - 3y,\quad \dfrac{dy}{dt} = 2x + y$
27. $\dfrac{dx}{dt} = 3x + 2y,\quad \dfrac{dy}{dt} = x + y$
28. $\dfrac{dx}{dt} = -x + y,\quad \dfrac{dy}{dt} = x - 3y$
29. $\dfrac{dx}{dt} = 4x + 2y,\quad \dfrac{dy}{dt} = x - 2y$

30. $\dfrac{dx}{dt} = 2x - y,\quad \dfrac{dy}{dt} = x + y$

31. $\dfrac{dx}{dt} = 3x + y,\quad \dfrac{dy}{dt} = x - 2y$

32. $\dfrac{dx}{dt} = 3x - 4y,\quad \dfrac{dy}{dt} = -x + y$

33. $\dfrac{dx}{dt} = 2x - 4y,\quad \dfrac{dy}{dt} = x + y$

34. $\dfrac{dx}{dt} = -2x + 3y,\quad \dfrac{dy}{dt} = x - 3y$

35. $\dfrac{dx}{dt} = 6x + y,\quad \dfrac{dy}{dt} = x - 2y$

In each of Problems 36 through 45, determine the nature of the critical point at the origin in those cases where Theorem 25 applies.

36. $\dfrac{dx}{dt} = x - 2y + x^2y^2,\quad \dfrac{dy}{dt} = x + y + xy$

37. $\dfrac{dx}{dt} = x + y - x^3,\quad \dfrac{dy}{dt} = 2x + 3y + 2xy$

38. $\dfrac{dx}{dt} = x + 2y - x^2y,\quad \dfrac{dy}{dt} = -x - y + xy^3$

39. $\dfrac{dx}{dt} = 2x + 4y - 3x^2,\quad \dfrac{dy}{dt} = x - 2y + xy$

40. $\dfrac{dx}{dt} = 3x - y + 2x^3y^2,\quad \dfrac{dy}{dt} = x - y - x^2$

41. $\dfrac{dx}{dt} = x + 3y - y^2,\quad \dfrac{dy}{dt} = -x + y + 3x^2y$

42. $\dfrac{dx}{dt} = x + 4y - x^2y^4,\quad \dfrac{dy}{dt} = -4x + y + x^3$

43. $\dfrac{dx}{dt} = 3x + y - x^3y,\quad \dfrac{dy}{dt} = x + 2y - y^3$

44. $\dfrac{dx}{dt} = 4x - y + x^2y,\quad \dfrac{dy}{dt} = x - 2y + x^2 - y^2$

45. $\dfrac{dx}{dt} = 3x + y - x^3,\quad \dfrac{dy}{dt} = x + 3y + x^2y^2$

CHAPTER EIGHT

Some Notes on the History of Differential Equations

Sir Isaac Newton (1642–1727) and Gottfried Wilhelm Leibniz (1646–1716) are jointly credited with independently discovering the calculus, although some of the basic ideas are implicit in earlier Greek and Indian manuscripts.

Particularly in Newton's case, the prime motivation for the development of the calculus was its use in describing motion, velocity, acceleration, and other physical phenomena. Dominating these considerations in turn were astronomy and mechanics (remember that Newton's greatest achievement lay in his derivation of the inverse square law of gravitational attraction to explain the motions of the planets as described by Kepler). Although his methods were sometimes analytic, much of Newton's writing was couched in geometrical terms, and it was left to others to consider differential equations from an analytic point of view.

James Bernoulli (1655–1705), one of a number of Swiss Bernoullis who reached prominence in science and mathematics, was among the first to explicitly use the calculus to solve a differential equation. In 1690 he solved the problem of finding the curve along which a pendulum takes the same time to make one complete oscillation, regardless of the length of arc through which it swings. The curve is a cycloid, and in deriving it Bernoulli made first use of the word *integral* in connection with the solution of a differential equation.

In 1691, James Bernoulli's brother John (1667–1748) solved the problem of the catenary, the curve assumed by a cable hanging from two pegged ends. In 1695, James Bernoulli considered the differential equation

$$y' = P(x)y + Q(x)y^n,$$

which is linear when n equals zero or one. Leibniz, in 1696, reduced this to a linear equation by the substitution $z = y^{1-n}$, valid for $n = 2, 3, 4, \ldots$.

Use of differential equations in finding curves intersecting a given family of curves in a given angle was due to John Bernoulli, and Leibniz, in about 1694. Orthogonal trajectories were obtained by Newton in 1715 in response to a challenge by Leibniz.

In 1733, Daniel Bernoulli (1700–1782), son of John, considered the problem of an oscillating, uniformly heavy hanging chain. The displacement $y(x)$ at a distance x from one end was found to be given by an equation of the form

$$A \frac{d}{dx}\left(x \frac{dy}{dx}\right) + y = 0,$$

in which A is constant. Bernoulli derived a series solution which in modern notation would be written

$$y = KJ_0\left(2\sqrt{\frac{x}{A}}\right).$$

Thus, Daniel Bernoulli actually used a zero order Bessel function, though it would not be known by this name for many years.

For nonuniform thickness, y is described by

$$A \frac{d}{dx}\left(g(x) \frac{dy}{dx}\right) + y \frac{dg}{dx} = 0,$$

in which $g(x)$ is the distribution of weight along the chain. Bernoulli obtained one solution of the form

$$y = 2K\sqrt{\frac{A}{2x}}\, J_1\left(2\sqrt{\frac{2x}{A}}\right),$$

involving what we now know as a Bessel function of the first kind of order one.

In 1734 the French mathematician Alexis-Claude Clairaut (1713–1765) studied the differential equation

$$y' = xy + f(y'),$$

now called Clairaut's equation. This was a break with tradition, as up to this time most work had concentrated on specific differential equations arising in specific contexts.

About 1727, the Swiss mathematician Leonard Euler (1707–1783) began a systematic study of certain second order differential equations. At the time, Euler was a resident mathematician in the St. Petersburg Academy in Russia. In 1734, Daniel Bernoulli wrote to him concerning a fourth order differential equation describing bending of a loaded beam. In response to this and his own previous work, Euler studied constant coefficient linear differential equations of arbitrary order n, publishing some results in about 1739 and the years that followed. Joseph-Louis Lagrange (1736–1813) later attempted to extend this work to the case of nonconstant coefficients.

Closely related to Lagrange's work was the study of what is now known as the Riccati equation,

$$y' = P(x) + Q(x)y + R(x)y^2,$$

introduced by Count Jacopo Francesco Riccati of Venice (1676–1754) in about 1724.

Lagrange, who was of French and Italian origin, is sometimes credited with developing the method of variation of parameters. He also formalized much of

Newton's work in his masterpiece *Mécanique analytique*, published in 1788. Reduction of order is sometimes credited to Jean Le Rond d'Alembert (1717–1783), who also contributed to the development of partial derivatives and partial differential equations.

Series solutions were used from Newton's time onward, often without regard to such details as convergence. But it was Euler who systematized their use, and even had in his possession an ad hoc form of the method of Frobenius. Georg Frobenius (1849–1917) formalized the method much later. Euler came in this way to a solution of what we now know as Bessel's equation,

$$y'' + \frac{1}{x}y' + \left(1 - \frac{n^2}{x^2}\right)y = 0,$$

obtaining series solutions we now recognize as $J_n(x)$. Euler also treated the hypergeometric differential equation

$$x(1 - x)y'' + [c - (a + b + 1)x]y' - aby = 0,$$

although the name was bestowed by his friend and colleague Johann Pfaff (1765–1825). Solutions of this equation were studied in the early nineteenth century by the great German mathematician Karl Friedrich Gauss (1777–1855).

Systems of ordinary differential equations arise naturally from putting Newton's laws (which deal with vectors) into component form. Much of the early work on systems appears in Lagrange's *Mécanique analytique*. There, Lagrange tackled the classical three-body problem of astronomy (which, in its general form, remains unsolved to this day). Lagrange also applied variation of parameters to nth order equations, though the method is hinted at in Newton's *Principia*. Further work on the n-body problem was done by Pierre Simon de Laplace (1749–1827) in his *Mécanique celeste*.

The nineteenth century saw an expanded use of series and a recognition of special functions. The first systematic study of Bessel's equation was undertaken by Friedrich Wilhelm Bessel (1784–1846), director of the observatory at Königsberg, in connection with a problem in astronomy. He derived various expressions for $J_n(x)$ as well as recurrence formulas and results on the zeros of Bessel functions.

Legendre polynomials, which arise in connection with potential theory in spherical coordinates, appeared in the writings of Laplace and Adrien-Marie Legendre (1752–1833), a mathematician at the Ecole Militaire. The first systematic treatise on Legendre polynomials was written by Robert Murphy of Cambridge in 1833.

Hermite polynomials are named for Charles Hermite (1822–1901), professor at the Ecole Polytechnique and Sorbonne in Paris. Laguerre polynomials are named after Edmond Laguerre (1834–1886) of the College de France.

Sturm-Liouville theory originated in the work of Charles Sturm (1803–1855), a professor of mechanics at the Sorbonne, and Joseph Liouville (1809–1852), professor of mathematics at the College de France. Problems associated with finding eigenvalues and making eigenfunction expansions date back to about 1750, arising in connection with special functions, which were just beginning to emerge, and with problems in heat conduction. Sturm and Liouville were the first to construct a general theory for second order differential equations.

Laplace transforms date back to writings of Laplace in the 1780s, Poisson in the 1820s, and Fourier's famous 1811 paper on heat conduction (which is discussed in the chapter on Fourier series). Their popularity as a computational tool in elementary differential equations and electrical engineering is due in large part to the work of Oliver Heaviside (1850–1925), a telephone and telegraph engineer who spent the later part of his career developing vector and transform methods for use in engineering problem-solving.

Stability and the qualitative theory of nonlinear systems, as introduced in Section 7.3, is a relatively new area of mathematical activity. Seminal work was done by Alexander Michailovich Liapunov (1857–1918), a Russian mathematician and mechanical engineer. His classic work appeared in 1892 and gave birth to an approach to stability through what are now known as Liapunov functions. Alfred Lienard (1869–1958), director of the School of Mines in Paris, worked on spiral approaches of trajectories in the phase plane, among other things. A classic result on closed trajectories in the phase plane is known as the Poincaré-Bendixson theorem. Ivar Bendixson (1861–1935) was a professor at the University of Stockholm in Sweden. Henri Poincaré (1854–1912) was perhaps the greatest mathematician of his time, and did important work in differential equations, dynamical systems, mathematical physics, stability, and an emerging field now known as topology. Finally, van der Pol's equation is due to Balthasar van der Pol (b. 1889), a Dutch radio engineer who derived his equation in the 1920s.

PART TWO

VECTORS AND MATRICES

CHAPTER NINE

Vectors and Vector Spaces

9.0 INTRODUCTION

Many quantities encountered in engineering and the sciences require two pieces of information for their complete description. For example, forces, velocity, and acceleration have both a magnitude and a direction. Such quantities are called *vectors*, a concept to be distinguished from, say, mass or temperature which can be completely specified by a single number. In this chapter we shall develop the algebra and geometry of vectors as they appear in real-world problems, and also generalize these to the notion of a vector space. This concept is useful in developing much of the material on matrices, the object of the next chapter.

9.1 ALGEBRA AND GEOMETRY OF VECTORS

In the context of vectors, it is customary to refer to real numbers as *scalars*. Thus, π, 2, and $\sqrt{41}$ are scalars. Quantities such as mass, temperature, and density, which can be measured by a single real number, are called *scalar quantities*.

Vectors, by contrast, must be specified by giving both a direction and a magnitude. If we push on an object, for example, not only the "amount," but also the direction of the push will determine the effect on the object.

One way of packing both direction and magnitude into a concept is to define a vector (in 3-space) as an *ordered triple* (a, b, c) of real numbers or scalars. Given a rectangular coordinate system, as in Figure 102, we think of the ordered triple (a, b, c) in two interchangeable ways: (1) as coordinates of a point, and (2) as an arrow from the origin to the point (a, b, c).

In (2), we have both a direction [standing at the origin, it is the direction from $(0, 0, 0)$ to (a, b, c)] and a magnitude [the length of the arrow or, equivalently, the distance from $(0, 0, 0)$ to (a, b, c), which is $\sqrt{a^2 + b^2 + c^2}$].

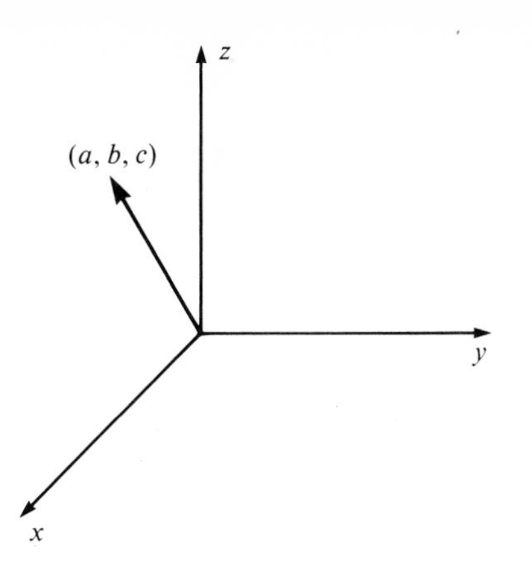

Figure 102. (a, b, c) as a geometric point and as a vector.

When thinking of (a, b, c) as a vector, we call a, b, and c the first, second, and third *component*, respectively. Two vectors are equal exactly when their respective components are equal.

Soon we shall develop a special notation to help us distinguish the emphasis on (a, b, c), as vector or as geometric point. In the meantime, to help identify scalars and vectors, we shall write scalars as Greek or Roman letters (α, β, a, b, A, W, ...) and vectors as Greek or Roman letters written in boldface type ($\boldsymbol{\alpha}$, $\boldsymbol{\beta}$, $\mathbf{a}$, $\mathbf{b}$, $\mathbf{A}$, $\mathbf{W}$, ...). The magnitude of a vector $\mathbf{F}$ will be written $\|\mathbf{F}\|$. Thus, if $\mathbf{F} = (a, b, c)$, then $\|\mathbf{F}\| = \sqrt{a^2 + b^2 + c^2}$.

In Figure 103, we show the vectors $(1, -1, 2)$ and $(-3, 2, 4)$. We reiterate the importance of realizing that, for example, $(1, -1, 2)$ is both a point in 3-space *and* a notation for an arrow from $(0, 0, 0)$ to $(1, -1, 2)$. This arrow represents a vector

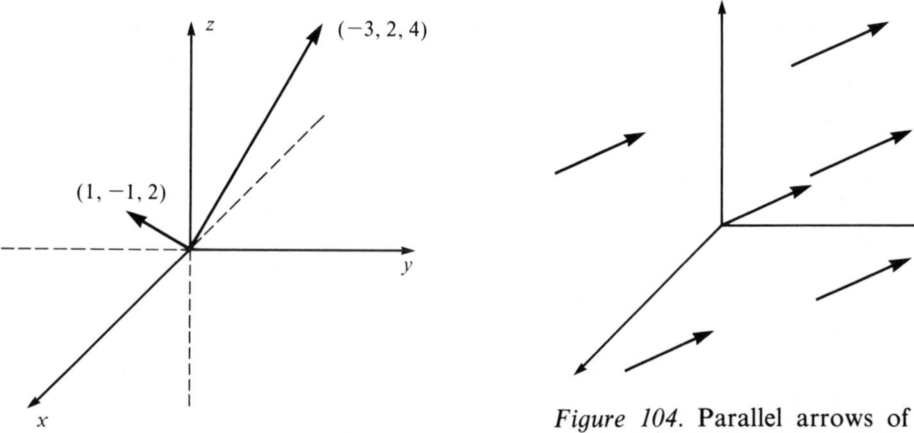

Figure 103

Figure 104. Parallel arrows of the same length representing the same vector.

having the direction (from the origin) of the arrow, and magnitude equal to the arrow's length (here, $\sqrt{6}$).

Actually, *any* arrow having the right length and direction will do just as well in representing a vector. Thus, in Figure 104, all the arrows are of the same length and direction and hence represent the same vector. In many applications it is convenient to represent a vector by an arrow from a point other than the origin. We shall return to this idea shortly.

There are two basic algebraic operations involving vectors. Let

$$\mathbf{F} = (a_1, b_1, c_1) \quad \text{and} \quad \mathbf{G} = (a_2, b_2, c_2).$$

We define the *vector sum* of **F** and **G** by

$$\mathbf{F} + \mathbf{G} = (a_1 + a_2, b_1 + b_2, c_1 + c_2).$$

Thus, we add vectors by adding respective components. For example,

$$(-1, 3, \pi) + (\sqrt{2}, 1, 6) = (-1 + \sqrt{2}, 4, \pi + 6).$$

Next, the *scalar product* of vector (a, b, c) with scalar α is defined by multiplying each component of (a, b, c) by α, that is,

$$\alpha(a, b, c) = (\alpha a, \alpha b, \alpha c).$$

For example,

$$\sqrt{3}(0, 1, -2) = (0, \sqrt{3}, -2\sqrt{3}).$$

Here are the basic properties of these operations:

1. $\mathbf{F} + \mathbf{G} = \mathbf{G} + \mathbf{F}$. This is called the *commutative law*.
2. $(\mathbf{F} + \mathbf{G}) + \mathbf{H} = \mathbf{F} + (\mathbf{G} + \mathbf{H})$. This is called the *associative law*.
3. $\mathbf{F} + (0, 0, 0) = \mathbf{F}$. For this reason we call $(0, 0, 0)$ the *zero vector*, denoted $\mathbf{0}$. It is the only vector we cannot represent as an arrow, as it has zero magnitude (length).
4. If $\mathbf{F} = (a, b, c)$, we denote by $-\mathbf{F}$ the vector $(-a, -b, -c)$. We call $-\mathbf{F}$ the *negative* of $\mathbf{F}$, and for any vector $\mathbf{G}$, we write $\mathbf{G} + (-\mathbf{F})$ as just $\mathbf{G} - \mathbf{F}$. If $\mathbf{F} = (a_1, b_1, c_1)$ and $\mathbf{G} = (a_2, b_2, c_2)$, then

$$\mathbf{G} - \mathbf{F} = (a_2 - a_1, b_2 - b_1, c_2 - c_1).$$

5. $\alpha(\mathbf{F} + \mathbf{G}) = \alpha\mathbf{F} + \alpha\mathbf{G}$
6. $(\alpha\beta)\mathbf{F} = \alpha(\beta\mathbf{F})$
7. $(\alpha + \beta)\mathbf{F} = \alpha\mathbf{F} + \beta\mathbf{F}$

Property 4 is mostly notation; the others can be proved easily. As an illustration, we prove property 5.

Proof of property 5 Let

$$\mathbf{F} = (a_1, b_1, c_1) \quad \text{and} \quad \mathbf{G} = (a_2, b_2, c_2).$$

Then

$$\begin{aligned}
\alpha(\mathbf{F} + \mathbf{G}) &= \alpha[(a_1, b_1, c_1) + (a_2, b_2, c_2)] \\
&= \alpha(a_1 + a_2, b_1 + b_2, c_1 + c_2) \\
&= (\alpha(a_1 + a_2), \alpha(b_1 + b_2), \alpha(c_1 + c_2)) \\
&= (\alpha a_1 + \alpha a_2, \alpha b_1 + \alpha b_2, \alpha c_1 + \alpha c_2) \\
&= (\alpha a_1, \alpha b_1, \alpha c_1) + (\alpha a_2, \alpha b_2, \alpha c_2) \\
&= \alpha(a_1, b_1, c_1) + \alpha(a_2, b_2, c_2) \\
&= \alpha\mathbf{F} + \alpha\mathbf{G}.
\end{aligned}$$

Besides the above, we have the following fact about the magnitude of a scalar multiple of a vector:

8. $\|\alpha\mathbf{F}\| = |\alpha|\ \|\mathbf{F}\|$. For, if $\mathbf{F} = (a, b, c)$, then $\alpha\mathbf{F} = (\alpha a, \alpha b, \alpha c)$, and

$$\begin{aligned}
\|\alpha\mathbf{F}\| &= \sqrt{(\alpha a)^2 + (\alpha b)^2 + (\alpha c)^2} \\
&= \sqrt{\alpha^2(a^2 + b^2 + c^2)} \\
&= |\alpha|\sqrt{a^2 + b^2 + c^2} \\
&= |\alpha|\ \|\mathbf{F}\|.
\end{aligned}$$

Since (a, b, c) is both a geometric point and a vector, it is natural to seek some geometric interpretation of vector addition and scalar multiplication. For vector addition, such an interpretation takes the form of the *parallelogram law*, which says the following:

Parallelogram law for addition: $\mathbf{F} + \mathbf{G}$ is represented by the diagonal of the parallelogram having $\mathbf{F}$ and $\mathbf{G}$ as sides.

This is depicted in Figure 105, which can be read in two ways: First, $\mathbf{F}$ and $\mathbf{G}$ can be drawn from the same point; or second, either vector may be represented by the parallel arrow drawn from the end of the other. In reading vector diagrams, the latter viewpoint is often useful.

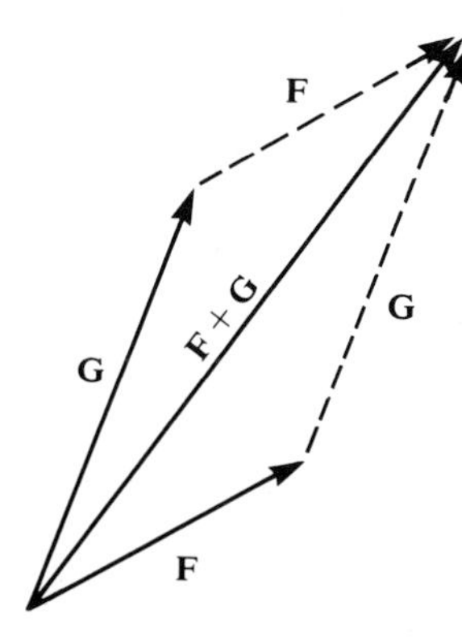

Figure 105. Parallelogram law for vector addition.

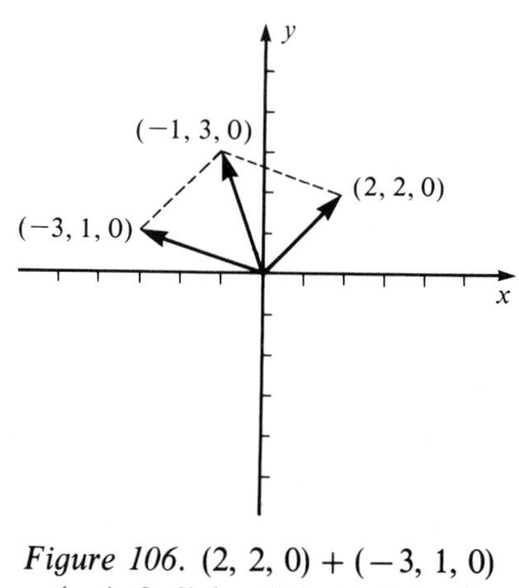

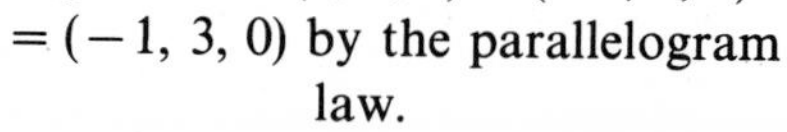

Figure 106. $(2, 2, 0) + (-3, 1, 0) = (-1, 3, 0)$ by the parallelogram law.

The parallelogram law is perhaps most easily visualized when **F** and **G** are in the xy-plane (zero third components) and we can count off coordinates on the axes. An illustration is given in Figure 106, where we have $(2, 2, 0) + (-3, 1, 0) = (-1, 3, 0)$.

The scalar product $\alpha\mathbf{F}$ also has a geometric interpretation, which one can guess from Figures 107 and 108. The tipoff is the relationship $\|\alpha\mathbf{F}\| = |\alpha|\,\|\mathbf{F}\|$, which suggests that $\alpha\mathbf{F}$ has length equal to $|\alpha|$ times the length of **F**. Thus, α acts as a scaling factor. Further, if α is negative, then $\alpha\mathbf{F}$ is in the opposite direction from **F**. In summary: If $\alpha > 0$, $\alpha\mathbf{F}$ is in the same direction as **F**, and is longer if $\alpha > 1$, shorter if $\alpha < 1$; and if $\alpha < 0$, $\alpha\mathbf{F}$ has direction opposite that of **F**, and is longer if $|\alpha| > 1$, shorter if $|\alpha| < 1$. In this connection, it is useful to remember that two vectors are parallel (in the same, or opposite, directions) exactly when each is a scalar multiple of the other.

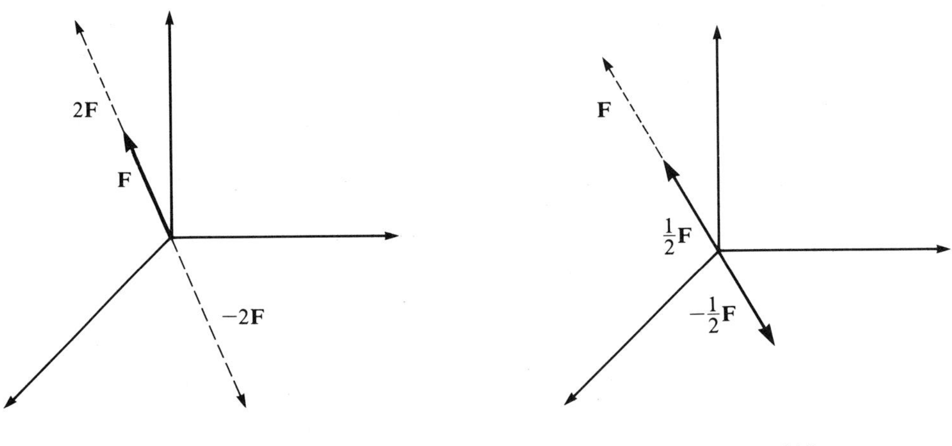

Figure 107 *Figure 108*

Our two operations enable us to decompose any vector into a sum of scalar multiples of "standardized" vectors as follows. Write

$$\mathbf{F} = (a, b, c) = a(1, 0, 0) + b(0, 1, 0) + c(0, 0, 1).$$

It is customary to denote

$$\mathbf{i} = (1, 0, 0), \qquad \mathbf{j} = (0, 1, 0), \qquad \mathbf{k} = (0, 0, 1).$$

These are *unit* (length 1) vectors along the positive x, y, and z axes, respectively, and are shown in Figure 109. We then have

$$\mathbf{F} = a\mathbf{i} + b\mathbf{j} + c\mathbf{k}.$$

When a component is zero, we usually just omit it in this notation. Thus, instead of $2\mathbf{i} + 0\mathbf{j} + 0\mathbf{k}$, we would write just $2\mathbf{i}$.

This gives us a handy way of distinguishing emphasis. Henceforth, we usually use (a, b, c) to denote a geometric point, and $a\mathbf{i} + b\mathbf{j} + c\mathbf{k}$ for the vector represented by the arrow from $(0, 0, 0)$ to (a, b, c), or by any parallel arrow of the same length and direction.

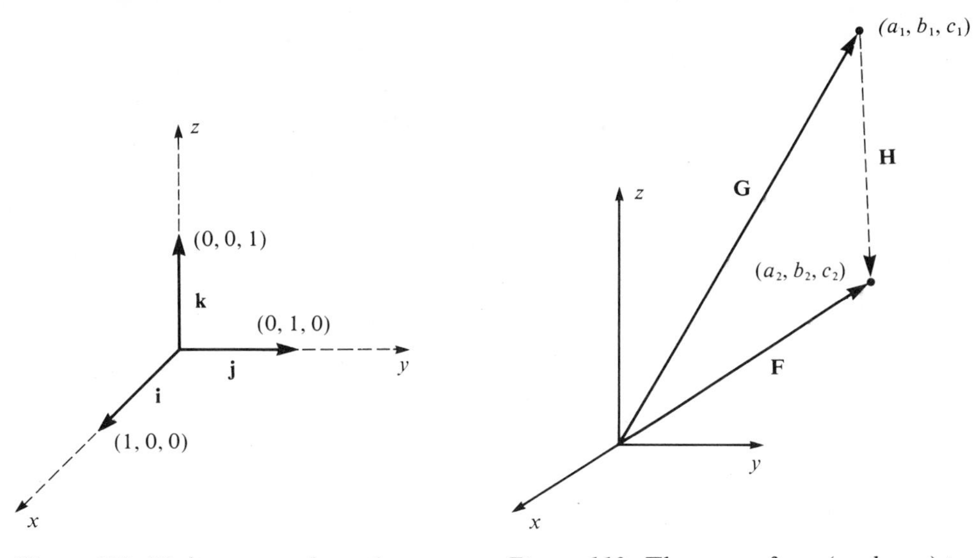

Figure 109. Unit vectors along the axes.

Figure 110. The arrow from (a_1, b_1, c_1) to (a_2, b_2, c_2) is

$$(a_2 - a_1)\mathbf{i} + (b_2 - b_1)\mathbf{j} + (c_2 - c_1)\mathbf{k}.$$

This leads to one last observation for this section. In Figure 110 we have points (a_1, b_1, c_1) and (a_2, b_2, c_2). We ask: What is the vector represented by the arrow from (a_1, b_1, c_1) to (a_2, b_2, c_2)? This arrow has a length and a direction, and hence does represent some vector—what are its components?

Let

$$\mathbf{F} = a_2\mathbf{i} + b_2\mathbf{j} + c_2\mathbf{k} \quad \text{and} \quad \mathbf{G} = a_1\mathbf{i} + b_1\mathbf{j} + c_1\mathbf{k}.$$

Call the unknown vector **H**. From the parallelogram law, $\mathbf{G} + \mathbf{H} = \mathbf{F}$; hence

$$\mathbf{H} = \mathbf{F} - \mathbf{G} = (a_2 - a_1)\mathbf{i} + (b_2 - b_1)\mathbf{j} + (c_2 - c_1)\mathbf{k}.$$

This fact will be so useful that we repeat it:

The arrow from (a_1, b_1, c_1) to (a_2, b_2, c_2) represents the vector

$$(a_2 - a_1)\mathbf{i} + (b_2 - b_1)\mathbf{j} + (c_2 - c_1)\mathbf{k}.$$

We conclude this section with several illustrations involving use of vector addition and scalar multiplication.

EXAMPLE 198

Find the equation of the straight line L through $(1, -2, 4)$ and $(6, 2, -3)$.

There are many ways to do this, but we shall illustrate the use of vectors here. Let (x, y, z) be any point on the line. Then

$$(x - 1)\mathbf{i} + (y + 2)\mathbf{j} + (z - 4)\mathbf{k}$$

is the arrow from $(1, -2, 4)$ to (x, y, z), and hence lies along L. But, similarly, the arrow from $(1, -2, 4)$ to $(6, 2, -3)$ is along L, and this is

$$5\mathbf{i} + 4\mathbf{j} - 7\mathbf{k}.$$

These two vectors are parallel; hence are scalar multiples of each other. For some scalar t,

$$(x - 1)\mathbf{i} + (y + 2)\mathbf{j} + (z - 4)\mathbf{k} = t(5\mathbf{i} + 4\mathbf{j} - 7\mathbf{k}) = 5t\mathbf{i} + 4t\mathbf{j} - 7t\mathbf{k}.$$

Then

$$x - 1 = 5t, \qquad y + 2 = 4t, \quad \text{and} \quad z - 4 = -7t.$$

As t varies over all real values, the point $(x, y, z) = (1 + 5t, -2 + 4t, 4 - 7t)$ thus determined varies over L. Note that when $t = 0$ we get $(1, -2, 4)$, and when $t = 1$ we get $(6, 2, -3)$.

These equations are called *parametric* equations for L, and t is the *parameter*. Eliminating t gives us

$$\frac{x - 1}{5} = \frac{y + 2}{4} = \frac{z - 4}{-7},$$

an equivalent expression for L.

We may also envision L as swept out by an arrow pivoted at the origin and extending to $(1 + 5t, -2 + 4t, 4 - 7t)$ as t varies from $-\infty$ to ∞.

EXAMPLE 199

Find a vector $\mathbf{F}$ of length 17 in the xy-plane making an angle of 42° with the positive x-axis.

By "find a vector," we mean determine its components. Figure 111 shows $\mathbf{F}$ and the numbers a and b such that $\mathbf{F} = a\mathbf{i} + b\mathbf{j}$. These numbers are the components of $\mathbf{F}$. From the right triangle, we read:

$$\cos(42°) = \frac{a}{17} \quad \text{and} \quad \sin(42°) = \frac{b}{17}.$$

Then,

$$a = 17\cos(42°) \quad \text{and} \quad b = 17\sin(42°).$$

Thus,

$$\mathbf{F} = [17\cos(42°)]\mathbf{i} + [17\sin(42°)]\mathbf{j}.$$

This is approximately $12.633\mathbf{i} + 11.375\mathbf{j}$.

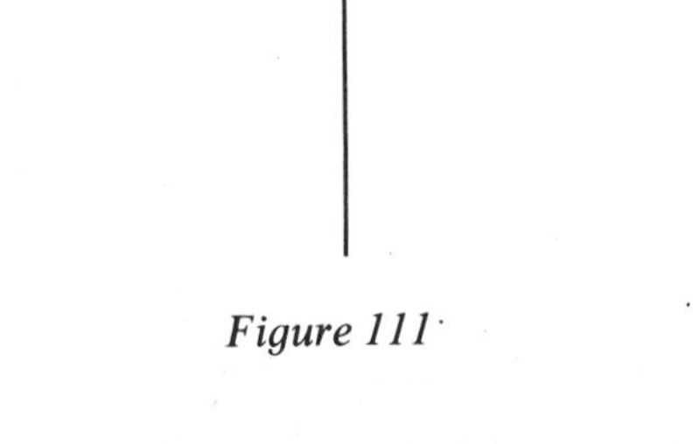

Figure 111

EXAMPLE 200

Proving geometry theorems is not the aim of this book. However, as practice in using vectors, we illustrate how vector algebra can make geometry proofs very neat.

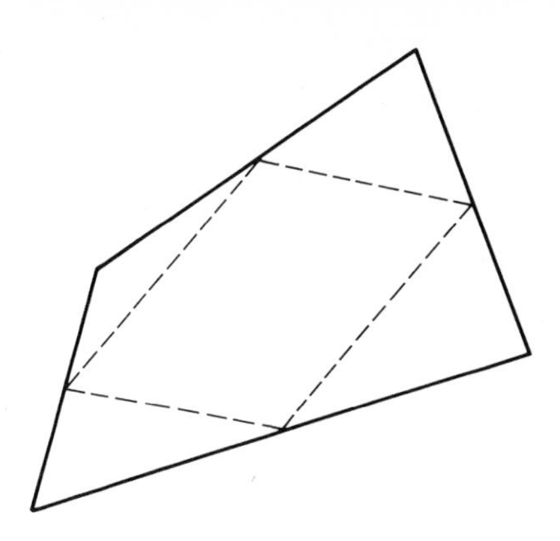

Figure 112. Parallelogram formed from midpoints of sides of a quadrilateral.

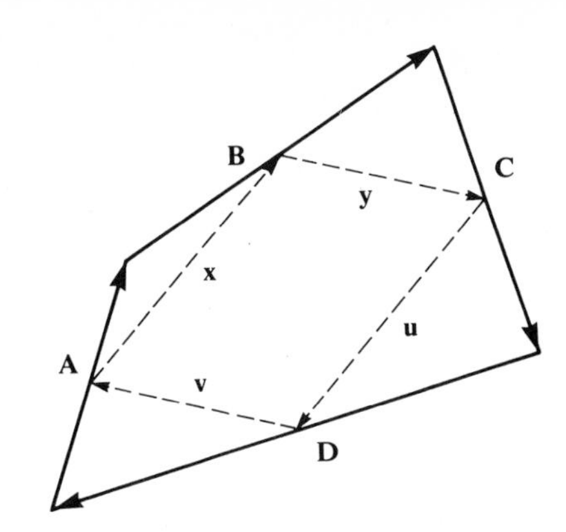

Figure 113

Consider this theorem: If successive midpoints of the sides of a quadrilateral are joined by straight lines, the resulting figure is a parallelogram. Referring to Figure 112, we must show that the dashed lines form a parallelogram.

Redraw the quadrilateral as in Figure 113, with vectors as sides. The dashed vectors **x**, **y**, **u**, and **v** connect the midpoints. We must show that **x** and **u** are parallel and of the same length, and similarly for **y** and **v**.

Now, from the parallelogram law,

$$\mathbf{x} = \tfrac{1}{2}\mathbf{A} + \tfrac{1}{2}\mathbf{B} = \tfrac{1}{2}(\mathbf{A} + \mathbf{B})$$

and

$$\mathbf{u} = \tfrac{1}{2}\mathbf{C} + \tfrac{1}{2}\mathbf{D} = \tfrac{1}{2}(\mathbf{C} + \mathbf{D}).$$

But,

$$\mathbf{A} + \mathbf{B} + \mathbf{C} + \mathbf{D} = \mathbf{0}$$

(see Problem 36). Hence,

$$\mathbf{A} + \mathbf{B} = -(\mathbf{C} + \mathbf{D}).$$

Then, $\mathbf{x} = -\mathbf{u}$, and so **x** and **u** are parallel. They are also of the same length, since

$$\|\mathbf{x}\| = \|-\mathbf{u}\| = |-1|\,\|\mathbf{u}\| = \|\mathbf{u}\|.$$

A similar argument holds for **y** and **v**.

PROBLEMS FOR SECTION 9.1

In Problems 1 through 5, compute $\mathbf{F} + \mathbf{G}$, $\mathbf{F} - \mathbf{G}$, $\|\mathbf{F}\|$, $\|\mathbf{G}\|$, $2\mathbf{F}$, and $3\mathbf{G}$.

1. $\mathbf{F} = 2\mathbf{i} - 3\mathbf{j} + 5\mathbf{k}$, $\mathbf{G} = \sqrt{2}\,\mathbf{i} + 6\mathbf{j} - 5\mathbf{k}$

2. $\mathbf{F} = \mathbf{i} - 3\mathbf{k}$, $\mathbf{G} = 4\mathbf{j}$

3. $\mathbf{F} = 2\mathbf{i} - 5\mathbf{j}$, $\mathbf{G} = \mathbf{i} + 5\mathbf{j} - \mathbf{k}$

4. $\mathbf{F} = \sqrt{2}\,\mathbf{i} + \mathbf{j} - 6\mathbf{k}$, $\mathbf{G} = 8\mathbf{i} + 2\mathbf{k}$

5. $\mathbf{F} = \mathbf{i} + \mathbf{j} + \mathbf{k}$, $\mathbf{G} = 2\mathbf{i} - 2\mathbf{j} + 2\mathbf{k}$

In Problems 6 through 10, all the vectors have zero third component to make them easier to draw. Calculate $\mathbf{F} + \mathbf{G}$ and $\mathbf{F} - \mathbf{G}$ in each case, and draw a diagram showing the parallelogram law for $\mathbf{F} + \mathbf{G}$ and $\mathbf{F} - \mathbf{G}$.

6. $\mathbf{F} = \mathbf{i}$, $\mathbf{G} = 6\mathbf{j}$
7. $\mathbf{F} = 2\mathbf{i} - \mathbf{j}$, $\mathbf{G} = \mathbf{i} - \mathbf{j}$
8. $\mathbf{F} = -3\mathbf{i} + \mathbf{j}$, $\mathbf{G} = 4\mathbf{j}$
9. $\mathbf{F} = \mathbf{i} - 2\mathbf{j}$, $\mathbf{G} = \mathbf{i} - 3\mathbf{j}$
10. $\mathbf{F} = -\mathbf{i} + 4\mathbf{j}$, $\mathbf{G} = -2\mathbf{i} - 3\mathbf{j}$

In each of Problems 11 through 15, compute $\alpha\mathbf{F}$. Draw diagrams of $\mathbf{F}$, $\alpha\mathbf{F}$, and $-\alpha\mathbf{F}$. (The vectors are in the plane to make this easier.)

11. $\mathbf{F} = \mathbf{i} + \mathbf{j}$, $\alpha = -\frac{1}{2}$
12. $\mathbf{F} = 6\mathbf{i} - 2\mathbf{j}$, $\alpha = 2$
13. $\mathbf{F} = -3\mathbf{j}$, $\alpha = -4$
14. $\mathbf{F} = 6\mathbf{i} + 6\mathbf{j}$, $\alpha = \frac{1}{2}$
15. $\mathbf{F} = -3\mathbf{i} + 2\mathbf{j}$, $\alpha = 3$

In each of Problems 16 through 25, find the equation of the straight line through the given points. You may leave the line in terms of a parameter t, or eliminate t as in Example 198.

16. (1, 0, 4), (2, 1, 1)
17. (3, 0, 0), (−3, 1, 0)
18. (2, 1, 1), (2, 1, −2)
19. (0, 1, 3), (0, 0, 1)
20. (1, 0, −4), (−2, −2, 5)
21. (2, −3, 6), (−1, 6, 4)
22. (−4, −2, 5), (1, 1, −5)
23. (3, 3, −5), (2, −6, 1)
24. (0, −3, 0), (1, −1, 5)
25. (4, −8, 1), (−1, 0, 0)

In each of Problems 26 through 35, find a vector in the xy-plane having the given length and making the given angle with the positive x-axis. Your answers may have sines and cosines in them. If you have a calculator, approximate these sines and cosines, and also any square roots encountered, to give the components to four decimal places.

26. $\sqrt{5}$, 45°
27. 6, 60°
28. 12, 135°
29. 1, 315°
30. 14, 90°
31. $\sqrt{2}$, 30°
32. 5, 140°
33. 15, 175°
34. 12, 225°
35. 25, 270°

36. Let P, Q, R, and S be any four points in space. Let $\mathbf{A}$ be the vector from P to Q, $\mathbf{B}$ from Q to R, $\mathbf{C}$ from R to S, and $\mathbf{D}$ from S to P. Determine $\mathbf{A} + \mathbf{B} + \mathbf{C} + \mathbf{D}$.

37. Let $\mathbf{A}$ be any vector except $\mathbf{0}$. Find a scalar t such that $\|t\mathbf{A}\| = 1$.

38. Prove, using vector methods, that the altitudes of any triangle intersect in a single point.

39. Let P, Q, and R be three given points, not on the same line. Let $\mathbf{A}$ be the vector from P to Q, and $\mathbf{B}$ the vector from P to R. Find the vector from P to the midpoint of the segment between R and Q. Your answer should be some vector combination of scalar multiples of $\mathbf{A}$ and $\mathbf{B}$.

40. Show, using vector methods, that lines drawn from a vertex of a parallelogram to the midpoints of the opposite sides trisect a diagonal of the parallelogram.

41. Suppose that $\alpha\mathbf{F} + \beta\mathbf{G} = \mathbf{0}$, and $\mathbf{F}$ and $\mathbf{G}$ are nonzero vectors which are not parallel. Show that α and β must be zero.

9.2 DOT PRODUCT OF VECTORS

In this section we define the first of several vector operations to be used later.

Suppose that

$$\mathbf{F} = a_1\mathbf{i} + b_1\mathbf{j} + c_1\mathbf{k} \quad \text{and} \quad \mathbf{G} = a_2\mathbf{i} + b_2\mathbf{j} + c_2\mathbf{k}.$$

The *dot product* of $\mathbf{F}$ with $\mathbf{G}$ is the *scalar*

$$\mathbf{F} \cdot \mathbf{G} = a_1 b_1 + a_2 b_2 + a_3 b_3.$$

For example,

$$(\sqrt{2}\mathbf{i} + \mathbf{j} + 3\mathbf{k}) \cdot (-2\mathbf{i} + 4\mathbf{j} + \pi\mathbf{k}) = -2\sqrt{2} + 4 + 3\pi.$$

Here are some basic properties of the dot product:

1. $\mathbf{F} \cdot \mathbf{G} = \mathbf{G} \cdot \mathbf{F}$.
2. $(\mathbf{F} + \mathbf{G}) \cdot \mathbf{H} = \mathbf{F} \cdot \mathbf{H} + \mathbf{G} \cdot \mathbf{H}$.
3. $\alpha(\mathbf{F} \cdot \mathbf{G}) = (\alpha\mathbf{F}) \cdot \mathbf{G} = \mathbf{F} \cdot (\alpha\mathbf{G})$.
4. $\mathbf{F} \cdot \mathbf{F} = \|\mathbf{F}\|^2$.
5. $\mathbf{F} \cdot \mathbf{F} = 0$ exactly when $\mathbf{F} = \mathbf{0}$.

For illustration, we shall prove properties 1, 4, and 5.

Proof of property 1 Let

$$\mathbf{F} = a_1\mathbf{i} + b_1\mathbf{j} + c_1\mathbf{k} \quad \text{and} \quad \mathbf{G} = a_2\mathbf{i} + b_2\mathbf{j} + c_2\mathbf{k}.$$

Then, using the commutativity of real-number multiplication,

$$\mathbf{F} \cdot \mathbf{G} = a_1 a_2 + b_1 b_2 + c_1 c_2 = a_2 a_1 + b_2 b_1 + c_2 c_1 = \mathbf{G} \cdot \mathbf{F}.$$

Proof of properties 4 and 5 If $\mathbf{F} = a\mathbf{i} + b\mathbf{j} + c\mathbf{k}$, then

$$\mathbf{F} \cdot \mathbf{F} = a^2 + b^2 + c^2 = \|\mathbf{F}\|^2.$$

Further, $a^2 + b^2 + c^2 = 0$ only when $a = b = c = 0$, in which case $\mathbf{F} = \mathbf{0}$.

To see a geometric side to dot product, consider $\mathbf{F}$ and $\mathbf{G}$ as in Figure 114, and let θ be the angle between the arrows as shown. The arrow from the end of $\mathbf{G}$ to the end of $\mathbf{F}$ is $\mathbf{F} - \mathbf{G}$. These arrows form a triangle with sides of length $\|\mathbf{F}\|$, $\|\mathbf{G}\|$, and $\|\mathbf{F} - \mathbf{G}\|$.

Now recall the law of cosines. For the triangle of Figure 115, it says that

$$a^2 + b^2 - 2ab\cos(\theta) = c^2.$$

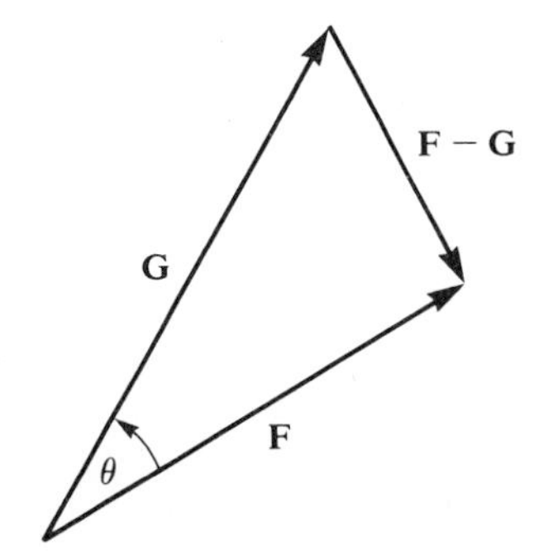

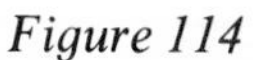
Figure 114

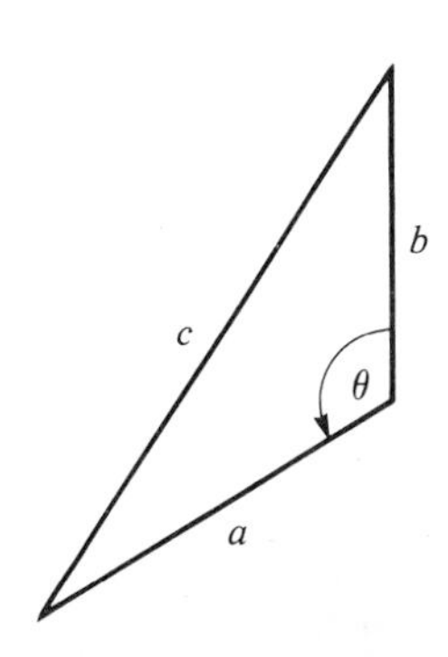

Figure 115. Law of cosines: $a^2 + b^2 - 2ab\cos(\theta) = c^2$.

Letting $a = \|\mathbf{F}\|$, $b = \|\mathbf{G}\|$, and $c = \|\mathbf{F} - \mathbf{G}\|$, we have

$$\|\mathbf{F}\|^2 + \|\mathbf{G}\|^2 - 2\|\mathbf{F}\|\,\|\mathbf{G}\|\cos(\theta) = \|\mathbf{F} - \mathbf{G}\|^2.$$

Writing

$$\mathbf{F} = a_2\mathbf{i} + b_2\mathbf{j} + c_2\mathbf{k} \quad \text{and} \quad \mathbf{G} = a_1\mathbf{i} + b_1\mathbf{j} + c_1\mathbf{k},$$

we have

$$a_2^2 + b_2^2 + c_2^2 + a_1^2 + b_1^2 + c_1^2 - 2\|\mathbf{F}\|\,\|\mathbf{G}\|\cos(\theta)$$
$$= (a_2 - a_1)^2 + (b_2 - b_1)^2 + (c_2 - c_1)^2.$$

After simplifying, this becomes

$$a_1a_2 + b_1b_2 + c_1c_2 = \|\mathbf{F}\|\,\|\mathbf{G}\|\cos(\theta).$$

If neither $\mathbf{F}$ nor $\mathbf{G}$ is $\mathbf{0}$, then $\|\mathbf{F}\| \neq 0$ and $\|\mathbf{G}\| \neq 0$, and we have

$$\cos(\theta) = \frac{\mathbf{F}\cdot\mathbf{G}}{\|\mathbf{F}\|\,\|\mathbf{G}\|}.$$

Thus, dot product gives us the cosine of the angle (hence also the angle) between two vectors or lines. This can be extremely useful, as the following examples show.

EXAMPLE 201

Let $\mathbf{F} = -\mathbf{i} + 3\mathbf{j} + \mathbf{k}$ and $\mathbf{G} = 2\mathbf{j} - 4\mathbf{k}$. The angle θ between $\mathbf{F}$ and $\mathbf{G}$ can be obtained from

$$\cos(\theta) = \frac{6 - 4}{\sqrt{11}\sqrt{20}} = \frac{2}{\sqrt{220}}.$$

Using radians, θ is that unique number in $[0, \pi]$ with $\cos(\theta) = 2/\sqrt{220}$. Since $2/\sqrt{220}$ is about 0.1348, θ is approximately 1.436 radians (about 82.25 degrees).

EXAMPLE 202

Find the angle between the lines

$$x = 1 + 6t, \qquad y = 2 - 4t, \qquad z = -1 + 3t$$

and

$$x = 4 - 3p, \qquad y = 2p, \qquad z = -5 + 4p$$

where they intersect.

Strictly speaking, this question is ambiguous. If two lines intersect, as in Figure 116, there are two angles between them. By convention, the smaller of these two angles is the angle between the lines.

The strategy in solving this problem by vector means is to take a vector $\mathbf{F}$ along one line, a vector $\mathbf{G}$ along the other, and find the cosine of the angle θ between them.

To find a vector along the first line, find two points on it by choosing two values of the parameter t, say 0 and 1. These give us $(1, 2, -1)$ and $(7, -2, 2)$, respectively. Thus, let

$$\mathbf{F} = (7 - 1)\mathbf{i} + (-2 - 2)\mathbf{j} + [2 - (-1)]\mathbf{k} = 6\mathbf{i} - 4\mathbf{j} + 3\mathbf{k}.$$

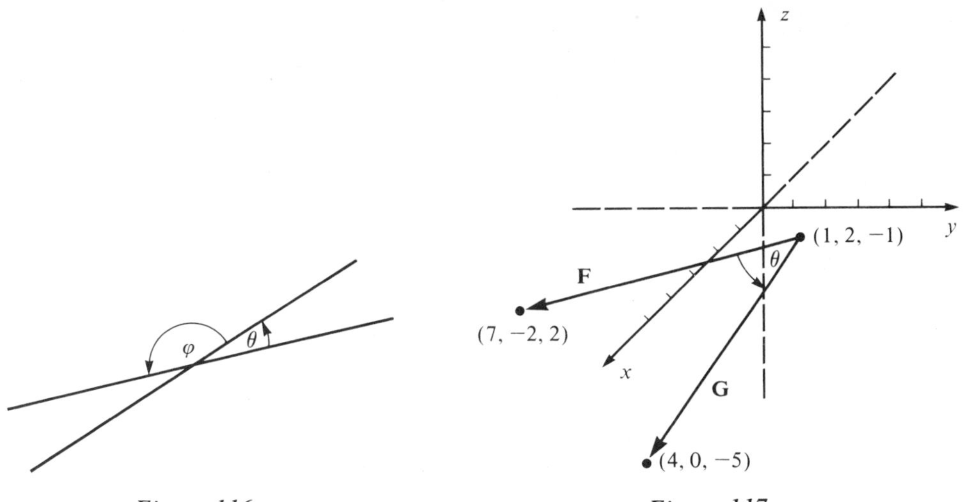

Figure 116

Figure 117

For the second line, choose two values of parameter p, say $p = 0$ and $p = 1$, to get $(4, 0, -5)$ and $(1, 2, -1)$. We may choose

$$\mathbf{G} = (4 - 1)\mathbf{i} + (0 - 2)\mathbf{j} + [-5 - (-1)]\mathbf{k} = 3\mathbf{i} - 2\mathbf{j} - 4\mathbf{k}.$$

These vectors are shown in Figure 117. Then,

$$\cos(\theta) = \frac{6(3) + (-4)(-2) + (3)(-4)}{\sqrt{6^2 + 4^2 + 3^2}\sqrt{3^2 + 2^2 + 4^2}} = \frac{14}{\sqrt{1769}}.$$

Thus, θ is about 1.23 radians, or about 70.56 degrees.

Here, we happened to find the point of intersection, $(1, 2, -1)$, but did not need it in the solution. **F** could have been *any* vector along the first line and **G** *any* vector along the second, and we would still have found the same result.

EXAMPLE 203

The points $A(1, -2, 1)$, $B(0, 1, 6)$, and $C(-3, 4, -2)$ form a triangle. Find the angle between the line AB and the line from A to the midpoint of the line BC. (Such a line is called a *median*.) The situation is shown in Figure 118.

To solve the problem using vectors, visualize the sides of the triangle as vectors, as shown in Figure 119. If P is the midpoint of BC, then $\mathbf{H}_1 = \mathbf{H}_2$ because they have the same length and direction. Further, from the coordinates of A, B, and C, we have

$$\mathbf{F} = -\mathbf{i} + 3\mathbf{j} + 5\mathbf{k} \quad \text{and} \quad \mathbf{G} = -4\mathbf{i} + 6\mathbf{j} - 3\mathbf{k}.$$

Now, we want the angle between **F** and **K**. We know **F**; to find **K**, note by the parallelogram law that

$$\mathbf{F} + \mathbf{H}_1 = \mathbf{K} \quad \text{and} \quad \mathbf{K} + \mathbf{H}_2 = \mathbf{G}.$$

Since $\mathbf{H}_1 = \mathbf{H}_2$, then $\mathbf{H}_1 = \mathbf{K} - \mathbf{F} = \mathbf{G} - \mathbf{K}$. Thus, $\mathbf{K} = \frac{1}{2}(\mathbf{F} + \mathbf{G})$, and we have

$$\mathbf{K} = \tfrac{1}{2}(-5\mathbf{i} + 9\mathbf{j} + 2\mathbf{k}).$$

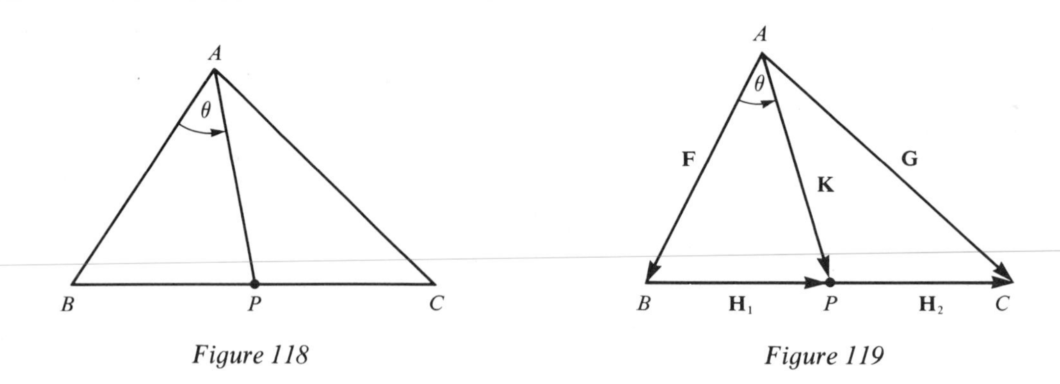

Figure 118

Figure 119

Then,

$$\cos(\theta) = \frac{\mathbf{F} \cdot \mathbf{K}}{\|\mathbf{F}\|\,\|\mathbf{K}\|} = \frac{42}{\sqrt{35}\sqrt{110}},$$

which is about 0.6769. Then θ is about 0.8273 radian or approximately 47.4 degrees.

The following theorem is often useful in doing computations with dot products (see Example 209).

THEOREM 29

For any scalars α and β,

$$\|\alpha\mathbf{F} + \beta\mathbf{G}\|^2 = \alpha^2\|\mathbf{F}\|^2 + 2\alpha\beta\mathbf{F} \cdot \mathbf{G} + \beta^2\|\mathbf{G}\|^2.$$

Proof Using the indicated properties of dot product, we have

$$\begin{aligned}
\|\alpha\mathbf{F} + \beta\mathbf{G}\|^2 &= (\alpha\mathbf{F} + \beta\mathbf{G}) \cdot (\alpha\mathbf{F} + \beta\mathbf{G}) && \text{(property 4)}\\
&= [(\alpha\mathbf{F} + \beta\mathbf{G}) \cdot (\alpha\mathbf{F})] + [(\alpha\mathbf{F} + \beta\mathbf{G}) \cdot (\beta\mathbf{G})] && \text{(property 2)}\\
&= (\alpha\mathbf{F}) \cdot (\alpha\mathbf{F}) + (\beta\mathbf{G}) \cdot (\alpha\mathbf{F}) + (\alpha\mathbf{F}) \cdot (\beta\mathbf{G}) + (\beta\mathbf{G}) \cdot (\beta\mathbf{G}) && \text{(property 2)}\\
&= \|\alpha\mathbf{F}\|^2 + 2(\alpha\mathbf{F}) \cdot (\beta\mathbf{G}) + \|\beta\mathbf{G}\|^2 && \text{(properties 1 and 4)}\\
&= |\alpha|^2\,\|\mathbf{F}\|^2 + 2(\alpha\beta)\mathbf{F} \cdot \mathbf{G} + |\beta|^2\|\mathbf{G}\|^2 && \text{(property 3)}\\
&= \alpha^2\|\mathbf{F}\|^2 + 2\alpha\beta\mathbf{F} \cdot \mathbf{G} + \beta^2\|\mathbf{G}\|^2.
\end{aligned}$$

Motivated by the geometry of arrow representations, we call **F** and **G** *orthogonal* (or *perpendicular*) if the angle θ between them is 90°. This happens exactly when $\cos(\theta) = 0$ or $\mathbf{F} \cdot \mathbf{G} = 0$. Thus, vanishing of the dot product is a test for orthogonality. Since $\mathbf{F} \cdot \mathbf{0} = 0$ for every **F**, it is customary to consider **0** as orthogonal to every vector.

EXAMPLE 204

Let

$$\mathbf{F} = -4\mathbf{i} + \mathbf{j} + 2\mathbf{k} \quad \text{and} \quad \mathbf{G} = 2\mathbf{i} + 4\mathbf{k}.$$

Then, $\mathbf{F} \cdot \mathbf{G} = -8 + 8 = 0$; so **F** and **G** are orthogonal.

However, $\mathbf{H} = 6\mathbf{i} - \mathbf{j} - 2\mathbf{k}$ and $\mathbf{L} = 3\mathbf{i} + \mathbf{j} + 4\mathbf{k}$ are not orthogonal, since $\mathbf{H} \cdot \mathbf{L} = 9$. If θ is the angle between $\mathbf{H}$ and $\mathbf{L}$, then $\cos(\theta) = 9/\sqrt{41}\sqrt{26}$. This is about 0.2757; so θ is about 1.29 radians (approximately 74 degrees).

EXAMPLE 205

The lines

$$L_1: \quad x = 2 - 4t, \quad y = 6 + t, \quad z = 3t$$

and

$$L_2: \quad x = -2 + p, \quad y = 7 + 2p, \quad z = 3 - 4p$$

intersect at $(-2, 7, 3)$. (This point is on L_1 when $t = 1$ and on L_2 when $p = 0$.) Are the lines perpendicular?

Letting $t = 0$, we find another point $(2, 6, 0)$ on L_1. The vector $[2 - (-2)]\mathbf{i} + (6 - 7)\mathbf{j} + (0 - 3)\mathbf{k}$, or $4\mathbf{i} - \mathbf{j} - 3\mathbf{k}$, is along L_1.

Letting $p = 1$, we get another point $(-1, 9, -1)$ on L_2. Then $[-1 - (-2)]\mathbf{i} + (9 - 7)\mathbf{j} + (-1 - 3)\mathbf{k}$, or $\mathbf{i} + 2\mathbf{j} - 4\mathbf{k}$, is along L_2.

The dot product of these vectors is $4 - 2 + 12 = 14 \neq 0$; so L_1 and L_2 are not perpendicular. (In fact, the angle θ between them is given by $\cos(\theta) = 14/\sqrt{26}\sqrt{21} \approx 0.599$; so θ is approximately 0.93 radian or 53.19 degrees.)

Using the dot product expression for the angle between two vectors, we can derive an important inequality due to Cauchy, Schwarz, and Bunjakowski.*

THEOREM 30 Cauchy-Schwarz Inequality

$$|\mathbf{F} \cdot \mathbf{G}| \leq \|\mathbf{F}\| \, \|\mathbf{G}\|.$$

Proof If $\mathbf{F}$ or $\mathbf{G}$ is $\mathbf{0}$, then the inequality becomes $0 \leq 0$, which is true. If $\mathbf{F} \neq \mathbf{0}$ and $\mathbf{G} \neq \mathbf{0}$, and θ is the angle between $\mathbf{F}$ and $\mathbf{G}$, then

$$-1 \leq \cos(\theta) = \frac{\mathbf{F} \cdot \mathbf{G}}{\|\mathbf{F}\| \, \|\mathbf{G}\|} \leq 1.$$

Hence,

$$-\|\mathbf{F}\| \, \|\mathbf{G}\| \leq \mathbf{F} \cdot \mathbf{G} \leq \|\mathbf{F}\| \, \|\mathbf{G}\|.$$

This is just another way of writing

$$|\mathbf{F} \cdot \mathbf{G}| \leq \|\mathbf{F}\| \, \|\mathbf{G}\|.$$

EXAMPLE 206

Let $\mathbf{F} = -3\mathbf{i} + 2\mathbf{j} - \mathbf{k}$ and $\mathbf{G} = 2\mathbf{i} - 4\mathbf{j} + \mathbf{k}$. Then $\mathbf{F} \cdot \mathbf{G} = -15$, $\|\mathbf{F}\| = \sqrt{14}$, and $\|\mathbf{G}\| = \sqrt{21}$. Thus, $\|\mathbf{F}\| \|\mathbf{G}\| = \sqrt{294}$, about 17.14, certainly at least as large as $|\mathbf{F} \cdot \mathbf{G}|$, which is 15.

---

* In the Soviet Union, this inequality is known as *Bunjakowski's inequality*; in the West, it is the *Cauchy-Schwarz inequality*.

We conclude this section with some additional examples showing uses of the dot product and the Cauchy-Schwarz inequality.

EXAMPLE 207

Here is an application of dot product to a problem commonly encountered with calculations involving forces. What is the effect of a given force **F** in a direction specified by a vector **r**?

Referring to Figure 120, we want the vector **H** which is the projection of **F** onto the direction of **r**. Let θ be the angle between **r** and **F**, and suppose first that $0 \leq \theta \leq \pi/2$. We have

$$\cos(\theta) = \frac{\mathbf{F} \cdot \mathbf{r}}{\|\mathbf{F}\| \, \|\mathbf{r}\|}.$$

But, by the way **H** is defined, **H** and **F** form two sides of a right triangle, and

$$\cos(\theta) = \frac{\|\mathbf{H}\|}{\|\mathbf{F}\|}.$$

Thus,

$$\frac{\|\mathbf{H}\|}{\|\mathbf{F}\|} = \frac{\mathbf{F} \cdot \mathbf{r}}{\|\mathbf{F}\| \, \|\mathbf{r}\|}.$$

Solving for $\|\mathbf{H}\|$ gives us

$$\|\mathbf{H}\| = \frac{\mathbf{F} \cdot \mathbf{r}}{\|\mathbf{r}\|}.$$

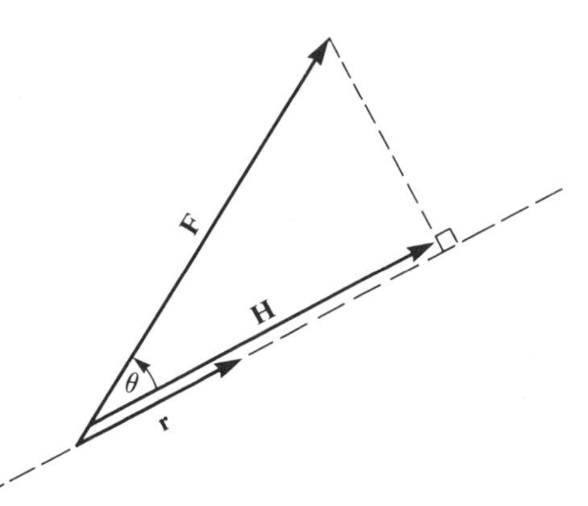

Figure 120. Projection of **F** onto the direction of **r**.

Now we know the magnitude of **H**, but we also know its direction, namely that of **r**. Thus, $\mathbf{H} = t\mathbf{r}$ for some scalar t, and choosing $t = \mathbf{F} \cdot \mathbf{r}/\|\mathbf{r}\|^2$ gives us the desired magnitude. Thus,

$$\mathbf{H} = \left(\frac{\mathbf{F} \cdot \mathbf{r}}{\|\mathbf{r}\|^2}\right)\mathbf{r},$$

and we call **H** the *projection of* **F** *in the direction of* **r**.

For example, let $\mathbf{r} = \mathbf{i} + \mathbf{j} + \mathbf{k}$ and $\mathbf{F} = 3\mathbf{i} - 2\mathbf{j} + 6\mathbf{k}$. Then the projection of **F** in the direction of **r** is

$$\mathbf{H} = \left(\frac{\mathbf{F} \cdot \mathbf{r}}{\|\mathbf{r}\|^2}\right) r = \left(\frac{(3 - 2 + 6)}{3}\right)\mathbf{r} = \frac{7}{3}(\mathbf{i} + \mathbf{j} + \mathbf{k}).$$

We have assumed thus far that $0 \leq \theta \leq \pi/2$; if $\pi/2 < \theta \leq \pi$, then $\cos(\theta) < 0$, and the above reasoning fails (the student should be able to spot where).

If $\pi/2 < \theta \leq \pi$, a modification of the above derivation gives us

$$\mathbf{H} = \left(\frac{-\mathbf{F} \cdot \mathbf{r}}{\|\mathbf{r}\|^2}\right)\mathbf{r}.$$

We leave the verification of this to the student.

EXAMPLE 208

Here is an application of dot product to a problem in analytic geometry. Find the equation of a plane passing through $(-6, 1, 1)$ and perpendicular to $-2\mathbf{i} + 4\mathbf{j} + \mathbf{k}$.

Reason as follows: Let (x, y, z) be any point on the desired plane. Then

$$(x + 6)\mathbf{i} + (y - 1)\mathbf{j} + (z - 1)\mathbf{k}$$

is a vector in this plane, and hence must be perpendicular to $-2\mathbf{i} + 4\mathbf{j} + \mathbf{k}$. Thus their dot product is zero:

$$-2(x + 6) + 4(y - 1) + (z - 1) = 0.$$

This reduces to

$$-2x + 4y + z = 17.$$

Thus far, any point (x, y, z) on the plane satisfies this equation. Conversely, by reversing the above reasoning, any (x, y, z) whose coordinates satisfy the equation lies on the plane. Thus, we can call $-2x + 4y + z = 17$ the *equation of the plane.*

Usually a vector perpendicular to a plane is called *normal* to the plane. Thus, in this example, $-2\mathbf{i} + 4\mathbf{j} + \mathbf{k}$ is normal to $-2x + 4y + z = 17$.

EXAMPLE 209

Here is an example from geometry. The triangle inequality says that the sum of the lengths of two sides of any triangle is at least as large as the length of the third side. This fact follows easily from the Cauchy-Schwarz inequality, as we now show.

Use vectors as sides of the triangle, as in Figure 121. By the parallelogram law, the sides are $\mathbf{F}$, $\mathbf{G}$, and $\mathbf{F} - \mathbf{G}$. Using Theorem 29 with $\alpha = 1$ and $\beta = -1$, we have

$$\begin{aligned}\|\mathbf{F} - \mathbf{G}\|^2 &= (\mathbf{F} - \mathbf{G}) \cdot (\mathbf{F} - \mathbf{G}) \\ &= \mathbf{F} \cdot \mathbf{F} - 2\mathbf{F} \cdot \mathbf{G} + \mathbf{G} \cdot \mathbf{G} \\ &= \|\mathbf{F}\|^2 - 2\mathbf{F} \cdot \mathbf{G} + \|\mathbf{G}\|^2 \\ &\leq \|\mathbf{F}\|^2 + 2|\mathbf{F} \cdot \mathbf{G}| + \|\mathbf{G}\|^2.\end{aligned}$$

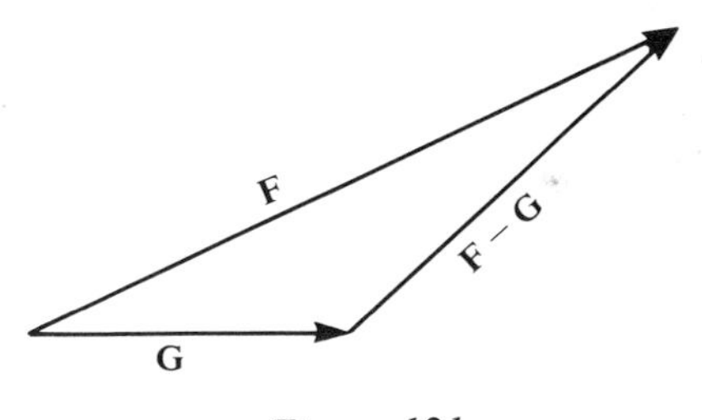

Figure 121

But, $|\mathbf{F} \cdot \mathbf{G}| \leq \|\mathbf{F}\| \, \|\mathbf{G}\|$. Hence,

$$\|\mathbf{F} - \mathbf{G}\|^2 \leq \|\mathbf{F}\|^2 + 2\|\mathbf{F}\| \, \|\mathbf{G}\| + \|\mathbf{G}\|^2 = (\|\mathbf{F}\| + \|\mathbf{G}\|)^2.$$

Hence,

$$\|\mathbf{F} - \mathbf{G}\| \leq \|\mathbf{F}\| + \|\mathbf{G}\|,$$

and this is what we wanted to show.

The point to many examples in this and the preceding section is that many calculations involving lines, planes, triangles and the like become easier to visualize and execute if such things as directions and sides are considered as vectors and the appropriate vector operations used. We shall see further examples of this in the next two sections when we consider cross products and scalar triple products.

PROBLEMS FOR SECTION 9.2

In each of Problems 1 through 10, find the dot product of the given vectors, the cosine of the angle between them, determine if they are orthogonal, and explicitly verify the Cauchy-Schwarz inequality. If you have a calculator, find the angle between the two vectors to four decimal places in both radians and degrees.

1. $\mathbf{i}$, $2\mathbf{i} - 3\mathbf{j} + \mathbf{k}$
2. $2\mathbf{i} - 6\mathbf{j} + \mathbf{k}$, $\mathbf{i} - \mathbf{j}$
3. $-4\mathbf{i} - 2\mathbf{j} + 3\mathbf{k}$, $6\mathbf{i} - 2\mathbf{j} - \mathbf{k}$
4. $\mathbf{i}$, $\mathbf{j}$
5. $-3\mathbf{i} + 2\mathbf{k}$, $6\mathbf{j}$
6. $8\mathbf{i} - 3\mathbf{j} + 2\mathbf{k}$, $-8\mathbf{i} - 3\mathbf{j} + \mathbf{k}$
7. $\mathbf{i} - 2\mathbf{j}$, $\mathbf{i} - 2\mathbf{j}$
8. $-5\mathbf{i} + 6\mathbf{j} - 3\mathbf{k}$, $5\mathbf{i} - 3\mathbf{j} - 3\mathbf{k}$
9. $\mathbf{i} - 3\mathbf{k}$, $2\mathbf{j} + 6\mathbf{k}$
10. $\mathbf{i} + \mathbf{j} + 2\mathbf{k}$, $\mathbf{i} - \mathbf{j} + 2\mathbf{k}$

In each of Problems 11 through 20, find the projection of **F** in the direction of **r**. In each case, drawing a diagram might be helpful. Also, remember to distinguish the cases $0 \le \theta \le \pi/2$ and $\pi/2 \le \theta \le \pi$.

11. $\mathbf{F} = -3\mathbf{i} + \mathbf{k}$, $\mathbf{r} = 2\mathbf{j} + \mathbf{k}$
12. $\mathbf{F} = \mathbf{i} + \mathbf{j} + \mathbf{k}$, $\mathbf{r} = \mathbf{i} - \mathbf{j} + 2\mathbf{k}$
13. $\mathbf{F} = 2\mathbf{i} - 6\mathbf{j} - 2\mathbf{k}$, $\mathbf{r} = 3\mathbf{i} - 4\mathbf{j} + 2\mathbf{k}$
14. $\mathbf{F} = 6\mathbf{i} - 2\mathbf{j} - 3\mathbf{k}$, $\mathbf{r} = 8\mathbf{i} + 2\mathbf{k}$
15. $\mathbf{F} = -\mathbf{j}$, $\mathbf{r} = 6\mathbf{i} - \mathbf{j} + \mathbf{k}$
16. $\mathbf{F} = 2\mathbf{i} - \mathbf{j} + 4\mathbf{k}$, $\mathbf{r} = \mathbf{i} + 3\mathbf{k}$
17. $\mathbf{F} = 8\mathbf{i} + 2\mathbf{k}$, $\mathbf{r} = -\mathbf{i} + \mathbf{j}$
18. $\mathbf{F} = \mathbf{i} - 3\mathbf{j} + 2\mathbf{k}$, $\mathbf{r} = \mathbf{i} + 2\mathbf{k}$
19. $\mathbf{F} = -3\mathbf{i} + \mathbf{j} - 4\mathbf{k}$, $\mathbf{r} = -2\mathbf{i} - \mathbf{j} + 2\mathbf{k}$
20. $\mathbf{F} = \mathbf{i} - 2\mathbf{j} + \mathbf{k}$, $\mathbf{r} = 3\mathbf{j} - \mathbf{k}$

In each of Problems 21 through 30, find the equation of a plane through the given point and normal to the given vector.

21. $(-1, 1, 2)$, $3\mathbf{i} - \mathbf{j} + 4\mathbf{k}$
22. $(-1, 0, 0)$, $\mathbf{i} - 2\mathbf{j}$
23. $(2, -3, 4)$, $8\mathbf{i} - 6\mathbf{j} + 4\mathbf{k}$
24. $(-1, -1, -5)$, $-3\mathbf{i} + 2\mathbf{j}$
25. $(0, -1, 4)$, $8\mathbf{i} - 3\mathbf{j} + 4\mathbf{k}$
26. $(-2, 3, 2)$, $4\mathbf{i} - 8\mathbf{j} + 6\mathbf{k}$
27. $(0, 2, 0)$, $-2\mathbf{i} + \mathbf{k}$
28. $(0, 0, 0)$, $3\mathbf{i} - \mathbf{j} - 5\mathbf{k}$
29. $(1, -1, 5)$, $6\mathbf{i} - 14\mathbf{j} - 2\mathbf{k}$
30. $(-2, 1, -1)$, $4\mathbf{i} + 3\mathbf{j} + \mathbf{k}$

In each of Problems 31 through 40, you are given three points A, B, and C, forming vertices of a triangle. Find the cosine of the angle between the line AB and the line from A to the midpoint of the line BC. If you have a calculator, find the angle in both radians and degrees to four decimal places.

	A	B	C
31.	$(1, -2, 6)$	$(3, 0, 1)$	$(4, 2, -7)$
32.	$(1, -2, 6)$	$(0, 4, -3)$	$(-3, -2, 7)$
33.	$(0, 0, -2)$	$(1, -3, 4)$	$(-2, 6, 1)$
34.	$(-1, 6, 2)$	$(3, 0, 4)$	$(0, 8, -2)$
35.	$(0, 0, -2)$	$(1, -2, 6)$	$(0, 4, 1)$
36.	$(3, -2, -3)$	$(-2, 0, 1)$	$(1, 1, 7)$
37.	$(0, 5, -1)$	$(1, -2, 5)$	$(7, 0, -1)$
38.	$(2, -2, 1)$	$(1, 1, 7)$	$(0, -2, 3)$
39.	$(-4, 1, 2)$	$(2, -8, 6)$	$(6, 0, -2)$
40.	$(-2, 1, 7)$	$(-3, 4, 0)$	$(-2, 0, 0)$

41. Suppose $\mathbf{F} \cdot \mathbf{X} = 0$ for every vector $\mathbf{X}$. What can you say about $\mathbf{F}$?

42. Suppose $\mathbf{F} \cdot \mathbf{i} = \mathbf{F} \cdot \mathbf{j} = \mathbf{F} \cdot \mathbf{k} = 0$.
 (a) What can you say about $\mathbf{F}$?
 (b) How can the question be answered by using facts about orthogonal vectors?

43. Let $\mathbf{F} \neq \mathbf{0}$. Prove that the unit vector $\mathbf{u}$ making $|\mathbf{F} \cdot \mathbf{u}|$ a maximum must be parallel to $\mathbf{F}$.

44. Prove that, for any vector $\mathbf{F}$,

$$\mathbf{F} = (\mathbf{F} \cdot \mathbf{i})\mathbf{i} + (\mathbf{F} \cdot \mathbf{j})\mathbf{j} + (\mathbf{F} \cdot \mathbf{k})\mathbf{k}.$$

45. Let $\mathbf{A}$ and $\mathbf{B}$ be given vectors, with $\mathbf{B} \neq \mathbf{0}$. Prove that $\mathbf{A}$ can be written as $\mathbf{F} + \mathbf{G}$, where $\mathbf{F}$ is parallel to $\mathbf{B}$ and $\mathbf{G}$ is orthogonal to $\mathbf{B}$.

46. Let **F**, **G**, and **H** be nonzero vectors, each orthogonal to the other two. Show that, for any vector **A**, there are scalars α, β, and γ such that

$$\mathbf{A} = \alpha\mathbf{F} + \beta\mathbf{G} + \gamma\mathbf{H}.$$

47. Suppose we defined

$$(a_1\mathbf{i} + b_1\mathbf{j} + c_1\mathbf{k}) \cdot (a_2\mathbf{i} + b_2\mathbf{j} + c_2\mathbf{k}) = 2a_1a_2 + b_1b_2 + c_1c_2.$$

Which properties of dot product would still hold, and which would fail?

48. Suppose we defined

$$(a_1\mathbf{i} + b_1\mathbf{j} + c_1\mathbf{k}) \cdot (a_2\mathbf{i} + b_2\mathbf{j} + c_2\mathbf{k}) = a_1a_2 - b_1b_2 + c_1c_2.$$

Which properties of dot product would still hold, and which would fail?

49. It is not necessary to use the law of cosines to prove the Cauchy-Schwarz inequality. Fill in the details of the following proof.

We want to prove that $|\mathbf{F} \cdot \mathbf{G}| \leq \|\mathbf{F}\|\,\|\mathbf{G}\|$. If $\mathbf{F} = \mathbf{0}$ or $\mathbf{G} = \mathbf{0}$, the inequality is true. Thus, suppose that $\mathbf{F} \neq \mathbf{0}$ and $\mathbf{G} \neq \mathbf{0}$. By Theorem 29,

$$0 \leq \|\alpha\mathbf{F} + \beta\mathbf{G}\|^2 = \alpha^2\|\mathbf{F}\|^2 + 2\alpha\beta\mathbf{F} \cdot \mathbf{G} + \beta^2\|\mathbf{G}\|^2$$

for *any* scalars α and β. Choose $\alpha = \|\mathbf{G}\|$ and $\beta = -\|\mathbf{F}\|$ and simplify the resulting expression to get the Cauchy-Schwarz inequality.

50. Prove property 2 of dot product.

51. Prove property 3 of dot product.

52. Prove that the diagonals of a rhombus must be perpendicular to each other. Recall that a rhombus is a parallelogram with all sides of equal length.

53. Let **F** and **G** be nonzero vectors. Show that the vector

$$\frac{1}{\|\mathbf{F}\|\,\|\mathbf{G}\|}(\|\mathbf{G}\|\mathbf{F} + \|\mathbf{F}\|\mathbf{G})$$

bisects the angle between **F** and **G** (that is, show that this vector makes the same angle with **F** as it does with **G**).

9.3 CROSS PRODUCT OF VECTORS

The dot product produces a *scalar* from two vectors; the cross product produces a *vector* from two vectors.

Let

$$\mathbf{F} = a_1\mathbf{i} + b_1\mathbf{j} + c_1\mathbf{k} \quad \text{and} \quad \mathbf{G} = a_2\mathbf{i} + b_2\mathbf{j} + c_2\mathbf{k}.$$

We define

$$\mathbf{F} \times \mathbf{G} = (b_1c_2 - b_2c_1)\mathbf{i} + (a_2c_1 - a_1c_2)\mathbf{j} + (a_1b_2 - a_2b_1)\mathbf{k}.$$

This is read "**F** cross **G**."

There is a notational device* which is helpful in remembering the components of $\mathbf{F} \times \mathbf{G}$. If you expand the "determinant"

$$\begin{vmatrix} \mathbf{i} & \mathbf{j} & \mathbf{k} \\ a_1 & b_1 & c_1 \\ a_2 & b_2 & c_2 \end{vmatrix}$$

by row 1, you get exactly $\mathbf{F} \times \mathbf{G}$. Thus, for example,

$$(\mathbf{i} + 2\mathbf{j} - 3\mathbf{k}) \times (-2\mathbf{i} + \mathbf{j} + 4\mathbf{k}) = \begin{vmatrix} \mathbf{i} & \mathbf{j} & \mathbf{k} \\ 1 & 2 & -3 \\ -2 & 1 & 4 \end{vmatrix}$$

$$= (8 + 3)\mathbf{i} + (6 - 4)\mathbf{j} + (1 + 4)\mathbf{k} = 11\mathbf{i} + 2\mathbf{j} + 5\mathbf{k}.$$

As a second example,

$$(\mathbf{i} - 2\mathbf{j}) \times (3\mathbf{i} + 2\mathbf{j} - 4\mathbf{k}) = \begin{vmatrix} \mathbf{i} & \mathbf{j} & \mathbf{k} \\ 1 & -2 & 0 \\ 3 & 2 & -4 \end{vmatrix}$$

$$= (8 - 0)\mathbf{i} + (0 + 4)\mathbf{j} + (2 + 6)\mathbf{k} = 8\mathbf{i} + 4\mathbf{j} + 8\mathbf{k}.$$

Here are some basic properties of the cross product:

1. $\mathbf{F} \times \mathbf{G}$ is orthogonal to both $\mathbf{F}$ and $\mathbf{G}$.
2. $\|\mathbf{F} \times \mathbf{G}\| = \|\mathbf{F}\| \, \|\mathbf{G}\| \sin(\theta)$, where θ is the angle between $\mathbf{F}$ and $\mathbf{G}$.
3. $\mathbf{F} \times (\mathbf{G} + \mathbf{H}) = (\mathbf{F} \times \mathbf{G}) + (\mathbf{F} \times \mathbf{H})$.
4. $(\alpha\mathbf{F}) \times \mathbf{G} = \alpha(\mathbf{F} \times \mathbf{G}) = \mathbf{F} \times (\alpha\mathbf{G})$.
5. $\mathbf{F} \times \mathbf{G} = -(\mathbf{G} \times \mathbf{F})$.

Proofs of these are for the most part routine calculation. We shall prove properties 1 and 2, and leave the others to the student. After these proofs we shall return to geometric aspects of the cross product.

Proof of property 1 Calculate $\mathbf{F} \cdot (\mathbf{F} \times \mathbf{G})$. If $\mathbf{F} = a_1\mathbf{i} + b_1\mathbf{j} + c_1\mathbf{k}$ and $\mathbf{G} = a_2\mathbf{i} + b_2\mathbf{j} + c_2\mathbf{k}$, then we find that

$$\mathbf{F} \cdot (\mathbf{F} \times \mathbf{G}) = a_1(b_1c_2 - b_2c_1) + b_1(a_2c_1 - a_1c_2) + c_1(a_1b_2 - a_2b_1),$$

and this is zero because all the terms cancel. Thus, $\mathbf{F}$ is orthogonal to $\mathbf{F} \times \mathbf{G}$. In similar fashion, $\mathbf{G}$ is orthogonal to $\mathbf{F} \times \mathbf{G}$.

* If you have never seen a 3×3 determinant, skip this notational device and go to the next paragraph.

Proof of property 2 Compute

$$
\begin{aligned}
\|\mathbf{F} \times \mathbf{G}\|^2 &= (b_1c_2 - b_2c_1)^2 + (a_2c_1 - a_1c_2)^2 + (a_1b_2 - a_2b_1)^2 \\
&= (a_1^2 + b_1^2 + c_1^2)(a_2^2 + b_2^2 + c_2^2) - (a_1a_2 + b_1b_2 + c_1c_2)^2 \\
&= \|\mathbf{F}\|^2\|\mathbf{G}\|^2 - (\mathbf{F} \cdot \mathbf{G})^2 \\
&= \|\mathbf{F}\|^2\|\mathbf{G}\|^2 - \|\mathbf{F}\|^2\|\mathbf{G}\|^2 \cos^2(\theta) \\
&= \|\mathbf{F}\|^2\|\mathbf{G}\|^2[1 - \cos^2(\theta)] \\
&= \|\mathbf{F}\|^2\|\mathbf{G}\|^2 \sin^2(\theta).
\end{aligned}
$$

Taking the square root of both sides now gives us

$$\|\mathbf{F} \times \mathbf{G}\| = \|\mathbf{F}\|\,\|\mathbf{G}\|\,|\sin(\theta)|.$$

Since $0 \leq \theta \leq \pi$, then $\sin(\theta) \geq 0$ and $|\sin(\theta)| = \sin(\theta)$, and we have proved property 2.

Using these properties, let us examine what $\mathbf{F} \times \mathbf{G}$ looks like. By property 2, $\mathbf{F} \times \mathbf{G}$ has length equal to the length of $\mathbf{F}$ times the length of $\mathbf{G}$ times the sine of the angle between $\mathbf{F}$ and $\mathbf{G}$. If $\mathbf{F}$ and $\mathbf{G}$ are parallel, then $\theta = 0$; so $\sin(\theta) = 0$ and $\mathbf{F} \times \mathbf{G} = \mathbf{0}$. If $\mathbf{F}$ and $\mathbf{G}$ are not parallel, then $\mathbf{F}$ and $\mathbf{G}$ determine a plane, as shown in the leftmost part of Figure 122. By property 1, $\mathbf{F} \times \mathbf{G}$ must be perpendicular to this plane; hence must point out of the plane from one side or the other. To see which side, we take a cue from the fact that $\mathbf{i} \times \mathbf{j} = \mathbf{k}$. In fact, the following right-hand rule, as shown in Figure 122, applies:

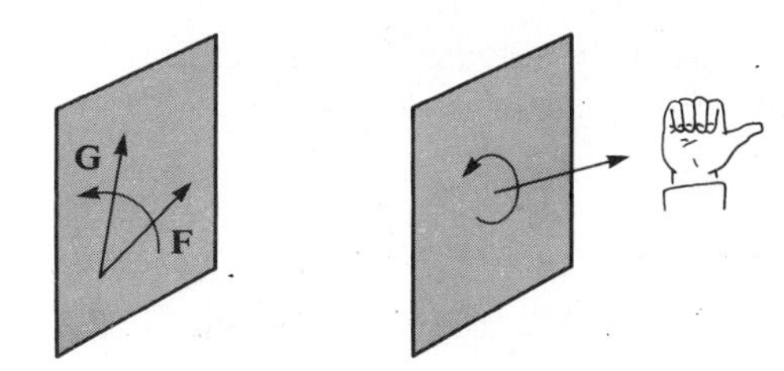

Figure 122. Right-hand rule for $\mathbf{F} \times \mathbf{G}$.

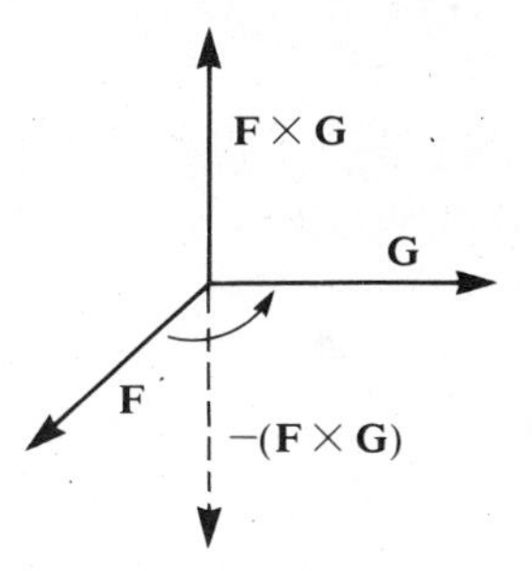

Figure 123. Figuring the direction of $\mathbf{F} \times \mathbf{G}$.

If you place your right hand so that the fingers curl in the direction of rotation *from* $\mathbf{F}$ *to* $\mathbf{G}$, then the thumb points (roughly) along $\mathbf{F} \times \mathbf{G}$.

Put another way, if you walk *from* $\mathbf{F}$ *to* $\mathbf{G}$, and look *up* perpendicular to the plane of $\mathbf{F}$ and $\mathbf{G}$, you should be looking along $\mathbf{F} \times \mathbf{G}$. If you look *down*, you will be looking in the direction of $-(\mathbf{F} \times \mathbf{G})$, which by property 5 is $\mathbf{G} \times \mathbf{F}$. This is shown in Figure 123.

These considerations emphasize the "antisymmetry" of the cross product, namely, $\mathbf{F} \times \mathbf{G}$ is minus $\mathbf{G} \times \mathbf{F}$. Reversing the order of the factors reverses the direction of the resulting vector. This can also be seen from the definition or from the determinant expression for $\mathbf{F} \times \mathbf{G}$. To compute $\mathbf{F} \times \mathbf{G}$, we put the components of $\mathbf{F}$ in the second row and those of $\mathbf{G}$ in the third row. For $\mathbf{G} \times \mathbf{F}$, we interchange these rows, resulting in a sign change.

EXAMPLE 210

Let $\mathbf{F} = 3\mathbf{i} - 2\mathbf{j} + 6\mathbf{k}$ and $\mathbf{G} = \mathbf{i} + 2\mathbf{j} - \mathbf{k}$. Then,

$$\mathbf{F} \times \mathbf{G} = \begin{vmatrix} \mathbf{i} & \mathbf{j} & \mathbf{k} \\ 3 & -2 & 6 \\ 1 & 2 & -1 \end{vmatrix} = -10\mathbf{i} + 9\mathbf{j} + 8\mathbf{k}$$

and

$$\mathbf{G} \times \mathbf{F} = \begin{vmatrix} \mathbf{i} & \mathbf{j} & \mathbf{k} \\ 1 & 2 & -1 \\ 3 & -2 & 6 \end{vmatrix} = 10\mathbf{i} - 9\mathbf{j} - 8\mathbf{k} = -(\mathbf{F} \times \mathbf{G}).$$

Note that the angle between $\mathbf{F}$ and $\mathbf{G}$ is given by

$$\cos(\theta) = \frac{\mathbf{F} \cdot \mathbf{G}}{\|\mathbf{F}\| \, \|\mathbf{G}\|} = \frac{-7}{\sqrt{49}\sqrt{6}}.$$

Thus, in this example,

$$\sin(\theta) = \sqrt{1 - \cos^2(\theta)} = \sqrt{\frac{245}{294}}.$$

Then,

$$\|\mathbf{F}\| \, \|\mathbf{G}\| \sin(\theta) = \sqrt{245}.$$

It is easy to check directly that indeed $\|\mathbf{F} \times \mathbf{G}\| = \sqrt{10^2 + 9^2 + 8^2} = \sqrt{245}$.

As with dot product, cross product is useful in solving a variety of problems which can be posed in terms of vectors. Here are some examples.

EXAMPLE 211

Find the equation of the plane containing the points $(1, 2, 1)$, $(-1, 1, 3)$, and $(-2, -2, -2)$.

If we can find a normal (perpendicular) vector to the proposed plane, then we can proceed as in Example 208. To find a normal vector, note that

$$\mathbf{F} = (-1 - 1)\mathbf{i} + (1 - 2)\mathbf{j} + (3 - 1)\mathbf{k} = -2\mathbf{i} - \mathbf{j} + 2\mathbf{k}$$

must be in the plane we seek, as it is represented by the arrow from $(1, 2, 1)$ to $(-1, 1, 3)$.

Similarly,

$$\mathbf{G} = (-2 - 1)\mathbf{i} + (-2 - 2)\mathbf{j} + (-2 - 1)\mathbf{k} = -3\mathbf{i} - 4\mathbf{j} - 3\mathbf{k}$$

is in the plane. Then $\mathbf{F} \times \mathbf{G}$ is perpendicular to this plane. Compute

$$\mathbf{F} \times \mathbf{G} = \begin{vmatrix} \mathbf{i} & \mathbf{j} & \mathbf{k} \\ -2 & -1 & 2 \\ -3 & -4 & -3 \end{vmatrix} = 11\mathbf{i} - 12\mathbf{j} + 5\mathbf{k}.$$

This is a normal vector to the plane we seek.

Now proceed as in Example 208. If (x, y, z) is a point in the plane, then $(x - 1)\mathbf{i} + (y - 2)\mathbf{j} + (z - 1)\mathbf{k}$ is a vector in the plane; hence must be perpendicular to $11\mathbf{i} - 12\mathbf{j} + 5\mathbf{k}$. Thus, the dot product must be zero, that is,

$$11(x - 1) - 12(y - 2) + 5(z - 1) = 0.$$

This works out to

$$11x - 12y + 5z = -8.$$

The student should check that each of the three given points is indeed on this plane.

EXAMPLE 212

Find the area of the parallelogram with incident sides extending from $(0, 1, -2)$ to $(1, 2, 2)$ and from $(0, 1, -2)$ to $(1, 4, 1)$, as shown in Figure 124.

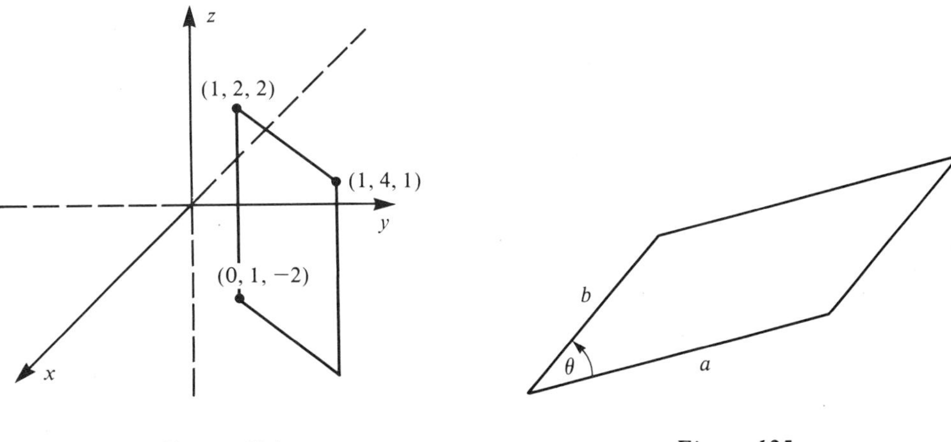

Figure 124

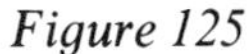

Figure 125

Recall that the area of a parallelogram with sides of length a and b incident with an angle θ, as shown in Figure 125, is $ab\sin(\theta)$. Thus, we need only put vectors $\mathbf{F}$ and $\mathbf{G}$ in for the sides of the parallelogram, and $\|\mathbf{F}\|\,\|\mathbf{G}\|\sin(\theta)$ is the area. This is exactly $\|\mathbf{F} \times \mathbf{G}\|$. Compute

$$\mathbf{F} = (1 - 0)\mathbf{i} + (2 - 1)\mathbf{j} + [2 - (-2)]\mathbf{k} = \mathbf{i} + \mathbf{j} + 4\mathbf{k}$$

and

$$\mathbf{G} = (1 - 0)\mathbf{i} + (4 - 1)\mathbf{j} + [1 - (-2)]\mathbf{k} = \mathbf{i} + 3\mathbf{j} + 3\mathbf{k}.$$

Then,

$$\mathbf{F} \times \mathbf{G} = \begin{vmatrix} \mathbf{i} & \mathbf{j} & \mathbf{k} \\ 1 & 1 & 4 \\ 1 & 3 & 3 \end{vmatrix} = -9\mathbf{i} + \mathbf{j} + 2\mathbf{k}.$$

The area of the parallelogram is $\|-9\mathbf{i} + \mathbf{j} + 2\mathbf{k}\|$ or $\sqrt{86}$.

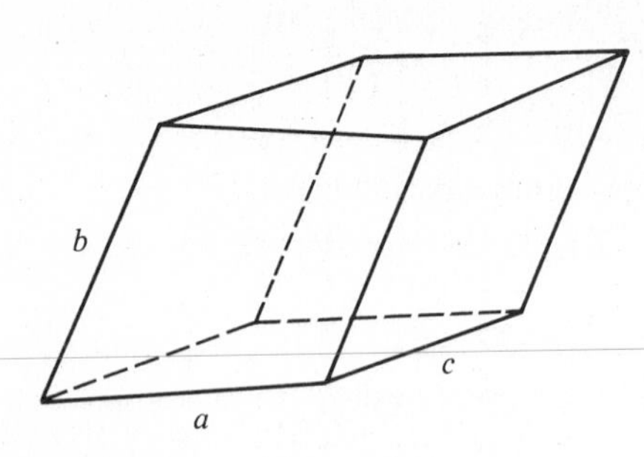

Figure 126. A parallelopiped.

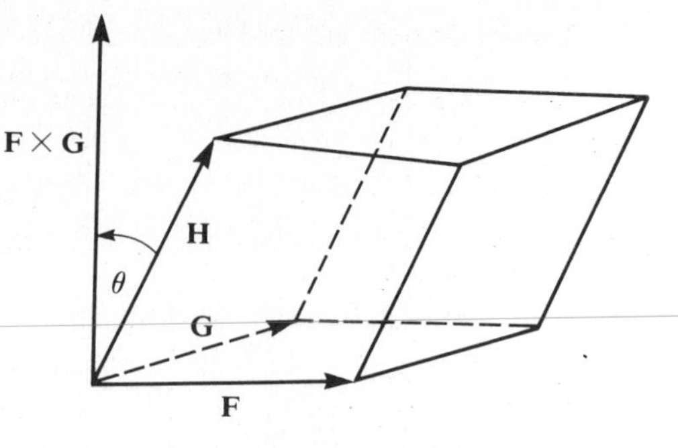

Figure 127. Volume of the parallelopiped is $|\mathbf{H} \cdot (\mathbf{F} \times \mathbf{G})|$.

EXAMPLE 213

Suppose we have a parallelopiped, as in Figure 126. What is the volume? If the sides were mutually perpendicular, the answer would be abc. Here, however, we allow that the box has been skewed out of plumb. Approach the problem as follows.

Imagine three incident sides of the parallelopiped to be vectors **F**, **G**, and **H**, as in Figure 127. We claim that the volume is

$$|\mathbf{H} \cdot (\mathbf{F} \times \mathbf{G})|.$$

To see this, note that

$$|\mathbf{H} \cdot (\mathbf{F} \times \mathbf{G})| = \|\mathbf{H}\| \, \|\mathbf{F} \times \mathbf{G}\| \cos(\theta).$$

where θ is the angle between **H** and $\mathbf{F} \times \mathbf{G}$. (If **F** and **G** are drawn in the xy-plane, then θ is the angle between **H** and the z-axis.)

Now,

$$\|\mathbf{F} \times \mathbf{G}\| = \text{area of the base of the parallelopiped}$$

and

$$\|\mathbf{H}\|\cos(\theta) = \text{height of the parallelopiped.}$$

Hence, $\|\mathbf{H}\| \, \|\mathbf{F} \times \mathbf{G}\| \cos(\theta)$ is the volume, as we wanted.

As a concrete example, suppose that one corner is at $(-1, 2, 2)$, and that three incident sides extend from this point to $(0, 1, 1)$, $(-4, 6, 8)$, and $(-3, -2, 4)$. To find the volume, form vectors along these sides:

$$\mathbf{F} = [0 - (-1)]\mathbf{i} + (1 - 2)\mathbf{j} + (1 - 2)\mathbf{k} = \mathbf{i} - \mathbf{j} - \mathbf{k},$$
$$\mathbf{G} = [-4 - (-1)]\mathbf{i} + (6 - 2)\mathbf{j} + (8 - 2)\mathbf{k} = -3\mathbf{i} + 4\mathbf{j} + 6\mathbf{k},$$
$$\mathbf{H} = [-3 - (-1)]\mathbf{i} + (-2 - 2)\mathbf{j} + (4 - 2)\mathbf{k} = -2\mathbf{i} - 4\mathbf{j} + 2\mathbf{k}.$$

Then $\mathbf{F} \times \mathbf{G} = -2\mathbf{i} - 3\mathbf{j} + \mathbf{k}$; so the volume is

$$|\mathbf{H} \cdot (\mathbf{F} \times \mathbf{G})| = |4 + 12 + 2| = 18 \text{ cubic units.}$$

Sometimes $\mathbf{H} \cdot (\mathbf{F} \times \mathbf{G})$ is called a *scalar triple product*. We shall pursue some facts about this and certain other vector combinations in the next section.

PROBLEMS FOR SECTION 9.3

In each of Problems 1 through 15, compute $\mathbf{F} \times \mathbf{G}$ and $\mathbf{G} \times \mathbf{F}$. In each case, use the dot product to compute $\cos(\theta)$, with θ the angle between $\mathbf{F}$ and $\mathbf{G}$. Then determine $\sin(\theta) = \sqrt{1 - \cos^2(\theta)}$, calculate $\|\mathbf{F}\|\,\|\mathbf{G}\|\sin(\theta)$, and verify directly that this equals $\|\mathbf{F} \times \mathbf{G}\|$.

1. $\mathbf{F} = -3\mathbf{i} + 6\mathbf{j} + \mathbf{k}$, $\mathbf{G} = -\mathbf{i} - 2\mathbf{j} + \mathbf{k}$

2. $\mathbf{F} = 6\mathbf{i} - \mathbf{k}$, $\mathbf{G} = \mathbf{j} + 2\mathbf{k}$

3. $\mathbf{F} = 2\mathbf{i} - 3\mathbf{j} + 4\mathbf{k}$, $\mathbf{G} = -3\mathbf{i} + 2\mathbf{j}$

4. $\mathbf{F} = 8\mathbf{i} + 6\mathbf{j}$, $\mathbf{G} = 14\mathbf{j}$

5. $\mathbf{F} = 5\mathbf{i} + 3\mathbf{j} + 4\mathbf{k}$, $\mathbf{G} = 20\mathbf{i} + 6\mathbf{k}$

6. $\mathbf{F} = 2\mathbf{k}$, $\mathbf{G} = 8\mathbf{i} - \mathbf{j}$

7. $\mathbf{F} = 18\mathbf{i} - 3\mathbf{j} + 4\mathbf{k}$, $\mathbf{G} = 22\mathbf{j} - \mathbf{k}$

8. $\mathbf{F} = \mathbf{i} - 3\mathbf{j} - \mathbf{k}$, $\mathbf{G} = 18\mathbf{i} - 21\mathbf{j}$

9. $\mathbf{F} = -4\mathbf{i} + 6\mathbf{k}$, $\mathbf{G} = \mathbf{i} - 2\mathbf{j} + 7\mathbf{k}$

10. $\mathbf{F} = -3\mathbf{i} + 2\mathbf{j} + \mathbf{k}$, $\mathbf{G} = 8\mathbf{i} - 6\mathbf{j} + 2\mathbf{k}$

11. $\mathbf{F} = \mathbf{i} + 3\mathbf{j} - \mathbf{k}$, $\mathbf{G} = 2\mathbf{i} + 6\mathbf{j} - \mathbf{k}$

12. $\mathbf{F} = 2\mathbf{j} - \mathbf{k}$, $\mathbf{G} = \mathbf{i} + 6\mathbf{j} + 3\mathbf{k}$

13. $\mathbf{F} = -\mathbf{i} + 8\mathbf{j} + 3\mathbf{k}$, $\mathbf{G} = 7\mathbf{i} + 2\mathbf{j}$

14. $\mathbf{F} = 2\mathbf{i} + 3\mathbf{j} - 5\mathbf{k}$, $\mathbf{G} = -2\mathbf{i} + 3\mathbf{j} - 4\mathbf{k}$

15. $\mathbf{F} = 5\mathbf{i} + 8\mathbf{j} - 3\mathbf{k}$, $\mathbf{G} = \mathbf{i} - \mathbf{j} - 3\mathbf{k}$

In each of Problems 16 through 30, (a) find a vector normal to the plane determined by the three given points, and (b) find the equation of the plane containing the three points.

16. $(-1, 1, 6)$, $(2, 0, 1)$, $(3, 0, 0)$

17. $(4, 1, 1)$, $(-2, -2, 3)$, $(6, 0, 1)$

18. $(1, 0, -2)$, $(0, 0, 0)$, $(5, 1, 1)$

19. $(0, 0, 2)$, $(-4, 1, 0)$, $(2, -1, 1)$

20. $(-4, 2, -6)$, $(1, 1, 3)$, $(-2, 4, 5)$

21. $(1, 0, 2)$, $(0, 1, 6)$, $(-7, 1, 9)$

22. $(-2, 1, 6)$, $(0, 0, 1)$, $(2, 1, -2)$

23. $(3, 3, -4)$, $(6, 1, 3)$, $(2, -4, 3)$

24. $(8, 0, 7)$, $(0, 0, 0)$, $(1, -2, 1)$

25. $(-3, -4, -7)$, $(8, 0, 2)$, $(-5, 6, 3)$

26. $(7, -1, 4)$, $(0, -1, 5)$, $(5, 7, 2)$

27. $(2, 3, -2)$, $(1, 1, -5)$, $(-2, -3, 1)$

28. $(8, 0, 4)$, $(6, 1, 1)$, $(7, -1, 0)$

29. $(9, 0, -5)$, $(4, 4, -4)$, $(-2, -1, -3)$

30. $(7, 0, -2)$, $(8, 1, 1)$, $(9, -2, 7)$

In each of Problems 31 through 40, find the volume of the parallelopiped with one corner at P and sides PQ, PR, and PS.

	P	Q	R	S
31.	(1, 1, 1)	(−2, 1, 6)	(3, 5, 7)	(0, 1, 6)
32.	(0, 1, −6)	(−3, 1, 4)	(1, 7, 2)	(−3, 0, 4)
33.	(1, 6, 1)	(−2, 1, 4)	(3, 0, 0)	(2, 2, −4)
34.	(0, 1, 7)	(9, 1, 3)	(−2, 4, 1)	(3, 0, −2)
35.	(1, 1, 1)	(2, 2, 2)	(6, 1, 3)	(−2, 4, 1)
36.	(0, 1, 1)	(2, 9, −7)	(6, 0, −5)	(−6, 0, 3)
37.	(0, 1, 1)	(8, −1, 3)	(2, 1, 6)	(3, 4, 4)
38.	(2, 0, 0)	(0, 1, 1)	(1, 3, 3)	(2, 4, 7)
39.	(1, 3, 1)	(2, 2, 5)	(6, 1, 7)	(9, 8, 8)
40.	(2, 0, 4)	(1, 1, 7)	(2, 8, 8)	(5, 4, 7)

In each of Problems 41 through 50, find the area of the parallelogram with one corner at P and sides PQ and PR.

	P	Q	R
41.	(1, −3, 7)	(2, 1, 1)	(6, −1, 2)
42.	(0, 1, 1)	(2, 0, −4)	(−3, −2, 1)
43.	(6, 1, 1)	(7, −2, 4)	(8, −4, 3)
44.	(−2, 1, 6)	(7, 1, 1)	(8, −1, 9)
45.	(−2, 1, 6)	(2, 1, −7)	(4, 1, 1)
46.	(0, 0, 3)	(1, 1, −2)	(2, −5, 3)
47.	(4, 2, −3)	(6, 2, −1)	(2, −6, 4)
48.	(−3, −1, 2)	(8, 0, 2)	(−2, −6, 4)
49.	(1, 1, −6)	(5, −3, 0)	(−2, 4, 1)
50.	(−4, 6, 3)	(1, 1, −5)	(2, 2, 2)

In each of Problems 51 through 60, find a vector normal to the given plane.

51. $8x - y + z = 12$ **52.** $x - y + 2z = 0$ **53.** $2x - 3y + 4z = 1$
54. $x - 3y + 2z = 9$ **55.** $7x + y - 7z = 7$ **56.** $x - y = 0$
57. $4x + 6y + 4z = -5$ **58.** $2x + y - 7z = -10$ **59.** $-3x + 2y - 8z = 21$
60. $4x + 2y - z = 13$
61. Prove that $\mathbf{F} \times (\mathbf{G} + \mathbf{H}) = \mathbf{F} \times \mathbf{G} + \mathbf{F} \times \mathbf{H}$.
62. Prove that $(\alpha\mathbf{F}) \times \mathbf{G} = \alpha(\mathbf{F} \times \mathbf{G}) = \mathbf{F} \times (\alpha\mathbf{G})$.
63. Prove that $\mathbf{F} \times (\mathbf{F} \times \mathbf{G}) = (\mathbf{F} \cdot \mathbf{G})\mathbf{F} - (\mathbf{F} \cdot \mathbf{F})\mathbf{G}$.
64. Prove *Lagrange's identity:*

$$(\mathbf{F} \times \mathbf{G}) \cdot (\mathbf{H} \times \mathbf{K}) = (\mathbf{F} \cdot \mathbf{H})(\mathbf{G} \cdot \mathbf{K}) - (\mathbf{F} \cdot \mathbf{K})(\mathbf{G} \cdot \mathbf{H}).$$

65. Suppose that **F**, **G**, **H**, and **K** lie in the same plane. What can you say about $(\mathbf{F} \times \mathbf{G}) \times (\mathbf{H} \times \mathbf{K})$?
66. Prove that $\mathbf{F} \cdot (\mathbf{G} \times \mathbf{H}) = \mathbf{G} \cdot (\mathbf{H} \times \mathbf{F}) = \mathbf{H} \cdot (\mathbf{F} \times \mathbf{G})$.
67. Prove that $(\mathbf{F} \times \mathbf{G}) \times (\mathbf{H} \times \mathbf{K}) = [\mathbf{H} \cdot (\mathbf{G} \times \mathbf{F})]\mathbf{G} - [\mathbf{H} \cdot (\mathbf{K} \times \mathbf{G})]\mathbf{F}$.
68. Prove that $\mathbf{F} \times \mathbf{G} = \mathbf{0}$ if and only if either $\mathbf{F} = \mathbf{0}$, $\mathbf{G} = \mathbf{0}$, or **F** and **G** are parallel.
69. Use vector operations to find a formula for the area of a triangle having vertices at (a_1, b_1, c_1), (a_2, b_2, c_2), and (a_3, b_3, c_3). What conditions must you place on the coordinates of the three points? (*Hint:* What happens if the three points all lie on a straight line?)

9.4 SCALAR TRIPLE PRODUCT AND VECTOR IDENTITIES

In the last example of the preceding section we encountered a volume expressed as the dot product of one vector with the cross product of two others. This is called a *scalar triple product.** More carefully, the scalar triple product of **F**, **G**, and **H** is defined to be

$$[\mathbf{F}, \mathbf{G}, \mathbf{H}] = \mathbf{F} \cdot (\mathbf{G} \times \mathbf{H}).$$

Note that the result is a *scalar*, not a vector. Scalar triple products arise in a variety of contexts. For example, the electromotive force dE induced in an element $d\mathbf{L}$ of a conducting wire moving with velocity **v** through a magnetic field having flux density **B** is given by $dE = [\mathbf{v}, \mathbf{B}, d\mathbf{L}]$.

Here are some basic properties of the scalar triple product:

1. If $\mathbf{F} = a_1\mathbf{i} + b_1\mathbf{j} + c_1\mathbf{k}$, $\mathbf{G} = a_2\mathbf{i} + b_2\mathbf{j} + c_2\mathbf{k}$, and $\mathbf{H} = a_3\mathbf{i} + b_3\mathbf{j} + c_3\mathbf{k}$, then $[\mathbf{F}, \mathbf{G}, \mathbf{H}]$ can be evaluated as a 3×3 determinant:

$$[\mathbf{F}, \mathbf{G}, \mathbf{H}] = \begin{vmatrix} a_1 & b_1 & c_1 \\ a_2 & b_2 & c_2 \\ a_3 & b_3 & c_3 \end{vmatrix}.$$

* Sometimes the scalar triple product is called the *box product*. The reason for this can be seen in property 5.

Expanded, this gives

$$[\mathbf{F}, \mathbf{G}, \mathbf{H}] = a_1b_2c_3 - a_1c_2b_3 + b_1c_2a_3 - b_1a_2c_3 + c_1a_2b_3 - c_1b_2a_3.$$

2. $[\mathbf{F}, \mathbf{G}, \mathbf{H}] = [\mathbf{G}, \mathbf{H}, \mathbf{F}] = [\mathbf{H}, \mathbf{F}, \mathbf{G}]$.
3. $[\mathbf{F}, \mathbf{G}, \mathbf{H}] = -[\mathbf{F}, \mathbf{H}, \mathbf{G}]$.
4. For any scalars α and β, we have:

$$[\alpha\mathbf{F} + \beta\mathbf{K}, \mathbf{G}, \mathbf{H}] = \alpha[\mathbf{F}, \mathbf{G}, \mathbf{H}] + \beta[\mathbf{K}, \mathbf{G}, \mathbf{H}],$$
$$[\mathbf{F}, \alpha\mathbf{G} + \beta\mathbf{K}, \mathbf{H}] = \alpha[\mathbf{F}, \mathbf{G}, \mathbf{H}] + \beta[\mathbf{F}, \mathbf{K}, \mathbf{H}],$$

and

$$[\mathbf{F}, \mathbf{G}, \alpha\mathbf{H} + \beta\mathbf{K}] = \alpha[\mathbf{F}, \mathbf{G}, \mathbf{H}] + \beta[\mathbf{F}, \mathbf{G}, \mathbf{K}].$$

5. If the vectors **F**, **G**, and **H** are drawn as arrows from a common point and considered as sides of a parallelopiped, then the volume of this parallelopiped is

$$|[\mathbf{F}, \mathbf{G}, \mathbf{H}]|.$$

6. If any one of **F**, **G**, or **H** is a sum of scalar multiples of the other two, then

$$[\mathbf{F}, \mathbf{G}, \mathbf{H}] = 0.$$

(For example, if $\mathbf{G} = 2\mathbf{F} + 3\mathbf{H}$, then $[\mathbf{F}, \mathbf{G}, \mathbf{H}] = 0$.)

Proofs of these are routine and are left to the exercises. Here we shall discuss the properties, make some observations, and give some examples.

Note first that $\mathbf{F} \cdot (\mathbf{G} \times \mathbf{H})$ can be written just $\mathbf{F} \cdot \mathbf{G} \times \mathbf{H}$, without parentheses, since $(\mathbf{F} \cdot \mathbf{G}) \times \mathbf{H}$ makes no sense. Property 2 can then be written

$$\mathbf{F} \cdot \mathbf{G} \times \mathbf{H} = \mathbf{G} \cdot \mathbf{H} \times \mathbf{F} = \mathbf{H} \cdot \mathbf{F} \times \mathbf{G}.$$

As long as the factors appear in cyclic order,

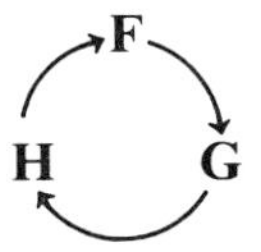

then the dot and cross may be inserted between any pairs. If, however, the vectors are interchanged out of this cyclic order, then a minus sign appears. Thus, in property 3, we have

$$\mathbf{F} \cdot \mathbf{G} \times \mathbf{H} = -\mathbf{F} \cdot \mathbf{H} \times \mathbf{G}.$$

Combining properties 2 and 3, the *absolute value* of $[\mathbf{F}, \mathbf{G}, \mathbf{H}]$ is the same regardless of the order of the vectors:

$$|[\mathbf{F}, \mathbf{G}, \mathbf{H}]| = |[\mathbf{G}, \mathbf{H}, \mathbf{F}]| = |[\mathbf{H}, \mathbf{G}, \mathbf{F}]| = \cdots.$$

This follows easily from property 1, since interchanging **F**, **G**, and **H** causes an interchange of rows of the determinant, which can cause only a change in sign.

EXAMPLE 214

Let $\mathbf{F} = 8\mathbf{i} - 2\mathbf{j} + 3\mathbf{k}$, $\mathbf{G} = 6\mathbf{i} + 4\mathbf{j} + 7\mathbf{k}$, and $\mathbf{H} = -7\mathbf{i} + 2\mathbf{j} - \mathbf{k}$. Then, from property 1,

$$[\mathbf{F}, \mathbf{G}, \mathbf{H}] = \begin{vmatrix} 8 & -2 & 3 \\ 6 & 4 & 7 \\ -7 & 2 & -1 \end{vmatrix} = 62.$$

As a check, note that

$$\mathbf{G} \times \mathbf{H} = \begin{vmatrix} \mathbf{i} & \mathbf{j} & \mathbf{k} \\ 6 & 4 & 7 \\ -7 & 2 & -1 \end{vmatrix} = -18\mathbf{i} - 43\mathbf{j} + 40\mathbf{k},$$

and so $\mathbf{F} \cdot \mathbf{G} \times \mathbf{H} = 8(-18) - 2(-43) + 3(40) = 62$.

The student can try these vectors in different order, and should always get $+62$ or -62, depending upon the changes made in the order. Note that, by property 5 and Example 213, $|[\mathbf{F}, \mathbf{G}, \mathbf{H}]|$ is the volume of the parallelopiped having sides **F**, **G**, and **H**, drawn as arrows from a common point, say, the origin.

Property 6 can be viewed in several ways. To be specific, suppose for example that $\mathbf{H} = \alpha\mathbf{F} + \beta\mathbf{G}$. Then, row 3 in the determinant of property 1 would be a sum of α times the first row plus β times the second row; hence the determinant must be zero. Geometrically, $\mathbf{H} = \alpha\mathbf{F} + \beta\mathbf{G}$ means that **H** is in the plane or line determined by **F** and **G**. In this case the "box" determined by **F**, **G**, and **H** is flat and has zero volume, again implying that $[\mathbf{F}, \mathbf{G}, \mathbf{H}] = 0$.

EXAMPLE 215

Let $\mathbf{F} = 2\mathbf{i} - 6\mathbf{j} + \mathbf{k}$, $\mathbf{G} = 8\mathbf{i} + 2\mathbf{j} - 4\mathbf{k}$, and $\mathbf{H} = -4\mathbf{i} - 14\mathbf{j} + 6\mathbf{k}$. Then, $\mathbf{H} = 2\mathbf{F} - \mathbf{G}$, and

$$[\mathbf{F}, \mathbf{G}, \mathbf{H}] = \begin{vmatrix} 2 & -6 & 1 \\ 8 & 2 & -4 \\ -4 & -14 & 6 \end{vmatrix} = 0.$$

Property 4 can be used to reduce calculations of $[\mathbf{F}, \mathbf{G}, \mathbf{H}]$ to calculations involving $[\mathbf{i}, \mathbf{j}, \mathbf{k}]$, which is just 1. Here is an illustration.

EXAMPLE 216

$$\begin{aligned}[][2\mathbf{i} - 4\mathbf{j}, 3\mathbf{i} + \mathbf{j}, \mathbf{k}] &= 2[\mathbf{i}, 3\mathbf{i} + \mathbf{j}, \mathbf{k}] - 4[\mathbf{j}, 3\mathbf{i} + \mathbf{j}, \mathbf{k}] \\ &= 2(3)[\mathbf{i}, \mathbf{i}, \mathbf{k}] + 2[\mathbf{i}, \mathbf{j}, \mathbf{k}] - 4(3)[\mathbf{j}, \mathbf{i}, \mathbf{k}] - 4[\mathbf{j}, \mathbf{j}, \mathbf{k}].\end{aligned}$$

Now, $[\mathbf{i}, \mathbf{i}, \mathbf{k}] = [\mathbf{j}, \mathbf{j}, \mathbf{k}] = 0$ by property 6, and

$$[\mathbf{i}, \mathbf{j}, \mathbf{k}] = 1 = -[\mathbf{j}, \mathbf{i}, \mathbf{k}].$$

Thus,

$$[2\mathbf{i} - 4\mathbf{j}, 3\mathbf{i} + \mathbf{j}, \mathbf{k}] = 2(1) - 12(-1) = 14.$$

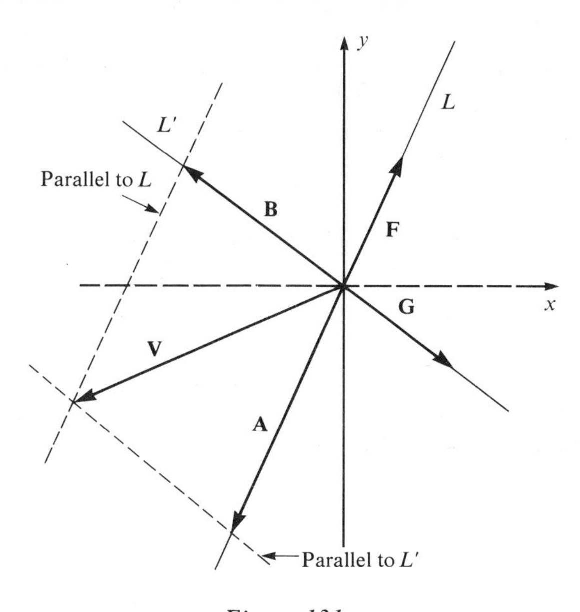

Figure 131

V is along L', then **V** is a scalar multiple of **G**; hence is in S. If **V** is along neither L nor L', then perform the following construction (see Figure 131).

Represent **V** as an arrow from the origin. Draw parallels to L and L' from the tip of the arrow representing **V**. As shown in Figure 131, these determine vectors **A** (along L) and **B** (along L') such that, by the parallelogram law, $\mathbf{A} + \mathbf{B} = \mathbf{V}$. Now, **A** is in S and **B** is in S; so **V** must be in S, because S is a subspace. Thus, the arbitrarily chosen vector **V** is in S; so in this case $S = R^2$.

By similar reasoning, but requiring more cases, one can prove the following:

THEOREM 35

The only subspaces of R^3 are (1) the trivial subspace, (2) R^3, (3) all vectors on a given straight line through the origin, and (4) all vectors on a given plane through the origin.

PROBLEMS FOR SECTION 9.5

In each of Problems 1 through 10, determine the vector sum of the two given n-vectors, and express the sum in terms of $\mathbf{e}_1, \ldots, \mathbf{e}_n$. Also determine the dot product of the vectors and the cosine of the angle between them.*

1. $(-1, 6, 2, 4, 0)$, $(6, -1, 4, 1, 1)$; in R^5

2. $(0, 1, 4, -3)$, $(2, 8, 6, -4)$; in R^4

3. $(1, -4, 3, 2)$, $(16, 0, 0, 4)$; in R^4

4. $(8, -3, 2)$, $(9, 1, 7)$; in R^3

5. $(-14, 6, 1, -1, 3, 2, 8)$, $(9, -3, 4, 4, 1, -2, 7)$; in R^7

6. $(3, -5, 2, 8)$, $(6, 1, -7, -2)$; in R^4

7. $(6, 1, 1, -1, 2)$, $(4, 3, 5, 1, -2)$; in R^5

8. $(5, -3, 2, 1, 6)$, $(1, -4, 5, 14, 12)$; in R^5

* If you have a calculator, give both radian and degree measures of the angle to four decimal places.

9. $(-3, 2, 1, -4, -5)$, $(18, 1, -5, 6, 7)$; in R^5
10. $(0, 1, 6)$, $(-4, 1, -2)$, $(14, 1, -7, 6, 1, 5)$; in R^6

In each of Problems 11 through 20, determine whether or not the given set S of vectors is a subspace of R^n for the given n.

11. S consists of all 5-vectors of the form (x, y, z, x, x).
12. S consists of all 4-vectors of the form $(x, 2x, 3x, y)$.
13. S consists of all vectors in R^6 of the form $(x, 0, 0, 1, 0, y)$.
14. S consists of all vectors in R^3 of the form $(0, x, y)$.
15. S consists of all vectors in R^4 of the form $(x, y, x + y, x - y)$.
16. S consists of all vectors in R^7 with third and fifth components zero.
17. S consists of all vectors in R^4 with first and second components equal.
18. S consists of all vectors in R^4 with third component 2.
19. S consists of all vectors in R^7 with seventh component the sum of the first six components.
20. S consists of all vectors in R^8 with zero first, second, and fourth components, and third component equal to the sixth.
21. Prove all ten properties listed in Theorem 32.
22. Prove the five properties of Theorem 33.
23. Let $\mathbf{F}$ and $\mathbf{G}$ be in R^n, and let α and β be any scalars. Prove that

$$(\alpha\mathbf{F} - \beta\mathbf{G}) \cdot (\alpha\mathbf{F} - \beta\mathbf{G}) = \alpha^2\|\mathbf{F}\|^2 - 2\alpha\beta\mathbf{F} \cdot \mathbf{G} + \beta^2\|\mathbf{G}\|^2.$$

24. Prove the Cauchy-Schwarz inequality for R^n: If $\mathbf{F}$ and $\mathbf{G}$ are in R^n, then

$$|\mathbf{F} \cdot \mathbf{G}| \leq \|\mathbf{F}\| \, \|\mathbf{G}\|.$$

Note: This was proved easily in Section 9.2 for the case $n = 3$ by using the fact that, in R^3, the law of cosines gave us

$$\cos(\theta) = \frac{\mathbf{F} \cdot \mathbf{G}}{\|\mathbf{F}\| \, \|\mathbf{G}\|}.$$

Since $-1 \leq \cos(\theta) \leq 1$, the Cauchy-Schwarz inequality followed. In R^n, however, when $n \geq 4$, we *defined* the angle between $\mathbf{F}$ and $\mathbf{G}$ by this formula, having no law of cosines to use as justification. Thus, one must prove the inequality without using the *definition* of $\cos(\theta)$ as $\mathbf{F} \cdot \mathbf{G}/\|\mathbf{F}\| \, \|\mathbf{G}\|$. In fact, the Cauchy-Schwarz inequality makes this definition possible.

Hint: Proceed as follows. Use the result of Problem 23, with the choices

$$\alpha = \|\mathbf{G}\| \quad \text{and} \quad \beta = \|\mathbf{F}\|.$$

If $\mathbf{F}$ or $\mathbf{G}$ is $\mathbf{0}$, the inequality becomes $0 \leq 0$, which is true. Hence, assume that $\mathbf{F}$ and $\mathbf{G}$ are both not $\mathbf{0}$. Then $\alpha \neq 0$ and $\beta \neq 0$. From Problem 23 and Theorem 33(4), we get

$$\begin{aligned} 0 \leq \|\alpha\mathbf{F} - \beta\mathbf{G}\|^2 &= (\alpha\mathbf{F} - \beta\mathbf{G}) \cdot (\alpha\mathbf{F} - \beta\mathbf{G}) \\ &= \|\mathbf{G}\|^2\|\mathbf{F}\|^2 - 2\|\mathbf{F}\| \, \|\mathbf{G}\|\mathbf{F} \cdot \mathbf{G} + \|\mathbf{F}\|^2\|\mathbf{G}\|^2. \end{aligned}$$

Show from this that

$$0 \leq \|\mathbf{F}\| \, \|\mathbf{G}\| - \mathbf{F} \cdot \mathbf{G}.$$

From this, complete the proof.

25. Prove the *parallelogram law* for R^n: If $\mathbf{F}$ and $\mathbf{G}$ are in R^n, then

$$\|\mathbf{F} + \mathbf{G}\|^2 + \|\mathbf{F} - \mathbf{G}\|^2 = 2(\|\mathbf{F}\|^2 + \|\mathbf{G}\|^2).$$

Hint: Use Theorem 33(4) to convert $\|\mathbf{F} + \mathbf{G}\|^2$ to $(\mathbf{F} + \mathbf{G}) \cdot (\mathbf{F} + \mathbf{G})$, and expand this using Problem 23 with $\alpha = 1, \beta = -1$. Similarly, expand $\|\mathbf{F} - \mathbf{G}\|^2$ by setting $\alpha = \beta = 1$ in Problem 23.

26. Prove *Pythagoras's theorem:* If **F** and **G** are orthogonal vectors in R^n, then

$$\|\mathbf{F} + \mathbf{G}\|^2 = \|\mathbf{F}\|^2 + \|\mathbf{G}\|^2.$$

Hint: Expand $\|\mathbf{F} + \mathbf{G}\|^2$ as $(\mathbf{F} + \mathbf{G}) \cdot (\mathbf{F} + \mathbf{G})$.

27. Prove the *triangle inequality* for R^n: If **F** and **G** are in R^n, then

$$\|\mathbf{F} + \mathbf{G}\| \le \|\mathbf{F}\| + \|\mathbf{G}\|.$$

9.6 LINEAR INDEPENDENCE AND DIMENSION

Suppose we are given m vectors $\mathbf{F}_1, \ldots, \mathbf{F}_m$ in R^n. Here we may have $m \le n$ or $m > n$. A *linear combination* of $\mathbf{F}_1, \ldots, \mathbf{F}_m$ is a sum of the form

$$\alpha_1 \mathbf{F}_1 + \cdots + \alpha_m \mathbf{F}_m,$$

in which each α_1 is a scalar.

For example, let $\mathbf{F}_1 = 2\mathbf{i} - 3\mathbf{j} + 4\mathbf{k}$ and $\mathbf{F}_2 = 8\mathbf{i} + 6\mathbf{j} + 4\mathbf{k}$ in R^3. Then $2\mathbf{F}_1 - 4\mathbf{F}_2$ is a linear combination of $\mathbf{F}_1$ and $\mathbf{F}_2$ (adding up to $-28\mathbf{i} - 30\mathbf{j} - 8\mathbf{k}$). Similarly, $\sqrt{3}\,\mathbf{F}_1 + 2\mathbf{F}_2$ is a linear combination of $\mathbf{F}_1$ and $\mathbf{F}_2$. Note that $\mathbf{F}_1$ is a linear combination of $\mathbf{F}_1$ and $\mathbf{F}_2$ ($\mathbf{F}_1 = 1\mathbf{F}_1 + 0\mathbf{F}_2$), as is $\mathbf{F}_2 = 0\mathbf{F}_1 + 1\mathbf{F}_2$.

Vectors $\mathbf{F}_1, \ldots, \mathbf{F}_m$ in R^n are called *linearly dependent* if one vector is a linear combination of the others. If this does not hold, then $\mathbf{F}_1, \ldots, \mathbf{F}_m$ are called *linearly independent.*

EXAMPLE 224

If $m = 1$, then $\mathbf{F}_1$ is linearly independent if $\mathbf{F}_1 \ne \mathbf{0}$. If $\mathbf{F}_1 = \mathbf{0}$, then $\mathbf{F}_1 = 2\mathbf{F}_1$, a linear combination of itself, and so is linearly dependent.

EXAMPLE 225

In R^2, the vectors $2\mathbf{i} + 3\mathbf{j}$, $8\mathbf{i} - 4\mathbf{j}$, and $-26\mathbf{i} + 25\mathbf{j}$ are linearly dependent, since

$$-26\mathbf{i} + 25\mathbf{j} = 3(2\mathbf{i} + 3\mathbf{j}) - 4(8\mathbf{i} - 4\mathbf{j}),$$

a linear combination of $2\mathbf{i} + 3\mathbf{j}$ and $8\mathbf{i} - 4\mathbf{j}$.

EXAMPLE 226

The vectors **i**, **j**, and **k** are linearly independent in R^3, as no one of these vectors is a linear combination of the other two.

EXAMPLE 227

In R^2, two nonzero vectors are linearly dependent exactly when one is a scalar multiple of the other, that is, exactly when they are parallel.

EXAMPLE 228

In R^3, three nonzero vectors are linearly dependent if (1) all three are parallel (hence, they differ only in length and may be opposite in direction), or (2) one lies in the plane determined by the other two.

EXAMPLE 229

Any collection of n-vectors containing the zero vector is linearly dependent in R^n. If, say, $\mathbf{F}_1, \ldots, \mathbf{F}_m$ are in R^n and $\mathbf{F}_1 = \mathbf{0}$, then

$$\mathbf{F}_1 = 0\mathbf{F}_2 + \cdots + 0\mathbf{F}_m.$$

Hence, $\mathbf{F}_1$ is a linear combination of the other $\mathbf{F}_j$'s.

In general, determining whether given vectors $\mathbf{F}_1, \ldots, \mathbf{F}_m$ are linearly dependent or independent in R^n can be difficult from a practical point of view. Here are two useful criteria. The first holds only for three vectors in R^3 and is quite easy to apply. The second holds in general, and is of primarily theoretical value (for example, it will be used in solving systems of equations in Chapter 10).

THEOREM 36

Vectors **F**, **G**, and **H** in R^3 are linearly dependent if and only if $[\mathbf{F}, \mathbf{G}, \mathbf{H}] = 0$.

Proof If **F**, **G**, and **H** are linearly dependent, then one is a linear combination of the others. By property 6 of scalar triple product, $[\mathbf{F}, \mathbf{G}, \mathbf{H}] = 0$.

Conversely, suppose that $[\mathbf{F}, \mathbf{G}, \mathbf{H}] = 0$. Then, $\mathbf{F} \cdot \mathbf{G} \times \mathbf{H} = 0$. This can happen in several ways, which we now consider.

First, **F**, **G**, or **H** might be **0**. Then, **F**, **G**, and **H** are linearly dependent, by Example 229.

Thus, suppose that **F**, **G**, and **H** are all nonzero. Then **F** is orthogonal to $\mathbf{G} \times \mathbf{H}$. If $\mathbf{G} \times \mathbf{H} = \mathbf{0}$, then **G** and **H** are parallel. Then each is a multiple of the other, say, $\mathbf{G} = \alpha\mathbf{H}$. In this event $\mathbf{G} = \alpha\mathbf{H} + 0\mathbf{F}$, and hence **F**, **G**, and **H** are linearly dependent.

The other possibility is that $\mathbf{G} \times \mathbf{H}$ is not **0**. In this event, **G** and **H** determine a plane. Now, $\mathbf{G} \times \mathbf{H}$ is orthogonal to this plane, and **F** is orthogonal to $\mathbf{G} \times \mathbf{H}$; hence **F** must lie in the plane determined by **G** and **H**. But then **F** is a linear combination of **G** and **H**, implying once again that **F**, **G**, and **H** are linearly dependent. This completes the proof.

EXAMPLE 230

Let $\mathbf{F} = 2\mathbf{i} - 4\mathbf{j} + 6\mathbf{k}$, $\mathbf{G} = 4\mathbf{i} + 8\mathbf{j} - 8\mathbf{k}$, and $\mathbf{H} = 3\mathbf{i} + 2\mathbf{j} - \mathbf{k}$. Then,

$$[\mathbf{F}, \mathbf{G}, \mathbf{H}] = \begin{vmatrix} 2 & -4 & 6 \\ 4 & 8 & -8 \\ 3 & 2 & -1 \end{vmatrix} = 0.$$

Thus, **F**, **G**, and **H** are linearly dependent. Here, any one of the vectors lies in the plane determined by the other two.

THEOREM 37

Let $m > 1$. Then, vectors $\mathbf{F}_1, \ldots, \mathbf{F}_m$ are linearly dependent in R^n if and only if there are scalars $\alpha_1, \ldots, \alpha_m$, not all zero, such that

$$\alpha_1\mathbf{F}_1 + \cdots + \alpha_m\mathbf{F}_m = \mathbf{0}.$$

Proof Suppose first that there are scalars $\alpha_1, \ldots, \alpha_m$, not all zero, such that

$$\alpha_1\mathbf{F}_1 + \cdots + \alpha_m\mathbf{F}_m = \mathbf{0}.$$

Look in order at $\alpha_1, \alpha_2, \ldots$ and choose the first nonzero one (this may be α_1). Say this is α_j. Then we can write

$$\mathbf{F}_j = -\frac{1}{\alpha_j}(\alpha_{j+1}\mathbf{F}_{j+1} + \cdots + \alpha_m\mathbf{F}_m) + 0\mathbf{F}_1 + \cdots + 0\mathbf{F}_{j-1},$$

and $\mathbf{F}_j$ is a linear combination of other $\mathbf{F}_i$'s, implying linear dependence.

This argument fails if $j = m$. But if $j = m$, then $\alpha_1 = \cdots = \alpha_{m-1} = 0$, and $\alpha_m \neq 0$. Then we have just

$$\alpha_m\mathbf{F}_m = \mathbf{0}.$$

Dividing by α_m gives us $\mathbf{F}_m = \mathbf{0}$. Then $\mathbf{F}_m$ is a linear combination of $\mathbf{F}_1, \ldots, \mathbf{F}_{m-1}$ (since $\mathbf{F}_m = 0\mathbf{F}_1 + \cdots + 0\mathbf{F}_{m-1}$); hence $\mathbf{F}_1, \ldots, \mathbf{F}_m$ are linearly dependent.

Conversely, suppose that $\mathbf{F}_1, \ldots, \mathbf{F}_m$ are linearly dependent. Then some $\mathbf{F}_j$ is a linear combination of the other given vectors, say,

$$\mathbf{F}_j = \alpha_1\mathbf{F}_1 + \cdots + \alpha_{j-1}\mathbf{F}_{j-1} + \alpha_{j+1}\mathbf{F}_{j+1} + \cdots + \alpha_m\mathbf{F}_m.$$

Then,

$$\alpha_1\mathbf{F}_1 + \cdots + \alpha_{j-1}\mathbf{F}_{j-1} - \mathbf{F}_j + \alpha_{j+1}\mathbf{F}_{j+1} + \cdots + \alpha_m\mathbf{F}_m = \mathbf{0},$$

and not all the coefficients are zero (the coefficient of $\mathbf{F}_j$ is -1). This completes the proof.

Another, sometimes useful, way of stating Theorem 37 is that for $\mathbf{F}_1, \ldots, \mathbf{F}_n$ to be linearly independent in R^n, the only way an equation

$$\alpha_1\mathbf{F}_1 + \cdots + \alpha_m\mathbf{F}_m = \mathbf{0}$$

can be true is for $\alpha_1 = \alpha_2 = \cdots = \alpha_m = 0$.

This somewhat abstract-looking statement is of considerable importance in the next chapter in the practical consideration of solving systems of equations. Linear independence is also important in applying matrix methods to the solution of systems of differential equations such as one encounters with electrical networks and mechanical systems.

Linear independence also underlies the notion of a basis. Suppose that $\mathbf{F}_1, \ldots, \mathbf{F}_m$ are vectors in a subspace S of R^n. We say that $\mathbf{F}_1, \ldots, \mathbf{F}_m$ form a *basis* for S if the following two conditions are met:

1. $\mathbf{F}_1, \ldots, \mathbf{F}_m$ are linearly independent.
2. Every vector in S is a linear combination of $\mathbf{F}_1, \ldots, \mathbf{F}_m$. That is, every vector in S has the form $\alpha_1\mathbf{F}_1 + \cdots + \alpha_m\mathbf{F}_m$ for some choice of scalars $\alpha_1, \ldots, \alpha_m$.

EXAMPLE 231

In R^2, consider the subspace S consisting of all vectors along the line $y = 2x$ (see Theorem 34). Every such vector is of the form $(\alpha, 2\alpha)$ or $\alpha(\mathbf{a} + 2\mathbf{b})$, where $\mathbf{a} = (1, 0)$ and $\mathbf{b} = (0, 1)$. Thus, $\mathbf{F}_1 = \mathbf{a} + 2\mathbf{b}$ alone constitutes a basis for S.

Note that $\mathbf{a}$ and $\mathbf{b}$ form a basis for all of R^2 because $\mathbf{a}$ and $\mathbf{b}$ are linearly independent and every vector in R^2 has the form $\alpha\mathbf{a} + \beta\mathbf{b}$, with α and β scalars.

EXAMPLE 232

In R^3, let S be the subspace of vectors on the plane $x + y + z = 0$. Thus, S consists of all 3-vectors of the form $(x, y, -x - y)$, with x and y arbitrary scalars.

In terms of $\mathbf{i}$, $\mathbf{j}$, and $\mathbf{k}$, a typical vector in S is $x\mathbf{i} + y\mathbf{j} + (-x - y)\mathbf{k}$. This can be written as

$$x(\mathbf{i} - \mathbf{k}) + y(\mathbf{j} - \mathbf{k}).$$

Thus, every vector in S is a linear combination of $\mathbf{i} - \mathbf{k}$ and $\mathbf{j} - \mathbf{k}$. These vectors are linearly independent, and hence form a basis for S.

EXAMPLE 233

In R^5, let S consist of all vectors $(0, x, y, 0, y)$, with x and y any scalars. Then every vector in S is of the form

$$x(0, 1, 0, 0, 0) + y(0, 0, 1, 0, 1),$$

and hence is a linear combination of $\mathbf{e}_2$ and $\mathbf{e}_3 + \mathbf{e}_5$ in R^5. Since $\mathbf{e}_2$ and $\mathbf{e}_3 + \mathbf{e}_5$ are linearly independent, they form a basis for S.

Given a subspace S of R^n, it is sometimes desirable to find a basis. Usually the description of S provides the clues one needs to do this. Here are two examples.

EXAMPLE 234

Find a basis for the subspace of R^6 consisting of all vectors of the form $(x, 0, y, x - y, x + y, 2x)$.

Begin by dissecting a typical vector in S. We can write

$$(x, 0, y, x - y, x + y, 2x) = x(1, 0, 0, 1, 1, 2) + y(0, 0, 1, -1, 1, 0).$$

Thus, every vector in S is a linear combination of just two vectors, $\mathbf{F} = (1, 0, 0, 1, 1, 2)$ and $\mathbf{G} = (0, 0, 1, -1, 1, 0)$. Further, $\mathbf{F}$ and $\mathbf{G}$ are linearly independent as neither is a scalar multiple of the other. Thus, $\mathbf{F}$ and $\mathbf{G}$ form a basis for S.

EXAMPLE 235

Find a basis for the subspace of R^3 consisting of all vectors in the plane $3x - 2y + z = 0$.

Every vector in S has the form $(x, y, -3x + 2y)$ or

$$x(1, 0, -3) + y(0, 1, 2).$$

Thus, every vector in S is a linear combination of $\mathbf{F} = (1, 0, -3)$ and $\mathbf{G} = (0, 1, 2)$. Since $\mathbf{F}$ and $\mathbf{G}$ are linearly independent, they form a basis for S.

In general, a vector space or subspace may have many different bases. For example, $\mathbf{i}$, $\mathbf{j}$, and $\mathbf{k}$ form a basis for R^3; so do $\mathbf{i}$, $\mathbf{j} + \mathbf{k}$, and $\mathbf{j} - \mathbf{k}$; so do $\mathbf{i} + \mathbf{j} + \mathbf{k}$, $2\mathbf{j}$, and $3\mathbf{k}$, and so on. In many courses on linear algebra, it is proved that every basis for a given vector space or subspace has the same *number* of vectors in it. This number is called the *dimension* of the vector space or subspace.

EXAMPLE 236

R^n has dimension n, a basis being $\mathbf{e}_1, \ldots, \mathbf{e}_n$.

EXAMPLE 237

The subspace of R^6 of Example 220 has dimension 4, a basis consisting of the 6-vectors (0, 1, 0, 0, 0, 0), (0, 0, 0, 1, 0, 0), (0, 0, 0, 0, 1, 0), and (0, 0, 0, 0, 0, 1). The subspace of R^4 in Example 221 has dimension 1, a basis being just the vector (1, 1, 1, 1). The subspace of R^2 in Example 231 has dimension 1, the subspace of R^3 in Example 232 has dimension 2, and the subspace of R^5 in Example 233 has dimension 2.

The notions of linear independence and basis are used to explain certain aspects of linear equations, which are treated in the next chapter. The next section places these notions in an abstract setting. This material is not essential in pursuing Chapter 10, but can be used to place certain other results in perspective. For example, the reason that an nth order linear homogeneous differential equation has a fundamental system of solutions consisting of exactly n functions is that the (abstract) vector space of solutions has dimension n.

PROBLEMS FOR SECTION 9.6

In each of Problems 1 through 10, determine whether the given vectors are linearly dependent or linearly independent in the given R^n.

1. $3\mathbf{i} + 2\mathbf{j}$, $\mathbf{i} - \mathbf{j}$; in R^3
2. $2\mathbf{i}$, $3\mathbf{j}$, $5\mathbf{i} - 12\mathbf{j}$; in R^3
3. $8\mathbf{e}_1 + 2\mathbf{e}_2$, $\mathbf{e}_5 - \mathbf{e}_6$; in R^7
4. $\mathbf{e}_1$, $\mathbf{e}_2 + \mathbf{e}_3$, $-4\mathbf{e}_1 + 6\mathbf{e}_2 + 6\mathbf{e}_3$; in R^4
5. $(1, 2, -3, 1)$, $(4, 0, 0, 2)$; in R^4
6. $\mathbf{e}_2 - \mathbf{e}_3$, $\mathbf{e}_3 - \mathbf{e}_4$, $\mathbf{e}_4 - \mathbf{e}_1$; in R^6
7. $8\mathbf{e}_1 - 6\mathbf{e}_2$, $-4\mathbf{e}_1 + 3\mathbf{e}_2$, $\mathbf{e}_1 + \mathbf{e}_4$; in R^5
8. $\mathbf{e}_2 - \mathbf{e}_1$, $\mathbf{e}_3 - \mathbf{e}_2$, $\mathbf{e}_3 - \mathbf{e}_1$; in R^8
9. $(-2, 0, 0, 1, 1)$, $(1, 0, 0, 0, 0)$, $(0, 0, 0, 0, 2)$; in R^5
10. $(3, 0, 0, 4)$, $(2, 0, 0, 8)$; in R^4

In each of Problems 11 through 20, use the scalar triple product and Theorem 35 to determine whether or not the three given vectors are linearly dependent in R^3.

11. $3\mathbf{i} + 6\mathbf{j} - \mathbf{k}$, $8\mathbf{i} + 2\mathbf{j} - 4\mathbf{k}$, $\mathbf{i} - \mathbf{j} + \mathbf{k}$
12. $\mathbf{i} + 6\mathbf{j} - 2\mathbf{k}$, $-\mathbf{i} + 4\mathbf{j} - 3\mathbf{k}$, $\mathbf{i} + 16\mathbf{j} - 7\mathbf{k}$
13. $4\mathbf{i} - 3\mathbf{j} + \mathbf{k}$, $10\mathbf{i} - 3\mathbf{j}$, $2\mathbf{i} - 6\mathbf{j} + 3\mathbf{k}$
14. $8\mathbf{i} + 6\mathbf{j}$, $2\mathbf{i} - 4\mathbf{j}$, $\mathbf{i} + \mathbf{k}$
15. $-\mathbf{i} + 3\mathbf{j} - 2\mathbf{k}$, $4\mathbf{i} - 6\mathbf{j} - \mathbf{k}$, $8\mathbf{i} - 2\mathbf{k}$
16. $3\mathbf{i} + 6\mathbf{j} - \mathbf{k}$, $-2\mathbf{i} + 4\mathbf{k}$, $-5\mathbf{i} - 6\mathbf{j} + 5\mathbf{k}$
17. $12\mathbf{i} - 3\mathbf{k}$, $\mathbf{i} + 2\mathbf{j} - \mathbf{k}$, $-3\mathbf{i} + 4\mathbf{j}$
18. $-\mathbf{i} + 2\mathbf{j} + 3\mathbf{k}$, $3\mathbf{i} - 4\mathbf{j} + 3\mathbf{k}$, $8\mathbf{i} + 2\mathbf{j} + \mathbf{k}$
19. $5\mathbf{i} + 2\mathbf{j} - \mathbf{k}$, $8\mathbf{i} + 7\mathbf{j} + \mathbf{k}$, $\mathbf{i} + 2\mathbf{j} - 3\mathbf{k}$
20. $2\mathbf{i} + 4\mathbf{j} + \mathbf{k}$, $10\mathbf{i} - 2\mathbf{j} + 6\mathbf{k}$, $3\mathbf{i} - 6\mathbf{j} + 2\mathbf{k}$

In each of Problems 21 through 30, you are given a subspace S of R^n. Determine whether the given vectors form a basis for S. Determine the dimension of the subspace in each problem.

21. S consists of all vectors $(x, y, -y, -x)$ in R^4; $(1, 0, 0, -1)$, $(0, 2, -2, 0)$.

22. S consists of all vectors $(x, y, 2x, 3y)$ in R^4; $(1, 0, 2, 0)$, $(0, 1, 0, 3)$, $(1, -1, 2, -3)$.

23. S consists of all vectors on the plane $2x - y + z = 0$ in R^3; $\mathbf{i} + 2\mathbf{j}$, $\mathbf{j} + \mathbf{k}$.

24. S consists of all vectors of the form $(x, y, -y, x - y, z)$ in R^5; $(1, 0, 0, 1, 0)$, $(0, -1, 1, 1, 0)$, $(0, 0, 0, 0, 3)$.

25. S consists of all vectors with second component zero in R^4; $(1, 0, 0, 0)$, $(0, 0, 2, 0)$, $(0, 0, 0, 3)$.

26. S consists of all vectors $(-x, x, y, 2y)$ in R^4; $(1, -1, 0, 0)$, $(1, -1, 2, 4)$.

27. S is all of R^3; $2\mathbf{i}$, $\mathbf{i} + \mathbf{j}$, $2\mathbf{i} - 3\mathbf{j} + \mathbf{k}$.

28. S is all of R^3; $2\mathbf{i}$, $4\mathbf{i} + 2\mathbf{j}$, $\mathbf{i} - \mathbf{j}$.

29. S consists of all vectors in R^2 on the line $y = 4x$; $(2, 8)$.

30. S consists of all vectors in R^3 on the plane $4x + 2y - z = 0$; $\mathbf{i} + 4\mathbf{k}$, $\mathbf{j} + 2\mathbf{k}$.

31. We know that the set of all vectors in R^2 along any given line through the origin is a subspace of R^2. Determine the dimension for such a subspace by finding a basis.

32. We know that the set of all vectors on a given plane through the origin in R^3 is a subspace of R^3. Determine the dimension of such a subspace of R^3.

33. Prove that any three distinct vectors in R^2 must be linearly dependent.

34. Prove that two nonzero vectors in R^3 are linearly dependent exactly when the arrows representing them are parallel.

35. Prove that three nonzero vectors in R^3 are linearly dependent exactly when one is parallel to a plane or line determined by the other two.

36. Let $\mathbf{F}_1, \ldots, \mathbf{F}_m$ be linearly independent vectors in R^n. Prove that $\mathbf{F}_1, \ldots, \mathbf{F}_r$ are linearly independent for any r with $1 \leq r \leq m$.

37. Let $\mathbf{F}_1, \ldots, \mathbf{F}_m$ form a basis for a subspace S of R^n. Let $\mathbf{G}$ be any other vector in S. Prove that $\mathbf{F}_1, \ldots, \mathbf{F}_m$, $\mathbf{G}$ are linearly dependent.

38. Let $\mathbf{F}_1, \ldots, \mathbf{F}_m$ be vectors in R^n. Suppose some $\mathbf{F}_j = \mathbf{0}$. Prove that $\mathbf{F}_1, \ldots, \mathbf{F}_m$ are linearly dependent.

39. Let $\mathbf{F}$, $\mathbf{G}$, and $\mathbf{H}$ be any three linearly independent vectors in R^3. Let $\mathbf{V}$ be any vector in R^3. Show that

$$\mathbf{V} = \frac{[\mathbf{V}, \mathbf{G}, \mathbf{H}]}{[\mathbf{F}, \mathbf{G}, \mathbf{H}]}\mathbf{F} + \frac{[\mathbf{V}, \mathbf{H}, \mathbf{F}]}{[\mathbf{F}, \mathbf{G}, \mathbf{H}]}\mathbf{G} + \frac{[\mathbf{V}, \mathbf{F}, \mathbf{G}]}{[\mathbf{F}, \mathbf{G}, \mathbf{H}]}\mathbf{H}.$$

40. Prove that any three linearly independent vectors in R^3 form a basis for R^3. (*Hint:* Use the result of Problem 39.)

9.7 CHAPTER SUPPLEMENT: ABSTRACT VECTOR SPACES

In Section 9.5 we gave the name *vector space* to the set R^n endowed with the operations of addition and multiplication by scalars. The algebraic properties of R^n were listed in Theorem 31. Using this list as a prototype, we can extend the notion of vector space to sets other than R^n as follows.

Let V be any set of objects. Suppose there are defined on V two algebraic operations:

a sum $\mathbf{a} + \mathbf{b}$ for each $\mathbf{a}$ and $\mathbf{b}$ in V

and

a product $\alpha\mathbf{a}$ of scalars α with objects $\mathbf{a}$ in V.

Suppose that these two operations satisfy the following properties:

1. $\mathbf{a} + \mathbf{b}$ is in V whenever $\mathbf{a}$ and $\mathbf{b}$ are in V.
2. $\mathbf{a} + \mathbf{b} = \mathbf{b} + \mathbf{a}$ for every $\mathbf{a}, \mathbf{b}$ in V.
3. $(\mathbf{a} + \mathbf{b}) + \mathbf{c} = \mathbf{a} + (\mathbf{b} + \mathbf{c})$ for every $\mathbf{a}, \mathbf{b}, \mathbf{c}$ in V.
4. There is some object $\boldsymbol{\theta}$ in V such that $\mathbf{a} + \boldsymbol{\theta} = \mathbf{a}$ for every $\mathbf{a}$ in V. We call $\boldsymbol{\theta}$ the *zero vector* of V.
5. For each $\mathbf{a}$ in V, there is some $\mathbf{b}$ in V such that $\mathbf{a} + \mathbf{b} = \boldsymbol{\theta}$. We denote such $\mathbf{b}$ as $-\mathbf{a}$.
6. $\alpha\mathbf{a}$ is in V whenever α is a scalar and $\mathbf{a}$ is in V.
7. $(\alpha + \beta)\mathbf{a} = \alpha\mathbf{a} + \beta\mathbf{a}$ for any $\mathbf{a}$ in V and scalars α and β.
8. $(\alpha\beta)\mathbf{a} = \alpha(\beta\mathbf{a})$ for $\mathbf{a}$ in V and α, β scalars.
9. $\alpha(\mathbf{a} + \mathbf{b}) = \alpha\mathbf{a} + \beta\mathbf{b}$ for $\mathbf{a}, \mathbf{b}$ in V and α scalar.
10. $\alpha\boldsymbol{\theta} = \boldsymbol{\theta}$ for every scalar α.

Then we call the set V, together with the two given algebraic operations, a *vector space*.

Strictly speaking, one must also say something about how the scalars are chosen. In many cases, it is agreed that the scalars are real numbers, in which case V is a *real vector space*. If they are complex numbers, then V is a *complex vector space*. In this section we shall assume that the scalars are real unless specific exception is noted.

This definition of vector space is quite general, but a comparison with Theorem 32 shows that R^n, with the usual addition and scalar multiplication, certainly satisfies all the requirements. We shall now give some examples of vector spaces other than the familiar R^n. In many examples, we shall drop the boldface notation when the objects in the vector space are familiar quantities usually denoted in other ways. Thus, in Examples 238, 239, and 240, the objects are functions which we denote as $f(x), g(x), \ldots$ in the usual way, rather than $\mathbf{f}(x), \mathbf{g}(x), \ldots$. We retain the boldface notation in general discussions in which the vector space is not given specifically.

EXAMPLE 238

Let V be the set of all functions continuous on $[0, 1]$ (in the sense of beginning calculus). We know that sums of continuous functions are continuous, and that products of real numbers with continuous functions are continuous. Thus, properties 1 and 6 of the definition hold.

Also, property 2 holds, since $f(x) + g(x) = g(x) + f(x)$ for any functions f and g in V and $0 \leq x \leq 1$; property 3 holds, since $[f(x) + g(x)] + h(x) = f(x) + [g(x) + h(x)]$ for $0 \leq x \leq 1$; property 4 holds, with $\boldsymbol{\theta}$ the zero function: $\theta(x) = 0$ for $0 \leq x \leq 1$; property 5 holds, since $f(x) - f(x) = \theta(x) = 0$ for $0 \leq x \leq 1$; property 7 holds, since $(\alpha + \beta)f(x) = \alpha f(x) + \beta f(x)$ for $0 \leq x \leq 1$; property 8 holds, since $(\alpha\beta)f(x) = \alpha[\beta f(x)]$ for $0 \leq x \leq 1$; property 9 holds, since $\alpha[f(x) + g(x)] = \alpha f(x) + \beta g(x)$ for $0 \leq x \leq 1$; and property 10 holds, since $\alpha\theta(x) = \alpha 0 = 0 = \theta(x)$ for $0 \leq x \leq 1$.

EXAMPLE 239

Let V consist of all functions $f(x)$ differentiable on the whole real line, with $f'(0) = 0$. With the usual addition of functions and multiplication by scalars, V is a vector space.

EXAMPLE 240

Let V consist of all solutions of the differential equation $y'' + 2xy' + x^2y = 0$. With the usual addition of functions and multiplication by scalars, V is a vector space. For example, sums of solutions are solutions, and scalar multiples of solutions are solutions. Further, $y(x) = 0$ is a solution and is the zero vector of V. Here, V is the *solution space* of the differential equation, a point of view useful in the analysis of certain differential equations.

We shall pursue more examples in the exercises. For now, it is enough to realize that the "natural" vector space R^n is really a special case of a general concept which applies to objects quite different from vectors in the traditional sense. Sometimes R^n is referred to as a *concrete vector space*, and vector spaces such as those of Examples 238, 239, and 240 are referred to as *abstract vector spaces*. Customarily, one also refers to objects in an abstract vector space as vectors, although here the term has lost its original connection with arrows representing forces. Thus, for example, $\cos(x)$ is a vector in the vector space of Example 239.

Many concepts which arise naturally in R^n carry over to abstract vector spaces. We shall illustrate this with linear independence and bases, and show the connection with general solutions to certain differential equations.

Suppose then that V is an (abstract) vector space. A *linear combination* of vectors $\mathbf{v}_1, \ldots, \mathbf{v}_n$ in V is a sum

$$\alpha_1\mathbf{v}_1 + \cdots + \alpha_n\mathbf{v}_n,$$

with $\alpha_1, \ldots, \alpha_n$ scalars.

If some $\mathbf{v}_j$ is a linear combination of the other $\mathbf{v}_i$'s, we call $\mathbf{v}_1, \ldots, \mathbf{v}_n$ *linearly dependent*. Otherwise, $\mathbf{v}_1, \ldots, \mathbf{v}_n$ are linearly independent.

For example,

$$\cos(2x), \qquad \sin^2(x), \quad \text{and} \quad \cos^2(x)$$

are linearly dependent in the vector space of Example 239 since, for all x,

$$\cos(2x) = \cos^2(x) - \sin^2(x).$$

In the vector space of Example 238, $f(x) = 3x^2$ and $g(x) = \cos(x)$ are linearly independent, as neither is a scalar multiple of the other for all x in $[0, 1]$.

Vectors $\mathbf{v}_1, \ldots, \mathbf{v}_n$ in a vector space V form a *basis* for V if $\mathbf{v}_1, \ldots, \mathbf{v}_n$ are linearly independent and every vector in V is a linear combination of $\mathbf{v}_1, \ldots, \mathbf{v}_n$.

EXAMPLE 241

Let V be the vector space of all solutions of

$$y'' + 4y = 0.$$

It is routine to check that V is a vector space; V is called the *solution space* of $y'' + 4y = 0$.

From Chapter 2 we know that $\sin(2x)$ and $\cos(2x)$ form a fundamental set of solutions of this differential equation. That is, every solution is of the form

$$\alpha \cos(2x) + \beta \sin(2x),$$

with α and β scalars.

In vector space terminology, this says that every solution is a linear combination of $\sin(2x)$ and $\cos(2x)$. These vectors of V are also linearly independent. Thus, $\sin(2x)$ and $\cos(2x)$ form a basis for the solution space of $y'' + 4y = 0$. The general solution of $y'' + 4y = 0$ is a linear combination of a basis for the solution space.

It is sometimes possible to imitate the notions of dot product and length in an abstract vector space. As an example, consider the vector space V of Example 238 consisting of all functions continuous on the interval $[0, 1]$.

We can define the *dot product* of two such functions by

$$f \cdot g = \int_0^1 f(x)g(x)\, dx.$$

Thus, for example, the dot product of x with e^x would be $\int_0^1 xe^x\, dx$, or 1.

The *length* of f would be defined by

$$\sqrt{f \cdot f} = \left[\int_0^1 f(x)^2\, dx\right]^{1/2}.$$

Thus, the length of e^x would be

$$\|e^{2x}\| = \sqrt{\int_0^1 e^{2x}\, dx} = \sqrt{\frac{1}{2}(e^2 - 1)}.$$

It is easy to check that this integral dot product has all the properties listed in Theorem 33 for the standard dot product in R^n. This enables us to translate geometrically motivated relationships from R^n to other settings where they might not be so obvious.

For example, the triangle inequality in R^n says that

$$\|\mathbf{F} - \mathbf{G}\| \leq \|\mathbf{F} - \mathbf{H}\| + \|\mathbf{H} - \mathbf{G}\|.$$

In the vector space of functions continuous on $[0, 1]$, with the integral definition of length, this becomes

$$\left\{\int_0^1 [f(x) - g(x)]^2\, dx\right\}^{1/2} \leq \left\{\int_0^1 [f(x) - h(x)]^2\, dx\right\}^{1/2} + \left\{\int_0^1 [h(x) - g(x)]^2\, dx\right\}^{1/2}.$$

As another example, the Cauchy-Schwarz inequality

$$|\mathbf{F} \cdot \mathbf{G}| \leq \|\mathbf{F}\|\, \|\mathbf{G}\|$$

becomes, in the vector space of functions continuous on $[0, 1]$,

$$\left|\int_0^1 f(x)g(x)\, dx\right| \leq \left[\int_0^1 f(x)^2\, dx\right]^{1/2}\left[\int_0^1 g(x)^2\, dx\right]^{1/2}.$$

These are useful integral inequalities in many parts of analysis.

These examples serve as a brief introduction to the subject of linear algebra, which is a study of vector spaces in the general setting. The problems at the end of this section develop some of the elementary theory. For the purpose of studying matrices and their applications in engineering, the vector space R^n will suffice, and so we shall not pursue abstract vector spaces beyond this brief exposure.

PROBLEMS FOR SECTION 9.7

In each of Problems 1 through 10, verify that the given set is a vector space with the given algebraic operations (as we did in Example 238). Scalars are understood to be real numbers in these problems.

1. Let P_n be the set of polynomials of degree $\leq n$. Define the sum of two polynomials and the product of a polynomial by a scalar in the usual way.
2. Let V consist of all functions f continuous on $[0, 1]$ such that $f(1) = 0$. Define addition of functions and multiplication by scalars in the usual way.
3. Let P consist of all polynomials with real coefficients, with the usual addition of polynomials and multiplication by scalars.
4. Let S consist of all solutions of $y'' - 8y = 0$, with the usual addition of functions and multiplication of functions by scalars. *Note:* Don't solve the differential equation to do this problem.
5. Let T consist of all 2-vectors (x, y), with addition defined by
$$(x_1, y_1) + (x_2, y_2) = (2x_1 + 2x_2, y_1 + y_2)$$
and scalar multiplication by $\alpha(x, y) = (\alpha x, \alpha y)$.
6. Let W consist of all 3-vectors (x, y, z) with addition defined by
$$(x_1, y_1, z_1) + (x_2, y_2, z_2) = (3x_1 + 3x_2, 2y_1 + 2y_2, z_1 + z_2)$$
and scalar multiplication by $\alpha(x, y, z) = (\alpha x, \alpha y, \alpha z)$.
7. Let V consist of all functions $f(x)$ continuous on $[0, 1]$ with $\int_0^1 f(x)\,dx = 0$. Use the usual addition of functions and multiplication of a function by a constant.
8. Let V consist of all polynomials of degree 2 or less, having 1 as a root. Use the usual addition of polynomials and multiplication of polynomials by scalars.
9. Let V consist of all $x\mathbf{i} + y\mathbf{j} + z\mathbf{k}$ such that $x - y + z = 0$ and $2x - 3y + 2z = 0$. Use the usual operations in R^3.
10. Let V consist of all functions continuous on $[-1, 1]$, such that $f(\frac{1}{2}) = f(1) = 0$. Use the usual addition of functions and multiplication of functions by constants.

In each of Problems 11 through 24, give all of the properties of the definition that are violated in showing that the given set is not a vector space with the given operations.

11. Q is the set of rational numbers, with the usual addition and multiplication by real numbers.
12. S is the set of all solutions of $y'' - 4y = 8$, with the usual addition of functions and multiplication by scalars.
13. P is the set of all polynomials of degree 4, with the usual addition of polynomials and multiplication by scalars.
14. V is the set of 2-vectors (x, y), with addition defined by
$$(x_1, y_1) + (x_2, y_2) = (x_1 + x_2, y_1 - y_2)$$
and scalar multiplication by $\alpha(x, y) = (\alpha x, y)$.
15. P consists of all polynomials of degree 5, with the usual addition of polynomials, but α times $(a_0 + a_1x + a_2x^2 + a_3x^3 + a_4x^4 + a_5x^5)$ defined to be 1.

16. V consists of all solutions of $y'' + 4y = x$, with the usual addition of functions and multiplication by constants.
17. V consists of all functions continuous on $[0, 1]$ with $f(\frac{1}{2}) = 3$, with the usual addition of functions and multiplication by scalars.
18. V consists of all vectors in R^2 parallel to the line $x + y = 1$, with the usual vector operations in R^2.
19. V consists of all vectors in R^3 parallel to the plane $x + 2y - z = 4$, with the usual vector operations in R^3.
20. V consists of all vectors in R^3 with second component -2, with the usual vector operations in R^3.
21. V consists of all polynomials of degree 3 with a zero at 4, with the usual addition of polynomials and multiplication by scalars.
22. V consists of all solutions of $x^2y'' - 2xy' + y = 0$ satisfying $y(1) = 6$, with the usual addition of functions and multiplication by scalars.
23. V consists of all functions continuous on $[-1, 1]$ with $\int_{-1}^{1} f(x)\,dx = 1$. Use the usual addition of functions and multiplication by scalars.
24. V consists of all vectors in R^3 of the form $x\mathbf{i} + y\mathbf{j} - \mathbf{k}$, with the usual vector operations in R^3.
25. Let V be a vector space. Prove that there can be only one zero vector. (*Hint:* Suppose that $\boldsymbol{\theta}$ and $\boldsymbol{\theta}'$ are zero vectors of V. Look at $\boldsymbol{\theta} + \boldsymbol{\theta}'$.)
26. Let V be a vector space, and let $\mathbf{a}$ be in V. Prove that there is only one $\mathbf{b}$ in V such that $\mathbf{a} + \mathbf{b} = \boldsymbol{\theta}$. (*Hint:* Suppose that $\mathbf{a} + \mathbf{b} = \mathbf{a} + \mathbf{b}' = \boldsymbol{\theta}$. To show that $\mathbf{b} = \mathbf{b}'$, add $\mathbf{b}'$ to both sides of $\boldsymbol{\theta} = \mathbf{a} + \mathbf{b}$.)
27. Let P_n be the vector space of polynomials of degree $\leq n$ (see Problem 1). Define $p \cdot q = \int_{-1}^{1} p(x)q(x)\,dx$ for p and q in P_n. Show that this dot product satisfies properties 1, 2, 3, and 5 of Theorem 32.
28. Let C be the vector space of functions continuous on $[0, \pi]$. Define $f \cdot g = \int_0^{\pi} x^2 f(x)g(x)\,dx$ for f, g in C. Prove that this dot product satisfies properties 1, 2, 3, and 5 of Theorem 32.
29. Let C be the vector space of functions continuous on $[0, \pi]$, and define $f \cdot g = \int_0^{\pi} f(x)g(x)\,dx$ for f, g in C. Prove that $\sin(x), \sin(2x), \ldots, \sin(nx)$ are mutually orthogonal (that is, the dot product of any two of these functions is zero). This fact is important in the theory of Fourier series. Here, n is any positive integer.

CHAPTER SUMMARY

This chapter had a dual purpose. The first four sections were devoted to the kinds of operations one encounters in working with 3-dimensional vectors, which might represent forces, velocity, acceleration, and the like. In Section 9.5 we saw a departure from the world of everyday experience in the sense that we cannot draw 4-vectors or 5-vectors. Nevertheless, such quantities are important in posing and solving engineering and physics problems. For example, in classical mechanics the description of a physical system with several degrees of freedom may be most conveniently achieved by using the language of n-tuples. One can also draw examples from other disciplines. For example, modern economists often use n-vectors and matrix notation to represent economic states in which many different economic factors are brought into account.

Introduction of lengths and angles into R^n by means of the dot product enables us to generalize familiar geometric concepts from three to higher dimensions. This provides valuable insight and also useful language for describing n-dimensional problems and solutions. For example, in solving certain partial differential equations in n variables, certain geometric configurations such as the n-dimensional cone become important. And in linear programming one can sometimes describe an optimal solution in terms of points inside an n-dimensional polyhedron.

Finally, in Section 9.7, we reached the notion of an abstract vector space. The generality of this concept makes it applicable to a wide variety of topics, and in a more advanced treatment would provide a natural setting for the development of differential equations, transforms, and Fourier analysis.

SUPPLEMENTARY PROBLEMS

In each of Problems 1 through 15, compute $\mathbf{F} \cdot \mathbf{G}$, $\mathbf{F} \times \mathbf{G}$, and the cosine of the angle between $\mathbf{F}$ and $\mathbf{G}$. If you have a calculator, determine this angle in both degrees and radians to four decimal places.

F	G	F	G
1. $2\mathbf{i} - 3\mathbf{j} + \mathbf{k}$	$\mathbf{i} + 6\mathbf{j} - 4\mathbf{k}$	**2.** $3\mathbf{i} + \mathbf{j}$	$\mathbf{i} - 2\mathbf{j} - \mathbf{k}$
3. $-2\mathbf{i} + \mathbf{j} - 6\mathbf{k}$	$4\mathbf{i} + 2\mathbf{j} + 3\mathbf{k}$	**4.** $7\mathbf{i} - 3\mathbf{k}$	$-\mathbf{i} + 4\mathbf{j}$
5. $8\mathbf{i} + 2\mathbf{j} + \mathbf{k}$	$\mathbf{i} - 2\mathbf{j} + 5\mathbf{k}$	**6.** $7\mathbf{i} + \mathbf{j} - 6\mathbf{k}$	$2\mathbf{i} - 3\mathbf{j} - 4\mathbf{k}$
7. $-8\mathbf{i} - 3\mathbf{j} + \mathbf{k}$	$\mathbf{j} - 4\mathbf{k}$	**8.** $2\mathbf{j} - 3\mathbf{k}$	$\mathbf{j} - 4\mathbf{k}$
9. $8\mathbf{i} - 4\mathbf{j} + 3\mathbf{k}$	$-2\mathbf{i} + 6\mathbf{j} - 2\mathbf{k}$	**10.** $-4\mathbf{i} + 5\mathbf{j} - \mathbf{k}$	$2\mathbf{i} + 4\mathbf{j} + 2\mathbf{k}$
11. $3\mathbf{i} - 4\mathbf{j}$	$\mathbf{i} - \mathbf{j} - 5\mathbf{k}$	**12.** $-3\mathbf{i} - 5\mathbf{j} + \mathbf{k}$	$8\mathbf{i} - 3\mathbf{j} + 2\mathbf{k}$
13. $5\mathbf{i} + 6\mathbf{j} - 4\mathbf{k}$	$2\mathbf{i} + 6\mathbf{j} + 3\mathbf{k}$	**14.** $\mathbf{j} - 5\mathbf{k}$	$2\mathbf{i} + 2\mathbf{j} - 3\mathbf{k}$
15. $14\mathbf{i} + 8\mathbf{j} + 7\mathbf{k}$	$-3\mathbf{i} + 7\mathbf{k}$		

In each of Problems 16 through 25, find the projection of $\mathbf{F}$ in the direction of $\mathbf{r}$.

F	r	F	r
16. $\mathbf{i} + 2\mathbf{j}$	$3\mathbf{k}$	**17.** $2\mathbf{i} + 3\mathbf{j} + \mathbf{k}$	$2\mathbf{i}$
18. $-\mathbf{i} - 2\mathbf{j} - 4\mathbf{k}$	$\mathbf{i} - \mathbf{j}$	**19.** $3\mathbf{i} - 4\mathbf{j} + \mathbf{k}$	$2\mathbf{i} + \mathbf{k}$
20. $2\mathbf{i} + 8\mathbf{j} - 5\mathbf{k}$	$\mathbf{i} + \mathbf{j} + \mathbf{k}$	**21.** $\mathbf{i} + 6\mathbf{j} + 7\mathbf{k}$	$2\mathbf{i} + \mathbf{k}$
22. $-3\mathbf{i} + 2\mathbf{j} - \mathbf{k}$	$\mathbf{i} - \mathbf{j} - \mathbf{k}$	**23.** $-3\mathbf{i} - 2\mathbf{j} + 8\mathbf{k}$	$3\mathbf{i} + 2\mathbf{j}$
24. $5\mathbf{i} + 3\mathbf{j} + 4\mathbf{k}$	$\mathbf{i} - 2\mathbf{j} + \mathbf{k}$	**25.** $-6\mathbf{i} + 3\mathbf{j} - \mathbf{k}$	$-2\mathbf{i} + \mathbf{j}$

In each of Problems 26 through 35, find the equation of a straight line through the two given points. Your answer may be in terms of just x, y, and z, or may have x, y, and z written in terms of another parameter.

26. $(1, -2, 4)$, $(6, 1, 1)$ **27.** $(0, 2, 3)$, $(-2, 4, 1)$ **28.** $(6, 5, 2)$, $(3, -5, 1)$
29. $(-2, 1, -5)$, $(6, 7, 2)$ **30.** $(0, 1, -4)$, $(8, -3, -5)$ **31.** $(2, 14, 1)$, $(7, 0, 0)$
32. $(-3, 1, -5)$, $(8, 10, -2)$ **33.** $(6, 4, -8)$, $(-5, 1, 1)$ **34.** $(-7, 8, -3)$, $(-1, -1, -1)$
35. $(8, 0, 0)$, $(-5, 1, 7)$

In each of Problems 36 through 45, find the equation of a plane containing the given point and normal to the given vector.

36. $(2, 1, -4)$, $3\mathbf{i} - 2\mathbf{j} + \mathbf{k}$ **37.** $(0, 0, 0)$, $4\mathbf{i} - 5\mathbf{j} - \mathbf{k}$ **38.** $(1, 1, -3)$, $-6\mathbf{i} + \mathbf{j} - 2\mathbf{k}$
39. $(4, 4, 7)$, $-4\mathbf{i} + 2\mathbf{j} + 3\mathbf{k}$ **40.** $(1, 5, -2)$, $5\mathbf{i} + 3\mathbf{j} - \mathbf{k}$ **41.** $(-3, -7, 0)$, $\mathbf{i} + 2\mathbf{k}$
42. $(2, 5, 3)$, $4\mathbf{i} - 6\mathbf{j} + \mathbf{k}$ **43.** $(4, 4, -7)$, $3\mathbf{i} + 2\mathbf{k}$ **44.** $(2, 2, -5)$, $4\mathbf{j} - 2\mathbf{k}$
45. $(5, 2, -6)$, $4\mathbf{j} - 3\mathbf{k}$

In each of Problems 46 through 55, find the equation of a plane containing the three given points.

46. $(2, 2, -6)$, $(0, 0, 0)$, $(1, 5, 3)$ **47.** $(-3, 2, 6)$, $(1, 1, -5)$, $(-4, 5, 0)$
48. $(-3, 8, 1)$, $(1, 1, -7)$, $(0, 2, 0)$ **49.** $(-5, 3, 8)$, $(1, -4, -4)$, $(6, 6, -2)$
50. $(3, 1, -7)$, $(2, 2, 4)$, $(0, 0, -2)$ **51.** $(6, -1, 3)$, $(2, 2, -6)$, $(4, 8, 2)$

52. $(-6, 4, 4)$, $(9, 0, 2)$, $(0, 0, -1)$

53. $(-3, -1, 0)$, $(2, -7, 4)$, $(10, 4, 7)$

54. $(4, -6, 0)$, $(1, 1, -2)$, $(0, 4, -3)$

55. $(2, 0, -1)$, $(4, 4, 6)$, $(-2, 3, -1)$

In each of Problems 56 through 65, find $[\mathbf{F}, \mathbf{G}, \mathbf{H}]$.

	F	G	H
56.	$3\mathbf{i} - \mathbf{j} + 4\mathbf{k}$	$\mathbf{i} - 2\mathbf{k}$	$\mathbf{i} + \mathbf{j}$
57.	$8\mathbf{i} + 3\mathbf{j} - \mathbf{k}$	$\mathbf{i} - 2\mathbf{j} - 5\mathbf{k}$	$3\mathbf{i} + 4\mathbf{j} - \mathbf{k}$
58.	$3\mathbf{i} + \mathbf{j} - \mathbf{k}$	$2\mathbf{i} + 4\mathbf{j} - 9\mathbf{k}$	$\mathbf{j} - 4\mathbf{k}$
59.	$\mathbf{i} + 4\mathbf{j} + 3\mathbf{k}$	$-3\mathbf{j} + 6\mathbf{k}$	$\mathbf{i} - 2\mathbf{j} + 8\mathbf{k}$
60.	$\mathbf{i} - 3\mathbf{j}$	$3\mathbf{i} + \mathbf{j} + 2\mathbf{k}$	$\mathbf{i} - 2\mathbf{j} + 5\mathbf{k}$
61.	$3\mathbf{i} - \mathbf{j} + 3\mathbf{k}$	$2\mathbf{i} + \mathbf{j} + \mathbf{k}$	$\mathbf{i} - 2\mathbf{j} + 5\mathbf{k}$
62.	$4\mathbf{i} + 6\mathbf{j} - 2\mathbf{k}$	$3\mathbf{i} + \mathbf{j} - \mathbf{k}$	$\mathbf{j} - 4\mathbf{k}$
63.	$\mathbf{i} - 2\mathbf{j} + \mathbf{k}$	$3\mathbf{i} + 2\mathbf{j}$	$\mathbf{i} - 5\mathbf{j} + 8\mathbf{k}$
64.	$3\mathbf{i} + 2\mathbf{j} + 7\mathbf{k}$	$3\mathbf{i} - \mathbf{j} + \mathbf{k}$	$\mathbf{i} - 2\mathbf{j} + 5\mathbf{k}$
65.	$\mathbf{i} - 3\mathbf{j}$	$4\mathbf{j} - \mathbf{k}$	$5\mathbf{k}$

In each of Problems 66 through 70, find the volume of the parallelopiped having incident sides AB, AC, and AD.

	A	B	C	D
66.	$(-3, 2, 2)$	$(1, 1, 0)$	$(0, -1, 0)$	$(4, 3, -7)$
67.	$(-2, 4, 4)$	$(7, 2, -3)$	$(5, 5, 8)$	$(-2, 4, 1)$
68.	$(6, -1, 4)$	$(0, -3, 0)$	$(-5, 7, 2)$	$(1, 1, -7)$
69.	$(4, 4, -2)$	$(0, 0, 0)$	$(4, -2, 8)$	$(5, 7, 1)$
70.	$(-2, 2, 2)$	$(0, 2, -1)$	$(5, 5, 7)$	$(-3, 5, -3)$

71. Show that the cancellation law fails for cross products. That is, give an example in which $\mathbf{F} \times \mathbf{G} = \mathbf{F} \times \mathbf{H}$, but $\mathbf{G} \neq \mathbf{H}$. You cannot use $\mathbf{0}$ in your example.

72. Following up on Problem 71, suppose that you are told that $\mathbf{F} \times \mathbf{G} = \mathbf{F} \times \mathbf{H}$ for all vectors $\mathbf{F}$. Show that $\mathbf{G} = \mathbf{H}$.

73. Let P be a parallelogram with adjacent sides $\mathbf{F}$ and $\mathbf{G}$. Prove that the area of P is the square root of

$$\begin{vmatrix} \|F\|^2 & \mathbf{F} \cdot \mathbf{G} \\ \mathbf{F} \cdot \mathbf{G} & \|G\|^2 \end{vmatrix}.$$

74. Let θ be the angle between $\mathbf{F}$ and $\mathbf{G}$. Show that

$$\sin(\theta) = \frac{\|\mathbf{F} \times \mathbf{G}\|}{\|\mathbf{F}\|\,\|\mathbf{G}\|}.$$

75. Let $P_1, \ldots, P_n$ be n points in 3-space. Form n vectors $\mathbf{F}_1, \ldots, \mathbf{F}_n$ by drawing arrows from P_1 to P_2, P_2 to P_3, $\ldots$, P_{n-1} to P_n, and P_n to P_1, as in Figure 132. Show that $\mathbf{F}_1 + \mathbf{F}_2 + \cdots + \mathbf{F}_n = \mathbf{0}$.

76. Show that

$$(\mathbf{F} \times \mathbf{G}) \cdot (\mathbf{H} \times \mathbf{K}) = \begin{vmatrix} \mathbf{F} \cdot \mathbf{H} & \mathbf{F} \cdot \mathbf{K} \\ \mathbf{G} \cdot \mathbf{H} & \mathbf{G} \cdot \mathbf{K} \end{vmatrix}.$$

77. Let the sides of a triangle be represented by vectors $\mathbf{F}$, $\mathbf{G}$, and $\mathbf{H}$, as shown in Figure 133. Show that

$$\mathbf{F} \times (\mathbf{G} \times \mathbf{H}) = \mathbf{G} \times (\mathbf{H} \times \mathbf{F}) = \mathbf{H} \times (\mathbf{F} \times \mathbf{G}) = \mathbf{0}.$$

(*Hint:* Use the result of Problem 75.)

78. Use vectors to find a formula for the angle between the diagonal of a cube and one of the edges of the cube.

79. Use vectors to show that the distance in the xy-plane between the point (x_0, y_0) and the line $ax + by + c = 0$ is

$$\frac{|ax_0 + by_0 + c|}{\sqrt{a^2 + b^2}},$$

assuming that $a^2 + b^2 \neq 0$.

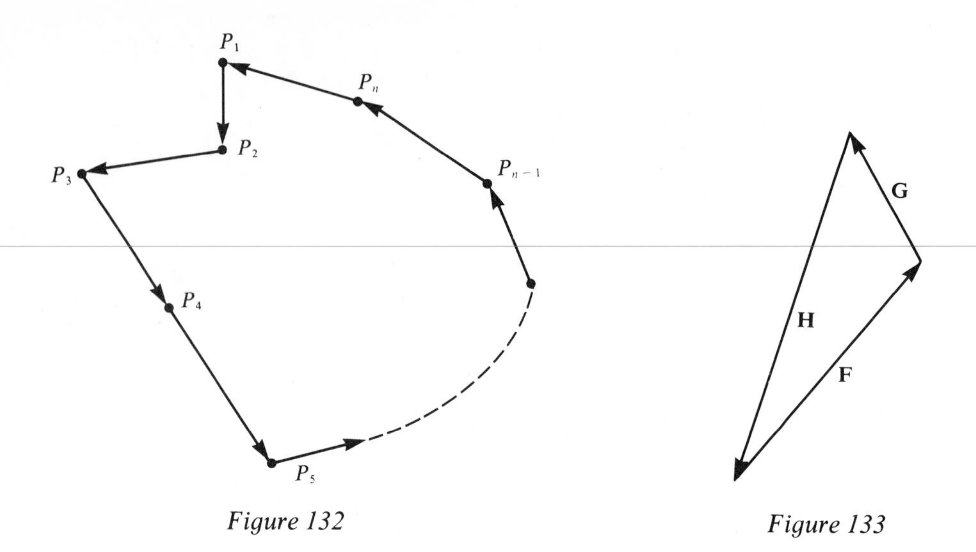

Figure 132

Figure 133

80. Use vectors to show that the distance in R^3 between the point (x_0, y_0, z_0) and the plane $ax + by + cz + d = 0$ is

$$\frac{|ax_0 + by_0 + cz_0 + d|}{\sqrt{a^2 + b^2 + c^2}},$$

assuming that the denominator is not zero.

81. Consider an n-sided polygon in the plane, as shown in Figure 134. The polygon is assumed to be *convex*, which means that a line drawn between *any two* points inside the polygon is entirely within the polygon. By contrast, the polygon of Figure 135 is not convex. Show that the area of the polygon is

$$\frac{1}{2}\left\{\begin{vmatrix} x_1 & y_1 \\ x_2 & y_2 \end{vmatrix} + \begin{vmatrix} x_2 & y_2 \\ x_3 & y_3 \end{vmatrix} + \cdots + \begin{vmatrix} x_{n-1} & y_{n-1} \\ x_n & y_n \end{vmatrix} + \begin{vmatrix} x_n & y_n \\ x_1 & y_1 \end{vmatrix}\right\}.$$

What role did convexity play in the argument?

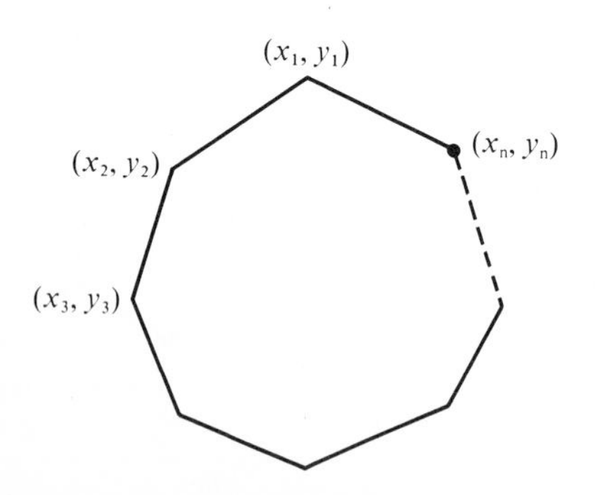

Figure 134. Convex n-sided polygon in the plane.

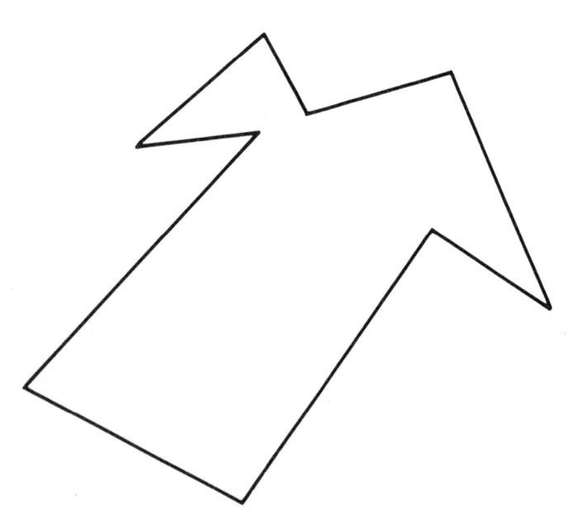

Figure 135. A nonconvex plane polygon.

82. Prove that the sum of the squares of the lengths of the diagonals of any quadrilateral in R^2 is twice the sum of the squares of the lengths of the lines joining the midpoints of opposite sides.

83. Prove that $\mathbf{i} - \mathbf{j} + \mathbf{k}$, $2\mathbf{i} + 3\mathbf{j}$, and $4\mathbf{i} - 6\mathbf{k}$ form a basis for R^3.

84. Prove that e^{2x} and e^{-2x} form a basis for the solution space of $y'' - 4y = 0$. What is the dimension of this solution space?

85. Show that the set of all vectors $\mathbf{F}$ in R^3 satisfying $\|\mathbf{F}\| = 1$ is not a subspace of R^3.

86. Give an example of a set S of vectors in R^3 such that (1) $\mathbf{F} + \mathbf{G}$ is in S whenever $\mathbf{F}$ and $\mathbf{G}$ are in S, but (2) for some $\mathbf{F}$ in S and some scalar α, $\alpha\mathbf{F}$ is not in S.

87. Give an example of a set S of vectors in R^3 such that (1) $\alpha\mathbf{F}$ is in S whenever $\mathbf{F}$ is in S and α is any scalar, but (2) for some $\mathbf{F}$ and $\mathbf{G}$ in S, $\mathbf{F} + \mathbf{G}$ is not in S.

88. Let S be a subspace of R^3, and suppose that S has dimension 3. Prove that $S = R^3$. That is, show that every vector in R^3 is in S.

89. Let S be a subspace of R^3 consisting of all vectors parallel to a given plane P through the origin. Show that S has dimension 2.

90. In the vector space of functions continuous on $[0, 1]$, show that $\sin(x)$, e^x, and e^{2x} are linearly independent.

CHAPTER TEN

Matrices and Determinants

10.0 INTRODUCTION

In the preceding chapter we studied vectors and vector spaces. In this chapter we shall center our attention upon objects called matrices and some of their applications. In the course of this study we shall also encounter determinants, which are numbers formed from matrices according to certain rules which have proved useful.

Consider, for example, a system of equations such as

$$\begin{aligned} x + 2y - z + 4w &= 0, \\ 3x - 4y + 2z - 6w &= 0, \\ x - 3y - 2z + w &= 0. \end{aligned}$$

The rectangular array

$$\begin{bmatrix} 1 & 2 & -1 & 4 \\ 3 & -4 & 2 & -6 \\ 1 & -3 & -2 & 1 \end{bmatrix}$$

in which the numerical coefficients of the unknowns are displayed in the pattern of the equations, carries most of the essential information about the system. Whether we call the unknowns x, y, z, and w or, say, s, t, u, and v, is not very important.

Such a rectangular array is called a *matrix*. Matrices may have numbers, functions, or other things in them. Their important feature is that they provide a grid in which we can store objects in named locations (for example, row 1, column 2), and from which these objects can be recalled.

If matrices were only useful as bookkeeping devices, they would not be worth much study. They possess, however, manipulative and algebraic properties which make them extremely useful in both theory and applications.

In this chapter we concentrate on certain properties of matrices which render them useful in a wide variety of contexts, such as electrical circuit theory, crystal physics, and solving systems of differential equations. These and various other applications will be treated along the way.

10.1 NOTATION AND ALGEBRA OF MATRICES

In this section we develop some of the basic notation and manipulative properties of matrices.

A *matrix* is a rectangular array of objects, written in rows and columns. These objects can be numbers, functions, or in fact anything you want. For example,

$$\begin{bmatrix} 2 & 1 & 3 \\ 1 & 4 & 2 \end{bmatrix}, \quad \begin{bmatrix} e^x & e^{2x} \\ 3 & 14 \end{bmatrix} \quad \text{and} \quad \begin{bmatrix} 1+i & 2 \\ 0 & -i \\ 0 & -2 \end{bmatrix}$$

are matrices. For our purposes, the objects will be real numbers, until specific exception is made in Section 10.15.

If a matrix A has n rows and m columns, we say that A is "n by m," written $n \times m$. A is *square* if $n = m$. Above, the first matrix on the left is 2×3, the second is 2×2 (square), and the third is 3×2. The entry in row i and column j of A is called the *i-j entry* and is denoted A_{ij}. The first subscript always gives row location, and the second subscript always gives column location in the array. Thus, if A is the matrix on the left above, then $A_{11} = 2$, $A_{12} = 1$, and $A_{23} = 2$. Similarly, the row 4, column 1 entry of a matrix B would be B_{41}; the row 3, column 2 entry of a matrix h would be h_{32}, and so on.

We shall use both upper- and lowercase letters for matrices. In some cases, a comma is inserted in the subscript to avoid ambiguity. For example, the row 1, column 12 entry of K would be written $K_{1,12}$, since K_{112} might be misread as the row 11, column 2 entry (which would be written $K_{11,2}$).

Two matrices A and B are *equal* if they have the same number of rows, the same number of columns, and the *i-j* entry of A equals that of B for each row i and column j. Thus,

$$A = \begin{bmatrix} 1 & 2 & 1 \\ 0 & 1 & 4 \\ 1 & 2 & 2 \end{bmatrix} \quad \text{and} \quad B = \begin{bmatrix} 1 & 2 & 1 \\ 0 & 1 & 4 \\ 1 & 2 & 3 \end{bmatrix}$$

are unequal because $A_{33} \neq B_{33}$.

The *sum* $A + B$ of two $n \times m$ matrices A and B is the $n \times m$ matrix having $A_{ij} + B_{ij}$ as its *i-j* entry. Thus, we add matrices of the same size by adding corresponding entries. For example,

$$\begin{bmatrix} 1 & 2 & -3 \\ 4 & 0 & 2 \end{bmatrix} + \begin{bmatrix} -1 & 6 & 3 \\ 8 & 12 & 14 \end{bmatrix} = \begin{bmatrix} 0 & 8 & 0 \\ 12 & 12 & 16 \end{bmatrix}.$$

We cannot add A and B if they differ in number of rows or number of columns.

We multiply a matrix A by a scalar* α by multiplying each entry of A by α. For example,

---

* *Scalar* for us means a real or complex number. For the most part, scalars will be real until we reach eigenvalues and eigenvectors in Section 10.15.

$$3\begin{bmatrix} 2 & 0 \\ 0 & 0 \\ 1 & 4 \\ 2 & 6 \end{bmatrix} = \begin{bmatrix} 6 & 0 \\ 0 & 0 \\ 3 & 12 \\ 6 & 18 \end{bmatrix}$$

and

$$\sqrt{2}\begin{bmatrix} -3 & 1 \\ 6 & \sqrt{2} \end{bmatrix} = \begin{bmatrix} -3\sqrt{2} & \sqrt{2} \\ 6\sqrt{2} & 2 \end{bmatrix}.$$

Besides these two rather obvious operations of addition and multiplication by a scalar, some (*but not all*) pairs of matrices can be multiplied. If A is an $n \times r$ matrix and B is $r \times m$, the *product* AB is defined to be the $n \times m$ matrix having

$$A_{i1}B_{1j} + A_{i2}B_{2j} + \cdots + A_{ir}B_{rj}$$

as its i-j entry. More compactly,

$$(AB)_{ij} = \sum_{k=1}^{r} A_{ik}B_{kj}.$$

Figure 136 shows how we fill in the i-j entry of the product AB. If we think of the rows of A and the columns of B as r-vectors, then the i-j entry of AB is the dot product of row i of A with column j of B:

$$i\text{-}j \text{ entry of } AB = (\text{row } i \text{ of } A) \cdot (\text{column } j \text{ of } B).$$

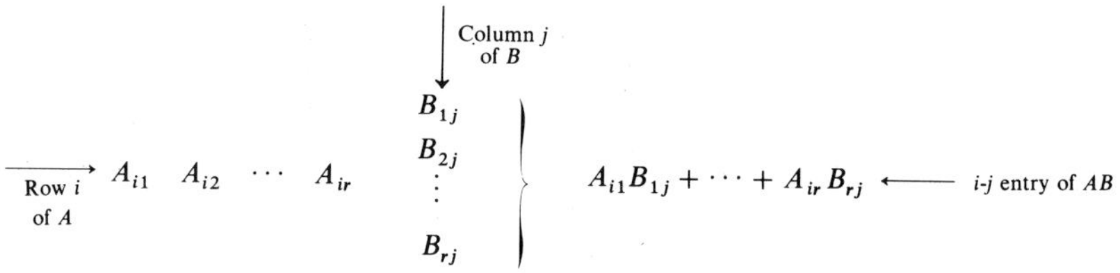

Figure 136

This again emphasizes that we can form the product AB exactly when

$$\text{number of columns of } A = \text{number of rows of } B.$$

Thus, not every pair of matrices can be multiplied. Further, even if AB is defined, BA might not be. For example, if A is 1×3 and B is 3×2, then AB is 1×2, but BA is not defined. And, even when AB and BA are both defined, they are probably not equal.

EXAMPLE 242

$$\begin{bmatrix} 1 & 3 \\ 2 & 5 \end{bmatrix}\begin{bmatrix} 1 & 1 & 3 \\ 2 & 1 & 4 \end{bmatrix} = \begin{bmatrix} 1\cdot1 + 3\cdot2 & 1\cdot1 + 3\cdot1 & 1\cdot3 + 3\cdot4 \\ 2\cdot1 + 5\cdot2 & 2\cdot1 + 5\cdot1 & 2\cdot3 + 5\cdot4 \end{bmatrix}$$
$$= \begin{bmatrix} 7 & 4 & 15 \\ 12 & 7 & 26 \end{bmatrix}.$$

But, $\begin{bmatrix} 1 & 1 & 3 \\ 2 & 1 & 4 \end{bmatrix}\begin{bmatrix} 1 & 3 \\ 2 & 5 \end{bmatrix}$ is not defined, because $\begin{bmatrix} 1 & 1 & 3 \\ 2 & 1 & 4 \end{bmatrix}$ is 2×3, $\begin{bmatrix} 1 & 3 \\ 2 & 5 \end{bmatrix}$ is 2×2, and $3 \neq 2$.

EXAMPLE 243

$$\begin{bmatrix} 1 & 1 & 2 & 1 \\ 4 & 1 & 6 & 2 \end{bmatrix}\begin{bmatrix} -1 & 8 \\ 2 & 1 \\ 1 & 1 \\ 12 & 6 \end{bmatrix}$$

$$= \begin{bmatrix} 1\cdot(-1) + 1\cdot 2 + 2\cdot 1 + 1\cdot 12 & 1\cdot 8 + 1\cdot 1 + 2\cdot 1 + 1\cdot 6 \\ 4\cdot(-1) + 1\cdot 2 + 6\cdot 1 + 2\cdot 12 & 4\cdot 8 + 1\cdot 1 + 6\cdot 1 + 2\cdot 6 \end{bmatrix}$$

$$= \begin{bmatrix} 15 & 17 \\ 28 & 51 \end{bmatrix}.$$

Here, the product in the other direction is also defined:

$$\begin{bmatrix} -1 & 8 \\ 2 & 1 \\ 1 & 1 \\ 12 & 6 \end{bmatrix}\begin{bmatrix} 1 & 1 & 2 & 1 \\ 4 & 1 & 6 & 2 \end{bmatrix}$$

$$= \begin{bmatrix} -1\cdot 1 + 8\cdot 4 & -1\cdot 1 + 8\cdot 1 & -1\cdot 2 + 8\cdot 6 & -1\cdot 1 + 8\cdot 2 \\ 2\cdot 1 + 1\cdot 4 & 2\cdot 1 + 1\cdot 1 & 2\cdot 2 + 1\cdot 6 & 2\cdot 1 + 1\cdot 2 \\ 1\cdot 1 + 1\cdot 4 & 1\cdot 1 + 1\cdot 1 & 1\cdot 2 + 1\cdot 6 & 1\cdot 1 + 1\cdot 2 \\ 12\cdot 1 + 6\cdot 4 & 12\cdot 1 + 6\cdot 1 & 12\cdot 2 + 6\cdot 6 & 12\cdot 1 + 6\cdot 2 \end{bmatrix}$$

$$= \begin{bmatrix} 31 & 7 & 46 & 15 \\ 6 & 3 & 10 & 4 \\ 5 & 2 & 8 & 3 \\ 36 & 18 & 60 & 24 \end{bmatrix}.$$

Note that in this example AB is 2×2, while BA is 4×4.

EXAMPLE 244

$$[1 \quad -1 \quad 0 \quad 1 \quad -2 \quad -4]\begin{bmatrix} 0 \\ 1 \\ 1 \\ 4 \\ 2 \\ 6 \end{bmatrix}$$

$$= [1 \cdot 0 + (-1) \cdot 1 + 0 \cdot 1 + 1 \cdot 4 + (-2) \cdot 2 + (-4) \cdot 6] = [-25].$$

This is a 1×1 matrix. Such a matrix $[\alpha]$ is usually identified with its only entry, and just written as the number α.

EXAMPLE 245

Often matrix multiplication can simplify otherwise bulky expressions. Consider the system of n equations in m unknowns:

$$\begin{aligned} a_{11}x_1 + a_{12}x_2 + \cdots + a_{1m}x_m &= b_1, \\ a_{21}x_1 + a_{22}x_2 + \cdots + a_{2m}x_m &= b_2, \\ &\vdots \\ a_{n1}x_1 + a_{n2}x_2 + \cdots + a_{nm}x_m &= b_n. \end{aligned}$$

Observe that the matrix equation

$$\begin{bmatrix} a_{11} & a_{12} & \cdots & a_{1m} \\ a_{21} & a_{22} & \cdots & a_{2m} \\ \vdots & \vdots & \cdots & \vdots \\ a_{n1} & a_{n2} & \cdots & a_{nm} \end{bmatrix} \begin{bmatrix} x_1 \\ x_2 \\ \vdots \\ x_m \end{bmatrix} = \begin{bmatrix} b_1 \\ b_2 \\ \vdots \\ b_n \end{bmatrix}$$

becomes, after carrying out the product on the left,

$$\begin{bmatrix} a_{11}x_1 + a_{12}x_2 + \cdots + a_{1m}x_m \\ a_{21}x_1 + a_{22}x_2 + \cdots + a_{2m}x_m \\ \vdots \qquad \vdots \qquad \cdots \qquad \vdots \\ a_{n1}x_1 + a_{n2}x_2 + \cdots + a_{nm}x_m \end{bmatrix} = \begin{bmatrix} b_1 \\ b_2 \\ \vdots \\ b_n \end{bmatrix}.$$

By definition of equality of matrices (here, both sides are $n \times 1$ matrices), this is exactly the original system of equations. Thus, the system can be written compactly as $AX = B$, where A is the $n \times m$ matrix of coefficients having a_{ij} as its i-j entry,

$$X = \begin{bmatrix} x_1 \\ x_2 \\ \vdots \\ x_m \end{bmatrix}, \quad \text{and} \quad B = \begin{bmatrix} b_1 \\ b_2 \\ \vdots \\ b_n \end{bmatrix}.$$

This matrix formulation will provide a starting point for a method of solving such systems of equations in Sections 10.6 and 10.7.

If A is square, say $n \times n$, then AA is defined and is also $n \times n$. Hence, $A(AA)$ is defined and is $n \times n$, and so on. We let $AA = A^2$, $AA^2 = A^3$, and, in general, $AA^{k-1} = A^k$.

Some of the rules for manipulating matrices by addition and multiplication are similar to those for real numbers, except that not just any two matrices can be added or multiplied and, even when AB and BA are both defined, they need not be equal.

THEOREM 38

1. $A + B = B + A$ if A and B are both $n \times m$.
2. $A(B + C) = AB + AC$ if A is $n \times k$ and B and C are $k \times m$.
3. $(B + C)A = BA + CA$ if B and C are $n \times k$ and A is $k \times m$.
4. $A(BC) = (AB)C$ if A is $n \times m$, B is $m \times k$, and C is $k \times r$.

Proof of (1) The *i-j* entry of $A + B$ is $A_{ij} + B_{ij}$. This is the same as $B_{ij} + A_{ij}$, the *i-j* entry of $B + A$.

Proof of (2) First, $B + C$ is $k \times m$; so $A(B + C)$ is defined and is $n \times m$. We must show that the *i-j* entry of $A(B + C)$ is the same as the *i-j* entry of $AB + AC$. For clarity, consider the special case that $n = k = 2$ and $m = 3$. Using the definition of matrix product, we have that

$$\begin{aligned} i\text{-}j \text{ entry of } A(B + C) &= A_{i1}(B + C)_{1j} + A_{i2}(B + C)_{2j} \\ &= A_{i1}(B_{1j} + C_{1j}) + A_{i2}(B_{2j} + C_{2j}) \\ &= (A_{i1}B_{1j} + A_{i2}B_{2j}) + (A_{i1}C_{1j} + A_{i2}C_{2j}) \\ &= (i\text{-}j \text{ entry of } AB) + (i\text{-}j \text{ entry of } AC). \end{aligned}$$

The proof in general follows the same lines of reasoning.

Proof of (3) Details are similar to (2), and are left to the student.

Proof of (4) First, BC is $m \times r$; so $A(BC)$ is $n \times r$. Similarly, AB is $n \times k$; so $(AB)C$ is $n \times r$. Thus, at least, $(AB)C$ and $A(BC)$ are the same size.

Again, for clarity, we shall look at the special case that $n = m = k = r = 2$. Then all the matrices and products involved are 2×2, and by using the definition of matrix product twice, we have

$$\begin{aligned} i\text{-}j \text{ entry of } A(BC) &= A_{i1}(BC)_{1j} + A_{i2}(BC)_{2j} \\ &= A_{i1}(B_{11}C_{1j} + B_{12}C_{2j}) + A_{i2}(B_{21}C_{1j} + B_{22}C_{2j}). \end{aligned}$$

Similarly,

$$\begin{aligned} i\text{-}j \text{ entry of } (AB)C &= (AB)_{i1}C_{1j} + (AB)_{i2}C_{2j} \\ &= (A_{i1}B_{11} + A_{i2}B_{21})C_{1j} + (A_{i1}B_{12} + A_{i2}B_{22})C_{2j}. \end{aligned}$$

By direct comparison, the *i-j* entry of $A(BC)$ is the same as that of $(AB)C$. The proof, in general, follows the same kind of reasoning.

Another similarity (but not an exact one) between matrix and number arithmetic is the existence of zero matrices. For real numbers, $x + 0 = 0 + x = 0$. Further, zero is the only number that has no effect upon addition to another number. For matrices, we have a *different* zero matrix for each size matrix. Define the $n \times m$ zero matrix 0_{nm}

as the $n \times m$ matrix whose entries are all zero. For example,

$$0_{12} = [0 \quad 0], \qquad 0_{22} = \begin{bmatrix} 0 & 0 \\ 0 & 0 \end{bmatrix}, \quad \text{and} \quad 0_{43} = \begin{bmatrix} 0 & 0 & 0 \\ 0 & 0 & 0 \\ 0 & 0 & 0 \\ 0 & 0 & 0 \end{bmatrix}.$$

THEOREM 39

If A is $n \times m$, then $A + 0_{nm} = 0_{nm} + A = A$.

A proof of this is left to the student. Note that the zero matrix we add to A must be the same size as A.

Given a matrix B, we denote by $-B$ the matrix obtained by replacing each entry of B by its negative. For example,

$$-\begin{bmatrix} 2 & 1 & -6 \\ -1 & 4 & 2 \end{bmatrix} = \begin{bmatrix} -2 & -1 & 6 \\ 1 & -4 & -2 \end{bmatrix}.$$

It is easy to check that, for $n \times m$ matrices A and B,

$$A + B = 0_{nm} \quad \text{exactly when} \quad A = -B.$$

Theorems 38 and 39 show some of the similarities between matrix operations and arithmetic of ordinary numbers. There are, however, differences which the student must understand. One difference we have seen is that it may not be possible to multiply two matrices in a given order (see Example 242, where AB works but BA does not). Here are some other differences.

Difference 1 Even if AB and BA are both defined, usually $AB \neq BA$. (That is, matrix multiplication is *noncommutative*.)

We have seen one instance of this in Example 243. Here is another example with A and B both the same size.

EXAMPLE 246

Let

$$A = \begin{bmatrix} 1 & 0 \\ -2 & 4 \end{bmatrix} \quad \text{and} \quad B = \begin{bmatrix} -2 & 6 \\ 1 & 3 \end{bmatrix}.$$

Then,

$$AB = \begin{bmatrix} -2 & 6 \\ 8 & 0 \end{bmatrix} \quad \text{but} \quad BA = \begin{bmatrix} -14 & 24 \\ -5 & 12 \end{bmatrix}.$$

Difference 2 Possibly $AB = AC$ with $B \neq C$ and A not a zero matrix.

Thus, there is in general no "cancellation" of A in an equation $AB = AC$.

EXAMPLE 247

$$\begin{bmatrix} 1 & 1 \\ 3 & 3 \end{bmatrix}\begin{bmatrix} 4 & 2 \\ 3 & 16 \end{bmatrix} = \begin{bmatrix} 1 & 1 \\ 3 & 3 \end{bmatrix}\begin{bmatrix} 2 & 7 \\ 5 & 11 \end{bmatrix} = \begin{bmatrix} 7 & 18 \\ 21 & 54 \end{bmatrix}$$

but

$$\begin{bmatrix} 4 & 2 \\ 3 & 16 \end{bmatrix} \neq \begin{bmatrix} 2 & 7 \\ 5 & 11 \end{bmatrix}.$$

Similarly, it is possible to have $BA = CA$ with $B \neq C$.

Difference 3 The product AB may be a zero matrix, with neither A nor B a zero matrix.

EXAMPLE 248

$$\begin{bmatrix} 1 & 2 \\ 0 & 0 \end{bmatrix}\begin{bmatrix} 6 & 4 \\ -3 & -2 \end{bmatrix} = \begin{bmatrix} 0 & 0 \\ 0 & 0 \end{bmatrix}.$$

As a second illustration,

$$\begin{bmatrix} 1 & 4 \\ 2 & 8 \end{bmatrix}\begin{bmatrix} 8 & -2 \\ -2 & \frac{1}{2} \end{bmatrix} = \begin{bmatrix} 0 & 0 \\ 0 & 0 \end{bmatrix}.$$

Differences 1, 2, and 3 have no analogues in ordinary number multiplication, and hence must be carefully considered by the student and recalled in the course of doing calculations with matrices.

The definition of matrix addition was a fairly obvious one, but matrix multiplication may at first seem unusual. In the next section we give an illustration of how the matrix product we defined can relate to a real-world problem. Problem 41 at the end of this section also gives a rationale for the definition of matrix multiplication.

PROBLEMS FOR SECTION 10.1

In each of Problems 1 through 10, find the sum of the given matrices.

1. $\begin{bmatrix} 1 & -1 & 3 \\ 2 & -4 & 6 \\ -1 & 1 & 2 \end{bmatrix}$ and $\begin{bmatrix} -4 & 0 & 0 \\ -2 & -1 & 6 \\ 8 & 15 & 4 \end{bmatrix}$

2. $\begin{bmatrix} -2 & 2 \\ 0 & 1 \\ 14 & 2 \\ 6 & 8 \end{bmatrix}$ and $\begin{bmatrix} 3 & 4 \\ 2 & 1 \\ 14 & 16 \\ 1 & 25 \end{bmatrix}$

3. $\begin{bmatrix} -22 & 1 & 6 & 4 & 5 \\ -3 & -2 & 14 & 2 & 25 \\ 18 & 1 & 16 & -4 & -6 \end{bmatrix}$ and $\begin{bmatrix} 0 & 1 & 3 & 1 & 14 \\ -8 & 6 & -10 & 4 & 10 \\ 21 & 6 & 17 & 3 & 2 \end{bmatrix}$

4. $[14]$ and $[-22]$

5. $[-4 \quad 8 \quad 6]$ and $[22 \quad 17 \quad -3]$

6. $\begin{bmatrix} -4 & 1 & 1 \\ 2 & 0 & 6 \end{bmatrix}$ and $\begin{bmatrix} -3 & -2 & 5 \\ 3 & 1 & 7 \end{bmatrix}$

7. $\begin{bmatrix} 1 \\ 0 \\ -4 \end{bmatrix}$ and $\begin{bmatrix} -3 \\ 1 \\ 5 \end{bmatrix}$

8. $\begin{bmatrix} 2 & 1 & 1 & 7 & 3 \\ 8 & 0 & 0 & 2 & 4 \end{bmatrix}$ and $\begin{bmatrix} -2 & 3 & 0 & 4 & 6 \\ -2 & 2 & 1 & 3 & 7 \end{bmatrix}$

9. $\begin{bmatrix} 2 & 1 \\ 0 & 0 \\ 2 & 5 \end{bmatrix}$ and $\begin{bmatrix} -2 & 4 \\ -5 & 1 \\ -6 & 3 \end{bmatrix}$

10. $\begin{bmatrix} 4 & -3 & 2 & 1 \\ 3 & 4 & 7 & 5 \\ 0 & 0 & 9 & 2 \end{bmatrix}$ and $\begin{bmatrix} 10 & 0 & 0 & 9 \\ -3 & 0 & 1 & 4 \\ 2 & 2 & -1 & -7 \end{bmatrix}$

In each of Problems 11 through 15, find the matrix obtained by multiplying the given matrix by the given scalar.

11. $\begin{bmatrix} -4 & -3 \\ 2 & 1 \\ 6 & 8 \end{bmatrix}$ by -4

12. $\begin{bmatrix} 18 & 14 & -2 & 3 \\ 12 & 25 & 3 & -6 \\ -4 & 0 & 0 & 1 \end{bmatrix}$ by 6

13. $\begin{bmatrix} 2 \\ 0 \\ -1 \end{bmatrix}$ by 22

14. $\begin{bmatrix} 8 & 1 & -3 \\ 4 & -2 & 1 \\ 7 & 7 & 4 \end{bmatrix}$, by -3

15. $\begin{bmatrix} 2 & -4 & 1 & 7 \\ 3 & 3 & -4 & 2 \\ 0 & 0 & -5 & 9 \end{bmatrix}$ by 8

In each of Problems 16 through 33, compute the products AB and BA where possible. Specify any products that are not defined, and explain why.

16. $A = \begin{bmatrix} -4 & 6 & 2 \\ -2 & -2 & 3 \\ 1 & 1 & 8 \end{bmatrix}$, $B = \begin{bmatrix} -2 & 4 & 6 & 12 & 5 \\ -3 & -3 & 1 & 1 & 4 \\ 0 & 0 & 1 & 6 & -9 \end{bmatrix}$

17. $A = \begin{bmatrix} -2 & -4 \\ 3 & -1 \end{bmatrix}$, $B = \begin{bmatrix} 6 & 8 \\ 1 & -4 \end{bmatrix}$

18. $A = [-1 \quad 6 \quad 2 \quad 14 \quad -22]$, $B = \begin{bmatrix} -3 \\ 2 \\ 6 \\ 0 \\ -4 \end{bmatrix}$

19. $A = \begin{bmatrix} -3 & 1 \\ 6 & 2 \\ 18 & -22 \\ 1 & 16 \end{bmatrix}$, $B = \begin{bmatrix} -16 & 0 & 0 & 28 \\ 0 & 1 & 1 & 26 \end{bmatrix}$

20. $A = [-26 \quad 1 \quad 13]$, $B = \begin{bmatrix} 22 \\ 1 \\ 6 \\ 14 \\ 2 \end{bmatrix}$

21. $A = \begin{bmatrix} -21 & 4 & 8 & -3 \\ 12 & 1 & 0 & 14 \\ 1 & 16 & 0 & -8 \\ 13 & 4 & 8 & 0 \end{bmatrix}, \quad B = \begin{bmatrix} -9 & 16 & 3 & 2 \\ 5 & 9 & 14 & 0 \end{bmatrix}$

22. $A = [-21 \quad 16], \quad B = \begin{bmatrix} 32 & 4 & 16 \\ -8 & 7 & 0 \end{bmatrix}$

23. $A = \begin{bmatrix} 4 & 6 \\ -3 & 1 \end{bmatrix}, \quad B = [1 \quad 16 \quad -7 \quad 9]$

24. $A = \begin{bmatrix} 2 & 7 & -5 & 6 & 10 \\ 4 & 6 & 0 & 0 & -5 \end{bmatrix}, \quad B = \begin{bmatrix} 3 \\ -4 \\ 7 \\ 3 \\ 9 \end{bmatrix}$

25. $A = \begin{bmatrix} -2 & 4 \\ 3 & 9 \end{bmatrix}, \quad B = \begin{bmatrix} 1 & -3 & 7 & 2 \\ -5 & 6 & 1 & 0 \end{bmatrix}$

26. $A = [1 \quad -3 \quad 4], \quad B = \begin{bmatrix} -4 & -2 & 0 \\ 0 & 5 & 3 \\ -3 & 1 & 1 \end{bmatrix}$

27. $A = \begin{bmatrix} 3 \\ 0 \\ -1 \\ 4 \end{bmatrix}, \quad B = [3 \quad -2 \quad 7]$

28. $A = \begin{bmatrix} 3 & -8 \\ 1 & 6 \end{bmatrix}, \quad B = \begin{bmatrix} 1 & -4 & 3 \\ -4 & 7 & 0 \end{bmatrix}$

29. $A = [-3 \quad 5 \quad 7 \quad 2], \quad B = \begin{bmatrix} -3 & 2 \\ 0 & -2 \\ 1 & 8 \\ 3 & -3 \end{bmatrix}$

30. $A = \begin{bmatrix} -4 & 2 & 8 \\ 1 & 6 & -4 \\ 2 & 2 & 0 \end{bmatrix}, \quad B = \begin{bmatrix} 2 & -4 & 6 \\ 1 & 9 & -5 \\ 0 & -5 & 1 \end{bmatrix}$

31. $A = \begin{bmatrix} 1 \\ 0 \\ -3 \end{bmatrix}, \quad B = [2 \quad -6]$

32. $A = \begin{bmatrix} 2 & -3 & 5 \\ 0 & 0 & -4 \end{bmatrix}$, $B = \begin{bmatrix} 0 \\ 3 \\ -4 \end{bmatrix}$

33. $A = \begin{bmatrix} 1 & -2 \\ 2 & 4 \end{bmatrix}$, $B = \begin{bmatrix} -1 & 3 & 2 & 9 & -4 & 6 \\ 0 & -1 & 6 & 0 & 9 & -4 \end{bmatrix}$

In each of Problems 34 through 39, determine if AB and BA are defined, and how many rows and columns each product has when it is defined.

34. A is 14×21, B is 21×14.

35. A is 18×4, B is 18×4.

36. A is 6×2, B is 4×6.

37. A is 1×3, B is 1×3.

38. A is 7×6, B is 6×7.

39. A is 8×6, B is 8×8.

40. Find nonzero matrices A, B, and C such that $BA = CA$ but $B \neq C$.

41. Consider the system of equations

$$\begin{aligned} a_1 x + b_1 y &= c, \\ a_2 x + b_2 y &= d. \end{aligned} \tag{1}$$

(a) Show that the system (1) can be written

$$AX = B,$$

where

$$A = \begin{bmatrix} a_1 & b_1 \\ a_2 & b_2 \end{bmatrix}, \quad X = \begin{bmatrix} x \\ y \end{bmatrix} \quad \text{and} \quad B = \begin{bmatrix} c \\ d \end{bmatrix}.$$

(b) Suppose that

$$\begin{aligned} x &= h_1 u + k_1 v, \\ y &= h_2 u + k_2 v. \end{aligned} \tag{2}$$

Show that (2) can be written

$$X = HU,$$

where

$$H = \begin{bmatrix} h_1 & k_1 \\ h_2 & k_2 \end{bmatrix} \quad \text{and} \quad U = \begin{bmatrix} u \\ v \end{bmatrix}.$$

(c) Substitute for x and y from (2) into (1). Show that the resulting system can be written

$$(AH)U = B,$$

where AH is the matrix product of A and H.

10.2 MATRIX MULTIPLICATION AND RANDOM WALKS IN CRYSTALS

In the last section we developed a nonintuitive concept of matrix multiplication. A more obvious definition might have been, for example, to simply multiply corresponding matrix entries. The reason we chose the definition we did was that it is a useful one. In the context of representing linear transformations by matrices, it is even natural, but we shall not deal with that here. Here, we shall describe a physical problem whose solution falls easily out of matrix multiplication.

In physics, a crystal can often be thought of as a lattice of sites occupied by atoms. In a defective crystal, one or more sites may be vacant at any time. An atom may jump from a given site to a neighboring, unoccupied one; after a time, another atom may in turn jump to the site just vacated, or the same atom may make a second jump to another empty site; and so on. What results is sometimes called a *random walk* of the atoms, in the sense that an atom describes a path from one point to another through the lattice, but does so randomly. In each time interval, it may have various choices of direction in which to move, and it may move in one of them or remain where it is.

A mathematical way of looking at this is to draw a lattice diagram in which points designate the positions to be occupied and lines connect neighboring points (an atom can move from one point to another if the two are joined by a line). Such a diagram is called a *graph*. Figure 137 shows a typical graph, having six points labeled $v_1, \ldots, v_6$. In it, v_3 has neighbors v_2, v_1, and v_4, while v_6 and v_4 are the only neighbors of v_5.

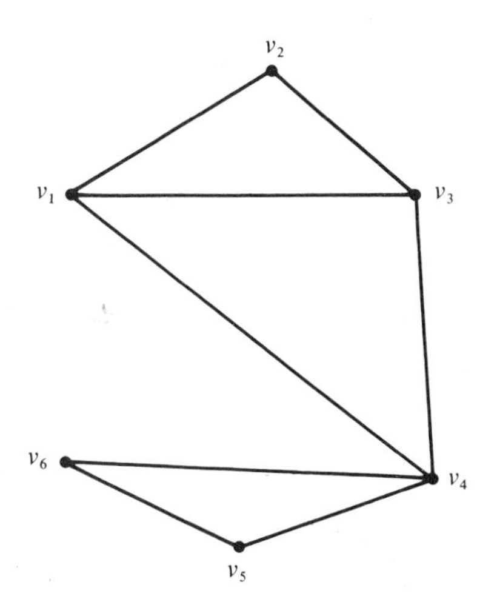

Figure 137. A typical graph.

A *walk of length n* in a graph is a sequence $t_1, \ldots, t_{n+1}$ of points (not necessarily all different), with t_{i-1} and t_i neighbors for $i = 1, \ldots, n + 1$. In Figure 137, $v_1, v_2, v_3, v_1, v_4, v_6$ is a walk of length 5 from v_1 to v_6 (a v_1—v_6 walk of length 5); $v_1, v_2, v_1, v_2, v_1, v_3, v_4, v_6, v_5, v_4, v_1, v_3, v_1$ is a v_1—v_1 walk of length 12. Walks represent possible paths of an atom through the crystal. We allow $t_{n+1} = t_1$ in a walk because an atom may return to its original site.

In studying any kind of random walk phenomenon, it is often important to know how many different walks there are from one site to another. Two walks $w_1, w_2, \ldots, w_m$ and $t_1, t_2, \ldots, t_n$ are different if $m \neq n$, or if $m = n$ and some t_i is different from w_i.

For example, v_1, v_3, v_2, v_1 and v_1, v_3, v_4, v_1 are different v_1—v_1 walks of length 3 in Figure 137. We now pose the following problem: Given a graph with points labeled $v_1, v_2, \ldots$, how many different v_i—v_j walks are there of length k?

In order to use matrix multiplication to solve this problem, we must first associate a matrix with a graph. The *adjacency matrix* of a graph G with n points $v_1, \ldots, v_n$ is the $n \times n$ matrix A with i-j entry 1 if v_1 and v_j are neighbors, and 0 otherwise. As an example, the graph of Figure 137 has a 6×6 adjacency matrix

$$\begin{bmatrix} 0 & 1 & 1 & 1 & 0 & 0 \\ 1 & 0 & 1 & 0 & 0 & 0 \\ 1 & 1 & 0 & 1 & 0 & 0 \\ 1 & 0 & 1 & 0 & 1 & 1 \\ 0 & 0 & 0 & 1 & 0 & 1 \\ 0 & 0 & 0 & 1 & 1 & 0 \end{bmatrix}.$$

We put $A_{ii} = 0$ because we do not think of a point as being its own neighbor.

The adjacency matrix provides a computational device for counting v_i—v_j walks of length k, even allowing the possibility that $i = j$.

THEOREM 40

Let A be the adjacency matrix of a graph G having n points labeled $v_1, \ldots, v_n$. Then the number of distinct v_i—v_j walks of length k is the i-j entry of A^k, for any positive integer k.

Proof We proceed by induction on k.

Begin with $k = 1$. Then, $A^k = A$. Now, A_{ii} does equal the number of v_i—v_i walks of length 1, as this number is 0. For $i \neq j$, $A_{ij} = 1$ if v_1 and v_j have a line connecting them, and in this case v_i, v_j is the one walk of length 1 from v_i to v_j. And $A_{ij} = 0$ if v_i and v_j are not neighbors, in which case there is no v_i—v_j walk of length 1. Thus, in all cases the i-j entry of A is the number of distinct v_i—v_j walks of length 1 in G.

Now assume that the statement of the theorem is true for some given k. We must prove it for $k + 1$. Hence, we are assuming that $(A^k)_{ij}$ = number of distinct v_i—v_j walks of length k, and we must show that $(A^{k+1})_{ij}$ = number of distinct v_i—v_j walks of length $k + 1$.

How do we get a v_i—v_j walk of length $k + 1$ in G? Clearly, by making a v_i—v_r walk of length 1 for some v_r neighboring on v_i, and then a v_r—v_j walk of length k (see Figure 138). Thus,

$$\begin{array}{l} \text{number of distinct} \\ \text{walks of length } k+1 \\ \text{from } v_i \text{ to } v_j \end{array} = \begin{array}{l} \text{sum of number of distinct walks} \\ \text{of length } k \text{ from } v_r \text{ to } v_j, \text{ for} \\ \text{each } v_r \text{ neighboring on } v_i. \end{array}$$

Now, $A_{ir} = 1$ when v_r neighbors on v_i, 0 otherwise. And, by the inductive hypothesis, the number of distinct walks of length k from v_r to v_j is $(A^k)_{rj}$.

Then, for $r = 1, \ldots, n$,

$$A_{ir}(A^k)_{rj} = \begin{cases} 0 & \text{if } v_r \text{ is not a neighbor of } v_i; \\ \text{number of distinct walks of length } k+1 \text{ from } v_i \\ \quad \text{to } v_j \text{ through } v_r \text{ when } v_r \text{ neighbors on } v_i. \end{cases}$$

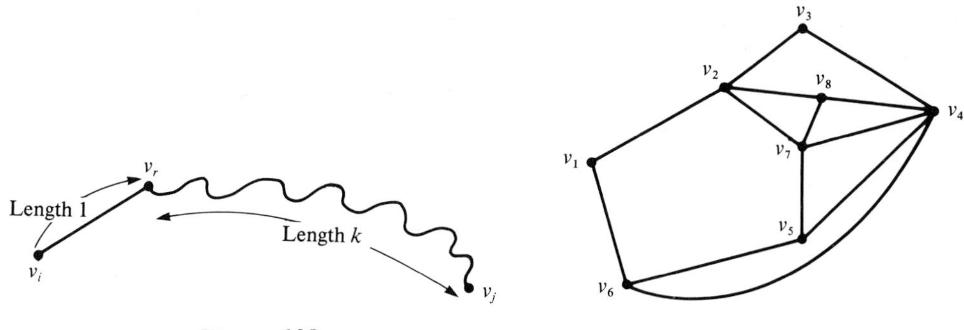

Figure 138 *Figure 139*

Hence,

$$A_{i1}(A^k)_{1j} + A_{i2}(A^k)_{2j} + \cdots + A_{in}(A^k)_{nj}$$

counts the number of v_i—v_j walks of length $k + 1$. But this number is exactly the *i-j* entry of the product AA^k or A^{k+1}, thus completing the proof.

Here is an example of the use of Theorem 40.

EXAMPLE 249

Let G be the graph of Figure 139. The adjacency matrix is

$$A = \begin{bmatrix} 0 & 1 & 0 & 0 & 0 & 1 & 0 & 0 \\ 1 & 0 & 1 & 0 & 0 & 0 & 1 & 1 \\ 0 & 1 & 0 & 1 & 0 & 0 & 0 & 0 \\ 0 & 0 & 1 & 0 & 1 & 1 & 1 & 1 \\ 0 & 0 & 0 & 1 & 0 & 1 & 1 & 0 \\ 1 & 0 & 0 & 1 & 1 & 0 & 0 & 0 \\ 0 & 1 & 0 & 1 & 1 & 0 & 0 & 1 \\ 0 & 1 & 0 & 1 & 0 & 0 & 1 & 0 \end{bmatrix}.$$

A simple but tedious calculation gives us

$$A^2 = \begin{bmatrix} 2 & 0 & 1 & 1 & 1 & 0 & 1 & 1 \\ 0 & 4 & 0 & 3 & 1 & 1 & 1 & 1 \\ 1 & 0 & 2 & 0 & 1 & 1 & 2 & 2 \\ 1 & 3 & 0 & 5 & 2 & 1 & 2 & 1 \\ 1 & 1 & 1 & 2 & 3 & 1 & 1 & 2 \\ 0 & 1 & 1 & 1 & 1 & 3 & 2 & 1 \\ 1 & 1 & 2 & 2 & 1 & 2 & 4 & 2 \\ 1 & 1 & 2 & 1 & 2 & 1 & 2 & 3 \end{bmatrix} \quad \text{and} \quad A^3 = \begin{bmatrix} 0 & 5 & 1 & 4 & 2 & 4 & 3 & 2 \\ 5 & 2 & 7 & 4 & 5 & 4 & 9 & 8 \\ 1 & 7 & 0 & 8 & 3 & 2 & 3 & 2 \\ 4 & 4 & 8 & 6 & 8 & 8 & 11 & 10 \\ 2 & 5 & 3 & 8 & 4 & 6 & 8 & 4 \\ 4 & 4 & 2 & 8 & 6 & 2 & 4 & 4 \\ 3 & 9 & 3 & 11 & 8 & 4 & 6 & 7 \\ 2 & 8 & 2 & 10 & 4 & 4 & 7 & 4 \end{bmatrix}.$$

From A^2, for example, we have in G:

two v_5—v_8 walks of length 2 (they are v_5, v_7, v_8 and v_5, v_4, v_8), and
three v_2—v_4 walks of length 2 (they are v_2, v_3, v_4; v_2, v_8, v_4; and v_2, v_7, v_4).

From A^3, we can count walks of length 3. For example, there are:

> eleven walks of length 3 from v_4 to v_7 (they are v_4, v_7, v_4, v_7; v_4, v_3, v_4, v_7; v_4, v_8, v_4, v_7; v_4, v_5, v_4, v_7; v_4, v_6, v_4, v_7; v_4, v_7, v_8, v_7; v_4, v_7, v_5, v_7; v_4, v_7, v_2, v_7; v_4, v_3, v_2, v_7; v_4, v_8, v_2, v_7; and v_4, v_6, v_5, v_7); and
> seven walks of length 3 from v_7 to v_8 (they are v_7, v_8, v_7, v_8; v_7, v_8, v_4, v_8; v_7, v_8, v_2, v_8; v_7, v_2, v_7, v_8; v_7, v_4, v_7, v_8; v_7, v_5, v_7, v_8; and v_7, v_5, v_4, v_8).

Even for walks of length two or three, it is easy to miss some if you attempt to enumerate them directly from the graph; hence the value of this computable matrix method. Of course, for large n and/or k, calculating A^k can be very time-consuming or costly, even on a computer. Here, the arithmetic is simplified by the symmetry of A. Since v_i is a neighbor of v_j exactly when v_j is a neighbor of v_i, then $A_{ij} = A_{ji}$. Such a matrix is called *symmetric*. Hence, in calculating A^k as AA^{k-1}, we need only fill in the i-j entries on and above the main diagonal: the rest can be assigned by symmetry. This reduces the arithmetic by almost half. There remains, however, a lot of work if n or k is large.

PROBLEMS FOR SECTION 10.2

1. Let G be as shown below. Compute A^4, where A is the adjacency matrix, and hence find the number of distinct walks of length 4 from v_1 to v_4 and from v_2 to v_4. From A^3, find the number of distinct walks of length 3 from v_2 to v_3 and from v_3 to v_4. List all the walks in these four cases.

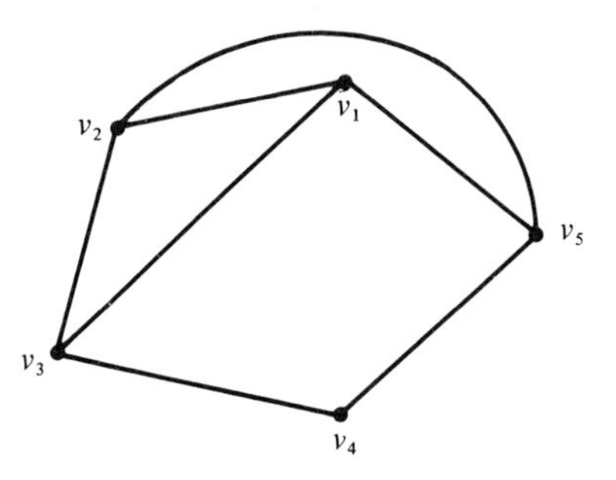

2. Let G be as shown below. How many distinct walks of length 5 are there from v_1 to v_4 in G? From v_2 to v_3? List all the walks in each case.

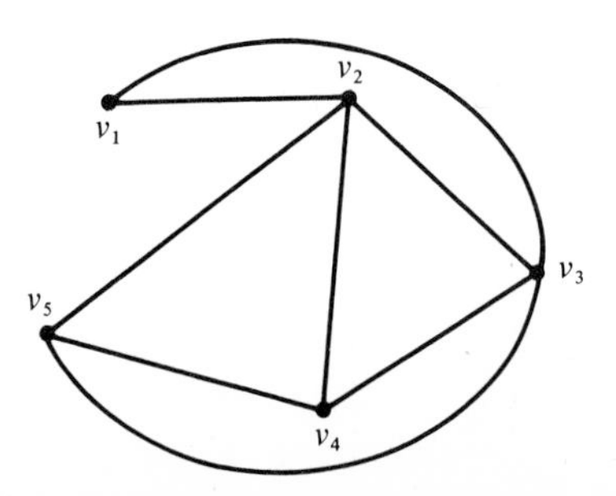

3. Let G be the graph shown below. How many distinct walks of length 5 are there in G from v_2 to v_2; from v_1 to v_4; from v_4 to v_5? How many distinct walks in G are there of length 4 from v_1 to v_2; from v_4 to v_5?

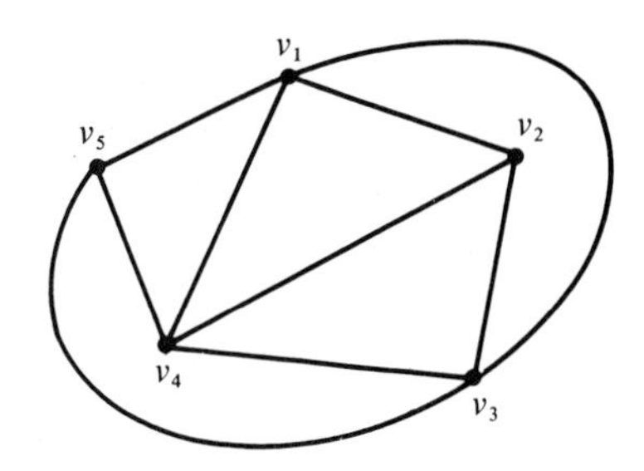

Let A be the adjacency matrix of a graph G.

4. Prove that the i-i entry of A^2 is the number of vertices neighboring on v_i in G.
5. Prove that the i-i entry of A^3 is the number of triangles in G containing v_i as a vertex (a triangle in G consists of three vertices of G, each neighboring on the others).

10.3 SOME SPECIAL KINDS OF MATRICES

In this chapter we consider certain kinds of matrices which are used often enough to warrant a special name. We have already seen as one example the $n \times m$ zero matrix 0_{nm}, which has n rows, m columns, and all entries zero.

Given any square matrix A, the *main diagonal* consists of the entries A_{11}, A_{22}, $\ldots$, A_{nn}. In Figure 140 the x's follow the main diagonal. In the matrix

$$\begin{bmatrix} 1 & -4 & 2 \\ 0 & 1 & 6 \\ 0 & -2 & 4 \end{bmatrix}$$

the main diagonal has three entries, namely 1, 1, and 4. Usually main diagonal entries are listed in this way, from upper left to lower right.

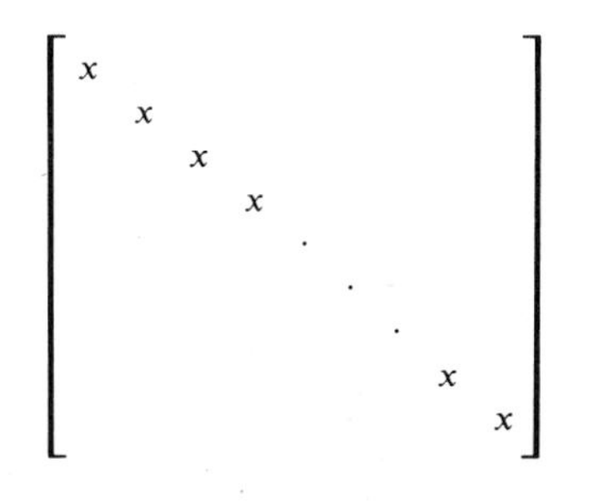

Figure 140. Main diagonal of a matrix.

A square matrix D is a *diagonal matrix* if every entry off the main diagonal is zero (the main diagonal may or may not have zeros on it).

For example,

$$\begin{bmatrix} 1 & 0 & 0 & 0 \\ 0 & -1 & 0 & 0 \\ 0 & 0 & 2 & 0 \\ 0 & 0 & 0 & 4 \end{bmatrix} \text{ and } \begin{bmatrix} -1 & 0 \\ 0 & 1 \end{bmatrix}$$

are diagonal matrices (4×4 and 2×2, respectively). The matrix

$$\begin{bmatrix} 1 & 0 & 0 \\ 0 & -2 & 4 \\ 0 & 0 & 3 \end{bmatrix}$$

is not a diagonal matrix.

One nice feature of such matrices is that sums and products of diagonal matrices are again diagonal.

THEOREM 41

Let A and B be $n \times n$ diagonal matrices, then

1. $A + B$ is a diagonal matrix.
2. AB and BA are diagonal and equal.

Proofs are left to the student. In the case of multiplication, it is easy to see that AB is obtained by multiplying the respective main diagonal entries of A and B. For example,

$$\begin{bmatrix} 2 & 0 & 0 \\ 0 & -1 & 0 \\ 0 & 0 & 3 \end{bmatrix}\begin{bmatrix} 4 & 0 & 0 \\ 0 & 2 & 0 \\ 0 & 0 & -2 \end{bmatrix} = \begin{bmatrix} 8 & 0 & 0 \\ 0 & -2 & 0 \\ 0 & 0 & -6 \end{bmatrix}.$$

This explains why we should expect to have $AB = BA$ when both are diagonal.

A diagonal matrix with all 1's down the main diagonal is called an *identity matrix*. The $n \times n$ identity matrix is denoted I_n. Thus,

$$I_2 = \begin{bmatrix} 1 & 0 \\ 0 & 1 \end{bmatrix}, \quad I_3 = \begin{bmatrix} 1 & 0 & 0 \\ 0 & 1 & 0 \\ 0 & 0 & 1 \end{bmatrix},$$

and so on. These matrices derive their name from the fact that they behave like the number 1 when multiplied by matrices of the appropriate dimensions.

THEOREM 42

If A is $m \times n$, then $AI_n = A$ and $I_m A = A$.

We leave the proof as an exercise.

EXAMPLE 250

Let

$$A = \begin{bmatrix} 1 & 0 \\ 2 & 1 \\ -1 & 8 \end{bmatrix}, \text{ a } 3 \times 2 \text{ matrix.}$$

Then,

$$I_3 A = \begin{bmatrix} 1 & 0 & 0 \\ 0 & 1 & 0 \\ 0 & 0 & 1 \end{bmatrix} \begin{bmatrix} 1 & 0 \\ 2 & 1 \\ -1 & 8 \end{bmatrix} = \begin{bmatrix} 1 & 0 \\ 2 & 1 \\ -1 & 8 \end{bmatrix} = A$$

and

$$AI_2 = \begin{bmatrix} 1 & 0 \\ 2 & 1 \\ -1 & 8 \end{bmatrix} \begin{bmatrix} 1 & 0 \\ 0 & 1 \end{bmatrix} = \begin{bmatrix} 1 & 0 \\ 2 & 1 \\ -1 & 8 \end{bmatrix} = A.$$

Note, however, that $I_2 A$ and AI_3 are not defined.

Given any $n \times m$ matrix A, the *transpose* A^t of A is the $m \times n$ matrix obtained by writing the rows of A as the columns of A^t. Thus, the i-j entry of A^t is the j-i entry of A:

$$(A^t)_{ij} = A_{ji} \quad \text{for} \quad i = 1, \ldots, m, \quad j = 1, \ldots, n.$$

Note also that, if A is $n \times m$, then A^t is $m \times n$.

EXAMPLE 251

If

$$A = \begin{bmatrix} 1 & -1 & 3 & 2 \\ 0 & 0 & 1 & 6 \\ 1 & -2 & 4 & 0 \end{bmatrix},$$

then

$$A^t = \begin{bmatrix} 1 & 0 & 1 \\ -1 & 0 & -2 \\ 3 & 1 & 4 \\ 2 & 6 & 0 \end{bmatrix}.$$

Here are some facts we shall need about transposes.

THEOREM 43

1. $(I_n)^t = I_n$.
2. For any matrix A, $(A^t)^t = A$.
3. Whenever AB is defined, $(AB)^t = B^t A^t$.

Part 1 of the theorem is obvious, since row k and column k of I_n are the same; hence interchanging them has no effect.

For part 2, observe that we get A^t by interchanging rows and columns of A and $(A^t)^t$ by interchanging rows and columns of A^t. This leaves everything exactly where it was.

More carefully, the i-j place of $(A^t)^t$ is

$$[(A^t)^t]_{ij} = (A^t)_{ji} = A_{ij}.$$

Thus, $(A^t)^t$ and A are identical.

Part 3 is more complicated, and is perhaps a surprise the first time you see it. In words, it says that the transpose of a product is the product of the transposes *in reverse order*. To illustrate the reasoning involved, here is a proof of part 3.

Proof of Part 3 Let A be $n \times k$, and B, $k \times m$. Then AB is $n \times m$, and $(AB)^t$ is $m \times n$. By definition of transpose and matrix product, the i-j entry of $(AB)^t$ is

$$[(AB)^t]_{ij} = (AB)_{ji} = A_{j1}B_{1i} + A_{j2}B_{2i} + \cdots + A_{jk}B_{ki}.$$

Next, B^t is $m \times k$ and A^t is $k \times n$; so B^tA^t is defined and is $m \times n$, the same as $(AB)^t$. The i-j entry of B^tA^t is

$$\begin{aligned}(B^tA^t)_{ij} &= B^t_{i1}A^t_{1j} + B^t_{i2}A^t_{2j} + \cdots + B^t_{ik}A^t_{kj}\\ &= B_{1i}A_{j1} + B_{2i}A_{j2} + \cdots + B_{ki}A_{jk}\\ &= A_{j1}B_{1i} + A_{j2}B_{2i} + \cdots + A_{jk}B_{ki},\end{aligned}$$

the same as the i-j entry of $(AB)^t$. Thus, $(AB)^t = B^tA^t$.

The last kind of matrix we shall treat for now is the *symmetric matrix*. We call A symmetric if it equals its own transpose, that is, if

$$A = A^t.$$

Another way of saying this is that each $A_{ij} = A_{ji}$.

If A is symmetric and $n \times m$, then A^t is $m \times n$, and $A = A^t$ forces $m = n$. Thus, only square matrices can be symmetric.

EXAMPLE 252

$$\begin{bmatrix} 1 & -1 & 3 \\ -1 & 0 & 2 \\ 3 & 2 & 4 \end{bmatrix} \quad \text{and} \quad \begin{bmatrix} -3 & 0 & 0 & 2 \\ 0 & 1 & -5 & 6 \\ 0 & -5 & 0 & -2 \\ 2 & 6 & -2 & 1 \end{bmatrix}$$

are symmetric, while

$$\begin{bmatrix} -2 & 1 & 0 \\ 1 & -1 & 5 \\ 0 & 6 & 0 \end{bmatrix}$$

is not (look at the 2-3 and 3-2 entries, which are different).

Given that A is symmetric, one only need know the main diagonal entries and the entries above (or below) the main diagonal. These can then be reflected across the main diagonal to fill in the rest of the matrix. When storing a matrix in computer memory, this can result in a significant savings.

PROBLEMS FOR SECTION 10.3

1. Let D be an $n \times n$ diagonal matrix with no zeros on the main diagonal. Find a diagonal matrix E such that $DE = ED = I_n$.
2. Suppose that A and B are $n \times n$ symmetric matrices. Prove that $A + B$ is symmetric.
3. Show that $A + A^t$ is symmetric for any square matrix A.
4. Let A and B be symmetric, $n \times n$ matrices.
   (a) Give an example to show that AB need not be symmetric.
   (b) Prove that, in general, AB is symmetric if and only if $AB = BA$.
5. Let A be an $n \times n$ matrix. Suppose that $AB = B$ for every $n \times n$ matrix B. Show that A must be I_n.
6. An $n \times n$ matrix A is *upper triangular* if all entries below the main diagonal are zero (i.e., $A_{ij} = 0$ if $i > j$).
   (a) Show that a sum of upper triangular matrices is upper triangular.
   (b) Show that a product of upper triangular matrices is upper triangular.
7. Let $A = \begin{bmatrix} a & b \\ c & d \end{bmatrix}$ be any 2×2 matrix, with $ad - bc \neq 0$.
   Find a matrix B such that $AB = BA = I_2$.
8. Let A and B be $n \times m$ matrices. Show that $(A + B)^t = A^t + B^t$.
9. Prove Theorem 41.
10. Prove Theorem 42.

10.4 ELEMENTARY ROW OPERATIONS AND ELEMENTARY MATRICES

In this section we shall discuss three row manipulations which play an important role in many uses of matrices. Because they are basic to much of what follows, they are called *elementary row operations*. There are three types:

Type I: Interchange of two rows.

Type II: Multiplication of a row by a nonzero scalar.

Type III: Addition of a scalar multiple of one row to another row.

In type II, we multiply a row by a number by multiplying each entry of the row by the number. In type III, we add a scalar multiple of one row to another row by adding the scalar multiple of each entry of the first row to the respective entry of the second row.

For example, let

$$A = \begin{bmatrix} -2 & 1 & 6 \\ 1 & 1 & 2 \\ 0 & 1 & 3 \\ 2 & -3 & 4 \end{bmatrix}.$$

Type I: If we interchange rows 1 and 3 of A, we obtain

$$\begin{bmatrix} 0 & 1 & 3 \\ 1 & 1 & 2 \\ -2 & 1 & 6 \\ 2 & -3 & 4 \end{bmatrix}.$$

Type II: If we multiply each entry of row 4 by $\sqrt{3}$, we obtain

$$\begin{bmatrix} -2 & 1 & 6 \\ 1 & 1 & 2 \\ 0 & 1 & 3 \\ 2\sqrt{3} & -3\sqrt{3} & 4\sqrt{3} \end{bmatrix}$$

Type III: If we add $\sqrt{5}$(row 3) to row 2, we obtain

$$\begin{bmatrix} -2 & 1 & 6 \\ 1 & 1+\sqrt{5} & 2+3\sqrt{5} \\ 0 & 1 & 3 \\ 2 & -3 & 4 \end{bmatrix}.$$

It is sometimes useful to be able to perform elementary row operations by matrix multiplication. To this end, we define an *elementary matrix* as one obtained by performing an elementary row operation on I_n. For example,

$$\begin{bmatrix} 0 & 1 \\ 1 & 0 \end{bmatrix} \text{(interchange rows 1 and 2 of } I_2\text{, a type-I operation)}$$

and

$$\begin{bmatrix} 1 & 0 & 0 \\ 0 & 1 & 0 \\ 0 & 3 & 1 \end{bmatrix} \quad \text{(add 3(row 2) to row 3 of } I_3\text{, a type-III operation)}$$

are elementary matrices. We call the elementary matrix type I, II, or III, according to the type of row operation used to obtain it from I_n.

Now, let A be any $n \times m$ matrix. We claim that, if E is an elementary matrix, then EA is exactly the matrix obtained by performing on A the elementary row operation used to obtain E from I_n.

We shall leave a proof of this to the student (see Problems 26, 27, and 28), but here are some examples.

EXAMPLE 253

Let A be the 4×3 matrix used above to illustrate the elementary row operations. Form E from I_4 by interchanging rows 1 and 3:

$$E = \begin{bmatrix} 0 & 0 & 1 & 0 \\ 0 & 1 & 0 & 0 \\ 1 & 0 & 0 & 0 \\ 0 & 0 & 0 & 1 \end{bmatrix}.$$

The student can verify that

$$EA = \begin{bmatrix} 0 & 1 & 3 \\ 1 & 1 & 2 \\ -2 & 1 & 6 \\ 2 & -3 & 4 \end{bmatrix},$$

which is exactly the matrix obtained from A by interchanging rows 1 and 3. Thus, the type-I elementary matrix E performs the given type-I elementary row operation on A by taking the product EA.

EXAMPLE 254

Form E from I_4 by multiplying row 4 by $\sqrt{3}$:

$$E = \begin{bmatrix} 1 & 0 & 0 & 0 \\ 0 & 1 & 0 & 0 \\ 0 & 0 & 1 & 0 \\ 0 & 0 & 0 & \sqrt{3} \end{bmatrix}.$$

With A as above, the student can check that

$$EA = \begin{bmatrix} -2 & 1 & 6 \\ 1 & 1 & 2 \\ 0 & 1 & 3 \\ 2\sqrt{3} & -3\sqrt{3} & 4\sqrt{3} \end{bmatrix},$$

which is the matrix A after row 4 has been multiplied by $\sqrt{3}$.

EXAMPLE 255

Form E from A by adding $\sqrt{5}$(row 3) to row 2:

$$E = \begin{bmatrix} 1 & 0 & 0 & 0 \\ 0 & 1 & \sqrt{5} & 0 \\ 0 & 0 & 1 & 0 \\ 0 & 0 & 0 & 1 \end{bmatrix}.$$

With A as above, we find that

$$EA = \begin{bmatrix} -2 & 1 & 6 \\ 1 & 1+\sqrt{5} & 2+3\sqrt{5} \\ 0 & 1 & 3 \\ 2 & -3 & 4 \end{bmatrix}.$$

This is the matrix we would get from A by adding $\sqrt{5}$(row 3) to row 2.

To reiterate: *Any elementary row operation on $A(n \times m)$ can be achieved by multiplying A on the left by the elementary matrix formed by performing the row operation on I_n.*

In fact, we can achieve any sequence of elementary row operations by a product of elementary matrices. Here is an example showing how to do this.

EXAMPLE 256

Suppose that we want to perform the following sequence of elementary row operations to produce a matrix B:

$$A = \begin{bmatrix} 2 & 1 & 0 \\ 0 & 1 & 2 \\ -1 & 3 & 2 \end{bmatrix} \xrightarrow[\text{rows 1 and 2 of } A]{\text{operation } S_1\text{: interchange}} \begin{bmatrix} 0 & 1 & 2 \\ 2 & 1 & 0 \\ -1 & 3 & 2 \end{bmatrix}$$

$$\xrightarrow[\text{row 3 by 2}]{\text{operation } S_2\text{: multiply}} \begin{bmatrix} 0 & 1 & 2 \\ 2 & 1 & 0 \\ -2 & 6 & 4 \end{bmatrix}$$

$$\xrightarrow[\text{2(row 1) to row 3}]{\text{operation } S_3\text{: add}} \begin{bmatrix} 0 & 1 & 2 \\ 2 & 1 & 0 \\ -2 & 8 & 8 \end{bmatrix} = B.$$

Now, each operation can be performed by an elementary matrix:

$$S_1 \text{ is performed by } E_1 = \begin{bmatrix} 0 & 1 & 0 \\ 1 & 0 & 0 \\ 0 & 0 & 1 \end{bmatrix} \text{ (perform } S_1 \text{ on } I_3\text{);}$$

$$S_2 \text{ is performed by } E_2 = \begin{bmatrix} 1 & 0 & 0 \\ 0 & 1 & 0 \\ 0 & 0 & 2 \end{bmatrix} \text{ (perform } S_2 \text{ on } I_3\text{);}$$

$$S_3 \text{ is performed by } E_3 = \begin{bmatrix} 1 & 0 & 0 \\ 0 & 1 & 0 \\ 2 & 0 & 1 \end{bmatrix} \text{ (perform } S_3 \text{ on } I_3\text{).}$$

Now observe that the end result is

$$B = \begin{bmatrix} 0 & 1 & 2 \\ 2 & 1 & 0 \\ -2 & 8 & 8 \end{bmatrix} = \begin{bmatrix} 1 & 0 & 0 \\ 0 & 1 & 0 \\ 2 & 0 & 1 \end{bmatrix}\begin{bmatrix} 1 & 0 & 0 \\ 0 & 1 & 0 \\ 0 & 0 & 2 \end{bmatrix}\begin{bmatrix} 0 & 1 & 0 \\ 1 & 0 & 0 \\ 0 & 0 & 1 \end{bmatrix}\begin{bmatrix} 2 & 1 & 0 \\ 0 & 1 & 2 \\ -1 & 3 & 2 \end{bmatrix}.$$

In the product on the right, we have

$$B = E_3 E_2 E_1 A,$$

where $E_1 A$ is the result of performing S_1 on A, $E_2(E_1 A)$ the result of performing S_2 on $E_1 A$, and $E_3(E_2 E_1 A)$ the final result of performing S_3 on $E_2 E_1 A$ to form B.

One must be careful in this to multiply the elementary matrices on the left of A in the proper order. Changing the order of the product changes the order of the operations performed, and may yield a different result.

If we obtain B from A by a sequence of elementary row operations, then we say that A is *row-equivalent* to B. In view of the remarks above, this means that $B = E_k E_{k-1} \cdots E_1 A$ for some elementary matrices $E_1, \ldots, E_k$ (which perform the row operations producing B from A). Row-equivalence of matrices has the following three properties:

THEOREM 44

1. Every matrix is row-equivalent to itself.
2. If A is row-equivalent to B, then B is row-equivalent to A.
3. If A is row-equivalent to B, and B to C, then A is row-equivalent to C.

To prove (1), note that A can always be obtained from A by an elementary row operation (e.g., multiply row 1 by 1).

For (3), if B is obtained from A by a sequence of elementary row operations, and then C is obtained from B by more elementary row operations, then we have in fact obtained C from A by elementary row operations.

A proof of (2) involves recognizing that we can undo any elementary row by an elementary row operation of the same type. For example, if we get B from A by multiplying row i by α, we can get A from B by multiplying row i by $1/\alpha$. Or, if we interchange rows i and j of A to get B, we can interchange rows i and j of B to get A. Or, if we obtain B from A by adding α(row i) to row j of A, then we obtain A from B by adding $-\alpha$(row i) to row j of B. Thus, however we get B from A by elementary row operations, we can start with B, unravel the operations, and come out with A again.

In view of this result, we shall just say that A and B are row-equivalent if A is row-equivalent to B (and hence also B to A). As an example, we have from Example 256 that

$$\begin{bmatrix} 2 & 1 & 0 \\ 0 & 1 & 2 \\ -1 & 3 & 2 \end{bmatrix} \text{ and } \begin{bmatrix} 0 & 1 & 2 \\ 2 & 1 & 0 \\ -2 & 8 & 8 \end{bmatrix}$$

are row-equivalent.

In the next section we shall show that every matrix is row-equivalent to a matrix having a form particularly convenient for later considerations, such as solving systems of linear equations.

PROBLEMS FOR SECTION 10.4

In each of Problems 1 through 15, perform the indicated elementary row operation on A to produce B. Then find an elementary matrix E such that $EA = B$.

1. $A = \begin{bmatrix} -2 & 1 & 4 & 2 \\ 0 & 1 & 16 & 3 \\ 1 & -2 & 4 & 8 \end{bmatrix}$; multiply row 2 by $\sqrt{3}$.

2. $A = \begin{bmatrix} 2 & 16 & -4 \\ 8 & 1 & 7 \\ 2 & 2 & 3 \end{bmatrix}$; add $\sqrt{3}$(row 3) to row 1.

3. $A = \begin{bmatrix} -2 & 1 & 1 & 7 & 13 \\ 3 & 0 & -1 & 14 & 2 \end{bmatrix}$; add 5(row 1) to row 2.

4. $A = \begin{bmatrix} 18 & 2 \\ 4 & 6 \\ -3 & 10 \end{bmatrix}$; multiply row 3 by $\sqrt{7}$.

5. $A = \begin{bmatrix} 2 & 2 & -6 \\ 1 & 0 & 3 \\ -2 & 4 & 3 \end{bmatrix}$; interchange rows 2 and 3.

6. $A = \begin{bmatrix} -3 & 5 & -7 & -8 & 5 \\ 14 & 5 & 1 & 3 & -6 \\ 2 & 0 & 4 & 3 & 9 \\ -5 & 4 & 8 & -2 & 4 \end{bmatrix}$; add 13(row 2) to row 3.

7. $A = \begin{bmatrix} -9 & 0 & 4 \\ 3 & 5 & 2 \\ 14 & 4 & 4 \end{bmatrix}$; multiply row 1 by 5.

8. $A = \begin{bmatrix} 14 & 6 & 3 & -7 \\ 3 & 2 & 7 & 0 \\ 2 & 2 & -4 & 5 \end{bmatrix}$; add 3(row 1) to row 2.

9. $A = \begin{bmatrix} 22 & 13 & -9 \\ 3 & 5 & 12 \\ -6 & 8 & 0 \end{bmatrix}$; interchange rows 1 and 3.

10. $A = \begin{bmatrix} 14 & 2 & -3 & 7 \\ 8 & 5 & 12 & -3 \end{bmatrix}$; add 3(row 2) to row 1.

11. $A = \begin{bmatrix} -2 & 3 & -6 & 3 & 1 \\ 0 & -5 & 3 & -5 & 9 \end{bmatrix}$; multiply row 2 by $\sqrt{3}$.

12. $A = \begin{bmatrix} -2 & 3 & 1 \\ 0 & 1 & 5 \\ 6 & -4 & 2 \end{bmatrix}$; multiply row 3 by $\sqrt{5}$.

13. $A = \begin{bmatrix} 0 & -2 & 7 & 1 & 3 \\ 0 & -1 & -6 & 3 & 7 \\ 1 & 4 & -3 & -2 & 4 \end{bmatrix}$; interchange rows 1 and 3.

14. $A = \begin{bmatrix} 1 & -3 & 4 & 1 & 8 \\ -2 & 7 & 3 & 8 & 8 \\ 2 & -5 & 7 & 2 & 0 \end{bmatrix}$; add $\sqrt{7}$(row 1) to row 2.

15. $A = \begin{bmatrix} 2 \\ 1 \\ 0 \\ -2 \\ 7 \end{bmatrix}$; add $\sqrt{5}$(row 3) to row 1.

In each of Problems 16 through 25, produce a matrix B from the given matrix A by performing the sequence of operations $S_1, \ldots, S_k$ (k may vary with each problem). Produce a matrix C such that $CA = B$. (*Hint:* C may be found as a product of elementary matrices—see Example 256 and the discussion following it.)

16.
$$A = \begin{bmatrix} -2 & 14 & 6 \\ 8 & 1 & -3 \\ 2 & 9 & 5 \end{bmatrix}$$

S_1: add $\sqrt{13}$(row 3) to row 1; S_2: interchange rows 2 and 3; S_3: multiply row 1 by 5.

17.
$$A = \begin{bmatrix} -4 & 6 & -3 \\ 12 & 4 & -4 \\ 1 & 3 & 0 \end{bmatrix}$$

S_1: interchange rows 2 and 3; S_2: add $-$(row 1) to row 2.

18.
$$A = \begin{bmatrix} -3 & 15 \\ 2 & 8 \end{bmatrix}$$

S_1: add $\sqrt{3}$(row 2) to row 1; S_2: multiply row 2 by 15; S_3: interchange rows 1 and 2.

19.
$$A = \begin{bmatrix} 3 & -4 & 5 & 9 \\ 2 & 1 & 3 & -6 \\ 1 & 13 & 2 & 6 \end{bmatrix}$$

S_1: add row 1 to row 3; S_2: add $\sqrt{3}$(row 1) to row 2; S_3: multiply row 3 by 4; S_4: add row 2 to row 3.

20.
$$A = \begin{bmatrix} -1 & 0 & 3 & 0 \\ 1 & 3 & 2 & 9 \\ -9 & 7 & -5 & 7 \end{bmatrix}$$

S_1: multiply row 3 by 4; S_2: add 14(row 1) to row 2; S_3: interchange rows 3 and 2.

21.
$$A = \begin{bmatrix} 0 & -9 & 14 \\ 1 & 5 & 2 \\ 9 & 15 & 0 \end{bmatrix}$$

S_1: interchange rows 2 and 3; S_2: add 3(row 2) to row 3; S_3: interchange rows 1 and 3; S_4: multiply row 3 by 5.

22.
$$A = \begin{bmatrix} -3 & 7 & 1 & 1 \\ 0 & 3 & 3 & -5 \\ 2 & 1 & -5 & 3 \end{bmatrix}$$

S_1: add 2(row 1) to row 3; S_2: multiply row 3 by -5; S_3: interchange rows 2 and 3.

23.
$$A = \begin{bmatrix} 2 & -6 & 5 & 8 \\ 0 & 1 & -3 & 5 \\ 0 & -4 & 2 & -6 \\ 1 & 7 & 3 & -3 \end{bmatrix}$$

S_1: multiply row 4 by -5; S_2: add $\sqrt{3}$(row 4) to row 1;
S_3: interchange rows 1 and 3; S_4: multiply row 3 by -2.

24.
$$A = \begin{bmatrix} 2 & -3 & 1 \\ 0 & 0 & 0 \\ 1 & -5 & 0 \end{bmatrix}$$

S_1: interchange rows 1 and 2; S_2: multiply row 2 by 5;
S_3: add -3(row 3) to row 1.

25.
$$A = \begin{bmatrix} -5 & 1 & -4 \\ 0 & 3 & -2 \\ 0 & 2 & 2 \end{bmatrix}$$

S_1: add 4(row 2) to row 3; S_2: interchange rows 1 and 3;
S_3: multiply row 1 by -2.

26. Suppose that B is obtained from A by a type-I elementary row operation. Form E from I_n by performing the same operation on I_n, where n = number of rows of A. Give a rigorous proof that $EA = B$.

27. Suppose that B is obtained from A by a type-II elementary row operation. Form E from I_n by performing the same operation on I_n, where n = number of rows of A. Give a rigorous proof that $EA = B$.

28. Suppose that B is obtained from A by a type-III elementary row operation. Form E from I_n by performing the same operation on I_n, where n = number of rows of A. Give a rigorous proof that $EA = B$.

29. Define the elementary column operations of types I, II, and III by substituting the word *column* for *row* in the corresponding definitions of the row operations. Prove that each elementary column operation on A can be achieved as a matrix product AE, where E is obtained from I_m (m = number of columns of A) by performing the column operation on I_m.

10.5 REDUCED FORM OF A MATRIX

As we shall see throughout the rest of this chapter, various matrix operations and applications are simplified if the matrix we are working with has a special appearance. A matrix having this appearance is called a *reduced matrix*, a term we define as follows:

Let A be any $n \times m$ matrix. Then A is a *reduced matrix* (or is in *reduced form*) if the following conditions hold:

1. The first nonzero entry of each row is 1.
2. If row r has its first nonzero entry in column c, then all other entries of column c are zero.

3. Each row having all zero entries (if there is such a row) lies below any row having a nonzero entry.
4. If the first nonzero entry in row r_1 lies in column c_1, and the first nonzero entry of row r_2 is in column c_2, and if $r_1 < r_2$, then $c_1 < c_2$.

As a matter of terminology, we often refer to a row's first nonzero entry (if it has one) as its *leading entry*. Then (1) says that the leading entry of any row (if it has one) must be 1; (2) says that all entries directly above and below any leading entry are zero; and (4) says that the leading entries move down and to the right as you look at A.

EXAMPLE 257

The following four matrices are all in reduced form:

$$\begin{bmatrix} 0 & 1 & 2 & 0 & 0 \\ 0 & 0 & 0 & 1 & 0 \\ 0 & 0 & 0 & 0 & 0 \\ 0 & 0 & 0 & 0 & 0 \\ 0 & 0 & 0 & 0 & 0 \end{bmatrix}, \quad \begin{bmatrix} 1 & 0 & 0 & 3 & 1 \\ 0 & 1 & 0 & -2 & 4 \\ 0 & 0 & 1 & 0 & 1 \\ 0 & 0 & 0 & 0 & 0 \end{bmatrix},$$

$$\begin{bmatrix} 0 & 1 & 3 & 2 & 0 \\ 0 & 0 & 0 & 0 & 1 \\ 0 & 0 & 0 & 0 & 0 \end{bmatrix}, \quad \text{and} \quad \begin{bmatrix} 1 & -4 & 1 & 0 \\ 0 & 0 & 0 & 1 \end{bmatrix}.$$

But

$$\begin{bmatrix} 0 & 1 & 2 & 0 & 0 \\ 0 & 0 & 1 & 0 & 0 \\ 0 & 0 & 0 & 1 & 0 \\ 0 & 0 & 0 & 0 & 1 \end{bmatrix}$$

is not reduced, as row 2 has its leading entry in column 3, and this column has another nonzero entry, violating (2) of the definition. Notice that this matrix is row-equivalent to a reduced matrix [add -2(row 2) to row 1].

Similarly,

$$\begin{bmatrix} 0 & 1 & 1 & 0 & 0 \\ 1 & 0 & 0 & 0 & 0 \\ 0 & 0 & 0 & 0 & 0 \end{bmatrix}$$

is not reduced, as (4) of the definition is violated. Again, this is row-equivalent to a reduced matrix (interchange rows 1 and 2).

Finally,

$$\begin{bmatrix} 2 & 0 & 0 \\ 0 & 1 & 0 \\ 1 & 0 & 1 \end{bmatrix}$$

is not reduced. The leading entry of row 1 is 2, violating (1) of the definition, and there is a nonzero entry in column 1 below row 1, violating (2). Note that we can multiply row 1 by $\frac{1}{2}$, then add $-$(row 1) to row 3 to obtain the reduced matrix

$$\begin{bmatrix} 1 & 0 & 0 \\ 0 & 1 & 0 \\ 0 & 0 & 1 \end{bmatrix}.$$

In each of these examples which was not in reduced form, we were able by one or more elementary row operations to transform the matrix to one in reduced form. Thus, each of the matrices in the example was either reduced, or row-equivalent to a reduced matrix. We shall now show that in fact *every* matrix is row-equivalent to a reduced matrix.

THEOREM 45

Every matrix is row-equivalent to a reduced matrix.

Proof Let A be any matrix. We shall outline a sequence of elementary row operations culminating in a reduced matrix.

Suppose that A is not itself reduced (otherwise we have nothing to do). Reading from left to right, locate the first column, c_1, having a nonzero entry. Let α be the top nonzero entry of this column, occurring, say, in row r_1. Multiply row r_1 by $1/\alpha$, producing a matrix B. By choice of r_1, column c_1 of B has only zeros above the 1 in row r_1. If any row below r_1 has a nonzero entry β in column c_1, add $-\beta(\text{row } r_1)$ to this row. Repetition of this ends in a matrix C having zeros above and below row r_1 in column c_1. Now interchange rows 1 and r_1 of C to produce D, having leading entry 1 in row 1 and column c_1, and zeros below this 1. Further, by choice of c_1, any column of D preceding column c_1 has only zero entries.

If D is reduced, stop. If not, locate the first column, say c_2, to the right of column c_1 of D and having a nonzero entry below row 1. Let r_2 be the first row below row 1 having a nonzero entry, γ, in column c_2. Divide row r_2 of D by γ to obtain a matrix E with 1 as its r_2-c_2 entry. If column c_2 of E has a nonzero entry δ above or below row r_2, add $-\delta(\text{row } r_2)$ to this row, producing a matrix F with zeros in column c_2 above and below the 1 in row r_2. Finally, interchange rows 2 and r_2 of F to form G.

If G is reduced, stop. If not, locate the first column to the right of column c_2 and having a nonzero entry below row 2, and repeat the process above. Since A has only finitely many columns, eventually we must obtain a reduced matrix. Further, we have used only elementary row operations; hence, this reduced matrix is row-equivalent to A.

The process of obtaining a reduced matrix row-equivalent to a given matrix A is called *reducing A*. Here are three examples in which we follow essentially the method of the proof of Theorem 45 to reduce the given matrix.

EXAMPLE 258

Let

$$A = \begin{bmatrix} -2 & 1 & 3 \\ 0 & 1 & 1 \\ 2 & 0 & 1 \end{bmatrix}.$$

To reduce A, begin with column 1, which has a nonzero entry in row 1, and proceed as follows:

$$A \xrightarrow[\text{by } -\frac{1}{2}]{\text{multiply row 1}} \begin{bmatrix} 1 & -\frac{1}{2} & -\frac{3}{2} \\ 0 & 1 & 1 \\ 2 & 0 & 1 \end{bmatrix} \xrightarrow[\text{to row 3}]{\text{add } -2(\text{row 1})} \begin{bmatrix} 1 & -\frac{1}{2} & -\frac{3}{2} \\ 0 & 1 & 1 \\ 0 & 1 & 4 \end{bmatrix}.$$

Now column 2 of the last matrix has a nonzero entry below row 1, the highest in the matrix being the 2-2 entry. It is already 1. We now want zeros above and below this entry. Add $\frac{1}{2}$(row 2) to row 1 and $-$(row 2) to row 3 to obtain

$$\begin{bmatrix} 1 & 0 & -1 \\ 0 & 1 & 1 \\ 0 & 0 & 3 \end{bmatrix}.$$

Multiply row 3 by $\frac{1}{3}$ to obtain

$$\begin{bmatrix} 1 & 0 & -1 \\ 0 & 1 & 1 \\ 0 & 0 & 1 \end{bmatrix};$$

then add row 3 to row 1, and $-$(row 3) to row 2, giving us the reduced matrix

$$\begin{bmatrix} 1 & 0 & 0 \\ 0 & 1 & 0 \\ 0 & 0 & 1 \end{bmatrix}.$$

This reduced matrix is row-equivalent to A.

EXAMPLE 259

Let

$$B = \begin{bmatrix} 0 & 0 & 0 & 0 & 0 \\ 0 & 0 & 2 & 0 & 0 \\ 0 & 1 & 0 & 1 & 1 \\ 0 & 4 & 3 & 4 & 0 \end{bmatrix}.$$

The first column having a nonzero entry is the second, and its first nonzero entry is 1 in the 3-2 place. Add -4(row 3) to row 4 to obtain

$$\begin{bmatrix} 0 & 0 & 0 & 0 & 0 \\ 0 & 0 & 2 & 0 & 0 \\ 0 & 1 & 0 & 1 & 1 \\ 0 & 0 & 3 & 0 & -4 \end{bmatrix};$$

then interchange rows 3 and 1, giving us

$$\begin{bmatrix} 0 & 1 & 0 & 1 & 1 \\ 0 & 0 & 2 & 0 & 0 \\ 0 & 0 & 0 & 0 & 0 \\ 0 & 0 & 3 & 0 & -4 \end{bmatrix}.$$

Now multiply row 2 by $\frac{1}{2}$ to obtain

$$\begin{bmatrix} 0 & 1 & 0 & 1 & 1 \\ 0 & 0 & 1 & 0 & 0 \\ 0 & 0 & 0 & 0 & 0 \\ 0 & 0 & 3 & 0 & -4 \end{bmatrix},$$

and add -3(row 2) to row 4, giving

$$\begin{bmatrix} 0 & 1 & 0 & 1 & 1 \\ 0 & 0 & 1 & 0 & 0 \\ 0 & 0 & 0 & 0 & 0 \\ 0 & 0 & 0 & 0 & -4 \end{bmatrix}.$$

Since column 4 of this matrix has only zeros below row 2, we move to column 5. Multiply row 4 by $-\frac{1}{4}$, and then add $-$(row 4) to row 1 and interchange rows 3 and 4 to obtain

$$\begin{bmatrix} 0 & 1 & 0 & 1 & 0 \\ 0 & 0 & 1 & 0 & 0 \\ 0 & 0 & 0 & 0 & 1 \\ 0 & 0 & 0 & 0 & 0 \end{bmatrix},$$

a reduced matrix row-equivalent to B.

EXAMPLE 260

Let

$$D = \begin{bmatrix} 0 & 1 & 1 & 0 \\ 1 & 0 & 2 & 0 \\ 2 & 0 & 1 & 0 \end{bmatrix}.$$

We proceed:

$$D \xrightarrow[\text{to row 3}]{\text{add } -2(\text{row 2})} \begin{bmatrix} 0 & 1 & 1 & 0 \\ 1 & 0 & 2 & 0 \\ 0 & 0 & -3 & 0 \end{bmatrix} \xrightarrow[\text{rows 1 and 2}]{\text{interchange}} \begin{bmatrix} 1 & 0 & 2 & 0 \\ 0 & 1 & 1 & 0 \\ 0 & 0 & -3 & 0 \end{bmatrix}$$

$$\xrightarrow[\text{by } -\frac{1}{3}]{\text{multiply row 3}} \begin{bmatrix} 1 & 0 & 2 & 0 \\ 0 & 1 & 1 & 0 \\ 0 & 0 & 1 & 0 \end{bmatrix} \xrightarrow[\text{to row 2}]{\text{add } -2(\text{row 3}) \text{ to row 1 and } -(\text{row 3})} \begin{bmatrix} 1 & 0 & 0 & 0 \\ 0 & 1 & 0 & 0 \\ 0 & 0 & 1 & 0 \end{bmatrix}.$$

This is a reduced matrix row-equivalent to D.

Beginning with a matrix A, it is usually possible to arrive at a row-equivalent reduced matrix by many different sequences of elementary row operations. For example, letting A be as in Example 258, we could proceed

$$A = \begin{bmatrix} -2 & 1 & 3 \\ 0 & 1 & 1 \\ 2 & 0 & 1 \end{bmatrix} \xrightarrow[\text{to row 1}]{\text{add row 3}} \begin{bmatrix} 0 & 1 & 4 \\ 0 & 1 & 1 \\ 2 & 0 & 1 \end{bmatrix}$$

$$\xrightarrow[\text{to row 1}]{\text{add } -(\text{row 2})} \begin{bmatrix} 0 & 0 & 3 \\ 0 & 1 & 1 \\ 2 & 0 & 1 \end{bmatrix} \xrightarrow[\text{row 1 by } \frac{1}{3}]{\text{multiply}} \begin{bmatrix} 0 & 0 & 1 \\ 0 & 1 & 1 \\ 2 & 0 & 1 \end{bmatrix}$$

$$\xrightarrow[\text{rows 2 and 3}]{\text{add } -(\text{row 1}) \text{ to}} \begin{bmatrix} 0 & 0 & 1 \\ 0 & 1 & 0 \\ 2 & 0 & 0 \end{bmatrix} \xrightarrow[\text{rows 1 and 3}]{\text{interchange}} \begin{bmatrix} 2 & 0 & 0 \\ 0 & 1 & 0 \\ 0 & 0 & 1 \end{bmatrix}$$

$$\xrightarrow[\text{row 1 by } \frac{1}{2}]{\text{multiply}} \begin{bmatrix} 1 & 0 & 0 \\ 0 & 1 & 0 \\ 0 & 0 & 1 \end{bmatrix},$$

which is a reduced matrix. Note that, even though we went by a different route than in Example 258, we concluded with the same reduced matrix. This will always happen, a fact we record as Theorem 46.

THEOREM 46

If A' and A'' are reduced matrices, both row-equivalent to A, then $A' = A''$.

We leave a proof of this to the student (see Problem 27).

The significance of Theorem 46 is that each matrix A has *exactly one* reduced matrix row-equivalent to it. We denote this reduced matrix A_R. Thus, in Example 258,

$$A = \begin{bmatrix} -2 & 1 & 3 \\ 0 & 1 & 1 \\ 2 & 0 & 1 \end{bmatrix} \quad \text{and} \quad A_R = \begin{bmatrix} 1 & 0 & 0 \\ 0 & 1 & 0 \\ 0 & 0 & 1 \end{bmatrix}.$$

In Example 259,

$$B = \begin{bmatrix} 0 & 0 & 0 & 0 & 0 \\ 0 & 0 & 2 & 0 & 0 \\ 0 & 1 & 0 & 1 & 1 \\ 0 & 4 & 3 & 4 & 0 \end{bmatrix} \quad \text{and} \quad B_R = \begin{bmatrix} 0 & 1 & 0 & 1 & 0 \\ 0 & 0 & 1 & 0 & 0 \\ 0 & 0 & 0 & 0 & 1 \\ 0 & 0 & 0 & 0 & 0 \end{bmatrix}.$$

In the next section we shall use the reduced form of a matrix to develop the important notion of *rank* of a matrix.

PROBLEMS FOR SECTION 10.5

In each of Problems 1 through 25, determine whether or not the matrix is in reduced form. If it is not, list all the requirements of the definition which are violated, and produce a reduced matrix row-equivalent to the given matrix by using elementary row operations.

1. $\begin{bmatrix} 1 & -1 & 3 \\ 0 & 1 & 2 \\ 0 & 0 & 0 \end{bmatrix}$

2. $\begin{bmatrix} 3 & 1 & 1 & 4 \\ 0 & 1 & 0 & 0 \end{bmatrix}$

3. $\begin{bmatrix} -1 & 4 & 1 & 1 \\ 0 & 0 & 0 & 0 \\ 0 & 0 & 0 & 0 \\ 0 & 0 & 0 & 1 \end{bmatrix}$

4. $\begin{bmatrix} 1 & 0 & 1 & 1 & -1 \\ 0 & 1 & 0 & 0 & 2 \end{bmatrix}$

5. $\begin{bmatrix} 6 & 1 \\ 0 & 0 \\ 1 & 3 \\ 0 & 1 \end{bmatrix}$

6. $\begin{bmatrix} 2 & 2 \\ 1 & 1 \end{bmatrix}$

7. $\begin{bmatrix} -1 & 4 & 6 \\ 2 & 3 & -5 \\ 7 & 1 & 1 \end{bmatrix}$

8. $\begin{bmatrix} -3 & 4 & 4 \\ 0 & 0 & 0 \end{bmatrix}$

9. $\begin{bmatrix} -1 & 2 & 3 & 1 \\ 1 & 0 & 0 & 0 \end{bmatrix}$

10. $\begin{bmatrix} 8 & 2 & 1 & 0 \\ 0 & 1 & 1 & 3 \\ 4 & 0 & 0 & -3 \end{bmatrix}$

11. $\begin{bmatrix} 4 & 1 & -7 \\ 2 & 2 & 0 \\ 0 & 1 & 0 \end{bmatrix}$

12. $\begin{bmatrix} 0 & 0 & 1 \\ 1 & 0 & 0 \\ 0 & 0 & 1 \end{bmatrix}$

13. $\begin{bmatrix} -5 & 1 & 0 & 0 \\ 2 & 0 & 0 & 0 \\ 0 & 1 & 1 & -1 \end{bmatrix}$

14. $\begin{bmatrix} 6 \\ -3 \\ 1 \\ 1 \end{bmatrix}$

15. $\begin{bmatrix} 0 & 0 & 2 & 1 & -1 \\ 1 & -1 & 3 & 0 & 0 \end{bmatrix}$

16. $\begin{bmatrix} 0 & 1 & 2 & 0 & 0 \\ 0 & 0 & 0 & 0 & 1 \end{bmatrix}$

17. $\begin{bmatrix} 1 & 0 & -4 & 0 & 6 \\ 5 & 1 & -3 & -3 & 9 \\ 6 & 3 & 7 & -3 & 1 \end{bmatrix}$

18. $\begin{bmatrix} 0 & -1 & 5 & 2 \\ 1 & 1 & -5 & 2 \\ -5 & 3 & 7 & 3 \\ 0 & 2 & 7 & 0 \end{bmatrix}$

19. $\begin{bmatrix} 1 & -5 & 3 & 0 & 8 \\ -10 & 3 & 7 & 3 & 6 \end{bmatrix}$

20. $\begin{bmatrix} -4 & 0 & 1 & 1 & -5 \end{bmatrix}$

21. $\begin{bmatrix} -12 & 9 & 1 & -2 & 4 \\ 5 & 9 & 1 & 7 & 0 \end{bmatrix}$

22. $\begin{bmatrix} 5 & -2 & 3 \\ 0 & 1 & -6 \\ -3 & 5 & 11 \end{bmatrix}$

23. $\begin{bmatrix} -3 & 6 & 1 \\ 0 & -6 & 4 \\ 1 & -1 & 7 \\ 9 & -6 & 4 \end{bmatrix}$

24. $\begin{bmatrix} -2 & 3 & 8 & 5 \\ 1 & -5 & 3 & 3 \end{bmatrix}$

25. $\begin{bmatrix} 5 & -2 & 1 & 5 \\ 0 & 3 & 3 & -7 \\ 7 & -4 & 1 & 5 \\ 9 & 5 & 3 & -8 \end{bmatrix}$

26. Prove the following theorem: For any matrix A, there is a matrix B such that $BA = A_R$. (*Hint:* Combine Theorem 45 with the idea contained in Example 256.)

27. Prove Theorem 46.

10.6 RANK OF A MATRIX

The number of nonzero rows of A_R will turn out to be a significant number in many applications. It is called the *rank of* A, denoted rank(A). In Example 258, A was 3×3, and rank(A) = 3; in Example 259, B was 4×5, and rank(B) = 3. Clearly, the rank of a matrix never exceeds the number of rows of the matrix.

The following somewhat technical lemma will be used when we consider matrix inverses and solutions of systems of equations.

LEMMA 1

If A is $n \times n$, then rank $(A) = n$ if and only if $A_R = I_n$.

Proof If $A_R = I_n$, then A_R has n nonzero rows; hence, rank(A) = n.

Conversely, suppose that rank(A) = n. Then A_R has exactly n nonzero rows; hence, no zero rows. By (1) of the definition of reduced form, every row of A_R has leading entry 1. By (4) of the definition, A_R has 1's down its main diagonal. By (2) of the definition, all entries of column j above and below the 1 on the main diagonal are zero, for $j = 1, \ldots, n$. Thus, $A_R = I_n$.

Rank can also be defined in terms of the vector space concepts developed in the previous chapter. To understand how this is done, consider the example

$$A = \begin{bmatrix} 1 & -1 & 4 & 2 \\ 0 & 1 & 3 & 2 \\ 3 & -2 & 15 & 8 \end{bmatrix}.$$

The rows may be thought of as vectors in R^4, namely

$$(1, -1, 4, 2), \qquad (0, 1, 3, 2) \quad \text{and} \quad (3, -2, 15, 8).$$

The set of all linear combinations

$$\alpha(1, -1, 4, 2) + \beta(0, 1, 3, 2) + \gamma(3, -2, 15, 8)$$

is a subspace of R^4, and is called the *row space* of A. In fact, every vector in this row space is a linear combination of just $(1, -1, 4, 2)$ and $(0, 1, 3, 2)$, since

$$(3, -2, 15, 8) = 3(1, -1, 4, 2) + (0, 1, 3, 2).$$

Further, $(1, -1, 4, 2)$ and $(0, 1, 3, 2)$ are linearly independent, and hence form a basis for the row space. Thus, the row space of A has dimension 2.

A routine calculation gives us

$$A_R = \begin{bmatrix} 1 & 0 & 7 & 4 \\ 0 & 1 & 3 & 2 \\ 0 & 0 & 0 & 0 \end{bmatrix};$$

hence, $\text{rank}(A) = 2$. At least for this example, then,

$$\text{rank}(A) = \text{dimension of the row space of } A.$$

This example illustrates ideas which hold in general. If A is $n \times m$, then the n rows of A may be thought of as vectors in R^m. The linear combinations of these n rows form a subspace of R^m, called the *row space of A*.

THEOREM 47

For any matrix A, the rank of A equals the dimension of the row space of A.

Proof Observe first the correspondence between elementary row operations on A, and vector operations on the rows of A thought of as vectors in R^m (if A is $n \times m$). Say these row vectors are $\mathbf{F}_1, \ldots, \mathbf{F}_n$.

Type I: Interchange two rows of A.

This simply means interchange $\mathbf{F}_i$ and $\mathbf{F}_j$ in the list of row vectors.

Type II: Multiply row i by a nonzero scalar α.

This corresponds to replacing $\mathbf{F}_i$ by $\alpha\mathbf{F}_i$.

Type III: Add α(row i) to row j.

This corresponds to replacing $\mathbf{F}_j$ by $\alpha\mathbf{F}_i + \mathbf{F}_j$.

Now look at the row space of A. It is the collection of all vectors in R^m of the form

$$\alpha_1\mathbf{F}_1 + \alpha_2\mathbf{F}_2 + \cdots + \alpha_m\mathbf{F}_m,$$

where $\alpha_1, \ldots, \alpha_m$ can be any scalars.

Now, type-I and -II operations do not really exert any influence on a sum $\alpha_1\mathbf{F}_1 + \cdots + \alpha_m\mathbf{F}_m$; hence on the dimension of the row space.

In reducing A, we may obtain a row of zeros through type-III operations. This corresponds to finding that some $\mathbf{F}_j$ is a linear combination of the other $\mathbf{F}_i$'s. If, say, $\mathbf{F}_n$ is a linear combination of $\mathbf{F}_1, \ldots, \mathbf{F}_{n-1}$, then the linear combinations of $\mathbf{F}_1, \ldots, \mathbf{F}_n$ are really linear combinations of just $\mathbf{F}_1, \ldots, \mathbf{F}_{n-1}$. In this way, each zero row of A_R corresponds to omitting some $\mathbf{F}_j$ from the list of row vectors without changing the row space. Eventually we get one $\mathbf{F}_i$ for each nonzero row of A_R, and these $\mathbf{F}_i$'s are linearly independent. Thus, the dimension of the row space is the number of nonzero rows of A_R.

Rank will play an important role in solving systems of linear equations in the following two sections. We conclude this section with one additional illustration of Theorem 47.

EXAMPLE 261

Let

$$A = \begin{bmatrix} -1 & 4 & 0 & 1 & 6 \\ -2 & 8 & 0 & 2 & 12 \end{bmatrix}.$$

The row space is the subspace of R^5 consisting of all linear combinations

$$\alpha(-1, 4, 0, 1, 6) + \beta(-2, 8, 0, 2, 12).$$

We can reduce A as follows:

$$A \xrightarrow[\text{to row 2}]{\text{add } -2(\text{row 1})} \begin{bmatrix} -1 & 4 & 0 & 1 & 6 \\ 0 & 0 & 0 & 0 & 0 \end{bmatrix}$$

$$\xrightarrow[\text{row 1 by } -1]{\text{multiply}} \begin{bmatrix} 1 & -4 & 0 & -1 & -6 \\ 0 & 0 & 0 & 0 & 0 \end{bmatrix}.$$

Now, the row vectors of A are

$$\mathbf{F}_1 = (-1, 4, 0, 1, 6) \quad \text{and} \quad \mathbf{F}_2 = (-2, 8, 0, 2, 12).$$

The first row operation on A [add -2(row 1) to row 2] corresponds to replacing $\mathbf{F}_2$ by $\mathbf{F}'_2 = (0, 0, 0, 0, 0)$. Since linear combinations $\alpha\mathbf{F}_1 + \beta\mathbf{F}'_2$ are really just $\alpha\mathbf{F}_1$, we can omit $\mathbf{F}'_2$ from our list of vectors whose linear combinations produce the row space of A. Finally, multiplying row 1 by -1 gives us $\mathbf{F}'_1 = (1, -4, 0, -1, -6)$. This forms a basis for the row space of A (as does just $\mathbf{F}_1$); hence this row space has dimension 1, the rank of A.

PROBLEMS FOR SECTION 10.6

In each of Problems 1 through 20, determine rank(A) first by finding A_R and counting the nonzero rows, and next by finding a basis for the row space of A.

1. $\begin{bmatrix} -4 & 1 & 3 \\ 2 & 2 & 0 \end{bmatrix}$

2. $\begin{bmatrix} 1 & -1 & 4 \\ 0 & 1 & 3 \\ 2 & -1 & 11 \end{bmatrix}$

3. $\begin{bmatrix} -3 & 1 \\ 2 & 2 \\ 4 & -3 \end{bmatrix}$

4. $\begin{bmatrix} 6 & 0 & 0 & 1 & 1 \\ 12 & 0 & 0 & 2 & 2 \\ 1 & -1 & 0 & 0 & 0 \end{bmatrix}$

5. $\begin{bmatrix} 8 & -4 & 3 & 2 \\ 1 & -1 & 1 & 0 \end{bmatrix}$

6. $\begin{bmatrix} 1 & 3 & 0 \\ 0 & 0 & 1 \end{bmatrix}$

7. $\begin{bmatrix} 2 & 2 & 1 \\ 1 & -1 & 3 \\ 0 & 0 & 1 \\ 4 & 0 & 7 \end{bmatrix}$

8. $\begin{bmatrix} 0 & -1 & 0 \\ 0 & 0 & -1 \\ 0 & 0 & 2 \end{bmatrix}$

9. $\begin{bmatrix} 0 & -4 & 3 \\ 6 & 1 & 0 \\ 2 & 2 & 2 \end{bmatrix}$

10. $\begin{bmatrix} 1 & 0 & 0 \\ 2 & 0 & 0 \\ 1 & 0 & -1 \\ 3 & 0 & 0 \end{bmatrix}$

11. $\begin{bmatrix} -3 & 2 & 2 \\ 1 & 0 & 5 \\ 0 & 0 & 2 \end{bmatrix}$

12. $\begin{bmatrix} -4 & -2 & 1 & 6 \\ 0 & 4 & -4 & 2 \\ 1 & 0 & 0 & 0 \end{bmatrix}$

13. $\begin{bmatrix} -2 & 5 & 7 \\ 0 & 1 & -3 \\ -4 & 11 & 11 \end{bmatrix}$

14. $\begin{bmatrix} -3 & 2 & 1 & 1 \\ 6 & -4 & -2 & -2 \end{bmatrix}$

15. $\begin{bmatrix} 7 & -2 & 1 & -2 \\ 0 & 2 & 6 & 3 \\ 7 & 2 & 13 & 4 \\ 7 & 0 & 7 & 1 \end{bmatrix}$

16. $\begin{bmatrix} -4 & 2 & 5 \\ 0 & 0 & 0 \\ 0 & 0 & 0 \end{bmatrix}$

17. $\begin{bmatrix} 4 & 1 & -3 & 5 \\ 2 & 0 & 0 & -2 \\ 13 & 2 & 0 & -1 \end{bmatrix}$

18. $\begin{bmatrix} -4 & 2 & 6 & 1 \\ 0 & 0 & 4 & 1 \\ 4 & -2 & -2 & 0 \end{bmatrix}$

19. $\begin{bmatrix} 5 & -2 & 5 & 6 & 1 \\ -2 & 0 & 1 & -1 & 3 \\ -1 & -2 & 8 & 3 & 10 \end{bmatrix}$

20. $\begin{bmatrix} 3 & -3 & 5 & 1 \\ 0 & 2 & 1 & -5 \\ 0 & 0 & 0 & 1 \end{bmatrix}$

21. Let A be any $n \times m$ matrix. The row space is the subspace of R^m consisting of all linear combinations of the n row vectors. Similarly, we can define the *column space* as the subspace of R^n consisting of all linear combinations of the m column vectors. It is a remarkable fact that the dimension of the row and column spaces of any given matrix are equal. Prove this.

Hint: Let the row vectors of A be $\mathbf{v}_1, \ldots, \mathbf{v}_n$. If the dimension of the row space is r, then exactly r of these vectors are linearly independent. Say, for convenience, that $\mathbf{v}_1, \ldots, \mathbf{v}_r$ are linearly independent, while $\mathbf{v}_{r+1}, \ldots, \mathbf{v}_n$ are all linear combinations of $\mathbf{v}_1, \ldots, \mathbf{v}_n$. Say,

$$\begin{aligned} \mathbf{v}_{r+1} &= a_{r+1,1}\mathbf{v}_1 + \cdots + a_{r+1,r}\mathbf{v}_r, \\ \mathbf{v}_{r+2} &= a_{r+2,1}\mathbf{v}_1 + \cdots + a_{r+2,r}\mathbf{v}_r, \\ &\vdots \\ \mathbf{v}_n &= a_{n1}\mathbf{v}_1 + \cdots + a_{nr}\mathbf{v}_r. \end{aligned}$$

Show from this that column j of A can be written

$$\begin{bmatrix} A_{1j} \\ A_{2j} \\ \vdots \\ A_{nj} \end{bmatrix} = A_{1j}\begin{bmatrix} 1 \\ 0 \\ \vdots \\ 0 \\ a_{r+1,1} \\ \vdots \\ a_{n1} \end{bmatrix} + A_{2j}\begin{bmatrix} 0 \\ 1 \\ 0 \\ \vdots \\ a_{r+1,2} \\ \vdots \\ a_{n2} \end{bmatrix} + \cdots + A_{rj}\begin{bmatrix} 0 \\ 0 \\ \vdots \\ 0 \\ 1 \\ a_{r+1,r} \\ \vdots \\ a_{nr} \end{bmatrix}.$$

Conclude that every column vector of A is a linear combination of r vectors in R^n. Hence, conclude that

$$\text{dimension of the column space} \leq r.$$

By reversing the roles of rows and column spaces above, show that

$$r \leq \text{dimension of the column space}.$$

In each of Problems 22 through 25, find a basis for the row space of A, and also a basis for the column space of A; thus showing that the two spaces have the same dimension.

22. $\begin{bmatrix} -1 & 4 & 2 \\ 0 & 1 & 6 \\ 2 & 2 & 0 \\ 0 & 0 & 1 \end{bmatrix}$

23. $\begin{bmatrix} 1 & 1 & -4 & 2 \\ 0 & 1 & 1 & 3 \end{bmatrix}$

24. $\begin{bmatrix} 1 & -1 & 3 \\ 0 & -1 & 2 \\ 1 & -1 & 3 \end{bmatrix}$

25. $\begin{bmatrix} 8 & 4 \\ 2 & 1 \\ 0 & 3 \end{bmatrix}$

26. Use the result of Problem 21 to prove $\text{rank}(A) = \text{rank}(A^t)$.

27. Show that the set V_R of all $n \times m$ matrices having real entries is a vector space, using the usual addition of matrices and multiplication by scalars. Find a basis for this vector space, and hence find its dimension.

28. Show that the set V_C of all $n \times m$ matrices having complex entries is a vector space, using the usual addition of matrices and multiplication by scalars (here, scalars are complex numbers). Find a basis for this vector space, and hence find its dimension.

10.7 SOLUTION OF SYSTEMS OF LINEAR EQUATIONS: THE HOMOGENEOUS CASE

In this section we shall begin showing how to use matrices to solve systems of linear equations. The theoretical and practical aspects of this problem will occupy this and the next section.

We want to consider a system of linear equations of the form

$$\begin{aligned} a_{11}x_1 + a_{12}x_2 + \cdots + a_{1m}x_m &= b_1, \\ a_{21}x_1 + a_{22}x_2 + \cdots + a_{2m}x_m &= b_2, \\ &\vdots \\ a_{n1}x_1 + a_{n2}x_2 + \cdots + a_{nm}x_m &= b_n. \end{aligned}$$

The term *linear* means that each unknown x_j appears to the first power only, and that there are no cross product terms $x_i x_j$ with $i \neq j$.

We have above a total of n equations in m unknowns. The object is to find *all* values of $x_1, \ldots, x_m$ satisfying all n equations simultaneously, given the numbers $a_{11}, \ldots, a_{1m}, \ldots, a_{n1}, \ldots, a_{nm}, b_1, \ldots, b_n$.

In order to bring matrix machinery to bear on this problem, write the system in matrix form as

$$AX = B,$$

where

$$A = \begin{bmatrix} a_{11} & a_{12} & \cdots & a_{1m} \\ a_{21} & a_{22} & \cdots & a_{2m} \\ \vdots & \vdots & & \vdots \\ a_{n1} & a_{n2} & \cdots & a_{nm} \end{bmatrix}$$

is the $n \times m$ *matrix of coefficients* of the system,

$$X = \begin{bmatrix} x_1 \\ \vdots \\ x_n \end{bmatrix} \quad \text{and} \quad B = \begin{bmatrix} b_1 \\ \vdots \\ b_n \end{bmatrix}.$$

The number of rows of A is the number of equations, while the number of columns of A is the number of unknowns. Row k of A and B gives us equation k, $a_{k1}x_1 + a_{k2}x_2 + \cdots + a_{km}x_m = b_k$, while column j of A gives the coefficients of x_j in equations 1 through n.

In keeping with this matrix formulation, we shall write a solution $x_1 = \alpha_1, x_2 = \alpha_2, \ldots, x_m = \alpha_m$ as an $m \times 1$ column matrix

$$\begin{bmatrix} \alpha_1 \\ \alpha_2 \\ \vdots \\ \alpha_m \end{bmatrix}$$

because X is written as an $m \times 1$ matrix in $AX = B$.

EXAMPLE 262

The system

$$\begin{aligned} x_1 - 2x_2 &= 3, \\ 4x_1 + 6x_2 &= -5, \end{aligned}$$

can be written

$$\begin{bmatrix} 1 & -2 \\ 4 & 6 \end{bmatrix}\begin{bmatrix} x_1 \\ x_2 \end{bmatrix} = \begin{bmatrix} 3 \\ -5 \end{bmatrix}.$$

It is easy to check that a solution is $x_1 = \frac{8}{14}$, $x_2 = -\frac{17}{14}$. We would write this solution in matrix notation as $\begin{bmatrix} \frac{8}{14} \\ -\frac{17}{14} \end{bmatrix}$. In this example, the solution is easily obtained by elimination (multiply equation 1 by -4, add to equation 2, and solve for x_2; then solve for x_1).

EXAMPLE 263

The system

$$x_1 + 2x_2 = 1$$

has one equation and 2 unknowns, and can be written

$$[1 \quad 2]\begin{bmatrix} x_1 \\ x_2 \end{bmatrix} = [1].$$

We can solve this by writing $x_1 = 1 - 2x_2$. For any choice of x_2, say $x_2 = \alpha$, we get a solution by letting $x_1 = 1 - 2\alpha$. Thus, we have many solutions. These can be written in matrix form as

$$\begin{bmatrix} 1 - 2\alpha \\ \alpha \end{bmatrix} \quad \text{or} \quad \begin{bmatrix} 1 \\ 0 \end{bmatrix} + \alpha\begin{bmatrix} -2 \\ 1 \end{bmatrix},$$

where α is any number.

In these two examples the systems were simple, and matrix notation did not contribute anything to the solution. We now consider methods for handling larger systems. To do this, we consider first the special case of *homogeneous systems of equations*.

The system $AX = B$ is called *homogeneous* if $b_1 = b_2 = \cdots = b_n = 0$. If some $b_j \neq 0$, the system is *nonhomogeneous*. Thus,

$$\begin{bmatrix} 1 & -1 & 2 \\ 0 & 1 & 6 \end{bmatrix}\begin{bmatrix} x_1 \\ x_2 \\ x_3 \end{bmatrix} = \begin{bmatrix} 0 \\ 0 \end{bmatrix}$$

is homogeneous, while

$$\begin{bmatrix} 1 & -1 & 2 \\ 0 & 1 & 6 \end{bmatrix}\begin{bmatrix} x_1 \\ x_2 \\ x_3 \end{bmatrix} = \begin{bmatrix} -4 \\ 0 \end{bmatrix}$$

is nonhomogeneous. For the remainder of this section, we concentrate on homogeneous systems. A homogeneous system of n equations in m unknowns is usually written $AX = 0$, where A is $n \times m$ and

$$0 = \begin{bmatrix} 0 \\ 0 \\ \vdots \\ 0 \end{bmatrix}, \quad \text{an } n \times 1 \text{ column of zeros.}$$

To solve $AX = 0$, with A $n \times m$ and 0 $n \times 1$, proceed as follows.

Step 1 Reduce A to A_R. The *reduced system* $A_R X = 0$ has the same solutions as the original system.

To see this, recall that we reduce A by applying elementary row operations. Since rows of A correspond to equations of the system, the row operations correspond to interchanging two equations, multiplying an equation by a nonzero constant, and adding a multiple of one equation to another. The result of any of these is a new system with the same solutions as the original system.

Step 2 In the system $A_R X = 0$, label each unknown as dependent or independent, according to the following test: If column j contains the leading entry of any row of A, call x_j *dependent*; otherwise, x_j is *independent*.

Step 3 Express each dependent unknown in terms of the independent ones. If, say, x_j is dependent, because column j contains the leading entry of row i, then we can solve for x_j in terms of independent unknowns using equation i.

Step 4 For the final solution, the independent unknowns can be assigned values arbitrarily. The dependent unknowns are then determined from step 3.

This process is called the *Gauss-Jordan reduction method.** Matrices provide a neat notational means of carrying out the method quite mechanically, as the following examples illustrate.

EXAMPLE 264

Consider the system

$$\begin{aligned} x_1 - 3x_2 + 2x_3 &= 0, \\ -2x_1 + x_2 - 3x_3 &= 0. \end{aligned}$$

This is easy to solve by high-school methods, but we want to begin with a simple system to illustrate the method. In matrix form, the system is $AX = 0$, where

$$A = \begin{bmatrix} 1 & -3 & 2 \\ -2 & 1 & -3 \end{bmatrix}, \quad X = \begin{bmatrix} x_1 \\ x_2 \\ x_3 \end{bmatrix}; \text{ and } \quad 0 = \begin{bmatrix} 0 \\ 0 \end{bmatrix}.$$

Step 1 Reduce A. Proceed

$$A \xrightarrow[\text{to row 2}]{\text{add 2(row 1)}} \begin{bmatrix} 1 & -3 & 2 \\ 0 & -5 & 1 \end{bmatrix} \xrightarrow[\text{row 2 by } -\frac{1}{5}]{\text{multiply}} \begin{bmatrix} 1 & -3 & 2 \\ 0 & 1 & -\frac{1}{5} \end{bmatrix}$$

$$\xrightarrow[\text{to row 1}]{\text{add 3(row 2)}} \begin{bmatrix} 1 & 0 & \frac{7}{5} \\ 0 & 1 & -\frac{1}{5} \end{bmatrix} = A_R.$$

Step 2 The leading entry of row 1 is in column 1, and so x_1 is dependent; similarly, x_2 is dependent. Finally, x_3 is independent.

Step 3 From row 1, $x_1 + \frac{7}{5}x_3 = 0$; so $x_1 = -\frac{7}{5}x_3$. From row 2, $x_2 - \frac{1}{5}x_3 = 0$; so $x_2 = \frac{1}{5}x_3$.

Step 4 x_3 can have *any* scalar value, say $x_3 = \alpha$. For a solution, choose $x_1 = -\frac{7}{5}\alpha$ and $x_2 = \frac{1}{5}\alpha$. In column matrix form, the solution is

$$\begin{bmatrix} -\frac{7}{5}\alpha \\ \frac{1}{5}\alpha \\ \alpha \end{bmatrix} \quad \text{or} \quad \alpha \begin{bmatrix} -\frac{7}{5} \\ \frac{1}{5} \\ 1 \end{bmatrix}.$$

* This method is also known as *complete pivoting*.

Since α is arbitrary, this system has infinitely many different solutions. For example, with $\alpha = 5$, we get $x_1 = -7$, $x_2 = 1$, $x_3 = 5$; with $\alpha = 1$, we get $x_1 = -\frac{7}{5}$, $x_2 = \frac{1}{5}$, $x_3 = 1$; and so on.

We call the expression $\alpha \begin{bmatrix} -\frac{7}{5} \\ \frac{1}{5} \\ 1 \end{bmatrix}$ the *general solution* of this system, in the sense that it contains all solutions as α is given different values.

The general solution may contain several arbitrary parameters, as the next example shows.

EXAMPLE 265

Solve the system

$$\begin{aligned} x_1 - 3x_2 + x_3 - 7x_4 + 4x_5 &= 0, \\ x_1 + 2x_2 - 3x_3 \qquad\qquad &= 0, \\ x_2 - 4x_3 \qquad + x_5 &= 0. \end{aligned}$$

This is $AX = 0$, with

$$A = \begin{bmatrix} 1 & -3 & 1 & -7 & 4 \\ 1 & 2 & -3 & 0 & 0 \\ 0 & 1 & -4 & 0 & 1 \end{bmatrix}.$$

We find that

$$A_R = \begin{bmatrix} 1 & 0 & 0 & -\frac{35}{16} & \frac{13}{16} \\ 0 & 1 & 0 & \frac{28}{16} & -\frac{20}{16} \\ 0 & 0 & 1 & \frac{7}{16} & -\frac{9}{16} \end{bmatrix}.$$

From A_R, we read that x_1, x_2, and x_3 are dependent, while x_4 and x_5 are independent. From row 1,

$$x_1 - \tfrac{35}{16}x_4 + \tfrac{13}{16}x_5 = 0.$$

Thus,

$$x_1 = \tfrac{35}{16}x_4 - \tfrac{13}{16}x_5.$$

Similarly, from row 2,

$$x_2 = -\tfrac{28}{16}x_4 + \tfrac{20}{16}x_5,$$

and from row 3,

$$x_3 = -\tfrac{7}{16}x_4 + \tfrac{9}{16}x_5,$$

with x_4 and x_5 arbitrary.

The solution, in matrix form, is then

$$\begin{bmatrix} x_1 \\ x_2 \\ x_3 \\ x_4 \\ x_5 \end{bmatrix} = \begin{bmatrix} \frac{35}{16}x_4 - \frac{13}{16}x_5 \\ -\frac{28}{16}x_4 + \frac{20}{16}x_5 \\ -\frac{7}{16}x_4 + \frac{9}{16}x_5 \\ x_4 \\ x_5 \end{bmatrix} = x_4 \begin{bmatrix} \frac{35}{16} \\ -\frac{28}{16} \\ -\frac{7}{16} \\ 1 \\ 0 \end{bmatrix} + x_5 \begin{bmatrix} -\frac{13}{16} \\ \frac{20}{16} \\ \frac{9}{16} \\ 0 \\ 1 \end{bmatrix}.$$

Since x_4 and x_5 can be assigned any values, say $x_4 = \alpha$, and $x_5 = \beta$, the general solution can be written

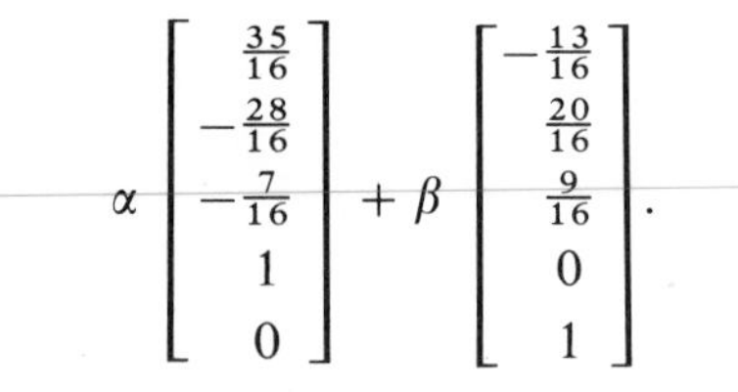

$$\alpha \begin{bmatrix} \frac{35}{16} \\ -\frac{28}{16} \\ -\frac{7}{16} \\ 1 \\ 0 \end{bmatrix} + \beta \begin{bmatrix} -\frac{13}{16} \\ \frac{20}{16} \\ \frac{9}{16} \\ 0 \\ 1 \end{bmatrix}.$$

Since α and β are arbitrary, we can replace them with the equally arbitrary $a = \frac{1}{16}\alpha$ and $b = \frac{1}{16}\beta$ and write more neatly

$$a \begin{bmatrix} 35 \\ -28 \\ -7 \\ 16 \\ 0 \end{bmatrix} + b \begin{bmatrix} -13 \\ 20 \\ 9 \\ 0 \\ 16 \end{bmatrix}.$$

EXAMPLE 266

Solve

$$\begin{aligned} -x_2 + 2x_3 + 4x_4 &= 0, \\ -x_3 + 3x_4 &= 0, \\ 2x_1 + x_2 + 3x_3 + 7x_4 &= 0, \\ 6x_1 + 2x_2 + 10x_3 + 28x_4 &= 0. \end{aligned}$$

Here we have $AX = 0$, with

$$A = \begin{bmatrix} 0 & -1 & 2 & 4 \\ 0 & 0 & -1 & 3 \\ 2 & 1 & 3 & 7 \\ 6 & 2 & 10 & 28 \end{bmatrix}.$$

We find that

$$A_R = \begin{bmatrix} 1 & 0 & 0 & 13 \\ 0 & 1 & 0 & -10 \\ 0 & 0 & 1 & -3 \\ 0 & 0 & 0 & 0 \end{bmatrix}.$$

Thus, x_1, x_2, and x_3 are dependent, while x_4 is independent. From rows 1, 2, and 3 of A_R, we have

$$\begin{aligned} x_1 &= -13x_4, \\ x_2 &= 10x_4, \\ x_3 &= 3x_4. \end{aligned}$$

Writing $x_4 = \alpha$, the general solution is

$$\alpha \begin{bmatrix} -13 \\ 10 \\ 3 \\ 1 \end{bmatrix},$$

with α arbitrary.

We pause here for a useful observation. Note that the number of arbitrary scalars in the general solution is the number of independent unknowns. If A is $n \times m$, this number is the total number of unknowns (m), minus the number of dependent unknowns. But the latter is the number of rows of A_R having leading entries. This in turn is the number of nonzero rows of A_R, or rank(A). This proves the following result:

THEOREM 48

If A is $n \times m$, then the number of arbitrary scalars in the general solution to $AX = 0$ is

$$m - \text{rank}(A).$$

Checking back, $m - \text{rank}(A)$ was $5 - 3 = 2$ in Example 265, and $4 - 3 = 1$ in Example 266.

We can look at these considerations from a vector space point of view. If A is $n \times m$, then solutions to $AX = 0$ can be thought of as vectors in R^m. If X_1 and X_2 are two solutions, then $A(X_1 + X_2) = AX_1 + AX_2 = 0 + 0 = 0$. Further, $A(\alpha X_1) = \alpha AX_1 = \alpha \cdot 0 = 0$. Thus, the totality of solutions of $AX = 0$ is a subspace of R^m. When we find the general solution, we are in fact expressing each vector in the solution space as a linear combination of certain linearly independent solutions. Thus, we are finding a basis for the solution space. The number of arbitrary scalars in the general solution is exactly the dimension of the solution space. Thus, Theorem 48 may be rephrased

$$m - \text{rank}(A) = \text{dimension of the solution space of the system } AX = 0.$$

For example, in Example 265 the solution space consists of all 5-vectors

$$a \begin{bmatrix} 35 \\ -28 \\ -7 \\ 16 \\ 0 \end{bmatrix} + b \begin{bmatrix} -13 \\ 20 \\ 9 \\ 0 \\ 16 \end{bmatrix},$$

with a and b arbitrary. Thus,

$$\begin{bmatrix} 35 \\ -28 \\ -7 \\ 16 \\ 0 \end{bmatrix} \quad \text{and} \quad \begin{bmatrix} -13 \\ 20 \\ 9 \\ 0 \\ 16 \end{bmatrix}$$

form a basis for the solution space of $AX = 0$ in this example. The dimension of this solution space is 2.

EXAMPLE 267

Solve

$$\begin{aligned}
-x_1 \qquad\quad + x_3 + x_4 + 2x_5 &= 0, \\
x_2 + 3x_3 \qquad + 4x_5 &= 0, \\
x_1 + 2x_2 + x_3 + x_4 + x_5 &= 0, \\
-3x_1 + x_2 \qquad\qquad + 4x_5 &= 0.
\end{aligned}$$

Here,

$$A = \begin{bmatrix} -1 & 0 & 1 & 1 & 2 \\ 0 & 1 & 3 & 0 & 4 \\ 1 & 2 & 1 & 1 & 1 \\ -3 & 1 & 0 & 0 & 4 \end{bmatrix}$$

and

$$A_R = \begin{bmatrix} 1 & 0 & 0 & 0 & -\frac{9}{8} \\ 0 & 1 & 0 & 0 & \frac{5}{8} \\ 0 & 0 & 1 & 0 & \frac{9}{8} \\ 0 & 0 & 0 & 1 & -\frac{2}{8} \end{bmatrix}.$$

Since $m = 5$ and $\operatorname{rank}(A) = 4$, the solution space has dimension 1, and there will be 1 arbitrary constant in the general solution. From rows 1 through 4 of A_R,

$$\begin{aligned}
x_1 &= \tfrac{9}{8}x_5, \\
x_2 &= -\tfrac{5}{8}x_5, \\
x_3 &= -\tfrac{9}{8}x_5, \\
x_4 &= \tfrac{2}{8}x_5,
\end{aligned}$$

with x_5 arbitrary. In matrix form, with $x_5 = \alpha$ (arbitrary), we have the general solution

$$\alpha \begin{bmatrix} \frac{9}{8} \\ -\frac{5}{8} \\ -\frac{9}{8} \\ \frac{2}{8} \\ 1 \end{bmatrix}.$$

A homogeneous system always has at least one solution, namely $x_1 = x_2 = \cdots = x_m = 0$. This is the *trivial solution.* In the above examples there were nontrivial solutions as well. However, it is possible for a homogeneous system to have only the trivial solution. The next theorem gives one circumstance in which this happens.

THEOREM 49

If A is square, say $n \times n$, then $AX = 0$ has only the trivial solution exactly when $\text{rank}(A) = n$.

Proof Suppose first that $\text{rank}(A) = n$. We get solutions of $AX = 0$ by solving $A_R X = 0$. But by Lemma 1, Section 10.6, $A_R = I_n$ when $\text{rank}(A) = n$. Thus, our system of equations reduces to $I_n X = 0$, or $X = 0$.

Conversely, suppose that $AX = 0$ has only the trivial solution. Then there can be no arbitrary constants in the general solution, as we could assign such a constant a nonzero value. By Theorem 48, $m - \text{rank}(A) = 0$. Since $m = n$ here, then $\text{rank}(A) = n$.

A useful alternative statement of Theorem 49 is the following.

Corollary 1

If A is $n \times n$, then $AX = 0$ has a nontrivial solution exactly when $\text{rank}(A) < n$.

EXAMPLE 268

Consider the system

$$\begin{aligned} 3x_1 - 11x_2 + 5x_3 &= 0, \\ 4x_1 + x_2 - 10x_3 &= 0, \\ 4x_1 + 9x_2 - 6x_3 &= 0. \end{aligned}$$

Here,

$$A = \begin{bmatrix} 3 & -11 & 5 \\ 4 & 1 & -10 \\ 4 & 9 & -6 \end{bmatrix} \quad \text{and} \quad A_R = \begin{bmatrix} 1 & 0 & 0 \\ 0 & 1 & 0 \\ 0 & 0 & 1 \end{bmatrix}.$$

From A_R, we read $x_1 = x_2 = x_3 = 0$, and the system has only the trivial solution. Note that here $n = 3 = \text{rank}(A)$.

There is an important situation in which there always exists a nontrivial solution.

THEOREM 50

If the homogeneous system $AX = 0$ has more unknowns than equations, then a nontrivial solution exists.

Proof Let A be $n \times m$. If there are more unknowns than equations, then $m > n$. Now, $\text{rank}(A)$ is the number of nonzero rows of A_R, and hence cannot exceed n. Then $m - \text{rank}(A) \geq m - n > 0$. Thus, there is at least one independent unknown which can be assigned values arbitrarily, and hence can be given nonzero values.

This is as much as we will say for now about homogeneous systems. In the next section we consider nonhomogeneous systems of linear equations.

PROBLEMS FOR SECTION 10.7

In each of Problems 1 through 20, find the general solution of the given homogeneous system, using the matrix methods of this section. In each case, also determine the dimension of the solution space.

1. $$\begin{aligned} x_1 + 2x_2 - x_3 + x_4 &= 0 \\ x_2 - x_3 + x_4 &= 0 \end{aligned}$$

2. $$\begin{aligned} -3x_1 + x_2 - x_3 + x_4 + x_5 &= 0 \\ x_2 + x_3 + 4x_5 &= 0 \\ -3x_3 + 2x_4 + x_5 &= 0 \end{aligned}$$

3. $$\begin{aligned} -2x_1 + x_2 + 2x_3 &= 0 \\ x_1 - x_2 &= 0 \\ x_1 + x_2 &= 0 \end{aligned}$$

4. $$\begin{aligned} 4x_1 + x_2 - 3x_3 + x_4 &= 0 \\ 2x_1 - x_3 &= 0 \end{aligned}$$

5. $$\begin{aligned} x_1 - x_2 + 3x_3 - x_4 + 4x_5 &= 0 \\ 2x_1 - 2x_2 + x_3 + x_4 &= 0 \\ x_1 - 2x_3 + x_5 &= 0 \\ x_3 + x_4 - x_5 &= 0 \end{aligned}$$

6. $$\begin{aligned} 6x_1 - x_2 + x_3 &= 0 \\ x_1 - x_4 + 2x_5 &= 0 \\ x_1 - 2x_5 &= 0 \end{aligned}$$

7. $$\begin{aligned} -10x_1 - x_2 + 4x_3 - x_4 + x_5 - x_6 &= 0 \\ x_2 - x_3 + 3x_4 &= 0 \\ 2x_1 - x_2 + x_5 &= 0 \\ x_2 - x_4 + x_6 &= 0 \end{aligned}$$

8. $$\begin{aligned} 8x_1 - 2x_3 + x_6 &= 0 \\ 2x_1 - x_2 + 3x_4 - x_6 &= 0 \\ x_2 + x_3 - 2x_5 - x_6 &= 0 \\ x_4 - 3x_5 + 2x_6 &= 0 \end{aligned}$$

9. $$\begin{aligned} x_2 - 3x_4 + x_5 &= 0 \\ 2x_1 - x_2 + 3x_4 &= 0 \\ 2x_1 - 3x_2 + 4x_5 &= 0 \end{aligned}$$

10. $$\begin{aligned} 4x_1 - 3x_2 + x_4 + x_5 - 3x_6 &= 0 \\ 2x_2 + 4x_4 - x_5 - 6x_6 &= 0 \\ 3x_1 - 2x_2 + 4x_5 - x_6 &= 0 \\ 2x_1 + x_2 - 3x_3 + 4x_4 &= 0 \end{aligned}$$

11. $$\begin{aligned} x_1 - 2x_2 + x_5 - x_6 + x_7 &= 0 \\ x_3 - x_4 + x_5 - 2x_6 + 3x_7 &= 0 \\ x_1 - x_5 + 2x_6 &= 0 \\ 2x_1 - 3x_4 + x_5 &= 0 \end{aligned}$$

12. $$\begin{aligned} 2x_1 - 4x_5 + x_7 + x_8 &= 0 \\ 2x_2 - x_6 + x_7 - x_8 &= 0 \\ x_3 - 4x_4 + x_8 &= 0 \\ x_2 - x_3 + x_4 &= 0 \\ x_2 - x_5 + x_6 - x_7 &= 0 \end{aligned}$$

For ease of recognition we have, up to this point, written the equations with the variables spaced so that the coefficients appear almost in matrix form. In Problems 13 through 20, the equations are written as they would probably appear in a normal encounter. Be careful when you write the coefficient matrix.

13. $$\begin{aligned} x_1 - 4x_3 + x_5 &= 0 \\ 2x_3 - 4x_4 &= 0 \\ x_2 - 5x_4 + 6x_5 &= 0 \end{aligned}$$

14. $$\begin{aligned} 12x_2 + 4x_2 - x_3 + x_4 &= 0 \\ 2x_3 - x_1 + 5x_3 - 5x_4 &= 0 \end{aligned}$$

15. $$\begin{aligned} -5x_1 + x_2 - 3x_3 + 4x_5 &= 0 \\ x_2 - 5x_3 + 7x_5 - x_4 &= 0 \end{aligned}$$

16. $$\begin{aligned} -3x_1 - 4x_2 + x_3 + x_4 &= 0 \\ x_2 - 4x_3 + 2x_4 &= 0 \\ -2x_1 + 4x_2 - 5x_3 + 2x_4 &= 0 \end{aligned}$$

17. $$\begin{aligned} 9x_2 + x_3 - 5x_4 &= 0 \\ x_1 + x_2 - 4x_4 &= 0 \\ x_3 + 8x_4 &= 0 \end{aligned}$$

18. $$\begin{aligned} -3x_1 + 3x_2 - 8x_3 + x_5 &= 0 \\ x_3 + 6x_4 - 2x_5 &= 0 \\ x_2 + x_4 + 5x_5 &= 0 \\ x_1 + x_2 + x_4 + 7x_5 &= 0 \end{aligned}$$

19. $$\begin{aligned} 5x_1 + x_2 - 4x_3 - x_6 &= 0 \\ x_2 + x_4 + 6x_5 - x_6 &= 0 \\ 2x_1 + x_3 - 5x_4 + 11x_6 &= 0 \end{aligned}$$

20. $$\begin{aligned} x_3 + 3x_4 - x_5 &= 0 \\ 2x_1 + 5x_2 + 3x_3 - x_5 &= 0 \end{aligned}$$

21. Suppose that we have a homogeneous system of n equations in m unknowns, and that $n \geq m$. Can there be nontrivial solutions?

10.8 SOLUTION OF NONHOMOGENEOUS SYSTEMS OF LINEAR EQUATIONS

We now consider systems $AX = B$, where A is $n \times m$, B is $n \times 1$, and at least one entry of B is nonzero.

In order to keep track of the entries in B as well as the coefficients of the unknowns, we introduce the *augmented matrix* $[A \mid B]$, an $n \times (m + 1)$ matrix formed by adjoining B to the right of A as an additional column. For example, the system

$$\begin{aligned} 2x_1 - x_2 + 3x_3 &= 4 \\ x_1 + 3x_2 - x_3 &= -2 \end{aligned}$$

can be written $AX = B$, with

$$A = \begin{bmatrix} 2 & -1 & 3 \\ 1 & 3 & -1 \end{bmatrix} \quad \text{and} \quad B = \begin{bmatrix} 4 \\ -2 \end{bmatrix}.$$

Here,

$$[A \mid B] = \left[\begin{array}{ccc:c} 2 & -1 & 3 & 4 \\ 1 & 3 & -1 & -2 \end{array}\right],$$

a 2×4 matrix. The dashed line helps us remember that the last column is really the column of matrix B in a system $AX = B$.

One dramatic difference between homogeneous and nonhomogeneous linear systems is that a nonhomogeneous system need not have a solution (here, the trivial solution $x_1 = x_2 = \cdots = x_n = 0$ is *not* a solution). For example, consider

$$\begin{aligned} 2x_1 - 3x_2 &= 6, \\ 4x_1 - 6x_2 &= 18. \end{aligned}$$

If $2x_1 - 3x_2 = 6$, then $4x_1 - 6x_2 = 2(2x_1 - 3x_2)$ must be 12, not 18. This system can have no solution.

Thus, we ought to worry about existence of a solution before tackling the problem of finding those solutions which do exist. Here is the main result on existence: It says that $AX = B$ can have a solution only if A and $[A \mid B]$ have the same rank, that is, the same number of nonzero rows in their reduced forms. This is stated as Theorem 51 below, but to get some feeling for it, look at the following example.

EXAMPLE 269

Consider

$$\begin{bmatrix} 2 & -3 \\ 4 & -6 \end{bmatrix} \begin{bmatrix} x_1 \\ x_2 \end{bmatrix} = \begin{bmatrix} 6 \\ 18 \end{bmatrix}.$$

Here,

$$A = \begin{bmatrix} 2 & -3 \\ 4 & -6 \end{bmatrix} \quad \text{and} \quad A_R = \begin{bmatrix} 1 & 0 \\ 0 & 0 \end{bmatrix}.$$

Thus, rank$(A) = 1$. But,

$$[A \mid B] = \begin{bmatrix} 2 & -3 & \vdots & 6 \\ 4 & -6 & \vdots & 18 \end{bmatrix}; \quad \text{and} \quad [A \mid B]_R = \begin{bmatrix} 1 & 0 & \vdots & 0 \\ 0 & 0 & \vdots & 1 \end{bmatrix},$$

with rank 2. The reduced system, read from $[A \mid B]_R$, is

$$x_1 + 0x_2 = 0,$$
$$0x_1 + 0x_2 = 1.$$

The second equation cannot be satisfied by any values of x_1 and x_2. The difficulty is that A_R has more zero rows than $[A \mid B]_R$, resulting in an equation in the system having all zero coefficients on one side, and a nonzero number on the other side. Such a system can have no solution.

THEOREM 51

The nonhomogeneous system $AX = B$ has a solution if and only if A and $[A \mid B]$ have the same rank.

Proof Let A be $n \times m$, and suppose first that rank$(A) =$ rank$[A \mid B] = r$. By Theorem 47 and Problem 21, Section 10.6, the column space of $[A \mid B]$ has dimension r, and certainly $r \leq m$. Thus, the $(m + 1)$th column of $[A \mid B]$ is a linear combination of the first m columns, say,

$$B = \alpha_1 \begin{bmatrix} A_{11} \\ A_{21} \\ \vdots \\ A_{n1} \end{bmatrix} + \alpha_2 \begin{bmatrix} A_{12} \\ A_{22} \\ \vdots \\ A_{n2} \end{bmatrix} + \cdots + \alpha_m \begin{bmatrix} A_{1m} \\ A_{2m} \\ \vdots \\ A_{nm} \end{bmatrix}.$$

This is precisely the same as writing

$$A \begin{bmatrix} \alpha_1 \\ \vdots \\ \alpha_m \end{bmatrix} = B,$$

implying that $AX = B$ has a solution.

Conversely, suppose that $AX = B$ has a solution, say, $X = \begin{bmatrix} \alpha_1 \\ \vdots \\ \alpha_m \end{bmatrix}$. Then

$$B = \alpha_1 \begin{bmatrix} A_{11} \\ A_{21} \\ \vdots \\ A_{n1} \end{bmatrix} + \alpha_2 \begin{bmatrix} A_{12} \\ A_{22} \\ \vdots \\ A_{n2} \end{bmatrix} + \cdots + \alpha_m \begin{bmatrix} A_{1m} \\ A_{2m} \\ \vdots \\ A_{nm} \end{bmatrix};$$

so B is a linear combination of columns of A. Thus, the column spaces of A and $[A \mid B]$ are the same. These column spaces then have the same dimensions, which are rank(A) and rank$[A \mid B]$, respectively. Thus, rank$(A) =$ rank$[A \mid B]$.

The next theorem tells us what to look for in solving $AX = B$.

THEOREM 52

Let U be any solution of $AX = B$. Then every solution of $AX = B$ is of the form $U + H$, where H is a solution of $AX = 0$.

Note: In writing $U + H$, we are considering solutions U and H as $m \times 1$ column matrices. Thus, the addition is performed as a matrix sum, one advantage of writing solutions as matrices. The student should also note the similarity between this theorem and Theorem 6, which treats the solution of nonhomogeneous second order linear differential equations.

Proof Let W be any solution of $AX = B$. Then $W - U$ is a solution of $AX = 0$, since

$$A(W - U) = AW - AU = B - B = 0.$$

Letting $H = W - U$, then H is a solution of $AX = 0$, and certainly

$$W = U + H.$$

Theorem 52 may appear almost circular, but is actually a very powerful result. It says:

To find an expression containing every solution of $AX = B$, we first find one (any one) solution of $AX = B$, and then add this to the general solution of $AX = 0$ (which we know how to find from the previous section). The resulting expression is called the *general solution* of $AX = B$.

Acting on this, we now present a procedure for finding all solutions of $AX = B$ by finding one solution of $AX = B$ and the general solution of $AX = 0$.

Step 1 Reduce $[A \,\vdots\, B]$ to $[A \,\vdots\, B]_R$. This matrix will be of the form $[A_R \,\vdots\, C]$, where C is $n \times 1$. Henceforth, work with the reduced system $A_R X = C$.

Step 2 If $\text{rank}[A \,\vdots\, B] \neq \text{rank}(A)$, stop; there is no solution. If these two ranks are equal, go to step 3.

Step 3 Identify the dependent variables. If column j contains the leading entry of row i, then use equation i to write x_j in terms of the independent variables *and* C_i (in the homogeneous case all the C_i's are zero).

Step 4 Write a column $\begin{bmatrix} x_1 \\ x_2 \\ \vdots \\ x_m \end{bmatrix}$ with each dependent x_j written in terms of independent unknowns and C_i.

Step 5 Rewrite this column as a sum of columns multiplied by the independent x_k's (as in the homogeneous case) *and* a column containing the C_i's appearing in the expressions for the dependent unknowns. This constant column is a particular solution of $A_R X = C$, and replacing the independent x_k's by arbitrary constants $\alpha, \beta, \ldots,$

gives us the general solution of $A_R X = 0$. Thus, we have the general solution of $A_R X = C$ (hence also of $AX = B$) written as a sum of a particular solution of $AX = B$ plus the general solution of $AX = 0$, as prescribed by Theorem 52.

EXAMPLE 270

Find the general solution of

$$\begin{aligned} -x_1 + x_2 + 3x_3 &= -2, \\ x_2 + 2x_3 &= 4. \end{aligned}$$

Step 1 Here,

$$[A \mid B] = \left[\begin{array}{ccc|c} -1 & 1 & 3 & -2 \\ 0 & 1 & 2 & 4 \end{array}\right]$$

and we find that

$$[A \mid B]_R = \left[\begin{array}{ccc|c} 1 & 0 & -1 & 6 \\ 0 & 1 & 2 & 4 \end{array}\right].$$

This is of the form $[A_R \mid C]$, with $A_R = \begin{bmatrix} 1 & 0 & -1 \\ 0 & 1 & 2 \end{bmatrix}$ and $C = \begin{bmatrix} 6 \\ 4 \end{bmatrix}$.

Step 2 Now, $\operatorname{rank}(A) = 2 = \operatorname{rank}[A \mid C]$; so a solution exists.

Step 3 We read from A_R that x_1 and x_2 are dependent, and x_3 independent. From $[A \mid B]_R$, we find

$$\begin{aligned} x_1 - x_3 &= 6, \\ x_2 + 2x_3 &= 4. \end{aligned}$$

Thus,

$$\begin{aligned} x_1 &= 6 + x_3, \\ x_2 &= 4 - 2x_3. \end{aligned}$$

Step 4 Write

$$\begin{bmatrix} x_1 \\ x_2 \\ x_3 \end{bmatrix} = \begin{bmatrix} 6 + x_3 \\ 4 - 2x_3 \\ x_3 \end{bmatrix}.$$

Step 5 Rewrite the result of step 4 as

$$\begin{bmatrix} 6 \\ 4 \\ 0 \end{bmatrix} + x_3 \begin{bmatrix} 1 \\ -2 \\ 1 \end{bmatrix}.$$

The general solution of $AX = B$ is then

$$\begin{bmatrix} 6 \\ 4 \\ 0 \end{bmatrix} + \alpha \begin{bmatrix} -1 \\ 2 \\ 1 \end{bmatrix},$$

with α any number.

Note that this is a sum of:

$$\begin{bmatrix} 6 \\ 4 \\ 0 \end{bmatrix}, \quad \text{a particular solution of } AX = B,$$

and

$$\alpha \begin{bmatrix} -1 \\ 2 \\ 1 \end{bmatrix}, \quad \text{the general solution of } AX = 0.$$

The student should verify by substitution that

$$x_1 = 6 + \alpha, \qquad x_2 = 4 - 2\alpha, \quad \text{and} \quad x_3 = \alpha$$

satisfy the given system for any α.

EXAMPLE 271

Solve

$$\begin{aligned} x_1 - x_2 + 2x_3 &= -1, \\ x_3 &= 0, \\ 3x_1 - 3x_2 + 7x_3 &= 1, \\ 10x_1 - 10x_2 + 24x_3 &= -2. \end{aligned}$$

Here,

$$[A \mid B] = \left[\begin{array}{ccc|c} 1 & -1 & 2 & -1 \\ 0 & 0 & 1 & 0 \\ 3 & -3 & 7 & 1 \\ 10 & -10 & 24 & -2 \end{array}\right]$$

and

$$[A \mid B]_R = \left[\begin{array}{ccc|c} 1 & -1 & 0 & 0 \\ 0 & 0 & 1 & 0 \\ 0 & 0 & 0 & 1 \\ 0 & 0 & 0 & 0 \end{array}\right].$$

Thus, $\text{rank}[A \mid B] = 3$ (there are 3 nonzero rows in $[A \mid B]_R$). But,

$$A_R = \begin{bmatrix} 1 & -1 & 0 \\ 0 & 0 & 1 \\ 0 & 0 & 0 \\ 0 & 0 & 0 \end{bmatrix};$$

so rank$(A) = 2$. By Theorem 51 this system has no solution. Note that, from $[A \mid B]_R$, the third equation of the reduced system would be $0x_1 + 0x_2 + 0x_3 = 1$, an impossibility.

EXAMPLE 272

Solve

$$\begin{aligned} x_1 \qquad\quad - x_3 + 2x_4 + x_5 + 6x_6 &= -3, \\ x_2 + x_3 + 3x_4 + 2x_5 + 4x_6 &= 1, \\ x_1 - 4x_2 + 3x_3 + x_4 \qquad + 2x_6 &= 0. \end{aligned}$$

Here,

$$[A \mid B] = \left[\begin{array}{cccccc|c} 1 & 0 & -1 & 2 & 1 & 6 & -3 \\ 0 & 1 & 1 & 3 & 2 & 4 & 1 \\ 1 & -4 & 3 & 1 & 0 & 2 & 0 \end{array}\right]$$

and

$$[A \mid B]_R = \left[\begin{array}{cccccc|c} 1 & 0 & 0 & \frac{27}{8} & \frac{15}{8} & \frac{60}{8} & -\frac{17}{8} \\ 0 & 1 & 0 & \frac{13}{8} & \frac{9}{8} & \frac{20}{8} & \frac{1}{8} \\ 0 & 0 & 1 & \frac{11}{8} & \frac{7}{8} & \frac{12}{8} & \frac{7}{8} \end{array}\right].$$

Now, rank$(A) = 3 =$ rank$[A \mid B]$; so at least one solution exists.

From $[A \mid B]_R$ we see that x_1, x_2, and x_3 are dependent, while x_4, x_5, and x_6 are independent. Further, from the first row of $[A \mid B]_R$ we read

$$x_1 + \tfrac{27}{8}x_4 + \tfrac{15}{8}x_5 + \tfrac{60}{8}x_6 = -\tfrac{17}{8}.$$

Hence,

$$x_1 = -\tfrac{17}{8} - \tfrac{27}{8}x_4 - \tfrac{15}{8}x_5 - \tfrac{60}{8}x_6.$$

Similarly, from the second and third rows,

$$\begin{aligned} x_2 &= \tfrac{1}{8} - \tfrac{13}{8}x_4 - \tfrac{9}{8}x_5 - \tfrac{20}{8}x_6, \\ x_3 &= \tfrac{7}{8} - \tfrac{11}{8}x_4 - \tfrac{7}{8}x_5 - \tfrac{12}{8}x_6. \end{aligned}$$

Thus,

$$\begin{bmatrix} x_1 \\ x_2 \\ x_3 \\ x_4 \\ x_5 \\ x_6 \end{bmatrix} = \begin{bmatrix} -\frac{17}{8} - \frac{27}{8}x_4 - \frac{15}{8}x_5 - \frac{60}{8}x_6 \\ \frac{1}{8} - \frac{13}{8}x_4 - \frac{9}{8}x_5 - \frac{20}{8}x_6 \\ \frac{7}{8} - \frac{11}{8}x_4 - \frac{7}{8}x_5 - \frac{12}{8}x_6 \\ x_4 \\ x_5 \\ x_6 \end{bmatrix}$$

$$= \begin{bmatrix} -\frac{17}{8} \\ \frac{1}{8} \\ \frac{7}{8} \\ 0 \\ 0 \\ 0 \end{bmatrix} + x_4 \begin{bmatrix} -\frac{27}{8} \\ -\frac{13}{8} \\ -\frac{11}{8} \\ 1 \\ 0 \\ 0 \end{bmatrix} + x_5 \begin{bmatrix} -\frac{15}{8} \\ -\frac{9}{8} \\ -\frac{7}{8} \\ 0 \\ 1 \\ 0 \end{bmatrix} + x_6 \begin{bmatrix} -\frac{60}{8} \\ -\frac{20}{8} \\ -\frac{12}{8} \\ 0 \\ 0 \\ 1 \end{bmatrix}.$$

The general solution can be written more neatly as

$$\frac{1}{8}\begin{bmatrix} -17 \\ 1 \\ 7 \\ 0 \\ 0 \\ 0 \end{bmatrix} + \alpha \begin{bmatrix} -27 \\ -13 \\ -11 \\ 8 \\ 0 \\ 0 \end{bmatrix} + \beta \begin{bmatrix} -15 \\ -9 \\ -7 \\ 0 \\ 8 \\ 0 \end{bmatrix} + \gamma \begin{bmatrix} -60 \\ -20 \\ -12 \\ 0 \\ 0 \\ 8 \end{bmatrix},$$

with α, β, and γ arbitrary scalars.

The student can check by substitution into the original system that the following satisfy the original system for any α, β, and γ:

$$\begin{aligned} x_1 &= -27\alpha - 15\beta - 60\gamma - \tfrac{17}{8}, \\ x_2 &= -13\alpha - 9\beta - 20\gamma + \tfrac{1}{8}, \\ x_3 &= -11\alpha - 7\beta - 12\gamma + \tfrac{7}{8}, \\ x_4 &= 8\alpha, \\ x_5 &= 8\beta, \\ x_6 &= 8\gamma. \end{aligned}$$

We conclude this section with a sufficient condition for $AX = B$ to have a unique solution when A is square.

THEOREM 53

Let A be $n \times n$. Then the nonhomogeneous system $AX = B$ has a unique solution if and only if $\text{rank}(A) = n$.

Proof Suppose first that $\text{rank}(A) = n$. Then $A_R = I_n$ by Lemma 1, Section 10.6. Then $[A \mid B]_R$ is of the form $[I_n \mid C]$ for some $n \times 1$ matrix C. This system has exactly one solution, namely $x_1 = C_1, x_2 = C_2, \ldots, x_n = C_n$, and this is then the only solution of $AX = B$.

Conversely, suppose that $AX = B$ has exactly one solution U. If $AX = 0$ has a solution H, then by Theorem 52, $U + H$ is a solution of $AX = B$. But then $U = U + H$; so $H = 0$. Thus, $AX = 0$ can have only the trivial solution. By Theorem 49, $\text{rank}(A) = n$.

EXAMPLE 273

Consider the system

$$\begin{bmatrix} 2 & -1 \\ 0 & 3 \end{bmatrix}\begin{bmatrix} x_1 \\ x_2 \end{bmatrix} = \begin{bmatrix} -1 \\ 4 \end{bmatrix}.$$

Here,

$$A = \begin{bmatrix} 2 & -1 \\ 0 & 3 \end{bmatrix} \quad \text{and} \quad A_R = \begin{bmatrix} 1 & 0 \\ 0 & 1 \end{bmatrix};$$

hence, rank(A) = 2 and the system has a unique solution. The solution is

$$\begin{bmatrix} \frac{1}{6} \\ \frac{4}{3} \end{bmatrix}.$$

In the next section we look at matrix inverses, which will return us again to questions involving solution of $AX = B$ when A is square.

PROBLEMS FOR SECTION 10.8

In each of Problems 1 through 20, either find the general solution or else show that the system of equations has no solution.

1.
$$\begin{aligned} 3x_1 - 2x_2 + x_3 &= 6 \\ x_1 + 10x_2 - x_3 &= 2 \\ -3x_1 - 2x_2 + x_3 &= 0 \end{aligned}$$

2.
$$\begin{aligned} 4x_1 - 2x_2 + 3x_3 + 10x_4 &= 1 \\ x_1 - 3x_4 &= 8 \\ 2x_1 - 3x_2 + x_4 &= 16 \end{aligned}$$

3.
$$\begin{aligned} 2x_1 - 3x_2 + x_4 - x_6 &= 0 \\ 3x_1 - 2x_3 + x_5 &= 1 \\ x_2 - x_4 + 6x_6 &= -3 \end{aligned}$$

4.
$$\begin{aligned} 2x_1 - 3x_2 &= 1 \\ -x_1 + 3x_2 &= 0 \\ x_1 - 4x_2 &= 3 \end{aligned}$$

5.
$$\begin{aligned} 3x_2 - 4x_4 &= 10 \\ x_1 - 3x_2 + 4x_5 - x_6 &= 8 \\ x_2 + x_3 - 6x_4 + x_6 &= -9 \\ x_1 - x_2 + x_6 &= 0 \end{aligned}$$

6.
$$\begin{aligned} 2x_1 - 3x_2 + x_4 &= 1 \\ 3x_2 + x_3 - x_4 &= 0 \\ 2x_1 - 3x_2 + 10x_3 &= 0 \end{aligned}$$

7.
$$\begin{aligned} 8x_2 - 4x_3 + 10x_6 &= 1 \\ x_3 + x_5 - x_6 &= 2 \\ x_4 - 3x_5 + 2x_6 &= 0 \end{aligned}$$

8.
$$\begin{aligned} 2x_1 - 3x_3 &= 1 \\ x_1 - x_2 + x_3 &= 1 \\ 2x_1 - 4x_2 + x_3 &= 2 \end{aligned}$$

9.
$$\begin{aligned} 14x_3 - 3x_5 + x_7 &= 2 \\ x_1 + x_2 + x_3 - x_4 + x_6 &= -4 \end{aligned}$$

10.
$$\begin{aligned} 3x_1 - 2x_2 &= -1 \\ 4x_1 + 3x_2 &= 4 \end{aligned}$$

11.
$$\begin{aligned} 7x_1 - 3x_2 + 4x_3 &= -7 \\ 2x_1 + x_2 - x_3 + 4x_4 &= 6 \\ x_2 - 3x_4 &= -5 \end{aligned}$$

12.
$$\begin{aligned} -4x_1 + 5x_2 - 6x_3 &= 2 \\ 2x_1 - 6x_2 + x_3 &= -5 \\ -6x_1 + 16x_2 - 11x_3 &= 1 \end{aligned}$$

13.
$$\begin{aligned} 4x_1 - x_2 + 4x_3 &= 1 \\ x_1 + x_2 - 5x_3 &= 0 \\ -2x_1 + x_2 + 7x_3 &= 4 \end{aligned}$$

14.
$$\begin{aligned} -6x_1 + 2x_2 - x_3 + x_4 &= 0 \\ x_1 + 4x_2 - x_4 &= -5 \\ x_2 + x_3 - 7x_4 &= 0 \end{aligned}$$

15.
$$\begin{aligned} 4x_1 - 3x_2 + x_3 &= -1 \\ -3x_1 + x_2 - 5x_3 &= 0 \\ -5x_1 - 14x_3 &= -10 \end{aligned}$$

16.
$$\begin{aligned} 9x_1 + x_2 - 4x_5 &= -1 \\ x_2 + 4x_3 - 4x_5 &= 2 \\ x_2 + x_4 - x_5 &= 0 \end{aligned}$$

17.
$$\begin{aligned} -5x_1 + 3x_2 - x_3 &= 0 \\ x_2 + x_4 &= -4 \\ x_1 - x_2 + 3x_3 - 6x_4 &= -11 \end{aligned}$$

18.
$$\begin{aligned} -6x_1 + x_2 - 4x_3 &= 1 \\ 2x_1 - x_2 - x_3 &= 8 \\ x_1 + 6x_2 - x_3 &= -3 \end{aligned}$$

19.
$$\begin{aligned} -5x_1 + 3x_2 + x_3 - x_4 &= -8 \\ 4x_1 + 3x_2 - x_4 &= 9 \\ 2x_1 + 3x_2 - 3x_3 + x_4 &= -7 \end{aligned}$$

20.
$$\begin{aligned} 3x_2 + x_4 - x_5 &= 15 \\ x_1 + 3x_2 + x_4 - 7x_5 &= 10 \\ -5x_1 + x_2 - 4x_3 + x_4 + 6x_5 &= 1 \\ 2x_1 + 4x_2 - x_3 + 10x_4 + 8x_5 &= 7 \end{aligned}$$

21. Let the $n \times m$ row reduced matrix A have rank k. Prove that the nonhomogeneous system of equations $AX = B$ has a solution if and only if $b_{k+1} = \cdots = b_n = 0$.

In each of Problems 22 through 27, show that the system of equations has a unique solution and find that solution.

22. $\begin{aligned} -2x_1 + 3x_2 &= -1 \\ -4x_2 + x_3 &= 0 \\ x_1 + 3x_2 &= 0 \end{aligned}$

23. $\begin{aligned} 2x_2 &= 1 \\ x_2 - 2x_3 &= 2 \\ x_1 - 2x_4 &= -10 \\ x_1 + x_4 &= 5 \end{aligned}$

24. $\begin{aligned} x_1 + 2x_3 &= -12 \\ x_1 - 3x_3 + x_4 &= 0 \\ x_2 - x_4 &= 1 \\ x_2 - x_3 &= 8 \end{aligned}$

25. $\begin{aligned} 4x_1 - x_2 &= 1 \\ x_2 + 3x_3 &= 0 \\ x_1 - 4x_3 &= -2 \end{aligned}$

26. $\begin{aligned} 2x_2 - x_4 &= 1 \\ x_1 - 3x_3 &= -2 \\ x_1 - 4x_2 &= 0 \\ 3x_2 &= 1 \end{aligned}$

27. $\begin{aligned} 2x_2 - 3x_3 &= 0 \\ 2x_1 - 3x_3 &= 0 \\ x_1 - x_2 + x_3 &= 1 \end{aligned}$

10.9 MATRIX INVERSES

As far as square matrices are concerned, the identity matrix behaves like the number 1. That is, if A is $n \times n$, then

$$AI_n = I_n A = A.$$

This is Theorem 42 when $n = m$. We now ask: If A is $n \times n$, can we find a matrix B such that $AB = BA = I_n$? If such a matrix exists, we call it an *inverse* of A (and, by symmetry, B is an inverse of A).

The following example shows that the question of existence of inverses is a delicate one.

EXAMPLE 274

It is easy to verify that $\begin{bmatrix} 1 & 0 \\ 0 & 0 \end{bmatrix}$ has no inverse. For suppose that

$$\begin{bmatrix} 1 & 0 \\ 0 & 0 \end{bmatrix}\begin{bmatrix} a & b \\ c & d \end{bmatrix} = \begin{bmatrix} 1 & 0 \\ 0 & 1 \end{bmatrix}.$$

Then, after carrying out the matrix multiplication on the left, we would have

$$\begin{bmatrix} a & b \\ 0 & 0 \end{bmatrix} = \begin{bmatrix} 1 & 0 \\ 0 & 1 \end{bmatrix}.$$

By definition of matrix equality, we would then have $0 = 1$ (compare the 2-2 entries).

On the other hand, it is easy to check that

$$\begin{bmatrix} 2 & 1 \\ 1 & 4 \end{bmatrix}\begin{bmatrix} \frac{4}{7} & -\frac{1}{7} \\ -\frac{1}{7} & \frac{2}{7} \end{bmatrix} = \begin{bmatrix} \frac{4}{7} & -\frac{1}{7} \\ -\frac{1}{7} & \frac{2}{7} \end{bmatrix}\begin{bmatrix} 2 & 1 \\ 1 & 4 \end{bmatrix} = \begin{bmatrix} 1 & 0 \\ 0 & 1 \end{bmatrix}.$$

Thus, each matrix in the product on the left is the inverse of the other.

A matrix is called *nonsingular* when it has an inverse, and *singular* otherwise. Thus, in Example 274, $\begin{bmatrix} 1 & 0 \\ 0 & 0 \end{bmatrix}$ was singular, while $\begin{bmatrix} 2 & 1 \\ 1 & 4 \end{bmatrix}$ was nonsingular. Here are some facts about nonsingular and singular matrices.

THEOREM 54

A nonsingular matrix has exactly one inverse.

Proof Let B and C be inverses of the nonsingular matrix A. Then $AB = BA = I_n$ and $AC = CA = I_n$. To show that $B = C$, we have

$$B = BI_n = B(AC) = (BA)C = I_n C = C.$$

In view of Theorem 54, we denote the unique inverse of a nonsingular matrix A as A^{-1}.

THEOREM 55

1. I_n is nonsingular.
2. If A and B are nonsingular $n \times n$ matrices, then AB is also nonsingular, and
$$(AB)^{-1} = B^{-1}A^{-1}.$$
3. If A is nonsingular, then so is A^{-1}, and $(A^{-1})^{-1} = A$.
4. If A is nonsingular, then so is A^t, and $(A^t)^{-1} = (A^{-1})^t$.
5. If A and B are $n \times n$ matrices, and either A or B is singular, then AB and BA are singular.

Proof of (1) Note that $I_n \cdot I_n = I_n$.

Proof of (2) Note that A^{-1} and B^{-1} exist because A and B are nonsingular. Now,

$$(AB)(B^{-1}A^{-1}) = A(BB^{-1})A^{-1} = AI_n A^{-1} = AA^{-1} = I_n.$$

Similarly, $(B^{-1}A^{-1})(AB) = I_n$. Thus, $B^{-1}A^{-1}$ is the inverse of AB, denoted $(AB)^{-1}$.

Proof of (3) $A^{-1}A = AA^{-1} = I_n$; so A^{-1} has inverse A. Thus, A is nonsingular and, by notation, $(A^{-1})^{-1} = A$.

Proof of (4) This is left to the student [note Theorem 43(3)].

A proof of (5) at this stage involves a subtlety (see Problems 38 and 39) which is easily circumvented when we have the use of determinants. Thus, we defer a proof of (5) until the end of Section 10.13.

We can now give a necessary and sufficient condition for a square matrix to be nonsingular.

THEOREM 56

Let A be $n \times n$. Then A is nonsingular if and only if $\text{rank}(A) = n$.

Proof Consider the equation $AB = I_n$, with B an $n \times n$ matrix of unknowns for which we want to solve. We need

$$\text{column } j \text{ of } AB = A \cdot (\text{column } j \text{ of } B)$$

$$= \text{column } j \text{ of } I_n = \begin{bmatrix} 0 \\ \vdots \\ 1 \\ \vdots \\ 0 \end{bmatrix} \leftarrow j\text{th place.}$$

Thus, column j of B must be found from the system of equations

$$A \begin{bmatrix} b_{1j} \\ b_{2j} \\ \vdots \\ b_{nj} \end{bmatrix} = \begin{bmatrix} 0 \\ \vdots \\ 1 \\ \vdots \\ 0 \end{bmatrix} \leftarrow j\text{th place}.$$

Now, if $\text{rank}(A) = n$, then this system has a unique solution by Theorem 53. In this case, there is a unique matrix B such that $AB = I_n$. But then A is nonsingular.

Conversely, if A is nonsingular, then the above system has a unique solution for $j = 1, \ldots, n$ because these solutions form the columns of A^{-1}. By Theorem 53, $\text{rank}(A) = n$.

With this much theory in hand, we turn to a method for computing A^{-1} when A is nonsingular, or for showing that A is singular.

Let A be any $n \times n$ matrix. Proceed as follows.

Step 1 Write an $n \times 2n$ matrix consisting of I_n placed on the left of A (as in Figure 141). This matrix is denoted $[I_n \mid A]$. The first n columns are called the left side, the second n columns, the right side.

Step 2 Reduce A to A_R. Each elementary row operation performed on the right side is also performed on the left side. This results in an $n \times 2n$ matrix $[C \mid A_R]$.

Step 3 If $A_R \neq I_n$, then A is singular and no inverse exists. If $A_R = I_n$, then $A^{-1} = C$.

$$\left[\begin{array}{ccccc|ccccc} 1 & 0 & 0 & \cdots & 0 & A_{11} & A_{12} & A_{13} & \cdots & A_{1n} \\ 0 & 1 & 0 & \cdots & 0 & A_{21} & A_{22} & A_{23} & \cdots & A_{2n} \\ 0 & 0 & 1 & \cdots & 0 & A_{31} & A_{32} & A_{33} & \cdots & A_{3n} \\ \vdots & & & & \vdots & \vdots & \vdots & \vdots & & \vdots \\ 0 & 0 & 0 & \cdots & 1 & A_{n1} & A_{n2} & A_{n3} & \cdots & A_{nn} \end{array}\right]$$

Figure 141. $n \times 2n$ matrix $[I_n \mid A]$.

Why does this work? By beginning on the left side with I_n, and performing exactly the elementary row operations used to reduce A, we produce in C a matrix that performs the complete sequence of reducing operations when multiplied on the left of A (recall Example 256, Section 10.4). That is, whether or not $A_R = I_n$, the matrix C in the final form $[C \mid A_R]$ has the property that $CA = A_R$. When $A_R = I_n$, then C must be A^{-1}.

EXAMPLE 275

Let

$$A = \begin{bmatrix} 2 & -1 & 3 \\ 1 & 0 & -2 \\ 4 & 0 & 2 \end{bmatrix}.$$

We want to determine whether A is nonsingular or singular. If A is nonsingular, we want A^{-1}.

Begin with the 3×6 matrix $[I_3 \mid A]$ and reduce A, performing the same operations on the left side:

$$\left[\begin{array}{ccc|ccc} 1 & 0 & 0 & 2 & -1 & 3 \\ 0 & 1 & 0 & 1 & 0 & -2 \\ 0 & 0 & 1 & 4 & 0 & 2 \end{array}\right] \xrightarrow[\text{by } \frac{1}{2}]{\text{multiply row 1}} \left[\begin{array}{ccc|ccc} \frac{1}{2} & 0 & 0 & 1 & -\frac{1}{2} & \frac{3}{2} \\ 0 & 1 & 0 & 1 & 0 & -2 \\ 0 & 0 & 1 & 4 & 0 & 2 \end{array}\right]$$

$$\xrightarrow[\text{add } -4(\text{row 1}) \text{ to row 3}]{\text{add } -(\text{row 1}) \text{ to row 2}} \left[\begin{array}{ccc|ccc} \frac{1}{2} & 0 & 0 & 1 & -\frac{1}{2} & \frac{3}{2} \\ -\frac{1}{2} & 1 & 0 & 0 & \frac{1}{2} & -\frac{7}{2} \\ -2 & 0 & 1 & 0 & 2 & -4 \end{array}\right]$$

$$\xrightarrow{\text{multiply row 2 by 2}} \left[\begin{array}{ccc|ccc} \frac{1}{2} & 0 & 0 & 1 & -\frac{1}{2} & \frac{3}{2} \\ -1 & 2 & 0 & 0 & 1 & -7 \\ -2 & 0 & 1 & 0 & 2 & -4 \end{array}\right]$$

$$\xrightarrow[\text{add } -2(\text{row 2}) \text{ to row 3}]{\text{add } \frac{1}{2}(\text{row 2}) \text{ to row 1}} \left[\begin{array}{ccc|ccc} 0 & 1 & 0 & 1 & 0 & -2 \\ -1 & 2 & 0 & 0 & 1 & -7 \\ 0 & -4 & 1 & 0 & 0 & 10 \end{array}\right]$$

$$\xrightarrow[\text{by } \frac{1}{10}]{\text{multiply row 3}} \left[\begin{array}{ccc|ccc} 0 & 1 & 0 & 1 & 0 & -2 \\ -1 & 2 & 0 & 0 & 1 & -7 \\ 0 & -\frac{4}{10} & \frac{1}{10} & 0 & 0 & 1 \end{array}\right]$$

$$\xrightarrow[\text{add } 7(\text{row 3}) \text{ to row 2}]{\text{add } 2(\text{row 3}) \text{ to row 1}} \left[\begin{array}{ccc|ccc} 0 & \frac{2}{10} & \frac{2}{10} & 1 & 0 & 0 \\ -1 & -\frac{8}{10} & \frac{7}{10} & 0 & 1 & 0 \\ 0 & -\frac{4}{10} & \frac{1}{10} & 0 & 0 & 1 \end{array}\right] = [C \mid I_n].$$

Since I_3 has appeared on the right side, A is nonsingular, and the left side is A^{-1}. Thus,

$$A^{-1} = \begin{bmatrix} 0 & \frac{1}{5} & \frac{1}{5} \\ -1 & -\frac{4}{5} & \frac{7}{10} \\ 0 & -\frac{2}{5} & \frac{1}{10} \end{bmatrix}.$$

EXAMPLE 276

Let

$$A = \begin{bmatrix} -3 & 1 & -1 \\ 1 & 0 & 1 \\ -2 & 2 & 2 \end{bmatrix}.$$

Proceed:

$$\left[\begin{array}{ccc:ccc} 1 & 0 & 0 & -3 & 1 & -1 \\ 0 & 1 & 0 & 1 & 0 & 1 \\ 0 & 0 & 1 & -2 & 2 & 2 \end{array}\right] \xrightarrow[\text{by } -\frac{1}{3}]{\text{multiply row 1}} \left[\begin{array}{ccc:ccc} -\frac{1}{3} & 0 & 0 & 1 & -\frac{1}{3} & \frac{1}{3} \\ 0 & 1 & 0 & 1 & 0 & 1 \\ 0 & 0 & 1 & -2 & 2 & 2 \end{array}\right]$$

$$\xrightarrow[\text{add 2(row 1) to row 3}]{\text{add }-\text{(row 1) to row 2}} \left[\begin{array}{ccc:ccc} -\frac{1}{3} & 0 & 0 & 1 & -\frac{1}{3} & \frac{1}{3} \\ \frac{1}{3} & 1 & 0 & 0 & \frac{1}{3} & \frac{2}{3} \\ -\frac{2}{3} & 0 & 1 & 0 & \frac{4}{3} & \frac{8}{3} \end{array}\right]$$

$$\xrightarrow{\text{multiply row 2 by 3}} \left[\begin{array}{ccc:ccc} -\frac{1}{3} & 0 & 0 & 1 & -\frac{1}{3} & \frac{1}{3} \\ 1 & 3 & 0 & 0 & 1 & 2 \\ -\frac{2}{3} & 0 & 1 & 0 & \frac{4}{3} & \frac{8}{3} \end{array}\right]$$

$$\xrightarrow[\text{add }-\frac{4}{3}\text{(row 2) to row 3}]{\text{add }\frac{1}{3}\text{(row 2) to row 1}} \left[\begin{array}{ccc:ccc} 0 & 1 & 0 & 1 & 0 & 1 \\ 1 & 3 & 0 & 0 & 1 & 2 \\ -\frac{6}{3} & -4 & 1 & 0 & 0 & 0 \end{array}\right].$$

We now have the reduced form of A on the right, and it is not I_3. Hence, A is singular.

Inverses bear on the problem of solving systems of equations in the following ways.

THEOREM 57

Let A be an $n \times n$ matrix. Then,

1. The nonhomogeneous system $AX = B$ has a unique solution if and only if A is nonsingular. In this case, the unique solution is $X = A^{-1}B$.
2. $AX = 0$ has a nontrivial solution exactly when A is singular.

Proof of (1) By Theorem 53, $AX = B$ has a unique solution exactly when $\text{rank}(A) = n$. By Theorem 56, this is equivalent to existence of A^{-1}. Thus, $AX = B$ has a unique solution exactly when A is nonsingular.

When A^{-1} exists, we can multiply both sides of $AX = B$ on the left by A^{-1} to get $X = A^{-1}B$.

Proof of (2) By Corollary 1, Section 10.7, $AX = 0$ has a nontrivial solution if and only if $\text{rank}(A) < n$. By Theorem 56, this is equivalent to the assertion that A is singular.

EXAMPLE 277

Solve the system

$$\begin{aligned} 2x_1 - x_2 + 3x_3 &= 4, \\ x_1 + 9x_2 - 2x_3 &= -8, \\ 4x_1 - 8x_2 + 11x_3 &= 15, \end{aligned}$$

or show that it has no solution.

The system is

$$\begin{bmatrix} 2 & -1 & 3 \\ 1 & 9 & -2 \\ 4 & -8 & 11 \end{bmatrix} \begin{bmatrix} x_1 \\ x_2 \\ x_3 \end{bmatrix} = \begin{bmatrix} 4 \\ -8 \\ 15 \end{bmatrix}.$$

We find that

$$A^{-1} = \tfrac{1}{53} \begin{bmatrix} 83 & -13 & -25 \\ -19 & 10 & 7 \\ -44 & 12 & 19 \end{bmatrix}.$$

Thus, the system has a unique solution, and it is

$$X = A^{-1}B = \begin{bmatrix} \frac{61}{53} \\ -\frac{51}{53} \\ \frac{13}{53} \end{bmatrix}.$$

PROBLEMS FOR SECTION 10.9

In each of Problems 1 through 25, find A^{-1}, or show that A is singular, using the method of this section.*

1. $\begin{bmatrix} -1 & 2 \\ 2 & 1 \end{bmatrix}$

2. $\begin{bmatrix} 1 & 1 & -3 \\ 2 & 16 & 1 \\ 0 & 0 & 4 \end{bmatrix}$

3. $\begin{bmatrix} -3 & 4 & 1 \\ 1 & 2 & 0 \\ 1 & 1 & 3 \end{bmatrix}$

4. $\begin{bmatrix} -2 & 1 & -5 \\ 1 & 1 & 4 \\ 0 & 3 & 3 \end{bmatrix}$

5. $\begin{bmatrix} -2 & 1 & 1 \\ 0 & 1 & 1 \\ -3 & 0 & 6 \end{bmatrix}$

6. $\begin{bmatrix} -1 & 1 & 1 & 0 \\ 1 & 0 & 2 & 0 \\ 1 & 1 & 1 & 1 \\ 3 & 0 & 0 & 1 \end{bmatrix}$

7. $\begin{bmatrix} 12 & 1 & 14 \\ -3 & 2 & 0 \\ 0 & 9 & 14 \end{bmatrix}$

8. $\begin{bmatrix} 0 & 0 & -1 \\ 1 & 12 & 0 \\ 1 & -2 & 4 \end{bmatrix}$

9. $\begin{bmatrix} -1 & 1 & 16 & 2 \\ 0 & 0 & 1 & 4 \\ 0 & 0 & 1 & 6 \\ 0 & 1 & 1 & -3 \end{bmatrix}$

* Some pocket calculators will find matrix inverses, and it is a good idea to learn to use these; but you should do these problems by the method of the section for practice.

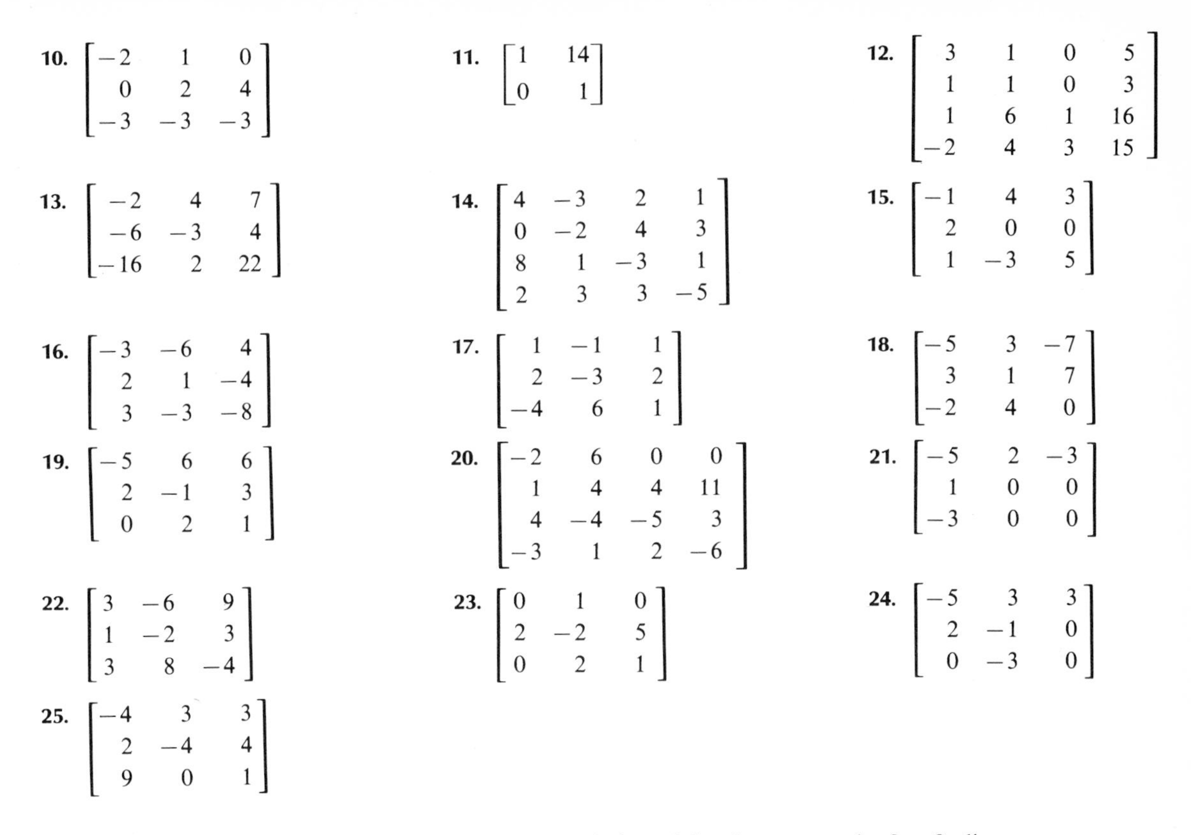

10. $\begin{bmatrix} -2 & 1 & 0 \\ 0 & 2 & 4 \\ -3 & -3 & -3 \end{bmatrix}$

11. $\begin{bmatrix} 1 & 14 \\ 0 & 1 \end{bmatrix}$

12. $\begin{bmatrix} 3 & 1 & 0 & 5 \\ 1 & 1 & 0 & 3 \\ 1 & 6 & 1 & 16 \\ -2 & 4 & 3 & 15 \end{bmatrix}$

13. $\begin{bmatrix} -2 & 4 & 7 \\ -6 & -3 & 4 \\ -16 & 2 & 22 \end{bmatrix}$

14. $\begin{bmatrix} 4 & -3 & 2 & 1 \\ 0 & -2 & 4 & 3 \\ 8 & 1 & -3 & 1 \\ 2 & 3 & 3 & -5 \end{bmatrix}$

15. $\begin{bmatrix} -1 & 4 & 3 \\ 2 & 0 & 0 \\ 1 & -3 & 5 \end{bmatrix}$

16. $\begin{bmatrix} -3 & -6 & 4 \\ 2 & 1 & -4 \\ 3 & -3 & -8 \end{bmatrix}$

17. $\begin{bmatrix} 1 & -1 & 1 \\ 2 & -3 & 2 \\ -4 & 6 & 1 \end{bmatrix}$

18. $\begin{bmatrix} -5 & 3 & -7 \\ 3 & 1 & 7 \\ -2 & 4 & 0 \end{bmatrix}$

19. $\begin{bmatrix} -5 & 6 & 6 \\ 2 & -1 & 3 \\ 0 & 2 & 1 \end{bmatrix}$

20. $\begin{bmatrix} -2 & 6 & 0 & 0 \\ 1 & 4 & 4 & 11 \\ 4 & -4 & -5 & 3 \\ -3 & 1 & 2 & -6 \end{bmatrix}$

21. $\begin{bmatrix} -5 & 2 & -3 \\ 1 & 0 & 0 \\ -3 & 0 & 0 \end{bmatrix}$

22. $\begin{bmatrix} 3 & -6 & 9 \\ 1 & -2 & 3 \\ 3 & 8 & -4 \end{bmatrix}$

23. $\begin{bmatrix} 0 & 1 & 0 \\ 2 & -2 & 5 \\ 0 & 2 & 1 \end{bmatrix}$

24. $\begin{bmatrix} -5 & 3 & 3 \\ 2 & -1 & 0 \\ 0 & -3 & 0 \end{bmatrix}$

25. $\begin{bmatrix} -4 & 3 & 3 \\ 2 & -4 & 4 \\ 9 & 0 & 1 \end{bmatrix}$

In each of Problems 26 through 31, find the unique solution of the given system by first finding A^{-1}, using the method of this section, and then calculating $X = A^{-1}B$.

26. $\begin{aligned} 3x_1 - 4x_2 + 6x_3 &= 0 \\ x_1 + x_2 - 3x_3 &= 4 \\ 2x_1 - x_2 + 6x_3 &= -1 \end{aligned}$

27. $\begin{aligned} x_1 - x_2 + 3x_3 - x_4 &= 1 \\ x_2 - 3x_3 + 5x_4 &= 2 \\ x_1 - x_3 + x_4 &= 0 \\ x_1 + 2x_2 - x_4 &= -5 \end{aligned}$

28. $\begin{aligned} 8x_1 - x_2 - x_3 &= 4 \\ x_1 + 2x_2 - 3x_3 &= 0 \\ 2x_1 - x_2 + 4x_3 &= 5 \end{aligned}$

29. $\begin{aligned} 2x_1 - 6x_2 + 3x_3 &= -4 \\ -x_1 + x_2 + x_3 &= 5 \\ 2x_1 + 6x_2 - 5x_3 &= 8 \end{aligned}$

30. $\begin{aligned} 12x_1 + x_2 - 3x_3 &= 4 \\ x_1 - x_2 + 3x_3 &= -5 \\ -2x_1 + x_2 + x_3 &= 0 \end{aligned}$

31. $\begin{aligned} 4x_1 + 6x_2 - 3x_3 &= 0 \\ 2x_1 + 3x_2 - 4x_3 &= 0 \\ x_1 - x_2 + 3x_3 &= -7 \end{aligned}$

32. Let A be an $n \times n$ matrix. Prove that if A is nonsingular, then every vector in R^n can be written as a linear combination of the columns of A, considered as vectors in R^n. (*Hint:* Use Theorem 56.)

33. Let A be an $n \times n$ matrix. Prove that A is nonsingular if and only if the rows of A, thought of as vectors in R^n, form a basis for R^n. (*Hint:* Use Theorem 55(4) and the result of Problem 32.)

34. Let A be a nonsingular matrix. Prove that A^t is nonsingular by showing that $(A^t)^{-1} = (A^{-1})^t$.

35. Let A be a nonsingular matrix. Prove that, for any positive integer k, A^k is nonsingular. (*Hint:* Prove that $(A^k)^{-1} = (A^{-1})^k$.)

36. Let A, B, and C be $n \times n$ matrices. Suppose that

$$BA = AC = I_n.$$

Prove that $B = C$.

37. Consider the following "proof" of Theorem 55(5): Suppose that AB is singular. If A and B are nonsingular, then so is AB by (2), a contradiction. Hence, A or B is singular. Why does this argument fail to establish (5)?

38. Try the following "proof" of Theorem 55(5): Suppose that A is singular. We want to show that AB is singular. If AB is nonsingular, then AB has an inverse, say C. Then $(AB)C = I_n$. Then $A(BC) = I_n$. Hence, BC is the inverse of A; so A must be nonsingular, a contradiction. Thus, AB is singular.

A similar argument shows that AB is singular if B is singular. Why does this argument fail?

10.10 DETERMINANTS: DEFINITION AND BASIC PROPERTIES

In this section we introduce the concept of determinant. The *determinant* of a square matrix A is a number, denoted $|A|$ or $\det(A)$, produced from the matrix in a way we shall now define, beginning with "small" matrices.

First, if A is 1×1, say $A = [a]$, then the determinant of A is defined by

$$|a| = a.$$

Thus, $|\pi| = \pi$ and $|-2| = -2$ (note that $|\ \ |$ denotes determinant here, not absolute value).

If A is 2×2, we define

$$\begin{vmatrix} A_{11} & A_{12} \\ A_{21} & A_{22} \end{vmatrix} = A_{11}A_{22} - A_{12}A_{21}.$$

For example,

$$\begin{vmatrix} 6 & -3 \\ 2 & 1 \end{vmatrix} = 6(1) - (-3)(2) = 12$$

and

$$\begin{vmatrix} 12 & \sqrt{2} \\ 3 & 4 \end{vmatrix} = 48 - 3\sqrt{2}.$$

For 3×3 matrices, we set

$$\begin{vmatrix} A_{11} & A_{12} & A_{13} \\ A_{21} & A_{22} & A_{23} \\ A_{31} & A_{32} & A_{33} \end{vmatrix} = A_{11}A_{22}A_{33} - A_{11}A_{23}A_{32} - A_{12}A_{21}A_{33} + A_{12}A_{23}A_{31} + A_{13}A_{21}A_{32} - A_{13}A_{22}A_{31}.$$

For example,

$$\begin{vmatrix} 8 & -2 & 4 \\ 1 & 0 & -3 \\ -2 & 8 & 3 \end{vmatrix} = 218.$$

In general, the determinant of an $n \times n$ matrix is called an *$n \times n$ determinant* or a *determinant of order n*. We shall now define $|A|$, for $n \geq 3$, as a sum of multiples of determinants of $(n-1) \times (n-1)$ matrices. Each of these is in turn a sum of multiples of $(n-2) \times (n-2)$ determinants, and we can (in theory) keep applying the definition until all determinants involved are 2×2 or 3×3, for which we have explicit formulas.

Some notation will help describe this process. If A is an $n \times n$ matrix, and $n \geq 2$, we let

$$M_{ij} = \text{determinant of the } (n-1) \times (n-1) \text{ matrix formed by deleting row } i \text{ and column } j \text{ of } A.$$

We call M_{ij} the *minor of A_{ij} in A*. The *cofactor of A_{ij} in A* is defined to be

$$(-1)^{i+j} M_{ij}.$$

For example, let

$$A = \begin{bmatrix} -1 & 6 & 2 \\ 0 & 1 & 4 \\ 8 & -3 & 7 \end{bmatrix}.$$

From the formula for 2×2 determinants, we have

$$M_{11} = \text{minor of } -1 \text{ in } A = \begin{vmatrix} 1 & 4 \\ -3 & 7 \end{vmatrix} = 19, \quad \text{(cover up row 1, column 1 of } A)$$

$$M_{23} = \text{minor of 4 in } A = \begin{vmatrix} -1 & 6 \\ 8 & -3 \end{vmatrix} = -45, \quad \text{(cover up row 2, column 3 of } A)$$

and

$$M_{12} = \text{minor of 6 in } A = \begin{vmatrix} 0 & 4 \\ 8 & 7 \end{vmatrix} = -32. \quad \text{(cover up row 1, column 2 of } A)$$

Then the cofactors of -1, 4, and 6 in A are, respectively,

$$(-1)^{1+1}(19) = 19, \qquad (-1)^{2+3}(-45) = 45, \quad \text{and} \quad (-1)^{1+2}(-32) = 32.$$

Now suppose that A is $n \times n$, and $n \geq 2$. The *cofactor* (or *Laplace*) *expansion of* $|A|$ *by row* k is defined to be the sum over row k of each row entry multiplied by its cofactor:

$$\sum_{j=1}^{n} (-1)^{k+j} A_{kj} M_{kj},$$

for any $k = 1, 2, \ldots, n$.

The *cofactor* (or *Laplace*) *expansion by column* k is the sum over column k of each column entry multiplied by its cofactor:

$$\sum_{i=1}^{n} (-1)^{i+k} A_{ik} M_{ik},$$

for any $k = 1, 2, \ldots, n$.

It is a remarkable fact that the n possible row cofactor expansions of $|A|$ and the n possible column cofactor expansions all yield exactly the same number. This is stated below as Theorem 58, and allows us to *define* $|A|$ for any $n \times n$ matrix A by the cofactor expansions. For example, if A is 4×4, then the cofactor expansion by any row or column gives $|A|$ as a sum of 3×3 determinants, which have been explicitly defined.

To get some feeling for cofactor expansions, look at the cases $n = 2$ and $n = 3$. If $n = 2$, then

$$A = \begin{bmatrix} A_{11} & A_{12} \\ A_{21} & A_{22} \end{bmatrix}.$$

The four possible row and column cofactor expansions of $|A|$ are:

by row 1:

$$\begin{aligned}\sum_{j=1}^{2} (-1)^{1+j} A_{1j} M_{1j} &= (-1)^{1+1} A_{11} M_{11} + (-1)^{1+2} A_{12} M_{12} \\ &= A_{11}A_{22} - A_{12}A_{21};\end{aligned}$$

by row 2:

$$\begin{aligned}\sum_{j=1}^{2} (-1)^{2+j} A_{2j} M_{2j} &= (-1)^{2+1} A_{21} A_{12} + (-1)^{2+2} A_{22} A_{11} \\ &= -A_{21}A_{12} + A_{22}A_{11};\end{aligned}$$

by column 1:

$$\begin{aligned}\sum_{i=1}^{2} (-1)^{i+1} A_{i1} M_{i1} &= (-1)^{1+1} A_{11} A_{22} + (-1)^{2+1} A_{21} A_{12} \\ &= A_{11}A_{22} - A_{21}A_{22};\end{aligned}$$

and by column 2:

$$\begin{aligned}\sum_{i=1}^{2} (-1)^{i+2} A_{i2} M_{i2} &= (-1)^{1+2} A_{12} A_{21} + (-1)^{2+2} A_{22} A_{11} \\ &= -A_{12}A_{21} + A_{22}A_{11}.\end{aligned}$$

These four cofactor expansions all yield exactly the same result, namely $|A|$.

When $n = 3$, there are nine row and column expansions possible. By way of illustration, here is the expansion of $|A|$ by column 2 when A is 3×3:

$$\begin{aligned}\sum_{i=1}^{3} (-1)^{2+i} A_{i2} M_{i2} &= -A_{12} \begin{vmatrix} A_{21} & A_{23} \\ A_{31} & A_{33} \end{vmatrix} + A_{22} \begin{vmatrix} A_{11} & A_{13} \\ A_{31} & A_{33} \end{vmatrix} - A_{32} \begin{vmatrix} A_{11} & A_{13} \\ A_{21} & A_{23} \end{vmatrix} \\ &= -A_{12}(A_{21}A_{33} - A_{23}A_{31}) + A_{22}(A_{11}A_{33} - A_{31}A_{13}) \\ &\quad - A_{32}(A_{11}A_{23} - A_{13}A_{21}) \\ &= |A|,\end{aligned}$$

as defined previously.

THEOREM 58

The cofactor expansions of $|A|$ by any row or column yield the same number for any given A.

We defer a proof of this to the end of this section in order to turn immediately to properties of determinants. These properties will be important in developing methods for evaluating determinants with a minimum of labor.

THEOREM 59

If B is formed from A by multiplying any row (or column) of A by a scalar α, then

$$|B| = \alpha|A|.$$

Proof Suppose we multiply row k by α. Expanding $|B|$ by row k, we get

$$|B| = \sum_{i=1}^{n} (-1)^{k+1} B_{ki} M_{ki}.$$

But, $B_{ki} = \alpha A_{ki}$. Further, the minor M_{ki} is the same for B as for A, since we obtain M_{ki} by deleting row k and column i, and A and B are the same except possibly in row k. Thus,

$$|B| = \sum_{i=1}^{n} (-1)^{k+i} \alpha A_{ki} M_{ki} = \alpha \sum_{i=1}^{n} (-1)^{k+i} A_{ki} M_{ki} = \alpha|A|.$$

A similar argument holds for the columns.

THEOREM 60

If A has a zero row or column, then $|A| = 0$.

Proof Form B from A by multiplying the zero row or column by 2. By Theorem 59, $|B| = 2|A|$. But $B = A$; so $|B| = |A|$. Then, $|A| = 2|A|$; hence, $|A| = 0$.

THEOREM 61

If B is obtained from A by interchanging two rows (or columns), then $|B| = -|A|$.

Proof We proceed by induction on the order of A. If A is 2×2, say,

$$A = \begin{bmatrix} A_{11} & A_{12} \\ A_{21} & A_{22} \end{bmatrix},$$

then

$$B = \begin{bmatrix} A_{21} & A_{22} \\ A_{11} & A_{12} \end{bmatrix} \quad \text{if we interchange two rows of } A, \text{ and}$$

$$B = \begin{bmatrix} A_{12} & A_{11} \\ A_{22} & A_{21} \end{bmatrix} \quad \text{if we interchange two columns of } A.$$

In either case, check directly that $|B| = -|A|$.

Now assume that Theorem 61 holds for $(n-1) \times (n-1)$ matrices, and let A be $n \times n$. Say we form B by interchanging rows i and j. Expand by another row k to get

$$|B| = \sum_{s=1}^{n} (-1)^{k+s} B_{ks} M_{ks}$$

and

$$|A| = \sum_{s=1}^{n} (-1)^{k+s} A_{ks} N_{ks},$$

where N_{ks} is the minor of A formed by deleting row k, column s of A. For $s = 1, \ldots, n$, we get N_{ks} from M_{ks} by interchanging rows i and j. By the induction hypothesis, recalling that N_{ks} and M_{ks} are $(n-1) \times (n-1)$ determinants, we have

$$N_{ks} = -M_{ks} \quad \text{for} \quad s = 1, \ldots, n.$$

Hence, $|B| = -|A|$.

A similar argument holds if columns are interchanged.

EXAMPLE 278

Expanding by any row or column, we get

$$\begin{vmatrix} 4 & -1 & 6 \\ 1 & 9 & 3 \\ 2 & 1 & 4 \end{vmatrix} = 28, \quad \text{while}$$

$$\begin{vmatrix} 2 & 1 & 4 \\ 1 & 9 & 3 \\ 4 & -1 & 6 \end{vmatrix} = -28 \quad \text{(switch rows 1 and 3)}$$

and

$$\begin{vmatrix} 4 & 6 & -1 \\ 1 & 3 & 9 \\ 2 & 4 & 1 \end{vmatrix} = -28 \quad \text{(switch columns 2 and 3).}$$

THEOREM 62

If A has two identical rows or columns, then $|A| = 0$.

Proof Suppose that A has two identical rows (or columns). Form B from A by interchanging the two identical rows (or columns). By Theorem 61, $|B| = -|A|$. But, $B = A$ because the rows (or columns) interchanged were identical. Thus, $|A| = -|A|$, and we must have $|A| = 0$.

EXAMPLE 279

$$\begin{vmatrix} -1 & 3 & 4 & 2 \\ 0 & 1 & 1 & -5 \\ 2 & 6 & 17 & 3 \\ 0 & 1 & 1 & -5 \end{vmatrix} = 0$$

immediately because rows 2 and 4 are identical.

THEOREM 63

If one row (or column) of A is a constant multiple of another, then $|A| = 0$.

Proof We shall give the argument for columns; a similar argument holds if one row is a constant multiple of another row.

Suppose then that column i of A is a constant multiple α of column j. If $\alpha = 0$, then $|A| = 0$ by Theorem 60. If $\alpha \neq 0$, then we can obtain from A a matrix B with two identical columns by multiplying column i of A by $1/\alpha$. By Theorem 62, $|A| = 0$.

EXAMPLE 280

$$\begin{vmatrix} -1 & 4 & 2 & 1 \\ 8 & \sqrt{2} & 4 & 0 \\ -2 & 8 & 4 & 2 \\ 1 & 19 & 0 & -4 \end{vmatrix} = 0, \quad \text{since row 3 is twice row 1.}$$

THEOREM 64

If we obtain B from A by adding a constant multiple of one row (or column) to another row (or column), then

$$|B| = |A|.$$

Proof Suppose we add α(row i) to row j of A to form B. Thus,

$$B = \begin{bmatrix} A_{11} & A_{12} & \cdots & A_{1n} \\ \vdots & \vdots & & \vdots \\ \alpha A_{i1} + A_{j1} & \alpha A_{i2} + A_{j2} & \cdots & \alpha A_{in} + A_{jn} \\ \vdots & \vdots & & \vdots \\ A_{n1} & A_{n2} & \cdots & A_{nn} \end{bmatrix} \leftarrow \text{Row } j.$$

Expand $|B|$ by row j to get

$$\begin{aligned} |B| &= \sum_{k=1}^{n} (-1)^{j+k}(\alpha A_{ik} + A_{jk})M_{jk} \\ &= \sum_{k=1}^{n} (-1)^{j+k}\alpha A_{ik} M_{jk} + \sum_{k=1}^{n} (-1)^{j+k} A_{jk} M_{jk}. \end{aligned}$$

Now, $\sum_{k=1}^{n} (-1)^{j+k} A_{jk} M_{jk} = |A|$, since we form M_{jk} by deleting row j, column k of B, and B is the same as A except for row j.

Next,

$$\sum_{k=1}^{n} (-1)^{j+k}\alpha A_{ik} M_{jk} = 0,$$

since this is the expansion by row j of the determinant formed from A by replacing row j by α(row i), and this is zero by Theorem 63.

A similar argument holds for columns.

EXAMPLE 281

It is easy to check that

$$\begin{vmatrix} 4 & -2 & 1 \\ 3 & 0 & -5 \\ 1 & -3 & -4 \end{vmatrix} = -83.$$

If we add -3(row 2) to row 3, we get

$$\begin{vmatrix} 4 & -2 & 1 \\ 3 & 0 & -5 \\ -8 & -3 & 11 \end{vmatrix},$$

and this determinant is also -83.

THEOREM 65

$|A| = |A^t|$ for any $n \times n$ matrix A.

Proof Recall that we form A^t, the transpose of A, by rewriting the rows of A as columns of A^t. Thus, the cofactor expansion of $|A^t|$ by any row is exactly the cofactor expansion of $|A|$ by the corresponding column. By Theorem 58, $|A^t| = |A|$.

THEOREM 66

If A and B are $n \times n$ matrices, then

$$|AB| = |A|\,|B|.$$

In words, the determinant of a product is the product of the determinants.

Proof The general argument is notationally messy; so we shall develop the details for the case $n = 2$ to explain the ideas involved. Suppose then that A and B are 2×2 matrices.

Begin by looking at the 4×4 matrix

$$P = \begin{bmatrix} A_{11} & A_{12} & 0 & 0 \\ A_{21} & A_{22} & 0 & 0 \\ -1 & 0 & B_{11} & B_{12} \\ 0 & -1 & B_{21} & B_{22} \end{bmatrix}.$$

We shall show that $|P| = |A|\,|B|$ and also that $|P| = |AB|$, and hence conclude that $|A|\,|B| = |AB|$ (at least for $n = 2$).

To show that $|P| = |A|\,|B|$, expand $|P|$ by row 1 to get

$$|P| = A_{11}\begin{vmatrix} A_{22} & 0 & 0 \\ 0 & B_{11} & B_{12} \\ -1 & B_{21} & B_{22} \end{vmatrix} - A_{12}\begin{vmatrix} A_{21} & 0 & 0 \\ -1 & B_{11} & B_{12} \\ 0 & B_{21} & B_{22} \end{vmatrix}.$$

Expanding each of these 3×3 determinants by row 1 gives us

$$|P| = A_{11}A_{22}\begin{vmatrix} B_{11} & B_{12} \\ B_{21} & B_{22} \end{vmatrix} - A_{12}A_{21}\begin{vmatrix} B_{11} & B_{12} \\ B_{21} & B_{22} \end{vmatrix}$$

$$= (A_{11}A_{22} - A_{12}A_{21})\begin{vmatrix} B_{11} & B_{12} \\ B_{21} & B_{22} \end{vmatrix} = |A|\,|B|.$$

To show that $|P| = |AB|$, first form a new matrix P' from P by

adding A_{11}(row 3) to row 1 and A_{12}(row 4) to row 1

and

adding A_{21}(row 3) to row 2 and A_{22}(row 4) to row 2.

Then,

$$P' = \begin{bmatrix} 0 & 0 & A_{11}B_{11} + A_{12}B_{21} & A_{11}B_{12} + A_{12}B_{22} \\ 0 & 0 & A_{21}B_{11} + A_{22}B_{21} & A_{21}B_{12} + A_{22}B_{22} \\ -1 & 0 & B_{11} & B_{12} \\ 0 & -1 & B_{21} & B_{22} \end{bmatrix}.$$

Recognize the 2×2 block in the upper right corner of P' as the entries of the product AB. Letting $C = AB$, then

$$P' = \begin{vmatrix} 0 & 0 & C_{11} & C_{12} \\ 0 & 0 & C_{21} & C_{22} \\ -1 & 0 & B_{11} & B_{12} \\ 0 & -1 & B_{21} & B_{22} \end{vmatrix}.$$

Now, by Theorem 64, $|P| = |P'|$. Expanding $|P'|$ by column 1 gives us

$$|P| = |P'| = (-1)\begin{vmatrix} 0 & C_{11} & C_{12} \\ 0 & C_{21} & C_{22} \\ -1 & B_{21} & B_{22} \end{vmatrix}.$$

Finally, expanding this 3×3 determinant by column 1 gives us

$$|P| = |P'| = \begin{vmatrix} C_{11} & C_{12} \\ C_{21} & C_{22} \end{vmatrix} = |AB|.$$

For $n \geq 3$, one proceeds in much the same way. Here is a sketch of the details. First, define the $2n \times 2n$ matrix

$$P = \left[\begin{array}{ccccc:ccc} & & & & & 0 & \cdots & 0 \\ & & & & & 0 & \cdots & 0 \\ & & A & & & \vdots & & \vdots \\ & & & & & 0 & \cdots & 0 \\ \hdashline -1 & 0 & 0 & \cdots & 0 & & & \\ 0 & -1 & 0 & \cdots & 0 & & B & \\ \vdots & \vdots & \vdots & & \vdots & & & \\ 0 & 0 & 0 & \cdots & -1 & & & \end{array}\right]$$

with A in the upper left corner, B in the lower right, zeros above B, and $-I_n$ below A. The strategy is to show that $|P| = |A|\,|B|$ and also that $|P| = |AB|$.

To show that $|P| = |A||B|$, expand $|P|$ by row 1, obtaining $|P|$ as a sum of n determinants of order $2n - 1$; expand each of these by row 1, and so on. After n such steps, we obtain an expression consisting of a sum of terms adding up to $|A|$, each with a factor of $|B|$. Thus, $|P| = |A||B|$.

To show that $|P| = |AB|$, form a new matrix P' from P as follows. First, add

$$\begin{gathered} A_{11}[\text{row}(n+1)] \text{ to row } 1, \\ A_{12}[\text{row}(n+2)] \text{ to row } 1, \\ \vdots \\ A_{1n}[\text{row}(2n)] \text{ to row } 1. \end{gathered}$$

The first row of the matrix formed thus far is

$$\underbrace{0 \quad 0 \quad \cdots \quad 0}_{n} \quad A_{11}B_{11} + A_{12}B_{21} + \cdots + A_{1n}B_{n1} \quad \cdots$$

$$A_{11}B_{1n} + A_{12}B_{2n} + \cdots + A_{1n}B_{nn}.$$

Now add

$$\begin{gathered} A_{21}[\text{row}(n+1)] \text{ to row } 2, \\ A_{22}[\text{row}(n+2)] \text{ to row } 2, \\ \vdots \\ A_{2n}[\text{row}(2n)] \text{ to row } 2, \end{gathered}$$

and so on. Eventually we obtain

$$P' = \left[\begin{array}{cccc:c} 0 & 0 & \cdots & 0 & \\ \vdots & & & \vdots & C \\ 0 & 0 & \cdots & 0 & \\ \hdashline -1 & 0 & \cdots & 0 & \\ \vdots & & & \vdots & B \\ 0 & 0 & \cdots & -1 & \end{array}\right];$$

where $C = AB$. By Theorem 64, $|P'| = |P|$. Expand $|P'|$ by column 1; then expand the resulting $(2n - 1) \times (2n - 1)$ determinant by its column 1, and so on. After n such expansions, we are left with

$$|P| = |P'| = |C| = |AB|.$$

EXAMPLE 282

Let

$$A = \begin{bmatrix} -4 & 6 & 3 \\ 8 & 1 & 1 \\ -2 & 0 & 7 \end{bmatrix}$$

and

$$B = \begin{bmatrix} 14 & 2 & -3 \\ -6 & 1 & -1 \\ 4 & 1 & 4 \end{bmatrix}.$$

Then

$$AB = \begin{bmatrix} -80 & 1 & 18 \\ 110 & 18 & -21 \\ 0 & 3 & 34 \end{bmatrix}.$$

One can check that $|A| = -370$, $|B| = 140$, and $|AB| = -51{,}800 = |A||B|$.

We conclude this section with a sketch of the proof of Theorem 58. The proof is by induction on n. When $n = 2$, we have already shown explicitly that both row and both column cofactor expansions of $|A|$ yield the same value, namely $A_{11}A_{22} - A_{12}A_{21}$. To complete the induction, we must show that the statement of Theorem 58 is true for $(n+1) \times (n+1)$ determinants if it holds for $n \times n$ determinants.

Thus assume that, for any $n \times n$ matrix, the cofactor expansions by all rows and columns are equal. Now, let A be any $(n+1) \times (n+1)$ matrix.

First look at the cofactor expansion of $|A|$ by rows i and j with, say, $i < j$.

By row i:

$$\sum_{t=1}^{n+1} (-1)^{i+t} A_{it} M_{it} \tag{a}$$

By row j:

$$\sum_{s=1}^{n+1} (-1)^{j+s} A_{js} M_{js} \tag{b}$$

Now, each M_{it} in the expansion by row i is an $n \times n$ determinant, and hence may be expanded by any row. Expand M_{it} by its row $(j-1)$, which contains the entries of row j of A (with column t deleted). The terms in the cofactor expansion of M_{it} by its row $(j-1)$ are of the form

$$(-1)^{j-1+s} A_{js} M_{itjs} \quad \text{if} \quad s < t$$

and

$$(-1)^{j+1+s-1} A_{js} M_{itjs} \quad \text{if} \quad s > t,$$

where M_{ijts} is the j, s minor of M_{it}. The reason for two cases, $s < t$ or $s > t$, is that column s of M_{ij} is column s of A (with A_{is} deleted) if $s < t$; but if $s > t$, then deletion of column t from A to form M_{it} makes column s of M_{ij} the $(s+1)$ column of A.

Thus, in (a), typical terms are of the form

$$(-1)^{i+t+j-1+s} A_{it} A_{js} M_{itjs} \quad \text{if} \quad s < t$$

and

$$(-1)^{i+t+j+s-2} A_{it} A_{js} M_{itjs} \quad \text{if} \quad s > t.$$

A similar argument leads us to the conclusion that typical terms in (b) are

$$(-1)^{j+s+i+t} A_{js} A_{it} M_{jsit} \quad \text{if} \quad t < s$$

and

$$(-1)^{j+s+i+t-1} A_{js} A_{it} M_{jsit} \quad \text{if} \quad t > s.$$

Now, $M_{itjs} = M_{jsit}$, since both are formed from $|A|$ by deleting rows i and j and columns s and t. Comparing terms, we see that the $s < t$ terms in (a) match exactly with those in (b); similarly, the $s > t$ terms match. Thus, the expansions by rows i and j result in the same number.

A similar argument by term by term comparison shows that the cofactor expansion by any column agrees with that by any row. We leave details of this to the student.

Using the properties of determinants derived in this section, it is possible to expand determinants much more efficiently than by straightforward row or column expansion. We shall discuss ideas along these lines in the next section.

PROBLEMS FOR SECTION 10.10

In each of Problems 1 through 20, evaluate the given determinant by a row cofactor expansion (choose any row) and by a column cofactor expansion (choose any column).

1. $\begin{vmatrix} -4 & 6 \\ 1 & 7 \end{vmatrix}$

2. $\begin{vmatrix} 8 & -3 \\ -1 & 4 \end{vmatrix}$

3. $\begin{vmatrix} 16 & 2 \\ -3 & -4 \end{vmatrix}$

4. $\begin{vmatrix} -8 & -4 & 2 \\ 0 & 1 & 1 \\ 4 & 1 & 3 \end{vmatrix}$

5. $\begin{vmatrix} 2 & -2 & 1 \\ 1 & 1 & 6 \\ 3 & -1 & 4 \end{vmatrix}$

6. $\begin{vmatrix} 7 & -3 & 1 \\ 1 & -2 & 4 \\ -3 & 1 & 0 \end{vmatrix}$

7. $\begin{vmatrix} -14 & -3 & 2 \\ 1 & -1 & 1 \\ 0 & 1 & -3 \end{vmatrix}$

8. $\begin{vmatrix} 5 & -1 & 7 \\ 0 & 0 & 2 \\ 1 & -4 & 3 \end{vmatrix}$

9. $\begin{vmatrix} 8 & 8 & 8 \\ 4 & 4 & 4 \\ -1 & 3 & 0 \end{vmatrix}$

10. $\begin{vmatrix} 2 & 20 & -5 \\ 7 & -9 & 3 \\ 1 & -1 & 4 \end{vmatrix}$

11. $\begin{vmatrix} -4 & 3 & 7 \\ 0 & 1 & 4 \\ -5 & 0 & 0 \end{vmatrix}$

12. $\begin{vmatrix} -15 & 12 & 0 \\ 1 & 14 & 1 \\ -2 & 0 & 3 \end{vmatrix}$

13. $\begin{vmatrix} 8 & -5 & 4 \\ 3 & -2 & 1 \\ 1 & -1 & 4 \end{vmatrix}$

14. $\begin{vmatrix} 8 & -8 & 4 \\ 2 & 3 & -7 \\ 1 & -1 & 2 \end{vmatrix}$

15. $\begin{vmatrix} -5 & 1 & 6 \\ 1 & -1 & 1 \\ 0 & -1 & 0 \end{vmatrix}$

16. $\begin{vmatrix} 5 & -4 & 3 \\ -1 & 1 & 6 \\ -2 & 2 & 4 \end{vmatrix}$

17. $\begin{vmatrix} -5 & 0 & 1 & 6 \\ 2 & -1 & 3 & 7 \\ 4 & 4 & -5 & -8 \\ 1 & -1 & 6 & 2 \end{vmatrix}$

18. $\begin{vmatrix} 4 & 3 & -5 & 6 \\ 1 & -5 & 15 & 2 \\ 0 & -5 & 1 & 7 \\ 8 & 9 & 0 & 15 \end{vmatrix}$

19. $\begin{vmatrix} 22 & -1 & 3 & 0 & 0 \\ 1 & 4 & -5 & 0 & 2 \\ -1 & 1 & 6 & 0 & -5 \\ 4 & 7 & 9 & 1 & -7 \\ 6 & 6 & -3 & 4 & 1 \end{vmatrix}$

20. $\begin{vmatrix} 5 & -3 & 0 & 1 & -4 \\ -5 & -3 & 2 & -3 & 1 \\ 6 & -3 & 1 & 4 & 0 \\ 0 & 0 & 5 & 5 & 0 \\ 0 & 0 & -3 & -2 & 0 \end{vmatrix}$

21. Write out complete details of the proof of Theorem 66 for the case $n = 3$.

22. Let A be any $n \times n$ matrix and α any scalar. Form B from A by multiplying every entry of A by α, that is, $B = \alpha A$. Prove that

$$|B| = \alpha^n |A|.$$

23. Let (x_1, y_1), (x_2, y_2), and $x_3, y_3)$ be points in the plane. Prove that these points lie on a straight line if and only if

$$\begin{vmatrix} 1 & x_1 & y_1 \\ 1 & x_2 & y_2 \\ 1 & x_3 & y_3 \end{vmatrix} = 0.$$

24. An $n \times n$ matrix A is called *skew-symmetric* if $A = -A^t$. Prove that $|A| = 0$ if A is skew-symmetric and n is odd.
25. In Section 10.13 we shall prove that A is nonsingular exactly when $|A| \neq 0$. Assuming this for now, prove that if A is nonsingular, then $|A^{-1}| = 1/|A|$.
26. An $n \times n$ matrix A is called *orthogonal* if $A^t = A^{-1}$. Prove that the determinant of an orthogonal matrix must be 1 or -1.
27. Let A be $n \times n$, and let α be any nonzero scalar. Multiply A_{ij} by α^{i-j} to form a new matrix B. How are $|A|$ and $|B|$ related?

10.11 PRACTICE IN EVALUATING DETERMINANTS

Evaluating determinants efficiently is in some sense more an art than a science. In this section we shall explain, primarily by example, a technique that is often useful, combining row and column cofactor expansions with the effects of row and column operations on determinants, as discussed in Section 10.10.

Look at the cofactor expansion of det A by any row (or column), say,

$$|A| = \sum_{j=1}^{n} (-1)^{k+j} A_{kj} M_{kj}.$$

If A is $n \times n$, this sum usually involves working out n determinants of $(n-1) \times (n-1)$ matrices. When n is large, this is not much help. However, observe that each j such that $A_{kj} = 0$ reduces the labor by one term. This leads to our first practical rule: Expand by a row (or column) having as many zeros as possible. For example,

$$A = \begin{bmatrix} -3 & 1 & 16 & -8 \\ 0 & 1 & 14 & 0 \\ 0 & 3 & 0 & 1 \\ 0 & 14 & 6 & 0 \end{bmatrix}$$

has only one nonzero entry in column 1. Expanding by this column, we have

$$|A| = -3 \begin{vmatrix} 1 & 14 & 0 \\ 3 & 0 & 1 \\ 14 & 6 & 0 \end{vmatrix},$$

and only one determinant appears in the cofactor expansion on the right because $A_{j1} = 0$ for $j = 2, 3, 4$. Next, the determinant above has only one nonzero entry in column 3; so we can expand it by that column to get

$$\begin{vmatrix} 1 & 14 & 0 \\ 3 & 0 & 1 \\ 14 & 6 & 0 \end{vmatrix} = (-1)^{2+3}(1)\begin{vmatrix} 1 & 14 \\ 14 & 6 \end{vmatrix}.$$

We are now down to a determinant that we can do by sight:

$$\begin{vmatrix} 1 & 14 \\ 14 & 6 \end{vmatrix} = 6 - 14^2 = -190.$$

Working back through the intermediate steps, we have

$$|A| = (-3)(-1)(-190) = -570.$$

Of course, this example was rigged to illustrate the use of the practical rule stated above. What if A has no rows or columns consisting mostly of zeros? *Then make some.* Use elementary row and column operations to get a row or column with as many zeros as possible. Then expand the determinant of the resulting matrix by cofactors, using this row or column, and go to work by the same method on the smaller determinant(s) appearing in the expansion. Eventually, we get everything in terms of 2×2 determinants (or 3×3 determinants if we stop the process there), and the rest is easy.

Here are some examples.

EXAMPLE 283

Let

$$A = \begin{bmatrix} -5 & 1 & 1 & -6 & 2 \\ 0 & 1 & 3 & 8 & -4 \\ -1 & 2 & 1 & 8 & 9 \\ 0 & 0 & 1 & 14 & 2 \\ 1 & 1 & 0 & 0 & 0 \end{bmatrix}.$$

Notice that row 5 has three zeros. One more would be nice, as then we could expand by that row and write $|A|$ as a multiple of one 4×4 determinant. Add $-$(column 1) to column 2 to get

$$\begin{bmatrix} -5 & 6 & 1 & -6 & 2 \\ 0 & 1 & 3 & 8 & -4 \\ -1 & 3 & 1 & 8 & 9 \\ 0 & 0 & 1 & 14 & 2 \\ 1 & 0 & 0 & 0 & 0 \end{bmatrix}.$$

By Theorem 64, this matrix has the same determinant as A, and expanding by row 5 gives us

$$|A| = (-1)^{1+5}(1)\begin{vmatrix} 6 & 1 & -6 & 2 \\ 1 & 3 & 8 & -4 \\ 3 & 1 & 8 & 9 \\ 0 & 1 & 14 & 2 \end{vmatrix}.$$

We can get zeros as the 4-3 and 4-4 entries of the last matrix by adding -14(column 2) to column 3 and -2(column 2) to column 4, giving us

$$\begin{vmatrix} 6 & 1 & -20 & 0 \\ 1 & 3 & -34 & -10 \\ 3 & 1 & -6 & 7 \\ 0 & 1 & 0 & 0 \end{vmatrix}.$$

By Theorem 64 (twice), we have not changed the value of the resulting determinant, and expansion by row 4 gives us

$$|A| = \begin{vmatrix} 6 & 1 & -20 & 0 \\ 1 & 3 & -34 & -10 \\ 3 & 1 & -6 & 7 \\ 0 & 1 & 0 & 0 \end{vmatrix} = (-1)^{4+2}(1)\begin{vmatrix} 6 & -20 & 0 \\ 1 & -34 & -10 \\ 3 & -6 & 7 \end{vmatrix}.$$

Now we have a 3×3 determinant, which can be evaluated readily enough directly. Or, we can add -6(row 2) to row 1 and -3(row 2) to row 3 (again applying Theorem 64) to get

$$|A| = \begin{vmatrix} 0 & 184 & 60 \\ 1 & -34 & -10 \\ 0 & 96 & 37 \end{vmatrix} = (-1)^{1+2}(1)\begin{vmatrix} 184 & 60 \\ 96 & 37 \end{vmatrix} = -1048.$$

Expand by column 1

EXAMPLE 284

Let

$$A = \begin{bmatrix} -6 & 0 & 1 & 3 & 2 \\ -1 & 5 & 0 & 1 & 7 \\ 8 & 3 & 2 & 1 & 7 \\ 0 & 1 & 5 & -3 & 2 \\ 1 & 15 & -3 & 9 & 4 \end{bmatrix}.$$

If we add -2(row 1) to row 3, -5(row 1) to row 4, and 3(row 1) to row 5, we get

$$|A| = \begin{vmatrix} -6 & 0 & 1 & 3 & 2 \\ -1 & 5 & 0 & 1 & 7 \\ 20 & 3 & 0 & -5 & 3 \\ 30 & 1 & 0 & -18 & -8 \\ -17 & 15 & 0 & 18 & 10 \end{vmatrix} = (-1)^{1+3}(1)\begin{vmatrix} -1 & 5 & 1 & 7 \\ 20 & 3 & -5 & 3 \\ 30 & 1 & -18 & -8 \\ -17 & 15 & 18 & 10 \end{vmatrix}.$$

Expand by column 3

Now add 5(column 1) to column 2, column 1 to column 3, and 7(column 1) to column 4 in the last matrix, giving us

$$|A| = \begin{vmatrix} -1 & 0 & 0 & 0 \\ 20 & 103 & 15 & 143 \\ 30 & 151 & 12 & 202 \\ -17 & -70 & 1 & -109 \end{vmatrix} \underset{\uparrow}{=} (-1)^{1+1}(-1) \begin{vmatrix} 103 & 15 & 143 \\ 151 & 12 & 202 \\ -70 & 1 & -109 \end{vmatrix}.$$

Expand by
row 1

Finally, exploit the 1 in the 2-3 place of the last matrix to get zeros in the 2-2 and 1-2 places: Add -15(row 3) to row 1 and -12(row 3) to row 2, giving us

$$|A| = -\begin{vmatrix} 1153 & 0 & 1778 \\ 991 & 0 & 1510 \\ -70 & 1 & -109 \end{vmatrix} \underset{\uparrow}{=} -(-1)^{2+3}(1) \begin{vmatrix} 1153 & 1778 \\ 991 & 1510 \end{vmatrix}$$

Expand by
column 2

$$= (1153)(1510) - (1778)(991) = -20{,}968.$$

EXAMPLE 285

Let

$$A = \begin{bmatrix} 14 & 0 & 3 & 5 & -7 & 1 \\ 2 & -6 & 0 & 18 & 0 & 3 \\ -6 & -5 & 7 & 20 & 18 & 0 \\ 0 & 4 & 8 & 0 & -3 & 2 \\ 9 & -5 & 4 & 3 & 3 & 0 \\ 10 & 0 & 5 & 3 & 21 & 10 \end{bmatrix}.$$

This is a 6×6 matrix, and evaluating its determinant will take some effort. One way to proceed is the following: In A, add -3(row 1) to row 2, -2(row 1) to row 4, and -10(row 1) to row 6, giving

$$|A| = \begin{vmatrix} 14 & 0 & 3 & 5 & -7 & 1 \\ -40 & -6 & -9 & 3 & 21 & 0 \\ -6 & -5 & 7 & 20 & 18 & 0 \\ -28 & 4 & 2 & -10 & 11 & 0 \\ 9 & -5 & 4 & 3 & 3 & 0 \\ -130 & 0 & -25 & -47 & 91 & 0 \end{vmatrix}$$

Expand by
column 6

$$\overset{\downarrow}{=} (-1)^{1+6}(1) \begin{vmatrix} -40 & -6 & -9 & 3 & 21 \\ -6 & -5 & 7 & 20 & 18 \\ -28 & 4 & 2 & -10 & 11 \\ 9 & -5 & 4 & 3 & 3 \\ -130 & 0 & -25 & -47 & 91 \end{vmatrix}$$

Multiply row 3 by $\frac{1}{2}$

$$= -2\begin{vmatrix} -40 & -6 & -9 & 3 & 21 \\ -6 & -5 & 7 & 20 & 18 \\ -14 & 2 & 1 & -5 & \frac{11}{2} \\ 9 & -5 & 4 & 3 & 3 \\ -130 & 0 & -25 & -47 & 91 \end{vmatrix}$$

Add 9(row 3) to row 1,
-7(row 3) to row 2,
-4(row 3) to row 4,
25(row 3) to row 5

$$= -2\begin{vmatrix} -166 & 12 & 0 & -42 & \frac{141}{2} \\ 92 & -19 & 0 & 55 & -\frac{41}{2} \\ -14 & 2 & 1 & -5 & \frac{11}{2} \\ 65 & -13 & 0 & 23 & -19 \\ -480 & 50 & 0 & -172 & \frac{457}{2} \end{vmatrix}$$

Expand by column 3 — Add row 1 to row 3

$$= -2\begin{vmatrix} -166 & 12 & -42 & \frac{141}{2} \\ 92 & -19 & 55 & -\frac{41}{2} \\ 65 & -13 & 23 & -19 \\ -480 & 50 & -172 & \frac{457}{2} \end{vmatrix} = -2\begin{vmatrix} -166 & 12 & -42 & \frac{141}{2} \\ 92 & -19 & 55 & -\frac{41}{2} \\ -101 & -1 & -19 & \frac{103}{2} \\ -480 & 50 & -172 & \frac{457}{2} \end{vmatrix}$$

Add 12(row 3) to row 1,
-19(row 3) to row 2,
50(row 3) to row 4

$$= -2\begin{vmatrix} -1378 & 0 & -270 & \frac{1377}{2} \\ 2011 & 0 & 416 & -999 \\ -101 & -1 & -19 & \frac{103}{2} \\ -5530 & 0 & -1122 & \frac{5607}{2} \end{vmatrix}$$

Expand by column 2

$$= -2(-1)^{3+2}(-1)\begin{vmatrix} -1378 & -270 & \frac{1377}{2} \\ 2011 & 416 & -999 \\ -5530 & -1122 & \frac{5607}{2} \end{vmatrix}$$

Expand by row 1

$$= -2\left\{-1378\begin{vmatrix} 416 & -999 \\ -1122 & \frac{5607}{2} \end{vmatrix} + 270\begin{vmatrix} 2011 & -999 \\ -5530 & \frac{5607}{2} \end{vmatrix} + \frac{1377}{2}\begin{vmatrix} 2011 & 416 \\ -5530 & -1122 \end{vmatrix}\right\}$$

$$= -2(-1{,}532{,}376) = 3{,}064{,}752.$$

Note: In the previous two examples, we continued to use row and column operations until we finally obtained a single 2×2 determinant to evaluate. Here, when we reached a 3×3 matrix, we evaluated its determinant directly as a sum of multiples of three 2 × 2 determinants. This was because the particular 3 × 3 matrix encountered in Example 285 contained fairly large numbers, and the arithmetic of evaluating its determinant directly looked just as easy as performing another set of row or column operations.

Example 285 makes the point that evaluating even a relatively small determinant generally involves a lot of calculation. A pocket calculator* is a great help, if available. By hand, determinants of order 4, 5, or 6 are very tedious, and as n increases the calculations get out of practical range very quickly.

We shall conclude this section with a general result that is often useful. Recall from Problem 6 of Section 10.3 that a matrix U is called *upper triangular* if every entry below the main diagonal is zero. That is, $U_{ij} = 0$ if $i > j$. Such a matrix has the appearance of Figure 142.

$$\begin{bmatrix} U_{11} & U_{12} & \cdots & U_{1n} \\ 0 & U_{22} & \cdots & U_{2n} \\ \vdots & \vdots & & \vdots \\ 0 & 0 & \cdots & U_{nn} \end{bmatrix}$$

Figure 142. Upper triangular matrix.

THEOREM 67

The determinant of an upper triangular matrix is the product of its main-diagonal entries.

Proof Proceed by induction on the order of the matrix.

If A is a 1 × 1 matrix, then $A = [A_{11}]$. Here, A is upper triangular by default and $|A| = A_{11}$.

Now suppose that the theorem holds for all $(n-1) \times (n-1)$ upper triangular matrices. If A is $n \times n$, then expanding $|A|$ by column 1 gives us

$$|A| = \sum_{j=1}^{n} (-1)^{j+1} A_{j1} M_{j1} = A_{11} M_{11},$$

since $A_{21} = A_{31} = \cdots = A_{n1} = 0$.

Now, M_{11} is upper triangular and $(n-1) \times (n-1)$, with main-diagonal entries $A_{22}, \ldots, A_{nn}$, as you can see by covering up row 1, column 1 of A in Figure 142. Thus, by the induction assumption, $M_{11} = A_{22} \cdots A_{nn}$, proving the theorem.

EXAMPLE 286

$$\begin{vmatrix} -1 & -3 & 4 \\ 0 & \pi & 16 \\ 0 & 0 & 14 \end{vmatrix} = -14\pi.$$

---

* The TI-58, TI-59, and HP-41C will do quite large determinants (up to 9 × 9 or, under some circumstances, even higher).

Clearly a similar result can be stated for lower triangular matrices (all entries above the main diagonal are zero).

PROBLEMS FOR SECTION 10.11

In each of Problems 1 through 25, evaluate the given determinant.

1. $$\begin{vmatrix} -3 & 1 & 14 \\ 0 & 1 & 6 \\ 2 & -3 & 4 \end{vmatrix}$$

2. $$\begin{vmatrix} 8 & 14 & 0 \\ 1 & -2 & 3 \\ 0 & 1 & 16 \end{vmatrix}$$

3. $$\begin{vmatrix} -5 & 2 & 4 \\ 1 & -3 & 4 \\ 0 & 1 & 3 \end{vmatrix}$$

4. $$\begin{vmatrix} -8 & -5 & 2 \\ 3 & -2 & -14 \\ 3 & 4 & -7 \end{vmatrix}$$

5. $$\begin{vmatrix} 2 & -3 & -5 \\ -7 & 4 & -4 \\ 1 & 3 & 5 \end{vmatrix}$$

6. $$\begin{vmatrix} 15 & 3 & -5 \\ 1 & 7 & -4 \\ -3 & 2 & -5 \end{vmatrix}$$

7. $$\begin{vmatrix} 8 & 0 & 0 & 4 \\ 9 & 1 & -7 & 2 \\ -8 & 1 & 14 & 2 \\ 0 & 0 & 1 & -3 \end{vmatrix}$$

8. $$\begin{vmatrix} -5 & 4 & 1 & 7 \\ -9 & 3 & 2 & -5 \\ -2 & 0 & -1 & 1 \\ 1 & 14 & 0 & 3 \end{vmatrix}$$

9. $$\begin{vmatrix} 14 & 13 & -2 & 5 \\ 7 & 1 & 1 & 7 \\ 0 & 2 & 12 & 3 \\ 1 & -6 & 5 & 2 \end{vmatrix}$$

10. $$\begin{vmatrix} -3 & 0 & 1 & -5 \\ 1 & 0 & 3 & 0 \\ 1 & 7 & 6 & -1 \\ 2 & 4 & 2 & -3 \end{vmatrix}$$

11. $$\begin{vmatrix} -8 & 5 & 1 & 7 & 2 \\ 0 & 1 & 3 & 5 & -6 \\ 2 & 2 & 1 & 5 & 3 \\ 0 & 4 & 3 & 7 & 2 \\ 1 & 1 & -7 & -6 & 5 \end{vmatrix}$$

12. $$\begin{vmatrix} 5 & 15 & 3 & 1 & 7 & 2 \\ 0 & 0 & 1 & 4 & -5 & 2 \\ 1 & 7 & -1 & 3 & 1 & 9 \\ 0 & 0 & 1 & -3 & -1 & 4 \\ 1 & 1 & 7 & -4 & 1 & 6 \\ 1 & 0 & 0 & 3 & -9 & -4 \end{vmatrix}$$

13. $$\begin{vmatrix} 1 & 3 & -9 & 5 \\ -6 & 2 & 1 & -1 \\ 0 & 3 & 2 & -6 \\ 8 & 5 & 3 & -8 \end{vmatrix}$$

14. $$\begin{vmatrix} 4 & 100 & -3 \\ 296 & -4 & 10 \\ 92 & 0 & -25 \end{vmatrix}$$

15. $$\begin{vmatrix} 203 & 13 & 693 \\ -12 & 1 & 10 \\ 0 & -5 & 64 \end{vmatrix}$$

16. $$\begin{vmatrix} 10 & 1023 & -3 \\ -3 & 3 & 21 \\ 0 & -5 & 14 \end{vmatrix}$$

17. $$\begin{vmatrix} -54 & 23 & 15 \\ 33 & 10 & -34 \\ 21 & -32 & 15 \end{vmatrix}$$

18. $$\begin{vmatrix} 23 & -12 & 15 & 9 \\ 3 & 9 & 11 & 14 \\ -5 & 6 & 17 & 12 \\ 17 & 13 & -7 & 8 \end{vmatrix}$$

19. $$\begin{vmatrix} 12 & 15 & 283 & -45 \\ 19 & -4 & 0 & 19 \\ -13 & 10 & 4 & 15 \\ -54 & 0 & 0 & 23 \end{vmatrix}$$

20. $$\begin{vmatrix} -2 & 4 & 15 & -12 \\ 20 & -5 & 0 & 23 \\ 4 & 6 & -2 & 14 \\ -9 & 0 & 0 & 13 \end{vmatrix}$$

21. $$\begin{vmatrix} 17 & 12 & -5 & 0 \\ -1 & 0 & 12 & 0 \\ 21 & -15 & -1 & 2 \\ 13 & -12 & 9 & 6 \end{vmatrix}$$

22. $$\begin{vmatrix} 46 & 2 & 1 & 0 \\ -3 & 13 & -3 & 6 \\ -5 & 12 & 17 & -7 \\ 1 & -1 & 0 & 0 \end{vmatrix}$$

23. $$\begin{vmatrix} 15 & -3 & 2 & 12 \\ 17 & 6 & 0 & 14 \\ -3 & 5 & 5 & -5 \\ 3 & 0 & 0 & -2 \end{vmatrix}$$

24. $$\begin{vmatrix} 2 & 4 & -6 & 3 & 5 \\ -6 & 16 & 15 & 4 & -6 \\ 0 & 14 & 12 & 9 & 5 \\ -4 & 0 & 22 & 6 & -8 \\ 4 & -4 & 15 & 8 & 10 \end{vmatrix}$$

25. $$\begin{vmatrix} -5 & 6 & 1 & -1 & 0 & 2 \\ 1 & 3 & -1 & -1 & 3 & 1 \\ 4 & 2 & -2 & 1 & 1 & 0 \\ 0 & 0 & 3 & 1 & -2 & 4 \\ 1 & 0 & 0 & -2 & 1 & 7 \\ 0 & 0 & 1 & 1 & -1 & 7 \end{vmatrix}$$

26. Redo Theorem 67, replacing upper triangular ($U_{ij} = 0$ if $i > j$) with lower triangular ($L_{ij} = 0$ if $i < j$).

27. Show that

$$\begin{vmatrix} 1 & \alpha & \alpha^2 \\ 1 & \beta & \beta^2 \\ 1 & \gamma & \gamma^2 \end{vmatrix} = (\alpha - \beta)(\gamma - \alpha)(\beta - \gamma).$$

(This is called *Vandermonde's determinant.*)

28. Show that

$$\begin{vmatrix} \alpha & \beta & \gamma & \delta \\ \beta & \gamma & \delta & \alpha \\ \gamma & \delta & \alpha & \beta \\ \delta & \alpha & \beta & \gamma \end{vmatrix} = (\alpha + \beta + \gamma + \delta)(\beta - \alpha + \delta - \gamma) \begin{vmatrix} 0 & 1 & -1 & 1 \\ 1 & \gamma & \delta & \alpha \\ 1 & \delta & \alpha & \beta \\ 1 & \alpha & \beta & \gamma \end{vmatrix}$$

29. Show that

$$\begin{vmatrix} 1 & 1 & 1 & 1 \\ \alpha & \beta & \gamma & \delta \\ \alpha^2 & \beta^2 & \gamma^2 & \delta^2 \\ \alpha^3 & \beta^3 & \gamma^3 & \delta^3 \end{vmatrix} = (\beta - \alpha)(\beta - \gamma)(\beta - \delta)(\gamma - \alpha)(\delta - \alpha)(\delta - \gamma).$$

30. An $n \times n$ matrix A is called *block diagonal* if it has the appearance of Figure 143, where each D_i is $k_i \times k_i$, $k_1 + k_2 + \cdots + k_r = n$, and every entry of A not in the D_i's is zero. Show that $|A| = |D_1|\,|D_2| \cdots |D_r|$.

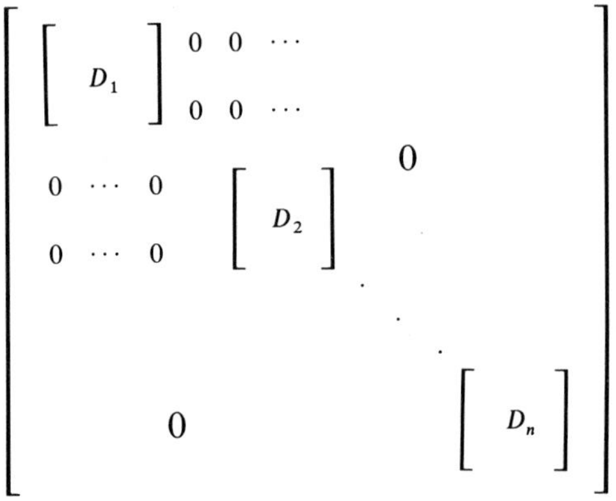

$$\begin{bmatrix} \begin{bmatrix} D_1 \end{bmatrix} & \begin{matrix} 0 & 0 & \cdots \\ 0 & 0 & \cdots \end{matrix} & & 0 \\ \begin{matrix} 0 & \cdots & 0 \\ 0 & \cdots & 0 \end{matrix} & \begin{bmatrix} D_2 \end{bmatrix} & & \\ & & \ddots & \\ & 0 & & \begin{bmatrix} D_n \end{bmatrix} \end{bmatrix}$$

Figure 143. Block diagonal matrix.

Use the result of Problem 30 to evaluate the following determinants.

31. $\begin{vmatrix} 22 & -4 & 3 & 0 & 0 \\ 1 & 0 & 6 & 0 & 0 \\ 2 & 2 & 0 & 0 & 0 \\ 0 & 0 & 0 & 1 & -1 \\ 0 & 0 & 0 & 0 & 4 \end{vmatrix}$

32. $\begin{vmatrix} 8 & 4 & 0 & 0 & 0 & 0 \\ -6 & 2 & 0 & 0 & 0 & 0 \\ 0 & 0 & 3 & 1 & 0 & 0 \\ 0 & 0 & 1 & 2 & 0 & 0 \\ 0 & 0 & 0 & 0 & -4 & 8 \\ 0 & 0 & 0 & 0 & 12 & 14 \end{vmatrix}$

33. $\begin{vmatrix} 16 & 14 & 0 & 2 & 0 & 0 \\ 1 & 1 & 3 & 2 & 0 & 0 \\ 0 & 0 & 1 & 4 & 0 & 0 \\ 1 & 3 & 6 & 2 & 0 & 0 \\ 0 & 0 & 0 & 0 & -5 & 3 \\ 0 & 0 & 0 & 0 & 4 & -9 \end{vmatrix}$

34. $\begin{vmatrix} 5 & 3 & 0 & 0 & 0 & 0 & 0 \\ 2 & -1 & 0 & 0 & 0 & 0 & 0 \\ 0 & 0 & 3 & 9 & 1 & 0 & 0 \\ 0 & 0 & 2 & 2 & -4 & 0 & 0 \\ 0 & 0 & 1 & -8 & 6 & 0 & 0 \\ 0 & 0 & 0 & 0 & 0 & -2 & -3 \\ 0 & 0 & 0 & 0 & 0 & 0 & 4 \end{vmatrix}$

10.12 AN APPLICATION OF DETERMINANTS TO ELECTRICAL CIRCUITS

In 1847, G. R. Kirchhoff published a classic paper in which he derived many of the electrical circuit laws which now bear his name. One of the less well known of these is the *matrix tree theorem*, which we shall now discuss.

Consider an electrical circuit, as in Figure 144. The underlying geometry of the network is shown in Figure 145. Such a diagram, consisting of points and connecting lines, is called a *graph*. (This notion arose in a different context in Section 10.2.) When the points are assigned labels, we have a *labeled graph*, as in Figure 146. Many of Kirchhoff's formulas depend upon geometric properties of the network's underlying graph; for example, the arrangements of the closed loops or circuits.

One of Kirchhoff's results requires knowing the number of *spanning trees* in a labeled graph. A *spanning tree* consists of all the points of the graph, together with some of the lines, subject to the following two conditions:

1. The lines of the spanning tree form no closed loops.
2. The spanning tree contains a path of lines between any pair of points.

For example, Figure 147 shows two different spanning trees in the labeled graph of Figure 146. The following result, the *matrix tree theorem*, tells how to compute the number of spanning trees in a labeled graph.

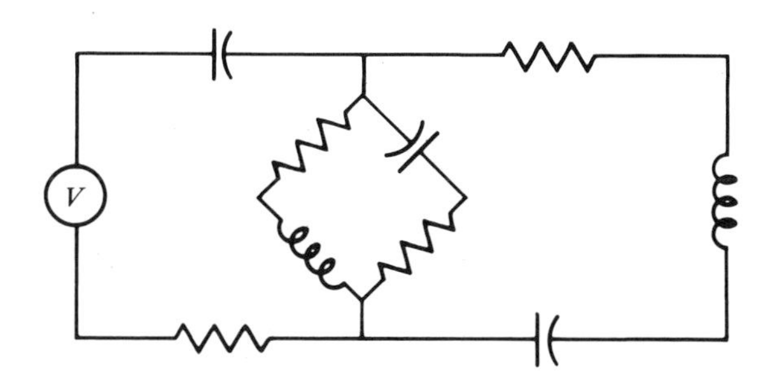

Figure 144

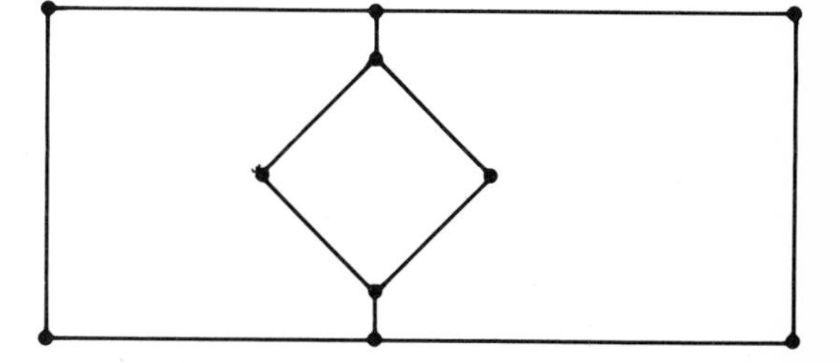
Figure 145

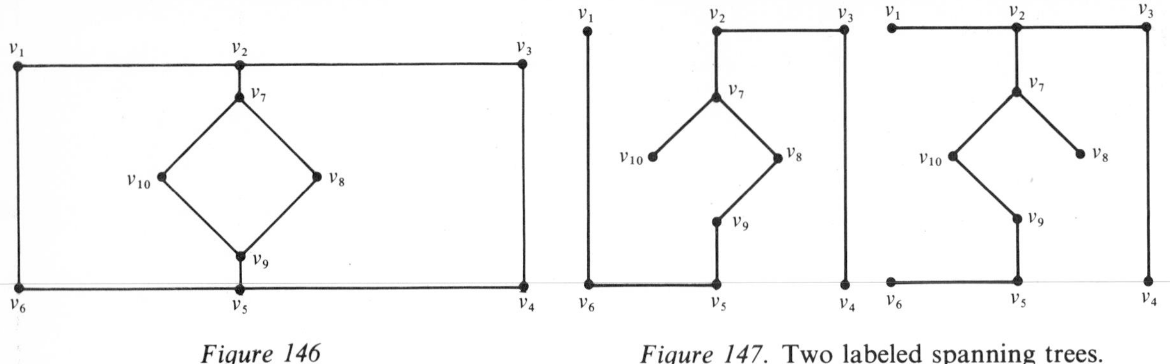

Figure 146 *Figure 147.* Two labeled spanning trees.

THEOREM 68

Let G be a graph with vertices labeled $1, \ldots, n$. Form an $n \times n$ matrix T (called the *tree matrix*) by setting

$$T_{ii} = \text{number of lines into point } i, \text{ for } i = 1, \ldots, n,$$

and

$$\text{for} \quad i = 1, \ldots, n \text{ and } j = 1, \ldots, n, \text{ but } i \neq j,$$

$$T_{ij} = \begin{cases} -1 & \text{if there is a line between points } i \text{ and } j \\ 0 & \text{otherwise.} \end{cases}$$

Then, all cofactors of T are equal, and their common value is the number of spanning trees in G.

A proof of this would take us farther afield than we wish to go, but we illustrate it below.

EXAMPLE 287

Consider the labeled graph shown in Figure 148. The 7×7 tree matrix is

$$T = \begin{bmatrix} 3 & -1 & 0 & 0 & 0 & -1 & -1 \\ -1 & 3 & -1 & -1 & 0 & 0 & 0 \\ 0 & -1 & 3 & -1 & 0 & -1 & 0 \\ 0 & -1 & -1 & 4 & -1 & 0 & -1 \\ 0 & 0 & 0 & -1 & 3 & -1 & -1 \\ -1 & 0 & -1 & 0 & -1 & 4 & -1 \\ -1 & 0 & 0 & -1 & -1 & -1 & 4 \end{bmatrix}.$$

Evaluate any cofactor of T. For example, $(-1)^{1+1}M_{11} = 386$. This is the number of spanning trees in this graph. Evaluation of any other cofactor of T would yield the same result.

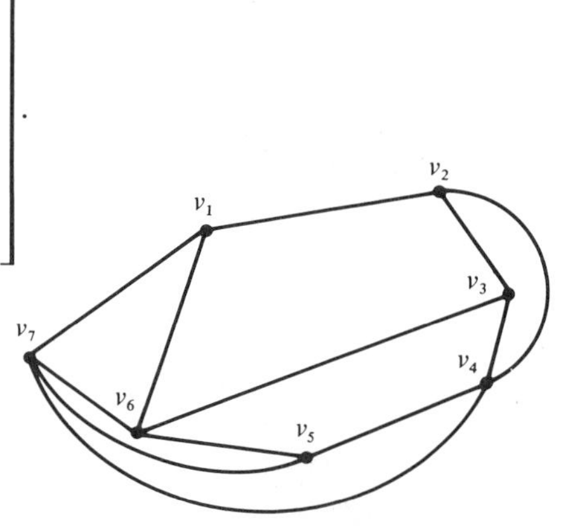

Figure 148

To appreciate the significance of the matrix tree theorem in calculating the number of spanning trees in a graph, try determining that there are 386 spanning trees in the graph of Figure 148 by some other means!

PROBLEMS FOR SECTION 10.12

In each of Problems 1 through 8, find the number of spanning trees in the given graph.

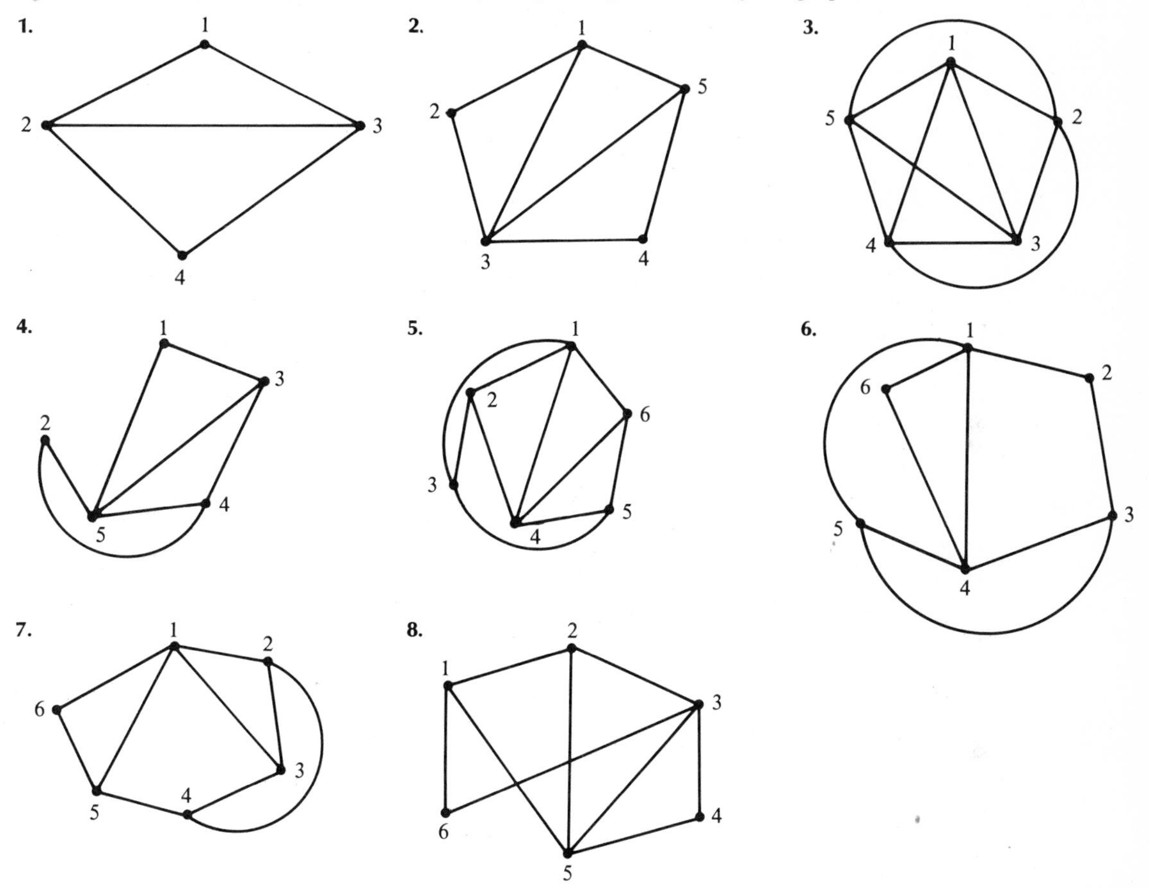

9. A complete graph on n points consists of n points and a line between each pair of distinct points. Such a graph is usually denoted K_n. Label the points of k_n, say $v_1, \ldots, v_n$, and use the matrix tree theorem and properties of determinants to show that the number of spanning trees in K_n is n^{n-2}.

In each of Problems 10 through 12, find the number of spanning trees in the given network.

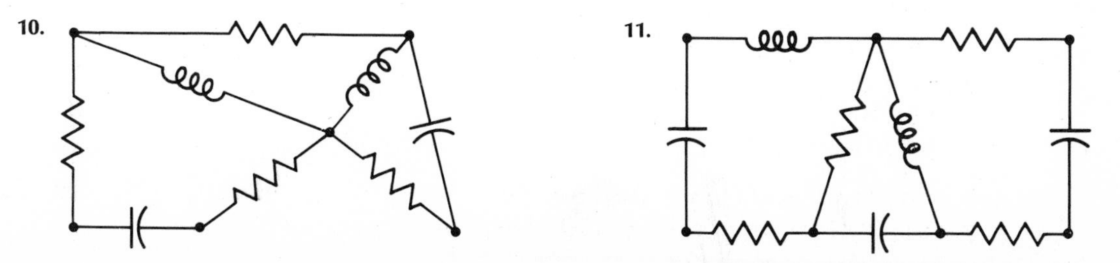

12.

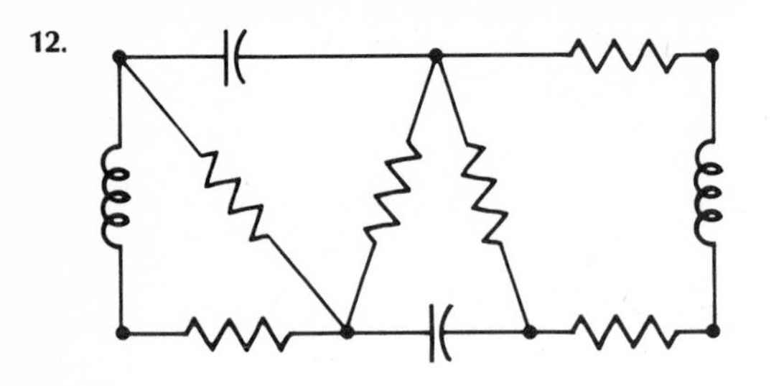

10.13 A DETERMINANT FORMULA FOR THE INVERSE OF A MATRIX

We now show how to use determinants to test for nonsingularity and find inverses. Recall that a square matrix may or may not have an inverse. If A has an inverse, A is nonsingular; otherwise, A is singular.

THEOREM 69

Let A be $n \times n$. Then,

$$A \text{ is nonsingular if and only if } |A| \neq 0.$$

Proof First recall that we obtain A_R from A by elementary row operations. By Theorems 59, 61, and 64, $|A|$ is a nonzero multiple of $|A_R|$, since in reducing A we never multiply any row by zero.

Now, if A is nonsingular, then $A_R = I_n$ by Lemma 1 (Section 10.6) and Theorem 56. Then $|A_R| = |I_n| = 1$ by Theorem 67; so $|A_R| \neq 0$.

Conversely, if $|A| \neq 0$, then $|A_R| \neq 0$. Then A_R must have no zero rows by Theorem 60; hence, rank$(A) = n$. By Theorem 56, A is nonsingular.

Using this result, we can produce an explicit formula for A^{-1} via $|A|$.

THEOREM 70

Let A be an $n \times n$ nonsingular matrix. Define an $n \times n$ matrix B by

$$B_{ij} = \frac{\text{cofactor of } A_{ji}}{|A|} = \frac{(-1)^{i+j}M_{ji}}{|A|} \quad \text{for } i = 1, \ldots, n \text{ and } j = 1, \ldots, n.$$

Then, $B = A^{-1}$.

Note: The i-j entry of B involves the j-i cofactor (*not* the i-j cofactor).

Proof Since $|A| \neq 0$ by Theorem 69, B is well defined. To show that $B = A^{-1}$, it suffices to show that $AB = I_n$. By definition of matrix product, the i-j entry of AB is

$$(AB)_{ij} = \sum_{k=1}^{n} A_{ik} B_{kj} = \sum_{k=1}^{n} A_{ij} \frac{(-1)^{k+j}M_{jk}}{|A|} = \frac{1}{|A|} \sum_{k=1}^{n} (-1)^{k+j} A_{ik} M_{jk}.$$

Now look at the sum $\sum_{k=1}^{n} (-1)^{k+j} A_{ik} M_{jk}$. When $i = j$, this is the cofactor expansion of $|A|$ by row i. Thus,

$$(AB)_{ii} = \frac{1}{|A|} |A| = 1,$$

and AB has 1's down the main diagonal.

Now suppose that $i \neq j$. Then $\sum_{k=1}^{n} (-1)^{k+j} A_{ij} M_{jk}$ is the cofactor expansion by row j of $|C|$, where C is obtained from A by replacing row j by row i. Now C has two identical rows; hence, $|C| = 0$ by Theorem 62.

Thus, the off-diagonal entries of AB are all zero, and we have $AB = I_n$.

EXAMPLE 288

Let

$$A = \begin{bmatrix} 8 & 0 & 1 \\ 3 & -2 & 1 \\ 1 & 4 & 0 \end{bmatrix}.$$

We find that $|A| = -18$. We can calculate the entries of A^{-1} using the formula in Theorem 70. We have

$$(A^{-1})_{11} = \frac{(-1)^{1+1} M_{11}}{-18} = -\tfrac{1}{18} \begin{vmatrix} -2 & 1 \\ 4 & 0 \end{vmatrix} = \tfrac{4}{18},$$

$$(A^{-1})_{12} = \frac{(-1)^{1+2} M_{21}}{-18} = \tfrac{1}{18} \begin{vmatrix} 0 & 1 \\ 4 & 0 \end{vmatrix} = -\tfrac{4}{18},$$

$$(A^{-1})_{13} = \frac{(-1)^{1+3} M_{31}}{-18} = \tfrac{1}{18} \begin{vmatrix} 0 & 1 \\ -2 & 1 \end{vmatrix} = -\tfrac{2}{18},$$

and so on. Continuing in this way, we find that

$$A^{-1} = \tfrac{1}{18} \begin{bmatrix} 4 & -4 & -2 \\ -1 & 1 & 5 \\ -14 & 32 & 16 \end{bmatrix}.$$

This is not the most efficient way to compute inverses—the method of Section 10.9 is much better in general. However, Theorem 70 has the value of giving an *explicit formula* for A^{-1}, as opposed to a *method* for finding it. Such a formula is sometimes useful in theoretical calculations.

In closing this section, we note that Theorem 69 gives us an easy way of proving Theorem 55(5). Argue as follows: Suppose that A and B are $n \times n$ matrices, and either A or B is singular. Then, by Theorem 69, $|A| = 0$ or $|B| = 0$. Then, $|AB| = |A|\,|B| = 0$; hence, by Theorem 69 again, AB is singular.

To make this a valid proof, the reader should go back and check that we have not used the conclusion of Theorem 55(5) to prove any of the results leading to Theorem 69.

PROBLEMS FOR SECTION 10.13

In each of Problems 1 through 15, find A^{-1} using the method of this section.

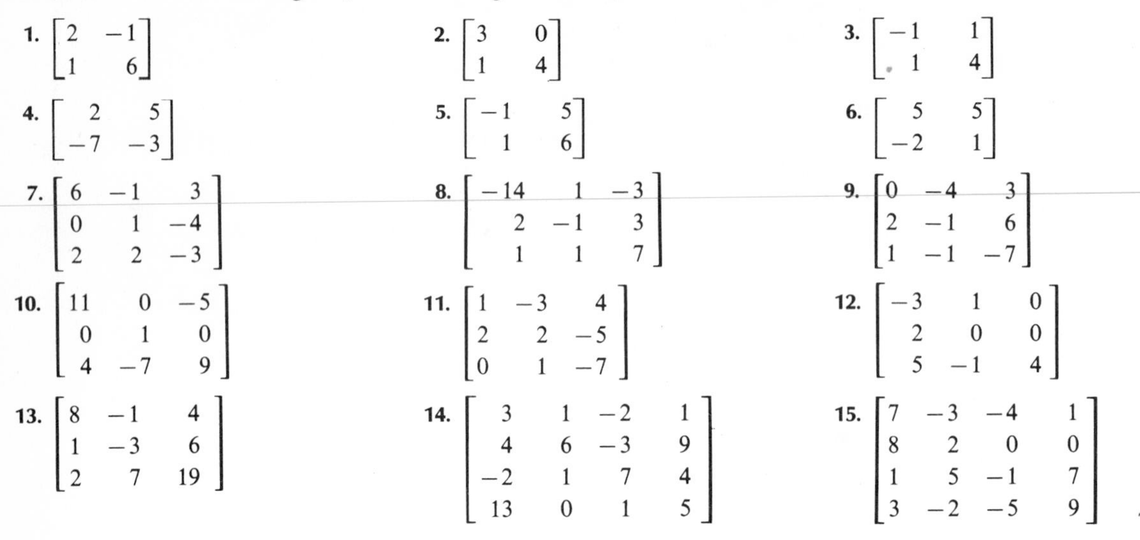

1. $\begin{bmatrix} 2 & -1 \\ 1 & 6 \end{bmatrix}$ **2.** $\begin{bmatrix} 3 & 0 \\ 1 & 4 \end{bmatrix}$ **3.** $\begin{bmatrix} -1 & 1 \\ 1 & 4 \end{bmatrix}$

4. $\begin{bmatrix} 2 & 5 \\ -7 & -3 \end{bmatrix}$ **5.** $\begin{bmatrix} -1 & 5 \\ 1 & 6 \end{bmatrix}$ **6.** $\begin{bmatrix} 5 & 5 \\ -2 & 1 \end{bmatrix}$

7. $\begin{bmatrix} 6 & -1 & 3 \\ 0 & 1 & -4 \\ 2 & 2 & -3 \end{bmatrix}$ **8.** $\begin{bmatrix} -14 & 1 & -3 \\ 2 & -1 & 3 \\ 1 & 1 & 7 \end{bmatrix}$ **9.** $\begin{bmatrix} 0 & -4 & 3 \\ 2 & -1 & 6 \\ 1 & -1 & -7 \end{bmatrix}$

10. $\begin{bmatrix} 11 & 0 & -5 \\ 0 & 1 & 0 \\ 4 & -7 & 9 \end{bmatrix}$ **11.** $\begin{bmatrix} 1 & -3 & 4 \\ 2 & 2 & -5 \\ 0 & 1 & -7 \end{bmatrix}$ **12.** $\begin{bmatrix} -3 & 1 & 0 \\ 2 & 0 & 0 \\ 5 & -1 & 4 \end{bmatrix}$

13. $\begin{bmatrix} 8 & -1 & 4 \\ 1 & -3 & 6 \\ 2 & 7 & 19 \end{bmatrix}$ **14.** $\begin{bmatrix} 3 & 1 & -2 & 1 \\ 4 & 6 & -3 & 9 \\ -2 & 1 & 7 & 4 \\ 13 & 0 & 1 & 5 \end{bmatrix}$ **15.** $\begin{bmatrix} 7 & -3 & -4 & 1 \\ 8 & 2 & 0 & 0 \\ 1 & 5 & -1 & 7 \\ 3 & -2 & -5 & 9 \end{bmatrix}$.

10.14 CRAMER'S RULE: DETERMINANT SOLUTION OF SYSTEMS OF EQUATIONS

If A is an $n \times n$, nonsingular matrix, then the nonhomogeneous system of equations $AX = B$ has a unique solution, namely

$$X = A^{-1}B.$$

Here is a determinant formula for this solution, known as *Cramer's rule.*

THEOREM 71

If A is an $n \times n$, nonsingular matrix, then the unique solution of the nonhomogeneous system $AX = B$ is given by

$$x_k = \frac{|A(k; B)|}{|A|} \quad \text{for} \quad k = 1, \ldots, n,$$

where $A(k; B)$ is the matrix obtained from A by replacing column k by B.

Proof Here is a heuristic argument which shows why Cramer's rule works. Begin with A, and multiply column k by x_k. This has the effect of multiplying $|A|$ by x_k, and we have

$$x_k|A| = \begin{vmatrix} A_{11} & \cdots & A_{1k}x_k & \cdots & A_{1n} \\ A_{21} & \cdots & A_{2k}x_k & \cdots & A_{2n} \\ \vdots & \cdots & \vdots & \cdots & \vdots \\ A_{n1} & \cdots & A_{nk}x_k & \cdots & A_{nn} \end{vmatrix}.$$

Next, for each $j \neq k$, add x_j (column j) to column k of the last matrix.

By Theorem 64, this does not change the value of the determinant. Thus,

$$x_k|A| = \begin{vmatrix} A_{11} & \cdots & A_{11}x_1 + A_{12}x_2 + \cdots + A_{1n}x_n & \cdots & A_{1n} \\ A_{21} & \cdots & A_{21}x_1 + A_{22}x_2 + \cdots + A_{2n}x_n & \cdots & A_{2n} \\ \vdots & \cdots & \vdots \quad\quad \vdots \quad \cdots \quad \vdots & \cdots & \vdots \\ A_{n1} & \cdots & \underbrace{A_{n1}x_1 + A_{n2}x_2 + \cdots + A_{nn}x_n}_{\text{column } k} & \cdots & A_{nn} \end{vmatrix}.$$

But,

$$\begin{bmatrix} A_{11}x_1 + \cdots + A_{1n}x_n \\ A_{21}x_1 + \cdots + A_{2n}x_n \\ \vdots \quad \cdots \quad \vdots \\ A_{n1}x_1 + \cdots + A_{nn}x_n \end{bmatrix} = AX = B = \begin{bmatrix} b_1 \\ b_2 \\ \vdots \\ b_n \end{bmatrix} \quad \text{if} \quad \begin{bmatrix} x_1 \\ x_2 \\ \vdots \\ x_n \end{bmatrix}$$

is a solution of $AX = B$. In this event,

$$x_k|A| = \begin{vmatrix} A_{11} & \cdots & b_1 & \cdots & A_{1n} \\ A_{21} & \cdots & b_2 & \cdots & A_{2n} \\ \vdots & \cdots & \vdots & \cdots & \vdots \\ A_{n1} & & b_n & & A_{nn} \end{vmatrix} = |A(k;\, B)|,$$

and we can solve for x_k to obtain $x_k = |A(k;\, B)|/|A|$, as we wanted to show.

EXAMPLE 289

Suppose that we want to solve the system

$$\begin{aligned} 2x_1 - 3x_2 &= 8, \\ x_1 + 4x_2 &= -3. \end{aligned}$$

Here, the matrix of coefficients is

$$A = \begin{bmatrix} 2 & -3 \\ 1 & 4 \end{bmatrix}$$

and $|A| = 11$. Then Cramer's rule applies, and we have

$$x_1 = \tfrac{1}{11}\begin{vmatrix} 8 & -3 \\ -3 & 4 \end{vmatrix} = \tfrac{23}{11}$$

and

$$x_2 = \tfrac{1}{11}\begin{vmatrix} 2 & 8 \\ 1 & -3 \end{vmatrix} = -\tfrac{14}{11}.$$

Note that in this example we obtain x_1 as $1/|A|$ times the determinant formed by replacing column 1 of A by $\begin{bmatrix} 8 \\ -3 \end{bmatrix}$; x_2 is $1/|A|$ times the determinant formed by replacing column 2 of A by $\begin{bmatrix} 8 \\ -3 \end{bmatrix}$.

In general, to solve $AX = B$ using Cramer's rule, we systematically take all the determinants formed by replacing column 1, then column 2, ..., then column n of A by B. Dividing these by $|A|$ gives us, in turn, $x_1, x_2, \ldots, x_n$.

EXAMPLE 290

Solve the system

$$\begin{aligned} x_1 - 3x_2 - 4x_3 &= 1, \\ -x_1 + x_2 - 3x_3 &= 14, \\ x_2 - 3x_3 &= 5. \end{aligned}$$

Here, the matrix of coefficients is

$$A = \begin{bmatrix} 1 & -3 & -4 \\ -1 & 1 & -3 \\ 0 & 1 & -3 \end{bmatrix}$$

and we find that $|A| = 13$. Cramer's rule applies, and we have

$$x_1 = \tfrac{1}{13} \begin{vmatrix} 1 & -3 & -4 \\ 14 & 1 & -3 \\ 5 & 1 & -3 \end{vmatrix} = -\tfrac{117}{13} = -9,$$

$$x_2 = \tfrac{1}{13} \begin{vmatrix} 1 & 1 & -4 \\ -1 & 14 & -3 \\ 0 & 5 & -3 \end{vmatrix} = -\tfrac{10}{13},$$

$$x_3 = \tfrac{1}{13} \begin{vmatrix} 1 & -3 & 1 \\ -1 & 1 & 14 \\ 0 & 1 & 5 \end{vmatrix} = -\tfrac{25}{13}.$$

As these two examples indicate, Cramer's rule is very pleasant for $n = 2$, and reasonably useful for $n = 3$. For $n \geq 4$, the Gauss-Jordan method is generally more efficient. Gauss-Jordan also has the advantage of not being restricted to nonsingular coefficient matrices, producing all solutions when there is more than one.

As with many other applications of determinants, the value of Cramer's rule lies in its presentation of a formula for the solution, as opposed to simply a method for finding it. A method is what is wanted for specific problems; a formula is useful for general discussions. A good illustration of this (which the student may not yet have seen) is the use of Cramer's rule in treating implicit partial differentiation through Jacobians in multivariable calculus. Another illustration is the Wronskian in determining linear independence or dependence of solutions of certain kinds of differential equations.

PROBLEMS FOR SECTION 10.14

In each of Problems 1 through 14, solve the system using Cramer's rule, or show that Cramer's rule does not apply.

1.
$$\begin{aligned} 8x_1 - 4x_2 + 3x_3 &= 0 \\ x_1 + 5x_2 - x_3 &= -5 \\ -2x_1 + 6x_2 + x_3 &= -4 \end{aligned}$$

2.
$$\begin{aligned} 15x_1 - 4x_2 &= 5 \\ 8x_1 + x_2 &= -4 \end{aligned}$$

3.
$$\begin{aligned} x_1 + 4x_2 &= 3 \\ x_1 + x_2 &= 0 \end{aligned}$$

4.
$$\begin{aligned} 5x_1 - 6x_2 + x_3 &= 4 \\ -x_1 + 3x_2 - 4x_3 &= 5 \\ 2x_1 + 3x_2 + x_3 &= -8 \end{aligned}$$

5.
$$\begin{aligned} x_1 + x_2 - 3x_3 &= 0 \\ x_2 - 4x_3 &= 0 \\ x_1 - x_2 - x_3 &= 5 \end{aligned}$$

6.
$$\begin{aligned} -8x_1 - 6x_2 + x_3 &= 4 \\ x_1 - x_2 + 3x_3 &= 2 \\ x_1 - x_3 &= 4 \end{aligned}$$

7.
$$\begin{aligned} -5x_1 + 6x_2 - 8x_3 + x_4 &= 5 \\ x_1 - 3x_3 + x_4 &= 6 \\ x_2 + 4x_3 - x_4 &= -8 \\ x_1 - 8x_3 + 5x_4 &= 0 \end{aligned}$$

8.
$$\begin{aligned} 6x_1 + 4x_2 - x_3 + 3x_4 - x_5 &= 7 \\ x_1 - 4x_2 + x_5 &= -5 \\ x_1 - 3x_2 + x_3 - 4x_5 &= 0 \\ -2x_1 + x_3 - 2x_5 &= 4 \\ x_3 - x_4 - x_5 &= 8 \end{aligned}$$

9.
$$\begin{aligned} 2x_1 - 4x_2 + x_3 - x_4 &= 6 \\ x_2 - 3x_3 &= 10 \\ x_1 - 4x_3 &= 0 \\ x_2 - x_3 + 2x_4 &= 4 \end{aligned}$$

10.
$$\begin{aligned} 2x_1 - 3x_2 + x_4 &= 2 \\ x_2 - x_3 + x_4 &= 2 \\ x_3 - 2x_4 &= 5 \\ x_1 - 3x_2 + 4x_3 &= 0 \end{aligned}$$

11.
$$\begin{aligned} 14x_1 - 3x_3 &= 5 \\ 2x_1 - 4x_3 + x_4 &= 2 \\ x_1 - x_2 + x_3 - x_4 &= 1 \\ x_3 - 4x_4 &= -5 \end{aligned}$$

12.
$$\begin{aligned} x_2 - 4x_4 &= 18 \\ x_1 - x_2 + 3x_3 &= -1 \\ x_1 + x_2 - 3x_3 + x_4 &= 5 \\ x_2 + 3x_4 &= 0 \end{aligned}$$

13.
$$\begin{aligned} 2x_1 - 3x_2 &= 10 \\ x_1 - 4x_3 + x_4 &= 5 \\ x_1 - x_2 + 3x_4 &= -2 \\ x_1 - x_2 + x_3 + x_4 &= 0 \end{aligned}$$

14.
$$\begin{aligned} x_3 - 4x_4 + x_5 &= 1 \\ x_1 - x_2 + 3x_4 &= 2 \\ x_2 - 3x_4 + 5x_5 &= -3 \\ x_2 + 3x_5 &= 1 \\ x_1 - x_2 + 3x_3 &= -4 \end{aligned}$$

15. Let A be a nonsingular matrix. What happens in Cramer's rule if we allow B to be the $n \times 1$ zero matrix in the system $AX = B$?

10.15 EIGENVALUES AND EIGENVECTORS

Let A be an $n \times n$ matrix. A real or complex number λ is called an *eigenvalue** of A if, for some nonzero $n \times 1$ matrix x, $Ax = \lambda x$. We usually think of x as a vector written as a column, and call it an *eigenvector* associated with λ.

---

* This word is an uncivilized mixture of German and English which has become fairly standard. Other commonly used terms are *proper value* and *characteristic value.* The word *eigen* means "proper" in German.

Eigenvalues and eigenvectors play a significant role in matrix theory and applications, but for the remainder of this section we shall concentrate on the concepts themselves. First, here are two examples which simply display some eigenvalues and eigenvectors. How these eigenvalues and eigenvectors were found will be discussed following the examples (see, in particular, Examples 293 and 294).

EXAMPLE 291

Let $A = \begin{bmatrix} 1 & 1 \\ 2 & -1 \end{bmatrix}$. Then $\sqrt{3}$ is an eigenvalue, and any vector $\begin{bmatrix} \alpha \\ (\sqrt{3}-1)\alpha \end{bmatrix}$ with $\alpha \neq 0$ is an eigenvector associated with $\sqrt{3}$. To verify this, calculate

$$\begin{bmatrix} 1 & 1 \\ 2 & -1 \end{bmatrix}\begin{bmatrix} \alpha \\ (\sqrt{3}-1) \end{bmatrix} = \begin{bmatrix} \alpha + (\sqrt{3}-1)\alpha \\ 2\alpha - (\sqrt{3}-1)\alpha \end{bmatrix} = \sqrt{3}\begin{bmatrix} \alpha \\ (\sqrt{3}-1)\alpha \end{bmatrix}.$$

Note in particular that the eigenvalue $\sqrt{3}$ has many eigenvectors associated with it.

Another eigenvalue of A is $-\sqrt{3}$, with associated eigenvectors $\begin{bmatrix} \alpha \\ (-\sqrt{3}-1)\alpha \end{bmatrix}$ for $\alpha \neq 0$. Again, there are many eigenvectors associated with eigenvalue $-\sqrt{3}$.

The numbers $\sqrt{3}$ and $-\sqrt{3}$ are the only eigenvalues of A in this example. A reason for this will be seen shortly.

EXAMPLE 292

Let

$$A = \begin{bmatrix} 1 & -1 & 0 \\ 0 & 1 & 1 \\ 0 & 0 & -1 \end{bmatrix}.$$

Then, 1 is an eigenvalue, with associated eigenvectors $\begin{bmatrix} \alpha \\ 0 \\ 0 \end{bmatrix}$, for $\alpha \neq 0$.

Another eigenvalue is -1, with associated eigenvectors $\begin{bmatrix} \alpha \\ 2\alpha \\ -4\alpha \end{bmatrix}$, for $\alpha \neq 0$. These are the only eigenvalues of A in this example.

There is a geometric way of looking at certain eigenvalues, but we shall defer this to the next chapter. For the remainder of this section we shall consider the following questions suggested by the examples above. Given an $n \times n$ matrix A:

1. Does A have eigenvalues?
2. How many eigenvalues does A have?
3. How do we compute the eigenvalues?
4. How do we find eigenvectors associated with eigenvalues?

A fifth question (what do we do with eigenvalues and eigenvectors once we find them) will be considered in later sections. Here, we shall answer the first four questions.

Suppose then that λ is an eigenvalue of A, and x an associated eigenvector. Rewrite the defining equation $Ax = \lambda x$ as $\lambda I_n x - Ax = 0$ or $(\lambda I_n - A)x = 0$. This equation says that the homogeneous system $(\lambda I_n - A)X = 0$ has a nontrivial solution (namely, x). Hence, by Theorems 57(2) and 69, we must have

$$|\lambda I_n - A| = 0.$$

This equation must be satisfied by λ if λ is to be an eigenvalue of A.

Conversely, if $|\lambda I_n - A| = 0$, then $\lambda I_n - A$ is singular, and there is a nontrivial solution x of the system $(\lambda I_n - A)X = 0$. Then $Ax = \lambda x$, and λ is an eigenvalue, with associated eigenvector x.

We summarize these important results as a theorem.

THEOREM 72

Let A be an $n \times n$ matrix. Then:

1. λ is an eigenvalue of A if and only if $|\lambda I_n - A| = 0$.
2. If λ is an eigenvalue of A, then any nontrivial solution of $(\lambda I_n - A)X = 0$ is an associated eigenvector.

This is the key to answering our first four questions. Given A, we form $\lambda I_n - A$:

$$\lambda I_n - A = \begin{bmatrix} \lambda - A_{11} & -A_{12} & \cdots & -A_{1n} \\ -A_{21} & \lambda - A_{22} & \cdots & -A_{2n} \\ \vdots & \vdots & \cdots & \vdots \\ -A_{n1} & -A_{n2} & \cdots & \lambda - A_{nn} \end{bmatrix}.$$

The determinant of this matrix involves powers of λ, multiplied by entries of A, and so is a polynomial in λ. We call this polynomial the *characteristic polynomial* of A. The term $(\lambda - A_{11})(\lambda - A_{22}) \cdots (\lambda - A_{nn})$ in the determinant expansion contributes the highest power of λ, namely λ^n; hence, this polynomial is of degree n. This product also gives the coefficient of λ^n, and it is 1. In general, then,

$$|\lambda I_n - A| = \lambda^n + a_{n-1}\lambda^{n-1} + a_{n-2}\lambda^{n-2} + \cdots + a_1\lambda + a_0.$$

The equation $|\lambda I_n - A| = 0$ is called the *characteristic equation* of A. Its solutions (roots of the characteristic polynomial) are exactly the eigenvalues of A. Associated with each such eigenvalue, the nontrivial solutions of $(\lambda I_n - A)X = 0$ are the eigenvectors.

The reason for allowing eigenvalues to be complex is inherent in (1) of the theorem. A polynomial with real coefficients may have complex roots. For example, $\begin{bmatrix} 1 & -2 \\ 2 & 0 \end{bmatrix}$ has characteristic polynomial $\lambda^2 - \lambda + 4$, with roots $(1 \pm i\sqrt{15})/2$. These are the eigenvalues of A, and A would have no eigenvalues at all if we allowed only real ones. For many applications, it is important not to exclude any eigenvalues.

Since the system $(\lambda I_n - A)X = 0$ will generally have solutions with complex entries if λ is complex, we must also now allow complex eigenvectors.

In light of the theorem, and the discussion following it, we shall reexamine the two previous examples, together with an example involving complex eigenvalues and eigenvectors.

EXAMPLE 293

Let

$$A = \begin{bmatrix} 1 & 1 \\ 2 & -1 \end{bmatrix},$$

as in Example 291. Then,

$$\lambda I_2 - A = \begin{bmatrix} \lambda - 1 & -1 \\ -2 & \lambda + 1 \end{bmatrix},$$

with determinant $(\lambda - 1)(\lambda + 1) - 2$ or $\lambda^2 - 3$. This is the characteristic polynomial of A. The characteristic equation is $\lambda^2 - 3 = 0$, with solutions $\sqrt{3}$ and $-\sqrt{3}$. These are exactly the eigenvalues of A.

Corresponding to $\lambda = \sqrt{3}$, we solve the system $(\sqrt{3}I_2 - A)X = 0$ to obtain $\begin{bmatrix} \alpha \\ (\sqrt{3} - 1)\alpha \end{bmatrix}$. Choosing $\alpha \neq 0$ gives us the eigenvectors associated with $\sqrt{3}$.

Corresponding to $\lambda = -\sqrt{3}$, we solve $(-\sqrt{3}I_2 - A)X = 0$ to obtain

$$\begin{bmatrix} \alpha \\ (-\sqrt{3} - 1)\alpha \end{bmatrix}.$$

Choosing $\alpha \neq 0$ gives us the eigenvectors associated with $-\sqrt{3}$.

EXAMPLE 294

Let

$$A = \begin{bmatrix} 1 & -1 & 0 \\ 0 & 1 & 1 \\ 0 & 0 & -1 \end{bmatrix},$$

as in Example 292. Then,

$$\lambda I_3 - A = \begin{bmatrix} \lambda - 1 & 1 & 0 \\ 0 & \lambda - 1 & -1 \\ 0 & 0 & \lambda + 1 \end{bmatrix},$$

with determinant $(\lambda + 1)(\lambda - 1)^2$. This is the characteristic polynomial of A, and its roots, 1 and -1, are the eigenvalues of A. There can be no other eigenvalues.

Corresponding to $\lambda = +1$, we solve $(I_3 - A)X = 0$ to obtain $\begin{bmatrix} \alpha \\ 0 \\ 0 \end{bmatrix}$. Each choice of $\alpha \neq 0$ gives us an eigenvector associated with 1.

Corresponding to $\lambda = -1$, we solve $(-I_3 - A)X = 0$ to obtain $\begin{bmatrix} \alpha \\ 2\alpha \\ -4\alpha \end{bmatrix}$.

Each $\alpha \neq 0$ gives us an eigenvector associated with -1.

EXAMPLE 295

Let

$$A = \begin{bmatrix} 1 & -2 \\ 2 & 0 \end{bmatrix}.$$

This has characteristic polynomial $\lambda^2 - \lambda + 4$, with eigenvalues $(1 + i\sqrt{15})/2$ and $(1 - i\sqrt{15})/2$. Corresponding to $(1 + i\sqrt{15})/2$, we solve

$$\left[\left(\frac{1 + i\sqrt{15}}{2}\right)I_2 - A\right]X = 0 \quad \text{to obtain} \quad \begin{bmatrix} \alpha \\ \left(\frac{1 - i\sqrt{15}}{4}\right)\alpha \end{bmatrix}.$$

These are associated eigenvectors when $\alpha \neq 0$.

Similarly, corresponding to $(1 - i\sqrt{15})/2$, we solve

$$\left[\left(\frac{1 - i\sqrt{15}}{2}\right)I_2 - A\right]X = 0 \quad \text{to obtain} \quad \begin{bmatrix} \alpha \\ \left(\frac{1 + i\sqrt{15}}{4}\right)\alpha \end{bmatrix}.$$

These are eigenvectors when $\alpha \neq 0$.

It may seem that we have now completely answered the four questions posed above. In the next section we shall explore the extent to which this is true.

PROBLEMS FOR SECTION 10.15

In each of Problems 1 through 23, find the characteristic polynomial and all the eigenvalues of the given matrix. Corresponding to each eigenvalue, find an eigenvector.

1. $\begin{bmatrix} 1 & 3 \\ 2 & 1 \end{bmatrix}$ **2.** $\begin{bmatrix} -2 & 0 \\ 1 & 4 \end{bmatrix}$ **3.** $\begin{bmatrix} -5 & 0 \\ 1 & 2 \end{bmatrix}$ **4.** $\begin{bmatrix} 6 & -2 \\ -3 & 4 \end{bmatrix}$

5. $\begin{bmatrix} 1 & -6 \\ 2 & 2 \end{bmatrix}$ **6.** $\begin{bmatrix} 0 & 1 \\ 0 & 0 \end{bmatrix}$ **7.** $\begin{bmatrix} -5 & 2 \\ 2 & -4 \end{bmatrix}$ **8.** $\begin{bmatrix} 15 & 1 \\ -2 & 6 \end{bmatrix}$

9. $\begin{bmatrix} -10 & 4 \\ 4 & -5 \end{bmatrix}$ **10.** $\begin{bmatrix} 5 & -4 \\ 4 & 1 \end{bmatrix}$ **11.** $\begin{bmatrix} 2 & 0 & 0 \\ 1 & 0 & 2 \\ 0 & 0 & 3 \end{bmatrix}$ **12.** $\begin{bmatrix} -2 & 1 & 0 \\ 1 & 3 & 0 \\ 0 & 0 & -1 \end{bmatrix}$

13. $\begin{bmatrix} -3 & 1 & 1 \\ 0 & 0 & 0 \\ 0 & 1 & 0 \end{bmatrix}$ **14.** $\begin{bmatrix} 0 & 0 & -1 \\ 0 & 0 & 1 \\ 2 & 0 & 0 \end{bmatrix}$ **15.** $\begin{bmatrix} -14 & 1 & 0 \\ 0 & 2 & 0 \\ 1 & 0 & 2 \end{bmatrix}$ **16.** $\begin{bmatrix} 3 & 0 & 0 \\ 1 & -2 & -8 \\ 0 & -5 & 1 \end{bmatrix}$

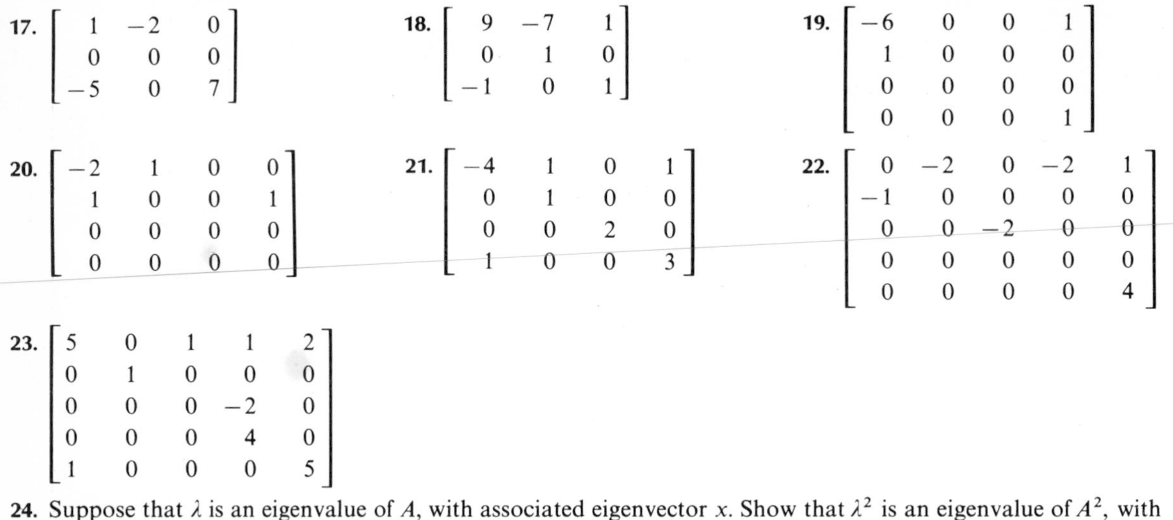

17. $\begin{bmatrix} 1 & -2 & 0 \\ 0 & 0 & 0 \\ -5 & 0 & 7 \end{bmatrix}$ **18.** $\begin{bmatrix} 9 & -7 & 1 \\ 0 & 1 & 0 \\ -1 & 0 & 1 \end{bmatrix}$ **19.** $\begin{bmatrix} -6 & 0 & 0 & 1 \\ 1 & 0 & 0 & 0 \\ 0 & 0 & 0 & 0 \\ 0 & 0 & 0 & 1 \end{bmatrix}$

20. $\begin{bmatrix} -2 & 1 & 0 & 0 \\ 1 & 0 & 0 & 1 \\ 0 & 0 & 0 & 0 \\ 0 & 0 & 0 & 0 \end{bmatrix}$ **21.** $\begin{bmatrix} -4 & 1 & 0 & 1 \\ 0 & 1 & 0 & 0 \\ 0 & 0 & 2 & 0 \\ 1 & 0 & 0 & 3 \end{bmatrix}$ **22.** $\begin{bmatrix} 0 & -2 & 0 & -2 & 1 \\ -1 & 0 & 0 & 0 & 0 \\ 0 & 0 & -2 & 0 & 0 \\ 0 & 0 & 0 & 0 & 0 \\ 0 & 0 & 0 & 0 & 4 \end{bmatrix}$

23. $\begin{bmatrix} 5 & 0 & 1 & 1 & 2 \\ 0 & 1 & 0 & 0 & 0 \\ 0 & 0 & 0 & -2 & 0 \\ 0 & 0 & 0 & 4 & 0 \\ 1 & 0 & 0 & 0 & 5 \end{bmatrix}$

24. Suppose that λ is an eigenvalue of A, with associated eigenvector x. Show that λ^2 is an eigenvalue of A^2, with eigenvector x. Is λ^k an eigenvalue of A^k for $k = 3, 4, \ldots$?

25. Show that the eigenvalues of an upper triangular matrix are exactly the main-diagonal entries.

26. Let λ_1 and λ_2 be distinct eigenvalues of A, with associated eigenvectors x_1 and x_2, respectively. Show that x_1 and x_2 are linearly independent.

27. Can a matrix with at least one complex, nonreal entry have only real eigenvalues? If not, give a proof; if yes, give an example.

28. Find the general form of all 2×2 matrices with real entries and eigenvalues 4 and -2.

10.16 COMPUTATIONAL ASPECTS OF EIGENVALUES AND EIGENVECTORS

We now know that the eigenvalues of A are exactly the roots of a certain polynomial. This raises several important points.

1. Real numbers are not enough.

We have seen that we must allow complex eigenvalues, or possibly have matrices with no eigenvalues at all. The latter course mitigates against many important applications; so in the last section we chose the former. This necessitates the following convention: *Henceforth, when dealing with eigenvalues and eigenvectors of matrices, "scalar" will mean complex number.*

Of course, any given scalar may still be real, since any real number α can be thought of as a complex number $\alpha + i \cdot 0$. When we want to specify that a scalar α is real, we shall explicitly say that α is real.

Matrix operations are performed the same for complex as for real matrices, using complex arithmetic. For a review of addition, subtraction, multiplication, and division of complex numbers, see Chapter 14.

2. Finding eigenvalues is usually difficult in practice.

This is because finding the eigenvalues of A is equivalent to finding the roots of the characteristic polynomial, and finding the roots of a polynomial is usually difficult. For example, the polynomial

$$x^4 - 1.371587913x^3 + 0.074844795x^2 + 0.224876581x - 0.0443323982$$

has roots approximately equal to 0.2666, -0.4114, 1.1714, and 0.3451. A computer or a good hand-held calculator* is useful in handling a polynomial like this. Further, this polynomial is more typical of those encountered in applied mathematics, physics, engineering, business, or economics than the easily solved characteristic polynomials of the last section's examples.

3. Eigenvalues may be repeated.

The characteristic polynomial of an $n \times n$ matrix has n roots, but some of these may be repeated. For example, $(\lambda - 1)^3(\lambda + 2)^2(\lambda - 4)$ has roots 1, 1, 1, -2, -2, and 4. Here, 1 has multiplicity 3; -2 has multiplicity 2; and 4 is a simple root, having multiplicity 1. It is convenient to be able to consistently speak of an $n \times n$ matrix as having n eigenvalues. This leads to a second convention: *We always list n eigenvalues for an n × n matrix. If some eigenvalue is a root of multiplicity k of the characteristic polynomial, then this eigenvalue is repeated k times in the list.*

For example,

$$\begin{bmatrix} 0 & 1 & -2 \\ 0 & 0 & 4 \\ 0 & 0 & 0 \end{bmatrix}$$

has characteristic polynomial λ^3; so the eigenvalues are 0, 0, and 0 (one root of multiplicity 3 of the characteristic polynomial).

In practice, finding the multiplicity of an eigenvalue can be tedious. One way is through use of derivatives. If λ is a root of $p(x)$, then λ has multiplicity k if $p(x)$ and its first $(k - 1)$ derivatives equal zero at $x = \lambda$, but the kth derivative is nonzero at λ.

Another way to determine multiplicity is through division. If λ is a root of $p(x)$, then $p(x)$ has $(x - \lambda)$ as a factor. Divide $(x - \lambda)$ into $p(x)$ to obtain a polynomial $q(x)$ of one lower degree. If $q(\lambda) = 0$, then we can divide $(x - \lambda)$ into $q(x)$, and so on. Continue this process until we have $p(x) = (x - \lambda)^k t(x)$, where $t(x)$ is a polynomial and $t(\lambda) \neq 0$. Then, k is the multiplicity of λ.

Of the three computational aspects we have discussed here, the first and third have led to agreements concerning scalars to be used and listing of eigenvalues. The second, determination of the eigenvalues, is difficult to handle. In Chapter 20, we shall describe a method of approximating the real eigenvalues of matrices having only real entries.

---

* For example, the HP-34C has a SOLVE key which can handle real zeros of polynomials of not too high a degree.

PROBLEMS FOR SECTION 10.16

In each of Problems 1 through 14, determine the characteristic polynomial and the eigenvalues of the given matrix. List the eigenvalues according to the convention of this section, including each eigenvalue in the list as many times as its multiplicity.

1. $\begin{bmatrix} -2 & -4 \\ 1 & 3 \end{bmatrix}$ **2.** $\begin{bmatrix} 3 & -4 \\ 2 & 1 \end{bmatrix}$ **3.** $\begin{bmatrix} 9 & -2 \\ -3 & 1 \end{bmatrix}$

4. $\begin{bmatrix} 1 & 7 \\ -2 & -6 \end{bmatrix}$ **5.** $\begin{bmatrix} -1 & 0 & 4 \\ 0 & 1 & 0 \\ 0 & 0 & -1 \end{bmatrix}$ **6.** $\begin{bmatrix} -2 & 0 & 0 \\ 1 & 1 & 2 \\ 0 & 1 & 0 \end{bmatrix}$

7. $\begin{bmatrix} 6 & 0 & 0 & 1 \\ 1 & 0 & 0 & 4 \\ 0 & 0 & 0 & 1 \\ 0 & 0 & 0 & 1 \end{bmatrix}$ **8.** $\begin{bmatrix} 0 & -1 & 1 & 2 \\ 0 & 0 & 1 & 1 \\ 0 & 0 & 0 & 0 \\ 0 & 0 & 0 & 1 \end{bmatrix}$ **9.** $\begin{bmatrix} -2 & 1 & 4 \\ 0 & 1 & 1 \\ 0 & 0 & 2 \end{bmatrix}$

10. $\begin{bmatrix} 5 & 1 & 0 & 9 \\ 0 & 1 & 0 & 9 \\ 0 & 0 & 0 & 9 \\ 0 & 0 & 0 & 0 \end{bmatrix}$ **11.** $\begin{bmatrix} 0 & 0 & 0 & 0 \\ 0 & 0 & 4 & 0 \\ 0 & 1 & 1 & 0 \\ 1 & 3 & 0 & 3 \end{bmatrix}$ **12.** $\begin{bmatrix} 1 & 0 & 0 & 2 \\ 4 & 0 & 2 & 0 \\ 0 & 2 & 0 & 0 \\ 0 & 0 & 0 & 0 \end{bmatrix}$

13. $\begin{bmatrix} -6 & 0 & 0 & 0 \\ 4 & 1 & 0 & 2 \\ 0 & 1 & 1 & 2 \\ 1 & 0 & 0 & -3 \end{bmatrix}$ **14.** $\begin{bmatrix} 0 & 0 & 1 & 2 \\ 1 & 0 & 1 & 3 \\ 0 & 0 & 1 & 4 \\ 0 & 0 & 0 & 1 \end{bmatrix}$

15. Show that the eigenvalues of any matrix of the form $\begin{bmatrix} \alpha & \beta \\ \beta & \gamma \end{bmatrix}$ are real if α, β, and γ are real.

16. Show that the eigenvalues of any matrix of the form $\begin{bmatrix} \alpha & \beta & \gamma \\ \beta & \varepsilon & \delta \\ \gamma & \delta & \phi \end{bmatrix}$ are real if all the entries are real.

10.17 APPLICATION OF EIGENVALUES TO SYSTEMS OF DIFFERENTIAL EQUATIONS

Eigenvalues are important in solving systems of differential equations as might arise, for example, in considering mass-spring systems or electrical circuits. The basic ideas are easily seen from an example, which we shall do in considerable detail.

Consider the mass-spring system of Figure 149, in which y_1 and y_2 measure displacements of masses m_1 and m_2, respectively, from equilibrium positions. The spring constants are k_1 and k_2 as shown, and we chose $m_1 = m_2 = 1$ for convenience.

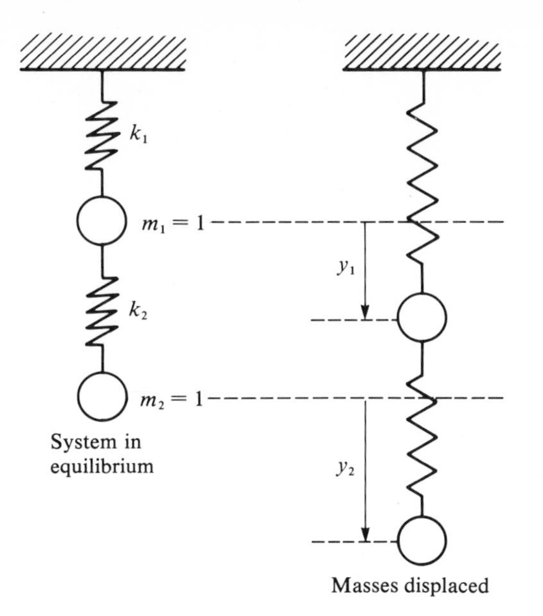

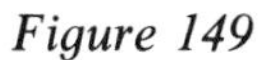

Figure 149

The motion is governed by the pair of differential equations

$$y_1'' = -k_1 y_1 + k_2(y_2 - y_1) = -(k_1 + k_2)y_1 + k_2 y_2,$$
$$y_2'' = -k_2(y_2 - y_1) = +k_2 y_1 - k_2 y_2,$$

where y_1 and y_2 are functions of time t and primes denote derivatives with respect to t. We assume no damping and no external driving forces.

Write this system in matrix form as follows. Let

$$Y = \begin{bmatrix} y_1 \\ y_2 \end{bmatrix}.$$

If we define the derivative of a matrix to be the matrix obtained by differentiating each matrix entry, then

$$Y'' = \begin{bmatrix} y_1'' \\ y_2'' \end{bmatrix}.$$

The system of differential equations is then

$$Y'' = \begin{bmatrix} -(k_1 + k_2) & k_2 \\ k_2 & -k_2 \end{bmatrix} Y.$$

To be specific, say $k_1 = 5$ and $k_2 = 6$. Then $Y'' = AY$, where

$$A = \begin{bmatrix} -11 & 6 \\ 6 & -6 \end{bmatrix}.$$

Now, $Y'' = AY$ is reminiscent of the single differential equation $x'' = ax$, with a constant. The latter always has an exponential solution. This suggests that we try

$$Y = e^{wt}C,$$

where w is to be determined and

$$C = \begin{bmatrix} c_1 \\ c_2 \end{bmatrix}$$

is an as yet unknown constant matrix. Substituting $Y = e^{wt}C$ into $Y'' = AY$ gives us

$$w^2 e^{wt}C = Ae^{wt}C.$$

Dividing out e^{wt} gives us

$$AC = \lambda C,$$

in which we wrote $\lambda = w^2$. This equation tells us that C must be an eigenvector corresponding to eigenvalue $\lambda = w^2$ in order for $Y = e^{wt}C$ to satisfy $Y'' = AY$.

Hence, we solve for the eigenvalues of A. We obtain

$$\lambda_1 = -2 \quad \text{and} \quad \lambda_2 = -15.$$

Corresponding to $\lambda_1 = -2$, we get the eigenvector $\begin{bmatrix} 6 \\ 9 \end{bmatrix}$. (Actually, any eigenvector corresponding to -2 will serve.) Thus, with $w_1 = \pm\sqrt{-2} = \pm\sqrt{2}\,i$, we get solutions of the form

$$Y_1 = \alpha e^{\sqrt{2}it}\begin{bmatrix} 6 \\ 9 \end{bmatrix} + \beta e^{-\sqrt{2}it}\begin{bmatrix} 6 \\ 9 \end{bmatrix},$$

where α and β are arbitrary scalars.

Similarly, with $\lambda_2 = -15$, we get eigenvector $\begin{bmatrix} 12 \\ -8 \end{bmatrix}$. (Again, *any* eigenvector corresponding to -15 will do.) With $w_2 = \pm\sqrt{\lambda_2} = \pm\sqrt{15}\,i$, we get solutions

$$Y_2 = \gamma e^{\sqrt{15}it}\begin{bmatrix} 12 \\ -8 \end{bmatrix} + \delta e^{-\sqrt{15}it}\begin{bmatrix} 12 \\ -8 \end{bmatrix},$$

where γ and δ are arbitrary scalars. Following the procedure of Chapter 2, Section 2.3, we can write these solutions as

$$Y_1 = [a\cos(\sqrt{2}\,t) + b\sin(\sqrt{2}\,t)]\begin{bmatrix} 6 \\ 9 \end{bmatrix}$$

and

$$Y_2 = [c\cos(\sqrt{15}\,t) + d\sin(\sqrt{15}\,t)]\begin{bmatrix} 12 \\ -8 \end{bmatrix},$$

with a, b, c, and d arbitrary scalars.

Thus, $Y'' = AY$ has as general solution

$$\begin{aligned} Y &= Y_1 + Y_2 \\ &= \begin{bmatrix} 6a\cos(\sqrt{2}\,t) + 6b\sin(\sqrt{2}\,t) + 12c\cos(\sqrt{15}\,t) + 12d\sin(\sqrt{15}\,t) \\ 9a\cos(\sqrt{2}\,t) + 9b\sin(\sqrt{2}\,t) - 8c\cos(\sqrt{15}\,t) - 8d\sin(\sqrt{15}\,t) \end{bmatrix}. \end{aligned}$$

In terms of components, we can write

$$\begin{aligned} y_1 &= 6a\cos(\sqrt{2}\,t) + 6b\sin(\sqrt{2}\,t) + 12c\cos(\sqrt{15}\,t) + 12d\sin(\sqrt{15}\,t), \\ y_2 &= 9a\cos(\sqrt{2}\,t) + 9b\sin(\sqrt{2}\,t) - 8c\cos(\sqrt{15}\,t) - 8d\sin(\sqrt{15}\,t). \end{aligned}$$

We can determine specific values for a, b, c, and d if we are given the initial position $Y(0)$ and the initial velocity $Y'(0)$. For example, say

$$Y(0) = \begin{bmatrix} 1 \\ 3 \end{bmatrix} \quad \text{and} \quad Y'(0) = \begin{bmatrix} 2 \\ 4 \end{bmatrix}.$$

Then,

$$Y(0) = \begin{bmatrix} 1 \\ 3 \end{bmatrix} = a\begin{bmatrix} 6 \\ 9 \end{bmatrix} + c\begin{bmatrix} 12 \\ -8 \end{bmatrix}$$

and

$$Y'(0) = \begin{bmatrix} 2 \\ 4 \end{bmatrix} = \sqrt{2}\,b\begin{bmatrix} 6 \\ 9 \end{bmatrix} + \sqrt{15}\,d\begin{bmatrix} 12 \\ -8 \end{bmatrix}.$$

Then,

$$\begin{aligned} 6a + 12c &= 1 \\ 9a - 8c &= 3 \end{aligned} \qquad \text{and} \qquad \begin{aligned} 6\sqrt{2}\,b + 12\sqrt{15}\,d &= 2 \\ 9\sqrt{2}\,b - 8\sqrt{15}\,d &= 4. \end{aligned}$$

Solve these for a, b, c, and d to get

$$a = \frac{11}{39}, \qquad b = \frac{16}{39\sqrt{2}}, \qquad c = \frac{-3}{52}, \quad \text{and} \quad d = \frac{-1}{26\sqrt{15}}.$$

Thus,

$$\begin{aligned} Y(t) = {} & \left[\frac{11}{39}\cos(\sqrt{2}\,t) + \frac{16}{39\sqrt{2}}\sin(\sqrt{2}\,t)\right]\begin{bmatrix} 6 \\ 9 \end{bmatrix} \\ & + \left[\frac{-3}{52}\cos(\sqrt{15}\,t) - \frac{1}{26\sqrt{15}}\sin(\sqrt{15}\,t)\right]\begin{bmatrix} 12 \\ -8 \end{bmatrix}. \end{aligned}$$

In terms of components, this is

$$\begin{aligned} y_1 &= \frac{66}{39}\cos(\sqrt{2}\,t) + \frac{96}{39\sqrt{2}}\sin(\sqrt{2}\,t) - \frac{36}{52}\cos(\sqrt{15}\,t) - \frac{12}{26\sqrt{15}}\sin(\sqrt{15}\,t), \\ y_2 &= \frac{99}{39}\cos(\sqrt{2}\,t) + \frac{144}{39\sqrt{2}}\sin(\sqrt{2}\,t) + \frac{24}{52}\cos(\sqrt{15}\,t) + \frac{8}{26\sqrt{15}}\sin(\sqrt{15}\,t). \end{aligned}$$

This section suggests the importance of matrix methods and the notions of eigenvalue and eigenvector in solving systems of differential equations. In fact, a more important application of matrices to systems can be seen once we have considered the concept of diagonalization in the next section.

PROBLEMS FOR SECTION 10.17

In each of Problems 1 through 8, solve the system $Y'' = AY$, with A the given matrix.

1. $A = \begin{bmatrix} 1 & 3 \\ 2 & 2 \end{bmatrix}$

2. $A = \begin{bmatrix} 0 & 1 \\ -4 & 0 \end{bmatrix}$

3. $A = \begin{bmatrix} -4 & 2 \\ 0 & 1 \end{bmatrix}$

4. $A = \begin{bmatrix} 6 & -3 \\ 2 & 0 \end{bmatrix}$

5. $A = \begin{bmatrix} 2 & -2 \\ 4 & 2 \end{bmatrix}$

6. $A = \begin{bmatrix} 5 & 1 \\ 1 & -1 \end{bmatrix}$

7. $\begin{bmatrix} -2 & 0 & 0 \\ 1 & 1 & 2 \\ 0 & 1 & 0 \end{bmatrix}$

8. $A = \begin{bmatrix} -1 & 0 & 4 \\ 0 & 1 & 0 \\ 0 & 1 & -1 \end{bmatrix}$

9. Set up and solve the equations of motion of the spring system of Figure 150 using methods of this section.
10. Set up and solve the equations of motion of the spring system of Figure 151 using the methods of this section.

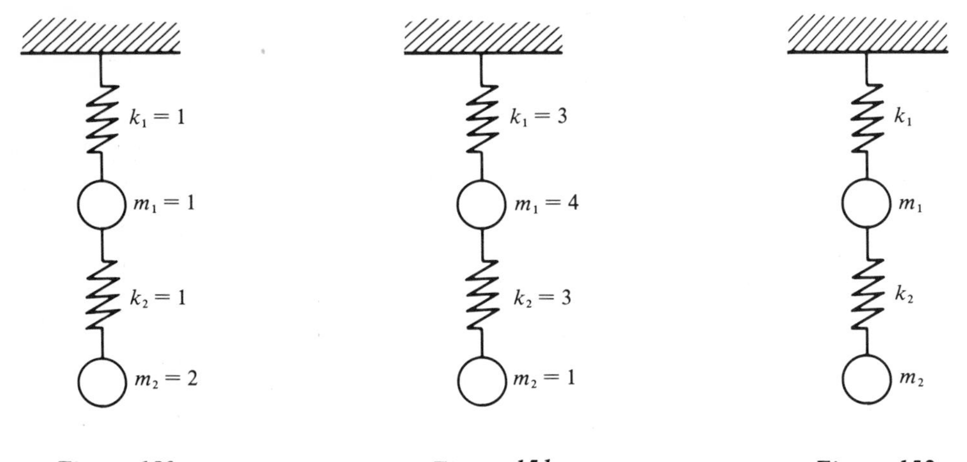

Figure 150 *Figure 151* *Figure 152*

11. Show that the motion of the spring system of Figure 152 is governed by $Y'' = AY$, where

$$A = \begin{bmatrix} -\dfrac{k_1 + k_2}{m_1} & \dfrac{k_2}{m_1} \\ \dfrac{k_2}{m_2} & -\dfrac{k_2}{m_2} \end{bmatrix}.$$

Show that the eigenvalues of A must be real and negative, no matter how k_1, k_2, m_1, and m_2 are chosen (they must of course be positive). What does this say about the nature of the resulting motion?

12. Consider the system of n first order differential equations $Y' = AY$, in which

$$Y = \begin{bmatrix} y_1(t) \\ \vdots \\ y_n(t) \end{bmatrix}$$

and A is an $n \times n$ matrix with real entries. Show that $Y = e^{\lambda t}C$ is a nontrivial solution exactly when λ is an eigenvalue of A and C is an associated eigenvector.

13. Use the result of Problem 12 to show that the general solution of $Y' = AY$ is

$$Y = \alpha_1 e^{\lambda_1 t} C_1 + \alpha_2 e^{\lambda_2 t} C_2 + \cdots + \alpha_n e^{\lambda_n t} C_n,$$

with $\alpha_1, \ldots, \alpha_n$ arbitrary real numbers; $\lambda_1, \ldots, \lambda_n$ the eigenvalues of A; and C_j any eigenvector associated with λ_j. Assume that A has n distinct eigenvalues.

14. Use the result of Problem 13 to find the general solution of

$$Y' = \begin{bmatrix} -1 & 6 \\ 1 & 1 \end{bmatrix} Y.$$

15. Use the result of Problem 13 to find the general solution of

$$y_1' = 2y_1 - 4y_2,$$
$$y_2' = 6y_1 + 7y_2.$$

16. Use the result of Problem 13 to find the solution of

$$Y' = \begin{bmatrix} 2 & -3 \\ 3 & 4 \end{bmatrix} Y, \qquad Y(0) = \begin{bmatrix} 1 \\ -1 \end{bmatrix}.$$

17. Find the general solution of $y'' + ay' + by = 0$ as follows: Let $z_1 = y$ and $z_2 = y'$. Write the second order differential equation as a system, solve the system, and then convert the answer back in terms of y. Compare the result with the general solution of $y'' + ay' + by = 0$ found in Chapter 2.

10.18 DIAGONALIZATION

A very pleasant type of matrix to work with is the diagonal matrix. Recall that an $n \times n$ matrix D is *diagonal* if $D_{ij} = 0$ whenever $i \neq j$. This means that every entry off the main diagonal is zero. It is convenient to write a diagonal matrix having diagonal entries $d_1, \ldots, d_n$ as

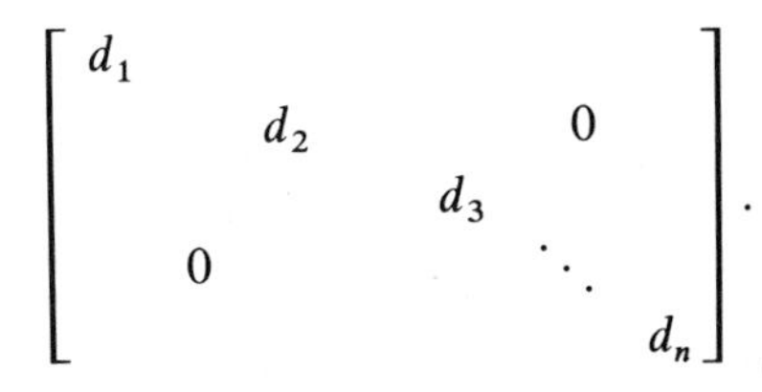

$$\begin{bmatrix} d_1 & & & & \\ & d_2 & & 0 & \\ & & d_3 & & \\ & 0 & & \ddots & \\ & & & & d_n \end{bmatrix}.$$

Here are some properties enjoyed by diagonal matrices:

1. A product of diagonal matrices is diagonal. In fact, it is easy to check that

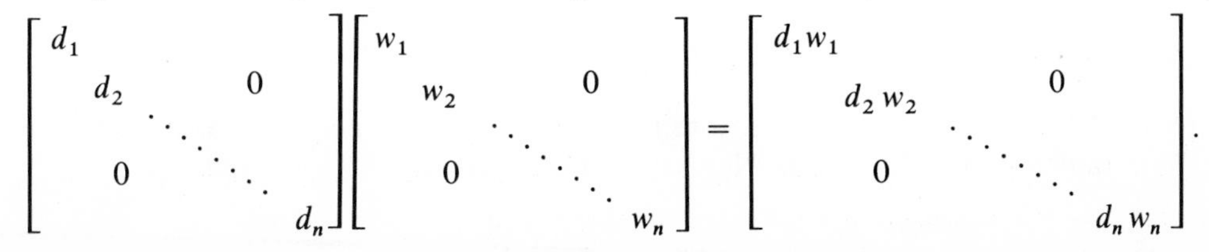

$$\begin{bmatrix} d_1 & & & \\ & d_2 & & 0 \\ & 0 & \ddots & \\ & & & d_n \end{bmatrix} \begin{bmatrix} w_1 & & & \\ & w_2 & & 0 \\ & 0 & \ddots & \\ & & & w_n \end{bmatrix} = \begin{bmatrix} d_1 w_1 & & & \\ & d_2 w_2 & & 0 \\ & 0 & \ddots & \\ & & & d_n w_n \end{bmatrix}.$$

2. The determinant of a diagonal matrix is just the product of its main-diagonal entries. (This is because a diagonal matrix is upper triangular.)
3. Hence, a diagonal matrix is nonsingular exactly when it has no zero on the main diagonal.
4. In view of (1) and (3), if D is diagonal and nonsingular, then

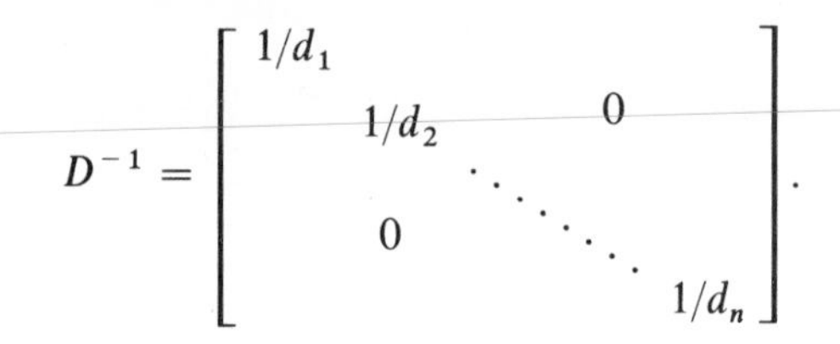

$$D^{-1} = \begin{bmatrix} 1/d_1 & & & \\ & 1/d_2 & & 0 \\ & & \ddots & \\ & 0 & & 1/d_n \end{bmatrix}.$$

5. Using (2), the eigenvalues of a diagonal matrix are exactly its main-diagonal entries.

Of course, most matrices are not diagonal. Sometimes, however, for a given $n \times n$ matrix A, one can find a matrix P such that $P^{-1}AP$ is diagonal. This will be useful in several applications of matrices. When such a P exists, we call A *diagonalizable*, and say that P *diagonalizes* A, and that A is *similar to* a diagonal matrix.

We shall see that not every matrix is diagonalizable. However, the following theorem gives a useful sufficient condition for A to be diagonalizable.

THEOREM 73

Let A be an $n \times n$ matrix, and let $v_1, \ldots, v_n$ be linearly independent eigenvectors corresponding, respectively, to eigenvalues $\lambda_1, \ldots, \lambda_n$. Then, there exists a matrix P such that

$$P^{-1}AP = \begin{bmatrix} \lambda_1 & & & \\ & \lambda_2 & & 0 \\ & & \ddots & \\ & 0 & & \lambda_n \end{bmatrix}.$$

This says that an $n \times n$ matrix with n linearly independent eigenvectors is diagonalizable; further, it is similar to a diagonal matrix having the eigenvalues of the original matrix down its main diagonal. These eigenvalues need not be distinct.

Here is a constructive proof which gives a recipe for P.

Proof Form an $n \times n$ matrix P by letting

$$\text{column } i \text{ of } P = v_i.$$

(*Note:* This may have complex entries.) Since $v_1, \ldots, v_n$ are linearly independent, then P has rank n, and so P^{-1} exists. Now compute the product $P^{-1}AP$. First,

$$\text{column } i \text{ of } AP = A \cdot (\text{column } i \text{ of } P) = Av_i = \lambda_i v_i.$$

Then,

$$P^{-1}AP = P^{-1}\begin{bmatrix} | & | & & | \\ \lambda_1 v_1 & \lambda_2 v_2 & \cdots & \lambda_n v_n \\ | & | & & | \end{bmatrix},$$

where the vertical lines indicate that column i is the $n \times 1$ matrix $\lambda_i v_i$.

Now, column i of this product is $P^{-1}(\lambda_i v_i)$, or $\lambda_i P^{-1} v_i$ (remember that v_i is $n \times 1$). But, $v_i =$ column i of P; so

$$P^{-1}v_i = \text{column } i \text{ of } P^{-1}P = \begin{bmatrix} 0 \\ 0 \\ \vdots \\ 1 \\ 0 \\ \vdots \\ 0 \end{bmatrix} \leftarrow i\text{th place}$$

Then

$$\text{column } i \text{ of } P^{-1}AP = \lambda_i \begin{bmatrix} 0 \\ \vdots \\ 1 \\ 0 \\ \vdots \\ 0 \end{bmatrix} = \begin{bmatrix} 0 \\ \vdots \\ \lambda_i \\ 0 \\ \vdots \\ 0 \end{bmatrix} \leftarrow i\text{th place}$$

Hence,

$$P^{-1}AP = \begin{bmatrix} \lambda_1 & 0 & 0 & \cdots & 0 \\ 0 & \lambda_2 & 0 & \cdots & 0 \\ \vdots & \vdots & \vdots & & \vdots \\ 0 & 0 & 0 & \cdots & \lambda_n \end{bmatrix}.$$

Here are several more examples. The fourth is particularly important, illustrating that a matrix with multiple eigenvalues may be diagonalizable.

EXAMPLE 296

Let

$$A = \begin{bmatrix} -1 & 4 \\ 0 & 3 \end{bmatrix}.$$

The eigenvalues of A are -1 and 3, and associated eigenvectors are, respectively,

$$\begin{bmatrix} 1 \\ 0 \end{bmatrix} \text{ and } \begin{bmatrix} 1 \\ 1 \end{bmatrix}.$$

These are linearly independent (thought of as vectors in R^2). Let

$$P = \begin{bmatrix} 1 & 1 \\ 0 & 1 \end{bmatrix}.$$

We claim that P diagonalizes A. To check this, we find that

$$P^{-1} = \begin{bmatrix} 1 & -1 \\ 0 & 1 \end{bmatrix},$$

and it is easy to verify that

$$P^{-1}AP = \begin{bmatrix} -1 & 0 \\ 0 & 3 \end{bmatrix}.$$

What would happen if we chose different eigenvectors corresponding to -1 and 3? According to the proof of the theorem, nothing: Any matrix whose columns are linearly independent eigenvectors will diagonalize A. For example,

$$\begin{bmatrix} 3 \\ 0 \end{bmatrix} \quad \text{and} \quad \begin{bmatrix} -2 \\ -2 \end{bmatrix}$$

are eigenvectors associated with -1 and 3, respectively.

Letting

$$Q = \begin{bmatrix} 3 & -2 \\ 0 & -2 \end{bmatrix},$$

then

$$Q^{-1} = \begin{bmatrix} \frac{1}{3} & -\frac{1}{3} \\ 0 & -\frac{1}{2} \end{bmatrix},$$

and it is easy to check that

$$Q^{-1}AQ = \begin{bmatrix} -1 & 0 \\ 0 & 3 \end{bmatrix}.$$

What would happen if we wrote the eigenvectors in a different order as columns of a new matrix S? Then S would still diagonalize A, but the eigenvalues appearing on the main diagonal of the resulting matrix would appear in the order corresponding to the eigenvectors in the columns of S. For example, let

$$S = \begin{bmatrix} 1 & 1 \\ 1 & 0 \end{bmatrix}.$$

This has the same columns as P, but in reverse order. Now S will diagonalize A, because it consists of linearly independent eigenvectors as columns, but we find that

$$S^{-1} = \begin{bmatrix} 0 & 1 \\ 1 & -1 \end{bmatrix} \quad \text{and} \quad S^{-1}AS = \begin{bmatrix} 3 & 0 \\ 0 & -1 \end{bmatrix}.$$

EXAMPLE 297

Let

$$A = \begin{bmatrix} 5 & 0 & 1 \\ 0 & 1 & 2 \\ 1 & 0 & -3 \end{bmatrix}.$$

Then A has eigenvalues $1, 1 + \sqrt{17}$, and $1 - \sqrt{17}$.

It is possible to find linearly independent eigenvectors associated with these eigenvalues; for example,

$$\begin{bmatrix} 0 \\ 1 \\ 0 \end{bmatrix}, \quad \begin{bmatrix} 1 \\ (2\sqrt{17} - 8)/\sqrt{17} \\ \sqrt{17} - 4 \end{bmatrix}, \text{ and } \begin{bmatrix} 1 \\ (2\sqrt{17} + 8)/\sqrt{17} \\ -4 - \sqrt{17} \end{bmatrix}, \text{ respectively.}$$

Then

$$P = \begin{bmatrix} 0 & 1 & 1 \\ 1 & (2\sqrt{17} - 8)/\sqrt{17} & (2\sqrt{17} + 8)/\sqrt{17} \\ 0 & \sqrt{17} - 4 & -4 - \sqrt{17} \end{bmatrix}$$

diagonalizes A:

$$P^{-1}AP = \begin{bmatrix} 1 & 0 & 0 \\ 0 & 1 + \sqrt{17} & 0 \\ 0 & 0 & 1 - \sqrt{17} \end{bmatrix}.$$

EXAMPLE 298

Sometimes we must use a complex matrix to diagonalize A, even though A itself may have only real entries. For example, let

$$A = \begin{bmatrix} -1 & -4 \\ 3 & -2 \end{bmatrix}.$$

The eigenvalues are $(-3 + \sqrt{47}\, i)/2$ and $(-3 - \sqrt{47}\, i)/2$. Linearly independent eigenvectors are, respectively,

$$\begin{bmatrix} 1 \\ (1 - \sqrt{47}\, i)/8 \end{bmatrix} \text{ and } \begin{bmatrix} 1 \\ (1 + \sqrt{47}\, i)/8 \end{bmatrix}.$$

A matrix that diagonalizes A is then

$$\begin{bmatrix} 1 & 1 \\ (1 - \sqrt{47}\, i)/8 & (1 + \sqrt{47}\, i)/8 \end{bmatrix}.$$

EXAMPLE 299

It is not necessary for A to have n distinct eigenvalues in order to be diagonalizable. For example, let

$$A = \begin{bmatrix} 5 & -4 & 4 \\ 12 & -11 & 12 \\ 4 & -4 & 5 \end{bmatrix}.$$

The eigenvalues are 1, 1, and -3. Associated with -3, we find an eigenvector $\begin{bmatrix} 1 \\ 3 \\ 1 \end{bmatrix}$.

Now consider the eigenvectors associated with 1. We must solve the system $(I_3 - A)X = 0$, or

$$\begin{bmatrix} -4 & 4 & -4 \\ -12 & 12 & -12 \\ -4 & 4 & -4 \end{bmatrix}\begin{bmatrix} x \\ y \\ z \end{bmatrix} = \begin{bmatrix} 0 \\ 0 \\ 0 \end{bmatrix}.$$

This has general solution $\begin{bmatrix} \alpha \\ \beta \\ \beta - \alpha \end{bmatrix}$, with α and β arbitrary. The key is that we can find two linearly independent eigenvectors associated with eigenvalue 1. For example,

$$\begin{bmatrix} 1 \\ 0 \\ -1 \end{bmatrix} \quad \text{and} \quad \begin{bmatrix} 0 \\ 1 \\ 1 \end{bmatrix}$$

are two such eigenvectors. The matrix

$$\begin{bmatrix} 1 & 0 & 1 \\ 0 & 1 & 3 \\ -1 & 1 & 1 \end{bmatrix}$$

then diagonalizes A.

We shall now show that the condition of Theorem 73 is necessary as well as sufficient: If A does not have n linearly independent eigenvectors, then A is not diagonalizable. But much more can be said. Above, we saw that

$$P^{-1}AP = \begin{bmatrix} \lambda_1 & & & 0 \\ & \lambda_2 & & \\ & & \ddots & \\ 0 & & & \lambda_n \end{bmatrix}$$

if column i of P is an eigenvector associated with λ_i and the n eigenvectors used are independent. We shall show that, whenever an equation

$$Q^{-1}AQ = \begin{bmatrix} d_1 & & & 0 \\ & d_2 & & \\ & & \ddots & \\ 0 & & & d_n \end{bmatrix}$$

holds, then the d_i's must necessarily be eigenvalues of A, and the columns of Q must be associated eigenvectors (linearly independent, because Q is nonsingular).

THEOREM 74

Let A be an $n \times n$ diagonalizable matrix. Then A must have n linearly independent eigenvectors. Further, if

$$Q^{-1}AQ = \begin{bmatrix} d_1 & & & 0 \\ & d_2 & & \\ & & \ddots & \\ 0 & & & d_n \end{bmatrix},$$

then $d_1, \ldots, d_n$ are the eigenvalues of A, and the column i of Q is an eigenvector associated with d_i. (*Note:* Again, this theorem does not imply that $d_1, \ldots, d_n$ are distinct.)

Proof Since A is diagonalizable, there is a nonsingular matrix that diagonalizes A. Let Q be any such matrix. Then,

$$Q^{-1}AQ = \begin{bmatrix} d_1 & & & 0 \\ & d_2 & & \\ & & \ddots & \\ 0 & & & d_n \end{bmatrix}.$$

Since Q is nonsingular, its columns are linearly independent. Denote column i of Q by v_i. We shall prove the theorem by showing that v_i is an eigenvector of A associated with d_i.

Let $D = Q^{-1}AQ$. Then, $QD = AQ$. Compute both of these products:

$$AQ = A\begin{bmatrix} | & | & & | \\ v_1 & v_2 & \cdots & v_n \\ | & | & & | \end{bmatrix} = \begin{bmatrix} | & | & & | \\ Av_1 & Av_2 & \cdots & Av_n \\ | & | & & | \end{bmatrix},$$

the $n \times n$ matrix with Av_i as column i, and

$$QD = \begin{bmatrix} | & | & & | \\ v_1 & v_2 & \cdots & v_n \\ | & | & & | \end{bmatrix}D = \begin{bmatrix} | & | & & | \\ d_1v_1 & d_2v_2 & \cdots & d_nv_n \\ | & | & & | \end{bmatrix},$$

the $n \times n$ matrix having $d_i v_i$ as column i. Comparing QD with AQ gives us $Av_i = d_i v_i$. Hence, d_i is an eigenvalue of A, and v_i an associated eigenvector.

The two theorems of this section not only say when a matrix can be diagonalized, but also describe the makeup of any diagonalizing matrix and the exact form of the resulting diagonal matrix.

Finally, here is an example of a matrix that cannot be diagonalized.

EXAMPLE 300

Let

$$B = \begin{bmatrix} 1 & -1 \\ 0 & 1 \end{bmatrix}.$$

The eigenvalues of B are 1 and 1. By solving $(I_2 - B)X = 0$, we find that the only eigenvectors associated with 1 are $\begin{bmatrix} \alpha \\ 0 \end{bmatrix}$, with $\alpha \neq 0$. While there are infinitely many such eigenvectors, no two are linearly independent. Hence, B fails to be diagonalizable.

In fact, if there were P such that

$$P^{-1}BP = \begin{bmatrix} d_1 & 0 \\ 0 & d_2 \end{bmatrix},$$

then by the last part of Theorem 74, P would have to have eigenvectors as columns. Then P would be of the form $\begin{bmatrix} \alpha & \beta \\ 0 & 0 \end{bmatrix}$, and this is a singular matrix (note that its determinant is zero).

If you go back and check examples in which eigenvalues and eigenvectors of a matrix have been found, you can notice that eigenvectors associated with different eigenvalues are linearly independent. This is true in general, and is sometimes useful to know.

THEOREM 75

Eigenvectors associated with distinct eigenvalues of a matrix are linearly independent.

Proof We proceed by induction on the number of distinct eigenvalues being considered.

If we are looking at just one eigenvalue, then any associated eigenvector is a nonzero vector, and hence is linearly independent.

Now suppose that any $(k - 1)$ eigenvectors associated with $(k - 1)$ distinct eigenvalues of A are linearly independent, and that we have k distinct eigenvalues $\lambda_1, \ldots, \lambda_k$, with associated eigenvectors $v_1, \ldots, v_k$.

If $v_1, \ldots, v_k$ are linearly dependent, there are scalars $\alpha_1, \ldots, \alpha_k$ (possibly complex), not all zero, such that

$$\alpha_1 v_1 + \alpha_2 v_2 + \cdots + \alpha_k v_k = \begin{bmatrix} 0 \\ 0 \\ \vdots \\ 0 \end{bmatrix}.$$

Now, some α_i are not zero. By relabeling the scalars if necessary, we may suppose that $\alpha_1 \neq 0$. Compute $(\lambda_1 I_n - A)$ times both sides of the preceding equation:

$$\begin{aligned}
&(\lambda_1 I_n - A)(\alpha_1 v_1 + \alpha_2 v_2 + \cdots + \alpha_k v_k) \\
&\quad = \alpha_1(\lambda_1 v_1 - Av_1) + \alpha_2(\lambda_1 v_2 - Av_2) + \cdots + \alpha_k(\lambda_1 v_k - Av_k) \\
&\quad = \alpha_1(\lambda_1 v_1 - \lambda_1 v_1) + \alpha_2(\lambda_1 v_2 - \lambda_2 v_2) + \cdots + \alpha_k(\lambda_1 v_k - \lambda_k v_k) \\
&\quad = \alpha_2(\lambda_1 - \lambda_2)v_2 + \cdots + \alpha_k(\lambda_1 - \lambda_k)v_k = \begin{bmatrix} 0 \\ 0 \\ \vdots \\ 0 \end{bmatrix}.
\end{aligned}$$

By the induction hypothesis, $v_2, \ldots, v_k$ are linearly independent; hence

$$\alpha_2(\lambda_1 - \lambda_2) = \cdots = \alpha_k(\lambda_1 - \lambda_k) = 0.$$

Since λ_1 is distinct from each of $\lambda_2, \ldots, \lambda_k$, then $\alpha_2 = \cdots = \alpha_k = 0$. But then we also have that

$$\alpha_1 v_1 = \begin{bmatrix} 0 \\ 0 \\ \vdots \\ 0 \end{bmatrix},$$

and hence that $\alpha_1 = 0$, a contradiction. Then $v_1, \ldots, v_k$ are linearly independent.

If we combine this result with Theorem 73 we immediately have the following:

COROLLARY 2

Any $n \times n$ matrix with n distinct eigenvalues is diagonalizable.

Here is an example to show how this corollary can be used to draw a quick conclusion about diagonalization.

EXAMPLE 301

Let

$$A = \begin{bmatrix} -4 & 0 & 1 & 0 \\ 0 & -2 & 0 & 0 \\ 0 & 0 & 1 & 2 \\ 0 & 0 & 0 & 0 \end{bmatrix}.$$

The eigenvalues are 0, 1, -2, and -4, and these are distinct; hence, A is diagonalizable. Using Corollary 2 and Theorem 73, there is a matrix P with $P^{-1}AP$ equal to

$$\begin{bmatrix} 0 & 0 & 0 & 0 \\ 0 & 1 & 0 & 0 \\ 0 & 0 & -2 & 0 \\ 0 & 0 & 0 & -4 \end{bmatrix}.$$

This conclusion can be drawn without actually finding any eigenvectors or finding a matrix which diagonalizes A.

In the next section we apply diagonalization to systems of differential equations.

PROBLEMS FOR SECTION 10.18

In each of Problems 1 through 12, a matrix is given. Either show that the matrix is not diagonalizable (as in Example 300) or produce a nonsingular matrix that diagonalizes the given matrix.

1. $\begin{bmatrix} 0 & -1 \\ 4 & 3 \end{bmatrix}$ **2.** $\begin{bmatrix} 5 & 2 \\ 1 & 3 \end{bmatrix}$ **3.** $\begin{bmatrix} 1 & 0 \\ -4 & 1 \end{bmatrix}$ **4.** $\begin{bmatrix} -5 & 3 \\ 0 & 9 \end{bmatrix}$

5. $\begin{bmatrix} 5 & 0 & 0 \\ 1 & 0 & 3 \\ 0 & 0 & -2 \end{bmatrix}$ **6.** $\begin{bmatrix} 0 & 0 & 0 \\ 1 & 0 & 2 \\ 0 & 1 & 3 \end{bmatrix}$ **7.** $\begin{bmatrix} -2 & 0 & 1 \\ 1 & 1 & 0 \\ 0 & 0 & -2 \end{bmatrix}$

8. $\begin{bmatrix} 2 & 0 & 0 \\ 0 & 2 & 1 \\ 0 & -1 & 2 \end{bmatrix}$ **9.** $\begin{bmatrix} 1 & 0 & 0 & 0 \\ 0 & 4 & 1 & 0 \\ 0 & 0 & -3 & 1 \\ 0 & 0 & 1 & -2 \end{bmatrix}$ **10.** $\begin{bmatrix} -2 & 0 & 0 & 0 \\ -4 & -2 & 0 & 0 \\ 0 & 0 & -2 & 0 \\ 0 & 0 & 0 & -2 \end{bmatrix}$

11. $\begin{bmatrix} 8 & -7 & 1 & 0 \\ 0 & 1 & 0 & 0 \\ 0 & 0 & 0 & 0 \\ 1 & 0 & 0 & 0 \end{bmatrix}$ **12.** $\begin{bmatrix} -7 & 0 & 1 & 0 \\ 0 & 1 & 1 & 0 \\ -4 & 0 & 2 & 0 \\ 0 & 0 & 0 & 0 \end{bmatrix}$

13. Let A be an $n \times n$ diagonalizable matrix. Prove that the constant term in the characteristic polynomial of A is $(-1)^n$ times the product of the eigenvalues of A. (This is true in general, but easiest to prove if A is diagonalizable.)

14. Suppose that A^2 is diagonalizable. Prove that A is diagonalizable.

15. Suppose that A has eigenvalues $\lambda_1, \ldots, \lambda_n$, and that P diagonalizes A. Prove that, for any positive integer k,

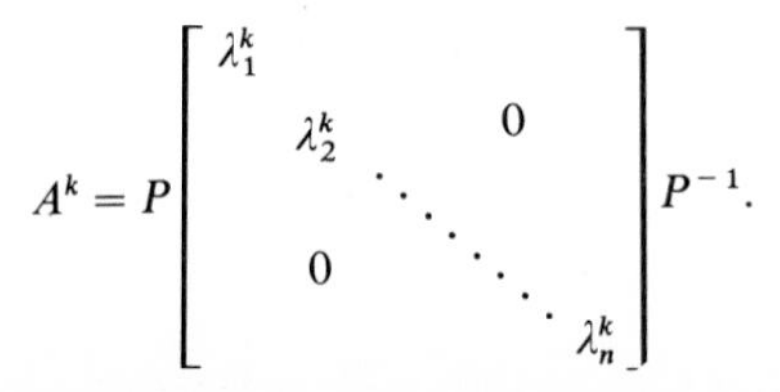

$$A^k = P \begin{bmatrix} \lambda_1^k & & & 0 \\ & \lambda_2^k & & \\ & & \ddots & \\ 0 & & & \lambda_n^k \end{bmatrix} P^{-1}.$$

In each of Problems 16 through 19, calculate the indicated power of the given matrix, using the result of Problem 15. In doing the calculation, a specific power k of a number d may be left written as d^k [for example, $(1+\sqrt{6})^{22}$ may be left as this, and need not be computed (it is about $6.77(10^{11})$].

16. $A = \begin{bmatrix} -1 & 0 \\ 1 & -5 \end{bmatrix}$; compute A^{18}.

17. $A = \begin{bmatrix} -3 & 2 \\ 1 & -4 \end{bmatrix}$; compute A^{16}.

18. $A = \begin{bmatrix} 0 & -2 \\ 1 & 0 \end{bmatrix}$; compute A^{43}.

19. $A = \begin{bmatrix} -2 & 3 \\ 3 & -4 \end{bmatrix}$; compute A^{31}.

20. Let A be any real, 2×2 matrix. Prove that there is some nonsingular matrix P such that $P^{-1}AP$ has one of the following forms:

$$\begin{bmatrix} \alpha & 0 \\ 0 & \beta \end{bmatrix}, \text{ with } \alpha \neq \beta; \text{ or } \begin{bmatrix} \alpha & 0 \\ 0 & \alpha \end{bmatrix}; \text{ or } \begin{bmatrix} \alpha & -1 \\ 0 & \alpha \end{bmatrix}.$$

21. Suppose that $L^{-1}AL = B$ for some nonsingular L. Prove that A and B are both diagonalizable, or neither is. If both are diagonalizable, and $P^{-1}AP = D_1$, a diagonal matrix, and $Q^{-1}BQ = D_2$, a diagonal matrix, how are D_1 and D_2 related?

10.19 APPLICATION OF DIAGONALIZATION TO SYSTEMS OF DIFFERENTIAL EQUATIONS

We shall briefly discuss how diagonalization can help us solve systems of differential equations. Consider a system

$$Y' = AY,$$

where A is an $n \times n$ matrix of scalars. Written out the long way, this system is

$$\begin{aligned} y_1' &= A_{11}y_1 + A_{12}y_2 + \cdots + A_{1n}y_n, \\ y_2' &= A_{21}y_1 + A_{22}y_2 + \cdots + A_{2n}y_n, \\ &\vdots \\ y_n' &= A_{n1}y_1 + A_{n2}y_2 + \cdots + A_{nn}y_n. \end{aligned}$$

We shall show how to solve this system easily if A is diagonalizable.

In the event that A is diagonalizable, there is a nonsingular matrix P such that $P^{-1}AP = D$, a diagonal matrix. Change variables by putting

$$Y = PZ,$$

where

$$Z = \begin{bmatrix} z_1 \\ \vdots \\ z_n \end{bmatrix}.$$

Substitution into the system $Y' = AY$ gives us

$$(PZ)' = A(PZ).$$

Since P is a constant matrix (its columns must be eigenvectors of A), then we obtain

$$PZ' = APZ$$

or

$$Z' = (P^{-1}AP)Z = DZ.$$

What have we gained by this? The point is that the system $Y' = AY$ is usually a *coupled system.* That is, each y_j' is given in terms of all of $y_1, \ldots, y_n$, not just y_j. However, since D is diagonal, the system $Z' = DZ$ is of the form

$$\begin{aligned} z_1' &= d_1 z_1, \\ z_2' &= d_2 z_2, \\ &\vdots \\ z_n' &= d_n z_n, \end{aligned}$$

an *uncoupled system* with each z_j' in terms of a constant times just z_j. Each equation $z_j' = d_j z_j$ is easily solved, giving us Z, hence, $Y = PZ$.

EXAMPLE 302

Consider the three by three system

$$\begin{aligned} y_1' &= y_1 - y_2 + 2y_3, \\ y_2' &= 3y_1 \qquad + 4y_3, \\ y_3' &= 2y_1 + y_2. \end{aligned}$$

This is $Y' = AY$ with

$$A = \begin{bmatrix} 1 & -1 & 2 \\ 3 & 0 & 4 \\ 2 & 1 & 0 \end{bmatrix}.$$

Eigenvalues of A are found to be 2, $(-1 + \sqrt{13})/2$, and $(-1 - \sqrt{13})/2$. Corresponding eigenvectors are, respectively,

$$\begin{bmatrix} 0 \\ 2 \\ 1 \end{bmatrix}, \quad \begin{bmatrix} -5 + \sqrt{13} \\ 2\sqrt{13} \\ 7 - \sqrt{13} \end{bmatrix} \quad \text{and} \quad \begin{bmatrix} -5 - \sqrt{13} \\ -2\sqrt{13} \\ 7 + \sqrt{13} \end{bmatrix}.$$

Thus, let

$$P = \begin{bmatrix} 0 & -5 + \sqrt{13} & -5 - \sqrt{13} \\ 2 & 2\sqrt{13} & -2\sqrt{13} \\ 1 & 7 - \sqrt{13} & 7 + \sqrt{13} \end{bmatrix}.$$

Then, P diagonalizes A, and setting $Y = PZ$ transforms $Y' = AY$ into $Z' = DZ$, where $D = P^{-1}AP$. We know that

$$D = \begin{bmatrix} 2 & 0 & 0 \\ 0 & \dfrac{-1 + \sqrt{13}}{2} & 0 \\ 0 & 0 & \dfrac{-1 - \sqrt{13}}{2} \end{bmatrix}.$$

Thus, $Z' = DZ$ can be written

$$z_1' = 2z_1,$$
$$z_2' = \left(\frac{-1+\sqrt{13}}{2}\right)z_2,$$
$$z_3' = \left(\frac{-1-\sqrt{13}}{2}\right)z_3.$$

These are easily solved to give

$$z_1 = ae^{2t},$$
$$z_2 = be^{(-1+\sqrt{13})t/2},$$
$$z_3 = ce^{(-1-\sqrt{13})t/2},$$

where a, b, and c are arbitrary constants. In matrix form,

$$Z = \begin{bmatrix} ae^{2t} \\ be^{(-1+\sqrt{13})t/2} \\ ce^{(-1-\sqrt{13})t/2} \end{bmatrix}.$$

Now solve for Y. Since $Y = PZ$, then

$$Y = \begin{bmatrix} 0 & -5+\sqrt{13} & -5-\sqrt{13} \\ 2 & 2\sqrt{13} & -2\sqrt{13} \\ 1 & 7-\sqrt{13} & 7+\sqrt{13} \end{bmatrix} \begin{bmatrix} ae^{2t} \\ be^{(-1+\sqrt{13})t/2} \\ ce^{(-1-\sqrt{13})t/2} \end{bmatrix}$$

$$= \begin{bmatrix} (-5+\sqrt{13})be^{(-1+\sqrt{13})t/2} + (-5-\sqrt{13})ce^{(-1-\sqrt{13})t/2} \\ 2ae^{2t} + 2\sqrt{13}\,be^{(-1+\sqrt{13})t/2} - 2\sqrt{13}\,ce^{(-1-\sqrt{13})t/2} \\ ae^{2t} + (7-\sqrt{13})be^{(-1+\sqrt{13})t/2} + (7+\sqrt{13})ce^{(-1-\sqrt{13})t/2} \end{bmatrix}.$$

In terms of components, the general solution is

$$y_1 = (-5+\sqrt{13})be^{(-1+\sqrt{13})t/2} + (-5-\sqrt{13})ce^{(-1-\sqrt{13})t/2},$$
$$y_2 = 2ae^{2t} + 2\sqrt{13}\,be^{(-1+\sqrt{13})t/2} - 2\sqrt{13}\,ce^{(-1-\sqrt{13})t/2},$$
$$y_3 = ae^{2t} + (7-\sqrt{13})be^{(-1+\sqrt{13})t/2} + (7+\sqrt{13})ce^{(-1-\sqrt{13})t/2}.$$

For many engineering purposes, we would be content to approximate $\sqrt{13}$ by 3.606. The solution is then approximately

$$Y_{\text{app}} = \begin{bmatrix} -1.394be^{1.303t} - 8.606ce^{-2.303t} \\ 2ae^{2t} + 7.211e^{1.303t} - 7.211ce^{-2.303t} \\ ae^{2t} + 3.394be^{1.303t} + 10.606ce^{-2.303t} \end{bmatrix}.$$

Note: One can see an important fact from the example: We never had to compute P^{-1}. This is because of two things. First, the transformation is $Y = PZ$; so after finding Z we get Y without needing P^{-1}. Second, $D = P^{-1}AP$, but we know D from Theorem 74 without actually forming the product $P^{-1}AP$; namely, D has to be the diagonal matrix with the eigenvalues of A down its diagonal (in the order corresponding to eigenvectors of A as columns of P).

Our ability to solve $Y' = AY$ without inverting P is important, as when n is large, finding P^{-1} can involve considerable labor.

PROBLEMS FOR SECTION 10.19

Using the methods of this section, solve the given systems of differential equations.

1. $y_1' = 3y_1 + 2y_2$, $y_2' = y_1 + 4y_2$
2. $y_1' = 6y_1 - y_2$, $y_2' = y_1 - 3y_2$
3. $y_1' = -4y_1 + 3y_2$, $y_2' = 2y_1 - 3y_2$
4. $y_1' = 7y_1 + 3y_2$, $y_2' = 8y_1 - 4y_2$
5. $y_1' = y_1 + 3y_2$, $y_2' = 4y_1 - 3y_2$
6. $y_1' = -7y_1 - 3y_2$, $y_2' = 5y_1 + 6y_2$
7. $y_1' = 5y_1 - y_2$, $y_2' = -2y_2$, $y_3' = 8y_1 + 7y_2$
8. $y_1' = 5y_1 - 4y_2 + 2y_3$, $y_2' = y_1 - 6y_2 + 3y_3$, $y_3' = 2y_3$

Note: In Problems 9 through 12, the numbers do not work out as nicely, and you should have calculator help.

9. $y_1' = 3y_1 - y_2 - y_3$, $y_2' = y_1 + 2y_2 - y_3$, $y_3' = 4y_2 - y_3$
(One eigenvalue is 0.1895.)
10. $y_1' = 4y_1 - 6y_2 + y_3$, $y_2' = y_1 + 3y_2 - 2y_2$, $y_3' = y_2 + 4y_3$
(One eigenvalue is 4.1229.)
11. $y_1' = -y_1 + 2y_2 - y_3$, $y_2' = 4y_2 + 3y_3$, $y_3' = -3y_1 + 6y_3$
(One eigenvalue is -1.7850.)
12. $y_1' = -3y_1 + 5y_3$, $y_2' = y_1 + 4y_2 + y_3$, $y_3' = -5y_2 + y_3$
(One eigenvalue is -3.6215.)

10.20 EIGENVALUES AND EIGENVECTORS OF REAL, SYMMETRIC MATRICES

Let A be a real, $n \times n$ matrix. In general, A may have real and/or complex eigenvalues. However, when A is *symmetric* (each $A_{ij} = A_{ji}$ or, equivalently, $A = A^t$), then we can make the following strong assertion.

THEOREM 76

The eigenvalues of a real, symmetric matrix are real.

Before proving the theorem, we recall that the complex conjugate of a number $a + ib$ is $a - ib$, and is denoted $\overline{a + ib}$. We take the conjugate of a matrix by taking the conjugate of each entry. The operations of taking the transpose of a matrix and taking the conjugate are interchangeable: $(\overline{M})^t = (\overline{M^t})$ for any matrix M. Finally, note that for any number $\alpha = a + ib$, we have $\alpha\bar{\alpha} = a^2 + b^2$, a real number; further, $\alpha = \bar{\alpha}$ exactly when $b = 0$. In this case, α is real. We can now prove the theorem.

Proof Let A be a real, $n \times n$ matrix, and let λ be an eigenvalue. Choose any eigenvector

$$X = \begin{bmatrix} x_1 \\ \vdots \\ x_n \end{bmatrix}$$

for λ. Then $AX = \lambda X$. Multiply by $\bar{X}^t$ to get

$$\bar{X}^t AX = \bar{X}^t \lambda X = \lambda \bar{X}^t X.$$

Now,

$$\bar{X}^t X = [\bar{x}_1 \quad \cdots \quad \bar{x}_n] \begin{bmatrix} x_1 \\ \vdots \\ x_n \end{bmatrix} = [\bar{x}_1 x_1 + \bar{x}_2 x_2 + \cdots + \bar{x}_n x_n].$$

Thus, $\bar{X}^t X$ is a 1×1 matrix whose single entry is real. Now look at the 1×1 matrix $\bar{X}^t AX$. Its conjugate is

$$\overline{\bar{X}^t AX} = (\overline{\bar{X}^t})\,\bar{A}\bar{X} = X^t \bar{A}\bar{X} = X^t A\bar{X},$$

since the hypothesis that A is real gives us $\bar{A} = A$. Since $X^t A\bar{X}$ is 1×1, its transpose is itself. Thus,

$$X^t A\bar{X} = (X^t A\bar{X})^t = \bar{X}^t A^t (X^t)^t = \bar{X}^t AX.$$

Here, we used the fact that A is symmetric to write $A^t = A$.

Comparing the last two equations (the last takes up where the one before ends), we have

$$\overline{\bar{X}^t AX} = \bar{X}^t AX.$$

Thus, $\bar{X}^t AX$ is real.

We now have shown that both $\bar{X}^t X$ and $\bar{X}^t AX$ are real. Since $\bar{X}^t AX = \lambda \bar{X}^t X$, then λ must be real also.

Of course, with any real eigenvalue λ, we can also associate real eigenvectors, since the eigenvectors are solutions of $(\lambda I_n - A)X = 0$, and are obtained from arithmetic operations applied to the entries of the real matrix $\lambda I_n - A$. Thus, in particular, a real, symmetric matrix has real eigenvalues and eigenvectors.

We can also derive the useful geometric result that eigenvectors corresponding to distinct eigenvalues of a real, symmetric matrix, are orthogonal in R^n. Recall that two vectors $(x_1, \ldots, x_n)$ and $(y_1, \ldots, y_n)$ in R^n are orthogonal if their dot product is zero, that is, if

$$(x_1, \ldots, x_n) \cdot (y_1, \ldots, y_n) = x_1 y_1 + \cdots + x_n y_n = 0.$$

It is often useful to observe that this dot product can be written in matrix form as follows. If

$$X = \begin{bmatrix} x_1 \\ \vdots \\ x_n \end{bmatrix} \quad \text{and} \quad Y = \begin{bmatrix} y_1 \\ \vdots \\ y_n \end{bmatrix},$$

then

$$X^tY = [x_1 \quad \cdots \quad x_n]\begin{bmatrix} y_1 \\ \vdots \\ y_n \end{bmatrix} = [x_1y_1 + \cdots + x_n y_n],$$

a 1×1 matrix whose entry is $(x_1, \ldots, x_n) \cdot (y_1, \ldots, y_n)$.

THEOREM 77

Let A be a real, symmetric matrix. Then eigenvectors associated with distinct eigenvalues of A are orthogonal.

Proof Let λ and μ be distinct eigenvalues of A. Let X be an eigenvector associated with λ, and Y an eigenvector associated with μ. Write

$$X = \begin{bmatrix} x_1 \\ \vdots \\ x_n \end{bmatrix} \quad \text{and} \quad Y = \begin{bmatrix} y_1 \\ \vdots \\ y_n \end{bmatrix}.$$

We want to show that $X \cdot Y = 0$.

Begin with

$$\begin{aligned}
\lambda X^tY &= (\lambda X^t)Y = (AX)^tY \\
&= \left[\sum_{j=1}^{n} A_{1j}x_j \quad \sum_{j=1}^{n} A_{2j}x_j \quad \cdots \quad \sum_{j=1}^{n} A_{nj}x_j\right]\begin{bmatrix} y_1 \\ \vdots \\ y_n \end{bmatrix} \\
&= \left[\sum_{j=1}^{n} A_{1j}y_1x_j + \sum_{j=1}^{n} A_{2j}y_2x_j + \cdots + \sum_{j=1}^{n} A_{nj}y_nx_j\right] \\
&= \left[x_1 \sum_{j=1}^{n} A_{1j}y_j + x_2 \sum_{j=1}^{n} A_{2j}y_j + \cdots + x_n \sum_{j=1}^{n} A_{nj}y_j\right] \\
&= [x_1 \quad \cdots \quad x_n]\begin{bmatrix} \sum_{j=1}^{n} A_{1j}y_j \\ \vdots \\ \sum_{j=1}^{n} A_{nj}y_j \end{bmatrix} \\
&= X^t(Y^tA)^t \\
&= X^t(A^tY) \\
&= X^t(AY) \\
&= X^t(\mu Y) \\
&= \mu(X^tY).
\end{aligned}$$

Then, $(\lambda - \mu)X^tY = 0$.

But $\lambda - \mu \neq 0$; hence, $X^tY = 0$, and this is just the matrix way of writing $(x_1, \ldots, x_n) \cdot (y_1, \ldots, y_n) = 0$. Thus, X and Y are orthogonal in R^n.

We call a collection of vectors in R^n *mutually orthogonal* if each vector is orthogonal to each of the others. By the previous theorem, eigenvectors of distinct eigenvalues of a real, symmetric matrix are mutually orthogonal.

EXAMPLE 303

Let

$$A = \begin{bmatrix} 3 & 0 & -2 \\ 0 & 2 & 0 \\ -2 & 0 & 0 \end{bmatrix}.$$

Then A is real and symmetric. We find that the eigenvalues are 2, -1, and 4. Associated eigenvectors are, respectively,

$$\begin{bmatrix} 0 \\ 1 \\ 0 \end{bmatrix}, \quad \begin{bmatrix} 1 \\ 0 \\ 2 \end{bmatrix} \quad \text{and} \quad \begin{bmatrix} 2 \\ 0 \\ -1 \end{bmatrix}.$$

These are mutually orthogonal in R^3.

We shall point out a fact that will be discussed in the next section. Notice that one of the eigenvectors above (the first one) happens to have length 1, while the others do not. Multiplying an eigenvector by a nonzero real number results in another eigenvector; so multiply $\begin{bmatrix} 1 \\ 0 \\ 2 \end{bmatrix}$ by the reciprocal of its length and $\begin{bmatrix} 2 \\ 0 \\ -1 \end{bmatrix}$ by the reciprocal of its length to obtain three unit eigenvectors:

$$\begin{bmatrix} 0 \\ 1 \\ 0 \end{bmatrix}, \quad \begin{bmatrix} 1/\sqrt{5} \\ 0 \\ 2/\sqrt{5} \end{bmatrix}, \quad \text{and} \quad \begin{bmatrix} 2/\sqrt{5} \\ 0 \\ -1/\sqrt{5} \end{bmatrix}.$$

The matrix

$$Q = \begin{bmatrix} 0 & 1/\sqrt{5} & 2/\sqrt{5} \\ 1 & 0 & 0 \\ 0 & 2/\sqrt{5} & -1/\sqrt{5} \end{bmatrix}$$

then diagonalizes A, but also has the remarkable property that $Q^{-1} = Q^t$. Such a matrix is called an *orthogonal matrix*.

In the next section we shall discuss how a real, symmetric matrix can always be diagonalized by an orthogonal matrix.

PROBLEMS FOR SECTION 10.20

In each of Problems 1 through 15, find the eigenvalues of the given real, symmetric matrix. Also find a matrix Q whose columns are mutually orthogonal unit vectors and which diagonalizes A (as in remarks following Example 303).

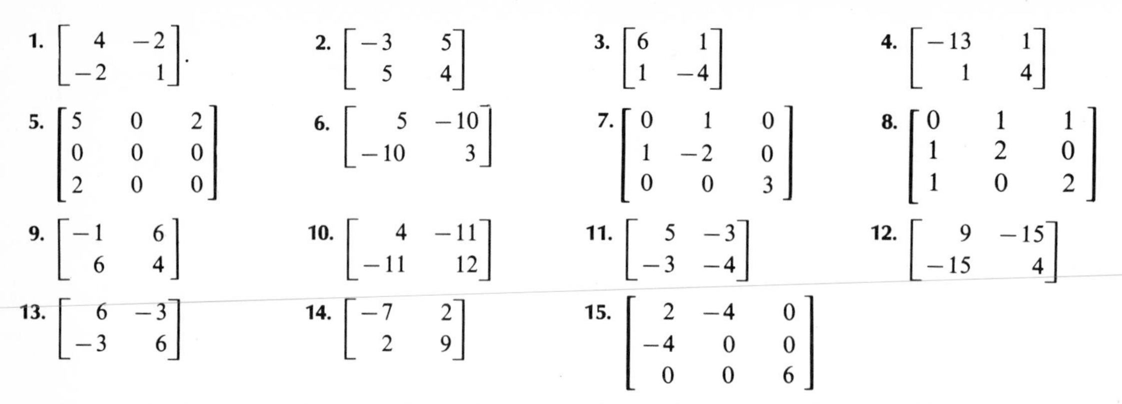

16. Let A and B be $n \times n$ real, symmetric matrices. Prove that AB is symmetric if and only if $AB = BA$.
17. Let A be any $n \times n$ matrix (not necessarily symmetric). Are the eigenvalues of A different in general from those of A^t?
18. Let A be a real, symmetric matrix and λ an eigenvalue of multiplicity k. Prove that there can be at most k mutually orthogonal eigenvectors associated with λ.

10.21 ORTHOGONAL MATRICES AND DIAGONALIZATION OF REAL, SYMMETRIC MATRICES

We shall now pursue the idea with which we ended the last section. Recall that a real, square matrix Q is called orthogonal if $Q^{-1} = Q^t$ (or, equivalently, $QQ^t = I_n$). The matrix Q following Example 303 is orthogonal. Here are some basic facts about orthogonal matrices, after which we shall tie them in with diagonalization.

Let Q be an $n \times n$, real matrix.

THEOREM 78

Q is orthogonal if and only if Q^t is orthogonal.

This is a simple consequence of the definition, and we leave the proof to the student.

Vectors $\mathbf{F}_1, \ldots, \mathbf{F}_r$ in R^n are said to be *orthonormal* if each $\mathbf{F}_i$ is orthogonal to each other $\mathbf{F}_j$ (so $\mathbf{F}_i \cdot \mathbf{F}_j = 0$ if $i \neq j$), and each $\mathbf{F}_i$ has length 1.

THEOREM 79

Q is orthogonal if and only if the rows of Q are orthonormal vectors in R^n.

Proof Suppose first that the rows of Q are orthonormal. Then,

$$\begin{aligned} i\text{-}j \text{ entry of } QQ^t &= (\text{row } i \text{ of } Q) \cdot (\text{column } j \text{ of } Q^t) \\ &= (\text{row } i \text{ of } Q) \cdot (\text{row } j \text{ of } Q) \\ &= \begin{cases} 0 & \text{if } i \neq j \\ 1 & \text{if } i = j \end{cases} \end{aligned}$$

by the orthonormality of the rows of Q. Then, $QQ^t = I_n$, and Q is orthogonal.

Conversely, suppose that Q is orthogonal. Then, $QQ^t = I_n$; so, for $i \neq j$,

$$(\text{row } i) \cdot (\text{row } j) = (QQ^t)_{ij} = (I_n)_{ij} = 0.$$

Hence, the rows of Q are mutually orthogonal. Further,

$$\|\text{row } i\|^2 = (\text{row } i) \cdot (\text{row } i) = (QQ^t)_{ii} = (I_n)_{ii} = 1;$$

so each row of Q is also a unit vector.

By combining the last two theorems, we have the following:

COROLLARY 3

Q is orthogonal if and only if the columns of Q are orthonormal vectors in R^n.

THEOREM 80

If Q is orthogonal, then $|Q|$ is either $+1$ or -1.

Proof Since $QQ^t = I_n$, then $|QQ^t| = |Q|\,|Q^t| = |Q|^2 = |I_n| = 1$.
Hence, $|Q|$ must be $+1$ or -1.

EXAMPLE 304

As an illustration of orthogonal matrices, we shall determine all such matrices of order 2.

Let

$$Q = \begin{bmatrix} a & b \\ c & d \end{bmatrix}$$

be an orthogonal matrix. What can be said about a, b, c, and d?

First,

$$QQ^t = \begin{bmatrix} a & b \\ c & d \end{bmatrix}\begin{bmatrix} a & c \\ b & d \end{bmatrix} = \begin{bmatrix} a^2 + b^2 & ac + bd \\ ac + bd & c^2 + d^2 \end{bmatrix} = \begin{bmatrix} 1 & 0 \\ 0 & 1 \end{bmatrix}.$$

Then,

$$a^2 + b^2 = c^2 + d^2 = 1 \quad \text{and} \quad ac + bd = 0.$$

Further, $|Q| = +1$ or -1; hence, $ad - bc = +1$ or -1. This leads us to consider two cases.

Case 1 $|Q| = +1$.
We now have four equations for a, b, c, and d, namely: $a^2 + b^2 = c^2 + d^2 = 1$, $ac + bd = 0$, and $ad - bc = 1$. We find (omitting the tedious algebraic details) that these force Q to have the form

$$\begin{bmatrix} a & -c \\ c & a \end{bmatrix} \text{ or } \begin{bmatrix} a & c \\ -c & a \end{bmatrix}.$$

We also find that $|a| \leq 1$ and $c^2 = 1 - a^2$. Hence, for some ϕ, $0 \leq \phi \leq \pi$, $a = \cos(\phi)$ and $c = \sin(\phi)$. Thus, Q must have the form

$$\begin{bmatrix} \cos(\phi) & -\sin(\phi) \\ \sin(\phi) & \cos(\phi) \end{bmatrix} \quad \text{or} \quad \begin{bmatrix} \cos(\phi) & \sin(\phi) \\ -\sin(\phi) & \cos(\phi) \end{bmatrix}.$$

Case 2 $|Q| = -1$.
Here we find that Q must have the form

$$\begin{bmatrix} \cos(\phi) & \sin(\phi) \\ \sin(\phi) & -\cos(\phi) \end{bmatrix}.$$

The student may recognize from analytic geometry that the 2×2 orthogonal matrices represent rotations of coordinate systems in the plane.

Further geometrical aspects of orthogonal matrices will be discussed in the next section, where we shall treat quadratic forms. For now, we want to relate orthogonal matrices to diagonalization.

THEOREM 81

Any real, symmetric matrix can be diagonalized by an orthogonal matrix.

This is a very strong theorem. We have seen a nondiagonalizable matrix in Example 300. It failed to have enough linearly independent eigenvectors. This theorem asserts that an $n \times n$ real, symmetric matrix must necessarily have n orthonormal eigenvectors, as these must form the columns of an orthogonal matrix that diagonalizes A.

A proof of this theorem is very tedious, and we omit it. The difficult part is in showing that n mutually orthogonal eigenvectors exist, even when some eigenvalues have multiplicity greater than 1.

In practice, diagonalizing a real, symmetric matrix by an orthogonal matrix is not difficult if the eigenvalues can be found. Here is an example.

EXAMPLE 305

Let

$$A = \begin{bmatrix} 1 & 0 & \sqrt{2} \\ 0 & 2 & 0 \\ \sqrt{2} & 0 & 0 \end{bmatrix}.$$

The eigenvalues are 2, 2, and -1. Corresponding to -1, we find eigenvectors

$$\begin{bmatrix} \alpha \\ 0 \\ -\sqrt{2}\,\alpha \end{bmatrix}, \quad \alpha \neq 0.$$

Corresponding to eigenvalue 2, we find eigenvectors

$$\begin{bmatrix} \sqrt{2}\gamma \\ \beta \\ \gamma \end{bmatrix},$$

with not both β and γ zero.

The important thing to notice is that we can find two linearly independent eigenvectors corresponding to eigenvalue 2, even though this is a multiple eigenvalue. By choosing α, β, and γ appropriately, we can find three linearly independent eigenvectors; for example,

$$\begin{bmatrix} 1 \\ 0 \\ -\sqrt{2} \end{bmatrix}, \quad \begin{bmatrix} 0 \\ 1 \\ 0 \end{bmatrix}, \quad \text{and} \quad \begin{bmatrix} \sqrt{2} \\ 0 \\ 1 \end{bmatrix}.$$

These are mutually orthogonal, but are not all of length 1. Multiply the first and third by $1/\sqrt{3}$ to obtain mutually orthogonal eigenvectors of unit length. These form the columns of an orthogonal matrix Q:

$$Q = \begin{bmatrix} 1/\sqrt{3} & 0 & \sqrt{2/3} \\ 0 & 1 & 0 \\ -\sqrt{2/3} & 0 & 1/\sqrt{3} \end{bmatrix}$$

which diagonalizes A:

$$Q^{-1}AQ = \begin{bmatrix} -1 & 0 & 0 \\ 0 & 2 & 0 \\ 0 & 0 & 2 \end{bmatrix}.$$

In the next section we apply Theorem 81 to the study of quadratic forms.

PROBLEMS FOR SECTION 10.21

In each of Problems 1 through 4, verify that the given matrix is orthogonal.

1. $\begin{bmatrix} \sqrt{3}/2 & -1/2 \\ 1/2 & \sqrt{3}/2 \end{bmatrix}$

2. $\begin{bmatrix} 1/\sqrt{2} & 0 & -1/\sqrt{2} \\ 0 & 1 & 0 \\ -1/\sqrt{2} & 0 & 1/\sqrt{2} \end{bmatrix}$

3. $\begin{bmatrix} 1/\sqrt{5} & 2/\sqrt{5} & 0 \\ -2/\sqrt{5} & 1/\sqrt{5} & 0 \\ 0 & 0 & 1 \end{bmatrix}$

4. $\begin{bmatrix} 1/\sqrt{3} & -\sqrt{2/3} & 0 \\ 1/\sqrt{3} & 1/\sqrt{6} & -1/\sqrt{2} \\ 1/\sqrt{3} & 1/\sqrt{6} & 1/\sqrt{2} \end{bmatrix}$

In each of Problems 5 through 12, find an orthogonal matrix that diagonalizes the given matrix.

5. $\begin{bmatrix} -2 & 1 \\ 1 & 3 \end{bmatrix}$

6. $\begin{bmatrix} 4 & -2 \\ -2 & 1 \end{bmatrix}$

7. $\begin{bmatrix} -2 & 6 \\ 6 & -2 \end{bmatrix}$

8. $\begin{bmatrix} 2 & -1 & 0 \\ -1 & 0 & 3 \\ 0 & 3 & 4 \end{bmatrix}$

9. $\begin{bmatrix} 0 & 0 & 0 \\ 0 & 1 & -2 \\ 0 & -2 & 0 \end{bmatrix}$

10. $\begin{bmatrix} 1 & 3 & 0 \\ 3 & 0 & 1 \\ 0 & 1 & 1 \end{bmatrix}$

11. $\begin{bmatrix} 0 & 0 & 0 & 0 \\ 0 & 1 & -2 & 0 \\ 0 & -2 & 1 & 0 \\ 0 & 0 & 0 & 0 \end{bmatrix}$

12. $\begin{bmatrix} 5 & 0 & 0 & 0 \\ 0 & 0 & -1 & 0 \\ 0 & -1 & 0 & 0 \\ 0 & 0 & 0 & 0 \end{bmatrix}$

13. A real, symmetric matrix is said to be *positive definite* if all its eigenvalues are positive.
    Let A be a real, symmetric matrix. Prove that A is positive definite if and only if there is a nonsingular matrix Q such that $A = Q^tQ$.
14. Use the result of Problem 13 to show that every real, nonsingular matrix A can be written $A = QS$, where S is real, symmetric, and positive definite, and Q is an orthogonal matrix.

10.22 APPLICATION OF ORTHOGONAL MATRICES TO REAL QUADRATIC FORMS

A *real quadratic form* in $x_1, \ldots, x_n$ is a polynomial

$$\sum_{j=1}^{n} \sum_{i=1}^{n} a_{ij} x_i x_j,$$

with all the a_{ij}'s real numbers. As examples, we might have

$$8x_1^2 + 4x_1x_2 - x_2^2 \qquad (n = 2)$$

or

$$-14x_1^2 + 8x_1x_2 + 3x_1x_3 - 4x_2x_3 + 5x_3^2 \qquad (n = 3).$$

Such expressions arise in a number of contexts. In physics, the kinetic energy of a particle is a quadratic form; and in analytic geometry, a conic such as $x^2 + y^2 = 14$ or $x^2/4 - y^2 = 22$ may be thought of as having a quadratic form on the left side (with x in place of x_1 and y in place of x_2).

In general, $\sum_{j=1}^{n} \sum_{i=1}^{n} a_{ij} x_i x_j$ will include square terms,

$$a_{11}x_1^2,\ a_{22}x_2^2,\ \ldots,\ a_{nn}x_n^2,$$

together with the mixed products

$$(a_{ij} + a_{ji})x_i x_j, \qquad i \neq j,$$

and can always be written as a matrix product X^tAX, where A is a real, symmetric matrix and

$$X = \begin{bmatrix} x_1 \\ \vdots \\ x_n \end{bmatrix}.$$

We need only put $A_{ii} = a_{ii}$ for $i = 1, \ldots, n$, and, for $i \neq j$,

$$A_{ij} = \frac{(a_{ij} + a_{ji})}{2}.$$

For example, $8x_1^2 + 4x_1x_2 - x_2^2$ can be written* as $8x_1^2 + 2x_1x_2 + 2x_2x_1 - x_2^2$, and this is exactly

$$[x_1 \quad x_2]\begin{bmatrix} 8 & 2 \\ 2 & -1 \end{bmatrix}\begin{bmatrix} x_1 \\ x_2 \end{bmatrix}.$$

Similarly, $-14x_1^2 + 8x_1x_2 + 3x_1x_3 - 4x_2x_3 + 5x_3^2$ can be written

$$-14x_1^2 + 4x_1x_2 + 4x_2x_1 + \tfrac{3}{2}x_1x_3 + \tfrac{3}{2}x_3x_1 - 2x_2x_3 - 2x_3x_2 + 0x_2^2 + 5x_3^2,$$

and this is

$$[x_1 \quad x_2 \quad x_3]\begin{bmatrix} -14 & 4 & \frac{3}{2} & x_1 \\ 4 & 0 & -2 & x_2 \\ \frac{3}{2} & -2 & 5 & x_3 \end{bmatrix}\begin{bmatrix} x_1 \\ x_2 \\ x_3 \end{bmatrix}.$$

A fundamental problem with real quadratic forms is to choose a coordinate system $(y_1, \ldots, y_n)$ in which the cross-product terms do not appear. That is, we want

$$\sum_{j=1}^{n} \sum_{i=1}^{n} a_{ij}x_ix_j = \alpha_1 y_1^2 + \alpha_2 y_2^2 + \cdots + \alpha_n y_n^2.$$

For example, put

$$x_1 = \frac{1}{\sqrt{2}} y_1 + \frac{1}{\sqrt{2}} y_2,$$

$$x_2 = \frac{1}{\sqrt{2}} y_1 - \frac{1}{\sqrt{2}} y_2$$

into $x_1^2 - 2x_1x_2 + x_2^2$, and we obtain just $2y_2^2$, a much simpler quadratic form.

We shall now show that such a coordinate system always exists, how to find it, and what values the α_i's will have.

THEOREM 82

Let A be a real, symmetric matrix with eigenvalues $\lambda_1, \ldots, \lambda_n$. Let Q be an orthogonal matrix that diagonalizes A. Then, the change of coordinates

$$X = QY \text{ transforms } \sum_{j=1}^{n} \sum_{i=1}^{n} a_{ij}x_ix_j \quad \text{into} \quad \lambda_1 y_1^2 + \lambda_2 y_2^2 + \cdots + \lambda_n y_n^2.$$

We note here that the eigenvalues of A need not be distinct, as the last theorem of the previous section ensures the existence of an orthogonal matrix diagonalizing A in any event.

---

* Note that X^tAX is a 1×1 matrix, and recall that we agreed to write such a matrix $[\alpha]$ as just α.

Proof The proof is a straightforward calculation. We have

$$\begin{aligned} X^tAX &= (QY)^tA(QY) \\ &= Y^t(Q^tAQ)Y \\ &= Y^t\begin{bmatrix} \lambda_1 & & 0 \\ & \ddots & \\ 0 & & \lambda_n \end{bmatrix}Y = \lambda_1 y_1^2 + \cdots + \lambda_n y_n^2. \end{aligned}$$

This result is sometimes called the *principal axis theorem* because it defines new axes with respect to which the quadratic form has a particularly simple appearance. One also sees $\lambda_1 y_1^2 + \cdots + \lambda_n y_n^2$ referred to as the *standard form* of $\sum_{j=1}^n \sum_{i=1}^n a_{ij}x_ix_j$.

Here are some examples, which include applications to analytic geometry.

EXAMPLE 306

Above, we saw the quadratic form $x_1^2 - 2x_1x_2 + x_2^2$ transformed into $2y_2^2$ by a change of variables. We shall now apply the last theorem to see how this change of variables was obtained.

Note first that the quadratic form is X^tAX, with $A = \begin{bmatrix} 1 & -1 \\ -1 & 1 \end{bmatrix}$.

A has eigenvalues 0 and 2, with corresponding orthonormal eigenvectors

$$\begin{bmatrix} 1/\sqrt{2} \\ 1/\sqrt{2} \end{bmatrix} \quad \text{and} \quad \begin{bmatrix} 1/\sqrt{2} \\ -1/\sqrt{2} \end{bmatrix}.$$

Thus,

$$Q = \begin{bmatrix} 1/\sqrt{2} & 1/\sqrt{2} \\ 1/\sqrt{2} & -1/\sqrt{2} \end{bmatrix}$$

is an orthogonal matrix that diagonalizes A. Putting $X = QY$ transforms $x_1^2 - 2x_1x_2 + x_2^2$ into $\lambda_1 y_1^2 + \lambda_2 y_2^2$, or just $2y_2^2$ in this case.

If we write $X = QY$ out in coordinate form, we have

$$x_1 = \frac{1}{\sqrt{2}}y_1 + \frac{1}{\sqrt{2}}y_2$$

and

$$x_2 = \frac{1}{\sqrt{2}}y_1 - \frac{1}{\sqrt{2}}y_2,$$

which is exactly the change of variables we pulled out of thin air before. Geometrically, this is a rotation of the plane through an angle of 45°.

Look at this example geometrically. For any constant $K > 0$, the points (x_1, x_2) satisfying $x_1^2 - 2x_1x_2 + x_2^2 = K$ form a conic in the plane. Such a conic is easier to visualize when put into standard form, and this is exactly what we have done by rotating the plane, as is often taught in elementary calculus courses. In this way, the

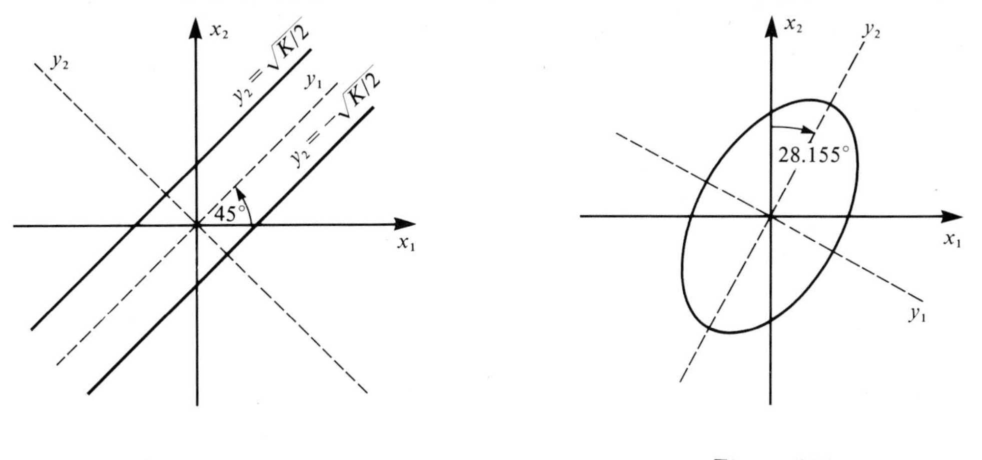

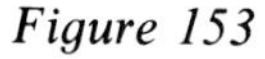
Figure 153

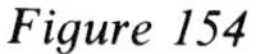
Figure 154

given conic becomes $2y_2^2 = K$, which we immediately identify as the two straight lines $y_2 = \sqrt{K/2}$ and $y_2 = -\sqrt{K/2}$. In Figure 153 we have drawn the original locus and the rotated coordinate system with a common origin so that the two can be compared.

EXAMPLE 307

Analyze the conic $4x_1^2 - 3x_1x_2 + 2x_2^2 = 8$.

First, observe that this is $X^tAX = 8$, with

$$A = \begin{bmatrix} 4 & -\frac{3}{2} \\ -\frac{3}{2} & 2 \end{bmatrix}.$$

The eigenvalues of A are $(6 + \sqrt{13})/2$ and $(6 - \sqrt{13})/2$. Without actually finding an orthogonal matrix Q that diagonalizes A, we know that one exists such that $X = QY$ transforms the given conic to $\lambda_1 y_1^2 + \lambda_2 y_2^2 = 8$. After a little rewriting, this is

$$y_1^2 + \frac{6 - \sqrt{13}}{6 + \sqrt{13}}\, y_2^2 = \frac{16}{6 + \sqrt{13}}.$$

This is in the standard form of an ellipse. The intercepts with the y_1 and y_2 axes are about $(\pm 1.29, 0)$ and $(0, \pm 2.59)$, respectively.

As an exercise, we shall find Q anyway. Orthonormal eigenvectors corresponding to $(6 + \sqrt{13})/2$ and $(6 - \sqrt{13})/2$, respectively, are

$$\begin{bmatrix} \dfrac{3}{\sqrt{26 - 4(13)^{1/2}}} \\ \dfrac{2 - \sqrt{13}}{\sqrt{26 - 4(13)^{1/2}}} \end{bmatrix} \quad \text{and} \quad \begin{bmatrix} \dfrac{3}{\sqrt{26 + 4(13)^{1/2}}} \\ \dfrac{2 + \sqrt{13}}{\sqrt{26 + 4(13)^{1/2}}} \end{bmatrix}.$$

These form the columns of Q. Since every 2×2 orthogonal matrix is a rotation of the plane, so should this one be. In fact, Q is approximately

$$\begin{bmatrix} 0.8816745988 & 0.4718579255 \\ -0.4718579255 & 0.8816745988 \end{bmatrix},$$

and this represents a clockwise rotation through about 28.155 degrees (about 0.49 radian). The graph of $4x_1^2 - 3x_1x_2 + 2x_2^2 = 8$ is shown in Figure 154, with the original x_1 and x_2 axes and the rotated y_1, y_2 coordinate system drawn for reference.

EXAMPLE 308

Here is a three-dimensional example, where things are not as easily visualized. The methods are the same, however. Consider $x_1^2 + 2x_2^2 + 2\sqrt{2}x_1x_3$, which is X^tAX with

$$A = \begin{bmatrix} 1 & 0 & \sqrt{2} \\ 0 & 2 & 0 \\ \sqrt{2} & 0 & 0 \end{bmatrix}.$$

Eigenvalues are $-1, 2$, and 2 (see Example 305).

Putting $X = QY$, where Q is an orthogonal matrix that diagonalizes A, gives us the standard form $-y_1^2 + 2y_2^2 + 2y_3^3$.

From Example 305,

$$Q = \begin{bmatrix} 1/\sqrt{3} & 0 & \sqrt{2/3} \\ 0 & 1 & 0 \\ -\sqrt{2/3} & 0 & 1/\sqrt{3} \end{bmatrix}.$$

In coordinate form, then, the change of variables is

$$x_1 = \frac{1}{\sqrt{3}}y_1 + \sqrt{\frac{2}{3}}y_3,$$

$$x_2 = y_2,$$

$$x_3 = -\sqrt{\frac{2}{3}}y_1 + \frac{1}{\sqrt{3}}y_3.$$

The student can check by direct substitution that this transforms $x_1^2 + 2x_2^2 + 2\sqrt{2}x_1x_3$ into $-y_1^2 + 2y_2^2 + 2y_3^2$.

PROBLEMS FOR SECTION 10.22

In each of Problems 1 through 6, find A such that the quadratic form can be written as X^tAX.

1. $x_1^2 + 2x_1x_2 + 6x_2^2$
2. $3x_1^2 + 3x_2^2 - 4x_1x_2 - 3x_1x_3 + 2x_2x_3 + x_3^2$
3. $x_1^2 - 4x_1x_2 + x_2^2$
4. $2x_1^2 - x_2^2 + 2x_1x_2$
5. $-x_1^2 + x_4^2 - 2x_1x_4 + 3x_2x_4 - x_1x_3 + 4x_2x_3$
6. $x_1^2 - x_2^2 + x_3^2 - x_1x_3 + 4x_2x_3$

In each of Problems 7 through 12, write out the quadratic form X^tAX defined by the given matrix.

7. $\begin{bmatrix} -2 & 1 \\ 1 & 6 \end{bmatrix}$ **8.** $\begin{bmatrix} 14 & -3 & 0 \\ -3 & 2 & 1 \\ 0 & 1 & 7 \end{bmatrix}$ **9.** $\begin{bmatrix} 6 & 1 & -7 \\ 1 & 2 & 0 \\ -7 & 0 & 1 \end{bmatrix}$ **10.** $\begin{bmatrix} 0 & -2 & 1 \\ -2 & 1 & 3 \\ 1 & 3 & 14 \end{bmatrix}$

11. $\begin{bmatrix} 8 & 0 & 0 \\ 0 & 2 & -4 \\ 0 & -4 & 3 \end{bmatrix}$ **12.** $\begin{bmatrix} 7 & 1 & -2 \\ 1 & 0 & -1 \\ -2 & -1 & 3 \end{bmatrix}$

In each of Problems 13 through 19, find the standard form of the given quadratic form.

13. $-5x_1^2 + 4x_1x_2 + 3x_2^2$ **14.** $4x_1^2 - 12x_1x_2 + x_2^2$ **15.** $-3x_1^2 + 4x_1x_2 + 7x_2^2$
16. $4x_1^2 - 4x_1x_2 + x_2^2$ **17.** $-6x_1x_2 + 4x_2^2$ **18.** $5x_1^2 + 4x_1x_3 + 2x_2^2$
19. $-2x_1x_2 + 2x_3^2$

In each of Problems 20 through 24, obtain the reduced form of the quadratic form, and identify the given locus as an ellipse, parabola, hyperbola, or straight line(s). Sketch the locus.

20. $x_1^2 - 2x_1x_2 + 4x_2^2 = 6$ **21.** $3x_1^2 + 5x_1x_2 - 3x_2^2 = 5$ **22.** $-2x_1^2 + 3x_2^2 + x_1x_2 = 5$
23. $4x_1^2 - 4x_2^2 + 6x_1x_2 = 8$ **24.** $6x_1^2 + 2x_1x_2 + 5x_2^2 = 14$

25. Let A be a real, symmetric matrix. Prove that there is a matrix P such that $X = PY$ transforms X^tAX into $y_1^2 + y_2^2 + \cdots + y_p^2 - y_{p+1}^2 - \cdots - y_r^2$, where $r = \text{rank}(A)$ and p = number of positive eigenvalues of A. *Note:* We do not require that P be orthogonal. This is called the *canonical form* of X^tAX.

26. Prove that the quadratic forms X^tAX and X^tBX have the same canonical form (see Problem 25) if and only if for some matrix P, $B = P^{-1}AP$.

27. Let A be a real, symmetric matrix. A is positive definite if all its eigenvalues are positive (see Problem 13 of Section 10.21). Prove that A is positive definite if and only if $X^tAX > 0$ for every nonzero vector X, where

$$X = \begin{bmatrix} x_1 \\ \vdots \\ x_n \end{bmatrix}.$$

10.23 UNITARY, HERMITIAN, AND SKEW-HERMITIAN MATRICES

In this section we shall discuss properties of some other special kinds of matrices which are frequently encountered.

Recall that an orthogonal matrix A is a real matrix having the property that $A^{-1} = A^t$. We saw that this is equivalent to the assertion that the row (or column) vectors are orthonormal. We shall now consider the complex analogue of this situation.

Let U be a square matrix with complex entries. We let $\overline{U}$ denote the matrix obtained by replacing each entry of U by its complex conjugate (thus, U is a real matrix if and only if $U = \overline{U}$). Note that $(\overline{U})^t = \overline{U^t}$ and, if U^{-1} exists, then $\overline{(U^{-1})} = (\overline{U})^{-1}$. Thus, we can without ambiguity write $\overline{U}^t$ and $\overline{U}^{-1}$.

We call U a *unitary matrix* if

$$U^t = \overline{U}^{-1}.$$

Note that an orthogonal matrix is thus a unitary matrix with all real entries.

In general, the row and column vectors of an $n \times n$ unitary matrix U are not in R^n, as the coordinates are complex. We can still define a kind of dot product (often called an *inner product*) by setting

$$(x_1, x_2, \ldots, x_n) \cdot (y_1, y_2, \ldots, y_n) = \sum_{j=1}^{n} \bar{x}_j y_j$$

when each x_i and y_i is complex. In terms of matrices, if

$$X = \begin{bmatrix} x_1 \\ \vdots \\ x_n \end{bmatrix} \quad \text{and} \quad Y = \begin{bmatrix} y_1 \\ \vdots \\ y_n \end{bmatrix},$$

then this inner product is $\overline{X}^t Y$.

Complex n-vectors $\mathbf{F}_1, \ldots, \mathbf{F}_n$ are said to form a *unitary system* if

$$\mathbf{F}_i \cdot \mathbf{F}_j = \begin{cases} 0 & \text{if } i \neq j, \\ 1 & \text{if } i = j. \end{cases}$$

Note that, when each $\mathbf{F}_i$ has real entries, unitary system is the same as orthonormal.

Here is the unitary matrix analogue of Theorem 79.

THEOREM 83

Let U be an $n \times n$ complex matrix. Then U is unitary if and only if the row (column) vectors form a unitary system.

Proof The argument is like that used to prove Theorem 79, and we leave details to the student.

Just as unitary matrices generalize orthogonal matrices to the complex case, the notion of a hermitian matrix generalizes the concept of a symmetric matrix. Recall that an $n \times n$ real matrix A is symmetric if $A = A^t$. We define an $n \times n$ complex matrix H to be *hermitian* if

$$\overline{H} = H^t.$$

If H is real, this means that $H = \overline{H}$, and H is then symmetric.

If $\overline{H} = -H^t$, we call H *skew-hermitian*. This generalizes the notion of skew-symmetric mentioned in Problem 24, Section 10.10.

EXAMPLE 309

$$H = \begin{bmatrix} 15 & 8i & 6-2i \\ -8i & 0 & -4+i \\ 6+2i & -4-i & -3 \end{bmatrix}$$

is hermitian, since

$$\overline{H} = \begin{bmatrix} 15 & -8i & 6+2i \\ 8i & 0 & -4-i \\ 6-2i & -4+i & -3 \end{bmatrix}$$

and $\overline{H}^t = H$.

The matrix

$$K = \begin{bmatrix} 0 & 8i & -6+2i \\ 8i & 0 & 4-i \\ 6+2i & -4-i & 0 \end{bmatrix}$$

is skew-hermitian.

Recall that a real quadratic form is an expression X^tAX, where A is real, symmetric and

$$X = \begin{bmatrix} x_1 \\ \vdots \\ x_n \end{bmatrix}$$

has real entries. By the same token, we can define an *hermitian form* as an expression

$$\overline{Z}^t HZ,$$

where H is hermitian and

$$Z = \begin{bmatrix} z_1 \\ \vdots \\ z_n \end{bmatrix}$$

is allowed to have complex entries. We can write

$$\overline{Z}^t HZ = \sum_{j=1}^{n} \sum_{k=1}^{n} H_{jk} \bar{z}_j z_k .$$

Similarly, $\overline{Z}^t SZ$ is a *skew-hermitian form* when S is skew-hermitian. Again, we allow Z to have complex entries.

The following is a rather remarkable result.

THEOREM 84

Let $Z = \begin{bmatrix} z_1 \\ \vdots \\ z_n \end{bmatrix}$ be any complex $n \times 1$ matrix. Then,

1. $\overline{Z}^t HZ$ is real if H is hermitian.
2. $\overline{Z}^t SZ$ is zero or pure imaginary if S is skew-hermitian.

Note: By *pure imaginary*, here we mean any number of the form αi, with α real.

Proof of (1) Let H be hermitian. Then $\overline{H}^t = H$; hence,

$$\overline{(\overline{Z}^t HZ)} = \overline{\overline{Z}}^t \overline{H} \overline{Z} = Z^t H \overline{Z}.$$

Now, $Z^t H \overline{Z}$ is a 1×1 matrix, and hence equals its own transpose. Continuing the above string of equalities, we now have

$$Z^t H \overline{Z} = (Z^t H \overline{Z})^t = \overline{Z}^t H^t (Z^t)^t = \overline{Z}^t H^t Z = \overline{Z}^t HZ.$$

In sum, then, we have shown that

$$\overline{(\overline{Z}^t H Z)} = \overline{Z}^t H Z.$$

Hence, $\overline{Z}^t H Z$ is real.

Proof of (2) Let S be skew-symmetric. Then $\overline{S}^t = -S$; hence [imitating the argument used in proving (1)],

$$\overline{(\overline{Z}^t S Z)} = -\overline{Z}^t S Z.$$

Writing $\overline{Z}^t S Z = \alpha + i\beta$, this says that

$$\alpha - i\beta = -\alpha - i\beta.$$

Then $\alpha = 0$, and $\overline{Z}^t S Z = \beta i$, which is zero if $\beta = 0$, and pure imaginary if $\beta \neq 0$.

This theorem enables us to draw interesting conclusions about the eigenvalues of hermitian and skew-hermitian matrices. To complete the picture, we include a statement about eigenvalues of unitary matrices.

THEOREM 85

1. The eigenvalues of any hermitian matrix are real.
2. The eigenvalues of any skew-hermitian matrix must be zero or pure imaginary.
3. The eigenvalues of any unitary matrix must have absolute value 1.

Note: (1) implies (but is stronger than) Theorem 76, which says that the eigenvalues of any real, symmetric matrix must be real. In (3), there is no implication that the eigenvalues of a unitary matrix must be real—only that their absolute value must be 1. Thus, $(1/\sqrt{2}) + (i/\sqrt{2})$ might be an eigenvalue of a unitary matrix, but 3 cannot be.

Proof of (1) and (2) Let A be any $n \times n$ matrix, real or complex. If λ is an eigenvalue of A, then for some nonzero $n \times 1$ matrix Z (possibly complex),

$$AZ = \lambda Z.$$

Then,

$$\overline{Z}^t A Z = \overline{Z}^t(\lambda Z) = \lambda \overline{Z}^t Z.$$

If

$$Z = \begin{bmatrix} z_1 \\ \vdots \\ z_n \end{bmatrix},$$

then

$$\overline{Z}^t Z = \bar{z}_1 z_1 + \cdots + \bar{z}_n z_n = |z_1|^2 + \cdots + |z_n|^2,$$

which is real and nonzero. Thus,

$$\lambda = \frac{\overline{Z}^t A Z}{\overline{Z}^t Z}.$$

Now, if A is hermitian, then $\overline{Z}^t AZ$ is real by Theorem 84(1); hence, λ is real, proving (1).

If A is skew-hermitian, then $\overline{Z}^t AZ$ is zero or pure imaginary by Theorem 84(2); hence, λ is zero or pure imaginary, proving (2).

Proof of (3) Let U be any unitary matrix. If λ is an eigenvalue, and Z an associated eigenvector, then

$$UZ = \lambda Z.$$

Hence,

$$\overline{UZ} = \overline{U}\,\overline{Z} = \overline{\lambda}\overline{Z},$$

and so

$$(\overline{U}\,\overline{Z})^t = (\overline{\lambda}\overline{Z})^t.$$

Now, $\overline{\lambda}$ is a scalar; so we have [remember Theorem 43(3)]

$$\overline{Z}^t\overline{U}^t = \overline{\lambda}\overline{Z}^t.$$

Now, since U is unitary, we have $\overline{U}^t = U^{-1}$. Hence,

$$\overline{Z}^t U^{-1} = \overline{\lambda}\overline{Z}^t.$$

Multiply both sides of this equation on the right by UZ. Then,

$$\overline{Z}^t U^{-1} UZ = \overline{\lambda}\overline{Z}^t UZ = \overline{\lambda}\overline{Z}^t \lambda Z,$$

since $UZ = \lambda Z$. Then,

$$\overline{Z}^t Z = \overline{\lambda}\lambda\overline{Z}^t Z.$$

Since $\overline{Z}^t Z \neq 0$, then $\overline{\lambda}\lambda = |\overline{\lambda}|^2 = 1$. Thus, $|\lambda| = 1$, as we wanted to show.

Graphically, we may represent Theorem 85 as in Figure 155. Eigenvalues of a unitary matrix may be real or complex but still have magnitude 1, and so lie on the unit circle about the origin in the complex plane. Eigenvalues of hermitian matrices lie on the horizontal real axis, and those of skew-symmetric matrices lie on the vertical axis.

Note that a real, symmetric matrix is hermitian; so Theorem 85(1) gives us Theorem 76 again. Also, an orthogonal matrix is unitary; so Theorem 85(3) gives us the previously unstated result that *the eigenvalues of any orthogonal matrix have magnitude 1.*

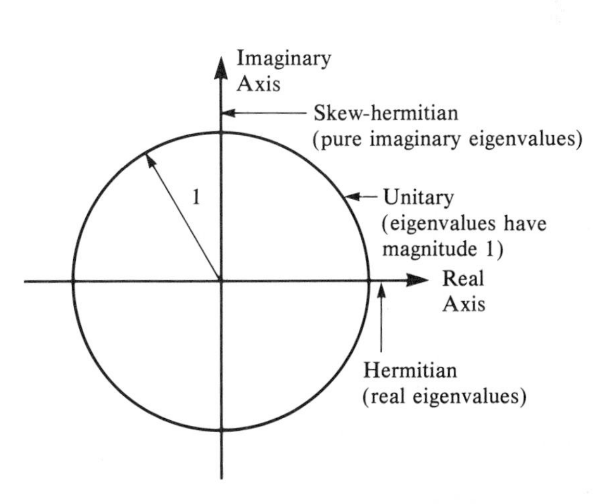

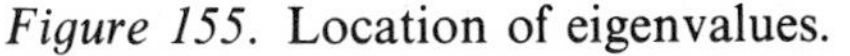
Figure 155. Location of eigenvalues.

PROBLEMS FOR SECTION 10.23

1. Let A be either hermitian, skew-hermitian, or unitary. Prove that $A\overline{A}^t = \overline{A}^t A$.

In each of Problems 2 through 10, determine whether the given matrix is unitary, hermitian, skew-hermitian, or none of these. Find the eigenvalues of the given matrices.

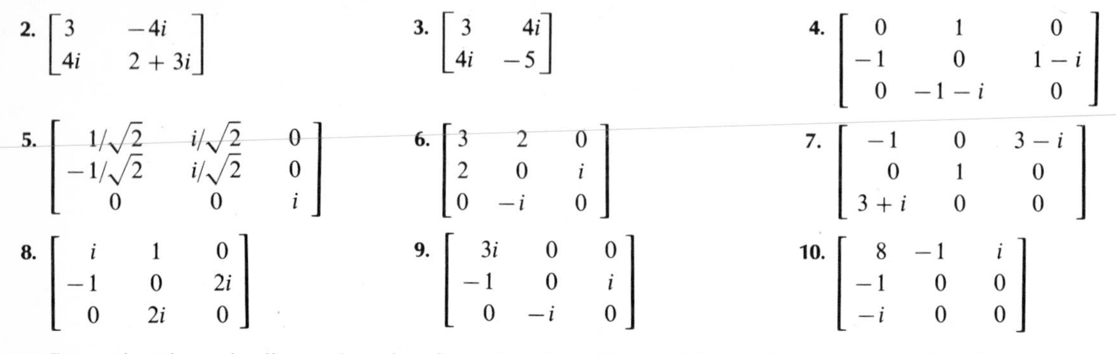

2. $\begin{bmatrix} 3 & -4i \\ 4i & 2+3i \end{bmatrix}$

3. $\begin{bmatrix} 3 & 4i \\ 4i & -5 \end{bmatrix}$

4. $\begin{bmatrix} 0 & 1 & 0 \\ -1 & 0 & 1-i \\ 0 & -1-i & 0 \end{bmatrix}$

5. $\begin{bmatrix} 1/\sqrt{2} & i/\sqrt{2} & 0 \\ -1/\sqrt{2} & i/\sqrt{2} & 0 \\ 0 & 0 & i \end{bmatrix}$

6. $\begin{bmatrix} 3 & 2 & 0 \\ 2 & 0 & i \\ 0 & -i & 0 \end{bmatrix}$

7. $\begin{bmatrix} -1 & 0 & 3-i \\ 0 & 1 & 0 \\ 3+i & 0 & 0 \end{bmatrix}$

8. $\begin{bmatrix} i & 1 & 0 \\ -1 & 0 & 2i \\ 0 & 2i & 0 \end{bmatrix}$

9. $\begin{bmatrix} 3i & 0 & 0 \\ -1 & 0 & i \\ 0 & -i & 0 \end{bmatrix}$

10. $\begin{bmatrix} 8 & -1 & i \\ -1 & 0 & 0 \\ -i & 0 & 0 \end{bmatrix}$

11. Prove that the main-diagonal entries of any skew-hermitian matrix must be zero or pure imaginary.
12. Prove that the main-diagonal entries of any hermitian matrix must be real.
13. Prove that a product of two unitary matrices is unitary.
14. Suppose that A is both unitary and hermitian. Prove that $A = A^{-1}$. Give a 3×3 example of such a matrix (different from I_3).
15. Let A be any $n \times n$ matrix. Prove that A is a sum of a hermitian and a skew-hermitian matrix.
16. Let H be hermitian, and α any scalar. Prove that if α is real, then αH is hermitian. Does this hold if α is complex but not real?
17. Let S be skew-hermitian, and α any scalar. Prove that if α is real, then αS is skew-hermitian. Does this hold if α is complex but not real?

In each of Problems 18 through 22, compute $\overline{Z}^t AZ$ for the given matrices A and Z.

18. $A = \begin{bmatrix} 2 & 2-i \\ 2+i & 5 \end{bmatrix}, \quad Z = \begin{bmatrix} -i \\ 2+i \end{bmatrix}$

19. $A = \begin{bmatrix} 5 & 3i & 4-i \\ -3i & 0 & 2-2i \\ 4+i & 2+2i & -3 \end{bmatrix}, \quad Z = \begin{bmatrix} 1 \\ -5 \\ 3-i \end{bmatrix}$

20. $A = \begin{bmatrix} i & 3-i \\ -3-i & 0 \end{bmatrix}, \quad Z = \begin{bmatrix} 2+5i \\ -3i \end{bmatrix}$

21. $A = \begin{bmatrix} 0 & -3i & 2+i \\ -3i & -4i & 0 \\ -2+i & 0 & 3i \end{bmatrix}, \quad Z = \begin{bmatrix} 0 \\ -2i \\ 1-5i \end{bmatrix}$

22. $A = \begin{bmatrix} 4 & -1 & 2i \\ -1 & 0 & 6-i \\ -2i & 6+i & -5 \end{bmatrix}, \quad Z = \begin{bmatrix} 3i \\ 4 \\ 0 \end{bmatrix}$

CHAPTER SUMMARY

In this chapter we have developed some of the basic notions and applications of matrices and determinants. As an overview, we might think of it as broken down into four major portions:

Basic matrix concepts and manipulation, Sections 1 through 6 and Section 9;

Solutions of systems of linear equations, Sections 7 and 8;

Determinants and some applications, Sections 10 through 14; and

Eigenvalues, eigenvectors, and ramifications, Sections 15 through 23.

While it is possible to treat these topics in an abstract way, we usually had numbers as matrix entries. The interplay between matrices, determinants, and n-vectors is important to the development of the subject, and contributes both to our understanding of matrices and to our skill in manipulating geometric quantities in 3 and higher dimensions.

As one example, the fact that each $n \times m$ matrix generates a column subspace of R^n and a row subspace of R^m enables us to use facts about these subspaces to determine what the general solution of a system of linear equations should look like.

The geometry of R^n is also inherent in such things as quadratic forms and rotation of axes, and diagonalization may be thought of as changing coordinates to a system in which coupled first order differential equations become uncoupled.

The many themes wound about matrices are the main reason why they are so useful today in posing and solving a variety of problems, ranging from neutron absorption in nuclear reactors to reconstruction of pictures sent back to Earth from our space probes.

SOME NOTES ON THE HISTORY OF MATRICES AND DETERMINANTS

Matrices and determinants grew out of geometric considerations and systems of linear equations in the eighteenth and nineteenth centuries. Perhaps surprisingly, the early emphasis was on the determinant, not the matrix. While many properties of determinants were discovered by talented amateurs and figures of minor importance in the overall history of mathematics, such mathematical giants as Gauss and Cauchy also contributed to their development. For example, Cauchy stated the "correct" definition of the product of two matrices, and showed that the determinant of a product is the product of the determinants. Cauchy also proved Laplace's formula for cofactor expansions. Gauss's main interest in determinants was in connection with quadratic forms.

A leading figure in the history of determinants was James Joseph Sylvester (1814–1897), who was denied a teaching post at Cambridge because he was a Jew. His varied career included periods at the University of Virginia and Johns Hopkins

University in the United States, as well as nonteaching posts as a lawyer and actuary in his native England. Sylvester's work touched upon determinant applications to solution of polynomial equations, as well as geometric aspects of quadratic forms. Out of this, and Lagrange's work on planetary motion, came the determinant for the characteristic equation of a matrix (then called the *secular equation*).

Carl Gustav Jacob Jacobi (1804–1851) and Eugene Catalan (1814–1894) are credited with discovering the determinant used in transforming multiple integrals. This determinant is today called the *Jacobian*. The *Wronskian*, a determinant involving solutions of differential equations, is named after the Polish mathematician H. Wronski (1778–1853).

In this connection, it may be noted that Lewis Carroll (1832–1898) was quite a talented amateur mathematician who was greatly interested in determinants. His books on Alice's adventures in Wonderland so delighted the Queen that she commanded him to send her his next work, which turned out to be a treatise on determinants. The exact wording of her reaction has not been recorded for history.

The term *matrix* was apparently coined by Sylvester around 1850. Arthur Cayley (1821–1895) pointed out the rationale for thinking of matrices as preceding determinants in the logical fabric of mathematics, but in fact the historical development was the reverse, and many properties of matrices were known before the subject was formalized. Cayley collaborated with Sylvester on a series of papers in which they developed matrices, with emphasis on the geometry of n dimensions, invariants of quadratic forms, and linear transformations. Cayley and Sylvester must have made an interesting pair—the latter was a lively, well-traveled man of many accomplishments, while the former was as calm and even as Sylvester was excitable. Cayley spent his entire career at Cambridge, except for a year (1892) visiting Sylvester at Johns Hopkins.

The fact that real, symmetric matrices have real eigenvalues was known to Cauchy and established later by Charles Hermite, who developed hermitian matrices and is also known for Hermite polynomials. Rudolph Friedrich Alfred Clebsch (1833–1872) showed that the nonzero eigenvalues of a real, skew-symmetric matrix are pure imaginary. Hermite had the idea of orthogonal matrices, but Georg Frobenius (1849–1917) gave the definition that A is orthogonal if $A^{-1} = A^t$. Frobenius is also credited with a method for solving differential equations by generalized power series (see Chapter 5).

Of the topics we have omitted for lack of space, two of the most important are the *Cayley-Hamilton theorem*, which states that a square matrix satisfies its own characteristic equation, and the *Jordan canonical form*. William Rowan Hamilton (1805–1865) was a major figure in the development of quaternions and vectors. A precocious child, Hamilton had mastered eight languages, including Persian and Sanskrit, by age fourteen. Camille Jordan (1838–1922) was a French mathematician who made fundamental contributions to the growing fields of modern algebra and topology.

SUPPLEMENTARY PROBLEMS

In each of Problems 1 through 10, find the reduced form of the given matrix and determine its rank.

1. $\begin{bmatrix} -4 & 5 & 8 & 1 \\ 3 & 0 & -5 & 2 \\ 1 & -1 & 5 & 8 \end{bmatrix}$ **2.** $\begin{bmatrix} -4 & 7 & 1 & -1 \\ 2 & -3 & 6 & 2 \end{bmatrix}$ **3.** $\begin{bmatrix} 3 & -1 \\ 2 & 5 \\ 5 & -3 \end{bmatrix}$

4. $\begin{bmatrix} -1 & 1 \\ 2 & -2 \end{bmatrix}$ **5.** $\begin{bmatrix} 8 & -3 & 2 & 1 & -5 \\ 0 & 0 & 0 & 0 & 0 \end{bmatrix}$ **6.** $\begin{bmatrix} 6 & -2 & 3 & 1 & 1 \\ -3 & 2 & 5 & 3 & -5 \\ 0 & 2 & 13 & 7 & -9 \end{bmatrix}$

7. $\begin{bmatrix} 3 & -2 & 1 \\ 0 & 0 & 0 \\ 0 & -1 & 1 \end{bmatrix}$ **8.** $\begin{bmatrix} -7 & 0 & 0 & 0 \\ 1 & 0 & 0 & 0 \\ 0 & 0 & -2 & 4 \end{bmatrix}$ **9.** $\begin{bmatrix} 3 & 4 & 1 & 0 \\ -3 & 2 & 2 & 0 \\ 0 & 0 & 0 & 1 \end{bmatrix}$

10. $\begin{bmatrix} -5 & 7 & 7 & 0 \\ 0 & 7 & -1 & 1 \\ 0 & 0 & 0 & 0 \\ 2 & -1 & 3 & 9 \\ 1 & 0 & 0 & 1 \end{bmatrix}$

In each of Problems 11 through 20, find the inverse of the given matrix, or else show that the matrix is nonsingular. Use any method you choose.

11. $\begin{bmatrix} -3 & 5 \\ 2 & -2 \end{bmatrix}$ **12.** $\begin{bmatrix} 13 & 1 \\ 2 & 0 \end{bmatrix}$ **13.** $\begin{bmatrix} 1 & -5 \\ -3 & 15 \end{bmatrix}$

14. $\begin{bmatrix} -4 & 2 \\ 0 & 4 \end{bmatrix}$ **15.** $\begin{bmatrix} -4 & 1 & -1 \\ 0 & 1 & 0 \\ 2 & -3 & 2 \end{bmatrix}$ **16.** $\begin{bmatrix} -3 & 1 & -2 \\ 0 & -20 & 28 \\ 1 & -7 & 10 \end{bmatrix}$

17. $\begin{bmatrix} 6 & -3 & 1 \\ 2 & -2 & 4 \\ 0 & 1 & -2 \end{bmatrix}$ **18.** $\begin{bmatrix} -5 & 1 & -1 \\ 0 & 7 & 1 \\ 2 & 2 & -2 \end{bmatrix}$ **19.** $\begin{bmatrix} -5 & 0 & 1 & 1 \\ 0 & -2 & 1 & -1 \\ 3 & -4 & 0 & 2 \\ 2 & -5 & 0 & 0 \end{bmatrix}$

20. $\begin{bmatrix} -3 & 1 & 5 & 0 \\ 2 & -1 & 5 & 3 \\ 1 & -5 & 4 & 4 \\ -3 & -4 & 19 & 7 \end{bmatrix}$

In each of Problems 21 through 30, evaluate the determinant.

21. $\begin{vmatrix} -3 & 6 \\ 2 & -4 \end{vmatrix}$ **22.** $\begin{vmatrix} 5 & -3 \\ 2 & 0 \end{vmatrix}$ **23.** $\begin{vmatrix} 3 & -2 & 4 \\ 1 & -1 & 4 \\ 3 & -5 & 6 \end{vmatrix}$

24. $\begin{vmatrix} 1 & -4 & 7 \\ 7 & -3 & 10 \\ -2 & 1 & 6 \end{vmatrix}$ **25.** $\begin{vmatrix} 1 & 0 & -3 & 6 \\ 2 & -1 & 5 & 3 \\ 0 & 1 & 0 & -4 \\ 1 & -4 & 6 & 5 \end{vmatrix}$ **26.** $\begin{vmatrix} -4 & 1 & 0 & 0 \\ 2 & -1 & 0 & -3 \\ 0 & 1 & -2 & 0 \\ 4 & 1 & -5 & 6 \end{vmatrix}$

27. $\begin{vmatrix} 6 & -2 & 1 & 0 \\ 1 & -4 & 3 & 0 \\ 2 & 0 & -6 & 2 \\ 0 & -2 & 1 & -5 \end{vmatrix}$ **28.** $\begin{vmatrix} 4 & -1 & 2 & 8 \\ 2 & -1 & 5 & 2 \\ 0 & 2 & 3 & -3 \\ 0 & 3 & -5 & 1 \end{vmatrix}$ **29.** $\begin{vmatrix} 3 & -2 & 3 & -6 & 1 \\ 0 & -2 & 0 & 0 & -2 \\ 1 & -1 & 0 & 0 & 5 \\ -2 & 4 & 0 & 1 & -5 \\ 0 & 0 & 1 & -1 & 0 \end{vmatrix}$

30. $\begin{vmatrix} 1 & -2 & 0 & 0 & 0 & 0 & 0 \\ 0 & -3 & 0 & 0 & 0 & 0 & 0 \\ 0 & 0 & -1 & 4 & 2 & 0 & 0 \\ 0 & 0 & 1 & -3 & 1 & 0 & 0 \\ 0 & 0 & 4 & -3 & 1 & 0 & 0 \\ 0 & 0 & 0 & 0 & 0 & -6 & 3 \\ 0 & 0 & 0 & 0 & 0 & 2 & 7 \end{vmatrix}$

In each of Problems 31 through 40, find the general solution of the given system, or else show that there is no solution.

31. $x_1 - 4x_2 + x_3 = 0$
$-2x_1 + x_2 - 4x_3 = 0$

32. $x_1 + 6x_2 + x_3 = 1$
$-2x_1 - x_2 + 7x_3 = -3$
$-3x_1 + 4x_2 + 15x_3 = 1$

33. $x_1 + x_2 - x_3 + 4x_4 = -5$
$2x_1 - 4x_2 + x_3 = -2$
$x_2 - x_3 + 7x_4 = -1$

34. $2x_1 - x_2 + 5x_3 = 0$
$-2x_1 + 3x_2 + 4x_3 = 0$
$x_1 - x_2 + 5x_3 = 0$
$3x_1 + 6x_2 - x_3 = 0$

35. $2x_1 + 3x_2 - x_4 = 5$
$x_1 - 4x_2 - 3x_3 - x_4 = -8$
$x_2 - 6x_3 + x_4 = 0$

36. $-3x_1 + x_2 - 5x_4 = 1$
$2x_1 + 3x_2 - 5x_3 = 0$
$-5x_1 + 4x_2 - 4x_3 - x_4 = -10$

37. $-4x_1 + x_2 - x_3 = 9$
$2x_1 + x_2 + 7x_3 = 1$
$x_1 - 4x_2 + x_3 = 0$

38. $6x_1 + x_2 - 4x_3 + x_4 = -6$
$x_1 - x_3 + 4x_5 = 9$

39. $2x_1 + 7x_2 - x_3 = 1$
$x_2 - x_3 + 4x_4 = 0$
$5x_1 + 5x_2 - 3x_3 + x_4 = -6$

40. $-3x_1 + x_2 - x_4 = 0$
$x_2 - x_3 + 5x_4 = 0$

In each of Problems 41 through 45, reduce the given quadratic form to standard form.

41. $2x_1^2 - 6x_1x_2 + x_2^2$

42. $-4x_1^2 + 5x_1x_2 + x_2^2$

43. $-6x_1^2 + 4x_1x_2 + 3x_2^2$

44. $2x_1^2 - x_1x_2 + 3x_2x_3 - 4x_1x_3 + 2x_2^2 - x_3^2$

45. $-4x_1^2 + 3x_2^2 + 2x_1x_2 - x_3^2$

In each of Problems 46 through 50, find an orthogonal matrix P which diagonalizes the given matrix.

46. $\begin{bmatrix} 1 & -3 \\ -3 & 4 \end{bmatrix}$.

47. $\begin{bmatrix} -2 & 0 \\ 0 & 1 \end{bmatrix}$

48. $\begin{bmatrix} -5 & 2 \\ 2 & 8 \end{bmatrix}$

49. $\begin{bmatrix} -1 & 2 & 0 \\ 2 & 5 & 0 \\ 0 & 0 & -4 \end{bmatrix}$

50. $\begin{bmatrix} -4 & 0 & -3 \\ 0 & 1 & 0 \\ -3 & 0 & 0 \end{bmatrix}$

In each of Problems 51 through 60, find the eigenvalues of the given matrix. Corresponding to each eigenvalue, find an eigenvector.

51. $\begin{bmatrix} 1 & -3 \\ -3 & 6 \end{bmatrix}$

52. $\begin{bmatrix} 1 & 0 & -1 \\ 0 & 0 & 0 \\ -1 & 0 & 4 \end{bmatrix}$

53. $\begin{bmatrix} i & 1 \\ 1 & 3i \end{bmatrix}$

54. $\begin{bmatrix} 2-3i & 0 \\ i & -5 \end{bmatrix}$

55. $\begin{bmatrix} 1-i & 2i \\ 0 & 4 \end{bmatrix}$

56. $\begin{bmatrix} i & 0 & -1 \\ -1 & 0 & 0 \\ 0 & 0 & 4i \end{bmatrix}$

57. $\begin{bmatrix} 3-2i & 0 \\ 1 & -4i \end{bmatrix}$ **58.** $\begin{bmatrix} -3 & 0 & 0 \\ 1 & -2 & 0 \\ 0 & 0 & -4 \end{bmatrix}$ **59.** $\begin{bmatrix} 1 & -5 & 3 \\ 0 & 0 & 0 \\ -2 & 0 & 0 \end{bmatrix}$ **60.** $\begin{bmatrix} -2 & 1 & -3 \\ 0 & 0 & 0 \\ 4 & 0 & 0 \end{bmatrix}$

In each of Problems 61 through 70, use the following definition: Let A and B be $n \times n$ matrices. Then, A is *similar* to B if, for some nonsingular P, $P^{-1}AP = B$.

61. Let A be similar to B. Prove that B is similar to A.

62. Prove that every matrix is similar to itself.

63. Let A be similar to B, and B similar to C. Prove that A is similar to C.

64. Let A be similar to B. Prove that A and B are either both singular or both nonsingular.

65. Let A be similar to B, and let A be nonsingular. Prove that A^{-1} is similar to B^{-1}.

66. Let A be similar to B. Prove that, for any positive integer n, A^n is similar to B^n.

67. Let A be similar to B. Prove that αA is similar to αB for any scalar α.

68. Let A be similar to B. Prove that A and B have the same rank.

69. Let A be similar to B. Prove that $|A| = |B|$.

70. Let A be similar to B. Prove that

$$\sum_{i=1}^{n} A_{ii} = \sum_{i=1}^{n} B_{ii}.$$

(The sum of the diagonal elements of a matrix is called its *trace*.)

71. Multiply A_{ij} by α^{i-j} to form a new matrix B from A. How are $|A|$ and $|B|$ related?

72. Suppose that $a + ib$ is an eigenvalue of A. Prove that $a - ib$ is also an eigenvalue, assuming that the entries of A are real numbers.

73. Let A be an $n \times n$ diagonalizable matrix. Prove that there are $n \times n$ matrices $P_1, \ldots, P_n$ such that

1. $A = \lambda_1 P_1 + \cdots + \lambda_n P_n$, with $\lambda_1, \ldots, \lambda_n$ the eigenvalues of A.
2. $P_j^2 = P_j$ for $j = 1, \ldots, n$.
3. $P_k P_j = 0$ if $k \neq j$ (here, 0 is the $n \times n$ zero matrix).
4. $P_1 + \cdots + P_n = I_n$.

(We call $\lambda_1 P_1 + \cdots + \lambda_n P_n$ a *spectral decomposition* of A, and the matrices P_j are called *projections*.)

Hint: First assert that, for some P, $P^{-1}AP = D$, a diagonal matrix. Write $D = \lambda_1 T_{11} + \cdots + \lambda_n T_{nn}$, where T_{jj} has 1 in the j-j place and 0 everywhere else. Then solve for A.

74. Let A be an $n \times n$ diagonalizable matrix with characteristic equation

$$\lambda^n + a_{n-1}\lambda^{n-1} + \cdots + a_1\lambda + a_0 = 0.$$

Prove that

$$A^n + a_{n-1}A^{n-1} + \cdots + a_1 A + a_0 I_n = 0,$$

the $n \times n$ zero matrix. This is a special case of the *Cayley-Hamilton theorem*, which asserts that a matrix satisfies its own characteristic equation. The theorem is true for any square matrix, but is easier to prove if A is diagonalizable.

75. Prove that every matrix is the sum of a symmetric and skew-symmetric matrix.

76. Prove that $\text{rank}(AB) \leq \text{rank}(A)$, assuming that AB is defined.

77. For each positive integer n, find an orthogonal matrix with determinant -1.

78. Prove that $|\alpha A| = \alpha^n |A|$ for A any $n \times n$ matrix and α any scalar.

79. Prove that AB and BA have the same eigenvalues for any $n \times n$ matrices A and B. Will they have the same eigenvectors?

80. An $n \times n$ matrix A is *nilpotent* if, for some positive integer k, $A^k = 0$. Prove that all the eigenvalues of a nilpotent matrix must be zero.

81. Let λ be an eigenvalue of the nonsingular matrix A.

(a) Prove that $\lambda \neq 0$.

(b) Show that $1/\lambda$ is an eigenvalue of A^{-1}.

82. A is an *involution* if $A^2 = I_n$. What can be said about the eigenvalues of an involution?

83. Prove *Gerschgorin's theorem*: Let A be an $n \times n$ matrix, and let

$$r_k = \sum_{\substack{j=1 \\ j \neq k}}^{n} A_{kj}.$$

Let C_k be a circle of radius r_k and center (α_k, β_k), where $A_{kk} = \alpha_k + i\beta_k$. Let λ be an eigenvalue of A. Then, plotted as a point in the plane, λ lies on or inside one of the circles $C_1, \ldots, C_n$. (This result can be used to obtain rough estimates on the magnitude of eigenvalues.)

PART THREE

VECTOR ANALYSIS

CHAPTER ELEVEN

Vector Analysis

11.0 INTRODUCTION

In Chapter 9 we developed the basic algebra and geometry of vectors in 3-space, including dot product, cross product, and scalar triple product. In this chapter we combine vector algebra and geometry with the processes of calculus, primarily the derivative and integral. The resulting field of study is called *vector analysis* (or sometimes *vector calculus*), a subject which has developed into a powerful tool for treating certain problems in engineering and physics.

We assume that the student has read, or was already familiar with, Chapter 9, Sections 9.1, 9.2, and 9.3, and begin with the notions of vector functions and derivatives of vector functions.

11.1 VECTOR FUNCTIONS OF ONE VARIABLE

It is often the case that we have a vector whose components are functions, either of one or several variables. In this section we treat the one-variable case, in which we have a vector of the form

$$\mathbf{R}(t) = x(t)\mathbf{i} + y(t)\mathbf{j} + z(t)\mathbf{k}.$$

For example, we might have

$$\mathbf{R}(t) = \cos(t)\mathbf{i} - 2t^2\mathbf{j} + e^t\mathbf{k}.$$

We call $\mathbf{R}(t)$ *continuous* wherever x, y, and z are continuous. Thus, $\mathbf{R}(t)$ above is continuous for all t, while $(1/t)\mathbf{i} + 2t\mathbf{j} - t^2\mathbf{k}$ is not continuous at 0, since the $\mathbf{i}$-component is not defined there.

We define the *derivative* $\mathbf{R}'(t)$ of $\mathbf{R}(t)$ by

$$\mathbf{R}'(t) = x'(t)\mathbf{i} + y'(t)\mathbf{j} + z'(t)\mathbf{k}$$

wherever x, y, and z are differentiable. Thus, we differentiate a vector by differentiating each component. The derivative of $\mathbf{R}(t)$, evaluated at t_0, is denoted

$$\mathbf{R}'(t_0) \quad \text{or} \quad \frac{d\mathbf{R}}{dt}(t_0).$$

For example, if

$$\mathbf{R}(t) = \cos(t)\mathbf{i} - 2t^2\mathbf{j} + e^t\mathbf{k},$$

then

$$\mathbf{R}'(t) = -\sin(t)\mathbf{i} - 4t\mathbf{j} + e^t\mathbf{k}.$$

In particular,

$$\mathbf{R}'(0) = \mathbf{k} \quad \text{and} \quad R'(\tfrac{1}{2}\pi) = -\mathbf{i} - 2\pi\mathbf{j} + e^{\pi/2}\mathbf{k}.$$

As a second example, let

$$\mathbf{K}(t) = |t|\mathbf{i} - 3t^2\mathbf{j}.$$

Then,

$$\mathbf{K}'(t) = \begin{cases} \mathbf{i} - 6t\mathbf{j} & \text{if} \quad t > 0, \\ -\mathbf{i} - 6t\mathbf{j} & \text{if} \quad t < 0. \end{cases}$$

Here, $\mathbf{K}'(t)$ does not exist at $t = 0$, since $|t|$ is not differentiable there.

Geometrically, we may think of $\mathbf{R}(t)$ as an adjustable arrow pivoted at the origin. As t varies, $\mathbf{R}(t)$ sweeps out a curve in space (see Figure 156). The curve is the locus of points $(x(t), y(t), z(t))$.

The vector $\mathbf{R}'(t_0)$ can be interpreted as the tangent vector to this curve at the

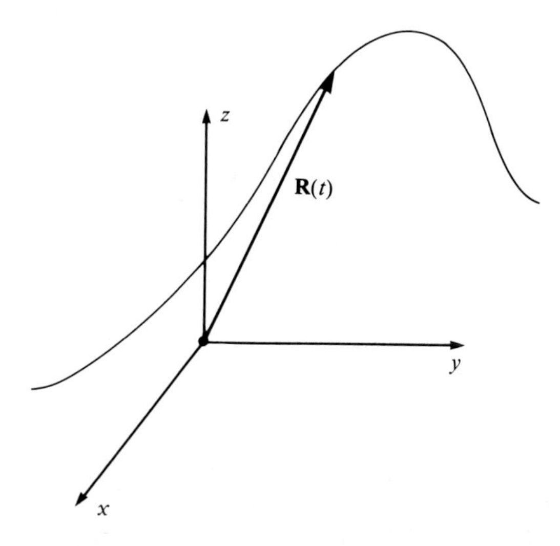

Figure 156. Vector sweeping out a curve.

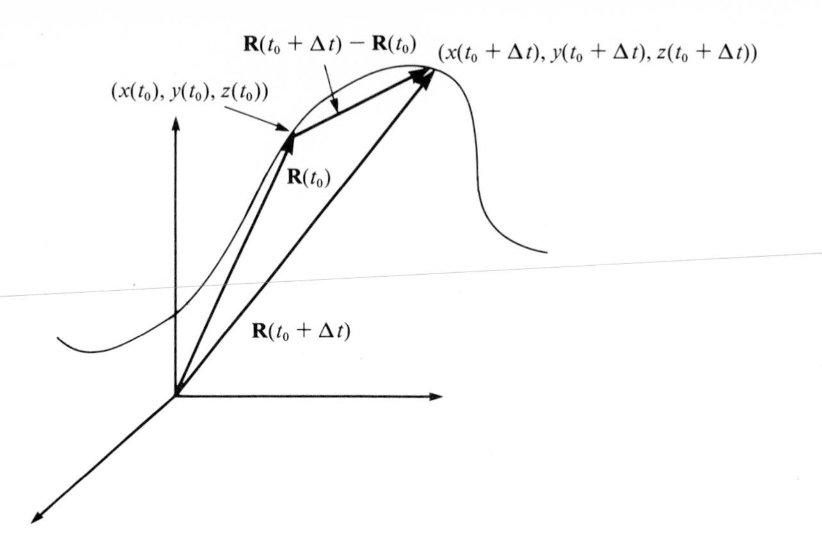

Figure 157

point $(x(t_0), y(t_0), z(t_0))$. To see this, note that $\mathbf{R}(t_0 + \Delta t) - \mathbf{R}(t_0)$ is the vector represented by the arrow from $(x(t_0), y(t_0), z(t_0))$ to $(x(t_0 + \Delta t), y(t_0 + \Delta t), z(t_0 + \Delta t))$, as shown in Figure 157. Now,

$$\frac{1}{\Delta t}(\mathbf{R}(t_0 + \Delta t) - \mathbf{R}(t_0))$$
$$= \left(\frac{x(t_0 + \Delta t) - x(t_0)}{\Delta t}\right)\mathbf{i} + \left(\frac{y(t_0 + \Delta t) - y(t_0)}{\Delta t}\right)\mathbf{j} + \left(\frac{z(t_0 + \Delta t) - z(t_0)}{\Delta t}\right)\mathbf{k}$$

is a vector along the line between $(x(t_0), y(t_0), z(t_0))$ and $(x(t_0 + \Delta t), y(t_0 + \Delta t), z(t_0 + \Delta t))$. In the limit as $\Delta t \to 0$, this vector slides into a position tangent to the curve at $(x(t_0), y(t_0), z(t_0))$, and we identify the right side in the above limit as $x'(t_0)\mathbf{i} + y'(t_0)\mathbf{j} + z'(t_0)\mathbf{k}$, or $\mathbf{R}'(t_0)$. This limit process is depicted in Figure 158.

The tangent vector figures prominently in many considerations, among them calculation of the length of a curve. Suppose we have a curve C, given parametrically as a locus of points $(x(t), y(t), z(t))$ as t varies from a to b. We shall assume that x', y', and z' are continuous on $[a, b]$. How can we attach a number to C which we can think of as length?

One way to proceed is to subdivide $[a, b]$ into subintervals

$$a = t_0 < t_1 < \cdots < t_{n-1} < t_n = b.$$

Corresponding to each t_j is a point $(x(t_j), y(t_j), z(t_j))$ on C, as shown in Figure 159. The line segments between successive points on C form a polygon "approximation" to C, and the sum of the lengths of these segments approximates the length of C. Thus,

$$\text{length of } C \simeq \sum_{j=1}^{n} \{[x(t_j) - x(t_{j-1})]^2 + [y(t_j) - y(t_{j-1})]^2 + [z(t_j) - z(t_{j-1})]^2\}^{1/2}.$$

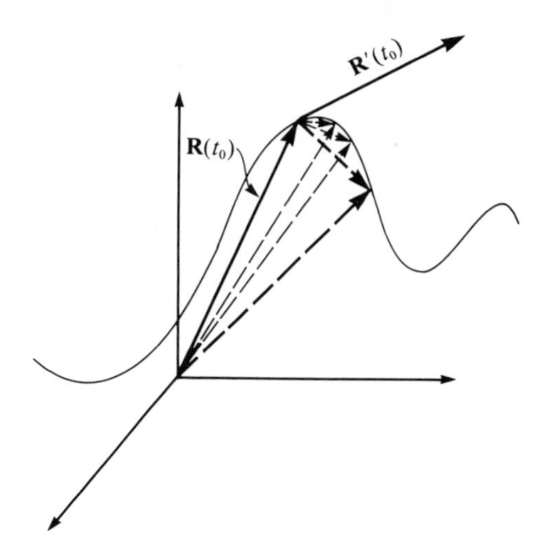

Figure 158. Vectors moving in the limit toward the tangent vector.

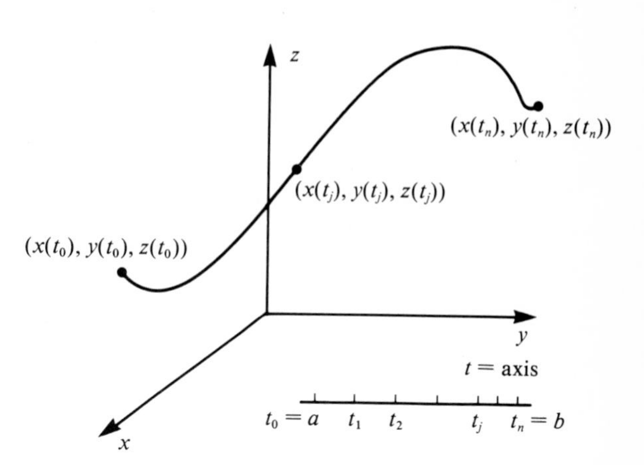

Figure 159. Points on C corresponding to partition of $[a, b]$ on the t-axis.

Now let $n \to \infty$, and each $t_j - t_{j-1} \to 0$. Thus, we choose more points on C, and choose them closer together, in forming the polygonal approximation. The polygon then fits C more closely, and in the limit the lengths of the polygons approach the number we call the length of C.*

This sounds fine, but how do we actually carry out this limit? To do this, look at a typical term in the above sum. Applying the Mean Value Theorem to $x(t)$ on the interval $[t_{j-1}, t_j]$, there is some α_j, $t_{j-1} < \alpha_j < t_j$, such that

$$x(t_j) - x(t_{j-1}) = x'(\alpha_j)(t_j - t_{j-1}).$$

Similarly, there are β_j and γ_j between t_{j-1} and t_j such that

$$y(t_j) - y(t_{j-1}) = y'(\beta_j)(t_j - t_{j-1})$$

and

$$z(t_j) - z(t_{j-1}) = z'(\gamma_j)(t_j - t_{j-1}).$$

Then, the approximating sum becomes

$$\sum_{j=1}^{n} [x'(\alpha_j)^2 + y'(\beta_j)^2 + z'(\gamma_j)^2]^{1/2}(t_j - t_{j-1}).$$

In the limit, this approaches

$$\int_a^b [x'(t)^2 + y'(t)^2 + z'(t)^2]^{1/2}\, dt.$$

---

* One can produce examples in which this limit of lengths of polygons is infinite. Such a curve is called *nonrectifiable*. An example is $y = \sin(1/x)$, $0 < x \leq \pi$. A curve with finite length is called *rectifiable*.

Letting $\mathbf{R}(t) = x(t)\mathbf{i} + y(t)\mathbf{j} + z(t)\mathbf{k}$, this gives us

$$\text{length of } C = \int_a^b \|\mathbf{R}'(t)\|\, dt.$$

In words, the length of C is the integral of the length of the tangent vector to C.

EXAMPLE 310

Let C be given by $x = 2\cos(t)$, $y = 2\sin(t)$, $z = t$, $0 \le t \le 2$. Then $\mathbf{R}(t) = 2\cos(t)\mathbf{i} + 2\sin(t)\mathbf{j} + t\mathbf{k}$ sweeps out C as t varies from 0 to 2, and the tangent to C is $\mathbf{R}'(t) = -2\sin(t)\mathbf{i} + 2\cos(t)\mathbf{j} + \mathbf{k}$.

Then,

$$\|\mathbf{R}'(t)\| = \sqrt{4\sin^2(t) + 4\cos^2(t) + 1} = \sqrt{5},$$

and C has length

$$\int_0^2 \sqrt{5}\, dt = 2\sqrt{5} \text{ units.}$$

In most instances, $\mathbf{R}'(t)$ is too complicated to integrate explicitly. For example, let

$$\mathbf{R}(t) = 2t\mathbf{i} - \cos(3t)\mathbf{j} + t^3\mathbf{k}, \qquad 0 \le t \le 1.$$

Then,

$$\mathbf{R}'(t) = 2\mathbf{i} + 3\sin(3t)\mathbf{j} + 3t^2\mathbf{k};$$

so

$$\|\mathbf{R}'(t)\| = \sqrt{4 + 9\sin^2(3t) + 9t^4}.$$

The length of C is

$$\int_0^1 \sqrt{4 + 9\sin^2(3t) + 9t^4}\, dt,$$

which cannot be integrated in closed form. One must then apply a numerical technique. For example, using the rectangular rule, with 200 subdivisions of [0, 1], we get 3.1960, and this is approximately the length of C.

Sometimes it is convenient to parametrize C by arc length. Let

$$s(t) = \int_a^t \|\mathbf{R}'(p)\|\, dp,$$

for $a \le t \le b$. Then $s(a) = 0$, $s(b) =$ length of C, and $s(t)$ for $a < t < b$ is the length of C from $(x(a), y(a), z(a))$ to $(x(t), y(t), z(t))$, as shown in Figure 160. Here, p is used as a

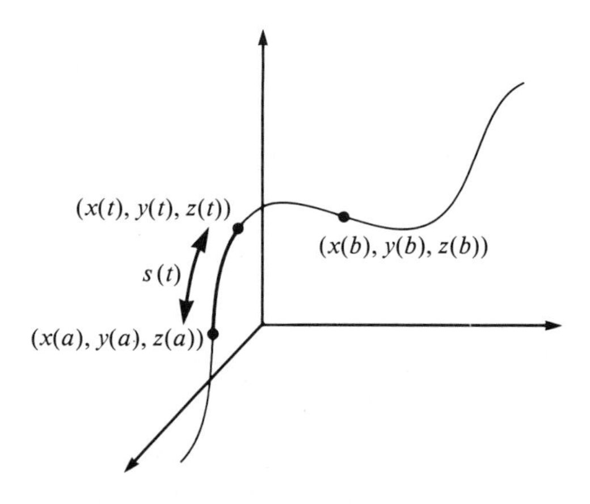

Figure 160. $s(t) =$ length of C from $(x(a), y(a), z(a))$ to $(x(t), y(t), z(t))$.

dummy variable of integration to avoid ambiguity. Assuming that $\mathbf{R}'(t)$ is continuous, then

$$\frac{ds}{dt} = \|\mathbf{R}'(t)\| = \sqrt{\left(\frac{dx}{dt}\right)^2 + \left(\frac{dy}{dt}\right)^2 + \left(\frac{dz}{dt}\right)^2}.$$

Sometimes this is written in differential form

$$ds = \sqrt{dx^2 + dy^2 + dz^2}.$$

Note that s is a strictly increasing function of t. In theory at least, we can invert $s = s(t)$ to solve for t in terms of s: $t = t(s)$. Substituting into the equations for C gives us a parametrization of C in terms of s:

$$x = x[t(s)] = X(s), \qquad y = y[t(s)] = Y(s), \qquad z = z[t(s)] = Z(s),$$

where s varies from 0 to L, the length of C.

This parametrization has an interesting feature. Letting

$$\mathbf{R}(s) = X(s)\mathbf{i} + Y(s)\mathbf{j} + Z(s)\mathbf{k},$$

the tangent vector is

$$\mathbf{R}'(s) = X'(s)\mathbf{i} + Y'(s)\mathbf{j} + Z'(s)\mathbf{k},$$

and this has length 1, since

$$\begin{aligned}\|\mathbf{R}'(s)\| &= \sqrt{\left(\frac{dX}{ds}\right)^2 + \left(\frac{dY}{ds}\right)^2 + \left(\frac{dZ}{ds}\right)^2}\\ &= \sqrt{\left(\frac{dx}{dt}\frac{dt}{ds}\right)^2 + \left(\frac{dy}{dt}\frac{dt}{ds}\right)^2 + \left(\frac{dz}{dt}\frac{dt}{ds}\right)^2}\\ &= \frac{1}{ds}\sqrt{dx^2 + dy^2 + dz^2} = 1.\end{aligned}$$

Thus, when arc length is used as parameter, the tangent vector always has length 1. This fact is convenient in many later calculations.

EXAMPLE 311

Let $x = 2\cos(t)$, $y = 2\sin(t)$, and $z = t$, for $0 \le t \le 2$, as in Example 310.

Using p as a dummy variable of integration, we have, for $0 \le t \le 2$,

$$s(t) = \int_0^t \sqrt{5}\, dp = \sqrt{5}\, t.$$

Thus, $t = (1/\sqrt{5})s$, and we can rewrite C as

$$x = 2\cos\left(\frac{s}{\sqrt{5}}\right), \qquad y = 2\sin\left(\frac{s}{\sqrt{5}}\right), \qquad z = \frac{s}{\sqrt{5}},$$

for $0 \le s \le 2\sqrt{5}$. In terms of s,

$$\mathbf{R}(s) = 2\cos\left(\frac{s}{\sqrt{5}}\right)\mathbf{i} + 2\sin\left(\frac{s}{\sqrt{5}}\right)\mathbf{j} + \left(\frac{s}{\sqrt{5}}\right)\mathbf{k}$$

and

$$\mathbf{R}'(s) = -\frac{2}{\sqrt{5}} \sin\left(\frac{s}{\sqrt{5}}\right)\mathbf{i} + \frac{2}{\sqrt{5}} \cos\left(\frac{s}{\sqrt{5}}\right)\mathbf{j} + \frac{1}{\sqrt{5}}\mathbf{k}.$$

It is easy to check that $\|\mathbf{R}'(s)\| = 1$ for $0 \le s \le 2\sqrt{5}$.

Henceforth, we always use s to denote arc length when discussing curves and vectors.

We now turn to the problem of differentiating various kinds of products involving vectors. Let $f(t)$ be any differentiable scalar-valued function of t, and let $\mathbf{F}(t) = F_1(t)\mathbf{i} + F_2(t)\mathbf{j} + F_3(t)\mathbf{k}$. Then, $f(t)\mathbf{F}(t)$ is a vector:

$$f(t)\mathbf{F}(t) = f(t)F_1(t)\mathbf{i} + f(t)F_2(t)\mathbf{j} + f(t)F_3(t)\mathbf{k}.$$

Suppressing t for convenience, we have

$$\begin{aligned}(f\mathbf{F})' &= (fF_1)'\mathbf{i} + (fF_2)'\mathbf{j} + (fF_3)'\mathbf{k}\\ &= (f'F_1 + fF_1')\mathbf{i} + (f'F_2 + fF_2')\mathbf{j} + (f'F_3 + fF_3')\mathbf{k}\\ &= f'(F_1\mathbf{i} + F_2\mathbf{j} + F_3\mathbf{k}) + f(F_1'\mathbf{i} + F_2'\mathbf{j} + F_3'\mathbf{k})\\ &= f'\mathbf{F} + f\mathbf{F}'.\end{aligned}$$

This is similar to the familiar rule for differentiating a product of two functions, except that here $\mathbf{F}$ is a vector.

For dot and cross product, let $\mathbf{F}$ be as above, and let

$$\mathbf{G}(t) = G_1(t)\mathbf{i} + G_2(t)\mathbf{j} + G_3(t)\mathbf{k}.$$

Then,

$$\mathbf{F} \cdot \mathbf{G} = F_1G_1 + F_2G_2 + F_3G_3$$

and

$$\begin{aligned}(\mathbf{F} \cdot \mathbf{G})' &= F_1'G_1 + F_1G_1' + F_2'G_2 + F_2G_2' + F_3'G_3 + F_3G_3'\\ &= F_1'G_1 + F_2'G_2 + F_3'G_3 + F_1G_1' + F_2G_2' + F_3G_3'\\ &= \mathbf{F}' \cdot \mathbf{G} + \mathbf{F} \cdot \mathbf{G}'.\end{aligned}$$

In summary,

$$(\mathbf{F} \cdot \mathbf{G})' = \mathbf{F}' \cdot \mathbf{G} + \mathbf{F} \cdot \mathbf{G}'.$$

Similarly, but omitting the details, we get

$$(\mathbf{F} \times \mathbf{G})' = \mathbf{F}' \times \mathbf{G} + \mathbf{F} \times \mathbf{G}'.$$

In all three situations, then, the rule for differentiating the product involved is reminiscent of the rule for differentiating a product of two functions, as learned in elementary calculus. Remember, however, that in cross product the order of the factors is crucial, since $\mathbf{F} \times \mathbf{G} = -(\mathbf{G} \times \mathbf{F})$.

In dealing with specific vectors, it is usually easier to first form $f\mathbf{F}$, $\mathbf{F} \cdot \mathbf{G}$, or $\mathbf{F} \times \mathbf{G}$ and then simply differentiate the result. However, in general discussions the above formulas are important to know.

EXAMPLE 312

Let

$$f(t) = 3e^t, \qquad \mathbf{F} = t\mathbf{i} - \mathbf{j} + 3t\mathbf{k}, \quad \text{and} \quad \mathbf{G} = t^2\mathbf{i} + 2t\mathbf{j} - \mathbf{k}.$$

Then,

$$f\mathbf{F} = 3te^t\mathbf{i} - 3e^t\mathbf{j} + 9te^t\mathbf{k}.$$

Hence, directly,

$$(f\mathbf{F})' = (3e^t + 3te^t)\mathbf{i} - 3e^t\mathbf{j} + (9e^t + 9te^t)\mathbf{k}.$$

By the product rule,

$$\begin{aligned} f'\mathbf{F} + f\mathbf{F}' &= 3e^t(t\mathbf{i} - \mathbf{j} + 3t\mathbf{k}) + 3e^t(\mathbf{i} + 3\mathbf{k}) \\ &= (3te^t + 3e^t)\mathbf{i} - 3e^t\mathbf{j} + (9te^t + 9e^t)\mathbf{k} \\ &= (f\mathbf{F})'. \end{aligned}$$

Next,

$$\mathbf{F} \cdot \mathbf{G} = t^3 - 2t - 3t = t^3 - 5t;$$

so

$$(\mathbf{F} \cdot \mathbf{G})' = 3t^2 - 5.$$

As a verification of the rule for differentiating dot products, we have

$$\begin{aligned} \mathbf{F}' \cdot \mathbf{G} + \mathbf{F} \cdot \mathbf{G}' &= (\mathbf{i} + 3\mathbf{k}) \cdot (t^2\mathbf{i} + 2t\mathbf{j} - \mathbf{k}) + (t\mathbf{i} - \mathbf{j} + 3t\mathbf{k}) \cdot (2t\mathbf{i} + 2\mathbf{j}) \\ &= t^2 - 3 + 2t^2 - 2 \\ &= 3t^2 - 5 \\ &= (\mathbf{F} \cdot \mathbf{G})'. \end{aligned}$$

Finally, look at the cross product. We have directly

$$\mathbf{F} \times \mathbf{G} = (1 - 6t^2)\mathbf{i} + (3t^3 + t)\mathbf{j} + 3t^2\mathbf{k}.$$

Hence,

$$(\mathbf{F} \times \mathbf{G})' = -12t\mathbf{i} + (9t^2 + 1)\mathbf{j} + 6t\mathbf{k}.$$

Using the formula, we have

$$\begin{aligned} \mathbf{F}' \times \mathbf{G} + \mathbf{F} \times \mathbf{G}' &= (\mathbf{i} + 3\mathbf{k}) \times (t^2\mathbf{i} + 2t\mathbf{j} - \mathbf{k}) + (t\mathbf{i} - \mathbf{j} + 3t\mathbf{k}) \times (2t\mathbf{i} + 2\mathbf{j}) \\ &= [-6t\mathbf{i} + (3t^2 + 1)\mathbf{j} + 2t\mathbf{k}] + (-6t\mathbf{i} + 6t^2\mathbf{j} + 4t\mathbf{k}) \\ &= -12t\mathbf{i} + (9t^2 + 1)\mathbf{j} + 6t\mathbf{k} \\ &= (\mathbf{F} \times \mathbf{G})'. \end{aligned}$$

Finally, the chain rule applies to vectors in the following form: If $t = h(p)$ and $h'(p)$ exists, then

$$\frac{d}{dp}\mathbf{F}[h(p)] = \frac{dt}{dp}\frac{d\mathbf{F}}{dt}.$$

This is because we differentiate by component, and the chain rule works for each component:

$$\begin{aligned}
\frac{d}{dp}\mathbf{F}[h(p)] &= \frac{d}{dp}\{F_1[h(p)]\mathbf{i} + F_2[h(p)]\mathbf{j} + F_3[h(p)]\mathbf{k}\} \\
&= \left(\frac{d}{dp}F_1[h(p)]\right)\mathbf{i} + \left(\frac{d}{dp}F_2[h(p)]\right)\mathbf{j} + \left(\frac{d}{dp}F_3[h(p)]\right)\mathbf{k} \\
&= \frac{dF_1}{dt}\frac{dh}{dp}\mathbf{i} + \frac{dF_2}{dt}\frac{dh}{dp}\mathbf{j} + \frac{dF_3}{dt}\frac{dh}{dp}\mathbf{k} \\
&= \frac{dh}{dp}\left(\frac{dF_1}{dt}\mathbf{i} + \frac{dF_2}{dt}\mathbf{j} + \frac{dF_3}{dt}\mathbf{k}\right) \\
&= \frac{dt}{dp}\frac{d\mathbf{F}}{dt}.
\end{aligned}$$

This result is particularly useful in differentiating a unit tangent $\mathbf{R}'(s)$, since we usually have $\mathbf{R}$ in terms of some other parameter t and may not even be able to find s in closed form as a function of t. Here is an example of this.

EXAMPLE 313

Let $\mathbf{R} = 3t^2\mathbf{i} - \sin(t)\mathbf{j} + 2t\mathbf{k}$, for $a \leq t \leq b$. We want to find $d\mathbf{R}/ds$, where s measures arc length along the curve described by $\mathbf{R}$.

Here,

$$s = \int_a^t \sqrt{36t^2 + \cos^2(t) + 4}\, dt.$$

We cannot integrate this, but we can differentiate to find

$$\frac{ds}{dt} = \sqrt{36t^4 + \cos^2(t) + 4}.$$

Then,

$$\begin{aligned}
\frac{d\mathbf{R}}{ds} = \frac{dt}{ds}\frac{d\mathbf{R}}{dt} &= \frac{1}{\frac{ds}{dt}}\frac{d\mathbf{R}}{dt} \\
&= \frac{1}{\sqrt{36t^4 + \cos^2(t) + 4}}[6t\mathbf{i} - \cos(t)\mathbf{j} + 2\mathbf{k}].
\end{aligned}$$

The kind of chain rule differentiation shown in the above example will be very important in the next section in treating velocity and acceleration along a curve, as well as in analyzing how the position vector $\mathbf{R}(t)$ can be used to determine certain useful facts about the curve it describes.

PROBLEMS FOR SECTION 11.1

In each of Problems 1 through 10, (a) determine the vector $f\mathbf{F}$ and compute $(f\mathbf{F})'$ directly, and (b) use the product rule to compute $(f\mathbf{F})'$ as $f'\mathbf{F} + f\mathbf{F}'$.

1. $f(t) = 4\cos(3t)$, $\mathbf{F}(t) = \mathbf{i} + 3t^2\mathbf{j} + 2t\mathbf{k}$

2. $f(t) = 1 - 2t^3$, $\mathbf{F}(t) = t\mathbf{i} - \cosh(t)\mathbf{j} + e^t\mathbf{k}$

3. $f(t) = \sin^2(t)$, $\mathbf{F}(t) = 4\mathbf{i} + t^5\mathbf{k}$

4. $f(t) = t^2 - 2t + 3$, $\mathbf{F}(t) = \ln(t)\mathbf{i} + e^t\mathbf{j} - t^2\mathbf{k}$

5. $f(t) = 2t + 3$, $\mathbf{F}(t) = (1 - 3t)\mathbf{i} + t^4\mathbf{j} - t\mathbf{k}$

6. $f(t) = 5t^2$, $\mathbf{F}(t) = \sin^2(3t)\mathbf{i} - t\mathbf{j} + 3t^2\mathbf{k}$

7. $f(t) = 2t^3 - t$, $\mathbf{F}(t) = 2\mathbf{i} - t^3\mathbf{j} + e^{-t}\mathbf{k}$

8. $f(t) = t + 2$, $\mathbf{F}(t) = t^2\mathbf{i} + \sinh(t)\mathbf{j} - \arctan(t)\mathbf{k}$

9. $f(t) = -3e^{5t}$, $\mathbf{F}(t) = \mathbf{i} + t^2\mathbf{j} - \mathbf{k}$

10. $f(t) = t - \cosh(t)$, $\mathbf{F}(t) = t^2\mathbf{i} - 2t\mathbf{j} + t\mathbf{k}$

In each of Problems 11 through 25, (a) determine $\mathbf{F} \cdot \mathbf{G}$ and, from this, $(\mathbf{F} \cdot \mathbf{G})'$; (b) determine $(\mathbf{F} \cdot \mathbf{G})'$ by using the product formula of this section; (c) determine $\mathbf{F} \times \mathbf{G}$ and, from this, $(\mathbf{F} \times \mathbf{G})'$; and (d) determine $(\mathbf{F} \times \mathbf{G})'$ by using the product formula of this section.

11. $\mathbf{F} = \cos(2t)\mathbf{i} + \sin(t)\mathbf{j} - e^{-t}\mathbf{k}$, $\mathbf{G} = 2t^2\mathbf{i} - 3t\mathbf{k}$

12. $\mathbf{F} = 8\mathbf{i} + 6t\mathbf{j} - t^3\mathbf{k}$, $\mathbf{G} = \ln(t)\mathbf{i} - t^2\mathbf{j}$

13. $\mathbf{F} = t^3\mathbf{i} + 18\mathbf{j} - e^{2t}\mathbf{k}$, $\mathbf{G} = \mathbf{i} + \mathbf{j} - 6t\mathbf{k}$

14. $\mathbf{F} = \arctan(1 - t)\mathbf{i} + 2t^2\mathbf{j}$, $\mathbf{G} = e^t\mathbf{j} - 6\mathbf{k}$

15. $\mathbf{F} = \sin(2\pi t)\mathbf{i} - 3t\mathbf{j} + t^3\mathbf{k}$, $\mathbf{G} = -8t^2\mathbf{i} + t^2\mathbf{j} - 3\cos(t)\mathbf{k}$

16. $\mathbf{F} = -2t^5\mathbf{i} + \cosh(t)\mathbf{k}$, $\mathbf{G} = \sqrt{t}\,\mathbf{i} + t^2\mathbf{j}$

17. $\mathbf{F} = -\frac{1}{2}t^2\mathbf{i} - 3\mathbf{j} + e^t\mathbf{k}$, $\mathbf{G} = 4t^4\mathbf{j}$

18. $\mathbf{F} = -\sinh(t)\mathbf{i} - \cosh(t)\mathbf{j} + \mathbf{k}$, $\mathbf{G} = 2t\mathbf{i} + 2t\mathbf{k}$

19. $\mathbf{F} = \sin(2\pi t)\mathbf{i} + \cos(2\pi t)\mathbf{j} + e^{2\pi}\mathbf{k}$, $\mathbf{G} = \sin(2\pi t)\mathbf{i} - \cos(2\pi t)\mathbf{j} + \mathbf{k}$

20. $\mathbf{F} = 2\sinh(t)\mathbf{i} + t^3\mathbf{k}$, $\mathbf{G} = -8t^3\mathbf{i}$

21. $\mathbf{F} = 8t\mathbf{i} - 2t\mathbf{j}$, $\mathbf{G} = \sin(2t)\mathbf{i} - \mathbf{k}$

22. $\mathbf{F} = -\mathbf{i} + 3t^2\mathbf{k}$, $\mathbf{G} = 2t^2\mathbf{i} - e^t\mathbf{j} + (1/t)\mathbf{k}$

23. $\mathbf{F} = e^{-t}\mathbf{i} + 2t\mathbf{j} - \mathbf{k}$, $\mathbf{G} = -3\cosh(2t)\mathbf{i} - 4t^2\mathbf{j} + e^{-t}\mathbf{k}$

24. $\mathbf{F} = -8\mathbf{i} + 7t\mathbf{j} + \mathbf{k}$, $\mathbf{G} = 4\sin(t)\mathbf{i} + 2t\mathbf{j} - 3\mathbf{k}$

25. $\mathbf{F} = \mathbf{i} + 17t\mathbf{j} + t^3\mathbf{k}$, $\mathbf{G} = -5t^2\mathbf{i} + 3e^t\mathbf{j} - \ln(t)\mathbf{k}$

In each of Problems 26 through 35, a curve is given parametrically. Find the tangent vector to the given curve.

26. $x = 1,\ y = 2t,\ z = 1;\quad 0 \le t \le 4$

27. $x = x,\ y = \sin(2\pi x),\ z = \cos(2\pi x);\quad 0 \le x \le 1$

28. $x = \cosh(t),\ y = \sinh(t),\ z = 4t;\quad 0 \le t \le 2$

29. $x = e^t\cos(t),\ y = e^t\sin(t),\ z = e^t;\quad 0 \le t \le \pi$

30. $x = y = t^2,\ z = 3t^2;\quad 0 \le t \le 1$

31. $x = y = z = t^3;\quad 1 \le t \le 2.$

32. $x = 2\cos(2t),\ y = 2\sin(2t),\ z = 1 - 3t;\quad 0 \le t \le 2\pi$

33. $x = 4\ln(2t + 1),\ y = 4\sinh(3t),\ z = 2;\quad 3 \le t \le 9$

34. $x = 2 - \sinh(t),\ y = \cosh(t),\ z = \ln(t);\quad 1 \le t \le 2$

35. $x = 4t^2 - 2,\ y = 3t,\ z = t;\quad -1 \le t \le 5$

In each of Problems 36 through 45, write an integral for the length of the given curve. Do not attempt to integrate the resulting expression.

36. $x = \cosh(t),\ y = \sinh(t),\ z = e^{2t};\quad 0 \le t \le \pi$

37. $x = 1 - 2t,\ y = \sin(t),\ z = 3t^3;\quad 1 \le t \le 4$

38. $x = 1 + 3t^2,\ y = 2t,\ z = t^4;\quad 0 \le t \le 5$

39. $x = 1 + 2e^t,\ y = \sinh(t),\ z = \ln(t);\quad 1 \le t \le 6$

40. $x = 3\cos(2t),\ y = 1,\ z = 1 + \cos(t);\quad 0 \le t \le \frac{1}{2}\pi$

41. $x = e^{-3t},\ y = 2t,\ z = 1 + \ln(3t);\quad 2 \le t \le \pi$

42. $x = 4 + 3t^2,\ y = 1 + t^3,\ z = t;\quad 1 \le t \le 3$

43. $x = 2t^2 - \cos(t),\ y = 1 - \sin(t),\ z = 3t;\quad 0 \le t \le 1$

44. $x = 2te^t,\ y = -2te^t,\ z = 3t^2 + 1;\quad 1 \le t \le \pi$

45. $x = \cosh(3t),\ y = |t|,\ z = 1 + t^2;\quad 1 \le t \le 4$

In each of Problems 46 through 60, use the given vector $\mathbf{R}(t)$ to compute $d\mathbf{R}/ds$ in terms of t, as in Example 313.

46. $\mathbf{R} = 3t^2\mathbf{i} - 2t\mathbf{j} + e^t\mathbf{k}$

47. $\mathbf{R} = 4\cos(t)\mathbf{i} + 2\sin(3t)\mathbf{j} - \mathbf{k}$

48. $\mathbf{R} = (3 - t)\mathbf{i} + 6t^3\mathbf{j} - \cosh(t)\mathbf{k}$

49. $\mathbf{R} = e^t\sin(t)\mathbf{i} + e^t\cos(t)\mathbf{j} + t^3\mathbf{k}$

50. $\mathbf{R} = (3t^2 - 1)\mathbf{i} + 2t\mathbf{j} - t^4\mathbf{k}$

51. $\mathbf{R} = [1 - \cos(2t)]\mathbf{i} + 6t^2\mathbf{j} - 4\mathbf{k}$

52. $\mathbf{R} = \sin^2(t)\mathbf{i} - e^t\mathbf{j} + 2t\mathbf{k}$

53. $\mathbf{R} = [1 + 2\sin(t)]\mathbf{i} + 2t^2\mathbf{j} + 4t^3\mathbf{k}$

54. $\mathbf{R} = 18t^3\mathbf{i} + t^2\mathbf{j} - \cos(3t)\mathbf{k}$

55. $\mathbf{R} = (t^2 + 2)\mathbf{i} - 3t\mathbf{j} + 4t^2\mathbf{k}$

56. $\mathbf{R} = \mathbf{i} - 16t^2\mathbf{j} + t^3\mathbf{k}$

57. $\mathbf{R} = (1 + e^t)\mathbf{i} + e^{-3t}\mathbf{j} + t\mathbf{k}$

58. $\mathbf{R} = -3t^2\mathbf{i} + 4t\mathbf{j} - 3t^3\mathbf{k}$

59. $\mathbf{R} = \cosh(2t)\mathbf{i} - t\mathbf{j} + \mathbf{k}$

60. $\mathbf{R} = 4\mathbf{i} - t^3\mathbf{j} + t\mathbf{k}$

In each of Problems 61 through 64, find the length of the given curve C. Also, find arc length s as a function of the given parameter of C, and express t explicitly in terms of s. Using this, find parametric equations of C using s as parameter.

61. $x = t,\ y = \cosh(t),\ z = 1;\quad 0 \le t \le \pi$

62. $x = \sin(t),\ y = \cos(t),\ z = 45;\quad 0 \le t \le \pi$

63. $x = y = z = t^3;\quad -1 \le t \le 1$

64. $x = 2t^2,\ y = 3t^2,\ z = 4t^2;\quad 1 \le t \le 3$

65. Let $\mathbf{R}(t) = x(t)\mathbf{i} + y(t)\mathbf{j} + z(t)\mathbf{k}$. Prove that

$$(\mathbf{R} \times \mathbf{R}')' = \mathbf{R} \times \mathbf{R}''.$$

66. Derive a formula for $[\mathbf{F} \cdot (\mathbf{G} \times \mathbf{H})]'$ in terms of $\mathbf{F}$, $\mathbf{G}$, $\mathbf{H}$, $\mathbf{F}'$, $\mathbf{G}'$, and $\mathbf{H}'$.

67. Suppose that C is given by $x = x$, $y = f(x)$, $z = 0$; $a \le x \le b$. Assuming that $f'(x)$ is continuous on $[a, b]$, show how our formula for length of C reduces to the elementary calculus result

$$\int_a^b \sqrt{1 + f'(x)^2}\, dx.$$

68. Let $\mathbf{R}(t) = x(t)\mathbf{i} + y(t)\mathbf{j} + z(t)\mathbf{k}$ be the position of a particle at time t, moving along a curve C. Suppose $\mathbf{R} \times \mathbf{R}' = \mathbf{0}$. Prove that $\mathbf{R}$ does not change direction.

69. Let $\mathbf{R}(t) = x\mathbf{i} + y\mathbf{j} + z\mathbf{k}$.
   (a) Show that $\|\mathbf{R}(t) \times [\mathbf{R}(t + \Delta t) - \mathbf{R}(t)]\|$ is twice the area of the sector bounded by C, $\mathbf{R}(t)$, and $\mathbf{R}(t + \Delta t)$.
   (b) Prove that if $\mathbf{R} \times \mathbf{R}'' = \mathbf{0}$, then the particle moves so that equal areas are swept out in equal times. (Note Kepler's laws of planetary motion.)

11.2 VELOCITY, ACCELERATION, CURVATURE, AND TORSION

In this section we shall use vector functions to analyze motion of a particle. In doing this, we shall encounter some elementary notions of the differential geometry of curves in 3-space.

Imagine a particle moving in 3-space (or in a plane), so that at time t, the particle is at $(x(t), y(t), z(t))$. We call

$$\mathbf{R}(t) = x(t)\mathbf{i} + y(t)\mathbf{j} + z(t)\mathbf{k}$$

the *position vector* of the particle. If $\mathbf{R}(t)$ is drawn as an arrow with blunt end at the origin, then the arrowhead sweeps out the path, or trajectory, C of the particle as time progresses, as shown in Figure 156 of the previous section. In any time interval $t_1 \le t \le t_2$, the particle moves a distance

$$\int_{t_1}^{t_2} \| R'(t) \|\, dt,$$

as we saw previously.

The *velocity* of the particle at any time t is defined to be

$$\mathbf{v}(t) = \mathbf{R}'(t) = \frac{dx}{dt}\mathbf{i} + \frac{dy}{dt}\mathbf{j} + \frac{dz}{dt}\mathbf{k}.$$

Note that this is in the direction of the tangent to the trajectory. The magnitude of the velocity is the *speed*, denoted just v:

$$v = \| \mathbf{v} \|.$$

The *acceleration* is the rate of change of the velocity:

$$\mathbf{a}(t) = \mathbf{v}'(t) = \frac{d^2x}{dt^2}\mathbf{i} + \frac{d^2y}{dt^2}\mathbf{j} + \frac{d^2z}{dt^2}\mathbf{k}.$$

EXAMPLE 314

Suppose that

$$\mathbf{R}(t) = \sin(t)\mathbf{i} + 2e^{-t}\mathbf{j} + t^2\mathbf{k}.$$

Then the velocity is

$$\mathbf{v}(t) = \cos(t)\mathbf{i} - 2e^{-t}\mathbf{j} + 2t\mathbf{k},$$

the speed is

$$v = \| \mathbf{v} \| = \sqrt{\cos^2(t) + 4e^{-2t} + 4t^2},$$

and the acceleration is

$$\mathbf{a} = -\sin(t)\mathbf{i} + 2e^{-t}\mathbf{j} + 2\mathbf{k}.$$

One objective of this section is to derive a decomposition of the acceleration vector into tangential and centripetal components. This will involve a scalar quantity called the radius of curvature of C, which we shall define in terms of a concept called the curvature. Roughly, the idea is this: Look at C at any point P. We think of the curvature of C at P as a measure of the amount the tangent is changing direction at P. Intuitively, a straight line should have zero curvature, and a circle should have the same curvature at each point. To this end, we define the *curvature* κ of C at P as the magnitude of the rate of change, with respect to arc length, of the unit tangent to C at P. If $\mathbf{T} = \mathbf{T}(s)$ is the unit tangent, then

$$\kappa = \left\| \frac{d\mathbf{T}}{ds} \right\|.$$

In most instances, $\mathbf{R}$ is a function of t, not s; hence, $d\mathbf{T}/ds$ is computed by chain rule from the relations

$$\mathbf{T} = \frac{d\mathbf{R}}{ds} \quad \text{and} \quad \frac{d\mathbf{R}}{ds} = \frac{dt}{ds}\frac{d\mathbf{R}}{dt}.$$

From the latter, we obtain

$$\frac{d\mathbf{R}}{ds} = \left(\frac{1}{\sqrt{(x')^2 + (y')^2 + (z')^2}} \right) \mathbf{R}'(t).$$

Omitting the details, one can show that

$$\kappa = \frac{\| \mathbf{R}' \times \mathbf{R}'' \|}{\| \mathbf{R}' \|^3}.$$

Another way of writing this is

$$\kappa = \frac{[\| \mathbf{R}' \|^2 \| \mathbf{R}'' \|^2 - (\mathbf{R}' \cdot \mathbf{R}'')^2]^{1/2}}{\| \mathbf{R}' \|^3}.$$

EXAMPLE 315

Let C be described by

$$\mathbf{R}(t) = t\mathbf{i} + (t - 2)\mathbf{j} + (3t - 1)\mathbf{k}.$$

Then, C is a straight line and $\mathbf{R}'' = 0$; hence, $\kappa = 0$ as expected.

EXAMPLE 316

Let C be the circle of radius 2 about the origin in the plane $z = 4$, given by

$$\mathbf{R}(t) = 2\cos(t)\mathbf{i} + 2\sin(t)\mathbf{j} + 4\mathbf{k}.$$

Then, one can apply any of the above formulas to get $\kappa = \frac{1}{2}$, a constant, as expected. In general, any circle of radius r has curvature $1/r$.

The quantity $\rho = 1/\kappa$ is called the *radius of curvature* of the curve. The motivation comes from Figure 161, in which we attempt to find a circle which is tangent to the concave side of C at P and is a "best fit" to C at P in the sense that near P the circle approximates C better than any other circle tangent to C at P. One can show that the circle of best fit should have radius $\rho = 1/\kappa$. This circle is called the *osculating circle* to C at P, and its center is the *center of curvature* of C at P.

We can now achieve our stated objective of decomposing $\mathbf{a}$ into tangential and centripetal components.* We assume throughout that $\kappa \neq 0$.

THEOREM 86

At any point P on C there are orthogonal unit vectors $\mathbf{T}$ and $\mathbf{N}$ such that $\mathbf{T}$ is tangent to C at P, $\mathbf{N}$ is normal (perpendicular) to C at P, and

$$\mathbf{a} = \frac{dv}{dt}\mathbf{T} + \frac{v^2}{\rho}\mathbf{N}.$$

Figure 162 shows $\mathbf{T}$ and $\mathbf{N}$ in a typical case.

---

* Here is a coincidence with a historical lesson. As I was preparing this page of the manuscript, the March 1981 issue of *Scientific American* arrived in the mail. In an article entitled "Newton's Discovery of Gravity," I. Bernard Cohen suggests that the decomposition given above as Theorem 86 was instrumental in leading Newton to postulate his universal law of gravitational attraction. The decomposition was communicated to Newton by Robert Hooke in November, 1679, and occurred in private correspondence between the two over the next several years.

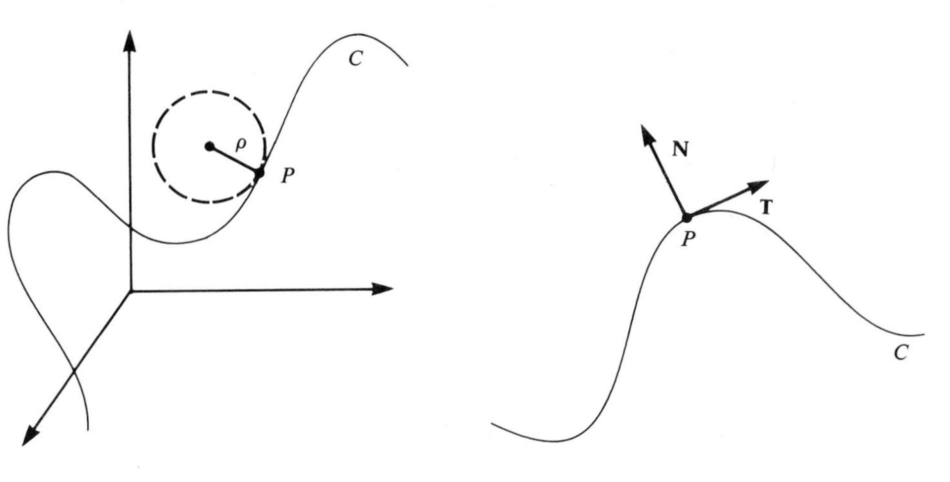

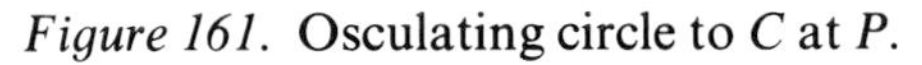
Figure 161. Osculating circle to C at P.

Figure 162

Proof Note first that

$$\mathbf{T} = \frac{\mathbf{R}'}{\|\mathbf{R}'\|} = \frac{1}{v}\,\mathbf{v} \qquad \text{(recall that } v = \|\mathbf{v}\|\text{);}$$

hence,

$$\mathbf{v} = v\mathbf{T}.$$

Then,

$$\mathbf{a} = \frac{d\mathbf{v}}{dt} = \frac{d}{dt}(v\mathbf{T}) = \frac{dv}{dt}\,\mathbf{T} + v\,\frac{d\mathbf{T}}{dt} = \frac{dv}{dt}\,\mathbf{T} + v\left(\frac{ds}{dt}\,\frac{d\mathbf{T}}{ds}\right).$$

Since $ds/dt = v$, then

$$\mathbf{a} = \frac{dv}{dt}\,\mathbf{T} + v^2\,\frac{d\mathbf{T}}{ds}.$$

This suggests that we define

$$\mathbf{N} = \rho\,\frac{d\mathbf{T}}{ds},$$

and we than have

$$\mathbf{a} = \frac{dv}{dt}\,\mathbf{T} + \frac{v^2}{\rho}\,\mathbf{N}.$$

There remains to show that $\mathbf{N}$ is a unit vector orthogonal to $\mathbf{T}$. To show that $\mathbf{N}$ is orthogonal to $\mathbf{T}$, recall that $\|\mathbf{T}\| = 1$; hence, $\mathbf{T} \cdot \mathbf{T} = 1$. Differentiating both sides with respect to s gives us

$$\frac{d\mathbf{T}}{ds} \cdot \mathbf{T} + \mathbf{T} \cdot \frac{d\mathbf{T}}{ds} = 2\mathbf{T} \cdot \frac{d\mathbf{T}}{ds} = 0.$$

Thus, **T** is orthogonal to $d\mathbf{T}/ds$, and hence also to **N**, which is a scalar multiple of $d\mathbf{T}/ds$.

Finally,

$$\| \mathbf{N} \| = \rho \left\| \frac{d\mathbf{T}}{ds} \right\| = \rho\kappa = 1,$$

and the proof is complete.

We call dv/dt the *tangential component of the acceleration* and v^2/ρ the *centripetal component*. Often, dv/dt is denoted a_t and v^2/ρ is denoted a_n; so that the decomposition is written

$$\mathbf{a} = a_t\mathbf{T} + a_n\mathbf{N}.$$

In practice, one usually computes a_t and a_n by using the Pythagorean theorem on the triangle of Figure 163. We have

$$\| \mathbf{a} \|^2 = a_t^2 + a_n^2.$$

Now, $\| \mathbf{a} \|^2$ is easy to compute from **R**, since

$$\| \mathbf{a} \|^2 = \| \mathbf{R}''(t) \|^2.$$

Next, a_t can be computed by

$$a_t = \frac{dv}{dt} = \frac{d}{dt}\frac{ds}{dt} = \frac{d}{dt} \| \mathbf{R}'(t) \| .$$

Finally, we get a_n^2, and hence a_n, from

$$a_n^2 = \| \mathbf{a} \|^2 - a_t^2.$$

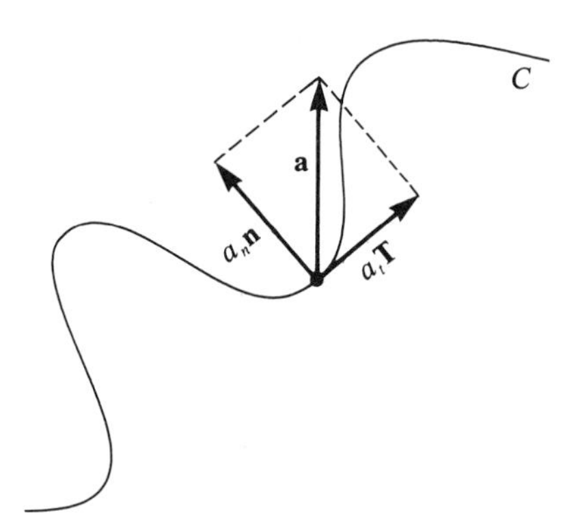

Figure 163. Computing a_t and a_n.

EXAMPLE 317

Let the position of a particle at time t be given by

$$\mathbf{R}(t) = [\cos(t) + t\sin(t)]\mathbf{i} + [\sin(t) - t\cos(t)]\mathbf{j} + t^2\mathbf{k}, \quad t > 0.$$

Then,

$$\mathbf{v} = t\cos(t)\mathbf{i} + t\sin(t)\mathbf{j} + 2t\mathbf{k},$$
$$v = \| \mathbf{v} \| = \sqrt{5}\, t,$$
$$\mathbf{a} = [\cos(t) - t\sin(t)]\mathbf{i} + [\sin(t) + t\cos(t)]\mathbf{j} + 2\mathbf{k},$$

and

$$\| \mathbf{a} \| = \sqrt{5 + t^2}.$$

The tangential component of the acceleration is

$$a_t = \frac{dv}{dt} = \sqrt{5}.$$

From this and $\| \mathbf{a} \|$, we get

$$a_n^2 = \| \mathbf{a} \|^2 - a_t^2 = t^2.$$

Hence,

$$a_n = |t| = t,$$

since $t > 0$ in this example, Thus,

$$\mathbf{a} = \sqrt{5}\,\mathbf{T} + t\mathbf{N}.$$

From a_n we easily get the radius of curvature of the trajectory. Since $a_n = v^2/\rho$, then

$$\rho = \frac{v^2}{a_n} = 5t.$$

Hence, the curvature is

$$\kappa = \frac{1}{\rho} = \frac{1}{5t}.$$

As $t \to 0$, $\mathbf{R}(t) \to \mathbf{i}$, and the initial point of the curve is (1, 0, 0). Near (1, 0, 0), the curve has very large curvature, since $\kappa \to \infty$ as $t \to 0$. As $t \to \infty$, $\kappa \to 0$; so the curve tends to straighten out after a long time.

We have not explicitly computed **T** and **N** in this example but can easily do so. We have

$$\begin{aligned}\mathbf{T} &= \frac{d\mathbf{R}}{ds} = \frac{dt}{ds}\frac{d\mathbf{R}}{dt} = \frac{1}{v}\,\mathbf{v} \\ &= \frac{1}{\sqrt{5}}\cos(t)\mathbf{i} + \frac{1}{\sqrt{5}}\sin(t)\mathbf{j} + \frac{2}{\sqrt{5}}\,\mathbf{k}.\end{aligned}$$

Next,

$$\begin{aligned}\mathbf{N} &= \rho\,\frac{d\mathbf{T}}{ds} = \rho\,\frac{dt}{ds}\frac{d\mathbf{T}}{dt} = \frac{\rho}{v}\frac{d\mathbf{T}}{dt} \\ &= \frac{5t}{\sqrt{5}t}\left(\frac{-1}{\sqrt{5}}\sin(t)\mathbf{i} + \frac{1}{\sqrt{5}}\cos(t)\mathbf{j}\right) \\ &= -\sin(t)\mathbf{i} + \cos(t)\mathbf{j}.\end{aligned}$$

Note: In this example, the various quantities (**v**, ρ, **T**, **N**, and so on) are easy to calculate because v and $\| \mathbf{a} \|$ turn out to be fairly simple expressions. Usually, these quantities are more complicated functions of t. This adds to the amount of work in doing the computations but does not change the basic ideas or methods.

In conjunction with **T** and **N**, the vector $\mathbf{B} = \mathbf{T} \times \mathbf{N}$ is called the *binormal* and is a unit vector perpendicular to the plane of **T** and **N**. The vectors **T**, **N**, and **B** form a right-handed coordinate system in the sense that, if one takes the x-axis along **T**, the y-axis along **N**, and the z-axis along **B**, then we obtain a system as shown in Figure 164. By contrast, a left-handed coordinate system is shown in Figure 165.

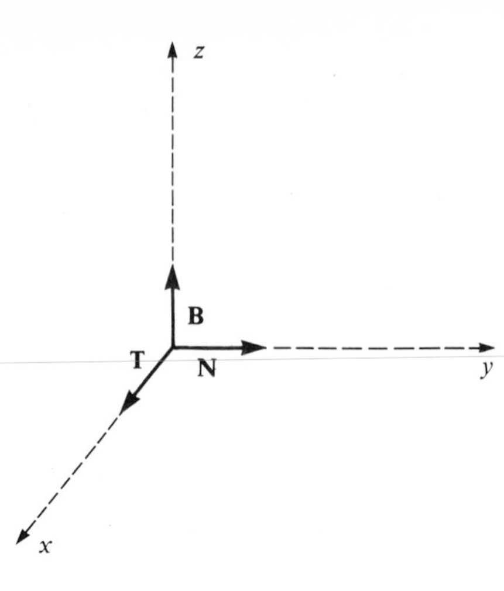

Figure 164. Right-handed coordinate system formed by tangent, normal, and binormal vectors.

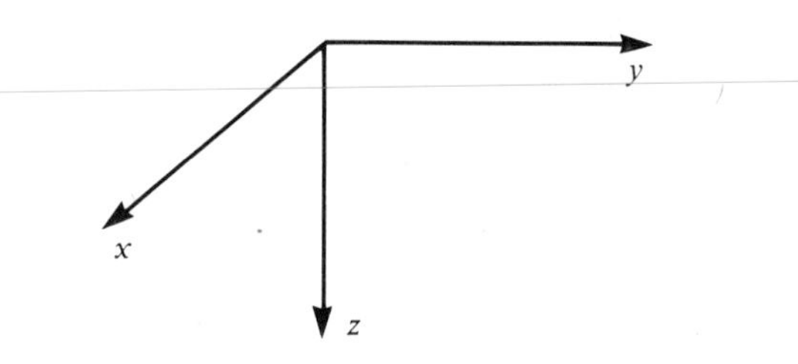

Figure 165. Left-handed coordinate system.

From previous information, we have

$$\frac{d\mathbf{T}}{ds} = \kappa \mathbf{N}.$$

One can also show that, for some scalar function $\tau(t)$,

$$\frac{d\mathbf{N}}{ds} = -\kappa \mathbf{T} + \tau \mathbf{B}$$

and

$$\frac{d\mathbf{B}}{ds} = -\tau \mathbf{N}.$$

Here, τ is called the *torsion* of the curve, and these three formulas are called the *Frenet formulas.* Roughly, torsion measures the amount a curve twists in the sense that the coordinate system formed by **T**, **N**, and **B** at each point seems to twist about C as one moves along the curve (see Figure 166).

We shall next turn our attention to vector fields and develop the three basic operations of vector calculus: gradient, divergence, and curl. These will occupy the next three sections.

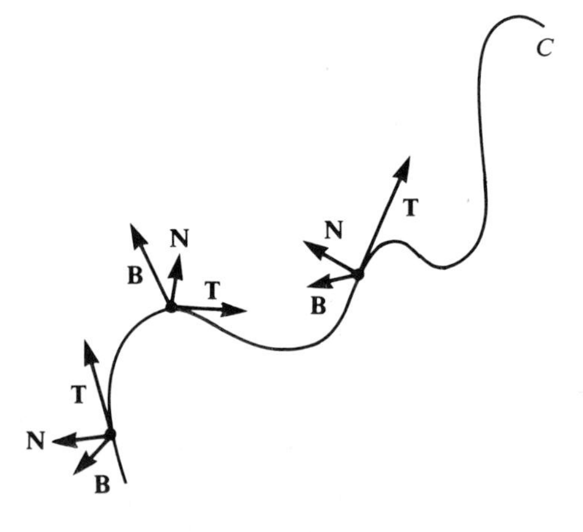

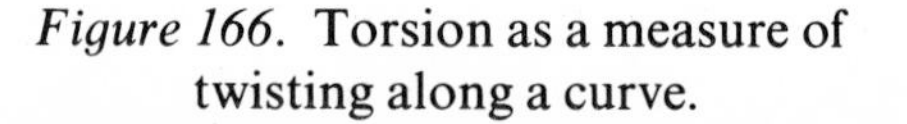

Figure 166. Torsion as a measure of twisting along a curve.

PROBLEMS FOR SECTION 11.2

In each of Problems 1 through 20, find $\mathbf{v}$, v, $\mathbf{a}$, a_t, and a_n. Use a_t and $\|\mathbf{a}\|$ to compute ρ and then κ. Finally, explicitly find $\mathbf{T}$ and $\mathbf{N}$.

1. $\mathbf{R} = 3t\mathbf{i} - 2\mathbf{j} + t^2\mathbf{k}$
2. $\mathbf{R} = t\sin(t)\mathbf{i} + t\cos(t)\mathbf{j} + \mathbf{k}$
3. $\mathbf{R} = 2t\mathbf{i} - 2t^2\mathbf{j} + \mathbf{k}$
4. $\mathbf{R} = e^t\sin(t)\mathbf{i} + e^t\cos(t)\mathbf{j} - 2\mathbf{k}$
5. $\mathbf{R} = 2\sin(t)\mathbf{i} + t\mathbf{j} + 2\cos(t)\mathbf{k}$
6. $\mathbf{R} = e^t[\sin(t)\mathbf{i} + \cos(t)\mathbf{j} + 3t\mathbf{k}]$
7. $\mathbf{R} = 2t^2\mathbf{i} + t\mathbf{j} + t\mathbf{k}$
8. $\mathbf{R} = e^{-t}(\mathbf{i} + \mathbf{j} - 2\mathbf{k})$
9. $\mathbf{R} = \alpha\cos(t)\mathbf{i} + \alpha\sin(t)\mathbf{j} + \beta t\mathbf{k}$, with α and β constant. (This curve is called a *circular helix*.)
10. $\mathbf{R} = 3t^2\mathbf{i} - 3t\mathbf{j} + 3t^3\mathbf{k}$
11. $\mathbf{R} = 2\sinh(t)\mathbf{i} - 2\cosh(t)\mathbf{k}$
12. $\mathbf{R} = 2\cos(t)\mathbf{i} + 2\sin(t)\mathbf{j} - t^2\mathbf{k}$
13. $\mathbf{R} = \alpha\cos(t)\mathbf{i} + \beta\sin(t)\mathbf{j}$
14. $\mathbf{R} = 2t\mathbf{i} - \cos(t)\mathbf{j} - \sin(t)\mathbf{k}$
15. $\mathbf{R} = \sin(\omega t)\mathbf{i} - 2t\mathbf{j} + \cos(\omega t)\mathbf{k}$
16. $\mathbf{R} = t^2\mathbf{i} - t^2\mathbf{j} + 3t\mathbf{k}$
17. $\mathbf{R} = (1 + 2t)\mathbf{i} - 2t\mathbf{j} + t^2\mathbf{k}$
18. $\mathbf{R} = e^{-t}(\mathbf{i} + \mathbf{j} - t\mathbf{k})$
19. $\mathbf{R} = \alpha t^2\mathbf{i} + \beta t^2\mathbf{j} + 2\gamma t\mathbf{k}$
20. $\mathbf{R} = 3t\cos(t)\mathbf{i} + 3t\sin(t)\mathbf{j}$
21. Let $\mathbf{R}(t)$ be the position vector defining a curve C. Suppose that $\mathbf{T}$ is a constant vector. Prove that C is a straight line.
22. Prove that

$$\kappa = \frac{\|\mathbf{R}' \times \mathbf{R}''\|}{\|\mathbf{R}'\|^3}.$$

23. Prove that

$$\kappa = \frac{\{\|\mathbf{R}'\|^2\,\|\mathbf{R}''\|^2 - (\mathbf{R}' \cdot \mathbf{R}'')^2]^{1/2}}{\|\mathbf{R}'\|^3}.$$

24. Prove that the torsion is given by

$$\tau(s) = -\mathbf{N}(s) \cdot \frac{d\mathbf{B}}{ds}.$$

Use as the definition of τ the relationship $d\mathbf{B}/ds = -\tau(s)\mathbf{N}(s)$.

25. Show that

$$\tau = [\mathbf{T}, \mathbf{N}, \mathbf{N}'].$$

26. Show that

$$\tau = \frac{1}{\kappa^2}[\mathbf{R}', \mathbf{R}'', \mathbf{R}'''].$$

27. Show that

$$\tau = \frac{[\mathbf{R}', \mathbf{R}'', \mathbf{R}''']}{(\mathbf{R}' \cdot \mathbf{R}')(\mathbf{R}'' \cdot \mathbf{R}'') - (\mathbf{R}' \cdot \mathbf{R}'')^2}.$$

28. Derive the Frenet formulas, as follows.
    (a) Show that $d\mathbf{T}/ds$ is parallel to $\mathbf{N}$, and use the implied relationship $d\mathbf{T}/ds = \kappa\mathbf{N}$ to define κ.
    (b) Show that $(d\mathbf{N}/ds) + \kappa\mathbf{T}$ is perpendicular to both $\mathbf{T}$ and $\mathbf{N}$, and hence is parallel to $\mathbf{B}$. Then, for some scalar-valued function which we call τ, $(d\mathbf{N}/ds) + \kappa\mathbf{T} = \tau\mathbf{B}$. Take this as the definition of τ.
    (c) Prove that $d\mathbf{B}/ds = -\tau\mathbf{N}$ by differentiating $\mathbf{T} \times \mathbf{N}$.
29. Let $\mathbf{R}(t) = \alpha[\cos(\omega t)\mathbf{i} + \sin(\omega t)\mathbf{j}]$ be the position vector of a particle moving in the xy-plane.
    (a) Show that the angular speed (speed divided by distance α from the center of the circular path) is ω.
    (b) Show that $\mathbf{a}$ is toward the origin, with constant magnitude $\omega^2\alpha$. (This is the *centripetal acceleration*. The *centripetal force* is $m\mathbf{a}$, and the *centrifugal force* is $-m\mathbf{a}$, where m is the mass of the particle.)

30. Imagine a disk rotating counterclockwise with constant angular speed ω, with $\mathbf{B} = \cos(\omega t)\mathbf{i} + \sin(\omega t)\mathbf{j}$ a unit vector rotating with the disk. A particle moves from the center toward the edge of the disk, with $\mathbf{R} = t\mathbf{B}$.

(a) Show that $\mathbf{v} = \mathbf{B} + t\mathbf{B}'$.

(b) Show that $\mathbf{a} = 2\mathbf{B}' + t\mathbf{B}'' = 2\mathbf{B}' - \omega^2 t\mathbf{B}$.

(c) In the above expression for $\mathbf{a}$, the term $2\mathbf{B}'$ is called the *Coriolis acceleration.* Show that a person of mass m walking along the position vector $\mathbf{R}(t)$ will feel a force of $-2m\mathbf{B}'$ against the sense of rotation of the disk. (This force is behind the fact that water emptying out of a tub in the northern hemisphere swirls counterclockwise, while in the southern hemisphere it swirls clockwise.)

11.3 VECTOR FIELDS

Up to now we have dealt with vector functions of one variable. Two other kinds of vector functions commonly encountered are those with components which are real-valued functions of two or three variables. Such vector functions in two variables usually have the appearance

$$\mathbf{G}(x, y) = G_1(x, y)\mathbf{i} + G_2(x, y)\mathbf{j}$$

and, in three variables,

$$\mathbf{F}(x, y, z) = F_1(x, y, z)\mathbf{i} + F_2(x, y, z)\mathbf{j} + F_3(x, y, z)\mathbf{k}.$$

Such expressions are called *vector fields*—**G** in the plane and **F** in 3-space. This terminology arises because we can think of **F** and **G** as defining vector arrows at each point for which they are defined.

For example, let

$$\mathbf{G}(x, y) = xy\mathbf{i} + (x - y)\mathbf{j}$$

for every (x, y). Thus,

$$\mathbf{G}(1, 1) = \mathbf{i}, \qquad \mathbf{G}(2, -3) = -6\mathbf{i} + 5\mathbf{j}, \qquad \mathbf{G}(0, 1) = -\mathbf{j},$$

and so on. We can get a pictorial representation of this vector field by drawing from each (x, y) the vector $xy\mathbf{i} + (x - y)\mathbf{j}$, as shown in Figure 167. Such diagrams are particularly helpful in picturing, for example, electrical or magnetic fields.

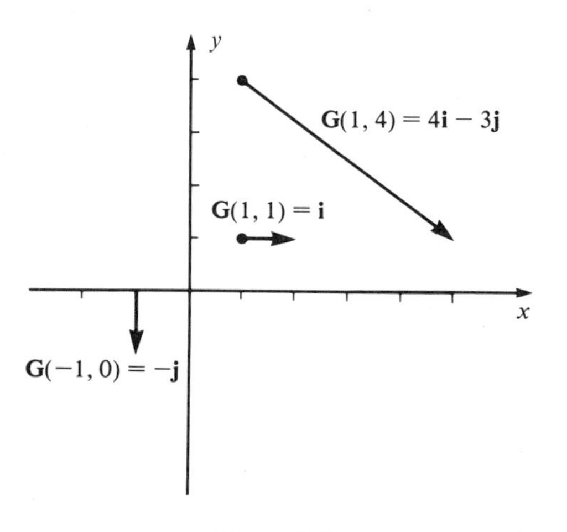

Figure 167

A vector field is called *continuous* if each of the components is continuous. We also define the *partial derivative* of a vector field to be the vector field obtained by taking the partial derivative of each component.

For example, let

$$\mathbf{F} = 3xz\mathbf{i} + 2e^z\cos(x)\mathbf{j} - y^2\mathbf{k}.$$

Then,

$$\frac{\partial \mathbf{F}}{\partial x} = 3z\mathbf{i} - 2e^z \sin(x)\mathbf{j},$$

$$\frac{\partial \mathbf{F}}{\partial y} = -2y\mathbf{k}, \quad \text{and} \quad \frac{\partial \mathbf{F}}{\partial z} = 3x\mathbf{i} + 2e^z \cos(x)\mathbf{j}.$$

Note that these partials are again vector fields.

Given a vector field $\mathbf{F}(x, y, z)$, suppose we draw from each point (x, y, z) the arrow representing the vector $\mathbf{F}(x, y, z)$. Often we can envision these arrows as tangents to curves in space. These curves are called *streamlines, lines of force*, or *flow lines of* $\mathbf{F}$. Terminology varies according to context. If, say, $\mathbf{F}$ is the velocity of a fluid, one usually refers to streamlines or flow lines; if $\mathbf{F}$ is a magnetic force, one usually speaks of lines of force. As an example, if you put iron filings on a piece of cardboard and hold a magnet underneath, the filings will arrange themselves along the magnetic lines of force, with more filings where the field strength is greater.

Given $\mathbf{F}$, how do we construct these lines of force? In effect, this is a problem of constructing curves having given tangents. To do this, suppose that C: $x = x(s)$, $y = y(s)$, $z = z(s)$ is a line of force of $\mathbf{F}$. Here, C is a curve in 3-space, which we parametrize in terms of arc length, measured along C from some point on C. As usual, we let

$$\mathbf{R}(s) = x(s)\mathbf{i} + y(s)\mathbf{j} + z(s)\mathbf{k}$$

denote the position vector. Then,

$$\mathbf{R}'(s) = \frac{dx}{ds}\mathbf{i} + \frac{dy}{ds}\mathbf{j} + \frac{dz}{ds}\mathbf{k}$$

is a unit tangent to C, and hence is aligned with $\mathbf{F}(x(s), y(s), z(s))$ at the point $(x(s), y(s), z(s))$. For some scalar t, then, $\mathbf{R}' = t\mathbf{F}$. Letting $\mathbf{F} = F_1\mathbf{i} + F_2\mathbf{j} + F_3\mathbf{k}$, we then have

$$\frac{dx}{ds} = tF_1, \qquad \frac{dy}{ds} = tF_2, \qquad \frac{dz}{ds} = tF_3.$$

This is a system of differential equations whose solution gives the coordinate functions x, y, and z of the lines of force.

If F_1, F_2, and F_3 are nonzero, we can eliminate t and write the system as

$$\frac{dx}{F_1} = \frac{dy}{F_2} = \frac{dz}{F_3},$$

since all three quotients are equal to $t\ ds$.

Here is an illustration of the construction of the lines of force of a vector field.

EXAMPLE 318

Let $\mathbf{F} = x^2\mathbf{i} + 2y\mathbf{j} - \mathbf{k}$.

The lines of force satisfy

$$\frac{dx}{ds} = tx^2, \qquad \frac{dy}{ds} = 2ty, \qquad \frac{dz}{ds} = -t.$$

These can also be written

$$\frac{dx}{x^2} = \frac{dy}{2y} = \frac{dz}{-1},$$

except at $x = 0$ or $y = 0$, and can be solved in pairs.

Begin, for example, with $dz = -dx/x^2$. Integrating both sides gives us

$$z = \frac{1}{x} + k_1,$$

where k_1 is a constant of integration. Next, $dz = -dy/2y$, which gives us

$$z = -\tfrac{1}{2}\ln(y) + k_2.$$

Here, it seems convenient to put everything in terms of z, and we get as the lines of force the curves

$$x = \frac{1}{z - k_1}, \qquad y = k_3 e^{-2z}, \qquad z = z,$$

where $k_3 = e^{2k_2}$.

This is in fact a family of curves, going through different points as k_1 and k_3 are chosen. For example, suppose we want the line of force through $(-1, 6, 2)$. Then,

$$x = -1 = \frac{1}{2 - k_1};$$

so $k_1 = 3$. Next,

$$y = 6 = k_3 e^{-4};$$

so $k_3 = 6e^4$. Thus, the line of force of $\mathbf{F}$ passing through $(-1, 6, 2)$ is given parametrically by

$$x = \frac{1}{z - 3}, \qquad y = 6e^{4-2z}, \qquad z = z.$$

EXAMPLE 319

Let $\mathbf{F} = -y\mathbf{j} + z\mathbf{k}$.

Here, $F_1 = 0$, and the lines of force satisfy

$$\frac{dx}{ds} = 0, \qquad \frac{dy}{ds} = -ty, \qquad \frac{dz}{ds} = tz.$$

Since $dx/ds = 0$, then $x = \text{constant} = k_1$. (Thus, the line of force is in a plane parallel to the yz-plane.)

The other two equations can be written

$$\frac{dy}{-y} = \frac{dz}{z}.$$

This is a separable differential equation, and integrating both sides gives us

$$-\ln(y) = \ln(z) + k_2 \quad \text{or} \quad \frac{1}{y} = k_3 z,$$

where $k_3 = e^{k_2}$.

If we want the line of force through, say, $(-4, 1, 7)$, then we need

$$x = k_1 = -4$$

and

$$\frac{1}{y} = 1 = k_3 z = 7k_3;$$

so $k_3 = \frac{1}{7}$. Thus, the line of force of $\mathbf{F}$ through $(-4, 1, 7)$ is

$$x = -4, \qquad \frac{1}{y} = \frac{z}{7}.$$

This is a hyperbola in the plane $x = -4$.

In the next section we shall consider the gradient, which is an important vector field generated by a scalar function.

PROBLEMS FOR SECTION 11.3

In each of Problems 1 through 10, you are given a vector field $\mathbf{G}(x, y)$ in the plane.

(a) Find $\dfrac{\partial \mathbf{G}}{\partial x}$ and $\dfrac{\partial \mathbf{G}}{\partial y}$.

(b) Make a diagram in which each indicated vector $\mathbf{G}(x, y)$ is drawn as an arrow with initial point (x, y).

1. $\mathbf{G} = 3x\mathbf{i} - 4xy\mathbf{j}$; $\mathbf{G}(0, 1), \mathbf{G}(1, 3), \mathbf{G}(1, 4), \mathbf{G}(-1, 2)$
2. $\mathbf{G} = e^x\mathbf{i} - 2x^2y\mathbf{j}$; $\mathbf{G}(0, 0), \mathbf{G}(1, 0), \mathbf{G}(0, 1), \mathbf{G}(3, -2)$
3. $\mathbf{G} = 4xy\mathbf{i} - y\mathbf{j}$; $\mathbf{G}(0, 0), \mathbf{G}(-1, 0), \mathbf{G}(0, 1), \mathbf{G}(1, 3)$
4. $\mathbf{G} = 3\mathbf{i} - 4x\mathbf{j}$; $\mathbf{G}(1, 0), \mathbf{G}(2, 0), \mathbf{G}(0, 1), \mathbf{G}(3, -1)$
5. $\mathbf{G} = 2xy\mathbf{i} + \cos(x)\mathbf{j}$; $\mathbf{G}(\frac{1}{2}\pi, 0), \mathbf{G}(0, 0), \mathbf{G}(\pi, 1), \mathbf{G}(3\pi, 0)$
6. $\mathbf{G} = (-5x + 1)\mathbf{i} + 2x^2\mathbf{j}$; $\mathbf{G}(0, 2), \mathbf{G}(-1, 1), \mathbf{G}(3, 1), \mathbf{G}(1, -2)$
7. $\mathbf{G} = e^{-x}x\mathbf{i} + 8xy\mathbf{j}$; $\mathbf{G}(1, -3), \mathbf{G}(0, 0), \mathbf{G}(-1, -2), \mathbf{G}(0, 2)$
8. $\mathbf{G} = \sin(2x)y\mathbf{i} + x^2\mathbf{j}$; $\mathbf{G}(0, 1), \mathbf{G}(-3, 1), \mathbf{G}(0, 0), \mathbf{G}(1, -4)$
9. $\mathbf{G} = 2\ln(x + 1)y^2\mathbf{i} + 8xy^3\mathbf{j}$; $\mathbf{G}(0, 1), \mathbf{G}(0, -2), \mathbf{G}(1, 1), \mathbf{G}(1, 0)$
10. $\mathbf{G} = 3x^2\mathbf{i} + 11xy\mathbf{j}$; $\mathbf{G}(1, -2), \mathbf{G}(0, -2), \mathbf{G}(3, 0), \mathbf{G}(-1, 2)$

In each of Problems 11 through 20, find $\dfrac{\partial \mathbf{F}}{\partial x}, \dfrac{\partial \mathbf{F}}{\partial y}$, and $\dfrac{\partial \mathbf{F}}{\partial z}$.

11. $\mathbf{F} = e^{xy}\mathbf{i} - 2x^2y\mathbf{j} + \cosh(z + y)\mathbf{k}$
12. $\mathbf{F} = 4z^2\cos(x)\mathbf{i} - x^3yz\mathbf{j} + x^3y\mathbf{k}$
13. $\mathbf{F} = 3xy^3\mathbf{i} + \ln(x + y + z)\mathbf{j} + \cosh(xyz)\mathbf{k}$
14. $\mathbf{F} = -\sin(xy)z^3\mathbf{i} + 3xy^3z\mathbf{j} + \cosh(z - x)\mathbf{k}$
15. $\mathbf{F} = (14x - 2y)\mathbf{i} + (x^2 - y^2 - z^2)\mathbf{j} + 3xy^3e^{zx}\mathbf{k}$
16. $\mathbf{F} = xy^3\sin(z)\mathbf{i} + e^{x-y+2z}\mathbf{j} + (x^3y - 2z)\mathbf{k}$
17. $\mathbf{F} = -3x^2ye^x\mathbf{i} + xy^3z^2\mathbf{j} + [3x - 2\sin(xy)]\mathbf{k}$
18. $\mathbf{F} = x\ln(y - x)\mathbf{i} + 3xyz^2\mathbf{j} - [3x + 2\sin(xy)]z\mathbf{k}$
19. $\mathbf{F} = 5z\mathbf{i} + (8x - 2y)\mathbf{j} + x^3z\mathbf{k}$
20. $\mathbf{F} = -3xy^3z^2\mathbf{i} + \cosh(x - y + 2z)\mathbf{j} + 14z^2\mathbf{k}$

In each of Problems 21 through 40, find the lines of force of the given vector field, and then find the particular line of force going through the given point.

21. $\mathbf{F} = 3x^2\mathbf{i} - y\mathbf{j} + z^3\mathbf{k}$; $(2, -1, 6)$

22. $\mathbf{F} = (1/x)\mathbf{i} + e^y\mathbf{j} - \mathbf{k}$; $(1, 2, 1)$

23. $\mathbf{F} = e^y\mathbf{i} + 2x\mathbf{j} - [1/\sin(z)]\mathbf{k}$; $(1, 1, \pi/4)$

24. $\mathbf{F} = \mathbf{i} - 2\mathbf{j} + \mathbf{k}$; $(0, 1, 1)$

25. $\mathbf{F} = 3x\mathbf{i} + y\mathbf{j} - z^2\mathbf{k}$; $(1, 6, 2)$

26. $\mathbf{F} = \cos(y)\mathbf{i} + \sin(x)\mathbf{j} - 2\mathbf{k}$; $(\pi/4, \pi/2, 1)$

27. $\mathbf{F} = 2e^z\mathbf{i} - e^z\mathbf{j} + e^x\mathbf{k}$; $(1, 0, 2)$

28. $\mathbf{F} = -3\mathbf{i} + 2e^z\mathbf{j} - \cos(y)\mathbf{k}$; $(2, \pi/6, 3)$

29. $\mathbf{F} = 2z^3\mathbf{j} - y^2\mathbf{k}$; $(0, 1, 1)$

30. $\mathbf{F} = x^2y\mathbf{i} - 2xy^3\mathbf{j} + e^z\mathbf{k}$; $(2, 1, 6)$

31. $\mathbf{F} = -3\mathbf{i} + 2x\mathbf{j} - y\mathbf{k}$; $(0, -2, 4)$

32. $\mathbf{F} = 8x\mathbf{i} - 3z\mathbf{j} + 4y^2\mathbf{k}$; $(1, 2, 2)$

33. $\mathbf{F} = 3e^z\mathbf{i} - x^2\mathbf{k}$; $(1, 3, -4)$

34. $\mathbf{F} = -2x^2\mathbf{i} + yz\mathbf{j} + z^2\mathbf{k}$; $(2, -4, 6)$

35. $\mathbf{F} = 3y^3\mathbf{i} - 4e^z\mathbf{j} + y^3\mathbf{k}$; $(0, 0, 5)$

36. $\mathbf{F} = 2xy\mathbf{i} + y^2\mathbf{j} - z\mathbf{k}$; $(2, -1, 3)$

37. $\mathbf{F} = 3y^3\mathbf{i} - 4e^z\mathbf{j} + y^3\mathbf{k}$; $(3, 1, 0)$

38. $\mathbf{F} = 11y\mathbf{i} + e^{2x}\mathbf{j} - x^2\mathbf{k}$; $(1, 2, 1)$

39. $\mathbf{F} = 3y\mathbf{i} + x^2\mathbf{j} + y\mathbf{k}$; $(1, 1, 4)$

40. $\mathbf{F} = x^3\mathbf{i} - 3xy\mathbf{j} + z^3\mathbf{k}$; $(1, -4, 2)$

41. Construct a vector field whose lines of force are straight lines.

42. Construct a vector field whose lines of force are circles $x^2 + y^2 = r^2$.

43. Is it possible for a vector field $\mathbf{F}(x, y, z)$ to have lines of force lying only in the xy-plane?

11.4 GRADIENT

We continue to move toward the calculus of vector functions by developing the first of three vector differential operations.

Let $\phi(x, y, z)$ be a real-valued function of three variables. Such a function is often called a *scalar field*. Then grad ϕ, or $\nabla\phi$, is the vector field

$$\nabla\phi = \frac{\partial\phi}{\partial x}\mathbf{i} + \frac{\partial\phi}{\partial y}\mathbf{j} + \frac{\partial\phi}{\partial z}\mathbf{k},$$

wherever these partials are defined. We read grad ϕ as "gradient of ϕ"; $\nabla\phi$ is often read "del ϕ," and is a commonly used symbol for the gradient. If we want to evaluate $\nabla\phi$ at a particular point P_0, we write $\nabla\phi(P_0)$ or grad $\phi(P_0)$.

EXAMPLE 320

Let $\phi = 2xy + e^z - x^2z$. Then,

$$\nabla\phi = (2y - 2xz)\mathbf{i} + 2x\mathbf{j} + (e^z - x^2)\mathbf{k}$$

and

$$\nabla\phi(1, 2, -1) = 6\mathbf{i} + 2\mathbf{j} + (e^{-1} - 1)\mathbf{k}.$$

Gradient has several important properties, which we now discuss.

Property 1 relates the gradient to a directional derivative, which we shall now review. Let $P_0(x_0, y_0, z_0)$ be a given point and $\mathbf{u} = u_1\mathbf{i} + u_2\mathbf{j} + u_3\mathbf{k}$ a given vector specifying a direction from P_0. As a convenience, imagine that $\mathbf{u}$ has length 1. The *directional derivative* of a real-valued function $\phi(x, y, z)$ at P_0 in the direction of $\mathbf{u}$ is the rate of change of $\phi(x, y, z)$ with respect to (x, y, z), as (x, y, z) varies from (x_0, y_0, z_0) in the direction of $\mathbf{u}$.

To derive an expression for this directional derivative, note from Figure 168 that any (x, y, z) on the line from (x_0, y_0, z_0) in the direction of $\mathbf{u}$ can be written as $(x_0 + tu_1, y_0 + tu_2, z_0 + tu_3)$. When $t = 0$, we are at (x_0, y_0, z_0), and as t increases from 0, (x, y, z) moves out from P_0 in the direction specified by $\mathbf{u}$. The rate of change of $\phi(x, y, z)$ at P_0 along this direction is the derivative of $\phi(x_0 + tu_1, y_0 + tu_2, z_0 + tu_3)$ with respect to t, evaluated at $t = 0$. This is, by the chain rule,

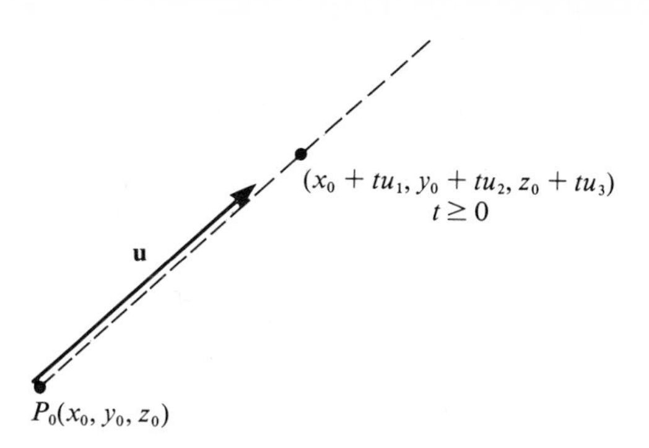

Figure 168

$$\frac{\partial \phi}{\partial x}\frac{\partial (x_0 + tu_1)}{\partial t} + \frac{\partial \phi}{\partial y}\frac{\partial (y_0 + tu_2)}{\partial t} + \frac{\partial \phi}{\partial z}\frac{\partial (z_0 + tu_3)}{\partial t},$$

with the partials evaluated at $t = 0$. This is equal to

$$\frac{\partial \phi}{\partial x} u_1 + \frac{\partial \phi}{\partial y} u_2 + \frac{\partial \phi}{\partial z} u_3,$$

with the partial derivatives evaluated at P_0. Now observe that this is exactly

$$\nabla\phi(P_0) \cdot \mathbf{u}.$$

This gives us the first property we want to state for the gradient.

1. The directional derivative of $\phi(x, y, z)$ at P_0 in the direction given by a unit vector $\mathbf{u}$ is the dot product of the gradient of ϕ at P_0 with $\mathbf{u}$.

Note: In using property 1, it is important that $\mathbf{u}$ be a unit vector. If in a given problem the direction is specified by a vector $\mathbf{v}$ of some other length, then replace $\mathbf{v}$ by the unit vector $\mathbf{v}/\|\mathbf{v}\|$, which has the direction of $\mathbf{v}$ but is of unit length.

EXAMPLE 321

Let $\phi(x, y, z) = x^2y - xe^z$. Let $P_0 = (2, -1, 0)$ and

$$\mathbf{u} = \frac{1}{\sqrt{21}}(2\mathbf{i} - 4\mathbf{j} + \mathbf{k}).$$

Then, $\nabla\phi(P_0) = -5\mathbf{i} + 4\mathbf{j} - 2\mathbf{k}$, and the directional derivative of ϕ at P_0 in the direction of $\mathbf{u}$ is

$$\nabla\phi(P_0) \cdot \mathbf{u} = \frac{-28}{\sqrt{21}}.$$

2. $\nabla\phi$ points in the direction in which ϕ increases at its greatest rate.

Let P_0 be any point around which ϕ is defined. If we look in various directions from P_0, we may see ϕ behaving differently—perhaps sometimes increasing, sometimes decreasing, or even remaining constant for some distance. We can ask, "In what direction does ϕ increase at the fastest rate?"

If $\mathbf{u}$ is a unit vector in this direction, then $\nabla\phi \cdot \mathbf{u}$ is the rate of change of ϕ along $\mathbf{u}$ at P_0 (here, $\nabla\phi$ is evaluated at P_0). This rate is also

$$\nabla\phi \cdot \mathbf{u} = \|\nabla\phi\| \, \|\mathbf{u}\| \cos(\theta),$$

where θ is the angle between $\nabla\phi$ and $\mathbf{u}$. Now, $\|\mathbf{u}\| = 1$ and $\|\nabla\phi(P_0)\|$ is some nonnegative number; so the right side is as large as possible when $\cos(\theta) = 1$, or $\theta = 0$. Then, $\mathbf{u}$ is parallel to and in the same direction as $\nabla\phi$; hence, $\nabla\phi(P_0)$ is in the direction of greatest increase of ϕ at P_0.

Note that $|\nabla\phi \cdot \mathbf{u}| = \|\nabla\phi\| \, \|\mathbf{u}\| \, |\cos(\theta)| = \|\nabla\phi\|$ if $\mathbf{u}$ is a unit vector in the direction of $\nabla\phi$. Hence:

3. $\nabla\phi(P_0)$ has magnitude equal to the maximum rate of increase of ϕ per unit distance at P_0.

4. If $\nabla\phi \neq \mathbf{0}$, then $\nabla\phi$ is normal (perpendicular) to the surface

$$\phi(x, y, z) = \text{constant}.$$

A proof of this is rather technical, but we can get some feeling for it by looking at examples.

EXAMPLE 322

Let $\phi(x, y, z) = x^2 + y^2 + z^2$. Then, $\nabla\phi = 2x\mathbf{i} + 2y\mathbf{j} + 2z\mathbf{k}$.

Now, the locus of points $x^2 + y^2 + z^2 = r^2$ (for any $r > 0$) is a sphere of radius r about the origin. At least intuitively, the normal to the sphere at any point (x, y, z) is on the line from $(0, 0, 0)$ through (x, y, z), and this is the direction of $\nabla\phi$.

For example, at $(0, 0, r)$, the "north pole," $\nabla\phi = 2r\mathbf{k}$, pointing straight up, consistent with intuition.

Usually we draw normal vectors to a surface from points on the surface instead of from the origin (see Figure 169).

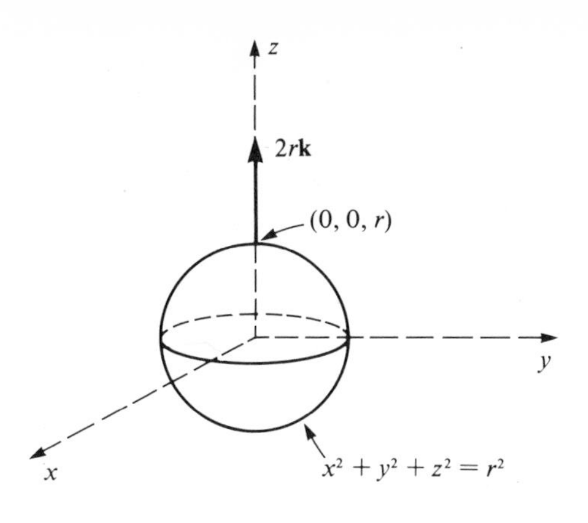

Figure 169. Normal to a sphere at (0, 0, r).

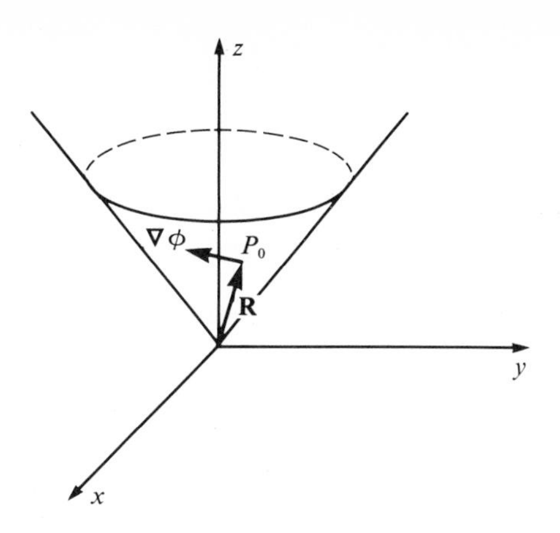

Figure 170. Cone $z = \sqrt{x^2 + y^2}$ and normal at P_0.

EXAMPLE 323

Let $\phi = z - \sqrt{x^2 + y^2}$. Then,

$$\nabla\phi = \frac{-x}{\sqrt{x^2 + y^2}}\mathbf{i} - \frac{y}{\sqrt{x^2 + y^2}}\mathbf{j} + \mathbf{k} = -\frac{x}{z}\mathbf{i} - \frac{y}{z}\mathbf{j} + \mathbf{k},$$

if $z \neq 0$. The surface $\phi = 0$, for example, is the cone $z = \sqrt{x^2 + y^2}$, as shown in Figure 170. At any point on the cone, $\nabla\phi$ points into the cone and is perpendicular to the position vector from the origin to that point, as shown in Figure 170.

If the picture is unconvincing, note that the vector from (0, 0, 0) to (x, y, z) on the cone is $\mathbf{R} = x\mathbf{i} + y\mathbf{j} + z\mathbf{k}$, and that

$$\mathbf{R}\cdot\left(-\frac{x}{z}\mathbf{i} - \frac{y}{z}\mathbf{j} + \mathbf{k}\right) = -\frac{x^2}{z} - \frac{y^2}{z} + z = 0$$

if $z = \sqrt{x^2 + y^2}$. Thus, $\mathbf{R}$, which is along the side of the cone, is perpendicular to $\nabla\phi$, as it should be if we are to call $\nabla\phi$ a normal vector to the cone.

Having determined a normal vector to ϕ = constant at any point P_0 where the gradient exists, we can now easily determine the *tangent plane at* P_0 (i.e., a plane through P_0 and normal to the normal vector). If (x, y, z) is on the tangent plane and $P_0 = (x_0, y_0, z_0)$, then

$$(x - x_0)\mathbf{i} + (y - y_0)\mathbf{j} + (z - z_0)\mathbf{k}$$

is orthogonal to $\nabla\phi(P_0)$. Thus, the tangent plane consists of all points (x, y, z) satisfying

$$\nabla\phi(P_0)\cdot[(x - x_0)\mathbf{i} + (y - y_0)\mathbf{j} + (z - z_0)\mathbf{k}] = 0.$$

In summary:

> At any point $P_0(x_0, y_0, z_0)$ on the surface $\phi(x, y, z) = \text{constant}$, the tangent plane is
>
> $$\frac{\partial \phi}{\partial x}(x - x_0) + \frac{\partial \phi}{\partial y}(y - y_0) + \frac{\partial \phi}{\partial z}(z - z_0) = 0,$$
>
> with the partial derivatives evaluated at P_0.

For example, consider again the cone $z = \sqrt{x^2 + y^2}$ at the point $(1, 1, \sqrt{2})$. Think of the cone as the surface $\phi = 0$, with $\phi(x, y, z) = z - \sqrt{x^2 + y^2}$. Then,

$$\nabla\phi(1, 1, \sqrt{2}) = \frac{-1}{\sqrt{2}}\mathbf{i} - \frac{1}{\sqrt{2}}\mathbf{j} + \mathbf{k}.$$

The tangent plane at $(1, 1, \sqrt{2})$ is then

$$-\frac{1}{\sqrt{2}}(x - 1) - \frac{1}{\sqrt{2}}(y - 1) + (z - \sqrt{2}) = 0$$

or

$$x + y - \sqrt{2}\,z = 0.$$

A straight line through P_0, and along the normal to the surface at P_0, is called the *normal line to the surface at* P_0. This is shown in Figure 171, which also suggests how to find the equation of the normal line. Begin with the normal vector $\nabla\phi(P_0)$. If (x, y, z) is on the normal line at P_0, then $(x - x_0)\mathbf{i} + (y - y_0)\mathbf{j} + (z - z_0)\mathbf{k}$ must be parallel to $\nabla\phi(P_0)$. Then, for some scalar t,

$$x - x_0 = t\,\frac{\partial \phi}{\partial x}(P_0),$$

$$y - y_0 = t\,\frac{\partial \phi}{\partial y}(P_0),$$

$$z - z_0 = t\,\frac{\partial \phi}{\partial z}(P_0).$$

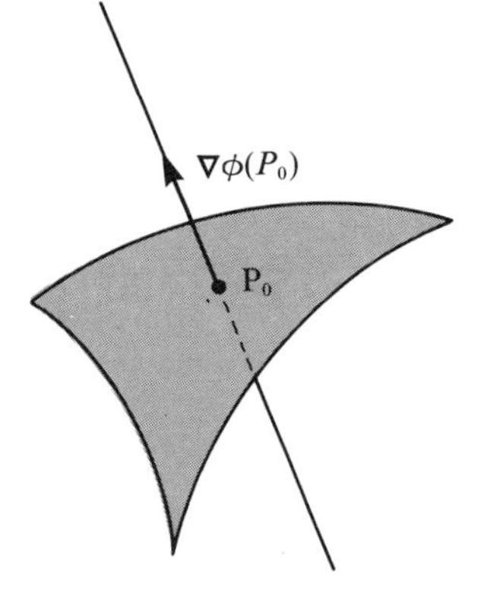

Figure 171. Normal line to a surface at P_0.

These are the parametric equations of the normal line at P_0.

As an example, consider again the cone $z = \sqrt{x^2 + y^2}$ at the point $(1, 1, \sqrt{2})$. We had that

$$\nabla\phi(1, 1, \sqrt{2}) = \frac{-1}{\sqrt{2}}\mathbf{i} - \frac{1}{\sqrt{2}}\mathbf{j} + \mathbf{k}.$$

Thus, the normal line to the cone at $(1, 1, \sqrt{2})$ is

$$x = 1 - \frac{t}{\sqrt{2}}, \qquad y = 1 - \frac{t}{\sqrt{2}}, \qquad z = \sqrt{2} + t.$$

Here are some examples showing use of the gradient.

EXAMPLE 324

Let $\phi(x, y, z) = 2xz + e^y z^2$.

(a) Find the rate of change of ϕ in the direction of $2\mathbf{i} + 3\mathbf{j} - \mathbf{k}$ at $(2, 1, 1)$.

To do this, use property 1 of gradient. First,

$$\nabla\phi = 2z\mathbf{i} + e^y z^2\mathbf{j} + (2x + 2e^y z)\mathbf{k}.$$

Then,

$$\nabla\phi(2, 1, 1) = 2\mathbf{i} + e\mathbf{j} + (4 + 2e)\mathbf{k}.$$

A unit vector in the direction of $2\mathbf{i} + 3\mathbf{j} - \mathbf{k}$ is

$$\mathbf{u} = \frac{1}{\sqrt{14}}(2\mathbf{i} + 3\mathbf{j} - \mathbf{k}).$$

Thus, the directional derivative is

$$\nabla\phi(2, 1, 1) \cdot \mathbf{u} = \frac{1}{\sqrt{14}}[4 + 3e - (4 + 2e)] = \frac{e}{\sqrt{14}}.$$

This is the rate of change of ϕ in the direction given.

(b) Find the direction in which ϕ has its greatest rate of change at $(2, 1, 1)$ and the magnitude of this rate of change.

Here, we use properties 2 and 3. The direction of greatest rate of change is the direction of $\nabla\phi$ at $(2, 1, 1)$, which is $2\mathbf{i} + e\mathbf{j} + (4 + 2e)\mathbf{k}$. The magnitude is

$$\|\nabla\phi(2, 1, 1)\| = \sqrt{5e^2 + 16e + 20}.$$

(c) Consider the surface $\phi(x, y, z) = 4 + e$. Find the normal vector and the tangent plane to this surface at $(2, 1, 1)$.

The normal vector is $\nabla\phi(2, 1, 1)$, and this is $2\mathbf{i} + e\mathbf{j} + (4 + 2e)\mathbf{k}$.

The tangent plane is

$$2(x - 2) + e(y - 1) + (4 + 2e)(z - 1) = 0$$

or

$$2x + ey + (4 + 2e)z = 8 + 3e.$$

(d) Find the normal line to the surface $2xz + e^y z^2 = 4 + e$ at $(2, 1, 1)$.

We already know that the normal vector at $(2, 1, 1)$ is $2\mathbf{i} + e\mathbf{j} + (4 + 2e)\mathbf{k}$. Thus, the normal line is

$$x - 2 = 2t, \qquad y - 1 = et, \qquad z - 1 = (4 + 2e)t.$$

This is the same as

$$\frac{x - 2}{2} = \frac{y - 1}{e} = \frac{z - 1}{4 + 2e}.$$

EXAMPLE 325

Find the tangent plane and normal line to $z = x^2 + y^2$ at $(2, -2, 8)$.

First, $(2, -2, 8)$ is on the surface since $8 = 2^2 + (-2)^2$.

Now, let $\phi(x, y, z) = z - x^2 - y^2$. The surface is then $\phi = 0$. A normal vector is $\nabla\phi = -2x\mathbf{i} - 2y\mathbf{j} + \mathbf{k}$. At $(2, -2, 8)$, the normal vector is $-4\mathbf{i} + 4\mathbf{j} + \mathbf{k}$. Thus, the tangent plane at $(2, -2, 8)$ is

$$-4(x-2) + 4(y+2) + (z-8) = 0$$

or

$$-4x + 4y + z = -8.$$

The normal line is

$$x - 2 = -4t, \qquad y + 2 = 4t, \qquad z - 8 = t.$$

Alternatively, this is

$$\frac{x-2}{-4} = \frac{y+2}{4} = \frac{z-8}{1}.$$

PROBLEMS FOR SECTION 11.4

In each of Problems 1 through 15, compute $\nabla\phi$ and $\nabla\phi(P_0)$ for the given point P_0.

1. $\phi = 2xyz$; $(1, 1, 1)$
2. $\phi = x^2y - \sin(zx)$; $(1, -1, \pi/4)$
3. $\phi = 2xy + e^z x$; $(-2, 1, 6)$
4. $\phi = \cos(xyz) + \arctan(y/x)$; $(-1, 1, \pi/2)$
5. $\phi = \cosh(2xy) - \sinh(z)$; $(-3, 1, 1)$
6. $\phi = \sqrt{x^2 + y^2 + z^2}$; $(1, 1, 1)$
7. $\phi = \ln(x + y + z)$; $(3, 1, 4)$
8. $\phi = -1/\|\mathbf{R}\|$, where $\mathbf{R} = x\mathbf{i} + y\mathbf{j} + z\mathbf{k}$; $(-2, 1, 1)$
9. $\phi = e^x \cos(y)\cos(z)$; $(0, \pi/4, \pi/4)$
10. $\phi = 2x^3y + ze^y$; $(1, 1, 2)$
11. $\phi = x^2y \cosh(xz)$; $(0, 0, 1)$
12. $\phi = e^{xy} + z^2x$; $(0, 0, 4)$
13. $\phi = x - 2\cos(y + z)$; $(1, 1, 0)$
14. $\phi = x^2 - 2y + z\ln(x)$; $(1, 3, 1)$
15. $\phi = \cosh(x - y + 2z)$; $(1, -1, 0)$

In each of Problems 16 through 25, find the tangent plane and normal line to the surface at the point.

16. $x^2 + y^2 + z^2 = 4$; $(1, 1, \sqrt{2})$
17. $z = x^2 + y$; $(-1, 1, 2)$
18. $z^2 = x^2 - y^2$; $(1, 1, 0)$
19. $\sinh(x + y + z) = 0$; $(0, 0, 0)$
20. $2x - 4y^2 + z^3 = 0$; $(-4, 0, 2)$
21. $x^2 - y^2 + z^2 = 0$; $(1, 1, 0)$
22. $2x - \cos(xyz) = 3$; $(1, \pi, 1)$
23. $3x^4 + 3y^4 + 6z^4 = 12$; $(1, 1, 1)$
24. $x^2 - 2y^2 + z^4 = 0$; $(1, 1, 1)$
25. $\cos(x) - \sin(y) + z = 1$; $(0, \pi, 0)$

In each of Problems 26 through 29, find the angle between the surfaces at the given point of intersection. (The angle between the surfaces is the smaller of the two angles between their tangent planes at the given point.)

26. $z = 3x^2 + 2y^2$, $-2x + 7y^2 - z = 0$; $(1, 1, 5)$
27. $x^2 + y^2 + z^2 = 4$, $z^2 + x^2 = 2$; $(1, \sqrt{2}, 1)$
28. $z = \sqrt{x^2 + y^2}$, $x^2 + y^2 = 4$; $(2, 2, \sqrt{8})$
29. $\frac{1}{2}x^2 + \frac{1}{2}y^2 + z^2 = 5$, $x + y + z = 5$; $(2, 2, 1)$

In each of Problems 30 through 40, find a unit vector from P_0 in the direction in which ϕ has its maximum rate of change, and also find the magnitude of this change.

30. $\phi = x^2 - 3xy + 2y^2$; $(0, 0, 1)$
31. $\phi = e^x \cos(yz)$; $(1, 1, \pi)$
32. $\phi = 2xy - 3xz^2$; $(1, 2, 1)$
33. $\phi = (1/x) - 3yz$; $(2, 1, 1)$
34. $\phi = 3x^2y - 2\sin(z)$; $(1, 3, \pi)$
35. $\phi = 14x - 3y^2 + 2xye^z$; $(0, 1, 0)$

36. $\phi = 3x\cos(z) - \sin(xyz)$; $(1, \pi, 1)$

37. $\phi = -3z^3 + e^x \sin(y) + x$; $(0, 1, -2)$

38. $\phi = -2xy^2 + x\ln(y + z)$; $(1, 3, -2)$

39. $\phi = \arctan(y/x) - \sin(x)$; $(3, -1, 1)$

40. $\phi = 4x^3y^2z^2 - 2\sin(x - y)$; $(1, -1, 1)$

41. Describe a function ϕ (not identically zero) such that $\nabla\phi = \mathbf{0}$. Can the surface $\phi =$ constant then have a normal vector?

42. Suppose $\nabla\phi = \mathbf{i} + \mathbf{k}$. What can you say about the surfaces $\phi =$ constant? Prove that the lines of force of $\nabla\phi$ are orthogonal to the surfaces $\phi =$ constant.

11.5 DIVERGENCE AND CURL

The gradient produces a vector field from a scalar field. We now develop the divergence and curl operations, which act on a vector field to produce, in turn, a scalar and a vector field. Let

$$\mathbf{F}(x, y, z) = F_1(x, y, z)\mathbf{i} + F_2(x, y, z)\mathbf{j} + F_3(x, y, z)\mathbf{k},$$

with $F_1(x, y, z)$, $F_2(x, y, z)$, and $F_3(x, y, z)$ real-valued functions of three real variables.

We define

$$\text{div } \mathbf{F} = \frac{\partial F_1}{\partial x} + \frac{\partial F_2}{\partial y} + \frac{\partial F_3}{\partial z} \qquad \text{(a scalar)}$$

and

$$\text{curl } \mathbf{F} = \left(\frac{\partial F_3}{\partial y} - \frac{\partial F_2}{\partial z}\right)\mathbf{i} + \left(\frac{\partial F_1}{\partial z} - \frac{\partial F_3}{\partial x}\right)\mathbf{j} + \left(\frac{\partial F_2}{\partial x} - \frac{\partial F_1}{\partial y}\right)\mathbf{k} \qquad \text{(a vector field).}$$

These are read, respectively, "divergence of **F**" and "curl of **F**."

For example, let $\mathbf{F} = y\mathbf{i} + 2xz\mathbf{j} + e^x z\mathbf{k}$. Then,

$$\text{div } \mathbf{F} = e^x$$

and

$$\text{curl } \mathbf{F} = -2x\mathbf{i} - e^x z\mathbf{j} + (2z - 1)\mathbf{k}.$$

Both divergence and curl have physical interpretations which we discuss briefly at the end of this section and more carefully later when we have the theorems of Gauss and Stokes.

Divergence and curl can be thought of as a dot and cross product, respectively, as follows. First, define the operator "vector":

$$\nabla = \frac{\partial}{\partial x}\mathbf{i} + \frac{\partial}{\partial y}\mathbf{j} + \frac{\partial}{\partial z}\mathbf{k}.$$

Here, ∇ is read "del," and is called the *del operator*. Taking a dot product of del with $F_1\mathbf{i} + F_2\mathbf{j} + F_3\mathbf{k}$ gives us

$$\nabla \cdot \mathbf{F} = \left(\frac{\partial}{\partial x}\mathbf{i} + \frac{\partial}{\partial y}\mathbf{j} + \frac{\partial}{\partial z}\mathbf{k}\right) \cdot (F_1\mathbf{i} + F_2\mathbf{j} + F_3\mathbf{k}) = \frac{\partial F_1}{\partial x} + \frac{\partial F_2}{\partial y} + \frac{\partial F_3}{\partial z} = \text{div } \mathbf{F}.$$

And, taking a cross product, we have

$$\nabla \times \mathbf{F} = \begin{vmatrix} \mathbf{i} & \mathbf{j} & \mathbf{k} \\ \frac{\partial}{\partial x} & \frac{\partial}{\partial y} & \frac{\partial}{\partial z} \\ F_1 & F_2 & F_3 \end{vmatrix}$$

$$= \left(\frac{\partial F_3}{\partial y} - \frac{\partial F_2}{\partial z}\right)\mathbf{i} + \left(\frac{\partial F_1}{\partial z} - \frac{\partial F_3}{\partial x}\right)\mathbf{j} + \left(\frac{\partial F_2}{\partial x} - \frac{\partial F_1}{\partial y}\right)\mathbf{k}$$

$$= \text{curl } \mathbf{F}.$$

Thus,

$$\nabla \cdot \mathbf{F} = \text{div } \mathbf{F}$$

and

$$\nabla \times \mathbf{F} = \text{curl } \mathbf{F}.$$

In words:

$$\text{del dot} = \text{div}$$

and

$$\text{del cross} = \text{curl}.$$

Use of the del operator gives us a compact way of writing divergence, curl, and the identities involving them, two of which appear as theorems below.

It is also used to define the "del squared," or Laplacian, operator: If ϕ is a scalar field, then $\nabla\phi$ is a vector, and the Laplacian of ϕ is defined to be the scalar field

$$\nabla \cdot (\nabla\phi) = \frac{\partial^2\phi}{\partial x^2} + \frac{\partial^2\phi}{\partial y^2} + \frac{\partial^2\phi}{\partial z^2}.$$

This is often denoted $\nabla^2\phi$, and the equation $\nabla^2\phi = 0$ is called *Laplace's equation.* It arises in heat and wave conduction, as we shall see later, and is one of the fundamental equations of physics and engineering.

Here are two basic relationships between gradient, divergence, and curl.

THEOREM 87

If ϕ is continuous with continuous first and second partial derivatives, then

$$\nabla \times (\nabla\phi) = \mathbf{0}.$$

[In words, this says that curl(grad ϕ) is the zero vector.]

Proof First, $\nabla\phi = \frac{\partial\phi}{\partial x}\mathbf{i} + \frac{\partial\phi}{\partial y}\mathbf{j} + \frac{\partial\phi}{\partial z}\mathbf{k}$. Then,

$$\nabla \times (\nabla\phi) = \begin{vmatrix} \mathbf{i} & \mathbf{j} & \mathbf{k} \\ \frac{\partial}{\partial x} & \frac{\partial}{\partial y} & \frac{\partial}{\partial z} \\ \frac{\partial\phi}{\partial x} & \frac{\partial\phi}{\partial y} & \frac{\partial\phi}{\partial z} \end{vmatrix}$$

$$= \left(\frac{\partial^2\phi}{\partial y\,\partial z} - \frac{\partial^2\phi}{\partial z\,\partial y}\right)\mathbf{i} + \left(\frac{\partial^2\phi}{\partial z\,\partial x} - \frac{\partial^2\phi}{\partial x\,\partial z}\right)\mathbf{j} + \left(\frac{\partial^2\phi}{\partial y\,\partial x} - \frac{\partial^2\phi}{\partial x\,\partial y}\right)\mathbf{k}.$$

Each component of $\nabla \times (\nabla\phi)$ is a difference of equal mixed partials, and hence is 0.

THEOREM 88

Let $\mathbf{F}$ be a continuous vector field with continuous first and second partials. Then,

$$\nabla \cdot (\nabla \times \mathbf{F}) = 0.$$

[In words, div(curl $\mathbf{F}$) is identically zero.]

Proof* Let $\mathbf{F} = F_1\mathbf{i} + F_2\mathbf{j} + F_3\mathbf{k}$. Referring to the definition of $\nabla \times \mathbf{F}$, we have

$$\nabla \cdot (\nabla \times \mathbf{F}) = \frac{\partial}{\partial x}\left(\frac{\partial F_3}{\partial y} - \frac{\partial F_2}{\partial z}\right) + \frac{\partial}{\partial y}\left(\frac{\partial F_1}{\partial z} - \frac{\partial F_3}{\partial x}\right) + \frac{\partial}{\partial z}\left(\frac{\partial F_2}{\partial x} - \frac{\partial F_1}{\partial y}\right).$$

After rearranging terms, we get

$$\nabla \cdot (\nabla \times \mathbf{F}) = \frac{\partial^2 F_3}{\partial x\,\partial y} - \frac{\partial^2 F_3}{\partial y\,\partial x} + \frac{\partial^2 F_1}{\partial z\,\partial y} - \frac{\partial^2 F_1}{\partial y\,\partial z} + \frac{\partial^2 F_2}{\partial x\,\partial z} - \frac{\partial^2 F_2}{\partial z\,\partial x} = 0.$$

To get some feeling for divergence, imagine that

$$\mathbf{F}(x, y, z, t) = F_1(x, y, z, t)\mathbf{i} + F_2(x, y, z, t)\mathbf{j} + F_3(x, y, z, t)\mathbf{k}$$

is the velocity of a fluid (say, oil or water) at point (x, y, z) and time t. We insert t to be more realistic, but it plays no role in calculating $\nabla \cdot \mathbf{F}$.

Now imagine a small rectangular box in the fluid, as shown in Figure 172. We are interested in measuring, in some sense, the rate per unit volume of fluid flow out of the box across its faces at time t. To do this, look at opposite pairs of faces.

First, look at the front and back faces, labeled II and I, parallel to the yz-plane. If Δx is small, then $\mathbf{F}$ on face II is approximately $\mathbf{F}(x + \Delta x, y, z, t)$. The outer normal to this face is $\mathbf{i}$, and the area is $\Delta y\,\Delta z$. Thus,

$$\mathbf{F}(x + \Delta x, y, z, t) \cdot \mathbf{i}\,\Delta y\,\Delta z = F_1(x + \Delta x, y, z, t)\,\Delta y\,\Delta z$$

= (normal component of fluid velocity outward across face II)(area of face II)

= (flux of flow out of the box across face II).

* One can also "derive" Theorem 88 by observing that $\nabla \cdot (\nabla \times \mathbf{F})$ is a scalar triple product $[\nabla, \nabla, \mathbf{F}]$ in which two terms are equal, and hence is zero.

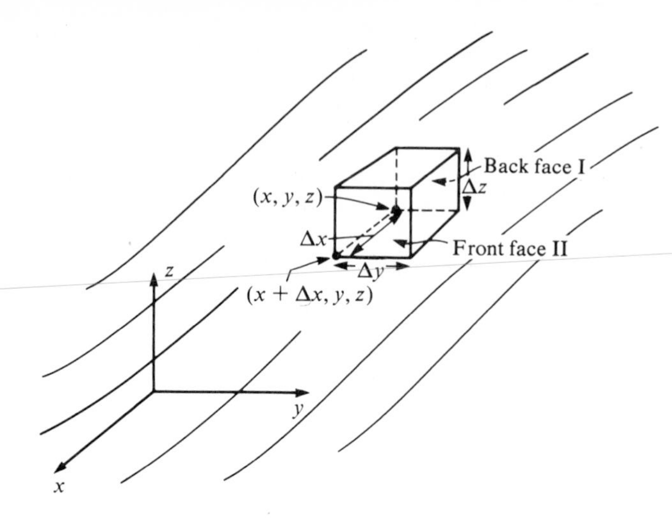

Figure 172. Fluid element.

On face I, **F** is about $\mathbf{F}(x, y, z. t)$ and the outer normal is $-\mathbf{i}$; hence, the flux out of the box across face I is approximately

$$\mathbf{F}(x, y, z, t) \cdot (-\mathbf{i})\ \Delta y\ \Delta z$$

or

$$-F_1(x, y, z, t)\ \Delta y\ \Delta z.$$

A similar calculation holds for the other sides. Summing, the total flux out of the box is

$$\begin{aligned}&[F_1(x + \Delta x, y, z, t) - F_1(x, y, z, t)]\ \Delta y\ \Delta z\\ &+ [F_2(x, y + \Delta y, z, t) - F_2(x, y, z, t)]\ \Delta x\ \Delta z\\ &+ [F_3(x, y, z + \Delta z, t) - F_3(x, y, z, t)]\ \Delta x\ \Delta y.\end{aligned}$$

The outward flux per unit volume is the outward flux divided by the volume, $\Delta x\ \Delta y\ \Delta z$. This is

$$\frac{F_1(x + \Delta x, y, z, t) - F_1(x, y, z, t)}{\Delta x} + \frac{F_2(x, y + \Delta y, z, t) - F_2(x, y, z, t)}{\Delta y} + \frac{F_3(x, y, z + \Delta z, t) - F_3(x, y, z, t)}{\Delta z}.$$

As Δx, Δy, and $\Delta z \to 0$, this gives us $\nabla \cdot \mathbf{F}$, which is then interpreted as the outward flux per unit volume of the flow at point (x, y, z) and time t. Thus, the divergence of **F** measures the expansion or "divergence away from the point" of the fluid at the point.

In the context of fluids, $\nabla \times \mathbf{F}$ measures the degree to which a fluid swirls, or rotates, about a given direction. We shall develop this interpretation in some detail when we have Stokes's theorem at our disposal. For now, we consider curl in connection with a rotating object.

Imagine that we have a body rotating with uniform angular speed about a line L (Figure 173). If ω is the angular speed, the angular velocity vector $\mathbf{\Omega}$ has magnitude ω and direction along L in the direction a right-handed screw would move if given the same rotation as the object. We suppose that L goes through the origin of our system and $\mathbf{R} = x\mathbf{i} + y\mathbf{j} + z\mathbf{k}$ for any point (x, y, z) on the rotating object.

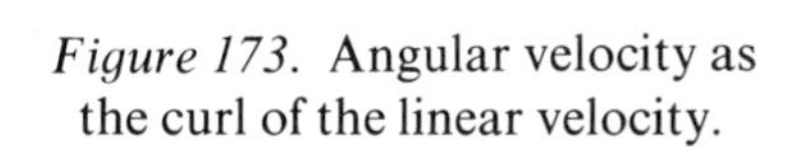

Figure 173. Angular velocity as the curl of the linear velocity.

Let $\mathbf{T}$ be the tangential (linear) velocity at any point. Then,

$$\|\mathbf{T}\| = \omega\|\mathbf{R}\|\,|\sin(\theta)| = \|\mathbf{\Omega} \times \mathbf{R}\|.$$

Since $\mathbf{T}$ is also in the direction of $\mathbf{\Omega} \times \mathbf{R}$, then, in fact, $\mathbf{T} = \mathbf{\Omega} \times \mathbf{R}$.

If you now write $\mathbf{\Omega} = \Omega_1\mathbf{i} + \Omega_2\mathbf{j} + \Omega_3\mathbf{k}$, then

$$\mathbf{\Omega} \times \mathbf{R} = (\Omega_2 z - \Omega_3 y)\mathbf{i} + (\Omega_3 x - \Omega_1 z)\mathbf{j} + (\Omega_1 y - \Omega_2 x)\mathbf{k}.$$

But we also have that

$$\nabla \times \mathbf{T} = \begin{vmatrix} \mathbf{i} & \mathbf{j} & \mathbf{k} \\ \dfrac{\partial}{\partial x} & \dfrac{\partial}{\partial y} & \dfrac{\partial}{\partial z} \\ \Omega_2 z - \Omega_3 y & \Omega_3 x - \Omega_1 z & \Omega_1 y - \Omega_2 x \end{vmatrix}$$

$$= 2\Omega_1\mathbf{i} + 2\Omega_2\mathbf{j} + 2\Omega_3\mathbf{k} = 2\mathbf{\Omega},$$

because Ω_1, Ω_2, and Ω_3 are constants.

Thus, we have that

$$\mathbf{\Omega} = \tfrac{1}{2}\nabla \times \mathbf{T},$$

which says that the angular velocity of the uniformly rotating body is a scalar multiple of the curl of the linear velocity. This is one motivation for the term *curl*. In fact, curl was once commonly known as *rotation*, and curl $\mathbf{F}$ was written rot $\mathbf{F}$, especially in British publications. It is also motivation for the term *irrotational* for a vector field whose curl is the zero vector.

PROBLEMS FOR SECTION 11.5

In each of Problems 1 through 20, compute $\nabla \cdot \mathbf{F}$ and $\nabla \times \mathbf{F}$, and verify explicitly that $\nabla \cdot (\nabla \times \mathbf{F}) = 0$.

1. $\mathbf{F} = x\mathbf{i} + y\mathbf{j} + 2z\mathbf{k}$

2. $\mathbf{F} = \sinh(xyz)\mathbf{j}$

3. $\mathbf{F} = 2xy\mathbf{i} + e^y\mathbf{j} + 2z\mathbf{k}$

4. $\mathbf{F} = x^2z\mathbf{i} - y\mathbf{j} + z^3\mathbf{k}$

5. $\mathbf{F} = -2e^z\mathbf{i} - y^2z\mathbf{j} + 2\mathbf{k}$

6. $\mathbf{F} = -yz\mathbf{j} - 6x^3\mathbf{k}$

7. $\mathbf{F} = 2x\mathbf{i} - 3y\mathbf{j} + \mathbf{k}$

8. $\mathbf{F} = \sinh(x)\mathbf{i} + \cosh(y)\mathbf{j} - xyz\mathbf{k}$

9. $\mathbf{F} = x^2\mathbf{i} + y^2\mathbf{j} + z^2\mathbf{k}$

10. $\mathbf{F} = -2xy\mathbf{i} - 2yz\mathbf{j} - 2xz\mathbf{k}$

11. $\mathbf{F} = 3x^2\mathbf{i} + yz\mathbf{j} + e^x\mathbf{k}$

12. $\mathbf{F} = \sinh(x - z)\mathbf{i} + 2y\mathbf{j} + z^2\mathbf{k}$

13. $\mathbf{F} = 2x^2y\mathbf{i} + zy\mathbf{j} - zy\mathbf{k}$

14. $\mathbf{F} = \cosh(2xz)\mathbf{i} + \sin(x)\mathbf{j} + e^z\mathbf{k}$

15. $\mathbf{F} = -3yz\mathbf{i} + xy^2\mathbf{j} - \ln(x + y + z)\mathbf{k}$
16. $\mathbf{F} = 8y^3x^2\mathbf{i} + 3xy\mathbf{j} - z^3\mathbf{k}$
17. $\mathbf{F} = \cos(xy)\mathbf{i} - z^2\mathbf{j} + (z + x)\mathbf{k}$
18. $\mathbf{F} = (14x - 2y)\mathbf{i} + 6y^3\mathbf{j} - zx\mathbf{k}$
19. $\mathbf{F} = 3y^2\mathbf{i} + 4xy^3\mathbf{j} - z^2\mathbf{k}$
20. $\mathbf{F} = \sin(x - y)\mathbf{i} + ye^{xz}\mathbf{j} - x^2z\mathbf{k}$

In each of Problems 21 through 35, compute $\nabla\phi$ and verify explicitly that $\nabla \times (\nabla\phi) = \mathbf{0}$.

21. $\phi = x - y + 2z^2$
22. $\phi = 18xyz - \sinh(yz) + e^x$
23. $\phi = -2x^3yz^2$
24. $\phi = \sin(xz) + \cos(yz)$
25. $\phi = -\arctan(yx/z)$
26. $\phi = 2e^x \ln(yz)$
27. $\phi = x^3y^2e^z$
28. $\phi = 2xy - 3z^2$
29. $\phi = -4xy^3 + z^2x$
30. $\phi = \cos(x + y + z)$
31. $\phi = -2\sin(x - y) + z^2$
32. $\phi = 4x - 3y^2z - 2$
33. $\phi = 3x^2y\cos(z) - 4x^2z - 5z$
34. $\phi = \ln(x - 4z^3) + \cos(xy^2)$
35. $\phi = 3z - \cos(e^{xyz})$
36. Find a vector field $\mathbf{F}$ with $\nabla \times \mathbf{F} = 3\mathbf{k}$.
37. Find a vector field $\mathbf{F}$ with $\nabla \cdot \mathbf{F} = 14xyz$.
38. Let ϕ be a scalar field, $\mathbf{F}$ a vector field. Derive expressions for $\nabla \cdot (\phi\mathbf{F})$ and $\nabla \times (\phi\mathbf{F})$ in terms of vector operations applied to ϕ and $\mathbf{F}$.
39. Prove that

$$\nabla(\mathbf{F} \cdot \mathbf{G}) = (\mathbf{F} \cdot \nabla)\mathbf{G} + (\mathbf{G} \cdot \nabla)\mathbf{F} + \mathbf{F} \times (\nabla \times \mathbf{G}) + \mathbf{G} \times (\nabla \times \mathbf{F}).$$

Here, $\mathbf{F} \cdot \nabla = F_1 \dfrac{\partial}{\partial x} + F_2 \dfrac{\partial}{\partial y} + F_3 \dfrac{\partial}{\partial z}$, and similarly for $\mathbf{G} \cdot \nabla$.

40. Prove that $\nabla \cdot (\mathbf{F} \times \mathbf{G}) = \mathbf{G} \cdot (\nabla \times \mathbf{F}) - \mathbf{F} \cdot (\nabla \times \mathbf{G})$.
41. Prove that $\nabla \cdot (\nabla\phi \times \nabla\psi) = 0$, where ϕ and ψ are scalar-valued functions of (x, y, z).
42. Let $\mathbf{R} = x\mathbf{i} + y\mathbf{j} + z\mathbf{k}$ and $r = \|\mathbf{R}\|$.
    (a) Prove that $\nabla r^n = nr^{n-2}\mathbf{R}$ for $n = 1, 2, \ldots$.
    (b) Prove that $\nabla \times [\phi(r)\mathbf{R}] = \mathbf{0}$, with ϕ scalar-valued.
43. Prove that $\nabla \times (\nabla \times \mathbf{F}) = \nabla(\nabla \cdot \mathbf{F}) - \nabla^2\mathbf{F}$, where

$$\nabla^2\mathbf{F} = \frac{\partial^2\mathbf{F}}{\partial x^2} + \frac{\partial^2\mathbf{F}}{\partial y^2} + \frac{\partial^2\mathbf{F}}{\partial z^2}.$$

44. Let $\mathbf{A}$ be constant and $\mathbf{R} = x\mathbf{i} + y\mathbf{j} + z\mathbf{k}$.
    (a) Show that $\nabla(\mathbf{R} \cdot \mathbf{A}) = \mathbf{A}$.
    (b) Show that $\nabla \cdot (\mathbf{R} - \mathbf{A}) = 3$.
    (c) Show that $\nabla \times (\mathbf{R} - \mathbf{A}) = \mathbf{0}$.
45. Prove that

$$\nabla \times (\mathbf{F} \times \mathbf{G}) = (\mathbf{G} \cdot \nabla)\mathbf{F} - (\mathbf{F} \cdot \nabla)\mathbf{G} + (\nabla \cdot \mathbf{G})\mathbf{F} - (\nabla \cdot \mathbf{F})\mathbf{G}.$$

11.6 LINE INTEGRALS

Thus far we have dealt with vector differential operations. We now develop the notion of an integral of vector fields over curves in space. First, we need some terminology concerning curves.

Usually, a curve C is given in parametric form as

$$x = x(t), \qquad y = y(t), \qquad z = z(t); \qquad a \le t \le b.$$

The geometric aspect consists of the graph of the curve, which consists of all points $(x(t), y(t), z(t))$ for $a \le t \le b$. The dynamic aspect consists of thinking of the curve as a pivoted vector $\mathbf{R}(t) = x(t)\mathbf{i} + y(t)\mathbf{j} + z(t)\mathbf{k}$ which sweeps through the graph as t varies

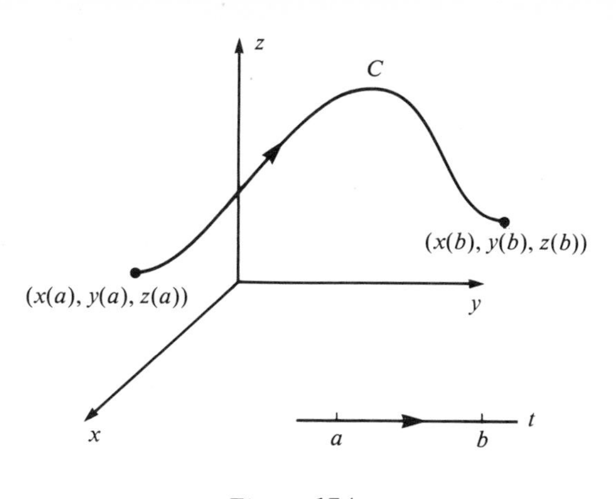

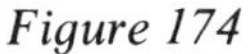
Figure 174

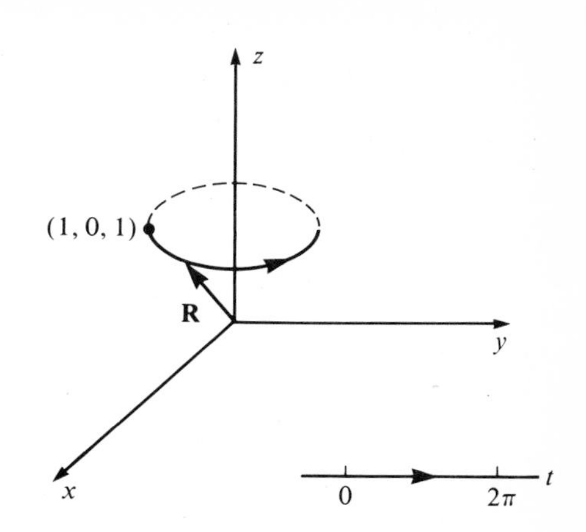

Figure 175. Circle swept out by $\mathbf{R} = \cos(t)\mathbf{i} + \sin(t)\mathbf{j} + \mathbf{k}$ as t varies from 0 to 2π.

from a to b. In this sense, the direction of the parameter assigns a direction to C. If t varies *from a to b*, we go from $(x(a), y(a), z(a))$ to $(x(b), y(b), z(b))$. We call $(x(a), y(a), z(a))$ the *initial point* of C and $(x(b), y(b), z(b))$ the *terminal point*. If we vary t *from b to a*, then we reverse direction on C, and the initial and terminal points reverse their roles. Often, we put an arrow on C to indicate direction (see Figure 174); and we write $t\colon a \to b$ to indicate that t varies from a to b or, similarly, $t\colon b \to a$ to say that t varies from b to a.

EXAMPLE 326

Let $x(t) = \cos(t)$, $y(t) = \sin(t)$, $z(t) = 1$, where $t\colon 0 \to 2\pi$.

This is the circle $x^2 + y^2 = 1$ in the plane $z = 1$ (see Figure 175). Here, t varies from 0 to 2π; so on C we go from initial point $(x(0), y(0), z(0)) = (1, 0, 1)$ to terminal point $(x(2\pi), y(2\pi), z(2\pi))$, which is also $(1, 0, 1)$.

As t varies from 0 to 2π, $(x(t), y(t), z(t))$ goes around C counterclockwise if viewed from a point above $(1, 0, 1)$ on the positive z-axis. If t varied from 2π to 0, we would reverse direction on C and go around clockwise.

Note that the vector $\mathbf{R}(t) = \cos(t)\mathbf{i} + \sin(t)\mathbf{j} + \mathbf{k}$ sweeps out C as $t\colon 0 \to 2\pi$.

We call C *closed* if the initial and terminal points are the same (as in Example 326). C is *continuous* if x, y, and z are continuous functions of t on $[a, b]$ and *simple* if it does not pass over the same point for different values of the parameter [i.e., $\mathbf{R}(t_1) \neq \mathbf{R}(t_2)$ if $t_1 \neq t_2$]. Technically, a closed curve cannot be simple, since $\mathbf{R}(a) = \mathbf{R}(b)$ in this case. However, we make an exception for closed curves and call a closed curve C simple if the end point is the only one visited at two different parameter values. In Example 326, C is a simple closed curved.

We call C *differentiable* if x, y, and z are differentiable functions of t for $a < t < b$. Recall that then C has a tangent vector $\mathbf{R}'(t) = x'(t)\mathbf{i} + y'(t)\mathbf{j} + z'(t)\mathbf{k}$. If x', y', and z' are themselves continuous and never all zero at the same time for $a < t < b$, then we call C *smooth*. In this case, C has a continuous tangent vector $\mathbf{R}'(t)$.

Finally, C is *piecewise smooth* if C consists of a finite number of smooth curves strung together. More carefully, C is smooth if x', y', and z' are continuous except for finitely many points on (a, b). To save writing, we call a curve *regular* if it is simple and piecewise smooth.

EXAMPLE 327

Let C be given by

$$x = 2t, \qquad y = -3t^2, \qquad z = t; \qquad t: 1 \to 2.$$

Here, $\mathbf{R}(t) = 2t\mathbf{i} - 3t^2\mathbf{j} + t\mathbf{k}$ sweeps out C as t varies from 1 to 2. The initial point is $(2, -3, 1)$, and the terminal point is $(4, -12, 2)$. This curve is smooth, as $\mathbf{R}'(t) = 2\mathbf{i} - 6t\mathbf{j} + \mathbf{k}$ is continuous on $(1, 2)$. It is not closed, as the terminal and initial points differ. Finally, it is simple. If $(2t_1, -3t_1^2, t_1) = (2t_2, -3t_2^2, t_2)$, then $z = t_1 = t_2$; so the curve visits each point only once. Here, C is regular.

As a second example, let H be given by

$$x = t, \qquad y = |t|, \qquad z = 0; \qquad t: -1 \to 1.$$

Here, we could eliminate t and write H as $y = |x|$, $-1 \le x \le 1$, in the xy-plane (as $z = 0$). The initial point is $(-1, 1, 0)$, and the terminal point is $(1, 1, 0)$; so H is not closed. It is simple but not smooth, as there is no tangent at $(0, 0, 0)$, which we get when $t = 0$. Nonexistence of the tangent here is due to the fact that $y = |t|$ has no derivative at $t = 0$. However, the coordinate functions x, y, and z have continuous derivatives except at $t = 0$; so H is piecewise smooth. Thus, H is regular.

As a third example, let D be given by

$$x = \cos(t), \qquad y = \sin(t), \qquad z = 1; \qquad t: 0 \to 4\pi.$$

Since $x^2 + y^2 = 1$, this is a circle of radius 1 about the origin in the plane $z = 1$. It is a closed curve, as the initial point $(1, 0, 1)$ coincides with the terminal point. While the graph of D is the same as that of C in Example 326, there is a considerable difference since D goes around the circle twice, once as $t: 0 \to 2\pi$, and once again as $t: 2\pi \to 4\pi$. Thus, D is not simple (e.g., we hit $(0, 1, 1)$ at $t = \pi/2$ and again at $t = 5\pi/2$). D is smooth, as the coordinate functions have continuous derivatives on $(0, 4\pi)$.

This example emphasizes that our curves are really more than just graphs. If you imagine yourself as a particle moving around the circle, you can see that going around twice, as with D, might expend more energy than going around just once, as with C.

We can now define the notion of the line integral of a vector field over a curve. Let

$$\mathbf{F}(x, y, z) = F_1(x, y, z)\mathbf{i} + F_2(x, y, z)\mathbf{j} + F_3(x, y, z)\mathbf{k}$$

be continuous at least on the graph of a regular curve C which is given by

$$x = x(t), \qquad y = y(t), \qquad z = z(t); \qquad t: a \to b.$$

Then, the *line integral of* $\mathbf{F}$ *over* C is denoted

$$\int_C \mathbf{F}$$

and is defined by

$$\int_C \mathbf{F} = \int_a^b \mathbf{F}(x(t), y(t), z(t)) \cdot \mathbf{R}'(t)\, dt.$$

Here, $\mathbf{R}(t) = x(t)\mathbf{i} + y(t)\mathbf{j} + z(t)\mathbf{k}$ as usual, and

$$\mathbf{F} \cdot \mathbf{R}' = F_1 x'(t) + F_2 y'(t) + F_3 z'(t).$$

Then,

$$\int_C \mathbf{F} = \int_a^b [F_1 x'(t) + F_2 y'(t) + F_3 z'(t)]\, dt.$$

In words, replace x, y, and z in F_1, F_2, and F_3 by the parametrized functions of t describing C, take the dot product with the tangent to C, and integrate the resulting function of t from a to b in the usual way.

EXAMPLE 328

Let $\mathbf{F}(x, y, z) = x^2\mathbf{i} - 2xyz\mathbf{j} + z^2\mathbf{k}$, and let C be given by

$$x = t, \qquad y = 2t, \qquad z = -4t; \qquad t: 0 \to 3.$$

Then $\mathbf{R}(t) = t\mathbf{i} + 2t\mathbf{j} - 4t\mathbf{k}$ and $\mathbf{R}' = \mathbf{i} + 2\mathbf{j} - 4\mathbf{k}$.

Replacing x by t, y by $2t$, and z by $-4t$ in $\mathbf{F}$ gives us

$$\mathbf{F} = t^2\mathbf{i} + 16t^3\mathbf{j} + 16t^2\mathbf{k}.$$

Then, $\mathbf{F} \cdot \mathbf{R}' = t^2 + 32t^3 - 64t^2 = 32t^3 - 63t^2$, and we have

$$\int_C \mathbf{F} = \int_0^3 (32t^3 - 63t^2)\, dt = 81.$$

EXAMPLE 329

Let $\mathbf{F} = 2x\mathbf{i} - e^{\sqrt{x}} \cos(z)\mathbf{j} + \mathbf{k}$, and let C be given by

$$x = t^2, \qquad y = 3t, \qquad z = 0; \qquad t: 1 \to 4.$$

Here, $\mathbf{R} = t^2\mathbf{i} + 3t\mathbf{j}$ and $\mathbf{R}' = 2t\mathbf{i} + 3\mathbf{j}$.

Putting $x = t^2$, $y = 3t$, and $z = 0$ into $\mathbf{F}$ gives us $\mathbf{F} = 2t^2\mathbf{i} - e^t\mathbf{j} + \mathbf{k}$. Then,

$$\mathbf{F} \cdot \mathbf{R}' = 4t^3 - 3e^t$$

and

$$\int_C \mathbf{F} = \int_1^4 (4t^3 - 3e^t)\, dt = 255 - 3(e^4 - e).$$

If C is regular, consisting of smooth pieces $C_1, \ldots, C_n$ (with the terminal point of C_i the same as the initial point of C_{i+1}), then we define

$$\int_C \mathbf{F} = \int_{C_1} \mathbf{F} + \cdots + \int_{C_n} \mathbf{F}.$$

EXAMPLE 330

Let $F = -\mathbf{i} + xyz\mathbf{j} - y^2\mathbf{k}$, and let C be given by

$$x = t, \qquad y = |t|, \qquad z = 1; \qquad t: -1 \rightarrow 1.$$

Then, C consists of smooth pieces C_1 and C_2, where

$$C_1: \ x = t, \qquad y = -t, \qquad z = 1; \qquad t: -1 \rightarrow 0$$

and

$$C_2: \ x = t, \qquad y = t, \qquad z = 1; \qquad t: 0 \rightarrow 1.$$

We shall work out $\int_{C_1} \mathbf{F}$ and $\int_{C_2} \mathbf{F}$ separately.

For $\int_{C_1} \mathbf{F}$: On C_1, $\mathbf{F} = -\mathbf{i} - t^2\mathbf{j} - t^2\mathbf{k}$, $\mathbf{R} = t\mathbf{i} - t\mathbf{j} + \mathbf{k}$, and $\mathbf{R}' = \mathbf{i} - \mathbf{j}$; so $\mathbf{F} \cdot \mathbf{R}' = -1 + t^2$. Then,

$$\int_{C_1} \mathbf{F} = \int_{-1}^{0} (-1 + t^2)\, dt = -\tfrac{2}{3}.$$

For $\int_{C_2} \mathbf{F}$: On C_2, $\mathbf{F} = -\mathbf{i} + t^2\mathbf{j} - t^2\mathbf{k}$, $\mathbf{R} = t\mathbf{i} + t\mathbf{j} + \mathbf{k}$, and $\mathbf{R}' = \mathbf{i} + \mathbf{j}$; so $\mathbf{F} \cdot \mathbf{R}' = -1 + t^2$. Then,

$$\int_{C_2} \mathbf{F} = \int_{0}^{1} (-1 + t^2)\, dt = -\tfrac{2}{3}.$$

Then,

$$\int_{C} \mathbf{F} = -\tfrac{2}{3} - \tfrac{2}{3} = -\tfrac{4}{3}.$$

EXAMPLE 331

Often, in working a line integral, the curve is described in words, and it is left to us to write one or more equations for it. Here is an illustration.

Evaluate $\int_C 2x\mathbf{i} - \mathbf{j} + z\mathbf{k}$ over the quarter-circle in the plane $z = 1$ from (2, 0, 1) to (0, 2, 1).

Here, we may use polar coordinates and write the curve C as

$$C: x = 2\cos(t), \qquad y = 2\sin(t), \qquad z = 1; \qquad t: 0 \rightarrow \tfrac{1}{2}\pi.$$

On C,

$$\mathbf{R} = 2\cos(t)\mathbf{i} + 2\sin(t)\mathbf{j} + \mathbf{k},$$
$$\mathbf{R}' = -2\sin(t)\mathbf{i} + 2\cos(t)\mathbf{j},$$

and

$$\mathbf{F} = 2[2\cos(t)]\mathbf{i} - \mathbf{j} + \mathbf{k}.$$

Then, on C,

$$\mathbf{F} \cdot \mathbf{R}' = -8\sin(t)\cos(t) - 2\cos(t),$$

and so

$$\int_C \mathbf{F} = \int_0^{\pi/2} [-8\sin(t)\cos(t) - 2\cos(t)]\,dt = -6.$$

There is another notation commonly used for line integrals. With $\mathbf{R} = x\mathbf{i} + y\mathbf{j} + z\mathbf{k}$, one often see $\int_C \mathbf{F}$ written as $\int_C \mathbf{F}\cdot d\mathbf{R}$.

To see how this is consistent with what we are doing, suppose, on C, $x = x(t)$, $y = y(t)$, and $z = z(t)$, for $t\colon a \to b$. Then,

$$d\mathbf{R} = dx\,\mathbf{i} + dy\,\mathbf{j} + dz\,\mathbf{k}.$$

If $\mathbf{F} = F_1\mathbf{i} + F_2\,\mathbf{j} + F_3\,\mathbf{k}$, then

$$\mathbf{F}\cdot d\mathbf{R} = F_1\,dx + F_2\,dy + F_3\,dz.$$

Then,

$$\int_a^b \mathbf{F}\cdot d\mathbf{R} = \int_a^b F_1\,dx + F_2\,dy + F_3\,dz$$

$$= \int_a^b \left(F_1\frac{dx}{dt} + F_2\frac{dy}{dt} + F_3\frac{dz}{dt}\right)dt = \int_a^b (\mathbf{F}\cdot\mathbf{R}')\,dt,$$

and this is how we defined $\int_C \mathbf{F}$.

There is also a differential notation which is useful in remembering how to evaluate line integrals. Since

$$\mathbf{F}\cdot d\mathbf{R} = F_1\,dx + F_2\,dy + F_3\,dz,$$

we often write $\int_C \mathbf{F}$ as $\int_C F_1\,dx + F_2\,dy + F_3\,dz$. In actually evaluating $\int_C F_1\,dx + F_2\,dy + F_3\,dz$, one simply substitutes $x = x(t)$, $y = y(t)$, and $z = z(t)$ from C and integrates over the parameter interval.

EXAMPLE 332

Evaluate $\int_C F_1\,dx + F_2\,dy + F_3\,dz$, where $\mathbf{F} = 2xy\mathbf{i} - y^2\mathbf{j} + e^x z\mathbf{k}$ and C is given by $C\colon x = -t,\ y = \sqrt{t},\ z = 3t;\ t\colon 1 \to 4$.

On C, $dx = -dt$, $dy = (1/2\sqrt{t})\,dt$, and $dz = 3\,dt$. Then,

$$\int_C F_1\,dx + F_2\,dy + F_3\,dz = \int_1^4 2(-t)(\sqrt{t})(-dt) - (\sqrt{t})^2\frac{1}{2\sqrt{t}}\,dt + e^{-t}(3t)(3\,dt)$$

$$= \int_1^4 (2t^{3/2} - \tfrac{1}{2}t^{1/2} + 9te^{-t})\,dt$$

$$= \tfrac{124}{5} - \tfrac{7}{3} + 18e^{-1} - 45e^{-4} \approx 28.2643.$$

EXAMPLE 333

Evaluate $\int_C F_1\,dx + F_2\,dy + F_3\,dz$, where $\mathbf{F} = 2xy\mathbf{i} + zy\mathbf{j} - e^z\mathbf{k}$ and C is the parabola $y = x^2$, $z = 0$, from $(0, 0, 0)$ to $(2, 4, 0)$ in the xy-plane.

Here we might as well use x as parameter. We have, on C,

$$dx = dx, \qquad dy = 2x\,dx, \qquad dz = 0\,dx.$$

Thus,

$$\int_C F_1\,dx + F_2\,dy + F_3\,dz = \int_0^2 [2x(x^2)\,dx + 0 \cdot x^2(2x\,dx) - e^0 \cdot 0\,dx]$$

$$= \int_0^2 2x^3\,dx = 8.$$

Just for illustration, we redo the problem using y as parameter. Write C as

$$C:\quad x = \sqrt{y}, \qquad y = y, \qquad z = 0; \qquad y: 0 \to 4.$$

On C,

$$dx = \frac{1}{2\sqrt{y}}\,dy, \qquad dy = dy \quad \text{and} \quad dz = 0\,dy.$$

Then,

$$\int_C F_1\,dx + F_2\,dy + F_3\,dz = \int_0^4 \left[2\sqrt{y}\,y\left(\frac{1}{2\sqrt{y}}\right)dy + 0 \cdot y\,dy - e^0 \cdot 0\,dy\right]$$

$$= \int_0^4 y\,dy = 8.$$

In general, two different parametrizations for C will lead to the same value for the line integral of $\mathbf{F}$ over C. In deciding which notation to use for a line integral, one often uses $\int_C \mathbf{F}$ for general discussions (as with Maxwell's equations later) and $\int_C F_1\,dx + F_2\,dy + F_3\,dz$ when we have a calculation or discussion in which the components of $\mathbf{F}$ have been written out.

Line integrals enjoy the properties one usually expects of integrals. In particular,

$$\int_C (\mathbf{F} + \mathbf{G}) = \int_C \mathbf{F} + \int_C \mathbf{G}$$

and, for any scalar α,

$$\int_C \alpha\mathbf{F} = \alpha \int_C \mathbf{F}.$$

Further, if we reverse direction on C to get a curve K (same graph, but initial and terminal points reversed), then

$$\int_C \mathbf{F} = -\int_K \mathbf{F}.$$

A line integral $\int_C \mathbf{F}$ has a physical interpretation as work done by the force $\mathbf{F}$ along C from initial to terminal point. To understand this, recall from elementary physics that $\mathbf{F} \cdot \mathbf{D}$ is by definition the work done by constant force $\mathbf{F}$ along a constant direction given by vector $\mathbf{D}$. Now, if $\mathbf{F}(x, y, z)$ is a (not necessarily constant) vector field and C is a regular curve (not necessarily a straight line), then at any point $(x(t), y(t), z(t))$ on C,

$$\mathbf{F}(x(t), y(t), z(t)) \cdot [x'(t)\mathbf{i} + y'(t)\mathbf{j} + z'(t)\mathbf{k}]$$

gives the work done by $\mathbf{F}$ at that point in the direction of C at that point (i.e., along its tangent direction). "Summing" over all the points on C is achieved by integrating $\mathbf{F} \cdot \mathbf{R}'$ over the parameter interval, and this gives us $\int_C \mathbf{F}$ as the total work done by $\mathbf{F}$ along C.

The differential notation for a line integral suggests a definition of line integrals of scalar fields which is sometimes useful. If $\phi(x, y, z)$ is a scalar field, and C is a regular curve parametrized by

$$x = x(t), \qquad y = y(t), \qquad z = z(t); \qquad t\colon a \to b,$$

then we define three kinds of line integral of ϕ:

$$\int_C \phi\, dx = \int_a^b \phi(x(t), y(t), z(t))x'(t)\, dt,$$

$$\int_C \phi\, dy = \int_a^b \phi(x(t), y(t), z(t))y'(t)\, dt,$$

and

$$\int_C \phi\, dz = \int_a^b \phi(x(t), y(t), z(t))z'(t)\, dt.$$

Finally, if s (arc length) is used to parametrize C, say,

$$x = x(s), \qquad y = y(s), \qquad z = z(s); \qquad s\colon 0 \to L,$$

then the line integral of ϕ with respect to s is defined by

$$\int_C \phi\, ds = \int_0^L \phi(x(s), y(s), z(s))\, ds.$$

If C is parametrized by t, and we want to evaluate $\int_C \phi\, ds$, we need not explicitly reparametrize C in terms of s, since

$$ds^2 = dx^2 + dy^2 + dz^2.$$

Thus, in terms of t,

$$\int_C \phi\, ds = \int_a^b \phi(x(t), y(t), z(t)) \cdot \sqrt{x'(t)^2 + y'(t)^2 + z'(t)^2}\, dt.$$

EXAMPLE 334

Let $\phi(x, y, z) = xyz^2$ and let C be given by

$$C\colon \quad x = 2t, \qquad y = t, \qquad z = \sqrt{t}; \qquad t\colon 1 \to 3.$$

Now, on C,

$$dx = 2\, dt, \qquad dy = dt, \quad \text{and} \quad dz = \frac{1}{2\sqrt{t}}\, dt.$$

Further,

$$\phi(x(t), y(t), z(t)) = (2t)(t)(\sqrt{t})^2 = 2t^3.$$

Then,

$$\int_C \phi\, dx = \int_1^3 2t^3(2\, dt) = 80,$$

$$\int_C \phi\, dy = \int_1^3 2t^3(dt) = 40,$$

$$\int_C \phi\, dz = \int_1^3 2t^3\left(\frac{1}{2\sqrt{t}}\right) dt = \int_1^3 t^{5/2}\, dt = \tfrac{2}{7}(3^{7/2} - 1).$$

Finally, in terms of the given parametrization,

$$\int_C \phi\, ds = \int_1^3 2t^3\sqrt{(2)^2 + (1)^2 + (1/2\sqrt{t})^2}\, dt = \int_1^3 t^{5/2}\sqrt{20t + 1}\, dt.$$

This cannot be done by elementary means, but can be approximated by any of the methods given in Chapter 20, Section 20.2, to yield 90.4062.

We shall pursue various interpretations of line integrals of scalar functions in some of the problems at the end of the following Problem Set. In the next section we shall develop an important relationship between double integrals and line integrals in the plane.

PROBLEMS FOR SECTION 11.6

In each of Problems 1 through 21, compute the line integral of the given vector field over the given curve.

1. $\mathbf{F} = 3x\mathbf{i} - y^2\mathbf{j} + \mathbf{k}$, $C: x = 2t,\ y = 1 - t,\ z = t^2 + 2;\ t: 1 \to 3$
2. $\mathbf{F} = \cos(xy)\mathbf{j}$, $C: x = 1,\ y = 2t - 1,\ z = t;\ t: 0 \to \pi$
3. $\mathbf{F} = -y\mathbf{i} + xy\mathbf{j} + x^2\mathbf{k}$, $C: x = \sqrt{t},\ y = 2t,\ z = t;\ t: 1 \to 4$
4. $\mathbf{F} = 8x^3\mathbf{i} + y^2\mathbf{j} - \sin(z)\mathbf{k}$, $C: x = y = t,\ z = 3t;\ t: 0 \to 4$
5. $\mathbf{F} = -\cos(x)\mathbf{i} - 2y^2\mathbf{k}$, $C: x = t,\ y = \pi,\ z = 3t^2;\ t: 1 \to 2$
6. $\mathbf{F} = 12\sqrt{x}\,\mathbf{i} + \sinh(y)\mathbf{j}$, $C: x = t^2,\ y = t,\ z = -t;\ t: -1 \to 1$
7. $\mathbf{F} = \mathbf{j} - 3x\mathbf{k}$, $C: x = t^2 + 1,\ y = -t,\ z = 1 + t;\ t: 2 \to 5$
8. $\mathbf{F} = 4\mathbf{i} - 6x\mathbf{j} + x^2\mathbf{k}$, $C: x = 3t,\ y = 1 - t,\ z = 3t^2;\ t: 1 \to 4$
9. $\mathbf{F} = \sin(x)\mathbf{i} + 2z\mathbf{j} - \mathbf{k}$, $C: x = 1,\ y = 3t^2,\ z = 4t;\ t: 0 \to 5$
10. $\mathbf{F} = 3e^x z\mathbf{i} - 4y^2\mathbf{k}$, $C: x = t,\ y = t^3,\ z = 2t;\ t: 0 \to 1$
11. $\mathbf{F} = 2\cosh(y)\mathbf{i} - 4z^2\mathbf{k}$, $C: x = e^t,\ y = 2t,\ z = \sqrt{t};\ t: 2 \to 4$
12. $\mathbf{F} = (1 - xy)\mathbf{i} + 5xz\mathbf{j}$, $C: x = 1,\ y = 3 - t,\ z = t^2;\ t: -1 \to 5$
13. $\mathbf{F} = 3\cos(x + y)\mathbf{i} - z\mathbf{j} + \mathbf{k}$, $C: x = y = 3t,\ z = t + 2;\ t: 0 \to 3$
14. $\mathbf{F} = 8e^{-x}\mathbf{i} + \cosh(z)\mathbf{j} - y^2\mathbf{k}$, $C: x = (1/\sqrt{2})t,\ y = 4t^2,\ z = 3t;\ t: 1 \to 8$
15. $\mathbf{F} = x\mathbf{i} - 4yz\mathbf{k}$, $C: x = 2t + 1,\ y = -t,\ z = 1 + t^2;\ t: 0 \to 6$
16. $\mathbf{F} = xy\mathbf{i} - yz\mathbf{j} + zx\mathbf{k}$, $C: y = \sqrt{1 - x^2},\ z = 2;\ x: 0 \to 1$
17. $\mathbf{F} = e^x\mathbf{i} - e^z\mathbf{k}$, $C: x = y = z;\ z: 1 \to 3$
18. $\mathbf{F} = 2\cos(x)\mathbf{i} + \sin(x)\mathbf{j} - y\mathbf{k}$, $C: y = 2x,\ z = 3y;\ y: 1 \to \pi$
19. $\mathbf{F} = -3xy\mathbf{i} + 2y\mathbf{k}$, C is the semicircle $x^2 + z^2 = 4$, $y = 1$, $z \geq 0$, oriented from $(2, 1, 0)$ to $(-2, 1, 0)$.

20. $\mathbf{F} = 3x\mathbf{i} - 2\mathbf{j} + z\mathbf{k}$, C consists of the straight line segment from $(1, 0, -2)$ to $(6, 1, 3)$ and then from $(6, 1, 3)$ to $(-1, -1, -4)$.

21. $\mathbf{F} = 4\mathbf{i} - 3x\mathbf{j} + z^2\mathbf{k}$, C is the line from $(1, 0, 3)$ to $(2, 1, 1)$.

In each of Problems 22 through 37, compute $\int_C \phi\, dx$, $\int_C \phi\, dy$, and $\int_C \phi\, dz$ over the given curve.

22. $\phi = x + y + z^2$, $C\colon x = 2y = \sqrt{z};\ x\colon 4 \to 8$

23. $\phi = x^2 - yz$, $C\colon x = t,\ y = \sqrt{t},\ z = \sqrt{t};\ t\colon 1 \to 4$

24. $\phi = 1 - \cosh(x + y + z)$, $C\colon x = y = z;\ z\colon 1 \to 3$

25. $\phi = xyz$, $C\colon x = y^3 = z^2;\ x\colon 1 \to 5$

26. $\phi = \cos(x) + \sin(z)$, $C\colon x = 1 + t,\ y = 1,\ z = 1 - t;\ t\colon 3 \to 4$

27. $\phi = e^{x^2} + y - e^{z^2}$, $C\colon x = \sqrt{t},\ y = t^2,\ z = \sqrt{t};\ t\colon 4 \to 16$

28. $\phi = -18xy + 6yz$, $C\colon x = t^{3/2},\ y = t^{5/2},\ z = t^{7/2};\ t\colon 1 \to 9$

29. $\phi = 3\sqrt{xyz}$, $C\colon x = t^2,\ y = t^3,\ z = t^{5/2};\ t\colon 2 \to 6$

30. $\phi = z^3 \sin(xy)$, $C\colon x = 3\sqrt{t},\ y = 2\sqrt{t},\ z = t^{2/3};\ t\colon 1 \to 4$

31. $\phi = -4x + 3yz$, C is the line segment from $(1, -2, 4)$ to $(1, 1, -6)$.

32. $\phi = xy^2 - z$, C is the line segment from $(0, 1, 0)$ to $(-2, 1, 1)$.

33. $\phi = 1 - xyz + z^3$, C consisting of the line segment from $(1, -1, 1)$ to $(0, 1, 3)$ and then to $(4, 0, 2)$.

34. $\phi = x^2 - y^2 + z^2$, C is the curve $y = x^2$, $x\colon 1 \to 3$, in the plane $z = 5$.

35. $\phi = 2\sin(x) + yz$, C is the line segment from $(1, -3, 1)$ to $(3, 2, 2)$.

36. $\phi = -x^2y + z - 4$, C is the circle $x^2 + y^2 = 4$ in the plane $z = 3$, oriented counterclockwise.

37. $\phi = 1 - 4yz$, C is the line segment from $(-1, 3, 3)$ to $(3, 3, 4)$.

In each of Problems 38 through 43, calculate the work done by $\mathbf{F}$ along C.

38. $\mathbf{F} = 3xy\mathbf{i} - 2\mathbf{j}$, C is the piece of the hyperbola $x^2 - y^2 = 1$, $z = 0$, from $(1, 0, 0)$ to $(2, \sqrt{3}, 0)$.

39. $\mathbf{F} = x^3\mathbf{i} - z\mathbf{j} + 2xy\mathbf{k}$, $C\colon x = t^2,\ y = z = \sqrt{t};\ t\colon 2 \to 4$

40. $\mathbf{F} = 2\mathbf{i}$, $C\colon x = 3t^2,\ y = \sin(t),\ z = t^3;\ t\colon 0 \to \pi$

41. $\mathbf{F} = x^3\mathbf{j} - yz\mathbf{k}$, $C\colon x = 1 - t,\ y = 1 + t,\ z = 2t;\ t\colon 1 \to 3$

42. $\mathbf{F} = -8z\mathbf{j} + \cosh(x)\mathbf{k}$, $C\colon x = t,\ y = 2t^2,\ z = e^t;\ t\colon 0 \to 5$

43. $\mathbf{F} = 16xz\mathbf{i} + y\mathbf{j}$, $C\colon x = \cos(t),\ y = \sin(t),\ z = 2t;\ t\colon 0 \to 2\pi$

In each of Problems 44 through 52, evaluate $\int_C \phi\, ds$.

44. $\phi = 4xy$, $C\colon x = y = t,\ z = 2t;\ t\colon 1 \to 2$

45. $\phi = 4\cosh(t)$, $C\colon x = y = 1,\ z = 3t^2;\ t\colon 0 \to 4$

46. $\phi = x + y$, $C\colon x = y = z = t^2;\ t\colon 0 \to 2$

47. $\phi = \sin(x) - xyz$, $C\colon x = t,\ y = 2t,\ z = 3t;\ t\colon 1 \to 5$

48. $\phi = 3xy + z$, $C\colon x = \sin(t),\ y = \cos(t),\ z = 4;\ t\colon 0 \to \pi/2$

49. $\phi = xz - y^2$, $C\colon x = y = z = 1 + t^2;\ t\colon 0 \to 1$

50. $\phi = 3y^3$, $C\colon x = z = t^2,\ y = 1;\ t\colon 1 \to 6$

51. $\phi = x - y + 3z$, $C\colon x = 3\cos(t),\ y = 1,\ z = 3\sin(t);\ t\colon 0 \to \pi/4$

52. $\phi = xz^3 + 2$, $C\colon x = 1,\ y = 2 + t,\ z = 3 - t;\ t\colon 1 \to 5$

53. Show that any ordinary Riemann integral $\int_a^b f(x)\, dx$ is a line integral $\int_C \mathbf{F}$ for appropriate choice of $\mathbf{F}$ and C.

54. Suppose a thin wire is in the shape of a curve $C\colon x = x(s),\ y = y(s),\ z = z(s);\ s\colon 0 \to L$. Let $\rho(x, y, z)$ be the density of the wire at point (x, y, z).

(a) Give a plausibility argument that the mass of the wire is $\int_C \rho\, ds$.

(b) Give an argument supporting the fact that the center of mass is at $(\bar{x}, \bar{y}, \bar{z})$, where

$$\bar{x} = \frac{\int_C x\rho\, ds}{\int_C \rho\, ds}, \qquad \bar{y} = \frac{\int_C y\rho\, ds}{\int_C \rho\, ds}, \qquad \bar{z} = \frac{\int_C z\rho\, ds}{\int_C \rho\, ds}.$$

11.7 GREEN'S THEOREM

In this section we confine our attention to vector fields in the xy-plane and relate line integrals over closed curves to double integrals over the enclosed region.

For this section, let C be a regular, closed curve in the plane (so $z = 0$ on C). We arbitrarily choose a direction, counterclockwise, and call it positive. Thus, if $(x(t), y(t))$ moves in a counterclockwise direction about C as t varies from a to b, we call C *positively oriented*; otherwise, C is *negatively oriented*. For example,

$$x = \cos(t), \qquad y = \sin(t); \qquad t: 0 \to 2\pi$$

is positively oriented, while

$$x = \cos(t), \qquad y = -\sin(t); \qquad t: 0 \to 2\pi$$

is negatively oriented.

Usually we indicate orientation by putting an arrow on the curve, counterclockwise for positive orientation (as in Figure 176) and clockwise for negative orientation (as in Figure 177).

A simple closed curve C in the plane encloses a region D; if you walk around C in the positive direction, D is over your left shoulder. For technical reasons, we include the points of C in D.

Often, when C is a positively oriented closed curve, we write $\int_C$ as $\oint_C$. This is simply a common notation.

We can now state Green's theorem.

THEOREM 89

Let C be a regular, closed, positively oriented curve in the plane, enclosing a region D. Let $\mathbf{F}(x, y) = F_1(x, y)\mathbf{i} + F_2(x, y)\mathbf{j}$, and assume that F_1, F_2, $\partial F_1/\partial y$, and $\partial F_2/\partial x$ are

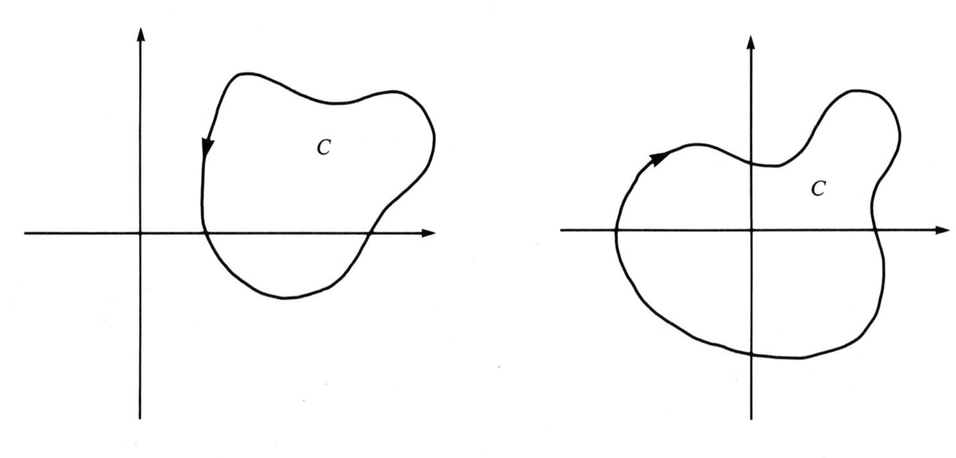

Figure 176. Positively oriented closed curve in the plane.

Figure 177. Negatively oriented closed curve in the plane.

continuous on D. Then,

$$\int_C F_1\,dx + F_2\,dy = \iint_D \left(\frac{\partial F_2}{\partial x} - \frac{\partial F_1}{\partial y}\right) dx\,dy.$$

In nondifferential notation, Green's formula reads:

$$\oint_C \mathbf{F} = \iint_D \left(\frac{\partial F_2}{\partial x} - \frac{\partial F_1}{\partial y}\right) dx\,dy.$$

A proof of this in general depends on delicate properties of curves which are beyond our intent here. We can give a proof when C has a special kind of description.

Proof for a Special Case Assume that C can be broken up into two curves, C_1 and C_2, where C_2 is parametrized by

$$y = f_2(x); \qquad x: a \to b$$

and C_1 is parametrized by

$$y = f_1(x); \qquad x: b \to a$$

as shown in Figure 178.

Assume that C can also be split up into C_3 and C_4, where

$$C_3: \quad x = g_2(y); \qquad y: c \to d,$$
$$C_4: \quad x = g_1(y); \qquad y: d \to c,$$

as shown in Figure 179.

We can now verify Green's theorem by evaluating both sides to obtain the same expression, using the assumed parametrizations for C.

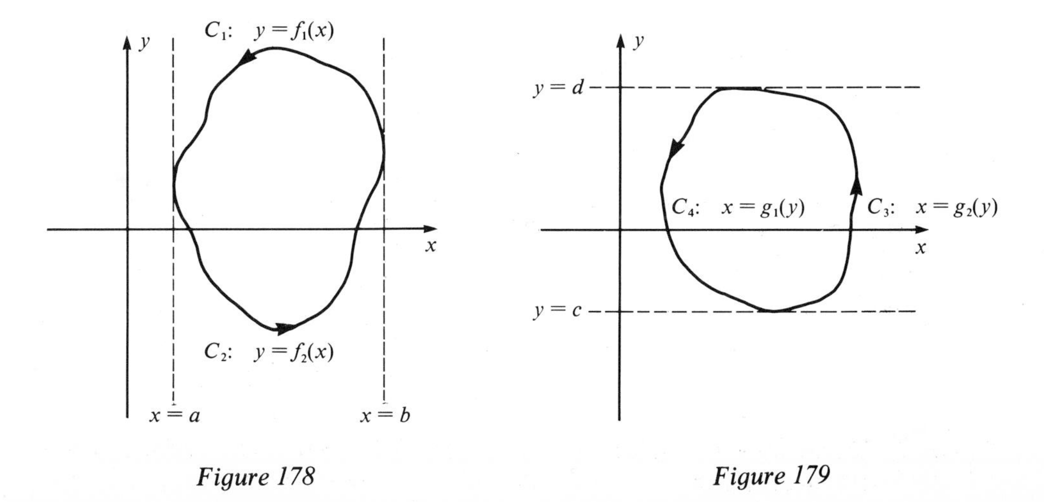

Figure 178 *Figure 179*

First, use x as parameter on C_1 and C_2 to calculate

$$\begin{aligned}\int_C F_1\,dx &= \int_{C_1} F_1(x, y)\,dx + \int_{C_2} F_1(x, y)\,dx\\ &= \int_b^a F_1(x, f_1(x))\,dx + \int_a^b F_1(x, f_2(x))\,dx\\ &= \int_a^b [F_1(x, f_2(x)) - F_1(x, f_1(x))]\,dx\end{aligned}$$

and

$$\begin{aligned}\iint_D -\frac{\partial F_1}{\partial y}\,dx\,dy &= \int_a^b \int_{f_2(x)}^{f_1(x)} -\frac{\partial F_1}{\partial y}\,dy\,dx\\ &= \int_a^b \left(\int_{f_2(x)}^{f_1(x)} -\frac{\partial F_1}{\partial y}\,dy\right)dx\\ &= \int_a^b [-F_1(x, f_1(x)) + F_1(x, f_2(x))]\,dx = \int_C F_1\,dx.\end{aligned}$$

Next, using y as parameter on C_3 and C_4, we get

$$\begin{aligned}\int_C F_2\,dy &= \int_{C_3} F_2\,dy + \int_{C_4} F_2\,dy\\ &= \int_c^d F_2(g_2(y), y)\,dy + \int_d^c F_2(g_1(y), y)\,dy\\ &= \int_c^d [F_2(g_2(y), y) - F_2(g_1(y), y)]\,dy\end{aligned}$$

and

$$\begin{aligned}\iint_D \frac{\partial F_2}{\partial x}\,dx\,dy &= \int_c^d \left(\int_{g_1(y)}^{g_2(y)} \frac{\partial F_2}{\partial x}\,dx\right)dy\\ &= \int_c^d [F_2(g_2(y), y) - F_2(g_1(y), y)]\,dy = \int_C F_2\,dy.\end{aligned}$$

Then,

$$\int_C F_1\,dx + F_2\,dy = \iint_D \left(\frac{\partial F_2}{\partial x} - \frac{\partial F_1}{\partial y}\right)dx\,dy,$$

as we wanted to show.

Here are two purely computational examples.

EXAMPLE 335

Let C be the unit circle

$$x = \cos(t), \qquad y = \sin(t); \qquad t\colon 0 \to 2\pi.$$

Let $\mathbf{F}(x, y) = x^2y\mathbf{i} + (x + y)\mathbf{j}$. Then, $F_1 = x^2y$ and $F_2 = x + y$.

Now, on C, $dx = -\sin(t)\,dt$ and $dy = \cos(t)\,dt$. Then,

$$\int_C F_1\,dx + F_2\,dy = \int_0^{2\pi} \{\cos^2(t)\sin(t)[-\sin(t)] + [\cos(t) + \sin(t)]\cos(t)\}\,dt$$

$$= \int_0^{2\pi} [-\cos^2(t)\sin^2(t) + \cos^2(t) + \sin(t)\cos(t)]\,dt$$

$$= \frac{3\pi}{4}.$$

Next, evaluate the double integral. Using polar coordinates $x = r\cos(\theta)$, $y = r\sin(\theta)$, where $0 \le \theta \le 2\pi$ and $0 \le r \le 1$, to describe D, we have:

$$\iint_D \left(\frac{\partial F_2}{\partial x} - \frac{\partial F_1}{\partial y}\right) dx\,dy = \iint_D (1 - x^2)\,dx\,dy$$

$$= \int_0^{2\pi}\int_0^1 [1 - r^2\cos^2(\theta)]r\,dr\,d\theta$$

$$= \int_0^{2\pi}\left[\frac{1}{2} - \frac{1}{4}\cos^2(\theta)\right] d\theta = \frac{3\pi}{4}.$$

EXAMPLE 336

Let $\mathbf{F} = y\mathbf{i} + 2xy\mathbf{j}$. Let C be the regular closed curve consisting of the line segments from $(0, 0)$ to $(1, 0)$, $(1, 0)$ to $(1, 1)$, and $(1, 1)$ to $(0, 0)$, as shown in Figure 180. Here, C consists of three smooth pieces, labeled C_1, C_2, and C_3. We may parametrize

$$C_1\colon\ x = x, \quad y = 0; \quad x\colon 0 \to 1,$$
$$C_2\colon\ x = 1, \quad y = y; \quad y\colon 0 \to 1,$$
$$C_3\colon\ x = y; \quad x\colon 1 \to 0$$

(we set the last condition, $x\colon 1 \to 0$, to keep the counterclockwise orientation).

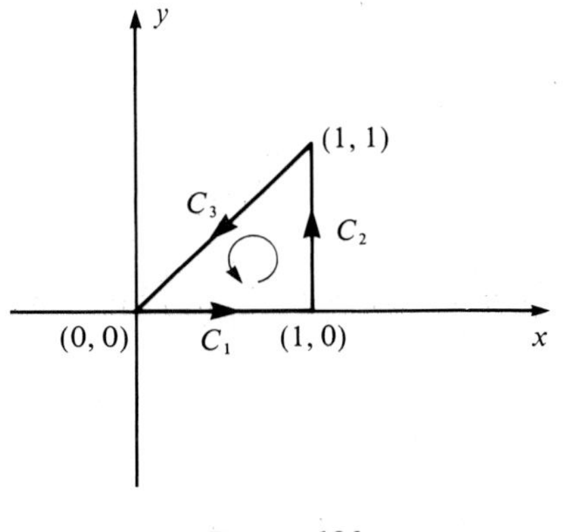

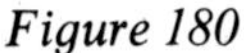

Figure 180

Then,

$$\int_{C_1} F_1\,dx + F_2\,dy = \int_0^1 0\,dx + 0\,d(0) = 0,$$

$$\int_{C_2} F_1\,dx + F_2\,dy = \int_0^1 y\,d(1) + 2y\,dy = \int_0^1 2y\,dy = 1,$$

and

$$\int_{C_3} F_1\,dx + F_2\,dy = \int_1^0 x\,dx + 2x(x)\,dx = \int_1^0 (x + 2x^2)\,dx = -\tfrac{7}{6}.$$

Then,

$$\int_C F_1\,dx + F_2\,dy = 1 - \tfrac{7}{6} = -\tfrac{1}{6}.$$

Next,

$$\iint_D \left(\frac{\partial F_2}{\partial x} - \frac{\partial F_1}{\partial y}\right) dx\,dy = \int_0^1 \int_0^x (2y - 1)\,dy\,dx = \int_0^1 (x^2 - x)\,dx = -\tfrac{1}{6}.$$

Obviously, Green's theorem would not be of much value if we simply went on computing both sides of the equation in special instances. In the next section we shall use it in discussing potential theory in the plane.

PROBLEMS FOR SECTION 11.7

In each of Problems 1 through 20, verify Green's theorem for the given **F** and C, as in Examples 335 and 336. All curves are oriented counterclockwise.

1. $\mathbf{F} = 2x\mathbf{i} - yx\mathbf{j}$, C is the circle $x^2 + y^2 = 9$.
2. $\mathbf{F} = y^2\mathbf{i} - \mathbf{j}$, C is the triangle from $(1, 0)$ to $(2, 0)$ to $(2, 4)$ to $(1, 0)$.
3. $\mathbf{F} = 8xy\mathbf{i} - 2xy\mathbf{j}$, C is the square with vertices at $(0, 0)$, $(1, 0)$, $(1, 1)$, and $(0, 1)$.
4. $\mathbf{F} = \cos(x)\mathbf{i} - \sin(x)\mathbf{j}$, C is the triangle with vertices at $(0, 0)$, $(2, 0)$, and $(1, 1)$.
5. $\mathbf{F} = ye^x\mathbf{i} - 2xy^2\mathbf{j}$, C is the square with vertices at $(1, 1)$, $(-1, 1)$, $(-1, -1)$, and $(1, -1)$.
6. $\mathbf{F} = -3y\mathbf{i} - 2x\mathbf{j}$, C consists of $y = x^3$ from $(0, 0)$ to $(1, 1)$ and the line segment from $(1, 1)$ to $(0, 0)$.
7. $\mathbf{F} = 8xy\mathbf{i}$, C is the ellipse $\frac{1}{8}x^2 + \frac{1}{16}y^2 = 1$.
8. $\mathbf{F} = 3y\mathbf{i} - 2xy\mathbf{j}$, C is the circle $(x - 3)^2 + (y - 2)^2 = 16$.
9. $\mathbf{F} = -16(y^2\mathbf{i} + x^2\mathbf{j})$, C is the square with vertices at $(1, 1)$, $(1, 2)$, $(2, 1)$, and $(2, 2)$.
10. $\mathbf{F} = 3e^x y\mathbf{j}$, C is the triangle with vertices at $(1, 1)$, $(2, 3)$, and $(1, 6)$.
11. $\mathbf{F} = 3y\mathbf{i}$, $C\colon x = \cos(2t)$, $y = \sin(2t)$; $t\colon 0 \to \pi$
12. $\mathbf{F} = x\mathbf{i} - 2y\mathbf{j}$, $C\colon x = \cos(t)/2$, $y = \sin(t)/2$; $t\colon 0 \to 2\pi$
13. $\mathbf{F} = xy\mathbf{i} + 2x\mathbf{j}$, C is the square with vertices at $(0, 0)$, $(1, 0)$, $(1, 1)$, and $(0, 1)$.
14. $\mathbf{F} = e^y\mathbf{i} - \cos(x)\mathbf{j}$, C is the square with vertices at $(0, 0)$, $(2, 2)$, $(2, 0)$, and $(0, 2)$.
15. $\mathbf{F} = 4\cos(y)\mathbf{i} - x^2\mathbf{j}$, C is the triangle with vertices at $(0, 0)$, $(1, 0)$, and $(1, 1)$.
16. $\mathbf{F} = xy^3\mathbf{i} - 3y\mathbf{j}$, C is the triangle with vertices at $(0, 0)$, $(3, 0)$, and $(3, 4)$.
17. $\mathbf{F} = 8y\mathbf{i} + xy\mathbf{j}$, C is the curve $\frac{1}{2}x^2 + \frac{1}{4}y^2 = 1$ (traversed once counterclockwise).
18. $\mathbf{F} = 4x^2\mathbf{i} - 3x^2y\mathbf{j}$, C is the circle $(x - 1)^2 + y^2 = 16$.
19. $\mathbf{F} = (x - y)\mathbf{i} + (3x - 2y)\mathbf{j}$, C is the circle $x^2 + y^2 = 4$.
20. $\mathbf{F} = e^y\mathbf{i} + 4x^2\mathbf{j}$, C is the triangle with vertices at $(0, 0)$, $(0, 4)$, and $(2, 0)$.
21. Let C be a regular closed curve bounding a region D in the xy-plane. Show that

$$\text{area of } D = \tfrac{1}{2}\int_C -y\,dx + x\,dy.$$

22. Let C be a regular simple closed curve in the xy-plane, enclosing a region D. At any point on C, let $\mathbf{T}$ be the unit tangent in the positive direction and $\mathbf{N}$ the unit outer normal (that is, $\mathbf{N}$ is orthogonal to $\mathbf{T}$ and points to the right as you traverse C counterclockwise, as shown in Figure 181). Let $\mathbf{F}$ satisfy the hypotheses of Green's theorem on D.

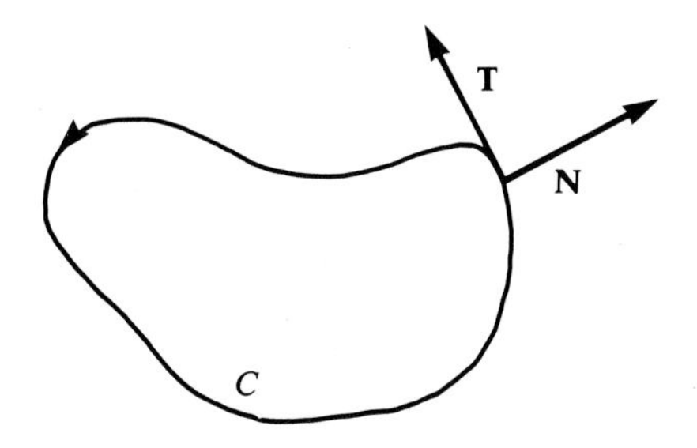

Figure 181. Tangent and outer normal to C.

(a) Prove that Green's theorem can be written

$$\iint_D (\nabla \cdot \mathbf{F})\, dx\, dy = \int_C \mathbf{F} \cdot \mathbf{N}\, ds.$$

(b) Prove that Green's theorem can also be written

$$\iint_D (\nabla \times \mathbf{F}) \cdot \mathbf{k}\, dx\, dy = \int_C \mathbf{F} \cdot \mathbf{T}\, ds.$$

23. (a) Derive the formula

$$\iint_D [f(\nabla \cdot \nabla g) + \nabla f \cdot \nabla g]\, dx\, dy = \int_C f \frac{\partial g}{\partial n}\, ds.$$

Here, D is a region in the xy-plane bounded by a regular curve C, $\partial g/\partial n$ is the directional derivative of g in the direction of the outer normal to D on C, and ∇ is the two-dimensional del operator

$$\nabla = \frac{\partial}{\partial x}\mathbf{i} + \frac{\partial}{\partial y}\mathbf{j}.$$

(b) From (a), show that

$$\iint_D [f(\nabla \cdot \nabla g) - g(\nabla \cdot \nabla f)]\, dx\, dy = \int_C \left(f \frac{\partial g}{\partial n} - g \frac{\partial f}{\partial n}\right) ds.$$

(c) Show that

$$\int_C \frac{\partial g}{\partial n}\, ds = 0 \quad \text{if} \quad \nabla \cdot \nabla g = 0.$$

(d) Prove that

$$\int_C f \frac{\partial g}{\partial n}\, ds = \int_C g \frac{\partial f}{\partial n}\, ds \quad \text{if} \quad \nabla \cdot \nabla f = \nabla \cdot \nabla g \text{ on } D.$$

The formulas of (a) and (b) are called *Green's identities.* They are used in treating the classical partial differential equations of physics and engineering, as well as in other contexts such as theoretical mechanics.

24. Let

$$\mathbf{F} = \frac{-y}{x^2 + y^2}\,\mathbf{i} + \frac{x}{x^2 + y^2}\,\mathbf{j}.$$

Let C be the unit circle $x = \cos(\theta)$, $y = \sin(\theta)$; θ: $0 \to 2\pi$, and let D be the enclosed region. Compute

$$\int_C \mathbf{F} \quad \text{and} \quad \iint_D \left(\frac{\partial F_2}{\partial x} - \frac{\partial F_1}{\partial y}\right) dx\, dy.$$

How do you explain the result?

11.8 POTENTIAL THEORY IN THE PLANE

A scalar function ϕ is called a *potential* for the vector field $\mathbf{F}$ if $\mathbf{F} = \nabla\phi$. (In physics, one often writes $\mathbf{F} = -\nabla\phi$.)

Potential functions have many ramifications—for example, thermodynamics has been described as a search for potentials. As a specific illustration,

$$\phi = \frac{gm}{r},$$

where $r = \sqrt{x^2 + y^2 + z^2}$ and g is the (constant) acceleration due to gravity, is called the *gravitational potential.* Here,

$$\mathbf{F} = \nabla\phi = -\frac{gm}{r^3}(x\mathbf{i} + y\mathbf{j} + z\mathbf{k}) = \frac{-gm}{\|\mathbf{R}\|^3}\mathbf{R}$$

is the gravitational force acting on a mass m at point (x, y, z) by a unit mass at the origin. The minus sign indicates that the force is attractive, and the magnitude is $\|\mathbf{F}\| = gm/r^2$, Newton's famous inverse square law.

In this section we shall look at potentials in the plane and how they are related to Green's theorem.

Suppose first that $\mathbf{F}(x, y) = F_1(x, y)\mathbf{i} + F_2(x, y)\mathbf{j}$ is defined in a region Ω of the plane and that $\mathbf{F} = \nabla\phi$ in Ω. We want to see the effect this has on $\int_C F_1\, dx + F_2\, dy$ for curves in Ω.

Parametrize C by $x = x(t)$, $y = y(t)$; t: $a \to b$. The initial point of C is $P_0 = (x(a), y(a))$ and the terminal point is $P_1 = (x(b), y(b))$. Now,

$$\int_C F_1\, dx + F_2\, dy = \int_a^b \left(F_1 \frac{dx}{dt} + F_2 \frac{dy}{dt}\right) dt.$$

But, from the chain rule for differentation (which we assume applies to ϕ),

$$\frac{d}{dt}\phi(x(t), y(t)) = \frac{\partial\phi}{\partial x}\frac{dx}{dt} + \frac{\partial\phi}{\partial y}\frac{dy}{dt} = F_1 \frac{dx}{dt} + F_2 \frac{dy}{dt}.$$

Thus,

$$\int_C F_1\,dx + F_2\,dy = \int_a^b \frac{d}{dt}\phi(x(t), y(t))\,dt$$
$$= \phi(x(b), y(b)) - \phi(x(a), y(a)) = \phi(P_1) - \phi(P_0).$$

This result depends *only on the endpoints of C, not on C itself.* If K were another regular curve in Ω from P_0 to P_1, we would have $\int_K F_1\,dx + F_2\,dy = \phi(P_1) - \phi(P_0)$ also.

We call the line integral of $\mathbf{F}$ *independent of path in* Ω if $\int_C \mathbf{F}$ depends only on the endpoints of C for any regular curve in Ω. We have therefore proved the following:

> If $\mathbf{F} = \nabla\phi$ in Ω, then the line integral of $\mathbf{F}$ is independent of path in Ω, and, in fact, $\int_C \mathbf{F}$ is the difference in ϕ evaluated at the endpoints of C.

In terms of work, when $\mathbf{F} = \nabla\phi$, we may say:

The work done by $\mathbf{F}$ along C equals the difference in potential (that is, the potential drop) at the ends of C.

We call $\mathbf{F}$ *conservative* over the region Ω if the work done around any closed curve is zero. If $\mathbf{F} = \nabla\phi$, then the potential drop around any closed curve is zero. Hence:

> If $\mathbf{F} = \nabla\phi$ in Ω, then $\mathbf{F}$ is conservative in Ω.

EXAMPLE 337

Let

$$\mathbf{F} = \frac{x}{x^2 + y^2}\,\mathbf{i} + \frac{y}{x^2 + y^2}\,\mathbf{j}.$$

Here, we can take Ω as the plane with the origin excluded, as $\mathbf{F}$ is not defined when $x = y = 0$. Observe that $\mathbf{F} = \nabla\phi$, where $\phi = \frac{1}{2}\ln(x^2 + y^2)$. As an illustration, let us test the results derived above.

First, look at independence of path. Choose two curves, say, from (1, 0) to (1, 1). Define

$$C_1:\quad x = 1, \qquad y = y; \qquad y: 0 \to 1$$

and

$$C_2:\quad x = 1 + \tfrac{1}{2}\cos(\theta), \quad y = \tfrac{1}{2} + \tfrac{1}{2}\sin(\theta); \qquad \theta: -\tfrac{1}{2}\pi \to \tfrac{1}{2}\pi.$$

These are shown in Figure 182. Compare

$$\int_{C_1} F_1\,dx + F_2\,dy = \int_0^1 \frac{y}{1 + y^2}\,dy = \left[\frac{1}{2}\ln(1 + y^2)\right]_0^1 = \frac{1}{2}\ln(2)$$

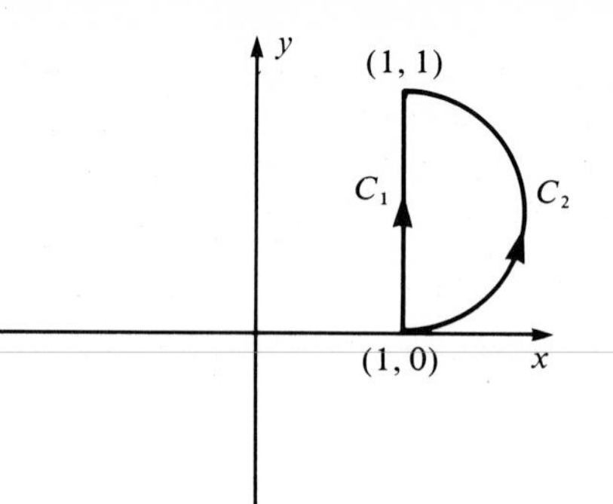

Figure 182. Two paths from (1, 0) to (1, 1).

and

$$\begin{aligned}\int_{C_2} F_1\,dx + F_2\,dy &= \int_{-\pi/2}^{\pi/2} \left(\frac{1+\frac{1}{2}\cos(\theta)}{[1+\frac{1}{2}\cos(\theta)]^2+[\frac{1}{2}+\frac{1}{2}\sin(\theta)]^2}\right)[-\tfrac{1}{2}\sin(\theta)\,d\theta] \\ &\quad + \int_{-\pi/2}^{\pi/2} \left(\frac{\frac{1}{2}+\frac{1}{2}\sin(\theta)}{[1+\frac{1}{2}\cos(\theta)]^2+[\frac{1}{2}+\frac{1}{2}\sin(\theta)]^2}\right)[\tfrac{1}{2}\cos(\theta)\,d\theta] \\ &= \int_{-\pi/2}^{\pi/2} \left(\frac{\frac{1}{4}\cos(\theta)-\frac{1}{2}\sin(\theta)}{\frac{3}{2}+\cos(\theta)+\frac{1}{2}\sin(\theta)}\right) d\theta \\ &= [\tfrac{1}{2}\ln[\tfrac{3}{2}+\cos(\theta)+\tfrac{1}{2}\sin(\theta)]]_{-\pi/2}^{\pi/2} = \tfrac{1}{2}\ln(2).\end{aligned}$$

This does not prove that the line integral of $\mathbf{F}$ is independent of path in Ω, but it does give a demonstration for two particular paths.

Observe that $\frac{1}{2}\ln(2)$ is exactly $\phi(1, 1) - \phi(1, 0)$, the potential drop between the endpoints of C_1 (and C_2).

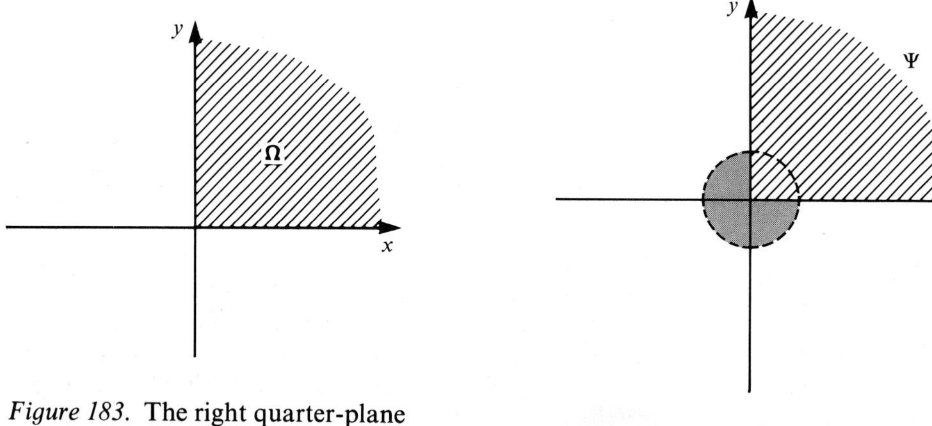

Figure 183. The right quarter-plane $x \geq 0$, $y \geq 0$, including the non-negative x- and y-axes.

Figure 184. Circles around (0, 0) extend outside of Ω.

Thus far, existence of a potential implies independence of path and a conservative vector field. The converse of this is not true in general, but it will be seen to hold with two additional assumptions about the region in which we are operating. We call a region Ω in the plane a *domain* if the following two conditions are satisfied:

1. About every point in Ω, we can draw a circle containing only points of Ω.
2. Between every pair of points in Ω, there is a regular curve lying completely in Ω.

EXAMPLE 338

(i) Let Ω consist of all points (x, y) with $x \geq 0$ and $y \geq 0$, as shown in Figure 183. Here, Ω satisfies (2) but not (1). For example, any circle you draw about (0, 0) contains points outside of Ω (the darkened area in Figure 184).

(ii) Let Ω consist of all points inside the circles $x^2 + y^2 = 1$ and $(x - 5)^2 + (y - 5)^2 = 1$. Then (1) holds but (2) fails. We cannot, for example, draw a regular curve from (0, 0) to (5, 5), both inside Ω, without passing outside the circles; hence, outside of Ω. This is shown in Figure 185.

(iii) The whole plane is a domain, as is the inside of the circle $x^2 + y^2 = 1$ (i.e., all points (x, y) with $x^2 + y^2 < 1$) and the right half-plane consisting of all (x, y) with $x > 0$.

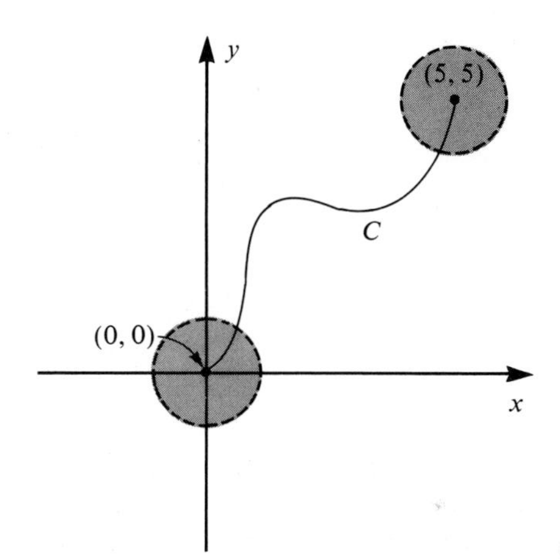

Figure 185. Curve C from (0, 0) to (5, 5) must pass outside Ω.

THEOREM 90

Let $\mathbf{F}$ be a continuous vector field defined over a domain Ω in the plane. Then, the following are equivalent:

1. $\mathbf{F}$ has a potential function on Ω.
2. $\int_C \mathbf{F} = 0$ for every regular closed curve in Ω.
3. The line integral of $\mathbf{F}$ is independent of path in Ω.

A proof of this theorem is outlined in Problem 32 at the end of this section [we have already shown that (1) implies both (2) and (3)].

This theorem is interesting, but it does not provide a practical test for existence of a potential, as each of the statements (1), (2), and (3) is difficult to verify directly. With one more condition on Ω, Green's theorem provides a simple test. We call a domain Ω *simply connected* if every regular closed curve in Ω encloses only points of Ω. For example, the plane with the origin removed is a domain but is not simply connected, as the unit circle about the origin encloses the "hole" formed by deleting the origin. Roughly speaking, a domain is simply connected if it has no "holes" in it. Figures 186 and 187 illustrate the contrast between simply connected and nonsimply connected.

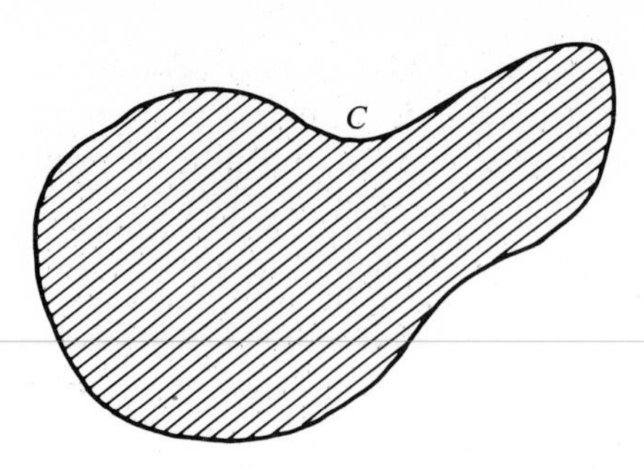

Figure 186. Region enclosed by C is simply connected.

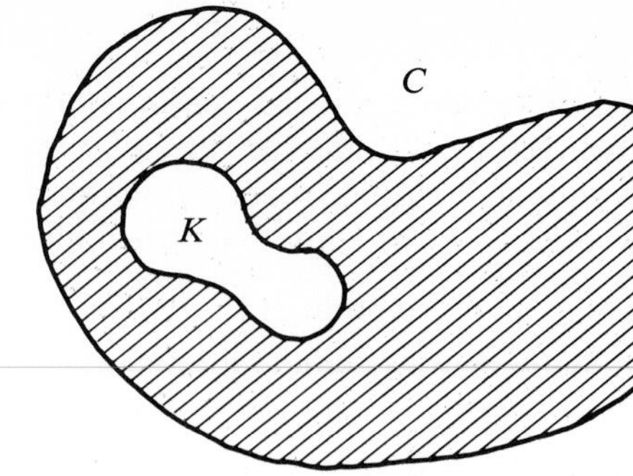

Figure 187. Region between C and K is not simply connected.

THEOREM 91

Let $\mathbf{F}$ be a vector field satisfying the conditions of Green's theorem on a simply connected domain Ω. Then, $\mathbf{F}$ has a potential function on Ω if and only if

$$\frac{\partial F_2}{\partial x} = \frac{\partial F_1}{\partial y}.$$

Proof If $\mathbf{F} = \nabla\phi$, then $F_1 = \partial\phi/\partial x$ and $F_2 = \partial\phi/\partial y$; so

$$\frac{\partial F_2}{\partial x} = \frac{\partial^2\phi}{\partial x\,\partial y} = \frac{\partial^2\phi}{\partial y\,\partial x} = \frac{\partial F_1}{\partial y}.$$

Conversely, suppose that $\partial F_2/\partial x = \partial F_1/\partial y$ on Ω. Let C be any simple regular closed curve in Ω, enclosing a region D. Since Ω is simply connected, Green's theorem applies to C and the region D it bounds. Then,

$$\int_C F_1\,dx + F_2\,dy = \iint_D \left(\frac{\partial F_2}{\partial x} - \frac{\partial F_1}{\partial y}\right) dx\,dy = 0.$$

Hence, $\mathbf{F}$ is conservative; so it has a potential on Ω, by the previous theorem.

If $\partial F_2/\partial x = \partial F_1/\partial y$ on Ω, but Ω is not simply connected, the above theorem's conclusion fails. An example is given in Problem 33 at the end of this section.

In practice, one uses the last theorem to see if a potential exists; if one does, it can be found by integration. Here are three examples.

EXAMPLE 339

Let $\mathbf{F} = (1 + 4x - y)\mathbf{i} + (e^y - 2y - x)\mathbf{j}$.

Here, F_1 and F_2 are defined everywhere; so we can let Ω be the entire xy-plane, certainly a simply connected domain. Now,

$$\frac{\partial F_2}{\partial x} = -1 = \frac{\partial F_1}{\partial y};$$

so $\mathbf{F}$ has a potential function. How can we produce one?

We want ϕ such that $\partial\phi/\partial x = F_1$ and $\partial\phi/\partial y = F_2$. Begin with either one of these, say,

$$\frac{\partial \phi}{\partial x} = 1 + 4x - y.$$

To reverse this partial derivative, integrate with respect to x, holding y fixed. We get

$$\phi(x, y) = x + 2x^2 - yx + f(y),$$

where $f(y)$ is the "constant" of integration. It may depend on y because y was treated as a constant in the integration.

Now we need

$$\frac{\partial \phi}{\partial y} = F_2 = e^y - 2y - x = \frac{\partial}{\partial y}[x + 2x^2 - yx + f(y)] = -x + f'(y).$$

Then we need

$$f'(y) = e^y - 2y;$$

so

$$f(y) = e^y - y^2 + K,$$

where K is any constant.

Thus, we may choose

$$\phi = x + 2x^2 - yx + e^y - y^2 + K,$$

for any scalar K. It is easy to check that indeed $\mathbf{F} = \nabla\phi$.

EXAMPLE 340

Let

$$\mathbf{F} = \frac{-y}{x^2 + y^2}\mathbf{i} + \frac{x}{x^2 + y^2}\mathbf{j}.$$

Here, we must be careful to work in a simply connected domain not containing the origin, as $\mathbf{F}$ is not defined if $x = y = 0$. We can, for example, let Ω be the right half-plane consisting of all (x, y) with $x > 0$ (other choices are possible). Here,

$$F_1 = \frac{-y}{x^2 + y^2} \quad \text{and} \quad F_2 = \frac{x}{x^2 + y^2},$$

and it is easy to check that $\partial F_2/\partial x = \partial F_1/\partial y$.

We now want ϕ such that $\mathbf{F} = \nabla\phi$. First, we need

$$\frac{\partial \phi}{\partial y} = F_2 = \frac{x}{x^2 + y^2}.$$

(We could have started with $\partial\phi/\partial x = F_1$.)

Integrate with respect to y, holding x fixed:

$$\phi = \arctan\left(\frac{y}{x}\right) + f(x).$$

Next,

$$\frac{\partial \phi}{\partial x} = \frac{-y}{x^2 + y^2} + f'(x) = F_1 = \frac{-y}{x^2 + y^2}.$$

Thus, we need $f'(x) = 0$; so $f(x) = K$, where K is any scalar. Then, $\phi = \arctan(y/x) + K$, for any K, will give us a potential for **F**.

EXAMPLE 341

Let $\mathbf{F} = 8xy\mathbf{i} - 16x^3\mathbf{j}$, with Ω the entire xy-plane.

Here,

$$\frac{\partial F_2}{\partial x} = -48x^2 \quad \text{and} \quad \frac{\partial F_1}{\partial y} = 8x,$$

and these are in general unequal. Hence, **F** has no potential.

What would happen if we tried to produce ϕ by integrating as in the previous two examples? Try and see. Begin with, say, $\partial\phi/\partial x = F_1 = 8xy$, and integrate with respect to x. We get

$$\phi = 4x^2y + f(y).$$

This would give us

$$\frac{\partial \phi}{\partial y} = 4x^2 + f'(y) = F_2 = -16x^3,$$

and we would need

$$f'(y) = -16x^3 - 4x^2,$$

an impossibility, as $f(y)$ was a "constant of integration" with respect to x and hence cannot involve x. On these grounds, we again conclude that **F** has no potential function.

Note that, if ϕ is a potential function for **F**, then so is $\phi + K$ for any constant K. This is consistent with the interpretation of $\int_C \mathbf{F}$ as work done by **F** along C, since the work done equals the potential *difference* at the ends of C, and this difference is the same for $\phi + K$ as it is for ϕ.

In the next section we introduce the notion of a surface integral in preparation for the theorems of Gauss and Stokes. Before going on, however, the student might try relating the concept of a potential function, as developed in this section, with the notion of an exact differential equation from Sections 1.4 and 1.5.

PROBLEMS FOR SECTION 11.8

In each of Problems 1 through 15, take Ω as the entire plane. Determine if **F** has a potential, using Theorem 91. If **F** has a potential, find one.

1. $\mathbf{F} = e^{xy}(y\mathbf{i} + x\mathbf{j})$

2. $\mathbf{F} = 2xy\mathbf{i} - 3\cos(yx)\mathbf{j}$

3. $\mathbf{F} = y\cos(xy)\mathbf{i} + x\cos(xy)\mathbf{j}$

4. $\mathbf{F} = (3x^2 - 2xy + 3e^x y)\mathbf{i} + (-x^2 + 3e^x)\mathbf{j}$

5. $\mathbf{F} = (-x^2 + y^2)\mathbf{i} + (x^2 - y)\mathbf{j}$

6. $\mathbf{F} = 2xye^x\mathbf{i} - 2ye^y\mathbf{j}$

7. $\mathbf{F} = [3x^2y + y\cos(xy)]\mathbf{i} + [x^3 + x\cos(xy)]\mathbf{j}$

8. $\mathbf{F} = \cosh(x)e^y\mathbf{i} + \sinh(x)e^y\mathbf{j}$

9. $\mathbf{F} = (3x^2 - 6yx + e^y)\mathbf{i} + (xe^y - 3x^2)\mathbf{j}$

10. $\mathbf{F} = (x^3y - 3y^2x^2)\mathbf{i} + (2xy^3 - x^2y)\mathbf{j}$

11. $\mathbf{F} = 3x^2y\mathbf{i} + (x^3 - 6y)\mathbf{j}$

12. $\mathbf{F} = (2xy^2 + 4y^3)\mathbf{i} + (2x^2y + 12xy)\mathbf{j}$

13. $\mathbf{F} = [\cos(xy) - xy\sin(xy)]\mathbf{i} - x^2\sin(xy)\mathbf{j}$

14. $\mathbf{F} = y^2e^{xy}\mathbf{i} + xye^{xy}\mathbf{j}$

15. $\mathbf{F} = 2x\mathbf{i} - 4xy\mathbf{j}$

In each of Problems 16 through 30, find a potential function for **F**, and use it to evaluate the line integral of **F** over the given regular curve.

16. $\mathbf{F} = e^x\cos(y)\mathbf{i} - e^x\sin(y)\mathbf{j}$, C is any regular curve from $(0, 0)$ to $(3, \pi)$.
17. $\mathbf{F} = 4x^3y^4\mathbf{i} + 4x^4y^3\mathbf{j}$, C is any regular curve from $(1, 2)$ to $(-3, 4)$.
18. $\mathbf{F} = 2xy\mathbf{i} + (x^2 + 1)\mathbf{j}$, C is the parabola $y = x^2$ from $(1, 1)$ to $(2, 4)$.
19. $\mathbf{F} = [y\cos(x) - xy\sin(x)]\mathbf{i} + x\cos(x)\mathbf{j}$, C is the straight line from $(\pi, 2)$ to $(3\pi/2, 4)$.
20. $\mathbf{F} = (16xy^2 + 4)\mathbf{i} + (16x^2y - 1)\mathbf{j}$, C is any regular curve from $(0, 0)$ to $(1, 2)$.
21. $\mathbf{F} = \cosh(x + y)(\mathbf{i} + \mathbf{j})$, C is the circle $x^2 + y^2 = r^2$ from $(r, 0)$ to $(r, 0)$, counterclockwise.
22. $\mathbf{F} = (2xy - e^y)\mathbf{i} + x(x - e^y)\mathbf{j}$, C is any regular curve from $(1, 0)$ to $(3, 16)$.
23. $\mathbf{F} = (-8y^3 + y)\mathbf{i} + (x - 24xy^2)\mathbf{j}$, C is any regular curve from $(2, 2)$ to $(-1, 6)$.
24. $\mathbf{F} = [y - 4\ln(y)]\mathbf{i} + [x - (4x/y)]\mathbf{j}$, C is any regular curve from $(2, 2)$ to $(4, 6)$ in the first quadrant.
25. $\mathbf{F} = y^2e^x\mathbf{i} + 2ye^x\mathbf{j}$, C is any regular curve from $(-1, 0)$ to $(3, 5)$.
26. $\mathbf{F} = (y^3e^{-x} - xy^3e^{-x})\mathbf{i} + (3y^2xe^{-x})\mathbf{j}$, C is any regular curve from $(-1, 0)$ to $(1, 1)$.
27. $\mathbf{F} = [2x - \sin(x - y)]\mathbf{i} + \sin(x - y)\mathbf{j}$, C is any regular curve from $(0, 0)$ to $(4, 4)$.
28. $\mathbf{F} = [1 - \cos(x)]\mathbf{i} - 12y^2\mathbf{j}$, C is any regular curve from $(-1, 3)$ to $(2, 5)$.
29. $\mathbf{F} = (1 - y^3)\mathbf{i} - 3xy^2\mathbf{j}$, C is any regular curve from $(-2, -2)$ to $(4, 6)$.
30. $\mathbf{F} = (y^3 - y)\mathbf{i} + (3xy^2 - x)\mathbf{j}$, C is any regular curve from $(0, 2)$ to $(3, 4)$.
31. Let the vector field **F** be continuous over a domain Ω in the plane. Prove that **F** is conservative if and only if the line integral of **F** is independent of path in Ω. (*Note:* **F** need not have a potential.)
32. Prove Theorem 90.

    Hint: (1) implies (2) and (1) implies (3) have already been done in the text. It is easy to show that (2) implies (3). To complete a logical chain, one can proceed as follows to show that (3) implies (1):

    Choose any point (x_0, y_0) in Ω. If (x, y) is any point in Ω, let $\phi(x, y) = \int_C \mathbf{F}$, where C is any regular curve in Ω from (x_0, y_0) to (x, y). Since the line integral of **F** is independent of path in Ω, ϕ is a function of x and y on Ω. There remains to show that $\mathbf{F} = \nabla\phi$. To prove that $\partial\phi/\partial x = F_1$, use the definition of domain to produce a point (x_1, y) in Ω with $x_1 < x$. Let K be a regular curve in Ω from (x_0, y_0) to (x_1, y) and L the straight line segment from (x_1, y) to (x, y). Write $\phi(x, y) = \int_C \mathbf{F} = \int_K \mathbf{F} + \int_L \mathbf{F}$ and evaluate $\partial\phi/\partial x$. Use a similar argument to show that $\partial\phi/\partial y = F_2$.
33. Let **F** be as in Example 340, but let Ω be the entire plane minus the origin.
    (a) Check that $\partial F_2/\partial x = \partial F_1/\partial y$ in all of Ω.
    (b) Show that $\int_C \mathbf{F}$ is not independent of path in Ω. One way to do this is to exhibit a closed path in Ω around which the line integral of **F** is not zero. Try $x = \cos(\theta)$, $y = \sin(\theta)$; $\theta\colon 0 \rightarrow 2\pi$.
    (c) How do you explain the results of (a) and (b) in light of Theorems 90 and 91?

11.9 SURFACES AND SURFACE INTEGRALS

Green's theorem has two generalizations to 3-space: they are Stokes's theorem and Gauss's divergence theorem [both of which are suggested by parts (a) and (b) of

Problem 22, Section 11.7]. To prepare the way for these, we must first develop some ideas about surfaces and the concept of a surface integral.

Usually, a surface is parametrized by equations of the form

$$x = x(u, v), \qquad y = y(u, v), \qquad z = z(u, v),$$

for (u, v) in some region D of the uv-plane. For example, using spherical coordinates θ and ϕ, a sphere of radius 1 about the origin is given by

$$x = \cos(\theta)\sin(\phi), \qquad y = \sin(\theta)\sin(\phi), \qquad z = \cos(\phi), \qquad 0 \leq \theta \leq 2\pi,\ 0 \leq \phi \leq \pi.$$

Figure 188 shows the surface, together with the range of values of θ and ϕ. Note that the graph of the surface is drawn in the usual xyz-coordinate system, while the parameters vary over a region of the uv-plane. This region is called the *parameter domain* and, in the case of the sphere above, is a rectangle, as shown in Figure 188.

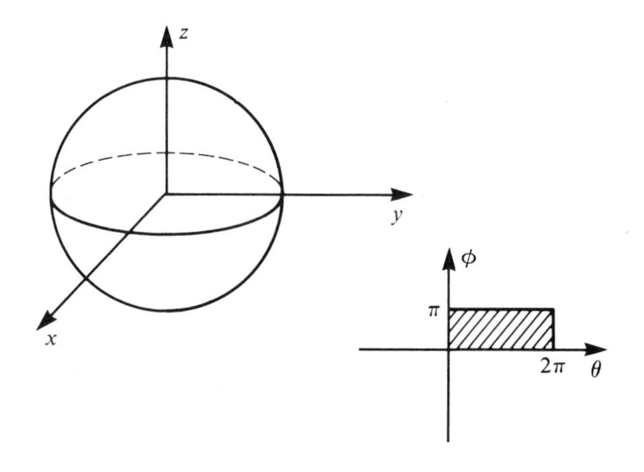

Figure 188

Often, a surface is described by an equation $f(x, y, z) = 0$. For example, the sphere of Figure 188 consists of all (x, y, z) satisfying

$$x^2 + y^2 + z^2 = 1,$$

and this is $f(x, y, z) = 0$ with $f(x, y, z) = x^2 + y^2 + z^2 - 1$.

A third way of describing a surface is in terms of a function $z = g(x, y)$. For example,

$$z = \sqrt{x^2 + y^2} \qquad \text{(for all } x \text{ and } y\text{)}$$

has as its graph the cone of Figure 189.

Suppose now that a surface is given in parametric form

$$x = x(u, v), \qquad y = y(u, v), \qquad z = z(u, v),$$

for (u, v) in some parameter domain D of the uv-plane. We can think of the position vector $\mathbf{R}(u, v) = x(u, v)\mathbf{i} + y(u, v)\mathbf{j} + z(u, v)\mathbf{k}$ as an arrow pivoted at the origin and

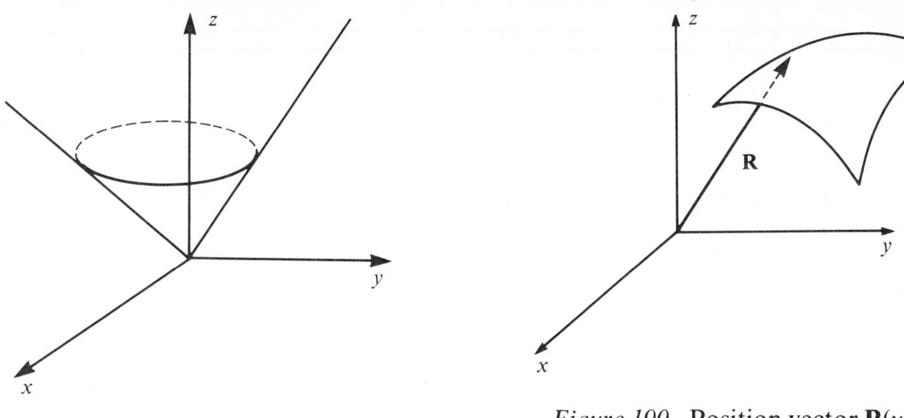

Figure 189. Cone $z = \sqrt{x^2 + y^2}$.

Figure 190. Position vector $\mathbf{R}(u, v)$ sweeping out a surface.

sweeping out the surface as (u, v) varies over D. This is shown in Figure 190 and is a straightforward generalization of the idea of a position vector sweeping out a curve when there is only a single parameter involved.

We call the surface S *simple* if $\mathbf{R}(u, v)$ does not return to the same point for different values of (u, v). This is analogous to a simple curve, which is one with no crossings.

On a surface, there is no clear way to distinguish a tangent vector at a point— there may, however, be a tangent plane containing many tangent vectors. In this event, we can define a normal vector at the point as a vector normal to the tangent plane there, as shown in Figure 191. To motivate a formal definition of a normal vector to a surface, in terms of the parameter functions defining the surface, consider Figure 192 in which P_0 is a point $(x(u_0, v_0), y(u_0, v_0), z(u_0, v_0))$. If we fix the parameter v at v_0 and let u vary, we have a curve

$$C_u: \quad x = x(u, v_0), \qquad y = y(u, v_0), \qquad z = z(u, v_0)$$

on the surface S.

Similarly, fixing $u = u_0$ and varying v gives us

$$C_v: \quad x = x(u_0, v), \qquad y = y(u_0, v), \qquad z = z(u_0, v).$$

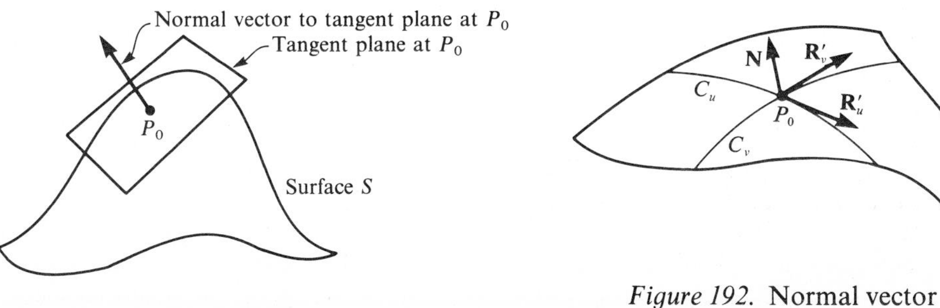

Figure 191

Figure 192. Normal vector 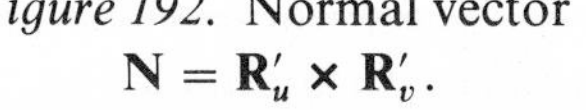
$\mathbf{N} = \mathbf{R}'_u \times \mathbf{R}'_v$.

These curves are defined by the position vectors

$$\mathbf{R}_u(u, v_0) = x(u, v_0)\mathbf{i} + y(u, v_0)\mathbf{j} + z(u, v_0)\mathbf{k}$$

and

$$\mathbf{R}_v(u_0, v) = x(u_0, v)\mathbf{i} + y(u_0, v)\mathbf{j} + z(u_0, v)\mathbf{k}.$$

The tangent vectors are, respectively,

$$\mathbf{R}'_u = \frac{\partial x}{\partial u}\mathbf{i} + \frac{\partial y}{\partial u}\mathbf{j} + \frac{\partial z}{\partial u}\mathbf{k}$$

and

$$\mathbf{R}'_v = \frac{\partial x}{\partial v}\mathbf{i} + \frac{\partial y}{\partial v}\mathbf{j} + \frac{\partial z}{\partial v}\mathbf{k}.$$

If we evaluate these at $u = u_0$ and $v = v_0$, we get two vectors in the tangent plane at P_0; hence, their cross product should be perpendicular to this tangent plane. This leads us to define the *normal* to S at P_0 as

$$\mathbf{N} = \mathbf{R}'_u \times \mathbf{R}'_v,$$

whenever this vector is not $\mathbf{0}$; if $\mathbf{N} = \mathbf{0}$, we say that S has no normal at P_0, as $\mathbf{0}$ does not specify any particular direction.

If you work out this cross product, you find that

$$\mathbf{N} = \begin{vmatrix} \dfrac{\partial y}{\partial u} & \dfrac{\partial y}{\partial v} \\ \dfrac{\partial z}{\partial u} & \dfrac{\partial z}{\partial v} \end{vmatrix} \mathbf{i} + \begin{vmatrix} \dfrac{\partial z}{\partial u} & \dfrac{\partial z}{\partial v} \\ \dfrac{\partial x}{\partial u} & \dfrac{\partial x}{\partial v} \end{vmatrix} \mathbf{j} + \begin{vmatrix} \dfrac{\partial x}{\partial u} & \dfrac{\partial x}{\partial v} \\ \dfrac{\partial y}{\partial u} & \dfrac{\partial y}{\partial v} \end{vmatrix} \mathbf{k}.$$

For those familiar with Jacobian notation, we can also write

$$\mathbf{N} = \frac{\partial(y, z)}{\partial(u, v)}\mathbf{i} + \frac{\partial(z, x)}{\partial(u, v)}\mathbf{j} + \frac{\partial(x, y)}{\partial(u, v)}\mathbf{k}.$$

Henceforth, we always mean this vector when we speak of the normal to a surface at a point. Of course, the partial derivatives are evaluated at the point.

The components of $\mathbf{N}$ can be easily remembered by their cyclic ordering. For the $\mathbf{i}$-component, the 2×2 determinant has partial derivatives of y and z with respect to u and v; for the $\mathbf{j}$-component, partials of z and x; and for the $\mathbf{k}$-component, partials of x and y. We can get these by reading around the cyclic chain:

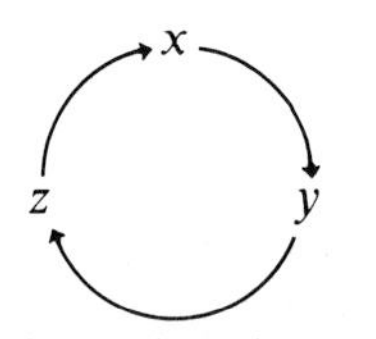

We call S *smooth* if it is simple and $\mathbf{N}$ is continuous on S except possibly at finitely many points; if S consists of finitely many smooth pieces, we call S *piecewise smooth* or *regular*. For example, a sphere is smooth, while a cube is regular, consisting of six smooth faces.

EXAMPLE 342

Consider the cone $z = \sqrt{x^2 + y^2}$, $x^2 + y^2 \leq 1$.

Here, S is given using x and y as parameters. In parametric form, we have

$$x = x, \qquad y = y, \qquad z = \sqrt{x^2 + y^2}, \quad \text{where} \quad x^2 + y^2 \leq 1.$$

Now,

$$\begin{vmatrix} \dfrac{\partial y}{\partial x} & \dfrac{\partial y}{\partial y} \\ \dfrac{\partial z}{\partial x} & \dfrac{\partial z}{\partial y} \end{vmatrix} = \begin{vmatrix} 0 & 1 \\ \dfrac{x}{\sqrt{x^2 + y^2}} & \dfrac{y}{\sqrt{x^2 + y^2}} \end{vmatrix} = \frac{-x}{\sqrt{x^2 + y^2}},$$

$$\begin{vmatrix} \dfrac{\partial z}{\partial x} & \dfrac{\partial z}{\partial y} \\ \dfrac{\partial x}{\partial x} & \dfrac{\partial x}{\partial y} \end{vmatrix} = \begin{vmatrix} \dfrac{x}{\sqrt{x^2 + y^2}} & \dfrac{y}{\sqrt{x^2 + y^2}} \\ 1 & 0 \end{vmatrix} = \frac{-y}{\sqrt{x^2 + y^2}},$$

and

$$\begin{vmatrix} \dfrac{\partial x}{\partial x} & \dfrac{\partial x}{\partial y} \\ \dfrac{\partial y}{\partial x} & \dfrac{\partial y}{\partial y} \end{vmatrix} = \begin{vmatrix} 1 & 0 \\ 0 & 1 \end{vmatrix} = 1.$$

Thus,

$$\mathbf{N} = \frac{-x}{\sqrt{x^2 + y^2}}\mathbf{i} - \frac{y}{\sqrt{x^2 + y^2}}\mathbf{j} + \mathbf{k} = \frac{-x}{z}\mathbf{i} - \frac{y}{z}\mathbf{j} + \mathbf{k},$$

as long as $x^2 + y^2 \neq 0$. At $x = y = 0$, $z = 0$; so $\mathbf{N}$ is not defined at the origin. This makes sense intuitively, as the cone has a sharp point there and hence no tangent plane or normal (see Figure 189). Here, S is smooth.

EXAMPLE 343

Let S be the paraboloid $z = x^2 + y^2$, with $x^2 + y^2 \leq 4$. The normal is found to be

$$\mathbf{N} = -2x\mathbf{i} - 2y\mathbf{j} + \mathbf{k},$$

and S is smooth. The surface is shown in Figure 193.

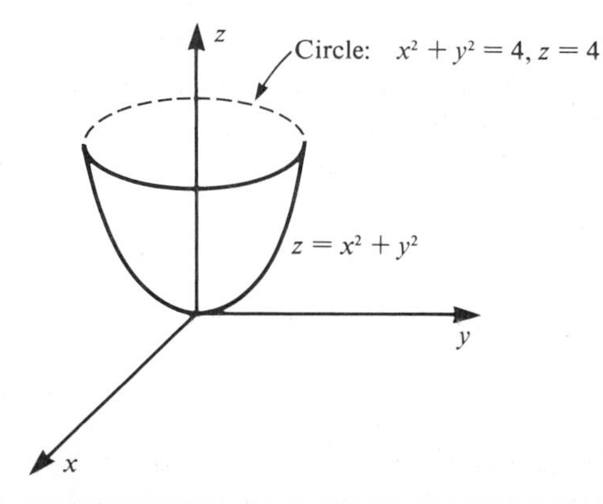

Figure 193. Paraboloid $z = x^2 + y^2$.

We can now define the surface integral, which is analogous to the line integral, with surfaces replacing curves and normals to surfaces playing the role previously played by tangents to curves.

Let $\mathbf{F}(x, y, z)$ be a continuous vector field defined over a region of space containing the smooth surface S, having normal vector $\mathbf{N}$. If S is parametrized

$$x = x(u, v), \qquad y = y(u, v), \qquad z = z(u, v),$$

for (u, v) in D, then $\mathbf{N}$ is also a function of u and v, and the *surface integral of* $\mathbf{F}$ *over* S is

$$\iint_S \mathbf{F} = \iint_D \mathbf{F}(x(u, v), y(u, v), z(u, v)) \cdot \mathbf{N}(u, v)\, du\, dv.$$

If S is regular, consisting of smooth pieces $S_1, \ldots, S_n$, then we define

$$\iint_S \mathbf{F} = \iint_{S_1} \mathbf{F} + \cdots + \iint_{S_n} \mathbf{F}.$$

EXAMPLE 344

Let S be the hemisphere

$$x = \cos(\theta)\sin(\phi), \qquad y = \sin(\theta)\sin(\phi), \qquad z = \cos(\phi),$$

$$0 \le \theta \le 2\pi, \qquad 0 \le \phi \le \frac{\pi}{2},$$

and let $\mathbf{F} = z\mathbf{i} + \mathbf{j} - xy\mathbf{k}$.

Substituting x, y, and z as functions of θ and ϕ into $\mathbf{F}$, we get, on S,

$$\mathbf{F} = \cos(\phi)\mathbf{i} + \mathbf{j} - \cos(\theta)\sin(\theta)\sin^2(\phi)\mathbf{k}.$$

Next, we compute (omitting the details)

$$\mathbf{N}(\theta, \phi) = -\sin(\phi)[\cos(\theta)\sin(\phi)\mathbf{i} + \sin(\theta)\sin(\phi)\mathbf{j} + \cos(\phi)\mathbf{k}].$$

Then, on S,

$$\mathbf{F} \cdot \mathbf{N} = -\cos(\theta)\sin^2(\phi)\cos(\phi) - \sin(\theta)\sin^2(\phi) - \cos(\theta)\sin(\theta)\sin^3(\phi)\cos(\phi).$$

Here, the parameter domain D is the rectangle $0 \le \theta \le 2\pi$, $0 \le \phi \le \pi/2$ in the $\theta\phi$-plane. Then,

$$\begin{aligned}\iint_S \mathbf{F} &= \int_0^{\pi/2}\int_0^{2\pi} [-\cos(\theta)\sin^2(\phi)\cos(\phi) - \sin(\theta)\sin^2(\phi) \\ &\qquad\qquad - \cos(\theta)\sin(\theta)\sin^3(\phi)\cos(\phi)]\, d\theta\, d\phi \\ &= 0.\end{aligned}$$

EXAMPLE 345

Let $\mathbf{F} = xyz\mathbf{i} - \mathbf{j} + 2\mathbf{k}$, and let S be the cone $z = \sqrt{x^2 + y^2}$, $x^2 + y^2 \le 1$, as shown in Figure 194. Here, x and y are the parameters. The parameter domain is the unit disk $x^2 + y^2 \le 1$ in the xy-plane, and

$$\mathbf{N} = \frac{-x}{\sqrt{x^2 + y^2}}\mathbf{i} - \frac{y}{\sqrt{x^2 + y^2}}\mathbf{j} + \mathbf{k},$$

for $x^2 + y^2 \ne 0$.

On S,

$$\mathbf{F} = xy\sqrt{x^2 + y^2}\,\mathbf{i} - \mathbf{j} + 2\mathbf{k}.$$

Then, on S,

$$\mathbf{F} \cdot \mathbf{N} = -x^2y + \frac{y}{\sqrt{x^2 + y^2}} + 2.$$

Then,

$$\iint_S \mathbf{F} = \iint_D \left(-x^2y + \frac{y}{\sqrt{x^2 + y^2}} + 2\right) dx\, dy.$$

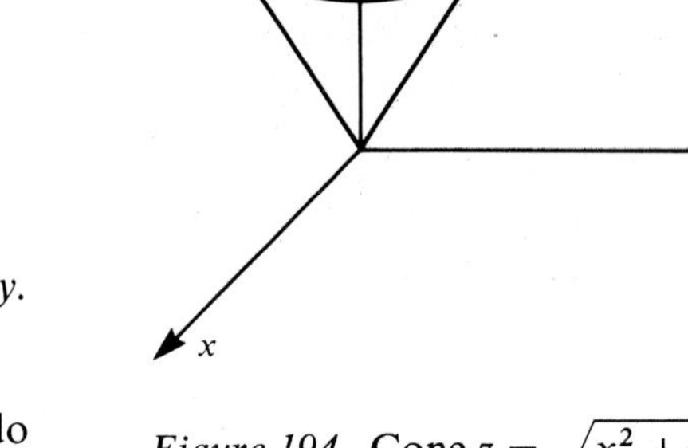

Figure 194. Cone $z = \sqrt{x^2 + y^2}$, $0 \leq x^2 + y^2 \leq 1$.

Since D is a disk, it is convenient to do this integral in polar coordinates. Let $x = r\cos(\theta)$, $y = r\sin(\theta)$, where $0 \leq r \leq 1$, $0 \leq \theta \leq 2\pi$. Then,

$$\iint_S \mathbf{F} = \int_0^{2\pi} \int_0^1 [-r^3 \cos^2(\theta)\sin(\theta) + \cos(\theta) + 2]r\, dr\, d\theta = 2\pi.$$

As with line integrals, surface integrals enjoy the following two properties:

1. If α is a scalar, then $\displaystyle\iint_S \alpha\mathbf{F} = \alpha \iint_S \mathbf{F}$.

2. $\displaystyle\iint_S (\mathbf{F} + \mathbf{G}) = \iint_S \mathbf{F} + \iint_S \mathbf{G}$.

Both of these follow easily from the definition of surface integral, and we leave it to the student to write out the details of proof.

Thus far, we have looked at computation of surface integrals. In the next section we state the two main theorems of vector integral calculus and then move into applications of these results.

PROBLEMS FOR SECTION 11.9

In each of Problems 1 through 10, compute the surface integral of each vector field **F** over each of the surfaces S in (a) through (f). (This is a total of 60 computations.) Label your solutions as 1-a [**F** as in 1, S as in (a)], and so on.

1. $\mathbf{F} = x\mathbf{i} + y\mathbf{j} + z\mathbf{k}$
2. $\mathbf{F} = 3\mathbf{i} - 2y\mathbf{j} + 6\mathbf{k}$
3. $\mathbf{F} = x^2\mathbf{i} - y\mathbf{j} + \mathbf{k}$
4. $\mathbf{F} = (x - y)\mathbf{i} + 6x\mathbf{j} - 3z\mathbf{k}$
5. $\mathbf{F} = 2y\mathbf{i} - 3z\mathbf{k}$
6. $\mathbf{F} = x^2\mathbf{i} - y^2\mathbf{j}$
7. $\mathbf{F} = z\mathbf{j}$
8. $\mathbf{F} = xy\mathbf{i} - 2y\mathbf{j} + (z - x)\mathbf{k}$
9. $\mathbf{F} = 2z\mathbf{i} + (x - y - z)\mathbf{k}$
10. $\mathbf{F} = 3yz\mathbf{i} + 8x\mathbf{j} - 2xyz\mathbf{k}$

(a) $z = x - 3y$; $0 \leq x \leq 1$, $0 \leq y \leq 2$
(b) $x - 3y + 2z = 0$; $1 \leq y \leq 3$, $2 \leq z \leq 5$
(c) $z = \sqrt{x^2 + y^2}$; $x^2 + y^2 \leq 3$
(d) $x = 2u + v$, $y = 2u - v$, $z = 4u$; $0 \leq u \leq 1$, $0 \leq v \leq 4$

(e) $z = x^2 + y^2;\quad x^2 + y^2 \leq 6$

(f) $x = 2u^2 - v,\quad y = u - v^2,\quad z = u + v;\quad 0 \leq u \leq 1,\quad 2 \leq v \leq 6$

11. Suppose that S is parametrized by x and y and given by $z = f(x, y)$, for (x, y) in some specified region D of the xy-plane.

Show that the normal to S is

$$\mathbf{N} = -\frac{\partial f}{\partial x}\mathbf{i} - \frac{\partial f}{\partial y}\mathbf{j} + \mathbf{k}$$

in two ways:

(a) By using the formula of this section for $\mathbf{N}$, with S given by

$$x = x, \qquad y = y, \qquad z = f(x, y).$$

(b) By looking at ∇G, where $G(x, y, z) = z - f(x, y)$ and the surface is given as the locus $G(x, y, z) = 0$.

12. Adapt the form of Green's theorem given in Problem 22(a), Section 11.7, to a three-dimensional formula in which the double integral becomes a triple integral, the curve a surface, and the line integral a surface integral.

13. Adapt the form of Green's theorem given in Problem 22(b), Section 11.7, to a three-dimensional formula in which the double integral is replaced by a surface integral (with S replacing D, and $\mathbf{N}$ replacing $\mathbf{k}$), and the line integral remains a line integral over an appropriate curve on S.

11.10 THEOREMS OF GAUSS AND STOKES: COMPUTATIONAL ASPECTS

We now develop the computational aspects of the two fundamental formulas of vector integral calculus.

For Gauss's divergence theorem, we need some terminology. Suppose that S is a regular surface. We call S *closed* if it bounds a finite, nonzero volume. For example, a sphere is closed, while a hemisphere is not. The surface consisting of a hemisphere and a flat disk covering the hemisphere would be closed. Intuitively, a surface is closed if it will hold water and not closed if the water could be poured out by moving or tipping the surface (as with the hemisphere).

The following terminology will be used in later applications: Given a closed surface S, we call the region enclosed by S the *interior* of S; the region outside of S is the *exterior* to S. The normal $\mathbf{N}$ to S is called an *inner normal* if it points toward the interior of S when drawn from points on S; otherwise, it is an *outer normal.* Figure 195 contrasts inner and outer normals in general. As a specific example, the normal to the cone computed in Example 342 is an inner normal, pointing into the volume bounded by the cone; $-\mathbf{N}$ is an outer normal. We call a surface *positively oriented* if the normal $\mathbf{N}$ is an outer

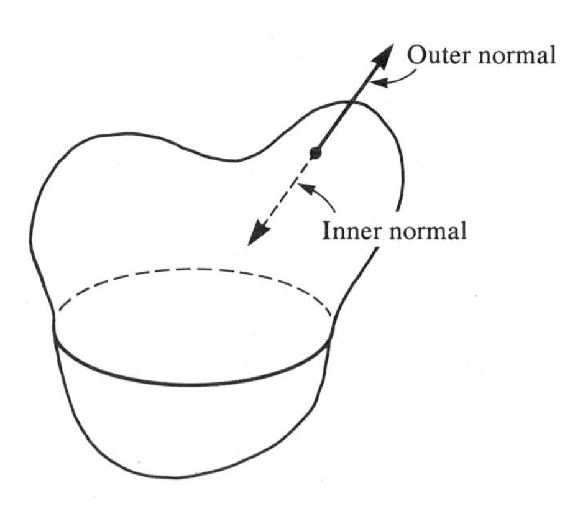

Figure 195. Inner and outer normal to a closed surface.

normal; otherwise, S is *negatively oriented.* Thus, in Example 342, the cone is negatively oriented. Note that the orientation is fixed once the surface is written in parametric form, since this determines $\mathbf{N}$ as defined in Section 11.9.

We are now in a position to state Gauss's divergence theorem.

THEOREM 92 Gauss's Divergence Theorem

Suppose that S is a regular, positively oriented closed surface, and that $\mathbf{F}$ and div $\mathbf{F}$ are continuous over S and the region V enclosed by S. Then,

$$\iint_S \mathbf{F} = \iiint_V \operatorname{div} \mathbf{F}\, dx\, dy\, dz.$$

This is often written

$$\iint_S \mathbf{F} = \iiint_V \nabla \cdot \mathbf{F}\, dx\, dy\, dz.$$

It is important to understand that S must be positively oriented for this to hold. In practice we calculate and evaluate $\iint_S \mathbf{F}$ in the usual way. If $\mathbf{N}$ is outer, the formula of Gauss's theorem holds; if $\mathbf{N}$ is inner, replace $\iint_S \mathbf{F}$ by $-\iint_S \mathbf{F}$ in Gauss's theorem.

A general proof of Gauss's theorem is beyond the techniques we have at our disposal, but one can do special cases. As one example, a proof when the surface is a cube is requested in the problems at the end of the section.

Here are two computational examples.

EXAMPLE 346

Let S be the closed parabolic bowl consisting of the two pieces

$$S_1: \quad z = x^2 + y^2; \qquad x^2 + y^2 \le 1$$

and

$$S_2: \quad x = r\cos(\theta), \qquad y = r\sin(\theta),$$

$$z = 1; \qquad 0 \le r \le 1, \qquad 0 \le \theta \le 2\pi.$$

Thus, S_1 is the bowl proper and S_2 is the circular cap on top, as shown in Figure 196. On S_1, x and y are the parameters; on S_2 we used polar coordinates.

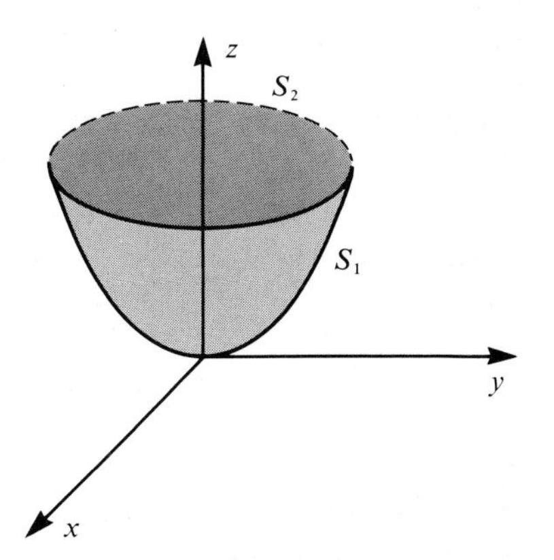

Figure 196. Parabolic bowl S_1 with top S_2.

Let $\mathbf{F} = (x - y + z)\mathbf{i} + 2x\mathbf{j} + \mathbf{k}$. We shall compute both sides of Gauss's formula and compare results.

First, $\nabla \cdot \mathbf{F} = 1$; so

$$\iiint_V \nabla \cdot \boldsymbol{F}\, dx\, dy\, dz = \iiint_V 1\, dx\, dy\, dz = \int_{-1}^{1} \int_{-\sqrt{1-x^2}}^{\sqrt{1-x^2}} \int_{x^2+y^2}^{1} dz\, dy\, dx.$$

Now,

$$\int_{x^2+y^2}^{1} dz = [z]_{x^2+y^2}^{1} = 1 - x^2 - y^2.$$

To evaluate

$$\int_{-1}^{1} \int_{-\sqrt{1-x^2}}^{\sqrt{1-x^2}} (1 - x^2 - y^2)\, dy\, dx,$$

switch to polar coordinates. The integral transforms to

$$\int_0^{2\pi} \int_0^1 (1 - r^2) r\, dr\, d\theta,$$

which works out to $\pi/2$.

Now evaluate $\iint_S \mathbf{F}$ as $\iint_{S_1} \mathbf{F} + \iint_{S_2} \mathbf{F}$.

On S_1, we find

$$\mathbf{N} = -2x\mathbf{i} - 2y\mathbf{j} + \mathbf{k} \qquad \text{(an } inner \text{ normal to } S_1)$$

and, replacing z by $x^2 + y^2$, we have

$$\begin{aligned} \mathbf{F} \cdot \mathbf{N} &= -2x(x - y + x^2 + y^2) - 2y(2x) + 1 \\ &= -2x^2 - 2xy - 2x(x^2 + y^2) + 1. \end{aligned}$$

Then,

$$\iint_{S_1} \mathbf{F} = \int_{-1}^{1} \int_{-\sqrt{1-x^2}}^{\sqrt{1-x^2}} [-2x^2 - 2xy - 2x(x^2 + y^2) + 1]\, dx\, dy.$$

To evaluate this, convert to polar coordinates, $x = r\cos(\theta)$, $y = r\sin(\theta)$, where r: $0 \rightarrow 1$ and $\theta: 0 \rightarrow 2\pi$. Then,

$$\iint_{S_1} \mathbf{F} = \int_0^{2\pi} \int_0^1 [-2r^2 \cos^2(\theta) - 2r^2 \cos(\theta)\sin(\theta) - 2r^3 \cos(\theta) + 1] r\, dr\, d\theta = \frac{\pi}{2}.$$

Since $\mathbf{N}$ was an inner normal here, we use $-\pi/2$ as the contribution of $\iint_{S_1} \mathbf{F}$ to $\iint_S \mathbf{F}$ in Gauss's theorem.

To evaluate $\iint_{S_2} \mathbf{F}$, we first compute $\mathbf{N} = r\mathbf{k}$ and observe that this is an *outer* normal. Next, on S_2,

$$\mathbf{F} = [r\cos(\theta) - r\sin(\theta) + 1]\mathbf{i} + 2r\cos(\theta)\mathbf{j} + \mathbf{k}.$$

Then,

$$\mathbf{F} \cdot \mathbf{N} = r.$$

Since the parameter domain here is the rectangle $0 \le r \le 1, 0 \le \theta \le 2\pi$, we have

$$\iint_{S_2} \mathbf{F} = \int_0^{2\pi} \int_0^1 r\, dr\, d\theta = \pi.$$

Then,

$$\iint_{S_1} \mathbf{F} + \iint_{S_2} \mathbf{F} = -\frac{\pi}{2} + \pi = \frac{\pi}{2}.$$

EXAMPLE 347

Let S be the sphere of radius r about the origin, given in spherical coordinates by

$$x = r\cos(\theta)\sin(\phi), \qquad y = r\sin(\theta)\sin(\phi), \qquad z = r\cos(\phi),$$
$$0 \le \theta \le 2\pi, \qquad 0 \le \phi \le \pi.$$

Let $\mathbf{F} = 4yz\mathbf{j}$. Then, $\nabla \cdot \mathbf{F} = 4z$, and, on S, $\nabla \cdot \mathbf{F} = 4r\cos(\phi)$.

Next, the normal to S is found to be

$$\mathbf{N} = -r^2\sin(\phi)[\cos(\theta)\sin(\phi)\mathbf{i} + \sin(\theta)\sin(\phi)\mathbf{j} + \cos(\phi)\mathbf{k}].$$

This is an inner normal. Now,

$$\begin{aligned} \mathbf{F} \cdot \mathbf{N} &= 4yz[-r^2\sin(\theta)\sin^2(\phi)] \\ &= 4[r\sin(\theta)\sin(\phi)][r\cos(\phi)][-r^2\sin(\theta)\sin^2(\phi)] \\ &= -4r^4\sin^2(\theta)\sin^3(\phi)\cos(\phi). \end{aligned}$$

Then,

$$\iint_S \mathbf{F} = \int_0^{\pi}\int_0^{2\pi} -4r^4\sin^2(\theta)\sin^3(\phi)\cos(\phi)\,d\theta\,d\phi = 0.$$

Next, $\nabla \cdot \mathbf{F} = 4z$; so

$$\iiint_V \nabla \cdot \mathbf{F}\,dx\,dy\,dz = \iiint_V 4z\,dx\,dy\,dz.$$

Transforming to spherical coordinates, this is

$$\int_0^{\pi}\int_0^{2\pi}\int_0^1 4r\cos(\phi)r^2\sin(\phi)\,dr\,d\theta\,d\phi = 0.$$

We now turn to the second main result of this section, Stokes's theorem, for which we require the notion of the boundary of a surface.

Let S be a regular surface, parametrized by

$$x = x(u, v), \qquad y = y(u, v), \qquad z = z(u, v),$$

for (u, v) in D. We assume that the parameter domain D is bounded by a regular closed curve C. As (u, v) traces over C in a counterclockwise direction, the point $(x(u, v), y(u, v), z(u, v))$ traces out a curve on S called the *boundary of* S, denoted ∂S. This process is shown in Figure 197. Here are two illustrations.

EXAMPLE 348

Let $z = x^2 + y^2$ for $x^2 + y^2 \le 2$. Here, x and y are the parameters, and the parameter domain is the solid disk $x^2 + y^2 \le 2$ in the xy-plane. The curve C bounding this disk

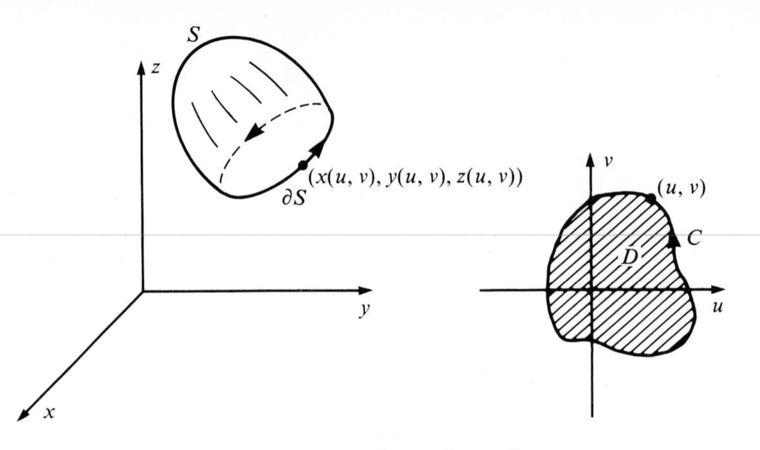

Figure 197. Boundary of a surface.

is $x^2 + y^2 = 2$. Substituting $x^2 + y^2 = 2$ into S gives us the curve $x^2 + y^2 = 2$, $z = 2$, and this is the boundary ∂S of S. It is the circular top of the bowl, as shown in Figure 198. As (x, y) moves around C counterclockwise, (x, y, z) moves around ∂S counterclockwise.

EXAMPLE 349

Let S be the hemisphere $x^2 + y^2 + z^2 = 1, z \geq 0$. Here we may write S as

$$z = \sqrt{1 - x^2 - y^2}, \qquad x^2 + y^2 \leq 1.$$

The curve C bounding the parameter domain is $x^2 + y^2 = 1$. Putting this into S, we have

$$\partial S\colon \; x^2 + y^2 = 1, \qquad z = \sqrt{1 - 1} = 0,$$

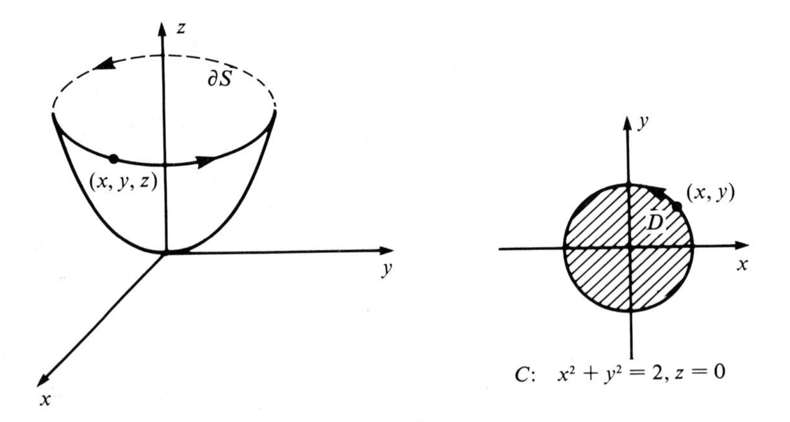

Figure 198. Boundary of the parabolic bowl $z = x^2 + y^2$, $x^2 + y^2 \leq 2$.

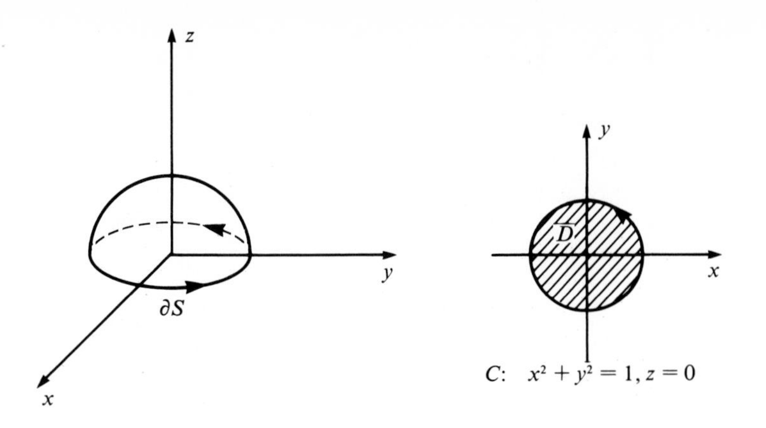

Figure 199. Boundary of the hemisphere $x^2 + y^2 + z^2 = 1, z \geq 0$.

which is the rim of the hemisphere in the xy-plane. As (x, y) goes around C counterclockwise, $(x, y, 0)$ goes around ∂S counterclockwise (see Figure 199).

Note that we always go around the boundary C of the parameter domain counterclockwise, and that this in turn determines a corresponding direction around ∂S.

With this as background, we can state Stokes's theorem.

THEOREM 93 Stokes's Theorem

Let S be a regular surface with boundary ∂S. Let $\mathbf{F}$ and curl $\mathbf{F}$ be continuous over S. Then,

$$\iint_S \text{curl } \mathbf{F} = \int_{\partial S} \mathbf{F}.$$

This may also be written

$$\iint_S \nabla \times \mathbf{F} = \int_{\partial S} \mathbf{F}.$$

As with Gauss's theorem, we shall treat applications later and be content here with a computational example. Note that S in Stokes's theorem need not be closed, nor do we require an outer normal in computing $\iint_S \nabla \times \mathbf{F}$; rather, we determine $\mathbf{N}$ in the usual way and use it in evaluating $\iint_S \nabla \times \mathbf{F}$ without regard to considerations of direction of $\mathbf{N}$. This makes sense, as S in Stokes's theorem may not be a closed surface, and hence may not bound a volume to determine an interior and an exterior region.

EXAMPLE 350

Let S be the cone $z = \sqrt{x^2 + y^2}$, $x^2 + y^2 \leq 4$. Here, the parameter domain D consists of the points on the disk $x^2 + y^2 \leq 4$; the curve C bounding D is the circle $x^2 + y^2 = 4$; and ∂S is the circle $x^2 + y^2 = 4$, $z = 2$ (the top rim of the cone). This is shown in Figure 200.

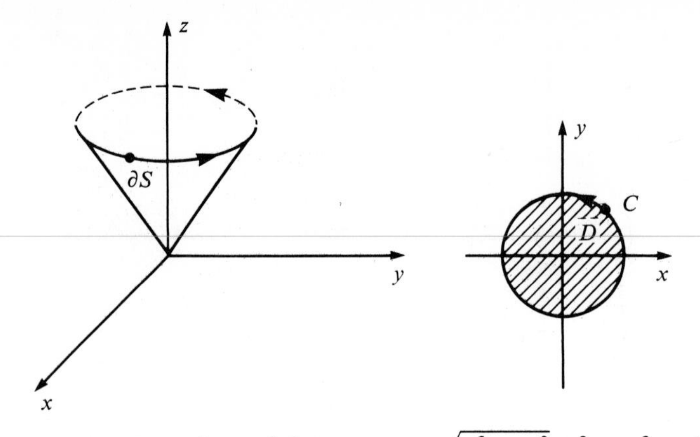

Figure 200. Boundary of the cone $z = \sqrt{x^2 + y^2}$, $x^2 + y^2 \leq 4$.

Let $\mathbf{F} = (x - y)\mathbf{i} + 2z\mathbf{j} + x^2\mathbf{k}$. We find that

$$\text{curl } \mathbf{F} = -2\mathbf{i} - 2x\mathbf{j} + \mathbf{k}.$$

Note: Since z does not appear in curl $\mathbf{F}$, replacing z by $\sqrt{x^2 + y^2}$ has no effect; so this is also curl $\mathbf{F}$ on S.

Next,

$$\mathbf{N} = \frac{-x}{\sqrt{x^2 + y^2}}\mathbf{i} - \frac{y}{\sqrt{x^2 + y^2}}\mathbf{j} + \mathbf{k},$$

for $x^2 + y^2 \neq 0$. Then,

$$\text{curl } \mathbf{F} \cdot \mathbf{N} = \frac{2x}{\sqrt{x^2 + y^2}} + \frac{2xy}{\sqrt{x^2 + y^2}} + 1.$$

Then,

$$\iint_S \text{curl } \mathbf{F} = \iint_D \left(\frac{2x}{\sqrt{x^2 + y^2}} + \frac{2xy}{\sqrt{x^2 + y^2}} + 1 \right) dx\, dy.$$

Switch to polar coordinates:

$$x = r\cos(\theta), \qquad y = r\sin(\theta), \qquad 0 \leq r \leq 2, \qquad 0 \leq \theta \leq 2\pi.$$

The last integral becomes

$$\int_0^{2\pi} \int_0^2 \left(\frac{2r\cos(\theta)}{r} + \frac{2r^2\cos(\theta)\sin(\theta)}{r} + 1 \right) r\, dr\, d\theta = 4\pi.$$

Now evaluate $\int_{\partial S} \mathbf{F}$. To do this, parametrize ∂S by:

$$x = 2\cos(\theta), \qquad y = 2\sin(\theta), \qquad z = 2, \qquad \theta\colon 0 \to 2\pi.$$

(We let θ vary from 0 to 2π to go around ∂S in the direction corresponding to going around C counterclockwise.)

On ∂S,

$$\mathbf{F} = [2\cos(\theta) - 2\sin(\theta)]\mathbf{i} + 4\mathbf{j} + 4\cos^2(\theta)\mathbf{k}.$$

Next, the tangent to ∂S is

$$\mathbf{R}' = -2\sin(\theta)\mathbf{i} + 2\cos(\theta)\mathbf{j}.$$

Then, on ∂S,

$$\mathbf{F} \cdot \mathbf{R}' = -4\cos(\theta)\sin(\theta) + 4\sin^2(\theta) + 8\cos(\theta).$$

Then,

$$\int_{\partial S} \mathbf{F} = \int_0^{2\pi} [-4\cos(\theta)\sin(\theta) + 4\sin^2(\theta) + 8\cos(\theta)]\, d\theta = 4\pi.$$

EXAMPLE 351

Let $\mathbf{F} = 3y\mathbf{i} - 2y\mathbf{j} + z^2\mathbf{k}$, and let S be the hemisphere $x^2 + y^2 + z^2 = 1$, $z \geq 0$, as shown in Figure 199. Here, the boundary of S is the circle $x^2 + y^2 = 1$, $z = 0$. Parametrize ∂S by $x = \cos(\theta)$, $y = \sin(\theta)$, $z = 0$, with θ: $0 \rightarrow 2\pi$. Calculate

$$\int_{\partial S} \mathbf{F} = \int_0^{2\pi} \{3\sin(\theta)[-\sin(\theta)] - 2\sin(\theta)\cos(\theta)\}\, d\theta$$

$$= \int_0^{2\pi} [-3\sin^2(\theta) - 2\sin(\theta)\cos(\theta)]\, d\theta = -3\pi.$$

Next, $\nabla \times \mathbf{F} = -3\mathbf{k}$; so $\iint_S \nabla \times \mathbf{F} = \iint_S -3\mathbf{k}$. To work this out, parametrize S by

$$x = \cos(\theta)\sin(\phi), \quad y = \sin(\theta)\sin(\phi), \quad z = \cos(\phi), \quad 0 \leq \theta \leq 2\pi, \quad 0 \leq \phi \leq \pi/2.$$

Now, from Example 344, we have

$$\mathbf{N} = -\cos(\theta)\sin^2(\phi)\mathbf{i} - \sin(\theta)\sin^2(\phi)\mathbf{j} + \sin(\phi)\cos(\phi)\mathbf{k}.$$

Then,

$$\iint_S \nabla \times \mathbf{F} = \int_0^{\pi/2} \int_0^{2\pi} (-3\mathbf{k}) \cdot \mathbf{N}\, d\theta\, d\phi$$

$$= \int_0^{\pi/2} \int_0^{2\pi} -3\sin(\phi)\cos(\phi)\, d\theta\, d\phi$$

$$= -6\pi \int_0^{\pi/2} \sin(\phi)\cos(\phi)\, d\phi = -3\pi.$$

In the next two sections we shall explore various applications of Gauss's and Stokes's theorems.

PROBLEMS FOR SECTION 11.10

In each of Problems 1 through 6, verify Gauss's theorem for each of the vector fields over each of the closed surfaces (a) through (d). Number your solutions 1-a, ..., 6-d.

1. $\mathbf{F} = x\mathbf{i} + y\mathbf{j} + z\mathbf{k}$ **2.** $\mathbf{F} = -y\mathbf{i} + x^2\mathbf{j} + z^2\mathbf{k}$ **3.** $\mathbf{F} = -xyz\mathbf{k}$
4. $\mathbf{F} = 2x\mathbf{i} - y^2\mathbf{j}$ **5.** $\mathbf{F} = xy\mathbf{i} - yz\mathbf{j} + 3\mathbf{k}$ **6.** $\mathbf{F} = 2yz\mathbf{i} + xy\mathbf{j} + 2z\mathbf{k}$

(a) S is the hemisphere $x^2 + y^2 + z^2 = 1, z \geq 0$, together with the base $x^2 + y^2 \leq 1, z = 0$.
(b) S is the cube with vertices at (0, 0, 0), (1, 1, 1), (1, 1, 0), (0, 1, 0), (1, 0, 0), (0, 0, 1), and (0, 1, 1).
(c) S is the cylinder $x^2 + y^2 = 4, 0 \leq z \leq 2$, together with the top $x^2 + y^2 \leq 4, z = 2$, and the base $x^2 + y^2 \leq 4, z = 0$.
(d) S is the paraboloid $z = x^2 + y^2, x^2 + y^2 \leq 9$, along with the top $x^2 + y^2 \leq 9, z = 9$.

In each of Problems 7 through 12, verify Stokes's theorem for the given vector field over each of the surfaces (e) through (h).

7. $\mathbf{F} = x\mathbf{i} - y\mathbf{j} + z\mathbf{k}$ **8.** $\mathbf{F} = 2xy\mathbf{i} + z^2\mathbf{k}$ **9.** $\mathbf{F} = (xy + z)\mathbf{i} - y\mathbf{j} + \mathbf{k}$
10. $\mathbf{F} = 2\mathbf{i} - 3y\mathbf{j} + (x + y + z)\mathbf{k}$ **11.** $\mathbf{F} = xy\mathbf{i} - yz\mathbf{j}$ **12.** $\mathbf{F} = -2y^2\mathbf{i} + z^2\mathbf{j} - x^2\mathbf{k}$

(e) S is the cone $x = \sqrt{y^2 + z^2}, y^2 + z^2 \leq 9$.
(f) S is the paraboloid $y = x^2 + z^2, x^2 + z^2 \leq 4$.
(g) S is given by $x = u, y = u + v, z = v, 0 \leq u \leq 1, 0 \leq v \leq 1$.
(h) S is the hemisphere $x^2 + y^2 + z^2 = 4, z \leq 0$.

13. Prove Gauss's theorem for a cube.
14. Show how Green's theorem may be thought of as a special case of Gauss's theorem.
15. Show how Green's theorem can be thought of as a special case of Stokes's theorem.

11.11 SOME APPLICATIONS OF GAUSS'S THEOREM

This section is devoted to various ways that Gauss's theorem can be used in physical settings.

1. DIVERGENCE OF A VECTOR

If $\mathbf{F} = F_1\mathbf{i} + F_2\mathbf{j} + F_3\mathbf{k}$, then the divergence of $\mathbf{F}$ is given by

$$\nabla \cdot \mathbf{F} = \frac{\partial F_1}{\partial x} + \frac{\partial F_2}{\partial y} + \frac{\partial F_3}{\partial z}.$$

In Section 11.5 we gave a rough argument showing a physical significance of divergence. Using Gauss's theorem, we can do things more neatly.

Let $\mathbf{F}(x, y, z, t)$ be the velocity of a fluid at point (x, y, z) and time t, as in Section 11.5. We include time to be more realistic, but t plays no role in computing $\nabla \cdot \mathbf{F}$.

Let $P_0(x_0, y_0, z_0)$ be any point in the fluid, and imagine a regular closed surface S about P_0 in the fluid (see Figure 201). By Gauss's theorem,

$$\iint_S \mathbf{F} = \iiint_V \nabla \cdot \mathbf{F}\, dx\, dy\, dz,$$

where V is the region enclosed by S. Let $\bar{V}$ denote the volume of V. By the mean value theorem for triple integrals,

$$\iiint_V \nabla \cdot \mathbf{F}\, dx\, dy\, dz = \bar{V}(\nabla \cdot \mathbf{F})|_P,$$

for some point P in V. Here, $(\nabla \cdot \mathbf{F})|_P$ is the divergence of $\mathbf{F}$ evaluated at P. Then,

$$(\nabla \cdot \mathbf{F})|_P = \frac{1}{\bar{V}} \iint_S \mathbf{F}.$$

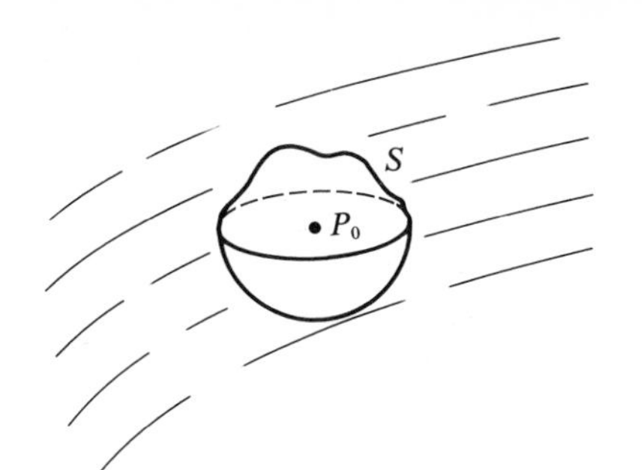

Figure 201. Infinitesimal volume element about P_0 in the fluid.

Now imagine that S contracts to P_0 (think of S as a balloon, and let the air out). Then P moves to P_0, $\bar{V} \to 0$, and we have

$$(\nabla \cdot \mathbf{F})|_{P_0} = \lim_{\bar{V} \to 0} \frac{1}{\bar{V}} \iint_S \mathbf{F},$$

giving the divergence of $\mathbf{F}$ at P_0 as a kind of limit. This expression has two interesting features.

First, the right side is independent of coordinate system (rectangular, spherical, cylindrical, or other) and can be used to find the divergence of a vector in other coordinate systems (see Section 11.13).

Second, we can interpret the right side physically. Since $\mathbf{N}$ is (presumably) an outer normal to S, $\mathbf{F} \cdot \mathbf{N}$ is the component of fluid velocity out across S at time t. Thus, $\iint_S \mathbf{F}$ is the flux of fluid flow out from V across S, and

$$\frac{1}{\bar{V}} \iint_S \mathbf{F} = \text{flux across } S \text{ per unit volume.}$$

Then,

$$\text{divergence of } \mathbf{F} \text{ at } P_0 = (\nabla \cdot \mathbf{F})|_{P_0}$$
$$= \text{flux of flow out from } P_0 \text{ per unit volume.}$$

Thus, divergence measures expansion of the fluid at any point. One may conjecture that this interpretation gives rise to the term *divergence*.

2. EQUATION OF HEAT CONDUCTION

Here we address the problem of deriving an equation governing heat conduction in a medium.

Suppose the medium has density $\rho(x, y, z)$, specific heat $\sigma(x, y, z)$, and coefficient of thermal conductivity $K(x, y, z)$. Let $u(x, y, z, t)$ be the temperature at time t and point (x, y, z).

To derive an equation for u, imagine a positively oriented closed surface S within the medium, enclosing a region V. By Fourier's law, the amount of heat energy leaving V across S in a time Δt equals $\iint_S (K\nabla u)\, \Delta t$.

Now, the change in temperature at (x, y, z) in V in the time interval Δt is approximately $(\partial u/\partial t)\,\Delta t$, and the resulting heat loss in V is

$$\left(\iiint_V \sigma\rho \frac{\partial u}{\partial t}\,dx\,dy\,dz\right)\Delta t.$$

In the absence of sources of heat within V, these must balance, and

$$\left(\iint_S K\nabla u\right)\Delta t = \left(\iiint_V \sigma\rho \frac{\partial u}{\partial t}\,dx\,dy\,dz\right)\Delta t.$$

Dividing out Δt, we have

$$\iint_S K\nabla u = \iiint_V \sigma\rho \frac{\partial u}{\partial t}\,dx\,dy\,dz.$$

Apply Gauss's theorem to the surface integral on the left to get

$$\iiint_V \nabla\cdot(K\nabla u)\,dx\,dy\,dz = \iiint_V \sigma\rho \frac{\partial u}{\partial t}\,dx\,dy\,dz.$$

Then,

$$\iiint_V \left(\nabla\cdot(K\nabla u) - \sigma\rho\frac{\partial u}{\partial t}\right)dx\,dy\,dz = 0.$$

But this must hold for *any* regular closed surface S bounding a region V in the medium. Hence,

$$\nabla\cdot(K\nabla u) - \sigma\rho\frac{\partial u}{\partial t} = 0.$$

This is the three-dimensional heat equation.

Note that

$$\begin{aligned}
\nabla\cdot(K\nabla u) &= \nabla\cdot\left(K\frac{\partial u}{\partial x}\mathbf{i} + K\frac{\partial u}{\partial y}\mathbf{j} + K\frac{\partial u}{\partial z}\mathbf{k}\right)\\
&= \frac{\partial}{\partial x}\left(K\frac{\partial u}{\partial x}\right) + \frac{\partial}{\partial y}\left(K\frac{\partial u}{\partial y}\right) + \frac{\partial}{\partial z}\left(K\frac{\partial u}{\partial z}\right)\\
&= K\left(\frac{\partial^2 u}{\partial x^2} + \frac{\partial^2 u}{\partial y^2} + \frac{\partial^2 u}{\partial z^2}\right) + \nabla K\cdot\nabla u\\
&= K\nabla^2 u + \nabla K\cdot\nabla u.
\end{aligned}$$

Thus, the heat equation is

$$K\nabla^2 u + \nabla K\cdot\nabla u = \sigma\rho\frac{\partial u}{\partial t}.$$

If K = constant, this reduces to

$$\frac{\partial u}{\partial t} = \frac{K}{\sigma\rho}\nabla^2 u.$$

The steady-state case occurs when u does not change with time. Then $\partial u/\partial t = 0$, and we have

$$\nabla^2 u = 0,$$

which is called *Laplace's equation.* This equation also arises in fluid flow, electricity and magnetism, astronomy, and other contexts as well, and hence is one of the fundamental equations of physical science.

3. SOME EQUATIONS OF HYDRODYNAMICS

Consider a fluid (say, water or mineral oil) flowing in some region of space. Recalling the trick used in deriving the heat equation, we insert an imaginary, positively oriented regular surface S into the fluid and observe the flow through S. Let $\rho(x, y, z, t)$ be the density of the fluid at time t and point (x, y, z), and let $\mathbf{F}(x, y, z, t)$ be the velocity.

Now, the amount of fluid flowing out of V across S in time Δt is $(\iint_S \rho\mathbf{F})\,\Delta t$, where V is the region bounded by S. This requires that the quantity of fluid inside S be changed by the amount

$$-\left(\iiint_V \frac{\partial\rho}{\partial t}\,dx\,dy\,dz\right)\Delta t$$

(minus sign for decrease). If, in addition, there are sources (or sinks) of fluid inside S, and we assume that fluid is created (or destroyed) at a rate proportional to ρ, say $K\rho$, then we must add a term

$$\left(\iiint_V K\rho\,dx\,dy\,dz\right)\Delta t$$

to the change in fluid inside S in time Δt.

Balancing loss from V through S on one side with decrease throughout V and fluid introduced into V by sources (or lost through sinks if $K < 0$), we have

$$\left(\iint_S \rho\mathbf{F}\right)\Delta t = -\left(\iiint_V \frac{\partial\rho}{\partial t}\,dx\,dy\,dz\right)\Delta t + \left(\iiint_V K\rho\,dx\,dy\,dz\right)\Delta t.$$

Dividing out Δt gives us

$$\iint_S \rho\mathbf{F} = \iiint_V \left(K\rho - \frac{\partial\rho}{\partial t}\right)dx\,dy\,dz.$$

Apply Gauss's theorem on the left to get

$$\iiint_V (\nabla\cdot\rho\mathbf{F})\,dx\,dy\,dz = \iiint_V \left(K\rho - \frac{\partial\rho}{\partial t}\right)dx\,dy\,dz.$$

Then,

$$\iiint_V \left(\nabla \cdot (\rho \mathbf{F}) + \frac{\partial \rho}{\partial t} - K\rho \right) dx\, dy\, dz = 0.$$

Again, by the arbitrary choice of S, we must have

$$\nabla \cdot (\rho \mathbf{F}) + \frac{\partial \rho}{\partial t} - K\rho = 0.$$

If $K = 0$ (no sources or sinks), this is

$$\nabla \cdot (\rho \mathbf{F}) + \frac{\partial \rho}{\partial t} = 0,$$

the *continuity equation* of fluid dynamics. It is, in fact, a law of conservation of matter.

In the above, $\partial \rho/\partial t$ = rate of change of ρ with time at a given point (x, y, z). If we imagine a particle of fluid moving along a path parametrized by

$$x = x(t), \qquad y = y(t), \qquad z = z(t),$$

then $\rho = \rho(x(t), y(t), z(t), t)$ and

$$\frac{d\rho}{dt} = \frac{\partial \rho}{\partial x}\frac{dx}{dt} + \frac{\partial \rho}{\partial y}\frac{dy}{dt} + \frac{\partial \rho}{\partial z}\frac{dz}{dt} + \frac{\partial \rho}{\partial t}.$$

This is the rate of change of ρ with time as the particle moves along the given path in the fluid. Now,

$$\mathbf{F} = \frac{dx}{dt}\mathbf{i} + \frac{dy}{dt}\mathbf{j} + \frac{dz}{dt}\mathbf{k}$$

is the velocity of the particle, and

$$\nabla \rho = \frac{\partial \rho}{\partial x}\mathbf{i} + \frac{\partial \rho}{\partial y}\mathbf{j} + \frac{\partial \rho}{\partial z}\mathbf{k}.$$

Thus,

$$\frac{d\rho}{dt} = \mathbf{F} \cdot \nabla \rho + \frac{\partial \rho}{\partial t}.$$

Solve this for $\partial \rho/\partial t$ and substitute into the continuity equation to get

$$\nabla \cdot (\rho \mathbf{F}) + \frac{d\rho}{dt} - \mathbf{F} \cdot \nabla \rho = 0$$

or

$$\frac{d\rho}{dt} = \mathbf{F} \cdot \nabla \rho - \nabla \cdot (\rho \mathbf{F}).$$

Now,

$$\nabla \cdot (\rho \mathbf{F}) = \mathbf{F} \cdot \nabla \rho + \rho \nabla \cdot \mathbf{F},$$

an identity derived in detail when treating the heat equation. Thus,

$$\frac{d\rho}{dt} = \mathbf{F} \cdot \nabla\rho - \mathbf{F} \cdot \nabla\rho - \rho\nabla \cdot \mathbf{F} = -\rho\nabla \cdot \mathbf{F},$$

which can be written

$$-\frac{1}{\rho}\frac{d\rho}{dt} = \nabla \cdot \mathbf{F}.$$

We call the fluid *incompressible* if $d\rho/dt = 0$ (for example, water under normal conditions). In this case, the simple equation

$$\nabla \cdot \mathbf{F} = 0$$

governs the fluid flow. In the special case that $\mathbf{F}$ has a potential function, say, $\mathbf{F} = \nabla\phi$, then $\nabla \cdot \mathbf{F} = 0$ becomes $\nabla \cdot (\nabla\phi) = 0$, and this is

$$\frac{\partial^2\phi}{\partial x^2} + \frac{\partial^2\phi}{\partial y^2} + \frac{\partial^2\phi}{\partial z^2} = 0,$$

which is Laplace's equation again.

4. GAUSS'S LAW AND POISSON'S EQUATION

An electric charge q at the origin produces an electric field

$$\mathbf{E} = \frac{q\mathbf{R}}{4\pi\varepsilon\|\mathbf{R}\|^3},$$

where $\mathbf{R} = x\mathbf{i} + y\mathbf{j} + z\mathbf{k}$, as usual, and ε is the electric permittivity of the medium.

Gauss's law says that

$$\iint_S \mathbf{E} = \begin{cases} \dfrac{q}{\varepsilon} & \text{if } S \text{ encloses the origin,} \\ 0 & \text{otherwise.} \end{cases}$$

To establish this result, let S be a regular closed surface, positively oriented, and consider two cases.

Case 1 S does not enclose the origin.
Then, Gauss's theorem applies, and

$$\iint_S \mathbf{E} = \iiint_V \nabla \cdot \mathbf{E}\, dx\, dy\, dz.$$

It is easy to check that $\nabla \cdot \mathbf{E} = 0$, establishing Gauss's law in this case.

Case 2 The origin is interior to S.
Now Gauss's theorem does not apply, as $\mathbf{E}$ is undefined at the origin inside S. Enclose $(0, 0, 0)$ in a sphere S' of sufficiently small radius that S' is entirely within S. By cutting

a small tube T from S to S', we get a new surface $\mathscr{S}$, consisting of S, S', and T, which does not contain the origin (see Figure 202). By case 1,

$$\iint_{\mathscr{S}} \mathbf{E} = 0 = \iint_S \mathbf{E} + \iint_{S'} \mathbf{E} + \iint_T \mathbf{E}.$$

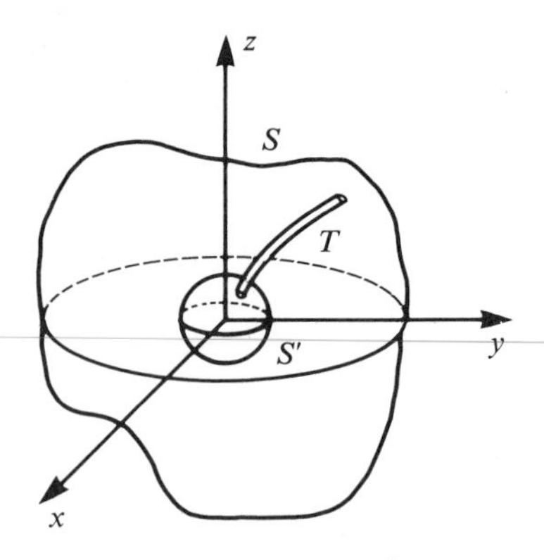

Figure 202. Tube T connects S to sphere S' about the origin.

Take a kind of limit as T contracts to a line. Then,

$$\iint_T \mathbf{E} \to 0$$

and we get

$$\iint_{\mathscr{S}} \mathbf{E} \to \iint_S \mathbf{E} + \iint_{S'} \mathbf{E} = 0.$$

Then, at least intuitively,

$$\iint_S \mathbf{E} = -\iint_{S'} \mathbf{E}.$$

But we can evaluate $\iint_{S'} \mathbf{E}$ explicitly, since S' is a sphere. If S' has radius r, then we can parametrize S' in terms of spherical coordinates

$$x = r\cos(\theta)\sin(\phi), \qquad y = r\sin(\theta)\sin(\phi),$$
$$z = r\cos(\phi),\ 0 \leq \theta \leq 2\pi, \qquad 0 \leq \phi \leq \pi.$$

Now,

$$\mathbf{N} = -\sin(\phi)[r^2\cos(\theta)\sin(\phi)\mathbf{i} + r^2\sin(\theta)\sin(\phi)\mathbf{j} + r^2\cos(\phi)\mathbf{k}].$$

(*Note:* This is an *inner* normal on S' considered by itself, but is *outer* on S' considered as part of $\mathscr{S}$.)

After a routine calculation, we find that, on S',

$$\mathbf{E}\cdot\mathbf{N} = -\frac{q}{4\pi\varepsilon}\sin(\phi),$$

and we get

$$\iint_{S'} \mathbf{E} = \int_0^{\pi}\int_0^{2\pi} \mathbf{E}\cdot\mathbf{N}\, d\theta\, d\phi = -\frac{q}{\varepsilon}.$$

Then,

$$\iint_S \mathbf{E} = -\left(-\frac{q}{\varepsilon}\right) = \frac{q}{\varepsilon},$$

finishing case 2.

We can use Gauss's law in an extended form to draw further conclusions. Let Q be the charge density. Then,

$$q = \iiint_V Q \, dx \, dy \, dz$$

is the distributed charge over the surface S bounding V. If S encloses the origin, then, by Gauss's law,

$$\iint_S \mathbf{E} = \frac{q}{\varepsilon} = \frac{1}{\varepsilon} \iiint_V Q \, dx \, dy \, dz.$$

Apply Gauss's theorem to $\iint_S \mathbf{E}$ (using an argument like that in case 2 above) to get

$$\iiint_V \nabla \cdot \mathbf{E} \, dx \, dy \, dz = \iiint_V \frac{1}{\varepsilon} Q \, dx \, dy \, dz.$$

Then,

$$\iiint_V \left(\nabla \cdot \mathbf{E} - \frac{1}{\varepsilon} Q \right) dx \, dy \, dz = 0.$$

Hence,

$$\nabla \cdot \mathbf{E} - \frac{1}{\varepsilon} Q = 0.$$

This is one of Maxwell's equations.

If $\mathbf{E} = \nabla \phi$, then we get

$$\frac{\partial^2 \phi}{\partial x^2} + \frac{\partial^2 \phi}{\partial y^2} + \frac{\partial^2 \phi}{\partial z^2} = \frac{1}{\varepsilon} Q.$$

This is called *Poisson's equation.* When $Q = 0$, we get Laplace's equation again.

5. GREEN'S IDENTITIES

There are several integral formulas which prove useful in handling partial differential equations of applied mathematics. *Green's first identity* is

$$\iint_S f \nabla g = \iiint_V (f \nabla^2 g + \nabla f \cdot \nabla g) \, dx \, dy \, dz.$$

Here, S is a regular closed surface bounding the region V, and f and g are assumed continuous with continuous first and second partials on S and in V. The formula is proved by using Gauss's theorem to write

$$\iint_S f \nabla g = \iiint_V \nabla \cdot (f \nabla g) \, dx \, dy \, dz$$

and then noting that

$$\nabla \cdot (f\nabla g) = f\nabla^2 g + \nabla f \cdot \nabla g.$$

Green's second identity is

$$\iint_S (f\nabla g - g\nabla f) = \iiint_V (f\nabla^2 g - g\nabla^2 f)\, dx\, dy\, dz.$$

To prove this, use Green's first identity with f and g interchanged to write:

$$\iint_S f\nabla g = \iiint_V (f\nabla^2 g + \nabla f \cdot \nabla g)\, dx\, dy\, dz$$

and

$$\iint_S g\nabla f = \iiint_V (g\nabla^2 f + \nabla g \cdot \nabla f)\, dx\, dy\, dz.$$

Subtracting the last two lines gives us Green's second identity.

Putting $f = 1$ into Green's second identity gives us

$$\iint_S \nabla g = \iiint_V \nabla^2 g\, dx\, dy\, dz.$$

Now, if g satisfies Laplace's equation in V, then $\nabla^2 g = 0$, and we conclude that

$$\iint_S \nabla g = 0.$$

There are many other applications of Gauss's theorem, but we shall stop with these for now and treat some applications of Stokes's theorem in the next section.

PROBLEMS FOR SECTION 11.11

1. Determine $\iint_S \nabla f$ over a piecewise smooth surface S if $\nabla^2 f = 0$ on S and in the region V enclosed by S.
2. Show that

$$\nabla^2 f(P_0) = \lim_{\bar{V}\to 0} \frac{1}{\bar{V}} \iint_S \nabla f,$$

where S is a piecewise smooth surface bounding a region V having volume $\bar{V}$, and $\lim_{\bar{V}\to 0}$ means that S contracts to P_0. Use this formula to give a physical interpretation of $\nabla^2 f$.

3. A *Dirichlet problem* consists of finding a function satisfying Laplace's equation $\nabla^2 u = 0$ for (x, y, z) in some region V enclosed by a piecewise smooth surface S, subject to the condition that u must take on given values on S:

$$u(x, y, z) = f(x, y, z) \quad \text{for} \quad (x, y, z) \text{ on } S.$$

Prove that there can be only one continuous solution having continuous first and second partials in V.

Hint: Suppose that $F(x, y, z)$ and $G(x, y, z)$ are solutions. Let $w(x, y, z) = F(x, y, z) - G(x, y, z)$. Show that $\nabla^2 w = 0$ in V and $w(x, y, z) = 0$ on S. Use Gauss's theorem to show that $w(x, y, z) = 0$ in V.

4. The partial differential equation governing heat conduction in a solid V bounded by a piecewise smooth surface S can be written

$$\frac{\partial u}{\partial t} = k\nabla^2 u + \phi(x, y, z, t),$$

where $u(x, y, z, t) =$ temperature at point (x, y, z) and time t. The temperature is given on the boundary S of V:

$$u(x, y, z, t) = f(x, y, z) \quad \text{for all} \quad (x, y, z) \text{ on } S \quad \text{and} \quad t > 0.$$

The initial temperature (i.e., at time 0) is given inside V:

$$u(x, y, z, 0) = g(x, y, z) \quad \text{for} \quad (x, y, z) \text{ in } V.$$

Show that there can be at most one temperature function $u(x, y, z, t)$, continuous in V and having continuous first and second partials, satisfying the partial differential equation and the initial and boundary conditions given.

Hint: Suppose that $A(x, y, z, t)$ and $B(x, y, z, t)$ are solutions. Let $w(x, y, z, t) = A(x, y, z, t) - B(x, y, z, t)$. Show that w satisfies $\partial w/\partial t = k\nabla^2 w$, that $w(x, y, z, t) = 0$ for (x, y, z) on S and $t > 0$, and that $w(x, y, z, 0) = 0$ for (x, y, z) in V.

Now let $I(t) = \frac{1}{2}\iiint_V w^2(x, y, z, t)\,dx\,dy\,dz$. Using the divergence theorem, show that

$$I'(t) = -k \iiint_V \left[\left(\frac{\partial w}{\partial x}\right)^2 + \left(\frac{\partial w}{\partial y}\right)^2 + \left(\frac{\partial w}{\partial z}\right)^2\right] dx\,dy\,dz.$$

Hence, conclude that $I'(t) \le 0$ for all times $t > 0$. Next, apply the mean value theorem to $I(t)$ on the interval $[0, t]$ to show that $I(t) \le 0$ for $t > 0$. Thus, show that $I(t) = 0$. Finally, show that $w(x, y, z, t) = 0$ for $t > 0$ and (x, y, z) in V, and hence that $A(x, y, z, t) = B(x, y, z, t)$.

5. Prove that the following problem can have only one solution which is continuous and has continuous first and second partials.

$$\frac{\partial u}{\partial t} = k\nabla^2 u + \phi(x, y, z, t) \quad \text{for} \quad (x, y, z) \text{ in } V, \quad t > 0,$$

$$\frac{\partial u}{\partial \eta} + hu = f(x, y, z, t) \quad \text{for} \quad (x, y, z) \text{ on } S \quad \text{and} \quad t > 0,$$

$$u(x, y, z, 0) = g(x, y, z) \quad \text{for} \quad (x, y, z) \text{ in } V.$$

Here, $\partial u/\partial \eta =$ normal derivative ($= \nabla u \cdot \mathbf{N}$, where $\mathbf{N} =$ normal to S) and h is a given positive constant. (*Hint:* Modify the proof in Problem 4.)

6. Let $f(x, y, z)$ and $g(x, y, z)$ satisfy Laplace's equation in a region V bounded by a piecewise smooth surface S. Suppose that $\partial f/\partial \eta = \partial g/\partial \eta$ at all points of S. Prove that for some constant K, $f(x, y, z) = g(x, y, z) + K$ for all (x, y, z) in V.

7. Express the moment of inertia I of a uniform solid about the z-axis as the flux of some vector field across the surface of the body. (Recall that $I = \iiint_V \rho(x^2 + y^2)\,dx\,dy\,dz$, where $\rho =$ constant density of the body.)

11.12 SOME APPLICATIONS OF STOKES'S THEOREM

1. CURL OF A VECTOR

In Section 11.5, we gave some indication of a physical significance of curl. Here we use Stokes's theorem to pursue the matter further.

Let $\mathbf{F}(x, y, z, t)$ be the velocity of a fluid. Imagine a plane (i.e., flat) surface S in the fluid about some point P_0, bounded by a regular closed curve C, as shown in Figure 203. By Stokes's theorem,

$$\iint_S \nabla \times \mathbf{F} = \int_C \mathbf{F}.$$

Now, on S, we can write, say, z in terms of x and y for (x, y) on the projection D of S into the xy-plane (see Figure 204). Then, by the mean value theorem for double integrals.

$$\iint_S \nabla \times \mathbf{F} = \iint_D (\nabla \times \mathbf{F}) \cdot \mathbf{N}\, dx\, dy = [(\nabla \times \mathbf{F}) \cdot \mathbf{N}]|_P \cdot (\text{area of } S),$$

for some point P on S. Then,

$$[(\nabla \times \mathbf{F}) \cdot \mathbf{N}]|_P = \left(\frac{1}{\text{area of } S}\right) \int_C \mathbf{F}.$$

Now let S contract to the point P_0, remaining in the same plane. This does not affect $\mathbf{N}$, which is constant, but forces P to approach P_0. Then we get

$$[(\nabla \times \mathbf{F}) \cdot \mathbf{N}]|_{P_0} = \lim_{S \to P_0} \left(\frac{1}{\text{area of } S}\right) \int_C \mathbf{F}.$$

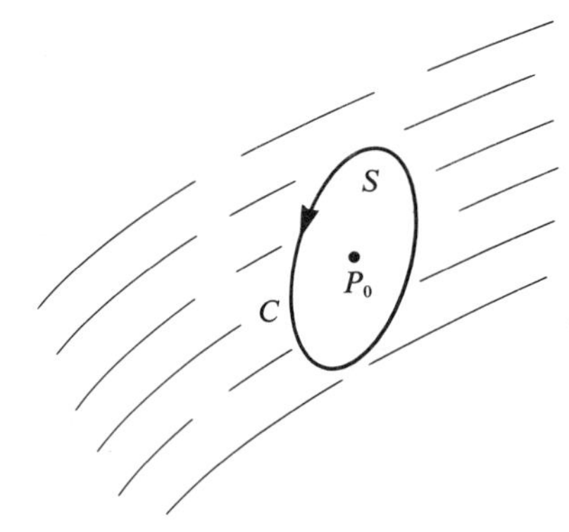

Figure 203. Flat surface element in fluid about P_0.

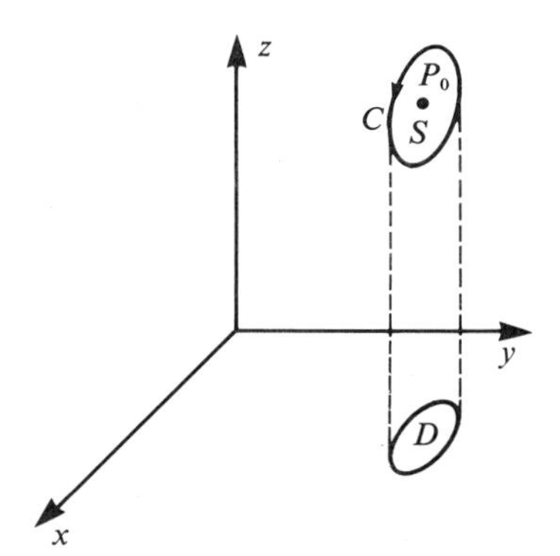

Figure 204. Projection of S onto xy-plane.

We may think of $\int_C \mathbf{F}$ as measuring the circulation of the fluid about C in the plane determined by $\mathbf{N}$. Then,

$$\left(\frac{1}{\text{area of } S}\right) \int_C \mathbf{F} = \text{circulation about } C \text{ per unit area enclosed by } C.$$

In the limit as S shrinks to P_0, we get

$$\begin{aligned} &\text{component of curl } \mathbf{F} \text{ at } P_0 \text{ normal to } S \\ &= \text{circulation per unit area in the plane normal to } S. \end{aligned}$$

2. MAXWELL'S EQUATIONS

We shall now apply Gauss's and Stokes's theorems, and some vector analysis, to a mathematical treatment of electric and magnetic fields.

As a starting point, define the following standard symbols:

$\mathbf{E}$ = electric intensity
$\mathbf{H}$ = magnetic intensity
$\mathbf{J}$ = current density
ε = permittivity of the medium
μ = permeability
σ = conductivity
Q = charge density
$\mathbf{D} = \varepsilon\mathbf{E}$ = electric flux density
$\mathbf{B} = \mu\mathbf{H}$ = magnetic flux density
$q = \iiint_V Q\, dx\, dy\, dz$ = total charge in a region V
$\phi = \iint_S \mathbf{B}$ = total magnetic flux through a closed surface S (toward the outside)
$i = \iint_S \mathbf{J}$ = total current flowing from inside to outside of S through S.

As with any real-world study, one begins with experimental observations and empirically derived relationships. The following have been observed:

Faraday's law:

$$\int_C \mathbf{E} = -\frac{\partial \phi}{\partial t}, \quad \text{around any closed curve } C.$$

This says that the tangential component of $\mathbf{E}$ measured around C is the negative of the rate of change with time of the magnetic flux through any surface bounded by C.

Ampère's law:

$$\int_C \mathbf{H} = i.$$

This says that the measure of the tangential component of magnetic intensity about C is the current flowing through any surface bounded by C.

Gauss's laws:

$$\iint_S \mathbf{D} = q \quad \text{and} \quad \iint_S \mathbf{B} = 0.$$

These say, respectively, that the measure of the normal component of electric flux density across a closed surface S is the total charge in the enclosed region, and that the measure of the normal component of magnetic flux density over any closed surface is zero.

We can now begin our analysis. Apply Stokes's theorem to Faraday's law to get

$$\int_C \mathbf{E} = \iint_S \nabla \times \mathbf{E} = -\frac{\partial \phi}{\partial t} = -\frac{\partial}{\partial t}\iint_S \mathbf{B} = \iint_S -\frac{\partial \mathbf{B}}{\partial t}.$$

Then,

$$\iint_S \left(\nabla \times \mathbf{E} + \frac{\partial \mathbf{B}}{\partial t}\right) = 0.$$

Since S is any regular closed surface in the medium, having C as boundary, then

$$\nabla \times \mathbf{E} + \frac{\partial \mathbf{B}}{\partial t} = \mathbf{0}.$$

This is often written

$$\boxed{\nabla \times \mathbf{E} = -\frac{\partial \mathbf{B}}{\partial t}.}$$

A similar analysis applied to Ampere's law gives us

$$\boxed{\nabla \times \mathbf{H} = \mathbf{J}.}$$

Maxwell observed that

$$\mathbf{J} = \sigma \mathbf{E} + \varepsilon \frac{\partial \mathbf{E}}{\partial t}.$$

Hence,

$$\boxed{\nabla \times \mathbf{H} = \sigma \mathbf{E} + \varepsilon \frac{\partial \mathbf{E}}{\partial t}.}$$

Now start on a new tack. Apply Gauss's theorem to Gauss's law $\iint_S \mathbf{D} = q$ to get

$$\iint_S \mathbf{D} = \iiint_V (\nabla \cdot \mathbf{D})\, dx\, dy\, dz = q = \iiint_V Q\, dx\, dy\, dz.$$

Then,

$$\boxed{\nabla \cdot \mathbf{D} = Q.}$$

Now go back to $\nabla \times \mathbf{E} = -\partial \mathbf{B}/\partial t$ and take the curl of both sides:

$$\nabla \times (\nabla \times \mathbf{E}) = \nabla \times \left(-\frac{\partial \mathbf{B}}{\partial t}\right) = -\frac{\partial}{\partial t}(\nabla \times \mathbf{B}).$$

We can interchange curl and $\partial/\partial t$ here because one involves only the space variables, the other only time. Since $\mathbf{B} = \mu \mathbf{H}$, then,

$$\nabla \times (\nabla \times \mathbf{E}) = -\frac{\partial}{\partial t}(\nabla \times \mu \mathbf{H}) = -\mu \frac{\partial}{\partial t}(\nabla \times \mathbf{H}).$$

It is routine to verify that this is the same as

$$\nabla(\nabla \cdot \mathbf{E}) - (\nabla \cdot \nabla)\mathbf{E} = -\mu \frac{\partial}{\partial t}(\nabla \times \mathbf{H}),$$

where

$$\nabla \cdot \nabla = \frac{\partial^2}{\partial x^2} + \frac{\partial^2}{\partial y^2} + \frac{\partial^2}{\partial z^2}.$$

Since

$$\nabla \times \mathbf{H} = \sigma \mathbf{E} + \varepsilon \frac{\partial \mathbf{E}}{\partial t},$$

we have finally

$$\boxed{\nabla(\nabla \cdot \mathbf{E}) - (\nabla \cdot \nabla)\mathbf{E} = -\mu \frac{\partial}{\partial t}\left(\sigma \mathbf{E} + \varepsilon \frac{\partial \mathbf{E}}{\partial t}\right).}$$

Now, in practice, we often have $Q = 0$. Then,

$$Q = \nabla \cdot \mathbf{D} = \nabla \cdot (\varepsilon \mathbf{E}) = \varepsilon \nabla \cdot \mathbf{E} = 0;$$

so

$$\nabla \cdot \mathbf{E} = 0$$

also. Then we have

$$\boxed{(\nabla \cdot \nabla)\mathbf{E} = \mu\sigma \frac{\partial \mathbf{E}}{\partial t} + \mu\varepsilon \frac{\partial^2 \mathbf{E}}{\partial t^2}.}$$

This is *Maxwell's equation for* **E**.

By a similar analysis, we can also obtain *Maxwell's equation for* **H**:

$$\boxed{(\nabla \cdot \nabla)\mathbf{H} = \mu\sigma \frac{\partial \mathbf{H}}{\partial t} + \mu\varepsilon \frac{\partial^2 \mathbf{H}}{\partial t^2}.}$$

As a special case, if $\sigma = 0$ (perfect dielectric), Maxwell's equations become

$$(\nabla \cdot \nabla)\mathbf{E} = \mu\varepsilon \frac{\partial^2 \mathbf{E}}{\partial t^2} \quad \text{and} \quad (\nabla \cdot \nabla)\mathbf{H} = \mu\varepsilon \frac{\partial^2 \mathbf{H}}{\partial t^2}.$$

This is a vector form of the three-dimensional wave equation.

If instead of $\sigma = 0$, we have $\varepsilon = 0$, then we have

$$(\nabla \cdot \nabla)\mathbf{E} = \mu\sigma \frac{\partial \mathbf{E}}{\partial t} \quad \text{and} \quad (\nabla \cdot \nabla)\mathbf{H} = \mu\sigma \frac{\partial \mathbf{E}}{\partial t}.$$

This is a vector form of the three-dimensional heat equation.

3. POTENTIAL THEORY IN THREE-SPACE

In Section 11.8, we discussed potential theory in the plane, making use of Green's theorem. We now wish to use Stokes's theorem to discuss potential theory in three-space.

Suppose that $\mathbf{F}$ is a vector field defined and continuous over a region Ω in three-space. Following the lead from the two-dimensional case, we say that:

1. $\mathbf{F}$ is *conservative* if $\int_C \mathbf{F} = 0$ for every regular closed curve in Ω.
2. The line integral of $\mathbf{F}$ is *independent of path* in Ω if $\int_C \mathbf{F} = \int_K \mathbf{F}$ whenever C and K are regular curves in Ω with the same initial point and the same terminal point.

If $\mathbf{F} = \nabla\phi$ in Ω, then, for any regular curve C from P_0 to P_1 in Ω,

$$\int_C \mathbf{F} = \phi(P_1) - \phi(P_0).$$

Thus, $\mathbf{F}$ is both conservative and independent of path in Ω. The proof is like that for the plane in Section 11.8.

The converse holds if Ω is a *domain* [i.e., (1) around any point in Ω is a sphere containing only points of Ω, and (2) between any two points in Ω there is a regular curve lying in Ω].

We can now ask for a simple test for existence of a potential function for a three-dimensional vector field. A hint is provided by the vector identity

$$\nabla \times (\nabla\phi) = \mathbf{0}.$$

If $\mathbf{F} = \nabla\phi$, then certainly $\nabla \times \mathbf{F} = \mathbf{0}$. Is the converse true? If $\nabla \times \mathbf{F} = \mathbf{0}$, does $\mathbf{F}$ have a potential function? The answer is yes—but only with an additional assumption on Ω. We call Ω *simply connected* if every closed curve in Ω is the boundary of a regular surface in Ω. For example, the region bounded by a doughnut-shaped surface is not simply connected (take C to go around the hole, as in Figure 205).

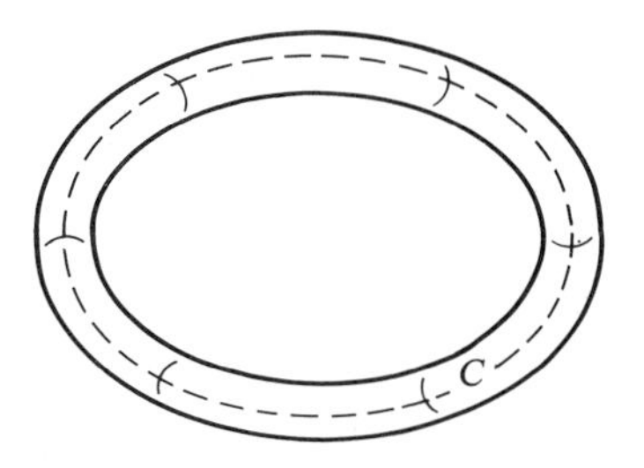

Figure 205. A nonsimply connected region in R^3.

Here is the test for a potential.

THEOREM 94

If $\mathbf{F}$ and $\nabla \times \mathbf{F}$ are continuous in a simply connected domain Ω, then $\mathbf{F}$ has a potential function if and only if $\nabla \times \mathbf{F} = \mathbf{0}$ in Ω.

Proof If $\mathbf{F} = \nabla\phi$ in Ω, then $\nabla \times \mathbf{F} = \nabla \times (\nabla\phi) = \mathbf{0}$.

Conversely, suppose that $\nabla \times \mathbf{F} = \mathbf{0}$ in Ω. Let C be any regular closed curve in Ω. Since Ω is simply connected, C is the boundary of a regular surface S lying in Ω. Then, by Stokes's theorem,

$$\int_C \mathbf{F} = \iint_S \nabla \times \mathbf{F} = \iint_S \mathbf{0} = 0.$$

Hence, $\mathbf{F}$ is independent of path in Ω, and so has a potential function.

In practice, we find potential functions in three dimensions just as we did in two—by integrating.

EXAMPLE 352

Let $\mathbf{F} = 2xy\mathbf{i} + z^2\mathbf{j} + (x - y + z)\mathbf{k}$.

Here, we can let Ω be all of three-space, as the components are continuous everywhere. Now, $\nabla \times \mathbf{F} = (-2z - 1)\mathbf{i} - \mathbf{j} - 2x\mathbf{k} \neq \mathbf{0}$; so $\mathbf{F}$ has no potential.

EXAMPLE 353

Let $\mathbf{F} = (yze^{xyz} - 4x)\mathbf{i} + (xze^{xyz} + z)\mathbf{j} + (xye^{xyz} + y)\mathbf{k}$, again with Ω all of three-space. It is easy to check that $\nabla \times \mathbf{F} = \mathbf{0}$; so $\mathbf{F}$ has a potential. Find one.

We want

$$\mathbf{F} = \nabla\phi = \frac{\partial \phi}{\partial x}\mathbf{i} + \frac{\partial \phi}{\partial y}\mathbf{j} + \frac{\partial \phi}{\partial z}\mathbf{k}.$$

Then we need

$$\frac{\partial \phi}{\partial x} = yze^{xyz} - 4x,$$

$$\frac{\partial \phi}{\partial y} = xze^{xyz} + z,$$

and

$$\frac{\partial \phi}{\partial z} = xye^{xyz} + y.$$

Choose one, say,

$$yze^{xyz} - 4x = \frac{\partial \phi}{\partial x}$$

and integrate with respect to x, treating y and z as constants. Then,

$$\phi(x, y, z) = e^{xyz} - 2x^2 + K(y, z),$$

where $K(y, z)$ is the "constant" of integration.

Next,

$$\frac{\partial \phi}{\partial y} = xze^{xyz} + \frac{\partial K}{\partial y} = xze^{xyz} + z.$$

Then,

$$\frac{\partial K}{\partial y} = z.$$

Integrating with respect to y, treating z as a constant, we get

$$K(y, z) = zy + A(z).$$

Thus far,

$$\phi = e^{xyz} - 2x^2 + zy + A(z).$$

Finally, we need

$$\frac{\partial \phi}{\partial z} = xye^{xyz} + y + A'(z) = xye^{xyz} + y.$$

Then, $A'(z) = 0$; so $A(z) =$ constant (arbitrary).

For any scalar c,

$$\phi = e^{xyz} - 2x^2 + zy + c$$

is a potential function for **F**.

In the next section we shall briefly consider curvilinear coordinates and show how such vector operations as divergence and curl can be written in terms of other coordinate systems.

PROBLEMS FOR SECTION 11.12

In each of Problems 1 through 10, take Ω as all of three-space and use Theorem 94 to determine whether or not **F** has a potential function. If a potential function exists, find one.

1. $\mathbf{F} = e^{xyz}[(1 + yzx)\mathbf{i} + x^2z\mathbf{j} + x^2y\mathbf{k}]$
2. $\mathbf{F} = \cosh(x + y)(\mathbf{i} + \mathbf{j})$
3. $\mathbf{F} = [\cos(x) + y\sin(xy)]\mathbf{i} + x\sin(xy)\mathbf{j} + \mathbf{k}$
4. $\mathbf{F} = 2x\mathbf{i} - 2y\mathbf{j} + 2z\mathbf{k}$
5. $\mathbf{F} = (2x^2 + 3y^2z)\mathbf{i} + 6yxz\mathbf{j} + 3y^2x\mathbf{k}$
6. $\mathbf{F} = \mathbf{i} - 2\mathbf{j} + 2z\mathbf{k}$
7. $\mathbf{F} = z\mathbf{i} + \mathbf{j} + x\mathbf{k}$
8. $\mathbf{F} = 2x\mathbf{i} - 2\mathbf{j} + \mathbf{k}$
9. $\mathbf{F} = yz\cos(x)\mathbf{i} + [z\sin(x) + 1]\mathbf{j} + y\sin(x)\mathbf{k}$
10. $\mathbf{F} = (x^2 - 2y)\mathbf{i} + xyz\mathbf{j} - yz^2\mathbf{k}$

In each of Problems 11 through 25, find a potential function for **F** and use it to evaluate $\int_C \mathbf{F}$, for C any smooth curve between the given endpoints.

11. $\mathbf{F} = \mathbf{i} - 9y^2z\mathbf{j} - 3y^3\mathbf{k}$; C from $(1, 1, 1)$ to $(0, 3, 5)$
12. $\mathbf{F} = [y\cos(xz) - xyz\sin(xz)]\mathbf{i} + x\cos(xz)\mathbf{j} - x^2y\sin(xz)\mathbf{k}$; C from $(1, 0, \pi)$ to $(1, 1, 7)$
13. $\mathbf{F} = 6x^2e^{yz}\mathbf{i} + 2x^3ze^{yz}\mathbf{j} + 2x^3ye^{yz}\mathbf{k}$; C from $(0, 0, 0)$ to $(1, 2, -1)$
14. $\mathbf{F} = -8y^2\mathbf{i} - (16xy + 4z)\mathbf{j} - 4y\mathbf{k}$; C from $(-2, 1, 1)$ to $(1, 3, 2)$
15. $\mathbf{F} = -\mathbf{i} + 2z^2\mathbf{j} + 4yz\mathbf{k}$; C from $(0, 0, -4)$ to $(1, 1, 6)$
16. $\mathbf{F} = (y - 4xz)\mathbf{i} + x\mathbf{j} + (3z^2 - 2x^2)\mathbf{k}$; C from $(1, 1, 1)$ to $(3, 1, 4)$

17. $\mathbf{F} = (4y^3 - 8z)\mathbf{i} + 12xy^2\mathbf{j} - 8x\mathbf{k}$; C from $(-1, 2, 2)$ to $(0, 1, 6)$

18. $\mathbf{F} = -\left(\frac{e^x}{yz}\right)\mathbf{i} + \left(\frac{e^x}{y^2z}\right)\mathbf{j} + \left(\frac{e^x}{yz^2}\right)\mathbf{k}$; C from $(1, 2, 2)$ to $(2, 3, 4)$

19. $\mathbf{F} = z\sin(yz)\mathbf{i} + xz^2\cos(yz)\mathbf{j} + [xyz\cos(yz) + x\sin(yz)]\mathbf{k}$; C from $(0, 1, 1)$ to $(1, 7, -2)$

20. $\mathbf{F} = \dfrac{1}{x - y + z}(\mathbf{i} - \mathbf{j} + \mathbf{k})$; C from $(1, -2, 2)$ to $(0, -1, 4)$

21. $\mathbf{F} = yz\cosh(xy)\mathbf{i} + xz\cosh(xy)\mathbf{j} + \sinh(xy)\mathbf{k}$; C from $(1, 1, 3)$ to $(1, -1, 4)$

22. $\mathbf{F} = (yz^2 - 2xz)\mathbf{i} + xz^2\mathbf{j} + (2xyz - x^2)\mathbf{k}$; C from $(-1, 4, 4)$ to $(0, 0, 0)$

23. $\mathbf{F} = zy\sin(xy)\mathbf{i} + zx\sin(xy)\mathbf{j} + (2z - \cos(xy))\mathbf{k}$; C from $(1, 7, 1)$ to $(1, 1, 3)$

24. $\mathbf{F} = (24x^2y - z^2)\mathbf{i} + 8x^3\mathbf{j} - 2xz\mathbf{k}$; C from $(0, 0, 0)$ to $(1, 1, 3)$

25. $\mathbf{F} = (3x^2 - 1)\mathbf{i} + 2yz\mathbf{j} + y^2\mathbf{k}$; C from $(1, -2, 1)$ to $(3, 3, -1)$

26. Prove that

$$\iint_S \nabla f \times \nabla g = \oint_C f\nabla g,$$

where C is a smooth closed curve bounding the smooth surface S.

27. Using Maxwell's equation $\nabla \times \mathbf{H} = \mathbf{0}$, and assuming that $\mathbf{E} = \mathbf{0}$, show that

$$\oint_C \mathbf{H} = \text{net electric current enclosed by the closed loop } C.$$

28. Assuming that $\nabla \times \mathbf{B} = \mu_0\mathbf{J}$, show that

$$\oint_C \mathbf{B} = \mu_0 i \text{ for any closed loop } C.$$

11.13 CURVILINEAR COORDINATES

Thus far, everything has been phrased in rectangular coordinates. While these are natural and appealing, some settings cry out for other systems—spheres for spherical coordinates, cylinders for cylindrical coordinates, and there are others as well.

To begin in general terms, suppose that we have a rectangular coordinate system with axes labeled x, y, and z, as usual. Imagine that we also have some other coordinate system, with coordinates labeled q_1, q_2, and q_3. We assume that the two systems are related by equations

$$x = x(q_1, q_2, q_3), \qquad y = y(q_1, q_2, q_3), \quad \text{and} \quad z = z(q_1, q_2, q_3),$$

and that these can be solved to get

$$q_1 = q_1(x, y, z), \qquad q_2 = q_2(x, y, z), \quad \text{and} \quad q_3 = q_3(x, y, z).$$

Further, we assume that each point has exactly one triple (q_1, q_2, q_3) which uniquely locates that point. We call (q_1, q_2, q_3) a *system of curvilinear coordinates.*

EXAMPLE 354

Spherical Coordinates. Any point (x, y, z) has spherical coordinates (r, θ, ϕ), where

$$r = \sqrt{x^2 + y^2 + z^2}, \qquad \theta = \arcsin\left(\frac{y}{\sqrt{x^2 + y^2}}\right),$$

$$\phi = \arccos\left(\frac{z}{\sqrt{x^2 + y^2 + z^2}}\right),$$

$$0 \leq r < \infty, \qquad 0 \leq \theta < 2\pi, \qquad 0 \leq \phi \leq \pi.$$

These are shown in Figure 206.

They invert to yield

$$x = r\cos(\theta)\sin(\phi), \qquad y = r\sin(\theta)\sin(\phi), \quad \text{and} \quad z = r\cos(\phi).$$

Here, $q_1 = r$, $q_2 = \theta$, and $q_3 = \phi$.

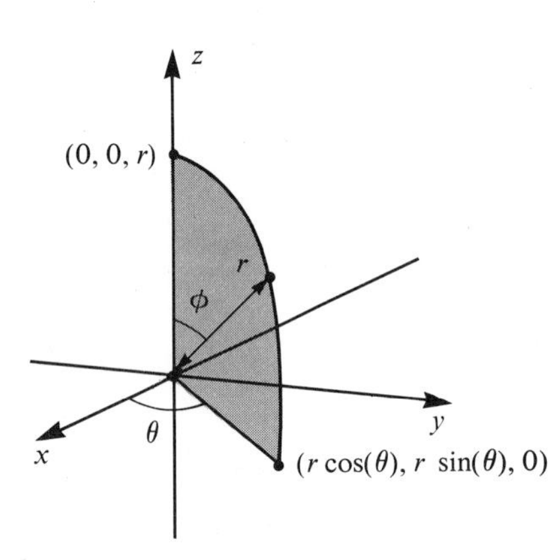

Figure 206. Spherical coordinates.

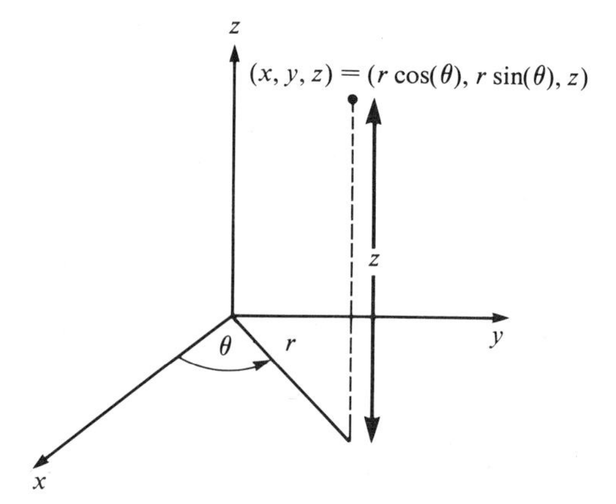

Figure 207. Cylindrical coordinates.

Cylindrical Coordinates. As shown in Figure 207, (x, y, z) is also given as (r, θ, z), where

$$r = \sqrt{x^2 + y^2}, \qquad \theta = \arctan\left(\frac{y}{x}\right), \qquad z = z,$$

$$0 \leq r < \infty, \qquad 0 \leq \theta < 2\pi, \qquad -\infty < z < \infty.$$

These invert to yield

$$x = r\cos(\theta), \qquad y = r\sin(\theta), \quad \text{and} \quad z = z.$$

Here, $q_1 = r$, $q_2 = \theta$, and $q_3 = z$.

If you think about rectangular coordinates for a moment, you can notice a very useful fact: Any point (K_1, K_2, K_3) is the intersection of the surfaces (here, planes) $x = K_1$, $y = K_2$, and $z = K_3$. Further, these surfaces are mutually orthogonal (any

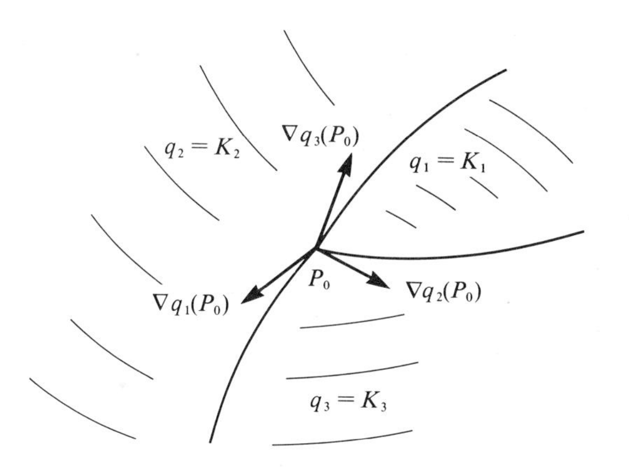

Figure 208. Orthogonal curvilinear coordinates.

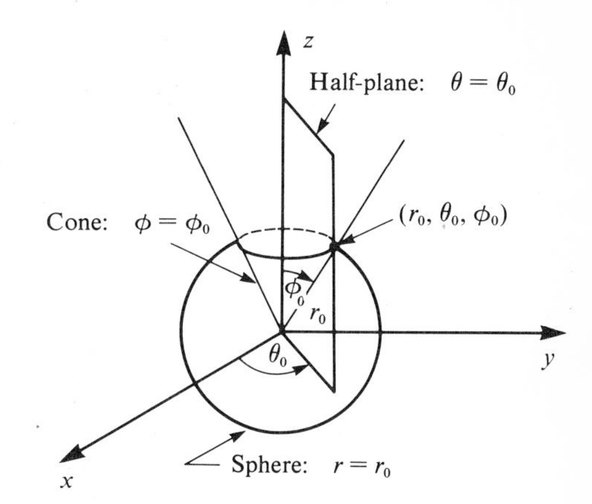

Figure 209. Spherical coordinates of a point at the intersection of mutually orthogonal coordinates surfaces.

two are perpendicular to one another). In curvilinear coordinates, we call the (not necessarily plane) surfaces $q_1 = K_1, q_2 = K_2, q_3 = K_3$ *coordinate surfaces.* If these are mutually orthogonal at each point, we call the coordinates (q_1, q_2, q_3) *orthogonal curvilinear coordinates.*

Specifically, at any point $P_0(K_1, K_2, K_3)$, the coordinate surfaces $q_1 = K_1, q_2 = K_2$, and $q_3 = K_3$ have normals

$$\nabla q_1(P_0), \qquad \nabla q_2(P_0), \quad \text{and} \quad \nabla q_3(P_0),$$

respectively. If the curvilinear coordinates are orthogonal, these vectors are mutually orthogonal (test by dot product in specific examples). In fact, we shall also suppose that $\nabla q_1(P_0)$, $\nabla q_2(P_0)$, and $\nabla q_3(P_0)$ form a right-handed system at P_0 for each point P_0, as shown in Figure 208.

EXAMPLE 355

Look at the preceding discussion in the case of spherical coordinates. A point (r_0, θ_0, ϕ_0) is the intersection of the sphere $r = r_0$, the cone $\phi = \phi_0$, and the half-plane $\theta = \theta_0$, as shown in Figure 209. Further, from Example 354, we calculate

$$\nabla r = \frac{1}{\sqrt{x^2 + y^2 + z^2}}(x\mathbf{i} + y\mathbf{j} + z\mathbf{k}),$$

$$\nabla \theta = \frac{-yx}{|x|(x^2 + y^2)}\mathbf{i} + \frac{x^2}{|x|(x^2 + y^2)}\mathbf{j},$$

and

$$\nabla \phi = \frac{-xz|z|}{\sqrt{x^2 + y^2}\sqrt{x^2 + y^2 + z^2}}\mathbf{i} - \frac{yz|z|}{\sqrt{x^2 + y^2}\sqrt{x^2 + y^2 + z^2}}\mathbf{j}$$
$$- \frac{|z|}{\sqrt{x^2 + y^2}}\left(\frac{\sqrt{x^2 + y^2 + z^2} - z^2/\sqrt{x^2 + y^2 + z^2}}{x^2 + y^2 + z^2}\right)\mathbf{k}.$$

It is easy to check that $\nabla r \cdot \nabla\theta = \nabla r \cdot \nabla\phi = \nabla\theta \cdot \nabla\phi = 0$ [except at (0, 0, 0), a "singular point" of spherical coordinates].

In rectangular coordinates, the differential element of arc length is given by

$$ds^2 = dx^2 + dy^2 + dz^2.$$

We shall assume that this is related to differentials in the q_i's by

$$ds^2 = \sum_{i=1}^{3} \sum_{j=1}^{3} h_{ij}{}^2 \, dq_i \, dq_j.$$

The numbers h_{ij} are called *scale factors*. We now seek to determine these so that we can calculate such quantities as area, volume, and arc length in curvilinear coordinates.

Write

$$dx = \frac{\partial x}{\partial q_1} dq_1 + \frac{\partial x}{\partial q_2} dq_2 + \frac{\partial x}{\partial q_3} dq_3,$$

$$dy = \frac{\partial y}{\partial q_1} dq_1 + \frac{\partial y}{\partial q_2} dq_2 + \frac{\partial y}{\partial q_3} dq_3,$$

and

$$dz = \frac{\partial z}{\partial q_1} dq_1 + \frac{\partial z}{\partial q_2} dq_2 + \frac{\partial z}{\partial q_3} dq_3.$$

Substitute these into the expression for ds^2 and we get

$$\begin{aligned}
ds^2 &= \left(\frac{\partial x}{\partial q_1}\right)^2 dq_1^2 + \left(\frac{\partial x}{\partial q_2}\right)^2 dq_2^2 + \left(\frac{\partial x}{\partial q_3}\right)^2 dq_3^2 \\
&\quad + 2\frac{\partial x}{\partial q_1}\frac{\partial x}{\partial q_2} dq_1\, dq_2 + 2\frac{\partial x}{\partial q_1}\frac{\partial x}{\partial q_3} dq_1\, dq_3 + 2\frac{\partial x}{\partial q_2}\frac{\partial x}{\partial q_3} dq_2\, dq_3 \\
&\quad + \left(\frac{\partial y}{\partial q_1}\right)^2 dq_1^2 + \left(\frac{\partial y}{\partial q_2}\right)^2 dq_2^2 + \left(\frac{\partial y}{\partial q_3}\right)^2 dq_3^2 \\
&\quad + 2\frac{\partial y}{\partial q_1}\frac{\partial y}{\partial q_2} dq_1\, dq_2 + 2\frac{\partial y}{\partial q_1}\frac{\partial y}{\partial q_3} dq_1\, dq_3 + 2\frac{\partial y}{\partial q_2}\frac{\partial y}{\partial q_3} dq_2\, dq_3 \\
&\quad + \left(\frac{\partial z}{\partial q_1}\right)^2 dq_1^2 + \left(\frac{\partial z}{\partial q_2}\right)^2 dq_2^2 + \left(\frac{\partial z}{\partial q_3}\right)^2 dq_3^2 \\
&\quad + 2\frac{\partial z}{\partial q_1}\frac{\partial z}{\partial q_2} dq_1\, dq_2 + 2\frac{\partial z}{\partial q_1}\frac{\partial z}{\partial q_3} dq_1\, dq_3 + 2\frac{\partial z}{\partial q_2}\frac{\partial z}{\partial q_3} dq_2\, dq_3 \\
&= \sum_{i=1}^{3} \sum_{j=1}^{3} h_{ij}{}^2 \, dq_i \, dq_j.
\end{aligned}$$

Solve for h_{ij} by equating like coefficients. Thus,

$$h_{11}^2 = \left(\frac{\partial x}{\partial q_1}\right)^2 + \left(\frac{\partial y}{\partial q_1}\right)^2 + \left(\frac{\partial z}{\partial q_1}\right)^2,$$

$$h_{22}^2 = \left(\frac{\partial x}{\partial q_2}\right)^2 + \left(\frac{\partial y}{\partial q_2}\right)^2 + \left(\frac{\partial z}{\partial q_2}\right)^2,$$

and

$$h_{33}^2 = \left(\frac{\partial x}{\partial q_3}\right)^2 + \left(\frac{\partial y}{\partial q_3}\right)^2 + \left(\frac{\partial z}{\partial q_3}\right)^2.$$

For $i \neq j$, $h_{ij} = 0$, assuming that we have an orthogonal system. For example,

$$\begin{aligned} h_{12}^2 &= 2\left(\frac{\partial x}{\partial q_1}\frac{\partial x}{\partial q_2} + \frac{\partial y}{\partial q_1}\frac{\partial y}{\partial q_2} + \frac{\partial z}{\partial q_1}\frac{\partial z}{\partial q_2}\right) \\ &= 2\left(\frac{\partial x}{\partial q_1}\mathbf{i} + \frac{\partial y}{\partial q_1}\mathbf{j} + \frac{\partial z}{\partial q_1}\mathbf{k}\right)\cdot\left(\frac{\partial x}{\partial q_2}\mathbf{i} + \frac{\partial y}{\partial q_2}\mathbf{j} + \frac{\partial z}{\partial q_2}\mathbf{k}\right) \\ &= 0 \quad \text{by orthogonality.} \end{aligned}$$

Rewrite h_{ii} as h_i for convenience. We then have

$$ds^2 = (h_1\, dq_1)^2 + (h_2\, dq_2)^2 + (h_3\, dq_3)^2.$$

EXAMPLE 356

In spherical coordinates, we get

$$h_r = \sqrt{\left(\frac{\partial x}{\partial r}\right)^2 + \left(\frac{\partial y}{\partial r}\right)^2 + \left(\frac{\partial z}{\partial r}\right)^2} = 1,$$

$$h_\theta = \sqrt{\left(\frac{\partial x}{\partial \theta}\right)^2 + \left(\frac{\partial y}{\partial \theta}\right)^2 + \left(\frac{\partial z}{\partial \theta}\right)^2} = r\sin(\phi),$$

and

$$h_\phi = \sqrt{\left(\frac{\partial x}{\partial \phi}\right)^2 + \left(\frac{\partial y}{\partial \phi}\right)^2 + \left(\frac{\partial z}{\partial \phi}\right)^2} = r.$$

Thus, in spherical coordinates,

$$ds^2 = dr^2 + r^2\sin^2(\phi)\, d\theta^2 + r^2\, d\phi^2.$$

Write ds_i = differential element of arc length in the q_i-direction. Then,

$$ds^2 = ds_1^2 + ds_2^2 + ds_3^2 = (h_1\, dq_1)^2 + (h_2\, dq_2)^2 + (h_3\, dq_3)^2.$$

Thus,

$$ds_i = h_i\, dq_i.$$

The differential elements of area are

$$ds_i\, ds_j = h_i h_j\, dq_i\, dq_j,$$

and the differential element of volume is

$$ds_1\, ds_2\, ds_3 = h_1 h_2 h_3\, dq_1\, dq_2\, dq_3.$$

In spherical coordinates, for example, this is

$$r^2 \sin(\phi)\, dr\, d\theta\, d\phi.$$

Here, $r^2 \sin(\phi)$ is recognized as the Jacobian

$$\frac{\partial(x, y, z)}{\partial(r, \theta, \phi)}.$$

Expressions for the differential elements of area and volume in curvilinear coordinates are at the basis of formulas for change of variables in double and triple integrals. For example, a triple integral $\iiint_V f(x, y, z)\, dx\, dy\, dz$ over a region V of 3-space transforms in spherical coordinates to

$$\iiint_V f[r\cos(\theta)\sin(\phi), r\sin(\theta)\sin(\phi), r\cos(\phi)]r^2 \sin(\phi)\, dr\, d\theta\, d\phi.$$

The appearance of the factor $r^2 \sin(\phi)$ in the transformed integral is due to the replacement of the differential element of volume $dx\, dy\, dz$ in rectangular coordinates by the corresponding $r^2 \sin(\phi)\, dr\, d\theta\, d\phi$ in spherical coordinates.

Similarly, if you go to cylindrical coordinates, you will find that the differential element of volume is $r\, dr\, d\theta\, dz$; hence, in cylindrical coordinates $\iiint_V f(x, y, z)\, dx\, dy\, dz$ transforms to

$$\iiint_V f[r\cos(\theta), r\sin(\theta), z]r\, dr\, d\theta\, dz.$$

Now let $\mathbf{u}_i$ be a unit vector in the direction of increasing q_i at any point $(x(q_1, q_2, q_3), y(q_1, q_2, q_3), z(q_1, q_2, q_3))$. For example, in spherical coordinates, these could be written in terms of **i**, **j**, and **k** as

$$\mathbf{u}_r = \cos(\theta)\sin(\phi)\mathbf{i} + \sin(\theta)\sin(\phi)\mathbf{j} + \cos(\phi)\mathbf{k},$$
$$\mathbf{u}_\theta = -\sin(\theta)\mathbf{i} + \cos(\theta)\mathbf{j},$$

and

$$\mathbf{u}_\phi = \cos(\theta)\cos(\phi)\mathbf{i} + \sin(\theta)\cos(\phi)\mathbf{j} - \sin(\phi)\mathbf{k}.$$

Note that, unlike rectangular coordinates where **i**, **j**, and **k** are fixed, in general, $\mathbf{u}_1$, $\mathbf{u}_2$, and $\mathbf{u}_3$ vary at each point.

Any vector field in curvilinear coordinates can be written

$$\mathbf{F}(q_1, q_2, q_3) = F_1(q_1, q_2, q_3)\mathbf{u}_1 + F_2(q_1, q_2, q_3)\mathbf{u}_2 + F_3(q_1, q_2, q_3)\mathbf{u}_3.$$

We now want to write expressions for gradient, divergence, and curl in curvilinear coordinates.

Gradient: Let $\psi = \psi(q_1, q_2, q_3)$ be scalar-valued. At any point (q_1, q_2, q_3), $\nabla\psi$ should be a vector normal to the surface $\psi =$ constant passing through that point, and it should have magnitude equal to the greatest rate of change of ψ from that point. Thus, the component of $\nabla\psi$ normal to $q_1 =$ constant is $\partial\psi/\partial s_1$ or $(1/h_1)\partial\psi/\partial q_1$. Arguing similarly for the other components, we have

$$\boxed{\nabla\psi(q_1, q_2, q_3) = \frac{1}{h_1}\frac{\partial\psi}{\partial q_1}\mathbf{u}_1 + \frac{1}{h_2}\frac{\partial\psi}{\partial q_2}\mathbf{u}_2 + \frac{1}{h_3}\frac{\partial\psi}{\partial q_3}\mathbf{u}_3.}$$

Divergence: We derive an expression for $\nabla \cdot \mathbf{F}(q_1, q_2, q_3)$ using the flux interpretation of divergence.

Let

$$\mathbf{F} = F_1\mathbf{u}_1 + F_2\mathbf{u}_2 + F_3\mathbf{u}_3.$$

Referring to Figure 210, the flux across face *abcd* is

$$\mathbf{F}(q_1 + ds_1, q_2, q_3) \cdot \mathbf{u}_1 h_2(q_1 + ds_1, q_2, q_3)h_3(q_1 + ds_1, q_2, q_3)\,dq_2\,dq_3.$$

Across *efgh*, it is

$$\mathbf{F}(q_1, q_2, q_3) \cdot \mathbf{u}_1 h_2(q_1, q_2, q_3)h_3(q_1, q_2, q_3)\,dq_2\,dq_3.$$

Across both faces, the total flux is approximately

$$\frac{\partial}{\partial q_1}(F_1 h_2 h_3)\,dq_1\,dq_2\,dq_3.$$

Similarly, the flux across the other pairs of opposite faces is

$$\frac{\partial}{\partial q_2}(F_2 h_1 h_3)\,dq_1\,dq_2\,dq_3$$

and

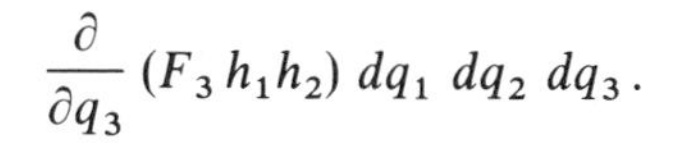

$$\frac{\partial}{\partial q_3}(F_3 h_1 h_2)\,dq_1\,dq_2\,dq_3.$$

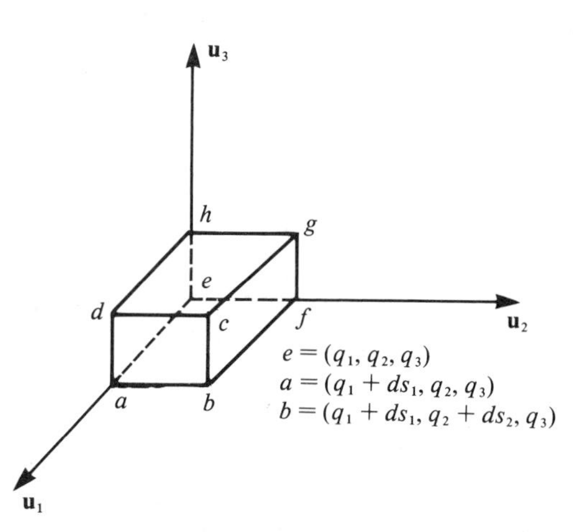

Figure 210. Calculating divergence in curvilinear coordinates.

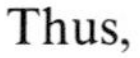

Thus,

$$\nabla \cdot \mathbf{F}(q_1, q_2, q_3) = \text{flux per unit volume}$$
$$= \frac{1}{h_1 h_2 h_3\, dq_1\, dq_2\, dq_3}$$
$$\times \left(\frac{\partial}{\partial q_1}(F_1 h_2 h_3) + \frac{\partial}{\partial q_2}(F_2 h_1 h_3) + \frac{\partial}{\partial q_3}(F_3 h_1 h_2)\right) dq_1\, dq_2\, dq_3$$
$$= \frac{1}{h_1 h_2 h_3}\left(\frac{\partial}{\partial q_1}(F_1 h_2 h_3) + \frac{\partial}{\partial q_2}(F_2 h_1 h_3) + \frac{\partial}{\partial q_3}(F_3 h_1 h_2)\right).$$

We now get the Laplacian free of charge. Recall that, in rectangular coordinates,

$$\nabla^2 f(x, y, z) = \nabla \cdot \nabla f = \frac{\partial^2 f}{\partial x^2} + \frac{\partial^2 f}{\partial y^2} + \frac{\partial^2 f}{\partial z^2}.$$

In orthogonal curvilinear coordinates,

$$\nabla^2 g(q_1, q_2, q_3) = \nabla \cdot [\nabla g(q_1, q_2, q_3)]$$
$$= \frac{1}{h_1 h_2 h_3}\left[\frac{\partial}{\partial q_1}\left(\frac{h_2 h_3}{h_1}\frac{\partial g}{\partial q_1}\right)\right.$$
$$\left. + \frac{\partial}{\partial q_2}\left(\frac{h_1 h_3}{h_2}\frac{\partial g}{\partial q_2}\right) + \frac{\partial}{\partial q_3}\left(\frac{h_1 h_2}{h_3}\frac{\partial g}{\partial q_3}\right)\right].$$

Curl: To derive an expression for curl in curvilinear coordinates, we use the interpretation of $(\nabla \times \mathbf{F}) \cdot \mathbf{N}$ as the swirl of a fluid with velocity $\mathbf{F}$ about a point in a plane normal to $\mathbf{N}$.

At point P, the component of $\nabla \times \mathbf{F}$ in the direction $\mathbf{u}_1$ is

$$\lim_{A \to 0} \frac{1}{A} \int_C \mathbf{F},$$

where C may be taken as a rectangle about P in the $\mathbf{u}_2$-$\mathbf{u}_3$ plane at P, as shown in Figure 211, and A is the area of this rectangle. Now,

$$A = h_2 h_3\, dq_2\, dq_3.$$

To compute the line integral, look at the line integral over each side a, b, c, and d of the rectangle. On side a, $\int_C \mathbf{F}$ is approximately

$$F_2(q_1, q_2, q_3) h_2(q_1, q_2, q_3)\, dq_2,$$

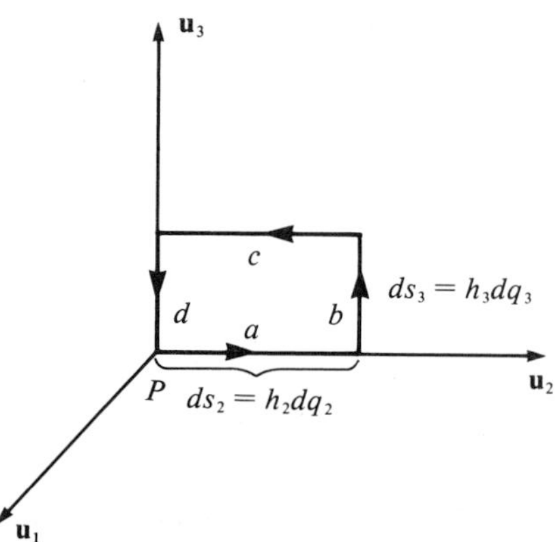

Figure 211. Calculating curl in curvilinear coordinates.

as $\mathbf{u}_2$ is tangent to a. On side c, $\int_C \mathbf{F}$ is approximately

$$-F_2(q_1, q_2, q_3 + dq_3)h_2(q_1, q_2, q_3 + dq_3)\,dq_2.$$

The net contribution from sides a and c is approximately

$$-\frac{\partial}{\partial q_3}(F_2 h_2)\,dq_2\,dq_3.$$

Similarly, from sides b and d, it is

$$\frac{\partial}{\partial q_2}(F_3 h_3)\,dq_2\,dq_3.$$

Then,

$$(\nabla \times \mathbf{F}) \cdot \mathbf{u}_1 = \frac{1}{h_2 h_3\,dq_2\,dq_3}\left(\frac{\partial}{\partial q_2}(F_3 h_3) - \frac{\partial}{\partial q_3}(F_2 h_2)\right)dq_2\,dq_3.$$

Similarly, we obtain the other components, giving us

$$\begin{aligned}\nabla \times \mathbf{F}(q_1, q_2, q_3) = {} & \frac{1}{h_2 h_3}\left(\frac{\partial}{\partial q_2}(F_3 h_3) - \frac{\partial}{\partial q_3}(F_2 h_2)\right)\mathbf{u}_1 \\ & + \frac{1}{h_1 h_3}\left(\frac{\partial}{\partial q_3}(F_1 h_1) - \frac{\partial}{\partial q_1}(F_3 h_3)\right)\mathbf{u}_2 \\ & + \frac{1}{h_1 h_2}\left(\frac{\partial}{\partial q_1}(F_2 h_2) - \frac{\partial}{\partial q_2}(F_1 h_1)\right)\mathbf{u}_3.\end{aligned}$$

This can be written

$$\nabla \times \mathbf{F} = \frac{1}{h_1 h_2 h_3}\begin{vmatrix} h_1 \mathbf{u}_1 & h_2 \mathbf{u}_2 & h_3 \mathbf{u}_3 \\ \dfrac{\partial}{\partial q_1} & \dfrac{\partial}{\partial q_2} & \dfrac{\partial}{\partial q_3} \\ F_1 h_1 & F_2 h_2 & F_3 h_3 \end{vmatrix}.$$

EXAMPLE 357

Returning to spherical coordinates, we had scale factors

$$h_r = 1, \qquad h_\theta = r\sin(\phi), \quad \text{and} \quad h_\phi = r.$$

(We use r, θ, and ϕ as subscripts in this standard system to help associate each scale factor with the appropriate coordinate.)

If $\mathbf{F} = F_r\,\mathbf{u}_r + F_\theta\,\mathbf{u}_\theta + F_\phi\,\mathbf{u}_\phi$, then we get

$$\nabla \cdot \mathbf{F} = \frac{1}{r^2}\frac{\partial}{\partial r}(r^2 F_r) + \frac{1}{r\sin(\phi)}\frac{\partial}{\partial \theta}(F_\theta) + \frac{1}{r\sin(\phi)}\frac{\partial}{\partial \phi}[F_\phi \sin(\phi)]$$

and

$$\nabla \times \mathbf{F} = \frac{1}{r^2 \sin(\phi)} \begin{vmatrix} \mathbf{u}_r & \mathbf{u}_\theta & \mathbf{u}_\phi \\ \dfrac{\partial}{\partial r} & \dfrac{\partial}{\partial \theta} & \dfrac{\partial}{\partial \phi} \\ F_r & r\sin(\phi)F_\theta & rF_\phi \end{vmatrix}.$$

If $f(r, \theta, \phi)$ is a scalar-valued function of spherical coordinates, then the gradient is given by

$$\nabla f = \frac{\partial f}{\partial r}\mathbf{u}_r + \frac{1}{r}\frac{\partial f}{\partial \phi}\mathbf{u}_\phi + \frac{1}{r\sin(\phi)}\frac{\partial f}{\partial \theta}\mathbf{u}_\theta.$$

From this, we get the Laplacian:

$$\nabla^2 f = \frac{1}{r^2}\frac{\partial}{\partial r}\left(r^2\frac{\partial f}{\partial r}\right) + \frac{1}{r^2\sin(\phi)}\frac{\partial}{\partial \phi}\left(\sin(\phi)\frac{\partial f}{\partial \phi}\right) + \frac{1}{r^2\sin^2(\phi)}\frac{\partial^2 f}{\partial \theta^2}.$$

This is as far as we shall go with curvilinear coordinates. In the exercises we ask the student to look at several systems which have found application in physics and engineering, including elliptic cylindrical coordinates, bipolar coordinates, and parabolic cylindrical coordinates. Expression of Laplace's equation in spherical and cylindrical coordinates will be particularly important in Chapter 13, where we consider solutions of certain partial differential equations.

PROBLEMS FOR SECTION 11.13

1. Compute the scale factors for cylindrical coordinates. Using these, compute $\nabla \cdot \mathbf{F}$ and $\nabla \times \mathbf{F}$ if $\mathbf{F}(r, \theta, z)$ is a vector field in cylindrical coordinates.

   If $g(r, \theta, z)$ is a scalar function in cylindrical coordinates, compute ∇g and $\nabla^2 g$.

2. *Elliptic cylindrical coordinates* are defined by

$$x = a\cosh(u)\cos(v), \qquad y = a\sinh(u)\sin(v), \quad \text{and} \quad z = z,$$

   where $0 \le u < \infty$, $0 \le v < 2\pi$, and $-\infty < z < \infty$.
   (a) Sketch the coordinate surfaces $u =$ constant, $v =$ constant, and $z =$ constant.
   (b) Determine the scale factors h_u, h_v, and h_z.
   (c) Determine $\nabla f(u, v, z)$ in this system.
   (d) Determine $\nabla \cdot \mathbf{F}(u, v, z)$ and $\nabla \times \mathbf{F}(u, v, z)$ in this system.
   (e) Determine $\nabla^2 f(u, v, z)$.

3. *Bipolar coordinates* are defined by

$$x = \frac{a\sinh(v)}{\cosh(v) - \cos(u)}, \qquad y = \frac{a\sin(u)}{\cosh(v) - \cos(u)}, \quad \text{and} \quad z = z,$$

   where $-\infty < u < \infty$, $0 \le v < 2\pi$, and $-\infty < z < \infty$.
   (a) Sketch the coordinate surfaces $u =$ constant, $v =$ constant, and $z =$ constant. Are these coordinates orthogonal?
   (b) Determine the scale factors h_u, h_v, and h_z.
   (c) Determine $\nabla f(u, v, z)$.
   (d) Determine $\nabla \cdot \mathbf{F}(u, v, z)$ and $\nabla \times \mathbf{F}(u, v, z)$.
   (e) Determine $\nabla^2 f(u, v, z)$.

4. *Parabolic cylindrical coordinates* are defined by

$$x = uv, \qquad y = \tfrac{1}{2}(u^2 - v^2), \quad \text{and} \quad z = z,$$

where $-\infty < v < \infty, 0 \leq u < \infty$, and $-\infty < z < \infty$.

(a) Sketch the coordinate surfaces u = constant, v = constant, and z = constant. Are these coordinates orthogonal?

(b) Determine the scale factors h_u, h_v, and h_z.

(c) Determine $\nabla f(u, v, z)$.

(d) Determine $\nabla \cdot \mathbf{F}(u, v, z)$ and $\nabla \times \mathbf{F}(u, v, z)$.

(e) Determine $\nabla^2 f(u, v, z)$.

5. Assume as given that

$$(\nabla^2 g)(P) = \lim_{S \to P} \frac{1}{\bar{V}} \iint_S \nabla g,$$

where P is any point, S is a regular closed surface containing P, V is the region bounded by S, and $\bar{V}$ = volume of V. Here, $\lim_{S \to P}$ means a limit as S shrinks to P.

Using this, and taking S as a rectangular box about P, derive the expression we obtained for $\nabla^2 g(q_1, q_2, q_3)$.

11.14 EXTENSIONS OF THE THEOREMS OF GREEN AND GAUSS

Green's formula

$$\int_C F_1\, dx + F_2\, dy = \iint_D \left(\frac{\partial F_2}{\partial x} - \frac{\partial F_1}{\partial y}\right) dx\, dy$$

assumes that F_1, F_2, $\partial F_2/\partial x$, and $\partial F_1/\partial y$ are continuous on C and in the enclosed region D. We shall now consider to what extent exceptions can be made to this and what effects these have on the formula.

Suppose that F_1, F_2, $\partial F_2/\partial x$, and $\partial F_1/\partial y$ are continuous on C, and in D, except perhaps at $P_1, \ldots, P_n$, where one or more of the functions may not be defined at all. Green's theorem no longer applies directly; so we proceed as follows.

Enclose each P_i by a circle K_i of sufficiently small radius so that K_i and the enclosed region D_i lie entirely within C and K_i and K_j do not intersect if $i \neq j$ (see Figure 212). Cut a channel from C to K_1, K_1 to $K_2, \ldots, K_{n-1}$ to K_n, as shown in Figure 213, to form a regular closed curve C^*, as shown in Figure 214. Now, C^* encloses a region D^* in which F_1, F_2, $\partial F_1/\partial y$, and $\partial F_2/\partial x$ are continuous (as $P_1, \ldots, P_n$ are *outside* D^*). Then,

$$\oint_{C^*} F_1\, dx + F_2\, dy = \iint_{D^*} \left(\frac{\partial F_2}{\partial x} - \frac{\partial F_1}{\partial y}\right) dx\, dy.$$

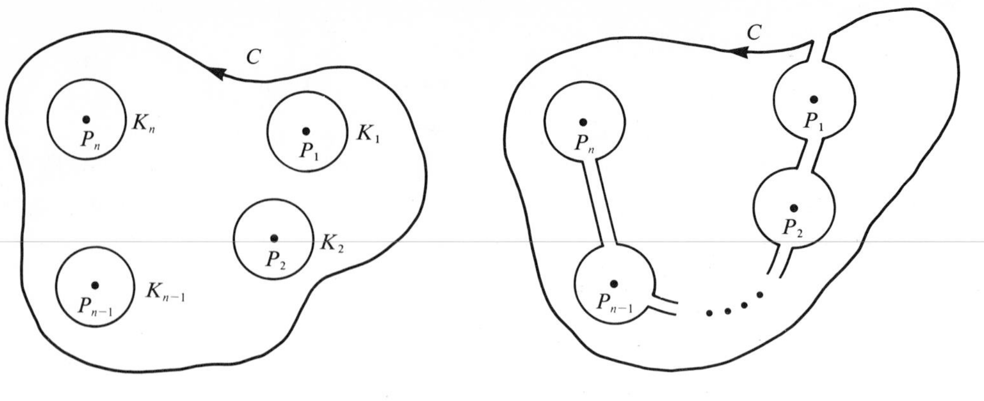

Figure 212 *Figure 213*

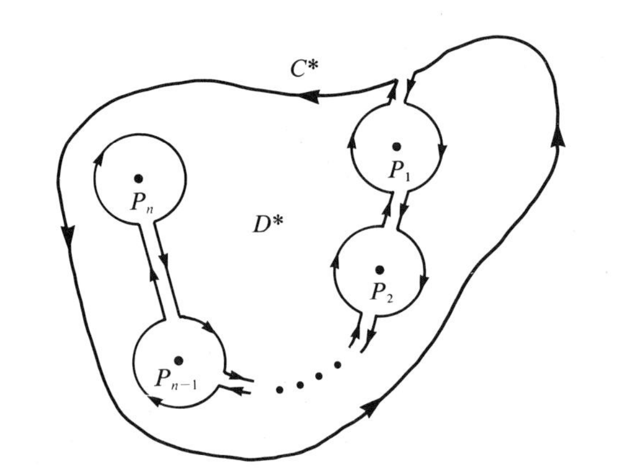

Figure 214. Points $P_1, \ldots, P_n$ lie *outside* the region D^* enclosed by C^*.

Now take a kind of limit as the channels become narrower, merging to single lines. On each channel/line, we integrate in both directions; hence, contributions to $\oint_{C^*}$ from these line segments cancel out. Further, as we shrink the channels, we restore the small pieces of C and the K_i's which were cut out. Finally, D^* approaches D', which is all of D with the interiors of the K_i's cut out. Then,

$$\oint_{C^*} F_1\,dx + F_2\,dy = \oint_C F_1\,dx + F_2\,dy + \sum_{j=1}^{n} \ointclockwise_{K_j} F_1\,dx + F_2\,dy$$

$$= \iint_{D'} \left(\frac{\partial F_2}{\partial x} - \frac{\partial F_1}{\partial y}\right) dx\,dy.$$

(Note that in the above integrals we go *clockwise* around the K_j's.)

Writing $\ointclockwise_{K_j} = -\oint_{K_j}$, we have the extended form of Green's theorem:

$$\oint_C F_1\,dx + F_2\,dy - \sum_{j=1}^{n} \oint_{K_j} F_1\,dx + F_2\,dy = \iint_{D'} \left(\frac{\partial F_2}{\partial x} - \frac{\partial F_1}{\partial y}\right) dx\,dy.$$

In particular, when $(\partial F_2/\partial x) - (\partial F_1/\partial y) = 0$ in D', then

$$\oint_C F_1\,dx + F_2\,dy = \sum_{j=1}^{n} \oint_{K_j} F_1\,dx + F_2\,dy,$$

with all integrals taken in the counterclockwise sense. This can be very useful in evaluating line integrals.

EXAMPLE 358

Evaluate

$$\int_C \frac{-y\,dx}{x^2+y^2} + \frac{x\,dy}{x^2+y^2},$$

where C is *any* regular closed curve about the origin.

Note that

$$\frac{\partial}{\partial x}\left(\frac{x}{x^2+y^2}\right) = \frac{\partial}{\partial y}\left(\frac{-y}{x^2+y^2}\right).$$

Thus,

$$\int_C \frac{-y\,dx}{x^2+y^2} + \frac{x\,dy}{x^2+y^2} = \int_K \frac{-y\,dx}{x^2+y^2} + \frac{x\,dy}{x^2+y^2},$$

where K is a circle about the origin and entirely within C. Parametrize C by

$$x = r\cos(\theta), \qquad y = r\sin(\theta); \qquad \theta: 0 \to 2\pi,$$

where r is the (constant) radius of K. Then,

$$\int_K \frac{-y\,dx}{x^2+y^2} + \frac{x\,dy}{x^2+y^2} = \int_0^{2\pi} \left(\frac{-r\sin(\theta)[-r\cos(\theta)]}{r^2} + \frac{r\cos(\theta)[r\cos(\theta)]}{r^2}\right) d\theta$$

$$= \int_0^{2\pi} d\theta = 2\pi.$$

A similar extension can be derived for Gauss's theorem, as in fact we did in treating Gauss's law, where we enclosed the trouble point (there, the origin) in a sphere. We leave this for the exercises.

PROBLEMS FOR SECTION 11.14

In each of Problems 1 through 5, evaluate $\oint_C \mathbf{F}$ with C any regular closed curve about the origin, using the method of Example 358.

1. $\mathbf{F} = \dfrac{x\mathbf{i}}{x^2+y^2} + \dfrac{y\mathbf{j}}{x^2+y^2}$

2. $\mathbf{F} = \left(\dfrac{1}{x^2+y^2}\right)^{3/2}(x\mathbf{i} + y\mathbf{j})$

3. $\mathbf{F} = \left(\dfrac{x}{x^2+y^2} + x^2\right)\mathbf{i} + \left(\dfrac{y}{x^2+y^2} - 2y\right)\mathbf{j}$

4. $\mathbf{F} = \left(\dfrac{-y}{x^2+y^2} + 2x^2\right)\mathbf{i} + \left(\dfrac{x}{x^2+y^2} - y\right)\mathbf{j}$

5. $\mathbf{F} = \left(\dfrac{x}{\sqrt{x^2 + y^2}} + 2x\right)\mathbf{i} + \left(\dfrac{y}{\sqrt{x^2 + y^2}} - 3y^2\right)\mathbf{j}$

6. Imitate the argument used in this section, to write $\iint_S \mathbf{F}$ as a sum of surface integrals $\iint_{S_1} \mathbf{F} + \cdots + \iint_{S_n} \mathbf{F}$, where $\mathbf{F}$ is a continuous vector field with continuous first partial derivatives on and inside the piecewise smooth surface S except at points $P_1, \ldots, P_n$, and $S_1, \ldots, S_n$ are piecewise smooth surfaces inside S, with S_i enclosing P_i, and no two of $S_1, \ldots, S_n$, S intersect. Assume that div $\mathbf{F} = 0$ in the region inside S and outside $S_1, \ldots, S_n$.

In each of Problems 7 through 10, evaluate $\iint_S \mathbf{F}$ as in this section, where S is any piecewise smooth surface enclosing a volume V about the origin.

7. $\mathbf{F} = \dfrac{1}{x^2 + y^2 + z^2}[(y - z)\mathbf{i} + (z - x)\mathbf{j} + (x - y)\mathbf{k}]$

8. $\mathbf{F} = \left(\dfrac{1}{y^2 + z^2}\right)\mathbf{i} + \left(\dfrac{1}{x^2 + z^2}\right)\mathbf{j} + \left(\dfrac{1}{y^2 + x^2}\right)\mathbf{k}$

9. $\mathbf{F} = y\mathbf{i} + z\mathbf{j} + x\mathbf{k}$

10. $\mathbf{F} = \left(\dfrac{1}{\sqrt{y^2 + z^2}}\right)\mathbf{i} + \left(\dfrac{1}{\sqrt{x^2 + z^2}}\right)\mathbf{j} + \left(\dfrac{1}{\sqrt{x^2 + y^2}}\right)\mathbf{k}$

SUPPLEMENTARY PROBLEMS

In each of Problems 1 through 10, compute $(\mathbf{F} \cdot \mathbf{G})'$ and $(\mathbf{F} \times \mathbf{G})'$, using any valid method.

1. $\mathbf{F} = 4t\mathbf{i} + t^2\mathbf{j} - \mathbf{k}, \quad \mathbf{G} = \cos(2t)\mathbf{i} - 4t\mathbf{k}$
2. $\mathbf{F} = -\sin(t)\mathbf{j} + 3\mathbf{k}, \quad \mathbf{G} = 2\mathbf{i} + t^2\mathbf{j} - \ln(t)\mathbf{k}$
3. $\mathbf{F} = 2\sqrt{t}\,\mathbf{i} + \sinh(t)\mathbf{j} + t\mathbf{k}, \quad \mathbf{G} = e^{-t}\mathbf{i} + \mathbf{j} - 2t\mathbf{k}$
4. $\mathbf{F} = te^t\mathbf{i} - \cos(t)\mathbf{j} + t\mathbf{k}, \quad \mathbf{G} = 2\mathbf{i} - t^3\mathbf{j} + e^{-t}\mathbf{k}$
5. $\mathbf{F} = 2\sinh(3t)\mathbf{j}, \quad \mathbf{G} = t^2\mathbf{i} - t\,\mathbf{j} + \mathbf{k}$
6. $\mathbf{F} = (1 - 2t)\mathbf{i} + t^3\mathbf{j} - 3t^2\mathbf{k}, \quad \mathbf{G} = 4t\mathbf{i} + (2/t)\mathbf{j} - \mathbf{k}$
7. $\mathbf{F} = (t^2 - 2)\mathbf{i} + (t - 3)\mathbf{j} - 4t\mathbf{k}, \quad \mathbf{G} = \mathbf{i} - \cosh(t)\mathbf{j} + t^3\mathbf{k}$
8. $\mathbf{F} = e^{t^2}\mathbf{i} - 2t\mathbf{j} + t^2\mathbf{k}, \quad \mathbf{G} = (1/t)(\mathbf{i} + \mathbf{j} + \mathbf{k})$
9. $\mathbf{F} = te^{-t}\mathbf{i} + 2t\mathbf{j} - \sqrt{t}\,\mathbf{k}, \quad \mathbf{G} = 2t^3\mathbf{i} + t^2\mathbf{j} - \cosh(t)\mathbf{k}$
10. $\mathbf{F} = 4t^3\mathbf{i} - 3t\mathbf{k}, \quad \mathbf{G} = 8t^2\mathbf{i} + e^{2t}\mathbf{j} - t\mathbf{k}$

In each of Problems 11 through 15, obtain an integral for the length of the given curve. You need not evaluate the integral.

11. $x = 2t^2,\ y = \cos(t),\ z = 1 + t,\ 0 \le t \le 1$
12. $x = 2 - \sin(t),\ y = 2 - \cos(t),\ z = 1,\ 0 \le t \le 2$
13. $x = t^2,\ y = 2 + t,\ z = e^t,\ 2 \le t \le 5$
14. $x = 4t,\ y = 2 + t^2,\ z = \sin(2t),\ -1 \le t \le 1$
15. $x = 3t^3,\ y = 2,\ z = t + e^{-t},\ 0 \le t \le 2\pi$

In each of Problems 16 through 20, compute the velocity, speed, and acceleration, and decompose the acceleration into tangential and centripetal components. Also find the curvature, radius of curvature, and torsion.

16. $\mathbf{R} = \mathbf{i} + t\mathbf{j} - 3t\mathbf{k}$
17. $\mathbf{R} = \sin(t)\mathbf{i} + \cos(t)\mathbf{j} - 4\mathbf{k}$
18. $\mathbf{R} = (1 + t)\mathbf{i} + (1 + t)\mathbf{j} - 2t\mathbf{k}$
19. $\mathbf{R} = t^2\mathbf{i} - t^2\mathbf{j} + t\mathbf{k}$
20. $\mathbf{R} = \sin(2t)\mathbf{i} - t\mathbf{j} + t\mathbf{k}$

In each of Problems 21 through 25, find the lines of force of the vector field.

21. $\mathbf{F} = \mathbf{i} - zy\mathbf{j} + z\mathbf{k}$
22. $\mathbf{F} = 4xy\mathbf{i} - x^2\mathbf{j} + z\mathbf{k}$
23. $\mathbf{F} = y\mathbf{i} - x\mathbf{j} + z\mathbf{k}$
24. $\mathbf{F} = z\mathbf{i} - 3y\mathbf{j} + x^3\mathbf{k}$
25. $\mathbf{F} = \mathbf{i} - y\mathbf{j} + z^2\mathbf{k}$

In each of Problems 26 through 30, find the direction in which ϕ has its greatest rate of change at the given point, and find the magnitude of this rate of change.

26. $\phi = x^2y - \cos(zx);\quad (1, 1, 1)$
27. $\phi = 2yz^3 + xy\cos(z);\quad (2, -1, 3)$

28. $\phi = -\ln(x + y + z) + x^2$; $(2, 1, 1)$
29. $\phi = e^{x-z}\cos(y + z)$; $(2, 2, 2)$
30. $\phi = 1 + \cos(zx) - x^3y$; $(2, -1, 3)$

In each of Problems 31 through 35, find the tangent plane and normal line to the given surface at the given point.

31. $z = x^2 + y^2 - 2$; $(2, 2, 6)$
32. $z^2 = x^2 + y^2$; $(1, 1, \sqrt{2})$
33. $x^2 + (y - 2)^2 + z^2 = 11$; $(1, 1, 3)$
34. $2x^2 - y^2 - z^2 = 0$; $(1, 1, 1)$
35. $4xyz + x^2 - y^2 = 0$; $(2, 4, \frac{3}{8})$

In each of Problems 36 through 40, compute div **F** and curl **F**.

36. $\mathbf{F} = xy\mathbf{i} - z^3\mathbf{j} + x\mathbf{k}$
37. $\mathbf{F} = (1 - z)\mathbf{i} + x^3\mathbf{j} - \cos(yz)\mathbf{k}$
38. $\mathbf{F} = \mathbf{i} - z^3\mathbf{j} + e^{xy}\mathbf{k}$
39. $\mathbf{F} = 2xy\mathbf{i} - z^2\mathbf{j} + x\mathbf{k}$
40. $\mathbf{F} = \sin(y - z)\mathbf{i} + e^z\mathbf{j} - x\mathbf{k}$

In each of Problems 41 through 50, compute the line integral of the given vector or scalar function over the given curve.

41. $\int_C x^2y\,dx - y\,dy$, C the straight line from $(-2, 3)$ to $(1, 1)$

42. $\int_C xz\mathbf{i} - y\mathbf{j} + xz\mathbf{k}$, C the straight line from $(1, 1, 0)$ to $(-2, 3, -1)$

43. $\int_C xy\,dz$, C the straight line from $(0, 0, 0)$ to $(1, 1, -2)$

44. $\int_C -4xy\,dx + zy\,dy - x\,dz$, C given by $x = 2t$, $y = 3t^2$, $z = t$, with t: $1 \to 4$

45. $\int_C x^2y\cos(z)\,dx$, C the straight line from $(1, 1, 1)$ to $(3, 1, 1)$

46. $\int_C x\mathbf{i} - y\mathbf{j} + xyz\mathbf{k}$, C the parabola $y = x^2$, $z = 4$, with x: $1 \to 3$

47. $\int_C 4xy\,dx - xz\,dy + e^{-x}\,dz$, C given by $x = y = t$, $z = -t$, with t: $-1 \to 4$

48. $\int_C -5yz\,dy - z^3\,dz$, C the straight line from $(1, 1, 1)$ to $(-4, 2, 5)$

49. $\int_C xz^2\,dx - y\,dy + zx\,dz$, C given by $x = 1 - 2t$, $y = t^3$, $z = 4t$, with t: $1 \to 3$

50. $\int_C x^3\,dx - 4yz\,dz$, C the straight line from $(0, 0, 1)$ to $(0, 2, 4)$

In each of Problems 51 through 55, verify Stokes's theorem for the given vector field and surface.

51. $\mathbf{F} = xy\mathbf{i} - \mathbf{j} + z\mathbf{k}$, S the hemisphere $x^2 + y^2 + z^2 = 4$, $z \geq 0$
52. $\mathbf{F} = 4xz\mathbf{i} - y\mathbf{j} + x^2\mathbf{k}$, S the cone $z = \sqrt{x^2 + y^2}$ with $0 \leq x^2 + y^2 \leq 4$
53. $\mathbf{F} = xyz\mathbf{i}$, S the frustum of a cone given by $z = \sqrt{x^2 + y^2}$, $1 \leq x^2 + y^2 \leq 4$
(*Hint*: Here, the boundary of S consists of the top and bottom circles about the frustum.)
54. $\mathbf{F} = 8\mathbf{i} - yz\mathbf{j} + x^3\mathbf{k}$, S the parabolic bowl $z = x^2 + y^2$, with $x^2 + y^2 \leq 9$
55. $\mathbf{F} = x\mathbf{i} - y\mathbf{j} + x\mathbf{k}$, S the disk $x^2 + z^2 \leq 5$, $y = 0$

In each of Problems 56 through 60, verify Gauss's theorem for the given vector field and surface.

56. $\mathbf{F} = -5x\mathbf{i} + y\mathbf{j} - z\mathbf{k}$, S the cube with vertices at $(0, 0, 0)$, $(1, 0, 0)$, $(0, 1, 0)$, $(0, 0, 1)$, $(1, 1, 0)$, $(1, 0, 1)$, $(0, 1, 1)$, and $(1, 1, 1)$

57. $\mathbf{F} = x^2\mathbf{i} - z\mathbf{j} + y\mathbf{k}$, S the sphere $(x-1)^2 + (y-1)^2 + z^2 = 4$

58. $\mathbf{F} = 4xy\mathbf{i} - z\mathbf{j} + 2\mathbf{k}$, S the cylinder $x^2 + y^2 = 1, 0 \le z \le 4$, together with the top and bottom disks

59. $\mathbf{F} = \mathbf{i} - y\mathbf{j} + xz\mathbf{k}$, S the cone $z = \sqrt{x^2 + y^2}, 0 \le x^2 + y^2 \le 4$, together with the disk $x^2 + y^2 = 4, z = 2$

60. $\mathbf{F} = x\mathbf{i} + y\mathbf{j} + z\mathbf{k}$, S the parabolic bowl $z = x^2 + y^2, 0 \le x^2 + y^2 \le 9$, together with the top disk $x^2 + y^2 = 9$, $z = 9$

In each of Problems 61 through 65, find a potential function for $\mathbf{F}$ and use this to evaluate the line integral of $\mathbf{F}$ over any smooth curve between the given points.

61. $\mathbf{F} = (z^2 - 2y)\mathbf{i} - 2x\mathbf{j} + 2zx\mathbf{k}$, C from $(0, 6, -2)$ to $(1, 1, -4)$

62. $\mathbf{F} = 2x\mathbf{i} - 3z\mathbf{j} + [-3y - \sin(z)]\mathbf{k}$, C from $(0, 0, 0)$ to $(1, 1, -4)$

63. $\mathbf{F} = -z\sin(xz)\mathbf{i} - ze^{yz}\mathbf{j} + [-x\sin(xz) - ye^{yz}]\mathbf{k}$, C from $(1, 0, 0)$ to $(-1, 0, -2)$

64. $\mathbf{F} = \mathbf{i} - 2y\mathbf{j} + \mathbf{k}$, C from $(-2, 2, 5)$ to $(0, 0, 0)$

65. $\mathbf{F} = 3x^2\mathbf{i} - z^2\mathbf{j} - 2yz\mathbf{k}$, C from $(-5, 2, 2)$ to $(1, 1, -4)$

In each of Problems 66 through 70, verify Green's theorem for the given vector field and curve.

66. $\mathbf{F} = xy\mathbf{i} - y\mathbf{j}$, C the circle $x^2 + y^2 = 9$

67. $\mathbf{F} = -xy\mathbf{i} - x\mathbf{j}$, C the triangle with vertices at $(1, 0)$, $(2, 0)$, and $(2, 4)$

68. $\mathbf{F} = xy^2\mathbf{i} - x\mathbf{j}$, C the square with vertices at $(0, 0)$, $(1, 0)$, $(0, 1)$, and $(1, 1)$

69. $\mathbf{F} = -2xy^2\mathbf{i} + xy\mathbf{j}$, C the semicircle $x^2 + y^2 = 4$ $(y \ge 0)$ and the x-axis $(-2 \le x \le 2)$

70. $\mathbf{F} = \mathbf{i} - y\mathbf{j}$, C the triangle with vertices at $(1, 2)$, $(-1, 3)$ and $(2, 2)$

71. Prove Stokes's theorem for the hemisphere $x^2 + y^2 + z^2 = a^2, z \ge 0$

72. Define the surface integral of a scalar-valued function $f(x, y, z)$ over a surface S parametrized by $x = x(u, v)$, $y = y(u, v)$, $z = z(u, v)$ for (u, v) in D, by setting

$$\iint_S f(x, y, z) = \iint_D f(x(u, v), y(u, v), z(u, v))\|\mathbf{N}\|\, du\, dv.$$

(a) Let S be a thin plastic shell, with $\rho(x, y, z)$ the density at point (x, y, z). Give a plausibility argument to support the definition of the mass of the shell as $\iint_S \rho(x, y, z)$.

(b) Let the center of mass of the shell be at $(\bar{x}, \bar{y}, \bar{z})$. Argue that

$$\bar{x} = \frac{1}{m}\iint_S x\rho, \quad \bar{y} = \frac{1}{m}\iint_S y\rho, \quad \text{and} \quad \bar{z} = \frac{1}{m}\iint_S z\rho.$$

Here, m is the mass of the shell.

(c) Give an argument to show that the moments of inertia of the shell about the x-, y-, and z-axes are given by

$$I_x = \iint_S \rho(y^2 + z^2), \quad I_y = \iint_S \rho(x^2 + z^2), \quad \text{and} \quad I_z = \iint_S \rho(x^2 + y^2).$$

73. Define the surface area of a surface S parametrized by $x = x(u, v)$, $y = y(u, v)$, $z = z(u, v)$, for (u, v) in D, by setting

$$\text{area} = \iint_D \|\mathbf{N}\|\, du\, dv.$$

Using this as a starting point, show that the area is also given by the integral

$$\iint_D \sqrt{EG - F^2}\, du\, dv,$$

where $E = x_u{}^2 + y_u{}^2 + z_u{}^2$, $G = x_v{}^2 + y_v{}^2 + z_v{}^2$, and $F = x_u x_v + y_u y_v + z_u z_v$. Here, a u or v subscript means a partial derivative with respect to u or v.

74. Use the result of Problem 73 to calculate the area of the cone $z = \sqrt{x^2 + y^2}, 0 \leq x^2 + y^2 \leq a^2$.

75. Use the result of Problem 73 to calculate the area of a sphere of radius a.

76. Use the result of Problem 73 to calculate the area of the frustum of a cone: $z = \sqrt{x^2 + y^2}, a^2 \leq x^2 + y^2 \leq b^2$.

A HISTORICAL SKETCH OF VECTORS AND VECTOR ANALYSIS

Vectors can be traced in their infancy to considerations of problems connected with complex numbers.

In the history of mathematics, complex numbers play a significant role, presenting philosophical as well as mathematical difficulties. Their necessity arises in solving such equations as $x^2 + 1 = 0$; however, the idea of a "number" whose square is negative was extremely distasteful. Complex numbers became more acceptable when they were interpreted geometrically. Caspar Wessell (1745–1818), a Norwegian surveyor, Jean-Robert Argand (1768–1822), a Swiss bookkeeper, and Karl Friedrich Gauss (1777–1855), one of the great mathematicians of all time, endowed complex numbers with a geometric interpretation. A complex number $x + iy$ can be thought of as a point in the plane (now often called the *Argand plane*). Multiplication by i corresponds to a 90° rotation counterclockwise, since $i(x + iy) = -y + ix$ is matched with $(-y, x)$, as shown in Figure 215. Thus, purely real numbers are graphed on the horizontal (real) axis, purely imaginary numbers (real multiples of i) on the vertical (imaginary) axis, and numbers $a + ib$ as points (a, b) or arrows from $(0, 0)$ to (a, b). It soon became apparent that this setting was useful in representing vectors in a plane, and, in fact, complex numbers can be added by the parallelogram law.

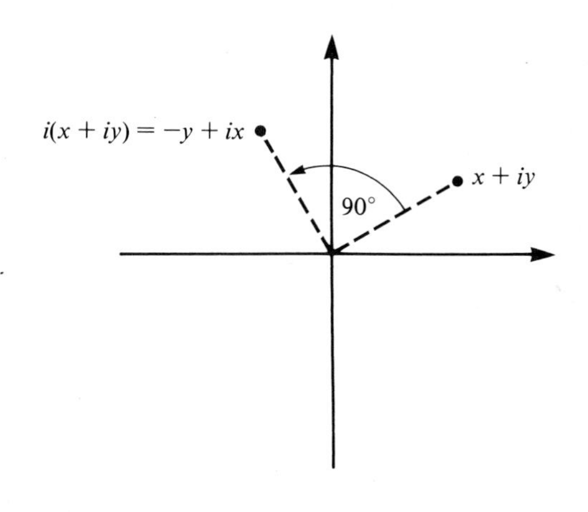

Figure 215. Multiplication by i as a 90° rotation.

In the early 1830s, vectors were routinely represented by complex numbers. Here, "vector" meant force. Since, however, forces are not necessarily confined to a plane, the question arose as to how to generalize to three-space. Due to the historical background, the emphasis was on the number aspect, and the problem became one of generalizing complex numbers to three dimensions. This problem was attacked by William Rowan Hamilton (1805–1865), who proposed a solution in the form of quaternions.

A *quaternion* is a "number" of the form

$$a + b\mathbf{i} + c\mathbf{j} + d\mathbf{k}$$

(using modern notation). One adds quaternions in the obvious way; for example,

$$(2 + 3\mathbf{i} + 4\mathbf{j} - 6\mathbf{k}) + (-1 + 2\mathbf{i} - 8\mathbf{j} + 16\mathbf{k}) = 1 + 5\mathbf{i} - 4\mathbf{j} + 10\mathbf{k}.$$

However, in "multiplying" quaternions, Hamilton used the rules

$$\mathbf{jk} = \mathbf{i}, \quad \mathbf{ki} = \mathbf{j}, \quad \mathbf{ij} = \mathbf{k}, \quad \mathbf{kj} = -\mathbf{i}, \quad \mathbf{ik} = -\mathbf{j}, \quad \mathbf{ji} = -\mathbf{k},$$

and

$$\mathbf{i}^2 = \mathbf{j}^2 = \mathbf{k}^2 = -1.$$

Today we recognize these as *cross products* of the unit vectors, except for the last three which reflect the equation $i^2 = -1$ for the imaginary unit i.

Hamilton believed that eventually quaternions would provide the natural vehicle for the mathematical expression of physical laws. In this he turned out to be wrong, and today quaternions are used primarily in abstract algebra.

Soon after Hamilton began to publicize his views on quaternions, there arose a debate among scientists over Hamilton's claims for them. James Clerk Maxwell (1831–1879), a professor of physics at Cambridge who derived the fundamental field equations of electricity and magnetism, separated in his thinking the scalar and vector parts of quaternions. Thus, attention centered on expressions $a\mathbf{i} + b\mathbf{j} + c\mathbf{k}$, or vectors as we know them.

Maxwell was the first to use the term *rotation* (curl) of a vector, in connection with swirl in fluid motion. He also introduced what he called *Laplace's operator*,

$$\nabla^2 = \frac{\partial^2}{\partial x^2} + \frac{\partial^2}{\partial y^2} + \frac{\partial^2}{\partial z^2}.$$

Maxwell further noted the identities div(curl) $= 0$ and curl(grad) $= \mathbf{0}$.

William Kingdon Clifford (1845–1879) of Queen's College, London, apparently originated the term *divergence* of a vector.

The outcome of the debate on quaternions versus vectors was in doubt until the work of Yale's Josiah Willard Gibbs (1839–1903). His book *Vector Analysis*, written in 1901 by E. B. Wilson but based on Gibbs's lectures, helped to popularize vector notation and algebra. A second book that was influential in establishing the use of vectors was Oliver Heaviside's *Electromagnetic Theory*, the first volume of which contained vector algebra. Heaviside (1850–1925) was primarily an electrical engineer, but he also helped develop the Laplace transform.

Today we recognize the main results of vector calculus in the integral theorems named for Gauss, Green, and Stokes. The divergence theorem, popularly named after Gauss, was published almost simultaneously and independently in 1831 by the Russian Michel Ostrogradsky (1801–1861). In Russia, the theorem is known as *Ostrogradsky's theorem.*

Stokes's theorem apparently first appeared in an 1850 letter addressed to Stokes by Lord Kelvin [Sir William Thompson (1824–1907), known for his work in electrostatics and heat conduction]. Sir George Gabriel Stokes (1819–1903), professor of mathematics at Cambridge, put the formula which bears his name as an examination question for the Smith prize. Thus, theorems are not always named after their discoverers, nor is it always clear who first stated a result.

Along the same lines, Green's theorem may be misnamed. George Green (1793–1841) was a self-taught mathematician primarily interested in electricity and magnetism. In the course of applying potential theory to this subject, he developed what are today known as *Green's formulas* or *Green's identities* (see Section 11.11). These were also presented to the St. Petersburg Academy of Sciences in 1828 by Ostrogradsky. The theorem we have named after Green (according to popular modern custom) is sometimes known as *Green's lemma.*

PART FOUR

FOURIER ANALYSIS AND BOUNDARY VALUE PROBLEMS

CHAPTER TWELVE

Fourier Series, Integrals, and Transforms

12.0 INTRODUCTION

The student has no doubt been exposed by this time to power series expansions about a point. Given $f(x)$ satisfying certain conditions, we can write a series

$$f(x) = \sum_{n=0}^{\infty} c_n(x - a)^n$$

for x in some interval about a. Note that, in effect, we are writing $f(x)$ as a sum of infinitely many polynomials, since $(x - a)^n$ is an nth degree polynomial in x.

Other kinds of series are also possible, in which one attempts to write $f(x)$ as a sum of functions which are not necessarily polynomials. One of the most useful is the *trigonometric series*, in which we attempt to write $f(x)$ as a series of sines and cosines of various amplitudes and frequencies.

The feasibility of doing this is not at all obvious but is suggested by the fact that a sound wave is a sum of fundamental harmonics, which are themselves represented by sine and cosine terms. In this chapter we shall show how, under fairly general conditions, familiar functions such as e^x or $3x^2 + 1$ can be decomposed into trigonometric sums valid on certain intervals. The resulting method forms one of the basic tools for solving certain important partial differential equations.

While most of this chapter is devoted to Fourier series, we also touch upon the related Fourier integral and Fourier transforms. In particular, the finite Fourier transform has become quite important in modern electrical engineering.

While we do not assume any familiarity with Sturm-Liouville theory, the student who has understood Section 6.4 will recognize that much of what we are doing in this chapter has its foundations in the theory of eigenfunction expansions. Having suggested this as an overview which may be meaningful to some students, we immediately retreat to basics and begin with the trigonometric Fourier series of a function.

12.1 THE FOURIER SERIES OF A FUNCTION

Suppose that $f(x)$ is defined on an interval $-L \leq x \leq L$. Using a little foresight which we shall explain later, we define the *Fourier series of* $f(x)$ on $[-L, L]$ to be the series

$$\frac{a_0}{2} + \sum_{n=1}^{\infty} a_n \cos\left(\frac{n\pi x}{L}\right) + b_n \sin\left(\frac{n\pi x}{L}\right),$$

where

$$a_0 = \frac{1}{L}\int_{-L}^{L} f(x)\,dx,$$

$$a_n = \frac{1}{L}\int_{-L}^{L} f(x)\cos\left(\frac{n\pi x}{L}\right)dx \quad \text{for} \quad n = 1, 2, 3, \ldots,$$

and

$$b_n = \frac{1}{L}\int_{-L}^{L} f(x)\sin\left(\frac{n\pi x}{L}\right)dx \quad \text{for} \quad n = 1, 2, 3, \ldots.$$

The numbers $a_0, a_1, a_2, a_3, \ldots, b_1, b_2, b_3, \ldots$ are the *Fourier coefficients* of $f(x)$ on $[-L, L]$. We shall explain how we got them in the next section. For now, we want to concentrate on just the mechanics of writing the Fourier series of a function on a given interval.

EXAMPLE 359

Let $f(x) = 2x + 1$, for $-3 \leq x \leq 3$.

Here, $L = 3$. The Fourier coefficients are

$$a_0 = \tfrac{1}{3}\int_{-3}^{3} (2x + 1)\,dx = 2,$$

$$a_n = \frac{1}{3}\int_{-3}^{3} (2x + 1)\cos\left(\frac{n\pi x}{3}\right)dx = 0 \quad \text{for} \quad n = 1, 2, 3, \ldots,$$

and

$$b_n = \frac{1}{3}\int_{-3}^{3} (2x + 1)\sin\left(\frac{n\pi x}{3}\right)dx = \frac{-12}{n\pi}\cos(n\pi) \quad \text{for} \quad n = 1, 2, 3, \ldots.$$

The Fourier series of $2x + 1$ on $[-3, 3]$ is

$$1 + \sum_{n=1}^{\infty} \frac{-12}{n\pi}\cos(n\pi)\sin\left(\frac{n\pi x}{3}\right).$$

Since $\cos(n\pi) = (-1)^n$, this can be written

$$1 + \frac{12}{\pi}\sum_{n=1}^{\infty} \frac{(-1)^{n+1}}{n}\sin\left(\frac{n\pi x}{3}\right).$$

EXAMPLE 360

Let

$$f(x) = \begin{cases} 0, & -2 \le x < 0, \\ 1, & 0 \le x \le 1, \\ 2, & 1 < x \le 2. \end{cases}$$

Calculate

$$a_0 = \tfrac{1}{2}\int_{-2}^{2} f(x)\,dx = \tfrac{1}{2}\int_{-2}^{0} 0\,dx + \tfrac{1}{2}\int_{0}^{1} 1\,dx + \tfrac{1}{2}\int_{1}^{2} 2\,dx = \tfrac{3}{2},$$

$$a_n = \frac{1}{2}\int_{-2}^{0} 0 \cdot \cos\left(\frac{n\pi x}{2}\right) dx + \frac{1}{2}\int_{0}^{1} 1 \cdot \cos\left(\frac{n\pi x}{2}\right) dx$$

$$+ \frac{1}{2}\int_{1}^{2} 2 \cdot \cos\left(\frac{n\pi x}{2}\right) dx = -\frac{1}{n\pi}\sin\left(\frac{n\pi}{2}\right),$$

and

$$b_n = \frac{1}{2}\int_{-2}^{0} 0 \cdot \sin\left(\frac{n\pi x}{2}\right) dx + \frac{1}{2}\int_{0}^{1} 1 \cdot \sin\left(\frac{n\pi x}{2}\right) dx + \frac{1}{2}\int_{1}^{2} 2 \cdot \sin\left(\frac{n\pi x}{2}\right) dx$$

$$= \frac{1}{n\pi}\left[\cos\left(\frac{n\pi}{2}\right) + 1 - 2\cos(n\pi)\right].$$

Thus, the Fourier series of $f(x)$ on $[-2, 2]$ is

$$\frac{3}{4} + \sum_{n=1}^{\infty}\left\{-\frac{1}{n\pi}\sin\left(\frac{n\pi}{2}\right)\cos\left(\frac{n\pi x}{2}\right) + \frac{1}{n\pi}\left[\cos\left(\frac{n\pi}{2}\right) + 1 - 2\cos(n\pi)\right]\sin\left(\frac{n\pi x}{2}\right)\right\}.$$

EXAMPLE 361

Let

$$f(x) = \begin{cases} 1, & 0 \le x \le 1, \\ -1, & -1 \le x < 0. \end{cases}$$

Then,

$$a_0 = \int_{-1}^{0} -1\,dx + \int_{0}^{1} 1\,dx = 0,$$

$$a_n = \int_{-1}^{0} -\cos(n\pi x)\,dx + \int_{0}^{1} \cos(n\pi x)\,dx = 0,$$

and

$$b_n = \int_{-1}^{0} -\sin(n\pi x)\,dx + \int_{0}^{1} \sin(n\pi x)\,dx = \frac{2}{n\pi}[1 - \cos(n\pi)].$$

The Fourier series of $f(x)$ on $[-1, 1]$ is

$$\sum_{n=1}^{\infty} \frac{2}{n\pi}[1 - \cos(n\pi)]\sin(n\pi x).$$

Note that $\cos(n\pi) = (-1)^n$; so

$$[1 - \cos(n\pi)] = [1 - (-1)^n] = \begin{cases} 0 & \text{if } n \text{ is even,} \\ 2 & \text{if } n \text{ is odd.} \end{cases}$$

Thus, the series can be written

$$\frac{4}{\pi} \sum_{n=1}^{\infty} \frac{1}{2n-1} \sin[(2n-1)\pi x].$$

Replacing n by $2n - 1$ in the summand omits all the even indices [for which $1 - \cos(n\pi)$ is zero] and replaces $1 - \cos(n\pi)$ by 2 for all the odd indices.

EXAMPLE 362

Let $f(x) = |x|$, $-\pi \leq x \leq \pi$.

Then,

$$a_0 = \frac{1}{\pi} \int_{-\pi}^{\pi} |x| \, dx = \frac{1}{\pi} \int_{-\pi}^{0} -x \, dx + \frac{1}{\pi} \int_{0}^{\pi} x \, dx = \pi,$$

$$a_n = \frac{1}{\pi} \int_{-\pi}^{0} -x \cos(nx) \, dx + \frac{1}{\pi} \int_{0}^{\pi} x \cos(nx) \, dx = \frac{2}{\pi n^2} [\cos(n\pi) - 1],$$

and

$$b_n = \frac{1}{\pi} \int_{-\pi}^{0} -x \sin(nx) \, dx + \frac{1}{\pi} \int_{0}^{\pi} x \sin(nx) \, dx = 0.$$

The Fourier series of $|x|$ on $[-\pi, \pi]$ is

$$\frac{\pi}{2} + \sum_{n=1}^{\infty} \frac{2}{\pi n^2} [\cos(n\pi) - 1] \cos(nx).$$

As above, since $\cos(n\pi) = (-1)^n$, this is

$$\frac{\pi}{2} + \sum_{n=1}^{\infty} \frac{-4}{(2n-1)^2 \pi} \cos[(2n-1)x].$$

Note that both the function and the interval must be specified in order to write a Fourier series; in particular, the same function will have a different Fourier series on two different intervals.

Thus far, we have written Fourier series for several functions, but we have left some questions unanswered. Among these are:

1. How did we choose the formulas for the Fourier coefficients?
2. What relationship does the Fourier series of a function on an interval have to the function there? In particular, to what does the Fourier series converge?

The second question is the more difficult. Look at Example 359, where $f(x) = 2x + 1$ for $-3 \leq x \leq 3$, and the Fourier series is

$$1 + \frac{12}{\pi} \sum_{n=1}^{\infty} \frac{(-1)^{n+1}}{n} \sin\left(\frac{n\pi x}{3}\right).$$

At $x = 0$, we have $f(0) = 1$, and, likewise, the series sums to 1, since all the sine terms vanish at zero. However, at, say, $x = 1$, $f(1) = 3$, while the series is

$$1 + \frac{12}{\pi} \sum_{n=1}^{\infty} \frac{(-1)^{n+1}}{n} \sin\left(\frac{n\pi}{3}\right).$$

This series is very complicated, and it is hard to tell whether it converges to 3 or not. Thus, without some further technique at our disposal, we cannot tell whether the series converges to the function values.

In fact, it is clear that sometimes the series does not converge to the function value. Putting $x = 3$ in Example 359, we have $f(3) = 7$. However, all the sine terms in the Fourier series are zero at $x = 3$, and the series reduces to just 1. Thus, $f(x)$ does not equal its Fourier series in this example.

These potentially disturbing illustrations have a nice resolution, which we give in the next section.

PROBLEMS FOR SECTION 12.1

In each of Problems 1 through 20, write the Fourier series of the given function on the given interval. Use integral tables to work out the Fourier coefficients where convenient.

1. $f(x) = x^2, \quad -\pi \le x \le \pi$

2. $f(x) = \cos(3x), \quad -2 \le x \le 2$

3. $f(x) = \frac{1}{2}e^{2x}, \quad -3 \le x \le 3$

4. $f(x) = \sinh(x), \quad -4 \le x \le 4$

5. $f(x) = x^2 - x, \quad -\pi \le x \le \pi$

6. $f(x) = \sin(x), \quad -\pi/2 \le x \le \pi/2$

7. $f(x) = -4, \quad -3 \le x \le 3$

8. $f(x) = \begin{cases} 0, & -\pi \le x \le \pi/2 \\ 1, & \pi/2 < x \le \pi \end{cases}$

9. $f(x) = \begin{cases} -1, & -2 \le x \le -1 \\ 0, & -1 < x \le 1 \\ 3, & 1 < x \le 2 \end{cases}$

10. $f(x) = \begin{cases} 2x, & -\pi \le x < 0 \\ \cos(x), & 0 < x \le \pi \end{cases}$

11. $f(x) = 2\sin(3x), \quad -\pi \le x \le \pi$

12. $f(x) = 1 - \cos(\pi x), \quad -1 \le x \le 1$

13. $f(x) = 1 - \cos(\pi x), \quad -\pi \le x \le \pi$

14. $f(x) = 2 + x^2, \quad -4 \le x \le 4$

15. $f(x) = 3x - e^x, \quad -3 \le x \le 3$

16. $f(x) = e^{-2x}, \quad -2 \le x \le 2$

17. $f(x) = 1 - |x|, \quad -3 \le x \le 3$

18. $f(x) = \begin{cases} 4, & -2 \le x < 0 \\ -4, & 0 \le x < 2 \end{cases}$

19. $f(x) = 2x^2 - 3x + 1, \quad -4 \le x \le 4$

20. $f(x) = \cos(x), \quad -5 \le x \le 5$

21. Suppose that $f(x)$ and $g(x)$ are defined on $[-L, L]$. Is the Fourier series of $f(x) + g(x)$ on $[-L, L]$ the sum of the Fourier series of $f(x)$ and $g(x)$ on $[-L, L]$?

22. Suppose that $f(x)$ and $g(x)$ are defined on $[-L, L]$. Let $-L < x_0 < L$, and let $f(x) = g(x)$ for $-L \le x \le L$, except at x_0, where $f(x_0) \ne g(x_0)$. How do the Fourier series of $f(x)$ and $g(x)$ differ on $[-L, L]$? What does this suggest about the relationship of a Fourier series to the given function on an interval?

12.2 FOURIER COEFFICIENTS AND CONVERGENCE OF FOURIER SERIES

How did we come to choose the Fourier coefficients in the last section? Here is a heuristic argument which indicates why the choice was, in a sense, the natural one.

Suppose that we have a series of sines and cosines which represents a given function on $[-L, L]$, say,

$$f(x) = \frac{a_0}{2} + \sum_{n=1}^{\infty} a_n \cos\left(\frac{n\pi x}{L}\right) + b_n \sin\left(\frac{n\pi x}{L}\right).$$

What would such an equation tell us about $a_0, a_1, a_2, \ldots, b_1, b_2, b_3, \ldots$?

To find a_0, integrate both sides, assuming that $\sum_{n=1}^{\infty}$ and $\int_{-L}^{L}$ can be interchanged:

$$\int_{-L}^{L} f(x)\,dx = \int_{-L}^{L} \frac{a_0}{2}\,dx + \sum_{n=1}^{\infty} a_n \int_{-L}^{L} \cos\left(\frac{n\pi x}{L}\right) dx + b_n \int_{-L}^{L} \sin\left(\frac{n\pi x}{L}\right) dx.$$

Now,

$$\int_{-L}^{L} \cos\left(\frac{n\pi x}{L}\right) dx = \int_{-L}^{L} \sin\left(\frac{n\pi x}{L}\right) dx = 0$$

(check this by direct integration).

Thus, we have

$$\int_{-L}^{L} f(x)\,dx = \frac{a_0}{2} \int_{-L}^{L} dx = a_0 L.$$

Then,

$$a_0 = \frac{1}{L} \int_{-L}^{L} f(x)\,dx,$$

exactly the choice made for a_0 in Section 12.1.

Now let m be any positive integer. Multiply the series for $f(x)$ by $\cos(m\pi x/L)$ to get

$$f(x) \cos\left(\frac{m\pi x}{L}\right) = \frac{a_0}{2} \cos\left(\frac{m\pi x}{L}\right) + \sum_{n=1}^{\infty} a_n \cos\left(\frac{n\pi x}{L}\right)\cos\left(\frac{m\pi x}{L}\right) + b_n \sin\left(\frac{n\pi x}{L}\right)\cos\left(\frac{m\pi x}{L}\right).$$

Integrate:

$$\int_{-L}^{L} f(x) \cos\left(\frac{m\pi x}{L}\right) dx = \frac{a_0}{2} \int_{-L}^{L} \cos\left(\frac{m\pi x}{L}\right) dx + \sum_{n=1}^{\infty} a_n \int_{-L}^{L} \cos\left(\frac{n\pi x}{L}\right)\cos\left(\frac{m\pi x}{L}\right) dx + b_n \int_{-L}^{L} \sin\left(\frac{n\pi x}{L}\right)\cos\left(\frac{m\pi x}{L}\right) dx.$$

Immediately, $\int_{-L}^{L} \cos(m\pi x/L)\,dx = 0$. Further, the student can verify by integration that

$$\int_{-L}^{L} \cos\left(\frac{n\pi x}{L}\right)\cos\left(\frac{m\pi x}{L}\right) dx = \begin{cases} 0 & \text{if } n \neq m \\ L & \text{if } n = m \end{cases}$$

and

$$\int_{-L}^{L} \sin\left(\frac{n\pi x}{L}\right)\cos\left(\frac{m\pi x}{L}\right) dx = 0 \quad \text{for} \quad n = 1, 2, 3, \ldots.$$

As n goes from 1 to ∞ in the last summation, every integral is zero except when $n = m$, and then the integral is $a_m L$. The series then collapses to

$$\int_{-L}^{L} f(x) \cos\left(\frac{m\pi x}{L}\right) dx = a_m L.$$

Hence,

$$a_m = \frac{1}{L}\int_{-L}^{L} f(x) \cos\left(\frac{m\pi x}{L}\right) dx \quad \text{for} \quad m = 1, 2, 3, \ldots,$$

again just as in Section 12.1.

To solve for the b_n's, go back to the original series and multiply by $\sin(m\pi x/L)$, where m is any positive integer:

$$f(x) \sin\left(\frac{m\pi x}{L}\right) = \frac{a_0}{2} \sin\left(\frac{m\pi x}{L}\right) + \sum_{n=1}^{\infty} a_n \cos\left(\frac{n\pi x}{L}\right)\sin\left(\frac{m\pi x}{L}\right) + b_n \sin\left(\frac{n\pi x}{L}\right)\sin\left(\frac{m\pi x}{L}\right).$$

Integrate:

$$\int_{-L}^{L} f(x) \sin\left(\frac{m\pi x}{L}\right) dx = \frac{a_0}{2}\int_{-L}^{L} \sin\left(\frac{m\pi x}{L}\right) dx + \sum_{n=1}^{\infty} a_n \int_{-L}^{L} \cos\left(\frac{n\pi x}{L}\right)\sin\left(\frac{m\pi x}{L}\right) dx + b_n \int_{-L}^{L} \sin\left(\frac{n\pi x}{L}\right)\sin\left(\frac{m\pi x}{L}\right) dx.$$

Now,

$$\int_{-L}^{L} \sin\left(\frac{m\pi x}{L}\right) dx = 0.$$

Further,

$$\int_{-L}^{L} \sin\left(\frac{n\pi x}{L}\right)\sin\left(\frac{m\pi x}{L}\right) dx = \begin{cases} 0 & \text{if} \quad n \neq m, \\ L & \text{if} \quad n = m. \end{cases}$$

Thus, the series collapses to

$$\int_{-L}^{L} f(x) \sin\left(\frac{m\pi x}{L}\right) dx = b_m L,$$

and we have

$$b_m = \frac{1}{L}\int_{-L}^{L} f(x) \sin\left(\frac{m\pi x}{L}\right) dx, \quad \text{for} \quad m = 1, 2, 3, \ldots.$$

The argument is not mathematically flawless, but it is fairly convincing in making a case for the choice of Fourier coefficients made in Section 12.1.

We now consider the second question posed in the last section: What is the relationship between a function and its Fourier series on a given interval $[-L, L]$?

If you look at Example 361, you can get some feeling for the complexity of the problem. There, $f(x) = 1$ for $0 \le x \le 1$, $f(x) = -1$ for $-1 \le x < 0$, and the Fourier series was

$$\frac{4}{\pi} \sum_{n=1}^{\infty} \frac{1}{2n-1} \sin[(2n-1)\pi x].$$

Clearly, this series converges to 0 at $x = 0$, 1, and -1, and hence certainly does not equal $f(x)$ at these points. At other points it is hard to tell what is happening. For example, $f(\frac{1}{2}) = 1$, while the series at $\frac{1}{2}$ is

$$\frac{4}{\pi} \sum_{n=1}^{\infty} \frac{1}{2n-1} \sin\left[\frac{(2n-1)\pi}{2}\right],$$

and who can tell if this adds up to 1?

In a situation such as this we need a theorem. The standard theorem on convergence of Fourier series requires that we introduce some terminology. This is done in outline form below.

1. The *right limit* of $f(x)$ at x_0 is defined as

$$\lim_{h \to 0+} f(x_0 + h).$$

In this notation, $\lim_{h \to 0^+}$ means that h approaches zero through positive values only. Similarly, the *left limit* of $f(x)$ at x_0 is defined to be

$$\lim_{h \to 0+} f(x_0 - h).$$

The right limit of $f(x)$ at x_0 is denoted $f(x_{0+})$, and the left limit is written $f(x_{0-})$, whenever these limits exist. Note that, if $\lim_{x \to x_0} f(x) = L$, then necessarily $f(x_{0+}) = f(x_{0-}) = L$; however, left and right limits may exist and differ, in which case $f(x)$ has no limit at x_0.

As an illustration, let

$$f(x) = \begin{cases} 2, & -\infty < x < 3, \\ \sqrt{x}, & 3 < x < \infty. \end{cases}$$

The graph of $f(x)$ is shown in Figure 216. Here, we have $f(3+) = \sqrt{3}$, and $f(3-) = 2$, while $\lim_{x \to 3} f(x)$ does not exist. Note that $f(3)$ is not defined, a fact which is irrelevant as far as existence of the limit at 3 is concerned. In this example, for any $x_0 < 3$, $f(x_{0+}) = f(x_{0-}) = \lim_{x \to x_0} f(x) = 2$; at any $x_0 > 3$, $f(x_{0+}) = f(x_{0-}) = \lim_{x \to x_0} f(x) = \sqrt{x_0}$.

2. A function $f(x)$ is *sectionally continuous* on $[-L, L]$ if (1) $f(x)$ is defined on $[-L, L]$, except possibly at finitely many points $-L, x_1, x_2, \ldots, x_n, L$, with $-L < x_1 < x_2 < \cdots < x_n < L$; (2) $f(x)$ is continuous on each subinterval $(-L, x_1)$, $(x_1, x_2), \ldots, (x_{n-1}, x_n), (x_n, L)$; and (3) $f(x)$ has a finite limit from the right at $-L$, from the left at L, and has both left and right limits at each $x_j, j = 1, 2, \ldots, n$.

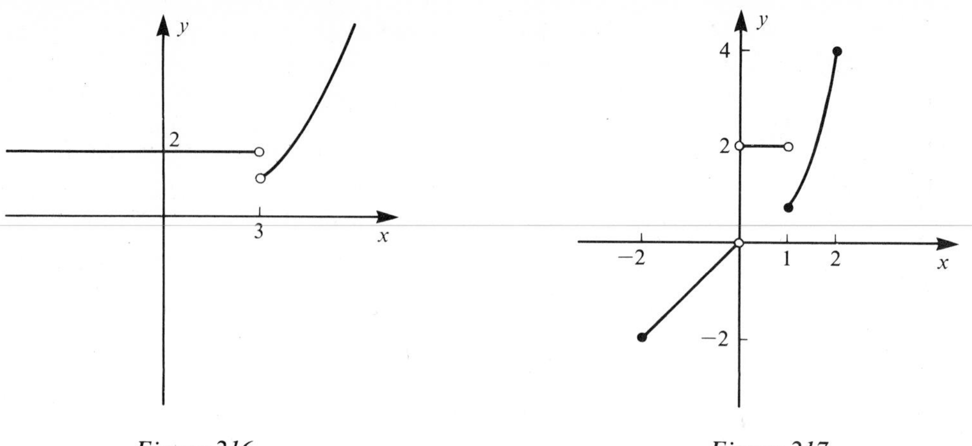

Figure 216

Figure 217

As an illustration, let

$$f(x) = \begin{cases} x & \text{for} \quad -2 \le x < 0, \\ 2 & \text{for} \quad 0 < x < 1, \\ x^2 & \text{for} \quad 1 \le x \le 2. \end{cases}$$

Then, $f(x)$ is continuous on $(-2, 0)$, $(0, 1)$, and $(1, 2)$. Further, $f(-2+) = -2$, $f(0+) = 2$, $f(0-) = 0$, $f(1+) = 1$, $f(1-) = 2$, and $f(2-) = 4$. Here, $f(x)$ is not defined at all at 0. From the definition, $f(x)$ is sectionally continuous on $[-2, 2]$. The graph of $f(x)$ is shown in Figure 217. Note that, if we wanted $\int_{-2}^{2} f(x)\, dx$, we would compute

$$\int_{-2}^{0} x\, dx + \int_{0}^{1} 2\, dx + \int_{1}^{2} x^2\, dx,$$

obtaining $\frac{7}{3}$.

3. Finally, we need the notions of right and left derivative of $f(x)$ at a point x_0. If $f(x)$ has a right limit at x_0, then the *right derivative* of $f(x)$ at x_0 is

$$f'_R(x_0) = \lim_{h \to 0+} \frac{f(x_0 + h) - f(x_{0+})}{h},$$

if this limit exists. Similarly, the *left derivative* of $f(x)$ at x_0 is

$$f'_L(x_0) = \lim_{h \to 0+} \frac{f(x_0 - h) - f(x_{0-})}{h}.$$

Wherever $f'(x_0)$ exists, then $f'_R(x_0) = f'_L(x_0) = f'(x_0)$. However, right and/or left derivatives may exist where $f(x)$ is not differentiable. With $f(x)$ as in Figure 217, we have, for example,

$$f'_R(1) = \lim_{h \to 0+} \frac{f(1 + h) - f(1+)}{h} = \lim_{h \to 0+} \frac{(1 + h)^2 - 1}{h} = 2,$$

while

$$f'_L(1) = \lim_{h\to 0+} \frac{f(1-h) - f(1-)}{h} = \lim_{h\to 0+} \frac{2-2}{h} = 0.$$

Intuitively, $f'_R(1)$ is the slope of $y = f(x)$ at 1 when viewed toward the right from 1 [where $f(x)$ looks like x^2]; $f'_L(1)$ is the slope at 1 when viewed from 1 toward the left, where $f(x)$ looks like the constant 2.

With these as preliminaries, we can now state the following convergence theorem.

THEOREM 95 Convergence of Fourier Series

Let $f(x)$ be sectionally continuous on $[-L, L]$. Then,

1. If $-L < x_0 < L$, and $f'_R(x_0)$ and $f'_L(x_0)$ both exist, then at x_0 the Fourier series of $f(x)$ on $[-L, L]$ converges to
$$\tfrac{1}{2}[f(x_{0+}) + f(x_{0-})].$$
2. At $-L$ and at L, if $f'_R(-L)$ and $f'_L(L)$ exist, then the Fourier series converges to
$$\tfrac{1}{2}[f(-L+) + f(L-)].$$

In words: At each point between $-L$ and L where the right and left derivatives exist, the Fourier series converges to the average of the left and right limits there; at both $-L$ and at L, the series converges to the average of the right limit at $-L$ and the left limit at L (assuming that the right derivative at $-L$ and the left derivative at L exist).

A proof of the above theorem is quite technical and is omitted in this treatment. Here is a restatement of the conclusions in a slightly different light:

1. Wherever $f(x)$ is continuous on $(-L, L)$ and both left and right derivatives exist, the series converges to $f(x)$ itself. This is because, if $f(x)$ is continuous at x_0, then the left and right limits at x_0 are exactly $f(x_0)$.
2. If $f(x)$ has a jump discontinuity at x_0, and both right and left derivatives exist at x_0, then, graphically, the Fourier series converges to the point midway between the gap in the graph of $y = f(x)$ at x_0. This is illustrated in Figure 218.
3. It should not be surprising that the Fourier series converges to the same value at L and at $-L$.

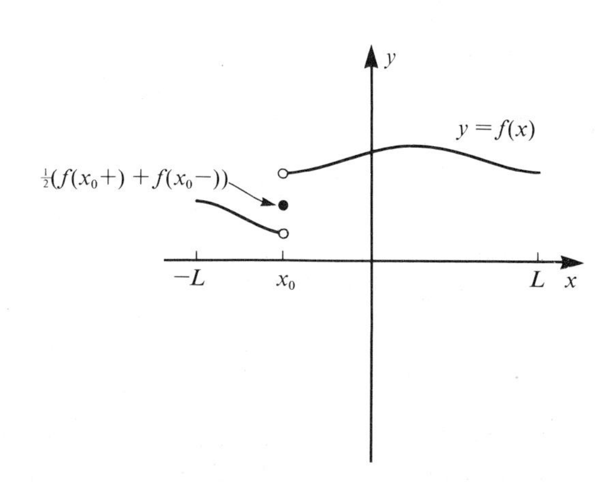

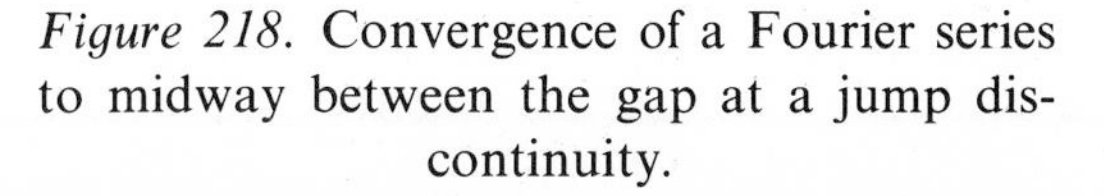
Figure 218. Convergence of a Fourier series to midway between the gap at a jump discontinuity.

The Fourier series of $f(x)$ on $[-L, L]$ is

$$\frac{a_0}{2} + \sum_{n=1}^{\infty} a_n \cos\left(\frac{n\pi x}{L}\right) + b_n \sin\left(\frac{n\pi x}{L}\right).$$

Substituting $x = L$ or $x = -L$ yields the same result, namely,

$$\frac{a_0}{2} + \sum_{n=1}^{\infty} a_n \cos(n\pi).$$

In view of these considerations, and Figure 218, one can often tell what a Fourier series converges to just by looking at the graph $y = f(x)$. In particular, for most functions one encounters in practice, it is obvious where $f(x)$ has right and left derivatives because most parts of the graph will be smooth. The Fourier series then converges to function values where the function is continuous and to midway between the gaps at jump discontinuities. At the endpoints, the series converges to midway between the gap between $f(L-)$ and $f(-L+)$. Here are three illustrations.

EXAMPLE 363

Let

$$f(x) = \begin{cases} 2, & -3 \le x < -1, \\ |x|, & -1 < x \le 0, \\ x^2, & 0 < x < 2, \\ x + 5, & 2 < x \le 3. \end{cases}$$

Figure 219 shows a graph of $y = f(x)$, and Figure 220 shows the graph of the Fourier series of $f(x)$ on $[-3, 3]$. Note that, where $f(x)$ is continuous, the series converges to $f(x)$ and the two graphs coincide. At -3 and at 3, the series converges to

$$\tfrac{1}{2}(2 + 8) = 5.$$

At -1, the series converges to

$$\tfrac{1}{2}(2 + |-1|) = \tfrac{3}{2}.$$

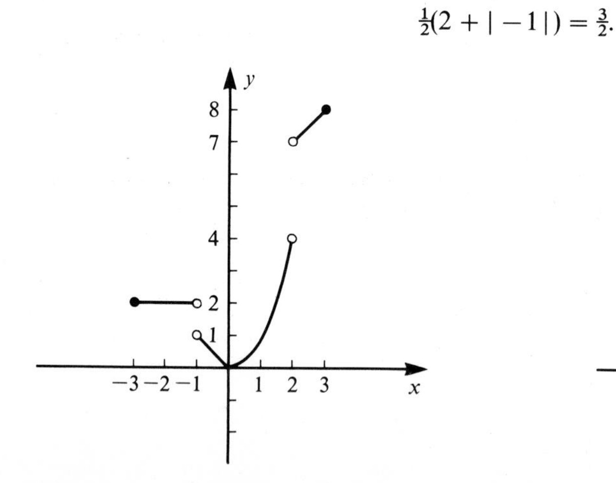

Figure 219. Graph of the function of example 363.

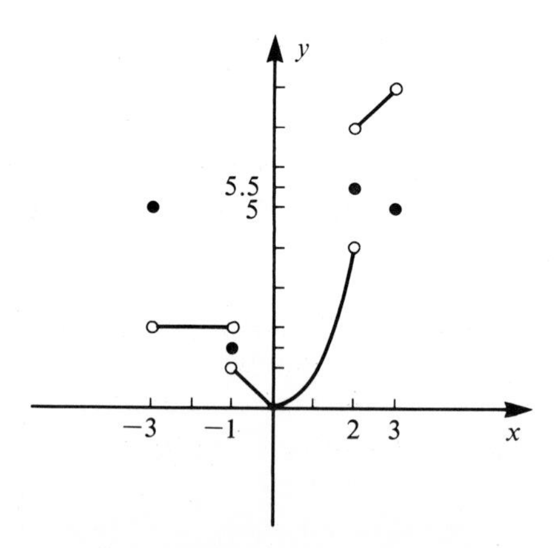

Figure 220. Fourier series of the function of example 363.

And, at 2, the series converges to

$$\tfrac{1}{2}[2^2 + (2 + 5)] = \tfrac{11}{2}.$$

It is obvious in this example that right and left derivatives exist where needed. For example, $f'_R(-3) = 0$, $f'_R(-1) = -1$, $f'_L(-1) = 0$, $f'_R(2) = 1$, $f'_L(2) = 4$, and $f'_L(3) = 1$; $f(x)$ has a derivative everywhere else on $(-3, 3)$.

EXAMPLE 364

Let

$$f(x) = \begin{cases} 1 - x, & -4 \le x < 0, \\ 2, & 0 < x < 3, \\ \sqrt{e^x}, & 3 < x < 4. \end{cases}$$

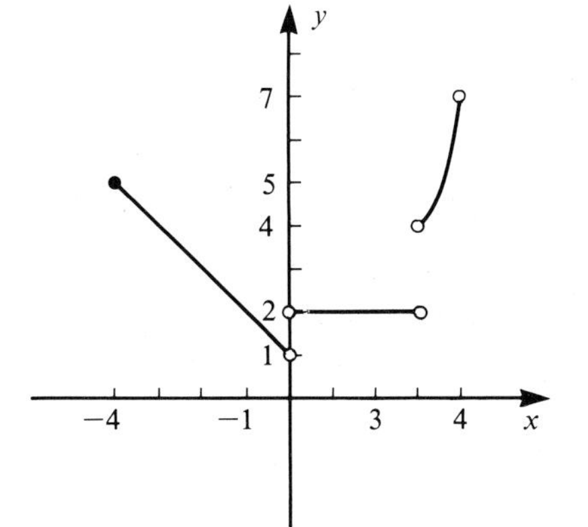

Figure 221. Graph of the function of example 364.

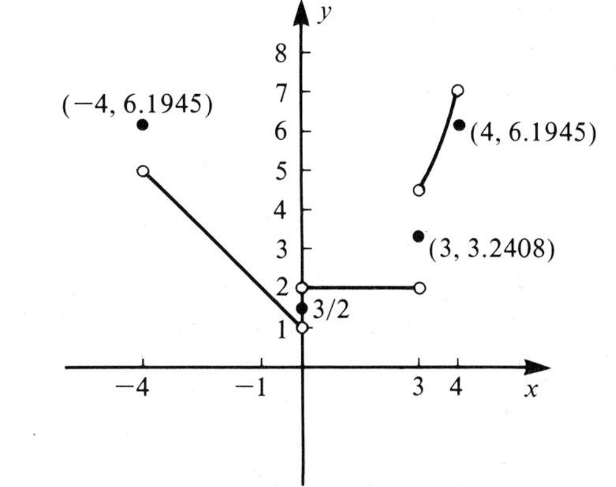

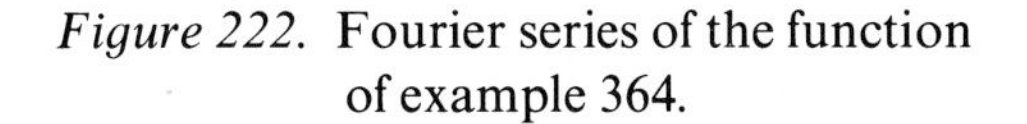

Figure 222. Fourier series of the function of example 364.

Figure 221 shows the graph of the function, and Figure 222 shows the graph of its Fourier series on $[-4, 4]$. Note that, at 4 and at -4, the series converges to

$$\tfrac{1}{2}(5 + \sqrt{e^4}),$$

which is about 6.1945. At 0, the series converges to

$$\tfrac{1}{2}(1 + 2) = \tfrac{3}{2},$$

and, at 3, the series converges to

$$\tfrac{1}{2}(2 + \sqrt{e^3}),$$

about 3.2408. On $(-4, 0)$, $(0, 3)$, and $(3, 4)$ the series converges to $f(x)$. Again, existence of right and left derivatives at the jump discontinuities is obvious in this example.

EXAMPLE 365

Let

$$f(x) = \begin{cases} 0, & -2 \le x < 0, \\ 1, & 0 \le x \le 1, \\ 2, & 1 < x \le 2, \end{cases}$$

as in Example 360. The graph of $y = f(x)$ is shown in Figure 223, and that of the Fourier series is shown in Figure 224. The jump discontinuities occur at 0 and 1; at 0 the series converges to $\frac{1}{2}$ and, at 1, to $\frac{3}{2}$. At -2 and at 2, the series converges to 1, which is midway between 0 and 2.

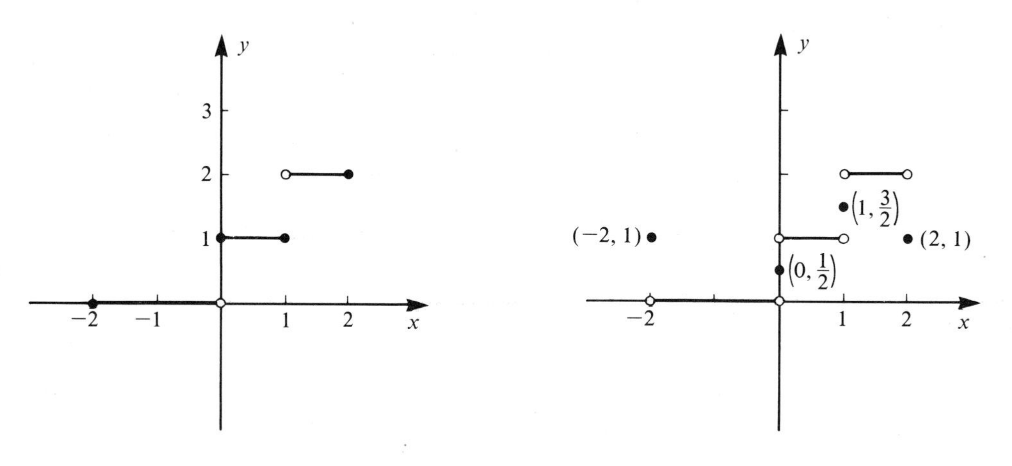

Figure 223. Graph of the function of example 365.

Figure 224. Fourier series of the function of example 365.

In general, Fourier series tend to converge slowly, in the following sense. Let $f(x)$ be defined on $[-L, L]$, with Fourier series

$$\frac{a_0}{2} + \sum_{n=1}^{\infty} a_n \cos\left(\frac{n\pi x}{L}\right) + b_n \sin\left(\frac{n\pi x}{L}\right).$$

Look at the partial sum

$$S_k(x) = \frac{a_0}{2} + \sum_{n=1}^{k} a_n \cos\left(\frac{n\pi x}{L}\right) + b_n \sin\left(\frac{n\pi x}{L}\right).$$

As $k \to \infty$, $S_k(x)$ approaches the Fourier series. By slow convergence of the Fourier series, we mean that usually k must be taken "large" for $S_k(x)$ to be "near" the entire infinite series.

Here are three somewhat untypical examples. The first two were chosen because they are simple and converge rapidly, and hence give an easy illustration of the idea of convergence. In the third example, we have to go out to fifty terms to get the kind of fit we got in the other examples.

EXAMPLE 366

Let $f(x)$ be the step function of Example 361. The partial sums are

$$S_1(x) = \frac{4}{\pi}\sin(\pi x),$$

$$S_2(x) = \frac{4}{\pi}\sin(\pi x) + \frac{4}{3\pi}\sin(3\pi x),$$

$$S_4(x) = \frac{4}{\pi}\sin(\pi x) + \frac{4}{3\pi}\sin(3\pi x) + \frac{4}{5\pi}\sin(5\pi x) + \frac{4}{7\pi}\sin(7\pi x),$$

and so on. In Figures 225, 226, and 227, we have a comparison of the graphs of $f(x)$ and several partial sums.

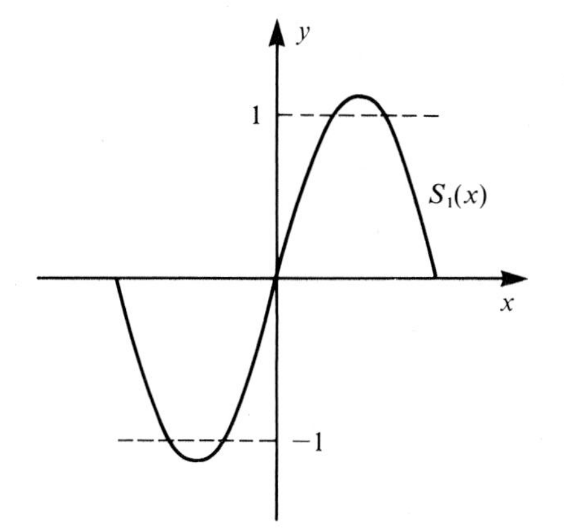

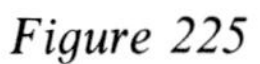
Figure 225

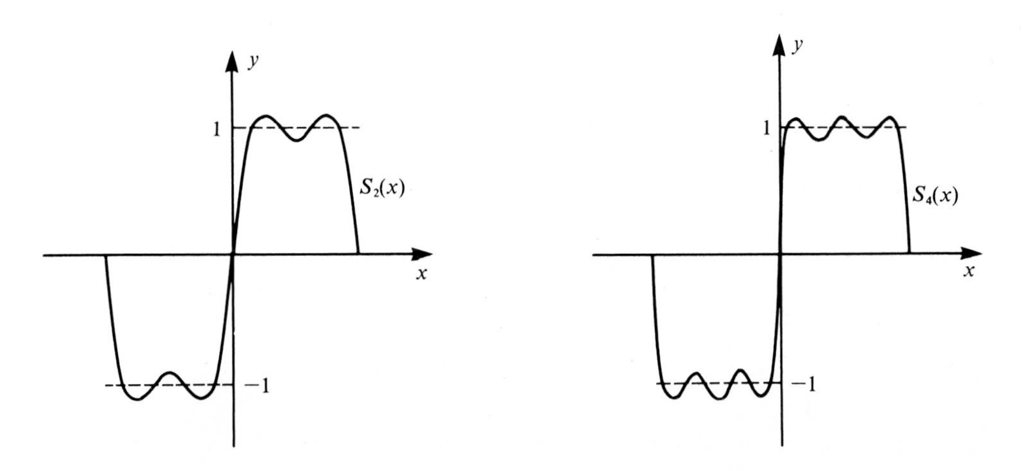

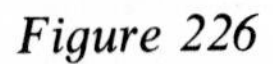
Figure 226

Figure 227

EXAMPLE 367

Let $f(x) = |x|$, $-\pi \leq x \leq \pi$, as in Example 362. Some partial sums are

$$S_1(x) = \frac{\pi}{2} - \frac{4}{\pi}\cos(x),$$

$$S_2(x) = \frac{\pi}{2} - \frac{4}{\pi}\cos(x) - \frac{4}{9\pi}\cos(3x),$$

$$S_3(x) = \frac{\pi}{2} - \frac{4}{\pi}\cos(x) - \frac{4}{9\pi}\cos(3x) - \frac{4}{25\pi}\cos(5x),$$

and so on. Figures 228, 229, and 230 compare the function and several of the partial sums.

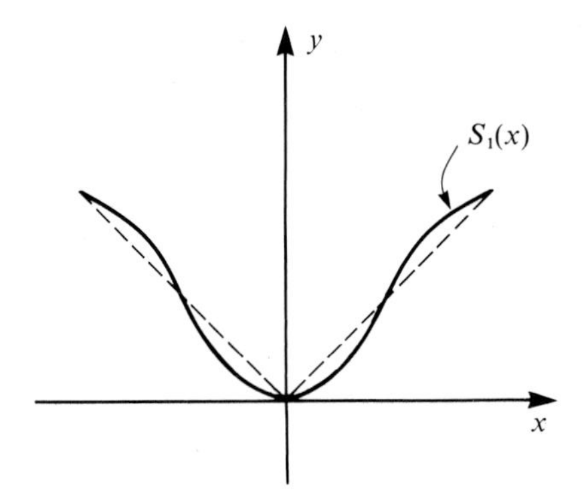

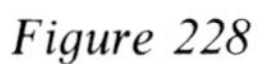

Figure 228

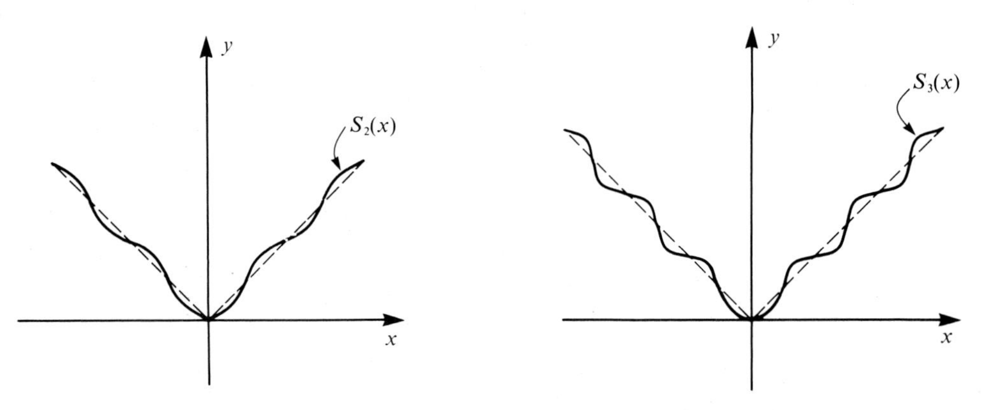

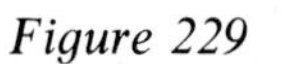

Figure 229

Figure 230

EXAMPLE 368

Let $f(x) = x$, $-\pi \leq x \leq \pi$. The Fourier series is

$$\sum_{n=1}^{\infty} \frac{2(-1)^{n+1}}{n} \sin(nx).$$

Partial sums are

$$S_1(x) = 2\sin(x),$$

$$S_2(x) = 2\sin(x) - \sin(2x),$$

$$S_3(x) = 2\sin(x) - \sin(2x) + \tfrac{2}{3}\sin(3x),$$

and so on. Figures 231, 232, and 233 compare the function with several partial sums. Note that convergence in this example is much slower than in the previous two.

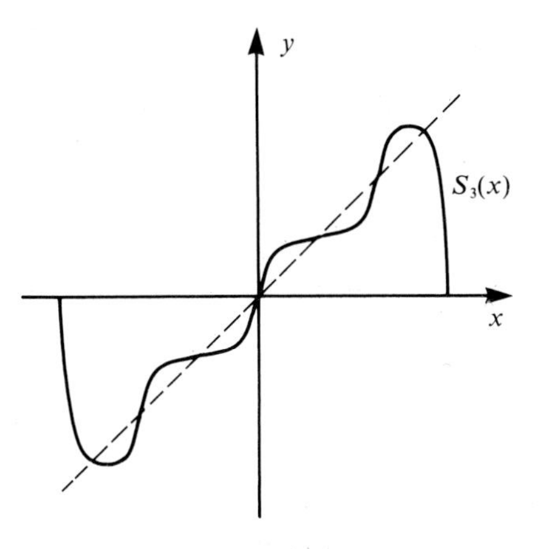

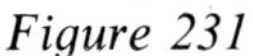

Figure 231

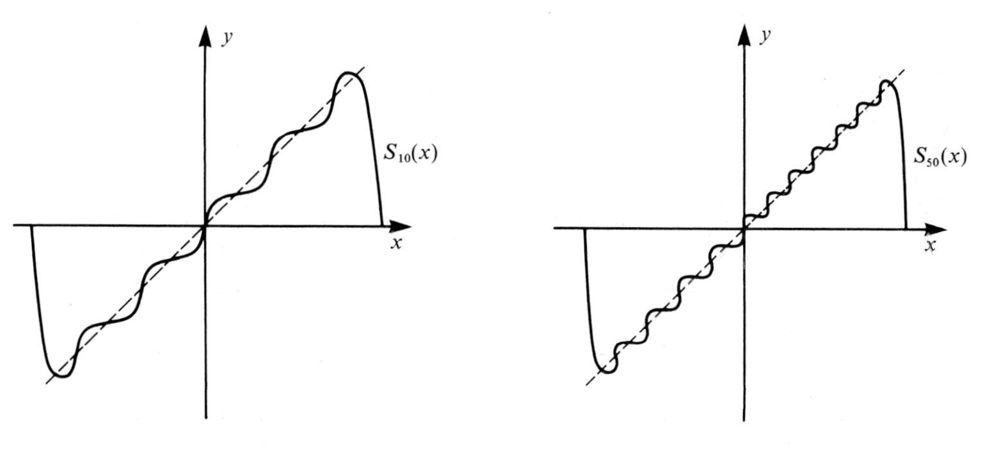

Figure 232 *Figure 233*

Often, in working with series, it is useful to integrate and differentiate term by term. Generally, this will cause difficulties. In the exercises we pursue some conditions under which Fourier series may be integrated and differentiated term by term.

The next section is devoted to Fourier series of functions defined just on $[0, L]$ instead of on interval $[-L, L]$. In this special circumstance, it is possible to arrange ahead of time to have just cosines or just sines appear in the series, a fact which is useful in solving many partial differential equations.

PROBLEMS FOR SECTION 12.2

In each of Problems 1 through 20, find $f(x_{0+}), f(x_{0-}), f'_R(x_0)$, and $f'_L(x_0)$ at each point x_0 in $(-L, L)$ where $f(x)$ has a discontinuity. Also find $f(-L+), f(L-), f_R(-L)$, and $f_L(L)$. Finally, determine what the Fourier series of $f(x)$ on $[-L, L]$ converges to for each point in $[-L, L]$. *Note:* You need not find the Fourier series to do this.

1. $f(x) = \begin{cases} x^2, & 1 \le x \le 3 \\ 0, & -2 \le x < 1 \\ 2x, & -3 \le x < -2 \end{cases}$

2. $f(x) = x^4, \quad -2 \le x \le 2$

3. $f(x) = x^{15}, \quad -3 \le x \le 3$

4. $f(x) = \begin{cases} 2x - 2, & -\pi \le x < \pi/2 \\ 3, & \pi/2 \le x \le \pi \end{cases}$

5. $f(x) = \begin{cases} x^2, & -\pi \le x < 0 \\ 2, & 0 \le x \le \pi \end{cases}$

6. $f(x) = e^{|x|}, \quad -3 \le x \le 3$

7. $f(x) = \begin{cases} \cos(x), & -\pi \le x \le 0 \\ \sin(x), & 0 < x \le \pi \end{cases}$

8. $f(x) = \begin{cases} 1/x, & -3 \le x < -1 \\ 2, & -1 \le x \le 2 \\ 3, & 2 < x \le 3 \end{cases}$

9. $f(x) = \begin{cases} 1 & 0 \le x \le 4 \\ -2, & -4 \le x < 0 \end{cases}$

10. $f(x) = \begin{cases} -2, & -3 \le x < -1 \\ 1 + x^2, & -1 < x \le 2 \\ x^3, & 2 < x \le 3 \end{cases}$

11. $f(x) = \begin{cases} x, & -2 \le x < 1 \\ e^x, & 1 \le x \le 2 \end{cases}$

12. $f(x) = \begin{cases} -1, & -1 \le x \le 0 \\ 4, & 0 < x \le 1 \end{cases}$

13. $f(x) = \begin{cases} 2 - x, & -3 \le x < -2 \\ 1, & -2 < x \le 0 \\ -1, & 0 < x < 1 \\ x^2 + 1, & 1 \le x \le 3 \end{cases}$

14. $f(x) = \begin{cases} \cos(\pi x), & -1 \le x < 0 \\ \sin(\pi x), & 0 < x \le 1 \end{cases}$

15. $f(x) = \begin{cases} x + 2, & -5 \le x \le 4 \\ 1, & 4 < x \le \frac{9}{2} \\ x - 4, & \frac{9}{2} < x \le 5 \end{cases}$

16. $f(x) = \begin{cases} x^2, & -1 \le x < \frac{1}{2} \\ 2x, & \frac{1}{2} < x \le \frac{2}{3} \\ 3, & \frac{2}{3} < x < \frac{3}{4} \\ 4x, & \frac{3}{4} \le x \le 1 \end{cases}$

17. $f(x) = \begin{cases} -2\sin(3x), & -\pi \le x < \pi/2 \\ x^2, & \pi/2 \le x \le \pi \end{cases}$

18. $f(x) = \begin{cases} |x|, & -1 \le x \le \frac{1}{2} \\ x^2, & \frac{1}{2} < x \le \frac{3}{4} \\ x - 4, & \frac{3}{4} < x \le 1 \end{cases}$

19. $f(x) = \begin{cases} x^2 - 2, & -3 \le x < 0 \\ e^x, & 0 < x \le 3 \end{cases}$

20. $f(x) = \begin{cases} -x^4, & -3 \le x < 0 \\ 2, & x = 0 \\ 4, & 0 < x \le 3 \end{cases}$

In each of Problems 21 through 30, (a) find the Fourier series of the function on the given interval, (b) draw a graph of the function, (c) draw a graph of the Fourier series, and (d) if you have calculator or computer help, compute the first four partial sums of the Fourier series and sketch the graphs of these partial sums on the same set of axes with a graph of the function.

21. $f(x) = e^{-|x|}, \quad -\pi \le x \le \pi$

22. $f(x) = x^2, \quad -\pi \le x \le \pi$

23. $f(x) = 2x - 4, \quad -6 \le x \le 6$

24. $f(x) = -2, \quad -\pi \le x \le \pi$

25. $f(x) = 3x, \quad -4 \le x \le 4$

26. $f(x) = \sin(x), \quad -\pi/2 \le x \le \pi/2$

27. $f(x) = \begin{cases} -2x, & -1 \le x < 0 \\ 0, & 0 \le x \le 1 \end{cases}$

28. $f(x) = 1 + \cos(2x), \quad -1 \le x \le 1$

29. $f(x) = 4x^2 - 1, \quad -2 \le x \le 2$

30. $f(x) = \begin{cases} -2\sin(\pi x), & -1 \le x < 0 \\ 2, & 0 < x \le 1 \end{cases}$

31. What is wrong with the following argument?

Theorem 95 doesn't make any sense because the sines and cosines in the Fourier series are periodic, while $f(x)$ need not be periodic. Hence, whenever $f(x)$ is not periodic, the Fourier series cannot converge to $f(x)$.

Problems 32 through 36 have to do with differentiating and integrating a Fourier series term by term.

32. Let $f(x) = x, \quad -\pi \le x \le \pi$.

(a) Write the Fourier series for $f(x)$ on $[-\pi, \pi]$.

(b) Use Theorem 95 to show that the Fourier series converges to $f(x)$ for $-\pi \le x \le \pi$.

(c) Note that $f'(x) = 1$, but that the series obtained by differentiating the Fourier series term by term diverges. Hence conclude that, in general, a Fourier series representation of a function cannot be differentiated term by term.

33. Prove the following theorem on differentiation of Fourier series.

Let $f(x)$ be continuous on $[-L, L]$, and suppose that $f(L) = f(-L)$. Let $f'(x)$ be sectionally continuous on $[-L, L]$. Then,

1. The Fourier series for $f(x)$ on $[-L, L]$ converges to $f(x)$ for $-L \le x \le L$.
2. On $(-L, L)$, wherever $f'(x)$ is continuous,

$$f'(x) = \sum_{n=1}^{\infty} \frac{n\pi}{L}\left[-a_n \sin\left(\frac{n\pi x}{L}\right) + b_n \cos\left(\frac{n\pi x}{L}\right)\right],$$

exactly the series we would obtain by differentiating term by term the Fourier series of $f(x)$ on $[-L, L]$.

Hint: Begin by writing the Fourier series for $f'(x)$ on $[-L, L]$; say this series is

$$\frac{A_0}{2} + \sum_{n=1}^{\infty} A_n \cos\left(\frac{n\pi x}{L}\right) + B_n \sin\left(\frac{n\pi x}{L}\right),$$

in which the A_n's and B_n's are the Fourier coefficients of $f'(x)$ on $[-L, L]$. Then, by integrating by parts, show that

$$A_n = \frac{n\pi}{L} b_n \quad \text{and} \quad B_n = -\frac{n\pi}{L} a_n \quad \text{for} \quad n = 1, 2, 3, \ldots.$$

Also show, by direct integration and the hypotheses of the theorem, that $A_0 = 0$.

34. Let $f(x) = |x|$ for $-1 \le x \le 1$.

(a) Write the Fourier series of $f(x)$ on $[-1, 1]$.

(b) Show that $f(x)$ satisfies the hypotheses of the theorem of Problem 33.

(c) By differentiating the result of (a) term by term, find a Fourier series for $f'(x)$, where

$$f'(x) = \begin{cases} -1, & -1 < x < 0, \\ 1, & 0 < x < 1. \end{cases}$$

(d) Verify this result directly by expanding the function of (c) in a Fourier series on $[-1, 1]$ and comparing with the series found in (c).

35. Prove the following theorem on integration of Fourier series term by term.

Let $f(x)$ be sectionally continuous on $[-L, L]$ with Fourier series

$$\frac{a_0}{2} + \sum_{n=1}^{\infty} a_n \cos\left(\frac{n\pi x}{L}\right) + b_n \sin\left(\frac{n\pi x}{L}\right).$$

Then, for $-L \le x \le L$,

$$\int_{-L}^{x} f(t)\,dt = \tfrac{1}{2}a_0(x + L)$$

$$+ \sum_{n=1}^{\infty} \frac{L}{n\pi}\left\{a_n \sin\left(\frac{n\pi x}{L}\right) - b_n\left[\cos\left(\frac{n\pi x}{L}\right) - \cos(n\pi)\right]\right\}.$$

This is exactly what we obtain by integrating the Fourier series of $f(x)$ term by term from $-L$ to x. Further, the result is valid even when this series does not converge to $f(x)$ at x.

Hint: Define

$$F(x) = \int_{-L}^{x} f(t)\,dt - \tfrac{1}{2}a_0\,x.$$

Verify that, by the Fourier convergence theorem, the Fourier series of $F(x)$ on $[-L, L]$ converges to $F(x)$ on the entire interval. Let the Fourier series of $F(x)$ on $[-L, L]$ be

$$F(x) = \frac{A_0}{2} + \sum_{n=1}^{\infty} A_n \cos\left(\frac{n\pi x}{L}\right) + B_n \sin\left(\frac{n\pi x}{L}\right).$$

Integrate by parts to show that, for $n = 1, 2, 3, \ldots,$

$$A_n = -\frac{L}{n\pi}\,b_n \quad \text{and} \quad B_n = \frac{L}{n\pi}\,a_n,$$

where the a_n's and b_n's are the Fourier coefficients of $f(x)$ on $[-L, L]$.

Next, show by substitution into the Fourier series that

$$F(L) = \frac{1}{2}A_0 + \sum_{n=1}^{\infty} -\frac{L}{n\pi}\,b_n \cos(n\pi),$$

and use this to solve for $\frac{1}{2}A_0$.

36. Let

$$f(x) = \begin{cases} 0, & -\pi \le x \le 0, \\ x, & 0 \le x \le \pi. \end{cases}$$

(a) Write the Fourier series of $f(x)$ on $[-\pi, \pi]$.

(b) Show that the Fourier series of $f(x)$ on $[-\pi, \pi]$ converges to $f(x)$ for $-\pi < x < \pi$.

(c) Use the result of Problem 35 to obtain a trigonometric series for

$$\int_{-\pi}^{x} f(t)\,dt,$$

for any x in $[-\pi, \pi]$.

In each of Problems 37 through 40, verify that $f(x)$ satisfies the hypotheses of the theorem given in Problem 33. Find the Fourier series of $f(x)$ on the given interval, and use Problem 33 to find the Fourier series of $f'(x)$ on the interval for those points where $f'(x)$ exists.

37. $f(x) = \begin{cases} x^2, & -1 \le x < 0 \\ x, & 0 < x \le 1 \end{cases}$

38. $f(x) = x^2 + 2, \quad -3 \le x \le 3$

39. $f(x) = (x - L)(x + L), \quad -L \le x \le L$

40. $f(x) = \begin{cases} x + 3, & -2 \le x \le 0 \\ x - 1, & 0 < x \le 2 \end{cases}$

In each of Problems 41 through 44, verify that $f(x)$ satisfies the hypotheses of the theorem of Problem 35. Find the Fourier series of $f(x)$ on the given interval, and use the result of Problem 35 to write the Fourier series of

$$\int_{-L}^{x} f(t)\,dt.$$

Next, verify this result directly by carrying out the integration to find $\int_{-L}^{x} f(t)\,dt$ explicitly as a function of x and expanding this function in a Fourier series on the given interval.

41. $f(x) = \begin{cases} -1, & -3 \le x \le 0 \\ 1, & 0 < x \le 3 \end{cases}$

42. $f(x) = e^{|x|}, \quad -4 \le x \le 4$

43. $f(x) = x^2 + 2, \quad -3 \le x \le 3$

44. $f(x) = \cos(2\pi x), \quad -2 \le x \le 2$

45. Earlier in this section we gave a heuristic argument to show why we chose the Fourier coefficients the way we did in Section 12.1. Here is another argument leading to the same conclusion, but with some additional dividends.

Suppose that $f(x)$ satisfies the hypotheses of Theorem 95 on an interval $[-L, L]$. We pose the following problem. We envision an approximation of $f(x)$ by a finite trigonometric series,

$$f(x) \approx \frac{\alpha_0}{2} + \sum_{n=1}^{k} \left[\alpha_n \cos\left(\frac{n\pi x}{L}\right) + \beta_n \sin\left(\frac{n\pi x}{L}\right) \right] = S_k(x).$$

The problem is to choose α_0 and the α_n's and β_n's to minimize the integral

$$I_k = \int_{-L}^{L} [f(x) - S_k(x)]^2\,dx.$$

With this choice, we call $S_k(x)$ a best *mean square approximation* of $f(x)$. Now solve the problem as follows.

First, show that

$$[f(x) - S_k(x)]^2 = f(x)^2 + \left\{ \frac{\alpha_0}{2} + \sum_{n=1}^{k} \left[\alpha_n \cos\left(\frac{n\pi x}{L}\right) + \beta_n \sin\left(\frac{n\pi x}{L}\right) \right] \right\}^2$$
$$- 2\left\{ \frac{\alpha_0}{2} f(x) + \sum_{n=1}^{k} \left[\alpha_n f(x) \cos\left(\frac{n\pi x}{L}\right) + \beta_n f(x) \sin\left(\frac{n\pi x}{L}\right) \right] \right\}$$

Next, show from this that

$$I_k = \int_{-L}^{L} f(x)^2\,dx + L\left[\frac{\alpha_0^2}{2} + \sum_{n=1}^{k} (\alpha_n{}^2 + \beta_n{}^2) \right]$$
$$-2L\left[\frac{\alpha_0 a_0}{2} + \sum_{n=1}^{k} (\alpha_n a_n + \beta_n b_n) \right].$$

In this expression, the a_n's and b_n's are the Fourier coefficients of $f(x)$ on $[-L, L]$.

The object now is to choose $\alpha_0, \alpha_1, \ldots, \alpha_k, \beta_1, \ldots, \beta_k$ to minimize the last equation for I_k. Think of I_k as a function of these $(2k + 1)$ variables. Set the partial derivative of I_k with respect to each of these variables equal to zero, and solve the resulting equations to show that the proper choice is

$$\alpha_0 = a_0, \quad \alpha_n = a_n \quad \text{for} \quad n = 1, \ldots, k, \quad \text{and} \quad \beta_n = b_n \quad \text{for} \quad n = 1, \ldots, k.$$

46. Prove *Bessel's inequality* for Fourier coefficients:

$$\frac{a_0^2}{2} + \sum_{n=1}^{\infty} (a_n{}^2 + b_n{}^2) \le \frac{1}{L} \int_{-L}^{L} f(x)^2\,dx.$$

Hint: In the outline for solving Problem 45, go to the last expression written for I_k and substitute $\alpha_0 = a_0$, $\alpha_n = a_n$, for $n = 1, \ldots, k$, and $\beta_n = b_n$ for $n = 1, \ldots, k$. Use the fact that I_k must be nonnegative, and let $k \to \infty$.

47. Continuing from Problem 46, prove that

$$\sum_{n=1}^{\infty} a_n^2 \quad \text{and} \quad \sum_{n=1}^{\infty} b_n^2$$

both converge.

48. Use the result of Problem 47 to show that

$$\lim_{n\to\infty} \int_{-L}^{L} f(x) \cos\left(\frac{n\pi x}{L}\right) dx = 0$$

and

$$\lim_{n\to\infty} \int_{-L}^{L} f(x) \sin\left(\frac{n\pi x}{L}\right) dx = 0$$

if $f(x)$ satisfies the hypotheses of Theorem 95 on $[-L, L]$. This is sometimes known as *Riemann's lemma.*

49. Let $f(x)$ satisfy the hypotheses of Theorem 95 on $[-L, L]$, and suppose that $f'(x)$ is sectionally continuous on $[-L, L]$. Suppose also that $f(L) = f(-L)$.

(a) Prove that $\sum_{n=1}^{\infty} \sqrt{a_n^2 + b_n^2}$ converges.

Hint: Expand $f'(x)$ in a Fourier series, and integrate by parts to relate the Fourier coefficients of $f'(x)$ to those of $f(x)$. You may also need Cauchy's inequality, which says that

$$\left(\sum_{n=1}^{k} A_n B_n\right)^2 \leq \left(\sum_{n=1}^{k} A_n^2\right)\left(\sum_{n=1}^{k} B_n^2\right),$$

for any numbers $A_1, \ldots, A_k, B_1, \ldots, B_k$.

(b) Use the result of (a) to show that the Fourier series of $f(x)$ on $[-L, L]$ converges uniformly.

(c) Prove that

$$\lim_{n\to\infty} na_n = \lim_{n\to\infty} nb_n = 0,$$

under the conditions assumed.

50. To do this problem, you have to understand the formula

$$e^{iA} = \cos(A) + i \sin(A).$$

Show that the Fourier series of $f(x)$ on $[-L, L]$ can be written in complex form as

$$\sum_{n=-\infty}^{\infty} k_n e^{in\pi x/L},$$

in which

$$k_n = \frac{1}{2L} \int_{-L}^{L} f(t) e^{-in\pi t/L}\, dt$$

for $n = 0, 1, 2, 3, \ldots, -1, -2, -3, \ldots$.

Hint: First, rewrite the Fourier series given in the text as

$$\frac{1}{L} \int_{-L}^{L} f(t) \left\{ \frac{1}{2} + \sum_{n=1}^{\infty} \left[\cos\left(\frac{n\pi t}{L}\right) \cos\left(\frac{n\pi x}{L}\right) + \sin\left(\frac{n\pi t}{L}\right) \sin\left(\frac{n\pi x}{L}\right) \right] \right\} dt.$$

Next, use an identity to rewrite this as

$$\frac{1}{L}\int_{-L}^{L} f(t)\left\{\frac{1}{2} + \sum_{n=1}^{\infty} \cos\left[\frac{n\pi(t-x)}{L}\right]\right\} dt.$$

Show that this can be written as

$$\frac{1}{2L}\int_{-L}^{L} f(t)\left\{\sum_{n=-\infty}^{\infty} \cos\left[\frac{n\pi(x-t)}{L}\right] + i\sin\left[\frac{n\pi(x-t)}{L}\right]\right\} dt.$$

Now convert things to complex exponentials.

12.3 FOURIER SERIES OF PERIODIC FUNCTIONS, WITH APPLICATIONS TO FORCED OSCILLATIONS AND RESONANCE

Thus far, we have considered the Fourier expansion of a function $f(x)$ defined on $[-L, L]$. Sometimes, however, we are interested in a function defined over the whole real line. Such a function can also be expanded in a Fourier series if it is periodic. Recall that $f(x)$ is periodic of period $2L$ if $f(x + 2L) = f(x)$ for all x. Thus, for example, $\sin(x)$ is periodic of period 2π. If $f(x)$ has period $2L$, then its graph from L to $3L$ duplicates its graph from $-L$ to L. Similarly, the graph from $3L$ to $5L$, $5L$ to $7L, \ldots$, $-3L$ to $-L$, $-5L$ to $-3L, \ldots$ is identical to the graph between $-L$ and L.

Suppose now that $f(x)$ is periodic of period $2L$. Then the Fourier expansion on $[-L, L]$ automatically extends to the intervals $[L, 3L]$, $[3L, 5L], \ldots, [-L, -3L]$, $[-5L, -3L], \ldots$. Convergence on each such interval reflects convergence on $[-L, L]$, which in turn is determined in many cases by our Fourier convergence theorem.

Here is an illustration.

EXAMPLE 369

Consider the square wave of Figure 234, given by

$$f(x) = \begin{cases} K & \text{for} \quad 2n\pi \le x < (2n+1)\pi \\ -K & \text{for} \quad (2n+1)\pi \le x < (2n+2)\pi \end{cases}$$

for $n = 0, \pm 1, \pm 2, \pm 3, \ldots$. Here, K is a given positive constant.

Note that the function is periodic of period 2π. Expand $f(x)$ in a Fourier series on the interval $[-\pi, \pi]$ in the usual way. Omitting the details, the Fourier series is found to be

$$\sum_{n=1}^{\infty} \frac{4K}{(2n-1)\pi} \sin[(2n-1)x].$$

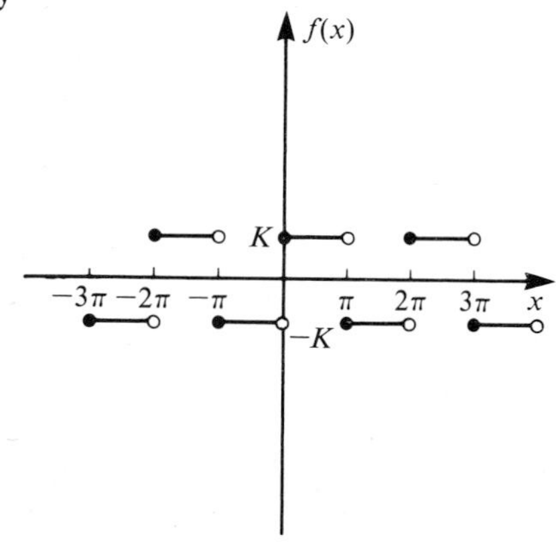

Figure 234

By Theorem 95, this series converges to

$$\begin{array}{rll} K & \text{if} & 0 < x < \pi, \\ -K & \text{if} & -\pi < x < 0, \text{ and} \\ 0 & \text{at} & x = 0,\ x = \pi, \text{ and } x = -\pi. \end{array}$$

These results now extend by the periodicity of $f(x)$ to the entire real line. For example, the series converges on the interval $[\pi, 3\pi]$ to

$$\begin{array}{rll} K & \text{if} & 2\pi < x < 3\pi, \\ -K & \text{if} & \pi < x < 2\pi, \text{ and} \\ 0 & \text{if} & x = \pi,\ 2\pi, \text{ or } 3\pi. \end{array}$$

In general, the Fourier series in this example converges to $f(x)$ at every value of x except integer multiples of π; there, the Fourier series converges to 0.

Fourier expansions of periodic functions enable us to explain phenomena which might otherwise appear somewhat mysterious. Here is an example.

EXAMPLE 370

Consider forced oscillations of a mass of one gram attached to a spring having spring modulus 49 grams per second squared and damping constant 0.01 grams per second. Related problems were discussed in Sections 2.6 and 2.9. We shall assume a periodic driving force given by

$$E(t) = \begin{cases} K & \text{for} \quad 2n\pi \le t < (2n+1)\pi \\ -K & \text{for} \quad (2n+1)\pi \le t < (2n+2)\pi \end{cases}$$

for $n = 0, 1, 2, 3, \ldots$. This is the periodic function of the previous example, with t (for time) replacing x as independent variable.

The differential equation governing the motion is

$$y'' + 0.01y' + 49y = E(t).$$

We shall examine the form of the steady-state solution, assuming zero initial conditions $y(0) = y'(0) = 0$.

Begin by expanding $E(t)$ in a Fourier series. From the preceding example, we have

$$y'' + 0.01y' + 49y = \sum_{n=1}^{\infty} \frac{4K}{(2n-1)\pi} \sin[(2n-1)t]$$

$$= \frac{4K}{\pi}\left[\sin(t) + \frac{1}{3}\sin(3t) + \frac{1}{5}\sin(5t) + \frac{1}{7}\sin(7t) + \cdots\right].$$

Since we are only going to look at the steady-state solution, disregard the homogeneous solution (in which the initial conditions play their role) and look in turn at the nonhomogeneous differential equations

$$y'' + 0.01y' + 49y = \frac{4K}{\pi}\sin(t),$$

$$y'' + 0.01y' + 49y = \frac{4K}{3\pi}\sin(3t),$$

$$y'' + 0.01y' + 49y = \frac{4K}{5\pi}\sin(5t),$$

$$\vdots$$

We shall solve each of these, and then sum the solutions to solve

$$y'' + 0.01y' + 49y = \sum_{n=1}^{\infty} \frac{4K}{(2n-1)\pi}\sin[(2n-1)t].$$

Thus, consider the nth equation,

$$y_n'' + 0.01y_n' + 49y_n = \frac{4K}{n\pi}\sin(nt),$$

in which n is an odd, positive integer. The subscript n was introduced just to remind us that we are working here on the nth term of the steady-state solution, which will be the sum of the y_n's.

Using the method of undetermined coefficients (Section 2.8), we attempt a solution of the form

$$y_n = A_n\cos(nt) + B_n\sin(nt).$$

Upon substituting into the differential equation for y_n, we find, with details omitted, that

$$A_n = \frac{0.04K}{\pi\,\Delta_n} \quad \text{and} \quad B_n = \frac{4K(49 - n^2)}{n\pi\,\Delta_n},$$

in which $\Delta_n = (49 - n^2)^2 + (0.01n)^2$.

The amplitude H_n of y_n is found, after some manipulation, to be

$$H_n = \sqrt{A_n^{\,2} + B_n^{\,2}} = \frac{4K}{n\pi\sqrt{\Delta_n}}.$$

Here are some computed values of these numbers:

n	Δ_n	A_n	B_n	H_n
1	2304.0001	$5.5262(10)^{-6}K$	$0.0265K$	$0.0265K$
3	1600.0009	$7.9577(10)^{-6}K$	$0.0106K$	$0.0106K$
5	576.0025	$2.2105(10)^{-5}K$	$0.0106K$	$0.0106K$
7	0.0049	$2.5984K$	0	$2.5984K$
9	1024.0081	$1.2434(10)^{-5}K$	$-0.0044K$	$0.0044K$
11	5184.0121	$2.4561(10)^{-6}K$	$-0.0016K$	$0.0016K$
13	14,400.0169	$8.8419(10)^{-7}K$	$-0.0008K$	$0.0008K$

Clearly, Δ_n decreases for $n = 1, 3, 5$, and then is near zero at $n = 7$; for $n = 9, 13, 15, \ldots, \Delta_n$ increases without bound. The effect this has on A_n is that A_n is negligibly small except at $n = 7$, while $A_7 = 2.5984K$. The B_n's are all fairly small, and as n increases from 9 onward, they steadily decrease in value. Thus, the amplitudes of the solutions $y_n = A_n \cos(nt) + B_n \sin(nt)$ are small except when $n = 7$. Here,

$$y_7 = 2.5984K \cos(7t).$$

Looking at the entire steady-state solution

$$y = y_1 + y_3 + y_5 + y_7 + y_9 + \cdots,$$

we see that the y_7 term dominates. Thus, we have the remarkable result that the steady-state solution is dominated by a cosine term, even though no such term is apparent in the driving force. The key here is that the frequency of this term is near the resonant frequency of the vibrating system.

The student should give some thought to justifying the method of obtaining the steady-state solution by adding the steady-state solutions of $y'' + 0.01y' + 49y = (4K/n)\sin(nt)$. This can be done by considering $y = y_1 + y_3 + \cdots$ and showing that term by term differentiation two times is valid, and hence that the solution can be verified by direct substitution back into the differential equation.

PROBLEMS FOR SECTION 12.3

In each of Problems 1 through 7, a periodic function is given by displaying part of the graph. Write a Fourier series for the function and determine, for all x, what the Fourier series converges to.

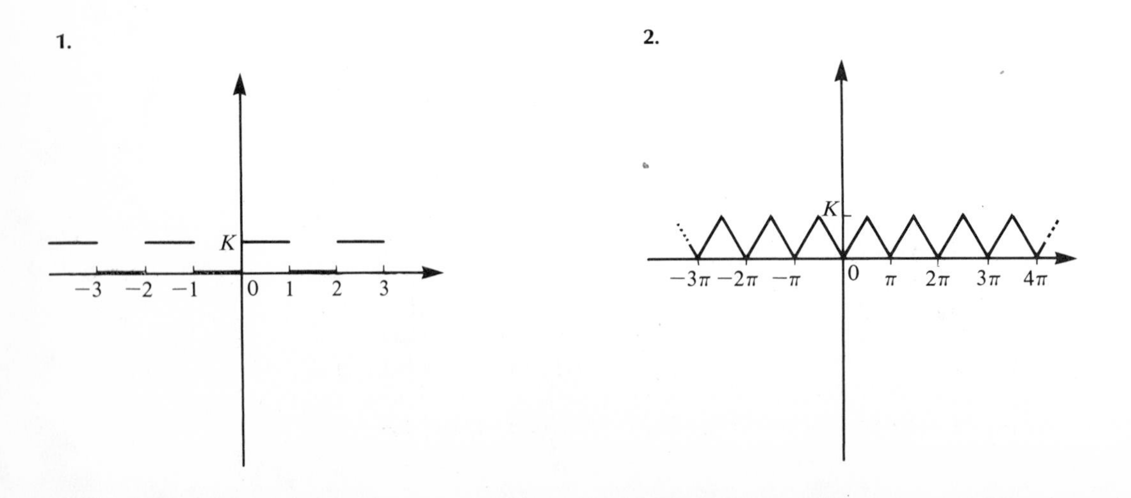

3.

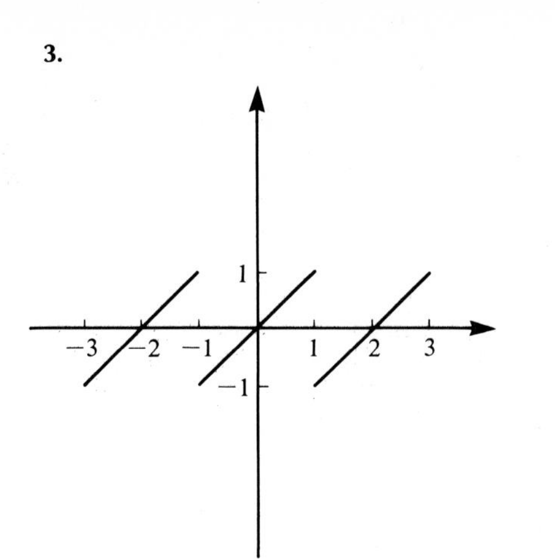

4.

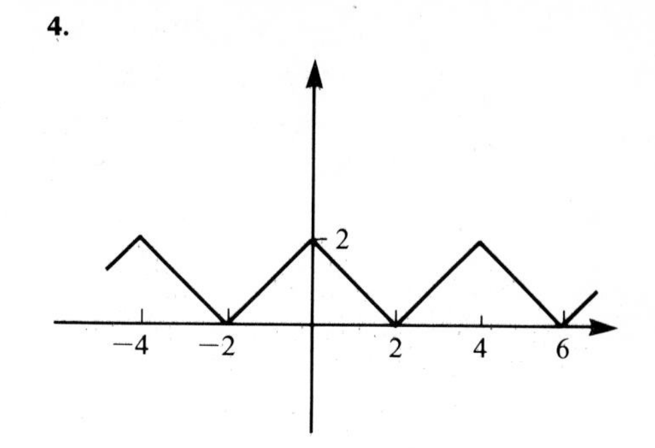

5.

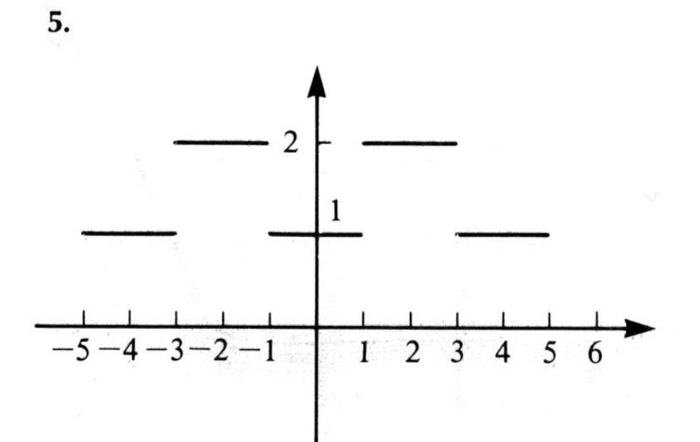

6.

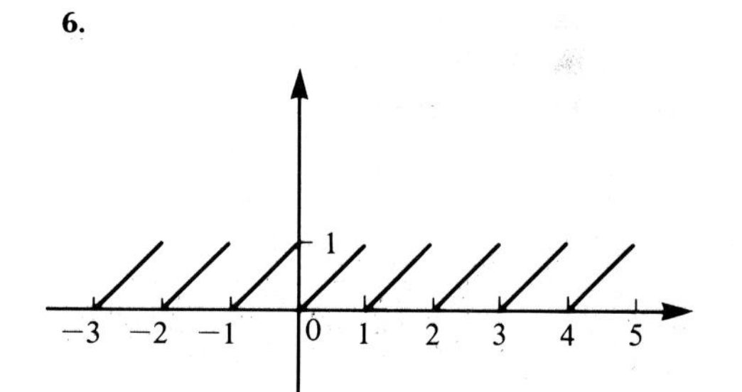

7.

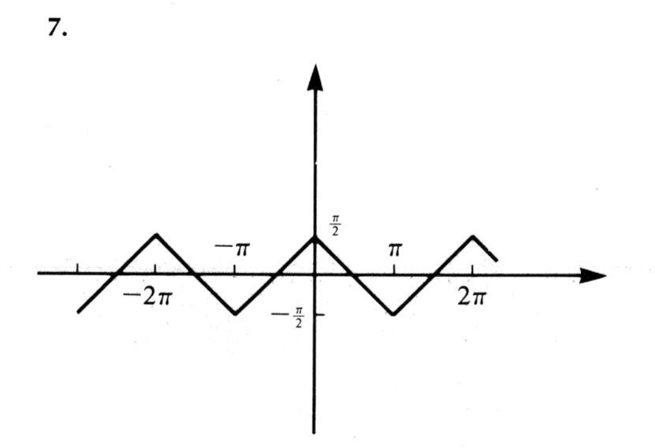

8. Find the steady-state solution of $y'' + 0.04y' + 25y = f(t)$ if $f(t)$ is as shown in Problem 1.

9. Find the steady-state solution of $y'' + 0.02y' + 12y = f(t)$ if $f(t)$ is the periodic function of Problem 3.

10. Find the steady-state solution of $y'' + 25y = f(t)$ if $f(t)$ is the periodic function of Problem 5.

11. Find the steady-state solution of $y'' + 8y = f(t)$ if $f(t)$ is the periodic function of Problem 7.

12.4 FOURIER SINE AND COSINE SERIES

Sometimes it is desirable to expand a function in a series containing just sines or just cosines. For example, we will need to do this in the next chapter to solve certain partial differential equations. On an interval $[-L, L]$, we have no control over this—we compute the Fourier coefficients and take what we get. On an interval $[0, L]$, however, we can choose to have just sines or just cosines appear in the series. We now show how to do this.

First, recall some facts about even and odd functions. On an interval $[-L, L]$, $g(x)$ is *even* if $g(-x) = g(x)$. Thus, the graph from $-L$ to 0 looks like that from 0 to L

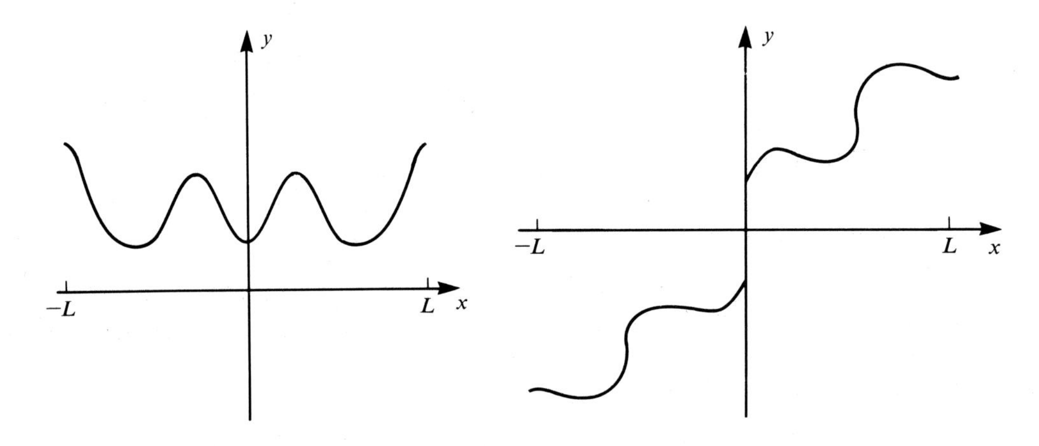

Figure 235. An even function. *Figure 236.* An odd function.

reflected back through the y-axis, as shown in Figure 235. We call $g(x)$ *odd* if $g(x) = -g(-x)$. In this case, the graph from $-L$ to 0 looks like that which we would obtain by taking the graph from 0 to L and reflecting down through the x-axis and then back across the y-axis, as shown in Figure 236. As some examples, we have:

$y = x^2$	even on $[-4, 4]$ (or any interval $[-L, L]$)
$y = x^3$	odd on $[-3, 3]$ (or any interval $[-L, L]$)
$y = \cos(x)$	even on $[-\pi, \pi]$
$y = \sin(x)$	odd on $[-\pi, \pi]$
$y = 2x^2 + x - 1$	neither even nor odd on, say, $[-4, 4]$

As the last example shows, a function need not be either even or odd.

If $g(x)$ is even on $[-L, L]$, then

$$\int_{-L}^{L} g(x)\,dx = 2\int_{0}^{L} g(x)\,dx.$$

Intuitively, this is because the "area" under $y = g(x)$ from $-L$ to 0 is exactly the same as that under the graph from 0 to L (see Figure 237).

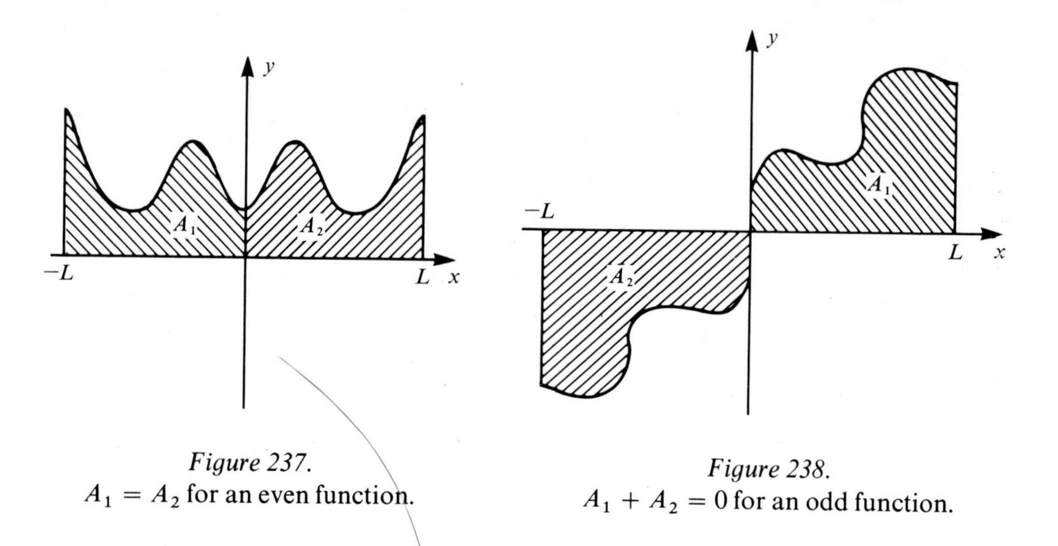

Figure 237.
$A_1 = A_2$ for an even function.

Figure 238.
$A_1 + A_2 = 0$ for an odd function.

Similarly, if $g(x)$ is odd on $[-L, L]$, then

$$\int_{-L}^{L} g(x)\, dx = 0.$$

This is because the "area" determined from $-L$ to 0 is the negative of that from 0 to L (see Figure 238).

Finally, it is easy to check that:

a product of two even or two odd functions is even,

while

a product of an even and an odd function is odd.

These facts can save some work in computing Fourier coefficients. For example, if $f(x)$ is odd on $[-L, L]$, then $f(x)\cos(n\pi x/L)$ is odd, and so each $a_n = 0$, while $f(x)\sin(n\pi x/L)$ is even, and so $b_n = (2/L)\int_0^L f(x)\sin(n\pi x/L)\,dx$. (For example, note that the Fourier series for $f(x) = x$, an odd function on $[-\pi, \pi]$, contains only sine terms in Example 368.)

If $f(x)$ is even, then $f(x)\sin(n\pi x/L)$ is odd, and so $b_n = 0$, and $f(x)\cos(n\pi x/L)$ is even, and so $a_n = (2/L)\int_0^L f(x)\cos(n\pi x/L)\,dx$. Thus, in Example 367, where $f(x) = |x|$, we have an even function on $[-\pi, \pi]$; so we expect to see only cosine terms and the constant term in the Fourier expansion.

For ease in remembering:

$f(x)$ even on $[-L, L]$	only cosine and constant terms in the Fourier expansion on $[-L, L]$
$f(x)$ odd on $[-L, L]$	only sine terms in the Fourier expansion on $[-L, L]$

We shall now show how to exploit these facts to get just sine or just cosine expansions of $f(x)$ on the "half" interval $[0, L]$.

COSINE SERIES

Suppose that $f(x)$ is defined on $[0, L]$, and we want a cosine series there. We extend $f(x)$ to a new function $g(x)$ defined on $[-L, L]$ by setting

$$g(x) = \begin{cases} f(x), & 0 \le x \le L, \\ f(-x), & -L \le x < 0. \end{cases}$$

This process is depicted in Figure 239. Note that, by design, $g(x)$ is *even* on $[-L, L]$ and agrees with $f(x)$ on $[0, L]$.

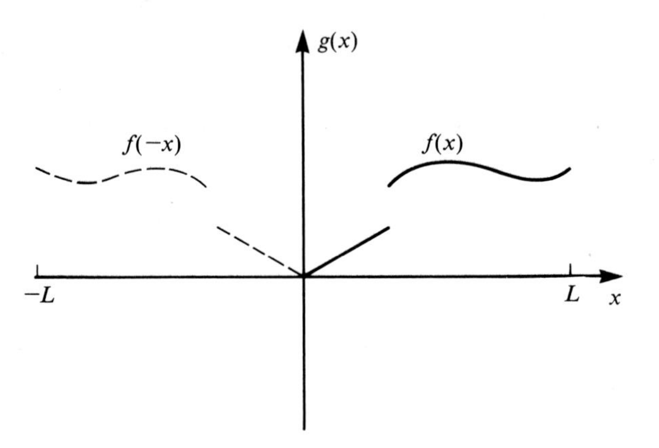

Figure 239. Extension of $f(x)$ defined on $[0, L]$ to an even function defined on $[-L, L]$.

Now expand $g(x)$ in a Fourier series on $[-L, L]$ in the normal way, but watch what happens to the coefficients. Because $g(x)$ is even and $g(x) = f(x)$ for $0 \le x \le L$, we have

$$a_0 = \frac{1}{L}\int_{-L}^{L} g(x)\,dx = \frac{2}{L}\int_0^L g(x)\,dx = \frac{2}{L}\int_0^L f(x)\,dx.$$

Next,

$$a_n = \frac{1}{L}\int_{-L}^{L} g(x)\cos\left(\frac{n\pi x}{L}\right)dx = \frac{2}{L}\int_0^L g(x)\cos\left(\frac{n\pi x}{L}\right)dx$$
$$= \frac{2}{L}\int_0^L f(x)\cos\left(\frac{n\pi x}{L}\right)dx,$$

because $g(x)$ and $\cos(n\pi x/L)$ are both even on $[-L, L]$; so their product is even as well.

Finally, $g(x)\sin(n\pi x/L)$ is odd on $[-L, L]$ because $\sin(n\pi x/L)$ is; so

$$b_n = \frac{1}{L}\int_{-L}^{L} g(x)\sin\left(\frac{n\pi x}{L}\right)dx = 0.$$

All the b_n's are 0, and we are left with a pure cosine series *for* $g(x)$ on $[-L, L]$.

Since $g(x) = f(x)$ on $[0, L]$, this is a cosine series for $f(x)$ on $[0, L]$.

In summary, the Fourier cosine series of $f(x)$ on $[0, L]$ is

$$\frac{a_0}{2} + \sum_{n=1}^{\infty} a_n \cos\left(\frac{n\pi x}{L}\right),$$

where

$$a_0 = \frac{2}{L}\int_0^L f(x)\,dx \quad \text{and} \quad a_n = \frac{2}{L}\int_0^L f(x)\cos\left(\frac{n\pi x}{L}\right)dx$$

$$\text{for} \quad n = 1, 2, 3, \ldots.$$

EXAMPLE 371

Let $f(x) = e^{2x}, 0 \leq x \leq 1$. Then,

$$a_0 = 2\int_0^1 e^{2x}\,dx = e^2 - 1$$

and

$$a_n = 2\int_0^1 e^{2x}\cos(n\pi x)\,dx = \frac{4}{4 + n^2\pi^2}[e^2\cos(n\pi) - 1].$$

The Fourier cosine series of e^{2x} on $[0, 1]$ is

$$\frac{e^2 - 1}{2} + \sum_{n=1}^{\infty} \frac{4}{4 + n^2\pi^2}[e^2\cos(n\pi) - 1]\cos(n\pi x).$$

EXAMPLE 372

Let $f(x) = \sin(x), 0 \leq x \leq \pi$. Then,

$$a_0 = \frac{2}{\pi}\int_0^{\pi} \sin(x)\,dx = \frac{4}{\pi}$$

and, for $n = 1, 2, 3, \ldots,$

$$a_n = \frac{2}{\pi}\int_0^{\pi}\sin(x)\cos(nx)\,dx = \begin{cases} 0 & \text{if } n = 1, \\ \dfrac{1}{\pi(1-n)}\{1 - \cos[(1-n)\pi]\} + \dfrac{1}{\pi(1+n)} & \\ \times \{1 - \cos[(1+n)\pi]\} & \text{if } n = 2, 3, \ldots. \end{cases}$$

Thus, the Fourier cosine series of $\sin(x)$ on $[0, \pi]$ is

$$\frac{2}{\pi} + \sum_{n=2}^{\infty}\left(\frac{1}{\pi(1-n)}\{1 - \cos[(1-n)\pi]\} + \frac{1}{\pi(1+n)}\{1 - \cos[(1+n)\pi]\}\right)\cos(nx).$$

Note that $\cos[(1-n)\pi] = (-1)^{1-n}$ and $\cos[(1+n)\pi] = (-1)^{1+n}$, since n is an integer. Further, $(-1)^{1-n} = (-1)^{1+n}$. Thus, the cosine series can be written

$$\frac{2}{\pi} + \sum_{n=2}^{\infty} \frac{2}{\pi}\left(\frac{1-(-1)^{n+1}}{1-n^2}\right)\cos(nx).$$

This can be further simplified by noting that

$$1-(-1)^{n+1} = \begin{cases} 0 & \text{if } n \text{ is odd,} \\ 2 & \text{if } n \text{ is even.} \end{cases}$$

Hence, the cosine series in this example becomes

$$\frac{2}{\pi} + \frac{4}{\pi}\sum_{n=1}^{\infty} \frac{\cos(2nx)}{1-4n^2}.$$

SINE SERIES

We now turn to Fourier sine expansions on an interval $[0, L]$. Suppose that $f(x)$ is defined on $[0, L]$. Extend $f(x)$ to an *odd* function $h(x)$ on $[-L, L]$ by setting

$$h(x) = \begin{cases} f(x), & 0 \le x \le L, \\ -f(-x), & -L \le x < 0. \end{cases}$$

This process is depicted in Figure 240.

Now expand $h(x)$ in a Fourier series on $[-L, L]$. We get

$$a_0 = \frac{1}{L}\int_{-L}^{L} h(x)\,dx = 0,$$

because $h(x)$ is odd.

For $n = 1, 2, 3, \ldots,$

$$a_n = \frac{1}{L}\int_{-L}^{L} h(x)\cos\left(\frac{n\pi x}{L}\right)dx = 0,$$

because $h(x)\cos(n\pi x/L)$ is odd (a product of an odd with an even function).

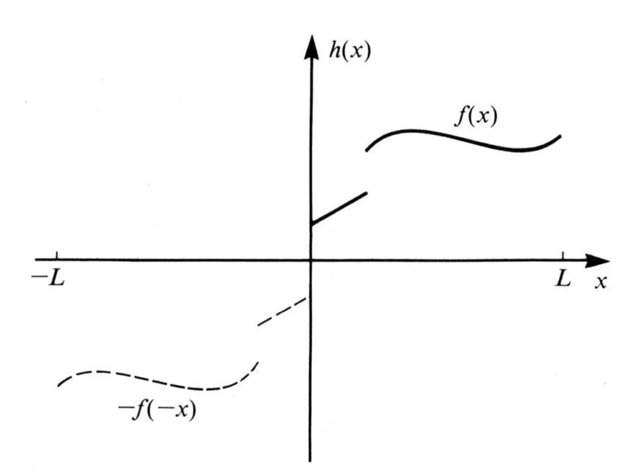

Figure 240. Extension of $f(x)$ defined on $[0, L]$ to an odd function defined on $[-L, L]$.

Finally,

$$b_n = \frac{1}{L}\int_{-L}^{L} h(x)\sin\left(\frac{n\pi x}{L}\right)dx = \frac{2}{L}\int_{0}^{L} h(x)\sin\left(\frac{n\pi x}{L}\right)dx$$
$$= \frac{2}{L}\int_{0}^{L} f(x)\sin\left(\frac{n\pi x}{L}\right)dx,$$

because a product of two odd functions is even and $h(x) = f(x)$ on $[0, L]$.

This gives us a Fourier series for $h(x)$ containing only sines, and hence a sine series for $f(x)$ on $[0, L]$.

In summary, the Fourier sine series for $f(x)$ on $[0, L]$ is

$$\sum_{n=1}^{\infty} b_n \sin\left(\frac{n\pi x}{L}\right),$$

where

$$b_n = \frac{2}{L}\int_0^L f(x)\sin\left(\frac{n\pi x}{L}\right)\,dx.$$

EXAMPLE 373

Let $f(x) = e^{2x}, 0 \leq x \leq 1$. Then,

$$b_n = 2\int_0^1 e^{2x}\sin(n\pi x)\,dx = \frac{2n\pi}{4 + n^2\pi^2}\,[1 - e^2\cos(n\pi)].$$

Thus, the sine expansion of e^{2x} on $[0, 1]$ is

$$\sum_{n=1}^{\infty} \frac{2n\pi}{4 + n^2\pi^2}\,[1 - e^2\cos(n\pi)]\sin(n\pi x).$$

EXAMPLE 374

Let $f(x) = \sin(x)$. We expand in a sine series on $[0, \pi]$. Calculate

$$b_n = \frac{2}{\pi}\int_0^{\pi} \sin(x)\sin(nx)\,dx = \begin{cases} 1 & \text{if } n = 1, \\ 0 & \text{if } n = 2, 3, 4, \ldots. \end{cases}$$

Thus, the sine series of $\sin(x)$ on $[0, \pi]$ is just $\sin(x)$.

Convergence of a sine or cosine series can be obtained from Theorem 95, but we have to be a little careful. To avoid ambiguities, we shall state the results separately for sine and cosine series.

THEOREM 96 Convergence of Fourier Cosine Series

Let $f(x)$ be sectionally continuous on $[0, L]$. Then,

1. If $0 < x_0 < L$ and both right and left derivatives exist at x_0, then at x_0 the Fourier cosine series of $f(x)$ converges to the average of the left and right limits at x_0. In particular, if $f(x)$ is continuous at x_0, then the series converges to $f(x_0)$.
2. At 0, if $f'_R(0)$ exists, then the series converges to $f(0^+)$; at L, if $f'_L(L)$ exists, then the series converges of $f(L-)$.

Note that (1) of the theorem is similar to Theorem 95; (2) is different. To understand (2), note that the cosine series of $f(x)$ on $[0, L]$ is actually the Fourier series on

$[-L, L]$ of the even extension $g(x)$ of $f(x)$. By Theorem 95, this converges at 0 to

$$\tfrac{1}{2}\left[\lim_{h\to 0+} g(0+h) + \lim_{h\to 0+} g(0-h)\right] = \tfrac{1}{2}\left[\lim_{h\to 0+} g(h) + \lim_{h\to 0+} g(-h)\right]$$

$$= \tfrac{1}{2}\left[\lim_{h\to 0+} f(h) + \lim_{h\to 0+} f(h)\right]$$

$$= \tfrac{1}{2}[f(0^+) + f(0^+)]$$

$$= f(0^+).$$

The result at L is derived similarly.

Here is the analogous result for sine series.

THEOREM 97 Convergence of Fourier Sine Series

Let $f(x)$ be sectionally continuous on $[0, L]$. Then,

1. If $0 < x_0 < L$ and both right and left derivatives exist at x_0, then at x_0 the Fourier sine series of $f(x)$ converges to the average of the left and right limits at x_0. In particular, if $f(x)$ is continuous at x_0, then the series converges to $f(x_0)$.
2. At 0 and at L, the sine series converges to 0.

For the sine series, (2) does not require a sophisticated argument; just put $x = 0$ or $x = L$ into the series and observe that all the terms are zero, regardless of the coefficients.

EXAMPLE 375

Let

$$f(x) = \begin{cases} 1, & 0 \le x \le \pi/2, \\ 2, & \pi/2 < x \le \pi. \end{cases}$$

To find the Fourier cosine series of $f(x)$ on $[0, \pi]$, compute

$$a_0 = \frac{2}{\pi}\int_0^{\pi} f(x)\,dx = \frac{2}{\pi}\int_0^{\pi/2} dx + \frac{2}{\pi}\int_{\pi/2}^{\pi} 2\,dx = 3$$

and

$$a_n = \frac{2}{\pi}\int_0^{\pi} f(x)\cos(nx)\,dx = \frac{2}{\pi}\int_0^{\pi/2} \cos(nx)\,dx + \frac{2}{\pi}\int_{\pi/2}^{\pi} 2\cos(nx)\,dx$$

$$= -\frac{2}{\pi n}\sin\left(\frac{n\pi}{2}\right).$$

The cosine series on $[0, \pi]$ is

$$\frac{3}{2} + \sum_{n=1}^{\infty} -\frac{2}{\pi n}\sin\left(\frac{n\pi}{2}\right)\cos(nx).$$

Since $\sin(n\pi/2) = 0$ when n is an even integer, this is

$$\frac{3}{2} + \sum_{n=1}^{\infty} \frac{-2}{(2n-1)\pi} \sin\left[\frac{(2n-1)\pi}{2}\right] \cos[(2n-1)x].$$

Since $\sin[(2n-1)\pi/2] = (-1)^n$, this series can be written

$$\frac{3}{2} + \sum_{n=1}^{\infty} \frac{2(-1)^{n+1}}{(2n-1)\pi} \cos[(2n-1)x].$$

Convergence is to:

$$\begin{array}{ll} 1, & 0 < x < \pi/2 \\ 2, & \pi/2 < x < \pi \end{array} \Bigg\} \text{[here, } f(x) \text{ is continuous]}$$

$$\begin{array}{ll} \frac{3}{2} & \text{at} \quad x = \pi/2 \text{ (the jump discontinuity)} \\ 1 & \text{at} \quad x = 0 \\ 2 & \text{at} \quad x = \pi \end{array}$$

For the sine series of $f(x)$ on $[0, \pi]$, compute

$$b_n = \frac{2}{\pi} \int_0^{\pi} f(x) \sin(nx)\, dx = \frac{2}{\pi} \int_0^{\pi/2} \sin(nx)\, dx + \frac{2}{\pi} \int_{\pi/2}^{\pi} 2 \sin(nx)\, dx$$

$$= \frac{2}{\pi n} \cos\left(\frac{n\pi}{2}\right) + \frac{2}{\pi n} - \frac{4}{\pi n} \cos(n\pi).$$

The sine series of $f(x)$ on $[0, \pi]$ is then

$$\frac{2}{\pi} \sum_{n=1}^{\infty} \frac{1}{n} \left[\cos\left(\frac{n\pi}{2}\right) + 1 - 2\cos(n\pi)\right] \sin(nx).$$

Since $\cos(n\pi) = (-1)^n$ for any integer n, this is

$$\frac{2}{\pi} \sum_{n=1}^{\infty} \frac{1}{n} \left[\cos\left(\frac{n\pi}{2}\right) + 1 - 2(-1)^n\right] \sin(nx).$$

This converges to:

$$\begin{array}{ll} 1, & 0 < x < \pi/2 \\ 2, & \pi/2 < x < \pi \\ \frac{3}{2} & \text{at} \quad x = \pi/2 \\ 0 & \text{at} \quad x = 0 \text{ and } x = \pi \end{array}$$

In the next section we make the transition from Fourier series, which represent a function on a finite interval, to Fourier integrals, which apply to functions defined over the whole real line.

PROBLEMS FOR SECTION 12.4

In each of Problems 1 through 25, compute both the sine and cosine series of the function on the given interval. In each case, tell what the series converges to.

1. $f(x) = 2x + x^2, \quad 0 \le x \le 3$

2. $f(x) = x^3 + 1, \quad 0 \le x \le 1$

3. $f(x) = x^3, \quad 0 \le x \le \pi$

4. $f(x) = -\sin(x), \quad 0 \le x \le \pi/4$

5. $f(x) = \cosh(x), \quad 0 \le x \le \pi$

6. $f(x) = x^2, \quad 0 \le x \le 3$

7. $f(x) = 3, \quad 0 \le x \le \pi$

8. $f(x) = \begin{cases} 1, & 0 \le x \le 1 \\ 2, & 1 < x \le 2 \end{cases}$

9. $f(x) = \begin{cases} e^x, & 0 \le x \le 2 \\ 1, & 2 < x \le 4 \end{cases}$

10. $f(x) = \begin{cases} 0, & 0 \le x \le 1 \\ \frac{1}{2}, & 1 < x \le 3 \\ 1, & 3 < x \le 4 \end{cases}$

11. $f(x) = \begin{cases} -x, & 0 \le x < 1 \\ x, & 1 \le x \le \pi \end{cases}$

12. $f(x) = \begin{cases} 1 - x, & 0 \le x \le \pi \\ 1 + x, & \pi < x \le 2\pi \end{cases}$

13. $f(x) = \begin{cases} 1, & 0 \le x < 1 \\ 2, & 1 \le x \le 3 \\ \cos(x), & 3 < x \le 5 \end{cases}$

14. $f(x) = \begin{cases} \cos(x), & 0 \le x < \frac{1}{2} \\ \sin(x), & \frac{1}{2} \le x \le \pi \end{cases}$

15. $f(x) = 1 - 2x, \quad 0 \le x \le 4$

16. $f(x) = x - \sin(2x), \quad 0 \le x \le \pi$

17. $f(x) = 3\cos(2x - 1), \quad 0 \le x \le 1$

18. $f(x) = \begin{cases} 1, & 0 \le x < \frac{1}{2} \\ -1, & \frac{1}{2} < x \le 1 \end{cases}$

19. $f(x) = e^{-x}, \quad 0 \le x \le 2$

20. $f(x) = 8x^2 + 1, \quad 0 \le x \le 3$

21. $f(x) = 4x + \sin(3x), \quad 0 \le x \le \pi$

22. $f(x) = \begin{cases} -2, & 0 \le x \le 1 \\ x, & 1 < x \le 2 \end{cases}$

23. $f(x) = \cos^2(\pi x), \quad 0 \le x \le 1$

24. $f(x) = \sinh(x), \quad 0 \le x \le 4$

25. $f(x) = 3x + 2, \quad 0 \le x \le 5$

26. Prove that any function defined on $[-L, L]$ can be written as a sum of an even and an odd function.

Sometimes Fourier series can be used to sum series of constants. Here are some illustrations in Problems 27 through 31.

27. Let $x = 0$ in the Fourier series for $f(x) = |x|$, $-\pi < x < \pi$, to obtain the sum of the series

$$\sum_{n=1}^{\infty} \frac{1}{(2n-1)^2}.$$

28. Write the Fourier cosine series for $\sin(x)$ on $[0, \pi]$ and set $x = \frac{1}{2}\pi$ to obtain the value of

$$\sum_{n=1}^{\infty} \frac{(-1)^n}{4n^2 - 1}.$$

29. Write the Fourier series for e^x on $[-\pi, \pi]$ and set $x = 0$ to obtain the sum of the series

$$\sum_{n=1}^{\infty} \frac{(-1)^n}{1 + n^2}.$$

30. Expand $f(x) = x$ in a Fourier series on $[-\pi, \pi]$ and choose x appropriately to sum the series

$$\sum_{n=1}^{\infty} \frac{1}{n^2}.$$

31. Let

$$f(x) = \begin{cases} 0, & -L \le x \le 0, \\ L, & 0 < x \le L. \end{cases}$$

Expand $f(x)$ in a Fourier series and choose x appropriately to sum the series

$$\sum_{n=1}^{\infty} \frac{(-1)^{n+1}}{2n-1}.$$

12.5 THE FOURIER INTEGRAL

In many applications we must deal with functions defined on the whole real line instead of just on an interval. If the function is periodic, a Fourier series may be considered. Otherwise, we go to the Fourier integral, which is the object of this section.

As a motivation (appealing, but not rigorous) for the Fourier integral, begin with the Fourier series of $f(x)$ on $[-L, L]$,

$$\frac{a_0}{2} + \sum_{n=1}^{\infty} a_n \cos\left(\frac{n\pi x}{L}\right) + b_n \sin\left(\frac{n\pi x}{L}\right),$$

where

$$a_0 = \frac{1}{L}\int_{-L}^{L} f(t)\,dt, \qquad a_n = \frac{1}{L}\int_{-L}^{L} f(t)\cos\left(\frac{n\pi t}{L}\right) dt;$$

and

$$b_n = \frac{1}{L}\int_{-L}^{L} f(t)\sin\left(\frac{n\pi t}{L}\right) dt.$$

(We use t as integration variable to avoid confusion with x in the series.)

Put $\lambda_n = n\pi/L$, and define

$$A(\lambda_n) = \frac{1}{\pi}\int_{-L}^{L} f(t)\cos(\lambda_n t)\,dt$$

and

$$B(\lambda_n) = \frac{1}{\pi}\int_{-L}^{L} f(t)\sin(\lambda_n t)\,dt.$$

Then, for $n = 1, 2, 3, \ldots,$

$$a_n = \frac{\pi}{L} A(\lambda_n) \quad\text{and}\quad b_n = \frac{\pi}{L} B(\lambda_n).$$

The Fourier series then becomes

$$\frac{a_0}{2} + \sum_{n=1}^{\infty} \left[\frac{\pi}{L} A(\lambda_n)\cos(\lambda_n x) + \frac{\pi}{L} B(\lambda_n)\sin(\lambda_n x)\right]$$

$$= \frac{a_0}{2} + \sum_{n=1}^{\infty} [A(\lambda_n)\cos(\lambda_n x) + B(\lambda_n)\sin(\lambda_n x)]\,\Delta\lambda,$$

where $\Delta\lambda = \lambda_{n+1} - \lambda_n = (n+1)(\pi/L) - (n\pi/L) = \pi/L.$

Now imagine that $L \to \infty$. Then, $\Delta\lambda \to 0$. Further, if $\int_{-\infty}^{\infty} |f(t)|\, dt$ is finite, then

$$a_0 = \frac{1}{L}\int_{-L}^{L} f(t)\, dt \to 0.$$

The last sum is suggestive of a Riemann sum for an integral, and, as $\Delta\lambda \to 0$,

$$\sum_{n=1}^{\infty} [A(\lambda_n)\cos(\lambda_n x) + B(\lambda_n)\sin(\lambda_n x)]\, \Delta\lambda \to \int_0^{\infty} [A(\lambda)\cos(\lambda x) + B(\lambda)\sin(\lambda x)]\, d\lambda,$$

where

$$A(\lambda) = \lim_{L\to\infty} \frac{1}{\pi}\int_{-L}^{L} f(t)\cos(\lambda_n t)\, dt = \frac{1}{\pi}\int_{-\infty}^{\infty} f(t)\cos(\lambda t)\, dt$$

and

$$B(\lambda) = \lim_{L\to\infty} \frac{1}{\pi}\int_{-L}^{L} f(t)\sin(\lambda_n t)\, dt = \frac{1}{\pi}\int_{-\infty}^{\infty} f(t)\sin(\lambda t)\, dt.$$

While much of this calculation was of a suspicious nature, it provides a good illustration of blundering into a good thing. In fact, we have the following theorem.

THEOREM 98

Suppose that $f(x)$ satisfies the hypotheses of Theorem 95 on every finite interval $[-L, L]$. In addition, suppose that

$$\int_{-\infty}^{\infty} |f(t)|\, dt$$

is finite.

Let

$$A(\lambda) = \frac{1}{\pi}\int_{-\infty}^{\infty} f(t)\cos(\lambda t)\, dt$$

and

$$B(\lambda) = \frac{1}{\pi}\int_{-\infty}^{\infty} f(t)\sin(\lambda t)\, dt,$$

for every $\lambda \geq 0$. Then, the *Fourier integral*

$$\int_0^{\infty} [A(\lambda)\cos(\lambda x) + B(\lambda)\sin(\lambda x)]\, d\lambda, \qquad -\infty < x < \infty,$$

converges:

1. To $f(x)$ wherever $f(x)$ is continuous.
2. To the average of the left and right limits of $f(x)$ wherever $f(x)$ has a discontinuity and the left and right derivatives exist.

We call $\int_0^{\infty} [A(\lambda)\cos(\lambda x) + B(\lambda)\sin(\lambda x)]\, d\lambda$ a *Fourier integral representation of* $f(x)$ on the whole real line.

EXAMPLE 376

Let

$$f(x) = \begin{cases} -2, & -1 \leq x \leq 0, \\ 1, & 0 < x \leq 1, \\ 0, & x < -1 \text{ or } x > 1. \end{cases}$$

Then,

$$\int_{-\infty}^{\infty} |f(x)|\, dx = \int_{-1}^{0} +2\, dx + \int_{0}^{1} dx = 3$$

is certainly finite, and $f(x)$ satisfies the hypotheses of Theorem 95 on every finite interval $[-L, L]$.

Compute, for $\lambda > 0$,

$$\begin{aligned} A(\lambda) &= \frac{1}{\pi} \int_{-\infty}^{\infty} f(t) \cos(\lambda t)\, dt \\ &= \frac{1}{\pi} \int_{-1}^{0} -2 \cos(\lambda t)\, dt + \frac{1}{\pi} \int_{0}^{1} \cos(\lambda t)\, dt \\ &= \frac{3}{\lambda \pi} \sin(\lambda). \end{aligned}$$

Note: $A(\lambda)$ can be assigned the value $3/\pi$ at $\lambda = 0$, since

$$\lim_{\lambda \to 0} \frac{3}{\pi \lambda} \sin(\lambda) = \frac{3}{\pi} \lim_{\lambda \to 0} \left(\frac{\sin(\lambda)}{\lambda} \right) = \frac{3}{\pi}.$$

Next, for $\lambda > 0$,

$$B(\lambda) = \frac{1}{\pi} \int_{-1}^{0} -2 \sin(\lambda t)\, dt + \frac{1}{\pi} \int_{0}^{1} \sin(\lambda t)\, dt = \frac{3}{\lambda \pi} [1 - \cos(\lambda)].$$

Again, $B(\lambda)$ can be assigned the value 0 at $\lambda = 0$, since

$$\lim_{\lambda \to 0} \frac{3}{\pi \lambda} [1 - \cos(\lambda)] = 0.$$

Thus, the Fourier integral for $f(x)$ is

$$\int_{0}^{\infty} \left\{ \frac{3}{\lambda \pi} \sin(\lambda) \cos(\lambda x) + \frac{3}{\lambda \pi} [1 - \cos(\lambda)] \sin(\lambda x) \right\} d\lambda.$$

This converges

to $f(x)$ for $x < -1$, $x > 1$, $0 < x < 1$, and $-1 < x < 0$,
to $\frac{1}{2}(-2 + 0) = -1$ at $x = -1$,
to $\frac{1}{2}(-2 + 1) = -\frac{1}{2}$ at $x = 0$,
and to $\frac{1}{2}(1 + 0) = \frac{1}{2}$ at $x = 1$.

EXAMPLE 377

Let

$$f(x) = \begin{cases} e^{-x} & \text{for } x \geq 0, \\ e^{x} & \text{for } x \leq 0. \end{cases}$$

Then,

$$\int_{-\infty}^{\infty} |f(x)|\, dx = \int_{-\infty}^{0} e^x\, dx + \int_0^{\infty} e^{-x}\, dx = 2,$$

which is finite.

Compute

$$A(\lambda) = \frac{1}{\pi}\int_{-\infty}^{\infty} f(t)\cos(\lambda t)\, dt$$

$$= \frac{1}{\pi}\int_{-\infty}^{0} e^t \cos(\lambda t)\, dt + \frac{1}{\pi}\int_0^{\infty} e^{-t}\cos(\lambda t)\, dt = \frac{2}{\pi}\left(\frac{1}{1+\lambda^2}\right)$$

and

$$B(\lambda) = \frac{1}{\pi}\int_{-\infty}^{\infty} f(t)\sin(\lambda t)\, dt$$

$$= \frac{1}{\pi}\int_{-\infty}^{0} e^t \sin(\lambda t)\, dt + \frac{1}{\pi}\int_0^{\infty} e^{-t}\sin(\lambda t)\, dt = 0.*$$

The Fourier integral for $f(x)$ is then

$$\frac{2}{\pi}\int_0^{\infty} \frac{1}{1+\lambda^2}\cos(\lambda x)\, d\lambda, \qquad -\infty < x < \infty.$$

By continuity of $f(x)$, this converges to $f(x)$ for every x. Thus, we have the equality

$$f(x) = \frac{2}{\pi}\int_0^{\infty} \frac{\cos(\lambda x)}{1+\lambda^2}\, d\lambda, \qquad -\infty < x < \infty.$$

In the next section we develop the Fourier integral analogues of the Fourier sine and cosine expansions.

PROBLEMS FOR SECTION 12.5

In each of Problems 1 through 15, compute the Fourier integral of $f(x)$, verifying first that $f(x)$ satisfies the hypotheses of Theorem 98. Determine what the Fourier integral converges to at each real number.

1. $f(x) = \begin{cases} x, & -\pi \leq x \leq \pi \\ 0, & |x| > \pi \end{cases}$

2. $f(x) = \begin{cases} C, & -10 \leq x \leq 10 \\ 0, & |x| > 10 \end{cases}$ (here, C is any constant)

* We could also observe that $f(t)$ is an even function on $(-\infty, \infty)$. Since $\sin(\lambda t)$ is odd, then $f(t)\sin(\lambda t)$ is odd; hence, $B(\lambda) = 0$.

3. $f(x) = \begin{cases} -1, & -\pi \le x < 0 \\ 1, & 0 \le x \le \pi \\ 0, & |x| > \pi \end{cases}$

4. $f(x) = \begin{cases} \sin(x), & -4 \le x < 0 \\ \cos(x), & 0 \le x \le 6 \\ 0, & x < -4 \text{ or } x > 6 \end{cases}$

5. $f(x) = \begin{cases} x^2, & -100 \le x \le 100 \\ 0, & |x| > 100 \end{cases}$

6. $f(x) = \begin{cases} |x|, & -\pi \le x \le \pi \\ 0, & |x| > \pi \end{cases}$

7. $f(x) = \begin{cases} \sin(x), & -\pi \le x \le \pi \\ 0, & |x| > \pi \end{cases}$

8. $f(x) = \begin{cases} e^x, & -5 \le x \le 5 \\ 0, & |x| > 5 \end{cases}$

9. $f(x) = \begin{cases} \sinh(x), & -5 \le x \le 9 \\ 0, & x < -5 \text{ or } x > 9 \end{cases}$

10. $f(x) = \begin{cases} 0, & |x| > 5 \\ \frac{1}{2}, & -5 \le x < 1 \\ 1, & 1 \le x \le 5 \end{cases}$

11. $f(x) = \begin{cases} \dfrac{\sin(x)}{x} & \text{if } x \ne 0 \\ 1 & \text{if } x = 0 \end{cases}$

12. $f(x) = \begin{cases} x^2 - 1, & 0 < x < 5 \\ 0, & x > 5 \text{ or } x < 0 \end{cases}$

13. $f(x) = \begin{cases} 1, & 0 \le x \le 5 \\ -1, & -5 \le x < 0 \\ 0, & |x| > 5 \end{cases}$

14. $f(x) = \begin{cases} \sin(\pi x), & 0 \le x \le 1 \\ 0, & x < 0 \text{ or } x > 1 \end{cases}$

15. $f(x) = \begin{cases} x^3, & -1 \le x \le 1 \\ 0, & |x| > 1 \end{cases}$

16. Show that the Fourier integral of $f(x)$ can be written

$$\lim_{\lambda \to \infty} \frac{1}{\pi} \int_{-\infty}^{\infty} f(t) \frac{\sin[\lambda(t - x)]\, dt}{t - x}.$$

17. Show that the Fourier integral of $f(x)$ can be written

$$\frac{1}{\pi} \int_0^{\infty} \int_{-\infty}^{\infty} f(t) \cos[\lambda(x - t)]\, dt\, d\lambda.$$

Hint: Write out the Fourier integral for $f(x)$, with the integral formulas for $A(\lambda)$ and $B(\lambda)$ written in, and observe that $\cos(\lambda t)\cos(\lambda x) + \sin(\lambda t)\sin(\lambda x) = \cos[\lambda(t - x)]$.

18. Use the result of Problem 17 to show that the Fourier integral of $f(x)$ can be written

$$\frac{1}{2\pi} \int_{-\infty}^{\infty} \int_{-\infty}^{\infty} f(t) \cos[\lambda(x - t)]\, dt\, d\lambda.$$

19. Use the result of Problem 18 to show that the Fourier integral of $f(x)$ can be written in complex exponential form as

$$\frac{1}{2\pi} \int_{-\infty}^{\infty} \int_{-\infty}^{\infty} e^{i\lambda(x-t)} f(t)\, dt\, d\lambda.$$

Hint: Recall that $e^{i\lambda(x-t)} = \cos[\lambda(x - t)] + i \sin[\lambda(x - t)]$. Now look at the formula in Problem 18. Is there any justification for replacing $\cos[\lambda(x - t)]$ by $\cos[\lambda(x - t)] + i \sin[\lambda(x - t)]$ without changing the value of the integral? Remember that sine is an odd function.

12.6 FOURIER SINE AND COSINE INTEGRALS

Analogous to Fourier sine and cosine series, we can develop sine and cosine integrals for certain functions defined on the half-line $0 \leq x < \infty$. The development mirrors that for series.

Suppose that $f(x)$ satisfies the hypotheses of Theorem 95 on every finite interval $[0, L]$. Suppose also that $\int_0^\infty |f(x)|\, dx$ is finite. Extend $f(x)$ to an even function $g(x)$ on the whole real line by putting

$$g(x) = \begin{cases} f(x), & x \geq 0, \\ f(-x), & x < 0. \end{cases}$$

Now,

$$\int_{-\infty}^{\infty} |g(x)|\, dx = \int_{-\infty}^{0} |g(x)|\, dx + \int_0^\infty |g(x)|\, dx = 2 \int_0^\infty |f(x)|\, dx$$

is finite; so we can expand $g(x)$ in a Fourier integral. Because $g(t)$ and $\cos(\lambda t)$ are even,

$$A(\lambda) = \frac{1}{\pi} \int_{-\infty}^{\infty} g(t) \cos(\lambda t)\, dt = \frac{2}{\pi} \int_0^\infty f(t) \cos(\lambda t)\, dt.$$

Because $\sin(\lambda t)$ is odd, $g(t) \sin(\lambda t)$ is odd, and

$$B(\lambda) = 0.$$

This gives us a Fourier cosine integral of $g(x)$ on the whole line, which is in turn the Fourier cosine integral of $f(x)$ for $x \geq 0$.

In summary, we let

$$A(\lambda) = \frac{2}{\pi} \int_0^\infty f(t) \cos(\lambda t)\, dt,$$

and the *Fourier cosine integral of* $f(x)$ is

$$\int_0^\infty A(\lambda) \cos(\lambda x)\, d\lambda, \qquad x \geq 0.$$

As the reader can no doubt guess by now, the *Fourier sine integral of* $f(x)$ is

$$\int_0^\infty B(\lambda) \sin(\lambda x)\, d\lambda, \qquad x \geq 0,$$

where

$$B(\lambda) = \frac{2}{\pi} \int_0^\infty f(t) \sin(\lambda t)\, dt.$$

EXAMPLE 378

Let $f(x) = e^{-x}$ for $x \geq 0$. Then $\int_0^\infty |f(x)|\,dx$ is finite, and we can proceed.

Fourier cosine integral. Compute

$$A(\lambda) = \frac{2}{\pi}\int_0^\infty e^{-t}\cos(\lambda t)\,dt = \frac{2}{\pi}\frac{1}{1+\lambda^2}.$$

Then, the cosine integral of e^{-x} on $x \geq 0$ is

$$\frac{2}{\pi}\int_0^\infty \frac{\cos(\lambda x)}{1+\lambda^2}\,d\lambda, \qquad x \geq 0.$$

Fourier sine integral. Compute

$$B(\lambda) = \frac{2}{\pi}\int_0^\infty e^{-t}\sin(\lambda t)\,dt = \frac{2}{\pi}\frac{\lambda}{1+\lambda^2}.$$

Then, the sine integral of e^{-x} on $x \geq 0$ is

$$\frac{2}{\pi}\int_0^\infty \frac{\lambda\sin(\lambda x)}{1+\lambda^2}\,d\lambda.$$

PROBLEMS FOR SECTION 12.6

In each of Problems 1 through 10, compute both the Fourier sine and cosine integrals of $f(x)$.

1. $f(x) = \begin{cases} x^2, & 0 \leq x \leq 10 \\ 0, & x > 10 \end{cases}$

2. $f(x) = \begin{cases} \sin(x), & 0 \leq x \leq 2\pi \\ 0, & x > 2\pi \end{cases}$

3. $f(x) = \begin{cases} 1, & 0 \leq x \leq 1 \\ 2, & 1 < x \leq 4 \\ 0, & x > 4 \end{cases}$

4. $f(x) = \begin{cases} \cosh(x), & 0 \leq x \leq 5 \\ 0, & x > 5 \end{cases}$

5. $f(x) = \begin{cases} 2x+1, & 0 \leq x \leq \pi \\ 2, & \pi < x \leq 3\pi \\ 1, & 3\pi < x \leq 10\pi \\ 0, & x > 10\pi \end{cases}$

6. $f(x) = \begin{cases} x, & 0 \leq x \leq 1 \\ x+1, & 1 < x \leq 2 \\ 0, & x > 2 \end{cases}$

7. $f(x) = e^{-x}\cos(x), \quad x \geq 0$

8. $f(x) = xe^{-x}, \quad x \geq 0$

9. $f(x) = \begin{cases} 1, & 0 \leq x \leq 10 \\ 0, & x > 10 \end{cases}$

10. $f(x) = \begin{cases} \sin(\pi x), & 0 \leq x \leq 1 \\ 0, & x > 1 \end{cases}$

11. State a theorem on convergence of Fourier cosine integrals and a theorem on convergence of Fourier sine integrals.

12.7 COMPUTER CALCULATION OF FOURIER COEFFICIENTS

The formulas for the Fourier coefficients are integrals, and, as the student has no doubt experienced, integrals can be difficult to evaluate. In some instances, evaluation by elementary means is impossible. For example, let $f(x) = \sqrt{\sin(x)}$, $0 \leq x \leq \pi$. If we want, say, a Fourier sine expansion on $[0, \pi]$, the coefficients are

$$b_n = \frac{2}{\pi} \int_0^{\pi} \sqrt{\sin(x)} \sin(nx)\, dx.$$

This cannot be evaluated without recourse to special functions and tables.

One thing to try in such a case is a numerical integration approximation. For example, we could use the rectangular rule, Simpson's rule, or Gaussian quadrature. These are generally impractical by hand, but, if one has some computer help, accuracy good enough for many engineering applications can be achieved. Here is an illustration.

EXAMPLE 379

Let $f(x) = \sqrt{\sin(x)}$, $0 \le x \le \pi$.

For a Fourier sine series on $[0, \pi]$, we have to evaluate

$$b_n = \frac{2}{\pi} \int_0^{\pi} \sqrt{\sin(x)} \sin(nx)\, dx.$$

Using the rectangular rule, in which the interval $[0, \pi]$ has been subdivided into 200 subintervals, we find that:

$$\begin{aligned} b_1 &= 1.11284 & b_3 &= 0.15898 & b_5 &= 0.07226 \\ b_7 &= 0.04336 & b_9 &= 0.02967 & b_{11} &= 0.02193 \\ b_{13} &= 0.01706 & b_{15} &= 0.01376 & b_{17} &= 0.01141 \\ b_{19} &= 0.00966 \end{aligned}$$

All the coefficients $b_2, b_4, \ldots$ with even subscripts are zero. Thus, for $0 < x < \pi$, $\sqrt{\sin(x)}$ is approximated by

$$\begin{aligned} &1.11284 \sin(x) + 0.15898 \sin(3x) + 0.07226 \sin(5x) + 0.04336 \sin(7x) \\ &+ 0.02967 \sin(9x) + 0.02193 \sin(11x) + 0.01706 \sin(13x) + 0.01376 \sin(15x) \\ &+ 0.01141 \sin(17x) + 0.00966 \sin(19x). \end{aligned}$$

Despite the slowness of convergence of Fourier series in general, the relatively "short" approximation above gives fairly good results. For example, $\sqrt{\sin(\pi/2)} = 1$, while the finite series sums to 0.98835; $\sqrt{\sin(\pi/4)} = \sqrt{\sqrt{2}/2}$, which is about 0.84090, while the finite series gives us 0.85224; and $\sqrt{\sin(\pi/3)}$ is about 0.93060, while the finite series gives us 0.92675.

In practice, the parameters of the problem being treated must determine the accuracy required. For example, greater accuracy is needed in measuring atomic spectra than in building bridges. Desired accuracy in turn dictates how far out in the series we have to go before stopping. In the case of $\sqrt{\sin(x)}$, the coefficients b_n are getting small quite fast.

PROBLEMS FOR SECTION 12.7

In each of Problems 1 through 20, use approximate integration to write the sum of the first ten terms of both the Fourier sine and cosine series of $f(x)$ on the given interval. Approximate the coefficients to four decimal places. Do not attempt these by hand.

1. $f(x) = e^{-x^2}, \quad 0 \le x \le \pi$
2. $f(x) = \sin(\sqrt{x}), \quad 0 \le x \le \pi$
3. $f(x) = \sqrt{x}\, e^x, \quad 0 \le x \le 2$
4. $f(x) = \cos(x^2), \quad 0 \le x \le 4$
5. $f(x) = \ln(1 + x)e^{-x}, \quad 0 \le x \le \pi$
6. $f(x) = \dfrac{2x + 1}{3x^2 + 4}, \quad 0 \le x \le 5$
7. $f(x) = \cosh[\sinh(x)], \quad 0 \le x \le 1$
8. $f(x) = x^{3/2}, \quad 0 \le x \le \pi$
9. $f(x) = \ln[1 + \sin(x)], \quad 0 \le x \le \pi$
10. $f(x) = \dfrac{1}{1 + e^x}, \quad 0 \le x \le 2$
11. $f(x) = \dfrac{1}{x^2 + 2}, \quad 0 \le x \le 3$
12. $f(x) = x^4 e^{-2x}, \quad 0 \le x \le 2$
13. $f(x) = e^{\sqrt{x}}, \quad 0 \le x \le 5$
14. $f(x) = (x - 2)\cos(x^2 + 1), \quad 0 \le x \le \pi$
15. $f(x) = 1 + \sin^2(4x^3), \quad 0 \le x \le 2$
16. $f(x) = -2e^{-x} + \sinh(\sqrt{x}), \quad 0 \le x \le 3$
17. $f(x) = 8\ln(2 + x^3), \quad 0 \le x \le \pi$
18. $f(x) = \sin^2(3x^2), \quad 0 \le x \le \pi$
19. $f(x) = 2\cosh(3x)e^{-\sqrt{x}}, \quad 0 \le x \le 0.3450$
20. $f(x) = 8x^4\sin(\sqrt{x}), \quad 0 \le x \le 1.3224$

12.8 MULTIPLE FOURIER SERIES

In many applications we need a Fourier expansion of a function of two or more variables. In this section we shall develop the basic ideas for functions of two variables, in preparation for solving wave propagation and heat conduction problems in two and three dimensions in the next chapter. We proceed informally with the main intention of clarifying how the coefficients should be chosen.

The easiest case to understand (because the details are neater) is the Fourier sine series of $f(x, y)$ on a rectangle $0 \le x \le L$, $0 \le y \le K$. First, imagine that y is fixed at some value in $[0, K]$, and consider the function of x,

$$g(x) = f(x, y), \qquad y \text{ fixed}, \qquad 0 \le x \le L.$$

Expand $g(x)$ in a Fourier sine series on $[0, L]$:

$$g(x) = \sum_{n=1}^{\infty} a_n(y) \sin\left(\frac{n\pi x}{L}\right),$$

where

$$a_n(y) = \frac{2}{L}\int_0^L f(x, y)\sin\left(\frac{n\pi x}{L}\right) dx \quad \text{for} \quad n = 1, 2, 3, \ldots.$$

Here, the Fourier coefficient is a function of y because y is frozen inside $f(x, y)$; x is the integration variable.

Now, each $a_n(y)$ can (we assume) be expanded in a Fourier sine series on $[0, K]$. We have

$$a_n(y) = \sum_{m=1}^{\infty} b_{nm} \sin\left(\frac{m\pi y}{K}\right),$$

where b_{nm} is the mth Fourier coefficient in the sine series of $a_n(y)$. Thus,

$$b_{nm} = \frac{2}{K} \int_0^K a_n(y) \sin\left(\frac{m\pi y}{K}\right) dy.$$

Plugging in the previous formula for $a_n(y)$, this gives us

$$b_{nm} = \frac{2}{K} \int_0^K \left[\frac{2}{L} \int_0^L f(x, y) \sin\left(\frac{n\pi x}{L}\right) dx\right] \sin\left(\frac{m\pi y}{K}\right) dy.$$

Now put everything together to get

$$\begin{aligned} f(x, y) &= \sum_{n=1}^{\infty} a_n(y) \sin\left(\frac{n\pi x}{L}\right) \\ &= \sum_{n=1}^{\infty} \sum_{m=1}^{\infty} b_{nm} \sin\left(\frac{n\pi x}{L}\right) \sin\left(\frac{m\pi y}{K}\right), \end{aligned}$$

where

$$b_{nm} = \frac{4}{LK} \int_0^K \int_0^L f(x, y) \sin\left(\frac{n\pi x}{L}\right) \sin\left(\frac{m\pi y}{K}\right) dx\, dy$$

for $n = 1, 2, 3, \ldots, m = 1, 2, 3, \ldots$.

This is the *double Fourier sine expansion* of $f(x, y)$ on $0 \le x \le L, 0 \le y \le K$. Of course, convergence of the series to $f(x, y)$ must be given by a two-variable analogue of Theorem 95; so the use of equal signs above between $f(x, y)$ and the series was only for notational convenience. However, we do get from this informal reasoning the form that a double Fourier sine series should take. Here is an illustration.

EXAMPLE 380

Let $f(x, y) = xy$ for $0 \le x \le 1, 0 \le y \le 2$.

Compute

$$\begin{aligned} b_{nm} &= \frac{4}{2} \int_0^2 \int_0^1 xy \sin(n\pi x) \sin\left(\frac{m\pi y}{2}\right) dx\, dy \\ &= 2 \int_0^2 y \sin\left(\frac{m\pi y}{2}\right) dy \int_0^1 x \sin(n\pi x)\, dx^* \\ &= 2\left(\frac{-4\cos(m\pi)}{m\pi}\right)\left(\frac{-\cos(n\pi)}{n\pi}\right) \\ &= \frac{8}{nm\pi^2}(-1)^{n+m}. \end{aligned}$$

* This step depends on the fact that $f(x, y)$ is *separable*, that is, $f(x, y)$ is a function of x times a function of y. Not all functions of two variables are separable in this sense; for example, e^{xy} is not. For nonseparable $f(x, y)$, computation of the double Fourier coefficients in explicit form is sometimes impossible. The student should try computing b_{nm} for e^{xy}.

Thus, the Fourier sine series of xy on $0 \le x \le 1, 0 \le y \le 2$ is

$$\frac{8}{\pi^2} \sum_{n=1}^{\infty} \sum_{m=1}^{\infty} \frac{(-1)^{n+m}}{nm} \sin(n\pi x) \sin\left(\frac{m\pi y}{2}\right).$$

In an analogous way, one can get a double series of just cosine terms representing $f(x, y)$ on $0 \le x \le L, 0 \le y \le K$. We find that the cosine series is

$$\frac{a_{oo}}{4} + \sum_{m=1}^{\infty} \frac{1}{2} a_{om} \cos\left(\frac{m\pi y}{K}\right) + \sum_{n=1}^{\infty} \sum_{m=1}^{\infty} a_{nm} \cos\left(\frac{n\pi x}{L}\right) \cos\left(\frac{m\pi y}{K}\right),$$

where

$$a_{nm} = \frac{4}{LK} \int_0^K \int_0^L f(x, y) \cos\left(\frac{n\pi x}{L}\right) \cos\left(\frac{m\pi y}{K}\right) dx\, dy$$

for $n = 0, 1, 2, 3, \ldots, m = 0, 1, 2, 3, \ldots$.

EXAMPLE 381

Consider again $f(x, y) = xy$, $0 \le x \le 1$, $0 \le y \le 2$. We shall find the Fourier cosine series of $f(x, y)$.

We have

$$a_{oo} = \frac{4}{2} \int_0^2 \int_0^1 xy\, dx\, dy = 2$$

and

$$a_{om} = \frac{4}{2} \int_0^2 \int_0^1 xy \cos\left(\frac{m\pi y}{2}\right) dx\, dy = \frac{4}{m^2\pi^2} [(-1)^m - 1]$$

for $m = 1, 2, 3, \ldots$. Also, for $n = 1, 2, 3, \ldots$, and $m = 1, 2, 3, \ldots$, we have

$$a_{nm} = \frac{4}{2} \int_0^2 \int_0^1 xy \cos(n\pi x) \cos\left(\frac{m\pi y}{2}\right) dx\, dy$$

$$= \frac{8}{m^2\pi^2} [(-1)^m - 1] \frac{1}{n^2\pi^2} [(-1)^n - 1].$$

Thus, the Fourier cosine expansion of xy on $0 \le x \le 1, 0 \le y \le 2$ is

$$\frac{1}{2} + \sum_{m=1}^{\infty} \frac{2}{m^2\pi^2} [(-1)^m - 1] \cos\left(\frac{m\pi y}{2}\right)$$

$$+ \sum_{n=1}^{\infty} \sum_{m=1}^{\infty} \frac{8}{n^2 m^2 \pi^4} [(-1)^m - 1][(-1)^n - 1] \cos(n\pi x) \cos\left(\frac{m\pi y}{2}\right).$$

This can be written

$$\frac{1}{2} - \frac{4}{\pi^2} \sum_{m=1}^{\infty} \frac{1}{(2m-1)^2} \cos\left[\frac{(2m-1)\pi y}{2}\right]$$

$$+ \frac{32}{\pi^4} \sum_{n=1}^{\infty} \sum_{m=1}^{\infty} \frac{1}{(2n-1)^2(2m-1)^2} \cos[(2n-1)\pi x] \cos\left[\frac{(2m-1)\pi y}{2}\right].$$

Finally, one can write the Fourier series of $f(x, y)$ on $-L \leq x \leq L$, $-K \leq y \leq K$, analogous to a Fourier series on $-L \leq x \leq L$. Again omitting the details, the series is

$$\frac{a_{oo}}{4} + \frac{1}{2}\sum_{n=1}^{\infty}\left[a_{no}\cos\left(\frac{n\pi x}{L}\right) + c_{no}\sin\left(\frac{n\pi x}{L}\right)\right]$$
$$+ \frac{1}{2}\sum_{m=1}^{\infty}\left[a_{om}\cos\left(\frac{m\pi y}{K}\right) + b_{om}\sin\left(\frac{m\pi y}{K}\right)\right]$$
$$+ \sum_{n=1}^{\infty}\sum_{m=1}^{\infty}\left[a_{nm}\cos\left(\frac{n\pi x}{L}\right)\cos\left(\frac{m\pi y}{K}\right) + b_{nm}\cos\left(\frac{n\pi x}{L}\right)\sin\left(\frac{m\pi y}{K}\right)\right.$$
$$\left. + c_{nm}\sin\left(\frac{n\pi x}{L}\right)\cos\left(\frac{m\pi y}{K}\right) + d_{nm}\sin\left(\frac{n\pi x}{L}\right)\sin\left(\frac{m\pi y}{K}\right)\right].$$

The coefficients are given by

$$a_{nm} = \frac{1}{LK}\int_{-K}^{K}\int_{-L}^{L} f(x, y)\cos\left(\frac{n\pi x}{L}\right)\cos\left(\frac{m\pi y}{K}\right)dx\,dy,$$

$$b_{nm} = \frac{1}{LK}\int_{-K}^{K}\int_{-L}^{L} f(x, y)\cos\left(\frac{n\pi x}{L}\right)\sin\left(\frac{m\pi y}{K}\right)dx\,dy,$$

$$c_{nm} = \frac{1}{LK}\int_{-K}^{K}\int_{-L}^{L} f(x, y)\sin\left(\frac{n\pi x}{L}\right)\cos\left(\frac{m\pi y}{K}\right)dx\,dy,$$

and

$$d_{nm} = \frac{1}{LK}\int_{-K}^{K}\int_{-L}^{L} f(x, y)\sin\left(\frac{n\pi x}{L}\right)\sin\left(\frac{m\pi y}{K}\right)dx\,dy,$$

for $n = 0, 1, 2, 3, \ldots, m = 0, 1, 2, 3, \ldots$.

In most of the applications we shall encounter in the next chapter, Fourier sine and cosine series play the key roles.

The final two sections of this chapter are devoted to transforms which fall somewhat naturally out of Fourier series and integrals. These are the Fourier transform and the finite Fourier transform.

PROBLEMS FOR SECTION 12.8

In each of Problems 1 through 12, find the Fourier sine series of $f(x, y)$ on the given rectangle.

1. $x^2y;\ 0 \leq x \leq \pi,\ 0 \leq y \leq \pi$
2. $x - y;\ 0 \leq x \leq 2,\ 0 \leq y \leq 2$
3. $3x + y^2;\ 0 \leq x \leq 1,\ 0 \leq y \leq 2$
4. $\sin(2x - y);\ 0 \leq x \leq 2,\ 0 \leq y \leq 2$
5. $e^{x+y};\ 0 \leq x \leq 4,\ 0 \leq y \leq 2$
6. $x^2y^2;\ 0 \leq x \leq 2,\ 0 \leq y \leq 3$
7. $x\sinh(y);\ 0 \leq x \leq 4,\ 0 \leq y \leq 2$
8. $x^2 + y^2;\ 0 \leq x \leq 2,\ 0 \leq y \leq 1$
9. $4y\sin(2x);\ 0 \leq x \leq 4,\ 0 \leq y \leq \pi$
10. $(x + y)^2;\ 0 \leq x \leq 3,\ 0 \leq y \leq 4$
11. $x - 4y;\ 0 \leq x \leq 1,\ 0 \leq y \leq 2$
12. $x^2 - y\cos(x);\ 0 \leq x \leq 4,\ 0 \leq y \leq 2$

In each of Problems 13 through 24, find the Fourier cosine series of the given function on the given rectangle.

13. $2xy^2$; $0 \le x \le 2\pi$, $0 \le y \le 2\pi$

14. $x^2 - 5y$; $0 \le x \le 1$, $0 \le y \le 3$

15. $y\cos(x)$; $0 \le x \le \pi$, $0 \le y \le 1$

16. $\cos(x)\sin(y)$; $0 \le x \le 2\pi$, $0 \le y \le 2$

17. $x(1-y)$; $0 \le x \le \pi$, $0 \le y \le 1$

18. $2xy - 1$; $0 \le x \le 4$, $0 \le y \le 1$

19. 3; $0 \le x \le 4$, $0 \le y \le 2$

20. $8e^{x-y}$; $0 \le x \le 2$, $0 \le y \le 4$

21. $1 - x^2y$; $0 \le x \le 3$, $0 \le y \le 1$

22. $2x\cos(\pi y)$; $0 \le x \le 1$, $0 \le y \le 1$

23. x^2y^2; $0 \le x \le 2$, $0 \le y \le 5$

24. $x + y - 2$; $0 \le x \le 2$, $0 \le y \le 4$

25. Find the Fourier series of xe^y on $-2 \le x \le 2$, $-1 \le y \le 1$.

26. Find the Fourier series of $x - y^2$ on $-3 \le x \le 3$, $-2 \le y \le 2$.

27. Give a heuristic argument for the choice of coefficients in the Fourier cosine series of $f(x, y)$ on $0 \le x \le L$, $0 \le y \le K$.

28. Give a heuristic argument for the choice of coefficients in the Fourier series of $f(x, y)$ on $-L \le x \le L$, $-K \le y \le K$.

12.9 FINITE FOURIER TRANSFORMS

In mathematics, a *transform* is a device (often an integral or series) which changes equations from one form to another. The most commonly used transforms (such as the Laplace transform of Chapter 4) change differential equations into algebraic equations, which are presumably easier to solve. The solution of the algebraic equation is then transformed back the other way to obtain a solution of the differential equation.

In this and the next section we shall treat two kinds of transforms arising out of Fourier analysis. The Fourier transform, which is developed in Section 12.10, is defined from the Fourier integral; finite Fourier sine and cosine transforms, which we consider now, have their origins in Fourier series.

Begin with a function $f(x)$ which is sectionally continuous on $[0, \pi]$.* We shall define the finite Fourier sine and cosine transforms of $f(x)$, and then we shall display some of the properties which will be of use to us in solving partial differential equations in Chapter 13.

FINITE FOURIER SINE TRANSFORM

The *finite Fourier sine transform* $f_S(n)$ of $f(x)$ is defined to be

$$f_S(n) = \int_0^\pi f(x)\sin(nx)\,dx \quad \text{for} \quad n = 1, 2, 3, \ldots.$$

Note that $f_S(n)$ is defined not for all x in some interval, but for the positive integers 1, 2, 3, ... This accounts for the name finite (or *discrete*) Fourier transform.

* We standardize the interval to have length π as a convenience. In applications, this amounts to a scaling of units of length.

The motivation for the definition comes from the Fourier sine series of $f(x)$ on $[0, \pi]$, which is

$$\frac{2}{\pi}\sum_{n=1}^{\infty}\left(\int_0^{\pi} f(t)\sin(nt)\,dt\right)\sin(nx).$$

In terms of $f_S(n)$, the Fourier sine series of $f(x)$ on $[0, \pi]$ is then

$$\frac{2}{\pi}\sum_{n=1}^{\infty} f_S(n)\sin(nx).$$

The alternate notation $S_n[f(x)]$ for $f_S(n)$ is also convenient and will be useful in developing the operational formulas for the sine transform.

EXAMPLE 382

Let $f(x) = K$, constant. Then,

$$f_S(n) = \int_0^{\pi} K\sin(nx)\,dx = \frac{K}{n}[1-(-1)^n] \quad \text{for} \quad n = 1, 2, 3, \ldots.$$

We could also write this as

$$S_n(K) = \frac{K}{n}[1-(-1)^n] = \begin{cases} \dfrac{2K}{n} & \text{if } n \text{ is odd,} \\ 0 & \text{if } n \text{ is even.} \end{cases}$$

As a second example, let $g(x) = x^2$. Then,

$$g_S(n) = \int_0^{\pi} x^2\sin(nx)\,dx = \frac{1}{n^3}[(n^2\pi^2 - 2)(-1)^{n+1} - 2].$$

This result could also be denoted $S_n(x^2)$.

As with Laplace transforms, there are tables of finite Fourier sine transforms which one consults when using them to solve problems. A table is included at the end of Chapter 13, where we pursue applications. Here, we are more concerned with basic properties, of which we shall state three.

THEOREM 99

1. $S_n[f(x) + g(x)] = S_n[f(x)] + S_n[g(x)]$ and
2. $S_n[\alpha f(x)] = \alpha S_n[f(x)]$ for any constant α,

whenever these transforms are defined.

In words, the transform of a sum is the sum of the transforms, and constants factor through the transform. Both of these facts follow easily from properties of integrals.

The following result is of prime importance in applying the transform to the solution of differential equations.

THEOREM 100

Let $f(x)$ and $f'(x)$ be continuous on $[0, \pi]$, and let $f''(x)$ be sectionally continuous there. Then,

$$S_n[f''(x)] = -n^2 f_S(n) + nf(0) - n(-1)^n f(\pi), \quad \text{for} \quad n = 1, 2, 3, \ldots.$$

Proof Integrate by parts twice as follows:

$$\begin{aligned} S_n[f''(x)] &= \int_0^\pi f''(x)\sin(nx)\,dx = [\sin(nx)f'(x)]_0^\pi - \int_0^\pi f'(x)n\cos(nx)\,dx \\ &= -n\left\{[f(x)\cos(nx)]_0^\pi + \int_0^\pi f(x)n\sin(nx)\,dx\right\} \\ &= -n[f(\pi)\cos(n\pi) - f(0) + nf_S(n)] \\ &= -n^2 f_S(n) + nf(0) - n(-1)^n f(\pi). \end{aligned}$$

With corresponding assumptions on $f^{(3)}(x)$ and $f^{(4)}(x)$, we can continue, to get

$$\begin{aligned} S_n[f^{(4)}(x)] &= -n^2 S_n[f''(x)] + nf''(0) - n(-1)^n f''(\pi) \\ &= -n^2[-n^2 f_S(n) + nf(0) - n(-1)^n f(\pi)] + nf''(0) - n(-1)^n f''(\pi) \\ &= n^4 f_S(n) - n^3[f(0) - (-1)^n f(\pi)] + n[f''(0) - (-1)^n f''(\pi)]. \end{aligned}$$

From this we can get $S_n[f^{(6)}(x)]$ and so on, as long as the required conditions on the derivatives apply.

Although the main use of Theorem 100 is in solving differential equations, it can also be used to find new transforms from previously derived ones.

EXAMPLE 383

To find $S_n(x^4)$, let $f(x) = x^4$ in Theorem 100. We get

$$S_n(12x^2) = -n^2 S_n(x^4) + n(0)^4 - n(-1)^n \pi^4.$$

Solve this for $S_n(x^4)$, giving

$$S_n(x^4) = -\frac{1}{n^2} S_n(12x^2) - \frac{1}{n}(-1)^n \pi^4.$$

But, from Theorem 98(2) and Example 382, we have

$$S_n(12x^2) = 12S_n(x^2) = \frac{12}{n^3}[(n^2\pi^2 - 2)(-1)^{n+1} - 2].$$

Thus,

$$S_n(x^4) = \frac{-12}{n^5}[(n^2\pi^2 - 2)(-1)^{n+1} - 2] - \frac{1}{n}(-1)^n \pi^4.$$

In order to state the next result most easily, we introduce the notation

$$S_n^{-1}[f_S(n)] = f(x) \quad \text{if} \quad S_n[f(x)] = f_S(n).$$

We call S_n^{-1} the *inverse finite Fourier sine transform*; it denotes the function whose finite Fourier sine transform is $f_S(n)$. Thus, for example,

$$S_n^{-1}\left(\frac{K}{n}[1-(-1)^n]\right) = K, \quad \text{because} \quad S_n(K) = \frac{K}{n}[1-(-1)^n].$$

THEOREM 101

Let $S_n[f(x)] = f_S(n)$, as usual. Then,

$$S_n^{-1}\left(\frac{f_S(n)}{n^2}\right) = \frac{x}{\pi}\int_0^\pi (\pi - t)f(t)\,dt - \int_0^x (x-t)f(t)\,dt.$$

The above theorem is proved by taking the transform of the right side and showing that it is $f_S(n)/n^2$; here is an illustration.

EXAMPLE 384

Find a function whose finite Fourier sine transform is $[1-(-1)^n]/n^3$.

To apply the theorem, write

$$\frac{1-(-1)^n}{n^3} = \left(\frac{1-(-1)^n}{n}\right)\left(\frac{1}{n^2}\right)$$

and choose $f_S(n) = [1-(-1)^n]/n$. From Example 382 with $K = 1, f(x) = 1$. Then,

$$S_n^{-1}\left(\frac{1-(-1)^n}{n^3}\right) = \frac{x}{\pi}\int_0^\pi (\pi - t)\,dt - \int_0^x (x-t)\,dt = \frac{1}{2}x(\pi - x).$$

We now turn to the finite Fourier cosine transform.

FINITE FOURIER COSINE TRANSFORM

The *finite Fourier cosine transform* $f_C(n)$ of $f(x)$ is defined by

$$f_C(n) = \int_0^\pi f(x)\cos(nx)\,dx, \quad \text{for} \quad n = 0, 1, 2, 3, \ldots.$$

The motivation for this is that the Fourier cosine series of $f(x)$ on $[0, \pi]$ is

$$\frac{1}{\pi}\int_0^\pi f(t)\,dt + \sum_{n=1}^{\infty}\left[\frac{2}{\pi}\int_0^\pi f(t)\cos(nt)\,dt\right]\cos(nx)$$

$$= \frac{2}{\pi}\left[\frac{1}{2}f_C(0) + \sum_{n=1}^{\infty} f_C(n)\cos(nx)\right].$$

As before, we also develop a notation

$$C_n[f(x)] = f_C(n),$$

and the *inverse finite Fourier cosine transform* is given by

$$C_n^{-1}[f_C(n)] = f(x).$$

EXAMPLE 385

$$C_n(1) = \int_0^\pi \cos(nx)\,dx = \begin{cases} \pi & \text{if } n = 0, \\ 0 & \text{if } n = 1, 2, 3, \ldots. \end{cases}$$

$$C_n(x) = \int_0^\pi x\cos(nx)\,dx = \begin{cases} \pi^2/2 & \text{if } n = 0, \\ \dfrac{(-1)^n - 1}{n^2} & \text{if } n = 1, 2, 3, \ldots. \end{cases}$$

Properties of the cosine transform are analogous to those of the sine transform.

THEOREM 102

1. $C_n[f(x) + g(x)] = C_n[f(x)] + C_n[g(x)]$ and
2. $C_n[\alpha f(x)] = \alpha C_n[f(x)]$ for any constant α,

whenever these transforms are defined.

THEOREM 103

If $f(x)$ and $f'(x)$ are continuous on $[0, \pi]$ and $f''(x)$ is sectionally continuous on $[0, \pi]$, then

$$C_n[f''(x)] = -n^2 f_C(n) - f'(0) + (-1)^n f'(\pi) \quad \text{for} \quad n = 1, 2, 3, \ldots.$$

The proof of the above theorem is by integration by parts, as with Theorem 100, and is omitted. As with its sine transform counterpart, this formula is of main interest when dealing with differential equations. However, here is an illustration of its use in calculating a transform.

EXAMPLE 386

Find $C_n(x^3)$. We could of course calculate this directly from the definition. However, if we let $f(x) = x^3$ in Theorem 103, we have

$$C_n(6x) = -n^2 C_n(x^3) + (-1)^n(3\pi^2), \quad \text{for} \quad n = 1, 2, 3, \ldots.$$

Then,

$$C_n(x^3) = -\frac{6}{n^2}\, C_n(x) + \frac{3(-1)^n \pi^2}{n^2},$$

and now Example 385 gives us

$$C_n(x^3) = \frac{6}{n^4}\,[1 - (-1)^n] + \frac{3\pi^2(-1)^n}{n^2} \quad \text{for} \quad n = 1, 2, 3, \ldots.$$

For $n = 0$, we have directly that

$$C_0(x^3) = \int_0^\pi x^3\,dx = \frac{\pi^4}{4}.$$

Finally, here is the cosine counterpart of Theorem 101.

THEOREM 104

If $C_n[f(x)] = f_C(n)$, then

$$C_n^{-1}\left(\frac{f_C(n)}{n^2}\right) = \int_0^x \int_t^\pi f(p)\,dp\,dt + \frac{f_C(0)}{2\pi}(x-\pi)^2.$$

A proof of the above theorem is left to the student. If you check the table of finite Fourier cosine transforms given at the end of Chapter 13, you will note that the transform of even the most familiar functions generally has an unusual appearance. For this reason, Theorem 104 is not of great practical importance in computing inverse transforms and has primarily theoretical value.

We shall pursue some other properties of finite Fourier sine and cosine transforms in the exercises.

PROBLEMS FOR SECTION 12.9

In each of Problems 1 through 7, obtain the finite Fourier sine transform of the given function, either by using the definition or by using Theorem 100 in conjunction with a previously derived result.

1. x^3 **2.** x^5 **3.** e^x **4.** $\sin(\alpha x)$ **5.** $\cos(\alpha x)$ **6.** x^6 **7.** e^{-x}

In each of Problems 8 through 14, find the finite Fourier cosine transform of the given function, either directly from the definition or by using Theorem 103 and a previously derived result.

8. $f(x) = \begin{cases} 1, & 0 \le x \le \frac{1}{2} \\ -1, & \frac{1}{2} < x \le \pi \end{cases}$

9. x^2

10. x^4

11. e^x

12. $\sin(\alpha x)$

13. $\cos(\alpha x)$

14. x^5

15. Prove Theorem 101.

16. Prove Theorem 103.

17. Prove Theorem 104.

18. Show that

$$S_n[f'(x)] = -nC_n[f(x)] \quad \text{for} \quad n = 1, 2, 3, \ldots$$

if $f(x)$ and $f'(x)$ are sectionally continuous on $[0, \pi]$.

19. Show that

$$C_n[f'(x)] = nS_n[f(x)] - f(0) + (-1)^n f(\pi) \quad \text{for} \quad n = 0, 1, 2, 3, \ldots$$

if $f(x)$ and $f'(x)$ are sectionally continuous on $[0, \pi]$.

20. Let m be a positive integer. Show that

$$f_S(n+m) = S_n[f(x)\cos(mx)] + C_n[f(x)\sin(mx)].$$

21. Let m be a positive integer. Show that

$$f_C(n+m) = C_n[f(x)\cos(mx)] - S_n[f(x)\sin(mx)].$$

22. Let m be a positive integer with $m < n$. Show that

$$C_n[f(x)\cos(mx)] = \tfrac{1}{2}[f_C(n-m) + f_C(n+m)].$$

23. Show that, if m is a positive integer and $m < n$, then

$$C_n[f(x)\sin(mx)] = \tfrac{1}{2}[f_S(n+m) - f_S(n-m)].$$

12.10 FOURIER TRANSFORMS

Just as finite Fourier transforms can be found in the coefficients of sine and cosine series, the Fourier transform arises out of the coefficients in a Fourier integral expansion.

We shall begin with Fourier sine and cosine transforms and then consider the "full" Fourier transform.

FOURIER SINE TRANSFORM

Suppose that $f(x)$ is defined on $[0, \infty)$. We define the *Fourier sine transform* to be

$$\hat{f}_S(\lambda) = \int_0^\infty f(x)\sin(\lambda x)\,dx,$$

whenever this integral converges.* This is also denoted $\mathscr{F}_S[f(x)]$. Thus, the Fourier sine integral representing $f(x)$ on $[0, \infty)$ is nothing more than

$$\frac{2}{\pi}\int_0^\infty \hat{f}_S(\lambda)\sin(\lambda x)\,d\lambda.$$

EXAMPLE 387

Let

$$f(x) = \begin{cases} 1, & 0 \leq x \leq K, \\ 0, & x > K. \end{cases}$$

Then,

$$\hat{f}_S(\lambda) = \int_0^\infty f(x)\sin(\lambda x)\,dx = \int_0^K \sin(\lambda x)\,dx = \frac{1}{\lambda}[1 - \cos(K\lambda)].$$

As a second example, a straightforward integration gives

$$\mathscr{F}_S(e^{-x}) = \frac{\lambda}{1 + \lambda^2}.$$

The following theorem is a direct consequence of the definition.

THEOREM 105

1. $\mathscr{F}_S[f(x) + g(x)] = \mathscr{F}_S[f(x)] + \mathscr{F}_S[g(x)]$ and
2. $\mathscr{F}_S[\alpha f(x)] = \alpha\mathscr{F}_S[f(x)]$ for any constant α,

whenever these transforms are defined.

The following is the fundamental operational property of the Fourier sine transform.

---

* For convergence, it is certainly sufficient that $\int_0^\infty |f(x)|\,dx$ converge, although this is certainly not necessary.

THEOREM 106

Let $f(x)$ and $f'(x)$ be sectionally continuous on $[0, \infty)$. Assume also that $\lim_{x \to \infty} f(x) = \lim_{x \to \infty} f'(x) = 0$. Then,

$$\mathscr{F}_S[f''(x)] = -\lambda^2 \hat{f}_S(\lambda) + \lambda f(0).$$

Proof The proof is a straightforward integration by parts. We have

$$\begin{aligned}\mathscr{F}_S[f''(x)] &= \int_0^\infty f''(x) \sin(\lambda x)\, dx \\ &= [f'(x) \sin(\lambda x)]_0^\infty - \int_0^\infty f'(x)[\lambda \cos(\lambda x)]\, dx \\ &= -\lambda \left\{ [\cos(\lambda x) f(x)]_0^\infty - \int_0^\infty f(x)[-\lambda \sin(\lambda x)]\, dx \right\} \\ &= -\lambda^2 \hat{f}_S(\lambda) + \lambda f(0).\end{aligned}$$

FOURIER COSINE TRANSFORM

Analogous to the sine transform, we define the *Fourier cosine transform* by

$$\hat{f}_C(\lambda) = \mathscr{F}_C[f(x)] = \int_0^\infty f(x) \cos(\lambda x)\, dx.$$

In terms of this transform, the Fourier cosine integral of $f(x)$ on $[0, \infty)$ is

$$\frac{2}{\pi} \int_0^\infty \hat{f}_C(\lambda) \cos(\lambda x)\, d\lambda.$$

Here are the cosine transform versions of the results stated for sine transforms.

THEOREM 107

1. $\mathscr{F}_C[f(x) + g(x)] = \mathscr{F}_C[f(x)] + \mathscr{F}_C[g(x)]$ and
2. $\mathscr{F}_C[\alpha f(x)] = \alpha \mathscr{F}_C[f(x)]$ for any constant α,

whenever these transforms exist.

THEOREM 108

If $f(x)$ and $f'(x)$ are sectionally continuous on $[0, \infty)$ and $\lim_{x \to \infty} f'(x) = 0$, then

$$\mathscr{F}_C[f''(x)] = -\lambda^2 \hat{f}_C(\lambda) - f'(0).$$

A proof of this is achieved by two integrations by parts, and we leave the details to the student.

A short table of Fourier sine and cosine transforms is included at the end of Chapter 13. We conclude this section with the Fourier transform and an application to band-limited signals.

FOURIER TRANSFORM

In defining the Fourier transform, we shall make use of the complex exponential function (which can be reviewed in Section 2.4) and the form of the Fourier integral given in Problem 19 of Section 12.5.

The *Fourier transform of* $f(x)$ is defined as

$$\hat{f}(\lambda) = \mathscr{F}[f(x)] = \int_{-\infty}^{\infty} f(x)e^{-i\lambda x}\,dx,$$

whenever this integral converges. Note that the Fourier integral representation of $f(x)$ can then be written

$$\frac{1}{2\pi}\int_{-\infty}^{\infty} \hat{f}(\lambda)e^{i\lambda x}\,d\lambda.$$

As with all the other transforms we have encountered thus far, we have

$$\mathscr{F}[f(x) + g(x)] = \mathscr{F}[f(x)] + \mathscr{F}[g(x)]$$

and

$$\mathscr{F}[\alpha f(x)] = \alpha\mathscr{F}[f(x)] \quad \text{for any constant } \alpha.$$

The main operational rule for the Fourier transform is the following.

THEOREM 109

If

$$\lim_{x\to\pm\infty} f(x) = \lim_{x\to\pm\infty} f'(x) = \cdots = \lim_{x\to\pm\infty} f^{(n-1)}(x) = 0 \quad \text{and} \quad f(x), \ldots, f^{(n-1)}(x)$$

are sectionally continuous on $[-\infty, \infty]$, then

$$\mathscr{F}[f^{(n)}(x)] = (i\lambda)^n\hat{f}(\lambda).$$

Proof Begin with

$$\begin{aligned}\mathscr{F}[f'(x)] &= \int_{-\infty}^{\infty} f'(x)e^{-i\lambda x}\,dx \\ &= [f(x)e^{-i\lambda x}]_{-\infty}^{\infty} - \int_{-\infty}^{\infty} f(x)(-i\lambda)e^{-i\lambda x}\,dx \\ &= (i\lambda)\hat{f}(\lambda).\end{aligned}$$

Continuing from this, we have next that

$$\mathscr{F}[f''(x)] = (i\lambda)\mathscr{F}[f'(x)] = (i\lambda)^2\hat{f}(\lambda),$$

and continued applications of the result first proved yield the final result.

Here is an application of Fourier transforms to reconstruction of signals from specific samples.

EXAMPLE 388 Band-Limited Signals

Consider a signal $f(t)$, with t for time used as the variable instead of x because of context. We assume that $f(t)$ has a Fourier integral representation

$$f(t) = \frac{1}{2\pi} \int_{-\infty}^{\infty} \int_{-\infty}^{\infty} f(\xi) e^{i\lambda(t-\xi)}\, d\xi\, d\lambda,$$

as developed in Problem 19 of Section 12.5. In view of the definition of the Fourier transform, this can be written

$$f(t) = \frac{1}{2\pi} \int_{-\infty}^{\infty} \hat{f}(\lambda) e^{i\lambda t}\, d\lambda.$$

A signal is said to be *band limited* if $\hat{f}(\lambda) = 0$ outside of some finite interval. That is, for some smallest number L,

$$\hat{f}(\lambda) = 0 \quad \text{if} \quad |x| > L.$$

In this case, the Fourier integral of $f(t)$ is

$$f(t) = \frac{1}{2\pi} \int_{-\infty}^{\infty} \hat{f}(\lambda) e^{i\lambda t}\, d\lambda = \frac{1}{2\pi} \int_{-L}^{L} \hat{f}(\lambda) e^{i\lambda t}\, d\lambda.$$

We shall derive a rather remarkable result: We will know $f(t)$ for all t if we know it just for $t = 0, \pm\pi/L, \pm 2\pi/L, \ldots$. This is called the *sampling theorem*, and it says that a band-limited signal can be reconstructed from samples of the signal at certain values of t. This result is used in telephone communications engineering.

To derive the sampling theorem, proceed as follows. Using the complex form of the Fourier series (see Problem 50, Section 12.2), write

$$\hat{f}(\lambda) = \sum_{n=-\infty}^{\infty} k_n e^{in\pi\lambda/L},$$

where

$$k_n = \frac{1}{2L} \int_{-L}^{L} \hat{f}(\lambda) e^{-in\pi\lambda/L}\, d\lambda.$$

Comparing this formula for k_n with the Fourier integral representation of $f(t)$, we see that

$$k_n = \frac{\pi}{L} f\left(\frac{-n\pi}{L}\right).$$

We can now write

$$\begin{aligned} \hat{f}(\lambda) &= \sum_{n=-\infty}^{\infty} k_n e^{in\pi\lambda/L} \\ &= \sum_{n=-\infty}^{\infty} \frac{\pi}{L} f\left(\frac{-n\pi}{L}\right) e^{in\pi\lambda/L} \\ &= \frac{\pi}{L} \sum_{n=-\infty}^{\infty} f\left(\frac{n\pi}{L}\right) e^{-in\pi\lambda/L}, \end{aligned}$$

since we may replace n by $-n$ in summing from $-\infty$ to ∞. Plug this back into the Fourier integral representation of $f(t)$ to get

$$\begin{aligned} f(t) &= \frac{1}{2\pi}\int_{-L}^{L} \hat{f}(\lambda)e^{i\lambda t}\,d\lambda \\ &= \frac{1}{2\pi}\int_{-L}^{L} \frac{\pi}{L}\sum_{n=-\infty}^{\infty} f\left(\frac{n\pi}{L}\right)e^{-in\pi\lambda/L}e^{i\lambda t}\,d\lambda \\ &= \frac{1}{2L}\sum_{n=-\infty}^{\infty} f\left(\frac{n\pi}{L}\right)\int_{-L}^{L} e^{\lambda[(-in\pi/L)+it]}\,d\lambda \\ &= \frac{1}{2}\sum_{n=-\infty}^{\infty} f\left(\frac{n\pi}{L}\right)\frac{1}{i(Lt-n\pi)}\left[e^{(Lt-n\pi)i} - e^{-(Lt-n\pi)i}\right] \\ &= \sum_{n=-\infty}^{\infty} f\left(\frac{n\pi}{L}\right)\frac{\sin(Lt-n\pi)}{Lt-n\pi}. \end{aligned}$$

In the last line we used the fact that

$$\frac{e^{(Lt-n\pi)i} - e^{-(Lt-n\pi)i}}{2i} = \sin(Lt - n\pi).$$

This expression for $f(t)$ says that we need only know $f(0)$, $f(\pi/L)$, $f(-\pi/L)$, $f(2\pi/L), f(-2\pi/L), \ldots$ in order to know $f(t)$ for all t.

PROBLEMS FOR SECTION 12.10

In each of Problems 1 through 12, find the Fourier cosine transform of the given function.

1. e^{-x}

2. $f(x) = \begin{cases} \cos(x), & 0 \le x \le K \\ 0, & x > K \end{cases}$

3. $f(x) = \begin{cases} x^2, & 0 \le x \le K \\ 0, & x > K \end{cases}$

4. $f(x) = \begin{cases} 1, & 0 \le x \le K \\ -1, & K < x \le 2K \\ 0, & x > 2K \end{cases}$

5. $f(x) = \begin{cases} 2x, & 0 \le x \le 5 \\ 0, & x > 5 \end{cases}$

6. $f(x) = \begin{cases} 0, & 0 \le x \le K \\ \sinh(x), & K < x \le 2K \\ 0, & x > 2K \end{cases}$

7. xe^{-ax}, a any positive constant

8. x^2e^{-ax}, a any positive constant

9. e^{-ax^2}, a any positive constant

10. $e^x \cos(x)$

11. $\dfrac{1}{1+x^2}$

12. $\dfrac{1}{(1+x^2)^2}$

In each of Problems 13 through 22, find the Fourier sine transform of the given function.

13. $f(x) = \begin{cases} 2, & 0 \le x \le K \\ -2, & K < x \le 2K \\ 0, & x > 2K \end{cases}$

14. e^{-x}

15. xe^{-x}

16. xe^{-ax^2}, a any positive constant

17. $e^{-x}\sin(x)$

18. $\cos(x)$

19. $\dfrac{\cos(x)}{x}$

20. $\dfrac{e^{-x}}{x}$

21. $f(x) = \begin{cases} \sin(Kx), & 0 \le x \le 1 \\ 0, & x > 1 \end{cases}$

22. $f(x) = \begin{cases} \sinh(Kx), & 0 \le x \le \pi \\ 0, & x > \pi \end{cases}$

In each of Problems 23 through 30, find the Fourier transform of the given function.

23. $f(x) = \begin{cases} 1, & 0 \le x \le 1 \\ -1, & -1 \le x < 0 \\ 0, & |x| > 1 \end{cases}$

24. $f(x) = \begin{cases} \sin(x), & -K \le x \le K \\ 0, & |x| > K \end{cases}$

25. $e^{-a|x|}$, a any positive constant

26. $xe^{-a|x|}$, a any positive constant

27. $f(x) = \begin{cases} \cosh(x), & -K \le x \le K \\ 0, & |x| > K \end{cases}$

28. $f(x) = \begin{cases} x^2, & -K \le x \le K \\ 0, & |x| > K \end{cases}$

29. $\dfrac{1}{1 + x^2}$

30. $\dfrac{x}{1 + x^2}$

31. Show that

$$\mathscr{F}_S[f^{(4)}(x)] = \lambda^4 \hat{f}_S(\lambda) - \lambda^3 f(0) + \lambda f''(0),$$

under appropriate assumptions on $f(x)$ and its derivatives.

32. Show that

$$\mathscr{F}_C[f^{(4)}(x)] = \lambda^4 \hat{f}_C(\lambda) + \lambda^2 f'(0) - f^{(3)}(0),$$

under appropriate assumptions on $f(x)$ and its derivatives.

SOME HISTORICAL NOTES ON FOURIER SERIES, INTEGRALS, AND TRANSFORMS

Fourier series are named after Joseph Fourier, a French mathematician who lived during the Napoleonic era from 1768 to 1830. Fourier is justly honored by having his name attached to these important series, but the germ of the idea goes back at least to the Swiss mathematician Leonard Euler (1707–1783), and therein lies an interesting story.

In the approximate period 1729–1753, Euler considered the problem of interpolation: Given values of $f(1), f(2), \ldots, f(n)$, find $f(x)$ for $1 \le x \le n$. The problem arose in connection with planetary perturbations and led Euler to a trigonometric series. Specifically, he found that the solution of an equation of the form

$$f(x) = f(x - 1) + F(x)$$

is

$$f(x) = \int_0^x F(t)\,dt + 2\sum_{n=1}^{\infty} \int_0^x (F(t)\cos(2n\pi t)\,dt)\cos(2n\pi x)$$

$$+ 2\sum_{n=1}^{\infty} \int_0^x (F(t)\sin(2n\pi t)\,dt)\sin(2n\pi x).$$

Soon thereafter, in 1754, Jean Le Rond d'Alembert (1717–1783) obtained a trigonometric cosine expansion for the reciprocal of the distance between two planets in terms of the angle between the vectors from the origin to the planets. In this work, the integral formulas for the Fourier coefficients appear.

Other expressions involving series of sines and cosines soon began to be noticed. Using geometric series, Euler obtained

$$\frac{a\cos(x) - a^2}{1 - 2a\cos(x) + a^2} = \sum_{n=1}^{\infty} a^n \cos(nx)$$

and

$$\frac{a\sin(x)}{1 - 2a\cos(x) + a^2} = \sum_{n=1}^{\infty} a^n \sin(nx).$$

These are valid if $|a| < 1$. Letting $a = 1$, Euler concluded that

$$\sum_{n=1}^{\infty} \cos(nx) = -\tfrac{1}{2},$$

an incorrect result (this series diverges). Integrating from x to π then gives

$$\sum_{n=1}^{\infty} \frac{1}{n} \sin(nx) = \frac{\pi - x}{2},$$

a correct Fourier sine expansion on $(0, \pi)$, derived, however, by incorrect reasoning. The result $\sum_{n=1}^{\infty} \cos(nx) = -\frac{1}{2}$ was also obtained by Joseph Louis Lagrange (1736–1813) in connection with a study of sound waves.

Thus, as early as the 1750s, trigonometric series were "in the air," in the sense that leading mathematicians were encountering them and calculating with them. Their significance was not yet understood, however, and they raised important questions. For example, how can a nonperiodic function $f(x)$ be represented by a series of sines and cosines which are periodic? Although the integral formulas for the coefficients were known, they were not entirely trusted, and men such as Euler preferred deriving trigonometric series in other ways.

Finally, there arose a heated and lengthy debate, involving Euler, Lagrange, d'Alembert, Daniel Bernoulli (1700–1782), and other leading mathematicians of the period, concerning which functions could be expanded in trigonometric series and, indeed, over the very meaning of the term *function*. Up to this time, the term *function* was often taken for granted, but it was now found to mean different things to different people. The debate was to continue for many years, involving the great mathematicians of the time, including Pierre-Simon de Laplace (1749–1827) who joined the battle in about 1779.

Joseph Fourier was born in 1768 at the height of this debate. As a student, he showed a good talent for mathematics but only turned to it as a profession when his common heritage (he was the son of a tailor) closed the doors to a military commission.

An outstanding problem at the turn of the nineteenth century was the mathematical formulation and solution of the problem of heat conduction. In 1807, Fourier submitted a paper on this subject to the prestigious French Academy of Sciences in Paris. Such mathematical giants as Laplace, Lagrange, and Legendre refereed the work and rejected it for lack of rigor, encouraging Fourier to continue his research and fill in the missing details. In 1811, Fourier submitted a revised version and was awarded a prize by the Academy, which nonetheless refused publication because of details which were still unclear. Finally, in 1822, Fourier published his now classic *Theorie Analytique de la Chaleur*, incorporating most of his 1811 results, together with some new ones.

Basically, Fourier considered the partial differential equation

$$\frac{\partial^2 u}{\partial x^2} + \frac{\partial^2 u}{\partial y^2} + \frac{\partial^2 u}{\partial z^2} = k^2 \frac{\partial u}{\partial t},$$

in which $u(x, y, z, t)$ is the temperature at time t and point (x, y, z) in a body conducting heat energy. By methods which we shall study in the next chapter, Fourier derived trigonometric series solutions with explicit use of the formula for the coefficients in the series which now bear his name.

How did Fourier's work really differ from that of Euler and other predecessors who worked with such series?

First, Fourier's general methods for attacking the heat equation were of fundamental and lasting importance in solving partial differential equations. The method is known today as *separation of variables* or (particularly in Russian texts) the *Fourier method.*

Second, while others had the Fourier coefficients, none really appreciated their importance. Fourier was the first (albeit for not entirely rigorous reasons) to use them confidently and extensively and to assert convincingly the generality with which arbitrary functions could be expanded in Fourier series. In certain technical details, Fourier was often wrong, as modern convergence theorems show. Nonetheless, he took the major steps upon which the others built and for this deserves the credit inherent in the designation *Fourier series.*

Finally, importance of a piece of mathematics is often judged by its ramifications, as seen much later from a historical perspective. Questions arising from the study of Fourier series have had a profound impact on the development of mathematics, including results on convergence of series, solution of partial differential equations, properties of the real line, set theory, measure theory, and harmonic analysis.

Sufficient conditions for convergence of Fourier series were first established by Peter Gustav Lejeune Dirichlet (1805–1859) in about 1829. These are essentially the conditions of Theorem 95. The result that $a_n \to 0$ and $b_n \to 0$ as $n \to \infty$, if $f(x)$ is bounded and integrable on $[-L, L]$, is due to Bernhard Riemann (1826–1866), a student of Gauss who worked with Dirichlet for a time.

In 1873, Paul du Bois-Reymond (1831–1889) gave an example of a function which is continuous on $(-\pi, \pi)$ but whose Fourier series fails to converge at a point. Many other results, leading into measure theory and modern harmonic analysis, have since evolved.

The Fourier integral appeared near the end of Fourier's 1811 paper, when he attempted to extend his results to functions defined over $[0, \infty)$. A separate treatment appeared in Cauchy's 1816 prize-winning paper on surface waves in a fluid. Still a third derivation appeared in 1816 in Poisson's paper on water waves.

The Fourier transform is found in some early writings of Cauchy and Laplace (from about 1782 on). A formula for the Fourier cosine transform and its inverse appears in Fourier's 1811 paper.

SUPPLEMENTARY PROBLEMS

In each of Problems 1 through 15, write the Fourier series of the given function on the given interval, and determine what the series converges to at each point of the interval.

1. $f(x) = 3x^2 - 1, \quad -4 \le x \le 4$

2. $f(x) = \cos(x) - \sin(x), \quad -1 \le x \le 1$

3. $f(x) = \cosh(2x), \quad -3 \le x \le 3$

4. $f(x) = x^4, \quad -5 \le x \le 5$

5. $f(x) = (x + 2)^2, \quad -2 \le x \le 2$

6. $f(x) = \begin{cases} -2, & -1 \le x < \frac{1}{2} \\ 3, & \frac{1}{2} < x \le 1 \end{cases}$

7. $f(x) = \begin{cases} 1 - \sin(\pi x), & -1 \le x < 0 \\ 4, & 0 \le x < 1 \end{cases}$

8. $f(x) = \begin{cases} e^{-x}, & 0 \le x < 1 \\ 2, & -1 < x \le 0 \end{cases}$

9. $f(x) = \begin{cases} 1, & -4 \le x < -\frac{1}{2} \\ 0, & -\frac{1}{2} < x < \frac{1}{2} \\ 1, & \frac{1}{2} < x \le 4 \end{cases}$

10. $f(x) = \begin{cases} \cosh(x), & -2 \le x \le 0 \\ \sinh(x), & 0 < x \le 2 \end{cases}$

11. $f(x) = \begin{cases} \frac{1}{2}, & -4 \le x < -1 \\ 0, & -1 < x \le 2 \\ x, & 2 < x \le 4 \end{cases}$

12. $f(x) = \begin{cases} x^2, & -3 \le x \le -2 \\ x, & -2 < x < 1 \\ e^{-x}, & 1 < x \le 3 \end{cases}$

13. $f(x) = \begin{cases} x^2, & -1 \le x \le 0 \\ -x^2, & 0 < x \le 1 \end{cases}$

14. $f(x) = \begin{cases} \cos(Kx), & -2 \le x \le 0 \\ \sin(Kx), & 0 < x \le 2 \end{cases}$

15. $f(x) = \begin{cases} 1 - e^{-x}, & -3 \le x < -2 \\ 0, & -2 \le x \le 3 \end{cases}$

In each of Problems 16 through 25, write the Fourier cosine series of the given function on the given interval. Also determine what the series converges to at each point of the interval.

16. $f(x) = \cosh(4x), \quad 0 \le x \le 2$

17. $f(x) = \sin(x), \quad 0 \le x \le \pi$

18. $f(x) = 1 - x^2, \quad 0 \le x \le 3$

19. $f(x) = -3e^{-2x}, \quad 0 \le x \le 4$

20. $f(x) = \begin{cases} -3, & 0 \le x < \frac{1}{2} \\ x, & \frac{1}{2} < x \le 1 \end{cases}$

21. $f(x) = \begin{cases} x^2, & 0 \le x < 1 \\ 2x, & 1 < x \le 4 \end{cases}$

22. $f(x) = \begin{cases} 0, & 0 \le x < 3 \\ 1, & 3 < x \le 5 \end{cases}$

23. $f(x) = \begin{cases} x, & 0 \le x \le 1 \\ x^2, & 1 < x < 2 \\ x^3, & 2 < x \le 3 \end{cases}$

24. $f(x) = \begin{cases} 2 + x^2, & 0 \le x < \pi \\ \sin(2x), & \pi < x \le 2\pi \end{cases}$

25. $f(x) = \begin{cases} -2\cos(3x), & 0 \le x \le 2\pi \\ \sin(4x), & 2\pi < x \le 5\pi \end{cases}$

In each of Problems 26 through 40, find the Fourier sine series of the given function on the given interval, and also determine what the series converges to at each point of the interval.

26. $f(x) = 2 - 3x, \quad 0 \le x \le 3$

27. $f(x) = \sinh(3x), \quad 0 \le x \le 1$

28. $f(x) = x - x^2, \quad 0 \le x \le 2$

29. $f(x) = e^{-3x}, \quad 0 \le x \le 5$

30. $f(x) = (x - 2)^3, \quad 0 \le x \le 1$

31. $f(x) = \begin{cases} 2, & 0 \le x \le 1 \\ -1, & 1 < x \le 10 \end{cases}$

32. $f(x) = \begin{cases} x^2 - 1, & 0 \le x \le 2 \\ 0, & 2 < x \le 4 \end{cases}$

33. $f(x) = \begin{cases} 1, & 0 \le x < \pi \\ \cos(x), & \pi < x < 2\pi \end{cases}$

34. $f(x) = \begin{cases} e^{-x}, & 0 \le x \le 2 \\ 1, & 2 < x \le 3 \end{cases}$

35. $f(x) = \begin{cases} -1, & 0 \le x < 3 \\ 2, & 3 < x \le 5 \end{cases}$

36. $f(x) = \begin{cases} 1, & 0 \le x < 1 \\ 0, & x = 1 \\ -1, & 1 < x \le 2 \end{cases}$

37. $f(x) = \begin{cases} \cosh(2x), & 0 \le x < \pi \\ 0, & \pi \le x \le 3\pi \end{cases}$

38. $f(x) = \cos(x), \quad 0 \le x \le 5$

39. $f(x) = \begin{cases} \sin(x) - \cos(x), & 0 \le x < 3 \\ 4x, & 3 < x \le 4 \end{cases}$

40. $f(x) = \begin{cases} 2 - x^2, & 0 \le x < 3 \\ 3x, & 3 \le x < 7 \end{cases}$

In each of Problems 41 through 45, compute the Fourier integral of the given function.

41. $f(x) = \begin{cases} \cosh(x), & -5 \le x \le 5 \\ 0, & |x| > 5 \end{cases}$

42. $f(x) = \begin{cases} e^{-x}, & -2 \le x \le 2 \\ 0, & |x| > 2 \end{cases}$

43. $f(x) = \begin{cases} -1, & -3 \le x \le 0 \\ 1, & 0 < x \le 3 \\ e^{-|x|}, & |x| > 3 \end{cases}$

44. $f(x) - \begin{cases} 1 - x, & -2 \le x < 0 \\ 1, & 0 < x \le 2 \\ 0, & |x| > 2 \end{cases}$

45. $f(x) = \begin{cases} \sin(x), & 0 \le x < \pi \\ \cos(x), & -\pi < x < 0 \\ 0, & |x| > \pi \end{cases}$

In each of Problems 46 through 50, compute the Fourier cosine integral and the Fourier sine integral of the given function.

46. $f(x) = \begin{cases} 0, & 0 \le x \le 5 \\ e^{-x}, & x > 5 \end{cases}$

47. $f(x) = \begin{cases} 1, & 0 \le x \le 10 \\ \cos(x), & 10 < x \le 20 \\ 0, & x > 20 \end{cases}$

48. $f(x) = \begin{cases} -1, & 0 \le x < 2 \\ 1, & 2 < x \le 4 \\ 0, & x > 4 \end{cases}$

49. $f(x) = \begin{cases} \cosh(2x), & 0 \le x \le 1 \\ 0, & 1 < x < 2 \\ \sinh(2x), & 2 \le x < 3 \\ 0, & x > 3 \end{cases}$

50. $f(x) = \begin{cases} x^2, & 0 \le x < 5 \\ 0, & x > 5 \end{cases}$

In each of Problems 51 through 60, write double Fourier sine series and cosine series for the given function on the given rectangle.

51. $x^2(y - 1); \quad 0 \le x \le 1, \quad 0 \le y \le 5$

52. $(x - a)(y - b); \quad 0 \le x \le a, \quad 0 \le y \le b$

53. $\sin(\pi x) \cos(\pi y); \quad 0 \le x \le 1, \quad 0 \le y \le 1$

54. $x^2 \cos(y); \quad 0 \le x \le 1, \quad 0 \le y \le \pi$

55. $e^{-x-y}; \quad 0 \le x \le 1, \quad 0 \le y \le 1$

56. $\cosh(x) \sinh(y); \quad 0 \le x \le 2, \quad 0 \le y \le 4$

57. $xe^{-y}; \quad 0 \le x \le 1, \quad 0 \le y \le 1$

58. $2x^2y; \quad 0 \le x \le 4, \quad 0 \le y \le 1$

59. $\sin(x) \cosh(3y); \quad 0 \le x \le \pi, \quad 0 \le y \le 1$

60. $(x + 2y)^2; \quad 0 \le x \le 3, \quad 0 \le y \le 2$

61. Find the finite Fourier cosine transform of $\cos(\pi - x)$.

62. Find the finite Fourier sine transform of $\sinh(\pi - x)$.

63. Find the finite Fourier sine transform of $\sin(\pi - x)$.

64. Find the finite Fourier sine transform of

$$\arctan\left(\frac{\sin(x)}{1 - \cos(x)}\right).$$

65. Find the finite Fourier cosine transform of $\cosh(\pi - x)$.

66. Let $f(x)$ be sectionally continuous on $[0, L]$. Let $b_1, b_2, \ldots$ be the coefficients in the Fourier sine series of $f(x)$ on $[0, L]$. Show that

$$\sum_{n=1}^{\infty} b_n{}^2 \le \frac{2}{L} \int_0^L f(x)^2 \, dx.$$

67. Let $f(x)$ be sectionally continuous on $[0, L]$. Let $a_0, a_1, \ldots$ be the coefficients in the Fourier cosine series of $f(x)$ on $[0, L]$. Show that

$$\frac{1}{2}a_0^2 + \sum_{n=1}^{\infty} a_n{}^2 \leq \frac{2}{L}\int_0^L f(x)^2\, dx.$$

68. Evaluate the series $\sum_{n=1}^{\infty} 1/n^2$ by setting $x = \pi$ in the Fourier series of $f(x) = x^2$ on $[-\pi, \pi]$.

69. Let $f(x)$ be sectionally continuous on $[0, \infty)$. Define the even and odd extensions of $f(x)$ to $(-\infty, \infty)$ by setting, respectively,

$$H(x) = f(|x|) \quad \text{and} \quad G(x) = \frac{x}{|x|}f(|x|) \quad \text{for} \quad -\infty < x < \infty.$$

(a) Prove that, for any constant K,

$$2f_S(\lambda)\sin(K\lambda) = \mathscr{F}_C[G(x + K) - G(x - K)].$$

(b) Prove that, for any constant K,

$$2f_S(\lambda)\cos(K\lambda) = \mathscr{F}_S[G(x + K) + G(x - K)].$$

(c) Prove that, for any constant K,

$$2f_C(\lambda)\sin(K\lambda) = \mathscr{F}_S[H(x - K) - H(x + K)].$$

(d) Prove that, for any constant K,

$$2f_C(\lambda)\cos(K\lambda) = \mathscr{F}_C[H(x - K) + H(x + K)].$$

70. Define $f(x)$ for all real x by setting $f(x) = x^2$, $-2 \leq x \leq 2$, and $f(x + 4) = f(x)$ for all x. Write the Fourier series for $f(x)$ on the whole real line, and determine, for each x, the sum of the Fourier series.

71. Let $f(x) = e^{-x}$ for $-1 \leq x \leq 1$, and let $f(x + 2) = f(x)$ for all x. Write the Fourier series for $f(x)$ on the whole real line, and determine, for each x, the sum of the Fourier series.

72. Let $f(x) = 3x - 1$ for $0 \leq x \leq 3$, and let $f(x) = -x$ for $-3 \leq x < 0$. Extend $f(x)$ to the whole real line by setting $f(x + 6) = f(x)$. Write the Fourier series for $f(x)$ on the whole real line, and determine the sum of this series for all x.

CHAPTER THIRTEEN

Partial Differential Equations

13.0 INTRODUCTION

A *partial differential equation* is an equation which contains one or more partial derivatives. Such equations arise in many natural phenomena involving the interaction of three or more variables, as we shall see in the next section.

A *solution* of a partial differential equation is a function which satisfies the equation. For example, it is easy to check by direct substitution that

$$u(x, t) = \cos(2x)e^{-4a^2t}$$

is a solution of

$$\frac{\partial u}{\partial t} = a^2 \frac{\partial^2 u}{\partial x^2}.$$

Occasionally a partial differential equation can be solved by inspection, though this is rare. As one example, consider

$$\frac{\partial u}{\partial x} = -4\frac{\partial u}{\partial y}.$$

There may be many solutions of this, but a simple one is easy to guess. If $\partial u/\partial x$ and $\partial u/\partial y$ were both constant, we could get a solution easily. We therefore try $u(x, y) = Ax + By$. Then, $\partial u/\partial x = A$ and $\partial u/\partial y = B$, and we have a solution if $A = -4B$. We can choose, say, $B = 1$ and $A = -4$ and then $u = -4x + y$ is a solution. Unfortunately, things are not usually this simple.

A partial differential equation is of *order n* if it contains an nth partial derivative, but none higher. Thus, for example, Laplace's equation,

$$\frac{\partial^2 u}{\partial x^2} + \frac{\partial^2 u}{\partial y^2} + \frac{\partial^2 u}{\partial z^2} = 0,$$

is second order, as is the heat equation

$$\frac{\partial u}{\partial t} = \frac{\partial^2 u}{\partial x^2},$$

while the simple equation we solved before was first order.

Just as with ordinary differential equations, one usually begins the study of partial differential equations by concentrating on the linear case. The nonlinear case is too difficult for openers.

The *general linear first order partial differential equation* in three variables (one dependent and two independent) is

$$a(x, y)\frac{\partial u}{\partial x} + b(x, y)\frac{\partial u}{\partial y} + g(x, y) = 0.$$

The *general linear second order partial differential equation* in three variables (one dependent and two independent) has the form

$$a(x, y)\frac{\partial^2 u}{\partial x^2} + b(x, y)\frac{\partial^2 u}{\partial x\,\partial y} + c(x, y)\frac{\partial^2 u}{\partial y^2} + d(x, y)\frac{\partial u}{\partial x}$$

$$+ e(x, y)\frac{\partial u}{\partial y} + f(x, y)u + g(x, y) = 0.$$

Most of the equations we encounter will be of this form. In both cases, the equation is *homogeneous* if $g(x, y) = 0$ for all (x, y) under consideration and *nonhomogeneous* if $g(x, y) \neq 0$ for some (x, y). In the homogeneous case, we always seek a nontrivial solution, that is, one which is not identically zero.

One reason for restricting attention to linear first and second order equations is that they occur so frequently, particularly the second order. Even these require a vast theory when treated very generally; so we shall spend most of our time on those particular equations governing vibrational and heat conduction phenomena most commonly encountered in engineering and the sciences. The main tools we will be using are Fourier series and integrals, with some attention devoted at the end of the chapter to Fourier and Laplace transform methods.

We begin in the next section with a derivation of some of the fundamental partial differential equations.

PROBLEMS FOR SECTION 13.0

In each of Problems 1 through 10, find a nontrivial solution of the given partial differential equation by inspection and integration. Here, u is a function of the two variables x and y.

1. $\dfrac{\partial u}{\partial x} = 0$ **2.** $\dfrac{\partial u}{\partial y} = 2y^2$ **3.** $\dfrac{\partial^2 u}{\partial x^2} = 0$ **4.** $\dfrac{\partial^2 u}{\partial x\,\partial y} = 0$ **5.** $\dfrac{\partial u}{\partial x} + \dfrac{\partial u}{\partial y} = 0$

6. $x\dfrac{\partial u}{\partial x} = 3$ **7.** $\dfrac{\partial^2 u}{\partial x\,\partial y} = y$ **8.** $\dfrac{\partial u}{\partial x} = \dfrac{\partial u}{\partial y}$ **9.** $\dfrac{\partial^4 u}{\partial x^4} - x = 0$ **10.** $\dfrac{\partial u}{\partial x} - 2\dfrac{\partial u}{\partial y} = 0$

In each of Problems 11 through 16, find a solution satisfying the given partial differential equation and the added condition.

11. $\dfrac{\partial u}{\partial x} - \dfrac{\partial u}{\partial y} = 0;\quad u(0, 1) = 2$ **12.** $\dfrac{\partial^2 u}{\partial x\,\partial y} = 1;\quad u(0, 0) = 0$ **13.** $\dfrac{\partial^2 u}{\partial x^2} = 0;\quad u(1, 1) = 1$

14. $\dfrac{\partial^2 u}{\partial y\,\partial x} = 0;\quad u(0, 2) = -3$ **15.** $2\dfrac{\partial u}{\partial x} + 3\dfrac{\partial u}{\partial y} = 0;\quad u(1, 4) = 7$ **16.** $y\dfrac{\partial u}{\partial y} = 2;\quad u(1, 2) = 4$

In each of Problems 17 through 25, determine if the given equation has the form of a linear partial differential equation. If not, explain why.

17. $\frac{\partial^2 u}{\partial x^2} + \frac{\partial^2 u}{\partial y^2} + \frac{\partial u}{\partial x}\frac{\partial u}{\partial y} = 0$

18. $x\frac{\partial^2 u}{\partial x\,\partial y} + u\frac{\partial^2 u}{\partial y^2} = 3x^2$

19. $\frac{\partial u}{\partial x} + \frac{\partial^2 u}{\partial y^2} = 2xy^2$

20. $\frac{\partial^2 u}{\partial x^2} + \left(\frac{\partial u}{\partial y}\right)^2 - 2e^{xy} = 0$

21. $\frac{\partial u}{\partial x} + 3\frac{\partial u}{\partial y} - y^2\frac{\partial^2 u}{\partial x^2} + x^2\frac{\partial^2 u}{\partial y\,\partial x} = 4x\cos(xy)$

22. $8x\frac{\partial u}{\partial y} - \frac{\partial u}{\partial x}\frac{\partial u}{\partial y} = 4y^2$

23. $2xy\frac{\partial^2 u}{\partial y^2} - 4\frac{\partial u}{\partial x} = \left(\frac{\partial u}{\partial y}\right)^2$

24. $x^2\frac{\partial^2 u}{\partial x\,\partial y} - 4\frac{\partial u}{\partial x} = 0$

25. $\frac{\partial u}{\partial y} - 8\frac{\partial u}{\partial x} - \frac{\partial^2 u}{\partial x^2} = \cos^2(x)$

In each of Problems 26 through 31, verify that the given function is a solution of the given partial differential equation.

26. $\frac{\partial^2 u}{\partial x^2} + \frac{\partial^2 u}{\partial y^2} = 0;\quad u = e^{-y}\cos(x)$

27. $\frac{\partial^2 y}{\partial t^2} = a^2\frac{\partial^2 y}{\partial x^2};\quad y = \sin\left(\frac{n\pi x}{L}\right)\cos\left(\frac{n\pi at}{L}\right)$ for $n = 1, 2, 3, \ldots$

28. $\frac{\partial^2 u}{\partial x^2} + \frac{\partial^2 u}{\partial y^2} = 0;\quad u = \arctan\left(\frac{y}{x}\right)$ for $x > 0$, $y > 0$

29. $\frac{\partial^2 y}{\partial t^2} = a^2\frac{\partial^2 y}{\partial x^2};\quad y = \frac{1}{2}[f(x + at) + f(x - at)]$ for any twice-differentiable function of one variable

30. $\frac{\partial^2 u}{\partial t^2} = a^2\left(\frac{\partial^2 u}{\partial x^2} + \frac{\partial^2 u}{\partial y^2}\right);\quad u(x, y, t) = \sin(nx)\sin(my)\cos(\sqrt{m^2 + n^2}\,at)$ for any positive integers n and m

31. $x\frac{\partial u}{\partial x} + y\frac{\partial u}{\partial y} = 0;\quad u = f\left(\frac{y}{x}\right)$, $x \neq 0$, for any differentiable function of one variable

32. Suppose that $u_1(x, y)$ and $u_2(x, y)$ are solutions of the general second order homogeneous linear partial differential equation.
(a) Prove that $u_1(x, y) + u_2(x, y)$ is also a solution.
(b) Prove that $\alpha u_1(x, y)$ is a solution for any constant α.

33. Show that, for a nonlinear partial differential equation, a sum of solutions need not be a solution.

13.1 DERIVATION OF THE WAVE AND HEAT EQUATIONS

In this section we shall derive several partial differential equations governing wave motion and heat conduction. One can take these equations for granted and proceed to the next section, but the derivations do convey a lesson, and that is the role of boundary and initial conditions. Usually a partial differential equation is in terms of a function of time and one or more space variables. In most physically motivated problems, one has data given at some time, say, at $t = 0$. These constitute the *initial conditions*. One also often has conditions specified at different extremes of the space variables, giving *boundary conditions*.

A *boundary value problem* consists of a partial differential equation, together with boundary conditions. Often, initial conditions are also present. The remainder of this section shows how such problems arise and exhibits some of the more important ones.

THE WAVE EQUATION

Imagine that we have a flexible elastic string stretched tightly between two pegs at the same level, as shown in Figure 241. If we lift the string in some fashion and then release it to vibrate in a vertical plane, what will be the nature of the resulting motion?

To model this situation, place the x-axis along the length of the string at rest, and let the y-axis be vertical, with, say, the origin at the left peg, as in Figure 241. At any time t and horizontal coordinate x, let $y(x, t)$ be the vertical displacement of the string. Thus, the graph of $y = y(x, t)$ at any given time t shows the shape of the string at that time, as shown in Figure 242. We want to know $y(x, t)$ for $0 \leq x \leq L$ (L is the distance between the pegs) and time $t > 0$.

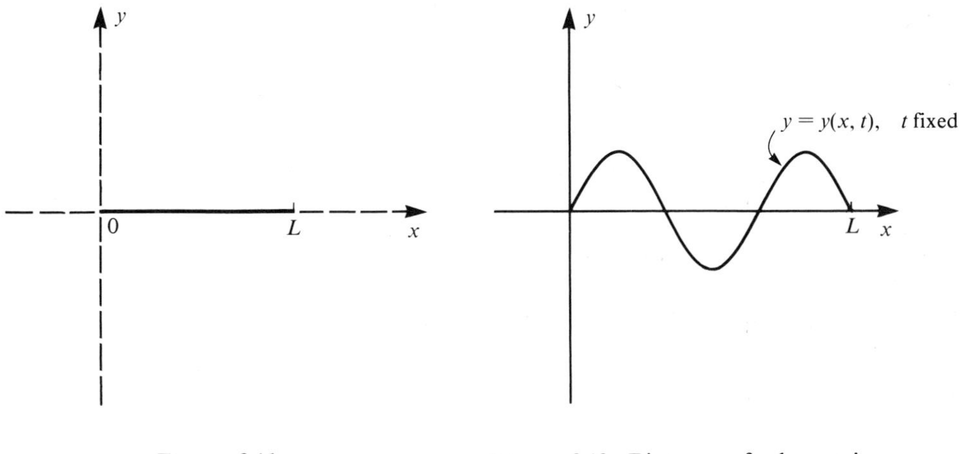

Figure 241. String stretched from 0 to L.

Figure 242. Picture of the string as a graph $y = y(x, t)$ at fixed time t.

To obtain a differential equation for y in a simple case, neglect damping forces such as air resistance and weight of the string, and assume that the tension $T(x, t)$ in the string always acts tangentially. Let ρ be the mass per unit length, and assume that ρ is constant.

Now apply Newton's law to the segment of string between x and $x + \Delta x$, as shown in Figure 243. We must have:

$$\text{net force due to tension} = (\text{segment mass}) \times (\text{segment acceleration}).$$

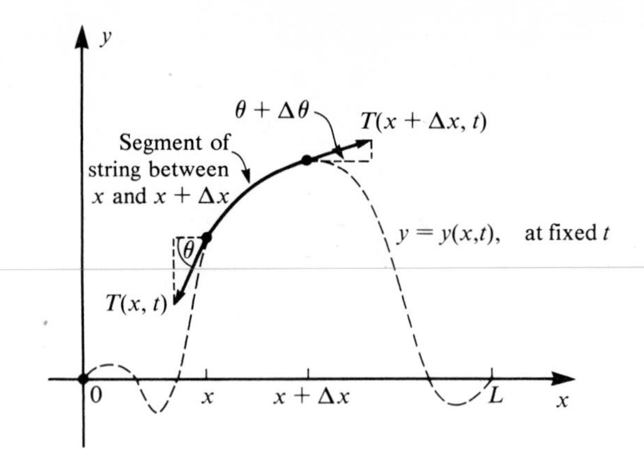

Figure 243. Forces acting on a segment of string between x and $x + \Delta x$.

For small Δx, consideration of the vertical component gives us approximately

$$T(x + \Delta x, t)\sin(\theta + \Delta\theta) - T(x, t)\sin(\theta) = \rho\,\Delta x\,\frac{\partial^2 y}{\partial t^2}(\bar{x}, t),$$

where $\bar{x}$ is the center of mass of the segment. Then,

$$\frac{T(x + \Delta x, t)\sin(\theta + \Delta\theta) - T(x, t)\sin(\theta)}{\Delta x} = \rho\,\frac{\partial^2 y}{\partial t^2}(\bar{x}, t).$$

As a convenience, write the vertical component of tension as $v(x, t)$. Then the last equation is

$$\frac{v(x + \Delta x, t) - v(x, t)}{\Delta x} = \rho\,\frac{\partial^2 y}{\partial t^2}(\bar{x}, t).$$

As $\Delta x \to 0$, we get

$$\frac{\partial v}{\partial x} = \rho\,\frac{\partial^2 y}{\partial t^2} \tag{a}$$

(note that $x < \bar{x} < x + \Delta x$; hence, $\bar{x} \to x$ as $\Delta x \to 0$).

Now look at the horizontal components in Newton's law. Because the horizontal component of the tension is zero, we get

$$T(x + \Delta x, t)\cos(\theta + \Delta\theta) - T(x, t)\cos(\theta) = 0.$$

Writing the horizontal component as $h(x, t)$, this says that

$$h(x + \Delta x, t) - h(x, t) = 0.$$

Hence, $h(x, t)$ is independent of x. But,

$$v = h\tan(\theta) = h\,\frac{\partial y}{\partial x}.$$

Substituting this for v into (a) gives us

$$\frac{\partial}{\partial x}\left(h\,\frac{\partial y}{\partial x}\right) = \rho\,\frac{\partial^2 y}{\partial t^2}$$

or

$$h\,\frac{\partial^2 y}{\partial x^2} = \rho\,\frac{\partial^2 y}{\partial t^2}.$$

This can be written in standard form

$$\frac{\partial^2 y}{\partial t^2} = a^2\,\frac{\partial^2 y}{\partial x^2},$$

where $a^2 = h/\rho$. This is the 1-dimensional wave equation (i.e., one independent space variable).

In order to uniquely determine a solution, we must know both the initial position and initial velocity of the string. Thus, we must be given initial conditions

$$y(x, 0) = f(x) \quad \text{for} \quad 0 \leq x \leq L \qquad \text{(initial position)}$$

and

$$\frac{\partial y}{\partial t}(x, 0) = g(x) \quad \text{for} \quad 0 \leq x \leq L \qquad \text{(initial velocity).}$$

Further, the data should reflect the fact that the string is fixed at both ends. Thus, we have boundary conditions

$$y(0, t) = y(L, t) = 0 \quad \text{for} \quad t \geq 0.$$

The boundary value problem for the vibrating string is then

$$\frac{\partial^2 y}{\partial t^2} = a^2\,\frac{\partial^2 y}{\partial x^2} \qquad (0 \leq x \leq L,\ t \geq 0),$$

$$y(0, t) = y(L, t) = 0 \qquad (t \geq 0),$$

$$y(x, 0) = f(x), \qquad \frac{\partial y}{\partial t}(x, 0) = g(x) \qquad (0 \leq x \leq L).$$

We shall solve this boundary value problem in the next section.

If we allowed an external force acting parallel to the y-axis, say, having magnitude F units per unit length, the wave equation would be

$$\frac{\partial^2 y}{\partial t^2} = a^2\,\frac{\partial^2 y}{\partial x^2} + \frac{F}{\rho}.$$

If F were just the weight of the string, we would have

$$\frac{\partial^2 y}{\partial t^2} = a^2\,\frac{\partial^2 y}{\partial x^2} - g.$$

If F were a damping force, say, proportional to the velocity of the string and having constant of proportionality (damping constant) α, then we would have

$$\frac{\partial^2 y}{\partial t^2} = a^2 \frac{\partial^2 y}{\partial x^2} - \alpha \frac{\partial y}{\partial t}.$$

In two dimensions (imagine that we are looking at vibrations in a stretched membrane or drum), then we would have (with no forcing)

$$\frac{\partial^2 z}{\partial t^2} = a^2 \left(\frac{\partial^2 z}{\partial x^2} + \frac{\partial^2 z}{\partial y^2} \right).$$

Here, the membrane is initially stretched over a frame in the xy-plane, and vertical displacement is measured as a function z of x, y, and t.

Similarly, we could add a forcing term to get

$$\frac{\partial^2 z}{\partial t^2} = a^2 \left(\frac{\partial^2 z}{\partial x^2} + \frac{\partial^2 z}{\partial y^2} \right) + \frac{F}{\rho}.$$

We would have to solve this subject to the given initial position and velocity, plus the boundary condition that the frame does not move.

THE HEAT EQUATION

We have already derived the heat equation under rather general conditions by means of Gauss's divergence theorem (see Section 11.11). Here, we shall derive a special case of the heat equation and examine forms that the boundary conditions can take.

Suppose we have a thin* homogeneous bar of uniform cross section placed along the x-axis from 0 to L, as shown in Figure 244. Assume that the sides of the bar are sufficiently well insulated that no heat energy is lost through them and that the bar is sufficiently thin that the temperature at any given time is constant on any given cross section perpendicular to the x-axis (although of course it may differ on different cross sections). Then, the temperature u is a function of x and t only.

To derive an equation for u, begin with the experimentally observed fact (called *Newton's law of cooling*) that the amount of heat energy per unit time passing between two parallel plates of area A, distance d apart, and temperatures T_1 and T_2, is proportional to $A|T_1 - T_2|/d$, and flows from the warmer to the cooler plate. Let k be the constant of proportionality. Then,

$$\begin{array}{l}\text{amount of heat energy per unit time flowing} \\ \text{from the warmer to the cooler plate}\end{array} = \frac{kA|T_1 - T_2|}{d}.$$

Here, k is the coefficient of thermal conductivity and depends upon the material in the plates.

Now, by conservation of energy, the rate at which heat flows into any portion of the bar (the *flux* term) must equal the rate at which that part of the bar absorbs heat energy (the *absorption* term). We shall obtain an equation for u by calculating the flux and absorption terms and setting them equal.

* "Thin" means that the thickness is much smaller than the length.

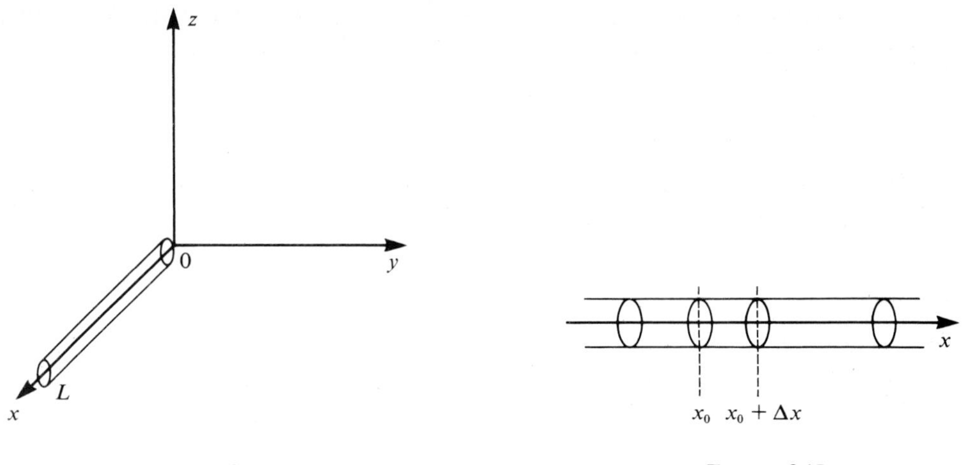

Figure 244 *Figure 245*

For the flux term, look at a portion of the bar between x_0 and $x_0 + \Delta x$, as shown in Figure 245. We imagine that Δx is very small. The instantaneous rate of energy transfer from left to right across the section at x_0 at time t is, from Newton's law of cooling,

$$R(x_0, t) = -\lim_{d \to 0} \frac{kA[u(x_0 + \frac{1}{2}d, t) - u(x_0 - \frac{1}{2}d, t)]}{d}.$$

The minus sign in front of the limit is due to the fact that energy flows from left to right exactly when the temperature left of x_0 is greater than that to the right of x_0.

Similarly, the rate of energy transfer from left to right at $x_0 + \Delta x$ and time t is

$$R(x_0 + \Delta x, t) = -\lim_{d \to 0} \frac{kA[u(x_0 + \Delta x + \frac{1}{2}d, t) - u(x_0 + \Delta x - \frac{1}{2}d, t)]}{d}.$$

We recognize these limits as

$$R(x_0, t) = -kA\frac{\partial u}{\partial x}(x_0, t)$$

and

$$R(x_0 + \Delta x, t) = -kA\frac{\partial u}{\partial x}(x_0 + \Delta x, t),$$

assuming that both k and the cross-sectional area A are constant.

The net rate F at which heat energy flows into the portion between x_0 and $x_0 + \Delta x$ is then

$$F = R(x_0, t) - R(x_0 + \Delta x, t) = kA\left(\frac{\partial u}{\partial x}(x_0 + \Delta x, t) - \frac{\partial u}{\partial x}(x_0, t)\right).$$

The amount of heat entering this portion of the bar in time Δt is then $F\ \Delta t$, or

$$kA\left(\frac{\partial u}{\partial x}(x_0 + \Delta x, t) - \frac{\partial u}{\partial x}(x_0, t)\right)\Delta t.$$

This is the flux term.

For the absorption term, the average change Δu in temperature over the time Δt is directly proportional to the flux $F\ \Delta t$ and inversely proportional to the mass Δm. Thus, for some constant s (called the *specific heat* of the bar),

$$\Delta u = \frac{F\ \Delta t}{s\ \Delta m}.$$

Since $\Delta m = \rho A\ \Delta x$, with ρ the density (assumed constant), then

$$\Delta u = \frac{F\ \Delta t}{s\rho A\ \Delta x}.$$

Now, the average change Δu is equal to the actual temperature change at some point $\bar{x}$ between x_0 and $x_0 + \Delta x$. Then,

$$\Delta u = u(\bar{x}, t + \Delta t) - u(\bar{x}, t) = \frac{F\ \Delta t}{s\rho A\ \Delta x}.$$

Then,

$$F\ \Delta t = s\rho A[u(\bar{x}, t + \Delta t) - u(\bar{x}, t)]\ \Delta x.$$

On the right in this equation is the absorption term.

Equating flux and absorption terms (both equal to $F\ \Delta t$), we have

$$kA\left(\frac{\partial u}{\partial x}(x_0 + \Delta x, t) - \frac{\partial u}{\partial x}(x_0, t)\right)\Delta t = s\rho A[u(\bar{x}, t + \Delta t) - u(\bar{x}, t)]\ \Delta x.$$

Upon dividing by $A\ \Delta x\ \Delta t$, we have

$$k\,\frac{\dfrac{\partial u}{\partial x}(x_0 + \Delta x, t) - \dfrac{\partial u}{\partial x}(x_0, t)}{\Delta x} = s\rho\left(\frac{u(\bar{x}, t + \Delta t) - u(\bar{x}, t)}{\Delta t}\right).$$

Let $\Delta x \to 0$ and $\Delta t \to 0$, noting that $\bar{x} \to x_0$ as $\Delta x \to 0$. Then,

$$k\,\frac{\partial^2 u}{\partial x^2} = s\rho\,\frac{\partial u}{\partial t}.$$

This is often written

$$\frac{\partial u}{\partial t} = a^2\,\frac{\partial^2 u}{\partial x^2},$$

with $a^2 = k/s\rho$. Sometimes a^2 is called the *thermal diffusivity* of the bar.

To determine $u(x, t)$ for all $t \geq 0$ and $0 \leq x \leq L$, we need boundary conditions (information at the ends of the bar) and initial data (temperature throughout the bar at time 0). As one example, we might have the boundary value problem

$$\frac{\partial u}{\partial t} = a^2 \frac{\partial^2 u}{\partial x^2} \qquad (0 < x < L,\ t > 0),$$

(boundary conditions) $\quad u(0, t) = u(L, t) = T_1 \quad (t > 0),$

(initial temperature) $\quad u(x, 0) = f(x) \quad (0 < x < L).$

This problem specifies that the ends of the bar are kept at constant temperature T_1 and that the temperature at time 0 at point x is $f(x)$.

Or, we might have *insulation conditions*, as in this boundary value problem

$$\frac{\partial u}{\partial t} = a^2 \frac{\partial^2 u}{\partial x^2} \qquad (0 < x < L,\ t > 0),$$

(insulated ends) $\quad \dfrac{\partial u}{\partial x}(0, t) = \dfrac{\partial u}{\partial x}(L, t) = 0 \quad (t > 0),$

(initial temperature) $\quad u(x, 0) = f(x) \quad (0 < x < L).$

Here, the boundary conditions specify no heat flow across the ends of the bar; hence the name *insulation conditions.*

In two dimensions, the heat equation is

$$\frac{\partial u}{\partial t} = a^2 \left(\frac{\partial^2 u}{\partial x^2} + \frac{\partial^2 u}{\partial y^2} \right);$$

in three dimensions it is

$$\frac{\partial u}{\partial t} = a^2 \left(\frac{\partial^2 u}{\partial x^2} + \frac{\partial^2 u}{\partial y^2} + \frac{\partial^2 u}{\partial z^2} \right).$$

Corresponding boundary and initial conditions would have to accompany these partial differential equations to determine a unique solution.

LAPLACE'S EQUATION

Another important partial differential equation is

$$\frac{\partial^2 u}{\partial x^2} + \frac{\partial^2 u}{\partial y^2} = 0 \qquad \text{(in 2 dimensions)}$$

or

$$\frac{\partial^2 u}{\partial x^2} + \frac{\partial^2 u}{\partial y^2} + \frac{\partial^2 u}{\partial z^2} = 0 \qquad \text{(in 3 dimensions).}$$

These are called *Laplace's equation.* Often, Laplace's equation is called the *steady-state heat equation* because it is exactly the two- or three-dimensional equation of heat conduction when $\partial u/\partial t = 0$ (that is, when u is constant with time). Laplace's equation occurs, however, in many other contexts. For example, in potential theory, a function satisfying Laplace's equation is called a *harmonic function.* In this context, one usually sees the vector-related notation $\nabla^2 u$ for $\partial^2 u/\partial x^2 + \partial^2 u/\partial y^2$ (or, in three dimensions, for $\partial^2 u/\partial x^2 + \partial^2 u/\partial y^2 + \partial^2 u/\partial z^2$). We read $\nabla^2 u$ as "del squared u," and ∇^2 is called the *Laplacian operator*, treated in some detail in the vector analysis chapter. In this notation, Laplace's equation becomes

$$\nabla^2 u = 0.$$

There are two types of problems which commonly arise involving Laplace's equation in two or three dimensions.

A *Dirichlet problem* consists of finding a function satisfying $\nabla^2 u = 0$ in a given region, assuming specified values $u = f$ on the boundary of the region. For example, in three dimensions, we might want to satisfy $\nabla^2 u = 0$ inside a sphere, with $u(x, y, z) = f(x, y, z)$ given on the surface of the sphere.

A *Neumann problem* consists of finding a function satisfying $\nabla^2 u = 0$ in a region, with given normal derivative $\partial u/\partial \eta = g$ on the boundary.

Dirichlet problems have unique solutions, with certain assumptions on the region and the data function f; Neumann problems have solutions uniquely specified to within an arbitrary added constant.

Finally, *Poisson's equation* has the form

$$\nabla^2 u = f,$$

where f is given. When f is identically zero, this reduces to Laplace's equation.

In many problems it is convenient to use cylindrical coordinates, spherical coordinates, or even some other kind of curvilinear coordinates. Laplace's equation in cylindrical and spherical coordinates is given in Section 11.13 and will be repeated later in this chapter for reference when we need it.

Notation Many treatments of partial differential equations indicate partial derivatives by subscripting the variable with respect to which the derivative is taken. In this notation, we would have, for example,

the one-dimensional wave equation: $\quad y_{tt} = a^2 y_{xx}$,

the one-dimensional heat equation: $\quad u_t = a^2 u_{xx}$,

and Laplace's equation:

$$u_{xx} + u_{yy} = 0 \qquad \text{(2 dimensions)}$$

or

$$u_{xx} + u_{yy} + u_{zz} = 0 \qquad \text{(3 dimensions)}.$$

The cylindrical coordinate form of $\nabla^2 u$ would be written

$$\nabla^2 u = \frac{1}{r}(ru_r)_r + \frac{1}{r^2} u_{\theta\theta} + u_{zz}.$$

Finally, the spherical coordinate form of $\nabla^2 u$ would be

$$\nabla^2 u = \frac{1}{r}(ru)_{rr} + \frac{1}{r^2 \sin^2(\phi)} u_{\theta\theta} + \frac{1}{r^2 \sin(\phi)} [u_\phi \sin(\phi)]_\phi.$$

As can be seen, the subscript notation is often easier to write and more compact in appearance. We shall use it whenever it seems convenient.

In the next section we shall begin solving boundary value problems.

PROBLEMS FOR SECTION 13.1

1. Consider the vibrating string problem:

$$\frac{\partial^2 y}{\partial t^2} = a^2 \frac{\partial^2 y}{\partial x^2} \qquad (0 < x < L,\ t > 0),$$

$$y(0, t) = y(L, t) = 0 \qquad (t > 0),$$

$$y(x, 0) = f(x), \qquad \frac{\partial y}{\partial t}(x, 0) = g(x) \qquad (0 < x < L).$$

(a) Put $x = \frac{1}{2}(\xi + \eta)$ and $t = (1/2a)(\xi - \eta)$ to change the partial differential equation to $\partial^2 Y/\partial\xi . \partial y = 0$, where $Y(\xi, \eta) = y[\frac{1}{2}(\xi + \eta), (1/2a)(\xi - \eta)]$.

(b) Show that $\partial^2 Y/\partial\xi\, \partial\eta = 0$ is satisfied by $Y(\xi, \eta) = F(\xi) + G(\eta)$, where F and G are any twice-differentiable functions of one variable.

(c) Conclude that $y(x, t) = F(x + at) + G(x - at)$ satisfies $\partial^2 y/\partial t^2 = a^2\, \partial^2 y/\partial x^2$ for any twice-differentiable functions F and G.

(d) Use the conditions $y(x, 0) = f(x)$, $(\partial y/\partial t)(x, 0) = g(x)$ to show that

$$F(\xi) = \frac{1}{2} f(\xi) + \frac{1}{2a} \int_0^{\xi} g(\alpha)\, d\alpha + k$$

and

$$G(\eta) = \frac{1}{2} f(\eta) - \frac{1}{2a} \int_0^{\eta} g(\alpha)\, d\alpha - k$$

for some constant k.

(e) Derive *d'Alembert's solution* to the vibrating string problem:

$$y(x, t) = \frac{1}{2}[f(x + at) + f(x - at)] + \frac{1}{2a} \int_{x - at}^{x + at} g(\alpha)\, d\alpha.$$

2. Write down the boundary value problem for heat conduction in a bar of length L with the left end kept at zero temperature, the right end insulated, and initial temperature throughout the bar given by $f(x)$. Assume that temperature is a function of x and t only. (Do not attempt to solve the problem yet.)

3. Write the boundary value problem for steady-state heat conduction in a thin plate placed over the rectangle $0 \le x \le \alpha$, $0 \le y \le \beta$. On the boundary, the top and bottom sides are kept at zero temperature, while the vertical sides are kept at constant temperature T.

4. Write the boundary value problem for vibration of a rectangular membrane occupying a region $0 \le x \le \alpha$, $0 \le y \le \beta$. The initial position is given as $f(x, y)$ and the initial velocity by $g(x, y)$. The membrane is fastened to a frame along the rectangle.

5. Write the boundary value problem for heat conduction in a right circular cylinder of height c, aligned along the z-axis. Use cylindrical coordinates, and assume that the temperature function depends on r, θ, z, and t. The initial temperature is $f(r, \theta, z)$, and the outer surface is kept at temperature 0.

6. Redo Problem 5, substituting the condition that the cylinder is insulated for the condition that it is kept at zero temperature.

7. Write the boundary value problem for heat conduction in a solid sphere of radius k, assuming initial temperature $f(\rho, \theta, \phi)$ (in spherical coordinates). Assume that the surface is kept at constant temperature T.
8. Redo Problem 7, with the condition that the surface of the sphere is insulated replacing the condition that the surface is kept at constant temperature.
9. Imagine an elastic string stretched between 0 and L and pegged down at the ends. It is released from rest from an initial displacement given by $y = f(x)$ and allowed to vibrate in a vertical plane. Its motion is opposed by air resistance which at each point is proportional to the square of the velocity at that point. Write a boundary value problem modeling this phenomenon.
10. Write the boundary value problem for steady-state heat conduction in a square plate placed over $0 \leq x \leq \alpha$, $0 \leq y \leq \alpha$. The vertical sides are kept at temperature T, and the top and bottom sides are insulated. The initial temperature in the plate is $f(x, y)$.
11. Show that $V(x, y, z) = (x^2 + y^2 + z^2)^{-1/2}$ satisfies Laplace's equation in any region not containing the origin.
12. Derive the damped wave equation

$$\frac{\partial^2 y}{\partial t^2} + K \frac{\partial y}{\partial t} = a^2 \frac{\partial^2 y}{\partial x^2},$$

assuming a damping force proportional to the velocity.

13. Assume, in addition to the damping force of Problem 12, a restoring force acting on the string proportional to the displacement. Show that the motion is then governed by the partial differential equation

$$\frac{\partial^2 y}{\partial t^2} + K \frac{\partial y}{\partial t} + Cy = a^2 \frac{\partial^2 y}{\partial x^2}.$$

This is called the *telegraph equation.*

14. Suppose that $Y_1(x, t)$ is a solution of

$$\frac{\partial^2 y}{\partial t^2} = a^2 \frac{\partial^2 y}{\partial x^2},$$

$$y(x, 0) = f(x), \qquad \frac{\partial y}{\partial t}(x, 0) = 0 \quad \text{for} \quad 0 < x < L,$$

$$y(0, t) = y(L, t) = 0 \quad \text{for} \quad t > 0.$$

Suppose that $Y_2(x, t)$ is a solution of

$$\frac{\partial^2 y}{\partial t^2} = a^2 \frac{\partial^2 y}{\partial x^2},$$

$$y(x, 0) = 0, \qquad \frac{\partial y}{\partial t}(x, 0) = g(x) \quad \text{for} \quad 0 < x < L,$$

$$y(0, t) = y(L, t) = 0 \quad \text{for} \quad t > 0.$$

Show that $Y_1(x, t) + Y_2(x, t)$ is a solution of

$$\frac{\partial^2 y}{\partial t^2} = a^2 \frac{\partial^2 y}{\partial x^2},$$

$$y(x, 0) = f(x), \qquad \frac{\partial y}{\partial t}(x, 0) = g(x), \qquad 0 < x < L,$$

$$y(0, t) = y(L, t) = 0, \qquad t > 0.$$

15. Show that

$$y(x, t) = \sin(x)\cos(at) + \frac{1}{a}\cos(x)\sin(at)$$

is a solution of the problem

$$\frac{\partial^2 y}{\partial t^2} = a^2 \frac{\partial^2 y}{\partial x^2},$$

$$y(x, 0) = \sin(x), \qquad \frac{\partial y}{\partial t}(x, 0) = \cos(x).$$

13.2 FOURIER SERIES SOLUTIONS OF THE WAVE EQUATION

In this section we shall use a classical method called *separation of variables* (or the *Fourier method*) to solve boundary value problems associated with the motion of a vibrating string. We shall consider several combinations of initial conditions, with a view toward teaching the method by illustration.

A. INITIALLY DISPLACED VIBRATING STRING WITH ZERO INITIAL VELOCITY

Consider the problem

$$\frac{\partial^2 y}{\partial t^2} = a^2 \frac{\partial^2 y}{\partial x^2} \qquad (0 < x < L,\ t > 0),$$

$$y(0, t) = y(L, t) = 0 \qquad (t > 0),$$

$$y(x, 0) = f(x) \quad \text{and} \quad \frac{\partial y}{\partial t}(x, 0) = 0 \qquad (0 < x < L).$$

This models vibration in an elastic string of length L pegged at the ends, picked up at time 0 to assume the configuration of $y = f(x)$, and released from rest (zero initial velocity).

Separation of variables consists of attempting a solution of the form

$$y(x, t) = X(x)T(t),$$

and determining $X(x)$ and $T(t)$ so that $y(x, t)$ satisfies the required conditions.

Substituting into the differential equation gives us

$$XT'' = a^2 X''T$$

or

$$\frac{X''}{X} = \frac{T''}{a^2 T}.$$

The left side depends only on x, the right side only on t, and x and t are independent. Thus, both sides must equal the same constant, say, λ:

$$\frac{X''}{X} = \frac{T''}{a^2 T} = \lambda.$$

(Often λ is called a *separation constant.**) Then,

$$X'' - \lambda X = 0 \quad \text{and} \quad T'' - \lambda a^2 T = 0.$$

Thus, we have separated the variables, cashing in one partial differential equation for two ordinary differential equations (and a new unknown, λ).

From the boundary conditions,

$$y(0, t) = X(0)T(t) = 0 \qquad (t > 0).$$

One way this can happen is that $T(t) = 0$ for all $t > 0$. Then, $y(x, t) = X(x)T(t) = 0$ for all x, $0 < x < L$, and all $t > 0$. This trivial solution is consistent with $y(x, 0) = f(x)$ only if $f(x) = 0$ for $0 \le x \le L$, in which case the string has not been displaced and, assuming zero initial velocity, never moves at all. This is physically realistic but not very interesting. Thus, we shall go to the case that the string has been initially displaced and that $T(t)$ is not identically zero. In this event, $X(0)T(t) = 0$ for all $t > 0$ implies that $X(0) = 0$. Similarly, $y(L, t) = X(L)T(t) = 0$ for all $t > 0$ now implies that $X(L) = 0$.

Concentrating on X for the moment, we have to determine values of λ for which there are functions X satisfying

$$X'' - \lambda X = 0 \qquad (0 < x < L),$$
$$X(0) = 0, \qquad X(L) = 0.$$

(This is a *Sturm-Liouville* or *eigenvalue* problem, discussed in Chapter 6, Section 4.)

From the setting of the problem, we expect λ to be real, not complex. Thus, we have three cases to consider.*

Case 1 $\lambda = 0$. Then $X'' = 0$; so $X = Ax + B$. But $X(0) = B = 0$. Further, $X(L) = AL = 0$; so $A = 0$ also. Then we get the trivial solution $X(x) = 0$, giving us the trivial solution $y(x, t) = 0$ for y. Assuming that the string is initially displaced and that motion occurs, we discard this solution. Thus, $\lambda = 0$ is not allowed.

Case 2 $\lambda > 0$. Say $\lambda = \alpha^2$, where $\alpha > 0$. Then $X'' - \alpha^2 X = 0$, with general solution

$$X(x) = Ae^{\alpha x} + Be^{-\alpha x}.$$

Now, $X(0) = A + B = 0$; so $A = -B$. Then,

$$X(L) = A(e^{\alpha L} - e^{-\alpha L}) = 0.$$

This can happen only if $A = 0$, again giving us $X(x) = 0$ and $y(x, t) = 0$. Thus, we discard this case, and there are no positive values of λ for this problem.

---

* For those familiar with Section 6.4, admissible values of λ are eigenvalues of a Sturm-Liouville problem.

* In Sturm-Liouville theory, it is shown that the λ's must be real under conditions such as prevail here. In most physically motivated problems, we expect λ to be real because the solution should be real-valued.

Case 3 $\lambda < 0$, say, $\lambda = -\alpha^2$ with $\alpha > 0$. Then, $X'' + \alpha^2 X = 0$, with general solution

$$X = A\cos(\alpha x) + B\sin(\alpha x).$$

Now, $X(0) = A = 0$ and $X(L) = B\sin(\alpha L) = 0$. If $B = 0$, we again have only the trivial solution. Thus, we are forced to consider

$$\sin(\alpha L) = 0.$$

Then αL must be an integer multiple of π. Since $\alpha > 0$, then $\alpha L = n\pi$, where $n = 1, 2, 3, \ldots$. Then,

$$\lambda = -\alpha^2 = -\frac{n^2\pi^2}{L^2}.$$

Thus, corresponding to each positive integer n, there is a value $\lambda_n = -n^2\pi^2/L^2$ for which we get a solution

$$X_n(x) = B_n \sin\left(\frac{n\pi x}{L}\right).$$

We use subscripts on λ_n, X_n, and the arbitrary constant B_n to indicate that we have a different solution for λ and X for each positive integer n. The λ_n's are called *eigenvalues* of the boundary value problem; the corresponding functions $B_n \sin(n\pi x/L)$ are *eigenfunctions.*

Now go back and solve for T. We had

$$T'' - \lambda a^2 T = 0.$$

Since λ can be any number $-n^2\pi^2/L^2$, we have

$$T'' + \frac{n^2\pi^2 a^2}{L^2} T = 0 \quad \text{for} \quad n = 1, 2, 3, \ldots.$$

This has general solution

$$T = A\cos\left(\frac{n\pi at}{L}\right) + B\sin\left(\frac{n\pi at}{L}\right).$$

Now recall that $(\partial y/\partial t)(x, 0) = 0$. Then, $X(x)T'(0) = 0$, forcing $T'(0) = 0$. Hence, $B = 0$ and, for $n = 1, 2, 3, \ldots$, we get

$$T_n(t) = A_n \cos\left(\frac{n\pi at}{L}\right).$$

Thus, for $n = 1, 2, 3, \ldots$, we have a function

$$y_n(x, t) = X_n(x)T_n(t) = C_n \sin\left(\frac{n\pi x}{L}\right)\cos\left(\frac{n\pi at}{L}\right),$$

where $C_n = B_n A_n$ is an arbitrary constant. For $n = 1, 2, 3, \ldots$, $y_n(x, t)$ satisfies the partial differential equation *and* the boundary conditions $y(0, t) = y(L, t) = 0$ *and one* of the initial conditions, $(\partial y/\partial t)(x, 0) = 0$. For any given n, $y_n(x, t)$ *may* or *may not* satisfy $y(x, 0) = f(x)$. In fact, we will have $y_n(x, 0) = C_n \sin(n\pi x/L) = f(x)$ for $0 \leq x \leq L$ only if $f(x)$ happens to be a constant multiple of $\sin(n\pi x/L)$, which need not be the case.

How do we satisfy $y(x, 0) = f(x)$ if $f(x)$ is not a multiple of $\sin(n\pi x/L)$? The key lies in recalling that many functions we encounter in practice can be expanded in a Fourier sine series on $[0, L]$. Thus, we may not be able to choose a *particular* C_n so that $C_n \sin(n\pi x/L) = f(x)$, but we may be able to choose *all* of $C_1, C_2, C_3, \ldots$ so that

$$\sum_{n=1}^{\infty} C_n \sin\left(\frac{n\pi x}{L}\right) = f(x).$$

This suggests that we try as $y(x, t)$ a sum, or *superposition*, of *all* the $y_n(x, t)$'s. That is, try a solution

$$y(x, t) = \sum_{n=1}^{\infty} C_n \sin\left(\frac{n\pi x}{L}\right)\cos\left(\frac{n\pi at}{L}\right).$$

This will still satisfy the partial differential equation and also the conditions $y(0, t) = y(L, t) = 0$ and $(\partial y/\partial t)(x, 0) = 0$. All we have to do to satisfy the fourth condition $y(x, 0) = f(x)$ is choose the C_n's so that

$$y(x, 0) = \sum_{n=1}^{\infty} C_n \sin\left(\frac{n\pi x}{L}\right) = f(x).$$

This is a Fourier sine series for $f(x)$ on $[0, L]$. If $f(x)$ is sectionally continuous* on $[0, L]$, then the work of the previous chapter tells us to choose

$$C_n = \frac{2}{L}\int_0^L f(x) \sin\left(\frac{n\pi x}{L}\right) dx.$$

We then satisfy the final condition $y(x, 0) = f(x)$ for $0 < x < L$.

Thus, the boundary value problem has solution

$$y(x, t) = \sum_{n=1}^{\infty} \left[\frac{2}{L}\int_0^L f(\xi) \sin\left(\frac{n\pi\xi}{L}\right) d\xi\right] \sin\left(\frac{n\pi x}{L}\right)\cos\left(\frac{n\pi at}{L}\right).$$

(Here, we used ξ as dummy integration variable to avoid confusion with x.)

As a specific example, suppose that

$$f(x) = \begin{cases} x, & 0 \leq x < L/2, \\ L - x, & L/2 \leq x \leq L, \end{cases}$$

as shown in Figure 246. That is, the string is picked up $L/2$ units at the center point and then released. Then,

$$C_n = \frac{2}{L}\int_0^{L/2} \xi \sin\left(\frac{n\pi\xi}{L}\right) d\xi + \frac{2}{L}\int_{L/2}^{L} (L - \xi) \sin\left(\frac{n\pi\xi}{L}\right) d\xi = \frac{4L}{n^2\pi^2} \sin\left(\frac{n\pi}{2}\right).$$

* In the real world, most functions describing an initial position of a stretched string will be sectionally continuous at the very least.

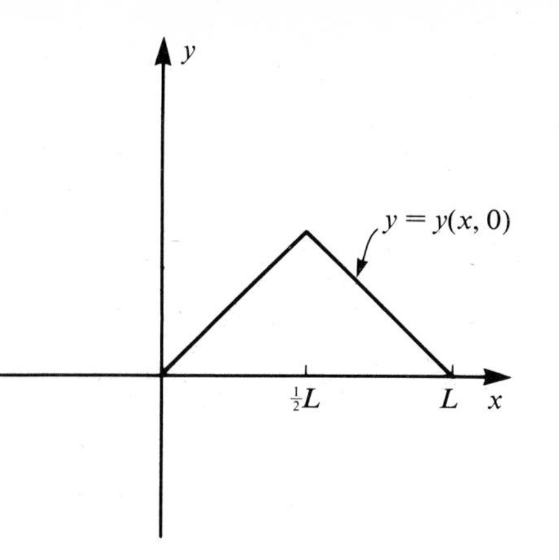

Figure 246. Initial position of string picked up $L/2$ units at its center point.

Thus, in this case, the string's motion is described by

$$y(x, t) = \sum_{n=1}^{\infty} \frac{4L}{n^2\pi^2} \sin\left(\frac{n\pi}{2}\right) \sin\left(\frac{n\pi x}{L}\right) \cos\left(\frac{n\pi at}{L}\right)$$

$$= \frac{4L}{\pi^2} \sum_{n=1}^{\infty} \frac{(-1)^{n+1}}{(2n-1)^2} \sin\left(\frac{(2n-1)\pi x}{L}\right) \cos\left(\frac{(2n-1)\pi at}{L}\right).$$

B. WAVE EQUATION WITH ZERO INITIAL DISPLACEMENT

Now consider the problem

$$\frac{\partial^2 y}{\partial t^2} = a^2 \frac{\partial^2 y}{\partial x^2} \qquad (0 < x < L,\ t > 0),$$

$$y(0, t) = y(L, t) = 0 \qquad (t > 0),$$

(zero initial displacement) $\quad y(x, 0) = 0 \qquad (0 < x < L),$

(given initial velocity) $\quad \dfrac{\partial y}{\partial t}(x, 0) = g(x) \qquad (0 < x < L).$

As in A, try $y(x, t) = X(x)T(t)$. Arguing as in A, we obtain

$$X'' - \lambda X = 0 \qquad T'' - \lambda a^2 T = 0$$

$$X(0) = X(L) = 0 \qquad T(0) = 0$$

Observe that the problem for X is exactly the same as that in problem A considered above. The difference between problem A and the current problem shows up only in $T(t)$. Thus, from A, we get eigenvalues

$$\lambda_n = -\frac{n^2\pi^2}{L^2}$$

and eigenfunctions

$$X_n(x) = B_n \sin\left(\frac{n\pi x}{L}\right), \qquad n = 1, 2, 3, \ldots.$$

Now solve for T. For $n = 1, 2, 3, \ldots$, the general solution of

$$T'' + \frac{n^2\pi^2a^2}{L^2} T = 0$$

is

$$T(t) = C \cos\left(\frac{n\pi at}{L}\right) + D \sin\left(\frac{n\pi at}{L}\right).$$

Since $T(0) = 0$, then $C = 0$. Hence, for $n = 1, 2, 3, \ldots$, we have

$$T_n(x, t) = D_n \sin\left(\frac{n\pi at}{L}\right).$$

Thus,

$$y_n(x, t) = D_n \sin\left(\frac{n\pi x}{L}\right) \sin\left(\frac{n\pi at}{L}\right).$$

For each n, this satisfies the wave equation and the conditions $y(0, t) = y(L, t) = 0$, $y(x, 0) = 0$. To satisfy $(\partial y/\partial t)(x, 0) = g(x)$, try a superposition

$$y(x, t) = \sum_{n=1}^{\infty} D_n \sin\left(\frac{n\pi x}{L}\right)\sin\left(\frac{n\pi at}{L}\right).$$

We need

$$\frac{\partial y}{\partial t}(x, 0) = g(x) = \sum_{n=1}^{\infty} \frac{n\pi a D_n}{L} \sin\left(\frac{n\pi x}{L}\right),$$

(obtained by differentiating the series term by term and putting $t = 0$).

This is a Fourier sine series for $g(x)$ on $[0, L]$. We must choose

$$\frac{n\pi a D_n}{L} = \frac{2}{L}\int_0^L g(x) \sin\left(\frac{n\pi x}{L}\right) dx.$$

Thus,

$$D_n = \frac{2}{n\pi a}\int_0^L g(x) \sin\left(\frac{n\pi x}{L}\right) dx.$$

The solution is then

$$y(x, t) = \sum_{n=1}^{\infty} \left[\frac{2}{n\pi a}\int_0^L g(\xi) \sin\left(\frac{n\pi \xi}{L}\right) d\xi\right] \sin\left(\frac{n\pi x}{L}\right)\sin\left(\frac{n\pi at}{L}\right).$$

For example, if $g(x) = 1$ for $0 < x < 1$, then we get

$$y(x, t) = \sum_{n=1}^{\infty} \frac{-2L}{n^2\pi^2 a}[(-1)^n - 1] \sin\left(\frac{n\pi x}{L}\right)\sin\left(\frac{n\pi at}{L}\right)$$

$$= \sum_{n=1}^{\infty} \frac{4L}{(2n-1)^2\pi^2 a} \sin\left(\frac{(2n-1)\pi x}{L}\right)\sin\left(\frac{(2n-1)\pi at}{L}\right),$$

since

$$(-1)^n - 1 = \begin{cases} 0 & \text{if } n \text{ is even,} \\ -2 & \text{if } n \text{ is odd.} \end{cases}$$

C. WAVE EQUATION WITH INITIAL VELOCITY AND DISPLACEMENT

Both A and B above are special cases of

$$\frac{\partial^2 y}{\partial t^2} = a^2 \frac{\partial^2 y}{\partial x^2} \qquad (0 < x < L,\ t > 0)$$

$$y(0, t) = y(L, t) = 0 \qquad (t > 0)$$

$$y(x, 0) = f(x), \quad \frac{\partial y}{\partial t}(x, 0) = g(x) \qquad (0 < x < L).$$

As an exercise, the student should try to solve this problem directly by separation of variables. However, we can also proceed by splitting this problem into two we have already solved.

Problem I	*Problem II*
$\frac{\partial^2 y}{\partial t^2} = a^2 \frac{\partial^2 y}{\partial x^2}$	$\frac{\partial^2 y}{\partial t^2} = a^2 \frac{\partial^2 y}{\partial x^2}$
$y(0, t) = y(L, t) = 0$	$y(0, t) = y(L, t) = 0$
$y(x, 0) = f(x)$	$y(x, 0) = 0$
$\frac{\partial y}{\partial t}(x, 0) = 0$	$\frac{\partial y}{\partial t}(x, 0) = g(x)$

Solve Problem I as in A to get a solution y_I. Solve Problem II as in B to get y_{II}. The reader can now check by direct substitution that $y = y_I + y_{II}$ solves the given problem.

For example, suppose the problem is

$$\frac{\partial^2 y}{\partial t^2} = a^2 \frac{\partial^2 y}{\partial x^2} \qquad (0 < x < L,\ t > 0),$$

$$y(0, t) = y(L, t) = 0 \qquad (t > 0),$$

$$y(x, 0) = \begin{cases} x, & 0 \le x \le L/2, \\ L - x, & L/2 < x \le L, \end{cases}$$

$$\frac{\partial y}{\partial t}(x, 0) = 1 \qquad (0 < x < L).$$

From A and B above, the solution is

$$y(x, t) = \frac{4L}{\pi^2} \sum_{n=1}^{\infty} \frac{(-1)^{n+1}}{(2n-1)^2} \sin\left(\frac{(2n-1)\pi x}{L}\right) \cos\left(\frac{(2n-1)\pi at}{L}\right)$$

$$+ \frac{4L}{\pi^2 a} \sum_{n=1}^{\infty} \frac{1}{(2n-1)^2} \sin\left(\frac{(2n-1)\pi x}{L}\right) \sin\left(\frac{(2n-1)\pi at}{L}\right)$$

$$= \frac{4L}{\pi^2} \sum_{n=1}^{\infty} \frac{1}{(2n-1)^2} \sin\left(\frac{(2n-1)\pi x}{L}\right)$$

$$\times \left[(-1)^{n+1} \cos\left(\frac{(2n-1)\pi at}{L}\right) + \frac{1}{a} \sin\left(\frac{(2n-1)\pi at}{L}\right)\right].$$

In the next section we shall use the method of separation of variables to solve some problems in heat conduction.

PROBLEMS FOR SECTION 13.2

In each of Problems 1 through 15, solve the boundary value problem.

1. $\dfrac{\partial^2 y}{\partial t^2} = a^2 \dfrac{\partial^2 y}{\partial x^2}$ $\quad (0 < x < 2,\ t > 0)$

   $y(0, t) = y(2, t) = 0$ $\quad (t > 0)$

   $y(x, 0) = 0, \quad \dfrac{\partial y}{\partial t}(x, 0) = 2x$ $\quad (0 < x < 2)$

2. $\dfrac{\partial^2 y}{\partial t^2} = 3 \dfrac{\partial^2 y}{\partial x^2}$ $\quad (0 < x < 4,\ t > 0)$

   $y(0, t) = y(4, t) = 0$ $\quad (t > 0)$

   $y(x, 0) = 2\sin(\pi x), \quad \dfrac{\partial y}{\partial t}(x, 0) = 0$ $\quad (0 < x < 4)$

3. $\dfrac{\partial^2 y}{\partial t^2} = 4 \dfrac{\partial^2 y}{\partial x^2}$ $\quad (0 < x < 3,\ t > 0)$

   $y(0, t) = y(3, t) = 0$ $\quad (t > 0)$

   $y(x, 0) = 0, \quad \dfrac{\partial y}{\partial t}(x, 0) = x$ $\quad (0 < x < 3)$

4. $\dfrac{\partial^2 y}{\partial t^2} = 9 \dfrac{\partial^2 y}{\partial x^2}$ $\quad (0 < x < \pi,\ t > 0)$

   $y(0, t) = y(\pi, t) = 0$ $\quad (t > 0)$

   $y(x, 0) = \sin(x), \quad \dfrac{\partial y}{\partial t}(x, 0) = 1$ $\quad (0 < x < \pi)$

5. $\dfrac{\partial^2 y}{\partial t^2} = 8\,\dfrac{\partial^2 y}{\partial x^2}$ $\quad (0 < x < 2\pi,\ t > 0)$

$y(0, t) = y(2\pi, t) = 0$ $\quad (t > 0)$

$\dfrac{\partial y}{\partial t}(x, 0) = 2, \quad y(x, 0) = 0$ $\quad (0 < x < 2\pi)$

6. $\dfrac{\partial^2 y}{\partial t^2} = 4\,\dfrac{\partial^2 y}{\partial x^2}$ $\quad (0 < x < 5,\ t > 0)$

$y(0, t) = y(5, t) = 0$ $\quad (t > 0)$

$$\frac{\partial y}{\partial t}(x, 0) = \begin{cases} x, & 0 < x < \frac{5}{2} \\ 5 - x, & \frac{5}{2} < x < 5 \end{cases}$$

$y(x, 0) = 0$ $\quad (t > 0)$

7. $\dfrac{\partial^2 y}{\partial t^2} = 9\,\dfrac{\partial^2 y}{\partial x^2}$ $\quad (0 < x < 2,\ t > 0)$

$y(0, t) = y(2, t) = 0$ $\quad (t > 0)$

$y(x, 0) = x(2 - x), \quad \dfrac{\partial y}{\partial t}(x, 0) = 4$ $\quad (0 < x < 2)$

8. $\dfrac{\partial^2 y}{\partial t^2} = 16\,\dfrac{\partial^2 y}{\partial x^2}$ $\quad (0 < x < 5,\ t > 0)$

$y(0, t) = y(5, t) = 0$ $\quad (t > 0)$

$y(x, 0) = x(x - 5), \quad \dfrac{\partial y}{\partial t}(x, 0) = 0$ $\quad (0 < x < 5)$

9. $\dfrac{\partial^2 y}{\partial t^2} = 8\,\dfrac{\partial^2 y}{\partial x^2}$ $\quad (0 < x < 4,\ t > 0)$

$y(0, t) = y(4, t) = 0$ $\quad (t > 0)$

$y(x, 0) = x^2(x - 4), \quad \dfrac{\partial y}{\partial t}(x, 0) = 1$ $\quad (0 < x < 4)$

10. $\dfrac{\partial^2 y}{\partial t^2} = 14\,\dfrac{\partial^2 y}{\partial x^2}$ $\quad (0 < x < 6,\ t > 0)$

$y(0, t) = y(6, t) = 0$ $\quad (t > 0)$

$y(x, 0) = 0, \quad \dfrac{\partial y}{\partial t}(x, 0) = e^x$ $\quad (0 < x < 6)$

11. $\dfrac{\partial^2 y}{\partial t^2} = \dfrac{\partial^2 y}{\partial x^2}$ $\quad (0 < x < \pi,\ t > 0)$

$y(0, t) = y(\pi, t) = 0$ $\quad (t > 0)$

$y(x, 0) = \cos(x)\sin(x), \quad \dfrac{\partial y}{\partial t}(x, 0) = \pi - x$ $\quad (0 < x < \pi)$

12. $\dfrac{\partial^2 y}{\partial t^2} = 4\dfrac{\partial^2 y}{\partial x^2}$ $\quad (0 < x < L, t > 0)$

$y(0, t) = y(L, t) = 0 \quad (t > 0)$

$$y(x, 0) = \begin{cases} \dfrac{3}{L}x, & 0 < x \le L/3 \\ 1, & L/3 < x \le 2L/3 \\ 1 - \dfrac{1}{L}(3x - 2L), & 2L/3 < x < L \end{cases}$$

$\dfrac{\partial y}{\partial t}(x, 0) = 0 \quad (0 < x < L)$

13. $\dfrac{\partial^2 y}{\partial t^2} = 3\dfrac{\partial^2 y}{\partial x^2} + 2x \quad (0 < x < 2, t > 0)$

$y(0, t) = y(2, t) = 0 \quad (t > 0)$

$y(x, 0) = \dfrac{\partial y}{\partial t}(x, 0) = 0 \quad (0 < x < 2)$

(*Hint:* Write $Y(x, t) = y(x, t) + f(x)$, substitute into the differential equation, and choose $f(x)$ so as to simplify things. Then solve for $Y(x, t)$ to get $y(x, t)$.)

14. $\dfrac{\partial^2 y}{\partial t^2} = 9\dfrac{\partial^2 y}{\partial x^2} + x^2 \quad (0 < x < 4, t > 0)$

$y(0, t) = y(4, t) = 0 \quad (t > 0)$

$y(x, 0) = \dfrac{\partial y}{\partial t}(x, 0) = 0 \quad (0 < x < 4)$

(*Hint:* Read the hint to Problem 13.)

15. $\dfrac{\partial^2 y}{\partial t^2} = \dfrac{\partial^2 y}{\partial x^2} - \cos(x) \quad (0 < x < 2\pi, t > 0)$

$y(0, t) = y(2\pi, t) = 0 \quad (t > 0)$

$y(x, 0) = \dfrac{\partial y}{\partial t}(x, 0) = 0 \quad (0 < x < 2\pi)$

16. Consider the problem

$$\frac{\partial^2 y}{\partial t^2} = a^2\frac{\partial^2 y}{\partial x^2} \quad (0 < x < L, t > 0),$$

$$y(0, t) = y(L, t) = 0 \quad (t > 0),$$

$$y(x, 0) = f(x), \quad \frac{\partial y}{\partial t}(x, 0) = 0 \quad (0 < x < L).$$

Previously, we derived the solution

$$y(x, t) = \sum_{n=1}^{\infty} C_n \sin\left(\frac{n\pi x}{L}\right) \cos\left(\frac{n\pi a t}{L}\right),$$

where $C_n = (2/L) \int_0^L f(\xi) \sin(n\pi\xi/L)\, d\xi$. Using the trigonometric identity

$$2 \sin\left(\frac{n\pi x}{L}\right) \cos\left(\frac{n\pi a t}{L}\right) = \sin\left(\frac{n\pi(x - at)}{L}\right) + \sin\left(\frac{n\pi(x + at)}{L}\right),$$

show that $y(x, t)$ can be written

$$y(x, t) = \tfrac{1}{2}[f(x + at) + f(x - at)],$$

under the assumption that the Fourier sine series for $f(x)$ converges to $f(x)$ for $0 \le x \le L$. (Note d'Alembert's solution of the wave equation, Problem 1, Section 13.1.)

17. Solve the vibrating string problem in which we allow a damping force proportional to the velocity. With zero initial velocity and initial position $y = f(x)$, the boundary value problem is

$$\frac{\partial^2 y}{\partial t^2} = a^2 \frac{\partial^2 y}{\partial x^2} - \alpha \frac{\partial y}{\partial t} \qquad (0 < x < L,\ t > 0),$$

$$y(0, t) = y(L, t) = 0 \qquad (t > 0),$$

$$y(x, 0) = f(x), \quad \frac{\partial y}{\partial t}(x, 0) = 0 \qquad (0 < x < L).$$

Assume that $\alpha < 2\pi a/L$.

18. Solve the vibrating string problem in which damping is proportional to velocity (as in Problem 17) but with initial position zero and initial velocity $g(x)$.

19. Consider the vibrating string problem when we allow for the weight of the string (but no damping). Assuming zero initial velocity, the boundary value problem is

$$\frac{\partial^2 y}{\partial t^2} = a^2 \frac{\partial^2 y}{\partial x^2} - g \qquad (0 < x < L,\ t > 0),$$

$$y(0, t) = y(L, t) = 0 \qquad (t > 0),$$

$$y(x, 0) = f(x), \quad \frac{\partial y}{\partial t}(x, 0) = 0 \qquad (0 < x < L).$$

(a) Try a separation of variables $y = X(x)T(t)$. What happens?

(b) Let $Y(x, t) = y(x, t) + h(x)$. Substitute $y = Y - h$ into the wave equation and determine a choice of h which results in a partial differential equation for Y which is separable.

(c) Rewrite the above boundary value problem as a new one in terms of $Y(x, t)$. Solve this problem for Y, and hence obtain the solution y of the original boundary value problem.

20. Try extending d'Alembert's solution of the one-dimensional wave equation (see Problem 16 above) to the two-dimensional problem of a vibrating membrane on a fixed rectangular frame:

$$\frac{\partial^2 z}{\partial t^2} = a^2\left(\frac{\partial^2 z}{\partial x^2} + \frac{\partial^2 z}{\partial y^2}\right) \qquad (0 < x < \alpha,\ 0 < y < \beta,\ t > 0)$$

$$z(x, 0) = z(x, \beta) = 0 \qquad (0 \le x \le \alpha)$$

$$z(0, y) = z(\alpha, y) = 0 \qquad (0 \le y \le \beta)$$

$$z(x, y, 0) = f(x, y) \qquad (0 < x < \alpha,\ 0 < y < \beta)$$

$$\frac{\partial z}{\partial t}(x, y, 0) = g(x, y) \qquad (0 < x < \alpha,\ 0 < y < \beta).$$

21. Solve the boundary value problem for the longitudinal displacements in a cylindrical bar of length L if at time 0 the bar is stretched by an amount AL (A constant) and then both ends are released and kept free. The problem is

$$\frac{\partial^2 y}{\partial t^2} = a^2 \frac{\partial^2 y}{\partial x^2} \qquad (0 < x < L,\ t > 0),$$

$$\frac{\partial y}{\partial x}(0, t) = \frac{\partial y}{\partial x}(L, t) = 0 \qquad (t > 0),$$

$$y(x, 0) = Ax, \quad \frac{\partial y}{\partial t}(x, 0) = 0 \qquad (0 < x < L).$$

(In this problem, $a^2 = E/\delta$, where E = modulus of elasticity of the material of the bar and δ = mass per unit volume.)

22. Transverse vibrations in a homogeneous rod are governed by

$$a^2 \frac{\partial^4 u}{\partial x^4} + \frac{\partial^2 u}{\partial t^2} = 0 \qquad (0 < x < \pi,\ t > 0).$$

In this, $u(x, t)$ is the displacement at time t of the cross section through x; a^2 is $EI/\rho A$, where E is Young's modulus, I is the moment of inertia of a cross section with respect to the x-axis, ρ is the density, and A is the cross-sectional area. Both ρ and A are assumed constant.

(a) Try a separation of variables $u(x, t) = X(x)T(t)$ and obtain differential equations for X and T.

(b) Solve for the eigenvalues and eigenfunctions of the problem in the case of free ends:

$$\frac{\partial^2 u}{\partial x^2}(0, t) = \frac{\partial^2 u}{\partial x^2}(\pi, t) = \frac{\partial^3 u}{\partial x^3}(0, t) = \frac{\partial^3 u}{\partial x^3}(\pi, t) = 0.$$

(c) Solve for the eigenvalues and eigenfunctions in the case of supported ends:

$$u(0, t) = u(\pi, t) = \frac{\partial^2 u}{\partial x^2}(0, t) = \frac{\partial^2 u}{\partial x^2}(\pi, t) = 0.$$

23. Solve the telegraph equation

$$\frac{\partial^2 u}{\partial t^2} + A\frac{\partial u}{\partial t} + Bu = a^2 \frac{\partial^2 u}{\partial x^2} \qquad (0 < x < L,\ t > 0),$$

with A and B positive constants, subject to

$$u(0, t) = u(L, t) = 0, \qquad u(x, 0) = f(x), \qquad \frac{\partial u}{\partial t}(x, 0) = 0.$$

Assume that $L^2 A^2 < 4(L^2 B + a^2\pi^2)$.

24. Consider torsional oscillations of a homogeneous cylindrical shaft. If $\theta(x, t)$ is the angular displacement at time t of the cross section at x, then

$$\frac{\partial^2 \theta}{\partial t^2} = a^2 \frac{\partial^2 \theta}{\partial x^2}.$$

Solve this if $\theta(x, 0) = f(x)$, $(\partial\theta/\partial t)(x, 0) = 0$, and the ends of the shaft are fixed elastically:

$$\frac{\partial \theta}{\partial x}(0, t) - \alpha\theta(0, t) = 0$$

$$\frac{\partial \theta}{\partial x}(L, t) + \alpha\theta(L, t) = 0, \text{ with } \alpha \text{ a positive constant.}$$

13.3 FOURIER SERIES SOLUTIONS OF THE HEAT EQUATION

In this section we give three examples of Fourier series solutions of boundary value problems associated with the one-dimensional heat equation.

A. ENDS OF THE BAR KEPT AT ZERO TEMPERATURE

Consider the problem of determining the temperature distribution $u(x, t)$ in a thin, homogeneous bar of length L, given the initial temperature throughout the bar and the temperature at both ends at all times. The boundary value problem is

$$\frac{\partial u}{\partial t} = a^2 \frac{\partial^2 u}{\partial x^2} \qquad (0 < x < L,\ t > 0),$$

(zero temperature at both ends) $\quad u(0, t) = u(L, t) = 0 \quad (t > 0)$,

(given initial temperature) $\quad u(x, 0) = f(x) \quad (0 < x < L)$.

To try a separation of variables, set

$$u(x, t) = X(x)T(t).$$

Substitution into the heat equation gives us

$$\frac{T'}{a^2 T} = \frac{X''}{X}.$$

Since the left side depends only on time and the right side only on x, and t and x are independent, both sides must be constant. For some λ, then,

$$\frac{T'}{a^2T} = \frac{X''}{X} = \lambda.$$

Then,

$$X'' - \lambda X = 0 \quad \text{and} \quad T' - a^2\lambda T = 0.$$

From $u(0, t) = u(L, t) = 0$ we get

$$X(0) = X(L) = 0.$$

Work with the problem for X first. It is exactly the same problem we encountered with the wave equation. Hence, we have eigenvalues

$$\lambda_n = -\frac{n^2\pi^2}{L^2}, \qquad n = 1, 2, 3, \ldots$$

and eigenfunctions

$$X_n = B_n \sin\left(\frac{n\pi x}{L}\right).$$

Now solve for T. For $n = 1, 2, 3, \ldots$, we have

$$T' + \frac{n^2\pi^2a^2}{L^2} T = 0.$$

Hence,

$$T_n = D_n \exp\left(\frac{-n^2\pi^2a^2t}{L^2}\right).$$

For $n = 1, 2, 3, \ldots$, then,

$$u_n(x, t) = C_n \sin\left(\frac{n\pi x}{L}\right) \exp\left(\frac{-n^2\pi^2a^2t}{L^2}\right)$$

satisfies the heat equation and the boundary conditions $u(0, t) = u(L, t) = 0$. To satisfy $u(x, 0) = f(x)$, try a superposition

$$u(x, t) = \sum_{n=1}^{\infty} C_n \sin\left(\frac{n\pi x}{L}\right) \exp\left(\frac{-n^2\pi^2a^2t}{L^2}\right).$$

We then need

$$u(x, 0) = f(x) = \sum_{n=1}^{\infty} C_n \sin\left(\frac{n\pi x}{L}\right).$$

Thus, choose

$$C_n = \frac{2}{L} \int_0^L f(x) \sin\left(\frac{n\pi x}{L}\right) dx.$$

The solution is then

$$u(x, t) = \sum_{n=1}^{\infty} \left[\frac{2}{L} \int_0^L f(\xi) \sin\left(\frac{n\pi\xi}{L}\right) d\xi \right] \sin\left(\frac{n\pi x}{L}\right) \exp\left(\frac{-n^2\pi^2a^2t}{L^2}\right).$$

For example, suppose that the initial temperature is

$$f(x) = \frac{4x}{L}\left(-\frac{x}{L} + 1\right) \qquad (0 \le x \le L),$$

as shown in Figure 247. This increases from 0 at 0 to 1 at $L/2$, and then decreases to 0 at L. Calculate

$$C_n = \frac{2}{L} \int_0^L \left[-\frac{4x^2}{L^2} \sin\left(\frac{n\pi x}{L}\right) + \frac{4x}{L} \sin\left(\frac{n\pi x}{L}\right) \right] dx$$

$$= -\frac{16}{n^3\pi^3}[\cos(n\pi) - 1]$$

$$= \begin{cases} 0 & \text{if } n \text{ is even,} \\ \dfrac{32}{n^3\pi^3} & \text{if } n \text{ is odd.} \end{cases}$$

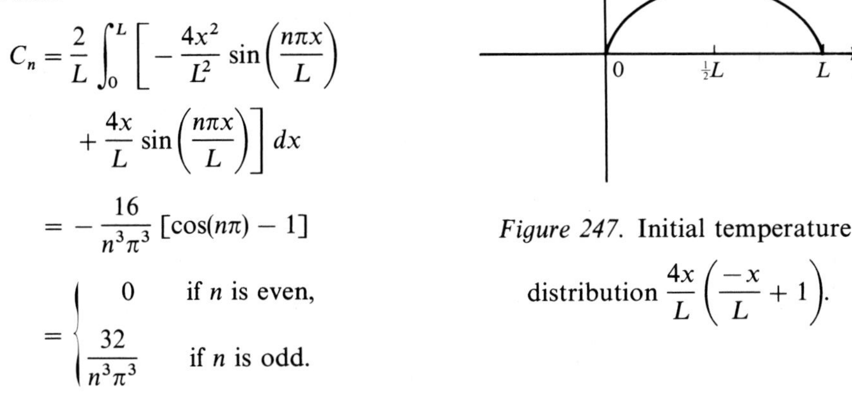

Figure 247. Initial temperature distribution $\dfrac{4x}{L}\left(\dfrac{-x}{L} + 1\right)$.

Thus, with the initial temperature of Figure 247, the solution is

$$u(x, t) = \sum_{n=1}^{\infty} \frac{32}{(2n-1)^3\pi^3} \sin\left(\frac{(2n-1)\pi x}{L}\right) \exp\left[\frac{-(2n-1)^2\pi^2a^2t}{L^2}\right].$$

B. TEMPERATURE IN A BAR WITH INSULATED ENDS

Consider the heat equation for a bar with no heat flow across the ends:

$$\frac{\partial u}{\partial t} = a^2 \frac{\partial^2 u}{\partial x^2} \qquad (0 < x < L, t > 0),$$

$$\frac{\partial u}{\partial x}(0, t) = \frac{\partial u}{\partial x}(L, t) = 0 \qquad (t > 0), \text{ (insulated ends)}$$

$$u(x, 0) = f(x) \qquad (0 < x < L), \text{ (given initial temperature)}$$

As usual, try $u(x, t) = X(x)T(t)$ to get

$$X'' - \lambda X = 0 \quad \text{and} \quad T' - \lambda a^2 T = 0.$$

Now, however, the insulation conditions give us

$$X'(0) = X'(L) = 0,$$

which are different boundary conditions than we had before. Consider cases:

Case 1 $\lambda = 0$. Then $X(x) = Ax + B$. Now, $X'(0) = A = 0$; so $X(x) = B$. This also satisfies $X'(L) = 0$. Thus, $\lambda = 0$ is an eigenvalue with eigenfunction X = constant.

Case 2 $\lambda > 0$. Say, $\lambda = \alpha^2$ with $\alpha > 0$. Now $X'' - \alpha^2 X = 0$ has general solution

$$X = Ae^{\alpha x} + Be^{-\alpha x}.$$

Since $X'(0) = \alpha(A - B) = 0$ and $\alpha > 0$, then $A = B$. Thus, $X = 2A\cosh(\alpha x)$. But then $X'(L) = 2\alpha A \sinh(\alpha L) = 0$ forces $A = 0$, and we get a trivial solution, which we discard. This problem has no eigenvalue $\lambda > 0$.

Case 3 $\lambda < 0$. Say, $\lambda = -\alpha^2$ with $\alpha > 0$. Now $X'' + \alpha^2 X = 0$ has general solution

$$X = A\cos(\alpha x) + B\sin(\alpha x).$$

First, $X'(0) = \alpha B = 0$; hence, $B = 0$. Next, $X'(L) = -\alpha A \sin(\alpha L) = 0$. To determine more possibilities for nontrivial solutions, we consider the case $A \neq 0$ and investigate how we can have $\sin(\alpha L) = 0$. This occurs when

$$\alpha L = \text{integer multiple of } \pi.$$

Since $\alpha > 0$ and $L > 0$, then $\alpha L = n\pi$, where $n = 1, 2, 3, \ldots$. The eigenvalues in this case are

$$\lambda_n = -\frac{n^2\pi^2}{L^2},$$

and the eigenfunctions are

$$X_n(x) = A_n \cos\left(\frac{n\pi x}{L}\right).$$

We can consolidate cases 1 and 3 to have

$$\text{eigenvalues } \lambda_n = -\frac{n^2\pi^2}{L^2}$$

and

$$\text{eigenfunctions } X_n = A_n \cos\left(\frac{n\pi x}{L}\right),$$

where $n = 0, 1, 2, 3, \ldots$.

With $\lambda_n = -n^2\pi^2/L^2$, the equation for T is

$$T' + \frac{n^2\pi^2 a^2}{L^2} T = 0,$$

with general solution

$$T_n = D_n \exp\left(\frac{-n^2\pi^2 a^2 t}{L^2}\right)$$

(when $n = 0$, this is T_0 = constant).

For $n = 0, 1, 2, \ldots$, then, writing $C_n = D_n A_n$, we have found that

$$u_n(x, t) = C_n \cos\left(\frac{n\pi x}{L}\right) \exp\left(\frac{-n^2\pi^2 a^2 t}{L^2}\right)$$

is a solution of

$$\frac{\partial u}{\partial t} = a^2 \frac{\partial^2 u}{\partial x^2} \qquad (0 < x < L,\ t > 0)$$

and

$$\frac{\partial u}{\partial x}(0, t) = \frac{\partial u}{\partial x}(L, t) = 0 \qquad (t > 0).$$

To satisfy $u(x, 0) = f(x)$, try a superposition

$$u(x, t) = \sum_{n=0}^{\infty} C_n \cos\left(\frac{n\pi x}{L}\right) \exp\left(\frac{-n^2\pi^2 a^2 t}{L^2}\right).$$

We now need

$$u(x, 0) = f(x) = \sum_{n=0}^{\infty} C_n \cos\left(\frac{n\pi x}{L}\right).$$

This is a Fourier cosine expansion of $f(x)$ on $[0, L]$. Thus, choose*

$$C_0 = \frac{1}{L}\int_0^L f(x)\,dx$$

and

$$C_n = \frac{2}{L}\int_0^L f(x)\cos\left(\frac{n\pi x}{L}\right) dx \quad \text{for} \quad n = 1, 2, 3, \ldots.$$

The solution is then

$$u(x, t) = \frac{1}{L}\int_0^L f(\xi)\,d\xi + \sum_{n=1}^{\infty}\left[\frac{2}{L}\int_0^L f(\xi)\cos\left(\frac{n\pi\xi}{L}\right) d\xi\right]\cos\left(\frac{n\pi x}{L}\right)\exp\left(\frac{-n^2\pi^2 a^2 t}{L^2}\right).$$

For example, if the initial temperature is given by

$$f(x) = x(L - x) \qquad (0 \le x \le L),$$

then

$$C_n = \frac{2}{L}\int_0^L x(L - x)\cos\left(\frac{n\pi x}{L}\right) dx = \begin{cases} \dfrac{-4L^2}{n^2\pi^2} & \text{if } n \text{ is even} \\ 0 & \text{if } n \text{ is odd} \end{cases}$$

* Here, the constant term is C_0, not $a_0/2$ as in the treatment of Fourier sine and cosine series. Here, then,

$$C_0 = \frac{1}{L}\int_0^L f(x)\,dx, \quad \text{while} \quad a_0 = \frac{2}{L}\int_0^L f(x)\,dx.$$

and

$$C_0 = \frac{1}{L}\int_0^L x(L - x)\,dx = \frac{L^2}{6}.$$

Hence, with this initial temperature, the temperature distribution is

$$u(x, t) = \frac{L^2}{6} - \frac{4L^2}{\pi^2}\sum_{n=1}^{\infty}\frac{1}{4n^2}\cos\left(\frac{2n\pi x}{L}\right)\exp\left(\frac{-4n^2\pi^2a^2t}{L^2}\right),$$

for $0 < x < L$ and $t > 0$.

Note: In all examples involving the wave equation, and in the first example of this section, we always got sines for our eigenfunctions, and the solution was a sum from $n = 1$ to ∞. In the last example (heat conduction in a bar with insulated ends) we got cosines for eigenfunctions and a zero eigenvalue as well, with the resulting sum from $n = 0$ to ∞.

It is important for the student to solve each problem by a logical progression of steps and take what comes, rather than presuppose that a certain kind of eigenfunction should occur. Of course, many times the same eigenvalue problem occurs over and over again in different contexts (as with $X'' - \lambda X = 0$, $X(0) = X(L) = 0$), and in this event previously derived solutions can be used without the need to repeat the derivation.

C. STEADY-STATE HEAT CONDUCTION IN A RECTANGULAR PLATE

Consider steady-state heat conduction in a flat plate having temperature values prescribed on the sides, as shown in Figure 248. The boundary value problem modeling this is

$$\frac{\partial^2 u}{\partial x^2} + \frac{\partial^2 u}{\partial y^2} = 0 \qquad (0 < x < \alpha,\ 0 < y < \beta),$$

$$u(x, 0) = u(x, \beta) = 0 \qquad (0 < x < \alpha),$$

$$u(0, y) = 0 \qquad (0 < y < \beta),$$

$$u(\alpha, y) = T \qquad (0 < y < \beta).$$

Put $u(x, y) = X(x)Y(y)$ into the differential equation to get

$$\frac{Y''}{Y} = -\frac{X''}{X} = \lambda,$$

for some constant λ.

Then,

$$X'' + \lambda X = 0 \quad \text{and} \quad Y'' - \lambda Y = 0.$$

Since $u(x, 0) = u(x, \beta) = 0$, then $Y(0) = Y(\beta) = 0$. Also, since $u(0, y) = 0$, then $X(0) = 0$.

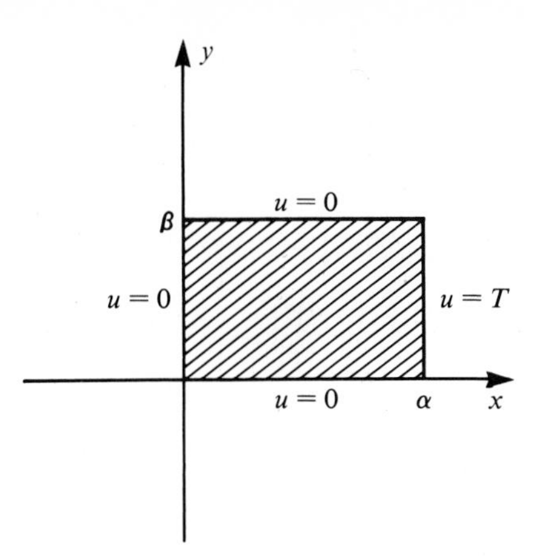

Figure 248. Flat plate with temperatures prescribed on the edges.

We have solved this eigenvalue problem for Y many times. We get

$$\lambda_n = -\frac{n^2\pi^2}{\beta^2} \quad \text{and} \quad Y_n(y) \doteq B_n \sin\left(\frac{n\pi y}{\beta}\right),$$

for $n = 1, 2, 3, \ldots$.

Now put $\lambda_n = -n^2\pi^2/\beta^2$ into the equation for X to get

$$X'' - \frac{n^2\pi^2}{\beta^2} X = 0.$$

This has general solution

$$X = Ae^{n\pi x/\beta} + Ce^{-n\pi x/\beta}.$$

Now, $X(0) = 0$; so $A + C = 0$. Thus,

$$X = D \sinh\left(\frac{n\pi x}{\beta}\right).$$

Thus far, for $n = 1, 2, \ldots, u_n(x, y)$ has the form

$$u_n(x, y) = C_n \sinh\left(\frac{n\pi x}{\beta}\right)\sin\left(\frac{n\pi y}{\beta}\right).$$

Now we need $u(\alpha, y) = T$. This cannot be satisfied for any fixed n; so we try a superposition

$$u(x, t) = \sum_{n=1}^{\infty} C_n \sinh\left(\frac{n\pi x}{\beta}\right)\sin\left(\frac{n\pi y}{\beta}\right).$$

Then we need

$$u(\alpha, y) = T = \sum_{n=1}^{\infty} C_n \sinh\left(\frac{n\pi \alpha}{\beta}\right)\sin\left(\frac{n\pi y}{\beta}\right).$$

This is a Fourier sine expansion of constant T on $(0, \beta)$. Thus, we should choose*

$$C_n \sinh\left(\frac{n\pi\alpha}{\beta}\right) = \frac{2}{\beta}\int_0^\beta T \sin\left(\frac{n\pi y}{\beta}\right) dy = \frac{2T}{n\pi}[1 - (-1)^n],$$

for $n = 1, 2, 3, \ldots$. Thus,

$$C_n = \frac{2T}{n\pi \sinh\left(\dfrac{n\pi\alpha}{\beta}\right)}[1 - (-1)^n] = \begin{cases} \dfrac{4T}{n\pi \sinh(n\pi\alpha/\beta)} & \text{if } n \text{ is odd,} \\ 0 & \text{if } n \text{ is even.} \end{cases}$$

Thus, the solution is

$$u(x, y) = \sum_{n=1}^{\infty} \frac{4T}{(2n-1)\pi \sinh\left(\dfrac{(2n-1)\pi\alpha}{\beta}\right)} \sinh\left(\frac{(2n-1)\pi x}{\beta}\right) \sin\left(\frac{(2n-1)\pi y}{\beta}\right).$$

In this section and the last, we have solved some classical boundary value problems dealing with wave motion of a string and heat conduction in a bar. For some approximation purposes, it is useful to consider strings and bars of infinite length. (As one example of a case in which the medium is considered infinite, the ocean is usually taken to be unbounded when studying sonar transmissions from submarines.) In such problems the Fourier integral replaces the Fourier series as the main tool. We shall consider typical examples in the next two sections.

PROBLEMS FOR SECTION 13.3

In each of Problems 1 through 5, solve the given boundary value problem.

1. $\dfrac{\partial u}{\partial t} = a^2 \dfrac{\partial^2 u}{\partial x^2}$ $\quad (0 < x < L,\ t > 0)$

$u(0, t) = u(L, t) = 0$ $\quad (t > 0)$

$u(x, 0) = x(L - x)$ $\quad (0 < x < L)$

2. $\dfrac{\partial u}{\partial t} = 4 \dfrac{\partial^2 u}{\partial x^2}$ $\quad (0 < x < L,\ t > 0)$

$u(0, t) = u(L, t) = 0$ $\quad (t > 0)$

$u(x, 0) = x^2(L - x)$ $\quad (0 < x < L)$

3. $\dfrac{\partial u}{\partial t} = 3 \dfrac{\partial^2 u}{\partial x^2}$ $\quad (0 < x < L,\ t > 0)$

$u(0, t) = u(L, t) = 0$ $\quad (t > 0)$

$u(x, 0) = L[1 - \cos(2\pi x/L)]$ $\quad (0 < x < L)$

4. $\dfrac{\partial u}{\partial t} = \dfrac{\partial^2 u}{\partial x^2}$ $\quad (0 < x < \pi,\ t > 0)$

$\dfrac{\partial u}{\partial x}(0, t) = \dfrac{\partial u}{\partial x}(\pi, t) = 0$ $\quad (t > 0)$

$u(x, 0) = \sin(x)$ $\quad (0 < x < \pi)$

5. $\dfrac{\partial u}{\partial t} = 4 \dfrac{\partial^2 u}{\partial x^2}$ $\quad (0 < x < 2\pi,\ t > 0)$

$\dfrac{\partial u}{\partial x}(0, t) = \dfrac{\partial u}{\partial x}(2\pi, t) = 0$ $\quad (t > 0)$

$u(x, 0) = x(2\pi - x)$ $\quad (0 < x < 2\pi)$

---

* *Important Note:* The *entire coefficient* of $\sin(n\pi y/\beta)$ must be set equal to the Fourier sine coefficient; thus, we must let $C_n \sinh(n\pi\alpha/\beta)$ equal the Fourier sine coefficient, not just C_n. We then solve for C_n by dividing out the term $\sinh(n\pi\alpha/\beta)$; hence its appearance in the denominator of the resulting solution.

6. A thin bar of length L has insulated ends and initial temperature equal to a constant T. Find the temperature function $u(x, t)$.

7. A thin bar of length L has initial temperature equal to a constant T, and the right end (at $x = L$) is insulated, while the left end is kept at temperature zero. Find the temperature distribution $u(x, t)$.

8. Imagine a thin bar of length L in which heat transfer between the bar and its surroundings is assumed to obey a linear transfer law. The heat equation is then

$$\frac{\partial u}{\partial t} = a^2 \frac{\partial^2 u}{\partial x^2} - hu \qquad (0 < x < L,\ t > 0),$$

where h is some given positive constant. Assume that the ends of the bar are insulated and that the initial temperature is $f(x)$.

After deriving the solution in general, find the solution when $f(x) = x(L - x)$.

9. Solve the boundary value problem

$$\frac{\partial u}{\partial t} = a^2 \frac{\partial^2 u}{\partial x^2} - hu \qquad (0 < x < L,\ t > 0),$$

$$u(0, t) = u(L, t) = 0 \qquad (t > 0),$$

$$u(x, 0) = f(x) \qquad (0 < x < L),$$

where h is a positive constant.

10. Solve for the steady-state temperature in a flat plate placed over $0 \le x \le \alpha$, $0 \le y \le \beta$, if the temperature on the vertical sides and bottom side is kept at 0, and on the top side at constant T.

11. Solve for the steady-state temperature in a flat plate placed over $0 \le x \le \alpha$, $0 \le y \le \beta$, if the temperature on the left side is T_1, on the right side T_2, and on the top and bottom sides zero. (*Hint:* Split this problem into two problems, in each of which the temperature is zero on three of the four sides.)

12. Solve for the steady-state temperature in a flat, circular disk, placed over $0 \le r \le k$ (in cylindrical coordinates), if the temperature on the boundary is $f(\theta) = T$ (constant) for $0 \le \theta \le 2\pi$.

13. Solve for the steady-state temperature in a flat, rectangular plate $0 \le x \le \alpha$, $0 \le y \le \beta$, if the top and bottom edges are insulated, the left side is kept at zero temperature, and the right side is kept at temperature $f(y)$.

14. Solve for the steady-state temperatures in a semiinfinite strip occupying $x \ge 0$, $0 \le y \le 1$. The temperature on the top and bottom sides is kept at 0, on the left side (the y-axis between 0 and 1) at constant temperature T. Here, there is no right boundary; so impose the condition (physically necessary) that only bounded solutions are acceptable.

15. Solve for the steady-state temperature in a semiinfinite plate over $0 \le x \le 1$, $y \ge 0$, if the temperature on the bottom side is $f(x)$, and both left and right sides are insulated. (*Hint:* Impose the condition that the solution must be bounded.)

16. Find the steady-state temperature in a wedge-shaped plate occupying the region $0 \le r \le k$, $0 \le \theta \le \alpha$ in polar coordinates. The edges $\theta = 0$ and $\theta = \alpha (0 \le r \le k)$ are kept at zero temperature; the arc $r = k (0 \le \theta \le \alpha)$ is kept at constant temperature T.

17. We have a thin bar of length L lying between $x = 0$ and $x = L$. Assuming a linear law of heat transfer from the surface of the bar to the surrounding medium, the partial differential equation for the temperature function $u(x, t)$ is

$$\frac{\partial u}{\partial t} = a^2 \frac{\partial^2 u}{\partial x^2} - Au \qquad (0 < x < L,\ t > 0),$$

where A is a positive constant. The temperature at the ends $x = 0$ and $x = L$ is kept at 0, and the initial temperature is $f(x)$.

(a) Try solving the resulting boundary value problem by separation of variables.
(b) Transform the above boundary value problem into a new one by setting

$$u(x, t) = e^{-At}v(x, t).$$

Solve the new problem for v, and hence find u.

18. Redo Problem 17, replacing the condition that the ends are kept at zero temperature with the condition that the left end is kept at zero temperature and the right end is insulated.

19. The current $I(x, t)$ and voltage $V(x, t)$ at time t and distance x from one end of a transmission line satisfy the system of equations

$$-\frac{\partial V}{\partial x} = RI + L\frac{\partial I}{\partial t}, \qquad -\frac{\partial I}{\partial x} = SV + K\frac{\partial V}{\partial t},$$

in which R = resistance, L = inductance, S = leakage conductance, and K = capacitance to ground, all per unit length. By differentiating appropriately and eliminating I and its partials (or V and its partials), show that V (or I) satisfies the telegraph equation of Problem 23, Section 13.2.

20. *Poisson's solution of a Dirichlet problem.* Consider a steady-state heat equation for a disk of radius 1. In polar coordinates, the temperature function $u(r, \theta)$ satisfies Laplace's equation

$$\nabla^2 u = \frac{1}{r}\frac{\partial}{\partial r}\left(r\frac{\partial u}{\partial r}\right) + \frac{1}{r^2}\frac{\partial^2 u}{\partial \theta^2} = 0$$

for $0 \le r < 1, 0 \le \theta \le 2\pi$. The temperature is given on the boundary:

$$u(1, \theta) = f(\theta).$$

(a) Assuming a solution of the form

$$u(r, \theta) = \frac{a_0}{2} + \sum_{n=1}^{\infty} r^n[a_n \cos(n\theta) + b_n \sin(n\theta)],$$

substitute in the formulas for the Fourier coefficients, interchange the integral and summation signs, and use a trigonometric identity to obtain

$$u(r, \theta) = \frac{1}{2\pi}\int_0^{2\pi} f(\phi)\left\{1 + 2\sum_{n=1}^{\infty} r^n \cos[n(\theta - \phi)]\right\} d\phi.$$

(b) Now use the complex exponential, defined by

$$e^{i\xi} = \cos(\xi) + i \sin(\xi).$$

Let $z = re^{i\xi}$ and use de Moivre's theorem to conclude that

$$z^n = r^n[\cos(n\xi) + i \sin(n\xi)].$$

Hence, conclude from (a) that

$$1 + 2\sum_{n=1}^{\infty} r^n \cos(n\xi) = \text{Re}\left(1 + 2\sum_{n=1}^{\infty} z^n\right),$$

where Re denotes the real part of a complex number.

(c) Use the geometric series

$$\frac{1}{1-z} = \sum_{n=0}^{\infty} z^n \quad \text{if} \quad |z| < 1$$

to show that

$$1 + 2\sum_{n=1}^{\infty} r^n \cos(n\xi) = \operatorname{Re}\left(\frac{1 + re^{i\xi}}{1 - re^{i\xi}}\right).$$

(d) Use the result of (c) to show that

$$1 + 2\sum_{n=1}^{\infty} r^n \cos(n\xi) = \frac{1 - r^2}{1 + r^2 - 2r\cos(\xi)}.$$

(e) Derive the solution

$$u(r, \theta) = \frac{1}{2\pi}\int_0^{2\pi} \frac{(1 - r^2)f(\phi)\,d\phi}{1 + r^2 - 2r\cos(\theta - \phi)}.$$

This is Poisson's integral solution of the Dirichlet problem for a unit disk. It is valid for $0 \le \theta \le 2\pi$ and $0 \le r < 1$. When $r = 1$, $u(1, \theta) = f(\theta)$.

21. Using the result of Problem 20, derive the integral solution

$$u(r, \theta) = \left(\frac{R^2 - r^2}{2\pi}\right)\int_0^{2\pi} \frac{f(\phi)\,d\phi}{R^2 + r^2 - 2rR\cos(\theta - \phi)},$$

for $0 \le r < R$, $0 \le \theta \le 2\pi$, of the Dirichlet problem for a disk of radius R about the origin. Here, the boundary data is $u(R, \theta) = f(\theta)$ for $0 \le \theta \le 2\pi$.

22. *Computer problem.* Using the result of Problem 20, show that the solution of

$$\nabla^2 u = 0 \qquad (0 \le r < 1,\ 0 \le \theta \le 2\pi),$$
$$u(1, \theta) = \theta - \sin(\theta)$$

is

$$u(r, \theta) = \begin{cases} \dfrac{1}{2\pi}\displaystyle\int_0^{2\pi} \frac{(1 - r^2)[\phi - \sin(\phi)]\,d\phi}{1 + r^2 - 2r\cos(\theta - \phi)} & \text{for} \quad 0 \le r < 1,\ 0 \le \theta \le 2\pi, \\ \theta - \sin(\theta) & \text{for} \quad r = 1. \end{cases}$$

This integral cannot be done in elementary terms. By numerical integration, calculate to four decimal places:

(a) $u(0, 0)$ (b) $u(0, \frac{1}{2}\pi)$ (c) $u(\frac{1}{4}, \frac{1}{2}\pi)$
(d) $u(\frac{1}{2}, \frac{1}{2}\pi)$ (e) $u(\frac{3}{4}, \frac{1}{2}\pi)$ (f) $u(\frac{9}{10}, \frac{1}{2}\pi)$

23. *Computer problem.* Using Poisson's integral formula (Problem 20), calculate the following values of $u(r, \theta)$, assuming that $u(r, \theta)$ is the solution of

$$\nabla^2 u = 0 \qquad (0 \le r < 1,\ 0 \le \theta \le 2\pi),$$
$$u(1, \theta) = 2\cos(3\theta) + \sin(\tfrac{1}{2}\theta).$$

(a) $u(\frac{1}{2}, \frac{1}{4}\pi)$ (b) $u(\frac{1}{3}, \frac{1}{6}\pi)$ (c) $u(\frac{1}{4}, \frac{2}{3}\pi)$ (d) $u(\frac{3}{4}, \frac{1}{2}\pi)$
(e) Calculate $u(r, \pi)$ for $r = 0, \frac{1}{4}, \frac{1}{3}, \frac{1}{2}, \frac{2}{3}, \frac{3}{4}, \frac{9}{10}, .95, .97, .99$. Do these values appear to approach $u(1, \pi)$?

24. Show that the partial differential equation

$$\frac{\partial u}{\partial t} = k\left(\frac{\partial^2 u}{\partial x^2} + A\frac{\partial u}{\partial x} + Bu\right)$$

can be reduced to

$$\frac{\partial v}{\partial t} = k\frac{\partial^2 v}{\partial x^2}$$

by setting

$$u(x, t) = e^{\alpha x + \beta t}v(x, t)$$

and choosing the constants α and β appropriately.

25. Use the method of Problem 24 to solve

$$\frac{\partial u}{\partial t} = k\left(\frac{\partial^2 u}{\partial x^2} - \frac{a}{L}\frac{\partial u}{\partial x}\right),$$

$$u(0, t) = u(L, t) = 0 \qquad (t \geq 0),$$

$$u(x, 0) = \frac{L}{ak}\left\{1 - \exp\left[-a\left(1 - \frac{x}{L}\right)\right]\right\} \qquad (0 \leq x \leq L).$$

Here, $u(x, t)$ measures the concentration of positive charge carriers on the base of length L of a transistor. In the problem, k and a are positive constants.

13.4 THE WAVE EQUATION FOR SEMIINFINITE AND INFINITE STRINGS

Often a physicist or engineer models a very long vibrating string by assuming infinite length. Imagine the string pegged at $x = 0$ and extending over all $x \geq 0$. Such a string has one end and is called *semiinfinite*. The resulting motion is governed by

$$\frac{\partial^2 y}{\partial t^2} = a^2\frac{\partial^2 y}{\partial x^2} \qquad (x > 0, t > 0),$$

(pegged left end) $\quad y(0, t) = 0 \quad (t > 0),$

(initial position) $\quad y(x, 0) = f(x) \quad (x > 0),$

(initial velocity) $\quad \frac{\partial y}{\partial t}(x, 0) = g(x) \quad (x > 0).$

For simplicity, we shall assume that $g(x) = 0$ [if $f(x)$ and $g(x)$ are both nonzero, proceed as in C of Section 13.2].

Begin as usual, putting $y = X(x)T(t)$ to get

$$X'' - \lambda X = 0, \qquad T'' - \lambda a^2 T = 0.$$

The condition $y(0, t) = 0$ becomes $X(0) = 0$. Here, we have no condition at L as there is no L, an important difference between this and previously considered problems.

Consider cases on λ.

Case 1 $\lambda = 0$. Then $X = Ax + B$; and $X(0) = B = 0$ gives us $X = Ax$, with A arbitrary.

Case 2 $\lambda > 0$. Say, $\lambda = \alpha^2$ with $\alpha > 0$. Then $X = Ae^{\alpha x} + Be^{-\alpha x}$. Now, $X(0) = A + B = 0$; so $A = -B$ and $X = C \sinh(\alpha x)$ with C arbitrary.

Case 3 $\lambda < 0$. Say, $\lambda = -\alpha^2$ with $\alpha > 0$. Then $X = A \cos(\alpha x) + B \sin(\alpha x)$. Since $X(0) = A = 0$, then $X = B \sin(\alpha x)$ for arbitrary B.

Thus, *every* real λ is an eigenvalue. The problem is that the condition $X(L) = 0$, which eliminated many possible values of λ before, is missing here. However, on physical grounds, we seek a bounded solution (waves should not have infinite amplitude). Thus, *we shall impose the following condition*:

$$\text{For some } M, \qquad |y(x, t)| \leq M \qquad \text{for all } x > 0,\ t > 0.$$

This eliminates cases 1 and 2, which produce unbounded X, and leaves case 3. Thus, we have

$$\text{eigenvalues:} \quad \lambda = -\alpha^2 \qquad (\alpha \text{ any positive number})$$

and

$$\text{eigenfunctions:} \quad X_\alpha = A_\alpha \sin(\alpha x) \qquad (A_\alpha \text{ arbitrary}).$$

With $\lambda = -\alpha^2$, $T_\alpha = A_\alpha \cos(\alpha a t) + B_\alpha \sin(\alpha a t)$. Since $(\partial y/\partial t)(x, 0) = 0$, then $T'_\alpha(0) = 0$ forces $B = 0$.

Thus, for $\alpha > 0$, we have

$$y_\alpha(x, t) = C_\alpha \sin(\alpha x) \cos(\alpha a t).$$

This probably will not satisfy $y(x, 0) = f(x)$ for any particular choice of α. Thus, we try a superposition, which must be an integral, not a sum, as α can be any positive number:

$$y(x, t) = \int_0^\infty y_\alpha(x, t)\, d\alpha = \int_0^\infty C_\alpha \sin(\alpha x) \cos(\alpha a t)\, d\alpha.$$

To have

$$y(x, 0) = f(x) = \int_0^\infty C_\alpha \sin(\alpha x)\, d\alpha,$$

we need a Fourier sine integral for $f(x)$ on $x > 0$. Thus, we should choose

$$C_\alpha = \frac{2}{\pi} \int_0^\infty f(\xi) \sin(\alpha \xi)\, d\xi.$$

The motion of the string is then described by an integral solution:

$$y(x, t) = \frac{2}{\pi} \int_0^\infty \int_0^\infty f(\xi) \sin(\alpha\xi) \sin(\alpha x) \cos(\alpha a t)\, d\xi\, d\alpha.$$

For example, suppose that

$$f(x) = \begin{cases} x, & 0 \le x \le 1, \\ 1, & 1 \le x \le 4, \\ 5 - x, & 4 \le x \le 5, \\ 0, & x \ge 5, \end{cases}$$

as shown in Figure 249. Then,

$$C_\alpha = \frac{2}{\pi}\left[\int_0^1 \xi \sin(\alpha\xi)\, d\xi + \int_1^4 \sin(\alpha\xi)\, d\xi + \int_4^5 (5 - \xi) \sin(\alpha\xi)\, d\xi\right]$$

$$= \frac{2}{\pi\alpha^2}[\sin(\alpha) + \sin(4\alpha) - \sin(5\alpha)].$$

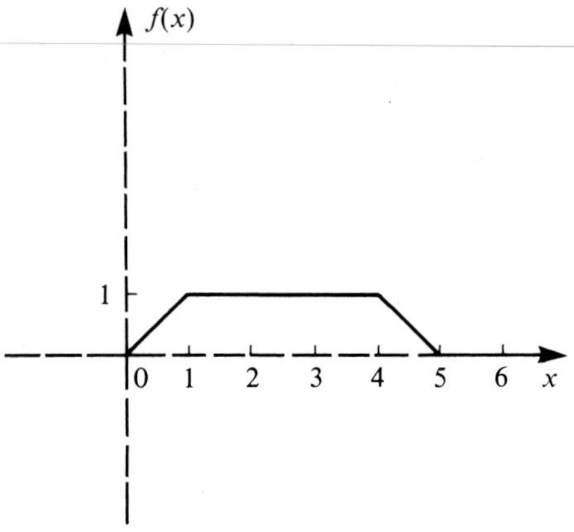

Figure 249

In this case, then,

$$y(x, t) = \frac{2}{\pi} \int_0^\infty \frac{1}{\alpha^2}[\sin(\alpha) + \sin(4\alpha) - \sin(5\alpha)] \sin(\alpha x) \cos(\alpha a t)\, d\alpha.$$

The above problem was fairly typical of boundary value problems on an infinite segment bounded at one end; usually one replaces the missing boundary condition by the physically plausible requirement that the solution be bounded, and this is enough to determine eigenvalues and eigenfunctions, and hence a Fourier integral solution. We illustrate this again for heat conduction problems in the next section.

As an illustration of wave motion in an unbounded medium, consider vibrations in an infinite string which extends from $-\infty$ to ∞. To be specific, look at the problem

$$\frac{\partial^2 y}{\partial t^2} = 4\frac{\partial^2 y}{\partial x^2} \qquad (-\infty < x < \infty,\ t > 0),$$

$$y(x, 0) = 0, \qquad \frac{\partial y}{\partial t}(x, 0) = e^{-|x|} = f(x).$$

Here, we release the string from the horizontal stretched position, but with initial velocity $e^{-|x|}$.

To solve the problem, assume that $y = XT$ and obtain, as usual,

$$X'' - \lambda X = 0 \quad \text{and} \quad T'' - 4\lambda T = 0.$$

Here, we have no boundary condition on X, and so must consider all cases on λ.

Case 1 $\lambda = 0$. Then $X = Ax + B$, which is unbounded for $-\infty < x < \infty$ unless $A = 0$. Thus, $\lambda = 0$ is an eigenvalue with constant eigenfunctions.

Case 2 $\lambda < 0$. Say, $\lambda = -\alpha^2$ with $\alpha > 0$. Then $X'' + \alpha^2 X = 0$; so $X = A\cos(\alpha x) + B\sin(\alpha x)$. This is bounded on $-\infty < x < \infty$ for all $\alpha > 0$. Thus, for any $\alpha > 0$, $\lambda = -\alpha^2$ is an eigenvalue with eigenfunctions $X_\alpha = A_\alpha \cos(\alpha x) + B_\alpha \sin(\alpha x)$.

Case 3 $\lambda > 0$. Say, $\lambda = \alpha^2$ with $\alpha > 0$. Then $X = Ce^{\alpha x} + De^{-\alpha x}$. On $-\infty < x < 0$, $e^{-\alpha x}$ is unbounded; so we must choose $D = 0$. But on $0 < x < \infty$, $e^{\alpha x}$ is unbounded; so we must also choose $C = 0$, giving us the trivial solution in this case. Thus, there are no positive eigenvalues for this problem.

Cases 1 and 2 can be combined as follows:

$$\text{eigenvalues:} \quad \lambda = -\alpha^2 \qquad (\alpha \geq 0)$$

and

$$\text{eigenfunctions:} \quad X_\alpha = A_\alpha \cos(\alpha x) + B_\alpha \sin(\alpha x).$$

Now we have $T'' + 4\alpha^2 T = 0$, with general solution

$$T_\alpha = C_\alpha \cos(2\alpha t) + D_\alpha \sin(2\alpha t).$$

Since $y(x, 0) = 0$, then $T(0) = 0$, requiring that $C_\alpha = 0$. Thus,

$$T_\alpha = D_\alpha \sin(2\alpha t).$$

For each $\alpha \geq 0$, then,

$$y_\alpha(x, t) = X_\alpha(x)T_\alpha(t) = [a_\alpha \cos(\alpha x) + b_\alpha \sin(\alpha x)] \sin(2\alpha t),$$

in which $a_\alpha = A_\alpha D_\alpha$ and $b_\alpha = B_\alpha D_\alpha$. To satisfy the condition $(\partial y/\partial t)(x, 0) = e^{-|x|}$, we try a superposition

$$\begin{aligned} y(x, t) &= \int_0^\infty y_\alpha(x, t)\, d\alpha \\ &= \int_0^\infty [a_\alpha \cos(\alpha x) + b_\alpha \sin(\alpha x)] \sin(2\alpha t)\, d\alpha, \end{aligned}$$

and attempt to choose the a_α's and b_α's appropriately. Now,

$$\frac{\partial y}{\partial t}(x, t) = \int_0^\infty [a_\alpha \cos(\alpha x) + b_\alpha \sin(\alpha x)] 2\alpha \cos(2\alpha t)\, d\alpha.$$

Thus, we require that

$$\frac{\partial y}{\partial t}(x, 0) = e^{-|x|} = \int_0^\infty [2\alpha a_\alpha \cos(\alpha x) + 2\alpha b_\alpha \sin(\alpha x)]\, d\alpha.$$

This is a Fourier integral expansion of $e^{-|x|}$ on $-\infty < x < \infty$. Thus, choose

$$2\alpha a_\alpha = \frac{1}{\pi} \int_{-\infty}^{\infty} e^{-|r|} \cos(\alpha r)\, dr$$

and

$$2\alpha b_\alpha = \frac{1}{\pi} \int_{-\infty}^{\infty} e^{-|r|} \sin(\alpha r)\, dr.$$

These are elementary integrations, and we obtain

$$2\alpha a_\alpha = \frac{1}{\pi}\frac{2}{1+\alpha^2} \quad \text{for} \quad \alpha > 0$$

and

$$2\alpha b_\alpha = 0.$$

Thus,

$$b_\alpha = 0 \quad \text{and} \quad a_\alpha = \frac{1}{\pi\alpha(1+\alpha^2)}.$$

The solution is then

$$y(x, t) = \frac{1}{\pi}\int_0^\infty \frac{1}{\alpha(1+\alpha^2)}\cos(\alpha x)\sin(2\alpha t)\,d\alpha.$$

Note that the integral is improper at 0. However, the integrand can be assigned the value $2t$ at $\alpha = 0$, since by L'Hospital's rule,

$$\lim_{\alpha\to 0}\frac{1}{\alpha(1+\alpha^2)}\cos(\alpha x)\sin(2\alpha t) = \lim_{\alpha\to 0}\frac{2t\cos(\alpha x)\cos(2\alpha t)}{1+3\alpha^2} = 2t.$$

In the next section we treat heat conduction problems in which the medium is infinite or semiinfinite.

PROBLEMS FOR SECTION 13.4

In each of Problems 1 through 8, solve the semiinfinite vibrating string problem

$$\frac{\partial^2 y}{\partial t^2} = a^2\frac{\partial^2 y}{\partial x^2} \qquad (x > 0,\ t > 0)$$

$$y(0, t) = 0 \qquad (t > 0)$$

$$y(x, 0) = f(x), \quad \frac{\partial y}{\partial t}(x, 0) = g(x) \qquad (x > 0)$$

$$y(x, t) \text{ is bounded for all } x,\ t > 0$$

for the following choices of $f(x)$ and $g(x)$.

1. $f(x) = \begin{cases} x(1-x), & 0 \le x \le 1 \\ 0, & x > 1 \end{cases}$ $\quad g(x) = 0$ for $x \ge 0$

2. $f(x) = e^{-x}$ and $g(x) = 0$ for $x \ge 0$.

3. $f(x) = \begin{cases} \sin(x), & 0 \le x \le \pi \\ 0, & x \ge \pi \end{cases}$ $\quad g(x) = 0$ for $x \ge 0$

4. $f(x) = 0,\quad g(x) = \begin{cases} 0, & 0 \le x < 1 \\ 2, & 1 \le x \le 4 \\ 0, & x > 4 \end{cases}$

5. $f(x) = 0, \qquad g(x) = \begin{cases} x, & 0 \le x \le 1 \\ 0, & x > 1 \end{cases}$

6. $f(x) = 0, \qquad g(x) = \begin{cases} x(1-x), & 0 \le x \le 1 \\ 0, & x \ge 1 \end{cases}$

7. $f(x) = \begin{cases} 1 - \cos(x), & 0 \le x \le \frac{1}{2}\pi \\ 0, & x > \frac{1}{2}\pi \end{cases} \qquad g(x) = \begin{cases} 0, & 0 \le x < 1 \\ 1, & 1 \le x \le 4 \\ 0, & x \ge 4 \end{cases}$

8. $f(x) = \begin{cases} x\cos(x), & 0 \le x \le \frac{1}{2}\pi \\ 0, & x > \frac{1}{2}\pi \end{cases} \qquad g(x) = \begin{cases} \sin(x), & 0 \le x \le \pi \\ 0, & x > \pi \end{cases}$

9. Set up and solve the boundary value problem for an infinite string (stretching from $-\infty$ to ∞) released from rest and given the initial displacement

$$y(x, 0) = f(x) = \begin{cases} x+1, & -1 \le x \le 0, \\ 1-x, & 0 \le x \le 1, \\ 0, & |x| > 1. \end{cases}$$

10. Set up and solve the boundary value problem for an infinite string released from rest and given the initial displacement

$$f(x) = \begin{cases} \sin(x), & -\pi \le x \le \pi, \\ 0, & |x| > \pi. \end{cases}$$

11. Set up and solve the boundary value problem for an infinite string initially stretched along the x-axis and released with an initial velocity

$$g(x) = \begin{cases} 1, & -2 \le x \le 2, \\ 0, & |x| > 2. \end{cases}$$

12. Set up and solve the boundary value problem for an infinite string with initial position

$$f(x) = \begin{cases} x\sin(x), & 0 \le x \le 4\pi, \\ 0, & x > 4\pi, \end{cases}$$

and initial velocity

$$g(x) = \begin{cases} e^{-x}, & -1 \le x \le 1, \\ 0, & |x| > 1. \end{cases}$$

13. Imagine a string stretched from 0 to ∞ and released from initial position $y = f(x)$ and with initial velocity $y = g(x)$.

Show how the solution for the resulting motion can be obtained from the motion of an infinite string (from $-\infty$ to ∞) having initial position $y = F(x)$ and initial velocity $y = G(x)$, assuming that

$$F(x) + F(-x) = -\frac{1}{a}\int_{-x}^{x} G(\xi)\, d\xi$$

for $x \ge 0$.

13.5 THE HEAT EQUATION IN SEMIINFINITE AND INFINITE REGIONS

In the spirit of Section 13.4, we now consider heat conduction in a bar extending from $x = 0$ to infinity. (Sometimes such a bar is said to be "semiinfinite.") We have as a typical problem:

$$\frac{\partial u}{\partial t} = a^2 \frac{\partial^2 u}{\partial x^2} \qquad (x > 0,\ t > 0),$$

$$u(x, 0) = f(x) \qquad (x > 0),$$

$$u(0, t) = 0 \qquad (t > 0).$$

We also require that $u(x, t)$ be bounded.

Putting $u = XT$, we get in the usual way

$$X'' - \lambda X = 0 \quad \text{and} \quad T' - \lambda a^2 T = 0.$$

We also have $X(0) = 0$.

As far as X is concerned, this problem is identical with that solved in Section 13.4; so we have without further labor

$$\text{eigenvalues:} \quad \lambda = -\alpha^2 \qquad (\alpha > 0)$$

and

$$\text{eigenfunctions:} \quad X_\alpha = A_\alpha \sin(\alpha x) \qquad (A_\alpha \text{ arbitrary}).$$

Next,

$$T' + \alpha^2 a^2 T = 0$$

has solution

$$T_\alpha = B_\alpha e^{-\alpha^2 a^2 t}.$$

Then,

$$u_\alpha = C_\alpha \sin(\alpha x) e^{-\alpha^2 a^2 t} \qquad (\alpha > 0)$$

with $C_\alpha = A_\alpha B_\alpha$. This leads us to try a solution

$$u(x, t) = \int_0^\infty C_\alpha \sin(\alpha x) e^{-\alpha^2 a^2 t}\, d\alpha.$$

We now need

$$u(x, 0) = f(x) = \int_0^\infty C_\alpha \sin(\alpha x)\, d\alpha.$$

Hence, choose

$$C_\alpha = \frac{2}{\pi} \int_0^\infty f(\xi) \sin(\alpha \xi)\, d\xi.$$

Then,

$$u(x, t) = \frac{2}{\pi} \int_0^\infty \int_0^\infty f(\xi) \sin(\alpha\xi) \sin(\alpha x) e^{-\alpha^2 a^2 t} \, d\xi \, d\alpha.$$

For example, suppose that

$$f(x) = \begin{cases} \pi - x, & 0 \leq x \leq \pi, \\ 0, & x > \pi. \end{cases}$$

Then,

$$C_\alpha = \frac{2}{\pi} \int_0^\infty f(\xi) \sin(\alpha\xi) \, d\xi$$

$$= \frac{2}{\pi} \int_0^\pi (\pi - \xi) \sin(\alpha\xi) \, d\xi$$

$$= \frac{2}{\alpha} \left(1 - \frac{\sin(\alpha\pi)}{\alpha\pi}\right).$$

Note that this has limit zero as $\alpha \to 0$. The temperature function is then

$$u(x, t) = \frac{2}{\pi} \int_0^\infty \frac{1}{\alpha} \left(1 - \frac{\sin(\alpha\pi)}{\alpha\pi}\right) \sin(\alpha x) e^{-\alpha^2 a^2 t} \, d\alpha.$$

The methods of this section extend readily to the case of an infinite bar (i.e., $-\infty < x < \infty$) or to the problem of finding steady-state temperatures in a flat plate extending over an unbounded region. Here are examples of these kinds of problems, which are also pursued in the exercises.

Suppose we want to find the temperature $u(x, t)$ in an infinitely long wire if the initial temperature distribution is

$$u(x, 0) = e^{-|x|}, \qquad (-\infty < x < \infty).$$

The problem is to solve

$$\frac{\partial u}{\partial t} = a^2 \frac{\partial^2 u}{\partial x^2} \qquad (-\infty < x < \infty, \, t > 0),$$

$$u(x, 0) = f(x) = e^{-|x|}, \qquad (-\infty < x < \infty).$$

Let $u = XT$ and obtain $X'' - \lambda X = 0$, $T' - \lambda a^2 T = 0$, as usual.

The problem for X is exactly that encountered with wave motion for an unbounded string, and we recall that the eigenvalues are

$$\lambda = -\alpha^2 \qquad (\alpha \geq 0)$$

and the eigenfunctions are

$$X_\alpha = A_\alpha \cos(\alpha x) + B_\alpha \sin(\alpha x).$$

Now look at the equation for T, which becomes $T' + \alpha^2 a^2 T = 0$. This has solution

$$T_\alpha = E_\alpha e^{-\alpha^2 a^2 t} \quad \text{for} \quad \alpha \geq 0.$$

Thus, for $\alpha \geq 0$, we have

$$u_\alpha(x, t) = [a_\alpha \cos(\alpha x) + b_\alpha \sin(\alpha x)]e^{-\alpha^2 a^2 t},$$

in which a_α and b_α are constants to be determined. Now we must satisfy

$$u(x, 0) = f(x).$$

To do this, we write a superposition

$$u(x, t) = \int_0^\infty [a_\alpha \cos(\alpha x) + b_\alpha \sin(\alpha x)]e^{-\alpha^2 a^2 t}\, d\alpha.$$

Then we need

$$u(x, 0) = f(x) = \int_0^\infty [a_\alpha \cos(\alpha x) + b_\alpha \sin(\alpha x)]\, d\alpha,$$

a Fourier integral expansion of $f(x)$ on $(-\infty, \infty)$. Thus, choose

$$a_\alpha = \frac{1}{\pi}\int_{-\infty}^{\infty} f(r)\cos(\alpha r)\, dr \quad \text{and} \quad b_\alpha = \frac{1}{\pi}\int_{-\infty}^{\infty} f(r)\sin(\alpha r)\, dr.$$

After integrating, we obtain

$$a_\alpha = \frac{1}{\pi}\int_{-\infty}^{\infty} e^{-|r|}\cos(\alpha r)\, dr = \frac{2}{\pi}\int_0^\infty e^{-r}\cos(\alpha r)\, dr = \frac{2}{\pi(1+\alpha^2)}$$

and

$$b_\alpha = \frac{1}{\pi}\int_{-\infty}^{\infty} e^{-|r|}\sin(\alpha r)\, dr = 0$$

[because $e^{-|r|}\sin(\alpha r)$ is odd on $(-\infty, \infty)$].

The solution is then

$$u(x, t) = \frac{2}{\pi}\int_0^\infty \frac{1}{1+\alpha^2}\cos(\alpha x)e^{-\alpha^2 a^2 t}\, d\alpha.$$

As another illustration of this type of problem we shall find the steady-state temperature $u(x, y)$ in a flat plate extending over the right quarter-plane, where the temperature on the side $x = 0$ is kept zero, and that on the bottom side $y = 0$ is kept at a constant 4 degrees Celsius for $0 \leq x \leq 2$, and zero degrees for $x > 2$. This is shown in Figure 250.

The boundary value problem is

$$\frac{\partial^2 u}{\partial x^2} + \frac{\partial^2 u}{\partial y^2} = 0 \qquad (x > 0,\ y > 0),$$

$$u(0, y) = 0 \quad \text{for} \quad y > 0$$

$$u(x, 0) = f(x) = \begin{cases} 4, & 0 \leq x \leq 2, \\ 0, & x > 2. \end{cases}$$

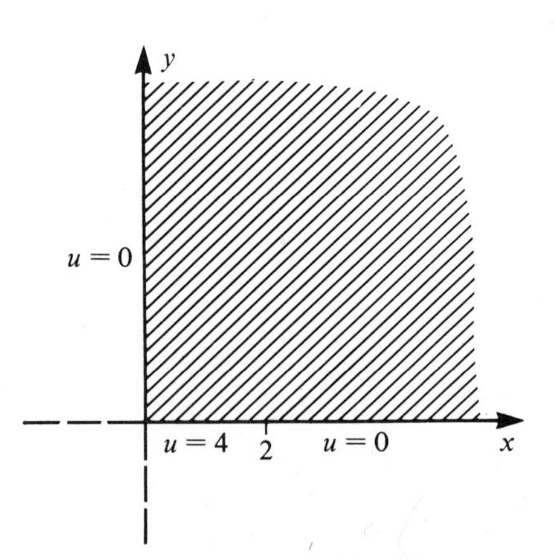

Figure 250. Temperatures prescribed on two sides of unbounded plate.

Note that the partial differential equation here is Laplace's equation in the two space variables x and y. Let $u(x, y) = X(x)Y(y)$, and obtain

$$X'' - \lambda X = 0, \qquad Y'' + \lambda Y = 0,$$

for λ constant, to be determined.

Now, $u(0, y) = 0$ for $y > 0$; hence $X(0) = 0$. Since we know more about X than we do about Y, begin with X. We have

$$X'' - \lambda X = 0; \qquad X(0) = 0.$$

We encountered this problem before with vibrations in a semiinfinite string, where we obtained

$$\text{eigenvalues:} \quad \lambda = -\alpha^2 \qquad (\alpha > 0)$$

and

$$\text{eigenfunctions:} \quad X_\alpha = A_\alpha \sin(\alpha x).$$

In using previous results, we are seeking a bounded solution, as usual.

Next, we have $Y'' + \lambda Y = 0$. Then, $Y'' - \alpha^2 Y = 0$; so

$$Y(y) = B_\alpha e^{\alpha y} + C_\alpha e^{-\alpha y}.$$

Since $\alpha > 0$, we must choose $B_\alpha = 0$ to have a bounded solution for $y > 0$. Thus, we are left with $Y_\alpha = C_\alpha e^{-\alpha y}$ and

$$u_\alpha(x, y) = D_\alpha \sin(\alpha x)e^{-\alpha y}.$$

To satisfy $u(x, 0) = f(x)$, we try a superposition

$$u(x, y) = \int_0^\infty D_\alpha \sin(\alpha x)e^{-\alpha y}\, d\alpha.$$

Then,

$$u(x, 0) = f(x) = \int_0^\infty D_\alpha \sin(\alpha x)\, d\alpha,$$

and we have a Fourier sine integral expansion of $f(x)$ on $[0, \infty)$. We have

$$D_\alpha = \frac{2}{\pi}\int_0^\infty f(r)\sin(\alpha r)\, dr = \frac{8}{\pi\alpha}[1 - \cos(2\alpha)].$$

Thus, the solution is

$$u(x, y) = \frac{8}{\pi}\int_0^\infty \frac{1 - \cos(2\alpha)}{\alpha}\sin(\alpha x)e^{-\alpha y}\, d\alpha.$$

In the next section we shall consider problems involving three independent variables. These will require multiple Fourier series for their solution.

PROBLEMS FOR SECTION 13.5

In each of Problems 1 through 5, solve the given boundary value problem.

1. $\dfrac{\partial u}{\partial t} = 4\dfrac{\partial^2 u}{\partial x^2} \quad (-\infty < x < \infty,\ t > 0)$

$$u(x, 0) = \begin{cases} -2, & -1 \le x < 0 \\ x, & 0 \le x \le 1 \\ 0, & |x| > 1 \end{cases}$$

2. $\dfrac{\partial u}{\partial t} = 16\dfrac{\partial^2 u}{\partial x^2} \quad (0 \le x < \infty,\ t > 0)$

$u(0, t) = 0, \quad u(x, 0) = e^{-x}\sin(x)$

3. $\nabla^2 u = 0 \quad (-\infty < x < \infty,\ y > 0)$

$$u(x, 0) = \begin{cases} -1, & -4 \le x < 0 \\ 0, & 0 \le x \le 4 \\ e^{-2|x|}, & |x| > 4 \end{cases}$$

4. $\nabla^2 u = 0, \quad (x > 0,\ y < 0)$

$$u(x, 0) = 0, \quad u(0, y) = \begin{cases} 0, & -5 \le y \le 0 \\ 2, & -7 \le y < -5 \\ 0, & y < -7 \end{cases}$$

5. $\nabla^2 u = 0, \quad (x > 0,\ -\infty < y < \infty)$

$$u(0, y) = \begin{cases} \cos(y), & -\pi \le y \le \pi \\ 0, & |y| > \pi \end{cases}$$

6. Find the temperature function for a thin, semiinfinite bar with temperature at $x = 0$ kept at zero, and initial temperature throughout the bar given by

$$f(x) = \begin{cases} x, & 0 \le x \le 4, \\ 0, & x > 4. \end{cases}$$

7. Solve for the temperature function in a thin, infinite bar (extending $-\infty < x < \infty$) if the initial temperature is $f(x)$. Assume as always that only bounded solutions are wanted.

After obtaining a formula for the temperature function in general, obtain the particular formula for the case that

$$f(x) = \begin{cases} 1 - |x|, & -1 \le x \le 1, \\ 0, & |x| > 1. \end{cases}$$

8. Find the temperature function for a thin, semiinfinite bar if the bar is initially kept at temperature zero, and for $t > 0$ the left end ($x = 0$) is kept at constant temperature A.

9. Find the temperature function for a thin, semiinfinite bar if the left end ($x = 0$) is insulated and the temperature at time zero is given by $f(x)$.

Find the particular case of this solution when $f(x)$ is given by $f(x) = e^{-x}$.

10. Find the steady-state temperature in a thin, flat plate extending over the region $x \ge 0$, $y \ge 0$, if the temperature on the side $x = 0$ is kept at e^{-y} and the temperature on the side $y = 0$ is kept at zero.

(*Hint:* Here, you are solving the two-variable Laplace equation $\partial^2 u/\partial x^2 + \partial^2 u/\partial y^2 = 0$ over the given region, subject to the given boundary conditions. This is a *Dirichlet problem* over an infinite region.)

11. Solve for the steady-state temperatures in the infinite flat plate covering the region $x \ge 0$ if the temperature on the boundary $x = 0$ is kept at $f(y)$, where

$$f(y) = \begin{cases} 1, & -1 \le y \le 1, \\ 0, & |y| > 1. \end{cases}$$

12. Solve for steady-state temperatures in the infinite flat plate covering the region $y \ge 0$ if the temperature on the boundary $y = 0$ is kept at 0 for $x < 4$, constant A for $4 \le x \le 8$, and 0 for $x > 8$.

13. Find the steady-state temperature in the infinite flat plate covering the region $0 \le y \le 1$, $x \ge 0$, if the temperature on the left and bottom sides is kept at 0 and on the top side is $f(x)$.

14. Redo Problem 13 under the boundary conditions that the temperature on the left side is kept at 1, on the top side at $f(x)$, and on the bottom side at 0. (*Hint:* Split this into two problems.)

15. Find the steady-state temperature in a flat plate covering the region $-1 \le x \le 1$, $y \ge 0$, if the temperature on the bottom side is 2, on the left side is e^{-y}, and on the right side is 3. (*Hint:* Split this into three problems.)

16. A thin bar is aligned along the positive x-axis. At time zero, the temperature at point x is $A - x$ for $0 \le x \le A$ (A is a positive constant) and zero for $x > A$. The left end is kept at temperature A at all times. Find the temperature distribution in the bar.

17. Find the temperature function for a semiinfinite bar if the left end ($x = 0$) is insulated and the initial temperature is $f(x)$.

18. Solve for the steady-state temperature in a flat plate extending over the region $x \ge 0$, $0 \le y \le 2$, if the left and bottom sides are insulated and the top side has temperature $f(x)$.

19. Solve for the steady-state temperature in a flat plate extending over the region $y \ge 0$, $0 \le x \le 2$, if the bottom side is insulated and the temperature on the left side is $f(y)$, on the right side, $g(y)$.

20. Find the steady-state temperature in a flat plate extending over the region $x \ge 0$, $y \ge 0$, if the temperature on the vertical side is given by e^{-y}, and the bottom side is insulated.

21. Find the steady-state temperature in a flat strip extending over the region $x \ge 0$, $0 \le y \le 1$, if the temperature on the two horizontal sides is kept at zero and that on the vertical side is kept at 2.

13.6 MULTIPLE FOURIER SERIES SOLUTIONS OF BOUNDARY VALUE PROBLEMS

Thus far, we have restricted the number of independent variables concerned to two. However, Fourier series and separation of variables do extend to higher-dimensional problems. Here is an example involving the wave equation in two dimensions, though the method is generally applicable to other problems as well. Imagine an elastic membrane stretched across a rectangular frame. If we put the origin at one corner and align the axes along the frame, as in Figure 251, we can think of the membrane as occupying the rectangle $0 \le x \le L$, $0 \le y \le K$. Vibrations in the membrane will be governed by the two-dimensional wave equation

$$\frac{\partial^2 z}{\partial t^2} = a^2\left(\frac{\partial^2 z}{\partial x^2} + \frac{\partial^2 z}{\partial y^2}\right) \qquad (0 < x < L,\ 0 < y < K,\ t > 0),$$

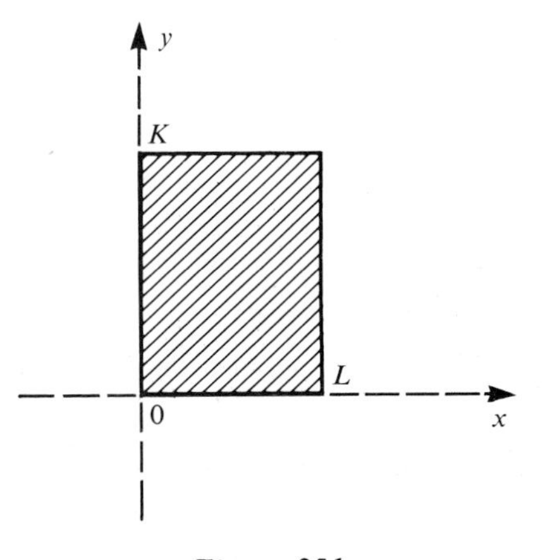

Figure 251

boundary conditions (the membrane is attached to the frame)

$$z(x, 0, t) = z(x, K, t) = 0 \quad \text{for} \quad 0 < x < L \quad \text{and} \quad t > 0,$$
$$z(0, y, t) = z(L, y, t) = 0 \quad \text{for} \quad 0 < y < K \text{ and } t > 0,$$

and initial conditions

$$z(x, y, 0) = f(x, y) \qquad \text{(initial position)},$$

$$\frac{\partial z}{\partial t}(x, y, 0) = g(x, y) \qquad \text{(initial velocity)}.$$

As an illustration, we shall solve this when $g(x, y) = 0$ (i.e., the membrane is released from rest).

Here we try a separation $z(x, y, t) = X(x)Y(y)T(t)$. Substituting into the differential equation gives us

$$XYT'' = a^2(X''YT + XY''T)$$

or

$$\frac{T''}{a^2T} - \frac{Y''}{Y} = \frac{X''}{X}.$$

The left side depends only on t and y, the right side only on x. Since x, y, and t are independent, both sides must equal some constant λ. Then,

$$\frac{X''}{X} = \lambda = \frac{T''}{a^2T} - \frac{Y''}{Y}.$$

Now look at

$$\frac{T''}{a^2T} - \frac{Y''}{Y} = \lambda.$$

This can be written

$$\frac{T''}{a^2T} - \lambda = \frac{Y''}{Y}.$$

Since y and t are independent, there must be a constant μ such that

$$\frac{T''}{a^2T} - \lambda = \frac{Y''}{Y} = \mu.$$

We now have a separation involving *two* separation constants:

$$X'' - \lambda X = 0,$$
$$Y'' - \mu Y = 0,$$

and

$$T'' - (\lambda + \mu)a^2T = 0.$$

From $z(0, y, t) = z(L, y, t) = 0$, we get

$$X(0) = X(L) = 0.$$

We have solved this eigenvalue problem for X many times, getting

$$\lambda_n = -\frac{n^2\pi^2}{L^2} \quad \text{for} \quad n = 1, 2, 3, \ldots$$

and

$$X_n = A_n \sin\left(\frac{n\pi x}{L}\right),$$

with A_n an arbitrary constant.

Similarly, $z(x, 0, t) = z(x, K, t) = 0$ gives us $Y(0) = Y(K) = 0$. Thus,

$$\mu_m = -\frac{m^2\pi^2}{K^2} \quad \text{for} \quad m = 1, 2, 3, \ldots$$

and

$$Y_m = B_m \sin\left(\frac{m\pi x}{K}\right),$$

with B_m an arbitrary constant.

Note: m and n are independent here; they are simply both positive integers.

Now the equation for T is

$$T'' + \left(\frac{n^2}{L^2} + \frac{m^2}{K^2}\right)\pi^2 a^2 T = 0,$$

with solution of the form

$$T = C \cos\left(\sqrt{\frac{n^2}{L^2} + \frac{m^2}{K^2}}\,\pi a t\right) + D \sin\left(\sqrt{\frac{n^2}{L^2} + \frac{m^2}{K^2}}\,\pi a t\right).$$

Since $T'(0) = 0$, then $D = 0$. Hence, for $n = 1, 2, \ldots$ and $m = 1, 2, \ldots$, we have

$$T_{nm} = C_{nm} \cos\left(\sqrt{\frac{n^2}{L^2} + \frac{m^2}{K^2}}\,\pi a t\right),$$

with C_{nm} an arbitrary constant.

Thus,

$$z_{nm} = k_{nm} \sin\left(\frac{n\pi x}{L}\right)\sin\left(\frac{m\pi y}{K}\right)\cos\left(\sqrt{\frac{n^2}{L^2} + \frac{m^2}{K^2}}\,\pi a t\right)$$

satisfies, for each m and n, all the conditions of the problem except perhaps $z(x, y, 0) = f(x, y)$. To satisfy this condition, we try a superposition (here, a double sum)

$$z(x, y, t) = \sum_{n=1}^{\infty}\sum_{m=1}^{\infty} k_{nm} \sin\left(\frac{n\pi x}{L}\right)\sin\left(\frac{m\pi y}{K}\right)\cos\left(\sqrt{\frac{n^2}{L^2} + \frac{m^2}{K^2}}\,\pi a t\right)$$

and attempt to choose the k_{nm}'s so that $z(x, y, 0) = f(x, y)$. Thus, we need

$$f(x, y) = \sum_{n=1}^{\infty}\sum_{m=1}^{\infty} k_{nm} \sin\left(\frac{n\pi x}{L}\right)\sin\left(\frac{m\pi y}{K}\right).$$

Now recognize (from Section 12.8) that what is required is a Fourier sine expansion of $f(x, y)$ on $0 \le x \le L$, $0 \le y \le K$. This tells us how to choose the k_{nm}'s, namely,

$$k_{nm} = \frac{4}{LK} \int_0^K \int_0^L f(x, y) \sin\left(\frac{n\pi x}{L}\right) \sin\left(\frac{m\pi y}{K}\right) dx\, dy,$$

and the solution is complete.

As a specific example, suppose that the initial position is given by

$$f(x, y) = x(x - L)y(y - K) \qquad (0 \le x \le L, 0 \le y \le K).$$

In this event,

$$\begin{aligned} k_{nm} &= \frac{4}{LK} \int_0^K \int_0^L x(x - L)y(y - K) \sin\left(\frac{n\pi x}{L}\right) \sin\left(\frac{m\pi y}{K}\right) dx\, dy \\ &= \frac{16L^2K^2}{(n\pi)^3(m\pi)^3} [\cos(n\pi) - 1][\cos(m\pi) - 1]. \end{aligned}$$

Now note that

$$\cos(n\pi) - 1 = (-1)^n - 1 = \begin{cases} -2 & \text{if } n \text{ is odd,} \\ 0 & \text{if } n \text{ is even.} \end{cases}$$

A similar result holds for $\cos(m\pi) - 1$, and the final solution can be written

$$\begin{aligned} z(x, y, t) = \frac{64L^2K^2}{\pi^6} \sum_{n=1}^{\infty} \sum_{m=1}^{\infty} \Bigg\{ &\frac{1}{(2n - 1)^3} \frac{1}{(2m - 1)^3} \sin\left(\frac{(2n - 1)\pi x}{L}\right) \\ &\times \sin\left(\frac{(2m - 1)\pi y}{K}\right) \cos\left(\sqrt{\frac{(2n - 1)^2}{L^2} + \frac{(2m - 1)^2}{K^2}}\, \pi a t\right)\Bigg\}. \end{aligned}$$

As a second illustration of the use of multiple Fourier series, solve for the temperature distribution in a square plate of side length 1, if the sides are kept at temperature 0, and the interior temperature at time 0 is given by $x(1 - x)y(1 - y)$. The boundary value problem is

$$\frac{\partial u}{\partial t} = a^2\left(\frac{\partial^2 u}{\partial x^2} + \frac{\partial^2 u}{\partial y^2}\right) \qquad (0 \le x \le 1, 0 \le y \le 1, t > 0),$$

$$u(x, 0, t) = u(x, 1, t) = u(0, y, t) = u(1, y, t) = 0,$$

$$u(x, y, 0) = x(1 - x)y(1 - y).$$

Let $u(x, y, t) = X(x)Y(y)T(t)$ and obtain, for separation constants λ and μ,

$$X'' - \mu X = 0, \qquad T' - \lambda a^2 T = 0, \quad \text{and} \quad Y'' - (\lambda - \mu)Y = 0.$$

We also have

$$X(0) = X(1) = 0 \qquad \text{and } Y(0) = Y(1) = 0.$$

The problem for X has been solved before, giving us

$$\mu = -n^2\pi^2$$

and

$$X_n = A_n \sin(n\pi x) \quad \text{for} \quad n = 1, 2, 3, \ldots.$$

Now look at the equation for Y, which becomes

$$Y'' - (\lambda + n^2\pi^2)Y = 0.$$

With $Y(0) = Y(1) = 0$, we again obtain from previous work that

$$\lambda + n^2\pi^2 = -m^2\pi^2 \quad \text{for} \quad m = 1, 2, 3, \ldots$$

and

$$Y_m = B_m \sin(m\pi y).$$

Thus,

$$\lambda = -(n^2 + m^2)\pi^2,$$

in which n and m can independently be any positive integers.

Finally, $T' - \lambda a^2 T = 0$ becomes

$$T' + (n^2 + m^2)a^2\pi^2 T = 0,$$

with solution

$$T = C \exp[-(n^2 + m^2)\pi^2 a^2 t].$$

For each n and m, then, products XYT have the form

$$u_{nm}(x, y, t) = k_{nm} \sin(n\pi x) \sin(m\pi y) \exp[-(n^2 + m^2)\pi^2 a^2 t].$$

Such a function will satisfy the partial differential equation and the boundary conditions for any positive integers n and m. To satisfy the initial condition, set

$$u(x, y, t) = \sum_{n=1}^{\infty} \sum_{m=1}^{\infty} k_{nm} \sin(n\pi x) \sin(m\pi y) \exp[-(n^2 + m^2)\pi^2 a^2 t].$$

We now need

$$u(x, y, 0) = \sum_{n=1}^{\infty} \sum_{m=1}^{\infty} k_{nm} \sin(n\pi x) \sin(m\pi y) = x(1 - x)y(1 - y).$$

This is a double Fourier sine expansion of $x(1 - x)y(1 - y)$ on $0 \leq x \leq 1, 0 \leq y \leq 1$. Accordingly, we choose

$$k_{nm} = 4 \int_0^1 \int_0^1 x(1 - x)y(1 - y) \sin(n\pi x) \sin(m\pi y)\, dx\, dy$$

$$= \begin{cases} \dfrac{64}{n^3 m^3 \pi^6} & \text{if } n \text{ and } m \text{ are both odd,} \\ 0 & \text{if } n \text{ is even or } m \text{ is even.} \end{cases}$$

The solution is

$$u(x, y, t) = \frac{64}{\pi^6} \sum_{n=1}^{\infty} \sum_{m=1}^{\infty} \left(\frac{1}{(2n-1)^3(2m-1)^3} \sin[(2n-1)\pi x] \right.$$

$$\left. \times \sin[(2m-1)\pi y] \exp\{-[(2n-1)^2 + (2m-1)^2]\pi^2 a^2 t\} \right).$$

In general, a problem involving n independent variables will have $(n-1)$ separation constants, and will, if solved by Fourier series methods, involve $(n-1)$ summations in the final solution. We pursue further examples in the exercises.

PROBLEMS FOR SECTION 13.6

In each of Problems 1 through 8, solve the given boundary value problem.

1. $\nabla^2 u(x, y, z) = 0 \qquad (0 \le x \le 1, 0 \le y \le 1, 0 \le z \le 1)$

$u(0, y, z) = u(1, y, z) = u(x, 0, z) = u(x, 1, z) = u(x, y, 0) = 0$

$u(x, y, 1) = xy$

2. $\dfrac{\partial u}{\partial t} = \dfrac{\partial^2 u}{\partial x^2} + \dfrac{\partial^2 u}{\partial y^2} \qquad (0 \le x \le \pi, 0 \le y \le \pi, t > 0)$

$u(0, y, t) = u(\pi, y, t) = u(x, 0, t) = u(x, \pi, t) = 0$

$u(x, y, 0) = x(\pi - x)y(\pi - y)^2$

3. $\dfrac{\partial^2 u}{\partial x^2} + \dfrac{\partial^2 u}{\partial y^2} = \dfrac{\partial^2 u}{\partial t^2} \qquad (0 \le x \le 2\pi, 0 \le y \le 2\pi, t > 0)$

$u(0, y, t) = u(2\pi, y, t) = u(x, 0, t) = u(x, 2\pi, t) = 0$

$u(x, y, 0) = x^2 \sin(y) \qquad (0 \le x \le 2\pi, 0 \le y \le 2\pi)$

$\dfrac{\partial u}{\partial t}(x, y, 0) = 0$

4. $\nabla^2 u = 0 \qquad (0 \le x \le \pi, 0 \le y \le 2\pi, 0 \le z \le 1)$

$u(0, y, z) = u(x, 0, z) = u(x, 2\pi, z) = u(x, y, 0) = u(x, y, 1) = 0$

$u(\pi, y, z) = 2$

5. $\dfrac{\partial^2 u}{\partial t^2} = 4\left(\dfrac{\partial^2 u}{\partial x^2} + \dfrac{\partial^2 u}{\partial y^2}\right) \qquad (0 \le x \le 2\pi, 0 \le y \le 2\pi, t > 0)$

$u(0, y, t) = u(2\pi, y, t) = u(x, 0, t) = u(x, 2\pi, t) = 0$

$u(x, y, 0) = 0, \qquad \dfrac{\partial u}{\partial t}(x, y, 0) = 1$

6. $$\frac{\partial^2 u}{\partial t^2} = 9\left(\frac{\partial^2 u}{\partial x^2} + \frac{\partial^2 u}{\partial y^2}\right) \qquad (0 \le x \le \pi,\ 0 \le y \le \pi,\ t > 0)$$

$$u(0, y, t) = u(\pi, y, t) = u(x, 0, t) = u(x, \pi, t) = 0$$

$$u(x, y, 0) = \sin(x)\cos(y)$$

$$\frac{\partial u}{\partial t}(x, y, 0) = xy$$

7. $$\nabla^2 u = 0 \qquad (0 \le x \le 1,\ 0 \le y \le 2\pi,\ 0 \le z \le \pi)$$

$$u(0, y, z) = u(1, y, z) = u(x, 0, z) = u(x, y, 0) = 0$$

$$u(x, 2\pi, z) = 2, \qquad u(x, y, \pi) = 1$$

8. $$\nabla^2 u = 0 \qquad (0 \le x \le 1,\ 0 \le y \le 1,\ 0 \le z \le 1)$$

$$u(0, y, z) = u(1, y, z) = u(x, 0, z) = u(x, 1, z) = 0$$

$$u(x, y, 0) = -1, \qquad u(x, y, 1) = 1$$

9. Find the temperature function for a cube $0 \le x \le L$, $0 \le y \le L$, $0 \le z \le L$, if all sides are insulated and the initial temperature is $f(x, y, z)$.

Hint: Here, you are solving

$$\frac{\partial u}{\partial t} = a^2\left(\frac{\partial^2 u}{\partial x^2} + \frac{\partial^2 u}{\partial y^2} + \frac{\partial^2 u}{\partial z^2}\right) \qquad (0 < x < L,\ 0 < y < L,\ 0 < z < L,\ t > 0),$$

$$\frac{\partial u}{\partial x}(0, y, z) = \frac{\partial u}{\partial x}(L, y, z) = 0 \qquad (0 < y < L,\ 0 < z < L),$$

$$\frac{\partial u}{\partial y}(x, 0, z) = \frac{\partial u}{\partial y}(x, L, z) = 0 \qquad (0 < x < L,\ 0 < z < L),$$

$$\frac{\partial u}{\partial z}(x, y, 0) = \frac{\partial u}{\partial z}(x, y, L) = 0 \qquad (0 < x < L,\ 0 < y < L),$$

$$u(x, y, z, 0) = f(x, y, z) \qquad (0 < x < L,\ 0 < y < L,\ 0 < z < L).$$

10. Solve the three-dimensional wave problem

$$\frac{\partial^2 u}{\partial t^2} = a^2\left(\frac{\partial^2 u}{\partial x^2} + \frac{\partial^2 u}{\partial y^2} + \frac{\partial^2 u}{\partial z^2}\right) \qquad (0 < x < A,\ 0 < y < B,\ 0 < z < C),$$

$$u(0, y, z, t) = u(A, y, z, t) = 0 \qquad (0 < y < B,\ 0 < z < C,\ t > 0),$$

$$u(x, 0, z, t) = u(x, B, z, t) = 0 \qquad (0 < x < A,\ 0 < z < C,\ t > 0),$$

$$u(x, y, 0, t) = u(x, y, C, t) = 0 \qquad (0 < x < A,\ 0 < y < B,\ t > 0),$$

$$u(x, y, z, 0) = f(x, y, z) \qquad (0 < x < A,\ 0 < y < B,\ 0 < z < C),$$

$$\frac{\partial u}{\partial t}(x, y, z, 0) = 0.$$

This corresponds to finding sound waves in a room, assuming a given initial disturbance with zero initial velocity.

11. Find an expression for sound waves in a cubical room, assuming an initial disturbance $f(x, y, z)$ and an initial velocity $g(x, y, z)$.

12. Solve for vibrations in a rectangular membrane attached to a fixed frame, if the initial position is given by $f(x, y, z)$ and the initial velocity is given by $g(x, y, z)$.

13. Solve the Dirichlet problem

$$\nabla^2 u = 0 \qquad (0 < x < 1, 0 < y < 1, 0 < z < 2),$$
$$u = 0 \qquad \text{(on all faces except the top)},$$
$$u(x, y, 2) = xy^2 \qquad (0 < x < 1, 0 < y < 1).$$

14. Solve the Dirichlet problem

$$\nabla^2 u = 0 \qquad (0 < x < 1, 0 < y < 4, 0 < z < 2),$$
$$u = 0 \qquad \text{(on the faces } x = 0, x = 1, z = 0, z = 2),$$
$$u(x, 0, z) = xz, \; u(x, 4, z) = \sin(z) \qquad (0 < x < 1, 0 < z < 2).$$

13.7 FOURIER-BESSEL SOLUTIONS OF BOUNDARY VALUE PROBLEMS

Sometimes the ideas used in obtaining Fourier series solutions of boundary value problems appear in other contexts. Here is an example involving Bessel functions.

Consider the problem of heat conduction in a homogeneous circular cylinder of radius k, assumed infinite in length. Here, it is convenient to use cylindrical coordinates, with the coordinate system as shown in Figure 252. We write the heat equation as

$$\frac{\partial u}{\partial t} = a^2\left(\frac{\partial^2 u}{\partial r^2} + \frac{1}{r}\frac{\partial u}{\partial r} + \frac{1}{r^2}\frac{\partial^2 u}{\partial \theta^2} + \frac{\partial^2 u}{\partial z^2}\right).$$

To simplify matters, we shall assume that the temperature at any point in the cylinder depends only on time t and the distance r from the axis of the cylinder (here, the z-axis). Then,

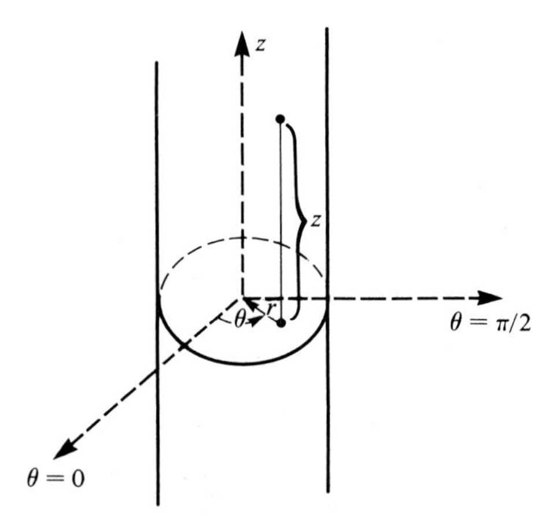

Figure 252.
Infinite cylinder aligned along z-axis.

$$\frac{\partial u}{\partial \theta} = \frac{\partial u}{\partial z} = 0$$

and the heat equation is

$$\frac{\partial u}{\partial t} = a^2\left(\frac{\partial^2 u}{\partial r^2} + \frac{1}{r}\frac{\partial u}{\partial r}\right) \qquad (0 \le r < k, t > 0).$$

We shall assume in this example that

$$u(k, t) = 0 \quad \text{for} \quad t > 0$$

(the surface is kept at zero temperature) and that

$$u(r, 0) = f(r) \quad \text{for} \quad 0 \le r < k$$

(interior temperature is given at time 0).

To separate variables, put

$$u(r, t) = R(r)T(t).$$

Substituting into the partial differential equation gives us

$$\frac{T'}{a^2 T} = \frac{R'' + \dfrac{1}{r} R'}{R} = \lambda,$$

for some constant λ (since r and t are independent). Then,

$$R'' + \frac{1}{r} R' - \lambda R = 0 \quad \text{and} \quad T' - \lambda a^2 T = 0.$$

Consider cases on λ.

Case 1 $\lambda = 0$. Then one easily solves the differential equation for R to get

$$R(r) = C \ln(r) + K.$$

Since $\ln(r)$ is unbounded as $r \to 0$, we must choose $C = 0$, getting $R =$ constant. Similarly, $T' - \lambda a^2 T = 0$ gives us $T = At + B$, and we must choose $A = 0$ to have T remain bounded. Thus, $T =$ constant also; hence, $u(r, t) = R(r)T(t) =$ constant in this case. This will satisfy $u(k, t) = 0$ for $t > 0$ only if this constant is zero, giving us the zero solution $u(r, t) = 0$. This is in fact the solution of the problem if $f(r) = 0$ for $0 \le r < k$. If, however, $f(r)$ is not identically zero (certainly the more interesting case), then $\lambda = 0$ does not give us a solution, and zero is not an eigenvalue.

Case 2 $\lambda > 0$. Then $T = e^{\lambda a^2 t}$ is unbounded, again violating our intuition that the temperature should not become infinite with time.

Case 3 $\lambda < 0$. Write $\lambda = -\alpha^2$, where $\alpha > 0$. Then $T = e^{-\alpha^2 a^2 t}$ and the equation for R is

$$r^2 R'' + rR' + \alpha^2 r^2 R = 0.$$

This is Bessel's equation of order zero, with general solution (see Chapter 6, Section 6.1)

$$R(r) = AJ_0(\alpha r) + BY_0(\alpha r).$$

Now, as $r \to 0$, $Y_0(r) \to -\infty$, which is physically unacceptable here. Thus, we must choose $B = 0$.

Thus far, for every $\alpha > 0$, we have a solution

$$u_\alpha(r, t) = A_\alpha J_0(\alpha r)e^{-\alpha^2 a^2 t}$$

of the heat equation. We now have to satisfy the given conditions. First, we need, for all $t > 0$,

$$u_\alpha(k, t) = A_\alpha J_0(\alpha k)e^{-\alpha^2 a^2 t} = 0.$$

To satisfy this with $A_\alpha \neq 0$, we need $J_0(\alpha k) = 0$. We know (from Chapter 6, Section 6.1) that $J_0(x)$ crosses the x-axis at infinitely many positive values, say, $q_1, q_2, \ldots$ of x. Thus, we choose $\alpha_n k = q_n$ or $\alpha_n = q_n/k$ for $n = 1, 2, 3, \ldots$. Then, for $n = 1, 2, 3, \ldots$, we have

$$u_n(r, t) = A_n J_0(\alpha_n r) \exp(-\alpha_n{}^2 a^2 t).$$

There remains to determine the A_n's. We have the remaining condition $u(r, 0) = f(r)$ for $0 \leq r \leq k$. Taking a cue from previous successes, we try a superposition

$$u(r, t) = \sum_{n=1}^{\infty} u_n(r, t) = \sum_{n=1}^{\infty} A_n J_0(\alpha_n r) \exp(-\alpha_n{}^2 a^2 t).$$

We now have to choose the A_n's so that

$$u(r, 0) = \sum_{n=1}^{\infty} A_n J_0(\alpha_n r) = f(r).$$

This is like a Fourier series, except that we have zero order Bessel functions instead of sines and cosines. For those who have read Section 6.4, we have a Fourier-Bessel expansion of $f(r)$, and it is clear from Sturm-Liouville theory how to choose the A_n's. Since we do not assume here that the reader is familiar with general eigenfunction expansions, we shall proceed from first principles.

The key to choosing the A_n's lies in the integral formula (noted in Section 6.1)

$$\int_0^k rJ_0(\alpha_n r)J_0(\alpha_m r)\, dr = 0 \quad \text{if} \quad n \neq m.$$

This is called an *orthogonality relationship*, and it can be used to find the A_n's here just as we used similar integral formulas involving sines and cosines to find the Fourier coefficients in Chapter 12.

Multiply both sides of the series $f(r) = \sum_{n=1}^{\infty} A_n J_0(\alpha_n r)$ by $rJ_0(\alpha_m r)$, with m any positive integer, to obtain

$$\sum_{n=1}^{\infty} A_n rJ_0(\alpha_n r)J_0(\alpha_m r) = rf(r)J_0(\alpha_m r).$$

Integrate both sides from 0 to k, interchanging the summation and integral, to obtain

$$\sum_{n=1}^{\infty} A_n \int_0^k rJ_0(\alpha_n r)J_0(\alpha_m r)\, dr = A_m \int_0^k rJ_0^2(\alpha_m r)\, dr$$

$$= \int_0^k rf(r)J_0(\alpha_m r)\, dr.$$

Then,

$$A_m = \frac{\int_0^k rf(r)J_0(\alpha_m r)\,dr}{\int_0^k rJ_0^2(\alpha_m r)\,dr} \quad \text{for} \quad m = 1, 2, 3, \ldots.$$

This gives us the constants in the series for $u(r, t)$. The solution of the problem is

$$u(r, t) = \sum_{n=1}^{\infty} \left(\frac{\int_0^k \rho f(\rho)J_0(\alpha_n \rho)\,d\rho}{\int_0^k \rho J_0^2(\alpha_n \rho)\,d\rho} \right) J_0(\alpha_n r) \exp(-\alpha_n{}^2 a^2 t),$$

with ρ used as dummy integration variable in place of r to avoid confusion.

It can be shown, using recurrence relations for J_0 given in Chapter 6, that

$$\int_0^k \rho J_0^2(\alpha_n \rho)\,d\rho = \frac{k^2}{2} J_1^2(\alpha_n k).$$

Thus, $u(r, t)$ is often written

$$u(r, t) = \frac{2}{k^2} \sum_{n=1}^{\infty} \left(\frac{\int_0^k \rho f(\rho)J_0(\alpha_n \rho)\,d\rho}{J_1^2(\alpha_n k)} \right) J_0(\alpha_n r) \exp(-\alpha_n{}^2 a^2 t).$$

Independent of the heat conduction problem we have been mainly concerned with, we have also derived an expansion of $f(r)$ in a series of Bessel functions:

$$f(r) = \sum_{n=1}^{\infty} A_n J_0(\alpha_n r) \qquad (0 < r < k),$$

with

$$A_n = \frac{2}{k^2} \frac{\int_0^k \rho f(\rho)J_0(\alpha_n \rho)\,d\rho}{J_1^2(\alpha_n k)}.$$

This is called a *Fourier-Bessel expansion for* $f(r)$, and one can prove a theorem on convergence very similar to the Fourier convergence theorem of Chapter 12, Section 2.

In the next section we shall consider problems in which Legendre polynomials play a role in deriving a solution. In particular, we shall see Fourier-Legendre series analogous to the Fourier-Bessel series of this section.

PROBLEMS FOR SECTION 13.7

1. Solve the boundary value problem

$$\frac{\partial u}{\partial t} = \frac{\partial^2 u}{\partial r^2} + \frac{1}{r}\frac{\partial u}{\partial r} - Au \qquad (0 \le r < 1,\ t > 0),$$

$$(A > 0,\ \text{constant}),$$

$$u(1, t) = 0, \qquad u(r, 0) = k \qquad (k > 0,\ \text{constant}).$$

This problem models temperatures $u(r, t)$ in a thin, flat, circular plate, allowing a linear law of heat transfer from the surface of the plate to the surrounding medium, which is at zero temperature.

2. Solve the following Dirichlet problem for $u(r, z)$ in cylindrical coordinates (with the assumption that u does not depend on θ).

$$\nabla^2 u = 0 \qquad (0 < r < 2,\ z > 0)$$
$$u(2, z) = 0 \qquad (z > 0)$$
$$u(r, 0) = k \qquad (k > 0,\ \text{constant})$$

3. An elastic membrane is stretched over a fixed circular frame of radius 1, given an initial displacement, and then released from rest. Solve for the resulting displacements.

Hint: If the frame is in the xy-plane, we have in cylindrical coordinates

$$\frac{\partial^2 z}{\partial t^2} = a^2\left(\frac{\partial^2 z}{\partial r^2} + \frac{1}{r}\frac{\partial z}{\partial r} + \frac{1}{r^2}\frac{\partial^2 z}{\partial \theta^2}\right) \qquad (0 \le r < 1,\ 0 \le \theta \le 2\pi,\ t > 0),$$

$$z(r, \theta, 0) = f(r, \theta),$$

$$\frac{\partial z}{\partial t}(r, \theta, 0) = 0,$$

$$z(1, \theta, 0) = 0, \qquad (0 \le \theta \le 2\pi).$$

To separate variables, put $z(r, \theta, t) = R(r)\Theta(\theta)T(t)$. Thus, we have a problem in multiple Fourier series, with one series a Fourier-Bessel series. Also impose the periodicity conditions

$$\Theta(0) = \Theta(2\pi) \quad \text{and} \quad \Theta'(0) = \Theta'(2\pi).$$

4. Solve the following boundary value problem (in cylindrical coordinates).

$$\frac{\partial u}{\partial t} = a^2\frac{\partial^2 u}{\partial r^2} + \frac{a^2}{r}\frac{\partial u}{\partial r} \qquad (0 < r < k,\ t > 0)$$

$$\frac{\partial u}{\partial r}(k, t) = -Au(k, t) \qquad (t > 0)$$

$$u(r, 0) = f(r) \qquad (0 \le r < k)$$

To separate the variables, put $u(r, t) = R(r)T(t)$. You will need to assume that an equation of the form

$$AJ_0(\alpha k) + BJ_0'(\alpha k) = 0$$

has infinitely many positive solutions for α, say, $\alpha_1, \alpha_2, \ldots$; these will enable you to find the eigenvalues of the problem.

5. Solve the heat conduction problem

$$\frac{\partial u}{\partial t} = \frac{\partial^2 u}{\partial r^2} + \frac{1}{r}\frac{\partial u}{\partial r} + \frac{1}{r^2}\frac{\partial^2 u}{\partial \theta^2} \qquad (0 \le r < 1,\ 0 \le \theta \le 2\pi,\ t > 0),$$

$$u(1, \theta, t) = 0 \qquad (t > 0),$$

$$u(r, \theta, 0) = f(r, \theta) \qquad (0 \le r < 1,\ 0 \le \theta \le 2\pi).$$

6. Solve the wave problem

$$\frac{\partial^2 u}{\partial t^2} = \frac{\partial^2 u}{\partial x^2} + \frac{\partial^2 u}{\partial y^2} + \cos(\omega t) \qquad (0 < x < L, 0 < y < L, t > 0),$$

$$u(x, L, t) = u(0, y, t) = 0 \qquad (0 < x < L, 0 < y < L),$$

$$\frac{\partial u}{\partial x}(L, y, t) = \frac{\partial u}{\partial y}(x, L, t) = 0 \qquad (0 < x < L, 0 < y < L),$$

$$u(x, y, 0) = \frac{\partial u}{\partial t}(x, y, 0) = 0 \qquad (0 < x < L, 0 < y < L).$$

7. A solid cylinder is bounded by $z = 0$, $z = L$, and $r = 1$. The cylindrical sides are insulated, while the top is kept at temperature 2 and the bottom at temperature 0. Find the steady-state temperature $u(r, z)$ inside the cylinder.

13.8 FOURIER-LEGENDRE SOLUTIONS OF BOUNDARY VALUE PROBLEMS

The ideas of the last section can be broadened to encompass Legendre polynomials. We shall discuss here an example in which this is a useful thing to do.

Consider a Dirichlet problem for a solid sphere of radius k centered at the origin. That is, we want a function satisfying Laplace's equation inside the sphere and having prescribed values on the surface of the sphere.

In spherical coordinates r, θ, and ϕ, as given in Section 11.13, we have

$$\frac{1}{r}\frac{\partial^2}{\partial r^2}(ru) + \frac{1}{r^2 \sin^2(\phi)}\frac{\partial^2 u}{\partial \theta^2} + \frac{1}{r^2 \sin(\phi)}\frac{\partial}{\partial \phi}\left(\frac{\partial u}{\partial \phi}\sin(\phi)\right) = 0.$$

Consider the special case that u is independent of θ. We then have

$$\frac{1}{r}\frac{\partial^2}{\partial r^2}(ru) + \frac{1}{r^2 \sin(\phi)}\frac{\partial}{\partial \phi}\left(\frac{\partial u}{\partial \phi}\sin(\phi)\right) = 0$$

for $0 \le r < k$, $0 \le \phi \le \pi$.

We suppose that u is given on the surface of the sphere:

$$u(k, \phi) = f(\phi) \qquad (0 < \phi < \pi).$$

This boundary value problem may be thought of as one of steady-state heat conduction in a solid sphere, with surface temperatures dependent only on the azimuthal angle ϕ. Alternatively, u is also the electrostatic potential inside the sphere if there are no charges inside, under the assumption that $u = f(\phi)$ is given on the surface of the sphere.

To solve the problem, substitute $u(r, \phi) = R(r)\Phi(\phi)$ into the partial differential equation to get

$$\frac{r(rR)''}{R} + \frac{1}{\sin(\phi)}\frac{[\Phi' \sin(\phi)]'}{\Phi} = 0.$$

Then, for some λ,

$$r^2R'' + 2rR' - \lambda R = 0 \qquad (0 \le r < k)$$

and

$$\frac{1}{\sin(\phi)}[\Phi' \sin(\phi)]' + \lambda\Phi = 0 \qquad (0 < \phi < \pi).$$

In the last equation, change variables by setting

$$x = \cos(\phi).$$

Since $0 < \phi < \pi$, then $-1 < x < 1$. We have

$$\Phi' \sin(\phi) = \frac{1 - \cos^2(\phi)}{\sin(\phi)} \frac{\partial\Phi}{d\phi} = -(1 - x^2)\frac{d\Phi}{dx}.$$

Hence, the differential equation for Φ becomes

$$\frac{d}{dx}\left((1 - x^2)\frac{\partial\Phi}{dx}\right) + \lambda\Phi(x) = 0 \qquad (-1 < x < 1).$$

When $\lambda = n(n + 1)$, for $n = 0, 1, 2, 3, \ldots$, this has the Legendre polynomial $P_n(x)$ as solution. Further, when $\lambda = n(n + 1)$, the equation for R is

$$r^2R'' + 2rR' - n(n + 1)R = 0,$$

an Euler equation (see Chapter 2, Section 12) with solution of the form

$$R = Ar^n + Br^{-n-1}.$$

Since $r^{-n-1} \to \infty$ as $r \to 0$, we must choose $B = 0$. Thus, for $n = 0, 1, 2, 3, \ldots$, we have

$$u_n(r, x) = C_n r^n P_n(x)$$

as a solution of the partial differential equation. In terms of ϕ,

$$u_n(r, \phi) = C_n r^n P_n[\cos(\phi)].$$

To satisfy the boundary condition $u(k, \phi) = f(\phi)$, we write a superposition

$$u(r, \phi) = \sum_{n=0}^{\infty} C_n r^n P_n[\cos(\phi)]$$

and try to choose the C_n's so that

$$u(k, \phi) = \sum_{n=0}^{\infty} C_n k^n P_n[\cos(\phi)] = f(\phi).$$

This is a Fourier-type expansion of $f(\phi)$ in a series of Legendre polynomials. The orthogonality of Legendre polynomials enables us to find the C_n's. Letting $x = \cos(\phi)$ again, we want

$$\sum_{n=0}^{\infty} C_n k^n P_n(x) = f[\arccos(x)].$$

Multiply through by $P_m(x)$ and integrate from -1 to 1 to get

$$\sum_{n=0}^{\infty} C_n k^n \int_{-1}^{1} P_n(x)P_m(x)\, dx = \int_{-1}^{1} f[\arccos(x)]P_m(x)\, dx.$$

Now recall (Chapter 6, Section 6.3) that

$$\int_{-1}^{1} P_n(x)P_m(x)\, dx = \begin{cases} 0 & \text{if } n \neq m, \\ \dfrac{2}{2m+1} & \text{if } n = m. \end{cases}$$

Thus, the summation in the previous equation collapses to one term, and we have

$$C_m k^m \left(\frac{2}{2m+1}\right) = \int_{-1}^{1} f[\arccos(x)]P_m(x)\, dx.$$

Then,

$$C_m = \left(\frac{2m+1}{2}\right)\frac{1}{k^m}\int_{-1}^{1} f[\arccos(x)]P_m(x)\, dx,$$

for $m = 0, 1, 2, \ldots$.

Thus, the solution of the boundary value problem is

$$u(r, \theta) = \sum_{n=0}^{\infty} \left\{\left(\frac{2n+1}{2}\right)\int_{-1}^{1} f[\arccos(x)]P_n(x)\, dx\right\}\left(\frac{r}{k}\right)^n P_n[\cos(\phi)],$$

for $0 \leq r < k, 0 < \phi < \pi$.

The series representation

$$f(x) = \sum_{n=0}^{\infty} \left[\left(\frac{2n+1}{2}\right)\int_{0}^{1} f[\arccos(t)]P_n(t)\, dt\right] P_n(x)$$

is called a *Fourier-Legendre series for* $f(x)$, and it is valid under conditions similar to those of the Fourier convergence theorem of Section 12.2.

PROBLEMS FOR SECTION 13.8

1. Consider heat conduction in a solid hemisphere, $0 \leq r \leq k$, $0 \leq \phi < \pi/2$, $0 \leq \theta \leq 2\pi$ (spherical coordinates). Solve for the steady-state temperature function in the hemisphere if the base is kept at zero temperature and the curved surface at constant temperature A. Assume that u is a function of r and ϕ only.
2. Redo Problem 1, with the condition that the base is at temperature zero replaced by the condition that the base is insulated.
3. Redo Problem 1 when the temperature on the hemispherical surface is $u(k, \phi) = f[\cos(\phi)]$, not necessarily the constant A.
4. Solve the boundary value problem

$$\frac{\partial u}{\partial t} = \frac{\partial}{\partial x}\left((1 - x^2)\frac{\partial u}{\partial x}\right) \qquad (-1 \leq x \leq 1,\ t > 0),$$

$$u(x, 0) = f(x) \qquad (-1 < x < 1).$$

This gives the temperature in a bar extending from -1 to 1, if the thermal conductivity is $1 - x^2$. Assume that the lateral surface is insulated.

5. Solve for the steady-state temperature in a hollowed-out sphere given in spherical coordinates by $a \leq r \leq b$. The inner surface $r = a$ is kept at constant temperature T, and the outer surface $r = b$ is kept at constant temperature 0. Assume that u depends only on r and ϕ.

13.9 LAPLACE TRANSFORM SOLUTIONS OF BOUNDARY VALUE PROBLEMS

Laplace transforms can sometimes be used to solve boundary value problems if one of the variables ranges over 0 to ∞.

As an illustration, consider a one-dimensional heat conduction problem. Imagine a semiinfinite bar of metal (placed with one end at $x = 0$ and extending out the x-axis of our coordinate system). At time 0, the temperature is a constant A; from time 0 to t_0, the surface temperature is kept at constant value B and, after t_0, is kept at 0. We want the temperature at all points $x > 0$ and times $t > 0$.

Thus, we want to solve the problem

$$\frac{\partial u}{\partial t} = a^2 \frac{\partial^2 u}{\partial x^2} \quad (x > 0,\ t > 0),$$

$$u(x, 0) = A \qquad (x > 0),$$

$$u(0, t) = \begin{cases} B, & \text{for } 0 < t < t_0, \\ 0, & \text{for } t > t_0. \end{cases}$$

Begin by taking the Laplace transform of the partial differential equation, thinking of t as the transformed variable and carrying x along as just a symbol. We have

$$\mathcal{L}\left(\frac{\partial u}{\partial t}\right) = a^2 \mathcal{L}\left(\frac{\partial^2 u}{\partial x^2}\right).$$

Now,

$$\mathcal{L}\left(\frac{\partial u}{\partial t}\right) = s\mathcal{L}(u) - u(x, 0)$$

and

$$\mathcal{L}\left(\frac{\partial^2 u}{\partial x^2}\right) = \int_0^\infty e^{-st} \frac{\partial^2 u}{\partial x^2}\, dt = \frac{\partial^2}{\partial x^2} \int_0^\infty e^{-st} u(x, t)\, dt = \frac{\partial^2}{\partial x^2} \mathcal{L}(u).$$

Here, $\partial^2/\partial x^2$ and $\int_0^\infty \cdots dt$ interchange because x and t are independent. For convenience, let $v(x, s) = \mathcal{L}[u(x, t)]$. We then have

$$sv - u(x, 0) = a^2 \frac{d^2 v}{dx^2}.$$

Since $u(x, 0) = A$, we have

$$\frac{d^2 v}{dx^2} - \frac{s}{a^2} v = -\frac{A}{a^2}.$$

Now think of this as a differential equation in terms of x, with s carried along as a parameter. The general solution is

$$v(x, s) = \alpha e^{\sqrt{s}\, x/a} + \beta e^{-\sqrt{s}\, x/a} + \frac{A}{s},$$

where α and β are to be determined and may involve s.

First, we must choose $\alpha = 0$ to get a bounded solution. Thus,

$$v(x, s) = \beta e^{-\sqrt{s}\, x/a} + \frac{A}{s}.$$

From this we have

$$v(0, s) = \beta + \frac{A}{s}. \tag{a}$$

But also $v(0, s) = \mathscr{L}[u(0, t)]$, and $u(0, t)$ is given as boundary data in the problem. We can write

$$u(0, t) = B[1 - h(t - t_0)],$$

where

$$h(t - t_0) = \begin{cases} 0, & t < t_0, \\ 1, & t > t_0, \end{cases}$$

[we called such a function $u(t - t_0)$ in the chapter on Laplace transforms, but that would be ambiguous here].

Then,

$$\mathscr{L}[u(0, t)] = B\mathscr{L}(1) - B\mathscr{L}[h(t - t_0)] = \frac{B}{s} - \frac{Be^{-t_0 s}}{s}. \tag{b}$$

Putting together the two equations (a) and (b) for $\mathscr{L}[u(0, t)]$, we have

$$\beta + \frac{A}{s} = \frac{B}{s} - \frac{Be^{-t_0 s}}{s}.$$

Then,

$$\beta = \frac{(B - A)}{s} - \frac{Be^{-t_0 s}}{s}.$$

Thus,

$$v(x, s) = \mathscr{L}[u(x, t)] = \left[\frac{(B - A)}{s} - \frac{Be^{-t_0 s}}{s}\right] e^{-\sqrt{s}\, x/a} + \frac{A}{s}.$$

The solution can now be read from a table of inverse Laplace transforms:

$$u(x, t) = (B - A)\mathscr{L}^{-1}\left(\frac{e^{-\sqrt{s}x/a}}{s}\right) - B\mathscr{L}^{-1}\left(\frac{e^{-\sqrt{s}x/a}}{s} e^{-t_0 s}\right) + A\mathscr{L}^{-1}\left(\frac{1}{s}\right)$$

$$= \begin{cases} A \operatorname{erf}\left(\dfrac{x}{2a\sqrt{t}}\right) + B \operatorname{erfc}\left(\dfrac{x}{2a\sqrt{t}}\right) & \text{for } x > 0,\ 0 < t < t_0, \\ A \operatorname{erf}\left(\dfrac{x}{2a\sqrt{t}}\right) + B \operatorname{erfc}\left(\dfrac{x}{2a\sqrt{t}}\right) - B \operatorname{erfc}\left(\dfrac{x}{2a\sqrt{t - t_0}}\right) & \text{for } x > 0,\ t > t_0. \end{cases}$$

Here

$$\operatorname{erf}(x) = \frac{2}{\sqrt{\pi}} \int_0^x e^{-\xi^2}\, d\xi$$

is called the *error function*,* and

$$\operatorname{erfc}(x) = 1 - \operatorname{erf}(x)$$

is the *complementary error function.*

Having solved the problem, we can step back and observe the process. Note that the Laplace transform converted the partial differential equation into an ordinary one containing s as a parameter. The solution of this ordinary differential equation involved "constants" of integration which could be functions of s. We solved for these constants by using the Laplace transform of the boundary data and the requirement that solutions be bounded. The inverse Laplace transform then gave the solution of the problem.

Note also that we chose a nontrivial example here, for which the Laplace transform is a relatively easy method. The student should test this remark by trying to solve this problem by Fourier methods.

Another advantage of the Laplace transform method is its effectiveness in treating systems of partial differential equations. We shall pursue this and other illustrations of the method in the problems which follow.

PROBLEMS FOR SECTION 13.9

1. Solve the following problem for a semiinfinite vibrating string, using the Laplace transform.

$$\frac{\partial^2 y}{\partial t^2} = a^2 \frac{\partial^2 y}{\partial x^2} \qquad (x > 0,\ t > 0)$$

$$y(x, 0) = 0 \qquad (x > 0)$$

$$\frac{\partial y}{\partial t}(x, 0) = A \qquad (x > 0)$$

$$y(0, t) = t \qquad (t > 0)$$

Also solve the problem by a previous method and compare results.

---

* So named because of its uses in probability and statistics.

2. Solve the following problem by Laplace transform.

$$\frac{\partial^2 y}{\partial t^2} = a^2 \frac{\partial^2 y}{\partial x^2} \qquad (x > 0,\ t > 0)$$

$$y(x, 0) = 0 \qquad (x > 0)$$

$$y(0, t) = \begin{cases} \sin(2\pi t), & 0 \le t \le 1 \\ 0, & t \ge 1 \end{cases}$$

$$\frac{\partial y}{\partial t}(x, 0) = 0 \qquad (x > 0)$$

This represents motion of a string initially at rest along the positive x-axis but with left end moving in the prescribed fashion. (Imagine, for example, a long jump-rope with the child at one end moving the rope in the given way.) Look for a solution which decays to zero with increasing x for all time [i.e., $y(x, t) \to 0$ as $x \to \infty$ for all $t \ge 0$].

3. Solve this heat conduction problem by Laplace transform.

$$\frac{\partial u}{\partial t} = a^2 \frac{\partial^2 u}{\partial x^2} \qquad (x > 0,\ t > 0)$$

$$u(x, 0) = e^{-x} \qquad (x > 0)$$

$$u(0, t) = 0 \qquad (t > 0)$$

Also solve the problem by separation of variables and compare results.

4. Solve this wave problem by Laplace transform.

$$\frac{\partial^2 y}{\partial t^2} = a^2 \frac{\partial^2 y}{\partial x^2} \qquad (x > 0,\ t > 0)$$

$$y(x, 0) = 0 \qquad (t > 0)$$

$$\frac{\partial y}{\partial t}(x, 0) = k \text{ (constant)} \qquad (t > 0)$$

$$y(0, t) = t^2 \qquad (t > 0)$$

We want $y(x, t)$ to remain finite for all $t > 0$ as $x \to \infty$.

5. Solve this boundary value problem by Laplace transform.

$$a \frac{\partial u}{\partial x} + \frac{\partial u}{\partial y} = y \qquad (x > 0,\ y > 0)$$

$$u(x, 0) = 0 \qquad (x > 0)$$

$$u(0, y) = y \qquad (y > 0)$$

Here, a is a positive constant.

6. Use the Laplace transform to solve

$$\frac{\partial u}{\partial x} - \frac{\partial u}{\partial y} = 3y^2 \qquad (x > 0,\ y > 0),$$

$$u(x, 0) = 0 \qquad (x > 0),$$

$$u(0, y) = -y \qquad (y > 0).$$

7. Use the Laplace transform to solve

$$\frac{\partial^2 y}{\partial t^2} = 4\frac{\partial^2 y}{\partial x^2} \qquad (x > 0,\ t > 0),$$

$$y(x, 0) = 0 \qquad (x > 0),$$

$$\frac{\partial y}{\partial t}(x, 0) = e^{-x} \qquad (x > 0),$$

$$y(0, t) = \sin(t) \qquad (t > 0).$$

We also want $y(x, t)$ to remain finite for each $t > 0$ as $x \to \infty$.

8. Use the Laplace transform to solve

$$\frac{\partial u}{\partial t} = 16\frac{\partial^2 u}{\partial x^2} \qquad (0 < x < 1,\ t > 0),$$

$$u(x, 0) = k \text{ (constant)} \qquad (0 < x < 1),$$

$$u(0, t) = 0 \qquad (t > 0),$$

$$u(1, t) = 0 \qquad (t > 0).$$

9. Solve the system

$$\frac{\partial u}{\partial t} + a\frac{\partial v}{\partial x} = 0,$$

$$\frac{\partial v}{\partial t} + b\frac{\partial u}{\partial x} = 0.$$

Here, a and b are positive constants, u and v are functions of x and t, and $t > 0$, $-\infty < x < \infty$. Assume initial conditions $u(x, 0) = 1$, $v(x, 0) = x$, and $v(0, t) = t$. Seek a solution with bounded Laplace transform for $s > 0$.

10. Solve the electrical circuit problem

$$\frac{\partial I}{\partial t} + \frac{1}{L}\frac{\partial E}{\partial x} + \frac{R}{L}I = 0,$$

$$\frac{\partial E}{\partial t} + \frac{1}{C}\frac{\partial I}{\partial x} + \frac{G}{C}E = 0,$$

where $-\infty < x < \infty$, $t > 0$, subject to $E(x, 0) = f(x)$ and $I(x, 0) = g(x)$.

13.10 FOURIER TRANSFORM SOLUTIONS OF BOUNDARY VALUE PROBLEMS

In this section we continue the theme begun in the previous section, except that we shall concentrate on the use of Fourier transforms instead of the Laplace transform. The general strategy followed in applying the Laplace transform to boundary value problems remains the same, however. Generally, we do the following:

1. Choose a transform appropriate for the range of variables in question. For example, if $-\infty < x < \infty$, then a Fourier transform in x may be appropriate; if $y > 0$, then a Fourier sine or cosine transform in y may be called for. In some problems, several transforms will work equally well in finding the solution; in other instances,

one transform turns out to be easier than the others, for example, in taking inverse transforms at the final stage.

2. The transform generally transforms the partial differential equation into an ordinary differential equation, with the boundary conditions incorporated through the appropriate operational formula.
3. The ordinary differential equation must be solved to obtain the transform of the solution of the boundary value problem.
4. The transformed solution found in (3) must be inverted to find the solution itself.

In (3), one must be alert to the various methods available for solving ordinary differential equations. For example, one might apply a Fourier transform in step (1), then use a Laplace transform in (3) to solve the ordinary differential equation. Be prepared at each stage to use *all* the tools you have available.

For the rest of this section we shall look at examples. For ease of reference, we first list the Fourier transforms, their inversion formulas, and the operational formulas we will be using. On pages 831–835 are some tables of the various Fourier transforms.

	Transform	Inversion formula	Operational formula
Finite Fourier cosine transform	$C_n[f(x)] = f_C(n) = \int_0^\pi f(x)\cos(nx)\,dx$ for $n = 0, 1, 2, 3, \ldots$	$f(x) = \frac{1}{\pi} f_C(0) + \frac{2}{\pi}\sum_{n=1}^{\infty} f_C(n)\cos(nx)$ for $0 < x < \pi$	$C_n[f''(x)] = -n^2 f_C(n) - f'(0) + (-1)^n f'(\pi)$
Finite Fourier sine transform	$S_n[f(x)] = f_S(n) = \int_0^\pi f(x)\sin(nx)\,dx$ for $n = 1, 2, 3, \ldots$	$f(x) = \frac{2}{\pi}\sum_{n=1}^{\infty} f_S(n)\sin(nx)$ for $0 < x < \pi$	$S_n[f''(x)] = -n^2 f_S(n) + nf(0) - n(-1)^n f(\pi)$
Fourier cosine transform	$\mathscr{F}_C[f(x)] = \hat{f}_C(\lambda) = \int_0^\infty f(x)\cos(\lambda x)\,dx$	$f(x) = \frac{2}{\pi}\int_0^\infty \hat{f}_C(\lambda)\cos(\lambda x)\,d\lambda$	$\mathscr{F}_C[f''(x)] = -\lambda^2 \hat{f}_C(\lambda) - f'(0)$
Fourier sine transform	$\mathscr{F}_S[f(x)] = \hat{f}_S(x) = \int_0^\infty f(x)\sin(\lambda x)\,dx$	$f(x) = \frac{2}{\pi}\int_0^\infty \hat{f}_S(\lambda)\sin(\lambda x)\,d\lambda$	$\mathscr{F}_S[f''(x)] = -\lambda^2 \hat{f}_S(\lambda) + \lambda f(0)$
Fourier transform	$\mathscr{F}[f(x)] = \hat{f}(\lambda) = \int_{-\infty}^\infty f(x)e^{-i\lambda x}\,dx$	$f(x) = \frac{1}{2\pi}\int_{-\infty}^\infty \hat{f}(\lambda)e^{i\lambda x}\,d\lambda$	$\mathscr{F}[f''(x)] = -\lambda^2 \hat{f}(\lambda)$

EXAMPLE 389

Solve the heat conduction problem in an infinite slab given by

$$\frac{\partial u}{\partial t} = \frac{\partial^2 u}{\partial x^2} \qquad (-\infty < x < \infty,\ t > 0),$$

$$u(x, 0) = f(x).$$

We can solve this by separation of variables, but here is a solution by Fourier transform for illustration. Since x ranges from $-\infty$ to ∞, a Fourier transform in x can be taken. We shall denote $\mathscr{F}[u(x, t)]$ as $\hat{u}(\lambda, t)$, with t untouched by the transform with respect to x.

First, take the Fourier transform of the partial differential equation. On the right side, we have

$$\mathscr{F}\left(\frac{\partial^2 u}{\partial x^2}\right) = -\lambda^2 \hat{u}(\lambda, t),$$

by the operational formula. For the left side, we have

$$\mathscr{F}\left(\frac{\partial u}{\partial t}\right) = \int_{-\infty}^{\infty} \frac{\partial u}{\partial t} e^{-i\lambda x}\, dx = \frac{\partial}{\partial t}\int_{-\infty}^{\infty} u(x, t)e^{-i\lambda x}\, dx = \frac{\partial}{\partial t}[\hat{u}(\lambda, t)].$$

Thus, the partial differential equation transforms to

$$\frac{d}{dt}\hat{u}(\lambda, t) + \lambda^2 \hat{u}(\lambda, t) = 0,$$

an ordinary differential equation in t with λ as parameter. This has general solution

$$\hat{u}(\lambda, t) = A_\lambda e^{-\lambda^2 t}.$$

To find A_λ, recall that $u(x, 0) = f(x)$ for $-\infty < x < \infty$. Then,

$$\mathscr{F}[u(x, 0)] = \hat{u}(\lambda, 0) = \hat{f}(\lambda) = A_\lambda.$$

Thus,

$$\hat{u}(\lambda, t) = \hat{f}(\lambda)e^{-\lambda^2 t}.$$

We now have the Fourier transform of the solution; to get $u(x, t)$, invert this by using the inversion formula. This gives us

$$u(x, t) = \frac{1}{2\pi}\int_{-\infty}^{\infty} \hat{f}(\lambda)e^{-\lambda^2 t}e^{i\lambda x}\, d\lambda.$$

Upon substituting the definition of $\hat{f}(\lambda)$ into this, we get

$$\begin{aligned} u(x, t) &= \frac{1}{2\pi}\int_{-\infty}^{\infty}\left(\int_{-\infty}^{\infty} f(r)e^{-i\lambda r}\, dr\right)e^{-\lambda^2 t}e^{i\lambda x}\, d\lambda \\ &= \frac{1}{2\pi}\int_{-\infty}^{\infty}\int_{-\infty}^{\infty} f(r)e^{-i\lambda(r-x)}e^{-\lambda^2 t}\, dr\, d\lambda. \end{aligned}$$

This can be simplified somewhat. Write

$$e^{i\lambda(r-x)} = \cos[\lambda(r-x)] + i\sin[\lambda(r-x)].$$

Then

$$u(x, t) = \frac{1}{2\pi}\int_{-\infty}^{\infty}\int_{-\infty}^{\infty} f(r)\cos[\lambda(r-x)]e^{-\lambda^2 t}\,dr\,d\lambda$$
$$+ \frac{i}{2\pi}\int_{-\infty}^{\infty}\int_{-\infty}^{\infty} f(r)\sin[\lambda(r-x)]e^{-\lambda^2 t}\,dr\,d\lambda.$$

Since $u(x, t)$ is real, the integral having coefficient i must vanish, and the solution is

$$u(x, t) = \frac{1}{2\pi}\int_{-\infty}^{\infty}\int_{-\infty}^{\infty} f(r)\cos[\lambda(r-x)]e^{-\lambda^2 t}\,dr\,d\lambda.$$

EXAMPLE 390

Solve

$$\frac{\partial u}{\partial t} = \frac{\partial^2 u}{\partial x^2} \qquad (0 < x < \infty,\ t > 0);$$

$$u(0, t) = 0, \qquad u(x, 0) = f(x).$$

This problem is similar to the one just solved, except that x varies from 0 to ∞ instead of from $-\infty$ to ∞. Thus, we try a Fourier sine transform in x. (We could also try a Fourier cosine transform in x.) We have

$$\mathscr{F}_S\left(\frac{\partial u}{\partial t}\right) = \mathscr{F}_S\left(\frac{\partial^2 u}{\partial x^2}\right).$$

Write

$$\mathscr{F}_S[u(x, t)] = \hat{u}_S(\lambda, t).$$

Since $\partial/\partial t$ and $\mathscr{F}_S$ are with respect to different variables, we have

$$\mathscr{F}_S\left(\frac{\partial u}{\partial t}\right) = \frac{d}{dt}\hat{u}_S(\lambda, t).$$

Next, using the operational formula and the condition $u(0, t) = 0$, we have

$$\mathscr{F}_S\left(\frac{\partial^2 u}{\partial x^2}\right) = -\lambda^2\hat{u}_S(\lambda, t) + \lambda u(0, t) = -\lambda^2\hat{u}_S(\lambda, t).$$

This gives us

$$\frac{d}{dt}\hat{u}_S(\lambda, t) = -\lambda^2 u_S(\lambda, t).$$

This has general solution

$$\hat{u}_S(\lambda, t) = A_\lambda e^{-\lambda^2 t}.$$

Now, $u(x, 0) = f(x)$; so $\hat{u}_S(\lambda, 0) = \hat{f}_S(\lambda) = A_\lambda$. Thus,

$$\hat{u}_S(\lambda, t) = \hat{f}_S(\lambda)e^{-\lambda^2 t}.$$

Then, by the inversion formula, the solution is

$$u(x, t) = \frac{2}{\pi}\int_0^\infty \hat{f}_S(\lambda)e^{-\lambda^2 t}\sin(\lambda x)\,d\lambda$$

$$= \frac{2}{\pi}\int_0^\infty \int_0^\infty f(r)\sin(\lambda r)e^{-\lambda^2 t}\sin(\lambda x)\,dr\,d\lambda.$$

Now try this example by cosine transform. What goes wrong?

EXAMPLE 391

Solve

$$\frac{\partial^2 u}{\partial x^2} + \frac{\partial^2 u}{\partial y^2} = -h \qquad (0 < x < \pi,\ y > 0).$$

This is Poisson's equation for a strip; we may think of $u(x, y)$ as the electrostatic potential in a space bounded by the planes $x = 0$, $x = \pi$, and $y = 0$, with a uniform distribution of charge with density $h/4\pi$.

We seek a solution satisfying

$$u(0, y) = u(x, 0) = 0,$$

$$u(\pi, y) = 1.$$

Because of the scale of length in the x-variable, we may use the finite Fourier sine transform in x. Writing $S_n[u(x, y)] = u_S(n, y)$, the partial differential equation transforms to

$$-n^2 u_S(n, y) - n(-1)^n u(\pi, y) + nu(0, y) + \frac{d^2 u_S}{dy^2}(n, y) = -hS_n(1).$$

Here we used the fact that d/dy and S_n commute because x and y are independent. Now put in the conditions $u(\pi, y) = 1$ and $u(0, y) = 0$ to get

$$\frac{d^2 u_S}{dy^2} - n^2 u_S = n(-1)^n - hS_n(1) = n(-1)^n - h\left[\frac{1-(-1)^n}{n}\right].$$

Since the right side does not involve y, this ordinary differential equation is easily solved to get

$$u_S(n, y) = \left[h\left(\frac{1-(-1)^n}{n^3}\right) + \frac{(-1)^{n+1}}{n}\right](1 - e^{-ny}).$$

The solution, by the inversion formula, is then

$$u(x, y) = \frac{2}{\pi}\sum_{n=1}^{\infty} u_S(n, y)\sin(nx)$$

$$= \frac{2}{\pi}\sum_{n=1}^{\infty}\left[h\left(\frac{1-(-1)^n}{n^3}\right) + \frac{(-1)^{n+1}}{n}\right](1 - e^{-ny})\sin(nx).$$

We should note that, in this particular problem, it is possible to invert $u_S(n, y)$ in closed form, thereby obtaining an explicit solution which is in a more convenient form than the above series solution. From the table of finite Fourier sine transforms, we have immediately that

$$S_n^{-1}\left(\frac{1-(-1)^n}{n^3}\right) = \frac{x}{2}(\pi - x),$$

$$S_n^{-1}\left(\frac{(-1)^{n+1}}{n}\right) = \frac{x}{\pi},$$

and

$$S_n^{-1}\left(\frac{(-1)^{n+1}e^{-ny}}{n}\right) = \frac{2}{\pi}\arctan\left(\frac{\sin(x)}{e^y + \cos(x)}\right).$$

The inverse of the remaining term in $u_S(n, y)$ does not appear in the table. However, we can use Theorem 101 (Section 12.9) and the table entry

$$S_n^{-1}\left(\frac{1-(-1)^n}{n}e^{-ny}\right) = \frac{2}{\pi}\arctan\left(\frac{\sin(x)}{\sinh(y)}\right)$$

to get

$$\begin{aligned} S_n^{-1}\left[\left(\frac{1-(-1)^n}{n^3}\right)e^{-ny}\right] &= \frac{x}{\pi}\int_0^\pi (\pi - r)\frac{2}{\pi}\arctan\left(\frac{\sin(r)}{\sinh(y)}\right) dr \\ &\quad - \int_0^x (x - r)\frac{2}{\pi}\arctan\left(\frac{\sin(r)}{\sinh(y)}\right) dr \\ &= A(x, y). \end{aligned}$$

The solution is then

$$u(x, y) = \frac{hx}{2}(\pi - x) + \frac{x}{\pi} - \frac{2}{\pi}\arctan\left(\frac{\sin(x)}{e^y + \cos(x)}\right) - hA(x, y).$$

This expression is easier to work with than the series for $u(x, y)$ because integrals are easy to approximate, and we can obtain $A(x, y)$, and hence $u(x, y)$, to a high degree of accuracy without much trouble.

PROBLEMS FOR SECTION 13.10

1. Solve

$$\nabla^2 u(x, y) = 0 \qquad (y \geq 0, \ -\infty < x < \infty),$$
$$u(x, 0) = f(x),$$

using the Fourier transform in x. Compare the solution with that obtained using separation of variables.

2. Try solving the boundary value problem in Problem 1 by using a Laplace transform in y.
3. Solve

$$\nabla^2 u(x, y) = 0 \qquad (0 < x < \pi, \ y > 0),$$
$$u(0, y) = 0, \qquad u(\pi, y) = 2, \qquad u(x, 0) = -4.$$

First, try a finite Fourier transform in x; then try a Fourier sine or cosine transform in y.

4. Solve

$$\frac{\partial u}{\partial t} = \nabla^2 u(x, y) \qquad (0 < x < \pi,\ y > 0,\ t > 0),$$

$$u(x, y, 0) = 0, \qquad \frac{\partial u}{\partial x}(0, y, t) = -5,$$

$$\frac{\partial u}{\partial x}(\pi, y, t) = 0, \qquad u(x, 0, t) = 0.$$

First, use a finite Fourier cosine transform in the x-variable; to solve the resulting ordinary differential equation, use a Laplace transform in the t-variable.

5. Solve

$$\nabla^2 u(x, y) = 0 \qquad (0 < x < \pi,\ 0 < y < 2),$$

$$u(0, y) = 0, \qquad u(\pi, y) = 4, \qquad \frac{\partial u}{\partial y}(x, 0) = u(x, 2) = 0,$$

using transform methods.

6. Solve

$$\frac{\partial u}{\partial t} = 9\,\frac{\partial^2 u}{\partial x^2} \qquad (0 < x < \infty,\ t > 0),$$

$$u(x, 0) = 0, \qquad u(0, t) = f(t),$$

using transform methods.

7. Solve

$$\nabla^2 u(x, y) = 0 \qquad (0 < x < \infty,\ 0 < y < 1),$$

$$u(0, y) = y^2(1 - y),$$

$$u(x, 0) = u(x, 1) = 0,$$

using transform methods.

8. Solve

$$\frac{\partial u}{\partial t} = \frac{\partial^2 u}{\partial x^2} - u \qquad (0 < x < \infty,\ t > 0),$$

$$\frac{\partial u}{\partial x}(0, t) = f(t),$$

$$u(x, 0) = 0,$$

using transform methods.

9. Solve

$$\frac{\partial u}{\partial t} - \frac{\partial^2 u}{\partial x^2} + tu = 0 \qquad (t > 0,\ x > 0),$$

$$u(x, 0) = xe^{-x}, \qquad \frac{\partial u}{\partial t}(0, t) = 0,$$

using transform methods.

10. Solve

$$\nabla^2 u = 0 \qquad (-\infty < x < \infty,\ 0 < y < 1),$$

$$u(x, 0) = \begin{cases} 0, & -\infty < x < 0 \\ e^{-ax}, & 0 < x < \infty \end{cases} \quad (a \text{ is a positive constant}),$$

$$u(x, 1) = 0,$$

using transform methods.

11. Solve

$$\nabla^2 u(x, y) = u(x, y) \qquad (-\infty < x < \infty,\ 0 < y < 1),$$

$$\frac{\partial u}{\partial y}(x, 0) = 0, \qquad u(x, 1) = e^{-x^2},$$

using transform methods.

Table of Fourier Transforms

$f(x)$		$\mathscr{F}[f(x)] = \hat{f}(\lambda)$
1		$2\pi\delta(\lambda)$
$\dfrac{1}{x}$		$i\ \mathrm{sgn}(\lambda)$, where $\mathrm{sgn}(\lambda) = \begin{cases} 1 & \text{if } \lambda > 0 \\ 0 & \text{if } \lambda = 0 \\ -1 & \text{if } \lambda < 0 \end{cases}$
e^{-ax}	$(a > 0)$	$\dfrac{2a}{a^2 + \lambda^2}$
xe^{-ax}	$(a > 0)$	$\dfrac{4ai}{(a^2 + \lambda^2)^2}$
$\lvert x \rvert e^{-ax}$	$(a > 0)$	$\dfrac{2(a^2 - \lambda^2)}{(a^2 + \lambda^2)^2}$
$e^{-a^2x^2}$	$(a > 0)$	$\dfrac{\sqrt{\pi}}{a} e^{-\lambda^2/4a^2}$
$\dfrac{1}{a^2 + x^2}$	$(a > 0)$	$\dfrac{\pi}{a} e^{-a\lvert\lambda\rvert}$
$\dfrac{x}{a^2 + x^2}$	$(a > 0)$	$-\dfrac{i}{2}\dfrac{\pi}{a} \lambda e^{-a\lvert\lambda\rvert}$

Table of Fourier Cosine Transforms

$f(x)$	$\mathscr{F}_C[f(x)] = \hat{f}_C(\lambda)$
$x^{r-1} \quad (0 < r < 1)$	$\lambda^{-r}\Gamma(r)\cos\left(\frac{r\pi}{2}\right)$
$e^{-ax} \quad (a > 0)$	$\frac{a}{a^2 + \lambda^2}$
$xe^{-ax} \quad (a > 0)$	$\frac{a^2 - \lambda^2}{(a^2 + \lambda^2)^2}$
$e^{-a^2x^2} \quad (a > 0)$	$\frac{\sqrt{\pi}}{2} a^{-1} e^{-\lambda^2/4a^2}$
$\frac{1}{a^2 + x^2} \quad (a > 0)$	$\frac{\pi}{2}\frac{1}{a} e^{-a\lambda}$
$\frac{1}{(a^2 + x^2)^2} \quad (a > 0)$	$\frac{\pi}{4} a^{-3} e^{-a\lambda}(1 + a\lambda)$
$\cos\left(\frac{x^2}{2}\right)$	$\frac{\sqrt{\pi}}{2}\left[\cos\left(\frac{\lambda^2}{2}\right) + \sin\left(\frac{\lambda^2}{2}\right)\right]$
$\sin\left(\frac{x^2}{2}\right)$	$\frac{\sqrt{\pi}}{2}\left[\cos\left(\frac{\lambda^2}{2}\right) - \sin\left(\frac{\lambda^2}{2}\right)\right]$

Table of Fourier Sine Transforms

$f(x)$	$\mathcal{F}_s[f(x)] = \hat{f}_s(\lambda)$
$\frac{1}{x}$	$\begin{cases} \pi/2 & \text{if} \quad \lambda > 0 \\ -\pi/2 & \text{if} \quad \lambda < 0 \end{cases}$
$x^{r-1} \quad (0 < r < 1)$	$\lambda^{-r}\Gamma(r) \sin\left(\frac{\pi r}{2}\right)$
$\frac{1}{\sqrt{x}}$	$\sqrt{\frac{\pi}{2}}\,\lambda^{-1/2}$
$e^{-ax} \quad (a > 0)$	$\frac{\lambda}{a^2 + \lambda^2}$
$xe^{-ax} \quad (a > 0)$	$\frac{2a\lambda}{(a^2 + \lambda^2)^2}$
$xe^{-a^2x^2} \quad (a > 0)$	$\frac{\sqrt{\pi}}{4} a^{-3} \lambda e^{-\lambda^2/4a^2}$
$x^{-1}e^{-ax} \quad (a > 0)$	$\arctan\left(\frac{\lambda}{a}\right)$
$\frac{x}{a^2 + x^2} \quad (a > 0)$	$\frac{\pi}{2} e^{-a\lambda}$
$\frac{x}{(a^2 + x^2)^2} \quad (a > 0)$	$2^{-3/2} a^{-1} \lambda e^{-a\lambda}$
$\frac{1}{x(a^2 + x^2)} \quad (a > 0)$	$\frac{\pi}{2} a^{-2}(1 - e^{-a\lambda})$

Table of Finite Fourier Cosine Transforms

$f(x)$	$C_n[f(x)] = f_C(n)$
1	$\begin{cases} 0 & \text{if} \quad n = 1, 2, 3, \ldots \\ \pi & \text{if} \quad n = 0 \end{cases}$
$\begin{cases} 1, & 0 \le x \le k \\ -1, & k < x \le \pi \end{cases}$	$\begin{cases} \frac{2}{n}\sin(nk) & \text{for} \quad n = 1, 2, 3, \ldots \\ 2k - \pi & \text{for} \quad n = 0 \end{cases}$
x	$\begin{cases} \frac{-[1-(-1)^n]}{n^2} & \text{for} \quad n = 1, 2, 3, \ldots \\ \pi^2/2 & \text{for} \quad n = 0 \end{cases}$
$\frac{x^2}{2\pi}$	$\begin{cases} (-1)^n/n & \text{for} \quad n = 1, 2, 3, \ldots \\ \pi^2/6 & \text{for} \quad n = 0 \end{cases}$
x^3	$\begin{cases} 3\pi^2 \frac{(-1)^n}{n^2} + 6\,\frac{[1-(-1)^n]}{n^4} & \text{if} \quad n = 1, 2, 3, \ldots \\ \pi^4/4 & \text{if} \quad n = 0 \end{cases}$
e^{cx}	$c\left(\frac{(-1)^n e^{c\pi} - 1}{n^2 + c^2}\right)$
$\sin(kx)$, k not an integer	$k\left(\frac{(-1)^n \cos(k\pi) - 1}{n^2 - k^2}\right)$
$\sin(px)$, $p = 0, 1, 2, \ldots$	$\begin{cases} p\left(\frac{(-1)^{n+p} - 1}{n^2 - p^2}\right) & \text{if} \quad n \ne p \\ 0 & \text{if} \quad n = p \end{cases}$
$\cos(px)$, $p = 1, 2, 3, \ldots$	$\begin{cases} 0 & \text{if} \quad n \ne p \\ \pi/2 & \text{if} \quad n = p \end{cases}$
$f(\pi - x)$	$(-1)^n f_C(n)$

Table of Finite Fourier Sine Transforms

$f(x)$	$S_n[f(x)] = f_S(n)$
$\dfrac{\pi - x}{\pi}$	$\dfrac{1}{n}$
$\dfrac{x}{\pi}$	$\dfrac{(-1)^{n+1}}{n}$
1	$\dfrac{1-(-1)^n}{n}$
$\dfrac{x(\pi^2 - x^2)}{6}$	$\dfrac{(-1)^{n+1}}{n^3}$
$\dfrac{x(\pi - x)}{2}$	$\dfrac{1-(-1)^n}{n^3}$
x^2	$\dfrac{\pi^2(-1)^{n+1}}{n} - \dfrac{2[1-(-1)^n]}{n^3}$
x^3	$\pi(-1)^n\left(\dfrac{6}{n^3} - \dfrac{\pi^2}{n}\right)$
e^{cx}	$\dfrac{n}{n^2 + c^2}[1 - (-1)^n e^{c\pi}]$
$\sin(px)$, $p = 1, 2, 3, \ldots$	$\begin{cases} 0, & n \neq p \\ \pi/2, & n = p \end{cases}$
$\cos(kx)$, $k \neq$ integer	$\dfrac{n}{n^2 - k^2}[1 - (-1)^n \cos(k\pi)]$
$\cos(px)$, $p = 1, 2, 3, \ldots$	$\begin{cases} \dfrac{n}{n^2 - p^2}[1 - (-1)^{n+p}], & n \neq p \\ \pi/2, & n = p \end{cases}$
$\dfrac{2}{\pi}\arctan\left(\dfrac{2b\sin(x)}{1 - b^2}\right)$ $(\lvert b \rvert < 1)$	$\left(\dfrac{1-(-1)^n}{n}\right)b^n$
$f(\pi - x)$	$(-1)^{n+1} f_S(n)$

13.11 SOME REMARKS ON EXISTENCE, UNIQUENESS, CLASSIFICATION, AND WELL-POSED PROBLEMS

Thus far, we have restricted our attention to specific kinds of problems in order to illustrate some of the techniques available for their solution. In this section we shall discuss some general facts about partial differential equations which can serve as a backdrop against which to view these specific results.

Although one can develop a general theory for nth order linear partial differential equations, the ideas are well illustrated by the important case that $n = 2$ (the heat equation, wave equation, and Laplace's equation are all second order—that is, each contains a second partial derivative but no higher derivative).

The *general linear second order partial differential equation* has the form

$$a(x, y)\frac{\partial^2 u}{\partial x^2} + 2b(x, y)\frac{\partial^2 u}{\partial x\,\partial y} + c(x, y)\frac{\partial^2 u}{\partial y^2} + d(x, y)\frac{\partial u}{\partial x} + e(x, y)\frac{\partial u}{\partial y} + f(x, y)u(x, y) + g(x, y) = 0.$$

(Letting $2b$ denote the coefficient of $\partial^2 u/\partial x\,\partial y$ is just a notational convenience.)

This partial differential equation is often written

$$au_{xx} + 2bu_{xy} + cu_{yy} + du_x + eu_y + fu + g = 0.$$

(See the remark on notation at the end of Section 13.1.)

The quantity $\Delta(x, y) = b^2 - ac$ is called the *discriminant* of the partial differential equation. The equation is said to be

hyperbolic at (x_0, y_0) if $\Delta(x_0, y_0) > 0$,
elliptic at (x_0, y_0) if $\Delta(x_0, y_0) < 0$,

and

parabolic at (x_0, y_0) if $\Delta(x_0, y_0) = 0$.

Thus, for example:

A wave-type equation $u_{xx} - u_{yy} = 0$ has $\Delta > 0$ at every point, and hence is hyperbolic everywhere;

a heat-type equation $u_x - u_{yy} = 0$ has $\Delta = 0$ at every point, and hence is parabolic everywhere;

and a Laplace equation $u_{xx} + u_{yy} = 0$ has $\Delta < 0$ everywhere, and hence is elliptic everywhere.

The equation $yu_{xx} + u_{yy} = 0$ (*Tricomi's equation*) has $\Delta = -y$, and hence is hyperbolic in the half-plane $y < 0$, elliptic in the half-plane $y > 0$, and parabolic on the x-axis, where $y = 0$.

The classification of linear second order partial differential equations as hyperbolic, elliptic, or parabolic (about a given point) has proved important for an understanding of the kinds of initial and boundary data one must furnish along with the partial differential equation in order to determine a unique solution. For example, the data furnished with the hyperbolic wave equation was of a different kind than that given with the parabolic heat equation, though the rationale previously given was the physical setting of the problem.

The type of equation also determines to a large extent the kind of technique one uses when attempting a numerical approximation of a solution. Methods suited to parabolic equations are not, for example, necessarily effective for hyperbolic or elliptic equations.

Given a problem involving a partial differential equation and some boundary and/or initial conditions, one usually asks three questions:

1. Does a solution exist?
2. Is there only one solution?
3. Does the solution depend continuously on the data given in the problem?

If the answer to all three is yes, the problem is said to be *well posed;* otherwise, it is *ill posed.*

While (1) and (2) are self-explanatory, (3) deserves some discussion. Intuitively, it means that small changes in the given data induce correspondingly small changes in the solution. For example, consider the Dirichlet problem

$$u_{xx} + u_{yy} = 0 \quad \text{in a rectangle } R,$$
$$u(x, y) = f(x, y) \quad \text{on the boundary of } R.$$

Here, the data is the function $f(x, y)$ defined on the boundary. Suppose that this problem has solution $F(x, y)$.

Now suppose that the problem

$$u_{xx} + u_{yy} = 0 \quad \text{in } R,$$
$$u(x, y) = g(x, y) \quad \text{on the boundary of } R,$$

has solution $G(x, y)$.

We say that this problem depends continuously on the data if, given $\varepsilon > 0$, there is some $\delta > 0$ such that

$$|F(x, y) - G(x, y)| < \varepsilon \quad \text{for } (x, y) \text{ in } R$$

whenever

$$|f(x, y) - g(x, y)| < \delta \quad \text{for } (x, y) \text{ on the boundary of } R.$$

That is, if the data functions are close together, then so are the solutions.

Similar definitions of continuous dependence on data can be written for problems with other kinds of data (for example, the vibrating string problem).

It may seem to the student that continuous dependence of the solution on the data ia a natural thing we ought to expect to happen. An example due to Hadamard shows that in fact it need not hold, even in simple problems involving analytic* functions. This example is given in the problems at the end of this section.

Concerning existence, there are some theorems which have been proved for various classes of problems, but we leave a treatment of them to a more rigorous

* *Analytic* here means that the function can be expanded in a power series about the origin.

course on partial differential equations. To give some flavor for the idea of an existence theorem, here is a classic one named after Augustin-Louis Cauchy and Sonya Kovalevski. The object of the theorem is the initial value first order problem (called a *Cauchy problem*):

$$u_t = F(x_1, \ldots, x_n, t, u, u_{x1}, \ldots, u_{x_n}),$$
$$u(x_1, \ldots, x_n, 0) = f(x_1, \ldots, x_n).$$

CAUCHY-KOVALEVSKI THEOREM

Suppose that F is analytic in a $(2n + 2)$-dimensional sphere about

$$(0, 0, \ldots, 0, F(0, \ldots, 0), f_{x1}(0, \ldots, 0), \ldots, f_{x_n}(0, \ldots, 0))$$

and that f is analytic in an n-dimensional sphere about $(0, \ldots, 0)$. Then the Cauchy problem has a unique, analytic solution in some $(n + 1)$-dimensional sphere about the origin in $(n + 1)$-space.

In many cases, the problem of proving uniqueness of a solution is easier than proving existence. Uniqueness is particularly important in problems modeling physical phenomena, where we know on physical grounds that there should be only one solution. As an illustration of a uniqueness proof, consider a problem in heat conduction:

$$u_t = a^2(u_{xx} + u_{yy} + u_{zz}) + \phi(x, y, z, t)$$
$$\text{for } (x, y, z) \text{ in a region } R, \text{ and } t > 0,$$
$$u(x, y, z, 0) = f(x, y, z) \quad \text{for } (x, y, z) \text{ in } R,$$

and

$$u(x, y, z, t) = g(x, y, z, t)$$
$$\text{for } t > 0 \text{ and } (x, y, z) \text{ on the boundary } S \text{ of } R.$$

Here, we may think of R as a solid occupying a region of 3-space and $u(x, y, z, t)$ as the temperature at point (x, y, z) and time t. The above problem specifies u in R at time 0 and on the boundary of R at all later times. The function ϕ allows for interior generation of heat energy inside R.

We would expect this problem to have a unique solution. To show that it does, suppose that u_1 and u_2 are solutions, and let $w = u_1 - u_2$. It is easy to check that

$$w_t = a^2(w_{xx} + w_{yy} + w_{zz}),$$
$$w(x, y, z, 0) = 0 \quad \text{throughout } R,$$

and

$$w(x, y, z, t) = 0 \quad \text{for } (x, y, z) \text{ on } S.$$

Define

$$I(t) = \tfrac{1}{2} \iiint_R w^2 \, dx \, dy \, dz.$$

Then,

$$I'(t) = \iiint_R w \frac{\partial w}{\partial t}\, dx\, dy\, dz = \iiint_R a^2 w\, \nabla^2 w\, dx\, dy\, dz,$$

where $\nabla^2 w = w_{xx} + w_{yy} + w_{zz}$ is the Laplacian of w. Apply the divergence theorem of Gauss to the last integral (see Chapter 11) to get

$$\iiint_R a^2 w\, \nabla^2 w\, dx\, dy\, dz = a^2 \iint_S w \frac{\partial w}{\partial \eta} - a^2 \iiint_R ({w_x}^2 + {w_y}^2 + {w_z}^2)\, dx\, dy\, dz.$$

Here, $\iint_S w\, \partial w/\partial \eta$ is a surface integral over the surface S bounding R, and $\partial w/\partial \eta$ is the directional derivative of w in the direction of the outer normal to S.

Now, $\iint_S w\, \partial w/\partial \eta = 0$ because $w = 0$ on S. Thus,

$$I'(t) = \iiint_R a^2 w\, \nabla^2 w\, dx\, dy\, dz = -a^2 \iiint_R ({w_x}^2 + {w_y}^2 + {w_z}^2)\, dx\, dy\, dz.$$

Hence, $I'(t) \leq 0$ for $t > 0$.

Apply the mean value theorem to $I(t)$ on any interval $[0, t]$. For some τ, $0 < \tau < t$ and

$$I(t) - I(0) = tI'(\tau).$$

But $I(0) = \frac{1}{2} \iiint_R w(x, y, z, 0)^2\, dx\, dy\, dz = 0$ because $w(x, y, z, 0) = 0$ in R.

Since $t > 0$ and $I'(\tau) \leq 0$, then

$$I(t) - I(0) = I(t) \leq 0 \quad \text{for all} \quad t > 0.$$

But

$$I(t) = \frac{1}{2} \iiint_R w^2\, dx\, dy\, dz \geq 0 \quad \text{for} \quad t > 0.$$

Thus,

$$I(t) = 0 \quad \text{for all} \quad t > 0.$$

Thus, w^2, and hence also w, must be zero in R. Then $u_1 = u_2$, and hence the solution is unique.

The uniqueness proof above is fairly typical of uniqueness proofs in several respects, namely:

1. It begins with the assumption that u_1 and u_2 are solutions, defines $w = u_1 - u_2$, and then tries to show that $w = 0$.
2. w satisfies a partial differential equation similar to (but often simpler than) that satisfied by u_1 and u_2; w also satisfies zero boundary and initial data.
3. The proof that $w = 0$ involves some clever trick, often requiring something like Gauss's theorem and an elementary vector identity tied into the use of Gauss's theorem.

We have included the considerations of this section simply to give the student some idea of what was omitted previously in studying particular problems in partial differential equations. Some extensions of these ideas are contained in the exercises.

PROBLEMS FOR SECTION 13.11

1. Write out a suitable definition for continuous dependence on data for the vibrating string problem

$$\begin{aligned} y_{tt} &= a^2 y_{xx} && (0 < x < L,\ t > 0),\\ y(0, t) &= y(L, t) = 0 && (t > 0),\\ y(x, 0) &= f(x) && (0 < x < L),\\ y_t(x, 0) &= g(x) && (0 < x < L). \end{aligned}$$

Prove that this problem does have continuous dependence on the data if $f(x)$ and $g(x)$ are bounded and continuous on $[0, L]$. (*Hint:* Use d'Alembert's solution to this problem.)

2. Write out a suitable definition for continuous dependence on data for the heat conduction problem

$$\begin{aligned} u_t &= a^2 u_{xx} && (0 < x < L,\ t > 0),\\ u(0, t) &= u(L, t) = 0 && (t > 0),\\ u(x, 0) &= f(x) && (0 < x < L). \end{aligned}$$

3. Here is Hadamard's example of a problem which does not depend continuously on the data. Consider:

$$\begin{aligned} u_{xx} + u_{yy} &= 0 && (-\infty < x < \infty,\ -\infty < y < \infty),\\ u(x, 0) &= 0 && (-\infty < x < \infty),\\ u_y(x, 0) &= \frac{1}{n}\sin(nx) && (-\infty < x < \infty). \end{aligned}$$

Here, n is any positive integer.
(a) Show that $u(x, y) = (1/n^2)\sinh(ny)\sin(nx)$ is a solution of the problem.
(b) Let $A(x, y)$ be a solution of

$$\begin{aligned} u_{xx} + u_{yy} &= 0,\\ u(x, 0) &= f(x),\\ u_y(x, 0) &= g(x). \end{aligned}$$

Let $B(x, y)$ be a solution of

$$\begin{aligned} u_{xx} + u_{yy} &= 0,\\ u(x, 0) &= f(x),\\ u_y(x, 0) &= g(x) + \frac{1}{n}\sin(nx). \end{aligned}$$

Prove that $B(x, y) - A(x, y) = (1/n^2)\sinh(ny)\sin(nx)$.

(c) Use the above results to show that

$$u_{xx} + u_{yy} = 0,$$
$$u(x, 0) = f(x),$$
$$u_y(x, 0) = g(x),$$

does not depend continuously on the data.

4. Consider the very general vibrating string problem:

$$y_{tt} = a^2 y_{tt} + F(x) \qquad (0 < x < L,\ t > 0),$$
$$y(0, t) = \alpha(t), \qquad y(L, t) = \beta(t) \qquad (t \geq 0),$$
$$y(x, 0) = f(x), \qquad y_t(x, 0) = g(x) \qquad (0 \leq x \leq L).$$

Assume that $\alpha(t)$ and $\beta(t)$ are continuous for $t \geq 0$ and that $f(x)$ and $g(x)$ are continuous for $0 \leq x \leq L$.

Prove that there can be at most one continuous solution $y(x, t)$ having continuous first and second partial derivatives.

Hint: Suppose that $y_1(x, t)$ and $y_2(x, t)$ are two such solutions. Let $u(x, t) = y_1(x, t) - y_2(x, t)$. First, derive the boundary value problem for $u(x, t)$. Then define

$$I(t) = \frac{1}{2} \int_0^L \left(u_x^2 + \frac{1}{a^2} u_t^2 \right) dx.$$

Prove that $I'(t) = 0$, for $t \geq 0$; hence, $I(t) =$ constant. By setting $t = 0$, prove that $I(t) = 0$ for all $t \geq 0$. Hence, show that $u(x, t) = 0$ for $0 \leq x \leq L, t \geq 0$.

5. In the boundary value problem of Problem 4, replace the conditions

$$y(0, t) = \alpha(t), \qquad y(L, t) = \beta(t),$$

by the conditions

$$y_t(0, t) = \alpha(t), \qquad y_t(L, t) = \beta(t).$$

Prove that there is still at most one continuous solution $y(x, t)$ having continuous first and second partials.

6. Consider the Dirichlet problem:

$$\nabla^2 u(x, y, z) = 0 \quad \text{for } (x, y, z) \text{ in a region } D \text{ bounded by a smooth surface } S,$$
$$u(x, y, z) = f(x, y, z) \quad \text{for } (x, y, z) \text{ on } S.$$

Prove that there is at most one solution $u(x, y, z)$ which is continuous and has continuous first partials in D.

Hint: Suppose that $u_1(x, y, z)$ and $u_2(x, y, z)$ are solutions of the problem. Let $w(x, y, z) = u_1(x, y, z) - u_2(x, y, z)$. Find the Dirichlet problem satisfied by $w(x, y, z)$. Then use the following form of Gauss's divergence theorem,

$$\iint_S w \frac{\partial w}{\partial \eta} = \iiint_D (w_x^2 + w_y^2 + w_z^2 + w\nabla^2 w)\, dx\, dy\, dz,$$

to conclude that $w(x, y, z) = 0$ throughout D.

7. Generalize Problem 6 as follows. Consider the problem

$$\nabla^2 u(x, y, z) = 0 \quad \text{for } (x, y, z) \text{ in a region } D \text{ bounded by a smooth surface } S,$$

$$a\frac{\partial u}{\partial \eta} + bu = f(x, y, z) \quad \text{for } (x, y, z) \text{ on } S.$$

Here, $\partial u/\partial \eta$ is the derivative of u in the direction of the outer normal to S.

Prove that there is at most one solution $u(x, y, z)$ of this problem which is continuous and has continuous first and second partial derivatives in D. Above, a and b are given constants.

13.12 SOME HISTORICAL NOTES ON PARTIAL DIFFERENTIAL EQUATIONS

As indicated in Section 13.11, we have not attempted to treat partial differential equations in general but have restricted our attention to certain important partial differential equations which arise frequently in engineering and physics. Thus, we shall restrict our historical remarks to the development of the wave and heat equations, Laplace's equation, and the separation of variables method.

Perhaps surprisingly, we may think of the wave equation as belonging to the eighteenth century and the heat equation to the nineteenth. In 1727, the Swiss mathematician John Bernoulli (1667–1748) treated the vibrating string problem by imagining the string as a weightless, flexible thread having a number of equally spaced weights, or beads, placed along it. His equation was time-independent, however, and hence not really a partial differential equation. The French mathematician Jean le Rond d'Alembert (1717–1783) introduced the time variable into the equation, and he let the number of weights in Bernoulli's treatment become infinite (and the distance between them go to zero) to derive the one-dimensional wave equation as we know it today. His work, which appeared in about 1746, also included the solution known today as d'Alembert's solution (Problem 1, Section 13.1).

In the 1750s, the Swiss mathematician Leonard Euler (1707–1783) considered what we would today call Fourier series solutions of the wave equation. Incidentally, Euler's work touched off a tremendous debate between the leading mathematicians of the day over the definition of the function concept.

In 1781, Euler used Fourier-Bessel series to solve the vibrating membrane problem for a circular drum. Of course, Bessel functions had not been named for Bessel at this point.

Laplace's equation, also known as the potential equation, arose in the context of gravitational attraction problems. The main figures were Pierre Simon de Laplace (1749–1827) and Adrien-Marie Legendre (1752–1833), both professors of mathematics at the Ecole Militaire in France.

In 1807, Joseph Fourier (1768–1830) submitted a paper to the Paris Academy. In it he derived the heat equation. The paper was rejected by Laplace, Lagrange, and Legendre, due to lack of rigor, but the results were promising enough for the Academy to set the problem of describing heat conduction for a prize in 1812. Fourier's 1811 revision of his earlier paper won the prize but suffered the same criticism. In

1822, Fourier published his *Theorie analytique de la chaleur*, laying the foundations not only for separation of variables and Fourier series, but for the Fourier integral as well.

Pioneering work on properties of harmonic functions (solutions to Laplace's equation) was done by George Green (1793–1841), a self-taught British mathematician whose main interest was in electricity and magnetism, and the Russian Michel Ostrogradsky (1801–1861), whose work was independent of Green's but paralleled Green's results almost exactly.

SUPPLEMENTARY PROBLEMS

1. Use the finite Fourier sine transform to solve the vibrating string problem:

$$\frac{\partial^2 y}{\partial t^2} = a^2 \frac{\partial^2 y}{\partial x^2} + x \qquad (0 < x < \pi,\ t > 0),$$

$$y(x, 0) = 0, \qquad \frac{\partial y}{\partial t}(x, 0) = 0, \qquad y(0, t) = 0, \qquad y(\pi, t) = 0.$$

Will separation of variables work in solving this problem?

2. Use the finite Fourier cosine transform to solve the problem of heat conduction in a bar with insulated ends in which heat is generated:

$$\frac{\partial u}{\partial t} = a^2 \frac{\partial^2 u}{\partial x^2} + h(x, t) \qquad (0 < x < \pi,\ t > 0),$$

$$u(x, 0) = f(x), \qquad \frac{\partial u}{\partial x}(0, t) = \frac{\partial u}{\partial x}(\pi, t) = 0.$$

3. Solve

$$\frac{\partial^2 u}{\partial t^2} + 2\frac{\partial u}{\partial t} + u(x, t) = 4\frac{\partial^2 u}{\partial x^2} \qquad (0 < x < \pi,\ t > 0),$$

$$u(x, 0) = 1, \qquad u(\pi, t) = 0, \qquad u_t(x, 0) = 0.$$

4. Solve

$$\frac{\partial^2 u}{\partial t^2} = a^2 \frac{\partial^2 u}{\partial x^2} + x^2 \qquad (0 < x < 4,\ t > 0),$$

$$u(x, 0) = 4 - x, \qquad u(0, t) = u(4, t) = 0, \qquad \frac{\partial u}{\partial t}(x, 0) = 0.$$

5. Solve

$$\frac{\partial u}{\partial t} = a^2 \frac{\partial^2 u}{\partial x^2} + \pi - x \qquad (0 < x < \pi,\ t > 0),$$

$$u(x, 0) = 4, \qquad u(\pi, t) = 0, \qquad u(0, t) = 0 \qquad (0 < x < \pi,\ t > 0).$$

6. Solve

$$\frac{\partial^2 u}{\partial t^2} = a^2 \frac{\partial^4 u}{\partial x^4} \qquad (0 < x < \pi,\ t > 0),$$

$$u(x, 0) = \frac{\partial u}{\partial t}(x, 0) = 0, \qquad u(0, t) = u(\pi, t) = 0,$$

$$\frac{\partial^2 u}{\partial x^2}(0, t) = 1 - t^2, \qquad \frac{\partial^2 u}{\partial x^2}(\pi, t) = 0.$$

7. Solve (cylindrical coordinates)

$$\nabla^2 u(r, \theta) = 0 \qquad (0 < r < 1,\ 0 < \theta < 2\pi),$$
$$u(r, 0) = u(r, 2\pi) = 0,$$
$$u(1, \theta) = \cos(2\theta).$$

8. Solve the three-dimensional Dirichlet problem:

$$\nabla^2 u(x, y, z) = 0 \qquad (0 < x < a,\ 0 < y < b,\ 0 < z < c),$$
$$u(a, y, z) = u(0, y, z) = u(x, b, z) = u(x, y, 0) = u(x, y, c) = 0,$$
$$u(x, 0, z) = x(1 - z).$$

9. Solve

$$\nabla^2 u(x, y) = 0 \qquad (x > 0,\ y > 0),$$
$$u(x, 0) = x, \qquad u(0, y) = g(y).$$

10. Solve

$$\nabla^2 u(x, y, z) = 0 \qquad (0 < x < a,\ 0 < y < b,\ 0 < z < c),$$

$$\frac{\partial u}{\partial x}(0, y, z) = y \sin\left(\frac{\pi z}{c}\right), \qquad \frac{\partial u}{\partial x}(a, y, z) = 0,$$

$$\frac{\partial u}{\partial y}(x, 0, z) = \frac{\partial u}{\partial y}(x, b, z) = \frac{\partial u}{\partial z}(x, y, 0) = \frac{\partial u}{\partial z}(x, y, c) = 0.$$

11. Solve

$$\frac{\partial^2 u}{\partial t^2} + \frac{\partial u}{\partial t} = \nabla^2 u(x, y, z, t) \qquad (0 < x < a,\ 0 < y < b,\ 0 < z < c,\ t > 0),$$

$$u(x, y, z, 0) = xyz, \qquad \frac{\partial u}{\partial t}(x, y, z, 0) = 0,$$

$$u(0, y, z, t) = u(a, y, z, t) = u(x, b, z, t) = u(x, 0, z, t)$$
$$= u(x, y, c, t) = u(x, y, 0, t) = 0.$$

12. Solve

$$\frac{\partial^2 u}{\partial t^2} = \frac{\partial^2 u}{\partial x^2} + \frac{\partial^3 u}{\partial x^2\, \partial t} \qquad (0 < x < \infty,\ t > 0),$$

$$u(x, 0) = x, \qquad \frac{\partial u}{\partial t}(x, 0) = 0, \qquad u(0, t) = 0.$$

13. Solve this Dirichlet problem in cylindrical coordinates:

$$\nabla^2 u(r, \theta, z) = 0 \qquad (0 < r < 2,\ 0 < \theta < 2\pi,\ 0 < z < L),$$

$$u(2, \theta, z) = u(r, \theta, L) = 0, \qquad u(r, \theta, 0) = f(r, \theta).$$

14. Solve the following Dirichlet problem in spherical coordinates:

$$\nabla^2 u(r, \theta, \phi) = 0 \qquad (0 < r < R,\ 0 < \theta < 2\pi,\ 0 < \phi < \pi),$$

$$u(R, \theta, \phi) = \sin(2\phi).$$

15. Solve the following wave problem in cylindrical coordinates:

$$\frac{\partial^2 y}{\partial t^2} = \nabla^2 y(r, \theta, z) \qquad (0 < r < 1,\ 0 < z < 1,\ 0 < \theta < 2\pi,\ t > 0),$$

$$y(r, \theta, z, 0) = z\sin(\theta), \qquad \frac{\partial y}{\partial t}(r, \theta, z, 0) = 0,$$

$$y(1, \theta, z, t) = y(r, \theta, 0, t) = y(r, \theta, 1, t) = 0.$$

16. Solve the three-dimensional wave propagation problem:

$$\frac{\partial^2 u}{\partial t^2} = \nabla^2 u(x, y, z, t) \qquad (0 < x < 1,\ 0 < y < 1,\ 0 < z < 1,\ t > 0),$$

$$u(x, y, z, 0) = 0, \qquad \frac{\partial u}{\partial t}(x, y, z, 0) = x(1 - y)(1 - z),$$

$$u(0, y, z, t) = u(1, y, z, t) = u(x, 0, z, t) = u(x, 1, z, t) = u(x, y, 0, t) = 0,$$

$$u(x, y, 1, t) = x^2.$$

17. Solve

$$\frac{\partial u}{\partial x} + x\frac{\partial u}{\partial t} = 0 \qquad (x > 0,\ t > 0),$$

$$u(x, 0) = 0, \qquad u(0, t) = 4t.$$

18. Solve

$$\frac{\partial^2 y}{\partial t^2} = a^2 \frac{\partial^2 y}{\partial x^2} \qquad (0 < x < L,\ t > 0),$$

$$y(x, 0) = 0, \qquad \frac{\partial y}{\partial t}(x, 0) = 0, \qquad y(0, t) = 0, \qquad K\frac{\partial y}{\partial x}(L, t) = f(t).$$

This problem models longitudinal vibrations in an elastic bar of length L which has the end $x = 0$ fixed and a force $f(t)$ per unit area acting on the end at $x = L$.

19. Solve

$$\frac{\partial^2 y}{\partial t^2} = -\frac{\partial^4 y}{\partial x^4} + f(x) \qquad (0 < x < \pi,\ t > 0),$$

$$y(0, t) = y(\pi, t) = 0, \qquad \frac{\partial^2 y}{\partial x^2}(0, t) = \frac{\partial^2 y}{\partial x^2}(\pi, t) = 0,$$

$$y(x, 0) = 0, \qquad \frac{\partial y}{\partial t}(x, 0) = 0.$$

This models transverse displacements in a beam hinged at ends 0 and π, with zero initial displacement and velocity.

20. Solve

$$\nabla^2 u(x, y) = 0 \qquad (0 < x < \pi,\ 0 < y < 1),$$

$$u(0, y) = \frac{\partial u}{\partial y}(x, 0) = u(x, 1) = 0, \qquad u(\pi, y) = A, \text{ constant.}$$

PART FIVE

COMPLEX ANALYSIS

CHAPTER FOURTEEN

Complex Numbers and Complex Functions

14.1 COMPLEX NUMBERS

Just about the time we become accustomed to finding roots of simple polynomial equations, we encounter examples which show the inadequacy of real numbers. One such is the equation

$$x^2 + 1 = 0,$$

which has no real-number solutions.

In order to solve such an equation, we must invent new numbers to complement the real numbers we already have. To do this, we let the letter i stand for a solution to $x^2 + 1 = 0$. Thus, i satisfies the rule

$$i^2 + 1 = 0$$

or

$$i^2 = -1.$$

We now consider symbols of the form

$$x + iy \qquad (\text{or} \quad x + yi),$$

where x and y are real numbers. Such a symbol is called a *complex number*. When $y = 0$, we consider iy to be zero; so $x + iy$ is just x, a real number. Thus, complex numbers contain the real numbers as special cases.

We call x the *real part* of $x + iy$ and y the *imaginary part*. These are denoted, respectively,

$$\text{Re}(x + iy) = x \quad \text{and} \quad \text{Im}(x + iy) = y.$$

Note that these are themselves real numbers.

EXAMPLE 392

$$\begin{aligned}
&\text{Re}(2 - 3i) = 2 && \text{and} && \text{Im}(2 - 3i) = -3.\\
&\text{Re}(\sqrt{2} + 5i) = \sqrt{2} && \text{and} && \text{Im}(\sqrt{2} + 5i) = 5.\\
&\text{Re}(7i) = 0 && \text{and} && \text{Im}(7i) = 7.\\
&\text{Re}(-4) = -4 && \text{and} && \text{Im}(-4) = 0.
\end{aligned}$$

Often, the letter z is used to refer to a complex number. If $z_1 = x_1 + iy_1$ and $z_2 = x_2 + iy_2$, then $z_1 = z_2$ exactly when $x_1 = x_2$ and $y_1 = y_2$. This fact is extremely important in solving equations involving complex quantities by equating the real parts of both sides and then the imaginary parts of both sides. For example, to solve $x^2 + (2x + y)i = 3 - 4i$, we need $x^2 = 3$ and $2x + y = -4$, from which we obtain $x = \sqrt{3}$ and $y = -4 - 2\sqrt{3}$, or $x = -\sqrt{3}$ and $y = -4 + 2\sqrt{3}$.

Arithmetic operations with complex numbers are defined as follows.

Addition:

$$(x_1 + iy_1) + (x_2 + iy_2) = (x_1 + x_2) + i(y_1 + y_2).$$

Subtraction:

$$(x_1 + iy_1) - (x_2 + iy_2) = (x_1 - x_2) + i(y_1 - y_2).$$

Multiplication:

$$(x_1 + iy_1) \cdot (x_2 + iy_2) = (x_1x_2 - y_1y_2) + i(x_1y_2 + x_2y_1).$$

Note that we multiply two complex numbers just as we would polynomials, using the rule $i^2 = -1$ to convert $i^2y_1y_2$ to $-y_1y_2$.

Division: Division is not so obvious. We have, if $x_2 + iy_2 \neq 0$,

$$\frac{x_1 + iy_1}{x_2 + iy_2} = \left(\frac{x_1x_2 + y_1y_2}{x_2^2 + y_2^2}\right) + i\left(\frac{x_2y_1 - x_1y_2}{x_2^2 + y_2^2}\right).$$

This can be verified by multiplying both sides by $x_2 + iy_2$.

The *complex conjugate* (or just *conjugate*) of $z = x + iy$ is denoted $\bar{z}$ and is defined by

$$\bar{z} = x - iy.$$

Thus,

$$\text{Re } z = \text{Re } \bar{z} \quad \text{and} \quad \text{Im } z = -\text{Im } \bar{z}.$$

Note that, if $z = x + iy$, then

$$z\bar{z} = x^2 + y^2,$$

a real number. Further,

$$\text{Re } z = \frac{1}{2}(z + \bar{z}) \quad \text{and} \quad \text{Im } z = \frac{1}{2i}(z - \bar{z}).$$

In carrying out a division of two complex numbers, one need not memorize the definition of division but can instead make use of conjugates as follows:

$$\frac{x_1 + iy_1}{x_2 + iy_2} = \left(\frac{x_1 + iy_1}{x_2 + iy_2}\right)\left(\frac{x_2 - iy_2}{x_2 - iy_2}\right) = \frac{x_1x_2 + y_1y_2}{x_2^2 + y_2^2} + i\left(\frac{x_2y_1 - x_1y_2}{x_2^2 + y_2^2}\right).$$

EXAMPLE 393

Here are some illustrations of these concepts.

Let $z = -3 - 4i$. Then, we have

$$\bar{z} = -3 + 4i,$$
$$\tfrac{1}{2}(z + \bar{z}) = \tfrac{1}{2}(-6) = -3 = \text{Re } z,$$

and

$$\frac{1}{2i}(z - \bar{z}) = \frac{1}{2i}(-4i - 4i) = -4 = \text{Im } z.$$

Letting $z_1 = 4 + 8i$, we have

$$z + z_1 = 1 + 4i,$$
$$z - z_1 = -7 - 12i,$$
$$zz_1 = (-12 + 32) + i(-24 - 16) = 20 - 40i,$$

and

$$\frac{z}{z_1} = \left(\frac{-3 - 4i}{4 + 8i}\right)\left(\frac{4 - 8i}{4 - 8i}\right) = \frac{-44}{80} + \frac{8}{80}i = \frac{-11}{20} + \frac{1}{10}i.$$

Complex numbers admit a convenient geometric interpretation. Draw a rectangular coordinate system, as in Figure 253, but label the axes real (horizontal) and imaginary (vertical). A number $x + iy$ may then be identified with the point (x, y) obtained by moving x units horizontally along the real axis and y units in the vertical direction along the imaginary axis. As usual, we move right if x is positive, left if x is negative, up if y is positive, and down if y is negative.

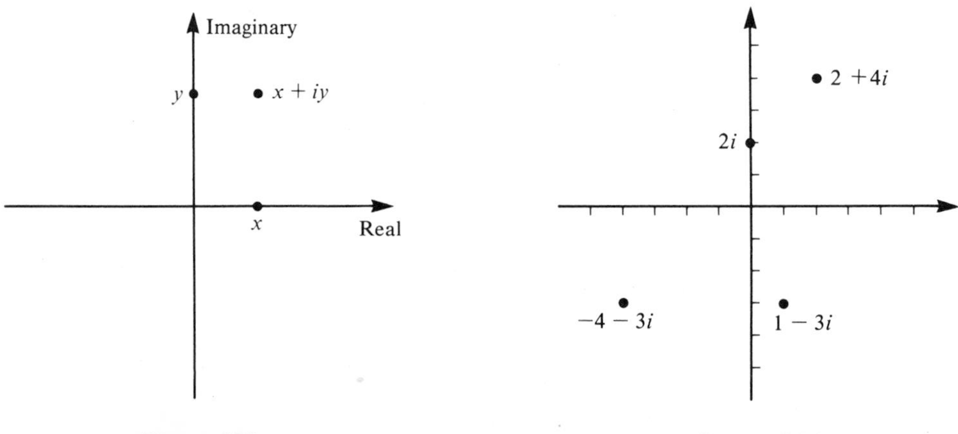

Figure 253.
Graphing complex numbers.

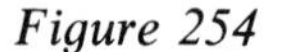

Figure 254

As examples, Figure 254 shows $2 + 4i$, $1 - 3i$, $2i$, and $-4 - 3i$. The plane, with points (x, y) thought of as complex numbers $x + iy$, is called the *complex plane.*

Note that the complex plane shows again how real numbers are special kinds of complex numbers (having zero imaginary part), since the real axis lies in the complex plane. Pure imaginary numbers iy, having no horizontal component, are graphed on the vertical (imaginary) axis.

It is easy to see that the parallelogram law holds for addition of complex numbers, as shown in Figure 255. (In fact, adding complex numbers $z_1 = x_1 + iy_1$ and $z_2 = x_2 + iy_2$ is exactly analogous to adding vectors $\mathbf{z}_1 = x_1\mathbf{i} + y_1\mathbf{j}$ and $\mathbf{z}_2 = x_2\mathbf{i} + y_2\mathbf{j}$, since we do both componentwise: $z_1 + z_2 = (x_1 + x_2) + i(y_1 + y_2)$, and, as vectors, $\mathbf{z}_1 + \mathbf{z}_2 = (x_1 + x_2)\mathbf{i} + (y_1 + y_2)\mathbf{j}$. Here, $\mathbf{i}$ is the unit horizontal vector, and i the complex number satisfying $i^2 = -1$.)

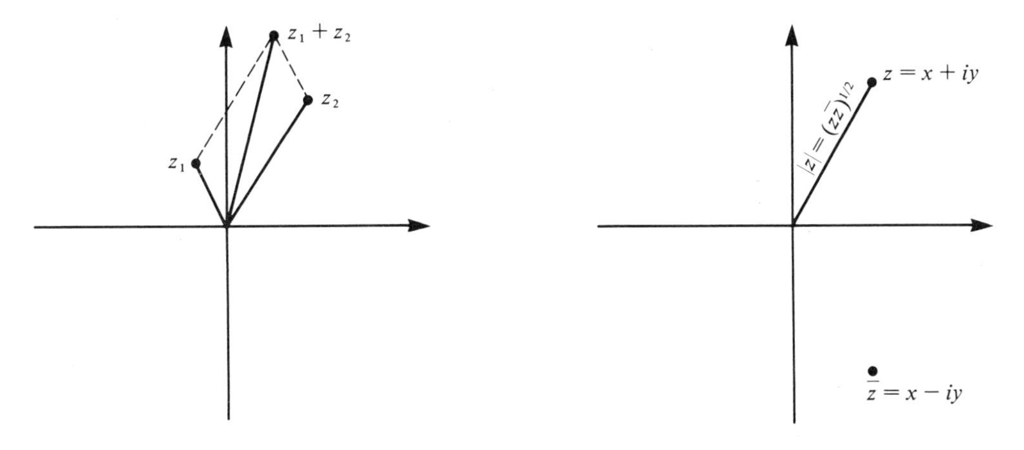

Figure 255. Parallelogram law for adding complex numbers.

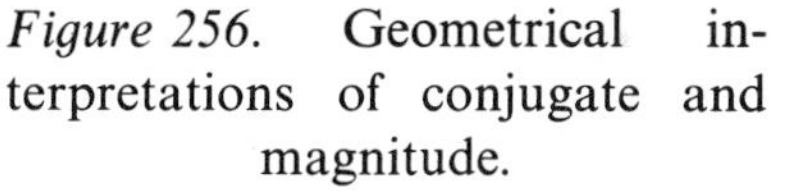

Figure 256. Geometrical interpretations of conjugate and magnitude.

The complex conjugate $\bar{z} = x - iy$ of $z = x + iy$ can be identified with the reflection of the point (x, y) across the real axis (see Figure 256). Further, $z\bar{z} = x^2 + y^2$ is the square of the distance from the origin to (x, y), also indicated in Figure 256. For this reason, the number $(z\bar{z})^{1/2}$, or $\sqrt{x^2 + y^2}$, is called the *magnitude* of z and is denoted $|z|$. The parallelogram law now gives us two important facts about magnitude:

1. $|z_1 + z_2| \leq |z_1| + |z_2|$. This is the *triangle inequality* and, as shown in Figure 257, says that the length of one side of a triangle cannot exceed the sum of the lengths of the other two sides.
2. $|z_1 - z_2|$ is the distance between z_1 and z_2. This is shown in Figure 258.

Thus, for example, the distance between $1 - 4i$ and $-2 - 6i$ is

$$|(1 - 4i) - (-2 - 6i)| = |3 + 2i| = \sqrt{13}.$$

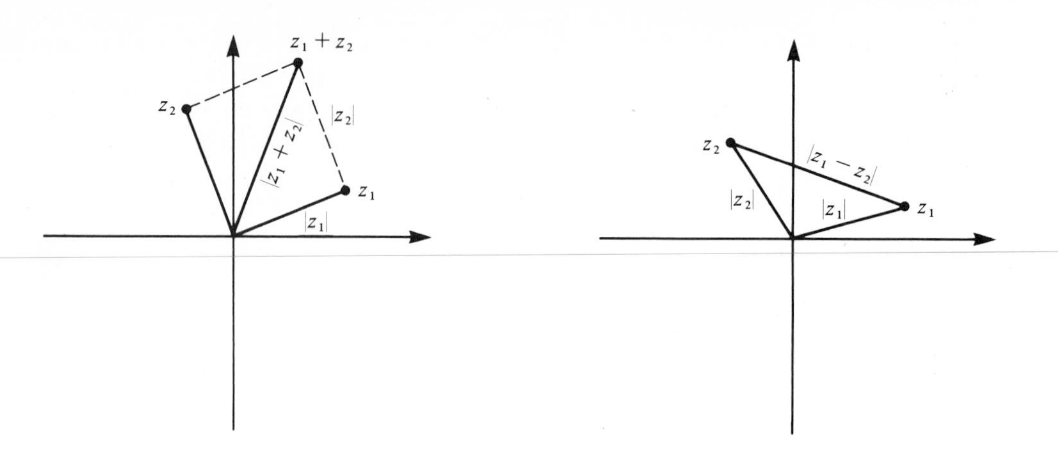

Figure 257. Triangle inequality.

Figure 258. Distance between z_1 and z_2.

Often, complex notation can be used to describe geometric loci in the plane, remembering that $x + iy$ is always identified with the point (x, y). Here are three examples.

EXAMPLE 394

Describe geometrically all numbers z with $|z| = 4$.

Since $|z|$ is the distance from the origin to z, then the equation $|z| = 4$ characterizes all points at distance 4 from the origin. Thus, $|z| = 4$ graphs as the circle of radius 4 about the origin in the complex plane, as shown in Figure 259.

Analytically, we could write $z = x + iy$, and then $|z| = 4$ gives us $|z|^2 = x^2 + y^2 = 16$, the equation of a circle of radius 4 about the origin.

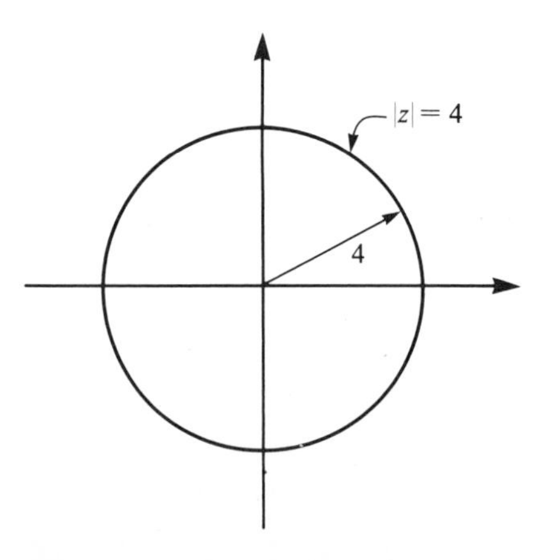

Figure 259. Locus of z with $|z| = 4$.

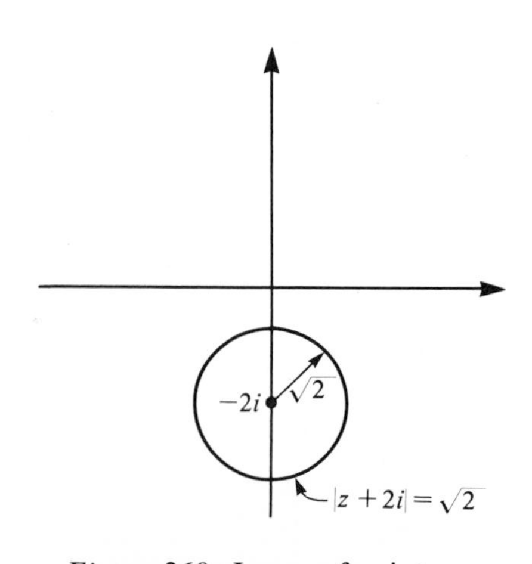

Figure 260. Locus of points z with $|z + 2i| = |1 + i|$.

EXAMPLE 395

Describe geometrically all numbers z with $|z + 2i| = |1 + i|$.

Since $|1 + i| = \sqrt{2}$, then we need $|z + 2i| = \sqrt{2}$. In words, we want all z whose distance from $-2i$ is fixed at $\sqrt{2}$. This is a circle of radius $\sqrt{2}$ and center $-2i$, as shown in Figure 260.

EXAMPLE 396

Describe the locus of points z such that

$$|z|^2 + 3\ \mathrm{Re}(z^2) = 4.$$

This one is more complicated than the previous two. Write $z = x + iy$, and we obtain

$$x^2 + y^2 + 3\ \mathrm{Re}(x^2 - y^2 + 2ixy) = 4.$$

Then,

$$x^2 + y^2 + 3(x^2 - y^2) = 4.$$

This can be written

$$x^2 - \tfrac{1}{2}y^2 = 1.$$

This is an hyperbola, as shown in Figure 261.

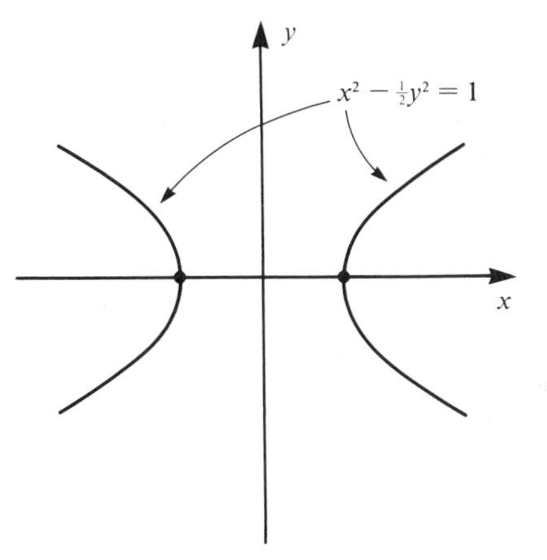

Figure 261. Locus of points z with $|z|^2 + 3\ \mathrm{Re}(z^2) = 4$.

We conclude this section by noting that complex arithmetic follows most of the "normal" rules we are accustomed to from experience with real numbers. In particular:

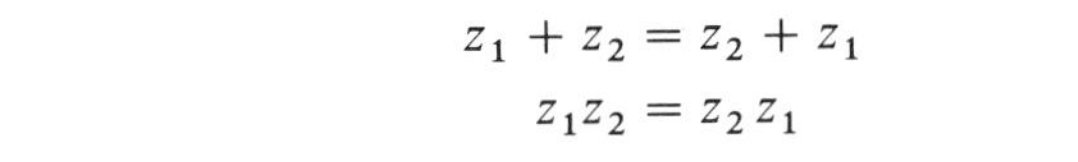

1.
$$z_1 + z_2 = z_2 + z_1$$
$$z_1 z_2 = z_2 z_1$$
(commutative laws)

2.
$$z_1 + (z_2 + z_3) = (z_1 + z_2) + z_3$$
$$z_1(z_2 z_3) = (z_1 z_2) z_3$$
(associative laws)

3.
$$z_1(z_2 + z_3) = z_1 z_2 + z_1 z_3$$
(distributive law)

4.
$$z + 0 = 0 + z = z$$
and
$$z \cdot 1 = 1 \cdot z = z$$

5. The equation $z_1 = w + z_2$ has a unique solution, given z_1 and z_2. If $z_1 = x_1 + iy_1$ and $z_2 = x_2 + iy_2$, this solution is

$$w = z_1 - z_2 = (x_1 - x_2) + i(y_1 - y_2).$$

6. The equation $z_1 = wz_2$ has a unique solution for w if $z_2 \neq 0$. In fact, $w = z_1/z_2$, as given previously.

An important difference between real and complex numbers is that real numbers are *ordered*, while complex numbers are not. If a and b are real, then either $a < b$,

$a = b$, or $a > b$. In fact, $a < b$ exactly when the point denoting a lies to the left of that denoting b on the real line. There is no comparable way of assigning a "less than" relationship for pairs of complex numbers. We shall pursue this idea in the exercises.

In the next section we consider another way of writing complex numbers which will prove useful.

PROBLEMS FOR SECTION 14.1

In each of Problems 1 through 25, perform the indicated arithmetic operations, writing the result in the form $a + ib$.

1. $(3 - 4i)(6 + 2i)$

2. $(1 - i) + (2 + 4i)$

3. $i(6 - 2i)$

4. $1/i$

5. $(2 - i)/(4 + i)$

6. $\left[2i + \left(\frac{3 - i}{2i}\right)\right](1 - i)$

7. $\frac{1 + i}{2 - i} - \frac{4}{3i}$

8. $[(i + 2) - (3 - 4i)]/(2 - i)(3 + i)$

9. $(2 + 4i)\overline{(6 - 3i)}$

10. $(-2 - 3i)\overline{(8 - 4i)}/\overline{(6 + 2i)}$

11. $8i/(6 - i)$

12. $i^3 - 4i^2 + 2$

13. $(3 + i)^3$

14. $(17 - 3i)/(2 + 4i)$

15. $\frac{(i + 2)(3 - i)}{(4 - i)(6 + i)}$

16. $\frac{(3 - 5i)(3 - 7i)}{i^3(3 + i)}$

17. $(i + 1)(2 - 4i)(4 + i)(6 - 2i)$

18. $(2 - i)^4$

19. $\frac{(9 - 2i)(2 + i)}{(-3 + 2i)^2}$

20. $\left(\frac{(-3 + i)(1 - 3i)^2}{-2 + i}\right)^2$

21. $(2 + 5i)(-4i)(-6 + i)$

22. $\frac{-4 - 7i}{(-4i)(2 - 5i)}$

23. $\frac{(6 + i)(2 - 5i)}{(1 + i)^3}$

24. $i(-3 + i)(5 + i)$

25. $\frac{2i(5 - 2i)(-8 + 4i)}{(-4 + i)(1 - 2i)}$

In each of Problems 26 through 30, let $z = x + iy$.

26. Find $\text{Re}(z^2)$ and $\text{Im}(z^2)$.

27. Find $|z + 2|$ and $|z - i|$.

28. Find $\text{Re}(2z - 3\bar{z} + 4)$.

29. Find $\text{Im}[(z^2 - 2x)/(z + 1)]$.

30. Find $\text{Im}(2\bar{z}/|z|)$.

31. Describe the locus of points in the complex plane satisfying $|z| = |z - i|$.

32. Describe the locus of points in the complex plane satisfying $|z|^2 + \text{Im}(z) = 16$.

33. Let z be any complex number except 0, i, or $-i$. Prove that $z/(1 + z^2)$ is real (i.e., has zero imaginary part) if and only if $z\bar{z} = 1$.

34. Prove that $|z_1 + z_2| \leq |1 + \bar{z}_1 z_2|$ if $|z_1| \leq 1$ and $|z_2| \leq 1$.

35. Suppose we try to define $x_1 + iy_1 < x_2 + iy_2$ to mean $x_1 < x_2$ and $y_1 < y_2$. Is it then true that, for every pair of complex numbers z_1 and z_2, either $z_1 < z_2$, $z_2 < z_1$, or $z_1 = z_2$?

36. Suppose we try to define $x_1 + iy_1 < x_2 + iy_2$ to mean that $x_1 < x_2$ or $y_1 < y_2$. Is it then true that, for every pair of complex numbers z_1 and z_2, either $z_1 < z_2$, $z_2 < z_1$, or $z_1 = z_2$?

37. Prove that z is real or pure imaginary if and only if $z^2 = (\bar{z})^2$.

38. Let z_1, z_2, and z_3 be complex numbers. Prove that z_1, z_2, and z_3 form the vertices of an equilateral triangle in the complex plane if and only if

$$z_1^2 + z_2^2 + z_3^2 = z_1 z_2 + z_2 z_3 + z_1 z_3.$$

39. Describe the locus of points z such that

$$z^2 + \bar{z}^2 = 4.$$

40. Describe the locus of points z such that

$$\operatorname{Im}(z - i) = \operatorname{Re}(z + 1).$$

41. Describe the locus of points z such that

$$|z - i| + |z + 8 - 2i| = 9.$$

(*Hint:* Think of $|z - \alpha|$ as the distance between z and α, and read in words what this equation says. This should remind you of something from analytic geometry.)

42. Prove the binomial theorem for complex numbers:

For any complex numbers z and w, and any positive integer n,

$$(z + w)^n = \sum_{j=0}^{n} \binom{n}{j} z^{n-j} w^j.$$

Here,

$$\binom{n}{j} = \frac{n!}{j!(n-j)!}.$$

The factorial notation is defined by $k! = 1 \cdot 2 \cdots k$ if k is a positive integer. For $k = 0$, we define $0! = 1$. Thus, for example, $1! = 1$, $2! = 2$, $3! = 6$, $4! = 24$, and so on.

43. Describe geometrically the locus of points z such that

$$|z - z_1|\,|z - z_2| \cdots |z - z_n| = K,$$

where $z_1, \ldots, z_n$ are given complex numbers and K is a given positive constant. (*Hint:* Start with $n = 1, 2, 3$ to get some feeling for the problem.)

14.2 POLAR FORM OF A COMPLEX NUMBER

Given $z = x + iy$, we can assign polar coordinates to the point (x, y) by setting

$$x = r\cos(\theta), \qquad y = r\sin(\theta).$$

Then,

$$z = r[\cos(\theta) + i\sin(\theta)].$$

This is the *polar form* of z, as shown in Figure 262. In this expression,

$$r = |z| = \sqrt{x^2 + y^2}$$

is the *magnitude*, or *modulus*, of z, and θ (in radians) is called the *argument* of z, denoted $\arg z$. Note that we may replace θ by $\theta + 2n\pi$ (n any integer) in the polar form of z; hence, $\arg z$ really denotes an infinity of numbers, differing from each other by integer multiples of 2π. The value θ of $\arg z$ lying in $-\pi < \theta \leq \pi$ is called the *principal argument* of z, denoted $\text{Arg } z$.

Geometrically, if $z \neq 0$, we obtain values of $\arg z$ by rotating the positive real axis to the line from (0, 0) to (x, y); a counterclockwise rotation gives positive values of $\arg z$, and a clockwise rotation gives negative values.

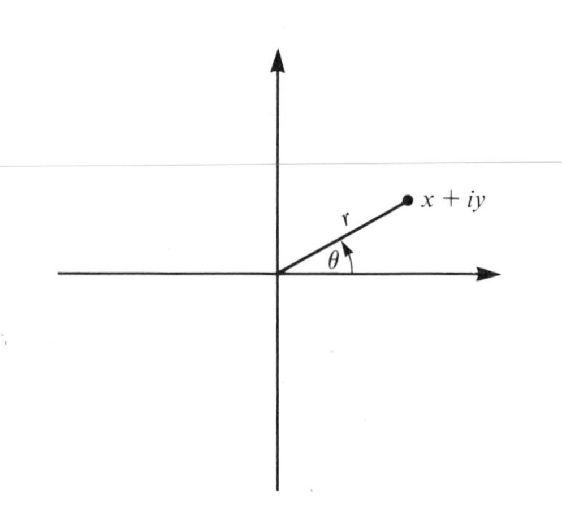

Figure 262.
Polar form of a complex number.

EXAMPLE 397

Let $z = 1 + i$. Then, $|z| = \sqrt{2}$ and $\arg z$ is any number $\pi/4 + 2n\pi$ (n any integer). Here, $\text{Arg } z = \pi/4$.

Note that we get negative values of $\arg z$ by clockwise rotations from the positive real axis. For example, with $n = -1$, $\arg(1 + i) = (\pi/4) - 2\pi = -7\pi/4$. This indicates a *clockwise* rotation of $7\pi/4$ from the positive real axis, as shown in Figure 263.

EXAMPLE 398

Let $z = 17 - 3i$. Then, $|z| = \sqrt{17^2 + 3^2} = \sqrt{298}$ (this is about 17.263).

To find $\arg(17 - 3i)$, note from Figure 264 that we can rotate the positive x-axis

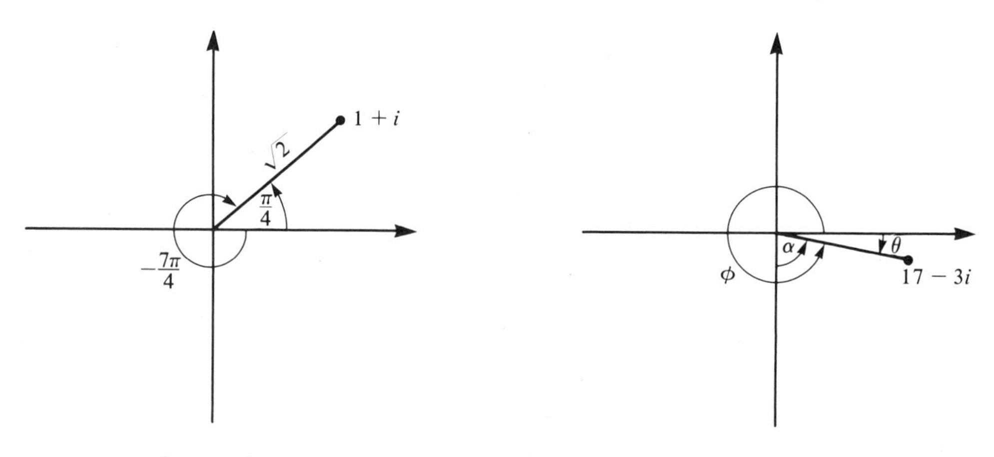

Figure 263.
Argument and modulus of $1 + i$.

Figure 264. Argument of $17 - 3i$.

clockwise through an angle θ, where $\tan(\theta) = \frac{3}{17}$, in order to arrive at the line from the origin to the point $17 - 3i$. Since a clockwise rotation is indicated by a minus sign, then

$$\arg(17 - 3i) = -\arctan(\tfrac{3}{17}) + 2n\pi, \qquad n \text{ any integer.}$$

If we want to use a counterclockwise rotation, then we want the angle ϕ of Figure 264. Here,

$$\phi = \frac{3\pi}{2} + \alpha = \frac{3\pi}{2} + \arctan\left(\frac{17}{3}\right).$$

Thus, we could also write

$$\arg(17 - 3i) = \frac{3\pi}{2} + \arctan\left(\frac{17}{3}\right) + 2k\pi, \qquad k \text{ any integer.}$$

As a check, recall that any two values of $\arg(17 - 3i)$ must differ by any integer multiple of 2π. Here, one can verify that indeed

$$\frac{3\pi}{2} + \arctan\left(\frac{17}{3}\right) - \left[-\arctan\left(\frac{3}{17}\right)\right] = 2\pi.$$

This can be done exactly using trigonometry, or we can use a calculator to obtain approximate values. We have $3\pi/2 \approx 4.71238898$, $\arctan(\frac{17}{3}) \approx 1.396124128$, $\arctan(\frac{3}{17}) \approx 0.174672199$, and $2\pi \approx 6.283185307$; and a little arithmetic shows the above equation to hold, at least to nine decimal places.

While properties of $\arg(z)$ are most conveniently derived by using complex exponentials, which we will develop later, one can also use trigonometric identities for this purpose. Let z_1 and z_2 have polar forms

$$z_1 = r_1[\cos(\theta_1) + i \sin(\theta_1)],$$
$$z_2 = r_2[\cos(\theta_2) + i \sin(\theta_2)].$$

Then,

$$\begin{aligned} z_1 z_2 &= r_1 r_2\{[\cos(\theta_1)\cos(\theta_2) - \sin(\theta_1)\sin(\theta_2)] \\ &\qquad + i[\cos(\theta_1)\sin(\theta_2) + \cos(\theta_2)\sin(\theta_1)]\} \\ &= r_1 r_2[\cos(\theta_1 + \theta_2) + i \sin(\theta_1 + \theta_2)]. \end{aligned}$$

This tells us two things. First,

$$|z_1 z_2| = r_1 r_2 = |z_1||z_2|,$$

a fact easily derived directly. Second,

$$\arg(z_1 z_2) = \theta_1 + \theta_2 + 2n\pi = \arg z_1 + \arg z_2 + 2n\pi,$$

for any integer n. That is, the argument of $z_1 z_2$ is the sum of the arguments of z_1 and z_2, *to within integer multiples of* 2π.

Similarly, we find that

$$\left|\frac{z_1}{z_2}\right| = \frac{|z_1|}{|z_2|},$$

and

$$\arg\left(\frac{z_1}{z_2}\right) = \arg z_1 - \arg z_2 + 2n\pi \qquad (n \text{ any integer}).$$

EXAMPLE 399

We illustrate the above results about modulus and argument.

Let $z_1 = 2i$ and $z_2 = 3 + 3i$. Then,

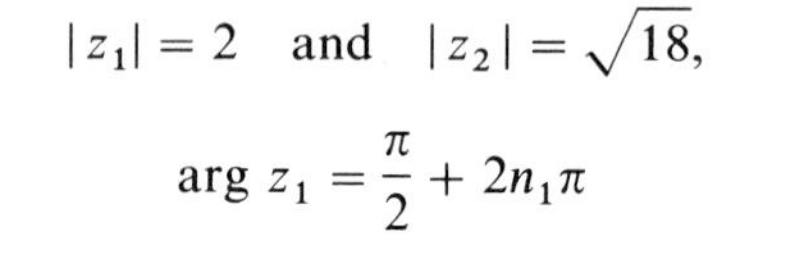

$$|z_1| = 2 \quad \text{and} \quad |z_2| = \sqrt{18},$$

$$\arg z_1 = \frac{\pi}{2} + 2n_1\pi$$

and

$$\arg z_2 = \frac{\pi}{4} + 2n_2\pi$$

(with n_1 and n_2 any integers), and

$$\text{Arg } z_1 = \frac{\pi}{2} \quad \text{and} \quad \text{Arg } z_2 = \frac{\pi}{4}$$

(see Figure 265). Now,

$$z_1 z_2 = -6 + 6i,$$

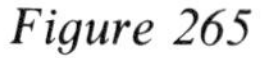

Figure 265

and we have

$$|z_1 z_2| = \sqrt{36 + 36}$$
$$= \sqrt{72} = |z_1||z_2|.$$

Finally,

$$\arg(z_1 z_2) - [\arg(z_1) + \arg(z_2)] = \left(\frac{3}{4}\pi + 2n\pi\right) - \left(\frac{\pi}{2} + 2n_1\pi + \frac{\pi}{4} + 2n_2\pi\right)$$
$$= 2(n - n_1 - n_2)\pi.$$

Thus, $\arg(z_1 z_2)$ is indeed equal to $\arg(z_1) + \arg(z_2)$ to within an integer multiple of 2π.

Now,

$$\frac{z_1}{z_2} = \frac{2i}{3 + 3i} = \frac{1}{3}(1 + i).$$

Then,

$$\left|\frac{z_1}{z_2}\right| = \sqrt{\frac{1}{9} + \frac{1}{9}} = \sqrt{\frac{2}{9}} = \frac{|z_1|}{|z_2|}$$

and

$$\arg\left(\frac{z_1}{z_2}\right) - [\arg(z_1) - \arg(z_2)] = \frac{\pi}{4} + 2n\pi - \left[\frac{\pi}{2} + 2n_1\pi - \left(\frac{\pi}{4} + 2n_2\pi\right)\right]$$
$$= 2(n - n_1 + n_2)\pi.$$

Thus, $\arg(z_1/z_2)$ differs from $\arg(z_1) - \arg(z_2)$ by an integer multiple of 2π.

There is a useful relationship due to de Moivre which can be proved easily using properties of argument and modulus.

LEMMA 1 de Moivre's Formula

For any positive integer n,

$$[\cos(\theta) + i\sin(\theta)]^n = \cos(n\theta) + i\sin(n\theta).$$

Proof Let $z = \cos(\theta) + i\sin(\theta)$. Then, $|z| = 1$. Then, $|z^2| = |z|^2 = 1$, and, in general, $|z|^n = 1$.

Next,

$$\arg(z^2) = \arg(z \cdot z) = \arg(z) + \arg(z) + 2k\pi$$
$$= 2\arg z + 2k\pi, \qquad k \text{ any integer}.$$

In general, a straightforward induction argument shows that

$$\arg(z^n) = n\arg z + 2k\pi.$$

If $\arg z = \theta$, then $\arg(z^n) = n\theta$ (to within $2k\pi$). Thus, the polar form of $z^n = [\cos(\theta) + i\sin(\theta)]^n$ is

$$\cos(n\theta) + i\sin(n\theta),$$

proving the lemma.

Here is an illustration of de Moivre's formula used to prove trigonometric identities.

EXAMPLE 400

Express $\cos(3\theta)$ and $\sin(3\theta)$ in terms of $\sin(\theta)$ and $\cos(\theta)$.

By de Moivre's formula,

$$[\cos(\theta) + i\sin(\theta)]^3 = \cos(3\theta) + i\sin(3\theta).$$

Then,

$$\cos^3(\theta) - 3\cos(\theta)\sin^2(\theta) + i[3\sin(\theta)\cos^2(\theta) - \sin^3(\theta)] = \cos(3\theta) + i\sin(3\theta).$$

Equating the real parts of both sides and then the imaginary parts of both sides, we have

$$\cos(3\theta) = \cos^3(\theta) - 3\cos(\theta)\sin^2(\theta)$$

and

$$\sin(3\theta) = 3\sin(\theta)\cos^2(\theta) - \sin^3(\theta).$$

This is a good illustration of a real-valued problem (the identities themselves do not involve complex numbers) being solved by introducing complex quantities and then separating out the real and imaginary parts. This theme recurs many times in complex analysis.

PROBLEMS FOR SECTION 14.2

In each of Problems 1 through 15, determine the polar form of the given complex number. You may leave θ in the form $\arctan(\alpha)$ for appropriate α.

1. $2 - 6i$ **2.** $1 + 4i$ **3.** $8 - 2i$ **4.** $-3 - 6i$ **5.** $-14i$
6. $-2 - 12i$ **7.** $3 + 9i$ **8.** $-4 - i$ **9.** $-8 - 3i$ **10.** $5 + i$

11. $(1 - i)(3 - 4i)$ **12.** $(9 - 3i)^2$ **13.** $\dfrac{2 - 4i}{4 + 3i}$ **14.** i^3 **15.** $(3 - i)(2 + 5i)(-4 + i)$

In each of Problems 16 through 25, a complex number is given in polar form. Write the number in the form $a + ib$.

16. $3[\cos(\pi/4) + i\sin(\pi/4)]$
17. $9[\cos(7\pi/4) + i\sin(7\pi/4)]$
18. $8[\cos(2\pi/3) + i\sin(2\pi/3)]$
19. $14[\cos(7\pi/6) + i\sin(7\pi/6)]$
20. $4[\cos(11\pi/4) + i\sin(11\pi/4)]$
21. $16[\cos(5\pi/6) + i\sin(5\pi/6)]$
22. $5[\cos(5\pi/4) + i\sin(5\pi/4)]$
23. $14[\cos(\pi/6) + i\sin(\pi/6)]$
24. $7[\cos(8\pi/3) + i\sin(8\pi/3)]$
25. $15[\cos(15\pi/4) + i\sin(15\pi/4)]$

26. Prove that if $z_2 \neq 0$, then

$$\arg\left(\frac{z_1}{z_2}\right) = \arg z_1 - \arg z_2 + 2n\pi,$$

where n is any integer.

27. Prove *Lagrange's identity:* If $z_1, \ldots, z_n$ and $w_1, \ldots, w_n$ are complex numbers, then

$$\left|\sum_{j=1}^{n} z_j w_j\right|^2 = \left(\sum_{j=1}^{n} |z_j|^2\right)\left(\sum_{j=1}^{n} |w_j|^2\right) - \sum_{1 \le k < j \le n} |z_k \bar{w}_j - z_j \bar{w}_k|^2.$$

28. Use the result of Problem 27 (whether or not you proved it) to show that

$$\left|\sum_{j=1}^{n} z_j w_j\right|^2 \le \left(\sum_{j=1}^{n} |z_j|^2\right)\left(\sum_{j=1}^{n} |w_j|^2\right).$$

29. Prove that

$$\big||z_1| - |z_2|\big| \le |z_1 - z_2|.$$

30. Prove that

$$|z_1 + z_2|^2 + |z_1 - z_2|^2 = 2(|z_1|^2 + |z_2|^2).$$

(*Hint:* In problems involving magnitude, it is often useful to recall that $|z|^2 = z\bar{z}$.)

31. Prove *Lagrange's trigonometric identity* (not related to Problem 27):

$$1 + 2\sum_{j=1}^{n} \cos(j\theta) = \frac{\sin[(n + \frac{1}{2})\theta]}{\sin\left(\dfrac{\theta}{2}\right)} \quad \text{if} \quad \sin\left(\frac{\theta}{2}\right) \neq 0.$$

(*Hint:* Begin with the algebraic identity

$$\sum_{j=1}^{n} z^j = \frac{z - z^{n+1}}{1 - z} \quad \text{if} \quad z \neq 1.$$

You may assume this identity. Let $z = \cos(\theta) + i\sin(\theta)$, use de Moivre's formula, and separate real and imaginary parts.)

32. Prove that, for any complex numbers a and b,

$$\left|\frac{az + b}{\bar{b}z + \bar{a}}\right| = 1$$

if $|z| = 1$.

33. Use the binomial theorem to expand $[\cos(\theta) + i\sin(\theta)]^n$ in powers of $\cos(\theta)$ and $\sin(\theta)$. Separate the real and imaginary parts of this expansion and use de Moivre's formula to write $\cos(n\theta)$ and $\sin(n\theta)$ in terms of powers of $\cos(\theta)$ and $\sin(\theta)$.

14.3 FUNCTIONS AND SETS IN THE COMPLEX PLANE

We are about to consider limits and derivatives of complex-valued functions. These will involve some technical aspects of functions and sets of complex numbers. This section is devoted to the terminology used to treat these technicalities.

The student has seen the notion of a function many times, particularly in calculus, where functions usually produce real numbers from real numbers. For example, the function

$$f(x) = 3x^2 + 2 \qquad (0 \le x \le 4)$$

produces numbers in the interval [2, 50] from numbers in [0, 4]. Specifically, $f(0) = 2$, $f(\frac{1}{2}) = \frac{11}{4}$, $f(2) = 14$, and so on. Other familiar functions are e^x for $-\infty < x < \infty$, $\ln(x)$ (defined only for $x > 0$), and trigonometric functions such as $\sin(x)$ and $\cos(x)$.

In complex analysis we treat functions which send complex numbers to complex numbers (with the understanding that real numbers are complex numbers of the form $x + 0i$). For example, we might have

$$f(z) = z^2 - 2z + 1 \quad \text{for all complex } z,$$

$$g(z) = \frac{2z - 1}{z^2 + 5} \quad \text{for all complex numbers with } z^2 + 5 \neq 0,$$

$$h(z) = \operatorname{Re}(z) - 5 \quad \text{for all } z \text{ with } |z| < 5,$$

or

$$p(z) = z\bar{z} - \text{Im } z + \frac{4z^3}{3} \quad \text{for all } z \text{ with } |z - 2| < 4.$$

Such functions are simply called *complex functions.*

Of course, it is easy to write complex functions such as these four, because they involve only arithmetic operations we have already defined for complex numbers. It is not so obvious how to define, say, e^z, $\ln(z)$, $z^{1/3}$, or $\cos(z)$ for complex z. We shall treat this problem in Section 14.5.

Given any complex function $f(z)$, we can always write $f(z) = \text{Re } f(z) + i \text{ Im } f(z)$. If $z = x + iy$, then $f(z) = \text{Re } f(x + iy) + i \text{ Im } f(x + iy)$. Now, $\text{Re } f(x + iy)$ is a real-valued function, say, $u(x, y)$, of the two real variables x and y. And $\text{Im } f(x + iy)$ is a real-valued function $v(x, y)$ of the real variables x and y. Thus, we may always write

$$f(z) = u(x, y) + iv(x, y),$$

with u and v real-valued functions of two real variables. This point of view will be very important later.

As an example, let $f(z) = z^2$. With $z = x + iy$, we have $z^2 = x^2 - y^2 + i(2xy)$; so $f(z) = (x^2 - y^2) + i(2xy)$. Here, $u(x, y) = x^2 - y^2$ and $v(x, y) = 2xy$.

In many later considerations, we shall have to emphasize certain properties of the sets of complex numbers under discussion. The following definitions are standard.

Given a complex number z_0 and a positive number δ, the *open disk* of radius δ about z_0 consists of all complex numbers z at distance less than δ from z_0. These are exactly those z *inside* the circle of radius δ about z_0, as shown in Figure 266, and are characterized by the inequality

$$|z - z_0| < \delta.$$

An open disk of radius δ about z_0 is also called a *δ-neighborhood* of z_0. The *closed disk*

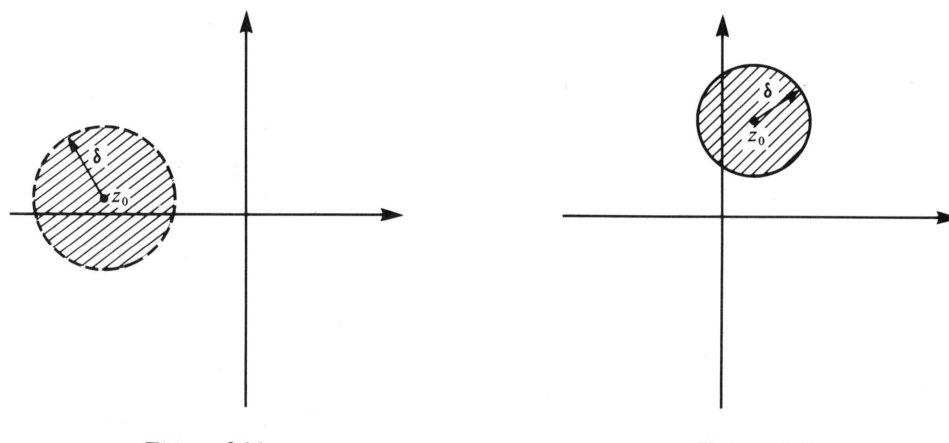

Figure 266.
Open disk of radius δ about z_0.

Figure 267.
Closed disk of radius δ about z_0.

of radius δ about z_0 consists of all z at distance less than or equal to δ from z_0. These are all z with

$$|z - z_0| \leq \delta,$$

as shown in Figure 267, and consist of all points on or inside the circle of radius δ about z_0.

Suppose now that S is any set of complex numbers and z_0 any particular complex number. Thinking of the complex plane, we usually refer to numbers in S as *points*.

We define:

1. z_0 is an *interior point* of S if all the points in *some* δ-neighborhood of z_0 are in S.
2. S is *open* if every point of S is an interior point.
3. z_0 is a *boundary point* of S if *every* δ-neighborhood of z_0 contains at least one point in S and one point not in S (z_0 itself may or may not be in S). The *boundary of S* consists of all boundary points of S.
4. S is *closed* if every boundary point of S actually belongs to S. Here are some examples.

EXAMPLE 401

Let S consist of all z with Re $z > 0$. Thus, S consists of all $x + iy$ with $x > 0$ and may be pictured as the right half-plane of Figure 268. We make several observations about S.

1. Every point of S is an interior point. Given any z_0 in S, let δ be one-half the distance from z_0 to the imaginary axis, as shown in Figure 268. Then this δ-neighborhood of z_0 is entirely within S.
2. As a result of (1), S is open.
3. The boundary points of S are exactly the pure imaginary numbers. For example, for *every* $\delta > 0$, the δ-neighborhood of i contains points in S and points not in S. The boundary of S is the vertical (imaginary) axis.
4. S does not contain all of its boundary points (in fact, it contains none of them). Hence, S is not closed.

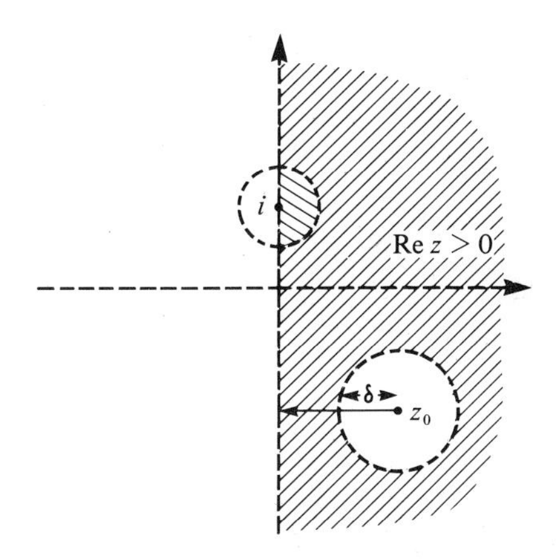

Figure 268.
Right half-plane Re $z > 0$.

EXAMPLE 402

Let S consist of all complex numbers. Then every point is an interior point; so S is open. Here, S has no boundary points, and so by default is also closed (there are no boundary points of S not in S). This example shows that closed and open are not exclusive concepts—here, S is both open and closed.

EXAMPLE 403

Let S consist of all $x + iy$ with $x \geq 0$ and $y > 0$ (see Figure 269). Then,

1. S is not open, as, for example, i is in S but is not an interior point (no δ-neighborhood of i contains only points of S).
2. S is not closed, since, for example, 1 is a boundary point but is not in S. The boundary of S consists of all $x + iy$ with $x = 0$ and $y \geq 0$, or $y = 0$ and $x \geq 0$.

This example shows that a set may be neither open nor closed.

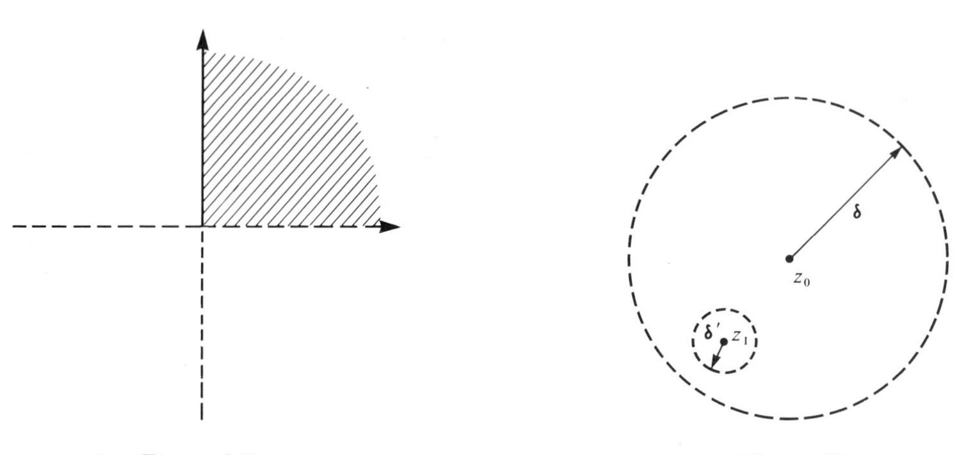

Figure 269.
Points $x + iy$ with $x \geq 0$ and $y > 0$.

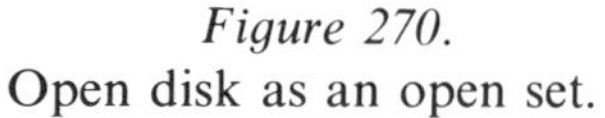

Figure 270.
Open disk as an open set.

EXAMPLE 404

Let S be the open disk of radius δ about z_0. Then S is an open set in the above sense. As indicated in Figure 270, about any z_1 in S we can draw a δ'-neighborhood containing only points of S. (Of course, as z_1 is chosen nearer the circle $|z - z_0| = \delta$, δ' must be chosen smaller.) The boundary points of S are those on the circle of radius δ about z_0. Thus, the boundary of S consists of all z with $|z - z_0| = \delta$. None of these points are in S.

Note that the closed disk $|z - z_0| \leq \delta$ thus contains all its boundary points, and hence is a closed set.

We shall also use the following terminology. Let S be any set of complex numbers.

1. S is *bounded* if, for some positive number M, $|z| \leq M$ for every z in S. This is equivalent to saying that one can draw a circle in the complex plane of sufficiently large radius to enclose all points of S.
2. S is *connected* if any two points in S can be joined by a polygonal line (a path consisting of finitely many straight line segments) consisting only of points of S.
3. S is a *domain* if S is both open and connected.

EXAMPLE 405

Let S consist of all z with $|z + i| < 4$. This is an open disk of radius 4 about $-i$ (see Figure 271). Then S is a domain. First, it is open because it is an open disk. It is connected because any two points in the disk can be joined by a polygonal line in the disk (in fact, by just a straight line in this example).

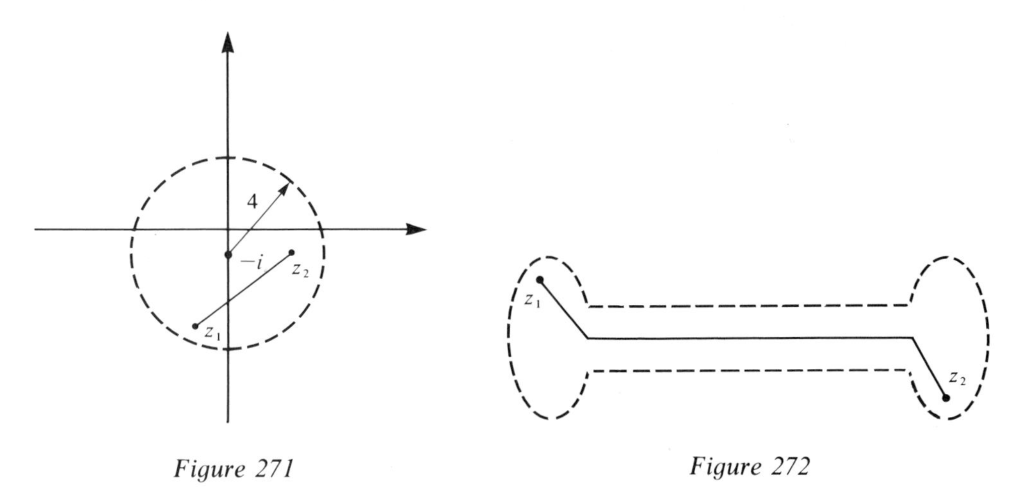

Figure 271 *Figure 272*

EXAMPLE 406

Let S be an open set consisting of the points inside the dumbbell-shaped outline of Figure 272. Then S is a domain because it is connected. For example, z_1 and z_2 as shown can be joined by the 3-section polygonal line of the diagram.

Finally, the term *region* is often used to refer to a set consisting of a domain together with some or all of its boundary points. Thus, for example, an open disk is a region, as is a closed disk.

PROBLEMS FOR SECTION 14.3

In each of Problems 1 through 10, sketch the indicated set in the complex plane. Determine whether it is open, closed, both open and closed, or neither open nor closed. Determine the boundary of the set, whether or not it is a domain, and whether or not it is bounded.

1. $|z| \geq 2$ **2.** $\text{Im}(z - 3i) \leq 4$ **3.** $1 < \text{Re}(z + 2) \leq 3$

4. $3 < |z - i| < 5$ **5.** $\text{Im}\left(\frac{z + i}{z + 2}\right) < 4$ **6.** $\left|\frac{z + i}{z - i}\right| < 9$

7. $\pi/2 \leq \text{Arg } z \leq 3\pi/2$ **8.** $0 < \text{Arg}(1 + 2z) < \pi/2$ **9.** $|1/z| \geq 4$

10. $\text{Re } z \leq 1$ and $0 < \text{Arg } z < \pi/4$

In each of Problems 11 through 20, write the given function in the form $u(x, y) + iv(x, y)$.

11. z^3 **12.** $2z^2 - \text{Im}(z) + 1$ **13.** $(z - i)/(2z + 1)$ **14.** $z + 3\bar{z} - |z|$

15. $(z^2 - i)/|z|$ **16.** $2z^2 - 5z + i$ **17.** $\text{Re}(z^2) - \text{Im}(z + 1) - 5 + 3i$ **18.** $|z^2 + 4i| - 3\bar{z}$

19. $2z + (\bar{z})^2 - 4$ **20.** $3\bar{z} + \text{Re}(\bar{z} - 2z) + 5\,\text{Im}(z^2)$

14.4 LIMITS AND DERIVATIVES OF COMPLEX FUNCTIONS

Let $f(z)$ be a complex function and z_0 a complex number. We want to define the limit of $f(z)$ as z approaches z_0. There is a subtlety here not encountered in the single-variable real case, however, and it is this: In "z approaches z_0" we want to allow z to move toward z_0 from *any* direction. This means that we want $f(z)$ to be defined at least for all z in any direction from z_0, though perhaps only very close to z_0. This is covered by the following definition.

We say that $f(z)$ *has limit* L *as* z *approaches* z_0, written

$$\lim_{z \to z_0} f(z) = L,$$

if:

1. $f(z)$ is defined in some open disk about z_0, except possibly at z_0 itself; and
2. given $\varepsilon > 0$, there exists $\delta > 0$ such that $|f(z) - L| < \varepsilon$ whenever $|z - z_0| < \delta$ and $z \neq z_0$.

Note that (2) may be reworded: Given any positive (hence real) number ε, there is a δ-neighborhood about z_0 such that $f(z)$ is within ε distance of L whenever z is within this neighborhood and different from z_0. We exclude z_0 itself from consideration, as with limits of real-valued functions, because the value $f(z_0)$ (if defined) does not influence the value, if any, which $f(z)$ *approaches* as z *approaches* z_0.

In elementary calculus, one usually spends some time doing ε-δ proofs of limits. Here, we will take the definition as a starting point for further discussion but will not concern ourselves with ε-δ proof techniques.

We call $f(z)$ *continuous* at z_0 if

$$\lim_{z \to z_0} f(z) = f(z_0).$$

This requires that $f(z)$ be defined for all z in some open disk about z_0, including at z_0 itself.

As in the real case, complex polynomials

$$\alpha_0 + \alpha_1 z + \alpha_2 z^2 + \cdots + \alpha_n z^n \qquad (\alpha_0, \ldots, \alpha_n \text{ complex})$$

are continuous everywhere. Further, *rational functions* (i.e., quotients of polynomials)

$$\frac{\alpha_0 + \alpha_1 z + \cdots + \alpha_n z^n}{\beta_0 + \beta_1 z + \cdots + \beta_m z^m}$$

are continuous wherever the denominator does not vanish.

We say that $f(z)$ is *differentiable* at z_0 if

$$\lim_{z \to z_0} \frac{f(z) - f(z_0)}{z - z_0}$$

exists (finite). We then call this limit the *derivative* of $f(z)$ at z_0 and denote it

$$f'(z_0) \quad \text{or} \quad \left.\frac{df}{dz}\right|_{z_0}.$$

EXAMPLE 407

Let $f(z) = 2z^2 - 1$. Then $f(z)$ is differentiable everywhere and $f'(z) = 4z$.
As a specific example, consider $f'(z_0)$ at $z_0 = 1 - i$. We have

$$\begin{aligned} f'(1-i) &= \lim_{z \to 1-i} \frac{(2z^2 - 1) - [2(1-i)^2 - 1]}{z - (1-i)} \\ &= \lim_{z \to 1-i} \frac{2[z - (1-i)][z + (1-i)]}{z - (1-i)} \\ &= \lim_{z \to 1-i} 2[z + (1-i)] \\ &= 2[(1-i) + (1-i)] = 4(1-i). \end{aligned}$$

If we write $\Delta z = z - z_0$ in the definition of derivative, then we have the equivalent formulation

$$f'(z_0) = \lim_{\Delta z \to 0} \frac{f(z_0 + \Delta z) - f(z_0)}{\Delta z}.$$

Again, we emphasize in these limits for $f'(z_0)$ that z must be allowed to approach z_0 (or Δz approach 0) from any direction. To illustrate the importance of this, here is an example in which $f'(z_0)$ does not exist.

EXAMPLE 408

Let $f(z) = \bar{z}$. Then $f'(i)$ does not exist.
To see this, we must examine

$$\lim_{z \to i} \frac{f(z) - f(i)}{z - i}.$$

Here,

$$\frac{f(z) - f(i)}{z - i} = \frac{\bar{z} - (-i)}{z - i} = \frac{\bar{z} + i}{z - i}.$$

Examine the behavior of this quotient as z approaches i along different paths. If $z \to i$ along the imaginary axis, then $z = \alpha i$ (α real), and

$$\frac{\bar{z} + i}{z - i} = \frac{-\alpha i + i}{\alpha i - i} = -1,$$

a constant which remains at -1 as $z \to i$ vertically (see Figure 273). If $z \to i$ along a horizontal line, then $z = \beta + i$, with β real, and

$$\frac{\bar{z} + i}{z - i} = 1,$$

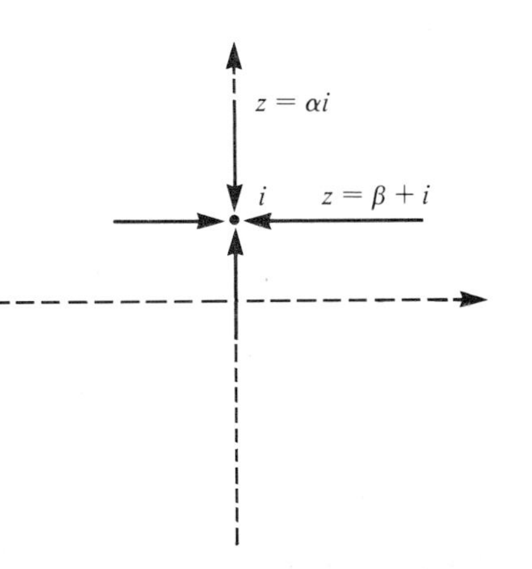

Figure 273. z approaching i along vertical and horizontal paths.

which remains at $+1$ as $\beta \to 0$. Thus, we get different values along different paths of approach, and so $f'(i)$ does not exist.

In general, the rules for differentiating sums, products, and quotients are the same for complex as for real-valued functions. Whenever $f'(z_0)$ and $g'(z_0)$ exist, then

1. $(\alpha f)'(z_0) = \alpha f'(z_0)$ for any complex constant α.
2. $(f + g)'(z_0) = f'(z_0) + g'(z_0)$.
3. $(fg)'(z_0) = f'(z_0)g(z_0) + f(z_0)g'(z_0)$.
4. $\left(\dfrac{f}{g}\right)'(z_0) = \dfrac{g(z_0)f'(z_0) - f(z_0)g'(z_0)}{g(z_0)^2}$ if $g(z_0) \neq 0$.
5. The chain rule holds: If $f'(z_0)$ exists, and $z_0 = g(w_0)$ and $g'(w_0)$ exists, then the derivative of $f[g(z)]$ at w_0 is

$$f'(z_0)g'(w_0).$$

It is also easy to check that differentiability of $f(z)$ at z_0 implies continuity of $f(z)$ at z_0.

A function $f(z)$ is said to be *analytic* at z_0 if $f'(z)$ exists for all z in some δ-neighborhood of z_0. Thus, analytic at z_0 means differentiable in some open disk about z_0. For example, $f(z) = z^2$ is analytic at any point z_0. By contrast, the student can use the definition to check that $f(z) = |z|^2$ is differentiable at 0 but not at any other point. Thus, $|z|^2$ is not analytic at 0.

We say that $f(z)$ is *analytic in a domain D* if $f(z)$ is analytic at each point of D.

The next section is devoted to a more thorough examination of the notion of analyticity.

PROBLEMS FOR SECTION 14.4

In each of Problems 1 through 10, use the definition of derivative to obtain $f'(z_0)$ at the given point or to show that $f'(z_0)$ does not exist.

1. $2z^2 + 1; \quad z_0 = 1 - i$
2. $z - 2\bar{z}; \quad z_0 = 3i$
3. $z/(z + 1); \quad z_0 = 4$
4. $(2z + 1)/z^2; \quad z_0 = -2 + i$
5. $(z - \bar{z})/|z|; \quad z_0 = 5i$
6. $\text{Re } z; \quad z_0 = 4 - i$
7. $\bar{z}; \quad z_0 = 1 + i$
8. $(z + 1)/2\bar{z}; \quad z_0 = 3i$
9. $\text{Re}(z) + 2i; \quad z_0 = 3 - 2i$
10. $(z + 2)/(z^2 + i); \quad z_0 = 2$

In each of Problems 11 through 20, use the usual rules for differentiation to find the derivative $f'(z)$ of the given function.

11. $8z^3 - 3z + 1$
12. $(z - 1)^2(z^3 + 2)$
13. $10(3z^2 - 4)$
14. $\dfrac{z + 1}{z - 1}$
15. $\dfrac{3z - 2}{z^2 + 1 - i}$
16. $4iz^2 - 8z + 2$
17. $\dfrac{iz^3}{z - 4 + 2i}$
18. $(iz - 2)^5$
19. $(4iz + 2)^3(8z^2 - iz + 1)$
20. $\dfrac{(z - 1)}{(iz - 2)^4}$

21. Prove that if $f(z)$ is differentiable at z_0, then it is continuous at z_0.

22. Give an example to show that a complex function may be continuous but not differentiable at a point.

23. Prove that a function can have only one limit as $z \to z_0$. That is, prove that, if $\lim_{z\to z_0} f(z) = A$ and $\lim_{z\to z_0} f(z) = B$, then $A = B$.

24. Suppose that $\lim_{z\to z_0} f(z) = L$. Prove that $\lim_{z\to z_0} \operatorname{Re} f(z) = \operatorname{Re} L$ and $\lim_{z\to z_0} \operatorname{Im} f(z) = \operatorname{Im} L$.

25. Prove that $\lim_{z\to z_0} [f(z) + g(z)] = \lim_{z\to z_0} f(z) + \lim_{z\to z_0} g(z)$ if both limits on the right exist.

26. Prove that $\lim_{z\to z_0} f(z)g(z) = [\lim_{z\to z_0} f(z)][\lim_{z\to z_0} g(z)]$ if both limits on the right exist.

27. Prove that $(f + g)'(z_0) = f'(z_0) + g'(z_0)$ if $f(z)$ and $g(z)$ are both differentiable at z_0.

28. Prove that $(fg)'(z_0) = f(z_0)g'(z_0) + g(z_0)f'(z_0)$ if $f(z)$ and $g(z)$ are both differentiable at z_0.

14.5 THE CAUCHY-RIEMANN EQUATIONS

There is an extremely important test for analyticity of $f(z)$ which can be stated in terms of $\operatorname{Re} f(z)$ and $\operatorname{Im} f(z)$. For notational convenience, write $z = x + iy$ and $f(z) = u(x, y) + iv(x, y)$, as discussed in Section 14.3.

THEOREM 110 Cauchy-Riemann Equations

Let $f(z) = u(x, y) + iv(x, y)$ be continuous in an open disk about $z_0 = x_0 + iy_0$ and differentiable at z_0. Then, at (x_0, y_0),

$$\frac{\partial u}{\partial x} = \frac{\partial v}{\partial y} \quad \text{and} \quad \frac{\partial u}{\partial y} = -\frac{\partial v}{\partial x}.$$

These are called the *Cauchy-Riemann equations* for u and v.

Proof Since $f'(z_0)$ exists, then

$$f'(z_0) = \lim_{\Delta z \to 0} \frac{f(z_0 + \Delta z) - f(z_0)}{\Delta z}.$$

Write $\Delta z = \Delta x + i\Delta y$, and write

$$\frac{f(z_0 + \Delta z) - f(z_0)}{\Delta z} = \frac{[u(x_0 + \Delta x, y_0 + \Delta y) + iv(x_0 + \Delta x, y_0 + \Delta y)]\,[u(x_0, y_0) + iv(x_0, y_0)]}{\Delta x + i\Delta y}$$

$$= \left(\frac{u(x_0 + \Delta x, y_0 + \Delta y) - u(x_0, y_0)}{\Delta x + i\,\Delta y}\right)$$

$$+\, i\left(\frac{v(x_0 + \Delta x, y_0 + \Delta y) - v(x_0, y_0)}{\Delta x + i\,\Delta y}\right).$$

Now, we must obtain the same value in the limit as Δz approaches zero along *any* path. Choose two specific paths as follows:

I. Let $\Delta z \to 0$ along the x-axis. Then, $\Delta y = 0$ and $\Delta z = \Delta x$. Then,

$$f'(z_0) = \lim_{\Delta x \to 0} \left\{ \left(\frac{u(x_0 + \Delta x, y_0) - u(x_0, y_0)}{\Delta x} \right) + i \left(\frac{v(x_0 + \Delta x, y_0) - v(x_0, y_0)}{\Delta x} \right) \right\}$$

$$= \frac{\partial u}{\partial x}(x_0, y_0) + i \frac{\partial v}{\partial x}(x_0, y_0).$$

II. Let $\Delta z \to 0$ along the y-axis. Then, $\Delta x = 0$ and $\Delta z = i\, \Delta y$. Then,

$$f'(z_0) = \lim_{\Delta y \to 0} \left\{ \left(\frac{u(x_0, y_0 + \Delta y) - u(x_0, y_0)}{i\, \Delta y} \right) + i \left(\frac{v(x_0, y_0 + \Delta y) - v(x_0, y_0)}{i\, \Delta y} \right) \right\}$$

$$= \frac{1}{i} \frac{\partial u}{\partial y}(x_0, y_0) + \frac{\partial v}{\partial y}(x_0, y_0)$$

$$= \frac{\partial v}{\partial y}(x_0, y_0) - i \frac{\partial u}{\partial y}(x_0, y_0).$$

Comparing the real and imaginary parts of the two expressions for $f'(z_0)$ from I and II, we have

$$\frac{\partial u}{\partial x}(x_0, y_0) = \frac{\partial v}{\partial y}(x_0, y_0)$$

and

$$\frac{\partial v}{\partial x}(x_0, y_0) = -\frac{\partial u}{\partial y}(x_0, y_0),$$

and these are the Cauchy-Riemann equations.

As an immediate consequence, we have the following corollary.

COROLLARY 1

If $f(z) = u(x, y) + iv(x, y)$ is analytic in a domain D, then $u(x, y)$ and $v(x, y)$ satisfy the Cauchy-Riemann equations at all points of D.

EXAMPLE 409

Let $f(z) = z^3$. Then $f'(z)$ exists for all z; in fact, $f'(z) = 3z^2$.

It is easy to check that

$$f(z) = (x^3 - 3xy^2) + i(3x^2y - y^3).$$

We have, at any (x, y),

$$\frac{\partial u}{\partial x} = 3x^2 - 3y^2 = \frac{\partial v}{\partial y}$$

and

$$\frac{\partial v}{\partial x} = 6xy = -\frac{\partial u}{\partial y}.$$

Thus, the Cauchy-Riemann equations hold in this example at all points (x, y).

The converse of Corollary 1 fails, as one can show by fairly complicated examples (see Problem 13). With an additional hypothesis, however, the Cauchy-Riemann equations imply analyticity. To prove the theorem, we need the following result from the calculus of real-valued functions of two variables: If $h(x, y)$ is continuous, with continuous partials $\partial h/\partial x$ and $\partial h/\partial y$, in some open disk about (x_0, y_0), then there exists a function $A(\Delta x, \Delta y)$ such that $A(\Delta x, \Delta y) \to 0$ as $(\Delta x, \Delta y) \to (0, 0)$ and

$$h(x_0 + \Delta x, y_0 + \Delta y) - h(x_0, y_0) = \frac{\partial h}{\partial x}(x_0, y_0)\,\Delta x + \frac{\partial h}{\partial y}(x_0, y_0)\,\Delta y + A(\Delta x, \Delta y)\sqrt{(\Delta x)^2 + (\Delta y)^2}.$$

THEOREM 111

Let $u(x, y)$ and $v(x, y)$ be continuous with continuous first partial derivatives satisfying the Cauchy-Riemann equations at all points in a domain D. Then the complex function $f(z) = u(x, y) + iv(x, y)$ is analytic in D.

Proof Let z_0 be any point in D. We want to consider

$$\lim_{\Delta z \to 0} \frac{f(z_0 + \Delta z) - f(z_0)}{\Delta z}.$$

Let $\Delta z = \Delta x + i\,\Delta y$. Noting the fact about real-valued functions of two real variables stated just before this theorem, we can write

$$\begin{aligned} f(z_0 + \Delta z) - f(z_0) &= [u(x_0 + \Delta x, y_0 + \Delta y) - u(x_0, y_0)] \\ &\quad + i[v(x_0 + \Delta x, y_0 + \Delta y) - v(x_0, y_0)] \\ &= \frac{\partial u}{\partial x}(x_0, y_0)\,\Delta x + \frac{\partial u}{\partial y}(x_0, y_0)\,\Delta y \\ &\quad + A(\Delta x, \Delta y)\sqrt{(\Delta x)^2 + (\Delta y)^2} \\ &\quad + i\left[\frac{\partial v}{\partial x}(x_0, y_0)\,\Delta x + \frac{\partial v}{\partial y}(x_0, y_0)\,\Delta y \right. \\ &\quad \left. + B(\Delta x, \Delta y)\sqrt{(\Delta x)^2 + (\Delta y)^2}\right], \end{aligned}$$

where $A(\Delta x, \Delta y) \to 0$ and $B(\Delta x, \Delta y) \to 0$ as $(\Delta x, \Delta y) \to (0, 0)$.

Now use the Cauchy-Riemann equations and some algebraic manipulation to write

$$f(z_0 + \Delta z) - f(z_0) = \left[\frac{\partial u}{\partial x}(x_0, y_0) + i\frac{\partial v}{\partial x}(x_0, y_0)\right](\Delta x + i\,\Delta y)$$
$$+ [A(\Delta x, \Delta y) + iB(\Delta x, \Delta y)]\sqrt{(\Delta x)^2 + (\Delta y)^2}.$$

Then

$$\frac{f(z_0 + \Delta z) - f(z_0)}{\Delta z} = \frac{\partial u}{\partial x}(x_0, y_0) + i\frac{\partial v}{\partial x}(x_0, y_0)$$
$$+ [A(\Delta x, \Delta y) + iB(\Delta x, \Delta y)]\frac{\sqrt{(\Delta x)^2 + (\Delta y)^2}}{\Delta x + i\,\Delta y}.$$

Now, observe that

$$\left|\frac{\sqrt{(\Delta x)^2 + (\Delta y)^2}}{\Delta x + i\,\Delta y}\right| = \frac{\sqrt{(\Delta x)^2 + (\Delta y)^2}}{\sqrt{(\Delta x)^2 + (\Delta y)^2}} = 1.$$

Then, as $\Delta z \to 0$, we have $(\Delta x, \Delta y) \to (0, 0)$ and

$$\lim_{\Delta z \to 0} \frac{f(z_0 + \Delta z) - f(z_0)}{\Delta z} = \frac{\partial u}{\partial x}(x_0, y_0) + i\frac{\partial v}{\partial x}(x_0, y_0).$$

Hence, the limit exists, and so $f'(z_0)$ exists.

Thus far, we have shown that $f(z)$ is differentiable at every point of D. To prove that $f(z)$ is analytic in D, let z_0 be any point of D. Since D is open, there is an open disk about z_0 containing only points in D. Then $f(z)$ is differentiable on this disk, proving analyticity of $f(z)$ at z_0.

Using the Cauchy-Riemann equations, we can derive another important fact about the real and imaginary parts of an analytic function.

THEOREM 112

Let $f(z) = u(x, y) + iv(x, y)$ be analytic in a domain D. Then,

$$\frac{\partial^2 u}{\partial x^2} + \frac{\partial^2 u}{\partial y^2} = 0$$

and

$$\frac{\partial^2 v}{\partial x^2} + \frac{\partial^2 v}{\partial y^2} = 0$$

for (x, y) in D.

Proof We shall see later that analyticity of $f(z)$ in D implies that $u(x, y)$ and $v(x, y)$ have continuous second partial derivatives in D. Assuming this for now, we can conclude that the mixed partials are equal, that is,

$$\frac{\partial^2 u}{\partial x\,\partial y} = \frac{\partial^2 u}{\partial y\,\partial x} \quad \text{and} \quad \frac{\partial^2 v}{\partial x\,\partial y} = \frac{\partial^2 v}{\partial y\,\partial x}.$$

Since $\partial u/\partial x = \partial v/\partial y$, then

$$\frac{\partial^2 u}{\partial x^2} = \frac{\partial^2 v}{\partial x\,\partial y} = \frac{\partial^2 v}{\partial y\,\partial x} = \frac{\partial}{\partial y}\left(\frac{\partial v}{\partial x}\right) = \frac{\partial}{\partial y}\left(-\frac{\partial u}{\partial y}\right) = -\frac{\partial^2 u}{\partial y^2}.$$

Thus,

$$\frac{\partial^2 u}{\partial x^2} + \frac{\partial^2 u}{\partial y^2} = 0.$$

Similarly,

$$\frac{\partial^2 v}{\partial x^2} + \frac{\partial^2 v}{\partial y^2} = 0.$$

In general, one often sees $\partial^2 u/\partial x^2 + \partial^2 u/\partial y^2$ written as $\nabla^2 u$ (read "del squared u").

The equation $\nabla^2 u = 0$ is called *Laplace's equation* and is of fundamental importance in such phenomena as heat conduction and wave propagation. A function satisfying Laplace's equation in D is said to be *harmonic* in D. Thus, Theorem 112 may be phrased:

The real and imaginary parts of a function analytic in D are harmonic in D.

EXAMPLE 410

Let $f(z) = z^2 + 2i$. Then $f(z) = (x^2 - y^2) + i(2xy + 2)$. Here, $u(x, y) = x^2 - y^2$ and $v(x, y) = 2xy + 2$.

For any (x, y),

$$\frac{\partial u}{\partial x} = 2x = \frac{\partial v}{\partial y}$$

and

$$\frac{\partial u}{\partial y} = -2y = -\frac{\partial v}{\partial x}.$$

Further, $u(x, y)$, $v(x, y)$, and their first partials are continuous in the entire complex plane (which is a domain). Thus, $f(z)$ is analytic everywhere.

Note also that

$$\frac{\partial^2 u}{\partial x^2} + \frac{\partial^2 u}{\partial y^2} = 2 - 2 = 0$$

and

$$\frac{\partial^2 v}{\partial x^2} + \frac{\partial^2 v}{\partial y^2} = 0 + 0 = 0;$$

so $u(x, y)$ and $v(x, y)$ are harmonic functions, as predicted by Theorem 112.

In concluding this section, we note one side effect of the proof of Theorem 112. If $f(z) = u(x, y) + iv(x, y)$ has a derivative at $z_0 = x_0 + iy_0$, then

$$f'(z_0) = \frac{\partial u}{\partial x}(x_0, y_0) + i\frac{\partial v}{\partial x}(x_0, y_0).$$

This gives us $f'(z_0)$ in terms of partials of $u(x, y)$ and $v(x, y)$. Using the Cauchy-Riemann equations, we have the equivalent expressions

$$\begin{aligned} f'(z_0) &= \frac{\partial u}{\partial x}(x_0, y_0) - i\frac{\partial u}{\partial y}(x_0, y_0) \\ &= \frac{\partial v}{\partial y}(x_0, y_0) + i\frac{\partial v}{\partial x}(x_0, y_0) \\ &= \frac{\partial v}{\partial y}(x_0, y_0) - i\frac{\partial u}{\partial y}(x_0, y_0). \end{aligned}$$

The remainder of this chapter is devoted to developing complex versions of functions such as $\ln(x)$, e^x, $\sin(x)$, x^α, and $\sinh(x)$, and in examining the properties of these functions.

PROBLEMS FOR SECTION 14.5

In each of Problems 1 through 12, find $u(x, y)$ and $v(x, y)$ such that $f(z) = u(x, y) + iv(x, y)$. Determine whether or not the Cauchy-Riemann equations hold in the given domain D, and use Theorems 110 and 111 to determine whether or not $f(z)$ is analytic in D. If $f(z)$ is analytic in D, verify by direct calculation that $u(x, y)$ and $v(x, y)$ are harmonic in D.

1. $f(z) = z - i$; D = complex plane

2. $f(z) = z^2 - iz$; D = complex plane

3. $f(z) = \dfrac{2z + 1}{z}$; D consists of all $z \neq 0$

4. $f(z) = i|z|^2$; D = complex plane

5. $f(z) = \dfrac{z}{\text{Re}(z)}$; D consists of all $x + iy$ with $x \neq 0$

6. $f(z) = \dfrac{\text{Im}(z) + 2}{|z|}$; D consists of all $z \neq 0$

7. $f(z) = (z + 2)^2$; D = complex plane

8. $f(z) = \bar{z} + \text{Im}(3z)$; D = complex plane

9. $f(z) = iz + |z|$; D = complex plane

10. $f(z) = z^3 + 2z + 1$; D = complex plane

11. $f(z) = \dfrac{z - 2i}{iz + 1}$; D = complex plane except the number i

12. $f(z) = i\,\text{Im}(z) - |z|^2$; D = complex plane

13. Let

$$f(z) = \begin{cases} \dfrac{z^5}{|z|^4} & \text{if } z \neq 0, \\ 0 & \text{if } z = 0. \end{cases}$$

Prove that $\text{Re} f(z)$ and $\text{Im} f(z)$ satisfy the Cauchy-Riemann equations at $z = 0$, but that $f'(0)$ does not exist. Why does this not contradict Theorem 111?

14. Let $f(z) = |z|^2$.
(a) Show that the real and imaginary parts of $f(z)$ satisfy the Cauchy-Riemann equations at (0, 0).
(b) Prove that $f(z)$ is not analytic at $z = 0$ but is differentiable at $z = 0$.
(c) Why does the result of (b) not contradict Theorem 111?

15. Suppose that $f(z)$ is analytic in a domain D and $f'(z) = 0$ for all z in D. Prove that $f(z)$ must be constant in D.

14.6 RATIONAL POWERS AND ROOTS

In this section we want to consider the meaning of z^r when r is any rational number (i.e., quotient of integers). We begin with simple cases and build up.

First, we have $z^2 = z \cdot z$, which was defined in Section 14.1. We let $z^3 = z^2 \cdot z$, $z^4 = z^3 \cdot z$, and so on. In general, for any positive integer n, $z^n = z^{n-1} \cdot z$. By convention, we let $z^0 = 1$.

If n is a negative integer and $z \neq 0$, we let

$$z^n = \frac{1}{z^{-n}}.$$

Now consider the problem of assigning a meaning to $z^{m/n}$, where m and n are integers. We can set this equal to $(z^m)^{1/n}$ once we have defined the $1/n$ power of a complex number. We now consider how to do this.

To assign a meaning to $z^{1/n}$ when n is a positive integer, let $w = z^{1/n}$. Then, $z = w^n$. Write z and w in polar form:

$$z = r[\cos(\theta) + i\sin(\theta)]$$

and

$$w = R[\cos(\phi) + i\sin(\phi)].$$

By de Moivre's formula,

$$w^n = R^n\cos(n\phi) + iR^n\sin(n\phi) = z = r\cos(\theta) + ir\sin(\theta).$$

Then,

$$R^n = r$$

and

$$n\phi = \theta + 2k\pi \qquad (k \text{ any integer}).$$

Thus,

$$|w| = R = r^{1/n}, \text{ the positive } n\text{th root of the positive number } r,$$

and

$$\arg w = \phi = \frac{\theta + 2k\pi}{n}, \qquad (k = 0, \pm 1, \pm 2, \pm 3, \ldots).$$

Thus,

$$w = z^{1/n} = r^{1/n}\left[\cos\left(\frac{\theta + 2k\pi}{n}\right) + i\sin\left(\frac{\theta + 2k\pi}{n}\right)\right],$$

where k is any integer. In fact, we obtain n distinct values of $z^{1/n}$ for $k = 0, 1, 2, \ldots, n - 1$; for $k = n, n + 1, \ldots$ or $k = -1, -2, -3, \ldots$, we repeat these n values. The n distinct values of $z^{1/n}$ thus obtained are called the nth *roots of* z. Note that, in the formula for $z^{1/n}$, θ may be any value of arg z.

EXAMPLE 411

Determine the cube roots of 8.

Letting $z = 8$, we have $r = |z| = 8$, and one value of arg z is 0. Here, $r^{1/3} = 2$, and the cube roots of 8 are

$$2\left[\cos\left(\frac{2k\pi}{3}\right) + i\sin\left(\frac{2k\pi}{3}\right)\right] \qquad (k = 0,\ 1,\ 2).$$

These are:

$$2 \qquad (k = 0),$$
$$-1 + \sqrt{3}\,i \qquad (k = 1),$$

and

$$-1 - \sqrt{3}\,i \qquad (k = 2).$$

EXAMPLE 412

Find the fourth roots of $1 - i$.

Here, $|1 - i| = \sqrt{2}$ and one value of arg z is $-\pi/4$. The four fourth roots of $1 - i$ are

$$2^{1/8}[\cos(-\pi/16) + i\sin(-\pi/16)] \qquad (k = 0),$$
$$2^{1/8}[\cos(7\pi/16) + i\sin(7\pi/16)] \qquad (k = 1),$$
$$2^{1/8}[\cos(15\pi/16) + i\sin(15\pi/16)] \qquad (k = 2),$$

and

$$2^{1/8}[\cos(23\pi/16) + i\sin(23\pi/16)] \qquad (k = 3).$$

These are approximately:

$$1.07 - 0.213i,$$
$$0.213 + 1.07i,$$
$$-1.07 + 0.213i,$$

and

$$-0.213 - 1.07i.$$

A sketch of these roots is shown in Figure 274.

EXAMPLE 413

The nth roots of 1 are often called nth *roots of unity*. They are

$$\cos\left(\frac{2k\pi}{n}\right) + i\sin\left(\frac{2k\pi}{n}\right) \qquad (k = 0,\ 1,\ 2,\ \ldots,\ n - 1).$$

These form the vertices of a regular polygon of n sides inscribed in the unit circle about the origin, oriented so that one vertex is at 1. Figure 275 shows this for $n = 5$, in which case the fifth roots of unity are 1, $\cos(2\pi/5) + i\sin(2\pi/5)$, $\cos(4\pi/5) + i\sin(4\pi/5)$, $\cos(6\pi/5) + i\sin(6\pi/5)$, and $\cos(8\pi/5) + i\sin(8\pi/5)$. These are approximately 1, $0.309 + 0.951i$, $-0.809 + 0.588i$, $-0.809 - 0.588i$, and $0.309 - 0.951i$.

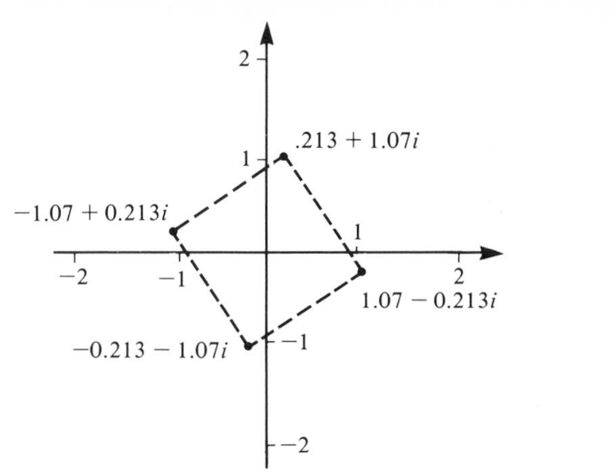

Figure 274. Fourth roots of $1 - i$ (forming vertices of a square).

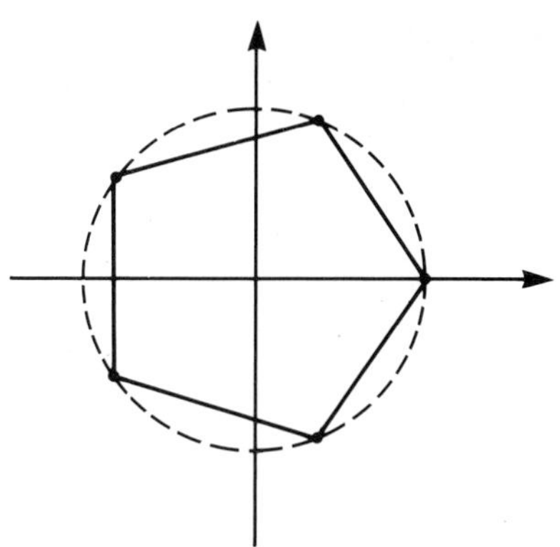

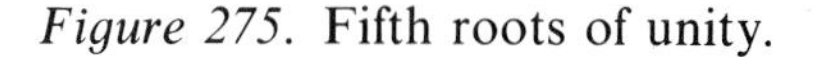

Figure 275. Fifth roots of unity.

Now we know how to compute all the values of $z^{1/n}$, and we can return to the general problem of $z^{m/n}$. Recall that we defined this to be $(z^m)^{1/n}$; that is, raise z to the integer power m, then find the nth roots of the resulting complex number. Here are three examples.

EXAMPLE 414

Find all possible values of $(2 - 2i)^{3/5}$.

First, calculate $(2 - 2i)^3 = -16 - 16i$. Thus, we want the fifth roots of $-16 - 16i$. This number has magnitude $\sqrt{512}$ and an argument of $\theta = 5\pi/4$. The fifth roots of $-16 - 16i$ are then

$$\sqrt{512}^{1/5}\left[\cos\left(\frac{5\pi}{20}\right) + i\sin\left(\frac{5\pi}{20}\right)\right],$$

$$\sqrt{512}^{1/5}\left[\cos\left(\frac{13\pi}{20}\right) + i\sin\left(\frac{13\pi}{20}\right)\right],$$

$$\sqrt{512}^{1/5}\left[\cos\left(\frac{21\pi}{20}\right) + i\sin\left(\frac{21\pi}{20}\right)\right],$$

$$\sqrt{512}^{1/5}\left[\cos\left(\frac{29\pi}{20}\right) + i\sin\left(\frac{29\pi}{20}\right)\right],$$

and

$$\sqrt{512}^{1/5}\left[\cos\left(\frac{37\pi}{20}\right) + i\sin\left(\frac{37\pi}{20}\right)\right].$$

To four decimal places, the $\frac{3}{5}$ powers of $2 - 2i$ are then, respectively,

$$1.3195 + 1.3195i,$$
$$-0.8472 + 1.6627i,$$
$$-1.8431 - 0.2919i,$$
$$-0.2919 - 1.8431i,$$

and

$$1.6627 - 0.8472i.$$

EXAMPLE 415

Calculate all values of $(3 + 5i)^{4/7}$.

First, we find that $(3 + 5i)^4 = -644 - 960i$. Thus, we need the seventh roots of $-644 - 960i$. This number has magnitude $\sqrt{644^2 + 960^2} = 1156$ and an argument $\pi + \arctan(\frac{960}{644})$, as shown in Figure 276. Thus, the seventh roots of $-644 - 960i$ are [with $\theta = \pi + \arctan(\frac{960}{644})$]

$$(1156)^{1/7}\left[\cos\left(\frac{\theta + 2k\pi}{7}\right) + i \sin\left(\frac{\theta + 2k\pi}{7}\right)\right]$$
$$(k = 0, 1, 2, 3, 4, 5, 6).$$

These give all values of $(3 + 5i)^{4/7}$.

We can approximate $1156^{1/7}$ as 2.7388 and θ as 4.1215 radians. The seven values of $(3 + 5i)^{4/7}$ are then approximately

$$2.2776 + 1.5210i \qquad (k = 0),$$
$$0.2309 + 2.7290i \qquad (k = 1),$$
$$-1.9879 + 1.8821i \qquad (k = 2),$$
$$-2.7120 - 0.3821i \qquad (k = 3),$$
$$-1.3921 - 2.3586i \qquad (k = 4),$$
$$0.9760 - 2.5590i \qquad (k = 5),$$

and

$$2.6092 - 0.8324i \qquad (k = 6).$$

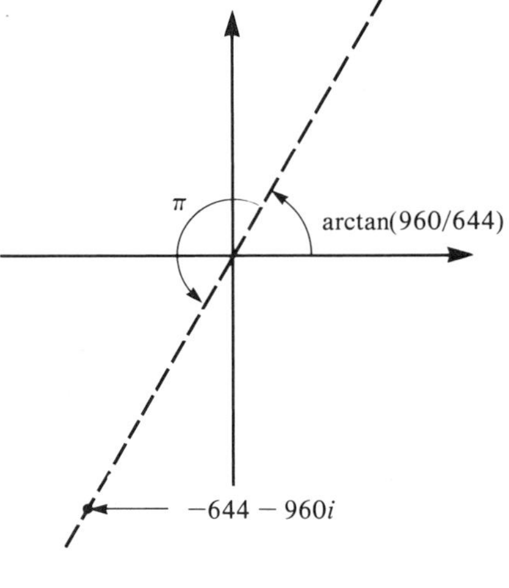

Figure 276.
Argument of $-644 - 960i$.

Of course, we could achieve the same result by first computing the seven seventh roots of $3 + 5i$ and then raising these to the fourth power. The seventh roots of $3 + 5i$ are given by

$$(\sqrt{34})^{1/7}\left[\cos\left(\frac{\phi + 2k\pi}{7}\right) + i \sin\left(\frac{\phi + 2k\pi}{7}\right)\right],$$

for $k = 0, 1, 2, 3, 4, 5, 6$. Here, $\phi = \arctan(\frac{5}{3})$ is an argument of $3 + 5i$.

The seven $\frac{4}{7}$-powers of $3 + 5i$ are then

$$(\sqrt{34})^{4/7}\left[\cos\left(\frac{\phi + 2k\pi}{7}\right) + i \sin\left(\frac{\phi + 2k\pi}{7}\right)\right]^4,$$

for $k = 0, 1, 2, 3, 4, 5, 6$. By de Moivre's theorem, these are

$$(\sqrt{34})^{4/7}\left[\cos\left(\frac{4\phi + 8k\pi}{7}\right) + i\sin\left(\frac{4\phi + 8k\pi}{7}\right)\right],$$

for $k = 0, 1, 2, 3, 4, 5, 6$. If you have a calculator, you can verify that these are the same numbers obtained previously.

The order in which one performs the computations in evaluating $z^{m/n}$ is dictated by convenience. In many cases it makes no difference, while in others, one way may be decidedly easier.

EXAMPLE 416

Calculate all values of $(2 - i)^{-2/3}$.

First, write

$$\begin{aligned}(2 - i)^{-2/3} &= \left(\frac{1}{2 - i}\right)^{2/3}\\ &= (\tfrac{2}{5} + \tfrac{1}{5}i)^{2/3}\\ &= (\tfrac{3}{25} + \tfrac{4}{25}i)^{1/3}.\end{aligned}$$

Thus, we need the cube roots of $\frac{3}{25} + \frac{4}{25}i$. This number has magnitude $\frac{1}{5}$ and an argument of $\arctan(\frac{4}{3})$. Thus, the $-\frac{2}{3}$ powers of $2 - i$ are

$$(\tfrac{1}{5})^{1/3}\left\{\cos\left[\frac{\arctan(\frac{4}{3}) + 2k\pi}{3}\right] + i\sin\left[\frac{\arctan(\frac{4}{3}) + 2k\pi}{3}\right]\right\}$$

for $k = 0, 1, 2$. To four decimal places, these are

$$0.5571 + 0.1179i,$$
$$-0.4326 + 0.3935i,$$

and

$$-0.1245 - 0.5714i.$$

Our next task is to develop the complex exponential and logarithm functions, after which we can define general powers z^α, with α not necessarily a rational number.

PROBLEMS FOR SECTION 14.6

In each of Problems 1 through 25, find all possible distinct values of the indicated power of the complex number. In some cases the argument must be left in terms of an inverse tangent function; if you have a calculator, you should calculate all the solutions to four decimal places, as in Examples 414, 415, and 416.

1. $(1 - i)^{1/2}$ **2.** $i^{1/4}$ **3.** $16^{1/4}$ **4.** $(1 + i)^{3/2}$

5. $(-16)^{1/4}$ **6.** $\left(\dfrac{1 + i}{1 - i}\right)^{1/3}$ **7.** $1^{1/10}$ **8.** $(-1)^{1/5}$

9. $(-2 - 2i)^{1/4}$ **10.** $(-i)^{1/3}$ **11.** $i^{1/10}$ **12.** $(2 + 2i)^{4/5}$

13. $\left(\dfrac{1-4i}{6+i}\right)^{2/3}$ **14.** $[(8+3i)(2-4i)]^{1/4}$ **15.** $(3-4i)^{-3/4}$ **16.** $(4i)^{-1/2}$

17. $[(1+2i)(3-i)]^{-3/4}$ **18.** $\left(\dfrac{2-4i}{4+i}\right)^{2/5}$ **19.** $(1-6i)^{1/5}$ **20.** $(8-i)^{-3/7}$

21. $(4-3i)^{1/6}$ **22.** $(-3-5i)^{-3/11}$ **23.** $(1-4i)^{2/5}$ **24.** $(5+3i)^{-4/5}$

25. $(-3+2i)^{2/3}$

26. Let $w_1, \ldots, w_n$ be the nth roots of 1. Prove that $w_1 + w_2 + \cdots + w_n = 0$.

27. Show that the numbers

$$\frac{-b \pm (b^2 - 4ac)^{1/2}}{2a}$$

satisfy $az^2 + bc + c = 0$ for complex a, b, and c.

28. Solve $z^2 + iz - 2 = 0$.

29. Solve $z^2 + (1-i)z + i = 0$.

30. Solve for z in

$$z^4 - 2z^2 = -2.$$

(*Hint*: Let $t = z^2$; then solve for t, and then z.)

31. Solve for z in $z^6 - (1+i)z^3 = 1 - i$.

32. Show that $(z^m)^{1/n} = (z^{1/n})^m$ for integers m and n, in the sense that both sides give exactly the same collection of complex numbers when the indicated operations are carried out in the given order.

14.7 THE COMPLEX EXPONENTIAL FUNCTION

This section is devoted to a development of the complex analogue of e^x. We begin with an argument which is not rigorous but nevertheless suggests a direction in which to proceed.

If a and b are real numbers, we know that

$$e^{a+b} = e^a \cdot e^b.$$

Proceed formally to write

$$e^{x+iy} = e^x \cdot e^{iy}.$$

Now, e^x causes no problem, but what does e^{iy} mean? Recall that, for real a,

$$e^a = \sum_{n=0}^{\infty} \frac{a^n}{n!} = 1 + a + \frac{a^2}{2!} + \frac{a^3}{3!} + \frac{a^4}{4!} + \frac{a^5}{5!} + \frac{a^6}{6!} + \cdots.$$

Put $a = iy$ into this, remembering that

$$i^{4k} = 1, \quad i^{4k+1} = i, \quad i^{4k+2} = -1 \quad \text{and} \quad i^{4k+3} = -i.$$

We get, by rearranging terms,

$$e^{iy} = 1 + iy + \frac{(iy)^2}{2!} + \frac{(iy)^3}{3!} + \frac{(iy)^4}{4!} + \frac{(iy)^5}{5!} + \frac{(iy)^6}{6!}$$

$$+ \frac{(iy)^7}{7!} + \frac{(iy)^8}{8!} + \cdots$$

$$= 1 - \frac{y^2}{2!} + \frac{y^4}{4!} - \frac{y^6}{6!} + \frac{y^8}{8!} + \cdots + i\left(y - \frac{y^3}{3!} + \frac{y^5}{5!} - \frac{y^7}{7!} + \cdots\right).$$

Recalling the Taylor series about 0 for $\sin(y)$ and $\cos(y)$, we obtain *Euler's formula:*

$$e^{iy} = \cos(y) + i\sin(y).$$

This suggests that we *define:*

$$e^{x+iy} = e^x[\cos(y) + i\sin(y)].$$

This is a *complex exponential function,* and it is defined for all complex z.

EXAMPLE 417

Here are some computed values of e^z.

1. $e^i = e^0[\cos(1) + i\sin(1)] = \cos(1) + i\sin(1)$.
2. $e^{2-3i} = e^2[\cos(-3) + i\sin(-3)] = e^2[\cos(3) - i\sin(3)]$.
3. $e^{4+\pi i} = e^4[\cos(\pi) + i\sin(\pi)] = -e^4$ (a real number!).
4. $e^{4+(\pi i/2)} = e^4\left[\cos\left(\frac{\pi}{2}\right) + i\sin\left(\frac{\pi}{2}\right)\right] = ie^4$ (pure imaginary).

Here are the basic properties of e^z.

THEOREM 113

1. e^z is analytic in the entire complex plane, and

$$\frac{d}{dz}e^z = e^z.$$

2. $e^{z_1}e^{z_2} = e^{z_1+z_2}$.
3. $e^{z_1}/e^{z_2} = e^{z_1-z_2}$.
4. For real y, $|e^{iy}| = 1$.
5. $|e^{x+iy}| = e^x$ whenever x and y are real.
6. $e^z \neq 0$ for all z.
7. $e^z = 1$ if and only if $z = 2n\pi i$, for some integer n.
8. $e^z = e^w$ if and only if $z = w + 2n\pi i$ for some integer n.

Proof of (1) Write $e^{x+iy} = e^x\cos(y) + ie^x\sin(y)$ and apply Theorem 111 with $u(x, y) = e^x\cos(y)$ and $v(x, y) = e^x\sin(y)$. It is easy to check that $u(x, y)$ and $v(x, y)$ are continuous with continuous first partials satisfying the Cauchy-Riemann equations. Further, from the end of Section 14.5, we have that

$$\frac{d}{dz}(e^z) = \frac{\partial}{\partial x}[e^x \cos(y)] - i\frac{\partial}{\partial y}[e^x \cos(y)] = e^x \cos(y) + ie^x \sin(y) = e^z.$$

Proof of (2) Let $z_1 = x_1 + iy_1$ and $z_2 = x_2 + iy_2$. Then,

$$\begin{aligned} e^{z_1}e^{z_2} &= \{e^{x_1}[\cos(y_1) + i\sin(y_1)]\}\{e^{x_2}[\cos(y_2) + i\sin(y_2)]\} \\ &= e^{x_1+x_2}\{[\cos(y_1)\cos(y_2) - \sin(y_1)\sin(y_2)] \\ &\quad + i[\cos(y_1)\sin(y_2) + \sin(y_1)\cos(y_2)]\} \\ &= e^{x_1+x_2}[\cos(y_1 + y_2) + i\sin(y_1 + y_2)] \\ &= e^{(x_1+x_2)+i(y_1+y_2)} = e^{z_1+z_2}. \end{aligned}$$

Part (3) is proved similarly and is left to the student. We also leave (4) and (5) to the student.

Proof of (7) First, if $z = 2n\pi i$, then

$$e^z = \cos(2n\pi) + i\sin(2n\pi) = 1.$$

Conversely, suppose that $e^z = 1$. Write $z = x + iy$. Then,

$$e^x[\cos(y) + i\sin(y)] = 1.$$

Then,

$$e^x \cos(y) = 1 \quad \text{and} \quad e^x \sin(y) = 0.$$

Since $e^x \neq 0$, then $\sin(y) = 0$; so $y = 2n\pi$ for some integer n. Then,

$$e^x \cos(y) = e^x \cos(2n\pi) = 1.$$

Now, $\cos(2n\pi) = 1$; so $e^x = 1$, and hence $x = 0$. This leaves $z = x + iy = 2n\pi i$.

Proof of (8) If $z = w + 2n\pi i$, then $e^z = e^{w+2n\pi i} = e^w \cdot e^{2n\pi i} = e^w$, by (2) and (7).

Conversely, suppose that $e^z = e^w$. Then $e^{z-w} = 1$ by (3). By (7), $z - w = 2n\pi i$ for some integer n, proving (8).

We conclude this section with the sometimes useful observation that the polar form of z can now be written

$$z = re^{i\theta},$$

where $r = |z|$ and $\theta = \arg(z)$. For example,

$$-1 - i = \sqrt{2}\,e^{i[(5\pi/4)+2n\pi]} = \sqrt{2}\,e^{5\pi i/4},$$

since $e^{2n\pi i} = 1$. Finally, note that $e^{\pi i} = -1$, $e^{2\pi i} = 1$, $e^{\pi i/2} = i$ and $e^{3\pi i/2} = -i$.

PROBLEMS FOR SECTION 14.7

In each of Problems 1 through 15, write e^z in the form $a + ib$, as in Example 417.

1. e^i **2.** e^{1-i} **3.** $e^{\pi-i}$ **4.** $e^{i+3\pi}$ **5.** e^{2-2i}

6. $e^{2-(\pi i/2)}$ 7. $e^{\pi(1+i)/4}$ 8. $e^{-3i+\pi}$ 9. e^{-5+7i} 10. $e^{2-7\pi i}$
11. e^{5i} 12. $e^{(1-i)(2-3i)}$ 13. $e^{4i(2+5i)}$ 14. $e^{(2+4i)2}$ 15. $e^{9\pi i}$

In each of Problems 16 through 25, write the given z in the form $re^{i\theta}$ for appropriate r and θ.

16. $3i$ 17. $2+i$ 18. $1-i$ 19. $-1-2i$ 20. $3+i$
21. $2-4i$ 22. $-3-9i$ 23. $6+12i$ 24. $4-2i$ 25. $-2+i$

26. Prove that $e^{-z} = 1/e^z$ for any complex number z.
27. Prove that, for any positive integer n, $(e^z)^n = e^{nz}$.
28. Use the result of Problem 27 to give a short proof of de Moivre's formula (Section 14.2).
29. Show that, for any positive integer n, the nth roots of unity are $e^{2k\pi i/n}$, with $k = 0, 1, 2, \ldots, n-1$.
30. Write e^{z^2} in the form $u(x, y) + iv(x, y)$, and show that the Cauchy-Riemann equations hold.
31. Write $e^{1/z}$ in the form $u(x, y) + iv(x, y)$, and show that the Cauchy-Riemann equations hold for $z \neq 0$.
32. Prove Theorem 113(3).
33. Prove Theorem 113(4).
34. Prove Theorem 113(5).
35. Prove Theorem 113(6).

14.8 THE COMPLEX LOGARITHM FUNCTION

In beginning calculus, $y = \ln(x)$ is taken to mean that $e^y = x$, an equation which makes sense when $x > 0$.

For the complex logarithm, we take the same approach and define $w = \ln(z)$ if $e^w = z$. We now ask: For what values of z can we find such w? To answer this question, write z in polar form:

$$z = x + iy = re^{i\theta}.$$

We want $w = u + iv$ such that $e^w = z$. Thus, we need

$$e^w = e^u \cdot e^{iv} = re^{i\theta}.$$

We solve this for w, thinking of z as given, as follows. First, take absolute values, recalling that $|e^{iy}| = 1$ for any real y. Then,

$$|e^u e^{iv}| = e^u = |re^{i\theta}| = r.$$

Thus, $r = e^u$. Hence, if $z \neq 0$, then $u = \ln(r)$, the real-valued natural logarithm of r. But then $e^{iv} = e^{i\theta}$, which means [by Theorem 113(8)] that

$$v = \theta + 2n\pi.$$

Then,

$$w = u + iv = \ln(r) + i(\theta + 2n\pi),$$

where n is any integer.

Let us summarize what we have found.

Given any complex $z \neq 0$, then all the solutions of $e^w = z$ are

$$w = \ln(r) + i(\theta + 2n\pi),$$

where $r = |z|$, $\theta = \arg z$, and $\ln(r)$ is the usual real natural logarithm of the positive number r. Since $\arg z$ is determined only to within integer multiples of 2π, this can be written

$$w = \ln|z| + i \arg z.$$

In view of this, we *define*, for $z \neq 0$,

$$\ln(z) = \ln|z| + i \arg z.$$

This is the *natural logarithm* of z, and it has infinitely many different values (because $\arg z$ does). The *principal logarithm function* is defined by

$$\mathrm{Ln}(z) = \ln|z| + i \operatorname{Arg} z.$$

EXAMPLE 418

Here are three examples of $\ln(z)$ and $\mathrm{Ln}(z)$.

1. $$\ln(i) = \ln|i| + i \arg(i) = i\left(\frac{\pi}{2} + 2n\pi\right) \qquad (n = 0, \pm 1, \pm 2, \dots)$$

   and

   $$\mathrm{Ln}(i) = i\left(\frac{\pi}{2}\right).$$

2. $$\ln(1 - i) = \ln|1 - i| + i \arg(1 - i) = \ln(\sqrt{2}) + i\left(-\frac{\pi}{4} + 2n\pi\right)$$

   and

   $$\mathrm{Ln}(1 - i) = \ln(\sqrt{2}) - i\left(\frac{\pi}{4}\right).$$

3. $$\ln(-4) = \ln|-4| + i \arg(-4) = \ln(4) + \pi i + 2n\pi i$$

   and

   $$\mathrm{Ln}(-4) = \ln|-4| + i \operatorname{Arg}(-4) = \ln(4) + \pi i.$$

Note: The above example emphasizes that we can now take a natural logarithm of a negative number—it turns out to be a complex number, explaining why the logarithm of a negative number makes no sense in real-variable calculus.

Here are the basic properties of $\ln(z)$ and $\mathrm{Ln}(z)$.

THEOREM 114

1. e^z and $\ln(z)$ are related by
   (a) $e^{\ln(z)} = z$
   and
   (b) $\ln(e^z) = z + 2n\pi i$ (n any integer).
2. $\ln(z_1 z_2) = \ln(z_1) + \ln(z_2) + 2n\pi i$ (n any integer).

3. $\ln\left(\frac{z_1}{z_2}\right) = \ln(z_1) - \ln(z_2) + 2n\pi i \qquad (n \text{ any integer}).$

4. $\mathrm{Ln}(z)$ is analytic in the domain consisting of all $z \neq 0$, and

$$\frac{d}{dz}\,\mathrm{Ln}(z) = \frac{1}{z}.$$

Proof of (1)-a

$$e^{\ln(z)} = e^{\ln|z| + i\arg(z)} = e^{\ln|z|}\{\cos[\arg(z)] + i\sin[\arg(z)]\},$$

and this is just z in polar form, since $e^{\ln|z|} = |z|$.

Proof of (1)-b Let $\ln(e^z) = w$. Then, by definition, $e^w = e^z$. Then, by Theorem 113(8), $w = z + 2n\pi i$ for any integer n. Thus,

$$\ln(e^z) = z + 2n\pi i.$$

Proof of (2) We have

$$\begin{aligned}\ln(z_1 z_2) &= \ln|z_1 z_2| + i\arg(z_1 z_2)\\ &= \ln|z_1| + \ln|z_2| + i[\arg(z_1) + \arg(z_2) + 2n\pi]\\ &= [\ln|z_1| + i\arg(z_1)] + [\ln|z_2| + i\arg(z_2)] + 2n\pi i\\ &= \ln(z_1) + \ln(z_2) + 2n\pi i.\end{aligned}$$

Part (3) is proved similarly. For (4), use Theorem 111; we leave the details to the student.

In closing this section, we caution the student to remember part (1) of the theorem when treating complex exponentials and logarithms. In the real case, $e^a = b$ and $a = \ln(b)$ are synonymous (when $b > 0$). In the complex case, $\ln(z)$ has infinitely many values, and $e^{z+2n\pi i} = e^z$ for any integer n; hence the need for more care.

PROBLEMS FOR SECTION 14.8

In each of Problems 1 through 20, find all possible values of $\ln(z)$; also find $\mathrm{Ln}(z)$.

1. $\ln(2i)$ **2.** $\ln(1+i)$ **3.** $\ln(-9)$ **4.** $\ln(3-2i)$
5. $\ln(1-2i)$ **6.** $\ln[(1+i)^{1/4}]$ **7.** $\ln(i^{1/3})$ **8.** $\ln[(2-2i)^{1/5}]$
9. $\ln(-4+2i)$ **10.** $\ln(-6-3i)$ **11.** $\ln(2+4i)$ **12.** $\ln[(1-i)(2+3i)]$
13. $\ln[(-1+2i)^2]$ **14.** $\ln(-5i)$ **15.** $\ln(7-2i)$ **16.** $\ln(-8)$
17. $\ln[(2i)^{1/2}]$ **18.** $\ln(6-18i)$ **19.** $\ln[(2+2i)^3]$ **20.** $\ln(-4+3i)$

21. Verify that

$$\mathrm{Ln}[(1-i)(1+i)] = \mathrm{Ln}(1-i) + \mathrm{Ln}(1+i)$$

by calculating both sides of the equation separately.

22. Verify that

$$\mathrm{Ln}\left(\frac{1-i}{1+i}\right) = \mathrm{Ln}(1-i) - \mathrm{Ln}(1+i)$$

by calculating both sides of the equation.

For each of Problems 23 through 27, verify Theorem 114(1), parts (a) and (b), by direct calculation for the given complex number.

23. $-1-i$ **24.** i **25.** -4 **26.** $1+3i$ **27.** $-2-4i$

28. Prove that

$$\tfrac{1}{2}\,\mathrm{Ln}(z\bar{z}) = \mathrm{Ln}\,|z| \qquad (z \neq 0).$$

29. Prove that, for any integers m and n,

$$\ln(z^{m/n}) = \frac{m}{n}\ln(z) + 2k\pi i \qquad (k \text{ any integer}).$$

30. Prove Theorem 114(4) when Re(z) is positive by writing

$$\mathrm{Ln}(z) = \mathrm{Ln}(x+iy) = \mathrm{Ln}\,|x+iy| + i\,\mathrm{Arg}(x+iy)$$
$$= \frac{1}{2}\ln(x^2+y^2) + i\arctan\left(\frac{y}{x}\right).$$

Now apply Theorem 111.

14.9 GENERAL POWERS

Using the exponential and logarithm functions, we can define z^α for any complex $z \neq 0$ and any complex α. Modeling the real case, we define

$$z^\alpha = e^{\alpha \ln(z)} \qquad (z \neq 0).$$

Since $\ln(z)$ has infinitely many values, then so will z^α. Here are some examples.

EXAMPLE 419

1.

$$\begin{aligned} 2^i &= e^{i\ln(2)} = e^{i[\ln|2| + i\arg(2)]} \\ &= e^{i[\ln(2)+2n\pi i]} = e^{i\ln(2)}e^{-2n\pi} \\ &= e^{-2n\pi}\{\cos[\ln(2)] + i\sin[\ln(2)]\} \qquad (n \text{ any integer}). \end{aligned}$$

2.

$$\begin{aligned} (1-i)^{1+i} &= e^{(1+i)\ln(1-i)} \\ &= \exp\left\{(1+i)\left[\ln(\sqrt{2}) + \left(-\frac{\pi}{4} + 2n\pi\right)\right]\right\} \\ &= \exp\{[\ln(\sqrt{2}) - \tfrac{1}{4}\pi - 2n\pi] + i[\ln(\sqrt{2}) - \tfrac{1}{4}\pi + 2n\pi]\} \\ &= \exp[\ln(\sqrt{2}) + \tfrac{1}{4}\pi - 2n\pi]\,\exp\{i[\ln(\sqrt{2}) - \tfrac{1}{4}\pi]\} \\ &= \exp[\ln(\sqrt{2}) + \tfrac{1}{4}\pi - 2n\pi] \\ &\quad \times \{\cos[\ln(\sqrt{2}) - \tfrac{1}{4}\pi] + i\sin[\ln(\sqrt{2}) - \tfrac{1}{4}\pi]\}. \end{aligned}$$

Both 2^i and $(1-i)^{1+i}$ have infinitely many values, as n can be any integer.

If we use $\mathrm{Ln}(z)$ in place of $\ln(z)$ in defining z^α, we get a single-valued function called the *principal value* of z^α, denoted $\Pr[z^\alpha]$. Thus,

$$\Pr[z^\alpha] = e^{\alpha\,\mathrm{Ln}(z)}.$$

EXAMPLE 420

From Example 419,

$$\Pr[2^i] = \cos[\ln(2)] + i\sin[\ln(2)] \qquad (\text{choose } n = 0)$$

and

$$\Pr[(1-i)^{1+i}] = e^{\ln(\sqrt{2}) + (\pi/4)}\left\{\cos\left[\ln(\sqrt{2}) - \frac{\pi}{4}\right] + i\sin\left[\ln(\sqrt{2}) - \frac{\pi}{4}\right]\right\}.$$

To four decimals, we have

$$\Pr[2^i] = 0.7692 + 0.6390i \quad \text{and} \quad \Pr[(1-i)^{1+i}] = 2.8079 - 1.3179i.$$

THEOREM 115

For $z \neq 0$, $\Pr[z^\alpha]$ is analytic, and

$$\frac{d}{dz}\Pr[z^\alpha] = \alpha\Pr[z^{\alpha-1}].$$

In simpler terms, $(z^\alpha)' = \alpha z^{\alpha-1}$ if we understand principal values in computing z^α and $z^{\alpha-1}$. We leave a proof of this to the student.

With arbitrary powers now defined, we can pick up some additional facts about e^z and $\ln(z)$.

THEOREM 116

1. For any z and α, $(e^z)^\alpha = e^{\alpha z}$.
2. For any $z \neq 0$ and any α, $\ln(z^\alpha) = \alpha\ln(z) + 2n\pi i$ (n any integer).

Proof of (2) By definition of z^α and by Theorem 114(1)-b, we have

$$\ln(z^\alpha) = \ln(e^{\alpha\ln(z)}) = \alpha\ln(z) + 2n\pi i.$$

We leave (1) to the student.

PROBLEMS FOR SECTION 14.9

In each of Problems 1 through 20, determine all possible values of z^α, and also determine $\Pr[z^\alpha]$.

1. i^{1+i}
2. $(1+i)^{2i}$
3. i^i
4. $(2-i)^{1+i}$
5. $(-1+i)^{-3i}$
6. $(-4)^{2-i}$
7. 6^{-2-3i}
8. $(7i)^{3i}$
9. $(1-i)^{-2-2i}$
10. i^{2-4i}
11. 2^{3-i}
12. $(-4)^{1+i}$
13. $(3-2i)^i$
14. $(4+i)^{1-i}$
15. $(-4i)^i$
16. $(i^{2/3})^i$
17. $[(1-i)^{1/2}]^{1-i}$
18. $(-3i)^{4+2i}$
19. $(3)^{i/2}$
20. $(2+5i)^{-10i}$
21. Prove Theorem 113.
22. Prove Theorem 114(1).

14.10 COMPLEX TRIGONOMETRIC AND HYPERBOLIC FUNCTIONS

Defining the trigonometric functions for complex values is relatively simple once we have the complex exponential function (so we could have put this section right after Section 14.7). To motivate the definition, remember Euler's formula

$$e^{iy} = \cos(y) + i \sin(y).$$

Then,

$$e^{-iy} = \cos(y) - i \sin(y).$$

These can be solved to yield

$$\cos(y) = \frac{1}{2}(e^{iy} + e^{-iy}) \quad \text{and} \quad \sin(y) = \frac{1}{2i}(e^{iy} - e^{-iy}).$$

This leads us to *define*, for any complex z,

$$\cos(z) = \frac{1}{2}(e^{iz} + e^{-iz})$$

and

$$\sin(z) = \frac{1}{2i}(e^{iz} - e^{-iz}).$$

The other trigonometric functions are defined in the usual way:

$$\tan(z) = \frac{\sin(z)}{\cos(z)}, \qquad \cos(z) = \frac{1}{\tan(z)},$$

$$\sec(z) = \frac{1}{\cos(z)}, \qquad \csc(z) = \frac{1}{\sin(z)},$$

whenever the denominators are not zero.

It is a simple matter to write these functions in the form $u(x, y) + iv(x, y)$. As an illustration, we give the details for $\sin(z)$. Write $z = x + iy$, and then

$$\begin{aligned}
\sin(z) &= \frac{1}{2i}(e^{iz} - e^{-iz}) \\
&= \frac{1}{2i}(e^{ix-y} - e^{-ix+y}) \\
&= \frac{1}{2i}(e^{-y}e^{ix} - e^{y}e^{-ix}) \\
&= \frac{e^{-y}}{2i}[\cos(x) + i \sin(x)] - \frac{e^{y}}{2i}[\cos(x) - i \sin(x)] \\
&= \sin(x)\left(\frac{e^{y} + e^{-y}}{2}\right) + i \cos(x)\left(\frac{e^{y} - e^{-y}}{2}\right) \\
&= \sin(x) \cosh(y) + i \cos(x) \sinh(y).
\end{aligned}$$

Similarly,

$$\cos(z) = \cos(x)\cosh(y) - i\sin(x)\sinh(y).$$

Using these, it is easy to show by Theorem 111 that $\sin(z)$ and $\cos(z)$ are analytic in the whole complex plane, and that

$$\frac{d}{dz}\cos(z) = -\sin(z)$$

and

$$\frac{d}{dz}\sin(z) = \cos(z).$$

It is a routine task to show that the complex trigonometric functions satisfy all the identities we are accustomed to from the real trigonometric functions. For example,

$$\sin^2(z) + \cos^2(z) = 1,$$
$$\sin(z_1 \pm z_2) = \sin(z_1)\cos(z_2) \pm \sin(z_2)\cos(z_1),$$
$$\cos(z_1 \pm z_2) = \cos(z_1)\cos(z_2) \mp \sin(z_1)\sin(z_2),$$
$$\sin(2z) = 2\sin(z)\cos(z),$$

and so on. We shall make use of such identities freely without pausing to prove them in the complex case.

Another important property also carries over: $\sin(z)$ and $\cos(z)$ are periodic of period 2π. Thus,

$$\sin(z + 2n\pi) = \sin(z) \quad \text{and} \quad \cos(z + 2n\pi) = \cos(z)$$

for all z and integer n.

Finally, $\sin(z)$ and $\cos(z)$ have the same zeros as the corresponding real-valued functions. Thus,

$$\sin(z) = 0 \quad \text{if and only if} \quad z = n\pi \qquad (n \text{ integer})$$

and

$$\cos(z) = 0 \quad \text{if and only if} \quad z = (2n+1)\frac{\pi}{2} \qquad (n \text{ integer}).$$

An important difference between the real and complex sine and cosine functions is that the real functions are bounded (between -1 and $+1$), while the complex functions can take on values arbitrarily large in magnitude. For example, if y is real, then $\cos(iy) = \frac{1}{2}(e^{-y} + e^{y}) \to \infty$ as $y \to \infty$ or $y \to -\infty$.

We conclude this section by mentioning briefly the *complex hyperbolic functions*, given by

$$\cosh(z) = \frac{e^z + e^{-z}}{2}, \qquad \sinh(z) = \frac{e^z - e^{-z}}{2},$$

$$\tanh(z) = \frac{\sinh(z)}{\cosh(z)}, \qquad \coth(z) = \frac{1}{\tanh(z)},$$

$$\operatorname{sech}(z) = \frac{1}{\cosh(z)} \quad \text{and} \quad \operatorname{csch}(z) = \frac{1}{\sinh(z)}$$

(whenever the denominators are not zero).

As with the trigonometric functions, basic identities and derivative formulas carry over in the same form to the complex hyperbolic functions (in effect, just replace x by z). We pursue some of these in the exercises.

A difference between the real and complex case is illustrated by a relationship between trigonometric and hyperbolic functions. It is easy to check directly from the definitions that

$$\cos(iz) = \cosh(z) \quad \text{and} \quad -i\sin(iz) = \sinh(z).$$

Such relationships are not possible when we are dealing only with real numbers.

Now that we have the basic functions and their derivative formulas, we can turn to a development of the complex integral, which forms the backbone of complex analysis.

PROBLEMS FOR SECTION 14.10

In each of Problems 1 through 20, compute the indicated function values.

1. $\sin(i)$ **2.** $\cosh(1-i)$ **3.** $\tan(2i)$ **4.** $\cos(-1-i)$
5. $\sinh(i^i)$ **6.** $\csc(2+i)$ **7.** $\cos(-2-4i)$ **8.** $\sin(\pi+i)$
9. $\tanh(\pi i)$ **10.** $\cot[1+(\pi/4)i]$ **11.** $\sin(e^i)$ **12.** $\cosh[\ln(i)]$
13. $\cos[(1+i)^i]$ **14.** $\sinh(2+i)$ **15.** $\cos[(1-i)^{1-i}]$ **16.** $\sinh[(1+i)^{1/3}]$
17. $e^{\sin(2-i)}$ **18.** $\ln[\cos(3i)]$ **19.** $[\sin(1-i)]^{3/4}$ **20.** $\sinh[\cos(i)]$

21. Prove that $\sin^2(z) + \cos^2(z) = 1$ for all z.

22. Prove that $\sin(z)$ and $\cos(z)$ are analytic in the entire complex plane.

23. Write $\cosh(z)$ and $\sinh(z)$ in the form $u(x, y) + iv(x, y)$. Use these and Theorem 111 to show that $\cosh(z)$ and $\sinh(z)$ are analytic in the entire complex plane.

24. Prove that $\sin(z + 2n\pi) = \sin(z)$ and $\cos(z + 2n\pi) = \cos(z)$ for any integer n.

25. For x real, $\cosh(x)$ is never zero. Is it possible for $\cosh(z)$ to be zero for any complex z?

26. Find $u(x, y)$ and $v(x, y)$ such that $\tan(z) = u(x, y) + iv(x, y)$. Determine where $\tan(z)$ is analytic.

27. Find all z such that $\sin(z) = \frac{1}{2}$.
(*Hint:* We want $(e^{iz} - e^{-iz})/2i = \frac{1}{2}$. Write this as $e^{2iz} - ie^{iz} - 1 = 0$, solve for e^{iz} by using the quadratic formula, and then solve for z.)

28. Find all z such that $\sin(z) = i$.

29. Let $z = x + iy$. Prove that

$$|\sinh(y)| \le |\sin(z)| \le \cosh(y).$$

30. Prove that

$$z = \tan\left\{\frac{1}{i}\ln\left[\left(\frac{1+iz}{1-iz}\right)^{1/2}\right]\right\}.$$

SUPPLEMENTARY PROBLEMS

In each of Problems 1 through 25, carry out the indicated operations and write the given quantity in the form $a + ib$. In each case find all possible values.

1. e^{4-i} **2.** $(1-2i)^2(3i)^{3/4}$ **3.** $\sinh(-5i)$ **4.** $\ln[(3-2i)^2]$
5. $\cos[\ln(1+i)]$ **6.** $3^{1/i}$ **7.** $(2+4i)/e^{1-i}$ **8.** $\sin(4i)$

9. $\cosh(-1 + i)$ **10.** $2i - e^{2-i}$ **11.** $(2 + i)^{4/7}$ **12.** $(i^i)^{2i}$
13. $\cosh(3^i)$ **14.** $\ln(-4 + 6i)$ **15.** $(1 + 3i)^e$ **16.** $(3 - i)^{\cos(i)}$
17. $(4 + i)^{-3/7}$ **18.** $1^{1/9}$ **19.** $(2 + i)^{3/8}$ **20.** $i^{1/5}$

21. $\dfrac{(2 + 3i)^{2/3}}{(1 - i)^{1/2}}$ **22.** $\ln(2 + 8i)$ **23.** $\sinh[(2 + 4i)^2]$ **24.** $\ln(e^{2-i})$

25. $\sin[\cos(2i)]$
26. Use de Moivre's formula to write $\cos(5x)$ in terms of $\cos(x)$ and $\sin(x)$.
27. Determine the locus of points z satisfying

$$\left|\frac{z - i}{z + 3 - 2i}\right| = 3.$$

28. Prove that $\sin(2z) = 2\sin(z)\cos(z)$.
29. Prove that $\cosh^2(z) - \sinh^2(z) = 1$.
30. Write $\sec(z)$ in the form $u(x, y) + iv(x, y)$. Determine from these where $\sec(z)$ is analytic.
31. Find $u(x, y)$ and $v(x, y)$ such that $\cot(z) = u(x, y) + iv(x, y)$.
32. Find $u(x, y)$ and $v(x, y)$ such that $e^{z^2} = u(x, y) + iv(x, y)$.
33. Find all solutions of the equation $z^5 = -4i$.
34. Prove that $f(z) = |z|$ is not analytic.
35. Prove that

$$z^{-1} = \frac{\bar{z}}{|z|^2}.$$

36. Prove that

$$\cosh(z_1 + z_2) = \cosh(z_1)\cosh(z_2) + \sinh(z_1)\sinh(z_2).$$

37. Show that the locus of points z such that

$$\left|\frac{z + 1}{1 - z}\right| = K \qquad (K \text{ constant})$$

is a straight line if $K = 1$ and a circle if $K \neq 1$.
38. Let z_1 and z_2 be given complex numbers such that

$$0 < \operatorname{Arg}(z_1) - \operatorname{Arg}(z_2) < \pi.$$

Show that the triangle with vertices 0, z_1, and z_2 has area $\frac{1}{2}\operatorname{Im}(\bar{z}_1 z_2)$.
39. Show that

$$\cos(\theta) + \cos(3\theta) + \cdots + \cos[(2n - 1)\theta] = \frac{\sin(2n\theta)}{2\sin(\theta)}$$

if $\sin(\theta) \neq 0$.
40. Suppose that $z = \sin(w)$. Show that

$$w = -i\ln[(1 - z^2)^{1/2} + iz].$$

41. Prove that

$$\tan(z) = \frac{\sin(2x) + i\sinh(2y)}{\cos(2x) + \cosh(2y)}.$$

CHAPTER FIFTEEN

Integration in the Complex Plane

15.0 INTRODUCTION

Differentiation of complex functions carries no surprises. At least in general appearance, derivative formulas for the standard complex functions resemble those of their real counterparts, with x replaced by z. We shall now turn to integrals of complex functions, where the differences between real and complex calculus become dramatic.

15.1 LINE INTEGRALS IN THE COMPLEX PLANE

Given $f(z)$, the symbol $\int f(z)\,dz$ stands for any function whose derivative is $f(z)$. This is the *antiderivative*, or *indefinite integral*, of $f(z)$. Thus, $\int z^2\,dz = \frac{1}{3}z^3 + K$ for any complex constant K. This notion is not essentially different from the real case.

The definite integral $\int_a^b g(x)\,dx$ of beginning calculus is replaced in complex analysis by $\int_C f(z)\,dz$, the integral of a complex function $f(z)$ over a curve C in the plane. We shall now define this notion.

Let C be a smooth curve in the plane. This concept was discussed in Section 11.6, but briefly the idea is this. We are given two real-valued functions $x(t)$ and $y(t)$ of a real parameter t, $a \leq t \leq b$. As t varies from a to b, the point $(x(t), y(t))$ describes a curve C in the plane. In the current context we denote $(x(t), y(t))$ as $z(t) = x(t) + iy(t)$. We call C *smooth* if $x'(t)$ and $y'(t)$ are continuous and never vanish simultaneously for $a < t < b$. In this case, the vector $x(t)\mathbf{i} + y(t)\mathbf{j}$ is tangent to C.

Note that C is *oriented* in the sense that, as t varies from a to b, $z(t)$ varies from $z(a)$, the *initial point*, to $z(b)$, the *terminal point*. If we want to go the other way, we can vary t from b to a. Then $z(b)$ is the initial point, and $z(a)$ is the terminal point. In drawing C as a graph in the plane, we often indicate direction by inserting an arrow on C, as shown in Figure 277.

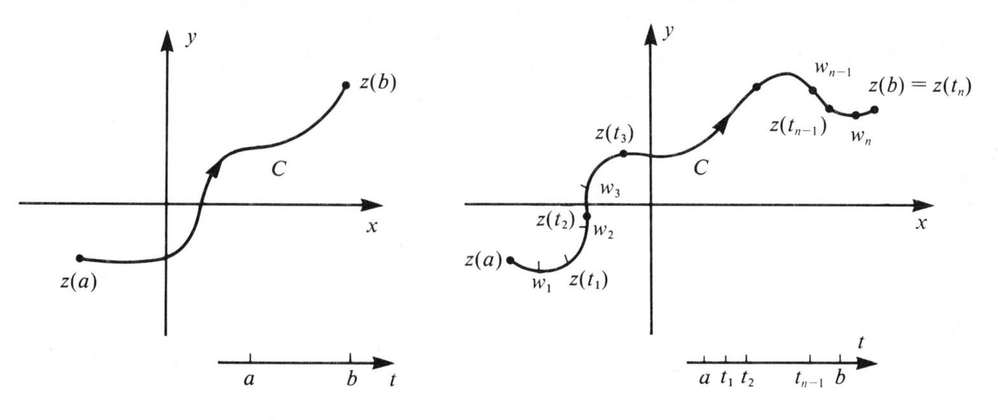

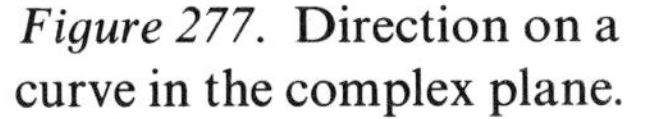

Figure 277. Direction on a curve in the complex plane.

Figure 278. Subdividing C to define the line integral.

We call C *piecewise smooth* if C consists of finitely many smooth pieces. This happens when $[a, b]$ can be broken up into finitely many intervals, $[a, t_1]$, $[t_1, t_2]$,..., $[t_{n-1}, b]$, such that $z(t)$ is smooth on each subinterval. For examples of smooth and piecewise-smooth curves, refer to Chapter 11, Section 6, or to Examples 421 through 426 just ahead.

Now let $f(z)$ be a complex function, defined at least on all points of C. Subdivide $[a, b]$ by inserting points

$$a = t_0 < t_1 < t_2 < \cdots < t_{n-1} < t_n = b.$$

Letting $z(t_j) = z_j$, we then have points $z_1, \ldots, z_{n-1}$ inserted on C between the initial and terminal points, as shown in Figure 278. Also insert on C a point w_j between z_{j-1} and z_j (possibly $w_j = z_{j-1}$ or $w_j = z_j$), and form the sum

$$\sum_{j=1}^{n} f(w_j)(z_j - z_{j-1}).$$

Now take a limit as $n \to \infty$ and as each $|z_j - z_{j-1}| \to 0$.

If the limit exists and has the same value no matter how the z_j's and w_j's are chosen (subject to the above conditions), then we call this limit the *integral of $f(z)$ over C* and denote it $\int_C f(z)\,dz$. Sometimes $\int_C f(z)\,dz$ is also called a *contour integral* (with contour C) or a *line integral* of $f(z)$.

THEOREM 117

If C is piecewise smooth and $f(z)$ is continuous on C, then $\int_C f(z)\,dz$ exists.

We shall not prove this theorem here, but use it as background in asserting the existence of various integrals.

One rarely uses the definition to evaluate $\int_C f(z)\,dz$. Before developing other methods, we give two examples that we shall use later.

EXAMPLE 421

Let k be any given complex number and C a piecewise-smooth curve from A to B. Then,

$$\int_C k\,dz = k(B - A).$$

To establish this, note that

$$\begin{aligned}\sum_{j=1}^{n} f(w_j)(z_j - z_{j-1}) &= \sum_{j=1}^{n} k(z_j - z_{j-1})\\ &= k[(z_1 - A) + (z_2 - z_1) + (z_3 - z_2) + \cdots\\ &\quad + (z_{n-1} - z_{n-2}) + (B - z_{n-1})]\\ &= k(B - A).\end{aligned}$$

Thus, in the limit, we get

$$\int_C k\,dz = k(B - A).$$

As a specific illustration, suppose that $k = 2i$ and that C goes from $1 - i$ to $3 + 2i$, as shown in Figure 279. Then,

$$\begin{aligned}\int_C k\,dz &= 2i[3 + 2i - (1 - i)]\\ &= -6 + 4i.\end{aligned}$$

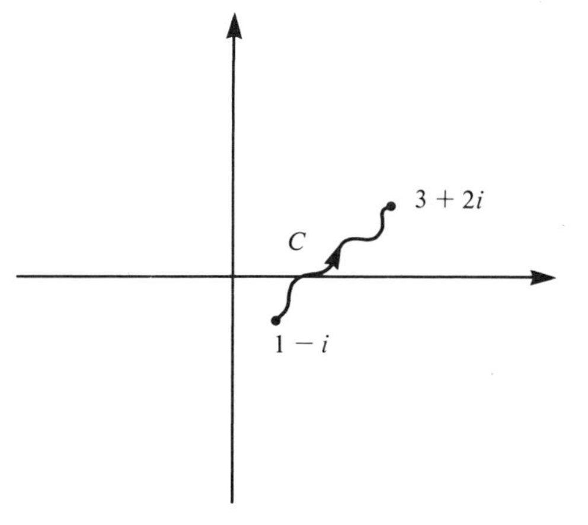

Figure 279

EXAMPLE 422

Let C be a piecewise-smooth curve from A to B. Then,

$$\int_C z\,dz = \tfrac{1}{2}(B^2 - A^2).$$

For this, choose $w_j = z_j$ in the above discussion to get

$$S_n = \sum_{j=1}^{n} f(w_j)(z_j - z_{j-1}) = \sum_{j=1}^{n} z_j(z_j - z_{j-1}).$$

Next, choose $w_j = z_{j-1}$ to get

$$T_n = \sum_{j=1}^{n} z_{j-1}(z_j - z_{j-1}).$$

By Theorem 117, both S_n and T_n approach the same limit $\int_C z\,dz$ as $n \to \infty$ and as $|z_j - z_{j-1}| \to 0$. Thus,

$$\lim_{\substack{n\to\infty\\ |z_j - z_{j-1}|\to 0}} (S_n + T_n) = 2\int_C z\,dz.$$

But, a direct calculation yields

$$\begin{aligned} S_n + T_n &= z_1(z_1 - A) + z_2(z_2 - z_1) + \cdots + z_{n-1}(z_{n-1} - z_{n-2}) + B(B - z_{n-1}) \\ &\quad + A(z_1 - A) + z_1(z_2 - z_1) + \cdots + z_{n-2}(z_{n-1} - z_{n-2}) + z_{n-1}(B - z_{n-1}) \\ &= B^2 - A^2, \end{aligned}$$

with all other terms cancelling.

Thus,

$$\int_C z\, dz = \tfrac{1}{2}(B^2 - A^2).$$

As a specific example, suppose that C goes from $1 - i$ to $3 + 2i$, as shown in Figure 279. Then,

$$\int_C z\, dz = \tfrac{1}{2}[(3 + 2i)^2 - (1 - i)^2] = \tfrac{1}{2}(5 + 14i).$$

As in real calculus, one rarely resorts to the definition to evaluate an integral, and we shall develop better ways soon. First, for the record, here are some basic properties enjoyed by the complex line integral. The person familiar with line integrals of vector functions will find none of them surprising.

1. $$\int_C [f(z) + g(z)]\, dz = \int_C f(z)\, dz + \int_C g(z)\, dz,$$

   whenever both integrals on the right exist.
2. $\int_C kf(z)\, dz = k \int_C f(z)\, dz$ for any constant k, whenever $\int_C f(z)\, dz$ exists.
3. Let $\int_C f(z)\, dz$ and $\int_K f(z)\, dz$ exist, and suppose that the terminal point of C is the initial point of K. Let L be the curve formed from C and K together, as shown in Figure 280. Then,

$$\int_L f(z)\, dz = \int_C f(z)\, dz + \int_K f(z)\, dz.$$

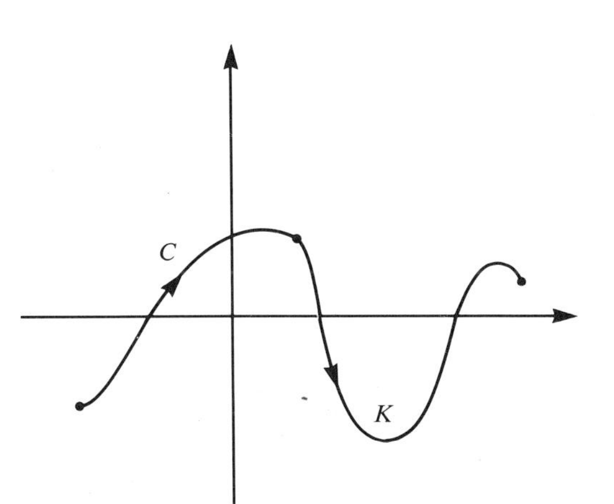

Figure 280. L consists of C and K "strung together."

4. If $\int_C f(z)\,dz$ exists, and if K is obtained from C by reversing direction, then

$$\int_K f(z)\,dz = -\int_C f(z)\,dz.$$

This completes the preliminaries, and we move to the practical business of evaluating $\int_C f(z)\,dz$ as efficiently as possible. We consider two methods in Theorems 118 and 119 of this section, along with other methods in subsequent sections.

THEOREM 118

Let $f(z)$ be continuous on a smooth curve C given by $z = z(t)$, $a \leq t \leq b$. Then,

$$\int_C f(z)\,dz = \int_a^b f[z(t)]z'(t)\,dt.$$

A proof of the above theorem is sketched in Problem 38 at the end of this section, and the student is asked to fill in the details. It is important, however, to understand what this result says and how to use it. In effect, it gives a method for converting a complex line integral into one or two Riemann integrals, as encountered in elementary calculus, but possibly involving complex constants which can be factored out of the integrals. In summary, to evaluate $\int_C f(z)\,dz$ using Theorem 118, proceed as follows:

1. Write the curve C as $z(t) = x(t) + iy(t)$ as t varies from a to b, using any parameter that is convenient.
2. Substitute $z(t)$ wherever you see z in $f(z)$, obtaining $f[z(t)]$.
3. Calculate $z'(t) = x'(t) + iy'(t)$.
4. Multiply $f[z(t)]$ by $x'(t) + iy'(t)$.
5. Integrate the resulting expression from a to b with respect to t.

Here are some examples.

EXAMPLE 423

Let $f(z) = \bar{z}$ and, on C, $z(t) = \cos(t) + i \sin(t)$, $0 \leq t \leq \pi/2$. Varying t from 0 to $\pi/2$, $z(t)$ moves over the arc of the unit circle from 1 to i, as shown in Figure 281.

On C,

$$f[z(t)]z'(t) = \overline{z(t)}z'(t) = [\cos(t) - i\sin(t)]$$

$$[-\sin(t) + i\cos(t)] = i.$$

Then,

$$\int_C \bar{z}\,dz = \int_0^{\pi/2} i\,dt = i\,\frac{\pi}{2}.$$

Figure 281. Graph of $z(t) = \cos(t) + i\sin(t)$, t from 0 to $\pi/2$.

EXAMPLE 424

Let $f(z) = z^2$, C as in Example 423.

Now, on C,

$$
\begin{aligned}
f[z(t)]z'(t) &= [\cos(t) + i\sin(t)]^2[-\sin(t) + i\cos(t)] \\
&= -3\cos^2(t)\sin(t) + \sin^3(t) \\
&\quad + i[-3\sin^2(t)\cos(t) + \cos^3(t)].
\end{aligned}
$$

Then,

$$
\begin{aligned}
\int_C z^2\,dz &= \int_0^{\pi/2} [-3\cos^2(t)\sin(t) + \sin^3(t)]\,dt \\
&\quad + i\int_0^{\pi/2} [-3\sin^2(t)\cos(t) + \cos^3(t)]\,dt \\
&= \frac{-1-i}{3}.
\end{aligned}
$$

EXAMPLE 425

Evaluate $\int_C (1 - z)\,dz$, where C is given by $z(t) = t - it^2$, t varying from 0 to 1.

Here,

$$
f[z(t)]z'(t) = [1 - (t - it^2)](1 - 2it) = (1 - t + 2t^3) + i(3t^2 - 2t).
$$

Then,

$$
\int_C (1 - z)\,dz = \int_0^1 (1 - t + 2t^3)\,dt + i\int_0^1 (3t^2 - 2t)\,dt = 1.
$$

EXAMPLE 426

Evaluate $\int_C \bar{z}\,dz$ on the straight line segment from 0 to $4i$.

Here, we may write C as $z(t) = it$, with t varying from 0 to 4. Now, on C,

$$
f[z(t)]z'(t) = \overline{it}(i) = (-it)i = t.
$$

Then,

$$
\int_C \bar{z}\,dz = \int_0^4 t\,dt = 8.
$$

EXAMPLE 427

Evaluate $\int_C (\operatorname{Re} z)\,dz$, with C the straight line segment from 1 to $2 + i$, as shown in Figure 282.

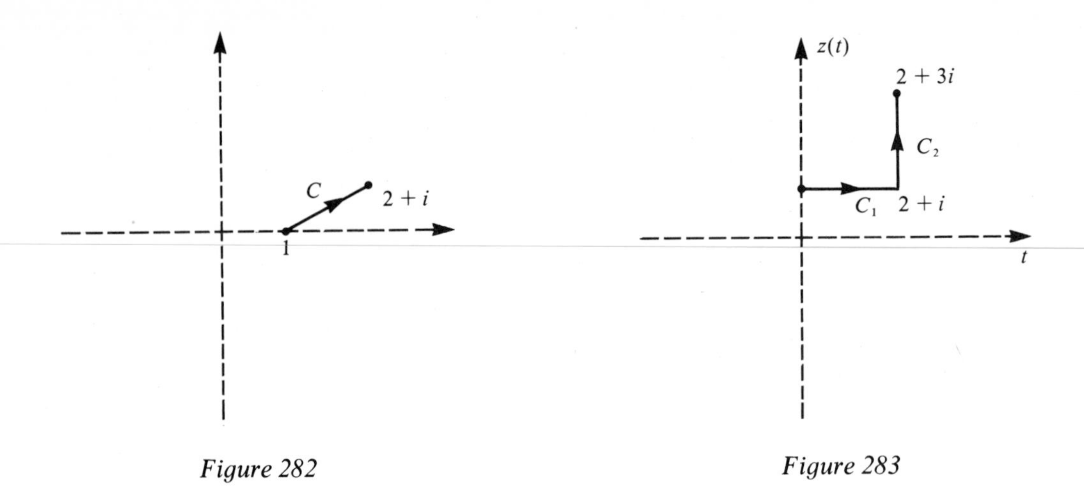

Figure 282

Figure 283

To write C in a usable form, note that the line through (1, 0) and (2, 1) is $y = x - 1$. Thus, C may be written $z(t) = t + (t - 1)i$, with t varying from 1 to 2. Here,

$$f[z(t)]z'(t) = \{\text{Re}[t + (t - 1)i]\}(1 + i) = (1 + i)t.$$

Then,

$$\int_C (\text{Re } z)\, dz = \int_1^2 (1 + i)t\, dt = \tfrac{3}{2}(1 + i).$$

If C is not smooth but is piecewise smooth, we can obtain $\int_C f(z)\, dz$ as the sum of the integrals over each smooth piece.

EXAMPLE 428

Let $f(z) = 2z + 1$, and consider C as shown in Figure 283, where C consists of smooth pieces C_1 and C_2.

Write C_1 as $z(t) = t + i$, with t varying from 0 to 2. On C_1,

$$f[z(t)]z'(t) = [2(t + i) + 1](1) = 2t + 1 + 2i.$$

Then,

$$\int_{C_1} (2z + 1)\, dz = \int_0^2 (2t + 1)\, dt + 2i \int_0^2 dt = 6 + 4i.$$

Write C_2 as $z(t) = 2 + it$, t varying from 1 to 3. On C_2,

$$f[z(t)]z'(t) = [2(2 + it) + 1](i) = -2t + 5i.$$

Then,

$$\int_{C_2} (2z + 1)\, dz = \int_1^3 -2t\, dt + 5i \int_1^3 dt = -8 + 10i.$$

Then,

$$\int_C (2z + 1)\, dz = (6 + 4i) + (-8 + 10i) = -2 + 14i.$$

The next theorem gives another way of evaluating some integrals. It is the complex version of the fundamental theorem of integral calculus.

THEOREM 119

Let $f(z)$ be continuous in a domain D, and let C be a piecewise-smooth curve in D from z_0 to z_1. Let $F(z)$ be analytic in D and $F'(z) = f(z)$. Then,

$$\int_C f(z)\, dz = F(z_1) - F(z_0).$$

Proof By the chain rule and Theorem 118, we have

$$\begin{aligned}\int_C f(z)\, dz &= \int_a^b f[z(t)]z'(t)\, dt \\ &= \int_a^b F'[z(t)]z'(t)\, dt \\ &= \int_a^b \frac{d}{dt} F[z(t)]\, dt \\ &= F[z(b)] - F[z(a)] \\ &= F(z_1) - F(z_0).\end{aligned}$$

EXAMPLE 429

Let $f(z) = z^2$ and let C be as in Example 423. There, $z_0 = 1$ and $z_1 = i$.

Now, $f(z)$ is continuous everywhere, and $F(z) = z^3/3$ is an analytic function whose derivative is z^2. Then,

$$\int_C f(z)\, dz = \frac{i^3}{3} - \frac{1^3}{3} = \frac{-1 - i}{3},$$

just as we had in Example 424.

EXAMPLE 430

Redo Example 428, using Theorem 119.

With $f(z) = 2z + 1$, an antiderivative is $F(z) = z^2 + z$. Then,

$$\int_C f(z)\, dz = F(2 + 3i) - F(i) = -2 + 14i.$$

Note: Example 423 cannot be redone using Theorem 119—there is no analytic $F(z)$ with $F'(z) = \bar{z}$ in any region of the plane.

EXAMPLE 431

Evaluate $\int_C 2\sin(3z)\,dz$ on any piecewise-smooth curve C from 1 to i.

Here, $F(z) = -\frac{2}{3}\cos(3z)$ is an antiderivative of $2\sin(3z)$; so

$$\int_C 2\sin(3z)\,dz = -\tfrac{2}{3}[\cos(3i) - \cos(3)].$$

Theorem 119 can also simplify some calculations. In Examples 423 and 424, we can write C as $z = e^{it}$, $0 \le t \le \pi/2$. Then, in Example 423, $f(z) = \bar{z} = e^{-it}$; so $f[z(t)]z'(t) = e^{-it}ie^{it} = i$ and $\int_C \bar{z}\,dz = \int_0^{\pi/2} i\,dt = i\pi/2$.

Similarly, in Example 424, $f(z) = z^2 = e^{2it}$ on C; so

$$f[z(t)]z'(t) = e^{2it}ie^{it} = ie^{3it}.$$

Then,

$$\int_C z^2\,dz = \int_0^{\pi/2} ie^{3it}\,dt = [\tfrac{1}{3}e^{3it}]_0^{\pi/2} = \tfrac{1}{3}(e^{3\pi i/2} - 1) = \frac{-1-i}{3}.$$

This avoids such calculations as $\int_0^{\pi/2} \sin^3(t)\,dt$, which arose in Example 424.

One important consequence of Theorem 119 is that, under the conditions of the theorem, $\int_C f(z)\,dz$ depends only on the endpoints of C and not on C itself, since the number $F(z_1) - F(z_0)$ uses only the endpoints.

EXAMPLE 432

$\int_C e^z\,dz = 0$ for any piecewise-smooth closed curve in the plane.

To establish this, note that we can let $f(z) = F(z) = e^z$ in Theorem 119. Since C is a closed curve, then $z_0 = z_1$; so $\int_C e^z\,dz = e^{z_1} - e^{z_1} = 0$.

We conclude this section with an estimate which is sometimes useful.

THEOREM 120

Let $f(z)$ be continuous on the piecewise-smooth curve C. If $|f(z)| \le M$ for all z on C, and if L is the length of C, then

$$\left|\int_C f(z)\,dz\right| \le ML.$$

Proof A typical sum in the limit definition of $\int_C f(z)\,dz$ is

$$S_n = \sum_{j=1}^{n} f(w_j)(z_j - z_{j-1}).$$

Now,

$$|S_n| \le \sum_{j=1}^{n} |f(w_j)|\,|z_j - z_{j-1}| \le M \sum_{j=1}^{n} |z_j - z_{j-1}|.$$

Note that $\sum_{j=1}^{n} |z_j - z_{j-1}|$ is the sum of the lengths of the line segments between z_0 and z_1, z_1 and $z_2, \ldots, z_{n-1}$ and z_n. As $n \to \infty$ and each $|z_j - z_{j-1}| \to 0$, the length of this polygonal path approaches the length of C:

$$\sum_{j=1}^{n} |z_j - z_{j-1}| \to L.$$

Since $|S_n| \to |\int_C f(z)\, dz|$, then $|\int_C f(z)\, dz| \le ML$.

EXAMPLE 433

Obtain an upper bound for $|\int_C e^{\text{Re}\, z}\, dz|$, where C is the circle of radius 2 about the origin.

Write z on C as $2\cos(t) + 2i\sin(t)$, t varying from 0 to 2π. Now, on C,

$$e^{\text{Re}\, z} = e^{2\cos(t)}.$$

This will be a maximum when $\cos(t)$ is a maximum, which is 1 at $t = 0$ or 2π. Thus, on C,

$$e^{\text{Re}\, z} \le e^2.$$

Since C has length 4π, then

$$\left| \int_C e^{\text{Re}\, z}\, dz \right| \le 4\pi e^2,$$

which is approximately 92.85.

The method only tells us that this number bounds the integral in magnitude, not how close the integral is to the bound.

PROBLEMS FOR SECTION 15.1

In each of Problems 1 through 20, evaluate $\int_C f(z)\, dz$ by using Theorem 118.

1. $f(z) = \text{Im}\, z$, C: $z(t) = t^2 - it$ as t varies from 1 to 3.
2. $f(z) = z^2 - iz$, C: $z(t) = t + it^2$, t varies from 0 to 4.
3. $f(z) = z\, \text{Re}\, z$, C: $z(t) = (1 + i)t$, t varies from 0 to 2.
4. $f(z) = 1/z$, C: $z(t) = it$, t varies from 1 to 3.
5. $f(z) = z - 1$, C: $z(t) = \cos(t) - i\sin(t)$, t varies from 0 to $\pi/3$.
6. $f(z) = iz^2$, C: $z(t) = [\cos(t)/2] + i[\sin(t)/4]$, t varies from $\pi/4$ to $\pi/2$.
7. $f(z) = \text{Re}(z^2 + 1)$, C: $z(t) = t - it^2$, t varies from 0 to 5.
8. $f(z) = (z - 1)^2$, C: $z(t) = t + 2$, t varies from 1 to 4.
9. $f(z) = 2z\cos(z)$, C: $z(t) = it$, t varies from 0 to 3.
10. $f(z) = z^2 e^z$, C: $z(t) = (1 - i)t + 1$, t varies from 1 to 4.
11. $f(z) = \bar{z}^2$, C is the straight line segment from $2 - i$ to $3 + 4i$.
12. $f(z) = 1 + z^2$, C is the semicircle of radius 3 from $3i$ to -3.
13. $f(z) = e^{-z}$, C is the line segment from 0 to $-2 - 3i$.
14. $f(z) = (1 - i)z + 2$, C consisting of the line segment from 1 to $2 - i$, and then the segment from $2 - i$ to -4.
15. $f(z) = -i\sin(z)$, C is the line segment from 1 to $-i$.
16. $f(z) = |z|^2$, C is the segment from $2i$ to 5.
17. $f(z) = (z - i)^3$, C is given by $z(t) = t - it^2$, t varying from 1 to 3.

18. $f(z) = 2z - i$, C is given by $z(t) = \cos(t) - 4i\sin(t)$, t varying from 0 to π.

19. $f(z) = e^{iz}$, C is the straight line segment from i to $1 - i$.

20. $f(z) = \sinh(z)$, C is the line segment from $2i$ to $5i$.

21. Evaluate $\int_C (2\bar{z} + iz)\,dz$, where C consists of the straight line segment from i to $3i$, and then the line segment from $3i$ to $5 + 6i$. Use any method.

22. Evaluate $\int_C 1/(z - a)\,dz$, where a is a given complex number and C is the circle of radius R centered at a, traversed counterclockwise.

First do the integral using Theorem 118 and the parametrization $C: z = a + Re^{it}$, $t: 0 \to 2\pi$. Next, try using Theorem 119. Why is this theorem not applicable?

In each of Problems 23 through 35, use Theorem 119 to evaluate the given integral. Assume that all curves considered are piecewise smooth.

23. $f(z) = 3e^{2z}$, C is any curve from $2i$ to $1 + i$.

24. $f(z) = z^2 + 2z - 5$, C is any curve from 0 to $3i$.

25. $f(z) = i\cos[(2 - i)z]$, C is any curve from 0 to π.

26. $f(z) = \cosh(2iz)$, C is any curve from 1 to $3 - i$.

27. $f(z) = 1/z$, C is the straight line segment from i to $2 + 4i$.

28. $f(z) = \sin(3i + z)$, C is the arc of the unit circle from i to -1, moving counterclockwise.

29. $f(z) = ze^{(1+i)z^2}$, C is the line segment from 2 to $8 - 4i$.

30. $f(z) = z^2\sin(5z^3)$, C is any curve from π to πi.

31. $f(z) = e^z\sin(z)$, C is any curve from 1 to $-5i$.

32. $f(z) = -iz\cos(3z^2)$, C is any curve from -1 to $3 + i$.

33. $f(z) = 2z^2e^{-z^3}$, C is any curve from $2 + i$ to $1 - 3i$.

34. $f(z) = 2z^2 - (2 + i)z$, C is any curve from 3 to $4 - i$.

35. $f(z) = e^{iz}$, C is any closed curve in the plane.

36. Use the limit definition of the integral to show that the integral of a sum is the sum of the integrals.

37. Use the limit definition of the integral to show that

$$\int_C \alpha f(z)\,dz = \alpha\int_C f(z)\,dz,$$

for any complex number α.

38. Prove Theorem 118.

Hint: Argue as follows. Form a sum $\sum_{j=1}^{n} f(w_j)(z_j - z_{j-1})$ as in the limit definition of integral. Write $z = x + iy$ and $f(z) = u(x, y) + iv(x, y)$, and let C be given by $z(t) = x(t) + iy(t)$, as t varies from a to b. Write $z_j - z_{j-1} = (x_j - x_{j-1}) + i(y_j - y_{j-1})$. If $w_j = a_j + ib_j$, show that S_n becomes

$$S_n = \sum_{j=1}^{n} [u(a_j, b_j)(x_j - x_{j-1}) - v(a_j, b_j)(y_j - y_{j-1})]$$

$$+ i\sum_{j=1}^{n} [v(a_j, b_j)(x_j - x_{j-1}) + u(a_j, b_j)(y_j - y_{j-1})].$$

Now refer to Chapter 11, Section 6, on real line integrals; this sum has limit

$$\int_C u(x, y)\,dx - v(x, y)\,dy + i\int_C v(x, y)\,dx + u(x, y)\,dy$$

as $n \to \infty$, $x_j - x_{j-1} \to 0$, and $y_j - y_{j-1} \to 0$. Both integrals here are real line integrals.

Finally, show that this expression can be written

$$\int_a^b \left[u(x(t), y(t)) \frac{dx}{dt} - v(x(t), y(t)) \frac{dy}{dt} \right] dt + i \int_a^b \left[v(x(t), y(t)) \frac{dx}{dt} + u(x(t), y(t)) \frac{dy}{dt} \right] dt,$$

and that this is the same as $\int_a^b f[z(t)]z'(t)\, dt$.

39. Find an upper bound for $|\int_C \cos(z^2)\, dz|$ if C is the circle of radius 4 about the origin.

40. Find an upper bound for $|\int_C 1/(1+z)\, dz|$ if C is the straight line between $2+i$ and $4+2i$.

15.2 THE CAUCHY INTEGRAL THEOREM

In this section we consider what may be the fundamental result of complex analysis, in the sense that it has many useful and important consequences. The theorem is named after Augustin Louis Cauchy, who first had the general idea but was only able to prove it under rather strict hypotheses. Later, Edward Goursat was able to prove it under much weaker hypotheses than those required in Cauchy's proof.

We will need some terminology. As a convenience, we shall define a *path* to be a piecewise-smooth curve. A path is *closed* if its terminal and initial points coincide and *simple* if it does not cross itself.

A domain D is called *simply connected* if every simple closed curve in D encloses only points of D. Thus, an open disk is a simply connected domain, while the domain consisting of all points between the circles in Figure 284 is not simply connected, because the simple closed curve C encloses points not in D, namely, those inside C_1. Similarly, if we remove one point (or a finite number of points) from a domain D, the resulting domain is not simply connected, as a simple closed curve in D around one of the removed points will enclose a point not in D. In dealing with integrals $\int_C f(z)\, dz$ of an analytic function over a closed curve C, we shall soon see that it makes a great deal of difference whether or not the domain over which $f(z)$ is analytic is simply connected or not.

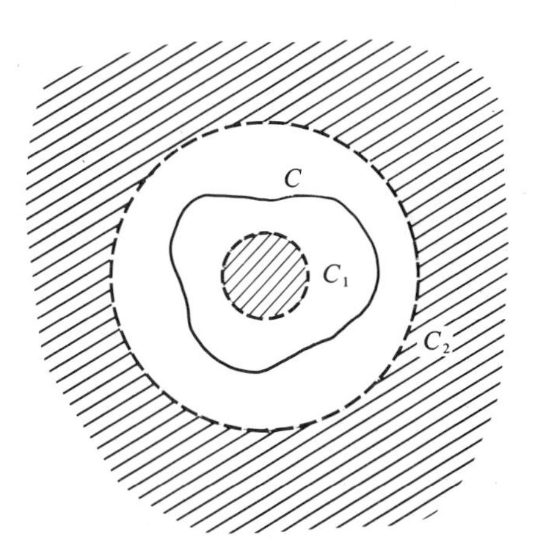

Figure 284. The region between C_1 and C_2 is not simply connected.

THEOREM 121 Cauchy Integral Theorem

Let $f(z)$ be analytic in a simply connected domain D. Then, for every simple closed path C in D,

$$\int_C f(z)\, dz = 0.$$

A rigorous proof is quite demanding, but we shall sketch the main ideas. Before doing this, we give two proofs under stricter hypotheses.

Proof I If $f(z)$ has an analytic antiderivative $F(z)$ in D, then $\int_C f(z)\,dz = 0$ by Theorem 119. This argument fails in general because we have not shown the existence of such an antiderivative.

Proof II Write $f(z) = u(x, y) + iv(x, y)$, and assume that u, v, and their first partials are continuous in D. [This is true if $f(z)$ is analytic in D, but we do not prove it here.] Using Green's theorem (see Section 11.7) and part of the argument outlined in Problem 38, Section 15.1, we have

$$\int_C f(z)\,dz = \int_C u\,dx - v\,dy + i\int_C v\,dx + u\,dy$$
$$= \iint_R \left(-\frac{\partial v}{\partial x} - \frac{\partial u}{\partial y}\right) dx\,dy + i\iint_R \left(\frac{\partial u}{\partial x} - \frac{\partial v}{\partial y}\right) dx\,dy,$$

where R is the region enclosed by C. Both these double integrals are zero by virtue of the Cauchy-Riemann equations.

We now outline Goursat's proof, which requires no assumptions beyond those stated in the theorem. This proof is not essential for later work, and the reader can omit it and go directly to the examples which follow it.

Step 1 C can be approximated arbitrarily closely by a polygon.

The idea of this is illustrated in Figure 285. Given any $\delta > 0$, we can draw a polygon P which is, at any point, at distance less than δ from C.

A proof of this requires a delicate topological argument and is beyond our scope.

Step 2 The region inside any polygon P can be subdivided into triangles, say, T_1, T_2, ..., T_n (n varies with the choice of polygon), as shown in Figure 286. If C_j is the path

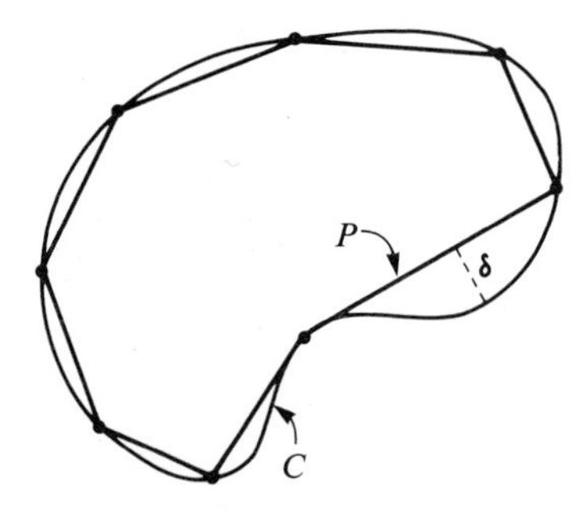

Figure 285. Approximating a curve C by a polygon P.

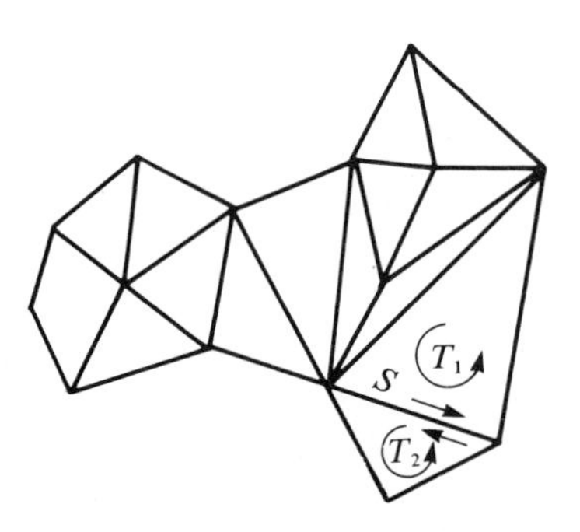

Figure 286. Triangulating a polygon: integrals over interior sides cancel.

around T_j, oriented counterclockwise, then

$$\int_P f(z)\,dz = \int_{C_1} f(z)\,dz + \int_{C_2} f(z)\,dz + \cdots + \int_{C_n} f(z)\,dz.$$

To understand this result, look at triangle T_1 in Figure 286. The arrows around C_1 indicate the direction of integration. Now, one side S of T_1 is common to exactly one other triangle, say, T_2. But in T_2, the integral around C_2 traverses S in the opposite direction than that taken on S while going around C_1. Thus, the total contribution to $\int_P f(z)\,dz$ from the interior side S is zero. Since each interior side is common to exactly two triangles, the integrals over these sides cancel in the sum $\int_{C_1} f(z)\,dz + \cdots + \int_{C_n} f(z)\,dz$, leaving just $\int_P f(z)\,dz$.

Step 3 If C is a triangle, then $\int_C f(z)\,dz = 0$.

This requires proof, which we give below after completing the sketch of the entire argument.

Step 4 $\int_C f(z)\,dz = 0$.

To prove this, note that we are done, by step 3, if C is a triangle. If C is not a triangle, let ε be any positive number. Using step 1, we find a polygon P_ε such that

$$\left| \int_C f(z)\,dz - \int_{P_\varepsilon} f(z)\,dz \right| < \varepsilon.$$

Using steps 2 and 3, $\int_{P_\varepsilon} f(z)\,dz = 0$. Thus,

$$\left| \int_C f(z)\,dz \right| < \varepsilon.$$

Since ε can be chosen arbitrarily small, then $\int_C f(z)\,dz = 0$ and the theorem is proved.

Now here are the details of step 3.

Proof of Step 3 Consider $\int_C f(z)\,dz$, with C a triangle as shown in Figure 287, oriented counterclockwise.

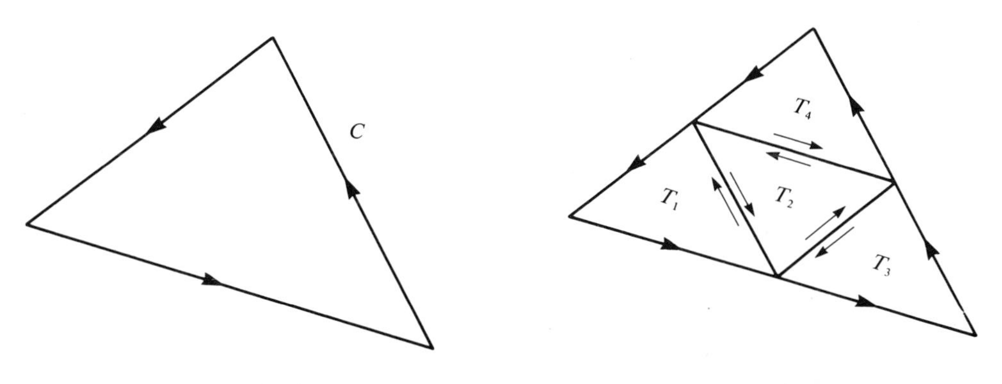

Figure 287

Figure 288

Subdivide the triangular region inside C into four congruent triangles, say, T_1, T_2, T_3, and T_4, as shown in Figure 288. Then (see step 2),

$$\int_C f(z)\,dz = \int_{C_1} f(z)\,dz + \int_{C_2} f(z)\,dz + \int_{C_3} f(z)\,dz + \int_{C_4} f(z)\,dz,$$

where C_j is the piecewise-smooth curve bounding T_j. Then, by the triangle inequality,

$$\left|\int_C f(z)\,dz\right| \leq \left|\int_{C_1} f(z)\,dz\right| + \left|\int_{C_2} f(z)\,dz\right| + \left|\int_{C_3} f(z)\,dz\right| + \left|\int_{C_4} f(z)\,dz\right|.$$

Then, for at least one value of $j = 1, 2, 3$, or 4,

$$\left|\int_C f(z)\,dz\right| \leq 4\left|\int_{C_j} f(z)\,dz\right|.$$

For notational convenience, denote this C_j as K_1.

Now center attention on K_1. Subdivide the region bounded by K_1 into four congruent triangles and argue, as above, that for at least one of these, say, K_2, we have

$$\left|\int_{K_1} f(z)\,dz\right| \leq 4\left|\int_{K_2} f(z)\,dz\right|.$$

Thus far,

$$\left|\int_C f(z)\,dz\right| \leq 4^2\left|\int_{K_2} f(z)\,dz\right|.$$

The relationship of K_1 and K_2 to C is shown in Figure 289. Continuing in this way, we generate a sequence of triangles $K_1, K_2, \ldots, K_n$ such that

$$\left|\int_C f(z)\,dz\right| \leq 4^n\left|\int_{K_n} f(z)\,dz\right|.$$

In the limit as $n \to \infty$, the triangular regions bounded by $K_1, K_2, \ldots, K_n$ collapse down to a single point, say, z_0. By assumption, $f(z)$ is analytic at z_0; hence, $f'(z_0)$ exists. Then we can write, for z sufficiently close to z_0,

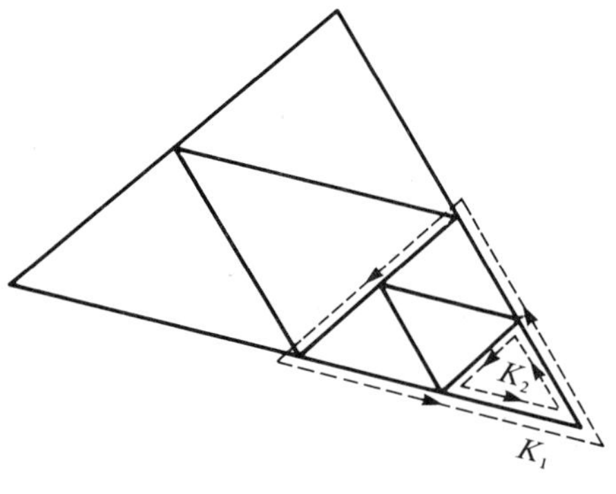

Figure 289

$$f(z) = f(z_0) + (z - z_0)f'(z_0) + h(z)(z - z_0),$$

where $\lim_{z \to z_0} h(z) = 0$. In fact,

$$h(z) = \frac{f(z) - f(z_0)}{z - z_0} - f'(z_0).$$

Then,

$$\int_{K_n} f(z)\,dz = \int_{K_n} f(z_0)\,dz + \int_{K_n} (z - z_0)f'(z_0)\,dz + \int_{K_n} h(z)(z - z_0)\,dz.$$

Since K_n is a closed curve, we obtain, by Example 421 or Theorem 119, that

$$\int_{K_n} f(z_0)\,dz = f(z_0)\int_{K_n} dz = 0$$

and

$$\int_{K_n} (z - z_0)f'(z_0)\,dz = f'(z_0)\int_{K_n} z\,dz - z_0 f'(z_0)\int_{K_n} dz = 0.$$

Thus,

$$\int_{K_n} f(z)\,dz = \int_{K_n} h(z)(z - z_0)\,dz.$$

Now examine the integral on the right. Let ε be any positive number. Since $\lim_{z\to z_0} h(z) = 0$, then there is some δ such that $|h(z)| < \varepsilon$ if $|z - z_0| < \delta$. Further, the triangular regions bounded by $K_1, K_2, \ldots, K_n$ shrink to z_0 as $n \to \infty$; so we can choose n so large, say, $n = N$, such that $|z - z_0| < \delta$ whenever z is on or inside K_N. If L_N is the length of K_N, then $|z - z_0| < L_N/2$ for all z on K_N, as indicated in Figure 290.

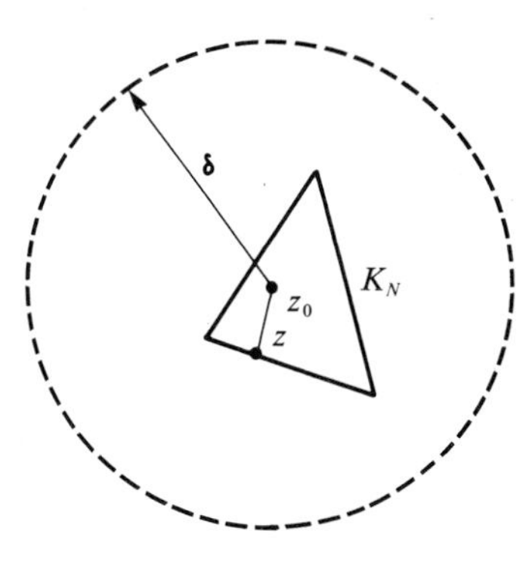

Figure 290

Then,

$$|h(z)(z - z_0)| \le \varepsilon\left(\frac{L_N}{2}\right)$$

for z on K_N. Hence, by Theorem 120,

$$\left|\int_{K_N} h(z)(z - z_0)\,dz\right| \le \left(\frac{\varepsilon L_N}{2}\right)(L_N) = \frac{\varepsilon L_N{}^2}{2}.$$

Now, if L is the length of the original triangle C, then K_1 has length $L/2$, K_2 has length $L/4, \ldots$, and K_N has length $L/2^N$. Thus,

$$\begin{aligned}\left|\int_C f(z)\,dz\right| &\le 4^N\left|\int_{K_N} f(z)\,dz\right|\\ &= 4^N\left|\int_{K_N} h(z)(z - z_0)\,dz\right|\\ &\le 4^N\varepsilon\,\frac{L_N{}^2}{2} = \frac{4^N\varepsilon}{2}\,\frac{L^2}{(2^N)^2} = \frac{\varepsilon L^2}{2}.\end{aligned}$$

Since ε is any positive number, then the nonnegative number $|\int_C f(z)\,dz|$ is arbitrarily small, and hence must be zero. Thus, $\int_C f(z)\,dz = 0$, completing the proof.

EXAMPLE 434

It is important to understand the hypotheses of Cauchy's integral theorem. These examples are designed to clarify them.

1. $\int_C z^n \, dz = 0$ for any nonnegative integer n and any simple closed path in the plane, because z^n is analytic everywhere, and the entire plane is a simply connected domain. [This result also follows easily from Theorem 119 with $F(z) = z^{n+1}/(n+1)$.]

2. Consider $\int_C 1/z \, dz$, where C is the unit circle about the origin, oriented counterclockwise.

   We may write C as $z(t) = \cos(t) + i\sin(t)$, $0 \le t \le 2\pi$. A little more compactly, we can write $z(t) = e^{it}$, $0 \le t \le 2\pi$. Then, $z'(t) = ie^{it}$, and

$$\int_C \frac{1}{z} \, dz = \int_0^{2\pi} \frac{1}{e^{it}} (ie^{it}) \, dt = i \int_0^{2\pi} dt = 2\pi i.$$

   We do not get zero here, even though $1/z$ is analytic on C. The reason the Cauchy integral theorem does not hold is that we cannot find a *simply connected* domain D containing C and in which $1/z$ is analytic. In fact, $1/z$ is not analytic at 0, which is inside the region bounded by C.

   In general, any time C goes around one or more points where $f(z)$ is not analytic, then Cauchy's integral theorem fails to hold. In such a case, $\int_C f(z) \, dz$ *may* or *may not* be zero—the theorem simply doesn't say.

3. To illustrate the last sentence of (2), consider $\int_C 1/z^2 \, dz$, with C as in (2). Now $1/z^2$ is not analytic at 0, and hence is not analytic in a simply connected domain containing C. But by direct calculation,

$$\int_C \frac{1}{z^2} \, dz = \int_0^{2\pi} \frac{1}{(e^{it})^2} ie^{it} \, dt = \int_0^{2\pi} ie^{-it} \, dt$$

$$= i \int_0^{2\pi} \cos(t) \, dt + \int_0^{2\pi} \sin(t) \, dt = 0.$$

   Thus, $\int_C 1/z^2 \, dz = 0$, although the conditions of the Cauchy integral theorem fail.

Cauchy's integral theorem is the springboard for many of the important results of complex analysis. We shall discuss several of these in the next section.

It will be convenient to introduce the notation $\oint_C f(z) \, dz$ for the integral of $f(z)$ counterclockwise around a simple closed path C. The notation $\int_C f(z) \, dz$ may still be used, but $\oint_C f(z) \, dz$ sometimes provides a reminder that C is closed. Unless specific exception is taken, all integrals around closed curves are counterclockwise. We conclude this section with two more examples.

EXAMPLE 435

Consider

$$\oint_C \frac{(2z + 1) \, dz}{z^3 - iz^2 + 6z},$$

with C the circle of radius $\frac{1}{3}$ about $3i$.

Write $z^3 - iz^2 + 6z = z(z + 2i)(z - 3i)$. The denominator of $f(z)$ thus vanishes at $0, -2i$, and $3i$; so $f(z)$ is not analytic at these points. As a quotient of polynomials, $f(z)$ is analytic everywhere else.

Use partial fractions to write

$$\frac{2z+1}{z^3 - iz^2 + 6z} = \left(\frac{1}{6}\right)\frac{1}{z} + \left(\frac{-1+4i}{10}\right)\frac{1}{z+2i} - \left(\frac{1+6i}{15}\right)\frac{1}{z-3i}.$$

Now look at Figure 291, where we have drawn C and the points 0, $-2i$, and $3i$ where $f(z)$ is not analytic. Note that only $3i$ lies inside C; note also that $1/z$ and $1/(z+2i)$ are analytic on and inside C. Thus,

$$\oint_C \frac{1}{z}\,dz = \oint_C \frac{1}{z+2i}\,dz = 0$$

by Cauchy's integral theorem. For the last integral, write $C: z = 3i + \frac{1}{3}e^{it}$, as t varies from 0 to 2π. Then,

$$\oint_C \frac{dz}{z-3i} = \int_0^{2\pi} \frac{1}{\frac{1}{3}e^{it}}\,\tfrac{1}{3}ie^{it}\,dt$$

$$= \int_0^{2\pi} i\,dz = 2\pi i.$$

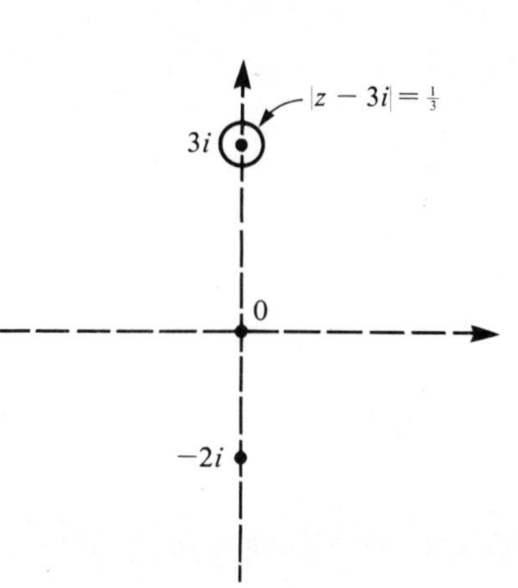

Figure 291

Thus,

$$\oint_C \frac{(2z+1)\,dz}{z^3 - iz^2 + 6z} = -\left(\frac{1+6i}{15}\right)2\pi i = \frac{\pi}{15}(12 - 2i).$$

EXAMPLE 436

Evaluate $\oint_c [z - \text{Re}(z)]\,dz$ on $C: |z| = 2$.

Write $\oint_C [z - \text{Re}(z)]\,dz = \oint_C z\,dz - \oint_C \text{Re}(z)\,dz$. Now, $\oint_C z\,dz = 0$ because z is analytic everywhere. But $\text{Re}(z)$ is not analytic (look at the Cauchy-Riemann equations) on and inside C (or anywhere else). Thus, we must do this integral directly.

Write $C: z = 2\cos(t) + 2i\sin(t)$, as t varies from 0 to 2π. Then,

$$\oint_C \text{Re}(z)\,dz = \int_0^{2\pi} 2\cos(t)[-2\sin(t) + 2i\cos(t)]\,dt = 4\pi i.$$

Thus,

$$\oint_C [z - \text{Re}(z)]\,dz = -4\pi i.$$

In the next section we pursue some important ramifications of Cauchy's integral theorem.

PROBLEMS FOR SECTION 15.2

In each of Problems 1 through 15, determine $\oint_C f(z)\,dz$. Use Cauchy's integral theorem when it applies; when it does not apply, use methods from Section 15.1. Remember that all closed curves are oriented counterclockwise.

Note: It is often helpful to sketch the curve and note any points where $f(z)$ fails to be analytic (if there are any).

1. $\oint_C \sin(3z)\, dz, \quad C: |z| = 4$

2. $\oint_C \frac{2z}{z - i}\, dz, \quad C: |z - i| = 4$

3. $\oint_C \frac{1}{(z - 2i)^3}\, dz, \quad C: |z - 2i| = 2$

4. $\oint_C z^2 \sin(z)\, dz$, C is the square with vertices at 0, 1, $1 + i$, and i.

5. $\oint_C \bar{z}\, dz, \quad C: |z| = 1$

6. $\oint_C \frac{1}{\bar{z}}\, dz, \quad C: |z| = 4$

7. $\oint_C ze^z\, dz, \quad C: |z - 3i| = 8$

8. $\oint_C (z^2 - 4z + 8)\, dz$, C is the rectangle with vertices at 1, 8, $8 + 4i$, and $1 + 4i$.

9. $\oint_C |z|^2\, dz, \quad C: |z| = 5$

10. $\oint_C \sin(1/z)\, dz, \quad C: |z - 1 + 2i| = 1$

11. $\oint_C z \operatorname{Re}(z)\, dz, \quad C: |z| = 4$

12. $\oint_C [z^2 + \operatorname{Im}(z)]\, dz$, C is the square with vertices at 0, $-2i$, $2 - 2i$, and 2.

13. $\oint_C (z^2 - |z|)\, dz, \quad C: |z| = 4$

14. $\oint_C \left(\frac{z^3 - 2z + 4}{z^8 - 2}\right) dz, \quad C: |z - 12| = 2$

15. $\oint_C \frac{(z^2 - 8z)}{\cos(z)}\, dz$, C is the rectangle with vertices at $\frac{-\pi i}{4}, \frac{\pi}{4} - \frac{\pi i}{4}, \frac{\pi}{4} + \frac{\pi i}{4}$, and $\frac{\pi i}{4}$.

16. Use partial fractions and Cauchy's integral theorem to evaluate

$$\oint_C \frac{1}{z^2 - 1}\, dz,$$

where C is given by $C: |z - i| = \frac{1}{4}$.

17. Use Cauchy's integral theorem to evaluate

$$\oint_C \frac{z\,dz}{z^2 + (1 - 2i)z - 2i},$$

where C is given by $C: |z + 1| = 1$.

15.3 SOME CONSEQUENCES OF THE CAUCHY INTEGRAL THEOREM

In this section we consider six important results which follow from Cauchy's integral theorem.

A. INDEPENDENCE OF PATH

Suppose that $f(z)$ is analytic in a simply connected domain D and that z_0 and z_1 are any two points of D. Let C_1 and C_2 be paths from z_0 to z_1, as shown in Figure 292. By reversing direction on C_2, we obtain a closed path K in D beginning and ending at z_0. By Cauchy's integral theorem,

$$\oint_K f(z)\,dz = 0.$$

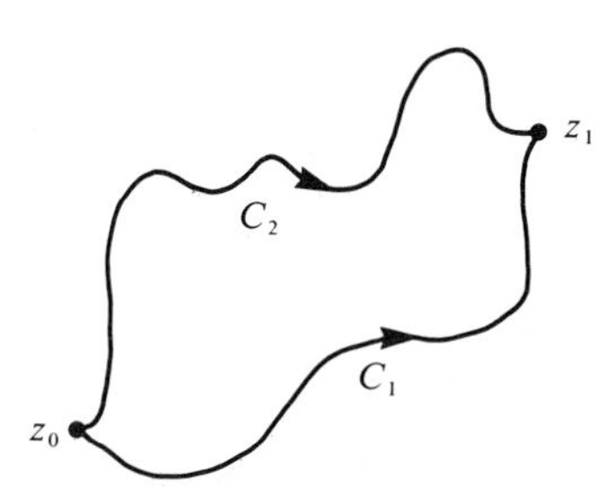

Figure 292

Decomposing K back into C_1 and C_2, and remembering that in K we reversed direction on C_2, we have

$$\int_{C_1} f(z)\,dz - \int_{C_2} f(z)\,dz = 0.$$

Thus,

$$\int_{C_1} f(z)\,dz = \int_{C_2} f(z)\,dz.$$

Thus, in D, $\int_C f(z)\,dz$ depends only on the endpoints of K and not on the path (in D) chosen between those endpoints. We say in this case that $\int_C f(z)\,dz$ is *independent of path in D*, and we summarize as follows.

THEOREM 122

If $f(z)$ is analytic in a simply connected domain D, then $\int_C f(z)\,dz$ is independent of path in D.

When $\int_C f(z)\,dz$ is independent of path in D, we usually write $\int_C f(z)\,dz$ as $\int_{z_0}^{z_1} f(z)\,dz$, where z_0 is the initial point of C and z_1 is the terminal point.

EXAMPLE 437

We know that $\sin(z)$ is analytic in the entire plane. Thus, $\int_C \sin(z)\,dz$ is independent of path. For example, suppose C is a path from $-i$ to $2+i$. Then we would write $\int_C \sin(z)\,dz$ as $\int_{-i}^{2+i} \sin(z)\,dz$. We know from Theorem 119 that

$$\int_{-i}^{2+i} \sin(z)\,dz = \Big[-\cos(z)\Big]_{-i}^{2+i}$$
$$= -\cos(2+i) + \cos(-i).$$

Note that this result is entirely dependent upon the endpoints of C and not on any other part of C at all.

B. EXISTENCE OF AN ANTIDERIVATIVE

According to Theorem 119, we can evaluate $\int_C f(z)\,dz$ easily if we know an antiderivative. However, not every function has an antiderivative—for example, $f(z) = \bar{z}$ does not. The next theorem gives sufficient conditions for an antiderivative to exist.

THEOREM 123

If $f(z)$ is analytic in a simply connected domain D, then $f(z)$ has an analytic antiderivative in D. That is, there is an analytic function $F(z)$ such that $F'(z) = f(z)$ for z in D.

Proof Choose any z_0 in D. By Theorem 122, $\int_C f(z)\,dz$ is independent of path in D. Thus, let $F(z) = \int_{z_0}^{z} f(z)\,dz$ for any z in D. We shall prove that $F(z)$ is analytic in D and that $F'(z) = f(z)$.

Let z_1 be any point of D. We have (see Figure 293)

$$F(z_1 + \Delta z) - F(z_1) = \int_{z_0}^{z_1+\Delta z} f(z)\,dz - \int_{z_0}^{z_1} f(z)\,dz = \int_{z_1}^{z_1+\Delta z} f(z)\,dz,$$

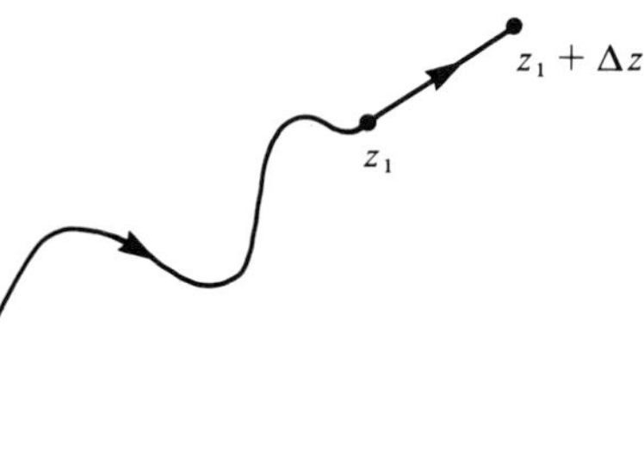

Figure 293

where $\Delta z = z - z_1$. Then,

$$\frac{F(z_1+\Delta z) - F(z_1)}{\Delta z} - f(z_1) = \frac{1}{\Delta z}\int_{z_1}^{z_1+\Delta z} f(z)\,dz - f(z_1)$$
$$= \frac{1}{\Delta z}\int_{z_1}^{z_1+\Delta z} [f(z) - f(z_1)]\,dz,$$

since $f(z_1)$ is constant.

Since $f(z)$ is analytic in D, then at least $f(z)$ is continuous at z_1. Given any $\varepsilon > 0$, there is some $\delta > 0$ such that

$$|f(z) - f(z_1)| < \varepsilon \quad \text{if} \quad |z - z_1| < \delta.$$

Remembering that $z - z_1 = \Delta z$, then, for $|\Delta z| < \delta$, we have

$$\left|\frac{F(z_1 + \Delta z) - F(z_1)}{\Delta z} - f(z_1)\right| = \left|\frac{1}{\Delta z}\int_{z_1}^{z_1+\Delta z} [f(z) - f(z_1)]\, dz\right|$$

$$= \left|\frac{1}{\Delta z}\right|\left|\int_{z_1}^{z_1+\Delta z} [f(z) - f(z_1)]\, dz\right|$$

$$\leq \frac{1}{|\Delta z|}\, \varepsilon\, |\Delta z| = \varepsilon.$$

Thus,

$$F'(z_1) = \lim_{\Delta z \to 0} \frac{F(z_1 + \Delta z) - F(z_1)}{\Delta z} = f(z_1).$$

This proves that $F(z)$ is differentiable at all points of D and that $F'(z) = f(z)$. Since D is open, we can find about any point of D an open neighborhood wholly contained in D, and hence in which $F(z)$ is differentiable. Thus, $F(z)$ is analytic in D.

EXAMPLE 438

Consider $\int_C 1/z\, dz$ under two different circumstances.

1. Let D be the domain consisting of all z with Re $z > 0$ and Im $z > 0$. This is the right quarter-plane (see Figure 294). In D, $\int_C 1/z\, dz$ is independent of path, since D is simply connected and $1/z$ is analytic in D. In fact, $Ln(z)$ is an antiderivative of $1/z$ in D; so

$$\int_{z_0}^{z_1} \frac{1}{z} = Ln(z_1) - Ln(z_0)$$

for any points z_0, z_1 in D.

2. Now let D be the entire plane, minus the origin. Again, $1/z$ is analytic in D, but D is not simply connected. Now, $Ln(z)$ is *not* an antiderivative of $1/z$ in all of D, because $Ln(z)$ is analytic only for $-\pi < \text{Arg}\, z < \pi$, which excludes the negative real axis. Note that, in fact, $\int_C 1/z\, dz = 2\pi i$, not zero, if C is the unit circle $|z| = 1$ traversed counterclockwise.

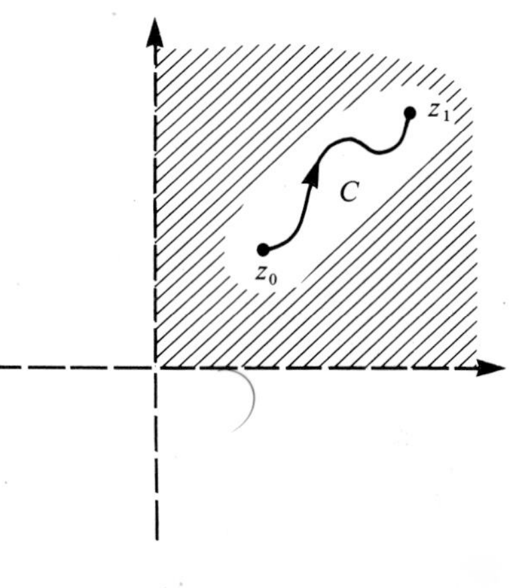

Figure 294

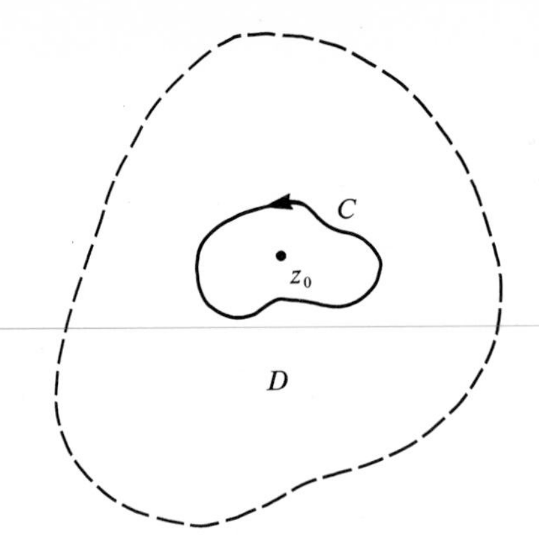

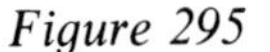
Figure 295

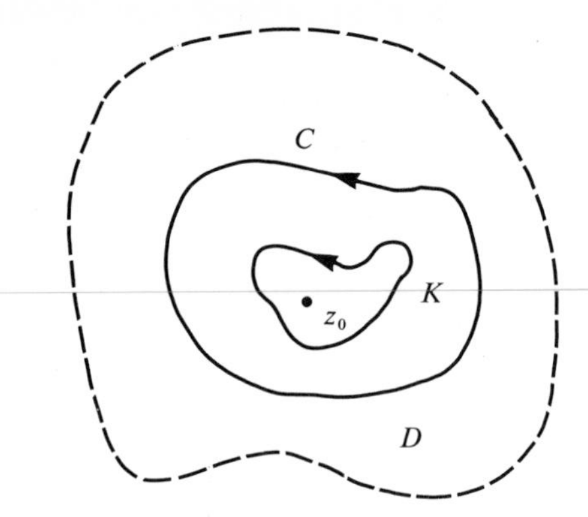

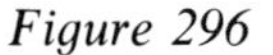
Figure 296

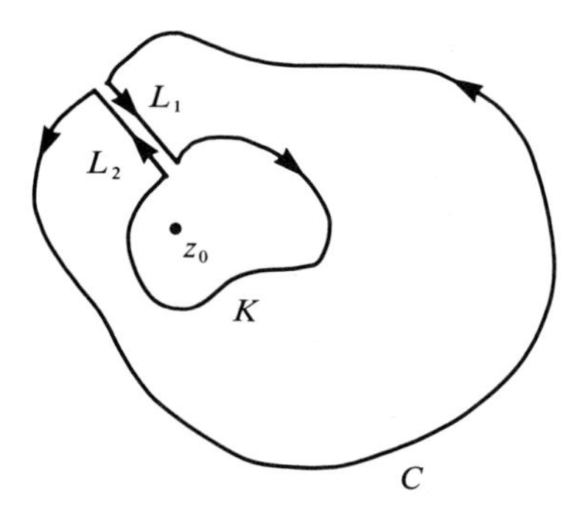

Figure 297

C. CHANGE OF PATH

Consider the problem of evaluating $\oint_C f(z)\,dz$, where $f(z)$ is analytic in some domain D, except at z_0, and C encloses z_0, as shown in Figure 295. By using Cauchy's integral theorem, we may replace C by any other simple closed path K in D enclosing z_0, as shown in Figure 296. This is often very convenient in evaluating integrals.

To see why this can be done, envision K as in Figure 296 and form a cut L between C and K, as indicated in Figure 297. Then, C, K, L_1, and L_2 form a new closed curve C^* which *does not* enclose z_0. By Cauchy's integral theorem,

$$\oint_{C^*} f(z)\,dz = 0.$$

Now, in integrating around C^* counterclockwise, we go around C counterclockwise, around K clockwise, and along L_1 and L_2. As L_1 and L_2 are taken closer together, they merge to a straight line, with the integral over L_1 cancelling that over L_2 because of the opposite directions. Thus, we have

$$\oint_C f(z)\,dz + \oint_K f(z)\,dz = 0.$$

Reversing direction on K to go around it counterclockwise, we have

$$\oint_C f(z)\,dz = \oint_K f(z)\,dz.$$

EXAMPLE 439

Evaluate $\displaystyle\oint_C \frac{dz}{z - z_0}$ over *any* simple closed path enclosing z_0.

Noting that $1/(z - z_0)$ is analytic everywhere except at z_0, we replace C by a circle about z_0 of sufficiently small radius r to be inside C (see Figure 298). We can write K as $z = z_0 + re^{it}$, with t varying from 0 to 2π. Then,

$$\oint_C \frac{dz}{z - z_0} = \oint_K \frac{dz}{z - z_0} = \int_0^{2\pi} \frac{1}{re^{it}}(ire^{it})\,dt = 2\pi i.$$

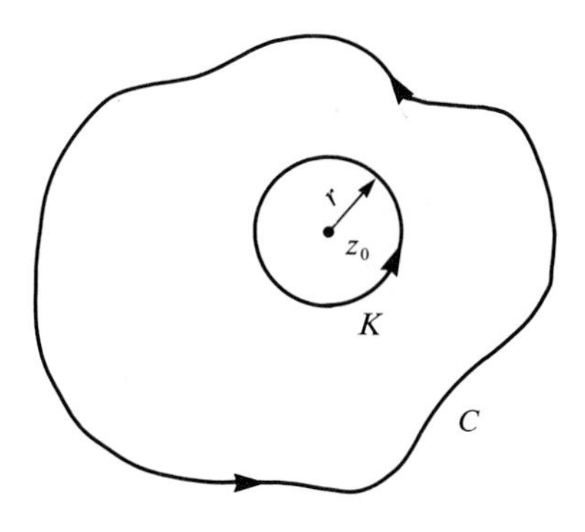

Figure 298

D. THE CAUCHY INTEGRAL FORMULA

The following result may be the most important consequence of Cauchy's integral theorem.

THEOREM 124 The Cauchy Integral Formula

Let $f(z)$ be analytic in a simply connected domain D. Let z_0 be any point in D, and let C be any simple closed path in D enclosing z_0. Then,

$$\oint_C \frac{f(z)}{z - z_0}\,dz = 2\pi i f(z_0).$$

Before proving the theorem, we give two examples of its use in evaluating integrals.

EXAMPLE 440

$$\oint_C \frac{e^{z^2}}{z - i}\,dz = 2\pi i e^{(i)^2} = \frac{2\pi i}{e}$$

for any simple closed curve C enclosing i, since e^{z^2} is analytic in the entire plane.

EXAMPLE 441

$$\oint_C \frac{e^{2z}\sin(z^2)}{z - 1}\,dz = 2\pi i e^2 \sin(1)$$

for any simple closed path C enclosing 1. If K is any simple closed path not enclosing 1 and not passing through 1, then

$$\oint_C \frac{e^{2z}\sin(z^2)}{z-1}\,dz = 0,$$

by Cauchy's integral theorem.

Proof of the Cauchy Integral Formula Since $f(z_0)$ is constant, we have

$$\begin{aligned}\oint_C \frac{f(z)\,dz}{z-z_0} &= \oint_C \left(\frac{f(z)-f(z_0)+f(z_0)}{z-z_0}\right) dz\\ &= f(z_0)\oint_C \frac{dz}{z-z_0} + \oint_C \left(\frac{f(z)-f(z_0)}{z-z_0}\right) dz\\ &= 2\pi i f(z_0) + \oint_C \left(\frac{f(z)-f(z_0)}{z-z_0}\right) dz.\end{aligned}$$

In the last line, we used Example 439.

Now consider the last integral. For Cauchy's integral formula to be true, we must show that this integral is zero. Replacing C by a circle K about z_0, we have

$$\oint_C \left(\frac{f(z)-f(z_0)}{z-z_0}\right) dz = \oint_K \left(\frac{f(z)-f(z_0)}{z-z_0}\right) dz.$$

Now, $f(z)$ is analytic at z_0. Thus, given $\varepsilon > 0$, there is some $\delta > 0$ such that

$$|f(z)-f(z_0)| \le \frac{\varepsilon}{2\pi} \quad \text{if} \quad |z-z_0| < \delta.$$

Choose K to have radius $\frac{1}{2}\delta$. Then, for z on K,

$$\left|\frac{f(z)-f(z_0)}{z-z_0}\right| \le \frac{\varepsilon}{2\pi}\left(\frac{1}{\frac{1}{2}\delta}\right) = \frac{\varepsilon}{\pi\delta}.$$

By Theorem 120,

$$\left|\oint_C \left(\frac{f(z)-f(z_0)}{z-z_0}\right) dz\right| \le \frac{\varepsilon}{\pi\delta}\left(2\pi\,\frac{1}{2}\,\delta\right) = \varepsilon.$$

Since ε is arbitrarily small, then

$$\oint_C \left(\frac{f(z)-f(z_0)}{z-z_0}\right) dz = 0,$$

proving the theorem.

E. HIGHER DERIVATIVES OF ANALYTIC FUNCTIONS

Using Cauchy's integral formula, we can show that an analytic function has derivatives of all orders, and we can derive formulas for these derivatives.

THEOREM 125 Cauchy's Integral Formula for Higher Derivatives

Let $f(z)$ be analytic at z_0. Then $f(z)$ has derivatives of all orders at z_0, and the nth derivative at z_0 is

$$f^{(n)}(z_0) = \frac{n!}{2\pi i} \oint_C \frac{f(z)}{(z - z_0)^{n+1}} \, dz,$$

where C is any simple closed path about z_0 such that $f(z)$ is analytic on and inside C.

Proof We proceed by induction on n. First, we prove the formula for $n = 1$. Begin with

$$f'(z_0) = \lim_{\Delta z \to 0} \frac{f(z_0 + \Delta z) - f(z_0)}{\Delta z}$$

$$= \lim_{\Delta z \to 0} \frac{1}{2\pi i \, \Delta z} \left[\oint_C \frac{f(z)}{z - (z_0 + \Delta z)} \, dz - \oint_C \frac{f(z)}{z - z_0} \, dz \right],$$

by applying Cauchy's integral formula to $f(z_0 + \Delta z)$ and to $f(z_0)$.

Thus,

$$f'(z_0) = \frac{1}{2\pi i} \lim_{\Delta z \to 0} \oint_C \frac{f(z)}{\Delta z} \left[\frac{1}{z - (z_0 + \Delta z)} - \frac{1}{z - z_0} \right] dz.$$

It is easy to check that

$$\frac{1}{\Delta z} \left[\frac{1}{z - (z_0 + \Delta z)} - \frac{1}{z - z_0} \right] = \frac{1}{(z - z_0)^2} + \frac{\Delta z}{(z - z_0 - \Delta z)(z - z_0)^2}.$$

Thus,

$$f'(z_0) = \frac{1}{2\pi i} \oint_C \frac{f(z)}{(z - z_0)^2} \, dz + \frac{1}{2\pi i} \lim_{\Delta z \to 0} \Delta z \oint_C \frac{f(z)}{(z - z_0 - \Delta z)(z - z_0)^2} \, dz.$$

We must show that the limit on the right is zero.

First, $f(z)$ is continuous, and hence bounded on C, say, $|f(z)| \leq M$. Now let δ be the smallest distance between z_0 and a point of C (see Figure 299). Then,

$$|z - z_0| \geq \delta$$

for all z on C; hence,

$$\frac{1}{|z - z_0|^2} \leq \frac{1}{\delta^2}.$$

Then, on C,

$$\frac{|f(z)|}{|z - z_0|^2} \leq \frac{M}{\delta^2}.$$

Now choose $|\Delta z| \leq \delta/2$. Then, $z_0 + \Delta z$ is at least $\delta/2$ units from any z on C:

$$|z - z_0 - \Delta z| \geq \frac{\delta}{2}.$$

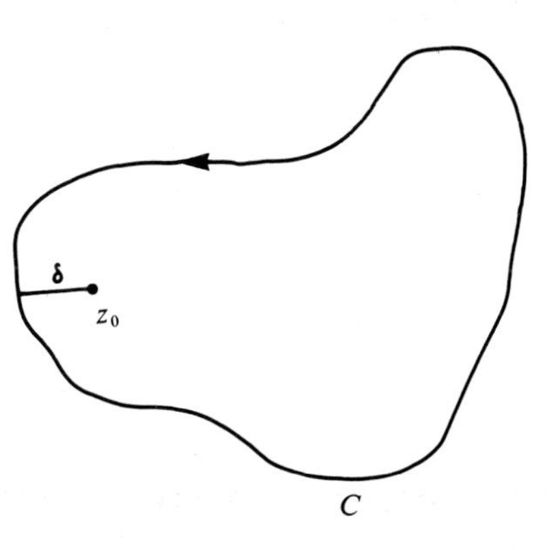

Figure 299. δ is the smallest distance between z_0 and C.

Then,

$$\frac{1}{|z - z_0 - \Delta z|} \le \frac{2}{\delta}.$$

Thus,

$$\left| \Delta z \oint_C \frac{f(z)}{(z - z_0)^2(z - z_0 - \Delta z)}\, dz \right| \le |\Delta z| \cdot \frac{M}{\delta^2} \cdot \frac{2}{\delta} \cdot (\text{length of } C)$$

for z on C and $|\Delta z| \le \delta/2$. As $\Delta z \to 0$, certainly $2|\Delta z| M/\delta^3 \to 0$. Thus,

$$\lim_{\Delta z \to 0} \Delta z \oint_C \frac{f(z)}{(z - z_0)^2(z - z_0 - \Delta z)}\, dz = 0$$

also, proving the formula for $f'(z_0)$.

We leave it for the student to complete the proof by establishing the formula for $f^{(n+1)}(z_0)$, assuming that for $f^{(n)}(z_0)$.

As an important consequence of Theorem 125, we have a bound on $f^{(n)}(z_0)$.

COROLLARY 1

Let $f(z)$ be analytic in a domain containing all points on and inside the circle C of radius r about z_0. Let $|f(z)| \le M$ for all z on C. Then,

$$|f^{(n)}(z_0)| \le \frac{Mn!}{r^n}.$$

Proof By Theorems 120 and 125,

$$|f^{(n)}(z_0)| = \frac{n!}{2\pi} \left| \oint_C \frac{f(z)}{(z - z_0)^{n+1}}\, dz \right| \le \frac{n!}{2\pi} \frac{M}{r^{n+1}} (2\pi r) = \frac{Mn!}{r^n}.$$

Sometimes Cauchy's integral formula for derivatives can be used to evaluate integrals. Here are two examples.

EXAMPLE 442

Evaluate

$$\oint_C \frac{4e^{z^2}}{(z - i)^3}\, dz,$$

where C is any simple closed path in the plane not passing through i.

Consider two cases:

1. C does not enclose i. Then the integral is zero by Cauchy's integral theorem, as $4e^{z^2}/(z - i)^3$ is analytic on and inside C.
2. C encloses i. Now Theorem 125 applies. Let $f(z) = 4e^{z^2}$. Then,

$$f^{(2)}(i) = \frac{2}{2\pi i} \oint_C \frac{4e^{z^2}}{(z - i)^3}\, dz.$$

Then,

$$\oint_C \frac{4e^{z^2}}{(z-i)^3}\,dz = \pi i f^{(2)}(i) = \pi i(-8e^{-1}).$$

EXAMPLE 443

Evaluate

$$\oint_C \frac{2\sin(z^3)}{(z-1)^4}\,dz,$$

where C is any simple closed path not passing through 1.

Again, we have two cases:

1. If 1 is not enclosed by C, then the integral is zero.
2. If 1 encloses C, then, letting $f(z) = 2\sin(z^3)$, we have

$$f^{(3)}(1) = \frac{6}{2\pi i}\oint_C \frac{2\sin(z^3)\,dz}{(z-1)^4}.$$

Then,

$$\oint_C \frac{2\sin(z^3)\,dz}{(z-1)^4} = \frac{\pi i}{3} f^{(3)}(1) = \frac{\pi i}{3}[-108\sin(1) - 42\cos(1)].$$

Finally, we note that the formula for $f^{(n)}(z_0)$ is easy to remember from Cauchy's integral formula. It is exactly what you would get by differentiating the integral formula n times with respect to z_0 under the integral sign.

F. A GENERALIZATION OF CAUCHY'S INTEGRAL FORMULA

Cauchy's integral formula can be extended in several ways. Here is one which will be useful in treating Laurent series in the next chapter.

Let w be any complex number, and suppose that $f(z)$ is analytic on and in the region between the two concentric circles $C_1: |z-w| = r_1$ and $C_2: |z-w| = r_2$ (see Figure 300). The domain $r_1 < |z-w| < r_2$ bounded by C_1 and C_2 is often called an *annulus* or *ring*.

We claim that, for any z_0 in this annulus,

$$f(z_0) = \frac{1}{2\pi i}\oint_{C_2} \frac{f(z)}{z-z_0}\,dz - \frac{1}{2\pi i}\oint_{C_1} \frac{f(z)\,dz}{z-z_0}.$$

Before proving this, we note that this formula is most commonly used when w is a point at which $f(z)$ is not analytic or perhaps not even defined.

For example, let $f(z) = 1/z$ and $w = 0$. Let z_0 be any nonzero complex number. Choose r_1 and r_2 so that $r_1 < |z_0| < r_2$, and let C_1 be the circle $|z| = r_1$ and C_2 the circle $|z| = r_2$ (see Figure 301). Then,

$$f(z_0) = \frac{1}{z_0} = \frac{1}{2\pi i}\oint_{C_2} \frac{\frac{1}{z}\,dz}{z-z_0} - \frac{1}{2\pi i}\oint_{C_1} \frac{\frac{1}{z}\,dz}{z-z_0}.$$

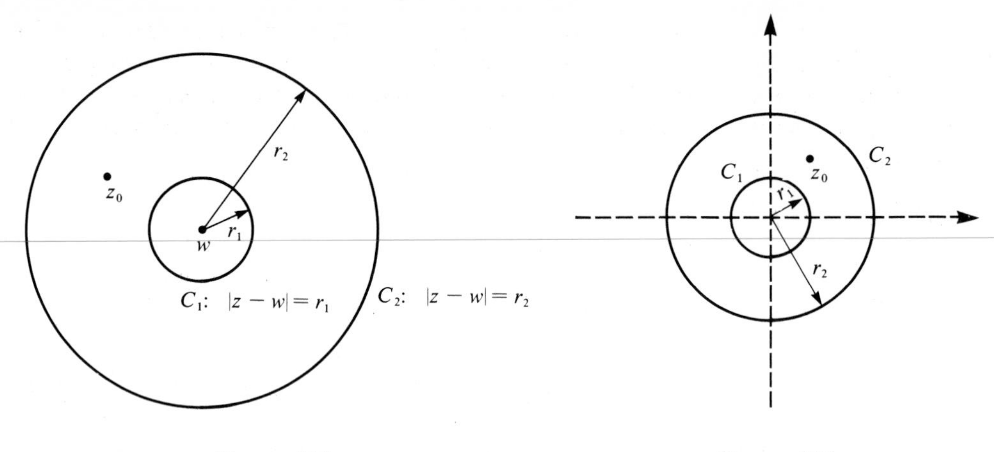

Figure 300

Figure 301

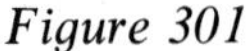

Now here is a proof of the formula in the general case. Form two cuts L_1 and L_2 between C_1 and C_2, and form a new simple closed path K enclosing z_0. This process is depicted in Figure 302(a), (b), and (c). Now, $f(z)$ is analytic on K and on the simply connected domain bounded by K. By Cauchy's integral formula,

$$f(z_0) = \frac{1}{2\pi i} \oint_K \frac{f(z)\, dz}{z - z_0}.$$

Now,

$$\oint_K = \int_{C_2} + \int_{C_1} + \int_{L_1} + \int_{L_2}.$$

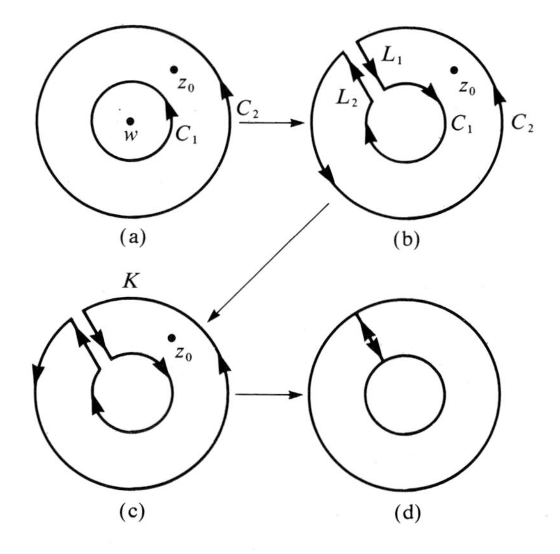

Figure 302

Note that counterclockwise orientation on K induces *clockwise* orientation on C_1. Further, if we merge L_1 and L_2, then the integrals along L_1 and L_2 cancel each other out, and we are left with

$$f(z_0) = \frac{1}{2\pi i} \oint_{C_2} \frac{f(z)\, dz}{z - z_0} + \frac{1}{2\pi i} \oint_{C_1} \frac{f(z)\, dz}{z - z_0}.$$

Reversing orientation on C_1 to have all integrals counterclockwise gives us the desired result.

In the next chapter we develop some results on complex sequences and series. Cauchy's integral formula, and the extension of it presented above, will play an important role in developing the complex version of the Taylor series, as well as an expansion called the *Laurent expansion*, which has no obvious counterpart for real functions.

PROBLEMS FOR SECTION 15.3

1. Evaluate $\int_{i}^{1+2i} z^3 \, dz$. **2.** Evaluate $\int_{\pi}^{\pi i} \sin(2z) \, dz$. **3.** Evaluate $\int_{1-2i}^{4+i} ze^{-z^2} \, dz$.

In each of Problems 4 through 15, use Cauchy's integral formula or the formula for $f^{(n)}(z_0)$ to evaluate the given integral.

4. $\oint_C \frac{z^4 \, dz}{z - 2i}$, C is any simple closed path enclosing $2i$.

5. $\oint_C \frac{\sin(z^2)}{z - 5} \, dz$, C is any simple closed path enclosing 5.

6. $\oint_C \frac{z^3 - 5z + i}{(z - 1 + 2i)} \, dz$, C is any simple closed path enclosing $1 - 2i$.

7. $\oint_C \frac{2z^3}{(z - 2)^2} \, dz$, C is any simple closed path about 2.

8. $\oint_C \frac{ie^{z^3} \, dz}{(z - 2 + i)^2}$, C is any simple closed path enclosing $2 - i$.

9. $\oint_C \frac{\cos(z - i)}{(z + 2i)^3} \, dz$, C is any simple closed path about $-2i$.

10. $\oint_C \frac{z \sin(3z) \, dz}{(z + 4)^3}$, C is any simple closed path about -4.

11. $\oint_C \frac{[z^2 - \cos(3z)] \, dz}{(z + 4i)^4}$, C is any simple closed path about $-4i$.

12. $\oint_C \frac{\sinh(z^3) \, dz}{(z - 8i)^3}$, C is any simple closed path about $8i$.

13. $\oint_C \frac{[\cos(z) - \sin(z)] \, dz}{(z + i)^4}$, C is any simple closed path about $-i$.

14. $\oint_C \frac{2z \cosh(z)}{(z - 2 + i)^3} \, dz$, C is any simple closed path about $2 - i$.

15. $\oint_C \frac{(z + i)^2}{(z - 2i)^4} \, dz$, C is any simple closed path about $2i$.

16. Let $f(z)$ be analytic in a domain D. Let z_0 be in D and suppose that the circle $|z - z_0| = r$ and all points inside this circle are in D. Prove that

$$f(z_0) = \frac{1}{2\pi} \int_0^{2\pi} f(z_0 + re^{i\theta})\, d\theta.$$

Thus, $f(z_0)$ is the average of the values of $f(z)$ on a circle centered at z_0.

17. Under the conditions of Problem 16, prove that

$$|f(z_0)| \leq M,$$

where M is the maximum of the values $|f(z)|$ for z on the circle $|z - z_0| = r$.

18. Prove *Morera's theorem*, which is nearly a converse of Cauchy's integral theorem:

If $f(z)$ is continuous in a domain D, and $\int_C f(z)\, dz = 0$ for every simple closed path C in D, then $f(z)$ is analytic in D.

(*Hint*: Prove that $f(z)$ has an antiderivative in D.)

19. Use Corollary 1 to prove *Liouville's theorem*:

If $f(z)$ is analytic in the entire plane, and, for some M, $|f(z)| \leq M$ for all z, then $f(z)$ must be a constant.

(*Hint*: Show that $f'(z) = 0$ for all z.)

20. Use Liouville's theorem (Problem 19) to prove the following:

If $f(z)$ is analytic for all z, and $f(z)$ is not a constant function, then, given positive numbers M and r, there exists z with $|z| > r$ and $|f(z)| > M$. [That is, such a function can be made to take on arbitrarily large values (in magnitude) at points arbitrarily far away from the origin.]

21. Use the result of Problem 20 to prove the *Fundamental Theorem of Algebra*:

If $p(z) = a_0 + a_1 z + \cdots + a_{n-1}z^{n-1} + z^n$, a polynomial with $a_0, a_1, \ldots, a_{n-1}$ complex, then there is some number z_0 such that $p(z_0) \neq 0$.

(*Hint*: Assume instead that $p(z) \neq 0$ for all z. Apply the result of Problem 20 to $f(z) = 1/p(z)$.)

SUPPLEMENTARY PROBLEMS

In each of Problems 1 through 25, evaluate the given integral. Use any of the methods of this chapter which apply to the problem.

1. $\oint_C \frac{e^{2z}}{(z-1)^3}\, dz$, C is any simple closed path enclosing 1.

2. $\int_{-2+3i}^{i} 4iz \sin(z^2)\, dz$

3. $\oint_C \frac{dz}{(z-i)^3}$, C is any simple closed path not going through i.

4. $\int_C 2i|z|\bar{z}\, dz$, C is the line from $-i$ to $4i$.

5. $\int_C \operatorname{Im}(z + 3z)\, dz$, C is the line from 1 to $-i$.

6. $\oint_C \frac{-(2+i)\sin(z^2)\, dz}{(z+4)^2}$, C is any simple closed path enclosing -4.

7. $\int_{3+i}^{2-5i} (z^3 - 3z^3 + i)\, dz$

8. $\oint_C 3iz^2\, dz$, C is the circle of radius 2 about $-4i$.

9. $\int_C [2z^2 - i\,\mathrm{Im}(z)]\, dz$, C is the straight line from $2 + 3i$ to $-1 + 2i$.

10. $\int_C (z - i)^2\, dz$, C is the semicircle of radius 1 about 0 from i to $-i$.

11. $\int_C \mathrm{Re}(z + 4)\, dz$, C is the line from $3 + i$ to $2 - 5i$.

12. $\oint_C \frac{3z^2 \cosh(z)\, dz}{(z + 2i)^2}$, C is any simple closed path not passing through $-2i$.

13. $\int_C (z + i|z|)\, dz$, C is the line from 0 to $-2i$.

14. $\oint_C \frac{2iz^2\, dz}{(z + 1 - 2i)^2}$, C is any simple closed path about $-1 + 2i$.

15. $\int_0^i -ze^{iz^2}\, dz$

16. $\int_{2+i}^{-1} \cosh(3iz)\, dz$

17. $\int_C [z - \mathrm{Re}(z^2)]\, dz$, C is the line from -2 to $3 + i$.

18. $\oint_C (1 - z^2)\, dz$, C is any simple closed path in the plane.

19. $\int_2^4 (z^2 + 3z^4 - i)\, dz$

20. $\oint_C \frac{-2\cos(z^3)\, dz}{z + 3i}$, C is any simple closed path not passing through $-3i$.

21. $\oint_C e^{-z^2}\, dz$, C is any simple closed path about $2 - i$.

22. $\int_C [z^2 - \mathrm{Re}(iz)]\, dz$, C is the line from -3 to $4 - i$.

23. $\int_C (3iz^2 - 1)\, dz$, C is the semicircle of radius 3 about 0 from $-3i$ to $3i$.

24. $\oint_C \frac{\cosh[\sinh(z)]\, dz}{z + 3}$, C is any simple closed path not passing through -3.

25. $\displaystyle\int_{-5}^{2i} [z^3 + iz^2 - \sin(z)]\, dz$

26. Examine the integral $\oint_C e^z/z\, dz$, with C the unit circle about the origin, to obtain the result that

$$\int_0^{\pi} e^{\cos(\theta)} \cos[\sin(\theta)]\, d\theta = \pi.$$

CHAPTER SIXTEEN

Complex Sequences and Series, and the Taylor and Laurent Expansions

16.0 INTRODUCTION

Two of the most important tools in complex analysis are the Taylor and Laurent series. In order to understand these, it is necessary to have some background on sequences and series of complex numbers and functions, which also have some interest in their own right.

In order to keep the treatment geared to the complex case, we assume a basic knowledge of real sequences and series. The main points we will need from the real case are compiled in the Appendix to this chapter. The student who feels a little hesitant in this area might benefit from reading the Appendix first and then proceeding to Section 1 of this chapter.

16.1 COMPLEX SEQUENCES

A *complex sequence* is an ordered list which assigns to each positive integer n a complex number z_n. One often denotes this by writing

$$z_1, z_2, \ldots, z_n, \ldots.$$

The number z_n is called the nth *term* of the sequence. For example, we might have

$$i, i^2, i^3, i^4, \ldots, i^n, \ldots,$$

in which the nth term is i^n, or

$$1 + i, \frac{1+i}{2}, \frac{1+i}{4}, \frac{1+i}{8}, \ldots,$$

in which the nth term is $(1 + i)/2^{n-1}$.

A sequence $z_1, z_2, \ldots$ is said to *converge to the number L* if, given $\varepsilon > 0$, there is some positive integer N such that $|z_n - L| < \varepsilon$ whenever $n \geq N$. Another way of putting this is that, given any ε-neighborhood of L, there is some index N such that all of $z_N, z_{N+1}, z_{N+2}, \ldots$ are in this neighborhood (see Figure 303). In this event, we write

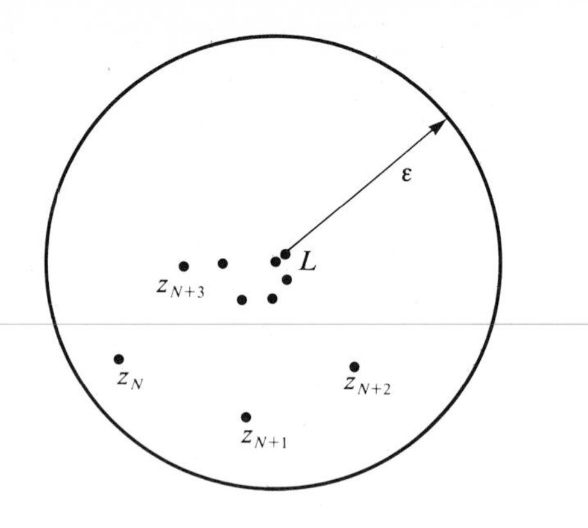

Figure 303. Convergence of a complex sequence to L.

$$\lim_{n\to\infty} z_n = L \quad \text{or} \quad z_n \to L.$$

When such a number L exists, we say that $z_1, z_2, \ldots$ *converges*, and we call L the *limit* of the sequence. If no such L exists, we say that $z_1, z_2, \ldots$ *diverges*.

The following theorem enables us to treat complex sequences $z_1, z_2, \ldots$ in terms of real sequences, in which each term is a real number.

THEOREM 126

Let $z_n = x_n + iy_n$. Then, $z_n \to A + iB$ if and only if $x_n \to A$ and $y_n \to B$.

Proof Suppose first that $z_n \to A + iB$.

Let $\varepsilon > 0$. Then, for some N, we have

$$|z_n - (A + iB)| < \varepsilon$$

if $n \geq N$. Then, for $n \geq N$,

$$\sqrt{(x_n - A)^2 + (y_n - B)^2} < \varepsilon.$$

Then $|x_n - A|$ and $|y_n - B|$ must both be less than ε if $n \geq N$; hence, $x_n \to A$ and $y_n \to B$.

Conversely, suppose that $x_n \to A$ and $y_n \to B$. Again, let $\varepsilon > 0$. For some N_1,

$$|x_n - A| < \frac{\varepsilon}{2} \quad \text{if} \quad n \geq N_1.$$

Similarly, for some N_2,

$$|y_n - b| < \frac{\varepsilon}{2} \quad \text{if} \quad n \geq N_2.$$

Let $N = N_1 + N_2$. If $n \geq N$, then $n \geq N_1$ and $n \geq N_2$; so

$$|z_n - (A + iB)| = \sqrt{(x_n - A)^2 + (y_n - B)^2} \leq \sqrt{\frac{\varepsilon^2}{4} + \frac{\varepsilon^2}{4}} = \frac{\varepsilon}{\sqrt{2}} < \varepsilon.$$

Thus, $z_n \to A + iB$.

In practice, this is how we usually treat a complex sequence $z_1, z_2, \ldots$. Here are some examples.

EXAMPLE 444

Let $z_n = \left(\dfrac{n^2 - 2n + 3}{3n^2 - 4}\right) + \left(\dfrac{2n - 1}{2n + 1}\right)i.$

Here, $z_n = x_n + iy_n$, with $x_n = \dfrac{n^2 - 2n + 3}{3n^2 - 4}$ and $y_n = \dfrac{2n - 1}{2n + 1}$.

First, write

$$x_n = \frac{1 - \dfrac{2}{n} + \dfrac{3}{n^2}}{3 - \dfrac{4}{n^2}}.$$

As $n \to \infty$, clearly $\dfrac{2}{n}, \dfrac{3}{n^2}$, and $\dfrac{4}{n^2} \to 0$; so $x_n \to \dfrac{1}{3}$.

Next, write $y_n = \dfrac{2 - \dfrac{1}{n}}{2 + \dfrac{1}{n}}$. As $n \to \infty$, $y_n \to \dfrac{2}{2} = 1$.

Thus, $z_n \to \frac{1}{3} + i$.

EXAMPLE 445

Let $z_n = \dfrac{3}{n} - \left(\dfrac{n^2 - 3}{n}\right)i.$

Here, $\dfrac{3}{n} \to 0$ as $n \to \infty$, but $\dfrac{n^2 - 3}{n} = n - \dfrac{3}{n}$ does not have a finite limit.

Thus, $z_1, z_2, \ldots$ diverges.

EXAMPLE 446

Let $z_n = \sin(in\pi/4)$.

Here,

$$z_n = \frac{1}{2i}(e^{-n\pi/4} - e^{n\pi/4}) = \left(\frac{e^{n\pi/4} - e^{-n\pi/4}}{2}\right)i = i \sinh\left(\frac{n\pi}{4}\right).$$

As n is chosen larger, $\sinh(n\pi/4)$ correspondingly grows larger, because of the $e^{n\pi/4}$ term. Thus, $z_1, z_2, \ldots$ diverges.

Here are some approximate values of z_n for various values of n:

n	z_n
1	$0.8687i$
10	$1287.9854i$
50	$5.6712(10^{16})i$
100	$6.4325(10^{33})i$

EXAMPLE 447

Let $z_n = i^n$. The sequence is then

$$i, \ -1, \ -i, \ 1, \ i, \ -1, \ -i, \ 1, \ i, \ -1, \ -i, \ 1, \ldots.$$

Note that the values i, -1, $-i$, and 1 repeat periodically in the sequence. This forces the sequence to diverge, since it is impossible for *all* values of z_n past some N to be within a distance of, say, $\frac{1}{2}$ from any given complex number. Thus, the definition of convergence cannot be satisfied with $\varepsilon = \frac{1}{2}$.

The main point to be remembered when working with a complex sequence z_n is that we can always write $z_n = x_n + iy_n$ and consider separately the two real sequences x_n and y_n. In this sense, complex sequences present us with no difficulties not already encountered with real sequences.

In the next section we develop the Cauchy convergence criterion for complex sequences. This will be a direct consequence of the Cauchy criterion for real sequences, and it is of theoretical value for later developments.

PROBLEMS FOR SECTION 16.1

In each of Problems 1 through 15, determine whether or not the sequence converges. If it converges, determine the limit.

1. $1 + \dfrac{2in}{n+1}$

2. i^{2n}

3. $\dfrac{n^2+1}{2n^2} - \left(\dfrac{n-1}{n}\right)i$

4. $e^{n\pi i/3}$

5. $(-1)^{in}$

6. $\left(1 + \dfrac{1}{n}\right)^n i$

7. $\dfrac{2n^2 i + 3}{n^2 + 1}$

8. $(1 - i)^{-n}$

9. $(-i)^{4n}$

10. $\dfrac{(-1)^n}{2n} + \left(\dfrac{3n-2}{n}\right)i$

11. $\left(\dfrac{-n^2}{n+i}\right)i$

12. $\tan(in)$

13. $(i)^{2/n}$

14. $\dfrac{1}{1-i}\left(\dfrac{n^2+4}{n^2-3n}\right)$

15. $\left(\dfrac{-1}{n}\right)^n + 3i\dfrac{n^2}{n^2+2}$

16. Suppose that $z_n \to L_1$ and $z_n \to L_2$. Prove that $L_1 = L_2$.

17. Prove that $z_n + w_n \to L + S$ if $z_n \to L$ and $w_n \to S$.

18. Prove that $z_n w_n \to LS$ if $z_n \to L$ and $w_n \to S$.

19. Prove that $cz_n \to cL$ if $z_n \to L$.

20. A sequence $z_1, z_2, \ldots$ is *bounded* if there is some number M such that $|z_n| \leq M$ for $n = 1, 2, 3, \ldots$. Prove that a convergent sequence must be bounded.

21. Continuing from Problem 20, can we say that every bounded sequence converges?

16.2 CAUCHY'S CONVERGENCE CRITERION FOR COMPLEX SEQUENCES

A complex sequence $z_1, z_2, \ldots$ is called a *Cauchy sequence* if, given $\varepsilon > 0$, there exists N such that $|z_n - z_m| < \varepsilon$ whenever $m \geq N$ and $n \geq N$. In words, this says that all terms of the sequence can be made arbitrarily close to one another by going far enough out in the sequence. The Cauchy convergence criterion is that this property characterizes the convergent sequences: $z_1, z_2, \ldots$ converges if and only if it is a Cauchy sequence.

This theorem is extremely important in many theoretical considerations, but it is not a major tool in treating specific sequences (Example 448 just ahead is an exception). Mainly, its importance stems from the fact that it implies existence of a limit solely in terms of "closeness" of the z_n's to one another and does not require that we first produce a candidate for the limit itself (a task which for some sequences is a practical impossibility).

The complex version of Cauchy's criterion is easy to prove if one already has in hand the real version, which depends on fairly deep properties of real numbers and is proved in the Appendix to this chapter.

THEOREM 127 The Cauchy Convergence Criterion

A complex sequence $z_1, z_2, \ldots$ converges if and only if it is a Cauchy sequence.

Proof Suppose first that $z_n \to L$. Let $\varepsilon > 0$. For some N, $|z_n - L| < \varepsilon/2$ whenever $n \geq N$. Then, for $m \geq N$ and $n \geq N$,

$$|z_n - z_m| = |(z_n - L) - (z_m - L)| \leq |z_n - L| + |z_m - L| < \frac{\varepsilon}{2} + \frac{\varepsilon}{2} = \varepsilon.$$

Conversely, suppose that $z_1, z_2, \ldots$ is a Cauchy sequence. Let $z_n = x_n + iy_n$. We shall prove that $x_1, x_2, \ldots$ and $y_1, y_2, \ldots$ are Cauchy sequences and apply the real version of the theorem to them. Let $\varepsilon > 0$. For some N, $|z_n - z_m| < \varepsilon$ if $n \geq N$. But then

$$|x_n - x_m| \leq |z_n - z_m| < \varepsilon$$

and

$$|y_n - y_m| \leq |z_n - z_m| < \varepsilon$$

if $n \geq N$. Thus, $x_1, x_2, \ldots$ is a real Cauchy sequence, and hence converges to some number A; and $y_1, y_2, \ldots$ is a real Cauchy sequence, and hence converges to some number B. Then, $z_n \to A + iB$ by Theorem 126.

Here is an illustration in which the Cauchy criterion plays a key role.

EXAMPLE 448

Let $z_1 = 2i$, $z_2 = 1 - i$, and, for $n \geq 3$, let

$$z_n = \tfrac{1}{2}(z_{n-1} + z_{n-2}).$$

Thus,

$$z_3 = \frac{1}{2}(1 + i), \qquad z_4 = \frac{1}{2}\left(\frac{3}{2} - \frac{i}{2}\right), \qquad z_5 = \frac{1}{2}\left(\frac{5}{4} + \frac{i}{4}\right),$$

and so on. We shall prove that this is a Cauchy sequence, and hence converges. In this example, Cauchy's criterion is useful because we don't know each z_n explicitly (for example, what is $z_{40,000}$?). Thus, guessing a possible limit does not seem feasible (although Problem 4 asks you to figure out a way to do it).

To show that this is a Cauchy sequence, begin by observing that

$$|z_2 - z_1| = |1 + 3i| = \sqrt{10}.$$

Next,

$$|z_3 - z_2| = \left|-\frac{1}{2} + \frac{3i}{2}\right| = \frac{\sqrt{10}}{2},$$

$$|z_4 - z_3| = \left|\frac{1}{4} - \frac{3i}{4}\right| = \frac{\sqrt{10}}{4},$$

and

$$|z_5 - z_4| = \left|-\frac{1}{8} + \frac{3i}{8}\right| = \frac{\sqrt{10}}{8}.$$

This suggests that, in general,

$$|z_n - z_{n-1}| = \frac{\sqrt{10}}{2^{n-2}},$$

and we leave it to the student to verify this.

Now, let m and n be any positive integers, say $n = m + t$ with $t > 0$. Then,

$$\begin{aligned}
|z_n - z_m| &= |(z_n - z_{n-1}) + (z_{n-1} - z_{n-2}) + \cdots + (z_{m+1} - z_m)| \\
&\leq |z_n - z_{n-1}| + |z_{n-1} - z_{n-2}| + \cdots + |z_{m+1} - z_m| \\
&= \frac{\sqrt{10}}{2^{n-2}} + \frac{\sqrt{10}}{2^{n-3}} + \cdots + \frac{\sqrt{10}}{z^{m-1}} \\
&= \sqrt{10}\left(\frac{1}{2^{m+t-2}} + \frac{1}{2^{m+t-3}} + \cdots + \frac{1}{2^{m-1}}\right) \\
&= \frac{\sqrt{10}}{2^{m-1}}\left(\frac{1}{2^{t-1}} + \frac{1}{2^{t-2}} + \cdots + 1\right).
\end{aligned}$$

But,

$$1 + \frac{1}{2} + \frac{1}{2^2} + \cdots + \frac{1}{2^{t-2}} + \frac{1}{2^{t-1}} = \frac{1 - (\frac{1}{2})^t}{1 - \frac{1}{2}} = 2\left[1 - \left(\frac{1}{2}\right)^t\right].$$

(This is a standard formula from geometric series. See Example 452 just ahead in the next section.)

Thus,

$$|z_n - z_m| \le \frac{\sqrt{10}}{2^{m-1}} 2\left[1 - \left(\frac{1}{2}\right)^t\right] = \frac{\sqrt{10}}{2^{m-2}}\left[1 - \left(\frac{1}{2}\right)^t\right] \le \frac{\sqrt{10}}{2^{m-2}}.$$

Clearly, $\sqrt{10}/2^{m-2} \to 0$ as $m \to \infty$. Given $\varepsilon > 0$, there is some N such that, for $m \ge N$, $\sqrt{10}/2^{m-2} < \varepsilon$. Then, for m and $n \ge N$, we have $|z_n - z_m| < \varepsilon$. Thus, $z_1, z_2, \ldots$ is a Cauchy sequence, and so must converge.

Note that it is not immediately obvious what this sequence converges to. However, once we know that the limit exists, the value of the limit can be determined (see Problems 1 and 4 at the end of this section).

For another application of the Cauchy criterion, note the proof of Theorem 130 in the next section.

PROBLEMS FOR SECTION 16.2

1. Having proved in Example 448 that the sequence converges, now show that its limit is $\frac{2}{3}$. Why was it important to use the Cauchy criterion to establish existence of the limit before its value could be determined? (*Hint:* How are the limits of z_n, z_{n-1}, and z_{n-2} related as $n \to \infty$?)
2. Use the Cauchy convergence criterion to prove that $|z_1|, |z_2|, \ldots$ converges if $z_1, z_2, \ldots$ converges. (*Hint:* You need an inequality involving $||z_n| - |z_m||$; look through the problems at the end of Chapter 14, Section 1.)
3. Is it true that convergence of $|z_1|, |z_2|, \ldots$ implies convergence of $z_1, z_2, \ldots$? Either prove that it does, or give a counterexample.
4. Generalize Example 448 and Problem 1 as follows. Let z_1 and z_2 be any complex numbers. For $n \ge 3$, define
$$z_n = \tfrac{1}{2}(z_{n-1} + z_{n-2}).$$
Prove that $z_1, z_2, \ldots$ converges and that the limit is $\frac{1}{3}(z_1 + 2z_2)$.
5. Suppose that $z_1, z_2, \ldots$ converges. What can be said about convergence of the real sequence $|z_1|, |z_2|, \ldots$?
6. As a converse of Problem 5, does convergence of $|z_1|, |z_2|, \ldots$ imply anything about convergence of $z_1, z_2, \ldots$?

16.3 COMPLEX SERIES

Given a complex sequence $z_1, z_2, \ldots$, we may consider the *complex series*

$$\sum_{n=1}^{\infty} z_n = z_1 + z_2 + \cdots.$$

We assign a meaning to such a sum as follows. Let

$$S_m = \sum_{n=1}^{m} z_n = z_1 + z_2 + \cdots + z_m.$$

Thus, $S_1 = z_1$, $S_2 = z_1 + z_2$, $S_3 = z_1 + z_2 + z_3$, and so on. We call $S_1, S_2, \ldots$ the *sequence of partial sums* of $\sum_{n=1}^{\infty} z_n$. If the sequence of partial sums converges, we say that $\sum_{n=1}^{\infty} z_n$ converges, and we define the sum of the series to be $\lim_{m\to\infty} S_m$. Thus,

$$\sum_{n=1}^{\infty} z_n = \lim_{m\to\infty} \sum_{n=1}^{m} z_n.$$

If the sequence of partial sums diverges, then we say that $\sum_{n=1}^{\infty} z_n$ diverges.

As with sequences, questions about complex series can always be reduced to questions about real series.

THEOREM 128

Let $z_n = x_n + iy_n$. Then, $\sum_{n=1}^{\infty} z_n = A + iB$ if and only if $\sum_{n=1}^{\infty} x_n = A$ and $\sum_{n=1}^{\infty} y_n = B$.

We leave a proof of this to the student.

In practice, it is usually impossible to determine the sum of a series explicitly. Usually, the best we can do is determine whether or not the series converges. In this, Theorem 128 is especially important, because it implies that $\sum_{n=1}^{\infty} z_n$ converges only when both real series $\sum_{n=1}^{\infty} x_n$ and $\sum_{n=1}^{\infty} y_n$ converge, and we have a number of tests for real series to apply to these. Among these tests are the comparison, ratio, root, and integral tests, which are reviewed in the Appendix to this chapter. Here are some examples.

EXAMPLE 449

$\sum_{n=1}^{\infty} \left(\frac{1}{n^2} - \frac{i}{n^3}\right)$ converges, because $\sum_{n=1}^{\infty} \frac{1}{n^2}$ and $\sum_{n=1}^{\infty} \frac{1}{n^3}$ converge (for example, by the integral test).

EXAMPLE 450

$\sum_{n=1}^{\infty} \left[\frac{n^2}{n^4+2} + i\left(\frac{n^2+1}{n^2}\right)\right]$ diverges. Here, $\sum_{n=1}^{\infty} \frac{n^2}{n^4+2}$ converges, but $\sum_{n=1}^{\infty} \frac{n^2+1}{n^2}$ diverges $\left(\text{for example, } \lim_{n\to\infty} \frac{n^2+1}{n^2} = 1 \neq 0\right)$.

EXAMPLE 451

$\sum_{n=1}^{\infty} \frac{2^n i}{n!}$ converges because $\sum_{n=1}^{\infty} \frac{2^n}{n!}$ converges (for example, by the ratio test).

EXAMPLE 452

$\sum_{n=0}^{\infty} z^n = 1/(1-z)$ if $|z| < 1$. (*Note:* This series begins at $n = 0$ instead of $n = 1$; this is just notational convenience.) This is a complex version of the geometric series and is one of the rare instances in which we can determine the sum of the series.

Let

$$S_m = \sum_{n=0}^{m} z_n = 1 + z + z^2 + \cdots + z^m.$$

Then,

$$zS_m = z + z^2 + z^3 + \cdots + z^{m+1}.$$

Then,

$$S_m - zS_m = 1 - z^{m+1}.$$

If $z \neq 1$, then

$$S_m = \frac{1 - z^{m+1}}{1 - z}.$$

As $m \to \infty$, $z^{m+1} \to 0$ if $|z| < 1$. Thus, for $|z| < 1$,

$$\sum_{n=0}^{\infty} z^n = \frac{1}{1 - z}.$$

For example,

$$\sum_{n=0}^{\infty} \left(\frac{i}{2}\right)^n = \frac{1}{1 - \dfrac{i}{2}} = \frac{2}{2 - i} = \frac{4 + 2i}{5},$$

since $|i/2| = \frac{1}{2} < 1$.

The following result is similar to the real case.

THEOREM 129

If $\sum_{n=1}^{\infty} z_n$ converges, then $\lim_{n\to\infty} z_n = 0$. (Thus, if $\lim_{n\to\infty} z_n \neq 0$, then $\sum_{n=1}^{\infty} z_n$ diverges.)

Proof Let $S_m = \sum_{n=1}^{m} z_n$. Then, $S_{m-1} = \sum_{n=1}^{m-1} z_n$ and $S_m - S_{m-1} = z_m$.

If $\sum_{n=1}^{\infty} z_n$ converges, then $\lim_{m\to\infty} S_m = \lim_{m\to\infty} S_{m-1} = \sum_{n=1}^{\infty} z_n$. Then, $\lim_{m\to\infty} (S_m - S_{m-1}) = \lim_{m\to\infty} z_m = 0$.

EXAMPLE 453

Continuing from Example 452, $\sum_{n=0}^{\infty} z^n$ diverges if $|z| \geq 1$, since in this case $\lim_{n\to\infty} z^n \neq 0$.

Combining both examples, we have $\sum_{n=0}^{\infty} z^n$ converging exactly when $|z| < 1$; in this case, $\sum_{n=0}^{\infty} z^n = 1/(1 - z)$.

The trick used to sum $\sum_{n=0}^{\infty} z^n$ can also be used to sum $\sum_{n=1}^{\infty} nz^{n-1}$. Let

$$S_m = \sum_{n=1}^{m} nz^{n-1} = 1 + 2z + 3z^2 + \cdots + mz^{m-1}.$$

Then,

$$zS_m = z + 2z^2 + 3z^3 + \cdots + mz^m;$$

so

$$S_m - zS_m = 1 + z + z^2 + \cdots + z^{m-1} - mz^m$$
$$= \left(\sum_{n=0}^{m-1} z^n\right) - mz^m.$$

Then,

$$S_m = \frac{1}{1-z} \sum_{n=0}^{m-1} z^n - \frac{m}{1-z} z^m.$$

As $m \to \infty$,

$$S_m \to \sum_{n=1}^{\infty} nz^{n-1} \quad \text{and} \quad \sum_{n=0}^{m-1} z^n \to \sum_{n=0}^{\infty} z^n = \frac{1}{1-z},$$

if $|z| < 1$. Further, if $|z| < 1$, then

$$\left|\frac{m}{1-z} z^m\right| \to 0.$$

Thus, for $|z| < 1$,

$$\sum_{n=1}^{\infty} nz^{n-1} = \left(\frac{1}{1-z}\right)\left(\frac{1}{1-z}\right) = \frac{1}{(1-z)^2}.$$

The series version of the Cauchy convergence criterion (Theorem 127) is given next. It is exactly the criterion for sequences applied to the sequence of partial sums of a series. However, it is worth stating separately for convenience.

THEOREM 130 Cauchy Convergence Criterion for Series

$\sum_{n=0}^{\infty} z_n$ converges if and only if, given $\varepsilon > 0$, there is some N such that, for every $n \geq N$ and every positive integer p,

$$|z_{n+1} + z_{n+2} + \cdots + z_{n+p}| < \varepsilon.$$

Proof The mth partial sum of $\sum_{n=1}^{\infty} z_n$ is $S_m = \sum_{n=1}^{m} z_n$.

Now, remember that $\sum_{n=1}^{\infty} z_n$ converges if and only if the sequence $S_1, S_2, \ldots$ converges, and apply Cauchy's criterion (Theorem 125); we leave the details to the student.

Here is an application of this theorem.

THEOREM 131

If $\sum_{n=1}^{\infty} |z_n|$ converges, then so does $\sum_{n=1}^{\infty} z_n$.

In this event, $|\sum_{n=1}^{\infty} z_n| \leq \sum_{n=1}^{\infty} |z_n|$.

Proof Note that, for any positive integers n and p,

$$|z_{n+1} + \cdots + z_{n+p}| \leq |z_{n+1}| + \cdots + |z_{n+p}|.$$

Given $\varepsilon > 0$, we can, by Theorem 130, find N such that the right side is less than ε if $n \geq N$. Then, $|z_{n+1} + \cdots + z_{n+p}| < \varepsilon$ if $n \geq N$; so $\sum_{n=1}^{\infty} z_n$ converges, by Theorem 130.

Finally, $|\sum_{n=1}^{N} z_n| \leq \sum_{n=1}^{N} |z_n|$ for each positive integer N, implying that

$$\lim_{N\to\infty} \left| \sum_{n=1}^{N} z_n \right| = \left| \sum_{n=1}^{\infty} z_n \right| \leq \lim_{N\to\infty} \sum_{n=1}^{N} |z_n| = \sum_{n=1}^{\infty} |z_n|.$$

EXAMPLE 454

$\sum_{n=1}^{\infty} i^{2n}/n$ converges. This is just the alternating harmonic series $-1 + \frac{1}{2} - \frac{1}{3} + \frac{1}{4} - \frac{1}{5} + \cdots$ in complex notation. But $|i^{2n}| = 1$; so $\sum_{n=1}^{\infty} |i^{2n}/n| = \sum_{n=1}^{\infty} 1/n$, the divergent harmonic series. Thus, $\sum_{n=1}^{\infty} i^{2n}/n$ converges conditionally.

In the next section we shall consider series in which each term is a function of z of a particular form. In treating these series, we will frequently want to use the ratio test as applied to complex series, and so we state it here. Given a series $\sum_{n=1}^{\infty} z_n$, we can test for absolute convergence by applying the ratio test to the real series $\sum_{n=1}^{\infty} |z_n|$. To do this, let $\lim_{n\to\infty} |z_{n+1}/z_n| = R$, assuming that this limit exists. Then,

$$\sum_{n=1}^{\infty} |z_n| \begin{cases} \text{converges if } 0 \leq R < 1, \\ \text{diverges if } R > 1 \text{ (and also if } R = \infty). \end{cases}$$

If $0 \leq R < 1$, then $\sum_{n=1}^{\infty} |z_n|$ converges, and hence $\sum_{n=1}^{\infty} z_n$ converges.

This test fails to distinguish conditionally convergent series, but, for the series we want to consider next, absolute convergence will turn out to be more important.

PROBLEMS FOR SECTION 16.3

1. Prove Theorem 128.

2. Write out a complete proof of Theorem 130.

In each of Problems 3 through 10, use the theorems of this section and results for real series to determine whether or not the given series converges.

3. $\sum_{n=1}^{\infty} \left(\frac{1}{n^2} - \frac{i}{2^n}\right)$

4. $\sum_{n=1}^{\infty} \left(\frac{2}{n} - \frac{i}{n^2}\right)$

5. $\sum_{n=1}^{\infty} \left(\frac{-i}{n}\right)^n$

6. $\sum_{n=1}^{\infty} \frac{(1+i)^n}{4^n}$

7. $\sum_{n=1}^{\infty} \tan(in)$

8. $\sum_{n=1}^{\infty} \left(\frac{\sin(n)}{n^2} + \frac{i^{3n}}{2n}\right)$

9. $\sum_{n=1}^{\infty} \frac{\cosh(in)}{e^n}$

10. $\sum_{n=1}^{\infty} \left(e^{-n} + \frac{i}{n^2+1}\right)$

In each of Problems 11 through 15, show that the series converges by showing that it converges absolutely.

11. $\sum_{n=1}^{\infty} \frac{(1+i)^n}{3^n}$

12. $\sum_{n=1}^{\infty} \frac{(n+2)^2}{(2i)^n}$

13. $\sum_{n=1}^{\infty} \frac{n - 2^n i}{n!(n+1)^2}$

14. $\sum_{n=1}^{\infty} \frac{e^{in}}{2n^3}$

15. $\sum_{n=1}^{\infty} \frac{(2-3i)^{2n}}{(18+2i)^{3n+1}}$

16. Suppose that $\sum_{n=1}^{\infty} z_n = A$ and $\sum_{n=1}^{\infty} w_n = B$. Prove that $\sum_{n=1}^{\infty} (z_n + w_n) = A + B$.

17. Let $\sum_{n=1}^{\infty} z_n = A$. Prove that $\sum_{n=1}^{\infty} az_n = aA$ for any complex number a.

16.4 COMPLEX POWER SERIES

We now turn to one of the fundamental tools of complex analysis—the power series. Let $a_0, a_1, a_2, \ldots$ be a given complex sequence and z_0 a complex number. The series

$$\sum_{n=0}^{\infty} a_n(z-z_0)^n$$

is called a *power series* with *center* z_0 and *coefficient sequence* $a_0, a_1, a_2, \ldots$.

Note that we start the series at $n = 0$ to allow a constant term. Thus,

$$\sum_{n=0}^{\infty} a_n(z-z_0)^n = a_0 + a_1(z-z_0) + a_2(z-z_0)^2 + a_3(z-z_0)^3 + \cdots$$

Given such a series, we first seek to determine all z such that the series converges. There are three possibilities.

1. Clearly, the series converges at $z = z_0$. Possibly this is the only number at which the series converges. (For example, $\sum_{n=0}^{\infty} n^n z^n$ converges only at $z = 0$. Here, $z_0 = 0$ and $a_n = n^n$.)
2. The second possibility is that $\sum_{n=0}^{\infty} a_n(z-z_0)^n$ converges for all z. This happens, for example, with $\sum_{n=0}^{\infty} z^n/n!$ (here, $z_0 = 0$, $a_n = 1/n!$).
3. The first two possibilities represent extremes. The third possibility is that the series converges for some $z \neq z_0$ but does not converge for all z. For example, from Examples 452 and 453, $\sum_{n=0}^{\infty} z^n$ converges if $|z| < 1$ and diverges if $|z| \geq 1$.

The following theorem gives a convenient test to determine in specific cases which of the above possibilities holds, and it also says a great deal more about the third possibility.

THEOREM 132 Convergence of Power Series

Given $\sum_{n=0}^{\infty} a_n(z-z_0)^n$, suppose that

$$R = \lim_{n\to\infty} \left| \frac{a_{n+1}}{a_n} \right|$$

exists finite or is equal to $+\infty$.

1. If $R = \infty$, then $\sum_{n=0}^{\infty} a_n(z-z_0)^n$ converges only at z_0 (the first possibility considered above).

2. If $R = 0$, then $\sum_{n=0}^{\infty} a_n(z - z_0)^n$ converges absolutely for all z. Thus, by Theorem 131, the series also converges for all z (this is the second of the above possibilities).
3. If $0 < R < \infty$, then $\sum_{n=0}^{\infty} a_n(z - z_0)^n$ converges absolutely (and hence also converges) for all z with $|z - z_0| < 1/R$, and it diverges for all z with $|z - z_0| > 1/R$. If $|z - z_0| = 1/R$, no general conclusion can be drawn about convergence of $\sum_{n=0}^{\infty} a_n(z - z_0)^n$.

Proof Apply the ratio test to the real series

$$\sum_{n=0}^{\infty} |a_n(z - z_0)^n| \qquad (z \neq z_0).$$

Thus, consider

$$\lim_{n\to\infty} \left| \frac{a_{n+1}(z - z_0)^{n+1}}{a_n(z - z_0)^n} \right| = \lim_{n\to\infty} \left| \frac{a_{n+1}}{a_n}(z - z_0) \right| = |z - z_0| \lim_{n\to\infty} \left| \frac{a_{n+1}}{a_n} \right|.$$

Now look at the possibilities for R as considered in the three conclusions of the theorem.

1. If $\lim_{n\to\infty} |a_{n+1}/a_n| = \infty$, then $\lim_{n\to\infty} |a_{n+1}(z - z_0)^{n+1}/a_n(z - z_0)^n| = \infty$. Then, $\lim_{n\to\infty} |a_n(z - z_0)^n| \neq 0$, and hence $\lim_{n\to\infty} a_n(z - z_0)^n \neq 0$. By Theorem 129, $\sum_{n=0}^{\infty} a_n(z - z_0)^n$ diverges for all $z \neq z_0$.
2. Suppose now that $R = 0$. Then, $\lim_{n\to\infty} |a_{n+1}(z - z_0)^{n+1}/a_n(z - z_0)^n| = |z - z_0|R = 0 < 1$. Thus, by the ratio test,

$$\sum_{n=0}^{\infty} |a_n(z - z_0)^n|$$

converges for all $z \neq z_0$. By Theorem 131, $\sum_{n=0}^{\infty} a_n(z - z_0)^n$ also converges for all $z \neq z_0$. Clearly, the series also converges at z_0, proving (2).
3. Now suppose that $0 < R < \infty$. By the ratio test,

$$\sum_{n=0}^{\infty} |a_n(z - z_0)^n| \begin{cases} \text{converges if } \lim_{n\to\infty} \left| \dfrac{a_{n+1}(z - z_0)^{n+1}}{a_n(z - z_0)^n} \right| < 1, \\ \text{diverges if } \lim_{n\to\infty} \left| \dfrac{a_{n+1}(z - z_0)^{n+1}}{a_n(z - z_0)^n} \right| > 1. \end{cases}$$

Thus,

$$\sum_{n=0}^{\infty} |a_n(z - z_0)^n| \begin{cases} \text{converges if } |z - z_0|R < 1, \\ \text{diverges if } |z - z_0|R > 1, \end{cases}$$

giving us the conclusion of (3). The ratio test fails if

$$\lim_{n\to\infty} \left| \frac{a_{n+1}(z - z_0)^{n+1}}{a_n(z - z_0)^n} \right| = 1,$$

that is, if $|z - z_0| = 1/R$, proving (3).

In practice, we let

$$r = \frac{1}{R} = \frac{1}{\lim_{n\to\infty} \left| \dfrac{a_{n+1}}{a_n} \right|}.$$

The number r is called the *radius of convergence* of the power series.

Paraphrasing Theorem 132, we have that the power series either converges only at z_0 ($r = 0$); or it converges for all z ($r = \infty$); or it converges for all z inside the circle of radius r about z_0, diverging for all z outside this circle.

The circle $|z - z_0| = r$ is called the *circle of convergence* (though the series converges for points *within* this circle, not necessarily for points *on* it). The disk $|z - z_0| < r$ is called the *disk of convergence*.

We can combine cases (2) and (3) by calling the entire plane the disk of convergence when $r = \infty$.

Theorem 132 does not apply to all power series, because it assumes that $\lim_{n\to\infty} |a_{n+1}/a_n|$ exists finite or is infinite, and this does not always happen. For example, let

$$a_n = \begin{cases} 1 & \text{if } n \text{ is odd,} \\ 0 & \text{if } n \text{ is even.} \end{cases}$$

Then we cannot write the ratio a_{n+1}/a_n. Thus, the theorem does not cover the series $z + z^3 + z^5 + z^7 + \cdots$. Similarly, consider the example

$$a_n = \begin{cases} 2 & \text{if } n \text{ is odd,} \\ 1 & \text{if } n \text{ is even.} \end{cases}$$

Now, a_{n+1}/a_n is defined for all n, but

$$\frac{a_{n+1}}{a_n} = \begin{cases} 2 & \text{if } n \text{ is even,} \\ \frac{1}{2} & \text{if } n \text{ is odd,} \end{cases}$$

and this ratio has no limit as $n \to \infty$. Thus, the theorem does not treat the power series $1 + 2z + z^2 + 2z^3 + z^4 + 2z^5 + \cdots$.

It is true in general that, for any power series $\sum_{n=0}^{\infty} a_n(z - z_0)^n$, there is some R, $0 \le R \le \infty$, such that the conclusion of Theorem 132 holds; then $1/R$ is the radius of convergence of the series. Theorem 132 gives a formula by which this radius of convergence can be computed in a special case. In the general case, while the radius of convergence certainly exists, we may have no way of explicitly computing it.

EXAMPLE 455

For the power series $\sum_{n=0}^{\infty} (n^n/n!)(z - i)^n$, compute

$$R = \lim_{n\to\infty} \left| \frac{a_{n+1}}{a_n} \right| = \lim_{n\to\infty} \left| \frac{(n+1)^{n+1}}{(n+1)!} \frac{n!}{n^n} \right| = \lim_{n\to\infty} \left(1 + \frac{1}{n}\right)^n = e.$$

Then $r = 1/e$ is the radius of convergence. The circle of convergence is $|z - i| = 1/e$, and the disk of convergence consists of all z with $|z - i| < 1/e$ (see Figure 304). The series converges if $|z - i| < 1/e$ and diverges if $|z - i| > 1/e$. Each point on the circle of convergence $|z - i| = 1/e$ would have to be treated separately.

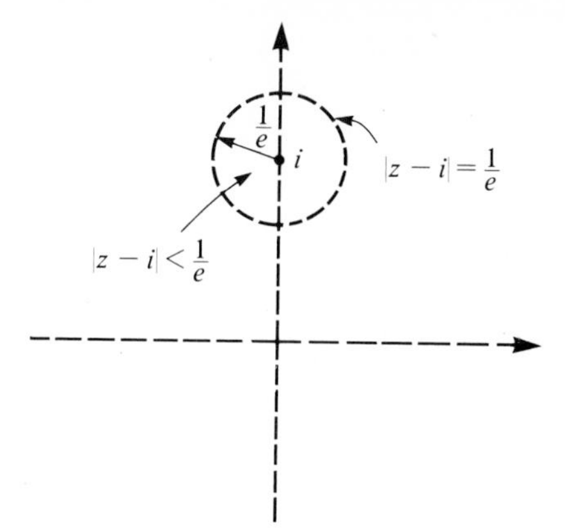

Figure 304. Circle and disk of convergence of $\sum_{n=0}^{\infty} (n^n/n!)(z - i)^n$.

Suppose now that we have a power series $\sum_{n=0}^{\infty} a_n(z - z_0)^n$ with radius of convergence $r > 0$ (we allow $r = \infty$). For each z with $|z - z_0| < r$, $\sum_{n=0}^{\infty} a_n(z - z_0)^n$ converges to some complex number, say, $f(z)$. What can we say about the resulting function $f(z)$?

THEOREM 133

Let $\sum_{n=0}^{\infty} a_n(z - z_0)^n$ have radius of convergence r, with $0 < r \leq \infty$. For $|z - z_0| < r$, let $f(z) = \sum_{n=0}^{\infty} a_n(z - z_0)^n$. Then,

1. $f(z)$ is analytic on $|z - z_0| < r$.
2. For $|z - z_0| < r$, $f'(z) = \sum_{n=1}^{\infty} na_n(z - z_0)^{n-1}$, and this power series also has radius of convergence r.
3. For $k = 2, 3, \ldots$, and $|z - z_0| < r$,

$$f^{(k)}(z) = \sum_{n=k}^{\infty} n(n - 1) \cdots (n - k + 1)a_n(z - z_0)^{n-k}.$$

In words, a power series represents an analytic function inside its circle of convergence, and the kth derivative is the power series obtained by differentiating the original series k times term by term.

Proof First, note that

$$\lim_{n\to\infty} \left| \frac{(n + 1)a_{n+1}}{na_n} \right| = \lim_{n\to\infty} \left(\frac{n + 1}{n}\right) \left| \frac{a_{n+1}}{a_n} \right| = \lim_{n\to\infty} \left| \frac{a_{n+1}}{a_n} \right|,$$

since $\lim_{n\to\infty} (n + 1)/n = 1$. Thus, the series $\sum_{n=0}^{\infty} a_n(z - z_0)^n$ and $\sum_{n=1}^{\infty} na_n(z - z_0)^{n-1}$ have the same radius of convergence, assuming that $\lim_{n\to\infty} |a_{n+1}/a_n|$ exists finite or equals infinity. If this limit does not exist, one can show by a deeper argument that the two series still have the same radius of convergence (see Problem 16).

Now set $g(z) = \sum_{n=1}^{\infty} na_n(z - z_0)^{n-1}$. We want to show that $f'(z) = g(z)$ if $|z - z_0| < r$. To make the details easier to follow, we shall write down a proof when $z_0 = 0$. Thus, $f(z) = \sum_{n=0}^{\infty} a_n z^n$ and $g(z) = \sum_{n=1}^{\infty} na_n z^{n-1}$. We have

$$\left| \frac{f(z + \Delta z) - f(z)}{\Delta z} - g(z) \right| = \left| \sum_{n=0}^{\infty} a_n \left(\frac{(z + \Delta z)^n - z^n}{\Delta z} - nz^{n-1} \right) \right|.$$

Now choose Δz so small that $|z| + |\Delta z| \leq \varepsilon < r$ (see Figure 305). By the binomial theorem, with $|z| = \rho$,

$$\left|\frac{(z+\Delta z)^n - z^n}{\Delta z} - nz^{n-1}\right|$$

$$= \left|\left[\frac{1}{\Delta z}\sum_{j=0}^{n}\binom{n}{j}z^{n-j}(\Delta z)^j\right] - \frac{z^n}{\Delta z} - nz^{n-1}\right|$$

$$= \left|\sum_{j=2}^{n}\binom{n}{j}z^{n-j}(\Delta z)^{j-1}\right|$$

$$\leq \sum_{j=2}^{n}\binom{n}{j}|z^{n-j}|\,|\Delta z|^{j-1}$$

$$= \sum_{j=2}^{n}\binom{n}{j}\rho^{n-j}|\Delta z|^{j-1}$$

$$= \frac{(\rho + |\Delta z|)^n - \rho^n}{|\Delta z|} - n\rho^{n-1}.$$

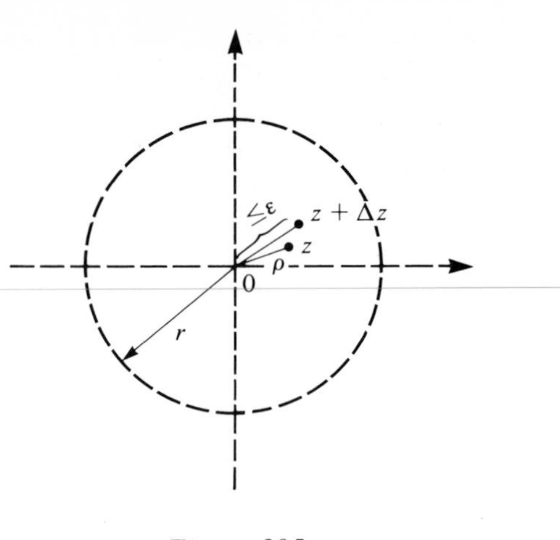

Figure 305

Thus,

$$\left|\frac{f(z+\Delta z) - f(z)}{\Delta z} - g(z)\right| \leq \sum_{n=0}^{\infty}|a_n|\left(\frac{(\rho + |\Delta z|)^n - \rho^n}{|\Delta z|} - n\rho^{n-1}\right)$$

$$= \sum_{n=0}^{\infty}|a_n|\varepsilon^n\left[\frac{\left(\dfrac{\rho + |\Delta z|}{\varepsilon}\right)^n - \left(\dfrac{\rho}{\varepsilon}\right)^n}{|\Delta z|} - \frac{n}{\rho}\left(\frac{\rho}{\varepsilon}\right)^n\right].$$

Since $\lim_{n\to\infty} a_n(\Delta z)^n = 0$, then, for some M, $|a_n(\Delta z)^n| < M$ for $n = 0, 1, 2, 3, \ldots$. Next, recall from Examples 452 and 453 that

$$\sum_{n=0}^{\infty} t^n = \frac{1}{1-t} \quad \text{and} \quad \sum_{n=1}^{\infty} nt^{n-1} = \frac{1}{(1-t)^2}$$

if $|t| < 1$. Then, if $\rho \leq \varepsilon$ and $|\Delta z| < \varepsilon$, we have

$$\left|\frac{f(z+\Delta z) - f(z)}{\Delta z} - g(z)\right| \leq M\left[\frac{1}{|\Delta z|}\left(\frac{r}{\varepsilon - \rho + |\Delta z|} - \frac{r}{\varepsilon - \rho}\right) - \frac{\varepsilon}{(\varepsilon - \rho)^2}\right]$$

$$= \frac{M|\Delta z|}{(\varepsilon - \rho)^2(\varepsilon - \rho - |\Delta z|)}.$$

As $\Delta z \to 0$, the quantity on the right goes to zero; hence,

$$\lim_{\Delta z \to 0}\frac{f(z+\Delta z) - f(z)}{\Delta z} = g(z),$$

as was to be proved.

Now use the fact that $\sum_{n=1}^{\infty} na_n z^{n-1}$ is again a power series with radius of convergence r to conclude that

$$f''(z) = g'(z) = \sum_{n=2}^{\infty} n(n-1)a_n z^{n-2},$$

and so on.

EXAMPLE 456

Look at Examples 452 and 453 again. We had

$$\sum_{n=0}^{\infty} z^n = \frac{1}{1-z} \quad \text{for} \quad |z| < 1.$$

By Theorem 133,

$$\sum_{n=1}^{\infty} nz^{n-1} = \frac{d}{dz}\left(\frac{1}{1-z}\right) = \frac{1}{(1-z)^2} \quad \text{for} \quad |z| < 1.$$

We derived this result in Example 453. Similarly,

$$\sum_{n=2}^{\infty} n(n-1)z^{n-2} = \frac{d}{dz}(1-z)^{-2} = \frac{2}{(1-z)^3},$$

$$\sum_{n=3}^{\infty} n(n-1)(n-2)z^{n-3} = \frac{d}{dz}[2(1-z)^{-3}] = \frac{6}{(1-z)^4},$$

and so on, provided $|z| < 1$.

The next theorem tells us that we can also integrate a power series term by term within its circle of convergence.

THEOREM 134

Let $f(z) = \sum_{n=0}^{\infty} a_n(z - z_0)^n$. If C is a simple path lying entirely within the disk of convergence, then

$$\int_C f(z)\, dz = \sum_{n=0}^{\infty} a_n \int_C (z - z_0)^n\, dz.$$

Proof First, there is some $\delta > 0$ such that the closed disk $|z - z_0| \leq \delta$ contains C and lies entirely within the circle of convergence (see Figure 306). For any positive

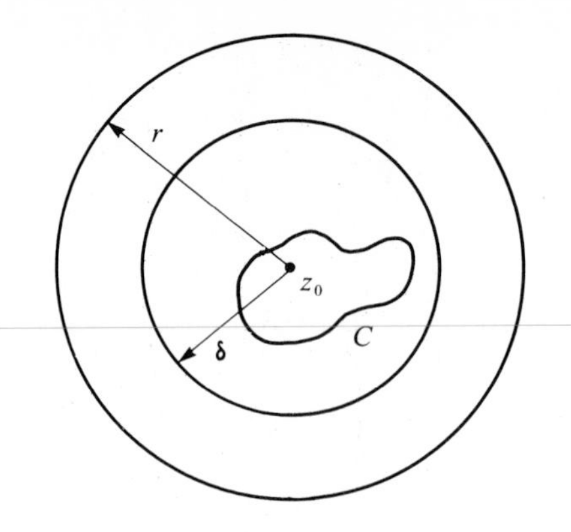

Figure 306

integer m,

$$\left| \int_C f(z)\, dz - \sum_{n=0}^{m} a_n \int_C (z - z_0)^n\, dz \right|$$

$$= \left| \int_C \sum_{n=0}^{\infty} a_n(z - z_0)^n\, dz - \int_C \sum_{n=0}^{m} a_n(z - z_0)^n\, dz \right|$$

$$= \left| \int_C \sum_{n=m+1}^{\infty} a_n(z - z_0)^n\, dz \right|$$

$$\leq (\text{length of } C) \cdot \left(\text{max. value of} \cdot \sum_{n=m+1}^{\infty} |a_n(z - z_0)^n| \text{ for } z \text{ on } C\right),$$

by Theorem 120. Now, choose any w with $|w - z_0| = \delta$. Since w is inside the circle of convergence, then $\sum_{n=0}^{\infty} |a_n(w - z_0)^n|$ converges. Hence, $\lim_{m\to\infty} \sum_{n=m+1}^{\infty} |a_n(w - z_0)^n| = 0$. Further, for each z on C, $|z - z_0| < |w - z_0|$. Then,

$$\left(\text{max. value of } \sum_{n=m+1}^{\infty} |a_n(z - z_0)^n|\right) \leq \sum_{n=m+1}^{\infty} |a_n(w - z_0)^n| \to 0$$

as $m \to \infty$, proving the theorem.

If C goes from w_0 to w_1, then

$$\int_C a_n(z - z_0)^n\, dz = \frac{a_n}{n+1}\,[(w_1 - z_0)^{n+1} - (w_0 - z_0)^{n+1}].$$

Thus, Theorem 134 may be rephrased:

$$\int_C \sum_{n=0}^{\infty} a_n(z - z_0)^n\, dz = \sum_{n=0}^{\infty} \frac{a_n}{n+1}\,[(w_1 - z_0)^{n+1} - (w_0 - z_0)^{n+1}].$$

EXAMPLE 457

Let C be any simple path from $\frac{1}{2}i$ to $-\frac{1}{2}$, lying inside the circle $|z| = 1$.
Let

$$f(z) = \sum_{n=0}^{\infty} \frac{z^n}{2n+1}.$$

It is easy to check that this series has radius of convergence 1; hence,

$$\int_C f(z)\,dz = \sum_{n=0}^{\infty} \int_{i/2}^{-1/2} \frac{z^n}{2n+1}\,dz$$

$$= \sum_{n=0}^{\infty} \frac{1}{(2n+1)(n+1)} \left[\left(-\frac{1}{2}\right)^{n+1} - \left(\frac{i}{2}\right)^{n+1}\right].$$

The sum of such a series is not obvious, and the result must be left in this infinite series form.

PROBLEMS FOR SECTION 16.4

In each of Problems 1 through 10, determine the center of the power series and the radius of convergence r. If $r > 0$, use Theorem 133 to find power series for $f'(z), f''(z)$, and $f'''(z)$, where $f(z)$ is the given power series inside the circle of convergence.

1. $\sum_{n=0}^{\infty} \frac{(n+1)}{2^n}(z+3i)^n$
2. $\sum_{n=0}^{\infty} \frac{(-1)^n(z-i)^{2n}}{(2n+1)!}$
3. $\sum_{n=0}^{\infty} \frac{n^n}{(n+1)^n}(z-1+2i)^n$
4. $\sum_{n=0}^{\infty} \left(\frac{2}{3i}\right)^n(z+1+4i)^n$
5. $\sum_{n=0}^{\infty} \left(\frac{i^n}{2^{n+1}}\right)(z+4-i)^n$
6. $\sum_{n=0}^{\infty} \frac{(1-i)^n}{n+2}(z-3i)^n$
7. $\sum_{n=0}^{\infty} \frac{n^2}{(2n+1)^2}(z+6+2i)^n$
8. $\sum_{n=0}^{\infty} \left(\frac{n^3}{4^n}\right)(z-3i)^{2n}$
9. $\sum_{n=0}^{\infty} \left(\frac{e^{in}}{2n+1}\right)(z+4)^n$
10. $\sum_{n=0}^{\infty} \left(\frac{1-i}{2+i}\right)^n(z-3i)^{4n}$
11. Suppose that $\sum_{n=0}^{\infty} a_n(z-z_0)^n$ has radius of convergence r_1 and that $\sum_{n=0}^{\infty} b_n(z-z_0)^n$ has radius of convergence r_2, where $r_1 > 0$ and $r_2 > 0$. What is the radius of convergence of $\sum_{n=0}^{\infty} (a_n + b_n)(z-z_0)^n$?
12. Let $\sum_{n=0}^{\infty} a_n(z-z_0)^n$ have radius of convergence r, where $0 < r < \infty$. Let $f(z) = \sum_{n=0}^{\infty} a_n(z-z_0)^n$ for $|z-z_0| < r$. Prove that r is the distance from z_0 to the nearest point at which $f(z)$ fails to be analytic.
13. Suppose that $\lim_{n\to\infty} |a_n|^{1/n}$ exists finite or equals ∞. Let $R = \lim_{n\to\infty} |a_n|^{1/n}$. Prove that:
    (a) $\sum_{n=0}^{\infty} a_n(z-z_0)^n$ converges only for $z = z_0$ if $R = \infty$.
    (b) $\sum_{n=0}^{\infty} a_n(z-z_0)^n$ converges for all z if $R = 0$.
    (c) If $0 < R < \infty$, then $\sum_{n=0}^{\infty} a_n(z-z_0)^n$ converges for $|z-z_0| < 1/R$ and diverges for $|z-z_0| > 1/R$.
14. Suppose that $\sum_{n=0}^{\infty} a_n(z-z_0)^n = \sum_{n=0}^{\infty} b_n(z-z_0)^n$ for all z with $|z-z_0| < r$, where $0 < r \le \infty$. Prove that $a_n = b_n$ for $n = 0, 1, 2, 3, \ldots$.

15. Let $f(z) = \sum_{n=0}^{\infty} a_n(z - z_0)^n$ and $g(z) = \sum_{n=0}^{\infty} b_n(z - z_0)^n$ converge in $|z - z_0| < r$, where $0 < r \leq \infty$. The *Cauchy product* of these series is defined to be

$$\sum_{n=0}^{\infty} c_n(z - z_0)^n,$$

where

$$c_n = \sum_{j=0}^{n} a_j b_{n-j}.$$

(Thus, $c_0 = a_0 b_0, c_1 = a_0 b_1 + a_1 b_0, c_2 = a_0 b_2 + a_1 b_1 + a_2 b_0$, and so on.)

It can be proved that the Cauchy product series converges to $f(z)g(z)$ for $|z - z_0| < r$. Assuming this, prove the following:

(a) $$\left(\frac{1}{1-z}\right)^2 = \sum_{n=0}^{\infty} (n+1)z^n \quad \text{for} \quad |z| < 1.$$

Use the fact that $1/(1 - z) = \sum_{n=0}^{\infty} z^n$ for $|z| < 1$.

(b) Let $f(z) = \sum_{n=0}^{\infty} z^n/n!$ for all z. Prove that $[f(z)]^2 = f(2z)$.

(c) Let $f(z) = \sum_{n=0}^{\infty} \frac{(-1)^n z^{2n}}{(2n)!}$ and $g(z) = \sum_{n=0}^{\infty} \frac{(-1)^n z^{2n+1}}{(2n+1)!}$ for all z.

Prove that $f(z)g(z) = \frac{1}{2}g(2z)$.

16. Prove that $\sum_{n=0}^{\infty} a_n(z - z_0)^n$ and $\sum_{n=1}^{\infty} na_n(z - z_0)^{n-1}$ have the same radius of convergence (without assuming that $\lim_{n\to\infty} |a_{n+1}/a_n|$ exists, finite or infinite).

Hint: Let $\sum_{n=0}^{\infty} a_n(z - z_0)^n$ have radius of convergence $r > 0$. Let $\sum_{n=0}^{\infty} a_n(z - z_0)^n$ have radius of convergence r'.

First, show that $r \leq r'$ by showing that $\sum_{n=1}^{\infty} na_n(z - z_0)^{n-1}$ converges whenever $\sum_{n=0}^{\infty} a_n(z - z_0)^n$ does. Argue that, if $\sum_{n=0}^{\infty} a_n(w - z_0)^n$ converges, then $|a_n(w - z_0)^n| \leq M$ for some M and n sufficiently large. Write

$$|na_n(z - z_0)^{n-1}| \leq \frac{Mn}{|w - z_0|}\left|\frac{z - z_0}{w - z_0}\right|^{n-1}$$

and use a comparison test to show that $\sum_{n=1}^{\infty} na_n(w - z_0)^{n-1}$ converges.

Next, assume that $r < r'$ and derive a contradiction.

17. Give an example of a power series $\sum_{n=0}^{\infty} a_n z^n$ with a finite, positive radius of convergence, but such that $\lim_{n\to\infty} |a_{n+1}/a_n|$ does not exist. Your example must have all the a_n's nonzero.

The following problems address series $\sum_{n=0}^{\infty} f_n(z)$ without the restriction that $f_n(z) = a_n(z - z_0)^n$. We say that $\sum_{n=0}^{\infty} f_n(z)$ converges to a function $f(z)$ on a region D if the series converges to $f(z)$ for each z in D. Such convergence is sometimes called *pointwise convergence*. The series converges *uniformly* to $f(z)$ on D if, given any $\varepsilon > 0$, there is some positive N such that

$$\left| f(z) - \sum_{j=0}^{n} f_j(z) \right| < \varepsilon$$

whenever $n \geq N$ and for all z in D. The following problems develop this concept.

18. Let $\sum_{n=0}^{\infty} f_n(z)$ converge uniformly to $f(z)$ on D. Prove that $\sum_{n=0}^{\infty} f_n(z)$ converges pointwise to $f(z)$ on D.

19. Let each $f_n(z)$ be continuous on a domain D. Let $\sum_{n=0}^{\infty} f_n(z)$ converge uniformly to $f(z)$ on D. Prove that $f(z)$ is continuous on D. (*Hint:* The proof is similar to that in the case of real-valued functions of a real variable and can be found in most advanced calculus books.)

20. Let each $f_n(z)$ be continuous on a domain D, and let $\sum_{n=0}^{\infty} f_n(z)$ converge uniformly to $f(z)$ on D. Prove that

$$\int_C f(z)\,dz = \sum_{n=0}^{\infty} \int_C f_n(z)\,dz$$

for any piecewise-smooth curve C in D.

21. Let each $f_n(z)$ be analytic in a simply connected domain D. Let $\sum_{n=0}^{\infty} f_n(z)$ converge uniformly to $f(z)$ on D. Show that $f(z)$ is also analytic on D. [*Hint:* Use the results of Problem 20 and Morera's theorem (Problem 18, Section 15.3).]

16.5 COMPLEX TAYLOR SERIES

Suppose that $\sum_{n=0}^{\infty} a_n(z-z_0)^n$ converges in $|z-z_0| < r$ (we allow $r = \infty$), say, to $f(z)$. Observe that

$$f(z_0) = a_0.$$

Next, $f'(z) = \sum_{n=1}^{\infty} na_n(z-z_0)^{n-1}$; so

$$f(z_0) = a_1.$$

Similarly, $f''(z) = \sum_{n=2}^{\infty} n(n-1)a_n(z-z_0)^{n-2}$; so

$$f''(z_0) = 2a_2,$$

and so on. In general, we find that $f^{(n)}(z_0) = n!\,a_n$, or

$$a_n = \frac{f^{(n)}(z_0)}{n!}.$$

Thus,

$$f(z) = \sum_{n=0}^{\infty} \frac{f^{(n)}(z_0)}{n!}(z-z_0)^n.$$

The series on the right is the *Taylor series* for $f(z)$ about z_0, and the number $f^{(n)}(z_0)/n!$ is the nth *Taylor coefficient*. Thus:

Inside its circle of convergence, any power series is in fact the Taylor series of the function defined by it.

Conversely, we shall now prove that any analytic function can be expanded in a Taylor series.

THEOREM 135

Let $f(z)$ be analytic at z_0. Then $f(z)$ has a Taylor series expansion

$$f(z) = \sum_{n=0}^{\infty} \frac{f^{(n)}(z_0)}{n!}(z-z_0)^n$$

in some disk about z_0.

Proof First, there is a circle C about z_0 such that $f(z)$ as analytic at all points on and inside C. Let z be any point inside C and w any point on C (see Figure 307). By an algebraic manipulation, we can write

$$\frac{1}{w-z}=\left(\frac{1}{w-z_0}\right)\left[\frac{1}{1-\left(\dfrac{z-z_0}{w-z_0}\right)}\right].$$

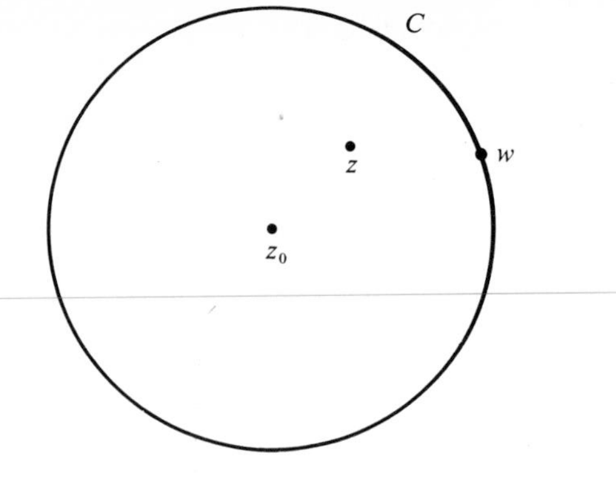

Figure 307

Now, $|(z-z_0)/(w-z_0)|<1$, since w is farther from z_0 than z is; so, by Example 452 (geometric series), we have

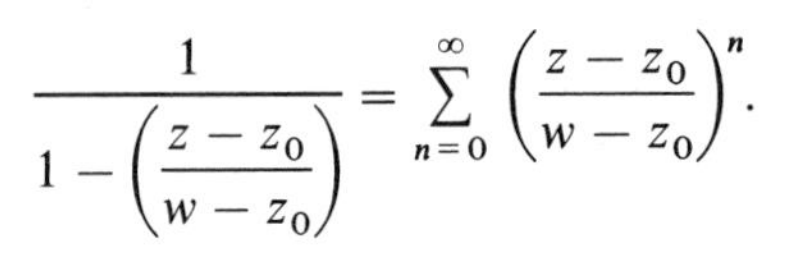

$$\frac{1}{1-\left(\dfrac{z-z_0}{w-z_0}\right)}=\sum_{n=0}^{\infty}\left(\frac{z-z_0}{w-z_0}\right)^n.$$

Thus,

$$\frac{1}{w-z}=\left(\frac{1}{w-z_0}\right)\sum_{n=0}^{\infty}\frac{(z-z_0)^n}{(w-z_0)^n}=\sum_{n=0}^{\infty}\frac{(z-z_0)^n}{(w-z_0)^{n+1}}.$$

By Cauchy's integral formula (Theorem 124), we have

$$f(z)=\frac{1}{2\pi i}\oint_C\frac{f(w)}{w-z}\,dw=\frac{1}{2\pi i}\oint_C\sum_{n=0}^{\infty}\frac{f(w)(z-z_0)^n}{(w-z_0)^{n+1}}\,dw.$$

One can show (see Problem 20, Section 16.4) that the last integral and summation can be interchanged. Thus, we have

$$f(z)=\sum_{n=0}^{\infty}\left(\frac{1}{2\pi i}\oint_C\frac{f(w)}{(w-z_0)^{n+1}}\,dw\right)(z-z_0)^n.$$

But, by Cauchy's integral formula for higher derivatives (Theorem 125), we have

$$\frac{1}{2\pi i}\oint_C\frac{f(w)}{(w-z_0)^{n+1}}\,dw=\frac{f^{(n)}(z_0)}{n!}.$$

Thus,

$$f(z)=\sum_{n=0}^{\infty}\frac{f^{(n)}(z_0)}{n!}(z-z_0)^n,$$

at least inside C.

In many instances it is important to be able to explicitly determine the Taylor series for a given $f(z)$ about a point z_0. This becomes a problem of computing $f^{(n)}(z_0)/n!$ for $n = 0, 1, 2, 3, \ldots$. In the case of many elementary functions, this can be done directly. In other cases one can often exploit known series to obtain the expansion. Here are some examples.

EXAMPLE 458

Let $f(z) = e^z$ and $z_0 = 0$. Since $f^{(n)}(z) = e^z$, then

$$\frac{f^{(n)}(0)}{n!} = \frac{1}{n!}.$$

Thus,

$$e^z = \sum_{n=0}^{\infty} \frac{z^n}{n!} \quad \text{for all } z.$$

Similarly, one can check that, for all z,

$$\sin(z) = \sum_{n=0}^{\infty} \frac{(-1)^n z^{2n+1}}{(2n+1)!}$$

and

$$\cos(z) = \sum_{n=0}^{\infty} \frac{(-1)^n z^{2n}}{(2n)!}.$$

Note that the above expansions would be the familiar real Taylor series of elementary calculus if we replaced z by x. The reverse of this is always true.

> If $f(x) = \sum_{n=0}^{\infty} a_n(x - x_0)^n$ is the real Taylor series for $f(x)$ about the real number x_0, and if $f(x)$ is analytic at x_0, then the complex Taylor series about x_0 is
>
> $$f(z) = \sum_{n=0}^{\infty} a_n(z - x_0)^n.$$
>
> In particular, if $f(x) = \sum_{n=0}^{\infty} a_n x^n$, then $f(z) = \sum_{n=0}^{\infty} a_n z^n$ in some disk about 0.

Often the geometric series can be used to obtain other expansions. Recall that $\sum_{n=0}^{\infty} z^n = 1/(1 - z)$ if $|z| < 1$. The next three examples show how to use this to obtain other series expansions.

EXAMPLE 459

Expand $1/(1 + z)$ about 0.

Here,

$$\frac{1}{1+z} = \frac{1}{1-(-z)} = \sum_{n=0}^{\infty} (-z)^n = \sum_{n=0}^{\infty} (-1)^n z^n \quad \text{if } |z| < 1.$$

EXAMPLE 460

Expand $1/(1+z^3)$ about 0.

Write, using Example 459,

$$\frac{1}{1+z^3}=\sum_{n=0}^{\infty}(-1)^n(z^3)^n=\sum_{n=0}^{\infty}(-1)^n z^{3n} \quad \text{for} \quad |z|<1.$$

EXAMPLE 461

Expand $1/(1+z)$ about $-2i$.

We want a series of the form $\sum_{n=0}^{\infty} a_n(z+2i)^n$. Some algebraic manipulation will help here. Write

$$\frac{1}{1+z}=\frac{1}{1+z+2i-2i}=\frac{1}{(1-2i)+(z+2i)}$$
$$=\frac{1}{(1-2i)}\left[\frac{1}{1+\left(\dfrac{z+2i}{1-2i}\right)}\right]$$
$$=\left(\frac{1}{1-2i}\right)\sum_{n=0}^{\infty}(-1)^n\left(\frac{z+2i}{1-2i}\right)^n$$
$$=\sum_{n=0}^{\infty}\frac{(-1)^n}{(1-2i)^{n+1}}(z+2i)^n,$$

provided that $|(z+2i)/(1-2i)|<1$ or, equivalently, $|z+2i|<\sqrt{5}$.

Sometimes a function satisfies a "simple" differential equation, and this can be used to compute the Taylor coefficients, as in the following.

EXAMPLE 462

Let $f(z)=e^{z^2}\int_0^z e^{-z^2}\,dz$. This defines a function analytic for all z, since $\int_C e^{-z^2}\,dz$ is independent of path in the entire plane. We want to expand $f(z)$ in a Taylor series about 0.

Note that

$$f'(z)=2ze^{z^2}\int_0^z e^{-z^2}\,dz+e^{z^2}(e^{-z^2})=2zf(z)+1.$$

Then,

$$f''(z)=2f(z)+2zf'(z),$$
$$f^{(3)}(z)=4f'(z)+2zf''(z),$$
$$f^{(4)}(z)=6f''(z)+2zf^{(3)}(z),$$
$$f^{(5)}(a)=8f^{(3)}(z)+2zf^{(4)}(z),$$

and, in general,

$$f^{(n)}(z)=(2n-2)f^{(n-2)}(z)+2zf^{(n-1)}(z)$$

for $n=2,3,4,\ldots$.

Obviously, $f(0) = 0$ and $f'(0) = 1$. Further, for $n = 2, 3, 4, \ldots,$

$$f^{(n)}(0) = (2n - 2)f^{(n-2)}(0).$$

This gives us $f^{(n)}(0)$ in terms of $f^{(n-2)}(0)$. In particular, $f(0) = 0$, $f^{(2)}(0) = 2f(0) = 0$, $f^{(4)}(0) = 6f^{(2)}(0) = 0$, and so on; so every even derivative of $f(z)$ at 0 is zero.

Now for the odd derivatives. If $n = 2k + 1$, then

$$\begin{aligned}
f^{(n)}(0) &= f^{(2k+1)}(0) = 4kf^{(2k-1)}(0)\\
&= (4k)(4k-4)f^{(2k-3)}(0)\\
&= (4k)(4k-4)(4k-8)f^{(2k-5)}(0)\\
&= (4k)(4k-4)(4k-8)(4k-12)f^{(2k-7)}(0)\\
&= (4k)(4k-4)(4k-8)\cdots[4k-4(k-1)]f^{[2k-(2k-1)]}(0)\\
&= 4^k(k)(k-1)\cdots(2)(1)f'(0)\\
&= 4^kk!\,f'(0) = 4^kk!
\end{aligned}$$

Thus,

$$a_{2k+1} = \frac{f^{(2k+1)}(0)}{(2k+1)!} = \frac{4^kk!}{(2k+1)!}.$$

The expansion is then

$$f(z) = \sum_{n=0}^{\infty} a_n z^n = \sum_{k=0}^{\infty} a_{2k+1}z^{2k+1} = \sum_{k=0}^{\infty} \frac{4^kk!}{(2k+1)!}z^{2k+1}.$$

The radius of convergence of this series cannot be found using our limit formula, since each $a_{2n} = 0$. However, from Problem 12, Section 16.4, the radius of convergence is infinite, since $f(z)$ is analytic in the entire plane.

Still another method for computing a power series is to multiply or divide known expansions. Here is an example.

EXAMPLE 463

Find the Taylor series for $\tan(z)$ about 0.

We know that

$$\tan(z) = \frac{\sin(z)}{\cos(z)},$$

and that

$$\sin(z) = z - \frac{z^3}{3!} + \frac{z^5}{5!} - \frac{z^7}{7!} + \cdots$$

and

$$\cos(z) = 1 - \frac{z^2}{2!} + \frac{z^4}{4!} - \frac{z^6}{6!} + \cdots.$$

Write $\tan(z) = \sum_{n=0}^{\infty} a_n z^n$. Then,

$$\tan(z)\cos(z) = \sin(z)$$

gives us

$$(a_0 + a_1 z + a_2 z^2 + \cdots)\left(1 - \frac{z^2}{2!} + \frac{z^4}{4!} - \frac{z^6}{6!} + \frac{z^8}{8!} - \cdots\right)$$
$$= z - \frac{z^3}{3!} + \frac{z^5}{5!} - \frac{z^7}{7!} + \frac{z^9}{9!} - \cdots.$$

Multiply a few terms, collecting the coefficient of each power of z, to get

$$a_0 + a_1 z + \left(a_2 - \frac{a_0}{2!}\right)z^2 + \left(a_3 - \frac{a_1}{2!}\right)z^3 + \left(\frac{a_0}{4!} - \frac{a_2}{2!} + a_4\right)z^4$$
$$+ \left(\frac{a_1}{4!} - \frac{a_3}{2!} + a_5\right)z^5 + \left(-\frac{a_0}{6!} + \frac{a_2}{4!} - \frac{a_4}{2!} + a_6\right)z^6 + \cdots$$
$$= z - \frac{z^3}{3!} + \frac{z^5}{5!} - \frac{z^7}{7!} + \cdots.$$

Comparing coefficients of like powers of z on both sides of the equation, we get

$$a_0 = 0$$
$$a_1 = 1$$
$$a_2 - \frac{a_0}{2} = 0$$
$$a_3 - \frac{a_1}{2} = -\frac{1}{6}$$
$$\frac{a_0}{24} - \frac{a_2}{2} + a_4 = 0$$
$$\frac{a_1}{24} - \frac{a_3}{2} + a_5 = \frac{1}{120}$$
$$-\frac{a_0}{720} + \frac{a_2}{24} - \frac{a_4}{2} + a_6 = 0$$

and so on.

Then $a_0 = 0$, $a_1 = 1$, $a_2 = 0$, $a_3 = \frac{1}{3}$, $a_4 = 0$, $a_5 = \frac{2}{15}$, $a_6 = 0$, and so on. Thus,

$$\tan(z) = z + \tfrac{1}{3}z^3 + \tfrac{2}{15}z^5 + \cdots.$$

This method does not give the entire expansion, but we can obtain as many terms as we have time and patience for. The method can also be easily programmed to produce a large number of the coefficients very quickly.

One can also do this example by long division, dividing the series for $\cos(z)$ into that for $\sin(z)$ to get $\tan(z)$.

We conclude this section with three observations:

1. The Taylor expansion of $f(z)$ about z_0 is unique. That is, if $f(z) = \sum_{n=0}^{\infty} a_n(z - z_0)^n$, then necessarily $a_n = f^{(n)}(0)/n!$. This follows from Theorem 133(3) upon substituting $z = z_0$ to get

$$f^{(k)}(z_0) = k!\, a_k.$$

The importance of this is that no matter how the a_n's in $\sum_{n=0}^{\infty} a_n(z - z_0)^n$ are derived, we know that the resulting power series is the Taylor series.

This justifies the methods we have seen in the worked examples. It also justifies the substitution method. For example, since

$$e^z = \sum_{n=0}^{\infty} \frac{1}{n!} z^n,$$

then $e^{z^2} = \sum_{n=0}^{\infty} (1/n!)z^{2n}$ is the Taylor series for e^{z^2} about 0.

2. For historical reasons, a Taylor series about the origin is often called a *Maclaurin series*. The series of Examples 458 and 459 are Maclaurin series; that of Example 461 is not.
3. One can often tell the radius of convergence of the Taylor series for $f(z)$ about z_0 just by looking at $f(z)$—it is the distance from z_0 to the nearest point at which $f(z)$ fails to be analytic. This is illustrated by

$$\frac{1}{1 - z} = \sum_{n=0}^{\infty} z^n,$$

with radius of convergence 1 (the distance from 0 to 1, where $f(z)$ fails to be analytic). Another illustration is

$$\frac{1}{1 + z} = \sum_{n=0}^{\infty} \frac{(-1)^n}{(1 - 2i)^{n+1}} (z + 2i)^n,$$

an expansion about $-2i$ (see Example 461). The radius of convergence is $\sqrt{5}$, exactly the distance from $-2i$ to the nearest point (in this case the only point), -1, where $f(z)$ fails to be analytic. One can envision this circumstance by thinking of the disk of convergence as expanding from z_0. If there is nothing to prevent it, the disk will expand until it takes up the entire plane, and the series converges for all z. If, however, the disk encounters a point where $f(z)$ fails to be analytic, then this point represents a barrier beyond which the disk cannot expand, for if it went beyond, then the disk of convergence would contain a point where $f(z)$ fails to be analytic, contradicting the fact that power series are analytic inside their disks of convergence.

In the next section we consider the Laurent expansion, which is concerned with points at which $f(z)$ fails to be analytic. The Laurent expansion will form the basis for the study of singularities, the residue theorem, and some important applications of complex integration, such as the summation of series and evaluation of real integrals.

PROBLEMS FOR SECTION 16.5

In each of Problems 1 through 25, derive the Taylor series for $f(z)$ about the given point. Determine in each case the radius of convergence of the resulting series.

1. $\cos(2z)$, 0 **2.** e^z, $-3i$ **3.** $\sin(z^2)/2$, 0

4. $[1 - \cos(2z)]/z$, 0 **5.** $\dfrac{1}{1-z}$, $4i$ **6.** $\dfrac{1}{2+z}$, $1 - 8i$

7. $\dfrac{3i}{1-z}$, 5 **8.** $\dfrac{1}{(1-z)^2}$, 0 **9.** $1 + \dfrac{1}{2+z^2}$, i

10. $e^z - \sin(z)$, 0 **11.** $\sinh(3z)$, 0 **12.** $z^2 - 3z + i$, $2 - i$

13. $\dfrac{3}{z-4i}$, -5

14. $\arctan(z)$, 0 [*Hint:* First look at the derivative of $\arctan(z)$.]

15. $\cosh(z^3)$, 0 **16.** $\dfrac{2+i}{z-1+3i}$, 0 **17.** $\cos(z^2) - \sin(z)$, 0

18. $\dfrac{1}{z-2-4i}$, $-2i$ **19.** $(z - 9i)^4$, $1 + i$ **20.** $\dfrac{\cos(z^2) - 1}{z}$, 0

21. $\cosh(z + 1)$, 0 **22.** $\dfrac{1}{z^2 + 4i}$, $-2 + 5i$ **23.** e^{3-z}, i

24. $\sin(z + i)$, $-i$ **25.** $\cosh(3z) - \sinh(3z)$, 0

26. Suppose that $f(z)$ satisfies $f''(z) = 2f(z) + 1$, $f'(0) = i$, and $f(0) = 1$. Find the Taylor series for $f(z)$ about 0.

27. The error function erf(z) is defined by

$$\operatorname{erf}(z) = \frac{2}{\sqrt{\pi}} \int_0^z e^{-w^2}\, dw.$$

It arises in probability and statistics. Derive a Taylor series for erf(z) about 0.

28. Find the first five terms in the Maclaurin series for $\sin^2(z)$

(a) By using the formula $a_n = f^{(n)}(0)/n!$, and then

(b) By multiplying the Maclaurin series for $\sin(z)$ by itself and determining the coefficients of $1, z, z^2, z^3$, and z^4 (as in Example 463).

29. Use the method of Example 463 to find the first six terms in the Maclaurin series for $\sec(z)$.

30. Use the method of Example 463 to find the first six terms in the Maclaurin series for $\tanh(z)$.

31. The *Bernoulli numbers* B_n are defined as the coefficients of $z^n/n!$ in the Maclaurin expansion of $f(z) = z/(e^z - 1)$ for $z \neq 0$ and $f(0) = 1$. Thus,

$$f(z) = B_0 + B_1 z + \frac{B_2}{2!} z^2 + \frac{B_3}{3!} z^3 + \frac{B_4}{4!} z^4 + \cdots.$$

(a) Find B_1, B_2, B_3, B_4, and B_5.

(b) Prove that

$$\binom{n+1}{0} B_0 + \binom{n+1}{1} B_1 + \cdots + \binom{n+1}{n} B_n = 0 \quad \text{for} \quad n \geq 1.$$

32. Determine the Maclaurin series for $1/(1 - z^2)$

(a) Directly from the geometric series, and

(b) By multiplying the geometric series for $1/(1 + z)$ and $1/(1 - z)$.

33. Find the Maclaurin series for $\operatorname{Ln}(1 + z)$ by integrating the geometric series for $1/(1 + z)$.

34. In the Maclaurin expansion $\sum_{n=0}^{\infty} a_n z^n$ of $1/(1 - z - z^2)$, prove that $a_0 = a_1 = 1$ and, for $n \geq 2$,

$$a_n = a_{n-1} + a_{n-2}.$$

The a_n's are called *Fibonacci numbers*, and there is an extensive literature on them, as well as a journal devoted to them. One can show that

$$a_n = \frac{1}{\sqrt{5}}\left[\left(\frac{1+\sqrt{5}}{2}\right)^n - \left(\frac{1-\sqrt{5}}{2}\right)^n\right].$$

35. Let the Maclaurin series for sec(z) be $\sum_{n=0}^{\infty} a_n z^n$. Since $\sec(-z) = \sec(z)$, it is easy to see that $a_1 = a_3 = a_5 = \cdots = 0$; thus, $\sec(z) = \sum_{n=0}^{\infty} a_{2n} z^{2n}$.

The *Euler numbers* $E_0, E_2, E_4, \ldots$ are defined by

$$E_{2n} = (-1)^n (2n)!\, a_{2n}.$$

(a) Solve explicitly for E_0, E_2, E_4, E_6, and E_8.
(b) Prove that

$$\frac{E_0}{(2n)!} + \frac{E_2}{(2n-2)!\,2!} + \frac{E_4}{(2n-4)!\,4!} + \cdots + \frac{E_{2n}}{(2n)!} = 0, \quad \text{for} \quad n = 1, 2, 3, \ldots.$$

36. The *Legendre polynomials* $P_n(z)$ are defined by

$$\frac{1}{\sqrt{1 - 2zw + w^2}} = \sum_{n=0}^{\infty} P_n(z) w^n.$$

Determine $P_0(z)$, $P_1(z)$, $P_2(z)$, $P_3(z)$, and $P_4(z)$. Check your results from Section 6.3.

16.6 LAURENT SERIES

If $f(z)$ is analytic at z_0, then we can expand $f(z)$ in a Taylor series about z_0. If $f(z)$ is not analytic at z_0, we may still be able to expand in a series if we allow negative powers of $z - z_0$. This is the idea behind the Laurent series.

THEOREM 136

Let $f(z)$ be analytic in the annulus $r_1 < |z - z_0| < r_2$. For z in this annulus, we can expand

$$f(z) = \sum_{n=-\infty}^{\infty} a_n (z - z_0)^n,$$

where

$$a_n = \frac{1}{2\pi i} \oint_C \frac{f(w)\, dw}{(w - z_0)^{n+1}} \quad \text{for} \quad n = 0, \pm 1, \pm 2, \ldots$$

and C is a circle $|z - z_0| = \rho$, with $r_1 < \rho < r_2$. [The series thus obtained is the *Laurent expansion of* $f(z)$ in the annulus $r_1 < |z - z_0| < r_2$.]

Proof Let z be in the annulus; so $r_1 < |z - z_0| < r_2$. Choose numbers R_1 and R_2 so that $r_1 < R_1 < |z - z_0| < R_2 < r_2$ (see Figure 308). Let C_2 be the circle $|z - z_0| = R_2$ and C_1 the circle $|z - z_0| = R_1$. By the generalized form of the Cauchy integral formula (see Part F of Section 3, Chapter 15), we have

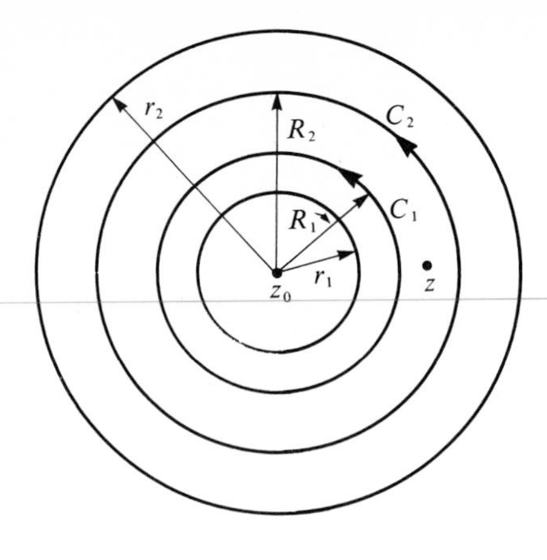

Figure 308

$$f(z) = \frac{1}{2\pi i}\oint_{C_2} \frac{f(w)\,dw}{w - z} - \frac{1}{2\pi i}\oint_{C_1} \frac{f(w)\,dw}{w - z}.$$

Here, we use w as integration variable to avoid confusion with the selected point z. Consider these integrals in turn.

For $\oint_{C_2}$: Write (as in the derivation of the Taylor expansion)

$$\frac{1}{w - z} = \sum_{n=0}^{\infty} \frac{(z - z_0)^n}{(w - z_0)^{n+1}}.$$

Then,

$$\begin{aligned}\oint_{C_2} \frac{f(w)\,dw}{w - z} &= \sum_{n=0}^{\infty} \left[\frac{1}{2\pi i}\oint_{C_2} \left(\frac{f(w)}{(w - z_0)^{n+1}}\right) dw\right](z - z_0)^n \\ &= \sum_{n=0}^{\infty} a_n(z - z_0)^n,\end{aligned}$$

where

$$a_n = \frac{1}{2\pi i}\oint_{C_2} \frac{f(w)}{(w - z_0)^{n+1}}\,dw \quad \text{for} \quad n = 0, 1, 2, \ldots .$$

For $\oint_{C_1}$: Write

$$\frac{1}{w - z} = \frac{1}{w - z_0 - (z - z_0)} = -\left(\frac{1}{z - z_0}\right)\frac{1}{1 - \left(\dfrac{w - z_0}{z - z_0}\right)}$$

and note that, for w on C_1, $|(w - z_0)/(z - z_0)| < 1$. By geometric series,

$$\begin{aligned}\left(\frac{1}{z - z_0}\right)\left[\frac{1}{1 - \left(\dfrac{w - z_0}{z - z_0}\right)}\right] &= \left(\frac{-1}{z - z_0}\right)\sum_{n=0}^{\infty}\left(\frac{w - z_0}{z - z_0}\right)^n \\ &= -\sum_{n=1}^{\infty} \frac{(w - z_0)^{n-1}}{(z - z_0)^n}.\end{aligned}$$

Then,

$$\oint_{C_1} \frac{f(w)\,dw}{w - z} = \sum_{n=1}^{\infty} \left(\frac{1}{2\pi i} \oint_{C_1} f(w)(w - z_0)^{n-1}\,dw \right) \left(\frac{-1}{(z - z_0)^n} \right)$$

$$= -\sum_{n=1}^{\infty} \frac{a_{-n}}{(z - z_0)^n},$$

where

$$a_{-n} = \frac{1}{2\pi i} \oint_{C_1} f(w)(z - z_0)^{n-1}\,dw \qquad (n = 1, 2, 3, \ldots).$$

Now, in both $\oint_{C_1}$ and $\oint_{C_2}$, we may replace C_1 and C_2 by C: $|z - z_0| = \rho$, where $r_1 < \rho < r_2$, by the change-of-path result of Section 15.3, Part C. We can then write the a_n's for $n = 0, 1, 2, \ldots$ and the a_{-n}'s for $n = 1, 2, \ldots$ in one formula:

$$a_n = \frac{1}{2\pi i} \oint_C \frac{f(w)\,dw}{(w - z_0)^{n+1}}.$$

We now obtain

$$f(z) = \sum_{n=0}^{\infty} a_n(z - z_0)^n - \left(-\sum_{n=1}^{\infty} \frac{a_{-n}}{(z - z_0)^n} \right) = \sum_{n=-\infty}^{\infty} a_n(z - z_0)^n.$$

This completes the proof.

Another way of looking at the Laurent expansion is suggested by writing, for $r_1 < |z - z_0| < r_2$,

$$f(z) = \sum_{n=-\infty}^{\infty} a_n(z - z_0)^n$$

$$= \sum_{n=0}^{\infty} a_n(z - z_0)^n + \sum_{n=1}^{\infty} \frac{a_{-n}}{(z - z_0)^n}$$

$$= f_1(z) + f_2(z).$$

Here, $f_1(z) = \sum_{n=0}^{\infty} a_n(z - z_0)^n$ is analytic at z_0; any problems experienced by $f(z)$ at z_0 are completely contained in $f_2(z) = \sum_{n=1}^{\infty} a_{-n}/(z - z_0)^n$. In fact, if $f(z)$ is analytic at z_0, then $f(z) = f_1(z)$, $f_2(z) = 0$, each $a_{-n} = 0$, and the Laurent expansion about z_0 reduces to the Taylor expansion about z_0.

In any given annulus $r_1 < |z - z_0| < r_2$, the Laurent expansion of $f(z)$ (if it has one) is unique. This is easy to show, and we omit the details.

In practice one almost never computes a Laurent expansion by evaluating the integral formula for the coefficients. On the contrary, one usually finds the series some other way and then uses certain of the coefficients (most often a_{-1}) to evaluate integrals, using the residue theorem of the next chapter. Here are some examples in which Laurent series are computed.

EXAMPLE 464

Often one can find a Laurent expansion from a Taylor expansion. Thus,

$$e^z = \sum_{n=0}^{\infty} \frac{z^n}{n!} \qquad \text{(Taylor expansion of } e^z \text{ about 0)}$$

gives us

$$e^{1/z} = \sum_{n=0}^{\infty} \frac{1}{n!\, z^n} \qquad \text{(Laurent expansion in } 0 < |z| < \infty\text{)}.$$

This fits Theorem 136 with $z_0 = 0$, $r_1 = 0$, and $r_2 = \infty$.

Similarly,

$$e^{1/z^2} = \sum_{n=0}^{\infty} \frac{1}{n!\, z^{2n}} \quad \text{in} \quad 0 < |z| < \infty.$$

The Taylor series for $\sin(z)$ about 0 is

$$\sin(z) = \sum_{n=0}^{\infty} \frac{(-1)^n z^{2n+1}}{(2n+1)!}.$$

Thus,

$$\sin\left(\frac{1}{z}\right) = \sum_{n=0}^{\infty} \frac{(-1)^n}{(2n+1)!\, z^{2n+1}} \quad \text{in} \quad 0 < |z| < \infty.$$

EXAMPLE 465

Expand $1/(1 + z^2)$ in a Laurent series about $-i$.

Here, we want a series in powers of $z + i$. Use partial fractions to write

$$\frac{1}{1+z^2} = \frac{i}{2}\left(\frac{1}{z+i}\right) - \frac{i}{2}\left(\frac{1}{z-i}\right).$$

The first term on the right is a power of $z + i$; so leave it alone and work on $1/(z - i)$. Write

$$\frac{1}{z-i} = \frac{1}{z-i+i-i}$$

$$= \frac{1}{-2i + (z+i)}$$

$$= \frac{1}{-2i\left(1 - \dfrac{z+i}{2i}\right)}$$

$$= -\frac{1}{2i} \sum_{n=0}^{\infty} \left(\frac{z+i}{2i}\right)^n$$

$$= \sum_{n=0}^{\infty} \frac{-(z+i)^n}{(2i)^{n+1}}$$

if $|(z+i)/2i| < 1$ or, equivalently, if $|z+i| < 2$. Thus,

$$\frac{1}{1+z^2} = \frac{i}{2}\left(\frac{1}{z+i}\right) - \frac{i}{2}\sum_{n=0}^{\infty} \frac{-(z+i)^n}{(2i)^{n+1}}$$

$$= \frac{i}{2}\left(\frac{1}{z+i}\right) + \frac{i}{2}\sum_{n=0}^{\infty} \frac{(z+i)^n}{(2i)^{n+1}}.$$

The expansion is valid in $0 < |z+i| < 2$. To understand this, note that the expansion is a sum of two terms, the first valid for all $z \neq -i$ (that is, $|z+i| > 0$), and the second a Taylor series valid for $|(z+i)/2i| < 1$ or, equivalently, $|z+i| < 2$. The sum is valid in the common region $|z+i| > 0$ and $|z+i| < 2$, or $0 < |z+i| < 2$.

EXAMPLE 466

Expand $1/(z+1)(z-3i)$ in a Laurent series about -1.

Begin by using partial fractions to write

$$\frac{1}{(z+1)(z-3i)} = \left(\frac{-1+3i}{10}\right)\left(\frac{1}{z+1}\right) + \left(\frac{1-3i}{10}\right)\left(\frac{1}{z-3i}\right),$$

and treat these one at a time.

First, $1/(z-3i)$ is analytic at -1, and so has a Taylor series about -1. Write

$$\frac{1}{z-3i} = \frac{1}{z-3i+1-1}$$

$$= \frac{1}{-1-3i+(z+1)}$$

$$= \frac{1}{-1-3i}\left[\frac{1}{1-\left(\frac{z+1}{1+3i}\right)}\right]$$

$$= \frac{-1}{1+3i}\sum_{n=0}^{\infty}\left(\frac{z+1}{1+3i}\right)^n$$

$$= -\sum_{n=0}^{\infty}\frac{(z+1)^n}{(1+3i)^{n+1}}$$

if $|(z+i)/(1-3i)| < 1$ or $|z+1| < \sqrt{10}$.

Next, $1/(z+1)$ is a Laurent series about -1, in $|z+1| > 0$. Thus, for $0 < |z+1| < \sqrt{10}$, we have

$$\frac{1}{(z+1)(z-3i)} = \left(\frac{-1+3i}{10}\right)\frac{1}{z+1} - \left(\frac{1-3i}{10}\right)\sum_{n=0}^{\infty}\frac{(z+1)^n}{(1+3i)^{n+1}}.$$

Again, note that the expansion is a sum of two terms, one valid in $|z+1| < \sqrt{10}$, the other in $0 < |z+1|$; the sum is valid in the common domain $0 < |z+1| < \sqrt{10}$.

In these examples, the Laurent expansion about z_0 was always valid in an annulus $0 < |z - z_0| < r_2$, that is, $r_1 = 0$ in Theorem 136. This is typical in most applications, including the residue theorem and classification of singularities which will occupy the next chapter.

PROBLEMS FOR SECTION 16.6

In each of Problems 1 through 20, determine the Laurent expansion in some annulus $0 < |z - z_0| < r$ about z_0. Determine in each case the largest r one can use here.

1. $\dfrac{2z}{1+z^2}$, $z_0 = i$

2. $\dfrac{\sin(z)}{z^2}$, $z_0 = 0$

3. $\dfrac{1-\cos(2z)}{z^2}$, $z_0 = 0$

4. $\cos\left(\dfrac{1}{z-i}\right)$, $z_0 = i$

5. $\dfrac{z^2}{1-z^4}$, $z_0 = 1$

6. $\dfrac{1}{z^2}e^{1/z}$, $z_0 = 0$

7. $\dfrac{z+i}{z-i}$, $z_0 = i$

8. $\dfrac{z^2+1}{2z-1}$, $z_0 = \frac{1}{2}$

9. $\dfrac{1}{z}\tan(z)$, $z_0 = 0$

10. $z^2\cos\left(\dfrac{i}{z}\right)$, $z_0 = 0$

11. $\dfrac{1+e^z}{\sin(z)+z\cos(z)}$, $z_0 = 0$

12. $\dfrac{z+z_0}{z-z_0}$, z_0 arbitrary

13. e^{1/z^3}, $z_0 = 0$

14. $\dfrac{\cos(z)}{z}$, $z_0 = 0$

15. $\dfrac{1}{z^2-i}$, $z_0 = i$

16. $\dfrac{\cosh(z)}{z^2}$, $z_0 = 0$

17. $\dfrac{2i}{z-1+i}$, $z_0 = 1-i$

18. $\sinh\left(\dfrac{1}{z^2}\right)$, $z_0 = 0$

19. $\dfrac{z}{z-2i}$, $z_0 = 2i$

20. $\sin\left(\dfrac{i}{2z}\right)$, $z_0 = 0$

21. The Bessel function $J_n(z)$ is defined by

$$\exp\left[z\left(w - \frac{1}{w}\right)\right] = \sum_{n=-\infty}^{\infty} J_n(z)w^n.$$

(a) Use the integral formula of Theorem 136 for the coefficients in the Laurent expansion to derive the expression

$$J_n(z) = \frac{1}{\pi}\int_0^{\pi} \cos[n\theta - z\sin(\theta)]\,d\theta.$$

(b) Write

$$\exp\left[z\left(w - \frac{1}{w}\right)\right] = e^{zw}e^{-z/w},$$

and multiply the expansions of e^{zw} and $e^{-z/w}$ about 0 to obtain the Maclaurin series

$$J_n(z) = \sum_{j=0}^{\infty} \frac{(-1)^j\left(\dfrac{z}{2}\right)^{n+2j}}{j!\,(n+j)!}.$$

22. (Refer to Problem 31, Section 16.5.) Show that the Laurent expansion of $1/(e^z - 1)$ about 0 is

$$\frac{1}{z} - \frac{1}{2} + \sum_{n=1}^{\infty} (-1)^{n-1} \frac{B_n}{(2n)!} z^{2n-1}.$$

APPENDIX: REAL SEQUENCES AND SERIES

In this section we review some facts about real sequences and series. We assume that the student has seen most of this material previously, and we therefore intend this section as a reference and reminder, not as a means of learning the material from the beginning.

We divide the material into two parts: tests for convergence of real series; and a sketch of a proof of the Cauchy convergence criterion for real sequences.

A. SOME TESTS FOR REAL SERIES

We assume that the reader has an understanding of what it means for a real series $\sum_{n=1}^{\infty} a_n$ to converge or diverge. Here are some of the more commonly used tests for convergence or divergence of $\sum_{n=1}^{\infty} a_n$.

1. If $\lim_{n\to\infty} a_n \neq 0$, then $\sum_{n=1}^{\infty} a_n$ diverges.

   Note: If $\lim_{n\to\infty} a_n = 0$, the series may converge or it may diverge—you need further testing to determine which is the case.

2. *Integral test.* Suppose each $a_n > 0$, and that $f(x)$ is a continuous function for $x \geq 1$ such that $a_n = f(n)$. Suppose that, for some x_0, $f(x)$ is a monotone nonincreasing function for $x \geq x_0$. Then, $\sum_{n=1}^{\infty} a_n$ converges if $\int_1^{\infty} f(x)\,dx$ converges and diverges if $\int_1^{\infty} f(x)\,dx$ diverges.

   By applying this test, one easily finds that the p-series $\sum_{n=1}^{\infty} 1/n^p$ converges if $p > 1$ and diverges if $p \leq 1$. Such series are handy to keep in mind when attempting to use comparison tests.

3. *Comparison test.* Let $0 \leq a_n \leq b_n$ for $n = 1, 2, 3, \ldots$. Then,

$$\text{if} \quad \sum_{n=1}^{\infty} b_n \quad \text{converges,} \quad \sum_{n=1}^{\infty} a_n \quad \text{converges;}$$

$$\text{if} \quad \sum_{n=1}^{\infty} a_n \quad \text{diverges,} \quad \sum_{n=1}^{\infty} b_n \quad \text{diverges;}$$

$$\text{if} \quad \sum_{n=1}^{\infty} a_n \quad \text{converges, no conclusion can be drawn about} \quad \sum_{n=1}^{\infty} b_n;$$

   and similarly,

$$\text{if} \quad \sum_{n=1}^{\infty} b_n \quad \text{diverges, no conclusion can be drawn about} \quad \sum_{n=1}^{\infty} a_n.$$

4. *Ratio test.* Let each $a_n > 0$, and let $\lim_{n\to\infty} a_{n+1}/a_n = L$. Then,

$$\text{if} \quad 0 \le L < 1, \quad \text{then} \quad \sum_{n=1}^{\infty} a_n \quad \text{converges;}$$

$$\text{if} \quad L > 1, \quad \text{then} \quad \sum_{n=1}^{\infty} a_n \quad \text{diverges; and}$$

$$\text{if} \quad L = 1, \quad \text{the test allows no conclusion.}$$

5. *Root test.* Let each $a_n > 0$, and let $\lim_{n\to\infty} (a_n)^{1/n} = L$. Then,

$$\text{if} \quad 0 \le L < 1, \quad \text{then} \quad \sum_{n=1}^{\infty} a_n \quad \text{converges;}$$

$$\text{if} \quad L > 1, \quad \text{then} \quad \sum_{n=1}^{\infty} a_n \quad \text{diverges; and}$$

$$\text{if} \quad L = 1, \quad \text{the test fails.}$$

6. *Absolute convergence.* If $\sum_{n=1}^{\infty} |a_n|$ converges, so does $\sum_{n=1}^{\infty} a_n$. However, $\sum_{n=1}^{\infty} a_n$ may converge while $\sum_{n=1}^{\infty} |a_n|$ diverges.

   When $\sum_{n=1}^{\infty} |a_n|$ converges, we say that $\sum_{n=1}^{\infty} a_n$ *converges absolutely.* When $\sum_{n=1}^{\infty} a_n$ converges, but $\sum_{n=1}^{\infty} |a_n|$ diverges, we say that $\sum_{n=1}^{\infty} a_n$ *converges conditionally.*

7. *Alternating series.* If each $a_n > 0$, then the series $\sum_{n=1}^{\infty} (-1)^{n+1} a_n = a_1 - a_2 + a_3 - a_4 + \cdots$ is called an *alternating series.*

   In order for such a series to converge, it is sufficient that

$$0 \le a_{n+1} \le a_n \quad \text{and} \quad \lim_{n\to\infty} a_n = 0.$$

   In the event that $\sum_{n=1}^{\infty} (-1)^{n+1} a_n$ converges, say, to S, then $\sum_{n=1}^{N} (-1)^{n+1} a_n$ differs from S in absolute value by no more than a_{N+1}. That is,

$$\left| \sum_{n=1}^{\infty} (-1)^{n+1} a_n - \sum_{n=1}^{N} (-1)^{n+1} a_n \right| \le a_{N+1}.$$

B. CAUCHY'S CONVERGENCE CRITERION FOR REAL SEQUENCES

The *Cauchy convergence criterion* says that a sequence $x_1, x_2, x_3, \ldots$ converges if and only if, given $\varepsilon > 0$, there exists N such that $|x_n - x_m| < \varepsilon$ whenever $n, m \ge N$.

A proof of this depends on the so-called completeness property of the real number system, which we shall now explain in terms of sequences. A sequence $x_1, x_2, \ldots$ of real numbers is said to be *bounded* if, for some real numbers A and B,

$$A \le x_j \le B$$

for $j = 1, 2, \ldots$. The *completeness property* of the real number system is that every bounded sequence contains a convergent subsequence. That is, from the bounded sequence $x_1, x_2, \ldots$, we must be able to form a sequence $y_1, y_2, \ldots$, with each y_j equal to some x_r with $r \ge j$, such that $y_1, y_2, \ldots$ converges.

For example, the sequence

$$1, \ -1, \ 1, \ -1, \ 1, \ -1, \ldots$$

with $x_j = (-1)^{j+1}$ is bounded. It does not itself converge. However, it contains a convergent subsequence, namely, 1, 1, 1, 1, ..., by choosing each $y_j = x_{2j-1}$. Thus, $y_1 = x_1 = 1$, $y_2 = x_3 = 1$, $y_3 = x_5 = 1$, and so on. Similarly, it also contains the convergent subsequence $-1, -1, -1, \ldots$, in which the jth term is x_{2j}.

Using this property, we can prove Cauchy's theorem as follows.

Proof of the Cauchy Convergence Criterion

Necessity

First suppose that $\lim_{n\to\infty} x_n = L$, and let $\varepsilon > 0$. For some N, $|x_n - L| < \varepsilon/2$ if $n \geq N$. Then, for m and $n \geq N$,

$$\begin{aligned} |x_n - x_m| &= |x_n - L + L - x_m| \\ &= |(x_n - L) + (-x_m + L)| \\ &\leq |x_n - L| + |L - x_m| \\ &= |x_n - L| + |x_m - L| \\ &< \frac{\varepsilon}{2} + \frac{\varepsilon}{2} = \varepsilon. \end{aligned}$$

Sufficiency

Suppose that, given $\varepsilon > 0$, there is some N such that $|x_n - x_m| < \varepsilon$ whenever n, $m \geq N$. We must show that there is some number L such that $\lim_{n\to\infty} x_n = L$. We now proceed in steps.

Step 1 The sequence $x_1, x_2, \ldots$ is bounded.

Choosing $\varepsilon = 1$, say, we have by assumption that, for some N, $|x_n - x_m| < 1$ if $m, n \geq N$. In particular, then,

$$|x_n - x_N| < 1 \quad \text{if} \quad n \geq N.$$

Then,

$$x_N - 1 < x_n < x_N + 1 \quad \text{for} \quad n \geq N.$$

Thus, the terms $x_N, x_{N+1}, x_{N+2}, \ldots$ of the sequence are bounded. Since there are only finitely many other terms $x_1, x_2, \ldots, x_{N-1}$ of the sequence, then the sequence is bounded.

Step 2 The sequence $x_1, x_2, \ldots$ has a convergent subsequence $y_1, y_2, \ldots$.

This is immediate by step 1 and the completeness property.

Step 3 Now, let $\lim_{n\to\infty} y_n = L$. Then, $\lim_{n\to\infty} x_n = L$ also.

To prove this, note that each y_n is some x_r. For notational convenience, say $y_n = x_{f(n)}$, with $f(n) > f(m)$ if $n > m$.

Let $\varepsilon > 0$. By hypothesis, for some N_1,

$$|x_n - x_m| < \tfrac{1}{2}\varepsilon \quad \text{if} \quad m, n \geq N_1.$$

Since $y_n \to L$, there is also some N_2 such that

$$|y_n - L| < \tfrac{1}{2}\varepsilon \quad \text{if} \quad n \geq N_2.$$

Now let N be the larger of the two numbers N_1 and N_2. Then, $N \geq N_1$ and $N \geq N_2$. We shall show that $x_n \to L$.

Let $n \geq N$. Since also $n \geq N_1$, and $f(n) \geq n \geq N_1$, we have

$$|x_n - x_{f(n)}| < \tfrac{1}{2}\varepsilon.$$

But also $n \geq N_2$; hence,

$$|y_n - L| = |x_{f(n)} - L| < \tfrac{1}{2}\varepsilon.$$

Then, for $n \geq N$,

$$\begin{aligned} |x_n - L| &= |x_n - x_{f(n)} + x_{f(n)} - L| \\ &\leq |x_n - x_{f(n)}| + |x_{f(n)} - L| \\ &< \tfrac{1}{2}\varepsilon + \tfrac{1}{2}\varepsilon = \varepsilon. \end{aligned}$$

This proves that $\lim_{n\to\infty} x_n = L$, completing the proof of the theorem.

SUPPLEMENTARY PROBLEMS

In each of Problems 1 through 20, find a Taylor expansion in some disk $|z - z_0| < r$ or a Laurent expansion in an annulus $0 < |z - z_0| < r$, whichever is appropriate. In each case, determine the largest possible value of r for which the expansion is valid. (In some cases, you may find $r = \infty$.)

1. $\dfrac{1}{z+4}$; $z_0 = 2 + i$

2. e^{2z}; $z_0 = -3i$

3. $\cos(z - 5i)$; $z_0 = 5i$

4. $\dfrac{1}{2z - 3 + i}$; $z_0 = -3$

5. $z^5 - iz^2 + (1 - i)z - 2$; $z_0 = 3 - i$

6. $\dfrac{\cos(2z)}{z^4}$; $z_0 = 0$

7. $\dfrac{e^{z^2}}{z}$; $z_0 = 0$

8. $\cosh(z - i)$; $z_0 = 0$

9. $\dfrac{1}{1 - iz}$; $z_0 = 4 + 5i$

10. $\dfrac{1}{i - z^2}$; $z_0 = 0$

11. $\dfrac{\sin(iz^3)}{z^2}$; $z_0 = 0$

12. $\dfrac{i - \cos(iz)}{z^4}$; $z_0 = 0$

13. $\dfrac{1}{2z - 2 + i}$; $z_0 = 1 + i$

14. $\dfrac{2 + z^2}{3 + z}$; $z_0 = 0$

15. $\dfrac{\cosh(iz^2)}{3z}$; $z_0 = 0$

16. $\dfrac{1}{4 - 2z^2}$; $z_0 = i$

17. $\cos(3z^2) - ie^z$; $z_0 = 0$

18. $\dfrac{1}{1 - z}$; $z_0 = -2 + 3i$

19. e^{2z+1}; $z_0 = 0$

20. $\cosh(z) - i\sinh(z)$; $z_0 = 0$

In each of Problems 21 through 30, find the first six terms in the Taylor series expansion of the given function about the given point.

21. $e^z \sin(iz)$; $z_0 = 0$
22. $\operatorname{sech}(z)$; $z_0 = 0$
23. $\cos(z^2)$; $z_0 = i$
24. $z^2 \sin(z)$; $z_0 = -3$
25. $\cosh(1/z)$; $z_0 = i$
26. e^{z^2}; $z_0 = 2i$
27. $\sin^2(2iz)$; $z_0 = 3 + i$
28. $\tanh(z)$; $z_0 = 0$
29. $e^{-z}\sinh(z)$; $z_0 = 0$
30. $3z^2\cosh(iz)$; $z_0 = -1$

31. Derive a complex form of Fourier series from the Taylor expansion as follows. Suppose that $f(z) = \sum_{n=0}^{\infty} a_n z^n$ in $|z| < R$. Write z in polar form as $z = re^{i\theta}$, with $r < R$, and show that

$$f(z) = \sum_{n=0}^{\infty} a_n r^n e^{in\theta},$$

where

$$a_n = \frac{1}{2\pi r^n} \int_0^{2\pi} f(re^{i\theta}) e^{-in\theta}\, d\theta.$$

32. Show in Problem 31 that

$$\sum_{n=0}^{\infty} |a_n|^2 r^{2n} = \frac{1}{2\pi} \int_0^{2\pi} |f(re^{i\theta})|^2\, d\theta.$$

This is called *Parseval's identity.*

33. Suppose that $f(z)$ is analytic in $0 < |z| < R$. Use the Laurent expansion to show that

$$\frac{1}{2\pi} \int_0^{2\pi} |f(re^{i\theta})|^2\, d\theta = \sum_{n=-\infty}^{\infty} |a_n|^2 r^{2n}.$$

34. Show that

$$\sum_{n=0}^{\infty} \left(\frac{z^n}{n!}\right)^2 = \frac{1}{2\pi} \int_0^{2\pi} e^{2z\cos(\theta)}\, d\theta.$$

(*Hint:* First show that

$$\left(\frac{z^k}{k!}\right)^2 = \frac{1}{2\pi i} \oint_C \frac{z^k e^{zw}\, dw}{k!\, w^{k+1}} \quad \text{for} \quad k = 0, 1, 2, \ldots,$$

with C the unit circle about the origin.)

CHAPTER SEVENTEEN

Singularities, Residues, and Applications to Real Integrals and Series

17.1 SINGULARITIES

A situation we encounter quite frequently is that $f(z)$ is analytic in an annulus $0 < |z - z_0| < r$ but not at z_0. In this case we say that $f(z)$ has an *isolated singularity* at z_0. We shall now see that isolated singularities fall naturally into three types.

Expand $f(z)$ in a Laurent expansion about z_0:

$$f(z) = \sum_{n=-\infty}^{\infty} a_n(z - z_0)^n \qquad (0 < |z - z_0| < r).$$

We call z_0:

a *removable singularity* of $f(z)$ if no negative powers of $z - z_0$ appear in this expansion;

an *essential singularity* if infinitely many negative powers of $z - z_0$ occur; and

a *pole* if only finitely many (but at least one) negative powers of $z - z_0$ appear.

EXAMPLE 467

Let $f(z) = \sin(z)/z$ for $z \neq 0$. Using the Taylor series for $\sin(z)$ about 0, from Example 458, we have

$$\frac{\sin(z)}{z} = 1 - \frac{z^2}{3!} + \frac{z^4}{5!} - \frac{z^6}{7!} + \frac{z^8}{9!} - \cdots.$$

This is the Laurent series of $\sin(z)/z$ in $0 < |z| < \infty$. Since this expansion has no negative powers of z, then 0 is a removable singularity of $\sin(z)/z$.

EXAMPLE 468

Let $f(z) = e^{1/(z-1)}$ for $z \neq 1$. Again using Example 458, we have

$$e^{1/(z-1)} = \sum_{n=0}^{\infty} \frac{1}{n!} \frac{1}{(z-1)^n} \qquad (0 < |z-1| < \infty).$$

This Laurent expansion about 1 has infinitely many negative powers of $z - 1$; hence, 1 is an essential singularity of $e^{1/(z-1)}$.

EXAMPLE 469

Let $f(z) = 1/(z+i)^3$. Here, $f(z)$ is already a Laurent expansion about $-i$, containing finitely many (exactly one) negative powers of $z + i$. Thus, $f(z)$ has a pole at $-i$.

If $f(z)$ has a pole at some point z_0, then the Laurent expansion of $f(z)$ in some annulus $0 < |z - z_0| < r$ has the form

$$f(z) = \frac{a_{-m}}{(z-z_0)^m} + \frac{a_{-m+1}}{(z-z_0)^{m-1}} + \cdots + \frac{a_{-1}}{z-z_0}$$
$$+ a_0 + a_1(z-z_0) + a_2(z-z_0)^2 + \cdots,$$

where $a_{-m} \neq 0$, but $a_{-n} = 0$ for every $n > m$. We call *m the order of the pole* at z_0. In Example 469, $f(z)$ has a pole of order 3 at $-i$. If $m = 1$, then $f(z)$ is said to have a *simple pole* at z_0. Such is the case with $\sin(z)/z^2$ at 0, since the Laurent expansion in $0 < |z| < \infty$ is

$$\frac{1}{z} - \frac{z}{3!} + \frac{z^3}{5!} - \frac{z^5}{7!} + \frac{z^7}{9!} - \cdots.$$

Sometimes poles are most easily understood in terms of zeros. We say that $f(z)$ has a *zero of order m at* z_0 if $f(z)$ is analytic at z_0 and

$$f(z_0) = f'(z_0) = \cdots = f^{(m-1)}(z_0) = 0,$$

while $f^{(m)}(z_0) \neq 0$. Thus, for example, z^m has a zero of order m at 0, while $\sin^2(z)$ has a zero of order 2 at π (and also at $0, 2\pi, 3\pi, \ldots$). As with poles, a zero of order 1 is called a *simple zero* [for example, $\sin(z)$ at π].

One can tell the order of a zero at z_0 from the Taylor expansion of $f(z)$ about z_0. Write

$$f(z) = \sum_{n=0}^{\infty} a_n(z-z_0)^n.$$

Now, $a_n = f^{(n)}(z_0)/n!$. If $f(z)$ has a zero of order m at z_0, then $a_0 = a_1 = \cdots = a_{m-1} = 0$, while $a_m \neq 0$. Then,

$$f(z) = \sum_{n=m}^{\infty} a_n(z-z_0)^n.$$

As a rule, then, a function $f(z)$ analytic at z_0 has a zero of order m at z_0 exactly when the lowest power of $z - z_0$ in the Taylor expansion of $f(z)$ about z_0 is m.

Zeros relate to poles in the following way.

LEMMA

Let $h(z)$ have a zero of order m at z_0. Let $g(z)$ be analytic, or have a removable singularity, at z_0, and let $g(z_0) \neq 0$. Then $f(z) = g(z)/h(z)$ has a pole of order m at z_0.

We leave a proof of this to the exercises.

Roughly speaking, the lemma says that poles of a quotient $g(z)/h(z)$ are determined by zeros of the denominator $h(z)$. As examples:

$1/(z^2 + 1)$ has simple poles at i and $-i$, as $z^2 + 1$ has simple zeros at i and $-i$;

and

$\sin(z)/z^3$ has a pole of order 2 at 0, since this is $[\sin(z)/z]1/z^2$, and $\sin(z)/z$ has a removable singularity at 0, while z^2 has a zero of order 2 there. In a sense, the zeros of $\sin(z)$ and z cancel each other out, leaving the z^2 in the denominator to give $\sin(z)/z^3$ a pole of order 2 at zero.

In the next section we shall see how singularities can be exploited to evaluate integrals.

PROBLEMS FOR SECTION 17.1

In each of Problems 1 through 20, determine where $f(z)$ has singularities, and classify each singularity as removable, essential, or a pole of appropriate order. Remember to use the lemma of this section where it applies.

1. $\dfrac{\cos(z)}{z^2}$

2. $\dfrac{1}{(z+i)^2(z-i)}$

3. $e^{1/z}(z-i)$

4. $\dfrac{\sin(z)}{z-\pi}$

5. $\dfrac{\cos(2z)}{(z-1)^2(z^2+1)}$

6. $\dfrac{z}{(z+1)^2}$

7. $\dfrac{z-i}{z^2+1}$

8. $\dfrac{\sin(z)}{\sinh(z)}$

9. $\dfrac{z}{z^4-1}$

10. $\tan(z)$

11. $\dfrac{1}{\cos^2(z)}$

12. $e^{1/(z+1)^2}$

13. $\dfrac{e^{iz}}{z^2}$

14. $\dfrac{\sin(z)}{z(z-\pi)(z-i)^2}$

15. $\dfrac{e^z}{(z+1)^4}$

16. $\dfrac{\sinh(z)}{z^4}$

17. $\coth(z)$

18. $\dfrac{2i-1}{(z^2+2z-3)^2}$

19. $\dfrac{\sin(z)}{(z+\pi)^2}$

20. $\dfrac{1}{\cos^2(z)}$

21. Prove the lemma of this section.

22. Let $f(z)$ have a pole at z_0. Prove that $|f(z)| \to \infty$ as $z \to z_0$ along any path.

23. Let $f(z)$ have a pole of order m at z_0. Let $P(z)$ be a polynomial of degree n. Prove that $P[f(z)]$ has a pole of order nm at z_0.

24. Let $f(z)$ be analytic in $0 < |z - z_0| < r$. Prove that $f(z)$ has a removable singularity at z_0 if and only if $\lim_{z \to z_0} (z - z_0)f(z) = 0$.

25. Prove that, in *every* neighborhood of 0, $e^{1/z}$ takes on every nonzero value infinitely often.

26. Let $R(z)$ be a rational function. Prove that $R(z)$ can have no essential singularity. (Recall that a rational function is a quotient of polynomials.)

17.2 RESIDUES AND THE RESIDUE THEOREM

We shall now prove a result which is useful in evaluating both real and complex integrals. To see the connection between isolated singularities of $f(z)$ and integrals of $f(z)$, suppose that $f(z)$ is analytic in $0 < |z - z_0| < r$ and that C is a simple closed path in this annulus enclosing z_0 (see Figure 309). The Laurent expansion of $f(z)$ in $0 < |z - z_0| < r$ has the form

$$\cdots \frac{a_{-3}}{(z - z_0)^3} + \frac{a_{-2}}{(z - z_0)^2} + \frac{a_{-1}}{z - z_0}$$
$$+ a_0 + a_1(z - z_0) + a_2(a - a_0)^2 + \cdots.$$

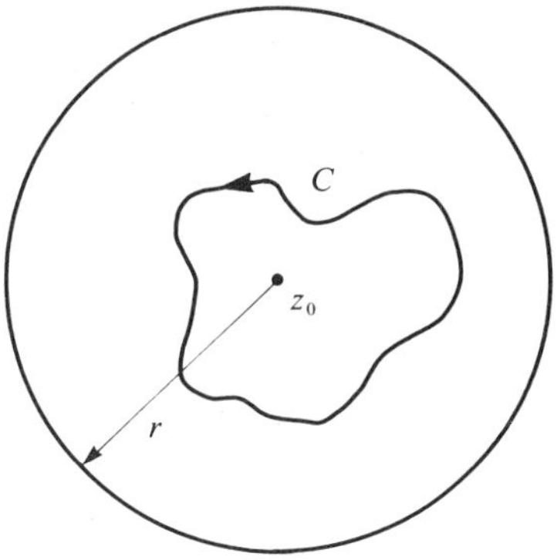

Figure 309

The formula for a_{-1} (from Theorem 136) is

$$a_{-1} = \frac{1}{2\pi i}\oint_C f(z)\,dz.$$

Thus,

$$\oint_C f(z)\,dz = 2\pi i a_{-1}.$$

The importance of this is that we can evaluate $\oint_C f(z)\,dz$ if we know the coefficient of $1/(z - z_0)$ in the Laurent expansion in $0 < |z - z_0| < r$. As shown by Examples 464, 465, and 466, often this number can be found without performing any integration, giving us a powerful method for evaluating $\oint_C f(z)\,dz$ under the conditions specified above.

The residue theorem expands this method to apply to the case that C encloses more than one isolated singularity of $f(z)$. The name of the theorem derives from the practice of calling a_{-1} the *residue of* $f(z)$ *at* z_0. More carefully:

If $f(z)$ has an isolated singularity at z_0, then the *residue of* $f(z)$ *at* z_0, denoted $\text{Res}_{z_0} f(z)$, is the coefficient of $1/(z - z_0)$ in the Laurent expansion of $f(z)$ in any annulus $0 < |z - z_0| < r$.

From Example 468, $e^{1/(z-1)}$ has residue 1 at 1; from Example 469, $1/(z + i)^3$ has residue 0 at $-i$. If we let $f(z) = i\cos(z)/3z$ for $z \neq 0$, then the Laurent expansion in $0 < |z| < \infty$ is

$$\frac{i}{3z} - \frac{iz}{6} + \frac{iz^3}{72} - \cdots.$$

Hence,

$$\underset{0}{\text{Res}}\, \frac{i\cos(z)}{3z} = \frac{i}{3}.$$

We can now state the main result.

THEOREM 137 The Residue Theorem

Let $f(z)$ be analytic in a domain D except at points $z_1, \dots, z_n$, where $f(z)$ has isolated singularities. Let C be a simple closed path in D enclosing $z_1, \dots, z_n$. Then,

$$\oint_C f(z)\, dz = 2\pi i \sum_{j=1}^{n} \operatorname*{Res}_{z_j} f(z).$$

In words, the integral of $f(z)$ around C equals $2\pi i$ times the sum of the residues of $f(z)$ at the isolated singularities enclosed by C. (*Note:* Isolated singularities of $f(z)$ *not* enclosed by C are irrelevant as far as $\oint_C f(z)\, dz$ is concerned, as long as C does not pass through any of them.)

Proof Enclose each z_j in a circle C_j of sufficiently small radius that no two of $C, C_1, \dots, C_n$ intersect. Next, cut a channel from C to C_1, C_1 to $C_2, \dots, C_{n-1}$ to C_n, and then back to C, as shown in Figure 310. Replace C by K as shown in Figure 311. Now, $f(z)$ is analytic on and inside K (note that each z_j is outside K); so

$$\oint_K f(z)\, dz = 0.$$

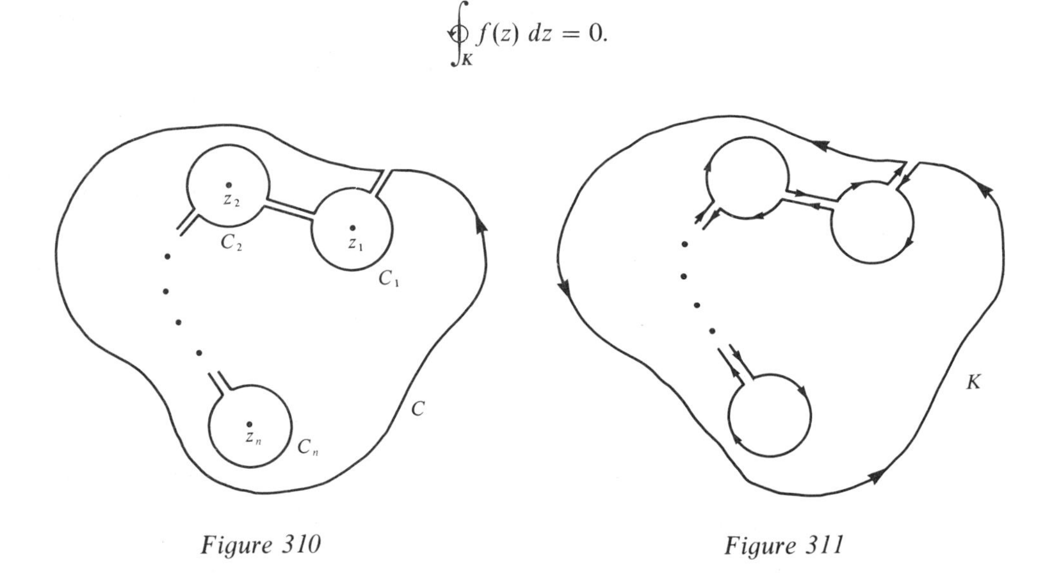

Figure 310 *Figure 311*

Now decompose K into $C, C_1, \dots, C_n$ and the lines connecting C to C_1, C_1 to C_2, and so on. As each pair of connecting lines is merged together, the integrals over them cancel, since the directions of integration are opposed. Noting that integration on the C_j's is clockwise in K, we have

$$\oint_C f(z)\, dz + \sum_{j=1}^{n} \oint_{C_j} f(z)\, dz = 0.$$

Then, taking all integrals counterclockwise, we have

$$\oint_C f(z)\, dz - \sum_{j=1}^{n} \oint_{C_j} f(z)\, dz = 0.$$

But, by Theorem 136 and the definition of residue,

$$\oint_{C_j} f(z)\, dz = 2\pi i \operatorname*{Res}_{z_j} f(z).$$

This proves the theorem.

Let us summarize the residue theorem in conjunction with Cauchy's integral theorem. To evaluate $\oint_C f(z)\, dz$:

1. Determine all singularities of $f(z)$ inside C.
2. If there are none, then the integral is zero by Cauchy's integral theorem.
3. If there are isolated singularities at $z_1, \ldots, z_n$, determine at each z_j individually the residue of $f(z)$ there. This can be done by inspecting the Laurent expansion in an annulus about z_j [remember that we only need the coefficient of the one term $1/(z - z_j)$] or by some other means (such as Theorem 138 just ahead).
4. In this case, the integral equals $2\pi i$ times the sum of the residues.

Here are two examples, after which we shall look at a relatively easy way to find residues, followed by some more examples.

EXAMPLE 470

Compute $\oint_C \sin(z)/z^2\, dz$, for C any simple closed path not passing through 0.

Here, the integrand has an isolated singularity (a simple pole) at 0, leading us to consider two cases:

Case 1 C does not enclose 0.

Then $\sin(z)/z^2$ is analytic on and inside C; so $\oint_C \sin(z)/z^2\, dz = 0$ by Cauchy's integral theorem.

Case 2 C encloses 0.

Now we must determine the residue of $\sin(z)/z^2$ at 0. The Laurent expansion of $\sin(z)/z^2$ about 0 is easily obtained by dividing the Maclaurin expansion of $\sin(z)$ about 0 by z^2. We get

$$\frac{\sin(z)}{z^2} = \frac{1}{z} - \frac{z}{6} + \frac{z^3}{120} - \cdots \quad \text{in} \quad 0 < |z| < \infty.$$

The coefficient of $1/z$ is 1, and this is the residue of $\sin(z)/z^2$ at 0. Thus, in this case,

$$\oint_C \frac{\sin(z)}{z^2}\, dz = 2\pi i(1) = 2\pi i.$$

EXAMPLE 471

Compute $\oint_C 1/(z^2 + 1)\, dz$ over any simple closed path C not passing through i or $-i$.

Since $1/(z^2 + 1)$ has isolated singularities at i and $-i$, we consider three cases:

Case 1 C does not enclose i or $-i$ (see Figure 312).

Then $f(z)$ is analytic on and inside C; so $\oint_C f(z)\, dz = 0$.

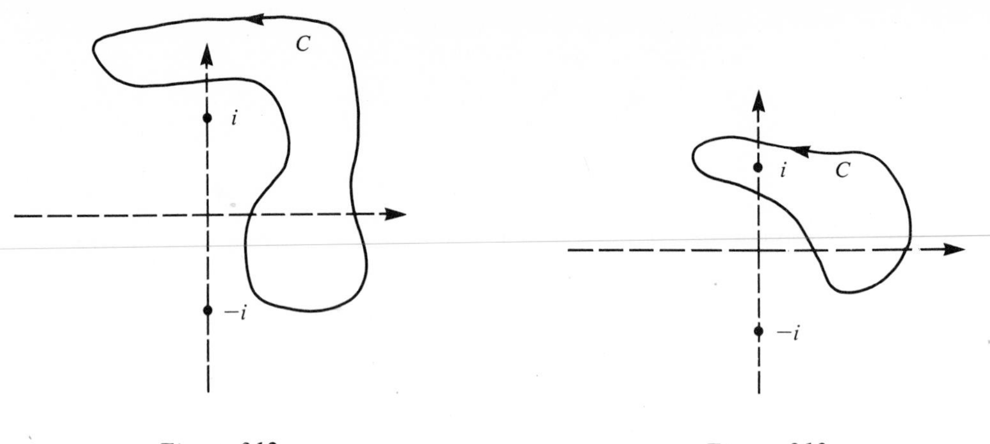

Figure 312

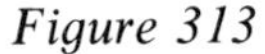

Figure 313

Case 2 C encloses exactly one of i or $-i$.

Say, C encloses i, but not $-i$, as shown in Figure 313. Then,

$$\frac{1}{1+z^2} = \frac{i}{2}\left(\frac{1}{z+i}\right) - \frac{i}{2}\left(\frac{1}{z-i}\right).$$

Now, $1/(z-i)$ is a Laurent expansion in $0 < |z-i| < \infty$. The term $1/(z+i)$ is analytic at i, and hence has a Taylor series about i and contributes no negative powers of $z-i$ to a Laurent expansion of $1/(z^2+1)$ about i. Thus,

$$\operatorname*{Res}_{i} \frac{1}{z^2+1} = -\frac{i}{2}.$$

Then,

$$\oint_C \frac{dz}{z^2+1} = 2\pi i\left(-\frac{i}{2}\right) = \pi.$$

The singularity at $-i$ is irrelevant here, since C does not enclose $-i$.

Similarly, one finds that, if C encloses $-i$ but not i, then

$$\oint_C \frac{dz}{z^2+1} = 2\pi i\left(\frac{i}{2}\right) = -\pi.$$

Case 3 C encloses both i and $-i$, as shown in Figure 314.

Now both singularities are inside C. We have

$$\oint_C \frac{dz}{z^2+1} = 2\pi i\left(\operatorname*{Res}_{i} \frac{1}{z^2+1} + \operatorname*{Res}_{-i} \frac{1}{z^2+1}\right) = 2\pi i\left(-\frac{i}{2} + \frac{i}{2}\right) = 0.$$

The effectiveness of the residue theorem is clearly a function of how easily one can calculate residues. In the case that z_0 is an essential singularity of $f(z)$, we usually have to look at the Laurent expansion; if, however, $f(z)$ has a pole at z_0, then there is an easier way.

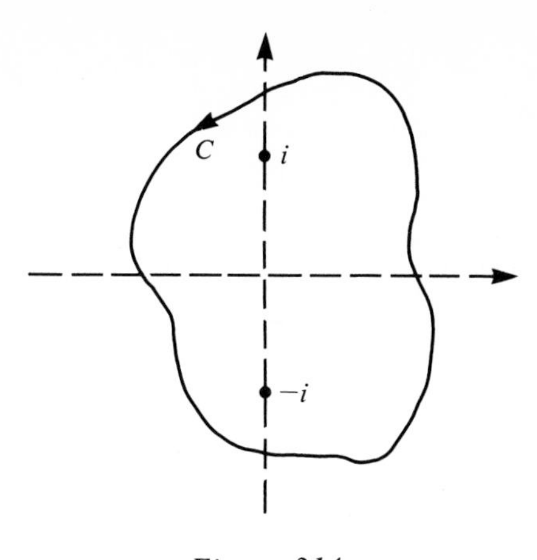

Figure 314

THEOREM 138

Let $f(z)$ have a pole of order m at z_0. Then,

$$\operatorname*{Res}_{z_0} f(z) = \frac{1}{(m-1)!} \lim_{z \to z_0} \frac{d^{m-1}}{dz^{m-1}} [(z - z_0)^m f(z)].$$

In particular,

1. If $m = 1$, this reduces to

$$\operatorname*{Res}_{z_0} f(z) = \lim_{z \to z_0} (z - z_0) f(z).$$

2. If $f(z) = g(z)/h(z)$, where $g(z)$ and $h(z)$ are analytic at z_0, $g(z_0) \neq 0$, and $h(z)$ has a simple zero at z_0, then

$$\operatorname*{Res}_{z_0} f(z) = \frac{g(z_0)}{h'(z_0)}.$$

Proof Expand $f(z)$ in a Laurent series in some annulus $0 < |z - z_0| < r$:

$$f(z) = \frac{a_{-m}}{(z - z_0)^m} + \frac{a_{-m+1}}{(z - z_0)^{m-1}} + \cdots + \frac{a_{-1}}{z - z_0}$$
$$+ a_0 + a_1(z - z_0) + a_2(z - z_0)^2 + \cdots.$$

Then,

$$(z - z_0)^m f(z) = a_{-m} + a_{-m+1}(z - z_0) + \cdots + a_{-1}(z - z_0)^{m-1}$$
$$+ (z - z_0)^m \sum_{n=0}^{\infty} a_n (z - z_0)^n.$$

Letting $g(z) = (z - z_0)^m f(z)$, we find that $g(z)$ is analytic at z_0, since it has a Taylor expansion there. Successive differentiation gives us

$$g^{(m-1)}(z) = (m-1)!\, a_{-1} + \frac{d^{m-1}}{dz^{m-1}} \left[(z - z_0)^m \sum_{n=0}^{\infty} a_n (z - z_0)^n \right].$$

Take a limit as z approaches z_0 and solve for a_{-1} [which is $\text{Res}_{z_0} f(z)$] to get

$$\operatorname*{Res}_{z_0} f(z) = \frac{1}{(m-1)!} \lim_{z \to z_0} g^{(m-1)}(z)$$

$$- \frac{1}{(m-1)!} \lim_{z \to z_0} \frac{d^{m-1}}{dz^{m-1}} \left[(z - z_0)^m \sum_{n=0}^{\infty} a_n (z - z_0)^n \right].$$

Now, differentiation of $\sum_{n=0}^{\infty} a_n(z - z_0)^{n+m}$ term by term $(m - 1)$ times yields a series having a $(z - z_0)$ factor in every term; hence, the second term on the right vanishes, proving the formula for $\text{Res}_{z_0} f(z)$.

For (1), suppose $m = 1$. Then $g(z) = (z - z_0)f(z)$, and

$$\frac{d^{m-1}}{dz^{m-1}} \left[(z - z_0)^m \sum_{n=0}^{\infty} a_n(z - z_0)^n \right] = (z - z_0) \sum_{n=0}^{\infty} a_n(z - z_0)^n,$$

which tends to zero as $z \to z_0$. Further, $(m - 1)! = 0! = 1$; so from the above proof we get

$$\operatorname*{Res}_{z_0} f(z) = \lim_{z \to z_0} (z - z_0)f(z).$$

Finally, for (2), note that $h(z_0) = 0$ but $h'(z_0) \neq 0$; so, by (1),

$$\begin{aligned} \operatorname*{Res}_{z_0} f(z) &= \lim_{z \to z_0} (z - z_0)f(z) \\ &= \lim_{z \to z_0} (z - z_0) \frac{g(z)}{h(z)} \\ &= \lim_{z \to z_0} g(z) \left[\frac{1}{\dfrac{h(z) - h(z_0)}{z - z_0}} \right] = \frac{g(z_0)}{h'(z_0)}. \end{aligned}$$

EXAMPLE 472

Find

$$\operatorname*{Res}_{i} \frac{\sin(2z)}{(z - i)^3}.$$

Since $\sin(2z)$ is analytic everywhere and $\sin(2i) \neq 0$, then the singularity at i lies completely in the denominator term $(z - i)^3$, which has a zero of order 3 at i. Thus, $\sin(2z)/(z - i)^3$ has a pole of order 3 at i. Then,

$$\begin{aligned} \operatorname*{Res}_{i} \frac{\sin(2z)}{(z - i)^3} &= \frac{1}{2!} \lim_{z \to i} \frac{d^2}{dz^2} \left((z - i)^3 \frac{\sin(2z)}{(z - i)^3} \right) \\ &= \lim_{z \to i} \frac{1}{2} \frac{d^2}{dz^2} \sin(2z) \\ &= \tfrac{1}{2} \lim_{z \to i} [-4 \sin(2z)] \\ &= -2 \sin(2i) \\ &= -\frac{2}{2i} (e^{-2} - e^2) = -2i \sinh(2). \end{aligned}$$

As a comparison, the student should try finding this residue by computing the coefficient of $1/(z - i)$ in the Laurent expansion of $\sin(2z)/(z - i)^3$ about i.

EXAMPLE 473

Find

$$\operatorname*{Res}_{-i} \frac{3z^2 + 1}{z^2 + 1}.$$

Here, we can use (2) of Theorem 138, with $g(z) = 3z^2 + 1$ and $h(z) = z^2 + 1$. Then, $g(-i) = -2$ and $h'(-i) = -2i$; so

$$\operatorname*{Res}_{-i} \frac{3z^2 + 1}{z^2 + 1} = \frac{-2}{-2i} = -i.$$

We could also use (1) of Theorem 138, as follows:

$$\operatorname*{Res}_{-i} \frac{3z^2 + 1}{z^2 + 1} = \operatorname*{Res}_{-i} \frac{3z^2 + 1}{(z + i)(z - i)} = \lim_{z \to -i} \frac{3z^2 + 1}{z - i} = -i.$$

EXAMPLE 474

Evaluate

$$\oint_C \frac{\sin(z)\,dz}{z^2(z^2 + 4)},$$

where C encloses 0, $2i$, and $-2i$ (see Figure 315).

Here, $f(z)$ has simple poles at 0, $2i$, and $-2i$. Using Theorem 138(1) three times, we get

$$\operatorname*{Res}_{0} f(z) = \tfrac{1}{4}, \qquad \operatorname*{Res}_{2i} f(z) = \frac{i \sin(2i)}{16},$$

$$\text{and} \quad \operatorname*{Res}_{-2i} f(z) = \frac{i \sin(2i)}{16}.$$

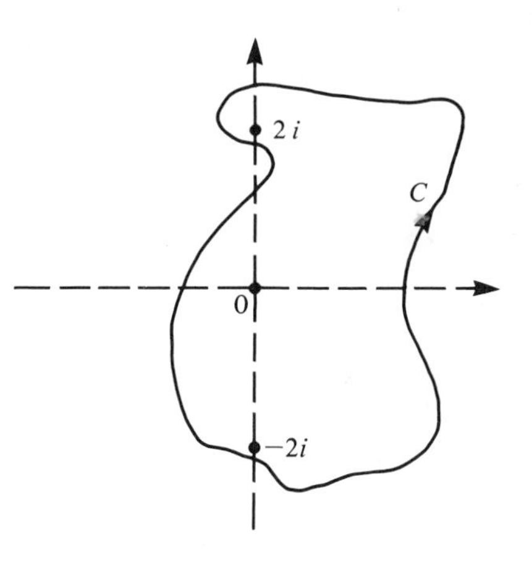

Figure 315

By the residue theorem,

$$\oint_C \frac{\sin(z)\,dz}{z^2(z^2 + 4)} = 2\pi i \left(\frac{1}{4} + \frac{i \sin(2i)}{16} + \frac{i \sin(2i)}{16} \right)$$

$$= \frac{\pi}{4}\,[2i - \sin(2i)].$$

EXAMPLE 475

Evaluate

$$\oint_C \frac{e^{z^2}\,dz}{z^3(z - i)},$$

where C encloses 0 and i.

Here, $f(z)$ has a pole of order three at 0 and a simple pole at i. Using Theorem 138, we calculate

$$\operatorname*{Res}_{0} f(z) = \frac{1}{2!} \lim_{z \to 0} \frac{d^2}{dz^2}\left(\frac{e^{z^2}}{z-i}\right) = 0$$

and

$$\operatorname*{Res}_{i} f(z) = \lim_{z \to i} \frac{e^{z^2}}{z^3} = \frac{e^{-1}}{-i} = \frac{i}{e}.$$

Then,

$$\oint_C \frac{e^{z^2}\,dz}{z^3(z-i)} = 2\pi i\left(\frac{i}{e}\right) = -\frac{2\pi}{e}.$$

EXAMPLE 476

Evaluate

$$\oint_C \frac{e^z \cos(z^2)\,dz}{(z-i)(z+4)^2},$$

where C encloses i and -4.

Here, $f(z)$ has a simple pole at i and a pole of order 2 at -4. Calculate

$$\operatorname*{Res}_{i} f(z) = \lim_{z \to i} \frac{e^z \cos(z^2)}{(z+4)^2} = \frac{e^i \cos(-1)}{(4+i)^2} = \frac{e^i \cos(1)}{(4+i)^2}$$

and

$$\begin{aligned}\operatorname*{Res}_{-4} f(z) &= \lim_{z \to -4} \frac{d}{dz}\left(\frac{e^z \cos(z^2)}{z-i}\right) \\ &= \frac{(-4-i)[e^{-4}\cos(16) + 8e^{-4}\sin(16)] - e^{-4}\cos(16)}{(-4-i)^2}.\end{aligned}$$

Then,

$$\begin{aligned}&\oint_C \frac{e^z \cos(z^2)}{(z-i)(z+4)^2}\,dz \\ &= 2\pi i\left\{\frac{e^i \cos(1)}{(4+i)^2} + \frac{(-4-i)[e^{-4}\cos(16) + 8e^{-4}\sin(16)] - e^{-4}\cos(16)}{(-4-i)^2}\right\}.\end{aligned}$$

This can be rewritten in a variety of ways, a chore we omit.

In the next section we show how the residue theorem can be used to evaluate certain kinds of real integrals which are otherwise intractable.

PROBLEMS FOR SECTION 17.2

For $j = 1, 2, \ldots, 20$, let Problem j be to find the residue of $f(z)$ at each singularity, with $f(z)$ as in Problem j of Section 17.1.

In each of Problems 21 through 35, use the residue theorem to evaluate $\int_C f(z)\,dz$. Assume in each case that C encloses all singularities of $f(z)$.

21. $\oint_C \frac{2z}{(z-i)^2}\,dz$ **22.** $\oint_C \frac{z^2+1}{(z-1)^2(z+2i)}\,dz$ **23.** $\oint_C \frac{\sin(z)}{z^2+4}\,dz$

24. $\oint_C \frac{e^z}{z}\,dz$ **25.** $\oint_C \frac{z-i}{2z+1}\,dz$ **26.** $\oint_C \frac{(z+i)}{z^2+6}\,dz$

27. $\oint_C \frac{\cos(2z)}{ze^z}\,dz$ **28.** $\oint_C \frac{z}{\sinh^2(z)}\,dz$ **29.** $\oint_C e^{2/z^2}\,dz$

30. $\oint_C \frac{iz\,dz}{(z^2+8)(z-i)^3}$ **31.** $\oint_C \frac{z^{1/2}}{z-1+2i}\,dz$ **32.** $\oint_C \frac{8z-4i+1}{\sin^3(z)}\,dz$

33. $\oint_C \frac{(1-z)^2}{z^3+2}\,dz$ **34.** $\oint_C \tanh(z)\,dz$ **35.** $\oint_C \frac{\mathrm{Ln}(z)}{z^2-1}\,dz$

36. Define the residue of $f(z)$ at ∞ to be the residue of $f(1/z)$ at 0. Prove that $\oint_C f(z)\,dz$ is $2\pi i$ times the sum of the residues of $f(z)$ *outside* C, including the one at infinity.

37. Recall the definition of residue at infinity from Problem 36. Let $R(z)$ be a rational function. Prove that the sum of the residues of $R(z)$ at all singularities of $R(z)$ and at ∞ is zero.

[*Hint:* Form $\oint_C R(z)\,dz$, where C is a circle of radius large enough to enclose all finite singularities of $R(z)$.]

38. Use the result of Problem 37 to prove the fundamental theorem of algebra, which says that every nonconstant polynomial with complex coefficients has a complex (possibly real) zero.

[*Hint:* Apply Problem 37 to the rational function $P'(z)/P(z)$, where $P(z)$ is any complex polynomial of degree at least one.]

17.3 APPLICATION OF THE RESIDUE THEOREM TO THE EVALUATION OF REAL INTEGRALS

The residue theorem is a powerful tool for evaluating many kinds of real integrals. We shall treat three classes of real integrals which can be done this way.

A. $\int_0^{2\pi} R[\cos(\theta), \sin(\theta)]\,d\theta$

Suppose that $R(x, y)$ is a quotient of polynomials in x and y. For example, we could have

$$R(x, y) = \frac{2xy^2 - x + y - xy}{2x^2 + 2y^2 - x}.$$

Then $R[\cos(\theta), \sin(\theta)]$ is a function of θ. For example, with $R(x, y)$ as above,

$$R[\cos(\theta), \sin(\theta)] = \frac{2\cos(\theta)\sin^2(\theta) - \cos(\theta) + \sin(\theta) - \sin(\theta)\cos(\theta)}{2 - \cos(\theta)}.$$

We shall assume that $R(x, y)$ has no poles on the unit circle $|z| = 1$ (this is the case in the above example). Then we can evaluate $\int_0^{2\pi} R[\cos(\theta), \sin(\theta)]\, d\theta$ as follows. Write $z = e^{i\theta}, 0 \leq \theta \leq 2\pi$. Then,

$$\cos(\theta) = \frac{1}{2}(e^{i\theta} + e^{-i\theta}) = \frac{1}{2}\left(z + \frac{1}{z}\right)$$

and

$$\sin(\theta) = \frac{1}{2i}(e^{i\theta} - e^{-i\theta}) = \frac{1}{2i}\left(z - \frac{1}{z}\right).$$

If $z = e^{i\theta}$, then $dz = ie^{i\theta}\, d\theta = iz\, d\theta$; so we obtain

$$\boxed{\int_0^{2\pi} R[\cos(\theta), \sin(\theta)]\, d\theta = \oint_{|z|=1} R\left[\frac{1}{2}\left(z + \frac{1}{z}\right), \frac{1}{2i}\left(z - \frac{1}{z}\right)\right]\frac{1}{iz}\, dz.}$$

This is an integral of a complex function around the unit circle and may be evaluated by any of our complex variable methods, including the residue theorem.

EXAMPLE 477

Evaluate

$$\int_0^{2\pi} \frac{\sin^2(\theta)\, d\theta}{2 + \cos(\theta)}.$$

The integrand is a rational function of $\sin(\theta)$ and $\cos(\theta)$, and $2 + \cos(\theta) \neq 0$ for $0 \leq \theta \leq 2\pi$; so the method will apply. Let

$$\sin(\theta) = \frac{1}{2i}\left(z - \frac{1}{z}\right) \quad \text{and} \quad \cos(\theta) = \frac{1}{2}\left(z + \frac{1}{z}\right).$$

We get:

$$\int_0^{2\pi} \frac{\sin^2(\theta)\, d\theta}{2 + \cos(\theta)} = \oint_{|z|=1} \frac{\left[\frac{1}{2i}\left(z - \frac{1}{z}\right)\right]^2}{2 + \frac{1}{2}\left(z + \frac{1}{z}\right)} \frac{1}{iz}\, dz$$

$$= \frac{-1}{2i}\oint_{|z|=1} \frac{(z^4 - 2z^2 + 1)}{z^2(z^2 + 4z + 1)}\, dz.$$

Use the residue theorem to evaluate this. The integrand has a pole of order 2 at 0 and simple poles at $-2 \pm \sqrt{3}$. But, the pole at $-2 - \sqrt{3}$ is *outside* $|z| = 1$, and hence is irrelevant; only 0 and $-2 + \sqrt{3}$ are *inside* $|z| = 1$.

Compute the residues at the poles, using Theorem 138. We get

$$\operatorname*{Res}_0 \frac{z^4 - 2z^2 + 1}{z^2(z^2 + 4z + 1)} = -4$$

and

$$\operatorname*{Res}_{-2+\sqrt{3}} \frac{z^4 - 2z^2 + 1}{z^2(z^2 + 4z + 1)} = \frac{42 - 24\sqrt{3}}{-12 + 7\sqrt{3}}.$$

Then, by the residue theorem,

$$\oint_{|z|=1} \frac{z^4 - 2z^2 + 1}{z^2(z^2 + 4z + 1)}\, dz = 2\pi i\left(-4 + \frac{42 - 24\sqrt{3}}{-12 + 7\sqrt{3}}\right) = 2\pi i\left(\frac{90 - 52\sqrt{3}}{-12 + 7\sqrt{3}}\right).$$

Then,

$$\int_0^{2\pi} \frac{\sin^2(\theta)\, d\theta}{2 + \cos(\theta)} = -\frac{1}{2i}(2\pi i)\left(\frac{90 - 52\sqrt{3}}{-12 + 7\sqrt{3}}\right) = \pi\left(\frac{52\sqrt{3} - 90}{7\sqrt{3} - 12}\right).$$

This is approximately 1.6836.

B. $\int_{-\infty}^{\infty} R(x)\, dx$

Suppose that $R(x)$ is a rational function of x, say, $R(x) = P(x)/Q(x)$, with $P(x)$ and $Q(x)$ polynomials. To ensure convergence of $\int_{-\infty}^{\infty} R(x)\, dx$, we shall assume that $Q(x)$ has no real zeros and that the degree of $Q(x)$ exceeds that of $P(x)$ by at least two.

To evaluate $\int_{-\infty}^{\infty} R(x)\, dx$, note first that $R(z) = P(z)/Q(z)$ has only poles as singularities and that these poles are exactly the zeros of $Q(z)$. Since we are assuming that none of the zeros of $Q(z)$ are real, they occur in complex conjugate pairs, say, $z_1, \bar{z}_1, z_2, \bar{z}_2, \ldots, z_n, \bar{z}_n$, with $z_1, \ldots, z_n$ in the upper half-plane. Let r be large enough that each $|z_j| < r$, and let C consist of the semicircle S: $z = re^{i\theta}$, $0 \le \theta \le \pi$, and the real axis L from $-r$ to r (see Figure 316). Then,

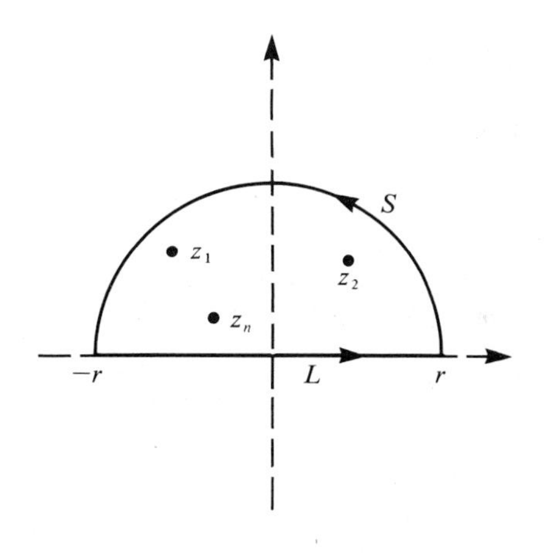

Figure 316

$$\oint_C R(z)\, dz = 2\pi i\left[\sum_{j=1}^{n} \operatorname*{Res}_{z_j} R(z)\right] = \int_S R(z)\, dz + \int_L R(z)\, dz.$$

Now, the degree of $Q(z)$ is at least as large as that of $z^2P(z)$ by assumption; so $|z^2P(z)/Q(z)|$ is bounded for $|z|$ large (say, $|z| > r$). For some M, then, $|P(z)/Q(z)| \le M/|z|^2$ for $|z|$ sufficiently large. Then,

$$\left|\int_S R(z)\, dz\right| \le \frac{M}{r^2}(\pi r) = \frac{\pi M}{r},$$

and this goes to zero as $r \to \infty$. Further, as $r \to \infty$,

$$\int_L R(z)\,dz \to \int_{-\infty}^{\infty} R(x)\,dx,$$

since $z = x$ on the real line. We thus obtain:

$$\int_{-\infty}^{\infty} R(x)\,dx = 2\pi i \cdot \begin{array}{l}\text{[sum of residues of } R(z) \text{ at} \\ \text{poles in the upper half-plane].}\end{array}$$

EXAMPLE 478

Evaluate

$$\int_{-\infty}^{\infty} \frac{dx}{x^6 + 64}.$$

The poles of $1/(z^6 + 64)$ are the zeros of $z^6 + 64$, which are:

$$2e^{\pi i/6}, \quad 2e^{i\pi/2}(= 2i), \quad 2e^{5\pi i/6}, \quad 2e^{7\pi i/6}, \quad 2e^{3\pi i/2} \quad \text{and} \quad 2e^{11\pi i/6}.$$

Three of these ($2e^{\pi i/6}$, $2i$, and $2e^{5\pi i/6}$) occur in the upper half-plane. Thus,

$$\int_{-\infty}^{\infty} \frac{dx}{x^6 + 64} = 2\pi i \left[\operatorname*{Res}_{2e^{\pi i/6}} \frac{1}{(z^6 + 64)} + \operatorname*{Res}_{2e^{\pi i/2}} \frac{1}{(z^6 + 64)} + \operatorname*{Res}_{2e^{5\pi i/6}} \frac{1}{(z^6 + 64)} \right].$$

Since the poles of $1/(z^6 + 64)$ are all simple zeros of the denominator, we can apply Theorem 138(2) with $g(z) = 1$ and $h(z) = z^6 + 64$. We get

$$\operatorname*{Res}_{2e^{\pi i/6}} \left(\frac{1}{z^6 + 64} \right) = \frac{1}{6(2e^{\pi i/6})^5} = \frac{1}{192} \left[\cos\left(\frac{5\pi}{6}\right) - i \sin\left(\frac{5\pi}{6}\right) \right],$$

$$\operatorname*{Res}_{2i} \left(\frac{1}{z^6 + 64} \right) = \frac{1}{6(2i)^5} = \frac{-i}{192},$$

and

$$\operatorname*{Res}_{2e^{5\pi i/6}} \left(\frac{1}{z^6 + 64} \right) = \frac{1}{6(2e^{5\pi i/6})^5} = \frac{1}{192} \left[\cos\left(\frac{\pi}{6}\right) - i \sin\left(\frac{\pi}{6}\right) \right].$$

Since $\cos(\pi/6) = -\cos(5\pi/6)$ and $\sin(\pi/6) = \sin(5\pi/6) = \frac{1}{2}$, we get

$$\int_{-\infty}^{\infty} \frac{dx}{x^6 + 64} = \frac{2\pi i}{192} \left(-\frac{i}{2} - i - \frac{i}{2} \right) = \frac{\pi}{48}.$$

C. $\displaystyle\int_{-\infty}^{\infty} R(x)\cos(ax)\,dx$ and $\displaystyle\int_{-\infty}^{\infty} R(x)\sin(ax)\,dx$

Using the method of B, it is straightforward to establish that

$$\int_{-\infty}^{\infty} R(x)\cos(ax)\,dx + i\int_{-\infty}^{\infty} R(x)\sin(ax)\,dx$$

$$= 2\pi i \cdot \begin{array}{l}\text{(sum of residues of } R(z)e^{iaz} \text{ at poles}\\ \text{of } R(z) \text{ in the upper half-plane).}\end{array}$$

Here, $R(x)$ is, as in B, a rational function whose denominator has no real zeros and has degree at least 2 greater than the degree of the numerator. In practice, one calculates the residues and evaluates the right side of the last equation; the real part of the right side is then $\int_{-\infty}^{\infty} R(x)\cos(ax)\,dx$ and the imaginary part of the right side is $\int_{-\infty}^{\infty} R(x)\sin(ax)\,dx$.

Here, a is any positive real constant.

EXAMPLE 479

Evaluate

$$\int_{-\infty}^{\infty} \frac{x\sin(\sqrt{3}\,x)\,dx}{x^6+64}.$$

Here, $R(x) = x/(x^6+64)$, and the complex function $R(z) = z/(z^6+64)$ has the same poles as $1/(z^6+64)$ (see Example 478). The poles in the upper half-plane are $2e^{\pi i/6}$, $2i$, and $2e^{5\pi i/6}$. Thus,

$$\int_{-\infty}^{\infty} \frac{\cos(\sqrt{3}\,x)\,dx}{x^6+64} + i\int_{-\infty}^{\infty} \frac{\sin(\sqrt{3}\,x)\,dx}{x^6+64}$$

$$= 2\pi i\left(\operatorname*{Res}_{2e^{\pi i/6}} \frac{ze^{i\sqrt{3}z}}{z^6+64} + \operatorname*{Res}_{2i} \frac{ze^{i\sqrt{3}z}}{z^6+64} + \operatorname*{Res}_{2e^{5\pi i/6}} \frac{ze^{i\sqrt{3}z}}{z^6+64}\right).$$

Again using Theorem 138, we get

$$\operatorname*{Res}_{2e^{\pi i/6}}\left(\frac{ze^{i\sqrt{3}z}}{z^6+64}\right) = \frac{e^{-\sqrt{3}}}{96}e^{i(3-2\pi/3)},$$

$$\operatorname*{Res}_{2i}\left(\frac{ze^{i\sqrt{3}z}}{z^6+64}\right) = \frac{1}{96}e^{-2\sqrt{3}},$$

and

$$\operatorname*{Res}_{2e^{5\pi i/6}}\left(\frac{ze^{i\sqrt{3}z}}{z^6+64}\right) = \frac{e^{-\sqrt{3}}}{96}e^{i(-3-10\pi/3)}.$$

Then, adding these residues together, we have

$$\int_{-\infty}^{\infty} \frac{\cos(\sqrt{3}\,x)\,dx}{x^6+64} + i\int_{-\infty}^{\infty} \frac{\sin(\sqrt{3}\,x)\,dx}{x^6+64}$$

$$= \frac{2\pi i e^{-\sqrt{3}}}{96}\left[e^{i(3-2\pi/3)} + e^{-\sqrt{3}} + e^{i(-3-10\pi/3)}\right]$$

$$= \frac{2\pi i e^{-\sqrt{3}}}{96}\left[\cos\left(3-\frac{2\pi}{3}\right) + i\sin\left(3-\frac{2\pi}{3}\right) + e^{-\sqrt{3}} + \cos\left(-3-\frac{10\pi}{3}\right) + i\sin\left(-3-\frac{10\pi}{3}\right)\right]$$

$$= \frac{\pi i e^{-\sqrt{3}}}{96}\left[2\cos\left(3-\frac{2\pi}{3}\right) + e^{-\sqrt{3}}\right],$$

since

$$\sin\left(3-\frac{2\pi}{3}\right) = \sin\left(3+\frac{10\pi}{3}\right) \quad \text{and} \quad \cos\left(3-\frac{2\pi}{3}\right) = \cos\left(-3-\frac{10\pi}{3}\right).$$

Now equate the real part of the left side with the real part of the right, followed by the imaginary parts, to get

$$\int_{-\infty}^{\infty} \frac{x\cos(\sqrt{3}\,x)\,dx}{x^6+64} = 0$$

(to be expected, as the integrand is an odd function) and

$$\int_{-\infty}^{\infty} \frac{x\sin(\sqrt{3}\,x)\,dx}{x^6+64} = \frac{\pi}{48}e^{-\sqrt{3}}\left[2\cos\left(3-\frac{2\pi}{3}\right) + e^{-\sqrt{3}}\right].$$

This is approximately 0.01634257.

Note that, although we are using complex methods, the final answers in evaluating these integrals must be real, because the integrals are real. If you obtain a real integral equal to a complex number, you have made a mistake somewhere.

There are many other kinds of real integrals which can be evaluated by the residue theorem. Some are pursued in the exercises. In the next section we show how the residue theorem can be used to sum certain kinds of real series.

PROBLEMS FOR SECTION 17.3

In each of Problems 1 through 20, evaluate the given integral using residue methods of this section.

1. $\displaystyle\int_0^{2\pi} \frac{d\theta}{2-\cos(\theta)}$

2. $\displaystyle\int_0^{2\pi} \frac{2\sin(\theta)\,d\theta}{\sin^2(\theta)+2}$

3. $\displaystyle\int_0^{2\pi} \frac{1}{4-\sin^2(\theta)}\,d\theta$

4. $\displaystyle\int_0^{2\pi} \frac{d\theta}{a^2\cos^2(\theta)+b^2\sin^2(\theta)}$ (a, b real, nonzero)

5. $\displaystyle\int_{-\infty}^{\infty} \frac{x^2 - x + 2}{x^4 + 10x^2 + 9}\,dx$

6. $\displaystyle\int_0^{\infty} \frac{x\sin(x)}{x^2 + a^2}\,dx \quad \left(\textit{Note: } 2\int_0^{\infty} = \int_{-\infty}^{\infty} \text{ here.}\right)$

7. $\displaystyle\int_{-\infty}^{\infty} \frac{dx}{x^4 + 1}$

8. $\displaystyle\int_{-\infty}^{\infty} \frac{\sin(3x)\,dx}{x^2 + 9}$

9. $\displaystyle\int_0^{\infty} \frac{\cos(2x)\,dx}{x^4 + 1}$

10. $\displaystyle\int_{-\infty}^{\infty} \frac{(x + 1)^2\,dx}{x^2 - 2x + 3}$

11. $\displaystyle\int_0^{\infty} \frac{\cos(ax)\,dx}{(x^2 + b^2)^2}$ (a, b real, nonzero)

12. $\displaystyle\int_0^{2\pi} \cos^{2n}(\theta)\,d\theta$ (n any positive integer)

Note: Observe that, after converting to a complex line integral, the integrand in Problem 12 has a pole of order $2n + 1$ at 0; the residue at 0 is the coefficient of z^{2n} in the binomial expansion of $(1 + z^2)^{2n}$.

13. $\displaystyle\int_{-\infty}^{\infty} \frac{dx}{x^6 + 3}$

14. $\displaystyle\int_0^{2\pi} \frac{d\theta}{a + b\cos(\theta)}$ $(0 < b < a)$

15. $\displaystyle\int_0^{2\pi} \frac{d\theta}{[a + b\cos(\theta)]^2}$ $(0 < b < a)$

16. $\displaystyle\int_{-\infty}^{\infty} \frac{dx}{(1 + x^2)^n}$

17. $\displaystyle\int_{-\infty}^{\infty} \frac{\cos(ax)\,dx}{(x^2 + b^2)(x^2 + c^2)}$ $(a, b, c > 0;\ b \neq c)$

18. $\displaystyle\int_{-\infty}^{\infty} \frac{x^4\,dx}{(1 + ax^2)^4}$ $(a > 0)$

19. $\displaystyle\int_{-\infty}^{\infty} \frac{x^2\cos^2(x)}{x^4 + 2}\,dx$

20. $\displaystyle\int_0^{2\pi} \frac{[\sin(\theta) + \cos(\theta)]\,d\theta}{2 - \cos(\theta)}$

21. Evaluate $\int_0^{\infty} \sin(x)/x\,dx$ by the residue theorem as follows. Consider $\int_C (e^{iz}/z)\,dz$, where C is the path shown in Figure 317. By Cauchy's integral theorem, $\int_C e^{iz}/z\,dz = 0$. Now break the line integral up into a sum of four integrals, over the two semicircular arcs of C and the two line segments on the real axis belonging to C. Show that, as $R \to \infty$ and $r \to 0$, $\int_{C_R} e^{iz}/z\,dz \to 0$ and $\int_{C_r} e^{iz}/z\,dz \to -\pi i$, and use this to obtain $\int_0^{\infty} \sin(x)/x\,dx = \pi/2$.

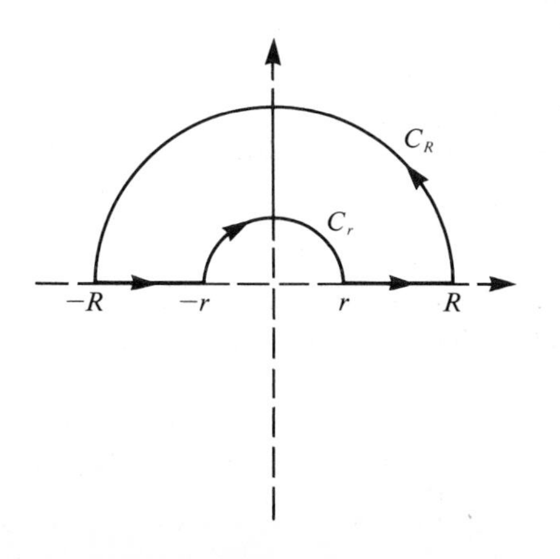

Figure 317

22. Using the same path as in Problem 21, consider $\int_C [\text{Ln}(z)]^2/(1 + z^2)\,dz$ to obtain

$$\int_0^\infty \frac{[\ln(x)]^2}{1 + x^2}\,dx = \frac{\pi^3}{8}.$$

23. Use the residue theorem to derive the formula

$$\int_{-\infty}^\infty \frac{dx}{(1 + x^2)^n} = \frac{\pi(2n - 2)!}{2^{2n-2}[(n - 1)!]^2}$$

for $n = 1, 2, 3, \ldots$.

24. Derive *Fresnel's integrals*:

$$\int_0^\infty \cos(x^2)\,dx = \int_0^\infty \sin(x^2)\,dx = \frac{1}{2}\sqrt{\frac{\pi}{2}}.$$

Hint: Consider $\int_C e^{-z^2}\,dz$, with C as shown in Figure 318 (C goes from 0 to r along the x-axis; then along the circle $|z| = r$ to $\frac{1}{2}\sqrt{2}(1 + i)r$; then along the straight line from $\frac{1}{2}\sqrt{2}(1 + i)r$ to the origin). Determine what happens along each part of C as $r \to \infty$.

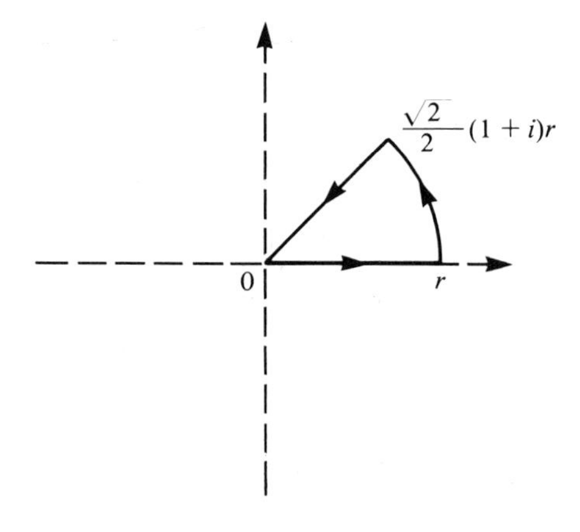

Figure 318

17.4 APPLICATION OF THE RESIDUE THEOREM TO THE SUMMATION OF SERIES

Usually, finding the sum of a convergent series is an impossible task. Using the residue theorem, however, we can often succeed with convergent series of the form

$$\sum_{n=-\infty}^{\infty} \frac{1}{P(n)} \quad \text{and} \quad \sum_{n=-\infty}^{\infty} \frac{(-1)^n}{P(n)},$$

where $P(z)$ is a polynomial of degree 2 or higher, having no integer zeros. In fact, one can show that, if $w_1, \dots, w_k$ are the zeros of $P(z)$, then

$$\sum_{n=-\infty}^{\infty} \frac{1}{P(n)} = -\pi \sum_{j=1}^{k} \operatorname*{Res}_{w_j} \left(\frac{\cot(\pi z)}{P(z)} \right)$$

and

$$\sum_{n=-\infty}^{\infty} \frac{(-1)^n}{P(n)} = -\pi \sum_{j=1}^{k} \operatorname*{Res}_{w_j} \left(\frac{\csc(\pi z)}{P(z)} \right).$$

We shall indicate a proof in Problems 11 and 12. Here are two illustrations of the use of these formulas.

EXAMPLE 480

Sum the series

$$\sum_{n=0}^{\infty} \frac{1}{n^2 + a^2},$$

where a is any positive constant.

Let

$$P(z) = z^2 + a^2.$$

Then $P(z)$ is a polynomial of degree 2, with zeros $\pm ai$, neither of which is an integer. Now note that $(-n)^2 + a^2 = n^2 + a^2$; so

$$\sum_{n=-\infty}^{\infty} \frac{1}{n^2 + a^2} = 2 \sum_{n=0}^{\infty} \frac{1}{n^2 + a^2} - \frac{1}{a^2}.$$

(The $-1/a^2$ term is needed because in $2\sum_{n=0}^{\infty}$ we counted the zero term $1/a^2$ twice.)

Thus,

$$\sum_{n=0}^{\infty} \frac{1}{n^2 + a^2} = \frac{1}{2a^2} + \frac{1}{2} \sum_{n=-\infty}^{\infty} \frac{1}{n^2 + a^2}.$$

But,

$$\sum_{n=-\infty}^{\infty} \frac{1}{n^2 + a^2} = -\pi \left(\operatorname*{Res}_{ai} \frac{\cot(\pi z)}{z^2 + a^2} + \operatorname*{Res}_{-ai} \frac{\cot(\pi z)}{z^2 + a^2} \right).$$

Since $\cot(\pi z)/(z^2 + a^2)$ has simple poles at $\pm ai$, we can use Theorem 138(2) to get

$$\operatorname*{Res}_{ai} \frac{\cot(\pi z)}{z^2 + a^2} = \frac{\cot(\pi ai)}{2ai}$$

and

$$\operatorname*{Res}_{-ai} \frac{\cot(\pi z)}{z^2 + a^2} = \frac{\cot(\pi a i)}{2ai}.$$

Thus,

$$\sum_{n=-\infty}^{\infty} \frac{1}{n^2 + a^2} = -\pi \frac{\cot(\pi a i)}{ai} = \frac{\pi}{a} i \cot(\pi a i) = \frac{\pi \coth(\pi a)}{a}.$$

Hence,

$$\sum_{n=0}^{\infty} \frac{1}{n^2 + a^2} = \frac{1}{2a^2} + \frac{\pi}{2a} \coth(\pi a).$$

Thus, for example,

$$\sum_{n=0}^{\infty} \frac{1}{n^2 + 1} = \frac{1}{2} + \frac{\pi}{2} \coth(\pi),$$

which is approximately 2.076674.

EXAMPLE 481

Find

$$\sum_{n=0}^{\infty} \frac{(-1)^n}{(2n+1)^3}.$$

Here, we let $P(z) = (2z + 1)^3$. Then $P(z)$ has degree 3 and only one zero at $z = -\frac{1}{2}$, which is not an integer. Since $\csc(-\pi/2) \neq 0$, then $\csc(\pi z)/(2z + 1)^3$ has a pole of order 3 at $-\frac{1}{2}$, with residue (by Theorem 138) equal to

$$\frac{1}{2} \lim_{z \to -1/2} \frac{d^2}{dz^2} \left[(z + \tfrac{1}{2})^3 \left(\frac{\csc(\pi z)}{(2z + 1)^3} \right) \right] = \frac{1}{2} \lim_{z \to -1/2} \frac{d^2}{dz^2} \left(\frac{\csc(\pi z)}{8} \right) = -\frac{\pi^2}{16}.$$

Then,

$$\sum_{n=-\infty}^{\infty} \frac{(-1)^n}{P(n)} = \sum_{n=-\infty}^{\infty} \frac{(-1)^n}{(2n+1)^3} = -\pi \left(\frac{-\pi^2}{16} \right) = \frac{\pi^3}{16}.$$

Next, the student can check that

$$\sum_{n=-\infty}^{\infty} \frac{(-1)^n}{(2n+1)^3} = 2 \sum_{n=0}^{\infty} \frac{(-1)^n}{(2n+1)^3}.$$

Then,

$$\sum_{n=0}^{\infty} \frac{(-1)^n}{(2n+1)^3} = \frac{1}{2} \left(\frac{\pi^3}{16} \right) = \frac{\pi^3}{32}.$$

This is approximately 0.968946.

PROBLEMS FOR SECTION 17.4

In each of Problems 1 through 10, sum the given series.

1. $\sum_{n=-\infty}^{\infty} \frac{1}{n^4 + a^4}$ (a any positive constant)

2. $\sum_{n=-\infty}^{\infty} \frac{(-1)^n}{n^4 + a^4}$ (a any positive constant)

3. $\sum_{n=-\infty}^{\infty} \frac{1}{(4n+3)^2}$

4. $\sum_{n=-\infty}^{\infty} \frac{(-1)^n}{(4n+3)^2}$

5. $\sum_{n=0}^{\infty} \frac{1}{(2n^2+1)^2}$

6. $\sum_{n=0}^{\infty} \frac{1}{(4n^2+3)^2}$

7. $\sum_{n=0}^{\infty} \frac{(-1)^n}{n^4 - 2}$

8. $\sum_{n=-\infty}^{\infty} \frac{(-1)^n}{3n^2 + 4}$

9. $\sum_{n=0}^{\infty} \frac{(-1)^n}{n^6 + 1}$

10. $\sum_{n=-\infty}^{\infty} \frac{1}{n^6 - 4}$

11. Derive the formula for $\sum_{n=-\infty}^{\infty} 1/P(n)$ as follows. Consider $\oint_{C_n} [\pi \cot(\pi z)/P(z)]\, dz$, where C_n is the path shown in Figure 319. Assume that $P(z)$ is a polynomial of degree ≥ 2, with zeros $w_1, \ldots, w_k$, with no w_j an integer. Begin by choosing n large enough so that all the w_j's are inside C_n. By the residue theorem,

$$\oint_{C_n} \frac{\pi \cot(\pi z)}{P(z)}\, dz = 2\pi i \cdot \text{(sum of residues of } \pi \cot(\pi z)/P(z) \text{ at singularities inside } C_n).$$

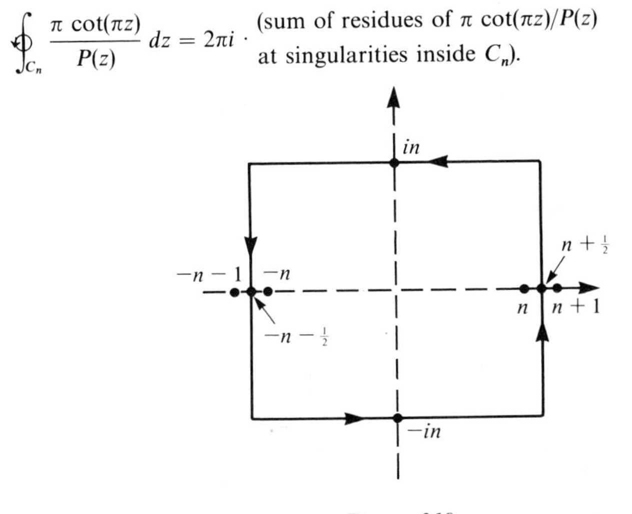

Figure 319

Evaluate these residues to obtain the right-hand side of this equation, then show that $\oint_{C_n} [\pi \cot(\pi z)/P(z)]\, dz \to 0$ as $n \to \infty$ to derive the desired formula.

12. Use an argument similar to that of Problem 11, with $\pi \csc(\pi z)/P(z)$ in place of $\pi \cot(\pi z)/P(z)$, to derive the formula

$$\sum_{n=-\infty}^{\infty} \frac{(-1)^n}{P(n)} = -\pi \sum_{j=1}^{k} \operatorname*{Res}_{w_j} \left(\frac{\csc(\pi z)}{P(z)} \right).$$

17.5 THE ARGUMENT PRINCIPLE

In this section we prove a result which is sometimes useful in estimating the number of zeros of a function and in evaluating integrals.

THEOREM 139 The Argument Principle

Let $f(z)$ be analytic in a domain D except at a finite number of poles. Let C be a simple closed path in D not passing through any zeros or poles of $f(z)$. Then,

$$\frac{1}{2\pi i}\oint_C \frac{f'(z)}{f(z)}\,dz = \begin{array}{l}\text{number of zeros of } f(z) \text{ inside } C\\ \text{minus number of poles of } f(z) \text{ inside } C.\end{array}$$

Here, each zero and pole is counted as many times as its order. Thus, for example, a pole of order 3 is counted three times.

Before proving the theorem, we show a typical use.

EXAMPLE 482

Consider $\oint_C \cot(z)\,dz$, with C as shown in Figure 320. Note that $\cot(z) = f'(z)/f(z)$, with $f(z) = \sin(z)$. Inside C, $f(z)$ has zeros of order 1 at -2π, $-\pi$, 0, π, and 2π and no poles. Then, according to the argument principle,

$$\oint_C \frac{f'(z)}{f(z)}\,dz = \oint_C \cot(z)\,dz$$

Number of zeros inside C

↓

$$= 2\pi i(5 - 0) = 10\pi i.$$

↑

Number of poles inside C

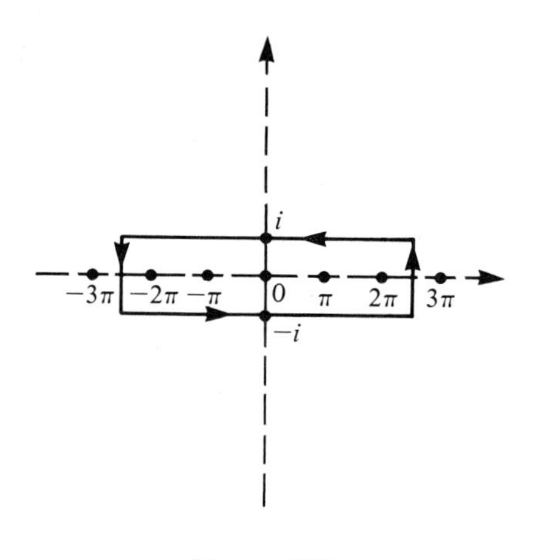

Figure 320

Proof of the Theorem Suppose first that $f(z)$ has a pole of order m at w inside C. Expand $f(z)$ in a Laurent series about w to get

$$f(z) = \frac{a_{-m}}{(z-w)^m} + \frac{a_{-m+1}}{(z-w)^{m+1}} + \cdots + \frac{a_{-1}}{z-w} + a_0 + a_1(z-w) + a_2(z-w)^2 + \cdots$$

in some annulus $0 < |z - w| < r$. Then,

$$(z-w)^m f(z) = a_{-m} + a_{-m+1}(z-w) + \cdots + a_{-1}(z-w)^{m-1} + a_0(z-w)^m + \cdots$$

is analytic at w.

Further, it is easy to check that

$$\frac{f'(z)}{f(z)} = \frac{-m}{z-w} + \frac{[(z-w)^m f(z)]'}{(z-w)^m f(z)}.$$

This is the Laurent series for $f'(z)/f(z)$ about w; hence,

$$\operatorname*{Res}_{w} \frac{f'(z)}{f(z)} = -m.$$

Similarly, if $f(z)$ has a zero of order n at w_0, then in some annulus about w_0,

$$\frac{f'(z)}{f(z)} = \frac{n}{z - w_0} + \frac{g'(z)}{g(z)},$$

where $g(z) = (z - w_0)^{-n} f(z)$.

Thus,

$$\operatorname*{Res}_{w_0} \frac{f'(z)}{f(z)} = n.$$

In sum, at a pole of $f(z)$ of order m inside C, $f'(z)/f(z)$ has residue $-m$; at a zero of $f(z)$ of order n inside C, $f'(z)/f(z)$ has residue $+n$. The argument principle now follows directly from the residue theorem.

PROBLEMS FOR SECTION 17.5

1. Let $P(z) = (z - w_1)(z - w_2) \cdots (z - w_n)$, where $w_1, \ldots, w_n$ are given complex numbers. Let C be a simple closed path enclosing $w_1, \ldots, w_n$. Prove that

$$\oint_C \frac{P'(z)}{P(z)}\, dz = 2\pi i n$$

   (a) By using the argument principle.
   (b) By the residue theorem.

2. Prove *Rouché's theorem:* Let $f(z)$ and $g(z)$ be analytic on and inside a closed path C, and suppose that $|f(z) - g(z)| < |f(z)|$ for all z on C. Then, $f(z)$ and $g(z)$ have the same number of zeros (counting multiplicities) enclosed by C.
   (*Hint:* Note that neither $f(z)$ nor $g(z)$ can have a zero on C, and apply the argument principle to $F(z) = g(z)/f(z)$.)

3. Use Rouché's theorem to prove the fundamental theorem of algebra.
   (*Hint:* Let $P(z) = z^n + a_{n-1}z^{n-1} + \cdots + a_1 z + a_0$, with $a_{n-1}, \ldots, a_0$ complex. We want to show that, for some z, $P(z) = 0$. Let $f(z) = z^n$ and $g(z) = P(z)$ in Rouché's theorem.)

4. Use Rouché's theorem to prove that if $f(z)$ is analytic on and inside a closed path C, and $f(z)$ is real for all z on C, then $f(z)$ must be constant.

5. Use Rouché's theorem to prove that $z^5 + 15z + 1 = 0$ has all its roots inside $|z| < 1$. (*Hint:* Let $f(z) = z^5$ in Rouché's theorem.)

SUPPLEMENTARY PROBLEMS

In each of Problems 1 through 30, evaluate the given integral.

1. $\displaystyle\oint_C \frac{z^2\, dz}{(z - i)^4}$, C any simple closed path enclosing i

2. $\displaystyle\int_0^{2\pi} \frac{3\,d\theta}{4+\cos^2(\theta)}$

3. $\displaystyle\int_{-\infty}^{\infty} \frac{dx}{(x^2+3)(x^2+2)}$

4. $\displaystyle\oint_C \frac{e^{2z}\,dz}{z(z-i)}$, C any simple closed path about 0 and i

5. $\displaystyle\int_0^{2\pi} \frac{\sin^2(z)\,dz}{2+\sin(z)}$

6. $\displaystyle\oint_C \frac{\sin(z)\cos(2z)\,dz}{z^2+1}$, C any simple closed path not passing through i or $-i$ (consider all cases)

7. $\displaystyle\oint_C \frac{\cosh(3z)\,dz}{z(z^2+3)}$, C any simple closed path not passing through 0, $\sqrt{3}\,i$, or $-\sqrt{3}\,i$

8. $\displaystyle\int_0^{2\pi} \frac{[\cos(\theta)-\sin(\theta)]\,d\theta}{2-\cos(\theta)}$

9. $\displaystyle\int_{-\infty}^{\infty} \frac{x^2\cos(x)\,dx}{x^6+1}$

10. $\displaystyle\oint_C \frac{(z-i)^2\,dz}{\sin^2(z)}$, C the square with vertices at $4+4i$, $4-4i$, $-4+4i$, and $-4-4i$

11. $\displaystyle\oint_C \frac{e^z(z^2-4)^2\,dz}{(z+i)^2}$, C the circle of radius 6 about 1

12. $\displaystyle\int_0^{2\pi} \frac{\sin(2\theta)\,d\theta}{\cos(\theta)+\sin(\theta)}$

13. $\displaystyle\oint_C \frac{\cosh(2z)\,dz}{(z-i)(z^2+1)}$, C the circle of radius 8 about 0

14. $\displaystyle\oint_C \frac{e^{z^2}\,dz}{z^4}$, C any simple closed path about the origin

15. $\displaystyle\oint_C \frac{\sin(z)\sinh(z)\,dz}{z^3}$, C any simple closed path about the origin

16. $\displaystyle\int_0^{2\pi} \frac{\sin^2(\theta)\,d\theta}{3+\cos^2(\theta)}$

17. $\displaystyle\oint_C \frac{ze^{-z}\,dz}{z^4+i}$, C the circle of radius 4 about the origin

18. $\displaystyle\oint_C \frac{dz}{z^8+1}$, C the circle of radius 1 about -1

19. $\displaystyle\int_{-\infty}^{\infty} \frac{x^4\,dx}{x^6+1}$

20. $\displaystyle\int_0^{2\pi} \frac{d\theta}{3+2\sin^2(\theta)}$

21. $\displaystyle\oint_C \frac{(z-2i)^3\,dz}{z^2-2z+2}$, C the circle of radius 8 about the origin

22. $\displaystyle\int_{-\infty}^{\infty} \frac{dx}{x^8+1}$

23. $\displaystyle\oint_C \frac{e^{\sin(z)}\,dz}{z^2}$, C any simple closed path about the origin

24. $\displaystyle\oint_C \frac{(z+1)\cosh(z)\,dz}{(z-i)(z+2i)}$, C the circle of radius 1 about $-2i$

25. $\displaystyle\oint_C \frac{(2z-3)\,dz}{(z+i)^3}$, C the circle of radius 3 about 0

26. $\displaystyle\int_0^{2\pi} \frac{d\theta}{\cos(\theta)+\sin^2(\theta)}$

27. $\displaystyle\int_{-\infty}^{\infty} \frac{(x^2-4)\,dx}{(x^2+4)^2}$

28. $\displaystyle\int_{-\infty}^{\infty} \frac{x\sin(x)\,dx}{(x^2+1)^2}$

29. $\displaystyle\oint_C \frac{(z^2 - z^4)\,dz}{(z-1)\cos(z)}$, C the circle of radius 8 about 0

30. $\displaystyle\oint_C \frac{[z - \cos(4iz)]\,dz}{(z^2+1)(z^2-1)}$, C the circle of radius 2 about the origin

31. Sum the series $\displaystyle\sum_{n=-\infty}^{\infty} \frac{1}{n^3+2}$.

32. Sum the series $\displaystyle\sum_{n=-\infty}^{\infty} \frac{1}{n^5+2}$.

33. Sum the series $\displaystyle\sum_{n=-\infty}^{\infty} \frac{(-1)^n}{n^6+4}$.

34. Suppose that $g(z)$ and $h(z)$ are analytic at z_0 and that $g(z_0) \neq 0$ and $h''(z_0) \neq 0$, while $h(z_0) = h'(z_0) = 0$. Show that $g(z)/h(z)$ has a second order pole at z_0 and that the residue there is

$$\frac{2g'(z_0)}{h''(z_0)} - \frac{2g(z_0)h^{(3)}(z_0)}{3[h''(z_0)]^2}.$$

35. Suppose that $g(z)$ and $h(z)$ are analytic at z_0 and that $g'(z_0)$ and $h^{(3)}(z_0)$ are not zero, while $g(z_0) = h(z_0) = h'(z_0) = h''(z_0) = 0$. Show that $g(z)/h(z)$ has a pole of order 3 at z_0 and that the residue there is

$$\frac{3g''(z_0)}{h^{(3)}(z_0)} - \frac{3g'(z_0)h^{(4)}(z_0)}{2[h^{(3)}(z_0)]^2}.$$

36. Suppose that $g(z)$ and $h(z)$ are analytic at z_0 and that $g(z_0)$ and $h^{(m)}(z_0)$ are not zero, while $h(z_0) = h'(z_0) = h''(z_0) = \cdots = h^{(m-1)}(z_0) = 0$. Show that $g(z)/h(z)$ has a pole of order m at z_0 and that the residue there is given by the $m \times m$ determinant

$$\begin{vmatrix} h^{(m)}(z_0) & 0 & 0 & \cdots & 0 & g(z_0) \\ \dfrac{h^{(m+1)}(z_0)}{(m+1)!} & \dfrac{h^{(m)}(z_0)}{m!} & 0 & \cdots & 0 & g'(z_0) \\ \dfrac{h^{(m+2)}(z_0)}{(m+2)!} & \dfrac{h^{(m+1)}(z_0)}{(m+1)!} & \dfrac{h^{(m)}(z_0)}{m!} & \cdots & 0 & \dfrac{g''(z_0)}{2!} \\ \vdots & \vdots & \vdots & & \vdots & \vdots \\ \dfrac{h^{(2m-1)}(z_0)}{(2m-1)!} & \dfrac{h^{(2m-2)}(z_0)}{(2m-2)!} & \dfrac{h^{(2m-3)}(z_0)}{(2m-3)!} & \cdots & \dfrac{h^{(m+1)}(z_0)}{(m+1)!} & \dfrac{g^{(m-1)}(z_0)}{(m-1)!} \end{vmatrix}.$$

CHAPTER EIGHTEEN

Conformal Mappings

18.0 INTRODUCTION

In elementary calculus one gains a certain insight from graphing functions $y = f(x)$ and visualizing their behavior by sketches. By the same token, we can often study complex functions from a geometric point of view. Given $w = f(z)$, we make two copies of the complex plane—one for the z-variable, and one for w. As z varies over curves or regions in the z-plane, we analyze how $w = f(z)$ varies in the w-plane, giving us a geometric description of the function. We shall give some examples of this in Section 18.1 below. When stressing this geometric aspect, we often refer to $w = f(z)$ as a *mapping*, and we call w the *image of* z under the mapping.

As one might expect, some mappings turn out to be more interesting than others. The most interesting ones are the *conformal mappings*, which preserve angles and sense of rotation. Conformal mappings and some of their applications will occupy the bulk of this and the next chapter.

18.1 SOME FAMILIAR FUNCTIONS AS MAPPINGS

Suppose we have a complex function $f(z)$. Write $w = f(z)$, and make two copies of the complex plane—one for the z-variable, and one for w. As z varies over curves or regions in the z-plane, we analyze how $w = f(z)$ varies in the w-plane, giving us a geometric description of the function. We shall give some examples of this below. When stressing this geometric aspect, we often refer to $w = f(z)$ as a *mapping*, and we call w the image of z under the mapping.

Now, imagine that we are given regions R in the z-plane and R^* in the w-plane. We say that $w = f(z)$ maps R *into* R^* if $f(z)$ is in R^* whenever z is in R. This is illustrated in Figure 321. The image of R in R^* may not take up all of R^*. If it does, we call the mapping *onto* R^*. In this case, given any w_0 in R^*, there is at least one z_0 in R such that $w_0 = f(z_0)$. This is illustrated in Figure 322.

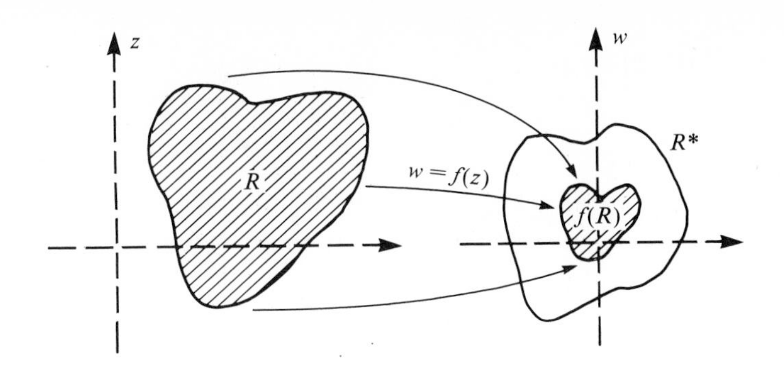

Figure 321. $w = f(z)$ mapping R into R^*.

Often, we write the image of R under $w = f(z)$ as $f(R)$. Thus, $f(z)$ maps R onto R^* exactly when $f(R) = R^*$.

We call $w = f(z)$ *one-to-one* on R if different values of z in R map to different values of w. That is, if $f(z_1) = f(z_2)$, then $z_1 = z_2$.

Here are some examples of mappings which illustrate these concepts.

EXAMPLE 483

Consider $w = z^2$. Note that

$$|w| = |z|^2 \quad \text{and} \quad \text{Arg } w = 2 \text{ Arg } z \qquad (\text{if } z \neq 0).$$

Thus, each nonzero z is mapped to a w having magnitude $|z|^2$ and twice the argument of z. Geometrically, the mapping rotates points counterclockwise to double their argument, while squaring the magnitude. For example, $1 + i$ maps to $(1 + i)^2 = 2i$. Note that $\text{Arg}(1 + i) = \pi/4$, while $\text{Arg } 2i = \pi/2$.

Considered as a mapping of the entire complex plane to the entire complex plane, $w = z^2$ is onto, but not one-to-one. It is onto because, given any w, there is some z with $w = z^2$. It is not one-to-one because, for any $w \neq 0$, there are two values of z with $w = z^2$.

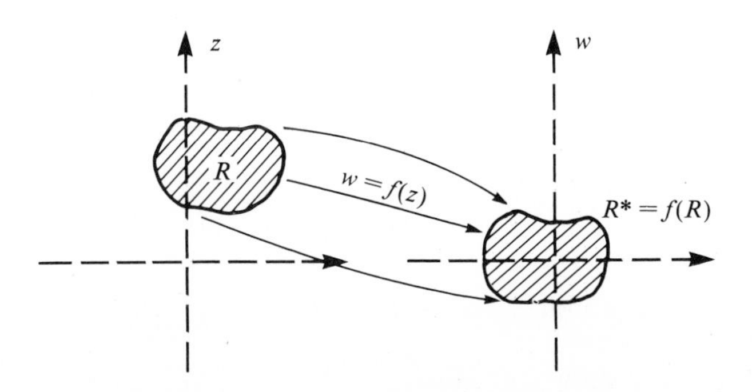

Figure 322. $w = f(z)$ mapping R onto R^*.

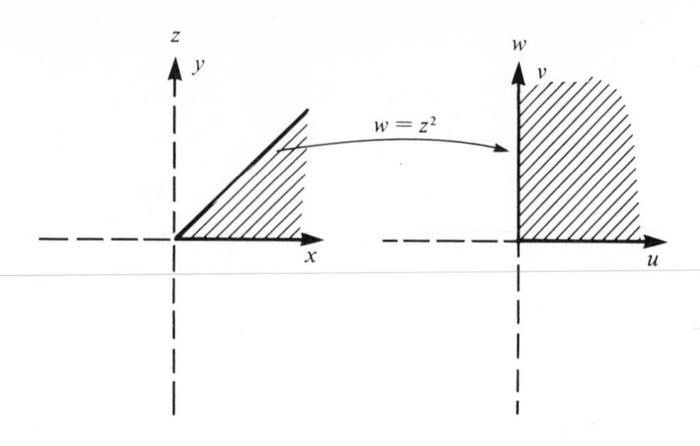

Figure 323. $w = z^2$ maps the sector $0 \leq \text{Arg } z \leq \pi/4$ one-to-one and onto the right quarter-plane $0 \leq \text{Arg } w \leq \pi/2$.

Suitably restricted, the mapping becomes one-to-one. For example, let R be the region

$$0 \leq \text{Arg } z \leq \frac{\pi}{4},$$

as shown in the z-plane in Figure 323. This maps onto the right quarter-plane $0 \leq \text{Arg } w \leq \pi/2$ in the w-plane. This mapping is onto, but also one-to-one, because for any w with $0 \leq \text{Arg } w \leq \pi/2$, there is only one z with $0 \leq \text{Arg } z \leq \pi/4$ which maps to w under $w = z^2$.

It is often instructive to examine the effects of a mapping on particular curves in the z-plane. Note that any circle $|z| = r$ is mapped by $w = z^2$ to a circle $|w| = r^2$; however, as z goes around $|z| = r$ once counterclockwise, w goes around $|w| = r^2$ *twice* counterclockwise, since arg w varies by 4π if arg z varies by 2π.

Straight lines parallel to the axes in the z-plane map to parabolas in the w-plane. To illustrate, consider a line Re $z = a$ (a constant). This may be written $z = a + it$, $-\infty < t < \infty$, and is parallel to the imaginary axis. The image in the w-plane consists of points $w = (a + it)^2$ or

$$w = a^2 - t^2 + 2ait.$$

Setting $w = u + iv$, then

$$u = a^2 - t^2, \qquad v = 2at.$$

Eliminating the parameter t gives us

$$v^2 = 4a^2(a^2 - u),$$

a parabola, as shown in Figure 324. This parabola has focus at the origin, vertex at $4a^4$, and opens to the left.

Similarly, one can show that a line Im $z = b$ maps to a parabola with focus at the origin, vertex on the real axis, and opening to the right.

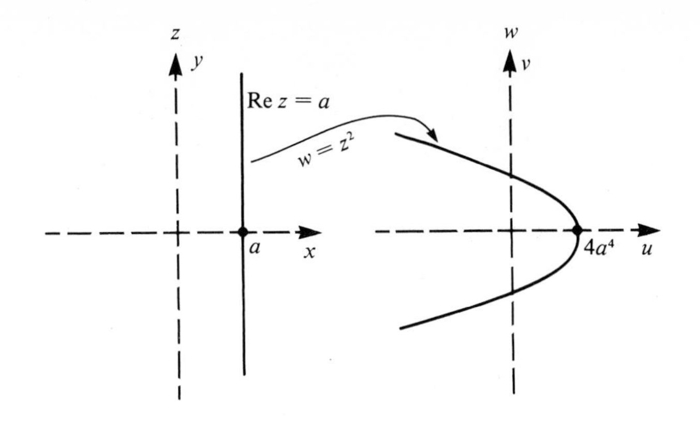

Figure 324. $w = z^2$ maps vertical lines to parabolas opening left.

EXAMPLE 484

Consider $w = e^z$.

If $z = x + iy$, then $w = e^x[\cos(y) + i\sin(y)]$; so

$$|w| = e^x \quad \text{and} \quad \arg w = y.$$

As a mapping of the z-plane to the w-plane, $w = e^z$ is not onto (no z maps to zero) and also not one-to-one [for example, $x + iy$ maps to the same point as $x + i(y + 2n\pi)$ for any integer n].

Consider a line $\text{Re } z = a$, parallel to the imaginary axis. Parametrically, the line is $z = a + it$, $-\infty < t < \infty$. The image is $w = e^{a+it} = e^a[\cos(t) + i\sin(t)]$. As t varies from $-\infty$ to ∞, $w = e^{a+it}$ runs over the circle $|w| = e^a$ infinitely often (one complete circuit every time t varies by 2π). This is shown in Figure 325.

Next, consider a line $\text{Im } z = b$, parallel to the real axis. This may be written $z = t + ib$, $-\infty < t < \infty$. Now,

$$w = e^{t+ib} = e^t[\cos(b) + i\sin(b)].$$

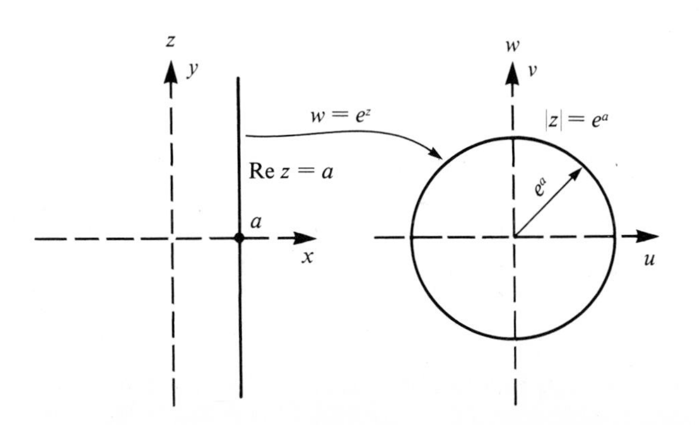

Figure 325. $w = e^z$ wraps a vertical line around a circle, covering the circle once for every interval of length 2π.

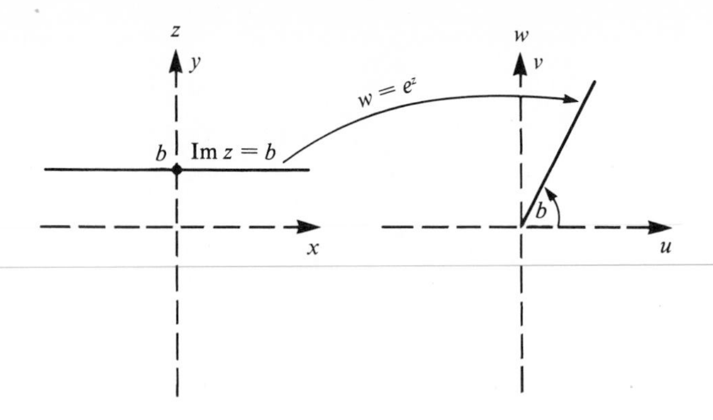

Figure 326. $w = e^z$ maps horizontal lines to half-rays from the origin.

Since b is constant, w now has fixed argument b and magnitude e^t, which varies from 0 to ∞ as t varies from $-\infty$ to ∞ (of course, e^t never actually achieves the value 0). Thus, w describes a half-line, or ray, emanating from the origin and making an angle of b radians with the positive real axis (see Figure 326).

Using these results, we can easily find the image of a rectangle R: $a \leq x \leq b$, $c \leq y \leq d$. We may think of R as the region bounded by the lines Re $z = a$, Re $z = b$, Im $z = c$, and Im $z = d$. The images of these lines bound a region R^*, which is then the image of R under the mapping (see Figure 327).

EXAMPLE 485

Consider the mapping $w = \cos(z)$. As a mapping from the z-plane to the w-plane, $w = \cos(z)$ is not one-to-one, since for a given z, all points $z + 2n\pi$ map to the same w for every integer n. The mapping is onto: Given any w, we can find z such that $w = \cos(z)$.

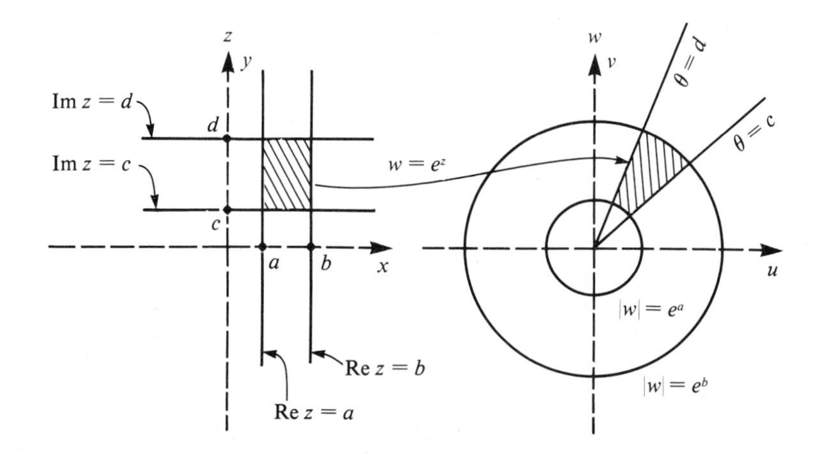

Figure 327. $w = e^z$ maps the rectangle shown to a wedge bounded by two half-rays and two circles.

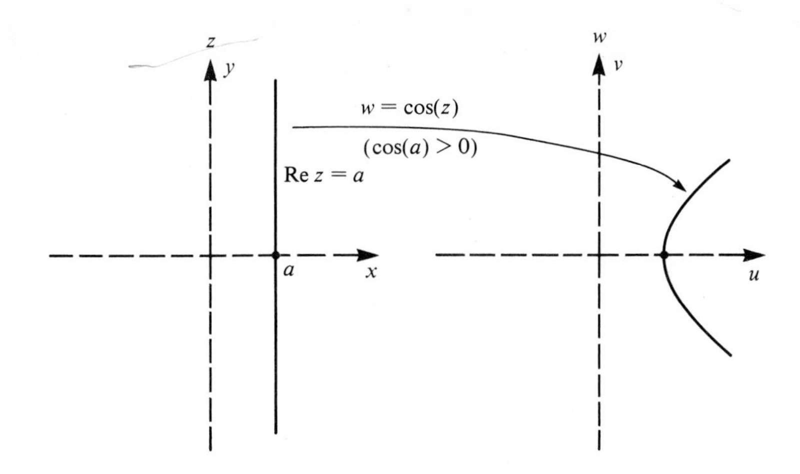

Figure 328. $w = \cos(z)$ maps a vertical line Re $z = a$ to the right branch of a hyperbola if $\cos(a) > 0$ and $\sin(a) \neq 0$.

Examine the effects of the mapping on lines parallel to the axes. If Re $z = a$, then $z = a + it$, $-\infty < t < \infty$. The image of this line consists of points

$$w = \cos(a + it) = \cos(a)\cosh(t) - i\sin(a)\sinh(t).$$

Writing $w = u + iv$, then

$$u = \cos(a)\cosh(t), \qquad v = -\sin(a)\sinh(t).$$

Eliminating the parameter t gives us

$$\frac{u^2}{\cos^2(a)} - \frac{v^2}{\sin^2(a)} = 1,$$

a family of hyperbolas with foci at $+1$ and -1. Actually, for a given a, the image of Re $z = a$ consists of only one branch of the hyperbola, since $\cosh(t) > 0$ for all t requires that u will be strictly positive [if $\cos(a) > 0$] or strictly negative [if $\cos(a) < 0$]. This is indicated in Figures 328 and 329.

These considerations are invalid if $\cos(a) = 0$ or $\sin(a) = 0$. If $\cos(a) = 0$, then $a = (2n + 1)\pi/2$ for some integer n. Then $u = 0$ and $v = (-1)^{n+1}\sinh(t)$. Now, $(-1)^{n+1}\sinh(t)$ takes on all real values as t varies between $-\infty$ and $+\infty$, whether n is even or odd. Thus, w traverses the entire imaginary axis as z varies over Re $z = a$, when $\cos(a) = 0$. This is shown in Figure 330.

When $\sin(a) = 0$, then $a = n\pi$ for some integer n. Then $u = (-1)^n\cosh(t)$ and $v = 0$. Since $\cosh(t) \geq 1$ for all t, then $u + iv$ varies over the portion $u \geq 1$ of the real line when n is even, and the portion $u \leq -1$ when n is odd. This is shown in Figures 331 and 332.

By a similar analysis, lines Im $z = b$, parallel to the real axis, map to $w = \cos(t + ib)$, $-\infty < t < \infty$. This is

$$w = \cos(t)\cosh(b) - i\sin(t)\sinh(b)$$

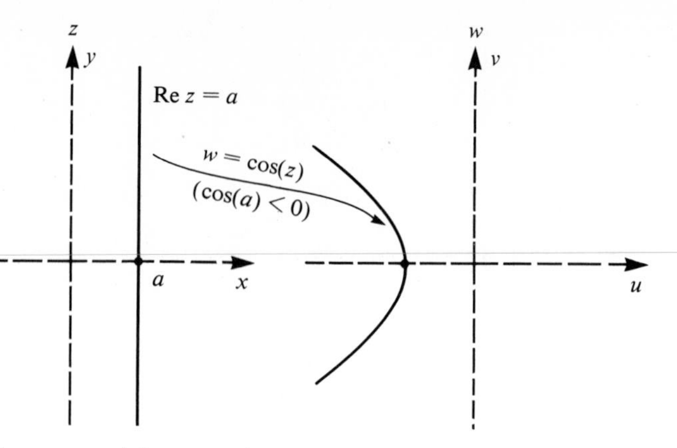

Figure 329. $w = \cos(z)$ maps the vertical line Re $z = a$ to the left branch of a hyperbola if $\cos(a) < 0$ and $\sin(a) \neq 0$.

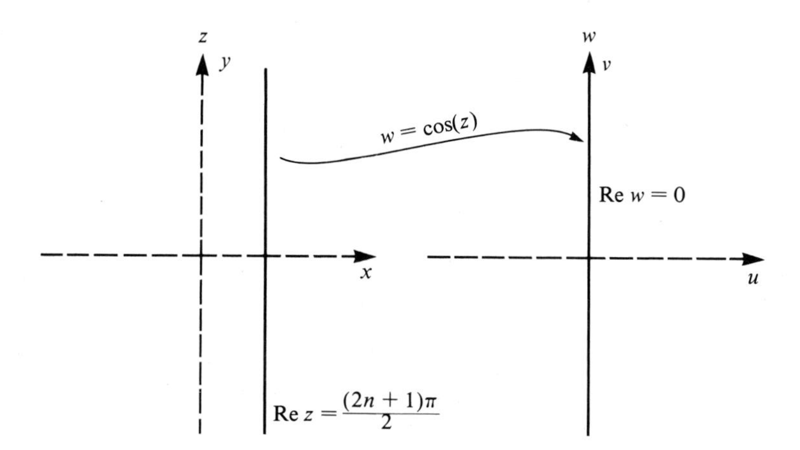

Figure 330. $w = \cos(z)$ maps vertical lines Re $z = (2n + 1)\pi/2$ to the imaginary axis.

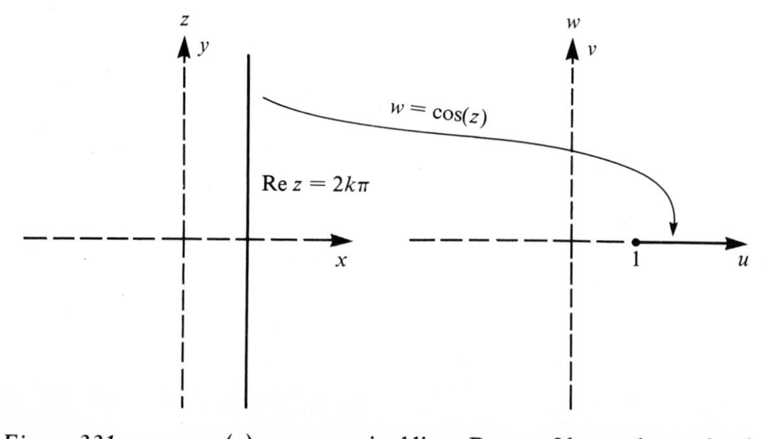

Figure 331. $w = \cos(z)$ maps vertical lines Re $z = 2k\pi$ to the real axis from 1 to ∞.

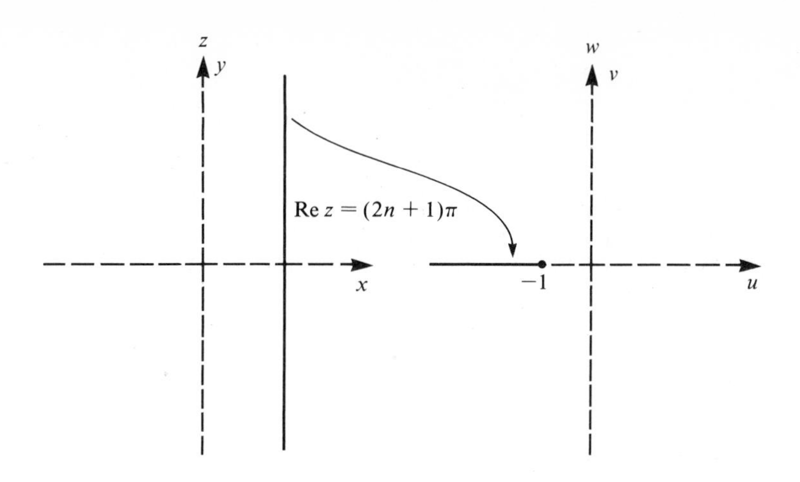

Figure 332. $w = \cos(z)$ maps vertical lines $\text{Re } z = (2k + 1)\pi$ to the real axis from $-\infty$ to -1.

or

$$u = \cos(t)\cosh(b), \qquad v = -\sin(t)\sinh(b).$$

If $b \neq 0$, we can eliminate t to write

$$\frac{u^2}{\cosh^2(b)} + \frac{v^2}{\sinh^2(b)} = 1,$$

ellipses with foci at $+1$ and -1 on the real axis. If $b = 0$, then $\text{Im } z = b$ is the real axis, and the image consists of $w = u + iv$ with

$$u = \cos(t), \qquad v = 0.$$

Then, w varies between -1 and $+1$ on the real axis as t varies from $-\infty$ to ∞.

The cases $b \neq 0$ and $b = 0$ are shown in Figures 333 and 334, respectively.

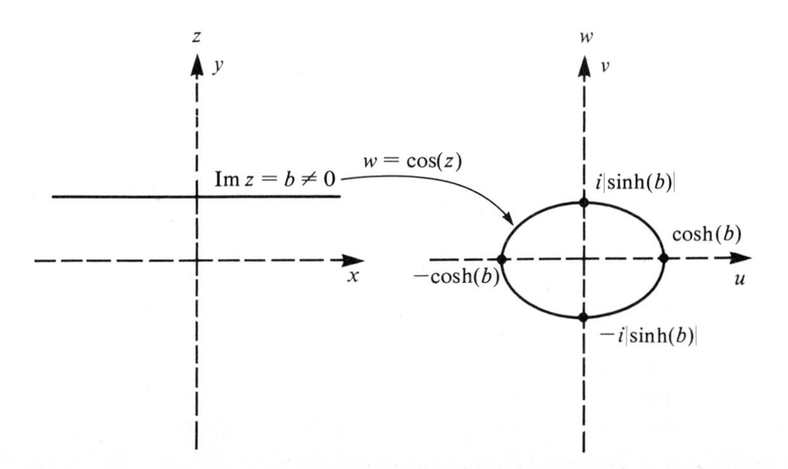

Figure 333. $w = \cos(z)$ maps horizontal lines $\text{Im } z = b \neq 0$ to ellipses.

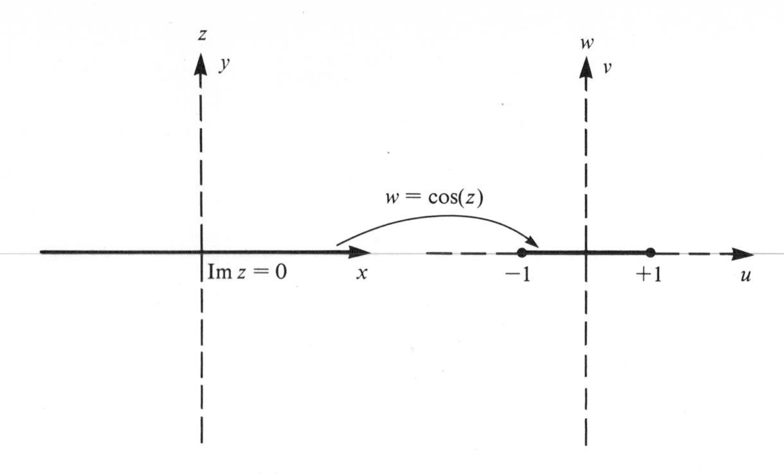

Figure 334. $w = \cos(z)$ maps the real axis Im $z = 0$ to the real interval $[-1, 1]$.

EXAMPLE 486

Determine the image under the mapping $w = \sin(z)$ of the strip

$$S: \frac{-\pi}{2} < \text{Re } z < \frac{\pi}{2}, \qquad \text{Im } z > 0.$$

The boundary of this strip consists of the lines Re $z = -\pi/2$, Re $z = \pi/2$, and the real interval $-\pi/2 \leq x \leq \pi/2$. This is shown in Figure 335.

The strategy is to examine the image of each boundary line of S, thus determining the boundary of the image of S.

The left boundary line of S is $z = -(\pi/2) + it$, $0 \leq t < \infty$. This maps to $w = \sin[-(\pi/2) + it] = -\cosh(t)$, which varies from -1 to $-\infty$ as t varies from 0 to ∞.

Similarly, the right boundary line $z = (\pi/2) + it$ maps to $w = \cosh(t)$, which varies from 1 to ∞ as t varies from $-\infty$ to ∞.

Finally, the bottom piece of boundary is $y = 0$, $-\pi/2 \leq x \leq \pi/2$, and this maps to $-1 \leq u \leq 1$, $v = 0$, in the w-plane.

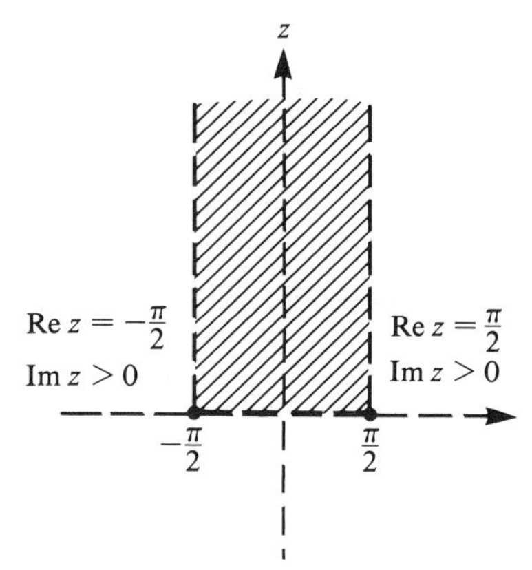

Figure 335. Strip bounded by vertical lines $x = -\pi/2$ and $x = \pi/2$, and the x-axis.

All told, then, the boundary of S maps to the real line in the w-plane (see Figure 336). This must be the boundary of the image of S. The image of S must then be the upper half-plane Im $w > 0$ or the lower half-plane Im $w < 0$. To see which, choose any point in S, say, $z = i$. Its image is $w = \sin(i) = i \sinh(1)$, which is in the upper half-plane. Thus, the image of S is the upper half-plane Im $w > 0$. This is shown in Figure 337.

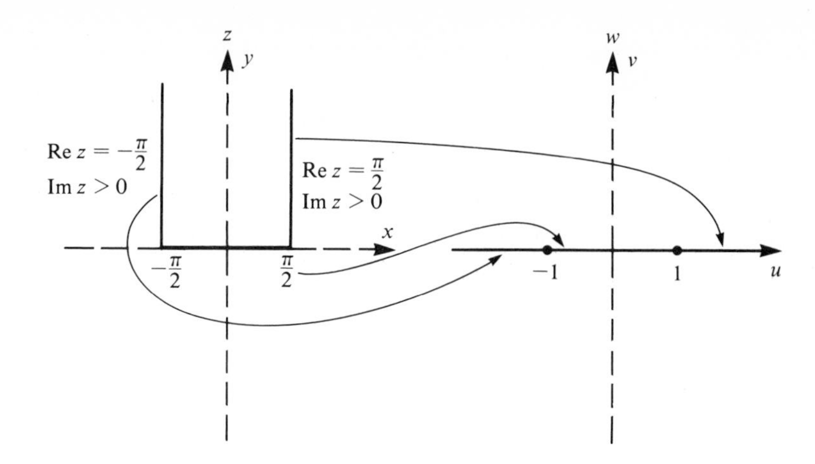

Figure 336. $w = \sin(z)$ maps $x = -\pi/2$, $y \geq 0$, to $u \leq -1$; $-\pi/2 \leq x \leq \pi/2$ to $-1 \leq u \leq 1$; and $x = \pi/2$, $y \geq 0$, to $u \geq 1$.

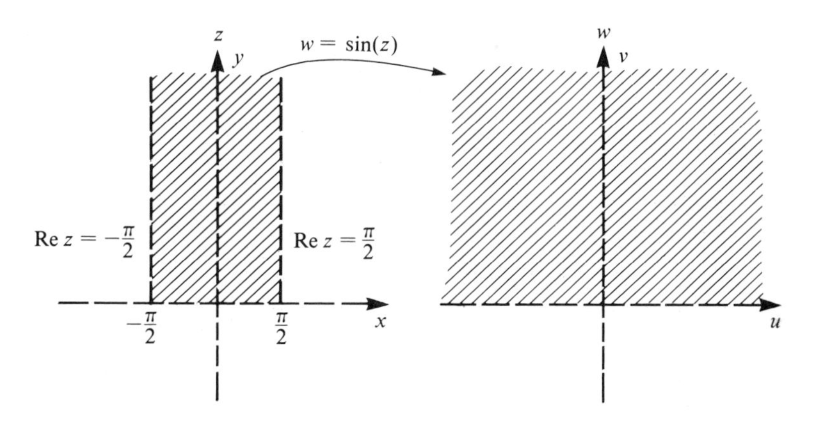

Figure 337. $w = \sin(z)$ maps the interior of the strip to the upper half-plane.

With these examples as an introduction to the notion of mapping, we shall be ready to begin the study of conformal mappings in the next section.

PROBLEMS FOR SECTION 18.1

1. In each of (a) through (e), find the image of the given region under the mapping $w = e^z$. In each case, draw a sketch of the given region and its image.
   (a) $0 < x < \pi, \quad 0 < y < \pi$
   (b) $-1 < x < 1, \quad -\pi/2 < y < \pi/2$
   (c) $0 < x < 1, \quad 0 < y < \pi/4$
   (d) $1 < x < 2, \quad 0 < y < \pi$
   (e) $-1 < x < 2, \quad -\pi/2 < y < \pi/2$

2. In each of (a) through (e), find the image of the given region under the mapping $w = \cos(z)$. In each case, sketch the region and its image.
   (a) $0 < x < 1, \quad 1 < y < 2$
   (b) $\pi/2 < x < \pi, \quad 1 < y < 3$
   (c) $0 < x < \pi, \quad \pi/2 < y < \pi$
   (d) $\pi < x < 2\pi, \quad 1 < y < 2$
   (e) $0 < x < \pi/2, \quad 0 < y < 1$
3. Determine the images of lines parallel to the axes under the mapping $w = \sinh(z)$.
4. Determine the images of lines parallel to the axes under the mapping $w = \sin(z)$.
5. Carry out the details omitted in Example 485 to determine images of lines parallel to the real axis under the mapping $w = \cos(z)$.
6. Determine the image under the mapping $w = z^3$ of the sector $\pi/6 < \text{Arg } z < \pi/3$. Sketch the region and its image.
7. Show that $w = \frac{1}{2}[z + (1/z)]$ maps circles $|z| =$ constant into ellipses with foci at $+1$ and -1 in the w-plane.
8. Show that $w = \frac{1}{2}[z + (1/z)]$ maps rays $\arg z =$ constant into hyperbolas with foci at $+1$ and -1 in the w-plane.
9. In Example 483, we showed that $w = z^2$ maps lines parallel to the axes to parabolas. Sketch in pencil the images of $\text{Im } z = 1$, $\text{Im } z = 2$, and $\text{Im } z = 4$. Sketch in ink the images of $\text{Re } z = 2$, $\text{Re } z = 3$, and $\text{Re } z = 5$. Show that each pencil parabola is perpendicular to each ink parabola wherever they intersect. (That is, at points of intersection, the tangents are perpendicular.)
10. Carry out an analysis similar to that of Example 483 for $w = (z - z_0)^2$, where z_0 is any given complex number.
11. Carry out an analysis similar to that of Example 483 for $w = z^n$ (n, any positive integer).
12. Consider the mapping $w = e^{\alpha z}$, where $\alpha = a + ib$ is constant, $a \neq 0$, and $b \neq 0$.
    (a) Show that a line $\text{Im } z = c$ maps to the curve (in polar coordinates)
    $$r = e^{-d}e^{a\theta/b},$$
    where
    $$d = -\frac{c}{b}(a^2 + b^2).$$
    (b) Show that $\text{Re } z = c$ maps to
    $$r = e^{d}e^{-a\theta/b}.$$
    The image curves in (a) and (b) are called *logarithmic spirals*. Sketch them for $a = b = c = 1$.
13. (a) Show that the map $w = 1/z$ maps circles about the origin to circles about the origin.
    (b) Find the image under $w = 1/z$ of lines $\text{Re } z = a$ and $\text{Im } z = b$.
14. Show that the mapping $w = 1/z$ maps every straight line to a circle or straight line, and every circle to a circle or straight line.
15. Determine the images of straight lines parallel to the axes under the mapping $w = \cosh(z)$.
16. Determine the image under the mapping $w = e^z$ of the infinite strip $0 < \text{Im } z < 2\pi$.
17. Determine the image under the mapping $w = \cos(z)$ of the rectangle with vertices $i\alpha$, $-i\alpha$, $i\alpha + \pi$, and $-i\alpha + \pi$, where α is any positive real number.

18.2 CONFORMAL MAPPINGS AND LINEAR FRACTIONAL TRANSFORMATIONS

We now center our attention upon mappings having two important properties.

Suppose that $w = f(z)$ is a given mapping. We say that the mapping *preserves angles* if curves intersecting at an angle θ in the z-plane map to curves intersecting at

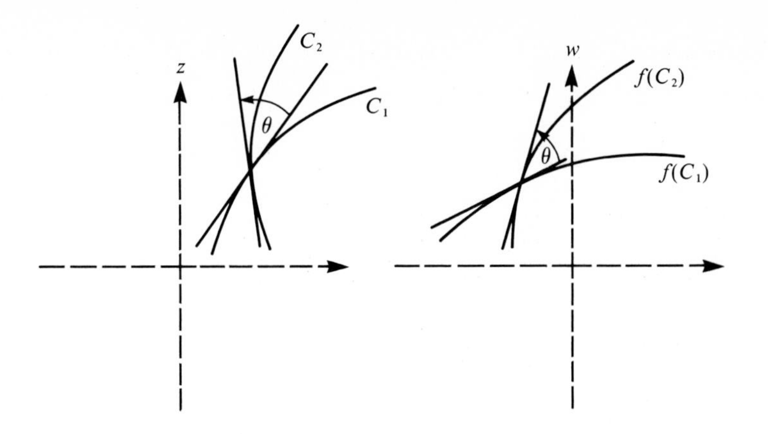

Figure 338. Angle-preserving mapping.

the same angle θ in the w-plane. We measure the angle between two curves at a point of intersection by the angle between their tangents there, as shown in Figure 338.

We say that $w = f(z)$ is *orientation-preserving* if a counterclockwise rotation in the z-plane maps to a counterclockwise rotation in the w-plane. That is, if a line L_1 is rotated counterclockwise in the z-plane to a new line L_2, then, in the w-plane, $f(L_2)$ is similarly counterclockwise from $f(L_1)$. This is illustrated in Figure 339. An orientation-preserving mapping need not preserve the actual *angle* between L_1 and L_2, just the direction of rotation from one to the other.

If $w = f(z)$ is both angle- and orientation-preserving, then we call it a *conformal* mapping.

Many examples will fall easily out of the following theorem.

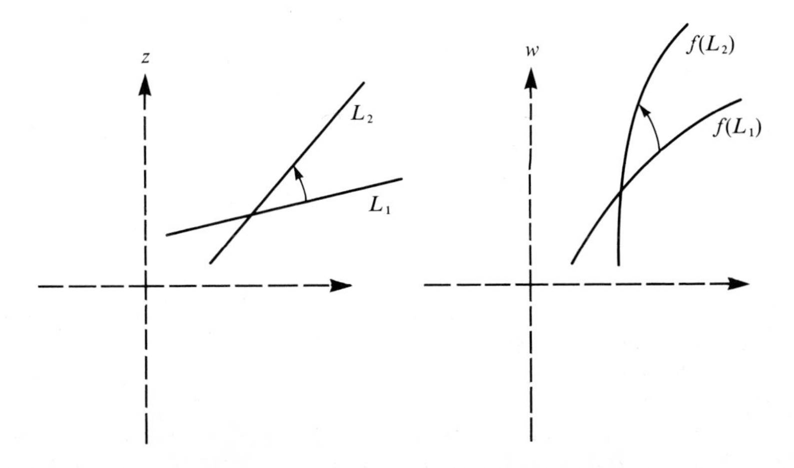

Figure 339. Orientation-preserving mapping.

THEOREM 140

Let $w = f(z)$ be a mapping of a domain D in the z-plane onto a domain D^* in the w-plane. If $f(z)$ is analytic in D, and $f'(z) \neq 0$ for every z in D, then $w = f(z)$ is a conformal mapping of D onto D^*.

Proof Let $z = z(t)$ be a smooth curve through z_0 in D, say, $z_0 = z(t_0)$. Writing $w_0 = f(z_0)$, we have

$$w - w_0 = \left(\frac{f(z) - f(z_0)}{z - z_0}\right)(z - z_0).$$

Hence,

$$\arg(w - w_0) = \arg\left(\frac{f(z) - f(z_0)}{z - z_0}\right) + \arg(z - z_0).$$

Note that $\arg(z - z_0)$ is the angle between the positive real axis and the line from z_0 to z (see Figure 340). Similarly, $\arg(w - w_0)$ is the angle between the positive real axis and the line from w_0 to w in the w-plane. Letting $z \to z_0$, then $\arg(z - z_0)$ approaches the angle θ between the positive real axis and the tangent to $z = z(t)$ at z_0 (Figure 341), and $\arg(w - w_0)$ approaches the angle ϕ between the positive real axis and the tangent to $w = f[z(t)]$ at w_0. Then,

$$\phi = f'(z_0) + \theta.$$

Thus,

$$\phi - \theta = f'(z_0).$$

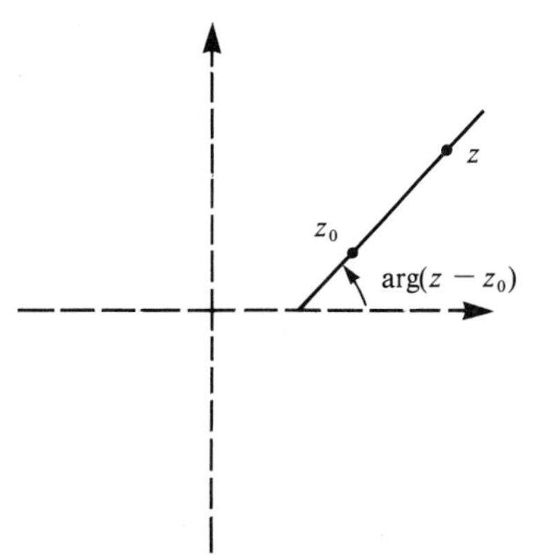

Figure 340

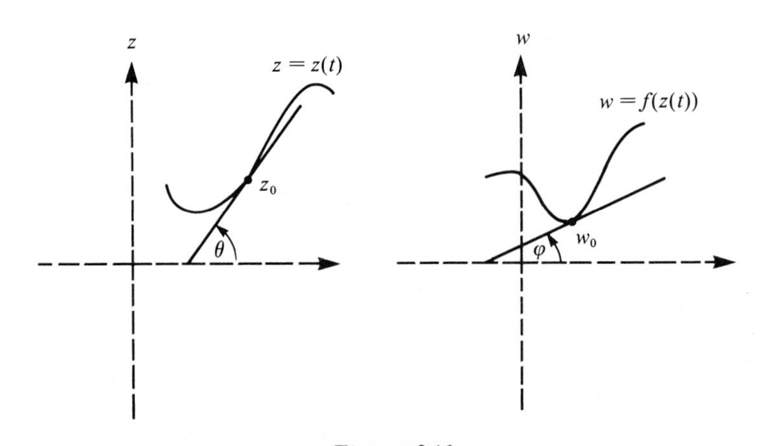

Figure 341

This result is independent of the curve $z(t)$. Given another curve $z = z_1(t)$ through z_0, we would obtain

$$\phi_1 - \theta_1 = f'(z_0),$$

where θ_1 is the angle between the positive real axis and the tangent to $z = z_1(t)$ at z_0, and ϕ_1 is the angle between the tangent to $w = f[z_1(t)]$ at w_0 and the positive real axis in the w-plane. Thus,

$$\phi - \theta = \phi_1 - \theta_1$$

or

$$\phi - \phi_1 = \theta - \theta_1.$$

Now, $\theta - \theta_1$ is the angle between $z = z(t)$ and $z = z_1(t)$ at z_0, and $\phi - \phi_1$ is the angle between $w = f[z(t)]$ and $w = f[z_1(t)]$ at w_0. Thus, $w = f(z)$ preserves angles.

From the same equation, $w = f(z)$ preserves orientation (otherwise we would get $\phi - \phi_1 = \theta_1 - \theta$).

Thus, $w = f(z)$ is conformal.

EXAMPLE 487

Consider the domain D consisting of the wedge $0 < \arg z < \pi/4$.

From Example 483, $w = z^2$ maps this to the wedge $0 < \arg w < \pi/2$ (see Figure 323). The mapping is conformal, as $f'(z) = 2z$, and this is zero only at $z = 0$, which is not in D.

Since $w = z^2$ is angle-preserving on any domain excluding 0, perpendicular lines should map to curves which intersect each other at right angles. Recall from Example 483 that the lines $\operatorname{Im} z = a$ and $\operatorname{Re} z = b$ map to parabolas. The reader can check that indeed these parabolas are perpendicular to each other where they intersect.

A particularly important class of conformal mappings consists of the *linear fractional transformations** which take the form

$$w = \frac{az + b}{cz + d},$$

where a, b, c, and d are constant, and $ad - bc \neq 0$. Note that

$$w' = \frac{ad - bc}{(cz + d)^2} \neq 0;$$

hence, this type of mapping is conformal on any domain if $z \neq -d/c$. Linear fractional transformations are important because we often want to construct a conformal mapping between two given domains, and sometimes this can be done by adjusting the constants a, b, c, and d.

To get some feeling for these mappings, consider some important special cases.

* Sometimes also called a *bilinear* or *Möbius* transformation.

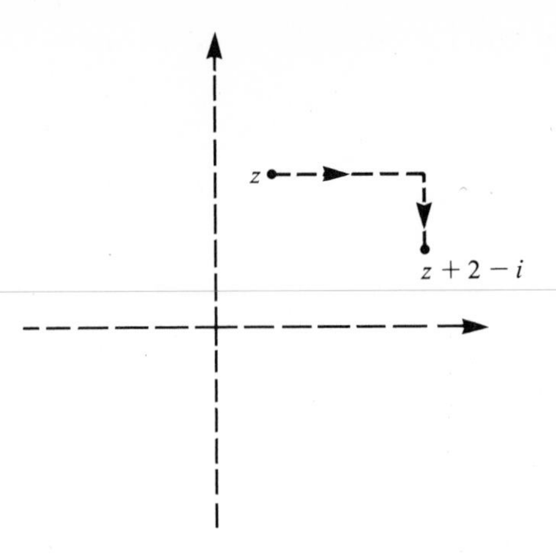

Figure 342. $w = z + 2 - i$ moves z two units right, one unit down.

Case 1 $w = z + b$. This is a *translation*, shifting z horizontally by Re b and vertically by Im b. For example, $w = z + 2 - i$ moves z right two units and down one unit (see Figure 342).

Case 2. $w = az$. This is a *rotation-magnification*. Note that $|w| = |a||z|$ and $\arg w = \arg z + \arg a$. Thus, $w = az$ magnifies $|z|$ by a factor of $|a|$, and rotates it counterclockwise by an angle $\arg a$.

In Figure 343, we consider $w = 2iz$. Here, $a = 2i$; so $|a| = 2$ and $\arg a = \pi/2$. Thus, w has twice the magnitude of z and is rotated 90 degrees from z. In Figure 344, we have $w = \frac{1}{4}(1 - i)z$. Here, $|w| = (\sqrt{2}/4)|z|$ and w is rotated $\arg \frac{1}{4}(1 - i) = 7\pi/4$ radians counterclockwise from z.

If $|a| = 1$, $w = az$ is a *pure rotation*, rotating z by $\arg a$ but preserving magnitude. Any rotation counterclockwise by θ radians has the form $w = e^{i\theta}z$.

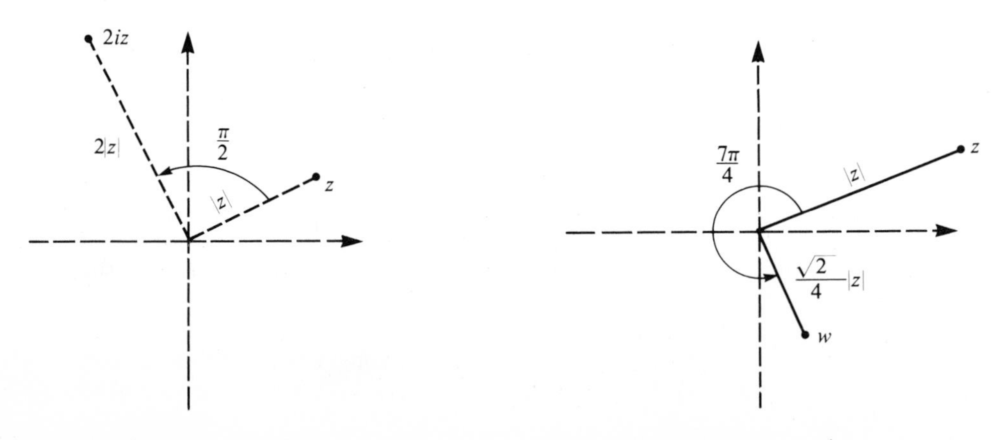

Figure 343. $w = 2iz$.

Figure 344. $w = \frac{1}{4}(1 - i)z$.

Case 3 $w = 1/z$, $z \neq 0$. This is called an *inversion*. We have

$$|w| = \frac{1}{|z|}$$

and

$$\arg w = \arg 1 - \arg z = -\arg z.$$

If z is outside the unit circle, we obtain $w = 1/z$ by moving toward the origin along the line from 0 to z until we are $1/|z|$ units from 0 and then reflecting across the real axis (see Figure 345). If z is inside the unit circle (but not zero), we move outward until we are $1/|z|$ units from the origin, and then reflect across the real axis.

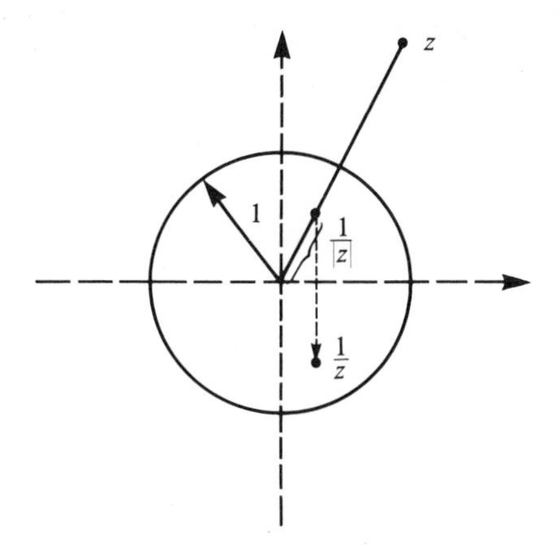

Figure 345. Inversion: $w = 1/z$.

Any linear fractional transformation can be envisioned as a sequence of mappings of these three kinds, the cumulative effect of which is the given linear fractional transformation. The steps are:

$$z \xrightarrow{\text{Rotation-magnification}} cz \xrightarrow{\text{Translation}} cz + d \xrightarrow{\text{Inversion}} \frac{1}{cz + d}$$

$$\xrightarrow[\text{Rotation-magnification}]{} \left(\frac{bc - ad}{c}\right)\frac{1}{cz + d}$$

$$\xrightarrow[\text{Translation}]{} \left(\frac{bc - ad}{c}\right)\left(\frac{1}{cz + d}\right) + \frac{a}{c} = \frac{az + b}{cz + d}.$$

Thus, in general, $w = (az + b)/(cz + d)$ is the end result of a rotation-magnification, translation, inversion, rotation-magnification, and translation, applied successively in this order.

We shall now establish some important facts about linear fractional transformations.

THEOREM 141

Given three distinct points z_1, z_2, z_3, and three distinct points w_1, w_2, w_3, then there is a linear fractional transformation mapping the z-plane to the w-plane such that z_1 maps to w_1, z_2 to w_2, and z_3 to w_3. In fact, such a mapping is obtained by solving for w in the equation

$$\left(\frac{w_1 - w}{w_1 - w_2}\right)\left(\frac{w_3 - w_2}{w_3 - w}\right) = \left(\frac{z_1 - z}{z_1 - z_2}\right)\left(\frac{z_3 - z_2}{z_3 - z}\right).$$

Proof We leave the details (which are algebraic manipulation) to the student.

EXAMPLE 488

Find a linear fractional transformation mapping $3 \to i$, $1 - i \to 4$, and $2 - i \to 6 + 2i$.

Let

$$z_1 = 3, \qquad z_2 = 1 - i, \qquad z_3 = 2 - i,$$
$$w_1 = i, \qquad w_2 = 4, \quad \text{and} \quad w_3 = 6 + 2i$$

in Theorem 141. We get

$$\left(\frac{i - w}{i - 4}\right)\left(\frac{2 + 2i}{6 + 2i - w}\right) = \left(\frac{3 - z}{2 + i}\right)\left(\frac{1}{2 - i - z}\right).$$

Solving for w in terms of z gives us

$$w = \frac{(20 + 4i)z - (16i + 68)}{(6 + 5i)z - (22 + 7i)}.$$

It is easy to check that this mapping does send each z_j to w_j.

When dealing with mappings, it is often convenient to consider the *extended complex plane*, formed by adjoining a point at infinity. This point is denoted ∞ and is the image of $z = -d/c$ under the linear fractional transformation $w = (az + b)/(cz + d)$. (Thus, 0 maps to ∞ under the inversion $w = 1/z$.) In Theorem 141 we can, if we choose, specify some z-value, say, z_3, to map to ∞. A slight modification of Theorem 141 gives us the following.

THEOREM 142

Given distinct z_1, z_2, z_3, and distinct w_1, w_2, a linear fractional transformation mapping $z_1 \to w_1$, $z_2 \to w_2$, and $z_3 \to \infty$ is obtained by solving for w in:

$$\frac{w_1 - w}{w_1 - w_2} = \left(\frac{z_1 - z}{z_1 - z_2}\right)\left(\frac{z_3 - z_2}{z_3 - z}\right).$$

We leave a proof of this to the student. Note, however, that in effect we get this equation defining w from the equation given in Theorem 141 by simply omitting the two factors involving w_3. This simplifies the algebra of finding w in terms of z.

EXAMPLE 489

Find a linear fractional transformation mapping

$$i \to 4i, \qquad 1 \to 3 - i, \quad \text{and} \quad 2 + i \to \infty.$$

Let $z_1 = i$, $z_2 = 1$, $z_3 = 2 + i$, $w_1 = 4i$, and $w_2 = 3 - i$ in Theorem 142. Then,

$$\frac{4i - w}{-3 + 5i} = \left(\frac{i - z}{i - 1}\right)\left(\frac{1 + i}{2 + i - z}\right).$$

Solving for w gives us

$$w = \frac{(-4 + 6i)z + (-2 - 4i)}{(1 - i)z + (-3 + i)}.$$

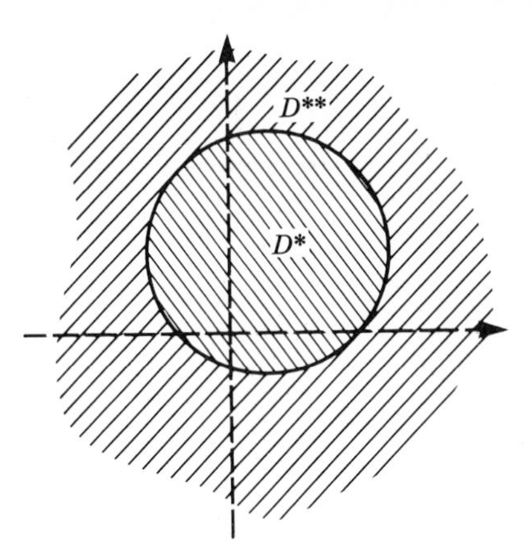

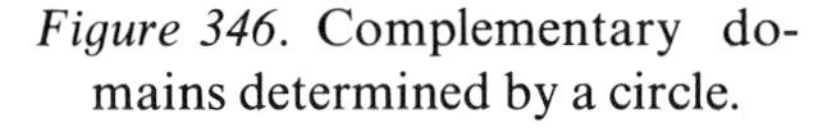

Figure 346. Complementary domains determined by a circle.

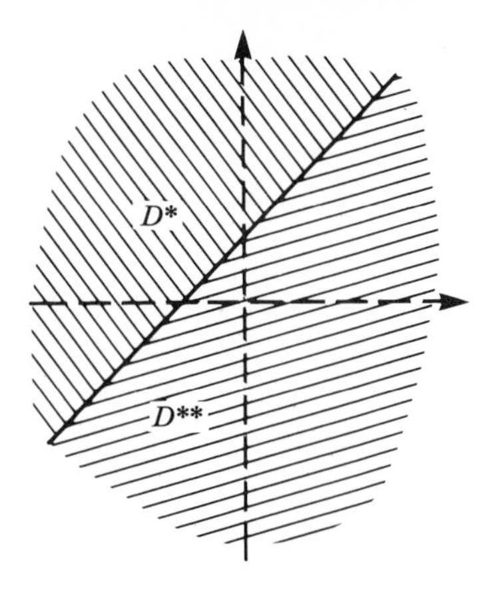

Figure 347. Complementary domains determined by a straight line.

The next theorem is sometimes useful in constructing linear fractional transformations between given domains. First, we observe that a circle or straight line in the plane separates the plane into two domains, D^* and D^{**}, both having the circle or straight line as boundary. These domains are called *complementary domains* of the circle or straight line and are illustrated in Figure 346 for a circle and Figure 347 for a straight line.

THEOREM 143

A linear fractional transformation always maps a circle to a circle or straight line, and it always maps a straight line to a circle or straight line.

Further, let K be a circle or straight line in the z-plane, having complementary domains D and D'. Let $w = (az + b)/(cz + d)$ map K to K^* in the w-plane. Then, $w = (az + b)/(cz + d)$ maps D onto one of the complementary domains of K^*, and it maps D' onto the other complementary domain of K^*.

Proof It is geometrically apparent that any translation or rotation-magnification maps a circle to a circle and a straight line to a straight line. Now look at inversions. Observe that any circle or straight line in the plane can be written

$$A(x^2 + y^2) + Bx + Cy + R = 0,$$

where A, B, C, and R are real. This gives a circle if $A \neq 0$ and a straight line if $A = 0$. In terms of z, this can be written

$$A|z|^2 + B\left(\frac{z + \bar{z}}{2}\right) + C\left(\frac{z - \bar{z}}{2i}\right) + R = 0.$$

If $w = 1/z$, then $z = 1/w$, and this becomes

$$A\left|\frac{1}{w}\right|^2 + \frac{B}{2}\left(\frac{1}{w} + \frac{1}{\bar{w}}\right) + \frac{C}{2i}\left(\frac{1}{w} - \frac{1}{\bar{w}}\right) + R = 0.$$

Recalling that $|w|^2 = w\bar{w}$, multiplying by $w\bar{w}$ gives us

$$R|w|^2 + B\left(\frac{w+\bar{w}}{2}\right) - C\left(\frac{w-\bar{w}}{2i}\right) + A = 0,$$

which is a circle if $R \neq 0$ and a straight line if $R = 0$.

We leave the proof of the rest of the theorem as an exercise. *Hint:* The complementary domains bounded by

$$A|z|^2 + B\left(\frac{z+\bar{z}}{2}\right) + C\left(\frac{z-\bar{z}}{2i}\right) + R = 0$$

are given by

$$A|z|^2 + B\left(\frac{z+\bar{z}}{2}\right) + C\left(\frac{z-\bar{z}}{2i}\right) + R > 0$$

and

$$A|z|^2 + B\left(\frac{z+\bar{z}}{2}\right) + C\left(\frac{z-\bar{z}}{2i}\right) + R < 0.$$

EXAMPLE 490

Consider $w = iz + (3 - 2i)$.

This is a rotation (by $\arg i = \pi/2$) and a translation (by $3 - 2i$).

Let K be the circle $(x-2)^2 + y^2 = 9$ of radius 3 about (2, 0) in the z-plane. This can be written

$$x^2 + y^2 - 4x - 5 = 0.$$

In complex notation, this is

$$|z|^2 - 4\left(\frac{z+\bar{z}}{2}\right) - 5 = 0.$$

Substituting $z = [w - (3 - 2i)]/i$, we get the image K^* of K in the w-plane. It is (after the substitution)

$$|w|^2 - 6\left(\frac{w+\bar{w}}{2}\right) = 0.$$

Writing $w = u + iv$, this is the same as

$$(u-3)^2 + v^2 = 9,$$

a circle of radius 3 about (3, 0) in the w-plane. On geometric grounds, this is what we should have expected from the transformation, which takes points on K, rotates them by 90 degrees, and then moves them three units to the right and two units down (see Figure 348).

Now look at the complementary domains D (enclosed by K) and D' (exterior to K), as shown in Figure 349. In view of what this mapping does to points, we would expect D to map to the interior of K^* and D' to map to the exterior of K^*, and in fact this is what happens. For example, $z = 2$, enclosed by K, maps to $w = 3$ inside K^*.

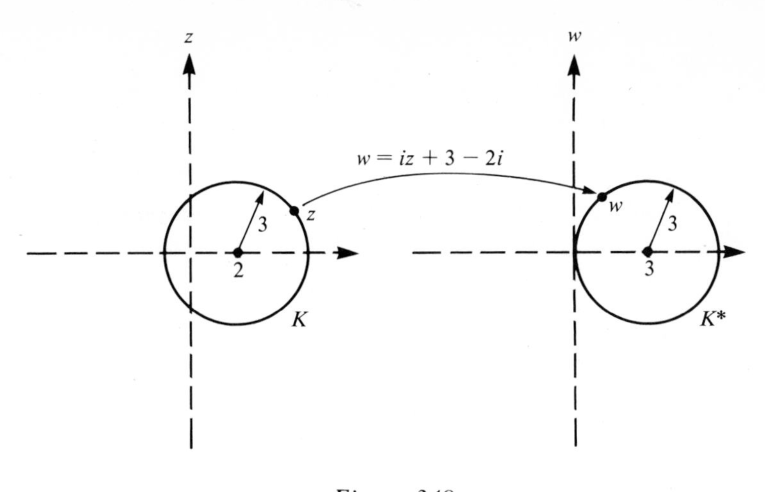

Figure 348

EXAMPLE 491

Examine the effects of inversion $w = 1/z$ on the line $\operatorname{Re} z = a$ (a constant, nonzero).

This line is $z = a + it$, $-\infty < t < \infty$. Then,

$$w = \frac{1}{z} = \frac{1}{a + it} = \frac{a}{a^2 + t^2} - i\,\frac{t}{a^2 + t^2}.$$

Writing $w = u + iv$, then

$$u = \frac{a}{a^2 + t^2} \quad \text{and} \quad v = \frac{-t}{a^2 + t^2}.$$

It is easy to check that

$$\left(u - \frac{1}{2a}\right)^2 + v^2 = \frac{1}{4a^2} \quad \text{or} \quad \left|w - \frac{1}{2a}\right| = \frac{1}{2a} \qquad (a \neq 0).$$

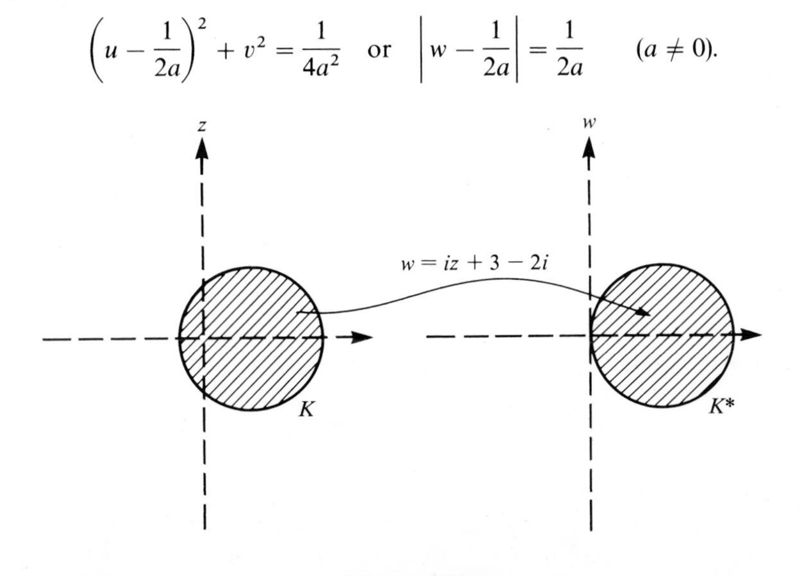

Figure 349

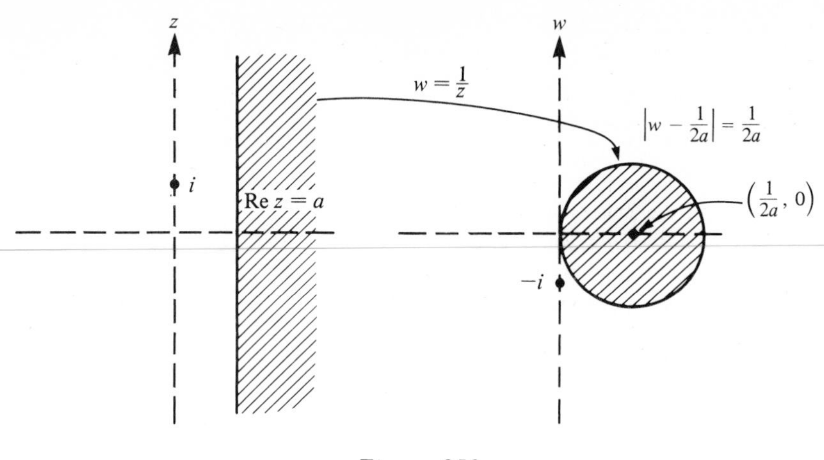

Figure 350

Thus, vertical lines $\text{Re } z = a \neq 0$ map to circles of radius $1/2a$ and center $(1/2a, 0)$, as shown in Figure 350.

Now, $\text{Re } z = a$ forms the boundary of two domains, D ($\text{Re } z < a$, to the left of the line) and D'($\text{Re } z > a$, to the right). Which maps to the interior, and which to the exterior, of the image of $\text{Re } z = a$? To determine this, choose a point i in, say, D (remember we are assuming that $a \neq 0$). Now, i maps to $-i$, which is exterior to the circle $[u - (1/2a)]^2 + v^2 = 1/4a^2$. Thus, the domain to the left of $\text{Re } z = a$ maps outside the circle $|w - (1/2a)| = 1/2a$, and the domain to the right maps inside the circle.

In the next section we shall take up the important problem of constructing a conformal mapping between two given domains.

PROBLEMS FOR SECTION 18.2

In each of Problems 1 through 10, find a linear fractional transformation mapping the given points to the indicated images.

1. $1 \to 1, \quad 2 \to -i, \quad 3 \to 1 + i$

2. $i \to i, \quad 1 \to -i, \quad 2 \to 0$

3. $1 \to 1 + i, \quad 2i \to 3 - i, \quad 4 \to \infty$

4. $-5 + 2i \to 1, \quad 3i \to 0, \quad -1 \to \infty$

5. $6 + i \to 2 - i, \quad i \to 3i, \quad 4 \to -1$

6. $10 \to 8 - i, \quad 3 \to 2i, \quad 4 \to 3 - i$

7. $1 \to 6 - 4i, \quad 1 + i \to 2, \quad 3 + 4i \to \infty$

8. $8 - i \to 1 - i, \quad i \to 1 + i, \quad 3i \to \infty$

9. $2i \to \frac{1}{2}, \quad 4i \to \frac{1}{3}, \quad i \to 6i$

10. $-1 \to -i, \quad i \to 2 - 3i, \quad 1 - 2i \to \infty$

In each of Problems 11 through 18, determine the image of the given circle or straight line under the given linear fractional transformation. Also determine the images of the complementary domains in the z-plane determined by the given circle or straight line.

11. $w = \dfrac{2z + i}{z - i}; \quad \text{Im } z = 2$

12. $w = \dfrac{2i}{z}; \quad \text{Re } z = -4$

13. $w = 2iz - 4; \quad \text{Re } z = 5$

14. $w = \dfrac{z - i}{iz}; \quad \left(\dfrac{z + \bar{z}}{2}\right) + \left(\dfrac{z - \bar{z}}{2i}\right) = 4$

15. $w = \dfrac{z - 1 + i}{2z + 1}$; $|z| = 4$

16. $w = 3z - i$; $|z - 4| = 3$

17. $w = \dfrac{2z - 5}{z + i}$; $(z + \bar{z}) - 3\left(\dfrac{z - \bar{z}}{2i}\right) = 5$

18. $w = \dfrac{(3i + 1)z - 2}{z}$; $|z - i| = 1$

19. Determine the image of any line $\operatorname{Im} z = a \neq 0$ under inversion $w = 1/z$. Determine the images of the complementary domains $\operatorname{Im} z < a$ and $\operatorname{Im} z > a$.

20. Let $w = f(z)$ be a conformal mapping of a domain D in the z-plane onto a domain S in the w-plane. Let $w^* = g(w)$ be a conformal mapping of S onto a domain T in the w^*-plane. Prove that $w^* = g[f(z)]$ is a conformal mapping of D onto T.

21. Prove that any linear fractional transformation

$$w = A\left(\frac{z - B}{z - \bar{B}}\right),$$

where $|A| = 1$ and $\operatorname{Im} B > 0$, maps the upper half-plane $\operatorname{Im} z > 0$ onto the unit disk $|w| < 1$. (*Hint:* First show that the boundary $\operatorname{Im} z = 0$ of the upper half-plane maps to the boundary $|w| = 1$ of the unit disk. Make use of Theorem 143.)

22. Show that any linear fractional transformation of the form

$$w = A\left(\frac{z - B}{\bar{B}z - 1}\right),$$

where $|B| < 1$ and $|A| = 1$, maps the domain $|z| < 1$ onto the domain $|w| < 1$. (*Hint:* Look at the boundaries $|z| = 1$ and $|w| = 1$ of these domains and use Theorem 143.)

23. Show that inversion maps the domain common to $|z - 1| < 1$ and $|z + i| < 1$ onto the quarter-plane $\operatorname{Re} w > \frac{1}{2}$, $\operatorname{Im} w > \frac{1}{2}$.

24. Prove that the mapping $w = \bar{z}$ is not a conformal mapping.

25. Prove that a composition of two linear fractional transformations is again a linear fractional transformation.

In Problems 26, 27, and 28, we are working in the extended complex plane, which includes a point at ∞.

26. We call z_0 a *fixed point* of $w = f(z)$ if $z_0 = f(z_0)$. Prove that every linear fractional transformation has either one or two fixed points. Give an example of a linear fractional transformation with exactly one fixed point and an example with two fixed points.

27. Prove that any linear fractional transformation with three fixed points must be the identity mapping $w = z$.

28. Prove that there cannot be two distinct linear fractional transformations mapping three given distinct points z_1, z_2, z_3 to three given distinct points w_1, w_2, w_3.

29. Let r be a given positive number, and let z_0 be any complex number with $|z_0| < r$. Show that the linear fractional transformation

$$w = A\left(\frac{z - z_0}{r^2 - \bar{z}_0 z}\right)$$

maps $|z| < r$ onto $|w| < r$ but maps z_0 to the center 0 if $|w| < r$. Here, A is any complex number with $|A| = r^2$.

30. Show that there is no linear fractional transformation mapping the disk $|z| < 1$ to the elliptical region $(u^2/4) + v^2 < \frac{1}{16}$. (*Hint:* What would the boundary of $|z| < 1$ have to map to?)

31. Show that every linear fractional transformation has an inverse, and then show that this inverse is also a linear fractional transformation. (*Hint:* Solve $w = (az + b)/(cz + d)$ for z in terms of w.)

18.3 CONSTRUCTING CONFORMAL MAPPINGS BETWEEN GIVEN DOMAINS

Often we want a conformal mapping between given domains. To give one illustration of why such a thing might be useful, consider the Dirichlet problem for a plane domain D, which is to find a function $u(x, y)$ such that

$$\frac{\partial^2 u}{\partial x^2} + \frac{\partial^2 u}{\partial y^2} = 0 \quad \text{for} \quad (x, y) \text{ in } D$$

and such that $u(x, y)$ takes on given values on the boundary of D. As we shall see in the next chapter, this problem is easy to solve if D is the unit disk $x^2 + y^2 < 1$. We can then construct a solution for another domain D^* by finding a conformal mapping between the unit disk and D^*, and then applying this mapping to the solution for the disk.

In this section we concentrate on the problem of constructing conformal mappings having prescribed properties. In general, this can be a very difficult problem; so we shall address it in three stages.

In Stage 1 we consider problems for which linear fractional transformations will do the job.

In Stage 2 we treat problems which can be solved using combinations of conformal mappings previously studied.

Finally, in Stage 3 we develop the Schwarz-Christoffel transformation, which can be used to map the upper half-plane to a polygonal domain.

STAGE 1

Linear fractional transformations will generally suffice to construct conformal mappings between domains whose boundaries are circles or straight lines. The key is Theorem 143. Here are three examples.

EXAMPLE 492

Map $|z| < 2$ onto $\text{Re } w > 0$.

The boundaries are $|z| = 2$ and $\text{Re } w = 0$. Choose three points on $|z| = 2$, say, $z_1 = 2$, $z_2 = 2i$, and $z_3 = -2$; and choose three points on $\text{Re } w = 0$, say, $w_1 = i$, $w_2 = 0$, and $w_3 = -i$, as shown in Figure 351. Now construct a linear fractional transformation mapping z_j to w_j. We find that it is

$$w = \frac{(1 - i)z + (-2 - 2i)}{(-1 + i)z + (-2 - 2i)}.$$

By Theorem 143, this maps $|z| = 2$ to a circle or straight line. But the three images w_1, w_2, and w_3 are on $\text{Re } w = 0$; so $|z| = 2$ is mapped to $\text{Re } w = 0$.

Again by Theorem 143, $|z| < 2$ is mapped to one of the complementary domains $\text{Re } w > 0$ or $\text{Re } w < 0$. To see which, choose any z in $|z| < 2$, say, $z = 0$. This is mapped to 1 in $\text{Re } w > 0$. Thus, the mapping sends $|z| < 2$ onto $\text{Re } w > 0$. Of course, there are also many other linear fractional transformations mapping $|z| < 2$ onto $\text{Re } w > 0$.

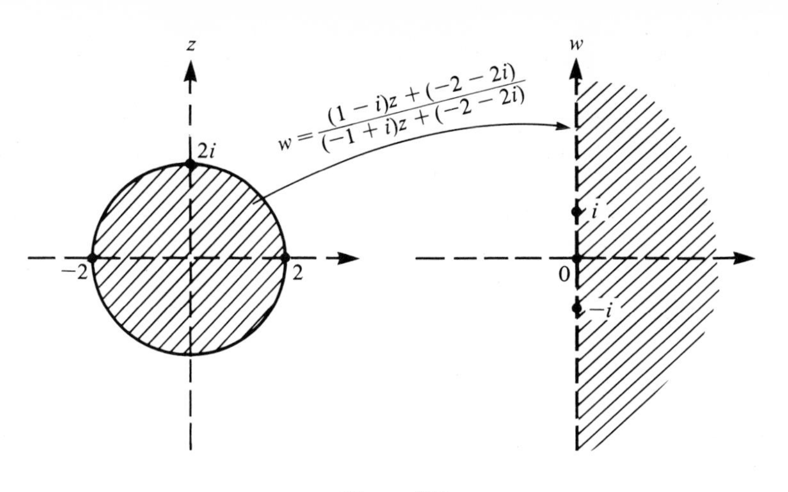

Figure 351

EXAMPLE 493

Map $|z| < 2$ onto Re $w < 0$.

From Example 492, we have a mapping of $|z| < 2$ onto Re $w > 0$. If we can now map Re $w > 0$ onto Re $w < 0$, then the two mappings in succession will take $|z| < 2$ onto Re $w < 0$ (see Figure 352). The mapping sending w to $-w$ maps Re $w > 0$ onto Re $w < 0$. Thus, replace w by $-w$ in Example 492 to get (after multiplying both sides by -1)

$$w = \frac{(-1+i)z + (2+2i)}{(-1+i)z + (-2-2i)}.$$

This maps $2 \to -i$, $2i \to 0$, and $-2 \to i$, and hence maps $|z| = 2$ onto Re $w = 0$. Since -1 in Re $w < 0$ maps to $(-3+4i)/5$ in Re $w < 0$, then this mapping takes $|z| < 2$ onto Re $w < 0$.

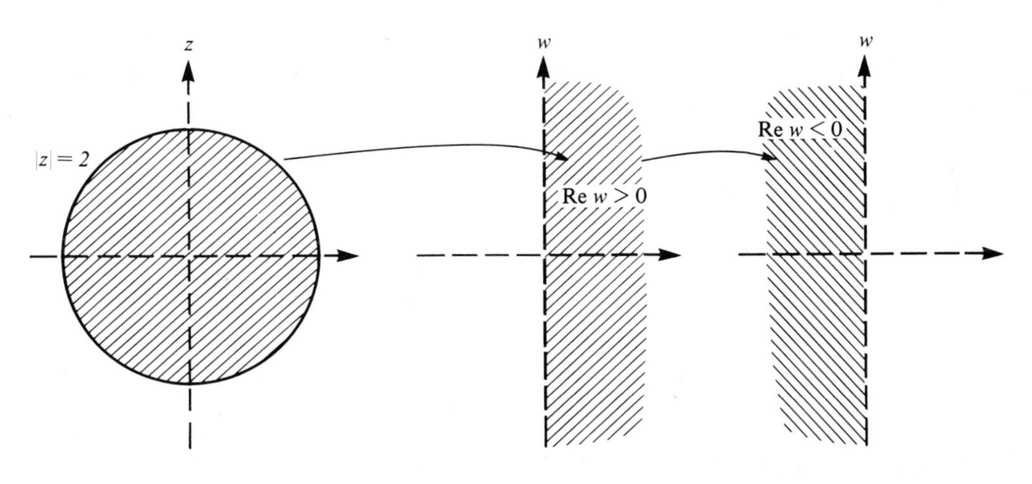

Figure 352

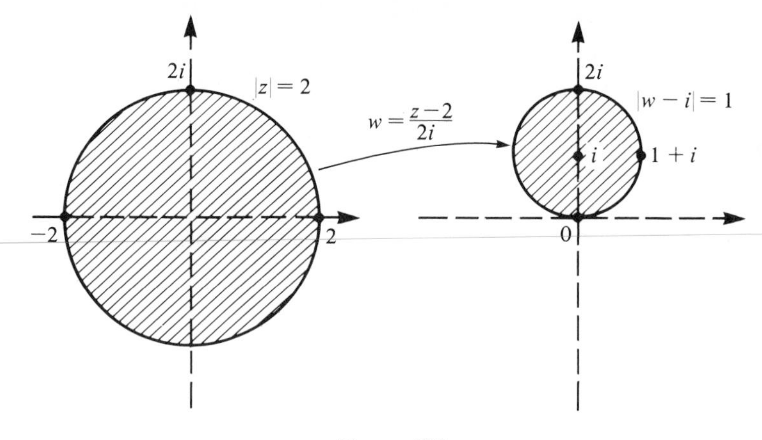

Figure 353

EXAMPLE 494

Map $|z| < 2$ onto $|w - i| < 1$.

Choose three points on the boundary of $|z| < 2$, say, $z_1 = 2$, $z_2 = 2i$, and $z_3 = -2$. Choose three on the boundary of $|w - i| < 1$, say, $w_1 = 0$, $w_2 = 1 + i$, and $w_3 = 2i$ (see Figure 353). The linear fractional mapping sending each z_j to w_j is

$$w = \frac{z - 2}{2i}.$$

This maps $|z| = 2$ onto $|w - i| = 1$ by Theorem 143. Hence, it also maps $|z| < 2$ onto either $|w - i| < 1$ or $|w - i| > 1$. Choose any point in $|z| < 2$, say, $z = 0$. This maps to $w = i$ inside $|w - i| < 1$. Thus, the mapping takes $|z| < 2$ onto $|w - i| < 1$. (Many other mappings will also do this.)

Actually, by choosing both the z_j's and w_j's counterclockwise around the respective boundary circles, we guaranteed that the inside of $|z| < 2$ would map to $|w - i| < 1$, since a conformal mapping preserves orientation.

If we wanted to map $|z| < 2$ to $|w - i| > 1$, we could proceed as follows. Choose three points on $|z| = 2$, say, $z_1 = 2$, $z_2 = 2i$, and $z_3 = -2$, as before. Choose three points on $|w - i| = 1$ in *clockwise* order, say, $w_1 = 2i$, $w_2 = 1 + i$, and $w_3 = 0$. The linear fractional transformation mapping each z_j to w_j is

$$w = \frac{iz + 2i}{z}.$$

This maps $|z| < 2$ onto $|w - i| > 1$ (see Figure 354).

STAGE 2

Often we can construct conformal mappings between given domains by composing two or more known conformal mappings. Here are two examples.

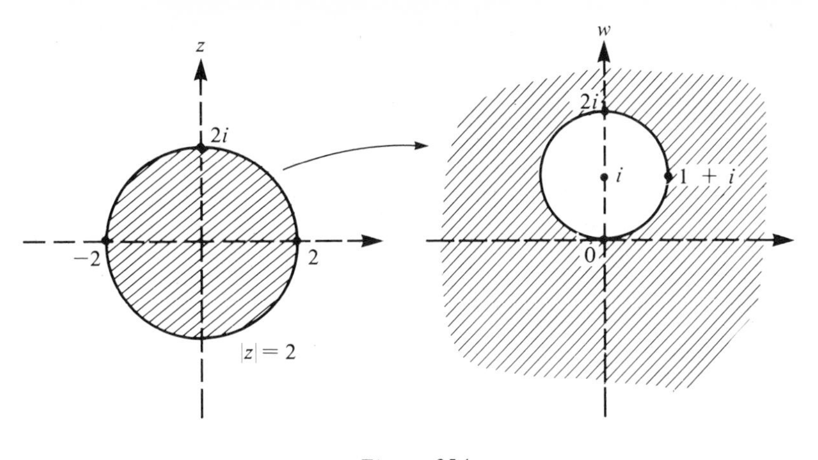

Figure 354

EXAMPLE 495

Map $|z| < 2$ onto the domain $u + v > 0$, as shown in Figure 355.

Here, the key is to recognize that $u + v > 0$ is the right half-plane $\text{Re } w > 0$ rotated counterclockwise by $\pi/4$. Thus, the final mapping can be thought of as a sequence of two mappings, sending in turn

$$|z| < 2 \rightarrow \text{Re } w > 0 \rightarrow u + v > 0,$$

as shown in Figure 356.

The first step, $|z| < 2 \rightarrow \text{Re } w > 0$, was done in Example 492, and we can use that mapping. To rotate by $\pi/4$, we multiply by $e^{i\pi/4}$. Thus, the complete effect is achieved by

$$w = \left(\frac{(1 - i)z + (-2 - 2i)}{(-1 + i)z + (-2 - 2i)} \right) e^{i\pi/4}.$$

For example, $z = 1$ is mapped to $\frac{1}{5}(3 + 4i)e^{i\pi/4}$, which is $(\sqrt{2}/10)(-1 + 7i)$. This is obtained by mapping 1 to $\frac{1}{5}(3 + 4i)$ by the mapping of Example 492, and then rotating counterclockwise by $\pi/4$ radians or 45 degrees. This is shown in Figure 357.

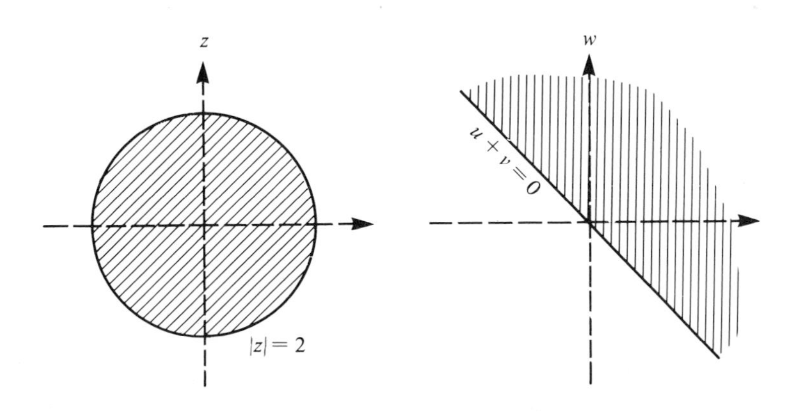

Figure 355

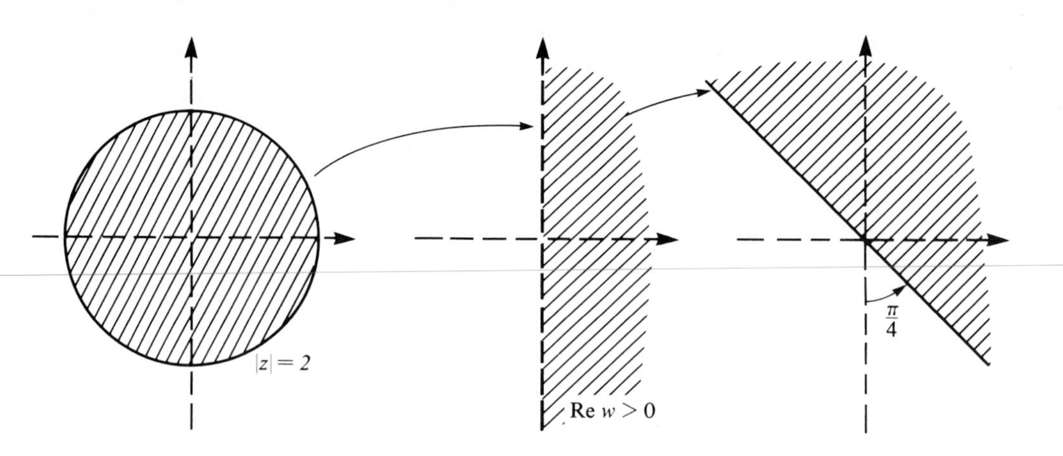

Figure 356

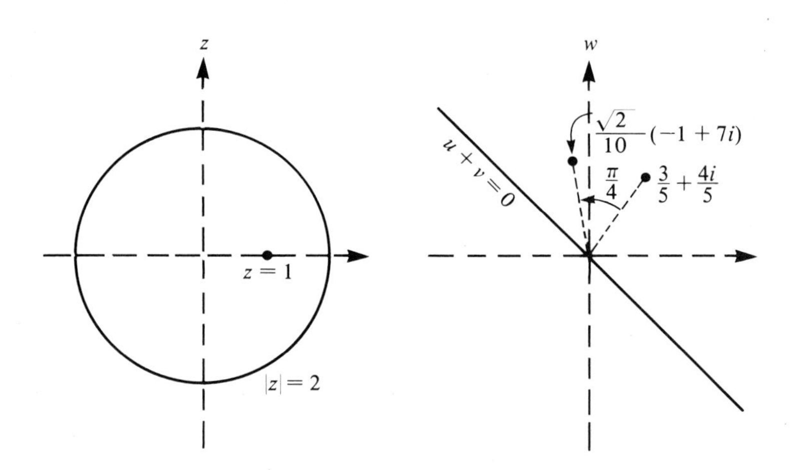

Figure 357

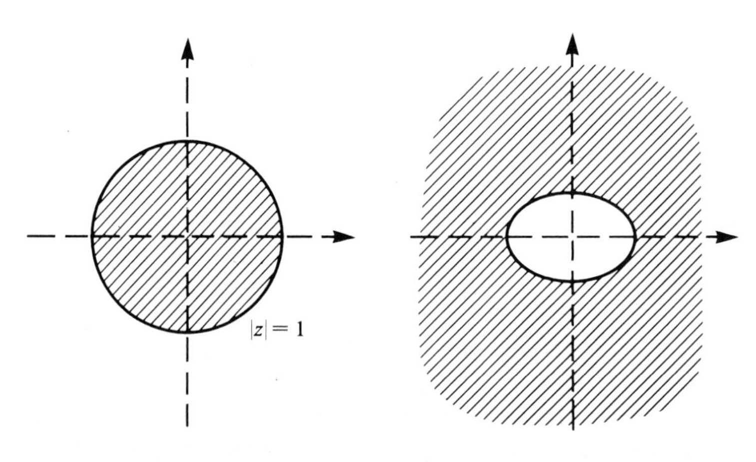

Figure 358

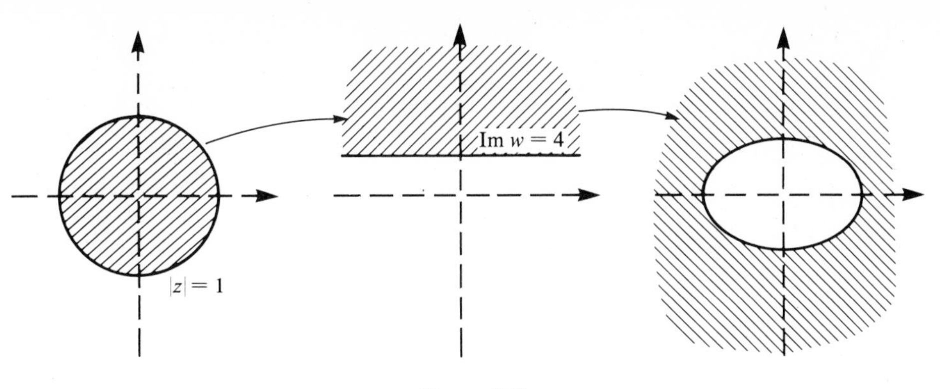

Figure 359

EXAMPLE 496

Map the domain $|z| < 1$ onto the domain *outside* the ellipse

$$\frac{u^2}{\cosh^2(4)} + \frac{v^2}{\sinh^2(4)} = 1,$$

as shown in Figure 358.

To do this, recall from Example 485 that $w = \cos(z)$ maps Im $z = 4$ to the ellipse

$$\frac{u^2}{\cosh^2(4)} + \frac{v^2}{\sinh^2(4)} = 1.$$

Thus, our strategy is to introduce an intermediate step and first map $|z| < 1$ onto Im $w > 4$, and then map this in turn onto the domain outside the ellipse, as indicated in Figure 359.

We can map $|z| < 1$ onto Im $w > 4$ by a linear fractional transformation. For example, by choosing $z_1 = 1$, $z_2 = i$, and $z_3 = -1$ on $|z| = 1$, and $w_1 = -1 + 4i$, $w_2 = 4i$, and $w_3 = 1 + 4i$ on Im $w = 4$, we can use Theorem 141 to construct the mapping

$$w = \frac{(3 + 3i)z + (-5 + 5i)}{(1 - i)z + (1 + i)}.$$

We want to apply cosine to this. Thus, the final result is

$$w = \cos\left(\frac{(3 + 3i)z + (-5 + 5i)}{(1 - i)z + (1 + i)}\right).$$

For example, $z = 0$ maps to $w = \cosh(5)$, outside the ellipse.

STAGE 3

Stage 1 was restricted to linear fractional transformations, which do not always succeed. Stage 2 required that we already know the right mappings to paste together for the final effect, an approach with obvious weaknesses.

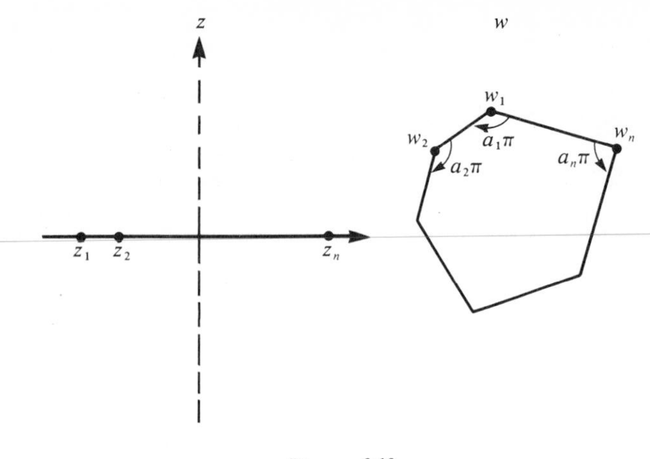

Figure 360

We shall now develop a mapping which is suited to sending the upper half-plane $\operatorname{Im} z > 0$ to the interior of a polygon in the w-plane. Again, one can always create problems this will not solve, but it will broaden our scope considerably.

Imagine then that we have a polygon P in the w-plane, oriented counterclockwise, with vertices at $w_1, \ldots, w_n$ and interior angles $a_1\pi, \ldots, a_n\pi$, as shown in Figure 360. Suppose there is a conformal mapping $w = f(z)$ taking $\operatorname{Im}(z) > 0$ onto the domain bounded by P. What can we say about $f(z)$?

First, let $z_1, \ldots, z_n$ be the points on the real axis (the boundary of $\operatorname{Im} z > 0$) which map to $w_1, \ldots, w_n$, respectively. As z moves from left to right passing over z_j, w moves along the polygon, counterclockwise, passing over w_j. Just before w_j, w is moving along a side of P, which is a straight line; hence, $\arg(dw/dz)$ is constant there. As w moves across w_j onto another side of P, $\arg(dw/dz)$ jumps by an amount $\pi(1 - a_j)$.

Thus, we want $f(z)$ to have the property that, as z moves from left to right across z_j, $\arg f'(z)$ is constant just to the left of z_j, increasing by $\pi(1 - a_j)$ just to the right of z_j. $\operatorname{Arg} f'(z)$ will then remain constant at the new value until z passes across z_{j+1}, where it will jump by $\pi(1 - a_{j+1})$, and so on.

Now, what kind of function behaves like this? Note that, as z moves from left to right across z_j on the real axis, $\arg(z - z_j)$ changes from π to 0, that is, it changes by $-\pi$. Then, $\arg(z - z_j)^{a_j - 1}$ changes by $-\pi(a_j - 1)$ or $\pi(1 - a_j)$. Thus, we would guess that $f'(z)$ consists of factors of the form $(z - z_j)^{a_j - 1}$, each such factor accounting for the change in $\arg f'(z)$ at z_j. This leads us to conjecture that $f'(z)$ has the form

$$f'(z) = A(z - z_1)^{a_1 - 1}(z - z_2)^{a_2 - 1} \cdots (z - z_n)^{a_n - 1}.$$

It can be shown that this is correct, and that the desired mapping is of the form

$$w = f(z) = A \int (z - z_1)^{a_1 - 1}(z - z_2)^{a_2 - 1} \cdots (z - z_n)^{a_n - 1}\, dz + B,$$

where A and B are constants determined by the lengths of the sides of the polygon and its orientation with respect to the axes.

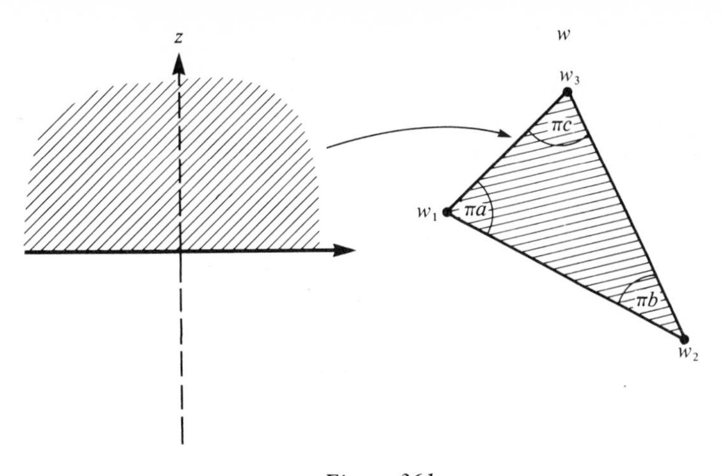

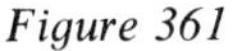
Figure 361

It is possible to adapt this formula to the case in which one of the z_j's, say, z_n is chosen as the point at infinity. Then $z_1, z_2, \ldots, z_{n-1}, \infty$ are mapped to the vertices of P. One can show that in this case we obtain the mapping by omitting the term involving $(z - z_n)$ from the above integral. Thus, when $z_n = \infty$, the mapping has the form

$$w = C + D \int (z - z_1)^{a_1 - 1} \cdots (z - z_{n-1})^{a_{n-1} - 1}\, dz,$$

where C and D are constants.

In practice, the Schwarz-Christoffel transformation is often difficult or impossible to find in explicit form because of the integration. The difficulty increases rapidly with the number of vertices of P. Here are some examples.

EXAMPLE 497

Map the upper half-plane onto a triangle with angles πa, πb, and πc.*

Let the vertex at angle πa be w_1, that at πb, w_2, and that at πc, w_3, as shown in Figure 361. First choose z_1, z_2, and z_3 on the real axis in the z-plane to map to w_1, w_2, and w_3, respectively. We choose $z_1 = 0$, $z_1 = 1$, and, to simplify the integral, $z_3 = \infty$. The mapping then has the form

$$w = C + D \int z^{a-1}(z - 1)^{b-1}\, dz.$$

The constants C and D represent a translation and rotation-magnification, respectively, and are determined by the size and position of the target triangle.

The integral in this transformation has the form of a beta function and cannot be integrated to give a simple, closed-form expression.

---

* Of course, we must have $\pi a + \pi b + \pi c = \pi$; hence, $a + b + c = 1$. We also assume that a, b, and c are positive.

EXAMPLE 498

To map the upper half-plane onto a rectangle, we may choose $z_1 = 0$, $z_2 = 1$, z_3 as any real number greater than 1, and $z_4 = \infty$ to correspond to the vertices of the rectangle. The Schwarz-Christoffel transformation then takes the form

$$w = C + D \int \frac{dz}{\sqrt{z(z-1)(z-z_2)}}.$$

The square root appears because each interior angle in a rectangle is $\pi/2$; hence, each $a_j = \frac{1}{2}$, and $1 - a_j = -\frac{1}{2}$. The resulting integral is an elliptic integral and cannot be evaluated in closed form.

The constants C and D, as before, are determined by the size of the rectangle and by its inclination with respect to the axes.

Lest these examples appear too discouraging, here is one in which the integration can be easily performed explicitly.

EXAMPLE 499

Map the upper half-plane onto the strip $\operatorname{Im} w > 0$, $-b < \operatorname{Re} w < b$ (see Figure 362).

As usual, begin with a choice of points on the x-axis to map to vertices of the "polygon." Here, we are stretching things a bit by thinking of the strip as a polygon with vertices at $-b$, $+b$, and ∞. Choose $z_1 = -1$ to map to $-b$, $z_2 = +1$ to map to $+b$, and let ∞ map to ∞. The interior angles are $\pi/2$ and $\pi/2$; so $a_1 = a_2 = \frac{1}{2}$. The mapping has the form

$$w = C + D \int (z+1)^{-1/2}(z-1)^{-1/2}\, dz.$$

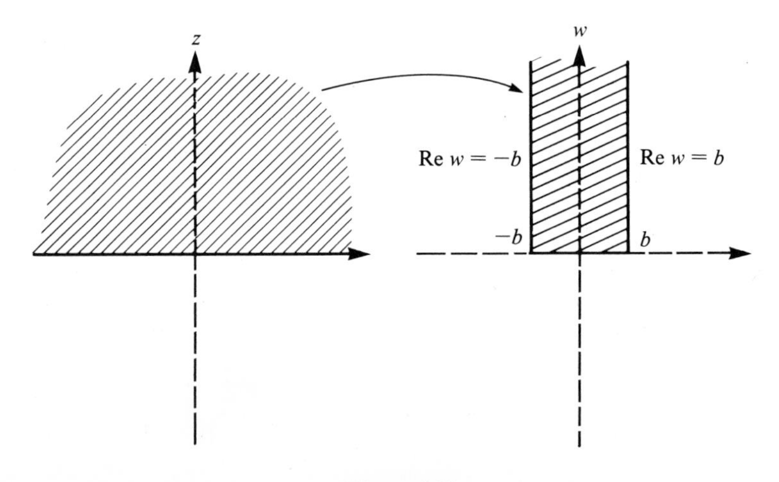

Figure 362

Now, C is a translation term, which we will choose to be zero. Next, write $(z-1)^{-1/2} = [-(1-z)]^{-1/2} = -i(1-z)^{-1/2}$, and absorb $-i$ into the constant D to write

$$w = E\int (1+z)^{-1/2}(1-z)^{-1/2}\,dz$$

$$= E\int \frac{dz}{(1-z^2)^{1/2}}.$$

Thus, w has the form

$$w = E\arcsin(z)$$

or, inverted,

$$z = \sin\left(\frac{w}{E}\right).$$

We must now choose E so that -1 maps to $-b$, and 1 maps to b. Thus, we need

$$-1 = \sin\left(\frac{-b}{E}\right) \quad\text{and}\quad 1 = \sin\left(\frac{b}{E}\right).$$

Thus, choose $b/E = \pi/2$ or $E = 2b/\pi$. The mapping is then

$$w = \frac{2b}{\pi}\arcsin(z).$$

Note that, with the choice $b = \pi/2$, this mapping is just $w = \arcsin(z)$, mapping the upper half-plane onto the strip $\text{Im } w > 0$, $-\pi/2 < \text{Re } w < \pi/2$. This is consistent with Example 486, which had z and w interchanged from their roles here but reached the same conclusion.

In the next chapter we shall consider some applications of conformal mappings.

PROBLEMS FOR SECTION 18.3

In each of Problems 1 through 10, find a linear fractional transformation between the given domains.

1. $|z| < 3$ onto $|w - i| < 4$
2. $|z| < 3$ onto $|w - i| > 4$
3. $|z + 2i| < 1$ onto $|w - 3| > 2$
4. $\text{Re } z > 1$ onto $\text{Im } w > 1$
5. $\text{Re } z < 0$ onto $|w| < 4$
6. $\text{Im } z > -4$ onto $|w - i| > 2$
7. $\text{Re } z > 0$ onto $\text{Im } z < 3$
8. $\text{Re } z < -4$ onto $|w + 1 - 2i| > 3$
9. $|z - 1| > 4$ onto $\text{Im } w < 2$
10. $|z - 1 + 3i| > 1$ onto $\text{Re } w < -5$
11. Find a conformal mapping which takes the upper half-plane $\text{Im } z > 0$ onto the wedge $0 < \arg w < \pi/3$.
12. Show that $w = \log(z)$ maps $\text{Re } z > 0$ onto the strip $0 < \text{Im } w < \pi$.
13. Show that $w = \log[(1+z)/(1-z)]$ maps $|z| < 1$ onto $-\pi/2 < \text{Im } z < \pi/2$.
14. Show that

$$w = \int (1 - z^n)^{-2/n}\,dz$$

maps $|z| < 1$ onto the interior of a regular n-sided polygon.

15. Consider the transformation

$$w = \frac{1}{2}\left(z + \frac{1}{z}\right).$$

(a) Show that this maps both $|z| < 1$ and $|z| > 1$ onto the complex w-plane with the interval $[-1, 1]$ on the real axis removed.

(b) Show that circles $|z| = r, 0 < r < 1$, map to ellipses.

SUPPLEMENTARY PROBLEMS

In each of Problems 1 through 10, find a linear fractional transformation mapping the three given points to the indicated images.

1. $1 \to 2i, \quad 3 \to -i, \quad 4 \to -7 + i$

2. $2 - 3i \to 4, \quad 1 + i \to 5 + i, \quad 4 - i \to 3 + 2i$

3. $i \to i, \quad 1 - 4i \to 2, \quad -3i \to 3$

4. $-2 + 3i \to -5 + i, \quad 2 - i \to 5 + 2i, \quad 3 \to \infty$

5. $3 \to 7i, \quad 4 \to 8i, \quad -1 \to 6 + 2i$

6. $-2 + 5i \to 3 - 2i, \quad 1 + i \to 3 - i, \quad 2 \to \infty$

7. $3 + 7i \to 3 + 7i, \quad 2 \to -2, \quad 4i \to 6$

8. $5i \to 4 - i, \quad 1 \to 3 + i, \quad 6i \to \infty$

9. $10i - 1 \to 3 + i, \quad i \to -i, \quad 9 \to 6 + 3i$

10. $1 - 2i \to 1 - 3i, \quad 5 \to 6i, \quad 2 \to 5 + 2i$

11. Find a conformal mapping which maps $|z| < 3$ onto $\text{Im}\, w > 0$.

12. Find a conformal mapping which maps $|z| > 1$ onto $\text{Re}\, w < -2$.

13. Find a conformal mapping which maps $\text{Im}\, z < 3$ onto $\text{Im}\, w > -4$.

14. Find a conformal mapping which maps $\text{Im}\, z < 0$ onto $u + v > 2$ in the $(w = u + iv)$-plane.

15. Find a conformal mapping which maps $\text{Re}\, z > 1$ onto $|w - i| < 3$.

16. Find a conformal mapping which maps $|z - 2 + i| < 1$ onto $\text{Re}\, w > -2$.

17. Find a conformal mapping which maps $|z + 2| < 1$ onto $u > v$.

18. Show that, for any given θ and z_0, the mapping

$$w = e^{i\theta}\left(\frac{z - z_0}{1 - \bar{z}_0 z}\right)$$

is a conformal mapping of the unit disk onto itself.

19. Show that, for any given θ and z_0, the mapping

$$w = e^{i\theta}\left(\frac{z - z_0}{z - \bar{z}_0}\right)$$

is a conformal mapping of $\text{Im}\, z > 0$ onto $|w| < 1$.

20. Define the *cross ratio* of z_1, z_2, z_3, and z_4 by

$$[z_1, z_2, z_3, z_4] = \left(\frac{z_4 - z_1}{z_4 - z_2}\right)\left(\frac{z_3 - z_2}{z_3 - z_1}\right).$$

Suppose that $w = T(z)$ is a linear fractional transformation, mapping each z_j to w_j, $j = 1, 2, 3, 4$. Prove that cross ratio is preserved; that is, show that

$$[z_1, z_2, z_3, z_4] = [w_1, w_2, w_3, w_4].$$

21. Continuing from Problem 20, prove that the cross ratio of z_1, z_2, z_3, and z_4 is the image of z_1 under the linear fractional transformation which maps $z_2 \to 1, z_3 \to 0$, and $z_4 \to \infty$.

22. Show that the cross ratio $[z_1, z_2, z_3, z_4]$ is real if and only if the four points z_1, z_2, z_3, and z_4 lie on a circle or straight line.

23. Two points z and z^* are said to be *symmetric with respect to a circle C* passing through points z_1, z_2, and z_3 if

$$[z^*, z_1, z_2, z_3] = \overline{[z, z_1, z_2, z_3]}.$$

Using the result of Problem 22, prove that the points which are symmetric to themselves (that is, $z = z^*$) are exactly those on C.

24. Continuing from Problem 23, prove the *Symmetry principle*: Suppose a linear fractional transformation $w = T(z)$ maps a circle C onto a circle K. Let z and z^* be symmetric with respect to C, and let $w = T(z)$, $w^* = T(z^*)$. Then w and w^* are symmetric with respect to K. (*Hint*: Note the result of Problem 20.)

CHAPTER NINETEEN

Some Applications of Complex Analysis

19.1 HARMONIC FUNCTIONS AND THE DIRICHLET PROBLEM FOR THE UNIT DISK

We have already used complex function methods to evaluate real integrals and sum infinite series. In this section we indicate a variety of other areas in which complex analysis plays an important role.

Given a domain D bounded by a simple closed curve C in the plane, the *Dirichlet problem for D* consists of finding a function $u(x, y)$ such that

$$\frac{\partial^2 u}{\partial x^2} + \frac{\partial^2 u}{\partial y^2} = 0 \quad \text{for} \quad (x, y) \text{ in } D$$

and

$$u(x, y) = g(x, y) \quad \text{for} \quad (x, y) \text{ on } C,$$

where $g(x, y)$ is a given function. Such problems arise in many contexts, among them heat conduction and vibrations in plates and membranes.

The differential equation $\partial^2 u/\partial x^2 + \partial^2 u/\partial y^2 = 0$ is called *Laplace's equation*, and any function satisfying it is called a *harmonic function.* For example, $u(x, y) = \ln(x^2 + y^2)$ is harmonic on any domain D not containing the origin.

Thus, we may rephrase the Dirichlet problem as one of finding a function which is harmonic in D and takes on prescribed values on the boundary of D.

Here is the connection between complex functions and harmonic functions.

THEOREM 144

Let D be a simply connected domain.

1. If $f(z) = u(x, y) + iv(x, y)$ is analytic in D, and $u(x, y)$ and $v(x, y)$ have continuous partial derivatives in D, then $u(x, y)$ and $v(x, y)$ are both harmonic in D.
2. If $u(x, y)$ is harmonic in D and has continuous first partials there, then, for some $v(x, y)$, $u(x, y) + iv(x, y)$ is analytic in D.

Proof of (1) By the Cauchy-Riemann equations,

$$\frac{\partial u}{\partial x} = \frac{\partial v}{\partial y} \quad \text{and} \quad \frac{\partial u}{\partial y} = -\frac{\partial v}{\partial x}.$$

Then,

$$\frac{\partial^2 u}{\partial x^2} + \frac{\partial^2 u}{\partial y^2} = \frac{\partial}{\partial x}\left(\frac{\partial v}{\partial y}\right) + \frac{\partial}{\partial y}\left(-\frac{\partial v}{\partial x}\right) = \frac{\partial^2 v}{\partial x\, \partial y} - \frac{\partial^2 v}{\partial y\, \partial x} = 0.$$

Similarly, $v(x, y)$ is also harmonic in D.

Proof of (2) Let

$$g(z) = \frac{\partial u}{\partial x} - i\frac{\partial u}{\partial y} \quad \text{for} \quad z = x + iy \text{ in } D.$$

It is easy to check from the Cauchy-Riemann equations that $g(z)$ is analytic in D. Since D is simply connected, there is some $G(z)$ analytic in D such that $G'(z) = g(z)$. Let $G(z) = A(x, y) + iB(x, y)$.

Now,

$$G'(z) = \frac{\partial A}{\partial x} + i\frac{\partial B}{\partial x} = \frac{\partial A}{\partial x} - i\frac{\partial A}{\partial y} = g(z) = \frac{\partial u}{\partial x} - i\frac{\partial u}{\partial y}.$$

Then,

$$\frac{\partial A}{\partial x} = \frac{\partial u}{\partial x} \quad \text{and} \quad \frac{\partial A}{\partial y} = \frac{\partial u}{\partial y}.$$

Thus, for some constant K,

$$A(x, y) = u(x, y) + K.$$

Define $f(z) = G(z) - K$ for z in D. Then, $f(z)$ is analytic in D. Further,

$$f(z) = G(z) - K = A(x, y) + iB(x, y) - K = u(x, y) + iB(x, y).$$

Thus, we may choose $v(x, y) = B(x, y)$, and (2) is proved.

A function $v(x, y)$ such that $u(x, u) + iv(x, y)$ is analytic in a domain D is called a *harmonic conjugate* of $u(x, y)$. One is rarely interested in computing a harmonic conjugate of a specific function $u(x, y)$. Rather, we use Theorem 144 to assert its existence in problems where $u(x, y)$ is really the function of interest. This enables us to treat $f(z) = u(x, y) + iv(x, y)$ instead of just $u(x, y)$, and complex function methods then become available to us. After analyzing $f(z)$ using complex techniques, the function $u(x, y)$ can always be retrieved as $\text{Re}\, f(z)$.

As an illustration of these ideas, we shall solve the Dirichlet problem for the unit disk. Thus, we want to find a function $u(x, y)$ such that

$$\frac{\partial^2 u}{\partial x^2} + \frac{\partial^2 u}{\partial y^2} = 0 \quad \text{for} \quad x^2 + y^2 < 1$$

and

$$u(x, y) = g(x, y), \quad \text{given for} \quad x^2 + y^2 = 1.$$

Note that the problem itself is a real-valued one; that is, complex numbers and complex functions have nothing to do with the problem, at least at first appearance. The idea is to introduce a harmonic conjugate of $u(x, y)$, examine the resulting complex function $f(z) = u(x, y) + iv(x, y)$, and then retrieve $u(x, y)$ later.

Begin by noting that a solution $u(x, y)$ will be the real part of a function $f(z) = u(x, y) + iv(x, y)$ analytic in $|z| < 1$. We may assume (by adding a constant if necessary) that $f(0) = u(0, 0)$. Expand $f(z)$ in a Taylor series about 0:

$$f(z) = \sum_{n=0}^{\infty} a_n z^n.$$

Then,

$$\begin{aligned} u(x, y) &= \operatorname{Re} f(z) = \tfrac{1}{2}[f(z) + \overline{f(z)}] \\ &= \sum_{n=0}^{\infty} \tfrac{1}{2} a_n z^n + \sum_{n=0}^{\infty} \tfrac{1}{2} \bar{a}_n \bar{z}^n \\ &= \tfrac{1}{2}(a_0 + \bar{a}_0) + \sum_{n=1}^{\infty} \tfrac{1}{2}(a_n z^n + \bar{a}_n \bar{z}^n) \\ &= u(0, 0) + \sum_{n=1}^{\infty} \tfrac{1}{2}(a_n z^n + \bar{a}_n \bar{z}^n), \end{aligned}$$

since

$$a_0 = f(0) = u(0, 0) = \bar{a}_0.$$

At this point we will assume that the expansion of $f(z)$ is valid in a disk of radius slightly larger than 1. If z is on the unit circle, then $|z| = 1$; so $\bar{z} = 1/z$. Then,

$$u(x, y) = u(0, 0) + \sum_{n=1}^{\infty} \frac{1}{2}\left(a_n z^n + \frac{\bar{a}_n}{z^n}\right).$$

We shall now determine the a_n's. If C is the unit circle, then

$$\frac{1}{2\pi i}\oint_C u(x, y) \cdot z^m \, dz = \frac{u(0, 0)}{2\pi i}\oint_C z^m \, dz + \sum_{n=1}^{\infty} \frac{1}{4\pi i}\oint_C (a_n z^{n+m} + \bar{a}_n z^{m-n}) \, dz.$$

But,

$$\frac{1}{2\pi i}\oint_C z^p \, dz = \begin{cases} 1 & \text{if } p = -1, \\ 0 & \text{if } p \neq -1, \end{cases}$$

for any integer p.

Thus, when $m = -1$, we get

$$\frac{1}{2\pi i}\oint_C u(x, y)\frac{1}{z} \, dz = u(0, 0)$$

and, when $m = -p - 1$, for $p = 1, 2, 3, \ldots$, we get

$$\frac{1}{2\pi i}\oint_C u(x, y) z^{-p-1} \, dz = \frac{a_p}{2}.$$

Thus, in general,

$$a_n = \frac{1}{\pi i} \oint_C u(x, y) z^{-n-1}\, dz \quad \text{for} \quad n = 1, 2, 3, \ldots.$$

We now have the coefficients in the Taylor series for $f(z)$, in terms of $u(x, y)$ on C, where $u(x, y) = g(x, y)$ is known. To avoid confusion in the next few lines, let $z = x + iy$ be a point inside the unit disk (so $|z| < 1$), and let $t = \alpha + i\beta$ be a point on the unit circle (so $|t| = 1$). Then,

$$f(z) = \sum_{n=0}^{\infty} a_n z^n = \frac{1}{2\pi i} \oint_C g(\alpha, \beta) \frac{1}{t}\, dt + \sum_{n=1}^{\infty} \left(\frac{1}{\pi i} \oint_C g(\alpha, \beta) \cdot t^{-n-1}\, dt \right) z^n$$

$$= \frac{1}{2\pi i} \oint_C g(\alpha, \beta) \left[1 + 2 \sum_{n=1}^{\infty} \left(\frac{z}{t} \right)^n \right] \frac{1}{t}\, dt.$$

But $|z/t| < 1$, since $|z| < 1$ and $|t| = 1$. Then, by geometric series, we have

$$\sum_{n=1}^{\infty} \left(\frac{z}{t} \right)^n = \sum_{n=0}^{\infty} \left(\frac{z}{t} \right)^n - 1 = \frac{1}{1 - \frac{z}{t}} - 1 = \frac{z}{t - z}.$$

Thus,

$$f(z) = \frac{1}{2\pi i} \oint_C g(\alpha, \beta) \left(1 + \frac{2z}{t - z} \right) \frac{1}{t}\, dt = \frac{1}{2\pi i} \oint_C g(\alpha, \beta) \left(\frac{t + z}{t - z} \right) \frac{1}{t}\, dt.$$

Finally, use this formula for $f(z)$ to retrieve $u(x, y)$, which was the main goal:

$$u(x, y) = \operatorname{Re} \left[\frac{1}{2\pi i} \oint_C g(\alpha, \beta) \left(\frac{t + z}{t - z} \right) \frac{1}{t}\, dt \right].$$

This gives a function harmonic in $x^2 + y^2 < 1$, strictly in terms of data given on $x^2 + y^2 = 1$. This formula for $u(x, y)$ is sometimes called the *Schwarz integral.*

We can obtain a more explicit integral by using polar coordinates. Let $t = e^{i\phi}$ for $0 \le \phi \le 2\pi$, and let $z = re^{i\theta}$ for $0 \le r \le 1$, $0 \le \theta \le 2\pi$. Then, $\alpha = \cos(\phi)$ and $\beta = \sin(\phi)$ in the Schwarz integral. Assuming that $u(x, y)$ transforms to $U(r, \theta)$ in polar coordinates, we get, after a lengthy calculation,

$$U(r, \theta) = \frac{1 - r^2}{2\pi} \int_0^{2\pi} g[\cos(\phi), \sin(\phi)] \left[\frac{1}{1 - 2r\cos(\theta - \phi) + r^2} \right] d\phi.$$

This is the *Poisson integral formula* for the solution of the Dirichlet problem for the unit disk. In general, this expression is impossible to integrate to a closed form. (An exception is given in Problem 12 at the end of this section.) However, it is still of value in providing some information about $u(x, y)$. For example, one can perform a numerical integration for specific values of r and θ.

In the next section we shall use conformal mappings to solve the Dirichlet problem for other regions.

PROBLEMS FOR SECTION 19.1

In each of Problems 1 through 6, use the proof of Theorem 144 to produce $v(x, y)$ such that $u(x, y) + iv(x, y)$ is analytic. Check that both $u(x, y)$ and $v(x, y)$ are harmonic.

1. $u(x, y) = x^2 - y^2$ **2.** $u(x, y) = x$ **3.** $u(x, y) = x + y$
4. $u(x, y) = e^x \cos(y)$ **5.** $u(x, y) = \sin(x)\cosh(y)$ **6.** $u(x, y) = x^3 - xy^2 - 2xy^2$

7. Show that the solution of the Dirichlet problem

$$\frac{\partial^2 u}{\partial x^2} + \frac{\partial^2 u}{\partial y^2} = 0 \quad \text{for} \quad (x, y) \text{ in } D,$$

$$u(x, y) = g(x, y) \quad \text{on the boundary of } D,$$

is unique (assuming existence). (*Hint:* Suppose that u_1 and u_2 are solutions; then show that $w = u_1 - u_2$ must be identically zero.)

8. Let $f(z)$ be analytic for $|z| < R$, and let $u(r, \theta) = \text{Re}\, f(z)$ for $z = re^{i\theta}$. Prove that, for $|z| < R$,

$$f(z) = \frac{1}{2\pi} \int_0^{2\pi} \left(\frac{Re^{i\phi} + z}{Re^{i\phi} - z} \right) u(R, \phi)\, d\phi.$$

9. Show that, for $|z| < 1$ and $|t| = 1$.

$$\frac{t + z}{t - z} = 1 + 2 \sum_{n=1}^{\infty} r^n e^{in(\theta - \phi)},$$

where $z = re^{i\theta}$ and $t = e^{i\phi}$.

Use this to show that Poisson's integral solution of the Dirichlet problem for the unit disk (in polar coordinates) is

$$U(r, \theta) = \frac{a_0}{2} + \sum_{n=1}^{\infty} r^n [a_n \cos(n\theta) + b_n \sin(n\theta)],$$

where

$$a_n = \frac{1}{\pi} \int_0^{2\pi} U(1, \phi) \cos(n\phi)\, d\phi$$

and

$$b_n = \frac{1}{\pi} \int_0^{2\pi} U(1, \phi) \sin(n\phi)\, d\phi.$$

This is in fact a Fourier series expression for the solution.

10. Use the result of Problem 9 to obtain

$$U(r, \theta) = \frac{4}{\pi} \text{Re}[\arctan(re^{i\theta})]$$

as a solution of the Dirichlet problem for the unit disk, with (in polar coordinates)

$$U(1, \theta) = \begin{cases} 1, & -\pi/2 < \theta < \pi/2, \\ -1, & \pi/2 < \theta < 3\pi/2. \end{cases}$$

11. Let $\phi(x, y)$ be harmonic in a domain D, and let $f(z) = u(x, y) + iv(x, y)$ be a conformal mapping of D onto a domain D' in the $(w = u + iv)$-plane. Let $\phi(x, y)$ transform into $\Phi(u, v)$ under this mapping. Prove that $\Phi(u, v)$ is harmonic in D'.

12. Use the Poisson integral formula to solve the Dirichlet problem for the unit disk, with (in polar coordinates)

$$U(1, \theta) = \begin{cases} 1, & 0 < \theta < \pi, \\ 0, & \pi < \theta < 2\pi. \end{cases}$$

13. Carry out the details omitted in deriving Poisson's integral to show that

$$\operatorname{Re}\left[\frac{1}{2\pi i}\left(\frac{t+z}{t-z}\right)\frac{1}{t}\right] dt = \frac{1-r^2}{2\pi}\left(\frac{dt}{1-2r\cos(\theta-\phi)+r^2}\right)$$

if $z = re^{i\theta}$ and $t = e^{i\phi}$.

19.2 CONFORMAL MAPPING SOLUTIONS OF DIRICHLET PROBLEMS

We have, at least in theory, solved the Dirichlet problem for the unit disk. Using this solution, we can construct solutions for other domains for which we can find conformal mappings onto the unit disk. Here are two examples.

EXAMPLE 500

We shall solve the Dirichlet problem for the upper half-plane $\operatorname{Im} z > 0$. Thus, we want $u(x, y)$ harmonic in $y > 0$ such that $u(x, 0) = g(x)$, a given function.

The strategy is to find a conformal mapping of $\operatorname{Im} z > 0$ onto $|w| < 1$, and then map the solution for the unit disk onto a solution for $\operatorname{Im} z > 0$ (note Problem 11, Section 19.1). Here, it is possible to find a linear fractional transformation, following the method of Section 18.3. Mapping $-1 \to 1$, $0 \to i$, and $1 \to -1$, we find that

$$w = \frac{(1+i)z + (1-i)}{(-1+i)z + (-1-i)}$$

maps $\operatorname{Im} z > 0$ onto $|w| < 1$ and $\operatorname{Im} z = 0$ onto $|w| = 1$.

Now we must keep our notation straight. We have a solution for $|w| < 1$. As a convenience, we shall write $w = X + iY$ and $z = x + iy$. On $\operatorname{Im} z = 0$, we shall use t as the variable, and on $|w| = 1$, we shall use T as the image of t. Thus, capitals denote points in the w-plane, and lowercase letters denote points in the z-plane (see Figure 363).

Let $U(X, Y)$ be the function obtained from $u(x, y)$ by the above linear fractional transformation. Then, $u(x, 0) = g(x)$ maps to $U(X, Y) = G(X, Y)$, given for $X^2 + Y^2 = 1$. The solution of

$$\frac{\partial^2 U}{\partial x^2} + \frac{\partial^2 U}{\partial y^2} = 0 \quad \text{for} \quad X^2 + Y^2 < 1,$$

$$U(X, Y) = G(X, Y) \quad \text{for} \quad X^2 + Y^2 = 1,$$

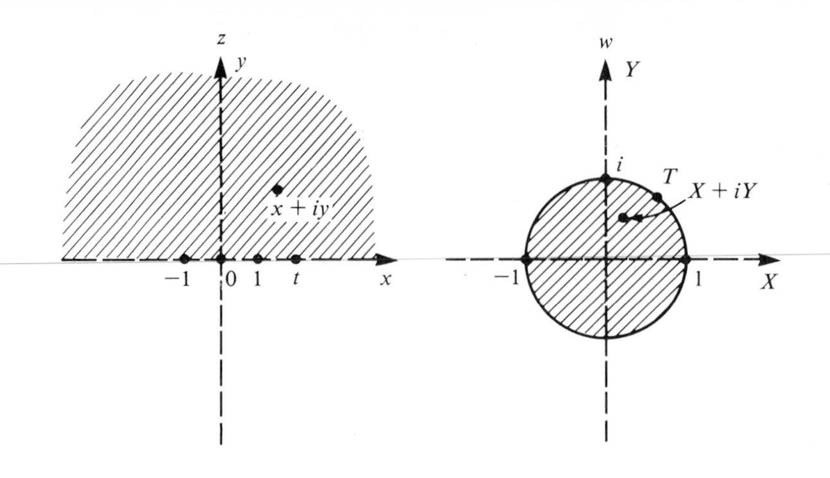

Figure 363

is given by the Schwarz integral

$$U(X, Y) = \text{Re}\left[\frac{1}{2\pi i}\int_C G(X, Y)\left(\frac{T+Z}{T-Z}\right)\frac{1}{T}\,dT\right],$$

where C is the unit circle $|w| = 1$.

Apply the change of variables

$$w = \frac{(1+i)z + (1-i)}{(-1+i)z + (-1-i)}.$$

Then, Im $z = 0$ maps to $|w| = 1$; so T on $|w| = 1$ maps to t on the real axis, given by

$$T = \frac{(1+i)t + (1-i)}{(-1+i)t + (-1-i)}.$$

Then we compute

$$dT = \frac{-2i\,dt}{-it^2 + 2t + i},$$

$$\frac{T+Z}{T-Z} = \frac{\dfrac{(1+i)t+(1-i)}{(-1+i)t+(-1-i)} + \dfrac{(1+i)z+(1-i)}{(-1+i)z+(-1-i)}}{\dfrac{(1+i)t+(1-i)}{(-1+i)t+(-1-i)} - \dfrac{(1+i)z+(1-i)}{(-1+i)z+(-1-i)}}$$

$$= \frac{2(x+iy)t + 2}{(1+i)t + (1-i)(x+iy) + (1-i)},$$

and

$$\frac{1}{T} = \frac{(-1+i)t + (-1-i)}{(1+i)t + (1-i)}.$$

Substitute these into the Schwarz integral [in effect, we are changing variables from (X, Y) to (x, y)]. Remember that $G(X, Y)$ converts to $g(x)$, $U(X, Y)$ to $u(x, y)$, and $|w| = 1$ to the real line. The transformed integral is

$$u(x, y) = \operatorname{Re}\left(\frac{1}{2\pi i}\int_{-\infty}^{\infty} g(t)\left[\frac{2(x+iy)t + 2}{(1+i)t + (1-i)(x+iy) + (1-i)}\right] \times \left[\frac{(-1+i)t + (-1-i)}{(1+i)t + (1-i)}\right]\left[\frac{-2i}{-it^2 + 2t + i}\right] dt\right).$$

After a good deal of calculation, which we omit, this reduces to

$$u(x, y) = \frac{y}{\pi}\int_{-\infty}^{\infty} \frac{g(t)\,dt}{(t-x)^2 + y^2} \quad \text{for} \quad -\infty < x < \infty, \quad y > 0.$$

This solves the Dirichlet problem for the upper half-plane. (Incidentally, one can derive the same result by Fourier integral or Fourier transform—see Chapter 13, Sections 5 and 10.)

EXAMPLE 501

Solve the Dirichlet problem for the right quarter-plane.

The problem is to find $u(x, y)$ such that

$$\frac{\partial^2 u}{\partial x^2} + \frac{\partial^2 u}{\partial y^2} = 0 \quad \text{if} \quad x > 0, \quad y > 0,$$

$$u(x, 0) = f(x) \quad \text{for} \quad x > 0,$$

and

$$u(0, y) = g(y) \quad \text{for} \quad y > 0.$$

First, we need a conformal mapping from the right quarter-plane to the unit disk. The easiest thing to do is to map the right quarter-plane to the upper half-plane by sending z to z^2, and then map the upper half-plane to the unit disk. Using the map of the last example, we get

$$w = \frac{(1+i)z^2 + (1-i)}{(-1+i)z^2 + (-1-i)},$$

sending the right quarter-plane to $|w| < 1$. To keep the orientation counterclockwise, we think of the positive direction on the boundary of the right quarter-plane as down the y-axis toward 0, and then out the x-axis toward infinity.

The Schwarz integral now gives us

$$u(x, y) = \text{Re}\left\{\frac{1}{2\pi i}\int_{\infty}^{0} g(t)\left[\frac{\dfrac{(1+i)t^2+(1-i)}{(-1+i)t^2+(-1-i)}+\dfrac{(1+i)z^2+(1-i)}{(-1+i)z^2+(-1-i)}}{\dfrac{(1+i)t^2+(1-i)}{(-1+i)t^2+(-1-i)}-\dfrac{(1+i)z^2+(1-i)}{(-1+i)z^2+(-1-i)}}\right]\right.$$

$$\times\left[\frac{(-1+i)t^2+(-1-i)}{(1+i)t^2+(1-i)}\right]\left[\frac{-2it\,dt}{-it^4+2t^2+i}\right]$$

$$+\frac{1}{2\pi i}\int_{0}^{\infty} f(t)\left[\frac{\dfrac{(1+i)t^2+(1-i)}{(-1+i)t^2+(-1-i)}+\dfrac{(1+i)z^2+(1-i)}{(-1+i)z^2+(-1-i)}}{\dfrac{(1+i)t^2+(1-i)}{(-1+i)t^2+(-1-i)}-\dfrac{(1+i)z^2+(1-i)}{(-1+i)z^2+(-1-i)}}\right]$$

$$\left.\times\left[\frac{(-1+i)t^2+(-1-i)}{(1+i)t^2+(1-i)}\right]\left[\frac{-2it\,dt}{-it^4+2t^2+i}\right]\right\}.$$

After some computation, we get, for the case $g(y) \equiv 0$,

$$u(x, y) = \frac{y}{\pi}\int_0^{\infty} f(t)\left[\frac{1}{y^2+(t-x)^2}-\frac{1}{y^2+(t+x)^2}\right]dt \qquad \text{for} \quad x > 0, \quad y > 0.$$

As with the last example, one could also solve this problem by separation of variables.

We pursue conformal mapping solutions of the Dirichlet problem for other domains in the exercises.

PROBLEMS FOR SECTION 19.2

1. Use the conformal mapping $w = Rz$ to solve the Dirichlet problem for a disk of radius R about the origin. Derive the solution (in polar coordinates)

$$U(r, \theta) = \frac{R^2 - r^2}{2\pi}\int_0^{2\pi} g[\cos(\phi), \sin(\phi)]\left[\frac{1}{R^2 - 2rR\cos(\theta - \phi) + r^2}\right]d\phi$$

for $0 \le r < R, 0 \le \theta \le 2\pi$.

2. Map the disk $|z - z_0| \le R$ onto $|w| \le 1$ to solve the Dirichlet problem for a disk of radius R centered at z_0.
3. Solve the Dirichlet problem for the domain Im $z < 0$.
4. Solve the Dirichlet problem for the domain Re $z > 0$.
5. Solve the Dirichlet problem for the domain Re $z < 0$, Im $z < 0$, assuming that the solution is zero for $z = y < 0$ and $g(x)$ for $z = x < 0$.
6. Solve the Dirichlet problem for the *exterior* of the circle $|z| = R$.
7. Solve the Dirichlet problem for the exterior of the ellipse $\frac{1}{2}x^2 + y^2 = 1$. (*Hint:* Choose a correctly in the mapping $w = z + (a^2/z)$ to send this ellipse to the unit circle.)
8. Solve the Dirichlet problem for the domain defined by $y > x$. (*Hint:* Use a conformal mapping which rotates this domain to one for which we know the solution.)

19.3 COMPLEX FUNCTIONS IN THE ANALYSIS OF FLUID FLOW

The fact that the real and imaginary parts of an analytic function are harmonic causes the theory of complex functions to be quite useful in studying the flow of fluids. We also assume here a familiarity with basic vector analysis (see Chapter 11).

Consider a two-dimensional flow of some fluid. The fluid may be air, water, oil, and so on. The flow is *two-dimensional* if the variables of the motion are functions of two space coordinates, x and y. They may also depend on time t. At each point (x, y) and time t, we assume that the fluid has:

velocity $\mathbf{V}(x, y, t)$ (a vector)

viscosity $\mu(x, y, t)$, temperature $T(x, y, t)$, density $\rho(x, y, t)$, pressure $p(x, y, t)$ — scalar quantities

An *incompressible fluid* is one in which the density remains constant. This is the case with water under normal conditions. The flow is *irrotational* if curl $\mathbf{V} = \mathbf{0}$.

We shall now consider irrotational, incompressible two-dimensional flows. In vector analysis, it is shown that curl $\mathbf{V} = \mathbf{0}$ implies existence of a *potential function* $u(x, y)$ such that $\mathbf{V} = -(\partial u/\partial x)\mathbf{i} - (\partial u/\partial y)\mathbf{j}$. Thus, $\mathbf{V}$ is minus the gradient of u. The minus sign is just a convention.

It is also convenient to introduce a *stream function* $v(x, y)$ as follows. Choose a point P_0 as a reference point. Consider curves C_1 and C_2 from P_0 to $P(x, y)$ enclosing a region D, as shown in Figure 364. At any time t, assuming conservation of mass in D, the mass of fluid per unit time entering D across C_1 must equal the amount leaving D across C_2. Thus, the mass per unit time of fluid flowing counterclockwise across C_1 is the same as that flowing across C_2; it is therefore a function of (x, y) (and possibly t) only. We denote this function $v(x, y)$ and call it the *stream function*.

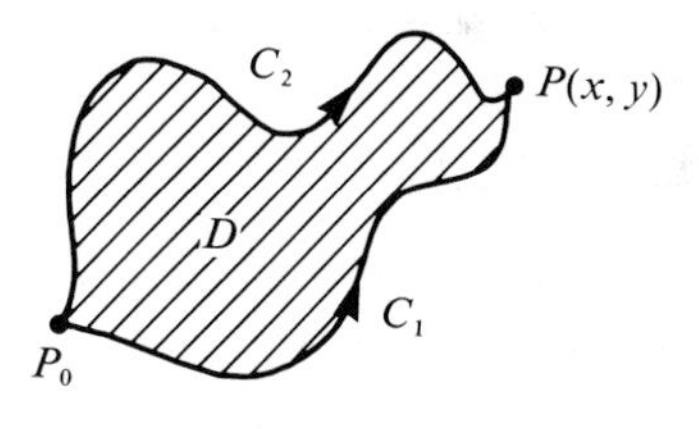

Figure 364

We shall now show that u and v have a surprising relationship—they satisfy the Cauchy-Riemann equations. To prove this, it is convenient to first suppose that units have been chosen so that $\rho = 1$ (recall that density is constant anyway in incompressible flows). Imagine points $P(x, y)$ and $P'(x + \Delta x, y + \Delta y)$ in the fluid. If C is a path in the fluid from P to P', then the mass per unit time crossing C counterclockwise is

$$\Delta v = v(x + \Delta x, y + \Delta y) - v(x, y),$$

which is approximately

$$\frac{\partial v}{\partial x}\Delta x + \frac{\partial v}{\partial y}\Delta y.$$

But, since $\mathbf{V} = -(\partial u/\partial x)\mathbf{i} - (\partial u/\partial y)\mathbf{j}$, we also have Δv approximated by

$$-\frac{\partial u}{\partial y}\,\Delta x + \frac{\partial u}{\partial x}\,\Delta y.$$

To see this, observe that $\rho(x, y)\,\Delta x$ is the mass per unit time moving across the line from P to $(x + \Delta x, y)$; that $\rho(x, y)\,\Delta y$ is the mass per unit time moving across the line from $(x + \Delta x, y)$ to $(x + \Delta x, y + \Delta y)$; and that we have chosen ρ as one.

We now have Δv approximated by $(\partial v/\partial x)\,\Delta x + (\partial v/\partial y)\,\Delta y$ and also by $-(\partial u/\partial y)\,\Delta x + (\partial u/\partial x)\,\Delta y$. Then, as we wanted to show,

$$\frac{\partial v}{\partial x} = -\frac{\partial u}{\partial y} \quad\text{and}\quad \frac{\partial v}{\partial y} = \frac{\partial u}{\partial x}.$$

Assuming continuity of these partials, we can now associate with the flow an analytic function

$$f(z) = u(x, y) + iv(x, y).$$

Streamlines of the flow are curves $v =$ constant, while *potential lines* are curves $u =$ constant. Sometimes potential lines are called *equipotential lines*, since along such lines the potential is constant.

Note that, along a streamline, $v =$ constant; hence, $dv = 0$. Then, $(\partial v/\partial x)\,dx + (\partial v/\partial y)\,dy = 0$, giving us the differential equation

$$\frac{dy}{dx} = -\frac{\dfrac{\partial v}{\partial x}}{\dfrac{\partial v}{\partial y}}.$$

Solutions of this differential equation give the streamlines of the flow.

One important feature of streamlines is that they are tangent to the flow, giving the direction of the flow at each point. Further, the flow cannot cross a streamline. Thus, for example, a streamline which is a circle may be interpreted as a cylindrical barrier in the flow, enabling us to study flow around a cylinder.

Note that the flow velocity $\mathbf{V}$ has magnitude

$$\sqrt{\left(\frac{\partial u}{\partial x}\right)^2 + \left(\frac{\partial u}{\partial y}\right)^2}$$

or, equivalently,

$$\sqrt{\left(\frac{\partial v}{\partial x}\right)^2 + \left(\frac{\partial v}{\partial y}\right)^2}.$$

Here are examples of flows represented by analytic functions.

EXAMPLE 502 Uniform Stream at an Angle θ

Consider the flow given by

$$f(z) = -Ke^{i\theta}z,$$

where K and θ are positive constants.

We have

$$f(z) = -K[x\cos(\theta) - y\sin(\theta)] - iK[y\cos(\theta) + x\sin(\theta)].$$

Thus,

$$u(x, y) = -Kx\cos(\theta) + Ky\sin(\theta)$$

and

$$v(x, y) = -Ky\cos(\theta) - Kx\sin(\theta).$$

The velocity of the flow is

$$\mathbf{V} = -\frac{\partial u}{\partial x}\mathbf{i} - \frac{\partial u}{\partial y}\mathbf{j} = K\cos(\theta)\mathbf{i} - K\sin(\theta)\mathbf{j}.$$

(In complex notation, $\mathbf{V} = Ke^{-i\theta}$.) The magnitude of $\mathbf{V}$ is K, a constant.

The streamlines are given by $v(x, y) = C$, where C is any real constant. This is equivalent to writing

$$-y\cos(\theta) - x\sin(\theta) = C$$

or

$$y = -x\tan(\theta) + a,$$

where a is any real constant. The streamlines are thus straight lines making an angle θ with the positive real axis. Since the streamlines determine the direction of the flow, we have a flow with uniform speed K making an angle θ with the positive real axis. This is indicated in Figure 365.

The equipotential lines are of the form

$$x\cos(\theta) - y\sin(\theta) = \text{constant}$$

or

$$y = x\cot(\theta) + b,$$

where b is any real constant. These lines are orthogonal to the streamlines of Figure 365.

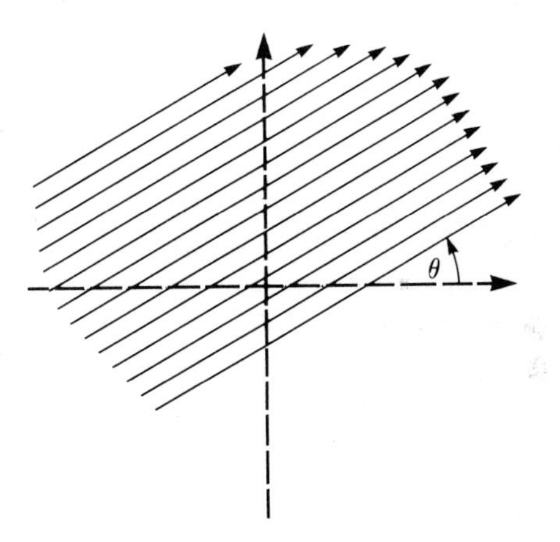

Figure 365. Streamlines of the flow $f(z) = -Ke^{i\theta}z$.

EXAMPLE 503 Flow Outside a Cylinder; Vortex at the Origin

We shall analyze the flow represented by

$$f(z) = \frac{iK}{2\pi}\ln(z),$$

where $K > 0$. We have

$$f(z) = -\frac{K}{2\pi}\arg(z) + i\frac{K}{4\pi}\ln(x^2 + y^2).$$

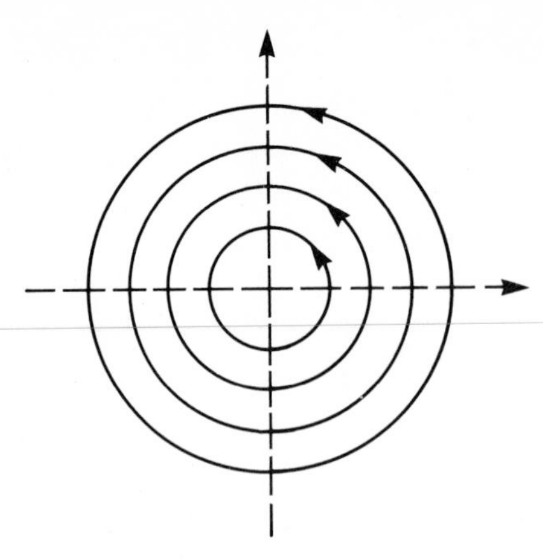

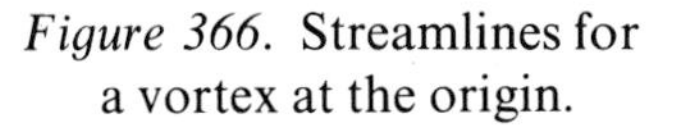

Figure 366. Streamlines for a vortex at the origin.

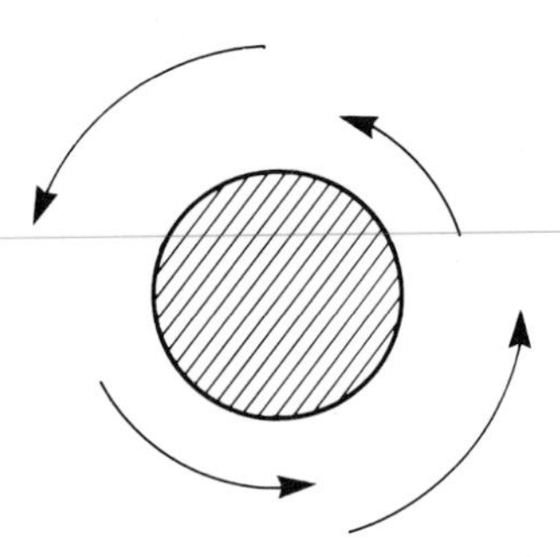

Figure 367. Flow about a cylinder.

Thus,

$$u(x, y) = -\frac{K}{2\pi}\arg(z) \quad \text{and} \quad v(x, y) = \frac{K}{4\pi}\ln(x^2 + y^2).$$

Streamlines are given by

$$\frac{K}{4\pi}\ln(x^2 + y^2) = \text{constant} \quad \text{or} \quad x^2 + y^2 = C.$$

Thus, the streamlines are circles about the origin. The equipotential lines are lines $\arg z = \text{constant}$, which are half-lines or rays emanating from the origin.

Thus, $f(z) = (iK/2\pi)\ln(z)$ represents flow in concentric circles about the origin. The number K is sometimes called the *circulation* of the flow. One also sees this flow referred to as a *vortex at the origin*, where K is called the *strength of the vortex*. The flow is illustrated in Figure 366.

Since flow cannot cross a streamline, we may also imagine a circular barrier $|z| = R$ inserted into the flow along the streamline $x^2 + y^2 = R^2$. We then have flow outside the cylinder with circulation K about the cylinder. This is illustrated in Figure 367.

EXAMPLE 504 Flow Outside an Elliptical Barrier

From the last example, we know that $f(z) = (iK/2\pi)\ln(z)$ for $|z| > R$ describes a flow with circulation K outside a cylinder of radius R. Using conformal mappings, we can replace the cylinder with an ellipse as follows.

The mapping

$$w = z + \frac{a^2}{z},$$

with a any positive (hence real) constant, is called a *Joukowski transformation*. Writing $z = x + iy$ and $w = X + iY$, it is easy to check that concentric circles

$$x^2 + y^2 = R^2$$

in the z-plane map to confocal ellipses

$$\frac{X^2}{\left(1+\frac{a^2}{R^2}\right)^2}+\frac{Y^2}{\left(1-\frac{a^2}{R^2}\right)^2}=R^2,$$

provided of course that $R \neq a$. If $R = a$, then $x^2 + y^2 = R^2 = a^2$ maps to the interval $[-2a, 2a]$ on the real axis.

Solve for z in the Joukowski transformation to get

$$z=\frac{w \pm \sqrt{w^2-4a^2}}{2}.$$

Choosing one root (say, take the $+$ sign with the radical), we transform $f(z) = (iK/2\pi)\ln(z)$ into

$$F(w)=\frac{iK}{2\pi}\ln\left(\frac{w+\sqrt{w^2-4a^2}}{2}\right).$$

This gives flow about an ellipse

$$\frac{X^2}{\left(1+\frac{a^2}{R^2}\right)^2}+\frac{Y^2}{\left(1-\frac{a^2}{R^2}\right)^2}=R^2$$

if $R > a$; and it gives flow about a flat plate $-2a \leq X \leq 2a$, $Y = 0$, if $R = a$.

As an example of an application of the residue theorem to fluid flow, we shall state without proof a theorem of Blasius. Suppose that $w = f(z)$ represents the flow about a barrier whose boundary is the closed curve C. Let the thrust of the fluid outside the cylinder be the vector $A\mathbf{i} + B\mathbf{j}$. Then A and B are given by

$$A - iB = \tfrac{1}{2}i\rho \int_C [f'(z)]^2\,dz.$$

Further, the moment of the thrust about the origin is

$$\operatorname{Re}\left\{-\tfrac{1}{2}\rho \int_C z[f'(z)]^2\,dz\right\}.$$

In terms of residues, these formulas become

$$A - iB = -\pi\rho \cdot \{\text{sum of residues of } [f'(z)]^2 \text{ at singularities inside } C\}$$

and

$$\text{thrust about the origin} = \operatorname{Re}\left\{-\pi i\rho \cdot \begin{matrix}[\text{sum of residues of } zf'(z)^2 \\ \text{at singularities inside } C]\end{matrix}\right\}.$$

PROBLEMS FOR SECTION 19.3

1. Draw the streamlines and equipotential lines for a flow with complex potential

$$f(z) = K\ln(z - z_0).$$

If $K > 0$, this flow is called a *source*; if $K < 0$, it is a *sink*. Why are these names appropriate?

2. Draw the streamlines and equipotential lines for a flow with complex potential

$$f(z) = K \ln\left(\frac{z + z_0}{z - z_0}\right).$$

(Such a flow is called a *dipole*.)

3. Draw streamlines and equipotential lines for a flow with complex potential

$$f(z) = K\left(z + \frac{a^2}{z}\right),$$

where a and K are positive constants. Show that this corresponds to flow around a cylindrical obstacle $|z| = a$.

4. Discuss the motion of a fluid having complex potential

$$f(z) = K\left(z + \frac{a^2}{z}\right) + \frac{ib}{2\pi}\log(z),$$

where a, b, and K are positive constants.

5. Use Blasius's theorem on the flow of Problem 3 to calculate the thrust on the cylinder. Show that the thrust takes the form of a lift. Also find the moment of the thrust about the origin.
6. Draw streamlines and equipotential lines for the flow with complex potential $f(z) = z^2$. Describe the flow physically.
7. A *stagnation point* of a flow is a point at which the velocity is zero.

   Calculate the stagnation points for the flows of Problems 1, 2, and 3, and also Examples 502 and 503.
8. Analyze the flow given by $f(z) = iz^2$. Show why this can be considered as flow around a corner.
9. Prove the theorem of Kutta and Joukowski: When a barrier bounded by a simple closed curve is placed in a uniform stream of speed K, then the resultant thrust on the barrier is at right angles to the stream, and its magnitude is $\alpha K\rho$ per unit length, where ρ is the density and α is the circulation of the flow.
10. Analyze the flow given by the complex potential

$$f(z) = iKa\sqrt{3}\ln\left(\frac{2z - ia\sqrt{3}}{2z + ia\sqrt{3}}\right).$$

Show that this potential represents irrotational flow around a cylinder $4x^2 + 4(y - a)^2 = a^2$, with an infinite plane boundary along the y-axis.

11. Use Blasius's theorem to show in Problem 10 that the force per unit width on the cylinder has y-component

$$2\sqrt{3}\,\pi\rho a K^2.$$

12. Prove Milne-Thomson's circle theorem: Suppose that a flow has complex potential $f(z)$ and that there are no singularities in the disk $|z| \le a$. Then, if a solid cylinder $|z| = a$ is placed in the flow, the resulting flow has complex potential $f(z) + \overline{f(a^2/\bar{z})}$.

19.4 COMPLEX FUNCTIONS AND ELECTROSTATIC POTENTIAL

Complex functions are often convenient in the mathematical treatment of electrostatic fields. Suppose we are given a charge distribution. If we place a unit positive charge at any point not already occupied by a charge, the resulting force due to this charge is the electric field intensity **E**. One can show that **E** is derivable from a potential function. That is, for some scalar function $u(x, y, z)$, **E** is minus the gradient of $u(x, y, z)$. (The minus is just a convention.) Then, $u(x, y, z)$ is the *electrostatic potential* for **E**.

If **E** is two-dimensional, then u is a function of just x and y. One can show that $u(x, y)$ is a harmonic function, and hence that there is some harmonic $v(x, y)$ such that $f(z) = u(x, y) + iv(x, y)$ is analytic. Then, $f(z)$ is called a *complex potential* for **E**. The curves $u(x, y) =$ constant and $v(x, y) =$ constant are called, respectively, *equipotential lines* and *flux lines* of the electrostatic field.

It is easy to see from the form of the complex potential that there is a direct analogy between complex electrostatics and the complex treatment of fluid flow. One simply replaces the electric field in electrostatics with the velocity vector in fluid flow.

Use of complex analysis in electrostatics is a convenience, not a necessity. As an illustration of finding a complex potential, imagine that we have two infinitely long, concentric cylinders of radii R_1 and R_2 ($R_1 < R_2$), bounding a region D. The inner cylinder is charged to a potential A, and the outer cylinder is charged to a potential B.

Note that

$$f(z) = a\ln(z) + b$$

is analytic in D. Letting $z = re^{i\theta}$ ($R_1 < r < R_2, 0 \leq \theta \leq 2\pi$), we have

$$f(z) = a\ln(r) + ia\theta + b.$$

Now, $u(r, \theta) = a\ln(r) + b$ is harmonic in D. To choose a and b so that $u(R_1, \theta) = A$ and $u(R_2, \theta) = B$, we solve for a and b in

$$a\ln(R_1) + b = A,$$
$$a\ln(R_2) + b = B.$$

We obtain

$$a = \frac{B - A}{\ln(R_2) - \ln(R_1)} \quad \text{and} \quad b = \frac{A\ln(R_2) - B\ln(R_1)}{\ln(R_2) - \ln(R_1)}.$$

Thus, the complex potential is

$$f(re^{i\theta}) = \left(\frac{B - A}{\ln(R_2) - \ln(R_1)}\right)[\ln(r) + i\theta] + \frac{A\ln(R_2) - B\ln(R_1)}{\ln(R_2) - \ln(R_1)}.$$

The real part of this is the potential function for the electrostatic field.

PROBLEMS FOR SECTION 19.4

1. Find flux lines and equipotential lines for an electrostatic field with complex potential $f(z) = 1/z$.
2. The complex potential of a pair of oppositely charged lines of the same strength q at $z = z_0$ and $z = \overline{z_0}$ is given by

$$f(z) = 2q\ln\left(\frac{z - \overline{z_0}}{z - z_0}\right).$$

Draw the equipotential and flux lines for this potential.

3. Two infinite, parallel plates, R units apart, are grounded, and a line charge q per unit length is located midway between the plates. Use the mapping $w = e^{\pi z/R}$ and the solution of Problem 2 to derive the potential

$$\phi(x, y) = 2q \operatorname{Re}\left(\frac{e^{\pi z/R} + i}{e^{\pi z/R} - i}\right).$$

19.5 INVERSES OF LAPLACE TRANSFORMS

The Laplace transform of a function $f(t)$ is given by

$$L[f(t)] = F(s) = \int_0^\infty e^{-st} f(t)\, dt,$$

for all s such that this integral converges. In Chapter 4 we discussed this transform and also the problem of finding the inverse transform: given $F(s)$, find $f(t)$ such that $L[f(t)] = F(s)$.

Of course, all these considerations were for real-valued functions. However, if we use complex functions, then a fairly general formula can be written for the inverse.

Suppose that $F(z)$ is analytic in the entire z-plane, except at a finite number of isolated singularities at $z_1, \ldots, z_n$. Assume that, for some real $a > 1$ and real b, $|z^a F(z)|$ remains bounded in the half-plane $\operatorname{Re} z > b$. Let C be the curve of Figure 368, consisting of part of a circle centered at the origin and part of the line $\operatorname{Re} z = c > b$. Then, the inverse Laplace transform of $F(s)$ is $f(t)$, where

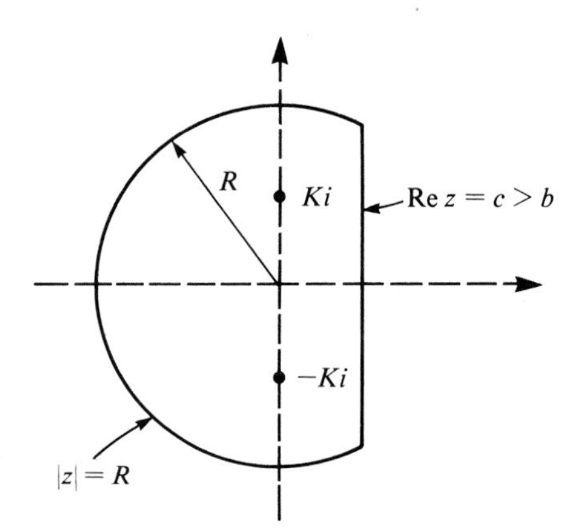

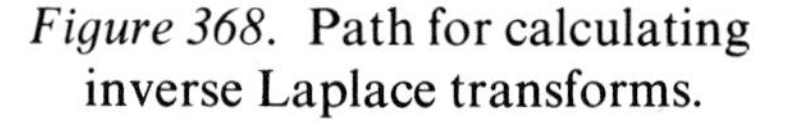
Figure 368. Path for calculating inverse Laplace transforms.

$$f(t) = \frac{1}{2\pi i} \oint_C e^{zt} F(z)\, dz.$$

Usually, this integral is evaluated by the residue theorem. This yields

$$f(t) = \text{sum of residues of } e^{zt}F(z) \text{ at singularities of } F(z).$$

Here is a simple example.

EXAMPLE 505

Find a function whose Laplace transform is $1/(s^2 + K^2)$, where K is any positive number.

Look at

$$\frac{1}{2\pi i} \oint_C \frac{1}{z^2 + K^2}\, dz.$$

Note that $|z^{3/2}/(z^2 + K^2)|$ remains bounded in, say, Re $z > 1$. Thus, we can let $F(z) = 1/(z^2 + K^2)$, $a = \frac{3}{2}$, and $b = 1$ in the above discussion. (Various other choices of a and b are also possible.)

Now, $F(z)$ is analytic except for simple poles at $z = \pm Ki$. Imagine that R in Figure 368 is chosen large enough that C encloses Ki and $-Ki$. Then, by the residue theorem, the inverse tranform is

$$\begin{aligned} f(t) &= \frac{1}{2\pi i}\int_C \frac{e^{tz}\,dz}{z^2 + K^2} \\ &= \operatorname*{Res}_{Ki} \frac{e^{tz}}{z^2 + K^2} + \operatorname*{Res}_{-Ki} \frac{e^{tz}}{z^2 + K^2} \\ &= \frac{e^{iKt}}{2iK} - \frac{e^{-iKt}}{2iK} \\ &= \frac{1}{K}\left(\frac{e^{iKt} - e^{-iKt}}{2i}\right) = \frac{\sin(Kt)}{K}. \end{aligned}$$

PROBLEMS FOR SECTION 19.5

In each of Problems 1 through 10, use the method of this section to find the inverse Laplace transform of the given function.

1. $\dfrac{s}{s^4 + K^4}$
2. $\dfrac{1}{(s-2)(s+6)}$
3. $\dfrac{2s+1}{s^3+5}$
4. $\dfrac{2s^2+1}{s^4+4}$
5. $\dfrac{s}{s^2-6}$
6. $\dfrac{3s-4}{s^4+2s^2-1}$
7. $\dfrac{3}{s^2(s^2-1)}$
8. $\dfrac{\sin(s)}{s^2-3}$
9. $\dfrac{1}{s^3+2}$
10. $\dfrac{s^2-3s+4}{s^6-64}$

19.6 COMPLEX FOURIER SERIES

Let $f(x)$ be a continuous real-valued function of the real variable x. Then, for $-L \le x \le L$, we can associate with $f(x)$ its Fourier series,

$$\frac{a_0}{2} + \sum_{n=1}^{\infty} a_n \cos\left(\frac{n\pi x}{L}\right) + b_n \sin\left(\frac{n\pi x}{L}\right),$$

as developed in Chapter 12, Section 1. The Fourier coefficients are

$$a_n = \frac{1}{L}\int_{-L}^{L} f(x)\cos\left(\frac{n\pi x}{L}\right) dx \qquad (n = 0, 1, 2, 3, \ldots)$$

and

$$b_n = \frac{1}{L}\int_{-L}^{L} f(x)\sin\left(\frac{n\pi x}{L}\right) dx \qquad (n = 1, 2, 3, \ldots).$$

There is a complex analog of this result. If $F(z)$ is analytic in the entire complex plane and has period 2π [so $F(z + 2\pi) = F(z)$ for all z], then

$$F(z) = \sum_{n=-\infty}^{\infty} c_n e^{inz},$$

where

$$c_n = \frac{1}{2\pi}\int_{-\pi}^{\pi} F(z)e^{-inz}\, dz \qquad (n = 0, \pm 1, \pm 2, \ldots).$$

(Note that, since $F(z)e^{-inz}$ is also analytic in the entire complex plane, its integral over any path depends only on the endpoints of the path. Thus, $\int_{-\pi}^{\pi} F(z)e^{-inz}\, dz$ makes sense.) We call $\sum_{n=-\infty}^{\infty} c_n e^{inz}$ the *complex Fourier series* for $F(z)$, and the c_n's are the *complex Fourier coefficients*.

The complex Fourier series can be put into a form resembling the real Fourier series when $L = 2\pi$ in the real series.

Write

$$\begin{aligned} F(z) &= \sum_{n=-\infty}^{\infty} c_n e^{inz} \\ &= \sum_{n=-\infty}^{\infty} c_n[\cos(nz) + i\sin(nz)] \\ &= c_0 + \sum_{n=1}^{\infty} [(c_n + c_{-n})\cos(nz) + i(c_n - c_{-n})\sin(nz)], \end{aligned}$$

where

$$c_0 = \frac{1}{2\pi}\int_{-\pi}^{\pi} F(z)\, dz,$$

$$c_n + c_{-n} = \frac{1}{\pi}\int_{-\pi}^{\pi} F(z)\left(\frac{e^{-inz} + e^{inz}}{2}\right) dz = \frac{1}{\pi}\int_{-\pi}^{\pi} F(z)\cos(nz)\, dz,$$

and

$$i(c_n - c_{-n}) = \frac{1}{\pi}\int_{-\pi}^{\pi} F(z)\left(\frac{e^{inz} - e^{-inz}}{2i}\right) dz = \frac{1}{\pi}\int_{-\pi}^{\pi} F(z)\sin(nz)\, dz.$$

These integrals resemble the integral formulas for the real Fourier coefficients.

EXAMPLE 506

$F(z) = e^z$ is analytic in the entire complex plane and has period 2π.

We shall compute its Fourier series. The Fourier coefficients are

$$\begin{aligned}
c_n &= \frac{1}{2\pi}\int_{-\pi}^{\pi} e^z e^{-inz}\,dz \\
&= \frac{1}{2\pi}\int_{-\pi}^{\pi} e^{(1-in)z}\,dz \\
&= \left[\frac{1}{2\pi(1-in)}e^{(1-in)z}\right]_{-\pi}^{\pi} \\
&= \frac{1}{\pi(1-in)}\left(\frac{e^{(1-in)\pi}-e^{-(1-in)\pi}}{2}\right) \\
&= \frac{1}{\pi(1-in)}\left(\frac{e^{\pi}-e^{-\pi}}{2}\right) \\
&= \frac{(-1)^n}{\pi(1-in)}\sinh(\pi).
\end{aligned}$$

Thus,

$$e^z = \sum_{n=-\infty}^{\infty}\frac{(-1)^n}{\pi(1-in)}\sinh(\pi)e^{inz}.$$

PROBLEMS FOR SECTION 19.6

1. Write the Fourier series for $\cos(2z)$.
2. Write the Fourier series for $\sin(4z)$.
3. Derive the formula for the complex Fourier coefficient c_K by multiplying both sides of $F(z) = \sum_{n=-\infty}^{\infty} c_n e^{inz}$ by e^{iKz} and integrating from $-\pi$ to π.
4. Derive a polar coordinate form of complex Fourier series as follows. Let $f(z)$ be analytic in $|z| < R$. Expand in a power series $f(z) = \sum_{n=0}^{\infty} a_n z^n$ for $|z| < R$. Putting $z = re^{i\theta}$, obtain $f(z) = \sum_{n=0}^{\infty} a_n r^n e^{in\theta}$, and show that
$$a_n = \frac{1}{2\pi r^n}\int_{-\pi}^{\pi} f(re^{i\phi})e^{-in\phi}\,d\phi \quad \text{for} \quad 0 < r < R,\ \ 0 \le \theta \le 2\pi.$$
5. Use the expansion $f(z) = \sum_{n=0}^{\infty} a_n r^n e^{in\theta}$ from Problem 4 to prove *Parseval's theorem*:
$$\frac{1}{2\pi}\int_{-\pi}^{\pi} |f(re^{i\theta})|^2\,d\theta = \sum_{n=0}^{\infty}|a_n|^2 r^{2n},$$
with the a_n's as in Problem 4.

In each of Problems 6 through 10, use the result of Problem 4 to obtain a series for $f(z)$ of the form $\sum_{n=0}^{\infty} a_n r^n e^{in\theta}$, valid in some disk $|z| < R$.

6. $f(z) = z^2 - z$
7. $f(z) = 3iz^{1/2}$
8. $f(z) = \cos^2(2z)$
9. $f(z) = (z + i)^2$
10. $f(z) = \sin(z) - \cos(2z)$
11. Let $z = x$, real, and substitute into the series derived in Example 506. Show how the resulting series gives the real Fourier series for e^x on $-\pi \le x \le \pi$.

SOME NOTES ON THE HISTORY OF COMPLEX ANALYSIS

The acceptance of complex numbers as a legitimate tool for algebra and analysis came only after a good deal of controversy and consternation among mathematicians. For example, in 1770, the great Swiss mathematician Leonard Euler (1707–1783) wrote, "Because all conceivable numbers are either greater than zero, or less than zero, or equal to zero, then it is clear that the square roots of negative numbers cannot be included among the possible numbers." He went on to refer to what we know as complex numbers as impossible, or fancied, numbers.

However, such numbers insisted on occurring in the solution of certain kinds of problems; for example, in finding the roots of polynomials. In 1799, Karl Friedrich Gauss (1777–1855) gave the first of his many proofs of the fundamental theorem of algebra, and this depended upon the existence of complex roots of polynomials.

The major breakthrough in gaining acceptance of complex numbers came in the nineteenth century with the gradual understanding of their geometric interpretation. This evolved slowly over many years and from many sources. Caspar Wessel (1745–1818) was a Norwegian surveyor who wrote a paper in 1797 called "On the Analytic Representation of Direction; an Attempt." In it, he grasped the basic ideas of representing complex numbers as points in the plane and of adding them by the parallelogram law. This work went largely unnoticed until 1897, but it indicates how the germ of the idea was in the air, and that its time was about to come.

In 1806, a Swiss bookkeeper, Jean-Robert Argand (1768–1822), wrote a small book on the geometrical representation of complex numbers. Despite the fact that we still sometimes refer to the complex plane as the Argand diagram, the major credit belongs to Gauss, who carried the prestige due the leading mathematician of his time. It is clear that by 1815 Gauss thoroughly understood the geometry of complex numbers and, by the 1830s, wrote freely about it.

Questions concerning complex functions and integrals began to be raised by Gauss and by Simeon-Denis Poisson (1781–1840). Poisson originally worked on the recently developed Fourier series, but he was also the first to integrate along paths in the complex plane. It was left to Cauchy, however, to formulate and analyze many of the properties of complex functions and integrals which today bear his name.

Augustin-Louis Cauchy (1789–1857) was born in Paris and grew to become a professor at the Ecole Polytechnique, the Sorbonne, and the College de France. He made important contributions in the mechanics of waves in elastic media and in the theory of light, but his most important work was in mathematics, where he authored over seven hundred papers (second only to Euler). His personal life was greatly influenced by his royalist sympathies, and he supported the Bourbons during a time of political upheaval in France. Napoleon III excused him from the oath of allegiance required at the time, probably in recognition of Cauchy's stature as one of the leading mathematicians of his day. Cauchy responded by donating his salary from the Sorbonne to the poor of Sceaux, the town in which he resided at the time.

In a series of papers from about 1814 to the early 1840s, Cauchy formulated his integral theorem and some of its consequences, including the notion of independence of path, and grasped the ideas of poles and residues and the residue theorem. He also

$$
\begin{aligned}
x_0 &= 1 \\
x_1 &= 1.499750125 \\
x_2 &= 1.416680519 \\
x_3 &= 1.414216580 \\
x_4 &= 1.414213563 \\
x_5 &= 1.414213562 \\
x_6 &= 1.414213562
\end{aligned}
\Bigg\} \text{ repetition}
$$

We have the same root as found before with $x_0 = 0$, but we found it in roughly one-third the time. This emphasizes the importance of a good initial guess, if possible.

In actually programming the method, one does not usually print out the intermediate x_n's, as we have done here, because they are of no interest. One programs the method to run until some $x_N = x_{N+1}$, and then prints out this approximate solution.

EXAMPLE 508

Solve

$$e^{x^2} - x^3 + 3x - 4 = 0.$$

With $x_0 = 0$ and $h = 0.001$, we get

$$0.851049094.$$

As a check, substitute this into the equation to get 0.000000001; thus, we have an approximate solution to a high degree of accuracy.

EXAMPLE 509

Find a solution of

$$\sin(x^3 + 2) = \frac{1}{x}.$$

With $x_0 = 1$ and $h = 0.001$, we get the approximate solution

$$2.226151190.$$

Substituting this into the function $\sin(x^3 + 2) - 1/x$, we obtain

$$-0.000000005,$$

giving some idea of the accuracy of the solution.

If you look at the graphs of $\sin(x^3 + 2)$ and $1/x$ in Figure 369, you expect there to be more points of intersection, and hence more solutions of the equation $\sin(x^3 + 2) = 1/x$. For exam-

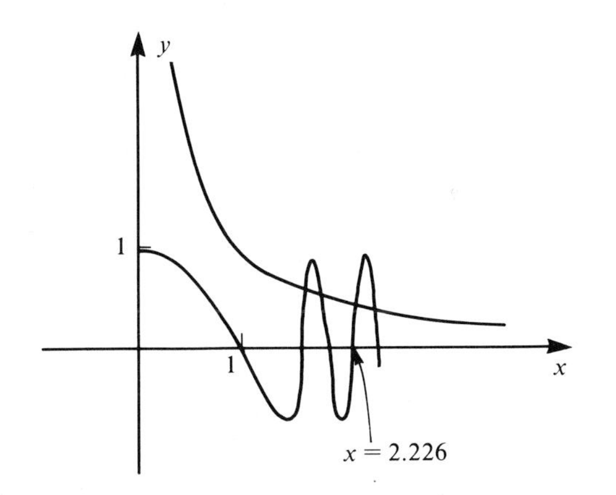

Figure 369. Estimating roots of $\sin(x^3 + 2) - 1/x = 0$ by the Newton-Raphson method.

ple, there appears to be a solution near $x = 1.7$. If you use $x_0 = 1.7$ in the Newton-Raphson method, then you find the approximate root

$$1.699896699.$$

Substituting this back into the equation yields -0.000000001; so we have a good approximation to another solution.

We also expect from the graphs that there will be solutions for larger values of x. For example, put $x_0 = 12$ into Newton-Raphson and you get 12.00216196. Upon substituting this into the equation, we get 0.000002097; so we have an approximation of another solution to within about five decimal places.

The use of graphs, illustrated in the above example, is very convenient for estimating how many solutions one might expect (if, in fact, there are any) and also for making an initial guess of a value of x_0.

PROBLEMS FOR SECTION 20.1

In each of Problems 1 through 20, use Newton-Raphson to find a solution of the given equation. If you do not find a solution, try showing by graphs that there is no solution.

1. $\cos(x^2) - (1/x) = 0$

2. $\ln(x^2 + x + 4) - \frac{1}{8}x = 0$

3. $x^5 - 0.25x^4 + 0.6x^3 - 3x + 4\sqrt{2} = 0$

4. $\cos(x^2) - \sin(x) = 0$

5. $\cosh(3x) - x^2 = 2$

6. $x^3 - 0.73x^2 + 2.42x - 5.63 = 0$

7. $\sin(3x - 4) + (4/3x) = 0$

8. $e^{-x^2} + x = 0$

9. $\tanh(2x) - x - 1 = 0$

10. $4x^7 - 3.2x^5 + 3x^3 - 2x^2 + \sqrt{3}x - 14.62 = 0$

11. $\tan(x + 1) - x^2 = 0$

12. $e^{-x} + x^3 - 2 = 0$

13. $\cosh(3x) - \sinh(2x) = 0$

14. $e^x = x^3$

15. $\tan(x) + \sin(3x + 2) = 0$

16. $\cos^2(x - 1) = x$

17. $\sin(x) + \cos(x) - \ln(x) = 0$

18. $(1/x^2) - e^{2x} = 0$

19. $\sin(x^3 + 3) = -x$

20. $\ln(x^2 + 2) - 6 + x = 0$

In each of Problems 21 through 30, use Newton-Raphson to find at least one solution of the given equation. Sketch graphs to help estimate x_0 and also to guess how many solutions (if any) one should look for.

21. $\sin(4x) = \cos(2x)$

22. $e^x = \tan(3x)$

23. $x^3 - 3x + 2 = \sin(2x)$

24. $\ln(x) = (1/x) + 4$

25. $\sqrt{x^2 + 3} = \cosh(3x)$

26. $\ln(x^2 + 4) = 8x + 7$

27. $e^{x+2} = \sin(x)$

28. $2/x = \tan(3x + 1)$

29. $\sin^2(x) = \cos(x)$

30. $x^3 - 3x + 5 = \cos(x) - \sin(x)$

20.2 NUMERICAL INTEGRATION

As the student may have experienced, most definite integrals cannot be done in closed form. In this section we give three schemes for approximating definite integrals.

METHOD 1: THE RECTANGULAR RULE

This rule is based on the definition of $\int_a^b f(x)\,dx$ as a limit of Riemann sums of the form

$$\sum_{j=1}^{n} f(c_j)(x_j - x_{j-1}),$$

where $x_{j-1} \le c_j \le x_j$ and $a = x_0 < x_1 < x_2 \cdots < x_{n-1} < x_n = b$.

One can obtain a good approximation to $\int_a^b f(x)\,dx$ by simply evaluating such a sum for large enough n. In practice, we usually subdivide $[a, b]$ into n equal subintervals, each of length $h = (b - a)/n$, and we can as a convenience choose c_j as x_j, though other choices are possible (for example, c_j as the center point of the jth subinterval). Choosing $c_j = x_{j-1}$ gives us

$$\int_a^b f(x)\,dx \approx h[f(x_0) + f(x_1) + \cdots + f(x_{n-1})].$$

As n increases, the quantity on the right approaches the exact value of the integral.

One can obtain bounds on the error in this method as follows. If m is the minimum value of $f''(x)$ on $[a, b]$ and M is the maximum value, then the error in approximating $\int_a^b f(x)\,dx$ by the rectangular rule is bounded by

$$\frac{m(b-a)^3}{12n^2} \le \varepsilon \le \frac{M(b-a)^3}{12n^2}.$$

Thus, as n increases, the error decreases rapidly (at least for twice-differentiable functions).

EXAMPLE 510

Consider $\int_1^4 x^2\,dx$, which we can do easily anyway to get 21. Here is what we get using the rectangular rule for various values of n:

n	Value of $\int_1^4 x^2\,dx$ by rectangular rule
2	20.43750000
4	20.85937500
10	20.97750000
20	20.99437500
30	20.99750000
50	20.99910000
100	20.99977500

EXAMPLE 511

Approximate

$$\int_0^{\pi/2} e^{-x^2} \sin(x^2 + 1)\,dx.$$

With $n = 100$, we get 0.748469604; with $n = 200$, we get 0.748468316. The two results agree to five decimal places.

EXAMPLE 512

Approximate

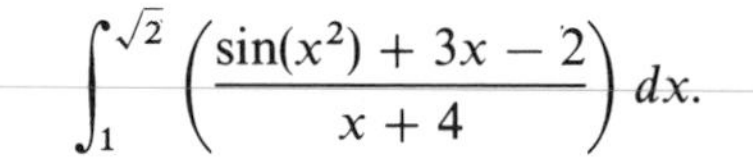

$$\int_1^{\sqrt{2}} \left(\frac{\sin(x^2) + 3x - 2}{x + 4}\right) dx.$$

With $n = 25$, we get 0.20427212; with $n = 50$, we get 0.20426771; and with $n = 100$, we get 0.20426662. For greater accuracy, we could choose a larger value of n.

METHOD 2: THE TRAPEZOIDAL RULE

The trapezoidal rule is motivated by Figure 370, in which $\int_a^b f(x)\,dx$ is approximated by the sum of the "areas" of the trapezoids shown. The area of a trapezoid as shown in Figure 371 is

$$\tfrac{1}{2}(a + b)h.$$

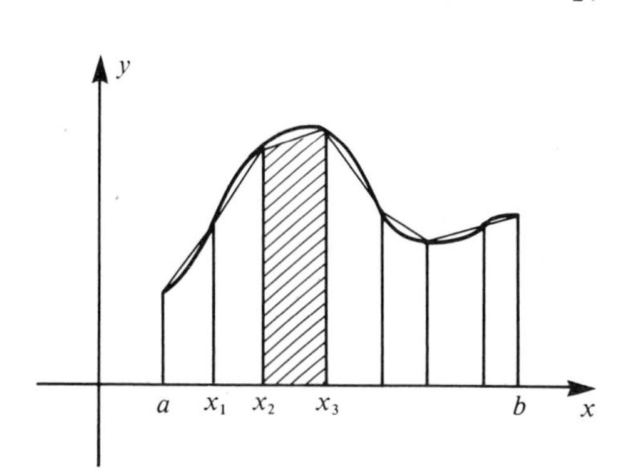

Figure 370. Trapezoidal approximation to $\int_a^b f(x)\,dx$.

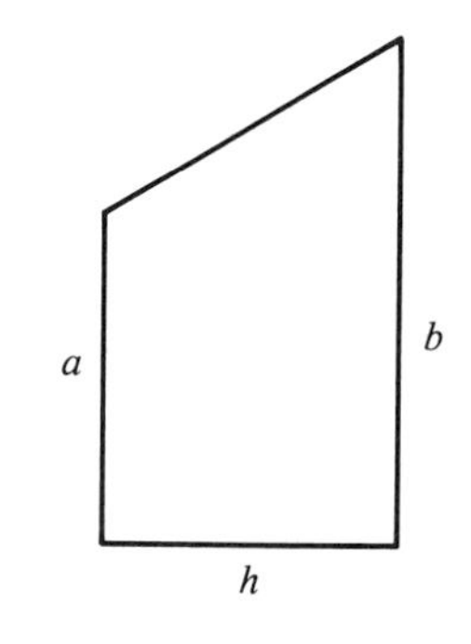

Figure 371. Area of a trapezoid is $\frac{1}{2}(a + b)h$.

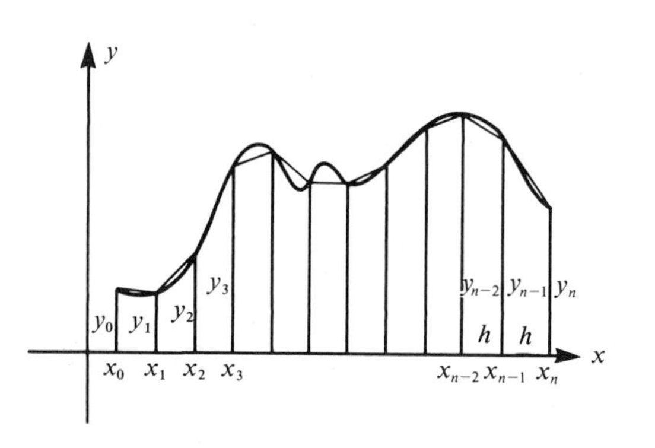

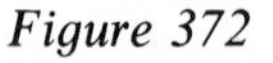

Figure 372

Thus, referring to Figure 372, we have the approximation

$$\int_a^b f(x)\,dx \approx \tfrac{1}{2}(y_0 + y_1)h + \tfrac{1}{2}(y_1 + y_2)h + \cdots + \tfrac{1}{2}(y_{n-1} + y_n)h,$$

in which $y_j = f(x_j)$ for the chosen partition $a = x_0 < x_1 < x_2 < \cdots < x_{n-1} < x_n = b$, and h is the spacing between successive partition points. If we set $h = (b - a)/n$, then the trapezoidal rule can be written

$$\int_a^b f(x)\,dx \approx \frac{(b-a)}{2n}(y_0 + 2y_1 + 2y_2 + \cdots + 2y_{n-1} + y_n).$$

It can be shown that the error in this method is

$$\frac{(b-a)}{12} f''(c)h^2,$$

for some c between a and b. Thus, the error is proportional to h^2.

EXAMPLE 513

Just for illustration, let us do $\int_1^4 x^2\,dx$ again and compare results with those obtained by the rectangular rule. The latter are reproduced here for ease of comparison.

n	Rectangular rule	Trapezoidal rule
2	20.43750000	22.12500000
4	20.85937500	21.28125000
10	20.97750000	21.04500000
20	20.99437500	21.01125000
30	20.99750000	21.00500000
50	20.99910000	21.00180000
100	20.99977500	21.00045000

EXAMPLE 514

Approximate

$$\int_0^{\pi/2} \sin(x^2)\,dx.$$

Here are comparisons using the rectangular and trapezoidal rules.

n	Rectangular rule	Trapezoidal rule
2	0.89293919	0.69947699
4	0.84420448	0.79620809
10	0.83064857	0.82305960
20	0.82874770	0.82685409
30	0.82839679	0.82755549
50	0.82821727	0.82791446
100	0.82814156	0.82806586
250	0.82812038	0.82810826

METHOD 3: SIMPSON'S RULE

Simpson's rule approximates $\int_a^b f(x)\,dx$ by fitting an arc of a parabola to $y = f(x)$ between x_j and x_{j+2} (j even) and using the fact that the area under the parabolic arc of Figure 373 is

$$\frac{h}{3}(y_0 + y_1 + y_2).$$

Thus,

$$\int_a^b f(x)\,dx \approx \frac{h}{3}(y_0 + 4y_1 + 2y_2 + 4y_3 + 2y_4 + \cdots + 2y_{n-2} + 4y_{n-1} + y_n),$$

with n even and $h = (b - a)/n$.

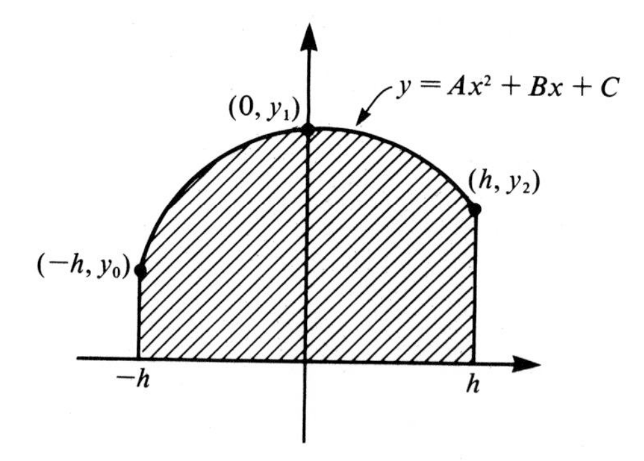

Figure 373

The error in Simpson's rule is

$$\frac{(b-a)}{180} f^{(4)}(c)h^4,$$

for some c between a and b. Thus, the error is proportional to h^4.

EXAMPLE 515

Consider again $\int_1^4 x^2\, dx$. With $n = 2$, Simpson's rule yields exactly 21.000000000. In fact, for any polynomial of degree three or less, Simpson's rule gives exactly the right answer with only $n = 2$, since $f^{(4)}(x) = 0$ for such a polynomial, making the error term zero.

EXAMPLE 516

Approximate

$$\int_0^1 \frac{dx}{\sqrt{2-\sin^2(x)}}.$$

For comparison, we use all three methods outlined thus far.

n	Rectangular rule	Trapezoidal rule	Simpson's rule
2	0.76260967	0.77253221	0.76555945
4	0.76513625	0.76757094	0.76591718
10	0.76582073	0.76620815	0.76594906
20	0.76591766	0.76601444	0.76594987
50	0.76594477	0.76596025	0.76594992
100	0.76594864	0.76595251	0.76594993
200	0.76594960	0.76595057	0.76594993

There are other methods of approximating integrals, but the above three are probably the simplest. Another method called *Gaussian quadrature* is very fast but more involved to implement. Many texts on numerical analysis have a discussion of this method, which utilizes the zeros of the Legendre polynomials discussed in Chapter 6, Section 3.

PROBLEMS FOR SECTION 20.2

In each of Problems 1 through 20, approximate the given integral using the rectangular rule, trapezoidal rule, and Simpson's rule, with $n = 2, 4, 10, 50, 100$, and 200. Carry out calculations to four decimal places.

1. $\displaystyle\int_0^1 \frac{dx}{\sqrt{x^2+2x+12}}$

2. $\displaystyle\int_0^{\pi/2} \sin(x^3)\, dx$

3. $\displaystyle\int_1^{\sqrt{2}} e^{-x^2}\cos(3x)\, dx$

4. $\displaystyle\int_0^{\sqrt{\pi}} \sin(x^2)\cos(x^2)\, dx$

5. $\displaystyle\int_1^3 \sinh(e^{-x})\, dx$

6. $\displaystyle\int_0^2 \frac{(1-2x)}{2+4x}\, dx$

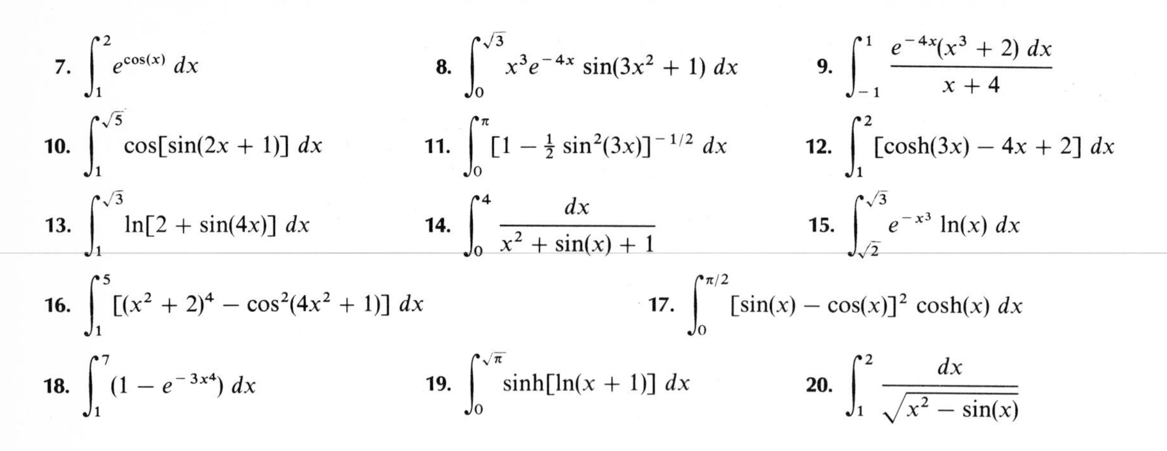

7. $\int_1^2 e^{\cos(x)}\,dx$

8. $\int_0^{\sqrt{3}} x^3 e^{-4x}\sin(3x^2+1)\,dx$

9. $\int_{-1}^{1} \frac{e^{-4x}(x^3+2)\,dx}{x+4}$

10. $\int_1^{\sqrt{5}} \cos[\sin(2x+1)]\,dx$

11. $\int_0^{\pi} [1-\frac{1}{2}\sin^2(3x)]^{-1/2}\,dx$

12. $\int_1^2 [\cosh(3x)-4x+2]\,dx$

13. $\int_1^{\sqrt{3}} \ln[2+\sin(4x)]\,dx$

14. $\int_0^4 \frac{dx}{x^2+\sin(x)+1}$

15. $\int_{\sqrt{2}}^{\sqrt{3}} e^{-x^3}\ln(x)\,dx$

16. $\int_1^5 [(x^2+2)^4-\cos^2(4x^2+1)]\,dx$

17. $\int_0^{\pi/2} [\sin(x)-\cos(x)]^2\cosh(x)\,dx$

18. $\int_1^7 (1-e^{-3x^4})\,dx$

19. $\int_0^{\sqrt{\pi}} \sinh[\ln(x+1)]\,dx$

20. $\int_1^2 \frac{dx}{\sqrt{x^2-\sin(x)}}$

20.3 POLYNOMIAL INTERPOLATION

The problem of interpolation dates back at least to Euler (eighteenth century), when attempts were made to construct elliptical orbits from tables of observed positions of planets.

In general, the interpolation problem we address here is the following: Given $(n+1)$ points $(x_0, y_0), (x_1, y_1), \ldots, (x_n, y_n)$, determine a smooth function $y = f(x)$ such that $y_j = f(x_j)$ for $j = 0, 1, 2, \ldots, n$. Here, *smooth* is usually taken to mean twice-differentiable, but in specific instances it may be understood to mean differentiable some set number of times.

For nth *degree polynomial interpolation*, we attempt to find an nth degree polynomial $p(x)$ passing through the given points and use it for $f(x)$. To find $p(x)$, write

$$p(x) = a_0 + a_1 x + a_2 x^2 + \cdots + a_n x^n.$$

We then require

$$\begin{aligned} p(x_0) &= a_0 + a_1 x_0 + a_2 x_0^2 + \cdots + a_n x_0{}^n = y_0 \\ p(x_1) &= a_0 + a_1 x_1 + a_2 x_1^2 + \cdots + a_n x_1{}^n = y_1 \\ &\vdots \\ p(x_n) &= a_0 + a_1 x_n + a_2 x_n{}^2 + \cdots + a_n x_n{}^n = y_n \end{aligned}$$

This provides us with $(n+1)$ equations to solve for the $(n+1)$ coefficients $a_0, a_1, \ldots, a_n$ of $p(x)$.

EXAMPLE 517

Suppose that the given points are

$(-1, 0)$, $(-0.8, 2)$, $(-0.6, 1)$, $(-0.4, -1)$, $(-0.2, 0)$ and $(0, -4)$.

Since we have six points, we use a fifth degree polynomial,

$$p(x) = a_0 + a_1 x + a_2 x^2 + a_3 x^3 + a_4 x^4 + a_5 x^5.$$

For this polynomial to pass through the six given points, we require

$$p(-1) = a_0 - a_1 + a_2 - a_3 + a_4 - a_5 = 0$$
$$p(-0.8) = a_0 - 0.8a_1 + 0.64a_2 - 0.512a_3 + 0.4096a_4 - 0.32768a_5 = 2$$
$$p(-0.6) = a_0 - 0.6a_1 + 0.36a_2 - 0.216a_3 + 0.1296a_4 - 0.07776a_5 = 1$$
$$p(-0.4) = a_0 - 0.4a_1 + 0.16a_2 - 0.064a_3 + 0.0256a_4 - 0.01024a_5 = -1$$
$$p(-0.2) = a_0 - 0.2a_1 + 0.04a_2 - 0.008a_3 + 0.0164a_4 - 0.00032a_5 = 0$$

and $$p(0) = a_0 = -4$$

Solve these to obtain

$$p(x) = -4 - 74.83333390x - 445.8333334x^2 - 1052.083333x^3 - 1041.666666x^4 - 364.5833331x^5.$$

For $-1 < x < 0$, we use this to calculate the values $f(x)$ of the interpolated function. For example, we can calculate $p(-0.9) = 1.6952$, and hence we take $f(-0.9)$ to be approximately 1.6952.

PROBLEMS FOR SECTION 20.3

In each of Problems 1 through 15, use nth degree polynomial interpolation [where $(n + 1)$ is the number of data points given] to obtain $f(x)$ for the requested values of x. Carry out calculations to four decimal places in finding the function values.

1. $(-1, 2)$, $(0, 3)$, $(1, 14)$, $(2, -2)$; $f(-0.5)$, $f(1.2)$, $f(\sqrt{2})$
2. $(0, 1)$, $(1, 4)$, $(2, \sqrt{5})$, $(3, -2)$; $f(3.5)$, $f(\frac{4}{3})$, $f(1.7)$
3. $(-2, 6)$, $(-1, 4)$, $(0, 1)$, $(1, 3)$; $f(-\sqrt{3})$, $f(1.2)$, $f(0.15)$
4. $(-3, 4)$, $(-2, 1)$, $(-1, 1)$, $(0, 4)$; $f(-2.6)$, $f(-1.3)$, $f(-0.75)$
5. $(1, -2)$, $(1.3, 4)$, $(1.6, -2)$, $(1.9, \sqrt{2})$; $f(1.1)$, $f(1.7)$, $f(1.8)$
6. $(2, -3)$, $(2.1, -3.4)$, $(2.2, -3.7)$, $(2.3, -3.9)$; $f(-3.14)$, $f(-3.27)$
7. $(2.3, 1.7)$, $(2.5, 3)$, $(2.7, -1)$, $(2.9, -3.2)$; $f(2.55)$, $f(2.8)$, $f(2.4)$
8. $(0, \sqrt{2})$, (1.14), $(2, -3)$, $(3, 1)$, $(4, 2)$; $f(1.5)$, $f(2.6)$, $f(\sqrt{3}/2)$
9. $(-1, 1)$, $(0, 2)$, $(1, -4)$, $(2, 6)$, $(3, 8)$; $f(2.3)$, $f(1.7)$, $f(\pi/2)$
10. $(-3.2, 4)$, $(-3.9, -4)$, $(-2.6, 1)$, $(-1.9, 0)$; $f(-\sqrt{2})$, $f(-1.5)$, $f(-3.6)$
11. $(-1, 0)$, $(0, 2)$, $(1, -3)$, $(2, 4)$, $(3, \sqrt{7})$; $f(\pi/4)$, $f(\sqrt{3})$, $f(-1.6)$
12. $(1, -7)$, $(1.5, 3)$, $(2, -1)$, $(2.5, -0.6)$, $(3, 1.2)$; $f(1.7)$, $f(2.3)$, $f(2.1)$
13. $(-3.2, 14)$, $(-2.6, 3)$, $(-2, \sqrt{3})$, $(-1.4, 1)$, $(-0.8, 12)$; $f(-1.7)$, $f(-2.3)$, $f(-3)$
14. $(1.4, 12)$, $(1.8, -3.2)$, $(2.2, 1.6)$, $(2.6, -3.2)$, $(3, 0)$; $f(1.7)$, $f(2)$, $f(2.3)$
15. $(0, -4)$, $(2, 3)$, $(4, -5)$, $(6, 1.2)$, $(8, 2.3)$, $(10, 1.9)$; $f(3)$, $f(\sqrt{7})$, $f(6.7)$

20.4 NUMERICAL DIFFERENTIATION

The problem we address in this section is that of computing $f'(x)$ when $f(x)$ is given in the form of a table of values at points $x_1, \ldots, x_n$. There are several methods available.

METHOD 1: DIFFERENCE-QUOTIENT APPROXIMATION

By definition,

$$f'(x) = \lim_{h \to 0} \frac{f(x+h) - f(x)}{h}.$$

Thus,

$$f'(x) \approx \frac{f(x+h) - f(x)}{h},$$

with greater accuracy the smaller h is chosen.

EXAMPLE 518

With $h = 0.01$, $f(3) = 4$, and $f(3.01) = 4.29$, we get

$$f'(3) \approx \frac{f(3+h) - f(3)}{h} = \frac{4.29 - 4}{0.01} = 29.$$

METHOD 2: THE THREE-POINT FORMULA

Suppose that we know $f(x_0)$, $f(x_1)$, and $f(x_2)$, where $x_1 = x_0 + h$ and $x_2 = x_0 + 2h$. Then,

$$f'(x_0) \approx \frac{1}{2h}[-3f(x_0) + 4f(x_1) - f(x_2)],$$

$$f'(x_1) \approx \frac{1}{2h}[-f(x_0) + f(x_2)],$$

and

$$f'(x_2) \approx \frac{1}{2h}[f(x_0) - 4f(x_1) + 3f(x_2)].$$

EXAMPLE 519

Suppose that $f(2) = 1.7$, $f(2.01) = 1.78$, and $f(2.02) = 2.1$. With $h = 0.01$, we have

$$f'(2) \approx \frac{1}{0.02}[-3(1.7) + 4(1.78) - 2.1] = -4,$$

$$f'(2.01) \approx \frac{1}{0.02}(-1.7 + 2.1) = 20,$$

and

$$f'(2.02) \approx \frac{1}{0.02}[1.7 - 4(1.78) + 3(2.1)] = 44.$$

METHOD 3: THE FIVE-POINT FORMULA

If we know $f(x_0)$, $f(x_0 + h)$, $f(x_0 + 2h)$, $f(x_0 + 3h)$, and $f(x_0 + 4h)$, then

$$f'(x_0 + 2h) \approx \frac{1}{12h}\,[f(x_0) - 8f(x_0 + h) + 8f(x_0 + 3h) - f(x_0 + 4h)].$$

EXAMPLE 520

Suppose that $f(1) = 0$, $f(1.01) = 0.02$, $f(1.02) = 0.03$, $f(1.03) = 0.076$, and $f(1.04) = 0.125$. Then,

$$f'(1.02) \approx \frac{1}{12(0.01)}\,[0 - 8(0.02) + 8(0.076) - 0.125] = 2.6917.$$

To get some feeling for these methods, let us use them in a situation involving a familiar function whose derivative is easily calculated.

EXAMPLE 521

Let $f(x) = x\cos(x)$. Then, $f'(x) = \cos(x) - x\sin(x)$.

Just for illustration, choose h equal to 0.01, and choose (to nine decimal places)

$$x_0 = \pi = 3.141592654$$
$$x_1 = \pi + h = 3.151592654$$
$$x_2 = \pi + 2h = 3.161592654$$
$$x_3 = \pi + 3h = 3.171592654$$
$$x_4 = \pi + 4h = 3.181592654$$

Using Method 1, we would have, for example,

$$f'(\pi) \approx \frac{f(\pi + 0.01) - f(\pi)}{0.01} = \frac{f(x_1) - f(x_0)}{x_1 - x_0}$$
$$= \frac{-3.151435076 - (-3.141592654)}{0.01} = -0.984242200.$$

In fact, $f'(\pi) = -1$; so the approximation in this case is in error by 0.015757800.

Now calculate $f'(x_2)$ using all three methods for purposes of comparison.

By Method 1,

$$f'(x_2) \approx \frac{f(x_2 + h) - f(x_2)}{h} = \frac{-3.170165544 + 3.160960357}{0.01} = -0.920518700.$$

By Method 2,

$$f'(x_2) \approx \frac{1}{2h}\,[-f(x_1) + f(x_3)] = -0.936523400.$$

By Method 3,

$$f'(x_2) \approx \frac{1}{12h}[f(x_0) - 8f(x_1) + 8f(x_3) - f(x_4)]$$

$$= \frac{1}{12(0.01)}[-3.141592654 - 8(-3.151435076)$$

$$+ 8(-3.170165544) - (-3.179047719)]$$

$$= -0.936572258.$$

From the exact formula for $f'(x)$, we compute (to nine decimal places)

$$f'(x_2) = -0.936572368.$$

Thus, the error in computing $f'(x_2)$ is:

-0.016053668	(using Method 1)
-0.000048968	(using Method 2)
-0.000000110	(using Method 3)

In general, as in the above example, Method 3 is the most accurate of the three methods for a given h, with Method 2 next, and then Method 1.

PROBLEMS FOR SECTION 20.4

In each of Problems 1 through 10, use the difference-quotient approximation to approximate the requested derivative, using the given information.

1. $f'(2.3)$; $f(2.4) = 1$, $f(2.3) = 1.08$
2. $f'(1.7)$; $f(1.8) = 0$, $f(1.7) = 0.02$
3. $f'(-2.4)$; $f(-2.3) = 1$, $f(-2.4) = 1.04$
4. $f'(1)$; $f(1) = 2.01$, $f(0.98) = 2.02$
5. $f'(0.9)$; $f(0.9) = 0.16$, $f(0.8) = 0.17$
6. $f'(3.16)$; $f(3.16) = 1.021$, $f(3.15) = 1.025$
7. $f'(4.26)$; $f(4.26) = -1.302$, $f(4.25) = -1.305$
8. $f'(6)$; $f(6) = 3.143$, $f(5.99) = 3.141$
9. $f'(-2)$; $f(-2) = -1.73$, $f(-2.1) = -1.71$
10. $f'(1.73)$; $f(1.73) = 0.0014$, $f(1.72) = 0.0017$

In each of Problems 11 through 20, find $f'(x_0)$, $f'(x_1)$, and $f'(x_2)$ from the given information using the three-point formula.

11. $f(1) = 2.01$, $f(1.01) = 2.04$, $f(1.02) = 2.09$
12. $f(4.7) = -1.02$, $f(4.8) = -1.13$, $f(4.9) = -1.14$
13. $f(-3.2) = 4.17$, $f(-3.1) = 4.21$, $f(-3.0) = 4.22$
14. $f(1.72) = 9.13$, $f(1.73) = 9.24$, $f(1.74) = 9.46$
15. $f(6.20) = -1.34$, $f(6.21) = -1.29$, $f(6.22) = -1.47$
16. $f(0.1) = -2.3$, $f(0.3) = -2.6$, $f(0.5) = -2.4$
17. $f(1.32) = 2.79$, $f(1.34) = 2.84$, $f(1.36) = 2.91$
18. $f(-3.52) = -1.73$, $f(-3.50) = -1.80$, $f(-3.48) = -1.77$
19. $f(5) = -2.71$, $f(5.3) = -2.76$, $f(5.6) = -2.80$
20. $f(2) = -1.76$, $f(2.01) = -1.74$, $f(2.02) = -1.73$

In each of Problems 21 through 30, use the five-point formula to approximate $f'(x_2)$, given the function values $f(x_0)$, $f(x_1)$, $f(x_3)$, and $f(x_4)$.

21. $f(2) = 1.73$, $f(2.1) = 1.74$, $f(2.3) = 1.79$, $f(2.4) = 1.85$
22. $f(-3.2) = 0.01$, $f(-3.1) = 0.02$, $f(-2.9) = 0.14$, $f(-2.8) = 0.13$

23. $f(4.37) = -2.16$, $f(4.38) = -2.17$, $f(4.40) = -2.21$, $f(4.41) = -2.19$
24. $f(6.21) = -1.01$, $f(6.22) = -0.99$, $f(6.24) = -0.88$, $f(6.25) = -0.85$
25. $f(-4.31) = 2.65$, $f(-4.30) = 2.67$, $f(-4.28) = 2.71$, $f(-4.27) = 2.70$
26. $f(1.91) = 4.63$, $f(1.92) = 4.65$, $f(1.94) = 4.62$, $f(1.95) = 4.61$
27. $f(2.71) = 1.63$, $f(2.74) = 1.63$, $f(2.80) = 1.67$, $f(2.83) = 1.74$
28. $f(0.92) = 2$, $f(0.94) = 2.31$, $f(0.98) = 2.27$, $f(1) = 2.25$
29. $f(1.87) = -2.14$, $f(1.90) = -2.01$, $f(1.96) = -2.16$, $f(1.99) = -2.25$
30. $f(6.21) = -1.81$, $f(6.25) = -1.74$, $f(6.33) = -1.83$, $f(6.37) = -1.64$

In each of Problems 31 through 40, you are given a differentiable function $f(x)$, a point x_0, and a value of h. Determine x_1, x_2, x_3, and x_4 (with $x_j = x_0 + jh$), and use all three methods to compute $f'(x_2)$. Make a table giving the result for $f'(x_2)$ using each method; the value of $f'(x_2)$ computed explicitly from the derivative of the function, calculated to four decimal places; and the error in the result produced by each approximate method.

31. $f(x) = x^2$; $x_0 = 2.3$; $h = 0.01$
32. $f(x) = x^3$; $x_0 = 1.74$; $h = 0.01$
33. $f(x) = x - \sin(2x)$; $x_0 = 3.1427$; $h = 0.1$
34. $f(x) = \sin(3x^2)$; $x_0 = 1$; $h = 0.001$
35. $f(x) = \cosh(2x) - x$; $x_0 = 0$; $h = 0.01$
36. $f(x) = 2x^3 - 4x$; $x_0 = 2$; $h = 0.002$
37. $f(x) = \sin^2(3x)$; $x_0 = 1$; $h = 0.03$
38. $f(x) = \tan(x) - x^2$; $x_0 = 0$; $h = 0.004$
39. $f(x) = e^{-2x} + x$; $x_0 = 4$; $h = 0.01$
40. $f(x) = x^2 - 2x + \cos(3x)$; $x_0 = 1$; $h = 0.002$

20.5 CUBIC SPLINES

Sometimes engineers use thin rods called *splines* to piece together smooth curves passing through given points. For this reason, the term *spline* is also applied to the mathematical curve-fitting problem which we now consider.

Suppose that we are given a function $f(x)$ for $a \le x \le b$. Suppose that $[a, b]$ is partitioned into subintervals by points

$$a = x_0 < x_1 < x_2 \cdots < x_{n-1} < x_n = b.$$

We seek to approximate $f(x)$ on each subinterval $[x_{j-1}, x_j]$ by a polynomial. It is permissible to use a different polynomial on each subinterval, but we must be careful to fit the polynomials used on successive intervals together in such a way that the resulting function is twice-differentiable. If the polynomials used are of degree three or less, then we call the function obtained by piecing together the polynomials a *cubic spline*.

More carefully, a *cubic spline approximation of* $f(x)$ on an interval $[a, b]$ partitioned into subintervals by points $a = x_0 < x_1 < x_2 < \cdots < x_{n-1} < x_n = b$ is a twice-differentiable function $P(x)$ such that

1. $P(x_j) = f(x_j)$ for $j = 0, 1, 2, \ldots, n$.
2. On each $[x_{j-1}, x_j]$, $P(x)$ is a polynomial of degree three or less.

One can show (as we shall see in the example below) that $P(x)$ is uniquely determined by $f(x)$, the interval and the chosen partition points, and the conditions $P(x_j) = f(x_j)$ for $j = 0, 1, 2, \ldots, n$ *if* we also specify values of $P'(a)$ and $P'(b)$. In practice, we usually specify that $P'(a) = f'(a)$ and $P'(b) = f'(b)$. Thus, to sum up, we determine $P(x)$ by the following conditions:

1. $P(x_j) = f(x_j)$ for $j = 0, 1, 2, \ldots, n$.
2. On each subinterval $[x_{j-1}, x_j]$, $P(x)$ has the form

$$p_j(x) = a_j + b_j x + c_j x^2 + d_j x^3 \quad \text{for} \quad j = 1, 2, \ldots, n.$$

3. $P'(a) = f'(a)$ and $P'(b) = f'(b)$.
4. For smoothness of fit at the "seams" $x_1, \ldots, x_{n-1}$, we require that

$$p_j(x_j) = p_{j+1}(x_j), \qquad p_j'(x_j) = p_{j+1}'(x_j) \quad \text{and} \quad p_j''(x_j) = p_{j+1}''(x_j)$$

for $j = 1, 2, \ldots, n-1$.

The last condition means that, at x_j, where $p_j(x)$ and $p_{j+1}(x)$ must be joined, we require that these two polynomials agree, that their first derivatives agree, and that their second derivatives agree.

These conditions yield equations which can then be solved for each a_j, b_j, c_j, and d_j, giving the part of $P(x)$ lying on each $[x_{j-1}, x_j]$.

EXAMPLE 522

Let $f(x) = e^{2x}$, for $0 \leq x \leq 3$. We shall insert partition points $x_1 = 1$ and $x_2 = 2$ and let $x_0 = 0$ and $x_3 = 3$. Here, $n = 3$.

Let

$$p_1(x) = a_1 + b_1 x + c_1 x^2 + d_1 x^3 \quad \text{for} \quad 0 \leq x \leq 1,$$
$$p_2(x) = a_2 + b_2 x + c_2 x^2 + d_2 x^3 \quad \text{for} \quad 1 \leq x \leq 2,$$

and

$$p_3(x) = a_3 + b_3 x + c_3 x^2 + d_3 x^3 \quad \text{for} \quad 2 \leq x \leq 3.$$

The cubic spline approximation will be the function

$$P(x) = \begin{cases} p_1(x) & \text{for} \quad 0 \leq x \leq 1, \\ p_2(x) & \text{for} \quad 1 \leq x \leq 2, \\ p_3(x) & \text{for} \quad 2 \leq x \leq 3. \end{cases}$$

We have twelve unknowns to solve for, namely, $a_1, b_1, \ldots, c_3, d_3$. First, we need

$$P(0) = f(0), \qquad P(1) = f(1), \qquad P(2) = f(2), \quad \text{and} \quad P(3) = f(3).$$

Thus,

$$p_1(0) = f(0), \qquad p_1(1) = p_2(1) = f(1), \qquad p_2(2) = p_3(2) = f(2),$$

and

$$p_3(3) = f(3).$$

These give us the following six equations:

$$
\begin{aligned}
a_1 &= 1 \\
a_1 + b_1 + c_1 + d_1 &= e^2 = 7.389056099 \\
a_2 + b_2 + c_2 + d_2 &= e^2 = 7.389056099 \\
a_2 + 2b_2 + 4c_2 + 8d_2 &= e^4 = 54.59815003 \\
a_3 + 2b_3 + 4c_3 + 8d_3 &= e^4 = 54.59815003 \\
a_3 + 3b_3 + 9c_3 + 27d_3 &= e^6 = 403.4287935
\end{aligned}
$$

Next, we shall require that

$$P'(0) = f'(0) \quad \text{and} \quad P'(3) = f'(3).$$

Thus,

$$p_1'(0) = f'(0) \quad \text{and} \quad p_3'(3) = f'(3),$$

giving us two more equations. They are:

$$
\begin{aligned}
b_1 &= 2 \\
b_3 &= 2e^6 = 806.8575870
\end{aligned}
$$

We have already used the conditions that $p_j(x_j) = p_{j+1}(x_j)$ for $j = 1, 2$. But we also want

$$p_j'(x_j) = p_{j+1}'(x_j) \quad \text{and} \quad p_j''(x_j) = p_{j+1}''(x_j)$$

for $j = 1, 2$. These give us four more equations:

$$
\begin{aligned}
b_1 + 2c_1 + 3d_1 &= b_2 + 2c_2 + 3d_2 \\
b_2 + 4c_2 + 12d_2 &= b_3 + 4c_3 + 12d_3 \\
2c_1 + 6d_1 &= 2c_2 + 6d_2 \\
2c_2 + 12d_2 &= 2c_3 + 12d_3
\end{aligned}
$$

We now have twelve linear equations in twelve unknowns. Actually, things are not really that bad, because we know a_1, b_1, and b_3 from these equations. Putting in these known values gives us nine equations in nine unknowns. They are:

$$
\begin{aligned}
c_1 + d_1 &= 4.389056099 \\
a_2 + b_2 + c_2 + d_2 &= 7.389056099 \\
a_2 + 2b_2 + 4c_2 + 8d_2 &= 54.59815003 \\
a_3 + 4c_3 + 8d_3 &= -752.2594370 \\
a_3 + 9c_3 + 27d_3 &= -2017.143968 \\
2c_1 + 3d_1 - b_2 - 2c_2 - 3d_2 &= -2 \\
b_2 + 4c_2 + 12d_2 - 4c_3 - 12d_3 &= 806.8575870 \\
2c_1 + 6d_1 - 2c_2 - 6d_2 &= 0 \\
2c_2 + 12d_2 - 2c_3 - 12d_3 &= 0
\end{aligned}
$$

These have solution

$$c_1 = -54.48222885 \qquad d_1 = 58.87128430 \qquad a_2 = 204.4432110$$
$$b_2 = -608.3296300 \qquad c_2 = 555.8474012 \qquad d_2 = -144.5719257$$
$$a_3 = 67.84265260 \qquad c_3 = -151.7462071 \qquad d_3 = -26.63965765$$

Thus, $P(x)$ is made up of the three cubic polynomials

$$p_1(x) = 1 + 2x - 54.48222885x^2 + 58.87128430x^3 \qquad (0 \le x \le 1),$$

$$p_2(x) = 204.4432110 - 608.3296300x + 555.8474012x^2 - 144.5719257x^3 \qquad (1 \le x \le 2),$$

and

$$p_3(x) = 67.84265260 + 806.8575870x - 151.7462071x^2 - 26.63965765x^3 \qquad (2 \le x \le 3).$$

PROBLEMS FOR SECTION 20.5

In each of Problems 1 through 10, find the cubic spline approximation of the given function on the given interval, using the indicated partition points of the interval.

1. $f(x) = 3x^4;\quad 1 < 2 < 3$
2. $f(x) = \cos(2x);\quad 0 < \pi/2 < \pi$
3. $f(x) = e^{-x};\quad 1 < 3 < 5$
4. $f(x) = \tan(x);\quad 0 < \pi/8 < \pi/4$
5. $f(x) = 4x^4 - 2;\quad 1 < 3 < 5 < 6$
6. $f(x) = \sin^2(x);\quad 0 < \pi/2 < \pi$
7. $f(x) = \ln(2x);\quad \frac{1}{2} < 1 < 2$
8. $f(x) = x^5 + x^3;\quad 0 < 1 < 2 < 3$
9. $f(x) = \cosh(x);\quad 0 < 1 < 2$
10. $f(x) = xe^{-x};\quad 0 < 1 < 2 < 3$
11. Let $P(x)$ be the cubic spline approximation of $f(x)$ on $[a, b]$ with partition $a = x_0 < x_1 < x_2 < \cdots < x_{n-1} < x_n = b$. Suppose that

$$P'(a) = f'(a) \quad \text{and} \quad P'(b) = f'(b).$$

Assume that $f'(x)$ and $f''(x)$ exist on $[a, b]$.
(a) Prove that

$$\int_a^b [P''(x)]^2\,dx \le \int_a^b [f''(x)]^2\,dx.$$

(b) Prove that there is equality in (a) if and only if $f(x) = P(x)$ on $[a, b]$.
Hint: Begin by showing that

$$\int_a^b P''(x)[f''(x) - P''(x)]\,dx = -\int_a^b P^{(3)}(x)[f'(x) - P'(x)]\,dx.$$

To do this, integrate the left side by parts and derive the right side. After doing this, use the fact that $P(x)$ is a polynomial of degree at most three on each subinterval to evaluate the integral on the right. Finally, look at

$$\int_a^b [f''(x) - P''(x)]^2\,dx.$$

12. Verify the assertion made in the text that the four conditions given for determining $P(x)$ determine $P(x)$ uniquely.

20.6 NUMERICAL SOLUTION OF INITIAL VALUE PROBLEMS

In this section we address ways of obtaining approximate solutions of the initial value problem

$$y' = f(x, y); \qquad y(x_0) = y_0,$$

with x_0 and y_0 given. The basic idea is to specify an increment h and obtain approximate values of a solution $y = y(x)$ at $x_0, x_0 + h, x_0 + 2h, x_0 + 3h, \ldots$.

Usually, we write $y(x_0 + nh)$ as y_n and $x_0 + nh$ as x_n. Thus,

$$\begin{aligned} y_0 &= y(x_0), \\ y_1 &= y(x_0 + h) = y(x_1), \\ y_2 &= y(x_0 + 2h) = y(x_2), \end{aligned}$$

and so on.

Here is a sketch of a simple and straightforward way of approximating a numerical solution $y = S(x)$ of $y' = f(x, y)$; $y(x_0) = y_0$. First, use the given information to construct the tangent line to $y = S(x)$ at x_0. Note that the tangent to the solution at (x_0, y_0) has slope $y'(x_0)$, or $f(x_0, y_0)$, which is known. Thus, we know the tangent to the solution at x_0, and we use this tangent as an approximation to $y = S(x)$ near x_0.

Now, if x_1 is near x_0, we move along the tangent line at x_0 until we reach x_1. Then, as one can see from Figure 374,

$$y_1 = y_0 + S'(x_0)(x_1 - x_0) = y_0 + f(x_0, y_0)(x_1 - x_0).$$

Now use this to approximate y_1, and compute $y_1' = f(x_1, y_1)$. Then, in turn, use this to approximate the slope of $y = S(x)$ at x_1. Then approximate y_2 by

$$y_2 = y_1 + S'(x_1)(x_2 - x_1) = y_1 + f(x_1, y_1)(x_2 - x_1).$$

Continuing in this way, we approximate y_3, y_4, and so on.

The beauty of this method is its simplicity. An unfortunate feature is that it is not very accurate, because successive uses of the tangent line at the approximated values

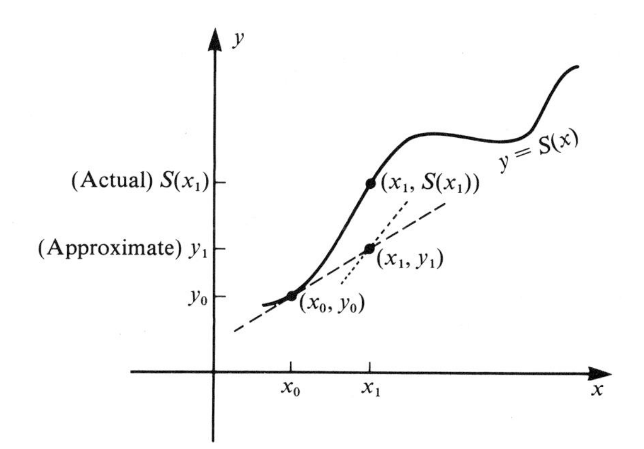

Figure 374. Euler, or tangent line, method.

$y_1, y_2, \ldots$ may lead one away from the actual solution $y = S(x)$. That is, the method can accumulate error, building it up from each step to the next. One partial remedy is to take each x_{j+1} closer to x_j, but a better means of improvement lies in modifying the data at each approximation to prevent a high degree of error propagation. The following three methods all employ a feedback scheme which yields more accurate results than the *Euler*, or *tangent line, method* we have just described.

All the methods discussed here are *iterative*—we obtain each value of y from one or more preceding ones.

METHOD 1: THE THREE-TERM TAYLOR SERIES METHOD

$$y_{n+1} = y_n + hf(x_n, y_n) + \frac{h^2}{2}\left[\frac{\partial f}{\partial x}(x_n, y_n) + \frac{\partial f}{\partial y}(x_n, y_n)f(x_n, y_n)\right].$$

The rationale for this method lies in the three-term Taylor expansion

$$y(x + h) = y(x) + y'(x)h + y''(x)\frac{h^2}{2} + y^{(3)}(\bar{x})\frac{h^3}{6},$$

where $x < \bar{x} < x + h$.

Here,

$$y'(x) = f[x, y(x)]$$

and

$$y''(x) = \frac{\partial f}{\partial x}[x, y(x)] + \frac{\partial f}{\partial y}[x, y(x)]y'(x)$$

$$= \frac{\partial f}{\partial x}[x, y(x)] + \frac{\partial f}{\partial y}[x, y(x)]f[x, y(x)].$$

Thus, we compute $y_{n+1} = y(x_n + h)$ from $y_n[=y(x_n)]$ by replacing x by x_n and omitting the term $y^{(3)}(\bar{x})h^3/6$ in the Taylor expansion. The error in the method is thus proportional to h^3.

EXAMPLE 523

Here is an example in which we can solve the initial value problem exactly and compare results.

Consider

$$y' = x + y; \qquad y(1) = -2.$$

Here, the differential equation is linear (see Chapter 1, Section 1.6) and we easily solve the problem to get

$$y = -x - 1,$$

an exact solution. We shall let $h = 0.1$ in the three-term Taylor method and obtain $y(1.1)$, $y(1.2)$, $y(1.3)$, ..., $y(1.10)$. Note that here $f(x, y) = x + y$, $\partial f/\partial x = \partial f/\partial y = 1$, and the formula becomes

$$y_{n+1} = y_n + h(x_n + y_n) + \frac{h^2}{2}(1 + x_n + y_n).$$

Form a table:

n	x_n	y_n	Exact value of $y(x_n + h)$ from $y(x) = -x - 1$	Approximate value of y_{n+1} from Method 1
0	1.0	−2.0	−2.1	−2.1
1	1.1	−2.1	−2.2	−2.2
2	1.2	−2.2	−2.3	−2.3
3	1.3	−2.3	−2.4	−2.4
4	1.4	−2.4	−2.5	−2.5
5	1.5	−2.5	−2.6	−2.6
6	1.6	−2.6	−2.7	−2.7
7	1.7	−2.7	−2.8	−2.8
8	1.8	−2.8	−2.9	−2.9
9	1.9	−2.9	−3.0	−3.0
10	2.0	−3.0	−3.1	−3.1

Here, the approximate values are exact because the problem was so simple. Try a more difficult problem.

EXAMPLE 524

Consider

$$y' = y\sin(x^2) - \cos(x^2) + 2y; \qquad y(3) = \sqrt{2}.$$

This is not solvable in elementary terms. With

$$f(x, y) = y\sin(x^2) - \cos(x^2) + 2y,$$

we have

$$\frac{\partial f}{\partial x} = 2xy\cos(x^2) + 2x\sin(x^2)$$

and

$$\frac{\partial f}{\partial y} = \sin(x^2) + 2.$$

Thus, the method is

$$y_{n+1} = y_n + h[y\sin(x^2) - \cos(x^2) + y]$$
$$+ \frac{h^2}{2}\{2xy\cos(x^2) + 2x\sin(x^2) + [\sin(x^2) + 2][y\sin(x^2) - \cos(x^2) + y]\}.$$

Now make a table of values. We shall use $h = 0.1$ again.

n	x_n	y_n	y_{n+1}
0	3.0	1.41421356	1.71381209
1	3.1	1.71381209	1.91561019
2	3.2	1.91561019	1.97866044
3	3.3	1.97866044	1.95118814
4	3.4	1.95118814	1.92379531
5	3.5	1.92379531	2.01755251
6	3.6	2.01755251	2.30748245
7	3.7	2.30748245	2.83071993
8	3.8	2.83071993	3.50413306
9	3.9	3.50413306	4.08336341
10	4.0	4.08336341	4.33505908
11	4.1	4.33505908	4.31521703
12	4.2	4.31521703	4.33209245
13	4.3	4.33209245	4.69393062
14	4.4	4.69393062	5.58318642
15	4.5	5.58318642	6.92468132
16	4.6	6.92468132	8.18946115
17	4.7	8.18946115	8.71906302
18	4.8	8.71906302	8.64414870
19	4.9	8.64414870	8.82779010
20	5.0	8.82779010	9.98776980

Note that one gets a fairly good picture of fluctuations in the solution over the interval $[3, 5]$ in this way. The error is of the order of h^3, or 0.001, since we chose $h = 0.1$. Figure 375 shows a graph of this approximate solution for $3 \leq x \leq 5$.

METHOD 2: THE RUNGE-KUTTA METHOD

$$y_{n+1} = y_n + \frac{h}{6}[A_n + 2B_n + 2C_n + D_n],$$

where

$$A_n = f(x_n, y_n),$$
$$B_n = f(x_n + \tfrac{1}{2}h, y_n + \tfrac{1}{2}hA_n),$$
$$C_n = f(x_n + \tfrac{1}{2}h, y_n + \tfrac{1}{2}hB_n),$$
$$D_n = f(x_n + h, y_n + hC_n).$$

The method is derived from taking weighted averages of $f(x, y)$ at certain points in $[x_n, x_{n+1}]$, and it has an error proportional to h^5. Thus, Runge-Kutta is in general more accurate than the three-term Taylor method. Further, it does not require computation of partial derivatives.

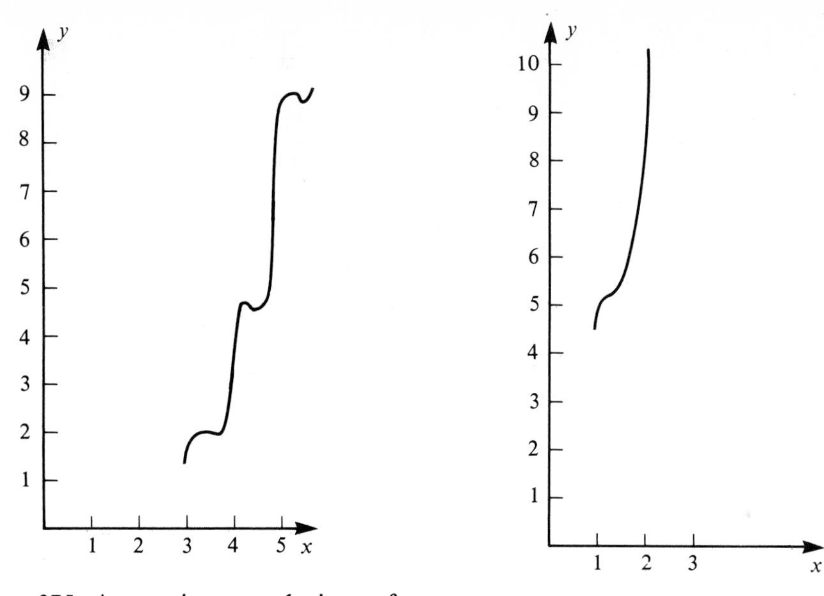

Figure 375. Approximate solution of $y' = y\sin(x^2) - \cos(x^2) + 2y$, $y(3) = \sqrt{2}$, on [3, 5].

Figure 376. Approximate solution of $y' = x^2 - \sin(y^2)$, $y(1) = 4.7$, on [1, 2.5].

EXAMPLE 525

Solve

$$y' = x^2 - \sin(y^2); \qquad y(1) = 4.7.$$

Using $h = 0.1$, we compute

n	x_n	y_n	y_{n+1}
0	1.0	4.70000000	4.87308189
1	1.1	4.87308189	5.04262649
2	1.2	5.04262649	5.13134592
3	1.3	5.13134592	5.21675174
4	1.4	5.21675174	5.40386478
5	1.5	5.40386478	5.67917650
6	1.6	5.67917650	5.87081208
7	1.7	5.87081208	6.21674655
8	1.8	6.21674655	6.52756575
9	1.9	6.52756575	6.87502843
10	2.0	6.87502843	7.29763507
11	2.1	7.29763507	7.75686697
12	2.2	7.75686697	8.23730331
13	2.3	8.23730331	8.85621401
14	2.4	8.85621401	9.39627580
15	2.5	9.39627580	10.04370098

A sketch of this approximate solution is shown in Figure 376.

METHOD 3: THE MILNE MULTISTEP PREDICTOR-CORRECTOR METHOD

1. Compute y_1, y_2, and y_3 by some other method, say, Runge-Kutta.
2. Compute

$$y_{n+1} = y_{n-3} + \frac{4h}{3}[2f(x_n, y_n) - f(x_{n-1}, y_{n-1}) + 2f(x_{n-2}, y_{n-2})].$$

   This is the *predictor formula.*
3. Recompute y_{n+1} by

$$y_{n+1} = y_{n-1} + \frac{h}{3}[f(x_{n-1}, y_{n-1}) + 4f(x_n, y_n) + f(x_{n+1}, y_{n+1})],$$

   using y_{n+1} from (2) in the right side of (3). This is the *corrector formula.*

Implement the method as follows. Once y_1, y_2, and y_3 are approximated, use these in the predictor formula to compute y_4. Use this in the corrector formula to recompute y_4. If the difference between the two values of y_4 is larger than the tolerance wanted, use the corrected value of y_4 in the corrector formula to correct again. Continue to correct until consecutive values of y_4 are within the desired tolerance.

Now repeat this process at each stage. Once y_{n-2}, y_{n-1}, and y_n have been approximated, use the predictor formula to approximate y_{n+1}; then correct as many times as needed.

In practice, if you need, say, four or more corrections at each step, then you should begin again with a smaller value of h.

EXAMPLE 526

Solve

$$y' = ye^{-x^2}; \qquad y(1) = 3.$$

We are given $x_0 = 1$ and $y_0 = 3$.

Using Runge-Kutta and $h = 0.1$, compute

$$y_1 = 3.10138741,$$
$$y_2 = 3.18525665,$$
$$y_3 = 3.25284852.$$

Now use predictor to compute

$$y_4 = 0.27973692 \qquad \text{(first approximation).}$$

Use this in corrector to get

$$y_4 = 3.30596390 \qquad \text{(second approximation).}$$

Use this value in corrector to get

$$y_4 = 3.30596348 \qquad \text{(third approximation).}$$

Correct again, getting

$$y_4 = 3.30596348 \qquad \text{(same as last value).}$$

We thus take

$$y_4 = 3.30596348.$$

Now use predictor to compute

$$y_5 = 0.22549273 \qquad \text{(first approximation).}$$

Use this in corrector to get

$$y_5 = 3.33573759 \qquad \text{(second approximation).}$$

Correct again, getting

$$y_5 = 3.34666483 \qquad \text{(third approximation).}$$

Correct again, getting

$$y_5 = 3.34670322 \qquad \text{(fourth approximation).}$$

Correct again, getting

$$y_5 = 3.34670336.$$

Correct again, getting

$$y_5 = 3.34670336 \qquad \text{(same as last value).}$$

Thus, we take

$$y_5 = 3.34670336.$$

We can continue in this way, computing and correcting y_6, y_7, and so on.

The above example was chosen to illustrate the idea of correcting values obtained from predictor. In practice, we would have gone back and started again with a smaller value of h because of the number of corrections needed at each stage.

PROBLEMS FOR SECTION 20.6

In each of Problems 1 through 20, use the three-term Taylor method and the Runge-Kutta method to approximate values of y_1 through y_{20}. Draw a graph of the approximate solution for each problem. Use $h = 0.1$ and carry out calculations to four decimal places.

1. $y' = xy^3 - \cos(x)\sin(y)$; $y(0) = -1$

2. $y' = xy + 1$; $y(1) = 2$

3. $y' = 4\sin(x - 1) + y$; $y(\pi) = \sqrt{2}$

4. $y' = \sinh(x) - y$; $y(1) = 5.6$

5. $y' = x\sin^2(y) - x^2 y$; $y(2.6) = 1.4$

6. $y' = -8x\cosh(y - x) + x^2$; $y(2) = -6$

7. $y' = 3y\sin[\cos(x)]$; $y(0) = 4$

8. $y' = (x - y)^3 + \cosh(x)$; $y(1) = -7$

9. $y' = x^2 + y^2$; $y(0) = 1$

10. $y' = 8x\sinh(y) + x^2 y$; $y(\sqrt{2}) = 15$

11. $y' = x - 4y\sin(x)$; $y(2) = 4$

12. $y' = -2\sin(xe^y)$; $y(2) = 4.162$

13. $y' = 4x^2y^2 - \sin(x)$; $y(1) = -4$

14. $y' = \sinh(x) - 2y + x^2$; $y(2) = 12.3$

15. $y' = -4xe^{-2y}\cos(x^2)$; $y(0) = 0.5$

16. $y' = xy^3 - e^x$; $y(2) = 7.31$

17. $y' = 4x\cos(y^2)$; $y(\pi) = 0.5$

18. $y' = 2x - y^3 + x^2y$; $y(1) = 2$

19. $y' = x - 4y\cos(y) - e^{3xy}$; $y(2.73) = 4.62$

20. $y' = \cosh(4x^2 - 2y^2)$; $y(0) = 14$

In each of Problems 21 through 30, use the Milne predictor-corrector method to approximate y_4 through y_{10}, having used Runge-Kutta to approximate y_1, y_2, and y_3. Use $h = 0.01$ and carry out calculations to four decimal places.

21. $y' = 4x^2 - 2y + 1$; $y(2) = \pi$

22. $y' = x^3 - \sin(xy)$; $y(1) = -0.26$

23. $y' = \ln[\cos(x) + 2y + 4]$; $y(\pi) = 3.42$

24. $y' = 8x^3ye^{xy}$; $y(1) = 0.42$

25. $y' = x - 2\cos(xy) + 4y^2$; $y(2) = 3.22$

26. $y' = -4xy^2 - e^y$; $y(3) = 0.38$

27. $y' = x + 2ye^x - \sin(y)$; $y(0) = 0.4$

28. $y' = y^3 - 3x^2y$; $y(0) = -0.112$

29. $y' = -8x + 3ye^x - \cos(2x)$; $y(0.3) = 1.04$

30. $y' = \cos(x - 2y + e^{-x})$; $y(2) = 7.3$

31. Use the *improved Euler method*

$$y_{n+1} = y_n + \frac{h}{2}\{f(x_n, y_n) + f[x_n + h, y_n + hf(x_n, y_n)]\}$$

to redo Problems 1 through 5. Compare with the results obtained by Taylor and Runge-Kutta.

32. Use the *second order Runge-Kutta method*

$$y_{n+1} = y_n + hf[x_n + \tfrac{1}{2}h, y_n + \tfrac{1}{2}hf(x_{n+1}, y_n)]$$

to redo Problems 6 through 10. Compare the results with those obtained by Taylor and Runge-Kutta. (This method is also known as the *modified Euler method* or the *improved polygon method*.)

20.7 NUMERICAL SOLUTION OF SECOND ORDER INITIAL VALUE PROBLEMS

In this section we discuss two methods for approximating the solution of the initial value problem

$$y'' = f(x, y, y'); \qquad y(x_0) = A, \qquad y'(x_0) = B,$$

with x_0, A, B, and $f(x, y, y')$ given.

For our usual notational convenience, we let

$$y_j = y(x_0 + jh),$$

where h is a fixed, positive number, and

$$x_j = x_0 + jh;$$

and we denote

$$y'(x_0 + jh) = y'_j$$

and

$$y''(x_0 + jh) = y''_j$$

for $j = 0, 1, 2, 3, \ldots$.

In this notation, $y_0 = A$ and $y'_0 = B$ are given numbers.

METHOD 1: THE TAYLOR APPROXIMATION

$$y(x+h) = y(x) + hy'(x) + \frac{h^2}{2}y''(x) + \frac{h^3}{6}y^{(3)}(x) + \cdots$$

and

$$y'(x+h) = y'(x) + hy''(x) + \frac{h^2}{2}y^{(3)}(x) + \cdots.$$

Thus,

$$y(x+h) \approx y(x) + hy'(x) + \frac{h^2}{2}y''(x)$$

and

$$y'(x+h) \approx y'(x) + hy''(x).$$

In subscript notation, we set

$$y_{j+1} = y_j + hy'_j + \frac{h^2}{2}y''_j$$

and

$$y'_{j+1} = y'_j + hy''_j.$$

We also have, from the differential equation, that

$$y''_j = f(x_j, y_j, y'_j).$$

Now proceed by setting, in turn, $j = 0$, computing y_1, y'_1, and y''_1; then setting $j = 1$ and computing y_2, y'_2, and y''_2, and so on. The results of each step enable us to compute the next step, eventually obtaining $y_1, y_2, y_3, \ldots$.

Here is an illustration.

EXAMPLE 527

Solve

$$y'' = x^2 - \cos(y) + 2e^{-x}y'; \qquad y(1) = -1, \qquad y'(1) = 3.$$

Here, $x_0 = 1$, $y_0 = -1$, $y'_0 = 3$, and $f(x, y, y') = x^2 - \cos(y) + 2e^{-x}y'$. We shall choose $h = 0.1$.

Step 1 With $j = 0$, we compute

$$y''_0 = f(x_0, y_0, y'_0) = 2.666974341.$$

Then,

$$y_1 = y_0 + hy'_0 + \frac{h^2}{2}y''_0 = -0.686665128$$

and

$$y_1' = y_0' + hy_0'' = 3.266697434.$$

Step 2 To compute y_2, use the results from step 1 to compute

$$y_1'' = f(x_1, y_1, y_1') = 2.611404959.$$

Then,

$$y_2 = y_1 + hy_1' + \frac{h^2}{2} y_1'' = -0.346938360$$

and

$$y_2' = y_1' + hy_1'' = 3.527837930.$$

Step 3 Compute

$$y_2'' = f(x_2, y_2, y_2') = 2.624710592.$$

Then,

$$y_3 = y_2 + hy_2' + \frac{h^2}{2} y_2'' = 0.018968986$$

and

$$y_3' = y_2' + hy_2'' = 3.79038989.$$

Step 4 Compute

$$y_3'' = f(x_3, y_3, y_3') = 2.756139316.$$

Then,

$$y_4 = y_3 + hy_3' + \frac{h^2}{2} y_3'' = 0.411780582$$

and

$$y_4' = y_3' + hy_3'' = 4.065922921.$$

Continuing in this way, we can next compute $y_4'' = f(x_4, y_4, y_4')$; hence, y_5 and y_5', and so on.

Thus far, we have

$$\begin{aligned} y(1) &= y_0 = -1 \qquad \text{(given)} \\ y(1.1) &= y_1 = -0.68665128 \\ y(1.2) &= y_2 = -0.346938360 \\ y(1.3) &= y_3 = 0.018968986 \\ y(1.4) &= y_4 = 0.411780582 \end{aligned}$$

METHOD 2: THE RUNGE-KUTTA-NYSTROM METHOD

In this method, we compute

$$A_j = \tfrac{1}{2}hf(x_j, y_j, y'_j),$$
$$B_j = \tfrac{1}{2}hf[x_j + \tfrac{1}{2}h, y_j + \tfrac{1}{2}h(y'_j + \tfrac{1}{2}A_j), y'_j + A_j],$$
$$C_j = \tfrac{1}{2}hf[x_j + \tfrac{1}{2}h, y_j + \tfrac{1}{2}h(y'_j + \tfrac{1}{2}A_j), y'_j + B_j],$$

and

$$D_j = \tfrac{1}{2}hf[x_j + h, y_j + h(y'_j + C_j), y'_j + 2C_j].$$

Then compute

$$y_{j+1} = y_j + h[y'_j + \tfrac{1}{3}(A_j + B_j + C_j)]$$

and

$$y'_{j+1} = y'_j + \tfrac{1}{3}[A_j + 2B_j + 2C_j + D_j].$$

Of course y'_{j+1} is only needed in computing y_{j+2}; the values of $y(x)$ are the main object of interest.

EXAMPLE 528

Solve

$$y'' = x - 4y^2 + 3e^{-y'}; \qquad y(2) = 3, \qquad y'(2) = -4.$$

Here, $x_0 = 2$, $y_0 = 3$, and $y'_0 = -4$. Choose $h = 0.1$, and compute the following values:

$x_0 = 1$	$y_0 = 3$	$y'_0 = -4$	(given)
$x_1 = 1.1$	$y_1 = 4.113939077$	$y'_1 = 21.19629843$	
$x_2 = 1.2$	$y_2 = 5.780895533$	$y'_2 = 11.09076746$	
$x_3 = 1.3$	$y_3 = 6.165990535$	$y'_3 = -0.932528890$	
$x_4 = 1.4$	$y_4 = 21.68109918$	$y'_4 = 150{,}138.3761$	

For comparison, the following example recomputes the values of y_1, y'_1, y_2, y'_2, y_3, y'_3, y_4, and y'_4 obtained in Example 527 by the Taylor series method.

EXAMPLE 529

Consider

$$y'' = x^2 - \cos(y) + 2e^{-x}y'; \qquad y(1) = -1, \qquad y'(1) = 3,$$

as in Example 527. Again, take $h = 0.1$, and construct the following table:

j	x_j	Taylor method y_j	Taylor method y'_j	Runge-Kutta-Nystrom method y_j	Runge-Kutta-Nystrom method y'_j
0	1.0	−1	3	−1	3 (given)
1	1.1	−0.686665128	3.266697434	−0.686780459	3.263427610
2	1.2	−0.346938360	3.527837930	−0.347408735	3.524225960
3	1.3	0.018968986	3.790308989	0.018291072	3.792004674
4	1.4	0.411780582	4.065922921	0.411699714	4.081153192

PROBLEMS FOR SECTION 20.7

In each of Problems 1 through 15, use $h = 0.1$ and solve for $y_1, y_2, \ldots, y_6$ using both methods of this section.

1. $y'' = x - 2y;\quad y(0) = 0,\quad y'(0) = 1$
2. $y'' = x^2 + 4y;\quad y(0) = 1,\quad y'(0) = -1$
3. $y'' = x - 8y + y';\quad y(0) = -1,\quad y'(0) = 4$
4. $y'' = 1 - \sin(x) + 2y + y';\quad y(1) = 2,\quad y'(1) = -3$
5. $y'' = \cos(x) - 3y + y';\quad y(2) = -1,\quad y'(2) = 0$
6. $y'' = x^2 - 4y + 2y';\quad y(3) = -\sqrt{2},\quad y'(3) = 0$
7. $y'' = 2 - x^2 + e^{-y} + y';\quad y(0) = y'(0) = 0$
8. $y'' = 1 + e^{-x} + y^2 - y';\quad y(-1) = 2,\quad y'(-1) = 3$
9. $y'' = 4x^3 - 6y + 2y';\quad y(1) = 0,\quad y'(1) = -4$
10. $y'' = 8x^2 - \cosh(y) + 3y';\quad y(2) = 1 = y'(2)$
11. $y'' = 4x - \sin(y');\quad y(\pi) = 0,\quad y'(\pi) = 0.1$
12. $y'' = 8x - 4y' + y^2;\quad y(0) = 1,\quad y'(0) = -2$
13. $y'' = 14x^2 - \sin(y) + y';\quad y(0) = -2,\quad y'(0) = 4$
14. $y'' = 2y - 4y';\quad y(0) = y'(0) = 1$
15. $y'' = 3x - 4y^2 + \cosh(y');\quad y(2) = 1,\quad y'(2) = 0$

20.8 NUMERICAL SOLUTION OF SECOND ORDER BOUNDARY VALUE PROBLEMS

In this section we give the *finite-difference method* for solving second order boundary value problems of the form

$$y'' = f(x, y, y');\qquad y(a) = \alpha,\qquad y(b) = \beta,$$

where $f(x, y, y')$, a, α, b, and β are given.

As with methods for treating initial value problems, we set

$$h = \frac{b-a}{n},\qquad x_j = a + jh,\quad \text{and}\quad y_j = y(x_j).$$

Note that $y(x_j) = y(a + jh) = y[a + (j-1)h + h] = y(x_{j-1} + h)$.

FINITE-DIFFERENCE METHOD

In the differential equation, replace x by x_j, y by y_j,

$$y' \text{ by } \frac{1}{2h}(y_{j+1} - y_{j-1}),$$

and

$$y'' \text{ by } \frac{1}{h^2}(y_{j+1} - 2y_j + y_{j-1}).$$

Write the resulting equation for $j = 1, 2, \ldots, n - 1$ and substitute in the boundary conditions $y(a) = \alpha$ and $y(b) = \beta$ where appropriate to obtain $(n - 1)$ equations in the $(n - 1)$ unknowns $y_1, y_2, \ldots, y_{n-1}$. (Note that $y_0 = \alpha$ and $y_n = \beta$ are known.)

EXAMPLE 530

Solve the problem

$$y'' = x - 2y + y'; \qquad y(0) = 1, \qquad y(1) = -3.$$

We shall let $n = 10$ and choose $h = \frac{1}{10} = 0.1000$, to four decimals. Replace x by x_j, y by y_j, y' by $(1/2h)(y_{j+1} - y_{j-1})$, and y'' by $(1/h^2)(y_{j+1} - 2y_j + y_{j-1})$. We get

$$\frac{1}{h^2}(y_{j+1} - 2y_j + y_{j-1}) = x_j - 2y_j + \frac{1}{2h}(y_{j+1} - y_{j-1}).$$

This can be written

$$\left(1 - \frac{h}{2}\right)y_{j+1} + (-2 + 2h^2)y_j + \left(1 + \frac{h}{2}\right)y_{j-1} = h^2x_j.$$

Since $h = 0.1000$, this is

$$0.9500y_{j+1} - 1.9800y_j + 1.0500y_{j-1} = 0.0100x_j. \tag{a}$$

Note that each x_j is known, since $x_j = a + jh = 0.1j$, as j varies from 1 to 9, inclusive. Substituting in turn $j = 1, 2, 3, \ldots, 9$ into (a) we get nine equations in the nine unknowns $y_1, \ldots, y_9$. These equations are:

$$\begin{aligned}
0.9500y_2 - 1.9800y_1 &= -1.0490 \\
0.9500y_3 - 1.9800y_2 + 1.0500y_1 &= 0.002 \\
0.9500y_4 - 1.9800y_3 + 1.0500y_2 &= 0.003 \\
0.9500y_5 - 1.9800y_4 + 1.0500y_3 &= 0.004 \\
0.9500y_6 - 1.9800y_5 + 1.0500y_4 &= 0.005 \\
0.9500y_7 - 1.9800y_6 + 1.0500y_5 &= 0.006 \\
0.9500y_8 - 1.9800y_7 + 1.0500y_6 &= 0.007 \\
0.9500y_9 - 1.9800y_8 + 1.0500y_7 &= 0.008 \\
-1.9800y_9 + 1.0500y_8 &= 2.8590
\end{aligned}$$

The first and last equations contain, respectively, the information that $y_0 = 1$ and $y_{10} = -3$. Solve these equations to get

$$\begin{array}{lll} y_1 = 0.7299 & y_4 = -0.3231 & y_7 = -1.6371 \\ y_2 = 0.4171 & y_5 = -0.7406 & y_8 = -2.0991 \\ y_3 = 0.0646 & y_6 = -1.1813 & y_9 = -2.5571 \end{array}$$

In this example, the boundary value problem can be solved exactly by the methods of Chapter 2. The solution is

$$y = e^{x/2}\left[\frac{3}{4}\cos\left(\frac{\sqrt{7}x}{2}\right) - 2.5361\sin\left(\frac{\sqrt{7}x}{2}\right)\right] + \frac{x}{2} + \frac{1}{4}.$$

(In this expression, the number -2.5361 is actually the quantity

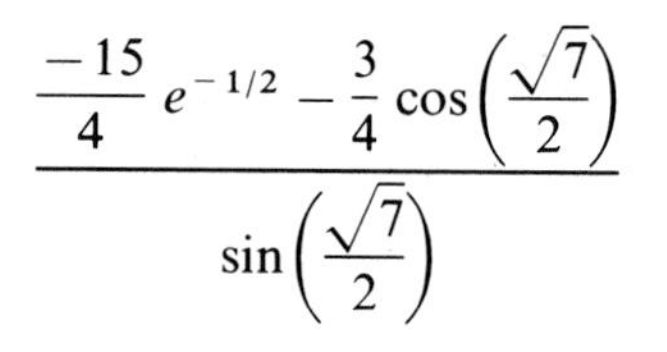

$$\frac{\dfrac{-15}{4}e^{-1/2} - \dfrac{3}{4}\cos\left(\dfrac{\sqrt{7}}{2}\right)}{\sin\left(\dfrac{\sqrt{7}}{2}\right)}$$

rounded off to four decimal places.) Using this solution, we can compute the exact values of $y_1, \ldots, y_9$ (to four decimal places) and compare with the approximate values we got from the finite-difference method. These rounded exact values are listed below:

$$\begin{array}{lll} y_1 = 0.7299 & y_4 = -0.3229 & y_7 = -1.6367 \\ y_2 = 0.4171 & y_5 = -0.7403 & y_8 = -2.0987 \\ y_3 = 0.0647 & y_6 = -1.1809 & y_9 = -2.5568. \end{array}$$

A comparison shows that the finite-difference results are quite close to the rounded exact results. For even better accuracy, one can choose h smaller.

EXAMPLE 531

Solve

$$y'' = x^2 - 4y + 4y'; \qquad y(1) = 2, \qquad y(2) = 4.$$

Just for illustration, choose $n = 10$, and let $h = (2 - 1)/10 = 0.1000$. As before, we shall carry out calculations to four decimal places.

Replace x by x_j, y by y_j, y' by $(1/2h)(y_{j+1} - y_{j-1})$ and y'' by $(1/h^2)(y_{j+1} - 2y_j + y_{j-1})$ as before. Since $h = 0.1$, we obtain

$$0.8y_{j+1} - 1.96y_j + 1.2y_{j-1} = 0.01x_j^{\,2}.$$

Substituting in turn $j = 1, 2, \ldots, 9$ into this equation gives us the system:

$$\begin{aligned}
0.8y_2 - 1.96y_1 &= 2.4121 \\
0.8y_3 - 1.96y_2 + 1.2y_1 &= 0.0144 \\
0.8y_4 - 1.96y_3 + 1.2y_2 &= 0.0169 \\
0.8y_5 - 1.96y_4 + 1.2y_3 &= 0.0196 \\
0.8y_6 - 1.96y_5 + 1.2y_4 &= 0.0225 \\
0.8y_7 - 1.96y_6 + 1.2y_5 &= 0.0256 \\
0.8y_8 - 1.96y_7 + 1.2y_6 &= 0.0289 \\
0.8y_9 - 1.96y_8 + 1.2y_7 &= 0.0324 \\
-1.96y_9 + 1.2y_8 &= -3.1639
\end{aligned}$$

These can be solved to yield approximate values of $y_1, \ldots, y_9$. In fact, as in the previous example, this boundary value problem can be solved exactly. With the coefficients rounded off to four decimal places, the exact solution is

$$y = -0.8756e^{2x} + 0.4527xe^{2x} + \tfrac{1}{4}x^2 + \tfrac{1}{2}x + \tfrac{3}{8}.$$

The table below gives the approximate values of $y_1, \ldots, y_9$ from the finite-difference method, together with the exact values (to four decimal places). As before, we can improve the approximation by choosing h smaller.

	Finite-difference value	Exact value
y_1	−2.2091	−2.1806
y_2	−2.3972	−2.3287
y_3	−2.5414	−2.4178
y_4	−2.6096	−2.4116
y_5	−2.5569	−2.2603
y_6	−2.1968	−1.8963
y_7	−1.5149	−1.2290
y_8	−0.3801	−0.1380
y_9	1.3815	1.5360

PROBLEMS FOR SECTION 20.8

In each of Problems 1 through 10, approximate a solution of the boundary value problem, using the value of n given with the problem. Sketch the resulting solution.

1. $y'' = x + 2y - 4y'$; $y(1) = 0$, $y(2) = -3$; $n = 4$
2. $y'' = 2x - 3y'$; $y(0) = -1$, $y(1) = 4$; $n = 4$
3. $y'' - 4y' = x - y$; $y(1) = 0$, $y(3) = -2$; $n = 6$
4. $y'' - x^2 = -2y$; $y(1) = y(2) = 3$; $n = 6$
5. $y'' = \sin(x) - 4y' + 2y$; $y(0) = 1$, $y(\pi) = 2$; $n = 6$
6. $y'' = 2x^2 - 4y + 2y'$; $y(-1) = 3$, $y(1) = \sqrt{2}$; $n = 6$
7. $y'' = 2\sqrt{x} - 3y'$; $y(1) = 4$, $y(2) = 6$; $n = 8$

8. $y'' = -3e^{-x} + x^2 - 2y + 7y'$; $y(1) = -1$, $y(2) = -3$; $n = 8$
9. $y'' = \cosh(2x) + 4y - 2y'$; $y(1) = 0$, $y(2) = -5$; $n = 10$
10. $y'' = 7x^3 - 4y' - 6y$; $y(2) = 4$, $y(4) = -7$; $n = 12$
11. Use the definition of y'' to give a rationale for replacing y'' by

$$\frac{y_{j+1} - 2y_j + y_{j-1}}{h^2}.$$

20.9 THE FINITE-DIFFERENCE METHOD FOR THE DIRICHLET PROBLEM

In two dimensions, the Dirichlet problem is to find a function $u(x, y)$ satisfying Laplace's equation

$$\frac{\partial^2 u}{\partial x^2} + \frac{\partial^2 u}{\partial y^2} = 0$$

for (x, y) interior to some given region R of the plane, subject to the condition that $u(x, y) = f(x, y)$, a given function, for (x, y) on the boundary of R.

Often, such a problem is approached through Fourier methods (series, integrals, or transforms, as shown in Sections 13.3 and 13.5) or through conformal mappings (see Sections 19.1 and 19.2). If the region R is fairly complicated, however, it may be difficult or impossible to obtain a closed-form solution which can be evaluated numerically at specific points by means of these methods. Thus, it is often useful in engineering applications to have some scheme for obtaining a numerical approximation of the solution.

The finite-difference method begins by placing a grid over R, say by vertical lines $x = jh$ and horizontal lines $y = jh$, with h some chosen positive number, as shown in Figure 377. We call h the *mesh size* of the grid, and j varies over integer values until R is covered. The points of intersection of these lines are called *grid points*. In practice, we are interested only in the grid points which fall inside R or on the boundary of R.

Figure 378 shows two typical grid points, P and Q. Note that P has four grid points adjacent to it, namely, E, N, W, and S, all lying inside R. In this case we call P an *interior grid point*. By contrast, not all the grid points neighboring Q are inside R: A and B are outside R, while D and C are inside. We call Q a *boundary grid point*. In general, a boundary grid point is a grid point inside or on the boundary of R which has an adjacent grid point outside R. Note that a boundary grid point need not itself lie on the boundary of R.

The basic strategy of the method is to use the boundary data of the problem [namely, $u(x, y) = f(x, y)$, given for (x, y) on the boundary of R] to assign values to $u(x, y)$ at the boundary grid points, and then use this information, together with relationships we shall establish between $u(x, y)$ at an interior grid point and its neighbors, to obtain approximate values of $u(x, y)$ at all points of the grid inside R or on the boundary of R.

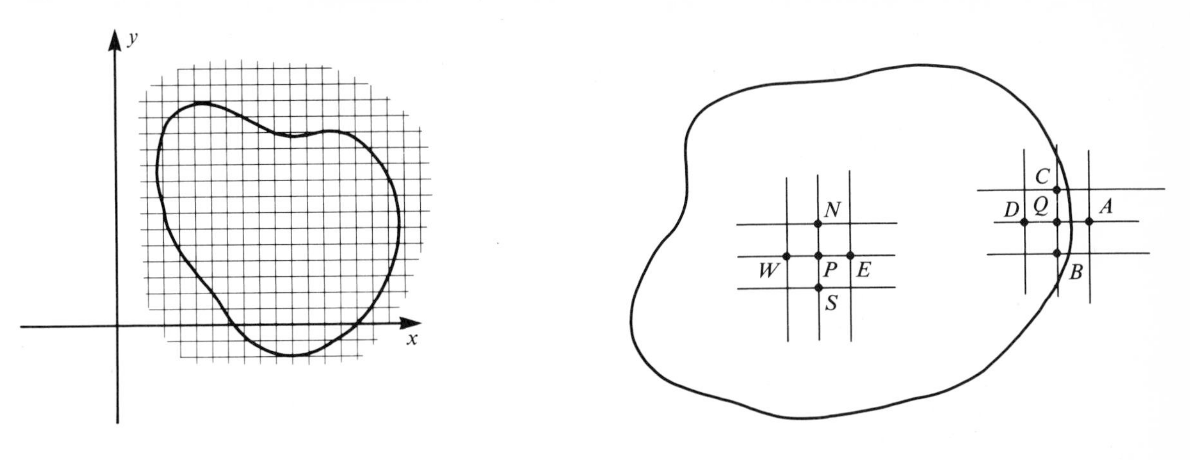

Figure 377. Grid placed over a region R.

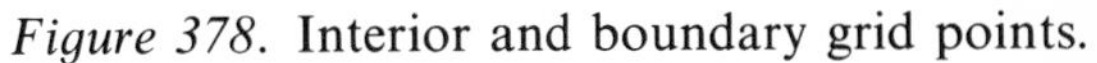

Figure 378. Interior and boundary grid points.

To carry out this scheme, we must approximate the partial derivatives of $u(x, y)$ by difference-quotients. We can approximate $\partial u/\partial x$ by

$$\frac{\partial u}{\partial x} \approx \frac{1}{2h}\left[u(x+h, y) - u(x-h, y)\right].$$

Then,

$$\frac{\partial^2 u}{\partial x^2} \approx \frac{1}{h^2}\left[u(x+h, y) - 2u(x, y) + u(x-h, y)\right].$$

Similarly,

$$\frac{\partial^2 u}{\partial y^2} \approx \frac{1}{h^2}\left[u(x, y+h) - 2u(x, y) + u(x, y-h)\right].$$

Then, Laplace's equation is approximated by

$$\frac{1}{h^2}\left[u(x+h, y) + u(x-h, y) + u(x, y+h) + u(x, y-h) - 4u(x, y)\right] = 0.$$

We can of course multiply out the factor of $1/h^2$ in this. Referring again to Figure 378, we have, at an interior grid point P,

$$u(E) + u(W) + u(N) + u(S) - 4u(P) = 0. \tag{a}$$

At a boundary grid point, such as Q in Figure 378, we assign $u(Q)$ the value of $f(x, y)$ at the point (x, y) nearest to Q and lying on the boundary of R. This gives us an equation of the form

$$u(Q) = f(x, y) \tag{b}$$

for each boundary grid point Q.

If the total number of interior grid points is N, then there will be N equations of the form (a). In some of these equations will occur terms involving values of $u(x, y)$ at boundary grid points. Thus, equations (b) will be substituted into some of the equations (a). The final result is a system of N linear equations in the N unknowns $u(P_1), \ldots, u(P_N)$, where $P_1, \ldots, P_N$ are the interior grid points. Solution of this system then yields values of the solution at the interior points, and hence gives us approximate values of $u(x, y)$ at all the grid points.

In general, this method does not admit "small" examples. Unless the grid mesh size is quite small relative to the diameter of the region R, the approximation can be expected to be very crude. Thus, one often encounters systems of equations with N on the order of 10^3, 10^4, or higher.

Note that, in the system of equations, each equation of the form (a) has at most five nonzero coefficients. Thus, the $N \times N$ matrix of the system is largely populated by zero entries. Such sparse matrices have for this reason become very important in modern numerical analysis.

In order to get some feeling for the method, here is an example.

EXAMPLE 532

Solve

$$\frac{\partial^2 u}{\partial x^2} + \frac{\partial^2 u}{\partial y^2} = 0 \quad \text{for } (x, y) \text{ in } R,$$

$$u(x, y) = x + y \quad \text{for } (x, y) \text{ on the boundary of } R,$$

with R the interior of the unit square $0 \le x \le 1$, $0 \le y \le 1$. Thus, the boundary of R consists of the sides of the square.

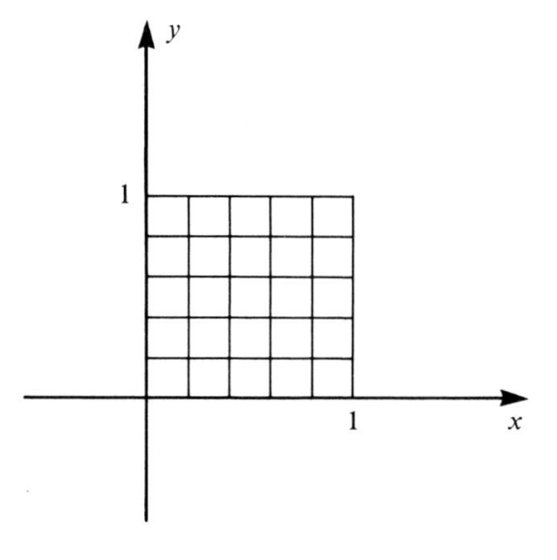

Figure 379

We shall choose $h = 0.2$, forming the grid shown in Figure 379. There are thirty-six grid points, of which sixteen are interior points and twenty are boundary grid points (which in this case actually lie on the boundary).

Since $u(x, y) = x + y$ on the boundary of R, we can assign values to $u(x, y)$ at the points on the boundary. These values are shown in Figure 380.

For convenience in writing the equations, label the sixteen interior grid points $P_1, \ldots, P_{16}$, as shown in Figure 381. Starting with P_1 and continuing on through P_{16}, equations (a) take the form:

$$u(P_2) + 0.2 + u(P_5) + 0.2 - 4u(P_1) = 0$$
$$u(P_3) + u(P_1) + u(P_6) + 0.4 - 4u(P_2) = 0$$
$$u(P_4) + u(P_2) + u(P_7) + 0.6 - 4u(P_3) = 0$$
$$0.2 + u(P_3) + u(P_8) + 0.8 - 4u(P_4) = 0$$
$$u(P_6) + 0.4 + u(P_9) + u(P_1) - 4u(P_5) = 0$$
$$u(P_7) + u(P_5) + u(P_{10}) + u(P_2) - 4u(P_6) = 0$$
$$u(P_8) + u(P_6) + u(P_{11}) + u(P_3) - 4u(P_7) = 0$$

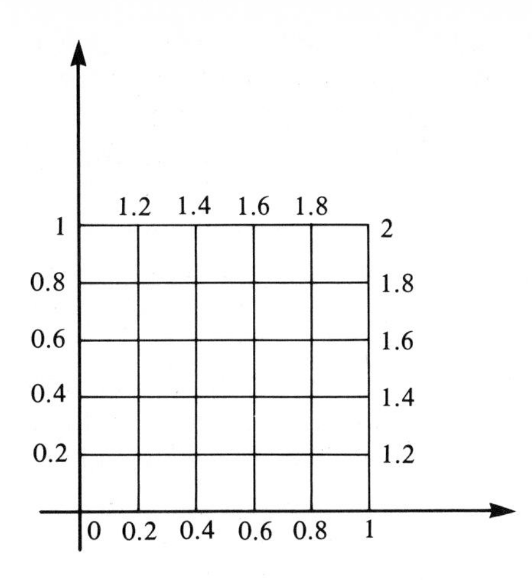

Figure 380. Values of $u(x, y)$ at boundary grid points.

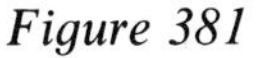

Figure 381

$$0.4 + u(P_7) + u(P_{12}) + u(P_4) - 4u(P_8) = 0$$
$$u(P_{10}) + 0.6 + u(P_{13}) + u(P_5) - 4u(P_9) = 0$$
$$u(P_{11}) + u(P_9) + u(P_{14}) + u(P_6) - 4u(P_{10}) = 0$$
$$u(P_{12}) + u(P_{10}) + u(P_{15}) + u(P_7) - 4u(P_{11}) = 0$$
$$0.6 + u(P_{11}) + u(P_{16}) + u(P_8) - 4u(P_{12}) = 0$$
$$u(P_{14}) + 0.8 + 1.2 + u(P_9) - 4u(P_{13}) = 0$$
$$u(P_{15}) + u(P_{13}) + 1.4 + u(P_{10}) - 4u(P_{14}) = 0$$
$$u(P_{16}) + u(P_{14}) + 1.6 + u(P_{11}) - 4u(P_{15}) = 0$$
$$1.8 + u(P_{15}) + 1.8 + u(P_{12}) - 4u(P_{16}) = 0$$

These sixteen equations in sixteen unknowns can be solved to yield:

$u(P_1) = u(0.2, 0.2) = 0.002241127$

$u(P_2) = u(0.4, 0.2) = 0.012017767$

$u(P_3) = u(0.6, 0.2) = 0.284586220$

$u(P_4) = u(0.8, 0.2) = 0.380737935$

$u(P_5) = u(0.2, 0.4) = -0.403052848$

$u(P_6) = u(0.4, 0.4) = -0.638756381$

$u(P_7) = u(0.6, 0.4) = 0.152669559$

$u(P_8) = u(0.8, 0.4) = 0.415621647$

$u(P_9) = u(0.2, 0.6) = 0.236515151$

$u(P_{10}) = u(0.4, 0.6) = 0.238365523$

$u(P_{11}) = u(0.6, 0.6) = 0.549226750$

$u(P_{12}) = u(0.8, 0.6) = 0.729079094$

$u(P_{13}) = u(0.2, 0.8) = 0.510747930$

$u(P_{14}) = u(0.4, 0.8) = 0.806476569$

$u(P_{15}) = u(0.6, 0.8) = 1.076792825$

$u(P_{16}) = u(0.8, 0.8) = 1.351467980$

Thus, we have approximate values of $u(x, y)$ at thirty-six points in and on the boundary of R. By choosing h smaller, we can increase accuracy and increase the number of points at which we have approximate solutions, but this means solving a larger system of equations.

PROBLEMS FOR SECTION 20.9

In each of Problems 1 through 10, use the finite-difference method with $h = 0.2$ to find an approximate solution of the Dirichlet problem for the given region.

1. R is given by $0 < x < 0.6, 0 < y < 0.6$; $u(x, y) = x - y$ on the boundary of R.
2. R is given by $0 < x < 0.6, 0 < y < 1$; $u(x, y) = 2x$ on the boundary of R.
3. R is given by $0 < x < 1, 0 < y < 0.6$; $u(x, y) = xy$ on the boundary of R.
4. R is given by $0 < x < 0.4, 0 < y < 2$; $u(x, y) = x^2 + 2y$ on the boundary of R.
5. R is given by $1.2 < x < 2, 0 < y < 0.4$; $u(x, y) = x^2 - y$ on the boundary of R.
6. R is given by $x^2 + y^2 < 0.36$; $u(x, y) = 2xy$ on the boundary of R.
7. R is given by $1.4 < x < 2, 2 < y < 3$; $u(x, y) = x^2 + 4xy$ on the boundary of R.
8. R is given by $-1 < x < -0.5, 1 < y < 2$; $u(x, y) = x - y^2$ on the boundary of R.
9. R is given by $1 < x < 2, -1 < y < -0.2$; $u(x, y) = -x + 6y^2$ on the boundary of R.
10. R is given by $-3 < x < -2.4, -1.2 < y < -1$; $u(x, y) = x\cos(y)$ on the boundary of R.
11. Using the definition of partial derivative, justify the approximation

$$\frac{\partial^2 u}{\partial x^2} \approx \frac{1}{h^2}[u(x + h, y) - 2u(x, y) + u(x - h, y)].$$

20.10 APPROXIMATION OF EIGENVALUES AND EIGENVECTORS

In attempting to approximate the eigenvalues of a matrix, there are basically two approaches one can take:

1. There are some results which locate the eigenvalues in certain regions of the plane.
2. Some results give numerical approximations of one or more eigenvalues.

As an example of (1), we state the following theorem.

THEOREM 145 Gerschgorin's Theorem

Let A be an $n \times n$ matrix having complex (possibly real) entries. Let

$$r_k = \sum_{\substack{j=1 \\ j \neq k}}^{n} |A_{kj}| \quad \text{for} \quad k = 1, 2, \ldots, n.$$

Let R be the region of the plane covered by the n closed disks of radius r_k about A_{kk}, $k = 1, \ldots, n$. Then the eigenvalues of A, when plotted as points in the complex plane, all lie in R.

In words, we form a disk corresponding to each row of A. The radius of the disk of row k is r_k, the sum of the absolute values of the entries of row k, omitting the diagonal entry. The center of this disk is the diagonal entry of row k. The collection of all these disks then covers all the eigenvalues of A, in the sense that each eigenvalue lies in at least one disk.

EXAMPLE 533

Let

$$A = \begin{bmatrix} 1 & -4 & 6 \\ 2 & 0 & 7 \\ 1 & -1 & 3 \end{bmatrix}.$$

From row 1, we get the disk D_1 of radius $|-4| + 6 = 10$, center 1; from row 2, the disk D_2 of radius 9 about 0; and from row 3, the disk D_3 of radius 2 about 3. These are shown in Figure 382.

The eigenvalues of A are found to be approximately

$$\lambda_1 = -0.6080, \qquad \lambda_2 = 2.3040 + 3.0811i,$$

and $\qquad \lambda_3 = 2.3040 - 3.0811i.$

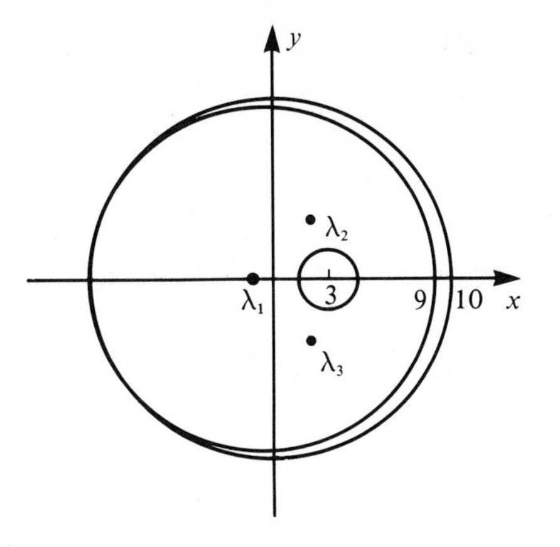

Figure 382. Location of eigenvalues by Gerschgorin's theorem.

These are plotted in Figure 382 and are seen to fall within the region covered by the disks.

The second kind of result consists of a scheme for actually estimating one or more of the eigenvalues. Here is one such scheme, called the *power method.*

POWER METHOD FOR FINDING A LARGEST (IN MAGNITUDE) EIGENVALUE

Step 1 Choose an initial vector (actually an $n \times 1$ matrix) v_0. Often, v_0 is just a guess.

Step 2 Compute $v_{n+1} = Av_n$ for $n = 0, 1, 2, \ldots$.

Step 3 For each v_n computed, calculate the *Rayleigh quotient*

$$R_{n+1} = \frac{v_n{}^t v_{n+1}}{v_n{}^t v_n}.$$

(Recall that u^t = transpose of u.)

Here, we are using matrix notation which can be unravelled as follows. The vector v_n is really an $n \times 1$ column vector, say,

$$v_n = \begin{bmatrix} v_{n1} \\ v_{n2} \\ \vdots \\ v_{nn} \end{bmatrix}.$$

Then,

$$v_n{}^t = [v_{n1} \quad v_{n2} \quad \cdots \quad v_{nn}].$$

Then,

$$v_n{}^t v_{n+1} = [v_{n1} \quad v_{n2} \quad \cdots \quad v_{nn}] \begin{bmatrix} v_{n+1,1} \\ v_{n+1,2} \\ \vdots \\ v_{n+1,n} \end{bmatrix}$$

$$= v_{n1}v_{n+1,1} + v_{n2}v_{n+1,2} + \cdots + v_{nn}v_{n+1,n},$$

the dot product of v_n and v_{n+1}.

Further,

$$v_n{}^t v_n = [v_{n1} \quad v_{n2} \quad \cdots \quad v_{nn}] \begin{bmatrix} v_{n1} \\ v_{n2} \\ \vdots \\ v_{nn} \end{bmatrix}$$

$$= v_{n1}{}^2 + v_{n2}{}^2 + \cdots + v_{nn}{}^2,$$

the square of the magnitude of v_n.

Thus,

$$R_{n+1} = \frac{v_{n1}v_{n+1,1} + v_{n2}v_{n+1,2} + \cdots + v_{nn}v_{n+1,n}}{v_{n1}{}^2 + v_{n2}{}^2 + \cdots + v_{nn}{}^2}.$$

Step 4 It can be shown that $\lim_{n\to\infty} R_n = \lambda_1$, the eigenvalue of A having largest magnitude (λ_1 is called the *dominant eigenvalue** of A). We compute R_n for enough values of n to get agreement between successive values for as many decimal places as we want, and then use this value of R_n as an approximation of λ_1. Thus, we proceed as follows.

Guess v_0, and compute $v_1 = Av_0$ and $R_1 = v_0{}^t v_1 / v_0{}^t v_0$. Then compute

$$v_2 = Av_1, \qquad R_2 = \frac{v_1{}^t v_2}{v_1{}^t v_1},$$

$$v_3 = Av_2, \qquad R_3 = \frac{v_2{}^t v_3}{v_2{}^t v_2},$$

$$\vdots \qquad\qquad \vdots$$

The method requires that A have a unique, real eigenvalue of largest magnitude; otherwise, the method fails.

EXAMPLE 534

Let

$$A = \begin{bmatrix} -1 & 6 \\ 0 & 4 \end{bmatrix}.$$

* Sometimes the magnitude of the dominant eigenvalue of A is also called the *spectral radius* of A.

The eigenvalues are easily found to be -1 and 4. To test the power method, choose some v_0 to initialize the process. Say,

$$v_0 = \begin{bmatrix} 1 \\ 1 \end{bmatrix}.$$

Then,

$$v_1 = \begin{bmatrix} -1 & 6 \\ 0 & 4 \end{bmatrix}\begin{bmatrix} 1 \\ 1 \end{bmatrix} = \begin{bmatrix} 5 \\ 4 \end{bmatrix}$$

and

$$R_1 = \frac{5+4}{1+1} = \frac{9}{2} = 4.5.$$

Next,

$$v_2 = \begin{bmatrix} -1 & 6 \\ 0 & 4 \end{bmatrix}\begin{bmatrix} 5 \\ 4 \end{bmatrix} = \begin{bmatrix} 19 \\ 16 \end{bmatrix}$$

and

$$R_2 = \frac{5(19) + 4(16)}{25 + 16} = 3.870.$$

Next,

$$v_3 = \begin{bmatrix} -1 & 6 \\ 0 & 4 \end{bmatrix}\begin{bmatrix} 19 \\ 16 \end{bmatrix} = \begin{bmatrix} 77 \\ 64 \end{bmatrix}$$

and

$$R_3 = \frac{(19)(77) + (16)(64)}{19^2 + 16^2} = 4.0308.$$

Next,

$$v_4 = \begin{bmatrix} -1 & 6 \\ 0 & 4 \end{bmatrix}\begin{bmatrix} 77 \\ 64 \end{bmatrix} = \begin{bmatrix} 307 \\ 256 \end{bmatrix}$$

and

$$R_4 = \frac{(77)(307) + (64)(256)}{77^2 + 64^2} = 3.9923.$$

Next,

$$v_5 = \begin{bmatrix} -1 & 6 \\ 0 & 4 \end{bmatrix}\begin{bmatrix} 307 \\ 256 \end{bmatrix} = \begin{bmatrix} 1229 \\ 1024 \end{bmatrix}$$

and

$$R_5 = \frac{(307)(1229) + (256)(1024)}{307^2 + 256^2} = 4.0019.$$

Next,

$$v_6 = \begin{bmatrix} -1 & 6 \\ 0 & 4 \end{bmatrix}\begin{bmatrix} 1229 \\ 1024 \end{bmatrix} = \begin{bmatrix} 4915 \\ 4096 \end{bmatrix}$$

and

$$R_6 = \frac{(1229)(4915) + (1024)(4096)}{1229^2 + 1024^2} = 3.9995.$$

The R_n's are converging toward 4, the eigenvalue of largest magnitude. More iterations can produce values of R_n still closer to 4.

One can compute

$$Av_6 = \begin{bmatrix} -1 & 6 \\ 0 & 4 \end{bmatrix}\begin{bmatrix} 4915 \\ 4096 \end{bmatrix} = \begin{bmatrix} 19{,}661 \\ 16{,}384 \end{bmatrix}$$
$$= (3.9995)\begin{bmatrix} 4915.8645 \\ 4096.5121 \end{bmatrix}.$$

Thus, v_6 is an approximate eigenvector associated with the approximate eigenvalue 3.9995. Actually, if we guess 4 as an eigenvalue, then

$$Av_6 = \begin{bmatrix} 19{,}661 \\ 16{,}384 \end{bmatrix} = 4\begin{bmatrix} 4915.25 \\ 4096 \end{bmatrix};$$

so $\begin{bmatrix} 4915 \\ 4096 \end{bmatrix}$ is a fair approximation of an eigenvector associated with eigenvalue 4.

PROBLEMS FOR SECTION 20.10

In each of Problems 1 through 10, use ten iterations to approximate the eigenvalue of largest magnitude and an associated eigenvector.

1. $\begin{bmatrix} -1 & 1 \\ 6 & 2 \end{bmatrix}$ **2.** $\begin{bmatrix} 6 & -1 \\ 3 & 4 \end{bmatrix}$ **3.** $\begin{bmatrix} 4 & 1 \\ -2 & 1 \end{bmatrix}$ **4.** $\begin{bmatrix} 1 & -7 \\ -3 & 1 \end{bmatrix}$

5. $\begin{bmatrix} -2 & 0 \\ 1 & 6 \end{bmatrix}$ **6.** $\begin{bmatrix} 1 & -2 & 1 \\ 4 & 0 & 3 \\ 2 & 1 & 7 \end{bmatrix}$ **7.** $\begin{bmatrix} 4 & -0.26 & 1 \\ -0.26 & 0 & 2 \\ 1 & 2 & -4 \end{bmatrix}$

8. $\begin{bmatrix} 1 & -4 & 1 \\ -4 & 0.2 & 3 \\ 1 & 3 & -7.4 \end{bmatrix}$ **9.** $\begin{bmatrix} -3.4 & 3.14 & 2.3 \\ 3.14 & 1 & 0 \\ 2.3 & 0 & 3.7 \end{bmatrix}$ **10.** $\begin{bmatrix} 2.6 & 1 & -7 \\ 1 & -4.3 & 0 \\ -7 & 0 & 4.2 \end{bmatrix}$

11. Prove Gerschgorin's theorem.

[*Hint*: Let λ_i be an eigenvalue of A and v_i an associated eigenvector. Use the equation $(A - \lambda_i I)v = 0$ to show that, if the jth component of v_i has largest magnitude of all the components of v_i, then λ_i is in the disk of radius r_j centered at A_{jj}.]

In each of Problems 12 through 20, (a) sketch the Gerschgorin disks for the given matrix, and (b) find the eigenvalues and verify explicitly that they lie in the region covered by disks. (*Note:* For problems 18, 19, and 20, you might use Newton-Raphson to find one real eigenvalue, and then find the other two eigenvalues by dividing out an appropriate factor of the characteristic polynomial and using the quadratic formula.)

12. $\begin{bmatrix} 1 & -1 \\ 6 & 4 \end{bmatrix}$ **13.** $\begin{bmatrix} 1.3 & -2.4 \\ 1 & -7.2 \end{bmatrix}$ **14.** $\begin{bmatrix} 0.16 & 2 \\ 1 & -4 \end{bmatrix}$

15. $\begin{bmatrix} -2.7 & 0 \\ 1 & 7.3 \end{bmatrix}$ **16.** $\begin{bmatrix} 1.7 & 0 \\ 1 & -6 \end{bmatrix}$ **17.** $\begin{bmatrix} -2.34 & 1.2 \\ 1.2 & 7.36 \end{bmatrix}$

18. $\begin{bmatrix} 1.73 & 0 & 1 \\ 2 & 1.6 & 0 \\ 1 & -1 & 4 \end{bmatrix}$ **19.** $\begin{bmatrix} 6.3 & -4.2 & -1 \\ 7 & 0 & 1.2 \\ -3 & -4.2 & 1.7 \end{bmatrix}$ **20.** $\begin{bmatrix} -4.7 & 0 & 1.2 \\ 3 & -4 & 1.7 \\ 3 & -4 & 2.6 \end{bmatrix}$

20.11 THE METHOD OF LEAST SQUARES

Suppose we are given n points

$$(x_1, y_1)\quad (x_2, y_2),\quad \ldots,\quad (x_n, y_n).$$

We seek to fit a straight line $y = a + bx$ to these points in the sense that the sum of the squares of the distances from the line to each point is a minimum. This is called a *least-squares fit*. [*Important note:* The distance from the line to a point is the distance measured in the vertical direction for this discussion of least squares (see Figure 383).]

To see how to choose the line $y = a + bx$, note from Figure 384 that the sum of the squares of the vertical distances from the points to the line is

$$S = \sum_{j=1}^{n} (y_j - a - bx_j)^2.$$

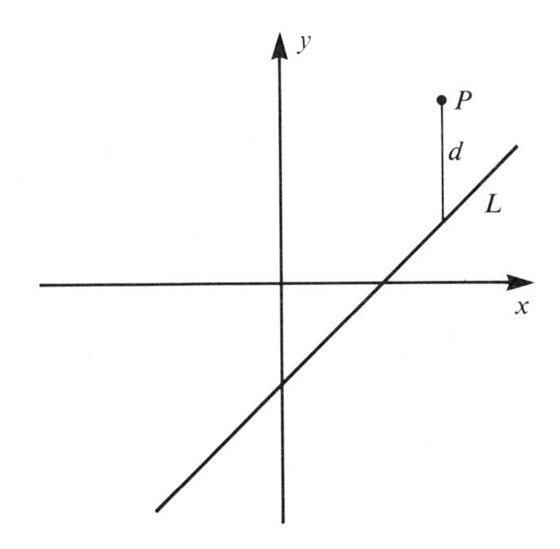

Figure 383. d is the vertical distance from point P to line L.

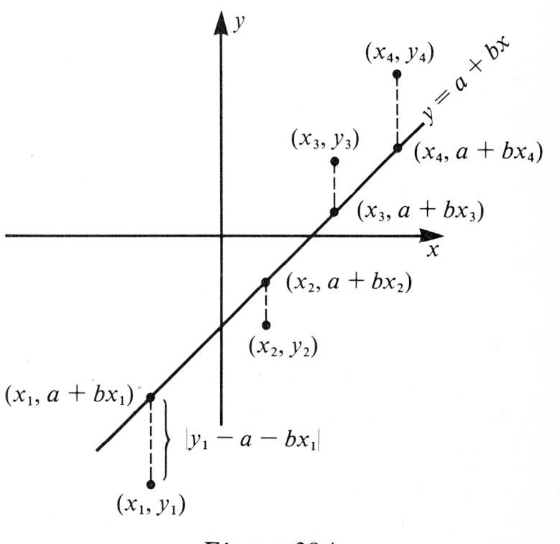

Figure 384

For a minimum, we need

$$\frac{\partial S}{\partial a} = -2\sum_{j=1}^{n}(y_j - a - bx_j) = 0$$

and

$$\frac{\partial S}{\partial b} = -2\sum_{j=1}^{n}x_j(y_j - a - bx_j) = 0.$$

Thus, we need

$$na + b\sum_{j=1}^{n}x_j = \sum_{j=1}^{n}y_j$$

and

$$a\sum_{j=1}^{n}x_j + b\sum_{j=1}^{n}x_j^2 = \sum_{j=1}^{n}x_j y_j.$$

Since we know the x_j's and y_j's, these are two equations for a and b.

EXAMPLE 535

Find a least squares, best-fitting straight line for the points

$$(1, 3), \quad (-2, 2), \quad (0, 4), \quad (3, 4), \quad \text{and} \quad (2, 7).$$

Here

$$\sum_{j=1}^{5}x_j = 1 - 2 + 0 + 3 + 2 = 4,$$

$$\sum_{j=1}^{5}x_j^2 = 1 + 4 + 0 + 9 + 4 = 18,$$

$$\sum_{j=1}^{5}y_j = 3 + 2 + 4 + 4 + 7 = 20,$$

and

$$\sum_{j=1}^{5}x_j y_j = 3 - 4 + 0 + 12 + 14 = 25.$$

Solve for a and b in

$$5a + 4b = 20,$$
$$4a + 18b = 25,$$

obtaining

$$a = \tfrac{260}{74}, \qquad b = \tfrac{45}{74}.$$

The line we seek is then

$$y = \tfrac{260}{74} + \tfrac{45}{74}x$$

or, approximately,

$$y = 3.5135 + 0.6081x.$$

The given points and this line are shown in Figure 385.

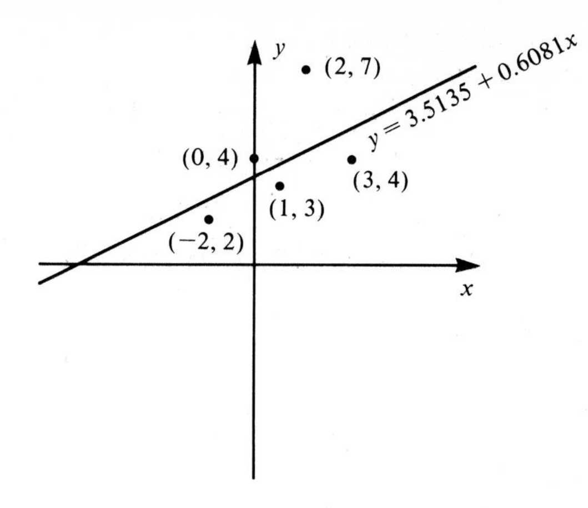

Figure 385. Least-squares straight-line fit to given points.

One can also have a least-squares fit with higher order polynomials. For example, suppose we want a parabola

$$y = a + bx + cx^2$$

such that the sum of the squares of the vertical distances to the points $(x_1, y_1), (x_2, y_2), \ldots, (x_n, y_n)$ is a minimum. One can check that a, b, and c can be obtained from the equations

$$na + b\sum_{j=1}^{n} x_j + c\sum_{j=1}^{n} x_j^2 = \sum_{j=1}^{n} y_j,$$

$$a\sum_{j=1}^{n} x_j + b\sum_{j=1}^{n} x_j^2 + c\sum_{j=1}^{n} x_j^3 = \sum_{j=1}^{n} x_j x_j,$$

and

$$a\sum_{j=1}^{n} x_j^2 + b\sum_{j=1}^{n} x_j^3 + c\sum_{j=1}^{n} x_j^4 = \sum_{j=1}^{n} x_j^2 y_j.$$

EXAMPLE 536

Find a parabola least-squares fit to

$$(-2, 4), \quad (-1, 3), \quad (2, 4), \quad (3, 1) \quad \text{and} \quad (4, 2).$$

Here,

$$\sum_{j=1}^{5} x_j = 6, \qquad \sum_{j=1}^{5} y_j = 14, \qquad \sum_{j=1}^{5} x_j^2 = 34, \qquad \sum_{j=1}^{5} x_j^3 = 90,$$

$$\sum_{j=1}^{5} x_j^4 = 370, \qquad \sum_{j=1}^{5} x_j y_j = 8, \quad \text{and} \quad \sum_{j=1}^{5} x_j^2 y_j = 76.$$

Now solve for a, b, and c in

$$\begin{aligned} 5a + 6b + 34c &= 14, \\ 6a + 34b + 90c &= 8, \\ 34a + 90b + 370c &= 76. \end{aligned}$$

We obtain

$$a = \frac{22{,}624}{6496}, \qquad b = \frac{-1384}{6496}, \quad \text{and} \quad c = \frac{-408}{6496}.$$

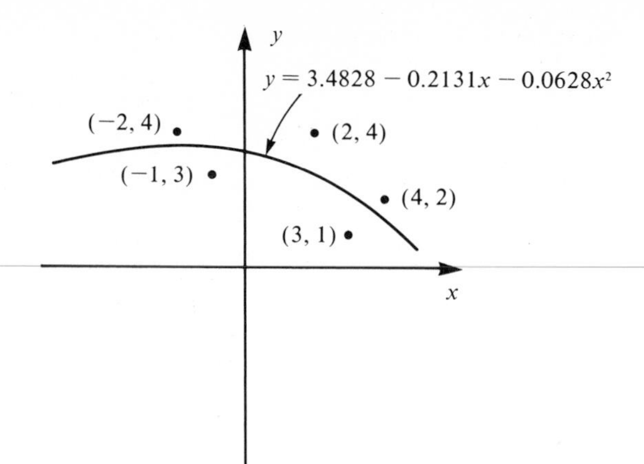

Figure 386. Parabola least-squares fit to given points.

The least-squares parabola for the given points is then

$$y = \frac{1}{6496}(22{,}624 - 1384x - 408x^2)$$

or, approximately,

$$y = 3.482758619 - 0.213054188x - 0.062807881x^2.$$

The points and the parabola are shown in Figure 386.

PROBLEMS FOR SECTION 20.11

In each of Problems 1 through 15, find a least-squares, straight-line fit to the given points. In each case, sketch the points and the straight line.

1. $(-1, 4)$, $(3, 0)$, $(2, 2)$

2. $(6, -2)$, $(1, 1)$, $(0, 3)$

3. $(1, -4)$, $(2, -5)$, $(3, -3)$

4. $(0, 1)$, $(1, 4)$, $(2, 2)$, $(-3, 4)$

5. $(1, 1)$, $(2, 1)$, $(3, 2)$, $(4, -2)$

6. $(0, 1)$, $(1, 5)$, $(-2, 3)$, $(-1, 5)$, $(3, 7)$

7. $(1, -4)$, $(2, -5)$, $(3, 3)$, $(4, 2)$, $(6, 1)$

8. $(-3, 1)$, $(-1, 2)$, $(-2, -2)$, $(4, 1)$

9. $(\sqrt{2}, 1)$, $(0, -2)$, $(1, \sqrt{3})$, $(2, 4)$, $(3, 5)$

10. $(-4, 1)$, $(-3, 3)$, $(2, -4)$, $(3, 5)$

11. $(6, 8)$, $(-5, 2)$, $(1, 7)$, $(3, -4)$, $(2, 9)$

12. $(-4, 2)$, $(\pi, \sqrt{3})$, $(1, 4)$, $(7, -2)$, $(3, 1)$

13. $(2, -4)$, $(3, 9)$, $(\pi/2, \sqrt{5})$, $(3, -3)$

14. $(1, 1)$, $(2, 2)$, $(3, 3)$, $(4, 7)$, $(5, -2)$

15. $(0, 0)$, $(1, -5)$, $(2, 4)$, $(3, 1)$, $(\sqrt{6}, 4)$

16. Derive the equations given in the section for finding a least-squares parabola fitting given points.

In each of Problems 17 through 25, find a parabola giving a least-squares fit for the given points. Sketch the points and the parabola.

17. $(6, -2)$, $(1, 4)$, $(0, 0)$

18. $(1, 1)$, $(-2, 3)$, $(4, 4)$, $(7, 6)$

19. $(0, -2)$, $(1, 4)$, $(6, 3)$, $(2, -5)$

20. $(1, 1)$, $(-2, 2)$, $(3, 3)$, $(4, 7)$

21. $(1, 5)$, $(2, -6)$, $(\pi, 14)$, $(1, 7)$, $(3, -5)$

22. $(8, -2)$, $(1, 4)$, $(2, 4)$, $(3, 4)$

23. $(0, -2)$, $(1, -4)$, $(2, 1)$, $(3, -2)$

24. $(0, 0)$, $(1, 1)$, $(2, -3)$, $(3, 5)$, $(-1, 4)$

25. $(1, -2)$, $(-1, 2)$, $(-2, 2)$, $(-3, 3)$, $(4, -5)$

SUPPLEMENTARY PROBLEMS

In each of Problems 1 through 10, use Newton-Raphson to find an approximate root of the given equation to four decimal places.

1. $x^5 - 4x^3 + 6x^2 - 2 = 0$ (find three real roots)
2. $\sin(x) = x - 0.012$
3. $e^{-x} + 4 = \tan(2x)$
4. $x^2 = \cosh(3x)$
5. $x^4 + 2x^2 - 3x + 2 = 0$
6. $\sin(2x) = \sqrt{\pi/8}$
7. $\sin(x)/x = e^{-x}$
8. $\sinh(x) + \cosh(2x) = 4$
9. $\ln(x) + x = 0$
10. $\cosh(1/x) - 4x^2 = 1$

11. Use the three-term Taylor method with $h = 0.1$, to approximate $y_1, \ldots, y_{10}$ for the problem

$$y' = xy - 2e^{-y}; \qquad y(0) = \pi.$$

12. Use Simpson's rule, with $n = 200$, to approximate

$$\int_0^{0.6} \frac{\cos(3x)\,dx}{\sqrt{x^4 + 6x^2 + 5}}.$$

13. Use Runge-Kutta with $h = 0.1$ to approximate $y_1, \ldots, y_{10}$ for the problem

$$y' = \cos(x - y) + x^2 - 2y; \qquad y(0) = 0.36.$$

14. Use a Taylor approximation with $h = 0.1$ to approximate $y_1, \ldots, y_9$ for the problem

$$y'' = x^2 + y^2; \qquad y(1) = 0, \qquad y'(1) = \sqrt{7}.$$

15. Use Runge-Kutta-Nystrom, with $h = 0.1$, to approximate $y_1, \ldots, y_9$ in the problem

$$y'' = \sin(x^2) - 2xe^{-y}; \qquad y(2) = 1, \qquad y'(2) = 7\sqrt{3}.$$

16. Use the finite-difference method, with $h = 0.1$, to approximate the solution of

$$y'' = 8x^2y - 4y + 2; \qquad y(0) = 1, \qquad y(1) = \tfrac{1}{2}.$$

17. Use the Milne predictor-corrector method, with $h = 0.01$, to approximate $y_4, \ldots, y_{12}$ for the problem $y' = x - 2y + 3x\sin(y)$. Use any method you wish to approximate y_1, y_2, and y_3, with $y(0) = 0.2$.

18. Use the trapezoidal rule, with $n = 200$, to approximate

$$\int_{-1}^{4} \frac{\sin(x^2)\,dx}{x^2 + 4}.$$

19. Use the finite-difference method, with $h = 0.2$, to approximate a solution of

$$y'' = y - 2xy'; \qquad y(0) = -0.36, \qquad y(1) = 0.43.$$

20. Find a root of $\tan(x) - \cot(x) + x = 2$, using Newton-Raphson with $h = 0.01$. Draw a sketch first to help guess an initial value x_0 with which to begin the algorithm.

21. Use Newton-Raphson to find the roots of the first six Legendre polynomials (note Figure 77 and Problem 5, Section 6.3).

22. Find a least-squares, best-fitting straight line for

$$(2, -1), \quad (1, 6), \quad (3, 4), \quad (4, 7), \quad \text{and} \quad (5, 2).$$

Sketch the points and the least-squares line.

23. Find a least-squares, best-fitting straight line for

$$(0, 6), \quad (2, 4), \quad (3, -5), \quad (0.2, 8), \quad (-\sqrt{7}, 3), \quad \text{and} \quad (1, 15).$$

Sketch the solution and the points.

24. Find a least-squares, best-fitting parabola for

$$(-1, 7), \quad (3, 4), \quad (1, 1), \quad (2, 9), \quad \text{and} \quad (5, 8).$$

Sketch the points and the parabola.

25. Derive a formula for finding a, b, c, and d to get a least-squares, best-fitting cubic $y = a + bx + cx^2 + dx^3$ to the points $(x_1, y_1), \ldots, (x_n, y_n)$.

26. Use the power method to approximate the largest eigenvalue of

$$\begin{bmatrix} -1 & 6 & 1 \\ 1 & 7 & 2 \\ 0 & 0 & 3 \end{bmatrix}.$$

Find the actual value of this eigenvalue by solving the characteristic equation. Also plot the eigenvalues and the Gerschgorin disks for this matrix.

27. Use the trapezoidal rule, with $n = 200$, to verify the integration of Example 477, Section 17.3.

28. Use the rectangular rule, with $n = 200$, to verify the answer for Problem 2, Section 17.3.

29. Use Simpson's rule, with $n = 200$, to evaluate the integral of Problem 4, Section 17.3, in the case that $a = \frac{1}{2}$ and $b = \frac{1}{4}$.

APPENDIX

A.1 REFERENCES

The following list of references makes no claim to being exhaustive but does, we hope, serve two purposes. First, it gives some standard and available texts which cover topics from this book, written at a level which should make them accessible to the student now. Second, it lists some references to specialized topics, such as Bessel functions and conformal mappings, which might be useful for further study.

ORDINARY DIFFERENTIAL EQUATIONS

Birkhoff, G. and Rota, G. C., *Ordinary Differential Equations.* 3rd ed. Wiley, New York, 1978.

Boyce, W. E. and DiPrima, R. C., *Elementary Differential Equations and Boundary Value Problems.* 3rd ed. Wiley, New York, 1977.

Spiegel, Murray R., *Applied Differential Equations.* 3rd ed. Prentice-Hall, Englewood Cliffs, N.J., 1981.

VECTORS AND MATRICES

Anton, Howard, *Elementary Linear Algebra.* 2nd ed. Wiley, New York, 1977.

Kolman, Bernard, *Elementary Linear Algebra.* 2nd ed. Macmillan, New York, 1977.

O'Neil, Peter V., *Introduction to Linear Algebra,* Wadsworth, Belmont, Calif., 1979.

Strang, Gilbert, *Linear Algebra and Its Applications.* Academic Press, New York, 1976.

VECTOR ANALYSIS

Davis, Harry F. and Snider, Arthur David, *Introduction to Vector Analysis.* 4th ed. Allyn & Bacon, Boston, 1979.

Schey, H. M., *Div, Grad, Curl and All That.* Norton, New York, 1973.

PARTIAL DIFFERENTIAL EQUATIONS AND FOURIER METHODS

Broman, Arne, *Introduction to Partial Differential Equations: From Fourier Series to Boundary Value Problems.* Addison-Wesley, Reading, Mass., 1968.

Churchill, R. V. and Brown, James Ward, *Fourier Series and Boundary Value Problems.* 3rd ed. McGraw-Hill, New York, 1978.

Davis, H. F., *Fourier Series and Orthogonal Functions.* Allyn & Bacon, Boston, 1963.

Franklin, Phillip, *An Introduction to Fourier Methods and the Laplace Transformation.* Dover, New York, 1949.

Jackson, D., *Fourier Series and Orthogonal Polynomials.* Carus Mathematical Monographs, no. 6. Mathematical Association of America, Washington, D.C., 1941.

Powers, D. L., *Boundary Value Problems.* Academic Press, New York, 1972.

Seeley, R. T., *An Introduction to Fourier Series and Integrals.* Benjamin, New York, 1966.

Weinberger, Hans, *A First Course in Partial Differential Equations.* Blaisdell Publishing Co., Waltham, Mass., 1965.

Young, Eutiquio C., *Partial Differential Equations, An Introduction.* Allyn & Bacon, Boston, 1972.

Zachmanoglou, E. C. and Thoe, Dale W., *Introduction to Partial Differential Equations with Applications.* Williams & Wilkins, Baltimore, 1976.

COMPLEX ANALYSIS

Dettman, J. W., *Applied Complex Variables.* Macmillan, New York, 1965.

Knopp, K., *Theory of Functions* (2 volumes). Dover, New York, 1945.

Markushevich, A. I., *Theory of Functions of a Complex Variable* (3 volumes). Translated from the Russian by Richard A. Silverman. Prentice-Hall, Englewood Cliffs, N.J., 1965.

Nehari, Z., *Introduction to Complex Analysis.* Allyn & Bacon, Boston, 1961.

Rothe, R. F., Ollendorf, F., and Pohlhausen, K., *Theory of Functions as Applied to Engineering Problems.* Dover, New York, 1961.

NUMERICAL METHODS

Henrici, Peter, *Elements of Numerical Analysis.* Wiley, New York, 1964.

———, *Computational Analysis with the HP-25 Pocket Calculator.* Wiley Interscience, New York, 1977.

Hildebrandt, F. B., *Introduction to Numerical Analysis.* McGraw-Hill, New York, 1956.

Noble, B., *Numerical Methods* (2 volumes). Wiley Interscience, New York, 1964.

Varga, R. S., *Matrix Iterative Analysis.* Prentice-Hall, Englewood Cliffs, N.J., 1962.

TRANSFORM METHODS

Churchill, R. V., *Operational Mathematics.* 3rd ed. McGraw-Hill, New York, 1972.

Jaeger, J. C., *An Introduction to the Laplace Transformation.* Wiley, New York, 1949.

Rainville, R. V., *The Laplace Transform.* Macmillan, New York, 1963.

Sneddon, Ian, *Fourier Transforms.* McGraw-Hill, New York, 1951.

———, *The Use of Integral Transforms.* McGraw-Hill, New York, 1972.

SPECIAL FUNCTIONS

Bowman, F., *Introduction to Bessel Functions.* Dover, New York, 1958.

Gray, Andrew and Mathews, G. B., *A Treatise on Bessel Functions and Their Application to Physics.* 2nd ed. Dover, New York, 1966.

Sneddon, Ian, *Special Functions of Mathematical Physics and Chemistry.* Oliver and Boyd, Ltd., New York, 1961.

Watson, G. N., *A Treatise on Bessel Functions.* 2nd ed. Cambridge Univ. Press, London, 1952.

CONFORMAL MAPPINGS

Bieberbach, L., *Conformal Mapping.* Chelsea Publishing Co., New York, 1953.

Kober, H., *Dictionary of Conformal Representations.* Dover, New York, 1957.

TABLES

Beyer, William H., ed., *CRC Standard Mathematical Tables*, 25th ed. CRC Press, Boca-Raton, Florida, 1978.

A.2 SOME USEFUL FORMULAS

Compiled here for reference are some frequently used facts from algebra, geometry, trigonometry, analytic geometry, and calculus.

CONSTANTS

$e = 2.71828\ 18284\ 59045\ 23536\ 02874\ 71352\ \cdots$

$\pi = 3.14159\ 26535\ 23846\ 26433\ 83279\ 50288\ \cdots$

$\gamma = 0.57721\ 56649\ 01532\ 86061\ \cdots$

(γ is Euler's constant.)

ALGEBRA

$(a+b)^2 = a^2 + 2ab + b^2$

$(a+b)^3 = a^3 + 3a^2b + 3ab^2 + b^3$

$(a+b)^4 = a^4 + 4a^3b + 6a^2b^2 + 4ab^3 + b^4$

Binomial formula: $(a+b)^n = \sum_{j=0}^{n} \binom{n}{j} a^{n-j}b^j$, where $\binom{n}{j} = \dfrac{n!}{j!(n-j)!}$.

Quadratic formula: The roots of $ax^2 + bx + c = 0$ are

$$x = \frac{-b \pm \sqrt{b^2 - 4ac}}{2a}, \qquad (a \neq 0).$$

GEOMETRY

Area of a triangle = $\frac{1}{2}$(base) × (height)

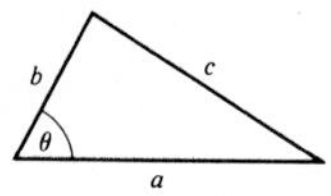

Law of cosines:

$$c^2 = a^2 + b^2 - 2ab\cos(\theta).$$

Area of a circle or radius $r = \pi r^2$.

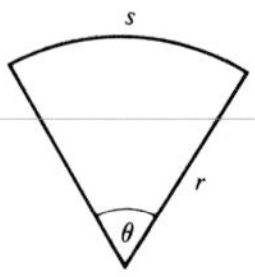

Circumference of a circle of radius $r = 2\pi r$.

Area of a circular sector of radius r and central angle θ (measured in radians) $= \frac{1}{2}r^2[\theta - \sin(\theta)]$.

Length of arc $s = r\theta$.

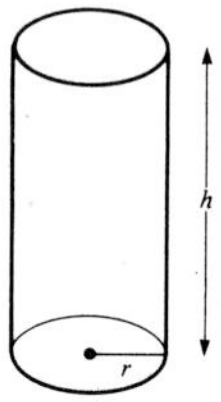

Volume of a right circular cylinder of radius r and height $h = \pi r^2 h$.

Curved surface area of cylinder $= 2\pi rh$.

Volume of a right circular cone of height h and base radius $r = \frac{1}{3}\pi r^2 h$.

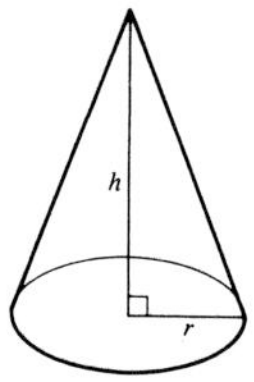

Lateral surface area (i.e., not including disk at bottom) $= \pi r\sqrt{r^2 + h^2}$.

Volume of a sphere of radius $r = \frac{4}{3}\pi r^3$.

Surface area of sphere $= 4\pi r^2$.

TRIGONOMETRY (angles are in radians)

$$\sin(\theta) = \cos\left(\theta - \frac{\pi}{2}\right) \qquad \sin(-\theta) = -\sin(\theta)$$

$$\cos(\theta) = -\sin\left(\theta - \frac{\pi}{2}\right) \qquad \cos(-\theta) = \cos(\theta)$$

$$\sin^2(\theta) + \cos^2(\theta) = 1$$

$$1 + \tan^2(\theta) = \sec^2(\theta)$$

$$1 + \cot^2(\theta) = \csc^2(\theta)$$

$$\sin(\alpha + \beta) = \sin(\alpha)\cos(\beta) + \sin(\beta)\cos(\alpha)$$

$$\cos(\alpha + \beta) = \cos(\alpha)\cos(\beta) - \sin(\alpha)\sin(\beta)$$

$$\tan(\alpha + \beta) = \frac{\tan(\alpha) + \tan(\beta)}{1 - \tan(\alpha)\tan(\beta)}$$

$$\sin(2\alpha) = 2\sin(\alpha)\cos(\alpha)$$

$$\cos(2\alpha) = \cos^2(\alpha) - \sin^2(\alpha) = 2\cos^2(\alpha) - 1$$

ANALYTIC GEOMETRY

The straight line through (x_1, y_1) and (x_2, y_2) is

$$(y - y_2) = \frac{(y_2 - y_1)}{(x_2 - x_1)}(x - x_2).$$

Here, the slope is $\dfrac{y_2 - y_1}{x_2 - x_1}$.

The straight line having slope m and passing through (x_0, y_0) is

$$y - y_0 = m(x - x_0).$$

The angle θ between two lines $a_1x + b_1y + c_1 = 0$ and $a_2x + b_2y + c_2 = 0$ is given by

$$\tan(\theta) = \frac{a_1b_2 - a_2b_1}{a_1a_2 + b_1b_2}.$$

The circle of radius r about (x_0, y_0) is

$$(x - x_0)^2 + (y - y_0)^2 = r^2.$$

CLASSIFICATION OF CONICS

Given the general quadratic,

$$ax^2 + 2hxy + by^2 + 2gx + 2fy + c = 0,$$

let

$$\Delta = \begin{vmatrix} a & h & g \\ h & b & f \\ g & f & c \end{vmatrix} \quad \text{and} \quad \sigma = \begin{vmatrix} a & h \\ h & b \end{vmatrix}.$$

Then,

1. If $\Delta = 0$, the quadratic represents one or two straight lines.
2. If $\Delta \neq 0$, $\sigma > 0$, and $\Delta/(a + b) < 0$, the quadratic is an ellipse.
3. If $\Delta \neq 0$ and $\sigma < 0$, the quadratic is an hyperbola.
4. If $\Delta \neq 0$ and $\sigma = 0$, the quadratic is a parabola.

CALCULUS

The following series expansions are commonly used.

Binomial: $$(1 + x)^\alpha = 1 + \alpha x + \frac{\alpha(\alpha - 1)}{2!}x^2 + \frac{\alpha(\alpha - 1)(\alpha - 2)}{3!}x^3 + \frac{\alpha(\alpha - 1)(\alpha - 2)(\alpha - 3)}{4!}x^4 + \cdots \quad \text{for} \quad |x| < 1.$$

$$e^x = \sum_{n=0}^{\infty} \frac{x^n}{n!} \quad \text{for all } x.$$

$$\ln(1 + x) = \sum_{n=1}^{\infty} \frac{(-1)^{n+1}}{n} x^n \quad \text{for all } x.$$

$$\sin(x) = \sum_{n=0}^{\infty} \frac{(-1)^n x^{2n+1}}{(2n + 1)!} \quad \text{for all } x.$$

$$\cos(x) = \sum_{n=0}^{\infty} \frac{(-1)^n x^{2n}}{(2n)!} \quad \text{for all } x.$$

$$\arcsin(x) = x + \frac{x^3}{2 \cdot 3} + \frac{1 \cdot 3}{2 \cdot 4 \cdot 5}x^5 + \frac{1 \cdot 3 \cdot 5}{2 \cdot 4 \cdot 6 \cdot 7}x^7 + \cdots \quad \text{for} \quad |x| < 1.$$

$$\arccos(x) = \frac{\pi}{2} - \arcsin(x) \quad \text{for} \quad |x| < 1.$$

$$\arctan(x) = \sum_{n=0}^{\infty} \frac{(-1)^{n+1}x^{2n+1}}{2n+1}.$$

A.3 GUIDE TO THEOREMS

The following guide is designed to enable the reader to locate theorems by the section in which they appear. Theorems are designated by name or a brief statement of content, not by complete formulation.

Number	*Section*	*Theorem*
1	1.9	Existence and uniqueness of solutions of $y' = f(x, y)$; $y(x_0) = y_0$
2	2.1	Existence and uniqueness of solutions of $y'' + P(x)y' + Q(x)y = 0; \quad y(x_0) = A, \quad y'(x_0) = B$
3	2.2	Linear combinations of solutions of $y'' + P(x)y' + Q(x)y = 0$ are solutions.
4	2.2	Wronskian test for linear independence of solutions of $y'' + P(x)y' + Q(x)y = 0$
5	2.3	General solution of $y'' + P(x)y' + Q(x)y = 0$
6	2.7	General solution of $y'' + P(x)y' + Q(x)y = F(x)$
7	3.1	Existence and uniqueness of solutions of $y^{(n)} + P_{n-1}(x)y^{(n-1)} + \cdots + P_1(x)y' + P_0(x)y = F(x)$; $y(x_0) = A_0, \quad y'(x_0) = A_1, \quad \ldots, \quad y^{(n-1)}(x_0) = A_{n-1}$
8	3.1	Linear combinations of solutions of $y^{(n)} + P_{n-1}(x)y^{(n-1)} + \cdots + P_1(x)y' + P_0(x)y = 0$ are solutions.
9	3.1	Wronskian test for linear independence of solutions of $y^{(n)} + P_{n-1}(x)y^{(n-1)} + \cdots + P_1(x)y' + P_0(x)y = 0$
10	3.1	General solution of $y^{(n)} + P_{n-1}(x)y^{(n-1)} + \cdots + P_1(x)y' + P_0(x) = 0$
11	3.1	General solution of $y^{(n)} + P_{n-1}(x)y^{(n-1)} + \cdots + P_1(x)y' + P_0(x)y = F(x)$
12	4.1	$\mathscr{L}[f(t) + g(t)] = \mathscr{L}[f(t)] + \mathscr{L}[g(t)]$ and $\mathscr{L}[af(t)] = a\mathscr{L}[f(t)]$.

Number	*Section*	*Theorem*
13	4.1	$\mathscr{L}^{-1}[f(t)+g(t)] = \mathscr{L}^{-1}[f(t)] + \mathscr{L}^{-1}[g(t)]$ and $\mathscr{L}^{-1}[af(t)] = a\mathscr{L}^{-1}[f(t)]$.
14	4.1	Existence of the Laplace transform
15	4.6	Convolution theorem for Laplace transforms
16	5.2	Power series solutions of $y' + q(x)y = r(x)$ and $y'' + P(x)y' + Q(x)y = F(x)$
17	5.3	Method of Frobenius
18	6.3	Rodrigues's formula for Legendre polynomials
19	6.3	Recurrence relations for Legendre polynomials
20	6.4	Sturm-Liouville theorem
21	6.4	Convergence of generalized Fourier series
22	6.5	Sturm separation theorem
23	6.5	Sturm comparison theorem
24	7.3	Classification of critical points of $\frac{dx}{dt} = ax + by$ and $\frac{dy}{dt} = cx + dy$
25	7.3	Critical points of $\frac{dx}{dt} = ax + by + P(x, y)$ and $\frac{dy}{dt} = cx + dy + Q(x, y)$
26	7.3	Liapunov's theorem
27	7.3	Poincaré-Bendixson theorem
28	7.3	Lienard's theorem
29	9.2	$\\|\alpha\mathbf{F} + \beta\mathbf{G}\\|^2 = \lvert\alpha\rvert^2\\|\mathbf{F}\\|^2 + 2\alpha\beta\mathbf{F}\cdot\mathbf{G} + \lvert\beta\rvert^2\\|\mathbf{G}\\|^2$
30	9.2	Cauchy-Schwarz inequality
31	9.4	Vector identities
32	9.5	R^n is a vector space.
33	9.5	Properties of dot product in R^n
34	9.5	Characterization of the subspaces of R^2
35	9.5	Characterization of the subspaces of R^3
36	9.6	$\mathbf{F}$, $\mathbf{G}$, and $\mathbf{H}$ in R^3 are linearly dependent if and only if $[\mathbf{F}, \mathbf{G}, \mathbf{H}] = 0$.
37	9.6	$\mathbf{F}_1, \ldots, \mathbf{F}_m$ in R^n are linearly dependent if and only if $\alpha_1\mathbf{F}_1 + \cdots + \alpha_m\mathbf{F}_m = \mathbf{0}$ for some scalars $\alpha_1, \ldots, \alpha_m$ not all zero.

Number	*Section*	*Theorem*
38	10.1	Elementary properties of matrix sums and products
39	10.1	$A + 0_{nm} = 0_{nm} + A = A$
40	10.2	Enumeration of walks in a graph
41	10.3	Sums and products of diagonal matrices are diagonal.
42	10.3	$AI_n = I_n A = A$
43	10.3	Properties of transposes
44	10.4	Properties of row-equivalence
45	10.5	Every matrix is row-equivalent to a reduced matrix.
46	10.5	The row-reduced form of A is unique.
	10.6	(Lemma 1) Rank $A = n$ if and only if $A_R = I_n$.
47	10.6	Rank of A equals the dimension of the row space.
48	10.7	The number of arbitrary scalars in the general solution of $AX = 0$ is the number of columns of A minus the rank of A.
49	10.7	If A is $n \times n$, then $AX = 0$ has only the trivial solution if and only if rank $A = n$.
	10.7	(Corollary 1) If A is $n \times n$, then $AX = 0$ has a nontrivial solution if and only if rank $A < n$.
50	10.7	$AX = 0$ has a nontrivial solution if the number of unknowns exceeds the number of equations.
51	10.8	$AX = B$ has a solution if and only if rank A = rank $[A \vdots B]$.
52	10.8	General solution of $AX = B$
53	10.8	Uniqueness of solution of $AX = B$
54	10.9	Uniqueness of matrix inverses
55	10.9	Properties of nonsingular matrices
56	10.9	If A is $n \times n$, then A is nonsingular if and only if rank $A = n$.
57	10.9	For $n \times n$ A, $AX = B$ has a unique solution if and only if A is nonsingular; $AX = 0$ has a nontrivial solution if and only if A is singular.
58	10.10	Cofactor expansions by any row or column of A are equal.
59	10.10	$\lvert B \rvert = \alpha \lvert A \rvert$ if B is formed from A by multiplying a row or column by α.
60	10.10	If A has a zero row or column, then $\lvert A \rvert = 0$.
61	10.10	If we interchange two rows or columns of A to get B, then $\lvert B \rvert = -\lvert A \rvert$.
62	10.10	If A has two identical rows or two identical columns, then $\lvert A \rvert = 0$.
63	10.10	If one row (or column) of A is a constant multiple of another row (or column), then $\lvert A \rvert = 0$.

Number	*Section*	*Theorem*
64	10.10	If we add a constant multiple of one row (or column) of A to another row (or column) to form B, then $\lvert B \rvert = \lvert A \rvert$.
65	10.10	$\lvert A \rvert = \lvert A^t \rvert$
66	10.10	$\lvert AB \rvert = \lvert A \rvert \lvert B \rvert$
67	10.11	The determinant of an upper triangular matrix is the product of its diagonal elements.
68	10.12	Matrix tree theorem
69	10.13	A is nonsingular if and only if $\lvert A \rvert \neq 0$.
70	10.13	A determinant formula for A^{-1}
71	10.14	Cramer's rule
72	10.15	Eigenvalues of A are the solutions of $\lvert \lambda I_n - A \rvert = 0$; eigenvectors associated with λ are nontrivial solutions of $(\lambda I_n - A)X = 0$.
73	10.18	A is diagonalizable if A has n linearly independent eigenvectors.
74	10.18	If A is diagonalizable, then A has n linearly independent eigenvectors.
75	10.18	Eigenvectors associated with distinct eigenvalues of A are linearly independent.
	10.18	(Corollary 2) Any $n \times n$ matrix with n distinct eigenvalues is diagonalizable.
76	10.20	The eigenvalues of a real, symmetric matrix are real.
77	10.20	Eigenvectors associated with distinct eigenvalues of a real, symmetric matrix are orthogonal.
78	10.21	Q is orthogonal if and only if Q^t is orthogonal.
79	10.21	Q is orthogonal if and only if the rows are orthonormal in R^n.
	10.21	(Corollary 3) Q is orthogonal if and only if its columns are orthonormal in R^n.
80	10.21	If Q is orthogonal, then $\lvert Q \rvert = 1$ or $\lvert Q \rvert = -1$.
81	10.21	Any real, symmetric matrix can be diagonalized by an orthogonal matrix.
82	10.22	Any real quadratic form can be transformed to $\lambda_1 y_1^2 + \cdots + \lambda_n y_n{}^2$ by an orthogonal matrix.
83	10.23	A square matrix U is unitary if and only if its rows (or columns) form a unitary system.
84	10.23	$\bar{Z}^t HZ$ is real if H is hermitian; it is zero or pure imaginary if H is skew-hermitian.
85	10.23	Eigenvalues of a hermitian matrix are real, of a skew-hermitian matrix are zero or pure imaginary, and of a unitary matrix are of absolute value one.

Number	*Section*	*Theorem*
86	11.2	Acceleration can be decomposed into $\mathbf{a} = \dfrac{dv}{dt}\mathbf{T} + \dfrac{v^2}{\rho}\mathbf{N}$.
87	11.5	$\nabla \times (\nabla \times \phi) = \mathbf{0}$
88	11.5	$\nabla \cdot (\nabla \times \mathbf{F}) = 0$
89	11.7	Green's theorem
90	11.8	$\mathbf{F}$ has a potential on Ω if and only if $\mathbf{F}$ is independent of path in Ω if and only if $\int_C \mathbf{F} = \mathbf{0}$ for every regular closed curve in Ω.
91	11.8	On a simply connected domain, $\mathbf{F}$ has a potential if and only if $$\frac{\partial F_2}{\partial x} = \frac{\partial F_1}{\partial y}.$$
92	11.10	Gauss's divergence theorem
93	11.10	Stokes's theorem
94	11.12	In a simply connected domain in R^3, $\mathbf{F}$ has a potential if and only if $\nabla \times \mathbf{F} = \mathbf{0}$.
95	12.2	Convergence of Fourier series
96	12.4	Convergence of Fourier cosine series
97	12.4	Convergence of Fourier sine series
98	12.5	Convergence of Fourier integrals
99	12.9	$S_n[f(x) + g(x)] = S_n[f(x)] + S_n[g(x)]$ and $S_n[\alpha f(x)] = \alpha S_n[f(x)]$.
100	12.9	$S_n[f''(x)] = -n^2 f_S(n) + nf(0) - n(-1)^n f(\pi)$
101	12.9	$S_n^{-1}\left[\dfrac{f_S(n)}{n^2}\right] = \dfrac{x}{\pi}\displaystyle\int_0^t (\pi - t) f(t)\, dt - \int_0^x (x - t) f(t)\, dt$
102	12.9	$C_n[f(x) + g(x)] = C_n[f(x)] + C_n[g(x)]$ and $C_n[\alpha f(x)] = \alpha C_n[f(x)]$.
103	12.9	$C_n[f''(x)] = -n^2 f_C(n) - f'(0) + (-1)^n f'(\pi)$
104	12.9	$C_n^{-1}\left[\dfrac{f_C(n)}{n^2}\right] = \displaystyle\int_0^x \int_t^\pi f(p)\, dp\, dt + \dfrac{f_C(0)}{2\pi}(x - \pi)^2$
105	12.10	$\mathscr{F}_S[f(x) + g(x)] = \mathscr{F}_S[f(x)] + \mathscr{F}_S[g(x)]$ and $\mathscr{F}_S[\alpha f(x)] = \alpha \mathscr{F}_S[f(x)]$.
106	12.10	$\mathscr{F}_S[f''(x)] = -\lambda^2 \hat{f}_S(\lambda) + \lambda f(0)$
107	12.10	$\mathscr{F}_C[f(x) + g(x)] = \mathscr{F}_C[f(x)] + \mathscr{F}_C[g(x)]$ and $\mathscr{F}_C[\alpha f(x)] = \alpha \mathscr{F}_C[f(x)]$.
108	12.10	$\mathscr{F}_C[f''(x)] = -\lambda^2 \hat{f}_C(\lambda) - f'(0)$
109	12.10	$\mathscr{F}[f^{(n)}(x)] = (i\lambda)^n \hat{f}(\lambda)$
110	14.5	Cauchy-Riemann equations

Number	*Section*	*Theorem*
	14.5	(Corollary 1) The real and imaginary parts of an analytic function satisfy the Cauchy-Riemann equations.
111	14.5	Continuity of $u(x, y)$ and $v(x, y)$ and their first partials, together with the Cauchy-Riemann equations, implies analyticity of $f(z) = u(x, y) + iv(x, y)$.
112	14.5	The real and imaginary parts of an analytic function are harmonic.
113	14.7	Properties of e^z
114	14.8	Properties of $\ln(z)$
115	14.9	$\mathrm{Pr}[z^\alpha]$ is analytic and has derivative $\alpha\ \mathrm{Pr}[z^{\alpha-1}]$.
116	14.9	$(e^z)^\alpha = e^{\alpha z}$ and $\ln(z^\alpha) = \alpha \ln(z) + 2n\pi i$.
117	14.10	Existence of $\int_C f(z)\, dz$
118	15.1	$\int_C f(z)\, dz = \int_a^b f[z(t)]z'(t)\, dt$
119	15.1	If $f(z) = F'(z)$ and $F(z)$ is analytic, then $\int_C f(z)\, dz = F(z_1) - F(z_0)$.
120	15.1	$\lvert \int_C f(z)\, dz \rvert \le$ (max. of $\lvert f(z) \rvert$ on C) $\cdot$ (length of C)
121	15.2	Cauchy integral theorem
122	15.3	An analytic function is independent of path in a simply connected domain.
123	15.3	An analytic function has an analytic antiderivative in a simply connected domain.
124	15.3	Cauchy's integral formula
125	15.3	Cauchy's integral formula for higher derivatives
	15.3	(Corollary 2) $\lvert f^{(n)}(z) \rvert \le \dfrac{Mn!}{r^n}$ if $\lvert f(z) \rvert \le M$ and $\lvert z - z_0 \rvert < r$.
126	16.1	$x_n + iy_n \to A + iB$ if and only if $x_n \to A$ and $y_n \to B$.
127	16.2	Cauchy convergence criterion for complex sequences
128	16.3	$\sum_{n=1}^{\infty} x_n + iy_n = A + iB$ if and only if $\sum_{n=1}^{\infty} x_n = A$ and $\sum_{n=1}^{\infty} y_n = B$.
129	16.3	If $\sum_{n=1}^{\infty} z_n$ converges, then $z_n \to 0$ as $n \to \infty$.
130	16.3	Cauchy convergence criterion for complex series
131	16.3	If $\sum_{n=1}^{\infty} \lvert z_n \rvert$ converges, then so does $\sum_{n=1}^{\infty} z_n$.
132	16.4	Convergence of power series
133	16.4	Analyticity of power series and term by term differentiation
134	16.4	Term by term integration of power series
135	16.5	Taylor expansion of an analytic function
136	16.6	Laurent expansion

Number	*Section*	*Theorem*
137	17.2	Residue theorem
138	17.2	At a pole of order m, $\operatorname{Res}_{z_0} f(z) = \frac{1}{(m-1)!} \frac{d^{m-1}}{dz^{m-1}} \lim_{z \to z_0} [(z - z_0)^m f(z)]$.
139	17.5	The argument principle
140	18.2	An analytic function with nonvanishing derivative is conformal.
141	18.2	There is a linear fractional transformation mapping three given points to three given images.
142	18.2	There is a linear fractional transformation mapping three given points to two given images and ∞.
143	18.2	A linear fractional transformation maps a circle to a circle or straight line, and a straight line to a circle or straight line.
144	19.1	If $u(x, y)$ is harmonic on a simply connected domain D, then, for some $v(x, y)$, $f(z) = u + iv$ is analytic on D.
145	20.10	Gerschgorin's theorem

Answers and Solutions to Selected Odd-Numbered Problems

Chapter 0

1. $\dfrac{dv}{dt} = -g - \dfrac{v^2}{1000\text{ m}}$

3. If $A(t)$ = amount in the account at time t, then $\dfrac{dA}{dt} = 0.06A$. We find that $A = 4000e^{0.06t}$. He becomes a millionaire in $\dfrac{1}{0.06}\ln(250)$ years, or slightly over 92 years.

5. If $P(t)$ is the population at time t, $B(t)$ the birth rate, and $D(t)$ the death rate, then $B = C_1 P$ and $D = C_2 P$ for some constants C_1 and C_2, and $\dfrac{dP}{dt} = (C_1 - C_2)P$

7. $\pi r^2\sqrt{2gh}\,dt = -\pi[R^2 - (R-h)^2]\,dh$; when empty, $h = 0$ and $t = \dfrac{14}{15r^2}\sqrt{\dfrac{R^5}{2g}}$

9. $\dfrac{dV}{dp} = -C\dfrac{V}{p}$, where C is a constant of proportionality. Solve to get $V = -Ap^{-C}$, with A a constant which can be determined only if more information is given.

11. $\dfrac{dI}{dh} = CI$, where $I(h)$ = intensity at depth h.

13. $\dfrac{dV}{dh} = CV^{2/3}$, where C is a constant and h is the distance fallen (from some reference point).

17. $m\dfrac{dv}{dt} = mg - \dfrac{4\rho\pi}{3}R^3 - 6\pi\mu Rv$, where ρ is the density of the grease.

Chapter 1

Section 1.0

1. Solution 3. Solution 5. Solution 7. Solution 9. Solution

Section 1.1

1. $\dfrac{y^3}{3} = \dfrac{x^2}{2} + \dfrac{x^4}{4} + C$ 3. Not separable 5. $e^y = 2e^{-x}(-x-1) + C$ 7. Not separable

9. $\frac{1}{2}(\ln|y|)^2 = \frac{3x^2}{2} + C$ **11.** Not separable **13.** Not separable **15.** Not separable

17. $\frac{\sin^2(y)}{2} = \ln|x| + C$ **19.** $(y-1)^2 = C\left(\frac{1+x}{1-x}\right)$

21. General solution, $\ln|y+2| = x^3 + C$; particular solution, $\ln|y+2| = x^3 + \ln(10) - 64$

23. General solution, $\frac{y^2}{2} = \frac{x^3}{3} + 2x + C$; particular solution, $\frac{y^2}{2} = \frac{x^3}{3} + 2x + \frac{133}{6}$

25. General solution, $(y+2)^2 = (x-1)^2 + C$; particular solution, $(y+2)^2 = (x-1)^2 + 60$

27. General solution, $-e^{-y} = \ln|x| + C$; particular solution, $-e^{-y} = \ln|x| - e^{-4}$

29. General solution, $(y+4)^2 = -3x^2 + K$; particular solution, $(y+4)^2 + 3x^2 = 133$

31. $\frac{dx}{dt} = Cx$, for some constant C.

(a) $x(t) = P_0 \exp \frac{t\ln(P/P_0)}{t_1}$ **(b)** $T = \frac{t_1 \ln(\frac{1}{2})}{\ln(P/P_0)}$ **(c)** $\tau = \frac{t_1\ln(0.1)}{\ln(P/P_0)}$

33. (a) At PG&E, the amount after n years is $(10{,}000)(1.0188)^{4n}\left(\text{here, } 1.0188 = 1 + \frac{0.075}{4}\right)$.

At PNT, after n years one has $10{,}000e^{0.075n}$. **(b)** About \$36.19.

35. At time t, the population is $P = 10{,}000e^{(1/3)\ln(2)(t-1)}$. At $t = 8$, P is about 50,397; $T = 3\ln(3)/\ln(2)$, about 4.755 years.

Section 1.2

1. $dt = \frac{m\,dv}{mg - kv}$; $v = \frac{mg}{k} - Ce^{-kt/m} \to \frac{mg}{k}$ as $t \to \infty$ **3.** $P\sqrt{\frac{P-1}{P+1}} = Ae^{kt}$ **5.** $p = Av^{-c_p/c_v}$, A constant

7. $\frac{1}{v} = \frac{t}{1000} + C$ **9.** $V = Ae^{K/p}$, with A and K constants **11.** $I = \frac{1}{L}(t + e^{-t})$

13. $r^{1/3} = Ch + K$ **15.** $T = Ct^{3/2} + K$ (T = thickness)

17. (b) $z(t) = x_0 - \cfrac{1}{kt + \cfrac{1}{x_0}}$ if $x_0 = y_0$; if $x_0 \neq y_0$, $z(t) = \frac{x_0 y_0 (e^{(x_0 - y_0)kt} - 1)}{y_0 e^{(x_0 - y_0)kt} - x_0}$

(c) $T_x = \frac{1}{k(x_0 - y_0)} \ln\left(\frac{2y_0 - x_0}{y_0}\right)$ if $x_0 < 2y_0$ (no solution if $x_0 \geq 2y_0$)

$T_y = \frac{1}{k(y_0 - x_0)} \ln\left(\frac{2x_0 - y_0}{x_0}\right)$ if $y_0 > 2x_0$

19. The resulting expression cannot be integrated in general.

Section 1.3

1. $\frac{2}{\sqrt{7}} \arctan\left(\frac{4(y/x) - 1}{\sqrt{7}}\right) = \ln|x| + C$ **3.** Not homogeneous

5. $\sqrt{2}\arctan\left(\frac{y}{\sqrt{2}x}\right) - \frac{1}{2}\ln\left(\frac{y^2}{x^2} + 2\right) = \ln|x| + C$

7. Not homogeneous **9.** Not homogeneous **11.** $-\frac{x}{y} = \ln|x| + C$ **13.** Not homogeneous

15. $x = C\left(1 - 4\frac{y}{x}\right)^{-1/4}$ **19.** $\frac{2}{\sqrt{3}}\arctan\left(\frac{(2y/x) - 1}{\sqrt{3}}\right) = \ln|x| + \frac{2}{\sqrt{3}}\arctan\left(\frac{7}{\sqrt{3}}\right)$

21. $y^2 + 2xy - x^2 = 73$ **23.** $\ln|2y^2 - 5xy + 8x^2| + \frac{2\sqrt{39}}{13}\arctan\left(\frac{\frac{4y}{x} - 5}{\sqrt{39}}\right) = \ln(200) + \frac{2\sqrt{39}}{13}\arctan\left(\frac{-19}{\sqrt{39}}\right)$

25. $\frac{2}{\sqrt{32}}\operatorname{arctanh}\left(\frac{-10}{\sqrt{32}}\right) - \frac{1}{2}\ln(17) - \ln(2) + \ln|x| = \frac{2}{\sqrt{32}}\operatorname{arctanh}\left(\frac{-(2y/x) - 4}{\sqrt{32}}\right) - \frac{1}{2}\ln\left|-\frac{y^2}{x^2} - 4\frac{y}{x} + 4\right|$

27. $\frac{-2}{\sqrt{5}}\operatorname{arctanh}\left(\frac{-(2y/x) - 1}{\sqrt{5}}\right) = \ln|x| - \frac{2}{\sqrt{5}}\operatorname{arctanh}\left(\frac{-9}{\sqrt{5}}\right) - \ln(2)$ **29.** $x^2 + y^2 = 13$

31. $\frac{1}{3}\frac{y^3}{x^3} = \ln|x| + \frac{64}{81} - \ln(3)$

33. $-\ln|x| = \frac{1}{\sqrt{6}}\arctan\left(\frac{\sqrt{6}y}{3x}\right) + \frac{1}{2}\ln\left|\frac{2y^2}{x^2} + 3\right| - \frac{1}{\sqrt{6}}\arctan(2\sqrt{6}) - \frac{1}{2}\ln(75)$

35. $\ln|x| - \ln(4) + \frac{2}{\sqrt{7}}\arctan\left(\frac{-9}{2\sqrt{7}}\right) = \frac{2}{\sqrt{7}}\arctan\left(\frac{(2y/x) - 1}{\sqrt{7}}\right)$

37. $\frac{6}{\sqrt{20}}\operatorname{arctanh}\left(\frac{2\left(\frac{y - \frac{1}{4}}{x + \frac{1}{4}}\right) + 4}{\sqrt{20}}\right) - \frac{1}{2}\ln\left|\left(\frac{y - \frac{1}{4}}{x + \frac{1}{4}}\right)^2 + 4\left(\frac{y - \frac{1}{4}}{x + \frac{1}{4}}\right) - 1\right| = \ln|x + \frac{1}{4}| + C$ **39.** $x + C = \frac{1}{2}(x - y + 3)^2$

41. $\ln|x - \frac{1}{11}| + C = -\frac{1}{2}\ln\left|\left(\frac{y - \frac{14}{11}}{x - \frac{1}{11}}\right)^2 + 5\left(\frac{y - \frac{14}{11}}{x - \frac{1}{11}}\right) - 2\right| + \frac{11}{\sqrt{33}}\operatorname{arctanh}\left(\frac{2\left(\frac{y - \frac{14}{11}}{x - \frac{1}{11}}\right) + 5}{\sqrt{33}}\right)$

43. $-8\ln|x - 2y + 4| + 2x - 6y = C$ **45.** $-\frac{2}{\sqrt{5}}\operatorname{arctanh}\left(\frac{2\left(\frac{y - 1}{x + 1}\right) - 3}{\sqrt{5}}\right) = \ln|x + 1| + C$

47. $-\frac{1}{2}\ln\left|\left(\frac{y - \frac{14}{5}}{x + \frac{16}{5}}\right)^2 + \left(\frac{y - \frac{14}{5}}{x + \frac{16}{5}}\right) + 3\right| + \frac{3}{\sqrt{11}}\arctan\left(\frac{2\left(\frac{y - \frac{14}{5}}{x + \frac{16}{5}}\right) + 1}{\sqrt{11}}\right) = \ln|x + \frac{16}{5}| + C$

49. $y^2 + xy - 2x^2 - 3x - 2y = C$

51. $-\frac{1}{2}\ln\left|\left(\frac{\theta + \frac{13}{3}}{r + \frac{8}{3}}\right)^2 + 2\left(\frac{\theta + \frac{13}{3}}{r + \frac{8}{3}}\right) - 2\right| = \ln|r + \frac{8}{3}| + C$ or $\left[\left(\frac{\theta + \frac{13}{3}}{r + \frac{8}{3}}\right)^2 + 2\left(\frac{\theta + \frac{13}{3}}{r + \frac{8}{3}}\right) - 2\right]^{-1/2} = A(r + \frac{8}{3})$

53. $\frac{5}{16}(-2u + z) + \frac{13}{128}\ln|16u - 8z| = -\frac{1}{2}u + C$

55. $\frac{1}{2}\ln\left|\frac{p^2}{(x - \frac{1}{2})^2} - 7\left(\frac{p}{x - \frac{1}{2}}\right) - 2\right| + \frac{1}{\sqrt{57}}\operatorname{arctanh}\left(\frac{2\left(\frac{p}{x - \frac{1}{2}}\right) - 7}{\sqrt{57}}\right) = \ln|x| + C$

57. $6\ln\left|3\left(\frac{q + \frac{17}{6}}{e - \frac{13}{24}}\right)^2 + 11\left(\frac{q + \frac{17}{6}}{e - \frac{13}{24}}\right) + 4\right| - \frac{108}{\sqrt{73}}\operatorname{arctanh}\left(\frac{6\left(\frac{q + \frac{17}{6}}{e - \frac{13}{24}}\right) + 11}{\sqrt{73}}\right) = \ln|e - \frac{13}{24}| + C$

59. $\left[5\left(\frac{x - 3}{t - 4}\right)^2 - 2\left(\frac{x - 3}{t - 4}\right) + 2\right]^{-1/2} = C(t - 4)$

Section 1.4

1. $2xy^2 + e^{xy} + y^2 = C$ **3.** $xy + e^x = C$ **5.** Not exact **7.** $x^2 + y^2 = C$ **9.** $\cosh(x)\sinh(y) = C$

11. $y^3 + xy + \ln|x| = C$ **13.** $y\ln|x| = C$ **15.** $\arctan\left(\frac{y}{x}\right) + x^2 = C$ **17.** Not exact

19. $x^2 = Cy^2$ **21.** $x^3 - y\sin(x) = C$ **23.** $y^2\cos(x) + 2y = C$ **25.** Not exact

27. $-4x^2 + e^{xy} + y^2 = 20 + e^{-12}$ **29.** $2xy - \tan(xy^2) = 4 - \tan(4)$ **31.** $xy^3 - y = -255$

33. $(1 - 3y^4)\,dx - 12xy^3\,dy = 0$; $-3xy^4 + x = C$ **35.** $e^y\,dx - (1 - xe^y)\,dy = 0$; $xe^y - y = C$

37. $(3y + 3x^2y^2)\,dx - (2x^3y - 3x)\,dy = 0$; not exact **39.** $[2x + \sin(x)]\,dx - [2y - y\cos(x)]\,dy = 0$; not exact

43. $r^2\cos(r\theta) = C$ **45.** $u^2 - 2uv = C$ **47.** $uv = C$ **49.** $x^3 + 3\sinh(ux) = C$

Section 1.5

1. $u = xe^{6x}$; $\frac{x^2}{2}ye^{6x} + \frac{e^{6x}}{9}(6x - 1) = C$ **3.** $u = y$; $x^3y^2 - y^2 = C$ **5.** $u = xy$; $x^7y^6 + x^2y^2\sin(x) = C$

7. $u = x^4y^3$; $x^6y^4 + x^5y^7 = C$ **9.** $u = e^x$; $(x + y^2)e^x = C$ **11.** $u = e^{-x}e^{2y}$; $(x^2 - 2xy)e^{-x+2y} = C$

17. $u = x^2y^3$; $x^3y^4 = 1296$ **19.** $u = y^2e^{x^2}$; $y^3e^{x^2} = 125e^9$ **21.** $u = xe^{x^2}$; $yx^2e^{x^2} = 12e^4$

23. $u = x$; $x^2y^3 - 2x^2 = -1143$ **25.** $u = e^{x^2}$; $xye^{x^2} = -2e$ **27.** Exact; $4x^2 + 3y^2 = 247$

29. $u = x^{-1/3}e^x$; $3x^{2/3}ye^x = -21e$ **33.** $e^{-x^2/2}\left(1 - \frac{1}{y}\right) = C$ **35.** $\frac{y^4x^8}{4} - \frac{x^6}{6} = C$ **37.** $\frac{1}{2}\frac{x^2}{y^2} + e^x = C$

39. $-\frac{3}{2}x^2y^{-2/3} - 2x = C$ **41.** $\frac{1}{2}\frac{1}{y^2x^2} + \ln|x| = C$ **43.** Yes (if $y \neq 0$)

45. Yes **47.** Yes **49.** No **51.** Yes

Section 1.6

1. $y = \cos(x) + \sin(x) + Ce^{-x}$ **3.** $y = \frac{x^3}{2} + 2x\ln(x) + Cx$ **5.** $y = 5(x - 1) + \frac{1}{3}e^{2x} + Ce^{-x}$

7. $y = Cx^3 + 2x^3\ln|x|$ **9.** $y = -4x - 12 + Ce^{x/3}$ **11.** $y = -\frac{1}{3}e^x + \frac{1}{17}[4\sin(x) + \cos(x)] + Ce^{4x}$

15. $R = \frac{kI}{c^2}\left(\frac{c}{I}x - 1\right) + Ke^{-cx/I}$ **17.** $y = \frac{2}{3}e^{4x} - \frac{11}{3}e^x$

19. $y = \frac{1}{x^3}[-x^4\cos(x) + (12x^2 - 24)\cos(x) + (4x^3 - 24x)\sin(x)] + \frac{1}{x^3}(-\pi^4 + 12\pi^2 - 24)$ **21.** $y = \frac{2}{5}x - \frac{22}{5x^4}$

23. $y = \frac{2}{11}x^4 - \frac{2}{5}x - \frac{359}{55}\sqrt{27}\,x^{-3/2}$ **25.** $y = \frac{27}{31}x^4 + \frac{9}{13}x^2 - \frac{982}{403}x^{5/9}$ **27.** $y = \frac{105}{26}e^{5x} - \frac{5}{26}\sin(x) - \frac{\cos(x)}{26}$

29. $\frac{1}{2}\frac{1}{x^2y^2} + 3x = C$ **31.** $\frac{2}{3}x^{-9/2}y^{-3/2} + \frac{4}{5}x^{-5/2} = C$ **33.** $-\frac{1}{3}x^{-12}y^{-3} + \frac{1}{10}x^{-10} = C$

35. $P(t) = -\frac{k}{B} - \frac{Ak}{A^2B^2 + a^2}[-AB\cos(at) + a\sin(at)] + Ce^{ABt}$, where A is the constant of proportionality between $P'(t)$ and $S(t) - D(t)$.

37. $P(t) = Akt - \frac{Ak}{a}\sin(at) - DAt$ (A as in solution to 35)

Section 1.7

1. $y = x + \left[\frac{C}{x} - \frac{1}{x}\ln(x)\right]^{-1}$; $y = x + \left[\frac{1}{2x} - \frac{1}{x}\ln(x)\right]^{-1}$

3. Using $S(x) = 2$, $y = 2 + [Cx^2 - \frac{1}{2}]^{-1}$; $y = 2 + (x^2 - \frac{1}{2})^{-1}$

5. $y = -e^x + \frac{1}{Ce^{-3x} + \frac{1}{2}e^{-x}}$; $y = -e^x + \frac{1}{-\frac{5}{14}e^{-3x} + \frac{1}{2}e^{-x}}$

7. Using $S(x) = 8x^2$, $y = 8x^2 + \left[C + \frac{1}{16x}\right]^{-1}$; $y = 8x^2 + \left[\frac{-27}{176} + \frac{1}{16x}\right]^{-1}$

9. Using $S(x) = 4x$, $y = 4x + \left[\frac{C}{x} - \frac{x^2}{3}\right]^{-1}$; $y = 4x + \left[\frac{5}{3x} - \frac{x^2}{3}\right]^{-1}$

11. Using $S(x) = 2x^2$, $y = 2x^2 + \left[\frac{C}{x^4e^x} + \frac{1 - x}{x^4}\right]^{-1}$

13. Using $S(x) = \sqrt{14}$, $y = \sqrt{14} + \frac{1}{z}$, where $z = \frac{C}{u(x)} + \frac{1}{u(x)}\int(1 - 2x)u(x)\,dx$ and $u(x) = \exp(2\sqrt{14}\,x - 2\sqrt{14}\,x^2 + \frac{1}{3}x^3)$

15. Using $S(x) = 3x$, $y = 3x + [Ce^{-5x^2/6} - \frac{1}{9}e^{-5x^2/6}\int e^{5x^2/6}\,dx]^{-1}$

Section 1.8

1. $I(t) = \dfrac{5}{49+25\omega}[(\frac{7}{5}-\omega)\cos(\omega t) + (\frac{7}{5}+\omega)\sin(\omega t)] + \left[2 - \dfrac{5}{49+25\omega}(\frac{7}{5}-\omega)\right]e^{-7t/5}$

3. $I(t) = \dfrac{1}{6A}[\frac{1}{48}\sin(\omega t) - \omega\cos(\omega t)] + \dfrac{\omega t}{6A}[\frac{1}{48}\cos(\omega t) + \omega\sin(\omega t)] - \dfrac{\omega}{288A^2}[\frac{1}{48}\cos(\omega t) + \omega\sin(\omega t)]$
$- \dfrac{\omega^2}{6A^2}[\frac{1}{48}\sin(\omega t) - \omega\cos(\omega t)] + Ke^{-t/48}$, where $A = \dfrac{1}{48^2} + \omega^2$ and $K = 2 + \dfrac{\omega}{6A} + \dfrac{\omega}{13{,}824A^2} - \dfrac{\omega^2}{6A^2}$.

5. **(a)** $I'(t) = \dfrac{E}{L}e^{-Rt/L} > 0$; hence, I increases with t **(b)** $t_0 = -\dfrac{L}{R}\ln(0.37)$, about $0.9943\dfrac{L}{R}$

(c) If $I(0) \neq 0$, $t_0 = -\dfrac{L}{R}\ln\left(\dfrac{0.37E}{E - I(0)}\right)$

7. $Q(t) = -\dfrac{A\omega RC^2}{1+R^2C^2\omega^2}\left[-\dfrac{1}{\omega RC}\cos(\omega t) - \sin(\omega t)\right] - RC\left(I_0 - \dfrac{A\omega^2RC^2}{1+R^2C^2\omega^2}\right)e^{-t/RC} + Q_0 + RC\left(I_0 - \dfrac{A\omega^2RC^2}{1+R^2C^2\omega^2}\right)$
$- \dfrac{AC}{1+R^2C^2\omega^2}$

9. $I(t) = -\dfrac{A}{23}e^{-t} + ke^{-2t/25}$

11. $I(t) = \dfrac{-\omega ARC^2}{1+R^2C^2\omega^2}\left[\dfrac{1}{RC}\sin(\omega t) - \omega\cos(\omega t)\right] - \dfrac{BC}{1-RC}e^{-t} + ke^{-t/RC}$ if $RC \neq 1$;

if $RC = 1$, then $I(t) = \dfrac{-\omega ARC^2}{1+R^2C^2\omega^2}\left[\dfrac{1}{RC}\sin(\omega t) - \omega\cos(\omega t)\right] - \dfrac{B}{R}te^{-t} + Ke^{-t}$

13. $I(t) = ke^{-Rt/L} + \dfrac{A}{R} + \dfrac{B}{L}e^{-R/L}e^{-Rx/L}\displaystyle\sum_{j=0}^{\alpha}\dfrac{(-1)^j j!\,x^{\alpha-j}}{(\alpha-j)!\left(\dfrac{R}{L}\right)^{j+1}}$

15. $I = \dfrac{2\omega RC^2}{1+R^2C^2\omega^2}e^{-t/RC} + \dfrac{2RC^2}{1+R^2C^2\omega^2}\left[\dfrac{1}{RC}\sin(\omega t) - \omega\cos(\omega t)\right]$

17. If $R \neq L$, then $Q(t) = \dfrac{t}{R} - \dfrac{1}{L}\left(\dfrac{1}{\dfrac{R}{L}-1}\right)e^{-t} - \dfrac{L}{R}\left(\dfrac{1}{L-R} - \dfrac{1}{R}\right)e^{-Rt/L} + \dfrac{1}{L}\left(\dfrac{1}{\dfrac{R}{L}-1}\right) + \dfrac{L}{R}\left(\dfrac{1}{L-R} - \dfrac{1}{R}\right)$;

if $R = L$, then $Q(t) = \dfrac{t}{R}(1 - e^{-t})$.

Section 1.9

1. $y = \dfrac{x}{2} - \dfrac{1}{4} + 11\dfrac{e^2e^{-2x}}{4}$ **3.** $\dfrac{y}{y+x} = \dfrac{6x}{7}$ **5.** $y = \dfrac{2x^4}{3} + \dfrac{x}{3}$ **7.** $y = -\frac{1}{2}e^{2x} - \frac{7}{2}e^{4x}$

9. $y = -\frac{52}{17}e^{4\pi}e^{-4x} + \frac{4}{17}\sin(x) - \frac{1}{17}\cos(x)$

11. $y_1 = 4x$, $y_2 = 4\left(x - \dfrac{x^2}{2}\right)$, $y_3 = 4\left(x - \dfrac{x^2}{2} + \dfrac{x^3}{6}\right)$, $y_4 = 4\left(x - \dfrac{x^2}{2} + \dfrac{x^3}{6} - \dfrac{x^4}{24}\right)$, $y_5 = 4\left(x - \dfrac{x^2}{2} + \dfrac{x^3}{6} - \dfrac{x^4}{24} + \dfrac{x^5}{120}\right)$

13. $y_1 = 1 + \dfrac{x^2}{2}$, $y_2 = 1 + \dfrac{x^2}{2} + \dfrac{x^4}{8}$, $y_3 = 1 + \dfrac{x^2}{2} + \dfrac{x^4}{8} + \dfrac{x^6}{48}$, $y_4 = 1 + \dfrac{x^2}{2} + \dfrac{x^4}{8} + \dfrac{x^6}{48} + \dfrac{x^8}{384}$,
$y_5 = 1 + \dfrac{x^2}{2} + \dfrac{x^4}{8} + \dfrac{x^6}{48} + \dfrac{x^8}{384} + \dfrac{x^{10}}{3840}$

15. $y_1 = -4 + 2x - \dfrac{x^2}{2} = y_2 = y_3 = y_4 = y_5$ (y_1 is the solution)

17. $y_1 = -\frac{1}{2} + 4x + \frac{x^2}{2}$, $y_2 = \frac{11}{6} - \frac{x}{2} + \frac{5x^2}{2} + \frac{x^3}{6}$

$y_3 = \frac{25}{24} + \frac{11x}{6} + \frac{x^2}{4} + \frac{5x^3}{6} + \frac{x^4}{24}$,

$y_4 = \frac{149}{120} + \frac{25x}{24} + \frac{17x^2}{12} + \frac{x^3}{12} + \frac{5x^4}{24} + \frac{x^5}{120}$

$y_5 = \frac{149}{120} + \frac{149}{120}x + \frac{49}{48}x^2 + \frac{17}{36}x^3 + \frac{x^4}{48} + \frac{x^5}{24} + \frac{x^6}{720}$

19. $y_1 = 1 - \cos(x)$; $y_2 = 1 - \cos(x) + \frac{1}{2}\cos^2(x)$;
$y_3 = 1 - \cos(x) + \frac{1}{2}\cos^2(x) - \frac{1}{6}\cos^3(x)$
$y_4 = 1 - \cos(x) + \frac{1}{2}\cos^2(x) - \frac{1}{6}\cos^3(x) + \frac{1}{24}\cos^4(x)$
$y_5 = 1 - \cos(x) + \frac{1}{2}\cos^2(x) - \frac{1}{6}\cos^3(x) + \frac{1}{24}\cos^4(x) - \frac{1}{120}\cos^5(x)$

31. Two solutions are $y(x) = 0$ and $y_2(x) = (\frac{2}{3}x)^{3/2}$; note that $\frac{\partial f}{\partial y} = \frac{1}{3}y^{-1/3}$ is not continuous in any rectangle about (0, 0).

Section 1.10

In 1–19, the exact solution is given.

1. $y = -1 - x + 5e^{x-2}$ **3.** $y = -\frac{1}{2}x^2 - \frac{1}{2}x - \frac{1}{4} + \frac{29}{4}e^{2x-4}$ **5.** $y = 4e^{x^2/2}$ **7.** $y = \frac{2x^2}{3} - \frac{9}{x}$

9. $y = \frac{2}{5}\sin(x) - \frac{1}{5}\cos(x) + \frac{9}{5}e^{-2(x-\pi)}$ **11.** $y = 3(x-1) - 4e^{-(x-1)}$ **13.** $y = \frac{-e^{-x}}{x}(x+1) - \frac{3}{x}$

15. $y = 3 - x - 6e^{2-x}$ **17.** $y = \frac{x^5}{2} - x - \frac{233}{54}x^3$ **19.** $y = \frac{1}{4}(2x+1) + \frac{9}{4}e^{2(x-1)}$

Section 1.11

1. $y = 2x + C$ **3.** $y = Cx^2$ **5.** $y = Cx^{-3}$ **7.** $y = Ce^{4x}$ **9.** $\frac{1}{2}\ln|2y^2 + xy + x^2| + \frac{1}{\sqrt{7}}\arctan\left(\frac{4y+x}{x\sqrt{7}}\right) = C$

11. $\frac{1}{2}y^2 = -x + C$ **13.** $\sqrt{1-y^2} - \ln\left|\frac{1+\sqrt{1-y^2}}{y}\right| = |x| + C$ **15.** $\frac{4}{3}y^{3/2} = -x + C$ **17.** $y = -\frac{3}{2}x + C$

19. $y^2 = \ln|x| - \frac{x^2}{2} + C$ **21.** $\ln|x| + C = \arctan\left(\frac{y}{x}\right) - \frac{1}{2}\ln\left|1 + \frac{y^2}{x^2}\right|$

23. $\sqrt{3}\arctan\left(\frac{y-1}{x}\right) - \frac{1}{2}\ln(x^2 + (y-1)^2) = C$ **25.** $\sqrt{3}\arctan\left(\frac{y}{1+x}\right) - \frac{1}{2}\ln((x+1)^2 + y^2) = C$

27. $-y + 2\ln|1+y| = x + C$ **29.** $y = \frac{4}{3}\ln|\sqrt{3}x - 1| + \frac{1}{\sqrt{3}}x + C$ **33.** $r = Ce^{\theta}$ **35.** $\frac{\theta^2}{2} = -2\ln|r| + C$

Supplementary Problems

1. $\ln|x| + C = \frac{1}{2}\arctan\left(\frac{y-\frac{3}{10}}{x-\frac{1}{5}} - 1\right) - \frac{1}{2}\ln\left|2\left(\frac{y-\frac{3}{10}}{x-\frac{1}{5}}\right) - 4\left(\frac{y-\frac{3}{10}}{x-\frac{1}{5}}\right) + 4\right|$ **3.** $-4ye^x + 3x^2e^x - 6e^x(x-1) = C$

5. $-y^2 + 3x^2 = C$ **7.** $\frac{1}{2}\ln|(y+1)^2 - 2(x-2)(y+1) - (x-2)^2| + \frac{1}{2\sqrt{2}}\ln\left|\frac{(y+1) - (x-2) - \sqrt{2}(x-2)}{(y+1) - (x-2) + \sqrt{2}(x-2)}\right| = C$

9. $4x^3e^y + \cos(x-y) = C$ **11.** $x - 2y + \ln|1 + x - y| = C$

13. $\frac{1}{2}\ln\left|\left(\frac{y+4}{x+2}\right)^2 + 3\left(\frac{y+4}{x+2}\right) - 4\right| + \frac{1}{5}\operatorname{arctanh}\left(\frac{2\left(\frac{y+4}{x+2}\right) + 3}{5}\right) = -\ln|x+2| + C$ **15.** $y = x + \frac{1}{C - \frac{x^2}{2}}$

17. $y = Ce^x - \frac{1}{5}[2\sin(2x) - \cos(2x)]$ **19.** $2\left(x - \frac{y}{2}\right)^2 + 12\left(x - \frac{y}{2}\right) = 11x + C$

21. $\dfrac{2}{\sqrt{3}}\arctan\left(\dfrac{2\frac{y}{x}-1}{\sqrt{3}}\right)=\ln|x|+\dfrac{2}{\sqrt{3}}\arctan\left(\dfrac{7}{\sqrt{3}}\right)$ **23.** $3x^3y^2+4x^2=38{,}416$

25. $y=(2-\pi^2)x^{-3}+\dfrac{2\sin(x)}{x^2}+\left(\dfrac{2-x^2}{x^3}\right)\cos(x)$ **27.** $8x^3y+e^x\sin(y)=8+e\sin(1)$

29. $y=\dfrac{1}{9}e^{-x}+\left(6-\dfrac{e}{9}\right)e^{8(x+1)}$

31. **(a)** If $C(t)$ is the number of cubic feet of carbon monoxide in the room at time t, then $C'(t)=0.2-\dfrac{0.2}{2000}C$; so $C(t)=2000(1-e^{-0.2t/2000})=2000(1-e^{-(0.0001)t})$. The concentration of carbon monoxide is $C(t)$ divided by the room volume, or $C(t)/2000$. **(b)** At about 130.85 minutes or 2.18 hours

33. $-\frac{1}{16}[\frac{1}{2}(1-4y)^2-2(1-4y)+\ln|1-4y|]-\frac{1}{8}[1-4y-\ln|1-4y|]+\ln|1-4y|=x+C$

35. **(a)** $2e^{(5/2)\ln(1.7/2)}$, about 1.3322 grams **(b)** In $\dfrac{80\ln(1/2)}{\ln(1.7/2)}$ years (about 341.2 years)

39. **(a)** $\dfrac{1}{\sqrt{8A^2-1}}\arctan\left(\dfrac{2A\frac{y}{x}-1}{\sqrt{8A^2-1}}\right)-\dfrac{1}{2}\ln\left|A\left(\dfrac{y}{x}\right)^2-\dfrac{y}{x}+2A\right|=\ln|x|+C$, where $A=\tan(30°)$, about 0.5774.

(c) $\arctan\dfrac{y}{x}-\dfrac{1}{2}\ln\left|\dfrac{y^2}{x^2}+1\right|=\ln|x|+C$

41. $I(t)=\dfrac{e^{-2t}}{L\left[\omega^2+\left(2+\frac{R}{L}\right)^2\right]}\left[-\left(2+\dfrac{R}{L}\right)\cos(\omega t)+\omega\sin(\omega t)\right]+Ke^{-Rt/L}$

43. **(a)** $y_1=-1-2x,\quad y_2=-3+2x-2x^2$
$y_3=-\frac{5}{3}-2x+2x^2-\frac{4}{3}x^3$
$y_4=-\frac{7}{3}+\frac{2}{3}x-2x^2+\frac{4}{3}x^3-\frac{2}{3}x^4$
$y_5=-\frac{31}{15}-\frac{2}{3}x+\frac{2}{3}x^2-\frac{4}{3}x^3+\frac{2}{3}x^4-\frac{4}{15}x^5$

45. $4x^2e^{-x}y=C$ **47.** $\ln|x|+C=-\ln\left|-6\dfrac{y^2}{x^2}-3\dfrac{y}{x}+4\right|-\dfrac{20}{\sqrt{105}}\operatorname{arctanh}\left(\dfrac{-12\frac{y}{x}-3}{\sqrt{105}}\right)$ **49.** $2x+\sin(xy)=C$

51. $-3x^{-1/3}y^{-1/3}-\frac{3}{8}x^{8/3}=C$ **53.** $y=3(x^2+2x+2)+Ce^x$ **55.** $y=Ce^x-2x^2-4x-4$

Chapter 2

Section 2.1

1. $c_1=\frac{8}{3},\ c_2=-\frac{5}{3}$ **3.** $c_1=3,\ c_2=15$ **5.** $c_1=\frac{1}{4},\ c_2=\frac{3}{4}$ **7.** $c_1=16,\ c_2=-32$

9. $c_1=0,\ c_2=e^{\pi/2}$ **11.** $c_1=\frac{4}{3},\ c_2=0$ **13.** $c_1=\dfrac{13+3\sqrt{13}}{26},\ c_2=\dfrac{13-3\sqrt{13}}{26}$ **15.** $c_1=\frac{20}{189},\ c_2=\frac{185}{224}$

Section 2.2

1. $y(x)=c_1\cosh(kx)+c_2\sinh(kx)$; $W=k\cosh(2kx)$ **3.** $y(x)=c_1e^{2x}+c_2e^{-6x}$; $W=-8e^{-4x}$

5. $y(x)=\dfrac{1}{x}[c_1\cos(\ln(x))+c_2\sin(\ln(x))]$; $W=\dfrac{1}{x^3}$ **7.** $y(x)=e^{5x}(c_1+c_2x)$; $W=e^{10x}$

9. $y(x)=\dfrac{1}{x}(c_1+c_2\ln(x))$; $W=\dfrac{1}{x^3}$ **11.** $y(x)=c_1e^{3x}+c_2e^{-14x}$; $W=-17e^{-14x}$

13. $y(x)=x^4(c_1+c_2\ln(x))$; $W=x^7$ **15.** $y(x)=\sqrt{\dfrac{2}{\pi x}}[c_1\sin(x)+c_2\cos(x)]$; $W=\dfrac{-2}{\pi x}$

17. $y=\cosh(k\pi)\cosh(kx)-\sinh(k\pi)\sinh(kx)$ **19.** $y=-\frac{2}{5}e^{-8x}+\frac{12}{5}e^{-3x}$ **21.** $y=\frac{12}{5}e^{3(x+1)}+\frac{3}{5}e^{-2(x+1)}$

23. $y = 7e^{-4(x-2)} - 4e^{-7(x-2)}$ **25.** $y = \sqrt{\frac{2}{\pi x}}\left[\left(\frac{5}{2\sqrt{2}} - \frac{8\pi}{\sqrt{2}}\right)\sin(x) + \frac{5\pi}{\sqrt{2}}\cos(x)\right]$

Section 2.3

1. $y = c_1 e^{(2+\sqrt{3})x} + c_2 e^{(2-\sqrt{3})x}$ **3.** $y = e^{8x}(c_1 + c_2 x)$ **5.** $y = c_1 e^{(-3+2\sqrt{3})x} + c_2 e^{(-3-2\sqrt{3})x}$
7. $y = c_1 e^{(-1+\sqrt{2})x} + c_2 e^{(-1-\sqrt{2})x}$ **9.** $y = c_1 e^{(-1+\sqrt{17})x} + c_2 e^{(-1-\sqrt{17})x}$ **11.** $y = e^{7x}(c_1 + c_2 x)$
13. $y = c_1 e^{(-7+\sqrt{69})x/2} + c_2 e^{(-7-\sqrt{69})x/2}$ **15.** $y = e^{-6x}(c_1 + c_2 x)$ **17.** $y = c_1 e^{(-3+\sqrt{17})x/2} + c_2 e^{(-3-\sqrt{17})x/2}$
19. $y = e^{5x}(c_1 + c_2 x)$ **21.** $y = c_1 e^{(7+\sqrt{47})x} + c_2 e^{(7-\sqrt{47})x}$ **23.** $y = c_1 e^{(-7+\sqrt{51})x} + c_2 e^{(-7-\sqrt{51})x}$
25. $y = e^{-9x}(c_1 + c_2 x)$ **27.** $y = e^{x-1}(5 - 4x)$ **29.** $y = e^x$ **31.** $y = e^{-6x}(-2 - 15x)$
33. $y = e^{(-3+\sqrt{17})x/2} + e^{(-3-\sqrt{17})x/2}$ **35.** $y = e^{3(x+1)}(5 + 4x)$

Section 2.4

1. $-e$ **3.** 1 **5.** -1 **7.** $e[\cos(4) + i\sin(4)]$ **9.** $\cos(1) + i\sin(1)$

Section 2.5

1. $y = e^{2x}[c_1\cos(2x) + c_2\sin(2x)]$ **3.** $y = e^{-x/2}\left[c_1\cos\left(\frac{\sqrt{3}x}{2}\right) + c_2\sin\left(\frac{\sqrt{3}x}{2}\right)\right]$ **5.** $y = e^{-11x}(c_1 + c_2 x)$
7. $y = c_1 + c_2 e^{4x}$ **9.** $y = c_1 e^{(2+\sqrt{2})x} + c_2 e^{(2-\sqrt{2})x}$ **11.** $y = c_1 e^{(-5+\sqrt{26})x} + c_2 e^{(-5-\sqrt{26})x}$
13. $y = e^{7x}(c_1 + c_2 x)$ **15.** $y = e^{3x/2}\left[c_1\cos\left(\frac{\sqrt{23}x}{2}\right) + c_2\sin\left(\frac{\sqrt{23}x}{2}\right)\right]$ **17.** $y = e^{-5x}(c_1 + c_2 x)$
19. $y = c_1 e^{(-11+\sqrt{113})x/2} + c_2 e^{(-11-\sqrt{113})x/2}$ **21.** $y = e^{-x}\left[\cos(\sqrt{3}x) + \frac{1}{\sqrt{3}}\sin(\sqrt{3}x)\right]$ **23.** $y = \frac{1}{2}e^{-3x} - \frac{1}{2}e^x$
25. $y = \left(\frac{-1+\sqrt{2}}{2\sqrt{2}}\right)e^{(2+\sqrt{2})x} + \left(\frac{1+\sqrt{2}}{2\sqrt{2}}\right)e^{(2-\sqrt{2})x}$
27. $y = e^{-x}\left\{\frac{e}{\sqrt{5}}[-\sin(\sqrt{5}) - 3\sqrt{5}\cos(\sqrt{5})]\cos(\sqrt{5}x) + \frac{e}{\sqrt{5}}[\cos(\sqrt{5}) - 3\sqrt{5}\sin(\sqrt{5})]\sin(\sqrt{5}x)\right\}$
29. $y = e^{2x}\left\{\frac{1}{\sqrt{3}e^{2\pi}}[\sin(\sqrt{3}\pi) + \sqrt{3}\cos(\sqrt{3}\pi)]\cos(\sqrt{3}x) + \frac{1}{\sqrt{3}e^{2\pi}}[-\cos(\sqrt{3}\pi) + \sqrt{3}\sin(\sqrt{3}\pi)]\sin(\sqrt{3}x)\right\}$
31. $\theta = c_1\cos\left(\sqrt{\frac{g}{L}}\,t\right) + c_2\sin\left(\sqrt{\frac{g}{L}}\,t\right)$
33. **(a)** The differential equation is $[(4xy') + (x^2 - 4)y]' = 0$, with general solution

$$y = Cxe^{-x^2/8}\int \frac{1}{x^2}e^{x^2/8}\,dx + Kxe^{-x^2/8}$$

(c) $y = K + Ce^{-3x}$ **(e)** $y = Cx^{-1} + Kx^{-2}$
37. $y = \left(\frac{3 + e^{-3-\sqrt{17}}}{e^{-3+\sqrt{17}} - e^{-3-\sqrt{17}}}\right)e^{(-3+\sqrt{17})x/2} + \left(\frac{3 + e^{-3+\sqrt{17}}}{e^{-3-\sqrt{17}} - e^{-3+\sqrt{17}}}\right)e^{(-3-\sqrt{17})x/2}$
39. $y = \left(\frac{1}{e^{20} - e^{24}}\right)e^{6x} + \left(\frac{1}{e^{20} - e^{16}}\right)e^{4x}$ **41.** $y = \frac{1}{\sin(2)}\sin(2x)$ **43.** $y = \left(\frac{5(1 - e^{-3})}{e^{-6} - e^{-3}}\right)e^{-2x} + \left(\frac{5(e^{-6} - 1)}{e^{-6} - e^{-3}}\right)e^{-x}$
45. $y = \left(\frac{3e^{-15-5\sqrt{11}} + 2e^{-6-2\sqrt{11}}}{e^{-21-3\sqrt{11}} - e^{-21+3\sqrt{11}}}\right)e^{(-3+\sqrt{11})x} + \left(\frac{2e^{-6+2\sqrt{11}} + 3e^{-15+5\sqrt{11}}}{e^{-21+3\sqrt{11}} - e^{-21-3\sqrt{11}}}\right)e^{(-3-\sqrt{11})x}$

Section 2.6

3. $\theta = c_1\cos\left(\sqrt{\frac{g}{L}}\,t\right) + c_2\sin\left(\sqrt{\frac{g}{L}}\,t\right)$ **5.** At most once (possibly none)

7. Undamped: $y = A\cos\left(\sqrt{\frac{k}{m}}\,t\right)$

Overdamped: $y = \left(\frac{Ar_2}{r_2 - r_1}\right)e^{r_1 t} - \left(\frac{Ar_1}{r_2 - r_1}\right)e^{r_2 t}$

Critically damped: $y = Ae^{-ct/2m}\left(1 + \frac{c}{2m}t\right)$

Underdamped: $y = de^{-ct/2m}\cos(\bar{\omega}t - \delta)$, where $\delta = \arctan(c/2m\bar{\omega})$ and $d = A\sqrt{1 + \frac{c^2}{4m^2\bar{\omega}^2}}$

9. (a) $y = \left(\frac{Ar_2}{r_2 - r_1}\right)e^{r_1 t} - \left(\frac{Ar_1}{r_2 - r_1}\right)e^{r_2 t}$ (b) $y = \left(\frac{A}{r_1 - r_2}\right)e^{r_1 t} + \left(\frac{A}{r_2 - r_1}\right)e^{r_2 t}$

11. $y(0) = A$ and $y'(0) = 0$ give us the underdamped solution of Problem 7;

$y(0) = 0$ and $y'(0) = A$ give us $y = \frac{A}{\bar{\omega}}e^{-ct/2m}\sin(\bar{\omega}t)$.

13. As $c \to \infty$, $y \to 0$ and the motion dies out.

15. Note that $\bar{\omega} \to 0$ as m gets larger; so the frequency $\bar{\omega}/2\pi$ becomes smaller with larger m.

23. $y = e^{-0.4t}[3.72\cos(0.6460t) + 5.0130\sin(0.6460t)]$ 25. $y = e^{-0.8847t}(3.24 + 4.3864t)$

27. $y = e^{-0.5727t}[2.36\cos(4.5248t) + 0.6081\sin(4.5248t)]$

Section 2.7

1. $y = c_1e^{(2+\sqrt{2})x} + c_2e^{(2-\sqrt{2})x} + y_p$ 3. $y = e^{-x}(c_1 + c_2x) + y_p$ 5. $y = c_1e^{(3+\sqrt{10})x} + c_2e^{(3-\sqrt{10})x} + y_p$

7. $y = c_1e^{(-4+3\sqrt{2})x} + c_2e^{(-4-3\sqrt{2})x} + y_p$ 9. $y = c_1e^{(-8+\sqrt{66})x} + c_2e^{(-8-\sqrt{66})x} + y_p$

11. $y = c_1e^{(2+\sqrt{6})x} + c_2e^{(2-\sqrt{6})x} + y_p$ 13. $y = c_1e^{2x} + c_2e^{x} + y_p$

15. $y = e^{x/2}\left[c_1\cos\left(\frac{\sqrt{39}\,x}{2}\right) + c_2\sin\left(\frac{\sqrt{39}\,x}{2}\right)\right] + y_p$

17. $y = -2\cos(x) + 9\sin(x) + x^3 - 6x$ 19. $y = \left(\frac{2\sqrt{7} - 1}{18\sqrt{7}}\right)e^{(2+\sqrt{7})x} - \left(\frac{2\sqrt{7} + 1}{18\sqrt{7}}\right)e^{(2-\sqrt{7})x} - \frac{1}{3}x + \frac{7}{9}$

21. $y = -\frac{29}{60}e^{-4x} - \frac{23}{5}e^{x} + \frac{4}{3}e^{2x} - \frac{1}{4}$ 23. $y = (\frac{1}{2}\pi^2 - \pi + \frac{1}{2})e^{\pi}e^{-x}\cos(x) + (\frac{1}{2}\pi^2 - \frac{1}{2})e^{\pi}e^{-x}\sin(x) + \frac{1}{2}x^2 + \frac{1}{2} - x$

25. $y = -\frac{7}{32}\cos(\sqrt{7}\,x)e^{-x} - \frac{19}{32\sqrt{7}}\sin(\sqrt{7}\,x)e^{-x} + \frac{3}{32}e^{4x} + \frac{1}{8}$

27. $y = \frac{3}{17}e^{x/2}\cos(\sqrt{15}\,x/2) + \frac{1}{17\sqrt{15}}e^{x/2}\sin(\sqrt{15}\,x/2) - \frac{3}{17}\cos(3x) + \frac{5}{17}\sin(3x)$

29. $y = \frac{e^{1-\sqrt{6}}}{2\sqrt{6}}(10 + 7\sqrt{6})e^{(-1+\sqrt{6})x} - \frac{e^{1+\sqrt{6}}}{2\sqrt{6}}(10 - 7\sqrt{6})e^{(-1-\sqrt{6})x} - 4$

Section 2.8

1. $y_h = c_1e^{(-3+\sqrt{14})x} + c_2e^{(-3-\sqrt{14})x}$; $y_p = -\frac{3}{5}x^2 - \frac{36}{25}x - \frac{146}{125}$

3. $y_h = c_1e^{(-1+\sqrt{29})x/2} + c_2e^{(-1-\sqrt{29})x/2}$; $y_p = -\frac{1}{7}x^3 - \frac{3}{49}x^2 - \frac{48}{343}x + \frac{1282}{2401}$

5. $y_h = e^{-x/2}\left[c_1\cos\left(\frac{\sqrt{55}\,x}{2}\right) + c_2\sin\left(\frac{\sqrt{55}\,x}{2}\right)\right]$; $y_p = \frac{1}{14}x + \frac{1}{196} - \frac{5}{17}\sin(3x) - \frac{3}{17}\cos(3x)$

7. $y_h = c_1e^{(-1+\sqrt{13})x} + c_2e^{(-1-\sqrt{13})x}$; $y_p = -\frac{1}{12}x^2 + \frac{1}{18}x - \frac{1}{216} - \frac{2}{9}e^{-3x}$

9. $y_h = c_1\cos(x) + c_2\sin(x)$; $y_p = x^2 - 6 + \frac{1}{10}\sinh(3x)$

11. $y_h = e^{2x}[c_1\cos(\sqrt{2}\,x) + c_2\sin(\sqrt{2}\,x)]$; $y_p = \frac{1}{2}e^{2x} - \frac{1}{2}e^{4x}$

13. $y_h = c_1e^{-x} + c_2xe^{-x}$; $y_p = -\frac{3}{2}x^2e^{-x} + \frac{4}{3}x^3e^{-x} + 1$

15. $y_h = c_1e^{-7x} + c_2e^{-x}$; $y_p = -\frac{4}{7} - \frac{66}{135}\cosh(2x) + \frac{96}{135}\sinh(2x)$

17. $y_h = c_1e^{3x} + c_2e^{-2x}$; $y_p = -\frac{4}{3}x^3 + \frac{2}{3}x^2 - \frac{14}{9}x + \frac{71}{54}$

19. $y_h = c_1\cosh(2x) + c_2\sinh(2x)$; $y_p = \frac{5}{4}x\cosh(2x)$

21. $y_h = c_1 e^{-2x} + c_2 e^{x};\quad y_p = -\dfrac{e^{2x}}{12}(2x-1)e^{-2x} + \dfrac{1}{3}e^{-x}(-x-1)e^{x} = -\dfrac{x}{2} - \dfrac{1}{4}$

23. $y_h = e^{x/2}\left[c_1 \cos\left(\dfrac{\sqrt{7}\,x}{2}\right) + c_2 \sin\left(\dfrac{\sqrt{7}\,x}{2}\right)\right];\quad y_p = \frac{4}{7}e^{x/2}$

25. $y_h = c_1 e^{(3+\sqrt{7})x} + c_2 e^{(3-\sqrt{7})x};\quad y_p = \frac{1}{37}\cos(x) - \frac{6}{37}\sin(x)$ **27.** $y_h = c_1 e^{2x} + c_2 e^{-3x};\quad y_p = -\dfrac{x}{6} - \dfrac{1}{36}$

29. $y_h = c_1 \cos(3x) + c_2 \sin(3x);\quad y_p = (\frac{1}{3}\ln|\cos(3x)|)\cos(3x) + x\sin(3x)$

31. $y_h = c_1 e^{-3x} + c_2 e^{4x};\quad y_p = -\frac{2}{15}\cosh(2x) + \frac{1}{30}\sinh(2x) - \frac{1}{12}$

33. $y_h = c_1 e^{2x} + c_2 e^{-4x};\quad y_p = \frac{1}{12}(e^{2x} + e^{-2x})e^{2x} - \frac{1}{12}(\frac{1}{4}e^{8x} - \frac{1}{2}e^{4x})e^{-4x} = \frac{1}{16}e^{4x} + \frac{1}{8}$

35. $y_h = c_1 e^{(-1+\sqrt{29})x/2} + c_2 e^{(-1-\sqrt{29})x/2};\quad y_p = -\frac{2}{7}x + \frac{19}{49}$ **37.** $y = -\frac{87}{25}e^{-x} - \frac{21}{5}xe^{-x} + \frac{12}{25}\cos(2x) + \frac{9}{25}\sin(2x)$

39. $y = \frac{7}{2} - 3e^{-x} - \sin(2x) - \frac{1}{2}\cos(2x) - 4x$ **41.** $y = \frac{111}{24}e^{-4x} + \frac{124}{189}e^{3x} + \frac{1}{12}x^2 + \frac{1}{72}x - \frac{131}{864}$

43. $y = \dfrac{-13{,}456}{925{,}100}e^{4x} - \dfrac{7}{638}e^{-7x} + \dfrac{37}{1450}\cos(3x) - \dfrac{9}{1450}\sin(3x)$ **45.** $y = (\frac{1}{4} - \pi)\sin(4x) + x\sin(4x)$

47. $W = 0$, and the formulas for u and v are invalid.

Section 2.9

In 1–9, y_T denotes the transient solution, and y_{SS} denotes the steady-state solution.

1. $y_T = c_1 e^{(-3+\sqrt{8})t} + c_2 e^{(-3-\sqrt{8})t};\quad y_{SS} = \dfrac{-18\omega}{(1-\omega^2)^2 + 36\omega^2}\cos(\omega t) + \dfrac{3(1-\omega^2)}{(1-\omega^2)^2 + 36\omega^2}\sin(\omega t)$

3. $y_T = c_1 e^{-2t} + c_2 e^{-t/2};\quad y_{SS} = -\frac{3}{17}\cos(2t) + \frac{5}{17}\sin(2t) - \frac{32}{481}\cos(3t) + \frac{30}{481}\sin(3t)$

5. $y_T = c_1 e^{(-3+\sqrt{3})x/2} + c_2 e^{(-3-\sqrt{3})x/2};\quad y_{SS} = \frac{2}{37}\cos(t) + \frac{12}{37}\sin(t) + \frac{24}{119}\cos(3t) - \frac{10}{119}\sin(3t)$

7. $y_T = e^{-t/5}(c_1 + c_2 t);\quad y_{SS} = -\frac{158}{1261}\cos(4t) + \frac{16}{1261}\sin(4t) - \frac{3}{190}\cos(3t) + \frac{11}{95}\sin(3t)$

9. $y_T = e^{-t/8}\left[c_1 \cos\left(\dfrac{\sqrt{31}\,t}{8}\right) + c_2 \sin\left(\dfrac{\sqrt{31}\,t}{8}\right)\right];\quad y_{SS} = 4 + \frac{142}{2525}\cos(6t) - \frac{6}{2525}\sin(6t)$

11. $y = e^{-t/2}\left[c_1 \cos\left(\dfrac{\sqrt{3}\,t}{2}\right) + c_2 \sin\left(\dfrac{\sqrt{3}\,t}{2}\right)\right] + \dfrac{1}{3}e^{-2t}$

13. $y = e^{-t}(c_1 + c_2 t) - \frac{3}{50}\cos(2t) + \frac{2}{25}\sin(2t) + \frac{4}{161}\cos(4t) - \frac{15}{322}\sin(4t)$ **15.** $y = e^{-t}(c_1 + c_2 t) - \frac{1}{2}e^{-t}\sin(t)$

17. $y = e^{-t/12}\left[c_1 \cos\left(\dfrac{\sqrt{23}\,t}{12}\right) + c_2 \sin\left(\dfrac{\sqrt{23}\,t}{12}\right)\right] - \frac{9}{404}e^{-t}\sin(2t) + \frac{11}{404}e^{-t}\cos(2t)$

19. $y = c_1 e^{-t} + c_2 e^{-2t/5} + \frac{1}{2} - \frac{1}{3}te^{-t}$

21. $y = \dfrac{3}{173}e^{-t/4}\cos\left(\dfrac{\sqrt{47}\,t}{4}\right) + \dfrac{1007}{173\sqrt{47}}e^{-t/4}\sin\left(\dfrac{\sqrt{47}\,t}{4}\right) - \dfrac{3}{173}\cos(4t) - \dfrac{39}{346}\sin(4t)$; underdamped

23. $y = \left(\dfrac{95{,}809 + 50{,}627\sqrt{7}}{193{,}214}\right)e^{(-3+\sqrt{7})t} + \left(\dfrac{95{,}809 - 50{,}626\sqrt{7}}{193{,}214}\right)e^{(-3-\sqrt{7})t} + \dfrac{1}{37}\cos(t) + \dfrac{6}{37}\sin(t) - \dfrac{7}{373}\cos(3t) + \dfrac{18}{373}\sin(3t)$; overdamped

25. $y = e^{-t}(\frac{73}{25} + \frac{41}{10}t) - \cos(t) + \frac{2}{25}\cos(3t) - \frac{3}{50}\sin(3t)$; critically damped

27. $y = \dfrac{1}{8155}e^{-3t/2}\left[6231\cos\left(\dfrac{\sqrt{19}\,t}{2}\right) + \dfrac{80{,}909}{\sqrt{19}}\sin\left(\dfrac{\sqrt{19}\,t}{2}\right)\right] + \dfrac{2}{7} - \dfrac{58}{1165}\cos(6t) + \dfrac{36}{1165}\sin(6t)$; underdamped

29. $y = e^{-t/6}\left[\dfrac{13}{5}\cos\left(\dfrac{\sqrt{59}\,t}{6}\right) + \dfrac{103}{5\sqrt{59}}\sin\left(\dfrac{\sqrt{59}\,t}{6}\right)\right] + \dfrac{2}{5}$; underdamped

31. $y = e^{-0.1068t}[0.2454\cos(1.8712t) + 1.8549\sin(1.8712t)] - 0.0154\cos(3t) + 0.0018\sin(3t)$

33. $y = e^{-0.0691t}[2.8136\cos(0.3492t) + 7.1043\sin(0.3492t)] + 0.1864e^{-t}$

35. $y = e^{-0.5667t}[-2.8512\cos(0.5879t) - 2.8393\sin(0.5879t)] + 1.7200 - 0.0688\cos(4t) + 0.0203\sin(4t)$

37. $y = e^{-0.3449t}[1.4946\cos(1.1516t) + 1.0012\sin(1.1516t)] + (0.2832t^2 + 0.8679t + 1.5054)e^{-t}$

39. $y = e^{-0.1454t}[0.3427\cos(0.8065t) + 1.3482\sin(0.8065t)] + 0.0433t^2 - 0.0375t - 0.1127$

Section 2.10

3. $I = -1.0004e^{-0.8337t} + 0.0004e^{-1999.1663t} - 0.0008\sin(40t) + \cos(40t)$
5. $I = 3.9844e^{-1.6699t} - 0.0071e^{-855.4730t} + 0.0227\cos(150t) + 0.0037\sin(150t)$
7. $I = 0.0044e^{-20.4168t} - 0.0042e^{-979.5832t} - 0.0002e^{-t}\cos(3t) - 0.0011e^{-t}\sin(3t)$
9. $I = 3.0190e^{-0.1137t} - 0.0065e^{-199.8863t} + 0.0001\sin(4t) - 0.0125\cos(4t)$
11. $I = 2.0130e^{-0.9184t} - 0.0197e^{-89.9907t} + 0.0067\cos(3t) + 0.0249\sin(3t)$
13. $I = 2.0035e^{-0.1334t} - 0.0017e^{-749.8666t} - 0.0033\cos(4t) + 0.0001\sin(4t) + 0.0015e^{-t}$
15. $I = 1.0214e^{-0.1137t} - 0.0125e^{-628.4577t} - 0.0051e^{-t} - 0.0038e^{-5t}$
17. $I = 1.0149e^{-4.8342t} - 0.0154e^{-318.2427t} - 0.0004\cos(2t+1) + 0.0009\sin(2t+1)$
19. $I = 1.0227e^{-1.7684t} - 0.0217e^{-221.7610t} - (0.0024t + 0.0010)e^{-4t}$

Section 2.11

1. $y = -2x + c_1x^2 + c_2$ 3. $x + c_2 = \sqrt{2}\ln\left|2\sqrt{\frac{y^2}{4} + \frac{y}{2} + c_1} + y + 1\right|$ 5. $y = c_1x^3 - \frac{x}{2} + c_2$

7. $y = -2x^2 + 5x + c_1e^{-(2/3)x} + c_2$ 9. $y = \frac{x^2}{6} + \frac{c_1}{x} + c_2$ 11. $y_2 = \frac{\sin(x)}{\sqrt{x}}$

13. $y_2 = xv$, where $v = \int x^{-2}e^{x^2/2}\,dx$ 15. $y_2 = x\int \frac{e^x}{x^2}\,dx$

17. $y_2 = (3x^2 - 1)\left[-\frac{1}{2}\ln|1 - x^2| + \frac{139}{270}\ln|3x^2 - 1| - \frac{13}{12\sqrt{3}}\ln\left|\frac{3x - \sqrt{3}}{3x + \sqrt{3}}\right| - \frac{112}{135}\frac{1}{3x^2 - 1} - \frac{11x}{6(3x^2 - 1)}\right]$

Section 2.12

1. $y = x^{3/2}(c_1x^{\sqrt{5}/2} + c_2x^{-\sqrt{5}/2})$ 3. $y = x^{3/2}\left[c_1\cos\left(\frac{\sqrt{7}\ln(x)}{2}\right) + c_2\sin\left(\frac{\sqrt{7}\ln(x)}{2}\right)\right]$

5. $y = x^{-3/2}\left[c_1\cos\left(\frac{\sqrt{31}\ln(x)}{2}\right) + c_2\sin\left(\frac{\sqrt{31}\ln(x)}{2}\right)\right]$ 7. $y = x^{-1/2}(c_1x^{\sqrt{21}/2} + c_2x^{-\sqrt{21}/2})$

9. $y = x^{-2}[c_1 + c_2\ln(x)]$ 11. $y = c_1x^3 + c_2x^2$ 13. $y = x^{5/2}\left[c_1\cos\left(\frac{\sqrt{23}\ln(x)}{2}\right) + c_2\sin\left(\frac{\sqrt{23}\ln(x)}{2}\right)\right]$

15. $y = x^{3/2}\left[c_1\cos\left(\frac{\sqrt{23}}{2}\ln(x)\right) + c_2\sin\left(\frac{\sqrt{23}}{2}\ln(x)\right)\right]$ 17. $y = x^{-3/2}\left[c_1\cos\left(\frac{\sqrt{11}\ln(x)}{2}\right) + c_2\sin\left(\frac{\sqrt{11}\ln(x)}{2}\right)\right]$

19. $y = x^{9/2}(c_1x^{\sqrt{41}/2} + c_2x^{-\sqrt{41}/2})$ 21. $y = x^{-1/2}(c_1x^{\sqrt{5}/2} + c_2x^{-\sqrt{5}/2})$ 23. $y = c_1x^2 + c_2x^{-2}$

25. $y = x^2[c_1 + c_2\ln(x)]$ 27. $y = \frac{1}{2}x^{-2}\sin[4\ln(x)]$ 29. $y = x^{-12}[-3 - 36\ln(x)]$

31. $y = -\frac{11}{192}x^5 + \frac{29}{3}x^{-1}$ 33. $y = x^{-1}\{\cos[6\ln(x)] + \frac{1}{6}\sin[6\ln(x)]\}$

35. $y = x^{9/2}\left[\frac{1}{\sqrt{41}}(-16 + 2\sqrt{41})x^{\sqrt{41}/2} + \frac{1}{\sqrt{41}}(16 + 2\sqrt{41})x^{-\sqrt{41}/2}\right]$

Section 2.13

1. $y = x^{3/2}\left[c_1\cos\left(\frac{\sqrt{19}\ln(x)}{2}\right) + c_2\sin\left(\frac{\sqrt{19}\ln(x)}{2}\right)\right]$

3. $y = e^{x/2}\left[c_1\cos\left(\frac{\sqrt{7}x}{2}\right) + c_2\sin\left(\frac{\sqrt{7}x}{2}\right)\right] - \frac{1}{8}e^{3x} + \frac{1}{2}x + \frac{1}{4}$

5. $y = x^{-1/2}(c_1x^{\sqrt{5}/2} + c_2x^{-\sqrt{5}/2})$ 7. $y = e^{15x}(c_1 + c_2x) + \frac{1}{196}e^x - \frac{60}{52{,}441}\cos(2x) - \frac{221}{52{,}441}\sin(2x)$

9. $y = e^{-2x}[c_1\cos(\sqrt{5}x) + c_2\sin(\sqrt{5}x)] + \frac{1}{14}e^x + \frac{1}{9}x^2 - \frac{8}{81}x + \frac{14}{729} - \frac{15}{89}\cos(2x) - \frac{24}{89}\sin(2x)$

11. $y = x^{3/2}(c_1x^{\sqrt{7}/2} + c_2x^{-\sqrt{7}/2})$

13. $y = e^x[c_1\cos(\sqrt{21}x) + c_2\sin(\sqrt{21}x)] - \frac{1}{22}x^2 - \frac{1}{121}x + \frac{186}{1331} + \frac{23}{525}\cosh(x) + \frac{2}{525}\sinh(x)$

15. $y = e^{-5x/2}(c_1 e^{\sqrt{33}x/2} + c_2 e^{-\sqrt{33}x/2}) - \frac{3}{68}\sin(2x) - \frac{5}{68}\cos(2x) - \frac{25}{181}\sin(4x) + \frac{45}{362}\cos(4x)$

17. $y = e^{-x}[c_1\cos(2x) + c_2\sin(2x)] + \frac{1}{5}x^3 - \frac{6}{25}x^2 - \frac{81}{125}x + \frac{472}{625}$ **19.** $y = c_1 e^{9x} + c_2 e^x - \frac{1}{8}xe^x + \frac{1}{9}x^2 + \frac{20}{81}x + \frac{182}{729}$

21. $y = 4x^{-2}\cos[5\ln(x)] + \frac{12}{5}x^{-2}\sin[5\ln(x)]$ **23.** $y = \frac{109}{144}e^{6x} + \frac{27}{144}e^{-2x} - \frac{1}{6}x + \frac{1}{18}$

25. $y = e^{-x} - \frac{918}{2401}e^{7x} - \frac{1}{7}x^3 + \frac{18}{49}x^2 - \frac{111}{343}x + \frac{918}{2401}$

Supplementary Problems

1. $y = x^{17/2}(c_1 x^{\sqrt{281}/2} + c_2 x^{-\sqrt{281}/2})$ **3.** $y = e^{-x}(c_1 e^{\sqrt{5}x} + c_2 e^{-\sqrt{5}x}) + e^x - \frac{13}{205}\cos(3x) + \frac{6}{205}\sin(3x)$

5. $y = e^{-x/2}(c_1 e^{\sqrt{5}x/2} + c_2 e^{-\sqrt{5}x/2}) + \frac{1}{5}e^{2x} - 2e^x$ **7.** $y = e^{4x}(c_1 + c_2 x) + \frac{1}{128}e^{-4x} + \frac{1}{4}x^2 e^{4x}$

9. $y = x^2(c_1 x^{\sqrt{3}} + c_2 x^{-\sqrt{3}})$ **11.** $y = \dfrac{x^3}{6} + \dfrac{x^2}{4} + \dfrac{x}{4} + c_1 e^{2x} + c_2$ **13.** $y = x^2(c_1 x^{\sqrt{2}} + c_2 x^{-\sqrt{2}})$

15. $y = e^{5x/2}\left[c_1\cos\left(\dfrac{\sqrt{11}\,x}{2}\right) + c_2\sin\left(\dfrac{\sqrt{11}\,x}{2}\right)\right] + \dfrac{1}{9}x^2 + \dfrac{10}{81}x + \dfrac{32}{729} - \dfrac{8}{89}\sin(x) - \dfrac{5}{89}\cos(x)$

17. $y = x^{3/2}\left[c_1\cos\left(\dfrac{\sqrt{47}\ln(x)}{2}\right) + c_2\sin\left(\dfrac{\sqrt{47}\ln(x)}{2}\right)\right]$ **19.** $y = e^{-13x/2}(c_1 e^{\sqrt{201}x/2} + c_2 e^{-\sqrt{201}x/2}) - \frac{1}{8}x^2 - \frac{13}{32}x - \frac{241}{1928}$

21. $y = \int \exp(c_1 e^{-x/2})\,dx + c_2$ **23.** $y = e^{x/2}(c_1 e^{\sqrt{33}x/2} + c_2 e^{-\sqrt{33}x/2}) - \frac{3}{8}x^2 + \frac{3}{32}x - \frac{187}{256}$

25. $y = e^{-3x/2}\left[c_1\cos\left(\dfrac{\sqrt{31}\,x}{2}\right) + c_2\sin\left(\dfrac{\sqrt{31}\,x}{2}\right)\right] + \dfrac{1}{10}x^3 - \dfrac{9}{100}x^2 - \dfrac{6}{1000}x + \dfrac{198}{10{,}000}$

27. $I = \dfrac{0.6}{4.25}[0.5\cos(2t) + 2\sin(2t)] - \dfrac{0.3}{4.25}e^{-0.5t}$; $Q = \dfrac{0.6}{4.25}[0.25\sin(2t) - \cos(2t)] + \dfrac{0.6}{4.25}e^{-0.5t}$

29. $y = -\dfrac{1}{3}e^{-t/2}\cos\left(\dfrac{\sqrt{11}\,t}{2}\right) + \dfrac{3}{\sqrt{11}}e^{-t/2}\sin\left(\dfrac{\sqrt{11}\,t}{2}\right) + \dfrac{1}{3}t - \dfrac{2}{3}$

31. $y = \dfrac{867}{251{,}600}e^{-12x} - \dfrac{3367}{251{,}600}e^{8x} + \dfrac{3}{20}xe^{8x} + \dfrac{25}{2516}\cos(2x) - \dfrac{1}{1258}\sin(2x)$

33. $y = e^x(\frac{11}{7} + \frac{8}{7}x) + \frac{3}{7}\cos(3x) - \frac{4}{7}\sin(3x)$ **35.** $y = -\frac{13}{6}e^x\cos(\sqrt{6}\,x) + \frac{1}{6}e^x$

37. $y = e^{-11x}(-\frac{633}{121} - \frac{6721}{121}x) + \frac{28}{121}$

Chapter 3

Section 3.0

3. (a)
$z_1' = z_2$
$z_2' = z_3$
$z_3' = z_4$
$z_4' = 2xz_2 + z_3 - 3z_4$

(c)
$z_1' = z_2$
$z_2' = z_3$
$z_3' = z_4$
$z_4' = 2z_3 - 6xz_1 + xe^x$

Section 3.1

5. $y = c_1 e^{4x} + c_2 e^{-x} + c_3 xe^{-x}$ **7.** $y = c_1 + c_2\cos(x) + c_3\sin(x)$ **9.** $y = c_1 e^{-3x} + c_2 xe^{-3x} + c_3 e^{2x}$

11. $y = c_1 e^{4x} + c_2 e^{2x} + c_3 e^{-x} + \frac{1}{64}(8x^2 - 4x + 11)$

13. $y = c_1 e^{4x} + c_2 e^{-x} + c_3 xe^{-x} + c_4 x^2 e^{-x} + \frac{1}{4}x^3 - \frac{33}{16}x^2 + \frac{255}{32}x - \frac{1729}{128}$

15. $y = c_1 e^{3x} + c_2 e^{-2x}\cos(x) + c_3 e^{-2x}\sin(x) + \frac{2}{15}\cos(3x) - \frac{1}{15}\sin(3x)$

17. $y = c_1 e^{-x} + c_2 xe^{-x} + c_3 x^2 e^{-x} + c_4 x^3 e^{-x} + \frac{1}{128}[17\cosh(3x) - 15\sinh(3x)]$

19. $y = \frac{8}{125}e^{4x} - \frac{11}{125}e^{-x} + \frac{10}{125}xe^{-x} + \frac{6}{125}\cos(2x) - \frac{33}{125}\sin(2x)$ **21.** $y = 1 + \frac{2}{5}\cos(x) + \frac{9}{5}\sin(x) - \frac{2}{5}e^{2x}$

23. $y = (-\frac{2792}{675} - \frac{269}{45}x)e^{-3x} - \frac{4}{25}e^{2x} - \frac{1}{9}x + \frac{8}{27}$

Section 3.2

1. $y = e^{2x}(c_1 + c_2 x + c_3 x^2 + c_4 x^3 + c_5 x^4)$ **3.** $y = c_1 e^{-x} + c_2 e^{4x} + c_3 e^x$

5. $y = c_1\cos(\sqrt{2}\,x) + c_2\sin(\sqrt{2}\,x) + e^{2x}[c_3\cos(x) + c_4\sin(x)]$ **7.** $y = c_1 e^{-2x} + e^x[c_2\cos(x) + c_3\sin(x)]$

9. $y = c_1 e^{-6x} + e^x(c_2 + c_3 x)$ **11.** $y = e^{2x}(c_1 + c_2 x) + c_3 e^{-6x} + c_4 e^{-x}$ **13.** $y = c_1 e^{2x} + c_2 e^{-2x} + c_3 e^x + c_4 e^{-x}$

15. $y = c_1 e^{2x} + e^{2x}[c_2 \cos(3x) + c_3 \sin(3x)]$ **17.** $y = e^{5x}(c_1 + c_2 x) + e^{-6x}(c_3 + c_4 x)$

19. $y = c_1 e^{7x} + c_2 e^{-x} + c_3 e^{-2x}$ **21.** $y = c_1 e^{7x} + c_2 e^{-2x} + c_3 e^{-11x}$ **23.** $y = c_1 e^{5x} + c_2 e^{-x} + c_3 x e^{-x} + c_4 x^2 e^{-x}$

25. $y = c_1 e^{-6x} + c_1 \cos(\sqrt{6}\,x) + c_3 \sin(\sqrt{6}\,x)$ **27.** $y = e^x - 2\cos(\sqrt{2}\,x) + \dfrac{1}{\sqrt{2}} \sin(\sqrt{2}\,x)$

29. $y = \frac{8}{61} e^{-4(x-\pi)} + e^{x-\pi}[-\frac{8}{61} \cos(6x) + \frac{20}{183} \sin(6x)]$ **31.** $y = \frac{179}{50} e^{-3x} + \frac{21}{50} e^{7x} - \frac{390}{50} x e^{7x}$

33. $y = -\frac{69}{45} e^{-2x} + \frac{1}{3} e^{7x} - \frac{36}{45} e^{3x}$ **35.** $y = -\frac{121}{169} e^{3x} - \frac{48}{169} e^{-10x} - \frac{286}{169} x e^{-10x}$

Section 3.3

1. $y = e^x(c_1 + c_2 x + c_3 x^2 + c_4 x^3) - \frac{1}{4} \cosh(2x)$

3. $y = c_1 e^{2x} + c_2 \cos(x) + c_3 \sin(x) - \frac{1}{2}x^2 + \frac{1}{2}x + \frac{3}{4} - \frac{3}{5}x \cos(x) + \frac{3}{10}x \sin(x)$

5. $y = e^{-2x}(c_1 + c_2 x + c_3 x^2) + \frac{1}{27} e^x + \frac{33}{125} \cos(x) - \frac{6}{125} \sin(x) - \frac{4}{3}x^3 e^{-2x}$

7. $y = c_1 + e^{2x}[c_1 \cos(4x) + c_2 \sin(4x)] + \frac{1}{60}x^3 + \frac{11}{100}x^2 - \frac{461}{1000}x$

9. $y = c_1 e^x + c_2 e^{-3x} \cos(4x) + c_3 e^{-3x} \sin(4x) - \frac{1}{25}x - \frac{19}{225} + \frac{5}{1664} \cosh(3x) - \frac{21}{1664} \sinh(2x)$

11. $y = c_1 + c_2 x + e^{5x}(c_1 + c_2 x) - \dfrac{2x^2}{25}$ **13.** $y = c_1 e^{2x} + e^{3x}[c_1 \cos(2x) + c_2 \sin(2x)] + \frac{7}{16} x e^{3x}[\cos(2x) + \sin(2x)]$

15. $y = c_1 e^{3x} + c_2 e^{-x} + c_3 e^{-2x} + \frac{5}{6} + \frac{11}{260} \cos(2x) - \frac{3}{260} \sin(2x)$

17. $y = c_1 e^{-2x} + e^{3x}[c_1 \cos(4x) + c_2 \sin(4x)] - \frac{33}{585} \cos(2x) - \frac{9}{585} \sin(2x)$

19. $y = c_1 e^{-3x} + e^x[c_1 \cos(x) + c_2 \sin(x)] + \frac{7}{30} e^{3x} - \frac{1}{3}x - \frac{4}{9}$

21. $y = c_1 e^x + c_2 e^{-x} + c_3 e^{2x} + \frac{1}{4} - \frac{1}{40} \cos(2x) + \frac{1}{40} \sin(2x)$

23. $y = c_1 e^{-4x} + c_2 e^{3x} + c_3 e^x - \frac{1}{15} \cosh(2x) - \frac{1}{10} \sinh(2x)$

25. $y = c_1 e^{-4x} + c_2 e^{-2x} + c_3 e^{-3x} + \frac{1}{24}x^3 - \frac{39}{288}x^2 + \frac{115}{576}x - \frac{865}{6912}$

27. $y = -\frac{25}{9} e^x - \frac{476}{5625} e^{-5x} - \frac{1120}{1875} x e^{-5x} - \frac{1}{25}x^2 - \frac{6}{125}x - \frac{86}{625}$

29. $y = \frac{1026}{1025} e^{3x} \cos(4x) - \frac{243}{4100} e^{3x} \sin(4x) + \frac{43}{50} + \frac{3}{82} e^{-2x}$

31. $y = \frac{11}{25} e^{-2x} + \frac{1081}{1350} e^{3x} - \frac{239}{180} x e^{3x} + \frac{1}{18}x + \frac{1}{108} - \frac{1}{4} e^{2x}$

33. $y = \frac{5}{2} e^x + e^{2x}[-\frac{3}{2} \cos(x) + \frac{1}{2} \sin(x)] - 1$ **35.** $y = \frac{69}{112} e^{-3x} + e^x(-\frac{293}{112} + \frac{299}{56}x) + \frac{1}{8} x e^{-3x} + \frac{1}{8}x^2 e^x$

Section 3.4

1. $y = c_1 x^{-1} + c_2 x^3 + c_3 x$ **3.** $y = c_1 x^2 + x^4\{c_2 \cos[2\ln(x)] + c_2 \sin[2\ln(x)]\}$

5. $y = c_1 x^3 + x^{-2}[c_1 + c_2 \ln(x)]$ **7.** $y = c_1 x^{-3} + x^4\{c_2 \cos[2\ln(x)] + c_3 \sin[2\ln(x)]\}$

9. $y = x^{-3}\{c_1 + c_2 \ln(x) + c_3[\ln(x)]^2\}$ **11.** $y = c_1 x^{-3} + x^2\{c_1 \cos[5\ln(x)] + c_2 \sin[5\ln(x)]\}$

13. $y = c_1 x^3 + x^4\{c_1 \cos[\ln(x)] + c_2 \sin[\ln(x)]\}$ **15.** $y = x^{-2} + x^6\{c_1 \cos[4\ln(x)] + c_2 \sin[4\ln(x)]\}$

17. $y = x\{c_1 + c_2 \ln(x) + c_3[\ln(x)]^2\}$ **19.** $y = x^8 + x^{-1}[c_1 + c_2 \ln(x)]$ **23.** $y = 8x^{-1} - [4\ln(2) + 2]x + 4x\ln(x)$

25. $y = -\frac{7}{17}x^{-3} + x\{-\frac{27}{17} \cos[\ln(x)] + \frac{57}{17} \sin[\ln(x)]\}$ **27.** $y = 2x^2 + 2x^{-2} \cos[\sqrt{3}\ln(x)] + \dfrac{4}{\sqrt{3}} \sin[\sqrt{3}\ln(x)]$

29. $y = \frac{144}{15}x^{-4} - \frac{1}{2}x^2 + \frac{6}{5}x$ **31.** $y = -\frac{5}{144}x^{-4} + \frac{5}{144}x^8 - \frac{60}{144}x^8 \ln(x)$

33. $y = \frac{5}{29}x^5 + \frac{24}{29} \cos[2\ln(x)] - \frac{56}{29} \sin[2\ln(x)]$ **35.** $y = -\frac{1}{24}x^7 - \frac{1}{4}x^{-3} + \frac{7}{24}x^{-5}$

Section 3.6

1. $-24\sin(2x) - 36\cos(2x)$ **3.** $-6\sin(2x) + 8\cos(2x) - 12 + 6x$ **5.** $32x^4 - 128x^3 + 96x^2 + 8x - 8$

7. $12x^4 - 68x^3 + 36x^2 + 72x$ **9.** $-3e^{4x}$ **11.** $(2D + 3)(D + 4)y = 0;\ e^{-4x},\ e^{-3x/2}$

13. $(D - 6)(D + 2)y = 0;\ e^{6x},\ e^{-2x}$ **15.** $(D - 7)(D + 5)y = 0;\ e^{7x},\ e^{-5x}$

17. $(8D + 1)(D - 2)y = 0;\ e^{-x/8},\ e^{2x}$ **19.** $(2D + 3)(D - 1)(D + 2) = 0;\ e^{-3x/2},\ e^x,\ e^{-2x}$

Supplementary Problems

1. $y = (\frac{49}{48} + \frac{769}{108}x + \frac{1681}{72}x^2)e^{-6x} + \frac{1}{216}x - \frac{9}{432}$ **3.** $y = \frac{25}{32} e^x + (\frac{39}{32} + \frac{31}{16}x)e^{-3x} - \frac{1}{16} x e^x - \frac{1}{2}x^2 e^{-3x}$

5. $y = -\frac{124}{65} e^{2x} + e^{3x}[-\frac{6}{25} \cos(8x) + \frac{71}{520} \sin(2x)] - 1$

7. $y = -\frac{1}{9}\left(\frac{17}{2}e^{-2} + \frac{4}{3}e^{-3}\right)e^{-2x} + \frac{1}{9}\left(\frac{10}{3} + 19e\right)e^{x} + \left(\frac{-12e+8}{9}\right)xe^{x} - \frac{3}{2} + \frac{2}{3}x^2e^x$

9. $y = \left(\frac{5+4\pi}{2e^{\pi}}\right)e^x - \left(\frac{5+4\pi}{2}\right)\cos(x) + \left(\frac{4\pi-3}{2}\right)\sin(x) - 4x - 4$

11. $y = -\frac{111}{45}e^{-2\pi/3}e^{4x} + e^x[\frac{24}{45}e^{-\pi/6}\cos(6x) - \frac{78}{45}e^{-\pi/6}\sin(6x)]$ **13.** $y = -5x^2 + 10x^2\ln(x) - 10x^2[\ln(x)]^2$

15. $y = -\frac{945}{243}x + \frac{1620}{243}x\ln(x) - \frac{27}{243}x\cos[3\ln(x)] - \frac{5}{243}x\sin[3\ln(x)]$

17. $y = \frac{3}{400}e^{-4x} - \frac{3}{400}e^{5x}\cos(2x) + \frac{2}{125}e^{5x}\sin(2x) + \frac{3}{85}xe^{-4x}$

19. $y = -\frac{6}{17}\cos(\sqrt{2}\,x) + \frac{8}{17\sqrt{2}}\sin(\sqrt{2}\,x) - \frac{4}{85}e^x\cos(2x) - \frac{18}{85}e^x\sin(2x) + \frac{2}{5}$

27. $82\sin(3x)$ **29.** $24x^2 + 120x$ **31.** $(2D+3)(D-4)y = 0$; e^{4x}, $e^{-3x/2}$

33. $(D-8)(D+3)y = 0$; e^{8x}, e^{-3x} **35.** $(D+1)(D-7)(D-2)y = 0$; e^{2x}, e^{7x}, e^{-x}

Chapter 4

Section 4.1

1. $\frac{6}{s^2-9} - \frac{4}{s}$ **3.** $\frac{16s}{(s^2+4)^2}$ **5.** $\frac{1}{s^2} - \frac{s}{s^2+25}$ **7.** $\frac{1}{s} - \frac{4}{s^2} + \frac{4}{(s+3)^3}$ **9.** $\frac{6s}{(s^2+9)^2} - \frac{1}{s} + \frac{4}{s^3}$

11. $\frac{24}{(s+1)^4}$ **13.** $\frac{14}{s^3} - \frac{1}{(s+4)^2}$ **15.** $\frac{-4s^4 + 144s^3 - 4s^2 - 48s}{(s^2+1)^4}$ **17.** $\frac{1}{2}\left(\frac{1}{s-6} + \frac{1}{s+6} + \frac{2}{s}\right)$

19. $\frac{3}{(s-3)^2} - \frac{3}{(s+3)^2} - \frac{54}{s^4}$ **21.** $\frac{4}{(s+1)^3} - \frac{1}{2s} + \frac{1}{2}\frac{s}{s^2+4}$ **23.** $\frac{1}{2}\frac{s}{s^2-16} - \frac{1}{2}\frac{s}{s^2+16}$

25. $\frac{s^2-9}{(s^2+9)^2} - \frac{s}{s^2+9} + \frac{5}{s+1}$ **27.** $4\cosh(\sqrt{14}\,t)$ **29.** $3e^{7t} + t$ **31.** $3\cosh(\sqrt{7}\,t) + \frac{17}{\sqrt{7}}\sinh(\sqrt{7}\,t)$

33. $\frac{1}{3^{3/2}}[\sin(\sqrt{3}\,t) - \sqrt{3}\,t\cos(\sqrt{3}\,t)] - \frac{1}{180}t^6$ **35.** $\frac{1}{6}e^{3t} - \frac{1}{6}e^{-3t} - 7\cos(\sqrt{15}\,t)$ **37.** $2te^{-7t}$

39. $2e^{-7t} - te^{-7t}$ **41.** $\frac{t^3}{6} - \frac{t^4}{12} + \frac{t^5}{30}$ **43.** $t^2 - t^3 + 2$ **45.** $\frac{2}{3}e^{3t} + \frac{1}{3}e^{-3t} - 1$

47. $\cos(\sqrt{6}\,t) - \frac{5}{\sqrt{6}}\sin(\sqrt{6}\,t)$ **49.** $-\frac{3}{2}t^2e^{-2t} + 4\cos(\sqrt{6}\,t)$

Section 4.2

1. $\frac{8}{s^3} - \frac{18}{s^4}$ **3.** $\frac{2}{s^3} - \frac{2}{s^2+4} + \frac{1}{s}$ **5.** $\frac{-4}{(s+3)^2+16} + \frac{3}{s}$ **7.** $\frac{2e^{-4s}}{s^2}$ **9.** $\frac{2}{(s+5)^3} + \frac{4s}{s^2+4} - \frac{6}{s^2-36}$

11. $\frac{e^{-4s}}{s^2} + \frac{4e^{-4s}}{s}$ **13.** $\frac{1}{s-3} - \frac{12}{s^4}$ **15.** $\frac{48e^{-2s}}{s^5}$ **17.** $\frac{1}{s} - \frac{6}{s^2-36} + \frac{s+1}{(s+1)^2-1}$ **19.** $\frac{s-2}{(s-2)^2+25} - \frac{18}{s^4}$

21. $\frac{6}{(s+4)^2+4} - \frac{s}{s^2+36} + \frac{6}{s}$ **23.** $\left(\frac{2}{s-3}\right)\left(\frac{1-e^{-(s-3)}}{1+e^{-(s-3)}}\right)$ **25.** $e^{-4s}\left\{\frac{128}{s} + \frac{16}{s^2} + \frac{48}{s^3} + \frac{12}{s^4}\right\}$ **27.** $\frac{s+4}{(s+4)^2-36}$

29. $\frac{1}{s}\frac{3}{(s+4)^2+9}$ **31.** $\frac{-e^{-5s}}{s^2} - \frac{3e^{-5s}}{s} + \frac{2}{s^2}$ **33.** $\frac{17e^{-6s}}{s} + \frac{3e^{-6s}}{s^2}$ **35.** $\frac{101e^{-7s}}{s} + \frac{28e^{-7s}}{s^2} + \frac{4e^{-7s}}{s^3}$

37. $\frac{h}{s}\left(\frac{1-e^{-4s}}{1+e^{-4s}}\right)$ **39.** $\frac{62e^{-2s}}{s} + \frac{96e^{-2s}}{s^2} + \frac{96}{s^3} + \frac{48}{s^4}$ **41.** $-\frac{4}{s^3} + e^{-9s}\left[\frac{6}{s^4} + \frac{64}{s^3} + \frac{336}{s^2} + \frac{1162}{s}\right]$

43. $\frac{1000e^{-5s}}{s} + \frac{600e^{-5s}}{s^2} + \frac{240e^{-5s}}{s^3} + \frac{48e^{-5s}}{s^4} - e^{-3(s+4)}\left(\frac{3s+13}{(s+4)^2}\right)$ **45.** $\frac{1}{s^2} - \frac{37e^{-s}}{s} - \frac{14e^{-s}}{s^2} - \frac{14e^{-s}}{s^3}$

47. $\frac{6320e^{-9s}}{s} + \frac{2862e^{-9s}}{s^2} + \frac{966e^{-9s}}{s^3} + \frac{216e^{-9s}}{s^4} + \frac{24e^{-9s}}{s^5}$ **49.** $\frac{1}{s} - \frac{4e^{-6}e^{-2s}}{s+3} - \frac{4e^{-6}e^{-2s}}{(s+3)^2} - \frac{2e^{-6}e^{-2s}}{(s+3)^3}$

51. $Y(s) = \frac{s-9}{s^2-6s+2}$ **53.** $Y = \frac{1}{(s+1)(s^2-3s-10)} + \frac{2s-10}{s^2-3s-10}$ **55.** $Y = \frac{1}{(s+4)(s^2+6s-18)} + \frac{-3s-16}{s^2+6s-18}$

57. $Y = \dfrac{8}{s(s^2+2s-7)} + \dfrac{s+2}{s^2+2s-7}$ **59.** $Y = \left(\dfrac{1}{s^2+8s-2}\right)\left(\dfrac{1}{s} - \dfrac{3s}{s^2-4} + s + 6\right)$

61. $Y = \left(\dfrac{1}{s^3-3s^2+4s-16}\right)\left(\dfrac{2e^{-3s}}{s} - s^2 + 2s - 1\right)$ **63.** $Y = \left(\dfrac{1}{s^4+12s^3-2}\right)\left(\dfrac{1}{s} - 4s^3 - 47s^2 + 12s\right)$

65. $Y = \left(\dfrac{1}{s^3-7s^2+s+1}\right)\left(\dfrac{8}{s^3} - \dfrac{3(s+2)}{(s+2)^2+1} + 2s^2 - 12s - 12\right)$

67. $Y = \left(\dfrac{1}{s^3+3s^2-2s+1}\right)\left(\dfrac{24}{(s+4)^4} + \dfrac{30}{s^4} + 5s^2 + 20s + 3\right)$

69. $Y = \left(\dfrac{1}{s^2-10s+8}\right)\left[\dfrac{3}{s^2+9} - e^{-4s}\left(\dfrac{4}{s} + \dfrac{1}{s^2}\right) + s - 16\right]$

Section 4.3

1. $\frac{1}{2}(e^{3t} - e^t)$ **3.** $2\cosh(2t)$ **5.** $e^{3t}\cos(3t)$ **7.** $\frac{4}{3}\sin(3t) - te^{3t}$ **9.** $\frac{4}{3}(e^{2t} - e^{-t})$

11. $e^t - 2te^t$ **13.** $e^{3t}[2\cos(2t) + 3\sin(2t)]$ **15.** $\dfrac{2}{3}e^t + \dfrac{1}{3}\cos(\sqrt{2}\,t) + \dfrac{1}{3\sqrt{2}}\sin(\sqrt{2}\,t)$

17. $e^{-2(t-4)}u(t-4)$ **19.** $3\cos[\sqrt{14}(t-2)]u(t-2)$ **21.** $e^{2(t-2)}\{\cos[2(t-2)] + 2\sin[2(t-2)]\}u(t-2)$

23. $e^{3(t-5)/2}\left\{2\cos\left[\dfrac{\sqrt{11}}{2}(t-5)\right] + \dfrac{8}{\sqrt{11}}\sin\left[\dfrac{\sqrt{11}}{2}(t-5)\right]\right\}u(t-5)$ **25.** $\frac{7}{3}e^{-2t} - \frac{1}{3}e^t$

27. $(-e^{-3(t-2)} + e^{-2(t-2)})u(t-2)$ **29.** $t^2e^t(1 - \frac{1}{3}t)$

31. $(-\frac{19}{9}e^{-2(t-4)} - \frac{4}{3}(t-4)e^{-2(t-4)} + \frac{1}{9}e^{t-4} + 2e^{-(t-4)})u(t-4)$

33. $\left\{-3\cos[\sqrt{5}(t-2)] - \dfrac{1}{\sqrt{5}}\sin[\sqrt{5}(t-2)]\right\}e^{t-2}u(t-2)$

35. $\frac{23}{169}e^{3t} - \frac{23}{169}\cos(2t) - \frac{69}{338}\sin(2t) + \frac{39}{676}t\sin(2t) + \frac{156}{1183}[\sin(2t) - 2t\cos(2t)]$

37. $(\frac{22}{3}e^{5(t-1)} - \frac{7}{3}e^{2(t-1)})u(t-1)$ **39.** $\frac{1}{9}\{1 - \cos[3(t-5)]\}u(t-5)$ **41.** $(\frac{4}{13}e^{-8(t-3)} + \frac{9}{13}e^{5(t-3)})u(t-3)$

43. $[1 + 5(t-6)]e^{5(t-6)}u(t-6)$ **45.** $e^{2t}\cos(\sqrt{15}\,t)$ **47.** $\dfrac{-7209}{882}e^{2t} - \dfrac{1890}{441}te^{2t} + \dfrac{983}{441}e^{-5t} + \dfrac{12{,}299}{882}e^{4t}$

49. $u(t-2)e^{-9(t-2)/2}\left\{\cosh\left[\dfrac{\sqrt{101}}{2}(t-2)\right] - \dfrac{15}{\sqrt{101}}\sinh\left[\dfrac{\sqrt{101}}{2}(t-2)\right]\right\}$ **51.** $u(t-3)\cos[2(t-3)]$

53. $e^{-2(t-3)}u(t-3)[\cos(t-3) - 7\sin(t-3)]$ **55.** $-2u(t-2)e^{5(t-2)}[(t-2) + \frac{1}{2}(t-2)^2]$

57. $0.2467u(t-1.27)[e^{0.5(t-1.27)} - e^{-(t-1.27)}]$ **59.** $-7.8e^{0.3t} + 9.8e^{-0.7t}$

61. $1.6725\cos(0.7746t) + 0.0198\sin(0.7746t) - 1.6725e^{-0.23t}$ **63.** $-3.8111e^{0.5974t} + 3.5111e^{-0.9374t}$

65. $0.0486u(t-4.72)\{\sin[2.1749(t-4.72)] - 2.1749(t-4.72)\cos[2.1749(t-4.72)]\}$ **67.** $y = 2 + t + (t-2)e^t$

69. $y = -\frac{1}{3} - \frac{2}{3}e^{-3t} + e^{-2t} + \frac{1}{3}u(t-3) + \frac{2}{3}u(t-3)e^{-3(t-3)} - u(t-3)e^{-2(t-3)}$

71. $y = 1 - \frac{1}{2}e^t + \frac{1}{4}te^t - \frac{1}{2}e^{-t} - \frac{1}{4}te^{-t}$

73. $y = \dfrac{11}{32}e^{-2t} + \dfrac{19}{32}e^t\cos(\sqrt{3}\,t) - \dfrac{1}{32\sqrt{3}}e^t\sin(\sqrt{3}\,t) + \dfrac{1}{16}\cos(2t) + \dfrac{1}{16}\sin(2t)$

75. $y = \frac{1}{4} + \frac{1}{4}t - \frac{9}{4}e^{2t} + \frac{21}{4}te^{2t} + \frac{1}{2}u(t-3) - \frac{1}{2}u(t-3)e^{2(t-3)} + u(t-3)(t-3)e^{2(t-3)}$

77. $y = \dfrac{1}{6} + \left(\dfrac{1}{6} - \dfrac{\sqrt{6}}{12}\right)e^{-\sqrt{6}t} + \left(\dfrac{1}{6} + \dfrac{\sqrt{6}}{12}\right)e^{\sqrt{6}t} - \dfrac{1}{2}e^{2t}$ **79.** $y = \frac{2}{21}e^{-3t} + \frac{4}{7}e^{4t} - \frac{2}{3}e^{3t}$

81. $y = -\dfrac{1}{10}\cos(t) + \dfrac{1}{20}\sin(t) + \dfrac{1}{10}e^{2t}\cos(\sqrt{5}\,t) - \dfrac{1}{4\sqrt{5}}e^{2t}\sin(\sqrt{5}\,t)$

83. $y = \dfrac{7}{23}e^{-3t} - \dfrac{7}{23}e^{3t/2}\cos\left(\dfrac{\sqrt{11}}{2}t\right) + \dfrac{63}{23\sqrt{11}}e^{3t/2}\sin\left(\dfrac{\sqrt{11}}{2}t\right)$

85. $y = e^{-4t}(\frac{54}{729} + \frac{108}{729}t + \frac{81}{729}t^2) + \frac{1404}{729}e^{-t} + \frac{54}{729}te^{-t} - 1$

Section 4.4

1. $\frac{10}{53}\cos(2t) - \frac{17}{106}\sin(2t) - \frac{10}{53}e^{-7t}$ **3.** $-\frac{1}{9}e^{t} + \frac{1}{9}e^{-2t} + \frac{10}{3}te^{-2t}$ **5.** $-3e^{-4t} + 10te^{-4t}$

7. $5e^{4t} + 4te^{4t} - 5e^{5t}$ **9.** $\frac{1}{3}\cos(\sqrt{2}\,t) + \frac{1}{3\sqrt{2}}\sin(\sqrt{2}\,t) + \frac{t}{2\sqrt{2}}\sin(2\sqrt{2}\,t) + \frac{3}{2\sqrt{2}}(\sin(\sqrt{2}\,t) + \sqrt{2}\,t\cos(\sqrt{2}\,t)) - \frac{1}{3}e^{t}$

11. $-\frac{1}{16}e^{2t} + \frac{1}{2}te^{2t} + \frac{1}{16}e^{-6t}$ **13.** $\frac{8}{25}e^{-2t} + \frac{42}{25}e^{3t} + \frac{18}{5}te^{3t}$ **15.** $\frac{41}{72}e^{t} + \frac{1}{6}te^{t} + \frac{14}{9}e^{-2t} - \frac{17}{8}e^{-3t}$

17. $\frac{41}{7}e^{-3t} - \frac{13}{7}e^{-3t/2}\cos\left(\frac{\sqrt{19}}{2}t\right) - \frac{129\sqrt{19}}{133}e^{-3t/2}\sin\left(\frac{\sqrt{19}}{2}t\right)$ **19.** $-27e^{-3t} - 27te^{-3t} + 28e^{-2t} - 8te^{-2t}$

21. $y = -\frac{1}{6}e^{-t} + \frac{1}{10}e^{t} + \frac{1}{15}e^{-4t}$ **23.** $y = -\frac{1}{16}e^{-3t} - \frac{1}{4}te^{-3t} + \frac{1}{16}e^{t}$

25. $y = \left(\frac{3+\sqrt{17}}{2\sqrt{17}}\right)e^{(-3+\sqrt{17})t} + \left(\frac{-3+\sqrt{17}}{2\sqrt{17}}\right)e^{(-3-\sqrt{17})t}$ **27.** $y = \frac{1}{21}e^{3t} - \frac{1}{21}e^{-4t} + \frac{1}{21}e^{-t} - \frac{1}{3}$

29. $y = -\frac{5}{52}\cos(2t) - \frac{1}{52}\sin(2t) + \frac{3}{65}e^{3t} + \frac{1}{20}e^{-2t}$

Section 4.5

1. $y = \frac{1}{120EI}u\left(x - \frac{L}{2}\right)\left(x - \frac{L}{2}\right)^5 + \frac{wL^2}{48EI}x^2 - \frac{wL}{24EI}x^3$ **3.** $y = \frac{wL^2}{48EI}x^2 - \frac{wL}{24EI}x^3 + \frac{x^6}{360} - \frac{Lx^5}{120}$

9. $I(t) = \frac{2}{R}u(t-5)[1 - e^{-R(t-5)/L}]$

11. If $R \neq L$, $I = \frac{Ae^{-4}}{R-L}u(t-4)[e^{-(t-4)} - e^{-R(t-4)/L}]$. If $R = L$, $I = \frac{Ae^{-4}}{R}u(t-4)(t-4)e^{-(t-4)}$

13. $I_1 = Au(t-k)\left[\frac{11}{8000} + \frac{61}{1600}\sqrt{\frac{5120}{59}}e^{-(t-k)/32}\sin\left(\sqrt{\frac{59}{5120}}(t-k)\right) - \frac{11}{8000}e^{-(t-k)/32}\cos\left(\sqrt{\frac{59}{5120}}(t-k)\right)\right]$

$I_2 = Au(t-k)\left[\frac{1}{10} - \frac{197}{80}\sqrt{\frac{5120}{59}}e^{-(t-k)/32}\sin\left(\sqrt{\frac{59}{5120}}(t-k)\right) - \frac{1}{10}e^{-(t-k)/32}\cos\left(\sqrt{\frac{59}{5120}}(t-k)\right)\right]$

$I_3 = Au(t-k)\left[1 - 25\sqrt{\frac{5120}{59}}e^{-(t-k)/32}\sin\left(\sqrt{\frac{59}{5120}}(t-k)\right) - e^{-(t-k)/32}\cos\left(\sqrt{\frac{59}{5120}}(t-k)\right)\right]$

Section 4.6

1. $\frac{1}{16}[\sinh(2t) - \sin(2t)]$ **3.** $(t-2)u(t-2)$

5. $\cos(at)\left[\frac{\sin[(a-b)t]}{2(a-b)} + \frac{\sin[(a+b)t]}{2(a+b)}\right] + \sin(at)\left[\frac{\sin[(a-b)t]}{2(a-b)} - \frac{\sin[(a+b)t]}{2(a+b)}\right]$ **7.** $\frac{1}{a^2+b^2}[\cosh(bt) - \cos(at)]$

9. $\frac{1}{4b}\left[\frac{1}{a-b}(e^{at} - e^{bt}) + \frac{1}{a+b}(e^{-bt} - e^{at}) + \frac{1}{a+b}(e^{bt} - e^{-at}) + \frac{1}{b-a}(e^{-bt} - e^{-at})\right]$

This can be written

$\frac{1}{2b}\left\{\frac{1}{a-b}[\cosh(at) - \cosh(bt)] + \frac{1}{a+b}[\cosh(bt) - \cosh(at)]\right\}$ or $\frac{\cosh(at) - \cosh(bt)}{a^2 - b^2}$

11. $\frac{1}{2}t - \frac{1}{2\sqrt{2}}\sin(\sqrt{2}\,t)$ **13.** $\frac{1}{a^4}\left(1 + \frac{t}{2}\cos(at)\right) - \frac{3}{2a^5}\sin(at)$ **15.** $-\frac{1}{5}e^{-2t} + \frac{1}{3}e^{3t} + \frac{1}{6}e^{-3t}$

17. $-\frac{1}{10}e^{2t}\cos(3t) + \frac{1}{120}e^{2t}\sin(3t) + \left(\frac{4\sqrt{12}-25}{80\sqrt{12}}\right)e^{-\sqrt{12}t} + \left(\frac{4\sqrt{12}+25}{80\sqrt{12}}\right)e^{\sqrt{12}t}$ **19.** $\frac{1}{2}u(t-4) - \frac{1}{2}u(t-4)e^{-2(t-4)}$

21. $u(t-3)(t-3)$ **23.** $-\frac{2}{25} + \frac{2}{5}t + \frac{2}{25}\cos(\sqrt{5}\,t)$ **25.** $\frac{5}{6}e^{5t} - \frac{5}{6}e^{3t}\cos(\sqrt{8}\,t) - \frac{5}{3\sqrt{8}}e^{3t}\sin(\sqrt{8}\,t)$

31. $y = -\frac{1}{2}f(t) * e^{-6t} + \frac{1}{2}f(t) * e^{-4t} - 2e^{-6t} + 3e^{-4t}$

33. $y = \frac{1}{42}f(t) * e^{3t} - \frac{1}{42}f(t) * e^{-3t} - \frac{\sqrt{2}}{28}f(t) * e^{\sqrt{2}t} - \frac{\sqrt{2}}{28}f(t) * e^{-\sqrt{2}t}$

35. $y = \frac{1}{4}f(t) * e^{6t} - \frac{1}{4}f(t) * e^{2t} + 2e^{6t} - 5e^{2t}$

37. $y = f(t) * \dfrac{2}{\sqrt{13}} e^{-t/2} \sinh\left(\dfrac{\sqrt{13}}{2} t\right) + 4e^{-t/2} \cosh\left(\dfrac{\sqrt{13}}{2} t\right) + \dfrac{12}{\sqrt{13}} e^{-t/2} \sinh\left(\dfrac{\sqrt{13}}{2} t\right)$

39. $y = \dfrac{1}{15} f(t) * e^{3t} - \dfrac{1}{15} f(t) * \cos(\sqrt{6}\, t) - \dfrac{1}{5\sqrt{6}} f(t) * \sin(\sqrt{6}\, t)$ **41.** $y = \frac{1}{10} f(t) * e^{9t} - \frac{1}{10} f(t) * e^{-t} - \frac{1}{5} e^{9t} + \frac{1}{5} e^{-t}$

43. $y = -\frac{1}{2} f(t) * e^{-6t} + \frac{1}{2} f(t) * e^{-4t} + e^{-6t} - e^{-4t}$ **45.** $y = \dfrac{2}{\sqrt{23}} f(t) * e^{3t/2} \sin\left(\dfrac{\sqrt{23}}{2} t\right)$

Section 4.7

1. $f(t) = -\frac{3}{2} + \frac{1}{2} e^{-2t}$ **3.** $f(t) = \cosh(t)$ **5.** $f(t) = -\frac{1}{2} e^{-t} + \frac{3}{2} \cos(t) + \frac{1}{2} \sin(t)$

7. $f(t) = \frac{1}{4} e^{-2t} + \frac{3}{4} e^{-6t}$ **9.** $f(t) = -1 + 2e^{-3t/2} \cosh\left(\dfrac{\sqrt{5}}{2} t\right)$ **11.** $f(t) = -2 \sinh(t)$

13. $f(t) = 3 + \dfrac{6}{\sqrt{15}} e^{t/2} \sin\left(\dfrac{\sqrt{15}}{2} t\right)$ **15.** $f(t) = -\frac{1}{8} + \frac{9}{4} t^2 + \frac{1}{8} \cosh(2\sqrt{3}\, t)$ **17.** $f(t) = 0$

19. $f(t) = \frac{5}{4} e^{t} - \frac{1}{4} e^{-t} - \frac{1}{2} t e^{-t}$ **21.** $f(t) = e^{-t/2} \cos\left(\dfrac{\sqrt{15}}{2} t\right) - \dfrac{1}{\sqrt{15}} e^{-t/2} \sin\left(\dfrac{\sqrt{15}}{2} t\right)$

23. $f(t) = \dfrac{4}{81} - \dfrac{4}{9} t + 2t^2 - \dfrac{4}{81} e^{t/2} \cosh\left(\dfrac{\sqrt{37}}{2} t\right) + \dfrac{76}{81\sqrt{37}} e^{t/2} \sinh\left(\dfrac{\sqrt{37}}{2} t\right)$

25. $f(t) = \left(\dfrac{3 + \sqrt{3}}{4}\right) e^{-\sqrt{3}t} + \left(\dfrac{3 - \sqrt{3}}{4}\right) e^{\sqrt{3}t} - \dfrac{1}{2} e^{-t}$ **27.** $f(t) = -\frac{1}{2} e^{t} + \frac{1}{2} e^{-t} - t e^{-t}$ **29.** $f(t) = e^{-t}$

31. $y = \dfrac{1}{4} + \dfrac{t}{4} + \left(A - \dfrac{1}{4}\right) e^{2t} + \left(-2A - \dfrac{7}{4}\right) t e^{2t}$, with $A = \dfrac{1}{2} e^{-2} - 2$

33. $y = -\dfrac{1}{6} - \dfrac{17}{6} e^{-2t} \cosh(\sqrt{10}\, t) + \left(\dfrac{B - 6}{\sqrt{10}}\right) e^{-2t} \sinh(\sqrt{10}\, t)$, with $B = \dfrac{e^2 + \dfrac{1}{3} \cosh(\sqrt{10}) + \left(\dfrac{17}{6} + \dfrac{12}{\sqrt{10}}\right) \sinh(\sqrt{10})}{\dfrac{2}{\sqrt{10}} \sinh(\sqrt{10}) - \cosh(\sqrt{10})}$

35. $y = \dfrac{-4}{k^2} + \dfrac{1}{k^2} t + \dfrac{4}{k^2} \cos(kt) - \dfrac{1}{k^3} \sin(kt) + A \cos(kt)$, with $A = \dfrac{1}{\cos(3k)} \left[-2 + \dfrac{1}{k^2} - \dfrac{4}{k^2} \cos(3k) + \dfrac{1}{k^3} \sin(3k)\right]$

37. $y = \left(\dfrac{1}{14} + \dfrac{B}{28}\right) e^{2t} + \left(\dfrac{1}{110} - \dfrac{B}{220}\right) e^{-2t} + \left(\dfrac{9}{462} + \dfrac{B}{462}\right) e^{9t} + \left(-\dfrac{1}{10} - \dfrac{B}{30}\right) e^{3t}$,

with $B = \dfrac{-\frac{1}{7} e^4 + \frac{1}{55} e^{-4} - \frac{81}{462} e^{18} + \frac{3}{10} e^6}{\frac{1}{14} e^4 + \frac{1}{110} e^{-4} + \frac{9}{462} e^{18} - \frac{1}{10} e^6}$

39. $y = -\frac{3}{4} e^{-2t} + \frac{1}{2} t e^{-2t} - \frac{1}{4} e^{-4t} + e^{-3t} + A(-3e^{-4t} + 4e^{-3t})$, with $A = \dfrac{-3 - \frac{1}{4} e^{-4} + \frac{1}{4} e^{-8} - e^{-6}}{-3e^{-8} + 4e^{-6}}$

41. $y = -5 + t + e^{-5t/2} \cosh\left(\dfrac{\sqrt{21}}{2} t\right) + \dfrac{2}{\sqrt{21}} e^{-5t/2} \sinh\left(\dfrac{\sqrt{21}}{2} t\right)\left(B + \dfrac{3}{2}\right)$,

with $B = \dfrac{-5 + e^{-5/2} \cosh\left(\dfrac{\sqrt{21}}{2}\right) + \left(\dfrac{15}{2\sqrt{21}} - \dfrac{\sqrt{21}}{2}\right) e^{-5/2} \sinh\left(\dfrac{\sqrt{21}}{2}\right)}{e^{-5/2} \cosh\left(\dfrac{\sqrt{21}}{2}\right) - \dfrac{5}{\sqrt{21}} e^{-5/2} \sinh\left(\dfrac{\sqrt{21}}{2}\right)}$

43. $y = \frac{1}{8} - \frac{1}{8} e^{-4t} - \frac{1}{2} t e^{-4t} + A e^{-4t} + (4A + B) t e^{-4t}$, with $A = \frac{1}{8}(1 + 27e^4)$ and $B = -13e^4$

45. $y = -\frac{8}{15} e^{t} + \frac{1}{5} e^{6t} + \frac{1}{3} e^{-2t} + \frac{1}{8} B(e^{6t} - e^{-2t})$, with $B = \dfrac{4}{3e^{24} + e^{-8}} \left(\frac{8}{15} e^4 - \frac{6}{5} e^{24} + \frac{2}{3} e^{-8}\right)$

47. $y = -\dfrac{1}{3} - t - \dfrac{1}{3} e^{-t} + \dfrac{2}{3} e^{t} + \dfrac{B}{\sqrt{2}} \sinh(\sqrt{2}\, t)$, with $B = \dfrac{1 - \frac{1}{3} e^{-2} - \frac{2}{3} e^2}{\cosh(2\sqrt{2})}$

49. $y = \frac{37}{432}e^{-t} + \frac{21}{108}te^{-t} + \frac{1}{6}t^2e^{-t} + \frac{1}{16}e^{3t} - \frac{4}{27}e^{2t} - 2Ae^{3t} + 3Ae^{2t}$, with $A = \dfrac{432 - 203e^{-1} - 27e^3 + 16e^2}{432(3e^2 - 2e^3)}$

51. $y = 2 + t - 2e^t + (B + 1)te^t$, with $B = \dfrac{-3}{2e}$

53. $y = \dfrac{1}{k^2 - 1}\left(\sin(t) + \dfrac{1}{k}\sin(kt)\right) + \cos(kt) + \dfrac{B}{k}\sin(kt)$, with $B = \dfrac{1}{\cos(2k)}\left(k\sin(2k) - \dfrac{1}{k^2 - 1}[\cos(2) + \cos(2k)]\right)$

55. $y = \frac{8}{81} - \frac{1}{9}t + \frac{1}{810}e^{9t} - \frac{1}{10}e^{-t} + \frac{1}{10}B(e^{9t} - e^{-t})$, with $B = \left(\dfrac{10}{9e^{18} + e^{-2}}\right)(\frac{1}{9} - \frac{1}{90}e^{18} - \frac{1}{10}e^{-2})$

Section 4.8

1. $y = -1 - \dfrac{c}{6}t^2$ **3.** $y = -7t$ **5.** $y = \frac{3}{2}t^2$ **7.** $y = 10t$ **9.** $y = ct^2e^{-t}$ (any constant)

Supplementary Problems

1. $f(t) = -1 - \frac{1}{2}e^{-\sqrt{2}t} - \frac{1}{2}e^{\sqrt{2}t} = -1 - \cosh(\sqrt{2}\,t)$

3. $y = -\dfrac{4}{9} + \dfrac{1}{3}t - \dfrac{1}{2}e^{6t} + \dfrac{7}{18}e^{-t} + \left(\dfrac{A - 2}{7}\right)e^{6t} + \left(\dfrac{6A + 2}{7}\right)e^{-t}$, with $A = \dfrac{\frac{1}{9} + \frac{11}{14}e^6 - \frac{85}{126}e^{-1}}{\frac{1}{7}e^6 + \frac{6}{7}e^{-1}}$

5. $y = -\frac{1}{8} + \frac{1}{24}e^{-4t} + \frac{1}{12}e^{2t} + \frac{1}{8}u(t - 3) - \frac{1}{24}u(t - 3)e^{-4(t-3)} - \frac{1}{12}u(t - 3)e^{2(t-3)}$

7. $y = -\dfrac{9}{125}\cos(t) + \dfrac{13}{125}\sin(t) - \dfrac{293}{42{,}875}e^{-7t} + \dfrac{3}{100}e^{-2t} + \dfrac{67}{1372} - \dfrac{63}{686}t + \dfrac{49}{686}t^2$

9. $y = \dfrac{1}{12}e^{-t} + \dfrac{1}{5}e^{-2t} + \dfrac{11}{120}e^{-7t} - \dfrac{9}{24}e^{-3t} - \dfrac{3A}{4}e^{-7t} + \dfrac{7A}{4}e^{-3t}$, with $A = \dfrac{3 - \frac{1}{12}e^{-1} - \frac{1}{5}e^{-2} - \frac{11}{120}e^{-7} + \frac{9}{24}e^{-3}}{\frac{7}{4}e^{-3} - \frac{3}{4}e^{-7}}$

11. $y = u(t - 6)[4e^{t-6} + 2(t - 6)e^{t-6} - 4e^{2(t-6)} + 2(t - 6)e^{2(t-6)}]$ **13.** $f(t) = \frac{1}{2} + t - \frac{1}{2}e^{-t}$

15. $y = -\frac{8}{9}e^{-2t} + \frac{4}{45}e^{7t} + \frac{4}{5}e^{-3t}$ **17.** $y = \frac{297}{15}\cos(5t) - \frac{149}{5}\sin(5t) + \frac{106}{5}e^{5t} - 43e^{-t}$

19. $y = \frac{147}{2030}\cos(5t) - \frac{63}{2030}\sin(5t) - \frac{87}{2030}e^{-5t} - \frac{6}{203}e^{2t}$

21. $y = u(t - 1)[\frac{1}{18} - \frac{1}{14}e^{-9(t-1)} + \frac{1}{63}e^{-2(t-2)}] - \frac{2}{7}e^{-9} + \frac{2}{7}e^{-2t}$

23. $y = -\frac{4}{21}u(t - 3) + \frac{2}{35}u(t - 3)e^{-7(t-3)} + \frac{2}{15}u(t - 3)e^{3(t-3)} - \frac{1}{87}\cos(3t) - \frac{5}{174}\sin(3t) - \frac{3}{580}e^{-7t} + \frac{1}{60}e^{3t} + \frac{1}{10}B(e^{3t} - e^{-7t})$,
with $B = \dfrac{-2 - \frac{3}{87}\sin(3) + \frac{5}{58}\cos(3) - \frac{21}{580}e^{-7} - \frac{1}{20}e^3}{\frac{1}{10}(3e^3 + 7e^{-7})}$

25. $y = \frac{43}{160}e^{-5t} + \frac{77}{96}e^{3t} - \frac{2}{15} + \frac{1}{16}e^{-t}$ **27.** $y = \frac{1}{2}\sin(t) + \frac{1}{4}e^t - \frac{1}{4}e^{-t}$ **29.** $y = \dfrac{6}{5(1 + 5e^{12})}e^{5t} + \dfrac{6e^{12}e^{-t}}{1 + 5e^{12}} - \dfrac{1}{5}$

39. $y_1 = -\frac{1}{2} + 2t + \frac{1}{2}e^{2t}$; $y_2 = \frac{1}{8} - \frac{1}{4}t - \frac{1}{8}e^{2t}$

47. **(a)** $y = 29 - 8t + t^2 - \left(\dfrac{29 + 50\sqrt{2}}{2}\right)e^{(\sqrt{2}-2)t/2} - \left(\dfrac{29 - 50\sqrt{2}}{2}\right)e^{-(\sqrt{2}-2)t/2}$

(b) $y = 2 + \left(\dfrac{3\sqrt{7} - 7}{7}\right)e^{(3+\sqrt{7})t} - \left(\dfrac{3\sqrt{7} + 7}{7}\right)e^{(3-\sqrt{7})t}$

(c) $y = \dfrac{1}{5}e^{-t} + \left(\dfrac{8\sqrt{5} - 5}{50}\right)e^{(3+\sqrt{5})t/2} - \left(\dfrac{8\sqrt{5} + 5}{50}\right)e^{(3-\sqrt{5})t/2}$

51. $-\dfrac{1}{k}\cos\left(\sqrt{\dfrac{k}{m}}(t - t_0)\right)u(t - t_0) + \dfrac{1}{k}\cos\left(\sqrt{\dfrac{k}{m}}(t - t_0 - \tau)\right)u(t - t_0 - \tau)$

53. **(a)** $x(t) = e^{4t} - 2e^t$; $y(t) = \frac{2}{3}e^{4t} - \frac{2}{3}e^t$
(c) $x(t) = 9e^{4t} + 5e^{3t}$; $y(t) = 6e^{4t} + 5e^{3t}$

Chapter 5

Section 5.1

1. 1 **3.** 1 **5.** 1 **7.** $\frac{2}{3}$ **9.** 1 **11.** 1 **13.** 1 **15.** 1 **17.** $\frac{3}{2}$ **19.** 0 **21.** 1

23. 1 **25.** 0 **27.** $\ln(2) + \frac{1}{2}x - \frac{1}{8}x^2 + \frac{1}{24}x^3 - \frac{1}{64}x^4 + \frac{1}{160}x^5$ **29.** $2x^2 - \frac{1}{3}x^4$ **31.** $1 + 4x + 6x^2 + 4x^3 + x^4$

33. $1 - 3x + 6x^2 - 10x^3 + 15x^4 - 21x^5$ **35.** $1 - x^2 + 2x^4$

37. $\sec(1) + \sec(1)\tan(1)\,x + \frac{1}{2}[\sec(1)\tan^2(1) + \sec^3(1)]x^2 + \frac{1}{6}[\sec(1)\tan^3(1) + 5\sec^3(1)\tan(1)]x^3$
$+ \frac{1}{24}[\sec(1)\tan^4(1) + 18\sec^3(1)\tan^2(1) + 5\sec(1)]x^4$
$+ \frac{1}{120}[\sec(1)\tan^5(1) + 58\sec^3(1)\tan^3(1) + 61\sec^5(1)\tan(1)]x^5$

39. $x + x^2 + \frac{1}{3}x^3 - \frac{7}{60}x^5$ **41.** $\sin(1) + \cos(1)x - \frac{1}{2}\sin(1)x^2 - \frac{1}{6}\cos(1)x^3 + \frac{1}{24}\sin(1)x^4 + \frac{1}{120}\cos(1)x^5$

43. $1 + \frac{\pi}{2} + \left(x - \frac{\pi}{2}\right) - 2\left(x - \frac{\pi}{2}\right)^2 + \frac{2}{3}\left(x - \frac{\pi}{2}\right)^4$ **45.** $\frac{9}{4} + 3(x - \frac{1}{2}) + (x - \frac{1}{2})^2$

47. $\frac{\sqrt{2}}{2} - \frac{3\sqrt{2}}{2}\left(x - \frac{\pi}{4}\right) - \frac{9\sqrt{2}}{4}\left(x - \frac{\pi}{4}\right)^2 + \frac{9\sqrt{2}}{4}\left(x - \frac{\pi}{4}\right)^3 + \frac{27\sqrt{2}}{16}\left(x - \frac{\pi}{4}\right)^4 - \frac{81\sqrt{2}}{80}\left(x - \frac{\pi}{4}\right)^5$

49. $\ln(\frac{4}{3}) + \frac{3}{4}(x - \frac{1}{3}) - \frac{9}{32}(x - \frac{1}{3})^2 + \frac{9}{64}(x - \frac{1}{3})^3 - \frac{81}{1024}(x - \frac{1}{3})^4 + \frac{243}{5120}(x - \frac{1}{3})^5$

51. $\sum_{n=4}^{\infty} \frac{(-1)^n x^n}{2n - 2}$ **53.** $\sum_{n=1}^{\infty} \frac{(2n + 5)x^n}{n + 1}$ **55.** $\sum_{n=2}^{\infty} \frac{2^{n-1}x^n}{(n - 1)^2 - 3}$ **57.** $\sum_{n=3}^{\infty} \frac{2(n - 3)}{n - 1}x^n$

59. $\sum_{n=-1}^{\infty} (n - 2)nx^n$ **61.** $1 + 2x + \sum_{n=2}^{\infty} (2^{n-1} + 1 + n)x^n$ **63.** $1 + \frac{1}{2}x + \sum_{n=2}^{\infty}\left(\frac{(n + 1)!}{(n + 1)^2} + 2^n\right)x^n$

65. $\sum_{n=1}^{\infty}\left(\frac{1}{2n} - (n + 2)^{n+2}\right)x^n$ **67.** $x + x^2 + x^3 + x^4 + \sum_{n=5}^{\infty} [1 + (n - 2)!]x^n$

69. $\frac{1}{2} + \frac{1}{3}x + \sum_{n=2}^{\infty}\left(\frac{1}{n + 2} + (-2)^{n-1}\right)x^n$ **71.** $1 - \frac{1}{2} + \frac{1}{8} - \frac{1}{48} + \frac{1}{384} - \frac{1}{3840} + \frac{1}{46{,}080}$
(about 0.60653212; the real value is 0.6065306...)

73. $\sum_{n=1}^{7} \frac{(-1)^{n+1}(\frac{1}{2})^n}{n}$ (about 0.40580357; the real value is 0.40546 ...)

Section 5.2

1. $(a_1 = 0)\ y = 1 - \frac{1}{6}x^3 + \frac{1}{180}x^6 - \frac{1}{12960}x^9 + \frac{1}{1610720}x^{12} + \cdots$
$(a_0 = 0)\ y = x - \frac{x^4}{12} + \frac{1}{504}x^7 - \frac{1}{45360}x^{10} - \frac{1}{7076160}x^{13} + \cdots$

3. $y = 1 + x^2 + \frac{1}{2}x^4 + \frac{1}{6}x^6 + \frac{1}{24}x^8 + \cdots$

5. $(a_1 = 0)\ y = 1 - \frac{1}{12}x^4 - \frac{1}{60}x^5 - \frac{1}{360}x^6 - \frac{1}{2520}x^7 + \cdots$ $(a_0 = 0)\ y = x + \frac{1}{2}x^2 + \frac{1}{6}x^3 + \frac{1}{24}x^4 - \frac{1}{24}x^5 + \cdots$

7. $(a_1 = 0)\ y = 1 - x^2 + \frac{1}{6}x^4 - \frac{1}{10}x^5 - \frac{1}{90}x^6 + \cdots$
$(a_0 = 0)\ y = x - \frac{1}{3}x^3 + \frac{1}{12}x^4 + \frac{1}{30}x^5 - \frac{7}{180}x^6 + \cdots$

9. $(a_1 = 0)\ y = 1 + 2x^2 - \frac{5}{6}x^3 + \frac{7}{8}x^4 - \frac{53}{120}x^5 + \cdots$ $(a_0 = 0)\ y = x - \frac{1}{2}x^2 + \frac{5}{6}x^3 - \frac{1}{3}x^4 + \frac{31}{120}x^5 + \cdots$

11. $(a_1 = 0)\ y = 1 - x^2 + \frac{1}{6}x^4 + \frac{1}{10}x^5 - \frac{1}{19}x^6 + \cdots$ $(a_0 = 0)\ y = x - \frac{1}{3}x^3 - \frac{1}{12}x^4 + \frac{1}{30}x^5 + \frac{7}{180}x^6 + \cdots$

13. $(a_1 = 0)\ y = 1 + \frac{1}{2}x^2 - \frac{1}{24}x^4 - \frac{1}{20}x^5 - \frac{13}{720}x^6 + \cdots$ $(a_0 = 0)\ y = x + \frac{1}{6}x^3 - \frac{1}{12}x^4 - \frac{1}{24}x^5 - \frac{7}{360}x^6 + \cdots$

15. $(a_1 = 0)\ y = 1 - 2x^2 - \frac{11}{6}x^3 - \frac{11}{6}x^4 - \frac{41}{30}x^5 + \cdots$ $(a_0 = 0)\ y = x + x^3 + \frac{11}{12}x^4 + \frac{11}{15}x^5 + \frac{41}{90}x^6 + \cdots$

17. $y = a_0 + a_1x + \frac{1}{3}a_1x^3 + \left(\frac{a_1 - 2a_0}{10}\right)x^5 - \frac{2}{15}a_1x^6 + \cdots$

19. $y = a_0 + a_1x + 6a_1x^2 + 24a_1x^3 + 72a_1x^4 + \cdots$ **21.** $y = a_0 + a_1x - \frac{2}{3}a_0x^3 - \frac{1}{3}a_1x^4 + \frac{3}{20}a_1x^5 + \cdots$

23. $y = a_0 + a_1x + \frac{1}{2}a_0x^2 - \frac{1}{6}a_1x^3 - \frac{1}{8}a_0x^4 + \cdots$ **25.** $y = a_0 + a_1x + 2a_0x^2 + \frac{2}{3}a_1x^3 + \left(\frac{7a_0 - a_1}{12}\right)x^4 + \cdots$

27. $y = a_0 + a_1x + a_2x^2 - \frac{1}{3}a_1x^3 + \left(\frac{4a_2 - a_0}{24}\right)x^4 + \cdots$ **29.** $y = a_0 + a_1x + a_2x^2 + a_3x^3 + \frac{1}{12}a_0x^4 + \cdots$

31. $y = a_0 + a_1x - \frac{1}{2}a_0x^2 + \left(\frac{a_0 - 3a_1}{6}\right)x^3 + \left(\frac{7a_0 + 2a_1}{24}\right)x^4 + \cdots$

33. $y = a_0 + a_1x - a_1x^2 + \left(\frac{1 + 4a_1}{6}\right)x^3 + \left(\frac{-1 - 3a_1}{6}\right)x^4 + \cdots$

35. $y = a_0(1 - x^2 + \frac{5}{12}x^4 - \frac{31}{360}x^6 - \frac{23}{5040}x^8 + \cdots)$

37. $y = a_0 + a_1 x + \left(\frac{1 - a_0}{2}\right)x^2 + \left(\frac{-a_0 - a_1}{6}\right)x^3 + \left(-\frac{1}{24} - \frac{a_1}{12}\right)x^4 + \cdots$

39. $y = a_0 + a_1 x - \frac{a_0}{6}x^3 - \frac{a_1}{12}x^4 + \left(\frac{3 - a_0}{60}\right)x^5 + \cdots$

41. $y = a_0 + a_1 x + \left(\frac{a_1 - 2}{4}\right)x^2 + \frac{a_1}{24}x^3 + \left(\frac{4 + a_1}{192}\right)x^4 + \cdots$

43. $y = a_0 + a_1 x + \left(\frac{1 - 3a_0}{2}\right)x^2 - \frac{a_1}{2}x^3 + \left(\frac{9a_0 - 2}{24}\right)x^4 + \cdots$

45. $y = a_0 + a_1 x - \left(\frac{4 + 3a_0}{2}\right)x^2 - \frac{1}{3}a_1 x^3 + \left(\frac{4 + 3a_0}{24}\right)x^4 + \cdots$

47. $y = a_0 + a_1 x + \left(\frac{3 + a_0 - a_1}{2}\right)x^2 + \left(\frac{-3 - a_0}{6}\right)x^3 + \left(\frac{-2a_0 - a_1 - 7}{2}\right)x^4 + \cdots$

49. $y = a_0 + a_1 x + \left(\frac{5 + 3a_0}{2}\right)x^2 + \left(\frac{a_1 - 1}{6}\right)x^3 + \left(\frac{-4 - 3a_0}{24}\right)x^4 + \cdots$

Section 5.3

1. $r^2 + r - 2 = 0;\quad a_n = \frac{(n + r - 1)a_{n-1}}{(n + r)(n + r + 1) - 2}$ for $n = 1, 2, \ldots$ $y_1 = \sum_{n=0}^{\infty} \frac{1}{(n+3)!} x^{n+1}$ (with $a_0 = \frac{1}{6}$)

$y_2 = \frac{1}{x^2} + \frac{1}{x} + \frac{1}{2}$ (with $A = 0, b_0 = 1, b_3 = 0$)

3. $r^2 = 0;\quad n^2 a_n + a_{n-2} = 0$ for $n = 2, 3, \ldots;\quad a_1 = 0\quad y_1 = a_0\left(1 - \frac{1}{2^2}x^2 + \frac{1}{2^2 4^2}x^4 - \frac{1}{2^2 4^2 6^2}x^6 + \cdots\right)$

With $a_0 = 1$ in y_1, $y_2 = y_1 \ln(x) + b_0 + \left(\frac{1 - b_0}{4}\right)x^2 + \left(\frac{2b_0 - 3}{128}\right)x^4 + \left(\frac{11 - 6b_0}{13{,}824}\right)x^6 + \cdots$

5. $r^2 = 0;\quad a_n(n + r)^2 + a_{n-3} = 0$ for $n = 3, 4, \ldots;\quad a_1 = a_2 = 0$

$y_1 = a_0\left(1 - \frac{1}{3^2}x^3 + \frac{1}{3^2 6^2}x^6 - \frac{1}{3^2 6^2 9^2}x^9 + \frac{1}{3^2 6^2 9^2 12^2}x^{12} + \cdots\right)$

With $b_0 = 0$, $y_2(x) = \frac{1}{2}y_1(x)\ln(x) + \frac{x^3}{27} - \frac{1}{3 \cdot 6^3}x^6 + \frac{99}{3^3 6^3 9^3}x^9 - \frac{1350}{3^3 6^3 9^3 12^3}x^{12} + \cdots$

7. $r^2 - 2r - 8 = 0; [(n + r)(n + r - 2) - 8]a_n + 2a_{n-1} = 0$ for $n = 1, 2, 3, \ldots$

$y_1 = a_0\left(x^4 - \frac{2}{7}x^5 + \frac{1}{4 \cdot 7}x^6 - \frac{1}{2 \cdot 3 \cdot 7 \cdot 9}x^7 + \frac{1}{3 \cdot 4 \cdot 7 \cdot 9 \cdot 10}x^8 - \frac{1}{2 \cdot 3 \cdot 5 \cdot 7 \cdot 9 \cdot 10 \cdot 11}x^9 + \cdots\right)$

With $a_0 = 1$, and $b_0 = b_6 = 0$,

$y_2(x) = y_1(x)\ln(x) - \frac{675}{x^2} - \frac{270}{x} - \frac{135}{2} - 15x - \frac{15}{4}x^2 - \frac{3}{2}x^3 + \frac{8}{49}x^5 - \frac{99}{448}x^6 + \cdots$

9. $r^2 - \frac{5}{4}r - \frac{1}{4} = 0;\quad a_n[(n + r)(n + r - \frac{5}{4}) - \frac{1}{4}] + a_{n-1}(n + r - 1) = 0$ for $n = 1, 2, 3, \ldots$

$y_1 = x^{(5+\sqrt{41})/8} \sum_{n=0}^{\infty} a_n x^n;$ with $a_n = \frac{-(8n - 3 + \sqrt{41})a_{n-1}}{n(8n + 2\sqrt{41})}$

and $y_2 = x^{(5-\sqrt{41})/8} \sum_{n=0}^{\infty} b_n x^n;$ with $b_n = \frac{-(8n - 3 - \sqrt{41})b_{n-1}}{n(8n - 2\sqrt{41})}$

11. $r^2 - \frac{7}{4}r - \frac{1}{8} = 0;\ a_n[(n + r)(n + r - \frac{7}{4}) - \frac{1}{8}] + a_{n-2}(n + r - 2) + a_{n-1} = 0$ for $n = 2, 3, 4, \ldots;\quad a_1 = \frac{-a_0}{r^2 + \frac{1}{4}r - \frac{7}{8}}$

$y_1(x) = x^{(7+\sqrt{57})/8} \sum_{n=0}^{\infty} a_n x^n,$ with $a_1 = \frac{-4a_0}{4 + \sqrt{57}},$ and $a_n = \frac{-[8a_{n-1} + (8n - 9 + \sqrt{57})a_{n-2}]}{n(8n + 2\sqrt{57})}$ for $n = 2, 3, 4, \ldots$

$y_2(x) = x^{(7-\sqrt{57})/8} \sum_{n=0}^{\infty} b_n x^n,$ with $b_1 = \frac{-16b_0}{9 - 4\sqrt{57}},$ and $b_n = \frac{-8b_{n-1} - (8n - 9 - \sqrt{57})b_{n-2}}{n(8n - 2\sqrt{57})}$ for $n = 2, 3, 4, \ldots$

$$y_2 = b_0 x^{(5-3\sqrt{5})/2}\left(1 + \sum_{n=0}^{\infty} b_n x^n\right); \quad \text{with} \quad b_n = \frac{(-1)^n}{n!(4-12\sqrt{5})(8-12\sqrt{5})\cdots(4n-12\sqrt{5})} \quad \text{for} \quad n = 1, 2, 3, \ldots$$

43. $$y_1 = a_0 x^{(-1+\sqrt{29})/2}\left(1 + \sum_{n=1}^{\infty} a_{2n} x^{2n}\right); \quad \text{with} \quad a_{2n} = \frac{(-1)^n}{2\cdot 4\cdots(2n)(4+\sqrt{29})(8+\sqrt{29})\cdots(4n+\sqrt{29})}, \ n = 1, 2, \ldots$$

$$y_2 = b_0 x^{(-1-\sqrt{29})/2}\left(1 + \sum_{n=1}^{\infty} b_{2n} x^{2n}\right); \quad \text{with} \quad b_{2n} = \frac{(-1)^n}{2\cdot 4\cdots(2n)(4-\sqrt{29})(8-\sqrt{29})\cdots(4n-\sqrt{29})} \quad \text{for} \quad n = 1, 2, 3, \ldots$$

45. $$y_1(x) = a_0 x^{5/4}\left(1 + \frac{x^2}{11} + \frac{x^4}{2!(11)(27)} + \frac{x^6}{3!(11)(27)(43)} + \frac{x^8}{4!(11)(27)(43)(59)} + \frac{x^{10}}{5!(11)(27)(43)(59)(75)} + \cdots\right.$$

$$y_2(x) = 1 + \frac{1}{6}x^2 + \frac{x^4}{6(4)(11)} + \frac{x^6}{6(4)(6)(11)(19)} + \frac{x^8}{6(4)(6)(8)(11)(19)(27)} + \frac{x^{10}}{6(4)(6)(8)(10)(11)(19)(27)(35)} + \cdots$$

47. $$y_1 = a_0 x^{1+\sqrt{3}}\left(1 + \sum_{n=1}^{\infty} a_{2n} x^{2n}\right); \quad \text{with} \quad a_{2n} = \frac{(-1)^n}{2\cdot 4\cdots(2n)(2+2\sqrt{3})(4+2\sqrt{3})\cdots(2n+2\sqrt{3})} \quad \text{for} \quad n = 1, 2, 3, \ldots$$

$$y_2 = b_0 x^{1-\sqrt{3}}\left(1 + \sum_{n=1}^{\infty} b_{2n} x^{2n}\right); \quad \text{with} \quad b_{2n} = \frac{(-1)^n}{2\cdot 4\cdots(2n)(2-2\sqrt{3})(4-2\sqrt{3})\cdots(2n-2\sqrt{3})} \quad \text{for} \quad n = 1, 2, 3, \ldots$$

49. $$y_1 = a_0 x^{\sqrt{2}}\left(1 + \sum_{n=1}^{\infty} a_{2n} x^{2n}\right); \quad \text{with} \quad a_{2n} = \frac{1}{2\cdot 4\cdots(2n)(2+2\sqrt{2})(4+2\sqrt{2})\cdots(2n+2\sqrt{2})} \quad \text{for} \quad n = 1, 2, 3, \ldots$$

$$y_2 = b_0 x^{-\sqrt{2}}\left(1 + \sum_{n=1}^{\infty} b_{2n} x^{2n}\right); \quad \text{with} \quad b_{2n} = \frac{1}{2\cdot 4\cdots(2n)(2-2\sqrt{2})(4-2\sqrt{2})\cdots(2n-2\sqrt{2})} \quad \text{for} \quad n = 1, 2, 3, \ldots$$

Chapter 6

Section 6.1

1. Write $\exp\left[\frac{x}{2}\left(t - \frac{1}{t}\right)\right] = \exp\left(\frac{xt}{2}\right)\exp\left(-\frac{x}{2t}\right) = \left[1 + \left(\frac{xt}{2}\right) + \frac{1}{2!}\left(\frac{xt}{2}\right)^2 + \cdots\right]\left[1 - \frac{x}{2t} + \frac{1}{2!}\left(\frac{x}{2t}\right)^2 + \cdots\right].$

Now, let n be a fixed nonnegative integer. In multiplying these two series, the terms involving t^n will be

$$\frac{1}{n!}\left(\frac{xt}{2}\right)^n(1) + \frac{1}{(n+1)!}\left(\frac{xt}{2}\right)^{n+1}\left(\frac{-x}{2t}\right) + \frac{1}{(n+2)!}\left(\frac{xt}{2}\right)^{n+2}\frac{1}{2!}\left(\frac{-x}{2t}\right)^2 + \frac{1}{(n+3)!}\left(\frac{xt}{2}\right)^{n+3}\frac{1}{3!}\left(\frac{-x}{2t}\right)^3 + \cdots,$$

which can be written $\sum_{j=0}^{\infty} \frac{1}{(n+j)!}\left(\frac{xt}{2}\right)^{n+j}\frac{1}{j!}\left(\frac{-x}{2t}\right)^j.$

Thus, the *coefficient* of t^n will be

$$\sum_{j=0}^{\infty} \frac{1}{j!(n+j)!}\left(\frac{x}{2}\right)^{n+j}\left(\frac{-x}{2}\right)^j = \sum_{j=0}^{\infty} \frac{(-1)^j}{j!(n+j)!}\left(\frac{x}{2}\right)^{n+2j} = \mathscr{J}_n(x).$$

3. Write $x^{-n}\mathscr{J}_n(x) = \frac{1}{2^n}\sum_{j=0}^{\infty}\frac{(-1)^j}{j!(n+j)!}\left(\frac{x}{2}\right)^{2j}$. Then, $[x^{-n}\mathscr{J}_n(x)]' = \frac{1}{2^n}\sum_{j=1}^{\infty}\frac{j(-1)^j}{j(j-1)!(n+j)!}\left(\frac{x}{2}\right)^{2j-1}$

$$= -x^{-n}\sum_{j=0}^{\infty}\frac{(-1)^j}{j!(n+1+j)!}\left(\frac{x}{2}\right)^{n+1+2j} = -x^{-n}\mathscr{J}_{n+1}(x).$$

A similar argument yields $[x^n\mathscr{J}_n(x)]' = x^n\mathscr{J}_{n-1}(x)$.

5. From Problem 4, $x\mathscr{J}_n'(x) = -n\mathscr{J}_n(x) + x\mathscr{J}_{n-1}(x)$ and $x\mathscr{J}_n'(x) = n\mathscr{J}_n(x) - x\mathscr{J}_{n+1}(x)$.
Add these to get $2x\mathscr{J}_n'(x) = x\mathscr{J}_{n-1}(x) - x\mathscr{J}_{n+1}(x)$. For $x \neq 0$, $2\mathscr{J}_n'(x) = \mathscr{J}_{n-1}(x) - \mathscr{J}_{n+1}(x)$, one of the results to be proved.
Now begin again with the results of Problem 4, but subtract to get

$0 = -2n\mathscr{J}_n(x) + x\mathscr{J}_{n-1}(x) + x\mathscr{J}_{n+1}(x)$. Then, $\left(\frac{2n}{x}\right)\mathscr{J}_n(x) = x\mathscr{J}_{n-1}(x) + x\mathscr{J}_{n+1}(x)$.

7. Put $n = 1$ into $x\mathcal{J}_n'(x) = -n\mathcal{J}_n(x) + x\mathcal{J}_{n-1}(x)$ to get $x\mathcal{J}_1'(x) = -\mathcal{J}_1(x) + x\mathcal{J}_0(x)$.

Then, $\int x^2\mathcal{J}_0(x)\,dx = \int x[x\mathcal{J}_1'(x) + \mathcal{J}_1(x)]dx$

$$= \int \underbrace{x^2}_{u} \underbrace{\mathcal{J}_1'(x)dx}_{dv} + \int x\mathcal{J}_1(x)dx$$

$$= x^2\mathcal{J}_1(x) - \int 2x\mathcal{J}_1(x)dx + \int x\mathcal{J}_1(x)dx$$

$$= x^2\mathcal{J}_1(x) - \int x\mathcal{J}_1(x)dx.$$

Next, put $n = 0$ into $x\mathcal{J}_n'(x) = n\mathcal{J}_n(x) - x\mathcal{J}_{n+1}(x)$ to get $x\mathcal{J}_0'(x) = -x\mathcal{J}_1(x)$.

Then, integration by parts again yields $\int x^2\mathcal{J}_0(x)dx = x^2\mathcal{J}_1(x) + \int x\mathcal{J}_0'(x)dx = x^2\mathcal{J}_1(x) + x\mathcal{J}_0(x) - \int \mathcal{J}_0(x)dx.$

9. From Problem 3, show that $\mathcal{J}_0'(x) = -\mathcal{J}_1(x)$. Then, upon setting $t = \alpha x$, we have

$$\int_0^1 \mathcal{J}_1(\alpha x)dx = \int_0^\alpha \mathcal{J}_1(t)\frac{1}{\alpha}\,dt = \frac{1}{\alpha}\int_0^\alpha -\mathcal{J}_0'(t)dt = -\frac{1}{\alpha}[\mathcal{J}_0(\alpha) - \mathcal{J}_0(0)].$$

But, $\mathcal{J}_0(\alpha) = 0$ and $\mathcal{J}_0(0) = 1$.

11. $y = c_1x\mathcal{J}_2(x) + c_2xY_2(x)$ **13.** $y = c_1x^2\mathcal{J}_2(\sqrt{x}) + c_2x^2Y_2(\sqrt{x})$ **15.** $y = c_1x^4\mathcal{J}_1(x) + c_2x^4Y_1(x)$

17. $y = c_1x^{-1}\mathcal{J}_2(\frac{1}{2}\sqrt{x}) + c_2x^{-1}Y_2(\frac{1}{2}\sqrt{x})$ **19.** $y = c_1x^2\mathcal{J}_4(x^2) + c_2x^2Y_4(x^2)$

Section 6.2

1. $\Gamma(\frac{1}{2}) = \int_0^\infty t^{-1/2}e^{-t}\,dx$. Let $x = t^{1/2}$. Then, $dx = \frac{1}{2}t^{-1/2}\,dt$; so $dt = 2x\,dx$, and

$$\int_0^\infty t^{-1/2}e^{-t}\,dt = \int_0^\infty \frac{1}{x}\,e^{-x^2}(2x)dx = 2\int_0^\infty e^{-x^2}\,dx.$$

Write $\Gamma(\frac{1}{2})^2 = \int_0^\infty\int_0^\infty 4e^{-x^2-y^2}\,dx\,dy$. Let $x = r\cos(\theta)$, $y = r\sin(\theta)$. Then,

$$[\Gamma(\tfrac{1}{2})]^2 = \int_0^\infty\int_0^{\pi/2} 4e^{-r^2}r\,d\theta\,dr = 4\left(\frac{\pi}{2}\right)\int_0^\infty re^{-r^2}\,dr = \left[2\pi(-\tfrac{1}{2}e^{-r^2})\right]_0^\infty = \pi. \quad \text{Thus,} \quad \Gamma(\tfrac{1}{2}) = \sqrt{\pi}.$$

3. **(a)** Let $u = 1 - t$ in $B(x, y)$ to get

$$B(x, y) = \int_1^0 (1-u)^{x-1}(u)^{y-1}(-1)du = \int_0^1 u^{y-1}(1-u)^{x-1}\,du = B(y, x).$$

(b) Let $t = \sin^2(\theta)$. Then, $1 - t = \cos^2(\theta)$ and $dt = 2\sin(\theta)\cos(\theta)d\theta$. Then,

$$B(x, y) = \int_0^{\pi/2} [\sin^2(\theta)]^{x-1}[\cos^2(\theta)]^{y-1}\cdot 2\sin(\theta)\cos(\theta)d\theta = 2\int_0^{\pi/2} \sin^{2x-1}(\theta)\cos^{2y-1}(\theta)d\theta.$$

(c) Put $t = u^2$ into $\Gamma(x)$ to get

$$\Gamma(x) = \int_0^\infty t^{x-1}e^{-t}\,dt = \int_0^\infty u^{2x-2}e^{-u^2}(2u)du = 2\int_0^\infty u^{2x-1}e^{-u^2}\,du.$$

Then, $\Gamma(x)\Gamma(y) = 4\int_0^\infty u^{2x-1}e^{-u^2}\,du\int_0^\infty u^{2y-1}e^{-u^2}\,du$

$$= 4\int_0^\infty u^{2x-1}e^{-u^2}\,du\int_0^\infty v^{2y-1}e^{-v^2}\,dv = 4\int_0^\infty\int_0^\infty u^{2x-1}v^{2y-1}e^{-u^2-v^2}\,du\,dv$$

$= 4\lim_{R\to\infty}\iint_{\Delta_R} u^{2x-1}v^{2y-1}e^{-u^2-v^2}\,du\,dv$. Let $x = r\cos(\theta)$ and $y = r\sin(\theta)$. Then,

$$\iint_{\Delta_R} u^{2x-1}v^{2y-1}e^{-u^2-v^2}\,du\,dv = \iint_{\Delta_R} [r\cos(\theta)]^{2x-1}[r\sin(\theta)]^{2y-1}e^{-r^2}r d\,r\,d\,\theta$$

$$= \iint_{\Delta_R} r^{2x+2y-1}\cos^{2x-1}(\theta)\sin^{2y-1}(\theta)e^{-r^2}\,dr\,d\theta = \int_0^R r^{2x+2y-1}e^{-r^2}\,dr \int_0^{\pi/2}\cos^{2x-1}(\theta)\sin^{2y-1}(\theta)d\theta.$$

Let $R \to \infty$. Using the result of part (b) and part (i) of (c), we have $\Gamma(x)\Gamma(y) = \Gamma(x+y)B(x, y)$.

5. $\frac{1}{2}B(\frac{2}{3}, \frac{3}{4})$ or $\frac{1}{2}\dfrac{\Gamma(\frac{2}{3})\Gamma(\frac{3}{4})}{\Gamma(\frac{17}{12})}$ **7.** $B(\frac{3}{2}, 1)$ or $\frac{2}{3}$

9. $\mathscr{J}_{1/2}(x) = \displaystyle\sum_{j=0}^{\infty} \frac{(-1)^j}{j!\,\Gamma(j+\frac{3}{2})}\left(\frac{x}{2}\right)^{2j+(1/2)}$ and $\sqrt{\dfrac{2}{\pi x}}\sin(x) = \sqrt{\dfrac{2}{\pi}}\displaystyle\sum_{j=0}^{\infty}\frac{(-1)^j}{(2j+1)!}x^{2j+(1/2)}$

These series are identical if, for $j = 0, 1, 2, \ldots$, we have $j!\,\Gamma(j+\frac{3}{2})2^{2j+(1/2)} = \sqrt{\dfrac{\pi}{2}}\,(2j+1)!$

By the result of Problem 2, $\Gamma(j+\frac{3}{2}) = \Gamma(j+\frac{1}{2}+1) = (j+\frac{1}{2})\Gamma(j+\frac{1}{2}) = (j+\frac{1}{2})\dfrac{(2j)!\sqrt{\pi}}{4^j j!}$.

Thus, $j!\Gamma(j+\frac{3}{2})2^{2j+(1/2)} = \dfrac{(j+\frac{1}{2})(2j)!\sqrt{\pi}}{4^j}2^{2j+(1/2)} = \dfrac{(2j+1)}{2}(2j)!\sqrt{\pi}\sqrt{2} = (2j+1)!\sqrt{\dfrac{\pi}{2}}$,

as we needed to show. The argument is similar for $\mathscr{J}_{-1/2}(x)$.

15. $y = A\mathscr{J}_{1/3}(x^2) + B\mathscr{J}_{-1/3}(x^2)$ **17.** $y = Ax^{-1}\mathscr{J}_{3/4}(2x^2) + Bx^{-1}\mathscr{J}_{3/4}(2x^2)$

19. $y = Ax^4\mathscr{J}_{3/4}(2x^3) + Bx^4\mathscr{J}_{-3/4}(2x^3)$ **21.** $y = Ax^{-2}\mathscr{J}_{1/2}(3x^3) + Bx^{-2}\mathscr{J}_{-1/2}(3x^3)$

23. $y = A\mathscr{J}_{3/5}(2x^3) + B\mathscr{J}_{-3/5}(2x^3)$

Section 6.3

3. Recall the binomial expansion

$$(1-z)^{-1/2} = 1 + \frac{1}{2}z + \frac{\frac{1}{2}(\frac{3}{2})}{2!}z^2 + \frac{\frac{1}{2}(\frac{3}{2})(\frac{5}{2})}{3!}z^3 + \frac{\frac{1}{2}(\frac{3}{2})(\frac{5}{2})(\frac{7}{2})}{4!}z^4 + \cdots.$$

Using this, $\dfrac{1}{(1-2xr+r^2)^{1/2}} = 1 + \frac{1}{2}(2xr-r^2) + \frac{3}{8}(2xr-r^2)^2 + \frac{15}{48}(2xr-r^2)^3$

$+\dfrac{105}{384}(2xr-r^2)^4 + \dfrac{945}{3840}(2xr-r^2)^5 + \cdots$. Carry out the powers of $2xr - r^2$ to write

$$\begin{aligned} H(x, r) &= 1 + \tfrac{1}{2}(2xr - r^2) + \tfrac{3}{8}(4x^2r^2 + r^4 - 4xr^3) + \tfrac{15}{48}(-r^6 + 6xr^5 - 12x^2r^4 + 8x^3r^3) \\ &\quad + \tfrac{105}{384}(r^8 - 8xr^7 + 24x^2r^6 - 32x^3r^5 + 16x^4r^4) \\ &\quad + \tfrac{945}{3840}(-r^{10} + 10xr^9 - 40x^2r^8 + 80x^3r^7 - 80x^4r^6 + 32x^5r^5) + \cdots \\ &= 1 + xr + (-\tfrac{1}{2} + \tfrac{3}{2}x^2)r^2 + (-\tfrac{3}{2}x + \tfrac{15}{6}x^3)r^3 \\ &\quad + (\tfrac{3}{8} - \tfrac{15}{4}x^2 + \tfrac{105}{24}x^4)r^4 + (\tfrac{15}{8}x - \tfrac{105}{12}x^3 + \tfrac{945}{120}x^5)r^5 + \cdots \\ &= P_0(x) + P_1(x)r + P_2(x)r^2 + P_3(x)r^3 + P_4(x)r^4 + P_5(x)r^5 + \cdots \end{aligned}$$

7. From Problem 2, $(n+1)P_{n+1}(x) + nP_{n-1}(x) = (2n+1)xP_n(x)$. Multiply by $P_{n-1}(x)$ and integrate from -1 to 1:

$$(n+1)\int_{-1}^{1} P_{n+1}(x)P_{n-1}(x) + n\int_{-1}^{1} P_{n-1}^2(x)dx = (2n+1)\int_{-1}^{1} xP_n(x)P_{n-1}(x)dx.$$

By the result of Problem 5, $\displaystyle\int_{-1}^{1} P_{n+1}(x)P_{n-1}(x)dx = 0$. Thus, $\displaystyle\int_{-1}^{1} P_{n-1}^2(x)dx = \left(\frac{2n+1}{n}\right)\int_{-1}^{1} xP_n(x)P_{n-1}(x)dx$.

Now use the same identity, with $(n-1)$ in place of n, and multiply through by $P_n(x)$ and integrate from -1 to 1 to get

$\displaystyle\int_{-1}^{1} P_n^2(x)dx = \left(\frac{2n-1}{n}\right)\int_{-1}^{1} xP_n(x)P_{n-1}(x)dx$. From the last two equations, $\displaystyle\int_{-1}^{1} P_n^2(x)dx = \left(\frac{2n-1}{2n+1}\right)\int_{-1}^{1} P_{n-1}^2(x)dx$.

We can use this to get $\displaystyle\int_{-1}^{1} P_n^2(x)dx$ in terms of $\displaystyle\int_{-1}^{1} P_0^2(x)dx$. Specifically,

$$\int_{-1}^{1} P_n^2(x)dx = \left(\frac{2n-1}{2n+1}\right)\int_{-1}^{1} P_{n-1}^2(x)dx = \left(\frac{2n-1}{2n+1}\right)\left(\frac{2n-3}{2n-1}\right)\int_{-1}^{1} P_{n-2}^2(x)dx = \cdots = \frac{1}{2n+1}\int_{-1}^{1} P_0(x)dx = \frac{2}{2n+1}.$$

Section 6.4

1. Periodic; $\lambda_n = \dfrac{n^2}{9}$ for $n = 1, 2, 3, \ldots$; $y_n = A_n \cos(n\pi x/3) + B_n \sin(n\pi x/3)$ with not both A_n and B_n zero.

3. Regular; $\lambda_n = C_n^2$, where $c_n > 0$ and c_n satisfies $c_n = A \cot(2\pi c_n)$; $y_n = A_n \cos(c_n x)$, $A_n \neq 0$, for $n = 1, 2, 3, \ldots$

5. Regular; $\lambda_n = c_n^2$, where $c_n > 0$ and satisfies $\tan(2c_n) = \frac{1}{2}$; $y_n = 2A_n c_n \cos(c_n x) + A_n \sin(c_n x)$, $A_n \neq 0$ for $n = 1, 2, 3, \ldots$

7. Regular: $\lambda_n = \dfrac{n^2\pi^2}{(\ln(2))^2}$, for $n = 0, 1, 2, 3, \ldots$; $y_n = A_n \cos\left(\dfrac{n\pi \ln(x)}{\ln(2)}\right)$, $A_n \neq 0$.

9. Regular; $\lambda_n = \dfrac{n^2\pi^2}{b^2}$ for $n = 0, 1, 2, \ldots$; $y_n = A_n \cos\left(\dfrac{n\pi x}{b}\right)$, with $A_n \neq 0$.

11. Regular; $\lambda_n = c_n^2$, where c_n satisfies $\tan(c_n \pi) = -2c_n$ and $c_n > 0$, $n = 1, 2, 3, \ldots$; $y_n = A_n \sin(c_n x)$, with $A_n \neq 0$.

13. Regular; $\lambda_n = \dfrac{n^2\pi^2 - 1}{2}$ for $n = 1, 2, 3, \ldots$; $y_n = A_n e^x \sin(n\pi x)$, with $A_n \neq 0$.

15. Regular; $\lambda_n = \left[\dfrac{n\pi}{\ln(2)}\right]^2 + 1$ for $n = 1, 2, 3, \ldots$; $y_n = \dfrac{A}{x} \sin\left(\dfrac{n\pi \ln(x)}{\ln(2)}\right)$ with $A_n \neq 0$.

17. Regular; $\lambda_n = c_n^2$, with c_n any positive solution of
$\tan(c_n \pi) = \dfrac{-c_n}{1 + 2c_n^2}$; and $y_n = A_n \cos(c_n x) + B_n \sin(c_n x)$, $A_n^2 + B_n^2 \neq 0$
One negative eigenvalue $\lambda_{-1} = -\alpha^2$, with α the positive soltuion of
$\tanh(\alpha\pi) = \dfrac{\alpha}{2\alpha^2 - 1}$, and $y_{-1} = A \cosh(\alpha x) + B \sinh(\alpha x)$, $A^2 + B^2 \neq 0$

19. $(xe^{-x}y')' + \lambda e^{-x}y = 0$. Here, $p(x) = e^{-x}$, and by Theorem 20(2), $\int_0^\infty e^{-x}\mathscr{L}_n(x)\mathscr{L}_m(x)\,dx = 0$ if $n \neq m$.

21. $\frac{1}{2} - \frac{3}{4}P_1(x) + \sum_{n=1}^{\infty} a_{2n+1}P_{2n+1}(x)$, with $a_{2n+1} = \dfrac{(-1)^{n+1}(4n+3)}{4n+4}\,\dfrac{(1)(3)\cdots(2n-1)}{2\cdot 4\cdots(2n)} = \dfrac{-(4n+3)}{(4n+4)} P_{2n}(0)$

23. $\frac{1}{2} + \sum_{n=2}^{\infty} a_n P_n(x)$, where $a_n = \left(\dfrac{n+1}{n+2}\right)P_n(0) + P_{n-2}(0)$ for $n = 2, 3, 4, \ldots$

25. $\frac{4}{3}P_0(x) + \frac{2}{3}P_2(x)$

Supplementary Problems

5. $y = Ax^2\mathscr{J}_1(\frac{1}{3}x^4) + Bx^2Y_1(\frac{1}{3}x^4)$

7. $y = Ax^3\mathscr{J}_{1/5}(3x^4) + Bx^3\mathscr{J}_{-1/5}(3x^4)$

31. Eigenvalues are values of λ with $\lambda > -\frac{3}{4}$ and satisfying $\tan\left(\dfrac{\sqrt{3+4\lambda}}{2}\right) = \sqrt{3+4\lambda}$; eigenfunctions are
$y = Ae^{-x/2}\sin\left(\dfrac{\sqrt{3+4\lambda}}{2}x\right)$ with $A \neq 0$.

33. $(\sqrt{1-x^2}\,y')' + \dfrac{\lambda}{\sqrt{1-x^2}}y = 0$

Chapter 7

Section 7.1

1. $x(t) = c_1 \cos(t/\sqrt{2}) + c_2 \sin(t/\sqrt{2}) + 1$ $\quad y(t) = (c_1 + \sqrt{2}c_2)\cos(t/\sqrt{2}) + (c_2 - \sqrt{2}c_1)\sin(t/\sqrt{2}) + 1$

3. $x(t) = c_1e^{t/2}$ $\quad y(t) = 3c_1e^{t/2}$

5. $x(t) = c_1e^{t/2}\cos(t/2) + c_2e^{t/2}\sin(t/2) - 1 - t$ $\quad y(t) = -c_2e^{t/2}\cos(t/2) + c_1e^{t/2}\sin(t/2) + 1$

7. $x(t) = c_1\cos(t) + c_2\sin(t)$ $\quad y(t) = \frac{1}{5}(c_1 - 3c_2)\cos(t) + \frac{1}{5}(3c_1 + c_2)\sin(t)$

9. $x(t) = c_1 + c_2e^t - t - \frac{1}{2}t^2 - \frac{3}{2}e^{-t}$ $\quad y(t) = c_2e^t - 1 - t + \frac{1}{2}e^{-t}$

11. $x(t) = c_1e^{\sqrt{3}t/2} + c_2e^{-\sqrt{3}t/2} - \frac{3}{7}\cos(t)$ $\quad y(t) = \dfrac{\sqrt{3}}{6}c_1e^{\sqrt{3}t/2} - \dfrac{\sqrt{3}}{6}c_2e^{-\sqrt{3}t/2} + \frac{3}{7}\sin(t) - 2$

13. $x(t) = c_1e^{t/\sqrt{2}} + c_2e^{-t/\sqrt{2}} + 4$ $\quad y(t) = (1 + \sqrt{2})c_1e^{t/\sqrt{2}} + (1 - \sqrt{2})c_2e^{-t/\sqrt{2}}$

15. $x(t) = c_1e^t\cos(t/\sqrt{2}) + c_2e^t\sin(t/\sqrt{2}) + \frac{8}{9} + \frac{2}{3}t$
$y(t) = (c_1 - \sqrt{2}c_2)e^t\cos(t/\sqrt{2}) + (\sqrt{2}c_1 + c_2)e^t\sin(t/\sqrt{2}) + \frac{4}{3} + 2t + t^2$

17. $x(t) = c_1 e^{(11+\sqrt{73})t/6} + c_2 e^{(11-\sqrt{73})t/6} + \frac{1}{6}e^{-t}$

$y(t) = \left(\frac{5+\sqrt{73}}{-2+2\sqrt{73}}\right)c_1 e^{(11+\sqrt{73})t/6} + \left(\frac{5-\sqrt{73}}{-2-2\sqrt{73}}\right)c_2 e^{(11-\sqrt{73})t/6} + \frac{1}{18}e^{-t}$

19. $x(t) = -3t^2 - t \quad y(t) = -3 - 3t^2 + 16(\frac{1}{4}t - 1) + ce^{-t/4}$

21. $x(t) = c_1 e^{-6t/7} + 8e^{-t} - \frac{8}{3}t + \frac{16}{9} \quad y(t) = \frac{3}{25}c_1 e^{-6t/7} + e^{-t} - \frac{2}{3}t + \frac{5}{18}$

23. $x(t) = c_1 e^{t/4}\cos\left(\frac{\sqrt{15}}{12}t\right) + c_2 e^{t/4}\sin\left(\frac{\sqrt{15}}{12}t\right) - 1$

$y(t) = \frac{1}{4}(\sqrt{15}\,c_2 - 5c_1)e^{t/4}\cos\left(\frac{\sqrt{15}}{12}t\right) + \frac{1}{4}(-\sqrt{15}\,c_1 - 5c_2)e^{t/4}\sin\left(\frac{\sqrt{15}}{12}t\right) - 1$

25. $x = c_1 e^{2t} + c_2 e^{4t/7} + \frac{1}{2} \quad y = -\frac{1}{2}c_1 e^{2t} - \frac{1}{8}c_2 e^{4t/7} + \frac{1}{4}t + \frac{3}{16}$

27. $x = c_1 t^{(1+\sqrt{33})/2} + c_2 t^{(1-\sqrt{33})/2} \quad y = \left(\frac{5-\sqrt{33}}{4}\right)c_1 t^{(1+\sqrt{33})/2} + \left(\frac{5+\sqrt{33}}{4}\right)c_2 t^{(1-\sqrt{33})/2}$

29. $x = c_1 t^{-1+\sqrt{8}} + c_2 t^{-1-\sqrt{8}} \quad y = \left(\frac{2-\sqrt{8}}{4}\right)c_1 t^{-1+\sqrt{8}} + \left(\frac{2+\sqrt{8}}{4}\right)c_2 t^{-1-\sqrt{8}}$

31. $x(t) = \frac{4}{\sqrt{15}}e^{-3t/4}\sin(\sqrt{15}\,t/4) \quad y(t) = -\frac{1}{\sqrt{15}}e^{-3t/4}\sin(\sqrt{15}\,t/4) + \frac{1}{3}e^{-3t/4}\cos(\sqrt{15}\,t/4) - \frac{1}{3}$

33. $x(t) = 24(1 - e^{-t/2}) + 2t^2 - 12t \quad y(t) = 12(1 - e^{-t/2}) + t^2 - 8t$

35. $x(t) = -4e^{t/2}\cos(t/2) + 8e^{t/2}\sin(t/2) + 8 \quad y(t) = -4e^{t/2}\sin(t/2)$

Section 7.2

1. $x(t) = 2 - 2e^{t/2} - t \quad y(t) = 1 - e^{t/2} - t$ **3.** $x(t) = \frac{4}{9} + \frac{t}{3} - \frac{4}{9}e^{3t/4} \quad y(t) = -\frac{2}{3} + \frac{2}{3}e^{3t/4}$

5. $x(t) = \frac{1}{\sqrt{5}}e^{(1+\sqrt{5})t/2} - \frac{1}{\sqrt{5}}e^{(1-\sqrt{5})t/2} \quad y(t) = \left(\frac{5+\sqrt{5}}{2\sqrt{5}}\right)e^{(1+\sqrt{5})t/2} + \left(\frac{-5+\sqrt{5}}{2\sqrt{5}}\right)e^{(1-\sqrt{5})t/2} - 1$

7. $x(t) = \frac{3}{4} - \frac{3}{4}e^{2t/3} + \frac{1}{2}t^2 + \frac{1}{2}t \quad y(t) = -\frac{3}{2}e^{2t/3} + t + \frac{3}{2}$

9. $x(t) = e^{-t}\cos(t) + t - 1 \quad y(t) = t^2 - t + e^{-t}\sin(t)$

11. $x(t) = \frac{1}{2}e^t - \frac{1}{4}t^2 - \frac{1}{2}t - \frac{1}{2} \quad y(t) = -\frac{1}{2}e^t + \frac{1}{2}t + \frac{1}{2}$

13. $x = \frac{9+\sqrt{11}}{(-2+2\sqrt{11})\sqrt{11}}e^{(1+\sqrt{11})t/5} + \frac{(9-\sqrt{11})}{(2+2\sqrt{11})\sqrt{11}}e^{(1-\sqrt{11})t/5}$

$y = \frac{1}{4\sqrt{11}}(-5+\sqrt{11})e^{(1+\sqrt{11})t/5} + \frac{1}{4\sqrt{11}}(5+\sqrt{11})e^{(1-\sqrt{11})t/5} + \frac{1}{2}$

15. $x = -e^{-x}\cos(2x) - 32e^{-x}\sin(2x) \quad y = 7e^{-x}\cos(2x) + 19e^{-x}\sin(2x) - 6$

Section 7.3

1. Unstable node **3.** Unstable node **5.** Stable, but not asymptotically stable, center
7. Unstable saddle point **9.** Asymptotically stable spiral point **11.** Unstable saddle point
13. Unstable node **15.** Asymptotically stable spiral point **17.** No conclusion **19.** node
21. No conclusion **23.** No conclusion

Supplementary Problems

1. $x(t) = -1 + c_1 e^{-t/3} \quad y(t) = t + c_1 e^{-t/3} + c_2$

3. $x(t) = c_1 + c_2 e^{-3t} + \frac{1}{6}t^2 + \frac{2}{9}t \quad y(t) = c_1 - \frac{7}{9} - 2c_2 e^{-3t} + \frac{5}{9}t + \frac{1}{6}t^2$

5. $x(t) = e^{3t}[c_1\cos(\sqrt{3}t) + c_2\sin(\sqrt{3}\,t)] \quad y(t) = 4 + \frac{e^{3t}}{19}[(-15c_1 - \sqrt{3}\,c_2)\cos(\sqrt{3}\,t) + (\sqrt{3}\,c_1 - 15c_2)\sin(\sqrt{3}\,t)]$

7. $x(t) = c_1 e^{t/\sqrt{7}} + c_2 e^{-t/\sqrt{7}} \quad y(t) = \left(-\frac{1}{3} + \frac{2}{\sqrt{7}}\right)c_1 e^{t/\sqrt{7}} - \left(\frac{1}{3} + \frac{2}{\sqrt{7}}\right)c_2 e^{-t/\sqrt{7}}$

9. $x(t) = c_1e^t + \frac{7}{3}c_2e^{-t/3} + 2t - 4 \qquad y(t) = c_1e^t + c_2e^{-t/3} + t - 2$

11. $x(t) = (\sqrt{3} - 3)c_1e^{(1+\sqrt{3})t/2} + (-\sqrt{3} - 3)c_2e^{(1-\sqrt{3})t/2} + t^2 - 8t + 16$
$y(t) = c_1e^{(1+\sqrt{3})t/2} + c_2e^{(1-\sqrt{3})t/2} + 2t - 4$

13. $x(t) = -c_1e^{-t} + \frac{1}{3}t^2 + \frac{1}{3}t \qquad y(t) = c_1e^{-t} + \frac{1}{3} - \frac{1}{3}t$

15. $x(t) = 1 + e^{3t/4}\left[\left(\frac{c_1 + \sqrt{7}c_2}{2}\right)\cos(\sqrt{7}\,t/4) + \left(\frac{c_2 - \sqrt{7}c_1}{2}\right)\sin(\sqrt{7}\,t/4)\right]$

$y(t) = c_1e^{3t/4}\cos(\sqrt{7}\,t/4) + c_2e^{3t/4}\sin(\sqrt{7}\,t/4) - 1$

17. $x(t) = \frac{1}{4}e^t[-\cos(t) + \sin(t)] + \frac{1}{4} \qquad y(t) = \frac{1}{4}t - \frac{1}{4}e^t\sin(t)$

19. $x(t) = -\frac{1}{2} + \frac{3}{4}e^t - \frac{1}{4}\cos(t) - \frac{1}{4}\sin(t) \qquad y(t) = \frac{1}{2} - \frac{3}{4}e^t + \frac{1}{4}\cos(t) - \frac{3}{4}\sin(t)$

21. $x(t) = \frac{2}{5}t^2 - \frac{t^3}{30} \qquad y(t) = \frac{1}{10}t^2$ **23.** $x(t) = 5 - 5e^{t/2} - \frac{1}{2}t \qquad y(t) = -\frac{5}{4} + \frac{5}{4}e^{t/2}$

25. $x(t) = \frac{9}{4}(t - 1 + e^{2t}) \qquad y(t) = \frac{3t^2}{8} + \frac{9}{2}(1 - t - e^{2t})$

27. Unstable node; $x(t) = c_1e^{(2+\sqrt{3})t} + c_2e^{(2-\sqrt{3})t} \qquad y(t) = \left(\frac{-1+\sqrt{3}}{2}\right)c_1e^{(2+\sqrt{3})t} + \left(\frac{-1-\sqrt{3}}{2}\right)c_2e^{(2-\sqrt{3})t}$

29. Unstable saddle point; $x(t) = c_1e^{(1+\sqrt{11})t} + c_2e^{(1-\sqrt{11})t} \qquad y(t) = \left(\frac{-3+\sqrt{11}}{2}\right)c_1e^{(1+\sqrt{11})t} + \left(\frac{-3-\sqrt{11}}{2}\right)c_2e^{(1-\sqrt{11})t}$

31. Unstable saddle point; $x(t) = c_1e^{(1+\sqrt{29})t/2} + c_2e^{(1-\sqrt{29})t/2}$
$y(t) = \left(\frac{\sqrt{29} - 5}{2}\right)c_1e^{(1+\sqrt{29})t/2} + \left(\frac{-\sqrt{29} - 5}{2}\right)c_2e^{(1-\sqrt{29})t/2}$

33. Unstable spiral point; $x(t) = e^{3t/2}[c_1\cos(\sqrt{15}\,t/2) + c_2\sin(\sqrt{15}\,t/2)]$
$y(t) = e^{3t/2}\left[\left(\frac{c_1 + \sqrt{15}c_2}{8}\right)\cos(\sqrt{15}\,t/2) + \left(\frac{\sqrt{15}c_1 + c_2}{8}\right)\sin(\sqrt{15}\,t/2)\right]$

35. Unstable saddle point; $x(t) = c_1e^{(2+\sqrt{17})t} + c_2e^{(2-\sqrt{17})t} \qquad y(t) = \left(\frac{-4+\sqrt{17}}{13}\right)c_1e^{(2+\sqrt{17})t} + \left(\frac{-4-\sqrt{17}}{13}\right)c_2e^{(2-\sqrt{17})t}$

37. Unstable node **39.** Unstable saddle point **41.** Unstable spiral point **43.** Unstable node **45.** Unstable node

Chapter 9

Section 9.1

1. $(2 + \sqrt{2})\mathbf{i} + 3\mathbf{j}; (2 - \sqrt{2})\mathbf{i} - 9\mathbf{j} + 10\mathbf{k}; \sqrt{38}; \sqrt{63}; 4\mathbf{i} - 6\mathbf{j} + 10\mathbf{k}; 3\sqrt{2}\mathbf{i} + 18\mathbf{j} - 15\mathbf{k}$

3. $3\mathbf{i} - \mathbf{k}; \mathbf{i} - 10\mathbf{j} + \mathbf{k}; \sqrt{29}; \sqrt{27}; 4\mathbf{i} - 10\mathbf{j}; 3\mathbf{i} + 15\mathbf{j} - 3\mathbf{k}$

5. $3\mathbf{i} - \mathbf{j} + 3\mathbf{k}; -\mathbf{i} + 3\mathbf{j} - \mathbf{k}; \sqrt{3}; \sqrt{12}; 2\mathbf{i} + 2\mathbf{j} + 2\mathbf{k}; 6\mathbf{i} - 6\mathbf{j} + 6\mathbf{k}$

7. $\mathbf{F} + \mathbf{G} = 3\mathbf{i} - 2\mathbf{j}; \mathbf{F} - \mathbf{G} = \mathbf{i}$ **9.** $2\mathbf{i} - 5\mathbf{j}; \mathbf{j}$ **11.** $-\frac{1}{2}\mathbf{i} - \frac{1}{2}\mathbf{j}$

13. $12\mathbf{j}$ **15.** $-9\mathbf{i} + 6\mathbf{j}$ **17.** $x = 3 - 6t, y = t, z = 0$ **19.** $x = 0, y = 1 - t, z = 3 - 2t$

21. $x = 2 + 3t, y = -3 - 9t, z = 6 + 2t$ **23.** $x = 3 + t, y = 3 + 9t, z = -5 - 6t$ **25.** $x = -1 + 5t, y = -8t, z = t$

27. $6\cos(60°)\mathbf{i} + 6\sin(60°)\mathbf{j}; 3\mathbf{i} + 5.1962\mathbf{j}$ **29.** $\sin(45°)\mathbf{i} - \cos(45°)\mathbf{j}; 0.7071\mathbf{i} - 0.7071\mathbf{j}$

31. $\sqrt{2}\cos(30°)\mathbf{i} + \sqrt{2}\sin(30°)\mathbf{j}; 1.2247\mathbf{i} + 0.7071\mathbf{j}$

33. $-15\cos(5°)\mathbf{i} + 15\sin(5°)\mathbf{j}; -14.9429\mathbf{i} + 1.3073\mathbf{j}$ **35.** $-25\mathbf{j}$ **37.** $t = \frac{1}{\|\mathbf{A}\|}$ **39.** $\frac{1}{2}(\mathbf{A} + \mathbf{B})$

Section 9.2

1. $2; \frac{2}{\sqrt{14}}$; not orthogonal; $2 \le \sqrt{1}\sqrt{14}$; 57.6885°, 1.0069 rad.

3. $-23;\ \dfrac{-23}{\sqrt{1189}}$; not orthogonal; $|-23| \le \sqrt{29}\sqrt{41} = \sqrt{1189}$ (about 34.4819); 131.8373°, 2.3010 rad.

5. $0; 0$; orthogonal; $0 \le 6\sqrt{13}$; 90°, $\pi/2$ rad. **7.** $5; 1$; not orthogonal; $5 \le \sqrt{5}\sqrt{5}$; 0°, 0 rad.

9. $-18; -\frac{9}{10}$; not orthogonal; $|-18| \le \sqrt{10}\sqrt{40} = 20$; 154.1581°, 2.6906 rad. **11.** $\frac{2}{5}\mathbf{i} + \frac{1}{5}\mathbf{k}$

13. $\frac{26}{29}\mathbf{r}$ **15.** $\frac{1}{38}\mathbf{r}$ **17.** $-4\mathbf{i} + 4\mathbf{j}$ **19.** $-\frac{2}{3}\mathbf{i} - \frac{1}{3}\mathbf{j} + \frac{2}{3}\mathbf{k}$ **21.** $3x - y + 4z = 4$

23. $8x - 6y + 4z = 50$ **25.** $8x - 3y + 4z = 19$ **27.** $-2x + z = 0$ **29.** $6x - 14y - 2z = 10$

31. $\cos(\theta) = \dfrac{112}{\sqrt{33}\sqrt{385}}$, about 0.9936; θ is 6.4635°, 0.1128 rad.

33. $\cos(\theta) = \dfrac{44}{\sqrt{91}\sqrt{46}}$, about 0.6801; θ is 47.1510°, 0.8229 rad.

35. $\cos(\theta) = \dfrac{85}{\sqrt{126}\sqrt{69}}$, about 0.9116; θ is 24.2712°, 0.4236 rad.

37. $\cos(\theta) = \dfrac{64}{\sqrt{86}\sqrt{61}}$, about 0.8836; θ is 27.9203°, 0.4873 rad.

39. $\cos(\theta) = \dfrac{93}{\sqrt{133}\sqrt{89}}$, about 0.8548; θ is 31.2629°, 0.5456 rad. **41.** $\mathbf{F} = \mathbf{0}$

Section 9.3

1. $\mathbf{F} \times \mathbf{G} = 8\mathbf{i} + 2\mathbf{j} + 12\mathbf{k} = -(\mathbf{G} \times \mathbf{F})$; $\cos(\theta) = \dfrac{-8}{\sqrt{276}}$; $\sin(\theta) = \sqrt{\dfrac{212}{276}}$; $\|\mathbf{F} \times \mathbf{G}\| = \sqrt{212}$

3. $-8\mathbf{i} - 12\mathbf{j} - 5\mathbf{k};\ \dfrac{-12}{\sqrt{377}};\ \sqrt{\dfrac{233}{377}};\ \sqrt{233}$ **5.** $18\mathbf{i} + 50\mathbf{j} - 60\mathbf{k};\ \dfrac{124}{\sqrt{21{,}800}};\ \sqrt{\dfrac{6424}{21{,}800}};\ \sqrt{6424}$

7. $-85\mathbf{i} - 18\mathbf{j} + 396\mathbf{k};\ \dfrac{-70}{\sqrt{169{,}265}};\ \sqrt{\dfrac{164{,}365}{169{,}265}};\ \sqrt{164{,}365}$ **9.** $12\mathbf{i} + 34\mathbf{j} + 8\mathbf{k};\ \dfrac{38}{\sqrt{2808}};\ \sqrt{\dfrac{1364}{2808}};\ \sqrt{1364}$

11. $3\mathbf{i} - \mathbf{j};\ \dfrac{21}{\sqrt{451}};\ \sqrt{\dfrac{10}{451}};\ \sqrt{10}$ **13.** $-6\mathbf{i} + 21\mathbf{j} - 58\mathbf{k};\ \dfrac{9}{\sqrt{3922}};\ \sqrt{\dfrac{3841}{3922}};\ \sqrt{3841}$

15. $-27\mathbf{i} + 12\mathbf{j} - 13\mathbf{k};\ \dfrac{6}{\sqrt{1078}};\ \sqrt{\dfrac{1042}{1078}};\ \sqrt{1042}$ **17.** $\mathbf{i} + 2\mathbf{j} + 6\mathbf{k}$; $x + 2y + 6z = 12$

19. $-3\mathbf{i} - 8\mathbf{j} + 2\mathbf{k}$; $-3x - 8y + 2z = 4$ **21.** $3\mathbf{i} - 25\mathbf{j} + 7\mathbf{k}$; $3x - 25y + 7z = 17$

23. $35\mathbf{i} - 28\mathbf{j} - 23\mathbf{k}$; $35x - 28y - 23z = 113$ **25.** $-25\mathbf{i} - 64\mathbf{j} + 59\mathbf{k}$; $25x + 64y - 59z = 82$

27. $-24\mathbf{i} + 15\mathbf{j} - 2\mathbf{k}$; $24x - 15y + 2z = -1$ **29.** $9\mathbf{i} - \mathbf{j} + 49\mathbf{k}$; $9x - y + 49z = -164$

31. 40 **33.** 129 **35.** 3 **37.** 126 **39.** 107 **41.** $\sqrt{1013}$ **43.** $\sqrt{98}$ **45.** 78 **47.** $\sqrt{836}$

49. $\sqrt{4232}$ **51.** $8\mathbf{i} - \mathbf{j} + \mathbf{k}$ **53.** $2\mathbf{i} - 3\mathbf{j} + 4\mathbf{k}$ **55.** $7\mathbf{i} + \mathbf{j} - 7\mathbf{k}$ **57.** $4\mathbf{i} + 6\mathbf{j} + 4\mathbf{k}$ **59.** $-3\mathbf{i} + 2\mathbf{j} - 8\mathbf{k}$

Section 9.4

1. -5 **3.** 38 **5.** 16 **7.** -304 **9.** -129 **11.** 6 **13.** 0 **15.** 4 **17.** -6 **19.** 0

25. $\frac{4}{3}$ **27.** $\frac{10}{3}$ **29.** $\frac{51}{2}$

Section 9.5

1. $5\mathbf{e}_1 + 5\mathbf{e}_2 + 6\mathbf{e}_3 + 5\mathbf{e}_4 + \mathbf{e}_5$; dot product $= 0$; $\cos(\theta) = 0$; θ is 90°, $\pi/2$ rad.

3. $17\mathbf{e}_1 - 4\mathbf{e}_2 + 3\mathbf{e}_3 + 6\mathbf{e}_4$; 24; $\cos(\theta) = \dfrac{24}{\sqrt{8160}}$; θ is 74.5924°, 1.3019 rad.

5. $-5\mathbf{e}_1 + 3\mathbf{e}_2 + 5\mathbf{e}_3 + 3\mathbf{e}_4 + 4\mathbf{e}_5 + 15\mathbf{e}_7$; -89; $\cos(\theta) = \dfrac{-89}{\sqrt{54{,}736}}$; θ is 112.3592°, 1.9610 rad.

7. $10\mathbf{e}_1 + 4\mathbf{e}_2 + 6\mathbf{e}_3$; 27; $\cos(\theta) = \dfrac{27}{\sqrt{2365}}$; θ is 56.2026°, 0.9822 rad.

9. $15\mathbf{e}_1 + 3\mathbf{e}_2 - 4\mathbf{e}_3 + 2\mathbf{e}_4 + 2\mathbf{e}_5$; -116; $\cos(\theta) = \dfrac{-116}{\sqrt{23{,}925}}$; θ is 138.5860°, 2.4188 rad.

11. Subspace **13.** Not a subspace **15.** Subspace **17.** Subspace **19.** Subspace

Section 9.6

1. Independent **3.** Independent **5.** Independent **7.** Dependent **9.** Independent
11. Independent **13.** Dependent **15.** Independent **17.** Independent **19.** Independent
21. Basis; dimension 2 **23.** Basis; dimension 2 **25.** Basis; dimension 3 **27.** Basis; dimension 3
29. Basis; dimension 1

Section 9.7

11. Part 6 is violated **13.** Parts 1, 4, 5, 6 are violated **15.** Parts 1, 4, 5, 6, 7, 9, 10 are violated
17. Parts 1, 4, 5, 6, 7, 10 are violated **19.** Parts 4, 5, 10 **21.** Parts 1, 4, 5, 10
23. Parts 1, 4, 5, 6, 10

Supplementary Problems

1. -20; $6\mathbf{i} + 9\mathbf{j} + 15\mathbf{k}$; $\cos(\theta) = \dfrac{-20}{\sqrt{742}}$; θ is 137.2416°, 2.3953 rad.

3. -24; $15\mathbf{i} - 18\mathbf{j} - 8\mathbf{k}$; $\cos(\theta) = \dfrac{-24}{\sqrt{1189}}$; θ is 134.1084°, 2.3406 rad.

5. 9; $12\mathbf{i} - 39\mathbf{j} - 18\mathbf{k}$; $\cos(\theta) = \dfrac{9}{\sqrt{2070}}$; θ is 78.5908°, 1.3717 rad.

7. -7; $11\mathbf{i} - 32\mathbf{j} - 8\mathbf{k}$; $\cos(\theta) = \dfrac{-7}{\sqrt{1258}}$; θ is 101.3826°, 1.7695 rad.

9. -46; $-10\mathbf{i} + 10\mathbf{j} + 40\mathbf{k}$; $\cos(\theta) = \dfrac{-46}{\sqrt{3916}}$; θ is 137.3142°, 2.3966 rad.

11. 7; $20\mathbf{i} + 15\mathbf{j} + \mathbf{k}$; $\cos(\theta) = \dfrac{7}{\sqrt{675}}$; θ is 74.3696°, 1.2980 rad.

13. 34; $42\mathbf{i} - 23\mathbf{j} + 18\mathbf{k}$; $\cos(\theta) = \dfrac{34}{\sqrt{3773}}$; θ is 56.3910°, 0.9842 rad.

15. 7; $56\mathbf{i} - 119\mathbf{j} + 24\mathbf{k}$; $\cos(\theta) = \dfrac{7}{\sqrt{17{,}922}}$; θ is 87.0027°, 1.5185 rad. **17.** $2\mathbf{i}$ **19.** $\frac{14}{5}\mathbf{i} + \frac{7}{5}\mathbf{k}$

21. $\frac{18}{5}\mathbf{i} + \frac{9}{5}\mathbf{k}$ **23.** $3\mathbf{i} + 2\mathbf{j}$ **25.** $-6\mathbf{i} + 3\mathbf{j}$ **27.** $x = -2t, y = 2 + 2t, z = 3 - 2t$
29. $x = -2 + 8t, y = 1 + 6t, z = -5 + 7t$ **31.** $x = 7 - 5t, y = 14t, z = t$
33. $x = 6 + 11t, y = 4 + 3t, z = -8 - 9t$ **35.** $x = 8 - 13t, y = t, z = 7t$ **37.** $4x - 5y - z = 0$
39. $-4x + 2y + 3z = 13$ **41.** $x + 2z = -3$ **43.** $3x + 2z = -2$ **45.** $4y - 3z = 26$
47. $39x + 35y + 11z = 19$ **49.** $106x - 72y + 95z = 14$ **51.** $78x + 14y - 30z = 364$
53. $-62x + 17y + 103z = 169$ **55.** $21x + 28y - 22z = 64$ **57.** 124 **59.** 21 **61.** 15 **63.** 47
65. 20 **67.** 69 **69.** 124

Chapter 10

Section 10.1

1. $\begin{bmatrix} -3 & -1 & 3 \\ 0 & -5 & 12 \\ 7 & 16 & 6 \end{bmatrix}$ **3.** $\begin{bmatrix} -22 & 2 & 9 & 5 & 19 \\ -11 & 4 & 4 & 6 & 35 \\ 39 & 7 & 33 & -1 & -4 \end{bmatrix}$ **5.** $[18 \quad 25 \quad 3]$ **7.** $\begin{bmatrix} -2 \\ 1 \\ 1 \end{bmatrix}$ **9.** $\begin{bmatrix} 0 & 5 \\ -5 & 1 \\ -4 & 8 \end{bmatrix}$

11. $\begin{bmatrix} 16 & 12 \\ -8 & -4 \\ -24 & -32 \end{bmatrix}$ **13.** $\begin{bmatrix} 44 \\ 0 \\ -22 \end{bmatrix}$ **15.** $\begin{bmatrix} 16 & -32 & 8 & 56 \\ 24 & 24 & -32 & 16 \\ 0 & 0 & -40 & 72 \end{bmatrix}$

17. $AB = \begin{bmatrix} -16 & 0 \\ 17 & 28 \end{bmatrix}$, $BA = \begin{bmatrix} 12 & -32 \\ -14 & 0 \end{bmatrix}$ **19.** $BA = \begin{bmatrix} 28 & 432 \\ 50 & 396 \end{bmatrix}$, $AB = \begin{bmatrix} 48 & 1 & 1 & -58 \\ -96 & 2 & 2 & 220 \\ -288 & -22 & -22 & -68 \\ -16 & 16 & 16 & 444 \end{bmatrix}$

21. AB is not defined; $BA = \begin{bmatrix} 410 & 36 & -56 & 227 \\ 17 & 253 & 40 & -1 \end{bmatrix}$ **23.** AB is not defined; BA is not defined

25. $AB = \begin{bmatrix} -22 & 30 & -10 & -4 \\ -42 & 45 & 30 & 6 \end{bmatrix}$; BA is not defined **27.** $AB = \begin{bmatrix} 9 & -6 & 21 \\ 0 & 0 & 0 \\ -3 & 2 & -7 \\ 12 & -8 & 28 \end{bmatrix}$; BA is not defined

29. $AB = [4 \quad 34]$; BA is not defined **31.** $AB = \begin{bmatrix} 2 & -6 \\ 0 & 0 \\ -6 & 18 \end{bmatrix}$; BA is not defined

33. $AB = \begin{bmatrix} -1 & 5 & -10 & 9 & -22 & 14 \\ -2 & 2 & 28 & 18 & 28 & -4 \end{bmatrix}$; BA is not defined **35.** AB is not defined; BA is not defined

37. AB is not defined; BA is not defined **39.** AB is not defined; BA is 8×6

Section 10.2

1. $A^3 = \begin{bmatrix} 4 & 5 & 7 & 2 & 7 \\ 5 & 4 & 7 & 2 & 7 \\ 7 & 7 & 2 & 6 & 2 \\ 2 & 2 & 6 & 0 & 6 \\ 7 & 7 & 2 & 6 & 2 \end{bmatrix}$ $A^4 = \begin{bmatrix} 19 & 18 & 11 & 14 & 11 \\ 18 & 19 & 11 & 14 & 11 \\ 11 & 11 & 20 & 4 & 20 \\ 14 & 14 & 4 & 12 & 4 \\ 11 & 11 & 20 & 4 & 20 \end{bmatrix}$

The number of distinct v_1—v_4 walks of length 4 is $(A^4)_{14} = 14$; they are 1—3—4—3—4, 1—3—2—3—4, 1—5—2—3—4, 1—2—1—3—4, 1—3—1—3—4, 1—5—1—3—4, 1—5—1—5—4, 1—2—1—5—4, 1—3—1—5—4, 1—3—4—5—4, 1—5—4—5—4, 1—5—2—5—4, 1—3—2—5—4, and 1—3—4—3—4.

The number of distinct v_2—v_4 walks of length 4 is $(A^4)_{24} = 14$; they are 2—5—4—5—4, 2—1—2—5—4, 2—3—1—5—4, 2—3—2—5—4, 2—3—4—5—4, 2—5—2—5—4, 2—5—1—5—4, 2—3—2—3—4, 2—3—4—3—4, 2—1—2—3—4, 2—5—2—3—4, 2—3—4—3—4, 2—5—4—3—4, and 2—5—1—3—4.

The number of distinct v_2—v_3 walks of length 3 is $(A^3)_{23} = 7$; they are 2—3—2—3, 2—3—1—3, 2—1—2—3, 2—5—2—3, 2—5—1—3, 2—5—4—3, and 2—3—4—3.

The number of distinct v_3—v_4 walks of length 3 is $(A^3)_{34} = 0$.

3. There are 96 walks of length 5 from v_2 to v_2; 147 of length 4 from v_1 to v_4; 124 of length 5 from v_4 to v_5; 32 of length 4 from v_1 to v_2; and 32 of length 4 from v_4 to v_5.

Section 10.3

1. Define E by $E_{ij} = 0$ if $i \neq j$, and $E_{ii} = 1/D_{ii}$. **7.** $B = \dfrac{1}{ad - bc}\begin{bmatrix} d & -c \\ -b & a \end{bmatrix}$

Section 10.4

1. $B = \begin{bmatrix} -2 & 1 & 4 & 2 \\ 0 & \sqrt{3} & 16\sqrt{3} & 3\sqrt{3} \\ 1 & -2 & 4 & 8 \end{bmatrix}$, $E = \begin{bmatrix} 1 & 0 & 0 \\ 0 & \sqrt{3} & 0 \\ 0 & 0 & 1 \end{bmatrix}$

3. $B = \begin{bmatrix} -2 & 1 & 1 & 7 & 13 \\ -7 & 5 & 4 & 49 & 67 \end{bmatrix}$, $E = \begin{bmatrix} 1 & 0 \\ 5 & 1 \end{bmatrix}$

5. $B = \begin{bmatrix} 2 & 2 & -6 \\ -2 & 4 & 3 \\ 1 & 0 & 3 \end{bmatrix}$, $E = \begin{bmatrix} 1 & 0 & 0 \\ 0 & 0 & 1 \\ 0 & 1 & 0 \end{bmatrix}$

7. $B = \begin{bmatrix} -45 & 0 & 20 \\ 3 & 5 & 2 \\ 14 & 4 & 4 \end{bmatrix}$, $E = \begin{bmatrix} 5 & 0 & 0 \\ 0 & 1 & 0 \\ 0 & 0 & 1 \end{bmatrix}$

9. $B = \begin{bmatrix} -6 & 8 & 0 \\ 3 & 5 & 12 \\ 22 & 13 & -9 \end{bmatrix}$, $E = \begin{bmatrix} 0 & 0 & 1 \\ 0 & 1 & 0 \\ 1 & 0 & 0 \end{bmatrix}$

11. $B = \begin{bmatrix} -2 & 3 & -6 & 3 & 1 \\ 0 & -5\sqrt{3} & 3\sqrt{3} & -5\sqrt{3} & 9\sqrt{3} \end{bmatrix}$, $E = \begin{bmatrix} 1 & 0 \\ 0 & \sqrt{3} \end{bmatrix}$

13. $B = \begin{bmatrix} 1 & 4 & -3 & -2 & 4 \\ 0 & -1 & -6 & 3 & 7 \\ 0 & -2 & 7 & 1 & 3 \end{bmatrix}$, $E = \begin{bmatrix} 0 & 0 & 1 \\ 0 & 1 & 0 \\ 1 & 0 & 0 \end{bmatrix}$

15. $B = \begin{bmatrix} 2 - 2\sqrt{5} \\ 1 \\ 0 \\ -2 \\ 7 \end{bmatrix}$, $E = \begin{bmatrix} 1 & 0 & \sqrt{5} & 0 & 0 \\ 0 & 1 & 0 & 0 & 0 \\ 0 & 0 & 1 & 0 & 0 \\ 0 & 0 & 0 & 1 & 0 \\ 0 & 0 & 0 & 0 & 1 \end{bmatrix}$

17. $B = \begin{bmatrix} -10 + 10\sqrt{13} & 70 + 45\sqrt{13} & 30 + 25\sqrt{13} \\ 2 & 9 & 5 \\ 8 & 1 & -3 \end{bmatrix}$, $C = \begin{bmatrix} 5 & 0 & 5\sqrt{13} \\ 0 & 0 & 1 \\ 0 & 1 & 0 \end{bmatrix}$

19. $B = \begin{bmatrix} 3 & -4 & 5 & 9 \\ 2 + 3\sqrt{3} & 1 - 4\sqrt{3} & 3 + 5\sqrt{3} & -6 + 9\sqrt{3} \\ 18 + 3\sqrt{3} & 37 - 4\sqrt{3} & 31 + 5\sqrt{3} & 54 + 9\sqrt{3} \end{bmatrix}$, $C = \begin{bmatrix} 1 & 0 & 0 \\ 0 & 1 & 0 \\ 0 & 1 & 1 \end{bmatrix}\begin{bmatrix} 1 & 0 & 0 \\ 0 & 1 & 0 \\ 0 & 0 & 4 \end{bmatrix}\begin{bmatrix} 1 & 0 & 0 \\ \sqrt{3} & 1 & 0 \\ 0 & 0 & 1 \end{bmatrix}\begin{bmatrix} 1 & 0 & 0 \\ 0 & 1 & 0 \\ 1 & 0 & 1 \end{bmatrix}$

21. $B = \begin{bmatrix} 28 & 50 & 2 \\ 9 & 15 & 0 \\ 0 & -45 & 70 \end{bmatrix}$, $C = \begin{bmatrix} 1 & 0 & 0 \\ 0 & 1 & 0 \\ 0 & 0 & 5 \end{bmatrix}\begin{bmatrix} 0 & 0 & 1 \\ 0 & 1 & 0 \\ 1 & 0 & 0 \end{bmatrix}\begin{bmatrix} 1 & 0 & 0 \\ 0 & 1 & 0 \\ 0 & 3 & 1 \end{bmatrix}\begin{bmatrix} 1 & 0 & 0 \\ 0 & 0 & 1 \\ 0 & 1 & 0 \end{bmatrix}$

23. $B = \begin{bmatrix} 0 & -4 & 2 & -6 \\ 0 & 1 & -3 & 5 \\ -4 + 10\sqrt{3} & 12 + 70\sqrt{3} & -10 + 30\sqrt{3} & -16 - 30\sqrt{3} \\ -5 & -35 & -15 & 15 \end{bmatrix}$,

$C = \begin{bmatrix} 1 & 0 & 0 & 0 \\ 0 & 1 & 0 & 0 \\ 0 & 0 & -2 & 0 \\ 0 & 0 & 0 & 1 \end{bmatrix}\begin{bmatrix} 0 & 0 & 1 & 0 \\ 0 & 1 & 0 & 0 \\ 1 & 0 & 0 & 0 \\ 0 & 0 & 0 & 1 \end{bmatrix}\begin{bmatrix} 1 & 0 & 0 & \sqrt{3} \\ 0 & 1 & 0 & 0 \\ 0 & 0 & 1 & 0 \\ 0 & 0 & 0 & 1 \end{bmatrix}\begin{bmatrix} 1 & 0 & 0 & 0 \\ 0 & 1 & 0 & 0 \\ 0 & 0 & 1 & 0 \\ 0 & 0 & 0 & -5 \end{bmatrix}$

25. $B = \begin{bmatrix} 0 & -28 & 12 \\ 0 & 3 & -2 \\ -5 & 1 & 4 \end{bmatrix}$, $C = \begin{bmatrix} -2 & 0 & 0 \\ 0 & 1 & 0 \\ 0 & 0 & 1 \end{bmatrix}\begin{bmatrix} 0 & 0 & 1 \\ 0 & 1 & 0 \\ 1 & 0 & 0 \end{bmatrix}\begin{bmatrix} 1 & 0 & 0 \\ 0 & 1 & 0 \\ 0 & 4 & 1 \end{bmatrix}$

Section 10.5

1. Not in reduced form (part 2 is violated); $\begin{bmatrix} 1 & 0 & 3 \\ 0 & 1 & 2 \\ 0 & 0 & 0 \end{bmatrix}$

3. Not in reduced form (parts 1 and 3) $\begin{bmatrix} 1 & 4 & 1 & 0 \\ 0 & 0 & 0 & 1 \\ 0 & 0 & 0 & 0 \\ 0 & 0 & 0 & 0 \end{bmatrix}$

5. Not in reduced form (parts 1, 2, 3, and 4) $\begin{bmatrix} 1 & 0 \\ 0 & 1 \\ 0 & 0 \\ 0 & 0 \end{bmatrix}$

7. Not in reduced form (parts 1, 2, and 4) $\begin{bmatrix} 1 & 0 & 0 \\ 0 & 1 & 0 \\ 0 & 0 & 1 \end{bmatrix}$

9. Not in reduced form (parts 1, 2, and 4) $\begin{bmatrix} 1 & 0 & 0 & 0 \\ 0 & 1 & \frac{3}{2} & \frac{1}{2} \end{bmatrix}$

11. Not in reduced form (parts 1, 2, and 4) $\begin{bmatrix} 1 & 0 & 0 \\ 0 & 1 & 0 \\ 0 & 0 & 1 \end{bmatrix}$

13. Not in reduced form (parts 1, 2, and 4) $\begin{bmatrix} 1 & 0 & 0 & 0 \\ 0 & 1 & 0 & 0 \\ 0 & 0 & 1 & -1 \end{bmatrix}$

15. Not in reduced form (parts 1, 2, and 4) $\begin{bmatrix} 1 & -1 & 0 & -\frac{3}{2} & \frac{3}{2} \\ 0 & 0 & 1 & \frac{1}{2} & -\frac{1}{2} \end{bmatrix}$

17. Not in reduced form (parts 1, 2, and 3) $\begin{bmatrix} 1 & 0 & 0 & -\frac{6}{5} & \frac{2}{5} \\ 0 & 1 & 0 & \frac{21}{10} & \frac{14}{5} \\ 0 & 0 & 1 & -\frac{3}{10} & -\frac{7}{5} \end{bmatrix}$

19. Not in reduced form (parts 1 and 2) $\begin{bmatrix} 1 & 0 & -\frac{44}{37} & -\frac{15}{47} & -\frac{54}{47} \\ 0 & 1 & -\frac{37}{47} & -\frac{3}{47} & -\frac{86}{47} \end{bmatrix}$

21. Not in reduced form (parts 1 and 2) $\begin{bmatrix} 1 & 0 & 0 & \frac{81}{153} & -\frac{36}{153} \\ 0 & 1 & \frac{17}{153} & \frac{74}{153} & \frac{20}{153} \end{bmatrix}$

23. Not in reduced form (parts 1, 2, and 4) $\begin{bmatrix} 1 & 0 & 0 \\ 0 & 1 & 0 \\ 0 & 0 & 1 \\ 0 & 0 & 0 \end{bmatrix}$

25. Not in reduced form (parts 1, 2, and 4) $\begin{bmatrix} 1 & 0 & 0 & 0 \\ 0 & 1 & 0 & 0 \\ 0 & 0 & 1 & 0 \\ 0 & 0 & 0 & 1 \end{bmatrix}$

Section 10.6

1. $A_R = \begin{bmatrix} 1 & 0 & -\frac{3}{5} \\ 0 & 1 & \frac{3}{5} \end{bmatrix}$; rank $= 2$; a basis consists of $(-4, 1, 3)$ and $(2, 2, 0)$

3. $A_R = \begin{bmatrix} 1 & 0 \\ 0 & 1 \\ 0 & 0 \end{bmatrix}$; rank $= 2$; $(-3, 1)$ and $(2, 2)$

5. $A_R = \begin{bmatrix} 1 & 0 & -\frac{1}{4} & \frac{1}{2} \\ 0 & 1 & -\frac{5}{4} & \frac{1}{2} \end{bmatrix}$; rank $= 2$; $(8, -4, -3, 2)$ and $(1, -1, 1, 0)$

7. $A_R = \begin{bmatrix} 1 & 0 & 0 \\ 0 & 1 & 0 \\ 0 & 0 & 1 \\ 0 & 0 & 0 \end{bmatrix}$; rank $= 3$; $(2, 2, 1)$, $(1, -1, 3)$, and $(0, 0, 1)$

9. $A_R = \begin{bmatrix} 1 & 0 & 0 \\ 0 & 1 & 0 \\ 0 & 0 & 1 \end{bmatrix}$; rank $= 3$; $(0, -4, 3)$, $(6, 1, 0)$ and $(2, 2, 2)$

11. $A_R = \begin{bmatrix} 1 & 0 & 0 \\ 0 & 1 & 0 \\ 0 & 0 & 1 \end{bmatrix}$; rank $= 3$; $(-3, 2, 2)$, $(1, 0, 5)$, and $(0, 0, 2)$

13. $A_R = \begin{bmatrix} 1 & 0 & -11 \\ 0 & 1 & -3 \\ 0 & 0 & 0 \end{bmatrix}$; rank $= 2$; $(-2, 5, 7)$ and $(0, 1, -3)$

15. $A_R = \begin{bmatrix} 1 & 0 & 1 & \frac{1}{7} \\ 0 & 1 & 3 & \frac{3}{2} \\ 0 & 0 & 0 & 0 \\ 0 & 0 & 0 & 0 \end{bmatrix}$; rank $= 2$; $(7, -2, 1, -2)$ and $(0, 2, 6, 3)$

17. $A_R = \begin{bmatrix} 1 & 0 & 0 & -1 \\ 0 & 1 & 0 & 6 \\ 0 & 0 & 1 & -1 \end{bmatrix}$; rank $= 3$; $(4, 1, -3, 5)$, $(2, 0, 0, -2)$, and $(13, 2, 0, -1)$

19. $A_R = \begin{bmatrix} 1 & 0 & -\frac{1}{2} & \frac{1}{2} & -\frac{3}{2} \\ 0 & 1 & -\frac{15}{4} & -\frac{7}{4} & -\frac{17}{4} \\ 0 & 0 & 0 & 0 & 0 \end{bmatrix}$; rank $= 2$; $(5, -2, 5, 6, 1)$ and $(-2, 0, 1, -1, 3)$

23. For the new space, $(1, 1, -4, 2)$, $(0, 1, 1, 3)$; for the column space, $(1, 0)$, $(1, 1)$

25. For the new space, $(8, 4)$, $(0, 3)$; for the column space, $(8, 2, 0)$, $(4, 1, 3)$

Section 10.7

1. $\alpha \begin{bmatrix} -1 \\ 1 \\ 1 \\ 0 \end{bmatrix} + \beta \begin{bmatrix} 1 \\ -1 \\ 0 \\ 1 \end{bmatrix}$; dimension 2

3. Only the trivial solution; dimension 0

5. $\alpha \begin{bmatrix} -\frac{9}{4} \\ -\frac{7}{4} \\ -\frac{5}{8} \\ \frac{13}{8} \\ 1 \end{bmatrix}$; dimension 1

7. $\alpha \begin{bmatrix} -\frac{5}{6} \\ -\frac{2}{3} \\ -\frac{8}{3} \\ -\frac{2}{3} \\ 1 \\ 0 \end{bmatrix} + \beta \begin{bmatrix} -\frac{5}{9} \\ -\frac{10}{9} \\ -\frac{13}{9} \\ -\frac{1}{9} \\ 0 \\ 1 \end{bmatrix}$; dimension 2

9. $\alpha \begin{bmatrix} 0 \\ 0 \\ 1 \\ 0 \\ 0 \end{bmatrix} + \beta \begin{bmatrix} -\frac{1}{2} \\ 1 \\ 0 \\ \frac{2}{3} \\ 1 \end{bmatrix}$; dimension 2

11. $\alpha \begin{bmatrix} 1 \\ 1 \\ 0 \\ 1 \\ 1 \\ 0 \\ 0 \end{bmatrix} + \beta \begin{bmatrix} -2 \\ -\frac{3}{2} \\ \frac{2}{3} \\ -\frac{4}{3} \\ 0 \\ 1 \\ 0 \end{bmatrix} + \gamma \begin{bmatrix} 0 \\ \frac{1}{2} \\ -3 \\ 0 \\ 0 \\ 0 \\ 1 \end{bmatrix}$; dimension 3

13. $\alpha \begin{bmatrix} -8 \\ -5 \\ 2 \\ 1 \\ 0 \end{bmatrix} + \beta \begin{bmatrix} -1 \\ -6 \\ 0 \\ 0 \\ 1 \end{bmatrix}$; dimension 2

15. $\alpha \begin{bmatrix} \frac{2}{5} \\ 5 \\ 1 \\ 0 \\ 0 \end{bmatrix} + \beta \begin{bmatrix} \frac{1}{5} \\ 1 \\ 0 \\ 1 \\ 0 \end{bmatrix} + \gamma \begin{bmatrix} -\frac{3}{5} \\ -7 \\ 0 \\ 0 \\ 1 \end{bmatrix}$; dimension 3

17. $\alpha \begin{bmatrix} -1 \\ 1 \\ -8 \\ 0 \end{bmatrix} + \beta \begin{bmatrix} 4 \\ 0 \\ -8 \\ 1 \end{bmatrix}$; dimension 2

19. $\alpha \begin{bmatrix} \frac{21}{13} \\ -1 \\ \frac{23}{13} \\ 1 \\ 0 \\ 0 \end{bmatrix} + \beta \begin{bmatrix} \frac{6}{13} \\ -6 \\ -\frac{12}{13} \\ 0 \\ 1 \\ 0 \end{bmatrix} + \gamma \begin{bmatrix} -\frac{44}{13} \\ 1 \\ -\frac{55}{13} \\ 0 \\ 0 \\ 1 \end{bmatrix}$; dimension 3

Section 10.8

1. Unique solution $\begin{bmatrix} 1 \\ \frac{1}{2} \\ 4 \end{bmatrix}$

3. $\alpha \begin{bmatrix} 1 \\ 1 \\ \frac{3}{2} \\ 1 \\ 0 \\ 0 \end{bmatrix} + \beta \begin{bmatrix} 0 \\ 0 \\ \frac{1}{2} \\ 0 \\ 1 \\ 0 \end{bmatrix} + \gamma \begin{bmatrix} -\frac{17}{2} \\ -6 \\ -\frac{51}{4} \\ 0 \\ 0 \\ 1 \end{bmatrix} + \begin{bmatrix} 0 \\ 0 \\ 0 \\ 0 \\ 1 \\ 0 \end{bmatrix}$

5. $\alpha \begin{bmatrix} 2 \\ 2 \\ 7 \\ \frac{3}{2} \\ 1 \\ 0 \end{bmatrix} + \beta \begin{bmatrix} -2 \\ -1 \\ -\frac{9}{2} \\ -\frac{3}{4} \\ 0 \\ 1 \end{bmatrix} + \begin{bmatrix} -2 \\ -2 \\ -31 \\ -4 \\ 1 \\ 0 \end{bmatrix}$

7. $\alpha \begin{bmatrix} 1 \\ 0 \\ 0 \\ 0 \\ 0 \\ 0 \end{bmatrix} + \beta \begin{bmatrix} 0 \\ -\frac{1}{2} \\ -1 \\ 3 \\ 1 \\ 0 \end{bmatrix} + \gamma \begin{bmatrix} 0 \\ -\frac{3}{4} \\ 1 \\ -2 \\ 0 \\ 1 \end{bmatrix} + \begin{bmatrix} 0 \\ \frac{9}{8} \\ 2 \\ 0 \\ 0 \\ 0 \end{bmatrix}$

9. $\alpha\begin{bmatrix}-1\\1\\0\\0\\0\\0\\0\end{bmatrix}+\beta\begin{bmatrix}1\\0\\0\\1\\0\\0\\0\end{bmatrix}+\gamma\begin{bmatrix}-3\\0\\3\\0\\14\\0\\0\end{bmatrix}+\gamma\begin{bmatrix}-1\\0\\0\\0\\0\\1\\0\end{bmatrix}+\varepsilon\begin{bmatrix}1\\0\\-1\\0\\0\\0\\14\end{bmatrix}+\begin{bmatrix}-\frac{29}{7}\\0\\\frac{1}{7}\\0\\0\\0\\0\end{bmatrix}$ **11.** $\alpha\begin{bmatrix}-\frac{19}{15}\\3\\\frac{67}{15}\\1\end{bmatrix}+\begin{bmatrix}\frac{22}{15}\\-5\\-\frac{121}{15}\\0\end{bmatrix}$

13. Unique solution $\begin{bmatrix}\frac{22}{87}\\\frac{93}{87}\\\frac{23}{87}\end{bmatrix}$ **15.** No solution **17.** $\alpha\begin{bmatrix}-1\\-1\\2\\1\end{bmatrix}+\begin{bmatrix}-\frac{3}{2}\\-4\\-\frac{63}{14}\\0\end{bmatrix}$ **19.** $\alpha\begin{bmatrix}\frac{2}{29}\\\frac{7}{29}\\\frac{18}{29}\\1\end{bmatrix}+\begin{bmatrix}\frac{67}{29}\\-\frac{7}{87}\\\frac{110}{29}\\0\end{bmatrix}$

23. $\begin{bmatrix}0\\\frac{1}{2}\\-\frac{3}{4}\\5\end{bmatrix}$ **25.** $\begin{bmatrix}-\frac{2}{19}\\-\frac{27}{19}\\\frac{9}{19}\end{bmatrix}$ **27.** $\begin{bmatrix}\frac{3}{2}\\\frac{3}{2}\\1\end{bmatrix}$

Section 10.9

1. $\frac{1}{5}\begin{bmatrix}-1&2\\2&1\end{bmatrix}$ **3.** $\frac{1}{31}\begin{bmatrix}-6&11&2\\3&10&-1\\1&-7&10\end{bmatrix}$ **5.** $\frac{1}{12}\begin{bmatrix}-6&6&0\\3&9&-2\\-3&3&2\end{bmatrix}$ **7.** Singular

9. $\frac{1}{2}\begin{bmatrix}-2&85&-55&2\\0&-9&7&2\\0&6&-4&0\\0&-1&1&0\end{bmatrix}$ **11.** $\begin{bmatrix}1&-14\\0&1\end{bmatrix}$ **13.** Singular **15.** $\begin{bmatrix}0&\frac{1}{2}&0\\\frac{5}{29}&\frac{4}{29}&-\frac{3}{29}\\\frac{3}{29}&-\frac{5}{29}&\frac{4}{29}\end{bmatrix}$

17. $\begin{bmatrix}3&-\frac{7}{5}&-\frac{1}{5}\\2&-1&0\\0&\frac{2}{5}&\frac{1}{5}\end{bmatrix}$ **19.** $\frac{1}{47}\begin{bmatrix}-7&6&24\\-2&-5&27\\4&10&-7\end{bmatrix}$ **21.** Singular **23.** $\begin{bmatrix}6&\frac{1}{2}&-\frac{5}{2}\\1&0&0\\-2&0&1\end{bmatrix}$ **25.** $\frac{1}{226}\begin{bmatrix}-4&-3&24\\34&-31&22\\36&27&10\end{bmatrix}$

27. $\begin{bmatrix}-\frac{21}{33}\\\frac{57}{33}\\\frac{9}{33}\\\frac{30}{33}\end{bmatrix}$ **29.** $\begin{bmatrix}\frac{22}{7}\\\frac{27}{7}\\\frac{30}{7}\end{bmatrix}$ **31.** $\begin{bmatrix}-3\\2\\0\end{bmatrix}$

Section 10.10

1. -34 **3.** -58 **5.** -12 **7.** -35 **9.** 0 **11.** -25 **13.** -5 **15.** -11
17. -773 **19.** -5987

Section 10.11

1. -82 **3.** 63 **5.** 96 **7.** -572 **9.** 4882 **11.** 3372 **13.** 1033 **15.** 74,706
17. 3855 **19.** 4,665,426 **21.** 27,246 **23.** -4132 **25.** -3144 **31.** -1224 **33.** 9636

Section 10.12

1. 8 **3.** 125 **5.** 130 **7.** 64 **11.** 40

Section 10.13

1. $\frac{1}{13}\begin{bmatrix}6&1\\-1&2\end{bmatrix}$ **3.** $\frac{1}{5}\begin{bmatrix}-4&1\\1&1\end{bmatrix}$ **5.** $\frac{1}{11}\begin{bmatrix}-6&5\\1&1\end{bmatrix}$ **7.** $\frac{1}{32}\begin{bmatrix}5&3&1\\-8&24&24\\2&14&6\end{bmatrix}$ **9.** $\frac{1}{83}\begin{bmatrix}-13&31&21\\-20&3&-6\\1&4&-8\end{bmatrix}$

11. $\frac{1}{43}\begin{bmatrix}9&17&-7\\-14&7&-13\\-2&1&-8\end{bmatrix}$ **13.** $\frac{1}{733}\begin{bmatrix}99&-47&-6\\7&-144&44\\-13&58&23\end{bmatrix}$ **15.** $\frac{1}{784}\begin{bmatrix}-52&131&-62&54\\208&-132&248&-216\\-496&360&-320&304\\-212&127&-102&190\end{bmatrix}$

Section 10.14

1. $x_1 = -\frac{66}{132}, x_2 = -\frac{114}{132}, x_3 = \frac{24}{132}$ **3.** $x_1 = -1, x_2 = 1$ **5.** $x_1 = \frac{5}{6}, x_2 = -\frac{20}{6}, x_3 = -\frac{5}{6}$

7. $x_1 = \frac{253}{128}, x_2 = -\frac{105}{128}, x_3 = -\frac{404}{128}, x_4 = -\frac{697}{128}$ **9.** $x_1 = -86, x_2 = -\frac{218}{4}, x_3 = -\frac{86}{4}, x_4 = \frac{74}{4}$

11. $x_1 = \frac{66}{186}, x_2 = -\frac{354}{186}, x_3 = -\frac{2}{186}, x_4 = \frac{232}{186}$ **13.** $x_1 = \frac{5}{16}, x_2 = -\frac{50}{16}, x_3 = -\frac{26}{16}, x_4 = -\frac{29}{16}$

Section 10.15

1. $\lambda^2 - 2\lambda - 5$; $1 + \sqrt{6}$ with $\begin{bmatrix} 3\alpha/\sqrt{6} \\ \alpha \end{bmatrix}$, $\alpha \neq 0$; $1 - \sqrt{6}$ with $\begin{bmatrix} -3\beta/\sqrt{6} \\ \beta \end{bmatrix}$, $\beta \neq 0$

3. $\lambda^2 + 3\lambda - 10$; 2 with $\begin{bmatrix} 0 \\ \alpha \end{bmatrix}$, $\alpha \neq 0$; -5 with $\begin{bmatrix} -7\beta \\ \beta \end{bmatrix}$, $\beta \neq 0$

5. $\lambda^2 - 3\lambda + 14$; $\dfrac{(3 + \sqrt{47}\,i)}{2}$ with $\begin{bmatrix} \alpha \\ [(-1 - \sqrt{47}\,i)/12]\alpha \end{bmatrix}$, $\alpha \neq 0$; $\dfrac{(3 - \sqrt{47}\,i)}{2}$ with $\begin{bmatrix} \beta \\ [(-1 + \sqrt{47}\,i)/12]\beta \end{bmatrix}$, $\beta \neq 0$

7. $\lambda^2 + 9\lambda + 16$; $\dfrac{(-9 + \sqrt{17})}{2}$ with $\begin{bmatrix} \alpha \\ [(1 + \sqrt{17})/4]\alpha \end{bmatrix}$, $\alpha \neq 0$; $\dfrac{(-9 - \sqrt{17})}{2}$ with $\begin{bmatrix} \beta \\ [(1 - \sqrt{17})/4]\beta \end{bmatrix}$, $\beta \neq 0$

9. $\lambda^2 + 15\lambda + 34$; $\dfrac{(-15 + \sqrt{89})}{2}$ with $\begin{bmatrix} \alpha \\ [(5 + \sqrt{89})/8]\alpha \end{bmatrix}$, $\alpha \neq 0$; $\dfrac{(-15 - \sqrt{89})}{2}$ with $\begin{bmatrix} \beta \\ [(5 - \sqrt{89})/8]\beta \end{bmatrix}$

11. $\lambda^3 - 5\lambda^2 + 6\lambda$; 0 with $\begin{bmatrix} 0 \\ \alpha \\ 0 \end{bmatrix}$, $\alpha \neq 0$; 2 with $\begin{bmatrix} 2\beta \\ \beta \\ 0 \end{bmatrix}$, $\beta \neq 0$; and 3 with $\begin{bmatrix} 0 \\ 2\gamma/3 \\ \gamma \end{bmatrix}$, $\gamma \neq 0$

13. $\lambda^3 + 2\lambda^2$; 0 with $\begin{bmatrix} \alpha/3 \\ 0 \\ \alpha \end{bmatrix}$, $\alpha \neq 0$; -3 with $\begin{bmatrix} \beta \\ 0 \\ 0 \end{bmatrix}$, $\beta \neq 0$ (only two distinct eigenvalues)

15. $\lambda^3 + 10\lambda^2 - 52\lambda + 56$; 2 with $\begin{bmatrix} 0 \\ 0 \\ \alpha \end{bmatrix}$, $\alpha \neq 0$; -14 with $\begin{bmatrix} -16\beta \\ 0 \\ \beta \end{bmatrix}$, $\beta \neq 0$ (only two distinct eigenvalues)

17. $\lambda^3 - 8\lambda^2 + 7\lambda$; 0 with $\begin{bmatrix} \alpha \\ \alpha/2 \\ 5\alpha/7 \end{bmatrix}$, $\alpha \neq 0$; 1 with $\begin{bmatrix} \beta \\ 0 \\ 5\beta/6 \end{bmatrix}$, $\beta \neq 0$; 7 with $\begin{bmatrix} 0 \\ 0 \\ \gamma \end{bmatrix}$, $\gamma \neq 0$

19. $\lambda^4 + 5\lambda^3 - 6\lambda^2$; 0 with $\begin{bmatrix} 0 \\ \alpha \\ \beta \\ 0 \end{bmatrix}$, $\alpha^2 + \beta^2 \neq 0$; 1 with $\begin{bmatrix} \gamma \\ \gamma \\ 0 \\ 7\gamma \end{bmatrix}$, $\gamma \neq 0$;

-6 with $\begin{bmatrix} -6\varepsilon \\ \varepsilon \\ 0 \\ 0 \end{bmatrix}$, $\varepsilon \neq 0$ (only three distinct eigenvalues)

21. $\lambda^4 - 2\lambda^3 - 14\lambda^2 + 41\lambda - 26$; 1 with $\begin{bmatrix} -2\alpha \\ -11\alpha \\ 0 \\ \alpha \end{bmatrix}$, $\alpha \neq 0$; 2 with $\begin{bmatrix} 0 \\ 0 \\ \beta \\ 0 \end{bmatrix}$, $\beta \neq 0$; $\dfrac{(-1 + \sqrt{53})}{2}$ with

$\begin{bmatrix} [(-7 + \sqrt{53})/2]\gamma \\ 0 \\ 0 \\ \gamma \end{bmatrix}$, $\gamma \neq 0$; $\dfrac{(-1 - \sqrt{53})}{2}$ with $\begin{bmatrix} \delta \\ 0 \\ 0 \\ [(7 - \sqrt{53})/2]\delta \end{bmatrix}$, $\delta \neq 0$

23. $(\lambda^3 - 5\lambda^2 + 4\lambda)(\lambda^2 - 10\lambda + 23)$; 0 with $\begin{bmatrix} -5\alpha \\ 0 \\ 23\alpha \\ 0 \\ \alpha \end{bmatrix}$, $\alpha \neq 0$; 1 with $\begin{bmatrix} 0 \\ \beta \\ 0 \\ 0 \\ 0 \end{bmatrix}$, $\beta \neq 0$; 4 with $\begin{bmatrix} 0 \\ 0 \\ 0 \\ \gamma \\ 0 \end{bmatrix}$, $\gamma \neq 0$;

$5 + \sqrt{2}$ with $\begin{bmatrix} \sqrt{2}\,\delta \\ 0 \\ 0 \\ 0 \\ \delta \end{bmatrix}$, $\delta \neq 0$; and $5 - \sqrt{2}$ with $\begin{bmatrix} -\sqrt{2}\,\varphi \\ 0 \\ 0 \\ 0 \\ \varphi \end{bmatrix}$, $\varphi \neq 0$

Section 10.16

1. $\lambda^2 - \lambda - 2;\quad -1,\ 2$ **3.** $\lambda^2 - 10\lambda + 3;\quad 5 + \sqrt{22},\ 5 - \sqrt{22}$ **5.** $\lambda^3 + \lambda^2 - \lambda - 1;\quad 1,\ -1,\ -1$
7. $\lambda^4 - 7\lambda^3 + 6\lambda^2;\quad 0,\ 0,\ 1,\ 6$ **9.** $\lambda^3 - \lambda^2 - 4\lambda + 4;\quad 1,\ 2,\ -2$
11. $\lambda^4 - 4\lambda^3 - \lambda^2 + 12\lambda;\quad 0,\ 3,\ (1 + \sqrt{17})/2,\ (1 - \sqrt{17})/2$ **13.** $\lambda^4 + 7\lambda^3 + \lambda^2 - 27\lambda + 18;\quad 1,\ 1,\ -3,\ -6$

Section 10.17

1. $Y = [a\cos(t) + b\sin(t)]\begin{bmatrix} 3 \\ -2 \end{bmatrix} + [ce^{2t} + de^{-2t}]\begin{bmatrix} 1 \\ 1 \end{bmatrix}$ **3.** $Y = [ae^{t} + be^{-t}]\begin{bmatrix} 2 \\ 5 \end{bmatrix} + [c\cos(2t) + d\sin(2t)]\begin{bmatrix} 1 \\ 0 \end{bmatrix}$

5. $Y = [ae^{\alpha t} + be^{-\alpha t}]\begin{bmatrix} 1 \\ -\sqrt{2} \end{bmatrix} + [c\cos(\beta t) + d\sin(\beta t)]\begin{bmatrix} 1 \\ \sqrt{2} \end{bmatrix}$, with $\alpha = (2 + 2\sqrt{2})^{1/2}$ and $\beta = (2\sqrt{2} - 2)^{1/2}$

7. $Y = (ae^{\sqrt{2}t} + be^{-\sqrt{2}t})\begin{bmatrix} 0 \\ 2 \\ 1 \end{bmatrix} + [c\cos(\sqrt{2}\,t) + d\sin(\sqrt{2}\,t)]\begin{bmatrix} 4 \\ -2 \\ 1 \end{bmatrix} + [e\cos(t) + h\sin(t)]\begin{bmatrix} 0 \\ -1 \\ 1 \end{bmatrix}$

15. $Y = \alpha_1 e^{(9+\sqrt{71}\,i)t/2}\begin{bmatrix} 8 \\ -5 - \sqrt{71}\,i \end{bmatrix} + \alpha_2 e^{(9-\sqrt{71}\,i)t/2}\begin{bmatrix} 8 \\ -5 + \sqrt{71}\,i \end{bmatrix}$

Section 10.18

1. $\begin{bmatrix} 2 & 2 \\ -3 - \sqrt{7}i & -3 + \sqrt{7}i \end{bmatrix}$ diagonalizes the given matrix **3.** Not diagonalizable

5. $\begin{bmatrix} 0 & 0 & 5 \\ 1 & -3 & 1 \\ 0 & 2 & 0 \end{bmatrix}$ diagonalizes the given matrix **7.** Not diagonalizable

9. $\begin{bmatrix} 1 & 0 & 0 & 0 \\ 0 & 1 & 2/(-13 + \sqrt{5}) & 2/(-13 - \sqrt{5}) \\ 0 & 0 & 1 & 1 \\ 0 & 0 & (1 + \sqrt{5})/2 & (1 - \sqrt{5})/2 \end{bmatrix}$ diagonalizes the given matrix **11.** Not diagonalizable

17. $A^{16} = \frac{1}{3}\begin{bmatrix} 2 & 1 \\ 1 & -1 \end{bmatrix}\begin{bmatrix} 2^{16} & 0 \\ 0 & 5^{16} \end{bmatrix}\begin{bmatrix} 1 & 1 \\ 1 & -2 \end{bmatrix} = \frac{1}{3}\begin{bmatrix} 2^{17} + 5^{16} & 2^{17} - 2(5^{16}) \\ 2^{16} - 5^{16} & 2^{16} + 2(5^{16}) \end{bmatrix}$

19. $A^{31} = \begin{bmatrix} 3 & 3 \\ -1 + \sqrt{10} & 1 + \sqrt{10} \end{bmatrix}\begin{bmatrix} (-3 + \sqrt{10})^{31} & 0 \\ 0 & (-3 - \sqrt{10})^{31} \end{bmatrix}\begin{bmatrix} (1 + \sqrt{10})/6 & -\frac{1}{2} \\ (1 - \sqrt{10})/6 & \frac{1}{2} \end{bmatrix}$

Section 10.19

1. $Y = \begin{bmatrix} -2 & 1 \\ 1 & 1 \end{bmatrix}\begin{bmatrix} ae^{2t} \\ be^{5t} \end{bmatrix}$ **3.** $Y = \begin{bmatrix} 1 & 3 \\ 1 & -2 \end{bmatrix}\begin{bmatrix} ae^{-t} \\ be^{-6t} \end{bmatrix}$ **5.** $Y = \begin{bmatrix} 3 & 1 \\ 2 & -2 \end{bmatrix}\begin{bmatrix} ae^{3t} \\ be^{-5t} \end{bmatrix}$

7. $Y = \begin{bmatrix} 0 & 1 & 5 \\ 0 & 7 & 0 \\ 1 & 0 & 8 \end{bmatrix}\begin{bmatrix} a \\ be^{-2t} \\ ce^{5t} \end{bmatrix}$

Section 10.20

1. $0,\ 5;\ Q = \begin{bmatrix} \frac{1}{\sqrt{5}} & \frac{-2}{\sqrt{5}} \\ \frac{2}{\sqrt{5}} & \frac{1}{\sqrt{5}} \end{bmatrix}$

3. $1+\sqrt{26},\ 1-\sqrt{26};\ Q = \begin{bmatrix} \frac{1}{\sqrt{52-10\sqrt{26}}} & \frac{1}{\sqrt{52+10\sqrt{26}}} \\ \frac{-5+\sqrt{26}}{\sqrt{52-10\sqrt{26}}} & \frac{-5-\sqrt{26}}{\sqrt{52+10\sqrt{26}}} \end{bmatrix}$

5. $0,\ \frac{5+\sqrt{41}}{2},\ \frac{5-\sqrt{41}}{2};\ Q = \begin{bmatrix} 0 & \frac{4}{\sqrt{82-10\sqrt{41}}} & \frac{4}{\sqrt{82+10\sqrt{41}}} \\ 1 & 0 & 0 \\ 0 & \frac{-5+\sqrt{41}}{\sqrt{82-10\sqrt{41}}} & \frac{-5-\sqrt{41}}{\sqrt{82+10\sqrt{41}}} \end{bmatrix}$

7. $3,\ -1+\sqrt{2},\ -1-\sqrt{2};\ Q = \begin{bmatrix} 0 & \frac{1}{\sqrt{4-2\sqrt{2}}} & \frac{1}{\sqrt{4+2\sqrt{2}}} \\ 0 & \frac{-1+\sqrt{2}}{\sqrt{4-2\sqrt{2}}} & \frac{-1-\sqrt{2}}{\sqrt{4+2\sqrt{2}}} \\ 1 & 0 & 0 \end{bmatrix}$

9. $8,\ -5;\ Q = \frac{1}{\sqrt{13}}\begin{bmatrix} 2 & 3 \\ 3 & -2 \end{bmatrix}$

11. $\frac{1+\sqrt{117}}{2},\ \frac{1-\sqrt{117}}{2};\ Q = \begin{bmatrix} \frac{6}{\sqrt{234-18\sqrt{117}}} & \frac{6}{\sqrt{234+18\sqrt{117}}} \\ \frac{9-\sqrt{117}}{\sqrt{234-18\sqrt{117}}} & \frac{9+\sqrt{117}}{\sqrt{234+18\sqrt{117}}} \end{bmatrix}$

13. $3,\ 9;\ Q = \frac{1}{\sqrt{2}}\begin{bmatrix} 1 & 1 \\ 1 & -1 \end{bmatrix}$

15. $0,\ 1+\sqrt{17},\ 1-\sqrt{17};\ Q = \begin{bmatrix} 0 & \frac{4}{\sqrt{34-2\sqrt{17}}} & \frac{4}{\sqrt{34+2\sqrt{17}}} \\ 0 & \frac{1-\sqrt{17}}{\sqrt{34-2\sqrt{17}}} & \frac{1+\sqrt{17}}{\sqrt{34+2\sqrt{17}}} \\ 1 & 0 & 0 \end{bmatrix}$

Section 10.21

5. $\begin{bmatrix} \frac{2}{\sqrt{58+10\sqrt{29}}} & \frac{2}{\sqrt{58-10\sqrt{29}}} \\ \frac{5+\sqrt{29}}{\sqrt{58+10\sqrt{29}}} & \frac{5-\sqrt{29}}{\sqrt{58-10\sqrt{29}}} \end{bmatrix}$

7. $\frac{1}{\sqrt{2}}\begin{bmatrix} 1 & 1 \\ -1 & 1 \end{bmatrix}$

9. $\begin{bmatrix} 1 & 0 & 0 \\ 0 & \frac{4}{\sqrt{34-2\sqrt{17}}} & \frac{4}{\sqrt{34+2\sqrt{17}}} \\ 0 & \frac{1-\sqrt{17}}{\sqrt{34-2\sqrt{17}}} & \frac{1+\sqrt{17}}{\sqrt{34+2\sqrt{17}}} \end{bmatrix}$

11. $\begin{bmatrix} 1 & 0 & 0 & 0 \\ 0 & 0 & \frac{1}{\sqrt{2}} & \frac{1}{\sqrt{2}} \\ 0 & 0 & \frac{1}{\sqrt{2}} & -\frac{1}{\sqrt{2}} \\ 0 & 1 & 0 & 0 \end{bmatrix}$

Section 10.22

1. $\begin{bmatrix} 1 & 1 \\ 1 & 6 \end{bmatrix}$ **3.** $\begin{bmatrix} 1 & -2 \\ -2 & 1 \end{bmatrix}$ **5.** $\begin{bmatrix} -1 & 0 & -\frac{1}{2} & -1 \\ 0 & 0 & 2 & \frac{3}{2} \\ -\frac{1}{2} & 2 & 0 & 0 \\ -1 & \frac{3}{2} & 0 & 1 \end{bmatrix}$

7. $-2x_1^2 + 2x_1x_2 + 6x_2^2$ **9.** $6x_1^2 + 2x_1x_2 - 14x_1x_3 + 2x_2^2 + x_3^2$ **11.** $8x_1^2 + 2x_2^2 - 8x_2x_3 + 3x_3^2$
13. $(-1+2\sqrt{5})y_1^2 + (-1-2\sqrt{5})y_2^2$ **15.** $(2+\sqrt{29})y_1^2 + (2-\sqrt{29})y_2^2$ **17.** $(2+\sqrt{13})y_1^2 + (2-\sqrt{13})y_2^2$
19. $y_1^2 + y_2^2 + 2y_3^2$ **21.** $y_1^2 - y_2^2 = 10/\sqrt{61}$ (hyperbola) **23.** $y_1^2 - y_2^2 = \frac{8}{5}$ (hyperbola)

Section 10.23

3. None; -1, -1 **5.** Unitary; i, $\dfrac{1+\sqrt{3}}{2\sqrt{2}} + i\left[\dfrac{1-\sqrt{3}}{2\sqrt{2}}\right]$, $\dfrac{1-\sqrt{3}}{2\sqrt{2}} + i\left[\dfrac{1+\sqrt{3}}{2\sqrt{2}}\right]$

7. Hermitian; 1, $\dfrac{-1+\sqrt{41}}{2}$, $\dfrac{-1-\sqrt{41}}{2}$ **9.** None; 1, -1, $3i$ **19.** -43 **21.** $62i$

Supplementary Problems

1. $\begin{bmatrix} 1 & 0 & 0 & \frac{271}{104} \\ 0 & 1 & 0 & \frac{44}{104} \\ 0 & 0 & 1 & \frac{121}{104} \end{bmatrix}$; 3 **3.** $\begin{bmatrix} 1 & 0 \\ 0 & 1 \\ 0 & 0 \end{bmatrix}$; 2 **5.** $\begin{bmatrix} 1 & -\frac{3}{8} & \frac{1}{4} & \frac{1}{8} & -\frac{5}{8} \\ 0 & 0 & 0 & 0 & 0 \end{bmatrix}$; 1 **7.** $\begin{bmatrix} 1 & 0 & -\frac{1}{3} \\ 0 & 1 & -1 \\ 0 & 0 & 0 \end{bmatrix}$; 2

9. $\begin{bmatrix} 1 & 0 & -\frac{1}{3} & 0 \\ 0 & 1 & \frac{1}{2} & 0 \\ 0 & 0 & 0 & 1 \end{bmatrix}$; 3 **11.** $\begin{bmatrix} \frac{1}{2} & \frac{5}{4} \\ \frac{1}{2} & \frac{3}{4} \end{bmatrix}$ **13.** Singular **15.** $\begin{bmatrix} -\frac{1}{3} & -\frac{1}{6} & -\frac{1}{6} \\ 0 & 1 & 0 \\ \frac{1}{3} & \frac{5}{3} & \frac{2}{3} \end{bmatrix}$ **17.** $\begin{bmatrix} 0 & \frac{1}{2} & 1 \\ -\frac{2}{5} & \frac{6}{5} & \frac{11}{5} \\ -\frac{1}{5} & \frac{3}{5} & \frac{3}{5} \end{bmatrix}$

19. $\frac{1}{56}\begin{bmatrix} -10 & 10 & 10 & -12 \\ -4 & 4 & 4 & -16 \\ -1 & 57 & 29 & -46 \\ 7 & -7 & 21 & -14 \end{bmatrix}$ **21.** 0 **23.** 22 **25.** 70 **27.** -664 **29.** 12

31. $\alpha\begin{bmatrix} -15 \\ -2 \\ 7 \end{bmatrix}$ **33.** $\alpha\begin{bmatrix} 3 \\ \frac{13}{3} \\ \frac{34}{3} \\ 1 \end{bmatrix} + \begin{bmatrix} -4 \\ -\frac{5}{3} \\ -\frac{2}{3} \\ 0 \end{bmatrix}$ **35.** $\alpha\begin{bmatrix} \frac{3}{4} \\ -\frac{1}{6} \\ \frac{5}{36} \\ 1 \end{bmatrix} + \begin{bmatrix} -\frac{1}{8} \\ \frac{7}{4} \\ \frac{7}{24} \\ 0 \end{bmatrix}$

37. $\frac{1}{21}\begin{bmatrix} -68 \\ -11 \\ 24 \end{bmatrix}$ **39.** $\alpha\begin{bmatrix} \frac{117}{30} \\ -\frac{19}{30} \\ \frac{101}{30} \\ 1 \end{bmatrix} + \begin{bmatrix} -\frac{36}{30} \\ \frac{17}{30} \\ \frac{17}{30} \\ 0 \end{bmatrix}$

41. $\left[\dfrac{3+\sqrt{37}}{2}\right]y_1^2 + \left[\dfrac{3-\sqrt{37}}{2}\right]y_2^2$ **43.** $\left[\dfrac{-3+\sqrt{97}}{2}\right]y_1^2 + \left[\dfrac{-3-\sqrt{97}}{2}\right]y_2^2$

45. $-y_1^2 + \left[\dfrac{-1+\sqrt{53}}{2}\right]y_2^2 + \left[\dfrac{-1-\sqrt{53}}{2}\right]y_3^2$ **47.** $\begin{bmatrix} 1 & 0 \\ 0 & 1 \end{bmatrix}$ (given matrix is already diagonal)

49. $\begin{bmatrix} 0 & \dfrac{2}{\sqrt{26+6\sqrt{13}}} & \dfrac{2}{\sqrt{26-6\sqrt{13}}} \\ 0 & \dfrac{3+\sqrt{13}}{\sqrt{26+6\sqrt{13}}} & \dfrac{3-\sqrt{13}}{\sqrt{26-6\sqrt{13}}} \\ 1 & 0 & 0 \end{bmatrix}$ **51.** $\dfrac{7+\sqrt{61}}{2}$, $\begin{bmatrix} 0 \\ -5-\sqrt{61} \end{bmatrix}$; $\dfrac{7-\sqrt{61}}{2}$, $\begin{bmatrix} 6 \\ -5+\sqrt{61} \end{bmatrix}$

53. $2i, \begin{bmatrix} 1 \\ i \end{bmatrix}$ (only one eigenvalue) **55.** $4, \begin{bmatrix} 2i \\ 3+i \end{bmatrix}$; $1-i, \begin{bmatrix} 1 \\ 0 \end{bmatrix}$ **57.** $3-2i, \begin{bmatrix} 3+2i \\ 1 \end{bmatrix}$; $-4i, \begin{bmatrix} 0 \\ 1 \end{bmatrix}$

59. $-3, \begin{bmatrix} 1 \\ -1 \\ 0 \end{bmatrix}$; $-2, \begin{bmatrix} 0 \\ 1 \\ 0 \end{bmatrix}$; $-4, \begin{bmatrix} 0 \\ 0 \\ 1 \end{bmatrix}$

Chapter 11

Section 11.1

1. $f\mathbf{F} = 4\cos(3t)\mathbf{i} + 12t^2\cos(3t)\mathbf{j} + 8t\cos(3t)\mathbf{k}$;
$(f\mathbf{F})' = -12\sin(3t)\mathbf{i} + [24t\cos(3t) - 36t^2\sin(3t)]\mathbf{j} + [8\cos(3t) - 24t\sin(3t)]\mathbf{k}$

3. $f\mathbf{F} = 4\sin^2(t)\mathbf{i} + t^5\sin^2(t)\mathbf{k}$
$(f\mathbf{F})' = 8\sin(t)\cos(t)\mathbf{i} + [5t^4\sin^2(t) + 2t^5\sin(t)\cos(t)]\mathbf{k}$

5. $f\mathbf{F} = (2t+3)(1-3t)\mathbf{i} + (2t^5 + 3t^4)\mathbf{j} - (2t^2 + 3t)\mathbf{k}$;
$(f\mathbf{F})' = (-12t - 7)\mathbf{i} + (10t^4 + 12t^3)\mathbf{j} - (4t + 3)\mathbf{k}$

7. $f\mathbf{F} = (4t^3 - 2t)\mathbf{i} + (t^4 - 2t^6)\mathbf{j} + (2t^3 - t)e^{-t}\mathbf{k}$;
$(f\mathbf{F})' = (12t^2 - 2)\mathbf{i} + (4t^3 - 12t^5)\mathbf{j} + (6t^2 - 1 - 2t^3 + t)e^{-t}\mathbf{k}$

9. $f\mathbf{F} = -3e^{5t}\mathbf{i} - 3t^2e^{5t}\mathbf{j} + 3e^{5t}\mathbf{k}$;
$(f\mathbf{F})' = -15e^{5t}\mathbf{i} + (-6t - 15t^2)e^{5t}\mathbf{j} + 15e^{5t}\mathbf{k}$

11. $\mathbf{F}\cdot\mathbf{G} = 2t^2\cos(2t) + 3te^{-t}$; $(\mathbf{F}\cdot\mathbf{G})' = 4t\cos(2t) - 4t^2\sin(2t) + 3e^{-t} - 3te^{-t}$;
$\mathbf{F}\times\mathbf{G} = -3t\sin(t)\mathbf{i} + (3t\cos(2t) - 2t^2e^{-t})\mathbf{j} - 2t^2\sin(t)\mathbf{k}$;
$(\mathbf{F}\times\mathbf{G})' = (-3\sin(t) - 3t\cos(t))\mathbf{i} + (3\cos(t) - 6t\sin(t) - 4te^{-t} + 2t^2e^{-t})\mathbf{j} + (-4t\sin(t) - 2t^2\cos(t))\mathbf{k}$

13. $\mathbf{F}\cdot\mathbf{G} = t^3 + 18 + 6te^{2t}$, $(\mathbf{F}\cdot\mathbf{G})' = 3t^2 + 6e^{2t} + 12te^{2t}$;
$\mathbf{F}\times\mathbf{G} = (-108t + e^{2t})\mathbf{i} + (-e^{2t} + 6t^4)\mathbf{j} + (t^3 - 18)\mathbf{k}$,
$(\mathbf{F}\times\mathbf{G})' = (-108 + 2e^{2t})\mathbf{i} + (-2e^{2t} + 24t^3)\mathbf{j} + 3t^2\mathbf{k}$

15. $\mathbf{F}\cdot\mathbf{G} = -8t^2\sin(2\pi t) - 3t^3 - 3t^3\cos(t)$,
$(\mathbf{F}\cdot\mathbf{G})' = -16t\sin(2\pi t) - 16\pi t^2\cos(2\pi t) - 9t^2 - 9t^2\cos(t) + 3t^3\sin(t)$;
$\mathbf{F}\times\mathbf{G} = [9t\cos(t) - t^5]\mathbf{i} + [-8t^5 + 3\cos(t)\sin(2\pi t)]\mathbf{j} + [t^2\sin(2\pi t) - 24t^3]\mathbf{k}$,
$(\mathbf{F}\times\mathbf{G})' = [9\cos(t) - 9t\sin(t) - 5t^4]\mathbf{i} + [-40t^4 - 3\sin(t)\sin(2\pi t) + 6\pi\cos(t)\cos(2\pi t)]\mathbf{j}$
$+ [2t\sin(2\pi t) + 2\pi t^2\cos(2\pi t) - 72t^2]\mathbf{k}$

17. $\mathbf{F}\cdot\mathbf{G} = -12t^4$, $(\mathbf{F}\cdot\mathbf{G})' = -48t^3$;
$\mathbf{F}\times\mathbf{G} = -4e^tt^4\mathbf{i} - 2t^6\mathbf{k}$, $(\mathbf{F}\times\mathbf{G})' = (-4e^tt^4 - 16e^tt^3)\mathbf{i} - 12t^5\mathbf{k}$

19. $\mathbf{F}\cdot\mathbf{G} = \sin^2(2\pi t) - \cos^2(2\pi t) + e^{2\pi}$, $(\mathbf{F}\cdot\mathbf{G})' = 8\pi\sin(2\pi t)\cos(2\pi t)$;
$\mathbf{F}\times\mathbf{G} = \cos(2\pi t)(e^{2\pi} + 1)\mathbf{i} + \sin(2\pi t)(e^{2\pi} - 1)\mathbf{j} - 2\sin(2\pi t)\cos(2\pi t)\mathbf{k}$,
$(\mathbf{F}\times\mathbf{G})' = -(e^{2\pi} + 1)2\pi\sin(2\pi t)\mathbf{i} + (e^{2\pi} - 1)2\pi\cos(2\pi t)\mathbf{j} - 4\pi\cos(4\pi t)\mathbf{k}$

21. $\mathbf{F}\cdot\mathbf{G} = 8t\sin(2t)$, $(\mathbf{F}\cdot\mathbf{G})' = 8\sin(2t) + 16t\cos(2t)$;
$\mathbf{F}\times\mathbf{G} = 2t\mathbf{i} + 8t\mathbf{j} + 2t\sin(2t)\mathbf{k}$, $(\mathbf{F}\times\mathbf{G})' = 2\mathbf{i} + 8\mathbf{j} + [2\sin(2t) + 4t\cos(2t)]\mathbf{k}$

23. $\mathbf{F}\cdot\mathbf{G} = -3e^{-t}\cosh(2t) - 8t^3 - e^{-t}$, $(\mathbf{F}\cdot\mathbf{G})' = 3e^{-t}\cosh(2t) - 6e^{-t}\sinh(2t) - 24t^2 + e^{-t}$;
$\mathbf{F}\times\mathbf{G} = (2te^{-t} - 4t^2)\mathbf{i} + [3\cosh(2t) - e^{-2t}]\mathbf{j} + [-4t^2e^{-t} + 6t\cosh(2t)]\mathbf{k}$,
$(\mathbf{F}\times\mathbf{G})' = (2e^{-t} - 2te^{-t} - 8t)\mathbf{i} + [6\sinh(2t) + 4e^{-2t}]\mathbf{j}$
$+ [-8te^{-t} + 4t^2e^{-t} + 6\cosh(2t) + 12t\sinh(2t)]\mathbf{k}$

25. $\mathbf{F}\cdot\mathbf{G} = -5t^2 + 51te^t - t^3\ln(t)$,
$(\mathbf{F}\cdot\mathbf{G})' = -10t + 51e^t + 51te^t - 3t^2\ln(t) - t^2$;
$\mathbf{F}\times\mathbf{G} = [-17t\ln(t) - 3t^3e^t]\mathbf{i} + [-5t^5 + \ln(t)]\mathbf{j} + (3e^t + 85t^3)\mathbf{k}$,
$(\mathbf{F}\times\mathbf{G})' = [-17\ln(t) - 17 - 9t^2e^t - 3t^3e^t]\mathbf{i} + [-25t^4 + (1/t)]\mathbf{j} + (3e^t + 255t^2)\mathbf{k}$

27. $\mathbf{i} + 2\pi\cos(2\pi x)\mathbf{j} - 2\pi\sin(2\pi x)\mathbf{k}$ **29.** $e^t[\cos(t) - \sin(t)]\mathbf{i} + e^t[\sin(t) + \cos(t)]\mathbf{j} + e^t\mathbf{k}$

31. $3t^2(\mathbf{i} + \mathbf{j} + \mathbf{k})$ **33.** $\dfrac{8}{2t+1}\mathbf{i} + 12\cosh(3t)\mathbf{j}$ **35.** $8t\mathbf{i} + 3\mathbf{j} + \mathbf{k}$

37. $\int_1^4 \sqrt{4 + \cos^2(t) + 81t^4}\,dt$ **39.** $\int_1^6 \sqrt{4e^{2t} + \cosh^2(t) + \dfrac{1}{t^2}}\,dt$ **41.** $\int_2^\pi \sqrt{9e^{-6t} + 4 + \dfrac{1}{t^2}}\,dt$

43. $\int_0^1 \sqrt{16t^2 + 10 + 8t\sin(t)}\,dt$ **45.** $\int_1^4 \sqrt{9\sinh^2(3t) + 1 + 4t^2}\,dt$

47. $\dfrac{1}{\sqrt{16\sin^2(t) + 36\cos^2(3t)}}[-4\sin(t)\mathbf{i} + 6\cos(3t)\mathbf{j}]$

49. $\dfrac{1}{\sqrt{9t^4 + 2e^{2t}}}\{e^t[\sin(t) + \cos(t)]\mathbf{i} + e^t[\sin(t) - \cos(t)]\mathbf{j} + 3t^2\mathbf{k}\}$

51. $\dfrac{1}{\sqrt{144t^2 + 4\sin^2(2t)}}[2\sin(2t)\mathbf{i} + 12t\mathbf{j}]$

53. $\dfrac{1}{\sqrt{4\cos^2(t) + 16t^2 + 144t^4}}[2\cos(t)\mathbf{i} + 4t\mathbf{j} + 12t^2\mathbf{k}]$

55. $\dfrac{1}{\sqrt{68t^2+9}}[2t\mathbf{i}-3\mathbf{j}+8t\mathbf{k}]$ **57.** $\dfrac{1}{\sqrt{e^{2t}+9e^{-6t}+1}}[e^t\mathbf{i}-3e^{-3t}\mathbf{j}+\mathbf{k}]$

59. $\dfrac{1}{\sqrt{4\sinh^2(2t)+1}}[2\sinh(2t)\mathbf{i}-\mathbf{j}]$

61. $L=\sinh(\pi)$; $s=\sinh(t)$, $t=\operatorname{arcsinh}(s)$; $x=\operatorname{arcsinh}(s)$, $y=\sqrt{1+s^2}$, $z=1$, $0\le s\le\sinh(\pi)$

63. $L=2\sqrt{3}$; $s=\sqrt{3}(t^3+1)$, $t=\left[\dfrac{s}{\sqrt{3}}-1\right]^{1/3}$; $x=y=z=\dfrac{s}{\sqrt{3}}-1$, $0\le s\le 2\sqrt{3}$

Section 11.2

1. $\mathbf{v}=3\mathbf{i}+2t\mathbf{k}$; $v=\sqrt{9+4t^2}$; $\mathbf{a}=2\mathbf{k}$; $a_t=\dfrac{4t}{\sqrt{9+4t^2}}$; $a_n=\dfrac{6}{\sqrt{9+4t^2}}$; $\rho=\dfrac{(9+4t^2)^{3/2}}{6}$,

$\kappa=\dfrac{6}{(9+4t^2)^{3/2}}$; $\mathbf{T}=\dfrac{1}{\sqrt{9+4t^2}}(3\mathbf{i}+2t\mathbf{k})$; $\mathbf{N}=\dfrac{1}{\sqrt{9+4t^2}}(-2t\mathbf{i}+3\mathbf{k})$

3. $\mathbf{v}=2\mathbf{i}-4t\mathbf{j}$; $v=2\sqrt{1+4t^2}$; $\mathbf{a}=-4\mathbf{j}$; $a_t=\dfrac{8t}{\sqrt{1+4t^2}}$; $a_n=\dfrac{4}{\sqrt{1+4t^2}}$; $\rho=(1+4t^2)^{3/2}$;

$\kappa=\dfrac{1}{(1+4t^2)^{3/2}}$; $\mathbf{T}=\dfrac{1}{\sqrt{1+4t^2}}(\mathbf{i}-2t\mathbf{j})$; $\mathbf{N}=\dfrac{1}{\sqrt{1+4t^2}}(-2t\mathbf{i}-\mathbf{j})$

5. $\mathbf{v}=2\cos(t)\mathbf{i}+\mathbf{j}-2\sin(t)\mathbf{k}$; $v=\sqrt{5}$; $\mathbf{a}=-2\sin(t)\mathbf{i}-2\cos(t)\mathbf{k}$; $a_t=0$; $a_n=2$; $\rho=\frac{5}{2}$; $\kappa=\frac{2}{5}$;
$\mathbf{T}=(1/\sqrt{5})[2\cos(t)\mathbf{i}+\mathbf{j}-2\sin(t)\mathbf{k}]$; $\mathbf{N}=-\sin(t)\mathbf{i}-\cos(t)\mathbf{k}$

7. $\mathbf{v}=4t\mathbf{i}+\mathbf{j}+\mathbf{k}$; $v=\sqrt{2+16t^2}$; $\mathbf{a}=4\mathbf{i}$; $a_t=\dfrac{16t}{\sqrt{2+16t^2}}$; $a_n=\dfrac{4}{\sqrt{1+8t^2}}$; $\rho=\dfrac{(1+8t^2)^{3/2}}{2}$;

$\kappa=\dfrac{2}{(1+8t^2)^{3/2}}$; $\mathbf{T}=\dfrac{1}{\sqrt{2+16t^2}}(4t\mathbf{i}+\mathbf{j}+\mathbf{k})$; $\mathbf{N}=\dfrac{1}{\sqrt{1+8t^2}}(\mathbf{i}-2t\mathbf{j}-2t\mathbf{k})$

9. $\mathbf{v}=-\alpha\sin(t)\mathbf{i}+\alpha\cos(t)\mathbf{j}+\beta\mathbf{k}$; $v=\sqrt{\alpha^2+\beta^2}$;

$\mathbf{a}=-\alpha\cos(t)\mathbf{i}-\alpha\sin(t)\mathbf{j}$; $a_t=0$; $a_n=\alpha$; $\rho=\dfrac{\alpha^2+\beta^2}{\alpha}$; $\kappa=\dfrac{\alpha}{\alpha^2+\beta^2}$;

$\mathbf{T}=\dfrac{1}{\sqrt{\alpha^2+\beta^2}}[-\alpha\sin(t)\mathbf{i}+\alpha\cos(t)\mathbf{j}+\beta\mathbf{k}]$; $\mathbf{N}=\dfrac{1}{\alpha}[-\alpha\cos(t)\mathbf{i}-\beta\sin(t)\mathbf{j}]$

11. $\mathbf{v}=2\cosh(t)\mathbf{i}-2\sinh(t)\mathbf{k}$; $v=2\sqrt{\cosh^2(t)+\sinh^2(t)}$
$=2\sqrt{\cosh(2t)}$; $\mathbf{a}=2\sinh(t)\mathbf{i}-2\cosh(t)\mathbf{k}$;

$a_t=\dfrac{4\sinh(t)\cosh(t)}{\sqrt{\cosh^2(t)+\sinh^2(t)}}$; $a_n=\dfrac{2}{\sqrt{\cosh^2(t)+\sinh^2(t)}}$

$\rho=2[\sinh^2(t)+\cosh^2(t)]^{3/2}$; $\kappa=\dfrac{1}{2[\sinh^2(t)+\cosh^2(t)]^{3/2}}$

$\mathbf{T}=\dfrac{1}{\sqrt{\sinh^2(t)+\cosh^2(t)}}[\cosh(t)\mathbf{i}-\sinh(t)\mathbf{k}]$;

$\mathbf{N}=\dfrac{1}{\sqrt{\sinh^2(t)+\cosh^2(t)}}[-\sinh(t)\mathbf{i}-\cosh(t)\mathbf{k}]$

13. $\mathbf{v}=-\alpha\sin(t)\mathbf{i}+\beta\cos(t)\mathbf{j}$; $v=\sqrt{\alpha^2\sin^2(t)+\beta^2\cos^2(t)}$;

$\mathbf{a}=-\alpha\cos(t)\mathbf{i}-\beta\sin(t)\mathbf{j}$; $a_t=\dfrac{(\alpha^2-\beta^2)\sin(t)\cos(t)}{\sqrt{\alpha^2\sin^2(t)+\beta^2\cos^2(t)}}$;

$a_n=\dfrac{|\alpha\beta|}{\sqrt{\alpha^2\sin^2(t)+\beta^2\cos^2(t)}}$; $\rho=\dfrac{[\alpha^2\sin^2(t)+\beta^2\cos^2(t)]^{3/2}}{|\alpha\beta|}$;

$$\kappa = \frac{|\alpha\beta|}{[\alpha^2 \sin^2(t) + \beta^2 \cos^2(t)]^{3/2}}; \quad \mathbf{T} = \frac{1}{\sqrt{\alpha^2 \sin^2(t) + \beta^2 \cos^2(t)}}[-\alpha \sin(t)\mathbf{i} + \beta \cos(t)\mathbf{j}]$$

$$\mathbf{N} = \frac{1}{\sqrt{\alpha^2 \sin^2(t) + \beta^2 \cos^2(t)}}[-\beta \cos(t)\mathbf{i} - \alpha \sin(t)\mathbf{j}]$$

15. $\mathbf{v} = \omega \cos(\omega t)\mathbf{i} - 2\mathbf{j} - \omega \sin(\omega t)\mathbf{k}; \quad v = \sqrt{\omega^2 + 4};$

$\mathbf{a} = -\omega^2 \sin(\omega t)\mathbf{i} - \omega^2 \cos(\omega t)\mathbf{k}; \quad a_t = 0; \quad a_n = \omega^2;$

$$\rho = \frac{\omega^2 + 4}{\omega^2}; \quad \kappa = \frac{\omega^2}{\omega^2 + 4};$$

$$\mathbf{T} = \frac{1}{\sqrt{\omega^2 + 4}}[\omega \cos(\omega t)\mathbf{i} - 2\mathbf{j} - \omega \sin(\omega t)\mathbf{k}];$$

$\mathbf{N} = -\sin(\omega t)\mathbf{i} - \cos(\omega t)\mathbf{k}$

17. $\mathbf{v} = 2\mathbf{i} - 2\mathbf{j} + 2t\mathbf{k}; \quad v = 2\sqrt{t^2 + 2};$

$$\mathbf{a} = 2\mathbf{k}; \quad a_t = \frac{2t}{\sqrt{t^2 + 2}}; \quad a_n = \frac{2\sqrt{2}}{\sqrt{t^2 + 2}};$$

$$\rho = \sqrt{2}(t^2 + 2)^{3/2}; \quad \kappa = \frac{1}{\sqrt{2}(t^2 + 2)^{3/2}};$$

$$\mathbf{T} = \frac{1}{\sqrt{t^2 + 2}}(\mathbf{i} - \mathbf{j} + t\mathbf{k});$$

$$\mathbf{N} = \frac{1}{\sqrt{2}\sqrt{t^2 + 2}}(-t\mathbf{i} + t\mathbf{j} + 2\mathbf{k})$$

19. $\mathbf{v} = 2\alpha t\mathbf{i} + 2\beta t\mathbf{j} + 2\gamma\mathbf{k}; \quad v = 2\sqrt{\alpha^2 t^2 + \beta^2 t^2 + \gamma^2};$

$$\mathbf{a} = 2\alpha\mathbf{i} + 2\beta\mathbf{j}; \quad a_t = \frac{2t(\alpha^2 + \beta^2)}{\sqrt{\alpha^2 t^2 + \beta^2 t^2 + \gamma^2}};$$

$$a_n = \frac{2|\gamma|\sqrt{\alpha^2 + \beta^2}}{\sqrt{\alpha^2 t^2 + \beta^2 t^2 + \gamma^2}};$$

$$\rho = \frac{2[(\alpha^2 + \beta^2)t^2 + \gamma^2]^{3/2}}{|\gamma|\sqrt{\alpha^2 + \beta^2}}; \qquad \kappa = \frac{|\gamma|\sqrt{\alpha^2 + \beta^2}}{2[(\alpha^2 + \beta^2)t^2 + \gamma^2]^{3/2}};$$

$$\mathbf{T} = \frac{1}{\sqrt{\alpha^2 t^2 + \beta^2 t^2 + \gamma^2}}(\alpha t\mathbf{i} + \beta t\mathbf{j} + \gamma\mathbf{k});$$

$$\mathbf{N} = \frac{1}{|\gamma|\sqrt{\alpha^2 + \beta^2}\sqrt{\alpha^2 t^2 + \beta^2 t^2 + \gamma^2}}[\alpha\gamma\mathbf{i} + \beta\gamma\mathbf{j} - (\alpha^2 t + \beta^2 t)\mathbf{k}]$$

Section 11.3

1. $\dfrac{\partial \mathbf{G}}{\partial x} = 3\mathbf{i} - 4y\mathbf{j}; \quad \dfrac{\partial \mathbf{G}}{\partial y} = -4x\mathbf{j}$ **3.** $4y\mathbf{i}; \quad 4x\mathbf{i} - \mathbf{j}$ **5.** $2y\mathbf{i} - \sin(x)\mathbf{j}; \quad 2x\mathbf{i}$

7. $(-e^{-x}x + e^{-x})\mathbf{i} + 8y\mathbf{j}; \quad 8x\mathbf{j}$ **9.** $\dfrac{2y^2}{x + 1}\mathbf{i} + 8y^3\mathbf{j}; \quad 4\ln(x + 1)y\mathbf{i} + 24xy^2\mathbf{j}$

11. $\dfrac{\partial \mathbf{F}}{\partial x} = ye^{xy}\mathbf{i} - 4xy\mathbf{j}; \quad \dfrac{\partial \mathbf{F}}{\partial y} = xe^{xy}\mathbf{i} - 2x^2\mathbf{j} + \sinh(z + y)\mathbf{k}; \quad \dfrac{\partial \mathbf{F}}{\partial z} = \sinh(z + y)\mathbf{k}$

13. $3y^3\mathbf{i} + \dfrac{1}{x + y + z}\mathbf{j} + yz \sinh(xyz)\mathbf{k};$

$9xy^2\mathbf{i} + \dfrac{1}{x + y + z}\mathbf{j} + xz \sinh(xyz)\mathbf{k}; \quad \dfrac{1}{x + y + z}\mathbf{j} + xy \sinh(xyz)\mathbf{k}$

15. $14\mathbf{i} + 2x\mathbf{j} + (3y^3 + 3xy^3 z)e^{zx}\mathbf{k}; \quad -2\mathbf{i} - 2y\mathbf{j} + 9xy^2 e^{zx}\mathbf{k}; \quad -2z\mathbf{j} + 3x^2 y^3 e^{zx}\mathbf{k}$

17. $(-6xy - 3x^2y)e^x\mathbf{i} + y^3z^2\mathbf{j} + (3 - 2y\cos(xy))\mathbf{k};\quad -3x^2e^x\mathbf{i} + 3xy^2z^2\mathbf{j} - 2x\cos(xy)\mathbf{k};\quad 2xy^3z\mathbf{j}$

19. $8\mathbf{j} + 3x^2z\mathbf{k};\quad -2\mathbf{j};\quad 5\mathbf{i} + x^3\mathbf{k}$

21. $\dfrac{2z^2}{3 + c_1z^2}\mathbf{i} + c_2\exp\left(\dfrac{1}{2z^2}\right)\mathbf{j} + z\mathbf{k};\quad \dfrac{24z^2}{36 + 11z^2}\mathbf{i} - \exp\left(\dfrac{36 - z^2}{72z^2}\right)\mathbf{j} + z\mathbf{k}$

23. $x\mathbf{i} + \ln(x^2 + c_1^2)\mathbf{j} + \arccos\left[\dfrac{1}{c_1}\arctan\left(\dfrac{x}{c_1}\right) + c_2\right]\mathbf{k};$

$x\mathbf{i} + \ln(x^2 + e - 1)\mathbf{j} + \arccos\left[\dfrac{1}{\sqrt{e-1}}\arctan\left(\dfrac{x}{\sqrt{e-1}}\right) - \arctan\left(\dfrac{1}{\sqrt{e-1}}\right) + \dfrac{\sqrt{2}}{2}\right]\mathbf{k}$

25. $c_1y^3\mathbf{i} + y\mathbf{j} + \dfrac{1}{\ln|y| + c_2}\mathbf{k};\quad \dfrac{1}{216}y^3\mathbf{i} + y\mathbf{j} + \dfrac{1}{1 + \ln\left(\dfrac{y^2}{36}\right)}\mathbf{k}$

27. $\ln(c_1 + 2e^z)\mathbf{i} + [c_2 - \frac{1}{2}\ln(2e^z + c_1)]\mathbf{j} + z\mathbf{k};$
$\ln(2e^z + e - 2e^2)\mathbf{i} + \frac{1}{2}[1 - \ln(2e^z + e - 2e^2)]\mathbf{j} + z\mathbf{k}$

29. $c_2\mathbf{i} + (c_1 - \frac{3}{2}z^4)^{1/3}\mathbf{j} + z\mathbf{k};\quad (-\frac{3}{2}z^4 + \frac{5}{2})^{1/3}\mathbf{j} + z\mathbf{k}$

31. $x\mathbf{i} + \left(c_1 - \dfrac{x^2}{3}\right)\mathbf{j} + \left(\dfrac{-x^3}{27} + \dfrac{c_1x}{3} + c_2\right)\mathbf{k};\quad x\mathbf{i} - \left(\dfrac{x^2}{3} - 2\right)\mathbf{j} + \left(-\dfrac{x^3}{27} - \dfrac{2x}{3} + 4\right)\mathbf{k}$

33. $-(9e^z + c_1)^{1/3}\mathbf{i} + c_2\mathbf{j} + z\mathbf{k};\quad -(9e^z - 9e^4 - 1)^{1/3}\mathbf{i} + 3\mathbf{j} + z\mathbf{k}$

35. $(c_1 + 3z)\mathbf{i} + |c_2 - 16e^z|^{1/4}\mathbf{j} + z\mathbf{k};\quad (3z - 5)\mathbf{i} + 2|e^5 - e^z|^{1/4}\mathbf{j} + z\mathbf{k}$

37. $(3z + c_1)\mathbf{i} + [-16e^z + c_2]\mathbf{j} + z\mathbf{k};\quad (3z + 3)\mathbf{i} + [-16e^z + 17]\mathbf{j} + z\mathbf{k}$

39. $x\mathbf{i} + \sqrt{\dfrac{2x^3}{9} + c_1}\,\mathbf{j} + (\frac{1}{3}x + c_2)\mathbf{k};\quad x\mathbf{i} + \sqrt{\dfrac{2x^3}{9} - \dfrac{7}{9}}\,\mathbf{j} + (\frac{1}{3}x + \frac{11}{3})\mathbf{k}$

41. One example is $\mathbf{i} + \mathbf{j} + \mathbf{k}$.

Section 11.4

1. $\nabla\varphi = 2yz\mathbf{i} + 2xz\mathbf{j} + 2xy\mathbf{k};\quad \nabla\varphi(1, 1, 1) = 2(\mathbf{i} + \mathbf{j} + \mathbf{k})$

3. $(2y + e^z)\mathbf{i} + 2x\mathbf{j} + xe^z\mathbf{k};\quad (2 + e^6)\mathbf{i} - 4\mathbf{j} - 2e^6\mathbf{k}$

5. $2y\sinh(2xy)\mathbf{i} + 2x\sinh(2xy)\mathbf{j} - \cosh(z)\mathbf{k};\quad -2\sinh(6)\mathbf{i} + 6\sinh(6)\mathbf{j} - \cosh(1)\mathbf{k}$

7. $\dfrac{1}{x + y + z}(\mathbf{i} + \mathbf{j} + \mathbf{k});\quad \frac{1}{8}(\mathbf{i} + \mathbf{j} + \mathbf{k})$

9. $e^x\cos(y)\cos(z)\mathbf{i} - e^x\sin(y)\cos(z)\mathbf{j} - e^x\cos(y)\sin(z)\mathbf{k};\quad \frac{1}{2}(\mathbf{i} - \mathbf{j} - \mathbf{k})$

11. $[2xy\cosh(xz) + x^2yz\sinh(xz)]\mathbf{i} + x^2\cosh(xz)\mathbf{j} + x^3y\sinh(xz)\mathbf{k};\quad \mathbf{0}$

13. $\mathbf{i} + 2\sin(y + z)\mathbf{j} + 2\sin(y + z)\mathbf{k};\quad \mathbf{i} + 2\sin(1)\mathbf{j} + 2\sin(1)\mathbf{k}$

15. $\sinh(x - y + 2z)(\mathbf{i} - \mathbf{j} + 2\mathbf{k});\quad \sinh(2)(\mathbf{i} - \mathbf{j} + 2\mathbf{k})$

17. $-2x + y - z = 1;\quad x = -1 + 2t,\quad y = 1 - t,\quad z = 2 + t$ **19.** $x + y + z = 0;\quad x = y = z$

21. $x = y;\quad x = 1 + 2t,\quad y = 1 - 2t,\quad z = 0$ **23.** $x + y + 2z = 4;\quad x = 1 + 12t,\quad y = 1 + 12t,\quad z = 1 + 24t$

25. $y + z = \pi;\quad x = 0,\quad y = \pi + t,\quad z = t$ **27.** $\theta = \pi/4[\cos(\theta) = 1/\sqrt{2}]$

29. $\cos(\theta) = 1;\quad \theta = 0$ radians **31.** Direction is given by $-\mathbf{i}$; magnitude is e.

33. $\frac{4}{17}(-\frac{1}{4}\mathbf{i} - 3\mathbf{j} - 3\mathbf{k});\quad \frac{17}{4}$ **35.** $\dfrac{1}{\sqrt{292}}(16\mathbf{i} - 6\mathbf{j});\quad \sqrt{292}$

37. $\dfrac{1}{\sqrt{1298 + 2\sin(1)}}\{[1 + \sin(1)]\mathbf{i} + \cos(1)\mathbf{j} - 36\mathbf{k}\};\quad \sqrt{1298 + 2\sin(1)}$

39. $\dfrac{1}{\sqrt{\frac{1}{10} + \cos^2(3) - \frac{1}{5}\cos(3)}}\{[\frac{1}{10} - \cos(3)]\mathbf{i} + \frac{3}{10}\mathbf{j}\};\quad \sqrt{\frac{1}{10} + \cos^2(3) - \frac{1}{5}\cos(3)}$

Section 11.5

1. $\nabla\cdot\mathbf{F} = 4;\quad \nabla\times\mathbf{F} = \mathbf{0}$ **3.** $2y + e^y + 2;\quad -2x\mathbf{k}$ **5.** $-2yz;\quad y^2\mathbf{i} - 2e^z\mathbf{j}$

7. $-1;\quad \mathbf{0}$ **9.** $2x + 2y + 2z;\quad \mathbf{0}$ **11.** $6x + z;\quad -y\mathbf{i} - e^x\mathbf{j}$

13. $4xy + z - y$; $(-z - y)\mathbf{i} - 2x^2\mathbf{k}$

15. $2xy - \dfrac{1}{x+y+z}$; $\left(-\dfrac{1}{x+y+z}\right)\mathbf{i} + \left(-3y + \dfrac{1}{x+y+z}\right)\mathbf{j} + (y^2 + 3z)\mathbf{k}$

17. $-y\sin(xy) + 1$; $2z\mathbf{i} - \mathbf{j} - x\sin(xy)\mathbf{k}$ **19.** $12xy^2 - 2z$; $(4y^3 - 6y)\mathbf{k}$

21. $\nabla\varphi = \mathbf{i} - \mathbf{j} + 4z\mathbf{k}$ **23.** $-6x^2yz^2\mathbf{i} - 2x^3z^2\mathbf{j} - 4x^3y\mathbf{k}$

25. $\dfrac{-zy}{z^2 + x^2y^2}\mathbf{i} - \dfrac{zx}{z^2 + x^2y^2}\mathbf{j} + \dfrac{yx}{z^2 + x^2y^2}\mathbf{k}$ **27.** $3x^2y^2e^z\mathbf{i} + 2x^3ye^z\mathbf{j} + x^3y^2e^z\mathbf{k}$

29. $(-4y^3 + z^2)\mathbf{i} - 12xy^2\mathbf{j} + 2zx\mathbf{k}$ **31.** $-2\cos(x - y)\mathbf{i} + 2\cos(x - y)\mathbf{j} + 2z\mathbf{k}$

33. $[6xy\cos(z) - 8xz]\mathbf{i} + 3x^2\cos(z)\mathbf{j} + [-3x^2y\sin(z) - 4x^2 - 5]\mathbf{k}$

35. $yze^{xyz}\sin(e^{xyz})\mathbf{i} + xze^{xyz}\sin(xyz)\mathbf{j} + [3 + xye^{xyz}\sin(xyz)]\mathbf{k}$

37. Any $\alpha x^2yz\mathbf{i} + \beta xy^2z\mathbf{j} + \gamma xyz^2\mathbf{k}$, with $\alpha + \beta + \gamma = 7$.

Section 11.6

1. $176/3$ **3.** $1573/30$ **5.** $-\sin(2) + \sin(1) - 18\pi^2$ **7.** -129 **9.** 1980 **11.** $\frac{1}{3}e^{12} - e^{-4} - \frac{32}{3} - \frac{1}{3}e^6 + e^{-2} + \frac{4}{3}(2^{3/2})$

13. $\frac{3}{2}\sin(18) - \frac{57}{2}$ **15.** $65{,}508/5$ **17.** 0 **19.** 0 **21.** $-55/6$ **23.** $\frac{81}{6}$; $\frac{58}{15}$; $\frac{58}{15}$

25. $\frac{6}{17}(5^{17/6} - 1)$; $\frac{2}{13}(5^{13/6} - 1)$; $\frac{3}{14}(5^{7/3} - 1)$ **27.** $992/5$; $32{,}640$; $992/5$

29. $\frac{24}{23}(6^{23/4} - 2^{23/4})$; $\frac{4}{3}(6^{27/4} - 2^{27/4})$; $\frac{6}{5}(6^{25/4} - 2^{25/4})$ **31.** 0; -30; 100

33. $\frac{307}{6}$; $\frac{81}{12}$; $\frac{81}{12}$ **35.** $2[\cos(1) - \cos(3) - \frac{1}{3}]$; $5[\cos(1) - \cos(3) - \frac{1}{3}]$; $\cos(1) - \cos(3) - \frac{1}{3}$

37. -164; 0; -41 **39.** $49{,}013/3$ **41.** $-164/3$ **43.** 16π **45.** $96\sinh(4) - 24\cosh(4) + 24$

47. $\sqrt{14}[\cos(1) - \cos(5) - 936]$ **49.** 0 **51.** $-\frac{3}{4}\pi + 27 - 9\sqrt{2}$

Section 11.7

In 1–19, answers give the common value of $\oint_C \mathbf{F}$ and $\iint_\Delta \left(\dfrac{\partial F_2}{\partial y} - \dfrac{\partial F_1}{\partial x}\right) dx\, dy$.

1. 0 **3.** -5 **5.** $-\frac{8}{3} - 2e + 2e^{-1}$ **7.** 0 **9.** 0 **11.** -3π **13.** $\frac{3}{2}$

15. $\frac{10}{3} - 4\sin(1)$ **17.** $-16\sqrt{2}\pi$ **19.** 16π

Section 11.8

In these solutions, K is any constant.

1. $e^{xy} + K$ **3.** $\sin(xy) + K$ **5.** No potential **7.** $x^3y + \sin(xy) + K$ **9.** $xe^y - 3x^2y + x^3 + K$

11. $x^3y - 3y^2 + K$ **13.** $x\cos(xy) + K$ **15.** No potential **17.** $\phi(x, y) = x^4y^4$; $20{,}720$

19. $\phi(x, y) = xy\cos(x)$; -2π **21.** $\phi(x, y) = \sinh(x + y)$; 0 **23.** $\phi(x, y) = xy - 8xy^3$; 1846

25. $\phi(x, y) = y^2e^x$; $25e^3$ **27.** $\phi(x, y) = x^2 + \cos(x - y)$; 16 **29.** $\phi(x, y) = x - xy^3$; -842

Section 11.9

For each odd-numbered **F**, the surface integral is given for the surfaces **(a)**, **(c)**, and **(e)**.

1. **(a)** 0 **(c)** 0 **(e)** -18π **3.** **(a)** $-\frac{14}{3}$ **(c)** $\pi(3 + \sqrt{3})$ **(e)** 24π

5. **(a)** 11 **(c)** $-2\pi(3)^{3/2}$ **(e)** -54π **7.** **(a)** -15 **(c)** 0 **(e)** 0

9. **(a)** 14 **(c)** $-2\pi\sqrt{3}$ **(e)** -18π

Section 11.10

For 1, 3, and 5, solutions are given for the surfaces of **(a)** and **(c)**.

1. **(a)** 2π **(c)** 24π **3.** **(a)** 0 **(c)** 0 **5.** **(a)** $-\pi/4$ **(c)** -8π

For 7, 9, and 11, solutions are given for the surfaces of **(e)** and **(g)**.

7. **(e)** 0 **(g)** 0 **9.** **(e)** 0 **(g)** $-\frac{3}{2}$ **11.** **(e)** 0 **(g)** $\frac{1}{2}$

Section 11.12

1. No potential 3. $\sin(x) - \cos(xy) + z + K$ 5. No potential 7. $xz + y + K$
9. $yz \sin(x) + y + K$ 11. $\phi = x - 3y^3z; \quad -403$ 13. $\phi = 2x^3e^{yz}; \quad 2e^{-2}$
15. $\phi = 2yz^2 - x; \quad 71$ 17. $\phi = 4xy^3 - 8xz; \quad 16$ 19. $\phi = xz \sin(yz); \quad 2\sin(14)$
21. $\phi = z \sinh(xy); \quad -7\sinh(1)$ 23. $\phi = x\cos(yz) - y + z^2; \quad 14 + \cos(3) - \cos(7)$
25. $\phi = x^3 - x + y^2z; \quad -11$

Section 11.13

1. $h_r = 1, \quad h_\theta = r, \quad h_z = 1$

$$\nabla \cdot \mathbf{F}(r, \theta, z) = \frac{1}{r}\left[\frac{2}{\partial r}(rF_r) + \frac{\partial}{\partial \theta}(F_\theta) + \frac{\partial}{\partial z}(rF_z)\right],$$

$$\nabla \times \mathbf{F}(r, \theta, z) = \frac{1}{r}\left[\frac{\partial}{\partial \theta}(F_z) - \frac{\partial}{\partial z}(rF_\theta)\right]\mathbf{u}_r + \left[\frac{\partial}{\partial z}(F_r) - \frac{\partial}{\partial r}(F_z)\right]\mathbf{u}_\theta + \frac{1}{r}\left[\frac{\partial}{\partial r}(rF_\theta) - \frac{\partial}{\partial \theta}(F_r)\right]\mathbf{u}_z$$

$$\nabla g = \frac{\partial g}{\partial r}\mathbf{u}_r + \frac{1}{r}\frac{\partial g}{\partial \theta}\mathbf{u}_\theta + \frac{\partial g}{\partial z}\mathbf{u}_z$$

$$\nabla^2 g = \frac{1}{r}\left[\frac{\partial}{\partial r}\left(r\frac{\partial g}{\partial r}\right) + \frac{\partial}{\partial \theta}\left(\frac{1}{r}\frac{\partial g}{\partial \theta}\right) + \frac{\partial}{\partial z}\left(r\frac{\partial g}{\partial z}\right)\right]$$

2. $h_u = \dfrac{a}{\cosh(v) - \cos(u)} = h_v, \; h_z = 1$

$$\nabla f = \left(\frac{\cosh(v) - \cos(u)}{a}\right)\frac{\partial f}{\partial u}\mathbf{u}_1 + \left(\frac{\cosh(v) - \cos(u)}{a}\right)\frac{\partial f}{\partial v}\mathbf{u}_2 + \frac{\partial f}{\partial z}\mathbf{u}_3$$

$$\nabla \cdot \mathbf{F} = \left(\frac{\cosh(v) - \cos(u)}{a}\right)^2\left[\frac{\partial}{\partial u}\left(\frac{a}{\cosh(v) - \cos(u)}F_1\right) + \frac{\partial}{\partial v}\left(\frac{a}{\cosh(v) - \cos(u)}F_2\right) + \frac{\partial}{\partial z}\left(\frac{a^2}{[\cosh(v) - \cos(u)]^2}F_3\right)\right]$$

$$\nabla \times \mathbf{F} = \left(\frac{\cosh(v) - \cos(u)}{a}\right)\left[\frac{\partial F_3}{\partial v} - \frac{\partial}{\partial z}\left(\frac{aF_2}{\cosh(v) - \cos(u)}\right)\right]\mathbf{u}_1 + \left(\frac{\cosh(v) - \cos(u)}{a}\right)\left[\frac{\partial}{\partial z}\left(\frac{aF_1}{\cosh(v) - \cos(u)}\right) - \frac{\partial F_3}{u}\right]\mathbf{u}_2 + \left(\frac{\cosh(v) - \cos(u)}{a}\right)^2\left[\frac{\partial}{\partial u}\left(\frac{aF_2}{\cosh(v) - \cos(u)}\right) - \frac{\partial}{\partial v}\left(\frac{aF_1}{\cosh(v) - \cos(u)}\right)\right]\mathbf{u}_3$$

$$\nabla^2 f = \left(\frac{\cosh(v) - \cos(u)}{a}\right)^2\left[\frac{\partial^2 f}{\partial u^2} + \frac{\partial^2 f}{\partial v^2} + \frac{\partial}{\partial z}\left(\frac{a^2}{[\cosh(v) - \cos(u)]^2}\frac{\partial f}{\partial z}\right)\right]$$

Section 11.14

1. 0 3. 0 5. 0 7. 0 9. $\nabla \cdot \mathbf{F} \neq 0$

Supplementary Problems

1. $(\mathbf{F} \cdot \mathbf{G})' = 4\cos(2t) - 8t\sin(2t) + 4;$
$(\mathbf{F} \times \mathbf{G})' = -12t^2\mathbf{i} + [32t + 2\sin(2t)]\mathbf{j} + [2t^2\sin(2t) - 2t\cos(2t)]\mathbf{k}$
3. $-2\sqrt{t}\,e^{-t} + t^{-1/2}e^{-t} + \cosh(t) - 4t;$
$[-1 - 2\sinh(t) - 2t\cosh(t)]\mathbf{i} + (e^{-t} - te^{-t} + 6\sqrt{t})\mathbf{j} + [t^{-1/2} + e^{-t}\sinh(t) - e^{-t}\cosh(t)]\mathbf{k}$
5. $-2\sinh(3t) - 6t\cosh(3t); \quad 6\cosh(3t)\mathbf{i} + [-4t\sinh(3t) - 6t^2\cosh(3t)]\mathbf{k}$
7. $2t - \cosh(t) - (t - 3)\sinh(t) - 16t^3;$
$[4t^3 - 9t^2 - 4\cosh(t) - 4t\sinh(t)]\mathbf{i} + (6t^2 - 5t^4 - 4)\mathbf{j} + [-2t\cosh(t) - (t^2 - 2)\sinh(t) - 1]\mathbf{k}$

9. $8t^3e^{-t} - 2t^4e^{-t} + 6t^2 + (1/2\sqrt{t})\cosh(t) + \sqrt{t}\sinh(t)$; $[-2\cosh(t) - 2t\sinh(t) + \frac{5}{2}t^{3/2}]\mathbf{i} + [-7t^{5/2} + e^{-t}\cosh(t) - te^{-t}\cosh(t) + te^{-t}\sinh(t)]\mathbf{j} + (3t^2 - e^{-t} - t^3e^{-t} - 16t^3)\mathbf{k}$

11. $\int_0^1 [16t^2 + \sin^2(t) + 1]^{1/2}\,dt$ **13.** $\int_2^5 [4t^2 + 1 + e^{2t}]^{1/2}\,dt$ **15.** $\int_0^{2\pi} [81t^4 + 1 + e^{-2t} - 2e^{-t}]^{1/2}\,dt$

17. $\mathbf{v} = \cos(t)\mathbf{i} - \sin(t)\mathbf{j}$; $v = 1$; $\mathbf{a} = -\sin(t)\mathbf{i} - \cos(t)\mathbf{j}$; $\mathbf{a} = 0\mathbf{T} + 1\mathbf{N}$ (here, $\mathbf{T} = \mathbf{v}$ and $\mathbf{N} = \mathbf{a}$); $\kappa = \rho = 1$; $\tau = 0$

19. $\mathbf{v} = 2t\mathbf{i} - 2t\mathbf{j} + \mathbf{k}$; $v = \sqrt{8t^2 + 1}$; $\mathbf{a} = 2\mathbf{i} - 2\mathbf{j}$;

$$\mathbf{a} = \left(\frac{8t}{8t^2+1}\right)(2t\mathbf{i} - 2t\mathbf{j} + \mathbf{k}) + \frac{1}{8t^2+1}(2\mathbf{i} - 2\mathbf{j} - 8t\mathbf{k});$$

$$\kappa = \frac{\sqrt{8}}{(8t^2+1)^{3/2}};\quad \rho = \frac{(8t^2+1)^{3/2}}{\sqrt{8}};\quad \tau = 0$$

21. $x = \ln|z| + c_1,\ \ y = c_2e^{-z}$ **23.** $y = \sqrt{c_1 - x^2},\ \ z = c_2\exp\left[\arcsin\left(\frac{x}{\sqrt{c_1}}\right)\right]$

25. $y = c_1e^{-x},\ \ z = \frac{1}{c_2 - x}$

27. $\nabla\phi(2, -1, 3) = -\cos(3)\mathbf{i} + [54 + 2\cos(3)]\mathbf{j} + [-54 + 2\sin(3)]\mathbf{k}$; magnitude $= \{5836 + \cos^2(3) + 216[\cos(3) - \sin(3)]\}^{1/2}$

29. $\nabla\phi(2, 2, 2) = \cos(4)\mathbf{i} - \sin(4)\mathbf{j} - [\cos(4) + \sin(4)]\mathbf{k}$; magnitude $= [2 - \sin(8)]^{1/2}$

31. $4x + 4y - z = 10$; $x = 2 - 4t$, $y = 2 - 4t$, $z = 6 + t$

33. $x - y + 3z = 9$; $x = 1 + t$, $y = 1 - t$, $z = 3(1 + t)$

35. $10x - 5y + 32z = 12$; $x = 2 + 10t$, $y = 4 - 5t$, $z = \frac{3}{8} + 32t$

37. $\nabla\cdot\mathbf{F} = y\sin(yz)$; $\nabla\times\mathbf{F} = z\sin(yz)\mathbf{i} - \mathbf{j} + 3x^2\mathbf{k}$

39. $\nabla\cdot\mathbf{F} = 2y$; $\nabla\times\mathbf{F} = 2z\mathbf{i} - \mathbf{j} - 2x\mathbf{k}$ **41.** $\frac{23}{2}$ **43.** $-\frac{2}{3}$ **45.** $\frac{26}{3}\cos(1)$ **47.** $\frac{325}{3} + e^{-4} - e$

49. $1276/3$ **51.** 0 **53.** 0 **55.** 5π **57.** $64\pi/3$ **59.** $-8\pi/3$ **61.** $\phi = z^2x - 2yx$; 14

63. $\phi = \cos(xz) - e^{yz}$; $\cos(2) - 1$ **65.** $\phi = x^3 - z^2y$; 118 **67.** $\frac{4}{3}$ **69.** $\frac{16}{3}$

CHAPTER 12

Section 12.1

1. $\frac{\pi^2}{3} + 4\sum_{n=1}^{\infty}\frac{(-1)^n}{n}\cos(nx)$

3. $\frac{\sinh(6)}{12} + \sum_{n=1}^{\infty}\frac{(-1)^n(e^6 - e^{-6})}{36 + n^2\pi^2}\left[3\cos\left(\frac{n\pi x}{3}\right) - \frac{n\pi}{6}\sin\left(\frac{n\pi x}{3}\right)\right]$ **5.** $\frac{2\pi^2}{3} + \sum_{n=1}^{\infty}\left[\frac{4(-1)^n}{n^2}\cos(nx) + \frac{2(-1)^n}{n}\sin(nx)\right]$

7. -4 [all coefficients of sine and cosine terms are zero; $f(x)$ *is* a Fourier series]

9. $\frac{1}{2} + \sum_{n=1}^{\infty}\left\{\frac{-2}{n\pi}\sin\left(\frac{n\pi}{2}\right)\cos\left(\frac{n\pi x}{2}\right) + \frac{4}{n\pi}\left[\cos\left(\frac{n\pi}{2}\right) - \cos(n\pi)\right]\sin\left(\frac{n\pi x}{2}\right)\right\}$

11. $2\sin(3x)$ ($f(x)$ is a Fourier series on $[-\pi, \pi]$) **13.** $1 - \frac{1}{\pi^2}\sin(\pi^2) + \sum_{n=1}^{\infty}\frac{2(-1)^n\sin(\pi^2)}{n^2 - \pi^2}\cos(nx)$

15. $-\frac{1}{3}\sinh(3) + \sum_{n=1}^{\infty}\left\{\frac{6(-1)^{n+1}\sinh(3)}{9 + n^2\pi^2}\cos\left(\frac{n\pi x}{3}\right) + \left[\frac{-18}{n\pi} + \frac{2n\pi}{9 + n^2\pi^2}\sinh(3)\right](-1)^n\sin\left(\frac{n\pi x}{3}\right)\right\}$

17. $-\frac{1}{2} + \sum_{n=1}^{\infty}\frac{6}{n^2\pi^2}[1 - \cos(n\pi)]\cos\left(\frac{n\pi x}{3}\right)$ or, equivalently, $-\frac{1}{2} + \sum_{n=1}^{\infty}\frac{12}{(2n-1)^2\pi^2}\cos\left(\frac{(2n-1)\pi x}{3}\right)$

19. $\frac{35}{3} + \sum_{n=1}^{\infty}\left[\frac{128}{n^2\pi^2}(-1)^n\cos\left(\frac{n\pi x}{4}\right) + \frac{24}{n\pi}(-1)^n\sin\left(\frac{n\pi x}{4}\right)\right]$

21. Yes (because the coefficients are integrals, and the integral of a sum is the sum of the integrals)

Section 12.2

1. $f(-2-)=-4,\ f(-2+)=0,\ f(1-)=0,\ f(1+)=1;\ f(-3+)=-6,\ f(3-)=9;\ f'_L(-2)=2,$ $f'_R(-2)=0;\ f'_L(1)=0,\ f'_R(1)=2;\ f'_R(-3)=2,\ f'_L(3)=6$. The Fourier series converges to x^2 for $1<x<3$; 0 for $-2<x<1$; $2x$ for $-3<x<-2$; $\frac{3}{2}$ at 3 and at -3; $\frac{1}{2}$ at 1; and to -2 at -2.

3. x^{15} has no discontinuities on $[-3,3]$; $f(-3+)=(-3)^{15},\ f(3-)=3^{15};\ f'_R(-3)=15(-3)^{14};\ f'_L(3)=15(3)^{14}$. The Fourier series converges to x^{15} for $-3<x<3$, and to $\frac{1}{2}[3^{15}+(-3)^{15}]=0$ at $x=3$ and $x=-3$.

5. $f(0-)=0,\ f(0+)=2;\ f'_R(0)=0=f'_L(0);\ f(-\pi+)=\pi^2,\ f(\pi-)=2;\ f'_R(-\pi)=-2\pi,\ f'_L(\pi)=0$. The Fourier series converges to x^2 for $-\pi<x<0$; to 2 for $0<x<\pi$; to 1 at $x=0$; and to $\frac{1}{2}(\pi^2+2)$ at $x=-\pi$ and at $x=\pi$.

7. $f(0-)=1,\ f(0+)=0;\ f'_R(0)=1,\ f'_L(0)=0;\ f(-\pi+)=-1,\ f(\pi-)=0;\ f'_R(-\pi)=0,\ f'_L(\pi)=-1$. The Fourier series converges to $\cos(x)$ for $-\pi<x<0$; to $\sin(x)$ for $0<x<\pi$; and to $\frac{1}{2}$ at $x=0$ and $-\frac{1}{2}$ at $x=-\pi$ and at $x=\pi$.

9. $f(0-)=-2,\ f(0+)=1;\ f'_R(0)=f'_L(0)=0;\ f(-4+)=-2,\ f(4-)=1;\ f'_R(-4)=f'_L(4)=0$. The Fourier series converges to -2 for $-4<x<0$; to 1 for $0<x<4$; to $-\frac{1}{2}$ at $x=0$; and to $-\frac{1}{2}$ at $x=-4$ and at $x=4$.

11. $f(1-)=1,\ f(1+)=e;\ f'_R(1)=e,\ f'_L(1)=1;\ f(-2+)=-2,\ f(2-)=e^2;\ f'_R(-2)=1,\ f'_L(2)=e^2$. The Fourier series converges to x for $-2<x<1$; to e^x for $1<x<2$; to $\frac{1}{2}(1+e)$ at $x=1$; and to $\frac{1}{2}(e^2-2)$ at $x=4$ and at $x=-4$.

13. $f(-2-)=4,\ f(-2+)=1;\ f'_R(-2)=0,\ f'_L(-2)=-1;\ f(0-)=1,\ f(0+)=-1;\ f'_R(0)=f'_L(0)=0;$ $f(1-)=-1,\ f(1+)=2;\ f'_R(1)=2,\ f'_L(1)=0;\ f(-3+)=5,\ f(3-)=10;\ f'_R(-3)=-1,\ f'_L(3)=6$. The Fourier series converges to $2-x$ for $-3<x<-2$; to 1 for $-2<x<1$; to -1 for $0<x<1$; to x^2+1 for $1<x<3$; to $\frac{5}{2}$ at $x=-2$; to 0 at $x=0$; to $\frac{1}{2}$ at $x=1$; and to $\frac{15}{2}$ at $x=-3$ and at $x=3$.

15. $f(4-)=6,\ f(4+)=1;\ f'_L(4)=1,\ f'_R(4)=0;\ f(\frac{9}{2}-)=1,\ f(\frac{9}{2}+)=\frac{1}{2};\ f'_R(\frac{9}{2})=1,\ f'_L(\frac{9}{2})=0;\ f(-5+)=-3,$ $f(5-)=1;\ f'_R(-5)=1,\ f'_L(5)=1$. The Fourier series converges to $x+2$ for $-5<x<4$; to 1 for $4<x<\frac{9}{2}$; to $x-4$ for $\frac{9}{2}<x<5$; to $\frac{7}{2}$ at $x=4$; to $\frac{3}{4}$ at $x=\frac{9}{2}$; and to -1 at $x=-5$ and at $x=5$.

17. $f\left(\frac{\pi}{2}-\right)=-2\sin\left(\frac{3\pi}{2}\right)=2,\ f\left(\frac{\pi}{2}+\right)=\frac{\pi^2}{4};\ f'_L\left(\frac{\pi}{2}\right)=-6\cos\left(\frac{3\pi}{2}\right)=0,\ f'_R\left(\frac{\pi}{2}\right)=\pi;\ f(-\pi+)=0,\ f(\pi-)=\pi^2;$

$f'_R(-\pi)=-6\cos(-3\pi)=6,\ f'_L(\pi)=2\pi$. The Fourier series converges to $-2\sin(3x)$ for $-\pi<x<\frac{\pi}{2}$; to x^2 for $\frac{\pi}{2}<x<\pi$; to $\frac{1}{2}\left(2+\frac{\pi^2}{4}\right)$ at $x=\frac{\pi}{2}$; and to $\frac{\pi^2}{2}$ at $x=\pi$ and at $x=-\pi$.

19. $f(0-)=-2,\ f(0+)=1;\ f'_L(0)=0,\ f'_R(0)=1;\ f(-3+)=7,\ f(3-)=e^3;\ f'_R(-3)=-6,\ f'_L(3)=e^3$. The Fourier series converges to x^2-2 for $-3<x<0$; to e^x for $0<x<3$; to $-\frac{1}{2}$ at $x=0$; and to $\frac{1}{2}(7+e^3)$ at $x=3$ and at $x=-3$.

21. **(a)** $\frac{1}{\pi}(1-e^{-\pi})+\frac{2}{\pi}\sum_{n=1}^{\infty}\left(\frac{1}{1+n^2}\right)[1+(-1)^{n+1}e^{-\pi}]\cos(nx)$

(c) The Fourier series converges to $e^{-|x|}$ for $-\pi\le x\le\pi$.

(d) The first four partial sums are

$$S_1(x)=\frac{1}{\pi}(1-e^{-\pi})+\frac{1}{\pi}(1+e^{-\pi})\cos(x)$$

$$S_2(x)=\frac{1}{\pi}(1-e^{-\pi})+\frac{1}{\pi}(1+e^{-\pi})\cos(x)+\frac{2}{5\pi}(1-e^{-\pi})\cos(2x)$$

$$S_3(x)=S_2(x)+\frac{1}{5\pi}(1+e^{-\pi})\cos(3x)$$

$$S_4(x)=S_3(x)+\frac{2}{17\pi}(1-e^{-\pi})\cos(4x)$$

23. **(a)** $-4+\frac{24}{\pi}\sum_{n=1}^{\infty}\frac{(-1)^{n+1}}{n}\sin\left(\frac{n\pi x}{6}\right)$

(c) The Fourier series converges to $2x-4$, for $-6<x<6$, and to -4 at $x=6$ and at $x=-6$.

(d) $S_1(x) = -4 + \frac{24}{\pi}\sin\left(\frac{\pi x}{6}\right)$

$S_2(x) = -4 + \frac{24}{\pi}\sin\left(\frac{\pi x}{6}\right) - \frac{12}{\pi}\sin\left(\frac{\pi x}{3}\right)$

$S_3(x) = -4 + \frac{24}{\pi}\sin\left(\frac{\pi x}{6}\right) - \frac{12}{\pi}\sin\left(\frac{\pi x}{3}\right) + \frac{8}{\pi}\sin\left(\frac{\pi x}{2}\right)$

$S_4(x) = -4 + \frac{24}{\pi}\sin\left(\frac{\pi x}{6}\right) - \frac{12}{\pi}\sin\left(\frac{\pi x}{3}\right) + \frac{8}{\pi}\sin\left(\frac{\pi x}{2}\right) - \frac{6}{\pi}\sin\left(\frac{2\pi x}{3}\right)$

25. (a) $\sum_{n=1}^{\infty} \frac{(-24)(-1)^n}{n\pi}\sin\left(\frac{n\pi x}{4}\right)$

(c) The Fourier series converges to $3x$ for $-4 < x < 4$, and to zero at $x = 4$ and $x = -4$.

(d) $S_1(x) = \frac{24}{\pi}\sin\left(\frac{\pi x}{4}\right)$

$S_2(x) = \frac{24}{\pi}\sin\left(\frac{\pi x}{4}\right) - \frac{12}{\pi}\sin\left(\frac{\pi x}{2}\right)$

$S_3(x) = \frac{24}{\pi}\sin\left(\frac{\pi x}{4}\right) - \frac{12}{\pi}\sin\left(\frac{\pi x}{2}\right) + \frac{8}{\pi}\sin\left(\frac{3\pi x}{4}\right)$

$S_4(x) = \frac{24}{\pi}\sin\left(\frac{\pi x}{4}\right) - \frac{12}{\pi}\sin\left(\frac{\pi x}{2}\right) + \frac{8}{\pi}\sin\left(\frac{3\pi x}{4}\right) - \frac{6}{\pi}\sin(\pi x)$

27. (a) $\frac{1}{2} + \sum_{n=1}^{\infty}\left\{\frac{2}{n^2\pi^2}[(-1)^n - 1]\cos(n\pi x) + \frac{2}{n\pi}(-1)^n\sin(n\pi x)\right\}$

(c) The Fourier series converges to $-2x$ for $-1 < x < 1$, and to -1 at $x = -1$ and at $x = +1$.

(d) $S_1(x) = \frac{1}{2} - \frac{4}{\pi^2}\cos(\pi x) - \frac{2}{\pi}\sin(\pi x)$

$S_2(x) = \frac{1}{2} - \frac{4}{\pi^2}\cos(\pi x) - \frac{2}{\pi}\sin(\pi x) + \frac{1}{\pi}\sin(2\pi x)$

$S_3(x) = \frac{1}{2} - \frac{4}{\pi^2}\cos(\pi x) - \frac{2}{\pi}\sin(\pi x) + \frac{1}{\pi}\sin(2\pi x) - \frac{4}{9\pi^2}\cos(3\pi x) - \frac{2}{3\pi}\sin(3\pi x)$

$S_4(x) = \frac{1}{2} - \frac{4}{\pi^2}\cos(\pi x) - \frac{2}{\pi}\sin(\pi x) + \frac{1}{\pi}\sin(2\pi x) - \frac{4}{9\pi^2}\cos(3\pi x) - \frac{2}{3\pi}\sin(3\pi x) + \frac{1}{2\pi}\sin(4\pi x)$

29. (a) $\frac{13}{3} + \sum_{n=1}^{\infty}\frac{64}{n^2\pi^2}(-1)^n\cos\left(\frac{n\pi x}{2}\right)$

(c) The Fourier series converges to $4x^2 - 1$ for $-2 \le x \le 2$.

(d) $S_1(x) = \frac{13}{3} - \frac{32}{\pi^2}\cos\left(\frac{\pi x}{2}\right)$

$S_2(x) = \frac{13}{3} - \frac{32}{\pi^2}\cos\left(\frac{\pi x}{2}\right) + \frac{8}{\pi^2}\cos(\pi x)$

$S_3(x) = \frac{13}{3} - \frac{32}{\pi^2}\cos\left(\frac{\pi x}{2}\right) + \frac{8}{\pi^2}\cos(\pi x) - \frac{32}{9\pi^2}\cos\left(\frac{3\pi x}{2}\right)$

$S_4(x) = \frac{13}{3} - \frac{32}{\pi^2}\cos\left(\frac{\pi x}{2}\right) + \frac{8}{\pi^2}\cos(\pi x) - \frac{32}{9\pi^2}\cos\left(\frac{3\pi x}{2}\right) + \frac{2}{\pi^2}\cos(2\pi x)$

31. The Fourier series only represents $f(x)$ on $[-L, L]$.

33. (a) By Theorem 95, the Fourier series converges to $f(x)$ for $-L < x < L$. At L and at $-L$, the Fourier series converges to $\frac{1}{2}[f(-L+) + f(L-)]$. But here $f(-L+) = f(-L)$ and $f(L-) = f(L)$ because $f(x)$ is continuous at L and at $-L$. Thus, the series converges at L and $-L$ to $\frac{1}{2}[f(L) + f(-L)]$. Finally, by hypothesis, $f(L) = f(-L)$. Hence, $\frac{1}{2}[f(L) + f(-L)] = f(L) = f(-L)$, and the Fourier series converges to $f(x)$ on all of $[-L, L]$.

(b) Write the Fourier series for $f'(x)$ on $[-L, L]$. The coefficients [using capital letters to distinguish the coefficients of $f'(x)$ from those of $f(x)$] are

$$A_0 = \frac{1}{L}\int_{-L}^{L} f'(x)\,dx, \quad A_n = \frac{1}{L}\int_{-L}^{L} f'(x)\cos\left(\frac{n\pi x}{L}\right)dx, \qquad n = 1, 2, 3, \ldots,$$

and $B_n = \frac{1}{L}\int_{-L}^{L} f'(x)\sin\left(\frac{n\pi x}{L}\right)dx, \qquad n = 1, 2, 3, \ldots.$

Since $f'(x)$ is sectionally continuous on $[-L, L]$, we can evaluate A_0 directly, and A_n and B_n by integration by parts.

Thus, $A_0 = \frac{1}{L}[f(x)]_{-L}^{L} = \frac{1}{L}[f(L) - f(-L)] = 0, \quad$ since $f(L) = f(-L)$,

$$A_n = \frac{1}{L}\left\{\left[f(x)\cos\left(\frac{n\pi x}{L}\right)\right]_{-L}^{L} - \int_{-L}^{L} f(x)\left(-\frac{n\pi}{L}\right)\sin\left(\frac{n\pi x}{L}\right)dx\right\}$$

$$= \frac{1}{L}[f(L)\cos(n\pi) - f(-L)\cos(n\pi)] + \frac{n\pi}{L^2}\int_{-L}^{L} f(x)\sin\left(\frac{n\pi x}{L}\right)dx = \frac{n\pi}{L}b_n,$$

and $B_n = \frac{1}{L}\left\{\left[f(x)\sin\left(\frac{n\pi x}{L}\right)\right]_{-L}^{L} - \int_{-L}^{L} f(x)\left(\frac{n\pi}{L}\right)\cos\left(\frac{n\pi x}{L}\right)dx\right\} = -\frac{n\pi}{L}\int_{-L}^{L} f(x)\cos\left(\frac{n\pi x}{L}\right)dx = -\frac{n\pi}{L}a_n.$

The Fourier series for $f'(x)$ on $[-L, L]$ is then $\frac{A_0}{2} + \sum_{n=1}^{\infty} A_n\cos\left(\frac{n\pi x}{L}\right) + B_n\sin\left(\frac{n\pi x}{L}\right)$

or $\sum_{n=1}^{\infty}\frac{n\pi}{L}\left[-a_n\sin\left(\frac{n\pi x}{L}\right) + b_n\cos\left(\frac{n\pi x}{L}\right)\right]$. By Theorem 95, this converges to $f'(x)$ at each point of $(-L, L)$ where $f'(x)$ is continuous.

Section 12.3

1. Consider $f(x)$ as given by $f(x) = \begin{cases} K, & 0 \le x < 1, \\ 0, & -1 < x < 0, \end{cases}$ with $f(x + 2) = f(x)$.

   The Fourier series on $[-1, 1]$ is $\frac{K}{2} + \frac{2K}{\pi}\sum_{n=1}^{\infty}\frac{1}{(2n-1)}\sin[(2n-1)\pi x]$.

   This converges to: K for $0 < x < 1$, $2 < x < 3$, $4 < x < 5$, $\ldots$, and for $-2 < x < -1$, $-4 < x < -3$, $\ldots$; 0 for $1 < x < 2$, $3 < x < 5$, $\ldots$, and for $-1 < x < 0$, $-3 < x < -2$, $\ldots$; $\frac{1}{2}K$ at $x = 0, \pm 1, \pm 2, \pm 3, \ldots$.

3. Here, $f(x) = x \quad (-1 < x < 1)$ and $f(x + 2) = f(x)$. The Fourier series is $\sum_{n=1}^{\infty}\frac{2}{n\pi}(-1)^{n+1}\sin(n\pi x)$.

   This converges to x for $x \neq$ integer, $\quad 0$ for $x = 0, \pm 1, \pm 2, \pm 3, \ldots$.

5. Here we may think of $f(x)$ as the periodic function defined by

   $f(x) = \begin{cases} 1, & -1 < x < 1, \\ 2, & 1 < x < 3 \end{cases}$ and $-3 < x < -1$, with $f(x + 6) = f(x)$.

   The Fourier series is $\frac{5}{3} - \sum_{n=1}^{\infty}\frac{2}{n\pi}\sin\left(\frac{n\pi}{3}\right)\cos\left(\frac{n\pi x}{3}\right)$

   This converges to:
   1 for $-1 < x < 1$, $3 < x < 4$, $\ldots$, $-5 < x < -3$, $\ldots$;
   2 for $1 < x < 3$, $5 < x < 7$, $\ldots$, $-3 < x < -1$, $\ldots$;
   $\frac{3}{2}$ at $x = 1, 3, 5, 7, \ldots, -1, -3, -5, \ldots$.

7. Here, $f(x) = \begin{cases} x + \frac{\pi}{2} & \text{for } -\pi \le x \le 0, \\ -x + \frac{\pi}{2} & \text{for } 0 < x \le \pi, \end{cases}$ and $f(x + 2\pi) = f(x)$.

   The Fourier series is $\sum_{n=1}^{\infty} -\frac{2}{\pi n^2}[\cos(n\pi) - 1]\cos(nx)$ or, equivalently, $\sum_{n=1}^{\infty}\frac{4}{(2n-1)^2\pi}\cos[(2n-1)x]$.

   This series converges to $f(x)$ for all x.

9. In Problem 3, the series for $f(t)$ is $\sum_{n=1}^{\infty} \frac{2}{n\pi}(-1)^{n+1}\sin(n\pi t)$. Thus, solve

$y'' + 0.02y' + 12y = \sum_{n=1}^{\infty} \frac{2}{n\pi}(-1)^{n+1}\sin(n\pi t)$. To do this by superposition, solve

$y_n'' + 0.02y_n' + 12y_n = \frac{2}{n\pi}(-1)^{n+1}\sin(n\pi t)$. The steady-state solution is found to be

$y_n = \frac{(-1)^{n+1}}{\Delta_n}\left[0.04\cos(n\pi t) + (12 - n^2\pi^2)\frac{2}{n\pi}\sin(n\pi t)\right]$, where $\Delta_n = (12 - n^2\pi^2)^2 + (0.02n\pi)^2$.

The steady-state solution is $y = \sum_{n=1}^{\infty} \frac{(-1)^{n+1}}{\Delta_n}\left[0.04\cos(n\pi t) + \frac{2(12 - n^2\pi^2)}{n\pi}\sin(n\pi t)\right]$.

11. In Problem 7, the series for $f(x)$ is $\sum_{n=1}^{\infty} \frac{4}{(2n-1)^2\pi}\cos[(2n-1)t]$. We attempt a solution $\sum_{n=1}^{\infty} y_{2n-1}$, where

$y_{2n-1}'' + 8y_{2n-1} = \frac{4}{(2n-1)^2\pi}\cos[(2n-1)t]$. This has steady-state solution

$y_{2n-1} = \frac{4}{(2n-1)^2\pi}\left(\frac{1}{8-(2n-1)^2}\right)\cos[(2n-1)t]$. Thus, the steady-state solution of $y'' + 8y = f(t)$ is

$\frac{4}{\pi}\sum_{n=1}^{\infty} \frac{1}{(2n-1)^2}\left(\frac{1}{8-(2n-1)^2}\right)\cos[(2n-1)t]$.

Section 12.4

1. $6 + \frac{12}{\pi^2}\sum_{n=1}^{\infty}\left(\frac{4(-1)^n - 1}{n^2}\right)\cos\left(\frac{n\pi x}{3}\right)$ This converges to $2x + x^2$ for $0 \le x \le 3$.

$\sum_{n=1}^{\infty}\left[\frac{12}{n\pi}(-1)^{n+1} + \frac{18}{n^3\pi^3}(n^2\pi^2 - 2)(-1)^{n+1} - \frac{54}{n^3\pi^3}\right]\sin\left(\frac{n\pi x}{3}\right)$ This converges to $2x + x^2$ for $0 < x < 3$, and to 0 at $x = 0$ and $x = 3$.

3. $\frac{\pi^3}{4} + \sum_{n=1}^{\infty}\frac{2}{\pi}\left\{\frac{3\pi^2}{n^2}(-1)^n + \frac{6}{n^4}[1 - (-1)^n]\right\}\cos(nx)$ This converges to x^3 for $0 \le x \le \pi$.

$\sum_{n=1}^{\infty}\frac{2}{n^3}(n^2\pi^2 - 6)(-1)^{n+1}\sin(nx)$ This converges to x^3 for $0 < x < \pi$, and to 0 at $x = 0$ and $x = \pi$.

5. $\frac{1}{\pi}\sinh(\pi) + \sum_{n=1}^{\infty}\frac{2}{\pi}\frac{1}{1+n^2}(-1)^n\sinh(\pi)\cos(nx)$. This converges to $\cosh(x)$ for $0 \le x \le \pi$.

$\frac{2}{\pi}\sum_{n=1}^{\infty}\frac{n}{1+n^2}[1 + (-1)^{n+1}\cosh(\pi)]\sin(nx)$ This converges to $\cosh(x)$ for $0 < x < \pi$, and to 0 at $x = 0$ and $x = \pi$.

7. The cosine series is just 3; it converges to 3 for all x. The sine series is $\sum_{n=1}^{\infty}\frac{12}{(2n-1)\pi}\sin[(2n-1)x]$.

This converges to 3 for $0 < x < \pi$, and to 0 at $x = 0$ and $x = \pi$.

9. $\frac{1}{4}(e^2 + 1) + \sum_{n=1}^{\infty}\left\{\frac{8e^2}{16 + n^2\pi^2}\left[\cos\left(\frac{n\pi}{2}\right) + \frac{n\pi}{4}\sin\left(\frac{n\pi}{2}\right)\right] - \frac{8}{16 + n^2\pi^2} - \frac{2}{n\pi}\sin\left(\frac{n\pi}{2}\right)\right\}\cos\left(\frac{n\pi x}{4}\right)$

This converges to e^x for $0 \le x < 2$, to 1 for $2 < x \le 4$, and to $\frac{1}{2}(1 + e^2)$ at $x = 2$.

$\sum_{n=1}^{\infty}\left\{\frac{8e^2}{16 + n^2\pi^2}\left[\sin\left(\frac{n\pi}{2}\right) - \frac{n\pi}{4}\cos\left(\frac{n\pi}{2}\right)\right] + \frac{2n\pi}{16 + n^2\pi^2} - \frac{2}{n\pi}\left[\cos(n\pi) - \cos\left(\frac{n\pi}{2}\right)\right]\right\}\sin\left(\frac{n\pi x}{4}\right)$

This converges to e^x for $0 < x < 2$, to 1 for $2 < x < 4$, to $\frac{1}{2}(e^2 + 1)$ at $x = 2$, and to 0 at $x = 0$ and $x = 4$.

11. $\frac{\pi^2 - 2}{2\pi} + \sum_{n=1}^{\infty}\left\{\frac{-4}{\pi n^2}\cos(n) - \frac{2}{\pi n}\sin(n) + \frac{2}{\pi n^2}[1 - (-1)^n]\right\}\cos(nx)$ This converges to $-x$ for $0 < x < 1$, to x for $1 < x < \pi$, and to 0 at $x = 1$, $x = 0$, and $x = \pi$. $\frac{2}{\pi}\sum_{n=1}^{\infty}\left[-\frac{2}{n^2}\sin(n) + \frac{2}{n}\cos(n) - \frac{\pi}{n}(-1)^n\right]\sin(nx)$

This converges to $-x$ for $0 < x < 1$, to x for $1 < x < \pi$, and to 0 at $x = 0$, $x = 1$, and $x = \pi$.

13. $1 + \frac{1}{5}[\sin(5) - \sin(3)] + \sum_{n=1}^{\infty}\left[\frac{4}{n\pi}\sin\left(\frac{3n\pi}{5}\right) - \frac{2}{n\pi}\sin\left(\frac{n\pi}{5}\right) + \frac{\sin(5 - n\pi) - \sin\left(\frac{15 - 3n\pi}{5}\right)}{5 - n\pi}\right.$

$\left. + \frac{\sin(5 + n\pi) - \sin\left(\frac{15 + 3n\pi}{5}\right)}{5 - n\pi}\right]\cos\left(\frac{n\pi x}{5}\right)$ This converges to 1 for $0 \le x < 1$, to 2 for $1 < x < 3$, to $\cos(x)$ for $3 < x \le 5$, to $\frac{3}{2}$ at $x = 1$, and to $\frac{1}{2}[2 + \cos(3)]$ at $x = 3$.

$$\sum_{n=1}^{\infty}\left[-\frac{4}{n\pi}\cos\left(\frac{3n\pi}{5}\right) + \frac{2}{n\pi}\cos\left(\frac{n\pi}{5}\right) + \frac{\cos\left(\frac{3n\pi - 15}{5}\right) - \cos(n\pi - 5)}{n\pi - 5} + \frac{\cos\left(\frac{3n\pi + 15}{5}\right) - \cos(n\pi + 5)}{n\pi + 5}\right]\sin\left(\frac{n\pi x}{5}\right)$$

This converges to 1 for $0 < x < 1$, to 2 for $1 < x < 3$, to $\cos(x)$ for $3 < x < 5$, to $\frac{3}{2}$ at $x = 1$, to $\frac{1}{2}[2 + \cos(3)]$ at $x = 3$, and to 0 at $x = 0$ and $x = 5$.

15. $-3 + \sum_{n=1}^{\infty}\frac{16}{n^2\pi^2}[1 - (-1)^n]\cos\left(\frac{n\pi x}{4}\right)$ or, equivalently, $-3 + \sum_{n=1}^{\infty}\frac{32}{(2n-1)^2\pi^2}\cos\left(\frac{(2n-1)\pi x}{4}\right)$

This converges to $1 - 2x$ for $0 \le x \le 4$. $\sum_{n=1}^{\infty}\frac{2}{n\pi}[7(-1)^n + 1]\sin\left(\frac{n\pi x}{4}\right)$

This converges to $1 - 2x$ for $0 < x < 4$, and to 0 at $x = 0$ and $x = 4$.

17. $3\sin(1) + \sum_{n=1}^{\infty}\sin(1)\left(\frac{12}{4 - n^2\pi^2}\right)[1 + (-1)^n]\cos(n\pi x)$ or, equivalently, $3\sin(1) + 6\sin(1)\sum_{n=1}^{\infty}\frac{1}{1 - n^2\pi^2}\cos(2nx)$;

$\sum_{n=1}^{\infty}\frac{6n\pi\cos(1)}{n^2\pi^2 - 4}[(-1)^{n+1} + 1]\sin(n\pi x)$ or, equivalently, $\sum_{n=1}^{\infty}\frac{12(2n-1)\pi\cos(1)}{(2n-1)^2\pi^2 - 4}\sin[(2n-1)\pi x]$.

This converges to $3\cos(2x - 1)$ for $0 < x < 1$, and to 0 at $x = 0$ and $x = 1$.

19. $\frac{1 - e^{-2}}{2} + \sum_{n=1}^{\infty}\frac{4}{n^2\pi^2 + 4}[1 + (-1)^{n+1}e^{-2}]\cos\left(\frac{n\pi x}{2}\right)$ This converges to e^{-x} for $0 \le x \le 2$.

$\sum_{n=1}^{\infty}\frac{2n\pi}{4 + n^2\pi^2}[1 + (-1)^{n+1}e^{-2}]\sin\left(\frac{n\pi x}{2}\right)$ This converges to e^{-x} for $0 < x < 2$, and to zero at $x = 0$ and $x = 2$.

21. $\frac{2}{\pi}\left(\pi^2 + \frac{1}{3}\right) - \frac{16}{9\pi}\cos(3x) + \sum_{\substack{n=1 \\ n \ne 3}}^{\infty}\left\{\frac{8}{\pi n^2}[(-1)^n - 1] + \frac{1}{\pi}[1 + (-1)^n]\left[\frac{6}{9 - n^2}\right]\right\}\cos(nx)$

This converges to $4x + \sin(3x)$ for $0 \le x \le \pi$. $\sin(3x) + \sum_{\substack{n=1 \\ n \ne 3}}^{\infty}\frac{8}{\pi}(-1)^{n+1}\sin(nx)$ This converges to $4x + \sin(3x)$ for $0 < x < \pi$, and to 0 at $x = 0$ and $x = \pi$.

23. $\frac{1}{2} + \frac{1}{2}\cos(2\pi x)$ [just use a trig identity for $\cos^2(\pi x)$] This converges to $\cos^2(\pi x)$ for $0 \le x \le 1$.

$$\sum_{\substack{n=1 \\ n \ne 2}}^{\infty}\left\{\frac{1}{n\pi}[1 - (-1)^n] + \frac{1}{(n^2 - 4)\pi}[1 - (-1)^n]\right\}\sin(n\pi x)$$

This converges to $\cos^2(\pi x)$ for $0 < x < 1$, and to 0 at $x = 0$ and $x = 1$.

25. $\frac{19}{2} - \sum_{n=1}^{\infty}\frac{60}{(2n-1)^2\pi^2}\cos\left(\frac{(2n-1)\pi x}{5}\right)$

This converges to $3x + 2$ for $0 \le x \le 5$. $\sum_{n=1}^{\infty}\frac{34(-1)^{n+1} + 4}{n\pi}\sin\left(\frac{n\pi x}{5}\right)$

This converges to $3x + 2$ for $0 < x < 5$, and to 0 at $x = 0$ and $x = 5$.

27. The Fourier series for $|x|$ on $[-\pi, \pi]$ is (from Example 362)

$\frac{\pi}{2} + \sum_{n=1}^{\infty}\frac{-4}{(2n-1)^2\pi}\cos[(2n-1)x]$. Let $x = 0$. Then, by Theorem 96, $0 = \frac{\pi}{2} + \sum_{n=1}^{\infty}\frac{-4}{(2n-1)^2\pi}$.

Then, $\sum_{n=1}^{\infty}\frac{1}{(2n-1)^2} = \frac{\pi^2}{8}$.

29. The Fourier series for e^x on $[-\pi, \pi]$ is $\frac{1}{\pi}\sinh(\pi) + \sum_{n=1}^{\infty}\left[\frac{2}{\pi}\frac{(-1)^n}{1+n^2}\sinh(\pi)\cos(nx) + \frac{2n(-1)^{n+1}}{\pi(1+n^2)}\sinh(\pi)\sin(nx)\right]$

Let $x = 0$. By Theorem 95, the series converges to e^0, or 1. Then, $1 = \frac{1}{\pi}\sinh(\pi) + \sum_{n=1}^{\infty}\frac{2}{\pi}\frac{(-1)^n}{1+n^2}\sinh(\pi)$.

Solve this to get $\sum_{n=1}^{\infty}\frac{(-1)^n}{1+n^2} = \frac{\pi}{2\sinh(\pi)} - \frac{1}{2}$ (approximately -0.3640).

Alternatively, this shows that $\sum_{n=2}^{\infty}\frac{(-1)^n}{1+n^2} = \frac{\pi}{2\sinh(\pi)}$ (approximately 0.1360).

31. The Fourier series is $\frac{L}{2} + \sum_{n=1}^{\infty}\frac{2L}{(2n-1)\pi}\sin\left(\frac{(2n-1)\pi x}{L}\right)$ Choose $x = L/2$. By Theorem 95, the series converges to L at $x = L/2$. Then, $L = \frac{L}{2} + \sum_{n=1}^{\infty}\frac{2L}{(2n-1)\pi}\sin\left[(2n-1)\frac{\pi}{2}\right]$.

But $\sin[(2n-1)\pi/2] = (-1)^{n+1}$ for $n = 1, 2, 3, \ldots$. Then, upon multiplying by $\pi/2L$, we have $\frac{\pi}{4} = \sum_{n=1}^{\infty}\frac{(-1)^{n+1}}{2n-1}$.

Section 12.5

1. $\int_0^{\infty}\left[\frac{2\sin(\lambda\pi)}{\pi\lambda^2} - \frac{2}{\lambda}\cos(\lambda\pi)\right]\sin(\lambda x)\,d\lambda$

This converges to x for $-\pi < x < \pi$; to 0 for $|x| > \pi$; to $\pi/2$ for $x = \pi$; and to $-\pi/2$ for $x = -\pi$.

3. $\int_0^{\infty}\frac{2}{\lambda}[1 - \cos(\lambda\pi)]\sin(\lambda x)\,d\lambda$

This converges to -1 for $-\pi < x < 0$; to 1 for $0 < x < \pi$; to $\frac{1}{2}$ at $x = \pi$; to $-\frac{1}{2}$ at $x = -\pi$; and to 0 for $|x| > \pi$.

5. $\int_0^{\infty}\left[\frac{400\cos(100\lambda)}{\pi\lambda^2} + \left(\frac{20{,}000\lambda^2 - 2}{\pi\lambda^3}\right)\sin(100\lambda)\right]\cos(\lambda x)\,d\lambda$

This converges to x^2 for $|x| \le 100$; to 0 for $|x| > 100$; and to $10^4/2$ at $x = \pm 100$.

7. $\int_0^{\infty}\frac{2}{\pi}\left(\frac{\sin(\pi\lambda)}{1-\lambda^2}\right)\sin(\lambda x)\,d\lambda$; this converges to $\sin(x)$ for $|x| \le \pi$ and to zero for $|x| > \pi$.

9. $\int_0^{\infty}\left\{\frac{1}{\pi(1+\lambda^2)}[\cos(9\lambda)\cosh(9) - \cos(5\lambda)\cosh(5)] + \frac{\lambda}{\pi(1+\lambda^2)}[\sin(9\lambda)\sinh(9) - \sin(5\lambda)\sinh(5)]\right\}\cos(\lambda x)\,d\lambda$

$+ \int_0^{\infty}\left\{\frac{-\lambda}{\pi(1+\lambda^2)}[\cos(9\lambda)\sinh(9) + \cos(5\lambda)\sinh(5)] + \frac{1}{\pi(1+\lambda^2)}[\sin(9\lambda)\cosh(9) + \sin(5\lambda)\cosh(5)]\right\}\sin(\lambda x)\,d\lambda$

This converges to $\sinh(x)$ for $-5 < x < 9$, to zero for $x > 9$ and for $x < -5$, and to $\frac{1}{2}\sinh(9)$ at 9 and $-\frac{1}{2}\sinh(5)$ at -5.

11. A good table of integrals will give

$$A(\lambda) = \frac{1}{\pi}\int_{-\infty}^{\infty}\frac{\sin(t)\cos(\lambda t)}{t}\,dt = \begin{cases} 0, & \lambda > 1 \\ 1, & 0 < \lambda < 1 \\ \frac{1}{2}, & \lambda = 1 \end{cases}$$

Further, $B(\lambda) = \frac{1}{\pi}\int_{-\infty}^{\infty}\frac{\sin(t)\sin(\lambda t)}{t}\,dt = 0$, since the integrand is an odd function of t. Thus, the Fourier integral is

$\int_0^{\infty}A(\lambda)\cos(\lambda x)\,d\lambda$, which is just $\int_0^1\cos(\lambda x)\,d\lambda$.

This converges to $\sin(x)/x$ for $x \ne 0$, and to 1 for $x = 0$, a fact which can be verified by direct integration.

13. $\int_0^{\infty}\frac{2}{\pi\lambda}[1 - \cos(5\lambda)]\sin(\lambda x)\,d\lambda$

This converges to 1 for $0 < x < 5$, to -1 for $-1 < x < 0$, to 0 for $|x| > 5$; to $\frac{1}{2}$ at $x = 5$; and to $-\frac{1}{2}$ at $x = -5$

15. $\int_0^{\infty}\frac{2}{\pi}\left[\left(\frac{3\lambda^2 - 6}{\lambda^4}\right)\sin(\lambda) - \left(\frac{\lambda^2 - 6}{\lambda^3}\right)\cos(\lambda)\right]\sin(\lambda x)\,d\lambda$

This converges to x^3 for $|x| < 1$, to 0 for $|x| > 1$; to $\frac{1}{2}$ at $x = 1$ and to $-\frac{1}{2}$ at $x = -1$

Section 12.6

1. $\int_0^\infty \frac{2}{\pi\lambda^3}\,[20\lambda\cos(10\lambda) + (100\lambda^2 - 2)\sin(10\lambda)]\cos(\lambda x)\,d\lambda$

   $\int_0^\infty \frac{2}{\pi\lambda^3}\,[20\lambda\sin(10\lambda) - (100\lambda^2 - 2)\cos(10\lambda) - 2]\sin(\lambda x)\,d\lambda$

3. $\int_0^\infty \left(\frac{4}{\pi\lambda}\sin(4\lambda) - \frac{2}{\pi\lambda}\sin(\lambda)\right)\cos(\lambda x)\,d\lambda$

   $\int_0^\infty \left(\frac{2}{\pi\lambda}\cos(\lambda) - \frac{4}{\pi\lambda}\cos(4\lambda) + \frac{2}{\pi\lambda}\right)\sin(\lambda x)\,d\lambda$

5. $\int_0^\infty \left[\frac{4}{\pi\lambda^2}\cos(\lambda\pi) + \frac{4}{\lambda}\sin(\lambda\pi) - \frac{4}{\pi\lambda^2} + \frac{2}{\pi\lambda}\sin(10\pi\lambda) - \frac{2}{\pi\lambda}\sin(\lambda\pi) - \frac{2}{\pi\lambda}\sin(3\lambda\pi)\right]\cos(\lambda x)\,d\lambda$

   $\int_0^\infty \left[\frac{4}{\pi\lambda^2}\sin(\lambda\pi) - \frac{4}{\lambda}\cos(\lambda\pi) + \frac{2}{\pi\lambda}\cos(\lambda\pi) + \frac{2}{\pi\lambda} - \frac{2}{\pi\lambda}\cos(3\lambda\pi) - \frac{2}{\pi\lambda}\cos(10\lambda\pi)\right]\sin(\lambda x)\,d\lambda$

7. $\int_0^\infty \frac{1}{\pi}\left[\frac{1}{1 + (1-\lambda)^2} + \frac{1}{1 + (1+\lambda)^2}\right]\cos(\lambda x)\,d\lambda$

   $\int_0^\infty \frac{1}{\pi}\left[\frac{\lambda - 1}{1 + (\lambda-1)^2} + \frac{\lambda + 1}{1 + (\lambda+1)^2}\right]\sin(\lambda x)\,d\lambda$

9. $\int_0^\infty \frac{2}{\pi}\frac{\sin(10\lambda)}{\lambda}\cos(\lambda x)\,d\lambda;\quad \int_0^\infty \frac{2}{\pi\lambda}\,[1 - \cos(10\lambda)]\sin(\lambda x)\,d\lambda$

11. Suppose that $\int_0^\infty |f(x)|\,dx$ converges and that $f(x)$ satisfies the hypotheses of Theorem 95 on each interval $[0, L]$, $L > 0$.

    Then, the Fourier cosine integral of $f(x)$ converges to $\frac{1}{2}[f(x+) + f(x-)]$ whenever $x > 0$ and the left and right derivatives exist at x; and it converges to $f(0+)$ at 0 if $f(x)$ has a right derivative at 0.

    The Fourier sine integral converges to $\frac{1}{2}[f(x+) + f(x-)]$ whenever $x > 0$ and $f(x)$ has left and right derivatives at x; and it converges to zero at $x = 0$.

Section 12.7

1. $0.2821 + 0.4394\cos(x) + 0.2075\cos(2x) + 0.0595\cos(3x) + 0.0103\cos(4x) + 0.0011\cos(5x) + 0.0001\cos(6x)$.
   (The other cosine coefficients are zero to four decimal places.)

   $0.02702\sin(x) + 0.3424\sin(2x) + 0.2725\sin(3x) + 0.1916\sin(4x) + 0.1418\sin(5x) + 0.1132\sin(6x) + 0.0949\sin(7x) + 0.0819\sin(8x) + 0.0722\sin(9x) + 0.0645\sin(10x)$

3. $3.5530 - 3.7865\cos(\pi x/2) + 1.0290\cos(\pi x) - 0.6339\cos(3\pi x/2) + 0.2733\cos(2\pi x) - 0.2438\cos(5\pi x/2) + 0.1204\cos(3\pi x) - 0.1296\cos(7\pi x/2) + 0.0666\cos(4\pi x) - 0.0813\cos(9\pi x/2) + 0.0418\cos(5\pi x)$

   $4.0547\sin(\pi x/2) - 2.8441\sin(\pi x) + 2.1129\sin(3\pi x/2) - 1.5669\sin(2\pi x) + 1.3114\sin(5\pi x/2) - 1.0601\sin(3\pi x) + 0.9383\sin(7\pi x/2) - 0.7943\sin(4\pi x) + 0.7244\sin(9\pi x/2) - 0.6305\sin(5\pi x)$

5. $0.1674 + 0.0667\cos(x) - 0.0478\cos(2x) - 0.0330\cos(3x) - 0.0287\cos(4x) - 0.0183\cos(5x) - 0.0157\cos(6x) - 0.0108\cos(7x) - 0.0096\cos(8x) - 0.0070\cos(9x) - 0.0065\cos(10x)$

   $0.2368\sin(x) - 0.0768\sin(2x) + 0.0537\sin(3x) + 0.0118\sin(4x) + 0.0196\sin(5x) + 0.0011\sin(6x) + 0.0103\sin(7x) - 0.0014\sin(8x) + 0.0066\sin(9x) - 0.0020\sin(10x)$.

7. $1.1173 - 0.2846\cos(\pi x) + 0.0977\cos(2\pi x) - 0.0472\cos(3\pi x) + 0.0275\cos(5\pi x) + 0.0126\cos(6\pi x) - 0.0094\cos(7\pi x) + 0.0072\cos(8\pi x) - 0.0058\cos(9\pi x) + 0.0047\cos(10\pi x)$

   $1.5047\sin(\pi x) - 0.2141\sin(2\pi x) + 0.5718\sin(3\pi x) - 0.1178\sin(4\pi x) + 0.3468\sin(5\pi x) - 0.0797\sin(6\pi x) + 0.2469\sin(7\pi x) - 0.0597\sin(8\pi x) + 0.1903\sin(9\pi x) - 0.0473\sin(10\pi x)$.

9. $0.4729 - 0.2737\cos(2x) - 0.0760\cos(4x) - 0.0349\cos(6x) - 0.0200\cos(8x) - 0.0130\cos(10x)$

   $0.7268\sin(x) + 0.0407\sin(3x) + 0.0095\sin(5x) + 0.0036\sin(7x) + 0.0017\sin(9x)$.

11. $0.2664 + 0.1804\cos(\pi x/3) + 0.0317\cos(2\pi x/3) + 0.0119\cos(\pi x) + 0.0002\cos(4\pi x/3) + 0.0016\cos(5\pi x/3) - 0.0007\cos(2\pi x) + 0.0006\cos(7\pi x/3) - 0.0005\cos(8\pi x/3) + 0.0004\cos(3\pi x) - 0.0003\cos(10\pi x/3)$

    $0.3256\sin(\pi x/3) + 0.1536\sin(2\pi x/3) + 0.1370\sin(\pi x) + 0.0712\sin(4\pi x/3) + 0.0775\sin(5\pi x/3) + 0.0446\sin(2\pi x) + 0.0538\sin(7\pi x/3) + 0.0325\sin(8\pi x/3) + 0.0411\sin(3\pi x) + 0.0255\sin(10\pi x/3)$.

13. $5.0251 - 3.2028\cos(\pi x/5) + 0.1887\cos(2\pi x/5) - 0.4030\cos(3\pi x/5) + 0.0297\cos(4\pi x/5) - 0.1569\cos(\pi x) + 0.0062\cos(6\pi x/5) - 0.0858\cos(7\pi x/5) - 0.0002\cos(8\pi x/5) - 0.0554\cos(9\pi x/5) - 0.0025\cos(2\pi x)$
$6.3133\sin(\pi x/5) - 2.4818\sin(2\pi x/5) + 2.2539\sin(3\pi x/5) - 1.2623\sin(4\pi x/5) + 1.3445\sin(\pi x) - 0.8430\sin(6\pi x/5) + 0.9505\sin(7\pi x/5) - 0.6294\sin(8\pi x/5) + 0.7298\sin(9\pi x/5) - 0.4990\sin(2\pi x)$

15. $1.4050 - 0.1956\cos(\pi x/2) - 0.1753\cos(\pi x) - 0.9179\cos(3\pi x/2) - 0.0341\cos(2\pi x) + 0.0627\cos(5\pi x/2) + 0.1163\cos(3\pi x) + 0.0535\cos(7\pi x/2) - 0.0524\cos(4\pi x) - 0.1055\cos(9\pi x/2) - 0.0139\cos(5\pi x)$
$1.8624\sin(\pi x/2) - 0.1064\sin(\pi x) + 0.4663\sin(3\pi x/2) - 0.2009\sin(2\pi x) + 0.2186\sin(5\pi x/2) - 0.0548\sin(3\pi x) + 0.3180\sin(7\pi x/2) + 0.0499\sin(4\pi x) + 0.1575\sin(9\pi x/2) - 0.1315\sin(5\pi x)$

17. $14.8582 - 10.4769\cos(x) + 1.0117\cos(2x) - 0.2426\cos(3x) + 0.4185\cos(4x) - 0.1337\cos(5x) + 0.1455\cos(6x) - 0.0877\cos(7x) + 0.0763\cos(8x) - 0.0557\cos(9x) + 0.0486\cos(10x)$
$18.4378\sin(x) - 8.7350\sin(2x) + 6.7990\sin(3x) - 3.6169\sin(4x) + 4.2458\sin(5x) - 2.3512\sin(6x) + 3.0051\sin(7x) - 1.7486\sin(8x) + 2.3093\sin(9x) - 1.3817\sin(10x)$

19. $1.5940 - 0.0027\cos(2.8986\pi x) + 0.1205\cos(5.7971\pi x) + 0.0238\cos(8.6957\pi x) + 0.0413\cos(11.5942\pi x) + 0.0149\cos(14.4928\pi x) + 0.0225\cos(17.3913\pi x) + 0.0104\cos(20.2899\pi x) + 0.0149\cos(23.1884\pi x) + 0.0080\cos(26.0870\pi x) + 0.0109\cos(28.9855\pi x)$
$1.9726\sin(2.8986\pi x) - 0.0176\sin(5.7971\pi x) + 0.7381\sin(8.6957\pi x) + 0.0046\sin(11.5942\pi x) + 0.4499\sin(14.4928\pi x) + 0.0070\sin(17.3913\pi x) + 0.3218\sin(20.2899\pi x) + 0.0069\sin(23.1884\pi x) + 0.2490\sin(26.0870\pi x) + 0.0064\sin(28.9855\pi x)$

Section 12.8

1. $\sum_{n=1}^{\infty}\sum_{m=1}^{\infty}\frac{4}{\pi}\left[\left(\frac{2-n^2\pi^2}{n^3}\right)(-1)^n - \frac{2}{n^3}\right]\left[\frac{(-1)^{m+1}}{m}\right]\sin(nx)\sin(my)$

3. $\sum_{n=1}^{\infty}\sum_{m=1}^{\infty}\left\{\frac{12}{nm\pi^2}(-1)^{n+1}[1-(-1)^n] + \frac{4}{nm^3\pi^4}[1-(-1)^n][1-(-1)^m][2(m^2\pi^2-8)]\right\}\sin(n\pi x)\sin\left(\frac{m\pi y}{2}\right)$

5. $\sum_{n=1}^{\infty}\sum_{m=1}^{\infty}4mn\pi^2\left(\frac{e^2(-1)^{m+1}+1}{4+m^2\pi^2}\right)\left(\frac{e^4(-1)^{n+1}+1}{16+n^2\pi^2}\right)\sin\left(\frac{n\pi x}{4}\right)\sin\left(\frac{m\pi y}{2}\right)$

7. $\sum_{n=1}^{\infty}\sum_{m=1}^{\infty}\left(\frac{16(-1)^n(-1)^m}{n(4+m^2\pi^2)}\right)[m\sinh(2)]\sin\left(\frac{n\pi x}{4}\right)\sin\left(\frac{m\pi y}{2}\right)$

9. $\sum_{n=1}^{\infty}\sum_{m=1}^{\infty}\frac{-8}{m}\left[\frac{\sin(8-n\pi)}{8-n\pi} - \frac{\sin(8+n\pi)}{8+n\pi}\right][-1)^m\sin\left(\frac{n\pi x}{4}\right)\sin(my)$

11. $\sum_{n=1}^{\infty}\sum_{m=1}^{\infty}\left\{\frac{4}{mn\pi^2}[(-1)^m-1](-1)^n - \frac{32}{mn\pi^2}[(-1)^n-1](-1)^m\right\}\sin(n\pi x)\sin\left(\frac{m\pi y}{2}\right)$

13. $\frac{8\pi^3}{3} + \sum_{m=1}^{\infty}\frac{32\pi}{m^2}(-1)^m\cos\left(\frac{my}{2}\right) + \sum_{n=1}^{\infty}\sum_{m=1}^{\infty}\frac{128}{\pi n^2m^2}[(-1)^n-1](-1)^m\cos\left(\frac{nx}{2}\right)\cos\left(\frac{my}{2}\right)$

15. $\sum_{m=1}^{\infty}\frac{2}{m^2\pi^2}[(-1)^m-1]\cos(x)\cos(m\pi y)$

17. $\frac{\pi}{4} + \sum_{m=1}^{\infty}\frac{1-(-1)^m}{m^2\pi}\cos(m\pi y) + \sum_{n=1}^{\infty}\sum_{m=1}^{\infty} - \frac{4}{n^2m^2\pi^3}[(-1)^n-1][(-1)^m-1]\cos(nx)\cos(m\pi y)$

19. 3

21. $-\frac{1}{2} + \sum_{m=1}^{\infty}\frac{12}{\pi^2}\frac{1}{(2m-1)^2}\cos[(2m-1)\pi y] + \sum_{n=1}^{\infty}\sum_{m=1}^{\infty}\frac{72}{n^2m^2\pi^4}(-1)^n[1-(-1)^m]\cos\left(\frac{n\pi x}{3}\right)\cos(m\pi y)$

23. $\frac{100}{9} + \sum_{m=1}^{\infty}\frac{400(-1)^m}{3m^2\pi^2}\cos\left(\frac{m\pi y}{5}\right) + \sum_{n=1}^{\infty}\sum_{m=1}^{\infty}\frac{1600}{n^2m^2\pi^4}(-1)^n(-1)^m\cos\left(\frac{n\pi x}{2}\right)\cos\left(\frac{m\pi y}{5}\right)$

25. $\sum_{n=1}^{\infty}\frac{4\sinh(1)}{n\pi}(-1)^{n+1}\sin\left(\frac{n\pi x}{2}\right) + \sum_{n=1}^{\infty}\sum_{m=1}^{\infty}\left(\frac{8}{n\pi}\frac{(-1)^n(-1)^m}{1+m^2\pi^2}\sinh(1)\right)\left[-\sin\left(\frac{n\pi x}{2}\right)\cos(m\pi y) + m\pi\sin\left(\frac{n\pi x}{2}\right)\sin(m\pi y)\right]$

Section 12.9

1. $\pi(-1)^n\left(\frac{6}{n^3} - \frac{\pi^2}{n}\right)$

3. $\frac{n}{n^2+1}[1-(-1)^n e^\pi]$

5. $\dfrac{n}{n^2-\alpha^2}[1-(-1)^n\cos(\alpha\pi)]$ for $\alpha \neq 1, 2, 3, \ldots$; 0 if $\alpha = 1, 2, 3, \ldots$.

7. $\dfrac{n}{n^2+1}[1-(-1)^n e^{-\pi}]$ **9.** $f_c(n) = \begin{cases} \dfrac{2\pi(-1)^n}{n^2} & \text{for } n = 1, 2, 3, \ldots \\ \dfrac{\pi^3}{3} & \text{for } n = 0 \end{cases}$ **11.** $\dfrac{(-1)^n e - 1}{1+n^2}$

13. If $\alpha = K = 1, 2, \ldots,$ $f_c(n) = \begin{cases} \pi/2 & \text{for } n = K \\ 0 & \text{for } n \neq K \end{cases}$; If $\alpha = 0$, $f_c(n) = \begin{cases} \pi, & n = 0 \\ 0, & n = 1, 2, \ldots \end{cases}$

If $\alpha \neq$ integer, then $f_c(n) = \dfrac{\alpha(-1)^n \sin(\alpha\pi)}{\alpha^2 - n^2}$

15. Let $y(x) = S_n^{-1}\left(\dfrac{f_s(n)}{n^2}\right)$. Then $y_s(n) = \dfrac{f_s(n)}{n^2}$; so $n^2 y_s(n) = f_s(n)$. Use this to show that $y(x)$ must satisfy $y''(x) = f(x)$; $y(0) = y(\pi) = 0$. Finally, check that $\dfrac{x}{\pi}\displaystyle\int_0^\pi (\pi - t)f(t)\,dt - \int_0^x (x-t)f(t)\,dt$ satisfies these conditions, and hence must be $y(x)$.

Section 12.10

1. $\dfrac{1}{1+\lambda^2}$ **3.** $\dfrac{2k\cos(\lambda k)}{\lambda^2} + \left(\dfrac{\lambda^2k^2-2}{\lambda^3}\right)\sin(\lambda k)$ **5.** $\dfrac{2}{\lambda^2}\cos(5\lambda) + \dfrac{10}{\lambda}\sin(5\lambda) - \dfrac{2}{\lambda^2}$ **7.** $\dfrac{a^2-\lambda^2}{(a^2+\lambda^2)^2}$

9. $\dfrac{\sqrt{\pi}}{2\sqrt{a}}e^{-\lambda^2/4a}$ **11.** $\dfrac{\pi}{2}e^{-\lambda}$ if $\lambda > 0$; $\dfrac{\pi}{2}e^{\lambda}$ if $\lambda < 0$ **13.** $\dfrac{2}{\lambda}[\cos(2\lambda k) - 2\cos(\lambda k) + 1]$ **15.** $\dfrac{2\lambda}{(1+\lambda^2)^2}$

17. $\dfrac{1}{2}\left[\dfrac{1}{1+(1-\lambda)^2} - \dfrac{1}{1+(1+\lambda)^2}\right]$ **19.** $\hat{f}(\lambda) = \begin{cases} 0 & \text{if } 0 < \lambda < 1 \\ \pi/4 & \text{if } \lambda = 1 \\ \pi/2 & \text{if } \lambda > 1 \end{cases}$

21. $\hat{f}(\lambda) = \begin{cases} \dfrac{1}{2} - \dfrac{1}{4\lambda}\sin(2\lambda) & \text{if } \lambda = k \\ \dfrac{\sin(k-\lambda)}{2(k-\lambda)} - \dfrac{\sin(k+\lambda)}{2(k+\lambda)} & \text{if } \lambda \neq k \end{cases}$ **23.** $-\dfrac{2i}{\lambda} - \dfrac{2}{\lambda}\sin(\lambda)$ **25.** $\dfrac{2a}{a^2+\lambda^2}$

27. $\dfrac{2}{1+\lambda^2}\sinh(k)\cos(k) + \dfrac{2\lambda}{1+\lambda^2}\cosh(k)\sin(k)$ [This uses the relations $\cos(A) = \frac{1}{2}(e^{iA} + e^{-iA})$ and $\sin(A) = \dfrac{1}{2i}(e^{iA} - e^{-iA})$.] **29.** $\pi e^{-|\lambda|}$

Supplementary Problems

1. $15 + \displaystyle\sum_{n=1}^{\infty} \frac{192}{n^2\pi^2}(-1)^n \cos\left(\frac{n\pi x}{4}\right)$ This converges to $3x^2 - 1$ for $-4 \leq x \leq 4$.

3. $\dfrac{1}{6}\sinh(6) + \displaystyle\sum_{n=1}^{\infty} \frac{12}{36+n^2\pi^2}(-1)^n \sinh(6)\cos\left(\frac{n\pi x}{3}\right)$ This converges to $\cosh(2x)$ for $-3 \leq x \leq 3$.

5. $\dfrac{16}{3} + \displaystyle\sum_{n=1}^{\infty}\left[\frac{16}{n^2\pi^2}(-1)^n\cos\left(\frac{n\pi x}{2}\right) - \frac{16}{n\pi}(-1)^n\sin\left(\frac{n\pi x}{2}\right)\right]$

This converges to $(x+2)^2$ for $-2 < x < 2$, and to 8 at $x = 2$ and $x = -2$.

7. $\dfrac{5}{2} + \dfrac{1}{\pi} + \left(\dfrac{6}{\pi} - \dfrac{1}{2}\right)\sin(\pi x) + \displaystyle\sum_{n=2}^{\infty}\frac{2}{\pi}\frac{1}{1-(2n-1)^2}\cos[(2n-1)\pi x] + \sum_{n=2}^{\infty}\frac{6}{(2n-1)\pi}\sin[(2n-1)\pi x]$ This converges to $1 - \sin(\pi x)$ for $-1 < x < 0$, to 4 for $0 < x < 1$, to $\frac{5}{2}$ at $x = 0$, and to $\frac{5}{2}$ at $x = -1$ and $x = 1$.

9. $\dfrac{7}{8} - \displaystyle\sum_{n=1}^{\infty}\frac{2}{n\pi}\sin\left(\frac{n\pi}{8}\right)\cos\left(\frac{n\pi x}{4}\right)$ This converges to 1 for $-4 \leq x < -\frac{1}{2}$ and for $\frac{1}{2} < x \leq 4$, to zero for $-\frac{1}{2} < x < \frac{1}{2}$, and to $\frac{1}{2}$ at $x = -\frac{1}{2}$ and $x = \frac{1}{2}$.

11. $\frac{15}{16} + \sum_{n=1}^{\infty} \left\{ -\frac{1}{2n\pi} \sin\left(\frac{n\pi}{4}\right) - \frac{2}{n\pi} \sin\left(\frac{n\pi}{2}\right) + \frac{4}{n^2\pi^2}\left[(-1)^n - \cos\left(\frac{n\pi}{2}\right)\right]\right\} \cos\left(\frac{n\pi x}{4}\right) + \sum_{n=1}^{\infty} \left\{\frac{1}{2n\pi}\left[(-1)^n - \cos\left(\frac{n\pi}{4}\right)\right] - \frac{4}{n^2\pi^2} \sin\left(\frac{n\pi}{2}\right) + \frac{2}{n\pi}\left[\cos\left(\frac{n\pi}{2}\right) - 2(-1)^n\right]\right\} \sin\left(\frac{n\pi x}{4}\right)$ This converges to $\frac{1}{2}$ for $-4 < x < -1$, to 0 for $-1 < x < 2$, to x for $2 < x < 4$, to $\frac{1}{4}$ at $x = -1$, to 1 at $x = 2$, and to $\frac{9}{4}$ at $x = -4$ and $x = 4$.

13. $\sum_{n=1}^{\infty} \frac{2}{n^3\pi^3} \{(n^2\pi^2 - 2)(-1)^n + 2\} \sin(n\pi x)$ This converges to x^2 for $-1 < x \le 0$, to $-x^2$ for $0 \le x < 1$, and to zero at $x = 0$, $x = -1$, and $x = 1$.

15. $\frac{1}{6}(1 + e^2 - e^3) + \sum_{n=1}^{\infty} \left\{ -\frac{1}{n\pi} \sin\left(\frac{2n\pi}{3}\right) + \frac{3e^2}{9 + n^2\pi^2}\left[\cos\left(\frac{2n\pi}{3}\right) + \frac{n\pi}{3}\sin\left(\frac{2n\pi}{3}\right)\right] - \frac{3e^3(-1)^n}{9 + n^2\pi^2}\right\} \cos\left(\frac{n\pi x}{3}\right) + \sum_{n=1}^{\infty} \left\{ -\frac{1}{n\pi}\left[\cos\left(\frac{2n\pi}{3}\right) - (-1)^n\right] - \frac{3e^2}{9 + n^2\pi^2}\left[\sin\left(\frac{2n\pi}{3}\right) - \frac{n\pi}{3}\cos\left(\frac{2n\pi}{3}\right)\right] - \frac{e^3 n\pi}{9 + n^2\pi^2}(-1)^n\right\} \sin\left(\frac{n\pi x}{3}\right)$
This converges to $1 - e^{-x}$ for $-3 < x < -2$, to zero for $-2 < x < 3$, to $1 - e^2$ at $x = -2$, and to $(1 - e^3)/2$ at $x = 3$ and $x = -3$.

17. $\frac{2}{\pi} + \frac{4}{\pi} \sum_{n=1}^{\infty} \frac{1}{1 - 4n^2} \cos(2nx)$ This converges to $\sin(x)$ for $0 \le x \le \pi$.

19. $\frac{3}{8}(e^{-8} - 1) - \sum_{n=1}^{\infty} \frac{48}{64 + n^2\pi^2} [1 - (-1)^n e^{-8}] \cos\left(\frac{n\pi x}{4}\right)$ This converges to $-3e^{-2x}$ for $0 \le x \le 4$.

21. $\frac{23}{12} + \sum_{n=1}^{\infty} \left\{\frac{16}{n^2\pi^2} \cos\left(\frac{n\pi}{4}\right) + \frac{(2n^2\pi^2 - 64)}{n^3\pi^3} \sin\left(\frac{n\pi}{4}\right) + \frac{16}{n^2\pi^2}[(-1)^n - 1] - \frac{4}{n\pi}\sin\left(\frac{n\pi}{4}\right)\right\} \cos\left(\frac{n\pi x}{4}\right)$
This converges to x^2 for $0 \le x < 1$, to $\frac{3}{2}$ at $x = 1$, and to $2x$ for $1 < x \le 4$.

23. $\frac{229}{36} + \sum_{n=1}^{\infty} \left\{\frac{-6}{n^2\pi^2} \cos\left(\frac{2n\pi}{3}\right) - \frac{6}{n^2\pi^2} + \frac{36}{n^3\pi^3}\sin\left(\frac{n\pi}{3}\right) + \frac{180}{n^3\pi^3}\sin\left(\frac{2n\pi}{3}\right) - \frac{48}{n^2\pi^2}\cos\left(\frac{2n\pi}{3}\right) + \frac{324}{n^4\pi^4}\left[\cos\left(\frac{2n\pi}{3}\right) - (-1)^n\right] + \frac{162}{n^2\pi^2}(-1)^n\right\} \cos\left(\frac{n\pi x}{3}\right)$ This converges to x for $0 \le x < 1$, to x^2 for $1 < x < 2$, to x^3 for $2 < x \le 3$, to 1 at $x = 1$, and to 6 at $x = 2$.

25. $\sum_{\substack{n=1 \\ n \ne 15 \\ n \ne 20}}^{\infty} \left\{\frac{4n}{\pi} \sin\left(\frac{2n\pi}{5}\right)\left(\frac{1}{225 - n^2}\right) + \frac{200}{\pi(400 - n^2)}\left[\cos\left(\frac{2n\pi}{5}\right) - (-1)^n\right]\right\} \cos\left(\frac{nx}{5}\right) + \frac{4}{5}\left(\frac{1}{\pi} - 1\right)\cos(3x)$

This converges to $-2\cos(3x)$ for $0 \le x \le 2\pi$, to $\sin(4x)$ for $2\pi < x \le 5\pi$, and to -1 at $x = 2\pi$.

27. $\sum_{n=1}^{\infty} \frac{2n\pi(-1)^{n+1}}{9 + n^2\pi^2} \sinh(3) \sin(n\pi x)$ This converges to $\sinh(3x)$ for $0 < x < 1$, and to zero at $x = 0$ and $x = 1$.

29. $\sum_{n=1}^{\infty} \left\{\frac{2n\pi}{225 + n^2\pi^2}[1 - e^{-15}(-1)^n]\right\} \sin\left(\frac{n\pi x}{5}\right)$ This converges to e^{-3x} for $0 < x < 5$, and to zero at $x = 0$ and $x = 5$.

31. $\sum_{n=1}^{\infty} \left\{\frac{4}{n\pi} + \frac{2}{n\pi}(-1)^n - \frac{6}{n\pi}\cos\left(\frac{n\pi}{10}\right)\right\} \sin\left(\frac{n\pi x}{10}\right)$ This converges to 2 for $0 < x < 1$, to -1 for $1 < x < 10$, to $\frac{1}{2}$ at $x = 1$, and to zero at $x = 0$ and $x = 10$.

33. $\frac{2}{\pi}\sin(x) + \sum_{\substack{n=1 \\ n \ne 2}}^{\infty} \left\{\frac{2}{n\pi}\left[1 - \cos\left(\frac{n\pi}{2}\right)\right] - \frac{2n}{\pi(n^2 - 4)}\left[(-1)^n + \cos\left(\frac{n\pi}{2}\right)\right]\right\} \sin\left(\frac{nx}{2}\right)$ This converges to 1 for $0 < x < \pi$, to $\cos(x)$ for $\pi < x < 2\pi$, and to zero at $x = 0$, $x = \pi$, and $x = 2\pi$.

35. $\sum_{n=1}^{\infty} \left[\frac{6}{n\pi}\cos\left(\frac{3n\pi}{5}\right) - \frac{2}{n\pi} - \frac{4}{n\pi}(-1)^n\right] \sin\left(\frac{n\pi x}{5}\right)$ This converges to -1 for $0 < x < 3$, to 2 for $3 < x < 5$, to $\frac{1}{2}$ at $x = 3$, and to zero at $x = 0$ and $x = 5$.

37. $\sum_{n=1}^{\infty} \left(\frac{1}{\pi}\frac{3}{36 + n^2}\right)\left[4\sin\left(\frac{n\pi}{3}\right)\sinh(2\pi) - \frac{2n}{3}\cos\left(\frac{n\pi}{3}\right)\cosh(2\pi) + \frac{2n}{3}\right] \sin\left(\frac{nx}{3}\right)$ This converges to $\cosh(2x)$ for $0 < x < \pi$, to $\frac{1}{2}\cosh(2\pi)$ at $x = \pi$, and to zero for $\pi < x \le 3\pi$ and $x = 0$.

39. $\displaystyle\sum_{n=1}^{\infty} \frac{\sin\left(\frac{12-3n\pi}{4}\right)}{4-n\pi} - \frac{\sin\left(\frac{12+3n\pi}{4}\right)}{4+n\pi} - \frac{32}{n\pi}(-1)^n + \frac{8}{16-n^2\pi^2} + \frac{12}{n\pi}\cos\left(\frac{3n\pi}{4}\right) - \frac{16}{n^2\pi^2}\sin\left(\frac{3n\pi}{4}\right) + \frac{1}{n\pi-4}$ $\displaystyle\times\left[\cos\left(\frac{3n\pi-12}{4}\right) + \cos\left(\frac{3n\pi+12}{4}\right)\right]\Bigg\}\sin\left(\frac{n\pi x}{4}\right)$ This converges to $\sin(x) - \cos(x)$ for $0 < x < 3$, to $4x$ for $3 < x < 4$, to $\frac{1}{2}[\sin(3) - \cos(3) + 12]$ at $x = 3$, and to zero at $x = 0$ and $x = 4$.

41. $\displaystyle\frac{2}{\pi}\int_0^{\infty} \frac{1}{1+\lambda^2}[\cos(5\lambda)\sinh(5) + \lambda\sin(5\lambda)\cosh(5)]\cos(\lambda x)$
This converges to $\cosh(x)$ for $-5 \le x \le 5$, and to zero for $|x| > 5$.

43. $\displaystyle\frac{2}{\pi}\int_0^{\infty}\left\{\frac{e^{-3}}{1+\lambda^2}[\cos(3\lambda) - \lambda\sin(3\lambda)]\cos(\lambda x) + \frac{2}{\pi\lambda}[1-\cos(3\lambda)]\sin(\lambda x)\,d\lambda\right.$ This converges to $e^{-|x|}$ for $|x| > 3$, to -1 for $-3 < x < 0$, to 1 for $0 < x < 3$, to $(-1 + e^{-3})/2$ at $x = -3$; to $(1 + e^{-3})/2$ at $x = 3$; and to 0 at $x = 0$.

45. $\displaystyle\frac{1}{2\pi}\int_0^{\infty}\left(\frac{1}{1-\lambda}\{\sin[(1-\lambda)\pi] - \cos[(1-\lambda)\pi] + 1\} + \frac{1}{1+\lambda}\{\sin[(1+\lambda)\pi] - \cos[(1+\lambda)\pi] + 1\}\right)\cos(\lambda x)\,d\lambda$
$\displaystyle+\frac{1}{2\pi}\int_0^{\infty}\left(\frac{1}{1-\lambda}\{\cos[(1-\lambda)x] + \sin[(1-\lambda)x] + 1\} + \frac{1}{1+\lambda}\{\cos[(1+\lambda)\pi] - \sin[(1+\lambda)\pi] - 1\}\right)\sin(\lambda x)\,d\lambda$
This converges to $\sin(x)$ for $0 < x < \pi$, to $\cos(x)$ for $-\pi < x < 0$, to zero for $|x| > \pi$, to $\frac{1}{2}$ at $x = 0$, to zero at $x = \pi$, and to $-\frac{1}{2}$ at $x = -\pi$.

47. $\displaystyle\int_0^{\infty}\left(\frac{2}{\pi\lambda}\sin(10\lambda) + \frac{1}{\pi(1-\lambda)}\{\sin[20(1-\lambda)] - \sin[10(1-\lambda)]\} + \frac{1}{\pi(1+\lambda)}\{\sin[20(1+\lambda)] - \sin[10(1+\lambda]\}\right)\cos(\lambda x)\,d\lambda$
This converges to 1 for $0 < x < 10$, to $\cos(x)$ for $10 < x < 20$, to zero for $x > 20$, to 1 at $x = 0$, to $\frac{1}{2}[1 + \cos(10)]$ at $x = 10$, and to $\frac{1}{2}\cos(20)$ at $x = 20$. $\displaystyle\int_0^{\infty}\left(\frac{2}{\pi\lambda}[1-\cos(10\lambda)] + \frac{1}{\pi(1-\lambda)}\{\cos[10(1-\lambda)] - \cos[20(1-\lambda)]\}\right.$ $\displaystyle\left.+\frac{1}{\pi(1+\lambda)}\{\cos[10(1+\lambda)] - \cos[20(1+\lambda)]\}\right)\sin(\lambda x)\,d\lambda$ This converges to 1 for $0 < x < 10$, to $\cos(x)$ for $10 < x < 20$, to 0 for $x > 20$, to 0 at $x = 0$, to $\frac{1}{2}[1 + \cos(10)]$ at $x = 10$, and to $\frac{1}{2}\cos(20)$ at $x = 20$.

49. $\displaystyle\frac{1}{\pi}\int_0^{\infty}\frac{2}{4+\lambda^2}[2\cos(\lambda)\sinh(2) + \lambda\sin(\lambda)\cosh(2) + 2\cos(3\lambda)\cosh(6) - 2\cos(2\lambda)\cosh(4) + \lambda\sin(3\lambda)\sinh(6)$ $-\lambda\sin(2\lambda)\sinh(4)]\cos(\lambda x)\,d\lambda$ This converges to $\cosh(2x)$ for $0 < x < 1$, to 0 for $1 < x < 2$, to $\sinh(2x)$ for $2 < x < 3$, to zero for $x > 3$, to 1 at $x = 0$, to $\frac{1}{2}\cosh(2)$ at $x = 1$, to $\frac{1}{2}\sinh(4)$ at $x = 2$, and to $\frac{1}{2}\sinh(6)$ at $x = 3$.
$\displaystyle\frac{1}{\pi}\int_0^{\infty}\frac{2}{4+\lambda^2}[\lambda - \lambda\cos(\lambda)\cosh(2) + 2\sin(3\lambda)\cosh(6) - \lambda\cos(3\lambda)\sinh(6) - 2\sin(2\lambda)\cosh(4) + \lambda\cos(2\lambda)\sinh(4)]$ $\sin(\lambda x)\,d\lambda$ This converges to $\cosh(2x)$ for $0 < x < 1$, to 0 for $1 < x < 2$ and for $x > 3$, to $\sinh(2x)$ for $2 < x < 3$, to zero at $x = 0$, to $\frac{1}{2}\cosh(2)$ at $x = 1$, to $\frac{1}{2}\sinh(4)$ at $x = 2$, and to $\frac{1}{2}\sinh(6)$ at $x = 3$.

51. $\displaystyle\sum_{n=1}^{\infty}\sum_{m=1}^{\infty} 4\left[\left(\frac{n^2\pi^2-2}{n^3\pi^3}\right)(-1)^n + \frac{2}{n^3\pi^3}\right]\left[\frac{4}{m\pi}(-1)^m + \frac{1}{m\pi}\right]\sin(n\pi x)\sin\left(\frac{m\pi y}{5}\right)$; $\displaystyle\frac{1}{2} + \sum_{m=1}^{\infty}\frac{10}{3}\left(\frac{(-1)^m - 1}{m^2\pi^2}\right)\cos\left(\frac{m\pi y}{5}\right)$ $\displaystyle+\sum_{n=1}^{\infty}\sum_{m=1}^{\infty}\frac{40}{n^2m^2\pi^4}(-1)^n[(-1)^m - 1]\cos(n\pi x)\cos\left(\frac{m\pi y}{5}\right)$

53. $\displaystyle\sum_{m=1}^{\infty}\frac{8}{\pi}\left(\frac{m}{4m^2-1}\right)\sin(\pi x)\sin(2m\pi y)$; $\displaystyle\frac{2}{\pi}\cos(\pi y) + \sum_{n=1}^{\infty}\frac{4}{\pi}\left(\frac{1}{1-4n^2}\right)\cos(2n\pi x)\cos(\pi y)$

55. $\displaystyle 4\pi^2\sum_{n=1}^{\infty}\sum_{m=1}^{\infty}\frac{nm}{(1+n^2\pi^2)(1+m^2\pi^2)}[1-(-1)^n e^{-1}][1-(-1)^m e^{-1}]\sin(n\pi x)\sin(m\pi y)$; $\displaystyle\left(1-\frac{1}{e}\right)^2 + \sum_{m=1}^{\infty}2\left(1-\frac{1}{e}\right)$ $\displaystyle\times\frac{1}{1+m^2\pi^2}[1-e^{-1}(-1)^m]\cos(m\pi y) + 4\sum_{n=1}^{\infty}\sum_{m=1}^{\infty}\frac{1}{(1+n^2\pi^2)(1+m^2\pi^2)}[1-e^{-1}(-1)^n][1-e^{-1}(-1)^m]\cos(n\pi x)\cos(m\pi y)$

57. $\sum_{n=1}^{\infty}\sum_{m=1}^{\infty} \frac{4}{n\pi}(-1)^{n+1}\left(\frac{m\pi}{1+m^2\pi^2}\right)[1-e^{-1}(-1)^m]\sin(n\pi x)\sin(m\pi y);\ \frac{1}{2}\left(1-\frac{1}{e}\right)+\sum_{m=1}^{\infty}\frac{1}{1+m^2\pi^2}[1-e^{-1}(-1)^m]$
$\times\cos(m\pi y)+\sum_{n=1}^{\infty}\sum_{m=1}^{\infty}\frac{4}{n^2\pi^2}[(-1)^n-1]\left(\frac{1}{1+m^2\pi^2}\right)[1-e^{-1}(-1)^m]\cos(n\pi x)\cos(m\pi y)$

59. $\sum_{m=1}^{\infty}\frac{2m}{9+m^2\pi^2}[1-(-1)^m\cosh(3)]\sin(x)\sin(m\pi y);\ \frac{2}{3\pi}\sinh(3)+\sum_{m=1}^{\infty}\frac{12}{\pi}\frac{(-1)^m}{9+m^2\pi^2}\sinh(3)\cos(m\pi y)$
$+\sum_{n=1}^{\infty}\sum_{m=1}^{\infty}\frac{6}{\pi(1-4n^2)}\frac{(-1)^m}{9+m^2\pi^2}\sinh(3)\cos(2nx)\cos(m\pi y)$

61. $f_c(n)=\begin{cases}-\frac{\pi}{2} & \text{if } n=1\\ 0 & \text{if } n\neq 1\end{cases}$ **63.** $f_s(n)=\begin{cases}\frac{\pi}{2} & \text{if } n=1\\ 0 & \text{if } n\neq 1\end{cases}$ **65.** $\frac{\sinh(\pi)}{1+n^2}$

71. $\sinh(1)+\sum_{n=1}^{\infty}\left[\frac{2(-1)^n}{1+n^2\pi^2}\sinh(1)\cos(n\pi x)+\frac{2n\pi(-1)^n}{1+n^2\pi^2}\sinh(1)\sin(n\pi x)\right]$ This converges to e^{-x} for $-1<x<1$; on any interval $2n+1<x<2n+3$, for $n=0,1,2,\ldots$, the series converges to $e^{x-2(n+1)}$; on any interval $2n+1<x<2n+3$, $n=-2,-3,-4,\ldots$, the series converges to $e^{x-2(n+1)}$. At $x=\pm 1,\pm 3,\pm 5,\ldots$, the series converges to $\cosh(1)$.

Chapter 13

Section 13.0

For most of these problems, many solutions are possible.

1. $u(x,y)=f(y)$ (any function of y) **3.** $u(x,y)=x+f(y)$ **5.** $u(x,y)=k(x-y)$, k constant

7. $u(x,y)=\frac{1}{2}xy^2$ **9.** $u(x,y)=\frac{x^5}{120}+f(y)$ **11.** $u(x,y)=2(x+y)$ **13.** $u(x,y)=x$ **15.** $u(x,y)=7$

17. Not linear $\left(\text{because of the } \frac{\partial u}{\partial x}\frac{\partial u}{\partial y} \text{ term}\right)$ **19.** Linear **21.** Linear

23. Not linear $\left[\text{because of the } \left(\frac{\partial u}{\partial y}\right)^2 \text{ term}\right]$ **25.** Linear

Section 13.1

3. $\frac{\partial^2 u}{\partial x^2}+\frac{\partial^2 u}{\partial y^2}=0,\quad 0<x<\alpha,\ 0<y<\beta$
$u(x,0)=u(x,\beta)=0,\quad 0<x<\alpha$
$u(0,y)=u(\alpha,y)=T,\quad 0<y<\beta$

5. $\frac{\partial u}{\partial t}=\frac{\partial^2 u}{\partial r^2}+\frac{1}{r}\frac{\partial u}{\partial r}+\frac{1}{r^2}\frac{\partial^2 u}{\partial\theta^2}+\frac{\partial^2 u}{\partial z^2},\quad 0\le r<R,\ 0<z<c,\ t>0$
(R is the radius of the cylinder) $0\le\theta\le 2\pi$
$u(r,\theta,z,0)=f(r,\theta,z)\quad 0\le r<R,\quad 0\le\theta\le 2\pi,\quad 0<z<c$
$u(R,\theta,z,t)=0\quad 0\le\theta\le 2\pi,\quad 0<z<c$

7. $\frac{\partial u}{\partial t}=\frac{\partial^2 u}{\partial r^2}+\frac{2}{r}\frac{\partial u}{\partial r}+\frac{1}{r^2\sin^2(\phi)}\frac{\partial^2 u}{\partial\theta^2}+\frac{1}{r^2}\frac{\partial^2 u}{\partial\phi^2}+\frac{\cot(\phi)}{r^2}\frac{\partial u}{\partial\phi},$
$0\le\theta\le 2\pi,\quad 0\le\phi\le\pi,\quad 0\le r\le k,\quad t>0$
$u(\rho,\theta,\phi,0)=f(\rho,\theta,\phi),\quad 0\le\rho<k,\quad 0\le\theta\le 2\pi,\quad 0\le\phi\le\pi$
$u(k,\theta,\phi,t)=T,\quad 0\le\theta\le 2\pi,\quad 0\le\phi\le\pi,\quad t>0$

9. $\frac{\partial^2 y}{\partial t^2}=a^2\frac{\partial^2 u}{\partial x^2}-K\left(\frac{\partial y}{\partial t}\right)^2$, for some constant K, $0<x<L$, $t>0$
$y(0,t)=y(L,t)=0,\quad t>0$
$\frac{\partial y}{\partial t}(x,0)=0,\quad 0<x<L$
$y(x,0)=f(x),\quad 0<x<L$

Section 13.2

1. $y(x, t) = \sum_{n=1}^{\infty} \frac{16}{n^2\pi^2 a}(-1)^{n+1} \sin\left(\frac{n\pi x}{2}\right)\sin\left(\frac{n\pi at}{2}\right)$ **3.** $y(x, t) = \sum_{n=1}^{\infty} \frac{9}{n^2\pi^2}(-1)^{n+1} \sin\left(\frac{n\pi x}{3}\right)\sin\left(\frac{2n\pi t}{3}\right)$

5. $y(x, t) = \sum_{n=1}^{\infty} \frac{8}{\sqrt{2}(2n-1)^2\pi} \sin\left(\frac{(2n-1)x}{2}\right)\sin(\sqrt{2}(2n-1)t)$

7. $y(x, t) = \sum_{n=1}^{\infty} \frac{32}{(2n-1)^3\pi^3} \sin\left(\frac{(2n-1)\pi x}{2}\right)\cos\left(\frac{3(2n-1)\pi t}{2}\right) + \sum_{n=1}^{\infty} \frac{32}{3\pi(2n-1)^2} \sin\left(\frac{(2n-1)\pi x}{2}\right)\sin\left(\frac{3(2n-1)\pi t}{2}\right)$

9. $y(x, t) = \sum_{n=1}^{\infty} \frac{256}{n^3\pi^3}[2(-1)^n + 1] \sin\left(\frac{n\pi x}{4}\right)\cos\left(\frac{n\pi t}{\sqrt{2}}\right) + \sum_{n=1}^{\infty} \frac{8}{\sqrt{2}(2n-1)^2\pi^2} \sin\left(\frac{(2n-1)\pi x}{4}\right)\sin\left(\frac{(2n-1)\pi t}{\sqrt{2}}\right)$

11. $y(x, t) = \frac{1}{2}\sin(2x)\cos(2t) + \sum_{n=1}^{\infty} \frac{2}{n^2}\sin(nx)\sin(nt)$

13. Choose $f(x)$ so that $-3f''(x) + 2x = 0$, $f(0) = 0$, and $f(2) = 0$. Then, $f(x) = \frac{x^3}{9} - \frac{4}{9}x = \frac{1}{9}x(x^2 - 4)$.

Further, $\frac{\partial^2 Y}{\partial t^2} = 3\frac{\partial^2 Y}{\partial x^2}$, $Y(0, t) = Y(2, t) = 0$, $Y(x, 0) = f(x)$, and $\frac{\partial Y}{\partial t}(x, 0) = 0$.

Solve this standard problem for $Y(x, t)$ to get

$Y(x, t) = \sum_{n=1}^{\infty} \frac{32}{3}\frac{1}{n^3\pi^3}(-1)^n \sin\left(\frac{n\pi x}{2}\right)\cos\left(\frac{\sqrt{3}n\pi t}{2}\right)$.

Then $y(x, t) = Y(x, t) - f(x)$.

15. Write $Y(x, t) = y(x, t) + f(x)$, and choose $f(x) = \cos(x) - 1$. Then, $\frac{\partial^2 Y}{\partial t^2} = \frac{\partial^2 Y}{\partial x^2}$, $Y(0, t) = Y(2\pi, t) = 0$,

$Y(x, 0) = \cos(x) - 1$, and $\frac{\partial Y}{\partial t}(x, 0) = 0$. Solve for $Y(x, t)$ to get

$Y(x, t) = \sum_{\substack{n=1 \\ n \neq 2}}^{\infty} \frac{8}{(n^2 - 4)\pi}[1 - (-1)^n] \sin\left(\frac{nx}{2}\right)\cos\left(\frac{nt}{2}\right)$; $y(x, t) = Y(x, t) + 1 - \cos(x)$.

Then, $y(x, t) = Y(x, t) - \cos(x) + 1$.

17. $y(x, t) = e^{-\alpha t/2}\sum_{n=1}^{\infty} A_n \sin\left(\frac{n\pi x}{L}\right)\cos\left(\frac{\sqrt{4n^2\pi^2a^2 - \alpha^2L^2}\,t}{2L}\right)$, with $A_n = \frac{2}{L}\int_0^L f(x)\sin\left(\frac{n\pi x}{L}\right)dx$.

19. **(a)** When you substitute $y = X(x)T(t)$ into the partial differential equation, you get $XT'' = a^2X''T - g$; and there is no way to isolate all functions of x on one side and all functions of t on the other.

(b) Choose $h(x)$ so that $a^2h''(x) + g = 0$. Thus, $h(x)$ has the form $h(x) = -\frac{gx^2}{2a^2} + Cx + D$.

(c) In terms of Y, the partial differential equation is $\frac{\partial^2 Y}{\partial t^2} = a^2\frac{\partial^2 Y}{\partial x^2}$. Next, $y(0, t) = Y(0, t) - h(0) = 0$

gives us $Y(0, t) = 0$ if we choose $D = 0$. Next, $y(L, t) = Y(L, t) - h(L) = Y(L, t) + \frac{gL^2}{2a^2} - CL = 0$

gives us $Y(L, t) = 0$ if we choose $C = \frac{gL}{2a^2}$.

Finally, $y(x, 0) = Y(x, 0) - h(x) = f(x)$ if $Y(x, 0) = f(x) + h(x)$ and $\frac{\partial y}{\partial t}(x, 0) = \frac{\partial Y}{\partial t}(x, 0) = 0$.

Thus, solve: $\frac{\partial^2 Y}{\partial t^2} = a^2\frac{\partial^2 Y}{\partial x^2}$ $0 < x < L, t > 0$, $Y(0, t) = Y(L, t) = 0$, $Y(x, 0) = f(x) + h(x)$, $\frac{\partial Y}{\partial t}(x, 0) = 0$,

with $h(x) = \frac{-gx^2}{2a^2} + \frac{gL}{2a^2}x = \frac{gx}{2a^2}(L - x)$. This has solution $Y(x, t) = \sum_{n=1}^{\infty} B_n \sin\left(\frac{n\pi x}{L}\right)\cos\left(\frac{n\pi at}{L}\right)$,

with $B_n = \frac{2}{L}\int_0^L \left[f(x) + \frac{gx}{2a^2}(L - x)\right]\sin\left(\frac{n\pi x}{L}\right)dx$. Then, $y(x, t) = Y(x, t) - \frac{gx}{2a^2}(L - x)$.

21. $y(x, t) = \dfrac{AL}{2} - \sum_{n=1}^{\infty} \dfrac{4AL}{(2n-1)^2\pi^2} \cos\left(\dfrac{(2n-1)\pi x}{L}\right)\cos\left(\dfrac{(2n-1)\pi at}{L}\right)$

23. $u(x, t) = \sum_{n=1}^{\infty} C_n e^{-At/2} \sin\left(\dfrac{n\pi x}{L}\right)\cos(rt)$, where $C_n = \dfrac{2}{L}\int_0^L f(x)\sin\left(\dfrac{n\pi x}{L}\right)dx$ and $r = \dfrac{1}{2L}\sqrt{4(BL^2 + a^2n^2\pi^2) - A^2L^2}$.

Section 13.3

1. $u(x, t) = \dfrac{8L^2}{\pi^3}\sum_{n=1}^{\infty} \dfrac{1}{(2n-1)^3}\sin\left(\dfrac{(2n-1)\pi x}{L}\right)\exp\left(\dfrac{-(2n-1)^2\pi^2a^2t}{L^2}\right)$

3. $u(x, t) = \sum_{n=1}^{\infty} \dfrac{-16L}{(2n-1)\pi[(2n-1)^2 - 4]}\sin\left(\dfrac{(2n-1)\pi x}{L}\right)\exp\left(\dfrac{-3(2n-1)^2\pi^2 t}{L^2}\right)$

5. $u(x, t) = \dfrac{2\pi^2}{3} - \sum_{n=1}^{\infty}\dfrac{4}{n^2}\cos(nx)\exp(-4n^2t)$

7. $u(x, t) = \sum_{n=1}^{\infty}\dfrac{4T}{(2n-1)\pi}\sin\left(\dfrac{(2n-1)\pi x}{2L}\right)\exp\left(\dfrac{-(2n-1)a^2\pi^2t}{4L^2}\right)$

9. $u(x, t) = e^{-ht}\sum_{n=1}^{\infty} C_n\sin\left(\dfrac{n\pi x}{L}\right)\exp\left(\dfrac{-n^2\pi^2a^2t}{L^2}\right)$, where $C_n = \dfrac{2}{L}\int_0^L f(x)\sin\left(\dfrac{n\pi x}{L}\right)dx$.

11. $$u(x, y) = \sum_{n=1}^{\infty}\frac{-4T_1}{(2n-1)\pi}\,\frac{\exp[-(2n-1)\pi\alpha/\beta]}{\sinh[(2n-1)\pi\alpha/\beta]}\sinh[(2n-1)\pi(x-\alpha)/\beta]\,\sin(n\pi y/\beta)$$
$$+\sum_{n=1}^{\infty}\frac{4T_2}{(2n-1)\pi}\,\frac{\sinh[(2n-1)\pi x/\beta]}{\sinh[(2n-1)\pi\alpha/\beta]}\sin\left(\frac{(2n-1)\pi y}{\beta}\right)$$

13. $u(x, y) = C_0x + \sum_{n=1}^{\infty} C_n\sinh\left(\dfrac{n\pi x}{\beta}\right)\cos\left(\dfrac{n\pi y}{\beta}\right)$, where $C_0 = \dfrac{1}{\alpha\beta}\int_0^\beta f(y)\,dy$ and, for $n = 1, 2, 3, \ldots,$

$C_n = \dfrac{2}{\beta\sinh(n\pi\alpha/\beta)}\int_0^\beta f(y)\cos\left(\dfrac{n\pi y}{\beta}\right)dy.$

15. $u(x, y) = C_0 + \sum_{n=1}^{\infty} C_n\cos(n\pi x)e^{-n\pi y}$, with $C_0 = \int_0^1 f(x)\,dx$ and, for $n = 1, 2, 3, \ldots,$ $C_n = 2\int_0^1 f(x)\cos(n\pi x)\,dx$.

17. **(a)** Using separation of variables, we get $u(x, t) = e^{-At}\sum_{n=1}^{\infty} C_n\sin\left(\dfrac{n\pi x}{L}\right)\exp\left(\dfrac{-n^2\pi^2a^2t}{L^2}\right)$,

in which $C_n = \dfrac{2}{L}\int_0^1 f(x)\sin\left(\dfrac{n\pi x}{L}\right)dx.$

(b) Setting $u = e^{-At}v$ gives us the following problem for v:

$$\frac{\partial v}{\partial t} = a^2\frac{\partial^2 v}{\partial x^2},\qquad v(0, t) = v(L, t) = 0,\qquad v(x, 0) = f(x).$$

Solve this to get $v(x, t) = \sum_{n=1}^{\infty} C_n\sin\left(\dfrac{n\pi x}{L}\right)\exp\left(\dfrac{-n^2\pi^2a^2t}{L^2}\right)$, with C_n as in **(a)**.

19. Begin with $-\dfrac{\partial v}{\partial x} = RI + L\dfrac{\partial I}{\partial t}$, and differentiate with respect to x to get

$$-\frac{\partial^2 v}{\partial x^2} = R\frac{\partial I}{\partial x} + L\frac{\partial^2 I}{\partial t\,\partial x}. \tag{1}$$

Now differentiate $-\dfrac{\partial I}{\partial x} = SV + K\dfrac{\partial V}{\partial t}$ with respect to t to get

$$-\frac{\partial^2 I}{\partial x\,\partial t} = S\frac{\partial V}{\partial t} + K\frac{\partial^2 V}{\partial t^2}. \tag{2}$$

Multiply (2) by $-L$ to get $L\dfrac{\partial^2 I}{\partial x\,\partial t} = -SL\dfrac{\partial V}{\partial t} - KL\dfrac{\partial^2 V}{\partial t^2}.$

Assuming that $\dfrac{\partial^2 I}{\partial x\,\partial t} = \dfrac{\partial^2 I}{\partial t\,\partial x}$, substitute the last equation into (1) to get

$$-\frac{\partial^2 V}{\partial x^2} = R\,\frac{\partial I}{\partial x} - LS\,\frac{\partial V}{\partial t} - LK\,\frac{\partial^2 V}{\partial t^2}. \qquad (3)$$

Now, in (3), set $\dfrac{\partial I}{\partial x} = -SV - K\,\dfrac{\partial V}{\partial t}$ to get $-\dfrac{\partial^2 V}{\partial x^2} = -SVR - KR\,\dfrac{\partial V}{\partial t} - LS\,\dfrac{\partial V}{\partial t} - LK\,\dfrac{\partial^2 V}{\partial t^2}$.

This can be written $LK\,\dfrac{\partial^2 V}{\partial t^2} + (LS + RK)\,\dfrac{\partial V}{\partial t} + RSV = \dfrac{\partial^2 V}{\partial x^2}$. Now compare with Problem 23, Section 13.2.

21. Change scale by setting $\bar{r} = r/R$, $0 \le r \le R$. Then from Problem 20 we get

$$u(r, \theta) = \frac{1}{2\pi}\int_0^{2\pi} \frac{\left(1 - \dfrac{r}{R}\right)^2 f(\phi)\,d\phi}{1 + \left(\dfrac{r}{R}\right)^2 - 2\,\dfrac{r}{R}\cos(\theta - \phi)}.$$ Some algebraic manipulation now gives the final result.

23. **(a)** 0.3144 **(b)** 0.5095 **(c)** 0.7230 **(d)** 0.6792
(e) 0.6366, 0.7066, 0.6957, 0.5812, 0.2972, 0.0744, −0.4902, −0.7429, −0.9356

25. Choose $\alpha = \dfrac{a}{2L}$ and $\beta = -\dfrac{1}{4}\dfrac{ka^2}{L^2}$. Then, $\dfrac{\partial v}{\partial t} = k\,\dfrac{\partial^2 v}{\partial x^2}$. Further, $v(0, t) = v(L, t) = 0$

and $v(x, 0) = \dfrac{1}{ak}\,e^{-ax/2L}\left\{1 - \exp\left[-a\left(1 - \dfrac{x}{L}\right)\right]\right\} = \dfrac{1}{ak}\,(e^{-ax/2L} - e^{-a}e^{ax/2L})$.

Solve this to get $v(x, t) = \displaystyle\sum_{n=1}^{\infty} \frac{4Ln\pi}{ak}\left(\frac{1 - e^{-a}}{a^2 + 4n^2\pi^2}\right)\sin\left(\frac{n\pi x}{L}\right)\exp\left(\frac{-kn^2\pi^2 t}{L^2}\right)$.

Then, $u(x, t) = \exp\left(\dfrac{ax}{2L} - \dfrac{ka^2 t}{4L^2}\right)v(x, t)$

Section 13.4

1. $y(x, t) = \dfrac{2}{\pi}\displaystyle\int_0^{\infty}\left(\frac{-\sin(\alpha)}{\alpha^2} + \frac{2[1 - \cos(\alpha)]}{\alpha^3}\right)\sin(\alpha x)\cos(\alpha a t)\,d\alpha$

3. $y(x, t) = \dfrac{2}{\pi}\displaystyle\int_0^{\infty}\frac{\sin(\alpha\pi)\sin(\alpha x)\cos(\alpha a t)}{1 - \alpha^2}\,d\alpha$

5. $y(x, t) = \dfrac{2}{\pi a}\displaystyle\int_0^{\infty}\left(\frac{\sin(\alpha) - \alpha\cos(\alpha)}{\alpha^3}\right)\sin(\alpha x)\sin(\alpha a t)\,d\alpha$

7. $y(x, t) = \displaystyle\int_0^{\infty}\left\{\frac{2}{\pi\alpha}\left[1 - \cos\left(\frac{\alpha\pi}{2}\right)\right] + \frac{1}{\pi(\alpha - 1)}\left[\cos\left((\alpha - 1)\frac{\pi}{2}\right) - 1\right] + \frac{1}{\pi(\alpha + 1)}\left[\cos\left((\alpha + 1)\frac{\pi}{2}\right) - 1\right]\right\}\sin(\alpha x)\cos(\alpha a t)\,d\alpha$

$\qquad + \displaystyle\int_0^{\infty}\frac{2}{\pi a\alpha^2}\,[\cos(\alpha) - \cos(4\alpha)]\sin(\alpha x)\sin(\alpha a t)\,d\alpha$

9. The problem is $\dfrac{\partial^2 y}{\partial t^2} = a^2\,\dfrac{\partial^2 y}{\partial x^2}$, $-\infty < x < \infty$, $t > 0$, $y(x, 0) = f(x)$, $\dfrac{\partial y}{\partial t}(x, 0) = 0$,

$y(x, t)$ bounded for $-\infty < x < \infty$, $t > 0$.

The solution is $y(x, t) = \dfrac{2}{\pi}\displaystyle\int_0^{\infty}\left(\frac{[1 - \cos(\alpha)]}{\alpha^2}\right)\cos(\alpha x)\cos(\alpha a t)\,d\alpha$.

11. The problem is $\dfrac{\partial^2 y}{\partial t^2} = a^2\,\dfrac{\partial^2 y}{\partial x^2}$, $-\infty < x < \infty$, $t > 0$, $y(x, 0) = 0$, $\dfrac{\partial y}{\partial t}(x, 0) = g(x)$.

The solution is $y(x, t) = \dfrac{2}{\pi a}\displaystyle\int_0^{\infty}\frac{\sin(2\alpha)}{\alpha^2}\cos(\alpha x)\sin(\alpha a t)\,d\alpha$.

Section 13.5

1. $u(x, t) = \dfrac{1}{\pi}\displaystyle\int_0^{\infty}\frac{1}{\alpha}\left(-\sin(\alpha) + \frac{\cos(\alpha) - 1}{\alpha}\right)\cos(\alpha x)\exp(-4\alpha^2 t)\,d\alpha + \frac{1}{\pi}\int_0^{\infty}\frac{1}{\alpha}\left(2 - 3\cos(\alpha) + \frac{\sin(\alpha)}{\alpha}\right)\sin(\alpha x)\exp(-4\alpha^2 t)\,d\alpha$

3. $u(x, y) = \frac{1}{\pi}\int_0^\infty \left\{\frac{1}{e^8(4+\alpha^2)}[4\cos(4\alpha) - 2\alpha\sin(4\alpha)] - \frac{1}{\alpha}\sin(4\alpha)\right\}\cos(\alpha x)e^{-\alpha y}\,d\alpha + \frac{1}{\pi}\int_0^\infty \frac{1}{\alpha}[1-\cos(4\alpha)]\sin(\alpha x)e^{-\alpha y}\,d\alpha$

5. $u(x, y) = \frac{1}{\pi}\int_0^\infty \left(\frac{\sin[(\alpha-1)\pi]}{\alpha-1} + \frac{\sin[(\alpha+1)\pi]}{\alpha+1}\right)\cos(\alpha y)e^{-\alpha x}\,d\alpha$

7. In general, $u(x, t) = \int_0^\infty [a_\alpha\cos(\alpha x) + b_\alpha\sin(\alpha x)]\exp(-\alpha^2 a^2 t)\,d\alpha$,

where $a_\alpha = \frac{1}{\pi}\int_{-\infty}^\infty f(x)\cos(\alpha x)\,dx$ and $b_\alpha = \frac{1}{\pi}\int_{-\infty}^\infty f(x)\sin(\alpha x)\,dx$.

With $f(x)$ as given, $u(x, t) = \frac{2}{\pi}\int_0^\infty \left(\frac{1-\cos(\alpha)}{\alpha^2}\right)\cos(\alpha x)\exp(-\alpha^2 a^2 t)\,d\alpha$.

9. $u(x, t) = \int_0^\infty b_\alpha\sin(\alpha x)\exp(-\alpha^2 a^2 t)\,d\alpha$, where $b_\alpha = \frac{2}{\pi}\int_0^\infty f(x)\sin(\alpha x)\,dx$.

If $f(x) = e^{-x}$, then $u(x, t) = \frac{2}{\pi}\int_0^\infty \frac{\alpha}{1+\alpha^2}\sin(\alpha x)\exp(-\alpha^2 a^2 t)\,d\alpha$.

11. $u(x, y) = \frac{2}{\pi}\int_0^\infty \frac{\sin(\alpha)}{\alpha}\cos(\alpha y)e^{-\alpha x}\,d\alpha$

13. $u(x, y) = \int_0^\infty b_\alpha\sin(\alpha x)\sinh(\alpha y)\,d\alpha$, where $b_\alpha = \frac{2}{\pi}\frac{1}{\sinh(\alpha)}\int_0^\infty f(x)\sin(\alpha x)\,dx$

15. $u(x, y) = \sum_{n=1}^\infty \frac{8}{(2n-1)\pi}\sin[(2n-1)\pi x]\exp[-(2n-1)y] + \int_0^\infty \frac{3}{\sinh(2\alpha)}\sin(\alpha y)\sinh[\alpha(x+1)]\,d\alpha$

$+ \frac{2}{\pi}\int_0^\infty \frac{\alpha}{1+\alpha^2}\frac{1}{\sinh(2\alpha)}\sin(\alpha y)\sinh[\alpha(1-x)]\,d\alpha$

17. $u(x, y) = \int_0^\infty a_\alpha\cos(\alpha x)\exp(-\alpha^2 a^2 t)\,d\alpha$, where $a_\alpha = \frac{2}{\pi}\int_0^\infty f(x)\cos(\alpha x)\,dx$.

19. $u(x, y) = \frac{2}{\pi}\int_0^\infty \frac{A_\alpha}{\sinh(2\alpha)}\cos(\alpha y)\sinh[\alpha(2-x)]\,d\alpha + \frac{2}{\pi}\int_0^\infty \frac{C_\alpha}{\sinh(2\alpha)}\cos(\alpha y)\sinh(\alpha x)\,d\alpha$,

with $A_\alpha = \int_0^\infty f(y)\cos(\alpha y)\,dy$ and $C_\alpha = \int_0^\infty g(y)\cos(\alpha y)\,dy$.

21. $u(x, y) = \sum_{n=1}^\infty \frac{8}{(2n-1)\pi}\sin\left(\frac{(2n-1)\pi y}{2}\right)\exp\left[\frac{-(2n-1)\pi x}{2}\right]$

Section 13.6

1. $u(x, y, z) = \sum_{n=1}^\infty\sum_{m=1}^\infty \frac{4(-1)^{n+m}}{nm\pi^2\sinh(\pi\sqrt{n^2+m^2})}\sin(n\pi x)\sin(m\pi y)\sinh(\pi\sqrt{n^2+m^2}\,z)$

3. $u(x, y, t) = \sum_{n=1}^\infty \frac{1}{\pi}\left\{\frac{8(-1)^{n+1}\pi^2}{n} + \frac{16}{n^3}[(-1)^n - 1]\right\}\sin\left(\frac{nx}{2}\right)\sin(y)\cos\left(\frac{\sqrt{n^2+4}}{2}t\right)$

5. $u(x, y, t) = \sum_{n=1}^\infty\sum_{m=1}^\infty \left[\frac{16}{\pi^2(2n-1)(2m-1)\sqrt{(2n-1)^2+(2m-1)^2}}\sin\left(\frac{(2n-1)x}{2}\right)\sin\left(\frac{(2m-1)y}{2}\right)\right]$

$\times\left\{\sin[\sqrt{(2n-1)^2+(2m-1)^2}\,t]\right\}$

7. $u(x, y, z) = \sum_{n=1}^\infty\sum_{m=1}^\infty \left\{\frac{32}{(2n-1)(2m-1)\pi^2}\frac{1}{\sinh[2\pi\sqrt{(2m-1)^2+\pi^2(2n-1)^2}]}\right\}$

$\times\{\sin[(2n-1)\pi x]\sin[(2m-1)z]\sinh[\sqrt{(2m-1)^2+\pi^2(2n-1)^2}\,y]\}$

$+\sum_{n=1}^\infty\sum_{m=1}^\infty \frac{16}{(2n-1)(2m-1)\pi^2}\frac{1}{\sinh\left(\pi\sqrt{\frac{(2m-1)^2}{4}+\pi^2(2n-1)^2}\right)}$

$\times\left[\sin[(2n-1)\pi x]\sin\left(\frac{(2m-1)y}{2}\right)\sinh\left(\sqrt{\frac{(2m-1)^2}{4}+\pi^2(2n-1)^2}\,z\right)\right]$

9. $$u(x, y, z, t) = \frac{a_{000}}{8} + \frac{1}{4}\sum_{n=1}^{\infty} a_{n00} \cos\left(\frac{n\pi z}{L}\right)\exp\left[\frac{-n^2\pi^2a^2t}{L^2}\right]$$
$$+ \frac{1}{4}\sum_{m=1}^{\infty} a_{0m0} \cos\left(\frac{m\pi y}{L}\right)\exp\left[\frac{-m^2\pi^2a^2t}{L^2}\right] + \frac{1}{4}\sum_{r=0}^{\infty} a_{00r} \cos\left(\frac{r\pi z}{L}\right)\exp\left[\frac{-r^2\pi^2a^2t}{L^2}\right]$$
$$+ \frac{1}{2}\sum_{n=1}^{\infty}\sum_{m=1}^{\infty} a_{nm0} \cos\left(\frac{n\pi x}{L}\right)\cos\left(\frac{m\pi y}{L}\right)\exp\left[\frac{(-n^2 - m^2)\pi^2a^2t}{L^2}\right]$$
$$+ \frac{1}{2}\sum_{n=1}^{\infty}\sum_{r=1}^{\infty} a_{n0r} \cos\left(\frac{n\pi x}{L}\right)\cos\left(\frac{r\pi z}{L}\right)\exp\left[\frac{(-n^2 - r^2)\pi^2a^2t}{L^2}\right]$$
$$+ \frac{1}{2}\sum_{m=1}^{\infty}\sum_{r=1}^{\infty} a_{0mr} \cos\left(\frac{m\pi y}{L}\right)\cos\left(\frac{r\pi z}{L}\right)\exp\left[\frac{(-m^2 - r^2)\pi^2a^2t}{L^2}\right]$$
$$+ \sum_{n=1}^{\infty}\sum_{m=1}^{\infty}\sum_{r=1}^{\infty} a_{nmr} \cos\left(\frac{n\pi x}{L}\right)\cos\left(\frac{m\pi y}{L}\right)\cos\left(\frac{r\pi z}{L}\right)\exp\left[\frac{(-n^2 - m^2 - r^2)\pi^2a^2t}{L^2}\right],$$

in which $a_{nmr} = \frac{8}{L^3}\int_0^L\int_0^L\int_0^L f(x, y, z)\cos\left(\frac{n\pi x}{L}\right)\cos\left(\frac{m\pi y}{L}\right)\cos\left(\frac{r\pi z}{L}\right) dx\, dy\, dz.$

11. Suppose the room occupies $0 \le x \le L$, $0 \le y \le L$, $0 \le z \le L$. We find that
$$u(x, y, z, t) = \sum_{n=1}^{\infty}\sum_{m=1}^{\infty}\sum_{r=1}^{\infty} b_{nmr} \sin\left(\frac{n\pi x}{L}\right)\sin\left(\frac{m\pi y}{L}\right)\sin\left(\frac{r\pi z}{L}\right)\cos\left(\frac{\sqrt{n^2 + m^2 + r^2}\,\pi at}{L}\right)$$
$$+ \sum_{n=1}^{\infty}\sum_{m=1}^{\infty}\sum_{r=1}^{\infty} c_{nmr} \sin\left(\frac{n\pi x}{L}\right)\sin\left(\frac{m\pi y}{L}\right)\sin\left(\frac{r\pi z}{L}\right)\sin\left(\frac{\sqrt{n^2 + m^2 + r^2}\,\pi at}{L}\right),$$
where $b_{nmr} = \frac{8}{L^3}\int_0^L\int_0^L\int_0^L f(x, y, z)\sin\left(\frac{n\pi x}{L}\right)\sin\left(\frac{m\pi y}{L}\right)\sin\left(\frac{r\pi z}{L}\right) dx\, dy\, dz$

and $c_{nmr} = \frac{8}{L^3}\int_0^L\int_0^L\int_0^L g(x, y, z)\sin\left(\frac{n\pi x}{L}\right)\sin\left(\frac{m\pi y}{L}\right)\sin\left(\frac{r\pi z}{L}\right) dx\, dy\, dz.$

13. $$u(x, y, z) = \sum_{n=1}^{\infty}\sum_{m=1}^{\infty} \frac{4}{\sinh[2\pi\sqrt{n^2 + m^2}]}\left(\frac{(-1)^{n+1}}{n\pi}\right)\left(\frac{(-1)^{m+1}}{m\pi} + \frac{2}{m^3\pi^3}[(-1)^m - 1]\right)$$
$$\times [\sin(n\pi x)\sin(m\pi y)\sinh(\pi\sqrt{n^2 + m^2}\, z)]$$

Section 13.7

1. $u(r, t) = \sum_{n=1}^{\infty} a_n J_0(\alpha_n r)e^{-At}\exp(-\alpha_n{}^2 t)$, where α_n is the nth positive zero of $J_0(x)$,

and $a_n = \frac{\int_0^1 rkJ_0(\alpha_n r)\, dr}{\int_0 rJ_0^2(\alpha_n r)\, dr} = \frac{2k}{\alpha_n}\frac{1}{J_1(\alpha_n)}.$

3. $u(r, \theta, t) = \sum_{n=1}^{\infty}\frac{1}{2}c_{n0}J_0(\sqrt{\lambda_n{}^{(0)}}\, r)\cos(a\sqrt{\lambda_n{}^{(0)}}\, t) + \sum_{n=1}^{\infty}\sum_{m=1}^{\infty}[c_{nm}\cos(m\theta) + d_{nm}\sin(m\theta)]J_m(\sqrt{\lambda_n{}^{(m)}}\, r)\cos(a\sqrt{\lambda_n{}^{(m)}}\, t),$

where J_m is the mth order Bessel function of the first kind, $\lambda_n{}^{(m)}$ is the nth positive zero of $J_m(x)$,

and $c_{nm} = \frac{1}{\pi\int_0^1 rJ_m(\sqrt{\lambda_n{}^{(m)}}\, r)^2\, dr}\int_0^1\int_{-\pi}^{\pi} rf(r, \theta)\cos(m\theta)J_m(\sqrt{\lambda_n{}^{(m)}}\, r)\, dr\, d\theta$

and $d_{nm} = \frac{1}{\pi\int_0^1 rJ_m(\sqrt{\lambda_n{}^{(m)}}\, r)^2\, dr}\int_0^1\int_{-\pi}^{\pi} rf(r, \theta)\sin(m\theta)J_m(\sqrt{\lambda_n{}^{(m)}}\, r)\, dr\, d\theta.$

5. $u(r, \theta, t) = \sum_{n=1}^{\infty}\frac{1}{2}c_{n0}J_0(\sqrt{\lambda_n{}^{(0)}}\, r)e^{-\lambda_n{}^{(0)}t} + \sum_{n=1}^{\infty}\sum_{m=1}^{\infty}[c_{nm}\cos(m\theta) + d_{nm}\sin(m\theta)]J_m(\sqrt{\lambda_n{}^{(m)}}\, r)e^{-\lambda_n{}^{(m)}t},$

where c_{nm}, d_{nm}, and $\lambda_n{}^{(m)}$ are as in the solution of Problem 3.

7. $u(r, z) = \frac{2z}{L}$

Section 13.8

1. $u(r, \phi) = A \sum_{n=1}^{\infty} (4n+1)\left(\frac{r}{k}\right)^{2n-1} \frac{(-1)^{n-1}(2n-2)!}{2^{2n-1}[(n-1)!]^2 n} P_{2n-1}[\cos(\phi)]$.

3. $u(r, \phi) = \sum_{n=0}^{\infty} a_{2n+1} r^{2n+1} P_{2n+1}[\cos(\phi)]$, where $a_{2n+1} = \left(\frac{4n+3}{k^{2n+1}}\right) \int_0^1 f(x) P_{2n+1}(x)\, dx$.

5. $u(r, \phi) = \left(\frac{b-r}{b-a}\right) T$

Section 13.9

1. $$y(x, t) = \begin{cases} At, & t < \dfrac{x}{a} \\ t + (A-1)\dfrac{x}{a}, & t > \dfrac{x}{a} \end{cases}$$

3. $u(x, t) = e^{-a^2 t}\{e^{-x} - \operatorname{erfc} x/2a\sqrt{t}\}$

5. $$u(x, y) = \begin{cases} \dfrac{y^2}{2} & \text{if } y < \dfrac{x}{a} \\ \dfrac{y^2}{2} + y - \dfrac{x}{a} - \dfrac{1}{2}\left(y - \dfrac{x}{a}\right)^2 & \text{if } y > \dfrac{x}{a} \end{cases}$$

7. $$y(x, t) = \begin{cases} \dfrac{1}{2} e^{-x} \sinh(2t) & \text{if } t < \dfrac{x}{2} \\ \dfrac{1}{2} e^{-x} \sinh(2t) + \sin\left(t - \dfrac{x}{2}\right) - \dfrac{1}{2} \sinh\left[2\left(t - \dfrac{x}{2}\right)\right] & \text{if } t > \dfrac{x}{2} \end{cases}$$

9. $$u(x, t) = \begin{cases} 1 - at & \text{if } t < \dfrac{x}{\sqrt{ba}} \\ 1 - at + \sqrt{\dfrac{a}{b}}\left(t - \dfrac{x}{\sqrt{ba}}\right) & \text{if } t > \dfrac{x}{\sqrt{ba}} \end{cases}$$

$$v(x, t) = \begin{cases} x & \text{if } t < \dfrac{x}{\sqrt{ba}} \\ x\left(1 - \dfrac{1}{\sqrt{ba}}\right) + t & \text{if } t > \dfrac{x}{\sqrt{ba}} \end{cases}$$

Section 13.10

1. $u(x, y) = \frac{1}{2\pi} \int_{-\infty}^{\infty} \hat{f}(\lambda) e^{-\lambda y} e^{-i\lambda x}\, d\lambda$

3. Using a finite Fourier sine transform in x, we get

$$u(x, y) = \sum_{n=1}^{\infty} \frac{2}{\pi}\left[\left(\frac{-4}{n} + \frac{6}{n}(-1)^n\right) e^{-ny} - \frac{2(-1)^n}{n}\right] \sin(nx)$$

Using a Fourier sine transform in y, we get

$$u(x, y) = \frac{2}{\pi} \int_0^{\infty} \left\{ \left[\frac{\frac{4}{\lambda} e^{-\lambda\pi} - 2 - \frac{4}{\lambda}}{e^{-\lambda\pi} - e^{\lambda\pi}}\right] e^{\lambda x} + \left[\frac{2 + \frac{4}{\lambda} - \frac{4}{\lambda} e^{\lambda\pi}}{e^{-\lambda\pi} - e^{\lambda\pi}}\right] e^{-\lambda x} - \frac{4}{\lambda} \right\} \sin(\lambda y)\, d\lambda.$$

5. Using a finite Fourier sine transform in x, we get

$$u(x, y) = \sum_{n=1}^{\infty} \frac{8}{\pi} \left\{ \frac{(-1)^n}{n} \left[\frac{\cosh(ny)}{\cosh(2n)} - 1\right] \right\} \sin(nx)$$

7. Using a Fourier sine transform in x, we get

$$u(x, y) = \frac{2}{\pi} \int_0^{\infty} \left[\frac{2}{\lambda^3}\left(\frac{2 + e^{-\lambda}}{e^{\lambda} - e^{-\lambda}}\right) e^{\lambda y} - \frac{2}{\lambda^3}\left(\frac{2 + e^{\lambda}}{e^{\lambda} - e^{-\lambda}}\right) e^{-\lambda y} - \frac{1}{\lambda} y^3 + \frac{1}{\lambda} y^2 - \frac{6}{\lambda^3} y + \frac{2}{\lambda^3} \right] \sin(\lambda x)\, d\lambda.$$

9. $u(x, t) = \dfrac{-4}{\pi}\displaystyle\int_0^{\infty} \frac{\lambda}{(1 + \lambda^2)^2} \sin(\lambda x) \exp(-\lambda^2 t - t^2/2)\, d\lambda$

11. Using a Fourier transform in x,

$$u(x, y) = \frac{1}{2\sqrt{\pi}} \int_{-\infty}^{\infty} \frac{\exp(-\lambda^2/4)}{\cosh(\sqrt{1 + \lambda^2})} \cosh(\sqrt{1 + \lambda^2}\, y) e^{i\lambda x}\, d\lambda.$$

Supplementary Problems

1. $y(x, t) = \displaystyle\sum_{n=1}^{\infty} \left[\frac{2(-1)^n}{a^2 n^3} \cos(ant) + \frac{2}{a^2 n^3} (-1)^{n+1}\right] \sin(nx)$

3. $u(x, t) = \dfrac{2}{\pi} \displaystyle\sum_{n=1}^{\infty} \left[\left(\frac{1 - (-1)^n}{n}\right)\cos(ant) + \left(\frac{1 - (-1)^n}{2n^2}\right)\sin(ant)\right] e^{-t} \sin(nx),$

using a finite Fourier transform in x.

5. Using a finite Fourier transform in x, we get

$$u(x, t) = \frac{2}{\pi} \sum_{n=1}^{\infty} \left[\left(\frac{4[1 - (-1)^n]}{n} - \frac{\pi}{a^2 n^3}\right) \exp(-a^2 n^2 t) + \frac{\pi}{a^2 n^3}\right] \sin(nx).$$

7. $u(r, \theta) = 4 \displaystyle\sum_{n=1}^{\infty} \frac{2n - 1}{(2n - 1)^2 - 16}\, r^{(2n-1)/2} \sin\left[\left(\frac{2n - 1}{2}\right)\theta\right]$

9. $u(x, y) = \dfrac{2}{\pi} \displaystyle\int_0^{\infty} \left\{\frac{x}{\lambda} + \left[\int_0^{\infty} g(t) \sin(\lambda t)\, dt\right] e^{-\lambda x}\right\} \sin(\lambda y)\, d\lambda$

11. $u(x, y, z, t) = \dfrac{8abc}{\pi^3} \displaystyle\sum_{n=1}^{\infty} \sum_{m=1}^{\infty} \sum_{k=1}^{\infty} \frac{(-1)^{n+m+k+1}}{nmk} e^{-t/2} \left\{\cos(\tfrac{1}{2} A_{nmk} t) + \frac{4}{A_{nmk}} \sin(\tfrac{1}{2} A_{nmk} t)\right\} \sin\left(\frac{n\pi x}{a}\right) \sin\left(\frac{m\pi y}{b}\right) \sin\left(\frac{k\pi z}{c}\right)$

where $A_{nmk} = \left[4\pi\left(\dfrac{n^2}{a^2} + \dfrac{m^2}{b^2} + \dfrac{k^2}{c^2}\right) - 1\right]^{1/2}$

13. $u(r, \theta, z) = \displaystyle\sum_{n=0}^{\infty} \sum_{m=1}^{\infty} J_n\left(\frac{\alpha_{mn} r}{2}\right) [A_{mn} \cos(n\theta) + B_{mn} \sin(n\theta)] \left[\sin\left(\frac{\alpha_{mn} L}{2}\right) \cos\left(\frac{\alpha_{mn} z}{2}\right) - \cos\left(\frac{\alpha_{mn} L}{2}\right) \sin\left(\frac{\alpha_{mn} z}{2}\right)\right]$

where α_{mn} is the mth root of $J_n(x)$, and

$$A_{m0} = \frac{\dfrac{1}{2\pi} \displaystyle\int_0^{2\pi} \int_0^2 f(r, \theta) J_0\left(\frac{\alpha_{m0} r}{2}\right) r\, dr\, d\theta}{\sin\left(\dfrac{\alpha_{m0} L}{2}\right) \displaystyle\int_0^2 r J_0^2\left(\frac{\alpha_{m0} r}{2}\right) dr},$$

$$A_{mn} = \frac{\dfrac{1}{\pi} \displaystyle\int_0^{2\pi} \int_0^2 r f(r, \theta) \cos(n\theta) J_n\left(\frac{\alpha_{nm} r}{2}\right) dr\, d\theta}{\sin\left(\dfrac{\alpha_{mn} L}{2}\right) \displaystyle\int_0^2 r J_n{}^2\left(\frac{\alpha_{mn} r}{2}\right) dr}$$

and $$B_{mn} = \frac{\dfrac{1}{\pi} \displaystyle\int_0^{2\pi} \int_0^2 r f(r, \theta) \sin(n\theta) J_n\left(\frac{\alpha_{mn} r}{2}\right) dr\, d\theta}{\sin\left(\dfrac{\alpha_{mn} L}{2}\right) \displaystyle\int_0^2 r J_n{}^2\left(\frac{\alpha_{mn} r}{2}\right) dr}$$

15. $y(r, \theta, z, t) = \dfrac{4}{\pi} \displaystyle\sum_{n=1}^{\infty} \sum_{m=1}^{\infty} \frac{(-1)^{m+1}}{nm} \frac{\int_0^1 r J_1(\alpha_n r)\, dr}{J_2^2(\alpha_n)} \cos(\sqrt{\alpha_n{}^2 + \pi^2}\, t) J_1(\alpha_n r) \sin(m\pi z) \sin(\theta)$

where α_n is the nth positive root of $J_1(x)$

17. $u(x, t) = \begin{cases} 0 & \text{if } t < \dfrac{x^2}{2} \\ 4t - 2x^2 & \text{if } t > \dfrac{x^2}{2} \end{cases}$

19. Using the finite Fourier sine transform,

$$y(x, t) = \frac{2}{\pi}\sum_{n=1}^{\infty}\frac{f_s(n)}{n^4}[1 - \cos(n^2 t)]\sin(nx)$$

Chapter 14

Section 14.1

1. $26 - 18i$ **3.** $12 + 6i$ **5.** $\frac{7}{17} - \frac{6}{17}i$ **7.** $\dfrac{3 + 29i}{15}$ **9.** $30i$ **11.** $-\frac{8}{37} + \frac{48}{37}i$ **13.** $18 + 26i$

15. $\frac{173}{629} + \frac{39}{629}i$ **17.** $152 - 64i$ **19.** $\frac{40}{169} + \frac{265}{169}i$ **21.** $-112 + 68i$ **23.** $-\frac{90}{8} + \frac{22}{8}i$ **25.** $-\frac{432}{85} + \frac{776}{85}i$

27. $\sqrt{(x + 2)^2 + y^2}$; $\sqrt{x^2 + (y - 1)^2}$ **29.** $\dfrac{x^2y + 4xy + y^3}{(x + 1)^2 + y^2}$

31. All points $z = x + \frac{1}{2}i$ (i.e., on the line $y = \frac{1}{2}$) **35.** No (e.g., try $z_1 = 1$ and $z_2 = i$)

39. The hyperbola $x^2 - y^2 = 2$ **41.** Ellipse with foci at (0, 1) and (−8, 2) and major axis of length 4

Section 14.2

1. $\sqrt{40}\,[\cos(\theta) + i\sin(\theta)]$, $\theta = \dfrac{3\pi}{2} + \arctan\left(\dfrac{1}{3}\right)$ **3.** $\sqrt{68}\,[\cos(\theta) + i\sin(\theta)]$, $\theta = -\arctan(\frac{1}{4})$

5. $14\left[-i\sin\left(\dfrac{\pi}{2}\right)\right]$ **7.** $\sqrt{90}\,[\cos(\theta) + i\sin(\theta)]$, $\theta = \arctan(3)$ **9.** $\sqrt{73}\,[\cos(\theta) + i\sin(\theta)]$, $\theta = \pi + \arctan(\frac{3}{8})$

11. $\sqrt{50}\,[\cos(\theta) + i\sin(\theta)]$, $\theta = \pi + \arctan(7)$ **13.** $\dfrac{\sqrt{500}}{25}[\cos(\theta) + i\sin(\theta)]$, $\theta = \pi + \arctan\left(\dfrac{11}{2}\right)$

15. $\sqrt{4930}\,[\cos(\theta) + i\sin(\theta)]$, $\theta = \pi + \arctan(\frac{41}{57})$ **17.** $\dfrac{9}{\sqrt{2}} - \dfrac{9}{\sqrt{2}}i$ **19.** $-7\sqrt{3} - 7i$

21. $-8\sqrt{3} + 8i$ **23.** $7\sqrt{3} + 7i$ **25.** $\dfrac{15}{\sqrt{2}} - \dfrac{15}{\sqrt{2}}i$

Section 14.3

1.

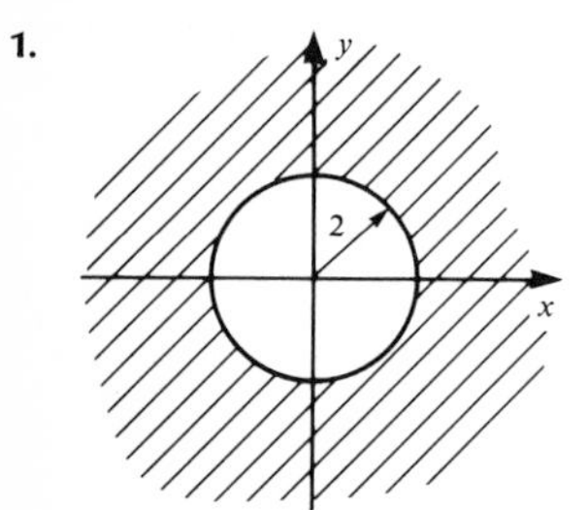

$|z| \geq 2$ consists of all points on and outside the circle of radius 2 about the origin; it is closed, with boundary the points on the circle $|z| = 2$; it is not a domain; it is not bounded.

3.

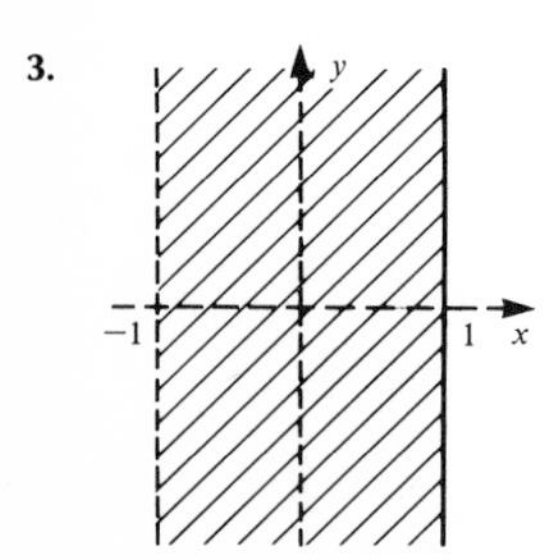

All points $x + iy$ with $-1 < x \leq 1$; not closed and not open; boundary consists of all points $x + iy$ with $x = 1$ or $x = -1$; not a domain; not bounded.

5. All z except $z = -2$; open set; the only boundary point is -2; the set is an unbounded domain.

7. Left half-plane, including the imaginary axis; closed, not open; boundary consists of all points on the imaginary axis; not a domain; not bounded.

9. All points on or inside the circle of radius 4 about the origin, with the origin excluded; not open; not closed; boundary consists of the origin and all points on $|z| = 4$; not a domain, but bounded.

11. $(x^3 - 3xy^2) + i(3x^2y - y^3)$ 13. $\dfrac{2x^2 + x + 2y^2 - 2y}{(2x+1)^2 + y^2} + i\left(\dfrac{-1 - 2x + y}{(2x+1)^2 + y^2}\right)$ 15. $\dfrac{x^2 - y^2}{x^2 + y^2} + i\left(\dfrac{2xy - 1}{x^2 + y^2}\right)$

17. $x^2 - y^2 - y - 5 + 3i$ 19. $x^2 - y^2 + 2x - 4 + i(2y - 2xy)$

Section 14.4

1. $4 - 4i$ 3. $\frac{1}{25}$ 5. $f'(5i)$ does not exist. 7. $f'(1 + i)$ does not exist. 9. $f'(3 - 2i)$ does not exist.

11. $24z^2 - 3$ 13. $60z$ 15. $\dfrac{-3z^2 + 4z + 3 - 3i}{(z^2 + 1 - i)^2}$ 17. $\dfrac{2iz^3 - (6 + 12i)z^2}{(z - 4 + 2i)^2}$

19. $3(4iz + 2)^2(4i)(8z^2 - iz + 1) + (4iz + 2)^3(16z - i)$

Section 14.5

1. $u = x, \quad v = y - 1$; analytic in D 3. $u = \dfrac{2x^2 + 2y^2 + x}{x^2 + y^2}, \quad v = -\dfrac{y}{x^2 + y^2}$; analytic in D

5. $u = 1, \quad v = \dfrac{y}{x}$; not analytic in D 7. $u = (x + 2)^2 - y^2, \quad v = 2xy + 4y$; analytic in D

9. $u = -y + \sqrt{x^2 + y^2}, \quad v = x$; not analytic in D 11. $u = \dfrac{x - 2x}{(1 - y)^2 + x^2}, \quad v = \dfrac{-y^2 - 2 + 3y - x^2}{(1 - y)^2 + x^2}$; analytic in D

Section 14.6

1. $w_1 = 2^{1/4}e^{7\pi i/8} = -1.0987 + 0.4551i$; $w_2 = 2^{1/4}e^{15\pi i/8} = 1.0987 - 0.4551i$

3. $w_1 = 2$; $w_2 = 2e^{\pi i/2} = 2i$; $w_3 = 2e^{\pi i} = -2$; $w_4 = 2e^{3\pi i/2} = -2i$

5. $w_1 = 2e^{\pi i/4} = 2[\cos(\pi/4) + i\sin(\pi/4)] = 1.4142 + 1.4142i$
$w_2 = 2e^{3\pi i/4} = 2[\cos(3\pi/4) + i\sin(3\pi/4)] = -1.4142 + 1.4142i$
$w_3 = 2e^{5\pi i/4} = 2[\cos(5\pi/4) + i\sin(5\pi/4)] = -1.4142 - 1.4142i$
$w_4 = 2e^{7\pi i/4} = 2[\cos(7\pi/4) + i\sin(7\pi/4)] = 1.4142 - 1.4142i$

7. $w_1 = 1$; $w_2 = e^{\pi i/5} = 0.8090 + 0.5878i$;
$w_3 = e^{2\pi i/5} = 0.3090 + 0.9511i$; $w_4 = e^{3\pi i/5} = -0.3090 + 0.9511i$;
$w_5 = e^{4\pi i/5} = -0.8090 + 0.5878i$; $w_6 = e^{\pi i} = -1$;
$w_7 = e^{6\pi i/5} = -0.8090 - 0.5878i$; $w_8 = e^{7\pi i/5} = -0.3090 - 0.9511i$;
$w_9 = e^{8\pi i/5} = 0.3090 - 0.9511i$; $w_{10} = e^{9\pi i/5} = 0.8090 - 0.5878i$

9. $w_1 = 8^{1/8}e^{5\pi i/16} = 0.7205 + 1.0782i$; $w_2 = 8^{1/8}e^{13\pi i/16} = -1.0782 + 0.7205i$;
$w_3 = 8^{1/8}e^{21\pi i/16} = -0.7205 - 1.0782i$; $w_4 = 8^{1/8}e^{29\pi i/16} = 1.0782 - 0.7205i$

11. $w_1 = e^{\pi i/20} = 0.9877 + 0.1564i$; $w_2 = e^{\pi i/4} = 0.7071 + 0.7071i$;
$w_3 = e^{9\pi i/20} = 0.1564 + 0.9877i$; $w_4 = e^{13\pi i/20} = -0.4540 + 0.8910i$;
$w_5 = e^{17\pi i/20} = -0.8910 + 0.4540i$; $w_6 = e^{21\pi i/20} = -0.9877 - 0.1564i$;
$w_7 = e^{25\pi i/20} = -0.7071 - 0.7071i$; $w_8 = e^{29\pi i/20} = -0.1564 - 0.9877i$;
$w_9 = e^{33\pi i/20} = 0.4540 - 0.8910i$; $w_{10} = e^{37\pi i/20} = 0.8910 - 0.4540i$

13. $w_1 = \left(\frac{629}{1369}\right)^{1/3} \exp\left\{\frac{i}{3}\left[\pi + \arctan\left(\frac{100}{621}\right)\right]\right\} = 0.3497 + 0.6878i$

$w_2 = \left(\frac{629}{1369}\right)^{1/3} \exp\left\{\frac{i}{3}\left[3\pi + \arctan\left(\frac{100}{621}\right)\right]\right\} = -0.7705 - 0.0410i$

$w_3 = \left(\frac{629}{1369}\right)^{1/3} \exp\left\{\frac{i}{3}\left[5\pi + \arctan\left(\frac{100}{621}\right)\right]\right\} = 0.4208 - 0.6468i$

15. $w_1 = \left(\frac{5}{625}\right)^{1/4} \exp\left\{\frac{i}{4}\left[\frac{\pi}{2} + \arctan\left(\frac{117}{44}\right)\right]\right\} = 0.2296 + 0.1916i$

$w_2 = \left(\frac{5}{625}\right)^{1/4} \exp\left\{\frac{i}{4}\left[\frac{5\pi}{2} + \arctan\left(\frac{117}{44}\right)\right]\right\} = -0.1916 + 0.2296i$

$w_3 = \left(\frac{5}{625}\right)^{1/4} \exp\left\{\frac{i}{4}\left[\frac{9\pi}{2} + \arctan\left(\frac{117}{44}\right)\right]\right\} = -0.2296 - 0.1916i$

$w_4 = \left(\frac{5}{625}\right)^{1/4} \exp\left\{\frac{i}{4}\left[\frac{13\pi}{2} + \arctan\left(\frac{117}{44}\right)\right]\right\} = 0.1916 - 0.2296i$

17. $w_1 = \frac{2^{1/8}}{(500)^{1/4}} e^{5\pi i/16} = 0.1281 + 0.1917i \qquad w_2 = \frac{2^{1/8}}{(500)^{1/4}} e^{13\pi i/16} = -0.1917 + 0.1281i$

$w_3 = \frac{2^{1/8}}{(500)^{1/4}} e^{21\pi i/16} = -0.1281 - 0.1917i \quad w_4 = \frac{2^{1/8}}{(500)^{1/4}} e^{29\pi i/16} = 0.1917 - 0.1281i$

19. $w_1 = (37)^{1/10} \exp\left\{\frac{i}{5}\left[\frac{3\pi}{2} + \arctan\left(\frac{1}{6}\right)\right]\right\} = 0.8046 + 1.1881i$

$w_2 = (37)^{1/10} \exp\left\{\frac{i}{5}\left[\frac{7\pi}{2} + \arctan\left(\frac{1}{6}\right)\right]\right\} = -0.8813 + 1.1324i$

$w_3 = (37)^{1/10} \exp\left\{\frac{i}{5}\left[\frac{11\pi}{2} + \arctan\left(\frac{1}{6}\right)\right]\right\} = -1.3493 - 0.4482i$

$w_4 = (37)^{1/10} \exp\left\{\frac{i}{5}\left[\frac{15\pi}{2} + \arctan\left(\frac{1}{6}\right)\right]\right\} = 0.0474 - 1.4341i$

$w_5 = (37)^{1/10} \exp\left\{\frac{i}{5}\left[\frac{19\pi}{2} + \arctan\left(\frac{1}{6}\right)\right]\right\} = 1.3786 - 0.3981i$

21. $w_1 = 5^{1/6} \exp\left\{\frac{i}{6}\left[\frac{3\pi}{2} + \arctan\left(\frac{4}{3}\right)\right]\right\} = 0.7713 + 1.0560i$

$w_2 = 5^{1/6} \exp\left\{\frac{i}{6}\left[\frac{7\pi}{2} + \arctan\left(\frac{4}{3}\right)\right]\right\} = -0.5288 + 1.1959i$

$w_3 = 5^{1/6} \exp\left\{\frac{i}{6}\left[\frac{11\pi}{2} + \arctan\left(\frac{4}{3}\right)\right]\right\} = -1.3001 + 0.1400i$

$w_4 = 5^{1/6} \exp\left\{\frac{i}{6}\left[\frac{15\pi}{2} + \arctan\left(\frac{4}{3}\right)\right]\right\} = -0.7713 - 1.0560i$

$w_5 = 5^{1/6} \exp\left\{\frac{i}{6}\left[\frac{19\pi}{2} + \arctan\left(\frac{4}{3}\right)\right]\right\} = 0.5288 - 1.1959i$

$w_6 = 5^{1/6} \exp\left\{\frac{i}{6}\left[\frac{23\pi}{2} + \arctan\left(\frac{4}{3}\right)\right]\right\} = 1.3001 - 0.1400i$

23. $w_1 = 17^{1/5} \exp\left\{\frac{i}{5}\left[\pi + \arctan\left(\frac{8}{15}\right)\right]\right\} = 1.3176 + 1.1704i$

$w_2 = 17^{1/5} \exp\left\{\frac{i}{5}\left[3\pi + \arctan\left(\frac{8}{15}\right)\right]\right\} = -0.7060 + 1.6148i$

$w_3 = 17^{1/5} \exp\left\{\frac{i}{5}\left[5\pi + \arctan\left(\frac{8}{15}\right)\right]\right\} = -1.7539 - 0.1724i$

$$w_4 = 17^{1/5}\exp\left\{\frac{i}{5}\left[7\pi + \arctan\left(\frac{8}{15}\right)\right]\right\} = -0.3780 - 1.7213i$$

$$w_5 = 17^{1/5}\exp\left\{\frac{i}{5}\left[7\pi + \arctan\left(\frac{8}{15}\right)\right]\right\} = 1.5203 - 0.8914i$$

25. $$w_1 = 13^{1/3}\exp\left\{\frac{i}{3}\left[\frac{3\pi}{2} + \arctan\left(\frac{5}{12}\right)\right]\right\} = -0.3085 + 2.3310i$$

$$w_2 = 13^{1/3}\exp\left\{\frac{i}{3}\left[\frac{7\pi}{2} + \arctan\left(\frac{5}{12}\right)\right]\right\} = -1.8644 - 1.4327i$$

$$w_3 = 13^{1/3}\exp\left\{\frac{i}{3}\left[\frac{11\pi}{2} + \arctan\left(\frac{5}{12}\right)\right]\right\} = 2.1730 - 0.8983i$$

29. $\dfrac{\sqrt{3} - 1 + i(1 - \sqrt{3})}{2}$, $\dfrac{-1 - \sqrt{3} + i(\sqrt{3} + 1)}{2}$

31. Roots are all possible values of $\left[\dfrac{1+i}{2} + \dfrac{1}{2}(4 - 2i)^{1/2}\right]^{1/3}$; the six values are $0.7228 + 0.6466i$, $-0.9214 + 0.3026i$, $0.1986 - 0.9493i$, $1.1557 + 0.0649i$, $-0.6341 + 0.9684i$, $-0.5216 - 1.0334i$.

Section 14.7

1. $\cos(1) + i\sin(1)$ (about $0.5403 + 0.8415i$) **3.** $e^{\pi}[\cos(1) - i\sin(1)]$ (about $12.5030 - 19.4722i$)

5. $e^2[\cos(2) - i\sin(2)]$ (about $-3.0749 + 6.7188i$) **7.** $e^{\pi/4}[\cos(\pi/4) + i\sin(\pi/4)]$ (about $1.5509 + 1.5509i$)

9. $e^{-5}[\cos(7) + i\sin(7)]$ (about $0.0051 + 0.0044i$) **11.** $\cos(5) + i\sin(5)$ (about $0.2837i - 0.9589i$)

13. $e^{-20}[\cos(8) + i\sin(8)]$ (zero to four decimal places) **15.** $\cos(9\pi) + i\sin(9\pi) = -1$

17. $\sqrt{5}\exp[i\arctan(\frac{1}{2})]$ **19.** $\sqrt{5}\exp\{i[\pi + \arctan(2)]\}$ **21.** $\sqrt{20}\exp\left\{i\left[\dfrac{3\pi}{2} + \arctan\left(\dfrac{1}{2}\right)\right]\right\}$

23. $\sqrt{180}\exp[i\arctan(2)]$ **25.** $\sqrt{5}\exp\left\{i\left[\dfrac{\pi}{2} + \arctan(2)\right]\right\}$

31. $u = \exp\left(\dfrac{x}{x^2+y^2}\right)\cos\left(\dfrac{y}{x^2+y^2}\right)$, $v = -\exp\left(\dfrac{x}{x^2+y^2}\right)\sin\left(\dfrac{y}{x^2+y^2}\right)$.

Section 14.8

In 1–19, n denotes any integer.

1. $\ln(2) + i\left(\dfrac{\pi}{2} + 2n\pi\right)$ **3.** $\ln(9) + 3n\pi i$ **5.** $\dfrac{1}{2}\ln(5) + i\left[\dfrac{3\pi}{2} + \arctan\left(\dfrac{1}{2}\right) + 2n\pi\right]$

7. $i\left[\dfrac{\pi}{6} + 2n\pi\right]$, $i\left[\dfrac{5\pi}{6} + 2n\pi\right]$, $i\left[\dfrac{3\pi}{2} + 2n\pi\right]$ for all three roots **9.** $\dfrac{1}{2}\ln(20) + i\left[\dfrac{\pi}{2} + \arctan(2) + 2n\pi\right]$

11. $\frac{1}{2}\ln(20) + i[\arctan(2) + 2n\pi]$ **13.** $\ln(5) + i[\pi + \arctan(\frac{4}{3}) + 2n\pi]$ **15.** $\dfrac{1}{2}\ln(53) + i\left[\dfrac{3\pi}{2} + \arctan\left(\dfrac{7}{2}\right) + 2n\pi\right]$

17. $\dfrac{1}{2}\ln(2) + i\left(\dfrac{\pi}{4} + 2n\pi\right)$, $\dfrac{1}{2}\ln(2) + i\left(\dfrac{3\pi}{4} + 2n\pi\right)$, for both roots **19.** $\dfrac{3}{2}\ln(8) + i\left(\dfrac{3\pi}{4} + 2n\pi\right)$

21. $\text{Ln}[(1 - i)(1 + i)] = \text{Ln}(2) = \ln(2)$

$\text{Ln}(1 - i) = \dfrac{1}{2}\ln(2) - \dfrac{\pi i}{4}$; $\text{Ln}(1 + i) = \dfrac{1}{2}\ln(2) + \dfrac{\pi i}{4}$

23. **(a)** $\ln(-1 - i) = \dfrac{1}{2}\ln(2) + i\left(\dfrac{5\pi}{4} + 2n\pi i\right)$; $e^{\ln(-1-i)} = \exp\left[\dfrac{1}{2}\ln(2)\right]\exp\left[i\left(\dfrac{5\pi}{4} + 2n\pi\right)\right] = \sqrt{2}\,e^{5\pi i/4} = -1 - i$

(b) $e^{-1-i} = e^{-1}e^{-i}$

$\ln(e^{-1-i}) = \ln(e^{-1}) + i(-1 + 2n\pi) = -1 - i + 2n\pi i$

25. **(a)** $\ln(-4) = \ln(4) + i(\pi + 2n\pi)$; so $e^{\ln(-4)} = e^{\ln(4)}e^{i\pi}e^{2n\pi i} = -4$

(b) $\ln(e^{-4}) = -4 + 2n\pi i$

27. (a) $\ln(-2-4i) = \frac{1}{2}\ln(20) + i[\pi + \arctan(2) + 2n\pi]$; so $e^{\ln(-2-4i)} = e^{\sqrt{20}} \cdot e^{i\pi} \cdot e^{i\arctan(2)} \cdot e^{2n\pi i}$ $= -e^{\sqrt{20}}e^{i\arctan(2)} = -(2+4i)$

(b) $e^{-2-4i} = e^{-2}e^{-4i}$; so $\ln(e^{-2-4i}) = \ln(e^{-2}) + i(-4+2n\pi) = -2-4i+2n\pi i$

Section 14.9

1. $i\exp\left\{-2n\pi - \frac{\pi}{2}\right\}$, n any integer; principal value $i\exp\{-\pi/2\}$

3. $\exp\left\{-2n\pi - \frac{\pi}{2}\right\}$, n any integer; principal value $\exp\{-\pi/2\}$

5. $\exp\left\{\frac{9\pi}{4} + 6n\pi\right\}\exp\left\{\frac{-3i}{2}\ln(2)\right\}$, n any integer; principal value $\exp\left\{\frac{9\pi}{4}\right\}\exp\left\{\frac{-3i}{2}\ln(2)\right\}$

7. $\exp\{-2\ln(6) + 6n\pi\}\exp\{-3i\ln(6)\}$, n any integer; principal value $\exp\{-2\ln(6)\}\exp\{-3i\ln(6)\}$

9. $\exp\left\{-\ln(2) - \frac{\pi}{2} + 4n\pi\right\}\exp\left\{i\left[\frac{\pi}{2} - \ln(2)\right]\right\}$, n any integer; principal value $\exp\{-\ln(2) - \pi/2\}\exp\left\{i\left[\frac{\pi}{2} - \ln(2)\right]\right\}$

11. $\exp\{3\ln(2) + 2n\pi\}\exp\{-i\ln(2)\}$, n any integer; principal value $\exp\{3\ln(2)\}\exp\{-i\ln(2)\}$

13. $\exp\{\arctan(\frac{2}{3}) - 2n\pi\}\exp\{i\ln(\sqrt{13})\}$, n any integer; principal value $\exp\{\arctan(\frac{2}{3})\}\exp\{i\ln(\sqrt{13})\}$

15. $\exp\left\{\frac{\pi}{2} - 2n\pi\right\}\exp\{i\ln(4)\}$, n any integer; principal value $\exp\left\{\frac{\pi}{2}\right\}\exp\{i\ln(4)\}$

17. $\exp\left\{\frac{1}{4}\ln(2) - \frac{\pi}{8} + n\pi\right\}\exp\left\{i\left[\frac{-\pi}{8} + n\pi - \frac{1}{4}\ln(2)\right]\right\}$, n any integer;

principal value $\exp\left\{\frac{1}{4}\ln(2) - \frac{\pi}{8}\right\}\exp\left\{i\left[-\frac{\pi}{8} - \frac{1}{4}\ln(2)\right]\right\}$

19. $\exp\{-n\pi\}\exp\left\{\frac{i}{2}\ln(3)\right\}$, n any integer; principal value $\exp\left\{\frac{i}{2}\ln(3)\right\}$

Section 14.10

1. $\frac{i}{2}\left(e - \frac{1}{e}\right)$ 3. $i\left(\frac{e^2 - e^{-2}}{e^2 + e^{-2}}\right)$; equivalently, $i\tanh(2)$ 5. $\sinh\left[\exp\left(-\frac{\pi}{2} - 2n\pi\right)\right]$

7. $\cos(2)\cosh(4) - i\sin(2)\sinh(4)$ 9. 0   11. $i\{\cos[\cos(1)]\sinh[\sin(1)] - \sin[\cos(1)]\cosh[\sin(1)]\}$

13. $\cos\left\{\exp\left(-\frac{\pi}{4} - 2n\pi\right)\cos[\frac{1}{2}\ln(2)]\right\}\cosh\left\{\exp\left(-\frac{\pi}{4} - 2n\pi\right)\sin[\frac{1}{2}\ln(2)]\right\}$

$-\frac{i}{2}\sin\left\{\exp\left(-\frac{\pi}{4} - 2n\pi\right)\cos[\frac{1}{2}\ln(2)]\right\}\sinh\left\{\exp\left(-\frac{\pi}{4} - 2n\pi\right)\sin[\frac{1}{2}\ln(2)]\right\}$, n any integer

15. $\cos[e^A\cos(B)]\cosh[e^A\sin(B)] + i\sin[e^A\cos(B)]\sinh[e^A\sin(B)]$, where $A = \frac{1}{2}\ln(2) + 2n\pi - \frac{\pi}{4}$ and $B = \frac{\pi}{4} + \frac{1}{2}\ln(2)$.

17. $\exp\{\sin(2)\cosh(1)\}\{\cos[\cos(2)\sinh(1)] - i\sin[\cos(2)\sinh(1)]\}$

19. The fourth roots of $[\sin(1)\cosh(1) - i\cos(1)\sinh(1)]^3$, approximately $1.2423 - 0.4410i$, $0.4410 + 1.2423i$, $-1.2423 + 0.4410i$, and $-0.4410 - 1.2423i$.

23. $\cosh(z) = \cos(y)\cosh(x) + i\sin(y)\sinh(x)$
$\sinh(z) = \cos(y)\sinh(x) + i\sin(y)\cosh(x)$

25. Yes; for example, $\cosh(z) = 0$ for any z of the form $z = \left(\frac{2n-1}{2}\right)\pi i$, n any integer.

27. $z = \frac{\pi}{6} + 2n\pi$ (n any integer) or $z = \frac{5\pi}{6} + 2m\pi$ (m any integer)

Supplementary Problems

1. $e^4\cos(1) - ie^4\sin(1)$ **3.** $-i\sin(5)$

5. $\cos\left[\frac{1}{2}\ln(2)\right]\cosh\left[\frac{\pi}{4} + 2n\pi\right] - i\sin\left[\frac{1}{2}\ln(2)\right]\sinh\left[\frac{\pi}{4} + 2n\pi\right]$, n any integer

7. $\{2\cos(1) - 4\sin(1) + i[2\sin(1) + 4\cos(1)]\}e^{-1}$ **9.** $\cosh(1)\cos(1) - i\sinh(1)\sin(1)$

11. These are the seventh roots of $-7 + 24i$, or $w_n = 25^{1/7}\exp\left\{\frac{i}{7}\left[\frac{\pi}{2} + \arctan\left(\frac{7}{24}\right) + 2n\pi\right]\right\}$, for $n = 0, 1, 2, 3, 4, 5, 6$.

These roots are approximately $1.5286 + 0.4147i$, $0.6288 + 1.4537i$, $-0.7445 + 1.3979i$, $-1.5571 + 0.2896i$, $-1.1972 - 1.0396i$, $0.0642 - 1.5825i$, $1.2773 - 0.9365i$.

13. $\cos\{e^{-2n\pi}\sin[\ln(3)]\}\cosh\{e^{-2n\pi}\cos[\ln(3)]\} + i\sin\{e^{-2n\pi}\sin[\ln(3)]\}\sinh\{e^{-2n\pi}\cos[\ln(3)]\}$, n any integer

15. $10^{e/2}\{\cos[e\arctan(3) + 2n\pi e] + i\sin[e\arctan(3) + 2n\pi e]\}$

17. $w_n = r^{1/7}\exp\left\{\frac{i}{7}\left[-\arctan\left(\frac{47}{52}\right) + 2n\pi\right]\right\}$, where $r = \frac{1}{17^3}\sqrt{52^2 + 47^2}$ and $n = 0, 1, 2, 3, 4, 5, 6$.

These roots are approximately $0.5421 - 0.0572i$, $0.3827 + 0.3882i$, $-0.0649 + 0.5412i$, $-0.4636 + 0.2867i$, $-0.5132 - 0.1837i$, $-0.1764 - 0.5158i$, and $0.2933 - 0.4595i$.

19. $w_n = (125)^{1/16}\exp\left\{\frac{i}{8}\left[\arctan\left(\frac{11}{2}\right) + 2n\pi\right]\right\}$, $n = 0, 1, 2, 3, 4, 5, 6, 7$. These are approximately $1.3319 + 0.2339i$, $0.7764 + 1.1072i$, $-0.2339 + 1.3319i$, $-1.1072 + 0.7764i$, $-1.3319 - 0.2339i$, $-0.7764 - 1.1072i$, $0.2339 - 1.3319i$, and $1.1072 - 0.7764i$.

21. There are six values, given by $\dfrac{13^{1/3}}{2^{1/4}}\dfrac{\exp\left\{\frac{i}{3}\left[\frac{\pi}{2} + \arctan\left(\frac{5}{12}\right) + 2n\pi\right]\right\}}{\exp\left\{\frac{i}{2}\left[-\frac{\pi}{4} + 2m\pi\right]\right\}}$, $n = 0, 1, 2$; $m = 0, 1$.

These values are approximately (for $n = 0$) $0.9874 + 1.7130i$ and $-0.9874 - 1.7130i$; (for $n = 1$) $-1.9772 - 0.0014i$ and $1.9772 + 0.0014i$; and (for $n = 2$), $0.9898 - 1.7116i$ and $-0.9898 + 1.7116i$.

23. $\cos(16)\sinh(12) + i\sin(16)\cosh(12)$ **25.** $\sin[\cosh(2)]$

27. This reduces to $(x + \frac{27}{8})^2 + (y - \frac{17}{8})^2 = \frac{45}{32}$, a circle of radius $\sqrt{45/32}$ about $(-\frac{27}{8}, \frac{17}{8})$.

33. The fifth roots of $-4i$, namely $w_n = 4^{1/5}\exp\left\{\frac{i}{5}\left(\frac{-\pi}{2} + 2n\pi\right)\right\}$, $n = 0, 1, 2, 3, 4$. These are approximately $1.2549 - 0.4078i$, $0.7756 + 1.0675i$, $-0.7756 + 1.0675i$, $-1.2549 - 0.4078i$, and $-1.3195i$.

Chapter 15

Section 15.1

1. $-\frac{52}{9} + 4i$ **3.** $\frac{16i}{3}$ **5.** $\frac{1 - \sqrt{3}}{2} + i\left(\frac{\sqrt{3}}{2} - \frac{\pi}{3}\right)$ **7.** $-\frac{1735}{3} - \frac{29{,}225}{6}i$ **9.** $2\cos(3i) - 2 + 6i\sin(3i)$

11. $\frac{131}{3} + \frac{5}{3}i$ **13.** $1 - e^2\cos(3) - e^2 i\sin(3)$ **15.** $i[\cos(i) - \cos(1)]$ **17.** $1172 + 2724i$

19. $e\sin(1) + i[e^{-1} - e\cos(1)]$ **21.** $\frac{228}{10} - 23i$ **23.** $\frac{3}{2}[e^2\cos(2) - \cos(4)] + \frac{3i}{2}[e^2\sin(2) - \sin(4)]$

25. $\frac{i}{2 - i}\sin[(2 - i)\pi]$ **27.** $\frac{1}{2}\ln(20) + i\arctan(2) - i\frac{\pi}{2}$ **29.** $\left(\frac{1 - i}{4}\right)(e^{112 - 16i} - e^{4 + 4i})$

31. $\frac{1}{2}\{-e^{-5i}[\sin(5i) + \cos(5i)] - e[\sin(1) - \cos(1)]\}$ **33.** $-\frac{2}{3}(e^{26}e^{-18i} - e^{-2}e^{-11i})$ **35.** 0

39. $8\pi\sqrt{\cosh(4)}$ (other bounds may be derived)

Section 15.2

1. 0 **3.** 0 **5.** $2\pi i$ **7.** 0 **9.** 0 **11.** 0 **13.** 0 **15.** 0 **17.** $\frac{(-4 - 2i)\pi}{5}$

Section 15.3

1. $-2-6i$ **3.** $-\frac{1}{2}(e^{-15-8i}-e^{3+4i})$ **5.** $2\pi i\sin(25)$ **7.** $48\pi i$ **9.** $-\pi i\cos(3i)$

11. $9\pi i\sin(12i)$ **13.** $\dfrac{\pi i}{3}[\cos(i)-\sin(i)]$ **15.** 0

Supplementary Problems

1. $4\pi ie^2$ **3.** 0 (whether or not C encloses i) **5.** $8+8i$ **7.** $-\frac{1}{2}+5i$ **9.** $\frac{71}{2}+\frac{1}{6}i$ **11.** $-\frac{13}{2}-39i$

13. 0 **15.** $\dfrac{i}{2}(e^{-i}-1)$ **17.** $-8+i$ **19.** $\frac{9208}{15}-2i$

21. 0 **23.** $54-6i$ **25.** $\dfrac{-1795}{12}+\cos(2i)+\dfrac{125i}{3}-\cos(5)$

Chapter 16

Section 16.1

1. Converges to $1+2i$ **3.** Converges to $\frac{1}{2}-i$ **5.** Converges to zero [taking the principal value of $(-1)^{in}$]

7. Converges to $2i$ **9.** Converges to 1 [taking the principal value of $(-i)^{4n}$] **11.** Diverges **13.** Converges to 1

15. Converges to $3i$ **21.** No; for example, $(-1)^n$ has no limit as $n\to\infty$.

Section 16.2

5. If $z_n\to L$, then $|z_n|\to|L|$.

Section 16.3

3. Converges **5.** Converges **7.** Diverges **9.** Converges

Section 16.4

1. Center $-3i$; radius of convergence 2;

$$f'(z)=\sum_{n=1}^{\infty}\frac{n(n+1)}{2^n}(z+3i)^{n-1};\quad f''(z)=\sum_{n=2}^{\infty}\frac{n(n-1)(n+1)}{2^n}(z+3i)^{n-2};$$

$$f'''(z)=\sum_{n=3}^{\infty}\frac{n(n-1)(n-2)(n+1)}{2^n}(z+3i)^{n-3}$$

3. Center $1-2i$; radius of convergence 1;

$$f'(z)=\sum_{n=1}^{\infty}\frac{n^{n+1}}{(n+1)^n}(z-1+2i)^{n-1};\quad f''(z)=\sum_{n=2}^{\infty}\frac{(n-1)n^{n+1}}{(n+1)^n}(z-1+2i)^{n-2};$$

$$f'''(z)=\sum_{n=3}^{\infty}\frac{(n-1)(n-2)n^{n+1}}{(n+1)^n}(z-1+2i)^{n-3}$$

5. Center $-4+i$; radius of convergence 2;

$$f'(z)=\sum_{n=1}^{\infty}\frac{ni^n}{2^{n+1}}(z+4-i)^{n-1};\quad f''(z)=\sum_{n=2}^{\infty}\frac{n(n-1)i^n}{2^{n+1}}(z+4-i)^{n-2};$$

$$f'''(z)=\sum_{n=3}^{\infty}\frac{n(n-1)(n-2)i^n}{2^{n+1}}(z+4-i)^{n-3}$$

7. Center $-6-2i$; radius of convergence 1;

$$f'(z)=\sum_{n=1}^{\infty}\frac{n^3}{(2n+1)^2}(z+6+2i)^{n-1};\quad f''(z)=\sum_{n=2}^{\infty}\frac{(n-1)n^3}{(2n+1)^2}(z+6+2i)^{n-2};$$

$$f'''(z)=\sum_{n=3}^{\infty}\frac{(n-1)(n-2)n^3}{(2n+1)^2}(z+6+2i)^{n-3}$$

9. Center -4; radius of convergence 1;

$$f'(z) = \sum_{n=1}^{\infty} \frac{ne^{in}}{2n+1}(z+4)^{n-1}; \quad f''(z) = \sum_{n=2}^{\infty} \frac{n(n-1)e^{in}}{2n+1}(z+4)^{n-2};$$

$$f'''(z) = \sum_{n=3}^{\infty} \frac{n(n-1)(n-2)e^{in}}{2n+1}(z+4)^{n-3}$$

Section 16.5

1. $\sum_{n=0}^{\infty} \frac{(-1)^n}{(2n)!} z^{2n}$; radius of convergence ∞ **3.** $\sum_{n=0}^{\infty} \frac{(-1)^n}{2(2n+1)!} z^{4n+2}$; radius of convergence ∞

5. $\sum_{n=0}^{\infty} \frac{(z-4i)^n}{(1-4i)^{n+1}}$, for $|z-4i| < \sqrt{17}$ **7.** $-3i \sum_{n=0}^{\infty} \frac{(-1)^n(z-5)^n}{4^{n+1}}$, for $|z-5| < 4$

9. $1 + \frac{1}{2\sqrt{2}(\sqrt{2}+1)} \sum_{n=0}^{\infty} \frac{(-1)^n(z-i)^n}{[i(\sqrt{2}+1)]^n} - \frac{1}{2\sqrt{2}(1-\sqrt{2})} \sum_{n=0}^{\infty} \frac{(-1)^n(z-i)^n}{[i(1-\sqrt{2})]^n}$, for $|z-i| < \sqrt{2}-1$

11. $\sum_{n=1}^{\infty} \frac{1}{(2n-1)!} 3^{2n-1}z^{2n-1}$, for all z **13.** $-3 \sum_{n=0}^{\infty} \frac{(z+5)^n}{(5+4i)^{n+1}}$, for $|z+5| < \sqrt{41}$

15. $\sum_{n=0}^{\infty} \frac{1}{(2n)!} z^{6n}$, for all z **17.** $\sum_{n=0}^{\infty} \frac{(-1)^n z^{4n}}{(2n)!} - \sum_{n=0}^{\infty} \frac{(-1)^n z^{2n+1}}{(2n+1)!}$, for all z

19. $3713 + 2016i + (-764 + 1952i)[z-(1+i)] + (-378 - 96i)[z-(1+i)]^2 + (4 - 32i)[z-(1+i)]^3 + [z-(1+i)]^4$, for all z

21. $\frac{e}{2} \sum_{n=0}^{\infty} \frac{z^n}{n!} + \frac{e^{-1}}{2} \sum_{n=0}^{\infty} \frac{(-1)^n z^n}{n!}$, for all z **23.** $\sum_{n=0}^{\infty} e^{3-i} \frac{(-1)^n(z-i)^n}{n!}$, for all z

25. $\sum_{n=0}^{\infty} \frac{(-1)^n 3^n z^n}{n!}$, for all z **27.** $\frac{2}{\sqrt{\pi}} \sum_{n=0}^{\infty} \frac{(-1)^n z^{2n+1}}{n!(2n+1)}$ **29.** $\sec(z) = 1 + \frac{z^2}{2} + \frac{5}{24}z^4 + \frac{61}{720}z^6 + \frac{277}{8064}z^8 + \cdots$

31. **(a)** $B_1 = -\frac{1}{2}$, $B_2 = \frac{1}{6}$, $B_3 = 0$, $B_4 = -\frac{1}{30}$, and $B_5 = 0$ **33.** $\sum_{n=0}^{\infty} \frac{(-1)^n z^{n+1}}{n+1}$

35. $E_0 = 1$, $E_2 = -1$, $E_4 = 5$, $E_6 = -61$, $E_8 = 1385$

Section 16.6

1. $\frac{1}{z-i} + \sum_{n=0}^{\infty} \frac{(-1)^n(z-i)^n}{(2i)^{n+1}}$, in $0 < |z-i| < 2$ **3.** $\frac{1}{z^2} + \sum_{n=0}^{\infty} \frac{(-1)^{n+1}2^{2n}z^{2n}}{(2n)!}$, for $0 < |z| < \infty$

5. $-\frac{1}{4}\frac{1}{z-1} - \frac{i}{4}\sum_{n=0}^{\infty} \frac{(-1)^n(z-1)^n}{(1+i)^{n+1}} + \frac{i}{4}\sum_{n=0}^{\infty} \frac{(-1)^n(z-1)^n}{(1-i)^{n+1}} + \frac{1}{4}\sum_{n=0}^{\infty} \frac{(-1)^n(z-1)^n}{2^{n+1}}$, in $0 < |z-1| < \sqrt{2}$

7. $1 + \frac{2i}{z-i}$, for $0 < |z| < \infty$ **9.** $1 + \frac{z^2}{3} + \frac{2z^4}{15} + \frac{17z^6}{315} + \frac{62z^8}{2835} + \cdots$ for $0 < |z| < \frac{\pi}{2}$

(The series can be written $\sum_{n=1}^{\infty} \frac{(-1)^{n-1}2^{2n}(2^{2n}-1)B_{2n}}{(2n)!} z^{2n-2}$, with B_{2n} the $(2n)$th Bernoulli number.)

11. $\frac{1}{z} + \frac{1}{2} + \frac{7}{12}z + \frac{1}{4}z^2 + \cdots$ in approximately $0 < |z| < 2.0288$ **13.** $\sum_{n=0}^{\infty} \frac{1}{n!\,z^{3n}}$, $0 < |z|$

15. $-\frac{(1-i)}{2\sqrt{2}} \sum_{n=0}^{\infty} \frac{(-1)^n(z-i)^n}{\left[\frac{\sqrt{2}}{2} + \left(\frac{\sqrt{2}}{2}+1\right)i\right]^{n+1}} + \frac{(1-i)}{2\sqrt{2}} \sum_{n=0}^{\infty} \frac{(-1)^n(z-i)^n}{\left[-\frac{\sqrt{2}}{2} + \left(1-\frac{\sqrt{2}}{2}\right)\right]^{n+1}}$, $|z-i| < \sqrt{2-\sqrt{2}}$

or, approximately, $|z-i| < 0.7654$. Note that $\frac{1}{z^2-i}$ is analytic at i.

17. $\frac{2i}{z-(1-i)}$, $0 < |z-(1-i)| < \infty$. (Here, $f(z)$ is a Laurent expansion.) **19.** $1 + \frac{2i}{z-2i}$, $0 < |z-2i| < \infty$

Supplementary Problems

1. $\displaystyle\sum_{n=0}^{\infty} \frac{(-1)^n[z-(2+i)]^n}{(6+i)^{n+1}}, \quad 0 < |z-(2+i)| < \sqrt{37}$ **3.** $\displaystyle\sum_{n=0}^{\infty} \frac{(-1)^n(z-5i)^{2n}}{(2n)!}$, for all z

5. $-18-328i+(139-487i)[z-(3-i)]+(180-261i)[z-(3-i)]^2+(80-60i)[z-(3-i)]^3+(15-5i)[z-(3-i)]^4 + [z-(3-i)]^5$, for all z

7. $\displaystyle\sum_{n=0}^{\infty} \frac{z^{2n-1}}{n!}, \quad 0<|z|<\infty$ **9.** $\displaystyle\sum_{n=0}^{\infty} \frac{(-1)^n i[z-(4+5i)]^n}{(4+6i)^{n+1}}, \quad |z-(4+5i)| < \sqrt{52}$

11. $\displaystyle\sum_{n=0}^{\infty} \frac{(-1)^n(i)^{2n+1}z^{6n+1}}{(2n+1)!}, \quad 0<|z|<\infty$ **13.** $\displaystyle\sum_{n=0}^{\infty} \frac{(-1)^n 2^n[z-(1+i)]^n}{(3i)^{n+1}}, \quad |z-(1+i)| < \frac{3}{2}$

15. $\displaystyle\frac{1}{6}\sum_{n=0}^{\infty} \frac{1}{n!}[i^n+(-i)^n]z^{2n-1}, \quad 0<|z|<\infty$ **17.** $\displaystyle\sum_{n=0}^{\infty} \frac{(-1)^n 3^{2n} z^{4n}}{(2n)!} - \sum_{n=0}^{\infty} \frac{iz^n}{n!}$, for all z

19. $\displaystyle\sum_{n=0}^{\infty} \frac{e\cdot 2^n z^n}{n!}$, for all z **21.** $\displaystyle iz+iz^2+\frac{2i}{3}z^3+\frac{i}{3}z^4+\frac{4i}{3}z^5$

23. $\displaystyle\cos(1)+2i\sin(1)(z-i)+[\sin(1)+2\cos(1)](z-i)^2+\left[-2i\cos(1)+\frac{4}{3}i\sin(1)\right](z-i)^3+\left[\frac{1}{6}\cos(1)+2\sin(1)\right](z-i)^4 - \left[\frac{11i}{15}\sin(1)+\frac{4i}{3}\cos(1)\right](z-i)^5$

25. $\cosh(i)-\sinh(i)(z-i)+[-i\sinh(i)+\frac{1}{2}\cosh(i)](z-i)^2+[\frac{5}{6}\sinh(i)+i\cosh(i)](z-i)^3+[\frac{1}{2}i\sinh(i)-\frac{31}{24}\cosh(i)](z-i)^4 + [-\frac{1}{24}\sinh(i)-\frac{49}{30}i\cosh(i)](z-i)^5$

27. $\sin^2(-2+6i)+2i\sin(-4+12i)(z-3-i)-4\cos(-4+12i)(z-3-i)^2+\frac{16}{3}i\sin(-4+12i)(z-3-i)^3 - \frac{16}{3}\cos(-4+12i)(z-3-i)^4+\frac{64}{15}i\sin(-4+12i)(z-3-i)^5$ [Here we used $\sin(4iz)=2\sin(2iz)\cos(2iz)$.]

29. $z-z^2+\frac{2}{3}z^3-\frac{1}{3}z^4+\frac{2}{15}z^5$

Chapter 17

Section 17.1

1. Pole of order 2 at 0 **3.** Essential singularity at 0 **5.** Poles of order 1 at i and $-i$; pole of order 2 at 1
7. Removable singularity at i; simple pole at $-i$ **9.** Simple poles at the fourth roots of 1
11. Poles of order 2 at odd-integer multiples of $\pi/2$ **13.** Pole of order 2 at 0 **15.** Pole of order 4 at -1
17. Simple poles at zeros of $\sinh(z)$, which are the numbers $n\pi i$, n any integer **19.** Simple pole at $-\pi$

Section 17.2

1. 0 at 0 **3.** $\frac{1}{2}-i$ at 0 **5.** $\operatorname{Res}_1 f(z) = -\sin(2)-\frac{1}{2}\cos(2)$; $\displaystyle\operatorname{Res}_i f(z) = \frac{\cos(2i)}{2i(i-1)^2}$; $\displaystyle\operatorname{Res}_{-i} f(z) = \frac{\cos(2i)}{-(i+1)^2(2i)}$

7. 1 at $-i$ **9.** $\displaystyle\operatorname{Res}_1 f(z) = \frac{1}{2(1-i)(1+i)} = \frac{1}{4}$; $\operatorname{Res}_{-1} f(z) = \frac{1}{4}$; $\operatorname{Res}_i f(z) = -\frac{1}{4}$; $\operatorname{Res}_{-i} f(z) = -\frac{1}{4}$

11. 0 at each double pole $\displaystyle\frac{(2n-1)\pi}{2}$, n any integer **13.** i at 0 **15.** $\frac{1}{6}e^{-1}$ at -1 **17.** 1 at each simple pole $n\pi i$

19. -1 at $-\pi$ **21.** $4\pi i$ **23.** $\pi\sin(2i)$ **25.** $\displaystyle\pi-\frac{\pi i}{2}$ **27.** $2\pi i$ **29.** 0 **31.** $2\pi i(1-2i)^{1/2}$

33. $\displaystyle 2\pi i\left[\frac{(1-w_0)^2}{(w_0-w_1)(w_0-w_2)}+\frac{(1-w_1)^2}{(w_1-w_0)(w_1-w_2)}+\frac{(1-w_2)^2}{(w_2-w_0)(w_2-w_1)}\right]$, where $\displaystyle w_j = 2^{1/3}\exp\left[\frac{\pi i+2j\pi i}{3}\right]$, $j=0, 1, 2.$

35. π^2

Section 17.3

1. $2\pi/\sqrt{3}$ **3.** $\dfrac{\pi}{\sqrt{3}}$ **5.** $\pi/3$ **7.** $\dfrac{\pi}{\sqrt{2}}$ **9.** $\dfrac{\pi e^{-\sqrt{2}}}{\sqrt{2}}[\sin(\sqrt{2})+\cos(\sqrt{2})]$ **11.** $\dfrac{\pi e^{-ab}}{b^3}(ab+1)$

13. $\dfrac{2\pi}{3^{11/6}}$ **15.** $\dfrac{2\pi a}{(a^2-b^2)^{3/2}}$ **17.** $\pi\left[\dfrac{e^{-ab}}{b(c^2-b^2)}+\dfrac{e^{-ac}}{c(b^2-c^2)}\right]$

19. $\dfrac{\pi}{2^{7/4}}+\dfrac{\pi}{2^{7/4}}\exp(1/2^{3/4})[\cos(2^{3/4})-\sin(2^{3/4})]$ (Let $\cos^2(x)=\frac{1}{2}[1+\cos(2x)]$.)

Section 17.4

1. $\dfrac{\pi}{4^{1/4}a^3}\left[\dfrac{\sinh(\alpha)+\sin(\alpha)}{\cosh(\alpha)-\cos(\alpha)}\right]$, where $\alpha=\dfrac{2\pi a}{4^{1/4}}$ **3.** $\dfrac{\pi^2}{8}$ **5.** $\dfrac{1}{2}+\dfrac{\pi}{8}\left[\pi\operatorname{csch}^2\left(\dfrac{\pi}{\sqrt{2}}\right)+\sqrt{2}\coth\left(\dfrac{\pi}{\sqrt{2}}\right)\right]$

7. $-\dfrac{1}{4}-\dfrac{\pi}{2^{11/4}}\left[\csc(\pi 2^{1/4})+\operatorname{csch}(\pi 2^{1/4})\right]$

Supplementary Problems

1. 0 **3.** $\pi\left[\dfrac{\sqrt{3}-\sqrt{2}}{\sqrt{6}}\right]$ **5.** $\pi\left[\dfrac{4-2\sqrt{3}}{4\sqrt{3}-6}\right]$

7. There are the following cases:

(a) C does not enclose 0, $\sqrt{3}\,i$, or $-\sqrt{3}\,i$; Integral $=0$

(b) C encloses $\sqrt{3}\,i$, but not 0 or $-\sqrt{3}\,i$; Integral $=\dfrac{\pi\cosh(\sqrt{3}\,i)}{3i}$

(c) C encloses $-\sqrt{3}\,i$, but not 0 or $\sqrt{3}\,i$; Integral $=\dfrac{\pi\cosh(\sqrt{3}\,i)}{3i}$

(d) C encloses 0, but not $\sqrt{3}\,i$ or $-\sqrt{3}\,i$; Integral $=\dfrac{2\pi i}{3}$

(e) C encloses 0 and $\sqrt{3}\,i$, but not $-\sqrt{3}\,i$; Integral $=\dfrac{2\pi i}{3}+\dfrac{\pi\cosh(\sqrt{3}\,i)}{3i}$

(f) C encloses 0 and $-\sqrt{3}\,i$, but not $\sqrt{3}\,i$; Integral $=\dfrac{2\pi i}{3}+\dfrac{\pi\cosh(\sqrt{3}\,i)}{3i}$

(g) C encloses $\sqrt{3}\,i$ and $-\sqrt{3}\,i$, but not 0; Integral $=\dfrac{2\pi\cosh(\sqrt{3}\,i)}{3i}$

(h) C encloses 0, $\sqrt{3}\,i$, and $-\sqrt{3}\,i$; Integral $=\dfrac{2\pi i}{3}+\dfrac{2\pi\cosh(\sqrt{3}\,i)}{3i}$

9. $\dfrac{\pi}{3}\left[e^{-1/2}\cos\left(\dfrac{\sqrt{3}}{2}\right)-e^{-1}\right]$ **11.** $(25+20i)\cos(1)+(20-25i)\sin(1)$ **13.** $-2\pi\sinh(2i)$ **15.** $2\pi i$

17. $\dfrac{\pi}{\sqrt{2}}(1-i)\cos\left[\sin\left(\dfrac{3\pi}{8}\right)+i\cos\left(\dfrac{3\pi}{8}\right)\right]+\dfrac{\pi}{\sqrt{2}}(-1+i)\cos\left[\sin\left(\dfrac{7\pi}{8}\right)+i\cos\left(\dfrac{7\pi}{8}\right)\right]$ **19.** $\dfrac{2\pi}{3}$ **21.** $(24-20i)\pi$

23. $2\pi i$ **25.** 0 **27.** 0

29. $2\pi i\left[\dfrac{\pi^2}{4}-\dfrac{\pi^4}{16}\right]\left[\dfrac{1}{1-\dfrac{\pi}{2}}-\dfrac{1}{1+\dfrac{\pi}{2}}\right]+2\pi i\left[\dfrac{9\pi^2}{4}-\dfrac{81\pi^4}{16}\right]\left[\dfrac{1}{\dfrac{3\pi}{2}-1}+\dfrac{1}{1+\dfrac{3\pi}{2}}\right]+2\pi i\left[\dfrac{25\pi^2}{4}-\dfrac{625\pi^4}{16}\right]\left[\dfrac{1}{1-\dfrac{5\pi}{2}}+\dfrac{1}{1+\dfrac{5\pi}{2}}\right]$

31. $\dfrac{-\pi(2^{4/3})}{24}\left\{(-1-\sqrt{3}\,i)\cot\left[\dfrac{\pi(1+\sqrt{3}\,i)}{2^{2/3}}\right]+2^{7/3}\cot(-2^{1/3}\pi)+(-1+\sqrt{3}\,i)\cot\left[\dfrac{\pi(1-\sqrt{3}\,i)}{2^{2/3}}\right]\right\}$

33. $\displaystyle\sum_{n=0}^{\infty}\frac{-\pi}{6}\exp\left[\frac{(-10n-5)\pi i}{6}\right]\csc\left[4^{1/6}\pi\exp\left(\frac{(2n+1)\pi i}{6}\right)\right]$

Chapter 18

Section 18.1

1. (a)

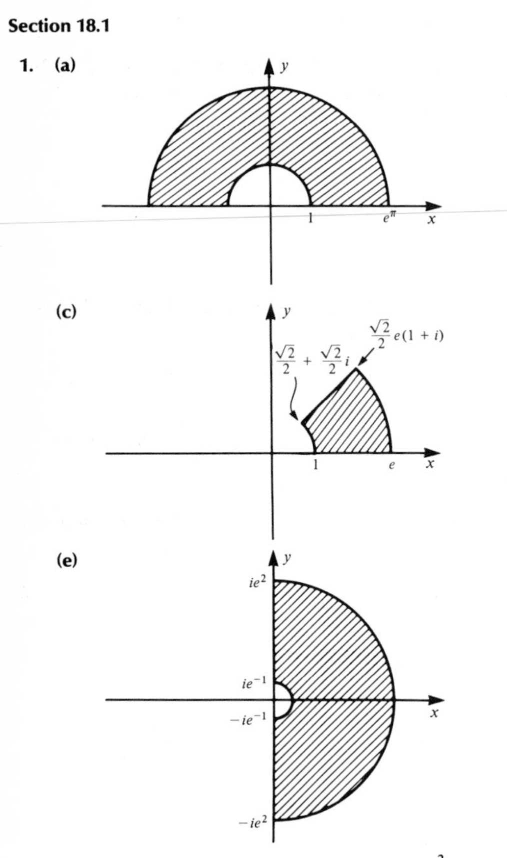

3. A line $x = k \neq 0$ maps to an ellipse $\dfrac{u^2}{\sinh^2(k)} + \dfrac{v^2}{\cosh^2(k)} = 1$. The line $x = 0$ maps to the interval from i to $-i$ on the imaginary axis (covering it infinitely many times).

   A line $y = C$ maps to one branch of the hyperbola $\dfrac{v^2}{\sin^2(C)} - \dfrac{u^2}{\cos^2(C)} = 1$, if $\sin(C) \neq 0$ and $\cos(C) \neq 0$. If $C = n\pi$ (n any integer), then $y = n\pi$ maps to the entire real axis; if $C = (2n-1)\pi/2$ (n any integer), then $y = C$ maps to the portion of the imaginary axis iv with $v \geq 1$ if n is odd and to iv with $v \leq -1$ if n is even.

5. On a line parallel to the real axis, $z = x + ik$ for some constant k. Then $w = \cos(x + ik) = \cosh(k)\cos(x) - i\sinh(k)\sin(x)$. Thus, $u = \cosh(k)\cos(x)$ and $v = -\sinh(k)\sin(x)$. This is an ellipse $\dfrac{u^2}{\cosh^2(k)} + \dfrac{v^2}{\sinh^2(k)} = 1$ if $\sinh(k) \neq 0$, that is, if $k \neq 0$. (Recall that $\cosh(k) \geq 1$ for all real k.) If $k = 0$, then $w = \cos(x)$, and the real axis maps to the interval $-1 \leq x \leq 1$ on the real axis in the w-plane.

15. A line $x = k \neq 0$ maps to an ellipse $\dfrac{u^2}{\cosh^2(k)} + \dfrac{v^2}{\sinh^2(k)} = 1$ if $k \neq 0$; the y-axis $x = 0$ maps to the interval $[-1, 1]$ on the real axis.

    A line $y = k$ maps to one branch of the hyperbola $\dfrac{u^2}{\cos^2(k)} - \dfrac{v^2}{\sin^2(k)} = 1$, unless $\cos(k) = 0$ or $\sin(k) = 0$. If $\cos(k) = 0$, then $k = (2n-1)\pi/2$ for some integer n, and $y = k$ maps to the entire imaginary axis. If $\sin(k) = 0$, then $k = n\pi$ for some integer n, and $y = k$ maps to the interval $u \geq 1$ on the real axis if n is even and to $u \leq -1$on the real axis if n is odd.

17. As z moves about the rectangle counterclockwise beginning at $-i\alpha$, w moves in the w-plane about the top of the ellipse $\dfrac{u^2}{\cosh^2(\alpha)} + \dfrac{v^2}{\sinh^2(\alpha)} = 1$, from $(\cosh(\alpha), 0)$ to $(-\cosh(\alpha), 0)$; then on the v-axis from $(-\cosh(\alpha), 0)$ to $(-1, 0)$ and back to $(-\cosh(\alpha), 0)$; then along the bottom of the ellipse from $(-\cosh(\alpha), 0)$ to $(\cosh(\alpha), 0)$; and finally on the real axis from $(\cosh(\alpha), 0)$ to $(1, 0)$ and back to $(\cosh(\alpha), 0)$. Points inside the rectangle map to points inside the ellipse.

Section 18.2

1. $w = \dfrac{(5+3i)z - 11 - 5i}{(5+i)z - 11 - 3i}$ 3. $w = \dfrac{(33+i)z - 48 - 16i}{5z - 20}$ 5. $w = \dfrac{z(-26 - 48i) + 78 + 224i}{-10z + 66 - 32i}$

7. $w = \dfrac{(-24 - 10i)z + 8 + 38i}{-iz - 4 + 3i}$ 9. $w = \dfrac{z(-6 + 54i) + 6i}{z(-21 + 216i) + 216 + 30i}$

11. $\operatorname{Im} z = 2$ maps to the circle $(u - \frac{7}{2})^2 + v^2 = \frac{9}{4}$; the half-plane $\operatorname{Im} z > 2$ maps to the inside of the circle; and $\operatorname{Im} z < 2$ maps to the exterior.

13. $\operatorname{Re} z = 5$ maps to the line $\operatorname{Im} w = 10$; the half-plane $\operatorname{Re} z > 5$ maps to the half-plane $\operatorname{Im} z > 10$; the half-plane $\operatorname{Re} z < 5$ maps to the half-plane $\operatorname{Im} z < 10$.

15. $|z| = 4$ maps to the circle $\left(u - \dfrac{11}{21}\right)^2 + \left(v + \dfrac{1}{63}\right)^2 = \dfrac{208}{3969}$, with center $\left(\dfrac{11}{21}, -\dfrac{1}{63}\right)$ and radius $\dfrac{4\sqrt{13}}{63}$. The interior $|z| < 4$ maps to the exterior of this circle.

17. The line $2x - 3y = 5$ maps to the circle $(u-1)^2 + (v + \frac{19}{4})^2 = \frac{377}{16}$. The half-plane $2x - 3y > 5$ maps to the exterior of the circle; the half-plane $2x - 5y < 5$ maps to the interior of the circle.

19. $\operatorname{Im} z = a \neq 0$ maps to the circle $u^2 + \left(v + \dfrac{1}{2a}\right)^2 = \left(\dfrac{1}{4a^2}\right)$; the region above $\operatorname{Im} z = a$ maps to the interior of this circle.

Section 18.3

In each of 1–9, only one of many possible answers is given.

1. $w = i + \dfrac{4z}{3}$ 3. $w = \dfrac{(3-3i)z + 8 + 4i}{(1-i)z + 2 + 2i}$ 5. $w = \dfrac{4z+4}{i(z-1)}$ 7. $w = i(3 - z)$ 9. $w = \dfrac{iz - 12 - i}{z - 1 + 4i}$

11. One such mapping sends $z \to z^{1/3}$ for $\operatorname{Im} z > 0$.

Supplementary Problems

1. $w = \dfrac{z(-11 - 49i) + 35 + 133i}{z(-14 + 7i) + 56 - 19i}$ 3. $w = \dfrac{z(4 - 4i) - (6 - 6i)}{z(1 - 4i) + 6 + i}$ 5. $w = \dfrac{z(32 + 72i) + 72 - 48i}{z(6 - i) + 6 - 21i}$

7. $w = \dfrac{z(-70 + 94i) - 20 - 8i}{z(23 + 25i) + 34 - 140i}$ 9. $w = \dfrac{z(81 - 31i) + (51 - 81i)}{z(11 - 19i) - (19 - 71i)}$

11. One such mapping is $w = \dfrac{z(-1 - i) - 3 + 3i}{z(1 - i) + 3 + 3i}$. 13. One such mapping is $w = -z - i$.

15. One such mapping is $w = \dfrac{z(2 - 4i) - 6 + 6i}{z(-1 + i)}$. 17. $w = e^{\pi i/4}\left(\dfrac{3(z+2)(-1-i) - 3 + 3i}{3(z+2)(1-i) + 3 + 3i}\right)$

Chapter 19

Section 19.1

1. $v = 2xy$ 3. $v = y - x$ 5. $v = \cos(x)\sinh(y)$

Section 19.2

3. $\displaystyle\int_{-\infty}^{\infty} \frac{yg(t)\,dt}{(t-x)^2 + 1}$ 5. $\displaystyle \operatorname{Re}\left\{\frac{1}{2\pi i}\int_{0}^{-\infty} \frac{g(t)[(t+z)(1+i) + 2i]\,dt}{(t-z)[(1+i)t + i]}\right\}$

Section 19.3

1. Streamlines are given by $\ln|z - z_0| = k$ (circles about z_0); equipotential lines are lines $\arg(z - z_0) = C$ (half-lines from z_0).
3. Streamlines are $x + \dfrac{a^2x}{x^2+y^2} = k$; equipotential lines are $y - \dfrac{a^2y}{x^2+y^2} = C$.
5. Thrust $= A\mathbf{i} + \beta\mathbf{j} = kb\rho\mathbf{j}$, acting vertically; Moment $= 0$
7. In 1, no stagnation point; in 3, $z = \pm a$

Section 19.4

1. Equipotential lines: $\dfrac{x}{x^2+y^2} = k$ (circles); Flux lines: $\dfrac{y}{x^2+y^2} = C$ (circles)

Section 19.5

1. $\dfrac{1}{k^2}\sin\left(\dfrac{kt}{\sqrt{2}}\right)\sinh\left(\dfrac{kt}{\sqrt{2}}\right)$
3. $\dfrac{1}{3\alpha}\left\{-e^{-\alpha t} + e^{\alpha t/2}\left[\cos\left(\dfrac{\sqrt{3}\,\alpha t}{2}\right) + \sin\left(\dfrac{\sqrt{3}\,\alpha t}{2}\right)\right]\right\} + \dfrac{1}{3\alpha^2}\left\{e^{-\alpha t} - e^{\alpha t/2}\left[\cos\left(\dfrac{\sqrt{3}\,\alpha t}{2}\right) - \sqrt{3}\sin\left(\dfrac{\sqrt{3}\,\alpha t}{2}\right)\right]\right\}$, where $\alpha = 5^{1/3}$
5. $\cosh(\sqrt{6}\,t)$ 7. $-3t + 3\sinh(t)$ 9. $\dfrac{1}{3\alpha^2}\left\{e^{-\alpha t} - e^{\alpha t/2}\left[\cos\left(\dfrac{\sqrt{3}\,\alpha t}{2}\right) - \sqrt{3}\sin\left(\dfrac{\sqrt{3}\,\alpha t}{2}\right)\right]\right\}$, with $\alpha = 2^{1/3}$

Section 19.6

1. $\cos(2z) = \frac{1}{2}(e^{2iz} + e^{-2iz})$ 7. $\displaystyle\sum_{n=0}^{\infty} \frac{6i\sqrt{r}}{\pi(1-2n)}\, e^{in\theta}$ 9. $-1 + 2ire^{i\theta} + r^2e^{2i\theta}$

Chapter 20

Section 20.1

We used $h = 0.001$ throughout; different choices of x_0 may result in different solutions.

1. 2.2733 (using $x_0 = 1$); 4.4940 (using $x_0 = 5$); 14.9897 (using $x_0 = 15$); other solutions exist
3. -1.4747 (other roots are complex, $1.0682 \pm 0.6423i$ and $-0.2058 \pm 1.5578i$)
5. 0.4803 (using $x_0 = 1$) 7. 58.9367 (using $x_0 = 1$) 9. -1.9993 (with $x_0 = -2$)
11. -0.6262 (with $x_0 = 1$) 13. No solutions 15. 3.4240 (with $x_0 = 1$) 17. 1.8895 (with $x_0 = 2$)
19. -0.1441 (with $x_0 = -0.5$) 21. -30.1069 (with $x_0 = 0.5$) 23. 1.3540 (with $x_0 = 1$)
25. 0.3922 (with $x_0 = 1$) 27. -3.3926 (with $x_0 = 1$) 29. 0.9046 (with $x_0 = 1$)

Section 20.2

		Rectangular	Trapezoidal	Simpsons
1.	$n = 2$	0.2744	0.2741	0.2743
	$n = 4$	0.2743	0.2742	0.2743
	$n = 10$	0.2743	0.2743	0.2743
	$n = 50$	0.2743	0.2743	0.2743
	$n = 100$	0.2743	0.2743	0.2743
	$n = 200$	0.2743	0.2743	0.2743
3.	$n = 2$	−0.0866	−0.0868	−0.0864
	$n = 4$	−0.0866	−0.0867	−0.0867
	$n = 10$	−0.0867	−0.0867	−0.0884
	$n = 50$	−0.0867	−0.0867	−0.0870
	$n = 100$	−0.0867	−0.0867	−0.0867
	$n = 200$	−0.0867	−0.0867	−0.0867

		Rectangular	Trapezoidal	Simpsons
5.	$n = 2$	0.3072	0.3488	0.3230
	$n = 4$	0.3173	0.3280	0.3210
	$n = 10$	0.3203	0.3220	0.3209
	$n = 50$	0.3208	0.3209	0.3209
	$n = 100$	0.3209	0.3209	0.3209
	$n = 200$	0.3209	0.3209	0.3209
7.	$n = 2$	1.1037	1.1307	1.1116
	$n = 4$	1.1106	1.1172	1.1127
	$n = 10$	1.1124	1.1135	1.1128
	$n = 50$	1.1128	1.1128	1.1128
	$n = 100$	1.1128	1.1128	1.1128
	$n = 200$	1.1128	1.1128	1.1128
9.	$n = 2$	4.0223	9.6052	6.7368
	$n = 4$	5.6957	6.8138	5.8833
	$n = 10$	6.0669	6.2003	6.0919
	$n = 50$	6.1114	6.1162	6.1130
	$n = 100$	6.1126	6.1138	6.1130
	$n = 200$	6.1129	6.1132	6.1130
11.	$n = 2$	3.6276	3.7922	4.0091
	$n = 4$	3.7064	3.7099	3.6825
	$n = 10$	3.7081	3.7081	3.9177
	$n = 50$	3.7081	3.7081	3.7500
	$n = 100$	3.7081	3.7081	3.7081
	$n = 200$	3.7081	3.7081	3.7186
13.	$n = 2$	0.2375	0.3021	0.2596
	$n = 4$	0.2556	0.2698	0.2591
	$n = 10$	0.2597	0.2619	0.2604
	$n = 50$	0.2604	0.2605	0.2651
	$n = 100$	0.2604	0.2604	0.2651
	$n = 200$	0.2604	0.2604	0.2627
15.	$n = 2$	0.0032	0.0033	0.0032
	$n = 4$	0.0032	0.0032	0.0032
	$n = 10$	0.0032	0.0032	0.0032
	$n = 50$	0.0032	0.0032	0.0032
	$n = 100$	0.0032	0.0032	0.0032
	$n = 200$	0.0032	0.0032	0.0032
17.	$n = 2$	0.6570	1.3781	0.9187
	$n = 4$	0.8376	1.0175	0.8974
	$n = 10$	0.8880	0.9168	0.8976
	$n = 50$	0.8972	0.8984	0.8976
	$n = 100$	0.8975	0.8978	0.8976
	$n = 200$	0.8976	0.8977	0.8976
19.	$n = 2$	1.1743	1.1352	1.1574
	$n = 4$	1.1652	1.1548	1.1613
	$n = 10$	1.1623	1.1606	1.1617
	$n = 50$	1.1618	1.1617	1.1903
	$n = 100$	1.1618	1.1617	1.1618
	$n = 200$	1.1618	1.1618	1.1618

Section 20.3

1. $f(-0.5) \approx -1.0625$; $f(1.2) \approx 14.1440$; $f(\sqrt{2}) \approx 12.7643$

3. $f(-\sqrt{3}) \approx 5.9019$; $f(1.2) \approx 4.5280$; $f(0.15) \approx 0.8346$

5. $f(1.1) \approx 2.6582$; $f(1.7) \approx -2.9582$; $f(1.8) \approx -2.0836$

7. $f(2.55) \approx 2.2593$; $f(2.8) \approx -2.7241$; $f(2.4) \approx 3.4931$
9. $f(2.3) \approx 9.8774$; $f(1.7) \approx 1.9243$; $f(\pi/2) \approx 0.2890$
11. $f(\pi/4) \approx -2.8892$; $f(\sqrt{3}) \approx 1.3940$; $f(-1.6) \approx -27.1979$
13. $f(-1.7) \approx 1.1905$; $f(-2.3) \approx 2.5187$; $f(-3) \approx 8.6503$
15. $f(3) \approx -2.8972$; $f(\sqrt{7}) \approx -1.0845$; $f(6.7) \approx 3.8683$

Section 20.4

1. -0.8 **3.** -0.4 **5.** -0.1 **7.** 0.3 **9.** -0.2 **11.** 2, 4, 6 **13.** 0.55, -0.05, 0.25
15. 16.5, -6.5, -29.5 **17.** 2, 3, 4 **19.** -0.1833, -0.15, -0.1167 **21.** 0.4333 **23.** -2.4167 **25.** 2.25
27. 0.5833 **29.** -3.0278

In 31–39, the answers are given in the following order: difference quotient approximation, three-point formula, five-point formula, and exact value (to four decimal places).

31. 4.6500, 4.6400, 4.6400, 4.6400 **33.** -0.7500, -0.8285, -0.8408, -0.8404
35. -0.9000, -0.9200, -0.9200, -0.9600 **37.** 0.4967, 0.2283, 0.2294, 0.2302
39. 1.0000, 1.0000, 1.0000, 0.9994

Section 20.5

1. $$P(x) = \begin{cases} -39 + 103.5x - 93x^2 + 31.5x^3 & 0 \le x \le 1 \\ -27 + 85.5x - 75x^2 + 25.5x^3 & 1 \le x \le 2 \end{cases}$$

3. $$P(x) = \begin{cases} 0.9088 - 0.7369x + 0.2189x^2 - 0.0229x^3 & 1 \le x \le 3 \\ 0.3277 - 0.1558x + 0.0251x^2 - 0.0014x^3 & 3 \le x \le 5 \end{cases}$$

5. $$P(x) = \begin{cases} 134.5870 - 247.8044x + 83.8478x^2 + 31.3696x^3 & 1 \le x \le 3 \\ -938.1956 + 992.6739x - 385.5435x^2 + 64.8913x^3 & 3 \le x \le 5 \\ -4337.6521 + 3032.3478x - 793.4783x^2 + 92.0870x^3 & 5 \le x \le 6 \end{cases}$$

7. $$P(x) = \begin{cases} -1.6931 + 5.1586x - 4.3172x^2 + 1.5448x^3 & \frac{1}{2} \le x \le 1 \\ -0.6593 + 1.8184x - 0.5342x^2 + 0.0682x^3 & 1 \le x \le 2 \end{cases}$$

9. $$P(x) = \begin{cases} 1 + 0.5x^2 + 0.0431x^3 & 0 \le x \le 1 \\ 0.2464 + 1.7821x - 0.9587x^2 + 0.4733x^3 & 1 \le x \le 2 \end{cases}$$

Section 20.6

1.

		Taylor		Runge-Kutta	
n	x_n	y_n	y_{n+1}	y_n	y_{n+1}
0	0	-1	-0.9231	-1	-0.9225
1	0.1	-0.9231	-0.8572	-0.9225	-0.8562
2	0.2	-0.8572	-0.8002	-0.8562	-0.7960
3	0.3	-0.8002	-0.7509	-0.7960	-0.7465
4	0.4	-0.7509	-0.7084	-0.7465	-0.7039
5	0.5	-0.7084	-0.6722	-0.7039	-0.6676
6	0.6	-0.6722	-0.6420	-0.6676	-0.6373
7	0.7	-0.6420	-0.6176	-0.6373	-0.6128
8	0.8	-0.6176	-0.5990	-0.6128	-0.5940
9	0.9	-0.5990	-0.5862	-0.5940	-0.5810
10	1.0	-0.5862	-0.5795	-0.5810	-0.5740

3.

		Taylor		Runge-Kutta	
n	x_n	y_n	y_{n+1}	y_n	y_{n+1}
0	3.1416	1.4142	1.9053	1.4142	1.9052
1	3.2416	1.9053	2.4219	1.9052	2.4217
2	3.3416	2.4219	2.9636	2.4217	2.9633
3	3.4416	2.9636	3.5301	2.9633	3.5297
4	3.5416	3.5301	4.1214	3.5297	4.1210

		Taylor		Runge-Kutta	
n	x_n	y_n	y_{n+1}	y_n	y_{n+1}
5	3.6416	4.1214	4.7380	4.1210	4.7376
6	3.7416	4.7380	5.3806	4.7376	5.3804
7	3.8416	5.3806	6.0506	5.3804	6.0506
8	3.9416	6.0506	6.8453	6.0506	6.7501
9	4.0416	6.8453	7.5861	6.7501	7.4814
10	4.1416	7.5861	8.3626	7.4814	8.2476

5.

		Taylor		Runge-Kutta	
n	x_n	y_n	y_{n+1}	y_n	y_{n+1}
0	2.6	1.4	0.8789	1.4	0.8364
1	2.7	0.8789	0.4889	0.8364	0.4565
2	2.8	0.4889	0.2435	0.4565	0.2221
3	2.9	0.2435	0.1151	0.2221	0.0978
4	3.0	0.1151	0.0535	0.0978	0.0398
5	3.1	0.0535	0.0248	0.0398	0.0151
6	3.2	0.0248	0.0115	0.0151	0.0054
7	3.3	0.0115	0.0054	0.0054	0.0018
8	3.4	0.0054	0.0026	0.0018	0.0016
9	3.5	0.0026	0.0013	0.0016	0.0005
10	3.6	0.0013	0.0007	0.0005	0.0001

7.

		Taylor		Runge-Kutta	
n	x_n	y_n	y_{n+1}	y_n	y_{n+1}
0	0	4	5.1372	4	5.1472
1	0.1	5.1372	6.5883	5.1472	6.6126
2	0.2	6.5883	8.4234	6.6126	8.4670
3	0.3	8.4234	10.7179	8.4670	10.7860
4	0.4	10.7179	10.6461	10.7860	13.6431
5	0.5	10.6461	13.3373	13.6431	17.0990
6	0.6	13.3373	16.5247	17.0990	21.1852
7	0.7	16.5247	20.1986	21.1852	25.8842
8	0.8	20.1986	24.2946	25.8842	31.1077
9	0.9	24.2946	28.6776	31.1077	36.6780
10	1.0	28.6776	33.1341	36.6780	42.3202

9.

		Taylor		Runge-Kutta	
n	x_n	y_n	y_{n+1}	y_n	y_{n+1}
0	0	1	1.1100	1	1.1115
1	0.1	1.1100	1.2490	1.1115	1.2531
2	0.2	1.2490	1.4310	1.2531	1.4398
3	0.3	1.4310	1.6784	1.4398	1.6963
4	0.4	1.6784	2.0301	1.6963	2.0673
5	0.5	2.0301	2.5610	2.0673	2.6444
6	0.6	2.5610	3.4361	2.6444	3.6532
7	0.7	3.4361	5.0953	3.6532	5.8445
8	0.8	5.0953	9.1190	5.8445	14.0354
9	0.9	9.1190	25.1815	14.0354	740.6277
10	1.0	25.1815	248.6320	740.6277	$4.1277(10^{26})$

11.

		Taylor		Runge-Kutta	
n	x_n	y_n	y_{n+1}	y_n	y_{n+1}
0	2	4	3.0016	4	2.9785
1	2.1	3.0016	2.3532	2.9785	2.3148
2	2.2	2.3532	1.9324	2.3148	1.8900
3	2.3	1.9324	1.6684	1.8900	1.6273

		Taylor		Runge-Kutta	
n	x_n	y_n	y_{n+1}	y_n	y_{n+1}
4	2.4	1.6684	1.5157	1.6273	1.4780
5	2.5	1.5157	1.4457	1.4780	1.4120
6	2.6	1.4457	1.4413	1.4120	1.4115
7	2.7	1.4413	1.4939	1.4115	1.4677
8	2.8	1.4939	1.6015	1.4677	1.5785
9	2.9	1.6015	1.7678	1.5785	1.7478
10	3.0	1.7678	2.0024	1.7478	1.9856

13.

		Taylor		Runge-Kutta	
n	x_n	y_n	y_{n+1}	y_n	y_{n+1}
0	1	−4	−7.1522	−4	−0.1594
1	1.1	−7.1522	−65.6315	−0.1594	−0.2299
2	1.2	−65.6315	−91,169.8170	−0.2299	−0.2827
3	1.3	—	—	−0.2827	−0.3142
4	1.4			−0.3142	−0.1517
5	1.5			−0.1517	−0.2174
6	1.6			−0.2174	−0.2547
7	1.7			−0.2547	−0.2681
8	1.8			−0.2681	−0.2658
9	1.9			−0.2658	−0.2551
10	2.0			−0.2551	−0.2405

15.

		Taylor		Runge-Kutta	
n	x_n	y_n	y_{n+1}	y_n	y_{n+1}
0	0	0.5	0.4926	0.5	0.4926
1	0.1	0.4926	0.4692	0.4926	0.4697
2	0.2	0.4692	0.4261	0.4697	0.4291
3	0.3	0.4261	0.3574	0.4291	0.3665
4	0.4	0.3574	0.2535	0.3665	0.2738
5	0.5	0.2535	0.0975	0.2738	0.1349
6	0.6	0.0975	−0.1443	0.1349	−0.0894
7	0.7	−0.1443	−0.5558	−0.0894	−0.5540
8	0.8	−0.5558	−1.4530	−0.5540	−502.3040
9	0.9	−1.4530	−6.0778	—	—
10	1.0	−6.0778	−36,777.5377	—	—

17.

		Taylor		Runge-Kutta	
n	x_n	y_n	y_{n+1}	y_n	y_{n+1}
0	3.1416	0.5	1.5477	0.5	1.0310
1	3.2416	1.5477	1.8780	1.0310	1.1914
2	3.3416	1.8780	−0.5477	1.1914	1.1248
3	3.4416	−0.5477	1.0796	1.1248	1.1603
4	3.5416	1.0796	0.8610	1.1603	1.1201
5	3.6416	0.8610	1.0402	1.1201	1.2665
6	3.7416	1.0402	0.7864	1.2665	1.3015
7	3.8416	0.7864	1.1776	1.3015	1.3676
8	3.9416	1.1776	0.9431	1.3676	1.2725
9	4.0416	0.9431	0.7681	1.2725	1.3417
10	14.1416	0.7681	1.1868	1.3417	1.2919

19.

		Taylor		Runge-Kutta	
n	x_n	y_n	y_{n+1}	y_n	y_{n+1}
0	2.73	4.62	4.6686	4.62	4.7823
1	2.74	4.6686	4.7006	4.7823	4.8340
2	2.75	4.7006	4.7007	4.8340	4.8511

		Taylor		Runge-Kutta	
n	x_n	y_n	y_{n+1}	y_n	y_{n+1}
3	2.76	4.7007	4.7243	4.8511	4.8571
4	2.78	4.7243	4.7428	4.8571	4.8597
5	2.79	4.7428	4.7579	4.8597	4.8609
6	2.80	4.7579	4.7707	4.8609	4.8616
7	2.81	4.7707	4.7818	4.8616	4.8622
8	2.82	4.7818	4.7916	4.8622	4.8627
9	2.83	4.7916	4.8004	4.8627	4.8632
10	2.84	4.8004	4.8084	4.8632	4.8637

21.

$x_0 = 2$	$y_0 = \pi$	
$x_1 = 2.01$	$y_1 = 3.2485$	Runge-Kutta
$x_2 = 2.02$	$y_2 = 3.3549$	Runge-Kutta
$x_3 = 2.03$	$y_3 = 3.4624$	Runge-Kutta
$x_4 = 2.04$	3.5660	Pred
	3.5661	Corr
	$y_4 = 3.5661$	
$x_5 = 2.05$	3.6711	Pred
	3.6727	Corr
	$y_5 = 3.6727$	
$x_6 = 2.06$	3.7753	Pred
	3.7754	Corr
	$y_6 = 3.7754$	
$x_7 = 2.07$	3.8812	Pred
	$y_7 = 3.8812$	Corr
$x_8 = 2.08$	3.9729	Pred
	3.9830	Corr
	$y_8 = 3.9830$	
$x_9 = 2.09$	4.0880	Pred
	$y_9 = 4.0880$	Corr
$x_{10} = 2.10$	4.1888	Pred
	4.1890	Corr
	$y_{10} = 4.1890$	

23.

$x_0 = 3.1416$	$y_0 = 3.42$	
$x_1 = 3.1516$	$y_1 = 3.4429$	Runge-Kutta
$x_2 = 3.1616$	$y_2 = 3.4658$	Runge-Kutta
$x_3 = 3.1716$	$y_3 = 3.4888$	Runge-Kutta
$x_4 = 3.1816$	3.5118	Pred
	$y_4 = 3.5118$	Corr
$x_5 = 3.1916$	3.5349	Pred
	$y_5 = 3.5349$	Corr
$x_6 = 3.2016$	3.5580	Pred
	$y_6 = 3.5580$	Corr
$x_7 = 3.2116$	3.5812	Pred
	$y_7 = 3.5812$	Corr
$x_8 = 3.2216$	3.6044	Pred
	$y_8 = 3.6044$	Corr
$x_9 = 3.2316$	3.6277	Pred
	$y_9 = 3.6277$	Corr
$x_{10} = 3.2416$	3.6509	Pred
	3.6510	Corr
	$y_{10} = 3.6510$	

25.

$x_0 = 2$	$y_0 = 3.22$	
$x_1 = 2.01$	$y_1 = 3.7010$	Runge-Kutta
$x_2 = 2.02$	$y_2 = 4.3722$	Runge-Kutta
$x_3 = 2.03$	$y_3 = 5.3435$	Runge-Kutta
$x_4 = 2.04$	6.7522	Pred
	6.8055	Corr
	$y_4 = 6.8055$	
$x_5 = 2.05$	9.2656	Pred
	9.3675	Corr
	$y_5 = 9.3675$	
$x_6 = 2.06$	14.3539	Pred
	$y_6 = 14.8663$	Corr
$x_7 = 2.07$	29.2339	Pred
	33.7402	Corr
	$y_7 = 33.7402$	
$x_8 = 2.08$	125.8192	Pred
	289.6214	Corr
	$y_8 = 289.6214$	
$x_9 = 2.09$	8919.5131	Pred
	1,065,292.090	Corr
	$y_9 = 1{,}065{,}292.090$	
$x_{10} = 2.10$	$12{,}105(10^{11})$	Pred

27.

$x_0 = 0$	$y_0 = 0.4$	
$x_1 = 0.01$	$y_1 = 0.4042$	Runge-Kutta
$x_2 = 0.02$	$y_2 = 0.4088$	Runge-Kutta
$x_3 = 0.03$	$y_3 = 0.4137$	Runge-Kutta
$x_4 = 0.04$	0.4183	Pred
	0.4184	Corr
	$y_4 = 0.4184$	
$x_5 = 0.05$	0.4234	Pred
	0.4238	Corr
	$y_5 = 0.4238$	
$x_6 = 0.06$	0.4290	Pred
	$y_6 = 0.4290$	Corr
$x_7 = 0.07$	0.4349	Pred
	$y_7 = 0.4349$	Corr
$x_8 = 0.08$	0.4406	Pred
	$y_8 = 0.4406$	Corr
$x_9 = 0.09$	0.4471	Pred
	$y_9 = 0.4471$	Corr
$x_{10} = 0.10$	0.4534	Pred
	0.4533	Corr
	$y_{10} = 0.4533$	

29.

x	y	
$x_0 = 0.3$	$y_0 = 1.0400$	
$x_1 = 0.31$	$y_1 = 1.0499$	Runge-Kutta
$x_2 = 0.32$	$y_2 = 1.0600$	Runge-Kutta
$x_3 = 0.33$	$y_3 = 1.0700$	Runge-Kutta
$x_4 = 0.34$	1.0807	Pred
	$y_4 = 1.0807$	Corr
$x_5 = 0.35$	1.0914	Pred
	1.0912	Corr
	$y_5 = 1.0912$	
$x_6 = 0.36$	1.1023	Pred
	$y_6 = 1.1023$	Corr
$x_7 = 0.37$	1.1133	Pred
	1.1134	Corr
	$y_7 = 1.1134$	
$x_8 = 0.38$	1.1251	Pred
	$y_8 = 1.1251$	Corr
$x_9 = 0.39$	1.1473	Pred
	1.1422	Corr
	$y_9 = 1.1422$	
$x_{10} = 0.40$	1.1498	Pred
	1.1496	Corr
	$y_{10} = 1.1496$	

Section 20.7

1.

		Taylor		Runge-Kutta-Nystrom	
j	x_j	y_j	y'_j	y_j	y'_j
0	0	0	1	0	1
1	0.1	0.1	1.0	0.0998	0.9950
2	0.2	0.1995	0.9900	0.1996	0.9801
3	0.3	0.2980	0.9801	0.2965	0.9555
4	0.4	0.3950	0.9605	0.3904	0.9218
5	0.5	0.4896	0.9315	0.4805	0.8796
6	0.6	0.5809	0.8936	0.5660	0.8299
7	0.7	0.6680	0.8474	0.6462	0.7736
8	0.8	0.7501	0.7938	0.7205	0.7118
9	0.9	0.8670	1.5438	0.7884	0.6458
10	1.0	1.0177	1.4704	0.8496	0.5769

3.

		Taylor		Runge-Kutta-Nystrom	
j	x_j	y_j	y'_j	y_j	y'_j
0	0	−1	4	−1	4
1	0.1	−0.5400	5.2000	−0.5437	5.0861
2	0.2	0.0281	6.1620	0.0071	5.8718
3	0.3	0.6750	6.7757	0.6173	6.2599
4	0.4	1.3609	6.9433	1.2950	6.1313
5	0.5	2.0375	6.5889	1.8803	5.4872
6	0.6	2.6504	5.6689	2.3752	4.3272
7	0.7	3.1426	4.1755	2.7299	2.6940
8	0.8	3.4588	2.1490	2.9010	0.6743
9	0.9	3.5501	−0.3231	2.8560	−1.6037
10	1.0	3.3787	−3.1055	2.5770	−3.9768

5.

		Taylor		Runge-Kutta-Nystrom	
j	x_j	y_j	y'_j	y_j	y'_j
0	2	−1	0	−1	0
1	2.1	−0.9871	0.2584	−0.9868	0.2658
2	2.2	−0.9477	0.5299	−0.9465	0.5421
3	2.3	−0.8808	0.8083	−0.8783	0.8219
4	2.4	−0.7860	1.0867	−0.7823	1.0975
5	2.5	−0.6638	1.3574	−0.6593	1.3604
6	2.6	−0.5153	1.6122	−0.5110	1.6019
7	2.7	−0.3426	1.8423	−0.3400	1.8130
8	2.8	−0.1485	2.0389	−0.1497	1.9848
9	2.9	0.0631	2.1931	0.0554	2.1086
10	3.0	0.2876	2.2964	0.2702	2.1766

7.

		Taylor		Runge-Kutta-Nystrom	
j	x_j	y_j	y'_j	y_j	y'_j
0	0	0	0	0	0
1	0.1	0.0150	0.3000	0.0155	0.3146
2	0.2	0.0614	0.6275	0.0639	0.6570
3	0.3	0.1418	0.9803	0.1497	1.0446
4	0.4	0.2586	1.3561	0.2736	1.4374
5	0.5	0.4141	1.7529	0.4379	1.8517
6	0.6	0.6102	2.1693	0.6446	2.2866
7	0.7	0.8489	2.6046	0.8959	2.7421
8	0.8	1.1321	3.0588	1.1938	3.2191
9	0.9	1.4617	3.5329	1.5405	3.7192
10	1.0	1.8398	4.0284	1.9385	4.2446

9.

		Taylor		Runge-Kutta-Nystrom	
j	x_j	y_j	y'_j	y_j	y'_j
0	1	0	−4	0	−4
1	1.1	−0.3814	−3.6281	−0.4149	−4.2425
2	1.2	−0.7256	−3.2562	−0.8356	−4.0986
3	1.3	−1.0326	−2.8843	−1.2182	−3.4620
4	1.4	−1.3024	−2.5124	−1.5082	−2.2326
5	1.5	−1.5350	−2.1405	−1.6420	−0.3248
6	1.6	−1.7305	−1.7686	−1.5484	2.3238
7	1.7	−1.8888	−1.3967	−1.1515	5.7431
8	1.8	−2.0099	−1.0248	−0.3744	9.9214
9	1.9	−2.0938	−0.6529	0.8561	14.7967
10	2.0	−2.1405	−0.2810	2.6043	20.2501

11.

		Taylor		Runge-Kutta-Nystrom	
j	x_j	y_j	y'_j	y_j	y'_j
0	3.1416	0	0.1	0	0.1
1	3.2416	0.0723	1.3467	0.0249	0.3946
2	3.3416	0.2669	2.5458	0.0784	0.6729
3	3.4416	0.5855	3.8263	0.1591	0.9400
4	3.5416	1.0401	5.2662	0.2662	1.2016
5	3.6416	1.6418	6.7679	0.3994	1.4639
6	3.7416	2.3891	8.1779	0.5592	1.7334
7	3.8416	3.2770	9.5797	0.7466	2.0174
8	3.9416	4.3126	11.1318	0.9635	2.3242
9	4.0416	5.5096	12.8075	1.2126	2.6630
10	4.1416	6.8700	14.4003	1.4975	3.0435

13.

		Taylor		Runge-Kutta-Nystrom	
j	x_j	y_j	y'_j	y_j	y'_j
0	0	−2	4	−2	4
1	0.1	−1.5755	4.4909	−1.5743	4.5272
2	0.2	−1.0989	5.0410	−1.0918	5.1388
3	0.3	−0.5649	5.6382	−0.5433	5.8484
4	0.4	0.0302	6.2646	0.0816	6.6691
5	0.5	0.6886	6.9040	0.7951	7.6264
6	0.6	1.4116	7.5559	1.6134	8.7789
7	0.7	2.2018	8.2488	2.5610	10.2348
8	0.8	3.0663	9.0419	3.6752	12.1339
9	0.9	4.0185	10.0026	5.0056	14.5627
10	1.0	5.0767	11.1607	6.6036	17.4689

15.

		Taylor		Runge-Kutta-Nystrom	
j	x_j	y_j	y'_j	y_j	y'_j
0	2	1	0	1	0
1	2.1	1.0150	0.3000	1.0154	0.3125
2	2.2	1.0611	0.6224	1.0630	0.6398
3	2.3	1.1398	0.9520	1.1433	0.9648
4	2.4	1.2510	1.2712	1.2552	1.2666
5	2.5	1.3924	1.5575	1.3949	1.5160
6	2.6	1.5593	1.7799	1.5552	1.6702
7	2.7	1.7429	1.8922	1.7237	1.6682
8	2.8	1.9288	1.8264	1.8813	1.4430
9	2.9	2.0950	1.4969	2.0038	0.9687
10	3.0	2.2121	0.8459	2.0687	0.3101

Section 20.8

1. $y_1 = -1.7110, \quad y_2 = -2.3718, \quad y_3 = -2.7272$

3. $y_1 = -0.1332, \quad y_2 = -0.3106, \quad y_3 = -0.5385, \quad y_4 = -0.8318, \quad y_5 = -1.2437$

5. $y_1 = -0.3646, \quad y_2 = 0.5439, \quad y_3 = 0.7846, \quad y_4 = 1.1231, \quad y_5 = 1.5321$

7. $y_1 = 4.4989, \quad y_2 = 4.8681, \quad y_3 = 5.1503, \quad y_4 = 5.3742, \quad y_5 = 5.5597, \quad y_6 = 5.7202, \quad y_7 = 5.8649$

9. $y_1 = -0.9544, \quad y_2 = -1.7285, \quad y_3 = -2.3741, \quad y_4 = -2.9272, \quad y_5 = -3.4111, \quad y_6 = -3.8395, \quad y_7 = -4.2180,$
$y_8 = -4.5446, \quad y_9 = -4.8107$

Section 20.9

1. $u(0.2, 0.2) = 0, \quad u(0.2, 0.4) = 0, \quad u(0.4, 0.2) = 0.1, \quad u(0.4, 0.4) = -0.1$

3. $u(0.2, 0.2) = 0.0270, \quad u(0.4, 0.2) = 0.0628, \quad u(0.6, 0.2) = 0.1111, \quad u(0.8, 0.2) = 0.1566,$
$u(0.2, 0.4) = 0.0450, \quad u(0.4, 0.4) = 0.1132, \quad u(0.6, 0.4) = 0.2249, \quad u(0.8, 0.4) = 0.3154$

5. $u(1.4, 0.2) = 1.7886, \quad u(1.6, 0.2) = 2.3943, \quad u(1.8, 0.2) = 3.0686$

7. $u(1.6, 2.2) = 17.0659, \quad u(1.8, 2.2) = 19.2388, \quad u(1.6, 2.4) = 18.1050, \quad u(1.8, 2.4) = 20.6492$
$u(1.6, 2.6) = 19.3047, \quad u(1.8, 2.6) = 22.0531, \quad u(1.6, 2.8) = 20.5408, \quad u(1.8, 2.8) = 23.4585$

9. $u(1.2, -0.4) = 1.0590, \quad u(1.4, -0.4) = 1.2108, \quad u(1.6, -0.4) = 1.3714, \quad u(1.8, -0.4) = 1.5593$
$u(1.2, -0.6) = 0.9280, \quad u(1.4, -0.6) = 1.0405, \quad u(1.6, -0.6) = 1.1476, \quad u(1.8, -0.6) = 1.2596$
$u(1.2, -0.8) = 0.7872, \quad u(1.4, -0.8) = 0.8758, \quad u(1.6, -0.8) = 0.9189, \quad u(1.8, -0.8) = 0.7877$

Section 20.10

1. With $v_0 = \begin{bmatrix} 1 \\ 1 \end{bmatrix}$, $R_{10} = 3.2863$ (actual value 3.3723)

3. With $v_0 = \begin{pmatrix} 1 \\ 0 \end{pmatrix}$, $R_{10} = 3.0199$ (actual value 2.6457)

5. With $v_0 = \begin{pmatrix} 0 \\ 1 \end{pmatrix}$, $R_1 = R_2 = \cdots = R_{10} = 6$ the actual value

7. With $v_0 = \begin{pmatrix} 1 \\ 0 \\ 0 \end{pmatrix}$, $R_{10} = 1.7895$ (actual value 4.9455)

9. With $v_0 = \begin{pmatrix} 0 \\ 1 \\ 0 \end{pmatrix}$, $R_{10} = -5.1870$ (actual value -5.4937)

In 13–19, the actual eigenvalues are given to four decimal places.

13. 1.0076, -6.9076 **15.** -2.7, 7.3 **17.** 7.5062, -2.4862 **19.** 4.0826, 2.5215, 0.7258

Section 20.11

1. $y = 3.2308 - 0.9231x$ **3.** $y = -5 + 0.5x$ **5.** $y = 2.5 - 0.8x$ **7.** $y = -4.6216 + 1.2568x$

9. $y = -1.8486 + 2.5593x$ **11.** $y = 3.9663 + 0.3098x$ **13.** $y = -4.0959 + 2.1544x$ **15.** $y = -1.8697 + 1.5798x$

17. $y = 4.8667x - 0.8667x^2$ **19.** $y = -1.3516 + 0.6560x - 0.0121x^2$ **21.** $y = 438.5062 - 536.6569x + 131.2859x^2$

23. $y = -2.7500 + 1.2500x - 0.2500x^2$ **25.** $y = 0.4266 - 1.0828x - 0.1037x^2$

Supplementary Problems

1. 0.7695, -0.5036, -2.5038 (other roots are $1.1189 \pm 0.8996i$) **3.** 11.6585 (with $x_0 = 0$) **5.** No real roots

7. 0 [with $\sin(x)/x$ defined to be 1 at $x = 0$] **9.** 0.5671

11.

n	x_n	y_n	y_{n+1}
0	0	π	3.1486
1	0.1	3.1486	3.1875
2	0.2	3.1875	3.2597
3	0.3	3.2597	3.3343
4	0.4	3.3343	3.4802
5	0.5	3.4802	3.6702
6	0.6	3.6702	3.9107
7	0.7	3.9107	4.2100
8	0.8	4.2100	4.5787
9	0.9	4.5787	5.0305
10	1.0	5.0305	5.5828

13.

n	x_n	y_n	y_{n+1}
0	0	0.3600	0.3811
1	0.1	0.3811	0.4022
2	0.2	0.4022	0.4245
3	0.3	0.4245	0.4491
4	0.4	0.4491	0.4769
5	0.5	0.4769	0.5086
6	0.6	0.5086	0.5449
7	0.7	0.5449	0.5869
8	0.8	0.5869	0.6343
9	0.9	0.6343	0.6883
10	1.0	0.6883	0.7495

15.

n	x_n	y_n	y'_n
0	2	1	0.5774
1	2.1	1.0462	0.3436
2	2.2	1.0682	0.0957
3	2.3	1.0653	-0.1520
4	2.4	1.0383	-0.3841
5	2.5	0.9894	-0.5890
6	2.6	0.9215	-0.7634
7	2.7	0.8374	-0.9165
8	2.8	0.7381	-1.0719
9	2.9	0.6200	-1.3057
10	3.0	0.4765	-1.5813

17.

n	x_n	y_n	
0	0	0.2	
1	0.01	0.1961	Runge-Kutta
2	0.02	0.1925	Runge-Kutta
3	0.03	0.1891	Runge-Kutta
4	0.04	-0.3313	Pred
		0.1859	Corr
5	0.05	-0.3096	Pred
		0.1829	Corr
6	0.06	-0.2885	Pred
		0.1801	Corr
7	0.07	-0.2680	Pred
		0.1775	Corr
8	0.08	-0.2479	Pred
		0.1751	Corr
9	0.09	-0.2284	Pred
		0.1729	Corr
10	0.10	-0.2093	Pred
		0.1709	Corr
11	0.11	-0.1908	Pred
		0.1691	Corr
12	0.12	-0.1727	Pred
		0.1674	Corr

19. $y_0 = -0.3600$ $y_1 = -0.2543$ $y_2 = -0.1532$ $y_3 = -0.0575$ $y_4 = 0.0320$ $y_5 = 0.1149$
$y_6 = 0.1911$ $y_7 = 0.2604$ $y_8 = 0.3231$ $y_9 = 0.3794$ $y_{10} = 0.4300$

21. $P_0(x)$: no roots
$P_1(x)$: 0
$P_2(x)$: 0.5774, −0.5774
$P_3(x)$: 0, 0.7746, −0.7746
$P_4(x)$: 0.8611, 0.3400, −0.3400, −0.8611
$P_5(x)$: 0, 0.9061, −0.9061, 0.5385, −0.5385

23. $y = 5.6891 - 0.8820x$ **27.** 1.6836 **29.** 50.2655

Index

Numbers in parentheses refer to problems.